# QUICK INDEX OF PAGE NUMBERS
# MATERIAL GROUPS AND MACHINING OPERATIONS




| MACHINING OPERATION / MATERIAL GROUP | 8.5 Cylindrical Grinding | 8.6 Cylindrical Grinding – CBN Wheels | 8.7 Cylindrical Grinding – Diamond Wheels | 8.8 Internal Grinding | 8.9 Internal Grinding – CBN Wheels | 8.10 Internal Grinding – Diamond Wheels | 8.11 Centerless Grinding | 8.13 Abrasive Belt Grinding | 8.14 Thread Grinding |
|---|---|---|---|---|---|---|---|---|---|
| 1. Free Machining Carbon Steels, Wrought | 8-61 | 8-93 | – | 8-101 | 8-133 | – | 8-143 | 8-177 | 8-179 |
| 2. Carbon Steels, Wrought | 8-61 | 8-93 | – | 8-101 | 8-133 | – | 8-143 | 8-177 | 8-179 |
| 3. Carbon and Ferritic Alloy Steels (High Temp. Service) | 8-61 | – | – | 8-101 | – | – | 8-143 | 8-177 | 8-179 |
| 4. Free Machining Alloy Steels, Wrought | 8-62 | 8-94 | – | 8-102 | 8-134 | – | 8-144 | 8-177 | 8-180 |
| 5. Alloy Steels, Wrought | 8-62 | 8-94 | – | 8-102 | 8-134 | – | 8-144 | 8-177 | 8-180 |
| 6. High Strength Steels, Wrought | 8-63 | 8-94 | – | 8-103 | 8-135 | – | 8-145 | – | 8-181 |
| 7. Maraging Steels, Wrought | 8-63 | 8-95 | – | 8-103 | 8-135 | – | 8-145 | – | 8-181 |
| 8. Tool Steels, Wrought | 8-63 | 8-95 | – | 8-103 | 8-135 | – | 8-145 | 8-177 | 8-181 |
| 9. Nitriding Steels, Wrought | 8-65 | 8-95 | – | 8-105 | 8-136 | – | 8-147 | – | 8-183 |
| 10. Armor Plate, Ship Plate, Aircraft Plate, Wrought | 8-65 | – | – | 8-105 | – | – | – | – | – |
| 11. Structural Steels, Wrought | – | – | – | – | – | – | – | – | – |
| 12. Free Machining Stainless Steels, Wrought | 8-66 | – | – | 8-106 | – | – | 8-147 | 8-177 | 8-183 |
| 13. Stainless Steels, Wrought | 8-66 | – | – | 8-106 | 8-136 | – | 8-148 | 8-177 | 8-184 |
| 14. Precipitation Hardening Stainless Steels, Wrought | 8-67 | – | – | 8-107 | – | – | 8-149 | 8-177 | 8-185 |
| 15. Carbon Steels, Cast | 8-68 | 8-95 | – | 8-108 | 8-136 | – | 8-150 | 8-177 | 8-185 |
| 16. Alloy Steels, Cast | 8-68 | 8-96 | – | 8-108 | 8-137 | – | 8-150 | 8-177 | 8-185 |
| 17. Tool Steels, Cast | 8-69 | 8-96 | – | 8-109 | 8-137 | – | 8-150 | 8-177 | 8-186 |
| 18. Stainless Steels, Cast | 8-70 | – | – | 8-110 | – | – | 8-151 | 8-177 | 8-187 |
| 19. Precipitation Hardening Stainless Steels, Cast | 8-71 | – | – | 8-111 | – | – | 8-152 | 8-177 | 8-188 |
| 20. Austenitic Manganese Steels, Cast | – | – | – | 8-111 | – | – | – | – | – |
| 21. Gray Cast Irons | 8-71 | – | – | 8-111 | – | – | 8-153 | 8-177 | 8-188 |
| 22. Compacted Graphite Cast Irons | 8-71 | – | – | 8-112 | – | – | 8-153 | – | 8-188 |
| 23. Ductile Cast Irons | 8-72 | – | – | 8-112 | – | – | 8-154 | 8-177 | 8-189 |
| 24. Malleable Cast Irons | 8-72 | – | – | 8-112 | – | – | 8-154 | 8-177 | 8-189 |
| 25. White Cast Irons (Abrasion Resistant) | 8-72 | – | – | 8-113 | – | – | 8-154 | – | – |
| 26. High Silicon Cast Irons | 8-73 | – | – | 8-113 | – | – | 8-155 | – | – |
| 27. Chromium-Nickel Alloy Castings | 8-73 | – | – | 8-113 | – | – | 8-155 | – | – |
| 28. Aluminum Alloys, Wrought | 8-73 | – | – | 8-113 | – | – | 8-155 | 8-177 | – |
| 29. Aluminum Alloys, Cast | 8-73 | – | – | 8-114 | – | – | 8-155 | 8-177 | – |
| 30. Magnesium Alloys, Wrought | 8-74 | – | – | 8-114 | – | – | 8-156 | 8-177 | – |
| 31. Magnesium Alloys, Cast | 8-74 | – | – | 8-114 | – | – | 8-156 | 8-177 | – |
| 32. Titanium Alloys, Wrought | 8-74 | – | – | 8-115 | – | – | 8-156 | 8-177 | – |
| 33. Titanium Alloys, Cast | 8-75 | – | – | 8-115 | – | – | 8-157 | 8-177 | – |
| 34. Copper Alloys, Wrought | 8-76 | – | – | 8-116 | – | – | 8-158 | 8-178 | – |
| 35. Copper Alloys, Cast | 8-77 | – | – | 8-117 | – | – | 8-159 | 8-178 | – |
| 36. Nickel Alloys, Wrought and Cast | 8-78 | – | – | 8-118 | – | – | 8-160 | – | 8-189 |
| 37. Beryllium Nickel Alloys, Wrought and Cast | 8-79 | – | – | 8-119 | – | – | 8-161 | – | 8-190 |
| 38. Nitinol Alloys, Wrought | 8-79 | – | – | 8-119 | – | – | 8-161 | – | – |
| 39. High Temperature Alloys, Wrought and Cast | 8-80 | – | – | 8-120 | – | – | 8-162 | 8-178 | 8-191 |
| 40. Refractory Alloys, Wrought, Cast, P/M | 8-82 | – | – | 8-123 | – | – | 8-164 | – | – |
| 41. Zinc Alloys, Cast | – | – | – | – | – | – | – | 8-178 | – |
| 42. Lead Alloys, Cast | – | – | – | – | – | – | – | – | – |
| 43. Tin Alloys, Cast | – | – | – | – | – | – | – | – | – |
| 44. Uranium, Wrought | 8-84 | – | – | – | – | – | 8-166 | – | – |
| 45. Zirconium Alloys, Wrought | 8-84 | – | – | 8-124 | – | – | 8-166 | – | – |
| 46. Manganese, Wrought | – | – | – | – | – | – | – | – | – |
| 47. Powder Metal Alloys | 8-84 | – | – | 8-124 | – | – | 8-166 | – | 8-193 |
| 48. Machinable Carbides | 8-86 | – | 8-97 | 8-127 | – | 8-139 | – | – | 8-194 |
| 49. Carbides | 8-87 | – | 8-97 | 8-127 | – | 8-139 | 8-168 | – | 8-194 |
| 50. Free Machining Magnetic Alloys | 8-87 | – | – | 8-127 | – | – | 8-169 | – | – |
| 51. Magnetic Alloys | 8-87 | – | 8-97 | 8-127 | – | 8-139 | 8-169 | – | 8-194 |
| 52. Free Machining Controlled Expansion Alloys | 8-88 | – | – | 8-128 | – | – | 8-169 | – | – |
| 53. Controlled Expansion Alloys | 8-88 | – | – | 8-128 | – | – | 8-170 | – | – |
| 54. Carbons and Graphites | 8-88 | – | 8-97 | 8-128 | – | – | 8-170 | – | – |
| 55. Glasses and Ceramics | 8-89 | – | 8-98 | 8-129 | – | 8-139 | 8-170 | – | – |
| 56. Plastics | 8-89 | – | – | 8-129 | – | – | 8-171 | – | – |
| 57. Composites | – | – | 8-98 | – | – | – | – | – | – |
| 58. Flame (Thermal) Sprayed Materials | 8-90 | – | 8-99 | 8-130 | – | 8-141 | 8-172 | – | – |
| 59. Plated Materials | 8-92 | – | – | 8-132 | – | – | 8-173 | – | – |
| 60. Precious Metals | 8-92 | – | – | – | – | – | – | – | – |
| 61. Rubber | 8-92 | – | – | 8-132 | – | – | 8-173 | – | – |

# MACHINING DATA HANDBOOK

# MACHINING DATA HANDBOOK

**3rd Edition**

## VOLUME TWO

Compiled by the
Technical Staff
of the
Machinability Data Center

**MACHINABILITY DATA CENTER**
Metcut Research Associates Inc.
3980 Rosslyn Drive
Cincinnati, Ohio 45209

1980

## MACHINABILITY DATA CENTER

MDC is a national information analysis center operated by Metcut Research Associates Inc. Metcut is an independent applied research, development and testing organization offering expertise and laboratory facilities in the fields of material removal, advanced manufacturing technology and materials engineering.

### Information Services Provided by MDC

MDC serves both government and industry by providing material removal data and information. MDC's services include providing analyzed data in response to technical inquiries, presenting seminars, and compiling and marketing data publications on subjects of current interest to the manufacturing industry. MDC also maintains a selected mailing list for providing notification of the availability of new information and services from the Center.

### Scope

The Machinability Data Center (MDC) collects, evaluates, stores, and disseminates material removal information, including specific and detailed machining data. Emphasis is given to engineering evaluation for the purpose of developing material removal parameters, such as speeds, feeds, depths of cut, tool material and geometry, cutting fluids and other significant variables. Data are being processed for all types of materials and for all kinds of material removal operations, such as turning, milling, drilling, tapping, grinding, electrical discharge machining, electrochemical machining, etc.

MDC has a data file of over thirty-two thousand selected documents pertaining to material removal technology. This data file is supported by a computer system for retrieval of information based upon the specific material (with definite chemical, physical, and mechanical properties) and the specific material removal operation being used. Computerized search techniques are employed utilizing a combination of search parameters to produce source data. Information retrieval can be refined to the extent necessary to satisfy the requirements of a specific inquiry by controlling the input search parameters.

**For information:** Contact the Machinability Data Center, 11240 Cornell Park Drive, Cincinnati, OH 45242. Telephone: (513) 489-6688. Telex II: 510-600-7802. FAX: (513) 489-3653.

ISBN 0-936974-00-1 (Set)
ISBN 0-936974-02-8 (Volume 2)
Library of Congress Catalog Card Number: 80-81480

Metcut Research Associates Inc.
3980 Rosslyn Drive
Cincinnati, OH 45209


First Edition published 1966. Second Edition 1972. Third Edition 1980. Seventh Printing 1989



Printed in the United States of America

Typeset in Helvetica and Bookman by Dayton Typographic Service Inc., Dayton, Ohio
Graphics by Repro Art Service Inc., Cincinnati, Ohio
Printed by The Otto Zimmerman & Son Company, Inc., Newport, Kentucky
Bound by C. J. Krehbiel Co., Cincinnati, Ohio

# CONTENTS
## Volume 2


**Section 7. Gear Cutting and Gear Grinding Operations** .......... **7–1**
- 7.1 Gear Hobbing .......... 7–3
- 7.2 Gear Cutting, Straight and Spiral Bevel .......... 7–31
- 7.3 Gear Shaping .......... 7–47
- 7.4 Gear Shaving .......... 7–69
- 7.5 Gear Grinding, Form .......... 7–89

**Section 8. Grinding Operations** .......... **8–1**
- 8.1 Surface Grinding—Horizontal Spindle, Reciprocating Table .......... 8–3
- 8.2 Surface Grinding—Horizontal Spindle, Reciprocating Table—Cubic Boron Nitride Wheels .......... 8–29
- 8.3 Surface Grinding—Horizontal Spindle, Reciprocating Table—Diamond Wheels .......... 8–39
- 8.4 Surface Grinding—Vertical Spindle, Rotary Table .......... 8–43
- 8.5 Cylindrical Grinding .......... 8–61
- 8.6 Cylindrical Grinding—Cubic Boron Nitride Wheels .......... 8–93
- 8.7 Cylindrical Grinding—Diamond Wheels .......... 8–97
- 8.8 Internal Grinding .......... 8–101
- 8.9 Internal Grinding—Cubic Boron Nitride Wheels .......... 8–133
- 8.10 Internal Grinding—Diamond Wheels .......... 8–139
- 8.11 Centerless Grinding .......... 8–143
- 8.12 Centerless Grinding—Work Traverse Rates .......... 8–175
- 8.13 Abrasive Belt Grinding .......... 8–177
- 8.14 Thread Grinding .......... 8–179

**Section 9. Introduction to Nontraditional Machining** .......... **9–1**

**Section 10. Mechanical Nontraditional Machining Operations** .......... **10–1**
- 10.1 Abrasive Flow Machining—AFM .......... 10–3
- 10.2 Abrasive Jet Machining—AJM .......... 10–15
- 10.3 Hydrodynamic Machining—HDM .......... 10–21
- 10.4 Low Stress Grinding—LSG .......... 10–37
- 10.5 Thermally Assisted Machining—TAM .......... 10–39
- 10.6 Total Form Machining—TFM .......... 10–41
- 10.7 Ultrasonic Machining—USM .......... 10–43
- 10.8 Water Jet Machining—WJM .......... 10–65

**Section 11. Electrical Nontraditional Machining Operations** .......... **11–1**
- 11.1 Electrochemical Deburring—ECD .......... 11–3
- 11.2 Electrochemical Discharge Grinding—ECDG .......... 11–5
- 11.3 Electrochemical Grinding—ECG .......... 11–9
- 11.4 Electrochemical Honing—ECH .......... 11–23
- 11.5 Electrochemical Machining—ECM .......... 11–25
- 11.6 Electrochemical Polishing—ECP .......... 11–63
- 11.7 Electrochemical Sharpening—ECS .......... 11–65
- 11.8 Electrochemical Turning—ECT .......... 11–67
- 11.9 Electro-stream™—ES .......... 11–69
- 11.10 Shaped Tube Electrolytic Machining—STEM™ .......... 11–71

# CONTENTS
## Volume 2


**Section 12. Thermal Nontraditional Machining Operations** .... **12–1**

12.1 Electron Beam Machining—EBM .... 12–3
12.2 Electrical Discharge Grinding—EDG .... 12–11
12.3 Electrical Discharge Machining—EDM .... 12–15
12.4 Electrical Discharge Sawing—EDS .... 12–47
12.5 Electrical Discharge Wire Cutting—EDWC .... 12–49
12.6 Laser Beam Machining—LBM .... 12–55
12.7 Laser Beam Torch—LBT .... 12–71
12.8 Plasma Beam Machining—PBM .... 12–97

**Section 13. Chemical Nontraditional Machining Operations** .... **13–1**

13.1 Chemical Machining—CHM .... 13–3
13.2 Electropolishing—ELP .... 13–17
13.3 Photochemical Machining—PCM .... 13–19
13.4 Thermochemical Machining—TCM .... 13–25

**Section 14. Tool Materials** .... **14–1**

14.1 General Guidelines for Selection of Tool Materials .... 14–3
14.2 High Speed Steels .... 14–4
14.3 Cast Alloys .... 14–6
14.4 Carbides .... 14–7
14.5 Micrograin Carbides .... 14–15
14.6 Coated Carbides .... 14–16
14.7 Ceramic Tool Materials .... 14–17
14.8 Diamond Tools .... 14–18

**Section 15. Tool Geometry** .... **15–1**

15.1 Turning and Boring Tools, Single Point .... 15–2
15.2 Threading Tools, Single Point .... 15–4
15.3 Die Threading Tools, Thread Chasers .... 15–6
15.4 Face Mills .... 15–8
15.5 Side and Slot Mills-Arbor Mounted .... 15–10
15.6 End Mills—Peripheral and Slotting, High Speed Steel .... 15–12
15.7 Drills, High Speed Steel Twist .... 15–14
15.8 Oil-Hole or Pressurized-Coolant Drills, High Speed Steel .... 15–16
15.9 Spade Drills, High Speed Steel .... 15–16
15.10 Gun Drills, Carbide .... 15–18
15.11 Reamers, High Speed Steel .... 15–19
15.12 Reamers, Carbide .... 15–20
15.13 Boring Tools, Carbide .... 15–21
15.14 Taps, High Speed Steel .... 15–23
15.15 Planing Tools .... 15–25
15.16 Broaches, High Speed Steel .... 15–27
15.17 Tool and Cutter Angles—Approximate Equivalents .... 15–28

**Section 16. Cutting Fluids*** .... **16–1**

16.1 Cutting Fluid Selection and Use .... 16–3
16.2 Cutting Fluid Recommendations .... 16–17
16.3 Cutting Fluid Key .... 16–65


*For the detailed contents of this section, see the title page of this section.

# CONTENTS
## Volume 2


**Section 17. Power and Force Requirements** .......... **17–1**
- 17.1 Determining Forces in Machining .......... 17–3
- 17.2 Determining Power Requirements in Machining .......... 17–7
- 17.3 Estimating Forces in Turning .......... 17–23
- 17.4 Estimating Torque and Thrust in Drilling .......... 17–27

**Section 18. Surface Technology*** .......... **18–1**
- 18.1 Introduction to Surface Technology .......... 18–3
- 18.2 Surface Texture .......... 18–5
- 18.3 Surface Integrity .......... 18–39

**Section 19. Machining Guidelines*** .......... **19–1**
- 19.1 General Machining Guidelines .......... 19–3
- 19.2 Guidelines for Drilling .......... 19–5
- 19.3 Tool Life .......... 19–11
- 19.4 Types of Machinability Data .......... 19–15

**Section 20. Grinding and Abrasive Machining*** .......... **20–1**
- 20.1 Grinding Wheels .......... 20–3
- 20.2 Grinding Guidelines .......... 20–13
- 20.3 Surface Roughness and Tolerances .......... 20–17
- 20.4 Grinding Formulas and Charts .......... 20–21
- 20.5 Abrasive Machining .......... 20–33

**Section 21. Economics in Machining and Grinding*** .......... **21–1**

**Section 22. Machine Chatter and Vibration*** .......... **22–1**

**Section 23. Numerical Control Machining*** .......... **23–1**
- 23.1 NC Machining Guidelines .......... 23–3
- 23.2 NC Vocabulary .......... 23–9

**Section 24. Computer-Aided Manufacturing Technologies*** .......... **24–1**

**Section 25. Machining Standards** .......... **25–1**
- 25.1 Machining Standards by Subject .......... 25–3
- 25.2 Alphabetical List of Machining Standards .......... 25–9
- 25.3 Sources of Machining Standards .......... 25–23

**Section 26. Materials Index** .......... **26–1**
- 26.1 Numerical List of Materials .......... 26–3
- 26.2 Alphabetical List of Materials .......... 26–17
- 26.3 Chemical Composition by Material Group .......... 26–37

**Section 27. Glossary** .......... **27–1**

**Section 28. Subject Index** .......... **28–1**


*For the detailed contents of this section, see the title page of this section.

# MACHINING DATA HANDBOOK

# SECTION 7

# GEAR CUTTING AND GEAR GRINDING OPERATIONS


7.1 Gear Hobbing ........ 7–3
7.2 Gear Cutting, Straight and Spiral Bevel ........ 7–31
7.3 Gear Shaping ........ 7–47
7.4 Gear Shaving ........ 7–69
7.5 Gear Grinding, Form ........ 7–89

| MATERIAL | HARD-NESS Bhn | CONDITION | DIAMETRAL PITCH / MODULE | NUMBER OF CUTS | FEED* PER REVOLUTION OF WORKPIECE in / mm | HOB SPEED fpm / m/min | HSS TOOL MATERIAL AISI / ISO |
|---|---|---|---|---|---|---|---|
| **1. FREE MACHINING CARBON STEELS, WROUGHT** **Low Carbon Resulfurized** 1116 1117 1118 1119 1211 1212 | 100 to 150 | Hot Rolled or Annealed | 1—2 | 2 | .060 | 220 | M2, M7 |
| | | | 3—10 | 1 | .060 | 230 | |
| | | | 11—19 | 1 | .060 | 240 | |
| | | | 20—48 | 1 | .050 | 250 | |
| | | | 48 & Finer | 1 | .030 | 280 | |
| | | | 25—13 | 2 | 1.50 | 67 | S4, S2 |
| | | | 12—2.5 | 1 | 1.50 | 70 | |
| | | | 2—1.5 | 1 | 1.50 | 73 | |
| | | | 1—.5 | 1 | 1.25 | 76 | |
| | | | .5 & Finer | 1 | .75 | 85 | |
| | 150 to 200 | Cold Drawn | 1—2 | 2 | .060 | 230 | M2, M7 |
| | | | 3—10 | 1 | .060 | 240 | |
| | | | 11—19 | 1 | .060 | 265 | |
| | | | 20—48 | 1 | .050 | 275 | |
| | | | 48 & Finer | 1 | .030 | 300 | |
| | | | 25—13 | 2 | 1.50 | 70 | S4, S2 |
| | | | 12—2.5 | 1 | 1.50 | 73 | |
| | | | 2—1.5 | 1 | 1.50 | 81 | |
| | | | 1—.5 | 1 | 1.25 | 84 | |
| | | | .5 & Finer | 1 | .75 | 90 | |
| **Low Carbon Resulfurized** 1213 1215 | 100 to 150 | Hot Rolled or Annealed | 1—2 | 2 | .060 | 240 | M2, M7 |
| | | | 3—10 | 1 | .060 | 250 | |
| | | | 11—19 | 1 | .060 | 265 | |
| | | | 20—48 | 1 | .050 | 275 | |
| | | | 48 & Finer | 1 | .030 | 300 | |
| | | | 25—13 | 2 | 1.50 | 73 | S4, S2 |
| | | | 12—2.5 | 1 | 1.50 | 76 | |
| | | | 2—1.5 | 1 | 1.50 | 81 | |
| | | | 1—.5 | 1 | 1.25 | 84 | |
| | | | .5 & Finer | 1 | .75 | 90 | |
| | 150 to 200 | Cold Drawn | 1—2 | 2 | .060 | 250 | M2, M7 |
| | | | 3—10 | 1 | .060 | 270 | |
| | | | 11—19 | 1 | .060 | 285 | |
| | | | 20—48 | 1 | .050 | 295 | |
| | | | 48 & Finer | 1 | .030 | 330 | |
| | | | 25—13 | 2 | 1.50 | 76 | S4, S2 |
| | | | 12—2.5 | 1 | 1.50 | 82 | |
| | | | 2—1.5 | 1 | 1.50 | 87 | |
| | | | 1—.5 | 1 | 1.25 | 90 | |
| | | | .5 & Finer | 1 | .75 | 100 | |
| **Low Carbon Resulfurized** 1108 1109 1110 1115 | 100 to 150 | Hot Rolled or Annealed | 1—2 | 2 | .060 | 190 | M2, M7 |
| | | | 3—10 | 1 | .060 | 205 | |
| | | | 11—19 | 1 | .060 | 220 | |
| | | | 20—48 | 1 | .050 | 240 | |
| | | | 48 & Finer | 1 | .030 | 260 | |
| | | | 25—13 | 2 | 1.50 | 58 | S4, S2 |
| | | | 12—2.5 | 1 | 1.50 | 62 | |
| | | | 2—1.5 | 1 | 1.50 | 67 | |
| | | | 1—.5 | 1 | 1.25 | 73 | |
| | | | .5 & Finer | 1 | .75 | 79 | |
| | 150 to 200 | Cold Drawn | 1—2 | 2 | .060 | 200 | M2, M7 |
| | | | 3—10 | 1 | .060 | 215 | |
| | | | 11—19 | 1 | .060 | 230 | |
| | | | 20—48 | 1 | .050 | 250 | |
| | | | 48 & Finer | 1 | .030 | 270 | |
| | | | 25—13 | 2 | 1.50 | 60 | S4, S2 |
| | | | 12—2.5 | 1 | 1.50 | 66 | |
| | | | 2—1.5 | 1 | 1.50 | 70 | |
| | | | 1—.5 | 1 | 1.25 | 76 | |
| | | | .5 & Finer | 1 | .75 | 82 | |

For hobbing Class 9 (per AGMA 390.03) or better gears, it may be necessary to reduce speeds and feeds by 50% and/or take 2 cuts.
To meet finish requirements it may be necessary to try both conventional and climb cutting.

See section 16 for Cutting Fluid Recommendations.
*Feeds are based on the largest standard recommended hob diameter. When using a smaller hob diameter, the Feed must be reduced proportionally.

# 7.1 Gear Hobbing

| MATERIAL | HARD-NESS Bhn | CONDITION | DIAMETRAL PITCH / MODULE | NUMBER OF CUTS | FEED* PER REVOLUTION OF WORKPIECE in / mm | HOB SPEED fpm / m/min | HSS TOOL MATERIAL AISI / ISO |
|---|---|---|---|---|---|---|---|
| **1. FREE MACHINING CARBON STEELS, WROUGHT (cont.)** **Medium Carbon Resulfurized** 1132, 1137, 1139, 1140, 1141, 1144, 1145, 1146, 1151 | 175 to 225 | Hot Rolled, Normalized, Annealed or Cold Drawn | 1—2 | 2 | .060 | 170 | |
| | | | 3—10 | 1 | .060 | 180 | |
| | | | 11—19 | 1 | .060 | 190 | M2, M7 |
| | | | 20—48 | 1 | .050 | 200 | |
| | | | 48 & Finer | 1 | .030 | 220 | |
| | | | 25—13 | 2 | 1.50 | 52 | |
| | | | 12—2.5 | 1 | 1.50 | 55 | |
| | | | 2—1.5 | 1 | 1.50 | 58 | S4, S2 |
| | | | 1—.5 | 1 | 1.25 | 60 | |
| | | | .5 & Finer | 1 | .75 | 67 | |
| | 275 to 325 | Quenched and Tempered | 1—2 | 2 | .060 | 120 | |
| | | | 3—10 | 1 | .060 | 140 | |
| | | | 11—19 | 1 | .060 | 145 | M2, M7 |
| | | | 20—48 | 1 | .050 | 150 | |
| | | | 48 & Finer | 1 | .030 | 165 | |
| | | | 25—13 | 2 | 1.50 | 37 | |
| | | | 12—2.5 | 1 | 1.50 | 43 | |
| | | | 2—1.5 | 1 | 1.50 | 44 | S4, S2 |
| | | | 1—.5 | 1 | 1.25 | 46 | |
| | | | .5 & Finer | 1 | .75 | 50 | |
| | 325 to 375 | Quenched and Tempered | 1—2 | 2 | .060 | 70 | |
| | | | 3—10 | 1 | .060 | 90 | |
| | | | 11—19 | 1 | .060 | 110 | M3, M42 |
| | | | 20—48 | 1 | .050 | 120 | |
| | | | 48 & Finer | 1 | .030 | 130 | |
| | | | 25—13 | 2 | 1.50 | 21 | |
| | | | 12—2.5 | 1 | 1.50 | 27 | |
| | | | 2—1.5 | 1 | 1.50 | 34 | S5, S11 |
| | | | 1—.5 | 1 | 1.25 | 37 | |
| | | | .5 & Finer | 1 | .75 | 40 | |
| **Low Carbon Leaded** 10L18, 11L17 | 100 to 150 | Hot Rolled, Normalized, Annealed or Cold Drawn | 1—2 | 2 | .065 | 230 | |
| | | | 3—10 | 1 | .065 | 250 | |
| | | | 11—19 | 1 | .060 | 270 | M2, M7 |
| | | | 20—48 | 1 | .050 | 290 | |
| | | | 48 & Finer | 1 | .030 | 300 | |
| | | | 25—13 | 2 | 1.65 | 70 | |
| | | | 12—2.5 | 1 | 1.65 | 76 | |
| | | | 2—1.5 | 1 | 1.50 | 82 | S4, S2 |
| | | | 1—.5 | 1 | 1.25 | 88 | |
| | | | .5 & Finer | 1 | .75 | 90 | |
| | 150 to 200 | Hot Rolled, Normalized, Annealed or Cold Drawn | 1—2 | 2 | .065 | 190 | |
| | | | 3—10 | 1 | .065 | 220 | |
| | | | 11—19 | 1 | .060 | 250 | M2, M7 |
| | | | 20—48 | 1 | .050 | 270 | |
| | | | 48 & Finer | 1 | .030 | 280 | |
| | | | 25—13 | 2 | 1.65 | 58 | |
| | | | 12—2.5 | 1 | 1.65 | 67 | |
| | | | 2—1.5 | 1 | 1.50 | 76 | S4, S2 |
| | | | 1—.5 | 1 | 1.25 | 82 | |
| | | | .5 & Finer | 1 | .75 | 85 | |
| | 200 to 250 | Hot Rolled, Normalized, Annealed or Cold Drawn | 1—2 | 2 | .060 | 140 | |
| | | | 3—10 | 1 | .050 | 180 | |
| | | | 11—19 | 1 | .050 | 200 | M2, M7 |
| | | | 20—48 | 1 | .050 | 225 | |
| | | | 48 & Finer | 1 | .030 | 250 | |
| | | | 25—13 | 2 | 1.50 | 43 | |
| | | | 12—2.5 | 1 | 1.25 | 55 | |
| | | | 2—1.5 | 1 | 1.25 | 60 | S4, S2 |
| | | | 1—.5 | 1 | 1.25 | 69 | |
| | | | .5 & Finer | 1 | .75 | 76 | |

For hobbing Class 9 (per AGMA 390.03) or better gears, it may be necessary to reduce speeds and feeds by 50% and/or take 2 cuts.
To meet finish requirements it may be necessary to try both conventional and climb cutting.

See section 16 for Cutting Fluid Recommendations.
*Feeds are based on the largest standard recommended hob diameter. When using a smaller hob diameter, the Feed must be reduced proportionally.

| MATERIAL | HARDNESS Bhn | CONDITION | DIAMETRAL PITCH / MODULE | NUMBER OF CUTS | FEED* PER REVOLUTION OF WORKPIECE in / mm | HOB SPEED fpm / m/min | HSS TOOL MATERIAL AISI / ISO |
|---|---|---|---|---|---|---|---|
| **Low Carbon Leaded** 12L13 12L14 12L15 | 100 to 150 | Hot Rolled, Normalized, Annealed or Cold Drawn | 1—2 | 2 | .065 | 255 | M2, M7 |
| | | | 3—10 | 1 | .065 | 265 | |
| | | | 11—19 | 1 | .060 | 295 | |
| | | | 20—48 | 1 | .050 | 305 | |
| | | | 48 & Finer | 1 | .030 | 340 | |
| | | | 25—13 | 2 | 1.65 | 78 | S4, S2 |
| | | | 12—2.5 | 1 | 1.65 | 81 | |
| | | | 2—1.5 | 1 | 1.50 | 90 | |
| | | | 1—.5 | 1 | 1.25 | 95 | |
| | | | .5 & Finer | 1 | .75 | 105 | |
| | 150 to 200 | Hot Rolled, Normalized, Annealed or Cold Drawn | 1—2 | 2 | .065 | 210 | M2, M7 |
| | | | 3—10 | 1 | .065 | 240 | |
| | | | 11—19 | 1 | .060 | 275 | |
| | | | 20—48 | 1 | .050 | 285 | |
| | | | 48 & Finer | 1 | .030 | 320 | |
| | | | 25—13 | 2 | 1.65 | 64 | S4, S2 |
| | | | 12—2.5 | 1 | 1.65 | 73 | |
| | | | 2—1.5 | 1 | 1.50 | 84 | |
| | | | 1—.5 | 1 | 1.25 | 87 | |
| | | | .5 & Finer | 1 | .75 | 100 | |
| | 200 to 250 | Hot Rolled, Normalized, Annealed or Cold Drawn | 1—2 | 2 | .060 | 165 | M2, M7 |
| | | | 3—10 | 1 | .060 | 190 | |
| | | | 11—19 | 1 | .060 | 225 | |
| | | | 20—48 | 1 | .050 | 240 | |
| | | | 48 & Finer | 1 | .030 | 275 | |
| | | | 25—13 | 2 | 1.50 | 50 | S4, S2 |
| | | | 12—2.5 | 1 | 1.50 | 58 | |
| | | | 2—1.5 | 1 | 1.50 | 69 | |
| | | | 1—.5 | 1 | 1.25 | 73 | |
| | | | .5 & Finer | 1 | .75 | 84 | |
| **Medium Carbon Leaded** 10L45 10L50 11L37 11L41 11L44 | 125 to 175 | Hot Rolled, Normalized, Annealed or Cold Drawn | 1—2 | 2 | .060 | 200 | M2, M7 |
| | | | 3—10 | 1 | .060 | 210 | |
| | | | 11—19 | 1 | .060 | 225 | |
| | | | 20—48 | 1 | .050 | 240 | |
| | | | 48 & Finer | 1 | .030 | 260 | |
| | | | 25—13 | 2 | 1.50 | 60 | S4, S2 |
| | | | 12—2.5 | 1 | 1.50 | 64 | |
| | | | 2—1.5 | 1 | 1.50 | 69 | |
| | | | 1—.5 | 1 | 1.25 | 73 | |
| | | | .5 & Finer | 1 | .75 | 79 | |
| | 175 to 225 | Hot Rolled, Normalized, Annealed or Cold Drawn | 1—2 | 2 | .060 | 190 | M2, M7 |
| | | | 3—10 | 1 | .060 | 200 | |
| | | | 11—19 | 1 | .060 | 210 | |
| | | | 20—48 | 1 | .050 | 225 | |
| | | | 48 & Finer | 1 | .030 | 240 | |
| | | | 25—13 | 2 | 1.50 | 58 | S4, S2 |
| | | | 12—2.5 | 1 | 1.50 | 60 | |
| | | | 2—1.5 | 1 | 1.50 | 64 | |
| | | | 1—.5 | 1 | 1.25 | 69 | |
| | | | .5 & Finer | 1 | .75 | 73 | |
| | 225 to 275 | Hot Rolled, Normalized, Annealed, Cold Drawn or Quenched and Tempered | 1—2 | 2 | .060 | 180 | M2, M7 |
| | | | 3—10 | 1 | .060 | 190 | |
| | | | 11—19 | 1 | .060 | 200 | |
| | | | 20—48 | 1 | .050 | 205 | |
| | | | 48 & Finer | 1 | .030 | 220 | |
| | | | 25—13 | 2 | 1.50 | 55 | S4, S2 |
| | | | 12—2.5 | 1 | 1.50 | 58 | |
| | | | 2—1.5 | 1 | 1.50 | 60 | |
| | | | 1—.5 | 1 | 1.25 | 62 | |
| | | | .5 & Finer | 1 | .75 | 67 | |

For hobbing Class 9 (per AGMA 390.03) or better gears, it may be necessary to reduce speeds and feeds by 50% and/or take 2 cuts.
To meet finish requirements it may be necessary to try both conventional and climb cutting.

See section 16 for Cutting Fluid Recommendations.
*Feeds are based on the largest standard recommended hob diameter. When using a smaller hob diameter, the Feed must be reduced proportionally.

# 7.1 Gear Hobbing

| MATERIAL | HARD-NESS<br>Bhn | CONDITION | DIAMETRAL PITCH<br>MODULE | NUMBER OF CUTS | FEED* PER REVOLUTION OF WORKPIECE<br>in<br>mm | HOB SPEED<br>fpm<br>m/min | HSS TOOL MATERIAL<br>AISI<br>ISO |
|---|---|---|---|---|---|---|---|
| **1. FREE MACHINING CARBON STEELS, WROUGHT (cont.)**<br>**Medium Carbon Leaded (cont.)**<br>(materials listed on preceding page) | 275 to 325 | Hot Rolled, Normalized, Annealed, Cold Drawn or Quenched and Tempered | 1—2 | 2 | .060 | 130 | M2, M7 |
| | | | 3—10 | 1 | .060 | 140 | |
| | | | 11—19 | 1 | .060 | 150 | |
| | | | 20—48 | 1 | .045 | 160 | |
| | | | 48 & Finer | 1 | .030 | 175 | |
| | | | 25—13 | 2 | 1.50 | 40 | S4, S2 |
| | | | 12—2.5 | 1 | 1.50 | 43 | |
| | | | 2—1.5 | 1 | 1.50 | 46 | |
| | | | 1—.5 | 1 | 1.15 | 49 | |
| | | | .5 & Finer | 1 | .75 | 53 | |
| | 325 to 375 | Quenched and Tempered | 1—2 | 2 | .060 | 80 | M3, M42 |
| | | | 3—10 | 1 | .045 | 90 | |
| | | | 11—19 | 1 | .045 | 105 | |
| | | | 20—48 | 1 | .025 | 120 | |
| | | | 48 & Finer | 1 | .025 | 140 | |
| | | | 25—13 | 2 | 1.50 | 24 | S5, S11 |
| | | | 12—2.5 | 1 | 1.15 | 27 | |
| | | | 2—1.5 | 1 | 1.15 | 32 | |
| | | | 1—.5 | 1 | .65 | 37 | |
| | | | .5 & Finer | 1 | .65 | 43 | |
| **2. CARBON STEELS, WROUGHT**<br>**Low Carbon**<br>1005 1019<br>1006 1020<br>1008 1021<br>1009 1022<br>1010 1023<br>1011 1025<br>1012 1026<br>1013 1029<br>1015 1513<br>1016 1518<br>1017 1522<br>1018 | 85 to 125 | Hot Rolled, Normalized, Annealed or Cold Drawn | 1—2 | 2 | .070 | 205 | M2, M7 |
| | | | 3—10 | 1 | .070 | 245 | |
| | | | 11—19 | 1 | .060 | 265 | |
| | | | 20—48 | 1 | .050 | 280 | |
| | | | 48 & Finer | 1 | .030 | 290 | |
| | | | 25—13 | 2 | 1.80 | 62 | S4, S2 |
| | | | 12—2.5 | 1 | 1.80 | 75 | |
| | | | 2—1.5 | 1 | 1.50 | 81 | |
| | | | 1—.5 | 1 | 1.25 | 85 | |
| | | | .5 & Finer | 1 | .75 | 88 | |
| | 125 to 175 | Hot Rolled, Normalized, Annealed or Cold Drawn | 1—2 | 2 | .070 | 175 | M2, M7 |
| | | | 3—10 | 1 | .070 | 190 | |
| | | | 11—19 | 1 | .060 | 200 | |
| | | | 20—48 | 1 | .050 | 210 | |
| | | | 48 & Finer | 1 | .030 | 220 | |
| | | | 25—13 | 2 | 1.80 | 53 | S4, S2 |
| | | | 12—2.5 | 1 | 1.80 | 58 | |
| | | | 2—1.5 | 1 | 1.50 | 60 | |
| | | | 1—.5 | 1 | 1.25 | 64 | |
| | | | .5 & Finer | 1 | .75 | 67 | |
| | 175 to 225 | Hot Rolled, Normalized, Annealed or Cold Drawn | 1—2 | 2 | .070 | 145 | M2, M7 |
| | | | 3—10 | 1 | .070 | 170 | |
| | | | 11—19 | 1 | .060 | 190 | |
| | | | 20—48 | 1 | .050 | 200 | |
| | | | 48 & Finer | 1 | .030 | 210 | |
| | | | 25—13 | 2 | 1.80 | 44 | S4, S2 |
| | | | 12—2.5 | 1 | 1.80 | 52 | |
| | | | 2—1.5 | 1 | 1.50 | 58 | |
| | | | 1—.5 | 1 | 1.25 | 60 | |
| | | | .5 & Finer | 1 | .75 | 64 | |
| | 225 to 275 | Annealed or Cold Drawn | 1—2 | 2 | .060 | 120 | M2, M7 |
| | | | 3—10 | 1 | .060 | 160 | |
| | | | 11—19 | 1 | .060 | 185 | |
| | | | 20—48 | 1 | .050 | 195 | |
| | | | 48 & Finer | 1 | .030 | 200 | |
| | | | 25—13 | 2 | 1.50 | 37 | S4, S2 |
| | | | 12—2.5 | 1 | 1.50 | 49 | |
| | | | 2—1.5 | 1 | 1.50 | 56 | |
| | | | 1—.5 | 1 | 1.25 | 59 | |
| | | | .5 & Finer | 1 | .75 | 60 | |

For hobbing Class 9 (per AGMA 390.03) or better gears, it may be necessary to reduce speeds and feeds by 50% and/or take 2 cuts. To meet finish requirements it may be necessary to try both conventional and climb cutting.

See section 16 for Cutting Fluid Recommendations.

*Feeds are based on the largest standard recommended hob diameter. When using a smaller hob diameter, the Feed must be reduced proportionally.

| MATERIAL | HARD-NESS Bhn | CONDITION | DIAMETRAL PITCH / MODULE | NUMBER OF CUTS | FEED* PER REVOLUTION OF WORKPIECE in / mm | HOB SPEED fpm / m/min | HSS TOOL MATERIAL AISI / ISO |
|---|---|---|---|---|---|---|---|
| **Medium Carbon**<br>1030 1044 1526<br>1033 1045 1527<br>1035 1046 1536<br>1037 1049 1541<br>1038 1050 1547<br>1039 1053 1548<br>1040 1055 1551<br>1042 1524 1552<br>1043 1525 | 125 to 175 | Hot Rolled, Normalized, Annealed or Cold Drawn | 1—2 | 2 | .070 | 165 | M2, M7 |
| | | | 3—10 | 1 | .070 | 175 | |
| | | | 11—19 | 1 | .060 | 185 | |
| | | | 20—48 | 1 | .050 | 200 | |
| | | | 48 & Finer | 1 | .030 | 210 | |
| | | | 25—13 | 2 | 1.80 | 50 | S4, S2 |
| | | | 12—2.5 | 1 | 1.80 | 53 | |
| | | | 2—1.5 | 1 | 1.50 | 56 | |
| | | | 1—.5 | 1 | 1.25 | 60 | |
| | | | .5 & Finer | 1 | .75 | 64 | |
| | 175 to 225 | Hot Rolled, Normalized, Annealed or Cold Drawn | 1—2 | 2 | .070 | 135 | M2, M7 |
| | | | 3—10 | 1 | .070 | 145 | |
| | | | 11—19 | 1 | .060 | 165 | |
| | | | 20—48 | 1 | .050 | 185 | |
| | | | 48 & Finer | 1 | .030 | 200 | |
| | | | 25—13 | 2 | 1.80 | 41 | S4, S2 |
| | | | 12—2.5 | 1 | 1.80 | 44 | |
| | | | 2—1.5 | 1 | 1.50 | 50 | |
| | | | 1—.5 | 1 | 1.25 | 56 | |
| | | | .5 & Finer | 1 | .75 | 60 | |
| | 225 to 275 | Hot Rolled, Normalized, Annealed, Cold Drawn or Quenched and Tempered | 1—2 | 2 | .060 | 110 | M2, M7 |
| | | | 3—10 | 1 | .050 | 125 | |
| | | | 11—19 | 1 | .050 | 150 | |
| | | | 20—48 | 1 | .050 | 165 | |
| | | | 48 & Finer | 1 | .030 | 175 | |
| | | | 25—13 | 2 | 1.50 | 34 | S4, S2 |
| | | | 12—2.5 | 1 | 1.25 | 38 | |
| | | | 2—1.5 | 1 | 1.25 | 46 | |
| | | | 1—.5 | 1 | 1.25 | 50 | |
| | | | .5 & Finer | 1 | .75 | 53 | |
| | 275 to 325 | Hot Rolled, Normalized, Annealed or Quenched and Tempered | 1—2 | 2 | .045 | 95 | M2, M7 |
| | | | 3—10 | 1 | .045 | 110 | |
| | | | 11—19 | 1 | .045 | 120 | |
| | | | 20—48 | 1 | .045 | 130 | |
| | | | 48 & Finer | 1 | .030 | 140 | |
| | | | 25—13 | 2 | 1.15 | 29 | S4, S2 |
| | | | 12—2.5 | 1 | 1.15 | 34 | |
| | | | 2—1.5 | 1 | 1.15 | 37 | |
| | | | 1—.5 | 1 | 1.15 | 40 | |
| | | | .5 & Finer | 1 | .75 | 43 | |
| | 325 to 375 | Quenched and Tempered | 1—2 | 2 | .045 | 70 | M3, M42 |
| | | | 3—10 | 1 | .045 | 80 | |
| | | | 11—19 | 1 | .025 | 100 | |
| | | | 20—48 | 1 | .025 | 105 | |
| | | | 48 & Finer | 1 | .025 | 120 | |
| | | | 25—13 | 2 | 1.15 | 21 | S5, S11 |
| | | | 12—2.5 | 1 | 1.15 | 24 | |
| | | | 2—1.5 | 1 | .65 | 30 | |
| | | | 1—.5 | 1 | .65 | 32 | |
| | | | .5 & Finer | 1 | .65 | 37 | |
| **High Carbon**<br>1060 1075 1090<br>1064 1078 1095<br>1065 1080 1561<br>1069 1084 1566<br>1070 1085 1572<br>1074 1086 | 175 to 225 | Hot Rolled, Normalized, Annealed or Cold Drawn | 1—2 | 2 | .070 | 125 | M2, M7 |
| | | | 3—10 | 1 | .070 | 135 | |
| | | | 11—19 | 1 | .060 | 155 | |
| | | | 20—48 | 1 | .050 | 175 | |
| | | | 48 & Finer | 1 | .030 | 190 | |
| | | | 25—13 | 2 | 1.80 | 38 | S4, S2 |
| | | | 12—2.5 | 1 | 1.80 | 41 | |
| | | | 2—1.5 | 1 | 1.50 | 47 | |
| | | | 1—.5 | 1 | 1.25 | 53 | |
| | | | .5 & Finer | 1 | .75 | 58 | |

For hobbing Class 9 (per AGMA 390.03) or better gears, it may be necessary to reduce speeds and feeds by 50% and/or take 2 cuts.
To meet finish requirements it may be necessary to try both conventional and climb cutting.

See section 16 for Cutting Fluid Recommendations.
*Feeds are based on the largest standard recommended hob diameter. When using a smaller hob diameter, the Feed must be reduced proportionally.

# 7.1 Gear Hobbing

| MATERIAL | HARDNESS Bhn | CONDITION | DIAMETRAL PITCH / MODULE | NUMBER OF CUTS | FEED* PER REVOLUTION OF WORKPIECE in / mm | HOB SPEED fpm / m/min | HSS TOOL MATERIAL AISI / ISO |
|---|---|---|---|---|---|---|---|
| **2. CARBON STEELS, WROUGHT (cont.) High Carbon (cont.)** (materials listed on preceding page) | 225 to 275 | Hot Rolled, Normalized, Annealed, Cold Drawn or Quenched and Tempered | 1—2 | 2 | .060 | 100 | M2, M7 |
| | | | 3—10 | 1 | .050 | 115 | |
| | | | 11—19 | 1 | .050 | 140 | |
| | | | 20—48 | 1 | .050 | 155 | |
| | | | 48 & Finer | 1 | .030 | 165 | |
| | | | 25—13 | 2 | 1.50 | 30 | S4, S2 |
| | | | 12—2.5 | 1 | 1.25 | 35 | |
| | | | 2—1.5 | 1 | 1.25 | 43 | |
| | | | 1—.5 | 1 | 1.25 | 47 | |
| | | | .5 & Finer | 1 | .75 | 50 | |
| | 275 to 325 | Hot Rolled, Normalized, Annealed or Quenched and Tempered | 1—2 | 2 | .045 | 85 | M2, M7 |
| | | | 3—10 | 1 | .045 | 100 | |
| | | | 11—19 | 1 | .045 | 110 | |
| | | | 20—48 | 1 | .045 | 120 | |
| | | | 48 & Finer | 1 | .030 | 130 | |
| | | | 25—13 | 2 | 1.15 | 26 | S4, S2 |
| | | | 12—2.5 | 1 | 1.15 | 30 | |
| | | | 2—1.5 | 1 | 1.15 | 34 | |
| | | | 1—.5 | 1 | 1.15 | 37 | |
| | | | .5 & Finer | 1 | .75 | 40 | |
| | 325 to 375 | Quenched and Tempered | 1—2 | 2 | .045 | 65 | M3, M42 |
| | | | 3—10 | 1 | .045 | 75 | |
| | | | 11—19 | 1 | .025 | 95 | |
| | | | 20—48 | 1 | .025 | 100 | |
| | | | 48 & Finer | 1 | .025 | 115 | |
| | | | 25—13 | 2 | 1.15 | 20 | S5, S11 |
| | | | 12—2.5 | 1 | 1.15 | 23 | |
| | | | 2—1.5 | 1 | .65 | 29 | |
| | | | 1—.5 | 1 | .65 | 30 | |
| | | | .5 & Finer | 1 | .65 | 35 | |
| **3. CARBON AND FERRITIC ALLOY STEELS (HIGH TEMPERATURE SERVICE)** ASTM A369: Grades FPA, FPB, FP1, FP2, FP12 | 150 to 200 | As Forged | 1—2 | 2 | .070 | 165 | M2, M7 |
| | | | 3—10 | 1 | .070 | 175 | |
| | | | 11—19 | 1 | .060 | 185 | |
| | | | 20—48 | 1 | .050 | 200 | |
| | | | 48 & Finer | 1 | .030 | 210 | |
| | | | 25—13 | 2 | 1.80 | 50 | S4, S2 |
| | | | 12—2.5 | 1 | 1.80 | 53 | |
| | | | 2—1.5 | 1 | 1.50 | 56 | |
| | | | 1—.5 | 1 | 1.25 | 60 | |
| | | | .5 & Finer | 1 | .75 | 64 | |
| ASTM A369: Grades FP3b, FP11 | 150 to 200 | Annealed or Normalized and Tempered | 1—2 | 2 | .070 | 135 | M2, M7 |
| | | | 3—10 | 1 | .070 | 145 | |
| | | | 11—19 | 1 | .060 | 165 | |
| | | | 20—48 | 1 | .050 | 185 | |
| | | | 48 & Finer | 1 | .030 | 200 | |
| | | | 25—13 | 2 | 1.80 | 41 | S4, S2 |
| | | | 12—2.5 | 1 | 1.80 | 44 | |
| | | | 2—1.5 | 1 | 1.50 | 50 | |
| | | | 1—.5 | 1 | 1.25 | 56 | |
| | | | .5 & Finer | 1 | .75 | 60 | |
| ASTM A369: Grades FP5, FP7, FP9, FP21, FP22 | 150 to 200 | Annealed or Normalized and Tempered | 1—2 | 2 | .060 | 110 | M2, M7 |
| | | | 3—10 | 1 | .050 | 125 | |
| | | | 11—19 | 1 | .050 | 150 | |
| | | | 20—48 | 1 | .050 | 165 | |
| | | | 48 & Finer | 1 | .030 | 175 | |
| | | | 25—13 | 2 | 1.50 | 34 | S4, S2 |
| | | | 12—2.5 | 1 | 1.25 | 38 | |
| | | | 2—1.5 | 1 | 1.25 | 46 | |
| | | | 1—.5 | 1 | 1.25 | 50 | |
| | | | .5 & Finer | 1 | .75 | 53 | |

For hobbing Class 9 (per AGMA 390.03) or better gears, it may be necessary to reduce speeds and feeds by 50% and/or take 2 cuts.
To meet finish requirements it may be necessary to try both conventional and climb cutting.

See section 16 for Cutting Fluid Recommendations.
*Feeds are based on the largest standard recommended hob diameter. When using a smaller hob diameter, the Feed must be reduced proportionally.

| MATERIAL | HARDNESS Bhn | CONDITION | DIAMETRAL PITCH / MODULE | NUMBER OF CUTS | FEED* PER REVOLUTION OF WORKPIECE in / mm | HOB SPEED fpm / m/min | HSS TOOL MATERIAL AISI / ISO |
|---|---|---|---|---|---|---|---|
| **4. FREE MACHINING ALLOY STEELS, WROUGHT**<br>**Medium Carbon Resulfurized**<br>4140<br>4140Se<br>4142Te<br>4145Se<br>4147Te<br>4150 | 150 to 200 | Hot Rolled, Normalized, Annealed or Cold Drawn | 1—2 | 2 | .060 | 145 | |
| | | | 3—10 | 1 | .060 | 155 | |
| | | | 11—19 | 1 | .060 | 165 | M2, M7 |
| | | | 20—48 | 1 | .050 | 180 | |
| | | | 48 & Finer | 1 | .030 | 200 | |
| | | | 25—13 | 2 | 1.50 | 44 | |
| | | | 12—2.5 | 1 | 1.50 | 47 | |
| | | | 2—1.5 | 1 | 1.50 | 50 | S4, S2 |
| | | | 1—.5 | 1 | 1.25 | 55 | |
| | | | .5 & Finer | 1 | .75 | 60 | |
| | 200 to 250 | Hot Rolled, Normalized, Annealed or Cold Drawn | 1—2 | 2 | .060 | 120 | |
| | | | 3—10 | 1 | .060 | 130 | |
| | | | 11—19 | 1 | .060 | 140 | M2, M7 |
| | | | 20—48 | 1 | .050 | 160 | |
| | | | 48 & Finer | 1 | .030 | 180 | |
| | | | 25—13 | 2 | 1.50 | 37 | |
| | | | 12—2.5 | 1 | 1.50 | 40 | |
| | | | 2—1.5 | 1 | 1.50 | 43 | S4, S2 |
| | | | 1—.5 | 1 | 1.25 | 49 | |
| | | | .5 & Finer | 1 | .75 | 55 | |
| | 275 to 325 | Quenched and Tempered | 1—2 | 2 | .045 | 90 | |
| | | | 3—10 | 1 | .045 | 100 | |
| | | | 11—19 | 1 | .045 | 110 | M2, M7 |
| | | | 20—48 | 1 | .045 | 120 | |
| | | | 48 & Finer | 1 | .030 | 130 | |
| | | | 25—13 | 2 | 1.15 | 27 | |
| | | | 12—2.5 | 1 | 1.15 | 30 | |
| | | | 2—1.5 | 1 | 1.15 | 34 | S4, S2 |
| | | | 1—.5 | 1 | 1.15 | 37 | |
| | | | .5 & Finer | 1 | .75 | 40 | |
| | 325 to 375 | Quenched and Tempered | 1—2 | 2 | .045 | 65 | |
| | | | 3—10 | 1 | .045 | 75 | |
| | | | 11—19 | 1 | .025 | 95 | M3, M42 |
| | | | 20—48 | 1 | .025 | 100 | |
| | | | 48 & Finer | 1 | .025 | 135 | |
| | | | 25—13 | 2 | 1.15 | 20 | |
| | | | 12—2.5 | 1 | 1.15 | 23 | |
| | | | 2—1.5 | 1 | .65 | 29 | S5, S11 |
| | | | 1—.5 | 1 | .65 | 30 | |
| | | | .5 & Finer | 1 | .65 | 41 | |
| **Medium and High Carbon Leaded**<br>41L30 43L40<br>41L40 51L32<br>41L45 52L100<br>41L47 86L20<br>41L50 86L40 | 150 to 200 | Hot Rolled, Normalized, Annealed or Cold Drawn | 1—2 | 2 | .060 | 170 | |
| | | | 3—10 | 1 | .060 | 180 | |
| | | | 11—19 | 1 | .060 | 190 | M2, M7 |
| | | | 20—48 | 1 | .050 | 200 | |
| | | | 48 & Finer | 1 | .030 | 220 | |
| | | | 25—13 | 2 | 1.50 | 52 | |
| | | | 12—2.5 | 1 | 1.50 | 55 | |
| | | | 2—1.5 | 1 | 1.50 | 58 | S4, S2 |
| | | | 1—.5 | 1 | 1.25 | 60 | |
| | | | .5 & Finer | 1 | .75 | 67 | |
| | 200 to 250 | Hot Rolled, Normalized, Annealed or Cold Drawn | 1—2 | 2 | .060 | 140 | |
| | | | 3—10 | 1 | .060 | 155 | |
| | | | 11—19 | 1 | .060 | 170 | M2, M7 |
| | | | 20—48 | 1 | .050 | 185 | |
| | | | 48 & Finer | 1 | .030 | 200 | |
| | | | 25—13 | 2 | 1.50 | 43 | |
| | | | 12—2.5 | 1 | 1.50 | 47 | |
| | | | 2—1.5 | 1 | 1.50 | 52 | S4, S2 |
| | | | 1—.5 | 1 | 1.25 | 56 | |
| | | | .5 & Finer | 1 | .75 | 60 | |

For hobbing Class 9 (per AGMA 390.03) or better gears, it may be necessary to reduce speeds and feeds by 50% and/or take 2 cuts.
To meet finish requirements it may be necessary to try both conventional and climb cutting.

See section 16 for Cutting Fluid Recommendations.
*Feeds are based on the largest standard recommended hob diameter. When using a smaller hob diameter, the Feed must be reduced proportionally.

# 7.1 Gear Hobbing

| MATERIAL | HARDNESS Bhn | CONDITION | DIAMETRAL PITCH / MODULE | NUMBER OF CUTS | FEED* PER REVOLUTION OF WORKPIECE in / mm | HOB SPEED fpm / m/min | HSS TOOL MATERIAL AISI / ISO |
|---|---|---|---|---|---|---|---|
| **4. FREE MACHINING ALLOY STEELS, WROUGHT (cont.) Medium and High Carbon Leaded (cont.)** (materials listed on preceding page) | 275 to 325 | Quenched and Tempered | 1—2 | 2 | .045 | 120 | M2, M7 |
| | | | 3—10 | 1 | .045 | 140 | |
| | | | 11—19 | 1 | .045 | 145 | |
| | | | 20—48 | 1 | .045 | 150 | |
| | | | 48 & Finer | 1 | .030 | 165 | |
| | | | 25—13 | 2 | 1.15 | 37 | S4, S2 |
| | | | 12—2.5 | 1 | 1.15 | 43 | |
| | | | 2—1.5 | 1 | 1.15 | 44 | |
| | | | 1—.5 | 1 | 1.15 | 46 | |
| | | | .5 & Finer | 1 | .75 | 50 | |
| | 325 to 375 | Quenched and Tempered | 1—2 | 2 | .045 | 70 | M3, M42 |
| | | | 3—10 | 1 | .045 | 90 | |
| | | | 11—19 | 1 | .045 | 110 | |
| | | | 20—48 | 1 | .045 | 120 | |
| | | | 48 & Finer | 1 | .030 | 130 | |
| | | | 25—13 | 2 | 1.15 | 21 | S5, S11 |
| | | | 12—2.5 | 1 | 1.15 | 27 | |
| | | | 2—1.5 | 1 | 1.15 | 34 | |
| | | | 1—.5 | 1 | 1.15 | 37 | |
| | | | .5 & Finer | 1 | .75 | 40 | |
| **5. ALLOY STEELS, WROUGHT Low Carbon** 4012 4615 4817 8617 4023 4617 4820 8620 4024 4620 5015 8622 4118 4621 5115 8822 4320 4718 5120 9310 4419 4720 6118 94B15 4422 4815 8115 94B17 | 125 to 175 | Hot Rolled, Annealed or Cold Drawn | 1—2 | 2 | .060 | 160 | M2, M7 |
| | | | 3—10 | 1 | .060 | 170 | |
| | | | 11—19 | 1 | .060 | 180 | |
| | | | 20—48 | 1 | .050 | 190 | |
| | | | 48 & Finer | 1 | .030 | 210 | |
| | | | 25—13 | 2 | 1.50 | 49 | S4, S2 |
| | | | 12—2.5 | 1 | 1.50 | 52 | |
| | | | 2—1.5 | 1 | 1.50 | 55 | |
| | | | 1—.5 | 1 | 1.25 | 58 | |
| | | | .5 & Finer | 1 | .75 | 64 | |
| | 175 to 225 | Hot Rolled, Annealed or Cold Drawn | 1—2 | 2 | .060 | 130 | M2, M7 |
| | | | 3—10 | 1 | .060 | 140 | |
| | | | 11—19 | 1 | .060 | 150 | |
| | | | 20—48 | 1 | .050 | 170 | |
| | | | 48 & Finer | 1 | .030 | 190 | |
| | | | 25—13 | 2 | 1.50 | 40 | S4, S2 |
| | | | 12—2.5 | 1 | 1.50 | 43 | |
| | | | 2—1.5 | 1 | 1.50 | 46 | |
| | | | 1—.5 | 1 | 1.25 | 52 | |
| | | | .5 & Finer | 1 | .75 | 58 | |
| | 225 to 275 | Hot Rolled, Normalized, Annealed or Cold Drawn | 1—2 | 2 | .060 | 105 | M2, M7 |
| | | | 3—10 | 1 | .050 | 115 | |
| | | | 11—19 | 1 | .050 | 130 | |
| | | | 20—48 | 1 | .050 | 150 | |
| | | | 48 & Finer | 1 | .030 | 165 | |
| | | | 25—13 | 2 | 1.50 | 32 | S4, S2 |
| | | | 12—2.5 | 1 | 1.25 | 35 | |
| | | | 2—1.5 | 1 | 1.25 | 40 | |
| | | | 1—.5 | 1 | 1.25 | 46 | |
| | | | .5 & Finer | 1 | .75 | 50 | |
| | 275 to 325 | Normalized or Quenched and Tempered | 1—2 | 2 | .045 | 90 | M2, M7 |
| | | | 3—10 | 1 | .045 | 100 | |
| | | | 11—19 | 1 | .045 | 110 | |
| | | | 20—48 | 1 | .045 | 120 | |
| | | | 48 & Finer | 1 | .030 | 130 | |
| | | | 25—13 | 2 | 1.15 | 27 | S4, S2 |
| | | | 12—2.5 | 1 | 1.15 | 30 | |
| | | | 2—1.5 | 1 | 1.15 | 34 | |
| | | | 1—.5 | 1 | 1.15 | 37 | |
| | | | .5 & Finer | 1 | .75 | 40 | |

For hobbing Class 9 (per AGMA 390.03) or better gears, it may be necessary to reduce speeds and feeds by 50% and/or take 2 cuts.
To meet finish requirements it may be necessary to try both conventional and climb cutting.

See section 16 for Cutting Fluid Recommendations.

*Feeds are based on the largest standard recommended hob diameter. When using a smaller hob diameter, the Feed must be reduced proportionally.

| MATERIAL | HARD-NESS Bhn | CONDITION | DIAMETRAL PITCH / MODULE | NUMBER OF CUTS | FEED* PER REVOLUTION OF WORKPIECE in / mm | HOB SPEED fpm / m/min | HSS TOOL MATERIAL AISI / ISO |
|---|---|---|---|---|---|---|---|
| **5. ALLOY STEELS, WROUGHT (cont.) Low Carbon (cont.)** (materials listed on preceding page) | 325 to 375 | Normalized or Quenched and Tempered | 1—2 | 2 | .045 | 65 | M3, M42 |
| | | | 3—10 | 1 | .045 | 70 | |
| | | | 11—19 | 1 | .025 | 75 | |
| | | | 20—48 | 1 | .025 | 90 | |
| | | | 48 & Finer | 1 | .025 | 100 | |
| | | | 25—13 | 2 | 1.15 | 20 | S5, S11 |
| | | | 12—2.5 | 1 | 1.15 | 21 | |
| | | | 2—1.5 | 1 | .65 | 23 | |
| | | | 1—.5 | 1 | .65 | 27 | |
| | | | .5 & Finer | 1 | .65 | 30 | |
| **Medium Carbon** 1330, 1335, 1340, 1345, 4027, 4028, 4032, 4037, 4042, 4047, 4130, 4135, 4137, 4140, 4142, 4145, 4147, 4150, 4161, 4340, 4427, 4626, 50B40, 50B44, 5046, 50B46, 50B50, 5060, 50B60, 5130, 5132, 5135, 5140, 5145, 5147, 5150, 5155, 5160, 51B60, 6150, 81B45, 8625, 8627, 8630, 8637, 8640, 8642, 8645, 86B45, 8650, 8655, 8660, 8740, 8742, 9254, 9255, 9260, 94B30 | 175 to 225 | Hot Rolled, Annealed or Cold Drawn | 1—2 | 2 | .060 | 125 | M2, M7 |
| | | | 3—10 | 1 | .060 | 135 | |
| | | | 11—19 | 1 | .050 | 145 | |
| | | | 20—48 | 1 | .050 | 160 | |
| | | | 48 & Finer | 1 | .030 | 180 | |
| | | | 25—13 | 2 | 1.50 | 38 | S4, S2 |
| | | | 12—2.5 | 1 | 1.50 | 41 | |
| | | | 2—1.5 | 1 | 1.25 | 44 | |
| | | | 1—.5 | 1 | 1.25 | 49 | |
| | | | .5 & Finer | 1 | .75 | 55 | |
| | 225 to 275 | Annealed, Normalized, Cold Drawn or Quenched and Tempered | 1—2 | 2 | .060 | 100 | M2, M7 |
| | | | 3—10 | 1 | .050 | 110 | |
| | | | 11—19 | 1 | .050 | 125 | |
| | | | 20—48 | 1 | .050 | 140 | |
| | | | 48 & Finer | 1 | .030 | 155 | |
| | | | 25—13 | 2 | 1.50 | 30 | S4, S2 |
| | | | 12—2.5 | 1 | 1.25 | 34 | |
| | | | 2—1.5 | 1 | 1.25 | 38 | |
| | | | 1—.5 | 1 | 1.25 | 43 | |
| | | | .5 & Finer | 1 | .75 | 47 | |
| | 275 to 325 | Normalized or Quenched and Tempered | 1—2 | 2 | .045 | 85 | M2, M7 |
| | | | 3—10 | 1 | .045 | 95 | |
| | | | 11—19 | 1 | .045 | 105 | |
| | | | 20—48 | 1 | .045 | 110 | |
| | | | 48 & Finer | 1 | .030 | 120 | |
| | | | 25—13 | 2 | 1.15 | 26 | S4, S2 |
| | | | 12—2.5 | 1 | 1.15 | 29 | |
| | | | 2—1.5 | 1 | 1.15 | 32 | |
| | | | 1—.5 | 1 | 1.15 | 34 | |
| | | | .5 & Finer | 1 | .75 | 37 | |
| | 325 to 375 | Normalized or Quenched and Tempered | 1—2 | 2 | .045 | 60 | M3, M42 |
| | | | 3—10 | 1 | .045 | 65 | |
| | | | 11—19 | 1 | .025 | 75 | |
| | | | 20—48 | 1 | .025 | 80 | |
| | | | 48 & Finer | 1 | .025 | 95 | |
| | | | 25—13 | 2 | 1.15 | 18 | S5, S11 |
| | | | 12—2.5 | 1 | 1.15 | 20 | |
| | | | 2—1.5 | 1 | .65 | 23 | |
| | | | 1—.5 | 1 | .65 | 24 | |
| | | | .5 & Finer | 1 | .65 | 29 | |
| **High Carbon** 50100, 51100, 52100, M-50 | 175 to 225 | Hot Rolled, Annealed or Cold Drawn | 1—2 | 2 | .060 | 115 | M2, M7 |
| | | | 3—10 | 1 | .060 | 125 | |
| | | | 11—19 | 1 | .050 | 135 | |
| | | | 20—48 | 1 | .050 | 150 | |
| | | | 48 & Finer | 1 | .030 | 170 | |
| | | | 25—13 | 2 | 1.50 | 35 | S4, S2 |
| | | | 12—2.5 | 1 | 1.50 | 38 | |
| | | | 2—1.5 | 1 | 1.25 | 41 | |
| | | | 1—.5 | 1 | 1.25 | 46 | |
| | | | .5 & Finer | 1 | .75 | 52 | |

For hobbing Class 9 (per AGMA 390.03) or better gears, it may be necessary to reduce speeds and feeds by 50% and/or take 2 cuts.
To meet finish requirements it may be necessary to try both conventional and climb cutting.

See section 16 for Cutting Fluid Recommendations.
*Feeds are based on the largest standard recommended hob diameter. When using a smaller hob diameter, the Feed must be reduced proportionally.

# 7.1 Gear Hobbing

| MATERIAL | HARDNESS Bhn | CONDITION | DIAMETRAL PITCH / MODULE | NUMBER OF CUTS | FEED* PER REVOLUTION OF WORKPIECE in / mm | HOB SPEED fpm / m/min | HSS TOOL MATERIAL AISI / ISO |
|---|---|---|---|---|---|---|---|
| **5. ALLOY STEELS, WROUGHT (cont.)** **High Carbon (cont.)** (materials listed on preceding page) | 225 to 275 | Normalized, Cold Drawn or Quenched and Tempered | 1—2 | 2 | .060 | 90 | M2, M7 |
| | | | 3—10 | 1 | .050 | 100 | |
| | | | 11—19 | 1 | .050 | 115 | |
| | | | 20—48 | 1 | .050 | 130 | |
| | | | 48 & Finer | 1 | .030 | 145 | |
| | | | 25—13 | 2 | 1.50 | 27 | S4, S2 |
| | | | 12—2.5 | 1 | 1.25 | 30 | |
| | | | 2—1.5 | 1 | 1.25 | 35 | |
| | | | 1—.5 | 1 | 1.25 | 40 | |
| | | | .5 & Finer | 1 | .75 | 44 | |
| | 275 to 325 | Normalized or Quenched and Tempered | 1—2 | 2 | .045 | 80 | M2, M7 |
| | | | 3—10 | 1 | .045 | 90 | |
| | | | 11—19 | 1 | .045 | 100 | |
| | | | 20—48 | 1 | .045 | 105 | |
| | | | 48 & Finer | 1 | .030 | 115 | |
| | | | 25—13 | 2 | 1.15 | 24 | S4, S2 |
| | | | 12—2.5 | 1 | 1.15 | 27 | |
| | | | 2—1.5 | 1 | 1.15 | 30 | |
| | | | 1—.5 | 1 | 1.15 | 32 | |
| | | | .5 & Finer | 1 | .75 | 35 | |
| | 325 to 375 | Normalized or Quenched and Tempered | 1—2 | 2 | .045 | 55 | M3, M42 |
| | | | 3—10 | 1 | .045 | 60 | |
| | | | 11—19 | 1 | .025 | 70 | |
| | | | 20—48 | 1 | .025 | 75 | |
| | | | 48 & Finer | 1 | .025 | 90 | |
| | | | 25—13 | 2 | 1.15 | 17 | S5, S11 |
| | | | 12—2.5 | 1 | 1.15 | 18 | |
| | | | 2—1.5 | 1 | .65 | 21 | |
| | | | 1—.5 | 1 | .65 | 23 | |
| | | | .5 & Finer | 1 | .65 | 27 | |
| **6. HIGH STRENGTH STEELS, WROUGHT** 300M, 98BV40, 4330V, D6ac, 4340, H11, 4340Si, H13 | 225 to 300 | Annealed | 1—2 | 2 | .060 | 80 | M2, M7 |
| | | | 3—10 | 1 | .050 | 95 | |
| | | | 11—19 | 1 | .050 | 105 | |
| | | | 20—48 | 1 | .050 | 110 | |
| | | | 48 & Finer | 1 | .030 | 120 | |
| | | | 25—13 | 2 | 1.50 | 24 | S4, S2 |
| | | | 12—2.5 | 1 | 1.25 | 29 | |
| | | | 2—1.5 | 1 | 1.25 | 32 | |
| | | | 1—.5 | 1 | 1.25 | 34 | |
| | | | .5 & Finer | 1 | .75 | 37 | |
| | 300 to 350 | Normalized | 1—2 | 2 | .045 | 65 | M2, M7 |
| | | | 3—10 | 1 | .045 | 80 | |
| | | | 11—19 | 1 | .045 | 90 | |
| | | | 20—48 | 1 | .045 | 95 | |
| | | | 48 & Finer | 1 | .030 | 105 | |
| | | | 25—13 | 2 | 1.15 | 20 | S4, S2 |
| | | | 12—2.5 | 1 | 1.15 | 24 | |
| | | | 2—1.5 | 1 | 1.15 | 27 | |
| | | | 1—.5 | 1 | 1.15 | 29 | |
| | | | .5 & Finer | 1 | .75 | 32 | |
| HP 9-4-20, HP 9-4-25, HP 9-4-30, HP 9-4-45 | 325 to 375 | Annealed | 1—2 | 2 | .045 | 60 | M3, M42 |
| | | | 3—10 | 1 | .045 | 70 | |
| | | | 11—19 | 1 | .025 | 80 | |
| | | | 20—48 | 1 | .025 | 85 | |
| | | | 48 & Finer | 1 | .025 | 95 | |
| | | | 25—13 | 2 | 1.15 | 18 | S5, S11 |
| | | | 12—2.5 | 1 | 1.15 | 21 | |
| | | | 2—1.5 | 1 | .65 | 24 | |
| | | | 1—.5 | 1 | .65 | 26 | |
| | | | .5 & Finer | 1 | .65 | 29 | |

For hobbing Class 9 (per AGMA 390.03) or better gears, it may be necessary to reduce speeds and feeds by 50% and/or take 2 cuts.
To meet finish requirements it may be necessary to try both conventional and climb cutting.

See section 16 for Cutting Fluid Recommendations.
*Feeds are based on the largest standard recommended hob diameter. When using a smaller hob diameter, the Feed must be reduced proportionally.

| MATERIAL | HARDNESS Bhn | CONDITION | DIAMETRAL PITCH / MODULE | NUMBER OF CUTS | FEED* PER REVOLUTION OF WORKPIECE in / mm | HOB SPEED fpm / m/min | HSS TOOL MATERIAL AISI / ISO |
|---|---|---|---|---|---|---|---|
| **7. MARAGING STEELS, WROUGHT** | 275 to 325 | Annealed | 1—2 | 2 | .045 | 120 | |
| 120 Grade | | | 3—10 | 1 | .045 | 130 | |
| 180 Grade | | | 11—19 | 1 | .045 | 145 | M2, M7 |
| | | | 20—48 | 1 | .045 | 155 | |
| | | | 48 & Finer | 1 | .030 | 170 | |
| | | | 25—13 | 2 | 1.15 | 37 | |
| | | | 12—2.5 | 1 | 1.15 | 40 | |
| | | | 2—1.5 | 1 | 1.15 | 44 | S4, S2 |
| | | | 1—.5 | 1 | 1.15 | 47 | |
| | | | .5 & Finer | 1 | .75 | 52 | |
| ASTM A538: Grades A, B, C | 275 to 325 | Annealed | 1—2 | 2 | .045 | 100 | |
| 200 Grade | | | 3—10 | 1 | .045 | 110 | |
| 250 Grade | | | 11—19 | 1 | .045 | 120 | M2, M7 |
| 300 Grade | | | 20—48 | 1 | .045 | 130 | |
| 350 Grade | | | 48 & Finer | 1 | .030 | 140 | |
| HY230 | | | 25—13 | 2 | 1.15 | 30 | |
| | | | 12—2.5 | 1 | 1.15 | 34 | |
| | | | 2—1.5 | 1 | 1.15 | 37 | S4, S2 |
| | | | 1—.5 | 1 | 1.15 | 40 | |
| | | | .5 & Finer | 1 | .75 | 43 | |
| **8. TOOL STEELS, WROUGHT** **High Speed** | 200 to 250 | Annealed | 1—2 | 2 | .045 | 40 | |
| M1 | | | 3—10 | 1 | .045 | 45 | |
| M2 | | | 11—19 | 1 | .045 | 55 | M2, M7 |
| M6 | | | 20—48 | 1 | .045 | 60 | |
| M10 | | | 48 & Finer | 1 | .030 | 80 | |
| T1 | | | 25—13 | 2 | 1.15 | 12 | |
| T2 | | | 12—2.5 | 1 | 1.15 | 14 | |
| T6 | | | 2—1.5 | 1 | 1.15 | 17 | S4, S2 |
| | | | 1—.5 | 1 | 1.15 | 18 | |
| | | | .5 & Finer | 1 | .75 | 24 | |
| **High Speed** | 225 to 275 | Annealed | 1—2 | 2 | .045 | 35 | |
| M3-1 M34 M46 | | | 3—10 | 1 | .045 | 40 | |
| M3-2 M36 M47 | | | 11—19 | 1 | .025 | 50 | M2, M7 |
| M4 M41 T4 | | | 20—48 | 1 | .025 | 55 | |
| M7 M42 T5 | | | 48 & Finer | 1 | .025 | 75 | |
| M30 M43 T8 | | | 25—13 | 2 | 1.15 | 11 | |
| M33 M44 T15 | | | 12—2.5 | 1 | 1.15 | 12 | |
| | | | 2—1.5 | 1 | .65 | 15 | S4, S2 |
| | | | 1—.5 | 1 | .65 | 17 | |
| | | | .5 & Finer | 1 | .65 | 23 | |
| **Hot Work** | 150 to 200 | Annealed | 1—2 | 2 | .060 | 70 | |
| H10 H22 | | | 3—10 | 1 | .060 | 75 | |
| H11 H23 | | | 11—19 | 1 | .060 | 85 | M2, M7 |
| H12 H24 | | | 20—48 | 1 | .050 | 90 | |
| H13 H25 | | | 48 & Finer | 1 | .030 | 110 | |
| H14 H26 | | | 25—13 | 2 | 1.50 | 21 | |
| H19 H42 | | | 12—2.5 | 1 | 1.50 | 23 | |
| H21 | | | 2—1.5 | 1 | 1.50 | 26 | S4, S2 |
| | | | 1—.5 | 1 | 1.25 | 27 | |
| | | | .5 & Finer | 1 | .75 | 34 | |
| | 200 to 250 | Annealed | 1—2 | 2 | .060 | 55 | |
| | | | 3—10 | 1 | .060 | 60 | |
| | | | 11—19 | 1 | .060 | 70 | M2, M7 |
| | | | 20—48 | 1 | .050 | 75 | |
| | | | 48 & Finer | 1 | .030 | 95 | |
| | | | 25—13 | 2 | 1.50 | 17 | |
| | | | 12—2.5 | 1 | 1.50 | 18 | |
| | | | 2—1.5 | 1 | 1.50 | 21 | S4, S2 |
| | | | 1—.5 | 1 | 1.25 | 23 | |
| | | | .5 & Finer | 1 | .75 | 29 | |

For hobbing Class 9 (per AGMA 390.03) or better gears, it may be necessary to reduce speeds and feeds by 50% and/or take 2 cuts.
To meet finish requirements it may be necessary to try both conventional and climb cutting.

See section 16 for Cutting Fluid Recommendations.
*Feeds are based on the largest standard recommended hob diameter. When using a smaller hob diameter, the Feed must be reduced proportionally.

# 7.1 Gear Hobbing

| MATERIAL | HARD-NESS Bhn | CONDITION | DIAMETRAL PITCH / MODULE | NUMBER OF CUTS | FEED* PER REVOLUTION OF WORKPIECE in / mm | HOB SPEED fpm / m/min | HSS TOOL MATERIAL AISI / ISO |
|---|---|---|---|---|---|---|---|
| 8. TOOL STEELS, WROUGHT (cont.) Hot Work (cont.) (materials listed on preceding page) | 325 to 375 | Quenched and Tempered | 1—2 | 2 | .045 | 45 | M3, M42 |
| | | | 3—10 | 1 | .045 | 50 | |
| | | | 11—19 | 1 | .025 | 60 | |
| | | | 20—48 | 1 | .025 | 65 | |
| | | | 48 & Finer | 1 | .025 | 85 | |
| | | | 25—13 | 2 | 1.15 | 14 | S5, S11 |
| | | | 12—2.5 | 1 | 1.15 | 15 | |
| | | | 2—1.5 | 1 | .65 | 18 | |
| | | | 1—.5 | 1 | .65 | 20 | |
| | | | .5 & Finer | 1 | .65 | 26 | |
| Cold Work A2, A3, A4, A6, A7, A8, A9, A10, D2, D3, D4, D5, D7, O1, O2, O6, O7 | 200 to 250 | Annealed | 1—2 | 2 | .060 | 55 | M2, M7 |
| | | | 3—10 | 1 | .060 | 60 | |
| | | | 11—19 | 1 | .060 | 70 | |
| | | | 20—48 | 1 | .050 | 75 | |
| | | | 48 & Finer | 1 | .030 | 95 | |
| | | | 25—13 | 2 | 1.50 | 17 | S4, S2 |
| | | | 12—2.5 | 1 | 1.50 | 18 | |
| | | | 2—1.5 | 1 | 1.50 | 21 | |
| | | | 1—.5 | 1 | 1.25 | 23 | |
| | | | .5 & Finer | 1 | .75 | 29 | |
| Shock Resisting S1, S2, S5, S6, S7 | 175 to 225 | Annealed | 1—2 | 2 | .060 | 70 | M2, M7 |
| | | | 3—10 | 1 | .060 | 75 | |
| | | | 11—19 | 1 | .060 | 85 | |
| | | | 20—48 | 1 | .050 | 90 | |
| | | | 48 & Finer | 1 | .030 | 110 | |
| | | | 25—13 | 2 | 1.50 | 21 | S4, S2 |
| | | | 12—2.5 | 1 | 1.50 | 23 | |
| | | | 2—1.5 | 1 | 1.50 | 26 | |
| | | | 1—.5 | 1 | 1.25 | 27 | |
| | | | .5 & Finer | 1 | .75 | 34 | |
| Mold P2, P4, P5, P6 | 100 to 150 | Annealed | 1—2 | 2 | .060 | 95 | M2, M7 |
| | | | 3—10 | 1 | .060 | 100 | |
| | | | 11—19 | 1 | .060 | 110 | |
| | | | 20—48 | 1 | .050 | 115 | |
| | | | 48 & Finer | 1 | .030 | 125 | |
| | | | 25—13 | 2 | 1.50 | 29 | S4, S2 |
| | | | 12—2.5 | 1 | 1.50 | 30 | |
| | | | 2—1.5 | 1 | 1.50 | 34 | |
| | | | 1—.5 | 1 | 1.25 | 35 | |
| | | | .5 & Finer | 1 | .75 | 38 | |
| Mold P20, P21 | 150 to 200 | Annealed | 1—2 | 2 | .060 | 70 | M2, M7 |
| | | | 3—10 | 1 | .060 | 75 | |
| | | | 11—19 | 1 | .060 | 85 | |
| | | | 20—48 | 1 | .050 | 90 | |
| | | | 48 & Finer | 1 | .030 | 110 | |
| | | | 25—13 | 2 | 1.50 | 21 | S4, S2 |
| | | | 12—2.5 | 1 | 1.50 | 23 | |
| | | | 2—1.5 | 1 | 1.50 | 26 | |
| | | | 1—.5 | 1 | 1.25 | 27 | |
| | | | .5 & Finer | 1 | .75 | 34 | |

For hobbing Class 9 (per AGMA 390.03) or better gears, it may be necessary to reduce speeds and feeds by 50% and/or take 2 cuts.
To meet finish requirements it may be necessary to try both conventional and climb cutting.

See section 16 for Cutting Fluid Recommendations.
*Feeds are based on the largest standard recommended hob diameter. When using a smaller hob diameter, the Feed must be reduced proportionally.

| MATERIAL | HARD-NESS Bhn | CONDITION | DIAMETRAL PITCH / MODULE | NUMBER OF CUTS | FEED* PER REVOLUTION OF WORKPIECE in / mm | HOB SPEED fpm / m/min | HSS TOOL MATERIAL AISI / ISO |
|---|---|---|---|---|---|---|---|
| **Special Purpose** L2, L6, L7 | 150 to 200 | Annealed | 1—2 | 2 | .060 | 65 | M2, M7 |
| | | | 3—10 | 1 | .060 | 70 | |
| | | | 11—19 | 1 | .060 | 80 | |
| | | | 20—48 | 1 | .050 | 85 | |
| | | | 48 & Finer | 1 | .030 | 105 | |
| | | | 25—13 | 2 | 1.50 | 20 | S4, S2 |
| | | | 12—2.5 | 1 | 1.50 | 21 | |
| | | | 2—1.5 | 1 | 1.50 | 24 | |
| | | | 1—.5 | 1 | 1.25 | 26 | |
| | | | .5 & Finer | 1 | .75 | 32 | |
| **Special Purpose** F1, F2 | 200 to 250 | Annealed | 1—2 | 2 | .060 | 50 | M2, M7 |
| | | | 3—10 | 1 | .060 | 55 | |
| | | | 11—19 | 1 | .060 | 65 | |
| | | | 20—48 | 1 | .050 | 70 | |
| | | | 48 & Finer | 1 | .030 | 90 | |
| | | | 25—13 | 2 | 1.50 | 15 | S4, S2 |
| | | | 12—2.5 | 1 | 1.50 | 17 | |
| | | | 2—1.5 | 1 | 1.50 | 20 | |
| | | | 1—.5 | 1 | 1.25 | 21 | |
| | | | .5 & Finer | 1 | .75 | 27 | |
| **Water Hardening** W1, W2, W5 SAE J438b: Types W108, W109, W110, W112, W209, W210, W310 | 150 to 200 | Annealed | 1—2 | 2 | .060 | 105 | M2, M7 |
| | | | 3—10 | 1 | .060 | 110 | |
| | | | 11—19 | 1 | .060 | 120 | |
| | | | 20—48 | 1 | .050 | 125 | |
| | | | 48 & Finer | 1 | .030 | 145 | |
| | | | 25—13 | 2 | 1.50 | 32 | S4, S2 |
| | | | 12—2.5 | 1 | 1.50 | 34 | |
| | | | 2—1.5 | 1 | 1.50 | 37 | |
| | | | 1—.5 | 1 | 1.25 | 38 | |
| | | | .5 & Finer | 1 | .75 | 44 | |
| **9. NITRIDING STEELS, WROUGHT** Nitralloy 125, Nitralloy 135, Nitralloy 135 Mod., Nitralloy 225, Nitralloy 230, Nitralloy EZ, Nitralloy N, Nitrex 1 | 200 to 250 | Annealed | 1—2 | 2 | .045 | 85 | M2, M7 |
| | | | 3—10 | 1 | .045 | 100 | |
| | | | 11—19 | 1 | .045 | 110 | |
| | | | 20—48 | 1 | .045 | 115 | |
| | | | 48 & Finer | 1 | .030 | 125 | |
| | | | 25—13 | 2 | 1.15 | 26 | S4, S2 |
| | | | 12—2.5 | 1 | 1.15 | 30 | |
| | | | 2—1.5 | 1 | 1.15 | 34 | |
| | | | 1—.5 | 1 | 1.15 | 35 | |
| | | | .5 & Finer | 1 | .75 | 38 | |
| | 300 to 350 | Normalized or Quenched and Tempered | 1—2 | 2 | .045 | 65 | M2, M7 |
| | | | 3—10 | 1 | .045 | 80 | |
| | | | 11—19 | 1 | .025 | 90 | |
| | | | 20—48 | 1 | .025 | 95 | |
| | | | 48 & Finer | 1 | .025 | 105 | |
| | | | 25—13 | 2 | 1.15 | 20 | S4, S2 |
| | | | 12—2.5 | 1 | 1.15 | 24 | |
| | | | 2—1.5 | 1 | .65 | 27 | |
| | | | 1—.5 | 1 | .65 | 29 | |
| | | | .5 & Finer | 1 | .65 | 32 | |
| **12. FREE MACHINING STAINLESS STEELS, WROUGHT** **Ferritic** 430F, 430F Se | 135 to 185 | Annealed | 1—2 | 2 | .060 | 175 | M2, M7 |
| | | | 3—10 | 1 | .060 | 205 | |
| | | | 11—19 | 1 | .060 | 220 | |
| | | | 20—48 | 1 | .050 | 235 | |
| | | | 48 & Finer | 1 | .030 | 250 | |
| | | | 25—13 | 2 | 1.50 | 53 | S4, S2 |
| | | | 12—2.5 | 1 | 1.50 | 62 | |
| | | | 2—1.5 | 1 | 1.50 | 67 | |
| | | | 1—.5 | 1 | 1.25 | 72 | |
| | | | .5 & Finer | 1 | .75 | 76 | |

For hobbing Class 9 (per AGMA 390.03) or better gears, it may be necessary to reduce speeds and feeds by 50% and/or take 2 cuts.
To meet finish requirements it may be necessary to try both conventional and climb cutting.

See section 16 for Cutting Fluid Recommendations.
*Feeds are based on the largest standard recommended hob diameter. When using a smaller hob diameter, the Feed must be reduced proportionally.

# 7.1 Gear Hobbing

| MATERIAL | HARDNESS<br>Bhn | CONDITION | DIAMETRAL PITCH<br>MODULE | NUMBER OF CUTS | FEED* PER REVOLUTION OF WORKPIECE<br>in<br>mm | HOB SPEED<br>fpm<br>m/min | HSS TOOL MATERIAL<br>AISI<br>ISO |
|---|---|---|---|---|---|---|---|
| **12. FREE MACHINING STAINLESS STEELS, WROUGHT (cont.)**<br>**Austenitic**<br>203EZ 303Pb<br>303 303 Plus X<br>303MA 303Se | 135 to 185 | Annealed | 1—2 | 2 | .060 | 125 | M2, M7 |
| | | | 3—10 | 1 | .060 | 155 | |
| | | | 11—19 | 1 | .060 | 170 | |
| | | | 20—48 | 1 | .050 | 185 | |
| | | | 48 & Finer | 1 | .030 | 200 | |
| | | | 25—13 | 2 | 1.50 | 38 | S4, S2 |
| | | | 12—2.5 | 1 | 1.50 | 47 | |
| | | | 2—1.5 | 1 | 1.50 | 52 | |
| | | | 1—.5 | 1 | 1.25 | 56 | |
| | | | .5 & Finer | 1 | .75 | 60 | |
| | 225 to 275 | Cold Drawn | 1—2 | 2 | .060 | 115 | M2, M7 |
| | | | 3—10 | 1 | .050 | 135 | |
| | | | 11—19 | 1 | .050 | 155 | |
| | | | 20—48 | 1 | .050 | 175 | |
| | | | 48 & Finer | 1 | .030 | 195 | |
| | | | 25—13 | 2 | 1.50 | 35 | S4, S2 |
| | | | 12—2.5 | 1 | 1.25 | 41 | |
| | | | 2—1.5 | 1 | 1.25 | 47 | |
| | | | 1—.5 | 1 | 1.25 | 53 | |
| | | | .5 & Finer | 1 | .75 | 59 | |
| **Martensitic**<br>416<br>416 Plus X<br>416Se<br>420F<br>420F Se<br>440F<br>440F Se | 135 to 185 | Annealed | 1—2 | 2 | .060 | 175 | M2, M7 |
| | | | 3—10 | 1 | .060 | 205 | |
| | | | 11—19 | 1 | .060 | 220 | |
| | | | 20—48 | 1 | .050 | 235 | |
| | | | 48 & Finer | 1 | .030 | 250 | |
| | | | 25—13 | 2 | 1.50 | 53 | S4, S2 |
| | | | 12—2.5 | 1 | 1.50 | 62 | |
| | | | 2—1.5 | 1 | 1.50 | 67 | |
| | | | 1—.5 | 1 | 1.25 | 72 | |
| | | | .5 & Finer | 1 | .75 | 76 | |
| | 185 to 240 | Annealed or Cold Drawn | 1—2 | 2 | .060 | 165 | M2, M7 |
| | | | 3—10 | 1 | .060 | 195 | |
| | | | 11—19 | 1 | .060 | 210 | |
| | | | 20—48 | 1 | .050 | 225 | |
| | | | 48 & Finer | 1 | .030 | 240 | |
| | | | 25—13 | 2 | 1.50 | 50 | S4, S2 |
| | | | 12—2.5 | 1 | 1.50 | 59 | |
| | | | 2—1.5 | 1 | 1.50 | 64 | |
| | | | 1—.5 | 1 | 1.25 | 69 | |
| | | | .5 & Finer | 1 | .75 | 73 | |
| | 275 to 325 | Quenched and Tempered | 1—2 | 2 | .045 | 100 | M2, M7 |
| | | | 3—10 | 1 | .045 | 135 | |
| | | | 11—19 | 1 | .045 | 155 | |
| | | | 20—48 | 1 | .045 | 160 | |
| | | | 48 & Finer | 1 | .030 | 175 | |
| | | | 25—13 | 2 | 1.15 | 30 | S4, S2 |
| | | | 12—2.5 | 1 | 1.15 | 41 | |
| | | | 2—1.5 | 1 | 1.15 | 47 | |
| | | | 1—.5 | 1 | 1.15 | 49 | |
| | | | .5 & Finer | 1 | .75 | 53 | |
| **13. STAINLESS STEELS, WROUGHT**<br>**Ferritic**<br>405 434<br>409 436<br>429 442<br>430 446 | 135 to 185 | Annealed | 1—2 | 2 | .060 | 125 | M2, M7 |
| | | | 3—10 | 1 | .060 | 135 | |
| | | | 11—19 | 1 | .060 | 150 | |
| | | | 20—48 | 1 | .050 | 160 | |
| | | | 48 & Finer | 1 | .030 | 170 | |
| | | | 25—13 | 2 | 1.50 | 38 | S4, S2 |
| | | | 12—2.5 | 1 | 1.50 | 41 | |
| | | | 2—1.5 | 1 | 1.50 | 46 | |
| | | | 1—.5 | 1 | 1.25 | 49 | |
| | | | .5 & Finer | 1 | .75 | 52 | |

For hobbing Class 9 (per AGMA 390.03) or better gears, it may be necessary to reduce speeds and feeds by 50% and/or take 2 cuts.
To meet finish requirements it may be necessary to try both conventional and climb cutting.

See section 16 for Cutting Fluid Recommendations.
*Feeds are based on the largest standard recommended hob diameter. When using a smaller hob diameter, the Feed must be reduced proportionally.

| MATERIAL | HARDNESS Bhn | CONDITION | DIAMETRAL PITCH / MODULE | NUMBER OF CUTS | FEED* PER REVOLUTION OF WORKPIECE in / mm | HOB SPEED fpm / m/min | HSS TOOL MATERIAL AISI / ISO |
|---|---|---|---|---|---|---|---|
| **Austenitic** | 135 to 185 | Annealed | 1—2 | 2 | .060 | 95 | |
| 201 308 | | | 3—10 | 1 | .060 | 105 | |
| 202 321 | | | 11—19 | 1 | .060 | 120 | M2, M7 |
| 301 347 | | | 20—48 | 1 | .050 | 130 | |
| 302 348 | | | 48 & Finer | 1 | .030 | 140 | |
| 304 384 | | | 25—13 | 2 | 1.50 | 29 | |
| 304L 385 | | | 12—2.5 | 1 | 1.50 | 32 | |
| 305 | | | 2—1.5 | 1 | 1.50 | 37 | S4, S2 |
| | | | 1—.5 | 1 | 1.25 | 40 | |
| | | | .5 & Finer | 1 | .75 | 43 | |
| | 225 to 275 | Cold Drawn | 1—2 | 2 | .060 | 90 | |
| | | | 3—10 | 1 | .050 | 100 | |
| | | | 11—19 | 1 | .050 | 115 | M2, M7 |
| | | | 20—48 | 1 | .050 | 125 | |
| | | | 48 & Finer | 1 | .030 | 135 | |
| | | | 25—13 | 2 | 1.50 | 27 | |
| | | | 12—2.5 | 1 | 1.25 | 30 | |
| | | | 2—1.5 | 1 | 1.25 | 35 | S4, S2 |
| | | | 1—.5 | 1 | 1.25 | 38 | |
| | | | .5 & Finer | 1 | .75 | 41 | |
| **Austenitic** | 135 to 185 | Annealed | 1—2 | 2 | .060 | 85 | |
| 302B 314 | | | 3—10 | 1 | .060 | 95 | |
| 309 316 | | | 11—19 | 1 | .060 | 110 | M2, M7 |
| 309S 316L | | | 20—48 | 1 | .050 | 120 | |
| 310 317 | | | 48 & Finer | 1 | .030 | 130 | |
| 310S 330 | | | 25—13 | 2 | 1.50 | 26 | |
| | | | 12—2.5 | 1 | 1.50 | 29 | |
| | | | 2—1.5 | 1 | 1.50 | 34 | S4, S2 |
| | | | 1—.5 | 1 | 1.25 | 37 | |
| | | | .5 & Finer | 1 | .75 | 40 | |
| | 225 to 275 | Cold Drawn | 1—2 | 2 | .060 | 70 | |
| | | | 3—10 | 1 | .050 | 80 | |
| | | | 11—19 | 1 | .050 | 90 | M2, M7 |
| | | | 20—48 | 1 | .050 | 105 | |
| | | | 48 & Finer | 1 | .030 | 120 | |
| | | | 25—13 | 2 | 1.50 | 21 | |
| | | | 12—2.5 | 1 | 1.25 | 24 | |
| | | | 2—1.5 | 1 | 1.25 | 27 | S4, S2 |
| | | | 1—.5 | 1 | 1.25 | 32 | |
| | | | .5 & Finer | 1 | .75 | 37 | |
| **Austenitic** | 210 to 250 | Annealed | 1—2 | 2 | .060 | 60 | |
| Nitronic 32 | | | 3—10 | 1 | .050 | 70 | |
| Nitronic 33 | | | 11—19 | 1 | .050 | 80 | M2, M7 |
| Nitronic 40 | | | 20—48 | 1 | .050 | 90 | |
| Nitronic 50 | | | 48 & Finer | 1 | .030 | 100 | |
| Nitronic 60 | | | 25—13 | 2 | 1.50 | 18 | |
| | | | 12—2.5 | 1 | 1.25 | 21 | |
| | | | 2—1.5 | 1 | 1.25 | 24 | S4, S2 |
| | | | 1—.5 | 1 | 1.25 | 27 | |
| | | | .5 & Finer | 1 | .75 | 30 | |
| **Martensitic** | 135 to 175 | Annealed | 1—2 | 2 | .060 | 110 | |
| 403 | | | 3—10 | 1 | .060 | 130 | |
| 410 | | | 11—19 | 1 | .060 | 145 | M2, M7 |
| 420 | | | 20—48 | 1 | .050 | 155 | |
| 422 | | | 48 & Finer | 1 | .030 | 170 | |
| 501 | | | 25—13 | 2 | 1.50 | 34 | |
| 502 | | | 12—2.5 | 1 | 1.50 | 40 | |
| | | | 2—1.5 | 1 | 1.50 | 44 | S4, S2 |
| | | | 1—.5 | 1 | 1.25 | 47 | |
| | | | .5 & Finer | 1 | .75 | 52 | |

For hobbing Class 9 (per AGMA 390.03) or better gears, it may be necessary to reduce speeds and feeds by 50% and/or take 2 cuts.
To meet finish requirements it may be necessary to try both conventional and climb cutting.

See section 16 for Cutting Fluid Recommendations.
*Feeds are based on the largest standard recommended hob diameter. When using a smaller hob diameter, the Feed must be reduced proportionally.

| MATERIAL | HARD-NESS Bhn | CONDITION | DIAMETRAL PITCH / MODULE | NUMBER OF CUTS | FEED* PER REVOLUTION OF WORKPIECE in / mm | HOB SPEED fpm / m/min | HSS TOOL MATERIAL AISI / ISO |
|---|---|---|---|---|---|---|---|
| **13. STAINLESS STEELS, WROUGHT (cont.) Martensitic (cont.)** (materials listed on preceding page) | 175 to 225 | Annealed | 1—2 | 2 | .060 | 100 | M2, M7 |
| | | | 3—10 | 1 | .060 | 120 | |
| | | | 11—19 | 1 | .060 | 135 | |
| | | | 20—48 | 1 | .050 | 145 | |
| | | | 48 & Finer | 1 | .030 | 160 | |
| | | | 25—13 | 2 | 1.50 | 30 | S4, S2 |
| | | | 12—2.5 | 1 | 1.50 | 37 | |
| | | | 2—1.5 | 1 | 1.50 | 41 | |
| | | | 1—.5 | 1 | 1.25 | 44 | |
| | | | .5 & Finer | 1 | .75 | 49 | |
| | 275 to 325 | Quenched and Tempered | 1—2 | 2 | .045 | 60 | M2, M7 |
| | | | 3—10 | 1 | .045 | 80 | |
| | | | 11—19 | 1 | .045 | 95 | |
| | | | 20—48 | 1 | .045 | 105 | |
| | | | 48 & Finer | 1 | .030 | 120 | |
| | | | 25—13 | 2 | 1.15 | 18 | S4, S2 |
| | | | 12—2.5 | 1 | 1.15 | 24 | |
| | | | 2—1.5 | 1 | 1.15 | 29 | |
| | | | 1—.5 | 1 | 1.15 | 32 | |
| | | | .5 & Finer | 1 | .75 | 37 | |
| **Martensitic** 414 431 Greek Ascoloy | 225 to 275 | Annealed | 1—2 | 2 | .060 | 70 | M2, M7 |
| | | | 3—10 | 1 | .050 | 90 | |
| | | | 11—19 | 1 | .050 | 105 | |
| | | | 20—48 | 1 | .050 | 115 | |
| | | | 48 & Finer | 1 | .030 | 130 | |
| | | | 25—13 | 2 | 1.50 | 21 | S4, S2 |
| | | | 12—2.5 | 1 | 1.25 | 27 | |
| | | | 2—1.5 | 1 | 1.25 | 32 | |
| | | | 1—.5 | 1 | 1.25 | 35 | |
| | | | .5 & Finer | 1 | .75 | 40 | |
| | 275 to 325 | Quenched and Tempered | 1—2 | 2 | .045 | 55 | M2, M7 |
| | | | 3—10 | 1 | .045 | 75 | |
| | | | 11—19 | 1 | .045 | 90 | |
| | | | 20—48 | 1 | .045 | 100 | |
| | | | 48 & Finer | 1 | .030 | 115 | |
| | | | 25—13 | 2 | 1.15 | 17 | S4, S2 |
| | | | 12—2.5 | 1 | 1.15 | 23 | |
| | | | 2—1.5 | 1 | 1.15 | 27 | |
| | | | 1—.5 | 1 | 1.15 | 30 | |
| | | | .5 & Finer | 1 | .75 | 35 | |
| **Martensitic** 440A 440B 440C | 225 to 275 | Annealed | 1—2 | 2 | .060 | 60 | M2, M7 |
| | | | 3—10 | 1 | .050 | 80 | |
| | | | 11—19 | 1 | .050 | 95 | |
| | | | 20—48 | 1 | .050 | 105 | |
| | | | 48 & Finer | 1 | .030 | 120 | |
| | | | 25—13 | 2 | 1.50 | 18 | S4, S2 |
| | | | 12—2.5 | 1 | 1.25 | 24 | |
| | | | 2—1.5 | 1 | 1.25 | 29 | |
| | | | 1—.5 | 1 | 1.25 | 32 | |
| | | | .5 & Finer | 1 | .75 | 37 | |
| | 275 to 325 | Quenched and Tempered | 1—2 | 2 | .045 | 50 | M2, M7 |
| | | | 3—10 | 1 | .045 | 70 | |
| | | | 11—19 | 1 | .045 | 85 | |
| | | | 20—48 | 1 | .045 | 95 | |
| | | | 48 & Finer | 1 | .030 | 110 | |
| | | | 25—13 | 2 | 1.15 | 15 | S4, S2 |
| | | | 12—2.5 | 1 | 1.15 | 21 | |
| | | | 2—1.5 | 1 | 1.15 | 26 | |
| | | | 1—.5 | 1 | 1.15 | 29 | |
| | | | .5 & Finer | 1 | .75 | 34 | |

For hobbing Class 9 (per AGMA 390.03) or better gears, it may be necessary to reduce speeds and feeds by 50% and/or take 2 cuts.
To meet finish requirements it may be necessary to try both conventional and climb cutting.

See section 16 for Cutting Fluid Recommendations.
*Feeds are based on the largest standard recommended hob diameter. When using a smaller hob diameter, the Feed must be reduced proportionally.

| MATERIAL | HARD-NESS Bhn | CONDITION | DIAMETRAL PITCH / MODULE | NUMBER OF CUTS | FEED* PER REVOLUTION OF WORKPIECE in / mm | HOB SPEED fpm / m/min | HSS TOOL MATERIAL AISI / ISO |
|---|---|---|---|---|---|---|---|
| **14. PRECIPITATION HARDENING STAINLESS STEELS, WROUGHT**<br>15-5 PH<br>16-6 PH<br>17-4 PH<br>17-7 PH<br>17-14 Cu Mo<br>AF-71<br>AFC-77<br>Almar 362 (AM-362)<br>AM-350<br>AM-355<br>AM-363<br>Custom 450<br>Custom 455<br>HNM<br>PH 13-8 Mo<br>PH 14-8 Mo<br>PH 15-7 Mo<br>Stainless W | 150 to 200 | Solution Treated | 1—2 | 2 | .060 | 65 | M2, M7 |
| | | | 3—10 | 1 | .060 | 85 | |
| | | | 11—19 | 1 | .060 | 100 | |
| | | | 20—48 | 1 | .050 | 110 | |
| | | | 48 & Finer | 1 | .030 | 125 | |
| | | | 25—13 | 2 | 1.50 | 20 | S4, S2 |
| | | | 12—2.5 | 1 | 1.50 | 26 | |
| | | | 2—1.5 | 1 | 1.50 | 30 | |
| | | | 1—.5 | 1 | 1.25 | 34 | |
| | | | .5 & Finer | 1 | .75 | 38 | |
| | 275 to 325 | Solution Treated or Hardened | 1—2 | 2 | .045 | 60 | M2, M7 |
| | | | 3—10 | 1 | .045 | 80 | |
| | | | 11—19 | 1 | .045 | 95 | |
| | | | 20—48 | 1 | .045 | 105 | |
| | | | 48 & Finer | 1 | .030 | 120 | |
| | | | 25—13 | 2 | 1.15 | 18 | S4, S2 |
| | | | 12—2.5 | 1 | 1.15 | 24 | |
| | | | 2—1.5 | 1 | 1.15 | 29 | |
| | | | 1—.5 | 1 | 1.15 | 32 | |
| | | | .5 & Finer | 1 | .75 | 37 | |
| | 325 to 375 | Solution Treated or Hardened | 1—2 | 2 | .045 | 50 | M3, M42 |
| | | | 3—10 | 1 | .045 | 70 | |
| | | | 11—19 | 1 | .025 | 85 | |
| | | | 20—48 | 1 | .025 | 95 | |
| | | | 48 & Finer | 1 | .025 | 110 | |
| | | | 25—13 | 2 | 1.15 | 15 | S5, S11 |
| | | | 12—2.5 | 1 | 1.15 | 21 | |
| | | | 2—1.5 | 1 | .65 | 26 | |
| | | | 1—.5 | 1 | .65 | 29 | |
| | | | .5 & Finer | 1 | .65 | 34 | |
| **15. CARBON STEELS, CAST**<br>**Low Carbon**<br>ASTM A426: Grade CP1<br>1010<br>1020 | 100 to 150 | Annealed, Normalized or Normalized and Tempered | 1—2 | 2 | .060 | 170 | M2, M7 |
| | | | 3—10 | 1 | .060 | 200 | |
| | | | 11—19 | 1 | .060 | 215 | |
| | | | 20—48 | 1 | .050 | 230 | |
| | | | 48 & Finer | 1 | .030 | 245 | |
| | | | 25—13 | 2 | 1.50 | 52 | S4, S2 |
| | | | 12—2.5 | 1 | 1.50 | 60 | |
| | | | 2—1.5 | 1 | 1.50 | 66 | |
| | | | 1—.5 | 1 | 1.25 | 70 | |
| | | | .5 & Finer | 1 | .75 | 75 | |
| **Medium Carbon**<br>ASTM A352: Grades LCA, LCB, LCC<br>ASTM A356: Grade 1<br>1030<br>1040<br>1050 | 125 tc 175 | Annealed, Normalized or Normalized and Tempered | 1—2 | 2 | .060 | 130 | M2, M7 |
| | | | 3—10 | 1 | .060 | 150 | |
| | | | 11—19 | 1 | .060 | 165 | |
| | | | 20—48 | 1 | .050 | 180 | |
| | | | 48 & Finer | 1 | .030 | 195 | |
| | | | 25—13 | 2 | 1.50 | 40 | S4, S2 |
| | | | 12—2.5 | 1 | 1.50 | 46 | |
| | | | 2—1.5 | 1 | 1.50 | 50 | |
| | | | 1—.5 | 1 | 1.25 | 55 | |
| | | | .5 & Finer | 1 | .75 | 59 | |
| | 175 to 225 | Annealed, Normalized or Normalized and Tempered | 1—2 | 2 | .060 | 110 | M2, M7 |
| | | | 3—10 | 1 | .060 | 125 | |
| | | | 11—19 | 1 | .060 | 145 | |
| | | | 20—48 | 1 | .050 | 150 | |
| | | | 48 & Finer | 1 | .030 | 165 | |
| | | | 25—13 | 2 | 1.50 | 34 | S4, S2 |
| | | | 12—2.5 | 1 | 1.50 | 38 | |
| | | | 2—1.5 | 1 | 1.50 | 44 | |
| | | | 1—.5 | 1 | 1.25 | 46 | |
| | | | .5 & Finer | 1 | .75 | 50 | |

For hobbing Class 9 (per AGMA 390.03) or better gears, it may be necessary to reduce speeds and feeds by 50% and/or take 2 cuts.
To meet finish requirements it may be necessary to try both conventional and climb cutting.

See section 16 for Cutting Fluid Recommendations.
*Feeds are based on the largest standard recommended hob diameter. When using a smaller hob diameter, the Feed must be reduced proportionally.

# 7.1 Gear Hobbing

| MATERIAL | HARD-NESS<br>Bhn | CONDITION | DIAMETRAL PITCH<br>MODULE | NUMBER OF CUTS | FEED* PER REVOLUTION OF WORKPIECE<br>in<br>mm | HOB SPEED<br>fpm<br>m/min | HSS TOOL MATERIAL<br>AISI<br>ISO |
|---|---|---|---|---|---|---|---|
| **15. CARBON STEELS, CAST (cont.)**<br>**Medium Carbon (cont.)**<br>(materials listed on preceding page) | 250 to 300 | Quenched and Tempered | 1—2 | 2 | .045 | 85 | M2, M7 |
| | | | 3—10 | 1 | .045 | 95 | |
| | | | 11—19 | 1 | .045 | 115 | |
| | | | 20—48 | 1 | .045 | 120 | |
| | | | 48 & Finer | 1 | .030 | 130 | |
| | | | 25—13 | 2 | 1.15 | 26 | S4, S2 |
| | | | 12—2.5 | 1 | 1.15 | 29 | |
| | | | 2—1.5 | 1 | 1.15 | 35 | |
| | | | 1—.5 | 1 | 1.15 | 37 | |
| | | | .5 & Finer | 1 | .75 | 40 | |
| **16. ALLOY STEELS, CAST**<br>**Low Carbon**<br>ASTM A217: Grade WC9<br>ASTM A352: Grades LC3, LC4<br>ASTM A426: Grades CP2, CP5, CP5b, CP11, CP12, CP15, CP21, CP22<br>1320 2320 4120 8020<br>2315 4110 4320 8620 | 150 to 200 | Annealed, Normalized or Normalized and Tempered | 1—2 | 2 | .060 | 130 | M2, M7 |
| | | | 3—10 | 1 | .060 | 150 | |
| | | | 11—19 | 1 | .060 | 165 | |
| | | | 20—48 | 1 | .050 | 175 | |
| | | | 48 & Finer | 1 | .030 | 190 | |
| | | | 25—13 | 2 | 1.50 | 40 | S4, S2 |
| | | | 12—2.5 | 1 | 1.50 | 46 | |
| | | | 2—1.5 | 1 | 1.50 | 50 | |
| | | | 1—.5 | 1 | 1.25 | 53 | |
| | | | .5 & Finer | 1 | .75 | 58 | |
| | 200 to 225 | Annealed, Normalized or Normalized and Tempered | 1—2 | 2 | .060 | 110 | M2, M7 |
| | | | 3—10 | 1 | .060 | 130 | |
| | | | 11—19 | 1 | .060 | 145 | |
| | | | 20—48 | 1 | .050 | 155 | |
| | | | 48 & Finer | 1 | .030 | 170 | |
| | | | 25—13 | 2 | 1.50 | 34 | S4, S2 |
| | | | 12—2.5 | 1 | 1.50 | 40 | |
| | | | 2—1.5 | 1 | 1.50 | 44 | |
| | | | 1—.5 | 1 | 1.25 | 47 | |
| | | | .5 & Finer | 1 | .75 | 52 | |
| | 250 to 300 | Quenched and Tempered | 1—2 | 2 | .045 | 90 | M2, M7 |
| | | | 3—10 | 1 | .045 | 100 | |
| | | | 11—19 | 1 | .045 | 110 | |
| | | | 20—48 | 1 | .045 | 120 | |
| | | | 48 & Finer | 1 | .030 | 130 | |
| | | | 25—13 | 2 | 1.15 | 27 | S4, S2 |
| | | | 12—2.5 | 1 | 1.15 | 30 | |
| | | | 2—1.5 | 1 | 1.15 | 34 | |
| | | | 1—.5 | 1 | 1.15 | 37 | |
| | | | .5 & Finer | 1 | .75 | 40 | |
| **Medium Carbon**<br>ASTM A27: Grades N1, N2, U-60-30, 60-30, 65-35, 70-36, 70-40<br>ASTM A148: Grades 80-40, 80-50, 90-60, 105-85, 120-95, 150-125, 175-145<br>ASTM A216: Grades WCA, WCB, WCC<br>ASTM A217: Grades WC1, WC4, WC5, WC6<br>ASTM A352: Grades LC1, LC2, LC2-1<br>ASTM A356: Grades 2, 5, 6, 8, 9, 10<br>ASTM A389: Grades C23, C24<br>ASTM A486: Classes 70, 90, 120<br>(materials continued on next page) | 175 to 225 | Annealed, Normalized, or Normalized and Tempered | 1—2 | 2 | .060 | 110 | M2, M7 |
| | | | 3—10 | 1 | .060 | 130 | |
| | | | 11—19 | 1 | .060 | 145 | |
| | | | 20—48 | 1 | .050 | 155 | |
| | | | 48 & Finer | 1 | .030 | 170 | |
| | | | 25—13 | 2 | 1.50 | 34 | S4, S2 |
| | | | 12—2.5 | 1 | 1.50 | 40 | |
| | | | 2—1.5 | 1 | 1.50 | 44 | |
| | | | 1—.5 | 1 | 1.25 | 47 | |
| | | | .5 & Finer | 1 | .75 | 52 | |
| | 225 to 250 | Normalized, Normalized and Tempered or Quenched and Tempered | 1—2 | 2 | .060 | 95 | M2, M7 |
| | | | 3—10 | 1 | .060 | 110 | |
| | | | 11—19 | 1 | .060 | 125 | |
| | | | 20—48 | 1 | .050 | 140 | |
| | | | 48 & Finer | 1 | .030 | 155 | |
| | | | 25—13 | 2 | 1.50 | 29 | S4, S2 |
| | | | 12—2.5 | 1 | 1.50 | 34 | |
| | | | 2—1.5 | 1 | 1.50 | 38 | |
| | | | 1—.5 | 1 | 1.25 | 43 | |
| | | | .5 & Finer | 1 | .75 | 47 | |

For hobbing Class 9 (per AGMA 390.03) or better gears, it may be necessary to reduce speeds and feeds by 50% and/or take 2 cuts.
To meet finish requirements it may be necessary to try both conventional and climb cutting.

See section 16 for Cutting Fluid Recommendations.
*Feeds are based on the largest standard recommended hob diameter. When using a smaller hob diameter, the Feed must be reduced proportionally.

| MATERIAL | HARD-NESS Bhn | CONDITION | DIAMETRAL PITCH / MODULE | NUMBER OF CUTS | FEED* PER REVOLUTION OF WORKPIECE in / mm | HOB SPEED fpm / m/min | HSS TOOL MATERIAL AISI / ISO |
|---|---|---|---|---|---|---|---|
| **16. ALLOY STEELS, CAST (cont.) Medium Carbon (cont.)** (materials continued from preceding page) ASTM A487: Classes 1N, 2N, 4N, 6N, 8N, 9N, 10N, DN, 1Q, 2Q, 4Q, 4QA, 6Q, 7Q, 8Q, 9Q, 10Q | 250 to 300 | Quenched and Tempered | 1—2 | 2 | .045 | 80 | M2, M7 |
| 1330 4130 80B30 8640 | | | 3—10 | 1 | .045 | 90 | |
| 1340 4140 8040 9525 | | | 11—19 | 1 | .045 | 105 | |
| 2325 4330 8430 9530 | | | 20—48 | 1 | .045 | 110 | |
| 2330 4340 8440 9535 | | | 48 & Finer | 1 | .030 | 120 | |
| 4125 8030 8630 | | | 25—13 | 2 | 1.15 | 24 | S4, S2 |
| | | | 12—2.5 | 1 | 1.15 | 27 | |
| | | | 2—1.5 | 1 | 1.15 | 32 | |
| | | | 1—.5 | 1 | 1.15 | 34 | |
| | | | .5 & Finer | 1 | .75 | 37 | |
| | 300 to 350 | Quenched and Tempered | 1—2 | 2 | .045 | 70 | M2, M7 |
| | | | 3—10 | 1 | .045 | 75 | |
| | | | 11—19 | 1 | .045 | 85 | |
| | | | 20—48 | 1 | .045 | 90 | |
| | | | 48 & Finer | 1 | .030 | 100 | |
| | | | 25—13 | 2 | 1.15 | 21 | S4, S2 |
| | | | 12—2.5 | 1 | 1.15 | 23 | |
| | | | 2—1.5 | 1 | 1.15 | 26 | |
| | | | 1—.5 | 1 | 1.15 | 27 | |
| | | | .5 & Finer | 1 | .75 | 30 | |
| | 350 to 400 | Quenched and Tempered | 1—2 | 2 | .045 | 55 | M3, M42 |
| | | | 3—10 | 1 | .045 | 60 | |
| | | | 11—19 | 1 | .025 | 70 | |
| | | | 20—48 | 1 | .025 | 75 | |
| | | | 48 & Finer | 1 | .025 | 85 | |
| | | | 25—13 | 2 | 1.15 | 17 | S5, S11 |
| | | | 12—2.5 | 1 | 1.15 | 18 | |
| | | | 2—1.5 | 1 | .65 | 21 | |
| | | | 1—.5 | 1 | .65 | 23 | |
| | | | .5 & Finer | 1 | .65 | 26 | |
| **18. STAINLESS STEELS, CAST Ferritic** ASTM A217: Grades C5, C12 ASTM A296: Grades CB-30, CC-50, CE-30, CA6N, CA-6NM, CD4MCu ASTM A297: Grade HC ASTM A487: Class CA6NM ASTM A608: Grade HC30 | 135 to 185 | Annealed | 1—2 | 2 | .060 | 120 | M2, M7 |
| | | | 3—10 | 1 | .060 | 130 | |
| | | | 11—19 | 1 | .060 | 145 | |
| | | | 20—48 | 1 | .050 | 155 | |
| | | | 48 & Finer | 1 | .030 | 165 | |
| | | | 25—13 | 2 | 1.50 | 37 | S4, S2 |
| | | | 12—2.5 | 1 | 1.50 | 40 | |
| | | | 2—1.5 | 1 | 1.50 | 44 | |
| | | | 1—.5 | 1 | 1.25 | 47 | |
| | | | .5 & Finer | 1 | .75 | 50 | |
| **Austenitic** ASTM A296: Grades CF-16F, CN-7M, CN-7MS ASTM A351: Grade CN-7M | 140 to 170 | Annealed or Normalized | 1—2 | 2 | .060 | 90 | M2, M7 |
| | | | 3—10 | 1 | .060 | 100 | |
| | | | 11—19 | 1 | .060 | 120 | |
| | | | 20—48 | 1 | .050 | 130 | |
| | | | 48 & Finer | 1 | .030 | 150 | |
| | | | 25—13 | 2 | 1.50 | 27 | S4, S2 |
| | | | 12—2.5 | 1 | 1.50 | 30 | |
| | | | 2—1.5 | 1 | 1.50 | 37 | |
| | | | 1—.5 | 1 | 1.25 | 40 | |
| | | | .5 & Finer | 1 | .75 | 46 | |

For hobbing Class 9 (per AGMA 390.03) or better gears, it may be necessary to reduce speeds and feeds by 50% and/or take 2 cuts.
To meet finish requirements it may be necessary to try both conventional and climb cutting.

See section 16 for Cutting Fluid Recommendations.
*Feeds are based on the largest standard recommended hob diameter. When using a smaller hob diameter, the Feed must be reduced proportionally.

# 7.1 Gear Hobbing

| MATERIAL | HARD-NESS Bhn | CONDITION | DIAMETRAL PITCH / MODULE | NUMBER OF CUTS | FEED* PER REVOLUTION OF WORKPIECE in / mm | HOB SPEED fpm / m/min | HSS TOOL MATERIAL AISI / ISO |
|---|---|---|---|---|---|---|---|
| **18. STAINLESS STEELS, CAST (cont.)** **Austenitic** ASTM A296: Grades CF-3, CF-8, CF-8C, CF-20 ASTM A351: Grades CF-3, CF-3A, CF-8, CF-8A, CF-8C ASTM A451: Grades CPF3, CPF3A, CPF8, CPF8A, CPF8C, CPF8C (Ta Max.) ASTM A452: Grades TP 304H, TP 347H | 135 to 185 | Annealed or Normalized | 1—2 | 2 | .060 | 90 | M2, M7 |
| | | | 3—10 | 1 | .060 | 100 | |
| | | | 11—19 | 1 | .060 | 115 | |
| | | | 20—48 | 1 | .050 | 125 | |
| | | | 48 & Finer | 1 | .030 | 135 | |
| | | | 25—13 | 2 | 1.50 | 27 | S4, S2 |
| | | | 12—2.5 | 1 | 1.50 | 30 | |
| | | | 2—1.5 | 1 | 1.50 | 35 | |
| | | | 1—.5 | 1 | 1.25 | 38 | |
| | | | .5 & Finer | 1 | .75 | 41 | |
| **Austenitic** ASTM A296: Grades CF-3M, CF-8M, CG-8M, CG-12, CH-20, CK-20 ASTM A351: Grades CF-3M, CF-3MA, CF-8M, CF-10MC, CH-8, CH-10, CH-20, CK-20, HK-30, HK-40, HT-30 ASTM A451: Grades CPF3M, CPF8M, CPF10MC, CPH8, CPH10, CPH20, CPK20 ASTM A452: Grade TP 316H | 135 to 185 | Annealed or Normalized | 1—2 | 2 | .060 | 80 | M2, M7 |
| | | | 3—10 | 1 | .060 | 90 | |
| | | | 11—19 | 1 | .060 | 105 | |
| | | | 20—48 | 1 | .050 | 115 | |
| | | | 48 & Finer | 1 | .030 | 125 | |
| | | | 25—13 | 2 | 1.50 | 24 | S4, S2 |
| | | | 12—2.5 | 1 | 1.50 | 27 | |
| | | | 2—1.5 | 1 | 1.50 | 32 | |
| | | | 1—.5 | 1 | 1.25 | 35 | |
| | | | .5 & Finer | 1 | .75 | 38 | |
| **Austenitic** ASTM A297: Grades HD, HE, HF, HH, HI, HK, HL, HN, HP, HT, HU ASTM A608: Grades HD50, HE35, HF30, HH30, HH33, HI35, HK30, HK40, HL30, HL40, HN40, HT50, HU50 | 160 to 210 | As Cast | 1—2 | 2 | .060 | 75 | M2, M7 |
| | | | 3—10 | 1 | .060 | 85 | |
| | | | 11—19 | 1 | .060 | 100 | |
| | | | 20—48 | 1 | .050 | 110 | |
| | | | 48 & Finer | 1 | .030 | 120 | |
| | | | 25—13 | 2 | 1.50 | 23 | S4, S2 |
| | | | 12—2.5 | 1 | 1.50 | 26 | |
| | | | 2—1.5 | 1 | 1.50 | 30 | |
| | | | 1—.5 | 1 | 1.25 | 34 | |
| | | | .5 & Finer | 1 | .75 | 37 | |
| **Martensitic** ASTM A217: Grade CA-15 ASTM A296: Grades CA-15, CA-15M, CA-40 ASTM A426: Grades CP7, CP9, CPCA15 ASTM A487: Classes CA15a, CA-15M | 135 to 175 | Annealed | 1—2 | 2 | .060 | 95 | M2, M7 |
| | | | 3—10 | 1 | .060 | 110 | |
| | | | 11—19 | 1 | .060 | 125 | |
| | | | 20—48 | 1 | .050 | 135 | |
| | | | 48 & Finer | 1 | .030 | 145 | |
| | | | 25—13 | 2 | 1.50 | 29 | S4, S2 |
| | | | 12—2.5 | 1 | 1.50 | 34 | |
| | | | 2—1.5 | 1 | 1.50 | 38 | |
| | | | 1—.5 | 1 | 1.25 | 41 | |
| | | | .5 & Finer | 1 | .75 | 44 | |

For hobbing Class 9 (per AGMA 390.03) or better gears, it may be necessary to reduce speeds and feeds by 50% and/or take 2 cuts.
To meet finish requirements it may be necessary to try both conventional and climb cutting.

See section 16 for Cutting Fluid Recommendations.
*Feeds are based on the largest standard recommended hob diameter. When using a smaller hob diameter, the Feed must be reduced proportionally.

| MATERIAL | HARD-NESS Bhn | CONDITION | DIAMETRAL PITCH / MODULE | NUMBER OF CUTS | FEED* PER REVOLUTION OF WORKPIECE in / mm | HOB SPEED fpm / m/min | HSS TOOL MATERIAL AISI / ISO |
|---|---|---|---|---|---|---|---|
| **18. STAINLESS STEELS, CAST (cont.) Martensitic (cont.)** (materials listed on preceding page) | 175 to 225 | Annealed, Normalized or Normalized and Tempered | 1—2 | 2 | .060 | 80 | M2, M7 |
| | | | 3—10 | 1 | .060 | 105 | |
| | | | 11—19 | 1 | .060 | 115 | |
| | | | 20—48 | 1 | .050 | 125 | |
| | | | 48 & Finer | 1 | .030 | 140 | |
| | | | 25—13 | 2 | 1.50 | 24 | S4, S2 |
| | | | 12—2.5 | 1 | 1.50 | 32 | |
| | | | 2—1.5 | 1 | 1.50 | 35 | |
| | | | 1—.5 | 1 | 1.25 | 38 | |
| | | | .5 & Finer | 1 | .75 | 43 | |
| | 275 to 325 | Quenched and Tempered | 1—2 | 2 | .045 | 50 | M2, M7 |
| | | | 3—10 | 1 | .045 | 70 | |
| | | | 11—19 | 1 | .045 | 80 | |
| | | | 20—48 | 1 | .045 | 90 | |
| | | | 48 & Finer | 1 | .030 | 105 | |
| | | | 25—13 | 2 | 1.15 | 15 | S4, S2 |
| | | | 12—2.5 | 1 | 1.15 | 21 | |
| | | | 2—1.5 | 1 | 1.15 | 24 | |
| | | | 1—.5 | 1 | 1.15 | 27 | |
| | | | .5 & Finer | 1 | .75 | 32 | |
| **19. PRECIPITATION HARDENING STAINLESS STEELS, CAST** ASTM A351: Grade CD-4MCu; ACI Grade CB-7Cu; ACI Grade CD-4MCu; 17-4 PH; AM-355 | 325 to 375 | Solution Treated | 1—2 | 2 | .045 | 45 | M3, M42 |
| | | | 3—10 | 1 | .045 | 60 | |
| | | | 11—19 | 1 | .025 | 70 | |
| | | | 20—48 | 1 | .025 | 75 | |
| | | | 48 & Finer | 1 | .025 | 85 | |
| | | | 25—13 | 2 | 1.15 | 14 | S5, S11 |
| | | | 12—2.5 | 1 | 1.15 | 18 | |
| | | | 2—1.5 | 1 | .65 | 21 | |
| | | | 1—.5 | 1 | .65 | 23 | |
| | | | .5 & Finer | 1 | .65 | 26 | |
| **21. GRAY CAST IRONS Ferritic** ASTM A48: Class 20; SAE J431c: Grade G1800 | 120 to 150 | Annealed | 1—2 | 1 | .060 | 155 | M2, M7 |
| | | | 3—10 | 1 | .050 | 180 | |
| | | | 11—19 | 1 | .030 | 210 | |
| | | | 20—48 | 1 | .015 | 220 | |
| | | | 48 & Finer | 1 | .015 | 240 | |
| | | | 25—13 | 1 | 1.50 | 47 | S4, S2 |
| | | | 12—2.5 | 1 | 1.25 | 55 | |
| | | | 2—1.5 | 1 | .75 | 64 | |
| | | | 1—.5 | 1 | .40 | 67 | |
| | | | .5 & Finer | 1 | .40 | 73 | |
| **Pearlitic- Ferritic** ASTM A48: Class 25; SAE J431c: Grade G2500 | 160 to 200 | As Cast | 1—2 | 1 | .060 | 100 | M2, M7 |
| | | | 3—10 | 1 | .050 | 110 | |
| | | | 11—19 | 1 | .030 | 130 | |
| | | | 20—48 | 1 | .015 | 135 | |
| | | | 48 & Finer | 1 | .015 | 150 | |
| | | | 25—13 | 1 | 1.50 | 30 | S4, S2 |
| | | | 12—2.5 | 1 | 1.25 | 34 | |
| | | | 2—1.5 | 1 | .75 | 40 | |
| | | | 1—.5 | 1 | .40 | 41 | |
| | | | .5 & Finer | 1 | .40 | 46 | |
| **Pearlitic** ASTM A48: Classes 30, 35, 40; SAE J431c: Grade G3000 | 190 to 220 | As Cast | 1—2 | 1 | .060 | 75 | M2, M7 |
| | | | 3—10 | 1 | .050 | 85 | |
| | | | 11—19 | 1 | .030 | 95 | |
| | | | 20—48 | 1 | .015 | 110 | |
| | | | 48 & Finer | 1 | .015 | 120 | |
| | | | 25—13 | 1 | 1.50 | 23 | S4, S2 |
| | | | 12—2.5 | 1 | 1.25 | 26 | |
| | | | 2—1.5 | 1 | .75 | 29 | |
| | | | 1—.5 | 1 | .40 | 34 | |
| | | | .5 & Finer | 1 | .40 | 37 | |

For hobbing Class 9 (per AGMA 390.03) or better gears, it may be necessary to reduce speeds and feeds by 50% and/or take 2 cuts.
To meet finish requirements it may be necessary to try both conventional and climb cutting.

See section 16 for Cutting Fluid Recommendations.
*Feeds are based on the largest standard recommended hob diameter. When using a smaller hob diameter, the Feed must be reduced proportionally.

# 7.1 Gear Hobbing

| MATERIAL | HARDNESS Bhn | CONDITION | DIAMETRAL PITCH / MODULE | NUMBER OF CUTS | FEED* PER REVOLUTION OF WORKPIECE in / mm | HOB SPEED fpm / m/min | HSS TOOL MATERIAL AISI / ISO |
|---|---|---|---|---|---|---|---|
| **21. GRAY CAST IRONS (cont.)** **Pearlitic + Free Carbides** ASTM A48: Classes 45, 50 SAE J431c: Grades G3500, G4000 | 220 to 260 | As Cast | 1—2 | 2 | .060 | 60 | M2, M7 |
| | | | 3—10 | 1 | .050 | 65 | |
| | | | 11—19 | 1 | .030 | 75 | |
| | | | 20—48 | 1 | .020 | 80 | |
| | | | 48 & Finer | 1 | .015 | 90 | |
| | | | 25—13 | 2 | 1.50 | 18 | S4, S2 |
| | | | 12—2.5 | 1 | 1.25 | 20 | |
| | | | 2—1.5 | 1 | .75 | 23 | |
| | | | 1—.5 | 1 | .50 | 24 | |
| | | | .5 & Finer | 1 | .40 | 27 | |
| **Pearlitic or Acicular + Free Carbides** ASTM A48: Classes 55, 60 | 250 to 320 | As Cast or Quenched and Tempered | 1—2 | 2 | .060 | 40 | M2, M7 |
| | | | 3—10 | 1 | .050 | 45 | |
| | | | 11—19 | 1 | .030 | 55 | |
| | | | 20—48 | 1 | .020 | 60 | |
| | | | 48 & Finer | 1 | .015 | 70 | |
| | | | 25—13 | 2 | 1.50 | 12 | S4, S2 |
| | | | 12—2.5 | 1 | 1.25 | 14 | |
| | | | 2—1.5 | 1 | .75 | 17 | |
| | | | 1—.5 | 1 | .50 | 18 | |
| | | | .5 & Finer | 1 | .40 | 21 | |
| **Austenitic (NI-RESIST)** ASTM A436: Types 1, 1b, 2, 2b, 3, 4, 5, 6 | 100 to 250 | As Cast | 1—2 | 2 | .060 | 50 | M2, M7 |
| | | | 3—10 | 1 | .050 | 55 | |
| | | | 11—19 | 1 | .030 | 65 | |
| | | | 20—48 | 1 | .020 | 70 | |
| | | | 48 & Finer | 1 | .015 | 80 | |
| | | | 25—13 | 2 | 1.50 | 15 | S4, S2 |
| | | | 12—2.5 | 1 | 1.25 | 17 | |
| | | | 2—1.5 | 1 | .75 | 20 | |
| | | | 1—.5 | 1 | .50 | 21 | |
| | | | .5 & Finer | 1 | .40 | 24 | |
| **23. DUCTILE CAST IRONS** **Ferritic** ASTM A536: Grades 60-40-18, 65-45-12 SAE J434c: Grades D4018, D4512 | 140 to 190 | Annealed | 1—2 | 1 | .060 | 135 | M2, M7 |
| | | | 3—10 | 1 | .050 | 155 | |
| | | | 11—19 | 1 | .030 | 170 | |
| | | | 20—48 | 1 | .020 | 185 | |
| | | | 48 & Finer | 1 | .015 | 200 | |
| | | | 25—13 | 1 | 1.50 | 41 | S4, S2 |
| | | | 12—2.5 | 1 | 1.25 | 47 | |
| | | | 2—1.5 | 1 | .75 | 52 | |
| | | | 1—.5 | 1 | .50 | 56 | |
| | | | .5 & Finer | 1 | .40 | 60 | |
| **Ferritic- Pearlitic** ASTM A536: Grade 80-55-06 SAE J434c: Grade D5506 | 190 to 225 | As Cast | 1—2 | 1 | .060 | 90 | M2, M7 |
| | | | 3—10 | 1 | .050 | 100 | |
| | | | 11—19 | 1 | .030 | 120 | |
| | | | 20—48 | 1 | .020 | 130 | |
| | | | 48 & Finer | 1 | .015 | 140 | |
| | | | 25—13 | 1 | 1.50 | 27 | S4, S2 |
| | | | 12—2.5 | 1 | 1.25 | 30 | |
| | | | 2—1.5 | 1 | .75 | 37 | |
| | | | 1—.5 | 1 | .50 | 40 | |
| | | | .5 & Finer | 1 | .40 | 43 | |
| | 225 to 260 | As Cast | 1—2 | 1 | .060 | 75 | M2, M7 |
| | | | 3—10 | 1 | .050 | 95 | |
| | | | 11—19 | 1 | .030 | 115 | |
| | | | 20—48 | 1 | .020 | 125 | |
| | | | 48 & Finer | 1 | .015 | 135 | |
| | | | 25—13 | 1 | 1.50 | 23 | S4, S2 |
| | | | 12—2.5 | 1 | 1.25 | 29 | |
| | | | 2—1.5 | 1 | .75 | 35 | |
| | | | 1—.5 | 1 | .50 | 38 | |
| | | | .5 & Finer | 1 | .40 | 41 | |

For hobbing Class 9 (per AGMA 390.03) or better gears, it may be necessary to reduce speeds and feeds by 50% and/or take 2 cuts. To meet finish requirements it may be necessary to try both conventional and climb cutting.

See section 16 for Cutting Fluid Recommendations.

*Feeds are based on the largest standard recommended hob diameter. When using a smaller hob diameter, the Feed must be reduced proportionally.

| MATERIAL | HARD-NESS<br>Bhn | CONDITION | DIAMETRAL PITCH / MODULE | NUMBER OF CUTS | FEED* PER REVOLUTION OF WORKPIECE<br>in / mm | HOB SPEED<br>fpm / m/min | HSS TOOL MATERIAL<br>AISI / ISO |
|---|---|---|---|---|---|---|---|
| **Pearlitic- Martensitic**<br>ASTM A536: Grade 100-70-03<br>SAE J434c: Grade D7003 | 240 to 300 | Normalized and Tempered | 1—2 | 1 | .060 | 60 | M2, M7 |
| | | | 3—10 | 1 | .050 | 65 | |
| | | | 11—19 | 1 | .030 | 70 | |
| | | | 20—48 | 1 | .020 | 80 | |
| | | | 48 & Finer | 1 | .015 | 90 | |
| | | | 25—13 | 1 | 1.50 | 18 | S4, S2 |
| | | | 12—2.5 | 1 | 1.25 | 20 | |
| | | | 2—1.5 | 1 | .75 | 21 | |
| | | | 1—.5 | 1 | .50 | 24 | |
| | | | .5 & Finer | 1 | .40 | 27 | |
| **Martensitic**<br>ASTM A536: Grade 120-90-02<br>SAE J434c: Grade DQ&T | 270 to 330 | Quenched and Tempered | 1—2 | 1 | .060 | 35 | M2, M7 |
| | | | 3—10 | 1 | .050 | 50 | |
| | | | 11—19 | 1 | .030 | 60 | |
| | | | 20—48 | 1 | .020 | 70 | |
| | | | 48 & Finer | 1 | .015 | 80 | |
| | | | 25—13 | 1 | 1.50 | 11 | S4, S2 |
| | | | 12—2.5 | 1 | 1.25 | 15 | |
| | | | 2—1.5 | 1 | .75 | 18 | |
| | | | 1—.5 | 1 | .50 | 21 | |
| | | | .5 & Finer | 1 | .40 | 24 | |
| **Austenitic (NI-RESIST Ductile)**<br>ASTM A439: Types D-2, D-2C, D-3A, D-5<br>ASTM A571: Type D-2M | 120 to 200 | Annealed | 1—2 | 1 | .035 | 45 | M2, M7 |
| | | | 3—10 | 1 | .030 | 50 | |
| | | | 11—19 | 1 | .020 | 60 | |
| | | | 20—48 | 1 | .020 | 65 | |
| | | | 48 & Finer | 1 | .015 | 75 | |
| | | | 25—13 | 1 | .90 | 14 | S4, S2 |
| | | | 12—2.5 | 1 | .75 | 15 | |
| | | | 2—1.5 | 1 | .50 | 18 | |
| | | | 1—.5 | 1 | .50 | 20 | |
| | | | .5 & Finer | 1 | .40 | 23 | |
| **Austenitic (NI-RESIST Ductile)**<br>ASTM A439: Types D-2B, D-3, D-4, D-5B | 140 to 275 | Annealed | 1—2 | 1 | .035 | 40 | M2, M7 |
| | | | 3—10 | 1 | .030 | 45 | |
| | | | 11—19 | 1 | .020 | 55 | |
| | | | 20—48 | 1 | .020 | 60 | |
| | | | 48 & Finer | 1 | .015 | 70 | |
| | | | 25—13 | 1 | .90 | 12 | S4, S2 |
| | | | 12—2.5 | 1 | .75 | 14 | |
| | | | 2—1.5 | 1 | .50 | 17 | |
| | | | 1—.5 | 1 | .50 | 18 | |
| | | | .5 & Finer | 1 | .40 | 21 | |
| **24. MALLEABLE CAST IRONS**<br>**Ferritic**<br>ASTM A47: Grades 32510, 35018<br>ASTM A602: Grade M3210<br>SAE J158: Grade M3210 | 110 to 160 | Malleablized | 1—2 | 1 | .060 | 200 | M2, M7 |
| | | | 3—10 | 1 | .050 | 225 | |
| | | | 11—19 | 1 | .030 | 260 | |
| | | | 20—48 | 1 | .020 | 270 | |
| | | | 48 & Finer | 1 | .015 | 300 | |
| | | | 25—13 | 1 | 1.50 | 60 | S4, S2 |
| | | | 12—2.5 | 1 | 1.25 | 69 | |
| | | | 2—1.5 | 1 | .75 | 79 | |
| | | | 1—.5 | 1 | .50 | 82 | |
| | | | .5 & Finer | 1 | .40 | 90 | |
| **Pearlitic**<br>ASTM A220: Grades 40010, 45006, 45008, 50005<br>ASTM A602: Grade M4504, M5003<br>SAE J158: Grades M4504, M5003 | 160 to 200 | Malleablized and Heat Treated | 1—2 | 1 | .060 | 125 | M2, M7 |
| | | | 3—10 | 1 | .050 | 150 | |
| | | | 11—19 | 1 | .030 | 175 | |
| | | | 20—48 | 1 | .020 | 180 | |
| | | | 48 & Finer | 1 | .015 | 200 | |
| | | | 25—13 | 1 | 1.50 | 38 | S4, S2 |
| | | | 12—2.5 | 1 | 1.25 | 46 | |
| | | | 2—1.5 | 1 | .75 | 53 | |
| | | | 1—.5 | 1 | .50 | 55 | |
| | | | .5 & Finer | 1 | .40 | 60 | |

For hobbing Class 9 (per AGMA 390.03) or better gears, it may be necessary to reduce speeds and feeds by 50% and/or take 2 cuts.
To meet finish requirements it may be necessary to try both conventional and climb cutting.

See section 16 for Cutting Fluid Recommendations.
*Feeds are based on the largest standard recommended hob diameter. When using a smaller hob diameter, the Feed must be reduced proportionally.

# 7.1 Gear Hobbing

| MATERIAL | HARD-NESS Bhn | CONDITION | DIAMETRAL PITCH / MODULE | NUMBER OF CUTS | FEED* PER REVOLUTION OF WORKPIECE in / mm | HOB SPEED fpm / m/min | HSS TOOL MATERIAL AISI / ISO |
|---|---|---|---|---|---|---|---|
| **24. MALLEABLE CAST IRONS (cont.)** **Pearlitic (cont.)** (materials listed on preceding page) | 200 to 240 | Malleablized and Heat Treated | 1—2 | 1 | .060 | 100 | M2, M7 |
| | | | 3—10 | 1 | .050 | 120 | |
| | | | 11—19 | 1 | .030 | 140 | |
| | | | 20—48 | 1 | .020 | 150 | |
| | | | 48 & Finer | 1 | .015 | 160 | |
| | | | 25—13 | 1 | 1.50 | 30 | S4, S2 |
| | | | 12—2.5 | 1 | 1.25 | 37 | |
| | | | 2—1.5 | 1 | .75 | 43 | |
| | | | 1—.5 | 1 | .50 | 46 | |
| | | | .5 & Finer | 1 | .40 | 49 | |
| **Tempered Martensite** ASTM A220: Grade 60004 ASTM A602: Grade M5503 SAE J158: Grade M5503 | 200 to 255 | Malleablized and Heat Treated | 1—2 | 1 | .060 | 90 | M2, M7 |
| | | | 3—10 | 1 | .050 | 110 | |
| | | | 11—19 | 1 | .030 | 130 | |
| | | | 20—48 | 1 | .020 | 140 | |
| | | | 48 & Finer | 1 | .015 | 150 | |
| | | | 25—13 | 1 | 1.50 | 27 | S4, S2 |
| | | | 12—2.5 | 1 | 1.25 | 34 | |
| | | | 2—1.5 | 1 | .75 | 40 | |
| | | | 1—.5 | 1 | .50 | 43 | |
| | | | .5 & Finer | 1 | .40 | 46 | |
| **Tempered Martensite** ASTM A220: Grade 70003 ASTM A602: Grade M7002 SAE J158: Grade M7002 | 220 to 260 | Malleablized and Heat Treated | 1—2 | 1 | .060 | 85 | M2, M7 |
| | | | 3—10 | 1 | .050 | 105 | |
| | | | 11—19 | 1 | .030 | 125 | |
| | | | 20—48 | 1 | .020 | 135 | |
| | | | 48 & Finer | 1 | .015 | 145 | |
| | | | 25—13 | 1 | 1.50 | 26 | S4, S2 |
| | | | 12—2.5 | 1 | 1.25 | 32 | |
| | | | 2—1.5 | 1 | .75 | 38 | |
| | | | 1—.5 | 1 | .50 | 41 | |
| | | | .5 & Finer | 1 | .40 | 44 | |
| **Tempered Martensite** ASTM A220: Grade 80002 ASTM A602: Grade M8501 SAE J158: Grade M8501 | 240 to 280 | Malleablized and Heat Treated | 1—2 | 1 | .060 | 80 | M2, M7 |
| | | | 3—10 | 1 | .050 | 100 | |
| | | | 11—19 | 1 | .030 | 120 | |
| | | | 20—48 | 1 | .020 | 130 | |
| | | | 48 & Finer | 1 | .015 | 140 | |
| | | | 25—13 | 1 | 1.50 | 24 | S4, S2 |
| | | | 12—2.5 | 1 | 1.25 | 30 | |
| | | | 2—1.5 | 1 | .75 | 37 | |
| | | | 1—.5 | 1 | .50 | 40 | |
| | | | .5 & Finer | 1 | .40 | 43 | |
| **Tempered Martensite** ASTM A220: Grade 90001 ASTM A602: Grade M8501 SAE J158: Grade M8501 | 250 to 320 | Malleablized and Heat Treated | 1—2 | 1 | .060 | 70 | M2, M7 |
| | | | 3—10 | 1 | .050 | 90 | |
| | | | 11—19 | 1 | .030 | 110 | |
| | | | 20—48 | 1 | .020 | 120 | |
| | | | 48 & Finer | 1 | .015 | 130 | |
| | | | 25—13 | 1 | 1.50 | 21 | S4, S2 |
| | | | 12—2.5 | 1 | 1.25 | 27 | |
| | | | 2—1.5 | 1 | .75 | 34 | |
| | | | 1—.5 | 1 | .50 | 37 | |
| | | | .5 & Finer | 1 | .40 | 40 | |

For hobbing Class 9 (per AGMA 390.03) or better gears, it may be necessary to reduce speeds and feeds by 50% and/or take 2 cuts.
To meet finish requirements it may be necessary to try both conventional and climb cutting.

See section 16 for Cutting Fluid Recommendations.
*Feeds are based on the largest standard recommended hob diameter. When using a smaller hob diameter, the Feed must be reduced proportionally.

| MATERIAL | HARDNESS Bhn | CONDITION | DIAMETRAL PITCH / MODULE | NUMBER OF CUTS | FEED* PER REVOLUTION OF WORKPIECE in / mm | HOB SPEED fpm / m/min | HSS TOOL MATERIAL AISI / ISO |
|---|---|---|---|---|---|---|---|
| **28. ALUMINUM ALLOYS, WROUGHT**<br>EC 2218 5252 6253<br>1060 2219 5254 6262<br>1100 2618 5454 6463<br>1145 3003 5456 6951<br>1175 3004 5457 7001<br>1235 3005 5652 7004<br>2011 4032 5657 7005<br>2014 5005 6053 7039<br>2017 5050 6061 7049<br>2018 5052 6063 7050<br>2021 5056 6066 7075<br>2024 5083 6070 7079<br>2025 5086 6101 7175<br>2117 5154 6151 7178 | 30 to 80<br>500kg | Cold Drawn | 1—2 | 1 | .060 | 850 | M2, M7 |
| | | | 3—10 | 1 | .050 | 900 | |
| | | | 11—19 | 1 | .030 | 1000 | |
| | | | 20—48 | 1 | .020 | 1050 | |
| | | | 48 & Finer | 1 | .015 | 1150 | |
| | | | 25—13 | 1 | 1.50 | 260 | S4, S2 |
| | | | 12—2.5 | 1 | 1.25 | 275 | |
| | | | 2—1.5 | 1 | .75 | 305 | |
| | | | 1—.5 | 1 | .50 | 320 | |
| | | | .5 & Finer | 1 | .40 | 350 | |
| | 75 to 150<br>500kg | Solution Treated and Aged | 1—2 | 1 | .060 | 800 | M2, M7 |
| | | | 3—10 | 1 | .050 | 850 | |
| | | | 11—19 | 1 | .030 | 950 | |
| | | | 20—48 | 1 | .020 | 1000 | |
| | | | 48 & Finer | 1 | .015 | 1100 | |
| | | | 25—13 | 1 | 1.50 | 245 | S4, S2 |
| | | | 12—2.5 | 1 | 1.25 | 260 | |
| | | | 2—1.5 | 1 | .75 | 290 | |
| | | | 1—.5 | 1 | .50 | 305 | |
| | | | .5 & Finer | 1 | .40 | 335 | |
| **29. ALUMINUM ALLOYS, CAST Sand and Permanent Mold**<br>A140 319.0 357.0 A712.0<br>201.0 328.0 359.0 D712.0<br>208.0 A332.0 B443.0 713.0<br>213.0 F332.0 514.0 771.0<br>222.0 333.0 A514.0 850.0<br>224.0 354.0 B514.0 A850.0<br>242.0 355.0 520.0 B850.0<br>295.0 C355.0 535.0<br>B295.0 356.0 705.0<br>308.0 A356.0 707.0<br>Hiduminium RR-350 | 40 to 100<br>500kg | As Cast | 1—2 | 1 | .060 | 700 | M2, M7 |
| | | | 3—10 | 1 | .050 | 750 | |
| | | | 11—19 | 1 | .030 | 850 | |
| | | | 20—48 | 1 | .020 | 900 | |
| | | | 48 & Finer | 1 | .015 | 1000 | |
| | | | 25—13 | 1 | 1.50 | 215 | S4, S2 |
| | | | 12—2.5 | 1 | 1.25 | 230 | |
| | | | 2—1.5 | 1 | .75 | 260 | |
| | | | 1—.5 | 1 | .50 | 275 | |
| | | | .5 & Finer | 1 | .40 | 305 | |
| | 70 to 125<br>500kg | Solution Treated and Aged | 1—2 | 1 | .060 | 700 | M2, M7 |
| | | | 3—10 | 1 | .050 | 750 | |
| | | | 11—19 | 1 | .030 | 850 | |
| | | | 20—48 | 1 | .020 | 900 | |
| | | | 48 & Finer | 1 | .015 | 1000 | |
| | | | 25—13 | 1 | 1.50 | 215 | S4, S2 |
| | | | 12—2.5 | 1 | 1.25 | 230 | |
| | | | 2—1.5 | 1 | .75 | 260 | |
| | | | 1—.5 | 1 | .50 | 275 | |
| | | | .5 & Finer | 1 | .40 | 305 | |
| **34. COPPER ALLOYS, WROUGHT**<br>145 332 360 482<br>147 335 365 485<br>173 340 366 544<br>187 342 367 623<br>191 349 368 624<br>314 350 370 638<br>316 353 377 642<br>330 356 385 782 | 10 $R_B$ to 70 $R_B$ | Annealed | 1—2 | 1 | .060 | 600 | M2, M7 |
| | | | 3—10 | 1 | .050 | 650 | |
| | | | 11—19 | 1 | .030 | 700 | |
| | | | 20—48 | 1 | .020 | 750 | |
| | | | 48 & Finer | 1 | .015 | 800 | |
| | | | 25—13 | 1 | 1.50 | 185 | S4, S2 |
| | | | 12—2.5 | 1 | 1.25 | 200 | |
| | | | 2—1.5 | 1 | .75 | 215 | |
| | | | 1—.5 | 1 | .50 | 230 | |
| | | | .5 & Finer | 1 | .40 | 245 | |
| | 60 $R_B$ to 100 $R_B$ | Cold Drawn | 1—2 | 1 | .060 | 650 | M2, M7 |
| | | | 3—10 | 1 | .050 | 700 | |
| | | | 11—19 | 1 | .030 | 750 | |
| | | | 20—48 | 1 | .020 | 800 | |
| | | | 48 & Finer | 1 | .015 | 850 | |
| | | | 25—13 | 1 | 1.50 | 200 | S4, S2 |
| | | | 12—2.5 | 1 | 1.25 | 215 | |
| | | | 2—1.5 | 1 | .75 | 230 | |
| | | | 1—.5 | 1 | .50 | 245 | |
| | | | .5 & Finer | 1 | .40 | 260 | |

For hobbing Class 9 (per AGMA 390.03) or better gears, it may be necessary to reduce speeds and feeds by 50% and/or take 2 cuts.
To meet finish requirements it may be necessary to try both conventional and climb cutting.

See section 16 for Cutting Fluid Recommendations.
*Feeds are based on the largest standard recommended hob diameter. When using a smaller hob diameter, the Feed must be reduced proportionally.

# 7.1 Gear Hobbing

| MATERIAL | HARDNESS Bhn | CONDITION | DIAMETRAL PITCH / MODULE | NUMBER OF CUTS | FEED* PER REVOLUTION OF WORKPIECE in / mm | HOB SPEED fpm / m/min | HSS TOOL MATERIAL AISI / ISO |
|---|---|---|---|---|---|---|---|
| **34. COPPER ALLOYS, WROUGHT (cont.)**<br>190 425 466 667<br>226 435 467 675<br>230 442 613 687<br>240 443 618 694<br>260 444 630 770<br>268 445 632<br>270 464 651<br>280 465 655 | 10 $R_B$ to 70 $R_B$ | Annealed | 1—2 | 1 | .060 | 250 | M2, M7 |
| | | | 3—10 | 1 | .050 | 300 | |
| | | | 11—19 | 1 | .030 | 350 | |
| | | | 20—48 | 1 | .020 | 400 | |
| | | | 48 & Finer | 1 | .015 | 450 | |
| | | | 25—13 | 1 | 1.50 | 76 | S4, S2 |
| | | | 12—2.5 | 1 | 1.25 | 90 | |
| | | | 2—1.5 | 1 | .75 | 105 | |
| | | | 1—.5 | 1 | .50 | 120 | |
| | | | .5 & Finer | 1 | .40 | 135 | |
| | 60 $R_B$ to 100 $R_B$ | Cold Drawn | 1—2 | 1 | .060 | 300 | M2, M7 |
| | | | 3—10 | 1 | .050 | 350 | |
| | | | 11—19 | 1 | .030 | 400 | |
| | | | 20—48 | 1 | .020 | 450 | |
| | | | 48 & Finer | 1 | .015 | 500 | |
| | | | 25—13 | 1 | 1.50 | 90 | S4, S2 |
| | | | 12—2.5 | 1 | 1.25 | 105 | |
| | | | 2—1.5 | 1 | .75 | 120 | |
| | | | 1—.5 | 1 | .50 | 135 | |
| | | | .5 & Finer | 1 | .40 | 150 | |
| 101 125 185 614<br>102 127 189 619<br>104 128 192 625<br>105 129 194 674<br>107 130 195 688<br>109 142 210 706<br>110 143 220 710<br>111 150 411 715<br>113 155 413 725<br>114 162 505 745<br>115 165 510 752<br>116 170 511 754<br>119 172 521 757<br>120 175 524<br>121 182 608<br>122 184 610 | 10 $R_B$ to 70 $R_B$ | Annealed | 1—2 | 1 | .060 | 100 | M2, M7 |
| | | | 3—10 | 1 | .050 | 110 | |
| | | | 11—19 | 1 | .030 | 130 | |
| | | | 20—48 | 1 | .020 | 135 | |
| | | | 48 & Finer | 1 | .015 | 150 | |
| | | | 25—13 | 1 | 1.50 | 30 | S4, S2 |
| | | | 12—2.5 | 1 | 1.25 | 34 | |
| | | | 2—1.5 | 1 | .75 | 40 | |
| | | | 1—.5 | 1 | .50 | 41 | |
| | | | .5 & Finer | 1 | .40 | 46 | |
| | 60 $R_B$ to 100 $R_B$ | Cold Drawn | 1—2 | 1 | .060 | 125 | M2, M7 |
| | | | 3—10 | 1 | .050 | 150 | |
| | | | 11—19 | 1 | .030 | 160 | |
| | | | 20—48 | 1 | .020 | 180 | |
| | | | 48 & Finer | 1 | .015 | 200 | |
| | | | 25—13 | 1 | 1.50 | 38 | S4, S2 |
| | | | 12—2.5 | 1 | 1.25 | 46 | |
| | | | 2—1.5 | 1 | .75 | 49 | |
| | | | 1—.5 | 1 | .50 | 55 | |
| | | | .5 & Finer | 1 | .40 | 60 | |
| **35. COPPER ALLOYS, CAST**<br>834 855 934 953<br>836 857 935 954<br>838 858 937 956<br>842 864 938 973<br>844 867 939 974<br>848 879 943 976<br>852 928 944 978<br>854 932 945 | 40 to 150 500kg | As Cast | 1—2 | 1 | .060 | 600 | M2, M7 |
| | | | 3—10 | 1 | .050 | 700 | |
| | | | 11—19 | 1 | .030 | 800 | |
| | | | 20—48 | 1 | .020 | 900 | |
| | | | 48 & Finer | 1 | .015 | 950 | |
| | | | 25—13 | 1 | 1.50 | 185 | S4, S2 |
| | | | 12—2.5 | 1 | 1.25 | 215 | |
| | | | 2—1.5 | 1 | .75 | 245 | |
| | | | 1—.5 | 1 | .50 | 275 | |
| | | | .5 & Finer | 1 | .40 | 290 | |
| 817 874 925<br>821 875 926<br>833 876 927<br>853 878 947<br>861 903 948<br>862 905 952<br>865 915 955<br>868 922 957<br>872 923 958 | 40 to 175 500kg | As Cast | 1—2 | 1 | .060 | 250 | M2, M7 |
| | | | 3—10 | 1 | .050 | 300 | |
| | | | 11—19 | 1 | .030 | 350 | |
| | | | 20—48 | 1 | .020 | 400 | |
| | | | 48 & Finer | 1 | .015 | 450 | |
| | | | 25—13 | 1 | 1.50 | 76 | S4, S2 |
| | | | 12—2.5 | 1 | 1.25 | 90 | |
| | | | 2—1.5 | 1 | .75 | 105 | |
| | | | 1—.5 | 1 | .50 | 120 | |
| | | | .5 & Finer | 1 | .40 | 135 | |

For hobbing Class 9 (per AGMA 390.03) or better gears, it may be necessary to reduce speeds and feeds by 50% and/or take 2 cuts.
To meet finish requirements it may be necessary to try both conventional and climb cutting.

See section 16 for Cutting Fluid Recommendations.
*Feeds are based on the largest standard recommended hob diameter. When using a smaller hob diameter, the Feed must be reduced proportionally.

| MATERIAL | HARD-NESS Bhn | CONDITION | DIAMETRAL PITCH / MODULE | NUMBER OF CUTS | FEED* PER REVOLUTION OF WORKPIECE in / mm | HOB SPEED fpm / m/min | HSS TOOL MATERIAL AISI / ISO |
|---|---|---|---|---|---|---|---|
| 801 815 828 916<br>803 818 863 917<br>805 820 902 962<br>807 822 907 963<br>809 824 909 964<br>811 825 910 966<br>813 826 911 993<br>814 827 913 | 40 to 200<br>500kg | As Cast | 1—2 | 1 | .060 | 100 | M2, M7 |
| | | | 3—10 | 1 | .050 | 110 | |
| | | | 11—19 | 1 | .030 | 130 | |
| | | | 20—48 | 1 | .020 | 135 | |
| | | | 48 & Finer | 1 | .015 | 150 | |
| | | | 25—13 | 1 | 1.50 | 30 | S4, S2 |
| | | | 12—2.5 | 1 | 1.25 | 34 | |
| | | | 2—1.5 | 1 | .75 | 40 | |
| | | | 1—.5 | 1 | .50 | 41 | |
| | | | .5 & Finer | 1 | .40 | 46 | |

For hobbing Class 9 (per AGMA 390.03) or better gears, it may be necessary to reduce speeds and feeds by 50% and/or take 2 cuts.
To meet finish requirements it may be necessary to try both conventional and climb cutting.

See section 16 for Cutting Fluid Recommendations.
*Feeds are based on the largest standard recommended hob diameter. When using a smaller hob diameter, the Feed must be reduced proportionally.

| MATERIAL | HARD-NESS Bhn | CONDITION | CUTTING SPEED ROUGHING fpm m/min | FINISHING fpm m/min | HSS TOOL MATERIAL AISI ISO |
|---|---|---|---|---|---|
| **1. FREE MACHINING CARBON STEELS, WROUGHT** **Low Carbon Resulfurized** 1116 1117 1118 1119 1211 1212 | 100 to 150 | Hot Rolled or Annealed | 125<br>38 | 250<br>76 | M2, M7<br>S4, S2 |
| | 150 to 200 | Cold Drawn | 140<br>43 | 275<br>84 | M2, M7<br>S4, S2 |
| **Low Carbon Resulfurized** 1213 1215 | 100 to 150 | Hot Rolled or Annealed | 140<br>43 | 275<br>84 | M2, M7<br>S4, S2 |
| | 150 to 200 | Cold Drawn | 145<br>44 | 290<br>88 | M2, M7<br>S4, S2 |
| **Low Carbon Resulfurized** 1108 1109 1110 1115 | 100 to 150 | Hot Rolled or Annealed | 110<br>34 | 225<br>69 | M2, M7<br>S4, S2 |
| | 150 to 200 | Cold Drawn | 125<br>38 | 250<br>76 | M2, M7<br>S4, S2 |
| **Medium Carbon Resulfurized** 1132 1137 1139 1140 1141 1144 1145 1146 1151 | 175 to 225 | Hot Rolled, Normalized, Annealed or Cold Drawn | 85<br>26 | 175<br>53 | M2, M7<br>S4, S2 |
| | 275 to 325 | Quenched and Tempered | 70<br>21 | 140<br>43 | M2, M7<br>S4, S2 |
| | 325 to 375 | Quenched and Tempered | 50<br>15 | 100<br>30 | M3<br>S5 |
| **Low Carbon Leaded** 10L18 11L17 | 100 to 150 | Hot Rolled, Normalized, Annealed or Cold Drawn | 140<br>43 | 275<br>84 | M2, M7<br>S4, S2 |
| | 150 to 200 | Hot Rolled, Normalized, Annealed or Cold Drawn | 130<br>40 | 265<br>81 | M2, M7<br>S4, S2 |

Cutting speed recommendations are for use with Alternate-Tooth Milling Cutters or Gear Generators.
For Planer-Type Generators use the recommended cutting speeds in section 7.3 Gear Shaping.
See section 16 for Cutting Fluid Recommendations.

# 7.2 Gear Cutting, Straight and Spiral Bevel

| MATERIAL | HARD-NESS Bhn | CONDITION | CUTTING SPEED ROUGHING fpm / m/min | CUTTING SPEED FINISHING fpm / m/min | HSS TOOL MATERIAL AISI / ISO |
|---|---|---|---|---|---|
| **1. FREE MACHINING CARBON STEELS, WROUGHT (cont.)** **Low Carbon Leaded (cont.)** (materials listed on preceding page) | 200 to 250 | Hot Rolled, Normalized, Annealed or Cold Drawn | 110 | 220 | M2, M7 |
| | | | 34 | 67 | S4, S2 |
| **Low Carbon Leaded** 12L13 12L14 12L15 | 100 to 150 | Hot Rolled, Normalized, Annealed or Cold Drawn | 145 | 290 | M2, M7 |
| | | | 44 | 88 | S4, S2 |
| | 150 to 200 | Hot Rolled, Normalized, Annealed or Cold Drawn | 145 | 290 | M2, M7 |
| | | | 44 | 88 | S4, S2 |
| | 200 to 250 | Hot Rolled, Normalized, Annealed or Cold Drawn | 110 | 225 | M2, M7 |
| | | | 34 | 69 | S4, S2 |
| **Medium Carbon Leaded** 10L45 10L50 11L37 11L41 11L44 | 125 to 175 | Hot Rolled, Normalized, Annealed or Cold Drawn | 110 | 225 | M2, M7 |
| | | | 34 | 69 | S4, S2 |
| | 175 to 225 | Hot Rolled, Normalized, Annealed or Cold Drawn | 100 | 200 | M2, M7 |
| | | | 30 | 60 | S4, S2 |
| | 225 to 275 | Hot Rolled, Normalized, Annealed, Cold Drawn or Quenched and Tempered | 90 | 175 | M2, M7 |
| | | | 27 | 53 | S4, S2 |
| | 275 to 325 | Hot Rolled, Normalized, Annealed, Cold Drawn or Quenched and Tempered | 80 | 155 | M2, M7 |
| | | | 24 | 47 | S4, S2 |
| | 325 to 375 | Quenched and Tempered | 60 | 110 | M3 |
| | | | 18 | 34 | S5 |
| **2. CARBON STEELS, WROUGHT** **Low Carbon** 1005 1006 1008 1009 1010 1011 1012 1013 1015 1016 1017 1018 1019 1020 1021 1022 1023 1025 1026 1029 1513 1518 1522 | 85 to 125 | Hot Rolled, Normalized, Annealed or Cold Drawn | 90 | 190 | M2, M7 |
| | | | 27 | 58 | S4, S2 |
| | 125 to 175 | Hot Rolled, Normalized, Annealed or Cold Drawn | 90 | 185 | M2, M7 |
| | | | 27 | 56 | S4, S2 |

Cutting speed recommendations are for use with Alternate-Tooth Milling Cutters or Gear Generators.
For Planer-Type Generators use the recommended cutting speeds in section 7.3 Gear Shaping.
See section 16 for Cutting Fluid Recommendations.

| MATERIAL | HARD-NESS Bhn | CONDITION | CUTTING SPEED ROUGHING fpm / m/min | FINISHING fpm / m/min | HSS TOOL MATERIAL AISI / ISO |
|---|---|---|---|---|---|
| **2. CARBON STEELS, WROUGHT (cont.)** **Low Carbon (cont.)** (materials listed on preceding page) | 175 to 225 | Hot Rolled, Normalized, Annealed or Cold Drawn | 80 | 165 | M2, M7 |
| | | | 24 | 50 | S4, S2 |
| | 225 to 275 | Annealed or Cold Drawn | 70 | 145 | M2, M7 |
| | | | 21 | 44 | S4, S2 |
| **Medium Carbon** 1030 1033 1035 1037 1038 1039 1040 1042 1043 1044 1045 1046 1049 1050 1053 1055 1524 1525 1526 1527 1536 1541 1547 1548 1551 1552 | 125 to 175 | Hot Rolled, Normalized, Annealed or Cold Drawn | 85 | 175 | M2, M7 |
| | | | 26 | 53 | S4, S2 |
| | 175 to 225 | Hot Rolled, Normalized, Annealed or Cold Drawn | 75 | 155 | M2, M7 |
| | | | 23 | 47 | S4, S2 |
| | 225 to 275 | Hot Rolled, Normalized, Annealed, Cold Drawn or Quenched and Tempered | 70 | 145 | M2, M7 |
| | | | 21 | 44 | S4, S2 |
| | 275 to 325 | Hot Rolled, Normalized, Annealed or Quenched and Tempered | 60 | 125 | M2, M7 |
| | | | 18 | 38 | S4, S2 |
| | 325 to 375 | Quenched and Tempered | 50 | 100 | M3 |
| | | | 15 | 30 | S5 |
| **High Carbon** 1060 1064 1065 1069 1070 1074 1075 1078 1080 1084 1085 1086 1090 1095 1561 1566 1572 | 175 to 225 | Hot Rolled, Normalized, Annealed or Cold Drawn | 70 | 145 | M2, M7 |
| | | | 21 | 44 | S4, S2 |
| | 225 to 275 | Hot Rolled, Normalized, Annealed, Cold Drawn or Quenched and Tempered | 65 | 130 | M2, M7 |
| | | | 20 | 40 | S4, S2 |
| | 275 to 325 | Hot Rolled, Normalized, Annealed or Quenched and Tempered | 60 | 120 | M2, M7 |
| | | | 18 | 37 | S4, S2 |
| | 325 to 375 | Quenched and Tempered | 45 | 95 | M3 |
| | | | 14 | 29 | S5 |

Cutting speed recommendations are for use with Alternate-Tooth Milling Cutters or Gear Generators.
For Planer-Type Generators use the recommended cutting speeds in section 7.3 Gear Shaping.
See section 16 for Cutting Fluid Recommendations.

# 7.2 Gear Cutting, Straight and Spiral Bevel

| MATERIAL | HARD-NESS Bhn | CONDITION | CUTTING SPEED ROUGHING fpm m/min | CUTTING SPEED FINISHING fpm m/min | HSS TOOL MATERIAL AISI ISO |
|---|---|---|---|---|---|
| **3. CARBON AND FERRITIC ALLOY STEELS (HIGH TEMPERATURE SERVICE)** ASTM A369: Grades FPA, FPB, FP1, FP2, FP12 | 150 to 200 | As Forged | 85<br>26 | 175<br>53 | M2, M7<br>S4, S2 |
| ASTM A369: Grades FP3b, FP11 | 150 to 200 | Annealed or Normalized and Tempered | 75<br>23 | 150<br>46 | M2, M7<br>S4, S2 |
| ASTM A369: Grades FP5, FP7, FP9, FP21, FP22 | 150 to 200 | Annealed or Normalized and Tempered | 70<br>21 | 140<br>43 | M2, M7<br>S4, S2 |
| **4. FREE MACHINING ALLOY STEELS, WROUGHT** **Medium Carbon Resulfurized** 4140, 4142Te, 4147Te, 4140Se, 4145Se, 4150 | 150 to 200 | Hot Rolled, Normalized, Annealed or Cold Drawn | 85<br>26 | 175<br>53 | M2, M7<br>S4, S2 |
| | 200 to 250 | Hot Rolled, Normalized, Annealed or Cold Drawn | 75<br>23 | 155<br>47 | M2, M7<br>S4, S2 |
| | 275 to 325 | Quenched and Tempered | 65<br>20 | 135<br>41 | M2, M7<br>S4, S2 |
| | 325 to 375 | Quenched and Tempered | 45<br>14 | 95<br>29 | M3<br>S5 |
| **Medium and High Carbon Leaded** 41L30, 41L50, 86L20, 41L40, 43L40, 86L40, 41L45, 51L32, 41L47, 52L100 | 150 to 200 | Hot Rolled, Normalized, Annealed or Cold Drawn | 85<br>26 | 175<br>53 | M2, M7<br>S4, S2 |
| | 200 to 250 | Hot Rolled, Normalized, Annealed or Cold Drawn | 75<br>23 | 155<br>47 | M2, M7<br>S4, S2 |
| | 275 to 325 | Quenched and Tempered | 65<br>20 | 135<br>41 | M2, M7<br>S4, S2 |
| | 325 to 375 | Quenched and Tempered | 45<br>14 | 90<br>27 | M3<br>S5 |

Cutting speed recommendations are for use with Alternate-Tooth Milling Cutters or Gear Generators.
For Planer-Type Generators use the recommended cutting speeds in section 7.3 Gear Shaping.
See section 16 for Cutting Fluid Recommendations.

| MATERIAL | HARDNESS Bhn | CONDITION | CUTTING SPEED ROUGHING fpm / m/min | CUTTING SPEED FINISHING fpm / m/min | HSS TOOL MATERIAL AISI / ISO |
|---|---|---|---|---|---|
| **5. ALLOY STEELS, WROUGHT**<br>**Low Carbon**<br>4012 4615 4817 8617<br>4023 4617 4820 8620<br>4024 4620 5015 8622<br>4118 4621 5115 8822<br>4320 4718 5120 9310<br>4419 4720 6118 94B15<br>4422 4815 8115 94B17 | 125 to 175 | Hot Rolled, Annealed or Cold Drawn | 85 | 175 | M2, M7 |
| | | | 26 | 53 | S4, S2 |
| | 175 to 225 | Hot Rolled, Annealed or Cold Drawn | 75 | 150 | M2, M7 |
| | | | 23 | 46 | S4, S2 |
| | 225 to 275 | Hot Rolled, Normalized, Annealed or Cold Drawn | 65 | 130 | M2, M7 |
| | | | 20 | 40 | S4, S2 |
| | 275 to 325 | Normalized or Quenched and Tempered | 60 | 120 | M2, M7 |
| | | | 18 | 37 | S4, S2 |
| | 325 to 375 | Normalized or Quenched and Tempered | 50 | 100 | M3 |
| | | | 15 | 30 | S5 |
| **Medium Carbon**<br>1330 4145 5132 8640<br>1335 4147 5135 8642<br>1340 4150 5140 8645<br>1345 4161 5145 86B45<br>4027 4340 5147 8650<br>4028 4427 5150 8655<br>4032 4626 5155 8660<br>4037 50B40 5160 8740<br>4042 50B44 51B60 8742<br>4047 5046 6150 9254<br>4130 50B46 81B45 9255<br>4135 50B50 8625 9260<br>4137 5060 8627 94B30<br>4140 50B60 8630<br>4142 5130 8637 | 175 to 225 | Hot Rolled, Annealed or Cold Drawn | 65 | 135 | M2, M7 |
| | | | 20 | 41 | S4, S2 |
| | 225 to 275 | Annealed, Normalized, Cold Drawn or Quenched and Tempered | 60 | 125 | M2, M7 |
| | | | 18 | 38 | S4, S2 |
| | 275 to 325 | Normalized or Quenched and Tempered | 55 | 110 | M2, M7 |
| | | | 17 | 34 | S4, S2 |
| | 325 to 375 | Normalized or Quenched and Tempered | 50 | 100 | M3 |
| | | | 15 | 30 | S5 |
| **High Carbon**<br>50100<br>51100<br>52100<br>M-50 | 175 to 225 | Hot Rolled, Annealed or Cold Drawn | 60 | 125 | M2, M7 |
| | | | 18 | 38 | S4, S2 |
| | 225 to 275 | Normalized, Cold Drawn or Quenched and Tempered | 50 | 100 | M2, M7 |
| | | | 15 | 30 | S4, S2 |

Cutting speed recommendations are for use with Alternate-Tooth Milling Cutters or Gear Generators.
For Planer-Type Generators use the recommended cutting speeds in section 7.3 Gear Shaping.
See section 16 for Cutting Fluid Recommendations.

## 7.2 Gear Cutting, Straight and Spiral Bevel

| MATERIAL | HARDNESS Bhn | CONDITION | CUTTING SPEED ROUGHING fpm / m/min | CUTTING SPEED FINISHING fpm / m/min | HSS TOOL MATERIAL AISI / ISO |
|---|---|---|---|---|---|
| **5. ALLOY STEELS, WROUGHT (cont.)** **High Carbon (cont.)** (materials listed on preceding page) | 275 to 325 | Normalized or Quenched and Tempered | 45 | 90 | M2, M7 |
| | | | 14 | 27 | S4, S2 |
| | 325 to 375 | Normalized or Quenched and Tempered | 40 | 80 | M3 |
| | | | 12 | 24 | S5 |
| **6. HIGH STRENGTH STEELS, WROUGHT** 300M, 4330V, 4340, 4340Si, 98BV40, D6ac, H11, H13 | 225 to 300 | Annealed | 45 | 90 | M2, M7 |
| | | | 14 | 27 | S4, S2 |
| | 300 to 350 | Normalized | 40 | 80 | M2, M7 |
| | | | 12 | 24 | S4, S2 |
| HP 9-4-20, HP 9-4-25, HP 9-4-30, HP 9-4-45 | 325 to 375 | Annealed | 40 | 80 | M2, M7 |
| | | | 12 | 24 | S4, S2 |
| **7. MARAGING STEELS, WROUGHT** 120 Grade, 180 Grade | 275 to 325 | Annealed | 60 | 120 | M2, M7 |
| | | | 18 | 37 | S4, S2 |
| ASTM A538: Grades A, B, C; 200 Grade, 250 Grade, 300 Grade, 350 Grade, HY230 | 275 to 325 | Annealed | 50 | 100 | M2, M7 |
| | | | 15 | 30 | S4, S2 |
| | | | | | |
| | | | | | |
| **8. TOOL STEELS, WROUGHT** **High Speed** M1, M2, M6, M10, T1, T2, T6 | 200 to 250 | Annealed | 50 | 100 | M2, M7 |
| | | | 15 | 30 | S4, S2 |
| **High Speed** M3-1, M3-2, M4, M7, M30, M33, M34, M36, M41, M42, M43, M44, M46, M47, T4, T5, T8, T15 | 225 to 275 | Annealed | 45 | 90 | M2, M7 |
| | | | 14 | 27 | S4, S2 |
| | | | | | |
| | | | | | |

Cutting speed recommendations are for use with Alternate-Tooth Milling Cutters or Gear Generators.
For Planer-Type Generators use the recommended cutting speeds in section 7.3 Gear Shaping.
See section 16 for Cutting Fluid Recommendations.

| MATERIAL | HARD-NESS Bhn | CONDITION | CUTTING SPEED ROUGHING fpm m/min | FINISHING fpm m/min | HSS TOOL MATERIAL AISI ISO |
|---|---|---|---|---|---|
| **Hot Work**<br>H10 H11 H12 H13 H14<br>H19 H21 H22 H23 H24<br>H25 H26 H42 | 150 to 200 | Annealed | 80<br>24 | 160<br>49 | M2, M7<br>S4, S2 |
| | 200 to 250 | Annealed | 65<br>20 | 130<br>40 | M2, M7<br>S4, S2 |
| | 325 to 375 | Quenched and Tempered | 45<br>14 | 95<br>29 | M3<br>S5 |
| **Cold Work**<br>A2 A3 A4 A6 A7 A8<br>A9 A10 D2 D3 D4 D5<br>D7 O1 O2 O6 O7 | 200 to 250 | Annealed | 65<br>20 | 130<br>40 | M2, M7<br>S4, S2 |
| | | | | | |
| **Shock Resisting**<br>S1 S2 S5<br>S6 S7 | 175 to 225 | Annealed | 70<br>21 | 145<br>44 | M2, M7<br>S4, S2 |
| **Mold**<br>P2 P4 P5 P6 | 100 to 150 | Annealed | 90<br>27 | 180<br>55 | M2, M7<br>S4, S2 |
| **Mold**<br>P20 P21 | 150 to 200 | Annealed | 80<br>24 | 165<br>50 | M2, M7<br>S4, S2 |
| **Special Purpose**<br>L2 L6 L7 | 150 to 200 | Annealed | 70<br>21 | 145<br>44 | M2, M7<br>S4, S2 |
| **Special Purpose**<br>F1 F2 | 200 to 250 | Annealed | 65<br>20 | 135<br>41 | M2, M7<br>S4, S2 |
| **Water Hardening**<br>W1, W2, W5<br>SAE J438b: Types W108, W109, W110, W112, W209, W210, W310 | 150 to 200 | Annealed | 90<br>27 | 180<br>55 | M2, M7<br>S4, S2 |

Cutting speed recommendations are for use with Alternate-Tooth Milling Cutters or Gear Generators.
For Planer-Type Generators use the recommended cutting speeds in section 7.3 Gear Shaping.
See section 16 for Cutting Fluid Recommendations.

# 7.2 Gear Cutting, Straight and Spiral Bevel

| MATERIAL | HARD-NESS Bhn | CONDITION | CUTTING SPEED ROUGHING fpm m/min | CUTTING SPEED FINISHING fpm m/min | HSS TOOL MATERIAL AISI ISO |
|---|---|---|---|---|---|
| **9. NITRIDING STEELS, WROUGHT** Nitralloy 125, Nitralloy 135, Nitralloy 135 Mod., Nitralloy 225, Nitralloy 230, Nitralloy EZ, Nitralloy N, Nitrex 1 | 200 to 250 | Annealed | 65 | 130 | M2, M7 |
| | | | 20 | 40 | S4, S2 |
| | 300 to 350 | Normalized or Quenched and Tempered | 55 | 110 | M2, M7 |
| | | | 17 | 34 | S4, S2 |
| **12. FREE MACHINING STAINLESS STEELS, WROUGHT** **Ferritic** 430F, 430F Se | 135 to 185 | Annealed | 85 | 175 | M2, M7 |
| | | | 26 | 53 | S4, S2 |
| **Austenitic** 203EZ, 303, 303MA, 303Pb, 303 Plus X, 303Se | 135 to 185 | Annealed | 60 | 150 | M2, M7 |
| | | | 18 | 46 | S4, S2 |
| | 225 to 275 | Cold Drawn | 65 | 135 | M2, M7 |
| | | | 20 | 41 | S4, S2 |
| **Martensitic** 416, 416 Plus X, 416Se, 420F, 420F Se, 440F, 440F Se | 135 to 185 | Annealed | 85 | 175 | M2, M7 |
| | | | 26 | 53 | S4, S2 |
| | 185 to 240 | Annealed or Cold Drawn | 75 | 155 | M2, M7 |
| | | | 23 | 47 | S4, S2 |
| | 275 to 325 | Quenched and Tempered | 60 | 120 | M2, M7 |
| | | | 18 | 37 | S4, S2 |
| **13. STAINLESS STEELS, WROUGHT** **Ferritic** 405, 409, 429, 430, 434, 436, 442, 446 | 135 to 185 | Annealed | 75 | 150 | M2, M7 |
| | | | 23 | 46 | S4, S2 |
| **Austenitic** 201, 202, 301, 302, 304, 304L, 305, 308, 321, 347, 348, 384, 385 | 135 to 185 | Annealed | 60 | 120 | M2, M7 |
| | | | 18 | 37 | S4, S2 |
| | 225 to 275 | Cold Drawn | 55 | 110 | M2, M7 |
| | | | 17 | 34 | S4, S2 |

Cutting speed recommendations are for use with Alternate-Tooth Milling Cutters or Gear Generators.
For Planer-Type Generators use the recommended cutting speeds in section 7.3 Gear Shaping.
See section 16 for Cutting Fluid Recommendations.

| MATERIAL | HARD-NESS Bhn | CONDITION | CUTTING SPEED ROUGHING fpm / m/min | CUTTING SPEED FINISHING fpm / m/min | HSS TOOL MATERIAL AISI / ISO |
|---|---|---|---|---|---|
| **Austenitic** 302B, 309, 309S, 310, 310S, 314, 316, 316L, 317, 330 | 135 to 185 | Annealed | 50 / 15 | 100 / 30 | M2, M7 / S4, S2 |
| | 225 to 275 | Cold Drawn | 45 / 14 | 90 / 27 | M2, M7 / S4, S2 |
| **Austenitic** Nitronic 32, Nitronic 33, Nitronic 40, Nitronic 50, Nitronic 60 | 210 to 250 | Annealed | 40 / 12 | 80 / 24 | M2, M7 / S4, S2 |
| **Martensitic** 403, 410, 420, 422, 501, 502 | 135 to 175 | Annealed | 75 / 23 | 150 / 46 | M2, M7 / S4, S2 |
| | 175 to 225 | Annealed | 70 / 21 | 140 / 43 | M2, M7 / S4, S2 |
| | 275 to 325 | Quenched and Tempered | 55 / 17 | 110 / 34 | M2, M7 / S4, S2 |
| **Martensitic** 414, 431, Greek Ascoloy | 225 to 275 | Annealed | 60 / 18 | 120 / 37 | M2, M7 / S4, S2 |
| | 275 to 325 | Quenched and Tempered | 55 / 17 | 110 / 34 | M2, M7 / S4, S2 |
| **Martensitic** 440A, 440B, 440C | 225 to 275 | Annealed | 55 / 17 | 115 / 35 | M2, M7 / S4, S2 |
| | 275 to 325 | Quenched and Tempered | 50 / 15 | 105 / 32 | M2, M7 / S4, S2 |

Cutting speed recommendations are for use with Alternate-Tooth Milling Cutters or Gear Generators.
For Planer-Type Generators use the recommended cutting speeds in section 7.3 Gear Shaping.
See section 16 for Cutting Fluid Recommendations.

# 7.2 Gear Cutting, Straight and Spiral Bevel

| MATERIAL | HARD-NESS Bhn | CONDITION | CUTTING SPEED ROUGHING fpm m/min | CUTTING SPEED FINISHING fpm m/min | HSS TOOL MATERIAL AISI ISO |
|---|---|---|---|---|---|
| **14. PRECIPITATION HARDENING STAINLESS STEELS, WROUGHT**<br>15-5 PH<br>16-6 PH<br>17-4 PH<br>17-7 PH<br>17-14 Cu Mo<br>AF-71<br>AFC-77<br>Almar 362 (AM-362)<br>AM-350<br>AM-355<br>AM-363<br>Custom 450<br>Custom 455<br>HNM<br>PH 13-8 Mo<br>PH 14-8 Mo<br>PH 15-7 Mo<br>Stainless W | 150 to 200 | Solution Treated | 55<br>17 | 110<br>34 | M2, M7<br>S4, S2 |
| | 275 to 325 | Solution Treated or Hardened | 45<br>14 | 95<br>29 | M2, M7<br>S4, S2 |
| | 325 to 375 | Solution Treated or Hardened | 40<br>12 | 85<br>26 | M3<br>S5 |
| **15. CARBON STEELS, CAST**<br>**Low Carbon**<br>ASTM A426: Grade CP1<br>1010<br>1020 | 100 to 150 | Annealed, Normalized or Normalized and Tempered | 95<br>29 | 190<br>58 | M2, M7<br>S4, S2 |
| **Medium Carbon**<br>ASTM A352: Grades LCA, LCB, LCC<br>ASTM A356: Grade 1<br>1030<br>1040<br>1050 | 125 to 175 | Annealed, Normalized or Normalized and Tempered | 90<br>27 | 185<br>56 | M2, M7<br>S4, S2 |
| | 175 to 225 | Annealed, Normalized or Normalized and Tempered | 75<br>23 | 155<br>47 | M2, M7<br>S4, S2 |
| | 250 to 300 | Quenched and Tempered | 65<br>20 | 135<br>41 | M2, M7<br>S4, S2 |
| **16. ALLOY STEELS, CAST**<br>**Low Carbon**<br>ASTM A217: Grade WC9<br>ASTM A352: Grades LC3, LC4<br>ASTM A426: Grades CP2, CP5, CP5b, CP11, CP12, CP15, CP21, CP22<br>1320 2320 4120 8020<br>2315 4110 4320 8620 | 150 to 200 | Annealed, Normalized or Normalized and Tempered | 85<br>26 | 170<br>52 | M2, M7<br>S4, S2 |
| | 200 to 225 | Annealed, Normalized or Normalized and Tempered | 70<br>21 | 145<br>44 | M2, M7<br>S4, S2 |
| | 250 to 300 | Quenched and Tempered | 65<br>20 | 130<br>40 | M2, M7<br>S4, S2 |

Cutting speed recommendations are for use with Alternate-Tooth Milling Cutters or Gear Generators.
For Planer-Type Generators use the recommended cutting speeds in section 7.3 Gear Shaping.
See section 16 for Cutting Fluid Recommendations.

| MATERIAL | HARD-NESS Bhn | CONDITION | CUTTING SPEED ROUGHING fpm m/min | CUTTING SPEED FINISHING fpm m/min | HSS TOOL MATERIAL AISI ISO |
|---|---|---|---|---|---|
| **Medium Carbon**<br>ASTM A27: Grades N1, N2, U-60-30, 60-30, 65-35, 70-36, 70-40<br>ASTM A148: Grades 80-40, 80-50, 90-60, 105-85, 120-95, 150-125, 175-145<br>ASTM A216: Grades WCA, WCB, WCC<br>ASTM A217: Grades WC1, WC4, WC5, WC6<br>ASTM A352: Grades LC1, LC2, LC2-1<br>ASTM A356: Grades 2, 5, 6, 8, 9, 10<br>ASTM A389: Grades C23, C24<br>ASTM A486: Classes 70, 90, 120<br>ASTM A487: Classes 1N, 2N, 4N, 6N, 8N, 9N, 10N, DN, 1Q, 2Q, 4Q, 4QA, 6Q, 7Q, 8Q, 9Q, 10Q<br>1330 4130 80B30 8640<br>1340 4140 8040 9525<br>2325 4330 8430 9530<br>2330 4340 8440 9535<br>4125 8030 8630 | 175 to 225 | Annealed, Normalized, or Normalized and Tempered | 70<br>21 | 145<br>44 | M2, M7<br>S4, S2 |
| | 225 to 250 | Normalized, Normalized and Tempered or Quenched and Tempered | 65<br>20 | 135<br>41 | M2, M7<br>S4, S2 |
| | 250 to 300 | Quenched and Tempered | 65<br>20 | 130<br>40 | M2, M7<br>S4, S2 |
| | 300 to 350 | Quenched and Tempered | 55<br>17 | 115<br>35 | M2, M7<br>S4, S2 |
| | 350 to 400 | Quenched and Tempered | 50<br>15 | 100<br>30 | M42, M3<br>S11, S5 |
| **18. STAINLESS STEELS, CAST**<br>**Ferritic**<br>ASTM A217: Grades C5, C12<br>ASTM A296: Grades CB-30, CC-50, CE-30, CA6N, CA-6NM, CD4MCu<br>ASTM A297: Grade HC<br>ASTM A487: Class CA6NM<br>ASTM A608: Grade HC30 | 135 to 185 | Annealed | 75<br>23 | 150<br>46 | M2, M7<br>S4, S2 |
| | | | | | |
| **Austenitic**<br>ASTM A296: Grades CF-16F, CN-7M, CN-7MS<br>ASTM A351: Grade CN-7M | 140 to 170 | Annealed or Normalized | 70<br>21 | 145<br>44 | M2, M7<br>S4, S2 |
| **Austenitic**<br>ASTM A296: Grades CF-3, CF-8, CF-8C, CF-20<br>ASTM A351: Grades CF-3, CF-3A, CF-8, CF-8A, CF-8C<br>ASTM A451: Grades CPF3, CPF3A, CPF8, CPF8A, CPF8C, CPF8C (Ta Max.)<br>ASTM A452: Grades TP 304H, TP 347H | 135 to 185 | Annealed or Normalized | 60<br>18 | 120<br>37 | M2, M7<br>S4, S2 |
| | | | | | |
| | | | | | |

Cutting speed recommendations are for use with Alternate-Tooth Milling Cutters or Gear Generators.
For Planer-Type Generators use the recommended cutting speeds in section 7.3 Gear Shaping.
See section 16 for Cutting Fluid Recommendations.

# 7.2 Gear Cutting, Straight and Spiral Bevel

| MATERIAL | HARD-NESS Bhn | CONDITION | CUTTING SPEED ROUGHING fpm m/min | FINISHING fpm m/min | HSS TOOL MATERIAL AISI ISO |
|---|---|---|---|---|---|
| **18. STAINLESS STEELS, CAST (cont.)** **Austenitic** ASTM A296: Grades CF-3M, CF-8M, CG-8M, CG-12, CH-20, CK-20 ASTM A351: Grades CF-3M, CF-3MA, CF-8M, CF-10MC, CH-8, CH-10, CH-20, CK-20, HK-30, HK-40, HT-30 ASTM A451: Grades CPF3M, CPF8M, CPF10MC, CPH8, CPH10, CPH20, CPK20 ASTM A452: Grade TP 316H | 135 to 185 | Annealed or Normalized | 50<br>15 | 100<br>30 | M2, M7<br>S4, S2 |
| | | | | | |
| **Austenitic** ASTM A297: Grades HD, HE, HF, HH, HI, HK, HL, HN, HP, HT, HU ASTM A608: Grades HD50, HE35, HF30, HH30, HH33, HI35, HK30, HK40, HL30, HL40, HN40, HT50, HU50 | 160 to 210 | As Cast | 50<br>15 | 100<br>30 | M2, M7<br>S4, S2 |
| | | | | | |
| **Martensitic** ASTM A217: Grade CA-15 ASTM A296: Grades CA-15, CA-15M, CA-40 ASTM A426: Grades CP7, CP9, CPCA15 ASTM A487: Classes CA15a, CA-15M | 135 to 175 | Annealed | 70<br>21 | 145<br>44 | M2, M7<br>S4, S2 |
| | 175 to 225 | Annealed, Normalized or Normalized and Tempered | 65<br>20 | 135<br>41 | M2, M7<br>S4, S2 |
| | 275 to 325 | Quenched and Tempered | 50<br>15 | 105<br>32 | M2, M7<br>S4, S2 |
| **19. PRECIPITATION HARDENING STAINLESS STEELS, CAST** ASTM A351: Grade CD-4MCu ACI Grade CB-7Cu ACI Grade CD-4MCu 17-4 PH AM-355 | 325 to 375 | Solution Treated | 35<br>11 | 75<br>23 | M3<br>S5 |
| | | | | | |
| **21. GRAY CAST IRONS** **Ferritic** ASTM A48: Class 20 SAE J431c: Grade G1800 | 120 to 150 | Annealed | 90<br>27 | 185<br>56 | M2, M7<br>S4, S2 |
| **Pearlitic- Ferritic** ASTM A48: Class 25 SAE J431c: Grade G2500 | 160 to 200 | As Cast | 75<br>23 | 150<br>46 | M2, M7<br>S4, S2 |

Cutting speed recommendations are for use with Alternate-Tooth Milling Cutters or Gear Generators.
For Planer-Type Generators use the recommended cutting speeds in section 7.3 Gear Shaping.
See section 16 for Cutting Fluid Recommendations.

| MATERIAL | HARD-NESS Bhn | CONDITION | CUTTING SPEED ROUGHING fpm m/min | CUTTING SPEED FINISHING fpm m/min | HSS TOOL MATERIAL AISI ISO |
|---|---|---|---|---|---|
| **Pearlitic**<br>ASTM A48: Classes 30, 35, 40<br>SAE J431c: Grade G3000 | 190 to 220 | As Cast | 60<br>18 | 120<br>37 | M2, M7<br>S4, S2 |
| **Pearlitic + Free Carbides**<br>ASTM A48: Classes 45, 50<br>SAE J431c: Grades G3500, G4000 | 220 to 260 | As Cast | 50<br>15 | 100<br>30 | M2, M7<br>S4, S2 |
| **Pearlitic or Acicular + Free Carbides**<br>ASTM A48: Classes 55, 60 | 250 to 320 | As Cast or Quenched and Tempered | 40<br>12 | 80<br>24 | M2, M7<br>S4, S2 |
| **Austenitic (NI-RESIST)**<br>ASTM A436: Types 1, 1b, 2, 2b, 3, 4, 5, 6 | 100 to 250 | As Cast | 40<br>12 | 85<br>26 | M2, M7<br>S4, S2 |
| **23. DUCTILE CAST IRONS**<br>**Ferritic**<br>ASTM A536: Grades 60-40-18, 65-45-12<br>SAE J434c: Grades D4018, D4512 | 140 to 190 | Annealed | 80<br>24 | 160<br>49 | M2, M7<br>S4, S2 |
| **Ferritic- Pearlitic**<br>ASTM A536: Grade 80-55-06<br>SAE J434c: Grade D5506 | 190 to 225 | As Cast | 70<br>21 | 140<br>43 | M2, M7<br>S4, S2 |
| | 225 to 260 | As Cast | 65<br>20 | 130<br>40 | M2, M7<br>S4, S2 |
| **Pearlitic- Martensitic**<br>ASTM A536: Grade 100-70-03<br>SAE J434c: Grade D7003 | 240 to 300 | Normalized and Tempered | 60<br>18 | 125<br>38 | M2, M7<br>S4, S2 |
| **Martensitic**<br>ASTM A536: Grade 120-90-02<br>SAE J434c: Grade DQ&T | 270 to 330 | Quenched and Tempered | 60<br>18 | 120<br>37 | M2, M7<br>S4, S2 |
| **Austenitic (NI-RESIST Ductile)**<br>ASTM A439: Types D-2, D-2C, D-3A, D-5<br>ASTM A571: Type D-2M | 120 to 200 | Annealed | 35<br>11 | 75<br>23 | M2, M7<br>S4, S2 |
| **Austenitic (NI-RESIST Ductile)**<br>ASTM A439: Types D-2B, D-3, D-4, D-5B | 140 to 275 | Annealed | 35<br>11 | 70<br>21 | M2, M7<br>S4, S2 |

Cutting speed recommendations are for use with Alternate-Tooth Milling Cutters or Gear Generators.
For Planer-Type Generators use the recommended cutting speeds in section 7.3 Gear Shaping.
See section 16 for Cutting Fluid Recommendations.

# 7.2 Gear Cutting, Straight and Spiral Bevel

| MATERIAL | HARD-NESS Bhn | CONDITION | CUTTING SPEED ROUGHING fpm m/min | CUTTING SPEED FINISHING fpm m/min | HSS TOOL MATERIAL AISI ISO |
|---|---|---|---|---|---|
| **24. MALLEABLE CAST IRONS** **Ferritic** ASTM A47: Grades 32510, 35018 ASTM A602: Grade M3210 SAE J158: Grade M3210 | 110 to 160 | Malleablized | 90 | 180 | M2, M7 |
| | | | 27 | 55 | S4, S2 |
| **Pearlitic** ASTM A220: Grades 40010, 45006, 45008, 50005 ASTM A602: Grade M4504, M5003 SAE J158: Grades M4504, M5003 | 160 to 200 | Malleablized and Heat Treated | 75 | 155 | M2, M7 |
| | | | 23 | 47 | S4, S2 |
| | 200 to 240 | Malleablized and Heat Treated | 70 | 145 | M2, M7 |
| | | | 21 | 44 | S4, S2 |
| **Tempered Martensite** ASTM A220: Grade 60004 ASTM A602: Grade M5503 SAE J158: Grade M5503 | 200 to 255 | Malleablized and Heat Treated | 65 | 130 | M2, M7 |
| | | | 20 | 40 | S4, S2 |
| **Tempered Martensite** ASTM A220: Grade 70003 ASTM A602: Grade M7002 SAE J158: Grade M7002 | 220 to 260 | Malleablized and Heat Treated | 60 | 125 | M2, M7 |
| | | | 18 | 38 | S4, S2 |
| **Tempered Martensite** ASTM A220: Grade 80002 ASTM A602: Grade M8501 SAE J158: Grade M8501 | 240 to 280 | Malleablized and Heat Treated | 60 | 120 | M2, M7 |
| | | | 18 | 37 | S4, S2 |
| **Tempered Martensite** ASTM A220: Grade 90001 ASTM A602: Grade M8501 SAE J158: Grade M8501 | 250 to 320 | Malleablized and Heat Treated | 55 | 110 | M2, M7 |
| | | | 17 | 34 | S4, S2 |
| **28. ALUMINUM ALLOYS, WROUGHT** EC 2218 5252 6253; 1060 2219 5254 6262; 1100 2618 5454 6463; 1145 3003 5456 6951; 1175 3004 5457 7001; 1235 3005 5652 7004; 2011 4032 5657 7005; 2014 5005 6053 7039; 2017 5050 6061 7049; 2018 5052 6063 7050; 2021 5056 6066 7075; 2024 5083 6070 7079; 2025 5086 6101 7175; 2117 5154 6151 7178 | 30 to 80 500kg | Cold Drawn | 145 | 290 | M2, M7 |
| | | | 44 | 88 | S4, S2 |
| | 75 to 150 500kg | Solution Treated and Aged | 145 | 275 | M2, M7 |
| | | | 44 | 84 | S4, S2 |

Cutting speed recommendations are for use with Alternate-Tooth Milling Cutters or Gear Generators.
For Planer-Type Generators use the recommended cutting speeds in section 7.3 Gear Shaping.
See section 16 for Cutting Fluid Recommendations.

| MATERIAL | HARDNESS Bhn | CONDITION | CUTTING SPEED ROUGHING fpm / m/min | CUTTING SPEED FINISHING fpm / m/min | HSS TOOL MATERIAL AISI / ISO |
|---|---|---|---|---|---|
| **29. ALUMINUM ALLOYS, CAST**<br>**Sand and Permanent Mold**<br>A140 319.0 357.0 A712.0<br>201.0 328.0 359.0 D712.0<br>208.0 A332.0 B443.0 713.0<br>213.0 F332.0 514.0 771.0<br>222.0 333.0 A514.0 850.0<br>224.0 354.0 B514.0 A850.0<br>242.0 355.0 520.0 B850.0<br>295.0 C355.0 535.0<br>B295.0 356.0 705.0<br>308.0 A356.0 707.0<br>Hiduminium RR-350 | 40 to 100 500kg | As Cast | 125 | 250 | M2, M7 |
| | | | 38 | 76 | S4, S2 |
| | 70 to 125 500kg | Solution Treated and Aged | 125 | 250 | M2, M7 |
| | | | 38 | 76 | S4, S2 |
| | | | | | |
| | | | | | |
| **34. COPPER ALLOYS, WROUGHT**<br>145 332 360 482<br>147 335 365 485<br>173 340 366 544<br>187 342 367 623<br>191 349 368 624<br>314 350 370 638<br>316 353 377 642<br>330 356 385 782 | 10 $R_B$ to 70 $R_B$ | Annealed | 140 | 285 | M2, M7 |
| | | | 43 | 87 | S4, S2 |
| | 60 $R_B$ to 100 $R_B$ | Cold Drawn | 155 | 310 | M2, M7 |
| | | | 47 | 95 | S4, S2 |
| 190 425 466 667<br>226 435 467 675<br>230 442 613 687<br>240 443 618 694<br>260 444 630 770<br>268 445 632<br>270 464 651<br>280 465 655 | 10 $R_B$ to 70 $R_B$ | Annealed | 115 | 230 | M2, M7 |
| | | | 35 | 70 | S4, S2 |
| | 60 $R_B$ to 100 $R_B$ | Cold Drawn | 125 | 255 | M2, M7 |
| | | | 38 | 78 | S4, S2 |
| 101 125 185 614<br>102 127 189 619<br>104 128 192 625<br>105 129 194 674<br>107 130 195 688<br>109 142 210 706<br>110 143 220 710<br>111 150 411 715<br>113 155 413 725<br>114 162 505 745<br>115 165 510 752<br>116 170 511 754<br>119 172 521 757<br>120 175 524<br>121 182 608<br>122 184 610 | 10 $R_B$ to 70 $R_B$ | Annealed | 100 | 200 | M2, M7 |
| | | | 30 | 60 | S4, S2 |
| | 60 $R_B$ to 100 $R_B$ | Cold Drawn | 105 | 215 | M2, M7 |
| | | | 32 | 66 | S4, S2 |
| | | | | | |
| | | | | | |
| | | | | | |
| | | | | | |

Cutting speed recommendations are for use with Alternate-Tooth Milling Cutters or Gear Generators.
For Planer-Type Generators use the recommended cutting speeds in section 7.3 Gear Shaping.
See section 16 for Cutting Fluid Recommendations.

# 7.2 Gear Cutting, Straight and Spiral Bevel

| MATERIAL | | | | HARDNESS Bhn | CONDITION | CUTTING SPEED ROUGHING fpm m/min | CUTTING SPEED FINISHING fpm m/min | HSS TOOL MATERIAL AISI ISO |
|---|---|---|---|---|---|---|---|---|
| **35. COPPER ALLOYS, CAST** | | | | | | | | |
| 834 | 855 | 934 | 953 | 40 to 150 500kg | As Cast | 125 | 250 | M2, M7 |
| 836 | 857 | 935 | 954 | | | 38 | 76 | S4, S2 |
| 838 | 858 | 937 | 956 | | | | | |
| 842 | 864 | 938 | 973 | | | | | |
| 844 | 867 | 939 | 974 | | | | | |
| 848 | 879 | 943 | 976 | | | | | |
| 852 | 928 | 944 | 978 | | | | | |
| 854 | 932 | 945 | | | | | | |
| 817 | 868 | 905 | 947 | 40 to 175 500kg | As Cast | 100 | 200 | M2, M7 |
| 821 | 872 | 915 | 948 | | | 30 | 60 | S4, S2 |
| 833 | 874 | 922 | 952 | | | | | |
| 853 | 875 | 923 | 955 | | | | | |
| 861 | 876 | 925 | 957 | | | | | |
| 862 | 878 | 926 | 958 | | | | | |
| 865 | 903 | 927 | | | | | | |
| 801 | 815 | 828 | 916 | 40 to 200 500kg | As Cast | 85 | 175 | M2, M7 |
| 803 | 818 | 863 | 917 | | | 26 | 53 | S4, S2 |
| 805 | 820 | 902 | 962 | | | | | |
| 807 | 822 | 907 | 963 | | | | | |
| 809 | 824 | 909 | 964 | | | | | |
| 811 | 825 | 910 | 966 | | | | | |
| 813 | 826 | 911 | 993 | | | | | |
| 814 | 827 | 913 | | | | | | |

Cutting speed recommendations are for use with Alternate-Tooth Milling Cutters or Gear Generators.
For Planer-Type Generators use the recommended cutting speeds in section 7.3 Gear Shaping.
See section 16 for Cutting Fluid Recommendations.

| MATERIAL | HARD-NESS Bhn | CONDITION | NUMBER OF CUTS | DIAMETRAL PITCH / MODULE | ROTARY FEED per cutter stroke (4″ P.D. cutter) in / mm | CUTTER SPEED fpm / m/min | HSS TOOL MATERIAL AISI / ISO |
|---|---|---|---|---|---|---|---|
| **1. FREE MACHINING CARBON STEELS, WROUGHT** **Low Carbon Resulfurized** 1116 1119 1117 1211 1118 1212 | 100 to 150 | Hot Rolled or Annealed | 4 | 1—4 | .022 | 85 | M2, M7 |
| | | | 2 | 5—10 | .016 | | |
| | | | 2 | 11—19 | .011 | | |
| | | | 2 | 20—48 | .008 | | |
| | | | 4 | 25—6 | .55 | 26 | S4, S2 |
| | | | 2 | 5.5—2.5 | .40 | | |
| | | | 2 | 2—1.5 | .28 | | |
| | | | 2 | 1—.5 | .20 | | |
| | 150 to 200 | Cold Drawn | 4 | 1—4 | .022 | 90 | M2, M7 |
| | | | 2 | 5—10 | .016 | | |
| | | | 2 | 11—19 | .011 | | |
| | | | 2 | 20—48 | .008 | | |
| | | | 4 | 25—6 | .55 | 27 | S4, S2 |
| | | | 2 | 5.5—2.5 | .40 | | |
| | | | 2 | 2—1.5 | .28 | | |
| | | | 2 | 1—.5 | .20 | | |
| **Low Carbon Resulfurized** 1213 1215 | 100 to 150 | Hot Rolled or Annealed | 4 | 1—4 | .022 | 100 | M2, M7 |
| | | | 2 | 5—10 | .016 | | |
| | | | 2 | 11—19 | .011 | | |
| | | | 2 | 20—48 | .008 | | |
| | | | 4 | 25—6 | .55 | 30 | S4, S2 |
| | | | 2 | 5.5—2.5 | .40 | | |
| | | | 2 | 2—1.5 | .28 | | |
| | | | 2 | 1—.5 | .20 | | |
| | 150 to 200 | Cold Drawn | 4 | 1—4 | .022 | 110 | M2, M7 |
| | | | 2 | 5—10 | .016 | | |
| | | | 2 | 11—19 | .011 | | |
| | | | 2 | 20—48 | .008 | | |
| | | | 4 | 25—6 | .55 | 34 | S4, S2 |
| | | | 2 | 5.5—2.5 | .40 | | |
| | | | 2 | 2—1.5 | .28 | | |
| | | | 2 | 1—.5 | .20 | | |
| **Low Carbon Resulfurized** 1108 1109 1110 1115 | 100 to 150 | Hot Rolled or Annealed | 4 | 1—4 | .022 | 80 | M2, M7 |
| | | | 2 | 5—10 | .016 | | |
| | | | 2 | 11—19 | .011 | | |
| | | | 2 | 20—48 | .008 | | |
| | | | 4 | 25—6 | .55 | 24 | S4, S2 |
| | | | 2 | 5.5—2.5 | .40 | | |
| | | | 2 | 2—1.5 | .28 | | |
| | | | 2 | 1—.5 | .20 | | |
| | 150 to 200 | Cold Drawn | 4 | 1—4 | .022 | 85 | M2, M7 |
| | | | 2 | 5—10 | .016 | | |
| | | | 2 | 11—19 | .011 | | |
| | | | 2 | 20—48 | .008 | | |
| | | | 4 | 25—6 | .55 | 26 | S4, S2 |
| | | | 2 | 5.5—2.5 | .40 | | |
| | | | 2 | 2—1.5 | .28 | | |
| | | | 2 | 1—.5 | .20 | | |
| **Medium Carbon Resulfurized** 1132 1144 1137 1145 1139 1146 1140 1151 1141 | 175 to 225 | Hot Rolled, Normalized, Annealed or Cold Drawn | 4 | 1—4 | .022 | 80 | M2, M7 |
| | | | 2 | 5—10 | .016 | | |
| | | | 2 | 11—19 | .011 | | |
| | | | 2 | 20—48 | .008 | | |
| | | | 4 | 25—6 | .55 | 24 | S4, S2 |
| | | | 2 | 5.5—2.5 | .40 | | |
| | | | 2 | 2—1.5 | .28 | | |
| | | | 2 | 1—.5 | .20 | | |
| | 275 to 325 | Quenched and Tempered | 4 | 1—4 | .018 | 55 | M2, M7 |
| | | | 2 | 5—10 | .012 | | |
| | | | 2 | 11—19 | .010 | | |
| | | | 2 | 20—48 | .008 | | |
| | | | 4 | 25—6 | .45 | 17 | S4, S2 |
| | | | 2 | 5.5—2.5 | .30 | | |
| | | | 2 | 2—1.5 | .25 | | |
| | | | 2 | 1—.5 | .20 | | |

For cutting gears of Class 9 (AGMA 390.03) or better, the Rotary Feed should be reduced and the number of cuts should be increased.
See section 16 for Cutting Fluid Recommendations.

# 7.3 Gear Shaping

| MATERIAL | HARD-NESS<br>Bhn | CONDITION | NUMBER OF CUTS | DIAMETRAL PITCH<br>MODULE | ROTARY FEED per cutter stroke (4″ P.D. cutter)<br>in<br>mm | CUTTER SPEED<br>fpm<br>m/min | HSS TOOL MATERIAL<br>AISI<br>ISO |
|---|---|---|---|---|---|---|---|
| **1. FREE MACHINING CARBON STEELS, WROUGHT (cont.)**<br>**Medium Carbon Resulfurized (cont.)**<br>(materials listed on preceding page) | 325<br>to<br>375 | Quenched<br>and<br>Tempered | 4<br>2<br>2<br>2 | 1—4<br>5—10<br>11—19<br>20—48 | .012<br>.010<br>.008<br>.008 | 40 | M3, M42 |
| | | | 4<br>2<br>2<br>2 | 25—6<br>5.5—2.5<br>2—1.5<br>1—.5 | .30<br>.25<br>.20<br>.20 | 12 | S5, S11 |
| **Low Carbon Leaded**<br>10L18<br>11L17 | 100<br>to<br>150 | Hot<br>Rolled,<br>Normalized,<br>Annealed<br>or Cold<br>Drawn | 4<br>2<br>2<br>2 | 1—4<br>5—10<br>11—19<br>20—48 | .022<br>.016<br>.011<br>.008 | 110 | M2, M7 |
| | | | 4<br>2<br>2<br>2 | 25—6<br>5.5—2.5<br>2—1.5<br>1—.5 | .55<br>.40<br>.28<br>.20 | 34 | S4, S2 |
| | 150<br>to<br>200 | Hot<br>Rolled,<br>Normalized,<br>Annealed<br>or Cold<br>Drawn | 4<br>2<br>2<br>2 | 1—4<br>5—10<br>11—19<br>20—48 | .022<br>.016<br>.011<br>.008 | 100 | M2, M7 |
| | | | 4<br>2<br>2<br>2 | 25—6<br>5.5—2.5<br>2—1.5<br>1—.5 | .55<br>.40<br>.28<br>.20 | 30 | S4, S2 |
| | 200<br>to<br>250 | Hot<br>Rolled,<br>Normalized,<br>Annealed<br>or Cold<br>Drawn | 4<br>2<br>2<br>2 | 1—4<br>5—10<br>11—19<br>20—48 | .022<br>.016<br>.011<br>.008 | 85 | M2, M7 |
| | | | 4<br>2<br>2<br>2 | 25—6<br>5.5—2.5<br>2—1.5<br>1—.5 | .55<br>.40<br>.28<br>.20 | 26 | S4, S2 |
| **Low Carbon Leaded**<br>12L13<br>12L14<br>12L15 | 100<br>to<br>150 | Hot<br>Rolled,<br>Normalized,<br>Annealed<br>or Cold<br>Drawn | 4<br>2<br>2<br>2 | 1—4<br>5—10<br>11—19<br>20—48 | .022<br>.016<br>.011<br>.008 | 125 | M2, M7 |
| | | | 4<br>2<br>2<br>2 | 25—6<br>5.5—2.5<br>2—1.5<br>1—.5 | .55<br>.40<br>.28<br>.20 | 38 | S4, S2 |
| | 150<br>to<br>200 | Hot<br>Rolled,<br>Normalized,<br>Annealed<br>or Cold<br>Drawn | 4<br>2<br>2<br>2 | 1—4<br>5—10<br>11—19<br>20—48 | .022<br>.016<br>.011<br>.008 | 130 | M2, M7 |
| | | | 4<br>2<br>2<br>2 | 25—6<br>5.5—2.5<br>2—1.5<br>1—.5 | .55<br>.40<br>.28<br>.20 | 40 | S4, S2 |
| | 200<br>to<br>250 | Hot<br>Rolled,<br>Normalized,<br>Annealed<br>or Cold<br>Drawn | 4<br>2<br>2<br>2 | 1—4<br>5—10<br>11—19<br>20—48 | .022<br>.016<br>.011<br>.008 | 90 | M2, M7 |
| | | | 4<br>2<br>2<br>2 | 25—6<br>5.5—2.5<br>2—1.5<br>1—.5 | .55<br>.40<br>.28<br>.20 | 27 | S4, S2 |
| **Medium Carbon Leaded**<br>10L45<br>10L50<br>11L37<br>11L41<br>11L44 | 125<br>to<br>175 | Hot<br>Rolled,<br>Normalized,<br>Annealed<br>or Cold<br>Drawn | 4<br>2<br>2<br>2 | 1—4<br>5—10<br>11—19<br>20—48 | .022<br>.016<br>.011<br>.008 | 90 | M2, M7 |
| | | | 4<br>2<br>2<br>2 | 25—6<br>5.5—2.5<br>2—1.5<br>1—.5 | .55<br>.40<br>.28<br>.20 | 27 | S4, S2 |

For cutting gears of Class 9 (AGMA 390.03) or better, the Rotary Feed should be reduced and the number of cuts should be increased.
See section 16 for Cutting Fluid Recommendations.

| MATERIAL | HARDNESS Bhn | CONDITION | NUMBER OF CUTS | DIAMETRAL PITCH / MODULE | ROTARY FEED per cutter stroke (4" P.D. cutter) in / mm | CUTTER SPEED fpm / m/min | HSS TOOL MATERIAL AISI / ISO |
|---|---|---|---|---|---|---|---|
| **1. FREE MACHINING CARBON STEELS, WROUGHT (cont.) Medium Carbon Leaded (cont.)** (materials listed on preceding page) | 175 to 225 | Hot Rolled, Normalized, Annealed or Cold Drawn | 4 | 1—4 | .022 | 85 | M2, M7 |
| | | | 2 | 5—10 | .016 | | |
| | | | 2 | 11—19 | .011 | | |
| | | | 2 | 20—48 | .008 | | |
| | | | 4 | 25—6 | .55 | 26 | S4, S2 |
| | | | 2 | 5.5—2.5 | .40 | | |
| | | | 2 | 2—1.5 | .28 | | |
| | | | 2 | 1—.5 | .20 | | |
| | 225 to 275 | Hot Rolled, Normalized, Annealed, Cold Drawn or Quenched and Tempered | 4 | 1—4 | .022 | 75 | M2, M7 |
| | | | 2 | 5—10 | .016 | | |
| | | | 2 | 11—19 | .011 | | |
| | | | 2 | 20—48 | .008 | | |
| | | | 4 | 25—6 | .55 | 23 | S4, S2 |
| | | | 2 | 5.5—2.5 | .40 | | |
| | | | 2 | 2—1.5 | .28 | | |
| | | | 2 | 1—.5 | .20 | | |
| | 275 to 325 | Hot Rolled, Normalized, Annealed, Cold Drawn or Quenched and Tempered | 4 | 1—4 | .018 | 60 | M2, M7 |
| | | | 2 | 5—10 | .012 | | |
| | | | 2 | 11—19 | .010 | | |
| | | | 2 | 20—48 | .008 | | |
| | | | 4 | 25—6 | .45 | 18 | S4, S2 |
| | | | 2 | 5.5—2.5 | .30 | | |
| | | | 2 | 2—1.5 | .25 | | |
| | | | 2 | 1—.5 | .20 | | |
| | 325 to 375 | Quenched and Tempered | 4 | 1—4 | .012 | 55 | M3, M42 |
| | | | 2 | 5—10 | .010 | | |
| | | | 2 | 11—19 | .008 | | |
| | | | 2 | 20—48 | .008 | | |
| | | | 4 | 25—6 | .30 | 17 | S5, S11 |
| | | | 2 | 5.5—2.5 | .25 | | |
| | | | 2 | 2—1.5 | .20 | | |
| | | | 2 | 1—.5 | .20 | | |
| **2. CARBON STEELS, WROUGHT Low Carbon** 1005 1006 1008 1009 1010 1011 1012 1013 1015 1016 1017 1018 1019 1020 1021 1022 1023 1025 1026 1029 1513 1518 1522 | 85 to 125 | Hot Rolled, Normalized, Annealed or Cold Drawn | 4 | 1—4 | .018 | 90 | M2, M7 |
| | | | 2 | 5—10 | .012 | | |
| | | | 2 | 11—19 | .010 | | |
| | | | 2 | 20—48 | .008 | | |
| | | | 4 | 25—6 | .45 | 27 | S4, S2 |
| | | | 2 | 5.5—2.5 | .30 | | |
| | | | 2 | 2—1.5 | .25 | | |
| | | | 2 | 1—.5 | .20 | | |
| | 125 to 175 | Hot Rolled, Normalized, Annealed or Cold Drawn | 4 | 1—4 | .018 | 80 | M2, M7 |
| | | | 2 | 5—10 | .012 | | |
| | | | 2 | 11—19 | .010 | | |
| | | | 2 | 20—48 | .008 | | |
| | | | 4 | 25—6 | .45 | 24 | S4, S2 |
| | | | 2 | 5.5—2.5 | .30 | | |
| | | | 2 | 2—1.5 | .25 | | |
| | | | 2 | 1—.5 | .20 | | |
| | 175 to 225 | Hot Rolled, Normalized, Annealed or Cold Drawn | 4 | 1—4 | .018 | 70 | M2, M7 |
| | | | 2 | 5—10 | .012 | | |
| | | | 2 | 11—19 | .010 | | |
| | | | 2 | 20—48 | .008 | | |
| | | | 4 | 25—6 | .45 | 21 | S4, S2 |
| | | | 2 | 5.5—2.5 | .30 | | |
| | | | 2 | 2—1.5 | .25 | | |
| | | | 2 | 1—.5 | .20 | | |
| | 225 to 275 | Annealed or Cold Drawn | 4 | 1—4 | .016 | 60 | M2, M7 |
| | | | 2 | 5—10 | .012 | | |
| | | | 2 | 11—19 | .010 | | |
| | | | 2 | 20—48 | .008 | | |
| | | | 4 | 25—6 | .40 | 18 | S4, S2 |
| | | | 2 | 5.5—2.5 | .30 | | |
| | | | 2 | 2—1.5 | .25 | | |
| | | | 2 | 1—.5 | .20 | | |

For cutting gears of Class 9 (AGMA 390.03) or better, the Rotary Feed should be reduced and the number of cuts should be increased.
See section 16 for Cutting Fluid Recommendations.

# 7.3 Gear Shaping

| MATERIAL | HARDNESS Bhn | CONDITION | NUMBER OF CUTS | DIAMETRAL PITCH<br>MODULE | ROTARY FEED per cutter stroke (4" P.D. cutter)<br>in<br>mm | CUTTER SPEED<br>fpm<br>m/min | HSS TOOL MATERIAL<br>AISI<br>ISO |
|---|---|---|---|---|---|---|---|
| **2. CARBON STEELS, WROUGHT (cont.)**<br>**Medium Carbon**<br>1030 1044 1526<br>1033 1045 1527<br>1035 1046 1536<br>1037 1049 1541<br>1038 1050 1547<br>1039 1053 1548<br>1040 1055 1551<br>1042 1524 1552<br>1043 1525 | 125 to 175 | Hot Rolled, Normalized, Annealed or Cold Drawn | 4<br>2<br>2<br>2 | 1—4<br>5—10<br>11—19<br>20—48 | .018<br>.012<br>.010<br>.008 | 75 | M2, M7 |
| | | | 4<br>2<br>2<br>2 | 25—6<br>5.5—2.5<br>2—1.5<br>1—.5 | .45<br>.30<br>.25<br>.20 | 23 | S4, S2 |
| | 175 to 225 | Hot Rolled, Normalized, Annealed or Cold Drawn | 4<br>2<br>2<br>2 | 1—4<br>5—10<br>11—19<br>20—48 | .018<br>.012<br>.010<br>.008 | 65 | M2, M7 |
| | | | 4<br>2<br>2<br>2 | 25—6<br>5.5—2.5<br>2—1.5<br>1—.5 | .45<br>.30<br>.25<br>.20 | 20 | S4, S2 |
| | 225 to 275 | Hot Rolled, Normalized, Annealed, Cold Drawn or Quenched and Tempered | 4<br>2<br>2<br>2 | 1—4<br>5—10<br>11—19<br>20—48 | .016<br>.012<br>.010<br>.008 | 60 | M2, M7 |
| | | | 4<br>2<br>2<br>2 | 25—6<br>5.5—2.5<br>2—1.5<br>1—.5 | .40<br>.30<br>.25<br>.20 | 18 | S4, S2 |
| | 275 to 325 | Hot Rolled, Normalized, Annealed or Quenched and Tempered | 4<br>2<br>2<br>2 | 1—4<br>5—10<br>11—19<br>20—48 | .016<br>.012<br>.010<br>.008 | 50 | M2, M7 |
| | | | 4<br>2<br>2<br>2 | 25—6<br>5.5—2.5<br>2—1.5<br>1—.5 | .40<br>.30<br>.25<br>.20 | 15 | S4, S2 |
| | 325 to 375 | Quenched and Tempered | 4<br>2<br>2<br>2 | 1—4<br>5—10<br>11—19<br>20—48 | .012<br>.010<br>.008<br>.008 | 40 | M3, M42 |
| | | | 4<br>2<br>2<br>2 | 25—6<br>5.5—2.5<br>2—1.5<br>1—.5 | .30<br>.25<br>.20<br>.20 | 12 | S5, S11 |
| **High Carbon**<br>1060 1075 1090<br>1064 1078 1095<br>1065 1080 1561<br>1069 1084 1566<br>1070 1085 1572<br>1074 1086 | 175 to 225 | Hot Rolled, Normalized, Annealed or Cold Drawn | 4<br>2<br>2<br>2 | 1—4<br>5—10<br>11—19<br>20—48 | .018<br>.012<br>.010<br>.008 | 65 | M2, M7 |
| | | | 4<br>2<br>2<br>2 | 25—6<br>5.5—2.5<br>2—1.5<br>1—.5 | .45<br>.30<br>.25<br>.20 | 20 | S4, S2 |
| | 225 to 275 | Hot Rolled, Normalized, Annealed, Cold Drawn or Quenched and Tempered | 4<br>2<br>2<br>2 | 1—4<br>5—10<br>11—19<br>20—48 | .016<br>.012<br>.010<br>.008 | 55 | M2, M7 |
| | | | 4<br>2<br>2<br>2 | 25—6<br>5.5—2.5<br>2—1.5<br>1—.5 | .40<br>.30<br>.25<br>.20 | 17 | S4, S2 |
| | 275 to 325 | Hot Rolled, Normalized, Annealed or Quenched and Tempered | 4<br>2<br>2<br>2 | 1—4<br>5—10<br>11—19<br>20—48 | .016<br>.012<br>.010<br>.008 | 45 | M2, M7 |
| | | | 4<br>2<br>2<br>2 | 25—6<br>5.5—2.5<br>2—1.5<br>1—.5 | .40<br>.30<br>.25<br>.20 | 14 | S4, S2 |

For cutting gears of Class 9 (AGMA 390.03) or better, the Rotary Feed should be reduced and the number of cuts should be increased.
See section 16 for Cutting Fluid Recommendations.

| MATERIAL | HARD-NESS Bhn | CONDITION | NUMBER OF CUTS | DIAMETRAL PITCH / MODULE | ROTARY FEED per cutter stroke (4″ P.D. cutter) in / mm | CUTTER SPEED fpm / m/min | HSS TOOL MATERIAL AISI / ISO |
|---|---|---|---|---|---|---|---|
| **2. CARBON STEELS, WROUGHT (cont.) High Carbon (cont.)** (materials listed on preceding page) | 325 to 375 | Quenched and Tempered | 4<br>2<br>2<br>2 | 1—4<br>5—10<br>11—19<br>20—48 | .012<br>.010<br>.008<br>.008 | 35 | M3, M42 |
| | | | 4<br>2<br>2<br>2 | 25—6<br>5.5—2.5<br>2—1.5<br>1—.5 | .30<br>.25<br>.20<br>.20 | 11 | S5, S11 |
| **3. CARBON AND FERRITIC ALLOY STEELS (HIGH TEMPERATURE SERVICE)** ASTM A369: Grades FPA, FPB, FP1, FP2, FP12 | 150 to 200 | As Forged | 4<br>2<br>2<br>2 | 1—4<br>5—10<br>11—19<br>20—48 | .018<br>.012<br>.010<br>.008 | 60 | M2, M7 |
| | | | 4<br>2<br>2<br>2 | 25—6<br>5.5—2.5<br>2—1.5<br>1—.5 | .45<br>.30<br>.25<br>.20 | 18 | S4, S2 |
| ASTM A369: Grades FP3b, FP11 | 150 to 200 | Annealed or Normalized and Tempered | 4<br>2<br>2<br>2 | 1—4<br>5—10<br>11—19<br>20—48 | .018<br>.012<br>.010<br>.008 | 55 | M2, M7 |
| | | | 4<br>2<br>2<br>2 | 25—6<br>5.5—2.5<br>2—1.5<br>1—.5 | .45<br>.30<br>.25<br>.20 | 17 | S4, S2 |
| ASTM A369: Grades FP5, FP7, FP9, FP21, FP22 | 150 to 200 | Annealed or Normalized and Tempered | 4<br>2<br>2<br>2 | 1—4<br>5—10<br>11—19<br>20—48 | .018<br>.012<br>.010<br>.008 | 50 | M2, M7 |
| | | | 4<br>2<br>2<br>2 | 25—6<br>5.5—2.5<br>2—1.5<br>1—.5 | .45<br>.30<br>.25<br>.20 | 15 | S4, S2 |
| **4. FREE MACHINING ALLOY STEELS, WROUGHT Medium Carbon Resulfurized** 4140 4145Se 4140Se 4147Te 4142Te 4150 | 150 to 200 | Hot Rolled, Normalized, Annealed or Cold Drawn | 4<br>2<br>2<br>2 | 1—4<br>5—10<br>11—19<br>20—48 | .022<br>.018<br>.011<br>.008 | 80 | M2, M7 |
| | | | 4<br>2<br>2<br>2 | 25—6<br>5.5—2.5<br>2—1.5<br>1—.5 | .55<br>.45<br>.28<br>.20 | 24 | S4, S2 |
| | 200 to 250 | Hot Rolled, Normalized, Annealed or Cold Drawn | 4<br>2<br>2<br>2 | 1—4<br>5—10<br>11—19<br>20—48 | .022<br>.018<br>.011<br>.008 | 75 | M2, M7 |
| | | | 4<br>2<br>2<br>2 | 25—6<br>5.5—2.5<br>2—1.5<br>1—.5 | .55<br>.45<br>.28<br>.20 | 23 | S4, S2 |
| | 275 to 325 | Quenched and Tempered | 4<br>2<br>2<br>2 | 1—4<br>5—10<br>11—19<br>20—48 | .018<br>.012<br>.010<br>.008 | 50 | M2, M7 |
| | | | 4<br>2<br>2<br>2 | 25—6<br>5.5—2.5<br>2—1.5<br>1—.5 | .45<br>.30<br>.25<br>.20 | 15 | S4, S2 |
| | 325 to 375 | Quenched and Tempered | 4<br>2<br>2<br>2 | 1—4<br>5—10<br>11—19<br>20—48 | .012<br>.010<br>.008<br>.008 | 35 | M3, M42 |
| | | | 4<br>2<br>2<br>2 | 25—6<br>5.5—2.5<br>2—1.5<br>1—.5 | .30<br>.25<br>.20<br>.20 | 11 | S5, S11 |

For cutting gears of Class 9 (AGMA 390.03) or better, the Rotary Feed should be reduced and the number of cuts should be increased.
See section 16 for Cutting Fluid Recommendations.

# 7.3 Gear Shaping

| MATERIAL | HARDNESS Bhn | CONDITION | NUMBER OF CUTS | DIAMETRAL PITCH / MODULE | ROTARY FEED per cutter stroke (4" P.D. cutter) in / mm | CUTTER SPEED fpm / m/min | HSS TOOL MATERIAL AISI / ISO |
|---|---|---|---|---|---|---|---|
| **4. FREE MACHINING ALLOY STEELS, WROUGHT (cont.) Medium and High Carbon Leaded** 41L30 41L47 51L32 86L40 41L40 41L50 52L100 41L45 43L40 86L20 | 150 to 200 | Hot Rolled, Normalized, Annealed or Cold Drawn | 4 | 1—4 | .022 | 80 | M2, M7 |
| | | | 2 | 5—10 | .016 | | |
| | | | 2 | 11—19 | .011 | | |
| | | | 2 | 20—48 | .008 | | |
| | | | 4 | 25—6 | .55 | 24 | S4, S2 |
| | | | 2 | 5.5—2.5 | .40 | | |
| | | | 2 | 2—1.5 | .28 | | |
| | | | 2 | 1—.5 | .20 | | |
| | 200 to 250 | Hot Rolled, Normalized, Annealed or Cold Drawn | 4 | 1—4 | .022 | 75 | M2, M7 |
| | | | 2 | 5—10 | .016 | | |
| | | | 2 | 11—19 | .011 | | |
| | | | 2 | 20—48 | .008 | | |
| | | | 4 | 25—6 | .55 | 23 | S4, S2 |
| | | | 2 | 5.5—2.5 | .40 | | |
| | | | 2 | 2—1.5 | .28 | | |
| | | | 2 | 1—.5 | .20 | | |
| | 275 to 325 | Quenched and Tempered | 4 | 1—4 | .018 | 50 | M2, M7 |
| | | | 2 | 5—10 | .012 | | |
| | | | 2 | 11—19 | .010 | | |
| | | | 2 | 20—48 | .008 | | |
| | | | 4 | 25—6 | .45 | 15 | S4, S2 |
| | | | 2 | 5.5—2.5 | .30 | | |
| | | | 2 | 2—1.5 | .25 | | |
| | | | 2 | 1—.5 | .20 | | |
| | 325 to 375 | Quenched and Tempered | 4 | 1—4 | .012 | 35 | M3, M42 |
| | | | 2 | 5—10 | .010 | | |
| | | | 2 | 11—19 | .008 | | |
| | | | 2 | 20—48 | .008 | | |
| | | | 4 | 25—6 | .30 | 11 | S5, S11 |
| | | | 2 | 5.5—2.5 | .25 | | |
| | | | 2 | 2—1.5 | .20 | | |
| | | | 2 | 1—.5 | .20 | | |
| **5. ALLOY STEELS, WROUGHT Low Carbon** 4012 4615 4817 8617 4023 4617 4820 8620 4024 4620 5015 8622 4118 4621 5115 8822 4320 4718 5120 9310 4419 4720 6118 94B15 4422 4815 8115 94B17 | 125 to 175 | Hot Rolled, Annealed or Cold Drawn | 4 | 1—4 | .018 | 70 | M2, M7 |
| | | | 2 | 5—10 | .012 | | |
| | | | 2 | 11—19 | .010 | | |
| | | | 2 | 20—48 | .008 | | |
| | | | 4 | 25—6 | .45 | 21 | S4, S2 |
| | | | 2 | 5.5—2.5 | .30 | | |
| | | | 2 | 2—1.5 | .25 | | |
| | | | 2 | 1—.5 | .20 | | |
| | 175 to 225 | Hot Rolled, Annealed or Cold Drawn | 4 | 1—4 | .018 | 65 | M2, M7 |
| | | | 2 | 5—10 | .012 | | |
| | | | 2 | 11—19 | .010 | | |
| | | | 2 | 20—48 | .008 | | |
| | | | 4 | 25—6 | .45 | 20 | S4, S2 |
| | | | 2 | 5.5—2.5 | .30 | | |
| | | | 2 | 2—1.5 | .25 | | |
| | | | 2 | 1—.5 | .20 | | |
| | 225 to 275 | Hot Rolled, Normalized, Annealed or Cold Drawn | 4 | 1—4 | .016 | 50 | M2, M7 |
| | | | 2 | 5—10 | .012 | | |
| | | | 2 | 11—19 | .010 | | |
| | | | 2 | 20—48 | .008 | | |
| | | | 4 | 25—6 | .40 | 15 | S4, S2 |
| | | | 2 | 5.5—2.5 | .30 | | |
| | | | 2 | 2—1.5 | .25 | | |
| | | | 2 | 1—.5 | .20 | | |
| | 275 to 325 | Normalized or Quenched and Tempered | 4 | 1—4 | .012 | 45 | M2, M7 |
| | | | 2 | 5—10 | .010 | | |
| | | | 2 | 11—19 | .008 | | |
| | | | 2 | 20—48 | .008 | | |
| | | | 4 | 25—6 | .30 | 14 | S4, S2 |
| | | | 2 | 5.5—2.5 | .25 | | |
| | | | 2 | 2—1.5 | .20 | | |
| | | | 2 | 1—.5 | .20 | | |

For cutting gears of Class 9 (AGMA 390.03) or better, the Rotary Feed should be reduced and the number of cuts should be increased.
See section 16 for Cutting Fluid Recommendations.

| MATERIAL | HARDNESS Bhn | CONDITION | NUMBER OF CUTS | DIAMETRAL PITCH / MODULE | ROTARY FEED per cutter stroke (4" P.D. cutter) in / mm | CUTTER SPEED fpm / m/min | HSS TOOL MATERIAL AISI / ISO |
|---|---|---|---|---|---|---|---|
| **5. ALLOY STEELS, WROUGHT (cont.)**<br>**Low Carbon (cont.)**<br>(materials listed on preceding page) | 325 to 375 | Normalized or Quenched and Tempered | 4 | 1—4 | .012 | 40 | M3, M42 |
| | | | 2 | 5—10 | .010 | | |
| | | | 2 | 11—19 | .008 | | |
| | | | 2 | 20—48 | .008 | | |
| | | | 4 | 25—6 | .30 | 12 | S5, S11 |
| | | | 2 | 5.5—2.5 | .25 | | |
| | | | 2 | 2—1.5 | .20 | | |
| | | | 2 | 1—.5 | .20 | | |
| **Medium Carbon**<br>1330 4427 81B45<br>1335 4626 8625<br>1340 50B40 8627<br>1345 50B44 8630<br>4027 5046 8637<br>4028 50B46 8640<br>4032 50B50 8642<br>4037 5060 8645<br>4042 50B60 86B45<br>4047 5130 8650<br>4130 5132 8655<br>4135 5135 8660<br>4137 5140 8740<br>4140 5145 8742<br>4142 5147 9254<br>4145 5150 9255<br>4147 5155 9260<br>4150 5160 94B30<br>4161 51B60<br>4340 6150 | 175 to 225 | Hot Rolled, Annealed or Cold Drawn | 4 | 1—4 | .018 | 60 | M2, M7 |
| | | | 2 | 5—10 | .012 | | |
| | | | 2 | 11—19 | .010 | | |
| | | | 2 | 20—48 | .008 | | |
| | | | 4 | 25—6 | .45 | 18 | S4, S2 |
| | | | 2 | 5.5—2.5 | .30 | | |
| | | | 2 | 2—1.5 | .25 | | |
| | | | 2 | 1—.5 | .20 | | |
| | 225 to 275 | Annealed, Normalized, Cold Drawn or Quenched and Tempered | 4 | 1—4 | .016 | 45 | M2, M7 |
| | | | 2 | 5—10 | .012 | | |
| | | | 2 | 11—19 | .010 | | |
| | | | 2 | 20—48 | .008 | | |
| | | | 4 | 25—6 | .40 | 14 | S4, S2 |
| | | | 2 | 5.5—2.5 | .30 | | |
| | | | 2 | 2—1.5 | .25 | | |
| | | | 2 | 1—.5 | .20 | | |
| | 275 to 325 | Normalized or Quenched and Tempered | 4 | 1—4 | .016 | 40 | M2, M7 |
| | | | 2 | 5—10 | .012 | | |
| | | | 2 | 11—19 | .010 | | |
| | | | 2 | 20—48 | .008 | | |
| | | | 4 | 25—6 | .40 | 12 | S4, S2 |
| | | | 2 | 5.5—2.5 | .30 | | |
| | | | 2 | 2—1.5 | .25 | | |
| | | | 2 | 1—.5 | .20 | | |
| | 325 to 375 | Normalized or Quenched and Tempered | 4 | 1—4 | .012 | 35 | M3, M42 |
| | | | 2 | 5—10 | .010 | | |
| | | | 2 | 11—19 | .008 | | |
| | | | 2 | 20—48 | .008 | | |
| | | | 4 | 25—6 | .30 | 11 | S5, S11 |
| | | | 2 | 5.5—2.5 | .25 | | |
| | | | 2 | 2—1.5 | .20 | | |
| | | | 2 | 1—.5 | .20 | | |
| **High Carbon**<br>50100<br>51100<br>52100<br>M-50 | 175 to 225 | Hot Rolled, Annealed or Cold Drawn | 4 | 1—4 | .018 | 55 | M2, M7 |
| | | | 2 | 5—10 | .012 | | |
| | | | 2 | 11—19 | .010 | | |
| | | | 2 | 20—48 | .008 | | |
| | | | 4 | 25—6 | .45 | 17 | S4, S2 |
| | | | 2 | 5.5—2.5 | .30 | | |
| | | | 2 | 2—1.5 | .25 | | |
| | | | 2 | 1—.5 | .20 | | |
| | 225 to 275 | Normalized, Cold Drawn or Quenched and Tempered | 4 | 1—4 | .016 | 40 | M2, M7 |
| | | | 2 | 5—10 | .012 | | |
| | | | 2 | 11—19 | .010 | | |
| | | | 2 | 20—48 | .008 | | |
| | | | 4 | 25—6 | .40 | 12 | S4, S2 |
| | | | 2 | 5.5—2.5 | .30 | | |
| | | | 2 | 2—1.5 | .25 | | |
| | | | 2 | 1—.5 | .20 | | |
| | 275 to 325 | Normalized or Quenched and Tempered | 4 | 1—4 | .016 | 35 | M2, M7 |
| | | | 2 | 5—10 | .012 | | |
| | | | 2 | 11—19 | .010 | | |
| | | | 2 | 20—48 | .008 | | |
| | | | 4 | 25—6 | .40 | 11 | S4, S2 |
| | | | 2 | 5.5—2.5 | .30 | | |
| | | | 2 | 2—1.5 | .25 | | |
| | | | 2 | 1—.5 | .20 | | |

For cutting gears of Class 9 (AGMA 390.03) or better, the Rotary Feed should be reduced and the number of cuts should be increased.
See section 16 for Cutting Fluid Recommendations.

## 7.3 Gear Shaping

| MATERIAL | HARD-NESS | CONDITION | NUMBER OF CUTS | DIAMETRAL PITCH / MODULE | ROTARY FEED per cutter stroke (4" P.D. cutter) | CUTTER SPEED | HSS TOOL MATERIAL |
|---|---|---|---|---|---|---|---|
| | Bhn | | | | in / mm | fpm / m/min | AISI / ISO |
| **5. ALLOY STEELS, WROUGHT (cont.)** | 325 | Normalized | 4 | 1—4 | .012 | 30 | M3, M42 |
| **High Carbon (cont.)** | to | or | 2 | 5—10 | .010 | | |
| (materials listed on preceding page) | 375 | Quenched | 2 | 11—19 | .008 | | |
| | | and | 2 | 20—48 | .008 | | |
| | | Tempered | 4 | 25—6 | .30 | 9 | S5, S11 |
| | | | 2 | 5.5—2.5 | .25 | | |
| | | | 2 | 2—1.5 | .20 | | |
| | | | 2 | 1—.5 | .20 | | |
| **6. HIGH STRENGTH STEELS, WROUGHT** | 225 | Annealed | 4 | 1—4 | .016 | 45 | M2, M7 |
| 300M 98BV40 | to | | 2 | 5—10 | .012 | | |
| 4330V D6ac | 300 | | 2 | 11—19 | .010 | | |
| 4340 H11 | | | 2 | 20—48 | .008 | | |
| 4340Si H13 | | | 4 | 25—6 | .40 | 14 | S4, S2 |
| | | | 2 | 5.5—2.5 | .30 | | |
| | | | 2 | 2—1.5 | .25 | | |
| | | | 2 | 1—.5 | .20 | | |
| | 300 | Normalized | 4 | 1—4 | .012 | 40 | M2, M7 |
| | to | | 2 | 5—10 | .010 | | |
| | 350 | | 2 | 11—19 | .008 | | |
| | | | 2 | 20—48 | .008 | | |
| | | | 4 | 25—6 | .30 | 12 | S4, S2 |
| | | | 2 | 5.5—2.5 | .25 | | |
| | | | 2 | 2—1.5 | .20 | | |
| | | | 2 | 1—.5 | .20 | | |
| HP 9-4-20 | 325 | Annealed | 4 | 1—4 | .012 | 45 | M3, M42 |
| HP 9-4-25 | to | | 2 | 5—10 | .010 | | |
| HP 9-4-30 | 375 | | 2 | 11—19 | .008 | | |
| HP 9-4-45 | | | 2 | 20—48 | .008 | | |
| | | | 4 | 25—6 | .30 | 14 | S5, S11 |
| | | | 2 | 5.5—2.5 | .25 | | |
| | | | 2 | 2—1.5 | .20 | | |
| | | | 2 | 1—.5 | .20 | | |
| **7. MARAGING STEELS, WROUGHT** | 275 | Annealed | 4 | 1—4 | .012 | 45 | M2, M7 |
| 120 Grade | to | | 2 | 5—10 | .010 | | |
| 180 Grade | 325 | | 2 | 11—19 | .008 | | |
| | | | 2 | 20—48 | .008 | | |
| | | | 4 | 25—6 | .30 | 14 | S4, S2 |
| | | | 2 | 5.5—2.5 | .25 | | |
| | | | 2 | 2—1.5 | .20 | | |
| | | | 2 | 1—.5 | .20 | | |
| ASTM A538: Grades A, B, C | 275 | Annealed | 4 | 1—4 | .012 | 35 | M2, M7 |
| 200 Grade | to | | 2 | 5—10 | .010 | | |
| 250 Grade | 325 | | 2 | 11—19 | .008 | | |
| 300 Grade | | | 2 | 20—48 | .008 | | |
| 350 Grade | | | 4 | 25—6 | .30 | 11 | S4, S2 |
| HY230 | | | 2 | 5.5—2.5 | .25 | | |
| | | | 2 | 2—1.5 | .20 | | |
| | | | 2 | 1—.5 | .20 | | |
| **8. TOOL STEELS, WROUGHT** | 200 | Annealed | 4 | 1—4 | .012 | 30 | M2, M7 |
| **High Speed** | to | | 2 | 5—10 | .011 | | |
| M1 T1 | 250 | | 2 | 11—19 | .009 | | |
| M2 T2 | | | 2 | 20—48 | .006 | | |
| M6 T6 | | | 4 | 25—6 | .30 | 9 | S4, S2 |
| M10 | | | 2 | 5.5—2.5 | .28 | | |
| | | | 2 | 2—1.5 | .23 | | |
| | | | 2 | 1—.5 | .15 | | |
| **High Speed** | 225 | Annealed | 4 | 1—4 | .012 | 20 | M2, M7 |
| M3-1 M34 M46 | to | | 2 | 5—10 | .011 | | |
| M3-2 M36 M47 | 275 | | 2 | 11—19 | .009 | | |
| M4 M41 T4 | | | 2 | 20—48 | .006 | | |
| M7 M42 T5 | | | 4 | 25—6 | .30 | 6 | S4, S2 |
| M30 M43 T8 | | | 2 | 5.5—2.5 | .28 | | |
| M33 M44 T15 | | | 2 | 2—1.5 | .23 | | |
| | | | 2 | 1—.5 | .15 | | |

For cutting gears of Class 9 (AGMA 390.03) or better, the Rotary Feed should be reduced and the number of cuts should be increased.
See section 16 for Cutting Fluid Recommendations.

| MATERIAL | HARD-NESS<br>Bhn | CONDITION | NUMBER OF CUTS | DIAMETRAL PITCH<br>MODULE | ROTARY FEED per cutter stroke (4" P.D. cutter)<br>in<br>mm | CUTTER SPEED<br>fpm<br>m/min | HSS TOOL MATERIAL<br>AISI<br>ISO |
|---|---|---|---|---|---|---|---|
| **Hot Work**<br>H10<br>H11<br>H12<br>H13<br>H14<br>H19<br>H21<br>H22<br>H23<br>H24<br>H25<br>H26<br>H42 | 150<br>to<br>200 | Annealed | 4<br>2<br>2<br>2 | 1—4<br>5—10<br>11—19<br>20—48 | .018<br>.015<br>.010<br>.008 | 50 | M2, M7 |
| | | | 4<br>2<br>2<br>2 | 25—6<br>5.5—2.5<br>2—1.5<br>1—.5 | .45<br>.40<br>.25<br>.20 | 15 | S4, S2 |
| | 200<br>to<br>250 | Annealed | 4<br>2<br>2<br>2 | 1—4<br>5—10<br>11—19<br>20—48 | .016<br>.012<br>.010<br>.008 | 35 | M2, M7 |
| | | | 4<br>2<br>2<br>2 | 25—6<br>5.5—2.5<br>2—1.5<br>1—.5 | .40<br>.30<br>.25<br>.20 | 11 | S4, S2 |
| | 325<br>to<br>375 | Quenched and Tempered | 4<br>2<br>2<br>2 | 1—4<br>5—10<br>11—19<br>20—48 | .012<br>.010<br>.008<br>.006 | 25 | M3, M42 |
| | | | 4<br>2<br>2<br>2 | 25—6<br>5.5—2.5<br>2—1.5<br>1—.5 | .30<br>.25<br>.20<br>.15 | 8 | S5, S11 |
| **Cold Work**<br>A2 D3<br>A3 D4<br>A4 D5<br>A6 D7<br>A7 O1<br>A8 O2<br>A9 O6<br>A10 O7<br>D2 | 200<br>to<br>250 | Annealed | 4<br>2<br>2<br>2 | 1—4<br>5—10<br>11—19<br>20—48 | .016<br>.012<br>.010<br>.008 | 35 | M2, M7 |
| | | | 4<br>2<br>2<br>2 | 25—6<br>5.5—2.5<br>2—1.5<br>1—.5 | .40<br>.30<br>.25<br>.20 | 11 | S4, S2 |
| | | | | | | | |
| | | | | | | | |
| **Shock Resisting**<br>S1<br>S2<br>S5<br>S6<br>S7 | 175<br>to<br>225 | Annealed | 4<br>2<br>2<br>2 | 1—4<br>5—10<br>11—19<br>20—48 | .018<br>.015<br>.010<br>.008 | 50 | M2, M7 |
| | | | 4<br>2<br>2<br>2 | 25—6<br>5.5—2.5<br>2—1.5<br>1—.5 | .45<br>.40<br>.25<br>.20 | 15 | S4, S2 |
| **Mold**<br>P2<br>P4<br>P5<br>P6 | 100<br>to<br>150 | Annealed | 4<br>2<br>2<br>2 | 1—4<br>5—10<br>11—19<br>20—48 | .018<br>.015<br>.010<br>.008 | 60 | M2, M7 |
| | | | 4<br>2<br>2<br>2 | 25—6<br>5.5—2.5<br>2—1.5<br>1—.5 | .45<br>.40<br>.25<br>.20 | 18 | S4, S2 |
| **Mold**<br>P20<br>P21 | 150<br>to<br>200 | Annealed | 4<br>2<br>2<br>2 | 1—4<br>5—10<br>11—19<br>20—48 | .018<br>.015<br>.010<br>.008 | 50 | M2, M7 |
| | | | 4<br>2<br>2<br>2 | 25—6<br>5.5—2.5<br>2—1.5<br>1—.5 | .45<br>.40<br>.25<br>.20 | 15 | S4, S2 |

For cutting gears of Class 9 (AGMA 390.03) or better, the Rotary Feed should be reduced and the number of cuts should be increased.
See section 16 for Cutting Fluid Recommendations.

# 7.3 Gear Shaping

| MATERIAL | HARD-NESS Bhn | CONDITION | NUMBER OF CUTS | DIAMETRAL PITCH / MODULE | ROTARY FEED per cutter stroke (4″ P.D. cutter) in / mm | CUTTER SPEED fpm / m/min | HSS TOOL MATERIAL AISI / ISO |
|---|---|---|---|---|---|---|---|
| **8. TOOL STEELS, WROUGHT (cont.)** **Special Purpose** L2, L6, L7 | 150 to 200 | Annealed | 4 | 1—4 | .018 | 45 | M2, M7 |
| | | | 2 | 5—10 | .015 | | |
| | | | 2 | 11—19 | .010 | | |
| | | | 2 | 20—48 | .008 | | |
| | | | 4 | 25—6 | .45 | 14 | S4, S2 |
| | | | 2 | 5.5—2.5 | .40 | | |
| | | | 2 | 2—1.5 | .25 | | |
| | | | 2 | 1—.5 | .20 | | |
| **Special Purpose** F1, F2 | 200 to 250 | Annealed | 4 | 1—4 | .016 | 35 | M2, M7 |
| | | | 2 | 5—10 | .012 | | |
| | | | 2 | 11—19 | .010 | | |
| | | | 2 | 20—48 | .008 | | |
| | | | 4 | 25—6 | .40 | 11 | S4, S2 |
| | | | 2 | 5.5—2.5 | .30 | | |
| | | | 2 | 2—1.5 | .25 | | |
| | | | 2 | 1—.5 | .20 | | |
| **Water Hardening** W1, W2, W5; SAE J438b: Types W108, W109, W110, W112, W209, W210, W310 | 150 to 200 | Annealed | 4 | 1—4 | .018 | 60 | M2, M7 |
| | | | 2 | 5—10 | .015 | | |
| | | | 2 | 11—19 | .010 | | |
| | | | 2 | 20—48 | .008 | | |
| | | | 4 | 25—6 | .45 | 18 | S4, S2 |
| | | | 2 | 5.5—2.5 | .40 | | |
| | | | 2 | 2—1.5 | .25 | | |
| | | | 2 | 1—.5 | .20 | | |
| **9. NITRIDING STEELS, WROUGHT** Nitralloy 125, Nitralloy 135, Nitralloy 135 Mod., Nitralloy 225, Nitralloy 230, Nitralloy EZ, Nitralloy N, Nitrex 1 | 200 to 250 | Annealed | 4 | 1—4 | .016 | 40 | M2, M7 |
| | | | 2 | 5—10 | .012 | | |
| | | | 2 | 11—19 | .010 | | |
| | | | 2 | 20—48 | .008 | | |
| | | | 4 | 25—6 | .40 | 12 | S4, S2 |
| | | | 2 | 5.5—2.5 | .30 | | |
| | | | 2 | 2—1.5 | .25 | | |
| | | | 2 | 1—.5 | .20 | | |
| | 300 to 350 | Normalized or Quenched and Tempered | 4 | 1—4 | .012 | 25 | M2, M7 |
| | | | 2 | 5—10 | .012 | | |
| | | | 2 | 11—19 | .010 | | |
| | | | 2 | 20—48 | .008 | | |
| | | | 4 | 25—6 | .30 | 8 | S4, S2 |
| | | | 2 | 5.5—2.5 | .30 | | |
| | | | 2 | 2—1.5 | .25 | | |
| | | | 2 | 1—.5 | .20 | | |
| **12. FREE MACHINING STAINLESS STEELS, WROUGHT** **Ferritic** 430F, 430F Se | 135 to 185 | Annealed | 4 | 1—4 | .020 | 60 | M2, M7 |
| | | | 2 | 5—10 | .015 | | |
| | | | 2 | 11—19 | .010 | | |
| | | | 2 | 20—48 | .008 | | |
| | | | 4 | 25—6 | .50 | 18 | S4, S2 |
| | | | 2 | 5.5—2.5 | .40 | | |
| | | | 2 | 2—1.5 | .25 | | |
| | | | 2 | 1—.5 | .20 | | |
| **Austenitic** 203EZ, 303, 303MA, 303Pb, 303 Plus X, 303Se | 135 to 185 | Annealed | 4 | 1—4 | .020 | 40 | M2, M7 |
| | | | 2 | 5—10 | .015 | | |
| | | | 2 | 11—19 | .010 | | |
| | | | 2 | 20—48 | .008 | | |
| | | | 4 | 25—6 | .50 | 12 | S4, S2 |
| | | | 2 | 5.5—2.5 | .40 | | |
| | | | 2 | 2—1.5 | .25 | | |
| | | | 2 | 1—.5 | .20 | | |
| | 225 to 275 | Cold Drawn | 4 | 1—4 | .020 | 30 | M2, M7 |
| | | | 2 | 5—10 | .015 | | |
| | | | 2 | 11—19 | .010 | | |
| | | | 2 | 20—48 | .008 | | |
| | | | 4 | 25—6 | .50 | 9 | S4, S2 |
| | | | 2 | 5.5—2.5 | .40 | | |
| | | | 2 | 2—1.5 | .25 | | |
| | | | 2 | 1—.5 | .20 | | |

For cutting gears of Class 9 (AGMA 390.03) or better, the Rotary Feed should be reduced and the number of cuts should be increased.
See section 16 for Cutting Fluid Recommendations.

| MATERIAL | HARD-NESS<br>Bhn | CONDITION | NUMBER OF CUTS | DIAMETRAL PITCH<br>MODULE | ROTARY FEED per cutter stroke (4″ P.D. cutter)<br>in<br>mm | CUTTER SPEED<br>fpm<br>m/min | HSS TOOL MATERIAL<br>AISI<br>ISO |
|---|---|---|---|---|---|---|---|
| **Martensitic**<br>416 420F Se<br>416 Plus X 440F<br>416Se 440F Se<br>420F | 135 to 185 | Annealed | 4 | 1—4 | .020 | 60 | M2, M7 |
| | | | 2 | 5—10 | .015 | | |
| | | | 2 | 11—19 | .010 | | |
| | | | 2 | 20—48 | .008 | | |
| | | | 4 | 25—6 | .50 | 18 | S4, S2 |
| | | | 2 | 5.5—2.5 | .40 | | |
| | | | 2 | 2—1.5 | .25 | | |
| | | | 2 | 1—.5 | .20 | | |
| | 185 to 240 | Annealed or Cold Drawn | 4 | 1—4 | .020 | 50 | M2, M7 |
| | | | 2 | 5—10 | .015 | | |
| | | | 2 | 11—19 | .010 | | |
| | | | 2 | 20—48 | .008 | | |
| | | | 4 | 25—6 | .50 | 15 | S4, S2 |
| | | | 2 | 5.5—2.5 | .40 | | |
| | | | 2 | 2—1.5 | .25 | | |
| | | | 2 | 1—.5 | .20 | | |
| | 275 to 325 | Quenched and Tempered | 4 | 1—4 | .016 | 25 | M2, M7 |
| | | | 2 | 5—10 | .012 | | |
| | | | 2 | 11—19 | .010 | | |
| | | | 2 | 20—48 | .008 | | |
| | | | 4 | 25—6 | .40 | 8 | S4, S2 |
| | | | 2 | 5.5—2.5 | .30 | | |
| | | | 2 | 2—1.5 | .25 | | |
| | | | 2 | 1—.5 | .20 | | |
| **13. STAINLESS STEELS, WROUGHT**<br>**Ferritic**<br>405 434<br>409 436<br>429 442<br>430 446 | 135 to 185 | Annealed | 4 | 1—4 | .020 | 40 | M2, M7 |
| | | | 2 | 5—10 | .015 | | |
| | | | 2 | 11—19 | .010 | | |
| | | | 2 | 20—48 | .008 | | |
| | | | 4 | 25—6 | .50 | 12 | S4, S2 |
| | | | 2 | 5.5—2.5 | .40 | | |
| | | | 2 | 2—1.5 | .25 | | |
| | | | 2 | 1—.5 | .20 | | |
| **Austenitic**<br>201 304L 348<br>202 305 384<br>301 308 385<br>302 321<br>304 347 | 135 to 185 | Annealed | 4 | 1—4 | .020 | 30 | M2, M7 |
| | | | 2 | 5—10 | .015 | | |
| | | | 2 | 11—19 | .010 | | |
| | | | 2 | 20—48 | .008 | | |
| | | | 4 | 25—6 | .50 | 9 | S4, S2 |
| | | | 2 | 5.5—2.5 | .40 | | |
| | | | 2 | 2—1.5 | .25 | | |
| | | | 2 | 1—.5 | .20 | | |
| | 225 to 275 | Cold Drawn | 4 | 1—4 | .020 | 25 | M2, M7 |
| | | | 2 | 5—10 | .015 | | |
| | | | 2 | 11—19 | .010 | | |
| | | | 2 | 20—48 | .008 | | |
| | | | 4 | 25—6 | .50 | 8 | S4, S2 |
| | | | 2 | 5.5—2.5 | .40 | | |
| | | | 2 | 2—1.5 | .25 | | |
| | | | 2 | 1—.5 | .20 | | |
| **Austenitic**<br>302B 314<br>309 316<br>309S 316L<br>310 317<br>310S 330 | 135 to 185 | Annealed | 4 | 1—4 | .020 | 30 | M2, M7 |
| | | | 2 | 5—10 | .015 | | |
| | | | 2 | 11—19 | .010 | | |
| | | | 2 | 20—48 | .008 | | |
| | | | 4 | 25—6 | .50 | 9 | S4, S2 |
| | | | 2 | 5.5—2.5 | .40 | | |
| | | | 2 | 2—1.5 | .25 | | |
| | | | 2 | 1—.5 | .20 | | |
| | 225 to 275 | Cold Drawn | 4 | 1—4 | .020 | 20 | M2, M7 |
| | | | 2 | 5—10 | .015 | | |
| | | | 2 | 11—19 | .010 | | |
| | | | 2 | 20—48 | .008 | | |
| | | | 4 | 25—6 | .50 | 6 | S4, S2 |
| | | | 2 | 5.5—2.5 | .40 | | |
| | | | 2 | 2—1.5 | .25 | | |
| | | | 2 | 1—.5 | .20 | | |

For cutting gears of Class 9 (AGMA 390.03) or better, the Rotary Feed should be reduced and the number of cuts should be increased.
See section 16 for Cutting Fluid Recommendations.

# 7.3 Gear Shaping

| MATERIAL | HARDNESS Bhn | CONDITION | NUMBER OF CUTS | DIAMETRAL PITCH / MODULE | ROTARY FEED per cutter stroke (4" P.D. cutter) in / mm | CUTTER SPEED fpm / m/min | HSS TOOL MATERIAL AISI / ISO |
|---|---|---|---|---|---|---|---|
| **13. STAINLESS STEELS, WROUGHT (cont.)** **Austenitic** Nitronic 32, Nitronic 50, Nitronic 33, Nitronic 60, Nitronic 40 | 210 to 250 | Annealed | 4 | 1—4 | .020 | 20 | M2, M7 |
| | | | 2 | 5—10 | .015 | | |
| | | | 2 | 11—19 | .010 | | |
| | | | 2 | 20—48 | .008 | | |
| | | | 4 | 25—6 | .50 | 6 | S4, S2 |
| | | | 2 | 5.5—2.5 | .40 | | |
| | | | 2 | 2—1.5 | .25 | | |
| | | | 2 | 1—.5 | .20 | | |
| **Martensitic** 403, 410, 420, 422, 501, 502 | 135 to 175 | Annealed | 4 | 1—4 | .020 | 40 | M2, M7 |
| | | | 2 | 5—10 | .015 | | |
| | | | 2 | 11—19 | .010 | | |
| | | | 2 | 20—48 | .008 | | |
| | | | 4 | 25—6 | .50 | 12 | S4, S2 |
| | | | 2 | 5.5—2.5 | .40 | | |
| | | | 2 | 2—1.5 | .25 | | |
| | | | 2 | 1—.5 | .20 | | |
| | 175 to 225 | Annealed | 4 | 1—4 | .020 | 35 | M2, M7 |
| | | | 2 | 5—10 | .015 | | |
| | | | 2 | 11—19 | .010 | | |
| | | | 2 | 20—48 | .008 | | |
| | | | 4 | 25—6 | .50 | 11 | S4, S2 |
| | | | 2 | 5.5—2.5 | .40 | | |
| | | | 2 | 2—1.5 | .25 | | |
| | | | 2 | 1—.5 | .20 | | |
| | 275 to 325 | Quenched and Tempered | 4 | 1—4 | .016 | 20 | M2, M7 |
| | | | 2 | 5—10 | .012 | | |
| | | | 2 | 11—19 | .010 | | |
| | | | 2 | 20—48 | .008 | | |
| | | | 4 | 25—6 | .40 | 6 | S4, S2 |
| | | | 2 | 5.5—2.5 | .30 | | |
| | | | 2 | 2—1.5 | .25 | | |
| | | | 2 | 1—.5 | .20 | | |
| **Martensitic** 414, 431, Greek Ascoloy | 225 to 275 | Annealed | 4 | 1—4 | .018 | 30 | M2, M7 |
| | | | 2 | 5—10 | .012 | | |
| | | | 2 | 11—19 | .010 | | |
| | | | 2 | 20—48 | .008 | | |
| | | | 4 | 25—6 | .45 | 9 | S4, S2 |
| | | | 2 | 5.5—2.5 | .30 | | |
| | | | 2 | 2—1.5 | .25 | | |
| | | | 2 | 1—.5 | .20 | | |
| | 275 to 325 | Quenched and Tempered | 4 | 1—4 | .016 | 20 | M2, M7 |
| | | | 2 | 5—10 | .012 | | |
| | | | 2 | 11—19 | .010 | | |
| | | | 2 | 20—48 | .008 | | |
| | | | 4 | 25—6 | .40 | 6 | S4, S2 |
| | | | 2 | 5.5—2.5 | .30 | | |
| | | | 2 | 2—1.5 | .25 | | |
| | | | 2 | 1—.5 | .20 | | |
| **Martensitic** 440A, 440B, 440C | 225 to 275 | Annealed | 4 | 1—4 | .018 | 30 | M2, M7 |
| | | | 2 | 5—10 | .012 | | |
| | | | 2 | 11—19 | .010 | | |
| | | | 2 | 20—48 | .008 | | |
| | | | 4 | 25—6 | .45 | 9 | S4, S2 |
| | | | 2 | 5.5—2.5 | .30 | | |
| | | | 2 | 2—1.5 | .25 | | |
| | | | 2 | 1—.5 | .20 | | |
| | 275 to 325 | Quenched and Tempered | 4 | 1—4 | .016 | 20 | M2, M7 |
| | | | 2 | 5—10 | .012 | | |
| | | | 2 | 11—19 | .010 | | |
| | | | 2 | 20—48 | .008 | | |
| | | | 4 | 25—6 | .40 | 6 | S4, S2 |
| | | | 2 | 5.5—2.5 | .30 | | |
| | | | 2 | 2—1.5 | .25 | | |
| | | | 2 | 1—.5 | .20 | | |

For cutting gears of Class 9 (AGMA 390.03) or better, the Rotary Feed should be reduced and the number of cuts should be increased.
See section 16 for Cutting Fluid Recommendations.

| MATERIAL | HARD-NESS | CONDITION | NUMBER OF CUTS | DIAMETRAL PITCH / MODULE | ROTARY FEED per cutter stroke (4" P.D. cutter) in / mm | CUTTER SPEED fpm / m/min | HSS TOOL MATERIAL AISI / ISO |
|---|---|---|---|---|---|---|---|
| | Bhn | | | | | | |
| **14. PRECIPITATION HARDENING STAINLESS STEELS, WROUGHT** 15-5 PH, 16-6 PH, 17-4 PH, 17-7 PH, 17-14 Cu Mo, AF-71, AFC-77, Almar 362 (AM-362), AM-350, AM-355, AM-363, Custom 450, Custom 455, HNM, PH 13-8 Mo, PH 14-8 Mo, PH 15-7 Mo, Stainless W | 150 to 200 | Solution Treated | 4<br>2<br>2<br>2 | 1—4<br>5—10<br>11—19<br>20—48 | .018<br>.012<br>.010<br>.008 | 35 | M2, M7 |
| | | | 4<br>2<br>2<br>2 | 25—6<br>5.5—2.5<br>2—1.5<br>1—.5 | .45<br>.30<br>.25<br>.20 | 11 | S4, S2 |
| | 275 to 325 | Solution Treated or Hardened | 4<br>2<br>2<br>2 | 1—4<br>5—10<br>11—19<br>20—48 | .018<br>.012<br>.010<br>.008 | 25 | M2, M7 |
| | | | 4<br>2<br>2<br>2 | 25—6<br>5.5—2.5<br>2—1.5<br>1—.5 | .45<br>.30<br>.25<br>.20 | 8 | S4, S2 |
| | 325 to 375 | Solution Treated or Hardened | 4<br>2<br>2<br>2 | 1—4<br>5—10<br>11—19<br>20—48 | .016<br>.012<br>.010<br>.008 | 20 | M3, M42 |
| | | | 4<br>2<br>2<br>2 | 25—6<br>5.5—2.5<br>2—1.5<br>1—.5 | .40<br>.30<br>.25<br>.20 | 6 | S5, S11 |
| **15. CARBON STEELS, CAST Low Carbon** ASTM A426: Grade CP1 1010 1020 | 100 to 150 | Annealed, Normalized or Normalized and Tempered | 4<br>2<br>2<br>2 | 1—4<br>5—10<br>11—19<br>20—48 | .018<br>.012<br>.010<br>.008 | 80 | M2, M7 |
| | | | 4<br>2<br>2<br>2 | 25—6<br>5.5—2.5<br>2—1.5<br>1—.5 | .45<br>.30<br>.25<br>.20 | 24 | S4, S2 |
| **Medium Carbon** ASTM A352: Grades LCA, LCB, LCC; ASTM A356: Grade 1; 1030 1040 1050 | 125 to 175 | Annealed, Normalized or Normalized and Tempered | 4<br>2<br>2<br>2 | 1—4<br>5—10<br>11—19<br>20—48 | .018<br>.012<br>.010<br>.008 | 75 | M2, M7 |
| | | | 4<br>2<br>2<br>2 | 25—6<br>5.5—2.5<br>2—1.5<br>1—.5 | .45<br>.30<br>.25<br>.20 | 23 | S4, S2 |
| | 175 to 225 | Annealed, Normalized or Normalized and Tempered | 4<br>2<br>2<br>2 | 1—4<br>5—10<br>11—19<br>20—48 | .018<br>.012<br>.010<br>.008 | 70 | M2, M7 |
| | | | 4<br>2<br>2<br>2 | 25—6<br>5.5—2.5<br>2—1.5<br>1—.5 | .45<br>.30<br>.25<br>.20 | 21 | S4, S2 |
| | 250 to 300 | Quenched and Tempered | 4<br>2<br>2<br>2 | 1—4<br>5—10<br>11—19<br>20—48 | .016<br>.012<br>.010<br>.008 | 55 | M2, M7 |
| | | | 4<br>2<br>2<br>2 | 25—6<br>5.5—2.5<br>2—1.5<br>1—.5 | .40<br>.30<br>.25<br>.20 | 17 | S4, S2 |

For cutting gears of Class 9 (AGMA 390.03) or better, the Rotary Feed should be reduced and the number of cuts should be increased.
See section 16 for Cutting Fluid Recommendations.

# 7.3 Gear Shaping

| MATERIAL | HARD-NESS<br>Bhn | CONDITION | NUMBER OF CUTS | DIAMETRAL PITCH<br>MODULE | ROTARY FEED per cutter stroke (4" P.D. cutter)<br>in<br>mm | CUTTER SPEED<br>fpm<br>m/min | HSS TOOL MATERIAL<br>AISI<br>ISO |
|---|---|---|---|---|---|---|---|
| **16. ALLOY STEELS, CAST**<br>**Low Carbon**<br>ASTM A217: Grade WC9<br>ASTM A352: Grades LC3, LC4<br>ASTM A426: Grades CP2, CP5, CP5b, CP11, CP12, CP15, CP21, CP22<br>1320 2320 4120 8020<br>2315 4110 4320 8620 | 150 to 200 | Annealed, Normalized or Normalized and Tempered | 4 | 1—4 | .018 | 65 | M2, M7 |
| | | | 2 | 5—10 | .012 | | |
| | | | 2 | 11—19 | .010 | | |
| | | | 2 | 20—48 | .008 | | |
| | | | 4 | 25—6 | .45 | 20 | S4, S2 |
| | | | 2 | 5.5—2.5 | .30 | | |
| | | | 2 | 2—1.5 | .25 | | |
| | | | 2 | 1—.5 | .20 | | |
| | 200 to 225 | Annealed, Normalized or Normalized and Tempered | 4 | 1—4 | .018 | 60 | M2, M7 |
| | | | 2 | 5—10 | .012 | | |
| | | | 2 | 11—19 | .010 | | |
| | | | 2 | 20—48 | .008 | | |
| | | | 4 | 25—6 | .45 | 18 | S4, S2 |
| | | | 2 | 5.5—2.5 | .30 | | |
| | | | 2 | 2—1.5 | .25 | | |
| | | | 2 | 1—.5 | .20 | | |
| | 250 to 300 | Quenched and Tempered | 4 | 1—4 | .016 | 50 | M2, M7 |
| | | | 2 | 5—10 | .012 | | |
| | | | 2 | 11—19 | .010 | | |
| | | | 2 | 20—48 | .008 | | |
| | | | 4 | 25—6 | .40 | 15 | S4, S2 |
| | | | 2 | 5.5—2.5 | .30 | | |
| | | | 2 | 2—1.5 | .25 | | |
| | | | 2 | 1—.5 | .20 | | |
| **Medium Carbon**<br>ASTM A27: Grades N1, N2, U-60-30, 60-30, 65-35, 70-36, 70-40<br>ASTM A148: Grades 80-40, 80-50, 90-60, 105-85, 120-95, 150-125, 175-145<br>ASTM A216: Grades WCA, WCB, WCC<br>ASTM A217: Grades WC1, WC4, WC5, WC6<br>ASTM A352: Grades LC1, LC2, LC2-1<br>ASTM A356: Grades 2, 5, 6, 8, 9, 10<br>ASTM A389: Grades C23, C24<br>ASTM A486: Classes 70, 90, 120<br>ASTM A487: Classes 1N, 2N, 4N, 6N, 8N, 9N, 10N, DN, 1Q, 2Q, 4Q, 4QA, 6Q, 7Q, 8Q, 9Q, 10Q<br>1330 4130 80B30 8640<br>1340 4140 8040 9525<br>2325 4330 8430 9530<br>2330 4340 8440 9535<br>4125 8030 8630 | 175 to 225 | Annealed, Normalized or Normalized and Tempered | 4 | 1—4 | .018 | 55 | M2, M7 |
| | | | 2 | 5—10 | .012 | | |
| | | | 2 | 11—19 | .010 | | |
| | | | 2 | 20—48 | .008 | | |
| | | | 4 | 25—6 | .45 | 17 | S4, S2 |
| | | | 2 | 5.5—2.5 | .30 | | |
| | | | 2 | 2—1.5 | .25 | | |
| | | | 2 | 1—.5 | .20 | | |
| | 225 to 250 | Normalized, Normalized and Tempered or Quenched and Tempered | 4 | 1—4 | .016 | 45 | M2, M7 |
| | | | 2 | 5—10 | .012 | | |
| | | | 2 | 11—19 | .010 | | |
| | | | 2 | 20—48 | .008 | | |
| | | | 4 | 25—6 | .40 | 14 | S4, S2 |
| | | | 2 | 5.5—2.5 | .30 | | |
| | | | 2 | 2—1.5 | .25 | | |
| | | | 2 | 1—.5 | .20 | | |
| | 250 to 300 | Quenched and Tempered | 4 | 1—4 | .016 | 40 | M2, M7 |
| | | | 2 | 5—10 | .012 | | |
| | | | 2 | 11—19 | .010 | | |
| | | | 2 | 20—48 | .008 | | |
| | | | 4 | 25—6 | .40 | 12 | S4, S2 |
| | | | 2 | 5.5—2.5 | .30 | | |
| | | | 2 | 2—1.5 | .25 | | |
| | | | 2 | 1—.5 | .20 | | |
| | 300 to 350 | Quenched and Tempered | 4 | 1—4 | .014 | 35 | M2, M7 |
| | | | 2 | 5—10 | .012 | | |
| | | | 2 | 11—19 | .010 | | |
| | | | 2 | 20—48 | .008 | | |
| | | | 4 | 25—6 | .36 | 11 | S4, S2 |
| | | | 2 | 5.5—2.5 | .30 | | |
| | | | 2 | 2—1.5 | .25 | | |
| | | | 2 | 1—.5 | .20 | | |
| | 350 to 400 | Quenched and Tempered | 4 | 1—4 | .012 | 30 | M3, M42 |
| | | | 2 | 5—10 | .010 | | |
| | | | 2 | 11—19 | .008 | | |
| | | | 2 | 20—48 | .008 | | |
| | | | 4 | 25—6 | .30 | 9 | S5, S11 |
| | | | 2 | 5.5—2.5 | .25 | | |
| | | | 2 | 2—1.5 | .20 | | |
| | | | 2 | 1—.5 | .20 | | |

For cutting gears of Class 9 (AGMA 390.03) or better, the Rotary Feed should be reduced and the number of cuts should be increased.
See section 16 for Cutting Fluid Recommendations.

| MATERIAL | HARD-NESS Bhn | CONDITION | NUMBER OF CUTS | DIAMETRAL PITCH / MODULE | ROTARY FEED per cutter stroke (4" P.D. cutter) in / mm | CUTTER SPEED fpm / m/min | HSS TOOL MATERIAL AISI / ISO |
|---|---|---|---|---|---|---|---|
| **18. STAINLESS STEELS, CAST**<br>**Ferritic**<br>ASTM A217: Grades C5, C12<br>ASTM A296: Grades CB-30, CC-50, CE-30, CA6N, CA-6NM, CD4MCu<br>ASTM A297: Grade HC<br>ASTM A487: Class CA6NM<br>ASTM A608: Grade HC30 | 135 to 185 | Annealed | 4<br>2<br>2<br>2 | 1—4<br>5—10<br>11—19<br>20—48 | .020<br>.015<br>.010<br>.008 | 40 | M2, M7 |
| | | | 4<br>2<br>2<br>2 | 25—6<br>5.5—2.5<br>2—1.5<br>1—.5 | .50<br>.40<br>.25<br>.20 | 12 | S4, S2 |
| **Austenitic**<br>ASTM A296: Grades CF-16F, CN-7M, CN-7MS<br>ASTM A351: Grade CN-7M | 140 to 170 | Annealed or Normalized | 4<br>2<br>2<br>2 | 1—4<br>5—10<br>11—19<br>20—48 | .018<br>.012<br>.010<br>.008 | 35 | M2, M7 |
| | | | 4<br>2<br>2<br>2 | 25—6<br>5.5—2.5<br>2—1.5<br>1—.5 | .45<br>.30<br>.25<br>.20 | 11 | S4, S2 |
| **Austenitic**<br>ASTM A296: Grades CF-3, CF-8, CF-8C, CF-20<br>ASTM A351: Grades CF-3, CF-3A, CF-8, CF-8A, CF-8C<br>ASTM A451: Grades CPF3, CPF3A, CPF8, CPF8A, CPF8C, CPF8C (Ta Max.)<br>ASTM A452: Grades TP 304H, TP 347H | 135 to 185 | Annealed or Normalized | 4<br>2<br>2<br>2 | 1—4<br>5—10<br>11—19<br>20—48 | .018<br>.012<br>.010<br>.008 | 30 | M2, M7 |
| | | | 4<br>2<br>2<br>2 | 25—6<br>5.5—2.5<br>2—1.5<br>1—.5 | .45<br>.30<br>.25<br>.20 | 9 | S4, S2 |
| **Austenitic**<br>ASTM A296: Grades CF-3M, CF-8M, CG-8M, CG-12, CH-20, CK-20<br>ASTM A351: Grades CF-3M, CF-3MA, CF-8M, CF-10MC, CH-8, CH-10, CH-20, CK-20, HK-30, HK-40, HT-30<br>ASTM A451: Grades CPF3M, CPF8M, CPF10MC, CPH8, CPH10, CPH20, CPK20<br>ASTM A452: Grade TP 316H | 135 to 185 | Annealed or Normalized | 4<br>2<br>2<br>2 | 1—4<br>5—10<br>11—19<br>20—48 | .018<br>.015<br>.010<br>.008 | 30 | M2, M7 |
| | | | 4<br>2<br>2<br>2 | 25—6<br>5.5—2.5<br>2—1.5<br>1—.5 | .45<br>.40<br>.25<br>.20 | 9 | S4, S2 |

For cutting gears of Class 9 (AGMA 390.03) or better, the Rotary Feed should be reduced and the number of cuts should be increased.
See section 16 for Cutting Fluid Recommendations.

# 7.3 Gear Shaping

| MATERIAL | HARDNESS<br>Bhn | CONDITION | NUMBER OF CUTS | DIAMETRAL PITCH<br>MODULE | ROTARY FEED per cutter stroke (4" P.D. cutter)<br>in<br>mm | CUTTER SPEED<br>fpm<br>m/min | HSS TOOL MATERIAL<br>AISI<br>ISO |
|---|---|---|---|---|---|---|---|
| **18. STAINLESS STEELS, CAST (cont.)**<br>**Austenitic**<br>ASTM A297: Grades HD, HE, HF, HH, HI, HK, HL, HN, HP, HT, HU<br>ASTM A608: Grades HD50, HE35, HF30, HH30, HH33, HI35, HK30, HK40, HL30, HL40, HN40, HT50, HU50 | 160 to 210 | As Cast | 4<br>2<br>2<br>2 | 1—4<br>5—10<br>11—19<br>20—48 | .018<br>.012<br>.010<br>.008 | 30 | M2, M7 |
| | | | 4<br>2<br>2<br>2 | 25—6<br>5.5—2.5<br>2—1.5<br>1—.5 | .45<br>.30<br>.25<br>.20 | 9 | S4, S2 |
| | | | | | | | |
| **Martensitic**<br>ASTM A217: Grade CA-15<br>ASTM A296: Grades CA-15, CA-15M, CA-40<br>ASTM A426: Grades CP7, CP9, CPCA15<br>ASTM A487: Classes CA15a, CA-15M | 135 to 175 | Annealed | 4<br>2<br>2<br>2 | 1—4<br>5—10<br>11—19<br>20—48 | .020<br>.015<br>.010<br>.008 | 40 | M2, M7 |
| | | | 4<br>2<br>2<br>2 | 25—6<br>5.5—2.5<br>2—1.5<br>1—.5 | .50<br>.40<br>.25<br>.20 | 12 | S4, S2 |
| | 175 to 225 | Annealed, Normalized or Normalized and Tempered | 4<br>2<br>2<br>2 | 1—4<br>5—10<br>11—19<br>20—48 | .020<br>.015<br>.010<br>.008 | 35 | M2, M7 |
| | | | 4<br>2<br>2<br>2 | 25—6<br>5.5—2.5<br>2—1.5<br>1—.5 | .50<br>.40<br>.25<br>.20 | 11 | S4, S2 |
| | 275 to 325 | Quenched and Tempered | 4<br>2<br>2<br>2 | 1—4<br>5—10<br>11—19<br>20—48 | .016<br>.012<br>.010<br>.007 | 20 | M2, M7 |
| | | | 4<br>2<br>2<br>2 | 25—6<br>5.5—2.5<br>2—1.5<br>1—.5 | .40<br>.30<br>.25<br>.18 | 6 | S4, S2 |
| **19. PRECIPITATION HARDENING STAINLESS STEELS, CAST**<br>ASTM A351: Grade CD-4MCu<br>ACI Grade CB-7Cu<br>ACI Grade CD-4MCu<br>17-4 PH<br>AM-355 | 325 to 375 | Solution Treated | 4<br>2<br>2<br>2 | 1—4<br>5—10<br>11—19<br>20—48 | .016<br>.012<br>.010<br>.008 | 20 | M3, M42 |
| | | | 4<br>2<br>2<br>2 | 25—6<br>5.5—2.5<br>2—1.5<br>1—.5 | .40<br>.30<br>.25<br>.20 | 6 | S5, S11 |
| | | | | | | | |
| **21. GRAY CAST IRONS**<br>**Ferritic**<br>ASTM A48: Class 20<br>SAE J431c: Grade G1800 | 120 to 150 | Annealed | 4<br>2<br>2<br>2 | 1—4<br>5—10<br>11—19<br>20—48 | .018<br>.012<br>.010<br>.008 | 100 | M2, M7 |
| | | | 4<br>2<br>2<br>2 | 25—6<br>5.5—2.5<br>2—1.5<br>1—.5 | .45<br>.30<br>.25<br>.20 | 30 | S4, S2 |

For cutting gears of Class 9 (AGMA 390.03) or better, the Rotary Feed should be reduced and the number of cuts should be increased.
See section 16 for Cutting Fluid Recommendations.

| MATERIAL | HARD-NESS Bhn | CONDITION | NUMBER OF CUTS | DIAMETRAL PITCH / MODULE | ROTARY FEED per cutter stroke (4" P.D. cutter) in / mm | CUTTER SPEED fpm / m/min | HSS TOOL MATERIAL AISI / ISO |
|---|---|---|---|---|---|---|---|
| **Pearlitic- Ferritic** ASTM A48: Class 25 SAE J431c: Grade G2500 | 160 to 200 | As Cast | 4 | 1—4 | .018 | 85 | M2, M7 |
| | | | 2 | 5—10 | .012 | | |
| | | | 2 | 11—19 | .010 | | |
| | | | 2 | 20—48 | .008 | | |
| | | | 4 | 25—6 | .45 | 26 | S4, S2 |
| | | | 2 | 5.5—2.5 | .30 | | |
| | | | 2 | 2—1.5 | .25 | | |
| | | | 2 | 1—.5 | .20 | | |
| **Pearlitic** ASTM A48: Classes 30, 35, 40 SAE J431c: Grade G3000 | 190 to 220 | As Cast | 4 | 1—4 | .018 | 70 | M2, M7 |
| | | | 2 | 5—10 | .012 | | |
| | | | 2 | 11—19 | .010 | | |
| | | | 2 | 20—48 | .008 | | |
| | | | 4 | 25—6 | .45 | 21 | S4, S2 |
| | | | 2 | 5.5—2.5 | .30 | | |
| | | | 2 | 2—1.5 | .25 | | |
| | | | 2 | 1—.5 | .20 | | |
| **Pearlitic + Free Carbides** ASTM A48: Classes 45, 50 SAE J431c: Grades G3500, G4000 | 220 to 260 | As Cast | 4 | 1—4 | .018 | 60 | M2, M7 |
| | | | 2 | 5—10 | .012 | | |
| | | | 2 | 11—19 | .010 | | |
| | | | 2 | 20—48 | .008 | | |
| | | | 4 | 25—6 | .45 | 18 | S4, S2 |
| | | | 2 | 5.5—2.5 | .30 | | |
| | | | 2 | 2—1.5 | .25 | | |
| | | | 2 | 1—.5 | .20 | | |
| **Pearlitic or Acicular + Free Carbides** ASTM A48: Classes 55, 60 | 250 to 320 | As Cast or Quenched and Tempered | 4 | 1—4 | .012 | 40 | M2, M7 |
| | | | 2 | 5—10 | .010 | | |
| | | | 2 | 11—19 | .008 | | |
| | | | 2 | 20—48 | .008 | | |
| | | | 4 | 25—6 | .30 | 12 | S4, S2 |
| | | | 2 | 5.5—2.5 | .25 | | |
| | | | 2 | 2—1.5 | .20 | | |
| | | | 2 | 1—.5 | .20 | | |
| **Austenitic (NI-RESIST)** ASTM A436: Types 1, 1b, 2, 2b, 3, 4, 5, 6 | 100 to 250 | As Cast | 4 | 1—4 | .018 | 50 | M2, M7 |
| | | | 2 | 5—10 | .012 | | |
| | | | 2 | 11—19 | .010 | | |
| | | | 2 | 20—48 | .008 | | |
| | | | 4 | 25—6 | .45 | 15 | S4, S2 |
| | | | 2 | 5.5—2.5 | .30 | | |
| | | | 2 | 2—1.5 | .25 | | |
| | | | 2 | 1—.5 | .20 | | |
| **23. DUCTILE CAST IRONS** **Ferritic** ASTM A536: Grades 60-40-18, 65-45-12 SAE J434c: Grades D4018, D4512 | 140 to 190 | Annealed | 4 | 1—4 | .018 | 100 | M2, M7 |
| | | | 2 | 5—10 | .012 | | |
| | | | 2 | 11—19 | .010 | | |
| | | | 2 | 20—48 | .008 | | |
| | | | 4 | 25—6 | .45 | 30 | S4, S2 |
| | | | 2 | 5.5—2.5 | .30 | | |
| | | | 2 | 2—1.5 | .25 | | |
| | | | 2 | 1—.5 | .20 | | |
| **Ferritic- Pearlitic** ASTM A536: Grade 80-55-06 SAE J434c: Grade D5506 | 190 to 225 | As Cast | 4 | 1—4 | .018 | 80 | M2, M7 |
| | | | 2 | 5—10 | .012 | | |
| | | | 2 | 11—19 | .010 | | |
| | | | 2 | 20—48 | .008 | | |
| | | | 4 | 25—6 | .45 | 24 | S4, S2 |
| | | | 2 | 5.5—2.5 | .30 | | |
| | | | 2 | 2—1.5 | .25 | | |
| | | | 2 | 1—.5 | .20 | | |
| | 225 to 260 | As Cast | 4 | 1—4 | .018 | 70 | M2, M7 |
| | | | 2 | 5—10 | .012 | | |
| | | | 2 | 11—19 | .010 | | |
| | | | 2 | 20—48 | .008 | | |
| | | | 4 | 25—6 | .45 | 21 | S4, S2 |
| | | | 2 | 5.5—2.5 | .30 | | |
| | | | 2 | 2—1.5 | .25 | | |
| | | | 2 | 1—.5 | .20 | | |

For cutting gears of Class 9 (AGMA 390.03) or better, the Rotary Feed should be reduced and the number of cuts should be increased.
See section 16 for Cutting Fluid Recommendations.

# 7.3 Gear Shaping

| MATERIAL | HARD-NESS Bhn | CONDITION | NUMBER OF CUTS | DIAMETRAL PITCH / MODULE | ROTARY FEED per cutter stroke (4" P.D. cutter) in / mm | CUTTER SPEED fpm / m/min | HSS TOOL MATERIAL AISI / ISO |
|---|---|---|---|---|---|---|---|
| **23. DUCTILE CAST IRONS (cont.)** **Pearlitic- Martensitic** ASTM A536: Grade 100-70-03 SAE J434c: Grade D7003 | 240 to 300 | Normalized and Tempered | 4<br>2<br>2<br>2 | 1—4<br>5—10<br>11—19<br>20—48 | .016<br>.012<br>.010<br>.008 | 60 | M2, M7 |
| | | | 4<br>2<br>2<br>2 | 25—6<br>5.5—2.5<br>2—1.5<br>1—.5 | .40<br>.30<br>.25<br>.20 | 18 | S4, S2 |
| **Martensitic** ASTM A536: Grade 120-90-02 SAE J434c: Grade DQ&T | 270 to 330 | Quenched and Tempered | 4<br>2<br>2<br>2 | 1—4<br>5—10<br>11—19<br>20—48 | .012<br>.010<br>.008<br>.008 | 45 | M2, M7 |
| | | | 4<br>2<br>2<br>2 | 25—6<br>5.5—2.5<br>2—1.5<br>1—.5 | .30<br>.25<br>.20<br>.20 | 14 | S4, S2 |
| **Austenitic (NI-RESIST Ductile)** ASTM A439: Types D-2, D-2C, D-3A, D-5 ASTM A571: Type D-2M | 120 to 200 | Annealed | 4<br>2<br>2<br>2 | 1—4<br>5—10<br>11—19<br>20—48 | .012<br>.010<br>.008<br>.008 | 40 | M2, M7 |
| | | | 4<br>2<br>2<br>2 | 25—6<br>5.5—2.5<br>2—1.5<br>1—.5 | .30<br>.25<br>.20<br>.20 | 12 | S4, S2 |
| **Austenitic (NI-RESIST Ductile)** ASTM A439: Types D-2B, D-3, D-4, D-5B | 140 to 275 | Annealed | 4<br>2<br>2<br>2 | 1—4<br>5—10<br>11—19<br>20—48 | .012<br>.010<br>.008<br>.008 | 35 | M2, M7 |
| | | | 4<br>2<br>2<br>2 | 25—6<br>5.5—2.5<br>2—1.5<br>1—.5 | .30<br>.25<br>.20<br>.20 | 11 | S4, S2 |
| **24. MALLEABLE CAST IRONS** **Ferritic** ASTM A47: Grades 32510, 35018 ASTM A602: Grade M3210 SAE J158: Grade M3210 | 110 to 160 | Malleablized | 4<br>2<br>2<br>2 | 1—4<br>5—10<br>11—19<br>20—48 | .018<br>.012<br>.010<br>.008 | 145 | M2, M7 |
| | | | 4<br>2<br>2<br>2 | 25—6<br>5.5—2.5<br>2—1.5<br>1—.5 | .45<br>.30<br>.25<br>.20 | 44 | S4, S2 |
| **Pearlitic** ASTM A220: Grades 40010, 45006, 45008, 50005 ASTM A602: Grade M4504, M5003 SAE J158: Grades M4504, M5003 | 160 to 200 | Malleablized and Heat Treated | 4<br>2<br>2<br>2 | 1—4<br>5—10<br>11—19<br>20—48 | .018<br>.012<br>.010<br>.008 | 100 | M2, M7 |
| | | | 4<br>2<br>2<br>2 | 25—6<br>5.5—2.5<br>2—1.5<br>1—.5 | .45<br>.30<br>.25<br>.20 | 30 | S4, S2 |
| | 200 to 240 | Malleablized and Heat Treated | 4<br>2<br>2<br>2 | 1—4<br>5—10<br>11—19<br>20—48 | .018<br>.012<br>.010<br>.008 | 70 | M2, M7 |
| | | | 4<br>2<br>2<br>2 | 25—6<br>5.5—2.5<br>2—1.5<br>1—.5 | .45<br>.30<br>.25<br>.20 | 21 | S4, S2 |
| **Tempered Martensite** ASTM A220: Grade 60004 ASTM A602: Grade M5503 SAE J158: Grade M5503 | 200 to 255 | Malleablized and Heat Treated | 4<br>2<br>2<br>2 | 1—4<br>5—10<br>11—19<br>20—48 | .018<br>.012<br>.010<br>.008 | 65 | M2, M7 |
| | | | 4<br>2<br>2<br>2 | 25—6<br>5.5—2.5<br>2—1.5<br>1—.5 | .45<br>.30<br>.25<br>.20 | 20 | S4, S2 |

For cutting gears of Class 9 (AGMA 390.03) or better, the Rotary Feed should be reduced and the number of cuts should be increased.
See section 16 for Cutting Fluid Recommendations.

| MATERIAL | HARD-NESS<br>Bhn | CONDITION | NUMBER OF CUTS | DIAMETRAL PITCH / MODULE | ROTARY FEED per cutter stroke (4″ P.D. cutter)<br>in / mm | CUTTER SPEED<br>fpm / m/min | HSS TOOL MATERIAL<br>AISI / ISO |
|---|---|---|---|---|---|---|---|
| **Tempered Martensite**<br>ASTM A220: Grade 70003<br>ASTM A602: Grade M7002<br>SAE J158: Grade M7002 | 220<br>to<br>260 | Malleablized<br>and<br>Heat<br>Treated | 4<br>2<br>2<br>2 | 1—4<br>5—10<br>11—19<br>20—48 | .018<br>.012<br>.010<br>.008 | 60 | M2, M7 |
| | | | 4<br>2<br>2<br>2 | 25—6<br>5.5—2.5<br>2—1.5<br>1—.5 | .45<br>.30<br>.25<br>.20 | 18 | S4, S2 |
| **Tempered Martensite**<br>ASTM A220: Grade 80002<br>ASTM A602: Grade M8501<br>SAE J158: Grade M8501 | 240<br>to<br>280 | Malleablized<br>and<br>Heat<br>Treated | 4<br>2<br>2<br>2 | 1—4<br>5—10<br>11—19<br>20—48 | .016<br>.012<br>.010<br>.008 | 50 | M2, M7 |
| | | | 4<br>2<br>2<br>2 | 25—6<br>5.5—2.5<br>2—1.5<br>1—.5 | .40<br>.30<br>.25<br>.20 | 15 | S4, S2 |
| **Tempered Martensite**<br>ASTM A220: Grade 90001<br>ASTM A602: Grade M8501<br>SAE J158: Grade M8501 | 250<br>to<br>320 | Malleablized<br>and<br>Heat<br>Treated | 4<br>2<br>2<br>2 | 1—4<br>5—10<br>11—19<br>20—48 | .012<br>.010<br>.008<br>.008 | 45 | M2, M7 |
| | | | 4<br>2<br>2<br>2 | 25—6<br>5.5—2.5<br>2—1.5<br>1—.5 | .30<br>.25<br>.20<br>.20 | 14 | S4, S2 |
| **28. ALUMINUM ALLOYS, WROUGHT**<br>EC 2218 5252 6253<br>1060 2219 5254 6262<br>1100 2618 5454 6463<br>1145 3003 5456 6951<br>1175 3004 5457 7001<br>1235 3005 5652 7004<br>2011 4032 5657 7005<br>2014 5005 6053 7039<br>2017 5050 6061 7049<br>2018 5052 6063 7050<br>2021 5056 6066 7075<br>2024 5083 6070 7079<br>2025 5086 6101 7175<br>2117 5154 6151 7178 | 30<br>to<br>80<br>500kg | Cold<br>Drawn | 4<br>2<br>2<br>2 | 1—4<br>5—10<br>11—19<br>20—48 | .028<br>.024<br>.015<br>.008 | 200 | M2, M7 |
| | | | 4<br>2<br>2<br>2 | 25—6<br>5.5—2.5<br>2—1.5<br>1—.5 | .70<br>.60<br>.40<br>.20 | 60 | S4, S2 |
| | 75<br>to<br>150<br>500kg | Solution<br>Treated<br>and<br>Aged | 4<br>2<br>2<br>2 | 1—4<br>5—10<br>11—19<br>20—48 | .028<br>.024<br>.015<br>.008 | 180 | M2, M7 |
| | | | 4<br>2<br>2<br>2 | 25—6<br>5.5—2.5<br>2—1.5<br>1—.5 | .70<br>.60<br>.40<br>.20 | 55 | S4, S2 |
| **29. ALUMINUM ALLOYS, CAST**<br>**Sand and Permanent Mold**<br>A140 319.0 357.0 A712.0<br>201.0 328.0 359.0 D712.0<br>208.0 A332.0 B443.0 713.0<br>213.0 F332.0 514.0 771.0<br>222.0 333.0 A514.0 850.0<br>224.0 354.0 B514.0 A850.0<br>242.0 355.0 520.0 B850.0<br>295.0 C355.0 535.0<br>B295.0 356.0 705.0<br>308.0 A356.0 707.0<br>Hiduminium RR-350 | 40<br>to<br>100<br>500kg | As Cast | 4<br>2<br>2<br>2 | 1—4<br>5—10<br>11—19<br>20—48 | .028<br>.024<br>.015<br>.008 | 180 | M2, M7 |
| | | | 4<br>2<br>2<br>2 | 25—6<br>5.5—2.5<br>2—1.5<br>1—.5 | .70<br>.60<br>.40<br>.20 | 55 | S4, S2 |
| | 70<br>to<br>125<br>500kg | Solution<br>Treated<br>and<br>Aged | 4<br>2<br>2<br>2 | 1—4<br>5—10<br>11—19<br>20—48 | .028<br>.024<br>.015<br>.008 | 160 | M2, M7 |
| | | | 4<br>2<br>2<br>2 | 25—6<br>5.5—2.5<br>2—1.5<br>1—.5 | .70<br>.60<br>.40<br>.20 | 49 | S4, S2 |

For cutting gears of Class 9 (AGMA 390.03) or better, the Rotary Feed should be reduced and the number of cuts should be increased.
See section 16 for Cutting Fluid Recommendations.

# 7.3 Gear Shaping

| MATERIAL | HARDNESS Bhn | CONDITION | NUMBER OF CUTS | DIAMETRAL PITCH / MODULE | ROTARY FEED per cutter stroke (4" P.D. cutter) in / mm | CUTTER SPEED fpm / m/min | HSS TOOL MATERIAL AISI / ISO |
|---|---|---|---|---|---|---|---|
| **34. COPPER ALLOYS, WROUGHT**<br>145 332 360 482<br>147 335 365 485<br>173 340 366 544<br>187 342 367 623<br>191 349 368 624<br>314 350 370 638<br>316 353 377 642<br>330 356 385 782 | 10 $R_B$ to 70 $R_B$ | Annealed | 4 | 1—4 | .018 | 100 | M2, M7 |
| | | | 2 | 5—10 | .012 | | |
| | | | 2 | 11—19 | .010 | | |
| | | | 2 | 20—48 | .008 | | |
| | | | 4 | 25—6 | .45 | 30 | S4, S2 |
| | | | 2 | 5.5—2.5 | .30 | | |
| | | | 2 | 2—1.5 | .25 | | |
| | | | 2 | 1—.5 | .20 | | |
| | 60 $R_B$ to 100 $R_B$ | Cold Drawn | 4 | 1—4 | .018 | 110 | M2, M7 |
| | | | 2 | 5—10 | .012 | | |
| | | | 2 | 11—19 | .010 | | |
| | | | 2 | 20—48 | .008 | | |
| | | | 4 | 25—6 | .45 | 34 | S4, S2 |
| | | | 2 | 5.5—2.5 | .30 | | |
| | | | 2 | 2—1.5 | .25 | | |
| | | | 2 | 1—.5 | .20 | | |
| 190 425 466 667<br>226 435 467 675<br>230 442 613 687<br>240 443 618 694<br>260 444 630 770<br>268 445 632<br>270 464 651<br>280 465 655 | 10 $R_B$ to 70 $R_B$ | Annealed | 4 | 1—4 | .016 | 70 | M2, M7 |
| | | | 2 | 5—10 | .012 | | |
| | | | 2 | 11—19 | .010 | | |
| | | | 2 | 20—48 | .008 | | |
| | | | 4 | 25—6 | .40 | 21 | S4, S2 |
| | | | 2 | 5.5—2.5 | .30 | | |
| | | | 2 | 2—1.5 | .25 | | |
| | | | 2 | 1—.5 | .20 | | |
| | 60 $R_B$ to 100 $R_B$ | Cold Drawn | 4 | 1—4 | .016 | 80 | M2, M7 |
| | | | 2 | 5—10 | .012 | | |
| | | | 2 | 11—19 | .010 | | |
| | | | 2 | 20—48 | .008 | | |
| | | | 4 | 25—6 | .40 | 24 | S4, S2 |
| | | | 2 | 5.5—2.5 | .30 | | |
| | | | 2 | 2—1.5 | .25 | | |
| | | | 2 | 1—.5 | .20 | | |
| 101 125 185 614<br>102 127 189 619<br>104 128 192 625<br>105 129 194 674<br>107 130 195 688<br>109 142 210 706<br>110 143 220 710<br>111 150 411 715<br>113 155 413 725<br>114 162 505 745<br>115 165 510 752<br>116 170 511 754<br>119 172 521 757<br>120 175 524<br>121 182 608<br>122 184 610 | 10 $R_B$ to 70 $R_B$ | Annealed | 4 | 1—4 | .012 | 30 | M2, M7 |
| | | | 2 | 5—10 | .012 | | |
| | | | 2 | 11—19 | .010 | | |
| | | | 2 | 20—48 | .008 | | |
| | | | 4 | 25—6 | .30 | 9 | S4, S2 |
| | | | 2 | 5.5—2.5 | .30 | | |
| | | | 2 | 2—1.5 | .25 | | |
| | | | 2 | 1—.5 | .20 | | |
| | 60 $R_B$ to 100 $R_B$ | Cold Drawn | 4 | 1—4 | .012 | 40 | M2, M7 |
| | | | 2 | 5—10 | .012 | | |
| | | | 2 | 11—19 | .010 | | |
| | | | 2 | 20—48 | .008 | | |
| | | | 4 | 25—6 | .30 | 12 | S4, S2 |
| | | | 2 | 5.5—2.5 | .30 | | |
| | | | 2 | 2—1.5 | .25 | | |
| | | | 2 | 1—.5 | .20 | | |

For cutting gears of Class 9 (AGMA 390.03) or better, the Rotary Feed should be reduced and the number of cuts should be increased.
See section 16 for Cutting Fluid Recommendations.

| MATERIAL | HARD-NESS Bhn | CONDITION | NUMBER OF CUTS | DIAMETRAL PITCH / MODULE | ROTARY FEED per cutter stroke (4" P.D. cutter) in / mm | CUTTER SPEED fpm / m/min | HSS TOOL MATERIAL AISI / ISO |
|---|---|---|---|---|---|---|---|
| **35. COPPER ALLOYS, CAST** | | | 4 | 1—4 | .018 | 90 | M2, M7 |
| 834 855 934 953 | 40 | | 2 | 5—10 | .012 | | |
| 836 857 935 954 | to | | 2 | 11—19 | .010 | | |
| 838 858 937 956 | 150 | As Cast | 2 | 20—48 | .008 | | |
| 842 864 938 973 | 500kg | | 4 | 25—6 | .45 | 27 | S4, S2 |
| 844 867 939 974 | | | 2 | 5.5—2.5 | .30 | | |
| 848 879 943 976 | | | 2 | 2—1.5 | .25 | | |
| 852 928 944 978 | | | 2 | 1—.5 | .20 | | |
| 854 932 945 | | | | | | | |
| 817 868 905 947 | 40 | | 4 | 1—4 | .016 | 60 | M2, M7 |
| 821 872 915 948 | to | | 2 | 5—10 | .012 | | |
| 833 874 922 952 | 175 | | 2 | 11—19 | .010 | | |
| 853 875 923 955 | 500kg | As Cast | 2 | 20—48 | .008 | | |
| 861 876 925 957 | | | 4 | 25—6 | .40 | 18 | S4, S2 |
| 862 878 926 958 | | | 2 | 5.5—2.5 | .30 | | |
| 865 903 927 | | | 2 | 2—1.5 | .25 | | |
| | | | 2 | 1—.5 | .20 | | |
| 801 815 828 916 | 40 | | 4 | 1—4 | .012 | 30 | M2, M7 |
| 803 818 863 917 | to | | 2 | 5—10 | .012 | | |
| 805 820 902 962 | 200 | | 2 | 11—19 | .010 | | |
| 807 822 907 963 | 500kg | As Cast | 2 | 20—48 | .008 | | |
| 809 824 909 964 | | | 4 | 25—6 | .30 | 9 | S4, S2 |
| 811 825 910 966 | | | 2 | 5.5—2.5 | .30 | | |
| 813 826 911 993 | | | 2 | 2—1.5 | .25 | | |
| 814 827 913 | | | 2 | 1—.5 | .20 | | |

For cutting gears of Class 9 (AGMA 390.03) or better, the Rotary Feed should be reduced and the number of cuts should be increased.
See section 16 for Cutting Fluid Recommendations.

| MATERIAL | HARD-NESS Bhn | CONDITION | DIAMETRAL PITCH / MODULE | FEED*† per rev. of gear in / mm | CUTTER PITCH LINE SPEED fpm / m/min | HSS TOOL MATERIAL AISI / ISO |
|---|---|---|---|---|---|---|
| **1. FREE MACHINING CARBON STEELS, WROUGHT**<br>**Low Carbon Resulfurized**<br>1116 1119<br>1117 1211<br>1118 1212 | 100 to 150 | Hot Rolled or Annealed | 1—4<br>5—10<br>11—19<br>20 & Finer | .012<br>.008<br>.005<br>.003 | 610 | M2, M7 |
| | | | 25—6<br>5—3<br>2—1.5<br>1 & Finer | .3<br>.2<br>.12<br>.07 | 185 | S4, S2 |
| | 150 to 200 | Cold Drawn | 1—4<br>5—10<br>11—19<br>20 & Finer | .012<br>.008<br>.005<br>.003 | 675 | M2, M7 |
| | | | 25—6<br>5—3<br>2—1.5<br>1 & Finer | .3<br>.2<br>.12<br>.07 | 205 | S4, S2 |
| **Low Carbon Resulfurized**<br>1213<br>1215 | 100 to 150 | Hot Rolled or Annealed | 1—4<br>5—10<br>11—19<br>20 & Finer | .012<br>.008<br>.005<br>.003 | 650 | M2, M7 |
| | | | 25—6<br>5—3<br>2—1.5<br>1 & Finer | .3<br>.2<br>.12<br>.07 | 200 | S4, S2 |
| | 150 to 200 | Cold Drawn | 1—4<br>5—10<br>11—19<br>20 & Finer | .012<br>.008<br>.005<br>.003 | 730 | M2, M7 |
| | | | 25—6<br>5—3<br>2—1.5<br>1 & Finer | .3<br>.2<br>.12<br>.07 | 225 | S4, S2 |
| **Low Carbon Resulfurized**<br>1108<br>1109<br>1110<br>1115 | 100 to 150 | Hot Rolled or Annealed | 1—4<br>5—10<br>11—19<br>20 & Finer | .012<br>.008<br>.005<br>.003 | 580 | M2, M7 |
| | | | 25—6<br>5—3<br>2—1.5<br>1 & Finer | .3<br>.2<br>.12<br>.07 | 175 | S4, S2 |
| | 150 to 200 | Cold Drawn | 1—4<br>5—10<br>11—19<br>20 & Finer | .012<br>.008<br>.005<br>.003 | 620 | M2, M7 |
| | | | 25—6<br>5—3<br>2—1.5<br>1 & Finer | .3<br>.2<br>.12<br>.07 | 190 | S4, S2 |
| **Medium Carbon Resulfurized**<br>1132 1144<br>1137 1145<br>1139 1146<br>1140 1151<br>1141 | 175 to 225 | Hot Rolled, Normalized, Annealed or Cold Drawn | 1—4<br>5—10<br>11—19<br>20 & Finer | .012<br>.008<br>.005<br>.003 | 500 | M2, M7 |
| | | | 25—6<br>5—3<br>2—1.5<br>1 & Finer | .3<br>.2<br>.12<br>.07 | 150 | S4, S2 |
| | 275 to 325 | Quenched and Tempered | 1—4<br>5—10<br>11—19<br>20 & Finer | .012<br>.008<br>.005<br>.003 | 350 | M2, M7 |
| | | | 25—6<br>5—3<br>2—1.5<br>1 & Finer | .3<br>.2<br>.12<br>.07 | 105 | S4, S2 |

See section 16 for Cutting Fluid Recommendations.

*Feed recommendations apply to CONVENTIONAL (axial-transverse) gear shaving.

†Feeds should be increased 100% for gears shaved by the DIAGONAL (angular-transverse) method.

# 7.4 Gear Shaving

| MATERIAL | HARD-NESS<br>Bhn | CONDITION | DIAMETRAL PITCH<br>MODULE | FEED*† per rev. of gear<br>in<br>mm | CUTTER PITCH LINE SPEED<br>fpm<br>m/min | HSS TOOL MATERIAL<br>AISI<br>ISO |
|---|---|---|---|---|---|---|
| **1. FREE MACHINING CARBON STEELS, WROUGHT (cont.)**<br>**Medium Carbon Resulfurized (cont.)**<br>(materials listed on preceding page) | 325<br>to<br>375 | Quenched<br>and<br>Tempered | 1—4<br>5—10<br>11—19<br>20 & Finer | .012<br>.008<br>.005<br>.003 | 275 | M2, M7 |
| | | | 25—6<br>5—3<br>2—1.5<br>1 & Finer | .3<br>.2<br>.12<br>.07 | 84 | S4, S2 |
| **Low Carbon Leaded**<br>10L18<br>11L17 | 100<br>to<br>150 | Hot Rolled, Normalized, Annealed or Cold Drawn | 1—4<br>5—10<br>11—19<br>20 & Finer | .012<br>.008<br>.005<br>.003 | 650 | M2, M7 |
| | | | 25—6<br>5—3<br>2—1.5<br>1 & Finer | .3<br>.2<br>.12<br>.07 | 200 | S4, S2 |
| | 150<br>to<br>200 | Hot Rolled, Normalized, Annealed or Cold Drawn | 1—4<br>5—10<br>11—19<br>20 & Finer | .012<br>.008<br>.005<br>.003 | 700 | M2, M7 |
| | | | 25—6<br>5—3<br>2—1.5<br>1 & Finer | .3<br>.2<br>.12<br>.07 | 215 | S4, S2 |
| | 200<br>to<br>250 | Hot Rolled, Normalized, Annealed or Cold Drawn | 1—4<br>5—10<br>11—19<br>20 & Finer | .012<br>.008<br>.005<br>.003 | 575 | M2, M7 |
| | | | 25—6<br>5—3<br>2—1.5<br>1 & Finer | .3<br>.2<br>.12<br>.07 | 175 | S4, S2 |
| **Low Carbon Leaded**<br>12L13<br>12L14<br>12L15 | 100<br>to<br>150 | Hot Rolled, Normalized, Annealed or Cold Drawn | 1—4<br>5—10<br>11—19<br>20 & Finer | .012<br>.008<br>.005<br>.003 | 700 | M2, M7 |
| | | | 25—6<br>5—3<br>2—1.5<br>1 & Finer | .3<br>.2<br>.12<br>.07 | 215 | S4, S2 |
| | 150<br>to<br>200 | Hot Rolled, Normalized, Annealed or Cold Drawn | 1—4<br>5—10<br>11—19<br>20 & Finer | .012<br>.008<br>.005<br>.003 | 780 | M2, M7 |
| | | | 25—6<br>5—3<br>2—1.5<br>1 & Finer | .3<br>.2<br>.12<br>.07 | 240 | S4, S2 |
| | 200<br>to<br>250 | Hot Rolled, Normalized, Annealed or Cold Drawn | 1—4<br>5—10<br>11—19<br>20 & Finer | .012<br>.008<br>.005<br>.003 | 600 | M2, M7 |
| | | | 25—6<br>5—3<br>2—1.5<br>1 & Finer | .3<br>.2<br>.12<br>.07 | 185 | S4, S2 |
| **Medium Carbon Leaded**<br>10L45<br>10L50<br>11L37<br>11L41<br>11L44 | 125<br>to<br>175 | Hot Rolled, Normalized, Annealed or Cold Drawn | 1—4<br>5—10<br>11—19<br>20 & Finer | .012<br>.008<br>.005<br>.003 | 600 | M2, M7 |
| | | | 25—6<br>5—3<br>2—1.5<br>1 & Finer | .3<br>.2<br>.12<br>.07 | 185 | S4, S2 |

See section 16 for Cutting Fluid Recommendations.

*Feed recommendations apply to CONVENTIONAL (axial-transverse) gear shaving.

†Feeds should be increased 100% for gears shaved by the DIAGONAL (angular-transverse) method.

| MATERIAL | HARDNESS Bhn | CONDITION | DIAMETRAL PITCH / MODULE | FEED*† per rev. of gear in / mm | CUTTER PITCH LINE SPEED fpm / m/min | HSS TOOL MATERIAL AISI / ISO |
|---|---|---|---|---|---|---|
| **1. FREE MACHINING CARBON STEELS, WROUGHT (cont.)**<br>**Medium Carbon Leaded (cont.)**<br>(materials listed on preceding page) | 175 to 225 | Hot Rolled, Normalized, Annealed or Cold Drawn | 1—4<br>5—10<br>11—19<br>20 & Finer | .012<br>.008<br>.005<br>.003 | 525 | M2, M7 |
| | | | 25—6<br>5—3<br>2—1.5<br>1 & Finer | .3<br>.2<br>.12<br>.07 | 160 | S4, S2 |
| | 225 to 275 | Hot Rolled, Normalized, Annealed, Cold Drawn or Quenched and Tempered | 1—4<br>5—10<br>11—19<br>20 & Finer | .012<br>.008<br>.005<br>.003 | 425 | M2, M7 |
| | | | 25—6<br>5—3<br>2—1.5<br>1 & Finer | .3<br>.2<br>.12<br>.07 | 130 | S4, S2 |
| | 275 to 325 | Hot Rolled, Normalized, Annealed, Cold Drawn or Quenched and Tempered | 1—4<br>5—10<br>11—19<br>20 & Finer | .012<br>.008<br>.005<br>.003 | 375 | M2, M7 |
| | | | 25—6<br>5—3<br>2—1.5<br>1 & Finer | .3<br>.2<br>.12<br>.07 | 115 | S4, S2 |
| | 325 to 375 | Quenched and Tempered | 1—4<br>5—10<br>11—19<br>20 & Finer | .012<br>.008<br>.005<br>.003 | 300 | M3 |
| | | | 25—6<br>5—3<br>2—1.5<br>1 & Finer | .3<br>.2<br>.12<br>.07 | 90 | S5 |
| **2. CARBON STEELS, WROUGHT**<br>**Low Carbon**<br>1005 1015 1023<br>1006 1016 1025<br>1008 1017 1026<br>1009 1018 1029<br>1010 1019 1513<br>1011 1020 1518<br>1012 1021 1522<br>1013 1022 | 85 to 125 | Hot Rolled, Normalized, Annealed or Cold Drawn | 1—4<br>5—10<br>11—19<br>20 & Finer | .012<br>.008<br>.005<br>.003 | 525 | M2, M7 |
| | | | 25—6<br>5—3<br>2—1.5<br>1 & Finer | .3<br>.2<br>.12<br>.07 | 160 | S4, S2 |
| | 125 to 175 | Hot Rolled, Normalized, Annealed or Cold Drawn | 1—4<br>5—10<br>11—19<br>20 & Finer | .012<br>.008<br>.005<br>.003 | 500 | M2, M7 |
| | | | 25—6<br>5—3<br>2—1.5<br>1 & Finer | .3<br>.2<br>.12<br>.07 | 150 | S4, S2 |
| | 175 to 225 | Hot Rolled, Normalized, Annealed or Cold Drawn | 1—4<br>5—10<br>11—19<br>20 & Finer | .012<br>.008<br>.005<br>.003 | 450 | M2, M7 |
| | | | 25—6<br>5—3<br>2—1.5<br>1 & Finer | .3<br>.2<br>.12<br>.07 | 135 | S4, S2 |
| | 225 to 275 | Annealed or Cold Drawn | 1—4<br>5—10<br>11—19<br>20 & Finer | .012<br>.008<br>.005<br>.003 | 375 | M2, M7 |
| | | | 25—6<br>5—3<br>2—1.5<br>1 & Finer | .3<br>.2<br>.12<br>.07 | 115 | S4, S2 |

See section 16 for Cutting Fluid Recommendations.

*Feed recommendations apply to CONVENTIONAL (axial-transverse) gear shaving.

†Feeds should be increased 100% for gears shaved by the DIAGONAL (angular-transverse) method.

# 7.4 Gear Shaving

| MATERIAL | HARDNESS Bhn | CONDITION | DIAMETRAL PITCH / MODULE | FEED*† per rev. of gear in / mm | CUTTER PITCH LINE SPEED fpm / m/min | HSS TOOL MATERIAL AISI / ISO |
|---|---|---|---|---|---|---|
| **2. CARBON STEELS, WROUGHT (cont.)**<br>**Medium Carbon**<br>1030 1044 1526<br>1033 1045 1527<br>1035 1046 1536<br>1037 1049 1541<br>1038 1050 1547<br>1039 1053 1548<br>1040 1055 1551<br>1042 1524 1552<br>1043 1525 | 125 to 175 | Hot Rolled, Normalized, Annealed or Cold Drawn | 1—4<br>5—10<br>11—19<br>20 & Finer | .012<br>.008<br>.005<br>.003 | 450 | M2, M7 |
| | | | 25—6<br>5—3<br>2—1.5<br>1 & Finer | .3<br>.2<br>.12<br>.07 | 135 | S4, S2 |
| | 175 to 225 | Hot Rolled, Normalized, Annealed or Cold Drawn | 1—4<br>5—10<br>11—19<br>20 & Finer | .012<br>.008<br>.005<br>.003 | 400 | M2, M7 |
| | | | 25—6<br>5—3<br>2—1.5<br>1 & Finer | .3<br>.2<br>.12<br>.07 | 120 | S4, S2 |
| | 225 to 275 | Hot Rolled, Normalized, Annealed, Cold Drawn or Quenched and Tempered | 1—4<br>5—10<br>11—19<br>20 & Finer | .012<br>.008<br>.005<br>.003 | 350 | M2, M7 |
| | | | 25—6<br>5—3<br>2—1.5<br>1 & Finer | .3<br>.2<br>.12<br>.07 | 105 | S4, S2 |
| | 275 to 325 | Hot Rolled, Normalized, Annealed or Quenched and Tempered | 1—4<br>5—10<br>11—19<br>20 & Finer | .012<br>.008<br>.005<br>.003 | 325 | M2, M7 |
| | | | 25—6<br>5—3<br>2—1.5<br>1 & Finer | .3<br>.2<br>.12<br>.07 | 100 | S4, S2 |
| | 325 to 375 | Quenched and Tempered | 1—4<br>5—10<br>11—19<br>20 & Finer | .012<br>.008<br>.005<br>.003 | 275 | M3 |
| | | | 25—6<br>5—3<br>2—1.5<br>1 & Finer | .3<br>.2<br>.12<br>.07 | 84 | S5 |
| **High Carbon**<br>1060 1075 1090<br>1064 1078 1095<br>1065 1080 1561<br>1069 1084 1566<br>1070 1085 1572<br>1074 1086 | 175 to 225 | Hot Rolled, Normalized, Annealed or Cold Drawn | 1—4<br>5—10<br>11—19<br>20 & Finer | .012<br>.008<br>.005<br>.003 | 375 | M2, M7 |
| | | | 25—6<br>5—3<br>2—1.5<br>1 & Finer | .3<br>.2<br>.12<br>.07 | 115 | S4, S2 |
| | 225 to 275 | Hot Rolled, Normalized, Annealed, Cold Drawn or Quenched and Tempered | 1—4<br>5—10<br>11—19<br>20 & Finer | .012<br>.008<br>.005<br>.003 | 325 | M2, M7 |
| | | | 25—6<br>5—3<br>2—1.5<br>1 & Finer | .3<br>.2<br>.12<br>.07 | 100 | S4, S2 |
| | 275 to 325 | Hot Rolled, Normalized, Annealed or Quenched and Tempered | 1—4<br>5—10<br>11—19<br>20 & Finer | .012<br>.008<br>.005<br>.003 | 300 | M2, M7 |
| | | | 25—6<br>5—3<br>2—1.5<br>1 & Finer | .3<br>.2<br>.12<br>.07 | 90 | S4, S2 |

See section 16 for Cutting Fluid Recommendations.

*Feed recommendations apply to CONVENTIONAL (axial-transverse) gear shaving.

†Feeds should be increased 100% for gears shaved by the DIAGONAL (angular-transverse) method.

| MATERIAL | HARDNESS Bhn | CONDITION | DIAMETRAL PITCH / MODULE | FEED*† per rev. of gear in / mm | CUTTER PITCH LINE SPEED fpm / m/min | HSS TOOL MATERIAL AISI / ISO |
|---|---|---|---|---|---|---|
| **2. CARBON STEELS, WROUGHT (cont.) High Carbon (cont.)** (materials listed on preceding page) | 325 to 375 | Quenched and Tempered | 1—4<br>5—10<br>11—19<br>20 & Finer | .012<br>.008<br>.005<br>.003 | 250 | M3 |
| | | | 25—6<br>5—3<br>2—1.5<br>1 & Finer | .3<br>.2<br>.12<br>.07 | 76 | S5 |
| **3. CARBON AND FERRITIC ALLOY STEELS (HIGH TEMPERATURE SERVICE)** ASTM A369: Grades FPA, FPB, FP1, FP2, FP12 | 150 to 200 | As Forged | 1—4<br>5—10<br>11—19<br>20 & Finer | .012<br>.008<br>.005<br>.003 | 450 | M2, M7 |
| | | | 25—6<br>5—3<br>2—1.5<br>1 & Finer | .3<br>.2<br>.12<br>.07 | 135 | S4, S2 |
| ASTM A369: Grades FP3b, FP11 | 150 to 200 | Annealed or Normalized and Tempered | 1—4<br>5—10<br>11—19<br>20 & Finer | .012<br>.008<br>.005<br>.003 | 400 | M2, M7 |
| | | | 25—6<br>5—3<br>2—1.5<br>1 & Finer | .3<br>.2<br>.12<br>.07 | 120 | S4, S2 |
| ASTM A369: Grades FP5, FP7, FP9, FP21, FP22 | 150 to 200 | Annealed or Normalized and Tempered | 1—4<br>5—10<br>11—19<br>20 & Finer | .012<br>.008<br>.005<br>.003 | 350 | M2, M7 |
| | | | 25—6<br>5—3<br>2—1.5<br>1 & Finer | .3<br>.2<br>.12<br>.07 | 105 | S4, S2 |
| **4. FREE MACHINING ALLOY STEELS, WROUGHT Medium Carbon Resulfurized**<br>4140 4145Se<br>4140Se 4147Te<br>4142Te 4150 | 150 to 200 | Hot Rolled, Normalized, Annealed or Cold Drawn | 1—4<br>5—10<br>11—19<br>20 & Finer | .012<br>.008<br>.005<br>.003 | 500 | M2, M7 |
| | | | 25—6<br>5—3<br>2—1.5<br>1 & Finer | .3<br>.2<br>.12<br>.07 | 150 | S4, S2 |
| | 200 to 250 | Hot Rolled, Normalized, Annealed or Cold Drawn | 1—4<br>5—10<br>11—19<br>20 & Finer | .012<br>.008<br>.005<br>.003 | 450 | M2, M7 |
| | | | 25—6<br>5—3<br>2—1.5<br>1 & Finer | .3<br>.2<br>.12<br>.07 | 135 | S4, S2 |
| | 275 to 325 | Quenched and Tempered | 1—4<br>5—10<br>11—19<br>20 & Finer | .012<br>.008<br>.005<br>.003 | 325 | M2, M7 |
| | | | 25—6<br>5—3<br>2—1.5<br>1 & Finer | .3<br>.2<br>.12<br>.07 | 100 | S4, S2 |
| | 325 to 375 | Quenched and Tempered | 1—4<br>5—10<br>11—19<br>20 & Finer | .012<br>.008<br>.005<br>.003 | 250 | M3 |
| | | | 25—6<br>5—3<br>2—1.5<br>1 & Finer | .3<br>.2<br>.12<br>.07 | 76 | S5 |

See section 16 for Cutting Fluid Recommendations.

*Feed recommendations apply to CONVENTIONAL (axial-transverse) gear shaving.

†Feeds should be increased 100% for gears shaved by the DIAGONAL (angular-transverse) method.

# 7.4 Gear Shaving

| MATERIAL | HARDNESS Bhn | CONDITION | DIAMETRAL PITCH / MODULE | FEED*† per rev. of gear in / mm | CUTTER PITCH LINE SPEED fpm / m/min | HSS TOOL MATERIAL AISI / ISO |
|---|---|---|---|---|---|---|
| **4. FREE MACHINING ALLOY STEELS, WROUGHT (cont.)**<br>**Medium and High Carbon Leaded**<br>41L30 41L50 86L20<br>41L40 43L40 86L40<br>41L45 51L32<br>41L47 52L100 | 150 to 200 | Hot Rolled, Normalized, Annealed or Cold Drawn | 1—4<br>5—10<br>11—19<br>20 & Finer | .012<br>.008<br>.005<br>.003 | 525 | M2, M7 |
| | | | 25—6<br>5—3<br>2—1.5<br>1 & Finer | .3<br>.2<br>.12<br>.07 | 160 | S4, S2 |
| | 200 to 250 | Hot Rolled, Normalized, Annealed or Cold Drawn | 1—4<br>5—10<br>11—19<br>20 & Finer | .012<br>.008<br>.005<br>.003 | 450 | M2, M7 |
| | | | 25—6<br>5—3<br>2—1.5<br>1 & Finer | .3<br>.2<br>.12<br>.07 | 135 | S4, S2 |
| | 275 to 325 | Quenched and Tempered | 1—4<br>5—10<br>11—19<br>20 & Finer | .012<br>.008<br>.005<br>.003 | 350 | M2, M7 |
| | | | 25—6<br>5—3<br>2—1.5<br>1 & Finer | .3<br>.2<br>.12<br>.07 | 105 | S4, S2 |
| | 325 to 375 | Quenched and Tempered | 1—4<br>5—10<br>11—19<br>20 & Finer | .012<br>.008<br>.005<br>.003 | 275 | M3 |
| | | | 25—6<br>5—3<br>2—1.5<br>1 & Finer | .3<br>.2<br>.12<br>.07 | 84 | S5 |
| **5. ALLOY STEELS, WROUGHT**<br>**Low Carbon**<br>4012 4615 4817 8617<br>4023 4617 4820 8620<br>4024 4620 5015 8622<br>4118 4621 5115 8822<br>4320 4718 5120 9310<br>4419 4720 6118 94B15<br>4422 4815 8115 94B17 | 125 to 175 | Hot Rolled, Annealed or Cold Drawn | 1—4<br>5—10<br>11—19<br>20 & Finer | .012<br>.008<br>.005<br>.003 | 475 | M2, M7 |
| | | | 25—6<br>5—3<br>2—1.5<br>1 & Finer | .3<br>.2<br>.12<br>.07 | 145 | S4, S2 |
| | 175 to 225 | Hot Rolled, Annealed or Cold Drawn | 1—4<br>5—10<br>11—19<br>20 & Finer | .012<br>.008<br>.005<br>.003 | 425 | M2, M7 |
| | | | 25—6<br>5—3<br>2—1.5<br>1 & Finer | .3<br>.2<br>.12<br>.07 | 130 | S4, S2 |
| | 225 to 275 | Hot Rolled, Normalized, Annealed or Cold Drawn | 1—4<br>5—10<br>11—19<br>20 & Finer | .012<br>.008<br>.005<br>.003 | 350 | M2, M7 |
| | | | 25—6<br>5—3<br>2—1.5<br>1 & Finer | .3<br>.2<br>.12<br>.07 | 105 | S4, S2 |
| | 275 to 325 | Normalized or Quenched and Tempered | 1—4<br>5—10<br>11—19<br>20 & Finer | .012<br>.008<br>.005<br>.003 | 300 | M2, M7 |
| | | | 25—6<br>5—3<br>2—1.5<br>1 & Finer | .3<br>.2<br>.12<br>.07 | 90 | S4, S2 |

See section 16 for Cutting Fluid Recommendations.

*Feed recommendations apply to CONVENTIONAL (axial-transverse) gear shaving.

†Feeds should be increased 100% for gears shaved by the DIAGONAL (angular-transverse) method.

| MATERIAL | HARD-NESS Bhn | CONDITION | DIAMETRAL PITCH / MODULE | FEED*† per rev. of gear in / mm | CUTTER PITCH LINE SPEED fpm / m/min | HSS TOOL MATERIAL AISI / ISO |
|---|---|---|---|---|---|---|
| **5. ALLOY STEELS, WROUGHT (cont.)** **Low Carbon (cont.)** (materials listed on preceding page) | 325 to 375 | Normalized or Quenched and Tempered | 1—4<br>5—10<br>11—19<br>20 & Finer | .012<br>.008<br>.005<br>.003 | 250 | M3 |
| | | | 25—6<br>5—3<br>2—1.5<br>1 & Finer | .3<br>.2<br>.12<br>.07 | 76 | S5 |
| **Medium Carbon**<br>1330 4145 5132 8640<br>1335 4147 5135 8642<br>1340 4150 5140 8645<br>1345 4161 5145 86B45<br>4027 4340 5147 8650<br>4028 4427 5150 8655<br>4032 4626 5155 8660<br>4037 50B40 5160 8740<br>4042 50B44 51B60 8742<br>4047 5046 6150 9254<br>4130 50B46 81B45 9255<br>4135 50B50 8625 9260<br>4137 5060 8627 94B30<br>4140 50B60 8630<br>4142 5130 8637 | 175 to 225 | Hot Rolled, Annealed or Cold Drawn | 1—4<br>5—10<br>11—19<br>20 & Finer | .012<br>.008<br>.005<br>.003 | 400 | M2, M7 |
| | | | 25—6<br>5—3<br>2—1.5<br>1 & Finer | .3<br>.2<br>.12<br>.07 | 120 | S4, S2 |
| | 225 to 275 | Annealed, Normalized, Cold Drawn or Quenched and Tempered | 1—4<br>5—10<br>11—19<br>20 & Finer | .012<br>.008<br>.005<br>.003 | 325 | M2, M7 |
| | | | 25—6<br>5—3<br>2—1.5<br>1 & Finer | .3<br>.2<br>.12<br>.07 | 100 | S4, S2 |
| | 275 to 325 | Normalized or Quenched and Tempered | 1—4<br>5—10<br>11—19<br>20 & Finer | .012<br>.008<br>.005<br>.003 | 275 | M2, M7 |
| | | | 25—6<br>5—3<br>2—1.5<br>1 & Finer | .3<br>.2<br>.12<br>.07 | 84 | S4, S2 |
| | 325 to 375 | Normalized or Quenched and Tempered | 1—4<br>5—10<br>11—19<br>20 & Finer | .012<br>.008<br>.005<br>.003 | 225 | M3 |
| | | | 25—6<br>5—3<br>2—1.5<br>1 & Finer | .3<br>.2<br>.12<br>.07 | 69 | S5 |
| **High Carbon**<br>50100<br>51100<br>52100<br>M-50 | 175 to 225 | Hot Rolled, Annealed or Cold Drawn | 1—4<br>5—10<br>11—19<br>20 & Finer | .012<br>.008<br>.005<br>.003 | 375 | M2, M7 |
| | | | 25—6<br>5—3<br>2—1.5<br>1 & Finer | .3<br>.2<br>.12<br>.07 | 115 | S4, S2 |
| | 225 to 275 | Normalized, Cold Drawn or Quenched and Tempered | 1—4<br>5—10<br>11—19<br>20 & Finer | .012<br>.008<br>.005<br>.003 | 300 | M2, M7 |
| | | | 25—6<br>5—3<br>2—1.5<br>1 & Finer | .3<br>.2<br>.12<br>.07 | 90 | S4, S2 |
| | 275 to 325 | Normalized or Quenched and Tempered | 1—4<br>5—10<br>11—19<br>20 & Finer | .012<br>.008<br>.005<br>.003 | 250 | M2, M7 |
| | | | 25—6<br>5—3<br>2—1.5<br>1 & Finer | .3<br>.2<br>.12<br>.07 | 76 | S4, S2 |

See section 16 for Cutting Fluid Recommendations.
*Feed recommendations apply to CONVENTIONAL (axial-transverse) gear shaving.
†Feeds should be increased 100% for gears shaved by the DIAGONAL (angular-transverse) method.

# 7.4 Gear Shaving

| MATERIAL | HARDNESS Bhn | CONDITION | DIAMETRAL PITCH / MODULE | FEED*† per rev. of gear in / mm | CUTTER PITCH LINE SPEED fpm / m/min | HSS TOOL MATERIAL AISI / ISO |
|---|---|---|---|---|---|---|
| **5. ALLOY STEELS, WROUGHT (cont.)**<br>**High Carbon (cont.)**<br>(materials listed on preceding page) | 325<br>to<br>375 | Normalized<br>or<br>Quenched<br>and<br>Tempered | 1—4<br>5—10<br>11—19<br>20 & Finer | .012<br>.008<br>.005<br>.003 | 200 | M3 |
| | | | 25—6<br>5—3<br>2—1.5<br>1 & Finer | .3<br>.2<br>.12<br>.07 | 60 | S5 |
| **6. HIGH STRENGTH STEELS, WROUGHT**<br>300M 98BV40<br>4330V D6ac<br>4340 H11<br>4340Si H13 | 225<br>to<br>300 | Annealed | 1—4<br>5—10<br>11—19<br>20 & Finer | .012<br>.008<br>.005<br>.003 | 300 | M2, M7 |
| | | | 25—6<br>5—3<br>2—1.5<br>1 & Finer | .3<br>.2<br>.12<br>.07 | 90 | S4, S2 |
| | 300<br>to<br>350 | Normalized | 1—4<br>5—10<br>11—19<br>20 & Finer | .012<br>.008<br>.005<br>.003 | 275 | M3 |
| | | | 25—6<br>5—3<br>2—1.5<br>1 & Finer | .3<br>.2<br>.12<br>.07 | 84 | S5 |
| HP 9-4-20<br>HP 9-4-25<br>HP 9-4-30<br>HP 9-4-45 | 325<br>to<br>375 | Annealed | 1—4<br>5—10<br>11—19<br>20 & Finer | .012<br>.008<br>.005<br>.003 | 200 | M3 |
| | | | 25—6<br>5—3<br>2—1.5<br>1 & Finer | .3<br>.2<br>.12<br>.07 | 60 | S5 |
| **7. MARAGING STEELS, WROUGHT**<br>120 Grade<br>180 Grade | 275<br>to<br>325 | Annealed | 1—4<br>5—10<br>11—19<br>20 & Finer | .012<br>.008<br>.005<br>.003 | 350 | M2, M7 |
| | | | 25—6<br>5—3<br>2—1.5<br>1 & Finer | .3<br>.2<br>.12<br>.07 | 105 | S4, S2 |
| ASTM A538: Grades A, B, C<br>200 Grade<br>250 Grade<br>300 Grade<br>350 Grade<br>HY230 | 275<br>to<br>325 | Annealed | 1—4<br>5—10<br>11—19<br>20 & Finer | .012<br>.008<br>.005<br>.003 | 300 | M2, M7 |
| | | | 25—6<br>5—3<br>2—1.5<br>1 & Finer | .3<br>.2<br>.12<br>.07 | 90 | S4, S2 |
| **8. TOOL STEELS, WROUGHT**<br>**High Speed**<br>M1 T1<br>M2 T2<br>M6 T6<br>M10 | 200<br>to<br>250 | Annealed | 1—4<br>5—10<br>11—19<br>20 & Finer | .012<br>.008<br>.005<br>.003 | 225 | M2, M7 |
| | | | 25—6<br>5—3<br>2—1.5<br>1 & Finer | .3<br>.2<br>.12<br>.07 | 69 | S4, S2 |
| **High Speed**<br>M3-1 M34 M46<br>M3-2 M36 M47<br>M4 M41 T4<br>M7 M42 T5<br>M30 M43 T8<br>M33 M44 T15 | 225<br>to<br>275 | Annealed | 1—4<br>5—10<br>11—19<br>20 & Finer | .012<br>.008<br>.005<br>.003 | 200 | M2, M7 |
| | | | 25—6<br>5—3<br>2—1.5<br>1 & Finer | .3<br>.2<br>.12<br>.07 | 60 | S4, S2 |

See section 16 for Cutting Fluid Recommendations.

*Feed recommendations apply to CONVENTIONAL (axial-transverse) gear shaving.

†Feeds should be increased 100% for gears shaved by the DIAGONAL (angular-transverse) method.

| MATERIAL | HARDNESS Bhn | CONDITION | DIAMETRAL PITCH / MODULE | FEED*† per rev. of gear in / mm | CUTTER PITCH LINE SPEED fpm / m/min | HSS TOOL MATERIAL AISI / ISO |
|---|---|---|---|---|---|---|
| **Hot Work** H10, H11, H12, H13, H14, H19, H21, H22, H23, H24, H25, H26, H42 | 150 to 200 | Annealed | 1—4<br>5—10<br>11—19<br>20 & Finer | .012<br>.008<br>.005<br>.003 | 325 | M2, M7 |
| | | | 25—6<br>5—3<br>2—1.5<br>1 & Finer | .3<br>.2<br>.12<br>.07 | 100 | S4, S2 |
| | 200 to 250 | Annealed | 1—4<br>5—10<br>11—19<br>20 & Finer | .012<br>.008<br>.005<br>.003 | 300 | M2, M7 |
| | | | 25—6<br>5—3<br>2—1.5<br>1 & Finer | .3<br>.2<br>.12<br>.07 | 90 | S4, S2 |
| | 325 to 375 | Quenched and Tempered | 1—4<br>5—10<br>11—19<br>20 & Finer | .012<br>.008<br>.005<br>.003 | 250 | M3 |
| | | | 25—6<br>5—3<br>2—1.5<br>1 & Finer | .3<br>.2<br>.12<br>.07 | 76 | S5 |
| **Cold Work** A2, A3, A4, A6, A7, A8, A9, A10, D2, D3, D4, D5, D7, O1, O2, O6, O7 | 200 to 250 | Annealed | 1—4<br>5—10<br>11—19<br>20 & Finer | .012<br>.008<br>.005<br>.003 | 300 | M2, M7 |
| | | | 25—6<br>5—3<br>2—1.5<br>1 & Finer | .3<br>.2<br>.12<br>.07 | 90 | S4, S2 |
| | | | | | | |
| | | | | | | |
| **Shock Resisting** S1, S2, S5, S6, S7 | 175 to 225 | Annealed | 1—4<br>5—10<br>11—19<br>20 & Finer | .012<br>.008<br>.005<br>.003 | 325 | M2, M7 |
| | | | 25—6<br>5—3<br>2—1.5<br>1 & Finer | .3<br>.2<br>.12<br>.07 | 100 | S4, S2 |
| **Mold** P2, P4, P5, P6 | 100 to 150 | Annealed | 1—4<br>5—10<br>11—19<br>20 & Finer | .012<br>.008<br>.005<br>.003 | 500 | M2, M7 |
| | | | 25—6<br>5—3<br>2—1.5<br>1 & Finer | .3<br>.2<br>.12<br>.07 | 150 | S4, S2 |
| **Mold** P20, P21 | 150 to 200 | Annealed | 1—4<br>5—10<br>11—19<br>20 & Finer | .012<br>.008<br>.005<br>.003 | 450 | M2, M7 |
| | | | 25—6<br>5—3<br>2—1.5<br>1 & Finer | .3<br>.2<br>.12<br>.07 | 135 | S4, S2 |

See section 16 for Cutting Fluid Recommendations.

*Feed recommendations apply to CONVENTIONAL (axial-transverse) gear shaving.

†Feeds should be increased 100% for gears shaved by the DIAGONAL (angular-transverse) method.

# 7.4 Gear Shaving

| MATERIAL | HARD-NESS<br>Bhn | CONDITION | DIAMETRAL PITCH / MODULE | FEED*† per rev. of gear<br>in / mm | CUTTER PITCH LINE SPEED<br>fpm / m/min | HSS TOOL MATERIAL<br>AISI / ISO |
|---|---|---|---|---|---|---|
| **8. TOOL STEELS, WROUGHT (cont.)**<br>**Special Purpose**<br>L2<br>L6<br>L7 | 150<br>to<br>200 | Annealed | 1—4<br>5—10<br>11—19<br>20 & Finer | .012<br>.008<br>.005<br>.003 | 425 | M2, M7 |
| | | | 25—6<br>5—3<br>2—1.5<br>1 & Finer | .3<br>.2<br>.12<br>.07 | 130 | S4, S2 |
| **Special Purpose**<br>F1<br>F2 | 200<br>to<br>250 | Annealed | 1—4<br>5—10<br>11—19<br>20 & Finer | .012<br>.008<br>.005<br>.003 | 375 | M2, M7 |
| | | | 25—6<br>5—3<br>2—1.5<br>1 & Finer | .3<br>.2<br>.12<br>.07 | 115 | S4, S2 |
| **Water Hardening**<br>W1, W2, W5<br>SAE J438b: Types W108, W109, W110, W112, W209, W210, W310 | 150<br>to<br>200 | Annealed | 1—4<br>5—10<br>11—19<br>20 & Finer | .012<br>.008<br>.005<br>.003 | 400 | M2, M7 |
| | | | 25—6<br>5—3<br>2—1.5<br>1 & Finer | .3<br>.2<br>.12<br>.07 | 120 | S4, S2 |
| **9. NITRIDING STEELS, WROUGHT**<br>Nitralloy 125<br>Nitralloy 135<br>Nitralloy 135 Mod.<br>Nitralloy 225<br>Nitralloy 230<br>Nitralloy EZ<br>Nitralloy N<br>Nitrex 1 | 200<br>to<br>250 | Annealed | 1—4<br>5—10<br>11—19<br>20 & Finer | .012<br>.008<br>.005<br>.003 | 375 | M2, M7 |
| | | | 25—6<br>5—3<br>2—1.5<br>1 & Finer | .3<br>.2<br>.12<br>.07 | 115 | S4, S2 |
| | 300<br>to<br>350 | Normalized<br>or<br>Quenched<br>and<br>Tempered | 1—4<br>5—10<br>11—19<br>20 & Finer | .012<br>.008<br>.005<br>.003 | 275 | M3 |
| | | | 25—6<br>5—3<br>2—1.5<br>1 & Finer | .3<br>.2<br>.12<br>.07 | 84 | S5 |
| **12. FREE MACHINING STAINLESS STEELS, WROUGHT**<br>**Ferritic**<br>430F<br>430F Se | 135<br>to<br>185 | Annealed | 1—4<br>5—10<br>11—19<br>20 & Finer | .012<br>.008<br>.005<br>.003 | 300 | M2, M7 |
| | | | 25—6<br>5—3<br>2—1.5<br>1 & Finer | .3<br>.2<br>.12<br>.07 | 90 | S4, S2 |
| **Austenitic**<br>203EZ<br>303<br>303MA<br>303Pb<br>303 Plus X<br>303Se | 135<br>to<br>185 | Annealed | 1—4<br>5—10<br>11—19<br>20 & Finer | .012<br>.008<br>.005<br>.003 | 250 | M2, M7 |
| | | | 25—6<br>5—3<br>2—1.5<br>1 & Finer | .3<br>.2<br>.12<br>.07 | 76 | S4, S2 |
| | 225<br>to<br>275 | Cold<br>Drawn | 1—4<br>5—10<br>11—19<br>20 & Finer | .012<br>.008<br>.005<br>.003 | 225 | M2, M7 |
| | | | 25—6<br>5—3<br>2—1.5<br>1 & Finer | .3<br>.2<br>.12<br>.07 | 69 | S4, S2 |

See section 16 for Cutting Fluid Recommendations.
*Feed recommendations apply to CONVENTIONAL (axial-transverse) gear shaving.
†Feeds should be increased 100% for gears shaved by the DIAGONAL (angular-transverse) method.

| MATERIAL | HARD-NESS Bhn | CONDITION | DIAMETRAL PITCH / MODULE | FEED*† per rev. of gear in / mm | CUTTER PITCH LINE SPEED fpm / m/min | HSS TOOL MATERIAL AISI / ISO |
|---|---|---|---|---|---|---|
| **Martensitic**<br>416, 416 Plus X, 416Se, 420F, 420F Se, 440F, 440F Se | 135 to 185 | Annealed | 1—4<br>5—10<br>11—19<br>20 & Finer | .012<br>.008<br>.005<br>.003 | 300 | M2, M7 |
| | | | 25—6<br>5—3<br>2—1.5<br>1 & Finer | .3<br>.2<br>.12<br>.07 | 90 | S4, S2 |
| | 185 to 240 | Annealed or Cold Drawn | 1—4<br>5—10<br>11—19<br>20 & Finer | .012<br>.008<br>.005<br>.003 | 275 | M2, M7 |
| | | | 25—6<br>5—3<br>2—1.5<br>1 & Finer | .3<br>.2<br>.12<br>.07 | 84 | S4, S2 |
| | 275 to 325 | Quenched and Tempered | 1—4<br>5—10<br>11—19<br>20 & Finer | .012<br>.008<br>.005<br>.003 | 225 | M2, M7 |
| | | | 25—6<br>5—3<br>2—1.5<br>1 & Finer | .3<br>.2<br>.12<br>.07 | 69 | S4, S2 |
| **13. STAINLESS STEELS, WROUGHT**<br>**Ferritic**<br>405, 409, 429, 430, 434, 436, 442, 446 | 135 to 185 | Annealed | 1—4<br>5—10<br>11—19<br>20 & Finer | .012<br>.008<br>.005<br>.003 | 275 | M2, M7 |
| | | | 25—6<br>5—3<br>2—1.5<br>1 & Finer | .3<br>.2<br>.12<br>.07 | 84 | S4, S2 |
| **Austenitic**<br>201, 202, 301, 302, 304, 304L, 305, 308, 321, 347, 348, 384, 385 | 135 to 185 | Annealed | 1—4<br>5—10<br>11—19<br>20 & Finer | .012<br>.008<br>.005<br>.003 | 225 | M2, M7 |
| | | | 25—6<br>5—3<br>2—1.5<br>1 & Finer | .3<br>.2<br>.12<br>.07 | 69 | S4, S2 |
| | 225 to 275 | Cold Drawn | 1—4<br>5—10<br>11—19<br>20 & Finer | .012<br>.008<br>.005<br>.003 | 200 | M2, M7 |
| | | | 25—6<br>5—3<br>2—1.5<br>1 & Finer | .3<br>.2<br>.12<br>.07 | 60 | S4, S2 |
| **Austenitic**<br>302B, 309, 309S, 310, 310S, 314, 316, 316L, 317, 330 | 135 to 185 | Annealed | 1—4<br>5—10<br>11—19<br>20 & Finer | .012<br>.008<br>.005<br>.003 | 200 | M2, M7 |
| | | | 25—6<br>5—3<br>2—1.5<br>1 & Finer | .3<br>.2<br>.12<br>.07 | 60 | S4, S2 |
| | 225 to 275 | Cold Drawn | 1—4<br>5—10<br>11—19<br>20 & Finer | .012<br>.008<br>.005<br>.003 | 175 | M2, M7 |
| | | | 25—6<br>5—3<br>2—1.5<br>1 & Finer | .3<br>.2<br>.12<br>.07 | 53 | S4, S2 |

See section 16 for Cutting Fluid Recommendations.
*Feed recommendations apply to CONVENTIONAL (axial-transverse) gear shaving.
†Feeds should be increased 100% for gears shaved by the DIAGONAL (angular-transverse) method.

# 7.4 Gear Shaving

| MATERIAL | HARDNESS Bhn | CONDITION | DIAMETRAL PITCH / MODULE | FEED*† per rev. of gear in / mm | CUTTER PITCH LINE SPEED fpm / m/min | HSS TOOL MATERIAL AISI / ISO |
|---|---|---|---|---|---|---|
| **13. STAINLESS STEELS, WROUGHT (cont.)**<br>**Austenitic**<br>Nitronic 32<br>Nitronic 33<br>Nitronic 40<br>Nitronic 50<br>Nitronic 60 | 210 to 250 | Annealed | 1—4<br>5—10<br>11—19<br>20 & Finer | .012<br>.008<br>.005<br>.003 | 175 | M2, M7 |
| | | | 25—6<br>5—3<br>2—1.5<br>1 & Finer | .3<br>.2<br>.12<br>.07 | 53 | S4, S2 |
| **Martensitic**<br>403<br>410<br>420<br>422<br>501<br>502 | 135 to 175 | Annealed | 1—4<br>5—10<br>11—19<br>20 & Finer | .012<br>.008<br>.005<br>.003 | 275 | M2, M7 |
| | | | 25—6<br>5—3<br>2—1.5<br>1 & Finer | .3<br>.2<br>.12<br>.07 | 84 | S4, S2 |
| | 175 to 225 | Annealed | 1—4<br>5—10<br>11—19<br>20 & Finer | .012<br>.008<br>.005<br>.003 | 250 | M2, M7 |
| | | | 25—6<br>5—3<br>2—1.5<br>1 & Finer | .3<br>.2<br>.12<br>.07 | 76 | S4, S2 |
| | 275 to 325 | Quenched and Tempered | 1—4<br>5—10<br>11—19<br>20 & Finer | .012<br>.008<br>.005<br>.003 | 200 | M2, M7 |
| | | | 25—6<br>5—3<br>2—1.5<br>1 & Finer | .3<br>.2<br>.12<br>.07 | 60 | S4, S2 |
| **Martensitic**<br>414<br>431<br>Greek Ascoloy | 225 to 275 | Annealed | 1—4<br>5—10<br>11—19<br>20 & Finer | .012<br>.008<br>.005<br>.003 | 225 | M2, M7 |
| | | | 25—6<br>5—3<br>2—1.5<br>1 & Finer | .3<br>.2<br>.12<br>.07 | 69 | S4, S2 |
| | 275 to 325 | Quenched and Tempered | 1—4<br>5—10<br>11—19<br>20 & Finer | .012<br>.008<br>.005<br>.003 | 200 | M2, M7 |
| | | | 25—6<br>5—3<br>2—1.5<br>1 & Finer | .3<br>.2<br>.12<br>.07 | 60 | S4, S2 |
| **Martensitic**<br>440A<br>440B<br>440C | 225 to 275 | Annealed | 1—4<br>5—10<br>11—19<br>20 & Finer | .012<br>.008<br>.005<br>.003 | 225 | M2, M7 |
| | | | 25—6<br>5—3<br>2—1.5<br>1 & Finer | .3<br>.2<br>.12<br>.07 | 69 | S4, S2 |
| | 275 to 325 | Quenched and Tempered | 1—4<br>5—10<br>11—19<br>20 & Finer | .012<br>.008<br>.005<br>.003 | 200 | M2, M7 |
| | | | 25—6<br>5—3<br>2—1.5<br>1 & Finer | .3<br>.2<br>.12<br>.07 | 60 | S4, S2 |

See section 16 for Cutting Fluid Recommendations.

*Feed recommendations apply to CONVENTIONAL (axial-transverse) gear shaving.

†Feeds should be increased 100% for gears shaved by the DIAGONAL (angular-transverse) method.

| MATERIAL | HARD-NESS Bhn | CONDITION | DIAMETRAL PITCH / MODULE | FEED*† per rev. of gear in / mm | CUTTER PITCH LINE SPEED fpm / m/min | HSS TOOL MATERIAL AISI / ISO |
|---|---|---|---|---|---|---|
| **14. PRECIPITATION HARDENING STAINLESS STEELS, WROUGHT**<br>15-5 PH<br>16-6 PH<br>17-4 PH<br>17-7 PH<br>17-14 Cu Mo<br>AF-71<br>AFC-77<br>Almar 362 (AM-362)<br>AM-350<br>AM-355<br>AM-363<br>Custom 450<br>Custom 455<br>HNM<br>PH 13-8 Mo<br>PH 14-8 Mo<br>PH 15-7 Mo<br>Stainless W | 150 to 200 | Solution Treated | 1—4<br>5—10<br>11—19<br>20 & Finer | .012<br>.008<br>.005<br>.003 | 225 | M2, M7 |
| | | | 25—6<br>5—3<br>2—1.5<br>1 & Finer | .3<br>.2<br>.12<br>.07 | 69 | S4, S2 |
| | 275 to 325 | Solution Treated or Hardened | 1—4<br>5—10<br>11—19<br>20 & Finer | .012<br>.008<br>.005<br>.003 | 200 | M2, M7 |
| | | | 25—6<br>5—3<br>2—1.5<br>1 & Finer | .3<br>.2<br>.12<br>.07 | 60 | S4, S2 |
| | 325 to 375 | Solution Treated or Hardened | 1—4<br>5—10<br>11—19<br>20 & Finer | .012<br>.008<br>.005<br>.003 | 175 | M3 |
| | | | 25—6<br>5—3<br>2—1.5<br>1 & Finer | .3<br>.2<br>.12<br>.07 | 53 | S5 |
| **15. CARBON STEELS, CAST**<br>**Low Carbon**<br>ASTM A426: Grade CP1<br>1010<br>1020 | 100 to 150 | Annealed, Normalized or Normalized and Tempered | 1—4<br>5—10<br>11—19<br>20 & Finer | .012<br>.008<br>.005<br>.003 | 500 | M2, M7 |
| | | | 25—6<br>5—3<br>2—1.5<br>1 & Finer | .3<br>.2<br>.12<br>.07 | 150 | S4, S2 |
| **Medium Carbon**<br>ASTM A352: Grades LCA, LCB, LCC<br>ASTM A356: Grade 1<br>1030<br>1040<br>1050 | 125 to 175 | Annealed, Normalized or Normalized and Tempered | 1—4<br>5—10<br>11—19<br>20 & Finer | .012<br>.008<br>.005<br>.003 | 425 | M2, M7 |
| | | | 25—6<br>5—3<br>2—1.5<br>1 & Finer | .3<br>.2<br>.12<br>.07 | 130 | S4, S2 |
| | 175 to 225 | Annealed, Normalized or Normalized and Tempered | 1—4<br>5—10<br>11—19<br>20 & Finer | .012<br>.008<br>.005<br>.003 | 375 | M2, M7 |
| | | | 25—6<br>5—3<br>2—1.5<br>1 & Finer | .3<br>.2<br>.12<br>.07 | 115 | S4, S2 |
| | 250 to 300 | Quenched and Tempered | 1—4<br>5—10<br>11—19<br>20 & Finer | .012<br>.008<br>.005<br>.003 | 325 | M2, M7 |
| | | | 25—6<br>5—3<br>2—1.5<br>1 & Finer | .3<br>.2<br>.12<br>.07 | 100 | S4, S2 |

See section 16 for Cutting Fluid Recommendations.
*Feed recommendations apply to CONVENTIONAL (axial-transverse) gear shaving.
†Feeds should be increased 100% for gears shaved by the DIAGONAL (angular-transverse) method.

# 7.4 Gear Shaving

| MATERIAL | HARD-NESS<br>Bhn | CONDITION | DIAMETRAL PITCH<br>MODULE | FEED*† per rev. of gear<br>in<br>mm | CUTTER PITCH LINE SPEED<br>fpm<br>m/min | HSS TOOL MATERIAL<br>AISI<br>ISO |
|---|---|---|---|---|---|---|
| **16. ALLOY STEELS, CAST**<br>**Low Carbon**<br>ASTM A217: Grade WC9<br>ASTM A352: Grades LC3, LC4<br>ASTM A426: Grades CP2, CP5, CP5b, CP11, CP12, CP15, CP21, CP22<br>1320 2320 4120 8020<br>2315 4110 4320 8620 | 150<br>to<br>200 | Annealed,<br>Normalized<br>or<br>Normalized<br>and<br>Tempered | 1—4<br>5—10<br>11—19<br>20 & Finer | .012<br>.008<br>.005<br>.003 | 425 | M2, M7 |
| | | | 25—6<br>5—3<br>2—1.5<br>1 & Finer | .3<br>.2<br>.12<br>.07 | 130 | S4, S2 |
| | 200<br>to<br>225 | Annealed,<br>Normalized<br>or<br>Normalized<br>and<br>Tempered | 1—4<br>5—10<br>11—19<br>20 & Finer | .012<br>.008<br>.005<br>.003 | 375 | M2, M7 |
| | | | 25—6<br>5—3<br>2—1.5<br>1 & Finer | .3<br>.2<br>.12<br>.07 | 115 | S4, S2 |
| | 250<br>to<br>300 | Quenched<br>and<br>Tempered | 1—4<br>5—10<br>11—19<br>20 & Finer | .012<br>.008<br>.005<br>.003 | 325 | M2, M7 |
| | | | 25—6<br>5—3<br>2—1.5<br>1 & Finer | .3<br>.2<br>.12<br>.07 | 100 | S4, S2 |
| **Medium Carbon**<br>ASTM A27: Grades N1, N2, U-60-30, 60-30, 65-35, 70-36, 70-40<br>ASTM A148: Grades 80-40, 80-50, 90-60, 105-85, 120-95, 150-125, 175-145<br>ASTM A216: Grades WCA, WCB, WCC<br>ASTM A217: Grades WC1, WC4, WC5, WC6<br>ASTM A352: Grades LC1, LC2, LC2-1<br>ASTM A356: Grades 2, 5, 6, 8, 9, 10<br>ASTM A389: Grades C23, C24<br>ASTM A486: Classes 70, 90, 120<br>ASTM A487: Classes 1N, 2N, 4N, 6N, 8N, 9N, 10N, DN, 1Q, 2Q, 4Q, 4QA, 6Q, 7Q, 8Q, 9Q, 10Q<br>1330 4130 80B30 8640<br>1340 4140 8040 9525<br>2325 4330 8430 9530<br>2330 4340 8440 9535<br>4125 8030 8630 | 175<br>to<br>225 | Annealed,<br>Normalized,<br>or<br>Normalized<br>and<br>Tempered | 1—4<br>5—10<br>11—19<br>20 & Finer | .012<br>.008<br>.005<br>.003 | 375 | M2, M7 |
| | | | 25—6<br>5—3<br>2—1.5<br>1 & Finer | .3<br>.2<br>.12<br>.07 | 115 | S4, S2 |
| | 225<br>to<br>250 | Normalized,<br>Normalized<br>and<br>Tempered<br>or<br>Quenched<br>and<br>Tempered | 1—4<br>5—10<br>11—19<br>20 & Finer | .012<br>.008<br>.005<br>.003 | 325 | M2, M7 |
| | | | 25—6<br>5—3<br>2—1.5<br>1 & Finer | .3<br>.2<br>.12<br>.07 | 100 | S4, S2 |
| | 250<br>to<br>300 | Quenched<br>and<br>Tempered | 1—4<br>5—10<br>11—19<br>20 & Finer | .012<br>.008<br>.005<br>.003 | 300 | M2, M7 |
| | | | 25—6<br>5—3<br>2—1.5<br>1 & Finer | .3<br>.2<br>.12<br>.07 | 90 | S4, S2 |
| | 300<br>to<br>350 | Quenched<br>and<br>Tempered | 1—4<br>5—10<br>11—19<br>20 & Finer | .012<br>.008<br>.005<br>.003 | 250 | M3 |
| | | | 25—6<br>5—3<br>2—1.5<br>1 & Finer | .3<br>.2<br>.12<br>.07 | 76 | S5 |
| | 350<br>to<br>400 | Quenched<br>and<br>Tempered | 1—4<br>5—10<br>11—19<br>20 & Finer | .012<br>.008<br>.005<br>.003 | 200 | M3 |
| | | | 25—6<br>5—3<br>2—1.5<br>1 & Finer | .3<br>.2<br>.12<br>.07 | 60 | S5 |

See section 16 for Cutting Fluid Recommendations.

*Feed recommendations apply to CONVENTIONAL (axial-transverse) gear shaving.

†Feeds should be increased 100% for gears shaved by the DIAGONAL (angular-transverse) method.

| MATERIAL | HARD-NESS Bhn | CONDITION | DIAMETRAL PITCH / MODULE | FEED*† per rev. of gear in / mm | CUTTER PITCH LINE SPEED fpm / m/min | HSS TOOL MATERIAL AISI / ISO |
|---|---|---|---|---|---|---|
| **18. STAINLESS STEELS, CAST**<br>**Ferritic**<br>ASTM A217: Grades C5, C12<br>ASTM A296: Grades CB-30, CC-50, CE-30, CA6N, CA-6NM, CD4MCu<br>ASTM A297: Grade HC<br>ASTM A487: Class CA6NM<br>ASTM A608: Grade HC30 | 135 to 185 | Annealed | 1—4<br>5—10<br>11—19<br>20 & Finer | .012<br>.008<br>.005<br>.003 | 250 | M2, M7 |
| | | | 25—6<br>5—3<br>2—1.5<br>1 & Finer | .3<br>.2<br>.12<br>.07 | 76 | S4, S2 |
| **Austenitic**<br>ASTM A296: Grades CF-16F, CN-7M, CN-7MS<br>ASTM A351: Grade CN-7M | 140 to 170 | Annealed or Normalized | 1—4<br>5—10<br>11—19<br>20 & Finer | .012<br>.008<br>.005<br>.003 | 200 | M2, M7 |
| | | | 25—6<br>5—3<br>2—1.5<br>1 & Finer | .3<br>.2<br>.12<br>.07 | 60 | S4, S2 |
| **Austenitic**<br>ASTM A296: Grades CF-3, CF-8, CF-8C, CF-20<br>ASTM A351: Grades CF-3, CF-3A, CF-8, CF-8A, CF-8C<br>ASTM A451: Grades CPF3, CPF3A, CPF8, CPF8A, CPF8C, CPF8C (Ta Max.)<br>ASTM A452: Grades TP 304H, TP 347H | 135 to 185 | Annealed or Normalized | 1—4<br>5—10<br>11—19<br>20 & Finer | .012<br>.008<br>.005<br>.003 | 175 | M2, M7 |
| | | | 25—6<br>5—3<br>2—1.5<br>1 & Finer | .3<br>.2<br>.12<br>.07 | 53 | S4, S2 |
| **Austenitic**<br>ASTM A296: Grades CF-3M, CF-8M, CG-8M, CG-12, CH-20, CK-20<br>ASTM A351: Grades CF-3M, CF-3MA, CF-8M, CF-10MC, CH-8, CH-10, CH-20, CK-20, HK-30, HK-40, HT-30<br>ASTM A451: Grades CPF3M, CPF8M, CPF10MC, CPH8, CPH10, CPH20, CPK20<br>ASTM A452: Grade TP 316H | 135 to 185 | Annealed or Normalized | 1—4<br>5—10<br>11—19<br>20 & Finer | .012<br>.008<br>.005<br>.003 | 175 | M2, M7 |
| | | | 25—6<br>5—3<br>2—1.5<br>1 & Finer | .3<br>.2<br>.12<br>.07 | 53 | S4, S2 |
| **Austenitic**<br>ASTM A297: Grades HD, HE, HF, HH, HI, HK, HL, HN, HP, HT, HU<br>ASTM A608: Grades HD50, HE35, HF30, HH30, HH33, HI35, HK30, HK40, HL30, HL40, HN40, HT50, HU50 | 160 to 210 | As Cast | 1—4<br>5—10<br>11—19<br>20 & Finer | .012<br>.008<br>.005<br>.003 | 175 | M2, M7 |
| | | | 25—6<br>5—3<br>2—1.5<br>1 & Finer | .3<br>.2<br>.12<br>.07 | 53 | S4, S2 |

See section 16 for Cutting Fluid Recommendations.

*Feed recommendations apply to CONVENTIONAL (axial-transverse) gear shaving.

†Feeds should be increased 100% for gears shaved by the DIAGONAL (angular-transverse) method.

# 7.4 Gear Shaving

| MATERIAL | HARD-NESS<br>Bhn | CONDITION | DIAMETRAL PITCH / MODULE | FEED*† per rev. of gear<br>in / mm | CUTTER PITCH LINE SPEED<br>fpm / m/min | HSS TOOL MATERIAL<br>AISI / ISO |
|---|---|---|---|---|---|---|
| **18. STAINLESS STEELS, CAST (cont.)**<br>**Martensitic**<br>ASTM A217: Grade CA-15<br>ASTM A296: Grades CA-15, CA-15M, CA-40<br>ASTM A426: Grades CP7, CP9, CPCA15<br>ASTM A487: Classes CA15a, CA-15M | 135<br>to<br>175 | Annealed | 1—4<br>5—10<br>11—19<br>20 & Finer | .012<br>.008<br>.005<br>.003 | 250 | M2, M7 |
| | | | 25—6<br>5—3<br>2—1.5<br>1 & Finer | .3<br>.2<br>.12<br>.07 | 76 | S4, S2 |
| | 175<br>to<br>225 | Annealed, Normalized or Normalized and Tempered | 1—4<br>5—10<br>11—19<br>20 & Finer | .012<br>.008<br>.005<br>.003 | 225 | M2, M7 |
| | | | 25—6<br>5—3<br>2—1.5<br>1 & Finer | .3<br>.2<br>.12<br>.07 | 69 | S4, S2 |
| | 275<br>to<br>325 | Quenched and Tempered | 1—4<br>5—10<br>11—19<br>20 & Finer | .012<br>.008<br>.005<br>.003 | 200 | M2, M7 |
| | | | 25—6<br>5—3<br>2—1.5<br>1 & Finer | .3<br>.2<br>.12<br>.07 | 60 | S4, S2 |
| **19. PRECIPITATION HARDENING STAINLESS STEELS, CAST**<br>ASTM A351: Grade CD-4MCu<br>ACI Grade CB-7Cu<br>ACI Grade CD-4MCu<br>17-4 PH<br>AM-355 | 325<br>to<br>375 | Solution Treated | 1—4<br>5—10<br>11—19<br>20 & Finer | .012<br>.008<br>.005<br>.003 | 175 | M2, M7 |
| | | | 25—6<br>5—3<br>2—1.5<br>1 & Finer | .3<br>.2<br>.12<br>.07 | 53 | S4, S2 |
| **21. GRAY CAST IRONS**<br>**Ferritic**<br>ASTM A48: Class 20<br>SAE J431c: Grade G1800 | 120<br>to<br>150 | Annealed | 1—4<br>5—10<br>11—19<br>20 & Finer | .012<br>.008<br>.005<br>.003 | 325 | M2, M7 |
| | | | 25—6<br>5—3<br>2—1.5<br>1 & Finer | .3<br>.2<br>.12<br>.07 | 100 | S4, S2 |
| **Pearlitic- Ferritic**<br>ASTM A48: Class 25<br>SAE J431c: Grade G2500 | 160<br>to<br>200 | As Cast | 1—4<br>5—10<br>11—19<br>20 & Finer | .012<br>.008<br>.005<br>.003 | 300 | M2, M7 |
| | | | 25—6<br>5—3<br>2—1.5<br>1 & Finer | .3<br>.2<br>.12<br>.07 | 90 | S4, S2 |
| **Pearlitic**<br>ASTM A48: Classes 30, 35, 40<br>SAE J431c: Grade G3000 | 190<br>to<br>220 | As Cast | 1—4<br>5—10<br>11—19<br>20 & Finer | .012<br>.008<br>.005<br>.003 | 230 | M2, M7 |
| | | | 25—6<br>5—3<br>2—1.5<br>1 & Finer | .3<br>.2<br>.12<br>.07 | 70 | S4, S2 |
| **Pearlitic + Free Carbides**<br>ASTM A48: Classes 45, 50<br>SAE J431c: Grades G3500, G4000 | 220<br>to<br>260 | As Cast | 1—4<br>5—10<br>11—19<br>20 & Finer | .012<br>.008<br>.005<br>.003 | 200 | M2, M7 |
| | | | 25—6<br>5—3<br>2—1.5<br>1 & Finer | .3<br>.2<br>.12<br>.07 | 60 | S4, S2 |

See section 16 for Cutting Fluid Recommendations.

*Feed recommendations apply to CONVENTIONAL (axial-transverse) gear shaving.

†Feeds should be increased 100% for gears shaved by the DIAGONAL (angular-transverse) method.

| MATERIAL | HARDNESS Bhn | CONDITION | DIAMETRAL PITCH / MODULE | FEED*† per rev. of gear in / mm | CUTTER PITCH LINE SPEED fpm / m/min | HSS TOOL MATERIAL AISI / ISO |
|---|---|---|---|---|---|---|
| **Pearlitic or Acicular + Free Carbides**<br>ASTM A48: Classes 55, 60 | 250<br>to<br>320 | As Cast<br>or<br>Quenched<br>and<br>Tempered | 1—4<br>5—10<br>11—19<br>20 & Finer | .012<br>.008<br>.005<br>.003 | 100 | M2, M7 |
| | | | 25—6<br>5—3<br>2—1.5<br>1 & Finer | .3<br>.2<br>.12<br>.07 | 30 | S4, S2 |
| **Austenitic (NI-RESIST)**<br>ASTM A436: Types 1, 1b, 2, 2b, 3, 4, 5, 6 | 100<br>to<br>250 | As Cast | 1—4<br>5—10<br>11—19<br>20 & Finer | .012<br>.008<br>.005<br>.003 | 150 | M2, M7 |
| | | | 25—6<br>5—3<br>2—1.5<br>1 & Finer | .3<br>.2<br>.12<br>.07 | 46 | S4, S2 |
| **23. DUCTILE CAST IRONS**<br>**Ferritic**<br>ASTM A536: Grades 60-40-18, 65-45-12<br>SAE J434c: Grades D4018, D4512 | 140<br>to<br>190 | Annealed | 1—4<br>5—10<br>11—19<br>20 & Finer | .012<br>.008<br>.005<br>.003 | 400 | M2, M7 |
| | | | 25—6<br>5—3<br>2—1.5<br>1 & Finer | .3<br>.2<br>.12<br>.07 | 120 | S4, S2 |
| **Ferritic- Pearlitic**<br>ASTM A536: Grade 80-55-06<br>SAE J434c: Grade D5506 | 190<br>to<br>225 | As Cast | 1—4<br>5—10<br>11—19<br>20 & Finer | .012<br>.008<br>.005<br>.003 | 350 | M2, M7 |
| | | | 25—6<br>5—3<br>2—1.5<br>1 & Finer | .3<br>.2<br>.12<br>.07 | 105 | S4, S2 |
| | 225<br>to<br>260 | As Cast | 1—4<br>5—10<br>11—19<br>20 & Finer | .012<br>.008<br>.005<br>.003 | 300 | M2, M7 |
| | | | 25—6<br>5—3<br>2—1.5<br>1 & Finer | .3<br>.2<br>.12<br>.07 | 90 | S4, S2 |
| **Pearlitic- Martensitic**<br>ASTM A536: Grade 100-70-03<br>SAE J434c: Grade D7003 | 240<br>to<br>300 | Normalized<br>and<br>Tempered | 1—4<br>5—10<br>11—19<br>20 & Finer | .012<br>.008<br>.005<br>.003 | 250 | M2, M7 |
| | | | 25—6<br>5—3<br>2—1.5<br>1 & Finer | .3<br>.2<br>.12<br>.07 | 76 | S4, S2 |
| **Martensitic**<br>ASTM A536: Grade 120-90-02<br>SAE J434c: Grade DQ&T | 270<br>to<br>330 | Quenched<br>and<br>Tempered | 1—4<br>5—10<br>11—19<br>20 & Finer | .012<br>.008<br>.005<br>.003 | 175 | M2, M7 |
| | | | 25—6<br>5—3<br>2—1.5<br>1 & Finer | .3<br>.2<br>.12<br>.07 | 53 | S4, S2 |
| **Austenitic (NI-RESIST Ductile)**<br>ASTM A439: Types D-2, D-2C, D-3A, D-5<br>ASTM A571: Type D-2M | 120<br>to<br>200 | Annealed | 1—4<br>5—10<br>11—19<br>20 & Finer | .012<br>.008<br>.005<br>.003 | 125 | M2, M7 |
| | | | 25—6<br>5—3<br>2—1.5<br>1 & Finer | .3<br>.2<br>.12<br>.07 | 38 | S4, S2 |

See section 16 for Cutting Fluid Recommendations.
*Feed recommendations apply to CONVENTIONAL (axial-transverse) gear shaving.
†Feeds should be increased 100% for gears shaved by the DIAGONAL (angular-transverse) method.

# 7.4 Gear Shaving

| MATERIAL | HARD-NESS<br>Bhn | CONDITION | DIAMETRAL PITCH<br>MODULE | FEED*† per rev. of gear<br>in<br>mm | CUTTER PITCH LINE SPEED<br>fpm<br>m/min | HSS TOOL MATERIAL<br>AISI<br>ISO |
|---|---|---|---|---|---|---|
| **23. DUCTILE CAST IRONS (cont.)**<br>**Austenitic (NI-RESIST Ductile)**<br>ASTM A439: Types D-2B, D-3, D-4, D-5B | 140 to 275 | Annealed | 1—4<br>5—10<br>11—19<br>20 & Finer | .012<br>.008<br>.005<br>.003 | 100 | M2, M7 |
| | | | 25—6<br>5—3<br>2—1.5<br>1 & Finer | .3<br>.2<br>.12<br>.07 | 30 | S4, S2 |
| **24. MALLEABLE CAST IRONS**<br>**Ferritic**<br>ASTM A47: Grades 32510, 35018<br>ASTM A602: Grade M3210<br>SAE J158: Grade M3210 | 110 to 160 | Malleablized | 1—4<br>5—10<br>11—19<br>20 & Finer | .012<br>.008<br>.005<br>.003 | 500 | M2, M7 |
| | | | 25—6<br>5—3<br>2—1.5<br>1 & Finer | .3<br>.2<br>.12<br>.07 | 150 | S4, S2 |
| **Pearlitic**<br>ASTM A220: Grades 40010, 45006, 45008, 50005<br>ASTM A602: Grade M4504, M5003<br>SAE J158: Grades M4504, M5003 | 160 to 200 | Malleablized and Heat Treated | 1—4<br>5—10<br>11—19<br>20 & Finer | .012<br>.008<br>.005<br>.003 | 400 | M2, M7 |
| | | | 25—6<br>5—3<br>2—1.5<br>1 & Finer | .3<br>.2<br>.12<br>.07 | 120 | S4, S2 |
| | 200 to 240 | Malleablized and Heat Treated | 1—4<br>5—10<br>11—19<br>20 & Finer | .012<br>.008<br>.005<br>.003 | 320 | M2, M7 |
| | | | 25—6<br>5—3<br>2—1.5<br>1 & Finer | .3<br>.2<br>.12<br>.07 | 100 | S4, S2 |
| **Tempered Martensite**<br>ASTM A220: Grade 60004<br>ASTM A602: Grade M5503<br>SAE J158: Grade M5503 | 200 to 255 | Malleablized and Heat Treated | 1—4<br>5—10<br>11—19<br>20 & Finer | .012<br>.008<br>.005<br>.003 | 320 | M2, M7 |
| | | | 25—6<br>5—3<br>2—1.5<br>1 & Finer | .3<br>.2<br>.12<br>.07 | 100 | S4, S2 |
| **Tempered Martensite**<br>ASTM A220: Grade 70003<br>ASTM A602: Grade M7002<br>SAE J158: Grade M7002 | 220 to 260 | Malleablized and Heat Treated | 1—4<br>5—10<br>11—19<br>20 & Finer | .012<br>.008<br>.005<br>.003 | 300 | M2, M7 |
| | | | 25—6<br>5—3<br>2—1.5<br>1 & Finer | .3<br>.2<br>.12<br>.07 | 90 | S4, S2 |
| **Tempered Martensite**<br>ASTM A220: Grade 80002<br>ASTM A602: Grade M8501<br>SAE J158: Grade M8501 | 240 to 280 | Malleablized and Heat Treated | 1—4<br>5—10<br>11—19<br>20 & Finer | .012<br>.008<br>.005<br>.003 | 250 | M2, M7 |
| | | | 25—6<br>5—3<br>2—1.5<br>1 & Finer | .3<br>.2<br>.12<br>.07 | 76 | S4, S2 |
| **Tempered Martensite**<br>ASTM A220: Grade 90001<br>ASTM A602: Grade M8501<br>SAE J158: Grade M8501 | 250 to 320 | Malleablized and Heat Treated | 1—4<br>5—10<br>11—19<br>20 & Finer | .012<br>.008<br>.005<br>.003 | 200 | M2, M7 |
| | | | 25—6<br>5—3<br>2—1.5<br>1 & Finer | .3<br>.2<br>.12<br>.07 | 60 | S4, S2 |

See section 16 for Cutting Fluid Recommendations.

*Feed recommendations apply to CONVENTIONAL (axial-transverse) gear shaving.

†Feeds should be increased 100% for gears shaved by the DIAGONAL (angular-transverse) method.

| MATERIAL | HARD-NESS<br>Bhn | CONDITION | DIAMETRAL PITCH<br>MODULE | FEED*† per rev. of gear<br>in<br>mm | CUTTER PITCH LINE SPEED<br>fpm<br>m/min | HSS TOOL MATERIAL<br>AISI<br>ISO |
|---|---|---|---|---|---|---|
| **28. ALUMINUM ALLOYS, WROUGHT**<br>EC 2218 5252 6253<br>1060 2219 5254 6262<br>1100 2618 5454 6463<br>1145 3003 5456 6951<br>1175 3004 5457 7001<br>1235 3005 5652 7004<br>2011 4032 5657 7005<br>2014 5005 6053 7039<br>2017 5050 6061 7049<br>2018 5052 6063 7050<br>2021 5056 6066 7075<br>2024 5083 6070 7079<br>2025 5086 6101 7175<br>2117 5154 6151 7178 | 30<br>to<br>80<br>500kg | Cold<br>Drawn | 1—4<br>5—10<br>11—19<br>20 & Finer | .012<br>.008<br>.005<br>.003 | 1200 | M2, M7 |
| | | | 25—6<br>5—3<br>2—1.5<br>1 & Finer | .3<br>.2<br>.12<br>.07 | 365 | S4, S2 |
| | 75<br>to<br>150<br>500kg | Solution<br>Treated<br>and<br>Aged | 1—4<br>5—10<br>11—19<br>20 & Finer | .012<br>.008<br>.005<br>.003 | 1100 | M2, M7 |
| | | | 25—6<br>5—3<br>2—1.5<br>1 & Finer | .3<br>.2<br>.12<br>.07 | 335 | S4, S2 |
| **29. ALUMINUM ALLOYS, CAST**<br>**Sand and Permanent Mold**<br>A140 319.0 357.0 A712.0<br>201.0 328.0 359.0 D712.0<br>208.0 A332.0 B443.0 713.0<br>213.0 F332.0 514.0 771.0<br>222.0 333.0 A514.0 850.0<br>224.0 354.0 B514.0 A850.0<br>242.0 355.0 520.0 B850.0<br>295.0 C355.0 535.0<br>B295.0 356.0 705.0<br>308.0 A356.0 707.0<br>Hiduminium RR-350 | 40<br>to<br>100<br>500kg | As Cast | 1—4<br>5—10<br>11—19<br>20 & Finer | .012<br>.008<br>.005<br>.003 | 1100 | M2, M7 |
| | | | 25—6<br>5—3<br>2—1.5<br>1 & Finer | .3<br>.2<br>.12<br>.07 | 335 | S4, S2 |
| | 70<br>to<br>125<br>500kg | Solution<br>Treated<br>and<br>Aged | 1—4<br>5—10<br>11—19<br>20 & Finer | .012<br>.008<br>.005<br>.003 | 1000 | M2, M7 |
| | | | 25—6<br>5—3<br>2—1.5<br>1 & Finer | .3<br>.2<br>.12<br>.07 | 305 | S4, S2 |
| **34. COPPER ALLOYS, WROUGHT**<br>145 332 360 482<br>147 335 365 485<br>173 340 366 544<br>187 342 367 623<br>191 349 368 624<br>314 350 370 638<br>316 353 377 642<br>330 356 385 782 | $10\ R_B$<br>to<br>$70\ R_B$ | Annealed | 1—4<br>5—10<br>11—19<br>20 & Finer | .012<br>.008<br>.005<br>.003 | 1000 | M2, M7 |
| | | | 25—6<br>5—3<br>2—1.5<br>1 & Finer | .3<br>.2<br>.12<br>.07 | 305 | S4, S2 |
| | $60\ R_B$<br>to<br>$100\ R_B$ | Cold<br>Drawn | 1—4<br>5—10<br>11—19<br>20 & Finer | .012<br>.008<br>.005<br>.003 | 1100 | M2, M7 |
| | | | 25—6<br>5—3<br>2—1.5<br>1 & Finer | .3<br>.2<br>.12<br>.07 | 335 | S4, S2 |
| 190 425 466 667<br>226 435 467 675<br>230 442 613 687<br>240 443 618 694<br>260 444 630 770<br>268 445 632<br>270 464 651<br>280 465 655 | $10\ R_B$<br>to<br>$70\ R_B$ | Annealed | 1—4<br>5—10<br>11—19<br>20 & Finer | .012<br>.008<br>.005<br>.003 | 600 | M2, M7 |
| | | | 25—6<br>5—3<br>2—1.5<br>1 & Finer | .3<br>.2<br>.12<br>.07 | 185 | S4, S2 |
| | $60\ R_B$<br>to<br>$100\ R_B$ | Cold<br>Drawn | 1—4<br>5—10<br>11—19<br>20 & Finer | .012<br>.008<br>.005<br>.003 | 700 | M2, M7 |
| | | | 25—6<br>5—3<br>2—1.5<br>1 & Finer | .3<br>.2<br>.12<br>.07 | 215 | S4, S2 |

See section 16 for Cutting Fluid Recommendations.
*Feed recommendations apply to CONVENTIONAL (axial-transverse) gear shaving.
†Feeds should be increased 100% for gears shaved by the DIAGONAL (angular-transverse) method.

# 7.4 Gear Shaving

| MATERIAL | HARD-NESS<br>Bhn | CONDITION | DIAMETRAL PITCH<br>MODULE | FEED*† per rev. of gear<br>in<br>mm | CUTTER PITCH LINE SPEED<br>fpm<br>m/min | HSS TOOL MATERIAL<br>AISI<br>ISO |
|---|---|---|---|---|---|---|
| **34. COPPER ALLOYS, WROUGHT (cont.)**<br>101 125 185 614<br>102 127 189 619<br>104 128 192 625<br>105 129 194 674<br>107 130 195 688<br>109 142 210 706<br>110 143 220 710<br>111 150 411 715<br>113 155 413 725<br>114 162 505 745<br>115 165 510 752<br>116 170 511 754<br>119 172 521 757<br>120 175 524<br>121 182 608<br>122 184 610 | 10 $R_B$<br>to<br>70 $R_B$ | Annealed | 1—4<br>5—10<br>11—19<br>20 & Finer | .012<br>.008<br>.005<br>.003 | 350 | M2, M7 |
| | | | 25—6<br>5—3<br>2—1.5<br>1 & Finer | .3<br>.2<br>.12<br>.07 | 105 | S4, S2 |
| | 60 $R_B$<br>to<br>100 $R_B$ | Cold Drawn | 1—4<br>5—10<br>11—19<br>20 & Finer | .012<br>.008<br>.005<br>.003 | 400 | M2, M7 |
| | | | 25—6<br>5—3<br>2—1.5<br>1 & Finer | .3<br>.2<br>.12<br>.07 | 120 | S4, S2 |
| | | | | | | |
| **35. COPPER ALLOYS, CAST**<br>834 855 934 953<br>836 857 935 954<br>838 858 937 956<br>842 864 938 973<br>844 867 939 974<br>848 879 943 976<br>852 928 944 978<br>854 932 945 | 40<br>to<br>150<br>500kg | As Cast | 1—4<br>5—10<br>11—19<br>20 & Finer | .012<br>.008<br>.005<br>.003 | 900 | M2, M7 |
| | | | 25—6<br>5—3<br>2—1.5<br>1 & Finer | .3<br>.2<br>.12<br>.07 | 275 | S4, S2 |
| | | | | | | |
| 817 868 905 947<br>821 872 915 948<br>833 874 922 952<br>853 875 923 955<br>861 876 925 957<br>862 878 926 958<br>865 903 927 | 40<br>to<br>175<br>500kg | As Cast | 1—4<br>5—10<br>11—19<br>20 & Finer | .012<br>.008<br>.005<br>.003 | 600 | M2, M7 |
| | | | 25—6<br>5—3<br>2—1.5<br>1 & Finer | .3<br>.2<br>.12<br>.07 | 185 | S4, S2 |
| 801 815 828 916<br>803 818 863 917<br>805 820 902 962<br>807 822 907 963<br>809 824 909 964<br>811 825 910 966<br>813 826 911 993<br>814 827 913 | 40<br>to<br>200<br>500kg | As Cast | 1—4<br>5—10<br>11—19<br>20 & Finer | .012<br>.008<br>.005<br>.003 | 350 | M2, M7 |
| | | | 25—6<br>5—3<br>2—1.5<br>1 & Finer | .3<br>.2<br>.12<br>.07 | 105 | S4, S2 |
| | | | | | | |

See section 16 for Cutting Fluid Recommendations.
*Feed recommendations apply to CONVENTIONAL (axial-transverse) gear shaving.
†Feeds should be increased 100% for gears shaved by the DIAGONAL (angular-transverse) method.

| MATERIAL | HARDNESS Bhn | CONDITION | DIAMETRAL PITCH / MODULE | WHEEL IDENTIFICATION ANSI / ISO | INFEED in/pass / mm/pass | WHEEL SPEED fpm / m/s | TRAVERSE RATE fpm / m/min |
|---|---|---|---|---|---|---|---|
| **1. FREE MACHINING CARBON STEELS, WROUGHT**<br>**Low Carbon Resulfurized**<br>1108 1115 1118 1212<br>1109 1116 1119 1213<br>1110 1117 1211 1215<br>**Medium Carbon Resulfurized**<br>1132 1140 1145<br>1137 1141 1146<br>1139 1144 1151<br>**Low Carbon Leaded**<br>10L18 12L13 12L15<br>11L17 12L14<br>**Medium Carbon Leaded**<br>10L45 11L37 11L44<br>10L50 11L41 | Over 48 $R_C$ | Carburized and/or Quenched and Tempered | 1 to 4<br>5 to 8<br>10 to 16<br>16 & finer | A60KV<br>A60KV<br>A80KV<br>A100LV | .0015<br>.0012<br>.0005<br>.0005 | 5500 to 6500 | 20-30 |
| | | | 25 to 6<br>5.5 to 3<br>2.5 to 2<br>1.5 & finer | A60KV<br>A60KV<br>A80KV<br>A100LV | .038<br>.030<br>.013<br>.013 | 28 to 33 | 6-9 |
| **2. CARBON STEELS, WROUGHT**<br>**Low Carbon**<br>1005 1012 1019 1026<br>1006 1013 1020 1029<br>1008 1015 1021 1513<br>1009 1016 1022 1518<br>1010 1017 1023 1522<br>1011 1018 1025<br>**Medium Carbon**<br>1030 1042 1053 1541<br>1033 1043 1055 1547<br>1035 1044 1524 1548<br>1037 1045 1525 1551<br>1038 1046 1526 1552<br>1039 1049 1527<br>1040 1050 1536<br>**High Carbon**<br>1060 1074 1085 1566<br>1064 1075 1086 1572<br>1065 1078 1090<br>1069 1080 1095<br>1070 1084 1561 | Over 48 $R_C$ | Carburized and/or Quenched and Tempered | 1 to 4<br>5 to 8<br>10 to 16<br>16 & finer | A60KV<br>A60KV<br>A80KV<br>A100LV | .0015<br>.0012<br>.0005<br>.0005 | 5500 to 6500 | 20-30 |
| | | | 25 to 6<br>5.5 to 3<br>2.5 to 2<br>1.5 & finer | A60KV<br>A60KV<br>A80KV<br>A100LV | .038<br>.030<br>.013<br>.013 | 28 to 33 | 6-9 |

*See section 16 for Cutting Fluid Recommendations.
See section 20 for additional information.

# 7.5 Gear Grinding, Form

| MATERIAL | HARDNESS Bhn | CONDITION | DIAMETRAL PITCH / MODULE | WHEEL IDENTIFICATION ANSI / ISO | INFEED in/pass / mm/pass | WHEEL SPEED fpm / m/s | TRAVERSE RATE fpm / m/min |
|---|---|---|---|---|---|---|---|
| **4. FREE MACHINING ALLOY STEELS, WROUGHT** **Medium Carbon Resulfurized** 4140 4142Te 4147Te 4140Se 4145Se 4150 **Medium and High Carbon Leaded** 41L30 41L47 51L32 86L40 41L40 41L50 52L100 41L45 43L40 86L20 | Over 48 $R_C$ | Carburized and/or Quenched and Tempered | 1 to 4<br>5 to 8<br>10 to 16<br>16 & finer | A60KV<br>A60KV<br>A80KV<br>A100LV | .0015<br>.0012<br>.0005<br>.0005 | 5500 to 6500 | 20-30 |
| | | | 25 to 6<br>5.5 to 3<br>2.5 to 2<br>1.5 & finer | A60KV<br>A60KV<br>A80KV<br>A100LV | .038<br>.030<br>.013<br>.013 | 28 to 33 | 6-9 |
| **5. ALLOY STEELS, WROUGHT** **Low Carbon** 4012 4615 4817 8617 4023 4617 4820 8620 4024 4620 5015 8622 4118 4621 5115 8822 4320 4718 5120 9310 4419 4720 6118 94B15 4422 4815 8115 94B17 **Medium Carbon** 1330 4145 5132 8640 1335 4147 5135 8642 1340 4150 5140 8645 1345 4161 5145 86B45 4027 4340 5147 8650 4028 4427 5150 8655 4032 4626 5155 8660 4037 50B40 5160 8740 4042 50B44 51B60 8742 4047 5046 6150 9254 4130 50B46 81B45 9255 4135 50B50 8625 9260 4137 5060 8627 94B30 4140 50B60 8630 4142 5130 8637 **High Carbon** 50100 51100 52100 M-50 | Over 48 $R_C$ | Carburized and/or Quenched and Tempered | 1 to 4<br>5 to 8<br>10 to 16<br>16 & finer | A60KV<br>A60KV<br>A80KV<br>A100LV | .0015<br>.0012<br>.0005<br>.0005 | 5500 to 6500 | 20-30 |
| | | | 25 to 6<br>5.5 to 3<br>2.5 to 2<br>1.5 & finer | A60KV<br>A60KV<br>A80KV<br>A100LV | .038<br>.030<br>.013<br>.013 | 28 to 33 | 6-9 |

*See section 16 for Cutting Fluid Recommendations.
See section 20 for additional information.

| MATERIAL | HARDNESS Bhn | CONDITION | DIAMETRAL PITCH / MODULE | WHEEL IDENTIFICATION ANSI / ISO | INFEED in/pass / mm/pass | WHEEL SPEED fpm / m/s | TRAVERSE RATE fpm / m/min |
|---|---|---|---|---|---|---|---|
| **6. HIGH STRENGTH STEELS, WROUGHT**<br>300M<br>4330V<br>4340<br>4340Si<br>98BV40<br>D6ac<br>H11<br>H13<br>HP 9-4-20<br>HP 9-4-25<br>HP 9-4-30<br>HP 9-4-45 | Over 48 $R_C$ | Quenched and Tempered | 1 to 4<br>5 to 8<br>10 to 16<br>16 & finer | A60KV<br>A60KV<br>A80KV<br>A100LV | .0015<br>.0012<br>.0005<br>.0005 | 5500 to 6500 | 20-30 |
| | | | 25 to 6<br>5.5 to 3<br>2.5 to 2<br>1.5 & finer | A60KV<br>A60KV<br>A80KV<br>A100LV | .038<br>.030<br>.013<br>.013 | 28 to 33 | 6-9 |
| **7. MARAGING STEELS, WROUGHT**<br>ASTM A538: Grades A, B, C<br>120 Grade<br>180 Grade<br>200 Grade<br>250 Grade<br>300 Grade<br>350 Grade<br>HY230 | Over 48 $R_C$ | Maraged | 1 to 4<br>5 to 8<br>10 to 16<br>16 & finer | A60KV<br>A60KV<br>A80KV<br>A100LV | .0015<br>.0012<br>.0005<br>.0005 | 5500 to 6500 | 20-30 |
| | | | 25 to 6<br>5.5 to 3<br>2.5 to 2<br>1.5 & finer | A60KV<br>A60KV<br>A80KV<br>A100LV | .038<br>.030<br>.013<br>.013 | 28 to 33 | 6-9 |
| **8. TOOL STEELS, WROUGHT**<br>**Group I**<br>A2 H13 L6 P20<br>A3 H14 L7 P21<br>A4 H19 O1 S1<br>A6 H21 O2 S2<br>A8 H22 O6 S5<br>A9 H23 O7 S6<br>A10 H24 P2 S7<br>H10 H25 P4 W1<br>H11 H26 P5 W2<br>H12 L2 P6 W5<br>SAE J438b: Types W108, W109, W110, W112, W209, W210, W310 | Over 48 $R_C$ | Quenched and Tempered | 1 to 4<br>5 to 8<br>10 to 16<br>16 & finer | A80JV<br>A80KV<br>A100KV<br>A120KV | .0015<br>.0012<br>.0005<br>.0005 | 5500 to 6500 | 20-30 |
| | | | 25 to 6<br>5.5 to 3<br>2.5 to 2<br>1.5 & finer | A80JV<br>A80KV<br>A100KV<br>A120KV | .038<br>.030<br>.013<br>.013 | 28 to 33 | 6-9 |
| **Group II**<br>D2 F2 M10 T4<br>D3 H42 M30 T8<br>D4 M1 M42<br>D5 M2 T1<br>F1 M3-1 T2 | Over 48 $R_C$ | Quenched and Tempered | 1 to 4<br>5 to 8<br>10 to 16<br>16 & finer | A80JV<br>A80KV<br>A100KV<br>A120KV | .0015<br>.0012<br>.0005<br>.0005 | 5500 to 6500 | 20-30 |
| | | | 25 to 6<br>5.5 to 3<br>2.5 to 2<br>1.5 & finer | A80JV<br>A80KV<br>A100KV<br>A120KV | .038<br>.030<br>.013<br>.013 | 28 to 33 | 6-9 |

*See section 16 for Cutting Fluid Recommendations.
See section 20 for additional information.

# 7.5 Gear Grinding, Form

| MATERIAL | HARDNESS Bhn | CONDITION | DIAMETRAL PITCH / MODULE | WHEEL IDENTIFICATION ANSI / ISO | INFEED in/pass / mm/pass | WHEEL SPEED fpm / m/s | TRAVERSE RATE fpm / m/min |
|---|---|---|---|---|---|---|---|
| **8. TOOL STEELS, WROUGHT (cont.)**<br>**Group III**<br>A7 M7 M41 M47<br>D7 M33 M43 T5<br>M3-2 M34 M44 T6<br>M6 M36 M46 | Over 48 $R_C$ | Quenched and Tempered | 1 to 4<br>5 to 8<br>10 to 16<br>16 & finer | A80JV<br>A80KV<br>A100KV<br>A120KV | .0015<br>.0012<br>.0005<br>.0005 | 5500 to 6500 | 20-30 |
| | | | 25 to 6<br>5.5 to 3<br>2.5 to 2<br>1.5 & finer | A80JV<br>A80KV<br>A100KV<br>A120KV | .038<br>.030<br>.013<br>.013 | 28 to 33 | 6-9 |
| **Group IV**<br>M4<br>T15 | Over 48 $R_C$ | Quenched and Tempered | 1 to 4<br>5 to 8<br>10 to 16<br>16 & finer | A80JV<br>A80KV<br>A100KV<br>A120KV | .0015<br>.0012<br>.0005<br>.0005 | 5500 to 6500 | 20-30 |
| | | | 25 to 6<br>5.5 to 3<br>2.5 to 2<br>1.5 & finer | A80JV<br>A80KV<br>A100KV<br>A120KV | .038<br>.030<br>.013<br>.013 | 28 to 33 | 6-9 |
| **17. TOOL STEELS, CAST**<br>ASTM A597: Grades CA-2, CD-2, CD-5, CH-12, CH-13, CO-1, CS-5 | Over 48 $R_C$ | Quenched and Tempered | 1 to 4<br>5 to 8<br>10 to 16<br>16 & finer | A80JV<br>A80KV<br>A100KV<br>A120KV | .0015<br>.0012<br>.0005<br>.0005 | 5500 to 6500 | 20-30 |
| | | | 25 to 6<br>5.5 to 3<br>2.5 to 2<br>1.5 & finer | A80JV<br>A80KV<br>A100KV<br>A120KV | .038<br>.030<br>.013<br>.013 | 28 to 33 | 6-9 |

*See section 16 for Cutting Fluid Recommendations.
See section 20 for additional information.

SECTION **8**

# GRINDING OPERATIONS


8.1 Surface Grinding—Horizontal Spindle, Reciprocating Table .......... 8–3
8.2 Surface Grinding—Horizontal Spindle, Reciprocating Table—
Cubic Boron Nitride Wheels .......... 8–29
8.3 Surface Grinding—Horizontal Spindle, Reciprocating Table—
Diamond Wheels .......... 8–39
8.4 Surface Grinding—Vertical Spindle, Rotary Table .......... 8–43
8.5 Cylindrical Grinding .......... 8–61
8.6 Cylindrical Grinding—Cubic Boron Nitride Wheels .......... 8–93
8.7 Cylindrical Grinding—Diamond Wheels .......... 8–97
8.8 Internal Grinding .......... 8–101
8.9 Internal Internal Grinding—Cubic Boron Nitride Wheels .......... 8–133
8.10 Internal Grinding—Diamond Wheels .......... 8–139
8.11 Centerless Grinding .......... 8–143
8.12 Centerless Grinding—Work Traverse Rates .......... 8–175
8.13 Abrasive Belt Grinding .......... 8–177
8.14 Thread Grinding .......... 8–179

| MATERIAL | HARD-NESS<br>Bhn | CONDITION | WHEEL SPEED<br>fpm<br>m/s | TABLE SPEED<br>fpm<br>m/min | DOWNFEED<br>in/pass<br>mm/pass | CROSSFEED<br>in/pass<br>mm/pass | WHEEL* IDENTIFI-CATION<br>ANSI<br>ISO |
|---|---|---|---|---|---|---|---|
| **1. FREE MACHINING CARBON STEELS, WROUGHT**<br>**Low Carbon Resulfurized**<br>1108 1115 1118 1212<br>1109 1116 1119 1213<br>1110 1117 1211 1215<br>**Medium Carbon Resulfurized**<br>1132 1140 1145<br>1137 1141 1146<br>1139 1144 1151<br>**Low Carbon Leaded**<br>10L18 12L13 12L15<br>11L17 12L14<br>**Medium Carbon Leaded**<br>10L45 11L37 11L44<br>10L50 11L41 | 50 $R_C$ max. | Hot Rolled, Normalized, Annealed, Cold Drawn or Quenched and Tempered | 5500 to 6500 | 50 to 100 | Rough: .003<br>Finish .001 max. | .050-.500<br>(Max: 1/4 of Wheel Width) | A46JV |
| | | | 28 to 33 | 15 to 30 | Rough: .075<br>Finish .025 max. | 1.25-12.5<br>(Max: 1/4 of Wheel Width) | A46JV |
| | Over 50 $R_C$ | Carburized and/or Quenched and Tempered | 5500 to 6500 | 50 to 100 | Rough: .002<br>Finish .0005 max. | .025-.250<br>(Max: 1/10 of Wheel Width) | A46IV |
| | | | 28 to 33 | 15 to 30 | Rough: .050<br>Finish .013 max. | .65-6.5<br>(Max: 1/10 of Wheel Width) | A46IV |
| **2. CARBON STEELS, WROUGHT**<br>**Low Carbon**<br>1005 1012 1019 1026<br>1006 1013 1020 1029<br>1008 1015 1021 1513<br>1009 1016 1022 1518<br>1010 1017 1023 1522<br>1011 1018 1025<br>**Medium Carbon**<br>1030 1042 1053 1541<br>1033 1043 1055 1547<br>1035 1044 1524 1548<br>1037 1045 1525 1551<br>1038 1046 1526 1552<br>1039 1049 1527<br>1040 1050 1536<br>**High Carbon**<br>1060 1074 1085 1566<br>1064 1075 1086 1572<br>1065 1078 1090<br>1069 1080 1095<br>1070 1084 1561 | 50 $R_C$ max. | Hot Rolled, Normalized, Annealed, Cold Drawn or Quenched and Tempered | 5500 to 6500 | 50 to 100 | Rough: .003<br>Finish .001 max. | .050-.500<br>(Max: 1/4 of Wheel Width) | A46JV |
| | | | 28 to 33 | 15 to 30 | Rough: .075<br>Finish .025 max. | 1.25-12.5<br>(Max: 1/4 of Wheel Width) | A46JV |
| | Over 50 $R_C$ | Carburized and/or Quenched and Tempered | 5500 to 6500 | 50 to 100 | Rough: .002<br>Finish .0005 max. | .025-.250<br>(Max: 1/10 of Wheel Width) | A46IV |
| | | | 28 to 33 | 15 to 30 | Rough: .050<br>Finish .013 max. | .65-6.5<br>(Max: 1/10 of Wheel Width) | A46IV |
| **3. CARBON AND FERRITIC ALLOY STEELS (HIGH TEMPERATURE SERVICE)**<br>ASTM A369: Grades FPA, FPB, FP1, FP2, FP3b, FP5, FP7, FP9, FP11, FP12, FP21, FP22 | 150 to 200 | As Forged, Annealed or Normalized and Tempered | 5500 to 6500 | 50 to 100 | Rough: .003<br>Finish .001 max. | .050-.500<br>(Max: 1/4 of Wheel Width) | A46JV |
| | | | 28 to 33 | 15 to 30 | Rough: .075<br>Finish .025 max. | 1.25-12.5<br>(Max: 1/4 of Wheel Width) | A46JV |

See section 16 for Cutting Fluid Recommendations.
*Wheel recommendations are for wet grinding. For DRY grinding—use a softer grade wheel. Also see section 20.2, Grinding Guidelines.

# 8.1 Surface Grinding-Horizontal Spindle, Reciprocating Table

| MATERIAL | HARD-NESS<br>Bhn | CONDITION | WHEEL SPEED<br>fpm<br>m/s | TABLE SPEED<br>fpm<br>m/min | DOWNFEED<br>in/pass<br>mm/pass | CROSSFEED<br>in/pass<br>mm/pass | WHEEL* IDENTIFI-CATION<br>ANSI<br>ISO |
|---|---|---|---|---|---|---|---|
| **4. FREE MACHINING ALLOY STEELS, WROUGHT**<br>**Medium Carbon Resulfurized**<br>4140 4142Te 4147Te<br>4140Se 4145Se 4150<br>**Medium and High Carbon Leaded**<br>41L30 41L47 51L32 86L40<br>41L40 41L50 52L100<br>41L45 43L40 86L20 | 50 $R_C$ max. | Hot Rolled, Normalized, Annealed, Cold Drawn or Quenched and Tempered | 5500 to 6500 | 50 to 100 | Rough: .003 Finish .001 max. | .050-.500 (Max: 1/4 of Wheel Width) | A46JV |
| | | | 28 to 33 | 15 to 30 | Rough: .075 Finish .025 max. | 1.25-12.5 (Max: 1/4 of Wheel Width) | A46JV |
| | Over 50 $R_C$ | Carburized and/or Quenched and Tempered | 5500 to 6500 | 50 to 100 | Rough: .002 Finish .0005 max. | .025-.250 (Max: 1/10 of Wheel Width) | A46IV |
| | | | 28 to 33 | 15 to 30 | Rough: .050 Finish .013 max. | .65-6.5 (Max: 1/10 of Wheel Width) | A46IV |
| **5. ALLOY STEELS, WROUGHT**<br>**Low Carbon**<br>4012 4615 4817 8617<br>4023 4617 4820 8620<br>4024 4620 5015 8622<br>4118 4621 5115 8822<br>4320 4718 5120 9310<br>4419 4720 6118 94B15<br>4422 4815 8115 94B17<br>**Medium Carbon**<br>1330 4145 5132 8640<br>1335 4147 5135 8642<br>1340 4150 5140 8645<br>1345 4161 5145 86B45<br>4027 4340 5147 8650<br>4028 4427 5150 8655<br>4032 4626 5155 8660<br>4037 50B40 5160 8740<br>4042 50B44 51B60 8742<br>4047 5046 6150 9254<br>4130 50B46 81B45 9255<br>4135 50B50 8625 9260<br>4137 5060 8627 94B30<br>4140 50B60 8630<br>4142 5130 8637<br>**High Carbon**<br>50100 51100 52100 M-50 | 50 $R_C$ max. | Hot Rolled, Normalized, Annealed, Cold Drawn or Quenched and Tempered | 5500 to 6500 | 50 to 100 | Rough: .003 Finish .0005 max. | .050-.500 (Max: 1/4 of Wheel Width) | A46JV |
| | | | 28 to 33 | 15 to 30 | Rough: .075 Finish .013 max. | 1.25-12.5 (Max: 1/4 of Wheel Width) | A46JV |
| | Over 50 $R_C$ | Carburized and/or Quenched and Tempered | 3000 to 4000 | 50 to 100 | Rough: .002 Finish .0005 max. | .025-.250 (Max: 1/10 of Wheel Width) | A46IV |
| | | | 15 to 20 | 15 to 30 | Rough: .050 Finish .013 max. | .65-6.5 (Max: 1/10 of Wheel Width) | A46IV |
| | | | | | | | |
| | | | | | | | |
| | | | | | | | |
| | | | | | | | |
| | | | | | | | |
| | | | | | | | |
| **6. HIGH STRENGTH STEELS, WROUGHT**<br>300M<br>4330V<br>4340<br>4340Si<br>98BV40<br>D6ac<br>H11<br>H13<br>HP 9-4-20<br>HP 9-4-25<br>HP 9-4-30<br>HP 9-4-45 | 50 $R_C$ max. | Annealed, Normalized or Quenched and Tempered | 4000 to 6000 | 50 to 100 | Rough: .002 Finish .0005 max. | .050-.250 (Max: 1/4 of Wheel Width) | A46JV |
| | | | 20 to 30 | 15 to 30 | Rough: .050 Finish .013 max. | 1.25-6.5 (Max: 1/4 of Wheel Width) | A46JV |
| | Over 50 $R_C$ | Quenched and Tempered | 3000 to 4000 | 50 to 100 | Rough: .002 Finish .0005 max. | .025-.250 (Max: 1/10 of Wheel Width) | A60HV |
| | | | 15 to 20 | 15 to 30 | Rough: .050 Finish .013 max. | .65-6.5 (Max: 1/10 of Wheel Width) | A60HV |
| | | | | | | | |
| | | | | | | | |

See section 16 for Cutting Fluid Recommendations.
*Wheel recommendations are for wet grinding. For DRY grinding—use a softer grade wheel. Also see section 20.2, Grinding Guidelines.

| MATERIAL | HARD-NESS Bhn | CONDITION | WHEEL SPEED fpm / m/s | TABLE SPEED fpm / m/min | DOWNFEED in/pass / mm/pass | CROSSFEED in/pass / mm/pass | WHEEL* IDENTIFI-CATION ANSI / ISO |
|---|---|---|---|---|---|---|---|
| **7. MARAGING STEELS, WROUGHT**<br>ASTM A538: Grades A, B, C<br>120 Grade<br>180 Grade<br>200 Grade<br>250 Grade<br>300 Grade<br>350 Grade<br>HY230 | 50 $R_C$ max. | Annealed or Maraged | 4000 to 6000 | 50 to 100 | Rough: .003 Finish .001 max. | .050-.500 (Max: 1/4 of Wheel Width) | A46JV |
| | | | 20 to 30 | 15 to 30 | Rough: .075 Finish .025 max. | 1.25-12.5 (Max: 1/4 of Wheel Width) | A46JV |
| | Over 50 $R_C$ | Maraged | 3000 to 4000 | 50 to 100 | Rough: .002 Finish .0005 max. | .025-.250 (Max: 1/10 of Wheel Width) | A60HV |
| | | | 15 to 20 | 15 to 30 | Rough: .050 Finish .013 max. | .65-6.5 (Max: 1/10 of Wheel Width) | A60HV |
| **8. TOOL STEELS, WROUGHT**<br>**Group I**<br>A2 H13 L6 P20<br>A3 H14 L7 P21<br>A4 H19 O1 S1<br>A6 H21 O2 S2<br>A8 H22 O6 S5<br>A9 H23 O7 S6<br>A10 H24 P2 S7<br>H10 H25 P4 W1<br>H11 H26 P5 W2<br>H12 L2 P6 W5<br>SAE J438b: Types W108, W109, W110, W112, W209, W210, W310 | 50 $R_C$ max. | Annealed or Quenched and Tempered | 5500 to 6500 | 50 to 100 | Rough: .003 Finish .0005 max. | .050-.500 (Max: 1/5 of Wheel Width) | A46JV |
| | | | 28 to 33 | 15 to 30 | Rough: .075 Finish .013 max. | 1.25-12.5 (Max: 1/5 of Wheel Width) | A46JV |
| | Over 50 $R_C$ | Quenched and Tempered | 3000 to 4000 | 50 to 100 | Rough: .002 Finish .0005 max. | .025-.250 (Max: 1/10 of Wheel Width) | A46IV |
| | | | 15 to 20 | 15 to 30 | Rough: .050 Finish .013 max. | .65-6.5 (Max: 1/10 of Wheel Width) | A46IV |
| **Group II**<br>D2 F2 M10 T4<br>D3 H42 M30 T8<br>D4 M1 M42<br>D5 M2 T1<br>F1 M3-1 T2 | 50 $R_C$ max. | Annealed or Quenched and Tempered | 5500 to 6500 | 50 to 100 | Rough: .003 Finish .0005 max. | .050-.500 (Max: 1/5 of Wheel Width) | A46JV |
| | | | 28 to 33 | 15 to 30 | Rough: .075 Finish .013 max. | 1.25-12.5 (Max: 1/5 of Wheel Width) | A46JV |
| | 50 $R_C$ to 58 $R_C$ | Quenched and Tempered | 3000 to 4000 | 50 to 100 | Rough: .002 Finish .0005 max. | .025-.250 (Max: 1/10 of Wheel Width) | A46IV |
| | | | 15 to 20 | 15 to 30 | Rough: .050 Finish .013 max. | .65-6.5 (Max: 1/10 of Wheel Width) | A46IV |
| | Over 58 $R_C$ | Quenched and Tempered | 3000 to 4000 | 50 to 100 | Rough: .001 Finish .0005 max. | .025-.250 (Max: 1/10 of Wheel Width) | A60HV |
| | | | 15 to 20 | 15 to 30 | Rough: .025 Finish .013 max. | .65-6.5 (Max: 1/10 of Wheel Width) | A60HV |
| **Group III**<br>A7 M7 M41 M47<br>D7 M33 M43 T5<br>M3-2 M34 M44 T6<br>M6 M36 M46 | 50 $R_C$ max. | Annealed or Quenched and Tempered | 4000 to 6000 | 50 to 100 | Rough: .002 Finish .0005 max. | .025-.250 (Max: 1/8 of Wheel Width) | A46IV |
| | | | 20 to 30 | 15 to 30 | Rough: .050 Finish .013 max. | .65-6.5 (Max: 1/8 of Wheel Width) | A46IV |
| | 50 $R_C$ to 58 $R_C$ | Quenched and Tempered | 3000 to 4000 | 75 to 150 | Rough: .002 Finish .0005 max. | .025-.250 (Max: 1/10 of Wheel Width) | A60IV |
| | | | 15 to 20 | 23 to 46 | Rough: .050 Finish .013 max. | .65-6.5 (Max: 1/10 of Wheel Width) | A60IV |

See section 16 for Cutting Fluid Recommendations.
*Wheel recommendations are for wet grinding. For DRY grinding—use a softer grade wheel. Also see section 20.2, Grinding Guidelines.

# 8.1 Surface Grinding-Horizontal Spindle, Reciprocating Table

| MATERIAL | HARD-NESS<br>Bhn | CONDITION | WHEEL SPEED<br>fpm<br>m/s | TABLE SPEED<br>fpm<br>m/min | DOWNFEED<br>in/pass<br>mm/pass | CROSSFEED<br>in/pass<br>mm/pass | WHEEL* IDENTIFI-CATION<br>ANSI<br>ISO |
|---|---|---|---|---|---|---|---|
| **8. TOOL STEELS, WROUGHT (cont.)**<br>**Group III (cont.)**<br>(materials listed on preceding page) | Over<br>58 $R_C$ | Quenched and Tempered | 3000<br>to<br>4000 | 75<br>to<br>150 | Rough: .001<br>Finish<br>.0003 max. | .020-.200<br>(Max: 1/12 of Wheel Width) | A60HV |
| | | | 15<br>to<br>20 | 23<br>to<br>46 | Rough: .025<br>Finish<br>.008 max. | .50-5.0<br>(Max: 1/12 of Wheel Width) | A60HV |
| **Group IV**<br>M4<br>T15 | 50 $R_C$<br>max. | Annealed or Quenched and Tempered | 4000<br>to<br>6000 | 50<br>to<br>100 | Rough: .002<br>Finish<br>.0005 max. | .020-.200<br>(Max: 1/10 of Wheel Width) | A46HV |
| | | | 20<br>to<br>30 | 15<br>to<br>30 | Rough: .050<br>Finish<br>.013 max. | .50-5.0<br>(Max: 1/10 of Wheel Width) | A46HV |
| | 50 $R_C$<br>to<br>58 $R_C$ | Quenched and Tempered | 3000<br>to<br>4000 | 75<br>to<br>150 | Rough: .001<br>Finish<br>.0003 max. | .020-.200<br>(Max: 1/12 of Wheel Width) | A60HV |
| | | | 15<br>to<br>20 | 23<br>to<br>46 | Rough: .025<br>Finish<br>.008 max. | .50-5.0<br>(Max: 1/12 of Wheel Width) | A60HV |
| | Over<br>58 $R_C$ | Quenched and Tempered | 3000<br>to<br>4000 | 75<br>to<br>150 | Rough: .001<br>Finish<br>.0003 max. | .020-.200<br>(Max: 1/12 of Wheel Width) | A60GV |
| | | | 15<br>to<br>20 | 23<br>to<br>46 | Rough: .025<br>Finish<br>.008 max. | .50-5.0<br>(Max: 1/12 of Wheel Width) | A60GV |
| **9. NITRIDING STEELS, WROUGHT**<br>Nitralloy 125<br>Nitralloy 135<br>Nitralloy 135 Mod.<br>Nitralloy 225<br>Nitralloy 230<br>Nitralloy EZ<br>Nitralloy N<br>Nitrex 1 | 200<br>to<br>350 | Annealed, Normalized or Quenched and Tempered | 5500<br>to<br>6500 | 50<br>to<br>100 | Rough: .003<br>Finish<br>.001 max. | .050-.500<br>(Max: 1/4 of Wheel Width) | A46JV |
| | | | 28<br>to<br>33 | 15<br>to<br>30 | Rough: .075<br>Finish<br>.025 max. | 1.25-12.5<br>(Max: 1/4 of Wheel Width) | A46JV |
| | 60 $R_C$<br>to<br>65 $R_C$ | Nitrided | 5500<br>to<br>6500 | 50<br>to<br>100 | Rough: .001<br>Finish<br>.0003 max. | .025-.250<br>(Max: 1/10 of Wheel Width) | A60HV |
| | | | 28<br>to<br>33 | 15<br>to<br>30 | Rough: .025<br>Finish<br>.008 max. | .65-6.5<br>(Max: 1/10 of Wheel Width) | A60HV |
| **10. ARMOR PLATE, SHIP PLATE, AIRCRAFT PLATE, WROUGHT**<br>HY80<br>MIL-S-12560<br>HY100<br>MIL-S-16216<br>HY180 | 45 $R_C$<br>max. | Annealed or Quenched and Tempered | 5500<br>to<br>6500 | 50<br>to<br>100 | Rough: .003<br>Finish<br>.001 max. | .050-.500<br>(Max: 1/4 of Wheel Width) | A46KV |
| | | | 28<br>to<br>33 | 15<br>to<br>30 | Rough: .075<br>Finish<br>.025 max. | 1.25-12.5<br>(Max: 1/4 of Wheel Width) | A46KV |
| | | | | | | | |
| | | | | | | | |
| **12. FREE MACHINING STAINLESS STEELS, WROUGHT**<br>**Ferritic**<br>430F 430F Se | 135<br>to<br>185 | Annealed | 5500<br>to<br>6500 | 50<br>to<br>100 | Rough: .002<br>Finish<br>.0005 max. | .050-.500<br>(Max: 1/4 of Wheel Width) | A46IV |
| | | | 28<br>to<br>33 | 15<br>to<br>30 | Rough: .050<br>Finish<br>.013 max. | 1.25-12.5<br>(Max: 1/4 of Wheel Width) | A46IV |
| **Austenitic**<br>203EZ 303Pb<br>303 303 Plus X<br>303MA 303Se | 135<br>to<br>275 | Annealed or Cold Drawn | 5500<br>to<br>6500 | 50<br>to<br>100 | Rough: .002<br>Finish<br>.0005 max. | .050-.500<br>(Max: 1/4 of Wheel Width) | C46JV |
| | | | 28<br>to<br>33 | 15<br>to<br>30 | Rough: .050<br>Finish<br>.013 max. | 1.25-12.5<br>(Max: 1/4 of Wheel Width) | C46JV |

See section 16 for Cutting Fluid Recommendations.
*Wheel recommendations are for wet grinding. For DRY grinding—use a softer grade wheel. Also see section 20.2, Grinding Guidelines.

| MATERIAL | HARDNESS | CONDITION | WHEEL SPEED | TABLE SPEED | DOWNFEED | CROSSFEED | WHEEL* IDENTIFICATION |
|---|---|---|---|---|---|---|---|
| | Bhn | | fpm | fpm | in/pass | in/pass | ANSI |
| | | | m/s | m/min | mm/pass | mm/pass | ISO |
| **Martensitic**<br>416 420F Se<br>416 Plus X 440 F<br>416Se 440F Se<br>420F | 135 to 240 | Annealed or Cold Drawn | 5500 to 6500 | 50 to 100 | Rough: .002<br>Finish .0005 max. | .050-.500 (Max: 1/4 of Wheel Width) | A46IV |
| | | | 28 to 33 | 15 to 30 | Rough: .050<br>Finish .013 max. | 1.25-12.5 (Max: 1/4 of Wheel Width) | A46IV |
| | Over 275 | Quenched and Tempered | 5500 to 6500 | 50 to 100 | Rough: .001<br>Finish .0005 max. | .025-.250 (Max: 1/10 of Wheel Width) | A46HV |
| | | | 28 to 33 | 15 to 30 | Rough: .025<br>Finish .013 max. | .65-6.5 (Max: 1/10 of Wheel Width) | A46HV |
| **13. STAINLESS STEELS, WROUGHT**<br>**Ferritic**<br>405 429 434 442<br>409 430 436 446 | 135 to 185 | Annealed | 5500 to 6500 | 50 to 100 | Rough: .002<br>Finish .0005 max. | .050-.500 (Max: 1/4 of Wheel Width) | A46IV |
| | | | 28 to 33 | 15 to 30 | Rough: .050<br>Finish .013 max. | 1.25-12.5 (Max: 1/4 of Wheel Width) | A46IV |
| **Austenitic**<br>201 304L 310S 330<br>202 305 314 347<br>301 308 316 348<br>302 309 316L 384<br>302B 309S 317 385<br>304 310 321 | 135 to 275 | Annealed or Cold Drawn | 5500 to 6500 | 50 to 100 | Rough: .002<br>Finish .0005 max. | .050-.500 (Max: 1/4 of Wheel Width) | C46JV |
| | | | 28 to 33 | 15 to 30 | Rough: .050<br>Finish .013 max. | 1.25-12.5 (Max: 1/4 of Wheel Width) | C46JV |
| | | | | | | | |
| | | | | | | | |
| **Austenitic**<br>Nitronic 32<br>Nitronic 33<br>Nitronic 40<br>Nitronic 50<br>Nitronic 60 | 210 to 250 | Annealed | 5500 to 6500 | 50 to 100 | Rough: .002<br>Finish .0005 max. | .050-.500 (Max: 1/4 of Wheel Width) | C46JV |
| | | | 28 to 33 | 15 to 30 | Rough: .050<br>Finish .013 max. | 1.25-12.5 (Max: 1/4 of Wheel Width) | C46JV |
| | 325 to 375 | Cold Drawn | 5500 to 6500 | 50 to 100 | Rough: .001<br>Finish .0005 max. | .025-.250 (Max: 1/10 of Wheel Width) | C46HV |
| | | | 28 to 33 | 15 to 30 | Rough: .025<br>Finish .013 max. | .65-6.5 (Max: 1/10 of Wheel Width) | C46HV |
| **Martensitic**<br>403 422 440C<br>410 431 501<br>414 440A 502<br>420 440B<br>Greek Ascoloy | 135 to 275 | Annealed | 5500 to 6500 | 50 to 100 | Rough: .002<br>Finish .0005 max. | .050-.500 (Max: 1/4 of Wheel Width) | A46IV |
| | | | 28 to 33 | 15 to 30 | Rough: .050<br>Finish .013 max. | 1.25-12.5 (Max: 1/4 of Wheel Width) | A46IV |
| | Over 275 | Quenched and Tempered | 5500 to 6500 | 50 to 100 | Rough: .001<br>Finish .0003 max. | .025-.250 (Max: 1/10 of Wheel Width) | A46HV |
| | | | 28 to 33 | 15 to 30 | Rough: .025<br>Finish .008 max. | .65-6.5 (Max: 1/10 of Wheel Width) | A46HV |
| | | | | | | | |
| | | | | | | | |

See section 16 for Cutting Fluid Recommendations.
*Wheel recommendations are for wet grinding. For DRY grinding—use a softer grade wheel. Also see section 20.2, Grinding Guidelines.

# 8.1 Surface Grinding-Horizontal Spindle, Reciprocating Table

| MATERIAL | HARD-NESS<br>Bhn | CONDITION | WHEEL SPEED<br>fpm<br>m/s | TABLE SPEED<br>fpm<br>m/min | DOWNFEED<br>in/pass<br>mm/pass | CROSSFEED<br>in/pass<br>mm/pass | WHEEL* IDENTIFI-CATION<br>ANSI<br>ISO |
|---|---|---|---|---|---|---|---|
| **14. PRECIPITATION HARDENING STAINLESS STEELS, WROUGHT**<br>15-5 PH<br>16-6 PH<br>17-4 PH<br>17-7 PH<br>17-14 Cu Mo<br>AF-71<br>AFC-77<br>Almar 362 (AM-362)<br>AM-350<br>AM-355<br>AM-363<br>Custom 450<br>Custom 455<br>HNM<br>PH 13-8 Mo<br>PH 14-8 Mo<br>PH 15-7 Mo<br>Stainless W | 150 to 200 | Solution Treated | 5500 to 6500 | 50 to 100 | Rough: .002 Finish .0005 max. | .050-.500 (Max: 1/4 of Wheel Width) | A46IV |
| | | | 28 to 33 | 15 to 30 | Rough: .050 Finish .013 max. | 1.25-12.5 (Max: 1/4 of Wheel Width) | A46IV |
| | 275 to 440 | Solution Treated or Hardened | 5500 to 6500 | 50 to 100 | Rough: .001 Finish .0005 max. | .025-.250 (Max: 1/10 of Wheel Width) | A46HV |
| | | | 28 to 33 | 15 to 30 | Rough: .025 Finish .013 max. | .65-6.5 (Max: 1/10 of Wheel Width) | A46HV |
| **15. CARBON STEELS, CAST**<br>**Low Carbon**<br>ASTM A426: Grade CP1<br>1010<br>1020<br>**Medium Carbon**<br>ASTM A352: Grades LCA, LCB, LCC<br>ASTM A356: Grade 1<br>1030 1040 1050 | 100 to 300 | Annealed, Normalized, Normalized and Tempered, or Quenched and Tempered | 5500 to 6500 | 50 to 100 | Rough: .003 Finish .001 max. | .050-.500 (Max: 1/4 of Wheel Width) | A46JV |
| | | | 28 to 33 | 15 to 30 | Rough: .075 Finish .025 max. | 1.25-12.5 (Max: 1/4 of Wheel Width) | A46JV |
| | Over 50 $R_C$ | Carburized and/or Quenched and Tempered | 5500 to 6500 | 50 to 100 | Rough: .002 Finish .0005 max. | .025-.250 (Max: 1/10 of Wheel Width) | A46IV |
| | | | 28 to 33 | 15 to 30 | Rough: .050 Finish .013 max. | .65-6.5 (Max: 1/10 of Wheel Width) | A46IV |
| **16. ALLOY STEELS, CAST**<br>**Low Carbon**<br>ASTM A217: Grade WC9<br>ASTM A352: Grades LC3, LC4<br>ASTM A426: Grades CP2, CP5, CP5b, CP11, CP12, CP15, CP21, CP22<br>1320 2320 4120 8020<br>2315 4110 4320 8620<br>**Medium Carbon**<br>ASTM A27: Grades N1, N2, U-60-30, 60-30, 65-35, 70-36, 70-40<br>ASTM A148: Grades 80-40, 80-50, 90-60, 105-85, 120-95, 150-125, 175-145<br>ASTM A216: Grades WCA, WCB, WCC<br>ASTM A217: Grades WC1, WC4, WC5, WC6<br>ASTM A352: Grades LC1, LC2, LC2-1<br>(materials continued on next page) | 150 to 400 | Annealed, Normalized, Normalized and Tempered, or Quenched and Tempered | 5500 to 6500 | 50 to 100 | Rough: .003 Finish .0005 max. | .050-.500 (Max: 1/4 of Wheel Width) | A46JV |
| | | | 28 to 33 | 15 to 30 | Rough: .075 Finish .013 max. | 1.25-12.5 (Max: 1/4 of Wheel Width) | A46JV |
| | Over 50 $R_C$ | Carburized and/or Quenched and Tempered | 3000 to 4000 | 50 to 100 | Rough: .002 Finish .0005 max. | .025-.250 (Max: 1/10 of Wheel Width) | A46IV |
| | | | 15 to 20 | 15 to 30 | Rough: .050 Finish .013 max. | .65-6.5 (Max: 1/10 of Wheel Width) | A46IV |

See section 16 for Cutting Fluid Recommendations.
*Wheel recommendations are for wet grinding. For DRY grinding—use a softer grade wheel. Also see section 20.2, Grinding Guidelines.

| MATERIAL | HARDNESS Bhn | CONDITION | WHEEL SPEED fpm m/s | TABLE SPEED fpm m/min | DOWNFEED in/pass mm/pass | CROSSFEED in/pass mm/pass | WHEEL* IDENTIFICATION ANSI ISO |
|---|---|---|---|---|---|---|---|
| **16. ALLOY STEELS, CAST (cont.) Medium Carbon (cont.)** (materials continued from preceding page) ASTM A356: Grades 2, 5, 6, 8, 9, 10 ASTM A389: Grades C23, C24 ASTM A486: Classes 70, 90, 120 ASTM A487: Classes 1N, 2N, 4N, 6N, 8N, 9N, 10N, DN, 1Q, 2Q, 4Q, 4QA, 6Q, 7Q, 8Q, 9Q, 10Q<br>1330 4130 80B30 8640<br>1340 4140 8040 9525<br>2325 4330 8430 9530<br>2330 4340 8440 9535<br>4125 8030 8630 | | | | | | | |
| **17. TOOL STEELS, CAST Group I** ASTM A597: Grades CA-2, CH-12, CH-13, CO-1, CS-5 | 50 $R_C$ max. | Annealed or Quenched and Tempered | 5500 to 6500 | 50 to 100 | Rough: .003 Finish .0005 max. | .050-.500 (Max: 1/5 of Wheel Width) | A46JV |
| | | | 28 to 33 | 15 to 30 | Rough: .075 Finish .013 max. | 1.25-12.5 (Max: 1/5 of Wheel Width) | A46JV |
| | 50 $R_C$ to 58 $R_C$ | Quenched and Tempered | 3000 to 4000 | 50 to 100 | Rough: .002 Finish .0005 max. | .025-.250 (Max: 1/10 of Wheel Width) | A46IV |
| | | | 15 to 20 | 15 to 30 | Rough: .050 Finish .013 max. | .65-6.5 (Max: 1/10 of Wheel Width) | A46IV |
| | Over 58 $R_C$ | Quenched and Tempered | 3000 to 4000 | 50 to 100 | Rough: .001 Finish .0005 max. | .025-.250 (Max: 1/10 of Wheel Width) | A60IV |
| | | | 15 to 20 | 15 to 30 | Rough: .025 Finish .013 max. | .65-6.5 (Max: 1/10 of Wheel Width) | A60IV |
| **Group II** ASTM A597: Grades CD-2, CD-5 | 200 to 250 | Annealed | 5500 to 6500 | 50 to 100 | Rough: .003 Finish .0005 max. | .050-.500 (Max: 1/5 of Wheel Width) | A46JV |
| | | | 28 to 33 | 15 to 30 | Rough: .075 Finish .013 max. | 1.25-12.5 (Max: 1/5 of Wheel Width) | A46JV |
| | 48 $R_C$ to 56 $R_C$ | Quenched and Tempered | 3000 to 4000 | 50 to 100 | Rough: .002 Finish .0005 max. | .025-.250 (Max: 1/10 of Wheel Width) | A46IV |
| | | | 15 to 20 | 15 to 30 | Rough: .050 Finish .013 max. | .65-6.5 (Max: 1/10 of Wheel Width) | A46IV |
| | Over 56 $R_C$ | Quenched and Tempered | 3000 to 4000 | 50 to 100 | Rough: .001 Finish .0005 max. | .025-.250 (Max: 1/10 of Wheel Width) | A60HV |
| | | | 15 to 20 | 15 to 30 | Rough: .025 Finish .013 max. | .65-6.5 (Max: 1/10 of Wheel Width) | A60HV |

See section 16 for Cutting Fluid Recommendations.
*Wheel recommendations are for wet grinding. For DRY grinding—use a softer grade wheel. Also see section 20.2, Grinding Guidelines.

# 8.1 Surface Grinding-Horizontal Spindle, Reciprocating Table

| MATERIAL | HARD-NESS<br>Bhn | CONDITION | WHEEL SPEED<br>fpm<br>m/s | TABLE SPEED<br>fpm<br>m/min | DOWNFEED<br>in/pass<br>mm/pass | CROSSFEED<br>in/pass<br>mm/pass | WHEEL* IDENTIFI-CATION<br>ANSI<br>ISO |
|---|---|---|---|---|---|---|---|
| **18. STAINLESS STEELS, CAST**<br>**Ferritic**<br>ASTM A217: Grades C5, C12<br>ASTM A296: Grades CB-30, CC-50, CE-30, CA6N, CA-6NM, CD4MCu<br>ASTM A297: Grade HC<br>ASTM A487: Class CA6NM<br>ASTM A608: Grade HC30 | 135<br>to<br>185 | Annealed | 5500<br>to<br>6500 | 50<br>to<br>100 | Rough: .002<br>Finish<br>.0005 max. | .050-.500<br>(Max: 1/4 of<br>Wheel Width) | A46IV |
| | | | 28<br>to<br>33 | 15<br>to<br>30 | Rough: .050<br>Finish<br>.013 max. | 1.25-12.5<br>(Max: 1/4 of<br>Wheel Width) | A46IV |
| | | | | | | | |
| | | | | | | | |
| **Austenitic**<br>ASTM A296: Grades CF-3, CF-3M, CF-8, CF-8C, CF-8M, CF-16F, CF-20, CG-8M, CG-12, CH-20, CK-20, CN-7M, CN-7MS<br>ASTM A297: Grades HD, HE, HF, HH, HI, HK, HL, HN, HP, HT, HU<br>ASTM A351: Grades CF-3, CF-3A, CF-3M, CF-3MA, CF-8, CF-8A, CF-8C, CF-8M, CF-10MC, CH-8, CH-10, CH-20, CK-20, CN-7M, HK-30, HK-40, HT-30<br>ASTM A451: Grades CPF8A, CPF3, CPF3A, CPF3M, CPF8, CPF8C, CPF8C (Ta Max.), CPF8M, CPF10MC, CPH8, CPH10, CPH20, CPK20<br>ASTM A452: Grades TP 304H, TP 316H, TP 347H<br>ASTM A608: Grades HD50, HE35, HF30, HH30, HH33, HI35, HK30, HK40, HL30, HL40, HN40, HT50, HU50 | 135<br>to<br>210 | Annealed,<br>Normalized<br>or<br>As Cast | 5500<br>to<br>6500 | 50<br>to<br>100 | Rough: .002<br>Finish<br>.0005 max. | .050-.500<br>(Max: 1/4 of<br>Wheel Width) | C46JV |
| | | | 28<br>to<br>33 | 15<br>to<br>30 | Rough: .050<br>Finish<br>.013 max. | 1.25-12.5<br>(Max: 1/4 of<br>Wheel Width) | C46JV |
| | | | | | | | |
| | | | | | | | |
| | | | | | | | |
| | | | | | | | |
| | | | | | | | |
| | | | | | | | |
| | | | | | | | |
| | | | | | | | |
| **Martensitic**<br>ASTM A217: Grade CA-15<br>ASTM A296: Grades CA-15, CA-15M, CA-40<br>ASTM A426: Grades CP7, CP9, CPCA15<br>ASTM A487: Classes CA15a, CA-15M | 135<br>to<br>225 | Annealed,<br>Normalized<br>or<br>Normalized<br>and<br>Tempered | 5500<br>to<br>6500 | 50<br>to<br>100 | Rough: .002<br>Finish<br>.0005 max. | .050-.500<br>(Max: 1/4 of<br>Wheel Width) | A46IV |
| | | | 28<br>to<br>33 | 15<br>to<br>30 | Rough: .050<br>Finish<br>.013 max. | 1.25-12.5<br>(Max: 1/4 of<br>Wheel Width) | A46IV |
| | Over<br>275 | Quenched<br>and<br>Tempered | 5500<br>to<br>6500 | 50<br>to<br>100 | Rough: .001<br>Finish<br>.0005 max. | .025-.250<br>(Max: 1/10 of<br>Wheel Width) | A46HV |
| | | | 28<br>to<br>33 | 15<br>to<br>30 | Rough: .025<br>Finish<br>.013 max. | .65-6.5<br>(Max: 1/10 of<br>Wheel Width) | A46HV |
| | | | | | | | |
| | | | | | | | |

See section 16 for Cutting Fluid Recommendations.
*Wheel recommendations are for wet grinding. For DRY grinding—use a softer grade wheel. Also see section 20.2, Grinding Guidelines.

| MATERIAL | HARDNESS<br>Bhn | CONDITION | WHEEL SPEED<br>fpm<br>m/s | TABLE SPEED<br>fpm<br>m/min | DOWNFEED<br>in/pass<br>mm/pass | CROSSFEED<br>in/pass<br>mm/pass | WHEEL* IDENTIFICATION<br>ANSI<br>ISO |
|---|---|---|---|---|---|---|---|
| **19. PRECIPITATION HARDENING STAINLESS STEELS, CAST**<br>ASTM A351: Grade CD-4MCu<br>ACI Grade CB-7Cu<br>ACI Grade CD-4MCu<br>17-4 PH<br>AM-355 | 325 to 450 | Solution Treated or Solution Treated and Aged | 5500 to 6500 | 50 to 100 | Rough: .001<br>Finish .0005 max. | .025-.250<br>(Max: 1/10 of Wheel Width) | A46HV |
| | | | 28 to 33 | 15 to 30 | Rough: .025<br>Finish .013 max. | .65-6.5<br>(Max: 1/10 of Wheel Width) | A46HV |
| | | | | | | | |
| | | | | | | | |
| **21. GRAY CAST IRONS**<br>**Ferritic**<br>ASTM A48: Class 20<br>SAE J431c: Grade G1800<br>**Pearlitic- Ferritic**<br>ASTM A48: Class 25<br>SAE J431c: Grade G2500<br>**Pearlitic**<br>ASTM A48: Classes 30, 35, 40<br>SAE J431c: Grade G3000<br>**Pearlitic + Free Carbides**<br>ASTM A48: Classes 45, 50<br>SAE J431c: Grades G3500, G4000<br>**Pearlitic or Acicular + Free Carbides**<br>ASTM A48: Classes 55, 60 | 45 $R_C$ max. | As Cast, Annealed or Quenched and Tempered | 5500 to 6500 | 50 to 100 | Rough: .003<br>Finish .001 max. | .050-.500<br>(Max: 1/3 of Wheel Width) | C36JV or A46JV |
| | | | 28 to 33 | 15 to 30 | Rough: .075<br>Finish .025 max. | 1.25-12.5<br>(Max: 1/3 of Wheel Width) | C36JV or A46JV |
| | 45 $R_C$ to 52 $R_C$ | As Cast, Annealed or Quenched and Tempered | 5500 to 6500 | 50 to 100 | Rough: .002<br>Finish .0005 max. | .050-.500<br>(Max: 1/5 of Wheel Width) | C36IV or A46IV |
| | | | 28 to 33 | 15 to 30 | Rough: .050<br>Finish .013 max. | 1.25-12.5<br>(Max: 1/5 of Wheel Width) | C36IV or A46IV |
| | 48 $R_C$ to 60 $R_C$ | Flame or Induction Hardened | 5500 to 6500 | 50 to 100 | Rough: .002<br>Finish .0005 max. | .025-.250<br>(Max: 1/10 of Wheel Width) | C46HV or A46HV |
| | | | 28 to 33 | 15 to 30 | Rough: .050<br>Finish .013 max. | .65-6.5<br>(Max: 1/10 of Wheel Width) | C46HV or A46HV |
| **Austenitic (NI-RESIST)**<br>ASTM A436: Types 1, 1b, 2, 2b, 3, 4, 5, 6 | 100 to 250 | As Cast | 5500 to 6500 | 50 to 100 | Rough: .005<br>Finish .001 max. | .050-.500<br>(Max: 1/3 of Wheel Width) | Rough: A46HV<br>Finish A80GV |
| | | | 28 to 33 | 15 to 30 | Rough: .13<br>Finish .025 max. | 1.25-12.5<br>(Max: 1/3 of Wheel Width) | Rough: A46HV<br>Finish A80GV |
| **22. COMPACTED GRAPHITE CAST IRONS** | 185 to 255 | As Cast | 5500 to 6500 | 50 to 100 | Rough: .003<br>Finish .001 max. | .050-.500<br>(Max: 1/3 of Wheel Width) | C36JV or A46IV |
| | | | 28 to 33 | 15 to 30 | Rough: .075<br>Finish .025 max. | 1.25-12.5<br>(Max: 1/3 of Wheel Width) | C36JV or A46IV |
| **23. DUCTILE CAST IRONS**<br>**Ferritic**<br>ASTM A536: Grades 60-40-18, 65-45-12<br>SAE J434c: Grades D4018, D4512<br>**Ferritic- Pearlitic**<br>ASTM A536: Grade 80-55-06<br>SAE J434c: Grade D5506<br>**Pearlitic- Martensitic**<br>ASTM A536: Grade 100-70-03<br>SAE J434c: Grade D7003<br>**Martensitic**<br>ASTM A536: Grade 120-90-02<br>SAE J434c: Grade DQ&T | 52 $R_C$ max. | Annealed, As Cast, Normalized and Tempered or Quenched and Tempered | 5500 to 6500 | 50 to 100 | Rough: .003<br>Finish .001 max. | .050-.500<br>(Max: 1/3 of Wheel Width) | C36IV or A46IV |
| | | | 28 to 33 | 15 to 30 | Rough: .075<br>Finish .025 max. | 1.25-12.5<br>(Max: 1/3 of Wheel Width) | C36IV or A46IV |
| | 53 $R_C$ to 60 $R_C$ | Flame or Induction Hardened | 5500 to 6500 | 50 to 100 | Rough: .002<br>Finish .0005 max. | .025-.250<br>(Max: 1/10 of Wheel Width) | C46HV or A46HV |
| | | | 28 to 33 | 15 to 30 | Rough: .050<br>Finish .013 max. | .65-6.5<br>(Max: 1/10 of Wheel Width) | C46HV or A46HV |
| | | | | | | | |
| | | | | | | | |

See section 16 for Cutting Fluid Recommendations.
*Wheel recommendations are for wet grinding. For DRY grinding—use a softer grade wheel. Also see section 20.2, Grinding Guidelines.

# 8.1 Surface Grinding-Horizontal Spindle, Reciprocating Table

| MATERIAL | HARD-NESS<br>Bhn | CONDITION | WHEEL SPEED<br>fpm<br>m/s | TABLE SPEED<br>fpm<br>m/min | DOWNFEED<br>in/pass<br>mm/pass | CROSSFEED<br>in/pass<br>mm/pass | WHEEL* IDENTIFI-CATION<br>ANSI<br>ISO |
|---|---|---|---|---|---|---|---|
| **23. DUCTILE CAST IRONS (cont.)**<br>**Austenitic (NI-RESIST Ductile)**<br>ASTM A439: Types D-2, D-2B, D-2C, D-3, D-3A, D-4, D-5, D-5B<br>ASTM A571: Type D-2M | 120<br>to<br>275 | Annealed | 5500<br>to<br>6500 | 50<br>to<br>100 | Rough: .005<br>Finish<br>.001 max. | .050-.500<br>(Max: 1/3 of Wheel Width) | Rough: A46HV<br>Finish<br>A80GV |
| | | | 28<br>to<br>33 | 15<br>to<br>30 | Rough: .13<br>Finish<br>.025 max. | 1.25-12.5<br>(Max: 1/3 of Wheel Width) | Rough: A46HV<br>Finish<br>A80GV |
| | | | | | | | |
| | | | | | | | |
| **24. MALLEABLE CAST IRONS**<br>**Ferritic**<br>ASTM A47: Grades 32510, 35018<br>ASTM A602: Grade M3210<br>SAE J158: Grade M3210<br>**Pearlitic**<br>ASTM A220: Grades 40010, 45006, 45008, 50005<br>ASTM A602: Grade M4504, M5003<br>SAE J158: Grades M4504, M5003<br>**Tempered Martensite**<br>ASTM A220: Grades 60004, 70003, 80002, 90001<br>ASTM A602: Grades M5503, M7002, M8501<br>SAE J158: Grades M5503, M7002, M8501 | 52 $R_C$<br>max. | Malleablized<br>or<br>Malleablized<br>and<br>Heat Treated | 5500<br>to<br>6500 | 50<br>to<br>100 | Rough: .003<br>Finish<br>.001 max. | .050-.500<br>(Max: 1/3 of Wheel Width) | C36IV<br>or<br>A46IV |
| | | | 28<br>to<br>33 | 15<br>to<br>30 | Rough: .075<br>Finish<br>.025 max. | 1.25-12.5<br>(Max: 1/3 of Wheel Width) | C36IV<br>or<br>A46IV |
| | Over<br>52 $R_C$ | Flame<br>or<br>Induction Hardened | 5500<br>to<br>6500 | 50<br>to<br>100 | Rough: .002<br>Finish<br>.0005 max. | .025-.250<br>(Max: 1/10 of Wheel Width) | C46HV<br>or<br>A46HV |
| | | | 28<br>to<br>33 | 15<br>to<br>30 | Rough: .050<br>Finish<br>.013 max. | .65-6.5<br>(Max: 1/10 of Wheel Width) | C46HV<br>or<br>A46HV |
| | | | | | | | |
| | | | | | | | |
| | | | | | | | |
| | | | | | | | |
| **25. WHITE CAST IRONS (ABRASION RESISTANT)**<br>ASTM A532:<br>Class I, Types A, B, C, D<br>Class II, Types A, B, C, D, E<br>Class III, Type A | 60 $R_C$<br>max. | As Cast,<br>Annealed<br>or<br>Hardened | 5500<br>to<br>6500 | 50<br>to<br>100 | Rough: .002<br>Finish<br>.0005 max. | .025-.250<br>(Max: 1/10 of Wheel Width) | C36IV |
| | | | 28<br>to<br>33 | 15<br>to<br>30 | Rough: .050<br>Finish<br>.013 max. | .65-6.5<br>(Max: 1/10 of Wheel Width) | C36IV |
| | | | | | | | |
| | | | | | | | |
| **26. HIGH SILICON CAST IRONS**<br>Duriron<br>Duriclor<br>ASTM A518 | 52 $R_C$ | As Cast | 5500<br>to<br>6500 | 50<br>to<br>100 | Rough: .002<br>Finish<br>.0005 max. | .025-.250<br>(Max: 1/10 of Wheel Width) | C46MB |
| | | | 28<br>to<br>33 | 15<br>to<br>30 | Rough: .050<br>Finish<br>.013 max. | .65-6.5<br>(Max: 1/10 of Wheel Width) | C46MB |
| **27. CHROMIUM-NICKEL ALLOY CASTINGS**<br>ASTM A560: Grades 50Cr-50Ni, 60Cr-40Ni | 275<br>to<br>375 | As Cast | 5500<br>to<br>6500 | 50<br>to<br>100 | Rough: .001<br>Finish<br>.0005 max. | .025-.250<br>(Max: 1/10 of Wheel Width) | C46HV |
| | | | 28<br>to<br>33 | 15<br>to<br>30 | Rough: .025<br>Finish<br>.013 max. | .65-6.5<br>(Max: 1/10 of Wheel Width) | C46HV |

See section 16 for Cutting Fluid Recommendations.

*Wheel recommendations are for wet grinding. For DRY grinding—use a softer grade wheel. Also see section 20.2, Grinding Guidelines.

| MATERIAL | HARDNESS Bhn | CONDITION | WHEEL SPEED fpm / m/s | TABLE SPEED fpm / m/min | DOWNFEED in/pass / mm/pass | CROSSFEED in/pass / mm/pass | WHEEL* IDENTIFICATION ANSI / ISO |
|---|---|---|---|---|---|---|---|
| **28. ALUMINUM ALLOYS, WROUGHT**<br>EC 2218 5252 6253<br>1060 2219 5254 6262<br>1100 2618 5454 6463<br>1145 3003 5456 6951<br>1175 3004 5457 7001<br>1235 3005 5652 7004<br>2011 4032 5657 7005<br>2014 5005 6053 7039<br>2017 5050 6061 7049<br>2018 5052 6063 7050<br>2021 5056 6066 7075<br>2024 5083 6070 7079<br>2025 5086 6101 7175<br>2117 5154 6151 7178 | 30 to 150 500kg | Cold Drawn or Solution Treated and Aged | 4000 to 5000 | 50 to 100 | Rough: .003 Finish .001 max. | .050-.500 (Max: 1/3 of Wheel Width) | C46JV† |
| | | | 20 to 25 | 15 to 30 | Rough: .075 Finish .025 max. | 1.25-12.5 (Max: 1/3 of Wheel Width) | C46JV† |
| **29. ALUMINUM ALLOYS, CAST**<br>**Sand and Permanent Mold**<br>A140 295.0 514.0 A712.0<br>201.0 B295.0 A514.0 D712.0<br>208.0 308.0 B514.0 713.0<br>213.0 319.0 520.0 771.0<br>222.0 355.0 535.0 850.0<br>224.0 C355.0 705.0 A850.0<br>242.0 B443.0 707.0 B850.0<br>Hiduminium RR-350<br>**Die Castings**<br>C443.0 518.0 | 40 to 125 500kg | As Cast or Solution Treated and Aged | 4000 to 5000 | 50 to 100 | Rough: .003 Finish .001 max. | .050-.500 (Max: 1/3 of Wheel Width) | C46JV† |
| | | | 20 to 25 | 15 to 30 | Rough: .075 Finish .025 max. | 1.25-12.5 (Max: 1/3 of Wheel Width) | C46JV† |
| **Sand and Permanent Mold**<br>328.0 333.0 A356.0<br>A332.0 354.0 357.0<br>F332.0 356.0 359.0<br>**Die Castings**<br>360.0 A380.0 390.0 A413.0<br>A360.0 383.0 392.0<br>380.0 A384.0 413.0 | 40 to 125 500kg | As Cast or Solution Treated and Aged | 4000 to 5000 | 50 to 100 | Rough: .003 Finish .001 max. | .050-.500 (Max: 1/3 of Wheel Width) | A46IV |
| | | | 20 to 25 | 15 to 30 | Rough: .075 Finish .025 max. | 1.25-12.5 (Max: 1/3 of Wheel Width) | A46IV |
| **30. MAGNESIUM ALLOYS, WROUGHT‡‡**<br>AZ21A AZ61A HM21A ZK60A<br>AZ31B AZ80A HM31A<br>AZ31C HK31A ZK40A | 50 to 90 500kg | Annealed, Cold Drawn or Solution Treated and Aged | 5500 to 6500 | 50 to 100 | Rough: .003 Finish .001 max. | .050-.500 (Max: 1/3 of Wheel Width) | C46KV or A46KV |
| | | | 28 to 33 | 15 to 30 | Rough: .075 Finish .025 max. | 1.25-12.5 (Max: 1/3 of Wheel Width) | C46KV or A46KV |

See section 16 for Cutting Fluid Recommendations.
*Wheel recommendations are for wet grinding. For DRY grinding—use a softer grade wheel. Also see section 20.2, Grinding Guidelines.

†Wax filled.
‡‡CAUTION: Potential Fire Hazard. Exercise caution in grinding and disposing of swarf. Do NOT use water or water-miscible cutting fluids for magnesium alloys.

# 8.1 Surface Grinding-Horizontal Spindle, Reciprocating Table

| MATERIAL | HARDNESS Bhn | CONDITION | WHEEL SPEED fpm | TABLE SPEED fpm | DOWNFEED in/pass | CROSSFEED in/pass | WHEEL* IDENTIFICATION ANSI |
|---|---|---|---|---|---|---|---|
| | | | m/s | m/min | mm/pass | mm/pass | ISO |
| **31. MAGNESIUM ALLOYS, CAST‡‡**<br>AM60A AZ91C ZE41A<br>AM100A AZ92A ZE63A<br>AS41A EZ33A ZH62A<br>AZ63A HK31A ZK51A<br>AZ81A HZ32A ZK61A<br>AZ91A K1A<br>AZ91B QE22A | 50 to 90 500kg | As Cast, Annealed or Solution Treated and Aged | 5500 to 6500 | 50 to 100 | Rough: .003 Finish .001 max. | .050-.500 (Max: 1/3 of Wheel Width) | C46KV or A46KV |
| | | | 28 to 33 | 15 to 30 | Rough: .075 Finish .025 max. | 1.25-12.5 (Max: 1/3 of Wheel Width) | C46KV or A46KV |
| **32. TITANIUM ALLOYS, WROUGHT‡‡**<br>**Commercially Pure**<br>99.5 99.0<br>99.2 98.9<br>Ti-0.2Pd<br>TiCODE-12 | 110 to 275 | Annealed | 3000 to 4000 | 60 | Rough: .001 Finish .0005 max. | .025-.250 (Max: 1/10 of Wheel Width) | C54JV‡ |
| | | | 15 to 20 | 18 | Rough: .025 Finish .013 max. | .65-6.5 (Max: 1/10 of Wheel Width) | C54JV‡ |
| **Alpha and Alpha-Beta Alloys**<br>Ti-8Mn<br>Ti-1Al-8V-5Fe<br>Ti-2Al-11Sn-5Zr-1Mo<br>Ti-3Al-2.5V<br>Ti-5Al-2Sn-2Zr-4Mo-4Cr (Ti-17)<br>Ti-5Al-2.5Sn<br>Ti-5Al-2.5Sn ELI<br>Ti-5Al-6Sn-2Zr-1Mo<br>Ti-6Al-2Cb-1Ta-0.8Mo<br>Ti-6Al-4V<br>Ti-6Al-4V ELI<br>Ti-6Al-6V-2Sn<br>Ti-6Al-2Sn-4Zr-2Mo<br>Ti-6Al-2Sn-4Zr-2Mo-.25Si<br>Ti-6Al-2Sn-4Zr-6Mo<br>Ti-7Al-4Mo<br>Ti-8Al-1Mo-1V | 300 to 380 | Annealed | 3000 to 4000 | 60 | Rough: .001 Finish .0005 max. | .025-.250 (Max: 1/10 of Wheel Width) | C54JV‡ |
| | | | 15 to 20 | 18 | Rough: .025 Finish .013 max. | .65-6.5 (Max: 1/10 of Wheel Width) | C54JV‡ |
| | 320 to 440 | Solution Treated and Aged | 3000 to 4000 | 60 | Rough: .001 Finish .0005 max. | .025-.250 (Max: 1/10 of Wheel Width) | C54JV‡ |
| | | | 15 to 20 | 18 | Rough: .025 Finish .013 max. | .65-6.5 (Max: 1/10 of Wheel Width) | C54JV‡ |
| **Beta Alloys**<br>Ti-3Al-8V-6Cr-4Mo-4Zr<br>Ti-8Mo-8V-2Fe-3Al<br>Ti-11.5Mo-6Zr-4.5Sn<br>Ti-10V-2Fe-3Al<br>Ti-13V-11Cr-3Al | 275 to 350 | Annealed or Solution Treated | 3000 to 4000 | 60 | Rough: .001 Finish .0005 max. | .025-.250 (Max: 1/10 of Wheel Width) | C54JV‡ |
| | | | 15 to 20 | 18 | Rough: .025 Finish .013 max. | .65-6.5 (Max: 1/10 of Wheel Width) | C54JV‡ |
| | 350 to 440 | Solution Treated and Aged | 3000 to 4000 | 60 | Rough: .001 Finish .0005 max. | .025-.250 (Max: 1/10 of Wheel Width) | C54JV‡ |
| | | | 15 to 20 | 18 | Rough: .025 Finish .013 max. | .65-6.5 (Max: 1/10 of Wheel Width) | C54JV‡ |

See section 16 for Cutting Fluid Recommendations.
*Wheel recommendations are for wet grinding. For DRY grinding—use a softer grade wheel. Also see section 20.2, Grinding Guidelines.

‡Use friable (green grit) silicon carbide.
‡‡CAUTION: Potential Fire Hazard. Exercise caution in grinding and disposing of swarf. Do NOT use water or water-miscible cutting fluids for magnesium alloys.

| MATERIAL | HARD-NESS Bhn | CONDITION | WHEEL SPEED fpm m/s | TABLE SPEED fpm m/min | DOWNFEED in/pass mm/pass | CROSSFEED in/pass mm/pass | WHEEL* IDENTIFI-CATION ANSI ISO |
|---|---|---|---|---|---|---|---|
| **33. TITANIUM ALLOYS, CAST‡‡**<br>**Commercially Pure**<br>99.0<br>Ti-0.2Pd<br>ASTM B367: Grades C-1, C-2, C-3, C-4, C-7A, C-7B, C-8A, C-8B | 150 to 250 | As Cast or As Cast and Annealed | 3000 to 4000 | 60 | Rough: .001 Finish .0005 max. | .025-.250 (Max: 1/10 of Wheel Width) | C54JV‡ |
| | | | 15 to 20 | 18 | Rough: .025 Finish .013 max. | .65-6.5 (Max: 1/10 of Wheel Width) | C54JV‡ |
| | | | | | | | |
| | | | | | | | |
| **Alpha and Alpha-Beta Alloys**<br>Ti-5Al-2.5Sn<br>Ti-6Al-4V<br>Ti-6Al-2Sn-4Zr-2Mo<br>Ti-8Al-1Mo-1V<br>ASTM B367: Grades C-5, C-6 | 300 to 350 | As Cast or As Cast and Annealed | 3000 to 4000 | 60 | Rough: .001 Finish .0005 max. | .025-.250 (Max: 1/10 of Wheel Width) | C54JV‡ |
| | | | 15 to 20 | 18 | Rough: .025 Finish .013 max. | .65-6.5 (Max: 1/10 of Wheel Width) | C54JV‡ |
| | | | | | | | |
| | | | | | | | |
| **34. COPPER ALLOYS, WROUGHT**<br>101 116 143 182<br>102 119 145 184<br>104 120 147 185<br>105 121 150 187<br>107 122 155 189<br>109 125 162 190<br>110 127 165 191<br>111 128 170†† 192<br>113 129 172†† 194<br>114 130 173†† 195<br>115 142 175†† | 10 $R_B$ to 70 $R_B$ | Annealed | 5500 to 6500 | 50 to 100 | Rough: .003 Finish .0005 max. | .050-.500 (Max: 1/3 of Wheel Width) | C36IV† |
| | | | 28 to 33 | 15 to 30 | Rough: .075 Finish .013 max. | 1.25-12.5 (Max: 1/3 of Wheel Width) | C36IV† |
| | 60 $R_B$ to 100 $R_B$ | Cold Drawn | 5500 to 6500 | 50 to 100 | Rough: .003 Finish .0005 max. | .050-.500 (Max: 1/3 of Wheel Width) | C36IV† |
| | | | 28 to 33 | 15 to 30 | Rough: .075 Finish .013 max. | 1.25-12.5 (Max: 1/3 of Wheel Width) | C36IV† |
| | | | | | | | |
| | | | | | | | |
| 210 332 368 464<br>220 335 370 465<br>226 340 377 466<br>230 342 385 467<br>240 349 411 482<br>260 350 413 485<br>268 353 425 667<br>270 356 435 687<br>280 360 442 688<br>314 365 443 694<br>316 366 444<br>330 367 445 | 10 $R_B$ to 70 $R_B$ | Annealed | 5500 to 6500 | 50 to 100 | Rough: .003 Finish .0005 max. | .050-.500 (Max: 1/3 of Wheel Width) | C36IV† |
| | | | 28 to 33 | 15 to 30 | Rough: .075 Finish .013 max. | 1.25-12.5 (Max: 1/3 of Wheel Width) | C36IV† |
| | 60 $R_B$ to 100 $R_B$ | Cold Drawn | 5500 to 6500 | 50 to 100 | Rough: .003 Finish .0005 max. | .050-.500 (Max: 1/3 of Wheel Width) | C36IV† |
| | | | 28 to 33 | 15 to 30 | Rough: .075 Finish .013 max. | 1.25-12.5 (Max: 1/3 of Wheel Width) | C36IV† |
| | | | | | | | |
| | | | | | | | |

See section 16 for Cutting Fluid Recommendations.
*Wheel recommendations are for wet grinding. For DRY grinding—use a softer grade wheel. Also see section 20.2, Grinding Guidelines.

†Wax filled.
‡Use friable (green grit) silicon carbide.
††CAUTION: Toxic Material, refer to National Institute for Occupational Safety and Health (NIOSH) for Precautions.
‡‡CAUTION: Potential Fire Hazard. Exercise caution in grinding and disposing of swarf.

# 8.1 Surface Grinding-Horizontal Spindle, Reciprocating Table

| MATERIAL | HARDNESS Bhn | CONDITION | WHEEL SPEED fpm / m/s | TABLE SPEED fpm / m/min | DOWNFEED in/pass / mm/pass | CROSSFEED in/pass / mm/pass | WHEEL* IDENTIFICATION ANSI / ISO |
|---|---|---|---|---|---|---|---|
| **34. COPPER ALLOYS, WROUGHT (cont.)**<br>505 608 623 642<br>510 610 624 651<br>511 613 625 655<br>521 614 630 674<br>524 618 632 675<br>544 619 638 | 10 $R_B$ to 70 $R_B$ | Annealed | 5500 to 6500 | 50 to 100 | Rough: .003 Finish .0005 max. | .050-.500 (Max: 1/3 of Wheel Width) | C36IV† |
| | | | 28 to 33 | 15 to 30 | Rough: .075 Finish .013 max. | 1.25-12.5 (Max: 1/3 of Wheel Width) | C36IV† |
| | 60 $R_B$ to 100 $R_B$ | Cold Drawn | 5500 to 6500 | 50 to 100 | Rough: .003 Finish .0005 max. | .050-.500 (Max: 1/3 of Wheel Width) | C36IV† |
| | | | 28 to 33 | 15 to 30 | Rough: .075 Finish .013 max. | 1.25-12.5 (Max: 1/3 of Wheel Width) | C36IV† |
| 706 752<br>710 754<br>715 757<br>725 770<br>745 782 | 10 $R_B$ to 70 $R_B$ | Annealed | 5500 to 6500 | 50 to 100 | Rough: .003 Finish .0005 max. | .050-.500 (Max: 1/3 of Wheel Width) | C36IV† |
| | | | 28 to 33 | 15 to 30 | Rough: .075 Finish .013 max. | 1.25-12.5 (Max: 1/3 of Wheel Width) | C36IV† |
| | 60 $R_B$ to 100 $R_B$ | Cold Drawn | 5500 to 6500 | 50 to 100 | Rough: .003 Finish .0005 max. | .050-.500 (Max: 1/3 of Wheel Width) | C36IV† |
| | | | 28 to 33 | 15 to 30 | Rough: .075 Finish .013 max. | 1.25-12.5 (Max: 1/3 of Wheel Width) | C36IV† |
| **35. COPPER ALLOYS, CAST**<br>801 811 818†† 825††<br>803 813 820†† 826††<br>805 814 821†† 827††<br>807 815 822†† 828††<br>809 817†† 824†† | 40 to 200 500kg | As Cast or Heat Treated | 5500 to 6500 | 50 to 100 | Rough: .003 Finish .0005 max. | .050-.500 (Max: 1/3 of Wheel Width) | C36IV† |
| | | | 28 to 33 | 15 to 30 | Rough: .075 Finish .013 max. | 1.25-12.5 (Max: 1/3 of Wheel Width) | C36IV† |
| | 34 $R_C$ to 45 $R_C$ | Heat Treated | 5500 to 6500 | 50 to 100 | Rough: .003 Finish .0005 max. | .050-.500 (Max: 1/3 of Wheel Width) | A46JV |
| | | | 28 to 33 | 15 to 30 | Rough: .075 Finish .013 max. | 1.25-12.5 (Max: 1/3 of Wheel Width) | A46JV |
| 833 852 862 874<br>834 853 863 875<br>836 854 864 876<br>838 855 865 878<br>842 857 867 879<br>844 858 868<br>848 861 872 | 35 to 200 500kg | As Cast | 5500 to 6500 | 50 to 100 | Rough: .003 Finish .0005 max. | .050-.500 (Max: 1/3 of Wheel Width) | C36IV† |
| | | | 28 to 33 | 15 to 30 | Rough: .075 Finish .013 max. | 1.25-12.5 (Max: 1/3 of Wheel Width) | C36IV† |
| | | | | | | | |
| | | | | | | | |
| 902 916 934 948<br>903 917 935 952<br>905 922 937 953<br>907 923 938 954<br>909 925 939 955<br>910 926 943 956<br>911 927 944 957<br>913 928 945 958<br>915 932 947 | 40 to 100 500kg | As Cast | 5500 to 6500 | 50 to 100 | Rough: .003 Finish .0005 max. | .050-.500 (Max: 1/3 of Wheel Width) | C36IV† |
| | | | 28 to 33 | 15 to 30 | Rough: .075 Finish .013 max. | 1.25-12.5 (Max: 1/3 of Wheel Width) | C36IV† |
| | Over 100 500kg | As Cast or Heat Treated | 5500 to 6500 | 50 to 100 | Rough: .003 Finish .0005 max. | .050-.500 (Max: 1/3 of Wheel Width) | A46JV |
| | | | 28 to 33 | 15 to 30 | Rough: .075 Finish .013 max. | 1.25-12.5 (Max: 1/3 of Wheel Width) | A46JV |

See section 16 for Cutting Fluid Recommendations.
*Wheel recommendations are for wet grinding. For DRY grinding—use a softer grade wheel. Also see section 20.2, Grinding Guidelines.

†Wax filled.
††CAUTION: Toxic Material, refer to National Institute for Occupational Safety and Health (NIOSH) for Precautions.

| MATERIAL | HARD-NESS Bhn | CONDITION | WHEEL SPEED fpm / m/s | TABLE SPEED fpm / m/min | DOWNFEED in/pass / mm/pass | CROSSFEED in/pass / mm/pass | WHEEL* IDENTIFI-CATION ANSI / ISO |
|---|---|---|---|---|---|---|---|
| 962 974<br>963 976<br>964 978<br>966†† 993<br>973 | 50 to 100 500kg | As Cast | 5500 to 6500 | 50 to 100 | Rough: .003 Finish .0005 max. | .050-.500 (Max: 1/3 of Wheel Width) | C36IV† |
| | | | 28 to 33 | 15 to 30 | Rough: .075 Finish .013 max. | 1.25-12.5 (Max: 1/3 of Wheel Width) | C36IV† |
| | Over 100 500kg | As Cast or Heat Treated | 5500 to 6500 | 50 to 100 | Rough: .003 Finish .0005 max. | .050-.500 (Max: 1/3 of Wheel Width) | A46JV |
| | | | 28 to 33 | 15 to 30 | Rough: .075 Finish .013 max. | 1.25-12.5 (Max: 1/3 of Wheel Width) | A46JV |
| **36. NICKEL ALLOYS, WROUGHT AND CAST**<br>Nickel 200<br>Nickel 201<br>Nickel 205<br>Nickel 211<br>Nickel 220<br>Nickel 230 | 80 to 170 | Annealed or Cold Drawn | 4500 to 5500 | 50 to 100 | Rough: .001 Finish .0005 max. | .040-.400 (Max: 1/6 of Wheel Width) | A46IV |
| | | | 23 to 28 | 15 to 30 | Rough: .025 Finish .013 max. | 1.0-10.0 (Max: 1/6 of Wheel Width) | A46IV |
| | | | | | | | |
| | | | | | | | |
| MONEL Alloy 400<br>MONEL Alloy 401<br>MONEL Alloy 404<br>MONEL Alloy R405<br>ASTM A296: Grades CZ-100, M-35<br>ASTM A494: Grades CZ-100, M-35 | 115 to 240 | Annealed, Cold Drawn or Cast | 5500 to 6500 | 50 to 100 | Rough: .001 Finish .0005 max. | .040-.400 (Max: 1/6 of Wheel Width) | A46IV |
| | | | 28 to 33 | 15 to 30 | Rough: .025 Finish .013 max. | 1.0-10.0 (Max: 1/6 of Wheel Width) | A46IV |
| | | | | | | | |
| | | | | | | | |
| DURANICKEL Alloy 301<br>MONEL Alloy 502<br>MONEL Alloy K500<br>NI-SPAN-C Alloy 902<br>PERMANICKEL Alloy 300 | 150 to 320 | Solution Treated | 5500 to 6500 | 50 to 100 | Rough: .001 Finish .0005 max. | .040-.400 (Max: 1/6 of Wheel Width) | A60IV |
| | | | 28 to 33 | 15 to 30 | Rough: .025 Finish .013 max. | 1.0-10.0 (Max: 1/6 of Wheel Width) | A60IV |
| | 330 to 360 | Aged | 5500 to 6500 | 50 to 100 | Rough: .001 Finish .0005 max. | .040-.400 (Max: 1/6 of Wheel Width) | A60IV |
| | | | 28 to 33 | 15 to 30 | Rough: .025 Finish .013 max. | 1.0-10.0 (Max: 1/6 of Wheel Width) | A60IV |
| **37. BERYLLIUM NICKEL ALLOYS, WROUGHT AND CAST††**<br>Berylco 440<br>Berylco 41C<br>Berylco 42C<br>Berylco 43C<br>Brush Alloy 200C<br>Brush Alloy 220C<br>Brush Alloy 260C | 200 to 250 | As Cast or Solution Treated | 5000 to 6000 | 35 | Rough: .001 Finish .0005 max. | .040-.400 (Max: 1/6 of Wheel Width) | A46IV |
| | | | 25 to 30 | 11 | Rough: .025 Finish .013 max. | 1.0-10.0 (Max: 1/6 of Wheel Width) | A46IV |
| | 283 to 425 | Hardened or Aged | 5000 to 6000 | 35 | Rough: .001 Finish .0005 max. | .040-.400 (Max: 1/6 of Wheel Width) | A46HV |
| | | | 25 to 30 | 11 | Rough: .025 Finish .013 max. | 1.0-10.0 (Max: 1/6 of Wheel Width) | A46HV |

See section 16 for Cutting Fluid Recommendations.
*Wheel recommendations are for wet grinding. For DRY grinding—use a softer grade wheel. Also see section 20.2, Grinding Guidelines.

†Wax filled.
††CAUTION: Toxic Material, refer to National Institute for Occupational Safety and Health (NIOSH) for Precautions.

# 8.1 Surface Grinding-Horizontal Spindle, Reciprocating Table

| MATERIAL | HARD-NESS<br>Bhn | CONDITION | WHEEL SPEED<br>fpm<br>m/s | TABLE SPEED<br>fpm<br>m/min | DOWNFEED<br>in/pass<br>mm/pass | CROSSFEED<br>in/pass<br>mm/pass | WHEEL* IDENTIFI-CATION<br>ANSI<br>ISO |
|---|---|---|---|---|---|---|---|
| **37. BERYLLIUM NICKEL ALLOYS, WROUGHT AND CAST†† (cont.)**<br>(materials listed on preceding page) | $47\ R_C$<br>to<br>$52\ R_C$ | Hardened<br>or<br>Aged | 5000<br>to<br>6000 | 35 | Rough: .001<br>Finish<br>.0005 max. | .040-.400<br>(Max: 1/6 of Wheel Width) | A60GV |
| | | | 25<br>to<br>30 | 11 | Rough: .025<br>Finish<br>.013 max. | 1.0-10.0<br>(Max: 1/6 of Wheel Width) | A60GV |
| **38. NITINOL ALLOYS, WROUGHT**<br>Nitinol 55Ni-45Ti<br>Nitinol 56Ni-44Ti<br>Nitinol 60Ni-40Ti | 210<br>to<br>340 | Wrought<br>or<br>Annealed | 5000 | 40 | Rough: .002<br>Finish<br>.0005 max. | .025-.250<br>(Max: 1/10 of Wheel Width) | C60IV |
| | | | 25 | 12 | Rough: .050<br>Finish<br>.013 max. | .65-6.5<br>(Max: 1/10 of Wheel Width) | C60IV |
| | $48\ R_C$<br>to<br>$52\ R_C$ | Quenched | 5000 | 40 | Rough: .002<br>Finish<br>.0005 max. | .025-.250<br>(Max: 1/10 of Wheel Width) | C60IV |
| | | | 25 | 12 | Rough: .050<br>Finish<br>.013 max. | .65-6.5<br>(Max: 1/10 of Wheel Width) | C60IV |
| **39. HIGH TEMPERATURE ALLOYS, WROUGHT AND CAST**<br>**Nickel Base, Wrought**<br>AF2-1DA<br>Astroloy<br>Haynes Alloy 263<br>IN-102<br>Incoloy Alloy 901<br>Incoloy Alloy 903<br>Inconel Alloy 617<br>Inconel Alloy 625<br>Inconel Alloy 700<br>Inconel Alloy 702<br>Inconel Alloy 706<br>Inconel Alloy 718<br>Inconel Alloy 721<br>Inconel Alloy 722<br>Inconel Alloy X-750<br>Inconel Alloy 751<br>M252<br>Nimonic 75<br>Nimonic 80<br>Nimonic 90<br>Nimonic 95<br>Rene 41<br>Rene 63<br>Rene 77<br>Rene 95<br>Udimet 500<br>Udimet 700<br>Udimet 710<br>Unitemp 1753<br>Waspaloy | 200<br>to<br>390 | Annealed<br>or<br>Solution Treated | 3000<br>to<br>3500 | 50<br>to<br>100 | Rough: .001<br>Finish<br>.0005 max. | .020-.200<br>(Max: 1/12 of Wheel Width) | A46HV |
| | | | 15<br>to<br>18 | 15<br>to<br>30 | Rough: .025<br>Finish<br>.013 max. | .50-5.0<br>(Max: 1/12 of Wheel Width) | A46HV |
| | 300<br>to<br>475 | Solution Treated and Aged | 3000<br>to<br>3500 | 50<br>to<br>100 | Rough: .001<br>Finish<br>.0005 max. | .020-.200<br>(Max: 1/12 of Wheel Width) | A46HV |
| | | | 15<br>to<br>18 | 15<br>to<br>30 | Rough: .025<br>Finish<br>.013 max. | .50-5.0<br>(Max: 1/12 of Wheel Width) | A46HV |

See section 16 for Cutting Fluid Recommendations.
*Wheel recommendations are for wet grinding. For DRY grinding—use a softer grade wheel. Also see section 20.2, Grinding Guidelines.

††CAUTION: Toxic Material, refer to National Institute for Occupational Safety and Health (NIOSH) for Precautions.

| MATERIAL | HARDNESS<br>Bhn | CONDITION | WHEEL SPEED<br>fpm<br>m/s | TABLE SPEED<br>fpm<br>m/min | DOWNFEED<br>in/pass<br>mm/pass | CROSSFEED<br>in/pass<br>mm/pass | WHEEL* IDENTIFICATION<br>ANSI<br>ISO |
|---|---|---|---|---|---|---|---|
| **Nickel Base, Wrought**<br>Hastelloy Alloy B<br>Hastelloy Alloy B-2<br>Hastelloy Alloy C<br>Hastelloy Alloy C-276<br>Hastelloy Alloy G<br>Hastelloy Alloy S<br>Hastelloy Alloy X<br>Incoloy Alloy 804<br>Incoloy Alloy 825<br>Inconel Alloy 600<br>Inconel Alloy 601<br>Refractaloy 26<br>Udimet 630 | 140 to 220 | Annealed or Solution Treated | 3000 to 4000 | 50 to 100 | Rough: .001 Finish .0005 max. | .025-.250 (Max: 1/10 of Wheel Width) | A46HV |
| | | | 15 to 20 | 15 to 30 | Rough: .025 Finish .013 max. | .65-6.5 (Max: 1/10 of Wheel Width) | A46HV |
| | 240 to 310 | Cold Drawn or Aged | 3000 to 4000 | 50 to 100 | Rough: .001 Finish .0005 max. | .025-.250 (Max: 1/10 of Wheel Width) | A46HV |
| | | | 15 to 20 | 15 to 30 | Rough: .025 Finish .013 max. | .65-6.5 (Max: 1/10 of Wheel Width) | A46HV |
| **Nickel Base, Wrought**<br>TD-Nickel††<br>TD-Ni-Cr†† | 180 to 200 | As Rolled | 5500 to 6500 | 50 to 100 | Rough: .002 Finish .0005 max. | .050-.500 (Max: 1/4 of Wheel Width) | A46IV |
| | | | 28 to 33 | 15 to 30 | Rough: .050 Finish .013 max. | 1.25-12.5 (Max: 1/4 of Wheel Width) | A46IV |
| **Nickel Base, Cast**<br>B-1900<br>GMR-235<br>GMR-235D<br>Hastelloy Alloy B<br>Hastelloy Alloy C<br>Hastelloy Alloy D<br>IN-100 (Rene 100)<br>IN-738<br>IN-792<br>Inconel Alloy 713C<br>Inconel Alloy 718<br>M252<br>MAR-M200<br>MAR-M246<br>MAR-M421<br>MAR-M432<br>Rene 80<br>Rene 125<br>SEL<br>SEL 15<br>TRW VI A<br>Udimet 500<br>Udimet 700<br>ASTM A296: Grades CW-12M, N-12M, CY-40<br>ASTM A297: Grades HW, HX<br>ASTM A494: Grades N-12M-1, N-12M-2, CY-40, CW-12M-1, CW-12M-2<br>ASTM A608: Grades HW50, HX50 | 200 to 425 | As Cast or Cast and Aged | 3000 to 3500 | 50 to 100 | Rough: .001 Finish .0005 max. | .020-.200 (Max: 1/12 of Wheel Width) | A46HV |
| | | | 15 to 18 | 15 to 30 | Rough: .025 Finish .013 max. | .50-5.0 (Max: 1/12 of Wheel Width) | A46HV |

See section 16 for Cutting Fluid Recommendations.

*Wheel recommendations are for wet grinding. For DRY grinding—use a softer grade wheel. Also see section 20.2, Grinding Guidelines.

††CAUTION: Toxic Material, refer to National Institute for Occupational Safety and Health (NIOSH) for Precautions.

# 8.1 Surface Grinding-Horizontal Spindle, Reciprocating Table

| MATERIAL | HARDNESS Bhn | CONDITION | WHEEL SPEED fpm / m/s | TABLE SPEED fpm / m/min | DOWNFEED in/pass / mm/pass | CROSSFEED in/pass / mm/pass | WHEEL* IDENTIFICATION ANSI / ISO |
|---|---|---|---|---|---|---|---|
| **39. HIGH TEMPERATURE ALLOYS, WROUGHT AND CAST (cont.)**<br>**Cobalt Base, Wrought**<br>AiResist 213<br>Haynes Alloy 25 (L605)<br>Haynes Alloy 188<br>J-1570<br>MAR-M905<br>MAR-M918<br>S-816<br>V-36 | 180 to 230 | Solution Treated | 3000 to 3500 | 50 to 100 | Rough: .001 Finish .0005 max. | .020-.200 (Max: 1/12 of Wheel Width) | A46HV |
| | | | 15 to 18 | 15 to 30 | Rough: .025 Finish .013 max. | .50-5.0 (Max: 1/12 of Wheel Width) | A46HV |
| | 270 to 320 | Solution Treated and Aged | 3000 to 3500 | 50 to 100 | Rough: .001 Finish .0005 max. | .020-.200 (Max: 1/12 of Wheel Width) | A46HV |
| | | | 15 to 18 | 15 to 30 | Rough: .025 Finish .013 max. | .50-5.0 (Max: 1/12 of Wheel Width) | A46HV |
| **Cobalt Base, Cast**<br>AiResist 13<br>AiResist 215<br>FSX-414<br>HS-6<br>HS-21<br>HS-31 (X-40)<br>HOWMET #3<br>MAR-M302<br>MAR-M322<br>MAR-M509<br>NASA Co-W-Re<br>WI-52<br>X-45 | 220 to 290 | As Cast or Cast and Aged | 3000 to 3500 | 50 to 100 | Rough: .001 Finish .0005 max. | .020-.200 (Max: 1/12 of Wheel Width) | A46HV |
| | | | 15 to 18 | 15 to 30 | Rough: .025 Finish .013 max. | .50-5.0 (Max: 1/12 of Wheel Width) | A46HV |
| | 290 to 425 | As Cast or Cast and Aged | 3000 to 3500 | 50 to 100 | Rough: .001 Finish .0005 max. | .020-.200 (Max: 1/12 of Wheel Width) | A46HV |
| | | | 15 to 18 | 15 to 30 | Rough: .025 Finish .013 max. | .50-5.0 (Max: 1/12 of Wheel Width) | A46HV |
| **Iron Base, Wrought**<br>A-286<br>Discaloy<br>Incoloy Alloy 800<br>Incoloy Alloy 800H<br>Incoloy Alloy 801<br>Incoloy Alloy 802<br>N-155<br>V-57<br>W-545<br>16-25-6<br>19-9DL | 180 to 230 | Solution Treated | 3000 to 4000 | 50 to 100 | Rough: .001 Finish .0005 max. | .025-.250 (Max: 1/10 of Wheel Width) | A46HV |
| | | | 15 to 20 | 15 to 30 | Rough: .025 Finish .013 max. | .65-6.5 (Max: 1/10 of Wheel Width) | A46HV |
| | 250 to 320 | Solution Treated and Aged | 3000 to 4000 | 50 to 100 | Rough: .001 Finish .0005 max. | .025-.250 (Max: 1/10 of Wheel Width) | A46HV |
| | | | 15 to 20 | 15 to 30 | Rough: .025 Finish .013 max. | .65-6.5 (Max: 1/10 of Wheel Width) | A46HV |
| **40. REFRACTORY ALLOYS, WROUGHT, CAST, P/M****<br>**Columbium**** †† ‡‡<br>C103<br>C129Y<br>Cb-1Zr<br>Cb-752<br>FS-85<br>FS-291<br>WC-3015 | 170 to 225 | Stress Relieved | 2000 to 4000 | 40 to 60 | Rough: .001 Finish .0005 max. | .020-.200 (Max: 1/12 of Wheel Width) | A46JV |
| | | | 10 to 20 | 12 to 18 | Rough: .025 Finish .013 max. | .50-5.0 (Max: 1/12 of Wheel Width) | A46JV |
| **Molybdenum****<br>Mo<br>Mo-50Re<br>TZC<br>TZM | 220 to 290 | Stress Relieved | 2000 to 4000 | 20 to 40 | Rough: .001 Finish .0005 max. | .020-.200 (Max: 1/12 of Wheel Width) | A46KV |
| | | | 10 to 20 | 6 to 12 | Rough: .025 Finish .013 max. | .50-5.0 (Max: 1/12 of Wheel Width) | A46KV |

See section 16 for Cutting Fluid Recommendations.
*Wheel recommendations are for wet grinding. For DRY grinding—use a softer grade wheel. Also see section 20.2, Grinding Guidelines.

**Due to the brittleness of refractory alloys, cracking, chipping, flaking and breakout tend to occur, particularly on the edges of the machined surfaces.
††CAUTION: Toxic Material, refer to National Institute for Occupational Safety and Health (NIOSH) for Precautions.
‡‡CAUTION: Potential Fire Hazard. Exercise caution in grinding and disposing of swarf.

| MATERIAL | HARD-NESS | CONDITION | WHEEL SPEED | TABLE SPEED | DOWNFEED | CROSSFEED | WHEEL* IDENTIFI-CATION |
|---|---|---|---|---|---|---|---|
| | | | fpm | fpm | in/pass | in/pass | ANSI |
| | Bhn | | m/s | m/min | mm/pass | mm/pass | ISO |
| **Tantalum****<br>ASTAR 811C<br>T-111<br>T-222<br>Ta-10W<br>Ta-Hf<br>Ta63 | 200 to 250 | Stress Relieved | 2000 to 4000 | 40 | Rough: .001<br>Finish<br>.0005 max. | .020-.200<br>(Max: 1/12 of Wheel Width) | A46JV |
| | | | 10 to 20 | 12 | Rough: .025<br>Finish<br>.013 max. | .50-5.0<br>(Max: 1/12 of Wheel Width) | A46JV |
| | | | | | | | |
| **Tungsten****<br>85% density<br>93% density<br>96% density<br>100% density | 180 to 320 | Pressed and Sintered, Forged, or Arc Cast | 2000 | 40 to 60 | Rough: .001<br>Finish<br>.0005 max. | .020-.200<br>(Max: 1/12 of Wheel Width) | A46NV or C36JV |
| | | | 10 | 12 to 18 | Rough: .025<br>Finish<br>.013 max. | .50-5.0<br>(Max: 1/12 of Wheel Width) | A46NV or C36JV |
| **Tungsten - 2 Thoria** ††** | 260 to 320 | Pressed and Sintered | 2000 | 40 to 60 | Rough: .001<br>Finish<br>.0005 max. | .020-.200<br>(Max: 1/12 of Wheel Width) | C60JV |
| | | | 10 | 12 to 18 | Rough: .025<br>Finish<br>.013 max. | .50-5.0<br>(Max: 1/12 of Wheel Width) | C60JV |
| **Tungsten Alloys****<br>GE-218<br>W-15Mo<br>W-5Re<br>W-25Re<br>W-25Re-30Mo | 260 to 320 | As Cast | 2000 | 40 to 60 | Rough: .001<br>Finish<br>.0005 max. | .020-.200<br>(Max: 1/12 of Wheel Width) | C60JV |
| | | | 10 | 12 to 18 | Rough: .025<br>Finish<br>.013 max. | .50-5.0<br>(Max: 1/12 of Wheel Width) | C60JV |
| | | | | | | | |
| **Tungsten Alloys****<br>Gyromet<br>Mallory 2000<br>W-10Ag<br>W-7Ni-4Cu | 290 to 320 | Pressed and Sintered | 2000 | 40 to 60 | Rough: .001<br>Finish<br>.0005 max. | .020-.200<br>(Max: 1/12 of Wheel Width) | C60JV |
| | | | 10 | 12 to 18 | Rough: .025<br>Finish<br>.013 max. | .50-5.0<br>(Max: 1/12 of Wheel Width) | C60JV |
| **Tungsten Alloys****<br>Anviloy 1100<br>Anviloy 1150<br>Anviloy 1200 | 290 to 320 | Pressed and Sintered | 4000 | 40 to 60 | Rough: .001<br>Finish<br>.0005 max. | .020-.200<br>(Max: 1/12 of Wheel Width) | C60JV |
| | | | 20 | 12 to 18 | Rough: .025<br>Finish<br>.013 max. | .50-5.0<br>(Max: 1/12 of Wheel Width) | C60JV |
| **44. URANIUM, WROUGHT†† ‡‡** | 56 $R_A$ to 58 $R_A$ | Annealed | 3000 to 5000 | 50 to 75 | Rough: .005<br>Finish<br>.002 max. | .025-.250<br>(Max: 1/10 of Wheel Width) | C46GV |
| | | | 15 to 25 | 15 to 23 | Rough: .13<br>Finish<br>.050 max. | .65-6.5<br>(Max: 1/10 of Wheel Width) | C46GV |
| | | | | | | | |

See section 16 for Cutting Fluid Recommendations.
*Wheel recommendations are for wet grinding. For DRY grinding—use a softer grade wheel. Also see section 20.2, Grinding Guidelines.

** Due to the brittleness of refractory alloys, cracking, chipping, flaking and breakout tend to occur, particularly on the edges of the machined surfaces.
†† CAUTION: Toxic Material, refer to National Institute for Occupational Safety and Health (NIOSH) for Precautions.
‡‡ CAUTION: Potential Fire Hazard. Exercise caution in grinding and disposing of swarf.

# 8.1 Surface Grinding-Horizontal Spindle, Reciprocating Table

| MATERIAL | HARD-NESS<br>Bhn | CONDITION | WHEEL SPEED<br>fpm<br>m/s | TABLE SPEED<br>fpm<br>m/min | DOWNFEED<br>in/pass<br>mm/pass | CROSSFEED<br>in/pass<br>mm/pass | WHEEL* IDENTIFI-CATION<br>ANSI<br>ISO |
|---|---|---|---|---|---|---|---|
| **45. ZIRCONIUM ALLOYS, WROUGHT‡‡**<br>Zr-2%Hf (Grade 11)<br>Zr-0.001%Hf (Grade 21)<br>Zircaloy 2 (Grade 32)<br>Zircaloy 4 (Grade 34) | 140<br>to<br>280 | Rolled,<br>Extruded<br>or Forged | 3000 | 40 | Rough: .001<br>Finish<br>.0005 max. | .040-.400<br>(Max: 1/6 of Wheel Width) | C46JV |
| | | | 15 | 12 | Rough: .025<br>Finish<br>.013 max. | 1.0-10.0<br>(Max: 1/6 of Wheel Width) | C46JV |
| **47. POWDER METAL ALLOYS****<br>**Copper**** | 50 $R_F$<br>to<br>70 $R_F$ | As Sintered | 5500<br>to<br>6500 | 50<br>to<br>100 | Rough: .003<br>Finish<br>.0005 max. | .050-.500<br>(Max: 1/3 of Wheel Width) | C36IV† |
| | | | 28<br>to<br>33 | 15<br>to<br>30 | Rough: .075<br>Finish<br>.013 max. | 1.25-12.5<br>(Max: 1/3 of Wheel Width) | C36IV† |
| **Brasses****<br>CZP-0218-T<br>CZP-0218-U<br>CZP-0218-W<br>90Cu-10Zn<br>90Cu-10Zn-0.5Pb<br>70Cu-30Zn<br>68.5Cu-30Zn-1.5Pb | 35 $R_H$<br>to<br>81 $R_H$ | As Sintered | 5500<br>to<br>6500 | 50<br>to<br>100 | Rough: .003<br>Finish<br>.0005 max. | .050-.500<br>(Max: 1/3 of Wheel Width) | C36IV† |
| | | | 28<br>to<br>33 | 15<br>to<br>30 | Rough: .075<br>Finish<br>.013 max. | 1.25-12.5<br>(Max: 1/3 of Wheel Width) | C36IV† |
| **Bronzes****<br>CT-0010-N<br>CT-0010-R<br>CT-0010-S<br>95Cu-5Al<br>77Cu-15Pb-7Sn-1Fe-1C | 30 $R_F$<br>to<br>75 $R_F$ | As Sintered | 5500<br>to<br>6500 | 50<br>to<br>100 | Rough: .003<br>Finish<br>.0005 max. | .050-.500<br>(Max: 1/3 of Wheel Width) | C36IV† |
| | | | 28<br>to<br>33 | 15<br>to<br>30 | Rough: .075<br>Finish<br>.013 max. | 1.25-12.5<br>(Max: 1/3 of Wheel Width) | C36IV† |
| **Copper-Nickel Alloys****<br>CZN-1818-T<br>CZN-1818-U<br>CZN-1818-W<br>CZNP-1618-U<br>CZNP-1618-W<br>90Cu-10Ni<br>62Cu-18Ni-18Zn-2Sn | 22 $R_H$<br>to<br>100 $R_H$ | As Sintered | 5500<br>to<br>6500 | 50<br>to<br>100 | Rough: .003<br>Finish<br>.0005 max. | .050-.500<br>(Max: 1/3 of Wheel Width) | C36IV† |
| | | | 28<br>to<br>33 | 15<br>to<br>30 | Rough: .075<br>Finish<br>.013 max. | 1.25-12.5<br>(Max: 1/3 of Wheel Width) | C36IV† |
| **Nickel**** | 45 $R_B$ | As Sintered | 4500<br>to<br>5500 | 50<br>to<br>100 | Rough: .001<br>Finish<br>.0005 max. | .040-.400<br>(Max: 1/6 of Wheel Width) | A60IV |
| | | | 23<br>to<br>28 | 15<br>to<br>30 | Rough: .025<br>Finish<br>.013 max. | 1.0-10.0<br>(Max: 1/6 of Wheel Width) | A60IV |

See section 16 for Cutting Fluid Recommendations.
*Wheel recommendations are for wet grinding. For DRY grinding—use a softer grade wheel. Also see section 20.2, Grinding Guidelines.

†Wax filled.
**Grinding of low-density parts is not recommended because surface porosity will be reduced or lost.
‡‡CAUTION: Potential Fire Hazard. Exercise caution in grinding and disposing of swarf.

| MATERIAL | HARD-NESS<br>Bhn | CONDITION | WHEEL SPEED<br>fpm<br>m/s | TABLE SPEED<br>fpm<br>m/min | DOWNFEED<br>in/pass<br>mm/pass | CROSSFEED<br>in/pass<br>mm/pass | WHEEL* IDENTIFI-CATION<br>ANSI<br>ISO |
|---|---|---|---|---|---|---|---|
| **Nickel Alloys****<br>67Ni-30Cu-3Fe | 34 $R_B$<br>to<br>50 $R_B$ | As Sintered | 5500<br>to<br>6500 | 50<br>to<br>100 | Rough: .001<br>Finish<br>.0005 max. | .040-.400<br>(Max: 1/6 of<br>Wheel Width) | A46IV |
| | | | 28<br>to<br>33 | 15<br>to<br>30 | Rough: .025<br>Finish<br>.013 max. | 1.0-10.0<br>(Max: 1/6 of<br>Wheel Width) | A46IV |
| **Refractory Metal Base****<br>87W-13Cu<br>85W-15Ag<br>74W-26Cu<br>72.5W-27.5Ag<br>65W-35Ag<br>56W+C-44Cu<br>55W-45Cu<br>51W-49Ag<br>50W+C-50Ag | 101<br>to<br>260 | As Sintered | 2000 | 50<br>to<br>100 | Rough: .001<br>Finish<br>.0005 max. | .020-.200<br>(Max: 1/12 of<br>Wheel Width) | C60JV |
| | | | 10 | 15<br>to<br>30 | Rough: .025<br>Finish<br>.013 max. | .50-5.0<br>(Max: 1/12 of<br>Wheel Width) | C60JV |
| | | | | | | | |
| | | | | | | | |
| **Refractory Metal Base****<br>61Mo-39Ag<br>50Mo-50Ag | 75 $R_B$<br>to<br>82 $R_B$ | As Sintered | 4000 | 40<br>to<br>80 | Rough: .001<br>Finish<br>.0005 max. | .020-.200<br>(Max: 1/12 of<br>Wheel Width) | A46KV |
| | | | 20 | 12<br>to<br>24 | Rough: .025<br>Finish<br>.013 max. | .50-5.0<br>(Max: 1/12 of<br>Wheel Width) | A46KV |
| **Irons****<br>F-0000-N<br>F-0000-P<br>F-0000-R<br>F-0000-S<br>F-0000-T | 50<br>to<br>67<br>500kg | As Sintered | 1700<br>to<br>2000 | 50<br>to<br>100 | Rough: .002<br>Finish<br>.0005 max. | .050-.500<br>(Max: 1/4 of<br>Wheel Width) | C36IV<br>or<br>A46IV |
| | | | 8.5<br>to<br>10 | 15<br>to<br>30 | Rough: .050<br>Finish<br>.013 max. | 1.25-12.5<br>(Max: 1/4 of<br>Wheel Width) | C36IV<br>or<br>A46IV |
| | | | | | | | |
| | | | | | | | |
| **Steels****<br>F-0008-P<br>F-0008-S<br>F-0005-S<br>FC-0205-S<br>FC-0208-P<br>FC-0208-S<br>FC-0508-P<br>FC-1000-N<br>FN-0205-S<br>FN-0205-T<br>FN-0405-R<br>FN-0405-S<br>FN-0405-T<br>FX-1005-T<br>FX-2008-T | 101<br>to<br>426 | As Sintered<br>or<br>Heat Treated | 5500<br>to<br>6500 | 50<br>to<br>100 | Rough: .003<br>Finish<br>.0005 max. | .050-.500<br>(Max: 1/4 of<br>Wheel Width) | A46JV |
| | | | 28<br>to<br>33 | 15<br>to<br>30 | Rough: .075<br>Finish<br>.013 max. | 1.25-12.5<br>(Max: 1/4 of<br>Wheel Width) | A46JV |
| | | | | | | | |
| | | | | | | | |
| | | | | | | | |
| | | | | | | | |
| **Stainless Steels****<br>SS-303-R<br>SS-304-R<br>SS-316-R<br>SS-410-R | 107<br>to<br>285 | As Sintered<br>or<br>Heat Treated | 5500<br>to<br>6500 | 50<br>to<br>100 | Rough: .002<br>Finish<br>.0005 max. | .050-.500<br>(Max: 1/4 of<br>Wheel Width) | C46JV |
| | | | 28<br>to<br>33 | 15<br>to<br>30 | Rough: .050<br>Finish<br>.013 max. | 1.25-12.5<br>(Max: 1/4 of<br>Wheel Width) | C46JV |

See section 16 for Cutting Fluid Recommendations.
*Wheel recommendations are for wet grinding. For DRY grinding—use a softer grade wheel. Also see section 20.2, Grinding Guidelines.

**Grinding of low-density parts is not recommended because surface porosity will be reduced or lost.

# 8.1 Surface Grinding-Horizontal Spindle, Reciprocating Table

| MATERIAL | HARDNESS<br>Bhn | CONDITION | WHEEL SPEED<br>fpm<br>m/s | TABLE SPEED<br>fpm<br>m/min | DOWNFEED<br>in/pass<br>mm/pass | CROSSFEED<br>in/pass<br>mm/pass | WHEEL* IDENTIFICATION<br>ANSI<br>ISO |
|---|---|---|---|---|---|---|---|
| **47. POWDER METAL ALLOYS** (cont.)**<br>**Aluminum Alloys****<br>90.5Al-5Sn-4Cu<br>88Al-5Sn-4Pb-3Cu<br>Al-1Mg-0.6Si-0.25Cu<br>Al-0.6Mg-0.4Si<br>Al-4.4Cu-0.8Si-0.4Mg | 55 $R_H$<br>to<br>98 $R_H$ | Solution Treated and Aged | 4000<br>to<br>5000 | 50<br>to<br>100 | Rough: .003<br>Finish<br>.001 max. | .050-.500<br>(Max: 1/3 of Wheel Width) | C46JV† |
| | | | 20<br>to<br>25 | 15<br>to<br>30 | Rough: .075<br>Finish<br>.025 max. | 1.25-12.5<br>(Max: 1/3 of Wheel Width) | C46JV† |
| **48. MACHINABLE CARBIDES**<br>Ferro-Tic | 40 $R_C$<br>to<br>51 $R_C$ | Annealed | 5500<br>to<br>6500 | 50<br>to<br>60 | Rough: .002<br>Finish<br>.001 max. | .010-.060<br>(Max: 1/12 of Wheel Width) | Rough: A46FV<br>Finish<br>A80GV |
| | | | 28<br>to<br>33 | 15<br>to<br>18 | Rough: .050<br>Finish<br>.025 max. | .25-1.5<br>(Max: 1/12 of Wheel Width) | Rough: A46FV<br>Finish<br>A80GV |
| | 68 $R_C$<br>to<br>70 $R_C$ | Hardened and Tempered | 5500<br>to<br>6500 | 50<br>to<br>60 | Rough: .002<br>Finish<br>.0005 max. | .010-.060<br>(Max: 1/12 of Wheel Width) | C46FV<br>or<br>A80GV |
| | | | 28<br>to<br>33 | 15<br>to<br>18 | Rough: .050<br>Finish<br>.013 max. | .25-1.5<br>(Max: 1/12 of Wheel Width) | C46FV<br>or<br>A80GV |
| **49. CARBIDES**<br>Titanium Carbide<br>Tungsten Carbide | 89 $R_A$<br>to<br>94 $R_A$ | — | 5000<br>to<br>6000 | 30<br>to<br>50 | Rough: .001<br>Finish<br>.0005 max. | .030-.060<br>(Max: 1/12 of Wheel Width) | Rough: C60GV<br>Finish<br>C100FV‡ |
| | | | 25<br>to<br>30 | 9<br>to<br>15 | Rough: .025<br>Finish<br>.013 max. | .75-1.5<br>(Max: 1/12 of Wheel Width) | Rough: C60GV<br>Finish<br>C100FV‡ |
| **50. FREE MACHINING MAGNETIC ALLOYS**<br>Magnetic Core Iron-FM (up to 2.5% Si) | 185<br>to<br>240 | Wrought | 5500<br>to<br>6500 | 50<br>to<br>100 | Rough: .003<br>Finish<br>.001 max. | .050-.500<br>(Max: 1/3 of Wheel Width) | A46JV |
| | | | 28<br>to<br>33 | 15<br>to<br>30 | Rough: .075<br>Finish<br>.025 max. | 1.25-12.5<br>(Max: 1/3 of Wheel Width) | A46JV |
| Hi Perm 49-FM | 185<br>to<br>240 | Wrought | 5500<br>to<br>6500 | 50<br>to<br>100 | Rough: .002<br>Finish<br>.0005 max. | .50-.500<br>(Max: 1/4 of Wheel Width) | C46JV |
| | | | 28<br>to<br>33 | 15<br>to<br>30 | Rough: .050<br>Finish<br>.013 max. | 1.25-12.5<br>(Max: 1/4 of Wheel Width) | C46JV |
| **51. MAGNETIC ALLOYS**<br>Magnetic Core Iron (up to 4% Si) | 185<br>to<br>240 | Wrought | 5500<br>to<br>6500 | 50<br>to<br>100 | Rough: .003<br>Finish<br>.001 max. | .050-.500<br>(Max: 1/3 of Wheel Width) | A46JV |
| | | | 28<br>to<br>33 | 15<br>to<br>30 | Rough: .075<br>Finish<br>.025 max. | 1.25-12.5<br>(Max: 1/3 of Wheel Width) | A46JV |
| Hi Perm 49<br>HyMu 80 | 185<br>to<br>240 | Wrought | 5500<br>to<br>6500 | 50<br>to<br>100 | Rough: .002<br>Finish<br>.0005 max. | .050-.500<br>(Max: 1/4 of Wheel Width) | C46JV |
| | | | 28<br>to<br>33 | 15<br>to<br>30 | Rough: .050<br>Finish<br>.013 max. | 1.25-12.5<br>(Max: 1/4 of Wheel Width) | C46JV |

See section 16 for Cutting Fluid Recommendations.
*Wheel recommendations are for wet grinding. For DRY grinding—use a softer grade wheel. Also see section 20.2, Grinding Guidelines.

†Wax filled.
‡Diamond wheel is preferable, see section 8.3.
**Grinding of low-density parts is not recommended because surface porosity will be reduced or lost.

| MATERIAL | HARDNESS Bhn | CONDITION | WHEEL SPEED fpm / m/s | TABLE SPEED fpm / m/min | DOWNFEED in/pass / mm/pass | CROSSFEED in/pass / mm/pass | WHEEL* IDENTIFICATION ANSI / ISO |
|---|---|---|---|---|---|---|---|
| Alnico I, Alnico II, Alnico III, Alnico IV, Alnico V, Alnico V-7, Alnico XII, Columax-5, Hyflux Alnico V-7 | 45 $R_C$ to 58 $R_C$ | As Cast | 5500 to 6500 | 30 | Rough: .002 Finish .0005 max. | .015 (Max: 1/10 of Wheel Width) | A54GV§ |
| | | | 28 to 33 | 9 | Rough: .050 Finish .013 max. | .40 (Max: 1/10 of Wheel Width) | A54GV§ |
| **52. FREE MACHINING CONTROLLED EXPANSION ALLOYS** Invar 36 | 125 to 220 | Annealed or Cold Drawn | 3000 to 4000 | 50 to 100 | Rough: .001 Finish .0005 max. | .040-.400 (Max: 1/6 of Wheel Width) | C46JV |
| | | | 15 to 20 | 15 to 30 | Rough: .025 Finish .013 max. | 1.0-10.0 (Max: 1/6 of Wheel Width) | C46JV |
| **53. CONTROLLED EXPANSION ALLOYS** Invar, Kovar | 125 to 250 | Annealed or Cold Drawn | 3000 to 4000 | 50 to 100 | Rough: .001 Finish .0005 max. | .040-.400 (Max: 1/6 of Wheel Width) | C46JV |
| | | | 15 to 20 | 15 to 30 | Rough: .025 Finish .013 max. | 1.0-10.0 (Max: 1/6 of Wheel Width) | C46JV |
| **54. CARBONS & GRAPHITES** Mechanical Grades | 40 to 100 Shore | Molded or Extruded | 5500 to 6500 | 50 to 60 | Rough: .010 Finish .003 max. | .025-.250 (Max: 1/10 of Wheel Width) | C36KV |
| | | | 28 to 33 | 15 to 18 | Rough: .25 Finish .075 max. | .65-6.5 (Max: 1/10 of Wheel Width) | C36KV |
| Brush Grade Carbon | 8 to 90 Shore | Molded or Extruded | 5500 to 6500 | 50 to 60 | Rough: .010 Finish .003 max. | .025-.250 (Max: 1/10 of Wheel Width) | C36KV |
| | | | 28 to 33 | 15 to 18 | Rough: .25 Finish .075 max. | .65-6.5 (Max: 1/10 of Wheel Width) | C36KV |
| Graphite | — | — | 5500 to 6500 | 50 to 60 | Rough: .010 Finish .003 max. | .025-.250 (Max: 1/10 of Wheel Width) | Rough: C36IV Finish C60IV |
| | | | 28 to 33 | 15 to 18 | Rough: .25 Finish .075 max. | .65-6.5 (Max: 1/10 of Wheel Width) | Rough: C36IV Finish C60IV |
| **55. GLASSES & CERAMICS** PYROCERAM | 700 Knoop 100g | — | 4000 to 6000 | 20 to 40 | Rough: .003 Finish .001 max. | .020-.100 (Max: 1/12 of Wheel Width) | C60JV |
| | | | 20 to 30 | 6 to 12 | Rough: .075 Finish .025 max. | .50-2.5 (Max: 1/12 of Wheel Width) | C60JV |
| Glass | 480 to 530 Knoop 100g | — | 4000 to 6000 | 20 to 40 | Rough: .003 Finish .001 max. | .020-.100 (Max: 1/10 of Wheel Width) | C100IV |
| | | | 20 to 30 | 6 to 12 | Rough: .075 Finish .025 max. | .50-2.5 (Max: 1/10 of Wheel Width) | C100IV |
| Porcelain Enamel | — | — | 5500 to 6500 | 35 | Rough: .001 Finish .0005 max. | .020-.100 (Max: 1/12 of Wheel Width) | C54IV |
| | | | 28 to 33 | 11 | Rough: .025 Finish .013 max. | .50-2.5 (Max: 1/12 of Wheel Width) | C54IV |

See section 16 for Cutting Fluid Recommendations.
*Wheel recommendations are for wet grinding. For DRY grinding—use a softer grade wheel. Also see section 20.2, Grinding Guidelines.

§Cubic boron nitride wheel is preferable, see section 8.2.

# 8.1 Surface Grinding-Horizontal Spindle, Reciprocating Table

| MATERIAL | HARD-NESS<br>Bhn | CONDITION | WHEEL SPEED<br>fpm<br>m/s | TABLE SPEED<br>fpm<br>m/min | DOWNFEED<br>in/pass<br>mm/pass | CROSSFEED<br>in/pass<br>mm/pass | WHEEL* IDENTIFI-CATION<br>ANSI<br>ISO |
|---|---|---|---|---|---|---|---|
| **55. GLASSES & CERAMICS (cont.)**<br>**Ceramics**<br>Alumina (Aluminum Oxide)<br>Alumina-Mullite<br>Aluminum Silicate<br>Beryllia (Beryllium Oxide)††<br>Magnesia (Magnesium Oxide)<br>Mullite<br>Silicon Carbide<br>Silicon Nitride<br>Thoria (Thorium Oxide)††<br>Titania (Titanium Oxide)<br>Titanium Diboride††<br>Zircon (Zirconium Silicate)<br>Zirconia (Zirconium Oxide) | Over 800 Knoop | Fired | 5000 to 5500 | 50 to 100 | Rough: .002 Finish .0005 max. | .025-.250 (Max: 1/10 of Wheel Width) | Rough: C46IV‡ Finish C60IV‡ |
| | | | 25 to 28 | 15 to 30 | Rough: .050 Finish .013 max. | .65-6.5 (Max: 1/10 of Wheel Width) | Rough: C46IV‡ Finish C60IV‡ |
| | | | | | | | |
| | | | | | | | |
| | | | | | | | |
| | | | | | | | |
| **56. PLASTICS**<br>**Thermoplastics** | — | Cast, Molded, Extruded, or Filled and Molded | 5500 to 6500 | 40 | Rough: .005 Finish .0005 max. | .050-.500 (Max: 1/3 of Wheel Width) | C54JV |
| | | | 28 to 33 | 12 | Rough: .13 Finish .013 max. | 1.25-12.5 (Max: 1/3 of Wheel Width) | C54JV |
| **Thermosetting Plastics** | — | Cast, Molded or Laminated | 5500 to 6500 | 40 | Rough: .005 Finish .0005 max. | .050-.500 (Max: 1/3 of Wheel Width) | A54GV |
| | | | 28 to 33 | 12 | Rough: .13 Finish .013 max. | 1.25-12.5 (Max: 1/3 of Wheel Width) | A54GV |
| **58. FLAME (THERMAL) SPRAYED MATERIALS**<br>**Sprayed Carbides**<br>Chromium Carbide<br>Chromium Carbide-Cobalt Blend<br>Columbium Carbide<br>Tantalum Carbide<br>Titanium Carbide<br>Tungsten Carbide<br>Tungsten Carbide-Cobalt<br>Tungsten Carbide (Cobalt)-Nickel Alloy Blend | — | — | 5500 to 6500 | 35 to 50 | Rough: .001 Finish .0005 max. | .010-.030 (Max: 1/12 of Wheel Width) | C80FV§ |
| | | | 28 to 33 | 11 to 15 | Rough: .025 Finish .013 max. | .25-.75 (Max: 1/12 of Wheel Width) | C80FV§ |
| | | | | | | | |
| | | | | | | | |
| **Inorganic Coating Materials**<br>Alumina (Pure)<br>Alumina (Grey) containing Titania<br>Alumina, Nickel-Aluminide Blends<br>Barium Titanate<br>Boron††<br>Calcium Titanate<br>Calcium Zirconate<br>Chromium Disilicide††<br>Chromium Oxide††<br>Cobalt (40%), Zirconia Blend<br>Columbium (Niobium)††<br>Glass (Kovar sealing)††<br>Hexaboron Silicide††<br>Magnesia Alumina Spinel<br>(materials continued on next page) | — | — | 5500 to 6500 | 35 to 50 | Rough: .001 Finish .0005 max. | .010-.030 (Max: 1/12 of Wheel Width) | C100GV§ |
| | | | 28 to 33 | 11 to 15 | Rough: .025 Finish .013 max. | .25-.75 (Max: 1/12 of Wheel Width) | C100GV§ |
| | | | | | | | |
| | | | | | | | |
| | | | | | | | |
| | | | | | | | |

See section 16 for Cutting Fluid Recommendations.
*Wheel recommendations are for wet grinding. For DRY grinding—use a softer grade wheel. Also see section 20.2, Grinding Guidelines.

‡Diamond wheel is preferable, see section 8.3
§Diamond wheel is preferable for flame sprayed carbides and ceramics, see section 8.3.
††CAUTION: Toxic Material, refer to National Institute for Occupational Safety and Health (NIOSH) for Precautions.

| MATERIAL | HARD-NESS | CONDITION | WHEEL SPEED | TABLE SPEED | DOWNFEED | CROSSFEED | WHEEL* IDENTIFI-CATION |
|---|---|---|---|---|---|---|---|
| | | | fpm | fpm | in/pass | in/pass | ANSI |
| | Bhn | | m/s | m/min | mm/pass | mm/pass | ISO |
| **58. FLAME (THERMAL) SPRAYED MATERIALS (cont.) Inorganic Coating Materials (cont.)**<br>(materials continued from preceding page)<br>Magnesium Zirconate<br>Molybdenum Disilicide<br>Mullite<br>Nickel (40%), Alumina Blend<br>Nickel Oxide<br>Rare Earth Oxides<br>Tantalum<br>Titania (50%), Alumina Blend<br>Titanium Oxide<br>Tungsten<br>Yttrium Zirconate<br>Zirconia (Lime Stabilized)<br>Zirconia, Nickel-Aluminide Blends<br>Zirconium Oxide (Hafnia Free, Lime Stabilized)<br>Zirconium Silicate | | | | | | | |
| **Sprayed Metals (Group I)**<br>Co-Cr-B Alloy (Self Fluxing)<br>Ni-Cr-B Alloy (Self Fluxing)<br>Nickel Chrome Steel (Special)<br>Stainless Steel | — | — | 5500 to 6500 | 35 to 75 | Rough: .001 Finish .0005 max. | .010-.100 (Max: 1/10 of Wheel Width) | C46HV |
| | | | 28 to 33 | 11 to 23 | Rough: .025 Finish .013 max. | .25-2.5 (Max: 1/10 of Wheel Width) | C46HV |
| **Sprayed Metals (Group II)**<br>Bronze<br>Chromium<br>Cobalt<br>Nickel<br>Molybdenum<br>Monel | — | — | 5500 to 6500 | 40 to 100 | Rough: .001 Finish .0005 max. | .015-.120 (Max: 1/10 of Wheel Width) | C60JV |
| | | | 28 to 33 | 12 to 30 | Rough: .025 Finish .013 max. | .40-3.0 (Max: 1/10 of Wheel Width) | C60JV |
| **Sprayed Metals (Group III)**<br>Carbon Steel<br>Iron<br>Precipitation Hardening Steel | — | — | 5500 to 6500 | 40 to 100 | Rough: .001 Finish .0005 max. | .015-.120 (Max: 1/10 of Wheel Width) | C46JV |
| | | | 28 to 33 | 12 to 30 | Rough: .025 Finish .013 max. | .40-3.0 (Max: 1/10 of Wheel Width) | C46JV |
| **59. PLATED MATERIALS**<br>Chromium Plate | — | — | 3000 to 4000 | 30 | Rough: .001 Finish .0005 max. | .025-.250 (Max: 1/10 of Wheel Width) | A60HV |
| | | | 15 to 20 | 9 | Rough: .025 Finish .013 max. | .65-6.5 (Max: 1/10 of Wheel Width) | A60HV |
| Nickel Plate<br>Silver Plate | — | — | 5500 to 6500 | 40 | Rough: .001 Finish .0005 max. | .040-.400 (Max: 1/6 of Wheel Width) | C60HV |
| | | | 28 to 33 | 12 | Rough: .025 Finish .013 max. | 1.0-10.0 (Max: 1/6 of Wheel Width) | C60HV |

See section 16 for Cutting Fluid Recommendations.
*Wheel recommendations are for wet grinding. For DRY grinding—use a softer grade wheel. Also see section 20.2, Grinding Guidelines.

# 8.1 Surface Grinding-Horizontal Spindle, Reciprocating Table

| MATERIAL | HARD-NESS | CONDITION | WHEEL SPEED | TABLE SPEED | DOWNFEED | CROSSFEED | WHEEL* IDENTIFI-CATION |
|---|---|---|---|---|---|---|---|
| | | | fpm | fpm | in/pass | in/pass | ANSI |
| | Bhn | | m/s | m/min | mm/pass | mm/pass | ISO |
| **61. RUBBER** | Soft | — | 5500 to 6500 | 50 to 100 | Rough: .003 Finish .001 max. | .050-.500 (Max: 1/4 of Wheel Width) | C24NB |
| | | | 28 to 33 | 15 to 30 | Rough: .075 Finish .025 max. | 1.25-12.5 (Max: 1/4 of Wheel Width) | C24NB |
| | Hard | — | 5500 to 6500 | 50 to 100 | Rough: .003 Finish .001 max. | .050-.500 (Max: 1/4 of Wheel Width) | C24LB |
| | | | 28 to 33 | 15 to 30 | Rough: .075 Finish .025 max. | 1.25-12.5 (Max: 1/4 of Wheel Width) | C24LB |

See section 16 for Cutting Fluid Recommendations.
*Wheel recommendations are for wet grinding. For DRY grinding—use a softer grade wheel. Also see section 20.2, Grinding Guidelines.

| MATERIAL | HARDNESS | CONDITION | WHEEL SPEED DRY | WHEEL SPEED WET* | TABLE SPEED | DOWNFEED | CROSSFEED | WHEEL IDENTIFICATION |
|---|---|---|---|---|---|---|---|---|
| | Bhn | | fpm<br>m/s | fpm<br>m/s | fpm<br>m/min | in/pass<br>mm/pass | in/pass<br>mm/pass | ANSI<br>ANSI |
| **1. FREE MACHINING CARBON STEELS, WROUGHT**<br>**Low Carbon Resulfurized**<br>1108 1118<br>1109 1119<br>1110 1211<br>1115 1212<br>1116 1213<br>1117 1215 | Over 50 $R_c$ | Carburized and/or Quenched and Tempered | 4000 | 5500 to 8000 | 50 to 125 | Rough: .001—.002<br>Finish: .0005 max. | .25—1.00<br>(Max: 1/2 of Wheel Width) | B100T75B |
| | | | 20 | 28 to 41 | 15 to 38 | Rough: .025—.050<br>Finish: .013 max. | 6.3—25<br>(Max: 1/2 of Wheel Width) | B100T75B |
| **Medium Carbon Resulfurized**<br>1132 1140 1145<br>1137 1141 1146<br>1139 1144 1151<br>**Low Carbon Leaded**<br>10L18 12L13 12L15<br>11L17 12L14<br>**Medium Carbon Leaded**<br>10L45 11L37 11L44<br>10L50 11L41 | | | | | | | | |
| **2. CARBON STEELS, WROUGHT**<br>**Low Carbon**<br>1005 1012 1019 1026<br>1006 1013 1020 1029<br>1008 1015 1021 1513<br>1009 1016 1022 1518<br>1010 1017 1023 1522<br>1011 1018 1025 | Over 50 $R_c$ | Carburized and/or Quenched and Tempered | 4000 | 5500 to 8000 | 50 to 125 | Rough: .001—.002<br>Finish: .0005 max. | .25—1.00<br>(Max: 1/2 of Wheel Width) | B100T75B |
| | | | 20 | 28 to 41 | 15 to 38 | Rough: .025—.050<br>Finish: .013 max. | 6.3—25<br>(Max: 1/2 of Wheel Width) | B100T75B |
| **Medium Carbon**<br>1030 1042 1053 1541<br>1033 1043 1055 1547<br>1035 1044 1524 1548<br>1037 1045 1525 1551<br>1038 1046 1526 1552<br>1039 1049 1527<br>1040 1050 1536<br>**High Carbon**<br>1060 1074 1085 1566<br>1064 1075 1086 1572<br>1065 1078 1090<br>1069 1080 1095<br>1070 1084 1561 | | | | | | | | |

*See section 16 for Cutting Fluid Recommendations.
See section 20 for additional information.

# 8.2 Surface Grinding-Horizontal Spindle, Reciprocating Table

## Cubic Boron Nitride Wheels

| MATERIAL | HARD-NESS | CONDITION | WHEEL SPEED DRY | WHEEL SPEED WET* | TABLE SPEED | DOWNFEED | CROSSFEED | WHEEL IDENTIFI-CATION |
|---|---|---|---|---|---|---|---|---|
| | Bhn | | fpm<br>m/s | fpm<br>m/s | fpm<br>m/min | in/pass<br>mm/pass | in/pass<br>mm/pass | ANSI<br>ANSI |
| **4. FREE MACHINING ALLOY STEELS, WROUGHT**<br>**Medium Carbon Resulfurized**<br>4140<br>4140Se<br>4142Te<br>4145Se<br>4147Te<br>4150<br>**Medium and High Carbon Leaded**<br>41L30 41L47 51L32 86L40<br>41L40 41L50 52L100<br>41L45 43L40 86L20 | Over 50 $R_c$ | Carburized and/or Quenched and Tempered | 4000 | 5500 to 8000 | 50 to 125 | Rough: .001—.002<br>Finish: .0005 max. | .25—1.00<br>(Max: 1/2 of Wheel Width) | B100T75B |
| | | | 20 | 28 to 41 | 15 to 38 | Rough: .025—.050<br>Finish: .013 max. | 6.3—25<br>(Max: 1/2 of Wheel Width) | B100T75B |
| | | | | | | | | |
| | | | | | | | | |
| **5. ALLOY STEELS, WROUGHT**<br>**Low Carbon**<br>4012 4615 4817 8617<br>4023 4617 4820 8620<br>4024 4620 5015 8622<br>4118 4621 5115 8822<br>4320 4718 5120 9310<br>4419 4720 6118 94B15<br>4422 4815 8115 94B17<br>**Medium Carbon**<br>1330 4145 5132 8640<br>1335 4147 5135 8642<br>1340 4150 5140 8645<br>1345 4161 5145 86B45<br>4027 4340 5147 8650<br>4028 4427 5150 8655<br>4032 4626 5155 8660<br>4037 50B40 5160 8740<br>4042 50B44 51B60 8742<br>4047 5046 6150 9254<br>4130 50B46 81B45 9255<br>4135 50B50 8625 9260<br>4137 5060 8627 94B30<br>4140 50B60 8630<br>4142 5130 8637<br>**High Carbon**<br>50100 51100 52100 M-50 | Over 50 $R_c$ | Carburized and/or Quenched and Tempered | 4000 | 5500 to 8000 | 50 to 125 | Rough: .001—.002<br>Finish: .0005 max. | .25—1.00<br>(Max: 1/2 of Wheel Width) | B100T75B |
| | | | 20 | 28 to 41 | 15 to 38 | Rough: .025—.050<br>Finish: .013 max. | 6.3—25<br>(Max: 1/2 of Wheel Width) | B100T75B |
| | | | | | | | | |
| | | | | | | | | |
| | | | | | | | | |
| | | | | | | | | |
| | | | | | | | | |
| | | | | | | | | |

*See section 16 for Cutting Fluid Recommendations.
See section 20 for additional information.

| MATERIAL | HARDNESS | CONDITION | WHEEL SPEED DRY | WHEEL SPEED WET* | TABLE SPEED | DOWNFEED | CROSSFEED | WHEEL IDENTIFICATION |
|---|---|---|---|---|---|---|---|---|
| | Bhn | | fpm<br>m/s | fpm<br>m/s | fpm<br>m/min | in/pass<br>mm/pass | in/pass<br>mm/pass | ANSI<br>ANSI |
| **6. HIGH STRENGTH STEELS, WROUGHT**<br>300M H11<br>4330V H13<br>4340 HP 9-4-20<br>4340Si HP 9-4-25<br>98BV40 HP 9-4-30<br>D6ac HP 9-4-45 | Over 50 $R_C$ | Quenched and Tempered | 4000 | 5500 to 8000 | 50 to 125 | Rough: .001—.002<br>Finish: .0005 max. | .25—1.00<br>(Max: 1/2 of Wheel Width) | B100T75B |
| | | | 20 | 28 to 41 | 15 to 38 | Rough: .025—.050<br>Finish: .013 max. | 6.3—25<br>(Max: 1/2 of Wheel Width) | B100T75B |
| **7. MARAGING STEELS, WROUGHT**<br>120 Grade 300 Grade<br>180 Grade 350 Grade<br>200 Grade HY230<br>250 Grade<br>ASTM A538: Grades A, B, C | Over 50 $R_C$ | Maraged | 4000 | 5500 to 8000 | 50 to 125 | Rough: .001—.002<br>Finish: .0005 max. | .25—1.00<br>(Max: 1/2 of Wheel Width) | B100T75B |
| | | | 20 | 28 to 41 | 15 to 38 | Rough: .025—.050<br>Finish: .013 max. | 6.3—25<br>(Max: 1/2 of Wheel Width) | B100T75B |
| **8. TOOL STEELS, WROUGHT**<br>**Group I**<br>A2 H13 L6 P20<br>A3 H14 L7 P21<br>A4 H19 O1 S1<br>A6 H21 O2 S2<br>A8 H22 O6 S5<br>A9 H23 O7 S6<br>A10 H24 P2 S7<br>H10 H25 P4 W1<br>H11 H26 P5 W2<br>H12 L2 P6 W5<br>SAE J438b: Types W108, W109, W110, W112, W209, W210, W310 | Over 50 $R_C$ | Quenched and Tempered | 4000 | 5500 to 8000 | 50 to 125 | Rough: .0005—.0015<br>Finish: .0005 max. | .12—1.00<br>(Max: 1/2 of Wheel Width) | B100T75B |
| | | | 20 | 28 to 41 | 15 to 38 | Rough: .013—.038<br>Finish: .013 max. | 3.0—25<br>(Max: 1/2 of Wheel Width) | B100T75B |

*See section 16 for Cutting Fluid Recommendations.
See section 20 for additional information.

# 8.2 Surface Grinding-Horizontal Spindle, Reciprocating Table

## Cubic Boron Nitride Wheels

| MATERIAL | HARD-NESS | CONDITION | WHEEL SPEED DRY | WHEEL SPEED WET* | TABLE SPEED | DOWNFEED | CROSSFEED | WHEEL IDENTIFI-CATION |
|---|---|---|---|---|---|---|---|---|
| | Bhn | | fpm / m/s | fpm / m/s | fpm / m/min | in/pass / mm/pass | in/pass / mm/pass | ANSI / ANSI |
| **8. TOOL STEELS, WROUGHT (cont.)** **Group II** D2 H42 M42 D3 M1 T1 D4 M2 T2 D5 M3-1 T4 F1 M10 T8 F2 M30 | Over 50 Rc | Quenched and Tempered | 4000 | 5500 to 8000 | 50 to 125 | Rough: .0005—.0015 Finish: .0005 max. | .12—1.00 (Max: 1/2 of Wheel Width) | B100T75B |
| | | | 20 | 28 to 41 | 15 to 38 | Rough: .013—.038 Finish: .013 max. | 3.0—25 (Max: 1/2 of Wheel Width) | B100T75B |
| **Group III** A7 M41 D7 M43 M3-2 M44 M6 M46 M7 M47 M33 T5 M34 T6 M36 | Over 50 Rc | Quenched and Tempered | 4000 | 5500 to 8000 | 50 to 125 | Rough: .0005—.0015 Finish: .0005 max. | .12—1.00 (Max: 1/2 of Wheel Width) | B100T75B |
| | | | 20 | 28 to 41 | 15 to 38 | Rough: .013—.038 Finish: .013 max. | 3.0—25 (Max: 1/2 of Wheel Width) | B100T75B |
| **Group IV** M4 T15 | Over 50 Rc | Quenched and Tempered | 4000 | 5500 to 8000 | 50 to 125 | Rough: .0005—.0010 Finish: .0002 max. | .06—1.00 (Max: 1/2 of Wheel Width) | B100T75B |
| | | | 20 | 28 to 41 | 15 to 38 | Rough: .013—.025 Finish: .005 max. | 1.5—25 (Max: 1/2 of Wheel Width) | B100T75B |
| **9. NITRIDING STEELS, WROUGHT** Nitralloy 125 Nitralloy 135 Nitralloy 135 Mod. Nitralloy 225 Nitralloy 230 Nitralloy EZ Nitralloy N Nitrex 1 | 60 Rc to 65 Rc | Nitrided | 4000 | 5500 to 8000 | 50 to 125 | Rough: .001—.002 Finish: .0005 max. | .25—1.00 (Max: 1/2 of Wheel Width) | B100T75B |
| | | | 20 | 28 to 41 | 15 to 38 | Rough: .025—.050 Finish: .013 max. | 6.3—25 (Max: 1/2 of Wheel Width) | B100T75B |

*See section 16 for Cutting Fluid Recommendations.
See section 20 for additional information.

| MATERIAL | HARD-NESS<br>Bhn | CONDITION | WHEEL SPEED DRY<br>fpm<br>m/s | WHEEL SPEED WET*<br>fpm<br>m/s | TABLE SPEED<br>fpm<br>m/min | DOWNFEED<br>in/pass<br>mm/pass | CROSSFEED<br>in/pass<br>mm/pass | WHEEL IDENTIFI-CATION<br>ANSI<br>ANSI |
|---|---|---|---|---|---|---|---|---|
| **13. STAINLESS STEELS, WROUGHT**<br>**Martensitic**<br>403 440A<br>410 440B<br>414 440C<br>420 501<br>422 502<br>431 Greek Ascoloy | Over 45 $R_c$ | Quenched and Tempered | 4000 | 5500 to 8000 | 50 to 125 | Rough: .001—.002<br>Finish: .0005 max. | .25—1.00<br>(Max: 1/2 of Wheel Width) | B100T75B |
| | | | 20 | 28 to 41 | 15 to 38 | Rough: .025—.050<br>Finish: .013 max. | 6.3—25<br>(Max: 1/2 of Wheel Width) | B100T75B |
| **15. CARBON STEELS, CAST**<br>**Low Carbon**<br>ASTM A426: Grade CP1<br>1010<br>1020<br>**Medium Carbon**<br>ASTM A352: Grades LCA, LCB, LCC<br>ASTM A356: Grade 1<br>1030 1040 1050 | Over 50 $R_c$ | Carburized and/or Quenched and Tempered | 4000 | 5500 to 8000 | 50 to 125 | Rough: .001—.002<br>Finish: .0005 max. | .25—1.00<br>(Max: 1/2 of Wheel Width) | B100T75B |
| | | | 20 | 28 to 41 | 15 to 38 | Rough: .025—.050<br>Finish: .013 max. | 6.3—25<br>(Max: 1/2 of Wheel Width) | B100T75B |
| | | | | | | | | |
| | | | | | | | | |
| **16. ALLOY STEELS, CAST**<br>**Low Carbon**<br>ASTM A217: Grade WC9<br>ASTM A352: Grades LC3, LC4<br>ASTM A426: Grades CP2, CP5, CP5b, CP11, CP12, CP15, CP21, CP22<br>1320 2320 4120 8020<br>2315 4110 4320 8620<br>**Medium Carbon**<br>ASTM A27: Grades N1, N2, U-60-30, 60-30, 65-35, 70-36, 70-40<br>ASTM A148: Grades 80-40, 80-50, 90-60, 105-85, 120-95, 150-125, 175-145<br>ASTM A216: Grades WCA, WCB, WCC<br>ASTM A217: Grades WC1, WC4, WC5, WC6<br>ASTM A352: Grades LC1, LC2, LC2-1<br>ASTM A356: Grades 2, 5, 6, 8, 9, 10<br>ASTM A389: Grades C23, C24<br>ASTM A486: Classes 70, 90, 120<br>(materials continued on next page) | Over 50 $R_c$ | Carburized and/or Quenched and Tempered | 4000 | 5500 to 8000 | 50 to 125 | Rough: .001—.002<br>Finish: .0005 max. | .25—1.00<br>(Max: 1/2 of Wheel Width) | B100T75B |
| | | | 20 | 28 to 41 | 15 to 38 | Rough: .025—.050<br>Finish: .013 max. | 6.3—25<br>(Max: 1/2 of Wheel Width) | B100T75B |
| | | | | | | | | |
| | | | | | | | | |
| | | | | | | | | |
| | | | | | | | | |

*See section 16 for Cutting Fluid Recommendations.
See section 20 for additional information.

# 8.2 Surface Grinding-Horizontal Spindle, Reciprocating Table
## Cubic Boron Nitride Wheels

| MATERIAL | HARD-NESS Bhn | CONDITION | WHEEL SPEED DRY fpm m/s | WHEEL SPEED WET* fpm m/s | TABLE SPEED fpm m/min | DOWNFEED in/pass mm/pass | CROSSFEED in/pass mm/pass | WHEEL IDENTIFI-CATION ANSI ANSI |
|---|---|---|---|---|---|---|---|---|
| **16. ALLOY STEELS, CAST (cont.) Medium Carbon (cont.)** (materials continued from preceding page) ASTM A487: Classes 1N, 2N, 4N, 6N, 8N, 9N, 10N, DN, 1Q, 2Q, 4Q, 4QA, 6Q, 7Q, 8Q, 9Q, 10Q; 1330 4130 80B30 8640 1340 4140 8040 9525 2325 4330 8430 9530 2330 4340 8440 9535 4125 8030 8630 | | | | | | | | |
| | | | | | | | | |
| | | | | | | | | |
| | | | | | | | | |
| **17. TOOL STEELS, CAST** ASTM A597: Grades CA-2, CD-2, CD-5, CH-12, CH-13, CO-1, CS-5 | Over 50 $R_C$ | Quenched and Tempered | 4000 | 5500 to 8000 | 50 to 125 | Rough: .001—.002 Finish: .0005 max. | .25—1.00 (Max: 1/2 of Wheel Width) | B100T75B |
| | | | 20 | 28 to 41 | 15 to 38 | Rough: .025—.050 Finish: .013 max. | 6.3—25 (Max: 1/2 of Wheel Width) | B100T75B |
| **21. GRAY CAST IRONS** **Ferritic** ASTM A48: Class 20 SAE J431c: Grade G1800 **Pearlitic- Ferritic** ASTM A48: Class 25 SAE J431c: Grade G2500 **Pearlitic** ASTM A48: Classes 30, 35, 40 SAE J431c: Grade G3000 **Pearlitic + Free Carbides** ASTM A48: Classes 45, 50 SAE J431c: Grades G3500, G4000 **Pearlitic or Acicular + Free Carbides** ASTM A48: Classes 55, 60 | 45 $R_C$ to 60 $R_C$ | As Cast, Annealed, Quenched and Tempered or Flame or Induction Hardened | 4000 | 5500 to 8000 | 50 to 125 | Rough: .001—.002 Finish: .0005 max. | .25—1.00 (Max: 1/2 of Wheel Width) | B100T75B |
| | | | 20 | 28 to 41 | 15 to 38 | Rough: .025—.050 Finish: .013 max. | 6.3—25 (Max: 1/2 of Wheel Width) | B100T75B |
| | | | | | | | | |
| | | | | | | | | |
| | | | | | | | | |
| | | | | | | | | |

*See section 16 for Cutting Fluid Recommendations.
See section 20 for additional information.

| MATERIAL | HARD-NESS Bhn | CONDITION | WHEEL SPEED DRY fpm m/s | WHEEL SPEED WET* fpm m/s | TABLE SPEED fpm m/min | DOWNFEED in/pass mm/pass | CROSSFEED in/pass mm/pass | WHEEL IDENTIFI-CATION ANSI ANSI |
|---|---|---|---|---|---|---|---|---|
| **23. DUCTILE CAST IRONS**<br>**Ferritic**<br>ASTM A536: Grades 60-40-18, 65-45-12<br>SAE J434c: Grades D4018, D4512<br>**Ferritic- Pearlitic**<br>ASTM A536: Grade 80-55-06<br>SAE J434c: Grade D5506<br>**Pearlitic- Martensitic**<br>ASTM A536: Grade 100-70-03<br>SAE J434c: Grade D7003<br>**Martensitic**<br>ASTM A536: Grade 120-90-02<br>SAE J434c: Grade DQ&T | Over 52 $R_C$ | Flame or Induction Hardened | 4000 | 5500 to 8000 | 50 to 125 | Rough: .001—.002<br>Finish: .0005 max. | .25—1.00<br>(Max: 1/2 of Wheel Width) | B100T75B |
| | | | 20 | 28 to 41 | 15 to 38 | Rough: .025—.050<br>Finish: .013 max. | 6.3—25<br>(Max: 1/2 of Wheel Width) | B100T75B |
| | | | | | | | | |
| | | | | | | | | |
| **24. MALLEABLE CAST IRONS**<br>**Ferritic**<br>ASTM A47: Grades 32510, 35018<br>ASTM A602: Grade M3210<br>SAE J158: Grade M3210<br>**Pearlitic**<br>ASTM A220: Grades 40010, 45006, 45008, 50005<br>ASTM A602: Grade M4504, M5003<br>SAE J158: Grades M4504, M5003<br>**Tempered Martensite**<br>ASTM A220: Grades 60004, 70003, 80002, 90001<br>ASTM A602: Grades M5503, M7002, M8501<br>SAE J158: Grades M5503, M7002, M8501 | Over 52 $R_C$ | Flame or Induction Hardened | 4000 | 5500 to 8500 | 50 to 125 | Rough: .001—.002<br>Finish: .0005 max. | .25—1.00<br>(Max: 1/2 of Wheel Width) | B100T75B |
| | | | 20 | 28 to 41 | 15 to 38 | Rough: .025—.050<br>Finish: .013 max. | 6.3—25<br>(Max: 1/2 of Wheel Width) | B100T75B |
| | | | | | | | | |
| | | | | | | | | |
| **25. WHITE CAST IRONS (ABRASION RESISTANT)**<br>ASTM A532:<br>Class I, Types A, B, C, D<br>Class II, Types A, B, C, D, E<br>Class III, Type A | Over 45 $R_C$ | As Cast, Annealed or Hardened | 4000 | 5500 to 8000 | 50 to 125 | Rough: .0005—.0010<br>Finish: .0002 max. | .06—1.00<br>(Max: 1/2 of Wheel Width) | B100T75B |
| | | | 20 | 28 to 41 | 15 to 38 | Rough: .013—.025<br>Finish: .005 max. | 1.5—25<br>(Max: 1/2 of Wheel Width) | B100T75B |
| | | | | | | | | |
| | | | | | | | | |

*See section 16 for Cutting Fluid Recommendations.
See section 20 for additional information.

## 8.2 Surface Grinding-Horizontal Spindle, Reciprocating Table
### Cubic Boron Nitride Wheels

| MATERIAL | HARD-NESS Bhn | CONDITION | WHEEL SPEED DRY fpm m/s | WHEEL SPEED WET* fpm m/s | TABLE SPEED fpm m/min | DOWNFEED in/pass mm/pass | CROSSFEED in/pass mm/pass | WHEEL IDENTIFI-CATION ANSI ANSI |
|---|---|---|---|---|---|---|---|---|
| **39. HIGH TEMPERATURE ALLOYS, WROUGHT AND CAST** **Nickel Base, Wrought** AF2-1DA, Astroloy, Haynes Alloy 263, IN-102, Incoloy Alloy 901, Incoloy Alloy 903, Inconel Alloy 617, Inconel Alloy 625, Inconel Alloy 700, Inconel Alloy 702, Inconel Alloy 706, Inconel Alloy 718, Inconel Alloy 721, Inconel Alloy 722, Inconel Alloy X-750, Inconel Alloy 751, M252, Nimonic 75, Nimonic 80, Nimonic 90, Nimonic 95, Rene 41, Rene 63, Rene 77, Rene 95, Udimet 500, Udimet 700, Udimet 710, Unitemp 1753, Waspaloy | 300 to 475 | Solution Treated or Solution Treated and Aged | See Note† | 5500 to 8000 | 50 to 125 | Rough: .001—.002 Finish: .0005 max. | .25—1.00 (Max: 1/2 of Wheel Width) | B100T75B |
| | | | See Note† | 28 to 41 | 15 to 38 | Rough: .025—.050 Finish: .013 max. | 6.3—25 (Max: 1/2 of Wheel Width) | B100T75B |
| **Nickel Base, Wrought** Hastelloy Alloy B, Hastelloy Alloy B-2, Hastelloy Alloy C, Hastelloy Alloy C-276, Hastelloy Alloy G, Hastelloy Alloy S, Hastelloy Alloy X, Incoloy Alloy 804, Incoloy Alloy 825, Inconel Alloy 600, Inconel Alloy 601, Refractaloy 26, Udimet 630 | 240 to 310 | Cold Drawn or Aged | See Note† | 5500 to 8000 | 50 to 125 | Rough: .001—.002 Finish: .0005 max. | .25—1.00 (Max: 1/2 of Wheel Width) | B100T75B |
| | | | See Note† | 28 to 41 | 15 to 38 | Rough: .025—.050 Finish: .013 max. | 6.3—25 (Max: 1/2 of Wheel Width) | B100T75B |

*See section 16 for Cutting Fluid Recommendations.
See section 20 for additional information.

†Not recommended.

| MATERIAL | HARDNESS Bhn | CONDITION | WHEEL SPEED DRY fpm m/s | WHEEL SPEED WET* fpm m/s | TABLE SPEED fpm m/min | DOWNFEED in/pass mm/pass | CROSSFEED in/pass mm/pass | WHEEL IDENTIFICATION ANSI ANSI |
|---|---|---|---|---|---|---|---|---|
| **Nickel Base, Cast**<br>B-1900<br>GMR-235<br>GMR-235D<br>Hastelloy Alloy B<br>Hastelloy Alloy C<br>Hastelloy Alloy D<br>IN-100 (Rene 100)<br>IN-738<br>IN-792<br>Inconel Alloy 713C<br>Inconel Alloy 718<br>M252<br>MAR-M200<br>MAR-M246<br>MAR-M421<br>MAR-M432<br>Rene 80<br>Rene 125<br>SEL<br>SEL 15<br>TRW VI A<br>Udimet 500<br>Udimet 700<br>ASTM A296: Grades CW-12M, N-12M, CY-40<br>ASTM A297: Grades HW, HX<br>ASTM A494: Grades N-12M-1, N-12M-2, CY-40, CW-12M-1, CW-12M-2<br>ASTM A608: Grades HW50, HX50 | 200 to 425 | As Cast or Cast and Aged | See Note† | 5500 to 8000 | 50 to 125 | Rough: .001—.002<br>Finish: .0005 max. | .25—1.00<br>(Max: 1/2 of Wheel Width) | B100T75B |
| | | | See Note† | 28 to 41 | 15 to 38 | Rough: .025—.050<br>Finish: .013 max. | 6.3—25<br>(Max: 1/2 of Wheel Width) | B100T75B |
| **Cobalt Base, Wrought**<br>AiResist 213<br>Haynes Alloy 25 (L605)<br>Haynes Alloy 188<br>J-1570<br>MAR-M905<br>MAR-M918<br>S-816<br>V-36 | 270 to 320 | Solution Treated and Aged | See Note† | 5500 to 8000 | 50 to 125 | Rough: .001—.002<br>Finish: .0005 max. | .25—1.00<br>(Max: 1/2 of Wheel Width) | B100T75B |
| | | | See Note† | 28 to 41 | 15 to 38 | Rough: .025—.050<br>Finish: .013 max. | 6.3—25<br>(Max: 1/2 of Wheel Width) | B100T75B |

*See section 16 for Cutting Fluid Recommendations.
See section 20 for additional information.

†Not Recommended.

# 8.2 Surface Grinding-Horizontal Spindle, Reciprocating Table

## Cubic Boron Nitride Wheels

| MATERIAL | HARD-NESS | CONDITION | WHEEL SPEED DRY | WHEEL SPEED WET* | TABLE SPEED | DOWNFEED | CROSSFEED | WHEEL IDENTIFI-CATION |
|---|---|---|---|---|---|---|---|---|
| | Bhn | | fpm | fpm | fpm | in/pass | in/pass | ANSI |
| | | | m/s | m/s | m/min | mm/pass | mm/pass | ANSI |
| **39. HIGH TEMPERATURE ALLOYS, WROUGHT AND CAST (cont.)**<br>**Cobalt Base, Cast**<br>AiResist 13, MAR-M302<br>AiResist 215, MAR-M322<br>FSX-414, MAR-M509<br>HS-6, NASA Co-W-Re<br>HS-21, WI-52<br>HS-31 (X-40), X-45<br>HOWMET #3 | 290 to 425 | As Cast or Cast and Aged | See Note† | 5500 to 8000 | 50 to 125 | Rough: .001—.002<br>Finish: .0005 max. | .25—1.00<br>(Max: 1/2 of Wheel Width) | B100T75B |
| | | | See Note† | 28 to 41 | 15 to 38 | Rough: .025—.050<br>Finish: .013 max. | 6.3—25<br>(Max: 1/2 of Wheel Width) | B100T75B |
| | | | | | | | | |
| **51. MAGNETIC ALLOYS**<br>Alnico I, Alnico V-7<br>Alnico II, Alnico XII<br>Alnico III, Columax-5<br>Alnico IV, Hyflux Alnico V-7<br>Alnico V | 45 $R_C$ to 58 $R_C$ | As Cast | — | 5500 to 8000 | 50 to 125 | Rough: .010<br>Finish: .002 max. | .020<br>(Max: 1/4 of Wheel Width) | B100T75B |
| | | | — | 28 to 41 | 15 to 38 | Rough: .25<br>Finish: .050 max. | .50<br>(Max: 1/4 of Wheel Width) | B100T75B |
| | | | | | | | | |
| | | | | | | | | |
| | | | | | | | | |

*See section 16 for Cutting Fluid Recommendations.
See section 20 for additional information.

†Not recommended.

| MATERIAL | HARD-NESS<br>Bhn | CONDITION | WHEEL SPEED DRY<br>fpm<br>m/s | WHEEL SPEED WET*<br>fpm<br>m/s | TABLE SPEED<br>fpm<br>m/min | DOWNFEED<br>in/pass<br>mm/pass | CROSSFEED<br>in/pass<br>mm/pass | WHEEL IDENTIFICATION<br>ANSI<br>ANSI |
|---|---|---|---|---|---|---|---|---|
| **5. ALLOY STEELS, WROUGHT**<br>**High Carbon**<br>50100<br>51100<br>52100<br>M-50 | 50 $R_C$<br>to<br>65 $R_C$ | Carburized<br>and/or<br>Quenched<br>and Tempered | — | 5000<br>to<br>6000 | 40 | Rough:<br>.003 | .080 | D80-N100-B-1/8† |
| | | | | | 50 | Finish:<br>.0005 max. | (Max: 1/4 of Wheel Width) | D100-N100-B-1/8† |
| | | | — | 25<br>to<br>30 | 12 | Rough:<br>.075 | 2.0 | D80-N100-B-1/8† |
| | | | | | 15 | Finish:<br>.013 max. | (Max: 1/4 of Wheel Width) | D100-N100-B-1/8† |
| **8. TOOL STEELS, WROUGHT**<br>**Group II**<br>D2 D5 H42 M3-1 M42 T4<br>D3 F1 M1 M10 T1 T8<br>D4 F2 M2 M30 T2<br>**Group III**<br>A7 M6 M34 M43 M47<br>D7 M7 M36 M44 T5<br>M3-2 M33 M41 M46 T6 | Over<br>58 $R_C$ | Quenched<br>and Tempered | — | 5000<br>to<br>6000 | 40 | Rough:<br>.0015 | .060 | D60-N100-B-1/8† |
| | | | | | 50 | Finish:<br>.0005 max. | (Max: 1/4 of Wheel Width) | D70-N100-B-1/8† |
| | | | — | 25<br>to<br>30 | 12 | Rough:<br>.038 | 1.50 | D60-N100-B-1/8† |
| | | | | | 15 | Finish:<br>.013 max. | (Max: 1/4 of Wheel Width) | D70-N100-B-1/8† |
| **Group IV**<br>M4<br>T15 | Over<br>58 $R_C$ | Quenched<br>and Tempered | — | 5000<br>to<br>6000 | 30 | Rough:<br>.0015 | .050 | D80-N100-M-1/8 |
| | | | | | 40 | Finish:<br>.0005 max. | (Max: 1/4 of Wheel Width) | D100-N100-M-1/8 |
| | | | — | 25<br>to<br>30 | 9 | Rough:<br>.038 | 1.25 | D80-N100-M-1/8 |
| | | | | | 12 | Finish:<br>.013 max. | (Max: 1/4 of Wheel Width) | D100-N100-M-1/8 |
| **13. STAINLESS STEELS, WROUGHT**<br>**Austenitic**<br>201 304L 310S 330<br>202 305 314 347<br>301 308 316 348<br>302 309 316L 384<br>302B 309S 317 385<br>304 310 321 | 135<br>to<br>275 | Annealed<br>or<br>Cold Drawn | — | 5000<br>to<br>6000 | 50 | Rough:<br>.0015 | .040 | D80-N100-M-1/8† |
| | | | | | 60 | Finish:<br>.0005 max. | (Max: 1/4 of Wheel Width) | D100-N100-M-1/8† |
| | | | — | 25<br>to<br>30 | 15 | Rough:<br>.038 | 1.0 | D80-N100-M-1/8† |
| | | | | | 18 | Finish:<br>.013 max. | (Max: 1/4 of Wheel Width) | D100-N100-M-1/8† |
| **21. GRAY CAST IRONS**<br>**Ferritic**<br>ASTM A48: Class 20<br>SAE J431c: Grade G1800<br>**Pearlitic- Ferritic**<br>ASTM A48: Class 25<br>SAE J431c: Grade G2500<br>**Pearlitic**<br>ASTM A48: Classes 30, 35, 40<br>SAE J431c: Grade G3000 | 48 $R_C$<br>to<br>60 $R_C$ | Flame<br>or<br>Induction<br>Hardened | — | 5000<br>to<br>6000 | 60 | Rough:<br>.002 | .060 | D80-N100-M-1/8 |
| | | | | | 65 | Finish:<br>.0005 max. | (Max: 1/4 of Wheel Width) | D100-N100-M-1/8 |
| | | | — | 25<br>to<br>30 | 18 | Rough:<br>.050 | 1.50 | D80-N100-M-1/8 |
| | | | | | 20 | Finish:<br>.013 max. | (Max: 1/4 of Wheel Width) | D100-N100-M-1/8 |
| **Pearlitic + Free Carbides**<br>ASTM A48: Classes 45, 50<br>SAE J431c: Grades G3500, G4000<br>**Pearlitic or Acicular + Free Carbides**<br>ASTM A48: Classes 55, 60 | | | | | | | | |

*See section 16 for Cutting Fluid Recommendations.
See section 20 for additional information.

†Metal-coated diamond.

# 8.3 Surface Grinding-Horizontal Spindle, Reciprocating Table

## Diamond Wheels

| MATERIAL | HARD-NESS<br>Bhn | CONDITION | WHEEL SPEED DRY<br>fpm<br>m/s | WHEEL SPEED WET*<br>fpm<br>m/s | TABLE SPEED<br>fpm<br>m/min | DOWNFEED<br>in/pass<br>mm/pass | CROSSFEED<br>in/pass<br>mm/pass | WHEEL IDENTIFICATION<br>ANSI<br>ANSI |
|---|---|---|---|---|---|---|---|---|
| **48. MACHINABLE CARBIDES**<br>Ferro-Tic | 68 $R_C$ to 72 $R_C$ | Hardened and Tempered | — | 1300 to 1500 | 20 | Rough: .002 | .030 | D100-L100-B-1/8 |
| | | | | | 30 | Finish: .0005 max. | (Max: 1/4 of Wheel Width) | D150-L100-B-1/8 |
| | | | — | 6.5 to 7.5 | 6 | Rough: .050 | .75 | D100-L100-B-1/8 |
| | | | | | 9 | Finish: .013 max. | (Max: 1/4 of Wheel Width) | D150-L100-B-1/8 |
| **49. CARBIDES**<br>Titanium Carbide<br>Tungsten Carbide | 89 $R_A$ to 94 $R_A$ | — | — | 5000 to 6000 | 40 | Rough: .001 | .050 | D150-R100-B-1/4 |
| | | | | | 50 | Finish: .0005 max. | (Max: 1/4 of Wheel Width) | D220-R75-B-1/8 |
| | | | — | 25 to 30 | 12 | Rough: .025 | 1.25 | D150-R100-B-1/4 |
| | | | | | 15 | Finish: .013 max. | (Max: 1/4 of Wheel Width) | D220-R75-B-1/8 |
| | 89 $R_A$ to 94 $R_A$ | — | 2500 to 3500 | — | 5 | Rough: .002 | .030 | D200-N100-M-1/4 |
| | | | | | 6 | Finish: .001 max. | (Max: 1/4 of Wheel Width) | D240-N75-M-1/8 |
| | | | 13 to 18 | — | 1.5 | Rough: .050 | .75 | D200-N100-M-1/4 |
| | | | | | 1.8 | Finish: .025 max. | (Max: 1/4 of Wheel Width) | D240-N75-M-1/8 |
| **51. MAGNETIC ALLOYS**<br>Magnetic Core Iron (up to 4% Si)<br>Hi Perm 49<br>HyMu 80 | 185 to 240 | Wrought | — | 5000 to 6000 | 30 | Rough: .005 | .030 | D120-N75-M-1/8 |
| | | | | | 40 | Finish: .001 max. | (Max: 1/8 of Wheel Width) | D150-N75-M-1/8 |
| | | | — | 25 to 30 | 9 | Rough: .13 | .75 | D120-N75-M-1/8 |
| | | | | | 12 | Finish: .025 max. | (Max: 1/8 of Wheel Width) | D150-N75-M-1/8 |
| Alnico I<br>Alnico II<br>Alnico III<br>Alnico IV<br>Alnico V<br>Alnico V-7<br>Alnico XII<br>Columax-5<br>Hyflux Alnico V-7 | 45 $R_C$ to 58 $R_C$ | As Cast | — | 2500 to 3500 | 8 | Rough: .002 | .030 | D150-R100-B-1/8 |
| | | | | | 10 | Finish: .0005 max. | (Max: 1/8 of Wheel Width) | D220-R75-B-1/8 |
| | | | — | 13 to 18 | 2.4 | Rough: .050 | .75 | D150-R100-B-1/8 |
| | | | | | 3 | Finish: .013 max. | (Max: 1/8 of Wheel Width) | D220-R75-B-1/8 |
| **54. CARBONS & GRAPHITES**<br>Mechanical Grades<br>Brush Grade Carbon | 8 to 100 Shore | Molded or Extruded | 5500 to 6500 | — | 40 | Rough: .005 | .050 | D60-N50-M-1/8 |
| | | | | | 50 | Finish: .001 max. | (Max: 1/4 of Wheel Width) | D150-N50-M-1/8 |
| | | | 28 to 33 | — | 12 | Rough: .13 | 1.25 | D60-N50-M-1/8 |
| | | | | | 15 | Finish: .025 max. | (Max: 1/4 of Wheel Width) | D150-N50-M-1/8 |

*See section 16 for Cutting Fluid Recommendations.
See section 20 for additional information.

| MATERIAL | HARDNESS | CONDITION | WHEEL SPEED DRY | WHEEL SPEED WET* | TABLE SPEED | DOWNFEED | CROSSFEED | WHEEL IDENTIFICATION |
|---|---|---|---|---|---|---|---|---|
| | Bhn | | fpm / m/s | fpm / m/s | fpm / m/min | in/pass / mm/pass | in/pass / mm/pass | ANSI / ANSI |
| Graphite | — | — | 5500 to 6500 | — | 30 | Rough: .003 | .050 | D60-N50-M-1/8 |
| | | | | | 40 | Finish: .001 max | (Max: 1/4 of Wheel Width) | D220-N50-M-1/8 |
| | | | 28 to 33 | — | 9 | Rough: .075 | 1.25 | D60-N50-M-1/8 |
| | | | | | 12 | Finish: .025 max. | (Max: 1/4 of Wheel Width) | D220-N50-M-1/8 |
| **55. GLASSES & CERAMICS** Machinable Glass-Ceramic (MACOR) | 250 Knoop 100g | Cast | — | 6500 to 7500 | 6 | Rough: .010 | .050 | D100-N100-M-1/8 |
| | | | | | 8 | Finish: .002 max. | (Max: 1/6 of Wheel Width) | D220-N100-M-1/8 |
| | | | — | 33 to 38 | 1.8 | Rough: .25 | 1.25 | D100-N100-M-1/8 |
| | | | | | 2.4 | Finish: .050 max. | (Max: 1/6 of Wheel Width) | D220-N100-M-1/8 |
| PYROCERAM | 700 Knoop 100g | — | — | 5500 to 6500 | 30 | Rough: .003 | .050 | D60-N100-M-1/8 |
| | | | | | 40 | Finish: .001 max. | (Max: 1/8 of Wheel Width) | D220-N100-M-1/8 |
| | | | — | 28 to 33 | 9 | Rough: .075 | 1.25 | D60-N100-M-1/8 |
| | | | | | 12 | Finish: .025 max. | (Max: 1/8 of Wheel Width) | D220-N100-M-1/8 |
| Glass | 480 to 530 Knoop 100g | — | — | 5500 to 6500 | 20 | Rough: .003 | .030 | D60-N100-M-1/8 |
| | | | | | 30 | Finish: .0005 max. | (Max: 1/8 of Wheel Width) | D220-N100-M-1/8 |
| | | | — | 28 to 33 | 6 | Rough: .075 | .75 | D60-N100-M-1/8 |
| | | | | | 9 | Finish: .013 max. | (Max: 1/8 of Wheel Width) | D220-N100-M-1/8 |
| **Ceramics** Alumina (Aluminum Oxide), Alumina-Mullite, Aluminum Silicate, Beryllia (Beryllium Oxide)††, Magnesia (Magnesium Oxide), Mullite, Silicon Carbide, Silicon Nitride, Thoria (Thorium Oxide)††, Titania (Titanium Oxide), Titanium Diboride††, Zircon (Zirconium Silicate), Zirconia (Zirconium Oxide) | Over 800 Knoop | Fired | — | 4500 to 5500 | 50 | Rough: .002 | .080 | D120-N100-M-1/8 |
| | | | | | 60 | Finish: .0005 max. | (Max: 1/4 of Wheel Width) | D220-N100-M-1/8 |
| | | | — | 23 to 28 | 15 | Rough: .050 | 2.00 | D120-N100-M-1/8 |
| | | | | | 18 | Finish: .013 max. | (Max: 1/4 of Wheel Width) | D220-N100-M-1/8 |

*See section 16 for Cutting Fluid Recommendations.
See section 20 for additional information.

††CAUTION: Toxic Material, refer to National Institute for Occupational Safety and Health (NIOSH) for Precautions.

# 8.3 Surface Grinding-Horizontal Spindle, Reciprocating Table
## Diamond Wheels

| MATERIAL | HARD-NESS | CONDITION | WHEEL SPEED DRY | WHEEL SPEED WET* | TABLE SPEED | DOWNFEED | CROSSFEED | WHEEL IDENTIFICATION |
|---|---|---|---|---|---|---|---|---|
| | Bhn | | fpm / m/s | fpm / m/s | fpm / m/min | in/pass / mm/pass | in/pass / mm/pass | ANSI / ANSI |
| **57. COMPOSITES**<br>Kevlar 49<br>Graphite Epoxy<br>Fiberglass Epoxy<br>(E) Glass<br>(S) Glass<br>Boron Epoxy | — | — | 5500 to 6500 | — | 5 | Rough: .003 | .030 | D180-R100-M-1/8 |
| | | | | | 6 | Finish: .001 max. | (Max: 1/8 of Wheel Width) | D220-N100-M-1/8 |
| | | | 28 to 33 | — | 1.5 | Rough: .075 | .75 | D180-R100-M-1/8 |
| | | | | | 1.8 | Finish: .025 max. | (Max: 1/8 of Wheel Width) | D220-N100-M-1/8 |
| **58. FLAME (THERMAL) SPRAYED MATERIALS**<br>**Sprayed Carbides**<br>Chromium Carbide<br>Chromium Carbide-Cobalt Blend<br>Columbium Carbide<br>Tantalum Carbide<br>Titanium Carbide<br>Tungsten Carbide<br>Tungsten Carbide-Cobalt<br>Tungsten Carbide (Cobalt)-Nickel Alloy Blend | — | — | 5500 to 6500 | 5500 to 6500 | 35 | Rough: .002 | .030 | D120-N100-B-1/8 |
| | | | | | 50 | Finish: .0005 max. | (Max: 1/8 of Wheel Width) | D320-R75-B-1/8 |
| | | | 28 to 33 | 28 to 33 | 11 | Rough: .050 | .75 | D120-N100-B-1/8 |
| | | | | | 15 | Finish: .013 max. | (Max: 1/8 of Wheel Width) | D320-R75-B-1/8 |
| **Inorganic Coating Materials**<br>Alumina (Pure)<br>Alumina (Grey) containing Titania<br>Alumina, Nickel-Aluminide Blends<br>Barium Titanate<br>Boron††<br>Calcium Titanate<br>Calcium Zirconate<br>Chromium Disilicide††<br>Chromium Oxide††<br>Cobalt (40%), Zirconia Blend<br>Columbium (Niobium)††<br>Glass (Kovar sealing)††<br>Hexaboron Silicide††<br>Magnesia Alumina Spinel<br>Magnesium Zirconate<br>Molybdenum Disilicide<br>Mullite<br>Nickel (40%), Alumina Blend<br>Nickel Oxide<br>Rare Earth Oxides<br>Tantalum<br>Titania (50%), Alumina Blend<br>Titanium Oxide<br>Tungsten<br>Yttrium Zirconate<br>Zirconia (Lime Stabilized)<br>Zirconia, Nickel-Aluminide Blends<br>Zirconium Oxide (Hafnia Free, Lime Stabilized)<br>Zirconium Silicate | — | — | 5500 to 6500 | 5500 to 6500 | 35 | Rough: .003 | .030 | D120-N100-M-1/8 |
| | | | | | 50 | Finish: .001 max. | (Max: 1/8 of Wheel Width) | D220-N75-M-1/8 |
| | | | 28 to 33 | 28 to 33 | 11 | Rough: .075 | .75 | D120-N100-M-1/8 |
| | | | | | 15 | Finish: .025 max. | (Max: 1/8 of Wheel Width) | D220-N75-M-1/8 |

*See section 16 for Cutting Fluid Recommendations.
See section 20 for additional information.

††CAUTION: Toxic Material, refer to National Institute for Occupational Safety and Health (NIOSH) for Precautions.

| MATERIAL | HARDNESS | CONDITION | WHEEL SPEED | TABLE* (WORK) SPEED | DOWNFEED per rev. of table | OPERATION | WHEEL IDENTIFICATION NARROW WORK AREA | BROAD WORK AREA |
|---|---|---|---|---|---|---|---|---|
| | | | fpm | fpm | in | | ANSI | ANSI |
| | Bhn | | m/s | m/min | mm | | ISO | ISO |
| **1. FREE MACHINING CARBON STEELS, WROUGHT**<br>**Low Carbon Resulfurized**<br>1108 1115 1118 1212<br>1109 1116 1119 1213<br>1110 1117 1211 1215<br>**Medium Carbon Resulfurized**<br>1132 1140 1145<br>1137 1141 1146<br>1139 1144 1151<br>**Low Carbon Leaded**<br>10L18 12L13 12L15<br>11L17 12L14<br>**Medium Carbon Leaded**<br>10L45 11L37 11L44<br>10L50 11L41 | 50 $R_C$ max. | Hot Rolled, Normalized, Annealed, Cold Drawn or Quenched and Tempered | 3500<br>to<br>6000 | 80<br>to<br>200 | .001<br>to<br>.005 | Roughing<br>Finishing | A30IV<br>A80IB | A30HV<br>A80HB |
| | | | 18<br>to<br>30 | 24<br>to<br>60 | .025<br>to<br>.13 | Roughing<br>Finishing | A30IV<br>A80IB | A30HV<br>A80HB |
| | Over 50 $R_C$ | Carburized and/or Quenched and Tempered | 3500<br>to<br>6000 | 100<br>to<br>250 | .0005<br>to<br>.001 | Roughing<br>Finishing | A46GV<br>A80GB | A46FV<br>A80FB |
| | | | 18<br>to<br>30 | 30<br>to<br>76 | .013<br>to<br>.025 | Roughing<br>Finishing | A46GV<br>A80GB | A46FV<br>A80FB |
| | | | | | | | | |
| | | | | | | | | |
| **2. CARBON STEELS, WROUGHT**<br>**Low Carbon**<br>1005 1012 1019 1026<br>1006 1013 1020 1029<br>1008 1015 1021 1513<br>1009 1016 1022 1518<br>1010 1017 1023 1522<br>1011 1018 1025<br>**Medium Carbon**<br>1030 1042 1053 1541<br>1033 1043 1055 1547<br>1035 1044 1524 1548<br>1037 1045 1525 1551<br>1038 1046 1526 1552<br>1039 1049 1527<br>1040 1050 1536<br>**High Carbon**<br>1060 1074 1085 1566<br>1064 1075 1086 1572<br>1065 1078 1090<br>1069 1080 1095<br>1070 1084 1561 | 50 $R_C$ max. | Hot Rolled, Normalized, Annealed, Cold Drawn or Quenched and Tempered | 3500<br>to<br>6000 | 80<br>to<br>200 | .001<br>to<br>.005 | Roughing<br>Finishing | A30IV<br>A80IB | A30HV<br>A80HB |
| | | | 18<br>to<br>30 | 24<br>to<br>60 | .025<br>to<br>.13 | Roughing<br>Finishing | A30IV<br>A80IB | A30HV<br>A80HB |
| | Over 50 $R_C$ | Carburized and/or Quenched and Tempered | 3500<br>to<br>6000 | 100<br>to<br>250 | .0005<br>to<br>.001 | Roughing<br>Finishing | A46GV<br>A80GB | A46FV<br>A80FB |
| | | | 18<br>to<br>30 | 30<br>to<br>76 | .013<br>to<br>.025 | Roughing<br>Finishing | A46GV<br>A80GB | A46FV<br>A80FB |
| | | | | | | | | |
| | | | | | | | | |
| | | | | | | | | |
| | | | | | | | | |
| **3. CARBON AND FERRITIC ALLOY STEELS (HIGH TEMPERATURE SERVICE)**<br>ASTM A369: Grades FPA, FPB, FP1, FP2, FP3b, FP5, FP7, FP9, FP11, FP12, FP21, FP22 | 150<br>to<br>200 | As Forged, Annealed or Normalized and Tempered | 3500<br>to<br>6000 | 80<br>to<br>200 | .001<br>to<br>.005 | Roughing<br>Finishing | A30IV<br>A80IB | A30HV<br>A80HB |
| | | | 18<br>to<br>30 | 24<br>to<br>60 | .025<br>to<br>.13 | Roughing<br>Finishing | A30IV<br>A80IB | A30HV<br>A80HB |
| | | | | | | | | |
| | | | | | | | | |
| | | | | | | | | |
| | | | | | | | | |

See section 16 for Cutting Fluid Recommendations.
*See section 20.4 for calculation of Table rpm and Downfeed in ipm.

# 8.4 Surface Grinding-Vertical Spindle, Rotary Table

| MATERIAL | HARD-NESS | CONDITION | WHEEL SPEED | TABLE* (WORK) SPEED | DOWNFEED per rev. of table | OPERATION | WHEEL IDENTIFICATION NARROW WORK AREA | WHEEL IDENTIFICATION BROAD WORK AREA |
|---|---|---|---|---|---|---|---|---|
| | Bhn | | fpm<br>m/s | fpm<br>m/min | in<br>mm | | ANSI<br>ISO | ANSI<br>ISO |
| **4. FREE MACHINING ALLOY STEELS, WROUGHT**<br>**Medium Carbon Resulfurized**<br>4140 4142Te 4147Te<br>4140Se 4145Se 4150<br>**Medium and High Carbon Leaded**<br>41L30 41L47 51L32 86L40<br>41L40 41L50 52L100<br>41L45 43L40 86L20 | 50 $R_C$ max. | Hot Rolled, Normalized, Annealed, Cold Drawn or Quenched and Tempered | 3500 to 6000 | 80 to 200 | .001 to .005 | Roughing<br>Finishing | A30IV<br>A80IB | A30HV<br>A80HB |
| | | | 18 to 30 | 24 to 60 | .025 to .13 | Roughing<br>Finishing | A30IV<br>A80IB | A30HV<br>A80HB |
| | Over 50 $R_C$ | Carburized and/or Quenched and Tempered | 3500 to 6000 | 100 to 250 | .0005 to .001 | Roughing<br>Finishing | A46GV<br>A80GB | A46FV<br>A80FB |
| | | | 18 to 30 | 30 to 76 | .013 to .025 | Roughing<br>Finishing | A46GV<br>A80GB | A46FV<br>A80FB |
| **5. ALLOY STEELS, WROUGHT**<br>**Low Carbon**<br>4012 4615 4817 8617<br>4023 4617 4820 8620<br>4024 4620 5015 8622<br>4118 4621 5115 8822<br>4320 4718 5120 9310<br>4419 4720 6118 94B15<br>4422 4815 8115 94B17<br>**Medium Carbon**<br>1330 4145 5132 8640<br>1335 4147 5135 8642<br>1340 4150 5140 8645<br>1345 4161 5145 86B45<br>4027 4340 5147 8650<br>4028 4427 5150 8655<br>4032 4626 5155 8660<br>4037 50B40 5160 8740<br>4042 50B44 51B60 8742<br>4047 5046 6150 9254<br>4130 50B46 81B45 9255<br>4135 50B50 8625 9260<br>4137 5060 8627 94B30<br>4140 50B60 8630<br>4142 5130 8637<br>**High Carbon**<br>50100 51100 52100 M-50 | 50 $R_C$ max. | Hot Rolled, Normalized, Annealed, Cold Drawn or Quenched and Tempered | 3500 to 6000 | 80 to 200 | .001 to .005 | Roughing<br>Finishing | A30IV<br>A80IB | A30HV<br>A80HB |
| | | | 18 to 30 | 24 to 60 | .025 to .13 | Roughing<br>Finishing | A30IV<br>A80IB | A30HV<br>A80HB |
| | Over 50 $R_C$ | Carburized and/or Quenched and Tempered | 3500 to 4500 | 100 to 250 | .0005 to .001 | Roughing<br>Finishing | A46GV<br>A80GB | A46FV<br>A80FB |
| | | | 18 to 23 | 30 to 76 | .013 to .025 | Roughing<br>Finishing | A46GV<br>A80GB | A46FV<br>A80FB |
| **6. HIGH STRENGTH STEELS, WROUGHT**<br>300M<br>4330V<br>4340<br>4340Si<br>98BV40<br>D6ac<br>H11<br>H13<br>HP 9-4-20<br>HP 9-4-25<br>HP 9-4-30<br>HP 9-4-45 | 50 $R_C$ max. | Annealed, Normalized or Quenched and Tempered | 3500 to 6000 | 80 to 200 | .001 to .003 | Roughing<br>Finishing | A30HV<br>A80HB | A30GV<br>A80GB |
| | | | 18 to 30 | 24 to 60 | .025 to .075 | Roughing<br>Finishing | A30HV<br>A80HB | A30GV<br>A80GB |
| | Over 50 $R_C$ | Quenched and Tempered | 3500 to 4500 | 100 to 250 | .0005 to .001 | Roughing<br>Finishing | A46GV<br>A100GB | A46FV<br>A100FB |
| | | | 18 to 23 | 30 to 76 | .013 to .025 | Roughing<br>Finishing | A46GV<br>A100GB | A46FV<br>A100FB |

See section 16 for Cutting Fluid Recommendations.
*See section 20.4 for calculation of Table rpm and Downfeed in ipm.

| MATERIAL | HARD-NESS<br>Bhn | CONDITION | WHEEL SPEED<br>fpm<br>m/s | TABLE* (WORK) SPEED<br>fpm<br>m/min | DOWNFEED per rev. of table<br>in<br>mm | OPERATION | WHEEL IDENTIFICATION<br>NARROW WORK AREA<br>ANSI<br>ISO | <br>BROAD WORK AREA<br>ANSI<br>ISO |
|---|---|---|---|---|---|---|---|---|
| **7. MARAGING STEELS, WROUGHT**<br>ASTM A538: Grades A, B, C<br>120 Grade<br>180 Grade<br>200 Grade<br>250 Grade<br>300 Grade<br>350 Grade<br>HY230 | 50 $R_C$ max. | Annealed or Maraged | 3500<br>to<br>6000 | 80<br>to<br>200 | .001<br>to<br>.005 | Roughing<br><br>Finishing | A30IV<br><br>A80IB | A30HV<br><br>A80HB |
| | | | 18<br>to<br>30 | 24<br>to<br>60 | .025<br>to<br>.13 | Roughing<br><br>Finishing | A30IV<br><br>A80IB | A30HV<br><br>A80HB |
| | Over 50 $R_C$ | Maraged | 3500<br>to<br>6000 | 100<br>to<br>250 | .0005<br>to<br>.001 | Roughing<br><br>Finishing | A46GV<br><br>A80GB | A46FV<br><br>A80FB |
| | | | 18<br>to<br>30 | 30<br>to<br>76 | .013<br>to<br>.025 | Roughing<br><br>Finishing | A46GV<br><br>A80GB | A46FV<br><br>A80FB |
| **8. TOOL STEELS, WROUGHT**<br>**Group I**<br>A2 H13 L6 P20<br>A3 H14 L7 P21<br>A4 H19 O1 S1<br>A6 H21 O2 S2<br>A8 H22 O6 S5<br>A9 H23 O7 S6<br>A10 H24 P2 S7<br>H10 H25 P4 W1<br>H11 H26 P5 W2<br>H12 L2 P6 W5<br>SAE J438b: Types W108, W109, W110, W112, W209, W210, W310 | 50 $R_C$ max. | Annealed or Quenched and Tempered | 3500<br>to<br>6000 | 80<br>to<br>200 | .001<br>to<br>.003 | Roughing<br><br>Finishing | A30IV<br><br>A80IB | A30HV<br><br>A80HB |
| | | | 18<br>to<br>30 | 24<br>to<br>60 | .025<br>to<br>.075 | Roughing<br><br>Finishing | A30IV<br><br>A80IB | A30HV<br><br>A80HB |
| | Over 50 $R_C$ | Quenched and Tempered | 3500<br>to<br>4500 | 100<br>to<br>250 | .0005<br>to<br>.001 | Roughing<br><br>Finishing | A46HV<br><br>A100HB | A46GV<br><br>A100GB |
| | | | 18<br>to<br>23 | 30<br>to<br>76 | .013<br>to<br>.025 | Roughing<br><br>Finishing | A46HV<br><br>A100HB | A46GV<br><br>A100GB |
| **Group II**<br>D2 F2 M10 T4<br>D3 H42 M30 T8<br>D4 M1 M42<br>D5 M2 T1<br>F1 M3-1 T2 | 50 $R_C$ max. | Annealed or Quenched and Tempered | 3500<br>to<br>6000 | 80<br>to<br>200 | .001<br>to<br>.003 | Roughing<br><br>Finishing | A30HV<br><br>A80HB | A30GV<br><br>A80GB |
| | | | 18<br>to<br>30 | 24<br>to<br>60 | .025<br>to<br>.075 | Roughing<br><br>Finishing | A30HV<br><br>A80HB | A30GV<br><br>A80GB |
| | 50 $R_C$ to 58 $R_C$ | Quenched and Tempered | 3500<br>to<br>4500 | 100<br>to<br>250 | .0005<br>to<br>.001 | Roughing<br><br>Finishing | A46GV<br><br>A100GB | A46FV<br><br>A100FB |
| | | | 18<br>to<br>23 | 30<br>to<br>76 | .013<br>to<br>.025 | Roughing<br><br>Finishing | A46GV<br><br>A100GB | A46FV<br><br>A100FB |
| | Over 58 $R_C$ | Quenched and Tempered | 3500<br>to<br>4500 | 100<br>to<br>250 | .0005<br>to<br>.001 | Roughing<br><br>Finishing | A46GV<br><br>A100GB | A46FV<br><br>A100FB |
| | | | 18<br>to<br>23 | 24<br>to<br>76 | .013<br>to<br>.025 | Roughing<br><br>Finishing | A46GV<br><br>A100GB | A46FV<br><br>A100FB |
| **Group III**<br>A7 M7 M41 M47<br>D7 M33 M43 T5<br>M3-2 M34 M44 T6<br>M6 M36 M46 | 50 $R_C$ max. | Annealed or Quenched and Tempered | 3500<br>to<br>6000 | 80<br>to<br>200 | .001<br>to<br>.003 | Roughing<br><br>Finishing | A30HV<br><br>A80HB | A30GV<br><br>A80GB |
| | | | 18<br>to<br>30 | 24<br>to<br>60 | .025<br>to<br>.075 | Roughing<br><br>Finishing | A30HV<br><br>A80HB | A30GV<br><br>A80GB |
| | 50 $R_C$ to 58 $R_C$ | Quenched and Tempered | 3500<br>to<br>4500 | 100<br>to<br>250 | .0005<br>to<br>.001 | Roughing<br><br>Finishing | A46GV<br><br>A100GB | A46FV<br><br>A100FB |
| | | | 18<br>to<br>23 | 30<br>to<br>76 | .013<br>to<br>.025 | Roughing<br><br>Finishing | A46GV<br><br>A100GB | A46FV<br><br>A100FB |

See section 16 for Cutting Fluid Recommendations.
*See section 20.4 for calculation of Table rpm and Downfeed in ipm.

# 8.4 Surface Grinding-Vertical Spindle, Rotary Table

| MATERIAL | HARD-NESS<br>Bhn | CONDITION | WHEEL SPEED<br>fpm<br>m/s | TABLE* (WORK) SPEED<br>fpm<br>m/min | DOWNFEED per rev. of table<br>in<br>mm | OPERATION | WHEEL IDENTIFICATION<br>NARROW WORK AREA<br>ANSI<br>ISO | <br>BROAD WORK AREA<br>ANSI<br>ISO |
|---|---|---|---|---|---|---|---|---|
| **8. TOOL STEELS, WROUGHT (cont.)**<br>**Group III (cont.)**<br>(materials listed on preceding page) | Over 58 $R_C$ | Quenched and Tempered | 3500 to 4500 | 100 to 250 | .0005 to .001 | Roughing<br>Finishing | A46GV<br>A100GB | A46FV<br>A100FB |
| | | | 18 to 23 | 30 to 76 | .013 to .025 | Roughing<br>Finishing | A46GV<br>A100GB | A46FV<br>A100FB |
| **Group IV**<br>M4<br>T15 | 50 $R_C$ max. | Annealed or Quenched and Tempered | 3500 to 6000 | 80 to 200 | .001 to .003 | Roughing<br>Finishing | A30HV<br>A80HB | A30GV<br>A80GB |
| | | | 18 to 30 | 24 to 60 | .025 to .075 | Roughing<br>Finishing | A30HV<br>A80HB | A30GV<br>A80GB |
| | 50 $R_C$ to 58 $R_C$ | Quenched and Tempered | 3500 to 4500 | 100 to 250 | .0005 to .001 | Roughing<br>Finishing | A46GV<br>A100GB | A46FV<br>A100FB |
| | | | 18 to 23 | 30 to 76 | .013 to .025 | Roughing<br>Finishing | A46GV<br>A100GB | A46FV<br>A100FB |
| | Over 58 $R_C$ | Quenched and Tempered | 3500 to 4500 | 100 to 250 | .0005 to .001 | Roughing<br>Finishing | A46GV<br>A100GB | A46FV<br>A100FB |
| | | | 18 to 23 | 30 to 76 | .013 to .025 | Roughing<br>Finishing | A46GV<br>A100GB | A46FV<br>A100FB |
| **9. NITRIDING STEELS, WROUGHT**<br>Nitralloy 125<br>Nitralloy 135<br>Nitralloy 135 Mod.<br>Nitralloy 225<br>Nitralloy 230<br>Nitralloy EZ<br>Nitralloy N<br>Nitrex 1 | 200 to 350 | Annealed, Normalized or Quenched and Tempered | 3500 to 6000 | 80 to 200 | .001 to .005 | Roughing<br>Finishing | A30HV<br>A80HB | A30GV<br>A80GB |
| | | | 18 to 30 | 24 to 60 | .025 to .13 | Roughing<br>Finishing | A30HV<br>A80HB | A30GV<br>A80GB |
| | 60 $R_C$ to 65 $R_C$ | Nitrided | 3500 to 4500 | 100 to 250 | .0005 to .001 | Roughing<br>Finishing | A46GV<br>A100GB | A46FV<br>A100FB |
| | | | 18 to 23 | 30 to 76 | .013 to .025 | Roughing<br>Finishing | A46GV<br>A100GB | A46FV<br>A100FB |
| **10. ARMOR PLATE, SHIP PLATE, AIRCRAFT PLATE, WROUGHT**<br>HY80<br>MIL-S-12560<br>HY100<br>MIL-S-16216<br>HY180 | 45 $R_C$ max. | Annealed or Quenched and Tempered | 3500 to 6000 | 80 to 200 | .001 to .003 | Roughing<br>Finishing | A30HV<br>A80HB | A30GV<br>A80GB |
| | | | 18 to 30 | 24 to 60 | .025 to .075 | Roughing<br>Finishing | A30HV<br>A80HB | A30GV<br>A80GB |
| | | | | | | | | |
| **11. STRUCTURAL STEELS, WROUGHT****<br>30 60 90 150<br>35 65 100 160<br>42 70 110 165<br>45 75 135 185<br>50 80 140 210<br>55 85 145 | 48 $R_C$ max. | Hot Rolled, Normalized, Stress Relieved or Quenched and Tempered | 3500 to 6000 | 80 to 200 | .001 to .005 | Roughing<br>Finishing | A24IV<br>A80IB | A24HV<br>A80HB |
| | | | 18 to 30 | 24 to 60 | .025 to .13 | Roughing<br>Finishing | A24IV<br>A80IB | A24HV<br>A80HB |
| | Over 48 $R_C$ | Quenched and Tempered | 3500 to 6000 | 100 to 250 | .0005 to .001 | Roughing<br>Finishing | A30IV<br>A80GB | A30HV<br>A80FB |
| | | | 18 to 30 | 30 to 76 | .013 to .025 | Roughing<br>Finishing | A30IV<br>A80GB | A30HV<br>A80FB |

See section 16 for Cutting Fluid Recommendations.
*See section 20.4 for calculation of Table rpm and Downfeed in ipm.

**In this handbook, Structural Steels are designated by yield strength in units of 1000 psi. For example, 50 means a steel with 50,000 psi yield strength. Similarly, 140 means 140,000 psi yield strength. Therefore, in order to select the appropriate block of machining data, it is necessary to know the yield strength or Bhn.

| MATERIAL | HARD-NESS | CONDITION | WHEEL SPEED | TABLE* (WORK) SPEED | DOWNFEED per rev. of table | OPERATION | WHEEL IDENTIFICATION NARROW WORK AREA | BROAD WORK AREA |
|---|---|---|---|---|---|---|---|---|
| | | | fpm | fpm | in | | ANSI | ANSI |
| | Bhn | | m/s | m/min | mm | | ISO | ISO |
| **12. FREE MACHINING STAINLESS STEELS, WROUGHT** **Ferritic** 430F 430F Se | 135 to 185 | Annealed | 3500 to 6000 | 100 to 350 | .001 to .003 | Roughing | A46IV | A46GV |
| | | | | | | Finishing | A80HB | A80GB |
| | | | 18 to 30 | 30 to 105 | .025 to .075 | Roughing | A46IV | A46GV |
| | | | | | | Finishing | A80HB | A80GB |
| **Austenitic** 203EZ 303Pb 303 303 Plus X 303MA 303Se | 135 to 275 | Annealed or Cold Drawn | 3500 to 6000 | 100 to 350 | .001 to .003 | Roughing | CA46IB | CA46GB |
| | | | | | | Finishing | CA80IB | CA80GB |
| | | | 18 to 30 | 30 to 105 | .025 to .075 | Roughing | CA46IB | CA46GB |
| | | | | | | Finishing | CA80IB | CA80GB |
| **Martensitic** 416 420F Se 416 Plus X 440F 416Se 440F Se 420F | 135 to 240 | Annealed or Cold Drawn | 3500 to 6000 | 100 to 350 | .001 to .003 | Roughing | A46IV | A46GV |
| | | | | | | Finishing | A80HB | A80GB |
| | | | 18 to 30 | 30 to 105 | .025 to .075 | Roughing | A46IV | A46GB |
| | | | | | | Finishing | A80HB | A80GB |
| | over 275 | Quenched and Tempered | 3500 to 6000 | 100 to 350 | .001 to .003 | Roughing | A46GV | A46FV |
| | | | | | | Finishing | A80GB | A80FB |
| | | | 18 to 30 | 30 to 105 | .025 to .075 | Roughing | A46GV | A46FV |
| | | | | | | Finishing | A80GB | A80FB |
| **13. STAINLESS STEELS, WROUGHT** **Ferritic** 405 429 434 442 409 430 436 446 | 135 to 185 | Annealed | 3500 to 6000 | 100 to 350 | .001 to .003 | Roughing | A46IV | A46GV |
| | | | | | | Finishing | A80HB | A80GB |
| | | | 18 to 30 | 30 to 105 | .025 to .075 | Roughing | A46IV | A46GV |
| | | | | | | Finishing | A80HB | A80GB |
| **Austenitic** 201 304L 310S 330 202 305 314 347 301 308 316 348 302 309 316L 384 302B 309S 317 385 304 310 321 | 135 to 275 | Annealed or Cold Drawn | 3500 to 6000 | 100 to 350 | .001 to .003 | Roughing | CA46IB | CA46GB |
| | | | | | | Finishing | CA80IB | CA80GB |
| | | | 18 to 30 | 30 to 105 | .025 to .075 | Roughing | CA46IB | CA46GB |
| | | | | | | Finishing | CA80IB | CA80GB |
| | | | | | | | | |
| **Austenitic** Nitronic 32 Nitronic 50 Nitronic 33 Nitronic 60 Nitronic 40 | 210 to 375 | Annealed or Cold Drawn | 3500 to 6000 | 100 to 350 | .001 to .003 | Roughing | CA46IB | CA46GB |
| | | | | | | Finishing | CA80IB | CA80GB |
| | | | 18 to 30 | 30 to 105 | .025 to .075 | Roughing | CA46IB | CA46GB |
| | | | | | | Finishing | CA80IB | CA80GB |
| **Martensitic** 403 422 440C 410 431 501 414 440A 502 420 440B Greek Ascoloy | 135 to 275 | Annealed | 3500 to 6000 | 100 to 350 | .001 to .003 | Roughing | A46IV | A46GV |
| | | | | | | Finishing | A80HB | A80GB |
| | | | 18 to 30 | 30 to 105 | .025 to .075 | Roughing | A46IV | A46GV |
| | | | | | | Finishing | A80HB | A80GB |
| | Over 275 | Quenched and Tempered | 3500 to 6000 | 100 to 350 | .001 to .003 | Roughing | A46GV | A46FV |
| | | | | | | Finishing | A80GB | A80FB |
| | | | 18 to 30 | 30 to 105 | .025 to .075 | Roughing | A46GV | A46FV |
| | | | | | | Finishing | A80GB | A80FB |

See section 16 for Cutting Fluid Recommendations.
*See section 20.4 for calculation of Table rpm and Downfeed in ipm.

# 8.4 Surface Grinding-Vertical Spindle, Rotary Table

| MATERIAL | HARD-NESS<br>Bhn | CONDITION | WHEEL SPEED<br>fpm<br>m/s | TABLE* (WORK) SPEED<br>fpm<br>m/min | DOWNFEED per rev. of table<br>in<br>mm | OPERATION | WHEEL IDENTIFICATION<br>NARROW WORK AREA<br>ANSI<br>ISO | WHEEL IDENTIFICATION<br>BROAD WORK AREA<br>ANSI<br>ISO |
|---|---|---|---|---|---|---|---|---|
| **14. PRECIPITATION HARDENING STAINLESS STEELS, WROUGHT**<br>15-5 PH<br>16-6 PH<br>17-4 PH<br>17-7 PH<br>17-14 Cu Mo<br>AF-71<br>AFC-77<br>Almar 362 (AM-362)<br>AM-350<br>AM-355<br>AM-363<br>Custom 450<br>Custom 455<br>HNM<br>PH 13-8 Mo<br>PH 14-8 Mo<br>PH 15-7 Mo<br>Stainless W | 150 to 440 | Solution Treated or Hardened | 3500 to 6000 | 100 to 350 | .001 to .003 | Roughing<br>Finishing | A46IV<br>A80HB | A46GV<br>A80GB |
| | | | 18 to 30 | 30 to 105 | .025 to .075 | Roughing<br>Finishing | A46IV<br>A80HB | A46GV<br>A80GB |
| **15. CARBON STEELS, CAST**<br>**Low Carbon**<br>ASTM A426: Grade CP1<br>1010<br>1020<br>**Medium Carbon**<br>ASTM A352: Grades LCA, LCB, LCC<br>ASTM A356: Grade 1<br>1030 1040 1050 | 100 to 300 | Annealed, Normalized, Normalized and Tempered, or Quenched and Tempered | 3500 to 6000 | 80 to 200 | .001 to .005 | Roughing<br>Finishing | A30IV<br>A80IB | A30HV<br>A80HB |
| | | | 18 to 30 | 24 to 60 | .025 to .13 | Roughing<br>Finishing | A30IV<br>A80IB | A30HV<br>A80HB |
| | Over 50 $R_C$ | Carburized and/or Quenched and Tempered | 3500 to 6000 | 100 to 250 | .0005 to .001 | Roughing<br>Finishing | A46GV<br>A80GB | A46FV<br>A80FB |
| | | | 18 to 30 | 30 to 76 | .013 to .025 | Roughing<br>Finishing | A46GV<br>A80GB | A46FV<br>A80FB |
| **16. ALLOY STEELS, CAST**<br>**Low Carbon**<br>ASTM A217: Grade WC9<br>ASTM A352: Grades LC3, LC4<br>ASTM A426: Grades CP2, CP5, CP5b, CP11, CP12, CP15, CP21, CP22<br>1320 2320 4120 8020<br>2315 4110 4320 8620<br>**Medium Carbon**<br>ASTM A27: Grades N1, N2, U-60-30, 60-30, 65-35, 70-36, 70-40<br>ASTM A148: Grades 80-40, 80-50, 90-60, 105-85, 120-95, 150-125, 175-145<br>ASTM A216: Grades WCA, WCB, WCC<br>ASTM A217: Grades WC1, WC4, WC5, WC6<br>ASTM A352: Grades LC1, LC2, LC2-1<br>(materials continued on next page) | 150 to 400 | Annealed, Normalized, Normalized and Tempered, or Quenched and Tempered | 3500 to 6000 | 80 to 200 | .001 to .005 | Roughing<br>Finishing | A30IV<br>A80IB | A30HV<br>A80HB |
| | | | 18 to 30 | 24 to 60 | .025 to .13 | Roughing<br>Finishing | A30IV<br>A80IB | A30HV<br>A80HB |
| | Over 50 $R_C$ | Carburized and/or Quenched and Tempered | 3500 to 6000 | 100 to 250 | .0005 to .002 | Roughing<br>Finishing | A46GV<br>A80GB | A46FV<br>A80FB |
| | | | 18 to 30 | 30 to 76 | .013 to .050 | Roughing<br>Finishing | A46GV<br>A80GB | A46FV<br>A80FB |

See section 16 for Cutting Fluid Recommendations.
*See section 20.4 for calculation of Table rpm and Downfeed in ipm.

| MATERIAL | HARD-NESS | CONDITION | WHEEL SPEED | TABLE* (WORK) SPEED | DOWNFEED per rev. of table | OPERATION | WHEEL IDENTIFICATION NARROW WORK AREA | WHEEL IDENTIFICATION BROAD WORK AREA |
|---|---|---|---|---|---|---|---|---|
| | Bhn | | fpm / m/s | fpm / m/min | in / mm | | ANSI / ISO | ANSI / ISO |
| **16. ALLOY STEELS, CAST (cont.) Medium Carbon (cont.)** (materials continued from preceding page) ASTM A356: Grades 2, 5, 6, 8, 9, 10 ASTM A389: Grades C23, C24 ASTM A486: Classes 70, 90, 120 ASTM A487: Classes 1N, 2N, 4N, 6N, 8N, 9N, 10N, DN, 1Q, 2Q, 4Q, 4QA, 6Q, 7Q, 8Q, 9Q, 10Q 1330 4130 80B30 8640 1340 4140 8040 9525 2325 4330 8430 9530 2330 4340 8440 9535 4125 8030 8630 | | | | | | | | |
| **17. TOOL STEELS, CAST Group I** ASTM A597: Grades CA-2, CH-12, CH-13, CO-1, CS-5 | 50 $R_C$ max. | Annealed or Quenched and Tempered | 3500 to 6000 | 80 to 200 | .001 to .003 | Roughing / Finishing | A30IV / A80IB | A30HV / A80HB |
| | | | 18 to 30 | 24 to 60 | .025 to .075 | Roughing / Finishing | A30IV / A80IB | A30HV / A80HB |
| | 50 $R_C$ to 58 $R_C$ | Quenched and Tempered | 3500 to 4500 | 100 to 250 | .0005 to .002 | Roughing / Finishing | A46HV / A100HB | A46GV / A100GB |
| | | | 18 to 23 | 30 to 76 | .013 to .050 | Roughing / Finishing | A46HV / A100HB | A46GV / A100GB |
| | Over 58 $R_C$ | Quenched and Tempered | 3500 to 4500 | 100 to 250 | .0005 to .001 | Roughing / Finishing | A46GV / A100GB | A46FV / A100FB |
| | | | 18 to 23 | 30 to 76 | .013 to .025 | Roughing / Finishing | A46GV / A100GB | A46FV / A100FB |
| **Group II** ASTM A597: Grades CD-2, CD-5 | 200 to 250 | Annealed | 3500 to 6000 | 80 to 200 | .001 to .003 | Roughing / Finishing | A30IV / A80IB | A30HV / A80HB |
| | | | 18 to 30 | 24 to 60 | .025 to .075 | Roughing / Finishing | A30IV / A80IB | A30HV / A80HB |
| | 48 $R_C$ to 56 $R_C$ | Quenched and Tempered | 3500 to 4500 | 100 to 250 | .0005 to .002 | Roughing / Finishing | A46HV / A100HB | A46GV / A100GB |
| | | | 18 to 23 | 30 to 76 | .013 to .050 | Roughing / Finishing | A46HV / A100HB | A46GV / A100GB |
| | Over 56 $R_C$ | Quenched and Tempered | 3500 to 4500 | 100 to 250 | .0005 to .001 | Roughing / Finishing | A46GV / A100GB | A46FV / A100FB |
| | | | 18 to 23 | 30 to 76 | .013 to .025 | Roughing / Finishing | A46GV / A100GB | A46FV / A100FB |

See section 16 for Cutting Fluid Recommendations.
*See section 20.4 for calculation of Table rpm and Downfeed in ipm.

# 8.4 Surface Grinding-Vertical Spindle, Rotary Table

| MATERIAL | HARDNESS | CONDITION | WHEEL SPEED | TABLE* (WORK) SPEED | DOWNFEED per rev. of table | OPERATION | WHEEL IDENTIFICATION NARROW WORK AREA | WHEEL IDENTIFICATION BROAD WORK AREA |
|---|---|---|---|---|---|---|---|---|
| | Bhn | | fpm<br>m/s | fpm<br>m/min | in<br>mm | | ANSI<br>ISO | ANSI<br>ISO |
| **18. STAINLESS STEELS, CAST**<br>**Ferritic**<br>ASTM A217: Grades C5, C12<br>ASTM A296: Grades CB-30, CC-50, CE-30, CA6N, CA-6NM, CD4MCu<br>ASTM A297: Grade HC<br>ASTM A487: Class CA6NM<br>ASTM A608: Grade HC30 | 135 to 185 | Annealed | 3500 to 6000 | 100 to 350 | .001 to .003 | Roughing<br>Finishing | A30IV<br>A80IB | A30HV<br>A80HB |
| | | | 18 to 30 | 30 to 105 | .025 to .075 | Roughing<br>Finishing | A30IV<br>A80IB | A30HV<br>A80HB |
| **Austenitic**<br>ASTM A296: Grades CF-3, CF-3M, CF-8, CF-8C, CF-8M, CF-16F, CF-20, CG-8M, CG-12, CH-20, CK-20, CN-7M, CN-7MS<br>ASTM A297: Grades HD, HE, HF, HH, HI, HK, HL, HN, HP, HT, HU<br>ASTM A351: Grades CF-3, CF-3A, CF-3M, CF-3MA, CF-8, CF-8A, CF-8C, CF-8M, CF-10MC, CH-8, CH-10, CH-20, CK-20, CN-7M, HK-30, HK-40, HT-30<br>ASTM A451: Grades CPF8A, CPF3, CPF3A, CPF3M, CPF8, CPF8C, CPF8C (Ta Max.), CPF8M, CPF10MC, CPH8, CPH10, CPH20, CPK20<br>ASTM A452: Grades TP 304H, TP 316H, TP 347H<br>ASTM A608: Grades HD50, HE35, HF30, HH30, HH33, HI35, HK30, HK40, HL30, HL40, HN40, HT50, HU50 | 135 to 210 | Annealed, Normalized or As Cast | 3500 to 6000 | 100 to 350 | .001 to .003 | Roughing<br>Finishing | A30IV<br>A80IB | A30HV<br>A80HB |
| | | | 18 to 30 | 30 to 105 | .025 to .075 | Roughing<br>Finishing | A30IV<br>A80IB | A30HV<br>A80HB |
| **Martensitic**<br>ASTM A217: Grade CA-15<br>ASTM A296: Grades CA-15, CA-15M, CA-40<br>ASTM A426: Grades CP7, CP9, CPCA15<br>ASTM A487: Classes CA15a, CA-15M | 135 to 225 | Annealed, Normalized or Normalized and Tempered | 3500 to 6000 | 100 to 350 | .001 to .003 | Roughing<br>Finishing | A30IV<br>A80IB | A30HV<br>A80HB |
| | | | 18 to 30 | 30 to 105 | .025 to .075 | Roughing<br>Finishing | A30IV<br>A80IB | A30HV<br>A80HB |
| | Over 275 | Quenched and Tempered | 3500 to 6000 | 100 to 350 | .0005 to .002 | Roughing<br>Finishing | A46GV<br>A80GB | A46FV<br>A80FB |
| | | | 18 to 30 | 30 to 105 | .013 to .050 | Roughing<br>Finishing | A46GV<br>A80GB | A46FV<br>A80FB |

See section 16 for Cutting Fluid Recommendations.
*See section 20.4 for calculation of Table rpm and Downfeed in ipm.

| MATERIAL | HARD-NESS | CONDITION | WHEEL SPEED | TABLE* (WORK) SPEED | DOWNFEED per rev. of table | OPERATION | WHEEL IDENTIFICATION NARROW WORK AREA | WHEEL IDENTIFICATION BROAD WORK AREA |
|---|---|---|---|---|---|---|---|---|
| | Bhn | | fpm<br>m/s | fpm<br>m/min | in<br>mm | | ANSI<br>ISO | ANSI<br>ISO |
| **19. PRECIPITATION HARDENING STAINLESS STEELS, CAST**<br>ASTM A351: Grade CD-4MCu<br>ACI Grade CB-7Cu<br>ACI Grade CD-4MCu<br>17-4 PH<br>AM-355 | 325 to 450 | Solution Treated or Solution Treated and Aged | 3500 to 6000 | 80 to 200 | .001 to .003 | Roughing<br>Finishing | A30IV<br>A80IB | A30HV<br>A80HB |
| | | | 18 to 30 | 24 to 60 | .025 to .075 | Roughing<br>Finishing | A30IV<br>A80IB | A30HV<br>A80HB |
| | | | | | | | | |
| **21. GRAY CAST IRONS**<br>**Ferritic**<br>ASTM A48: Class 20<br>SAE J431c: Grade G1800<br>**Pearlitic- Ferritic**<br>ASTM A48: Class 25<br>SAE J431c: Grade G2500<br>**Pearlitic**<br>ASTM A48: Classes 30, 35, 40<br>SAE J431c: Grade G3000<br>**Pearlitic + Free Carbides**<br>ASTM A48: Classes 45, 50<br>SAE J431c: Grades G3500, G4000<br>**Pearlitic or Acicular + Free Carbides**<br>ASTM A48: Classes 55, 60 | 52 $R_C$ max. | As Cast, Annealed or Quenched and Tempered | 3500 to 6000 | 100 to 350 | .003 to .008 | Roughing<br>Finishing | A30IV<br>C80IB | A30HV<br>C80HB |
| | | | 18 to 30 | 30 to 105 | .075 to .20 | Roughing<br>Finishing | A30IV<br>C80IB | A30HV<br>C80HB |
| | | | | | | | | |
| | | | | | | | | |
| **Austenitic (NI-RESIST)**<br>ASTM A436: Types 1, 1b, 2, 2b, 3, 4, 5, 6 | 100 to 250 | As Cast | 3500 to 6000 | 100 to 350 | .001 to .005 | Roughing<br>Finishing | A30HV<br>C80HB | A30GV<br>C80GB |
| | | | 18 to 30 | 30 to 105 | .025 to .13 | Roughing<br>Finishing | A30HV<br>C80HB | A30GV<br>C80GB |
| **22. COMPACTED GRAPHITE CAST IRONS** | 185 to 225 | As Cast | 3500 to 6000 | 100 to 350 | .001 to .005 | Roughing<br>Finishing | A30IV<br>C80IB | A30HV<br>C80HB |
| | | | 18 to 30 | 30 to 105 | .025 to .13 | Roughing<br>Finishing | A30IV<br>C80IB | A30HV<br>C80HB |
| **23. DUCTILE CAST IRONS**<br>**Ferritic**<br>ASTM A536: Grades 60-40-18, 65-45-12<br>SAE J434c: Grades D4018, D4512<br>**Ferritic- Pearlitic**<br>ASTM A536: Grade 80-55-06<br>SAE J434c: Grade D5506<br>**Pearlitic- Martensitic**<br>ASTM A536: Grade 100-70-03<br>SAE J434c: Grade D7003<br>**Martensitic**<br>ASTM A536: Grade 120-90-02<br>SAE J434c: Grade DQ&T | 52 $R_C$ max. | Annealed, As Cast, Normalized and Tempered or Quenched and Tempered | 3500 to 6000 | 100 to 350 | .001 to .005 | Roughing<br>Finishing | A24HV<br>A80HB | A24GV<br>A80GB |
| | | | 18 to 30 | 30 to 105 | .025 to .13 | Roughing<br>Finishing | A24HV<br>A80HB | A24GV<br>A80GB |
| | | | | | | | | |
| | | | | | | | | |

See section 16 for Cutting Fluid Recommendations.
*See section 20.4 for calculation of Table rpm and Downfeed in ipm.

# 8.4 Surface Grinding-Vertical Spindle, Rotary Table

| MATERIAL | HARD-NESS | CONDITION | WHEEL SPEED | TABLE* (WORK) SPEED | DOWNFEED per rev. of table | OPERATION | WHEEL IDENTIFICATION NARROW WORK AREA | WHEEL IDENTIFICATION BROAD WORK AREA |
|---|---|---|---|---|---|---|---|---|
| | Bhn | | fpm<br>m/s | fpm<br>m/min | in<br>mm | | ANSI<br>ISO | ANSI<br>ISO |
| **23. DUCTILE CAST IRONS (cont.)**<br>**Austenitic (NI-RESIST Ductile)**<br>ASTM A439: Types D-2, D-2B, D-2C, D-3, D-3A, D-4, D-5, D-5B<br>ASTM A571: Type D-2M | 120<br>to<br>275 | Annealed | 3500<br>to<br>6000 | 100<br>to<br>350 | .001<br>to<br>.005 | Roughing<br>Finishing | A30HV<br>A80HB | A30GV<br>A80GB |
| | | | 18<br>to<br>30 | 30<br>to<br>105 | .025<br>to<br>.13 | Roughing<br>Finishing | A30HV<br>A80HB | A30GV<br>A80GB |
| **24. MALLEABLE CAST IRONS**<br>**Ferritic**<br>ASTM A47: Grades 32510, 35018<br>ASTM A602: Grade M3210<br>SAE J158: Grade M3210<br>**Pearlitic**<br>ASTM A220: Grades 40010, 45006, 45008, 50005<br>ASTM A602: Grade M4504, M5003<br>SAE J158: Grades M4504, M5003<br>**Tempered Martensite**<br>ASTM A220: Grades 60004, 70003, 80002, 90001<br>ASTM A602: Grades M5503, M7002, M8501<br>SAE J158: Grades M5503, M7002, M8501 | 52 $R_C$<br>max. | Malleablized<br>or<br>Malleablized<br>and<br>Heat Treated | 3500<br>to<br>6000 | 100<br>to<br>350 | .001<br>to<br>.005 | Roughing<br>Finishing | A30HV<br>A80HB | A30GV<br>A80GB |
| | | | 18<br>to<br>30 | 30<br>to<br>105 | .025<br>to<br>.13 | Roughing<br>Finishing | A30HV<br>A80HB | A30GV<br>A80GB |
| **25. WHITE CAST IRONS (ABRASION RESISTANT)**<br>ASTM A532:<br>Class I, Types A, B, C, D<br>Class II, Types A, B, C, D, E<br>Class III, Type A | 60 $R_C$<br>max. | As Cast,<br>Annealed or<br>Hardened | 3500<br>to<br>6000 | 100<br>to<br>350 | .001<br>to<br>.005 | Roughing<br>Finishing | C24HV<br>C80HB | C24GV<br>C80GB |
| | | | 18<br>to<br>30 | 30<br>to<br>105 | .025<br>to<br>.13 | Roughing<br>Finishing | C24HV<br>C80HB | C24GV<br>C80GB |
| **28. ALUMINUM ALLOYS, WROUGHT**<br>EC 2218 5252 6253<br>1060 2219 5254 6262<br>1100 2618 5454 6463<br>1145 3003 5456 6951<br>1175 3004 5457 7001<br>1235 3005 5652 7004<br>2011 4032 5657 7005<br>2014 5005 6053 7039<br>(materials continued on next page) | 30<br>to<br>150<br>500kg | Cold Drawn or<br>Solution Treated<br>and Aged | 3500<br>to<br>6000 | 100<br>to<br>250 | .001<br>to<br>.004 | Roughing<br>Finishing | CA46IB<br>CA60IB | CA46GB<br>CA60GB |
| | | | 18<br>to<br>30 | 30<br>to<br>76 | .025<br>to<br>.100 | Roughing<br>Finishing | CA46IB<br>CA60IB | CA46GB<br>CA60GB |

See section 16 for Cutting Fluid Recommendations.
*See section 20.4 for calculation of Table rpm and Downfeed in ipm.

| MATERIAL | HARD-NESS | CONDITION | WHEEL SPEED | TABLE* (WORK) SPEED | DOWNFEED per rev. of table | OPERATION | WHEEL IDENTIFICATION NARROW WORK AREA | BROAD WORK AREA |
|---|---|---|---|---|---|---|---|---|
| | Bhn | | fpm<br>m/s | fpm<br>m/min | in<br>mm | | ANSI<br>ISO | ANSI<br>ISO |
| **28. ALUMINUM ALLOYS, WROUGHT (cont.)**<br>(materials continued from preceding page)<br>2017 5050 6061 7049<br>2018 5052 6063 7050<br>2021 5056 6066 7075<br>2024 5083 6070 7079<br>2025 5086 6101 7175<br>2117 5154 6151 7178 | | | | | | | | |
| **29. ALUMINUM ALLOYS, CAST**<br>**Sand and Permanent Mold**<br>A140 295.0 514.0 A712.0<br>201.0 B295.0 A514.0 D712.0<br>208.0 308.0 B514.0 713.0<br>213.0 319.0 520.0 771.0<br>222.0 355.0 535.0 850.0<br>224.0 C355.0 705.0 A850.0<br>242.0 B443.0 707.0 B850.0<br>Hiduminium RR-350<br>**Die Castings**<br>C443.0 518.0 | 40<br>to<br>125<br>500kg | As Cast<br>or<br>Solution Treated<br>and Aged | 3500<br>to<br>6000 | 100<br>to<br>250 | .001<br>to<br>.004 | Roughing<br><br>Finishing | CA46IB<br><br>CA60IB | CA46GB<br><br>CA60GB |
| | | | 18<br>to<br>30 | 30<br>to<br>76 | .025<br>to<br>.100 | Roughing<br><br>Finishing | CA46IB<br><br>CA60IB | CA46GB<br><br>CA60GB |
| **Sand and Permanent Mold**<br>328.0 333.0 A356.0<br>A332.0 354.0 357.0<br>F332.0 356.0 359.0<br>**Die Castings**<br>360.0 A380.0 390.0 A413.0<br>A360.0 383.0 392.0<br>380.0 A384.0 413.0 | 40<br>to<br>125<br>500kg | As Cast<br>or<br>Solution Treated<br>and Aged | 3500<br>to<br>6000 | 100<br>to<br>250 | .001<br>to<br>.004 | Roughing<br><br>Finishing | CA46IB<br><br>CA60IB | CA46GB<br><br>CA60GB |
| | | | 18<br>to<br>30 | 30<br>to<br>76 | .025<br>to<br>.100 | Roughing<br><br>Finishing | CA46IB<br><br>CA60IB | CA46GB<br><br>CA60GB |
| **30. MAGNESIUM ALLOYS, WROUGHT‡‡**<br>AZ21A AZ61A HM21A ZK60A<br>AZ31B AZ80A HM31A<br>AZ31C HK31A ZK40A | 50<br>to<br>90<br>500kg | Annealed,<br>Cold Drawn<br>or<br>Solution Treated<br>and Aged | 3500<br>to<br>6000 | 100<br>to<br>250 | .001<br>to<br>.004 | Roughing<br><br>Finishing | CA46IB<br><br>CA60IB | CA46GB<br><br>CA60GB |
| | | | 18<br>to<br>30 | 30<br>to<br>76 | .025<br>to<br>.100 | Roughing<br><br>Finishing | CA46IB<br><br>CA60IB | CA46GB<br><br>CA60GB |
| **31. MAGNESIUM ALLOYS, CAST‡‡**<br>AM60A AZ91A HK31A ZE63A<br>AM100A AZ91B HZ32A ZH62A<br>AS41A AZ91C K1A ZK51A<br>AZ63A AZ92A QE22A ZK61A<br>AZ81A EZ33A ZE41A | 50<br>to<br>90<br>500kg | As Cast,<br>Annealed<br>or<br>Solution Treated<br>and Aged | 3500<br>to<br>6000 | 100<br>to<br>250 | .001<br>to<br>.004 | Roughing<br><br>Finishing | CA46IB<br><br>CA60IB | CA46GB<br><br>CA60GB |
| | | | 18<br>to<br>30 | 30<br>to<br>76 | .025<br>to<br>.100 | Roughing<br><br>Finishing | CA46IB<br><br>CA60IB | CA46GB<br><br>CA60GB |

See section 16 for Cutting Fluid Recommendations.
*See section 20.4 for calculation of Table rpm and Downfeed in ipm.

‡‡CAUTION: Potential Fire Hazard. Exercise caution in grinding and disposing of swarf. Do NOT use water or water-miscible cutting fluids for magnesium alloys.

# 8.4 Surface Grinding-Vertical Spindle, Rotary Table

| MATERIAL | HARD-NESS | CONDITION | WHEEL SPEED | TABLE* (WORK) SPEED | DOWNFEED per rev. of table | OPERATION | WHEEL IDENTIFICATION | |
|---|---|---|---|---|---|---|---|---|
| | | | | | | | NARROW WORK AREA | BROAD WORK AREA |
| | Bhn | | fpm<br>m/s | fpm<br>m/min | in<br>mm | | ANSI<br>ISO | ANSI<br>ISO |
| **32. TITANIUM ALLOYS, WROUGHT‡‡**<br>**Commercially Pure**<br>99.5 99.0<br>99.2 98.9<br>Ti-0.2Pd<br>TiCODE-12 | 110 to 275 | Annealed | 3500 to 6000 | 50 to 100 | .0005 to .001 | Roughing<br>Finishing | C36HV<br>C60IV | C36GV<br>C60HV |
| | | | 18 to 30 | 15 to 30 | .013 to .025 | Roughing<br>Finishing | C36HV<br>C60IV | C36GV<br>C60HV |
| **Alpha and Alpha-Beta Alloys**<br>Ti-8Mn<br>Ti-1Al-8V-5Fe<br>Ti-2Al-11Sn-5Zr-1Mo<br>Ti-3Al-2.5V<br>Ti-5Al-2Sn-2Zr-4Mo-4Cr (Ti-17)<br>Ti-5Al-2.5Sn<br>Ti-5Al-2.5Sn ELI<br>Ti-5Al-6Sn-2Zr-1Mo<br>Ti-6Al-2Cb-1Ta-0.8Mo<br>Ti-6Al-4V<br>Ti-6Al-4V ELI<br>Ti-6Al-6V-2Sn<br>Ti-6Al-2Sn-4Zr-2Mo<br>Ti-6Al-2Sn-4Zr-2Mo-.25Si<br>Ti-6Al-2Sn-4Zr-6Mo<br>Ti-7Al-4Mo<br>Ti-8Al-1Mo-1V | 300 to 440 | Annealed or Solution Treated and Aged | 3500 to 6000 | 50 to 100 | .0005 to .001 | Roughing<br>Finishing | C36HV<br>C60IV | C36GV<br>C60HV |
| | | | 18 to 30 | 15 to 30 | .013 to .025 | Roughing<br>Finishing | C36HV<br>C60IV | C36GV<br>C60HV |
| **Beta Alloys**<br>Ti-3Al-8V-6Cr-4Mo-4Zr<br>Ti-8Mo-8V-2Fe-3Al<br>Ti-11.5Mo-6Zr-4.5Sn<br>Ti-10V-2Fe-3Al<br>Ti-13V-11Cr-3Al | 275 to 440 | Annealed, Solution Treated or Solution Treated and Aged | 3500 to 6000 | 50 to 100 | .0005 to .001 | Roughing<br>Finishing | C36HV<br>C60IV | C36GV<br>C60HV |
| | | | 18 to 30 | 15 to 30 | .013 to .025 | Roughing<br>Finishing | C36HV<br>C60IV | C36GV<br>C60HV |
| **33. TITANIUM ALLOYS, CAST‡‡**<br>**Commercially Pure**<br>99.0<br>Ti-0.2Pd<br>ASTM B367: Grades C-1, C-2, C-3, C-4, C-7A, C-7B, C-8A, C-8B | 150 to 250 | As Cast or As Cast and Annealed | 3500 to 6000 | 50 to 100 | .0005 to .001 | Roughing<br>Finishing | C36HV<br>C60IV | C36GV<br>C60HV |
| | | | 18 to 30 | 15 to 30 | .013 to .025 | Roughing<br>Finishing | C36HV<br>C60IV | C36GV<br>C60HV |

See section 16 for Cutting Fluid Recommendations.
*See section 20.4 for calculation of Table rpm and Downfeed in ipm.

‡‡CAUTION: Potential Fire Hazard. Exercise caution in grinding and disposing of swarf.

| MATERIAL | HARDNESS Bhn | CONDITION | WHEEL SPEED fpm / m/s | TABLE* (WORK) SPEED fpm / m/min | DOWNFEED per rev. of table in / mm | OPERATION | WHEEL IDENTIFICATION NARROW WORK AREA ANSI / ISO | WHEEL IDENTIFICATION BROAD WORK AREA ANSI / ISO |
|---|---|---|---|---|---|---|---|---|
| **Alpha and Alpha-Beta Alloys**<br>Ti-5Al-2.5Sn<br>Ti-6Al-4V<br>Ti-6Al-2Sn-4Zr-2Mo<br>Ti-8Al-1Mo-1V<br>ASTM B367: Grades C-5, C-6 | 300 to 350 | As Cast or As Cast and Annealed | 3500 to 6000 | 50 to 100 | .0005 to .001 | Roughing<br>Finishing | C36HV<br>C60IV | C36GV<br>C60HV |
| | | | 18 to 30 | 15 to 30 | .013 to .025 | Roughing<br>Finishing | C36HV<br>C60IV | C36GV<br>C60HV |
| **34. COPPER ALLOYS, WROUGHT**<br>101 116 143 182<br>102 119 145 184<br>104 120 147 185<br>105 121 150 187<br>107 122 155 189<br>109 125 162 190<br>110 127 165 191<br>111 128 170†† 192<br>113 129 172†† 194<br>114 130 173†† 195<br>115 142 175†† | 10 $R_B$ to 100 $R_B$ | Annealed or Cold Drawn | 3500 to 6000 | 100 to 250 | .001 to .004 | Roughing<br>Finishing | CA46IB<br>CA60IB | CA46GB<br>CA60GB |
| | | | 18 to 30 | 30 to 76 | .025 to .100 | Roughing<br>Finishing | CA46IB<br>CA60IB | CA46GB<br>CA60GB |
| 210 332 368 464<br>220 335 370 465<br>226 340 377 466<br>230 342 385 467<br>240 349 411 482<br>260 350 413 485<br>268 353 425 667<br>270 356 435 687<br>280 360 442 688<br>314 365 443 694<br>316 366 444<br>330 367 445 | 10 $R_B$ to 100 $R_B$ | Annealed or Cold Drawn | 3500 to 6000 | 100 to 250 | .001 to .004 | Roughing<br>Finishing | CA46IB<br>CA60IB | CA46GB<br>CA60GB |
| | | | 18 to 30 | 30 to 76 | .025 to .100 | Roughing<br>Finishing | CA46IB<br>CA60IB | CA46GB<br>CA60GB |
| 505 608 623 642<br>510 610 624 651<br>511 613 625 655<br>521 614 630 674<br>524 618 632 675<br>544 619 638 | 10 $R_B$ to 100 $R_B$ | Annealed or Cold Drawn | 3500 to 6000 | 100 to 250 | .001 to .004 | Roughing<br>Finishing | CA46IB<br>CA60IB | CA46GB<br>CA60GB |
| | | | 18 to 30 | 30 to 76 | .025 to .100 | Roughing<br>Finishing | CA46IB<br>CA60IB | CA46GB<br>CA60GB |

See section 16 for Cutting Fluid Recommendations.
*See section 20.4 for calculation of Table rpm and Downfeed in ipm.

††CAUTION: Toxic Material, refer to National Institute for Occupational Safety and Health (NIOSH) for Precautions.

# 8.4 Surface Grinding-Vertical Spindle, Rotary Table

| MATERIAL | HARD-NESS | CONDITION | WHEEL SPEED | TABLE* (WORK) SPEED | DOWNFEED per rev. of table | OPERATION | WHEEL IDENTIFICATION NARROW WORK AREA | WHEEL IDENTIFICATION BROAD WORK AREA |
|---|---|---|---|---|---|---|---|---|
| | | | fpm | fpm | in | | ANSI | ANSI |
| | Bhn | | m/s | m/min | mm | | ISO | ISO |
| **34. COPPER ALLOYS, WROUGHT (cont.)**<br>706 725 754 782<br>710 745 757<br>715 752 770 | 10 $R_B$<br>to<br>100 $R_B$ | Annealed or Cold Drawn | 3500<br>to<br>6000 | 100<br>to<br>250 | .001<br>to<br>.004 | Roughing<br>Finishing | CA46IB<br>CA60IB | CA46GB<br>CA60GB |
| | | | 18<br>to<br>30 | 30<br>to<br>76 | .025<br>to<br>.100 | Roughing<br>Finishing | CA46IB<br>CA60IB | CA46GB<br>CA60GB |
| **35. COPPER ALLOYS, CAST**<br>801 811 818†† 825††<br>803 813 820†† 826††<br>805 814 821†† 827††<br>807 815 822†† 828††<br>809 817†† 824†† | 40<br>to<br>200<br>500kg | As Cast<br>or<br>Heat Treated | 3500<br>to<br>6000 | 100<br>to<br>250 | .001<br>to<br>.004 | Roughing<br>Finishing | CA46IB<br>CA60IB | CA46GB<br>CA60GB |
| | | | 18<br>to<br>30 | 30<br>to<br>76 | .025<br>to<br>.100 | Roughing<br>Finishing | CA46IB<br>CA60IB | CA46GB<br>CA60GB |
| | | | | | | | | |
| 833 852 862 874<br>834 853 863 875<br>836 854 864 876<br>838 855 865 878<br>842 857 867 879<br>844 858 868<br>848 861 872 | 35<br>to<br>200<br>500kg | As Cast | 3500<br>to<br>6000 | 100<br>to<br>250 | .001<br>to<br>.004 | Roughing<br>Finishing | CA46IB<br>CA60IB | CA46GB<br>CA60GB |
| | | | 18<br>to<br>30 | 30<br>to<br>76 | .025<br>to<br>.100 | Roughing<br>Finishing | CA46IB<br>CA60IB | CA46GB<br>CA60GB |
| | | | | | | | | |
| 902 916 934 948<br>903 917 935 952<br>905 922 937 953<br>907 923 938 954<br>909 925 939 955<br>910 926 943 956<br>911 927 944 957<br>913 928 945 958<br>915 932 947 | 40<br>to<br>100<br>500kg | As Cast or Heat Treated | 3500<br>to<br>6000 | 100<br>to<br>250 | .001<br>to<br>.004 | Roughing<br>Finishing | CA46IB<br>CA60IB | CA46GB<br>CA60GB |
| | | | 18<br>to<br>30 | 30<br>to<br>76 | .025<br>to<br>.100 | Roughing<br>Finishing | CA46IB<br>CA60IB | CA46GB<br>CA60GB |
| | | | | | | | | |
| 962 974<br>963 976<br>964 978<br>966†† 993<br>973 | 50<br>to<br>100<br>500kg | As Cast or Heat Treated | 3500<br>to<br>6000 | 100<br>to<br>250 | .001<br>to<br>.004 | Roughing<br>Finishing | CA46IB<br>CA60IB | CA46GB<br>CA60GB |
| | | | 18<br>to<br>30 | 30<br>to<br>76 | .025<br>to<br>.100 | Roughing<br>Finishing | CA46IB<br>CA60IB | CA46GB<br>CA60GB |
| **36. NICKEL ALLOYS, WROUGHT AND CAST**<br>Nickel 200<br>Nickel 201<br>Nickel 205<br>Nickel 211<br>Nickel 220<br>Nickel 230 | 80<br>to<br>170 | Annealed<br>or<br>Cold Drawn | 3500<br>to<br>4500 | 100<br>to<br>200 | .0005<br>to<br>.001 | Roughing<br>Finishing | A36HV<br>A60HV | A36GV<br>A60GV |
| | | | 18<br>to<br>23 | 30<br>to<br>60 | .013<br>to<br>.025 | Roughing<br>Finishing | A36HV<br>A60HV | A36GV<br>A60GV |
| | | | | | | | | |

See section 16 for Cutting Fluid Recommendations.
*See section 20.4 for calculation of Table rpm and Downfeed in ipm.

††CAUTION: Toxic Material, refer to National Institute for Occupational Safety and Health (NIOSH) for Precautions.

| MATERIAL | HARD-NESS | CONDITION | WHEEL SPEED | TABLE* (WORK) SPEED | DOWNFEED per rev. of table | OPERATION | WHEEL IDENTIFICATION NARROW WORK AREA | BROAD WORK AREA |
|---|---|---|---|---|---|---|---|---|
| | Bhn | | fpm / m/s | fpm / m/min | in / mm | | ANSI / ISO | ANSI / ISO |
| MONEL Alloy 400<br>MONEL Alloy 401<br>MONEL Alloy 404<br>MONEL Alloy R405<br>ASTM A296: Grades CZ-100, M-35<br>ASTM A494: Grades CZ-100, M-35 | 115 to 240 | Annealed, Cold Drawn or Cast | 3500 to 4500 | 100 to 200 | .0005 to .001 | Roughing<br>Finishing | A36HV<br>A60HV | A36GV<br>A60GV |
| | | | 18 to 23 | 30 to 60 | .013 to .025 | Roughing<br>Finishing | A36HV<br>A60HV | A36GV<br>A60GV |
| | | | | | | | | |
| DURANICKEL Alloy 301<br>MONEL Alloy 502<br>MONEL Alloy K500<br>NI-SPAN-C Alloy 902<br>PERMANICKEL Alloy 300 | 150 to 320 | Solution Treated | 3500 to 4500 | 100 to 200 | .0005 to .001 | Roughing<br>Finishing | A36HV<br>A60HV | A36GV<br>A60GV |
| | | | 18 to 23 | 30 to 60 | .013 to .025 | Roughing<br>Finishing | A36HV<br>A60HV | A36GV<br>A60GV |
| | 330 to 360 | Aged | 3500 to 4500 | 100 to 200 | .0005 to .001 | Roughing<br>Finishing | A36GV<br>A60GV | A36FV<br>A60FV |
| | | | 18 to 23 | 30 to 60 | .013 to .025 | Roughing<br>Finishing | A36GV<br>A60GV | A36FV<br>A60FV |
| **37. BERYLLIUM NICKEL ALLOYS, WROUGHT AND CAST††**<br>Berylco 440<br>Berylco 41C<br>Berylco 42C<br>Berylco 43C<br>Brush Alloy 200C<br>Brush Alloy 220C<br>Brush Alloy 260C | 200 to 250 | As Cast or Solution Treated | 3000 to 3500 | 100 to 250 | .0005 to .001 | Roughing<br>Finishing | A30HV<br>A80HV | A30GV<br>A80GV |
| | | | 15 to 18 | 30 to 76 | .013 to .025 | Roughing<br>Finishing | A30HV<br>A80HV | A30GV<br>A80GV |
| | 283 to 425 | Hardened or Aged | 3000 to 3500 | 100 to 250 | .0005 to .001 | Roughing<br>Finishing | A36GV<br>A80GV | A36FV<br>A80FV |
| | | | 15 to 18 | 30 to 76 | .013 to .025 | Roughing<br>Finishing | A36GV<br>A80GV | A36FV<br>A80FV |
| | 47 $R_C$ to 52 $R_C$ | Hardened or Aged | 3000 to 3500 | 100 to 250 | .0005 to .001 | Roughing<br>Finishing | A36FV<br>A80FV | A36EV<br>A80EV |
| | | | 15 to 18 | 30 to 76 | .013 to .025 | Roughing<br>Finishing | A36FV<br>A80FV | A36EV<br>A80EV |

See section 16 for Cutting Fluid Recommendations.
*See section 20.4 for calculation of Table rpm and Downfeed in ipm.

††CAUTION: Toxic Material, refer to National Institute for Occupational Safety and Health (NIOSH) for Precautions.

# 8.4 Surface Grinding-Vertical Spindle, Rotary Table

| MATERIAL | HARD-NESS<br>Bhn | CONDITION | WHEEL SPEED<br>fpm<br>m/s | TABLE* (WORK) SPEED<br>fpm<br>m/min | DOWNFEED per rev. of table<br>in<br>mm | OPERATION | WHEEL IDENTIFICATION<br>NARROW WORK AREA<br>ANSI<br>ISO | <br>BROAD WORK AREA<br>ANSI<br>ISO |
|---|---|---|---|---|---|---|---|---|
| **39. HIGH TEMPERATURE ALLOYS, WROUGHT AND CAST**<br>**Nickel Base, Wrought**<br>AF2-1DA<br>Astroloy<br>Haynes Alloy 263<br>IN-102<br>Incoloy Alloy 901<br>Incoloy Alloy 903<br>Inconel Alloy 617<br>Inconel Alloy 625<br>Inconel Alloy 700<br>Inconel Alloy 702<br>Inconel Alloy 706<br>Inconel Alloy 718<br>Inconel Alloy 721<br>Inconel Alloy 722<br>Inconel Alloy X-750<br>Inconel Alloy 751<br>M252<br>Nimonic 75<br>Nimonic 80<br>Nimonic 90<br>Nimonic 95<br>Rene 41<br>Rene 63<br>Rene 77<br>Rene 95<br>Udimet 500<br>Udimet 700<br>Udimet 710<br>Unitemp 1753<br>Waspaloy | 200 to 390 | Annealed or Solution Treated | 3500 to 4500 | 100 to 200 | .0005 to .001 | Roughing<br>Finishing | A36GV<br>A60GV | A36FV<br>A60FV |
| | | | 18 to 23 | 30 to 60 | .013 to .025 | Roughing<br>Finishing | A36GV<br>A60GV | A36FV<br>A60FV |
| | 300 to 475 | Solution Treated and Aged | 3500 to 4500 | 100 to 200 | .0005 to .001 | Roughing<br>Finishing | A36GV<br>A60GV | A36FV<br>A60FV |
| | | | 18 to 23 | 30 to 60 | .013 to .025 | Roughing<br>Finishing | A36GV<br>A60GV | A36FV<br>A60FV |
| **Nickel Base, Wrought**<br>Hastelloy Alloy B<br>Hastelloy Alloy B-2<br>Hastelloy Alloy C<br>Hastelloy Alloy C-276<br>Hastelloy Alloy G<br>Hastelloy Alloy S<br>Hastelloy Alloy X<br>Incoloy Alloy 804<br>Incoloy Alloy 825<br>Inconel Alloy 600<br>Inconel Alloy 601<br>Refractaloy 26<br>Udimet 630 | 140 to 220 | Annealed or Solution Treated | 3500 to 4500 | 100 to 200 | .0005 to .001 | Roughing<br>Finishing | A36GV<br>A60GV | A36FV<br>A60FV |
| | | | 18 to 23 | 30 to 60 | .013 to .025 | Roughing<br>Finishing | A36GV<br>A60GV | A36FV<br>A60FV |
| | 240 to 310 | Cold Drawn or Aged | 3500 to 4500 | 100 to 200 | .0005 to .001 | Roughing<br>Finishing | A36GV<br>A60GV | A36FV<br>A60FV |
| | | | 18 to 23 | 30 to 60 | .013 to .025 | Roughing<br>Finishing | A36GV<br>A60GV | A36FV<br>A60FV |

See section 16 for Cutting Fluid Recommendations.
*See section 20.4 for calculation of Table rpm and Downfeed in.

| MATERIAL | HARD-NESS | CONDITION | WHEEL SPEED | TABLE* (WORK) SPEED | DOWNFEED per rev. of table | OPERATION | WHEEL IDENTIFICATION NARROW WORK AREA | BROAD WORK AREA |
|---|---|---|---|---|---|---|---|---|
| | Bhn | | fpm | fpm | in | | ANSI | ANSI |
| | | | m/s | m/min | mm | | ISO | ISO |
| **Nickel Base, Cast**<br>B-1900<br>GMR-235<br>GMR-235D<br>Hastelloy Alloy B<br>Hastelloy Alloy C<br>Hastelloy Alloy D<br>IN-100 (Rene 100)<br>IN-738<br>IN-792<br>Inconel Alloy 713C<br>Inconel Alloy 718<br>M252<br>MAR-M200<br>MAR-M246<br>MAR-M421<br>MAR-M432<br>Rene 80<br>Rene 125<br>SEL<br>SEL 15<br>TRW VI A<br>Udimet 500<br>Udimet 700<br>ASTM A296: Grades CW-12M, N-12M, CY-40<br>ASTM A297: Grades HW, HX<br>ASTM A494: Grades N-12M-1, N-12M-2, CY-40, CW-12M-1, CW-12M-2<br>ASTM A608: Grades HW50, HX50 | 200 to 425 | As Cast or Cast and Aged | 3500 to 4500 | 100 to 200 | .0005 to .001 | Roughing<br>Finishing | A36GV<br>A60GV | A36FV<br>A60FV |
| | | | 18 to 23 | 30 to 60 | .013 to .025 | Roughing<br>Finishing | A36GV<br>A60GV | A36FV<br>A60FV |
| **Cobalt Base, Wrought**<br>AiResist 213<br>Haynes Alloy 25 (L605)<br>Haynes Alloy 188<br>J-1570<br>MAR-M905<br>MAR-M918<br>S-816<br>V-36 | 180 to 230 | Solution Treated | 3000 to 4000 | 100 to 200 | .0005 to .001 | Roughing<br>Finishing | A30GV<br>A60GV | A30FV<br>A60FV |
| | | | 15 to 20 | 30 to 60 | .013 to .025 | Roughing<br>Finishing | A30GV<br>A60GV | A30FV<br>A60FV |
| | 270 to 320 | Solution Treated and Aged | 3000 to 4000 | 100 to 200 | .0005 to .001 | Roughing<br>Finishing | A30GV<br>A60GV | A30FV<br>A60FV |
| | | | 15 to 20 | 30 to 60 | .013 to .025 | Roughing<br>Finishing | A30GV<br>A60GV | A30FV<br>A60FV |

See section 16 for Cutting Fluid Recommendations.
*See section 20.4 for calculation of Table rpm and Downfeed in ipm.

# 8.4 Surface Grinding-Vertical Spindle, Rotary Table

| MATERIAL | HARD-NESS | CONDITION | WHEEL SPEED | TABLE* (WORK) SPEED | DOWNFEED per rev. of table | OPERATION | WHEEL IDENTIFICATION NARROW WORK AREA | WHEEL IDENTIFICATION BROAD WORK AREA |
|---|---|---|---|---|---|---|---|---|
| | Bhn | | fpm<br>m/s | fpm<br>m/min | in<br>mm | | ANSI<br>ISO | ANSI<br>ISO |
| **39. HIGH TEMPERATURE ALLOYS, WROUGHT AND CAST (cont.)**<br>**Cobalt Base, Cast**<br>AiResist 13<br>AiResist 215<br>FSX-414<br>HS-6<br>HS-21<br>HS-31 (X-40)<br>HOWMET #3<br>MAR-M302<br>MAR-M322<br>MAR-M509<br>NASA Co-W-Re<br>WI-52<br>X-45 | 220 to 290 | As Cast or Cast and Aged | 3000 to 4000 | 100 to 200 | .0005 to .001 | Roughing<br>Finishing | A30GV<br>A60GV | A30FV<br>A60FV |
| | | | 15 to 20 | 30 to 60 | .013 to .025 | Roughing<br>Finishing | A30GV<br>A60GV | A30FV<br>A60FV |
| | 290 to 425 | As Cast or Cast and Aged | 3000 to 4000 | 100 to 200 | .0005 to .001 | Roughing<br>Finishing | A30GV<br>A60GV | A30FV<br>A60FV |
| | | | 15 to 20 | 30 to 60 | .013 to .025 | Roughing<br>Finishing | A30GV<br>A60GV | A30FV<br>A60FV |
| **Iron Base, Wrought**<br>A-286<br>Discaloy<br>Incoloy Alloy 800<br>Incoloy Alloy 800H<br>Incoloy Alloy 801<br>Incoloy Alloy 802<br>N-155<br>V-57<br>W-545<br>16-25-6<br>19-9DL | 180 to 230 | Solution Treated | 3500 to 4500 | 100 to 200 | .0005 to .001 | Roughing<br>Finishing | A36GV<br>A60GV | A36FV<br>A60FV |
| | | | 18 to 23 | 30 to 60 | .013 to .025 | Roughing<br>Finishing | A36GV<br>A60GV | A36FV<br>A60FV |
| | 250 to 320 | Solution Treated and Aged | 3500 to 4500 | 100 to 200 | .0005 to .001 | Roughing<br>Finishing | A36GV<br>A60GV | A36FV<br>A60FV |
| | | | 18 to 23 | 30 to 60 | .013 to .025 | Roughing<br>Finishing | A36GV<br>A60GV | A36FV<br>A60FV |

See section 16 for Cutting Fluid Recommendations.
*See section 20.4 for calculation of Table rpm and Downfeed in ipm.

| MATERIAL | HARD-NESS<br>Bhn | CONDITION | WHEEL SPEED<br>fpm<br>m/s | WORK SPEED<br>fpm<br>m/min | INFEED on dia.<br>in/pass<br>mm/pass | TRAVERSE<br>Wheel width per rev. of work | WHEEL IDENTIFI-CATION*<br>ANSI<br>ISO |
|---|---|---|---|---|---|---|---|
| **1. FREE MACHINING CARBON STEELS, WROUGHT**<br>**Low Carbon Resulfurized**<br>1108 1118<br>1109 1119<br>1110 1211<br>1115 1212<br>1116 1213<br>1117 1215<br>**Medium Carbon Resulfurized**<br>1132 1140 1145<br>1137 1141 1146<br>1139 1144 1151<br>**Low Carbon Leaded**<br>10L18 12L13 12L15<br>11L17 12L14<br>**Medium Carbon Leaded**<br>10L45 11L37 11L44<br>10L50 11L41 | 50 $R_C$ max. | Hot Rolled, Normalized, Annealed, Cold Drawn or Quenched and Tempered | 5500 to 6500 | 70 to 100 | Rough: .002<br>Finish: .0005 max. | 1/2<br>1/6 | A60LV |
| | | | 28 to 33 | 21 to 30 | Rough: .050<br>Finish: .013 max. | 1/2<br>1/6 | A60LV |
| | Over 50 $R_C$ | Carburized and/or Quenched and Tempered | 5500 to 6500 | 70 to 100 | Rough: .002<br>Finish: .0005 max. | 1/4<br>1/6 | A60KV |
| | | | 28 to 33 | 21 to 30 | Rough: .050<br>Finish: .013 max. | 1/4<br>1/6 | A60KV |
| **2. CARBON STEELS, WROUGHT**<br>**Low Carbon**<br>1005 1012 1019 1026<br>1006 1013 1020 1029<br>1008 1015 1021 1513<br>1009 1016 1022 1518<br>1010 1017 1023 1522<br>1011 1018 1025<br>**Medium Carbon**<br>1030 1042 1053 1541<br>1033 1043 1055 1547<br>1035 1044 1524 1548<br>1037 1045 1525 1551<br>1038 1046 1526 1552<br>1039 1049 1527<br>1040 1050 1536<br>**High Carbon**<br>1060 1074 1085 1566<br>1064 1075 1086 1572<br>1065 1078 1090<br>1069 1080 1095<br>1070 1084 1561 | 50 $R_C$ max. | Hot Rolled, Normalized, Annealed, Cold Drawn or Quenched and Tempered | 5500 to 6500 | 70 to 100 | Rough: .002<br>Finish: .0005 max. | 1/2<br>1/6 | A60LV |
| | | | 28 to 33 | 21 to 30 | Rough: .050<br>Finish: .013 max. | 1/2<br>1/6 | A60LV |
| | Over 50 $R_C$ | Carburized and/or Quenched and Tempered | 5500 to 6500 | 70 to 100 | Rough: .002<br>Finish: .0005 max. | 1/4<br>1/8 | A60KV |
| | | | 28 to 33 | 21 to 30 | Rough: .050<br>Finish: .013 max. | 1/4<br>1/8 | A60KV |
| **3. CARBON AND FERRITIC ALLOY STEELS (HIGH TEMPERATURE SERVICE)**<br>ASTM A369: Grades FPA, FPB, FP1, FP2, FP3b, FP5, FP7, FP9, FP11, FP12, FP21, FP22 | 150 to 200 | As Forged, Annealed or Normalized and Tempered | 5500 to 6500 | 70 to 100 | Rough: .002<br>Finish: .0005 max. | 1/2<br>1/6 | A60LV |
| | | | 28 to 33 | 21 to 30 | Rough: .050<br>Finish: .013 max. | 1/2<br>1/6 | A60LV |

See section 16 for Cutting Fluid Recommendations.

*Wheel recommendations are for wet grinding of 2- to 4-inch [50 to 100 mm] diameter work. For DRY grinding—use a softer grade wheel. For LARGER diameter work—use a softer and/or coarser grit wheel. For SMALLER diameter work—use a harder grade wheel. Wheel recommendations also apply to plunge grinding applications. Also see section 20.2, Grinding Guidelines.

# 8.5 Cylindrical Grinding

| MATERIAL | HARD-NESS | CONDITION | WHEEL SPEED | WORK SPEED | INFEED on dia. | TRAVERSE Wheel width per rev. of work | WHEEL IDENTIFI-CATION* |
|---|---|---|---|---|---|---|---|
| | Bhn | | fpm / m/s | fpm / m/min | in/pass / mm/pass | | ANSI / ISO |
| **4. FREE MACHINING ALLOY STEELS, WROUGHT**<br>**Medium Carbon Resulfurized**<br>4140<br>4140Se<br>4142Te<br>4145Se<br>4147Te<br>4150<br>**Medium and High Carbon Leaded**<br>41L30 41L47 51L32 86L40<br>41L40 41L50 52L100<br>41L45 43L40 86L20 | $50\ R_C$ max. | Hot Rolled, Normalized, Annealed, Cold Drawn or Quenched and Tempered | 5500 to 6500 | 70 to 100 | Rough: .002<br>Finish: .0005 max. | 1/2<br>1/6 | A60LV |
| | | | 28 to 33 | 21 to 30 | Rough: .050<br>Finish: .013 max. | 1/2<br>1/6 | A60LV |
| | Over $50\ R_C$ | Carburized and/or Quenched and Tempered | 5500 to 6500 | 70 to 100 | Rough: .002<br>Finish: .0005 max. | 1/4<br>1/8 | A60KV |
| | | | 28 to 33 | 21 to 30 | Rough: .050<br>Finish: .013 max. | 1/4<br>1/8 | A60KV |
| **5. ALLOY STEELS, WROUGHT**<br>**Low Carbon**<br>4012 4615 4817 8617<br>4023 4617 4820 8620<br>4024 4620 5015 8622<br>4118 4621 5115 8822<br>4320 4718 5120 9310<br>4419 4720 6118 94B15<br>4422 4815 8115 94B17<br>**Medium Carbon**<br>1330 4145 5132 8640<br>1335 4147 5135 8642<br>1340 4150 5140 8645<br>1345 4161 5145 86B45<br>4027 4340 5147 8650<br>4028 4427 5150 8655<br>4032 4626 5155 8660<br>4037 50B40 5160 8740<br>4042 50B44 51B60 8742<br>4047 5046 6150 9254<br>4130 50B46 81B45 9255<br>4135 50B50 8625 9260<br>4137 5060 8627 94B30<br>4140 50B60 8630<br>4142 5130 8637<br>**High Carbon**<br>50100 51100 52100 M-50 | $50\ R_C$ max. | Hot Rolled, Normalized, Annealed, Cold Drawn or Quenched and Tempered | 5500 to 6500 | 70 to 100 | Rough: .002<br>Finish: .0005 max. | 1/2<br>1/6 | A60LV |
| | | | 28 to 33 | 21 to 30 | Rough: .050<br>Finish: .013 max. | 1/2<br>1/6 | A60LV |
| | Over $50\ R_C$ | Carburized and/or Quenched and Tempered | 4000 to 6000 | 70 to 100 | Rough: .002<br>Finish: .0004 max. | 1/4<br>1/8 | A60KV |
| | | | 20 to 30 | 21 to 30 | Rough: .050<br>Finish: .010 max. | 1/4<br>1/8 | A60KV |
| | | | | | | | |
| | | | | | | | |

See section 16 for Cutting Fluid Recommendations.

*Wheel recommendations are for wet grinding of 2- to 4-inch [50 to 100 mm] diameter work. For DRY grinding—use a softer grade wheel. For LARGER diameter work—use a softer and/or coarser grit wheel. For SMALLER diameter work—use a harder grade wheel. Wheel recommendations also apply to plunge grinding applications. Also see section 20.2, Grinding Guidelines.

| MATERIAL | HARDNESS<br>Bhn | CONDITION | WHEEL SPEED<br>fpm<br>m/s | WORK SPEED<br>fpm<br>m/min | INFEED on dia.<br>in/pass<br>mm/pass | TRAVERSE Wheel width per rev. of work | WHEEL IDENTIFICATION*<br>ANSI<br>ISO |
|---|---|---|---|---|---|---|---|
| **6. HIGH STRENGTH STEELS, WROUGHT**<br>300M<br>4330V<br>4340<br>4340Si<br>98BV40<br>D6ac<br>H11<br>H13<br>HP 9-4-20<br>HP 9-4-25<br>HP 9-4-30<br>HP 9-4-45 | 50 $R_C$ max. | Annealed, Normalized or Quenched and Tempered | 4000 to 5500 | 70 to 100 | Rough: .002<br>Finish: .0005 max. | 1/2<br>1/6 | A60KV |
| | | | 20 to 28 | 21 to 30 | Rough: .050<br>Finish: .013 max. | 1/2<br>1/6 | A60KV |
| | Over 50 $R_C$ | Quenched and Tempered | 4000 to 5500 | 70 to 100 | Rough: .002<br>Finish: .0004 max. | 1/4<br>1/8 | A80IV |
| | | | 20 to 28 | 21 to 30 | Rough: .050<br>Finish: .010 max. | 1/4<br>1/8 | A80IV |
| **7. MARAGING STEELS, WROUGHT**<br>ASTM A538: Grades A, B, C<br>120 Grade<br>180 Grade<br>200 Grade<br>250 Grade<br>300 Grade<br>350 Grade<br>HY230 | 50 $R_C$ max. | Annealed or Maraged | 5500 to 6500 | 70 to 100 | Rough: .002<br>Finish: .0005 max. | 1/2<br>1/6 | A60KV |
| | | | 28 to 33 | 21 to 30 | Rough: .050<br>Finish: .013 max. | 1/2<br>1/6 | A60KV |
| | Over 50 $R_C$ | Maraged | 4000 to 5500 | 70 to 100 | Rough: .002<br>Finish: .0005 max. | 1/4<br>1/8 | A80IV |
| | | | 20 to 28 | 21 to 30 | Rough: .050<br>Finish: .013 max. | 1/4<br>1/8 | A80IV |
| **8. TOOL STEELS, WROUGHT**<br>**Group I**<br>A2 H13 L6 P20<br>A3 H14 L7 P21<br>A4 H19 O1 S1<br>A6 H21 O2 S2<br>A8 H22 O6 S5<br>A9 H23 O7 S6<br>A10 H24 P2 S7<br>H10 H25 P4 W1<br>H11 H26 P5 W2<br>H12 L2 P6 W5<br>SAE J438b: Types W108, W109, W110, W112, W209, W210, W310 | 50 $R_C$ max. | Annealed or Quenched and Tempered | 5500 to 6500 | 60 to 100 | Rough: .002<br>Finish: .0005 max. | 1/2<br>1/6 | A60LV |
| | | | 28 to 33 | 18 to 30 | Rough: .050<br>Finish: .013 max. | 1/2<br>1/6 | A60LV |
| | Over 50 $R_C$ | Quenched and Tempered | 4000 to 6000 | 60 to 100 | Rough: .002<br>Finish: .0004 max. | 1/4<br>1/8 | A60KV |
| | | | 20 to 30 | 18 to 30 | Rough: .050<br>Finish: .010 max. | 1/4<br>1/8 | A60KV |

See section 16 for Cutting Fluid Recommendations.

*Wheel recommendations are for wet grinding of 2- to 4-inch [50 to 100 mm] diameter work. For DRY grinding—use a softer grade wheel. For LARGER diameter work—use a softer and/or coarser grit wheel. For SMALLER diameter work—use a harder grade wheel. Wheel recommendations also apply to plunge grinding applications. Also see section 20.2, Grinding Guidelines.

# 8.5 Cylindrical Grinding

| MATERIAL | HARD-NESS<br>Bhn | CONDITION | WHEEL SPEED<br>fpm<br>m/s | WORK SPEED<br>fpm<br>m/min | INFEED on dia.<br>in/pass<br>mm/pass | TRAVERSE<br>Wheel width per rev. of work | WHEEL IDENTIFI-CATION*<br>ANSI<br>ISO |
|---|---|---|---|---|---|---|---|
| **8. TOOL STEELS, WROUGHT (cont.)**<br>**Group II**<br>D2 H42 M42<br>D3 M1 T1<br>D4 M2 T2<br>D5 M3-1 T4<br>F1 M10 T8<br>F2 M30 | 50 $R_C$ max. | Annealed or Quenched and Tempered | 5500 to 6500 | 60 to 100 | Rough: .002<br>Finish: .0005 max. | 1/2<br>1/6 | A60KV |
| | | | 28 to 33 | 18 to 30 | Rough: .050<br>Finish: .013 max. | 1/2<br>1/6 | A60KV |
| | 50 $R_C$ to 58 $R_C$ | Quenched and Tempered | 4000 to 6000 | 60 to 100 | Rough: .002<br>Finish: .0004 max. | 1/3<br>1/8 | A60JV |
| | | | 20 to 30 | 18 to 30 | Rough: .050<br>Finish: .010 max. | 1/3<br>1/8 | A60JV |
| | Over 58 $R_C$ | Quenched and Tempered | 4000 to 5500 | 60 to 100 | Rough: .002<br>Finish: .0004 max. | 1/4<br>1/8 | A80IV |
| | | | 20 to 28 | 18 to 30 | Rough: .050<br>Finish: .010 max. | 1/4<br>1/8 | A80IV |
| **Group III**<br>A7 M41<br>D7 M43<br>M3-2 M44<br>M6 M46<br>M7 M47<br>M33 T5<br>M34 T6<br>M36 | 50 $R_C$ max. | Annealed or Quenched and Tempered | 4000 to 6000 | 60 to 100 | Rough: .001<br>Finish: .0004 max. | 1/2<br>1/6 | A60KV |
| | | | 20 to 30 | 18 to 30 | Rough: .025<br>Finish: .010 max. | 1/2<br>1/6 | A60KV |
| | 50 $R_C$ to 58 $R_C$ | Quenched and Tempered | 4000 to 5500 | 60 to 100 | Rough: .001<br>Finish: .0003 max. | 1/3<br>1/8 | A80JV |
| | | | 20 to 28 | 18 to 30 | Rough: .025<br>Finish: .008 max. | 1/3<br>1/8 | A80JV |
| | Over 58 $R_C$ | Quenched and Tempered | 4000 to 5500 | 60 to 100 | Rough: .001<br>Finish: .0003 max. | 1/4<br>1/8 | A80IV |
| | | | 20 to 28 | 18 to 30 | Rough: .025<br>Finish: .008 max. | 1/4<br>1/8 | A80IV |

See section 16 for Cutting Fluid Recommendations.

*Wheel recommendations are for wet grinding of 2- to 4-inch [50 to 100 mm] diameter work. For DRY grinding—use a softer grade wheel. For LARGER diameter work—use a softer and/or coarser grit wheel. For SMALLER diameter work—use a harder grade wheel. Wheel recommendations also apply to plunge grinding applications. Also see section 20.2, Grinding Guidelines.

| MATERIAL | HARD-NESS<br>Bhn | CONDITION | WHEEL SPEED<br>fpm<br>m/s | WORK SPEED<br>fpm<br>m/min | INFEED on dia.<br>in/pass<br>mm/pass | TRAVERSE Wheel width per rev. of work | WHEEL IDENTIFI-CATION*<br>ANSI<br>ISO |
|---|---|---|---|---|---|---|---|
| **Group IV**<br>M4<br>T15 | 50 $R_C$ max. | Annealed or Quenched and Tempered | 4000 to 5500 | 60 to 100 | Rough: .001<br>Finish: .0004 max. | 1/2<br>1/6 | A60JV |
| | | | 20 to 28 | 18 to 30 | Rough: .025<br>Finish: .010 max. | 1/2<br>1/6 | A60JV |
| | 50 $R_C$ to 58 $R_C$ | Quenched and Tempered | 4000 to 5500 | 60 to 100 | Rough: .001<br>Finish: .0003 max. | 1/3<br>1/8 | A80IV |
| | | | 20 to 28 | 18 to 30 | Rough: .025<br>Finish: .008 max. | 1/3<br>1/8 | A80IV |
| | Over 58 $R_C$ | Quenched and Tempered | 4000 to 5000 | 60 to 100 | Rough: .001<br>Finish: .0003 max. | 1/4<br>1/8 | A80HV |
| | | | 20 to 25 | 18 to 30 | Rough: .025<br>Finish: .008 max. | 1/4<br>1/8 | A80HV |
| **9. NITRIDING STEELS, WROUGHT**<br>Nitralloy 125<br>Nitralloy 135<br>Nitralloy 135 Mod.<br>Nitralloy 225<br>Nitralloy 230<br>Nitralloy EZ<br>Nitralloy N<br>Nitrex 1 | 200 to 350 | Annealed, Normalized or Quenched and Tempered | 5500 to 6500 | 70 to 100 | Rough: .001<br>Finish: .0003 max. | 1/2<br>1/6 | A60KV |
| | | | 28 to 33 | 21 to 30 | Rough: .025<br>Finish: .008 max. | 1/2<br>1/6 | A60KV |
| | 60 $R_C$ to 65 $R_C$ | Nitrided | 4000 to 5500 | 60 to 100 | Rough: .001<br>Finish: .0003 max. | 1/4<br>1/8 | A80IV |
| | | | 20 to 28 | 18 to 30 | Rough: .025<br>Finish: .008 max. | 1/4<br>1/8 | A80IV |
| **10. ARMOR PLATE, SHIP PLATE, AIRCRAFT PLATE, WROUGHT**<br>HY80<br>MIL-S-12560<br>HY100<br>MIL-S-16216<br>HY180 | 45 $R_C$ max. | Annealed or Quenched and Tempered | 5500 to 6500 | 70 to 100 | Rough: .002<br>Finish: .0005 max. | 1/2<br>1/6 | A60JV |
| | | | 28 to 33 | 21 to 30 | Rough: .050<br>Finish: .013 max. | 1/2<br>1/6 | A60JV |

See section 16 for Cutting Fluid Recommendations.

*Wheel recommendations are for wet grinding of 2- to 4-inch [50 to 100 mm] diameter work. For DRY grinding—use a softer grade wheel. For LARGER diameter work—use a softer and/or coarser grit wheel. For SMALLER diameter work—use a harder grade wheel. Wheel recommendations also apply to plunge grinding applications. Also see section 20.2, Grinding Guidelines.

# 8.5 Cylindrical Grinding

| MATERIAL | HARDNESS Bhn | CONDITION | WHEEL SPEED fpm / m/s | WORK SPEED fpm / m/min | INFEED on dia. in/pass / mm/pass | TRAVERSE Wheel width per rev. of work | WHEEL IDENTIFICATION* ANSI / ISO |
|---|---|---|---|---|---|---|---|
| **12. FREE MACHINING STAINLESS STEELS, WROUGHT**<br>**Ferritic**<br>430F<br>430F Se | 135 to 185 | Annealed | 5500 to 6500 | 50 to 100 | Rough: .002<br>Finish: .0005 max. | 1/2<br>1/6 | A60JV |
| | | | 28 to 33 | 15 to 30 | Rough: .050<br>Finish: .013 max. | 1/2<br>1/6 | A60JV |
| **Austenitic**<br>203EZ<br>303<br>303MA<br>303Pb<br>303 Plus X<br>303Se | 135 to 275 | Annealed or Cold Drawn | 5500 to 6500 | 50 to 100 | Rough: .002<br>Finish: .0005 max. | 1/2<br>1/6 | C54JV |
| | | | 28 to 33 | 15 to 30 | Rough: .050<br>Finish: .013 max. | 1/2<br>1/6 | C54JV |
| **Martensitic**<br>416<br>416 Plus X<br>416Se<br>420F<br>420F Se<br>440F<br>440F Se | 135 to 240 | Annealed or Cold Drawn | 5500 to 6500 | 50 to 100 | Rough: .002<br>Finish: .0005 max. | 1/4<br>1/8 | A60JV |
| | | | 28 to 33 | 15 to 30 | Rough: .050<br>Finish: .013 max. | 1/4<br>1/8 | A60JV |
| | Over 275 | Quenched and Tempered | 5500 to 6500 | 50 to 100 | Rough: .002<br>Finish: .0005 max. | 1/4<br>1/8 | A60IV |
| | | | 28 to 33 | 15 to 30 | Rough: .050<br>Finish: .013 max. | 1/4<br>1/8 | A60IV |
| **13. STAINLESS STEELS, WROUGHT**<br>**Ferritic**<br>405 434<br>409 436<br>429 442<br>430 446 | 135 to 185 | Annealed | 5500 to 6500 | 50 to 100 | Rough: .002<br>Finish: .0005 max. | 1/2<br>1/6 | A60JV |
| | | | 28 to 33 | 15 to 30 | Rough: .050<br>Finish: .013 max. | 1/2<br>1/6 | A60JV |
| **Austenitic**<br>201 308 317<br>202 309 321<br>301 309S 330<br>302 310 347<br>302B 310S 348<br>304 314 384<br>304L 316 385<br>305 316L | 135 to 275 | Annealed or Cold Drawn | 5500 to 6500 | 50 to 100 | Rough: .002<br>Finish: .0005 max. | 1/2<br>1/6 | C54JV |
| | | | 28 to 33 | 15 to 30 | Rough: .050<br>Finish: .013 max. | 1/2<br>1/6 | C54JV |

See section 16 for Cutting Fluid Recommendations.

*Wheel recommendations are for wet grinding of 2- to 4-inch [50 to 100 mm] diameter work. For DRY grinding—use a softer grade wheel. For LARGER diameter work—use a softer and/or coarser grit wheel. For SMALLER diameter work—use a harder grade wheel. Wheel recommendations also apply to plunge grinding applications. Also see section 20.2, Grinding Guidelines.

| MATERIAL | HARDNESS | CONDITION | WHEEL SPEED | WORK SPEED | INFEED on dia. | TRAVERSE Wheel width per rev. of work | WHEEL IDENTIFICATION* |
|---|---|---|---|---|---|---|---|
| | Bhn | | fpm<br>m/s | fpm<br>m/min | in/pass<br>mm/pass | | ANSI<br>ISO |
| **Austenitic**<br>Nitronic 32<br>Nitronic 33<br>Nitronic 40<br>Nitronic 50<br>Nitronic 60 | 210 to 250 | Annealed | 5500 to 6500 | 50 to 100 | Rough: .002<br>Finish: .0005 max. | 1/2<br>1/6 | C54JV |
| | | | 28 to 33 | 15 to 30 | Rough: .050<br>Finish: .013 max. | 1/2<br>1/6 | C54JV |
| | 325 to 375 | Cold Drawn | 5500 to 6500 | 50 to 100 | Rough: .002<br>Finish: .0005 max. | 1/4<br>1/8 | C54IV |
| | | | 28 to 33 | 15 to 30 | Rough: .050<br>Finish: .013 max. | 1/4<br>1/8 | C54IV |
| **Martensitic**<br>403 440A<br>410 440B<br>414 440C<br>420 501<br>422 502<br>431 Greek Ascoloy | 135 to 275 | Annealed | 5500 to 6500 | 50 to 100 | Rough: .002<br>Finish: .0005 max. | 1/4<br>1/8 | A60JV |
| | | | 28 to 33 | 15 to 30 | Rough: .050<br>Finish: .013 max. | 1/4<br>1/8 | A60JV |
| | Over 275 | Quenched and Tempered | 5500 to 6500 | 50 to 100 | Rough: .002<br>Finish: .0005 max. | 1/4<br>1/8 | A60IV |
| | | | 28 to 33 | 15 to 30 | Rough: .050<br>Finish: .013 max. | 1/4<br>1/8 | A60IV |
| **14. PRECIPITATION HARDENING STAINLESS STEELS, WROUGHT**<br>15-5 PH<br>16-6 PH<br>17-4 PH<br>17-7 PH<br>17-14 Cu Mo<br>AF-71<br>AFC-77<br>Almar 362 (AM-362)<br>AM-350<br>AM-355<br>AM-363<br>Custom 450<br>Custom 455<br>HNM<br>PH 13-8 Mo<br>PH 14-8 Mo<br>PH 15-7 Mo<br>Stainless W | 150 to 200 | Solution Treated | 5500 to 6500 | 70 to 100 | Rough: .002<br>Finish: .0005 max. | 1/4<br>1/8 | A60JV |
| | | | 28 to 33 | 21 to 30 | Rough: .050<br>Finish: .013 max. | 1/4<br>1/8 | A60JV |
| | 275 to 440 | Solution Treated or Hardened | 5500 to 6500 | 70 to 100 | Rough: .002<br>Finish: .0005 max. | 1/4<br>1/8 | A60IV |
| | | | 28 to 33 | 21 to 30 | Rough: .050<br>Finish: .013 max. | 1/4<br>1/8 | A60IV |

See section 16 for Cutting Fluid Recommendations.

*Wheel recommendations are for wet grinding of 2- to 4-inch [50 to 100 mm] diameter work. For DRY grinding—use a softer grade wheel. For LARGER diameter work—use a softer and/or coarser grit wheel. For SMALLER diameter work—use a harder grade wheel. Wheel recommendations also apply to plunge grinding applications. Also see section 20.2, Grinding Guidelines.

# 8.5 Cylindrical Grinding

| MATERIAL | HARDNESS<br>Bhn | CONDITION | WHEEL SPEED<br>fpm<br>m/s | WORK SPEED<br>fpm<br>m/min | INFEED on dia.<br>in/pass<br>mm/pass | TRAVERSE Wheel width per rev. of work | WHEEL IDENTIFICATION*<br>ANSI<br>ISO |
|---|---|---|---|---|---|---|---|
| **15. CARBON STEELS, CAST**<br>**Low Carbon**<br>ASTM A426: Grade CP1<br>1010<br>1020<br>**Medium Carbon**<br>ASTM A352: Grades LCA, LCB, LCC<br>ASTM A356: Grade 1<br>1030 1040 1050 | 100<br>to<br>300 | Annealed,<br>Normalized,<br>Normalized and Tempered,<br>or Quenched<br>and Tempered | 5500<br>to<br>6500 | 70<br>to<br>100 | Rough:<br>.002<br>Finish:<br>.0005 max. | 1/2<br>1/6 | A60LV |
| | | | 28<br>to<br>33 | 21<br>to<br>30 | Rough:<br>.050<br>Finish:<br>.013 max. | 1/2<br>1/6 | A60LV |
| | Over<br>50 $R_C$ | Carburized<br>and/or<br>Quenched and<br>Tempered | 5500<br>to<br>6500 | 70<br>to<br>100 | Rough:<br>.002<br>Finish:<br>.0005 max. | 1/4<br>1/8 | A60KV |
| | | | 28<br>to<br>33 | 21<br>to<br>30 | Rough:<br>.050<br>Finish:<br>.013 max. | 1/4<br>1/8 | A60KV |
| **16. ALLOY STEELS, CAST**<br>**Low Carbon**<br>ASTM A217: Grade WC9<br>ASTM A352: Grades LC3, LC4<br>ASTM A426: Grades CP2, CP5, CP5b, CP11, CP12, CP15, CP21, CP22<br>1320 4110 8020<br>2315 4120 8620<br>2320 4320 | 150<br>to<br>400 | Annealed,<br>Normalized,<br>Normalized and Tempered,<br>or Quenched<br>and Tempered | 5500<br>to<br>6500 | 70<br>to<br>100 | Rough:<br>.002<br>Finish:<br>.0005 max. | 1/2<br>1/6 | A60LV |
| | | | 28<br>to<br>33 | 21<br>to<br>30 | Rough:<br>.050<br>Finish:<br>.013 max. | 1/2<br>1/6 | A60LV |
| **Medium Carbon**<br>ASTM A27: Grades N1, N2, U-60-30, 60-30, 65-35, 70-36, 70-40<br>ASTM A148: Grades 80-40, 80-50, 90-60, 105-85, 120-95, 150-125, 175-145<br>ASTM A216: Grades WCA, WCB, WCC<br>ASTM A217: Grades WC1, WC4, WC5, WC6<br>ASTM A352: Grades LC1, LC2, LC2-1<br>ASTM A356: Grades 2, 5, 6, 8, 9, 10<br>ASTM A389: Grades C23, C24<br>ASTM A486: Classes 70, 90, 120<br>ASTM A487: Classes 1N, 2N, 4N, 6N, 8N, 9N, 10N, DN, 1Q, 2Q, 4Q, 4QA, 6Q, 7Q, 8Q, 9Q, 10Q<br>1330 4130 80B30 8640<br>1340 4140 8040 9525<br>2325 4330 8430 9530<br>2330 4340 8440 9535<br>4125 8030 8630 | Over<br>50 $R_C$ | Carburized<br>and/or<br>Quenched and<br>Tempered | 4000<br>to<br>6000 | 70<br>to<br>100 | Rough:<br>.002<br>Finish:<br>.0005 max. | 1/4<br>1/8 | A60KV |
| | | | 20<br>to<br>30 | 21<br>to<br>30 | Rough:<br>.050<br>Finish:<br>.013 max. | 1/4<br>1/8 | A60KV |

See section 16 for Cutting Fluid Recommendations.

*Wheel recommendations are for wet grinding of 2- to 4-inch [50 to 100 mm] diameter work. For DRY grinding—use a softer grade wheel. For LARGER diameter work—use a softer and/or coarser grit wheel. For SMALLER diameter work—use a harder grade wheel. Wheel recommendations also apply to plunge grinding applications. Also see section 20.2, Grinding Guidelines.

| MATERIAL | HARDNESS | CONDITION | WHEEL SPEED | WORK SPEED | INFEED on dia. | TRAVERSE Wheel width per rev. of work | WHEEL IDENTIFICATION* |
|---|---|---|---|---|---|---|---|
| | Bhn | | fpm / m/s | fpm / m/min | in/pass / mm/pass | | ANSI / ISO |
| **17. TOOL STEELS, CAST** **Group I** ASTM A597: Grades CA-2, CH-12, CH-13, CO-1, CS-5 | 50 $R_C$ max. | Annealed or Quenched and Tempered | 4000 to 6000 | 60 to 100 | Rough: .002 Finish: .0005 max. | 1/2 1/6 | A60LV |
| | | | 20 to 30 | 18 to 30 | Rough: .050 Finish: .013 max. | 1/2 1/6 | A60LV |
| | 50 $R_C$ to 58 $R_C$ | Quenched and Tempered | 4000 to 5500 | 60 to 100 | Rough: .002 Finish: .0004 max. | 1/3 1/8 | A60KV |
| | | | 20 to 28 | 18 to 30 | Rough: .050 Finish: .010 max. | 1/3 1/8 | A60KV |
| | Over 58 $R_C$ | Quenched and Tempered | 4000 to 5500 | 60 to 100 | Rough: .002 Finish: .0004 max. | 1/4 1/8 | A60JV |
| | | | 20 to 28 | 18 to 30 | Rough: .050 Finish: .010 max. | 1/4 1/8 | A60JV |
| **Group II** ASTM A597: Grades CD-2, CD-5 | 200 to 250 | Annealed | 4000 to 6000 | 60 to 100 | Rough: .002 Finish: .0005 max. | 1/2 1/6 | A60LV |
| | | | 20 to 30 | 18 to 30 | Rough: .050 Finish: .013 max. | 1/2 1/6 | A60LV |
| | 48 $R_C$ to 56 $R_C$ | Quenched and Tempered | 4000 to 5500 | 60 to 100 | Rough: .002 Finish: .0004 max. | 1/3 1/8 | A60KV |
| | | | 20 to 28 | 18 to 30 | Rough: .050 Finish: .010 max. | 1/3 1/8 | A60KV |
| | Over 56 $R_C$ | Quenched and Tempered | 4000 to 5500 | 60 to 100 | Rough: .002 Finish: .0003 max. | 1/4 1/8 | A80JV |
| | | | 20 to 28 | 18 to 30 | Rough: .050 Finish: .008 max. | 1/4 1/8 | A80JV |

See section 16 for Cutting Fluid Recommendations.

*Wheel recommendations are for wet grinding of 2- to 4-inch [50 to 100 mm] diameter work. For DRY grinding—use a softer grade wheel. For LARGER diameter work—use a softer and/or coarser grit wheel. For SMALLER diameter work—use a harder grade wheel. Wheel recommendations also apply to plunge grinding applications. Also see section 20.2, Grinding Guidelines.

## 8.5 Cylindrical Grinding

| MATERIAL | HARD-NESS Bhn | CONDITION | WHEEL SPEED fpm m/s | WORK SPEED fpm m/min | INFEED on dia. in/pass mm/pass | TRAVERSE Wheel width per rev. of work | WHEEL IDENTIFI-CATION* ANSI ISO |
|---|---|---|---|---|---|---|---|
| **18. STAINLESS STEELS, CAST**<br>**Ferritic**<br>ASTM A217: Grades C5, C12<br>ASTM A296: Grades CB-30, CC-50, CE-30, CA6N, CA-6NM, CD4MCu<br>ASTM A297: Grade HC<br>ASTM A487: Class CA6NM<br>ASTM A608: Grade HC30 | 135 to 185 | Annealed | 5500 to 6500 | 50 to 100 | Rough: .002<br>Finish: .0005 max. | 1/2<br>1/6 | A60JV |
| | | | 28 to 33 | 15 to 30 | Rough: .050<br>Finish: .013 max. | 1/2<br>1/6 | A60JV |
| **Austenitic**<br>ASTM A296: Grades CF-3, CF-3M, CF-8, CF-8C, CF-8M, CF-16F, CF-20, CG-8M, CG-12, CH-20, CK-20, CN-7M, CN-7MS<br>ASTM A297: Grades HD, HE, HF, HH, HI, HK, HL, HN, HP, HT, HU<br>ASTM A351: Grades CF-3, CF-3A, CF-3M, CF-3MA, CF-8, CF-8A, CF-8C, CF-8M, CF-10MC, CH-8, CH-10, CH-20, CK-20, CN-7M, HK-30, HK-40, HT-30<br>ASTM A451: Grades CPF8A, CPF3, CPF3A, CPF3M, CPF8, CPF8C, CPF8C (Ta Max.), CPF8M, CPF10MC, CPH8, CPH10, CPH20, CPK20<br>ASTM A452: Grades TP 304H, TP 316H, TP 347H<br>ASTM A608: Grades HD50, HE35, HF30, HH30, HH33, HI35, HK30, HK40, HL30, HL40, HN40, HT50, HU50 | 135 to 210 | Annealed, Normalized or As Cast | 5500 to 6500 | 50 to 100 | Rough: .002<br>Finish: .0005 max. | 1/2<br>1/6 | C54JV |
| | | | 28 to 33 | 15 to 30 | Rough: .050<br>Finish: .013 max. | 1/2<br>1/6 | C54JV |
| **Martensitic**<br>ASTM A217: Grade CA-15<br>ASTM A296: Grades CA-15, CA-15M, CA-40<br>ASTM A426: Grades CP7, CP9, CPCA15<br>ASTM A487: Classes CA15a, CA-15M | 135 to 225 | Annealed, Normalized or Normalized and Tempered | 5500 to 6500 | 50 to 100 | Rough: .002<br>Finish: .0005 max. | 1/4<br>1/8 | A60JV |
| | | | 28 to 33 | 15 to 30 | Rough: .050<br>Finish: .013 max. | 1/4<br>1/8 | A60JV |
| | Over 275 | Quenched and Tempered | 5500 to 6500 | 50 to 100 | Rough: .002<br>Finish: .0005 max. | 1/4<br>1/8 | A60IV |
| | | | 28 to 33 | 15 to 30 | Rough: .050<br>Finish: .013 max. | 1/4<br>1/8 | A60IV |

See section 16 for Cutting Fluid Recommendations.

*Wheel recommendations are for wet grinding of 2- to 4-inch [50 to 100 mm] diameter work. For DRY grinding—use a softer grade wheel. For LARGER diameter work—use a softer and/or coarser grit wheel. For SMALLER diameter work—use a harder grade wheel. Wheel recommendations also apply to plunge grinding applications. Also see section 20.2, Grinding Guidelines.

| MATERIAL | HARDNESS<br>Bhn | CONDITION | WHEEL SPEED<br>fpm<br>m/s | WORK SPEED<br>fpm<br>m/min | INFEED on dia.<br>in/pass<br>mm/pass | TRAVERSE Wheel width per rev. of work | WHEEL IDENTIFICATION*<br>ANSI<br>ISO |
|---|---|---|---|---|---|---|---|
| **19. PRECIPITATION HARDENING STAINLESS STEELS, CAST**<br>ASTM A351: Grade CD-4MCu<br>ACI Grade CB-7Cu<br>ACI Grade CD-4MCu<br>17-4 PH<br>AM-355 | 325<br>to<br>450 | Solution Treated<br>or<br>Solution Treated<br>and Aged | 5500<br>to<br>6500 | 70<br>to<br>100 | Rough:<br>.002<br>Finish:<br>.0005 max. | 1/4<br>1/8 | A60IV |
| | | | 28<br>to<br>33 | 21<br>to<br>30 | Rough:<br>.050<br>Finish:<br>.013 max. | 1/4<br>1/8 | A60IV |
| **21. GRAY CAST IRONS**<br>**Ferritic**<br>ASTM A48: Class 20<br>SAE J431c: Grade G1800<br>**Pearlitic- Ferritic**<br>ASTM A48: Class 25<br>SAE J431c: Grade G2500<br>**Pearlitic**<br>ASTM A48: Classes 30, 35, 40<br>SAE J431c: Grade G3000<br>**Pearlitic + Free Carbides**<br>ASTM A48: Classes 45, 50<br>SAE J431c: Grades G3500, G4000<br>**Pearlitic or Acicular + Free Carbides**<br>ASTM A48: Classes 55, 60 | 45 $R_C$<br>max. | As Cast,<br>Annealed<br>or Quenched<br>and Tempered | 5500<br>to<br>6500 | 70<br>to<br>100 | Rough:<br>.002<br>Finish:<br>.001 max. | 1/2<br>1/6 | C46KV |
| | | | 28<br>to<br>33 | 21<br>to<br>30 | Rough:<br>.050<br>Finish:<br>.025 max. | 1/2<br>1/6 | C46KV |
| | 45 $R_C$<br>to<br>52 $R_C$ | As Cast,<br>Annealed<br>or Quenched<br>and Tempered | 5500<br>to<br>6500 | 70<br>to<br>100 | Rough:<br>.002<br>Finish:<br>.0005 max. | 1/3<br>1/6 | C54JV |
| | | | 28<br>to<br>33 | 21<br>to<br>30 | Rough:<br>.050<br>Finish:<br>.013 max. | 1/3<br>1/6 | C54JV |
| | 48 $R_C$<br>to<br>60 $R_C$ | Flame<br>or<br>Induction Hardened | 5500<br>to<br>6500 | 70<br>to<br>100 | Rough:<br>.002<br>Finish:<br>.0005 max. | 1/4<br>1/8 | C60IV |
| | | | 28<br>to<br>33 | 21<br>to<br>30 | Rough:<br>.050<br>Finish:<br>.013 max. | 1/4<br>1/8 | C60IV |
| **Austenitic (NI-RESIST)**<br>ASTM A436: Types 1, 1b, 2, 2b, 3, 4, 5, 6 | 100<br>to<br>250 | As Cast | 5500<br>to<br>6500 | 40<br>to<br>65 | Rough:<br>.002<br>Finish:<br>.0005 max. | 1/2<br>1/6 | A60KV |
| | | | 28<br>to<br>33 | 12<br>to<br>20 | Rough:<br>.050<br>Finish:<br>.013 max. | 1/2<br>1/6 | A60KV |
| **22. COMPACTED GRAPHITE CAST IRONS** | 185<br>to<br>225 | As Cast | 5500<br>to<br>6500 | 70<br>to<br>100 | Rough:<br>.002<br>Finish:<br>.001 max. | 1/2<br>1/6 | C46KV |
| | | | 28<br>to<br>33 | 21<br>to<br>30 | Rough:<br>.050<br>Finish:<br>.025 max. | 1/2<br>1/6 | C46KV |

See section 16 for Cutting Fluid Recommendations.

*Wheel recommendations are for wet grinding of 2- to 4-inch [50 to 100 mm] diameter work. For DRY grinding—use a softer grade wheel. For LARGER diameter work—use a softer and/or coarser grit wheel. For SMALLER diameter work—use a harder grade wheel. Wheel recommendations also apply to plunge grinding applications. Also see section 20.2, Grinding Guidelines.

# 8.5 Cylindrical Grinding

| MATERIAL | HARDNESS<br>Bhn | CONDITION | WHEEL SPEED<br>fpm<br>m/s | WORK SPEED<br>fpm<br>m/min | INFEED on dia.<br>in/pass<br>mm/pass | TRAVERSE Wheel width per rev. of work | WHEEL IDENTIFICATION*<br>ANSI<br>ISO |
|---|---|---|---|---|---|---|---|
| **23. DUCTILE CAST IRONS**<br>**Ferritic**<br>ASTM A536: Grades 60-40-18, 65-45-12<br>SAE J434c: Grades D4018, D4512<br>**Ferritic- Pearlitic**<br>ASTM A536: Grade 80-55-06<br>SAE J434c: Grade D5506<br>**Pearlitic- Martensitic**<br>ASTM A536: Grade 100-70-03<br>SAE J434c: Grade D7003<br>**Martensitic**<br>ASTM A536: Grade 120-90-02<br>SAE J434c: Grade DQ&T | 52 $R_C$ max. | Annealed, As Cast<br>Normalized and Tempered<br>or Quenched<br>and Tempered | 5500 to 6500 | 70 to 100 | Rough: .002<br>Finish: .0005 max. | 1/4<br>1/8 | C54JV |
| | | | 28 to 33 | 21 to 30 | Rough: .050<br>Finish: .013 max. | 1/4<br>1/8 | C54JV |
| | 53 $R_C$ to 60 $R_C$ | Flame or Induction Hardened | 5500 to 6500 | 70 to 100 | Rough: .002<br>Finish: .0005 max. | 1/4<br>1/8 | C60IV |
| | | | 28 to 33 | 21 to 30 | Rough: .050<br>Finish: .013 max. | 1/4<br>1/8 | C60IV |
| **Austenitic (NI-RESIST Ductile)**<br>ASTM A439: Types D-2, D-2B, D-2C, D-3, D-3A, D-4, D-5, D-5B<br>ASTM A571: Type D-2M | 120 to 275 | Annealed | 5500 to 6500 | 40 to 65 | Rough: .002<br>Finish: .0005 max. | 1/2<br>1/6 | A60LV |
| | | | 28 to 33 | 12 to 20 | Rough: .050<br>Finish: .013 max. | 1/2<br>1/6 | A60LV |
| **24. MALLEABLE CAST IRONS**<br>**Ferritic**<br>ASTM A47: Grades 32510, 35018<br>ASTM A602: Grade M3210<br>SAE J158: Grade M3210<br>**Pearlitic**<br>ASTM A220: Grades 40010, 45006, 45008, 50005<br>ASTM A602: Grade M4504, M5003<br>SAE J158: Grades M4504, M5003<br>**Tempered Martensite**<br>ASTM A220: Grades 60004, 70003, 80002, 90001<br>ASTM A602: Grades M5503, M7002, M8501<br>SAE J158: Grades M5503, M7002, M8501 | 52 $R_C$ max. | Malleablized or Malleablized and Heat Treated | 5500 to 6500 | 70 to 100 | Rough: .002<br>Finish: .0005 max. | 1/3<br>1/6 | C54JV |
| | | | 28 to 33 | 21 to 30 | Rough: .050<br>Finish: .013 max. | 1/3<br>1/6 | C54JV |
| | Over 52 $R_C$ | Flame or Induction Hardened | 5500 to 6500 | 70 to 100 | Rough: .002<br>Finish: .0005 max. | 1/4<br>1/8 | C60IV |
| | | | 28 to 33 | 21 to 30 | Rough: .050<br>Finish: .013 max. | 1/4<br>1/8 | C60IV |
| **25. WHITE CAST IRONS (ABRASION RESISTANT)**<br>ASTM A532:<br>Class I, Types A, B, C, D<br>Class II, Types A, B, C, D, E<br>Class III, Type A | 60 $R_C$ max. | As Cast, Annealed or Hardened | 5500 to 6500 | 70 to 100 | Rough: .002<br>Finish: .0005 max. | 1/4<br>1/8 | C60IV |
| | | | 28 to 33 | 21 to 30 | Rough: .050<br>Finish: .013 max. | 1/4<br>1/8 | C60IV |

See section 16 for Cutting Fluid Recommendations.

*Wheel recommendations are for wet grinding of 2- to 4-inch [50 to 100 mm] diameter work. For DRY grinding—use a softer grade wheel. For LARGER diameter work—use a softer and/or coarser grit wheel. For SMALLER diameter work—use a harder grade wheel. Wheel recommendations also apply to plunge grinding applications. Also see section 20.2, Grinding Guidelines.

| MATERIAL | HARDNESS Bhn | CONDITION | WHEEL SPEED fpm / m/s | WORK SPEED fpm / m/min | INFEED on dia. in/pass / mm/pass | TRAVERSE Wheel width per rev. of work | WHEEL IDENTIFICATION* ANSI / ISO |
|---|---|---|---|---|---|---|---|
| **26. HIGH SILICON CAST IRONS**<br>Duriron<br>Duriclor<br>ASTM A518 | 52 $R_C$ | As Cast | 5500 to 6500 | 70 to 100 | Rough: .002<br>Finish: .0005 max. | 1/4<br>1/8 | Rough: C36KB<br>Finish: C60LB |
| | | | 28 to 33 | 21 to 30 | Rough: .050<br>Finish: .013. max. | 1/4<br>1/8 | Rough: C36KB<br>Finish: C60LB |
| **27. CHROMIUM-NICKEL ALLOY CASTINGS**<br>ASTM A560: Grades 50Cr-50Ni, 60Cr-40Ni | 275 to 375 | As Cast | 5500 to 6500 | 50 to 100 | Rough: .002<br>Finish: .0005 max. | 1/4<br>1/8 | C54IV |
| | | | 28 to 33 | 15 to 30 | Rough: .050<br>Finish: .013 max. | 1/4<br>1/8 | C54IV |
| **28. ALUMINUM ALLOYS, WROUGHT**<br>EC 2218 5252 6253<br>1060 2219 5254 6262<br>1100 2618 5454 6463<br>1145 3003 5456 6951<br>1175 3004 5457 7001<br>1235 3005 5652 7004<br>2011 4032 5657 7005<br>2014 5005 6053 7039<br>2017 5050 6061 7049<br>2018 5052 6063 7050<br>2021 5056 6066 7075<br>2024 5083 6070 7079<br>2025 5086 6101 7175<br>2117 5154 6151 7178 | 30 to 150 500kg | Cold Drawn or Solution Treated and Aged | 5500 to 6500 | 50 to 150 | Rough: .002<br>Finish: .0005 max. | 1/3<br>1/6 | C54JV† |
| | | | 28 to 33 | 15 to 46 | Rough: .050<br>Finish: .013 max. | 1/3<br>1/6 | C54JV† |
| **29. ALUMINUM ALLOYS, CAST**<br>**Sand and Permanent Mold**<br>A140 295.0 514.0 A712.0<br>201.0 B295.0 A514.0 D712.0<br>208.0 308.0 B514.0 713.0<br>213.0 319.0 520.0 771.0<br>222.0 355.0 535.0 850.0<br>224.0 C355.0 705.0 A850.0<br>242.0 B443.0 707.0 B850.0<br>Hiduminium RR-350<br>**Die Castings**<br>C443.0 518.0 | 40 to 125 500kg | As Cast or Solution Treated and Aged | 5500 to 6500 | 50 to 150 | Rough: .002<br>Finish: .0005 max. | 1/3<br>1/6 | C54JV† |
| | | | 28 to 33 | 15 to 46 | Rough: .050<br>Finish: .013 max. | 1/3<br>1/6 | C54JV† |

See section 16 for Cutting Fluid Recommendations.

*Wheel recommendations are for wet grinding of 2- to 4-inch [50 to 100 mm] diameter work. For DRY grinding—use a softer grade wheel. For LARGER diameter work—use a softer and/or coarser grit wheel. For SMALLER diameter work—use a harder grade wheel. Wheel recommendations also apply to plunge grinding applications. Also see section 20.2, Grinding Guidelines.

†Wax filled.

# 8.5 Cylindrical Grinding

| MATERIAL | HARD-NESS<br>Bhn | CONDITION | WHEEL SPEED<br>fpm<br>m/s | WORK SPEED<br>fpm<br>m/min | INFEED on dia.<br>in/pass<br>mm/pass | TRAVERSE<br>Wheel width per rev. of work | WHEEL IDENTIFI-CATION*<br>ANSI<br>ISO |
|---|---|---|---|---|---|---|---|
| **29. ALUMINUM ALLOYS, CAST (cont.)**<br>**Sand and Permanent Mold**<br>328.0 356.0<br>A332.0 A356.0<br>F332.0 357.0<br>333.0 359.0<br>354.0<br>**Die Castings**<br>360.0 A380.0 390.0 A413.0<br>A360.0 383.0 392.0<br>380.0 A384.0 413.0 | 40<br>to<br>125<br>500kg | As Cast<br>or<br>Solution Treated<br>and Aged | 5500<br>to<br>6500 | 50<br>to<br>150 | Rough:<br>.002<br>Finish:<br>.0005 max. | 1/3<br>1/6 | A54IV |
| | | | 28<br>to<br>33 | 15<br>to<br>46 | Rough:<br>.050<br>Finish:<br>.013 max. | 1/3<br>1/6 | A54IV |
| **30. MAGNESIUM ALLOYS, WROUGHT‡‡**<br>AZ21A HK31A<br>AZ31B HM21A<br>AZ31C HM31A<br>AZ61A ZK40A<br>AZ80A ZK60A | 50<br>to<br>90<br>500kg | Annealed,<br>Cold Drawn<br>or<br>Solution Treated and Aged | 5500<br>to<br>6500 | 70<br>to<br>150 | Rough:<br>.002<br>Finish:<br>.0005 max. | 1/3<br>1/6 | C60KV |
| | | | 28<br>to<br>33 | 21<br>to<br>46 | Rough:<br>.050<br>Finish:<br>.013 max. | 1/3<br>1/6 | C60KV |
| **31. MAGNESIUM ALLOYS, CAST‡‡**<br>AM60A AZ91C ZE41A<br>AM100A AZ92A ZE63A<br>AS41A EZ33A ZH62A<br>AZ63A HK31A ZK51A<br>AZ81A HZ32A ZK61A<br>AZ91A K1A<br>AZ91B QE22A | 50<br>to<br>90<br>500kg | As Cast,<br>Annealed<br>or<br>Solution Treated and Aged | 5500<br>to<br>6500 | 70<br>to<br>150 | Rough:<br>.002<br>Finish:<br>.0005 max. | 1/3<br>1/6 | C60KV |
| | | | 28<br>to<br>33 | 21<br>to<br>46 | Rough:<br>.050<br>Finish:<br>.013 max. | 1/3<br>1/6 | C60KV |
| **32. TITANIUM ALLOYS, WROUGHT‡‡**<br>**Commercially Pure**<br>99.5<br>99.2<br>99.0<br>98.9<br>Ti-0.2Pd<br>TiCODE-12 | 110<br>to<br>275 | Annealed | 3000<br>to<br>4000 | 50<br>to<br>100 | Rough:<br>.001<br>Finish:<br>.0005 max. | 1/5<br>1/10 | C60JV‡ |
| | | | 15<br>to<br>20 | 15<br>to<br>30 | Rough:<br>.025<br>Finish:<br>.013 max. | 1/5<br>1/10 | C60JV‡ |

See section 16 for Cutting Fluid Recommendations.

*Wheel recommendations are for wet grinding of 2- to 4-inch [50 to 100 mm] diameter work. For DRY grinding—use a softer grade wheel. For LARGER diameter work—use a softer and/or coarser grit wheel. For SMALLER diameter work—use a harder grade wheel. Wheel recommendations also apply to plunge grinding applications. Also see section 20.2, Grinding Guidelines.

‡Use friable (green grit) silicon carbide.

‡‡CAUTION: Potential Fire Hazard. Exercise caution in grinding and disposing of swarf. Do NOT use water or water-miscible cutting fluids for magnesium alloys.

| MATERIAL | HARD-NESS | CONDITION | WHEEL SPEED | WORK SPEED | INFEED on dia. | TRAVERSE Wheel width per rev. of work | WHEEL IDENTIFI-CATION* |
|---|---|---|---|---|---|---|---|
| | Bhn | | fpm | fpm | in/pass | | ANSI |
| | | | m/s | m/min | mm/pass | | ISO |
| **Alpha and Alpha-Beta Alloys**<br>Ti-8Mn<br>Ti-1Al-8V-5Fe<br>Ti-2Al-11Sn-5Zr-1Mo<br>Ti-3Al-2.5V<br>Ti-5Al-2Sn-2Zr-4Mo-4Cr (Ti-17)<br>Ti-5Al-2.5Sn<br>Ti-5Al-2.5Sn ELI<br>Ti-5Al-6Sn-2Zr-1Mo<br>Ti-6Al-2Cb-1Ta-0.8Mo<br>Ti-6Al-4V<br>Ti-6Al-4V ELI<br>Ti-6Al-6V-2Sn<br>Ti-6Al-2Sn-4Zr-2Mo<br>Ti-6Al-2Sn-4Zr-2Mo-.25Si<br>Ti-6Al-2Sn-4Zr-6Mo<br>Ti-7Al-4Mo<br>Ti-8Al-1Mo-1V | 300 to 380 | Annealed | 3000 to 4000 | 50 to 100 | Rough: .001<br>Finish: .0005 max. | 1/5<br>1/10 | C60JV‡ |
| | | | 15 to 20 | 15 to 30 | Rough: .025<br>Finish: .013 max. | 1/5<br>1/10 | C60JV‡ |
| | 320 to 440 | Solution Treated and Aged | 3000 to 4000 | 50 to 100 | Rough: .001<br>Finish: .0005 max. | 1/5<br>1/10 | C60JV‡ |
| | | | 15 to 20 | 15 to 30 | Rough: .025<br>Finish: .013 max. | 1/5<br>1/10 | C60JV‡ |
| **Beta Alloys**<br>Ti-3Al-8V-6Cr-4Mo-4Zr<br>Ti-8Mo-8V-2Fe-3Al<br>Ti-11.5Mo-6Zr-4.5Sn<br>Ti-10V-2Fe-3Al<br>Ti-13V-11Cr-3Al | 275 to 350 | Annealed or Solution Treated | 3000 to 4000 | 50 to 100 | Rough: .001<br>Finish: .0005 max. | 1/5<br>1/10 | C60JV‡ |
| | | | 15 to 20 | 15 to 30 | Rough: .025<br>Finish: .013 max. | 1/5<br>1/10 | C60JV‡ |
| | 350 to 440 | Solution Treated and Aged | 3000 to 4000 | 50 to 100 | Rough: .001<br>Finish: .0005 max. | 1/5<br>1/10 | C60JV‡ |
| | | | 15 to 20 | 15 to 30 | Rough: .025<br>Finish: .013 max. | 1/5<br>1/10 | C60JV‡ |
| **33. TITANIUM ALLOYS, CAST‡‡**<br>**Commercially Pure**<br>99.0<br>Ti-0.2Pd<br>ASTM B367: Grades C-1, C-2, C-3, C-4, C-7A, C-7B, C-8A, C-8B | 150 to 250 | As Cast or As Cast and Annealed | 3000 to 4000 | 50 to 100 | Rough: .001<br>Finish: .0005 max. | 1/5<br>1/10 | C60JV‡ |
| | | | 15 to 20 | 15 to 30 | Rough: .025<br>Finish: .013 max. | 1/5<br>1/10 | C60JV‡ |
| **Alpha and Alpha-Beta Alloys**<br>Ti-5Al-2.5Sn<br>Ti-6Al-4V<br>Ti-6Al-2Sn-4Zr-2Mo<br>Ti-8Al-1Mo-1V<br>ASTM B367: Grades C-5, C-6 | 300 to 350 | As Cast or As Cast and Annealed | 3000 to 4000 | 50 to 100 | Rough: .001<br>Finish: .0005 max. | 1/5<br>1/10 | C60JV‡ |
| | | | 15 to 20 | 15 to 30 | Rough: .025<br>Finish: .013 max. | 1/5<br>1/10 | C60JV‡ |

See section 16 for Cutting Fluid Recommendations.

*Wheel recommendations are for wet grinding of 2- to 4-inch [50 to 100 mm] diameter work. For DRY grinding—use a softer grade wheel. For LARGER diameter work—use a softer and/or coarser grit wheel. For SMALLER diameter work—use a harder grade wheel. Wheel recommendations also apply to plunge grinding applications. Also see section 20.2, Grinding Guidelines.

‡Use friable (green grit) silicon carbide.

‡‡CAUTION: Potential Fire Hazard. Exercise caution in grinding and disposing of swarf.

# 8.5 Cylindrical Grinding

| MATERIAL | HARDNESS Bhn | CONDITION | WHEEL SPEED fpm / m/s | WORK SPEED fpm / m/min | INFEED on dia. in/pass / mm/pass | TRAVERSE Wheel width per rev. of work | WHEEL IDENTIFICATION* ANSI / ISO |
|---|---|---|---|---|---|---|---|
| **34. COPPER ALLOYS, WROUGHT**<br>101 116 143 182<br>102 119 145 184<br>104 120 147 185<br>105 121 150 187<br>107 122 155 189<br>109 125 162 190<br>110 127 165 191<br>111 128 170†† 192<br>113 129 172†† 194<br>114 130 173†† 195<br>115 142 175†† | $10\ R_B$ to $70\ R_B$ | Annealed | 5500 to 6500 | 70 to 100 | Rough: .002<br>Finish: .0005 max. | 1/3<br>1/6 | C46JV |
| | | | 28 to 33 | 21 to 30 | Rough: .050<br>Finish: .013 max. | 1/3<br>1/6 | C46JV |
| | $60\ R_B$ to $100\ R_B$ | Cold Drawn | 5500 to 6500 | 70 to 100 | Rough: .002<br>Finish: .0005 max. | 1/3<br>1/6 | C46JV |
| | | | 28 to 33 | 21 to 30 | Rough: .050<br>Finish: .013 max. | 1/3<br>1/6 | C46JV |
| 210 332 368 464<br>220 335 370 465<br>226 340 377 466<br>230 342 385 467<br>240 349 411 482<br>260 350 413 485<br>268 353 425 667<br>270 356 435 687<br>280 360 442 688<br>314 365 443 694<br>316 366 444<br>330 367 445 | $10\ R_B$ to $70\ R_B$ | Annealed | 5500 to 6500 | 70 to 100 | Rough: .002<br>Finish: .0005 max. | 1/3<br>1/6 | C46KV |
| | | | 28 to 33 | 21 to 30 | Rough: .050<br>Finish: .013 max. | 1/3<br>1/6 | C46KV |
| | $60\ R_B$ to $100\ R_B$ | Cold Drawn | 5500 to 6500 | 70 to 100 | Rough: .002<br>Finish: .0005 max. | 1/3<br>1/6 | C46KV |
| | | | 28 to 33 | 21 to 30 | Rough: .050<br>Finish: .013 max. | 1/3<br>1/6 | C46KV |
| 505 613 632<br>510 614 638<br>511 618 642<br>521 619 651<br>524 623 655<br>544 624 674<br>608 625 675<br>610 630 | $10\ R_B$ to $70\ R_B$ | Annealed | 5500 to 6500 | 70 to 100 | Rough: .002<br>Finish: .0005 max. | 1/3<br>1/6 | C46KV |
| | | | 28 to 33 | 21 to 30 | Rough: .050<br>Finish: .013 max. | 1/3<br>1/6 | C46KV |
| | $60\ R_B$ to $100\ R_B$ | Cold Drawn | 5500 to 6500 | 70 to 100 | Rough: .002<br>Finish: .0005 max. | 1/3<br>1/6 | C46KV |
| | | | 28 to 33 | 21 to 30 | Rough: .050<br>Finish: .013 max. | 1/3<br>1/6 | C46KV |

See section 16 for Cutting Fluid Recommendations.

*Wheel recommendations are for wet grinding of 2- to 4-inch [50 to 100 mm] diameter work. For DRY grinding—use a softer grade wheel. For LARGER diameter work—use a softer and/or coarser grit wheel. For SMALLER diameter work—use a harder grade wheel. Wheel recommendations also apply to plunge grinding applications. Also see section 20.2, Grinding Guidelines.

††CAUTION: Toxic Material, refer to National Institute for Occupational Safety and Health (NIOSH) for Precautions.

| MATERIAL | HARDNESS<br>Bhn | CONDITION | WHEEL SPEED<br>fpm<br>m/s | WORK SPEED<br>fpm<br>m/min | INFEED on dia.<br>in/pass<br>mm/pass | TRAVERSE Wheel width per rev. of work | WHEEL IDENTIFICATION*<br>ANSI<br>ISO |
|---|---|---|---|---|---|---|---|
| 706 710 715 725 745 752 754 757 770 782 | 10 $R_B$ to 70 $R_B$ | Annealed | 5500 to 6500 | 70 to 100 | Rough: .002<br>Finish: .0005 max. | 1/3<br>1/6 | C46KV |
| | | | 28 to 33 | 21 to 30 | Rough: .050<br>Finish: .013 max. | 1/3<br>1/6 | C46KV |
| | 60 $R_B$ to 100 $R_B$ | Cold Drawn | 5500 to 6500 | 70 to 100 | Rough: .002<br>Finish: .0005 max. | 1/3<br>1/6 | C46KV |
| | | | 28 to 33 | 21 to 30 | Rough: .050<br>Finish: .013 max. | 1/3<br>1/6 | C46KV |
| **35. COPPER ALLOYS, CAST**<br>801 803 805 807 809 811 813 814 815 817†† 818†† 820†† 821†† 822†† 824†† 825†† 826†† 827†† 828†† | 40 to 200 500kg | As Cast or Heat Treated | 5500 to 6500 | 70 to 100 | Rough: .002<br>Finish: .0005 max. | 1/3<br>1/6 | C46JV |
| | | | 28 to 33 | 21 to 30 | Rough: .050<br>Finish: .013 max. | 1/3<br>1/6 | C46JV |
| | 34 $R_C$ to 45 $R_C$ | Heat Treated | 5500 to 6500 | 70 to 100 | Rough: .002<br>Finish: .0005 max. | 1/3<br>1/6 | A54LV |
| | | | 28 to 33 | 21 to 30 | Rough: .050<br>Finish: .013 max. | 1/3<br>1/6 | A54LV |
| 833 834 836 838 842 844 848 852 853 854 855 857 858 861 862 863 864 865 867 868 872 874 875 876 878 879 | 35 to 200 500kg | As Cast | 5500 to 6500 | 70 to 100 | Rough: .002<br>Finish: .0005 max. | 1/3<br>1/6 | C46KV |
| | | | 28 to 33 | 21 to 30 | Rough: .050<br>Finish: .013 max. | 1/3<br>1/6 | C46KV |
| 902 903 905 907 909 910 911 913 915 916 917 922 923 925 926 927 928 932 934 935 937 938 939 943 944 945 947 948 952 953 954 955 956 957 958 | 40 to 100 500kg | As Cast | 5500 to 6500 | 70 to 100 | Rough: .002<br>Finish: .0005 max. | 1/3<br>1/6 | C46KV |
| | | | 28 to 33 | 21 to 30 | Rough: .050<br>Finish: .013 max. | 1/3<br>1/6 | C46KV |

See section 16 for Cutting Fluid Recommendations.

*Wheel recommendations are for wet grinding of 2- to 4-inch [50 to 100 mm] diameter work. For DRY grinding—use a softer grade wheel. For LARGER diameter work—use a softer and/or coarser grit wheel. For SMALLER diameter work—use a harder grade wheel. Wheel recommendations also apply to plunge grinding applications. Also see section 20.2, Grinding Guidelines.

††CAUTION: Toxic Material, refer to National Institute for Occupational Safety and Health (NIOSH) for Precautions.

# 8.5 Cylindrical Grinding

| MATERIAL | HARDNESS Bhn | CONDITION | WHEEL SPEED fpm / m/s | WORK SPEED fpm / m/min | INFEED on dia. in/pass / mm/pass | TRAVERSE Wheel width per rev. of work | WHEEL IDENTIFICATION* ANSI / ISO |
|---|---|---|---|---|---|---|---|
| **35. COPPER ALLOYS, CAST (cont.)**<br>(materials listed on preceding page) | Over 100 500kg | As Cast or Heat Treated | 5500 to 6500 | 70 to 100 | Rough: .002<br>Finish: .0005 max. | 1/3<br>1/6 | A54LV |
| | | | 28 to 33 | 21 to 30 | Rough: .050<br>Finish: .013 max. | 1/3<br>1/6 | A54LV |
| 962<br>963<br>964<br>966††<br>973<br>974<br>976<br>978<br>993 | 50 to 100 500kg | As Cast | 5500 to 6500 | 70 to 100 | Rough: .002<br>Finish: .0005 max. | 1/3<br>1/6 | C46KV |
| | | | 28 to 33 | 21 to 30 | Rough: .050<br>Finish: .013 max. | 1/3<br>1/6 | C46KV |
| | Over 100 500kg | As Cast or Heat Treated | 5500 to 6500 | 70 to 100 | Rough: .002<br>Finish: .0005 max. | 1/3<br>1/6 | A54LV |
| | | | 28 to 33 | 21 to 30 | Rough: .050<br>Finish: .013 max. | 1/3<br>1/6 | A54LV |
| **36. NICKEL ALLOYS, WROUGHT AND CAST**<br>Nickel 200<br>Nickel 201<br>Nickel 205<br>Nickel 211<br>Nickel 220<br>Nickel 230 | 80 to 170 | Annealed or Cold Drawn | 5500 to 6500 | 50 to 100 | Rough: .002<br>Finish: .0005 max. | 1/5<br>1/10 | C54JV |
| | | | 28 to 33 | 15 to 30 | Rough: .050<br>Finish: .013 max. | 1/5<br>1/10 | C54JV |
| MONEL Alloy 400<br>MONEL Alloy 401<br>MONEL Alloy 404<br>MONEL Alloy R405<br>ASTM A296: Grades CZ-100, M-35<br>ASTM A494: Grades CZ-100, M-35 | 115 to 240 | Annealed, Cold Drawn or Cast | 5500 to 6500 | 50 to 100 | Rough: .002<br>Finish: .0005 max. | 1/5<br>1/10 | C54JV |
| | | | 28 to 33 | 15 to 30 | Rough: .050<br>Finish: .013 max. | 1/5<br>1/10 | C54JV |
| DURANICKEL Alloy 301<br>MONEL Alloy 502<br>MONEL Alloy K500<br>NI-SPAN-C Alloy 902<br>PERMANICKEL Alloy 300 | 150 to 320 | Solution Treated | 5500 to 6500 | 50 to 100 | Rough: .002<br>Finish: .0005 max. | 1/5<br>1/10 | C60JV |
| | | | 28 to 33 | 15 to 30 | Rough: .050<br>Finish: .013 max. | 1/5<br>1/10 | C60JV |

See section 16 for Cutting Fluid Recommendations.

*Wheel recommendations are for wet grinding of 2- to 4-inch [50 to 100 mm] diameter work. For DRY grinding—use a softer grade wheel. For LARGER diameter work—use a softer and/or coarser grit wheel. For SMALLER diameter work—use a harder grade wheel. Wheel recommendations also apply to plunge grinding applications. Also see section 20.2, Grinding Guidelines.

††CAUTION: Toxic Material, refer to National Institute for Occupational Safety and Health (NIOSH) for Precautions.

| MATERIAL | HARDNESS Bhn | CONDITION | WHEEL SPEED fpm / m/s | WORK SPEED fpm / m/min | INFEED on dia. in/pass / mm/pass | TRAVERSE Wheel width per rev. of work | WHEEL IDENTIFICATION* ANSI / ISO |
|---|---|---|---|---|---|---|---|
| **36. NICKEL ALLOYS, WROUGHT AND CAST (cont.)** (materials listed on preceding page) | 330 to 360 | Aged | 5500 to 6500 | 50 to 100 | Rough: .002 Finish: .0005 max. | 1/5 1/10 | C60JV |
| | | | 28 to 33 | 15 to 30 | Rough: .050 Finish: .013 max. | 1/5 1/10 | C60JV |
| **37. BERYLLIUM NICKEL ALLOYS, WROUGHT AND CAST††** Berylco 440, Berylco 41C, Berylco 42C, Berylco 43C, Brush Alloy 200C, Brush Alloy 220C, Brush Alloy 260C | 200 to 250 | As Cast or Solution Treated | 5500 to 6500 | 50 to 100 | Rough: .002 Finish: .0005 max. | 1/5 1/10 | A60KV |
| | | | 28 to 33 | 15 to 30 | Rough: .050 Finish: .013 max. | 1/5 1/10 | A60KV |
| | 283 to 425 | Hardened or Aged | 5500 to 6500 | 50 to 100 | Rough: .002 Finish: .0005 max. | 1/5 1/10 | A60JV |
| | | | 28 to 33 | 15 to 30 | Rough: .050 Finish: .013 max. | 1/5 1/10 | A60JV |
| | 47 $R_C$ to 52 $R_C$ | Hardened or Aged | 5500 to 6500 | 50 to 100 | Rough: .002 Finish: .0005 max. | 1/5 1/10 | A60IV |
| | | | 28 to 33 | 15 to 30 | Rough: .050 Finish: .013 max. | 1/5 1/10 | A60IV |
| **38. NITINOL ALLOYS, WROUGHT** Nitinol 55Ni-45Ti, Nitinol 56Ni-44Ti, Nitinol 60Ni-40Ti | 210 to 340 | Wrought or Annealed | 5000 | 50 to 100 | Rough: .002 Finish: .0005 max. | 1/5 1/10 | C60JV |
| | | | 25 | 15 to 30 | Rough: .050 Finish: .013 max. | 1/5 1/10 | C60JV |
| | 48 $R_C$ to 52 $R_C$ | Quenched | 5000 | 50 to 100 | Rough: .002 Finish: .0005 max. | 1/5 1/10 | C60IV |
| | | | 25 | 15 to 30 | Rough: .050 Finish: .013 max. | 1/5 1/10 | C60IV |

See section 16 for Cutting Fluid Recommendations.

*Wheel recommendations are for wet grinding of 2- to 4-inch [50 to 100 mm] diameter work. For DRY grinding—use a softer grade wheel. For LARGER diameter work—use a softer and/or coarser grit wheel. For SMALLER diameter work—use a harder grade wheel. Wheel recommendations also apply to plunge grinding applications. Also see section 20.2, Grinding Guidelines.

††CAUTION: Toxic Material, refer to National Institute for Occupational Safety and Health (NIOSH) for Precautions.

# 8.5 Cylindrical Grinding

| MATERIAL | HARDNESS<br>Bhn | CONDITION | WHEEL SPEED<br>fpm<br>m/s | WORK SPEED<br>fpm<br>m/min | INFEED on dia.<br>in/pass<br>mm/pass | TRAVERSE<br>Wheel width per rev. of work | WHEEL IDENTIFICATION*<br>ANSI<br>ISO |
|---|---|---|---|---|---|---|---|
| **39. HIGH TEMPERATURE ALLOYS, WROUGHT AND CAST**<br>**Nickel Base, Wrought**<br>AF2-1DA<br>Astroloy<br>Haynes Alloy 263<br>IN-102<br>Incoloy Alloy 901<br>Incoloy Alloy 903<br>Inconel Alloy 617<br>Inconel Alloy 625<br>Inconel Alloy 700<br>Inconel Alloy 702<br>Inconel Alloy 706<br>Inconel Alloy 718<br>Inconel Alloy 721<br>Inconel Alloy 722<br>Inconel Alloy X-750<br>Inconel Alloy 751<br>M252<br>Nimonic 75<br>Nimonic 80<br>Nimonic 90<br>Nimonic 95<br>Rene 41<br>Rene 63<br>Rene 77<br>Rene 95<br>Udimet 500<br>Udimet 700<br>Udimet 710<br>Unitemp 1753<br>Waspaloy | 200<br>to<br>390 | Annealed<br>or<br>Solution Treated | 3000<br>to<br>3500 | 50<br>to<br>100 | Rough:<br>.001<br>Finish:<br>.0002 max. | 1/5<br>1/10 | A60JV |
| | | | 15<br>to<br>18 | 15<br>to<br>30 | Rough:<br>.025<br>Finish:<br>.005 max. | 1/5<br>1/10 | A60JV |
| | 300<br>to<br>475 | Solution Treated<br>and Aged | 3000<br>to<br>3500 | 50<br>to<br>100 | Rough:<br>.001<br>Finish:<br>.0002 max. | 1/5<br>1/10 | A60JV |
| | | | 15<br>to<br>18 | 15<br>to<br>30 | Rough:<br>.025<br>Finish:<br>.005 max. | 1/5<br>1/10 | A60JV |
| | | | | | | | |
| | | | | | | | |
| | | | | | | | |
| | | | | | | | |
| **Nickel Base, Wrought**<br>Hastelloy Alloy B<br>Hastelloy Alloy B-2<br>Hastelloy Alloy C<br>Hastelloy Alloy C-276<br>Hastelloy Alloy G<br>Hastelloy Alloy S<br>Hastelloy Alloy X<br>Incoloy Alloy 804<br>Incoloy Alloy 825<br>Inconel Alloy 600<br>Inconel Alloy 601<br>Refractaloy 26<br>Udimet 630 | 140<br>to<br>220 | Annealed<br>or<br>Solution Treated | 3000<br>to<br>3500 | 50<br>to<br>100 | Rough:<br>.001<br>Finish:<br>.0002 max. | 1/5<br>1/10 | A60JV |
| | | | 15<br>to<br>18 | 15<br>to<br>30 | Rough:<br>.025<br>Finish:<br>.005 max. | 1/5<br>1/10 | A60JV |
| | 240<br>to<br>310 | Cold Drawn<br>or<br>Aged | 3000<br>to<br>3500 | 50<br>to<br>100 | Rough:<br>.001<br>Finish:<br>.0002 max. | 1/5<br>1/10 | A60JV |
| | | | 15<br>to<br>18 | 15<br>to<br>30 | Rough:<br>.025<br>Finish:<br>.005 max. | 1/5<br>1/10 | A60JV |

See section 16 for Cutting Fluid Recommendations.

*Wheel recommendations are for wet grinding of 2- to 4-inch [50 to 100 mm] diameter work. For DRY grinding—use a softer grade wheel. For LARGER diameter work—use a softer and/or coarser grit wheel. For SMALLER diameter work—use a harder grade wheel. Wheel recommendations also apply to plunge grinding applications. Also see section 20.2, Grinding Guidelines.

| MATERIAL | HARDNESS Bhn | CONDITION | WHEEL SPEED fpm / m/s | WORK SPEED fpm / m/min | INFEED on dia. in/pass / mm/pass | TRAVERSE Wheel width per rev. of work | WHEEL IDENTIFICATION* ANSI / ISO |
|---|---|---|---|---|---|---|---|
| **Nickel Base, Wrought**<br>TD-Nickel††<br>TD-Ni-Cr†† | 180 to 200 | As Rolled | 5500 to 6500 | 50 to 100 | Rough: .002<br>Finish: .0005 max. | 1/2<br>1/6 | A60JV |
| | | | 28 to 33 | 15 to 30 | Rough: .050<br>Finish: .013 max. | 1/2<br>1/6 | A60JV |
| **Nickel Base, Cast**<br>B-1900<br>GMR-235<br>GMR-235D<br>Hastelloy Alloy B<br>Hastelloy Alloy C<br>Hastelloy Alloy D<br>IN-100 (Rene 100)<br>IN-738<br>IN-792<br>Inconel Alloy 713C<br>Inconel Alloy 718<br>M252<br>MAR-M200<br>MAR-M246<br>MAR-M421<br>MAR-M432<br>Rene 80<br>Rene 125<br>SEL<br>SEL 15<br>TRW VI A<br>Udimet 500<br>Udimet 700<br>ASTM A296: Grades CW-12M, N-12M, CY-40<br>ASTM A297: Grades HW, HX<br>ASTM A494: Grades N-12M-1, N-12M-2, CY-40, CW-12M-1, CW-12M-2<br>ASTM A608: Grades HW50, HX50 | 200 to 425 | As Cast or Cast and Aged | 3000 to 3500 | 50 to 100 | Rough: .001<br>Finish: .0002 max. | 1/5<br>1/10 | A60JV |
| | | | 15 to 18 | 15 to 30 | Rough: .025<br>Finish: .005 max. | 1/5<br>1/10 | A60JV |
| | | | | | | | |
| | | | | | | | |
| | | | | | | | |
| | | | | | | | |
| **Cobalt Base, Wrought**<br>AiResist 213<br>Haynes Alloy 25 (L605)<br>Haynes Alloy 188<br>J-1570<br>MAR-M905<br>MAR-M918<br>S-816<br>V-36 | 180 to 230 | Solution Treated | 3000 to 3500 | 50 to 100 | Rough: .001<br>Finish: .0002 max. | 1/5<br>1/10 | A60JV |
| | | | 15 to 18 | 15 to 30 | Rough: .025<br>Finish: .005 max. | 1/5<br>1/10 | A60JV |
| | 270 to 320 | Solution Treated and Aged | 3000 to 3500 | 50 to 100 | Rough: .001<br>Finish: .0002 max. | 1/5<br>1/10 | A60JV |
| | | | 15 to 18 | 15 to 30 | Rough: .025<br>Finish: .005 max. | 1/5<br>1/10 | A60JV |

See section 16 for Cutting Fluid Recommendations.

*Wheel recommendations are for wet grinding of 2- to 4-inch [50 to 100 mm] diameter work. For DRY grinding—use a softer grade wheel. For LARGER diameter work—use a softer and/or coarser grit wheel. For SMALLER diameter work—use a harder grade wheel. Wheel recommendations also apply to plunge grinding applications. Also see section 20.2, Grinding Guidelines.

††CAUTION: Toxic Material, refer to National Institute for Occupational Safety and Health (NIOSH) for Precautions.

# 8.5 Cylindrical Grinding

| MATERIAL | HARDNESS | CONDITION | WHEEL SPEED | WORK SPEED | INFEED on dia. | TRAVERSE Wheel width per rev. of work | WHEEL IDENTIFICATION* |
|---|---|---|---|---|---|---|---|
| | Bhn | | fpm | fpm | in/pass | | ANSI |
| | | | m/s | m/min | mm/pass | | ISO |
| **39. HIGH TEMPERATURE ALLOYS, WROUGHT AND CAST (cont.)**<br>**Cobalt Base, Cast**<br>AiResist 13<br>AiResist 215<br>FSX-414<br>HS-6<br>HS-21<br>HS-31 (X-40)<br>HOWMET #3<br>MAR-M302<br>MAR-M322<br>MAR-M509<br>NASA Co-W-Re<br>WI-52<br>X-45 | 220 to 290 | As Cast or Cast and Aged | 3000 to 3500 | 50 to 100 | Rough: .001<br>Finish: .0002 max. | 1/5<br>1/10 | A60JV |
| | | | 15 to 18 | 15 to 30 | Rough: .025<br>Finish: .005 max. | 1/5<br>1/10 | A60JV |
| | 290 to 425 | As Cast or Cast and Aged | 3000 to 3500 | 50 to 100 | Rough: .001<br>Finish: .0002 max. | 1/5<br>1/10 | A60JV |
| | | | 15 to 18 | 15 to 30 | Rough: .025<br>Finish: .005 max. | 1/5<br>1/10 | A60JV |
| **Iron Base, Wrought**<br>A-286<br>Discaloy<br>Incoloy Alloy 800<br>Incoloy Alloy 800H<br>Incoloy Alloy 801<br>Incoloy Alloy 802<br>N-155<br>V-57<br>W-545<br>16-25-6<br>19-9DL | 180 to 230 | Solution Treated | 3000 to 4000 | 50 to 100 | Rough: .001<br>Finish: .0002 max. | 1/5<br>1/10 | A60JV |
| | | | 15 to 20 | 15 to 30 | Rough: .025<br>Finish: .005 max. | 1/5<br>1/10 | A60JV |
| | 250 to 320 | Solution Treated and Aged | 3000 to 4000 | 50 to 100 | Rough: .001<br>Finish: .0002 max. | 1/5<br>1/10 | A60JV |
| | | | 15 to 20 | 15 to 30 | Rough: .025<br>Finish: .005 max. | 1/5<br>1/10 | A60JV |
| **40. REFRACTORY ALLOYS, WROUGHT, CAST, P/M****<br>**Columbium**** †† ‡‡<br>C103<br>C129Y<br>Cb-1Zr<br>Cb-752<br>FS-85<br>FS-291<br>WC-3015 | 170 to 225 | Stress Relieved | 4000 | 50 to 100 | Rough: .001<br>Finish: .0003 max. | 1/5<br>1/10 | A60KV |
| | | | 20 | 15 to 30 | Rough: .025<br>Finish: .008 max. | 1/5<br>1/10 | A60KV |

See section 16 for Cutting Fluid Recommendations.

*Wheel recommendations are for wet grinding of 2- to 4-inch [50 to 100 mm] diameter work. For DRY grinding—use a softer grade wheel. For LARGER diameter work—use a softer and/or coarser grit wheel. For SMALLER diameter work—use a harder grade wheel. Wheel recommendations also apply to plunge grinding applications. Also see section 20.2, Grinding Guidelines.

**Due to the brittleness of refractory alloys, cracking, chipping, flaking and breakout tend to occur, particularly on the edges of the machined surfaces.

††CAUTION: Toxic Material, refer to National Institute for Occupational Safety and Health (NIOSH) for Precautions.

‡‡CAUTION: Potential Fire Hazard. Exercise caution in grinding and disposing of swarf.

| MATERIAL | HARDNESS | CONDITION | WHEEL SPEED | WORK SPEED | INFEED on dia. | TRAVERSE Wheel width per rev. of work | WHEEL IDENTIFICATION* |
|---|---|---|---|---|---|---|---|
| | Bhn | | fpm | fpm | in/pass | | ANSI |
| | | | m/s | m/min | mm/pass | | ISO |
| **Molybdenum****<br>Mo<br>Mo-50Re<br>TZC<br>TZM | 220 to 290 | Stress Relieved | 4000 | 50 to 100 | Rough: .001<br>Finish: .0003 max. | 1/5<br>1/10 | A60LV |
| | | | 20 | 15 to 30 | Rough: .025<br>Finish: .008 max. | 1/5<br>1/10 | A60LV |
| **Tantalum****<br>ASTAR 811C<br>T-111<br>T-222<br>Ta-10W<br>Ta-Hf<br>Ta63 | 200 to 250 | Stress Relieved | 4000 | 50 to 100 | Rough: .001<br>Finish: .0003 max. | 1/5<br>1/10 | A60KV |
| | | | 20 | 15 to 30 | Rough: .025<br>Finish: .008 max. | 1/5<br>1/10 | A60KV |
| **Tungsten****<br>85% density<br>93% density<br>96% density<br>100% density | 180 to 320 | Pressed and Sintered, Forged, or Arc Cast | 2000 | 50 to 100 | Rough: .001<br>Finish: .0003 max. | 1/5<br>1/10 | C60KV |
| | | | 10 | 15 to 30 | Rough: .025<br>Finish: .008 max. | 1/5<br>1/10 | C60KV |
| **Tungsten - 2 Thoria**** †† | 260 to 320 | Pressed and Sintered | 2000 | 50 to 100 | Rough: .001<br>Finish: .0003 max. | 1/5<br>1/10 | C60KV |
| | | | 10 | 15 to 30 | Rough: .025<br>Finish: .008 max. | 1/5<br>1/10 | C60KV |
| **Tungsten Alloys****<br>GE-218<br>W-15Mo<br>W-5Re<br>W-25Re<br>W-25Re-30Mo | 260 to 320 | As Cast | 2000 | 50 to 100 | Rough: .001<br>Finish: .0003 max. | 1/5<br>1/10 | C60KV |
| | | | 10 | 15 to 30 | Rough: .025<br>Finish: .008 max. | 1/5<br>1/10 | C60KV |
| **Tungsten Alloys****<br>Gyromet<br>Mallory 2000<br>W-10Ag<br>W-7Ni-4Cu | 290 to 320 | Pressed and Sintered | 2000 | 50 to 100 | Rough: .001<br>Finish: .0003 max. | 1/5<br>1/10 | C60KV |
| | | | 10 | 15 to 30 | Rough: .025<br>Finish: .008 max. | 1/5<br>1/10 | C60KV |

See section 16 for Cutting Fluid Recommendations.

*Wheel recommendations are for wet grinding of 2- to 4-inch [50 to 100 mm] diameter work. For DRY grinding—use a softer grade wheel. For LARGER diameter work—use a softer and/or coarser grit wheel. For SMALLER diameter work—use a harder grade wheel. Wheel recommendations also apply to plunge grinding applications. Also see section 20.2, Grinding Guidelines.

**Due to the brittleness of refractory alloys, cracking, chipping, flaking and breakout tend to occur, particularly on the edges of the machined surfaces.

††CAUTION: Toxic Material, refer to National Institute for Occupational Safety and Health (NIOSH) for Precautions.

# 8.5 Cylindrical Grinding

| MATERIAL | HARDNESS Bhn | CONDITION | WHEEL SPEED fpm / m/s | WORK SPEED fpm / m/min | INFEED on dia. in/pass / mm/pass | TRAVERSE Wheel width per rev. of work | WHEEL IDENTIFICATION* ANSI / ISO |
|---|---|---|---|---|---|---|---|
| **40. REFRACTORY ALLOYS, WROUGHT, CAST, P/M** (cont.)** **Tungsten Alloys**** Anviloy 1100, Anviloy 1150, Anviloy 1200 | 290 to 320 | Pressed and Sintered | 2000 | 50 to 100 | Rough: .001 Finish: .0003 max. | 1/5 1/10 | C60KV |
| | | | 10 | 15 to 30 | Rough: .025 Finish: .008 max. | 1/5 1/10 | C60KV |
| **44. URANIUM, WROUGHT†† ‡‡** | 56 $R_A$ to 58 $R_A$ | Annealed | 3000 to 5000 | 40 to 60 | Rough: .002 Finish: .0005 max. | 1/5 1/10 | C46HV |
| | | | 15 to 25 | 12 to 18 | Rough: .050 Finish: .013 max. | 1/5 1/10 | C46HV |
| **45. ZIRCONIUM ALLOYS, WROUGHT‡‡** Zr-2%Hf (Grade 11), Zr-0.001%Hf (Grade 21), Zircaloy 2 (Grade 32), Zircaloy 4 (Grade 34) | 140 to 280 | Rolled, Extruded or Forged | 3000 | 60 to 100 | Rough: .001 Finish: .0005 max. | 1/6 1/12 | C60KV or CA60PB |
| | | | 15 | 18 to 30 | Rough: .025 Finish: .013 max. | 1/6 1/12 | C60KV or CA60PB |
| **47. POWDER METAL ALLOYS§§** **Copper§§** | 50 $R_F$ to 70 $R_F$ | As Sintered | 5500 to 6500 | 70 to 100 | Rough: .002 Finish: .0005 max. | 1/3 1/6 | C46JV |
| | | | 28 to 33 | 21 to 30 | Rough: .050 Finish: .013 max. | 1/3 1/6 | C46JV |
| **Brasses§§** CZP-0218-T, CZP-0218-U, CZP-0218-W, 90Cu-10Zn, 90Cu-10Zn-0.5Pb, 70Cu-30Zn, 68.5Cu-30Zn-1.5Pb | 35 $R_H$ to 81 $R_H$ | As Sintered | 5500 to 6500 | 70 to 100 | Rough: .002 Finish: .0005 max. | 1/3 1/6 | C46KV |
| | | | 28 to 33 | 21 to 30 | Rough: .050 Finish: .013 max. | 1/3 1/6 | C46KV |
| **Bronzes§§** CT-0010-N, CT-0010-R, CT-0010-S, 95Cu-5Al, 77Cu-15Pb-7Sn-1Fe-1C | 30 $R_F$ to 75 $R_F$ | As Sintered | 5500 to 6500 | 70 to 100 | Rough: .002 Finish: .0005 max. | 1/3 1/6 | C46KV |
| | | | 28 to 33 | 21 to 30 | Rough: .050 Finish: .013 max. | 1/3 1/6 | C46KV |

See section 16 for Cutting Fluid Recommendations.

*Wheel recommendations are for wet grinding of 2- to 4-inch [50 to 100 mm] diameter work. For DRY grinding—use a softer grade wheel. For LARGER diameter work—use a softer and/or coarser grit wheel. For SMALLER diameter work—use a harder grade wheel. Wheel recommendations also apply to plunge grinding applications. Also see section 20.2, Grinding Guidelines.

**Due to the brittleness of refractory alloys, cracking, chipping, flaking and breakout tend to occur, particularly on the edges of the machined surfaces.

††CAUTION: Toxic Material, refer to National Institute for Occupational Safety and Health (NIOSH) for Precautions.

‡‡CAUTION: Potential Fire Hazard. Exercise caution in grinding and disposing of swarf.

§§Grinding of low-density parts is not recommended because surface porosity will be reduced or lost.

| MATERIAL | HARDNESS<br>Bhn | CONDITION | WHEEL SPEED<br>fpm<br>m/s | WORK SPEED<br>fpm<br>m/min | INFEED on dia.<br>in/pass<br>mm/pass | TRAVERSE Wheel width per rev. of work | WHEEL IDENTIFICATION*<br>ANSI<br>ISO |
|---|---|---|---|---|---|---|---|
| **Copper-Nickel Alloys§§**<br>CZN-1818-T<br>CZN-1818-U<br>CZN-1818-W<br>CZNP-1618-U<br>CZNP-1618-W<br>90Cu-10Ni<br>62Cu-18Ni-18Zn-2Sn | $22\ R_H$ to $100\ R_H$ | As Sintered | 5500 to 6500 | 70 to 100 | Rough: .002<br>Finish: .0005 max. | 1/3<br>1/6 | C46KV |
| | | | 28 to 33 | 21 to 30 | Rough: .050<br>Finish: .013 max. | 1/3<br>1/6 | C46KV |
| **Nickel§§** | $45\ R_B$ | As Sintered | 5500 to 6500 | 50 to 100 | Rough: .002<br>Finish: .0005 max. | 1/5<br>1/10 | A60JV |
| | | | 28 to 33 | 15 to 30 | Rough: .050<br>Finish: .013 max. | 1/5<br>1/10 | A60JV |
| **Nickel Alloys§§**<br>67Ni-30Cu-3Fe | $34\ R_B$ to $50\ R_B$ | As Sintered | 5500 to 6500 | 50 to 100 | Rough: .002<br>Finish: .0005 max. | 1/5<br>1/10 | C60IV |
| | | | 28 to 33 | 15 to 30 | Rough: .050<br>Finish: .013 max. | 1/5<br>1/10 | C60IV |
| **Refractory Metal Base§§**<br>87W-13Cu<br>85W-15Ag<br>74W-26Cu<br>72.5W-27.5Ag<br>65W-35Ag<br>56W + C-44Cu<br>55W-45Cu<br>51W-49Ag<br>50W + C-50Ag | 101 to 260 | As Sintered | 2000 | 50 to 100 | Rough: .001<br>Finish: .0003 max. | 1/5<br>1/10 | C60KV |
| | | | 10 | 15 to 30 | Rough: .025<br>Finish: .008 max. | 1/5<br>1/10 | C60KV |
| | | | | | | | |
| **Refractory Metal Base§§**<br>61Mo-39Ag<br>50Mo-50Ag | $75\ R_B$ to $82\ R_B$ | As Sintered | 4000 | 50 to 100 | Rough: .001<br>Finish: .0003 max. | 1/5<br>1/10 | A60LV |
| | | | 20 | 15 to 30 | Rough: .025<br>Finish: .008 max. | 1/5<br>1/10 | A60LV |

See section 16 for Cutting Fluid Recommendations.

*Wheel recommendations are for wet grinding of 2- to 4-inch [50 to 100 mm] diameter work. For DRY grinding—use a softer grade wheel. For LARGER diameter work—use a softer and/or coarser grit wheel. For SMALLER diameter work—use a harder grade wheel. Wheel recommendations also apply to plunge grinding applications. Also see section 20.2, Grinding Guidelines.

§§Grinding of low-density parts is not recommended because surface porosity will be reduced or lost.

# 8.5 Cylindrical Grinding

| MATERIAL | HARDNESS Bhn | CONDITION | WHEEL SPEED fpm m/s | WORK SPEED fpm m/min | INFEED on dia. in/pass mm/pass | TRAVERSE Wheel width per rev. of work | WHEEL IDENTIFICATION* ANSI ISO |
|---|---|---|---|---|---|---|---|
| **47. POWDER METAL ALLOYS§§(cont.)**<br>**Irons§§**<br>F-0000-N<br>F-0000-P<br>F-0000-R<br>F-0000-S<br>F-0000-T | 50 to 67 500kg | As Sintered | 5500 to 6500 | 70 to 100 | Rough: .002<br>Finish: .001 max. | 1/2<br>1/6 | C46KV |
| | | | 28 to 33 | 21 to 30 | Rough: .050<br>Finish: .025 max. | 1/2<br>1/6 | C46KV |
| **Steels§§**<br>F-0008-P<br>F-0008-S<br>F-0005-S<br>FC-0205-S<br>FC-0208-P<br>FC-0208-S<br>FC-0508-P<br>FC-1000-N<br>FN-0205-S<br>FN-0205-T<br>FN-0405-R<br>FN-0405-S<br>FN-0405-T<br>FX-1005-T<br>FX-2008-T | 101 to 426 | As Sintered or Heat Treated | 5500 to 6500 | 70 to 100 | Rough: .002<br>Finish: .0005 max. | 1/2<br>1/6 | A60MV |
| | | | 28 to 33 | 21 to 30 | Rough: .050<br>Finish: .013 max. | 1/2<br>1/6 | A60MV |
| | | | | | | | |
| | | | | | | | |
| **Stainless Steels§§**<br>SS-303-R<br>SS-304-R<br>SS-316-R<br>SS-410-R | 107 to 285 | As Sintered or Heat Treated | 5500 to 6500 | 50 to 100 | Rough: .002<br>Finish: .0005 max. | 1/2<br>1/6 | C54JV |
| | | | 28 to 33 | 15 to 30 | Rough: .050<br>Finish: .013 max. | 1/2<br>1/6 | C54JV |
| **Aluminum Alloys§§**<br>90.5Al-5Sn-4Cu<br>88Al-5Sn-4Pb-3Cu<br>Al-1Mg-0.6Si-0.25Cu<br>Al-0.6Mg-0.4Si<br>Al-4.4Cu-0.8Si-0.4Mg | 55 $R_H$ to 98 $R_H$ | Solution Treated and Aged | 5500 to 6500 | 50 to 150 | Rough: .002<br>Finish: .0005 max. | 1/2<br>1/6 | C54JV† |
| | | | 28 to 33 | 15 to 46 | Rough: .050<br>Finish: .013 max. | 1/2<br>1/6 | C54JV† |
| **48. MACHINABLE CARBIDES**<br>Ferro-Tic | 40 $R_C$ to 51 $R_C$ | Annealed | 4000 to 6000 | 40 to 60 | Rough: .001<br>Finish: .0003 max. | 1/6<br>1/12 | A60HV |
| | | | 20 to 30 | 12 to 18 | Rough: .025<br>Finish: .008 max. | 1/6<br>1/12 | A60HV |

See section 16 for Cutting Fluid Recommendations.

*Wheel recommendations are for wet grinding of 2- to 4-inch [50 to 100 mm] diameter work. For DRY grinding—use a softer grade wheel. For LARGER diameter work—use a softer and/or coarser grit wheel. For SMALLER diameter work—use a harder grade wheel. Wheel recommendations also apply to plunge grinding applications. Also see section 20.2, Grinding Guidelines.

†Wax filled.

§§Grinding of low-density parts is not recommended because surface porosity will be reduced or lost.

| MATERIAL | HARD-NESS<br>Bhn | CONDITION | WHEEL SPEED<br>fpm<br>m/s | WORK SPEED<br>fpm<br>m/min | INFEED on dia.<br>in/pass<br>mm/pass | TRAVERSE Wheel width per rev. of work | WHEEL IDENTIFI-CATION*<br>ANSI<br>ISO |
|---|---|---|---|---|---|---|---|
| **48. MACHINABLE CARBIDES (cont.)**<br>(materials listed on preceding page) | 68 $R_C$ to 70 $R_C$ | Hardened and Tempered | 4000 to 6000 | 40 to 60 | Rough: .001<br>Finish: .0003 max. | 1/6<br>1/12 | A60HV‡ |
| | | | 20 to 30 | 12 to 18 | Rough: .025<br>Finish: .008 max. | 1/6<br>1/12 | A60HV‡ |
| **49. CARBIDES**<br>Titanium Carbide<br>Tungsten Carbide | 89 $R_A$ to 94 $R_A$ | — | 5500 to 6500 | 50 to 60 | Rough: .001<br>Finish: .0003 max. | 1/5<br>1/10 | Rough: C60JV<br>Finish: C120IV‡ |
| | | | 28 to 33 | 15 to 18 | Rough: .025<br>Finish: .008 max. | 1/5<br>1/10 | Rough: C60JV<br>Finish: C120IV‡ |
| **50. FREE MACHINING MAGNETIC ALLOYS**<br>Magnetic Core Iron-FM (up to 2.5% Si) | 185 to 240 | Wrought | 5500 to 6500 | 70 to 100 | Rough: .002<br>Finish: .001 max. | 1/2<br>1/6 | A46KV |
| | | | 28 to 33 | 21 to 30 | Rough: .050<br>Finish: .025 max. | 1/2<br>1/6 | A46KV |
| Hi Perm 49-FM | 185 to 240 | Wrought | 5500 to 6500 | 50 to 100 | Rough: .002<br>Finish: .0005 max. | 1/5<br>1/10 | C54KV |
| | | | 28 to 33 | 15 to 30 | Rough: .050<br>Finish: .013 max. | 1/5<br>1/10 | C54KV |
| **51. MAGNETIC ALLOYS**<br>Magnetic Core Iron (up to 4% Si) | 185 to 240 | Wrought | 5500 to 6500 | 70 to 100 | Rough: .002<br>Finish: .001 max. | 1/2<br>1/6 | A46KV |
| | | | 28 to 33 | 21 to 30 | Rough: .050<br>Finish: .025 max. | 1/2<br>1/6 | A46KV |
| Hi Perm 49<br>HyMu 80 | 185 to 240 | Wrought | 5500 to 6500 | 50 to 100 | Rough: .002<br>Finish: .0005 max. | 1/5<br>1/10 | C54KV |
| | | | 28 to 33 | 15 to 30 | Rough: .050<br>Finish: .013 max. | 1/5<br>1/10 | C54KV |

See section 16 for Cutting Fluid Recommendations.

*Wheel recommendations are for wet grinding of 2- to 4-inch [50 to 100 mm] diameter work. For DRY grinding—use a softer grade wheel. For LARGER diameter work—use a softer and/or coarser grit wheel. For SMALLER diameter work—use a harder grade wheel. Wheel recommendations also apply to plunge grinding applications. Also see section 20.2, Grinding Guidelines.

‡Diamond wheel is preferable, see section 8.7.

# 8.5 Cylindrical Grinding

| MATERIAL | HARD-NESS<br>Bhn | CONDITION | WHEEL SPEED<br>fpm<br>m/s | WORK SPEED<br>fpm<br>m/min | INFEED on dia.<br>in/pass<br>mm/pass | TRAVERSE Wheel width per rev. of work | WHEEL IDENTIFI-CATION*<br>ANSI<br>ISO |
|---|---|---|---|---|---|---|---|
| **51. MAGNETIC ALLOYS (cont.)**<br>Alnico I<br>Alnico II<br>Alnico III<br>Alnico IV<br>Alnico V<br>Alnico V-7<br>Alnico XII<br>Columax-5<br>Hyflux Alnico V-7 | 45 $R_C$<br>to<br>58 $R_C$ | As Cast | 5500<br>to<br>6500 | 50<br>to<br>75 | Rough:<br>.002<br>Finish:<br>.0005 max. | 1/5<br>1/10 | A60JV |
| | | | 28<br>to<br>33 | 15<br>to<br>23 | Rough:<br>.050<br>Finish:<br>.013 max. | 1/5<br>1/10 | A60JV |
| | | | | | | | |
| **52. FREE MACHINING CONTROLLED EXPANSION ALLOYS**<br>Invar 36 | 125<br>to<br>220 | Annealed<br>or<br>Cold Drawn | 5500<br>to<br>6500 | 50<br>to<br>100 | Rough:<br>.002<br>Finish:<br>.0005 max. | 1/5<br>1/10 | C54KV |
| | | | 28<br>to<br>33 | 15<br>to<br>30 | Rough:<br>.050<br>Finish:<br>.013 max. | 1/5<br>1/10 | C54KV |
| **53. CONTROLLED EXPANSION ALLOYS**<br>Invar<br>Kovar | 125<br>to<br>250 | Annealed<br>or<br>Cold Drawn | 5500<br>to<br>6500 | 50<br>to<br>100 | Rough:<br>.002<br>Finish:<br>.0005 max. | 1/5<br>1/10 | C54KV |
| | | | 28<br>to<br>33 | 15<br>to<br>30 | Rough:<br>.050<br>Finish:<br>.013 max. | 1/5<br>1/10 | C54KV |
| **54. CARBONS & GRAPHITES**<br>Mechanical Grades | 40<br>to<br>100<br>Shore | Molded<br>or<br>Extruded | 5500<br>to<br>6500 | 50<br>to<br>75 | Rough:<br>.002<br>Finish:<br>.0005 max. | 1/2<br>1/6 | C36LV |
| | | | 28<br>to<br>33 | 15<br>to<br>23 | Rough:<br>.050<br>Finish:<br>.013 max. | 1/2<br>1/6 | C36LV |
| Brush Grade Carbon | 8<br>to<br>90<br>Shore | Molded<br>or<br>Extruded | 5500<br>to<br>6500 | 50<br>to<br>75 | Rough:<br>.002<br>Finish:<br>.0005 max. | 1/2<br>1/6 | C36LV |
| | | | 28<br>to<br>33 | 15<br>to<br>23 | Rough:<br>.050<br>Finish:<br>.013 max. | 1/2<br>1/6 | C36LV |

See section 16 for Cutting Fluid Recommendations.

*Wheel recommendations are for wet grinding of 2- to 4-inch [50 to 100 mm] diameter work. For DRY grinding—use a softer grade wheel. For LARGER diameter work—use a softer and/or coarser grit wheel. For SMALLER diameter work—use a harder grade wheel. Wheel recommendations also apply to plunge grinding applications. Also see section 20.2, Grinding Guidelines.

| MATERIAL | HARD-NESS<br>Bhn | CONDITION | WHEEL SPEED<br>fpm<br>m/s | WORK SPEED<br>fpm<br>m/min | INFEED on dia.<br>in/pass<br>mm/pass | TRAVERSE Wheel width per rev. of work | WHEEL IDENTIFI-CATION*<br>ANSI<br>ISO |
|---|---|---|---|---|---|---|---|
| Graphite | — | — | 5000 to 6500 | 50 to 75 | Rough: .002<br>Finish: .0005 max. | 1/2<br>1/6 | Rough: C36JV<br>Finish: C60JV |
| | | | 25 to 33 | 15 to 23 | Rough: .050<br>Finish: .013 max. | 1/2<br>1/6 | Rough: C36JV<br>Finish: C60JV |
| **55. GLASSES & CERAMICS**<br>Glass | 480 to 530 Knoop 100g | — | 5500 to 6500 | 70 to 100 | Rough: .001<br>Finish: .0005 max. | 1/5<br>1/10 | Rough: C80GV<br>Finish: A180KV |
| | | | 28 to 33 | 21 to 30 | Rough: .025<br>Finish: .013 max. | 1/5<br>1/10 | Rough: C80GV<br>Finish: A180KV |
| Porcelain Enamel | — | — | 5500 to 6500 | 70 to 100 | Rough: .001<br>Finish: .0005 max. | 1/5<br>1/10 | C70KV |
| | | | 28 to 33 | 21 to 30 | Rough: .025<br>Finish: .013 max. | 1/5<br>1/10 | C70KV |
| **Ceramics**<br>Alumina (Aluminum Oxide)<br>Alumina-Mullite<br>Aluminum Silicate<br>Beryllia (Beryllium Oxide)††<br>Magnesia (Magnesium Oxide)<br>Mullite<br>Silicon Carbide<br>Silicon Nitride<br>Thoria (Thorium Oxide)††<br>Titania (Titanium Oxide)<br>Titanium Diboride††<br>Zircon (Zirconium Silicate)<br>Zirconia (Zirconium Oxide) | Over 800 Knoop | Fired | 5500 to 6500 | 70 to 100 | Rough: .001<br>Finish: .0002 max. | 1/5<br>1/10 | Rough: C46KV†<br>Finish: C80KV† |
| | | | 28 to 33 | 21 to 30 | Rough: .025<br>Finish: .005 max. | 1/5<br>1/10 | Rough: C46KV†<br>Finish: C80KV† |
| | | | | | | | |
| **56. PLASTICS**<br>**Thermoplastics** | — | Cast, Molded, Extruded, or Filled and Molded | 5500 to 6500 | 70 to 100 | Rough: .005<br>Finish: .0005 max. | 1/3<br>1/6 | C60KV |
| | | | 28 to 33 | 21 to 30 | Rough: .13<br>Finish: .013 max. | 1/3<br>1/6 | C60KV |

See section 16 for Cutting Fluid Recommendations.

*Wheel recommendations are for wet grinding of 2- to 4-inch [50 to 100 mm] diameter work. For DRY grinding—use a softer grade wheel. For LARGER diameter work—use a softer and/or coarser grit wheel. For SMALLER diameter work—use a harder grade wheel. Wheel recommendations also apply to plunge grinding applications. Also see section 20.2, Grinding Guidelines.

†Diamond wheel is preferable, see section 8.7.

††CAUTION: Toxic Material, refer to National Institute for Occupational Safety and Health (NIOSH) for Precautions.

# 8.5 Cylindrical Grinding

| MATERIAL | HARD-NESS<br>Bhn | CONDITION | WHEEL SPEED<br>fpm<br>m/s | WORK SPEED<br>fpm<br>m/min | INFEED on dia.<br>in/pass<br>mm/pass | TRAVERSE Wheel width per rev. of work | WHEEL IDENTIFI-CATION*<br>ANSI<br>ISO |
|---|---|---|---|---|---|---|---|
| **56. PLASTICS (cont.)**<br>**Thermosetting Plastics** | — | Cast | 5500 to 6500 | 70 to 100 | Rough: .005<br>Finish: .0005 max. | 1/3<br>1/6 | C46IV |
| | | | 28 to 33 | 21 to 30 | Rough: .13<br>Finish: .013 max. | 1/3<br>1/6 | C46IV |
| | — | Molded | 5500 to 6500 | 70 to 100 | Rough: .005<br>Finish: .0005 max. | 1/3<br>1/6 | C46KV |
| | | | 28 to 33 | 21 to 30 | Rough: .13<br>Finish: .013 max. | 1/3<br>1/6 | C46KV |
| | — | Laminated | 5500 to 6500 | 70 to 100 | Rough: .005<br>Finish: .0005 max. | 1/3<br>1/6 | C36LV |
| | | | 28 to 33 | 21 to 30 | Rough: .13<br>Finish: .013 max. | 1/3<br>1/6 | C36LV |
| **58. FLAME (THERMAL) SPRAYED MATERIALS**<br>**Sprayed Carbides**<br>Chromium Carbide<br>Chromium Carbide-Cobalt Blend<br>Columbium Carbide<br>Tantalum Carbide<br>Titanium Carbide<br>Tungsten Carbide<br>Tungsten Carbide-Cobalt<br>Tungsten Carbide (Cobalt)-Nickel Alloy Blend | — | — | 5500 to 6500 | 70 to 100 | Rough: .001<br>Finish: .0005 max. | 1/4-1/2<br>1/12-1/6 | C80FV§ |
| | | | 28 to 33 | 21 to 30 | Rough: .025<br>Finish: .013 max. | 1/4-1/2<br>1/12-1/6 | C80FV§ |

See section 16 for Cutting Fluid Recommendations.

*Wheel recommendations are for wet grinding of 2- to 4-inch [50 to 100 mm] diameter work. For DRY grinding—use a softer grade wheel. For LARGER diameter work—use a softer and/or coarser grit wheel. For SMALLER diameter work—use a harder grade wheel. Wheel recommendations also apply to plunge grinding applications. Also see section 20.2, Grinding Guidelines.

§Diamond wheel is preferable for flame sprayed carbides and ceramics; see section 8.7.

| MATERIAL | HARD-NESS Bhn | CONDITION | WHEEL SPEED fpm / m/s | WORK SPEED fpm / m/min | INFEED on dia. in/pass / mm/pass | TRAVERSE Wheel width per rev. of work | WHEEL IDENTIFI-CATION* ANSI / ISO |
|---|---|---|---|---|---|---|---|
| **Inorganic Coating Materials**<br>Alumina (Pure)<br>Alumina (Grey) containing Titania<br>Alumina, Nickel-Aluminide Blends<br>Barium Titanate<br>Boron††<br>Calcium Titanate<br>Calcium Zirconate<br>Chromium Disilicide††<br>Chromium Oxide††<br>Cobalt (40%), Zirconia Blend<br>Columbium (Niobium)††<br>Glass (Kovar sealing)††<br>Hexaboron Silicide††<br>Magnesia Alumina Spinel<br>Magnesium Zirconate<br>Molybdenum Disilicide<br>Mullite<br>Nickel (40%), Alumina Blend<br>Nickel Oxide<br>Rare Earth Oxides<br>Tantalum<br>Titania (50%), Alumina Blend<br>Titanium Oxide<br>Tungsten<br>Yttrium Zirconate<br>Zirconia (Lime Stabilized)<br>Zirconia, Nickel-Aluminide Blends<br>Zirconium Oxide (Hafnia Free, Lime Stabilized)<br>Zirconium Silicate | — | — | 5500 to 6500 | 50 to 100 | Rough: .001<br>Finish: .0005 max. | 1/4-1/2<br>1/12-1/6 | C100GV§ |
| | | | 28 to 33 | 15 to 30 | Rough: .025<br>Finish: .013 max. | 1/4-1/2<br>1/2-1/6 | C100GV§ |
| **Sprayed Metals (Group I)**<br>Co-Cr-B Alloy (Self Fluxing)<br>Ni-Cr-B Alloy (Self Fluxing)<br>Nickel Chrome Steel (Special)<br>Stainless Steel | — | — | 5500 to 6500 | 70 to 100 | Rough: .001<br>Finish: .0005 max. | 1/4-1/2<br>1/12-1/6 | C60JV |
| | | | 28 to 33 | 21 to 30 | Rough: .025<br>Finish: .013 max. | 1/4-1/2<br>1/12-1/6 | C60JV |
| **Sprayed Metals (Group II)**<br>Bronze<br>Chromium<br>Cobalt<br>Nickel<br>Molybdenum<br>Monel | — | — | 5500 to 6500 | 50 to 100 | Rough: .001<br>Finish: .0005 max. | 1/4-1/2<br>1/12-1/6 | C60JV |
| | | | 28 to 33 | 15 to 30 | Rough: .025<br>Finish: .013 max. | 1/4-1/2<br>1/12-1/6 | C60JV |
| **Sprayed Metals (Group III)**<br>Carbon Steel<br>Iron<br>Precipitation Hardening Steel | — | — | 5500 to 6500 | 50 to 100 | Rough: .001<br>Finish: .0005 max. | 1/4-1/2<br>1/12-1/6 | C46KV |
| | | | 28 to 33 | 15 to 30 | Rough: .025<br>Finish: .013 max. | 1/4-1/2<br>1/12-1/6 | C46KV |

See section 16 for Cutting Fluid Recommendations.

*Wheel recommendations are for wet grinding of 2- to 4-inch [50 to 100 mm] diameter work. For DRY grinding—use a softer grade wheel. For LARGER diameter work—use a softer and/or coarser grit wheel. For SMALLER diameter work—use a harder grade wheel. Wheel recommendations also apply to plunge grinding applications. Also see section 20.2, Grinding Guidelines.

§Diamond wheel is preferable for flame sprayed carbides and ceramics; see section 8.7.

††CAUTION: Toxic Material, refer to National Institute for Occupational Safety and Health (NIOSH) for Precautions.

## 8.5 Cylindrical Grinding

| MATERIAL | HARD-NESS Bhn | CONDITION | WHEEL SPEED fpm m/s | WORK SPEED fpm m/min | INFEED on dia. in/pass mm/pass | TRAVERSE Wheel width per rev. of work | WHEEL IDENTIFI-CATION* ANSI ISO |
|---|---|---|---|---|---|---|---|
| **59. PLATED MATERIALS** Chromium Plate | — | — | 3000 to 4000 | 70 to 100 | Rough: .002 Finish: .0004 max. | 1/4 1/8 | A80IV† A120IV‡ |
| | | | 15 to 20 | 21 to 30 | Rough: .050 Finish: .010 max. | 1/4 1/8 | A80IV† A120IV‡ |
| Nickel Plate Silver Plate | — | — | 5500 to 6500 | 70 to 100 | Rough: .002 Finish: .0005 max. | 1/5 1/10 | A60KV |
| | | | 28 to 33 | 21 to 30 | Rough: .050 Finish: .013 max. | 1/5 1/10 | A60KV |
| **60. PRECIOUS METALS** Silver | — | — | 5500 to 6500 | 70 to 100 | Rough: .002 Finish: .0005 max. | 1/5 1/10 | A100JV |
| | | | 28 to 33 | 21 to 30 | Rough: .050 Finish: .013 max. | 1/5 1/10 | A100JV |
| **61. RUBBER** | Soft | — | 5500 to 6500 | 70 to 150 | Rough: .005 Finish: .001 max. | 1/2 1/6 | C30JB |
| | | | 28 to 33 | 21 to 46 | Rough: .13 Finish: .025 max. | 1/2 1/6 | C30JB |
| | Hard | — | 5500 to 6500 | 70 to 150 | Rough: .005 Finish: .001 max. | 1/2 1/6 | C36KB |
| | | | 28 to 33 | 21 to 46 | Rough: .13 Finish: .025 max. | 1/2 1/6 | C36KB |

See section 16 for Cutting Fluid Recommendations.

*Wheel recommendations are for wet grinding of 2- to 4-inch [50 to 100 mm] diameter work. For DRY grinding—use a softer grade wheel. For LARGER diameter work—use a softer and/or coarser grit wheel. For SMALLER diameter work—use a harder grade wheel. Wheel recommendations also apply to plunge grinding applications. Also see section 20.2, Grinding Guidelines.

†For commercial finish.

‡For high finish.

| MATERIAL | HARD-NESS Bhn | CONDITION | WHEEL SPEED fpm m/s | WORK SPEED fpm m/min | INFEED on dia. in/pass mm/pass | TRAVERSE Wheel width per rev. of work | WHEEL IDENTIFI-CATION ANSI ANSI |
|---|---|---|---|---|---|---|---|
| **1. FREE MACHINING CARBON STEELS, WROUGHT**<br>**Low Carbon Resulfurized**<br>1108 1118<br>1109 1119<br>1110 1211<br>1115 1212<br>1116 1213<br>1117 1215<br>**Medium Carbon Resulfurized**<br>1132 1140 1145<br>1137 1141 1146<br>1139 1144 1151<br>**Low Carbon Leaded**<br>10L18 12L13 12L15<br>11L17 12L14<br>**Medium Carbon Leaded**<br>10L45 11L37 11L44<br>10L50 11L41 | Over 50 $R_c$ | Carburized and/or Quenched and Tempered | 5000 to 7500 | 50 to 100 | Rough: .001<br>Finish: .0002 max. | 1/4<br><br>1/8 | B100T100B |
| | | | 25 to 38 | 15 to 30 | Rough: .025<br>Finish: .005 max. | 1/4<br><br>1/8 | B100T100B |
| | | | | | | | |
| | | | | | | | |
| **2. CARBON STEELS, WROUGHT**<br>**Low Carbon**<br>1005 1012 1019 1026<br>1006 1013 1020 1029<br>1008 1015 1021 1513<br>1009 1016 1022 1518<br>1010 1017 1023 1522<br>1011 1018 1025<br>**Medium Carbon**<br>1030 1042 1053 1541<br>1033 1043 1055 1547<br>1035 1044 1524 1548<br>1037 1045 1525 1551<br>1038 1046 1526 1552<br>1039 1049 1527<br>1040 1050 1536<br>**High Carbon**<br>1060 1074 1085 1566<br>1064 1075 1086 1572<br>1065 1078 1090<br>1069 1080 1095<br>1070 1084 1561 | Over 50 $R_c$ | Carburized and/or Quenched and Tempered | 5000 to 7500 | 50 to 100 | Rough: .001<br>Finish: .0002 max. | 1/4<br><br>1/8 | B100T100B |
| | | | 25 to 38 | 15 to 30 | Rough: .025<br>Finish: .005 max. | 1/4<br><br>1/8 | B100T100B |
| | | | | | | | |
| | | | | | | | |
| | | | | | | | |
| | | | | | | | |
| | | | | | | | |
| | | | | | | | |

See section 16 for Cutting Fluid Recommendations.
See section 20.1 for additional information.

# 8.6 Cylindrical Grinding
## Cubic Boron Nitride Wheels

| MATERIAL | HARDNESS | CONDITION | WHEEL SPEED | WORK SPEED | INFEED on dia. | TRAVERSE Wheel width per rev. of work | WHEEL IDENTIFICATION |
|---|---|---|---|---|---|---|---|
| | Bhn | | fpm | fpm | in/pass | | ANSI |
| | | | m/s | m/min | mm/pass | | ANSI |
| **4. FREE MACHINING ALLOY STEELS, WROUGHT**<br>**Medium Carbon Resulfurized**<br>4140<br>4140Se<br>4142Te<br>4145Se<br>4147Te<br>4150 | Over 50 $R_C$ | Carburized and/or Quenched and Tempered | 5000 to 7500 | 50 to 100 | Rough: .001<br>Finish: .0002 max. | 1/4<br>1/8 | B100T100B |
| | | | 25 to 38 | 15 to 30 | Rough: .025<br>Finish: .005 max. | 1/4<br>1/8 | B100T100B |
| **Medium and High Carbon Leaded**<br>41L30 41L47 51L32 86L40<br>41L40 41L50 52L100<br>41L45 43L40 86L20 | | | | | | | |
| | | | | | | | |
| **5. ALLOY STEELS, WROUGHT**<br>**Low Carbon**<br>4012 4615 4817 8617<br>4023 4617 4820 8620<br>4024 4620 5015 8622<br>4118 4621 5115 8822<br>4320 4718 5120 9310<br>4419 4720 6118 94B15<br>4422 4815 8115 94B17 | Over 50 $R_C$ | Carburized and/or Quenched and Tempered | 5000 to 7500 | 50 to 100 | Rough: .001<br>Finish: .0002 max. | 1/4<br>1/8 | B100T100B |
| | | | 25 to 38 | 15 to 30 | Rough: .025<br>Finish: .005 max. | 1/4<br>1/8 | B100T100B |
| **Medium Carbon**<br>1330 4145 5132 8640<br>1335 4147 5135 8642<br>1340 4150 5140 8645<br>1345 4161 5145 86B45<br>4027 4340 5147 8650<br>4028 4427 5150 8655<br>4032 4626 5155 8660<br>4037 50B40 5160 8740<br>4042 50B44 51B60 8742<br>4047 5046 6150 9254<br>4130 50B46 81B45 9255<br>4135 50B50 8625 9260<br>4137 5060 8627 94B30<br>4140 50B60 8630<br>4142 5130 8637 | | | | | | | |
| | | | | | | | |
| **High Carbon**<br>50100 51100 52100 M-50 | | | | | | | |
| | | | | | | | |
| **6. HIGH STRENGTH STEELS, WROUGHT**<br>300M H11<br>4330V H13<br>4340 HP 9-4-20<br>4340Si HP 9-4-25<br>98BV40 HP 9-4-30<br>D6ac HP 9-4-45 | Over 50 $R_C$ | Quenched and Tempered | 5000 to 7500 | 50 to 100 | Rough: .001<br>Finish: .0002 max. | 1/4<br>1/8 | B100T100B |
| | | | 25 to 38 | 15 to 30 | Rough: .025<br>Finish: .005 max. | 1/4<br>1/8 | B100T100B |

See section 16 for Cutting Fluid Recommendations.
See section 20.1 for additional information.

| MATERIAL | HARD-NESS | CONDITION | WHEEL SPEED | WORK SPEED | INFEED on dia. | TRAVERSE Wheel width per rev. of work | WHEEL IDENTIFI-CATION |
|---|---|---|---|---|---|---|---|
| | | | fpm | fpm | in/pass | | ANSI |
| | Bhn | | m/s | m/min | mm/pass | | ANSI |
| **7. MARAGING STEELS, WROUGHT**<br>120 Grade 300 Grade<br>180 Grade 350 Grade<br>200 Grade HY230<br>250 Grade<br>ASTM A538: Grades A, B, C | Over 50 $R_C$ | Maraged | 5000 to 7500 | 50 to 100 | Rough: .001<br>Finish: .0002 max. | 1/4<br>1/8 | B100T100B |
| | | | 25 to 38 | 15 to 30 | Rough: .025<br>Finish: .005 max. | 1/4<br>1/8 | B100T100B |
| **8. TOOL STEELS, WROUGHT**<br>A2 H14 M10 P20<br>A3 H19 M30 P21<br>A4 H21 M33 S1<br>A6 H22 M34 S2<br>A7 H23 M36 S5<br>A8 H24 M41 S6<br>A9 H25 M42 S7<br>A10 H26 M43 T1<br>D2 H42 M44 T2<br>D3 L2 M46 T4<br>D4 L6 M47 T5<br>D5 L7 O1 T6<br>D7 M1 O2 T8<br>F1 M2 O6 T15<br>F2 M3-1 O7 W1<br>H10 M3-2 P2 W2<br>H11 M4 P4 W5<br>H12 M6 P5<br>H13 M7 P6<br>SAE J438b: Types W108, W109, W110, W112, W209, W210, W310 | Over 50 $R_C$ | Quenched and Tempered | 5000 to 7500 | 50 to 100 | Rough: .001<br>Finish: .0002 max. | 1/4<br>1/8 | B100T100B |
| | | | 25 to 38 | 15 to 30 | Rough: .025<br>Finish: .005 max. | 1/4<br>1/8 | B100T100B |
| | | | | | | | |
| | | | | | | | |
| | | | | | | | |
| | | | | | | | |
| **9. NITRIDING STEELS, WROUGHT**<br>Nitralloy 125 Nitralloy 230<br>Nitralloy 135 Nitralloy EZ<br>Nitralloy 135 Mod. Nitralloy N<br>Nitralloy 225 Nitrex 1 | 60 $R_C$ to 65 $R_C$ | Nitrided | 5000 to 7500 | 50 to 100 | Rough: .001<br>Finish: .0002 max. | 1/4<br>1/8 | B100T100B |
| | | | 25 to 38 | 15 to 30 | Rough: .025<br>Finish: .005 max. | 1/4<br>1/8 | B100T100B |
| **15. CARBON STEELS, CAST**<br>**Low Carbon**<br>ASTM A426: Grade CP1<br>1010 1020<br>**Medium Carbon**<br>ASTM A352: Grades LCA, LCB, LCC<br>ASTM A356: Grade 1<br>1030 1040 1050 | Over 50 $R_C$ | Carburized and/or Quenched and Tempered | 5000 to 7500 | 50 to 100 | Rough: .001<br>Finish: .0002 max. | 1/4<br>1/8 | B100T100B |
| | | | 25 to 38 | 15 to 30 | Rough: .025<br>Finish: .005 max. | 1/4<br>1/8 | B100T100B |

See section 16 for Cutting Fluid Recommendations.
See section 20.1 for additional information.

# 8.6 Cylindrical Grinding
## Cubic Boron Nitride Wheels

| MATERIAL | HARDNESS Bhn | CONDITION | WHEEL SPEED fpm / m/s | WORK SPEED fpm / m/min | INFEED on dia. in/pass / mm/pass | TRAVERSE Wheel width per rev. of work | WHEEL IDENTIFICATION ANSI / ANSI |
|---|---|---|---|---|---|---|---|
| **16. ALLOY STEELS, CAST**<br>**Low Carbon**<br>ASTM A217: Grade WC9<br>ASTM A352: Grades LC3, LC4<br>ASTM A426: Grades CP2, CP5, CP5b, CP11, CP12, CP15, CP21, CP22<br>1320 4110 8020<br>2315 4120 8620<br>2320 4320<br>**Medium Carbon**<br>ASTM A27: Grades N1, N2, U-60-30, 60-30, 65-35, 70-36, 70-40<br>ASTM A148: Grades 80-40, 80-50, 90-60, 105-85, 120-95, 150-125, 175-145<br>ASTM A216: Grades WCA, WCB, WCC<br>ASTM A217: Grades WC1, WC4, WC5, WC6<br>ASTM A352: Grades LC1, LC2, LC2-1<br>ASTM A356: Grades 2, 5, 6, 8, 9, 10<br>ASTM A389: Grades C23, C24<br>ASTM A486: Classes 70, 90, 120<br>ASTM A487: Classes 1N, 2N, 4N, 6N, 8N, 9N, 10N, DN, 1Q, 2Q, 4Q, 4QA, 6Q, 7Q, 8Q, 9Q, 10Q<br>1330 4130 80B30 8640<br>1340 4140 8040 9525<br>2325 4330 8430 9530<br>2330 4340 8440 9535<br>4125 8030 8630 | Over 50 Rc | Carburized and/or Quenched and Tempered | 5000 to 7500 | 50 to 100 | Rough: .001<br>Finish: .0002 max. | 1/4<br>1/8 | B100T100B |
| | | | 25 to 38 | 15 to 30 | Rough: .025<br>Finish: .005 max. | 1/4<br>1/8 | B100T100B |
| **17. TOOL STEELS, CAST**<br>ASTM A597: Grades CA-2, CD-2, CD-5, CH-12, CH-13, CO-1, CS-5 | Over 50 Rc | Quenched and Tempered | 5000 to 7500 | 50 to 100 | Rough: .001<br>Finish: .0002 max. | 1/4<br>1/8 | B100T100B |
| | | | 25 to 38 | 15 to 30 | Rough: .025<br>Finish: .005 max. | 1/4<br>1/8 | B100T100B |

See section 16 for Cutting Fluid Recommendations.
See section 20.1 for additional information.

| MATERIAL | HARDNESS Bhn | CONDITION | WHEEL SPEED fpm / m/s | WORK SPEED fpm / m/min | INFEED on dia. in/pass / mm/pass | TRAVERSE Wheel width per rev. of work | WHEEL IDENTIFICATION ANSI |
|---|---|---|---|---|---|---|---|
| **48. MACHINABLE CARBIDES**<br>Ferro-Tic | 68 $R_C$ to 72 $R_C$ | Hardened and Tempered | 5500 to 6500 | 40 to 50 | Rough: .001<br>Finish: .0002 max. | 1/5<br>1/10 | D100-L100-B-1/8 |
| | | | 28 to 33 | 12 to 15 | Rough: .025<br>Finish: .005 max. | 1/5<br>1/10 | D100-L100-B-1/8 |
| **49. CARBIDES**<br>Titanium Carbide<br>Tungsten Carbide | 89 $R_A$ to 94 $R_A$ | — | 5500 to 6500 | 50 to 60 | Rough: .002<br>Finish: .0002 max. | 1/5<br>1/10 | D120-R100-B-1/8<br>D320-H75-B-1/8 |
| | | | 28 to 33 | 15 to 18 | Rough: .050<br>Finish: .005 max. | 1/5<br>1/10 | D120-R100-B-1/8<br>D320-H75-B-1/8 |
| **51. MAGNETIC ALLOYS**<br>Magnetic Core Iron (up to 4% Si)<br>Hi Perm 49<br>HyMu 80 | 185 to 240 | Wrought | 5500 to 6500 | 50 to 60 | Rough: .005<br>Finish: .001 max. | 1/4<br>1/8 | D120-N75-M-1/8<br>D220-N75-M-1/8 |
| | | | 28 to 33 | 15 to 18 | Rough: .13<br>Finish: .025 max. | 1/4<br>1/8 | D120-N75-M-1/8<br>D220-N75-M-1/8 |
| Alnico I<br>Alnico II<br>Alnico III<br>Alnico IV<br>Alnico V<br>Alnico V-7<br>Alnico XII<br>Columax-5<br>Hyflux Alnico V-7 | 45 $R_C$ to 58 $R_C$ | As Cast | 5500 to 6500 | 40 to 50 | Rough: .002<br>Finish: .0005 max. | 1/6<br>1/12 | D120-N75-M-1/8<br>D220-N75-M-1/8 |
| | | | 28 to 33 | 12 to 15 | Rough: .050<br>Finish: .013 max. | 1/6<br>1/12 | D120-N75-M-1/8<br>D220-N75-M-1/8 |
| **54. CARBONS & GRAPHITES**<br>Mechanical Grades<br>Brush Grade Carbon | 8 to 100 Shore | Molded or Extruded | 5500 to 6500 | 50 to 60 | Rough: .005<br>Finish: .001 max. | 1/6<br>1/12 | D60E*<br>D220E* |
| | | | 28 to 33 | 15 to 18 | Rough: .13<br>Finish: .025 max. | 1/6<br>1/12 | D60E*<br>D220E* |
| Graphite | — | — | 5500 to 6500 | 30 to 40 | Rough: .003<br>Finish: .001 max. | 1/5<br>1/10 | D60E*<br>D220E* |
| | | | 28 to 33 | 9 to 12 | Rough: .075<br>Finish: .025 max. | 1/5<br>1/10 | D60E*<br>D220E* |

See section 16 for Cutting Fluid Recommendations.
See section 20.1 for additional information.

*Electroplated.

# 8.7 Cylindrical Grinding
## Diamond Wheels

| MATERIAL | HARD-NESS | CONDITION | WHEEL SPEED | WORK SPEED | INFEED on dia. | TRAVERSE Wheel width per rev. of work | WHEEL IDENTIFI-CATION |
|---|---|---|---|---|---|---|---|
| | Bhn | | fpm / m/s | fpm / m/min | in/pass / mm/pass | | ANSI / ANSI |
| **55. GLASSES & CERAMICS**<br>Machinable Glass-Ceramic (MACOR) | 250 Knoop 100g | Cast | 5500 to 6500 | 50 to 60 | Rough: .005 | 1/4 | D100-N100-M-1/8 |
| | | | | | Finish: .001 max. | 1/8 | D220-N100-M-1/8 |
| | | | 28 to 33 | 15 to 18 | Rough: .13 | 1/4 | D100-N100-M-1/8 |
| | | | | | Finish: .025 max. | 1/8 | D220-N100-M-1/8 |
| PYROCERAM | 700 Knoop 100g | — | 5500 to 6500 | 20 to 40 | Rough: .002 | 1/5 | D120-N75-M-1/8 |
| | | | | | Finish: .001 max. | 1/10 | D220-N75-M-1/8 |
| | | | 28 to 33 | 6 to 12 | Rough: .050 | 1/5 | D120-N75-M-1/8 |
| | | | | | Finish: .025 max. | 1/10 | D220-N75-M-1/8 |
| Glass | 480 to 530 Knoop 100g | — | 4500 to 6000 | 20 to 40 | Rough: .002 | 1/6 | D120-N75-M-1/8 |
| | | | | | Finish: .0005 max. | 1/12 | D220-N75-M-1/8 |
| | | | 23 to 30 | 6 to 12 | Rough: .050 | 1/6 | D120-N75-M-1/8 |
| | | | | | Finish: .013 max. | 1/12 | D220-N75-M-1/8 |
| **Ceramics**<br>Alumina (Aluminum Oxide)<br>Alumina-Mullite<br>Aluminum Silicate<br>Beryllia (Beryllium Oxide)††<br>Magnesia (Magnesium Oxide)<br>Mullite<br>Silicon Carbide<br>Silicon Nitride<br>Thoria (Thorium Oxide)††<br>Titania (Titanium Oxide)<br>Titanium Diboride††<br>Zircon (Zirconium Silicate)<br>Zirconia (Zirconium Oxide) | Over 800 Knoop | Fired | 5500 to 6500 | 30 to 40 | Rough: .002 | 1/6 | D120-N100-M-1/8 |
| | | | | | Finish: .0005 max. | 1/12 | D220-N100-M-1/8 |
| | | | 28 to 33 | 9 to 12 | Rough: .050 | 1/6 | D120-N100-M-1/8 |
| | | | | | Finish: .013 max. | 1/12 | D220-N100-M-1/8 |
| **57. COMPOSITES**<br>Kevlar 49 | — | — | 4500 to 6500 | 75 to 150 | Rough: .001 | 1/8 | D100-N100-B-1/8 |
| | | | | | Finish: .0005 max. | 1/12 | D600-N100-B-1/8 |
| | | | 23 to 33 | 23 to 46 | Rough: .025 | 1/8 | D100-N100-B-1/8 |
| | | | | | Finish: .013 max. | 1/12 | D600-N100-B-1/8 |

See section 16 for Cutting Fluid Recommendations.
See section 20.1 for additional information.

††CAUTION: Toxic Material, refer to National Institute for Occupational Safety and Health (NIOSH) for Precautions.

| MATERIAL | HARDNESS Bhn | CONDITION | WHEEL SPEED fpm / m/s | WORK SPEED fpm / m/min | INFEED on dia. in/pass / mm/pass | TRAVERSE Wheel width per rev. of work | WHEEL IDENTIFICATION ANSI / ANSI |
|---|---|---|---|---|---|---|---|
| Graphite Epoxy | — | — | 4500 to 6500 | 50 to 60 | Rough: .002 | 1/8 | D120-N100-B-1/8 |
| | | | | | Finish: .0005 max. | 1/12 | D320-N100-B-1/8 |
| | | | 23 to 33 | 15 to 18 | Rough: .050 | 1/8 | D120-N100-B-1/8 |
| | | | | | Finish: .013 max. | 1/12 | D320-N100-B-1/8 |
| Fiberglass Epoxy (E) Glass (S) Glass | — | — | 3000 to 4000 | 50 to 60 | Rough: .002 | 1/8 | D80-M100-B-1/8 |
| | | | | | Finish: .0005 max. | 1/12 | D120-N100-B-1/8 |
| | | | 15 to 20 | 15 to 18 | Rough: .050 | 1/8 | D80-M100-B-1/8 |
| | | | | | Finish: .013 max. | 1/12 | D120-N100-B-1/8 |
| Boron Epoxy | — | — | 5500 to 6500 | 50 to 60 | Rough: .002 | 1/8 | D120-N100-B-1/8 |
| | | | | | Finish: .0005 max. | 1/12 | D320-N100-B-1/8 |
| | | | 28 to 33 | 15 to 60 | Rough: .050 | 1/8 | D120-N100-B-1/8 |
| | | | | | Finish: .013 max. | 1/12 | D320-N100-B-1/8 |
| **58. FLAME (THERMAL) SPRAYED MATERIALS** **Sprayed Carbides** Chromium Carbide, Chromium Carbide-Cobalt Blend, Columbium Carbide, Tantalum Carbide, Titanium Carbide, Tungsten Carbide, Tungsten Carbide-Cobalt, Tungsten Carbide (Cobalt)-Nickel Alloy Blend | — | — | 5500 to 6500 | 70 to 100 | Rough: .002 | 1/6 | D120-N100-B-1/8 |
| | | | | | Finish: .0005 max. | 1/12 | D220-N100-B-1/8 |
| | | | 28 to 33 | 21 to 30 | Rough: .050 | 1/6 | D120-N100-B-1/8 |
| | | | | | Finish: .013 max. | 1/12 | D220-N100-B-1/8 |

See section 16 for Cutting Fluid Recommendations.
See section 20.1 for additional information.

# 8.7 Cylindrical Grinding
## Diamond Wheels

| MATERIAL | HARD-NESS<br>Bhn | CONDITION | WHEEL SPEED<br>fpm<br>m/s | WORK SPEED<br>fpm<br>m/min | INFEED on dia.<br>in/pass<br>mm/pass | TRAVERSE Wheel width per rev. of work | WHEEL IDENTIFI-CATION<br>ANSI<br>ANSI |
|---|---|---|---|---|---|---|---|
| **58. FLAME (THERMAL) SPRAYED MATERIALS (cont.)**<br>**Inorganic Coating Materials**<br>Alumina (Pure)<br>Alumina (Grey) containing Titania<br>Alumina, Nickel-Aluminide Blends<br>Barium Titanate<br>Boron††<br>Calcium Titanate<br>Calcium Zirconate<br>Chromium Disilicide††<br>Chromium Oxide††<br>Cobalt (40%), Zirconia Blend<br>Columbium (Niobium)††<br>Glass (Kovar sealing)††<br>Hexaboron Silicide††<br>Magnesia Alumina Spinel<br>Magnesium Zirconate<br>Molybdenum Disilicide<br>Mullite<br>Nickel (40%), Alumina Blend<br>Nickel Oxide<br>Rare Earth Oxides<br>Tantalum<br>Titania (50%), Alumina Blend<br>Titanium Oxide<br>Tungsten<br>Yttrium Zirconate<br>Zirconia (Lime Stabilized)<br>Zirconia, Nickel-Aluminide Blends<br>Zirconium Oxide (Hafnia Free, Lime Stabilized)<br>Zirconium Silicate | — | — | 5500 to 6500 | 70 to 100 | Rough: .002<br>Finish: .0005 max. | 1/6<br>1/12 | D120-N100-B-1/8<br>D220-N100-B-1/8 |
| | | | 28 to 33 | 21 to 30 | Rough: .050<br>Finish: .013 max. | 1/6<br>1/12 | D120-N100-B-1/8<br>D220-N100-B-1/8 |

See section 16 for Cutting Fluid Recommendations.
See section 20.1 for additional information.

††CAUTION: Toxic Material, refer to National Institute for Occupational Safety and Health (NIOSH) for Precautions.

| MATERIAL | HARD-NESS | CONDITION | WHEEL SPEED | WORK SPEED | INFEED on dia. | TRAVERSE Wheel width per rev. of work | WHEEL IDENTIFI-CATION* |
|---|---|---|---|---|---|---|---|
| | Bhn | | fpm<br>m/s | fpm<br>m/min | in/pass<br>mm/pass | | ANSI<br>ISO |
| **1. FREE MACHINING CARBON STEELS, WROUGHT**<br>**Low Carbon Resulfurized**<br>1108 1118<br>1109 1119<br>1110 1211<br>1115 1212<br>1116 1213<br>1117 1215<br>**Medium Carbon Resulfurized**<br>1132 1140 1145<br>1137 1141 1146<br>1139 1144 1151<br>**Low Carbon Leaded**<br>10L18 12L13 12L15<br>11L17 12L14<br>**Medium Carbon Leaded**<br>10L45 11L37 11L44<br>10L50 11L41 | 50 $R_C$ max. | Hot Rolled, Normalized, Annealed, Cold Drawn or Quenched and Tempered | 5000 to 6500 | 75 to 200 | Rough: .0005<br>Finish: .0002 max. | 1/3<br>1/6 | A54MV |
| | | | 25 to 33 | 23 to 60 | Rough: .013<br>Finish: .005 max. | 1/3<br>1/6 | A54MV |
| | Over 50 $R_C$ | Carburized and/or Quenched and Tempered | 5000 to 6500 | 75 to 200 | Rough: .0005<br>Finish: .0002 max. | 1/3<br>1/6 | A60KV |
| | | | 25 to 33 | 23 to 60 | Rough: .013<br>Finish: .005 max. | 1/3<br>1/6 | A60KV |
| **2. CARBON STEELS, WROUGHT**<br>**Low Carbon**<br>1005 1012 1019 1026<br>1006 1013 1020 1029<br>1008 1015 1021 1513<br>1009 1016 1022 1518<br>1010 1017 1023 1522<br>1011 1018 1025<br>**Medium Carbon**<br>1030 1042 1053 1541<br>1033 1043 1055 1547<br>1035 1044 1524 1548<br>1037 1045 1525 1551<br>1038 1046 1526 1552<br>1039 1049 1527<br>1040 1050 1536<br>**High Carbon**<br>1060 1074 1085 1566<br>1064 1075 1086 1572<br>1065 1078 1090<br>1069 1080 1095<br>1070 1084 1561 | 50 $R_C$ max. | Hot Rolled, Normalized, Annealed, Cold Drawn or Quenched and Tempered | 5000 to 6500 | 75 to 200 | Rough: .0005<br>Finish: .0002 max. | 1/3<br>1/6 | A54MV |
| | | | 25 to 33 | 23 to 60 | Rough: .013<br>Finish: .005 max. | 1/3<br>1/6 | A54MV |
| | Over 50 $R_C$ | Carburized and/or Quenched and Tempered | 5000 to 6500 | 75 to 200 | Rough: .0005<br>Finish: .0002 max. | 1/3<br>1/6 | A60KV |
| | | | 25 to 33 | 23 to 60 | Rough: .013<br>Finish: .005 max. | 1/3<br>1/6 | A60KV |
| **3. CARBON AND FERRITIC ALLOY STEELS (HIGH TEMPERATURE SERVICE)**<br>ASTM A369: Grades FPA, FPB, FP1, FP2, FP3b, FP5, FP7, FP9, FP11, FP12, FP21, FP22 | 150 to 200 | As Forged, Annealed or Normalized and Tempered | 5000 to 6500 | 75 to 200 | Rough: .0005<br>Finish: .0002 max. | 1/3<br>1/6 | A54MV |
| | | | 25 to 33 | 23 to 60 | Rough: .013<br>Finish: .005 max. | 1/3<br>1/6 | A54MV |

Maximum hole length is 2.5 times hole diameter.
Maximum wheel width is 1.5 times wheel diameter.
See section 16 for Cutting Fluid Recommendations.
*Wheel recommendations are for wet grinding 0.8- to 2-inch [20 to 50 mm] diameter holes. For LARGER holes—use the same or softer grade wheel. For SMALLER holes—use a harder grade wheel. Also see section 20.2, Grinding Guidelines.

# 8.8 Internal Grinding

| MATERIAL | HARD-NESS<br>Bhn | CONDITION | WHEEL SPEED<br>fpm<br>m/s | WORK SPEED<br>fpm<br>m/min | INFEED on dia.<br>in/pass<br>mm/pass | TRAVERSE Wheel width per rev. of work | WHEEL IDENTIFI-CATION*<br>ANSI<br>ISO |
|---|---|---|---|---|---|---|---|
| **4. FREE MACHINING ALLOY STEELS, WROUGHT**<br>**Medium Carbon Resulfurized**<br>4140<br>4140Se<br>4142Te<br>4145Se<br>4147Te<br>4150<br>**Medium and High Carbon Leaded**<br>41L30 41L47 51L32 86L40<br>41L40 41L50 52L100<br>41L45 43L40 86L20 | 50 $R_C$ max. | Hot Rolled, Normalized, Annealed, Cold Drawn or Quenched and Tempered | 5000 to 6500 | 75 to 200 | Rough: .0005<br>Finish: .0002 max. | 1/3<br>1/6 | A60MV |
| | | | 25 to 33 | 23 to 60 | Rough: .013<br>Finish: .005 max. | 1/3<br>1/6 | A60MV |
| | Over 50 $R_C$ | Carburized and/or Quenched and Tempered | 5000 to 6500 | 75 to 200 | Rough: .0005<br>Finish: .0002 max. | 1/3<br>1/6 | A60KV |
| | | | 25 to 33 | 23 to 60 | Rough: .013<br>Finish: .005 max. | 1/3<br>1/6 | A60KV |
| **5. ALLOY STEELS, WROUGHT**<br>**Low Carbon**<br>4012 4615 4817 8617<br>4023 4617 4820 8620<br>4024 4620 5015 8622<br>4118 4621 5115 8822<br>4320 4718 5120 9310<br>4419 4720 6118 94B15<br>4422 4815 8115 94B17<br>**Medium Carbon**<br>1330 4145 5132 8640<br>1335 4147 5135 8642<br>1340 4150 5140 8645<br>1345 4161 5145 86B45<br>4027 4340 5147 8650<br>4028 4427 5150 8655<br>4032 4626 5155 8660<br>4037 50B40 5160 8740<br>4042 50B44 51B60 8742<br>4047 5046 6150 9254<br>4130 50B46 81B45 9255<br>4135 50B50 8625 9260<br>4137 5060 8627 94B30<br>4140 50B60 8630<br>4142 5130 8637<br>**High Carbon**<br>50100 51100 52100 M-50 | 50 $R_C$ max. | Hot Rolled, Normalized, Annealed, Cold Drawn or Quenched and Tempered | 5000 to 6500 | 75 to 200 | Rough: .0005<br>Finish: .0002 max. | 1/3<br>1/6 | A60MV |
| | | | 25 to 33 | 23 to 60 | Rough: .013<br>Finish: .005 max. | 1/3<br>1/6 | A60MV |
| | Over 50 $R_C$ | Carburized and/or Quenched and Tempered | 4000 to 6000 | 75 to 200 | Rough: .0005<br>Finish: .0002 max. | 1/3<br>1/6 | A60KV |
| | | | 20 to 30 | 23 to 60 | Rough: .013<br>Finish: .005 max. | 1/3<br>1/6 | A60KV |

Maximum hole length is 2.5 times hole diameter.
Maximum wheel width is 1.5 times wheel diameter.
See section 16 for Cutting Fluid Recommendations.
*Wheel recommendations are for wet grinding 0.8- to 2-inch [20 to 50 mm] diameter holes. For LARGER holes—use the same or softer grade wheel. For SMALLER holes—use a harder grade wheel. Also see section 20.2, Grinding Guidelines.

| MATERIAL | HARDNESS<br>Bhn | CONDITION | WHEEL SPEED<br>fpm<br>m/s | WORK SPEED<br>fpm<br>m/min | INFEED on dia.<br>in/pass<br>mm/pass | TRAVERSE Wheel width per rev. of work | WHEEL IDENTIFICATION*<br>ANSI<br>ISO |
|---|---|---|---|---|---|---|---|
| **6. HIGH STRENGTH STEELS, WROUGHT**<br>300M<br>4330V<br>4340<br>4340Si<br>98BV40<br>D6ac<br>H11<br>H13<br>HP 9-4-20<br>HP 9-4-25<br>HP 9-4-30<br>HP 9-4-45 | 50 $R_C$ max. | Annealed, Normalized or Quenched and Tempered | 4000 to 6000 | 75 to 200 | Rough: .0005<br>Finish: .0002 max. | 1/3<br>1/6 | A60KV |
| | | | 20 to 30 | 23 to 60 | Rough: .013<br>Finish: .005 max. | 1/3<br>1/6 | A60KV |
| | Over 50 $R_C$ | Quenched and Tempered | 4000 to 5500 | 75 to 200 | Rough: .0005<br>Finish: .0002 max. | 1/3<br>1/6 | A80IV |
| | | | 20 to 28 | 23 to 60 | Rough: .013<br>Finish: .005 max. | 1/3<br>1/6 | A80IV |
| **7. MARAGING STEELS, WROUGHT**<br>ASTM A538: Grades A, B, C<br>120 Grade<br>180 Grade<br>200 Grade<br>250 Grade<br>300 Grade<br>350 Grade<br>HY230 | 50 $R_C$ max. | Annealed or Maraged | 5000 to 6500 | 75 to 200 | Rough: .0005<br>Finish: .0002 max. | 1/3<br>1/6 | A60KV |
| | | | 25 to 33 | 23 to 60 | Rough: .013<br>Finish: .005 max. | 1/3<br>1/6 | A60KV |
| | Over 50 $R_C$ | Maraged | 4000 to 5500 | 75 to 200 | Rough: .0005<br>Finish: .0002 max. | 1/3<br>1/6 | A80IV |
| | | | 20 to 28 | 23 to 60 | Rough: .013<br>Finish: .005 max. | 1/3<br>1/6 | A80IV |
| **8. TOOL STEELS, WROUGHT**<br>**Group I**<br>A2 H13 L6 P20<br>A3 H14 L7 P21<br>A4 H19 O1 S1<br>A6 H21 O2 S2<br>A8 H22 O6 S5<br>A9 H23 O7 S6<br>A10 H24 P2 S7<br>H10 H25 P4 W1<br>H11 H26 P5 W2<br>H12 L2 P6 W5<br>SAE J438b: Types W108, W109, W110, W112, W209, W210, W310 | 50 $R_C$ max. | Annealed or Quenched and Tempered | 5000 to 6500 | 75 to 200 | Rough: .0005<br>Finish: .0002 max. | 1/3<br>1/6 | A60MV |
| | | | 25 to 33 | 23 to 60 | Rough: .013<br>Finish: .005 max. | 1/3<br>1/6 | A60MV |
| | Over 50 $R_C$ | Quenched and Tempered | 4000 to 6000 | 75 to 200 | Rough: .0005<br>Finish: .0002 max. | 1/3<br>1/6 | A60KV |
| | | | 20 to 30 | 23 to 60 | Rough: .013<br>Finish: .005 max. | 1/3<br>1/6 | A60KV |

Maximum hole length is 2.5 times hole diameter.
Maximum wheel width is 1.5 times wheel diameter.
See section 16 for Cutting Fluid Recommendations.
*Wheel recommendations are for wet grinding 0.8- to 2-inch [20 to 50 mm] diameter holes. For LARGER holes—use the same or softer grade wheel. For SMALLER holes—use a harder grade wheel. Also see section 20.2, Grinding Guidelines.

# 8.8 Internal Grinding

| MATERIAL | HARDNESS<br>Bhn | CONDITION | WHEEL SPEED<br>fpm<br>m/s | WORK SPEED<br>fpm<br>m/min | INFEED on dia.<br>in/pass<br>mm/pass | TRAVERSE Wheel width per rev. of work | WHEEL IDENTIFICATION*<br>ANSI<br>ISO |
|---|---|---|---|---|---|---|---|
| **8. TOOL STEELS, WROUGHT (cont.)**<br>**Group II**<br>D2 H42 M42<br>D3 M1 T1<br>D4 M2 T2<br>D5 M3-1 T4<br>F1 M10 T8<br>F2 M30 | 50 $R_C$ max. | Annealed or Quenched and Tempered | 5000 to 6500 | 75 to 200 | Rough: .0005<br>Finish: .0002 max. | 1/3<br>1/6 | A60LV |
| | | | 25 to 33 | 23 to 60 | Rough: .013<br>Finish: .005 max. | 1/3<br>1/6 | A60LV |
| | 50 $R_C$ to 58 $R_C$ | Quenched and Tempered | 4000 to 6000 | 75 to 200 | Rough: .0005<br>Finish: .0002 max. | 1/3<br>1/6 | A60KV |
| | | | 20 to 30 | 23 to 60 | Rough: .013<br>Finish: .005 max. | 1/3<br>1/6 | A60KV |
| | Over 58 $R_C$ | Quenched and Tempered | 4000 to 5500 | 75 to 200 | Rough: .0005<br>Finish: .0002 max. | 1/3<br>1/6 | A80JV |
| | | | 20 to 28 | 23 to 60 | Rough: .013<br>Finish: .005 max. | 1/3<br>1/6 | A80JV |
| **Group III**<br>A7 M41<br>D7 M43<br>M3-2 M44<br>M6 M46<br>M7 M47<br>M33 T5<br>M34 T6<br>M36 | 50 $R_C$ max. | Annealed or Quenched and Tempered | 4000 to 6000 | 75 to 200 | Rough: .0005<br>Finish: .0002 max. | 1/3<br>1/6 | A60KV |
| | | | 20 to 30 | 23 to 60 | Rough: .013<br>Finish: .005 max. | 1/3<br>1/6 | A60KV |
| | 50 $R_C$ to 58 $R_C$ | Quenched and Tempered | 4000 to 6000 | 75 to 200 | Rough: .0005<br>Finish: .0002 max. | 1/3<br>1/6 | A70KV |
| | | | 20 to 30 | 23 to 60 | Rough: .013<br>Finish: .005 max. | 1/3<br>1/6 | A70KV |
| | Over 58 $R_C$ | Quenched and Tempered | 4000 to 5500 | 75 to 200 | Rough: .0005<br>Finish: .0002 max. | 1/3<br>1/6 | A80JV |
| | | | 20 to 28 | 23 to 60 | Rough: .013<br>Finish: .005 max. | 1/3<br>1/6 | A80JV |

Maximum hole length is 2.5 times hole diameter.
Maximum wheel width is 1.5 times wheel diameter.
See section 16 for Cutting Fluid Recommendations.
*Wheel recommendations are for wet grinding 0.8- to 2-inch [20 to 50 mm] diameter holes. For LARGER holes—use the same or softer grade wheel. For SMALLER holes—use a harder grade wheel. Also see section 20.2, Grinding Guidelines.

| MATERIAL | HARD-NESS Bhn | CONDITION | WHEEL SPEED fpm m/s | WORK SPEED fpm m/min | INFEED on dia. in/pass mm/pass | TRAVERSE Wheel width per rev. of work | WHEEL IDENTIFI-CATION* ANSI ISO |
|---|---|---|---|---|---|---|---|
| **Group IV**<br>M4<br>T15 | 50 $R_C$ max. | Annealed or Quenched and Tempered | 4000 to 6000 | 75 to 200 | Rough: .0005<br>Finish: .0002 max. | 1/3<br>1/6 | A60JV |
| | | | 20 to 30 | 23 to 60 | Rough: .013<br>Finish: .005 max. | 1/3<br>1/6 | A60JV |
| | 50 $R_C$ to 58 $R_C$ | Quenched and Tempered | 4000 to 5500 | 75 to 200 | Rough: .0005<br>Finish: .0002 max. | 1/3<br>1/6 | A80JV |
| | | | 20 to 28 | 23 to 60 | Rough: .013<br>Finish: .005 max. | 1/3<br>1/6 | A80JV |
| | Over 58 $R_C$ | Quenched and Tempered | 4000 to 5500 | 75 to 200 | Rough: .0005<br>Finish: .0002 max. | 1/3<br>1/6 | A80IV |
| | | | 20 to 28 | 23 to 60 | Rough: .013<br>Finish: .005 max. | 1/3<br>1/6 | A80IV |
| **9. NITRIDING STEELS, WROUGHT**<br>Nitralloy 125<br>Nitralloy 135<br>Nitralloy 135 Mod.<br>Nitralloy 225<br>Nitralloy 230<br>Nitralloy EZ<br>Nitralloy N<br>Nitrex 1 | 200 to 350 | Annealed, Normalized or Quenched and Tempered | 5000 to 6500 | 75 to 200 | Rough: .0005<br>Finish: .0002 max. | 1/3<br>1/6 | A60KV |
| | | | 25 to 33 | 23 to 60 | Rough: .013<br>Finish: .005 max. | 1/3<br>1/6 | A60KV |
| | 60 $R_C$ to 65 $R_C$ | Nitrided | 4000 to 5500 | 75 to 200 | Rough: .0005<br>Finish: .0002 max. | 1/3<br>1/6 | A80IV |
| | | | 20 to 28 | 23 to 60 | Rough: .013<br>Finish: .005 max. | 1/3<br>1/6 | A80IV |
| **10. ARMOR PLATE, SHIP PLATE, AIRCRAFT PLATE, WROUGHT**<br>HY80<br>MIL-S-12560<br>HY100<br>MIL-S-16216<br>HY180 | 45 $R_C$ max. | Annealed or Quenched and Tempered | 5000 to 6500 | 75 to 200 | Rough: .0005<br>Finish: .0002 max. | 1/3<br>1/6 | A60JV |
| | | | 25 to 33 | 23 to 60 | Rough: .013<br>Finish: .005 max. | 1/3<br>1/6 | A60JV |

Maximum hole length is 2.5 times hole diameter.
Maximum wheel width is 1.5 times wheel diameter.
See section 16 for Cutting Fluid Recommendations.
*Wheel recommendations are for wet grinding 0.8- to 2-inch [20 to 50 mm] diameter holes. For LARGER holes—use the same or softer grade wheel. For SMALLER holes—use a harder grade wheel. Also see section 20.2, Grinding Guidelines.

# 8.8 Internal Grinding

| MATERIAL | HARD-<br>NESS<br><br>Bhn | CONDITION | WHEEL SPEED<br><br>fpm<br>m/s | WORK SPEED<br><br>fpm<br>m/min | INFEED on dia.<br><br>in/pass<br>mm/pass | TRAVERSE Wheel width per rev. of work | WHEEL IDENTIFI-CATION*<br><br>ANSI<br>ISO |
|---|---|---|---|---|---|---|---|
| **12. FREE MACHINING STAINLESS STEELS, WROUGHT**<br>**Ferritic**<br>430F<br>430F Se | 135<br>to<br>185 | Annealed | 5000<br>to<br>6500 | 75<br>to<br>200 | Rough:<br>.0005<br>Finish:<br>.0002 max. | 1/3<br><br>1/6 | A60KV |
| | | | 25<br>to<br>33 | 23<br>to<br>60 | Rough:<br>.013<br>Finish:<br>.005 max. | 1/3<br><br>1/6 | A60KV |
| **Austenitic**<br>203EZ<br>303<br>303MA<br>303Pb<br>303 Plus X<br>303Se | 135<br>to<br>275 | Annealed or<br>Cold Drawn | 5000<br>to<br>6500 | 75<br>to<br>200 | Rough:<br>.0005<br>Finish:<br>.0002 max. | 1/3<br><br>1/6 | C60KV |
| | | | 25<br>to<br>33 | 23<br>to<br>60 | Rough:<br>.013<br>Finish:<br>.005 max. | 1/3<br><br>1/6 | C60KV |
| **Martensitic**<br>416<br>416 Plus X<br>416Se<br>420F<br>420F Se<br>440F<br>440F Se | 135<br>to<br>240 | Annealed or<br>Cold Drawn | 5000<br>to<br>6500 | 75<br>to<br>200 | Rough:<br>.0005<br>Finish:<br>.0002 max. | 1/3<br><br>1/6 | A60KV |
| | | | 25<br>to<br>33 | 23<br>to<br>60 | Rough:<br>.013<br>Finish:<br>.005 max. | 1/3<br><br>1/6 | A60KV |
| | Over<br>275 | Quenched<br>and Tempered | 5000<br>to<br>6500 | 50<br>to<br>150 | Rough:<br>.0005<br>Finish:<br>.0002 max. | 1/3<br><br>1/6 | A60JV |
| | | | 25<br>to<br>33 | 15<br>to<br>46 | Rough:<br>.013<br>Finish:<br>.005 max. | 1/3<br><br>1/6 | A60JV |
| **13. STAINLESS STEELS, WROUGHT**<br>**Ferritic**<br>405 434<br>409 436<br>429 442<br>430 446 | 135<br>to<br>185 | Annealed | 5000<br>to<br>6500 | 75<br>to<br>200 | Rough:<br>.0005<br>Finish:<br>.0002 max. | 1/3<br><br>1/6 | A60KV |
| | | | 25<br>to<br>33 | 23<br>to<br>60 | Rough:<br>.013<br>Finish:<br>.005 max. | 1/3<br><br>1/6 | A60KV |
| **Austenitic**<br>201 308 317<br>202 309 321<br>301 309S 330<br>302 310 347<br>302B 310S 348<br>304 314 384<br>304L 316 385<br>305 316L | 135<br>to<br>275 | Annealed<br>or<br>Cold Drawn | 5000<br>to<br>6500 | 75<br>to<br>200 | Rough:<br>.0005<br>Finish:<br>.0002 max. | 1/3<br><br>1/6 | C60KV |
| | | | 25<br>to<br>33 | 23<br>to<br>60 | Rough:<br>.013<br>Finish:<br>.005 max. | 1/3<br><br>1/6 | C60KV |

Maximum hole length is 2.5 times hole diameter.
Maximum wheel width is 1.5 times wheel diameter.
See section 16 for Cutting Fluid Recommendations.
*Wheel recommendations are for wet grinding 0.8- to 2-inch [20 to 50 mm] diameter holes. For LARGER holes—use the same or softer grade wheel. For SMALLER holes—use a harder grade wheel. Also see section 20.2, Grinding Guidelines.

| MATERIAL | HARD-NESS<br>Bhn | CONDITION | WHEEL SPEED<br>fpm<br>m/s | WORK SPEED<br>fpm<br>m/min | INFEED on dia.<br>in/pass<br>mm/pass | TRAVERSE Wheel width per rev. of work | WHEEL IDENTIFI-CATION*<br>ANSI<br>ISO |
|---|---|---|---|---|---|---|---|
| **Austenitic**<br>Nitronic 32<br>Nitronic 33<br>Nitronic 40<br>Nitronic 50<br>Nitronic 60 | 210 to 250 | Annealed | 5000 to 6500 | 75 to 200 | Rough: .0005<br>Finish: .0002 max. | 1/3<br>1/6 | C60KV |
| | | | 25 to 33 | 23 to 60 | Rough: .013<br>Finish: .005 max. | 1/3<br>1/6 | C60KV |
| | 325 to 375 | Cold Drawn | 5000 to 6500 | 50 to 150 | Rough: .0005<br>Finish: .0002 max. | 1/3<br>1/6 | C60JV |
| | | | 25 to 33 | 15 to 46 | Rough: .013<br>Finish: .005 max. | 1/3<br>1/6 | C60JV |
| **Martensitic**<br>403 440A<br>410 440B<br>414 440C<br>420 501<br>422 502<br>431 Greek Ascoloy | 135 to 275 | Annealed | 5000 to 6500 | 75 to 200 | Rough: .0005<br>Finish: .0002 max. | 1/3<br>1/6 | A60KV |
| | | | 25 to 33 | 23 to 60 | Rough: .013<br>Finish: .005 max. | 1/3<br>1/6 | A60KV |
| | Over 275 | Quenched and Tempered | 5000 to 6500 | 50 to 150 | Rough: .0005<br>Finish: .0002 max. | 1/3<br>1/6 | A60JV |
| | | | 25 to 33 | 15 to 46 | Rough: .013<br>Finish: .005 max. | 1/3<br>1/6 | A60JV |
| **14. PRECIPITATION HARDENING STAINLESS STEELS, WROUGHT**<br>15-5 PH AM-355<br>16-6 PH AM-363<br>17-4 PH Custom 450<br>17-7 PH Custom 455<br>17-14 Cu Mo HNM<br>AF-71 PH 13-8 Mo<br>AFC-77 PH 14-8 Mo<br>Almar 362 (AM-362) PH 15-7 Mo<br>AM-350 Stainless W | 150 to 200 | Solution Treated | 5000 to 6500 | 75 to 200 | Rough: .0005<br>Finish: .0002 max. | 1/3<br>1/6 | A60KV |
| | | | 25 to 33 | 23 to 60 | Rough: .013<br>Finish: .005 max. | 1/3<br>1/6 | A60KV |
| | 275 to 440 | Solution Treated or Hardened | 5000 to 6500 | 50 to 150 | Rough: .0005<br>Finish: .0002 max. | 1/3<br>1/6 | A60JV |
| | | | 25 to 33 | 15 to 46 | Rough: .013<br>Finish: .005 max. | 1/3<br>1/6 | A60JV |

Maximum hole length is 2.5 times hole diameter.
Maximum wheel width is 1.5 times wheel diameter.
See section 16 for Cutting Fluid Recommendations.
*Wheel recommendations are for wet grinding 0.8- to 2-inch [20 to 50 mm] diameter holes. For LARGER holes—use the same or softer grade wheel. For SMALLER holes—use a harder grade wheel. Also see section 20.2, Grinding Guidelines.

# 8.8 Internal Grinding

| MATERIAL | HARD-NESS<br>Bhn | CONDITION | WHEEL SPEED<br>fpm<br>m/s | WORK SPEED<br>fpm<br>m/min | INFEED on dia.<br>in/pass<br>mm/pass | TRAVERSE Wheel width per rev. of work | WHEEL IDENTIFI-CATION*<br>ANSI<br>ISO |
|---|---|---|---|---|---|---|---|
| **15. CARBON STEELS, CAST**<br>**Low Carbon**<br>ASTM A426: Grade CP1<br>1010<br>1020<br>**Medium Carbon**<br>ASTM A352: Grades LCA, LCB, LCC<br>ASTM A356: Grade 1<br>1030 1040 1050 | 100<br>to<br>300 | Annealed,<br>Normalized,<br>Normalized and Tempered,<br>or Quenched<br>and Tempered | 5000<br>to<br>6500 | 75<br>to<br>200 | Rough:<br>.0005<br>Finish:<br>.0002 max. | 1/3<br>1/6 | A54MV |
| | | | 25<br>to<br>33 | 23<br>to<br>60 | Rough:<br>.013<br>Finish:<br>.005 max. | 1/3<br>1/6 | A54MV |
| | Over<br>50 $R_C$ | Carburized<br>and/or<br>Quenched and<br>Tempered | 5000<br>to<br>6500 | 75<br>to<br>200 | Rough:<br>.0005<br>Finish:<br>.0002 max. | 1/3<br>1/6 | A60KV |
| | | | 25<br>to<br>33 | 23<br>to<br>60 | Rough:<br>.013<br>Finish:<br>.005 max. | 1/3<br>1/6 | A60KV |
| **16. ALLOY STEELS, CAST**<br>**Low Carbon**<br>ASTM A217: Grade WC9<br>ASTM A352: Grades LC3, LC4<br>ASTM A426: Grades CP2, CP5, CP5b, CP11, CP12, CP15, CP21, CP22<br>1320 4110 8020<br>2315 4120 8620<br>2320 4320<br>**Medium Carbon**<br>ASTM A27: Grades N1, N2, U-60-30, 60-30, 65-35, 70-36, 70-40<br>ASTM A148: Grades 80-40, 80-50, 90-60, 105-85, 120-95, 150-125, 175-145<br>ASTM A216: Grades WCA, WCB, WCC<br>ASTM A217: Grades WC1, WC4, WC5, WC6<br>ASTM A352: Grades LC1, LC2, LC2-1<br>ASTM A356: Grades 2, 5, 6, 8, 9, 10<br>ASTM A389: Grades C23, C24<br>ASTM A486: Classes 70, 90, 120<br>ASTM A487: Classes 1N, 2N, 4N, 6N, 8N, 9N, 10N, DN, 1Q, 2Q, 4Q, 4QA, 6Q, 7Q, 8Q, 9Q, 10Q<br>1330 4130 80B30 8640<br>1340 4140 8040 9525<br>2325 4330 8430 9530<br>2330 4340 8440 9535<br>4125 8030 8630 | 150<br>to<br>400 | Annealed,<br>Normalized,<br>Normalized and Tempered,<br>or Quenched<br>and Tempered | 5000<br>to<br>6500 | 75<br>to<br>200 | Rough:<br>.0005<br>Finish:<br>.0002 max. | 1/3<br>1/6 | A60MV |
| | | | 25<br>to<br>33 | 23<br>to<br>60 | Rough:<br>.013<br>Finish:<br>.005 max. | 1/3<br>1/6 | A60MV |
| | Over<br>50 $R_C$ | Carburized<br>and/or<br>Quenched and<br>Tempered | 4000<br>to<br>6000 | 75<br>to<br>200 | Rough:<br>.0005<br>Finish:<br>.0002 max. | 1/3<br>1/6 | A60KV |
| | | | 20<br>to<br>30 | 23<br>to<br>60 | Rough:<br>.013<br>Finish:<br>.005 max. | 1/3<br>1/6 | A60KV |
| | | | | | | | |
| | | | | | | | |

Maximum hole length is 2.5 times hole diameter.
Maximum wheel width is 1.5 times wheel diameter.
See section 16 for Cutting Fluid Recommendations.
*Wheel recommendations are for wet grinding 0.8- to 2-inch [20 to 50 mm] diameter holes. For LARGER holes—use the same or softer grade wheel. For SMALLER holes—use a harder grade wheel. Also see section 20.2, Grinding Guidelines.

| MATERIAL | HARDNESS Bhn | CONDITION | WHEEL SPEED fpm m/s | WORK SPEED fpm m/min | INFEED on dia. in/pass mm/pass | TRAVERSE Wheel width per rev. of work | WHEEL IDENTIFICATION* ANSI ISO |
|---|---|---|---|---|---|---|---|
| **17. TOOL STEELS, CAST** **Group I** ASTM A597: Grades CA-2, CH-12, CH-13, CO-1, CS-5 | 50 $R_C$ max. | Annealed or Quenched and Tempered | 5000 to 6500 | 75 to 200 | Rough: .0005 Finish: .0002 max. | 1/3 1/6 | A60MV |
| | | | 25 to 33 | 23 to 60 | Rough: .013 Finish: .005 max. | 1/3 1/6 | A60MV |
| | 50 $R_C$ to 58 $R_C$ | Quenched and Tempered | 4000 to 6000 | 75 to 200 | Rough: .0005 Finish: .0002 max. | 1/3 1/6 | A80KV |
| | | | 20 to 30 | 23 to 60 | Rough: .013 Finish: .005 max. | 1/3 1/6 | A80KV |
| | Over 58 $R_C$ | Quenched and Tempered | 4000 to 5500 | 75 to 200 | Rough: .0005 Finish: .0002 max. | 1/3 1/6 | A80JV |
| | | | 20 to 28 | 23 to 60 | Rough: .013 Finish: .005 max. | 1/3 1/6 | A80JV |
| **Group II** ASTM A597: Grades CD-2, CD-5 | 200 to 250 | Annealed | 5000 to 6500 | 75 to 200 | Rough: .0005 Finish: .0002 max. | 1/3 1/6 | A60MV |
| | | | 25 to 33 | 23 to 60 | Rough: .013 Finish: .005 max. | 1/3 1/6 | A60MV |
| | 48 $R_C$ to 56 $R_C$ | Quenched and Tempered | 4000 to 6000 | 75 to 200 | Rough: .0005 Finish: .0002 max. | 1/3 1/6 | A80KV |
| | | | 20 to 30 | 23 to 60 | Rough: .013 Finish: .005 max. | 1/3 1/6 | A80KV |
| | Over 56 $R_C$ | Quenched and Tempered | 4000 to 5500 | 75 to 200 | Rough: .0005 Finish: .0002 max. | 1/3 1/6 | A80JV |
| | | | 20 to 28 | 23 to 60 | Rough: .013 Finish: .005 max. | 1/3 1/6 | A80JV |

Maximum hole length is 2.5 times hole diameter.
Maximum wheel width is 1.5 times wheel diameter.
See section 16 for Cutting Fluid Recommendations.
*Wheel recommendations are for wet grinding 0.8- to 2-inch [20 to 50 mm] diameter holes. For LARGER holes—use the same or softer grade wheel. For SMALLER holes—use a harder grade wheel. Also see section 20.2, Grinding Guidelines.

# 8.8 Internal Grinding

| MATERIAL | HARD-NESS Bhn | CONDITION | WHEEL SPEED fpm m/s | WORK SPEED fpm m/min | INFEED on dia. in/pass mm/pass | TRAVERSE Wheel width per rev. of work | WHEEL IDENTIFI-CATION* ANSI ISO |
|---|---|---|---|---|---|---|---|
| **18. STAINLESS STEELS, CAST**<br>**Ferritic**<br>ASTM A217: Grades C5, C12<br>ASTM A296: Grades CB-30, CC-50, CE-30, CA6N, CA-6NM, CD4MCu<br>ASTM A297: Grade HC<br>ASTM A487: Class CA6NM<br>ASTM A608: Grade HC30 | 135 to 185 | Annealed | 5000 to 6500 | 75 to 200 | Rough: .0005<br>Finish: .0002 max. | 1/3<br>1/6 | A60KV |
| | | | 25 to 33 | 23 to 60 | Rough: .013<br>Finish: .005 max. | 1/3<br>1/6 | A60KV |
| **Austenitic**<br>ASTM A296: Grades CF-3, CF-3M, CF-8, CF-8C, CF-8M, CF-16F, CF-20, CG-8M, CG-12, CH-20, CK-20, CN-7M, CN-7MS<br>ASTM A297: Grades HD, HE, HF, HH, HI, HK, HL, HN, HP, HT, HU<br>ASTM A351: Grades CF-3, CF-3A, CF-3M, CF-3MA, CF-8, CF-8A, CF-8C, CF-8M, CF-10MC, CH-8, CH-10, CH-20, CK-20, CN-7M, HK-30, HK-40, HT-30<br>ASTM A451: Grades CPF8A, CPF3, CPF3A, CPF3M, CPF8, CPF8C, CPF8C (Ta Max.), CPF8M, CPF10MC, CPH8, CPH10, CPH20, CPK20<br>ASTM A452: Grades TP 304H, TP 316H, TP 347H<br>ASTM A608: Grades HD50, HE35, HF30, HH30, HH33, HI35, HK30, HK40, HL30, HL40, HN40, HT50, HU50 | 135 to 210 | Annealed, Normalized or As Cast | 5000 to 6500 | 75 to 200 | Rough: .0005<br>Finish: .0002 max. | 1/3<br>1/6 | C60KV |
| | | | 25 to 33 | 23 to 60 | Rough: .013<br>Finish: .005 max. | 1/3<br>1/6 | C60KV |
| | | | | | | | |
| | | | | | | | |
| | | | | | | | |
| | | | | | | | |
| **Martensitic**<br>ASTM A217: Grade CA-15<br>ASTM A296: Grades CA-15, CA-15M, CA-40<br>ASTM A426: Grades CP7, CP9, CPCA15<br>ASTM A487: Classes CA15a, CA-15M | 135 to 225 | Annealed, Normalized or Normalized and Tempered | 5000 to 6500 | 75 to 200 | Rough: .0005<br>Finish: .0002 max. | 1/3<br>1/6 | A60KV |
| | | | 25 to 33 | 23 to 60 | Rough: .013<br>Finish: .005 max. | 1/3<br>1/6 | A60KV |
| | Over 275 | Quenched and Tempered | 5000 to 6500 | 50 to 150 | Rough: .0005<br>Finish: .0002 max. | 1/3<br>1/6 | A60JV |
| | | | 25 to 33 | 15 to 46 | Rough: .013<br>Finish: .005 max. | 1/3<br>1/6 | A60JV |

Maximum hole length is 2.5 times hole diameter.
Maximum wheel width is 1.5 times wheel diameter.
See section 16 for Cutting Fluid Recommendations.
*Wheel recommendations are for wet grinding 0.8- to 2-inch [20 to 50 mm] diameter holes. For LARGER holes—use the same or softer grade wheel. For SMALLER holes—use a harder grade wheel. Also see section 20.2, Grinding Guidelines.

| MATERIAL | HARD-NESS<br>Bhn | CONDITION | WHEEL SPEED<br>fpm<br>m/s | WORK SPEED<br>fpm<br>m/min | INFEED on dia.<br>in/pass<br>mm/pass | TRAVERSE Wheel width per rev. of work | WHEEL IDENTIFI-CATION*<br>ANSI<br>ISO |
|---|---|---|---|---|---|---|---|
| **19. PRECIPITATION HARDENING STAINLESS STEELS, CAST**<br>ASTM A351: Grade CD-4MCu<br>ACI Grade CB-7Cu<br>ACI Grade CD-4MCu<br>17-4 PH<br>AM-355 | 325 to 450 | Solution Treated or Solution Treated and Aged | 5000 to 6500 | 50 to 150 | Rough: .0005<br>Finish: .0002 max. | 1/3<br>1/6 | A60JV |
| | | | 25 to 33 | 15 to 46 | Rough: .013<br>Finish: .005 max. | 1/3<br>1/6 | A60JV |
| **20. AUSTENITIC MANGANESE STEELS, CAST**<br>ASTM A128: Grades A, B-1, B-2, B-3, B-4, C, D, E-1, E-2, F | 150 to 220 | Annealed | 5000 to 6500 | 75 to 200 | Rough: .0005<br>Finish: .0002 max. | 1/3<br>1/6 | A46JV |
| | | | 25 to 33 | 23 to 60 | Rough: .013<br>Finish: .005 max. | 1/3<br>1/6 | A46JV |
| **21. GRAY CAST IRONS**<br>**Ferritic**<br>ASTM A48: Class 20<br>SAE J431c: Grade G1800<br>**Pearlitic- Ferritic**<br>ASTM A48: Class 25<br>SAE J431c: Grade G2500<br>**Pearlitic**<br>ASTM A48: Classes 30, 35, 40<br>SAE J431c: Grade G3000<br>**Pearlitic + Free Carbides**<br>ASTM A48: Classes 45, 50<br>SAE J431c: Grades G3500, G4000<br>**Pearlitic or Acicular + Free Carbides**<br>ASTM A48: Classes 55, 60 | 45 $R_C$ max. | As Cast, Annealed or Quenched and Tempered | 5000 to 6500 | 75 to 200 | Rough: .002<br>Finish: .0002 max. | 1/3<br>1/6 | C46JV or A46KV |
| | | | 25 to 33 | 23 to 60 | Rough: .050<br>Finish: .005 max. | 1/3<br>1/6 | C46JV or A46KV |
| | 45 $R_C$ to 52 $R_C$ | As Cast, Annealed or Quenched and Tempered | 5000 to 6500 | 75 to 200 | Rough: .002<br>Finish: .0002 max. | 1/3<br>1/6 | C54JV or A60JV |
| | | | 25 to 33 | 23 to 60 | Rough: .050<br>Finish: .005 max. | 1/3<br>1/6 | C54JV or A60JV |
| | 48 $R_C$ to 60 $R_C$ | Flame or Induction Hardened | 5000 to 6500 | 75 to 200 | Rough: .001<br>Finish: .0002 max. | 1/3<br>1/6 | C60IV or A60IV |
| | | | 25 to 33 | 23 to 60 | Rough: .025<br>Finish: .005 max. | 1/3<br>1/6 | C60IV or A60IV |
| **Austenitic (NI-RESIST)**<br>ASTM A436: Types 1, 1b, 2, 2b, 3, 4, 5, 6 | 100 to 250 | As Cast | 5000 to 6500 | 75 to 200 | Rough: .0006<br>Finish: .0003 max. | 1/3<br>1/6 | Rough: A36HV<br>Finish: A60HV |
| | | | 25 to 33 | 23 to 60 | Rough: .015<br>Finish: .008 max. | 1/3<br>1/6 | Rough: A36HV<br>Finish: A60HV |

Maximum hole length is 2.5 times hole diameter.
Maximum wheel width is 1.5 times wheel diameter.
See section 16 for Cutting Fluid Recommendations.
*Wheel recommendations are for wet grinding 0.8- to 2-inch [20 to 50 mm] diameter holes. For LARGER holes—use the same or softer grade wheel. For SMALLER holes—use a harder grade wheel. Also see section 20.2, Grinding Guidelines.

# 8.8 Internal Grinding

| MATERIAL | HARD-NESS<br>Bhn | CONDITION | WHEEL SPEED<br>fpm<br>m/s | WORK SPEED<br>fpm<br>m/min | INFEED on dia.<br>in/pass<br>mm/pass | TRAVERSE Wheel width per rev. of work | WHEEL IDENTIFI-CATION*<br>ANSI<br>ISO |
|---|---|---|---|---|---|---|---|
| **22. COMPACTED GRAPHITE CAST IRONS** | 185 to 225 | As Cast | 5000 to 6500 | 75 to 200 | Rough: .002 | 1/3 | C46JV or A46KV |
| | | | | | Finish: .0002 max. | 1/6 | |
| | | | 25 to 33 | 23 to 60 | Rough: .050 | 1/3 | C46JV or A46KV |
| | | | | | Finish: .005 max. | 1/6 | |
| **23. DUCTILE CAST IRONS**<br>**Ferritic**<br>ASTM A536: Grades 60-40-18, 65-45-12<br>SAE J434c: Grades D4018, D4512<br>**Ferritic- Pearlitic**<br>ASTM A536: Grade 80-55-06<br>SAE J434c: Grade D5506<br>**Pearlitic- Martensitic**<br>ASTM A536: Grade 100-70-03<br>SAE J434c: Grade D7003<br>**Martensitic**<br>ASTM A536: Grade 120-90-02<br>SAE J434c: Grade DQ&T | $52\ R_C$ max. | Annealed, As Cast, Normalized and Tempered or Quenched and Tempered | 5000 to 6500 | 75 to 200 | Rough: .002 | 1/3 | C54JV or A60JV |
| | | | | | Finish: .0002 max. | 1/6 | |
| | | | 25 to 33 | 23 to 60 | Rough: .050 | 1/3 | C54JV or A60JV |
| | | | | | Finish: .005 max. | 1/6 | |
| | $53\ R_C$ to $60\ R_C$ | Flame or Induction Hardened | 5000 to 6500 | 75 to 200 | Rough: .001 | 1/3 | C60IV or A60IV |
| | | | | | Finish: .0002 max. | 1/6 | |
| | | | 25 to 33 | 23 to 60 | Rough: .025 | 1/3 | C60IV or A60IV |
| | | | | | Finish: .005 max. | 1/6 | |
| **Austenitic (NI-RESIST Ductile)**<br>ASTM A439: Types D-2, D-2B, D-2C, D-3, D-3A, D-4, D-5, D-5B<br>ASTM A571: Type D-2M | 120 to 275 | Annealed | 5000 to 6500 | 75 to 200 | Rough: .0006 | 1/3 | Rough: A36HV |
| | | | | | Finish: .0003 max. | 1/6 | Finish: A60HV |
| | | | 25 to 33 | 23 to 60 | Rough: .015 | 1/3 | Rough: A36HV |
| | | | | | Finish: .008 max. | 1/6 | Finish: A60HV |
| **24. MALLEABLE CAST IRONS**<br>**Ferritic**<br>ASTM A47: Grades 32510, 35018<br>ASTM A602: Grade M3210<br>SAE J158: Grade M3210<br>**Pearlitic**<br>ASTM A220: Grades 40010, 45006, 45008, 50005<br>ASTM A602: Grade M4504, M5003<br>SAE J158: Grades M4504, M5003<br>**Tempered Martensite**<br>ASTM A220: Grades 60004, 70003, 80002, 90001<br>ASTM A602: Grades M5503, M7002, M8501<br>SAE J158: Grades M5503, M7002, M8501 | $52\ R_C$ max. | Malleablized or Malleablized and Heat Treated | 5000 to 6500 | 75 to 200 | Rough: .002 | 1/3 | C54JV or A60JV |
| | | | | | Finish: .0002 max. | 1/6 | |
| | | | 25 to 33 | 23 to 60 | Rough: .050 | 1/3 | C54JV or A60JV |
| | | | | | Finish: .005 max. | 1/6 | |
| | Over $52\ R_C$ | Flame or Induction Hardened | 5000 to 6500 | 75 to 200 | Rough: .001 | 1/3 | C60IV or A60IV |
| | | | | | Finish: .0002 max. | 1/6 | |
| | | | 25 to 33 | 23 to 60 | Rough: .025 | 1/3 | C60IV or A60IV |
| | | | | | Finish: .005 max. | 1/6 | |

Maximum hole length is 2.5 times hole diameter.
Maximum wheel width is 1.5 times wheel diameter.
See section 16 for Cutting Fluid Recommendations.
*Wheel recommendations are for wet grinding 0.8- to 2-inch [20 to 50 mm] diameter holes. For LARGER holes—use the same or softer grade wheel. For SMALLER holes—use a harder grade wheel. Also see section 20.2, Grinding Guidelines.

| MATERIAL | HARD-NESS Bhn | CONDITION | WHEEL SPEED fpm / m/s | WORK SPEED fpm / m/min | INFEED on dia. in/pass / mm/pass | TRAVERSE Wheel width per rev. of work | WHEEL IDENTIFI-CATION* ANSI / ISO |
|---|---|---|---|---|---|---|---|
| **25. WHITE CAST IRONS (ABRASION RESISTANT)**<br>ASTM A532:<br>Class I, Types A, B, C, D<br>Class II, Types A, B, C, D, E<br>Class III, Type A | 60 $R_C$ max. | As Cast, Annealed or Hardened | 5000 to 6500 | 75 to 200 | Rough: .001<br>Finish: .0002 max. | 1/3<br>1/6 | C60IV |
| | | | 25 to 33 | 23 to 60 | Rough: .025<br>Finish: .005 max. | 1/3<br>1/6 | C60IV |
| **26. HIGH SILICON CAST IRONS**<br>Duriron<br>Duriclor<br>ASTM A518 | 52 $R_C$ | As Cast | 5000 to 6500 | 75 to 200 | Rough: .001<br>Finish: .0002 max. | 1/3<br>1/6 | A46LV |
| | | | 25 to 33 | 23 to 60 | Rough: .025<br>Finish: .005 max. | 1/3<br>1/6 | A46LV |
| **27. CHROMIUM-NICKEL ALLOY CASTINGS**<br>ASTM A560: Grades 50Cr-50Ni, 60Cr-40Ni | 275 to 375 | As Cast | 5000 to 6500 | 50 to 150 | Rough: .0005<br>Finish: .0002 max. | 1/3<br>1/6 | C60JV |
| | | | 25 to 33 | 15 to 46 | Rough: .013<br>Finish: .005 max. | 1/3<br>1/6 | C60JV |
| **28. ALUMINUM ALLOYS, WROUGHT**<br>EC 2218 5252 6253<br>1060 2219 5254 6262<br>1100 2618 5454 6463<br>1145 3003 5456 6951<br>1175 3004 5457 7001<br>1235 3005 5652 7004<br>2011 4032 5657 7005<br>2014 5005 6053 7039<br>2017 5050 6061 7049<br>2018 5052 6063 7050<br>2021 5056 6066 7075<br>2024 5083 6070 7079<br>2025 5086 6101 7175<br>2117 5154 6151 7178 | 30 to 150 500kg | Cold Drawn or Solution Treated and Aged | 5000 to 6500 | 75 to 200 | Rough: .003<br>Finish: .0002 max. | 1/3<br>1/6 | C46JV |
| | | | 25 to 33 | 23 to 60 | Rough: .075<br>Finish: .005 max. | 1/3<br>1/6 | C46JV |

Maximum hole length is 2.5 times hole diameter.
Maximum wheel width is 1.5 times wheel diameter.
See section 16 for Cutting Fluid Recommendations.
*Wheel recommendations are for wet grinding 0.8- to 2-inch [20 to 50 mm] diameter holes. For LARGER holes—use the same or softer grade wheel. For SMALLER holes—use a harder grade wheel. Also see section 20.2, Grinding Guidelines.

# 8.8 Internal Grinding

| MATERIAL | HARD-NESS<br>Bhn | CONDITION | WHEEL SPEED<br>fpm<br>m/s | WORK SPEED<br>fpm<br>m/min | INFEED on dia.<br>in/pass<br>mm/pass | TRAVERSE<br>Wheel width per rev. of work | WHEEL IDENTIFI-CATION*<br>ANSI<br>ISO |
|---|---|---|---|---|---|---|---|
| **29. ALUMINUM ALLOYS, CAST**<br>**Sand and Permanent Mold**<br>A140 295.0 514.0 A712.0<br>201.0 B295.0 A514.0 D712.0<br>208.0 308.0 B514.0 713.0<br>213.0 319.0 520.0 771.0<br>222.0 355.0 535.0 850.0<br>224.0 C355.0 705.0 A850.0<br>242.0 B443.0 707.0 B850.0<br>Hiduminium RR-350<br>**Die Castings**<br>C443.0 518.0 | 40 to 125 500kg | As Cast or Solution Treated and Aged | 5000 to 6500 | 75 to 200 | Rough: .003<br>Finish: .0002 max. | 1/3<br>1/6 | C46JV |
| | | | 25 to 33 | 23 to 60 | Rough: .075<br>Finish: .005 max. | 1/3<br>1/6 | C46JV |
| | | | | | | | |
| | | | | | | | |
| **Sand and Permanent Mold**<br>328.0 356.0<br>A332.0 A356.0<br>F332.0 357.0<br>333.0 359.0<br>354.0<br>**Die Castings**<br>360.0 A380.0 390.0 A413.0<br>A360.0 383.0 392.0<br>380.0 A384.0 413.0 | 40 to 125 500kg | As Cast or Solution Treated and Aged | 5000 to 6500 | 75 to 200 | Rough: .003<br>Finish: .0002 max. | 1/3<br>1/6 | A60HV |
| | | | 25 to 33 | 23 to 60 | Rough: .075<br>Finish: .005 max. | 1/3<br>1/6 | A60HV |
| | | | | | | | |
| | | | | | | | |
| **30. MAGNESIUM ALLOYS, WROUGHT‡‡**<br>AZ21A HK31A<br>AZ31B HM21A<br>AZ31C HM31A<br>AZ61A ZK40A<br>AZ80A ZK60A | 50 to 90 500kg | Annealed, Cold Drawn or Solution Treated and Aged | 5000 to 6500 | 75 to 200 | Rough: .003<br>Finish: .0002 max. | 1/3<br>1/6 | C46JV |
| | | | 25 to 33 | 23 to 60 | Rough: .075<br>Finish: .005 max. | 1/3<br>1/6 | C46JV |
| **31. MAGNESIUM ALLOYS, CAST‡‡**<br>AM60A AZ91C ZE41A<br>AM100A AZ92A ZE63A<br>AS41A EZ33A ZH62A<br>AZ63A HK31A ZK51A<br>AZ81A HZ32A ZK61A<br>AZ91A K1A<br>AZ91B QE22A | 50 to 90 500kg | As Cast, Annealed or Solution Treated and Aged | 5000 to 6500 | 75 to 200 | Rough: .003<br>Finish: .0002 max. | 1/3<br>1/6 | C46JV |
| | | | 25 to 33 | 23 to 60 | Rough: .075<br>Finish: .005 max. | 1/3<br>1/6 | C46JV |

Maximum hole length is 2.5 times hole diameter.
Maximum wheel width is 1.5 times wheel diameter.
See section 16 for Cutting Fluid Recommendations.
*Wheel recommendations are for wet grinding 0.8- to 2-inch [20 to 50 mm] diameter holes. For LARGER holes—use the same or softer grade wheel. For SMALLER holes—use a harder grade wheel. Also see section 20.2, Grinding Guidelines.

‡‡CAUTION: Potential Fire Hazard. Exercise caution in grinding and disposing of swarf. Do NOT use water or water-miscible cutting fluids for magnesium alloys.

| MATERIAL | HARD-NESS<br>Bhn | CONDITION | WHEEL SPEED<br>fpm<br>m/s | WORK SPEED<br>fpm<br>m/min | INFEED on dia.<br>in/pass<br>mm/pass | TRAVERSE Wheel width per rev. of work | WHEEL IDENTIFI-CATION*<br>ANSI<br>ISO |
|---|---|---|---|---|---|---|---|
| **32. TITANIUM ALLOYS, WROUGHT‡‡**<br>**Commercially Pure**<br>99.5<br>99.2<br>99.0<br>98.9<br>Ti-0.2Pd<br>TiCODE-12 | 110<br>to<br>275 | Annealed | 4000<br>to<br>5000 | 50<br>to<br>150 | Rough:<br>.0005<br>Finish:<br>.0002 max. | 1/3<br>1/6 | C60JV‡ |
| | | | 20<br>to<br>25 | 15<br>to<br>46 | Rough:<br>.013<br>Finish:<br>.005 max. | 1/3<br>1/6 | C60JV‡ |
| **Alpha and Alpha-Beta Alloys**<br>Ti-8Mn<br>Ti-1Al-8V-5Fe<br>Ti-2Al-11Sn-5Zr-1Mo<br>Ti-3Al-2.5V<br>Ti-5Al-2Sn-2Zr-4Mo-4Cr (Ti-17)<br>Ti-5Al-2.5Sn<br>Ti-5Al-2.5Sn ELI<br>Ti-5Al-6Sn-2Zr-1Mo<br>Ti-6Al-2Cb-1Ta-0.8Mo<br>Ti-6Al-4V<br>Ti-6Al-4V ELI<br>Ti-6Al-6V-2Sn<br>Ti-6Al-2Sn-4Zr-2Mo<br>Ti-6Al-2Sn-4Zr-2Mo-.25Si<br>Ti-6Al-2Sn-4Zr-6Mo<br>Ti-7Al-4Mo<br>Ti-8Al-1Mo-1V | 300<br>to<br>380 | Annealed | 4000<br>to<br>5000 | 50<br>to<br>150 | Rough:<br>.0005<br>Finish:<br>.0002 max. | 1/3<br>1/6 | C60JV‡ |
| | | | 20<br>to<br>25 | 15<br>to<br>46 | Rough:<br>.013<br>Finish:<br>.005 max. | 1/3<br>1/6 | C60JV‡ |
| | 320<br>to<br>440 | Solution Treated and Aged | 4000<br>to<br>5000 | 50<br>to<br>150 | Rough:<br>.0005<br>Finish:<br>.0002 max. | 1/3<br>1/6 | C60JV‡ |
| | | | 20<br>to<br>25 | 15<br>to<br>46 | Rough:<br>.013<br>Finish:<br>.005 max. | 1/3<br>1/6 | C60JV‡ |
| **Beta Alloys**<br>Ti-3Al-8V-6Cr-4Mo-4Zr<br>Ti-8Mo-8V-2Fe-3Al<br>Ti-11.5Mo-6Zr-4.5Sn<br>Ti-10V-2Fe-3Al<br>Ti-13V-11Cr-3Al | 275<br>to<br>350 | Annealed<br>or<br>Solution Treated | 4000<br>to<br>5000 | 50<br>to<br>150 | Rough:<br>.0005<br>Finish:<br>.0002 max. | 1/3<br>1/6 | C60JV‡ |
| | | | 20<br>to<br>25 | 15<br>to<br>46 | Rough:<br>.013<br>Finish:<br>.005 max. | 1/3<br>1/6 | C60JV‡ |
| | 350<br>to<br>440 | Solution Treated and Aged | 4000<br>to<br>5000 | 50<br>to<br>150 | Rough:<br>.0005<br>Finish:<br>.0002 max. | 1/3<br>1/6 | C60JV‡ |
| | | | 20<br>to<br>25 | 15<br>to<br>46 | Rough:<br>.013<br>Finish:<br>.005 max. | 1/3<br>1/6 | C60JV‡ |
| **33. TITANIUM ALLOYS, CAST‡‡**<br>**Commercially Pure**<br>99.0<br>Ti-0.2Pd<br>ASTM B367: Grades C-1, C-2, C-3, C-4, C-7A, C-7B, C-8A, C-8B | 150<br>to<br>250 | As Cast<br>or<br>As Cast and Annealed | 4000<br>to<br>5000 | 50<br>to<br>150 | Rough:<br>.0005<br>Finish:<br>.0002 max. | 1/3<br>1/6 | C60JV‡ |
| | | | 20<br>to<br>25 | 15<br>to<br>46 | Rough:<br>.013<br>Finish:<br>.005 max. | 1/3<br>1/6 | C60JV‡ |

Maximum hole length is 2.5 times hole diameter.
Maximum wheel width is 1.5 times wheel diameter.
See section 16 for Cutting Fluid Recommendations.
*Wheel recommendations are for wet grinding 0.8- to 2-inch [20 to 50 mm] diameter holes. For LARGER holes—use the same or softer grade wheel. For SMALLER holes—use a harder grade wheel. Also see section 20.2, Grinding Guidelines.

‡Use friable (green grit) silicon carbide.
‡‡CAUTION: Potential Fire Hazard. Exercise caution in grinding and disposing of swarf.

| MATERIAL | HARDNESS<br>Bhn | CONDITION | WHEEL SPEED<br>fpm<br>m/s | WORK SPEED<br>fpm<br>m/min | INFEED on dia.<br>in/pass<br>mm/pass | TRAVERSE Wheel width per rev. of work | WHEEL IDENTIFICATION*<br>ANSI<br>ISO |
|---|---|---|---|---|---|---|---|
| **33. TITANIUM ALLOYS, CAST‡‡ (cont.)**<br>**Alpha and Alpha-Beta Alloys**<br>Ti-5Al-2.5Sn<br>Ti-6Al-4V<br>Ti-6Al-2Sn-4Zr-2Mo<br>Ti-8Al-1Mo-1V<br>ASTM B367: Grades C-5, C-6 | 300 to 350 | As Cast or As Cast and Annealed | 4000 to 5000 | 50 to 150 | Rough: .0005 | 1/3 | C60JV‡ |
| | | | | | Finish: .0002 max. | 1/6 | |
| | | | 20 to 25 | 15 to 46 | Rough: .013 | 1/3 | C60JV‡ |
| | | | | | Finish: .005 max. | 1/6 | |
| **34. COPPER ALLOYS, WROUGHT**<br>101 116 143 182<br>102 119 145 184<br>104 120 147 185<br>105 121 150 187<br>107 122 155 189<br>109 125 162 190<br>110 127 165 191<br>111 128 170†† 192<br>113 129 172†† 194<br>114 130 173†† 195<br>115 142 175†† | 10 $R_B$ to 70 $R_B$ | Annealed | 5000 to 6500 | 75 to 200 | Rough: .002 | 1/3 | C46JV |
| | | | | | Finish: .0002 max. | 1/6 | |
| | | | 25 to 33 | 23 to 60 | Rough: .050 | 1/3 | C46JV |
| | | | | | Finish: .005 max. | 1/6 | |
| | 60 $R_B$ to 100 $R_B$ | Cold Drawn | 5000 to 6500 | 75 to 200 | Rough: .002 | 1/3 | C46JV |
| | | | | | Finish: .0002 max. | 1/6 | |
| | | | 25 to 33 | 23 to 60 | Rough: .050 | 1/3 | C46JV |
| | | | | | Finish: .005 max. | 1/6 | |
| 210 332 368 464<br>220 335 370 465<br>226 340 377 466<br>230 342 385 467<br>240 349 411 482<br>260 350 413 485<br>268 353 425 667<br>270 356 435 687<br>280 360 442 688<br>314 365 443 694<br>316 366 444<br>330 367 445 | 10 $R_B$ to 70 $R_B$ | Annealed | 5000 to 6500 | 75 to 200 | Rough: .002 | 1/3 | C46JV |
| | | | | | Finish: .0002 max. | 1/6 | |
| | | | 25 to 33 | 23 to 60 | Rough: .050 | 1/3 | C46JV |
| | | | | | Finish: .005 max. | 1/6 | |
| | 60 $R_B$ to 100 $R_B$ | Cold Drawn | 5000 to 6500 | 75 to 200 | Rough: .002 | 1/3 | C46JV |
| | | | | | Finish: .0002 max. | 1/6 | |
| | | | 25 to 33 | 23 to 60 | Rough: .050 | 1/3 | C46JV |
| | | | | | Finish: .005 max. | 1/6 | |
| 505 613 632<br>510 614 638<br>511 618 642<br>521 619 651<br>524 623 655<br>544 624 674<br>608 625 675<br>610 630 | 10 $R_B$ to 70 $R_B$ | Annealed | 5000 to 6500 | 75 to 200 | Rough: .002 | 1/3 | C46JV |
| | | | | | Finish: .0002 max. | 1/6 | |
| | | | 25 to 33 | 23 to 60 | Rough: .050 | 1/3 | C46JV |
| | | | | | Finish: .005 max. | 1/6 | |

Maximum hole length is 2.5 times hole diameter.
Maximum wheel width is 1.5 times wheel diameter.
See section 16 for Cutting Fluid Recommendations.
*Wheel recommendations are for wet grinding 0.8- to 2-inch [20 to 50 mm] diameter holes. For LARGER holes—use the same or softer grade wheel. For SMALLER holes—use a harder grade wheel. Also see section 20.2, Grinding Guidelines.

‡Use friable (green grit) silicon carbide.
††CAUTION: Toxic Material, refer to National Institute for Occupational Safety and Health (NIOSH) for Precautions.
‡‡CAUTION: Potential Fire Hazard. Exercise caution in grinding and disposing of swarf.

| MATERIAL | HARDNESS<br>Bhn | CONDITION | WHEEL SPEED<br>fpm<br>m/s | WORK SPEED<br>fpm<br>m/min | INFEED on dia.<br>in/pass<br>mm/pass | TRAVERSE Wheel width per rev. of work | WHEEL IDENTIFICATION*<br>ANSI<br>ISO |
|---|---|---|---|---|---|---|---|
| **34. COPPER ALLOYS, WROUGHT (cont.)**<br>(materials listed on preceding page) | 60 $R_B$<br>to<br>100 $R_B$ | Cold Drawn | 5000<br>to<br>6500 | 75<br>to<br>200 | Rough:<br>.002<br>Finish:<br>.0002 max. | 1/3<br>1/6 | C46JV |
| | | | 25<br>to<br>33 | 23<br>to<br>60 | Rough:<br>.050<br>Finish:<br>.005 max. | 1/3<br>1/6 | C46JV |
| 706 752<br>710 754<br>715 757<br>725 770<br>745 782 | 10 $R_B$<br>to<br>70 $R_B$ | Annealed | 5000<br>to<br>6500 | 75<br>to<br>200 | Rough:<br>.002<br>Finish:<br>.0002 max. | 1/3<br>1/6 | C46JV |
| | | | 25<br>to<br>33 | 23<br>to<br>60 | Rough:<br>.050<br>Finish:<br>.005 max. | 1/3<br>1/6 | C46JV |
| | 60 $R_B$<br>to<br>100 $R_B$ | Cold Drawn | 5000<br>to<br>6500 | 75<br>to<br>200 | Rough:<br>.002<br>Finish:<br>.0002 max. | 1/3<br>1/6 | C46JV |
| | | | 25<br>to<br>33 | 23<br>to<br>60 | Rough:<br>.050<br>Finish:<br>.005 max. | 1/3<br>1/6 | C46JV |
| **35. COPPER ALLOYS, CAST**<br>801 814 824††<br>803 815 825††<br>805 817†† 826††<br>807 818†† 827††<br>809 820†† 828††<br>811 821††<br>813 822†† | 40<br>to<br>200<br>500kg | As Cast<br>or<br>Heat Treated | 5000<br>to<br>6500 | 75<br>to<br>200 | Rough:<br>.002<br>Finish:<br>.0002 max. | 1/3<br>1/6 | C46JV |
| | | | 25<br>to<br>33 | 23<br>to<br>60 | Rough:<br>.050<br>Finish:<br>.005 max. | 1/3<br>1/6 | C46JV |
| | 34 $R_C$<br>to<br>45 $R_C$ | Heat Treated | 5000<br>to<br>6500 | 75<br>to<br>200 | Rough:<br>.002<br>Finish:<br>.0002 max. | 1/3<br>1/6 | A54KV |
| | | | 25<br>to<br>33 | 23<br>to<br>60 | Rough:<br>.050<br>Finish:<br>.005 max. | 1/3<br>1/6 | A54KV |
| 833 854 867<br>834 855 868<br>836 857 872<br>838 858 874<br>842 861 875<br>844 862 876<br>848 863 878<br>852 864 879<br>853 865 | 35<br>to<br>200<br>500kg | As Cast | 5000<br>to<br>6500 | 75<br>to<br>200 | Rough:<br>.002<br>Finish:<br>.0002 max. | 1/3<br>1/6 | C46JV |
| | | | 25<br>to<br>33 | 23<br>to<br>60 | Rough:<br>.050<br>Finish:<br>.005 max. | 1/3<br>1/6 | C46JV |

Maximum hole length is 2.5 times hole diameter.
Maximum wheel width is 1.5 times wheel diameter.
See section 16 for Cutting Fluid Recommendations.
*Wheel recommendations are for wet grinding 0.8- to 2-inch [20 to 50 mm] diameter holes. For LARGER holes—use the same or softer grade wheel. For SMALLER holes—use a harder grade wheel. Also see section 20.2, Grinding Guidelines.

††CAUTION: Toxic Material, refer to National Institute for Occupational Safety and Health (NIOSH) for Precautions.

# 8.8 Internal Grinding

| MATERIAL | HARDNESS Bhn | CONDITION | WHEEL SPEED fpm / m/s | WORK SPEED fpm / m/min | INFEED on dia. in/pass / mm/pass | TRAVERSE Wheel width per rev. of work | WHEEL IDENTIFICATION* ANSI / ISO |
|---|---|---|---|---|---|---|---|
| **35. COPPER ALLOYS, CAST (cont.)** 902, 903, 905, 907, 909, 910, 911, 913, 915, 916, 917, 922, 923, 925, 926, 927, 928, 932, 934, 935, 937, 938, 939, 943, 944, 945, 947, 948, 952, 953, 954, 955, 956, 957, 958 | 40 to 100 500kg | As Cast | 5000 to 6500 | 75 to 200 | Rough: .002 | 1/3 | C46JV |
| | | | | | Finish: .0002 max. | 1/6 | |
| | | | 25 to 33 | 23 to 60 | Rough: .050 | 1/3 | C46JV |
| | | | | | Finish: .005 max. | 1/6 | |
| | Over 100 500kg | As Cast or Heat Treated | 5000 to 6500 | 75 to 200 | Rough: .002 | 1/3 | A54KV |
| | | | | | Finish: .0002 max. | 1/6 | |
| | | | 25 to 33 | 23 to 60 | Rough: .050 | 1/3 | A54KV |
| | | | | | Finish: .005 max. | 1/6 | |
| 962, 963, 964, 966††, 973, 974, 976, 978, 993 | 50 to 100 500kg | As Cast | 5000 to 6500 | 75 to 200 | Rough: .002 | 1/3 | C46JV |
| | | | | | Finish: .0002 max. | 1/6 | |
| | | | 25 to 33 | 23 to 60 | Rough: .050 | 1/3 | C46JV |
| | | | | | Finish: .005 max. | 1/6 | |
| | Over 100 500kg | As Cast or Heat Treated | 5000 to 6500 | 75 to 200 | Rough: .002 | 1/3 | A54KV |
| | | | | | Finish: .0002 max. | 1/6 | |
| | | | 25 to 33 | 23 to 60 | Rough: .050 | 1/3 | A54KV |
| | | | | | Finish: .005 max. | 1/6 | |
| **36. NICKEL ALLOYS, WROUGHT AND CAST** Nickel 200, Nickel 201, Nickel 205, Nickel 211, Nickel 220, Nickel 230 | 80 to 170 | Annealed or Cold Drawn | 5000 to 6000 | 50 to 150 | Rough: .0005 | 1/3 | C54JV |
| | | | | | Finish: .0002 max. | 1/6 | |
| | | | 25 to 30 | 15 to 46 | Rough: .013 | 1/3 | C54JV |
| | | | | | Finish: .005 max. | 1/6 | |
| MONEL Alloy 400, MONEL Alloy 401, MONEL Alloy 404, MONEL Alloy R405, ASTM A296: Grades CZ-100, M-35, ASTM A494: Grades CZ-100, M-35 | 115 to 240 | Annealed, Cold Drawn or Cast | 5000 to 6000 | 50 to 150 | Rough: .0005 | 1/3 | C54JV |
| | | | | | Finish: .0002 max. | 1/6 | |
| | | | 25 to 30 | 15 to 46 | Rough: .013 | 1/3 | C54JV |
| | | | | | Finish: .005 max. | 1/6 | |

Maximum hole length is 2.5 times hole diameter.
Maximum wheel width is 1.5 times wheel diameter.
See section 16 for Cutting Fluid Recommendations.
*Wheel recommendations are for wet grinding 0.8- to 2-inch [20 to 50 mm] diameter holes. For LARGER holes—use the same or softer grade wheel. For SMALLER holes—use a harder grade wheel. Also see section 20.2, Grinding Guidelines.

††CAUTION: Toxic Material, refer to National Institute for Occupational Safety and Health (NIOSH) for Precautions.

| MATERIAL | HARDNESS<br>Bhn | CONDITION | WHEEL SPEED<br>fpm<br>m/s | WORK SPEED<br>fpm<br>m/min | INFEED on dia.<br>in/pass<br>mm/pass | TRAVERSE Wheel width per rev. of work | WHEEL IDENTIFICATION*<br>ANSI<br>ISO |
|---|---|---|---|---|---|---|---|
| DURANICKEL Alloy 301<br>MONEL Alloy 502<br>MONEL Alloy K500<br>NI-SPAN-C Alloy 902<br>PERMANICKEL Alloy 300 | 150 to 320 | Solution Treated | 5000 to 6000 | 50 to 150 | Rough: .0005<br>Finish: .0002 max. | 1/3<br>1/6 | C60JV |
| | | | 25 to 30 | 15 to 46 | Rough: .013<br>Finish: .005 max. | 1/3<br>1/6 | C60JV |
| | 330 to 360 | Aged | 5000 to 6000 | 50 to 150 | Rough: .0005<br>Finish: .0002 max. | 1/3<br>1/6 | C60JV |
| | | | 25 to 30 | 15 to 46 | Rough: .013<br>Finish: .005 max. | 1/3<br>1/6 | C60JV |
| **37. BERYLLIUM NICKEL ALLOYS, WROUGHT AND CAST††**<br>Berylco 440<br>Berylco 41C<br>Berylco 42C<br>Berylco 43C<br>Brush Alloy 200C<br>Brush Alloy 220C<br>Brush Alloy 260C | 200 to 250 | As Cast or Solution Treated | 5000 to 6000 | 50 to 150 | Rough: .0005<br>Finish: .0002 max. | 1/3<br>1/6 | A46KV |
| | | | 25 to 30 | 15 to 46 | Rough: .013<br>Finish: .005 max. | 1/3<br>1/6 | A46KV |
| | 283 to 425 | Hardened or Aged | 5000 to 6000 | 50 to 150 | Rough: .0005<br>Finish: .0002 max. | 1/3<br>1/6 | A60JV |
| | | | 25 to 30 | 15 to 46 | Rough: .013<br>Finish: .005 max. | 1/3<br>1/6 | A60JV |
| | 47 $R_C$ to 52 $R_C$ | Hardened or Aged | 5000 to 6000 | 50 to 150 | Rough: .0005<br>Finish: .0002 max. | 1/3<br>1/6 | A60IV |
| | | | 25 to 30 | 15 to 46 | Rough: .013<br>Finish: .005 max. | 1/3<br>1/6 | A60IV |
| **38. NITINOL ALLOYS, WROUGHT**<br>Nitinol 55Ni-45Ti<br>Nitinol 56Ni-44Ti<br>Nitinol 60Ni-40Ti | 210 to 340 | Wrought or Annealed | 5000 | 50 to 150 | Rough: .0005<br>Finish: .0002 max. | 1/3<br>1/6 | C60JV |
| | | | 25 | 15 to 46 | Rough: .013<br>Finish: .005 max. | 1/3<br>1/6 | C60JV |

Maximum hole length is 2.5 times hole diameter.
Maximum wheel width is 1.5 times wheel diameter.
See section 16 for Cutting Fluid Recommendations.
*Wheel recommendations are for wet grinding 0.8- to 2-inch [20 to 50 mm] diameter holes. For LARGER holes—use the same or softer grade wheel. For SMALLER holes—use a harder grade wheel. Also see section 20.2, Grinding Guidelines.

††CAUTION: Toxic Material, refer to National Institute for Occupational Safety and Health (NIOSH) for Precautions.

| MATERIAL | HARD-NESS | CONDITION | WHEEL SPEED | WORK SPEED | INFEED on dia. | TRAVERSE Wheel width per rev. of work | WHEEL IDENTIFI-CATION* |
|---|---|---|---|---|---|---|---|
| | Bhn | | fpm<br>m/s | fpm<br>m/min | in/pass<br>mm/pass | | ANSI<br>ISO |
| **38. NITINOL ALLOYS, WROUGHT (cont.)**<br>(materials listed on preceding page) | 48 $R_C$ to 52 $R_C$ | Quenched | 5000 | 50 to 150 | Rough: .0005<br>Finish: .0002 max. | 1/3<br>1/6 | C60JV |
| | | | 25 | 15 to 46 | Rough: .013<br>Finish: .005 max. | 1/3<br>1/6 | C60JV |
| **39. HIGH TEMPERATURE ALLOYS, WROUGHT AND CAST**<br>**Nickel Base, Wrought**<br>AF2-1DA<br>Astroloy<br>Haynes Alloy 263<br>IN-102<br>Incoloy Alloy 901<br>Incoloy Alloy 903<br>Inconel Alloy 617<br>Inconel Alloy 625<br>Inconel Alloy 700<br>Inconel Alloy 702<br>Inconel Alloy 706<br>Inconel Alloy 718<br>Inconel Alloy 721<br>Inconel Alloy 722<br>Inconel Alloy X-750<br>Inconel Alloy 751<br>M252<br>Nimonic 75<br>Nimonic 80<br>Nimonic 90<br>Nimonic 95<br>Rene 41<br>Rene 63<br>Rene 77<br>Rene 95<br>Udimet 500<br>Udimet 700<br>Udimet 710<br>Unitemp 1753<br>Waspaloy | 200 to 390 | Annealed or Solution Treated | 3000 to 4000 | 50 to 150 | Rough: .0005<br>Finish: .0002 max. | 1/3<br>1/6 | A60JV |
| | | | 15 to 20 | 15 to 46 | Rough: .013<br>Finish: .005 max. | 1/3<br>1/6 | A60JV |
| | 300 to 475 | Solution Treated and Aged | 3000 to 4000 | 50 to 150 | Rough: .0005<br>Finish: .0002 max. | 1/3<br>1/6 | A60JV |
| | | | 15 to 20 | 15 to 46 | Rough: .013<br>Finish: .005 max. | 1/3<br>1/6 | A60JV |

Maximum hole length is 2.5 times hole diameter.
Maximum wheel width is 1.5 times wheel diameter.
See section 16 for Cutting Fluid Recommendations.
*Wheel recommendations are for wet grinding 0.8- to 2-inch [20 to 50 mm] diameter holes. For LARGER holes—use the same or softer grade wheel. For SMALLER holes—use a harder grade wheel. Also see section 20.2, Grinding Guidelines.

| MATERIAL | HARD-NESS<br>Bhn | CONDITION | WHEEL SPEED<br>fpm<br>m/s | WORK SPEED<br>fpm<br>m/min | INFEED on dia.<br>in/pass<br>mm/pass | TRAVERSE Wheel width per rev. of work | WHEEL IDENTIFI-CATION*<br>ANSI<br>ISO |
|---|---|---|---|---|---|---|---|
| **Nickel Base, Wrought**<br>Hastelloy Alloy B<br>Hastelloy Alloy B-2<br>Hastelloy Alloy C<br>Hastelloy Alloy C-276<br>Hastelloy Alloy G<br>Hastelloy Alloy S<br>Hastelloy Alloy X<br>Incoloy Alloy 804<br>Incoloy Alloy 825<br>Inconel Alloy 600<br>Inconel Alloy 601<br>Refractaloy 26<br>Udimet 630 | 140<br>to<br>220 | Annealed<br>or<br>Solution Treated | 3000<br>to<br>4000 | 50<br>to<br>150 | Rough:<br>.0005<br>Finish:<br>.0002 max. | 1/3<br>1/6 | A60JV |
| | | | 15<br>to<br>20 | 15<br>to<br>46 | Rough:<br>.013<br>Finish:<br>.005 max. | 1/3<br>1/6 | A60JV |
| | 240<br>to<br>310 | Cold Drawn<br>or<br>Aged | 3000<br>to<br>4000 | 50<br>to<br>150 | Rough:<br>.0005<br>Finish:<br>.0002 max. | 1/3<br>1/6 | A60JV |
| | | | 15<br>to<br>20 | 15<br>to<br>46 | Rough:<br>.013<br>Finish:<br>.005 max. | 1/3<br>1/6 | A60JV |
| **Nickel Base, Wrought**<br>TD-Nickel††<br>TD-Ni-Cr†† | 180<br>to<br>200 | As Rolled | 5000 | 50<br>to<br>150 | Rough:<br>.0005<br>Finish:<br>.0002 max. | 1/3<br>1/6 | A60JV |
| | | | 25 | 15<br>to<br>46 | Rough:<br>.013<br>Finish:<br>.005 max. | 1/3<br>1/6 | A60JV |
| **Nickel Base, Cast**<br>B-1900<br>GMR-235<br>GMR-235D<br>Hastelloy Alloy B<br>Hastelloy Alloy C<br>Hastelloy Alloy D<br>IN-100 (Rene 100)<br>IN-738<br>IN-792<br>Inconel Alloy 713C<br>Inconel Alloy 718<br>M252<br>MAR-M200<br>MAR-M246<br>MAR-M421<br>MAR-M432<br>Rene 80<br>Rene 125<br>SEL<br>SEL 15<br>TRW VI A<br>Udimet 500<br>Udimet 700<br>ASTM A296: Grades CW-12M, N-12M, CY-40<br>ASTM A297: Grades HW, HX<br>ASTM A494: Grades N-12M-1, N-12M-2, CY-40, CW-12M-1, CW-12M-2<br>ASTM A608: Grades HW50, HX50 | 200<br>to<br>425 | As Cast<br>or<br>Cast and Aged | 3000<br>to<br>4000 | 50<br>to<br>150 | Rough:<br>.0005<br>Finish:<br>.0002 max. | 1/3<br>1/6 | A60JV |
| | | | 15<br>to<br>20 | 15<br>to<br>46 | Rough:<br>.013<br>Finish:<br>.005 max. | 1/3<br>1/6 | A60JV |

Maximum hole length is 2.5 times hole diameter.
Maximum wheel width is 1.5 times wheel diameter.
See section 16 for Cutting Fluid Recommendations.
*Wheel recommendations are for wet grinding 0.8- to 2-inch [20 to 50 mm] diameter holes. For LARGER holes—use the same or softer grade wheel. For SMALLER holes—use a harder grade wheel. Also see section 20.2, Grinding Guidelines.

††CAUTION: Toxic Material, refer to National Institute for Occupational Safety and Health (NIOSH) for Precautions.

# 8.8 Internal Grinding

| MATERIAL | HARD-NESS<br>Bhn | CONDITION | WHEEL SPEED<br>fpm<br>m/s | WORK SPEED<br>fpm<br>m/min | INFEED on dia.<br>in/pass<br>mm/pass | TRAVERSE Wheel width per rev. of work | WHEEL IDENTIFI-CATION*<br>ANSI<br>ISO |
|---|---|---|---|---|---|---|---|
| **39. HIGH TEMPERATURE ALLOYS, WROUGHT AND CAST (cont.)**<br>**Cobalt Base, Wrought**<br>AiResist 213<br>Haynes Alloy 25 (L605)<br>Haynes Alloy 188<br>J-1570<br>MAR-M905<br>MAR-M918<br>S-816<br>V-36 | 180 to 230 | Solution Treated | 3000 to 4000 | 50 to 150 | Rough: .0005<br>Finish: .0002 max. | 1/3<br>1/6 | A60JV |
| | | | 15 to 20 | 15 to 46 | Rough: .013<br>Finish: .005 max. | 1/3<br>1/6 | A60JV |
| | 270 to 320 | Solution Treated and Aged | 3000 to 4000 | 50 to 150 | Rough: .0005<br>Finish: .0002 max. | 1/3<br>1/6 | A60JV |
| | | | 15 to 20 | 15 to 46 | Rough: .013<br>Finish: .005 max. | 1/3<br>1/6 | A60JV |
| **Cobalt Base, Cast**<br>AiResist 13<br>AiResist 215<br>FSX-414<br>HS-6<br>HS-21<br>HS-31 (X-40)<br>HOWMET #3<br>MAR-M302<br>MAR-M322<br>MAR-M509<br>NASA Co-W-Re<br>WI-52<br>X-45 | 220 to 290 | As Cast or Cast and Aged | 3000 to 4000 | 50 to 150 | Rough: .0005<br>Finish: .0002 max. | 1/3<br>1/6 | A60JV |
| | | | 15 to 20 | 15 to 46 | Rough: .013<br>Finish: .005 max. | 1/3<br>1/6 | A60JV |
| | 290 to 425 | As Cast or Cast and Aged | 3000 to 4000 | 50 to 150 | Rough: .0005<br>Finish: .0002 max. | 1/3<br>1/6 | A60JV |
| | | | 15 to 20 | 15 to 46 | Rough: .013<br>Finish: .005 max. | 1/3<br>1/6 | A60JV |
| **Iron Base, Wrought**<br>A-286<br>Discaloy<br>Incoloy Alloy 800<br>Incoloy Alloy 800H<br>Incoloy Alloy 801<br>Incoloy Alloy 802<br>N-155<br>V-57<br>W-545<br>16-25-6<br>19-9DL | 180 to 230 | Solution Treated | 3000 to 4000 | 50 to 150 | Rough: .0005<br>Finish: .0002 max. | 1/3<br>1/6 | A60JV |
| | | | 15 to 20 | 15 to 46 | Rough: .013<br>Finish: .005 max. | 1/3<br>1/6 | A60JV |
| | 250 to 320 | Solution Treated and Aged | 3000 to 4000 | 50 to 150 | Rough: .0005<br>Finish: .0002 max. | 1/3<br>1/6 | A60JV |
| | | | 15 to 20 | 15 to 46 | Rough: .013<br>Finish: .005 max. | 1/3<br>1/6 | A60JV |

Maximum hole length is 2.5 times hole diameter.
Maximum wheel width is 1.5 times wheel diameter.
See section 16 for Cutting Fluid Recommendations.
*Wheel recommendations are for wet grinding 0.8- to 2-inch [20 to 50 mm] diameter holes. For LARGER holes—use the same or softer grade wheel. For SMALLER holes—use a harder grade wheel. Also see section 20.2, Grinding Guidelines.

| MATERIAL | HARD-NESS | CONDITION | WHEEL SPEED | WORK SPEED | INFEED on dia. | TRAVERSE Wheel width per rev. of work | WHEEL IDENTIFI-CATION* |
|---|---|---|---|---|---|---|---|
| | Bhn | | fpm / m/s | fpm / m/min | in/pass / mm/pass | | ANSI / ISO |
| **40. REFRACTORY ALLOYS, WROUGHT, CAST, P/M**** **Columbium**** †† ‡‡ C103 C129Y Cb-1Zr Cb-752 FS-85 FS-291 WC-3015 | 170 to 225 | Stress Relieved | 4000 | 50 to 150 | Rough: .001 Finish: .0005 max. | 1/3 1/6 | A60JV |
| | | | 20 | 15 to 46 | Rough: .025 Finish: .013 max. | 1/3 1/6 | A60JV |
| | | | | | | | |
| **Molybdenum**** Mo Mo-50Re TZC TZM | 220 to 290 | Stress Relieved | 4000 | 50 to 150 | Rough: .001 Finish: .0005 max. | 1/3 1/6 | A60JV |
| | | | 20 | 15 to 46 | Rough: .025 Finish: .013 max. | 1/3 1/6 | A60JV |
| **Tantalum**** ASTAR 811C T-111 T-222 Ta-10W Ta-Hf Ta63 | 200 to 250 | Stress Relieved | 4000 | 50 to 150 | Rough: .001 Finish: .0005 max. | 1/3 1/6 | A60JV |
| | | | 20 | 15 to 46 | Rough: .025 Finish: .013 max. | 1/3 1/6 | A60JV |
| **Tungsten**** 85% density 93% density 96% density 100% density | 180 to 320 | Pressed and Sintered, Forged, or Arc Cast | 2000 | 50 to 150 | Rough: .0005 Finish: .0002 max. | 1/3 1/6 | C60HV |
| | | | 10 | 15 to 46 | Rough: .013 Finish: .005 max. | 1/3 1/6 | C60HV |
| **Tungsten - 2 Thoria**** †† | 260 to 320 | Pressed and Sintered | 2000 | 50 to 150 | Rough: .0005 Finish: .0002 max. | 1/3 1/6 | C60HV |
| | | | 10 | 15 to 46 | Rough: .013 Finish: .005 max. | 1/3 1/6 | C60HV |

Maximum hole length is 2.5 times hole diameter.
Maximum wheel width is 1.5 times wheel diameter.
See section 16 for Cutting Fluid Recommendations.
*Wheel recommendations are for wet grinding 0.8- to 2-inch [20 to 50 mm] diameter holes. For LARGER holes—use the same or softer grade wheel. For SMALLER holes—use a harder grade wheel. Also see section 20.2, Grinding Guidelines.

**Due to the brittleness of refractory alloys, cracking, chipping, flaking and breakout tend to occur, particularly on the edges of the machined surfaces.
††CAUTION: Toxic Material, refer to National Institute for Occupational Safety and Health (NIOSH) for Precautions.
‡‡CAUTION: Potential Fire Hazard. Exercise caution in grinding and disposing of swarf.

# 8.8 Internal Grinding

| MATERIAL | HARD-NESS<br>Bhn | CONDITION | WHEEL SPEED<br>fpm<br>m/s | WORK SPEED<br>fpm<br>m/min | INFEED on dia.<br>in/pass<br>mm/pass | TRAVERSE Wheel width per rev. of work | WHEEL IDENTIFI-CATION*<br>ANSI<br>ISO |
|---|---|---|---|---|---|---|---|
| **40. REFRACTORY ALLOYS, WROUGHT, CAST, P/M** (cont.)**<br>**Tungsten Alloys****<br>GE-218<br>W-15Mo<br>W-5Re<br>W-25Re<br>W-25Re-30Mo | 260<br>to<br>320 | As Cast | 2000 | 50<br>to<br>150 | Rough:<br>.0005<br>Finish:<br>.0002 max. | 1/3<br><br>1/6 | C60HV |
| | | | 10 | 15<br>to<br>46 | Rough:<br>.013<br>Finish:<br>.005 max. | 1/3<br><br>1/6 | C60HV |
| **Tungsten Alloys****<br>Gyromet<br>Mallory 2000<br>W-10Ag<br>W-7Ni-4Cu | 290<br>to<br>320 | Pressed<br>and<br>Sintered | 2000 | 50<br>to<br>150 | Rough:<br>.0005<br>Finish:<br>.0002 max. | 1/3<br><br>1/6 | C60HV |
| | | | 10 | 15<br>to<br>46 | Rough:<br>.013<br>Finish:<br>.005 max. | 1/3<br><br>1/6 | C60HV |
| **Tungsten Alloys****<br>Anviloy 1100<br>Anviloy 1150<br>Anviloy 1200 | 290<br>to<br>320 | Pressed<br>and<br>Sintered | 3000 | 50<br>to<br>150 | Rough:<br>.0005<br>Finish:<br>.0002 max. | 1/3<br><br>1/6 | C60HV |
| | | | 15 | 15<br>to<br>46 | Rough:<br>.013<br>Finish:<br>.005 max. | 1/3<br><br>1/6 | C60HV |
| **45. ZIRCONIUM ALLOYS, WROUGHT‡‡**<br>Zr-2%Hf (Grade 11)<br>Zr-0.001%Hf (Grade 21)<br>Zircaloy 2 (Grade 32)<br>Zircaloy 4 (Grade 34) | 140<br>to<br>280 | Rolled,<br>Extruded<br>or Forged | 3000 | 50<br>to<br>150 | Rough:<br>.0005<br>Finish:<br>.0002 max. | 1/3<br><br>1/6 | C80KV |
| | | | 15 | 15<br>to<br>46 | Rough:<br>.013<br>Finish:<br>.005 max. | 1/3<br><br>1/6 | C80KV |
| **47. POWDER METAL ALLOYS§§**<br>**Copper§§** | 50 $R_F$<br>to<br>70 $R_F$ | As Sintered | 5000<br>to<br>6500 | 75<br>to<br>200 | Rough:<br>.002<br>Finish:<br>.0002 max. | 1/3<br><br>1/6 | C46JV |
| | | | 25<br>to<br>33 | 23<br>to<br>60 | Rough:<br>.050<br>Finish:<br>.005 max. | 1/3<br><br>1/6 | C46JV |
| **Brasses§§**<br>CZP-0218-T<br>CZP-0218-U<br>CZP-0218-W<br>90Cu-10Zn<br>90Cu-10Zn-0.5Pb<br>70Cu-30Zn<br>68.5Cu-30Zn-1.5Pb | 35 $R_H$<br>to<br>81 $R_H$ | As Sintered | 5000<br>to<br>6500 | 75<br>to<br>200 | Rough:<br>.002<br>Finish:<br>.0002 max. | 1/3<br><br>1/6 | C46JV |
| | | | 25<br>to<br>33 | 23<br>to<br>60 | Rough:<br>.050<br>Finish:<br>.005 max. | 1/3<br><br>1/6 | C46JV |

Maximum hole length is 2.5 times hole diameter.
Maximum wheel width is 1.5 times wheel diameter.
See section 16 for Cutting Fluid Recommendations.

*Wheel recommendations are for wet grinding 0.8- to 2-inch [20 to 50 mm] diameter holes. For LARGER holes—use the same or softer grade wheel. For SMALLER holes—use a harder grade wheel. Also see section 20.2, Grinding Guidelines.

**Due to the brittleness of refractory alloys, cracking, chipping, flaking and breakout tend to occur, particularly on the edges of the machined surfaces.

‡‡CAUTION: Potential Fire Hazard. Exercise caution in grinding and disposing of swarf.

§§Grinding of low-density parts is not recommended because surface porosity will be reduced or lost.

| MATERIAL | HARDNESS | CONDITION | WHEEL SPEED | WORK SPEED | INFEED on dia. | TRAVERSE Wheel width per rev. of work | WHEEL IDENTIFICATION* |
|---|---|---|---|---|---|---|---|
| | Bhn | | fpm | fpm | in/pass | | ANSI |
| | | | m/s | m/min | mm/pass | | ISO |
| **Bronzes**§§<br>CT-0010-N<br>CT-0010-R<br>CT-0010-S<br>95Cu-5Al<br>77Cu-15Pb-7Sn-1Fe-1C | 30 $R_F$ to 75 $R_F$ | As Sintered | 5000 to 6500 | 75 to 200 | Rough: .002<br>Finish: .0002 max. | 1/3<br>1/6 | C46JV |
| | | | 25 to 33 | 23 to 60 | Rough: .050<br>Finish: .005 max. | 1/3<br>1/6 | C46JV |
| **Copper-Nickel Alloys**§§<br>CZN-1818-T<br>CZN-1818-U<br>CZN-1818-W<br>CZNP-1618-U<br>CZNP-1618-W<br>90Cu-10Ni<br>62Cu-18Ni-18Zn-2Sn | 22 $R_H$ to 100 $R_H$ | As Sintered | 5000 to 6500 | 75 to 200 | Rough: .002<br>Finish: .0002 max. | 1/3<br>1/6 | C46JV |
| | | | 25 to 33 | 23 to 60 | Rough: .050<br>Finish: .005 max. | 1/3<br>1/6 | C46JV |
| **Nickel**§§ | 45 $R_B$ | As Sintered | 5000 to 6500 | 50 to 150 | Rough: .0005<br>Finish: .0002 max. | 1/3<br>1/6 | C60KV |
| | | | 25 to 33 | 15 to 46 | Rough: .013<br>Finish: .005 max. | 1/3<br>1/6 | C60KV |
| **Nickel Alloys**§§<br>67Ni-30Cu-3Fe | 34 $R_B$ to 50 $R_B$ | As Sintered | 5000 to 6500 | 50 to 150 | Rough: .0005<br>Finish: .0002 max. | 1/3<br>1/6 | C60IV |
| | | | 25 to 33 | 15 to 46 | Rough: .013<br>Finish: .005 max. | 1/3<br>1/6 | C60IV |
| **Refractory Metal Base**§§<br>87W-13Cu<br>85W-15Ag<br>74W-26Cu<br>72.5W-27.5Ag<br>65W-35Ag<br>56W + C-44Cu<br>55W-45Cu<br>51W-49Ag<br>50W + C-50Ag | 101 to 260 | As Sintered | 2000 | 50 to 150 | Rough: .0005<br>Finish: .0002 max. | 1/3<br>1/6 | C60HV |
| | | | 10 | 15 to 46 | Rough: .013<br>Finish: .005 max. | 1/3<br>1/6 | C60HV |

Maximum hole length is 2.5 times hole diameter.
Maximum wheel width is 1.5 times wheel diameter.
See section 16 for Cutting Fluid Recommendations.
*Wheel recommendations are for wet grinding 0.8- to 2-inch [20 to 50 mm] diameter holes. For LARGER holes—use the same or softer grade wheel. For SMALLER holes—use a harder grade wheel. Also see section 20.2, Grinding Guidelines.

§§Grinding of low-density parts is not recommended because surface porosity will be reduced or lost.

# 8.8 Internal Grinding

| MATERIAL | HARDNESS<br>Bhn | CONDITION | WHEEL SPEED<br>fpm<br>m/s | WORK SPEED<br>fpm<br>m/min | INFEED on dia.<br>in/pass<br>mm/pass | TRAVERSE Wheel width per rev. of work | WHEEL IDENTIFICATION*<br>ANSI<br>ISO |
|---|---|---|---|---|---|---|---|
| **47. POWDER METAL ALLOYS§§ (cont.)**<br>**Refractory Metal Base§§**<br>61Mo-39Ag<br>50Mo-50Ag | 75 $R_B$<br>to<br>82 $R_B$ | As Sintered | 4000 | 50<br>to<br>150 | Rough:<br>.001<br>Finish:<br>.0005 max. | 1/3<br>1/6 | A60JV |
| | | | 20 | 15<br>to<br>46 | Rough:<br>.025<br>Finish:<br>.013 max. | 1/3<br>1/6 | A60JV |
| **Irons§§**<br>F-0000-N<br>F-0000-P<br>F-0000-R<br>F-0000-S<br>F-0000-T | 50<br>to<br>67<br>500kg | As Sintered | 5000<br>to<br>6500 | 75<br>to<br>200 | Rough:<br>.002<br>Finish:<br>.0002 max. | 1/3<br>1/6 | C46JV<br>or<br>A46KV |
| | | | 25<br>to<br>33 | 23<br>to<br>60 | Rough:<br>.050<br>Finish:<br>.005 max. | 1/3<br>1/6 | C46JV<br>or<br>A46KV |
| **Steels§§**<br>F-0008-P<br>F-0008-S<br>F-0005-S<br>FC-0205-S<br>FC-0208-P<br>FC-0208-S<br>FC-0508-P<br>FC-1000-N<br>FN-0205-S<br>FN-0205-T<br>FN-0405-R<br>FN-0405-S<br>FN-0405-T<br>FX-1005-T<br>FX-2008-T | 101<br>to<br>426 | As Sintered<br>or<br>Heat Treated | 5000<br>to<br>6500 | 75<br>to<br>200 | Rough:<br>.0005<br>Finish:<br>.0002 max. | 1/3<br>1/6 | A54MV |
| | | | 25<br>to<br>33 | 23<br>to<br>60 | Rough:<br>.013<br>Finish:<br>.005 max. | 1/3<br>1/6 | A54MV |
| | | | | | | | |
| | | | | | | | |
| **Stainless Steels§§**<br>SS-303-R<br>SS-304-R<br>SS-316-R<br>SS-410-R | 107<br>to<br>285 | As Sintered<br>or<br>Heat Treated | 5000<br>to<br>6500 | 75<br>to<br>200 | Rough:<br>.0005<br>Finish:<br>.0002 max. | 1/3<br>1/6 | C60KV |
| | | | 25<br>to<br>33 | 23<br>to<br>60 | Rough:<br>.013<br>Finish:<br>.005 max. | 1/3<br>1/6 | C60KV |
| **Aluminum Alloys§§**<br>90.5Al-5Sn-4Cu<br>88Al-5Sn-4Pb-3Cu<br>Al-1Mg-0.6Si-0.25Cu<br>Al-0.6Mg-0.4Si<br>Al-4.4Cu-0.8Si-0.4Mg | 55 $R_H$<br>to<br>98 $R_H$ | Solution Treated<br>and Aged | 5000<br>to<br>6500 | 75<br>to<br>200 | Rough:<br>.003<br>Finish:<br>.0002 max. | 1/3<br>1/6 | C46JV |
| | | | 25<br>to<br>33 | 23<br>to<br>60 | Rough:<br>.075<br>Finish:<br>.005 max. | 1/3<br>1/6 | C46JV |

Maximum hole length is 2.5 times hole diameter.
Maximum wheel width is 1.5 times wheel diameter.
See section 16 for Cutting Fluid Recommendations.
*Wheel recommendations are for wet grinding 0.8- to 2-inch [20 to 50 mm] diameter holes. For LARGER holes—use the same or softer grade wheel. For SMALLER holes—use a harder grade wheel. Also see section 20.2, Grinding Guidelines.

§§Grinding of low-density parts is not recommended because surface porosity will be reduced or lost.

| MATERIAL | HARD-NESS<br>Bhn | CONDITION | WHEEL SPEED<br>fpm<br>m/s | WORK SPEED<br>fpm<br>m/min | INFEED on dia.<br>in/pass<br>mm/pass | TRAVERSE Wheel width per rev. of work | WHEEL IDENTIFI-CATION*<br>ANSI<br>ISO |
|---|---|---|---|---|---|---|---|
| **48. MACHINABLE CARBIDES**<br>Ferro-Tic | $40\ R_C$ to $51\ R_C$ | Annealed | 4000 to 6000 | 50 to 150 | Rough: .0005<br>Finish: .0002 max. | 1/3<br>1/6 | A60JV |
| | | | 20 to 30 | 15 to 46 | Rough: .013<br>Finish: .005 max. | 1/3<br>1/6 | A60JV |
| | $68\ R_C$ to $70\ R_C$ | Hardened and Tempered | — | — | — | — | † |
| | | | — | — | — | — | † |
| **49. CARBIDES**<br>Titanium Carbide<br>Tungsten Carbide | $89\ R_A$ to $94\ R_A$ | — | 5000 to 6000 | 50 to 100 | Rough: .0005<br>Finish: .0002 max. | 1/3<br>1/6 | Rough: C60KV<br>Finish: C120JV‡ |
| | | | 25 to 30 | 5 to 30 | Rough: .013<br>Finish: .005 max. | 1/3<br>1/6 | Rough: C60KV<br>Finish: C120JV‡ |
| **50. FREE MACHINING MAGNETIC ALLOYS**<br>Magnetic Core Iron-FM (up to 2.5% Si) | 185 to 240 | Wrought | 5000 to 6500 | 75 to 200 | Rough: .001<br>Finish: .0002 max. | 1/3<br>1/6 | A60LV |
| | | | 25 to 33 | 23 to 60 | Rough: .025<br>Finish: .005 max. | 1/3<br>1/6 | A60LV |
| Hi Perm 49-FM | 185 to 240 | Wrought | 5000 to 6000 | 50 to 150 | Rough: .0005<br>Finish: .0002 max. | 1/3<br>1/6 | C60KV |
| | | | 25 to 30 | 15 to 46 | Rough: .013<br>Finish: .005 max. | 1/3<br>1/6 | C60KV |
| **51. MAGNETIC ALLOYS**<br>Magnetic Core Iron (up to 4% Si) | 185 to 240 | Wrought | 5000 to 6500 | 75 to 200 | Rough: .001<br>Finish: .0002 max. | 1/3<br>1/6 | A60LV |
| | | | 25 to 33 | 23 to 60 | Rough: .025<br>Finish: .005 max. | 1/3<br>1/6 | A60LV |

Maximum hole length is 2.5 times hole diameter.
Maximum wheel width is 1.5 times wheel diameter.
See section 16 for Cutting Fluid Recommendations.
*Wheel recommendations are for wet grinding 0.8- to 2-inch [20 to 50 mm] diameter holes. For LARGER holes—use the same or softer grade wheel. For SMALLER holes—use a harder grade wheel. Also see section 20.2, Grinding Guidelines.

†Use diamond wheel, see section 8.10.
‡Diamond wheel is preferable, see section 8.10.

# 8.8 Internal Grinding

| MATERIAL | HARDNESS Bhn | CONDITION | WHEEL SPEED fpm / m/s | WORK SPEED fpm / m/min | INFEED on dia. in/pass / mm/pass | TRAVERSE Wheel width per rev. of work | WHEEL IDENTIFICATION* ANSI / ISO |
|---|---|---|---|---|---|---|---|
| **51. MAGNETIC ALLOYS (cont.)**<br>Hi Perm 49<br>HyMu 80 | 185 to 240 | Wrought | 5000 to 6000 | 50 to 150 | Rough: .0005<br>Finish: .0002 max. | 1/3<br>1/6 | C60KV |
| | | | 25 to 30 | 15 to 46 | Rough: .013<br>Finish: .005 max. | 1/3<br>1/6 | C60KV |
| Alnico I<br>Alnico II<br>Alnico III<br>Alnico IV<br>Alnico V<br>Alnico V-7<br>Alnico XII<br>Columax-5<br>Hyflux Alnico V-7 | 45 $R_C$ to 58 $R_C$ | As Cast | 5000 to 6000 | 50 to 150 | Rough: .0005<br>Finish: .0002 max. | 1/3<br>1/6 | A60JV |
| | | | 25 to 30 | 15 to 46 | Rough: .013<br>Finish: .002 max. | 1/3<br>1/6 | A60JV |
| **52. FREE MACHINING CONTROLLED EXPANSION ALLOYS**<br>Invar 36 | 125 to 220 | Annealed or Cold Drawn | 5000 to 6000 | 50 to 150 | Rough: .0005<br>Finish: .0002 max. | 1/3<br>1/6 | C60KV |
| | | | 25 to 30 | 15 to 46 | Rough: .013<br>Finish: .005 max. | 1/3<br>1/6 | C60KV |
| **53. CONTROLLED EXPANSION ALLOYS**<br>Invar<br>Kovar | 125 to 250 | Annealed or Cold Drawn | 5000 to 6000 | 50 to 150 | Rough: .0005<br>Finish: .0002 max. | 1/3<br>1/6 | C60KV |
| | | | 25 to 30 | 15 to 46 | Rough: .013<br>Finish: .005 max. | 1/3<br>1/6 | C60KV |
| **54. CARBONS & GRAPHITES**<br>Mechanical Grades | 40 to 100 Shore | Molded or Extruded | 5000 to 6500 | 50 | Rough: .0005<br>Finish: .0002 max. | 1/3<br>1/6 | C36KV |
| | | | 25 to 33 | 15 | Rough: .013<br>Finish: .005 max. | 1/3<br>1/6 | C36KV |
| Brush Grade Carbon | 8 to 90 Shore | Molded or Extruded | 5000 to 6500 | 50 | Rough: .0005<br>Finish: .0002 max. | 1/3<br>1/6 | C36KV |
| | | | 25 to 33 | 15 | Rough: .013<br>Finish: .005 max. | 1/3<br>1/6 | C36KV |

Maximum hole length is 2.5 times hole diameter.
Maximum wheel width is 1.5 times wheel diameter.
See section 16 for Cutting Fluid Recommendations.
*Wheel recommendations are for wet grinding 0.8- to 2-inch [20 to 50 mm] diameter holes. For LARGER holes—use the same or softer grade wheel. For SMALLER holes—use a harder grade wheel. Also see section 20.2, Grinding Guidelines.

| MATERIAL | HARD-NESS (Bhn) | CONDITION | WHEEL SPEED (fpm / m/s) | WORK SPEED (fpm / m/min) | INFEED on dia. (in/pass / mm/pass) | TRAVERSE Wheel width per rev. of work | WHEEL IDENTIFI-CATION* (ANSI / ISO) |
|---|---|---|---|---|---|---|---|
| Graphite | — | — | 5000 to 6500 | 50 | Rough: .0005 | 1/3 | Rough: C36IV |
| | | | | | Finish: .0002 max. | 1/6 | Finish: C60IV |
| | | | 25 to 33 | 15 | Rough: .013 | 1/3 | Rough: C36IV |
| | | | | | Finish: .005 max. | 1/6 | Finish: C60IV |
| **55. GLASSES & CERAMICS** Glass | 480 to 530 Knoop 100g | — | 5000 to 6500 | 50 to 150 | Rough: .0005 | 1/3 | C80JV |
| | | | | | Finish: .0002 max. | 1/6 | |
| | | | 25 to 33 | 15 to 46 | Rough: .013 | 1/3 | C80JV |
| | | | | | Finish: .005 max. | 1/6 | |
| Porcelain Enamel | — | — | 5000 to 6500 | 50 to 150 | Rough: .0005 | 1/3 | C60KV |
| | | | | | Finish: .0002 max. | 1/6 | |
| | | | 25 to 33 | 15 to 46 | Rough: .013 | 1/3 | C60KV |
| | | | | | Finish: .005 max. | 1/6 | |
| **Ceramics** Alumina (Aluminum Oxide) Alumina-Mullite Aluminum Silicate Beryllia (Beryllium Oxide)†† Magnesia (Magnesium Oxide) Mullite Silicon Carbide Silicon Nitride Thoria (Thorium Oxide)†† Titania (Titanium Oxide) Titanium Diboride†† Zircon (Zirconium Silicate) Zirconia (Zirconium Oxide) | Over 800 Knoop | Fired | 5000 to 6500 | 50 to 100 | Rough: .0005 | 1/3 | Rough: C46JV‡ |
| | | | | | Finish: .0002 max. | 1/6 | Finish: C60JV‡ |
| | | | 25 to 33 | 15 to 30 | Rough: .013 | 1/3 | Rough: C46JV‡ |
| | | | | | Finish: .005 max. | 1/6 | Finish: C60JV‡ |
| **56. PLASTICS** **Thermoplastics** | — | Cast, Molded, Extruded, or Filled and Molded | 5000 to 6500 | 50 to 150 | Rough: .001 | 1/3 | A60KV |
| | | | | | Finish: .0005 max. | 1/6 | |
| | | | 25 to 33 | 15 to 46 | Rough: .025 | 1/3 | A60KV |
| | | | | | Finish: .013 max. | 1/6 | |

Maximum hole length is 2.5 times hole diameter.
Maximum wheel width is 1.5 times wheel diameter.
See section 16 for Cutting Fluid Recommendations.
*Wheel recommendations are for wet grinding 0.8- to 2-inch [20 to 50 mm] diameter holes. For LARGER holes—use the same or softer grade wheel. For SMALLER holes—use a harder grade wheel. Also see section 20.2, Grinding Guidelines.

‡Diamond wheel is preferable, see section 8.10.
††CAUTION: Toxic Material, refer to National Institute for Occupational Safety and Health (NIOSH) for Precautions.

# 8.8 Internal Grinding

| MATERIAL | HARD-NESS<br>Bhn | CONDITION | WHEEL SPEED<br>fpm<br>m/s | WORK SPEED<br>fpm<br>m/min | INFEED on dia.<br>in/pass<br>mm/pass | TRAVERSE Wheel width per rev. of work | WHEEL IDENTIFI-CATION*<br>ANSI<br>ISO |
|---|---|---|---|---|---|---|---|
| **56. PLASTICS (cont.)**<br>**Thermosetting Plastics** | — | Cast | 5000 to 6500 | 50 to 150 | Rough: .001<br>Finish: .0005 max. | 1/3<br>1/6 | C46IV |
| | | | 25 to 33 | 15 to 46 | Rough: .025<br>Finish: .013 max. | 1/3<br>1/6 | C46IV |
| | — | Molded | 5000 to 6500 | 50 to 150 | Rough: .001<br>Finish: .0005 max. | 1/3<br>1/6 | C46KV |
| | | | 25 to 33 | 15 to 46 | Rough: .025<br>Finish: .013 max. | 1/3<br>1/6 | C46KV |
| | — | Laminated | 5000 to 6500 | 50 to 150 | Rough: .001<br>Finish: .0005 max. | 1/3<br>1/6 | C36LV |
| | | | 25 to 33 | 15 to 46 | Rough: .025<br>Finish: .013 max. | 1/3<br>1/6 | C36LV |
| **58. FLAME (THERMAL) SPRAYED MATERIALS**<br>**Sprayed Carbides**<br>Chromium Carbide<br>Chromium Carbide-Cobalt Blend<br>Columbium Carbide<br>Tantalum Carbide<br>Titanium Carbide<br>Tungsten Carbide<br>Tungsten Carbide-Cobalt<br>Tungsten Carbide (Cobalt)-Nickel Alloy Blend | — | — | 5500 to 6500 | 50 to 100 | Rough: .0005<br>Finish: .0002 max. | 1/3<br>1/6 | C80HV§ |
| | | | 28 to 33 | 15 to 30 | Rough: .013<br>Finish: .005 max. | 1/3<br>1/6 | C80HV§ |

Maximum hole length is 2.5 times hole diameter.
Maximum wheel width is 1.5 times wheel diameter.
See section 16 for Cutting Fluid Recommendations.
*Wheel recommendations are for wet grinding 0.8- to 2-inch [20 to 50 mm] diameter holes. For LARGER holes—use the same or softer grade wheel. For SMALLER holes—use a harder grade wheel. Also see section 20.2, Grinding Guidelines.

§Diamond wheel is preferable for flame sprayed carbides and ceramics; see section 8.10.

| MATERIAL | HARD-NESS Bhn | CONDITION | WHEEL SPEED fpm / m/s | WORK SPEED fpm / m/min | INFEED on dia. in/pass / mm/pass | TRAVERSE Wheel width per rev. of work | WHEEL IDENTIFI-CATION* ANSI / ISO |
|---|---|---|---|---|---|---|---|
| **Inorganic Coating Materials**<br>Alumina (Pure)<br>Alumina (Grey) containing Titania<br>Alumina, Nickel-Aluminide Blends<br>Barium Titanate<br>Boron††<br>Calcium Titanate<br>Calcium Zirconate<br>Chromium Disilicide††<br>Chromium Oxide††<br>Cobalt (40%), Zirconia Blend<br>Columbium (Niobium)††<br>Glass (Kovar sealing)††<br>Hexaboron Silicide††<br>Magnesia Alumina Spinel<br>Magnesium Zirconate<br>Molybdenum Disilicide<br>Mullite<br>Nickel (40%), Alumina Blend<br>Nickel Oxide<br>Rare Earth Oxides<br>Tantalum<br>Titania (50%), Alumina Blend<br>Titanium Oxide<br>Tungsten<br>Yttrium Zirconate<br>Zirconia (Lime Stabilized)<br>Zirconia, Nickel-Aluminide Blends<br>Zirconium Oxide (Hafnia Free, Lime Stabilized)<br>Zirconium Silicate | — | — | 5500 to 6500 | 50 to 100 | Rough: .0005<br>Finish: .0002 max. | 1/3<br>1/6 | C80IV§ |
| | | | 28 to 33 | 15 to 30 | Rough: .013<br>Finish: .005 max. | 1/3<br>1/6 | C80IV§ |
| **Sprayed Metals (Group I)**<br>Co-Cr-B Alloy (Self Fluxing)<br>Ni-Cr-B Alloy (Self Fluxing)<br>Nickel Chrome Steel (Special)<br>Stainless Steel | — | — | 5500 to 6500 | 50 to 150 | Rough: .0005<br>Finish: .0002 max. | 1/3<br>1/6 | C60JV |
| | | | 28 to 33 | 15 to 46 | Rough: .013<br>Finish: .005 max. | 1/3<br>1/6 | C60JV |
| **Sprayed Metals (Group II)**<br>Bronze<br>Chromium<br>Cobalt<br>Nickel<br>Molybdenum<br>Monel | — | — | 5500 to 6500 | 50 to 150 | Rough: .0005<br>Finish: .0002 max. | 1/3<br>1/6 | C80KV |
| | | | 28 to 33 | 15 to 46 | Rough: .013<br>Finish: .005 max. | 1/3<br>1/6 | C80KV |
| **Sprayed Metals (Group III)**<br>Carbon Steel<br>Iron<br>Precipitation Hardening Steel | — | — | 5500 to 6500 | 50 to 150 | Rough: .0005<br>Finish: .0002 max. | 1/3<br>1/6 | C60LV |
| | | | 28 to 33 | 15 to 46 | Rough: .013<br>Finish: .005 max. | 1/3<br>1/6 | C60LV |

Maximum hole length is 2.5 times hole diameter.
Maximum wheel width is 1.5 times wheel diameter.
See section 16 for Cutting Fluid Recommendations.
*Wheel recommendations are for wet grinding 0.8- to 2-inch [20 to 50 mm] diameter holes. For LARGER holes—use the same or softer grade wheel. For SMALLER holes—use a harder grade wheel. Also see section 20.2, Grinding Guidelines.

§Diamond wheel is preferable for flame sprayed carbides and ceramics; see section 8.10.
††CAUTION: Toxic Material, refer to National Institute for Occupational Safety and Health (NIOSH) for Precautions.

# 8.8 Internal Grinding

| MATERIAL | HARD-NESS | CONDITION | WHEEL SPEED | WORK SPEED | INFEED on dia. | TRAVERSE Wheel width per rev. of work | WHEEL IDENTIFI-CATION* |
|---|---|---|---|---|---|---|---|
| | Bhn | | fpm<br>m/s | fpm<br>m/min | in/pass<br>mm/pass | | ANSI<br>ISO |
| **59. PLATED MATERIALS**<br>Chromium Plate | — | — | 3000 to 4000 | 50 to 150 | Rough: .0005<br>Finish: .0002 max. | 1/3<br>1/6 | A80HV |
| | | | 15 to 20 | 15 to 46 | Rough: .013<br>Finish: .005 max. | 1/3<br>1/6 | A80HV |
| Nickel Plate<br>Silver Plate | — | — | 5000 to 6500 | 50 to 150 | Rough: .0005<br>Finish: .0002 max. | 1/3<br>1/6 | A60KV |
| | | | 25 to 33 | 15 to 46 | Rough: .013<br>Finish: .005 max. | 1/3<br>1/6 | A60KV |
| **61. RUBBER** | Soft | — | 5000 to 6500 | 50 to 150 | Rough: .001<br>Finish: .0005 max. | 1/3<br>1/6 | C30IV |
| | | | 25 to 33 | 15 to 46 | Rough: .025<br>Finish: .013 max. | 1/3<br>1/6 | C30IV |
| | Hard | — | 5000 to 6500 | 50 to 150 | Rough: .001<br>Finish: .0005 max. | 1/3<br>1/6 | C36JV |
| | | | 25 to 33 | 15 to 46 | Rough: .025<br>Finish: .013 max. | 1/3<br>1/6 | C36JV |

Maximum hole length is 2.5 times hole diameter.
Maximum wheel width is 1.5 times wheel diameter.
See section 16 for Cutting Fluid Recommendations.
*Wheel recommendations are for wet grinding 0.8- to 2-inch [20 to 50 mm] diameter holes. For LARGER holes—use the same or softer grade wheel. For SMALLER holes—use a harder grade wheel. Also see section 20.2, Grinding Guidelines.

| MATERIAL | HARDNESS Bhn | CONDITION | WHEEL SPEED fpm / m/s | WORK SPEED fpm / m/min | INFEED on dia. in/pass / mm/pass | TRAVERSE Wheel width per rev. of work | WHEEL IDENTIFICATION ANSI / ANSI |
|---|---|---|---|---|---|---|---|
| **1. FREE MACHINING CARBON STEELS, WROUGHT**<br>**Low Carbon Resulfurized**<br>1108 1118<br>1109 1119<br>1110 1211<br>1115 1212<br>1116 1213<br>1117 1215 | Over 50 $R_c$ | Carburized and/or Quenched and Tempered | 5,000 to 10,000 | 80 to 150 | Rough: .0005<br>Finish: .0002 max. | 1/3<br>1/8 | B120TB<br>B180TV |
| | | | 25 to 51 | 24 to 46 | Rough: .013<br>Finish: .005 max. | 1/3<br>1/8 | B120TB<br>B180TV |
| **Medium Carbon Resulfurized**<br>1132 1140 1145<br>1137 1141 1146<br>1139 1144 1151<br>**Low Carbon Leaded**<br>10L18 12L13 12L15<br>11L17 12L14<br>**Medium Carbon Leaded**<br>10L45 11L37 11L44<br>10L50 11L41 | | | | | | | |
| **2. CARBON STEELS, WROUGHT**<br>**Low Carbon**<br>1005 1012 1019 1026<br>1006 1013 1020 1029<br>1008 1015 1021 1513<br>1009 1016 1022 1518<br>1010 1017 1023 1522<br>1011 1018 1025 | Over 50 $R_c$ | Carburized and/or Quenched and Tempered | 5,000 to 10,000 | 80 to 150 | Rough: .0005<br>Finish: .0002 max. | 1/3<br>1/8 | B120TB<br>B180TV |
| | | | 25 to 51 | 24 to 46 | Rough: .013<br>Finish: .005 max. | 1/3<br>1/8 | B120TB<br>B180TV |
| **Medium Carbon**<br>1030 1042 1053 1541<br>1033 1043 1055 1547<br>1035 1044 1524 1548<br>1037 1045 1525 1551<br>1038 1046 1526 1552<br>1039 1049 1527<br>1040 1050 1536<br>**High Carbon**<br>1060 1074 1085 1566<br>1064 1075 1086 1572<br>1065 1078 1090<br>1069 1080 1095<br>1070 1084 1561 | | | | | | | |

See section 16 for Cutting Fluid Recommendations.
See section 20.1 for additional information.

# 8.9 Internal Grinding

## Cubic Boron Nitride Wheels

| MATERIAL | HARD-NESS Bhn | CONDITION | WHEEL SPEED fpm m/s | WORK SPEED fpm m/min | INFEED on dia. in/pass mm/pass | TRAVERSE Wheel width per rev. of work | WHEEL IDENTIFI-CATION ANSI ANSI |
|---|---|---|---|---|---|---|---|
| **4. FREE MACHINING ALLOY STEELS, WROUGHT**<br>**Medium Carbon Resulfurized**<br>4140<br>4140Se<br>4142Te<br>4145Se<br>4147Te<br>4150<br>**Medium and High Carbon Leaded**<br>41L30 41L47 51L32 86L40<br>41L40 41L50 52L100<br>41L45 43L40 86L20 | Over 50 $R_C$ | Carburized and/or Quenched and Tempered | 5,000 to 10,000 | 80 to 150 | Rough: .0005<br>Finish: .0002 max. | 1/3<br>1/8 | B120TB<br>B180TV |
| | | | 25 to 51 | 24 to 46 | Rough: .013<br>Finish: .005 max. | 1/3<br>1/8 | T120TB<br>T180TV |
| **5. ALLOY STEELS, WROUGHT**<br>**Low Carbon**<br>4012 4615 4817 8617<br>4023 4617 4820 8620<br>4024 4620 5015 8622<br>4118 4621 5115 8822<br>4320 4718 5120 9310<br>4419 4720 6118 94B15<br>4422 4815 8115 94B17<br>**Medium Carbon**<br>1330 4145 5132 8640<br>1335 4147 5135 8642<br>1340 4150 5140 8645<br>1345 4161 5145 86B45<br>4027 4340 5147 8650<br>4028 4427 5150 8655<br>4032 4626 5155 8660<br>4037 50B40 5160 8740<br>4042 50B44 51B60 8742<br>4047 5046 6150 9254<br>4130 50B46 81B45 9255<br>4135 50B50 8625 9260<br>4137 5060 8627 94B30<br>4140 50B60 8630<br>4142 5130 8637<br>**High Carbon**<br>50100 51100 52100 M-50 | Over 50 $R_C$ | Carburized and/or Quenched and Tempered | 5,000 to 10,000 | 80 to 150 | Rough: .0005<br>Finish: .0002 max. | 1/3<br>1/8 | B120TB<br>B180TV |
| | | | 25 to 51 | 24 to 46 | Rough: .013<br>Finish: .005 max. | 1/3<br>1/8 | B120TB<br>B180TV |

See section 16 for Cutting Fluid Recommendations.
See section 20.1 for additional information.

| MATERIAL | HARDNESS Bhn | CONDITION | WHEEL SPEED fpm m/s | WORK SPEED fpm m/min | INFEED on dia. in/pass mm/pass | TRAVERSE Wheel width per rev. of work | WHEEL IDENTIFICATION ANSI ANSI |
|---|---|---|---|---|---|---|---|
| **6. HIGH STRENGTH STEELS, WROUGHT**<br>300M H11<br>4330V H13<br>4340 HP 9-4-20<br>4340Si HP 9-4-25<br>98BV40 HP 9-4-30<br>D6ac HP 9-4-45 | Over 50 $R_C$ | Quenched and Tempered | 5,000 to 10,000 | 80 to 150 | Rough: .0005 | 1/3 | B120TB |
| | | | | | Finish: .0002 max. | 1/8 | B180TV |
| | | | 25 to 51 | 24 to 46 | Rough: .013 | 1/3 | B120TB |
| | | | | | Finish: .005 max. | 1/8 | B180TV |
| **7. MARAGING STEELS, WROUGHT**<br>120 Grade<br>180 Grade<br>200 Grade<br>250 Grade<br>300 Grade<br>350 Grade<br>HY230<br>ASTM A538: Grades A, B, C | Over 50 $R_C$ | Maraged | 5,000 to 10,000 | 80 to 150 | Rough: .0005 | 1/3 | B120TB |
| | | | | | Finish: .0002 max. | 1/8 | B180TV |
| | | | 25 to 51 | 24 to 46 | Rough: .013 | 1/3 | B120TB |
| | | | | | Finish: .005 max. | 1/8 | B180TV |
| | | | | | | | |
| | | | | | | | |
| **8. TOOL STEELS, WROUGHT**<br>A2 H14 M10 P20<br>A3 H19 M30 P21<br>A4 H21 M33 S1<br>A6 H22 M34 S2<br>A7 H23 M36 S5<br>A8 H24 M41 S6<br>A9 H25 M42 S7<br>A10 H26 M43 T1<br>D2 H42 M44 T2<br>D3 L2 M46 T4<br>D4 L6 M47 T5<br>D5 L7 O1 T6<br>D7 M1 O2 T8<br>F1 M2 O6 T15<br>F2 M3-1 O7 W1<br>H10 M3-2 P2 W2<br>H11 M4 P4 W5<br>H12 M6 P5<br>H13 M7 P6<br>SAE J438b: Types W108, W109, W110, W112, W209, W210, W310 | Over 50 $R_C$ | Quenched and Tempered | 5,000 to 10,000 | 80 to 150 | Rough: .0005 | 1/3 | B120TB |
| | | | | | Finish: .0002 max. | 1/8 | B180TV |
| | | | 25 to 51 | 24 to 46 | Rough: .013 | 1/3 | B120TB |
| | | | | | Finish: .005 max. | 1/8 | B180TV |
| | | | | | | | |
| | | | | | | | |
| | | | | | | | |
| | | | | | | | |

See section 16 for Cutting Fluid Recommendations.
See section 20.1 for additional information.

# 8.9 Internal Grinding
## Cubic Boron Nitride Wheels

| MATERIAL | HARD-NESS<br>Bhn | CONDITION | WHEEL SPEED<br>fpm<br>m/s | WORK SPEED<br>fpm<br>m/min | INFEED on dia.<br>in/pass<br>mm/pass | TRAVERSE Wheel width per rev. of work | WHEEL IDENTIFI-CATION<br>ANSI<br>ANSI |
|---|---|---|---|---|---|---|---|
| **9. NITRIDING STEELS, WROUGHT**<br>Nitralloy 125<br>Nitralloy 135<br>Nitralloy 135 Mod.<br>Nitralloy 225<br>Nitralloy 230<br>Nitralloy EZ<br>Nitralloy N<br>Nitrex 1 | 60 Rc to 65 Rc | Nitrided | 5,000 to 10,000 | 80 to 150 | Rough: .0005<br>Finish: .0002 max. | 1/3<br>1/8 | B120TB<br>B180TV |
| | | | 25 to 51 | 24 to 46 | Rough: .013<br>Finish: .005 max. | 1/3<br>1/8 | B120TB<br>B180TV |
| | | | | | | | |
| | | | | | | | |
| **13. STAINLESS STEELS, WROUGHT**<br>**Martensitic**<br>403 440A<br>410 440B<br>414 440C<br>420 501<br>422 502<br>431<br>Greek Ascoloy | Over 275 | Quenched and Tempered | 5,000 to 10,000 | 80 to 150 | Rough: .0005<br>Finish: .0002 max. | 1/3<br>1/8 | B120TB<br>B180TV |
| | | | 25 to 51 | 24 to 46 | Rough: .013<br>Finish: .005 max. | 1/3<br>1/8 | B120TB<br>B180TV |
| | | | | | | | |
| | | | | | | | |
| **15. CARBON STEELS, CAST**<br>**Low Carbon**<br>ASTM A426: Grade CP1<br>1010<br>1020<br>**Medium Carbon**<br>ASTM A352: Grades LCA, LCB, LCC<br>ASTM A356: Grade 1<br>1030 1040 1050 | Over 50 Rc | Carburized and/or Quenched and Tempered | 5,000 to 10,000 | 80 to 150 | Rough: .0005<br>Finish: .0002 max. | 1/3<br>1/8 | B120TB<br>B180TV |
| | | | 25 to 51 | 24 to 46 | Rough: .013<br>Finish: .005 max. | 1/3<br>1/8 | B120TB<br>B180TV |
| | | | | | | | |
| | | | | | | | |

See section 16 for Cutting Fluid Recommendations.
See section 20.1 for additional information.

| MATERIAL | HARD-NESS | CONDITION | WHEEL SPEED | WORK SPEED | INFEED on dia. | TRAVERSE Wheel width per rev. of work | WHEEL IDENTIFI-CATION |
|---|---|---|---|---|---|---|---|
| | Bhn | | fpm / m/s | fpm / m/min | in/pass / mm/pass | | ANSI / ANSI |
| **16. ALLOY STEELS, CAST** **Low Carbon** ASTM A217: Grade WC9; ASTM A352: Grades LC3, LC4; ASTM A426: Grades CP2, CP5, CP5b, CP11, CP12, CP15, CP21, CP22; 1320 2320 4120 8020 2315 4110 4320 8620 | Over 50 $R_c$ | Carburized and/or Quenched and Tempered | 5,000 to 10,000 | 80 to 150 | Rough: .0005 | 1/3 | B120TB |
| | | | | | Finish: .0002 max. | 1/8 | B180TV |
| | | | 25 to 51 | 24 to 46 | Rough: .013 | 1/3 | B120TB |
| | | | | | Finish: .005 max. | 1/8 | B180TV |
| **Medium Carbon** ASTM A27: Grades N1, N2, U-60-30, 60-30, 65-35, 70-36, 70-40; ASTM A148: Grades 80-40, 80-50, 90-60, 105-85, 120-95, 150-125, 175-145; ASTM A216: Grades WCA, WCB, WCC; ASTM A217: Grades WC1, WC4, WC5, WC6; ASTM A352: Grades LC1, LC2, LC2-1; ASTM A356: Grades 2, 5, 6, 8, 9, 10; ASTM A389: Grades C23, C24; ASTM A486: Classes 70, 90, 120; ASTM A487: Classes 1N, 2N, 4N, 6N, 8N, 9N, 10N, DN, 1Q, 2Q, 4Q, 4QA, 6Q, 7Q, 8Q, 9Q, 10Q; 1330 4130 80B30 8640 1340 4140 8040 9525 2325 4330 8430 9530 2330 4340 8440 9535 4125 8030 8630 | | | | | | | |
| **17. TOOL STEELS, CAST** ASTM A597: Grades CA-2, CD-2, CD-5, CH-12, CH-13, CO-1, CS-5 | Over 50 $R_c$ | Quenched and Tempered | 5,000 to 10,000 | 80 to 150 | Rough: .0005 | 1/3 | B120TB |
| | | | | | Finish: .0002 max. | 1/8 | B180TV |
| | | | 25 to 51 | 24 to 46 | Rough: .013 | 1/3 | B120TB |
| | | | | | Finish: .005 max. | 1/8 | B180TV |

See section 16 for Cutting Fluid Recommendations.
See section 20.1 for additional information.

| MATERIAL | HARDNESS Bhn | CONDITION | WHEEL SPEED fpm / m/s | WORK SPEED fpm / m/min | INFEED on dia. in/pass / mm/pass | TRAVERSE Wheel width per rev. of work | WHEEL IDENTIFICATION ANSI / ANSI |
|---|---|---|---|---|---|---|---|
| **48. MACHINABLE CARBIDES**<br>Ferro-Tic | 68 $R_C$ to 72 $R_C$ | Hardened and Tempered | 1200 to 1500 | 30 to 50 | Rough: .001 | 3 | ASD100-R100-B* |
| | | | | | Finish: .0003 max. | 6 | ASD150-R100-B* |
| | | | 6 to 7.5 | 9 to 15 | Rough: .025 | 3 | ASD100-R100-B* |
| | | | | | Finish: .008 max. | 6 | ASD150-R100-B* |
| **49. CARBIDES**<br>Titanium Carbide<br>Tungsten Carbide | 89 $R_A$ to 94 $R_A$ | — | 5000 to 6000 | 50 to 100 | Rough: .001 | 3 | D150-R75-B |
| | | | | | Finish: .0002 max. | 6 | D220-R75-B |
| | | | 25 to 30 | 15 to 30 | Rough: .025 | 3 | D150-R75-B |
| | | | | | Finish: .005 max. | 6 | D220-R75-B |
| **51. MAGNETIC ALLOYS**<br>Magnetic Core Iron (up to 4% Si)<br>Hi Perm 49<br>HyMu 80 | 185 to 240 | Wrought | 5000 to 6000 | 50 to 60 | Rough: .003 | 3 | D100-N100-M |
| | | | | | Finish: .001 max. | 6 | D220-N100-M |
| | | | 25 to 30 | 15 to 18 | Rough: .075 | 3 | D100-N100-M |
| | | | | | Finish: .025 max. | 6 | D220-N100-M |
| Alnico I<br>Alnico II<br>Alnico III<br>Alnico IV<br>Alnico V<br>Alnico V-7<br>Alnico XII<br>Columax-5<br>Hyflux Alnico V-7 | 45 $R_C$ to 58 $R_C$ | As Cast | 5000 to 6000 | 30 to 40 | Rough: .003 | 3 | D100-N75-M |
| | | | | | Finish: .001 max. | 6 | D220-N75-M |
| | | | 25 to 30 | 9 to 12 | Rough: .075 | 3 | D100-N75-M |
| | | | | | Finish: .025 max. | 6 | D220-N75-M |
| **55. GLASSES & CERAMICS**<br>Machinable Glass-Ceramic (MACOR) | 250 Knoop 100g | Cast | 5000 to 5500 | 50 to 100 | Rough: .003 | 3 | D100-N100-M |
| | | | | | Finish: .001 max. | 6 | D120-N100-M |
| | | | 25 to 28 | 15 to 30 | Rough: .075 | 3 | D100-N100-M |
| | | | | | Finish: .025 max. | 6 | D120-N100-M |
| PYROCERAM | 700 Knoop 100g | — | 4000 to 4500 | 50 to 100 | Rough: .001 | 3 | D100-N100-M |
| | | | | | Finish: .0002 max. | 6 | D220-N100-M |
| | | | 20 to 23 | 15 to 30 | Rough: .025 | 3 | D100-N100-M |
| | | | | | Finish: .005 max. | 6 | D220-N100-M |

See section 16 for Cutting Fluid Recommendations.
See section 20.1 for additional information.

*Metal-coated diamond.

# 8.10 Internal Grinding
## Diamond Wheels

| MATERIAL | HARDNESS Bhn | CONDITION | WHEEL SPEED fpm / m/s | WORK SPEED fpm / m/min | INFEED on dia. in/pass / mm/pass | TRAVERSE Wheel width per rev. of work | WHEEL IDENTIFICATION ANSI / ANSI |
|---|---|---|---|---|---|---|---|
| **55. GLASSES & CERAMICS (cont.)** Glass | 480 to 530 Knoop 100g | — | 4000 to 4500 | 50 to 100 | Rough: .001 | 3 | D100-R50-M |
| | | | | | Finish: .0002 max. | 6 | D220-R50-M |
| | | | 20 to 23 | 15 to 30 | Rough: .025 | 3 | D100-R50-M |
| | | | | | Finish: .005 max. | 6 | D220-R50-M |
| Mica | — | — | 5000 to 6000 | 50 to 150 | Rough: .003 | 3 | D100-N75-B |
| | | | | | Finish: .001 max. | 6 | D220-N75-B |
| | | | 25 to 30 | 15 to 30 | Rough: .075 | 3 | D100-N75-B |
| | | | | | Finish: .025 max. | 6 | D220-N75-B |
| Porcelain Enamel | — | — | 4000 to 4500 | 50 to 100 | Rough: .001 | 3 | D100-R100-M |
| | | | | | Finish: .0005 max. | 6 | D220-R100-M |
| | | | 20 to 23 | 15 to 30 | Rough: .025 | 3 | D100-R100-M |
| | | | | | Finish: .013 max. | 6 | D220-R100-M |
| **Ceramics** Alumina (Aluminum Oxide) Alumina-Mullite Aluminum Silicate Beryllia (Beryllium Oxide)†† Magnesia (Magnesium Oxide) Mullite Silicon Carbide Silicon Nitride Thoria (Thorium Oxide)†† Titania (Titanium Oxide) Titanium Diboride†† Zircon (Zirconium Silicate) Zirconia (Zirconium Oxide) | Over 800 Knoop | Fired | 4000 to 4500 | 50 to 100 | Rough: .002 | 3 | D80-N100-M |
| | | | | | Finish: .0005 max. | 6 | D100-N100-M |
| | | | 20 to 23 | 15 to 30 | Rough: .050 | 3 | D80-N100-M |
| | | | | | Finish: .013 max. | 6 | D100-N100-M |

See section 16 for Cutting Fluid Recommendations.
See section 20.1 for additional information.

††CAUTION: Toxic Material, refer to National Institute for Occupational Safety and Health (NIOSH) for Precautions.

| MATERIAL | HARD-NESS | CONDITION | WHEEL SPEED | WORK SPEED | INFEED on dia. | TRAVERSE Wheel width per rev. of work | WHEEL IDENTIFI-CATION |
|---|---|---|---|---|---|---|---|
| | Bhn | | fpm / m/s | fpm / m/min | in/pass / mm/pass | | ANSI / ANSI |
| **58. FLAME (THERMAL) SPRAYED MATERIALS** **Sprayed Carbides** Chromium Carbide, Chromium Carbide-Cobalt Blend, Columbium Carbide, Tantalum Carbide, Titanium Carbide, Tungsten Carbide, Tungsten Carbide-Cobalt, Tungsten Carbide (Cobalt)-Nickel Alloy Blend | — | — | 5000 to 6000 | 50 to 100 | Rough: .002 | 3 | D150-N100-B |
| | | | | | Finish: .0005 max. | 6 | D220-N100-B |
| | | | 25 to 30 | 15 to 30 | Rough: .050 | 3 | D150-N100-B |
| | | | | | Finish: .013 max. | 6 | D220-N100-B |
| **Inorganic Coating Materials** Alumina (Pure), Alumina (Grey) containing Titania, Alumina, Nickel-Aluminide Blends, Barium Titanate, Boron††, Calcium Titanate, Calcium Zirconate, Chromium Disilicide††, Chromium Oxide††, Cobalt (40%), Zirconia Blend, Columbium (Niobium)††, Glass (Kovar sealing)††, Hexaboron Silicide††, Magnesia Alumina Spinel, Magnesium Zirconate, Molybdenum Disilicide, Mullite, Nickel (40%), Alumina Blend, Nickel Oxide, Rare Earth Oxides, Tantalum, Titania (50%), Alumina Blend, Titanium Oxide, Tungsten, Yttrium Zirconate, Zirconia (Lime Stabilized), Zirconia, Nickel-Aluminide Blends, Zirconium Oxide (Hafnia Free, Lime Stabilized), Zirconium Silicate | — | — | 5000 to 6000 | 50 to 100 | Rough: .002 | 3 | D120-N100-B |
| | | | | | Finish: .0005 max. | 6 | D220-N100-B |
| | | | 25 to 30 | 15 to 30 | Rough: .050 | 3 | D120-N100-B |
| | | | | | Finish: .013 max. | 6 | D220-N100-B |

See section 16 for Cutting Fluid Recommendations.
See section 20.1 for additional information.

††CAUTION: Toxic Material, refer to National Institute for Occupational Safety and Health (NIOSH) for Precautions.

| MATERIAL | HARDNESS Bhn | CONDITION | WHEEL SPEED fpm / m/s | THRUFEED OF WORK* in/min / m/min | INFEED on dia. in/pass / mm/pass | WHEEL IDENTIFICATION† ANSI / ISO |
|---|---|---|---|---|---|---|
| **1. FREE MACHINING CARBON STEELS, WROUGHT**<br>**Low Carbon Resulfurized**<br>1108 1109 1110 1115 1116 1117 1118 1119 1211 1212 1213 1215<br>**Medium Carbon Resulfurized**<br>1132 1137 1139 1140 1141 1144 1145 1146 1151<br>**Low Carbon Leaded**<br>10L18 11L17 12L13 12L14 12L15<br>**Medium Carbon Leaded**<br>10L45 10L50 11L37 11L41 11L44 | 50 $R_C$ max. | Hot Rolled, Normalized, Annealed, Cold Drawn or Quenched and Tempered | 5500 to 6500 | 50 to 150 | Rough: .005 Finish: .0015 max. | A60MV |
| | | | 28 to 33 | 1.3 to 3.8 | Rough: .13 Finish: .038 max. | A60MV |
| | Over 50 $R_C$ | Carburized and/or Quenched and Tempered | 5500 to 6500 | 50 to 150 | Rough: .005 Finish: .0015 max. | A60LV |
| | | | 28 to 33 | 1.3 to 3.8 | Rough: .13 Finish: .038 max. | A60LV |
| **2. CARBON STEELS, WROUGHT**<br>**Low Carbon**<br>1005 1006 1008 1009 1010 1011 1012 1013 1015 1016 1017 1018 1019 1020 1021 1022 1023 1025 1026 1029 1513 1518 1522<br>**Medium Carbon**<br>1030 1033 1035 1037 1038 1039 1040 1042 1043 1044 1045 1046 1049 1050 1053 1055 1524 1525 1526 1527 1536 1541 1547 1548 1551 1552<br>**High Carbon**<br>1060 1064 1065 1069 1070 1074 1075 1078 1080 1084 1085 1086 1090 1095 1561 1566 1572 | 50 $R_C$ max. | Hot Rolled, Normalized, Annealed, Cold Drawn or Quenched and Tempered | 5500 to 6500 | 50 to 150 | Rough: .005 Finish: .0015 max. | A60MV |
| | | | 28 to 33 | 1.3 to 3.8 | Rough: .13 Finish: .038 max. | A60MV |
| | Over 50 $R_C$ | Carburized and/or Quenched and Tempered | 5500 to 6500 | 50 to 150 | Rough: .005 Finish: .0015 max. | A60LV |
| | | | 28 to 33 | 1.3 to 3.8 | Rough: .13 Finish: .038 max. | A60LV |
| **3. CARBON AND FERRITIC ALLOY STEELS (HIGH TEMPERATURE SERVICE)**<br>ASTM A369: Grades FPA, FPB, FP1, FP2, FP3b, FP5, FP7, FP9, FP11, FP12, FP21, FP22 | 150 to 200 | As Forged, Annealed or Normalized and Tempered | 5500 to 6500 | 50 to 150 | Rough: .005 Finish: .0015 max. | A60MV |
| | | | 28 to 33 | 1.3 to 3.8 | Rough: .13 Finish: .038 max. | A60MV |

See section 16 for Cutting Fluid Recommendations.
*See section 8.12 for Work Traverse Rates. As recommended starting conditions—use a regulating wheel angle with a positive inclination of 3° and a regulating wheel speed of 25 to 40 rpm.
†Wheel recommendations are for wet grinding 0.8- to 2-inch [20 to 50 mm] diameter work. For LARGER diameter work—use a softer grade and/or coarser grit wheel. For SMALLER diameter work—use a harder grade wheel. Also see section 20.2, Grinding Guidelines.

# 8.11 Centerless Grinding

| MATERIAL | HARD-NESS<br>Bhn | CONDITION | WHEEL SPEED<br>fpm<br>m/s | THRUFEED OF WORK*<br>in/min<br>m/min | INFEED on dia.<br>in/pass<br>mm/pass | WHEEL IDENTIFI-CATION†<br>ANSI<br>ISO |
|---|---|---|---|---|---|---|
| **4. FREE MACHINING ALLOY STEELS, WROUGHT**<br>**Medium Carbon Resulfurized**<br>4140<br>4140Se<br>4142Te<br>4145Se<br>4147Te<br>4150<br>**Medium and High Carbon Leaded**<br>41L30 41L47 51L32 86L40<br>41L40 41L50 52L100<br>41L45 43L40 86L20 | 50 $R_C$ max. | Hot Rolled, Normalized, Annealed, Cold Drawn or Quenched and Tempered | 5500 to 6500 | 50 to 150 | Rough: .005<br>Finish: .0015 max. | A60MV |
| | | | 28 to 33 | 1.3 to 3.8 | Rough: .13<br>Finish: .038 max. | A60MV |
| | Over 50 $R_C$ | Carburized and/or Quenched and Tempered | 5500 to 6500 | 50 to 100 | Rough: .005<br>Finish: .0015 max. | A80LV |
| | | | 28 to 33 | 1.3 to 3.8 | Rough: .13<br>Finish: .038 max. | A80LV |
| **5. ALLOY STEELS, WROUGHT**<br>**Low Carbon**<br>4012 4615 4817 8617<br>4023 4617 4820 8620<br>4024 4620 5015 8622<br>4118 4621 5115 8822<br>4320 4718 5120 9310<br>4419 4720 6118 94B15<br>4422 4815 8115 94B17<br>**Medium Carbon**<br>1330 4145 5132 8640<br>1335 4147 5135 8642<br>1340 4150 5140 8645<br>1345 4161 5145 86B45<br>4027 4340 5147 8650<br>4028 4427 5150 8655<br>4032 4626 5155 8660<br>4037 50B40 5160 8740<br>4042 50B44 51B60 8742<br>4047 5046 6150 9254<br>4130 50B46 81B45 9255<br>4135 50B50 8625 9260<br>4137 5060 8627 94B30<br>4140 50B60 8630<br>4142 5130 8637<br>**High Carbon**<br>50100 51100 52100 M-50 | 50 $R_C$ max. | Hot Rolled, Normalized, Annealed, Cold Drawn or Quenched and Tempered | 5500 to 6500 | 50 to 150 | Rough: .005<br>Finish: .0015 max. | A60MV |
| | | | 28 to 33 | 1.3 to 3.8 | Rough: .13<br>Finish: .038 max. | A60MV |
| | Over 50 $R_C$ | Carburized and/or Quenched and Tempered | 4000 to 6000 | 50 to 150 | Rough: .005<br>Finish: .0015 max. | A80LV |
| | | | 20 to 30 | 1.3 to 3.8 | Rough: .13<br>Finish: .038 max. | A80LV |

See section 16 for Cutting Fluid Recommendations.
*See section 8.12 for Work Traverse Rates. As recommended starting conditions—use a regulating wheel angle with a positive inclination of 3° and a regulating wheel speed of 25 to 40 rpm.
†Wheel recommendations are for wet grinding 0.8- to 2-inch [20 to 50 mm] diameter work. For LARGER diameter work—use a softer grade and/or coarser grit wheel. For SMALLER diameter work—use a harder grade wheel. Also see section 20.2, Grinding Guidelines.

| MATERIAL | HARDNESS Bhn | CONDITION | WHEEL SPEED fpm / m/s | THRUFEED OF WORK* in/min / m/min | INFEED on dia. in/pass / mm/pass | WHEEL IDENTIFICATION† ANSI / ISO |
|---|---|---|---|---|---|---|
| **6. HIGH STRENGTH STEELS, WROUGHT**<br>300M<br>4330V<br>4340<br>4340Si<br>98BV40<br>D6ac<br>H11<br>H13<br>HP 9-4-20<br>HP 9-4-25<br>HP 9-4-30<br>HP 9-4-45 | 50 $R_C$ max. | Annealed, Normalized or Quenched and Tempered | 5500 to 6500 | 50 to 150 | Rough: .005<br>Finish: .0015 max. | A60MV |
| | | | 28 to 33 | 1.3 to 3.8 | Rough: .13<br>Finish: .038 max. | A60MV |
| | Over 50 $R_C$ | Quenched and Tempered | 4000 to 6000 | 50 to 150 | Rough: .003<br>Finish: .001 max. | A80KV |
| | | | 20 to 30 | 1.3 to 3.8 | Rough: .075<br>Finish: .025 max. | A80KV |
| **7. MARAGING STEELS, WROUGHT**<br>ASTM A538: Grades A, B, C<br>120 Grade<br>180 Grade<br>200 Grade<br>250 Grade<br>300 Grade<br>350 Grade<br>HY230 | 50 $R_C$ max. | Annealed or Maraged | 5500 to 6500 | 50 to 150 | Rough: .005<br>Finish: .0015 max. | A60MV |
| | | | 28 to 33 | 1.3 to 3.8 | Rough: .13<br>Finish: .038 max. | A60MV |
| | Over 50 $R_C$ | Maraged | 4000 to 6000 | 50 to 150 | Rough: .003<br>Finish: .001 max. | A80KV |
| | | | 20 to 30 | 1.3 to 3.8 | Rough: .075<br>Finish: .025 max. | A80KV |
| **8. TOOL STEELS, WROUGHT**<br>**Group I**<br>A2 H13 L6 P20<br>A3 H14 L7 P21<br>A4 H19 O1 S1<br>A6 H21 O2 S2<br>A8 H22 O6 S5<br>A9 H23 O7 S6<br>A10 H24 P2 S7<br>H10 H25 P4 W1<br>H11 H26 P5 W2<br>H12 L2 P6 W5<br>SAE J438b: Types W108, W109, W110, W112, W209, W210, W310 | 50 $R_C$ max. | Annealed or Quenched and Tempered | 5500 to 6500 | 50 to 150 | Rough: .005<br>Finish: .0015 max. | A60MV |
| | | | 28 to 33 | 1.3 to 3.8 | Rough: .13<br>Finish: .038 max. | A60MV |
| | Over 50 $R_C$ | Quenched and Tempered | 4000 to 6000 | 50 to 150 | Rough: .005<br>Finish: .0015 max. | A60LV |
| | | | 20 to 30 | 1.3 to 3.8 | Rough: .13<br>Finish: .038 max. | A60LV |

See section 16 for Cutting Fluid Recommendations.

*See section 8.12 for Work Traverse Rates. As recommended starting conditions—use a regulating wheel angle with a positive inclination of 3° and a regulating wheel speed of 25 to 40 rpm.

†Wheel recommendations are for wet grinding 0.8- to 2-inch [20 to 50 mm] diameter work. For LARGER diameter work—use a softer grade and/or coarser grit wheel. For SMALLER diameter work—use a harder grade wheel. Also see section 20.2, Grinding Guidelines.

# 8.11 Centerless Grinding

| MATERIAL | HARD-NESS Bhn | CONDITION | WHEEL SPEED fpm / m/s | THRUFEED OF WORK* in/min / m/min | INFEED on dia. in/pass / mm/pass | WHEEL IDENTIFI-CATION† ANSI / ISO |
|---|---|---|---|---|---|---|
| **8. TOOL STEELS, WROUGHT (cont.)** **Group II** D2, D3, D4, D5, F1, F2, H42, M1, M2, M3-1, M10, M30, M42, T1, T2, T4, T8 | 50 $R_C$ max. | Annealed or Quenched and Tempered | 5500 to 6500 | 50 to 150 | Rough: .005 Finish: .0015 max. | A60MV |
| | | | 28 to 33 | 1.3 to 3.8 | Rough: .13 Finish: .038 max. | A60MV |
| | 50 $R_C$ to 58 $R_C$ | Quenched and Tempered | 4000 to 6000 | 50 to 150 | Rough: .004 Finish: .001 max. | A60LV |
| | | | 20 to 30 | 1.3 to 3.8 | Rough: .102 Finish: .025 max. | A60LV |
| | Over 58 $R_C$ | Quenched and Tempered | 4000 to 6000 | 50 to 150 | Rough: .003 Finish: .001 max. | A80KV |
| | | | 20 to 30 | 1.3 to 3.8 | Rough: .075 Finish: .025 max. | A80KV |
| **Group III** A7, D7, M3-2, M6, M7, M33, M34, M36, M41, M43, M44, M46, M47, T5, T6 | 50 $R_C$ max. | Annealed or Quenched and Tempered | 5500 to 6500 | 50 to 150 | Rough: .005 Finish: .0015 max. | A60LV |
| | | | 28 to 33 | 1.3 to 3.8 | Rough: .13 Finish: .038 max. | A60LV |
| | 50 $R_C$ to 58 $R_C$ | Quenched and Tempered | 4000 to 6000 | 50 to 150 | Rough: .004 Finish: .001 max. | A80KV |
| | | | 20 to 30 | 1.3 to 3.8 | Rough: .102 Finish: .025 max. | A80KV |
| | Over 58 $R_C$ | Quenched and Tempered | 4000 to 5500 | 50 to 150 | Rough: .003 Finish: .001 max. | A80JV |
| | | | 20 to 28 | 1.3 to 3.8 | Rough: .075 Finish: .025 max. | A80JV |

See section 16 for Cutting Fluid Recommendations.

*See section 8.12 for Work Traverse Rates. As recommended starting conditions—use a regulating wheel angle with a positive inclination of 3° and a regulating wheel speed of 25 to 40 rpm.

†Wheel recommendations are for wet grinding 0.8- to 2-inch [20 to 50 mm] diameter work. For LARGER diameter work—use a softer grade and/or coarser grit wheel. For SMALLER diameter work—use a harder grade wheel. Also see section 20.2, Grinding Guidelines.

| MATERIAL | HARD-NESS Bhn | CONDITION | WHEEL SPEED fpm / m/s | THRUFEED OF WORK* in/min / m/min | INFEED on dia. in/pass / mm/pass | WHEEL IDENTIFI-CATION† ANSI / ISO |
|---|---|---|---|---|---|---|
| **Group IV** M4 T15 | 50 $R_C$ max. | Annealed or Quenched and Tempered | 4000 to 6000 | 50 to 150 | Rough: .004 Finish: .001 max. | A60KV |
| | | | 20 to 30 | 1.3 to 3.8 | Rough: .102 Finish: .025 max. | A60KV |
| | 50 $R_C$ to 58 $R_C$ | Quenched and Tempered | 4000 to 5500 | 50 to 150 | Rough: .003 Finish: .001 max. | A80JV |
| | | | 20 to 28 | 1.3 to 3.8 | Rough: .075 Finish: .025 max. | A80JV |
| | Over 58 $R_C$ | Quenched and Tempered | 4000 to 5500 | 50 to 150 | Rough: .002 Finish: .001 max. | A80IV |
| | | | 20 to 28 | 1.3 to 3.8 | Rough: .050 Finish: .025 max. | A80IV |
| **9. NITRIDING STEELS, WROUGHT** Nitralloy 125, Nitralloy 135, Nitralloy 135 Mod., Nitralloy 225, Nitralloy 230, Nitralloy EZ, Nitralloy N, Nitrex 1 | 200 to 350 | Annealed, Normalized or Quenched and Tempered | 5500 to 6500 | 50 to 150 | Rough: .005 Finish: .0015 max. | A60LV |
| | | | 28 to 33 | 1.3 to 3.8 | Rough: .13 Finish: .038 max. | A60LV |
| | 60 $R_C$ to 65 $R_C$ | Nitrided | 4000 to 6000 | 50 to 150 | Rough: .003 Finish: .001 max. | A80JV |
| | | | 20 to 30 | 1.3 to 3.8 | Rough: .075 Finish: .025 max. | A80JV |
| **12. FREE MACHINING STAINLESS STEELS, WROUGHT** **Ferritic** 430F, 430F Se | 135 to 185 | Annealed | 5500 to 6500 | 50 to 150 | Rough: .005 Finish: .0015 max. | A60LV |
| | | | 28 to 33 | 1.3 to 3.8 | Rough: .13 Finish: .038 max. | A60LV |

See section 16 for Cutting Fluid Recommendations.

*See section 8.12 for Work Traverse Rates. As recommended starting conditions—use a regulating wheel angle with a positive inclination of 3° and a regulating wheel speed of 25 to 40 rpm.

†Wheel recommendations are for wet grinding 0.8- to 2-inch [20 to 50 mm] diameter work. For LARGER diameter work—use a softer grade and/or coarser grit wheel. For SMALLER diameter work—use a harder grade wheel. Also see section 20.2, Grinding Guidelines.

# 8.11 Centerless Grinding

| MATERIAL | HARD-NESS Bhn | CONDITION | WHEEL SPEED fpm / m/s | THRUFEED OF WORK* in/min / m/min | INFEED on dia. in/pass / mm/pass | WHEEL IDENTIFI-CATION† ANSI / ISO |
|---|---|---|---|---|---|---|
| **12. FREE MACHINING STAINLESS STEELS, WROUGHT (cont.)** **Austenitic** 203EZ, 303, 303MA, 303Pb, 303 Plus X, 303Se | 135 to 275 | Annealed or Cold Drawn | 5500 to 6500 | 50 to 150 | Rough: .005 Finish: .0015 max. | C60LV |
| | | | 28 to 33 | 1.3 to 3.8 | Rough: .13 Finish: .038 max. | C60LV |
| **Martensitic** 416, 416 Plus X, 416Se, 420F, 420F Se, 440F, 440F Se | 135 to 240 | Annealed or Cold Drawn | 5500 to 6500 | 50 to 150 | Rough: .005 Finish: .0015 max. | A60LV |
| | | | 28 to 33 | 1.3 to 3.8 | Rough: .13 Finish: .038 max. | A60LV |
| | Over 275 | Quenched and Tempered | 5500 to 6500 | 50 to 150 | Rough: .005 Finish: .0015 max. | A60KV |
| | | | 28 to 33 | 1.3 to 3.8 | Rough: .13 Finish: .038 max. | A60KV |
| **13. STAINLESS STEELS, WROUGHT** **Ferritic** 405, 409, 429, 430, 434, 436, 442, 446 | 135 to 185 | Annealed | 5500 to 6500 | 50 to 150 | Rough: .005 Finish: .0015 max. | A60LV |
| | | | 28 to 33 | 1.3 to 3.8 | Rough: .13 Finish: .038 max. | A60LV |
| **Austenitic** 201, 202, 301, 302, 302B, 304, 304L, 305, 308, 309, 309S, 310, 310S, 314, 316, 316L, 317, 321, 330, 347, 348, 384, 385 | 135 to 275 | Annealed or Cold Drawn | 5500 to 6500 | 50 to 150 | Rough: .005 Finish: .0015 max. | C60LV |
| | | | 28 to 33 | 1.3 to 3.8 | Rough: .13 Finish: .038 max. | C60LV |
| **Austenitic** Nitronic 32, Nitronic 33, Nitronic 40, Nitronic 50, Nitronic 60 | 210 to 250 | Annealed | 5500 to 6500 | 50 to 150 | Rough: .005 Finish: .0015 max. | C60LV |
| | | | 28 to 33 | 1.3 to 3.8 | Rough: .13 Finish: .038 max. | C60LV |

See section 16 for Cutting Fluid Recommendations.

*See section 8.12 for Work Traverse Rates. As recommended starting conditions—use a regulating wheel angle with a positive inclination of 3° and a regulating wheel speed of 25 to 40 rpm.

†Wheel recommendations are for wet grinding 0.8- to 2-inch [20 to 50 mm] diameter work. For LARGER diameter work—use a softer grade and/or coarser grit wheel. For SMALLER diameter work—use a harder grade wheel. Also see section 20.2, Grinding Guidelines.

| MATERIAL | HARD-NESS Bhn | CONDITION | WHEEL SPEED fpm / m/s | THRUFEED OF WORK* in/min / m/min | INFEED on dia. in/pass / mm/pass | WHEEL IDENTIFI-CATION† ANSI / ISO |
|---|---|---|---|---|---|---|
| **13. STAINLESS STEELS, WROUGHT (cont.)** **Austenitic (cont.)** (materials listed on preceding page) | 325 to 375 | Cold Drawn | 5500 to 6500 | 50 to 100 | Rough: .005 Finish: .0015 max. | C60KV |
| | | | 28 to 33 | 1.3 to 2.6 | Rough: .13 Finish: .038 max. | C60KV |
| **Martensitic** 403, 410, 414, 420, 422, 431, 440A, 440B, 440C, 501, 502, Greek Ascoloy | 135 to 275 | Annealed | 5500 to 6500 | 50 to 150 | Rough: .005 Finish: .0015 max. | A60LV |
| | | | 28 to 33 | 1.3 to 3.8 | Rough: .13 Finish: .038 max. | A60LV |
| | Over 275 | Quenched and Tempered | 5500 to 6500 | 50 to 150 | Rough: .005 Finish: .0015 max. | A60KV |
| | | | 28 to 33 | 1.3 to 3.8 | Rough: .13 Finish: .038 max. | A60KV |
| **14. PRECIPITATION HARDENING STAINLESS STEELS, WROUGHT** 15-5 PH, 16-6 PH, 17-4 PH, 17-7 PH, 17-14 Cu Mo, AF-71, AFC-77, Almar 362 (AM-362), AM-350, AM-355, AM-363, Custom 450, Custom 455, HNM, PH 13-8 Mo, PH 14-8 Mo, PH 15-7 Mo, Stainless W | 150 to 200 | Solution Treated | 5500 to 6500 | 50 to 150 | Rough: .005 Finish: .0015 max. | A60LV |
| | | | 28 to 33 | 1.3 to 3.8 | Rough: .13 Finish: .038 max. | A60LV |
| | 275 to 440 | Solution Treated or Hardened | 5500 to 6500 | 50 to 150 | Rough: .005 Finish: .0015 max. | A60KV |
| | | | 28 to 33 | 1.3 to 3.8 | Rough: .13 Finish: .038 max. | A60KV |

See section 16 for Cutting Fluid Recommendations.

*See section 8.12 for Work Traverse Rates. As recommended starting conditions—use a regulating wheel angle with a positive inclination of 3° and a regulating wheel speed of 25 to 40 rpm.

†Wheel recommendations are for wet grinding 0.8- to 2-inch [20 to 50 mm] diameter work. For LARGER diameter work—use a softer grade and/or coarser grit wheel. For SMALLER diameter work—use a harder grade wheel. Also see section 20.2, Grinding Guidelines.

# 8.11 Centerless Grinding

| MATERIAL | HARDNESS<br>Bhn | CONDITION | WHEEL SPEED<br>fpm<br>m/s | THRUFEED OF WORK*<br>in/min<br>m/min | INFEED on dia.<br>in/pass<br>mm/pass | WHEEL IDENTIFICATION†<br>ANSI<br>ISO |
|---|---|---|---|---|---|---|
| **15. CARBON STEELS, CAST**<br>**Low Carbon**<br>ASTM A426: Grade CP1<br>1010<br>1020<br>**Medium Carbon**<br>ASTM A352: Grades LCA, LCB, LCC<br>ASTM A356: Grade 1<br>1030 1040 1050 | 100<br>to<br>300 | Annealed, Normalized, Normalized and Tempered, or Quenched and Tempered | 5500<br>to<br>6500 | 50<br>to<br>150 | Rough:<br>.005<br>Finish:<br>.0015 max. | A60MV |
| | | | 28<br>to<br>33 | 1.3<br>to<br>3.8 | Rough:<br>.13<br>Finish:<br>.038 max. | A60MV |
| | Over<br>50 $R_C$ | Carburized and/or Quenched and Tempered | 5500<br>to<br>6500 | 50<br>to<br>150 | Rough:<br>.005<br>Finish:<br>.0015 max. | A60LV |
| | | | 28<br>to<br>33 | 1.3<br>to<br>3.8 | Rough:<br>.13<br>Finish:<br>.038 max. | A60LV |
| **16. ALLOY STEELS, CAST**<br>**Low Carbon**<br>ASTM A217: Grade WC9<br>ASTM A352: Grades LC3, LC4<br>ASTM A426: Grades CP2, CP5, CP5b, CP11, CP12, CP15, CP21, CP22<br>1320 4110 8020<br>2315 4120 8620<br>2320 4320<br>**Medium Carbon**<br>ASTM A27: Grades N1, N2, U-60-30, 60-30, 65-35, 70-36, 70-40<br>ASTM A148: Grades 80-40, 80-50, 90-60, 105-85, 120-95, 150-125, 175-145<br>ASTM A216: Grades WCA, WCB, WCC<br>ASTM A217: Grades WC1, WC4, WC5, WC6<br>ASTM A352: Grades LC1, LC2, LC2-1<br>ASTM A356: Grades 2, 5, 6, 8, 9, 10<br>ASTM A389: Grades C23, C24<br>ASTM A486: Classes 70, 90, 120<br>ASTM A487: Classes 1N, 2N, 4N, 6N, 8N, 9N, 10N, DN, 1Q, 2Q, 4Q, 4QA, 6Q, 7Q, 8Q, 9Q, 10Q<br>1330 4130 80B30 8640<br>1340 4140 8040 9525<br>2325 4330 8430 9530<br>2330 4340 8440 9535<br>4125 8030 8630 | 150<br>to<br>400 | Annealed, Normalized, Normalized and Tempered, or Quenched and Tempered | 5500<br>to<br>6500 | 50<br>to<br>150 | Rough:<br>.005<br>Finish:<br>.0015 max. | A60MV |
| | | | 28<br>to<br>33 | 1.3<br>to<br>3.8 | Rough:<br>.13<br>Finish:<br>.038 max. | A60MV |
| | Over<br>50 $R_C$ | Carburized and/or Quenched and Tempered | 4000<br>to<br>6000 | 50<br>to<br>150 | Rough:<br>.005<br>Finish:<br>.0015 max. | A80LV |
| | | | 20<br>to<br>30 | 1.3<br>to<br>3.8 | Rough:<br>.13<br>Finish:<br>.038 max. | A80LV |
| | | | | | | |
| | | | | | | |
| **17. TOOL STEELS, CAST**<br>**Group I**<br>ASTM A597: Grades CA-2, CH-12, CH-13, CO-1, CS-5 | 50 $R_C$<br>max. | Annealed or Quenched and Tempered | 5500<br>to<br>6500 | 50<br>to<br>150 | Rough:<br>.005<br>Finish:<br>.0015 max. | A60MV |
| | | | 28<br>to<br>33 | 1.3<br>to<br>3.8 | Rough:<br>.13<br>Finish:<br>.038 max. | A60MV |

See section 16 for Cutting Fluid Recommendations.

*See section 8.12 for Work Traverse Rates. As recommended starting conditions—use a regulating wheel angle with a positive inclination of 3° and a regulating wheel speed of 25 to 40 rpm.

†Wheel recommendations are for wet grinding 0.8- to 2-inch [20 to 50 mm] diameter work. For LARGER diameter work—use a softer grade and/or coarser grit wheel. For SMALLER diameter work—use a harder grade wheel. Also see section 20.2, Grinding Guidelines.

| MATERIAL | HARD-NESS Bhn | CONDITION | WHEEL SPEED fpm / m/s | THRUFEED OF WORK* in/min / m/min | INFEED on dia. in/pass / mm/pass | WHEEL IDENTIFI-CATION† ANSI / ISO |
|---|---|---|---|---|---|---|
| **17. TOOL STEELS, CAST (cont.)** **Group I (cont.)** (materials listed on preceding page) | 50 $R_C$ to 58 $R_C$ | Quenched and Tempered | 4000 to 6000 | 50 to 150 | Rough: .004 Finish: .001 max. | A60LV |
| | | | 20 to 30 | 1.3 to 3.8 | Rough: .102 Finish: .025 max. | A60LV |
| | Over 58 $R_C$ | Quenched and Tempered | 4000 to 6000 | 50 to 150 | Rough: .003 Finish: .001 max. | A60KV |
| | | | 20 to 30 | 1.3 to 3.8 | Rough: .075 Finish: .025 max. | A60KV |
| **Group II** ASTM A597: Grades CD-2, CD-5 | 200 to 250 | Annealed | 5500 to 6500 | 50 to 150 | Rough: .005 Finish: .0015 max. | A60MV |
| | | | 28 to 33 | 1.3 to 3.8 | Rough: .13 Finish: .038 max. | A60MV |
| | 48 $R_C$ to 56 $R_C$ | Quenched and Tempered | 4000 to 6000 | 50 to 150 | Rough: .004 Finish: .001 max. | A60LV |
| | | | 20 to 30 | 1.3 to 3.8 | Rough: .102 Finish: .025 max. | A60LV |
| | Over 56 $R_C$ | Quenched and Tempered | 4000 to 6000 | 50 to 150 | Rough: .003 Finish: .001 max. | A80KV |
| | | | 20 to 30 | 1.3 to 3.8 | Rough: .075 Finish: .025 max. | A80KV |
| **18. STAINLESS STEELS, CAST** **Ferritic** ASTM A217: Grades C5, C12; ASTM A296: Grades CB-30, CC-50, CE-30, CA6N, CA-6NM, CD4MCu; ASTM A297: Grade HC; ASTM A487: Class CA6NM; ASTM A608: Grade HC30 | 135 to 185 | Annealed | 5500 to 6500 | 50 to 150 | Rough: .005 Finish: .0015 max. | A60LV |
| | | | 28 to 33 | 1.3 to 3.8 | Rough: .13 Finish: .038 max. | A60LV |

See section 16 for Cutting Fluid Recommendations.

*See section 8.12 for Work Traverse Rates. As recommended starting conditions—use a regulating wheel angle with a positive inclination of 3° and a regulating wheel speed of 25 to 40 rpm.

†Wheel recommendations are for wet grinding 0.8- to 2-inch [20 to 50 mm] diameter work. For LARGER diameter work—use a softer grade and/or coarser grit wheel. For SMALLER diameter work—use a harder grade wheel. Also see section 20.2, Grinding Guidelines.

| MATERIAL | HARD-NESS Bhn | CONDITION | WHEEL SPEED fpm / m/s | THRUFEED OF WORK* in/min / m/min | INFEED on dia. in/pass / mm/pass | WHEEL IDENTIFI-CATION† ANSI / ISO |
|---|---|---|---|---|---|---|
| **18. STAINLESS STEELS, CAST (cont.)**<br>**Austenitic**<br>ASTM A296: Grades CF-3, CF-3M, CF-8, CF-8C, CF-8M, CF-16F, CF-20, CG-8M, CG-12, CH-20, CK-20, CN-7M, CN-7MS<br>ASTM A297: Grades HD, HE, HF, HH, HI, HK, HL, HN, HP, HT, HU<br>ASTM A351: Grades CF-3, CF-3A, CF-3M, CF-3MA, CF-8, CF-8A, CF-8C, CF-8M, CF-10MC, CH-8, CH-10, CH-20, CK-20, CN-7M, HK-30, HK-40, HT-30<br>ASTM A451: Grades CPF8A, CPF3, CPF3A, CPF3M, CPF8, CPF8C, CPF8C (Ta Max.), CPF8M, CPF10MC, CPH8, CPH10, CPH20, CPK20<br>ASTM A452: Grades TP 304H, TP 316H, TP 347H<br>ASTM A608: Grades HD50, HE35, HF30, HH30, HH33, HI35, HK30, HK40, HL30, HL40, HN40, HT50, HU50 | 135 to 210 | Annealed, Normalized or As Cast | 5500 to 6500 | 50 to 150 | Rough: .005<br>Finish: .0015 max. | C60LV |
| | | | 28 to 33 | 1.3 to 3.8 | Rough: .13<br>Finish: .038 max. | C60LV |
| **Martensitic**<br>ASTM A217: Grade CA-15<br>ASTM A296: Grades CA-15, CA-15M, CA-40<br>ASTM A426: Grades CP7, CP9, CPCA15<br>ASTM A487: Classes CA15a, CA-15M | 135 to 225 | Annealed, Normalized or Normalized and Tempered | 5500 to 6500 | 50 to 150 | Rough: .005<br>Finish: .0015 max. | A60LV |
| | | | 28 to 33 | 1.3 to 3.8 | Rough: .13<br>Finish: .038 max. | A60LV |
| | Over 275 | Quenched and Tempered | 5500 to 6500 | 50 to 150 | Rough: .005<br>Finish: .0015 max. | A60KV |
| | | | 28 to 33 | 1.3 to 3.8 | Rough: .13<br>Finish: .038 max. | A60KV |
| **19. PRECIPITATION HARDENING STAINLESS STEELS, CAST**<br>ASTM A351: Grade CD-4MCu<br>ACI Grade CB-7Cu<br>ACI Grade CD-4MCu<br>17-4 PH<br>AM-355 | 325 to 450 | Solution Treated or Solution Treated and Aged | 5500 to 6500 | 50 to 150 | Rough: .005<br>Finish: .0015 max. | A60KV |
| | | | 28 to 33 | 1.3 to 3.8 | Rough: .13<br>Finish: .038 max. | A60KV |

See section 16 for Cutting Fluid Recommendations.

*See section 8.12 for Work Traverse Rates. As recommended starting conditions—use a regulating wheel angle with a positive inclination of 3° and a regulating wheel speed of 25 to 40 rpm.

†Wheel recommendations are for wet grinding 0.8- to 2-inch [20 to 50 mm] diameter work. For LARGER diameter work—use a softer grade and/or coarser grit wheel. For SMALLER diameter work—use a harder grade wheel. Also see section 20.2, Grinding Guidelines.

| MATERIAL | HARDNESS Bhn | CONDITION | WHEEL SPEED fpm / m/s | THRUFEED OF WORK* in/min / m/min | INFEED on dia. in/pass / mm/pass | WHEEL IDENTIFICATION† ANSI / ISO |
|---|---|---|---|---|---|---|
| **21. GRAY CAST IRONS**<br>**Ferritic**<br>ASTM A48: Class 20<br>SAE J431c: Grade G1800<br>**Pearlitic- Ferritic**<br>ASTM A48: Class 25<br>SAE J431c: Grade G2500<br>**Pearlitic**<br>ASTM A48: Classes 30, 35, 40<br>SAE J431c: Grade G3000<br>**Pearlitic + Free Carbides**<br>ASTM A48: Classes 45, 50<br>SAE J431c: Grades G3500, G4000<br>**Pearlitic or Acicular + Free Carbides**<br>ASTM A48: Classes 55, 60 | 45 $R_C$ max. | As Cast, Annealed or Quenched and Tempered | 5500 to 6500 | 50 to 150 | Rough: .005<br>Finish: .002 max. | C46LV or A60KV |
| | | | 28 to 33 | 1.3 to 3.8 | Rough: .13<br>Finish: .050 max. | C46LV or A60KV |
| | 45 $R_C$ to 52 $R_C$ | As Cast, Annealed or Quenched and Tempered | 5500 to 6500 | 50 to 150 | Rough: .005<br>Finish: .0015 max. | C54KV or A60JV |
| | | | 28 to 33 | 1.3 to 3.8 | Rough: .13<br>Finish: .038 max. | C54KV or A60JV |
| | 48 $R_C$ to 60 $R_C$ | Flame or Induction Hardened | 5500 to 6500 | 50 to 150 | Rough: .004<br>Finish: .001 max. | C60JV |
| | | | 28 to 33 | 1.3 to 3.8 | Rough: .102<br>Finish: .025 max. | C60JV |
| **Austenitic (NI-RESIST)**<br>ASTM A436: Types 1, 1b, 2, 2b, 3, 4, 5, 6 | 100 to 250 | As Cast | 5500 to 6500 | 50 to 150 | Rough: .005<br>Finish: .0015 max. | A60LV |
| | | | 28 to 33 | 1.3 to 3.8 | Rough: .13<br>Finish: .038 max. | A60LV |
| **22. COMPACTED GRAPHITE CAST IRONS** | 185 to 255 | As Cast | 5500 to 6500 | 50 to 150 | Rough: .005<br>Finish: .002 max. | C46LV |
| | | | 28 to 33 | 1.3 to 3.8 | Rough: .13<br>Finish: .050 max. | C46LV |
| | | | | | | |

See section 16 for Cutting Fluid Recommendations.

*See section 8.12 for Work Traverse Rates. As recommended starting conditions—use a regulating wheel angle with a positive inclination of 3° and a regulating wheel speed of 25 to 40 rpm.

†Wheel recommendations are for wet grinding 0.8- to 2-inch [20 to 50 mm] diameter work. For LARGER diameter work—use a softer grade and/or coarser grit wheel. For SMALLER diameter work—use a harder grade wheel. Also see section 20.2, Grinding Guidelines.

# 8.11 Centerless Grinding

| MATERIAL | HARDNESS<br>Bhn | CONDITION | WHEEL SPEED<br>fpm<br>m/s | THRUFEED OF WORK*<br>in/min<br>m/min | INFEED on dia.<br>in/pass<br>mm/pass | WHEEL IDENTIFICATION†<br>ANSI<br>ISO |
|---|---|---|---|---|---|---|
| **23. DUCTILE CAST IRONS**<br>**Ferritic**<br>ASTM A536: Grades 60-40-18, 65-45-12<br>SAE J434c: Grades D4018, D4512<br>**Ferritic- Pearlitic**<br>ASTM A536: Grade 80-55-06<br>SAE J434c: Grade D5506<br>**Pearlitic- Martensitic**<br>ASTM A536: Grade 100-70-03<br>SAE J434c: Grade D7003<br>**Martensitic**<br>ASTM A536: Grade 120-90-02<br>SAE J434c: Grade DQ&T | 52 $R_C$ max. | Annealed, As Cast, Normalized and Tempered or Quenched and Tempered | 5500 to 6500 | 50 to 150 | Rough: .005<br>Finish: .0015 max. | C54KV |
| | | | 28 to 33 | 1.3 to 3.8 | Rough: .13<br>Finish: .038 max. | C54KV |
| | 53 $R_C$ to 60 $R_C$ | Flame or Induction Hardened | 5500 to 6500 | 50 to 150 | Rough: .004<br>Finish: .001 max. | C60JV |
| | | | 28 to 33 | 1.3 to 3.8 | Rough: .102<br>Finish: .025 max. | C60JV |
| **Austenitic (NI-RESIST Ductile)**<br>ASTM A439: Types D-2, D-2B, D-2C, D-3, D-3A, D-4, D-5, D-5B<br>ASTM A571: Type D-2M | 120 to 275 | Annealed | 5500 to 6500 | 50 to 150 | Rough: .005<br>Finish: .0015 max. | A60LV |
| | | | 28 to 33 | 1.3 to 3.8 | Rough: .13<br>Finish: .038 max. | A60LV |
| **24. MALLEABLE CAST IRONS**<br>**Ferritic**<br>ASTM A47: Grades 32510, 35018<br>ASTM A602: Grade M3210<br>SAE J158: Grade M3210<br>**Pearlitic**<br>ASTM A220: Grades 40010, 45006, 45008, 50005<br>ASTM A602: Grade M4504, M5003<br>SAE J158: Grades M4504, M5003<br>**Tempered Martensite**<br>ASTM A220: Grades 60004, 70003, 80002, 90001<br>ASTM A602: Grades M5503, M7002, M8501<br>SAE J158: Grades M5503, M7002, M8501 | 52 $R_C$ max. | Malleablized or Malleablized and Heat Treated | 5500 to 6500 | 50 to 150 | Rough: .005<br>Finish: .0015 max. | C54KV |
| | | | 28 to 33 | 1.3 to 3.8 | Rough: .13<br>Finish: .038 max. | C54KV |
| | Over 52 $R_C$ | Flame or Induction Hardened | 5500 to 6500 | 50 to 150 | Rough: .004<br>Finish: .001 max. | C60JV |
| | | | 28 to 33 | 1.3 to 3.8 | Rough: .102<br>Finish: .025 max. | C60JV |
| **25. WHITE CAST IRONS (ABRASION RESISTANT)**<br>ASTM A532:<br>Class I, Types A, B, C, D<br>Class II, Types A, B, C, D, E<br>Class III, Type A | 60 $R_C$ max. | As Cast, Annealed or Hardened | 5500 to 6500 | 50 to 150 | Rough: .004<br>Finish: .001 max. | C60JV |
| | | | 28 to 33 | 1.3 to 3.8 | Rough: .102<br>Finish: .025 max. | C60JV |

See section 16 for Cutting Fluid Recommendations.

*See section 8.12 for Work Traverse Rates. As recommended starting conditions—use a regulating wheel angle with a positive inclination of 3° and a regulating wheel speed of 25 to 40 rpm.

†Wheel recommendations are for wet grinding 0.8- to 2-inch [20 to 50 mm] diameter work. For LARGER diameter work—use a softer grade and/or coarser grit wheel. For SMALLER diameter work—use a harder grade wheel. Also see section 20.2, Grinding Guidelines.

| MATERIAL | HARD-NESS Bhn | CONDITION | WHEEL SPEED fpm m/s | THRUFEED OF WORK* in/min m/min | INFEED on dia. in/pass mm/pass | WHEEL IDENTIFI-CATION† ANSI ISO |
|---|---|---|---|---|---|---|
| **26. HIGH SILICON CAST IRONS**<br>Duriron<br>Duriclor<br>ASTM A518 | 52 $R_C$ | As Cast | 5500 to 6500 | 50 to 150 | Rough: .004 Finish: .001 max. | C60JV |
| | | | 28 to 33 | 1.3 to 3.8 | Rough: .102 Finish: .025 max. | C60JV |
| **27. CHROMIUM-NICKEL ALLOY CASTINGS**<br>ASTM A560: Grades 50Cr-50Ni, 60Cr-40Ni | 275 to 375 | As Cast | 5500 to 6500 | 50 to 150 | Rough: .005 Finish: .0015 max. | C60KV |
| | | | 28 to 33 | 1.3 to 3.8 | Rough: .13 Finish: .038 max. | C60KV |
| **28. ALUMINUM ALLOYS, WROUGHT**<br>EC 2218 5252 6253<br>1060 2219 5254 6262<br>1100 2618 5454 6463<br>1145 3003 5456 6951<br>1175 3004 5457 7001<br>1235 3005 5652 7004<br>2011 4032 5657 7005<br>2014 5005 6053 7039<br>2017 5050 6061 7049<br>2018 5052 6063 7050<br>2021 5056 6066 7075<br>2024 5083 6070 7079<br>2025 5086 6101 7175<br>2117 5154 6151 7178 | 30 to 150 500kg | Cold Drawn or Solution Treated and Aged | 5500 to 6500 | 50 to 150 | Rough: .005 Finish: .0015 max. | C54JV |
| | | | 28 to 33 | 1.3 to 3.8 | Rough: .13 Finish: .038 max. | C54JV |
| | | | | | | |
| | | | | | | |
| **29. ALUMINUM ALLOYS, CAST**<br>**Sand and Permanent Mold**<br>A140 295.0 514.0 A712.0<br>201.0 B295.0 A514.0 D712.0<br>208.0 308.0 B514.0 713.0<br>213.0 319.0 520.0 771.0<br>222.0 355.0 535.0 850.0<br>224.0 C355.0 705.0 A850.0<br>242.0 B443.0 707.0 B850.0<br>Hiduminium RR-350<br>**Die Castings**<br>C443.0 518.0 | 40 to 125 500kg | As Cast or Solution Treated and Aged | 5500 to 6500 | 50 to 150 | Rough: .005 Finish: .0015 max. | C54JV |
| | | | 28 to 33 | 1.3 to 3.8 | Rough: .13 Finish: .038 max. | C54JV |
| | | | | | | |
| | | | | | | |

See section 16 for Cutting Fluid Recommendations.

*See section 8.12 for Work Traverse Rates. As recommended starting conditions—use a regulating wheel angle with a positive inclination of 3° and a regulating wheel speed of 25 to 40 rpm.

†Wheel recommendations are for wet grinding 0.8- to 2-inch [20 to 50 mm] diameter work. For LARGER diameter work—use a softer grade and/or coarser grit wheel. For SMALLER diameter work—use a harder grade wheel. Also see section 20.2, Grinding Guidelines.

## 8.11 Centerless Grinding

| MATERIAL | HARDNESS Bhn | CONDITION | WHEEL SPEED fpm / m/s | THRUFEED OF WORK* in/min / m/min | INFEED on dia. in/pass / mm/pass | WHEEL IDENTIFICATION† ANSI / ISO |
|---|---|---|---|---|---|---|
| **29. ALUMINUM ALLOYS, CAST (cont.)** **Sand and Permanent Mold** 328.0, 356.0, A332.0, A356.0, F332.0, 357.0, 333.0, 359.0, 354.0 **Die Castings** 360.0, A380.0, 390.0, A413.0, A360.0, 383.0, 392.0, 380.0, A384.0, 413.0 | 40 to 125 500kg | As Cast or Solution Treated and Aged | 5500 to 6500 | 50 to 150 | Rough: .005 Finish: .0015 max. | A60IV |
| | | | 28 to 33 | 1.3 to 3.8 | Rough: .13 Finish: .038 max. | A60IV |
| **30. MAGNESIUM ALLOYS, WROUGHT‡‡** AZ21A, HK31A, AZ31B, HM21A, AZ31C, HM31A, AZ61A, ZK40A, AZ80A, ZK60A | 50 to 90 500kg | Annealed, Cold Drawn or Solution Treated and Aged | 5500 to 6500 | 50 to 150 | Rough: .005 Finish: .0015 max. | C60KV |
| | | | 28 to 33 | 1.3 to 3.8 | Rough: .13 Finish: .038 max. | C60KV |
| **31. MAGNESIUM ALLOYS, CAST‡‡** AM60A, AZ91C, ZE41A, AM100A, AZ92A, ZE63A, AS41A, EZ33A, ZH62A, AZ63A, HK31A, ZK51A, AZ81A, HZ32A, ZK61A, AZ91A, K1A, AZ91B, QE22A | 50 to 90 500kg | As Cast, Annealed or Solution Treated and Aged | 5500 to 6500 | 50 to 150 | Rough: .005 Finish: .0015 max. | C60KV |
| | | | 28 to 33 | 1.3 to 3.8 | Rough: .13 Finish: .038 max. | C60KV |
| **32. TITANIUM ALLOYS, WROUGHT‡‡** **Commercially Pure** 99.5, 99.2, 99.0, 98.9, Ti-0.2Pd, TiCODE-12 | 110 to 275 | Annealed | 4000 to 5500 | 50 to 150 | Rough: .001 Finish: .0005 max. | C60KV‡ |
| | | | 20 to 28 | 1.3 to 3.8 | Rough: .025 Finish: .013 max. | C60KV‡ |

See section 16 for Cutting Fluid Recommendations.

*See section 8.12 for Work Traverse Rates. As recommended starting conditions—use a regulating wheel angle with a positive inclination of 3° and a regulating wheel speed of 25 to 40 rpm.

†Wheel recommendations are for wet grinding 0.8- to 2-inch [20 to 50 mm] diameter work. For LARGER diameter work—use a softer grade and/or coarser grit wheel. For SMALLER diameter work—use a harder grade wheel. Also see section 20.2, Grinding Guidelines.

‡Use friable (green grit) silicon carbide.

‡‡CAUTION: Potential Fire Hazard. Exercise caution in grinding and disposing of swarf. Do NOT use water or water-miscible cutting fluids for magnesium alloys.

| MATERIAL | HARD-NESS Bhn | CONDITION | WHEEL SPEED fpm / m/s | THRUFEED OF WORK* in/min / m/min | INFEED on dia. in/pass / mm/pass | WHEEL IDENTIFI-CATION† ANSI / ISO |
|---|---|---|---|---|---|---|
| **Alpha and Alpha-Beta Alloys**<br>Ti-8Mn<br>Ti-1Al-8V-5Fe<br>Ti-2Al-11Sn-5Zr-1Mo<br>Ti-3Al-2.5V<br>Ti-5Al-2Sn-2Zr-4Mo-4Cr (Ti-17)<br>Ti-5Al-2.5Sn<br>Ti-5Al-2.5Sn ELI<br>Ti-5Al-6Sn-2Zr-1Mo<br>Ti-6Al-2Cb-1Ta-0.8Mo | 300 to 380 | Annealed | 4000 to 5500 | 50 to 150 | Rough: .001 Finish: .0005 max. | C60KV‡ |
| | | | 20 to 28 | 1.3 to 3.8 | Rough: .025 Finish: .013 max. | C60KV‡ |
| Ti-6Al-4V<br>Ti-6Al-4V ELI<br>Ti-6Al-6V-2Sn<br>Ti-6Al-2Sn-4Zr-2Mo<br>Ti-6Al-2Sn-4Zr-2Mo-.25Si<br>Ti-6Al-2Sn-4Zr-6Mo<br>Ti-7Al-4Mo<br>Ti-8Al-1Mo-1V | 320 to 440 | Solution Treated and Aged | 4000 to 5500 | 50 to 150 | Rough: .001 Finish: .0005 max. | C60KV‡ |
| | | | 20 to 28 | 1.3 to 3.8 | Rough: .025 Finish: .013 max. | C60KV‡ |
| **Beta Alloys**<br>Ti-3Al-8V-6Cr-4Mo-4Zr<br>Ti-8Mo-8V-2Fe-3Al<br>Ti-11.5Mo-6Zr-4.5Sn<br>Ti-10V-2Fe-3Al<br>Ti-13V-11Cr-3Al | 275 to 350 | Annealed or Solution Treated | 4000 to 5500 | 50 to 150 | Rough: .001 Finish: .0005 max. | C60KV‡ |
| | | | 20 to 28 | 1.3 to 3.8 | Rough: .025 Finish: .013 max. | C60KV‡ |
| | 350 to 440 | Solution Treated and Aged | 4000 to 5500 | 50 to 150 | Rough: .001 Finish: .0005 max. | C60KV‡ |
| | | | 20 to 28 | 1.3 to 3.8 | Rough: .025 Finish: .013 max. | C60KV‡ |
| **33. TITANIUM ALLOYS, CAST‡‡**<br>**Commercially Pure**<br>99.0<br>Ti-0.2Pd<br>ASTM B367: Grades C-1, C-2, C-3, C-4, C-7A, C-7B, C-8A, C-8B | 150 to 250 | As Cast or As Cast and Annealed | 4000 to 5500 | 50 to 150 | Rough: .001 Finish: .0005 max. | C60KV‡ |
| | | | 20 to 28 | 1.3 to 3.8 | Rough: .025 Finish: .013 max. | C60KV‡ |
| **Alpha and Alpha-Beta Alloys**<br>Ti-5Al-2.5Sn<br>Ti-6Al-4V<br>Ti-6Al-2Sn-4Zr-2Mo<br>Ti-8Al-1Mo-1V<br>ASTM B367: Grades C-5, C-6 | 300 to 350 | As Cast or As Cast and Annealed | 4000 to 5500 | 50 to 150 | Rough: .001 Finish: .0005 max. | C60KV‡ |
| | | | 20 to 28 | 1.3 to 3.8 | Rough: .025 Finish: .013 max. | C60KV‡ |

See section 16 for Cutting Fluid Recommendations.
*See section 8.12 for Work Traverse Rates. As recommended starting conditions—use a regulating wheel angle with a positive inclination of 3° and a regulating wheel speed of 25 to 40 rpm.
†Wheel recommendations are for wet grinding 0.8- to 2-inch [20 to 50 mm] diameter work. For LARGER diameter work—use a softer grade and/or coarser grit wheel. For SMALLER diameter work—use a harder grade wheel. Also see section 20.2, Grinding Guidelines.

‡Use friable (green grit) silicon carbide.
‡‡CAUTION: Potential Fire Hazard. Exercise caution in grinding and disposing of swarf.

# 8.11 Centerless Grinding

| MATERIAL | HARDNESS Bhn | CONDITION | WHEEL SPEED fpm / m/s | THRUFEED OF WORK* in/min / m/min | INFEED on dia. in/pass / mm/pass | WHEEL IDENTIFICATION† ANSI / ISO |
|---|---|---|---|---|---|---|
| **34. COPPER ALLOYS, WROUGHT**<br>101 116 143 182<br>102 119 145 184<br>104 120 147 185<br>105 121 150 187<br>107 122 155 189<br>109 125 162 190<br>110 127 165 191<br>111 128 170†† 192<br>113 129 172†† 194<br>114 130 173†† 195<br>115 142 175†† | 10 $R_B$ to 70 $R_B$ | Annealed | 5500 to 6500 | 50 to 150 | Rough: .005 Finish: .0015 max. | C54KV |
| | | | 28 to 33 | 1.3 to 3.8 | Rough: .13 Finish: .038 max. | C54KV |
| | 60 $R_B$ to 100 $R_B$ | Cold Drawn | 5500 to 6500 | 50 to 150 | Rough: .005 Finish: .0015 max. | C54KV |
| | | | 28 to 33 | 1.3 to 3.8 | Rough: .13 Finish: .038 max. | C54KV |
| 210 332 368 464<br>220 335 370 465<br>226 340 377 466<br>230 342 385 467<br>240 349 411 482<br>260 350 413 485<br>268 353 425 667<br>270 356 435 687<br>280 360 442 688<br>314 365 443 694<br>316 366 444<br>330 367 445 | 10 $R_B$ to 70 $R_B$ | Annealed | 5500 to 6500 | 50 to 150 | Rough: .005 Finish: .0015 max. | C54KV |
| | | | 28 to 33 | 1.3 to 3.8 | Rough: .13 Finish: .038 max. | C54KV |
| | 60 $R_B$ to 100 $R_B$ | Cold Drawn | 5500 to 6500 | 50 to 150 | Rough: .005 Finish: .0015 max. | C54KV |
| | | | 28 to 33 | 1.3 to 3.8 | Rough: .13 Finish: .038 max. | C54KV |
| 505 613 632<br>510 614 638<br>511 618 642<br>521 619 651<br>524 623 655<br>544 624 674<br>608 625 675<br>610 630 | 10 $R_B$ to 70 $R_B$ | Annealed | 5500 to 6500 | 50 to 150 | Rough: .005 Finish: .0015 max. | C54KV |
| | | | 28 to 33 | 1.3 to 3.8 | Rough: .13 Finish: .038 max. | C54KV |
| | 60 $R_B$ to 100 $R_B$ | Cold Drawn | 5500 to 6500 | 50 to 150 | Rough: .005 Finish: .0015 max. | C54KV |
| | | | 28 to 33 | 1.3 to 3.8 | Rough: .13 Finish: .038 max. | C54KV |

See section 16 for Cutting Fluid Recommendations.

*See section 8.12 for Work Traverse Rates. As recommended starting conditions—use a regulating wheel angle with a positive inclination of 3° and a regulating wheel speed of 25 to 40 rpm.

†Wheel recommendations are for wet grinding 0.8- to 2-inch [20 to 50 mm] diameter work. For LARGER diameter work—use a softer grade and/or coarser grit wheel. For SMALLER diameter work—use a harder grade wheel. Also see section 20.2, Grinding Guidelines.

††CAUTION: Toxic Material, refer to National Institute for Occupational Safety and Health (NIOSH) for Precautions.

| MATERIAL | HARD-NESS Bhn | CONDITION | WHEEL SPEED fpm m/s | THRUFEED OF WORK* in/min m/min | INFEED on dia. in/pass mm/pass | WHEEL IDENTIFI-CATION† ANSI ISO |
|---|---|---|---|---|---|---|
| 706 752<br>710 754<br>715 757<br>725 770<br>745 782 | $10\ R_B$ to $70\ R_B$ | Annealed | 5500 to 6500 | 50 to 150 | Rough: .005<br>Finish: .0015 max. | C54KV |
| | | | 28 to 33 | 1.3 to 3.8 | Rough: .13<br>Finish: .038 max. | C54KV |
| | $60\ R_B$ to $100\ R_B$ | Cold Drawn | 5500 to 6500 | 50 to 150 | Rough: .005<br>Finish: .0015 max. | C54KV |
| | | | 28 to 33 | 1.3 to 3.8 | Rough: .13<br>Finish: .038 max. | C54KV |
| **35. COPPER ALLOYS, CAST**<br>801 814 824††<br>803 815 825††<br>805 817†† 826††<br>807 818†† 827††<br>809 820†† 828††<br>811 821††<br>813 822†† | 40 to 200 500kg | As Cast or Heat Treated | 5500 to 6500 | 50 to 150 | Rough: .005<br>Finish: .0015 max. | C54KV |
| | | | 28 to 33 | 1.3 to 3.8 | Rough: .13<br>Finish: .038 max. | C54KV |
| | $34\ R_C$ to $45\ R_C$ | Heat Treated | 5500 to 6500 | 50 to 150 | Rough: .005<br>Finish: .0015 max. | A60LV |
| | | | 28 to 33 | 1.3 to 3.8 | Rough: .13<br>Finish: .038 max. | A60LV |
| 833 854 867<br>834 855 868<br>836 857 872<br>838 858 874<br>842 861 875<br>844 862 876<br>848 863 878<br>852 864 879<br>853 865 | 35 to 200 500kg | As Cast | 5500 to 6500 | 50 to 150 | Rough: .005<br>Finish: .0015 max. | C54KV |
| | | | 28 to 33 | 1.3 to 3.8 | Rough: .13<br>Finish: .038 max. | C54KV |
| 902 916 934 948<br>903 917 935 952<br>905 922 937 953<br>907 923 938 954<br>909 925 939 955<br>910 926 943 956<br>911 927 944 957<br>913 928 945 958<br>915 932 947 | 40 to 100 500kg | As Cast | 5500 to 6500 | 50 to 150 | Rough: .005<br>Finish: .0015 max. | C54KV |
| | | | 28 to 33 | 1.3 to 3.8 | Rough: .13<br>Finish: .038 max. | C54KV |

See section 16 for Cutting Fluid Recommendations.

*See section 8.12 for Work Traverse Rates. As recommended starting conditions—use a regulating wheel angle with a positive inclination of 3° and a regulating wheel speed of 25 to 40 rpm.

†Wheel recommendations are for wet grinding 0.8- to 2-inch [20 to 50 mm] diameter work. For LARGER diameter work—use a softer grade and/or coarser grit wheel. For SMALLER diameter work—use a harder grade wheel. Also see section 20.2, Grinding Guidelines.

††CAUTION: Toxic Material, refer to National Institute for Occupational Safety and Health (NIOSH) for Precautions.

# 8.11 Centerless Grinding

| MATERIAL | HARD-NESS<br>Bhn | CONDITION | WHEEL SPEED<br>fpm<br>m/s | THRUFEED OF WORK*<br>in/min<br>m/min | INFEED on dia.<br>in/pass<br>mm/pass | WHEEL IDENTIFI-CATION†<br>ANSI<br>ISO |
|---|---|---|---|---|---|---|
| **35. COPPER ALLOYS, CAST (cont.)**<br>(materials listed on preceding page) | Over<br>100<br>500kg | As Cast<br>or<br>Heat Treated | 5500<br>to<br>6500 | 50<br>to<br>150 | Rough:<br>.005<br>Finish:<br>.0015 max. | A60LV |
| | | | 28<br>to<br>33 | 1.3<br>to<br>3.8 | Rough:<br>.13<br>Finish:<br>.038 max. | A60LV |
| 962<br>963<br>964<br>966††<br>973<br>974<br>976<br>978<br>993 | 50<br>to<br>100<br>500kg | As Cast | 5500<br>to<br>6500 | 50<br>to<br>150 | Rough:<br>.005<br>Finish:<br>.0015 max. | C54KV |
| | | | 28<br>to<br>33 | 1.3<br>to<br>3.8 | Rough:<br>.13<br>Finish:<br>.038 max. | C54KV |
| | Over<br>100<br>500kg | As Cast<br>or<br>Heat Treated | 5500<br>to<br>6500 | 50<br>to<br>150 | Rough:<br>.005<br>Finish:<br>.0015 max. | A60LV |
| | | | 28<br>to<br>33 | 1.3<br>to<br>3.8 | Rough:<br>.13<br>Finish:<br>.038 max. | A60LV |
| **36. NICKEL ALLOYS, WROUGHT AND CAST**<br>Nickel 200<br>Nickel 201<br>Nickel 205<br>Nickel 211<br>Nickel 220<br>Nickel 230 | 80<br>to<br>170 | Annealed<br>or<br>Cold Drawn | 5500<br>to<br>6500 | 50<br>to<br>150 | Rough:<br>.005<br>Finish:<br>.0015 max. | C60KV |
| | | | 28<br>to<br>33 | 1.3<br>to<br>3.8 | Rough:<br>.13<br>Finish:<br>.038 max. | C60KV |
| MONEL Alloy 400<br>MONEL Alloy 401<br>MONEL Alloy 404<br>MONEL Alloy R405<br>ASTM A296: Grades CZ-100, M-35<br>ASTM A494: Grades CZ-100, M-35 | 115<br>to<br>240 | Annealed,<br>Cold Drawn<br>or<br>Cast | 5500<br>to<br>6500 | 50<br>to<br>150 | Rough:<br>.005<br>Finish:<br>.0015 max. | C60KV |
| | | | 28<br>to<br>33 | 1.3<br>to<br>3.8 | Rough:<br>.13<br>Finish:<br>.038 max. | C60KV |
| DURANICKEL Alloy 301<br>MONEL Alloy 502<br>MONEL Alloy K500<br>NI-SPAN-C Alloy 902<br>PERMANICKEL Alloy 300 | 150<br>to<br>320 | Solution Treated | 5500<br>to<br>6500 | 50<br>to<br>150 | Rough:<br>.005<br>Finish:<br>.0015 max. | C60KV |
| | | | 28<br>to<br>33 | 1.3<br>to<br>3.8 | Rough:<br>.13<br>Finish:<br>.038 max. | C60KV |

See section 16 for Cutting Fluid Recommendations.

*See section 8.12 for Work Traverse Rates. As recommended starting conditions—use a regulating wheel angle with a positive inclination of 3° and a regulating wheel speed of 25 to 40 rpm.

†Wheel recommendations are for wet grinding 0.8- to 2-inch [20 to 50 mm] diameter work. For LARGER diameter work—use a softer grade and/or coarser grit wheel. For SMALLER diameter work—use a harder grade wheel. Also see section 20.2, Grinding Guidelines.

††CAUTION: Toxic Material, refer to National Institute for Occupational Safety and Health (NIOSH) for Precautions.

| MATERIAL | HARD-NESS Bhn | CONDITION | WHEEL SPEED fpm / m/s | THRUFEED OF WORK* in/min / m/min | INFEED on dia. in/pass / mm/pass | WHEEL IDENTIFI-CATION† ANSI / ISO |
|---|---|---|---|---|---|---|
| **36. NICKEL ALLOYS, WROUGHT AND CAST (cont.)** (materials listed on preceding page) | 330 to 360 | Aged | 5500 to 6500 | 50 to 150 | Rough: .003 Finish: .001 max. | C60KV |
| | | | 28 to 33 | 1.3 to 3.8 | Rough: .075 Finish: .025 max. | C60KV |
| **37. BERYLLIUM NICKEL ALLOYS, WROUGHT AND CAST††** Berylco 440, Berylco 41C, Berylco 42C, Berylco 43C, Brush Alloy 200C, Brush Alloy 220C, Brush Alloy 260C | 200 to 250 | As Cast or Solution Treated | 5500 to 6500 | 50 to 150 | Rough: .005 Finish: .0015 max. | A60LV |
| | | | 28 to 33 | 1.3 to 3.8 | Rough: .13 Finish: .038 max. | A60LV |
| | 283 to 425 | Hardened or Aged | 5500 to 6500 | 50 to 150 | Rough: .003 Finish: .001 max. | A60KV |
| | | | 28 to 33 | 1.3 to 3.8 | Rough: .075 Finish: .025 max. | A60KV |
| | 47 $R_C$ to 52 $R_C$ | Hardened or Aged | 5500 to 6500 | 50 to 150 | Rough: .002 Finish: .001 max. | A60JV |
| | | | 28 to 33 | 1.3 to 3.8 | Rough: .050 Finish: .025 max. | A60JV |
| **38. NITINOL ALLOYS, WROUGHT** Nitinol 55Ni-45Ti, Nitinol 56Ni-44Ti, Nitinol 60Ni-40Ti | 210 to 340 | Wrought or Annealed | 4000 to 6000 | 50 to 150 | Rough: .003 Finish: .001 max. | C60KV or CA60PB |
| | | | 20 to 30 | 1.3 to 3.8 | Rough: .075 Finish: .025 max. | C60KV or CA60PB |
| | 48 $R_C$ to 52 $R_C$ | Quenched | 4000 to 6000 | 50 to 150 | Rough: .001 Finish: .0005 max. | C60JV or CA60PB |
| | | | 20 to 30 | 1.3 to 3.8 | Rough: .025 Finish: .013 max. | C60JV or CA60PB |

See section 16 for Cutting Fluid Recommendations.
*See section 8.12 for Work Traverse Rates. As recommended starting conditions—use a regulating wheel angle with a positive inclination of 3° and a regulating wheel speed of 25 to 40 rpm.
†Wheel recommendations are for wet grinding 0.8- to 2-inch [20 to 50 mm] diameter work. For LARGER diameter work—use a softer grade and/or coarser grit wheel. For SMALLER diameter work—use a harder grade wheel. Also see section 20.2, Grinding Guidelines.

††CAUTION: Toxic Material, refer to National Institute for Occupational Safety and Health (NIOSH) for Precautions.

# 8.11 Centerless Grinding

| MATERIAL | HARDNESS Bhn | CONDITION | WHEEL SPEED fpm / m/s | THRUFEED OF WORK* in/min / m/min | INFEED on dia. in/pass / mm/pass | WHEEL IDENTIFICATION† ANSI / ISO |
|---|---|---|---|---|---|---|
| **39. HIGH TEMPERATURE ALLOYS, WROUGHT AND CAST**<br>**Nickel Base, Wrought**<br>AF2-1DA<br>Astroloy<br>Haynes Alloy 263<br>IN-102<br>Incoloy Alloy 901<br>Incoloy Alloy 903<br>Inconel Alloy 617<br>Inconel Alloy 625<br>Inconel Alloy 700<br>Inconel Alloy 702<br>Inconel Alloy 706<br>Inconel Alloy 718<br>Inconel Alloy 721<br>Inconel Alloy 722<br>Inconel Alloy X-750<br>Inconel Alloy 751<br>M252<br>Nimonic 75<br>Nimonic 80<br>Nimonic 90<br>Nimonic 95<br>Rene 41<br>Rene 63<br>Rene 77<br>Rene 95<br>Udimet 500<br>Udimet 700<br>Udimet 710<br>Unitemp 1753<br>Waspaloy | 200 to 390 | Annealed or Solution Treated | 3000 to 4000 | 50 to 150 | Rough: .003<br>Finish: .001 max. | A60KV |
| | | | 15 to 20 | 1.3 to 3.8 | Rough: .075<br>Finish: .025 max. | A60KV |
| | 300 to 475 | Solution Treated and Aged | 3000 to 4000 | 50 to 150 | Rough: .003<br>Finish: .001 max. | A60KV |
| | | | 15 to 20 | 1.3 to 3.8 | Rough: .075<br>Finish: .025 max. | A60KV |
| **Nickel Base, Wrought**<br>Hastelloy Alloy B<br>Hastelloy Alloy B-2<br>Hastelloy Alloy C<br>Hastelloy Alloy C-276<br>Hastelloy Alloy G<br>Hastelloy Alloy S<br>Hastelloy Alloy X<br>Incoloy Alloy 804<br>Incoloy Alloy 825<br>Inconel Alloy 600<br>Inconel Alloy 601<br>Refractaloy 26<br>Udimet 630 | 140 to 220 | Annealed or Solution Treated | 3000 to 4000 | 50 to 150 | Rough: .003<br>Finish: .001 max. | A60KV |
| | | | 15 to 20 | 1.3 to 3.8 | Rough: .075<br>Finish: .025 max. | A60KV |
| | 240 to 310 | Cold Drawn or Aged | 3000 to 4000 | 50 to 150 | Rough: .003<br>Finish: .001 max. | A60KV |
| | | | 15 to 20 | 1.3 to 3.8 | Rough: .075<br>Finish: .025 max. | A60KV |

See section 16 for Cutting Fluid Recommendations.

*See section 8.12 for Work Traverse Rates. As recommended starting conditions—use a regulating wheel angle with a positive inclination of 3° and a regulating wheel speed of 25 to 40 rpm.

†Wheel recommendations are for wet grinding 0.8- to 2-inch [20 to 50 mm] diameter work. For LARGER diameter work—use a softer grade and/or coarser grit wheel. For SMALLER diameter work—use a harder grade wheel. Also see section 20.2, Grinding Guidelines.

| MATERIAL | HARD-NESS Bhn | CONDITION | WHEEL SPEED fpm / m/s | THRUFEED OF WORK* in/min / m/min | INFEED on dia. in/pass / mm/pass | WHEEL IDENTIFI-CATION† ANSI / ISO |
|---|---|---|---|---|---|---|
| **Nickel Base, Wrought**<br>TD-Nickel††<br>TD-Ni-Cr†† | 180 to 200 | As Rolled | 5500 to 6500 | 50 to 150 | Rough: .005<br>Finish: .0015 max. | A60KV |
| | | | 28 to 33 | 1.3 to 3.8 | Rough: .13<br>Finish: .038 max. | A60KV |
| **Nickel Base, Cast**<br>B-1900<br>GMR-235<br>GMR-235D<br>Hastelloy Alloy B<br>Hastelloy Alloy C<br>Hastelloy Alloy D<br>IN-100 (Rene 100)<br>IN-738<br>IN-792<br>Inconel Alloy 713C<br>Inconel Alloy 718<br>M252<br>MAR-M200<br>MAR-M246<br>MAR-M421<br>MAR-M432<br>Rene 80<br>Rene 125<br>SEL<br>SEL 15<br>TRW VI A<br>Udimet 500<br>Udimet 700<br>ASTM A296: Grades CW-12M, N-12M, CY-40<br>ASTM A297: Grades HW, HX<br>ASTM A494: Grades N-12M-1, N-12M-2, CY-40, CW-12M-1, CW-12M-2<br>ASTM A608: Grades HW50, HX50 | 200 to 425 | As Cast or Cast and Aged | 3000 to 4000 | 50 to 150 | Rough: .003<br>Finish: .001 max. | A60KV |
| | | | 15 to 20 | 1.3 to 3.8 | Rough: .075<br>Finish: .025 max. | A60KV |
| | | | | | | |
| | | | | | | |
| | | | | | | |
| | | | | | | |
| **Cobalt Base, Wrought**<br>AiResist 213<br>Haynes Alloy 25 (L605)<br>Haynes Alloy 188<br>J-1570<br>MAR-M905<br>MAR-M918<br>S-816<br>V-36 | 180 to 230 | Solution Treated | 3000 to 4000 | 50 to 150 | Rough: .003<br>Finish: .001 max. | A60KV |
| | | | 15 to 20 | 1.3 to 3.8 | Rough: .075<br>Finish: .025 max. | A60KV |
| | 270 to 320 | Solution Treated and Aged | 3000 to 4000 | 50 to 150 | Rough: .003<br>Finish: .001 max. | A60KV |
| | | | 15 to 20 | 1.3 to 3.8 | Rough: .075<br>Finish: .025 max. | A60KV |

See section 16 for Cutting Fluid Recommendations.

*See section 8.12 for Work Traverse Rates. As recommended starting conditions—use a regulating wheel angle with a positive inclination of 3° and a regulating wheel speed of 25 to 40 rpm.

†Wheel recommendations are for wet grinding 0.8- to 2-inch [20 to 50 mm] diameter work. For LARGER diameter work—use a softer grade and/or coarser grit wheel. For SMALLER diameter work—use a harder grade wheel. Also see section 20.2, Grinding Guidelines.

††CAUTION: Toxic Material, refer to National Institute for Occupational Safety and Health (NIOSH) for Precautions.

# 8.11 Centerless Grinding

| MATERIAL | HARDNESS Bhn | CONDITION | WHEEL SPEED fpm / m/s | THRUFEED OF WORK* in/min / m/min | INFEED on dia. in/pass / mm/pass | WHEEL IDENTIFICATION† ANSI / ISO |
|---|---|---|---|---|---|---|
| **39. HIGH TEMPERATURE ALLOYS, WROUGHT AND CAST (cont.)**<br>**Cobalt Base, Cast**<br>AiResist 13<br>AiResist 215<br>FSX-414<br>HS-6<br>HS-21<br>HS-31 (X-40)<br>HOWMET #3<br>MAR-M302<br>MAR-M322<br>MAR-M509<br>NASA Co-W-Re<br>WI-52<br>X-45 | 220 to 290 | As Cast or Cast and Aged | 3000 to 4000 | 50 to 150 | Rough: .003<br>Finish: .001 max. | A60KV |
| | | | 15 to 20 | 1.3 to 3.8 | Rough: .075<br>Finish: .025 max. | A60KV |
| | 290 to 425 | As Cast or Cast and Aged | 3000 to 4000 | 50 to 150 | Rough: .003<br>Finish: .001 max. | A60KV |
| | | | 15 to 20 | 1.3 to 3.8 | Rough: .075<br>Finish: .025 max. | A60KV |
| **Iron Base, Wrought**<br>A-286<br>Discaloy<br>Incoloy Alloy 800<br>Incoloy Alloy 800H<br>Incoloy Alloy 801<br>Incoloy Alloy 802<br>N-155<br>V-57<br>W-545<br>16-25-6<br>19-9DL | 180 to 230 | Solution Treated | 3000 to 4000 | 50 to 150 | Rough: .003<br>Finish: .001 max. | A60KV |
| | | | 15 to 20 | 1.3 to 3.8 | Rough: .075<br>Finish: .025 max. | A60KV |
| | 250 to 320 | Solution Treated and Aged | 3000 to 4000 | 50 to 150 | Rough: .003<br>Finish: .001 max. | A60KV |
| | | | 15 to 20 | 1.3 to 3.8 | Rough: .075<br>Finish: .025 max. | A60KV |
| **40. REFRACTORY ALLOYS, WROUGHT, CAST, P/M****<br>**Columbium**** †† ‡‡<br>C103<br>C129Y<br>Cb-1Zr<br>Cb-752<br>FS-85<br>FS-291<br>WC-3015 | 170 to 225 | Stress Relieved | 4000 | 50 to 150 | Rough: .001<br>Finish: .0005 max. | A60LV |
| | | | 20 | 1.3 to 3.8 | Rough: .025<br>Finish: .013 max. | A60LV |
| | | | | | | |
| | | | | | | |

See section 16 for Cutting Fluid Recommendations.

*See section 8.12 for Work Traverse Rates. As recommended starting conditions—use a regulating wheel angle with a positive inclination of 3° and a regulating wheel speed of 25 to 40 rpm.

†Wheel recommendations are for wet grinding 0.8- to 2-inch [20 to 50 mm] diameter work. For LARGER diameter work—use a softer grade and/or coarser grit wheel. For SMALLER diameter work—use a harder grade wheel. Also see section 20.2, Grinding Guidelines.

**Due to the brittleness of refractory alloys, cracking, chipping, flaking and breakout tend to occur, particularly on the edges of the machined surfaces.

††CAUTION: Toxic Material, refer to National Institute for Occupational Safety and Health (NIOSH) for Precautions.

‡‡CAUTION: Potential Fire Hazard. Exercise caution in grinding and disposing of swarf.

| MATERIAL | HARD-NESS Bhn | CONDITION | WHEEL SPEED fpm / m/s | THRUFEED OF WORK* in/min / m/min | INFEED on dia. in/pass / mm/pass | WHEEL IDENTIFI-CATION† ANSI / ISO |
|---|---|---|---|---|---|---|
| **Molybdenum****<br>Mo<br>Mo-50Re<br>TZC<br>TZM | 220 to 290 | Stress Relieved | 2000 to 4000 | 50 to 150 | Rough: .001<br>Finish: .0005 max. | A60MV |
| | | | 10 to 20 | 1.3 to 3.8 | Rough: .025<br>Finish: .013 max. | A60MV |
| **Tantalum****<br>ASTAR 811C<br>T-111<br>T-222<br>Ta-10W<br>Ta-Hf<br>Ta63 | 200 to 250 | Stress Relieved | 2000 to 4000 | 50 to 150 | Rough: .001<br>Finish: .0005 max. | A60KV |
| | | | 10 to 20 | 1.3 to 3.8 | Rough: .025<br>Finish: .013 max. | A60KV |
| **Tungsten****<br>85% density<br>93% density<br>96% density<br>100% density | 180 to 320 | Pressed and Sintered, Forged, or Arc Cast | 2000 | 50 to 150 | Rough: .001<br>Finish: .0003 max. | C70LV |
| | | | 10 | 1.3 to 3.8 | Rough: .025<br>Finish: .008 max. | C70LV |
| **Tungsten - 2 Thoria**** †† | 260 to 320 | Pressed and Sintered | 2000 | 50 to 150 | Rough: .001<br>Finish: .0005 max. | C70LV |
| | | | 10 | 1.3 to 3.8 | Rough: .025<br>Finish: .013 max. | C70LV |
| **Tungsten Alloys****<br>GE-218<br>W-15Mo<br>W-5Re<br>W-25Re<br>W-25Re-30Mo | 260 to 320 | As Cast | 2000 | 50 to 150 | Rough: .001<br>Finish: .0005 max. | C70LV |
| | | | 10 | 1.3 to 3.8 | Rough: .025<br>Finish: .013 max. | C70LV |
| **Tungsten Alloys****<br>Gyromet<br>Mallory 2000<br>W-10Ag<br>W-7Ni-4Cu | 290 to 320 | Pressed and Sintered | 2000 | 50 to 150 | Rough: .001<br>Finish: .0005 max. | C70LV |
| | | | 10 | 1.3 to 3.8 | Rough: .025<br>Finish: .013 max. | C70LV |

See section 16 for Cutting Fluid Recommendations.

*See section 8.12 for Work Traverse Rates. As recommended starting conditions—use a regulating wheel angle with a positive inclination of 3° and a regulating wheel speed of 25 to 40 rpm.

†Wheel recommendations are for wet grinding 0.8- to 2-inch [20 to 50 mm] diameter work. For LARGER diameter work—use a softer grade and/or coarser grit wheel. For SMALLER diameter work—use a harder grade wheel. Also see section 20.2, Grinding Guidelines.

**Due to the brittleness of refractory alloys, cracking, chipping, flaking and breakout tend to occur, particularly on the edges of the machined surfaces.

††CAUTION: Toxic Material, refer to National Institute for Occupational Safety and Health (NIOSH) for Precautions.

# 8.11 Centerless Grinding

| MATERIAL | HARD-NESS<br>Bhn | CONDITION | WHEEL SPEED<br>fpm<br>m/s | THRUFEED OF WORK*<br>in/min<br>m/min | INFEED on dia.<br>in/pass<br>mm/pass | WHEEL IDENTIFI-CATION†<br>ANSI<br>ISO |
|---|---|---|---|---|---|---|
| **40. REFRACTORY ALLOYS, WROUGHT, CAST, P/M** (cont.)**<br>**Tungsten Alloys****<br>Anviloy 1100<br>Anviloy 1150<br>Anviloy 1200 | 290<br>to<br>320 | Pressed<br>and<br>Sintered | 2000 | 50<br>to<br>150 | Rough:<br>.001<br>Finish:<br>.0005 max. | C70LV |
| | | | 10 | 1.3<br>to<br>3.8 | Rough:<br>.025<br>Finish:<br>.013 max. | C70LV |
| **44. URANIUM, WROUGHT†† ‡‡** | 56 $R_A$<br>to<br>58 $R_A$ | Annealed | 4000<br>to<br>6000 | 75<br>to<br>125 | Rough:<br>.005<br>Finish:<br>.002 max. | C46JB<br>or<br>C46KV |
| | | | 20<br>to<br>30 | 1.9<br>to<br>3.2 | Rough:<br>.13<br>Finish:<br>.050 max. | C46JB<br>or<br>C46KV |
| **45. ZIRCONIUM ALLOYS, WROUGHT‡‡**<br>Zr-2%Hf (Grade 11)<br>Zr-0.001%Hf (Grade 21)<br>Zircaloy 2 (Grade 32)<br>Zircaloy 4 (Grade 34) | 140<br>to<br>280 | Rolled,<br>Extruded<br>or Forged | 3000 | 50<br>to<br>150 | Rough:<br>.001<br>Finish:<br>.0005 max. | Rough: A60TB or CA60RB<br>Finish:<br>CA80PB |
| | | | 15 | 1.3<br>to<br>3.8 | Rough:<br>.025<br>Finish:<br>.013 max. | Rough: A60TB or CA60RB<br>Finish:<br>CA80PB |
| **47. POWDER METAL ALLOY §§**<br>**Copper§§** | 50 $R_F$<br>to<br>70 $R_F$ | As Sintered | 5500<br>to<br>6500 | 50<br>to<br>150 | Rough:<br>.005<br>Finish:<br>.0015 max. | C54KV |
| | | | 28<br>to<br>33 | 1.3<br>to<br>3.8 | Rough:<br>.13<br>Finish:<br>.038 max. | C54KV |
| **Brasses§§**<br>CZP-0218-T<br>CZP-0218-U<br>CZP-0218-W<br>90Cu-10Zn<br>90Cu-10Zn-0.5Pb<br>70Cu-30Zn<br>68.5Cu-30Zn-1.5Pb | 35 $R_H$<br>to<br>81 $R_H$ | As Sintered | 5500<br>to<br>6500 | 50<br>to<br>150 | Rough:<br>.005<br>Finish:<br>.0015 max. | C54KV |
| | | | 28<br>to<br>33 | 1.3<br>to<br>3.8 | Rough:<br>.13<br>Finish:<br>.038 max. | C54KV |
| **Bronzes§§**<br>CT-0010-N<br>CT-0010-R<br>CT-0010-S<br>95Cu-5Al<br>77Cu-15Pb-7Sn-1Fe-1C | 30 $R_F$<br>to<br>75 $R_F$ | As Sintered | 5500<br>to<br>6500 | 50<br>to<br>150 | Rough:<br>.005<br>Finish:<br>.0015 max. | C54KV |
| | | | 28<br>to<br>33 | 1.3<br>to<br>3.8 | Rough:<br>.13<br>Finish:<br>.038 max. | C54KV |

See section 16 for Cutting Fluid Recommendations.

*See section 8.12 for Work Traverse Rates. As recommended starting conditions—use a regulating wheel angle with a positive inclination of 3° and a regulating wheel speed of 25 to 40 rpm.

†Wheel recommendations are for wet grinding 0.8- to 2-inch [20 to 50 mm] diameter work. For LARGER diameter work—use a softer grade and/or coarser grit wheel. For SMALLER diameter work—use a harder grade wheel. Also see section 20.2, Grinding Guidelines.

**Due to the brittleness of refractory alloys, cracking, chipping, flaking and breakout tend to occur, particularly on the edges of the machined surfaces.

††CAUTION: Toxic Material, refer to National Institute for Occupational Safety and Health (NIOSH) for Precautions.

‡‡CAUTION: Potential Fire Hazard. Exercise caution in grinding and disposing of swarf.

§§Grinding of low-density parts is not recommended because surface porosity will be reduced or lost.

| MATERIAL | HARD-NESS Bhn | CONDITION | WHEEL SPEED fpm m/s | THRUFEED OF WORK* in/min m/min | INFEED on dia. in/pass mm/pass | WHEEL IDENTIFI-CATION† ANSI ISO |
|---|---|---|---|---|---|---|
| **Copper-Nickel Alloys§§**<br>CZN-1818-T<br>CZN-1818-U<br>CZN-1818-W<br>CZNP-1618-U<br>CZNP-1618-W<br>90Cu-10Ni<br>62Cu-18Ni-18Zn-2Sn | $22\ R_H$ to $100\ R_H$ | As Sintered | 5500 to 6500 | 50 to 150 | Rough: .005<br>Finish: .0015 max. | C54KV |
| | | | 28 to 33 | 1.3 to 3.8 | Rough: .13<br>Finish: .038 max. | C54KV |
| **Nickel§§** | $45\ R_B$ | As Sintered | 5500 to 6500 | 50 to 150 | Rough: .005<br>Finish: .0015 max. | A46KV |
| | | | 28 to 33 | 1.3 to 3.8 | Rough: .13<br>Finish: .038 max. | A46KV |
| **Nickel Alloys§§**<br>67Ni-30Cu-3Fe | $34\ R_B$ to $50\ R_B$ | As Sintered | 5500 to 6500 | 50 to 150 | Rough: .005<br>Finish: .0015 max. | C60JV |
| | | | 28 to 33 | 1.3 to 3.8 | Rough: .13<br>Finish: .038 max. | C60JV |
| **Refractory Metal Base§§**<br>87W-13Cu<br>85W-15Ag<br>74W-26Cu<br>72.5W-27.5Ag<br>65W-35Ag<br>56W+C-44Cu<br>55W-45Cu<br>51W-49Ag<br>50W+C-50Ag | 101 to 260 | As Sintered | 2000 | 50 to 150 | Rough: .001<br>Finish: .0005 max. | C70LV |
| | | | 10 | 1.3 to 3.8 | Rough: .025<br>Finish: .013 max. | C70LV |
| **Refractory Metal Base§§**<br>61Mo-39Ag<br>50Mo-50Ag | $75\ R_B$ to $82\ R_B$ | As Sintered | 2000 to 4000 | 50 to 150 | Rough: .001<br>Finish: .0005 max. | A60MV |
| | | | 10 to 20 | 1.3 to 3.8 | Rough: .025<br>Finish: .013 max. | A60MV |

See section 16 for Cutting Fluid Recommendations.

*See section 8.12 for Work Traverse Rates. As recommended starting conditions—use a regulating wheel angle with a positive inclination of 3° and a regulating wheel speed of 25 to 40 rpm.

†Wheel recommendations are for wet grinding 0.8- to 2-inch [20 to 50 mm] diameter work. For LARGER diameter work—use a softer grade and/or coarser grit wheel. For SMALLER diameter work—use a harder grade wheel. Also see section 20.2, Grinding Guidelines.

§§Grinding of low-density parts is not recommended because surface porosity will be reduced or lost.

# 8.11 Centerless Grinding

| MATERIAL | HARD-NESS Bhn | CONDITION | WHEEL SPEED fpm / m/s | THRUFEED OF WORK* in/min / m/min | INFEED on dia. in/pass / mm/pass | WHEEL IDENTIFI-CATION† ANSI / ISO |
|---|---|---|---|---|---|---|
| **47. POWDER METAL ALLOYS§§ (cont.)**<br>**Irons§§**<br>F-0000-N<br>F-0000-P<br>F-0000-R<br>F-0000-S<br>F-0000-T | 50 to 67 500kg | As Sintered | 5500 to 6500 | 50 to 150 | Rough: .005<br>Finish: .002 max. | C46LV or A60KV |
| | | | 28 to 33 | 1.3 to 3.8 | Rough: .13<br>Finish: .050 max. | C46LV or A60KV |
| **Steels§§**<br>F-0008-P<br>F-0008-S<br>F-0005-S<br>FC-0205-S<br>FC-0208-P<br>FC-0208-S<br>FC-0508-P<br>FC-1000-N<br>FN-0205-S<br>FN-0205-T<br>FN-0405-R<br>FN-0405-S<br>FN-0405-T<br>FX-1005-T<br>FX-2008-T | 101 to 426 | As Sintered or Heat Treated | 5500 to 6500 | 50 to 150 | Rough: .005<br>Finish: .0015 max. | A60NV |
| | | | 28 to 33 | 1.3 to 3.8 | Rough: .13<br>Finish: .038 max. | A60NV |
| | | | | | | |
| | | | | | | |
| **Stainless Steels§§**<br>SS-303-R<br>SS-304-R<br>SS-316-R<br>SS-410-R | 107 to 285 | As Sintered or Heat Treated | 5500 to 6500 | 50 to 150 | Rough: .005<br>Finish: .0015 max. | C60LV |
| | | | 28 to 33 | 1.3 to 3.8 | Rough: .13<br>Finish: .038 max. | C60LV |
| **Aluminum Alloys§§**<br>90.5Al-5Sn-4Cu<br>88Al-5Sn-4Pb-3Cu<br>Al-1Mg-0.6Si-0.25Cu<br>Al-0.6Mg-0.4Si<br>Al-4.4Cu-0.8Si-0.4Mg | 55 $R_H$ to 98 $R_H$ | Solution Treated and Aged | 5500 to 6500 | 50 to 150 | Rough: .005<br>Finish: .0015 max. | C54JV |
| | | | 28 to 33 | 1.3 to 3.8 | Rough: .13<br>Finish: .038 max. | C54JV |
| **49. CARBIDES**<br>Titanium Carbide<br>Tungsten Carbide | 89 $R_A$ to 94 $R_A$ | — | 5500 to 6500 | 50 to 60 | Rough: .001<br>Finish: .0003 max. | C60KV |
| | | | 28 to 33 | 1.3 to 1.5 | Rough: .025<br>Finish: .008 max. | C60KV |

See section 16 for Cutting Fluid Recommendations.

*See section 8.12 for Work Traverse Rates. As recommended starting conditions—use a regulating wheel angle with a positive inclination of 3° and a regulating wheel speed of 25 to 40 rpm.

†Wheel recommendations are for wet grinding 0.8- to 2-inch [20 to 50 mm] diameter work. For LARGER diameter work—use a softer grade and/or coarser grit wheel. For SMALLER diameter work—use a harder grade wheel. Also see section 20.2, Grinding Guidelines.

§§Grinding of low-density parts is not recommended because surface porosity will be reduced or lost.

| MATERIAL | HARD-NESS Bhn | CONDITION | WHEEL SPEED fpm / m/s | THRUFEED OF WORK* in/min / m/min | INFEED on dia. in/pass / mm/pass | WHEEL IDENTIFI-CATION† ANSI / ISO |
|---|---|---|---|---|---|---|
| **50. FREE MACHINING MAGNETIC ALLOYS**<br>Magnetic Core Iron-FM (up to 2.5% Si) | 185 to 240 | Wrought | 5500 to 6500 | 50 to 150 | Rough: .005<br>Finish: .0015 max. | A60LV |
| | | | 28 to 33 | 1.3 to 3.8 | Rough: .13<br>Finish: .038 max. | A60LV |
| Hi Perm 49-FM | 185 to 240 | Wrought | 5500 to 6500 | 50 to 150 | Rough: .005<br>Finish: .0015 max. | C60LV |
| | | | 28 to 33 | 1.3 to 3.8 | Rough: .13<br>Finish: .038 max. | C60LV |
| **51. MAGNETIC ALLOYS**<br>Magnetic Core Iron (up to 4% Si) | 185 to 240 | Wrought | 5500 to 6500 | 50 to 150 | Rough: .005<br>Finish: .0015 max. | A60MV |
| | | | 28 to 33 | 1.3 to 3.8 | Rough: .13<br>Finish: .038 max. | A60MV |
| Hi Perm 49<br>HyMu 80 | 185 to 240 | Wrought | 5500 to 6500 | 50 to 150 | Rough: .005<br>Finish: .0015 max. | C60LV |
| | | | 28 to 33 | 1.3 to 3.8 | Rough: .13<br>Finish: .038 max. | C60LV |
| Alnico I<br>Alnico II<br>Alnico III<br>Alnico IV<br>Alnico V<br>Alnico V-7<br>Alnico XII<br>Columax-5<br>Hyflux Alnico V-7 | 45 $R_C$ to 58 $R_C$ | As Cast | 5500 to 6500 | 50 to 150 | Rough: .003<br>Finish: .001 max. | A60KV |
| | | | 28 to 33 | 1.3 to 3.8 | Rough: .075<br>Finish: .025 max. | A60KV |
| **52. FREE MACHINING CONTROLLED EXPANSION ALLOYS**<br>Invar 36 | 125 to 220 | Annealed or Cold Drawn | 5500 to 6500 | 50 to 150 | Rough: .005<br>Finish: .0015 max. | C60LV |
| | | | 28 to 33 | 1.3 to 3.8 | Rough: .13<br>Finish: .038 max. | C60LV |

See section 16 for Cutting Fluid Recommendations.

*See section 8.12 for Work Traverse Rates. As recommended starting conditions—use a regulating wheel angle with a positive inclination of 3° and a regulating wheel speed of 25 to 40 rpm.

†Wheel recommendations are for wet grinding 0.8- to 2-inch [20 to 50 mm] diameter work. For LARGER diameter work—use a softer grade and/or coarser grit wheel. For SMALLER diameter work—use a harder grade wheel. Also see section 20.2, Grinding Guidelines.

## 8.11 Centerless Grinding

| MATERIAL | HARD-NESS<br>Bhn | CONDITION | WHEEL SPEED<br>fpm<br>m/s | THRUFEED OF WORK*<br>in/min<br>m/min | INFEED on dia.<br>in/pass<br>mm/pass | WHEEL IDENTIFI-CATION†<br>ANSI<br>ISO |
|---|---|---|---|---|---|---|
| **53. CONTROLLED EXPANSION ALLOYS**<br>Invar<br>Kovar | 125 to 250 | Annealed or Cold Drawn | 5500 to 6500 | 50 to 150 | Rough: .005<br>Finish: .0015 max. | C60LV |
| | | | 28 to 33 | 1.3 to 3.8 | Rough: .13<br>Finish: .038 max. | C60LV |
| **54. CARBONS & GRAPHITES**<br>Mechanical Grades | 40 to 100 Shore | Molded or Extruded | 5000 to 6500 | 50 | Rough: .005<br>Finish: .001 max. | C36NV |
| | | | 25 to 33 | 1.3 | Rough: .13<br>Finish: .025 max. | C36NV |
| Brush Grade Carbon | 8 to 90 Shore | Molded or Extruded | 5000 to 6500 | 50 | Rough: .005<br>Finish: .001 max. | C36NV |
| | | | 25 to 33 | 1.3 | Rough: .13<br>Finish: .025 max. | C36NV |
| Graphite | — | — | 5000 to 6500 | 50 | Rough: .005<br>Finish: .001 max. | Rough: C36JV<br>Finish: C60JV |
| | | | 25 to 33 | 1.3 | Rough: .13<br>Finish: .025 max. | Rough: C36JV<br>Finish: C60JV |
| **55. GLASSES & CERAMICS**<br>Glass | 480 to 530 Knoop 100g | — | 5500 to 6500 | 50 to 150 | Rough: .002<br>Finish: .001 max. | C100LV |
| | | | 28 to 33 | 1.3 to 3.8 | Rough: .050<br>Finish: .025 max. | C100LV |
| Porcelain Enamel | — | — | 5500 to 6500 | 50 to 150 | Rough: .002<br>Finish: .001 max. | C80LV |
| | | | 28 to 33 | 1.3 to 3.8 | Rough: .050<br>Finish: .025 max. | C80LV |

See section 16 for Cutting Fluid Recommendations.

*See section 8.12 for Work Traverse Rates. As recommended starting conditions—use a regulating wheel angle with a positive inclination of 3° and a regulating wheel speed of 25 to 40 rpm.

†Wheel recommendations are for wet grinding 0.8- to 2-inch [20 to 50 mm] diameter work. For LARGER diameter work—use a softer grade and/or coarser grit wheel. For SMALLER diameter work—use a harder grade wheel. Also see section 20.2, Grinding Guidelines.

| MATERIAL | HARDNESS Bhn | CONDITION | WHEEL SPEED fpm / m/s | THRUFEED OF WORK* in/min / m/min | INFEED on dia. in/pass / mm/pass | WHEEL IDENTIFICATION† ANSI / ISO |
|---|---|---|---|---|---|---|
| **Ceramics**<br>Alumina (Aluminum Oxide)<br>Alumina-Mullite<br>Aluminum Silicate<br>Beryllia (Beryllium Oxide)††<br>Magnesia (Magnesium Oxide)<br>Mullite<br>Silicon Carbide<br>Silicon Nitride<br>Thoria (Thorium Oxide)††<br>Titania (Titanium Oxide)<br>Titanium Diboride††<br>Zircon (Zirconium Silicate)<br>Zirconia (Zirconium Oxide) | Over 800 Knoop | Fired | 5500 to 6500 | 50 to 150 | Rough: .001<br>Finish: .0002 max. | Rough: C46KV<br>Finish: C80KV |
| | | | 28 to 33 | 1.3 to 3.8 | Rough: .025<br>Finish: .005 max. | Rough: C46KV<br>Finish: C80KV |
| | | | | | | |
| | | | | | | |
| **56. PLASTICS**<br>**Thermoplastics** | — | Cast, Molded, Extruded, or Filled and Molded | 5000 to 6000 | 5 to 30 | Rough: .005-.020<br>Finish: .003 max. | C60KV |
| | | | 25 to 30 | .13 to .75 | Rough: .13-.50<br>Finish: .075 max. | C60KV |
| **Thermosetting Plastics** | — | Cast | 5500 to 6500 | 5 to 30 | Rough: .005-.020<br>Finish: .003 max. | C46KV |
| | | | 28 to 33 | .13 to .75 | Rough: .13-.50<br>Finish: .075 max. | C46KV |
| | — | Molded | 5500 to 6500 | 5 to 30 | Rough: .005-.020<br>Finish: .003 max. | C46MV |
| | | | 28 to 33 | .13 to .75 | Rough: .13-.50<br>Finish: .075 max. | C46MV |
| | — | Laminated | 5500 to 6500 | 5 to 30 | Rough: .005-.020<br>Finish: .003 max. | C36NV |
| | | | 28 to 33 | .13 to .75 | Rough: .13-.50<br>Finish: .075 max. | C36NV |

See section 16 for Cutting Fluid Recommendations.

*See section 8.12 for Work Traverse Rates. As recommended starting conditions—use a regulating wheel angle with a positive inclination of 3° and a regulating wheel speed of 25 to 40 rpm.

†Wheel recommendations are for wet grinding 0.8- to 2-inch [20 to 50 mm] diameter work. For LARGER diameter work—use a softer grade and/or coarser grit wheel. For SMALLER diameter work—use a harder grade wheel. Also see section 20.2, Grinding Guidelines.

††CAUTION: Toxic Material, refer to National Institute for Occupational Safety and Health (NIOSH) for Precautions.

# 8.11 Centerless Grinding

| MATERIAL | HARD-NESS Bhn | CONDITION | WHEEL SPEED fpm m/s | THRUFEED OF WORK* in/min m/min | INFEED on dia. in/pass mm/pass | WHEEL IDENTIFI-CATION† ANSI ISO |
|---|---|---|---|---|---|---|
| **58. FLAME (THERMAL) SPRAYED MATERIALS** **Sprayed Carbides** Chromium Carbide, Chromium Carbide-Cobalt Blend, Columbium Carbide, Tantalum Carbide, Titanium Carbide, Tungsten Carbide, Tungsten Carbide-Cobalt, Tungsten Carbide (Cobalt)-Nickel Alloy Blend | — | — | 5500 to 6500 | 20 to 90 | Rough: .001 Finish: .0005 max. | C80HV |
| | | | 28 to 33 | .5 to 2.3 | Rough: .025 Finish: .013 max. | C80HV |
| | | | | | | |
| | | | | | | |
| **Inorganic Coating Materials** Alumina (Pure), Alumina (Grey) containing Titania, Alumina, Nickel-Aluminide Blends, Barium Titanate, Boron††, Calcium Titanate, Calcium Zirconate, Chromium Disilicide††, Chromium Oxide††, Cobalt (40%), Zirconia Blend, Columbium (Niobium)††, Glass (Kovar sealing)††, Hexaboron Silicide††, Magnesia Alumina Spinel, Magnesium Zirconate, Molybdenum Disilicide, Mullite, Nickel (40%), Alumina Blend, Nickel Oxide, Rare Earth Oxides, Tantalum, Titania (50%), Alumina Blend, Titanium Oxide, Tungsten, Yttrium Zirconate, Zirconia (Lime Stabilized), Zirconia, Nickel-Aluminide Blends, Zirconium Oxide (Hafnia Free, Lime Stabilized), Zirconium Silicate | — | — | 5500 to 6500 | 20 to 90 | Rough: .001 Finish: .0005 max. | C80IV |
| | | | 28 to 33 | .5 to 2.3 | Rough: .025 Finish: .013 max. | C80IV |
| | | | | | | |
| | | | | | | |
| | | | | | | |
| | | | | | | |
| **Sprayed Metals (Group I)** Co-Cr-B Alloy (Self Fluxing), Ni-Cr-B Alloy (Self Fluxing), Nickel Chrome Steel (Special), Stainless Steel | — | — | 5500 to 6500 | 20 to 90 | Rough: .001 Finish: .0005 max. | C60JV |
| | | | 28 to 33 | .5 to 2.3 | Rough: .025 Finish: .013 max. | C60JV |

See section 16 for Cutting Fluid Recommendations.

*See section 8.12 for Work Traverse Rates. As recommended starting conditions—use a regulating wheel angle with a positive inclination of 3° and a regulating wheel speed of 25 to 40 rpm.

†Wheel recommendations are for wet grinding 0.8- to 2-inch [20 to 50 mm] diameter work. For LARGER diameter work—use a softer grade and/or coarser grit wheel. For SMALLER diameter work—use a harder grade wheel. Also see section 20.2, Grinding Guidelines.

††CAUTION: Toxic Material, refer to National Institute for Occupational Safety and Health (NIOSH) for Precautions.

| MATERIAL | HARD-NESS Bhn | CONDITION | WHEEL SPEED fpm m/s | THRUFEED OF WORK* in/min m/min | INFEED on dia. in/pass mm/pass | WHEEL IDENTIFI-CATION† ANSI ISO |
|---|---|---|---|---|---|---|
| **Sprayed Metals (Group II)**<br>Bronze<br>Chromium<br>Cobalt<br>Nickel<br>Molybdenum<br>Monel | — | — | 5500 to 6500 | 20 to 90 | Rough: .001 Finish: .0005 max. | C60KV |
| | | | 28 to 33 | .5 to 2.3 | Rough: .025 Finish: .013 max. | C60KV |
| **Sprayed Metals (Group III)**<br>Carbon Steel<br>Iron<br>Precipitation Hardening Steel | — | — | 5500 to 6500 | 20 to 90 | Rough: .001 Finish: .0005 max. | C60LV |
| | | | 28 to 33 | .5 to 2.3 | Rough: .025 Finish: .013 max. | C60LV |
| **59. PLATED MATERIALS**<br>Chromium Plate | — | — | 3000 to 4000 | 50 to 150 | Rough: .002 Finish: .0005 max. | A150JV |
| | | | 15 to 30 | 1.3 to 3.8 | Rough: .050 Finish: .013 max. | A150JV |
| Nickel Plate<br>Silver Plate | — | — | 5500 to 6500 | 50 to 150 | Rough: .003 Finish: .001 max. | A60KV |
| | | | 28 to 33 | 1.3 to 3.8 | Rough: .075 Finish: .025 max. | A60KV |
| **61. RUBBER** | Hard | — | 5500 to 6500 | 50 to 150 | Rough: .005 Finish: .0015 max. | C30KV |
| | | | 28 to 33 | 1.3 to 3.8 | Rough: .13 Finish: .038 max. | C30KV |

See section 16 for Cutting Fluid Recommendations.
*See section 8.12 for Work Traverse Rates. As recommended starting conditions—use a regulating wheel angle with a positive inclination of 3° and a regulating wheel speed of 25 to 40 rpm.
†Wheel recommendations are for wet grinding 0.8- to 2-inch [20 to 50 mm] diameter work. For LARGER diameter work—use a softer grade and/or coarser grit wheel. For SMALLER diameter work—use a harder grade wheel. Also see section 20.2, Grinding Guidelines.

| REG. WHEEL ANGLE | REG. WHEEL DIA. | APPROXIMATE THRUFEED WORK TRAVERSE RATES inches per minute* | | | | | | | | | | |
|---|---|---|---|---|---|---|---|---|---|---|---|---|
| | | Regulating Wheel Speeds | | | | | | | | | | |
| | inches† | 10 rpm | 15 rpm | 20 rpm | 25 rpm | 30 rpm | 40 rpm | 50 rpm | 70 rpm | 90 rpm | 125 rpm | 160 rpm |
| | 8.0 | 4.4 | 6.6 | 8.8 | 11.0 | 13.2 | 17.5 | 21.9 | 30.7 | 39.5 | 54.8 | 70.2 |
| | 8.5 | 4.7 | 7.0 | 9.3 | 11.6 | 14.0 | 18.6 | 23.3 | 32.6 | 41.9 | 58.2 | 74.6 |
| | 9.0 | 4.9 | 7.4 | 9.9 | 12.3 | 14.8 | 19.7 | 24.7 | 34.5 | 44.4 | 61.7 | 78.9 |
| | 9.5 | 5.2 | 7.8 | 10.4 | 13.0 | 15.6 | 20.8 | 26.0 | 36.5 | 46.9 | 65.1 | 83.3 |
| | 10.0 | 5.5 | 8.2 | 11.0 | 13.7 | 16.4 | 21.9 | 27.4 | 38.4 | 49.3 | 68.5 | 87.7 |
| | 10.5 | 5.8 | 8.6 | 11.5 | 14.4 | 17.3 | 23.0 | 28.8 | 40.3 | 51.8 | 72.0 | 92.1 |
| 1° | 11.0 | 6.0 | 9.0 | 12.1 | 15.1 | 18.1 | 24.1 | 30.2 | 42.2 | 54.3 | 75.4 | 96.5 |
| | 11.5 | 6.3 | 9.5 | 12.6 | 15.8 | 18.9 | 25.2 | 31.5 | 44.1 | 56.7 | 78.8 | 100.9 |
| | 12.0 | 6.6 | 9.9 | 13.2 | 16.4 | 19.7 | 26.3 | 32.9 | 46.0 | 59.2 | 82.2 | 105.3 |
| | 12.5 | 6.9 | 10.3 | 13.7 | 17.1 | 20.5 | 27.4 | 34.2 | 47.9 | 61.6 | 85.6 | 109.7 |
| | 13.0 | 7.1 | 10.7 | 14.3 | 17.8 | 21.3 | 28.5 | 35.6 | 49.8 | 64.1 | 89.0 | 114.1 |
| | 13.5 | 7.4 | 11.1 | 14.8 | 18.5 | 22.1 | 29.6 | 36.9 | 51.7 | 66.5 | 92.4 | 118.5 |
| | 14.0 | 7.6 | 11.5 | 15.4 | 19.2 | 22.9 | 30.7 | 38.3 | 53.6 | 69.0 | 95.8 | 122.9 |
| | 8.0 | 8.8 | 13.2 | 17.5 | 21.9 | 26.3 | 35.1 | 43.9 | 61.4 | 78.9 | 109.6 | 140.3 |
| | 8.5 | 9.3 | 14.0 | 18.6 | 23.3 | 28.0 | 37.3 | 46.6 | 65.2 | 83.9 | 116.5 | 149.1 |
| | 9.0 | 9.9 | 14.8 | 19.7 | 24.7 | 29.6 | 39.5 | 49.3 | 69.1 | 88.8 | 123.3 | 157.9 |
| | 9.5 | 10.4 | 15.6 | 20.8 | 26.0 | 31.2 | 41.7 | 52.1 | 72.9 | 93.7 | 130.2 | 166.7 |
| | 10.0 | 11.0 | 16.4 | 21.9 | 27.4 | 32.9 | 43.9 | 54.8 | 76.7 | 98.7 | 137.1 | 175.4 |
| | 10.5 | 11.5 | 17.3 | 23.0 | 28.8 | 34.5 | 46.0 | 57.6 | 80.6 | 103.6 | 143.9 | 184.2 |
| 2° | 11.0 | 12.1 | 18.1 | 24.1 | 30.2 | 36.2 | 48.2 | 60.3 | 84.4 | 108.5 | 150.8 | 193.0 |
| | 11.5 | 12.6 | 18.9 | 25.2 | 31.5 | 37.8 | 50.4 | 63.0 | 88.3 | 113.5 | 157.6 | 201.7 |
| | 12.0 | 13.2 | 19.7 | 26.3 | 32.9 | 39.5 | 52.6 | 65.8 | 92.1 | 118.4 | 164.5 | 210.5 |
| | 12.5 | 13.7 | 20.5 | 27.4 | 33.2 | 41.1 | 54.8 | 68.5 | 96.0 | 123.4 | 171.3 | 219.2 |
| | 13.0 | 14.3 | 21.3 | 28.5 | 33.6 | 42.8 | 57.0 | 71.3 | 99.8 | 128.3 | 174.2 | 228.5 |
| | 13.5 | 14.8 | 22.1 | 29.6 | 33.9 | 44.4 | 59.2 | 74.0 | 103.7 | 133.3 | 177.0 | 236.7 |
| | 14.0 | 15.4 | 22.9 | 30.7 | 34.3 | 46.1 | 61.4 | 76.8 | 107.5 | 138.2 | 183.9 | 245.5 |
| | 8.0 | 13.2 | 19.7 | 26.3 | 32.9 | 39.5 | 52.6 | 65.8 | 92.1 | 118.4 | 164.4 | 210.5 |
| | 8.5 | 14.0 | 21.0 | 28.0 | 34.9 | 41.9 | 55.9 | 69.9 | 97.8 | 125.8 | 174.7 | 223.6 |
| | 9.0 | 14.8 | 22.2 | 29.6 | 37.0 | 44.4 | 59.2 | 74.0 | 103.6 | 133.2 | 185.0 | 236.8 |
| | 9.5 | 15.6 | 23.4 | 31.2 | 39.1 | 46.9 | 62.5 | 78.1 | 109.3 | 140.6 | 195.3 | 249.9 |
| | 10.0 | 16.4 | 24.7 | 32.9 | 41.1 | 49.3 | 65.8 | 82.2 | 115.1 | 148.0 | 205.5 | 263.1 |
| | 10.5 | 17.3 | 25.9 | 34.5 | 43.2 | 51.8 | 69.1 | 86.3 | 120.9 | 155.4 | 215.8 | 276.2 |
| 3° | 11.0 | 18.1 | 27.1 | 36.2 | 45.2 | 54.3 | 72.3 | 90.4 | 126.6 | 162.8 | 226.1 | 289.4 |
| | 11.5 | 18.9 | 28.4 | 37.8 | 47.3 | 56.7 | 75.6 | 94.5 | 132.4 | 170.2 | 236.4 | 302.6 |
| | 12.0 | 19.7 | 29.6 | 39.5 | 49.3 | 59.2 | 78.9 | 98.7 | 138.1 | 177.6 | 246.6 | 315.7 |
| | 12.5 | 20.5 | 30.9 | 41.1 | 51.4 | 61.6 | 82.2 | 102.8 | 143.9 | 185.0 | 256.9 | 328.8 |
| | 13.0 | 21.3 | 32.1 | 42.8 | 53.4 | 64.1 | 85.5 | 106.9 | 149.6 | 192.4 | 267.2 | 341.9 |
| | 13.5 | 22.1 | 33.4 | 44.4 | 55.5 | 66.5 | 88.8 | 111.0 | 155.4 | 199.8 | 277.5 | 355.0 |
| | 14.0 | 22.9 | 34.6 | 46.1 | 57.5 | 69.0 | 92.1 | 115.1 | 161.1 | 207.2 | 287.8 | 368.1 |
| | 8.0 | 17.5 | 26.3 | 35.1 | 43.8 | 52.6 | 70.1 | 87.7 | 122.7 | 157.8 | 219.2 | 280.5 |
| | 8.5 | 18.6 | 27.9 | 37.3 | 46.6 | 55.9 | 74.5 | 93.1 | 130.4 | 167.7 | 232.9 | 298.1 |
| | 9.0 | 19.7 | 29.6 | 39.4 | 49.3 | 59.2 | 78.9 | 98.6 | 138.1 | 177.5 | 246.6 | 315.6 |
| | 9.5 | 20.8 | 31.2 | 41.6 | 52.1 | 62.5 | 83.3 | 104.1 | 145.7 | 187.4 | 260.3 | 333.1 |
| | 10.0 | 21.9 | 32.9 | 43.8 | 54.8 | 65.7 | 87.7 | 109.6 | 153.4 | 197.2 | 273.9 | 350.7 |
| | 10.5 | 23.0 | 34.5 | 46.0 | 57.5 | 69.0 | 92.0 | 115.1 | 161.1 | 207.1 | 287.6 | 368.2 |
| 4° | 11.0 | 24.1 | 36.2 | 48.2 | 60.3 | 72.3 | 96.4 | 120.5 | 168.8 | 217.0 | 301.3 | 385.7 |
| | 11.5 | 25.2 | 37.8 | 50.4 | 63.0 | 75.6 | 100.8 | 126.0 | 176.4 | 226.8 | 315.0 | 403.3 |
| | 12.0 | 26.3 | 39.4 | 52.6 | 65.7 | 78.9 | 105.2 | 131.5 | 184.1 | 236.7 | 328.7 | 420.8 |
| | 12.5 | 27.4 | 41.0 | 54.8 | 68.5 | 82.2 | 109.6 | 136.9 | 191.7 | 246.5 | 342.4 | 438.4 |
| | 13.0 | 28.5 | 42.6 | 57.0 | 71.3 | 85.5 | 114.0 | 142.4 | 199.4 | 256.4 | 356.1 | 456.0 |
| | 13.5 | 29.6 | 44.2 | 59.2 | 74.1 | 88.8 | 118.4 | 147.8 | 207.0 | 266.2 | 369.8 | 473.6 |
| | 14.0 | 30.7 | 45.8 | 61.4 | 76.9 | 92.1 | 122.8 | 153.3 | 214.7 | 276.1 | 383.5 | 491.2 |

*To convert inches per minute to millimeters per minute multiply by 25.4.

†To convert inches to millimeters multiply by 25.4.

# 8.12 Centerless Grinding—Work Traverse Rates

| REG. WHEEL ANGLE | REG WHEEL DIA. | APPROXIMATE THRUFEED WORK TRAVERSE RATES inches per minute* — Regulating Wheel Speeds | | | | | | | | | | |
|---|---|---|---|---|---|---|---|---|---|---|---|---|
| | inches† | 10 rpm | 15 rpm | 20 rpm | 25 rpm | 30 rpm | 40 rpm | 50 rpm | 70 rpm | 90 rpm | 125 rpm | 160 rpm |
| | 8.0 | 21.9 | 32.9 | 43.8 | 54.8 | 65.7 | 87.6 | 109.5 | 153.3 | 197.2 | 273.8 | 350.5 |
| | 8.5 | 23.3 | 34.9 | 46.5 | 58.2 | 69.8 | 93.1 | 116.4 | 162.9 | 209.5 | 290.9 | 372.4 |
| | 9.0 | 24.6 | 37.0 | 49.3 | 61.6 | 73.9 | 98.6 | 123.2 | 172.5 | 221.8 | 308.0 | 394.3 |
| | 9.5 | 26.0 | 39.0 | 52.0 | 65.0 | 78.0 | 104.1 | 130.1 | 182.1 | 234.1 | 325.2 | 416.2 |
| | 10.0 | 27.4 | 41.1 | 54.8 | 68.5 | 82.1 | 109.5 | 136.9 | 191.7 | 246.4 | 342.3 | 438.1 |
| | 10.5 | 28.8 | 43.1 | 57.5 | 71.9 | 86.3 | 115.0 | 143.8 | 201.3 | 258.8 | 359.4 | 460.0 |
| 5° | 11.0 | 30.1 | 45.2 | 60.2 | 75.3 | 90.4 | 120.5 | 150.6 | 210.8 | 271.1 | 376.5 | 481.9 |
| | 11.5 | 31.5 | 47.2 | 63.0 | 78.7 | 94.5 | 126.0 | 157.4 | 220.4 | 283.4 | 393.6 | 503.8 |
| | 12.0 | 32.9 | 49.3 | 65.7 | 82.1 | 98.6 | 131.4 | 164.3 | 230.0 | 295.7 | 410.7 | 525.7 |
| | 12.5 | 34.3 | 51.3 | 68.5 | 85.5 | 102.7 | 136.9 | 171.2 | 239.6 | 308.0 | 427.8 | 547.6 |
| | 13.0 | 35.7 | 53.4 | 71.2 | 88.9 | 106.8 | 142.4 | 178.1 | 249.2 | 320.3 | 444.9 | 569.5 |
| | 13.5 | 37.1 | 55.4 | 74.0 | 92.3 | 110.9 | 147.9 | 185.0 | 258.8 | 332.6 | 462.0 | 591.4 |
| | 14.0 | 38.5 | 57.5 | 76.7 | 95.7 | 115.0 | 153.4 | 191.9 | 268.4 | 344.9 | 479.1 | 613.3 |
| | 8.0 | 26.3 | 39.4 | 52.5 | 65.7 | 78.8 | 105.1 | 131.4 | 183.9 | 236.4 | 328.4 | 420.3 |
| | 8.5 | 27.9 | 41.9 | 55.8 | 69.8 | 83.7 | 111.7 | 139.6 | 195.4 | 251.2 | 348.9 | 446.6 |
| | 9.0 | 29.6 | 44.3 | 59.1 | 73.9 | 88.7 | 118.2 | 147.8 | 206.9 | 266.0 | 369.4 | 472.9 |
| | 9.5 | 31.2 | 46.8 | 62.4 | 78.0 | 93.6 | 124.8 | 156.0 | 218.4 | 280.8 | 390.0 | 499.2 |
| | 10.0 | 32.8 | 49.3 | 65.7 | 82.1 | 98.5 | 131.4 | 164.2 | 229.9 | 295.6 | 410.5 | 525.4 |
| | 10.5 | 34.5 | 51.7 | 69.0 | 86.2 | 103.4 | 137.9 | 172.4 | 241.4 | 310.3 | 431.0 | 551.7 |
| 6° | 11.0 | 36.1 | 54.2 | 72.2 | 90.3 | 108.4 | 144.5 | 180.6 | 252.9 | 325.1 | 451.5 | 578.0 |
| | 11.5 | 37.8 | 56.6 | 75.5 | 94.4 | 113.3 | 151.1 | 188.8 | 264.4 | 339.9 | 472.1 | 604.2 |
| | 12.0 | 39.4 | 59.1 | 78.8 | 98.5 | 118.2 | 157.6 | 197.0 | 275.8 | 354.7 | 492.6 | 630.5 |
| | 12.5 | 41.0 | 61.6 | 82.1 | 102.6 | 123.1 | 164.3 | 205.2 | 287.4 | 369.5 | 513.1 | 656.8 |
| | 13.0 | 42.6 | 64.1 | 85.4 | 106.7 | 128.0 | 170.9 | 213.4 | 298.9 | 384.3 | 533.6 | 683.1 |
| | 13.5 | 44.2 | 66.6 | 88.7 | 110.8 | 132.9 | 177.5 | 221.6 | 310.4 | 399.1 | 554.1 | 709.4 |
| | 14.0 | 45.8 | 69.1 | 92.0 | 114.9 | 137.8 | 184.1 | 229.8 | 321.9 | 413.9 | 574.6 | 735.7 |
| | 8.0 | 30.6 | 45.9 | 61.3 | 76.6 | 91.9 | 122.5 | 153.1 | 214.4 | 275.7 | 382.9 | 490.1 |
| | 8.5 | 32.5 | 48.8 | 65.1 | 81.4 | 97.6 | 130.2 | 162.7 | 227.8 | 292.9 | 406.8 | 520.7 |
| | 9.0 | 34.5 | 51.7 | 68.9 | 86.1 | 103.4 | 137.8 | 172.3 | 241.2 | 310.1 | 430.7 | 551.3 |
| | 9.5 | 36.4 | 54.6 | 72.7 | 90.9 | 109.1 | 145.5 | 181.9 | 254.6 | 327.4 | 454.7 | 582.0 |
| | 10.0 | 38.3 | 57.4 | 76.6 | 95.7 | 114.9 | 153.1 | 191.4 | 268.0 | 344.6 | 478.6 | 612.6 |
| | 10.5 | 40.2 | 60.3 | 80.4 | 100.5 | 120.6 | 160.8 | 201.0 | 281.4 | 361.8 | 502.5 | 643.2 |
| 7° | 11.0 | 42.1 | 63.2 | 84.2 | 105.3 | 126.3 | 168.5 | 210.6 | 294.8 | 379.0 | 526.4 | 673.8 |
| | 11.5 | 44.0 | 66.0 | 88.1 | 110.1 | 132.1 | 176.1 | 220.1 | 308.2 | 396.3 | 550.4 | 704.5 |
| | 12.0 | 45.9 | 68.9 | 91.9 | 114.9 | 137.8 | 183.8 | 229.7 | 321.6 | 413.5 | 574.3 | 735.1 |
| | 12.5 | 47.8 | 71.8 | 95.7 | 119.7 | 143.5 | 191.5 | 239.3 | 335.0 | 430.7 | 598.2 | 765.7 |
| | 13.0 | 49.7 | 74.7 | 99.5 | 124.5 | 149.2 | 199.2 | 248.9 | 348.4 | 447.9 | 622.1 | 796.3 |
| | 13.5 | 51.6 | 77.6 | 103.3 | 129.3 | 154.9 | 206.9 | 258.5 | 361.8 | 465.1 | 646.0 | 826.9 |
| | 14.0 | 53.5 | 80.5 | 107.1 | 134.1 | 160.6 | 214.6 | 268.1 | 375.2 | 482.3 | 669.9 | 857.5 |
| | 8.0 | 35.0 | 52.5 | 70.0 | 87.4 | 104.9 | 139.9 | 174.9 | 244.8 | 314.8 | 437.2 | 559.6 |
| | 8.5 | 37.2 | 55.7 | 74.3 | 92.9 | 111.5 | 148.7 | 185.8 | 260.1 | 334.5 | 464.5 | 594.6 |
| | 9.0 | 39.3 | 59.0 | 78.7 | 98.4 | 118.0 | 157.4 | 196.7 | 275.4 | 354.1 | 491.9 | 629.6 |
| | 9.5 | 41.5 | 62.3 | 83.1 | 103.8 | 124.6 | 166.1 | 207.7 | 290.7 | 373.8 | 519.2 | 664.6 |
| | 10.0 | 43.7 | 65.6 | 87.4 | 109.3 | 131.2 | 174.9 | 218.6 | 306.1 | 393.5 | 546.5 | 699.5 |
| | 10.5 | 45.9 | 68.9 | 91.8 | 114.8 | 137.7 | 183.6 | 229.5 | 321.4 | 413.2 | 573.8 | 734.5 |
| 8° | 11.0 | 48.1 | 72.1 | 96.2 | 120.2 | 144.3 | 192.4 | 240.5 | 336.7 | 432.8 | 601.2 | 769.5 |
| | 11.5 | 50.3 | 75.4 | 100.6 | 125.7 | 150.8 | 201.1 | 251.4 | 352.0 | 452.5 | 628.5 | 804.5 |
| | 12.0 | 52.5 | 78.7 | 104.9 | 131.2 | 157.4 | 209.9 | 262.3 | 367.3 | 472.2 | 655.8 | 839.5 |
| | 12.5 | 54.7 | 82.0 | 109.3 | 136.7 | 164.0 | 218.6 | 273.2 | 382.6 | 491.9 | 683.1 | 874.5 |
| | 13.0 | 56.9 | 85.3 | 113.6 | 142.2 | 170.6 | 227.4 | 284.1 | 397.9 | 511.6 | 710.4 | 909.5 |
| | 13.5 | 59.1 | 88.6 | 118.0 | 147.7 | 177.2 | 236.1 | 295.0 | 413.2 | 531.3 | 737.7 | 944.5 |
| | 14.0 | 61.3 | 91.9 | 122.3 | 153.2 | 183.8 | 244.9 | 305.9 | 428.5 | 551.0 | 765.0 | 979.5 |

*To convert inches per minute to millimeters per minute multiply by 25.4.
†To convert inches to millimeters multiply by 25.4.

| MATERIAL | HARDNESS Bhn | CONDITION | OPERATION | ABRASIVE TYPE | GRAIN SIZE | BELT SPEED* fpm / m/s | CONTACT WHEEL TYPE† | CONTACT WHEEL HARDNESS Durometer / IRHD‡ |
|---|---|---|---|---|---|---|---|---|
| CARBON STEELS, ALLOY STEELS, AND TOOL STEELS; WROUGHT AND CAST | 100 to 45 $R_C$ | Hot Rolled, Normalized, Annealed, Cold Drawn or Quenched and Tempered | Roughing | $Al_2O_3$, $Al_2O_3$-$ZrO_2$ | 24-60 | 4000-5500 | SR | 70-95 |
| | | | Polishing | $Al_2O_3$ | 80-320 | 4500-6500 | SR, SFR | 20-60 |
| | | | Roughing | $Al_2O_3$, $Al_2O_3$-$ZrO_2$ | 24-60 | 20-28 | SR | 70-95 |
| | | | Polishing | $Al_2O_3$ | 80-320 | 23-33 | SR, SFR | 20-60 |
| | 45 $R_C$ to 56 $R_C$ | Quenched and Tempered | Roughing | $Al_2O_3$, $Al_2O_3$-$ZrO_2$ | 40-80 | 2500-4500 | SR | 70-95 |
| | | | Polishing | $Al_2O_3$ | 100-320 | 3000-4500 | SR, SFR | 20-60 |
| | | | Roughing | $Al_2O_3$, $Al_2O_3$-$ZrO_2$ | 40-80 | 13-23 | SR | 70-95 |
| | | | Polishing | $Al_2O_3$ | 100-320 | 15-23 | SR, SFR | 20-60 |
| STAINLESS STEELS, WROUGHT AND CAST | 135 to 56 $R_C$ | Annealed, Cold Drawn or Quenched and Tempered | Roughing | $Al_2O_3$, $Al_2O_3$-$ZrO_2$ | 36-60 | 2000-4000 | SR | 70-95 |
| | | | Polishing | $Al_2O_3$ | 80-320 | 3000-5500 | SR, SFR | 20-60 |
| | | | Roughing | $Al_2O_3$, $Al_2O_3$-$ZrO_2$ | 36-60 | 10-20 | SR | 70-95 |
| | | | Polishing | $Al_2O_3$ | 80-320 | 15-28 | SR, SFR | 20-60 |
| GRAY CAST IRONS DUCTILE CAST IRONS MALLEABLE CAST IRONS | 100 to 400 | All | Roughing | $Al_2O_3$, $Al_2O_3$-$ZrO_2$ | 24-60 | 3000-4500 6000-9000 | SR, SFR | 70-95 |
| | | | Polishing | $Al_2O_3$ | 80-180 | 4000-5500 | SR, SFR | 20-60 |
| | | | Roughing | $Al_2O_3$, $Al_2O_3$-$ZrO_2$ | 24-60 | 15-23 30-46 | SR, SFR | 70-95 |
| | | | Polishing | $Al_2O_3$ | 80-180 | 20-28 | SR, SFR | 20-60 |
| ALUMINUM ALLOYS, WROUGHT AND CAST MAGNESIUM ALLOYS,‡‡ WROUGHT AND CAST | 30 to 150 500kg | Cold Drawn, As Cast or Solution Treated and Aged | Roughing | $Al_2O_3$, SiC | 24-80 | 5000-6500 | SR, SFR | 70-95 |
| | | | Polishing | $Al_2O_3$, SiC | 100-240 | 5000-8000 | SR, SFR, B | 20-60 |
| | | | Roughing | $Al_2O_3$, SiC | 24-80 | 25-33 | SR, SFR | 70-95 |
| | | | Polishing | $Al_2O_3$, SiC | 100-240 | 25-40 | SR, SFR, B | 20-60 |
| TITANIUM ALLOYS,‡‡ WROUGHT AND CAST | 110 to 440 | Annealed, Solution Treated or Solution Treated and Aged | Roughing | SiC, $Al_2O_3$, $Al_2O_3$-$ZrO_2$ | 50-80 | 1000-2500 | SR | 70-95 |
| | | | Polishing | SiC | 100-240 | 1200-2500 | SR, SFR | 20-60 |
| | | | Roughing | SiC, $Al_2O_3$, $Al_2O_3$-$ZrO_2$ | 50-80 | 5-13 | SR | 70-95 |
| | | | Polishing | SiC | 100-240 | 6-13 | SR, SFR | 20-60 |

See section 16 for Cutting Fluid Recommendations.

*Use lower values of BELT SPEED when good surface integrity is required.

†SR—Serrated Rubber, SFR—Smooth Face Rubber, B—Buff Type Serrations are usually at 45° angle, although some heavy stock removal operations use 60°. Widths of lands and grooves vary from narrow lands and wide grooves for fast aggressive cuts to wide lands and narrow grooves for intermediate and finishing operations, depending upon workpiece shape and operating conditions.

‡IRHD (International Rubber Hardness degrees) is approximately equal to the durometer reading.

‡‡CAUTION: Potential Fire Hazard. Exercise caution in grinding and disposing of swarf. Do NOT use water or water-miscible cutting fluids for magnesium alloys.

# 8.13 Abrasive Belt Grinding

| MATERIAL | HARDNESS Bhn | CONDITION | OPERATION | ABRASIVE TYPE | GRAIN SIZE | BELT SPEED* fpm / m/s | CONTACT WHEEL TYPE† | CONTACT WHEEL HARDNESS Durometer / IRHD‡ |
|---|---|---|---|---|---|---|---|---|
| **COPPER ALLOYS,†† WROUGHT AND CAST** | 10 $R_B$ to 100 $R_B$ | Annealed or Cold Drawn | Roughing | $Al_2O_3$, SiC | 24-80 | 2000-4500 | SR, SFR | 70-95 |
| | | | Polishing | $Al_2O_3$, SiC | 100-240 | 4000-6000 | SR, SFR, B | 20-60 |
| | | | Roughing | $Al_2O_3$, SiC | 24-80 | 10-23 | SR, SFR | 70-95 |
| | | | Polishing | $Al_2O_3$, SiC | 100-240 | 20-30 | SR, SFR, B | 20-60 |
| **HIGH TEMPERATURE ALLOYS, WROUGHT AND CAST** **Nickel, Cobalt and Iron Base** | 140 to 475 | All | Roughing | $Al_2O_3$, $Al_2O_3$-$ZrO_2$ | 40-80 | 2000-4500 | SR | 70-95 |
| | | | Polishing | $Al_2O_3$ | 100-240 | 3000-5500 | SR, SFR | 20-60 |
| | | | Roughing | $Al_2O_3$, $Al_2O_3$-$ZrO_2$ | 40-80 | 10-23 | SR | 70-95 |
| | | | Polishing | $Al_2O_3$ | 100-240 | 15-28 | SR, SFR | 20-60 |
| **ZINC ALLOYS, CAST** | 80 to 100 | As Cast | Roughing | $Al_2O_3$, SiC | 150-220 | 5000-10,000 | SR, B | 40-95 |
| | | | Polishing | $Al_2O_3$, SiC | 220-320 | 5000-10,000 | SR, B | 20-60 |
| | | | Roughing | $Al_2O_3$, SiC | 150-220 | 25-50 | SR, B | 40-95 |
| | | | Polishing | $Al_2O_3$, SiC | 220-320 | 25-50 | SR, B | 20-60 |

See section 16 for Cutting Fluid Recommendations.

*Use lower values of BELT SPEED when good surface integrity is required.

†SR—Serrated Rubber, SFR—Smooth Face Rubber, B—Buff Type Serrations are usually at 45° angle, although some heavy stock removal operations use 60°. Widths of lands and grooves vary from narrow lands and wide grooves for fast aggressive cuts to wide lands and narrow grooves for intermediate and finishing operations, depending upon workpiece shape and operating conditions.

‡IRHD (International Rubber Hardness degrees) is approximately equal to the durometer reading.

††CAUTION: Beryllium-copper alloys are Toxic Materials; refer to National Institute for Occupational Safety and Health (NIOSH) for Precautions. See section 8.11 for specific beryllium-copper alloys indicated by ††.

| MATERIAL | HARDNESS Bhn | CONDITION | THREADS PER INCH / PITCH (P) mm | WHEEL IDENTIFICATION ANSI / ISO | WHEEL SPEED fpm / m/s | WORK SPEED fpm / m/min |
|---|---|---|---|---|---|---|
| **1. FREE MACHINING CARBON STEELS, WROUGHT**<br>**Low Carbon Resulfurized**<br>1108 1116 1211<br>1109 1117 1212<br>1110 1118 1213<br>1115 1119 1215<br>**Medium Carbon Resulfurized**<br>1132 1140 1145<br>1137 1141 1146<br>1139 1144 1151<br>**Low Carbon Leaded**<br>10L18 12L13 12L15<br>11L17 12L14<br>**Medium Carbon Leaded**<br>10L45 11L37 11L44<br>10L50 11L41 | 50 $R_C$ max. | Hot Rolled, Normalized, Annealed, Cold Drawn or Quenched and Tempered | 4 to 12<br>13 to 27<br>28 to 63<br>64 to 80 | A90RB<br>A120SB<br>A180TB<br>A220UB | 8500<br>9000<br>9500<br>10000 | 4.0 |
| | | | 6 to 2.5<br>2 to 1<br>.8 to .45<br>.4 to .35 | A90RB<br>A120SB<br>A180TB<br>A220UB | 43<br>46<br>48<br>51 | 1.2 |
| | Over 50 $R_C$ | Carburized and/or Quenched and Tempered | 4 to 12<br>13 to 27<br>28 to 63<br>64 to 80 | A100KV<br>A150JV<br>A220MV<br>A240PV | 7000<br>7500<br>8500<br>9000 | 6.0 |
| | | | 6 to 2.5<br>2 to 1<br>.8 to .45<br>.4 to .35 | A100KV<br>A150JV<br>A220MV<br>A240PV | 36<br>38<br>43<br>46 | 1.8 |
| **2. CARBON STEELS, WROUGHT**<br>**Low Carbon**<br>1005 1012 1019 1026<br>1006 1013 1020 1029<br>1008 1015 1021 1513<br>1009 1016 1022 1518<br>1010 1017 1023 1522<br>1011 1018 1025<br>**Medium Carbon**<br>1030 1042 1053 1541<br>1033 1043 1055 1547<br>1035 1044 1524 1548<br>1037 1045 1525 1551<br>1038 1046 1526 1552<br>1039 1049 1527<br>1040 1050 1536<br>**High Carbon**<br>1060 1074 1085 1566<br>1064 1075 1086 1572<br>1065 1078 1090<br>1069 1080 1095<br>1070 1084 1561 | 50 $R_C$ max. | Hot Rolled, Normalized, Annealed, Cold Drawn or Quenched and Tempered | 4 to 12<br>13 to 27<br>28 to 63<br>64 to 80 | A90RB<br>A120SB<br>A180TB<br>A220UB | 8500<br>9000<br>9500<br>10000 | 4.0 |
| | | | 6 to 2.5<br>2 to 1<br>.8 to .45<br>.4 to .35 | A90RB<br>A120SB<br>A180TB<br>A220UB | 43<br>46<br>48<br>51 | 1.2 |
| | Over 50 $R_C$ | Carburized and/or Quenched and Tempered | 4 to 12<br>13 to 27<br>28 to 63<br>64 to 80 | A100KV<br>A150JV<br>A220MV<br>A240PV | 7000<br>7500<br>8500<br>9000 | 6.0 |
| | | | 6 to 2.5<br>2 to 1<br>.8 to .45<br>.4 to .35 | A100KV<br>A150JV<br>A220MV<br>A240PV | 36<br>38<br>43<br>46 | 1.8 |
| **3. CARBON AND FERRITIC ALLOY STEELS (HIGH TEMPERATURE SERVICE)**<br>ASTM A369: Grades FPA, FPB, FP1, FP2, FP3b, FP5, FP7, FP9, FP11, FP12, FP21, FP22 | 150 to 200 | As Forged, Annealed or Normalized and Tempered | 4 to 12<br>13 to 27<br>28 to 63<br>64 to 80 | A90RB<br>A120SB<br>A180TB<br>A220UB | 8500<br>9000<br>9500<br>10000 | 4.0 |
| | | | 6 to 2.5<br>2 to 1<br>.8 to .45<br>.4 to .35 | A90RB<br>A120SB<br>A180TB<br>A220UB | 43<br>46<br>48<br>51 | 1.2 |

See section 16 for Cutting Fluid Recommendations.
See section 20 for additional information.

# 8.14 Thread Grinding

| MATERIAL | HARDNESS Bhn | CONDITION | THREADS PER INCH / PITCH (P) mm | WHEEL IDENTIFICATION ANSI / ISO | WHEEL SPEED fpm / m/s | WORK SPEED fpm / m/min |
|---|---|---|---|---|---|---|
| **4. FREE MACHINING ALLOY STEELS, WROUGHT**<br>**Medium Carbon Resulfurized**<br>4140 4145Se<br>4140Se 4147Te<br>4142Te 4150<br>**Medium and High Carbon Leaded**<br>41L30 41L47 51L32 86L40<br>41L40 41L50 52L100<br>41L45 43L40 86L20 | 50 $R_C$ max. | Hot Rolled, Normalized, Annealed, Cold Drawn or Quenched and Tempered | 4 to 12 | A90RB | 8500 | 4.0 |
| | | | 13 to 27 | A120SB | 9000 | |
| | | | 28 to 63 | A180TB | 9500 | |
| | | | 64 to 80 | A220UB | 10000 | |
| | | | 6 to 2.5 | A90RB | 43 | 1.2 |
| | | | 2 to 1 | A120SB | 46 | |
| | | | .8 to .45 | A180TB | 48 | |
| | | | .4 to .35 | A220UB | 51 | |
| | Over 50 $R_C$ | Carburized and/or Quenched and Tempered | 4 to 12 | A100KV | 7000 | 6.0 |
| | | | 13 to 27 | A150JV | 7500 | |
| | | | 28 to 63 | A220MV | 8500 | |
| | | | 64 to 80 | A240PV | 9000 | |
| | | | 6 to 2.5 | A100KV | 36 | 1.8 |
| | | | 2 to 1 | A150JV | 38 | |
| | | | .8 to .45 | A220MV | 43 | |
| | | | .4 to .35 | A240PV | 46 | |
| **5. ALLOY STEELS, WROUGHT**<br>**Low Carbon**<br>4012 4615 4817 8617<br>4023 4617 4820 8620<br>4024 4620 5015 8622<br>4118 4621 5115 8822<br>4320 4718 5120 9310<br>4419 4720 6118 94B15<br>4422 4815 8115 94B17<br>**Medium Carbon**<br>1330 4145 5132 8640<br>1335 4147 5135 8642<br>1340 4150 5140 8645<br>1345 4161 5145 86B45<br>4027 4340 5147 8650<br>4028 4427 5150 8655<br>4032 4626 5155 8660<br>4037 50B40 5160 8740<br>4042 50B44 51B60 8742<br>4047 5046 6150 9254<br>4130 50B46 81B45 9255<br>4135 50B50 8625 9260<br>4137 5060 8627 94B30<br>4140 50B60 8630<br>4142 5130 8637<br>**High Carbon**<br>50100 51100 52100 M-50 | 50 $R_C$ max. | Hot Rolled, Normalized, Annealed, Cold Drawn or Quenched and Tempered | 4 to 12 | A90RB | 8500 | 4.0 |
| | | | 13 to 27 | A120SB | 9000 | |
| | | | 28 to 63 | A180TB | 9500 | |
| | | | 64 to 80 | A220UB | 10000 | |
| | | | 6 to 2.5 | A90RB | 43 | 1.2 |
| | | | 2 to 1 | A120SB | 46 | |
| | | | .8 to .45 | A180TB | 48 | |
| | | | .4 to .35 | A220UB | 51 | |
| | Over 50 $R_C$ | Carburized and/or Quenched and Tempered | 4 to 12 | A100KV | 7000 | 6.0 |
| | | | 13 to 27 | A150JV | 7500 | |
| | | | 28 to 63 | A220MV | 8500 | |
| | | | 64 to 80 | A240PV | 9000 | |
| | | | 6 to 2.5 | A100KV | 36 | 1.8 |
| | | | 2 to 1 | A150JV | 38 | |
| | | | .8 to .45 | A220MV | 43 | |
| | | | .4 to .35 | A240PV | 46 | |

See section 16 for Cutting Fluid Recommendations.
See section 20 for additional information.

| MATERIAL | HARDNESS<br>Bhn | CONDITION | THREADS PER INCH<br>PITCH (P)<br>mm | WHEEL IDENTIFICATION<br>ANSI<br>ISO | WHEEL SPEED<br>fpm<br>m/s | WORK SPEED<br>fpm<br>m/min |
|---|---|---|---|---|---|---|
| **6. HIGH STRENGTH STEELS, WROUGHT**<br>300M<br>4330V<br>4340<br>4340Si<br>98BV40<br>D6ac<br>H11<br>H13<br>HP 9-4-20<br>HP 9-4-25<br>HP 9-4-30<br>HP 9-4-45 | 50 $R_C$ max. | Annealed, Normalized or Quenched and Tempered | 4 to 12<br>13 to 27<br>28 to 63<br>64 to 80 | A90RB<br>A120SB<br>A180TB<br>A220UB | 8500<br>9000<br>9500<br>10000 | 4.0 |
| | | | 6 to 2.5<br>2 to 1<br>.8 to .45<br>.4 to .35 | A90RB<br>A120SB<br>A180TB<br>A220UB | 43<br>46<br>48<br>51 | 1.2 |
| | Over 50 $R_C$ | Quenched and Tempered | 4 to 12<br>13 to 27<br>28 to 63<br>64 to 80 | A100KV<br>A150JV<br>A220MV<br>A240PV | 7000<br>7500<br>8500<br>9000 | 6.0 |
| | | | 6 to 2.5<br>2 to 1<br>.8 to .45<br>.4 to .35 | A100KV<br>A150JV<br>A220MV<br>A240PV | 36<br>38<br>43<br>46 | 1.8 |
| **7. MARAGING STEELS, WROUGHT**<br>ASTM A538: Grades A, B, C<br>120 Grade<br>180 Grade<br>200 Grade<br>250 Grade<br>300 Grade<br>350 Grade<br>HY230 | 50 $R_C$ max. | Annealed or Maraged | 4 to 12<br>13 to 27<br>28 to 63<br>64 to 80 | A90RB<br>A120SB<br>A180TB<br>A220UB | 8500<br>9000<br>9500<br>10000 | 4.0 |
| | | | 6 to 2.5<br>2 to 1<br>.8 to .45<br>.4 to .35 | A90RB<br>A120SB<br>A180TB<br>A220UB | 43<br>46<br>48<br>51 | 1.2 |
| | Over 50 $R_C$ | Maraged | 4 to 12<br>13 to 27<br>28 to 63<br>64 to 80 | A100KV<br>A150JV<br>A220MV<br>A240PV | 7000<br>7500<br>8500<br>9000 | 6.0 |
| | | | 6 to 2.5<br>2 to 1<br>.8 to .45<br>.4 to .35 | A100KV<br>A150JV<br>A220MV<br>A240PV | 36<br>38<br>43<br>46 | 1.8 |
| **8. TOOL STEELS, WROUGHT**<br>**Group I**<br>A2 H13 L6 P20<br>A3 H14 L7 P21<br>A4 H19 O1 S1<br>A6 H21 O2 S2<br>A8 H22 O6 S5<br>A9 H23 O7 S6<br>A10 H24 P2 S7<br>H10 H25 P4 W1<br>H11 H26 P5 W2<br>H12 L2 P6 W5<br>SAE J438b: Types W108, W109, W110, W112, W209, W210, W310 | 50 $R_C$ max. | Annealed or Quenched and Tempered | 4 to 12<br>13 to 27<br>28 to 63<br>64 to 80 | A90RB<br>A120SB<br>A180TB<br>A220UB | 8500<br>9000<br>9500<br>10000 | 4.0 |
| | | | 6 to 2.5<br>2 to 1<br>.8 to .45<br>.4 to .35 | A90RB<br>A120SB<br>A180TB<br>A220UB | 43<br>46<br>48<br>51 | 1.2 |
| | Over 50 $R_C$ | Quenched and Tempered | 4 to 12<br>13 to 27<br>28 to 63<br>64 to 80 | A100KV<br>A150JV<br>A220MV<br>A240PV | 7000<br>7500<br>8500<br>9000 | 6.0 |
| | | | 6 to 2.5<br>2 to 1<br>.8 to .45<br>.4 to .35 | A100KV<br>A150JV<br>A220MV<br>A240PV | 36<br>38<br>43<br>46 | 1.8 |
| **Group II**<br>D2 H42 M42<br>D3 M1 T1<br>D4 M2 T2<br>D5 M3-1 T4<br>F1 M10 T8<br>F2 M30 | 50 $R_C$ max. | Annealed or Quenched and Tempered | 4 to 12<br>13 to 27<br>28 to 63<br>64 to 80 | A90RB<br>A120SB<br>A180TB<br>A220UB | 8500<br>9000<br>9500<br>10000 | 4.0 |
| | | | 6 to 2.5<br>2 to 1<br>.8 to .45<br>.4 to .35 | A90RB<br>A120SB<br>A180TB<br>A220UB | 43<br>46<br>48<br>51 | 1.2 |

See section 16 for Cutting Fluid Recommendations.
See section 20 for additional information.

# 8.14 Thread Grinding

| MATERIAL | HARDNESS Bhn | CONDITION | THREADS PER INCH / PITCH (P) mm | WHEEL IDENTIFICATION ANSI / ISO | WHEEL SPEED fpm / m/s | WORK SPEED fpm / m/min |
|---|---|---|---|---|---|---|
| **8. TOOL STEELS, WROUGHT (cont.)**<br>**Group II (cont.)**<br>(materials listed on preceding page) | 50 $R_C$ to 58 $R_C$ | Quenched and Tempered | 4 to 12 | A100KV | 7000 | 6.0 |
| | | | 13 to 27 | A150JV | 7500 | |
| | | | 28 to 63 | A220MV | 8500 | |
| | | | 64 to 80 | A240PV | 9000 | |
| | | | 6 to 2.5 | A100KV | 36 | 1.8 |
| | | | 2 to 1 | A150JV | 38 | |
| | | | .8 to .45 | A220MV | 43 | |
| | | | .4 to .35 | A240PV | 46 | |
| | Over 58 $R_C$ | Quenched and Tempered | 4 to 12 | A100KV | 7000 | 6.0 |
| | | | 13 to 27 | A150JV | 7500 | |
| | | | 28 to 63 | A240MV | 8500 | |
| | | | 64 to 80 | A280PV | 9000 | |
| | | | 6 to 2.5 | A100KV | 36 | 1.8 |
| | | | 2 to 1 | A150JV | 38 | |
| | | | .8 to .45 | A240MV | 43 | |
| | | | .4 to .35 | A280PV | 46 | |
| **Group III**<br>A7 M33 M44<br>D7 M34 M46<br>M3-2 M36 M47<br>M6 M41 T5<br>M7 M43 T6 | 50 $R_C$ max. | Annealed or Quenched and Tempered | 4 to 12 | A90RB | 8500 | 4.0 |
| | | | 13 to 27 | A120SB | 9000 | |
| | | | 28 to 63 | A180TB | 9500 | |
| | | | 64 to 80 | A220UB | 10000 | |
| | | | 6 to 2.5 | A90RB | 43 | 1.2 |
| | | | 2 to 1 | A120SB | 46 | |
| | | | .8 to .45 | A180TB | 48 | |
| | | | .4 to .35 | A220UB | 51 | |
| | 50 $R_C$ to 58 $R_C$ | Quenched and Tempered | 4 to 12 | A100KV | 7000 | 4.0 |
| | | | 13 to 27 | A150JV | 7500 | 4.0 |
| | | | 28 to 63 | A220MV | 8500 | 6.0 |
| | | | 64 to 80 | A240PV | 9000 | 6.0 |
| | | | 6 to 2.5 | A100KV | 36 | 1.2 |
| | | | 2 to 1 | A150JV | 38 | 1.2 |
| | | | .8 to .45 | A220MV | 43 | 1.8 |
| | | | .4 to .35 | A240PV | 46 | 1.8 |
| | Over 58 $R_C$ | Quenched and Tempered | 4 to 12 | A100KV | 7000 | 4.0 |
| | | | 13 to 27 | A150JV | 7500 | 4.0 |
| | | | 28 to 63 | A240MV | 8500 | 6.0 |
| | | | 64 to 80 | A280PV | 9000 | 6.0 |
| | | | 6 to 2.5 | A100KV | 36 | 1.2 |
| | | | 2 to 1 | A150JV | 38 | 1.2 |
| | | | .8 to .45 | A240MV | 43 | 1.8 |
| | | | .4 to .35 | A280PV | 46 | 1.8 |
| **Group IV**<br>M4<br>T15 | 50 $R_C$ max. | Annealed or Quenched and Tempered | 4 to 12 | A90RB | 8500 | 4.0 |
| | | | 13 to 27 | A120SB | 9000 | |
| | | | 28 to 63 | A180TB | 9500 | |
| | | | 64 to 80 | A220UB | 10000 | |
| | | | 6 to 2.5 | A90RB | 43 | 1.2 |
| | | | 2 to 1 | A120SB | 46 | |
| | | | .8 to .45 | A180TB | 48 | |
| | | | .4 to .35 | A220UB | 51 | |
| | 50 $R_C$ to 58 $R_C$ | Quenched and Tempered | 4 to 12 | A100KV | 7000 | 4.0 |
| | | | 13 to 27 | A150JV | 7500 | 4.0 |
| | | | 28 to 63 | A220MV | 8500 | 6.0 |
| | | | 64 to 80 | A240PV | 9000 | 6.0 |
| | | | 6 to 2.5 | A100KV | 36 | 1.2 |
| | | | 2 to 1 | A150JV | 38 | 1.2 |
| | | | .8 to .45 | A220MV | 43 | 1.8 |
| | | | .4 to .35 | A240PV | 46 | 1.8 |

See section 16 for Cutting Fluid Recommendations.
See section 20 for additional information.

| MATERIAL | HARDNESS<br>Bhn | CONDITION | THREADS PER INCH<br>PITCH (P)<br>mm | WHEEL IDENTIFICATION<br>ANSI<br>ISO | WHEEL SPEED<br>fpm<br>m/s | WORK SPEED<br>fpm<br>m/min |
|---|---|---|---|---|---|---|
| **8. TOOL STEELS, WROUGHT (cont.)**<br>**Group IV (cont.)**<br>(materials listed on preceding page) | Over 58 $R_C$ | Quenched and Tempered | 4 to 12 | A100KV | 7000 | 4.0 |
| | | | 13 to 27 | A150JV | 7500 | 4.0 |
| | | | 28 to 63 | A240MV | 8500 | 6.0 |
| | | | 64 to 80 | A280PV | 9000 | 6.0 |
| | | | 6 to 2.5 | A100KV | 36 | 1.2 |
| | | | 2 to 1 | A150JV | 38 | 1.2 |
| | | | .8 to .45 | A240MV | 43 | 1.8 |
| | | | .4 to .35 | A280PV | 46 | 1.8 |
| **9. NITRIDING STEELS, WROUGHT**<br>Nitralloy 125<br>Nitralloy 135<br>Nitralloy 135 Mod.<br>Nitralloy 225<br>Nitralloy 230<br>Nitralloy EZ<br>Nitralloy N<br>Nitrex 1 | 200 to 350 | Annealed, Normalized or Quenched and Tempered | 4 to 12 | A90RB | 8500 | 4.0 |
| | | | 13 to 27 | A120SB | 9000 | |
| | | | 28 to 63 | A180TB | 9500 | |
| | | | 64 to 80 | A220UB | 10000 | |
| | | | 6 to 2.5 | A90RB | 43 | 1.2 |
| | | | 2 to 1 | A120SB | 46 | |
| | | | .8 to .45 | A180TB | 48 | |
| | | | .4 to .35 | A220UB | 51 | |
| | 60 $R_C$ to 65 $R_C$ | Nitrided | 4 to 12 | A100JV | 7000 | 6.0 |
| | | | 13 to 27 | A150KV | 7500 | |
| | | | 28 to 63 | A220MV | 8500 | |
| | | | 64 to 80 | A240PV | 9000 | |
| | | | 6 to 2.5 | A100JV | 36 | 1.8 |
| | | | 2 to 1 | A150KV | 38 | |
| | | | .8 to .45 | A220MV | 43 | |
| | | | .4 to .35 | A240PV | 46 | |
| **12. FREE MACHINING STAINLESS STEELS, WROUGHT**<br>**Ferritic**<br>430F<br>430F Se | 135 to 185 | Annealed | 4 to 12 | A90RB | 8500 | 4.0 |
| | | | 13 to 27 | A120SB | 9000 | |
| | | | 28 to 63 | A180TB | 9500 | |
| | | | 64 to 80 | A220UB | 10000 | |
| | | | 6 to 2.5 | A90RB | 43 | 1.2 |
| | | | 2 to 1 | A120SB | 46 | |
| | | | .8 to .45 | A180TB | 48 | |
| | | | .4 to .35 | A220UB | 51 | |
| **Austenitic**<br>203EZ<br>303<br>303MA<br>303Pb<br>303 Plus X<br>303Se | 135 to 275 | Annealed or Cold Drawn | 4 to 12 | A90RB | 8500 | 4.0 |
| | | | 13 to 27 | A120SB | 9000 | |
| | | | 28 to 63 | A180TB | 9500 | |
| | | | 64 to 80 | A220UB | 10000 | |
| | | | 6 to 2.5 | A90RB | 43 | 1.2 |
| | | | 2 to 1 | A120SB | 46 | |
| | | | .8 to .45 | A180TB | 48 | |
| | | | .4 to .35 | A220UB | 51 | |
| **Martensitic**<br>416<br>416 Plus X<br>416Se<br>420F<br>420F Se<br>440F<br>440F Se | 135 to 240 | Annealed or Cold Drawn | 4 to 12 | A90RB | 8500 | 4.0 |
| | | | 13 to 27 | A120SB | 9000 | |
| | | | 28 to 63 | A180TB | 9500 | |
| | | | 64 to 80 | A220UB | 10000 | |
| | | | 6 to 2.5 | A90RB | 43 | 1.2 |
| | | | 2 to 1 | A120SB | 46 | |
| | | | .8 to .45 | A180TB | 48 | |
| | | | .4 to .35 | A220UB | 51 | |
| | Over 275 | Quenched and Tempered | 4 to 12 | A100KV | 7000 | 6.0 |
| | | | 13 to 27 | A150KV | 7500 | |
| | | | 28 to 63 | A220MV | 8500 | |
| | | | 64 to 80 | A240PV | 9000 | |
| | | | 6 to 2.5 | A100KV | 36 | 1.8 |
| | | | 2 to 1 | A150KV | 38 | |
| | | | .8 to .45 | A220MV | 43 | |
| | | | .4 to .35 | A240PV | 46 | |

See section 16 for Cutting Fluid Recommendations.
See section 20 for additional information.

# 8.14 Thread Grinding

| MATERIAL | HARD-NESS Bhn | CONDITION | THREADS PER INCH / PITCH (P) mm | WHEEL IDENTIFI-CATION ANSI / ISO | WHEEL SPEED fpm / m/s | WORK SPEED fpm / m/min |
|---|---|---|---|---|---|---|
| **13. STAINLESS STEELS, WROUGHT** **Ferritic** 405, 434, 409, 436, 429, 442, 430, 446 | 135 to 185 | Annealed | 4 to 12<br>13 to 27<br>28 to 63<br>64 to 80 | A90RB<br>A120KV<br>A180TB<br>A220UB | 8500<br>9000<br>9500<br>10000 | 4.0 |
| | | | 6 to 2.5<br>2 to 1<br>.8 to .45<br>.4 to .35 | A90RB<br>A120KV<br>A180TB<br>A220UB | 43<br>46<br>48<br>51 | 1.2 |
| **Austenitic** 201, 304L, 310S, 330, 202, 305, 314, 347, 301, 308, 316, 348, 302, 309, 316L, 384, 302B, 309S, 317, 385, 304, 310, 321 | 135 to 275 | Annealed or Cold Drawn | 4 to 12<br>13 to 27<br>28 to 63<br>64 to 80 | A90RB<br>A120SB<br>A180TB<br>A220UB | 8500<br>9000<br>9500<br>10000 | 4.0 |
| | | | 6 to 2.5<br>2 to 1<br>.8 to .45<br>.4 to .35 | A90RB<br>A120SB<br>A180TB<br>A220UB | 43<br>46<br>48<br>51 | 1.2 |
| **Austenitic** Nitronic 32, Nitronic 33, Nitronic 40, Nitronic 50, Nitronic 60 | 210 to 250 | Annealed | 4 to 12<br>13 to 27<br>28 to 63<br>64 to 80 | A90RB<br>A120SB<br>A180TB<br>A220UB | 8500<br>9000<br>9500<br>10000 | 4.0 |
| | | | 6 to 2.5<br>2 to 1<br>.8 to .45<br>.4 to .35 | A90RB<br>A120SB<br>A180TB<br>A220UB | 43<br>46<br>48<br>51 | 1.2 |
| | 325 to 375 | Cold Drawn | 4 to 12<br>13 to 27<br>28 to 63<br>64 to 80 | A90RB<br>A120SB<br>A180TB<br>A220UB | 8500<br>9000<br>9500<br>10000 | 4.0 |
| | | | 6 to 2.5<br>2 to 1<br>.8 to .45<br>.4 to .35 | A90RB<br>A120SB<br>A180TB<br>A220UB | 43<br>46<br>48<br>51 | 1.2 |
| **Martensitic** 403, 440A, 410, 440B, 414, 440C, 420, 501, 422, 502, 431, Greek Ascoloy | 135 to 275 | Annealed | 4 to 12<br>13 to 27<br>28 to 63<br>64 to 80 | A90RB<br>A120SB<br>A180TB<br>A220UB | 8500<br>9000<br>9500<br>10000 | 4.0 |
| | | | 6 to 2.5<br>2 to 1<br>.8 to .45<br>.4 to .35 | A90RB<br>A120SB<br>A180TB<br>A220UB | 43<br>46<br>48<br>51 | 1.2 |
| | Over 275 | Quenched and Tempered | 4 to 12<br>13 to 27<br>28 to 63<br>64 to 80 | A100KV<br>A150KV<br>A220MV<br>A240PV | 7000<br>7500<br>8500<br>9000 | 6.0 |
| | | | 6 to 2.5<br>2 to 1<br>.8 to .45<br>.4 to .35 | A100KV<br>A150KV<br>A220MV<br>A240PV | 36<br>38<br>43<br>46 | 1.8 |
| | | | | | | |

See section 16 for Cutting Fluid Recommendations.
See section 20 for additional information.

| MATERIAL | HARD-NESS<br>Bhn | CONDITION | THREADS PER INCH<br>PITCH (P)<br>mm | WHEEL IDENTIFI-CATION<br>ANSI<br>ISO | WHEEL SPEED<br>fpm<br>m/s | WORK SPEED<br>fpm<br>m/min |
|---|---|---|---|---|---|---|
| **14. PRECIPITATION HARDENING STAINLESS STEELS, WROUGHT**<br>15-5 PH; AM-355<br>16-6 PH; AM-363<br>17-4 PH; Custom 450<br>17-7 PH; Custom 455<br>17-14 Cu Mo; HNM<br>AF-71; PH 13-8 Mo<br>AFC-77; PH 14-8 Mo<br>Almar 362 (AM-362); PH 15-7 Mo<br>AM-350; Stainless W | 150<br>to<br>200 | Solution Treated | 4 to 12<br>13 to 27<br>28 to 63<br>64 to 80 | A90RB<br>A120SB<br>A180TB<br>A220UB | 8500<br>9000<br>9500<br>10000 | 4.0 |
| | | | 6 to 2.5<br>2 to 1<br>.8 to .45<br>.4 to .35 | A90RB<br>A120SB<br>A180TB<br>A220UB | 43<br>46<br>48<br>51 | 1.2 |
| | 275<br>to<br>440 | Solution Treated<br>or<br>Hardened | 4 to 12<br>13 to 27<br>28 to 63<br>64 to 80 | A100KV<br>A150KV<br>A220MV<br>A240PV | 7000<br>7500<br>8500<br>9000 | 6.0 |
| | | | 6 to 2.5<br>2 to 1<br>.8 to .45<br>.4 to .35 | A100KV<br>A150KV<br>A220MV<br>A240PV | 36<br>38<br>43<br>46 | 1.8 |
| **15. CARBON STEELS, CAST**<br>**Low Carbon**<br>ASTM A426: Grade CP1<br>1010<br>1020<br>**Medium Carbon**<br>ASTM A352: Grades LCA, LCB, LCC<br>ASTM A356: Grade 1<br>1030 1040 1050 | 100<br>to<br>300 | Annealed, Normalized, Normalized and Tempered, or Quenched and Tempered | 4 to 12<br>13 to 27<br>28 to 63<br>64 to 80 | A90RB<br>A120SB<br>A180TB<br>A220UB | 8500<br>9000<br>9500<br>10000 | 4.0 |
| | | | 6 to 2.5<br>2 to 1<br>.8 to .45<br>.4 to .35 | A90RB<br>A120SB<br>A180TB<br>A220UB | 43<br>46<br>48<br>51 | 1.2 |
| | Over<br>50 $R_C$ | Carburized and/or Quenched and Tempered | 4 to 12<br>13 to 27<br>28 to 63<br>64 to 80 | A100KV<br>A150KV<br>A220MV<br>A240PV | 7000<br>7500<br>8500<br>9000 | 6.0 |
| | | | 6 to 2.5<br>2 to 1<br>.8 to .45<br>.4 to .35 | A100KV<br>A150KV<br>A220MV<br>A240PV | 36<br>38<br>43<br>46 | 1.8 |
| **16. ALLOY STEELS, CAST**<br>**Low Carbon**<br>ASTM A217: Grade WC9<br>ASTM A352: Grades LC3, LC4<br>ASTM A426: Grades CP2, CP5, CP5b, CP11, CP12, CP15, CP21, CP22<br>1320 2320 4120 8020<br>2315 4110 4320 8620<br>**Medium Carbon**<br>ASTM A27: Grades N1, N2, U-60-30, 60-30, 65-35, 70-36, 70-40<br>ASTM A148: Grades 80-40, 80-50, 90-60, 105-85, 120-95, 150-125, 175-145<br>ASTM A216: Grades WCA, WCB, WCC<br>ASTM A217: Grades WC1, WC4, WC5, WC6<br>ASTM A352: Grades LC1, LC2, LC2-1<br>ASTM A356: Grades 2, 5, 6, 8, 9, 10<br>ASTM A389: Grades C23, C24<br>ASTM A486: Classes 70, 90, 120<br>(materials continued on next page) | 150<br>to<br>400 | Annealed, Normalized, Normalized and Tempered, or Quenched and Tempered | 4 to 12<br>13 to 27<br>28 to 63<br>64 to 80 | A90RB<br>A120SB<br>A180TB<br>A220UB | 8500<br>9000<br>9500<br>10000 | 4.0 |
| | | | 6 to 2.5<br>2 to 1<br>.8 to .45<br>.4 to .35 | A90RB<br>A120SB<br>A180TB<br>A220UB | 43<br>46<br>48<br>51 | 1.2 |
| | Over<br>50 $R_C$ | Carburized and/or Quenched and Tempered | 4 to 12<br>13 to 27<br>28 to 63<br>64 to 80 | A100KV<br>A150JV<br>A220MV<br>A240PV | 7000<br>7500<br>8500<br>9000 | 6.0 |
| | | | 6 to 2.5<br>2 to 1<br>.8 to .45<br>.4 to .35 | A100KV<br>A150JV<br>A220MV<br>A240PV | 36<br>38<br>43<br>46 | 1.8 |

See section 16 for Cutting Fluid Recommendations.
See section 20 for additional information.

# 8.14 Thread Grinding

| MATERIAL | HARD-NESS Bhn | CONDITION | THREADS PER INCH / PITCH (P) mm | WHEEL IDENTIFI-CATION ANSI / ISO | WHEEL SPEED fpm / m/s | WORK SPEED fpm / m/min |
|---|---|---|---|---|---|---|
| **16. ALLOY STEELS, CAST (cont.)** **Medium Carbon (cont.)** (materials continued from preceding page) ASTM A487: Classes 1N, 2N, 4N, 6N, 8N, 9N, 10N, DN, 1Q, 2Q, 4Q, 4QA, 6Q, 7Q, 8Q, 9Q, 10Q; 1330, 4130, 80B30, 8640, 1340, 4140, 8040, 9525, 2325, 4330, 8430, 9530, 2330, 4340, 8440, 9535, 4125, 8030, 8630 | | | | | | |
| | | | | | | |
| **17. TOOL STEELS, CAST** **Group I** ASTM A597: Grades CA-2, CH-12, CH-13, CO-1, CS-5 | 50 $R_C$ max. | Annealed or Quenched and Tempered | 4 to 12<br>13 to 27<br>28 to 63<br>64 to 80 | A90RB<br>A120SB<br>A180TB<br>A220UB | 8500<br>9000<br>9500<br>10000 | 4.0 |
| | | | 6 to 2.5<br>2 to 1<br>.8 to .45<br>.4 to .35 | A90RB<br>A120SB<br>A180TB<br>A220UB | 43<br>46<br>48<br>51 | 1.2 |
| | 50 $R_C$ to 58 $R_C$ | Quenched and Tempered | 4 to 12<br>13 to 27<br>28 to 63<br>64 to 80 | A100KV<br>A150JV<br>A220MV<br>A240PV | 7000<br>7500<br>8500<br>9000 | 6.0 |
| | | | 6 to 2.5<br>2 to 1<br>.8 to .45<br>.4 to .35 | A100KV<br>A150JV<br>A220MV<br>A240PV | 36<br>38<br>43<br>46 | 1.8 |
| | Over 58 $R_C$ | Quenched and Tempered | 4 to 12<br>13 to 27<br>28 to 63<br>64 to 80 | A100KV<br>A150JV<br>A240MV<br>A280PV | 7000<br>7500<br>8500<br>9000 | 6.0 |
| | | | 6 to 2.5<br>2 to 1<br>.8 to .45<br>.4 to .35 | A100KV<br>A150JV<br>A240MV<br>A280PV | 36<br>38<br>43<br>46 | 1.8 |
| **Group II** ASTM A597: Grades CD-2, CD-5 | 200 to 250 | Annealed | 4 to 12<br>13 to 27<br>28 to 63<br>64 to 80 | A90RB<br>A120SB<br>A180TB<br>A220UB | 8500<br>9000<br>9500<br>10000 | 4.0 |
| | | | 6 to 2.5<br>2 to 1<br>.8 to .45<br>.4 to .35 | A90RB<br>A120SB<br>A180TB<br>A220UB | 43<br>46<br>48<br>51 | 1.2 |
| | 48 $R_C$ to 56 $R_C$ | Quenched and Tempered | 4 to 12<br>13 to 27<br>28 to 63<br>64 to 80 | A100KV<br>A150JV<br>A220MV<br>A240PV | 7000<br>7500<br>8500<br>9000 | 6.0 |
| | | | 6 to 2.5<br>2 to 1<br>.8 to .45<br>.4 to .35 | A100KV<br>A150JV<br>A220MV<br>A240PV | 36<br>38<br>43<br>46 | 1.8 |

See section 16 for Cutting Fluid Recommendations.
See section 20 for additional information.

| MATERIAL | HARDNESS | CONDITION | THREADS PER INCH | WHEEL IDENTIFICATION | WHEEL SPEED | WORK SPEED |
|---|---|---|---|---|---|---|
| | Bhn | | PITCH (P) mm | ANSI / ISO | fpm / m/s | fpm / m/min |
| **17. TOOL STEELS, CAST (cont.)** | Over 56 $R_C$ | Quenched and Tempered | 4 to 12 | A100KV | 7000 | 6.0 |
| **Group II (cont.)** | | | 13 to 27 | A150JV | 7500 | |
| (materials listed on preceding page) | | | 28 to 63 | A240MV | 8500 | |
| | | | 64 to 80 | A280PV | 9000 | |
| | | | 6 to 2.5 | A100KV | 36 | 1.8 |
| | | | 2 to 1 | A150JV | 38 | |
| | | | .8 to .45 | A240MV | 43 | |
| | | | .4 to .35 | A280PV | 46 | |
| **18. STAINLESS STEELS, CAST** | 135 to 185 | Annealed | 4 to 12 | A90RB | 8500 | 4.0 |
| **Ferritic** | | | 13 to 27 | A120SB | 9000 | |
| ASTM A217: Grades C5, C12 | | | 28 to 63 | A180TB | 9500 | |
| ASTM A296: Grades CB-30, CC-50, CE-30, CA6N, CA-6NM, CD4MCu | | | 64 to 80 | A220UB | 10000 | |
| ASTM A297: Grade HC | | | 6 to 2.5 | A90RB | 43 | 1.2 |
| ASTM A487: Class CA6NM | | | 2 to 1 | A120SB | 46 | |
| ASTM A608: Grade HC30 | | | .8 to .45 | A180TB | 48 | |
| | | | .4 to .35 | A220UB | 51 | |
| | | | | | | |
| **Austenitic** | 135 to 210 | Annealed, Normalized or As Cast | 4 to 12 | A90RB | 8500 | 4.0 |
| ASTM A296: Grades CF-3, CF-3M, CF-8, CF-8C, CF-8M, CF-16F, CF-20, CG-8M, CG-12, CH-20, CK-20, CN-7M, CN-7MS | | | 13 to 27 | A120SB | 9000 | |
| ASTM A297: Grades HD, HE, HF, HH, HI, HK, HL, HN, HP, HT, HU | | | 28 to 63 | A180TB | 9500 | |
| ASTM A351: Grades CF-3, CF-3A, CF-3M, CF-3MA, CF-8, CF-8A, CF-8C, CF-8M, CF-10MC, CH-8, CH-10, CH-20, CK-20, CN-7M, HK-30, HK-40, HT-30 | | | 64 to 80 | A220UB | 10000 | |
| ASTM A451: Grades CPF8A, CPF3, CPF3A, CPF3M, CPF8, CPF8C, CPF8C (Ta Max.), CPF8M, CPF10MC, CPH8, CPH10, CPH20, CPK20 | | | 6 to 2.5 | A90RB | 43 | 1.2 |
| ASTM A452: Grades TP 304H, TP 316H, TP 347H | | | 2 to 1 | A120SB | 46 | |
| ASTM A608: Grades HD50, HE35, HF30, HH30, HH33, HI35, HK30, HK40, HL30, HL40, HN40, HT50, HU50 | | | .8 to .45 | A180TB | 48 | |
| | | | .4 to .35 | A220UB | 51 | |
| | | | | | | |
| | | | | | | |
| **Martensitic** | 135 to 225 | Annealed, Normalized or Normalized and Tempered | 4 to 12 | A90RB | 8500 | 4.0 |
| ASTM A217: Grade CA-15 | | | 13 to 27 | A120SB | 9000 | |
| ASTM A296: Grades CA-15, CA-15M, CA-40 | | | 28 to 63 | A180TB | 9500 | |
| ASTM A426: Grades CP7, CP9, CPCA15 | | | 64 to 80 | A220UB | 10000 | |
| ASTM A487: Classes CA15a, CA-15M | | | 6 to 2.5 | A90RB | 43 | 1.2 |
| | | | 2 to 1 | A120SB | 46 | |
| | | | .8 to .45 | A180TB | 48 | |
| | | | .4 to .35 | A220UB | 51 | |

See section 16 for Cutting Fluid Recommendations.
See section 20 for additional information.

# 8.14 Thread Grinding

| MATERIAL | HARDNESS Bhn | CONDITION | THREADS PER INCH / PITCH (P) mm | WHEEL IDENTIFICATION ANSI / ISO | WHEEL SPEED fpm / m/s | WORK SPEED fpm / m/min |
|---|---|---|---|---|---|---|
| **18. STAINLESS STEELS, CAST (cont.)**<br>**Martensitic (cont.)**<br>(materials listed on preceding page) | Over 275 | Quenched and Tempered | 4 to 12 | A100KV | 7000 | 6.0 |
| | | | 13 to 27 | A150JV | 7500 | |
| | | | 28 to 63 | A220MV | 8500 | |
| | | | 64 to 80 | A240PV | 9000 | |
| | | | 6 to 2.5 | A100KV | 36 | 1.8 |
| | | | 2 to 1 | A150JV | 38 | |
| | | | .8 to .45 | A220MV | 43 | |
| | | | .4 to .35 | A240PV | 46 | |
| **19. PRECIPITATION HARDENING STAINLESS STEELS, CAST**<br>ASTM A351: Grade CD-4MCu<br>ACI Grade CB-7Cu<br>ACI Grade CD-4MCu<br>17-4 PH<br>AM-355 | 325 to 450 | Solution Treated or Solution Treated and Aged | 4 to 12 | A90RB | 8500 | 4.0 |
| | | | 13 to 27 | A120SB | 9000 | |
| | | | 28 to 63 | A180TB | 9500 | |
| | | | 64 to 80 | A220UB | 10000 | |
| | | | 6 to 2.5 | A90RB | 43 | 1.2 |
| | | | 2 to 1 | A120SB | 46 | |
| | | | .8 to .45 | A180TB | 48 | |
| | | | .4 to .35 | A220UB | 51 | |
| **21. GRAY CAST IRONS**<br>**Ferritic**<br>ASTM A48: Class 20<br>SAE J431c: Grade G1800<br>**Pearlitic- Ferritic**<br>ASTM A48: Class 25<br>SAE J431c: Grade G2500<br>**Pearlitic**<br>ASTM A48: Classes 30, 35, 40<br>SAE J431c: Grade G3000<br>**Pearlitic + Free Carbides**<br>ASTM A48: Classes 45, 50<br>SAE J431c: Grades G3500, G4000<br>**Pearlitic or Acicular + Free Carbides**<br>ASTM A48: Classes 55, 60 | 45 $R_C$ max. | As Cast, Annealed or Quenched and Tempered | 4 to 12 | A90RB | 8500 | 4.0 |
| | | | 13 to 27 | A120SB | 9000 | |
| | | | 28 to 63 | A180TB | 9500 | |
| | | | 64 to 80 | A220UB | 10000 | |
| | | | 6 to 2.5 | A90RB | 43 | 1.2 |
| | | | 2 to 1 | A120SB | 46 | |
| | | | .8 to .45 | A180TB | 48 | |
| | | | .4 to .35 | A220UB | 51 | |
| | 45 $R_C$ to 52 $R_C$ | As Cast, Annealed or Quenched and Tempered | 4 to 12 | A90RB | 8500 | 4.0 |
| | | | 13 to 27 | A120SB | 9000 | |
| | | | 28 to 63 | A180TB | 9500 | |
| | | | 64 to 80 | A220UB | 10000 | |
| | | | 6 to 2.5 | A90RB | 43 | 1.2 |
| | | | 2 to 1 | A120SB | 46 | |
| | | | .8 to .45 | A180TB | 48 | |
| | | | .4 to .35 | A220UB | 51 | |
| **Austenitic (NI-RESIST)**<br>ASTM A436: Types 1, 1b, 2, 2b, 3, 4, 5, 6 | 100 to 250 | As Cast | 4 to 12 | A90RB | 8500 | 4.0 |
| | | | 13 to 27 | A120SB | 9000 | |
| | | | 28 to 63 | A180TB | 9500 | |
| | | | 64 to 80 | A220UB | 10000 | |
| | | | 6 to 2.5 | A90RB | 43 | 1.2 |
| | | | 2 to 1 | A120SB | 46 | |
| | | | .8 to .45 | A180TB | 48 | |
| | | | .4 to .35 | A220UB | 51 | |
| **22. COMPACTED GRAPHITE CAST IRONS** | 185 to 255 | As Cast | 4 to 12 | A90RB | 8500 | 4.0 |
| | | | 13 to 27 | A120SB | 9000 | |
| | | | 28 to 63 | A180TB | 9500 | |
| | | | 64 to 80 | A220UB | 10000 | |
| | | | 6 to 2.5 | A90RB | 43 | 1.2 |
| | | | 2 to 1 | A120SB | 46 | |
| | | | .8 to .45 | A180TB | 48 | |
| | | | .4 to .35 | A220UB | 51 | |

See section 16 for Cutting Fluid Recommendations.
See section 20 for additional information.

| MATERIAL | HARD-NESS Bhn | CONDITION | THREADS PER INCH / PITCH (P) mm | WHEEL IDENTIFI-CATION ANSI / ISO | WHEEL SPEED fpm / m/s | WORK SPEED fpm / m/min |
|---|---|---|---|---|---|---|
| **23. DUCTILE CAST IRONS**<br>**Ferritic**<br>ASTM A536: Grades 60-40-18, 65-45-12<br>SAE J434c: Grades D4018, D4512<br>**Ferritic- Pearlitic**<br>ASTM A536: Grade 80-55-06<br>SAE J434c: Grade D5506<br>**Pearlitic- Martensitic**<br>ASTM A536: Grade 100-70-03<br>SAE J434c: Grade D7003<br>**Martensitic**<br>ASTM A536: Grade 120-90-02<br>SAE J434c: Grade DQ&T | $52 R_C$ max. | Annealed, As Cast, Normalized and Tempered or Quenched and Tempered | 4 to 12 | A90RB | 8500 | 4.0 |
| | | | 13 to 27 | A120SB | 9000 | |
| | | | 28 to 63 | A180TB | 9500 | |
| | | | 64 to 80 | A220UB | 10000 | |
| | | | 6 to 2.5 | A90RB | 43 | 1.2 |
| | | | 2 to 1 | A120SB | 46 | |
| | | | .8 to .45 | A180TB | 48 | |
| | | | .4 to .35 | A220UB | 51 | |
| **Austenitic (NI-RESIST Ductile)**<br>ASTM A439: Types D-2, D-2B, D-2C, D-3, D-3A, D-4, D-5, D-5B<br>ASTM A571: Type D-2M | 120 to 275 | Annealed | 4 to 12 | A90RB | 8500 | 4.0 |
| | | | 13 to 27 | A120SB | 9000 | |
| | | | 28 to 63 | A180TB | 9500 | |
| | | | 64 to 80 | A220UB | 10000 | |
| | | | 6 to 2.5 | A90RB | 43 | 1.2 |
| | | | 2 to 1 | A120SB | 46 | |
| | | | .8 to .45 | A180TB | 48 | |
| | | | .4 to .35 | A220UB | 51 | |
| **24. MALLEABLE CAST IRONS**<br>**Ferritic**<br>ASTM A47: Grades 32510, 35018<br>ASTM A602: Grade M3210<br>SAE J158: Grade M3210<br>**Pearlitic**<br>ASTM A220: Grades 40010, 45006, 45008, 50005<br>ASTM A602: Grade M4504, M5003<br>SAE J158: Grades M4504, M5003<br>**Tempered Martensite**<br>ASTM A220: Grades 60004, 70003, 80002, 90001<br>ASTM A602: Grades M5503, M7002, M8501<br>SAE J158: Grades M5503, M7002, M8501 | $52 R_C$ max. | Malleablized or Malleablized and Heat Treated | 4 to 12 | A90RB | 8500 | 4.0 |
| | | | 13 to 27 | A120SB | 9000 | |
| | | | 28 to 63 | A180TB | 9500 | |
| | | | 64 to 80 | A220UB | 10000 | |
| | | | 6 to 2.5 | A90RB | 43 | 1.2 |
| | | | 2 to 1 | A120SB | 46 | |
| | | | .8 to .45 | A180TB | 48 | |
| | | | .4 to .35 | A220UB | 51 | |
| **36. NICKEL ALLOYS, WROUGHT AND CAST**<br>Nickel 200<br>Nickel 201<br>Nickel 205<br>Nickel 211<br>Nickel 220<br>Nickel 230 | 80 to 170 | Annealed or Cold Drawn | 4 to 12 | A90RB | 8500 | 4.0 |
| | | | 13 to 27 | A120SB | 9000 | |
| | | | 28 to 63 | A180TB | 9500 | |
| | | | 64 to 80 | A220UB | 10000 | |
| | | | 6 to 2.5 | A90RB | 43 | 1.2 |
| | | | 2 to 1 | A120SB | 46 | |
| | | | .8 to .45 | A180TB | 48 | |
| | | | .4 to .35 | A220UB | 51 | |
| MONEL Alloy 400<br>MONEL Alloy 401<br>MONEL Alloy 404<br>MONEL Alloy R405<br>ASTM A296: Grades CZ-100, M-35<br>ASTM A494: Grades CZ-100, M-35 | 115 to 240 | Annealed, Cold Drawn or Cast | 4 to 12 | A90RB | 8500 | 4.0 |
| | | | 13 to 27 | A120SB | 9000 | |
| | | | 28 to 63 | A180TB | 9500 | |
| | | | 64 to 80 | A220UB | 10000 | |
| | | | 6 to 2.5 | A90RB | 43 | 1.2 |
| | | | 2 to 1 | A120SB | 46 | |
| | | | .8 to .45 | A180TB | 48 | |
| | | | .4 to .35 | A220UB | 51 | |

See section 16 for Cutting Fluid Recommendations.
See section 20 for additional information.

# 8.14 Thread Grinding

| MATERIAL | HARD-NESS Bhn | CONDITION | THREADS PER INCH / PITCH (P) mm | WHEEL IDENTIFI-CATION ANSI / ISO | WHEEL SPEED fpm / m/s | WORK SPEED fpm / m/min |
|---|---|---|---|---|---|---|
| **36. NICKEL ALLOYS, WROUGHT AND CAST (cont.)**<br>DURANICKEL Alloy 301<br>MONEL Alloy 502<br>MONEL Alloy K500<br>NI-SPAN-C Alloy 902<br>PERMANICKEL Alloy 300 | 150 to 320 | Solution Treated | 4 to 12<br>13 to 27<br>28 to 63<br>64 to 80 | A90RB<br>A120SB<br>A180TB<br>A220UB | 8500<br>9000<br>9500<br>10000 | 4.0 |
| | | | 6 to 2.5<br>2 to 1<br>.8 to .45<br>.4 to .35 | A90RB<br>A120SB<br>A180TB<br>A220UB | 43<br>46<br>48<br>51 | 1.2 |
| | 330 to 360 | Aged | 4 to 12<br>13 to 27<br>28 to 63<br>64 to 80 | A90RB<br>A120SB<br>A180TB<br>A220UB | 8500<br>9000<br>9500<br>10000 | 4.0 |
| | | | 6 to 2.5<br>2 to 1<br>.8 to .45<br>.4 to .35 | A90RB<br>A120SB<br>A180TB<br>A220UB | 43<br>46<br>48<br>51 | 1.2 |
| **37. BERYLLIUM NICKEL ALLOYS, WROUGHT AND CAST††**<br>Berylco 440<br>Berylco 41C<br>Berylco 42C<br>Berylco 43C<br>Brush Alloy 200C<br>Brush Alloy 220C<br>Brush Alloy 260C | 200 to 250 | As Cast or Solution Treated | 4 to 12<br>13 to 27<br>28 to 63<br>64 to 80 | A90RB<br>A120SB<br>A180TB<br>A220UB | 8500<br>9000<br>9500<br>10000 | 4.0 |
| | | | 6 to 2.5<br>2 to 1<br>.8 to .45<br>.4 to .35 | A90RB<br>A120SB<br>A180TB<br>A220UB | 43<br>46<br>48<br>51 | 1.2 |
| | 283 to 425 | Hardened or Aged | 4 to 12<br>13 to 27<br>28 to 63<br>64 to 80 | A90RB<br>A120SB<br>A180TB<br>A220UB | 8500<br>9000<br>9500<br>10000 | 4.0 |
| | | | 6 to 2.5<br>2 to 1<br>.8 to .45<br>.4 to .35 | A90RB<br>A120SB<br>A180TB<br>A220UB | 43<br>46<br>48<br>51 | 1.2 |
| | $47 R_C$ to $52 R_C$ | Hardened or Aged | 4 to 12<br>13 to 27<br>28 to 63<br>64 to 80 | A120JV<br>A180KV<br>A240MV<br>A320PV | 7000<br>7500<br>8500<br>9000 | 6.0 |
| | | | 6 to 2.5<br>2 to 1<br>.8 to .45<br>.4 to .35 | A120JV<br>A180KV<br>A240MV<br>A320PV | 36<br>38<br>43<br>46 | 1.8 |

See section 16 for Cutting Fluid Recommendations.
See section 20 for additional information.

††CAUTION: Toxic Material, refer to National Institute for Occupational Safety and Health (NIOSH) for Precautions.

| MATERIAL | HARDNESS Bhn | CONDITION | THREADS PER INCH<br>PITCH (P) mm | WHEEL IDENTIFICATION<br>ANSI<br>ISO | WHEEL SPEED<br>fpm<br>m/s | WORK SPEED<br>fpm<br>m/min |
|---|---|---|---|---|---|---|
| **39. HIGH TEMPERATURE ALLOYS, WROUGHT AND CAST**<br>**Nickel Base, Wrought**<br>AF2-1DA, Astroloy, Haynes Alloy 263, IN-102, Incoloy Alloy 901, Incoloy Alloy 903, Inconel Alloy 617, Inconel Alloy 625, Inconel Alloy 700, Inconel Alloy 702, Inconel Alloy 706, Inconel Alloy 718, Inconel Alloy 721, Inconel Alloy 722, Inconel Alloy X-750, Inconel Alloy 751, M252, Nimonic 75, Nimonic 80, Nimonic 90, Nimonic 95, Rene 41, Rene 63, Rene 77, Rene 95, Udimet 500, Udimet 700, Udimet 710, Unitemp 1753, Waspaloy | 200 to 390 | Annealed or Solution Treated | 4 to 12<br>13 to 27<br>28 to 63<br>64 to 80 | A90SB<br>A100TB<br>A180TB<br>A280TB | 9000 | 1.5 to 4.0 |
| | | | 6 to 2.5<br>2 to 1<br>.8 to .45<br>.4 to .35 | A90SB<br>A100TB<br>A180TB<br>A280TB | 46 | .45 to 1.2 |
| | 300 to 475 | Solution Treated and Aged | 4 to 12<br>13 to 27<br>28 to 63<br>64 to 80 | A90SB<br>A100TB<br>A180TB<br>A280TB | 9000 | 1.5 to 4.0 |
| | | | 6 to 2.5<br>2 to 1<br>.8 to .45<br>.4 to .35 | A90SB<br>A100TB<br>A180TB<br>A280TB | 46 | .45 to 1.2 |
| | | | | | | |
| **Nickel Base, Wrought**<br>Hastelloy Alloy B, Hastelloy Alloy B-2, Hastelloy Alloy C, Hastelloy Alloy C-276, Hastelloy Alloy G, Hastelloy Alloy S, Hastelloy Alloy X, Incoloy Alloy 804, Incoloy Alloy 825, Inconel Alloy 600, Inconel Alloy 601, Refractaloy 26, Udimet 630 | 140 to 220 | Annealed or Solution Treated | 4 to 12<br>13 to 27<br>28 to 63<br>64 to 80 | A90SB<br>A100TB<br>A180TB<br>A280TB | 9000 | 1.5 to 4.0 |
| | | | 6 to 2.5<br>2 to 1<br>.8 to .45<br>.4 to .35 | A90SB<br>A100TB<br>A180TB<br>A280TB | 46 | .45 to 1.2 |
| | 240 to 310 | Cold Drawn or Aged | 4 to 12<br>13 to 27<br>28 to 63<br>64 to 80 | A90SB<br>A100TB<br>A180TB<br>A280TB | 9000 | 1.5 to 4.0 |
| | | | 6 to 2.5<br>2 to 1<br>.8 to .45<br>.4 to .35 | A90SB<br>A100TB<br>A180TB<br>A280TB | 46 | .45 to 1.2 |
| **Nickel Base, Wrought**<br>TD-Nickel††<br>TD-Ni-Cr†† | 180 to 200 | As Rolled | 4 to 12<br>13 to 27<br>28 to 63<br>64 to 80 | A90SB<br>A100TB<br>A180TB<br>A280TB | 9000 | 1.5 to 4.0 |
| | | | 6 to 2.5<br>2 to 1<br>.8 to .45<br>.4 to .35 | A90SB<br>A100TB<br>A180TB<br>A280TB | 46 | .45 to 1.2 |
| | | | | | | |

See section 16 for Cutting Fluid Recommendations.
See section 20 for additional information.

††CAUTION: Toxic Material, refer to National Institute for Occupational Safety and Health (NIOSH) for Precautions.

# 8.14 Thread Grinding

| MATERIAL | HARD-NESS Bhn | CONDITION | THREADS PER INCH / PITCH (P) mm | WHEEL IDENTIFI-CATION ANSI / ISO | WHEEL SPEED fpm / m/s | WORK SPEED fpm / m/min |
|---|---|---|---|---|---|---|
| **39. HIGH TEMPERATURE ALLOYS, WROUGHT AND CAST (cont.)**<br>**Nickel Base, Cast**<br>B-1900<br>GMR-235<br>GMR-235D<br>Hastelloy Alloy B<br>Hastelloy Alloy C<br>Hastelloy Alloy D<br>IN-100 (Rene 100)<br>IN-738<br>IN-792<br>Inconel Alloy 713C<br>Inconel Alloy 718<br>M252<br>MAR-M200<br>MAR-M246<br>MAR-M421<br>MAR-M432<br>Rene 80<br>Rene 125<br>SEL<br>SEL 15<br>TRW VI A<br>Udimet 500<br>Udimet 700<br>ASTM A296: Grades CW-12M, N-12M, CY-40<br>ASTM A297: Grades HW, HX<br>ASTM A494: Grades N-12M-1, N-12M-2, CY-40, CW-12M-1, CW-12M-2<br>ASTM A608: Grades HW50, HX50 | 200 to 425 | As Cast or Cast and Aged | 4 to 12<br>13 to 27<br>28 to 63<br>64 to 80 | A90SB<br>A100TB<br>A180TB<br>A280TB | 9000 | 1.5 to 4.0 |
| | | | 6 to 2.5<br>2 to 1<br>.8 to .45<br>.4 to .35 | A90SB<br>A100TB<br>A180TB<br>A280TB | 46 | .45 to 1.2 |
| **Cobalt Base, Wrought**<br>AiResist 213<br>Haynes Alloy 25 (L605)<br>Haynes Alloy 188<br>J-1570<br>MAR-M905<br>MAR-M918<br>S-816<br>V-36 | 180 to 230 | Solution Treated | 4 to 12<br>13 to 27<br>28 to 63<br>64 to 80 | A90SB<br>A100TB<br>A180TB<br>A280TB | 9000 | 1.5 to 4.0 |
| | | | 6 to 2.5<br>2 to 1<br>.8 to .45<br>.4 to .35 | A90SB<br>A100TB<br>A180TB<br>A280TB | 46 | .45 to 1.2 |
| | 270 to 320 | Solution Treated and Aged | 4 to 12<br>13 to 27<br>28 to 63<br>64 to 80 | A90SB<br>A100TB<br>A180TB<br>A280TB | 9000 | 1.5 to 4.0 |
| | | | 6 to 2.5<br>2 to 1<br>.8 to .45<br>.4 to .35 | A90SB<br>A100TB<br>A180TB<br>A280TB | 46 | .45 to 1.2 |

See section 16 for Cutting Fluid Recommendations.
See section 20 for additional information.

| MATERIAL | HARD-NESS<br>Bhn | CONDITION | THREADS PER INCH<br>PITCH (P)<br>mm | WHEEL IDENTIFI-CATION<br>ANSI<br>ISO | WHEEL SPEED<br>fpm<br>m/s | WORK SPEED<br>fpm<br>m/min |
|---|---|---|---|---|---|---|
| **Cobalt Base, Cast**<br>AiResist 13<br>AiResist 215<br>FSX-414<br>HS-6<br>HS-21<br>HS-31 (X-40)<br>HOWMET #3<br>MAR-M302<br>MAR-M322<br>MAR-M509<br>NASA Co-W-Re<br>WI-52<br>X-45 | 220<br>to<br>290 | As Cast<br>or<br>Cast and Aged | 4 to 12<br>13 to 27<br>28 to 63<br>64 to 80 | A90SB<br>A100TB<br>A180TB<br>A280TB | 9000 | 1.5<br>to<br>4.0 |
| | | | 6 to 2.5<br>2 to 1<br>.8 to .45<br>.4 to .35 | A90SB<br>A100TB<br>A180TB<br>A280TB | 46 | .45<br>to<br>1.2 |
| | 290<br>to<br>425 | As Cast<br>or<br>Cast and Aged | 4 to 12<br>13 to 27<br>28 to 63<br>64 to 80 | A90SB<br>A100TB<br>A180TB<br>A280TB | 9000 | 1.5<br>to<br>4.0 |
| | | | 6 to 2.5<br>2 to 1<br>.8 to .45<br>.4 to .35 | A90SB<br>A100TB<br>A180TB<br>A280TB | 46 | .45<br>to<br>1.2 |
| **Iron Base, Wrought**<br>A-286<br>Discaloy<br>Incoloy Alloy 800<br>Incoloy Alloy 800H<br>Incoloy Alloy 801<br>Incoloy Alloy 802<br>N-155<br>V-57<br>W-545<br>16-25-6<br>19-9DL | 180<br>to<br>230 | Solution Treated | 4 to 12<br>13 to 27<br>28 to 63<br>64 to 80 | A90SB<br>A100TB<br>A180TB<br>A280TB | 9000 | 1.5<br>to<br>4.0 |
| | | | 6 to 2.5<br>2 to 1<br>.8 to .45<br>.4 to .35 | A90SB<br>A100TB<br>A180TB<br>A280TB | 46 | .45<br>to<br>1.2 |
| | 250<br>to<br>320 | Solution Treated<br>and Aged | 4 to 12<br>13 to 27<br>28 to 63<br>64 to 80 | A90SB<br>A100TB<br>A180TB<br>A280TB | 9000 | 1.5<br>to<br>4.0 |
| | | | 6 to 2.5<br>2 to 1<br>.8 to .45<br>.4 to .35 | A90SB<br>A100TB<br>A180TB<br>A280TB | 46 | .45<br>to<br>1.2 |
| **47. POWDER METAL ALLOYS**<br>**Irons**<br>F-0000-N<br>F-0000-P<br>F-0000-R<br>F-0000-S<br>F-0000-T | 50<br>to<br>67<br>500kg | As Sintered | 4 to 12<br>13 to 27<br>28 to 63<br>64 to 80 | A90RB<br>A120SB<br>A180TB<br>A220UB | 8500<br>9000<br>9500<br>10000 | 4.0 |
| | | | 6 to 2.5<br>2 to 1<br>.8 to .45<br>.4 to .35 | A90RB<br>A120SB<br>A180TB<br>A220UB | 43<br>46<br>48<br>51 | 1.2 |
| **Steels**<br>F-0008-P<br>F-0008-S<br>F-0005-S<br>FC-0205-S<br>FC-0208-P<br>FC-0208-S<br>FC-0508-P<br>FC-1000-N<br>FN-0205-S<br>FN-0205-T<br>FN-0405-R<br>FN-0405-S<br>FN-0405-T<br>FX-1005-T<br>FX-2008-T | 101<br>to<br>426 | As Sintered<br>or<br>Heat Treated | 4 to 12<br>13 to 27<br>28 to 63<br>64 to 80 | A90RB<br>A120SB<br>A180TB<br>A220UB | 8500<br>9000<br>9500<br>10000 | 4.0 |
| | | | 6 to 2.5<br>2 to 1<br>.8 to .45<br>.4 to .35 | A90RB<br>A120SB<br>A180TB<br>A220UB | 43<br>46<br>48<br>51 | 1.2 |
| | | | | | | |
| | | | | | | |

See section 16 for Cutting Fluid Recommendations.
See section 20 for additional information.

# 8.14 Thread Grinding

| MATERIAL | HARD-NESS<br>Bhn | CONDITION | THREADS PER INCH<br>PITCH (P) mm | WHEEL IDENTIFI-CATION<br>ANSI<br>ISO | WHEEL SPEED<br>fpm<br>m/s | WORK SPEED<br>fpm<br>m/min |
|---|---|---|---|---|---|---|
| **47. POWDER METAL ALLOYS (cont.)**<br>**Stainless Steels**<br>SS-303-R<br>SS-304-R<br>SS-316-R<br>SS-410-R | 107<br>to<br>285 | As Sintered<br>or<br>Heat Treated | 4 to 12<br>13 to 27<br>28 to 63<br>64 to 80 | A90RB<br>A120SB<br>A180TB<br>A220UB | 8500<br>9000<br>9500<br>10000 | 4.0 |
| | | | 6 to 2.5<br>2 to 1<br>.8 to .45<br>.4 to .35 | A90RB<br>A120SB<br>A180TB<br>A220UB | 43<br>46<br>48<br>51 | 1.2 |
| **48. MACHINABLE CARBIDES**<br>Ferro-Tic | 68 $R_C$<br>to<br>70 $R_C$ | Hardened<br>and<br>Tempered | 4 to 12<br>13 to 27<br>28 to 63<br>64 to 80 | D240R100B1/8<br>D320T100B1/8<br>D400V100B1/8<br>D400V100B1/8 | 4500<br>5000<br>5500<br>6000 | 1.5 |
| | | | 6 to 2.5<br>2 to 1<br>.8 to .45<br>.4 to .35 | D240R100B1/8<br>D320T100B1/8<br>D400V100B1/8<br>D400V100B1/8 | 23<br>25<br>28<br>30 | .45 |
| **49. CARBIDES**<br>Titanium Carbide<br>Tungsten Carbide | 89 $R_A$<br>to<br>94 $R_A$ | — | 4 to 12<br>13 to 27<br>28 to 63<br>64 to 80 | D240R100B1/8<br>D320T100B1/8<br>D400V100B1/8<br>D400V100B1/8 | 4500<br>5000<br>5500<br>6000 | 1.5 |
| | | | 6 to 2.5<br>2 to 1<br>.8 to .45<br>.4 to .35 | D240R100B1/8<br>D320T100B1/8<br>D400V100B1/8<br>D400V100B1/8 | 23<br>25<br>28<br>30 | .45 |
| **51. MAGNETIC ALLOYS**<br>Alnico I<br>Alnico II<br>Alnico III<br>Alnico IV<br>Alnico V<br>Alnico V-7<br>Alnico XII<br>Columax-5<br>Hyflux Alnico V-7 | 45 $R_C$<br>to<br>58 $R_C$ | As Cast | 4 to 12<br>13 to 27<br>28 to 63<br>64 to 80 | A120JV<br>A180KV<br>A240MV<br>A320PV | 7000<br>7500<br>8500<br>9000 | 6.0 |
| | | | 6 to 2.5<br>2 to 1<br>.8 to .45<br>.4 to .35 | A120JV<br>A180KV<br>A240MV<br>A320PV | 36<br>38<br>43<br>46 | 1.8 |

See section 16 for Cutting Fluid Recommendations.
See section 20 for additional information.

SECTION 9

# INTRODUCTION TO NONTRADITIONAL MACHINING

NONTRADITIONAL MACHINING is a generic designation applied to those material removal processes that have recently emerged, have not been used extensively heretofore, or are new to the user. These processes are sometimes labeled nonconventional, layless or nonmechanical. The designation "nontraditional" reflects a high degree of personal bias, and its use depends upon the experiences of the individual.

Thirty-one nontraditional machining processes have been selected for inclusion in this handbook. They have been grouped for discussion according to their primary energy mode; that is, Mechanical, Electrical, Thermal or Chemical, as shown in table 9-1. While many more nontraditional machining processes exist in laboratory use, these thirty-one represent those processes which have emerged from the laboratory since the early 1940's, are available or are in use commercially, and have machinability data and information which are available in the public domain.

Nontraditional processes provide manufacturing engineers with additional choices or alternatives to the traditional or conventional mechanical and abrasive material removal processes. Figure 9-1 and table 9-2 demonstrate the relationships among the conventional and the nontraditional machining processes with respect to surface roughness, dimensional tolerance and metal removal rate. It is important to note that carefully selected and properly applied nontraditional machining processes offer some unique capabilities and fresh opportunities for cost improvement in the field of material removal.

The state of the art for nontraditional machining processes is still in a period of rapid change. It can be expected, therefore, that many of the values, ranges and limitations expressed in the data presented can be exceeded. The data comprise publicly available information circa 1979. A check with knowledgeable process vendors is recommended before proceeding with or rejecting any of these material removal processes. The Machinability Data Center monitors developments and collects data for these processes. Inquiries are welcomed.

**TABLE 9-1 Current Commercially Available Nontraditional Material Removal Processes**

**MECHANICAL**
AFM — Abrasive Flow Machining
AJM — Abrasive Jet Machining
HDM — Hydrodynamic Machining
LSG — Low Stress Grinding
RUM — Rotary Ultrasonic Machining
TAM — Thermally Assisted Machining
TFM — Total Form Machining
USM — Ultrasonic Machining
WJM — Water Jet Machining

**ELECTRICAL**
ECD — Electrochemical Deburring
ECDG — Electrochemical Discharge Grinding
ECG — Electrochemical Grinding
ECH — Electrochemical Honing
ECM — Electrochemical Machining
ECP — Electrochemical Polishing
ECS — Electrochemical Sharpening
ECT — Electrochemical Turning
ES — Electro-stream™
STEM™ — Shaped Tube Electrolytic Machining

**THERMAL**
EBM — Electron Beam Machining
EDG — Electrical Discharge Grinding
EDM — Electrical Discharge Machining
EDS — Electrical Discharge Sawing
EDWC — Electrical Discharge Wire Cutting
LBM — Laser Beam Machining
LBT — Laser Beam Torch
PBM — Plasma Beam Machining

**CHEMICAL**
CHM — Chemical Machining
ELP — Electropolish
PCM — Photochemical Machining
TCM — Thermochemical Machining (or TEM—Thermal Energy Method)

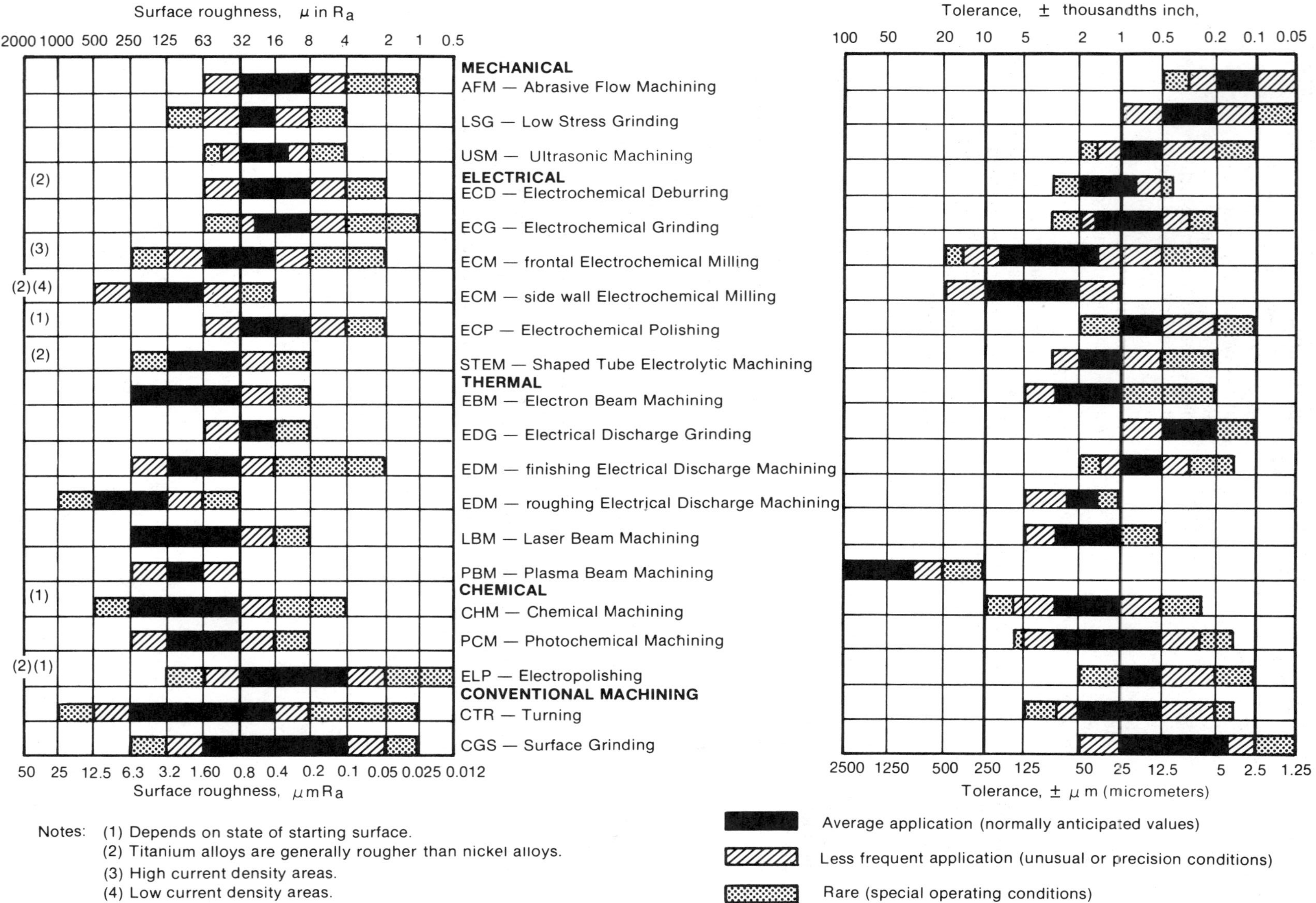


**Figure 9–1** Typical surface roughnesses and tolerances produced by nontraditional material removal processes.

**TABLE 9-2 Comparison of Material Removal Rates and Dimensional Tolerances for Conventional and Nontraditional Machining Processes**

| PROCESS | MAXIMUM RATE OF MATERIAL REMOVAL | TYPICAL POWER CONSUMPTION | WORK SPEED | TYPICAL PENETRATION RATE PER MINUTE | ACCURACY ± ATTAINABLE | ACCURACY ± AT MAXIMUM MATERIAL REMOVAL RATE | TYPICAL MACHINE INPUT |
|---|---|---|---|---|---|---|---|
| | $in^3/min$ | $hp/in^3/min$ | fpm | in | in | in | hp |
| | $cm^3/min$ | $kW/cm^3/min$ | m/min | mm | mm | mm | kW |
| Conventional Turning | 200 | 1 | 250 | — | 0.0002 | 0.005 | 30 |
| | 3300 | 0.046 | 76 | — | 0.005 | 0.13 | 22 |
| Conventional Grinding | 50 | 10 | 100 | — | 0.0001 | 0.002 | 25 |
| | 820 | 0.46 | 30 | — | 0.0025 | 0.05 | 20 |
| CHM | 30 | — | — | 0.001 | 0.0005 | 0.003 | — |
| | 490 | — | — | 0.025 | 0.013 | 0.075 | — |
| PBM | 10 | 20 | 50 | 10 | 0.02 | 0.1 | 200 |
| | 164 | 0.91 | 15 | 254 | 0.5 | 2.54 | 150 |
| ECG | 2 | 2 | 0.25 | — | 0.0002 | 0.0025 | 4 |
| | 33 | 0.091 | 0.08 | — | 0.005 | 0.063 | 3 |
| ECM | 1 | 160 | — | 0.5 | 0.0005 | 0.006 | 200 |
| | 16.4 | 7.28 | — | 12.7 | 0.013 | 0.15 | 150 |
| EDM | 0.3 | 40 | — | 0.5 | 0.00015 | 0.002 | 15 |
| | 4.9 | 1.82 | — | 12.7 | 0.004 | 0.05 | 11 |
| USM | 0.05 | 200 | — | 0.02 | 0.0002 | 0.0015 | 15 |
| | 0.82 | 9.10 | — | 0.50 | 0.005 | 0.040 | 11 |
| EBM | 0.0005 | 10,000 | 200 | 6 | 0.0002 | 0.002 | 10 |
| | 0.0082 | 455 | 60 | 150 | 0.005 | 0.050 | 7.5 |
| LBM, LBT | 0.0003 | 60,000 | — | 4 | 0.0005 | 0.005 | 2 |
| | 0.0049 | 2,731 | — | 102 | 0.013 | 0.13 | 1.5 |

## SECTION 10

# MECHANICAL NONTRADITIONAL MACHINING OPERATIONS

10.1 Abrasive Flow Machining—AFM ........ 10–3
10.2 Abrasive Jet Machining—AJM ........ 10–15
10.3 Hydrodynamic Machining—HDM ........ 10–21
10.4 Low Stress Grinding—LSG ........ 10–37
10.5 Thermally Assisted Machining—TAM ........ 10–39
10.6 Total Form Machining—TFM ........ 10–41
10.7 Ultrasonic Machining—USM and Rotary Ultrasonic Machining—RUM ........ 10–43
10.8 Water Jet Machining—WJM ........ 10–65

## PROCESS SUMMARY

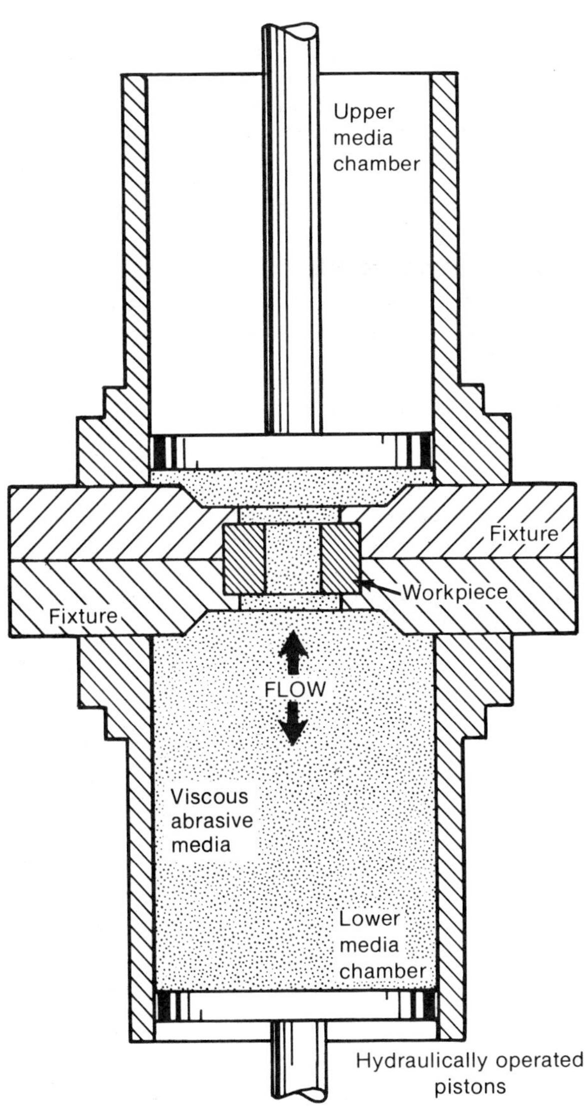


**Figure 10.1–1** AFM schematic.

### Principles

Abrasive flow machining (AFM) is the removal of material by a viscous, abrasive media flowing, under pressure, through or across a workpiece. Figure 10.1–1 contains a schematic presentation of the AFM process. The grit-loaded media of polymeric-base material is selected for its viscosity and type and proportion of grit, in order to suit the part shape and the intended action—deburring, polishing, radiusing. Generally, the putty-like media is extruded through or over the workpiece with motion usually in both directions using from one to one hundred flow reversals per fixture load. Aluminum oxide, silicon carbide, boron carbide or diamond abrasives are used. The velocity of the extruded media is dependent upon the principal parameters of viscosity, pressure, passage size and length.

### Practical Applications

Edge finishing, radiusing, deburring, polishing and minor surface material removal are accomplished with AFM. It is not a mass material removal process, but it is particularly useful for polishing or deburring inaccessible internal passages. Materials from soft aluminum to tough nickel alloys are being processed with AFM. Removal of undesired layers produced by thermal processes such as EDM, LBM, or nitriding is also achieved with AFM. Polishing can quickly improve 30- to 300-microinch $R_a$ [0.8 to 7.6 $\mu$m] finishes to one-tenth of the original roughness. Radii from 0.001 inch [0.025 mm] to 0.060 inch [1.5 mm] can be produced. Holes smaller than 1/64-inch [0.4 mm] diameter are sometimes difficult to process with AFM. Blind-hole polishing is impractical because AFM requires flowing media.

### Operating Parameters

**TABLE 10.1-1 Typical Values for AFM Operating Parameters**

| | |
|---|---|
| Media | |
| viscosity: | Stiff to fluid (see tables 10.1-2 and 10.1-4) |
| grit size: | #8 to #1000 (see table 10.1-3) |
| starting temperature: | 90° to 125°F [32° to 52°C] |
| grit types: | See table 10.1-2 |
| Flow | |
| pressure: | 100 to 3000 psi [700 to 20,000 kPa] |
| volume: | 3 to 100 fluid ounces [100 to 3000 ml] |
| rate: | 2 to 60 gal/min [7 to 225 L/min] |
| Strokes | 1 to 100 |

### Material Removal Rates and Tolerances

The greatest material removal action will occur at the point(s) of maximum flow restriction. AFM processing times frequently range from one to five minutes. Stock removal can be uniform within 10 percent of the stock removed and uniform within each passageway, but out-of-roundness will not be corrected. Production rates of 1000 pieces per hour for multiple-fixtured small parts have been achieved. Usually only a few thousandths inch of material are removed by AFM, as demonstrated in tables 10.1–5 and 10.1–6. Where passageways of dissimilar size are adjacent, stock removal will also be dissimilar, but fixture design can sometimes provide compensation.

# 10.1 Abrasive Flow Machining—AFM

## Surface Technology

Surface roughness to 2 microinches $R_a$ [0.05 $\mu$m] can be obtained and dimensional tolerances to a few ten-thousandths inch can be achieved. Control of the volume of media flow is of major importance to quality control. Surfaces are typically smoothed to 1/10 the prior roughness, in terms of microinches $R_a$. Unidirectional lay is evident after a few strokes of the machine. No metal smearing results from the AFM process. Action occurs on both the peaks and the valleys on the surface. The abrading action can leave residual stresses under 0.001-inch deep [0.025 mm] in the surface layers. Surface layers altered by prior processes are smoothly removed by AFM.

## Equipment and Tools

Low-, medium- and high-pressure systems are available in a range from 100 to 3000 psi [700 to 20,000 kPa]. Tooling is usually needed to direct, confine or sometimes control the flow paths. Hardened steel tooling directs the flow so that the major restriction occurs where material removal is desired. Ceramic or urethane inserts can minimize the tool wear that is experienced in some cases.

Many types of abrasives and viscosities are available as media. Media viscosity relates directly to the size of the restricting passage to be abraded. Fundamentally, viscosity must be high enough to maintain "extrusion-type" flow and high enough to hold the abrasive grains at the outermost surface of the extruding "slug" with sufficient force to allow the abrasive to cut the edges and/or surfaces of the restricting passage. At the same time, the media must be soft enough to flow at a reasonable rate, at the extrusion pressure available, in order to perform the finishing operation in an acceptable cycle time.

## Machining Characteristics

More uniform deburring than can be accomplished with hand tools is a significant quality advantage for AFM. Careful post-operation cleaning is recommended. This cleaning can readily be accomplished by an air blast or immersion in an agitated solvent.

Machining action is gentle and continuous, and the burrs or "chips" are retained in the media. The media can tolerate as much as 10 percent of its volume in "chips" or foreign material. The media life is limited at the point at which a substantial number of the grains become dull. Reconditioning of the media can be accomplished by measured additions of abrasive grit and lubricant.

## SELECTED DATA

**TABLE 10.1-2 Abrasive Flow Media Designations**

Viscosity Grade or Base*:

- LV or 500 — low
- LMV or 300 — low-medium
- MV or 200 — medium
- HMV or 100 — high-medium
- HV or 50 — high

Grit Type†:

- A — Aluminum oxide
- B — Boron carbide
- C — Silicon carbide (in ANSI B74.13 standard)
- D — Diamond
- S — Silicon carbide (in some manufacturers' nomenclatures)

Media Nomenclature (One system):

D075-20A(30)-36A(40)-700A(30)

- Manufacturer
- Media base (viscosity)
- Abrasive grit size
- Abrasive grit type
- Parts by weight to base (this amount of abrasive in 100 parts of base)

*Number and letter codes are not necessarily of equal viscosity—see manufacturers recommendations.

†See section 3.14 for grit hardness.

**TABLE 10.1-3 Grit Sizes Used in AFM**

| GRIT SIZE | AVERAGE PARTICLE SIZE | |
|---|---|---|
| | in | μm |
| 8 | 0.1817 | 4,615 |
| 10 | 0.1366 | 3,470 |
| 12 | 0.1003 | 2,550 |
| 14 | 0.0830 | 2,110 |
| 16 | 0.0655 | 1,660 |
| 20 | 0.0528 | 1,340 |
| 25 | 0.0408 | 1,035 |
| 30 | 0.0365 | 925 |
| 35 | 0.0280 | 710 |
| 45 | 0.0200 | 508 |
| 50 | 0.0170 | 432 |
| 60 | 0.0160 | 406 |
| 70 | 0.0131 | 328 |
| 80 | 0.0105 | 267 |
| 90 | 0.0085 | 216 |
| 100 | 0.0068 | 173 |
| 120 | 0.0056 | 142 |
| 140 | 0.0048 | 122 |
| 170 | 0.0034 | 86 |
| 200 | 0.0026 | 66 |
| 230 | 0.00248 | 63 |
| 270 | 0.00175 | 44 |
| 325 | 0.00128 | 33 |
| 400 | 0.00090 | 23 |
| 500 | 0.00073 | 19 |
| 600 | 0.00057 | 14 |
| 800 | 0.00044 | 11 |
| 900 | 0.00038 | 10 |
| 1,000 | 0.00032 | 8 |

# 10.1 Abrasive Flow Machining—AFM

**TABLE 10.1-4 Guide to Media Selection**
(Viscosity grades for passageways with 2:1 length-to-width ratio)

| PASSAGE WIDTH | | EXTRUSION PRESSURE | | | |
|---|---|---|---|---|---|
| in | mm | Low | | to | High |
| 1/64 | 0.4 | | | | LV* |
| 1/32 | 0.8 | | | LV | LMV |
| 1/16 | 1.6 | | LV | LMV | MV |
| 1/8 | 3.2 | LV | LMV | MV | HMV |
| 1/4 | 6.4 | LMV | MV | HMV | HV |
| 1/2 | 12.8 | MV | HMV | HV | |
| 1 | 25.4 | HMV | HV | | |
| 2 | 50.8 | HV | | | |

SOURCE: Extrude Hone Corporation.

NOTES: As usable media extrusion pressure increases, higher viscosity media grades may be used.

If passage length is substantially shorter than two times passage width, higher viscosity media or lower extrusion pressure may be used.

If passage length is substantially longer than two times passage width, lower viscosity media or higher extrusion pressures may be required.

Media *slug flow* rates affect uniformity of stock removal and edge radius size.

Slow *slug flow* rates (speed of abrasive *slug* passing through restricting passage) are best for uniformly removing material on close tolerance parts.

High *slug flow* rates produce larger edge radii.

*See table 10.1-2 for explanation of viscosity grade codes.

**TABLE 10.1-5 AFM Cutting Test Results for Cold Rolled Steel Discs—10 Full-Reverse Cycles**

Conditions:
- hole length .......... 0.25 inch [6.25 mm]
- hole diameter .......... 0.125 inch [3.18 mm] nominal—initially
- number full reverse cycles .......... 10
- equipment .......... Dynetics HL15
- initial media temperature .......... 30°C (86°F)

Media (moderately fast):
- base polymer 100 (high-medium viscosity) .......... 56.5%
- 36-grit size, white aluminum oxide (A) or black silicon carbide (S) or boron carbide (B) .......... 37.8%
- 700-grit size, black silicon carbide (S) .......... 5.7%

| MEDIA GRIT TYPE | PRESSURE psi | TOTAL TIME min | INITIAL DIAMETER inch | FINAL DIAMETER inch | INCREASE inch | RADIUS inch | STOCK REMOVAL milligrams |
|---|---|---|---|---|---|---|---|
| Aluminum oxide | 300 | 18.0 | 0.127 | 0.141 | 0.014 | 0.020 | 118 |
| | 400 | 3.8 | 0.127 | 0.145 | 0.018 | 0.020 | 176 |
| | 500 | 1.8 | 0.127 | 0.153 | 0.026 | 0.040 | 293 |
| Silicon carbide | 300 | 22.7 | 0.126 | 0.144 | 0.018 | 0.020 | 168 |
| | 400 | 4.2 | 0.127 | 0.149 | 0.022 | 0.025 | 239 |
| | 500 | 2.0 | 0.127 | 0.153 | 0.026 | 0.030 | 293 |
| Boron carbide | 300 | 19.3 | 0.126 | 0.160 | 0.034 | 0.020 | 328 |
| | 400 | 3.8 | 0.125 | 0.164 | 0.039 | 0.020 | 344 |
| | 500 | 1.6 | 0.126 | 0.173 | 0.047 | 0.040 | 540 |

SOURCE: W. B. Perry, Properties and capabilities of low pressure abrasive flow mediae, Technical paper MR75-831, Society of Manufacturing Engineers, 1975.

**TABLE 10.1-6 AFM Cutting Test Results for Cold Rolled Steel Discs—30 Full-Reverse Cycles**

Conditions:
- hole length .......... 0.25 inch [6.35 mm]
- hole diameter .......... 0.125 inch [3.18 mm] nominal—initially
- number of full reverse cycles .......... 30
- equipment .......... Dynetics HL15
- initial media temperature .......... 25°C (77°F)
- media pressure .......... 450 psi [3100 k Pa]

Media:
- base polymer .......... 64.5%
- 60-grit size, black silicon carbide (S) .......... 21.3%
- 700-grit size, black silicon carbide (S) .......... 14.2%
- base polymer .......... 100, 200, 300 (see table 10.1-2 for viscosities)

or

- base polymer .......... 56.5%
- 36-grit size, black silicon carbide (S) .......... 37.8%
- 700-grit size, black silicon carbide (S) .......... 5.7%
- base polymer .......... 100, 050 (see table 10.1-2 for viscosities)

| MEDIA | TOTAL TIME min | INITIAL DIAMETER inch | FINAL DIAMETER inch | RADIUS inch |
|---|---|---|---|---|
| Size 60 grit, Base 300 | 4.0 | 0.125 | 0.136 | 0.045 |
| Size 60 grit, Base 200 | 4.4 | 0.125 | 0.138 | 0.040 |
| Size 60 grit, Base 100 | 7.7 | 0.126 | 0.153 | 0.020 |
| Size 36 grit, Base 100* | 4.1 | 0.125 | 0.189 | 0.040 |
| Size 36 grit, Base 100* | 1.6 | 0.249 | 0.267 | 0.055 |
| Size 36 grit, Base 050 | 22.0 | 0.373 | 0.408 | 0.040 |

SOURCE: W. B. Perry, Technical paper MR75-831, Society of Manufacturing Engineers, 1975.
* Same media as in table 10.1-5.

## Stock Removal and Flow Rates[1]

One of the attractive features of AFM is the wide variety of available and easily adjustable process parameters. The effects of changing some of these parameters for specific media compositions are illustrated in figures 10.1-2 to 10.1-7. These data come from simple laboratory tests on cold-rolled-steel samples and are intended only to provide "guidelines" for use in actual production situations. The interrelationships of various parameters depend greatly on the characteristics of a given media composition. A few widely different media compositions were examined in order to formulate general rules that can be applied to AFM. Parameter relationships are also affected by the size and length of the flow passages. The data in figures 10.1-2 to 10.1-7 are limited to the examination of round holes machined normal to the test piece surfaces. The general rules which emerged are as follows:

1. Stock removal per unit time increases substantially with increasing extrusion pressure (figures 10.1-2 to 10.2-4).
2. Media flow rate increases with increasing extrusion pressure. Note: Flow rate increases proportionally with extrusion pressure, or logarithmically, as indicated in figure 10.1-5.
3. Flow rate increases as media temperature increases (figure 10.1-6). Note: In the tests conducted in preparing this graph, stock removal per unit volume of flow did not change with changes in flow velocity over the range covered by the graph.
4. Increasing hole length reduces both flow rate and diameter change per unit flow (figure 10.1-7).

These general rules apply for each of the medias examined, but the degree of application varies. These variances reflect differences in the content of base materials, lubricants and/or plasticizers as well as differences in the size and percentage of grit. The relative size and length of the flow passage also is likely to have an effect on these relationships.

[1]Based on text and figures from L. J. Rhoades, Extrude hone—edge and surface finishing—capabilities and costs, Technical paper MR77-366, Society of Manufacturing Engineers, Dearborn, MI, 1977

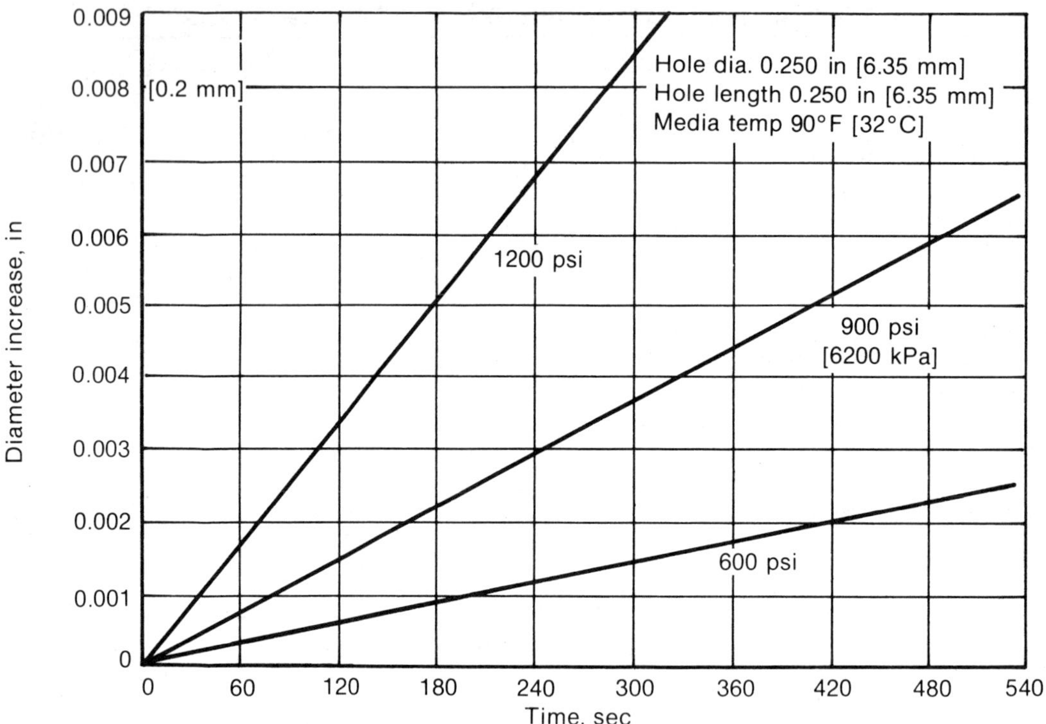


**Figure 10.1-2** Stock removal rates for AFM of cold-rolled-steel samples using media 70 (633) at three extrusion pressures. (From L. J. Rhoades, Extrude hone-edge and surface finishing-capabilities and costs, Technical paper MR77-366, Society of Manufacturing Engineers, Dearborn, MI, 1977, p. 14)

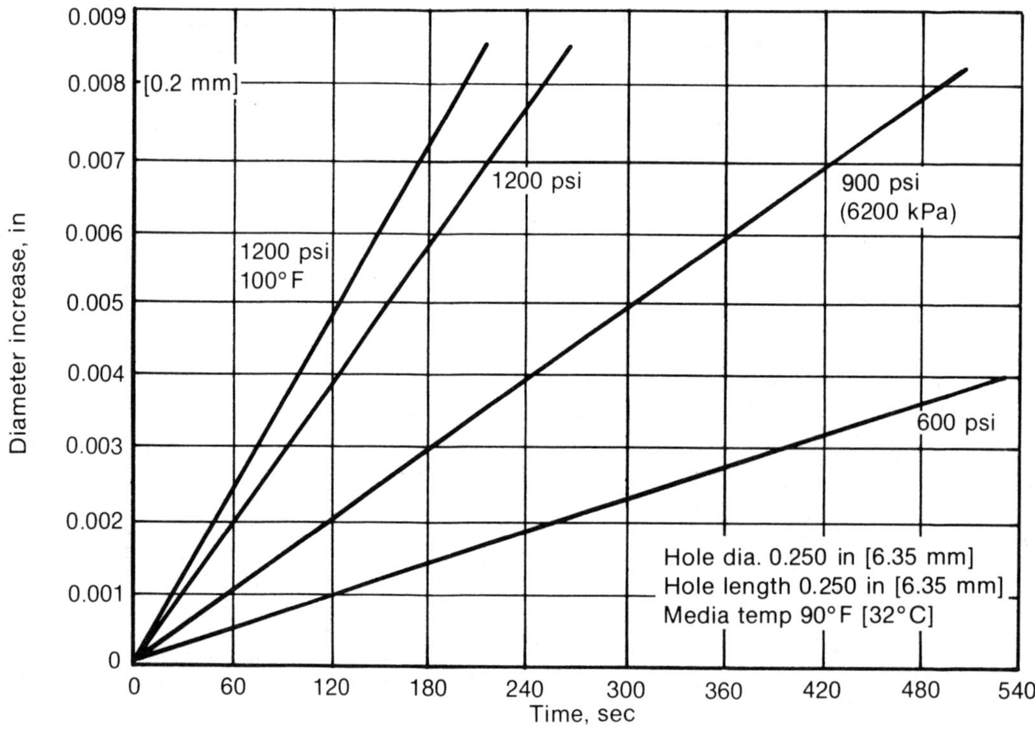


**Figure 10.1-3** Stock removal rates for AFM of cold-rolled-steel samples using media 36 (564) at four extrusion pressures. (L. J. Rhoades, p. 14)

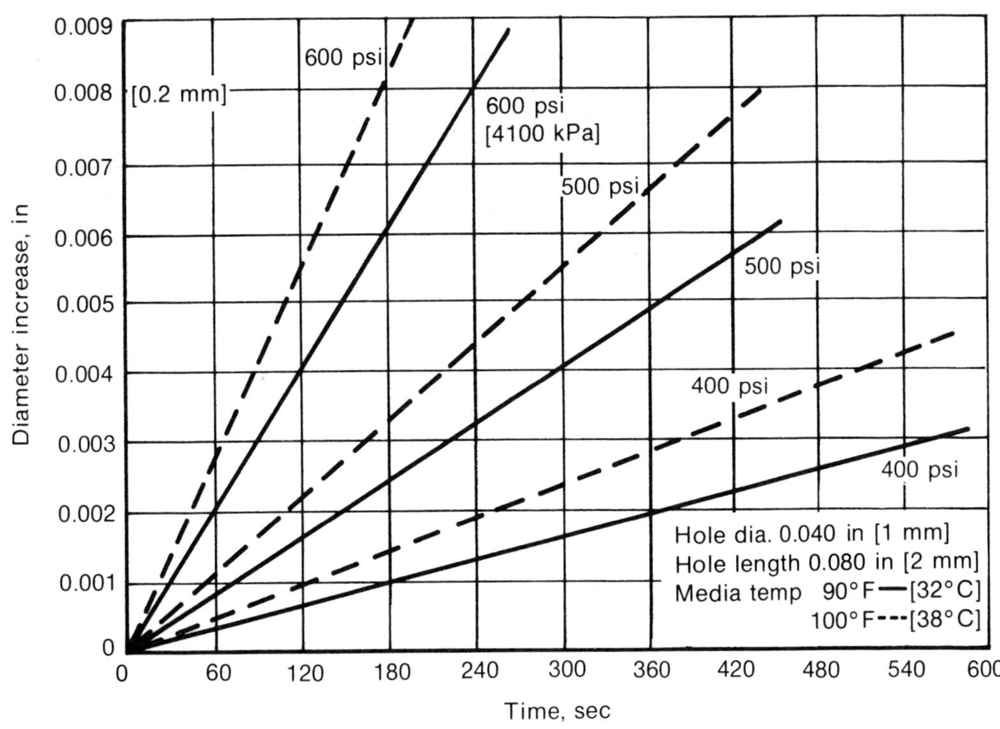


**Figure 10.1-4** Stock removal rates for AFM of cold-rolled-steel samples using media 100 (356) at three extrusion pressures. (L. J. Rhoades, p. 15)

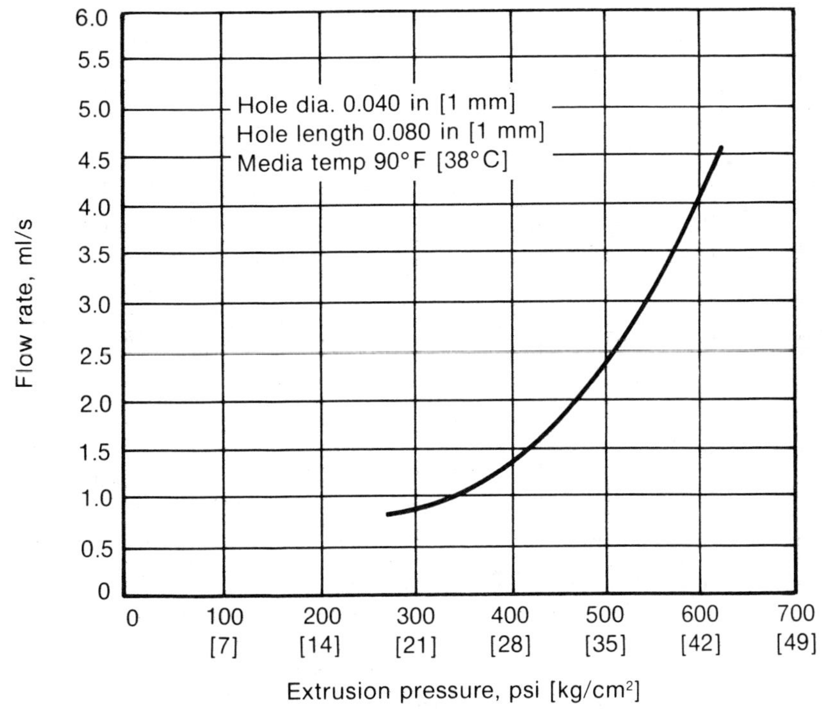


**Figure 10.1-5** Media flow rate versus extrusion pressure. Test conditions same as figure 10.1-4. (L. J. Rhoades, p. 16)

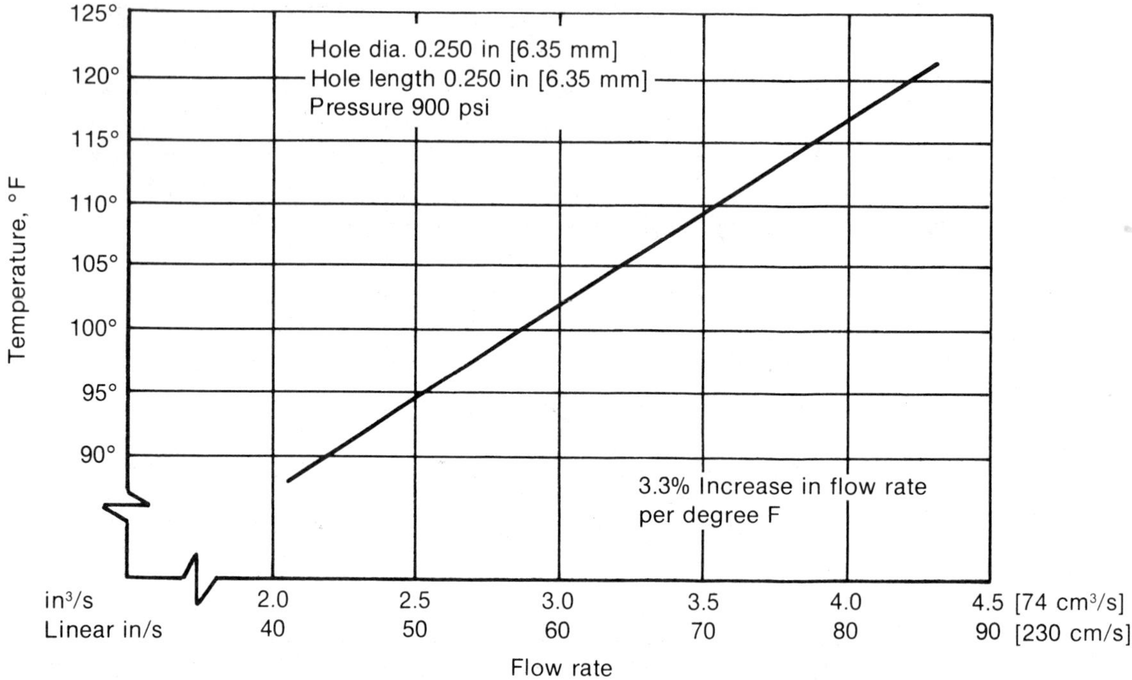


**Figure 10.1–6** Media flow rate versus temperature for AFM of cold-rolled-steel samples using media 70 (633). (L. J. Rhoades, p.-16)

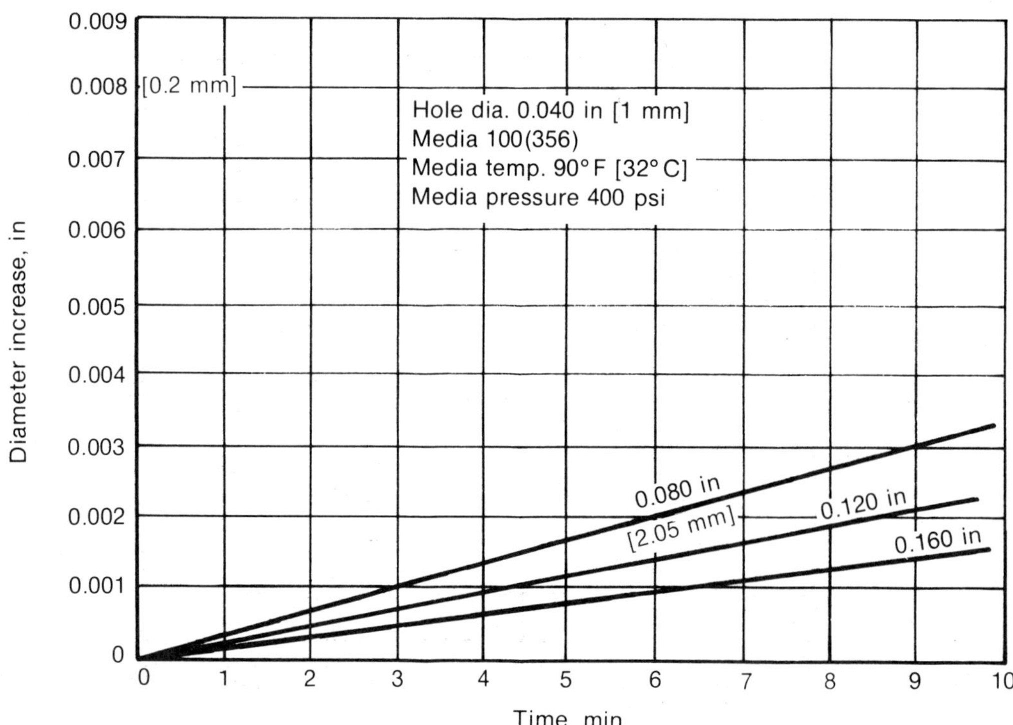


**Figure 10.1–7** Stock removal rates for AFM of cold-rolled-steel samples with three hole lengths. (L. J. Rhoades, p. 15)

## Hole Enlargement

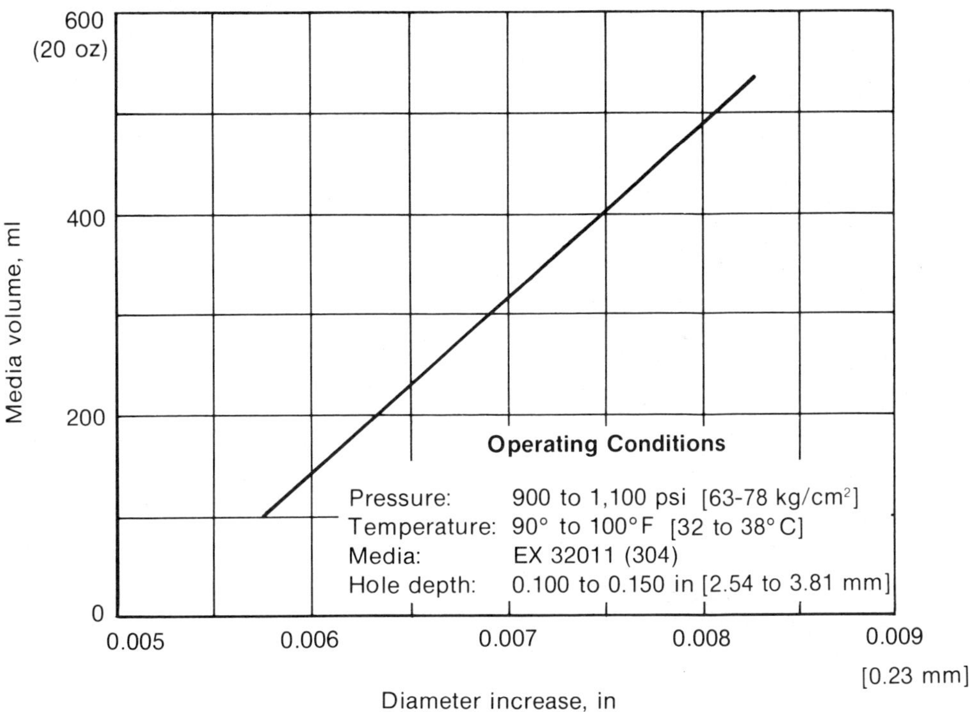


**Figure 10.1–8** Hole diameter increase as a function of media volume. Data from enlarging 0.015-inch diameter laser drilled holes in a nickel alloy. (Courtesy of General Electric Company)

# 10.1 Abrasive Flow Machining—AFM

## SPECIFIC APPLICATIONS DATA

**TABLE 10.1–7 Specific Applications Data for Abrasive Flow Machining**

| Work material: | **Stainless Steel** | **Hardened Die Steel** | **Tungsten Carbide** |
|---|---|---|---|
| Workpiece configuration: | 0.152 in<br>Deburr 12 drilled hole intersections | Radius & remove recast layer from extrusion die | Draw die polishing |
| Media type: | Dynetics D080-20S(50)-36S(50)-700S(40) | Extrude Hone MV70S | Extrude Hone HV30$\mu$D |
| grit type: | Silicon carbide | Silicon carbide | Diamond |
| grit size: | 50 parts-20 grit; 50 parts-36 grit; 40 parts-700 grit | 70 grit | 30 micron |
| volume: | 10-inch stroke | 3 passages, 6,000 $in^3$ | 1,500 $in^3$ |
| Pressure, psi: | 450 | 450 | 800 |
| Media temperature, °F: | | 80 | 110 |
| Process time: | 90 s | 4 min | 3 to 8 min |
| Number strokes (one way): | 6 | 20 | 30 |
| Pieces per fixture: | 1 | 1 | 1 |
| Material removed (Dimension changed): | 0.001 inch per surface | 0.00075 inch per surface | 0.001 to 0.002 inch |
| Surface roughness change, $\mu$in $R_a$: | 63 to 16 | 75 to 7 | 25-35 to 6-8 |

**TABLE 10.1-7—Continued**

| Work material: | Cast Stellite | High Strength Steel | AISI Type O2 Tool Steel |
|---|---|---|---|
| Workpiece configuration: | Jet fuel swirler polish, radius & deburr | Thread deburr | Polished retainer seat pockets |
| Media type: | Dynetics D060-20B(35)-36B(35)-700(45) | Extrude Hone MV70S | Extrude Hone MV24S |
| grit type: | Boron carbide | Silicon carbide | Silicon carbide |
| grit size: | 35 parts-20 grit; 35 parts-36 grit; 45 parts-700 grit | 70 grit | 24 grit |
| volume: | 12-inch stroke | 1,000 $in^3$ | 5,600 $in^3$ |
| Pressure, psi: | 450 | 600 | 1,400 |
| Media temperature, °F: | | 80 | |
| Process time: | 3 min | 2 min | 7 min |
| Number strokes (one way): | 6 | 4 | 14 |
| Pieces per fixture: | 3 | 1 | 1 (6 to 8 practical) |
| Material removed (Dimension changed): | Polished | 0.0002 inch per surface | 0.002 inch per surface |
| Surface roughness change, $\mu$in $R_a$: | 63 to 12 | 40 to 8 | 100 to 10 |

**TABLE 10.1-7—Continued**

| Work material: | **Cast Nickel Alloy** | **Cast Nickel Alloy** |
|---|---|---|
| Workpiece configuration: | 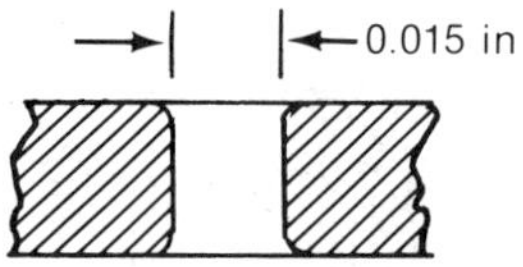 0.015 in<br>(section)<br>Remove recast from small holes | 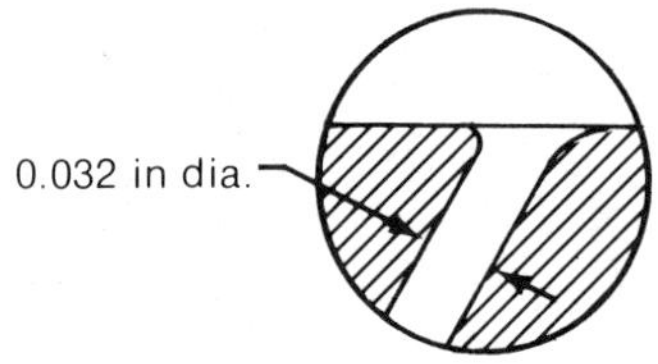 0.032 in dia.<br>Removal of recast & add radii to hole |
| Media type: | Dynetics D250-220S(50) | Extrude Hone VLV320S |
| grit type: | Silicon carbide | Silicon carbide |
| grit size: | 50 parts-220 grit | 320 grit |
| volume: | 1,000 ml | 480 ml |
| Pressure, psi: | 475 | 1,100 |
| Media temperature, °F: | 95 | 110 |
| Process time: | Controlled by volume, not strokes and time | Controlled by volume, not strokes and time |
| Number strokes (one way): | Controlled by volume, not strokes and time | Controlled by volume, not strokes and time |
| Pieces per fixture: | 1 | 1 |
| Material removed (Dimension changed): | 0.004 inch on diameter | 0.005 inch on diameter |
| Surface roughness change, $\mu$in $R_a$: | 50 to 16 | — |

## PROCESS SUMMARY

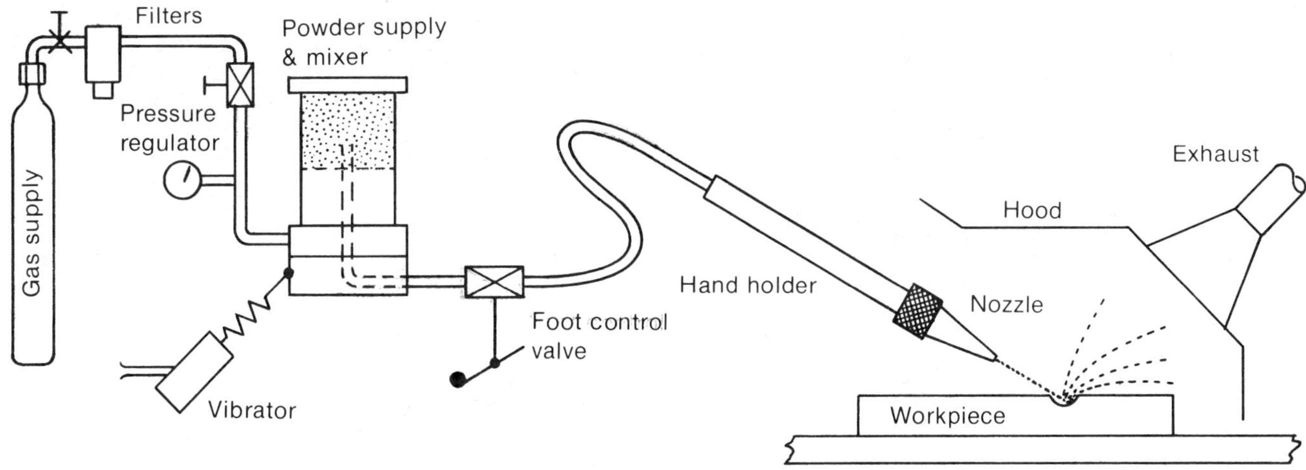


**Figure 10.2–1** AJM schematic.

### Principles

Abrasive jet machining (AJM) is the removal of material through the action of a focused, high velocity stream of fine grit or powder-loaded gas. The gas should be dry, clean and under modest pressure. Figure 10.2–1 shows a schematic of the AJM process. The mixing chamber sometimes uses a vibrator to promote a uniform flow of grit. The hard nozzle is directed close to the workpiece at a slight angle. The operation should be enclosed in a dust hood or be near an adequate exhaust collector. Cutting is controlled by type of abrasive, gas pressure, rate of abrasive flow, angle of and closeness of nozzle to workpiece, and duration of application. Masks of copper, glass or rubber are sometimes used to control overspray or the etching pattern.

### Practical Applications

Cutting, deburring, cleaning, deflashing and etching of glass, ceramics, or hard metals can be accomplished with AJM. It may also be used for deburring of cross holes, slots, and threads in small precision parts that require a burr-free finish, such as hydraulic valves, aircraft fuel systems and medical appliances. Parts machined from nylon, Teflon®, and Delrin® may also be deburred. The cleaning of irregular surfaces, such as holes that break through threads, is one advantage of the ability of the abrasive stream to follow contours. Drilling holes in metal is not a viable application because the walls will have significant taper, and the drilling speed is very slow.

Hand-held nozzles, template guides, pantograph guides, or motorized fixtures are used for simple and intricate shapes or holes in sensitive, brittle, thin or difficult materials. Discs as small as 1/16-inch [1.6 mm] diameter and discs up to 1/4-inch [6.35 mm] thick have been cut in glass without the occurrence of chipping. Other good applications for AJM are insulation stripping and wire cleaning without affecting the conductor, and micro-deburring, as on hypodermic needles. Frosting glass, trimming circuit boards, and trimming silicon and gallium are routine. Removal of films and delicate cleaning, such as the removal of smudges from antique documents, is also possible. Hybrid circuit resistor and capacitor trimming is a typical production application in the electronics industry.

®Trademark E. I. Du Pont de Nemours & Co., Inc.

# 10.2 Abrasive Jet Machining—AJM

## Operating Parameters

**TABLE 10.2-1 Typical Values for AJM Operating Parameters**

| | |
|---|---|
| **Abrasive** | |
| types: | Aluminum oxide for aluminum and brass<br>Silicon carbide for stainless steel and ceramic<br>Bicarbonate of soda for nylon, Teflon®, Delrin® and light cleaning<br>Glass beads for polishing |
| size: | 10 to 150 $\mu$m, well classified and clean |
| quantity: | 1 to 5 g/min for fine work<br>5 to 10 g/min for usual cuts<br>10 to 20 g/min for heavier cuts |
| **Carrier gas** | |
| types: | Dry air, carbon dioxide, nitrogen, nitrous oxide, helium (Do *not* use oxygen) |
| quantity: | Up to 1.0 ft³/min [28 L/min] |
| pressure: | 30 to 190 psi [207 to 1,310 kPa] |
| velocity: | 500 to 1,100 ft/s [152 to 335 m/s] |
| **Nozzle** | |
| material: | Tungsten carbide or sapphire |
| tip distance (ntd) to work: | 0.10 to 3.0 inches [2.54 to 76 mm] (see figures 10.2-2 and 10.2-3 |
| opening: | Diameter 0.005 to 0.046 inch [0.13 to 1.2 mm]<br>Rectangular size 0.003 x 0.020 to 0.026 x 0.026 inch [0.075 x 0.50 to 0.65 x 0.65 mm] |
| operating angle: | Vertical to 60° off vertical |

## Material Removal Rates and Tolerances

AJM is not a mass material removal process; it is a finishing process. The typical removal rate for plate glass is 0.001 cubic inch per minute [16.4 mm³/min]. The speed of drilling in glass is impressive; however, higher hardness ceramics cut at approximately 50 percent higher rates and metals cut at lesser rates, as low as 10 to 25 percent that of glass depending on the metal properties. Tumbling, vibratory or larger recirculating blast cabinets may be more practical for large parts. The rate of material removal is affected by several factors, including abrasive type and particle size, nozzle opening, nozzle-to-work-surface distance, nozzle angle, and particle velocity. Generally, the larger the particle and the higher the velocity, the faster the material removal.

Figure 10.2-4 illustrates the factors that can influence material removal. These data were taken on plate glass of 450 to 510 Knoop hardness using a 0.018-inch [0.46 mm] diameter nozzle, air as the gas at 75 psi [517 kPa], and 1/32-inch [0.79 mm] nozzle tip distance (ntd) from the workpiece (unless otherwise expressed). Silicon carbide abrasive would cut at approximately the same values.

Because of the small amount of abrasive flowing through the nozzle, it is possible to be very selective in material removal. The radius on holes, slots, grooves, and threads that are deburred abrasively will vary from sharp to 0.010 inch [0.25 mm] depending on abrasive used and blast time. It is common for micro-abrasive blasting to be a secondary deburring operation after rechasing threads, reaming holes, or using a wheel to remove the heaviest burrs. A practical minimum cut width is 0.005 inch [0.13 mm]. Tolerances are typically ±0.005 [±0.13 mm] with ±0.002 inch [±0.05 mm] possible with good fixturing and motion control. Taper is present in deep cuts.

## Surface Technology

Surfaces generated by AJM have a random "grainness" or matte texture. Surface roughness ranges from 6 to 63 microinches $R_a$ [0.15 to 1.6 $\mu$m] depending on grit size, as demonstrated in tables 10.2-2 and 10.2-3.

The AJM process is cool cutting; heat sensitive alloys are not affected. Constant motion of the nozzle is necessary to prevent unwanted grooves on the surface. Cleaning of the surfaces of softer materials may be necessary to remove any imbedded particles of grit. Cleaning is also required to remove grit from crevices or internal passageways in the workpiece.

## Equipment and Tooling

The equipment consists of four parts: (1) an apparatus which stores, mixes and propels the abrasive; (2) a work chamber; (3) a dust collector; and (4) a dry gas supply. A complete system is available for a modest investment. Single- or dual-tank, bench-top apparatus is available. The work station should provide sufficient illumination for the operator. The dust collector should be a commercial unit of at least 3/4 horsepower with a cloth-bag filter that can handle fine powder particles as small as 40 microinches [1 $\mu$m]. The gas supply of at least 100 to 125 pounds per square inch [690 to 862 kPa] must be dried to a moisture content of less than 50 parts per million. It is possible to use bottled carbon dioxide or nitrogen as a dry gas supply.

The nozzle tips are normally made of tungsten carbide. Their life will vary depending on the type of abrasive and the operating pressure used. Tip life of 8 to 15 hours with silicon carbide and 20 to 35 hours with aluminum oxide is reasonable.

Abrasive powder must be clean, dry, and very well classified. It is not practical to reuse abrasive powder because contamination and worn grit will cause a decline in cutting rates.

### Machining Chacteristics

The powders used in AJM are nontoxic, but adequate dust control is needed. Silica dust produced during AJM may be a health hazard. Commercial grades of abrasive powders are not suitable for AJM because they are not classified closely enough to assure proper flow and they may contain some silica dust. Ordinary shop air should not be used without adequate filtration to remove moisture and oil. Oxygen should never be used as the propelling gas because it may produce a violent chemical reaction when combined with the workpiece chips or the abrasive.

## SELECTED DATA

**TABLE 10.2-2 Surface Roughness for Glass Processed with AJM**

| ABRASIVE TYPE | GRIT SIZE | | SURFACE ROUGHNESS | |
|---|---|---|---|---|
| | μin | μm | μin $R_a$ | μm $R_a$ |
| Aluminum oxide | 400 | 10 | 6 to 8 | 0.15 to 0.20 |
| | 1100 | 28 | 14 to 20 | 0.36 to 0.51 |
| | 2000 | 50 | 38 to 55 | 0.97 to 1.40 |

**TABLE 10.2-3 Surface Roughness for 316 Stainless Steel (Annealed) Processed with AJM**
(Initial surface was ground to 18.5 μin $R_a$ [0.47 μm])

| ABRASIVE TYPE | GRIT SIZE | | SURFACE ROUGHNESS | |
|---|---|---|---|---|
| | μin | μm | μin $R_a$ | μm $R_a$ |
| Aluminum oxide | 400 | 10 | 8 to 20 | 0.20 to 0.50 |
| | 1000 | 25 | 10 to 21 | 0.25 to 0.53 |
| | 2000 | 50 | 15 to 38 | 0.38 to 0.96 |
| Silicon carbide | 800 | 20 | 12 to 20 | 0.30 to 0.50 |
| | 2000 | 50 | 17 to 34 | 0.43 to 0.86 |
| Glass bead | 2000 | 50 | 12 to 38 | 0.30 to 0.96 |

SOURCE: Comco Incorporated.

| ORIFICE mm | NOZZLE CUTTING PATTERN, mm | | | | | | |
|---|---|---|---|---|---|---|---|
| | Nozzle Tip Distance, mm | | | | | | |
| | 0.38 | 0.79 | 1.59 | 2.38 | 3.18 | 4.76 | 6.35 |
| 0.46 dia | 0.48 dia | 0.81 dia | 0.89 dia | 1.22 dia | 1.32 dia | 1.90 dia | 2.24 dia |
| 0.51 x 0.15 | 0.66 x 0.33 | 0.76 x 0.48 | 0.96 x 0.71 | 1.12 x 0.76 | 1.22 x 1.22 | 1.42 x 1.40 | 1.80 x 1.70 |
| 1.02 x 0.15 | 1.22 x 0.38 | 1.40 x 0.61 | 1.60 x 1.02 | 1.73 x 1.19 | 2.06 x 1.60 | 2.49 x 2.18 | 2.72 x 2.64 |
| 1.52 x 0.15 | 1.62 x 0.25 | 1.73 x 0.41 | 1.88 x 0.64 | 2.11 x 0.91 | 2.21 x 1.04 | 2.49 x 2.18 | 2.72 x 2.64 |
| 2.54 x 0.15 | 2.67 x 0.38 | 2.82 x 0.64 | 3.12 x 1.02 | 3.20 x 1.14 | 3.30 x 1.17 | 3.53 x 1.52 | 3.89 x 1.88 |
| 3.81 x 0.15 | 3.96 x 0.41 | 4.01 x 0.56 | 4.24 x 0.81 | 4.37 x 1.01 | 4.57 x 1.37 | 4.93 x 1.78 | 5.23 x 2.18 |

**Figure 10.2-2** Cutting pattern of nozzle versus nozzle tip distance. (Data courtesy of S. S. White Industrial Products)

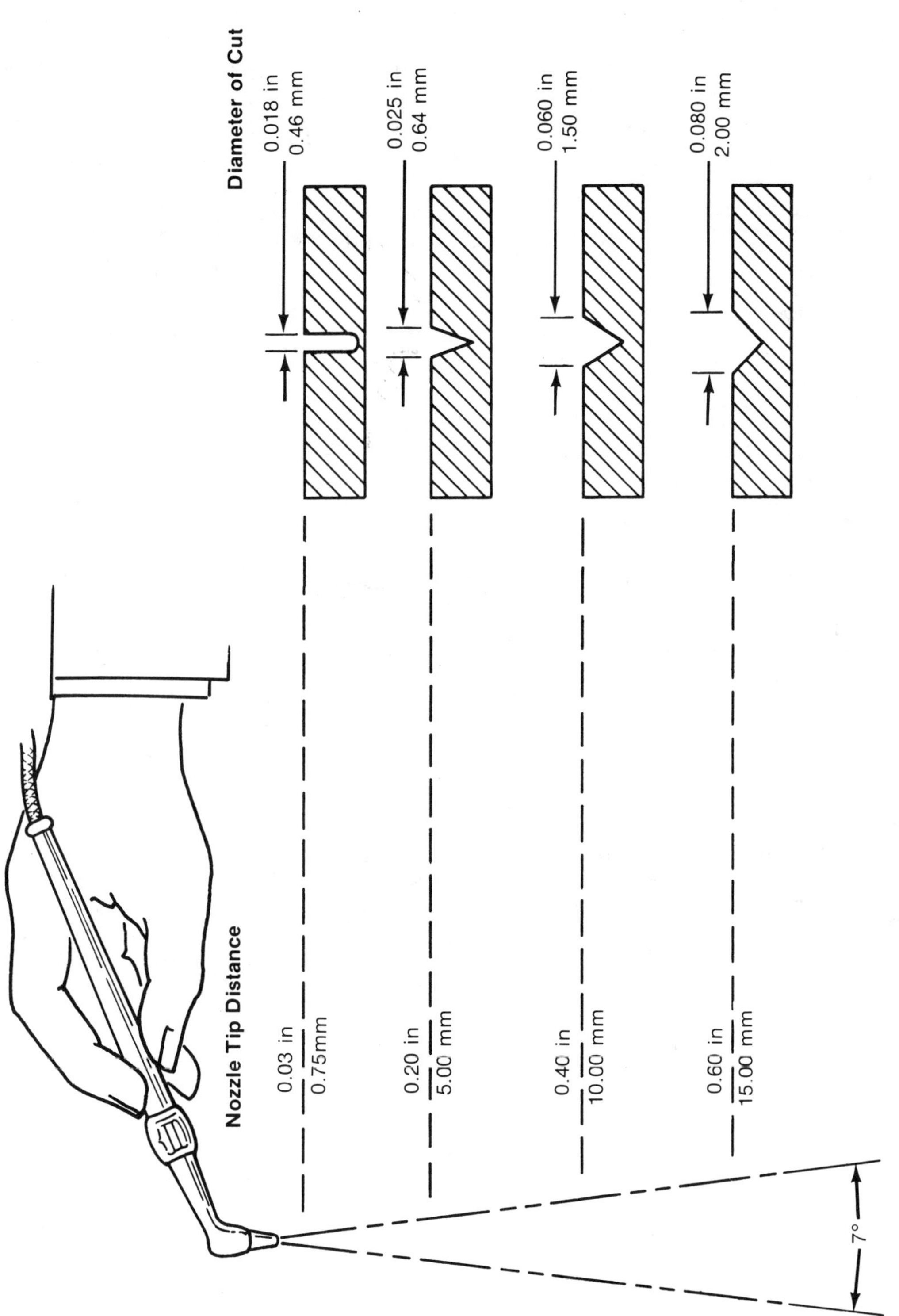


**Figure 10.2–3** Typical cutting action of 0.018 inch [0.46 mm] diameter nozzle. (Data courtesy of S. S. White Industrial Products)

## 10.2 Abrasive Jet Machining—AJM

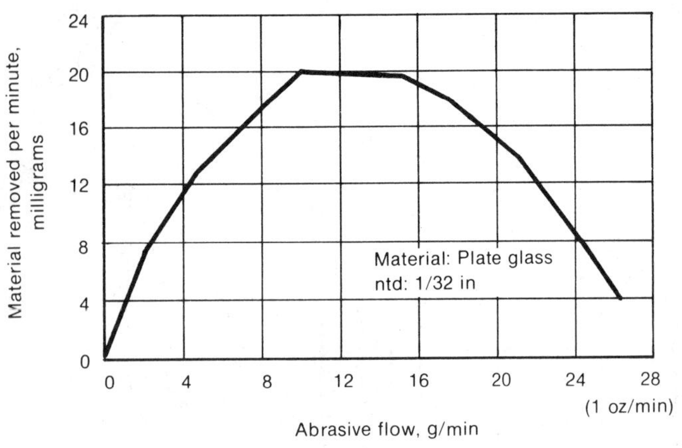


A. Abrasive flow effects on cutting speed

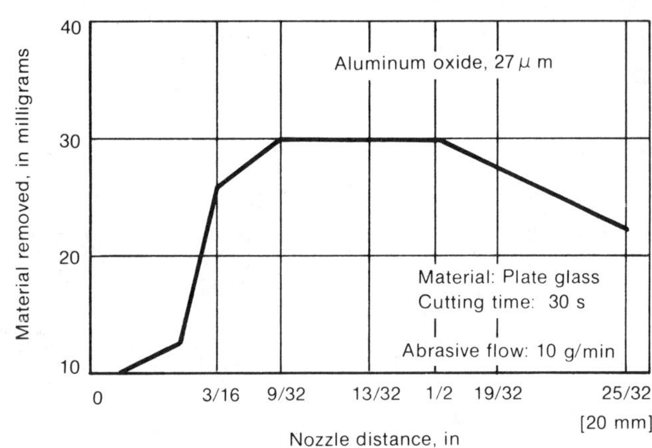


B. Cutting rate as function of nozzle tip distance (ntd)

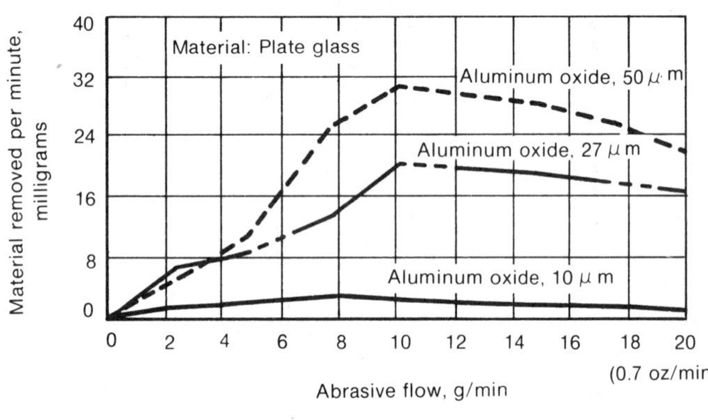


C. Effect of grit size on cutting speed

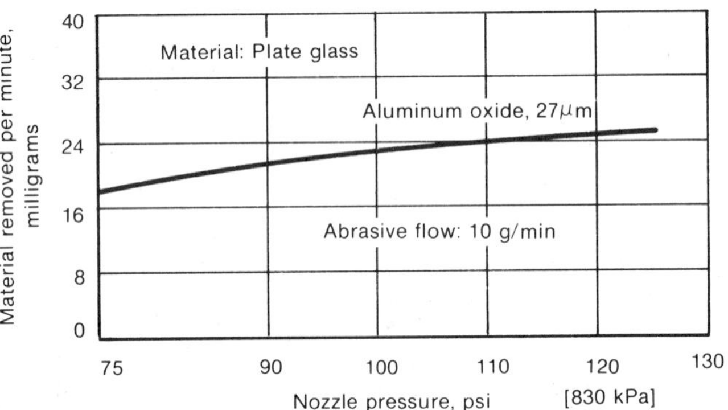


D. Effect of gas pressure on cutting speed

**Figure 10.2–4** Factors affecting material removal rates for AJM. (Adapted from W. Kulischenko, Abrasive jet machining techniques and related AJM equipment, Technical paper MR68-524, Society of Manufacturing Engineers, Dearborn, MI, 1968, pp. 4–5)

## PROCESS SUMMARY

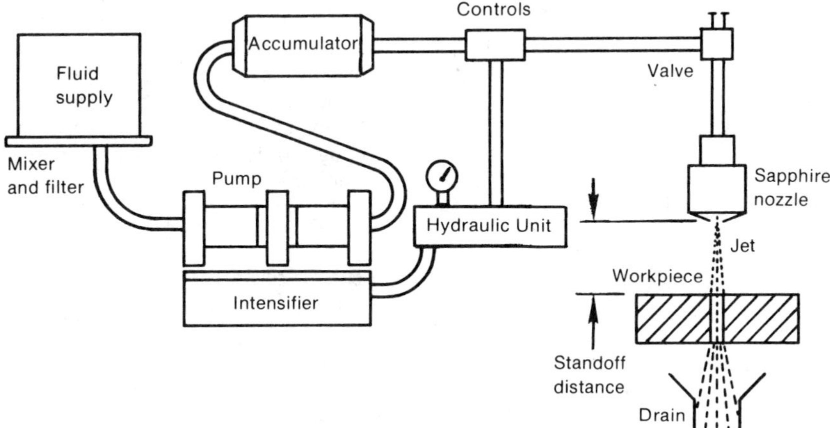


**Figure 10.3–1** HDM schematic.

### Principles

Hydrodynamic machining (HDM) removes material by the impingement of a high-velocity fluid against the workpiece (see figure 10.3–1). The coherent jet of water or water with additives (to aid coherence or prevent freezing) is propelled at speeds up to Mach 3 thereby cutting or shearing the workpiece. A synthetic-sapphire nozzle controls the jet stream and establishes the standoff distance which is important for controlling the cutting depth. Relatively small volumes of fluid are used. The fluid is pumped to the desired operating pressure aided by a hydraulically driven intensifier and an accumulator tank smooths out the pulses. The cut can be initiated at any point on the workpiece and the kerf (width of cut) is insensitive to dwell of the jet. The "chips" enter the exit stream. The power density in the stream can reach 6.5 x $10^8$ watts per square inch [$10^{10}$ watts/mm$^2$].

### Practical Applications

The ability of HDM to cut very thin, soft metals or nonmetallic materials in any position with a very narrow kerf leads to many form-cutting applications, such as those described in tables 10.3–3, 10.3–4 and 10.3–5. A cutting or deburring action can be initiated at any point on the workpiece. HDM can be substituted for a carbide slitting saw. Since it is a non-dulling tool, HDM can produce a better edge cut. The narrow kerf can result in material and cost savings. The low level cutting temperatures allow the processing of wood and paper products, such as cutting 3/4-inch [19 mm] acoustic ceiling tile at 250 feet per minute [76 m/min] using a 45,000 pounds per square inch [310 MPa] jet. HDM has been used to cut furniture forms of 1/2-inch [12.5 mm] laminated paperboard, 1/2-inch [12.5 mm] asbestos brake shoe linings and 3/4-inch [19 mm] thick shoe sole material. Some work has been done with thin, mild steel; however, cutting of hard metals is still experimental using pressures above 100,000 pounds per square inch [700 MPa]. Scribing of maskants for chemical machining is a productive application.

### Operating Parameters

**TABLE 10.3–1 Typical Values for HDM Operating Parameters**

| | |
|---|---|
| Fluid | |
| type: | Water or water plus additives—well filtered |
| additives: | Glycerine, polyethylene oxide, long-chain polymers |
| pressure: | 10 to 60 ksi [69 to 415 MPa] |
| jet velocity: | 1,000 to 3,000 ft/s [305 to 915 m/min] (figure 10.3–4) |
| flow: | Up to 2 gal/min [7.5 L/min] (figure 10.3–5) |
| jet force on workpiece: | 1 to 30 lb [4.5 to 134 N] (figure 10.3–4) |
| Power: | Up to 50 hp [38 kW] available; figures 10.3–4, 10.3–5, 10.3–6, 10.3–14 |
| Cutting rates—traverse rates: | See table 10.3–2 and figures 10.3–7, 10.3–9, 10.3–10, 10.3–11 |
| Nozzle | |
| material: | Synthetic sapphire most common, also hardened steels, 17-4 stainless |
| diameter: | 0.003 to 0.015 inch [0.075 to 0.38 mm] |
| angle: | Perpendicular to positive rake angles of 30° (figures 10.3–2 and 10.3–18) |
| Kerf width: | 0.003 to 0.016 inch [0.075 to 0.41 mm] |
| Standoff distance: | 0.1 to several inches [2.5 to 50 mm]; ⅛ inch [3 mm] is typical (see figures 10.3–16 and 10.3–17) |

# 10.3 Hydrodynamic Machining—HDM

### General Observations

- Increasing nozzle diameter improves performance and increases horsepower requirements (figure 10.3-12).
- Increasing pressure increases penetration, quality of cut and horsepower requirements.
- Lower pressure reduces edge quality and delamination effects.
- Traverse rates are directly related to cost and rates of production.
- Increased standoff distance decreases penetration and increases noise (figures 10.3-16 and 10.3-17).
- Additives can increase cutting performance and reduce cut width (figures 10.3-16 and 10.3-17).
- Smaller nozzles produce finer cuts.
- Larger nozzles are needed for slicing thicker materials.
- Lower traverse rates improve the quality of cut.
- Cutting performance can improve with positive rake angles.

### Material Removal Rates and Tolerances

Cutting rates depend on the work material (table 10.3-4) and vary directly with the horsepower applied and inversely with the material thickness. Rates up to 6,000 feet per minute [1,830 m/min] have been attained on paper products. Slots and through cuts are common. Removal rates for various materials are shown in figures 10.3-7 to 10.3-13. The lack of sensitivity to dwell allows contours to be controlled to an accuracy limited only by the machine elements. The kerf is about 0.001-inch [0.025 mm] larger than the orifice utilized. For composite materials, the best combination is a high-pressure jet from a small nozzle operating at a slight rake angle and close to the workpiece. Taper can be a factor if too large of a standoff distance is used.

### Surface Technology

Edge quality depends upon how easily the workpiece fractures. Soft materials cut smoothly and crushable materials can be slit with high quality edges. Test cuts are recommended before selecting the final operating parameters. The high energy density of the jet stream causes a slight rise in temperature which may melt some types of plastics at low traverse rates, but the temperature is not high enough to be a concern for paper-type materials. Tendency to delaminate layers of composite materials during cutting can be eliminated by reducing the cutting rate.

### Equipment and Tools

Machine and hydraulic components are available for HDM; however, a general-purpose machine tool is not regularly available. Each application is engineered to meet the requirements as found by sample test cuts made to determine the exact values for the key operating parameters. Equipment generating pressures up to 60,000 pounds per square inch [415 MPa] is commercially available with controls, filters and hydraulic seals of reasonable durability. A drain and sump need to be provided close to the workpiece. The jet stream does not dull and nozzle life is respectable, provided there is good filtration of the fluid. Filtration down to 0.5 micrometer (20 $\mu$in) followed by demineralization and deionization is recommended to reduce erosion of the nozzle. There is little frictional drag in the cut and virtually no forces at right angles to the jet; therefore, holding-fixture requirements are minimal. Multiple jets have been arranged for multiple cuts.

Use of NC or CNC is helpful with HDM. Recent equipment has reached 4-by-5-foot [1.2 x 1.5 m] table size with speeds of 15 inches per second [380 mm/s] and acceleration to top speed in 0.2 second.

### Machining Characteristics

The "chips" mix in the fluid so dust is eliminated; therefore, explosion and fire hazards are reduced. Air entrapped in the jet stream can create considerable noise with some work materials. The noise level increases with standoff distance but can be reduced by using certain fluid additives or certain rake angles. The noise levels are generally below the OSHA requirements. Periodic preventive maintenance is very desirable for equipment operating at these pressures.

## SELECTED DATA

There are several valid techniques for selecting operating parameters for HDM. The steps in table 10.3-2 are recommended as one logical sequence. Supporting data come from the other figures and tables in accordance with the terms as explained in figure 10.3-3.

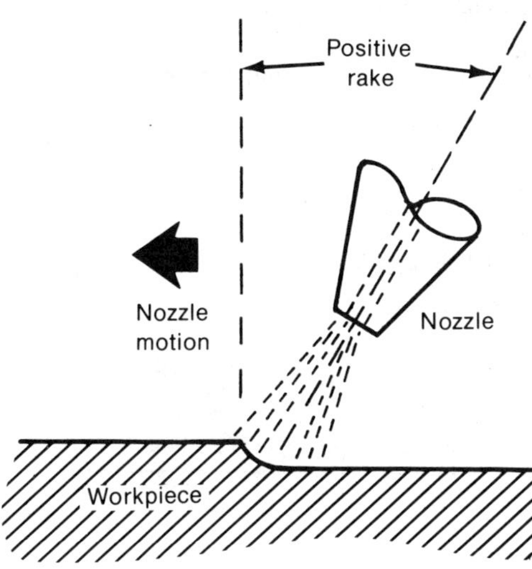


**Figure 10.3-2** HDM nozzle angles.

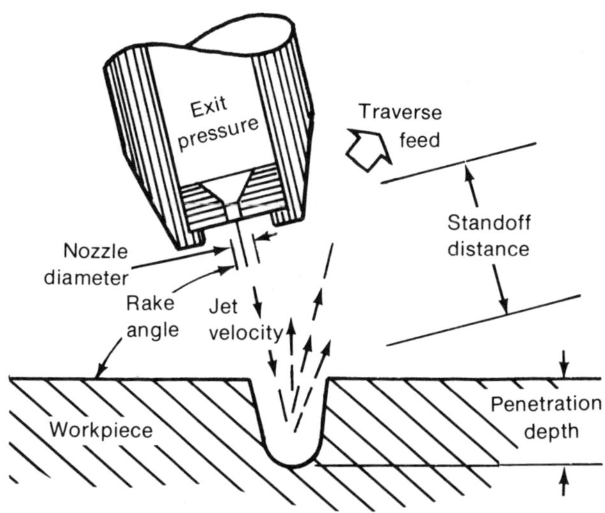


**Figure 10.3-3** HDM parameter terminology.

**TABLE 10.3-2 Steps To Select HDM Operating Parameters**

| Step | Source |
|---|---|
| 1. Can workpiece material be machined by HDM? | Tables 10.3-3, 10.3-4 and 10.3-5 |
| 2. Cutting rates required | User's process or production requirements |
| 3. Nozzle size and pressure | Test cuts or table 10.3-4, figures 10.3-7, 10.3-8, 10.3-13<br>Pressure is usually less than 65 ksi [450 MPa].<br>Nozzle size is approximately 0.001 inch [0.025 mm] less than cut width. |
| 4. Cutting rates | Table 10.3-4 and figures 10.3-7, 10.3-9, 10.3-10, 10.3-11<br>Rate is approximately inversely proportional to thickness being cut. |
| 5. Flow rate | Figure 10.3-5 |
| 6. Power | Figure 10.3-5 or 10.3-6 |
| 7. Standoff distance | 0 to 0.5 inch [0 to 0.13 mm], smallest distance is best |
| 8. Rake angle (straight-line cutting only) | Figures 10.3-2, 10.3-16, 10.3-17, 10.3-18 |
| 9. Fluid additives | Figures 10.3-16 and 10.3-17 |
| 10. Confirm parameters selected. Check surface quality, wetting, and kerf width for acceptability. | Trial cut of specimen material. |

**TABLE 10.3–3 Partial List of Materials Successfully Cut with HDM**
(With some maximum thickness of cut)

Abrasives—coated papers, cloth
Acoustic tile
Acrylic plastics
Aluminum
Aluminum honeycomb
Asbestos brake linings
Asphalt base materials

Bakery products
Boron-aluminum composite
Boron-epoxy composite
Boron-polyester composite
Box board
Brake linings
Brass

Carbon bricks and electrodes
Carpets
Ceramic foam
Coal
Concrete
Copper
Cork

Delrin

Epoxy—Kevlar

Fabrics, automotive
Fabrics, "soft" materials (except doubleknit fabrics)
Fiberglass-epoxy composite
Fiberglass insulation—to 5-inch [125 mm] thickness
Fiberglass—Nomex core sandwich—to 0.75-inch [19 mm] thickness
Fiberglass reinforced phenolic—to 0.5-inch [12.5 mm] thickness
Firebrick
Food (frozen meats, vegetables)
Furniture components

Glass—polyimide
Glass fiber reinforced polyester
Graphite
Graphite-epoxy composite—to 0.625-inch [15.9 mm] thickness
Gypsum boards

Insulating materials

Kevlar and epoxy—to 0.5-inch [12.5 mm] thickness

Laminated plastics
Lead
Leathers
Lucite plate—to 0.5 inch-[12.5 mm] thickness

Maple
Micarta—to 0.37 inch [9.5 mm] thickness
Mild steel
Mineral fibers (acoustic tile)

Neoprene rubber, 50 Shore hardness—to 2-inch [50 mm] thickness
Nomex honeycomb—to 3-inch [75 mm] thickness
Nylon (brushes, carpets)

Oak

PVC
Paper—corrugated board, filters, impregnated, coated, laminated newsprint
Plastics
Plexiglass
Plywood—to 1-inch [25 mm] thickness
Poplar
Polycarbonate—to 0.25-inch [6 mm] thickness
Polyester fabrics and carpets
Polyethylene sheets
Polypropylene tubes
Polystyrene
Polyurethane foam—to 1-inch [25 mm] thickness

Rocks
Roofing (composition, asbestos, rolls, shingles)
Rubber (foams, reinforced shoe soling, tire tubes)

Shoe leather
Silicone rubber—to 2-inch [50 mm] thickness
Slag
Steel (mild and thin)
Styrofoam
Styrene foams
Suits
Synthetic rubber
Teflon—to 0.37-inch [9.5 mm] thickness

Urethane foams

Vinyl sheets
Vinyl edge foam—to 1-inch [25 mm] thickness

White granite
Wood (fiber tiles, fiber padding, particle board, paneling veneer)

**TABLE 10.3–4 HDM Cutting Rates for Several Materials**

| MATERIAL | THICKNESS | | NOZZLE DIAMETER | | PRESSURE | | CUTTING SPEED | |
|---|---|---|---|---|---|---|---|---|
| | in | mm | in | mm | bar | MPa | in/s | m/s |
| ABS plastic | 0.11 | 2.8 | 0.003 | 0.075 | 2580 | 258 | 0.55 | 0.0142 |
| Acoustic tile | 0.75 | 19 | 0.010 | 0.25 | 3100 | 310 | 50 | 1.27 |
| Aluminum honeycomb | 1 | 25 | — | — | 4200 | 420 | 33 | 0.85 |
| Automotive carpet | 1 layer | 1 layer | — | — | 4200 | 420 | 118 | 3 |
| Automotive vinyl | 13 layers | 13 layers | — | — | 4200 | 420 | 165 | 4.25 |
| Boxboard (white lined chipboard) | 0.016 | 0.4 | 0.005 | 0.125 | 1770 | 177 | 37.5 | 0.955 |
| Boxboard (white lined chipboard) | 0.025 | 0.635 | 0.005 | 0.125 | 1740 | 174 | 31.5 | 0.805 |
| Boxboard (white lined chipboard) | 0.035 | 0.89 | 0.005 | 0.125 | 2440 | 244 | 24 | 0.61 |
| Brake shoe material | 0.5 | 13 | — | — | 4200 | 420 | 16.4 | 0.42 |
| Corduroy | 14 layers | 14 layers | — | — | 4200 | 420 | 74 | 1.9 |
| Cotton nylon fabric | 50 layers | 50 layers | — | — | 4200 | 420 | 126 | 3.2 |
| Crepe | 100 layers | 100 layers | — | — | 4200 | 420 | 166 | 4.25 |
| Delrin | 0.1 | 2.5 | 0.015 | 0.38 | 4120 | 412 | 8 | 0.2 |
| Ethyl polyvinyl chloride | 0.125 | 3.2 | — | — | 3500 | 350 | 8 | 0.2 |
| Fiberglass epoxy | 0.139 | 3.55 | 0.010 | 0.25 | 4120 | 412 | 0.1 | 0.0025 |
| Graphite epoxy | 0.067 | 1.7 | 0.008 | 0.2 | 3800 | 380 | 1 | 0.025 |
| Graphite epoxy | 0.136 | 3.45 | 0.010 | 0.25 | 4120 | 412 | 0.5 | 0.0125 |
| Graphite epoxy | 0.273 | 6.9 | 0.014 | 0.35 | 4120 | 412 | 1.1 | 0.0275 |
| Graphite epoxy prepreg. | 1 ply | 1 ply | 0.003 | 0.075 | 3800 | 380 | 39 | 1 |
| Hardwood pulp sheet | — | — | 0.004 | 0.1 | 2480 | 248 | 27.5 | 0.7 |
| Kevlar epoxy | 0.125 | 3.2 | 0.010 | 0.25 | 3800 | 380 | 0.5 | 0.0125 |
| Kevlar epoxy | 0.25 | 6.3 | 0.014 | 0.35 | 4120 | 412 | 0.8 | 0.0205 |
| Leather | 0.175 | 4.45 | 0.002 | 0.05 | 3030 | 303 | 0.35 | 0.0091 |
| Newsprint | — | — | 0.004 | 0.1 | 585 | 58.5 | 420 | 10.6 |
| Nylon tire fabric | 0.032 | 0.8 | — | — | 4200 | 420 | 166 | 4.25 |
| Plywood | 0.25 | 6.4 | — | — | 4200 | 420 | 67 | 1.7 |
| Plexiglass | 0.4 | 10 | 0.015 | 0.38 | 4120 | 412 | 2.7 | 0.07 |
| Polycarbonate | 0.2 | 5 | 0.015 | 0.38 | 4120 | 412 | 4 | 0.1 |
| Polyethylene (high density) | 0.12 | 3 | 0.002 | 0.05 | 2860 | 286 | 0.36 | 0.0092 |
| Polypropylene gasket | 1.6 | 40 | — | — | 4200 | 420 | 100 | 2.5 |
| Polypropylene (40%) glass fiber | 0.25 | 6.4 | — | — | 4200 | 420 | 16.6 | 0.42 |
| Rubber belt (cotton reinforced) | — | — | 0.002 | 0.05 | 2960 | 296 | 0.9 | 0.0232 |
| Styrene (high impact) | 0.12 | 3 | 0.003 | 0.075 | 2480 | 248 | 0.25 | 0.0064 |
| Tissue paper | 2 ply | 2 ply | — | — | 1000 | 100 | 1200 | 30.5 |
| Transite (asbestos cement) | 0.75 | 19 | 0.010 | 0.25 | 3800 | 380 | 0.71 | 0.018 |
| Wood veneer | 0.1 | 2.5 | — | — | 4200 | 420 | 166 | 4.25 |

**TABLE 10.3-5 Parameters for Hydrodynamic Machining of Various Nonmetallic Materials**

| MATERIAL | THICKNESS | | NOZZLE OPENING | | FEED RATE | |
|---|---|---|---|---|---|---|
| | in | cm | in | mm | in/min | m/min |
| Plywood | 0.25 | 0.6 | 0.012 | 0.3 | 200 | 5 |
| | 0.37 | 0.95 | 0.012 | 0.3 | 100 | 2.5 |
| | 1.0 | 2.5 | 0.012 | 0.3 | 60 | 1.5 |
| Foam rubber | 0.5 | 1.27 | 0.008 | 0.2 | 1000 est. | 25.4 |
| | 0.5 | 1.27 | 0.012 | 0.3 | 1000 est. | 25.4 |
| | 4.0 | 10.16 | 0.012 | 0.3 | 1000 est. | 25.4 |
| Neoprene, 50 Shore | 0.5 | 1.27 | 0.012 | 0.3 | 600 est. | 15.2 |
| | 2.0 | 6.08 | 0.012 | 0.3 | 100 | 2.5 |
| Grey silicone rubber, 40 Shore | 0.6 | 1.52 | 0.012 | 0.3 | 500 est. | 12.7 |
| Fiberglass and epoxy | 0.25 | 0.6 | 0.012 | 0.3 | 120 | 3 |
| Fiberglass and phenolic | 0.25 | 0.6 | 0.012 | 0.3 | 150 | 3.8 |
| Circuit board | 0.090 | 0.23 | 0.008 | 0.2 | 60 | 1.5 |
| Fiberglass and Nomex core | 0.5 | 1.27 | 0.012 | 0.3 | 360 | 9.1 |
| | 0.75 | 1.9 | 0.012 | 0.3 | 340 | 9 |
| Nomex core only | 2.0 | 5.08 | 0.012 | 0.3 | 300 | 7.6 |
| Kevlar and epoxy | 0.25 | 0.6 | 0.012 | 0.3 | 100 | 2.5 |
| Graphite and epoxy | 0.25 | 0.6 | 0.012 | 0.3 | 40 | 1 |
| ABS | 0.090 | 0.23 | 0.012 | 0.3 | 300 | 7.6 |
| Polycarbonate | 0.092 | 0.24 | 0.012 | 0.3 | 340 | 8.6 |

These feed rates can be extrapolated inversely on a nearly linear basis as the material thickness is reduced. That is, if a 0.25 inch [0.6 cm] thickness can be cut at 200 in/min [5 m/min], a 0.125 inch [0.3 cm] thickness can be cut at 400 in/min [10 m/min].

SOURCE: E. J. Buck and D. L. Zeulow, New techniques in water jet cutting, Technical paper MR79-376, Society of Manufacturing Engineers, Dearborn, MI, 1979, p. 5.

NOTE: The above materials were cut at a water pressure of 55,000 psi [3800 kg/cm$^2$].

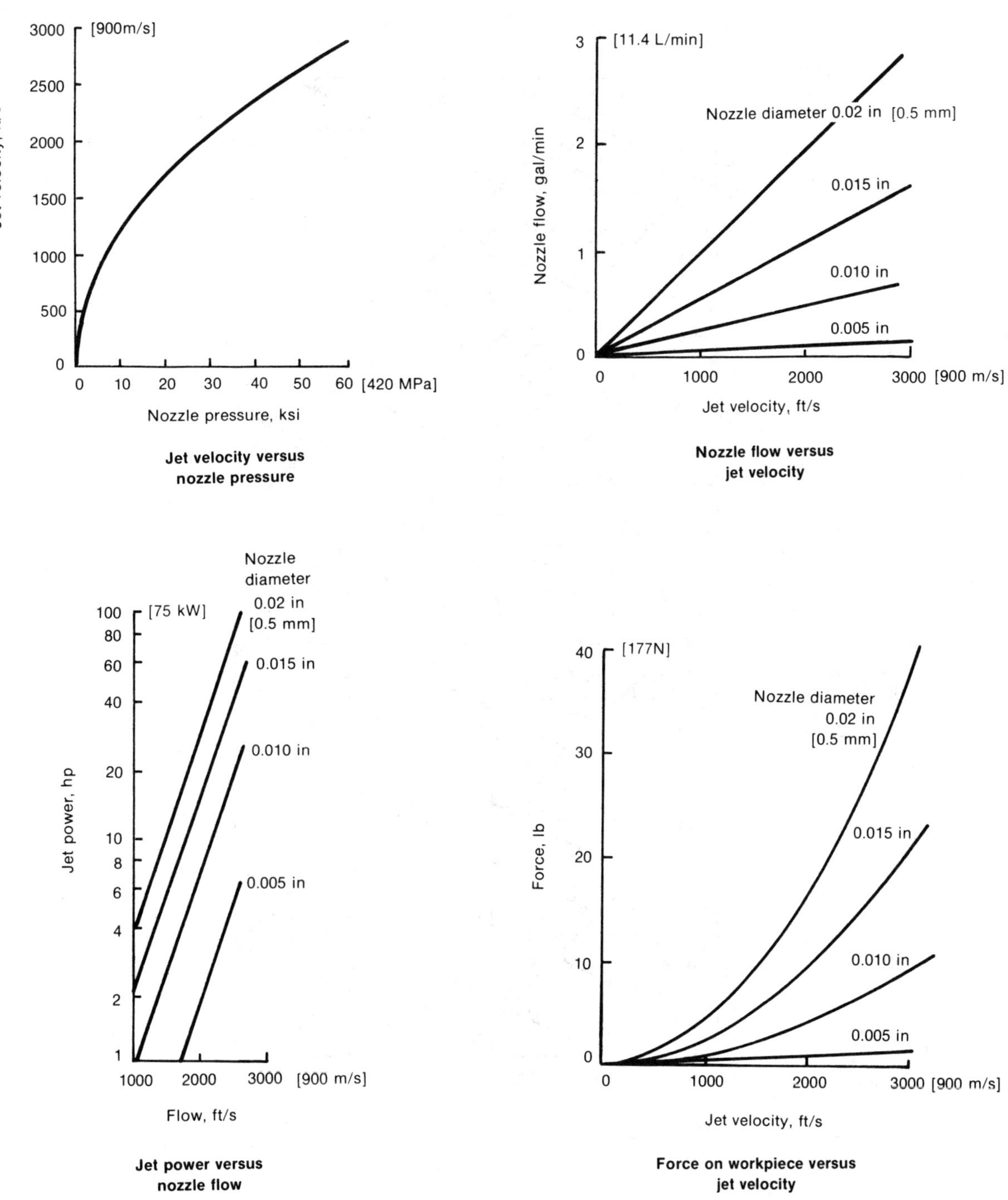


**Figure 10.3–4** Operating parameter relationships in HDM. (From K. F. Neusen, High velocity fluid jet cutting and slotting, Technical paper MR76-691, Society of Manufacturing Engineers, Dearborn, MI, 1976, pp. 8–10)

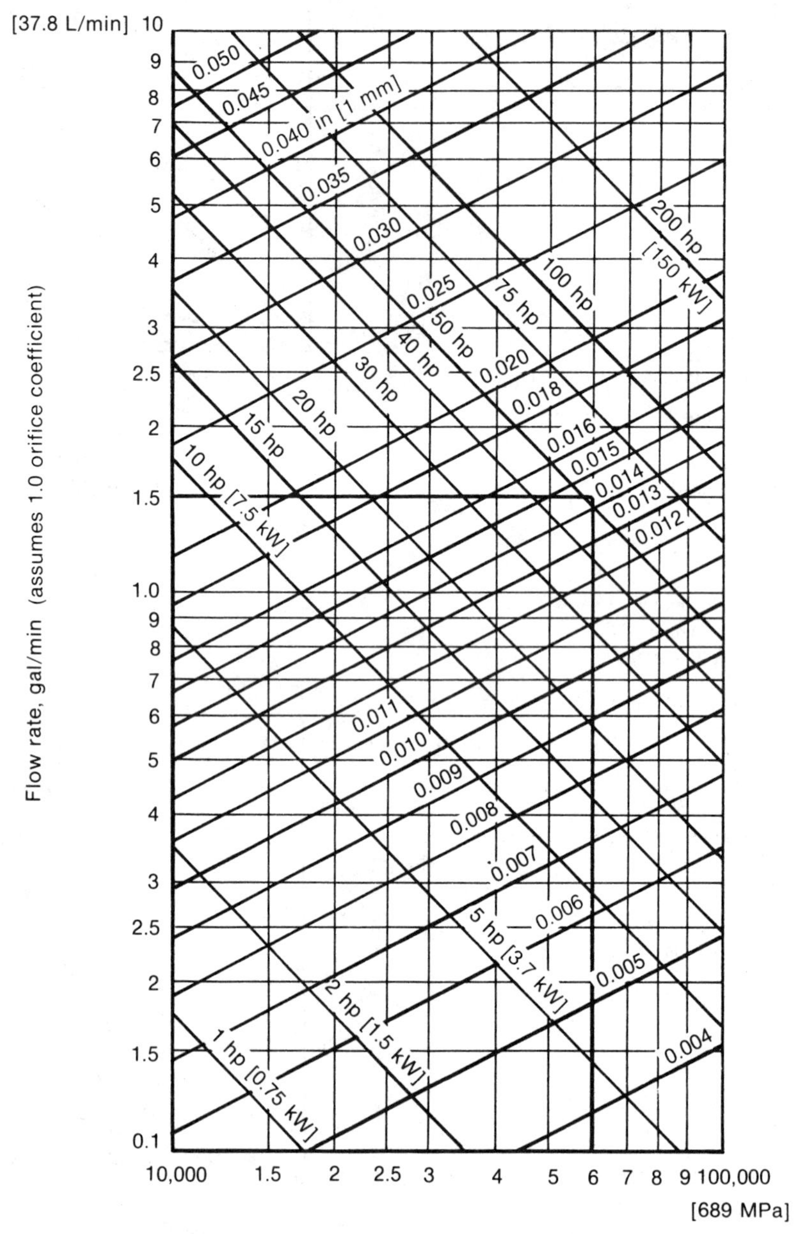


**Figure 10.3–5** Pressure versus flow rate with various nozzle sizes. Knowledge of two variables enables determination of two other variables. For example: (Note heavy lines) if a cutting pressure to 60,000 psi is required, it can be obtained using a flow rate of approximately 1.5 gal/min with a 0.014-inch diameter stream. (Courtesy of Flow Equipment Company)

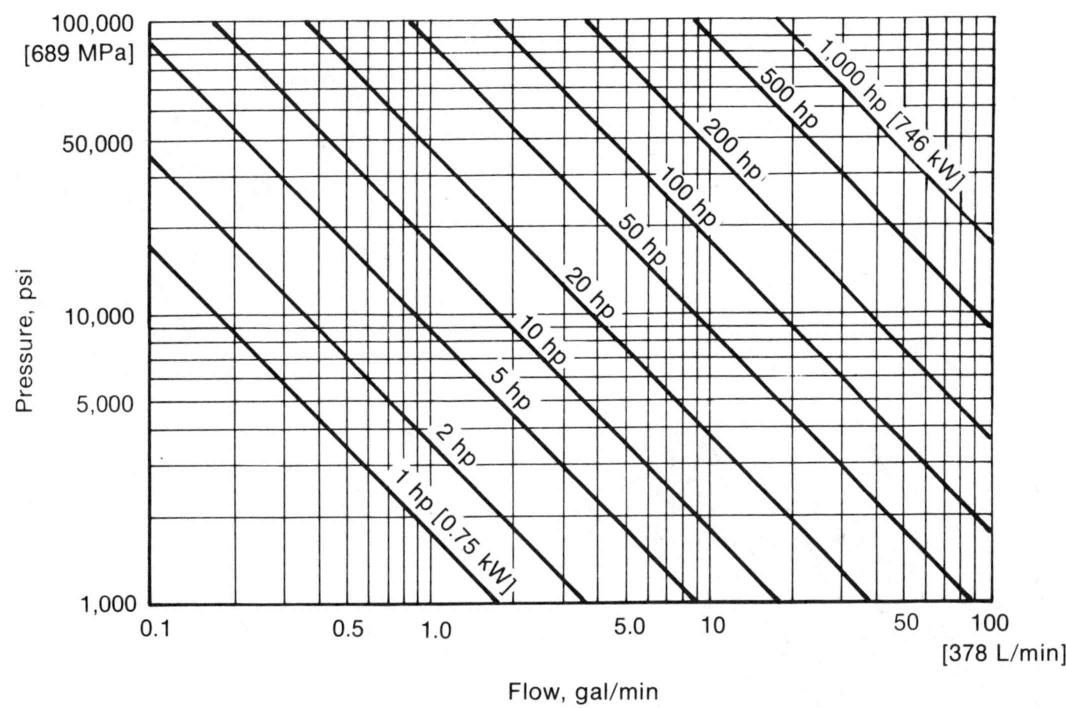


**Figure 10.3–6** Hydraulic horsepower requirements of HDM for various pressures and flow rates. Example: 60,000-psi gallons at 0.7 gal/min requires 20 hp. (Courtesy of Flow Equipment Company)

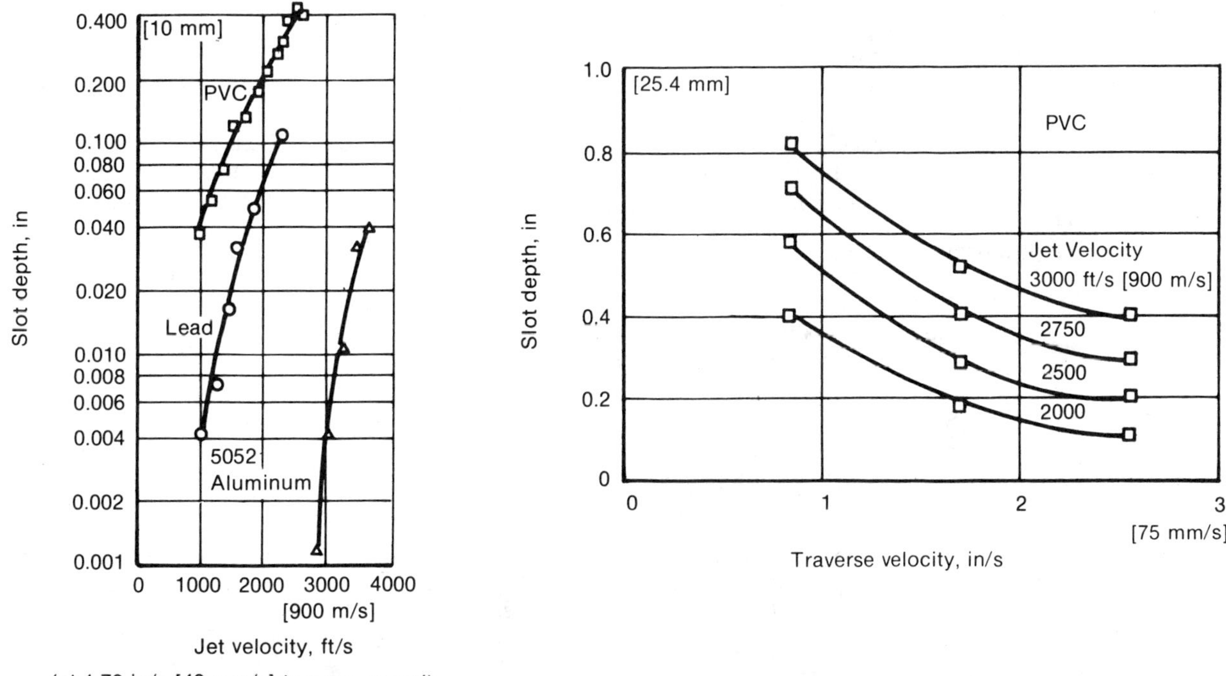


**Figure 10.3–7** Slot depth for various materials (fixed nozzle 0.0094-inch diameter of 17-4 stainless steel, 60,000 psi pressure, plexol-201 fluid). (From K. F. Neusen, pp. 10–11)

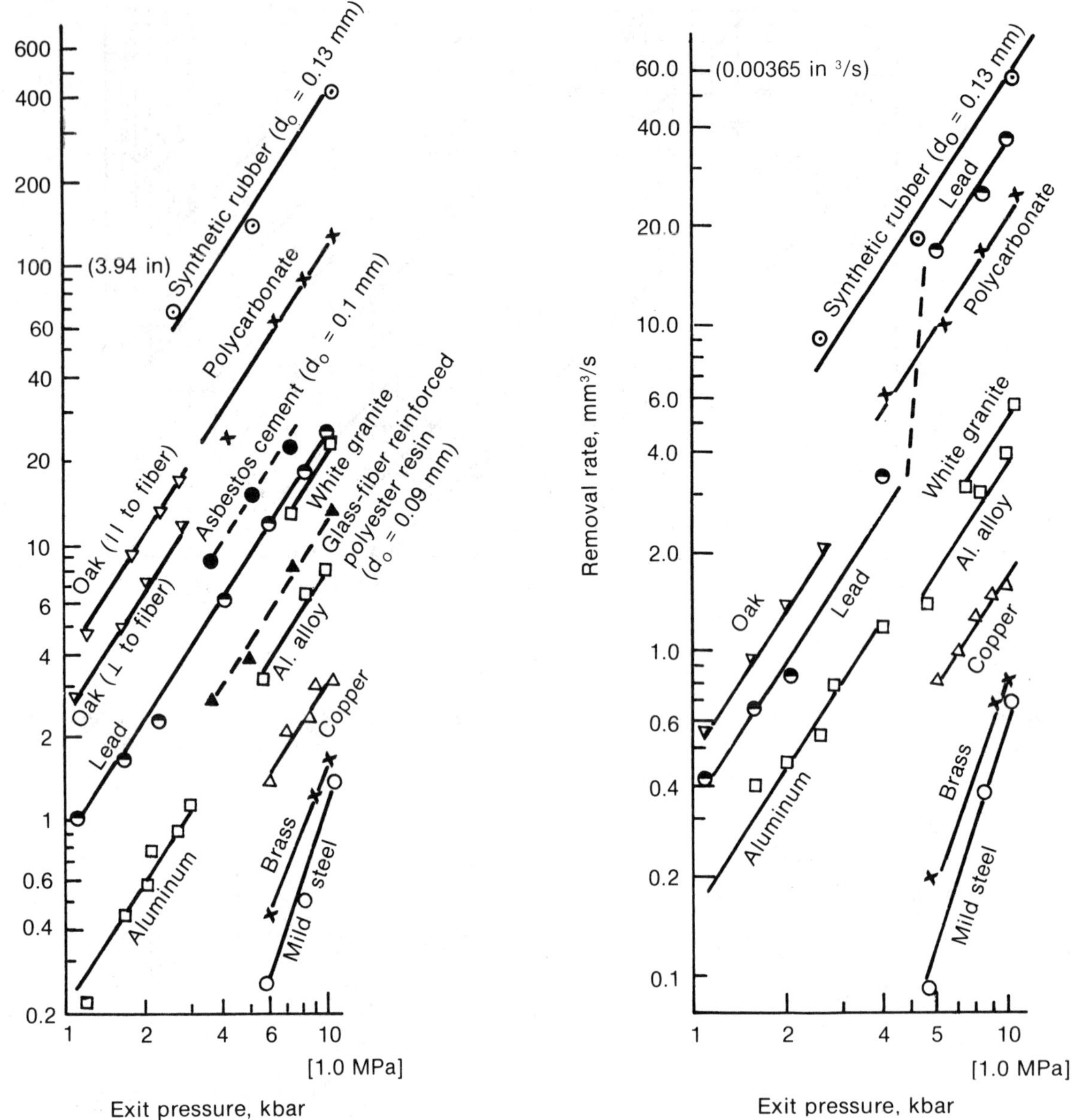


Fluid: 1:1 mix $H_2O$ and glycerine
Nozzle: Hardened steel, 0.15 mm (0.006 in) diameter
(unless $d_o$ otherwise designated)
Standoff distance: 3 mm (0.12 in)
Angle: 90° to workpiece
Feed rate: 1.0 mm/s (0.039 in/s)

**Figure 10.3–8** Depth of cut and material removal rate at different HDM pressures. (From O. Imanaka et al, Machining with continuous liquid jets at pressures up to 10 kbars, *Proceedings of the international conference on production engineering*, 1974 (Part I), p. 28. Japan Society of Precision Engineering, 1974)

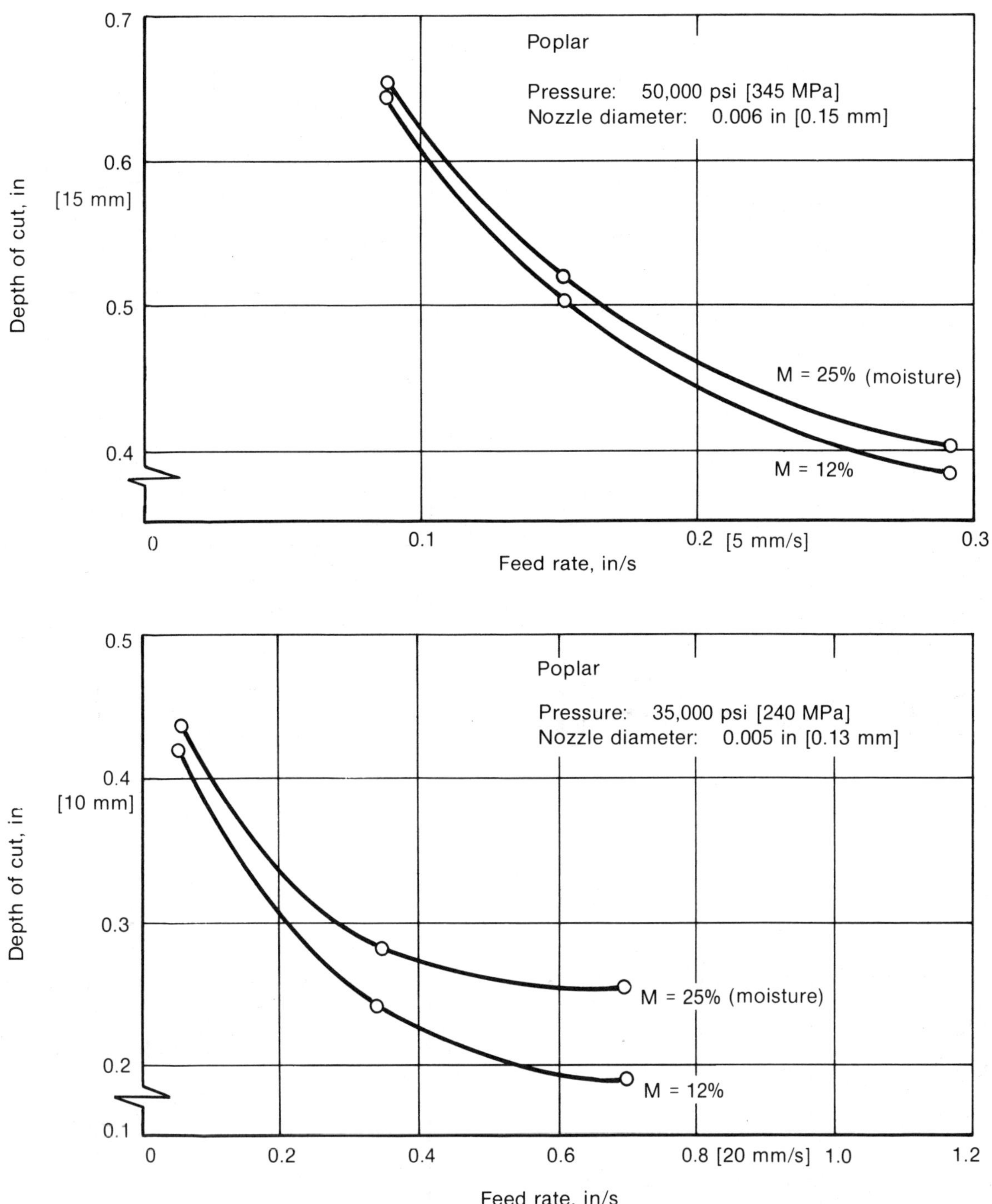


**Figure 10.3–9** Feed rate and depth of cut for poplar at two levels of moisture content and two levels of HDM pressure (cut perpendicular to grain, 2 mm standoff distance, nozzle at 90°, fluid = $H_2O$). (M. P. dePlessis and M. Hashish, High energy water jet cutting equations for wood, Concordia, Univ., Montreal, Quebec, 1976)

# 10.3 Hydrodynamic Machining—HDM

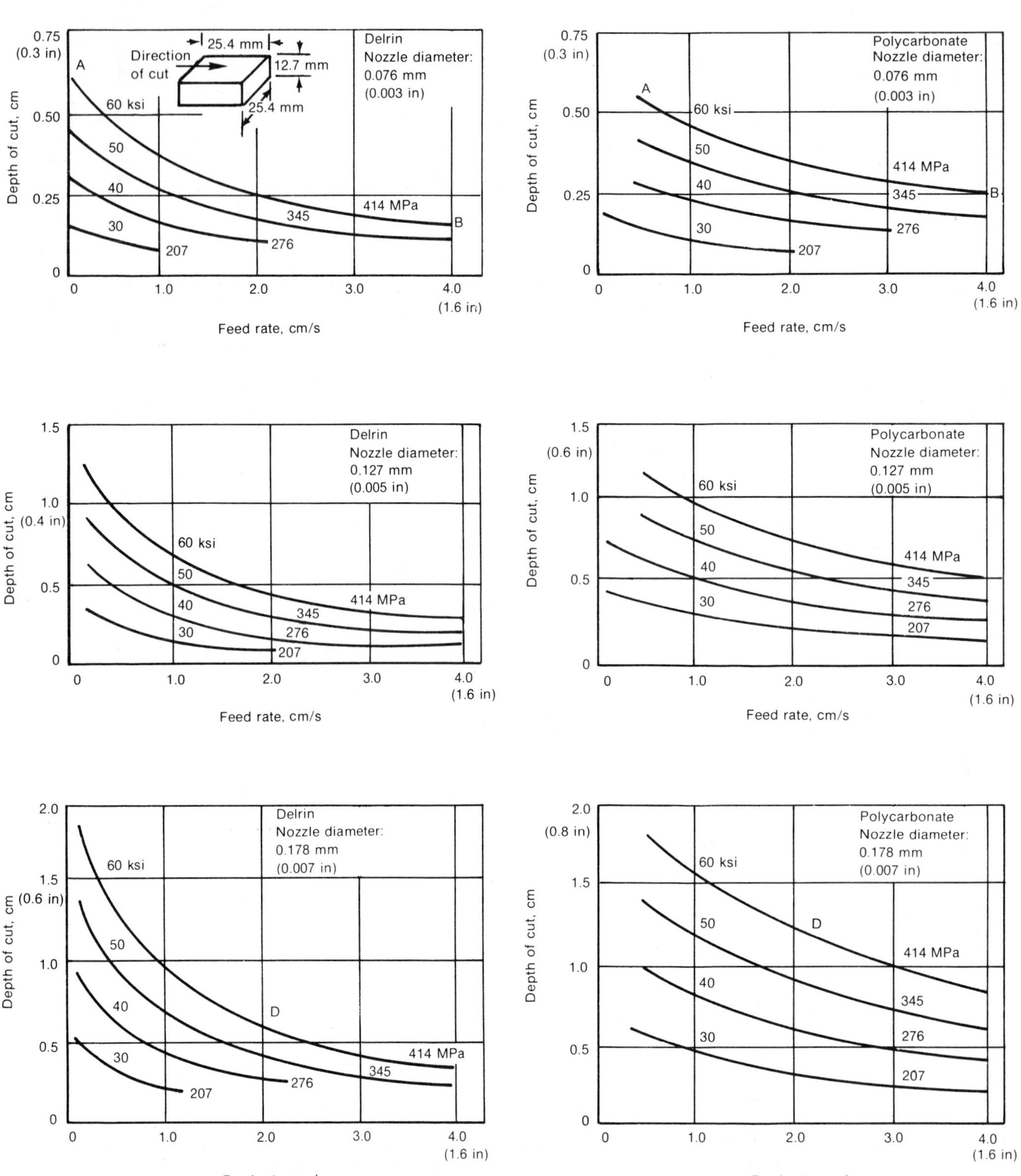


**Figure 10.3–10** Depth of cut versus feed rate at various pressures. Delrin-Polycarbonate; saphire nozzle, 0.38 mm above surface; fluid = $H_2O$. (U. H. Mohaupt and D. J. Burns, Machining unreinforced polymers with high-velocity water jets, *Experimental Mechanics* 14 (April 1974): pp. 154–155)

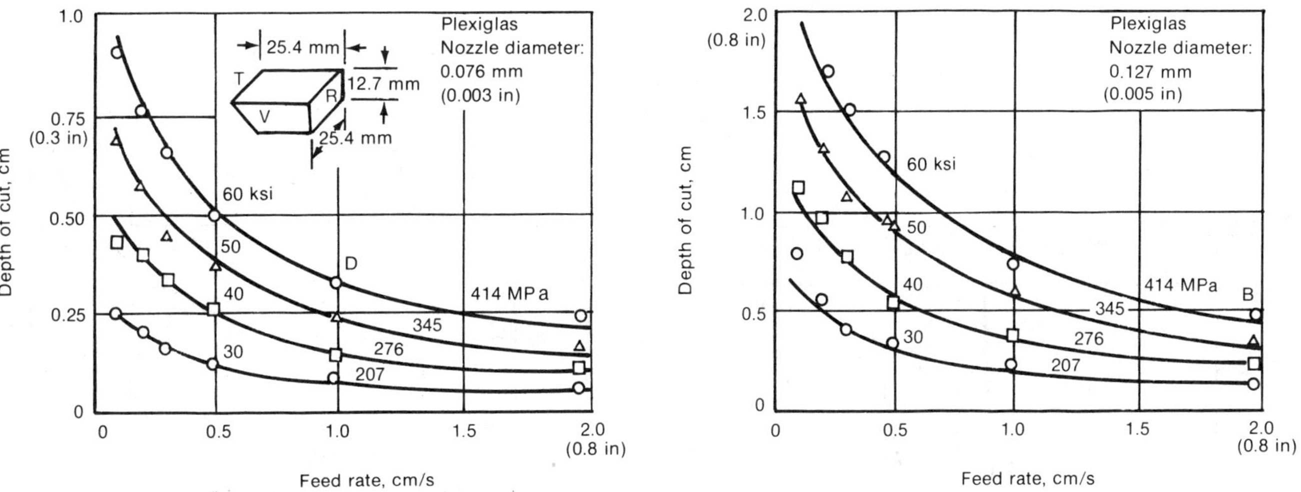


**Figure 10.3–11** Depth of cut versus feed rate at various pressures. Plexiglas. (U. H. Mohaupt and D. J. Burns, p. 155.)

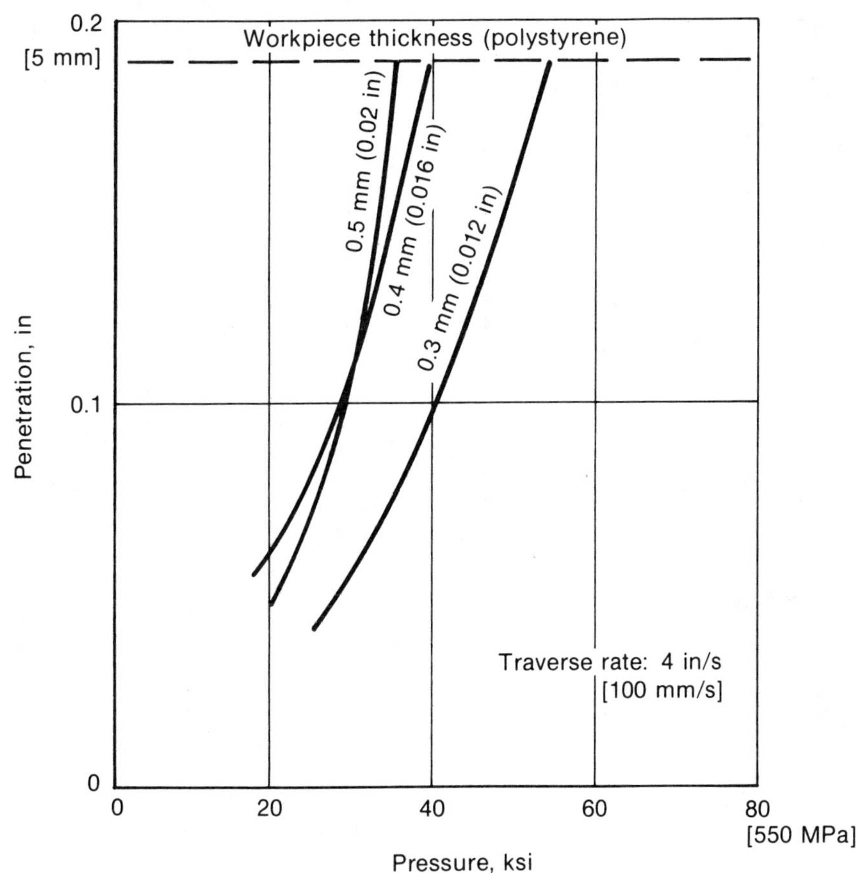


**Figure 10.3–12** Effect of nozzle oriface diameter on workpiece penetration. (Courtesy of Grumman Aerospace Corporation)

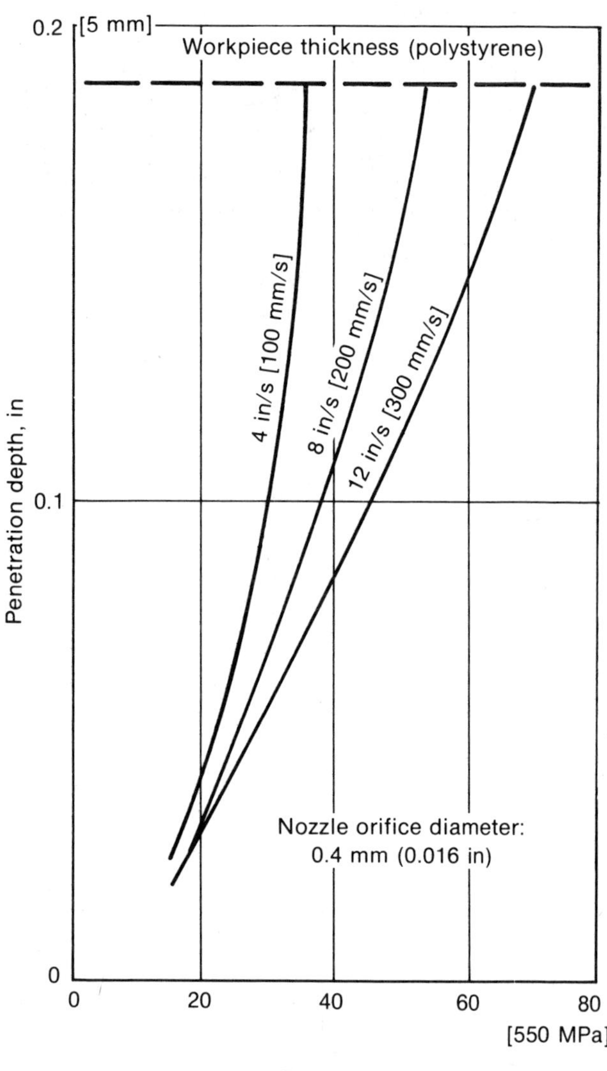


**Figure 10.3–13** Effect of jet pressure on workpiece penetration. (Courtesy of Grumman Aerospace Corporation)

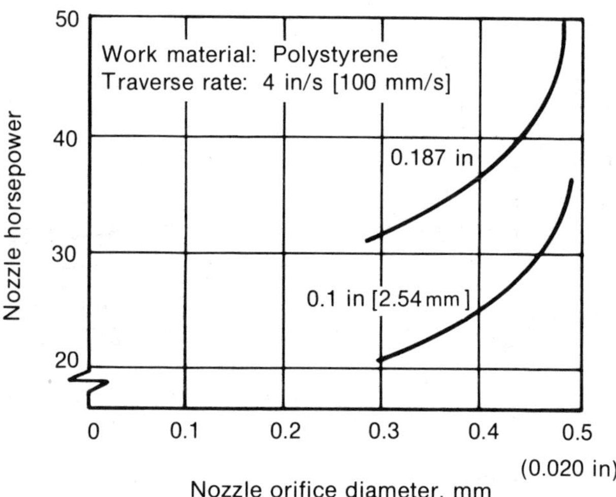


**Figure 10.3–14** Nozzle horsepower versus nozzle diameter for plastic casing at 4 in/s traverse velocity for fixed penetration. (Courtesy of Grumman Aerospace Corporation)

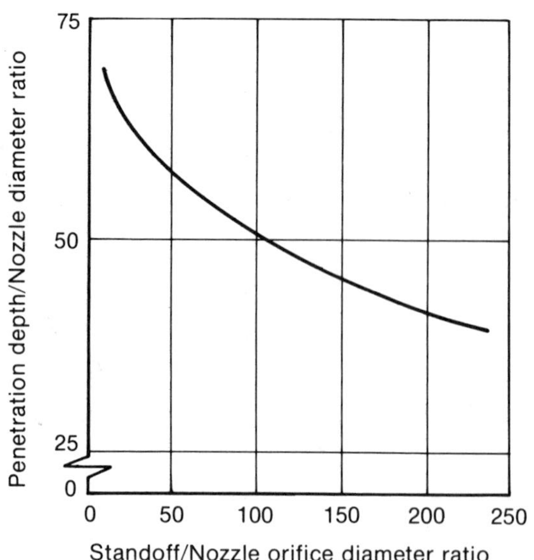


**Figure 10.3–15** Effect of standoff distance on penetration depth. (Courtesy of Grumman Aerospace Corporation)

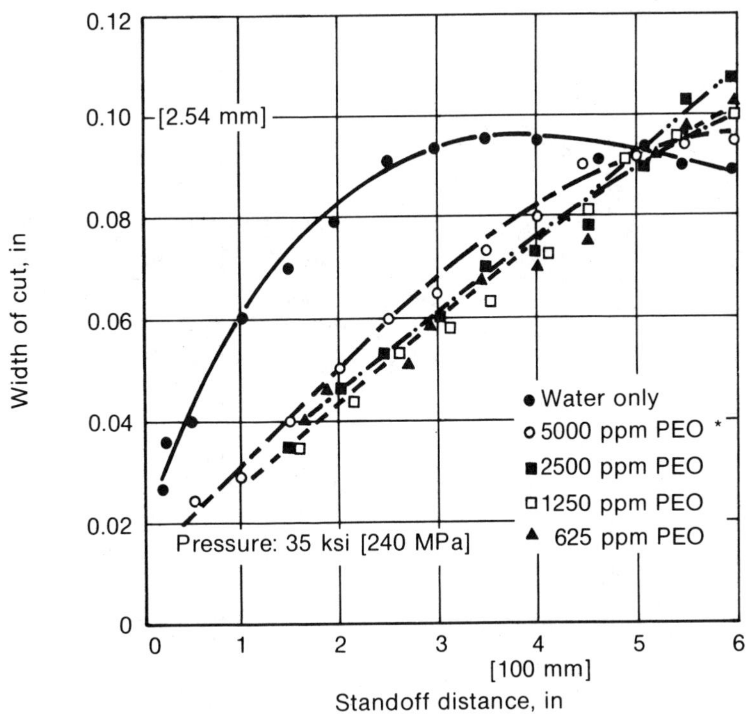


**Figure 10.3–16** Width of cut versus standoff distance for aluminum. *PEO = Polyethylene oxide. (Courtesy of Grumman Aerospace Corporation)

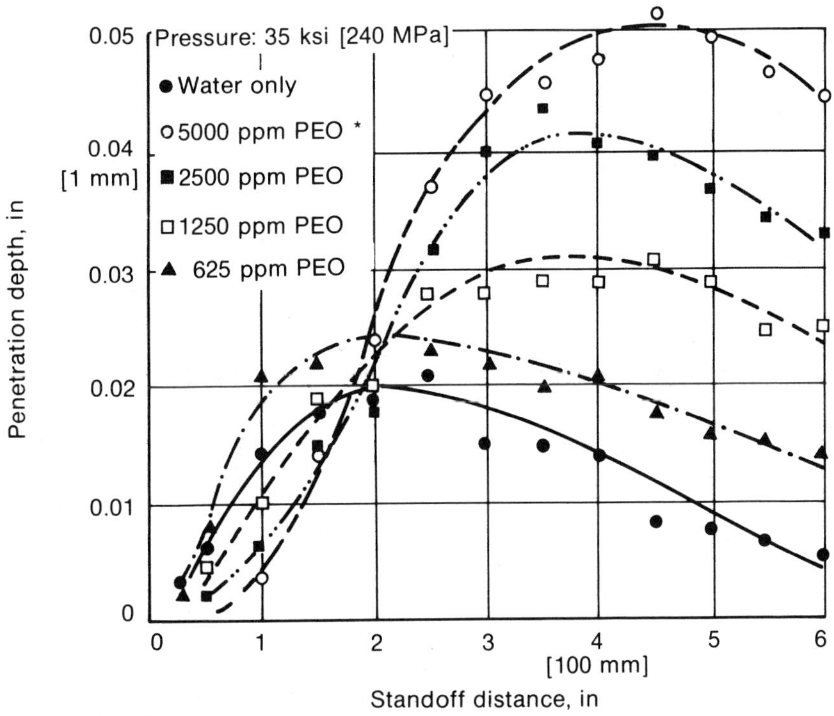


**Figure 10.3–17** Penetration depth versus standoff distance for aluminum. *PEO = Polyethylene oxide. (Courtesy of Grumman Aerospace Corporation)

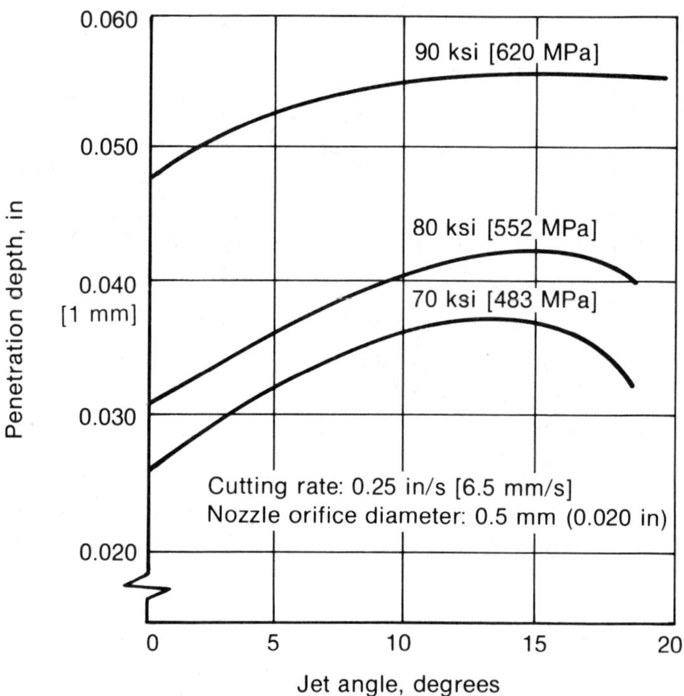


**Figure 10.3–18** Effect of jet angle on penetration depth for steel cut under water. (Courtesy of Grumman Aerospace Corporation)

## PROCESS SUMMARY

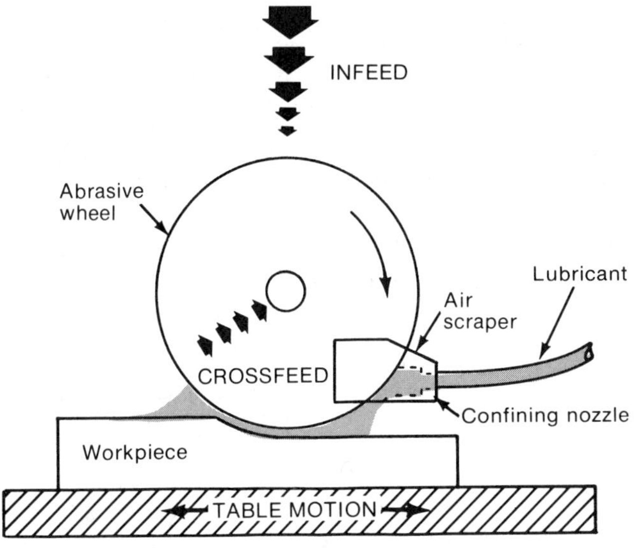


**Figure 10.4–1** LSG schematic.

### Principles

Low stress grinding (LSG) is an abrasive material removal process that leaves a low magnitude, generally compressive, residual stress in the surface of the workpiece. Figure 10.4–1 shows a schematic of the LSG process. The thermal effects from conventional grinding can produce high tensile stress in the workpiece surface. By careful selection and control of the operating parameters, the thermal effects are minimized, thereby reducing the residual stresses, the attendant distortion and the possible reduction in high-cycle fatigue strength. The process parameter guidelines can be applied to any of the grinding modes: surface, cylindrical, centerless, internal, etc. Additional description, details and data on LSG are provided in section 18.3, Surface Integrity, in the discussion of abrasive material removal processes.

### Practical Applications

Grinding burn, distortion and cracking always have been associated with the improper use of abrasive processes, particularly when grinding high strength steels, high temperature alloys, titanium alloys and other sensitive alloys. Low stress grinding eliminates these effects by reducing the specific energy input. LSG should be used only on critical or highly stressed areas of a workpiece, such as those areas subjected to high cyclical stress or stress-corrosion environments. The enhanced surface integrity, minimal distortion and minimal residual stress in surfaces produced by LSG are important on highly loaded structural components. As always, careful attention to the entire sequence of manufacturing processes is necessary to avoid reduction in the surface integrity produced in a component. Note that it is possible to reduce or eliminate the need for LSG by careful consideration of the total processing sequence and the proper placement of LSG within it.

To maintain an acceptable level of productivity, LSG parameters should be applied only to the removal of the last few thousandths inch of material from the workpiece (see table 10.4–2). LSG should *not* be applied "all over" the workpiece. It should be considered for use only on critical areas, and then only if it is the final process affecting the "as shipped" surface. Note that heat treatment, shot peening or other post-grinding processing may affect the ground surface or may make LSG unnecessary.

Productivity loss may also be relieved by an experimental investigation to determine what increase in the downfeed or infeed rates can be tolerated above those rates specified in table 10.4–2. Table 18.3–17 Grinding Parameters for IN-100 Turbine Blades (see section 18.3) provides data which show that no cracks were produced using a feed rate of 0.002 inch [0.05 mm] after certain low-stress-grinding parameters were adopted. Feed rates should not be increased, however, without doing the necessary experimental work for each particular alloy and situation. Generalizations about the grinding of sensitive alloys are dangerous.

### Operating Parameters

**TABLE 10.4–1 Typical Values for LSG Operating Parameters**

(Parameters are listed in descending order of significance for low stress generation in the workpiece surface.)

| PROCESS PARAMETERS | GUIDELINES |
|---|---|
| 1. Grinding wheel dressing technique | Frequent and coarse to maintain sharpness. Maintain sharpness of dressing tool. Avoid dwell in using crush, roll, or single-point dressing tools. |
| 2. Wheel speed | Low, under 3,500 fpm [1,050 m/min]. |
| 3. Downfeed (or infeed) rate | 0.0002 to 0.0005 inch/pass [0.005 to 0.013 mm/pass] with programmed reduction from conventional rates. |
| 4. Grinding fluid | Oil-base fluid is preferred. |
| 5. Wheel classification | Soft grade (G, H or I*). Open structure (6 or more). Grain size (60 or coarser). |
| 6. Table (workpiece) speed | High, 50 fpm [15 m/min] and up. Crossfeed is preferred to plunge motion. |
| 7. Grinding fluid flow control | Adequate to high fluid flow. Assure placement of fluid between wheel and workpiece. Flow controlling nozzle design. Reduce air film on wheel. |

*Cylindrical grinding frequently requires use of harder wheels (with J grade prevalent); however, the other parameters must be selected to compensate for this extra hardness.

# 10.4 Low Stress Grinding—LSG

### Material Removal Rates and Tolerances

Material removal rates are low, but *good productivity can be attained* by successively reducing conventional material removal rates to LSG levels as the finished dimensions are approached. Typical tolerances that can be obtained are in the ±0.0002 inch [±0.005 mm] range.

### Surface Technology

Surface roughness values are typical of those attainable by fine grinding and range from 16 to 32 microinches $R_a$ [0.4 to 0.8 μm]. Plastically deformed debris, distortion and residual stress are minimal or absent. The low residual-stress levels frequently are compressive, which further enhances the high-cycle-fatigue strength. The absence of microcracks also contributes to high surface integrity. A test specimen or check of the results from LSG is desirable. Either the distortion produced in a ground, thin test strip or a simple metallographic examination of the surface taken parallel to the lay can be a useful quality assurance test.

### Equipment and Tools

Conventional grinding equipment can be used to produce LSG results. It may be necessary to modify the equipment to obtain the desired lower wheel speeds and higher table speeds. The low infeed rates usually are attainable on a well maintained machine. If it is not practical to modify some machine condition to secure the desired operating parameter, it is still possible to attain LSG results. In this case all of the other parameters should be at their best levels to compensate for a deficiency in one of them.

### Machining Characteristics

Results from LSG are typical of those attained by any fine *gentle* grinding operation. Cleanliness of the lubricant, workpiece and machine are important, and it is most desirable to assure that a continuous, full stream of lubricant reaches the interface between the wheel and the workpiece. A closely fitted nozzle box is helpful.

## SELECTED DATA

**TABLE 10.4–2 Typical Infeed Schedule for Producing Surfaces by Low Stress Grinding**

| STEP | WORKPIECE STOCK REMOVAL* | INFEED RATE | |
|---|---|---|---|
| | | inch/pass | mm/pass |
| Rough Grinding | From raw material dimension to within 0.010 inch [0.25 mm] of desired final dimension | up to 0.002† | up to 0.050† |
| Dress wheel | | | |
| Semi-finish Grinding | From 0.010 inch [0.25 mm] to 0.001 inch [0.025 mm] of final size | 0.0003 to 0.0006 | 0.008 to 0.015 |
| Dress wheel | Optional | | |
| Finish Grinding (to low stress levels) | From 0.001 inch [0.025 mm] to final size | 0.0001 to 0.0002 | 0.0025 to 0.005 |
| | Optional: at final size, with continued application of grinding fluid | 2 to 4 spark-out passes | |

NOTE: For additional data and description of low stress grinding, see abrasive machining discussion in section 18.3, Surface Integrity.

*Data apply to the radius for cylindrical grinding and to the depth for surface grinding.

†This roughing rate should be such that roughness, tears, cracks, or microcracks in the surface are of lesser magnitude than the material remaining to be removed by the finishing and semi-finishing cuts.

## PROCESS SUMMARY

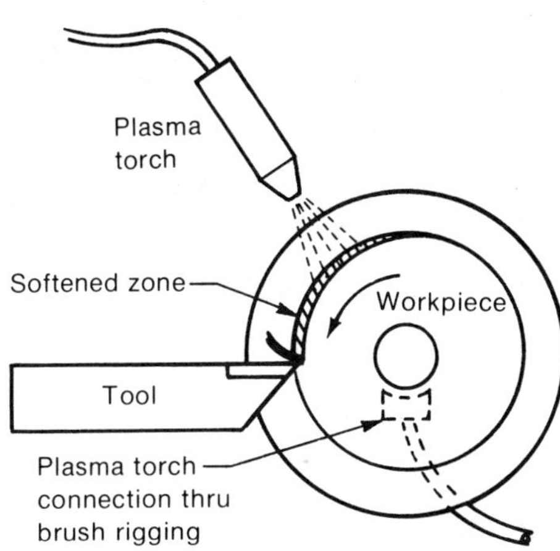


**Figure 10.5-1** TAM schematic.

### Principles

Thermally assisted machining (TAM) is the addition of significant amounts of heat to the workpiece immediately prior to single-point cutting so that the material is softened, but the strength of the tool bit is umimpaired (see figure 10.5-1). A steep temperature gradient is necessary, so the heat transfer properties of the work material are important. High energy density in the heating source is a prime necessity. Table 10.5-1 lists the energy densities for several heat sources. While resistive heating and induction heating offer possibilities, the plasma arc offers the most practical and effective method for external heating at the required rates. The plasma arc has a core temperature of 14500°F [8000°C] and a surface temperature of 6500°F [3600°C]. The torch can produce 2000°F [1100°C] in the workpiece in approximately one quarter revolution of the workpiece between the point of application of the torch and the cutting tool. This temperature will soften many high temperature alloys. The depth of heat penetration must be sufficient to secure material softening to the depth expected to be used by the cutting tool. Warm surfaces are not helpful—the temperature must reach the softening region for the work material. The residual temperature in the workpiece usually is only 100° to 200°F [35° to 100°C]. Temperatures in thin wall sections may be higher.

**TABLE 10.5-1 Energy Density of Heat Sources**

| HEAT SOURCE | DENSITY | |
|---|---|---|
| | W/in$^2$ | W/mm$^2$ |
| Electron beam | 6.5 x $10^9$ | 1 x $10^7$ |
| Laser beam | 6.5 x $10^9$ | 1 x $10^7$ |
| Plasma arc | 6.5 x $10^6$ | 1 x $10^4$ |
| Electric arc | 6.5 x $10^4$ | 1 x $10^2$ |
| Oxyacetylene torch | 6.5 x $10^3$ | 10 |

### Practical Applications

Hard-to-machine and high-temperature alloys are all candidates for "hot machining" or TAM. Work materials with hardness ranges of 40 to 70 $R_C$ should be considered for TAM. Poor thermal conductivity in the material can reduce the heat loss between the time of heat application and the time of cutting. Aluminum, copper, and titanium with their relatively high heat conductivities are not good candidate materials for TAM. Most of the applications have been for turning; however, experimental work has involved milling and slotting. Cuts of long duration have the greatest potential for economic benefit from the addition of TAM.

### Operating Parameters

The heat source (plasma torch) parameters are added to the conventional turning parameters. The standoff distance, angular position and shieldings for the torch are usually fixed by adapters to the machine tool. A careful balance between heat input, cutting depth and speed are needed to be sure that adequate, but not excessive, temperatures are achieved at the necessary depth in the workpiece surface at the instant of the cut.

### Material Removal Rates and Tolerances

Conventional turning tolerances apply equally to thermally assisted machining, but material removal rates increase because the workpiece is softened. The shear forces are reduced; however, the depth of the softening zone governs the maximum depth of cut at which improved rates can be attained. Table 10.5-2 lists improved rates of cutting attained.

### Surface Technology

When the heat input and cutting rates are properly balanced, the work material below the cut is usually not heated enough to alter the metallurgical structure. The surface roughness, plastic deformation, residual stresses, etc., associated with conventional turning apply equally to TAM.

# 10.5 Thermally Assisted Machining—TAM

**TABLE 10.5–2 Comparison of Conventional Turning with TAM**

| MATERIAL | CONVENTIONAL TURNING | | TAM | |
|---|---|---|---|---|
| | fpm | m/s | fpm | m/s |
| Nimonic 115 | 35 | 0.18 | 500 | 2.5 |
| Austenitic chrome/manganese | 10 | 0.05 | 500 | 2.5 |
| Stellite, soft | 20-50 | 0.10-0.25 | 400 | 2.0 |
| Stellite, hard | grind only | grind only | 350 | 1.8 |
| Hard steel, 50 $R_c$ | grind only | grind only | 500 | 2.5 |
| Inconel 718, STA | 100 | 0.5 | 400 | 2.0 |
| 18% Mn steel | 35 | 0.18 | 200 | 1.0 |
| Waspaloy | 90 | 0.45 | 400 | 2.0 |
| Rene 41 | 28 | 0.14 | 415 | 2.1 |

## Equipment and Tooling

Retrofitting adapters are available for applying plasma torch TAM to single-point, heavy-duty lathes Capacities to 4 kW in heating are available. Shrouds or screens are desirable to contain the very hot chips. Standard tool bits are applicable; however, consideration should be given to using cutters with higher temperature capabilities, such as tungsten carbide, cermets, Borazon®-CBN or ceramics. The machine tool and fixtures should be rigid enough to accept the higher cutting speeds used.

## Machining Characteristics

TAM has characteristics similar to conventional single-point cutting. A significant amount of noise may be produced by the plasma torch, so suitable muffling may be necessary. Protection from the chips is desirable because they may reach the 1000°F [550°C] temperature range. Eye protection from the arc is recommended.

## PROCESS SUMMARY

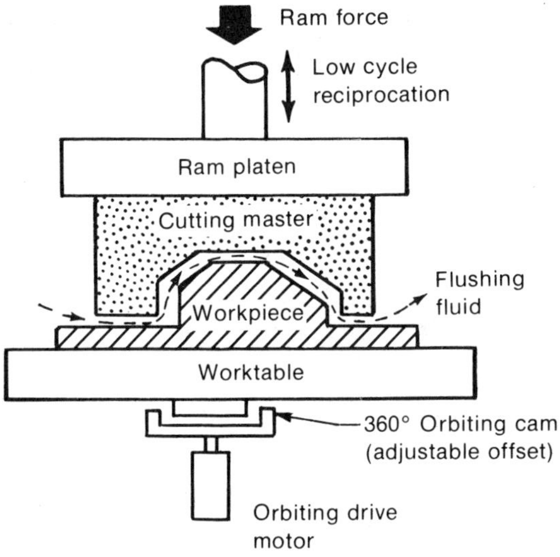


**Figure 10.6–1** TFM schematic.

### Principles

Total form machining (TFM) is a process in which an abrasive master abrades its full three-dimensional shape into the workpiece by the application of force while a full-circle, orbiting motion is applied to the workpiece via the worktable (see figure 10.6–1). The cutting master is advanced into the work until the desired depth of cut is achieved. Uniformity of cutting is promoted by the fluid which continuously transports the abraded particles out of the working gap. Adjustment of the orbiting cam drive controls the precision of the overcut from the cutting master. Cutting action takes place simultaneously over the full surface of abrasive contact.

### Practical Applications

Easily abradable, chalky, granular or friable materials are candidates for three-dimensional (or two-dimensional) forming or cavity-sinking operations by TFM. TFM can rapidly prepare graphite electrodes for EDM, and both the roughing and finishing electrodes can be made from the same master by adjusting the orbiting cam action. Redressing of worn electrodes can be accomplished in a few minutes. Using special fluids, TFM can polish surfaces produced by EDM. Foundry cores can be generated with TFM.

### Operating Parameters

**TABLE 10.6–1 Typical Values for TFM Operating Parameters**

| | |
|---|---|
| Abrasive master: | Steel die cavity with an untouched EDM surface or epoxy loaded with silicon carbide grit. |
| Grit size: | #180 to #20 (#100 provides a good balance between cutting rate and reasonable finish.) |
| Ram force: | Up to 1000 psi [6.9 MPa]. Normal operation is below 500 psi [3.5 MPa]. |
| Ram reciprocation: | Once per second. ±0.10-inch [2.5 mm] amplitude. |
| Orbit offset: | 0.020 to 0.180 inch [0.5 to 4.5 mm] is typical; up to 0.25 or 0.50 inch [6.5 or 12.5 mm] is available. |
| Fluid: | Filtered oil type comparable to EDM oils for machining graphite electrodes. |

### Material Removal Rates and Tolerances

Rates of material removal are determined by the ram force, the reciprocating offset and to some extent by the grit size and fluid flow rate. For three-dimensional forms, a comparison with profile milling of graphite shows an average saving in machining time of 90 percent with TFM. A 0.10-inch [2.5 mm] redressing of a graphite EDM electrode can be accomplished in 15 minutes with TFM. The sharpness of an inside corner is limited to the radius of the orbit setting. Outside corners can be as sharp as the master. Tolerances are only slightly greater than those of the master.

### Surface Technology

The characteristics of the machined surfaces are related directly to the fineness of the abrasive grit used in the cutting master and to some degree to the extent of the orbiting offset. For graphite, the surface roughness produced is often sufficiently good so that the electrode can go to the EDM machine without any benchwork being necessary.

# 10.6 Total Form Machining—TFM

## Equipment and Tooling

Equipment is regularly available in several sizes, with table sizes up to three feet [900 mm] square and up to 18 inches [460 mm] of ram travel. Orbiting offsets up to 0.50 inch [12.5 mm] are available. Filtered fluid systems accompany the machine or are built into the base.

The preparation of the abrasive cutting master, with appropriate overcut allowances, can start from a full-scale model and use a double-reverse, plastic-molding technique to secure the mirror image needed. Several other techniques for preparing the cutting master are available from the equipment builder.[1]

## Machining Characteristics

The fluid flow in TFM eliminates the dust associated with conventional profile milling of graphite or other friable materials.

[1]C. Krauter, 1976. EDM and TFM, The ideal marriage in a production die sinking center, Technical paper MR76-702, Society of Manufacturing Engineers, Dearborn, MI.

## PROCESS SUMMARY

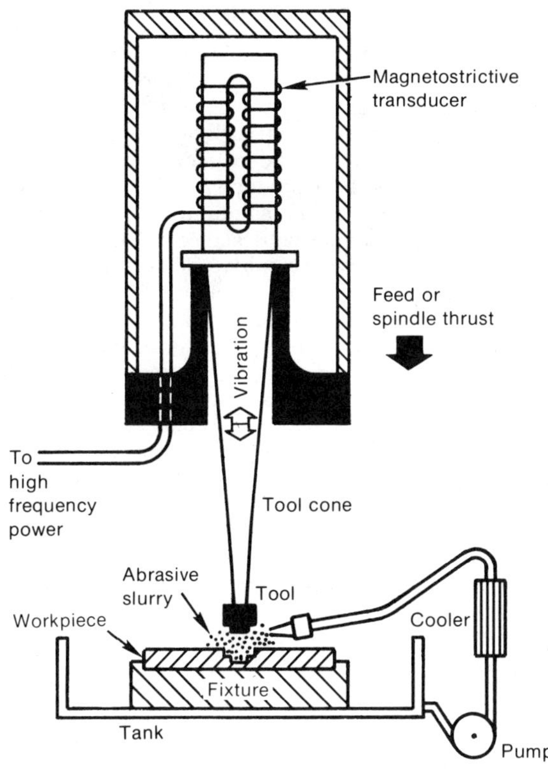


**Figure 10.7–1** USM schematic.

### Principles

Ultrasonic machining (USM) is the removal of material by the abrading action of a grit-loaded liquid slurry circulating between the workpiece and a tool vibrating perpendicular to the workface at a frequency above the audible range (see figure 10.7–1). A high-frequency power source activates a stack of magnetostrictive material, which produces a low amplitude vibration of the toolholder. This motion is transmitted under light pressure to the slurry which abrades the workpiece into a conjugate image of the tool form. A constant flow of slurry (usually cooled) is necessary to carry away the chips from the workface. The process is sometimes called ultrasonic abrasive machining (UAM) or impact machining.

A prime variation of USM is the addition of ultrasonic vibration to a rotating tool—usually a diamond-plated drill. Rotary ultrasonic machining (RUM) substantially increases the drilling efficiency. A piezoelectric device built into the rotating head provides the needed vibration. Milling-, drilling-, threading-, and grinding-type operations are performed with RUM (table 10.7–15 and figure 10.7–15).

### Practical Applications

While USM can cut any material, conductive or nonconductive, metallic, ceramic or composite, it is most effective on materials harder than 40 $R_C$. Holes, slots and irregular shapes can be produced in delicate ceramics. Tool wear and taper in the cut can be limiting factors. A practical depth-to-width ratio is 2.5:1. Holes as small as 0.003-inch [0.076 mm] diameter have been made. Current practical limits are 3.5-inch [89 mm] diameter tools operating in 2.5-inch [64 mm] deep cavities. Trepanning is preferred to solid drilling. Threading of ceramics can be accomplished with a rotary tool or part. Slicing, coining, dicing, lapping, engraving, deburring, broaching, boring and trepanning operations are also practical. An ultrasonic assist is sometimes helpful in the performance of conventional drilling, grinding and drawing operations.

### Operating Parameters

**TABLE 10.7–1 Typical Values for USM Operating Parameters**

| | |
|---|---|
| Power: | 200 to 4000 watts |
| Frequency: | 10 to 40 kHz (most frequently 20 kHz) |
| Abrasive | |
| type: | Boron carbide (most frequently) (see tables 10.7–5 and 10.7–6) |
| size: | 100 to 2000 (see table 10.7–10) |
| concentration: | 20% to 60% by volume in water (use lower percentage for the larger tools). Oil is used sometimes for finishing. |
| flow: | Ample, sharp and cool (replenish as becomes worn) (35° to 40°F [2° to 5°C] desirable) |
| Vibration amplitude: | 0.0005 to 0.0025 inch [0.013 to 0.063 mm] (mean diameter of grit should be approximately equal to the vibration amplitude) |
| Tool tip force: | 1 to 100 lb. [0.45 to 45 kg] (generally less than 10 lb [4.5 kg]) |
| Tool material: | Mild steel, 303 stainless steel, Monel, 52100 steel, molybdenum |
| Overcut and cutting gap: | Approximately 2 times grit size |
| Wear ratio: | 0.7:1 to 1000:1 (see table 10.7–16) |
| Feed rate: | 0.002 to 0.10 in/min [0.05 to 2.5 mm/min] (sensitive to debris accumulation in gap, to sharpness of grit and depth of cut) |
| Depth of cut: | Up to 2.5 inches [64 mm] |
| Area of cut: | Up to 3.5-inch [89 mm] diameter |
| Accuracy: | ±0.001 inch [±0.025 mm] typical; ±0.0002 inch [±0.005 mm] possible |
| Taper: | Up to 0.005 inch/inch [0.005 mm/mm] |
| Surface roughness, $R_a$: | 20 to 30 μin [0.51 to 0.76 μm] |

# 10.7 Ultrasonic Machining—USM and Rotary Ultrasonic Machining—RUM

## Material Removal Rates and Tolerances

Respectable machining rates combined with accuracy and good finishes can be achieved for very hard work materials. Vibration amplitude and frequency, area of cut, and static load are principal considerations; however, abrasive size and type, work material brittleness, and depth of cut are also important. The rate of penetration is inversely proportional to the area of the cut and proportional to the grit size and the square of the amplitude of vibration. Trepanning of cavities or holes is preferable to the use of solid tools. Cutting rates increase with the hardness of the abrasive grit. It is essential to maintain a good flow of fresh, sharp grit at the cutting interface and to flush this face free from chips and accumulations of debris. Excess static load can compact the grit and slow the penetration rate.

The delicate touch of USM, while slower in volume removal rate, can be especially valuable for reducing manufacturing losses of fragile workpieces. Breakout and chipping of exits of cuts can be a problem. Thin parts often are cemented to a sacrifice plate.

Accuracy is typically ±0.001 inch [±0.025 mm] and can achieve ±0.0002 inch [±0.005 mm] with special care. Typical operating conditions are listed in table 10.7-3. The rate of removal and other specific operating conditions are dependent upon the work material characteristics, as demonstrated in table 10.7-20.

## Surface Technology

Surface roughness decreases and finish improves with smaller size grit and smaller vibration amplitudes or higher frequencies (see figures 10.7-3 to 10.7-5). USM typically achieves roughness values of 20 to 30 microinches $R_a$ [0.51 to 0.76 μm]. Side walls of deep cavities can be rougher than the bottom face. There is no heat-affected zone, and there are no chemical or electrical alterations in the surfaces produced. The multitude of small impacts imparts a shallow, compressive residual stress to the surface. This residual stress promotes an increase in the high cycle fatigue strength of the work material after USM (see section 18.3, table 18.3-7). Side-wall channels can occur from the action of cavitation but will only be approximately the size of the abrasive grit.

## Equipment and Tools

USM equipment for cavity sinking or piercing is available in a power range from 200 to 2400 watts and, with special engineering order, up to 4000 watts. The most common frequency is 20 kHz which can be adjusted about plus or minus 10 percent to attain better tuning for specific tooling. Special attachments or equipment adaptations provide ultrasonic assist for conventional drilling, grinding, forming and drawing. Tool material and grit are matched to the application, and for the best productivity, great care should be exercised in selecting and maintaining the amplitude of vibration. The tool should be designed to resonate at the desired frequency for best results, and it must be strong enough to resist fatigue failure. Tool materials are selected on the basis of ductility and toughness rather than hardness. The overcut between the tool form and workpiece shape is approximately equal to twice the size of the abrasive grit. Tool storage needs attention in order to reduce nicks and scratches that can cause early fatigue failure. Most forces are low, so workpiece clamping can be minimal. Fixturing should be positive so as to eliminate vibration of the workpiece. The slurry recirculation system should have cooling capability.

RUM equipment is available in a range of sizes and accessories as well as in portable units. Practical limit to tool weight is 1.4 oz [40 g]. The rotary heads provide capability for threading.

## Machining Characteristics

There are no unusual safety considerations in USM, and the simplicity of the operation reduces the training or skill requirements. A good flow of slurry is desirable in order to maintain the cutting rate. Replenishment of the grit is needed at regular intervals. With water-based slurries, a rust-preventive treatment is desirable following the washout of the slurry. The proper choice of tool material and abrasive grit will result in satisfactory wear ratios between the workpiece and the tool. Taper in deep cuts can be relieved by the use of roughing and finishing tools. Because tool wear occurs mostly on the bottom face of the tool, profile precision is maintained.

## SELECTED DATA

There are several valid techniques for selecting operating parameters for USM. The steps in table 10.7-2 are recommended as one logical sequence. Supporting data come from the other figures and tables in accordance with the terms as explained in figure 10.7-2.

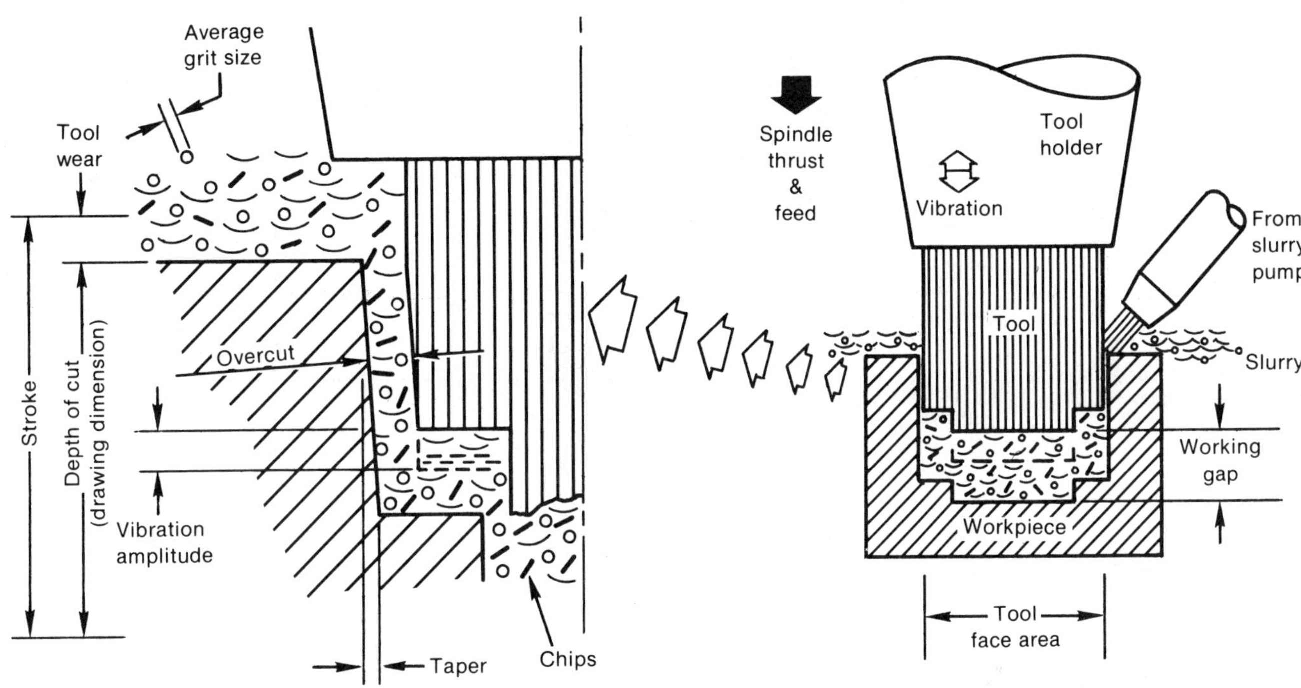


**Figure 10.7-2** USM parameter terminology

# 10.7 Ultrasonic Machining—USM and Rotary Ultrasonic Machining—RUM

**TABLE 10.7–2 Steps to Select USM Operating Parameters**

| ITEM | DATA |
|---|---|
| 1. Workpiece material: type, hardness, metallurgical state | From engineering drawings or specifications.<br>Preferably hardness greater than 40 $R_C$.<br>See table 10.7–4. |
| 2. Configuration, finish and tolerance | From engineering drawing:<br>Maximum contours depth-to-width = 2.5:1<br>Maximum dimension 4 inch [100 mm]<br>Minimum width cut 0.003 inch [0.076 mm]<br>Limited by tool type and shape. Consult machine vendors. |
| 3. Area of cut | From engineering drawing:<br>Calculate from part configuration.<br>Minimize area by use of trepanning where possible. |
| 4. Select type abrasive | Tables 10.7–5 to 10.7–10 |
| 5. Select abrasive grit size | Table 10.7–10; figures 10.7–3 to 10.7–5.<br>Base on surface roughness and tolerance needs. |
| 6. Select % concentration of slurry | Figure 10.7–6<br>Lower %'s for larger tool areas.<br>Water base generally; oil or alcohol sometimes used when water will dissolve workpiece. |
| 7. Select tool material | Tables 10.7–11 and 10.7–12 |
| 8. Estimate:<br>penetration rate<br>vibration amplitude<br>static load | Tables 10.7–13 to 10.7–15; figures 10.7–7 to 10.7–11 |
| 9. Select tool design | Allow overcut of 2 times grit size.<br>Figure 10.7–12 |
| 10. Select equipment size | Table 10.7–19 |
| 11. Verify vibration frequency and amplitude. Refine other operating settings. | Trial cut and tuning of system.<br>Establish wear rate. |
| 12. Select depth of cut | Depth of cut or stroke equals drawing requirements plus tool wear allowance.<br>Measure or estimate wear ratios.<br>Tables 10.7–16, 10.7–17 and 10.7–18; figures 10.7–13 and 10.7–14 |

**TABLE 10.7–3 Typical USM Operating Conditions**

| PARAMETER | ROUGHING | SEMI-FINISHING | FINISHING |
|---|---|---|---|
| Tolerance, inch | ±0.001 | ±0.0005 | ±0.00025 |
| Roughness, microinch $R_a$ | 30 | 25 | 20 |
| Grit size number | 230 | 400 | 800 |
| Grit average size, inch | 0.00248 | 0.00090 | 0.00044 |
| Grit average size, micron | 63 | 23 | 11.2 |
| Abrasive concentration, % | 55 | 45 | 35 |
| Overcut, inch | 0.005 | 0.002 | 0.0005 |
| Machine stroke | 0.0025 | 0.0015 | 0.0005 |

**TABLE 10.7-4 Materials That Have Been Successfully Machined Ultrasonically**

| | | | |
|---|---|---|---|
| Agate | Composites | Limestone | Silicon carbide |
| Alumina | Cold rolled steel | Lithium fluoride | Silicon nitrite |
| Aluminum oxide | Ebony | Micarta | Stainless steel—hardened |
| Barium titanate | Ferrite | Molybdenum | Steatite |
| Beryllium oxide | Formica | Molybdenum disilicate | Steel-hardened tools |
| Boron carbide (fused) | Garnet | Mother of pearl | Ti-6Al-4V |
| Boron composites | Germanium | Plaster of paris | Tungsten |
| Brass | Glass | Pyrolytic graphite | Tungsten carbide |
| Calcium | Glass-bonded mica | Quartz | Thorium oxide |
| Carbides | Graphite | Ruby | Uranium oxide |
| Carbon | Hardened 1095 steel | Sapphire | Uranium carbide |
| Ceramics | High pressure laminates | Silicon | Zirconium oxide |

**TABLE 10.7-5 Abrasives Used in USM**

| ABRASIVE | KNOOP HARDNESS | RELATIVE CUTTING POWER |
|---|---|---|
| Diamond | 6500-7000 | 1.0 |
| Cubic boron nitride (CBN) | 4700 | 0.95 |
| Boron carbide ($B_4C$) | 2800 | 0.50-0.60 |
| Silicon carbide (SiC) | 2480-2500 | 0.25-0.45 |
| Aluminum oxide ($Al_2O_3$) | 2000-2100 | 0.14-0.16 |

**TABLE 10.7-6 Selection of Abrasive**

| ABRASIVE | WORK MATERIAL |
|---|---|
| Boron carbide | Tungsten carbide, metals, high density ceramics, minerals, semi and precious stones. |
| Silicon carbide | Low density ceramics, glass silicon, germanium, mineral stones. |
| Aluminum oxide | Glass, low density, sintered or hard powder compounds. |

SOURCE: G.E. Littleford, Machining by ultrasonics, 1971.

**TABLE 10.7-7 USM Cutting Speed Index for Brittle Work Materials**
(Soda glass, using 100 mesh $B_4C$ abrasive = 100)

| WORK MATERIAL | ABRASIVE | | | |
|---|---|---|---|---|
| | Boron Carbide | | | Silicon Carbide |
| | 100 mesh | 200 mesh | 400 mesh | 100 mesh |
| Soda glass | 100 | 90 | 77 | 85 |
| Hysil | 73 | 66 | 54 | |
| B9 borosilicate glass | 86 | | | |
| Ferroxcube IIIC | 37 | | | 34 |
| Ferroxdure (demagnetized) | (32) | | | |
| Quartz crystal | (57) | | | |
| Fused alumina | 19 | | | |
| Synthetic sapphire | 19 | | | |
| Synthetic ruby | 18 | | | |
| Flint stone | (72) | | | |
| Barium-titanate ceramic | 110 | | | 109 |
| Ceramic 507 | (38) | | | 35 |
| Garnet | (58) | | | |
| Feldspar | (40) | | | |
| Spinel | (48) | | | |
| Slate | 67 | | | |
| Mycalex | (240) | | | (200) |

SOURCE: E. A. Neppiras and R. D. Foskett, Ultrasonic machining, *Philips Technical Review* 18 (1956–57), p. 372.

NOTES: Tool used was mild steel, H-form, 1/2-inch [12.7 mm] square, limbs 1/16-inch [1.59 mm] thick, see figure 10.7-9.

Numbers in parentheses are estimates, interpolated from measurements under somewhat different conditions.

**TABLE 10.7-8 USM Cutting Speed Index for Metals**
(Soda glass, using 100 mesh $B_4C$ abrasive = 100)

| WORK MATERIAL | ABRASIVE | | | | | | |
|---|---|---|---|---|---|---|---|
| | Boron Carbide | | | Silicon Carbide | Alumina | Sand | Diamond Powder |
| | 100 mesh | 220 mesh | 400 mesh | 100 mesh | 220 mesh | (grit size 0.012 in) | (grit size 0.001 in) |
| Soda glass | 100 | 90 | 77 | 85 | 65 | 47 | 90 |
| Brass (common yellow) | 6.6 | 5.6 | | | | | |
| Die steels | | | | | | | |
| K.E. 672 (approx. AISI O1), R. 66 | 1.4 | 1.3 | | | | | |
| C.S.K., R. 62 | 3.9 | 3.6 | | | | | |
| K.E. 672 (approx. AISI O1), R. 61 | 2.2 | 2.1 | | 1.48 | 0.1 | 0.1 | |
| K.E. 672 (approx. AISI O1), R. 58 | | 1.7 | | | | | |
| Stainless steels | | | | | | | |
| 18% Cr, 8% Ni, 0.1% C | 2.1 | 1.9 | | | | | |
| 3.5% Cr, 8.4% W, 0.35% V, 0.3% C | 1.2 | 1.1 | | | | | |
| Carbon-chrome bearing steel (heat-treated) | 1.4 | | | | | | |
| Sintered tungsten carbide, R. 76 | 4.1 | 3.5 | | 2.55 | 0.2 | 0.2 | 4.3 |
| Tungsten | 4.8 | 4.3 | | | | | |
| Stellite | 4.0 | 3.7 | | | | | |
| Germanium single crystal | (31) | | | (28) | | | |
| Titanium | (4.0) | | | | | | |
| Beryllium | (7) | | | | | | |

SOURCE: E. A. Neppiras and R. D. Foskett, p. 371.
NOTE: Same tool used as in table 10.7–7. Numbers in parentheses are estimates interpolated from measurements under somewhat different conditions.

**TABLE 10.7-9 USM Cutting Speed Index for Ceramic Work Materials**
(Soda glass, using 100 mesh $B_4C$ abrasive = 100)

| WORK MATERIAL | ABRASIVE | | | | | |
|---|---|---|---|---|---|---|
| | Boron Carbide | | Silicon Carbide | | | Alumina |
| | 100 mesh | 400 mesh | 100 mesh | 400 mesh | 600 mesh | 220 mesh |
| Chemical porcelain | 70 | 14 | 53.5 | 11 | 2 | 25 |
| Temperadex* | 60 | 10 | 45 | 7.5 | — | 20 |
| Faradex* | 55 | 7 | 41 | 5 | — | 2 |
| Vulcanex* | 190 | 51 | 168 | 44 | 2 | 62 |
| Z.Z. porcelain | 90 | 32.5 | 75 | 26 | 2 | 10 |
| H.T. porcelain | 45 | 24 | 41 | 22 | — | 8 |
| Frequentite* | 47.5 | 27.5 | 45 | 26 | — | 20 |
| Soda glass | 100 | 45 | 80 | 37 | 6 | 70 |

SOURCE: E. A. Neppiras and R. D. Foskett, p. 371.
NOTE: Tool used was mild steel, circular, 1/4-inch [6.5 mm] diameter.
*Trade names of proprietary materials.

**TABLE 10.7-10 Grit Sizes Used in USM**

| GRIT SIZE | PARTICLE SIZE | |
|---|---|---|
| | in | μm |
| Roughing | | |
| 120 | 0.0056 | 142 |
| 140 | 0.0048 | 122 |
| 170 | 0.0034 | 86 |
| 200 | 0.0026 | 66 |
| 230 | 0.00248 | 63 |
| 270 | 0.00175 | 44 |
| 325 | 0.00128 | 33 |
| 400 | 0.00090 | 23 |
| Finishing | | |
| 500 | 0.00073 | 19 |
| 600 | 0.00057 | 14 |
| 800 | 0.00044 | 11 |
| 900 | 0.00038 | 10 |
| 1000 | 0.00032 | 8 |
| 1200 | 0.00022 | 6 |

NOTE: In estimating tool size with respect to tolerances required, the abrasive particle diameter can be considered as an overcut factor. For example, if a 1/4 inch [6.3 mm] diameter hole was being drilled with USM (using 180 grit), the diameter of the hole in the workpiece would be equal to 0.250 [6.350 mm] + 2 times 0.0034 [86 μm x 10³] or approximately 0.257 inch [6.53 mm].

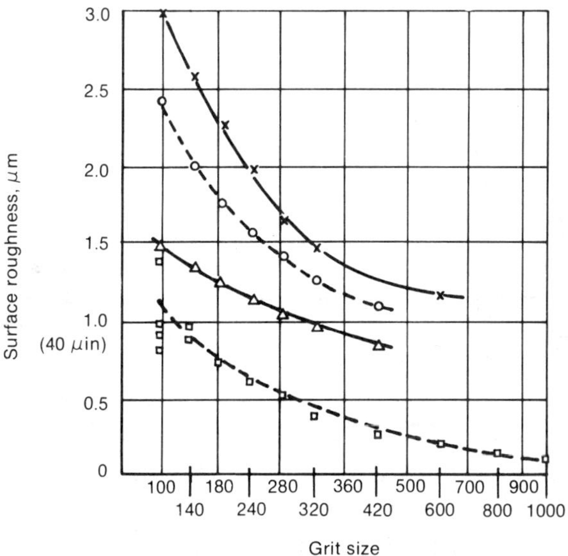


**Figure 10.7-3** Surface roughness versus grit size of boron carbide abrasive. Work material key: x = glass; o = silicon-semi-conductor; △ = mineralo-ceramic; □ = hard alloy steel. (D. C. Kennedy and R. J. Grieve, Ultrasonic machining—a review, *Production Engineer* 54 (1975), p. 485)

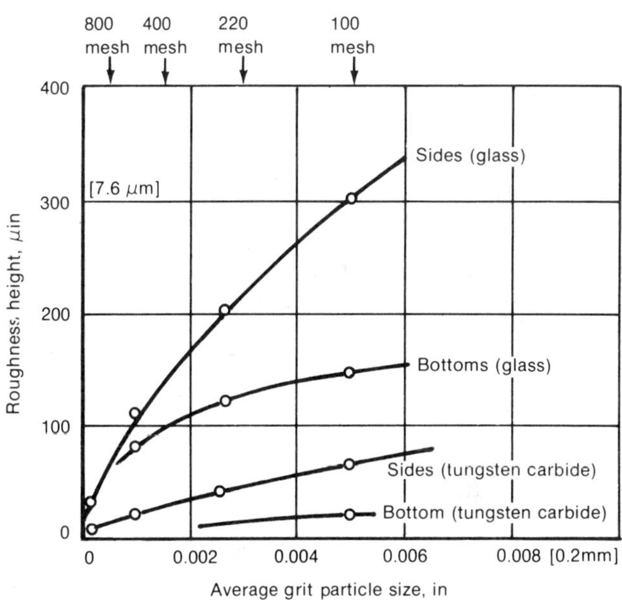


**Figure 10.7-4** Roughness height versus grit particle size for holes in glass and tungsten carbide. (E. A. Neppiras and R. D. Foskett, p. 378)

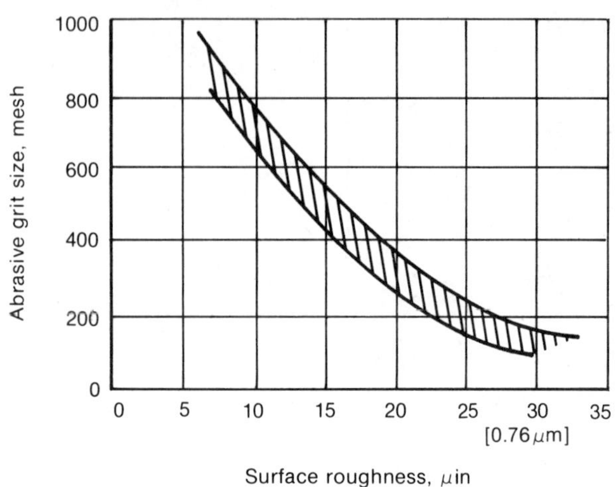


**Figure 10.7-5** Grit size versus surface roughness. Smaller particles produce smoother finishes; larger particles cut faster. (Data modified from A. L. Roses, Techniques of ultrasonic machining, *Tool and Manufacturing Engineer* 46 (1969): 71–75)

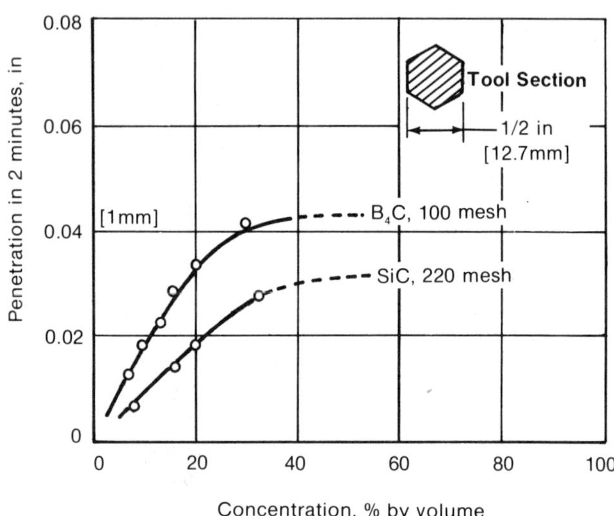


**Figure 10.7-6** Penetration in glass as a function of slurry concentration. (E. A. Neppiras and R. D. Foskett, p. 373)

**TABLE 10.7-11 Effect of Tool Material on Cutting Rate**
(Soda glass work material; $B_4C$ abrasive; optimum static load at 20 kHz)

| TOOL MATERIAL | TOOL SHAPE | CUTTING RATE,* in/min | | | |
|---|---|---|---|---|---|
| | | Amplitude = 0.002 in | | Amplitude = 0.001 in | |
| | | Abrasive 100 mesh | Abrasive 400 mesh | Abrasive 100 mesh | Abrasive 400 mesh |
| Copper | Circular, 1/2 inch dia. | 0.063 | 0.020 | 0.016 | 0.006 |
| Brass (BSS 251) | Circular, 1/2 inch dia. | 0.090 | 0.028 | 0.022 | 0.008 |
| Mild steel (EN2) | Circular, 1/2 inch dia. | 0.086 | 0.022 | 0.021 | 0.007 |
| Silver steel† | Circular, 1/2 inch dia. | 0.074 | 0.022 | 0.020 | 0.008 |
| Stainless steel (18% Cr, 8% Ni, 0.1% C) | Circular, 1/2 inch dia. | 0.071 | 0.020 | 0.018 | 0.006 |
| Tungsten carbide | Triangular, 1/8 inch base | 0.156 | 0.100 | 0.038 | 0.030 |
| Mild steel (EN2) | Triangular, 1/8 inch base | 0.150 | 0.102 | 0.036 | 0.029 |
| | | **CUTTING RATE,* mm/min** | | | |
| | | **Amplitude = 0.051 mm** | | **Amplitude = 0.025 mm** | |
| | | **Abrasive 100 mesh** | **Abrasive 400 mesh** | **Abrasive 100 mesh** | **Abrasive 400 mesh** |
| Copper | Circular, 12.7 mm dia. | 1.60 | 0.51 | 0.41 | 0.15 |
| Brass (BSS 251) | Circular, 12.7 mm dia. | 2.28 | 0.71 | 0.56 | 0.20 |
| Mild steel (EN2) | Circular, 12.7 mm dia. | 2.18 | 0.56 | 0.53 | 0.18 |
| Silver steel† | Circular, 12.7 mm dia. | 1.88 | 0.56 | 0.51 | 0.20 |
| Stainless steel (18% Cr, 8% Ni, 0.1% C) | Circular, 12.7 mm dia. | 1.80 | 0.51 | 0.46 | 0.15 |
| Tungsten carbide | Triangular, 3.2 mm base | 3.96 | 2.54 | 0.97 | 0.76 |
| Mild steel (EN2) | Triangular, 3.2 mm base | 3.81 | 2.59 | 0.91 | 0.74 |

SOURCE: Adapted from E. A. Neppiras and R. D. Foskett, p. 370.

*Averaged over about 0.1 inch [2.5 mm] penetration.

†0.87% C, 0.020% S, 0.025% P, 0.28% Mn; tempered in air after oil-quenching from 800°C.

**TABLE 10.7–12 Tool Wear for Various Tool Materials Cutting with 100 Mesh $B_4C$ Abrasive**
(Amplitude = 0.002 inch [0.051 mm]; optimum static load at 20 kHz)

| TOOL MATERIAL | TOOL SHAPE | TOOL WEAR* | | | | | | | | | |
|---|---|---|---|---|---|---|---|---|---|---|---|
| | | Soda Glass Work Material | | | | | Tungsten Carbide Work Material | | | | |
| | | Longitudinal tool wear | | Total penetration in workpiece | | Tool wear as % of stock removal | Longitudinal tool wear | | Total penetration in workpiece | | Tool wear as % of stock removal |
| | | in | mm | in | mm | | in | mm | in | mm | |
| Copper | Circular, 1/2 in [12.7 mm] dia. | 0.0025 | 0.063 | 0.520 | 13.2 | 0.48 | — | — | — | — | — |
| Mild steel (EN2) | Circular, 1/2 in [12.7 mm] dia. | 0.018 | 0.46 | 1.850 | 42.0 | 1.0 | 0.110 | 2.79 | 0.125 | 3.18 | 88 |
| Silver steel† | Circular, 1/2 in [12.7 mm] dia. | 0.0025 | 0.063 | 0.546 | 13.9 | 0.46 | 0.012 | 0.30 | 0.046 | 1.17 | 26 |
| Stainless steel (18% Cr, 8% Ni, 0.1% C) | Circular, 1/2 in [12.7 mm] dia. | 0.008 | 0.20 | 1.150 | 29.2 | 0.7 | 0.016 | 0.41 | 0.045 | 1.14 | 35 |
| Brass (BSS 251) | Circular, 1/2 in [12.7 mm] dia. | 0.021 | 0.53 | 1.250 | 31.8 | 1.68 | 0.175 | 4.45 | 0.125 | 3.18 | 140 |
| Sintered tungsten carbide | Triangular, 1/8 in [3.2 mm] base | 0.0015 | 0.038 | 1.510 | 38.4 | 0.1 | 0.138 | 3.51 | 0.125 | 3.18 | 110 |
| | | Tool Steel | | | | | Tungsten Carbide | | | | |
| Mild steel | Circular, hollow, int. dia. = 1/8 in [3.2 mm], ext. dia. = 1/4 in [6.4 mm] | | | | | | 0.156 | 3.96 | 0.125 | 3.18 | 125 |
| Mild steel | H-form, 1/2 in [12.7 mm] square, limbs, 1/16 in [1.6 mm] thick | 0.053 | 1.35 | 0.024 | 0.6 | 220 | | | | | |
| Mild steel | Circular, hollow, int. dia. = 0.33 in [8.4 mm], ext. dia. = 0.39 in [9.9 mm] | | | | | | 0.170 | 4.32 | 0.075 | 1.91 | 222 |
| Brass | Extrusion shape T-form, height of T = 16 mm (0.63 in) | | | | | | 0.256 (avg.) | 6.5 (avg.) | 0.125 | 3.18 | 205 |

SOURCE: Adapted from E. A. Neppiras and R. D. Foskett, p. 373.
*Average over 0.1 inch [2.5 mm] penetration.
†0.87% C, 0.020% S, 0.025% P, 0.28% Mn; tempered in air after oil-quenching from 800°C.

# 10.7 Ultrasonic Machining—USM and Rotary Ultrasonic Machining—RUM

**TABLE 10.7-13 Material Removal Rates for Various Work Materials**

| WORK MATERIAL | MATERIAL REMOVAL RATE | |
|---|---|---|
| | $in^3/min$ | $mm^3/min$ |
| Carbon | 0.015 | 246 |
| Glass | 0.016 | 262 |
| Ceramics | 0.005 | 82 |
| Silicon | 0.005 | 82 |
| Germanium | 0.006 | 98 |
| Quartz (crystal) | 0.007 | 115 |
| Ferrite | 0.014 | 230 |
| Tungsten carbide | 0.00025 | 4 |
| Tool steel (hardened) | 0.0002 | 3 |
| Stainless steel (hardened) | 0.0002 | 3 |
| Boron carbide (fused) | 0.00025 | 4 |
| Pyrolytic graphite | 0.010 | 16 |

**TABLE 10.7-14 USM Material Removal Rate Index**
(Cutting soda glass at 0.060 in/min [1.52 mm/min] = 100)

| WORK MATERIAL | INDEX |
|---|---|
| Soda glass | 100 |
| Ceramic | 70 |
| Silicon | 60 |
| Tungsten carbide | 4 |
| Stainless steel | 2 |

SOURCE: G. E. Littleford, Machining by ultrasonics, 1971.

**TABLE 10.7-15 Rotary Ultrasonic Machining (RUM) with Rotating Diamond Impregnated Tool**

| WORK MATERIAL | DRILL DIAMETER | | DEPTH OF CUT | | MACHINING TIME | CONDITIONS |
|---|---|---|---|---|---|---|
| | in | mm | in | mm | seconds | |
| 99.9 Alumina | 0.042 | 1.07 | 0.250 | 6.35 | 130 | 20 kHz frequency |
| | 0.125 | 3.18 | 0.250 | 6.35 | 20 | 3000 rpm |
| | 0.250 | 6.35 | 0.250 | 6.35 | 25 | 120 grit core drill diamond tool |
| | 0.375 | 9.53 | 0.250 | 6.35 | 30 | 12 lb. [5.5 Kg] pressure |
| | 0.500 | 12.70 | 0.250 | 6.35 | 38 | |
| Glass | 0.042 | 1.07 | 0.500 | 12.7 | 120 | |
| | 0.250 | 6.35 | 5.000 | 127 | 360 | |
| Ferrite | 0.080 | 2.03 | 0.250 | 6.35 | 70 | |
| | 0.125 | 3.18 | 0.250 | 6.35 | 30 | |
| | 0.250 | 6.35 | 0.250 | 6.35 | 35 | |
| Boron composite | 0.500 | 12.70 | 0.700 | 17.8 | 48 | |
| | 0.250 | 6.35 | 0.500 | 12.7 | 19 | |
| | 0.125 | 3.18 | 0.500 | 12.7 | 26 | |

SOURCE: W. R. Tyrrell, A new method for machining hard and brittle materials, *SAMPE Quarterly* 1 (January 1970): 55-59.

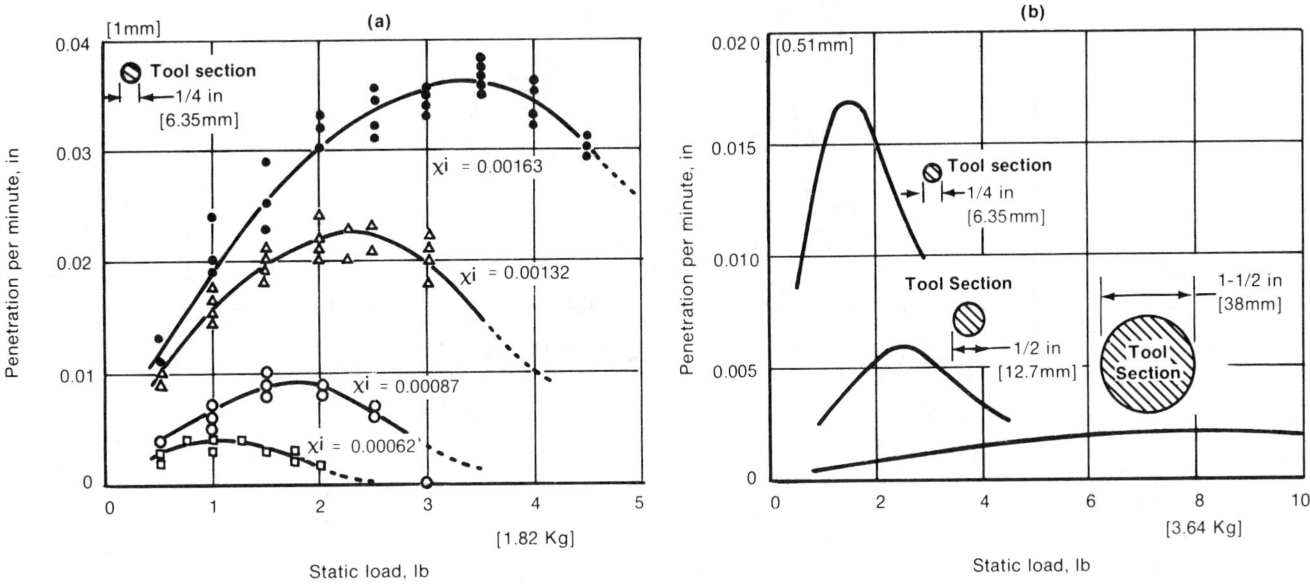


**Figure 10.7–7** Penetration rate as a function of static load for (a) various amplitudes ($\xi$) and (b) various tool areas. (E. A. Neppiras and R. D. Foskett, p. 369)

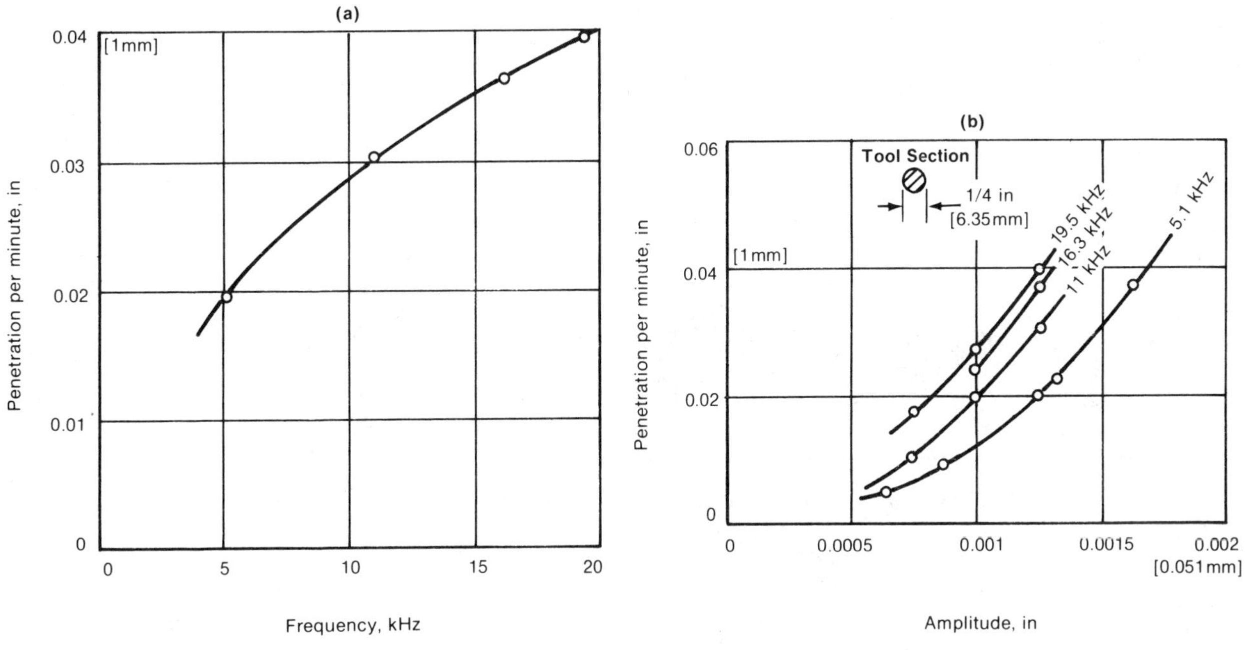


**Figure 10.7–8** Penetration rates in glass as a function of (a) frequency, for a peak to peak amplitude of 0.00125 inch [0.038 mm] and constant static load; and (b) amplitude, for four operating frequencies and constant static load. (E. A. Neppiras and R. D. Foskett, p. 369)

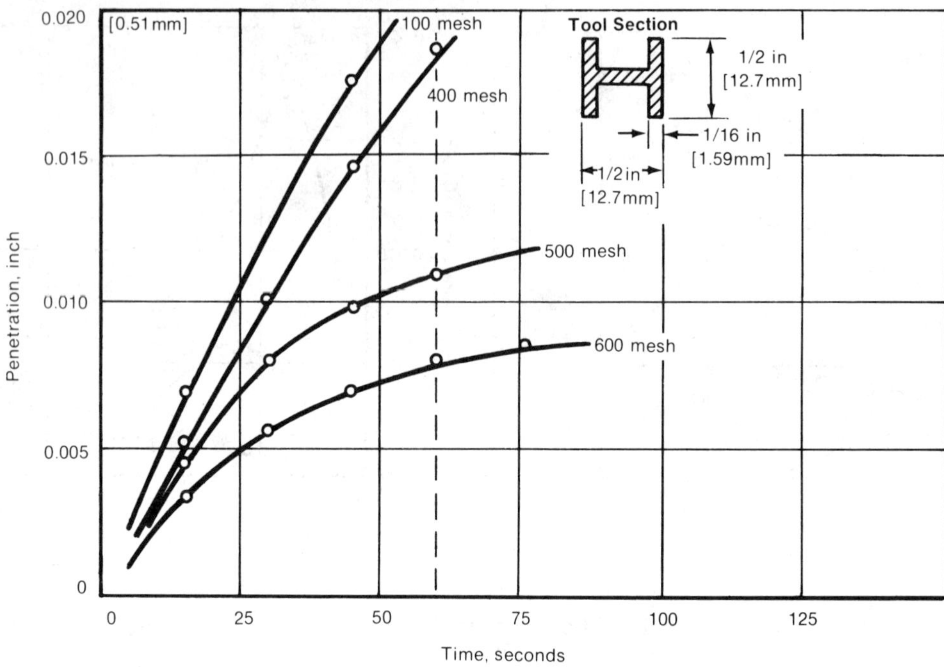


**Figure 10.7–9** Penetration rate as a function of time for various grit sizes. Grit sizes given in British Standard Sieves. (American Mesh sizes are essentially the same.) (E. A. Neppiras and R. D. Foskett, p. 370)

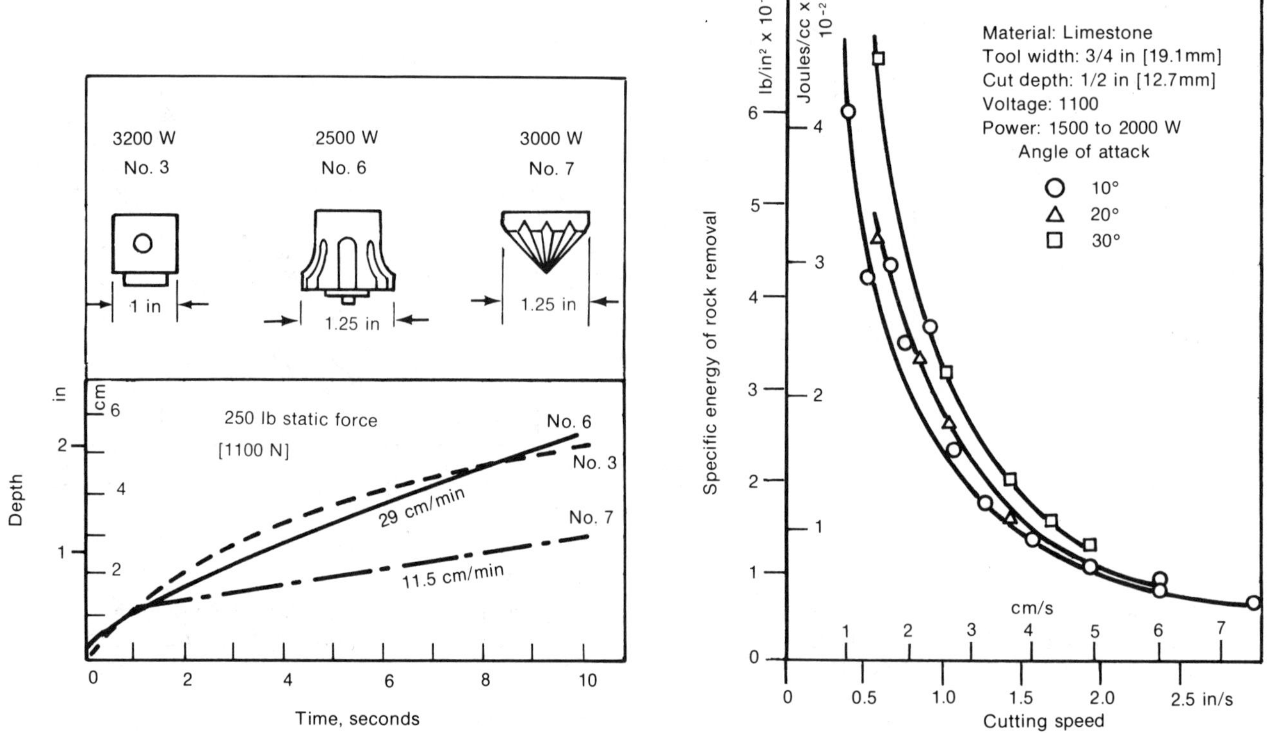


**Figure 10.7–10** Drilling and layer removal rates for medium Indiana limestone processed with USM. *Left*, depth of drilling versus time for three types of tools operated at 10 kHz; *right*, specific energy at various cutting speeds and angles of attack at 10 kHz. (From K. F. Graff, Application of sonic power to rock cutting, *Ultrasonics international 1973 conference proceedings*, pp. 31, 33)

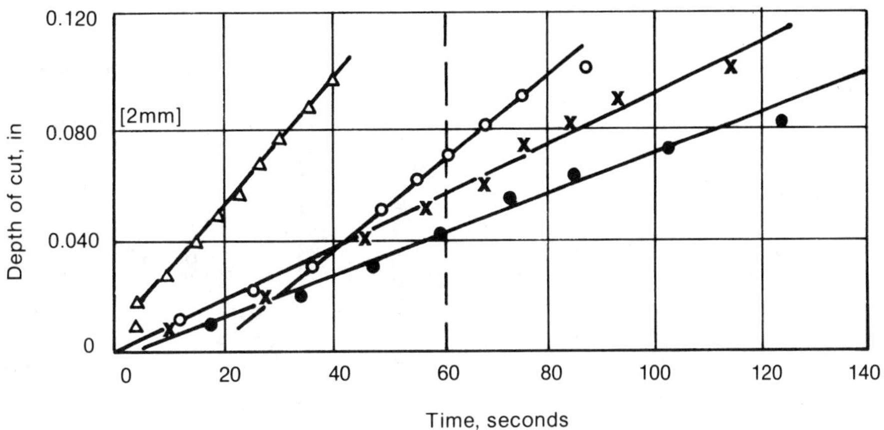


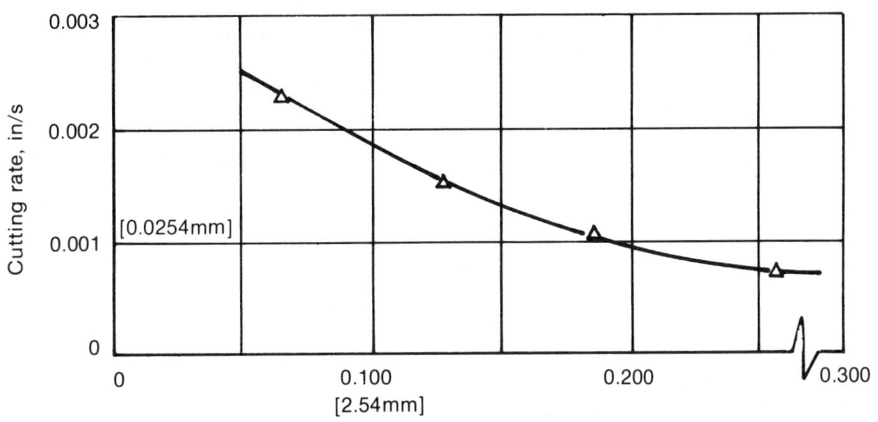


| KEY | DIAMETER OF TOOL | | CUTTING RATE | |
|---|---|---|---|---|
| | in | mm | in/s | mm/s |
| △ | 0.0625 | 1.59 | 2.3 | 58 |
| ○ | 0.125 | 3.18 | 1.58 | 40 |
| X | 0.180 | 4.57 | 1.1 | 28 |
| ● | 0.250 | 6.35 | 0.7 | 18 |

**Figure 10.7-11** Cutting rates for stainless steel tools into ceramic work material. *Top*, determining rates for various size tools; *bottom*, rates as a function of tool diameter. (J. Krawczyk, Ultrasonic grinding techniques in microminiaturization, Report No. TR-958, Diamond Ordnance Fuze Laboratories, Washington, DC, 1961)

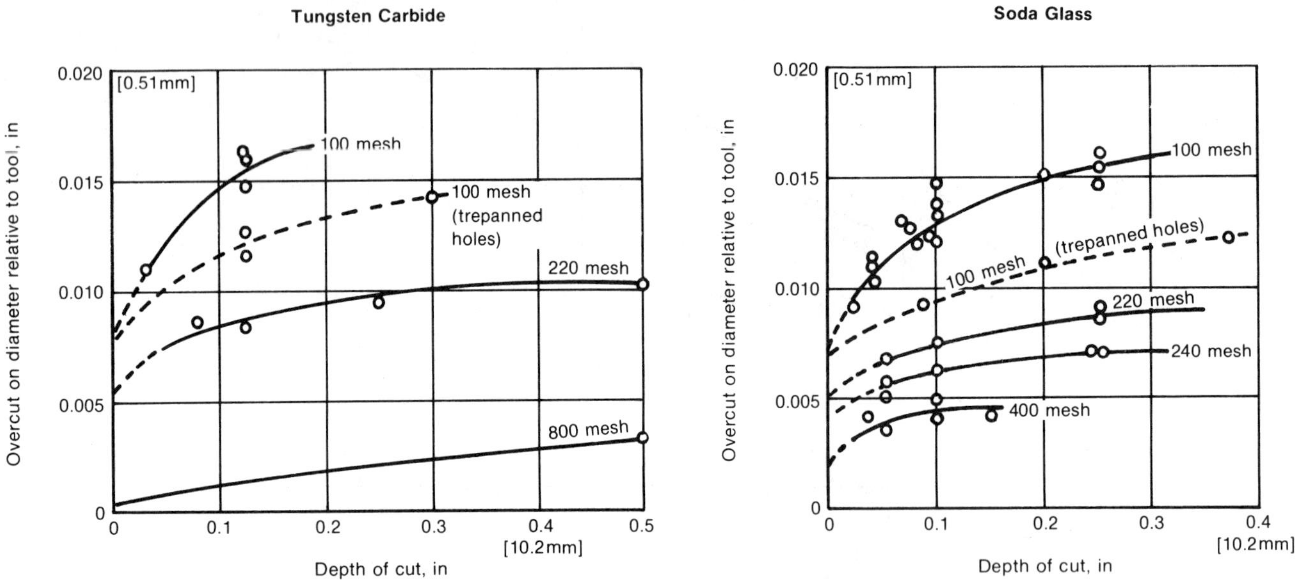


**Figure 10.7–12** Overcut values for tungsten carbide and soda glass work materials cut with $B_4C$ abrasive at various depths. (E. A. Neppiras and R. D. Foskett, p. 374)

**TABLE 10.7–16 General Wear Ratios**

| TOOL MATERIAL | APPROXIMATE WEAR RATIO (ratio of depth of cut to loss of tool size) — Workpiece Materials: Glass | Steatite | Tungsten carbide | Ceramic |
|---|---|---|---|---|
| Brass | 40-50 | 10 | 0.7 | 40 |
| Annealed low C steel | 100 | | 1.1 | 75 |
| Cold rolled steel | 100 | 35 | 3.9 | 75 |
| Stainless steel | 150 | 40 | 2.8 | 100 |
| Tungsten carbide | 1000 | | 0.9 | |

SOURCE: J. Krawczyk, Report No. TR-958, Diamond Ordnance Fuze Laboratories, 1961.

NOTE: General cutting conditions: 1/4-inch diameter mild steel tool; 20 kHz; 320 grit $B_4C$; 1-1/2 lb pressure.

**TABLE 10.7-17 Representative USM Penetrating and Tool Wear Rates at 700 Watts Input**

| MATERIAL | RATIO STOCK REMOVED TO TOOL WEAR | MAXIMUM PRACTICAL MACHINING AREA | | AVERAGE PENETRATING RATE* | |
|---|---|---|---|---|---|
| | | in² | cm² | in/min | mm/min |
| Glass | 100:1 | 4.0 | 25.8 | 0.150 | 3.81 |
| Ceramic | 75:1 | 3.0 | 19.4 | 0.060 | 1.52 |
| Germanium | 100:1 | 3.5 | 22.6 | 0.085 | 2.16 |
| Tungsten carbide | 1.5:1 | 1.2 | 7.7 | 0.010 | 0.25 |
| Tool steel | 1:1 | 0.875 | 5.6 | 0.005 | 0.13 |
| Mother of pearl | 100:1 | 4.0 | 25.8 | 0.150 | 3.81 |
| Synthetic ruby | 2:1 | 0.875 | 5.6 | 0.020 | 0.51 |
| Carbon-graphite | 100:1 | 3.0 | 19.4 | 0.080 | 2.00 |
| Ferrite | 100:1 | 3.5 | 22.6 | 0.125 | 3.18 |
| Quartz | 50:1 | 3.0 | 19.4 | 0.065 | 1.65 |
| Boron carbide | 2:1 | 0.875 | 5.6 | 0.008 | 0.20 |
| Glass-bonded mica | 100:1 | 3.5 | 22.6 | 0.125 | 3.18 |

SOURCE: Data from Raytheon Company, Impact grinders for ultrasonic machining, 1961.

NOTE: Tool material was cold rolled steel in all cases; #320 mesh Boron Carbide was used in all cases.

*1/2-inch [12.7 mm] diameter tool; 1/2-inch [12.7 mm] deep.

**TABLE 10.7-18 Representative USM Penetrating and Tool Wear Rates at 100 Watts Input**

| MATERIAL | SOLID TOOL DIAMETER | | FORCE | | BEST CUTTING RATE | | ABRASIVE/WATER RATIO BY VOLUME | STOCK REMOVAL RATIO WORKPIECE (vs) TOOL |
|---|---|---|---|---|---|---|---|---|
| | in | mm | lb | kg | in/min | mm/min | | |
| Glass | 0.25 | 6.35 | 1 | 0.45 | 0.075 | 1.91 | 1.4:1 | 100:1 |
| Aluminum oxide | 0.25 | 6.35 | 5 | 2.27 | 0.010 | 0.25 | 1.4:1 | 13:1 |
| Tungsten carbide | 0.25 | 6.35 | 1 | 0.45 | 0.002 | 0.05 | 1.4:1 | 1:1 |
| Carbon | 0.25 | 6.35 | 0.5 | 0.27 | 0.100 | 2.50 | 1.4:1 | 100:1 |
| Steatite | 0.25 | 6.35 | 1 | 0.45 | 0.090 | 2.29 | 1.4:1 | 200:1 |
| Germanium | 0.25 | 6.35 | 1 | 0.45 | 0.110 | 2.79 | 1.4:1 | 100:1 |
| Silicon | 0.25 | 6.35 | 1 | 0.45 | 0.110 | 2.79 | 1.4:1 | 100:1 |
| Quartz | 0.25 | 6.35 | 3 | 1.36 | 0.070 | 1.79 | 1.4:1 | 50:1 |
| Glass | 0.125 | 3.18 | 1 | 0.45 | 0.125 | 3.18 | 1.4:1 | 100:1 |
| Aluminum oxide | 0.125 | 3.18 | 1 | 0.45 | 0.030 | 0.76 | 0.25:1 | 10:1 |
| Carbon | 0.125 | 3.18 | 0.5 | 0.27 | 0.250 | 6.35 | 0.25:1 | 100:1 |
| Germanium | 0.125 | 3.18 | 1 | 0.45 | 0.125 | 3.18 | 0.25:1 | 100:1 |
| Silicon | 0.125 | 3.18 | 1 | 0.45 | 0.125 | 3.18 | 0.25:1 | 100:1 |
| Tungsten carbide | 0.125 | 3.18 | 1 | 0.45 | 0.004 | 0.10 | 0.25:1 | 1:1 |

SOURCE: Data from Raytheon Company, Impact grinders for ultrasonic machining, 1961.

NOTE: Tool material was cold rolled steel in all cases; #320 mesh Boron Carbide was used in all cases.

**TABLE 10.7–19 Equipment Size**
(Estimate of watts required for 20 kHz equipment using 120 mesh $B_4C$ abrasive and optimum parameters)

| CUTTING RATE | | | | | | WATTS |
|---|---|---|---|---|---|---|
| Glass | | Hard Alloy | | Germanium | | |
| in³/min x 10⁻⁴ | mm³/min | in³/min x 10⁻⁴ | mm³/min | in³/min x 10⁻⁴ | mm³/min | |
| 92 | 150 | 3.1 | 5 | 31 | 50 | 250 |
| 214 | 350 | 4.9 | 8 | 61 | 100 | 500-600 |
| 490 | 800 | 12.2 | 20 | 122 | 200 | 1000 |
| 900 | 1500 | 18.3 | 30 | 183 | 300 | 1500 |
| 1800 | 3000 | 45.8 | 75 | 427 | 700 | 4000 |

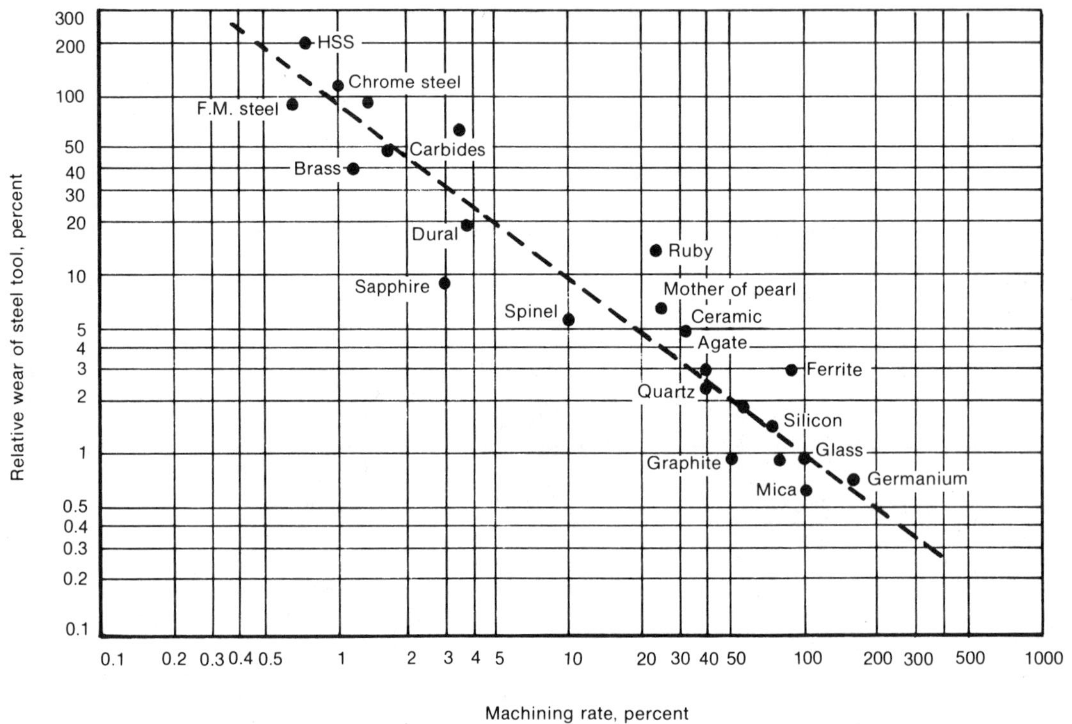


**Figure 10.7–13** Tool wear versus machining rate. Relative wear rate of steel tool in various work materials. 100% is 5 mm/min (0.20 in/min) in glass. (M. Adithan and V. C. Venkatesh, Tool wear phenomenon in ultrasonic drilling, *Proceedings of 5th all India machine tool design and research conference*, University of Roorkee, 1972, p. 625)

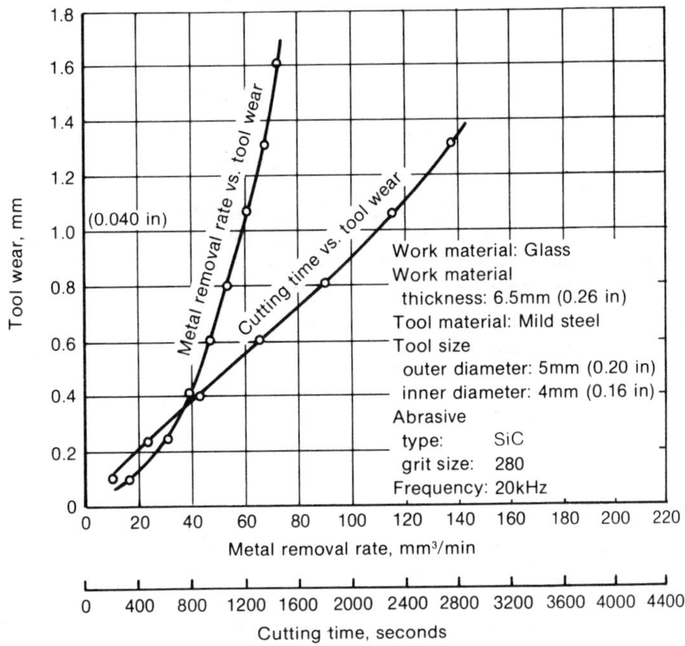


**Figure 10.7–14** Tool wear versus machining rate and cutting time. (M. Adithan and V. C. Venkatesh, p. 627)

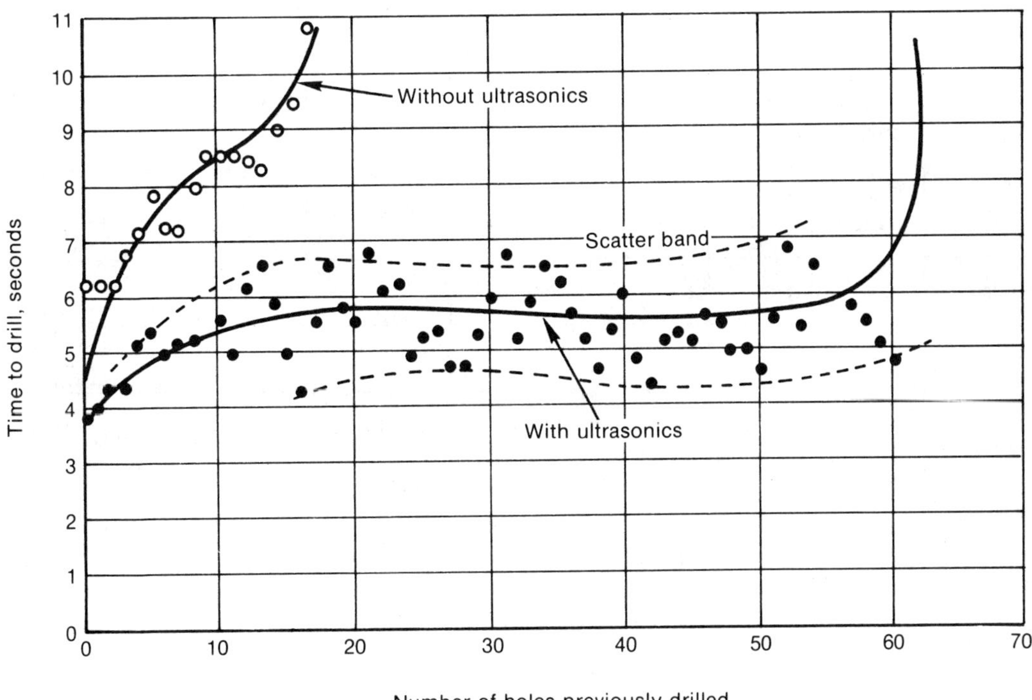


**Figure 10.7–15** Improvements in drilling time and number of holes drilled using rotary ultrasonic machining. Test drilling of 1/16-inch [1.59 mm] Ti-6Al-4V with #30 cobalt twist drill at 1390 rpm and 54 lb [24.5 kg] force in a Branson UMT3. (W. Tyrrell, Rotary ultrasonic machining, Technical paper MR70-516, Society of Manufacturing Engineers, Dearborn, MI, 1970)

# 10.7 Ultrasonic Machining—USM
## and Rotary Ultrasonic Machining—RUM

**TABLE 10.7–20 General Recommendations of Conservative Starting Conditions for USM**

| PARAMETER | WORK MATERIAL | | | | | | | |
|---|---|---|---|---|---|---|---|---|
| | Glass | | Ceramics | | Hard Metals (40-60 $R_c$) | Composites (e.g., glass epoxy) | Tungsten Carbide | Stone |
| | Roughing | Finishing | Large area | Small area | | | | |
| Tool material | mild steel | mild steel | stainless steel | stainless steel | tool steel | mild steel | mild steel | tool steel |
| Abrasive<br>Type<br>Size<br>Percent concentration<br>Carrier fluid | <br>$B_4C$<br>320<br>20<br>$H_2O$ | <br>$B_4C$<br>500<br>60<br>light oil | <br>$B_4C$<br>320<br>20<br>$H_2O$ | <br>$B_4C$<br>180<br>40<br>$H_2O$ | <br>SiC<br>240<br>50<br>$H_2O$ | <br>$B_4C$<br>320<br>40<br>$H_2O$ | <br>$B_4C$<br>240<br>50<br>$H_2O$ | <br>$Al_2O_3$<br>180<br>20<br>$H_2O$ |
| Power<br>Frequency, kHz<br>Watts<br>Amplitude,<br>in<br>[mm] | <br>20<br>700<br><br>0.002<br>[0.051] | <br>20<br>100<br><br>0.001<br>[0.025] | <br>20<br>400<br><br>0.002<br>[0.051] | <br>20<br>200<br><br>0.001<br>[0.025] | <br>20<br>500<br><br>0.0005<br>[0.013] | <br>20<br>300<br><br>0.0015<br>[0.038] | <br>20<br>500<br><br>0.0005<br>[0.013] | <br>10<br>1000<br><br>0.004<br>[0.102] |
| Spindle thrust, lb<br>[kg] | 5<br>[2.27] | 2<br>[0.91] | 2<br>[0.91] | 1<br>[0.45] | 4<br>[1.82] | 2<br>[0.91] | 4<br>[1.82] | 10<br>[4.54] |
| Material removal<br>Rate,<br>in³/min<br>[mm³/min]<br>Penetration,<br>in/min<br>[mm/min]<br>Relative percent* | <br><br>0.030<br>[491]<br><br>0.150<br>[3.81]<br>100 | <br><br>0.015<br>[246]<br><br>0.075<br>[1.90]<br>50 | <br><br>0.004<br>[65.5]<br><br>0.020<br>[0.51]<br>15 | <br><br>0.005<br>[81.9]<br><br>0.025<br>[0.64]<br>20 | <br><br>0.0002<br>[32.7]<br><br>0.001<br>[0.025]<br>6 | <br><br>0.016<br>[262]<br><br>0.080<br>[2.0]<br>50 | <br><br>0.0003<br>[4.9]<br><br>0.0015<br>[0.038]<br>4 | <br><br>0.15<br>[2458]<br><br>0.150<br>[3.81]<br>500 |
| Depth of cut, in<br>[mm] | 0.5<br>[12.7] | 0.5<br>[12.7] | 0.1<br>[2.5] | 0.1<br>[2.5] | 0.25<br>[6.4] | 0.1<br>[2.5] | 0.1<br>[2.5] | 1.0<br>[25.4] |
| Cutting time, minutes | 3.4 | 6.8 | 5.0 | 4.0 | 250.0 | 12.5 | 66.0 | 6.6 |
| Wear ratio (work to tool) | 100 | 200 | 75 | 75 | 75 | 100 | 2 | 150 |
| Tolerance, ±in<br>[±mm] | 0.0010<br>[0.025] | 0.0005<br>[0.013] | 0.0010<br>[0.025] | 0.0010<br>[0.025] | 0.0015<br>[0.038] | 0.0010<br>[0.025] | 0.0005<br>[0.013] | 0.0020<br>[0.051] |
| Surface roughness, $R_a$,<br>μin<br>[μm] | <br>40-60<br>[1-1.5] | <br>20-40<br>[0.5-1] | <br>40-60<br>[1-1.5] | <br>40-60<br>[1-1.5] | <br>10-20<br>[0.25-0.5] | <br>40-60<br>[1-1.5] | <br>10-20<br>[0.25-0.5] | <br>60-80<br>[1.5-2] |

NOTE: Based generally on a tool face area of 0.2 in² [1.29 cm²].
*Based on soda glass as 100%.

## SPECIFIC APPLICATIONS DATA

**TABLE 10.7–21 Specific Applications Data for Ultrasonic Machining**

| Work material: | **Tungsten Carbide** | | **Phenolic Molding** | **Carbon** |
|---|---|---|---|---|
| Workpiece configuration: | 0.50 in<br>Draw die | | 0.020 in<br>0.40 in | 4 in<br>3 in<br>2176 holes, 0.029 x 0.035 in |
| Abrasive type: | $B_4C$ | $B_4C$ | $Al_2O_3$ | SiC |
| grit size: | 320 | 600 | | 400 |
| percent concentration: | | | | 50 |
| Power supply frequency, kHz: | 20 | 20 | 20 | 20 |
| power, W: | | | | 600 |
| amplitude, in: | | | | 0.0015 |
| Spindle thrust, lb: | | | | Controlled feed rate |
| Depth cut, in: | 0.187 | 0.187 | 0.190 | 0.040 |
| Cutting time, min: | 15 | 10 | 2 | 10 |
| Wear ratio (work to tool): | | | | 0 |
| Tolerance, in: | | ±0.0002 | ±0.001 for size;<br>±0.0025 for depth | ±0.0005 |
| Surface roughness, $\mu$in $R_a$: | 22 | 15 | | 16 |
| Remarks: | Roughing cut, trepanned, 7/16-inch ID. Through hole. | Finishing cut, 0.4992-inch OD tool size. Through hole. | Bottom hole. | Very delicate part. |

**TABLE 10.7-21—Continued**

| Work material: | **Silicon Nitrite** | **Glass-Graphite Epoxy Composite** | **Glass** |
|---|---|---|---|
| Workpiece configuration: |  1.0 in OD<br>0.9 in ID<br>Rings — trepanned from block | 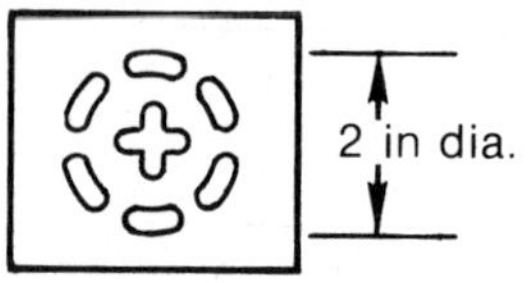 2 in dia.<br>23, slots 0. 025 x 0.060 in to 0.025 x 0.375 in | 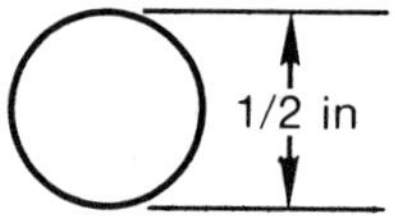 1/2 in |
| Abrasive type: | $B_4C$ | $B_4C$ | $B_4C$ |
| grit size: | 240 | 280 | 200 |
| percent concentration: | 50 | 40 | 35 |
| Power supply frequency, kHz: | 20 | 20 | 25 |
| power, W: | 600 | 300 | |
| amplitude, in: | 0.002 | 0.0011 | 0.0015 |
| Spindle thrust, lb: | 1.5 | 4 | |
| Depth cut, in: | 0.30 | 0.060 | 0.150 |
| Cutting time, min: | 12 | 2 | 1 |
| Wear ratio (work to tool): | 10:1 | 80:1 | 100:1 |
| Tolerance, in: | ±0.001 | ±0.001 | |
| Surface roughness, $\mu$in $R_a$: | 15 | 60 (est.) | |
| Remarks: | Hardest ceramic. | No burrs on through slots. | Low carbon steel tool for engraving. |

**TABLE 10.7-21—Continued**

| Work material: | **Glass** | **Sintered Tungsten Carbide** |
|---|---|---|
| Workpiece configuration: | 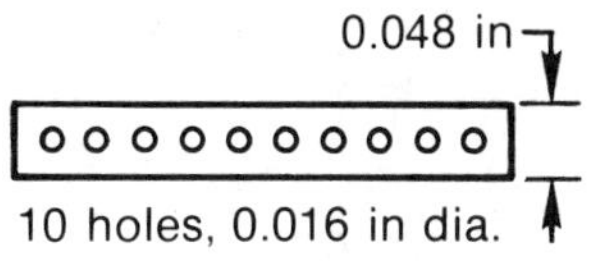 0.048 in<br>10 holes, 0.016 in dia. | 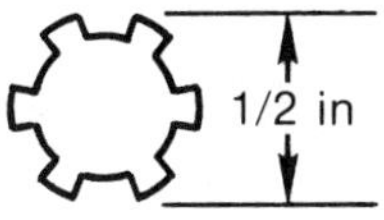 1/2 in<br>Spline trepanning |
| Abrasive type: | $B_4C$ | $B_4C$ |
| grit size: | 600 | 220 |
| percent concentration: | | |
| Power supply frequency, kHz: | 20 | 20 |
| power, W: | | |
| amplitude, in: | | 0.0025 |
| Spindle thrust, lb: | 0.33 | 3.5 |
| Depth cut, in: | 0.47 | 0.25 |
| Cutting time, min: | 0.0125 in/min | 42 |
| Wear ratio (work to tool): | 9:1 | |
| Tolerance, in: | | |
| Surface roughness, $\mu$in $R_a$: | | |
| Remarks: | 0.014-inch OD, 0.007-inch ID stainless steel tube used for tool. | Mild steel tool used. |

## PROCESS SUMMARY

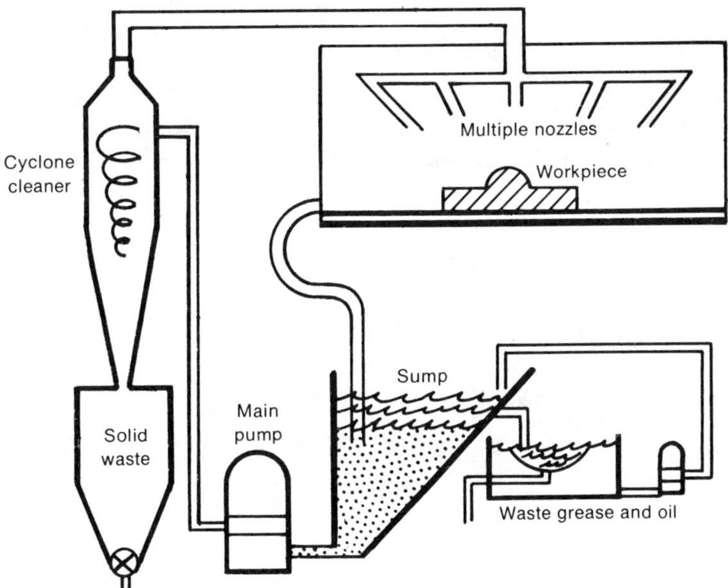


**Figure 10.8–1** WJM schematic.

### Principles

Water jet machining (WJM) is low pressure hydrodynamic machining. The pressure range for WJM is an order of magnitude below that used in HDM. There are two versions of WJM: one for mining, tunneling and large pipe cleaning that operates in the region from 250 to 1000 psi [1.7 to 6.9 MPa]; and one, for smaller parts and production shop situations that uses pressures below 250 psi [1.7 MPa]. The first version, or high-pressure range, is characterized by use of a pumped water supply with hoses and nozzles that generally are hand-directed. In the second version, more production-oriented and controlled equipment, such as that shown in figure 10.8–1, is involved. In some instances, abrasives are added to the fluid flow to promote rapid cutting. Single- or multiple-nozzle approachs to the workpiece depend on the size and number of parts per load. The principle is that WJM is high volume—not high pressure.

### Practical Applications

Production applications for WJM include descaling, deburring, surface cleaning, decorative finishing, nuclear decontamination, food utensil cleaning, internal deburring and degreasing. Some polishing activities and preparations for precise inspection also have used WJM. Pipe cleaning and casting cleaning are good applications.

### Operating Parameters

**TABLE 10.8–1 Typical Values for WJM Operating Parameters**

| | |
|---|---|
| Pressure: | 30 to 250 psi [0.2 to 1.7 MPa] production<br>250 to 1000 psi [1.7 to 6.9 MPa] civil engineering |
| Fluid: | Water<br>Water plus grit additives |

## 10.8 Water Jet Machining—WJM

### Material Removal Rates and Tolerances

Low pressure WJM is not a mass material removal process, it is a finishing, cleaning and deburring process. The higher pressure uses in civil engineering applications are outside the scope of a machinability handbook.

### Surface Technology

A polished, scoured clean surface is typical. When abrasives are used, a light peening texture may occur. Motion in the workpiece or the nozzle is desirable because dwelling on one spot can result in grooving. Oiling of iron bearing alloys is necessary to reduce rust formation after cleaning. Shake-out or cleaning of crevices in the workpiece is necessary when grit loaded fluids are used.

### Equipment and Tooling

Production equipment is available which offers adjustable nozzles in a splash chamber, pumps, filters and associated items in a single package. Portable equipment with flexible nozzles and lances also is commercially available.

### Machining Characteristics

Suitable for cleaning large areas.

SECTION **11**

# ELECTRICAL NONTRADITIONAL MACHINING OPERATIONS


11.1 Electrochemical Deburring—ECD ..... 11-3
11.2 Electrochemical Discharge Grinding—ECDG ..... 11-5
11.3 Electrochemical Grinding—ECG ..... 11-9
11.4 Electrochemical Honing—ECH ..... 11-23
11.5 Electrochemical Machining—ECM ..... 11-25
11.6 Electrochemical Polishing—ECP ..... 11-63
11.7 Electrochemical Sharpening—ECS ..... 11-65
11.8 Electrochemical Turning—ECT ..... 11-67
11.9 Electro-stream™—ES ..... 11-69
11.10 Shaped Tube Electrolytic Machining—STEM™ ..... 11-71

## PROCESS SUMMARY

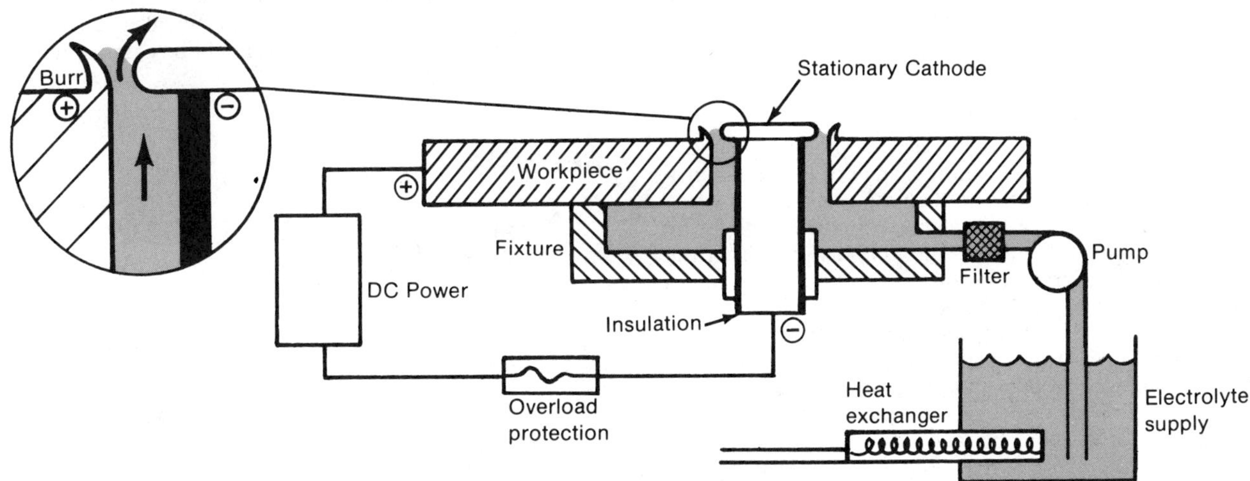


**Figure 11.1-1** ECD schematic.

### Principles

Electrochemical deburring (ECD) is a special version of ECM (see figure 11.1-1). ECD was developed to remove burrs and fins or to round sharp corners. Anodic dissolution occurs on the workpiece burrs in the presence of a closely placed cathodic tool whose configuration matches the burred edge. Normally, only a small portion of the cathode is electrically exposed so a maximum concentration of the electrolytic action is attained. The electrolyte flow usually is arranged to carry away any burrs that may break loose from the workpiece during the cycle. Voltages are low, current densities are high, electrolyte flow rate is modest, and electrolyte types are similar to those used for ECM. The electrode (tool) is stationary, so equipment is simpler than that used for ECM. Cycle time is short for deburring. Longer cycle time produces a natural radiusing action.

### Practical Applications

Almost any conductive material can be deburred or polished electrolytically. Most ECD is performed in seconds and is many times faster than hand deburring. The cycle time is limited by the load and unload parts of the operation. Drilling-burr removal is routine and involves only simple electrodes. Interior burrs, intersecting holes and internal fins can be deburred using precisely located electrodes. Applications include automotive connecting rods, gear teeth, blanking dies, valve parts, punch press blankings and machined pipe fittings. One 20-second ECD operation on automotive pistons simultaneously deburrs two oil ring slots, six piston wall oil holes, two piston pin oil holes and all weight bosses. The absence of mechanical contact or residual stress effects permits the use of ECD on thin parts or sections without fear of distortion or damage. Shallow reliefs (up to 0.030-inch [0.76 mm] deep) can also be produced.

### Operating Parameters

**TABLE 11.1-1 Typical Values for ECD Operating Parameters**

|  |  |
|---|---|
| Power supply | |
| type: | Direct current |
| voltage: | 4 to 25 V |
| current: | 50 to 500 A (approximately 10 to 15 A per linear inch of edge) |
| Gap: | 0.005 to 0.050 inch (generally 0.030 inch) [0.13 to 1.27mm—generally 0.76 mm] |
| Electrolyte | |
| types: | |
| frequently used: | 1 to 2 lb/gal [120 to 240 g/L] NaCl |
| most used: | 2 to 3 lb/gal [240 to 360 g/L] $NaNO_3$ |
| rarely used: | 2 to 4 lb/gal [240 to 480 g/L] $NaNO_2$ |
| rarely used: | 2 to 4 lb/gal [240 to 480 g/L] KCl |
| temperature: | 90° to 110°F [32° to 42°C] |
| pressure: | 15 to 50 psi [100 to 350 kPa] |
| flow: | 1 to 4 gal/min [4 to 15 L/min] per 100 A |
| filters: | 10 to 75 $\mu$m (microns) |
| Electrode material: | Brass or copper |

### Material Removal Rates and Tolerances

Most deburring occurs in 5 to 30 seconds, as shown in figures 11.1-2 and 11.1-3. The rate depends on work material, voltage, electrolyte, flow rate and tool design. Longer "on" times can produce a rounded corner in addition to deburring. Tolerances are in the range of ±0.0007 to ±0.002

inch [±0.018 to ±0.050 mm]. The larger tolerances are associated with lower current densities and large operating gaps; however, the principal factor controlling accuracy is the "on" time. Automatic controls are needed to hold the best values.

### Surface Technology

The highly focused electrolytic action produces smooth radii and surface roughness in the range of 8 to 16 microinches $R_a$ [0.2 to 0.4 μm]. With high current densities, surface roughness can be 2 microinches $R_a$ [0.05 μm]. Stray etching will occur at some considerable distance from the surface being deburred. Shallow striations (less than 0.0001 inch [0.0025 mm]) can occur where turbulent flow, gas bubbles or insulation discontinuities are present. Some alloys with large, distinct grains can exhibit selective etching on a microscale, particularly in low current density areas adjacent to the principal focal point of the electrode. ECD is not stress-inducing and can remove some previously strained surface layers. Alloys with nonconductive grains, for example, silicon grains in aluminums alloys, can exhibit a pock-marked surface where these grains have fallen out.

### Equipment and Tools

ECD equipment is smaller and less complex than that for ECM because feed motion is not required. Plastic tables or shallow tanks are arranged to accept the fixtures and to allow quick connections for electricity and electrolytes. Stainless steel fixtures and fittings are used to resist the corrosiveness of the electrolytes. Tooling and electrodes are tailored to conform to the configuration of the workpiece and the edges to be deburred. By properly insulating the cathode, effects on other exposed areas of the workpiece are reduced to negligible amounts. Power supplies range from 100 to 500 amperes, with multiples available if several parts are to be run concurrently. Automatic controls are desirable to secure repeatable results, particularly with the very short machining cycles. Multiple workpiece fixtures and automation of the ECD cycle are frequent. The cycle begins with closure of the lid or splash guard and placement of the electrode in correct relation to the workpiece. The cycle continues automatically from electrolyte-on, to current-on, through the cutting time to current-off, flushing, and opening of the lid. Hand-held tools for ECD also are available (see discussion of ECS, section 11.7). Electrolyte supplies, pumps and filters are sized to handle the flow rates. Flow rate depends on component size and fixtures, but usually ranges up to only a few hundred gallons because cycle times are so short.

ECD equipment for the 20-second operation on automotive pistons cited under "Practical Applications" was integrated into a transfer line that included cleaning, rinsing and oiling.

### Machining Characteristics

It is essential that burr size and location be consistent. Burrs of excessive size or random location can cause shorts in the electrical circuits unless they are knocked off before the workpiece is placed in the close gaps with the electrode. Uniform results are typical with the fixed electrode positions. Variations in material properties will sometimes be revealed by the more uniform ECD action. The metal hydroxides in the used electrolyte need to be filtered out before the electrolyte is reused. This action minimizes metal hydroxide deposition on the workpiece or tools and promotes more uniform material removal rates. Disposal of sludge from the used electrolyte should be carried out in an environmentally sound manner.

## SELECTED DATA

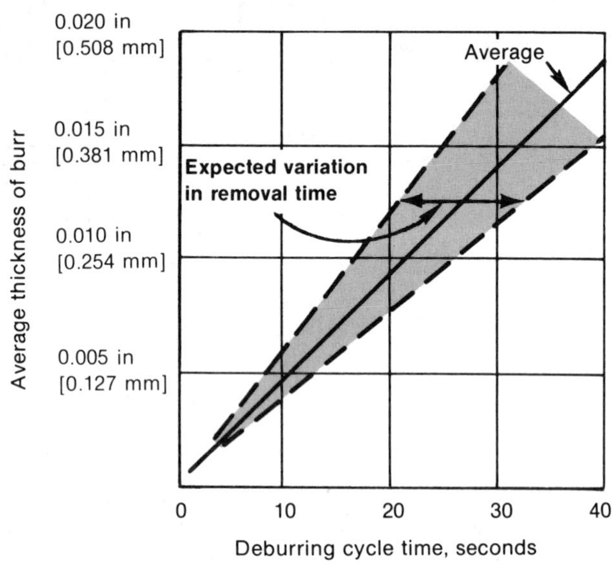


**Figure 11.1-2** Burr removal rates for aluminum alloys. Thickness of burr is measured at root of burr. (D. W. Sickels, Electrochemical deburring—some capabilities and limitations, Technical paper MR76-133, Society of Manufacturing Engineers, Dearborn, MI, 1976, p. 10)

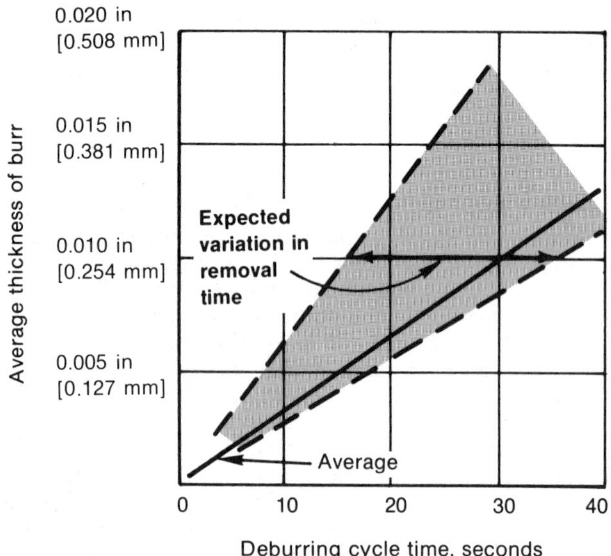


**Figure 11.1-3** Burr removal rates for iron base alloys. Thickness of burr is measured at root of burr. (D. W. Sickels, p. 9)

## PROCESS SUMMARY

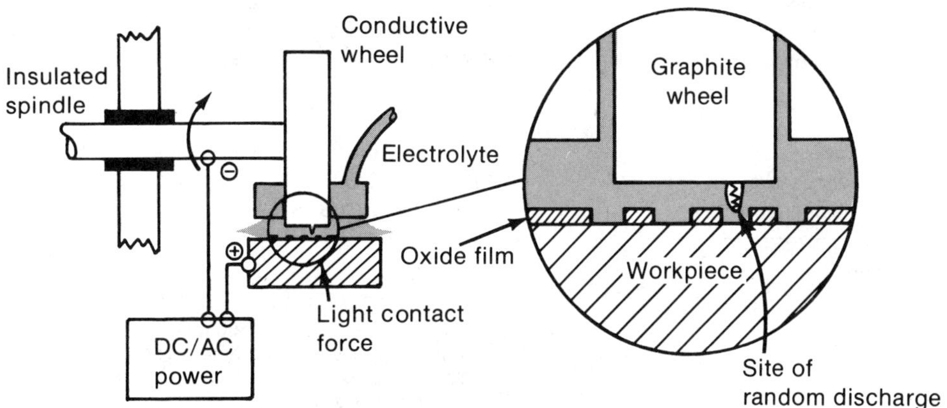


**Figure 11.2-1** ECDG schematic.

### Principles

Electrochemical discharge grinding (ECDG) combines the features of both electrochemical and electrical discharge methods of material removal (see figure 11.2-1). ECDG has the arrangement and electrolytes of electrochemical grinding (ECG), but uses a graphite wheel without abrasive grains. The random spark discharge is generated through the insulating oxide film on the workpiece by the power generated in an alternating-current source or by a pulsating direct-current source. The principal material removal comes from the electrolytic action of the low level direct-current voltages. The spark discharges erode the anodic films to allow the electrolytic action to continue. At increased operating levels, electrical discharge contributes to the material removal, as demonstrated in figure 11.2-2. There is no mechanical contact between the wheel and the workpiece; however, there are separating forces developed by the compression of the electrolyte in the space between the wheel and the workpiece. Careful electrolyte control is required to assure its placement between the workpiece and the wheel and to prevent boiling and possible severe arcing. Sometimes the process is called ECDM, electrochemical discharge machining.

### Practical Applications

The workpiece must be an electrically conductive material. As with ECM, the hardness of the workpiece does not affect the material removal rates, and almost any conductive material can be cut. Single-point carbide disposable-insert grinding or resharpening is done by ECDG. Plunge, face and surface grinding are all performed. Thin or delicate profiles can be form ground. The electrolytic action produces a burr-free surface. Circular forms that need to be stress free are good applications, as is the machining of honeycomb materials. The low cost wheel and the ease with which it can be formed for intricate profile grinding are distinct advantages. Careful economic comparisons between ECDG and ECG should be made before completing an application study.

# 11.2 Electrochemical Discharge Grinding—ECDG

## Operating Parameters

**TABLE 11.2–1 Typical Values for ECDG Operating Parameters**

| | | |
|---|---|---|
| Power supply | | |
| type: | Pulsating dc | ac |
| frequency: | 120 Hz | 60 Hz |
| voltage: | 4 to 12 V (8 optimum) | 8 to 12 V |
| current: | 200 to 1000 A | 200 to 500 A |
| Electrolyte | | |
| type: | NaCl, $NaNO_3$, proprietary neutral salts | |
| concentration: | 1½ to 2 lb/gal [180 to 240 g/L] | |
| temperature (inlet): | 80° to 100°F [27° to 38°C] | |
| Wheel | | |
| type: | Graphite, typically 300 mesh | |
| speed: | 4000 to 6000 fpm [1200 to 1800 m/min] (Lower speeds do not promote good electrolyte flow and permit gas bubbles to become too large, thus inhibiting maximum current densities—see figure 11.2–3.) | |
| Operating gap: | 0.0005 to 0.0015 inch [0.013 to 0.038 mm] (figure 11.2–4) | |
| Pressure between wheel and workpiece: | 5 to 20 psi [35 to 140 kPa] | |
| Current density for | | |
| carbides: | 600 A/in$^2$ [1 A/mm$^2$] maximum | |
| other metals: | 800 A/in$^2$ [1.3 A/mm$^2$] maximum | |
| steel, using pulsed dc: | 1200 A/in$^2$ [2.0 A/mm$^2$] maximum (Excessive current density can overheat workpiece and/or wheel.) | |
| Feed rates* | | |
| plunge grinding, carbide: | 0.020 in/min [0.5 mm/min] | |
| plunge grinding, steel: | 0.060 in/min [1.5 mm/min] | |
| surface grinding, carbide: | 0.15 in/min [3.8 mm/min] | |
| surface grinding, steel: | 0.50 in/min [12.7 mm/min] | |
| Wheel wear ratios | | |
| ac power: | 7:1 (carbide or steel) | |
| dc power: | 40:1 (steel) | |

*For 575 A/in$^2$ [0.9 A/mm$^2$] on an 8-inch [200 mm] diameter wheel.
Surface grinding at 0.10 inch [2.5 mm] depth of cut.

### Material Removal Rates and Tolerances

ECDG will remove about one cubic inch [16 $cm^3$] of tool steel per hour—about 5 times faster than EDG while using 10 to 15 times as much current. Dimensional accuracy is less than that from EDG and is typically ±0.0005 inch [±0.013 mm] with close control and ±0.001 inch [±0.025 mm] for routine production. While a typical depth of cut is one-half inch [12.5 mm] when machining at 200 amperes, the cubic inches per minute metal removal rate is 0.006 [0.1 $cm^3$/min] for carbide workpieces and 0.015 [0.25 $cm^3$/min] for steel workpieces. Single-point carbide disposable inserts have been ground to a depth of 0.010 inch [0.25 mm] at the rate of one insert per minute. Carbide thread chasers have been ground with a mean thread height deviation of ±0.0003 inch [±0.008 mm] and pitch deviation of ±0.00015 inch [±0.004 mm]. The accuracy is primarily a function of the degree to which the electrolyte film is uniform. Special scrapers are needed to control the electrolyte flow and to prevent air films.

### Surface Technology

Surface roughness values of 5 to 15 microinches $R_a$ [0.13 to 0.38 μm] for carbide workpieces and 15 to 30 microinches $R_a$ [0.38 to 0.76 μm] for steel workpieces can be expected. The surfaces are sometimes pitted with minute craters from the electrical discharge if the voltages used are too high. (Too-high voltage also deteriorates the wear ratio). The electrolytic dissolution promotes stress-free material removal action and burr-free edges.

### Equipment and Tools

Commercial equipment in the range from 20 to 1000 amperes is available on special engineering order. Tooling involves the electrolyte flow control so as to obtain a uniformly thick film between the workpiece and the wheel. Conforming graphite or plastic film scrapers are essential to tolerance and repeatability control. A sufficient flood of filtered electrolyte is essential to high material removal rates and to the prevention of excessive temperatures, gas bubbles or boiling in the working gap. Fixtures should be made of stainless steel, plastic, or materials compatible with the corrosiveness of the electrolyte.

### Machining Characteristics

Wheel-to-workpiece wear ratios are dependent upon the type of power supply, the voltage, the grade of graphite and the uniformity of electrolyte film. Figure 11.2-5 demonstrates the relationship between wheel wear and voltage.

## SELECTED DATA

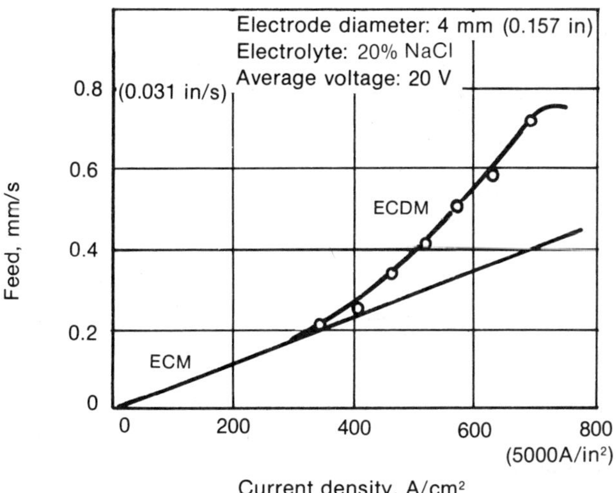


**Figure 11.2-2** Feed rate versus current density in drilling steel plate with ECDM. (M. Kubota, Metal removal in ECDM, *ISEM-5 Proceedings, International Symposium for Electromachining*, Zurich, Switzerland, 1977, p. 218)

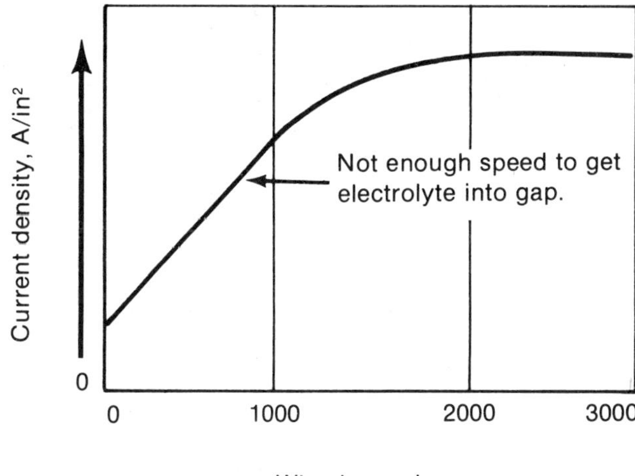


**Figure 11.2-3** Relationship between current density and wheel speed. (W. G. Voorhees, Electrochemical discharge machining, Technical paper MR67-165, Society of Manufacturing Engineers, Dearborn, MI, 1967, p. 7)

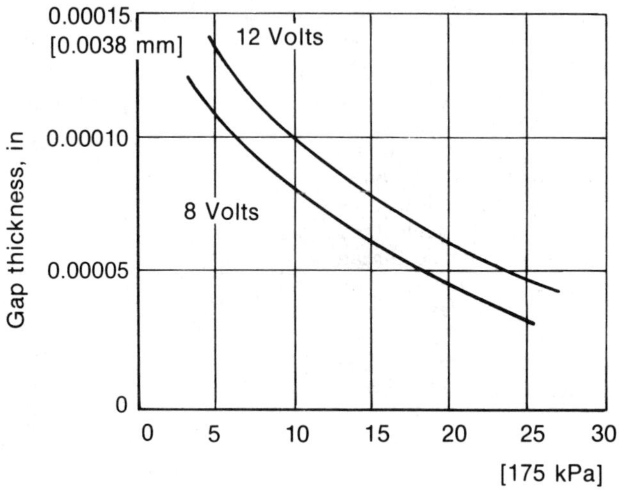


**Figure 11.2-4** Relationship between gap thickness, workpiece pressure and gap voltage. (W. G. Voorhees, p. 7)

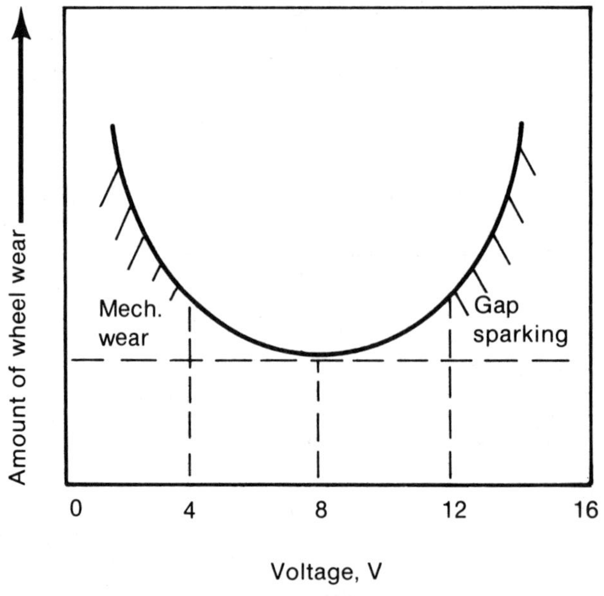


**Figure 11.2-5** Effect of gap voltage on wheel wear. (W. G. Voorhees, p. 7)

## PROCESS SUMMARY

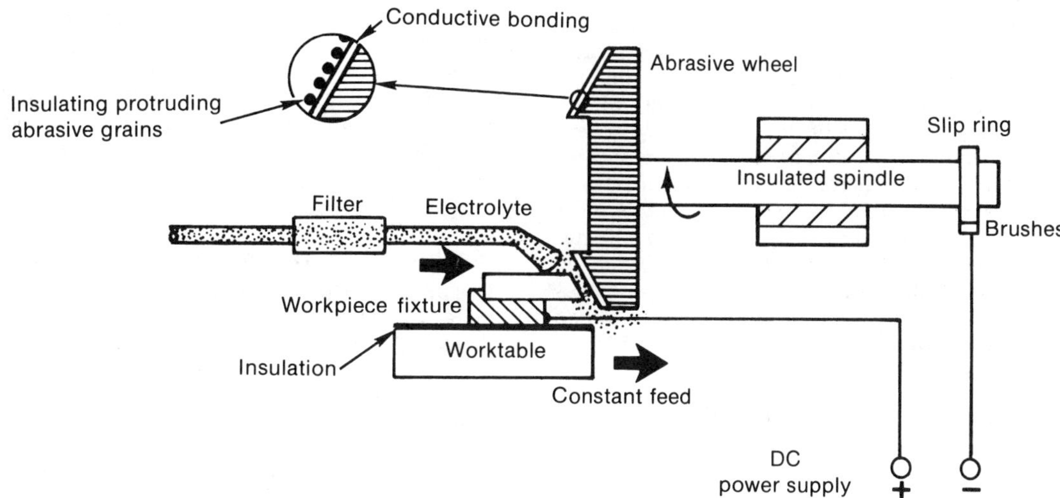


**Figure 11.3–1** ECG schematic.

### Principles

Electrochemical grinding (ECG) is a special form of electrochemical machining in which the conductive workpiece material is dissolved by anodic action, and any resulting films are removed by a rotating, conductive, abrasive wheel (see figure 11.3–1). The abrasive grains protruding from the wheel form the insulating electrical gap between the wheel and the workpiece. This gap must be filled with the electrolyte at all times. The conductive wheel uses conventional abrasives—aluminum oxide (because it is nonconductive) or diamond (for intricate shapes)—but lasts substantially longer than wheels used in conventional grinding. The reason for this is that the bulk of the material removal (95 to 98 percent) occurs by deplating, while only a small amount (2 to 5 percent) occurs by abrasive mechanical action. Maximum wheel contact arc lengths are about 3/4 to 1 inch [19 to 25 mm] to prevent overheating the electrolyte. The fastest material removal is obtained by using the highest attainable current densities without boiling the electrolyte. The corrosive salts used as electrolytes should be filtered, and flow rate should be controlled for the best process control.

### Practical Applications

Face, plunge, cone or contour grinding are usual ECG applications on high-strength, conductive materials, regardless of hardness. Almost any conductive material can be ground electrochemically. Low wheel wear, infrequent dressing, minimal forces on the workpiece and automatic deburring are characteristic. Sharpening, shaping or resharpening tungsten carbide tool bits, contouring fragile honeycomb parts, and cutting surgical needles and thin-wall tubing are practical. Grinding hard, tough materials without heat checks or burrs is frequently done. Inner corner radii are limited to 0.010 to 0.015 inch [0.25 to 0.38 mm]. Precision and fine finishes are combined with more rapid cutting, freedom from burrs and absence of residual stress or distortion. ECG is performed with 1/7 or less of the grinding wheel wear of conventional grinding. It also compares favorably with conventional milling and grinding (see table 11.3–5).

### Operating Parameters

**TABLE 11.3–1 Typical Values for ECG Operating Parameters**

| | |
|---|---|
| Power supply | |
| type: | Direct current |
| voltage: | 4 to 14 V |
| current: | 50 to 3,000 A |
| Gap: | 0.001 inch [0.025 mm] (about equal to protrusion of abrasive grains) |
| Electrolyte | |
| types: | |
| frequently used: | $NaNO_3$, 1 to 2 lb/gal [120 to 240 g/L] |
| rarely used: | $KNO_3$, 1½ to 2 lb/gal [180 to 240 g/L] |
| rarely used: | NaCl, ½ to 1½ lb/gal [60 to 180 g/L] |
| temperature: | 90° to 110°F [26° to 42°C] |
| pressure: | 2 to 10 psi [14 to 70 kPa] |
| filters: | 50 to 100 micrometers |
| flow control: | Close-fitting nozzle with air scraper |
| Abrasive wheel | |
| type: | Conductive |
| abrasive: | $Al_2O_3$ or diamond grit |
| speed: | 4,000 to 7,000 fpm [1,200 to 2,100 m/min] |
| Feed rate and depth cut: | Figure 11.3–3 |
| Length of contact: | Up to 0.75 inch [19 mm]. Limit length to prevent boiling in electrolyte. See also figure 11.3–3. |

# 11.3 Electrochemical Grinding—ECG

## Material Removal Rates and Tolerances

The material removal rate is proportional to the current density. The current density is limited by the anodic dissolution rate for a particular alloy-electrolyte system and the boiling point of the electrolyte. A convenient value for most common materials is 0.010 cubic inch per minute [0.16 $cm^3$/min] per 100 amperes of current. A plunge grinding feed rate of 0.050 inch per minute [1.27 mm/min] is attained at 500 amperes per square inch [77.5 A/$cm^2$], while surface grinding operates at 1.0 inch per minute [25.4 mm/min] with a 0.010 inch [0.25 mm] depth of cut at 1000 amperes per square inch [155 A/$cm^2$]. Size control is typically $\pm$0.001 inch [$\pm$0.025 mm] and is attainable in a one-pass cut. Accuracies to $\pm$0.0002 inch [$\pm$0.005 mm] can be achieved with the final pass without electrolytic action. Outside corners usually have a radius about 0.001 inch [0.025 mm]. Inside corners have a limit of 0.010- to 0.015-inch [0.25 to 0.38 mm] radius. Removal rates are 5 to 10 times those for conventional grinding on hard materials (Rockwell C45 and up).

## Surface Technology

Surface finishes improve with increasing current densities and higher feed rates. Roughness values of 8 to 32 microinches $R_a$ [0.2 to 0.8 $\mu$m] are common, with 1 to 4 microinches $R_a$ [0.025 to 0.1 $\mu$m] attainable. The surface texture is similar to that obtained with a metallographic polish. Workpiece hardness is not a factor; however, the material type can affect the range of roughness values. The surface is free from process-induced residual stress and there is no heat-affected zone. Both of these effects contribute to the production of workpieces with high surface integrity.

## Equipment and Tools

Grinding machine construction must be rigid enough to maintain precision under the deflecting forces which can reach 150 psi [1 MPa] between the workpiece and the wheel. The spindle must be insulated and capable of conducting the low-voltage, high-current d-c power to the wheel. Associated equipment is required to store, pressurize, filter and withstand the corrosive electrolytes. Controls, fixtures and mechanical and electrical systems must be made of suitable material or coated so as to be capable of operating in the salt mist environment. Conversion kits to convert mechanical grinders are available. Wheels must be conductive and must use an insulating grain. Electrolyte nozzles should be close fitting and should be made of Plexiglas or equivalent dielectric material (see table 11.3–4). Workpiece fixtures should be made of copper, copper-base alloys or 300-series stainless steel and should be constructed such that all cathodic components remain electrically insulated from anodic components during the ECG operation. The wheels must have the correct abrasive qualities of the insulating grit (silicon carbide and some forms of Borazon cannot be used because they are electrically conductive.)

## Machining Characteristics and Precautions

It is necessary to "condition" any ECG wheel to remove the metal matrix and to allow the nonconductive abrasive particles to protrude (see table 11.3–4). To accommodate the usually longer runs between wheel dressings, truing of the wheel should be more precise than that for conventional grinding—0.0002 to 0.0005 inch TIR [0.005 to 0.013 mm].

Handling and storage of the electrolytes should be done with care and strict attention to the cautions pertaining to the particular chemicals (see table 11.5–6, ECM). A moderate fire hazard exists when dry $NaNO_3$ is mixed with organic materials. With their heavy metal and salt content, the disposal of the used electrolytes needs careful, environmentally sound planning.

Areas adjacent to the current gap are exposed to the conductive spray and may be slightly eroded or pitted by the low current densities present there. Appropriate insulation in the tooling is required to avoid bimetal electrolytic corrosion. NaCl electrolytes are rarely used in ECG because they have strong corrosive effects on machine components. Citrates, tartrates or proprietary chemicals frequently are added to the electrolytes and can aid machining while reducing corrosiveness. A thorough washdown of the equipment and tools at the end of each day or each period of use is a good preventive maintenance practice. Workpieces should also be well rinsed after ECG, and some materials should be checked for possible adverse metallurgical reaction to the electrolytes.

Spindle current for the a-c drive motor is a good indication of the amount of mechanical grinding during ECG. An increase in current to 125 percent of the no-load value for roughing ECG or to 115 percent of the no-load value for finishing ECG is a reasonable upper limit for the amount of grinding. The d-c operating current should be selected such that resistive overheating does not occur in the workpiece.

## SELECTED DATA

There are several valid techniques for selecting operating parameters for ECG. The steps in table 11.3-2 are recommended as one logical sequence. Supporting data come from the other figures and tables in accordance with the terms as explained in figure 11.3-2.

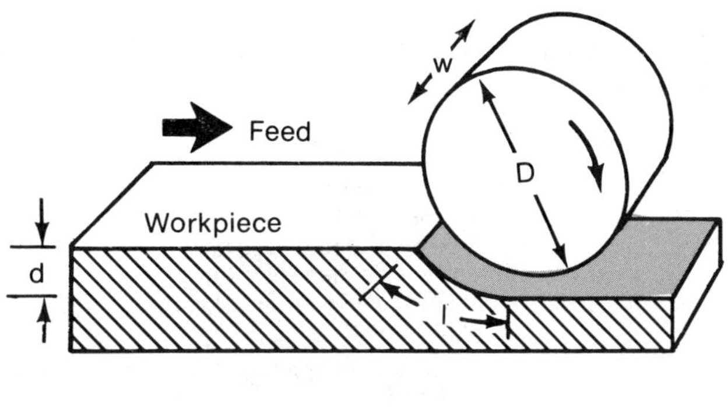


SURFACE ECG

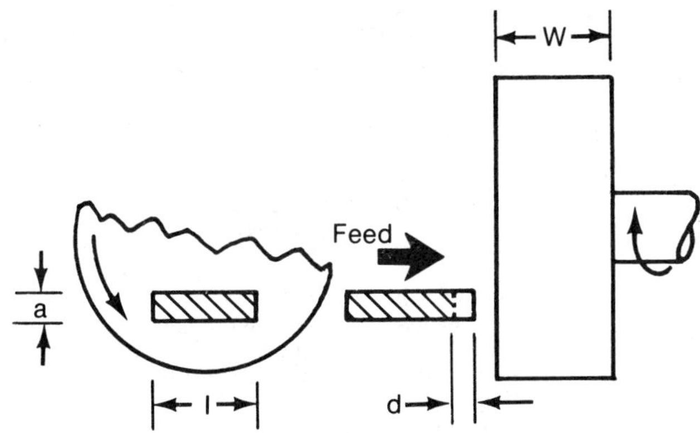


FACE PLUNGE ECG

**Figure 11.3-2** Parameter terminology for surface and face plunge ECG.

**TABLE 11.3-2 Steps To Select ECG Operating Parameters**

| ITEM | SURFACE ECG | FACE PLUNGE ECG |
|---|---|---|
| 1. Select wheel type to suit work material | Table 11.3-3 | Table 11.3-3 |
| 2. Select size of wheel | Largest possible diameter for particular part shape | Largest possible diameter for particular part shape |
| 3. Select wheel speed | 4,000 to 7,000 fpm [1,200 to 2,100 m/min]. Too low, electrolyte not inserted as well; too high, no extra benefits. | 4,000 to 7,000 fpm [1,200 to 2,100 m/min]. Too low, electrolyte not inserted as well; too high, no extra benefits. |
| 4. Select electrolyte, mix | Table 11.3-3 | Table 11.3-3 |
| 5. Select electrolyte specific gravity test value and operating temperature | Figures 11.5-10, 11.5-11, 11.5-12 in ECM section | Figures 11.5-10, 11.5-11, 11.5-12 in ECM section |
| 6. Select current density for particular material (A/in$^2$) [A/cm$^2$] | Table 11.3-3 (or from trial or laboratory testing) (or work from generalized material removal rate—see item 13). The current density in the workpiece must be low enough to prevent overheating. | Table 11.3-3 (or from trial or laboratory testing) (or work from generalized material removal rate—see item 13). The current density in the workpiece must be low enough to prevent overheating. |
| 7. Determine depth of cut (d) | Maximum to suit part configuration and wheel diameter | Not applicable |
| 8. Determine contact length (l) (maximum ¾ to 1 inch [19 to 25 mm] electrolyte flow path) | Figure 11.3-3 | Figure 11.3-2. |
| 9. Calculate contact area | l times w (If slot and side cutting also, then area on side of wheels must be included—see figure 11.3-8.) | l times a |
| 10. Calculate current required | Current density times contact area | Current density times contact area |
| 11. Select voltage setting (4 to 12 V dc approx.) | Set to attain required current during a trial cut after full contact made. Do not increase to point any sparking occurs. Final setting to be ½ volt less than value that gives sparking with the correct value of spindle current, item 12. (See also figures 11.3-6 and 11.3-7.) | Set to attain required current during a trial cut after full contact made. Do not increase to point any sparking occurs. Final setting to be ½ volt less than value that gives sparking with the correct value of spindle current, item 12. (See also figures 11.3-4 and 11.3-5.) |
| 12. Check spindle current | Full contact spindle current should be only a little larger than the no-load current so as to keep the mechanical grinding less than 5% of the total metal removal rate. A value of 115% for finishing ECG or 125% for roughing ECG is a practical limit. | Full contact spindle current should be only a little larger than the no-load current so as to keep the mechanical grinding less than 5% of the total metal removal rate. A value of 115% for finishing ECG or 125% for roughing ECG is a practical limit. |
| 13. Material removal rate (MRR) | Use generalized value of 0.010 in$^3$/min/100 A [164 mm$^3$/min/100 A] as a starting point or refine as in ECM to specific alloy dissolution rate and multiply by current. | Use generalized value of 0.010 in$^3$/min/100 A [164 mm$^3$/min/100 A] as a starting point or refine as in ECM to specific alloy dissolution rate and multiply by current. |
| 14. Volume metal to be removed, in$^3$ [mm$^3$] | d times w times length cut | l times a times d |
| 15. Time for cut or per pass | Volume divided by MRR | Volume divided by MRR |
| 16. Feed rate | Figure 11.3-3 | Current density times MRR |

**TABLE 11.3–3 Recommended Parameters for ECG of Various Materials**

| WORK MATERIAL | WHEEL TYPE | ELECTROLYTE* | | | MAXIMUM CURRENT DENSITY† | |
|---|---|---|---|---|---|---|
| | | Base Chemical | lb/gal $H_2O$ | g/L $H_2O$ | A/in² | A/cm² |
| Straight tungsten carbide grades | Diamond | $KNO_3$ | 1.5-1.7 | 180-200 | 500 | 78 |
| Tantalum or titanium carbide grades | Diamond | $KNO_3$ | 1.5-1.7 | 180-200 | 500 | 78 |
| High speed steel | Diamond | $NaNO_3$ | 1.0-1.5 | 120-180 | 500 | 78 |
| Tungsten | Diamond | KOH or NaOH | 1.0-1.5 | 120-180 | 500 | 78 |
| Low carbon steel | Aluminum oxide | $KNO_3$ (90%) + $KNO_2$ (10%) | 0.5-1.0 | 60-120 | 1000 | 155 |
| High carbon steel | Aluminum oxide | $NaNO_3$ | 1.0-1.5 | 120-180 | 1000 | 155 |
| Stainless steel | Aluminum oxide | $NaNO_3$ | 1.5-1.7 | 180-200 | 500 | 78 |
| Silicon iron | Aluminum oxide | NaCl | 1.0-1.5 | 120-180 | 500 | 78 |
| Copper alloys | Aluminum oxide | $NaNO_3$ or $KNO_3$ | 1.5-1.7 | 180-200 | 1500 | 233 |
| Aluminum alloys | Aluminum oxide | $NaNO_3$ | 1.0-1.2 | 120-140 | 1500 | 233 |
| Titanium alloys | Aluminum oxide | $NaNO_3$ | 1.0-1.2 | 120-140 | 1000 | 155 |
| A-286 | Aluminum oxide | $NaNO_3$ | 1.0-1.2 | 120-140 | 750 | 116 |
| Hastelloy alloy X | Aluminum oxide | $NaNO_3$ | 1.0-1.2 | 120-140 | 750 | 116 |
| M252 | Aluminum oxide | $NaNO_3$ | 1.0-1.2 | 120-140 | 750 | 116 |
| Udimet 500, 700 | Aluminum oxide | $NaNO_3$ or NaCl* | 0.9-1.0 | 110-120 | 750 | 116 |
| Waspaloy | Aluminum oxide | $NaNO_3$ | 1.0-1.2 | 120-140 | 750 | 116 |
| Inconel | Aluminum oxide | $NaNO_3$ | 1.0-1.2 | 120-140 | 750 | 116 |
| Rene 41 | Aluminum oxide | $NaNO_3$ | 1.8-2.0 | 180-230 | 500 | 78 |
| Rene 80 | Aluminum oxide | $NaNO_3$ | 1.5-1.9 | 120-140 | 500 | 78 |
| Nickel alloys | Aluminum oxide | $NaNO_3$ | 1.0-1.2 | 120-140 | 750 | 116 |
| Cobalt alloys | Aluminum oxide | $NaNO_3$ | 1.0-1.2 | 60- 80 | 500 | 78 |
| HS-31 (X-40) | Aluminum oxide | $NaNO_3$ + NaCl* | 0.5-0.7 | 60- 80 | 500 | 78 |
| Zirconium alloys | Aluminum oxide | $NaNO_3$ | 1.0-1.5 | 120-180 | 750 | 116 |
| Stellite | Aluminum oxide | $NaNO_3$ | 1.8-2.0 | 210-240 | 500 | 78 |

*There are many proprietary mixtures or additives to the base chemicals that have desirable effects on specific alloys and restrain the corrosion of equipment. These are recommended for initial trials. Note that NaCl is especially corrosive.

†Density may be limited by possibility of overheating workpiece.

**TABLE 11.3-4 Recommended Procedures for Conditioning the Wheel, Maintaining the Electrolyte, and Fitting the Nozzle in ECG Operations**

ECG wheels require "conditioning" after dressing to remove copper binder and allow nonconductive particles of aluminum oxide or diamond to protrude. Attention to electrolyte condition and flow nozzle on a regular basis is also recommended.

Wheel mounting procedure recommendations:

1. After removing worn-out wheel, clean flanges and spindle with warm, clean water and dry thoroughly. Check for any burrs on bearing surfaces of flanges.
2. Clean bearing surfaces and holes of flanges with an abrasive fabric; then blow clean with air to remove any abrasive particles. This is important for proper electrical conductivity between flanges and wheel.
3. Mount wheel. At this point, only hand-tighten flange nut.
4. Indicate wheel face to within 0.001 inch [0.025 mm] TIR (total indicated runout); this is accomplished by trial and error. Use piece of wood to tap wheel into position. Tighten flanges with wrench and re-check TIR. If wheel has moved, repeat procedure.

Conditioning cutting face of wheel:

1. Set source voltage at 4 to 5 volts.
2. Set spark suppress at maximum setting (if such control exists on power supply).
3. Start wheel rotation and full electrolyte flow.
4. Plunge grind into a piece of scrap steel. Plunge into steel slowly until ampere meter stabilizes.
5. Increase voltage and feed until sparking occurs.
6. Continue until cutting face and wheel has a dull gray or charcoal appearance.
7. Wheel is now ready to use.
8. Return spark control to normal setting.

Electrolyte checks:

If machine is not equipped with a coolant filter, electrolyte should be changed once a week, or when mounting a new wheel. Check electrolyte nozzles to ensure proper flow on all areas of wheel.

For cutoff wheels, electrolyte should flow evenly under full available pressure on each side of wheel and on wheel periphery.

Nozzle fitting:

The fitting of the nozzle is very important since it will help to confine and direct the electrolyte into the working gap between the wheel and workpiece contact area. A properly selected and fitted nozzle will also help to reduce "stray machining" caused by a combination of excessive electrolyte spray and uncontrolled current flow.

The point at which the wheel enters into the nozzle must be tightly fitted against the O.D. surface of the wheel. This will break up the air curtain around the periphery of the "spinning" wheel. The air curtain would seriously restrict the current flow between the anode and cathode in the working gap. The wheel will pick up the electrolyte by traveling through the supply line. It is also good practice to relieve the inside of the nozzle chamber to eliminate any possible wiping action between wheel and nozzle.

**TABLE 11.3–5 Comparison of ECG to Precision Conventional Milling and Grinding**

| MATERIAL | STOCK REMOVAL (MILLING) | STOCK REMOVAL (GRINDING) | TOOLING COSTS & REPLACEMENT | SIZE CONTROL | PRODUCTION OF FRAGILE PARTS | POTENTIAL OF HEAT DAMAGE | QUALITY OF SURFACE FINISH | PROBLEMS WITH BURRS |
|---|---|---|---|---|---|---|---|---|
| Machinery steel | – | = | – | – | + | = | = | + |
| Tool steel, soft | – | = | – | – | + | + | = | + |
| Tool steel, hard | + | + | – | – | + | + | = | + |
| Cast iron | – | – | – | – | + | = | – | = |
| Copper | – | + | – | = | + | = | + | + |
| Brass | – | + | – | = | + | = | = | + |
| Aluminum | – | – | – | – | + | = | – | + |
| Tungsten | – | + | – | = | + | + | + | + |
| Tungsten carbide | + | + | + | = | + | + | + | = |
| Beryllium | – | = | – | – | = | = | – | + |
| 300 stainless | – | + | + | = | + | + | + | + |
| 400 stainless | – | + | + | = | + | + | + | + |
| Titanium | – | = | = | = | + | + | = | + |
| A-236 | – | + | + | = | + | + | = | + |
| Waspaloy | + | + | + | = | + | + | + | + |
| Inconel 718 | + | + | + | = | + | + | + | + |
| Inconel X | + | + | + | = | + | + | + | + |
| Rene 41 | + | + | + | = | + | + | + | + |
| HS-21 | + | + | + | + | + | + | = | = |
| Hastelloy alloy X | + | + | + | = | + | + | = | + |
| PWA 1004 | + | + | + | = | + | + | = | + |
| PWA 689 | + | + | + | = | + | + | = | + |
| AMS 5668 | = | + | + | = | + | + | = | + |
| Udimet 500 | + | + | + | = | + | + | + | + |
| Udimet 700 | + | + | + | = | + | + | + | + |
| Greek Ascoloy | – | = | + | = | + | + | = | + |
| PWA 90 | + | + | + | = | + | + | + | + |

Legend: = Processes about equal
– ECG inferior to conventional machining
+ ECG superior to conventional machining

SOURCE: R. R. Brandi, Basics of electrochemical grinding, *American Machinist* 118 (April 24, 1974), p. 48.

NOTE: In general, ECG superiority increases with hardness or toughness of material.

# 11.3 Electrochemical Grinding—ECG

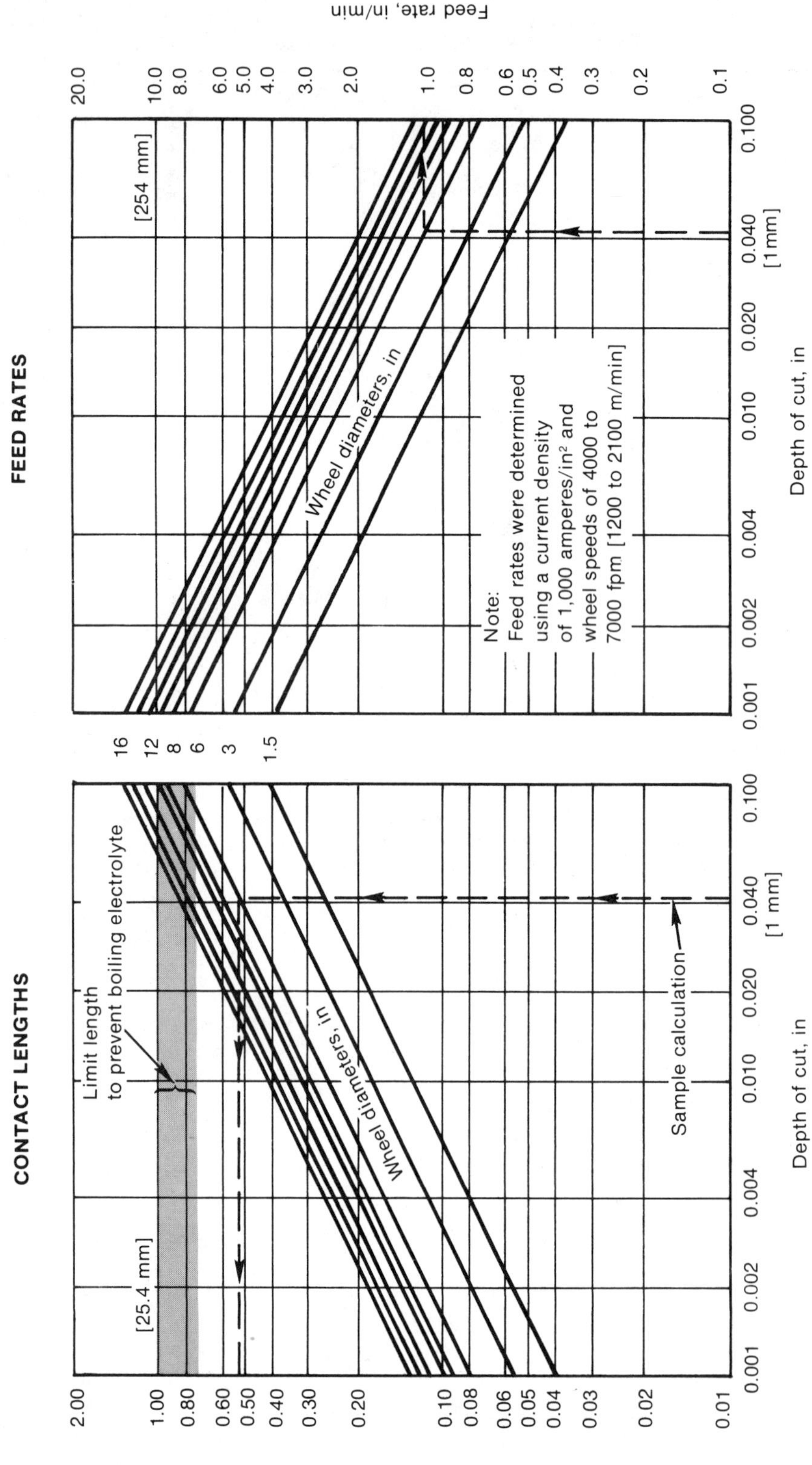


**Figure 11.3-3** Parameter charts for surface ECG. (Courtesy of Cincinnati Milacron)

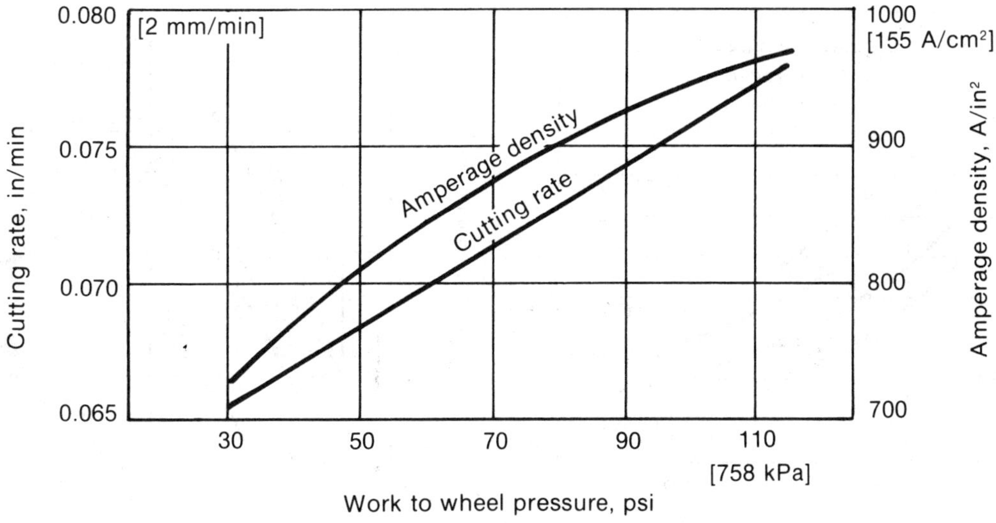


**Figure 11.3-4** Effects of work to wheel pressure in face ECG on M2 tool steel with aluminum oxide wheel. (J. R. Thompson, Operating parameters in electrochemical grinding with aluminum oxide wheels, *New Developments in Grinding*, Pittsburgh, PA: Carnegie Press, 1972, p. 793–812)

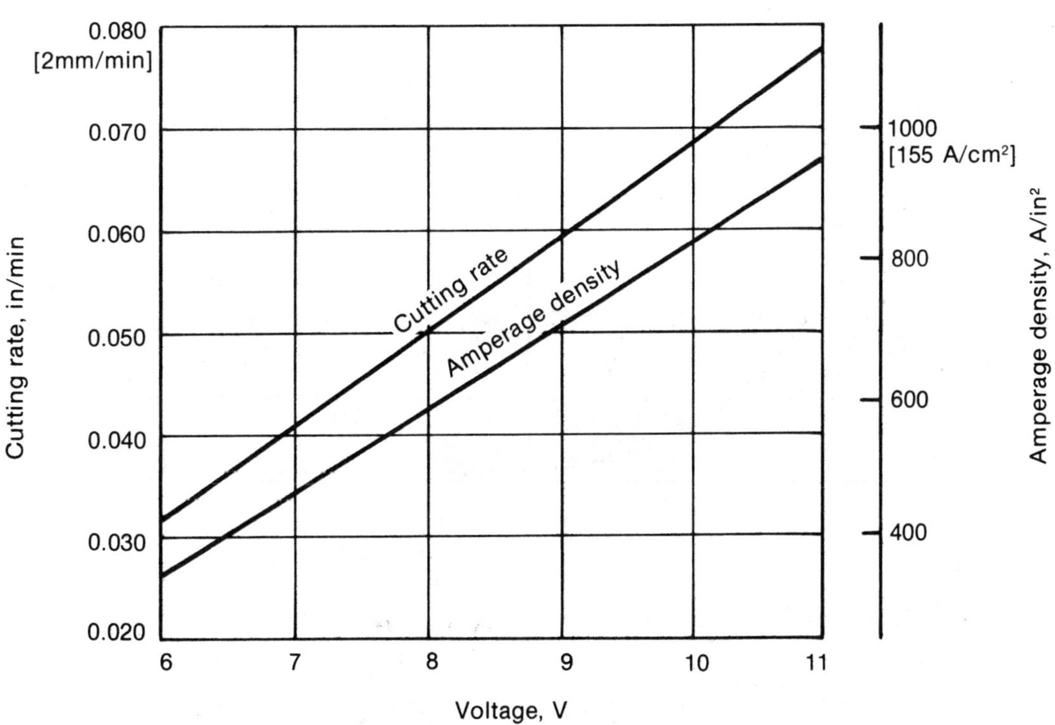


**Figure 11.3-5** Effects of voltage in face ECG on M2 tool steel with aluminum oxide wheel. (J. R. Thompson, p. 811)

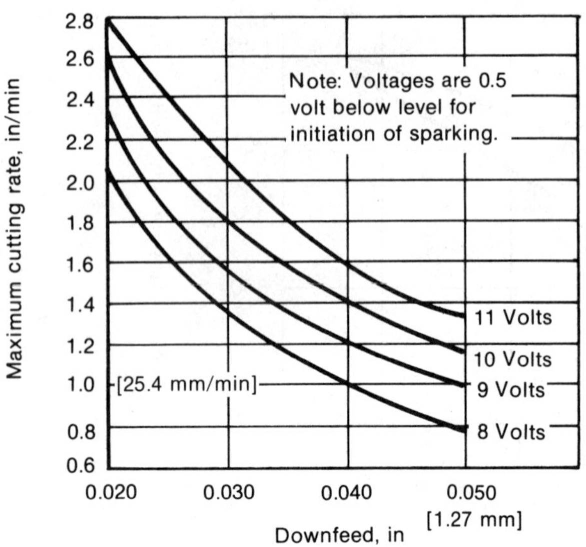


**Figure 11.3-6** Maximum cutting rate as a function of voltage and downfeed for surface ECG on M2 tool steel with aluminum oxide wheel. (J. R. Thompson, p. 812)

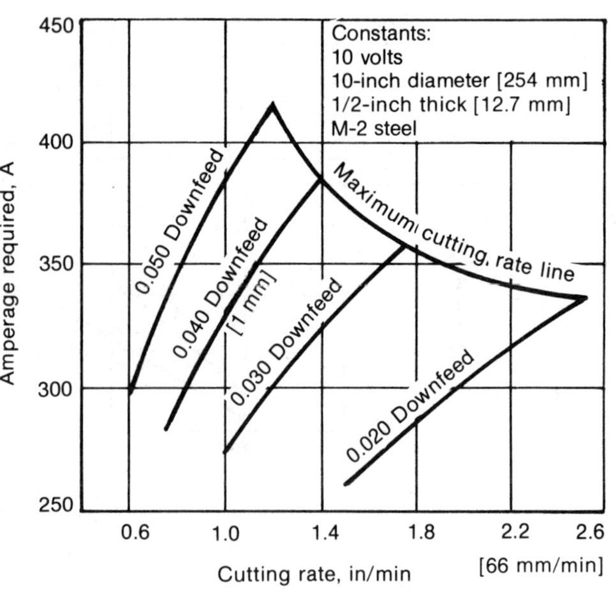


**Figure 11.3-7** Family of operating curves and maximum cutting rates for surface ECG on M2 tool steel with aluminum oxide wheels. (J. R. Thompson, p. 812)

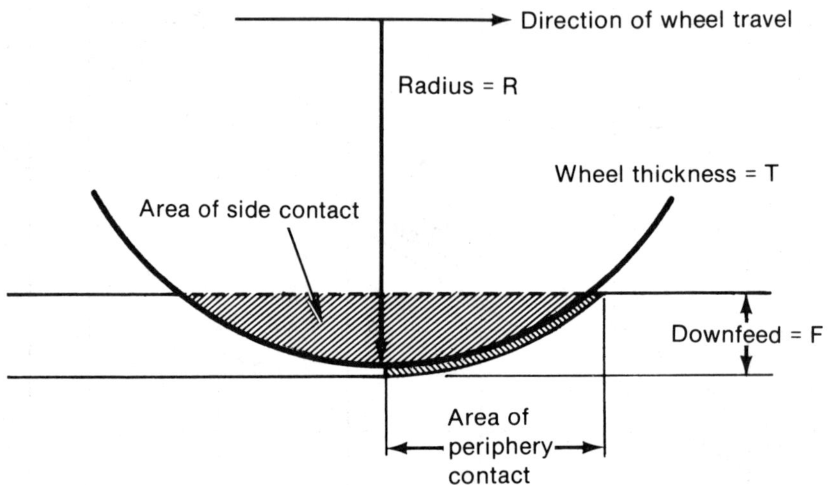


Total area of contact = Area of periphery contact + Area of side contact

$$\text{Area} = \frac{T\pi R}{180}\arccos\left(\frac{R-F}{R}\right) + 2\left[\frac{\pi R^2}{180}\arccos\left(\frac{R-F}{R}\right) - (R-F)\sqrt{2RF-F^2}\right]$$

(arccos is in degrees, not radians)

**Figure 11.3-8** Calculation of contact area of wheel and workpiece for surface ECG of slots. (J. R. Thompson, p. 810)

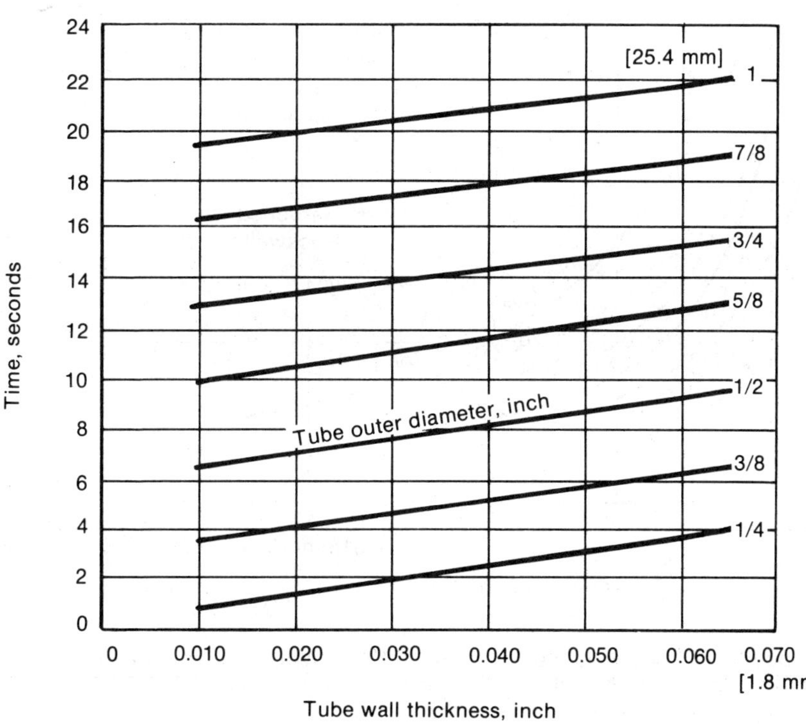


Burr-free, distortionless tubing cut off by ECG is easily accomplished for type 304 stainless steel. The data also are applicable to many other alloys to obtain rough production rates, even though the electrolytic dissolution rates will vary slightly with composition (see ECM section 11.5). For well radiused cut ends, a spindle dwell time of 2 to 5 seconds can be added at the end of the charted cutting time. These cutting times are the fastest burr-free rate without regard for wheel wear.

**Figure 11.3-9** ECG cutoff operation for type 304 stainless steel tubing. Maximum cutting rates for various tube diameters and wall thicknesses. (Courtesy of Everite Machine Products Company)

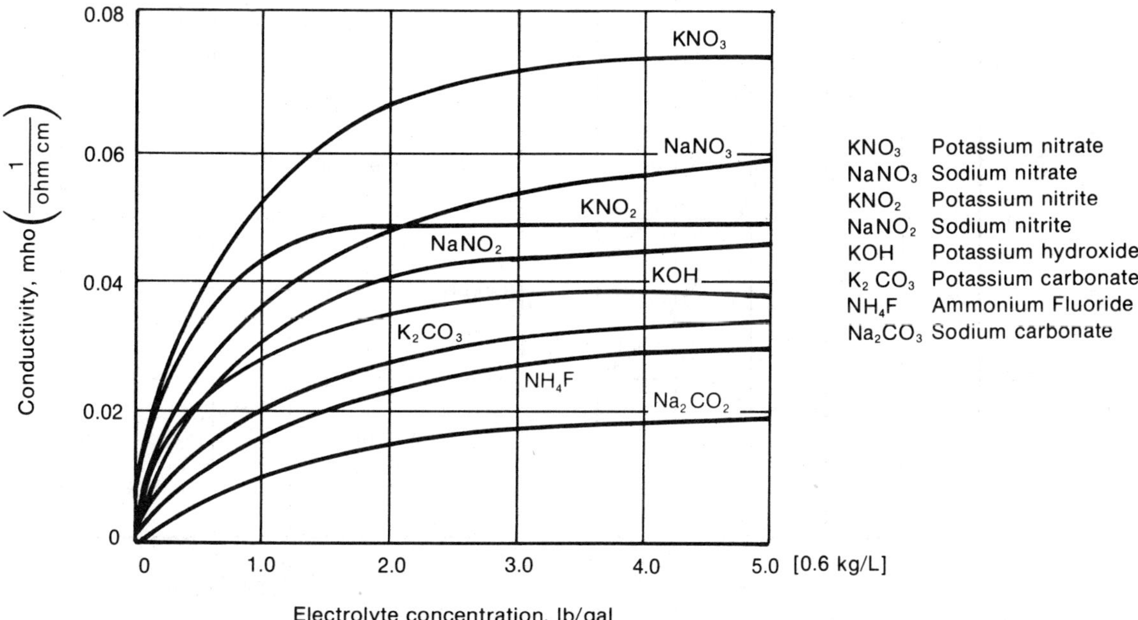


**Figure 11.3-10** Conductivity versus concentration for ECG electrolytes. (Courtesy of Anocut Engineering Company)

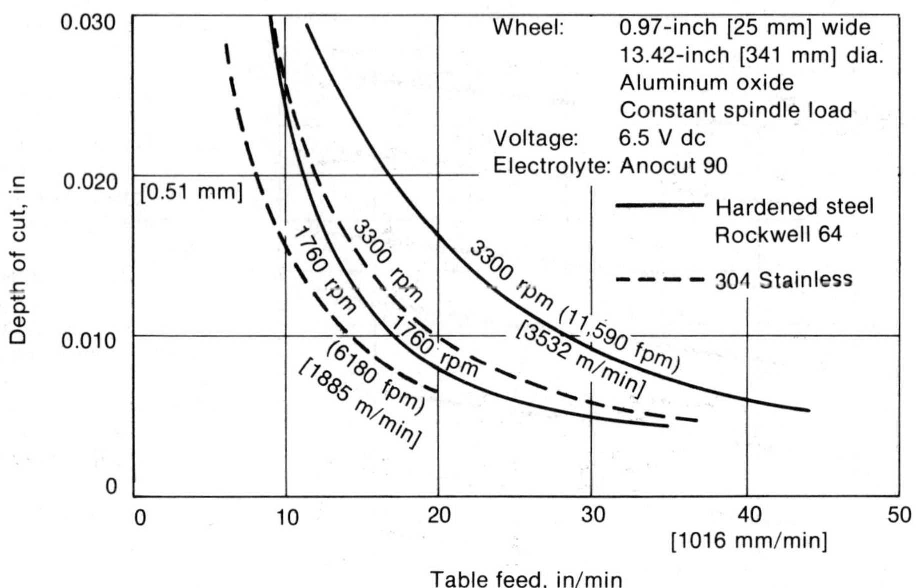


**Figure 11.3–11** Effect of wheel speed in ECG of two work materials. Note: Wheel speeds are somewhat higher than the typical 4000 to 7000 fpm. (L. A. Williams, III, Electrochemical grinding, Technical paper MR67-649, Society of Manufacturing Engineers, Dearborn, MI, 1967, p. 17)

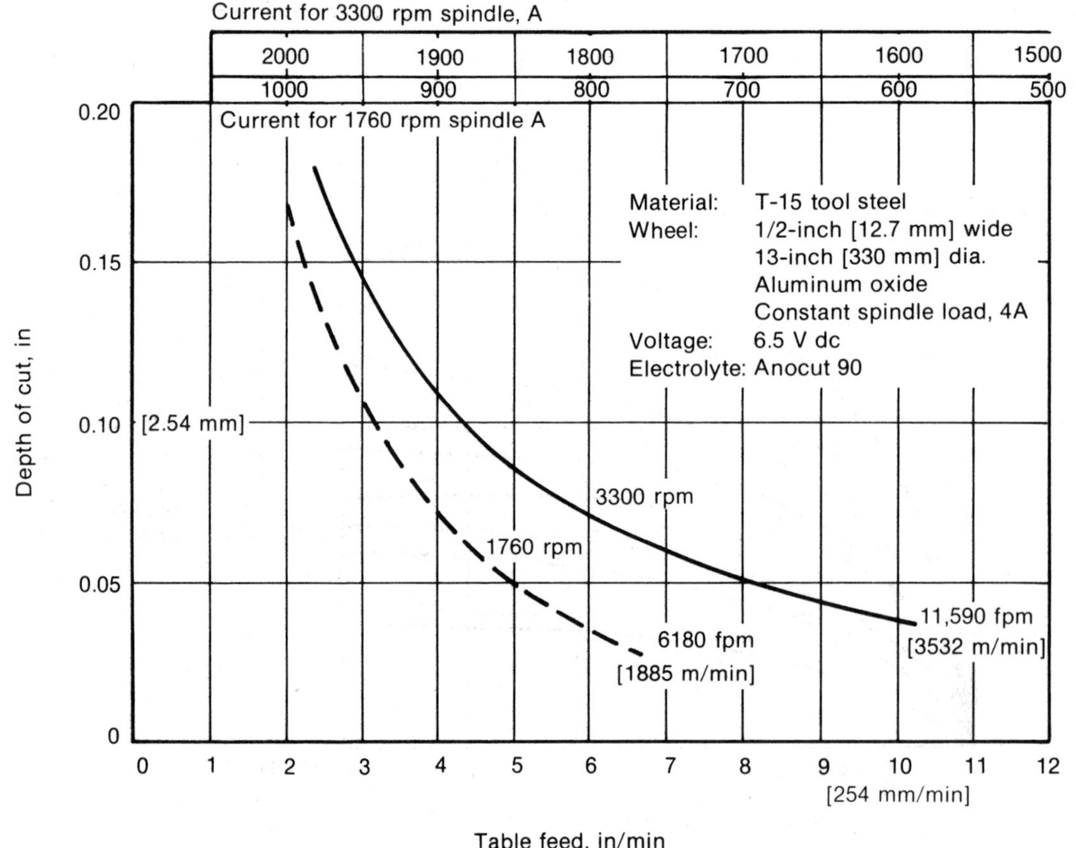


**Figure 11.3–12** Effect of wheel speed in ECG of T15 tool steel. Note: Wheel speeds are somewhat higher than the typical 4000 to 7000 fpm. (L. H. Williams, III, p. 17)

## SPECIFIC APPLICATIONS DATA

**TABLE 11.3–6 Specific Applications Data for Electrochemical Grinding**

| Work material: | **Fully Annealed Low Carbon Steel and Medium Carbon Steel Hardened to 60 $R_c$** | **Tungsten Carbide** | **Stellite Repair Weld on Waspaloy** |
|---|---|---|---|
| Workpiece configuration: | 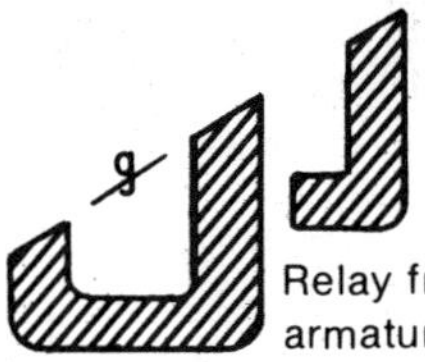 Relay frame & armature |  40 blades<br>Face mill cutters |  |
| Wheel type: | $Al_2O_3$ | Diamond | $Al_2O_3$ |
| diameter, in: | 14 | | 8 |
| speed, fpm: | 6,400 | 4,000 to 6,000 | 6,000 |
| Electrolyte type: | NaCl | $KNO_3$ | $NaNO_3$ |
| concentration, lb/gal: | 0.5 | 1.0 | 2.0 |
| Type of cut: | Face | Face | Contour |
| Depth cut, in: | 0.030 | 0.010 | 0.030 to 0.050 |
| Feed rate, in/min: | 0.105 | 3 faces in 80 min | 4.0 |
| Voltage, V: | 8 to 9 | | 8 |
| Current, A: | 1,200 | 300 | 125 |
| Wheel contact area, $in^2$: | 0.25 | | |
| Tolerance, finish, benefits and remarks: | Three parts per load.<br>17-second cutting cycle.<br>50 wheel oscillations per minute. | Eliminated rough grinding.<br>One wheel per year in place of several per week.<br>No heat checks—33% more parts per wheel dressing.<br>Accuracy was ±0.00025 inch.<br>30 wheel oscillations per minute. | Tolerance was ±0.0005. |

TABLE 11.3-6—Continued

| Work material: | **Inconel Alloy X** | **High-Nickel Magnetic Lamination** | **Stainless** |
|---|---|---|---|
| Workpiece configuration: | 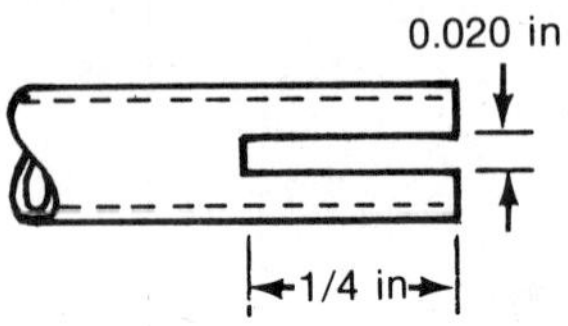 0.020 in; 1/4 in; 1/8-inch dia. 1/64-inch wall tubing | 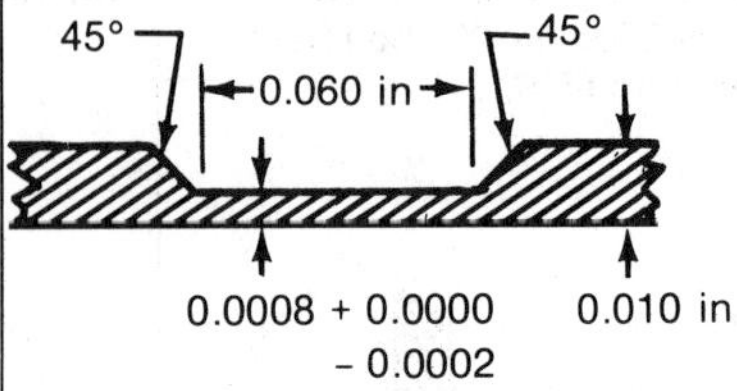 45°; 0.060 in; 45°; 0.0008 + 0.0000 - 0.0002; 0.010 in | 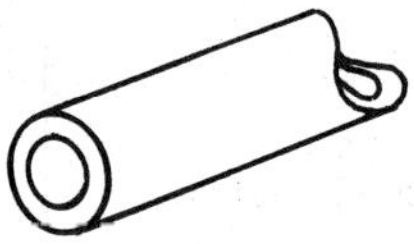 1/2-inch dia. x 0.015-inch thick wall |
| Wheel type: | $Al_2O_3$ | $Al_2O_3$ | $Al_2O_3$ |
| diameter, in: | 5 | 8 | 10 |
| speed, fpm: | 1,900 | 6,000 | 6,000 |
| Electrolyte type: | $NaNO_3$ | $NaNO_3$ | $NaNO_3 + NaNO_2$ |
| concentration, lb/gal: | 2.0 | 2.0 | 1.7 |
| Type of cut: | Surface | Contour | Surface cutoff |
| Depth cut, in: | 0.25 | 0.0092 | |
| Feed rate, in/min: | 2.0 | 3.0 | 3.0 |
| Voltage, V: | 4 | 6 | 8 |
| Current, A: | 25 | 120 | 25 |
| Wheel contact area, $in^2$: | | | |
| Tolerance, finish, benefits and remarks: | Burr-free and no distortion.<br>120 pieces per hour in place of 100 pieces per day.<br>0.007-inch radius at bottom of slot.<br>Tolerance on width was ±0.001 inch. | No distortion.<br>No nicks.<br>No wrinkles. | No burrs.<br>Less than 0.001-inch runout. |

## PROCESS SUMMARY

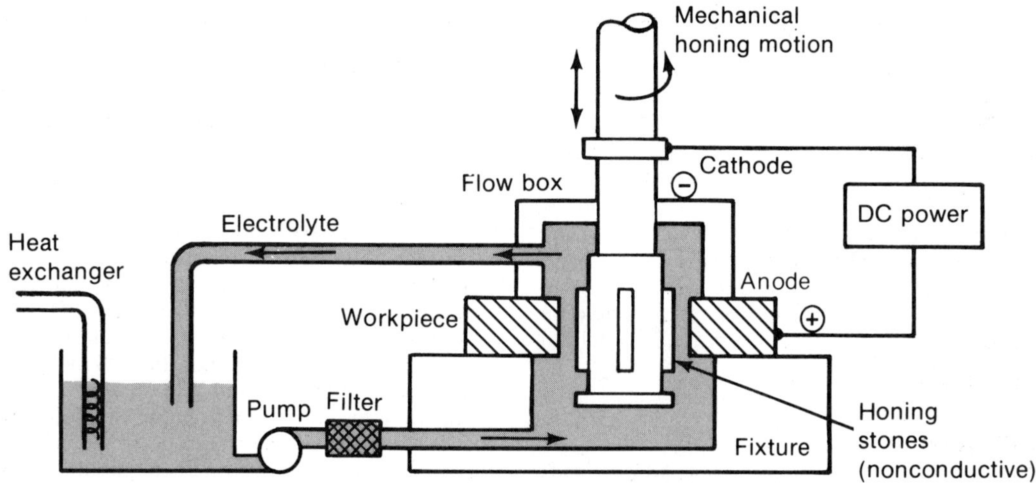


**Figure 11.4-1** ECH schematic.

### Principles

Electrochemical honing (ECH) is the removal of material by anodic dissolution combined with mechanical abrasion from a rotating and reciprocating abrasive stone (carried on a spindle which is the cathode) separated from the workpiece by a rapidly flowing electrolyte (see figure 11.4-1). The principal material removal action comes from the electrolytic dissolution. The abrasive stones are used to maintain size and to clean the surfaces to expose fresh metal to the electrolytic action. The small electrical gap is maintained by the nonconducting stones which are bonded to the expandable arbor with cement. The cement must be compatible with the electrolyte and the low direct-current voltage. The mechanical honing action uses materials, speeds and pressures typical of conventional honing.

### Practical Applications

The advantages of ECH are most pronounced when honing hard metals. The workpiece must be conductive. The absence of heavy mechanical action keeps the workpiece cool and free from heat distortion. The electrolytic action introduces no stresses and deburrs automatically. Blind holes in cast tool steel components and pinion gears of 62 $R_C$ high alloy steels have been honed using ECH.

### Operating Parameters

**TABLE 11.4-1 Typical Values for ECH Operating Parameters**

| | |
|---|---|
| Power supply | |
| type: | Direct current |
| voltage: | 6 to 30 V |
| current: | 100 to 3000 A |
| current density: | 100 to 3000 A/in$^2$ [15.5 to 465 A/cm$^2$] |
| Electrolyte | |
| types and concentrations: | $NaNO_3$, 2 lb/gal [240 g/L]<br>NaCl, 1 lb/gal [120 g/L] |
| temperature: | 100°F [38°C] |
| pressure: | 75 to 150 psi [500 to 1000 kPa] |
| flow: | Up to 25 gal/min [95 L/min] |
| gap: | 0.003 to 0.010 inch [0.076 to 0.25 mm] |

# 11.4 Electrochemical Honing—ECH

## Material Removal Rates and Tolerances

Material is removed 3 to 5 times faster by ECH than by conventional honing and 4 times faster than by internal grinding. Removal of 0.008 inch [0.20 mm] of stock from the bore of 8620 steel in 30 seconds has been reported. Tolerances up to ±0.0001 inch [±0.0025 mm] are possible. Production tolerances achieve ±0.0005 inch [±0.013 mm] on the diameter and 0.0002 inch [0.005 mm] on straightness.

## Surface Technology

Surface roughness of 8 to 32 microinches $R_a$ [0.2 to 0.8 μm] is routine, and roughness of 2 microinches $R_a$ [0.05 μm] is attainable. For a controlled surface roughness, the stones sometimes are allowed to cut for a few seconds after the current has been turned off. This sequence also will leave a light, compressive residual stress in the surface. For a stress-free surface, the last few seconds of action should be pure electrolytic material removal. This practice is helpful for controlling surface integrity.

## Equipment and Tooling

Commercial ECH equipment is made to special engineering order only for internal cylindrical honing. It comes with suitable pumps, tanks, filters, power packs and controls for the combined material removal modes. Automatic size gaging is sometimes incorporated. Tooling for control of the electrolyte flow path and for positioning should be fabricated from stainless steels or plastics that are resistant to the corrosiveness of the electrolyte. Sizes of equipment range from 3/8 inch to 6 inches [10 to 150 mm] in diameter with power packs up to 3000 amperes.

## Machining Characteristics

The electrolyte selection and control is not as critical as in ECM because of the cleaning action of the stones.

## Hone-Forming

A reversed-polarity version of ECH combines honing with plating. This "hone-forming" process uses plating solutions in place of deplating electrolytes and can achieve rapid, accurate metal deposition. This highly focused plating has achieved deposition rates of 0.0001 inch per minute [0.0025 mm/min] for copper, 0.0005 inch per minute [0.013 mm/min] for nickel and 0.0001 inch per minute [0.0025 mm/min] for chromium. The mechanical (abrasive) honing action prepares a clean surface for the plating and sizes the bores to close tolerances. The equipment is quite similar to ECH equipment, and sometimes the acronym ECF is applied.

## PROCESS SUMMARY

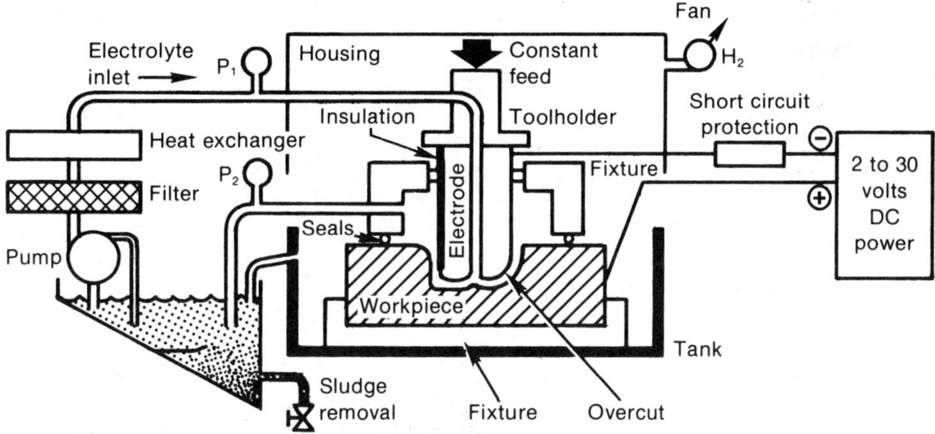


**Figure 11.5–1** ECM schematic.

### Principles

Electrochemical machining (ECM) is the removal of electrically conductive material by anodic dissolution in a *rapidly* flowing electrolyte which separates the workpiece from a shaped electrode (see figure 11.5–1). The filtered electrolyte is pumped under pressure and at controlled temperature to bring a controlled-conductivity fluid into the narrow gap of the cutting area. The shape imposed on the workpiece is nearly a mirror or conjugate image of the shape of the cathodic electrode. The electrode is advanced into the workpiece at a constant feed rate that exactly matches the rate of dissolution of the work material. The current density is the chief factor in setting feed rates and in attaining smoothness. Higher current densities and feed rates create better finishes and higher material removal rates. The hydraulic pressures and tool/workpiece separating forces increase with smaller gaps and can affect workpiece tolerances. Fixtures with epoxies or other securely fastened plastics for insulation locate and control the electrolyte flow so that it is continuous and uniformly turbulent. Materials and design must accommodate the high currents without distortion or overheating. Fixtures must be compatible with the organic salts or acids that are used as electrolytes. Metal dissolved by salt electrolytes forms a sludge of considerable volume, and excess sludge must be removed by filtration, centrifuge or settling. Conductivity control of the electrolyte is maintained by temperature control, usually cooling, and by periodic checks on chemical composition. Both inlet and outlet pressure controls provide flow control. The electrode design must include compensation or correction for the variable current density that comes with variations in the shape of the workpiece and the metal content of the electrolyte. Exact control of all critical parameters is needed to ensure the best results. Cutting action occurs on all exposed faces simultaneously, and there is no mechanical contact or wear of the electrodes. While metal ions are removed from the workpiece surface, hydrogen is generated at the electrode. Adequate provision is required to safely vent this hydrogen.

### Practical Applications

ECM is best suited for repetitive production of complex shapes in high strength, hard, or difficult-to-machine materials where high surface quality is needed. The work material must be electrically conductive. Small, odd-shaped, difficult or deep holes as small as 1/8-inch [3.2 mm] across can be "drilled" individually or in multiples. Electrodes do not wear; consequently, long tool life is possible. The stress-free material removal eliminates distortion from the ECM operation (but not necessarily from prior stress-inducing operations). Concentration of current density at the edges on the workpiece provides automatic rounding and the absence of burrs. Anodic dissolution generally proceeds independently of material hardness, so finish cuts in the heat-treated state are practical.

The characteristics of an application that make it a prime candidate for ECM are as follows:

- Material is hard or tough.
- Difficult contours are involved.
- Machining without cold working is needed.
- Repetitive production is involved (more than 30 pieces).
- Distortionless, burr-free parts are required.
- High surface quality is required.

# 11.5 Electrochemical Machining—ECM

## Operating Parameters

**TABLE 11.5–1 Typical Values for ECM Operating Parameters**

| | |
|---|---|
| Power supply | |
| type: | Direct current |
| voltage: | 5 to 30 V |
| current: | 50 to 40,000 A |
| Current density: | 50 to 3,000 A/in$^2$* [8 to 465 A/cm$^2$] |
| Electrolyte | |
| type and concentration: | |
| most used: | NaCl at 1/2 to 2 lb/gal [60 to 240 g/L] |
| frequently used: | $NaNO_3$ at 1 to 4 lb/gal [120 to 480 g/L] |
| less frequently used: | Proprietary mixtures |
| temperature: | 90° to 125°F [26° to 46°C] |
| flow rate: | 0.25 gal/min/100 A [0.95 L/min/100 A] |
| velocity: | 5,000 to 10,000 fpm [1,500 to 3,000 m/min] |
| inlet pressure: | 20 to 300 psi [137 to 2,060 kPa] |
| outlet pressure: | 0 to 45 psi [0 to 310 kPa] |
| Frontal working gap: | 0.003 to 0.030 inch [0.076 to 0.76 mm] |
| Side overcut: | 0.005 to 0.040 inch [0.127 to 1 mm] |
| Feed rate: | 0.020 to 0.50 in/min [0.51 to 13 mm/min] |
| Electrode material: | Brass, copper, bronze (table 11.5–10) |
| Tolerance | |
| 2-dimensional shapes: | ±0.001 inch [±0.025 mm] |
| 3-dimensional shapes: | ±0.002 inch [±0.051 mm] |
| Surface roughness, $R_a$, frontal: | 8 to 63 microinches [0.2 to 1.6 μm] |

*Limited by overtemperature in workpiece or electrolyte.

### Material Removal Rates and Tolerances

Material removal rate is independent of material hardness and for most common metals is approximately 0.10 cubic inch per minute [1.64 cm$^3$/min] per 1000 amperes. The cutting occurs simultaneously on all exposed surfaces, and this aids productivity. Accuracy up to ±0.004 inch [±0.1 mm] is usual for contoured cavities, and accuracy up to ±0.0010 inch [±0.025 mm] is common for frontal cuts or cuts made with refined tools. (These tolerances can be cut in half with well-developed tools and experienced operators.) Deep cavities will have a taper of up to 0.001 inch per inch [0.001 mm/mm] with 0.005 inch [0.13 mm] overcut gap unless a special tool design is used. Internal radii of 0.007 inch [0.18 mm] and external radii of 0.002 inch [0.051 mm] are attainable. Tolerance capabilities are dependent upon part geometry, tool design and particular shop practices and require careful checking with experienced people for particular applications.

### Surface Technology

Surfaces roughness of 16 to 32 microinches $R_a$ [0.4 to 0.8 μm] is normal. Roughness values decrease with the increase in cutting rates that accompanies higher current densities. The faster the cutting, the better the finish. Mirror finishes (8 microinches $R_a$ [0.2 μm] or better) are easily obtained in frontal cuts of stainless steels or nickel alloys. The side-gap areas generally are much rougher because current density is lower in these areas. Incorrect electrolyte, different work material heat treatment, low current density, or stray current can produce selective etching pits or intergranular attack at the grain boundaries. The absence of residual stress (in contrast with the presence of compressive stresses from many mechanical processes) produces a reduced high cycle fatigue strength (by comparison), but these values more nearly represent the true or unaffected material values. Any physical blemish in the tool will be reproduced on the workpiece, and poor flow conditions can produce striations in the surface. Hydrogen embrittlement is not a problem because the hydrogen is liberated at the electrode (as a gas) and is carried away by the flowing electrolyte as shown in figure 11.5–2.

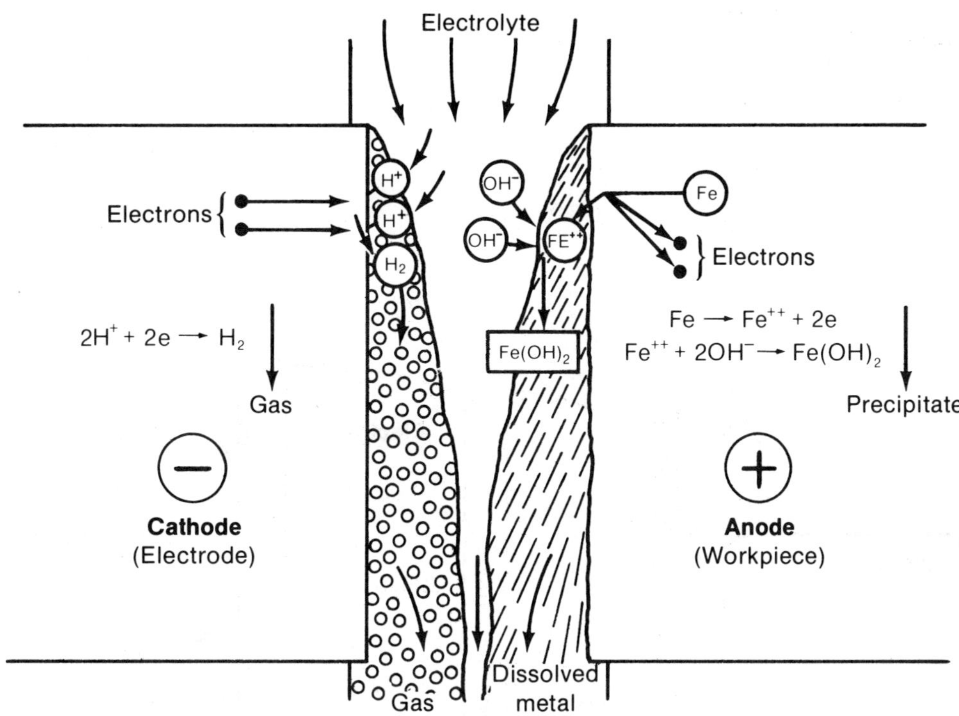


**Figure 11.5-2** Electrochemical reactions in the working gap during ECM.

### Equipment and Tooling

The available equipment ranges from small 50-ampere bench models to 40,000-ampere models with a five-foot-cube working space. Power sources providing from 4 to 30 volts direct current are available with special, fast-acting, short-circuit protective devices. Ancillary equipment, such as pumps, tanks, plastic piping, mixing vats and sludge-disposal devices, should be engineered as a complete system. Equipment and tools should be rigid enough to withstand the forces from the high hydraulic pressures. The fixtures, the insulation and the tool material must be chemically compatible with the electrolyte. Electrode materials are high-conductivity coppers, brasses or bronzes (see table 11.5-10).

Tooling trials and start-up costs should be carefully assessed and included in development plans. The electrode design and correction during tryout may require several "cut and try" cycles for complex shapes. Filtration of the electrolyte to the 7.5-micrometer level or better is most desirable. Placement of the filter immediately ahead of the electrode is good practice. With salt solutions, the volume of metal hydroxide or metal hydrate can be 100 to 500 times the volume of the metal removed, as noted in table 11.5-13. Settling or filtration for removal of these "chips" is necessary, and disposal requires environmentally satisfactory planning.

### Machining Characteristics

Process control and the sequence of operating steps must be exact and precisely repetitive for good production. The workpiece should be thoroughly cleaned and/or oiled after ECM to prevent corrosion. Periodic washdown of the equipment can reduce maintenance expenses. Special safety precautions should be instituted when the electrolyte used is a chlorate or a nitrite (see table 11.5-6).

# 11.5 Electrochemical Machining—ECM

## SELECTED DATA

There are several valid techniques for selecting operating parameters for ECM. The steps in table 11.5-2 are recommended as one logical sequence. Supporting data come from the other figures and tables in accordance with the terms as explained in figure 11.5-3.

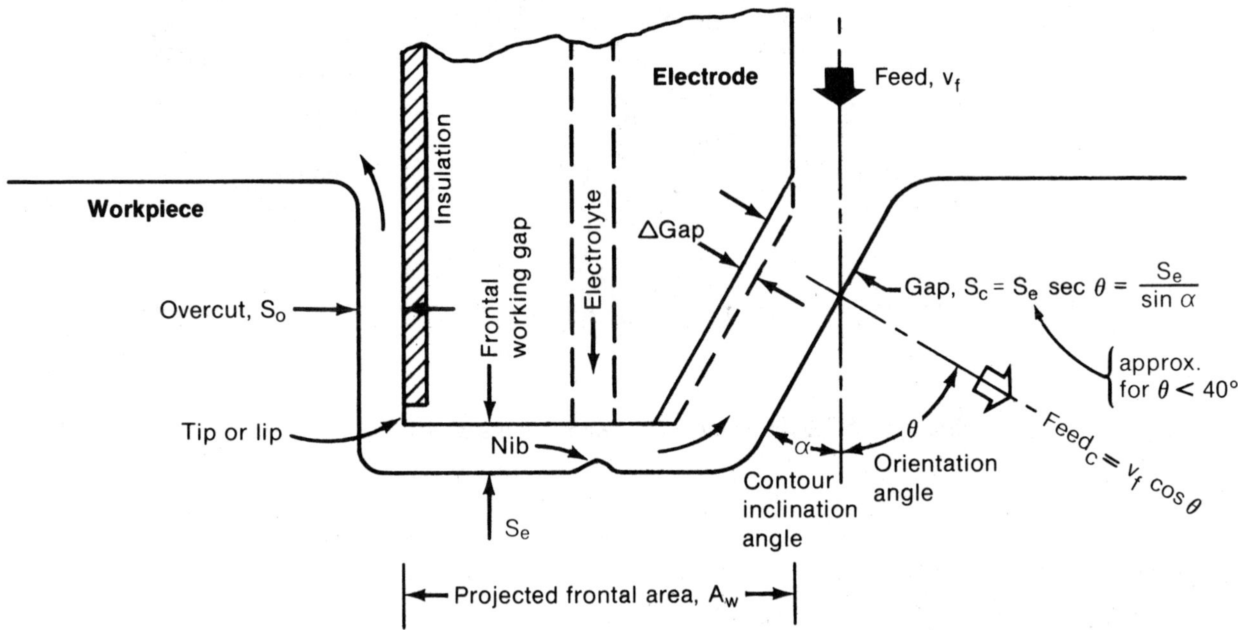


**Figure 11.5-3** ECM parameter terminology.

**TABLE 11.5–2 Steps To Select ECM Operating Parameters**

| ITEM | DATA |
|---|---|
| 1. Workpiece material | From engineering drawings obtain material type, composition, and metallurgical state. Must be electrically conductive material. |
| 2. Material removal rate, MRR, (at 100% efficiency) | A. For planning or rough estimating, good approximation can be had from the following values which apply to most common conductive materials and alloys:<br>0.1 $in^3$/min/1,000 A [1.64 $cm^3$/min/1,000 A]<br>1.0 $in^3$/min/10,000 A [16.4 $cm^3$/min/10,000 A]<br>B. For pure metals, removal rate depends on valence and the relationships established in Faraday's second law:<br>$\text{specific removal rate} = \frac{\text{atomic weight}}{\text{valence}} \times \frac{1}{\text{density}} \times \frac{1}{96{,}494}$<br>where 96,494 is Faraday's constant representing the number of coulombs (or ampere-seconds) of electricity required to remove one gram equivalent weight of material. Gram equivalent weight is atomic weight divided by valence. Values in table 11.5–3 represent theoretical metal removal rates assuming 100% current efficiency (all current goes to dissolving metal and none goes to gas generation at the anode).<br>C. For alloys, rates of material removal can be calculated from alloy chemical composition using Faraday's law (table 11.5–4). |
| 3. Assume efficiency, h, of electrochemical action | With complex tools, long electrolytic path, higher voltage or current density, valence variation or chemical activity, and appreciable stray current or spray, the efficiency can decline to 80 to 90%. For most estimating prior to a pilot run, 90% is a fair assumption. |
| 4. Determine projected area, $A_w$, of electrode perpendicular to feed direction | See figure 11.5–3 |
| 5. Determine maximum current density available | Rated amperes of equipment ÷ $A_w$ |
| 6. Establish feed rate, $v_f$, that exactly matches rate of material dissolution. Current density and feed rate will be in balance. | A. $\frac{\text{material removal rate}}{\text{projected area}} = \frac{\text{MRR times h}}{A_w} = v_f$<br>B. See figure 11.5–7 and 11.5–8 for quick approximations. (Figure 11.5–9 for one specific alloy-electrolyte combination.)<br>C. The current density that balances the feed rate is limited by (1) boiling of the electrolyte, see table 11.5–7, (2) overtemperature in the workpiece from resistive heating (and possible distortion or dimensional inaccuracies), or (3) available power supply (item 5). See also table 11.5–9. |
| 7. Electrolyte Selection<br>Composition<br>Concentration<br>Temperature | Table 11.5–5 and 11.5–6. Many applications try simple NaCl first, then try more complex electrolytes only if necessary. Keep concentrations as low as compatible with productivity so as to reduce accumulations of salt deposits on equipment and tooling.<br>Selection should be based upon compatibility with particular metallurgical state of alloy being machined, feed rate, surface roughness (both frontal and side) and surface integrity.*<br>Temperature control should be within ±1°F [±1.8°C] to attain consistent electrolyte conductivity at the inlet to the electrode, see table 11.5–7, figures 11.5–14 and 11.5–15. |
| 8. Electrolyte Testing<br>Select a measure for electrolyte condition or control | |
| Specific Gravity | Nomographs, figures 11.5–10 to 11.5–12. |
| Salimeter | Table 11.5–8, figure 11.5–13. |
| Conductivity | Figures 11.5–14 and 11.5–15. |
| Titration | Standard analytic chemical test. |

*For detail on interrelationships, see J. A. Cross, ECM machinability data, Technical paper MR73-228, Society of Manufacturing Engineers, Dearborn, MI, 1973, or J. A. Cross and A. U. Jollis, Electrochemical machining (ECM), Technical report AFML-TR-72-188, General Electric Company, 1972.

**TABLE 11.5-2—Continued**

| ITEM | DATA |
|---|---|
| 9. Voltage | A minimum voltage, ΔE, is required to initiate full electrochemical action, see figure 11.5-17 and table 11.5-9. The usual operating range is between 5 and 30 volts. The value is set to attain the desired current density that matches the feed rate. Typical setting ranges are:<br>Steel alloys 5 to 10 volts dc<br>High strength alloys 10 to 20<br>Titanium alloys 15 to 25<br>Superalloys 10 to 25 |
| 10. Tool design, cutting gaps and corrections<br>Frontal gap, $S_e$<br>Side gap, $S_c$<br>Orientation, Δ gap | The equilibrium frontal gap is a compromise depending on tool design and prior parameter selections, see figures 11.5-16 and 11.5-17. It is not independently adjustable. Electrode shape should be adjusted depending on orientation angle, see figure 11.5-18. |
| 11. Current | Cannot be set independently—it is a result of equilibrium among prior parameters to just balance feed rate and MRR. Adjusted by setting feed rate. Note that current increases as starting gap reaches operating or equilibrium gap during full engagement when selected current density (item 6) is attained. |
| 12. Electrolyte Pressures | The tool design and operating gaps will determine pressures required to get rapid turbulent flow. Inlet pressures range from 20 to 300 psi. Outlet pressures range from 0 to 45 psi (and sometimes to 75). Adequate provision to contain the considerable hydraulic forces is needed in the plumbing, fixtures, electrodes and machine if good accuracy is to be attained (figure 11.5-19). |
| 13. Starting Gap and Depth of Feed | These parameters can be set independently and are a function of the shape to be generated and the shape of the raw material stock. Adequate cutting stock is required to permit penetration of electrode to point of electrochemical equilibrium. |
| 14. Tool Construction | The insulation and tool materials must be compatible with the electrolyte being used. Typical materials are listed on tables 11.5-10 to 11.5-12. |
| 15. Sludge | The "chips" from ECM are in the form of a sludge in the electrolyte composed of metal hydroxides or hydrates. The volume of sludge is considerable with salt electrolytes, see table 11.5-13. |
| 16. Trial cut | The extensive interrelations among the ECM parameters make a trial cut highly desirable. If it is performed on a sample of the workpiece in the metallurgical state equal to the production item, a simple metallographic section of the ECM surface can reveal possible surface integrity imperfections. |

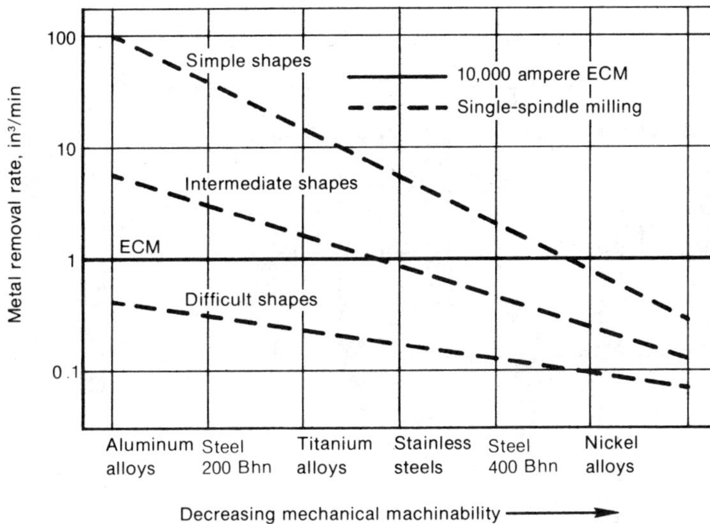


**Figure 11.5-4** Comparison of metal removal rates for 10,000 ampere ECM and single-spindle NC conventional milling of equal investment cost. (Courtesy of Lockheed-Burbank)

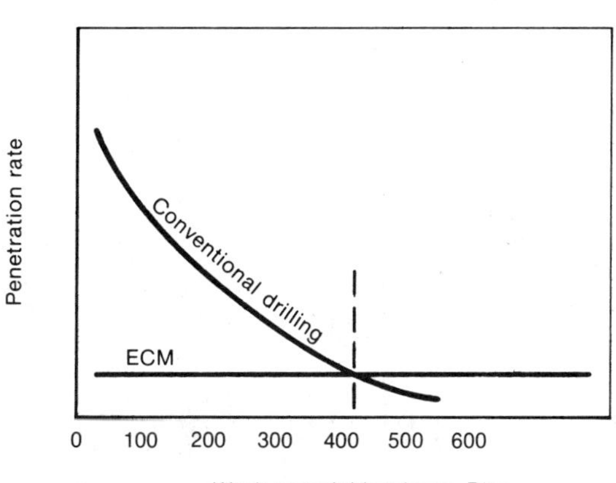


**Figure 11.5-5** Relative penetration rates of ECM and conventional drilling for increasing work material hardness. (Courtesy of Cincinnati Milacron)

## 11.5 Electrochemical Machining—ECM

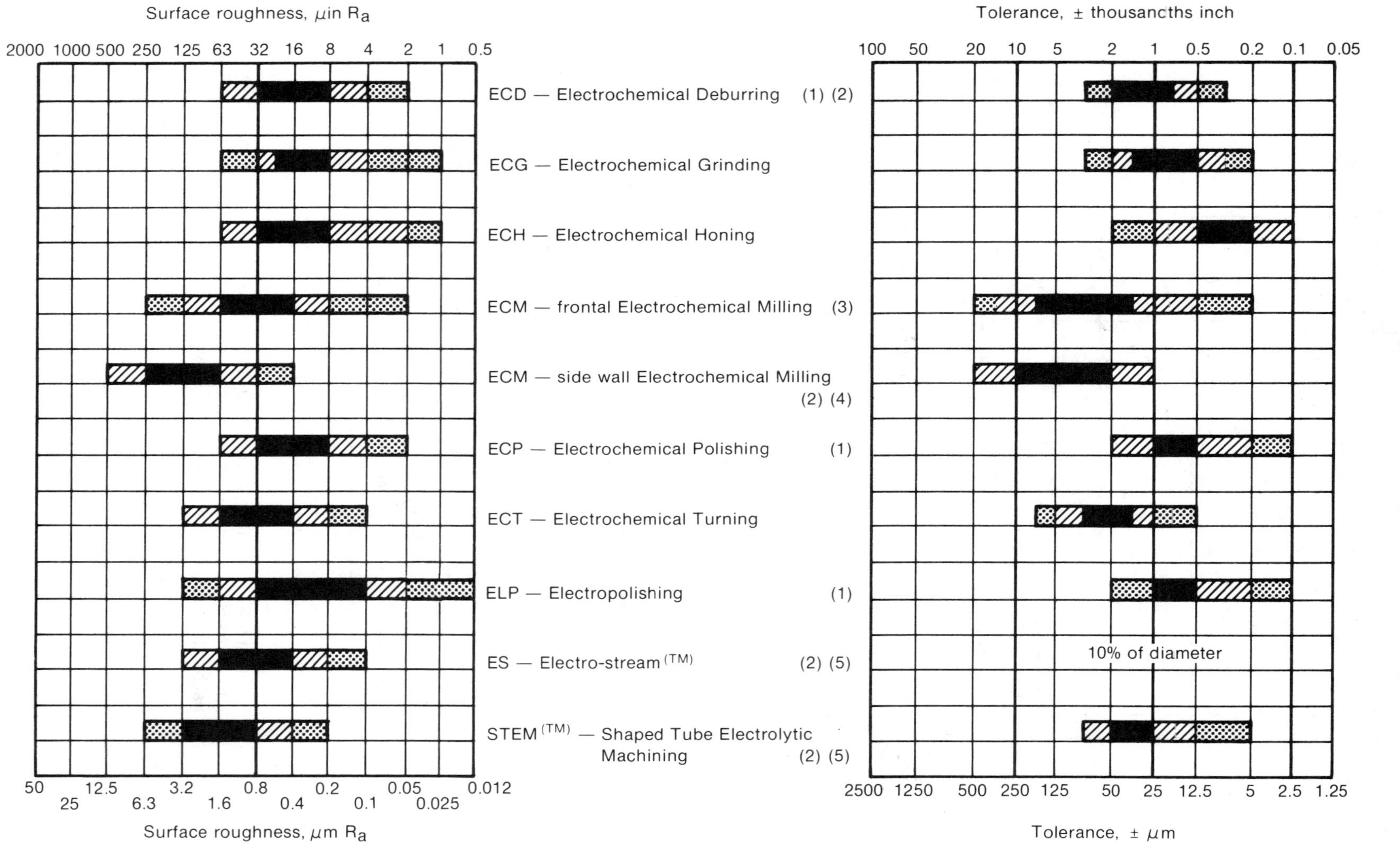


**Figure 11.5–6** Typical surface roughness values and typical tolerances from various forms of electrolytic machining.

**TABLE 11.5-3 Material Removal Rates for ECM of Pure Metals Assuming 100 Percent Current Efficiency**

| METAL | ATOMIC WEIGHT | VALENCE | DENSITY | | REMOVAL RATES AT 1000 AMPERES | | | |
|---|---|---|---|---|---|---|---|---|
| | g | | lb/in³ | g/cm³ | lb/hr | in³/min | g/min | cm³/min |
| Aluminum | 26.98 | 3 | 0.098 | 2.71 | 0.74 | 0.126 | 5.59 | 2.06 |
| Antimony | 121.75 | 3<br>5 | 0.239 | 6.62 | 3.33<br>2.00 | 0.23<br>0.14 | 25.17<br>15.12 | 3.77<br>2.30 |
| Arsenic | 74.92 | 3<br>5 | 0.207 | 5.73 | 2.05<br>1.23 | 0.17<br>0.10 | 15.50<br>9.30 | 2.79<br>1.64 |
| Beryllium | 9.012 | 2 | 0.067 | 1.86 | 0.37 | 0.092 | 2.80 | 1.5 |
| Bismuth | 208.98 | 3<br>5 | 0.354 | 9.80 | 5.73<br>3.44 | 0.27<br>0.16 | 43.32<br>26.01 | 4.43<br>2.62 |
| Cadmium | 112.40 | 2 | 0.313 | 8.67 | 4.62 | 0.25 | 34.93 | 4.1 |
| Chromium | 51.896 | 2<br>3<br>6 | 0.260 | 7.20 | 2.14<br>1.43<br>0.71 | 0.137<br>0.092<br>0.046 | 16.18<br>10.81<br>5.37 | 2.25<br>1.51<br>0.75 |
| Cobalt | 58.93 | 2<br>3 | 0.322 | 8.92 | 2.42<br>1.62 | 0.125<br>0.084 | 18.30<br>12.25 | 2.05<br>1.38 |
| Columbium (Niobium) | 92.906 | 3<br>4<br>5 | 0.310 | 8.59 | 2.55<br>1.92<br>1.53 | 0.132<br>0.103<br>0.08 | 19.28<br>14.52<br>11.57 | 2.16<br>1.69<br>1.34 |
| Copper | 63.546 | 1<br>2 | 0.324 | 8.97 | 5.22<br>2.61 | 0.268<br>0.134 | 39.46<br>19.73 | 4.39<br>2.20 |
| Germanium | 72.59 | 4 | 0.192 | 5.32 | 1.49 | 0.13 | 11.26 | 2.13 |
| Gold | 196.967 | 1<br>3 | 0.698 | 19.33 | 16.22<br>5.40 | 0.39<br>0.13 | 122.62<br>40.82 | 6.40<br>2.13 |
| Hafnium | 178.49 | 4 | 0.473 | 13.10 | 3.68 | 0.13 | 27.82 | 2.13 |
| Indium | 114.82 | 1<br>2<br>3 | 0.264 | 7.31 | 9.43<br>4.71<br>3.14 | 0.60<br>0.30<br>0.20 | 71.29<br>35.61<br>23.74 | 9.84<br>4.92<br>3.28 |
| Iridium | 192.20 | 3<br>4 | 0.813 | 22.52 | 5.27<br>3.96 | 0.11<br>0.08 | 39.84<br>29.94 | 1.80<br>1.31 |
| Iron | 55.847 | 2<br>3 | 0.284 | 7.86 | 2.30<br>1.53 | 0.14<br>0.09 | 17.39<br>11.57 | 2.21<br>1.47 |
| Lead | 207.19 | 2<br>4 | 0.410 | 11.36 | 8.52<br>4.26 | 0.35<br>0.17 | 64.41<br>32.21 | 5.74<br>2.79 |
| Magnesium | 24.312 | 2 | 0.063 | 1.75 | 1.00 | 0.27 | 7.56 | 4.43 |
| Manganese | 54.938 | 2<br>3<br>4<br>6<br>7 | 0.270 | 7.48 | 2.26<br>1.50<br>1.13<br>0.75<br>0.65 | 0.14<br>0.09<br>0.07<br>0.05<br>0.04 | 17.09<br>11.34<br>8.54<br>5.67<br>4.91 | 2.28<br>1.48<br>1.15<br>0.77<br>0.65 |
| Molybdenum | 95.94 | 3<br>4<br>6 | 0.369 | 10.22 | 2.63<br>1.97<br>1.32 | 0.12<br>0.09<br>0.06 | 19.88<br>14.89<br>9.98 | 1.95<br>1.47<br>0.98 |
| Nickel | 58.71 | 2<br>3 | 0.322 | 8.92 | 2.41<br>1.61 | 0.13<br>0.08 | 16.22<br>12.17 | 2.11<br>1.36 |
| Osmium | 109.20 | 2<br>3<br>4<br>8 | 0.815 | 22.58 | 7.82<br>5.22<br>3.91<br>1.95 | 0.16<br>0.10<br>0.08<br>0.04 | 59.12<br>39.46<br>29.56<br>14.74 | 2.62<br>1.64<br>1.31<br>0.66 |

**TABLE 11.5-3—Continued**

| METAL | ATOMIC WEIGHT | VALENCE | DENSITY | | REMOVAL RATES AT 1000 AMPERES | | | |
|---|---|---|---|---|---|---|---|---|
| | g | | lb/in³ | g/cm³ | lb/hr | in³/min | g/min | cm³/min |
| Palladium | 106.40 | 2<br>4<br>6 | 0.434 | 12.02 | 4.38<br>2.19<br>1.46 | 0.17<br>0.08<br>0.06 | 33.11<br>16.56<br>11.04 | 2.79<br>1.31<br>0.98 |
| Platinum | 195.09 | 2<br>4 | 0.775 | 21.47 | 8.02<br>4.01 | 0.17<br>0.09 | 60.63<br>30.32 | 2.79<br>1.47 |
| Rhenium | 186.20 | 3<br>4<br>5<br>6<br>7 | 0.756 | 20.94 | 7.60<br>3.82<br>3.07<br>2.55<br>2.19 | 0.17<br>0.08<br>0.07<br>0.06<br>0.05 | 57.46<br>28.88<br>23.21<br>19.28<br>16.56 | 2.79<br>1.31<br>1.15<br>0.98<br>0.82 |
| Rhodium | 102.905 | 3 | 0.447 | 12.38 | 2.82 | 0.11 | 21.32 | 1.80 |
| Silver | 107.868 | 1 | 0.379 | 10.50 | 8.87 | 0.39 | 67.06 | 6.39 |
| Tantalum | 180.948 | 5 | 0.600 | 16.62 | 2.98 | 0.08 | 22.53 | 1.31 |
| Thalium | 204.37 | 1<br>3 | 0.428 | 11.86 | 16.80<br>5.60 | 0.65<br>0.22 | 127.01<br>42.34 | 10.66<br>3.61 |
| Thorium | 232.038 | 4 | 0.421 | 11.66 | 4.76 | 0.19 | 35.99 | 3.12 |
| Tin | 118.69 | 2<br>4 | 0.264 | 7.31 | 4.88<br>2.44 | 0.31<br>0.15 | 36.89<br>18.45 | 5.05<br>2.52 |
| Titanium | 47.90 | 3<br>4 | 0.163 | 4.52 | 1.31<br>0.99 | 0.13<br>0.10 | 9.90<br>7.48 | 2.19<br>1.65 |
| Tungsten | 183.85 | 6<br>8 | 0.697 | 19.31 | 2.52<br>1.89 | 0.06<br>0.05 | 19.05<br>14.29 | 0.98<br>0.74 |
| Uranium | 238.03 | 4<br>6 | 0.689 | 19.09 | 4.90<br>3.27 | 0.12<br>0.08 | 37.04<br>24.72 | 1.92<br>1.29 |
| Vanadium | 50.942 | 3<br>5 | 0.220 | 6.09 | 1.40<br>0.84 | 0.11<br>0.06 | 10.58<br>6.35 | 1.74<br>1.05 |
| Zinc | 65.37 | 2 | 0.258 | 7.15 | 2.69 | 0.17 | 20.34 | 2.85 |
| Zirconium | 91.22 | 4 | 0.234 | 6.48 | 1.87 | 0.13 | 14.14 | 2.13 |

**TABLE 11.5-4 Material Removal Rates for ECM of Alloys Assuming 100 Percent Current Efficiency**

| ALLOY | THEORETICAL REMOVAL RATES FOR 1,000 AMPERES PER SQUARE INCH | |
|---|---|---|
| | in³/min | cm³/min |
| 4340 steel | 0.133 | 2.18 |
| 17-4 PH | 0.123 | 2.02 |
| A-286 | 0.117 | 1.92 |
| M252 | 0.110 | 1.80 |
| Rene 41 | 0.108 | 1.77 |
| Udimet 500 | 0.110 | 1.80 |
| Udimet 700 | 0.108 | 1.77 |
| L605 | 0.107 | 1.75 |

NOTE: Rates listed were calculated using Faraday's law and valences as follows:

| | | | | | |
|---|---|---|---|---|---|
| Aluminum | 3 | Copper | 2 | Silicon | 0 |
| Carbon | 0 | Iron | 2 | Titanium | 4 |
| Columbium | 3 | Manganese | 3 | Tungsten | 6 |
| Cobalt | 2 | Molybdenum | 4 | Vanadium | 5 |
| Chromium | 3 | Nickel | 2 | | |

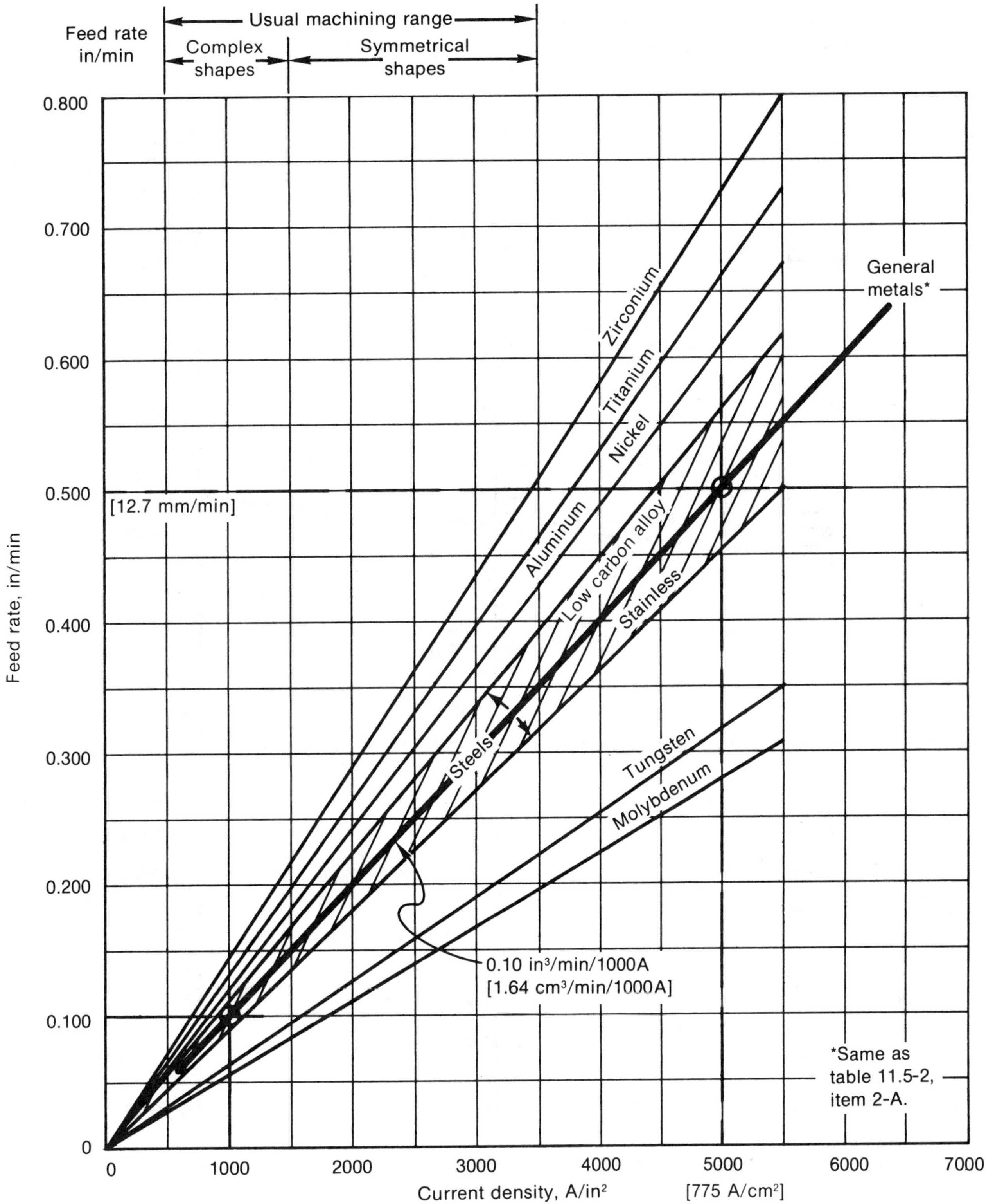


**Figure 11.5–7** Feed rates versus current density for ECM of various materials. (Courtesy of Cincinnati Milacron)

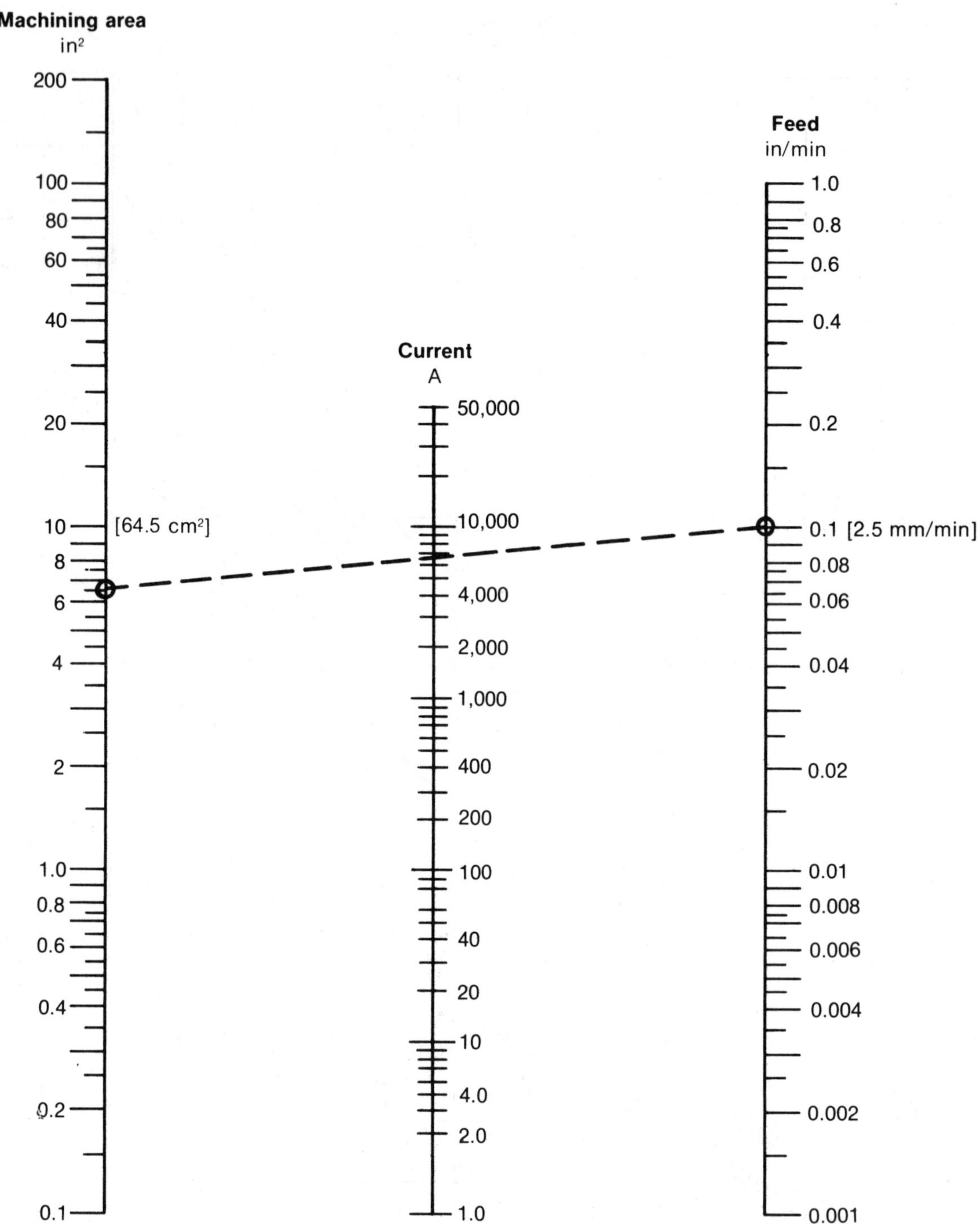


**Figure 11.5-8** Nomogram for approximating ECM feed rates and currents. (Courtesy of Anocut Inc.)

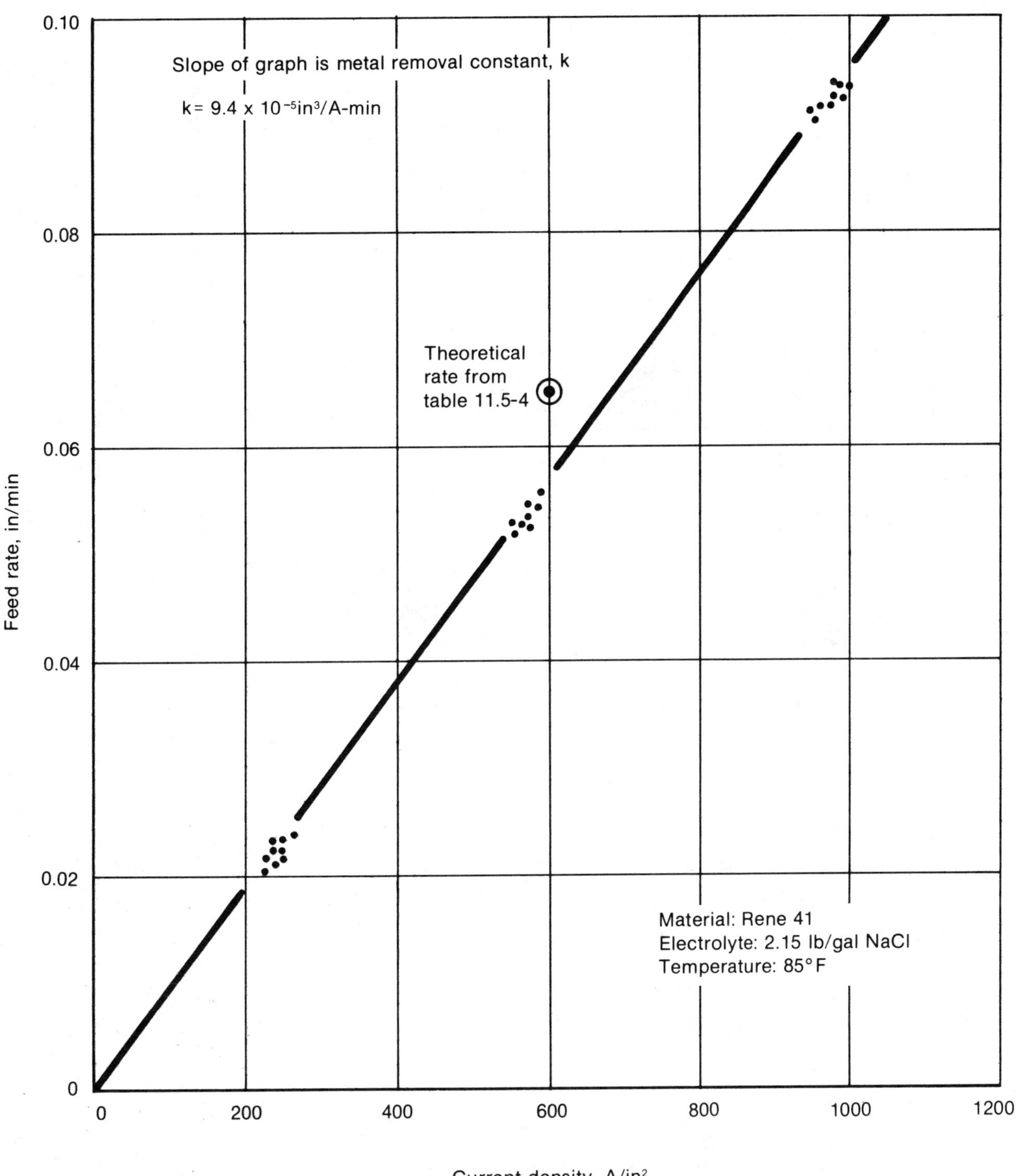


**Figure 11.5–9** Typical material removal rate data for a specific alloy-electrolyte combination with data points from ECM machinability data test rig. (G. Bellows, ECM machinability data and ratings, Technical paper MR67-710, Society of Manufacturing Engineers, Dearborn, MI, 1967)

**TABLE 11.5-5 Electrolyte Selection Guide**

| WORK MATERIAL* | ELECTROLYTE | | | | | REMARKS |
|---|---|---|---|---|---|---|
| | Composition† | Concentration‡ | | Temperature at Inlet | | |
| | | lb/gal | g/L | °F | °C | |
| ALUMINUM & ALUMINUM ALLOYS | NaCl | up to 1 | 120 | 90 to 125 | 32 to 52 | Generally better finishes. |
| | $NaNO_3$ | up to 2 | 240 | 90 to 125 | 32 to 52 | |
| 2024-T4 | $NaNO_3$ | 1.25 | 150 | 100 | 38 | |
| 7075 | NaCl + $NaNO_3$ | 0.9 + 0.2 | 108 + 24 | 100 | 38 | |
| BERYLLIUM | $NaNO_3$ | 4 | 480 | 100 | 38 | |
| COBALT ALLOYS | NaCl | up to 2 | 240 | 90 to 125 | 32 to 52 | |
| MAR-M509 | NaCl | 1.0 | 120 | 100 | 38 | |
| HS-21 | NaCl + $NaNO_3$ | 2 + 0.25 | 240 + 30 | 100 | 38 | |
| Haynes alloy 25 (L605) | NaCl + NaF | 0.25 + 0.02 | 30 + 2.4 | 100 | 38 | |
| HS-31 (X-40) | | | | | | |
| COLUMBIUM ALLOYS | — | — | — | — | — | Difficult to ECM. Special electrolytes and special flow control tooling required. |
| COPPER & COPPER ALLOYS | $NaNO_3$ | up to 4 | 480 | 90 to 125 | 32 to 52 | Generally better finishes. |
| | NaCl | up to 2 | 240 | 90 to 125 | 32 to 52 | |
| IRON BASE ALLOYS | NaCl | up to 2.5 | 300 | 90 to 125 | 32 to 52 | $NaClO_3$ for specialized applications. |
| Low carbon steels | | | | | | |
| 1018 | NaCl | 0.25 | 30 | 100 | 38 | |
| Medium carbon steels | | | | | | |
| 4130 | NaCl | 1 | 120 | 100 | 38 | |
| 5160 | NaCl | 0.4 to 1.0 | 48 to 120 | 90 | 32 | |
| High strength steels | | | | | | |
| Grade 250 maraging | NaCl | 1.25 (0.25 to 1.75) | 150 (30 to 210) | 110 | 42 | |
| 300M | NaCl or | 1 | 120 | 100 | 38 | |
| | NaCl + $NaNO_3$ | 1 + 2 | 120 + 240 | 100 | 38 | |
| 4340 | NaCl or | 0.9 to 1.0 | 108 to 120 | 95 to 100 | 35 to 38 | |
| | NaCl + $NaNO_3$ | 1 + 0.5 | 120 + 60 | 100 | 38 | Also 1.8 to 1.0 lb/gal $NaNO_3$. |
| A-286 | $NaNO_3$ | 2.0 | 240 | 100 | 38 | |
| D6ac | NaCl | 1.5 | 180 | 90 | 32 | |

*Caution: Particular metallurgical state must be known because IGA, excessive pitting, or excessive roughness can occur with some heat-treat states. A metallographic check on ECM'd test specimen is recommended.

†For safety and corrosiveness, see table 11.5-6. There are many proprietary mixtures or additives to the base composition that have desirable effects on finish, corrosiveness and sometimes feed rates. These should be explored through regular vendors or ECM operators.

‡A wide range of concentrations (and compositions) will permit ECM to continue, so the final selection will depend on a balanced relationship among all factors. The lower concentrations will reduce the salt deposits on equipment and tools. Note that about 80 percent of applications circa 1979 use NaCl (in various concentrations), 18 percent use $NaNO_3$, and all others comprise only 2 percent of total.

**TABLE 11.5-5—Continued**

| WORK MATERIAL* | ELECTROLYTE | | | | | REMARKS |
|---|---|---|---|---|---|---|
| | Composition† | Concentration‡ | | Temperature at Inlet | | |
| | | lb/gal | g/L | °F | °C | |
| Tool steels | | | | | | |
| M50 | $NaClO_3$§ | 2 | 240 | 90 | 32 | |
| L6 | $NaClO_3$§ | 3.75 | 450 | 86 | 25 | |
| D2 | NaCl | 1.0 | 120 | 100 | 38 | |
| Carpenter 20 | NaCl | 0.67 | 80 | 85 | 25 | |
| Stainless steels | | | | | | |
| Austenitic | | | | | | |
| 302 | NaCl + NaF | 0.25 + 0.02 | 30 + 2.4 | 100 | 38 | Wrought alloys with nonconductive inclusion can yield erratic results. |
| 303 | NaCl + $NaNO_3$ | 1.0 + 0.17 | 120 + 20 | 70 | 21 | |
| 316 | NaCl | 1 | 120 | 100 | 38 | |
| Martensitic | | | | | | |
| 410 | NaCl or | 0.8 | 96 | 80 | 27 | |
| | NaCl + $NaNO_3$ | 1.6 + 0.2 | 192 + 24 | 115 | 46 | |
| Precipitation hardening | | | | | | |
| 17-4 PH | NaCl or | 0.8 to 1 | 96 to 120 | 80 | 26 | |
| | $NaNO_3$ | 2 (0.25 to 4) | 240 (60 to 480) | 100 | 38 | |
| Gray cast iron | NaCl | up to 2.5 | 300 | 90 to 125 | 32 to 52 | Feed rates limited by size of graphite particles. |
| 12% Cr white cast iron | $NaNO_3$ or | up to 4 | 480 | 90 to 125 | 32 to 52 | Rougher finishes. |
| | NaCl | 1.2 | 144 | 120 | 49 | |
| MOLYBDENUM ALLOYS | NaOH | up to 1.5 | 180 | 100 | 38 | Electrolyte becomes acid with use. NaOH may be added to maintain pH. |
| TZM | NaCl | 1 | 120 | 95 | 35 | |
| NONMETALLICS $B_4C$, ZrC, $TiB_2$, $ZrB_2$ | NaCl | up to 2.5 | 300 | 100 | 38 | |
| NICKEL BASE ALLOYS | NaCl or | up to 2.5 | 300 | 100 | 38 | Side walls and low current density areas subject to IGA and/or rough finishes. |
| | $NaNO_3$ | up to 4 | 480 | 100 | 38 | |

*Caution: Particular metallurgical state must be known because IGA, excessive pitting, or excessive roughness can occur with some heat-treat states. A metallographic check on ECM'd test specimen is recommended.

†For safety and corrosiveness, see table 11.5-6. There are many proprietary mixtures or additives to the base composition that have desirable effects on finish, corrosiveness and sometimes feed rates. These should be explored through regular vendors or ECM operators.

‡A wide range of concentrations (and compositions) will permit ECM to continue, so the final selection will depend on a balanced relationship among all factors. The lower concentrations will reduce the salt deposits on equipment and tools. Note that about 80 percent of applications circa 1979 use NaCl (in various concentrations), 18 percent use $NaNO_3$, and all others comprise only 2 percent of total.

§See precautions, table 11.5-6.

**TABLE 11.5-5—Continued**

| WORK MATERIAL* | ELECTROLYTE | | | | | REMARKS |
|---|---|---|---|---|---|---|
| | Composition† | Concentration‡ | | Temperature at Inlet | | |
| | | lb/gal | g/L | °F | °C | |
| Wrought | | | | | | |
| M252 | NaCl | 2 | 240 | 105 | 41 | ST (not aged condition) gives better results. |
| Waspaloy | NaCl | 1 | 120 | 95 | 35 | |
| Hastelloy alloy X | NaCl | 2 | 240 | 75 | 24 | |
| Astroloy | $NaNO_3$ or | 1 to 2 | 120 to 240 | 100 | 38 | Cuts easily and rapidly. |
| | NaCl + NaF | 0.75 + 0.25 | 90 + 30 | 100 | 38 | |
| Inconel alloy 700 | NaCl | 1 | 120 | 100 | 38 | |
| Inconel 706 | NaCl | 0.8 | 96 | 75 | 24 | |
| Inconel 718 | NaCl or | 0.9 to 1.0 | 108 to 120 | 95 to 100 | 35 to 38 | Better results in solution treated and aged condition. |
| | $NaNO_3$ | 1.8 to 2.0 | 216 to 240 | 95 to 100 | 35 to 38 | |
| Rene 41 | NaCl | 2.15 (or 1.0) | 258 (or 120) | 75 (or 95) | 24 (or 35) | |
| Udimet 500 | NaCl | 1 | 120 | 100 | 38 | |
| Udimet 700 | NaCl | 1 | 120 | 100 | 38 | |
| Inconel alloy X | NaCl + $NaNO_3$ | 2 + 0.5 | 240 + 60 | 100 | 38 | |
| Cast | | | | | | |
| IN-100 | NaCl | 1 | 120 | 100 | 38 | |
| Rene 80 | NaCl | 1 | 120 | 100 | 38 | |
| Rene 125 | NaCl | 1 | 120 | 100 | 38 | |
| AF2-1DA | $NaNO_3$ | 2 | 240 | 100 | 38 | |
| Rene 95 | $NaNO_3$ or | 1.8 to 2.0 | 216 to 240 | 95 to 100 | 35 to 38 | |
| | NaCl | 1.0 | 120 | 100 | 38 | |
| TANTALUM | — | — | — | — | — | Difficult to ECM. Special electrolytes and flow controlled tooling required. |
| TITANIUM ALLOYS | NaCl | up to 1.5 | 180 | 95 to 105 | 35 to 41 | Use voltages greater than 12. |
| Commercially pure | — | — | — | — | — | Not recommended. Forms inhibiting films. Try ECG. |
| Alpha alloys | | | | | | |
| Ti-5Al-2.5Sn (A110) | NaCl | 0.9 to 1.0 | 108 to 120 | 95 to 100 | 35 to 38 | |
| Ti-6Al-2Sn-4Zr-2Mo | NaCl or | 1 | 120 | 90 | 32 | |
| | NaCl + NaF | 3/4 + 0.01 | 90 + 1.2 | 100 | 38 | |
| Alpha-beta alloys | | | | | | |
| Ti-6Al-4V | NaCl | 1 (0.25 to 1.75) | 120 (30 to 210) | 100 | 38 | |
| Ti-8Al-1Mo-1V | NaCl | 0.8 | 96 | 105 | 41 | |
| Ti-6Al-6V-2Sn | NaCl | 1 (0.25 to 1.75) | 120 (30 to 210) | 100 | 38 | 3:1 mixture NaCl:$NaNO_3$ up to 1.0 lb/gal. Proprietary additives helpful to reduce film formation. |

*Caution: Particular metallurgical state must be known because IGA, excessive pitting, or excessive roughness can occur with some heat-treat states. A metallographic check on ECM'd test specimen is recommended.

†For safety and corrosiveness, see table 11.5-6. There are many proprietary mixtures or additives to the base composition that have desirable effects on finish, corrosiveness and sometimes feed rates. These should be explored through regular vendors or ECM operators.

‡A wide range of concentrations (and compositions) will permit ECM to continue, so the final selection will depend on a balanced relationship among all factors. The lower concentrations will reduce the salt deposits on equipment and tools. Note that about 80 percent of applications circa 1979 use NaCl (in various concentrations), 18 percent use $NaNO_3$, and all others comprise only 2 percent of total.

**TABLE 11.5-5—Continued**

| WORK MATERIAL* | ELECTROLYTE | | | | | REMARKS |
|---|---|---|---|---|---|---|
| | Composition† | Concentration‡ | | Temperature at Inlet | | |
| | | lb/gal | g/L | °F | °C | |
| TUNGSTEN | NaOH<br>NaOH | 0.25 to 1.5<br>2 | 30 to 180<br>240 | 100<br>75 | 38<br>24 | Must constantly be replenished. (Personnel hazard and hard on pumps as nonlubricating.) |
| URANIUM | NaCl | up to 2 | 240 | 100 | 38 | |
| ZINC ALLOYS | NaCl | up to 2 | 240 | 100 | 38 | |
| ZIRCONIUM | NaCl | up to 2 | 240 | 100 | 38 | |

*Caution: Particular metallurgical state must be known because IGA, excessive pitting, or excessive roughness can occur with some heat-treat states. A metallographic check on ECM'd test specimen is recommended.

†For safety and corrosiveness, see table 11.5-6. There are many proprietary mixtures or additives to the base composition that have desirable effects on finish, corrosiveness and sometimes feed rates. These should be explored through regular vendors or ECM operators.

‡A wide range of concentrations (and compositions) will permit ECM to continue, so the final selection will depend on a balanced relationship among all factors. The lower concentrations will reduce the salt deposits on equipment and tools. Note that about 80 percent of applications circa 1979 use NaCl (in various concentrations), 18 percent use $NaNO_3$, and all others comprise only 2 percent of total.

**TABLE 11.5-6 Corrosiveness and Safety Considerations for ECM Electrolytes**

| ELECTROLYTE | COST RATIO* | SAFETY | CORROSIVENESS |
|---|---|---|---|
| Sodium chloride ($NaCl$) | 1.0 | Safe. | Highly corrosive. Brass or stainless steel required for tooling. |
| Sodium nitrate ($NaNO_3$) | 3.0† | Safe in solution; moderate fire hazard when dry and mixed with organic materials. | Noncorrosive—mild steel can be used for tooling. |
| Sodium chlorate ($NaClO_3$) | 7.5 | Safe in solution; very highly combustible when dry. Special precautions required. | Noncorrosive. |
| Sodium hydroxide ($NaOH$) | 2.0 | Caution—will cause severe burns. | Mildly corrosive on ferrous alloys and more corrosive on aluminum, zinc and cadmium. |
| Sodium fluoride ($NaF$) | 8.0 | Poisonous and needs check of medical hazards. | Highly corrosive. |
| Potassium chloride ($KCl$) | 2.3 | Safe. | Highly corrosive. |
| Sodium nitrite ($NaNO_2$) | 6.7 | Similar to $NaNO_3$, but less stable. Special fire precautions advised. | Noncorrosive. |
| Sodium sulphate ($NaSO_4$) | 3.0 | Safe. | Highly corrosive. |
| Potassium nitrate ($KNO_3$) | 8.0 | Similar to $NaNO_3$, but less stable. | Noncorrosive. |

*Based on salt as 1.0 and market circa 1972.

†As it takes twice as much $NaNO_3$ for conductivity equivalent to NaCl, the cost ratio for equal conductivity is 1:6. Similar factors affect other electrolytes.

**TABLE 11.5-7 Boiling Point of Water at Various Pressures**

| GAUGE PRESSURE | | ABSOLUTE PRESSURE | | BOILING TEMPERATURE | |
|---|---|---|---|---|---|
| in Hg | mm Hg | in Hg | mm Hg | °F | °C |
| 29.5 | 749.3 | 0.5 | 12.7 | 58.8 | 15 |
| 29.0 | 736.6 | 1.0 | 25.4 | 79.0 | 26 |
| 20.0 | 508.0 | 10 | 254 | 161.5 | 72 |
| 10.0 | 254.0 | 20 | 508 | 192.4 | 89 |
| psi | kPa | psi | kPa | | |
| 0 | 0 | 14.7 | 101 | 212.0 | 100 |
| 35.3 | 243 | 50 | 345 | 281.0 | 138 |
| 85.3 | 588 | 100 | 689 | 327.8 | 164 |
| 135.3 | 933 | 150 | 1,034 | 358.4 | 181 |
| 185.3 | 1,278 | 200 | 1,379 | 381.8 | 194 |

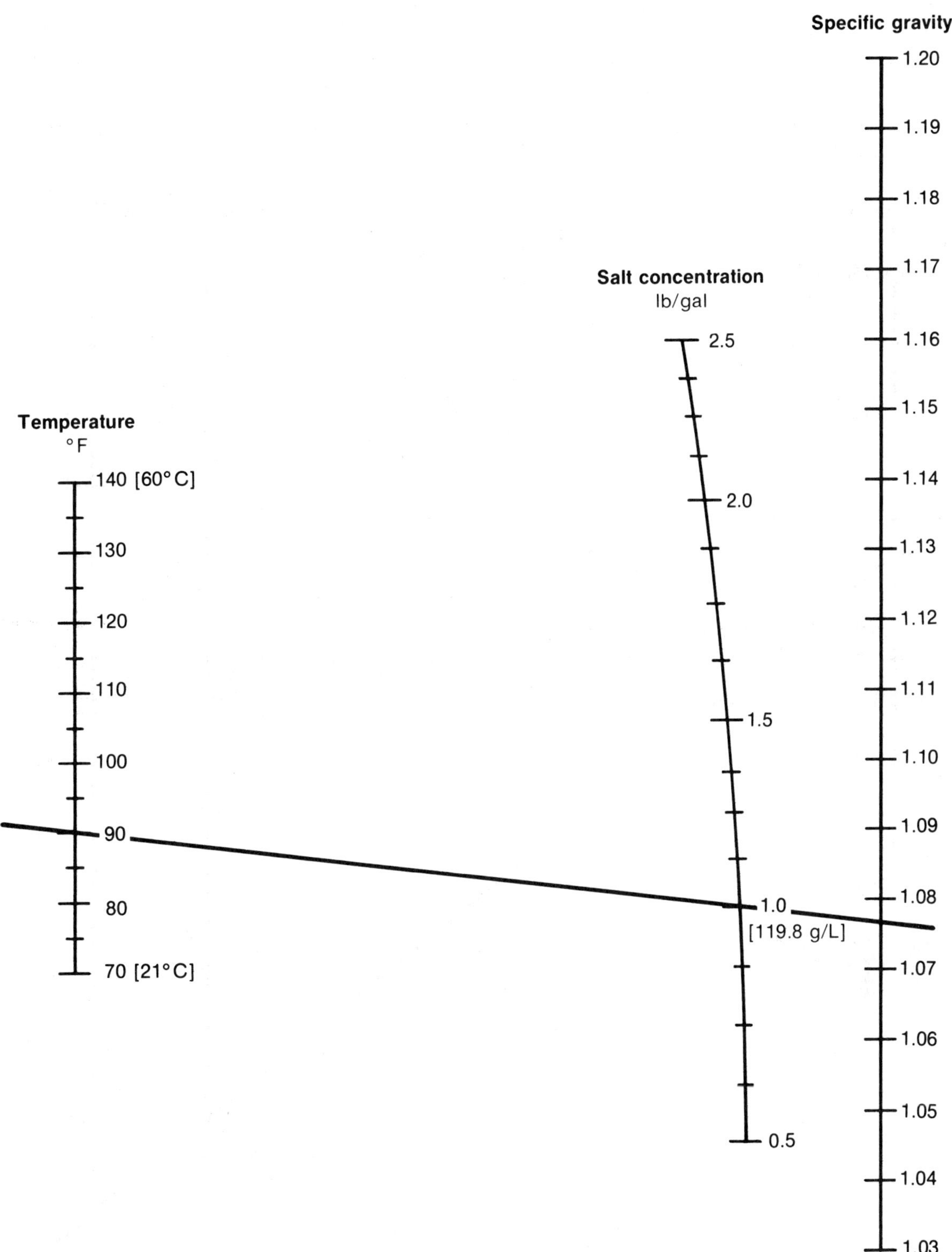


**Figure 11.5–10** Nomogram to determine the specific gravity of sodium chloride solutions. Specific gravity measured by 60°F/60°F hydrometer. (Courtesy of Anocut Engineering Company)

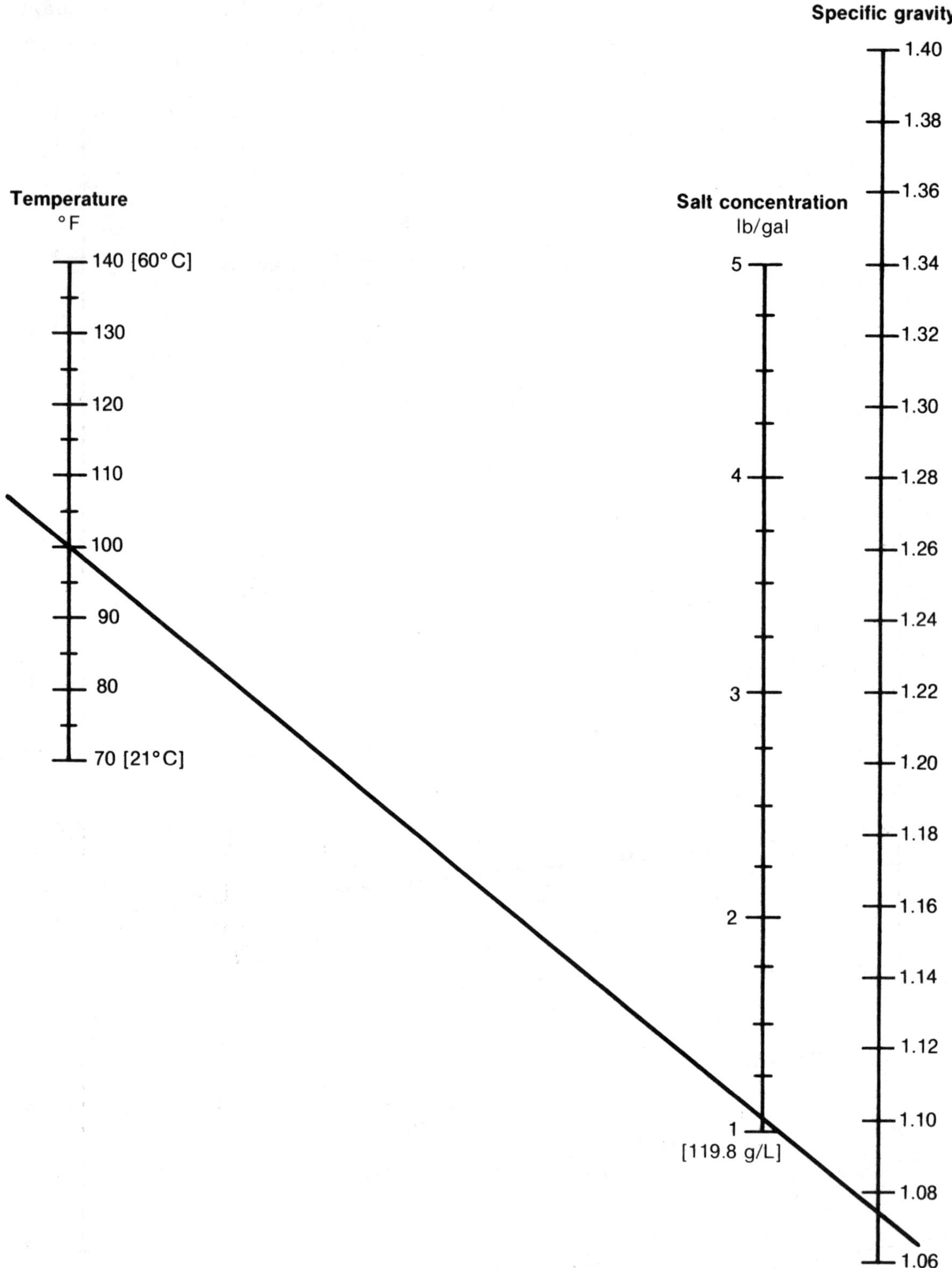


**Figure 11.5–11** Nomogram to determine the specific gravity of sodium nitrate solutions. Specific gravity measured by 60°F/60°F hydrometer. (Courtesy of Anocut Engineering Company)

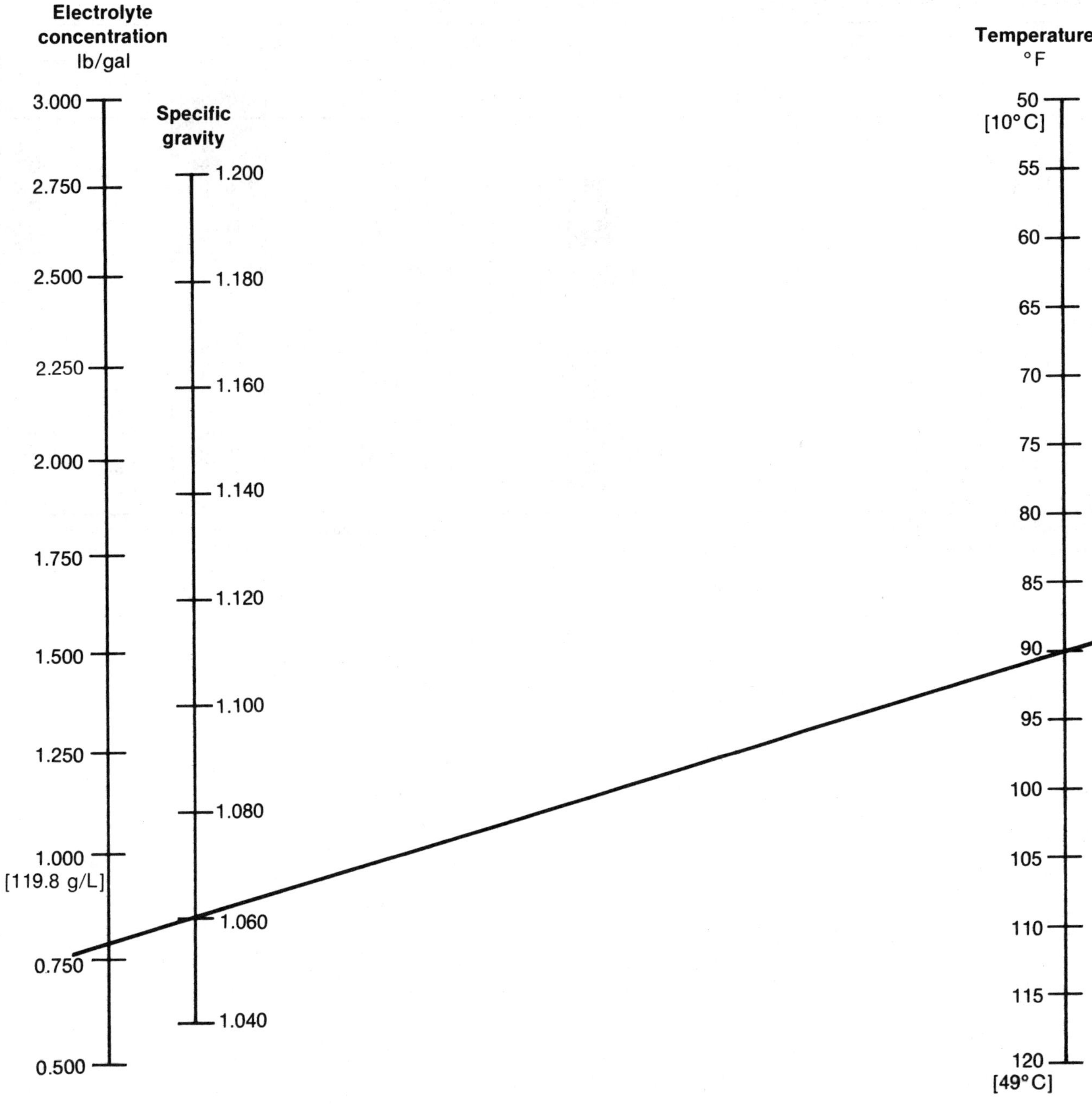


**Figure 11.5–12** Nomogram to determine the specific gravity of 50 percent sodium chloride, 50 percent sodium nitrate solutions. (Courtesy of General Electric Company)

**TABLE 11.5–8 Sodium Chloride (NaCl) Electrolyte Concentration Conversion Chart**

| HYDROMETER READING | | | CONCENTRATION/TEMPERATURE | | | | | | | |
|---|---|---|---|---|---|---|---|---|---|---|
| °Baume* | Specific Gravity | °Salimeter* | 100°F lb/gal | 38°C g/L | 110°F lb/gal | 43°C g/L | 120°F lb/gal | 49°C g/L | 130°F lb/gal | 54°C g/L |
| 8 | 1.06 | 30 | 0.7 | 83.9 | 0.6 | 71.9 | 0.6 | 71.9 | 0.6 | 71.9 |
| 9 | 1.07 | 34 | 0.8 | 95.8 | 0.7 | 83.9 | 0.7 | 83.9 | 0.7 | 83.9 |
| 10 | 1.07 | 38 | 0.9 | 107.8 | 0.8 | 95.8 | 0.8 | 95.8 | 0.8 | 95.8 |
| 11 | 1.08 | 42 | 1.0 | 119.8 | 0.9 | 107.8 | 0.9 | 107.8 | 0.9 | 107.8 |
| 12 | 1.09 | 46 | 1.1 | 131.8 | 1.0 | 119.8 | 1.0 | 119.8 | 1.0 | 119.8 |
| 13 | 1.10 | 50 | 1.2 | 143.8 | 1.2 | 143.8 | 1.1 | 131.8 | 1.1 | 131.8 |
| 14 | 1.11 | 54 | 1.3 | 155.7 | 1.3 | 155.7 | 1.3 | 155.7 | 1.2 | 143.8 |
| 15 | 1.11 | 58 | 1.5 | 179.7 | 1.4 | 167.7 | 1.4 | 167.7 | 1.3 | 155.7 |
| 16 | 1.12 | 63 | 1.6 | 191.7 | 1.5 | 179.7 | 1.5 | 179.7 | 1.5 | 179.7 |
| 17 | 1.13 | 67 | 1.7 | 203.7 | 1.7 | 203.7 | 1.6 | 191.7 | 1.6 | 191.7 |
| 18 | 1.14 | 71 | 1.9 | 227.6 | 1.8 | 215.6 | 1.8 | 215.6 | 1.7 | 203.7 |
| 19 | 1.15 | 76 | 2.0 | 239.6 | 2.0 | 239.6 | 1.9 | 227.6 | 1.9 | 227.6 |
| 20 | 1.16 | 80 | 2.1 | 251.6 | 2.1 | 251.6 | 2.0 | 239.6 | 2.0 | 239.6 |
| 21 | 1.17 | 84 | 2.3 | 275.5 | 2.2 | 263.6 | 2.2 | 263.6 | 2.1 | 251.6 |
| 22 | 1.18 | 88 | 2.5 | 299.5 | 2.4 | 287.5 | 2.4 | 287.5 | 2.3 | 275.5 |
| 23 | 1.19 | 93 | 2.6 | 311.5 | 2.6 | 311.5 | 2.5 | 299.5 | 2.5 | 299.5 |
| 24 | 1.20 | 97 | 2.8 | 335.4 | 2.8 | 335.4 | 2.7 | 323.5 | 2.6 | 311.5 |
| 25 | 1.20 | 100 | 3.0 | 359.4 | 2.9 | 347.4 | 2.9 | 347.4 | 2.8 | 335.4 |

SOURCE: Cincinnati Milacron.

*Generally, a reading on the Baume scale is approximately the percentage of salt by weight. A reading on the Salimeter scale represents the percentage of saturation in a sodium chloride solution.

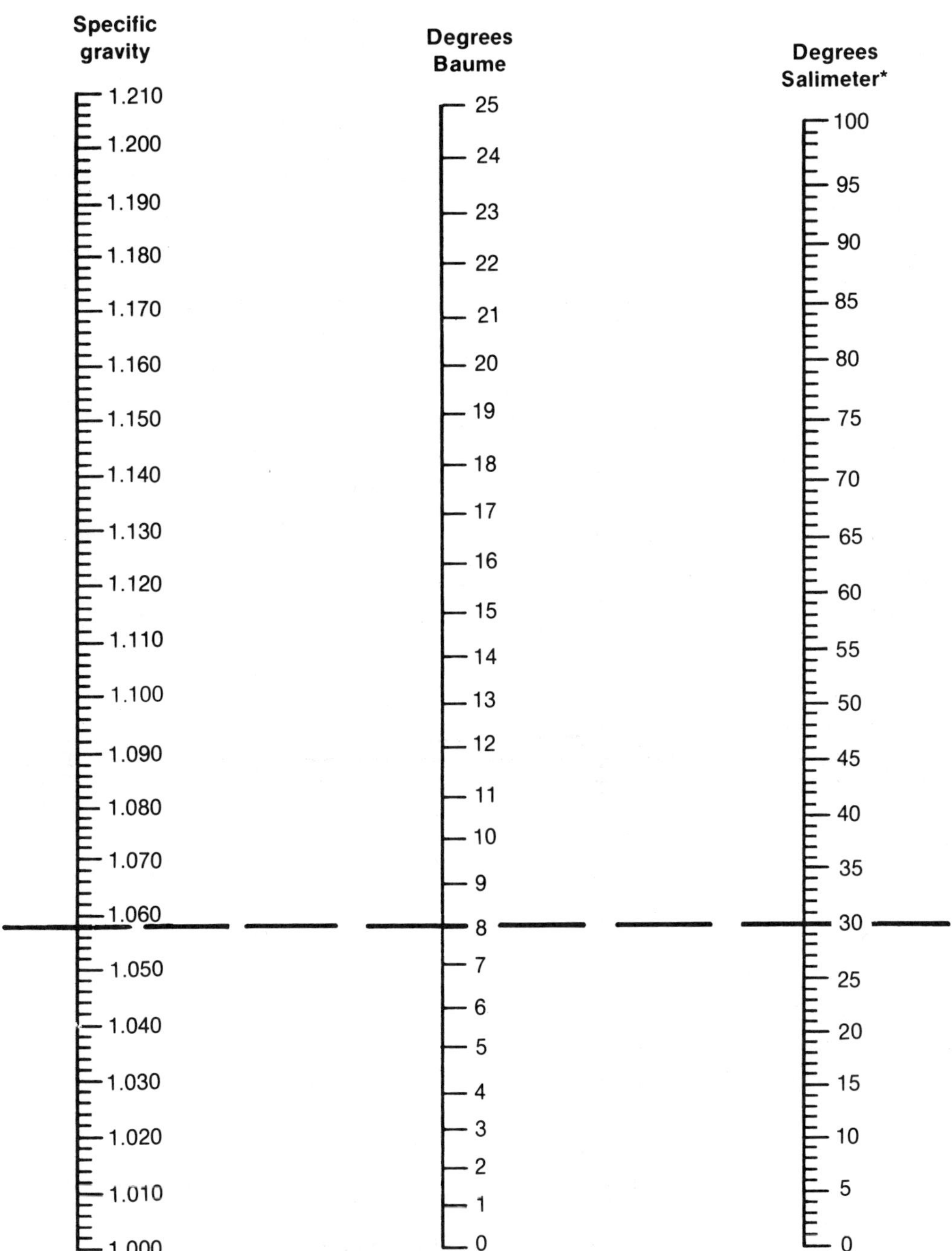


*Degrees Salimeter = 0.26395% salt by weight per degree salimeter.

**Figure 11.5-13** Nomogram for salt concentration. Hydrometer scale comparison at 68°F. (Courtesy of International Salt Company)

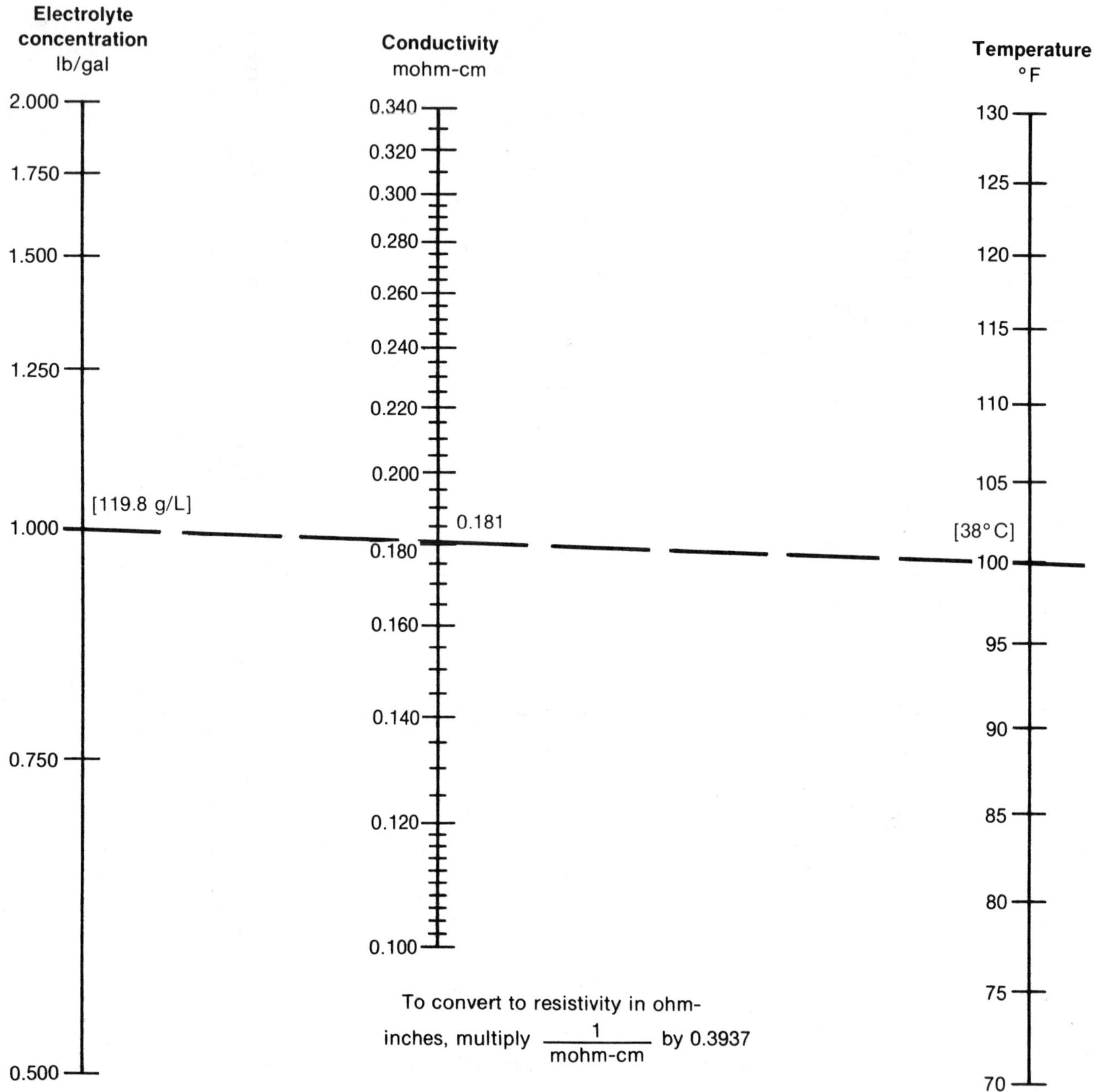


**Figure 11.5-14** Nomogram to determine the conductivity of sodium chloride electrolytes. (J. A. Cross and A. U. Jollis, Electrochemical machining (ECM), Technical report AFML-TR-72-188, Vol. II, General Electric Company, 1972, p. 83)

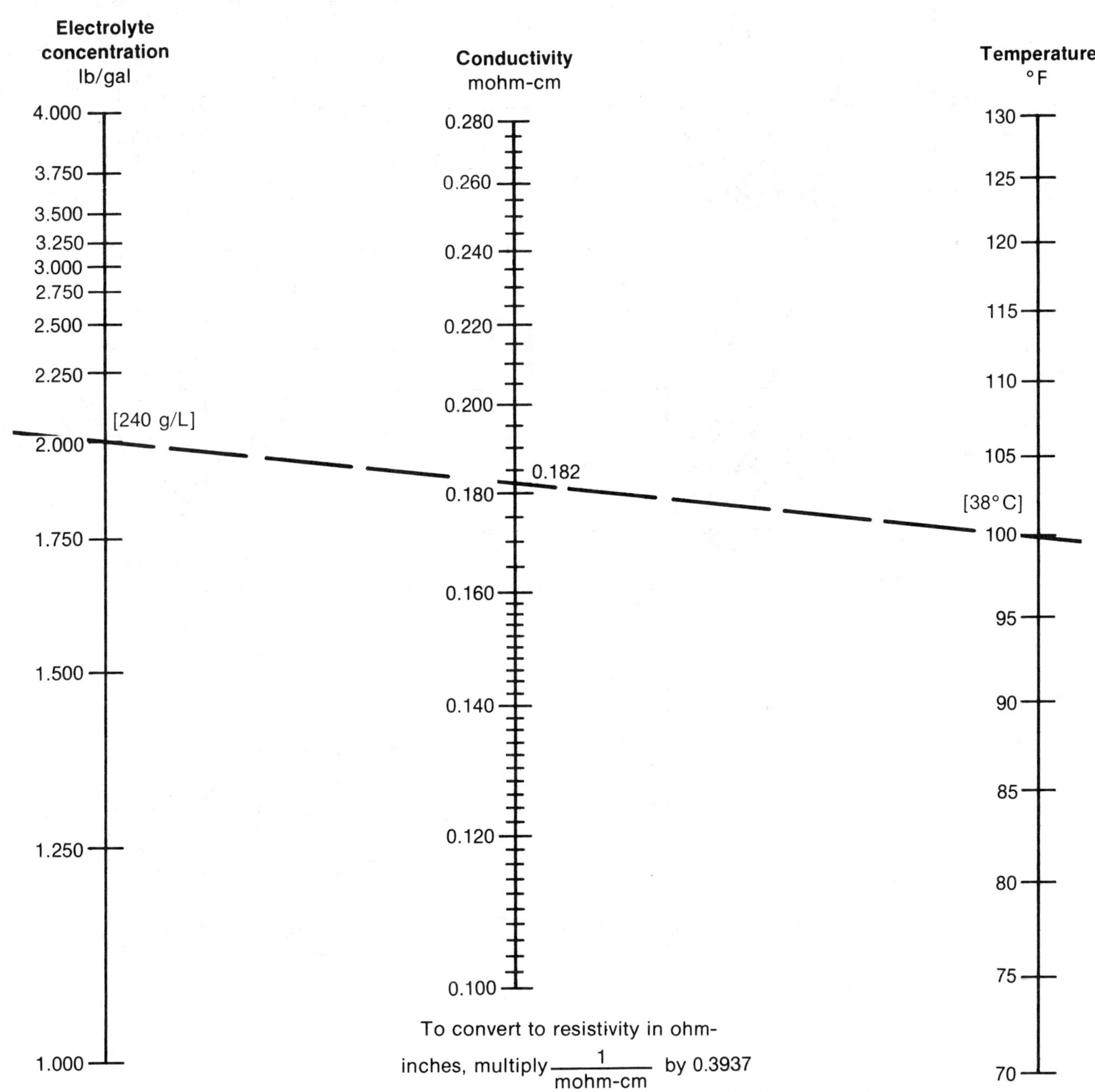


**Figure 11.5–15** Nomogram to determine the conductivity of sodium nitrate electrolytes. (J. A. Cross and A. U. Jollis, p. 84)

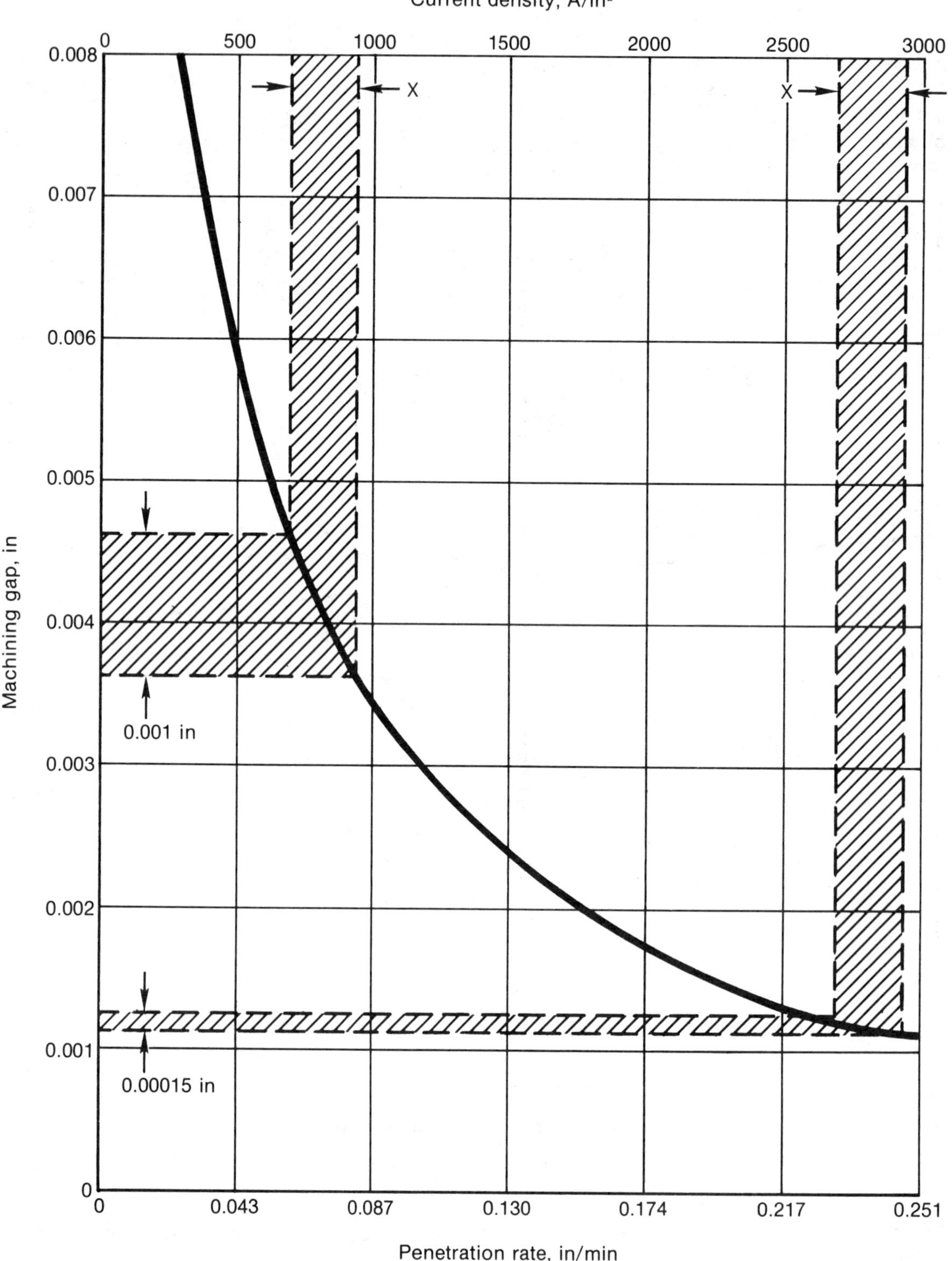


**Figure 11.5–16** ECM of steel in a neutral electrolyte at 120°F and 15 volts. The "X" bands represent the variability expected in normal ECM. (Courtesy of Chemform Inc.)

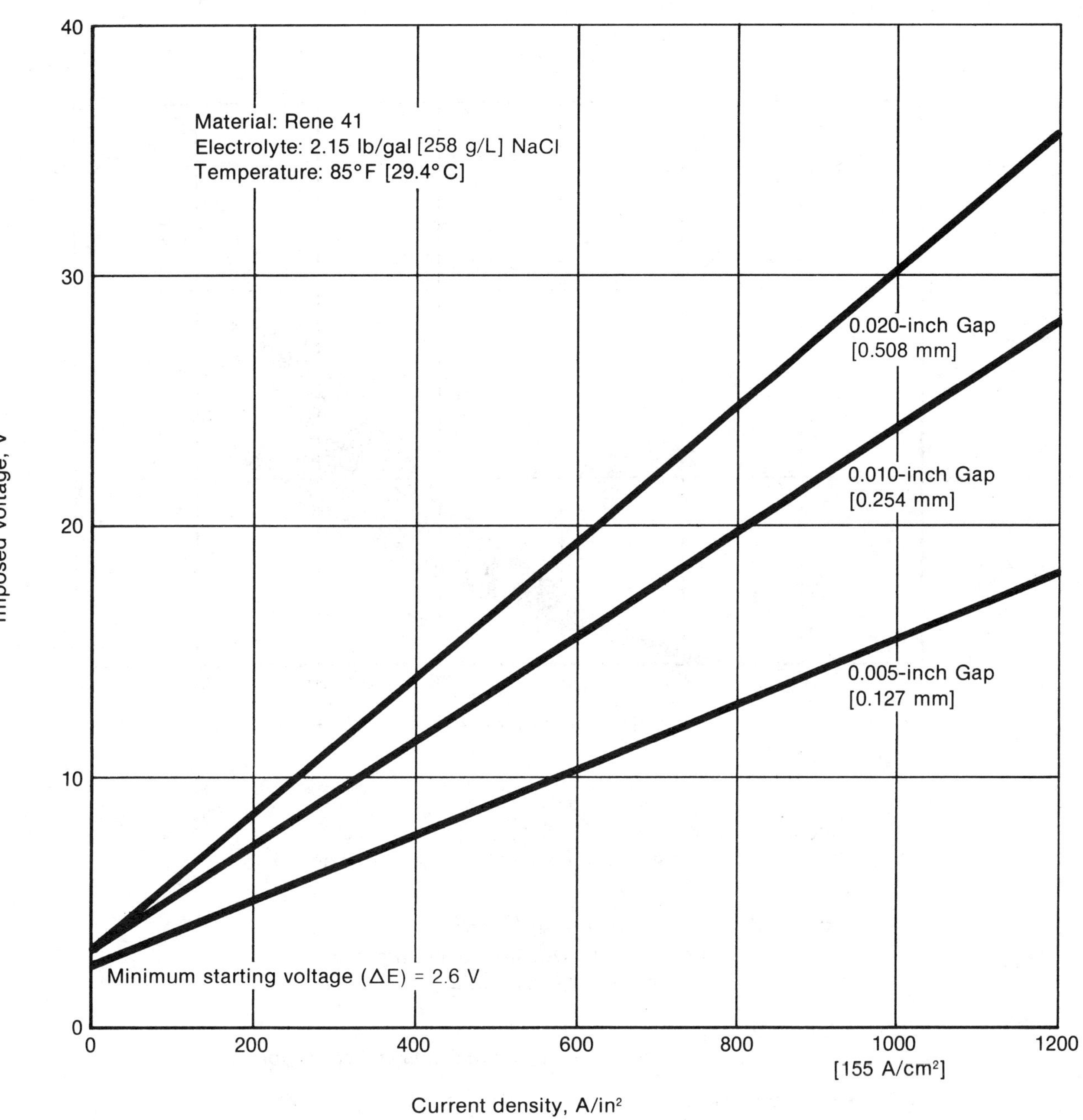


**Figure 11.5–17** Specific gap selection data for specific alloy-electrolyte combination. (Note that figure 11.5–9 and 11.5–17 are the same data set.) (G. Bellows, ECM machinability data and ratings, Technical paper MR67-710, Society of Manufacturing Engineers, Dearborn, MI, 1967)

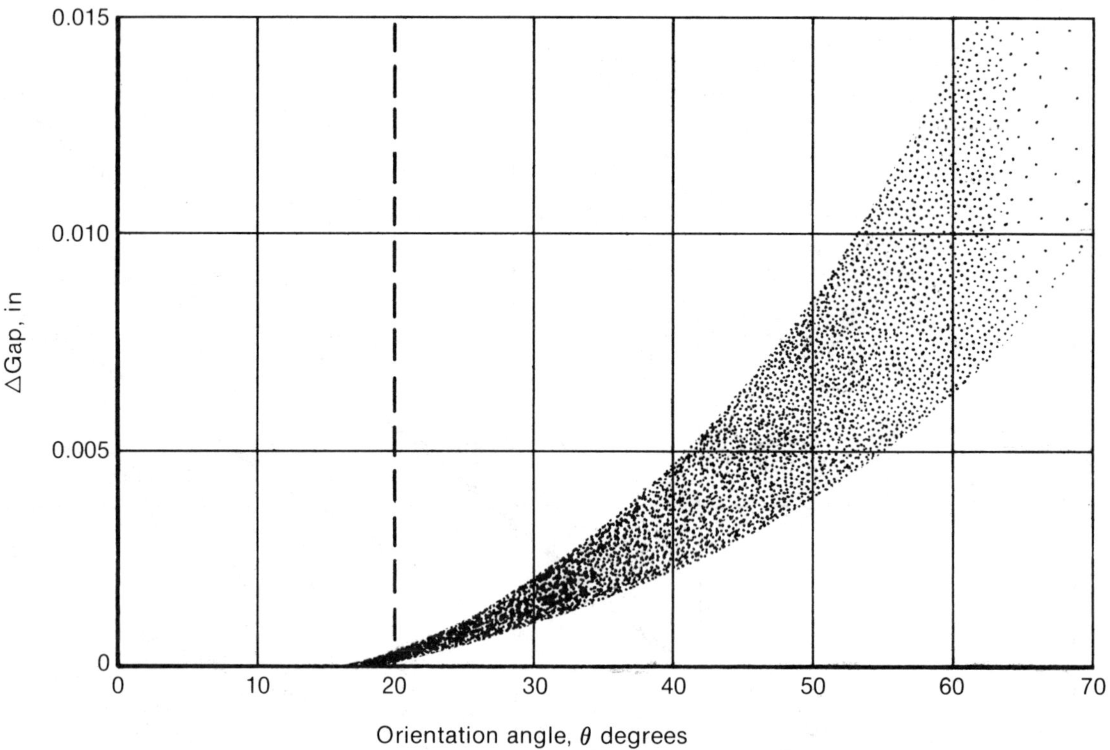


- ANGLE $\theta$ — $0°$ to $20°$ ($\alpha = 70°$ to $90°$)
  Use mirror or conjugate image of part and uniform gap, $S_c = S_e$.
- ANGLE $\theta$ — $20°$ to $40°$ ($\alpha = 50°$ to $70°$)
  Correct tool shape from uniform gap to
  $S_c = S_e \sec\theta$ or $S_c = \dfrac{Se}{\sin\alpha}$.
- ANGLE $\theta$ — greater than $40°$ ($\alpha$ less than $50°$)
  Develop tool shape by trial cuts and correction.
- An Approximation:
  $\Delta Gap \cong S_e (\sec\theta - \cos\theta)$

**Figure 11.5–18** Design of contour cutting electrodes in frontal ECM cutting under steady-state (equilibrium) process conditions.

**TABLE 11.5-9 Typical Operating Conditions for ECM of Aerospace Alloys**

| WORKPIECE MATERIAL | ELECTROLYTE | | | MINIMUM STARTING VOLTAGE (ΔE) | METAL REMOVAL CONSTANT (k)* | |
|---|---|---|---|---|---|---|
| | Type | Concentration | | | | |
| | | oz/gal | g/L | | $in^3$/A-min x $10^{-5}$ | $cm^3$/A-min x $10^{-3}$ |
| Inconel alloy 718, STA | NaCl | 16 | 120 | 3.3 | 8.8 | 1.44 |
| AF 95, STA (Rene 95) | $NaNO_3$† | 36 | 270 | 4.3 | 10.3 | 1.69 |
| Ti-6Al-6V-2Sn, Annealed | NaCl | 16 | 120 | 3.2 | 10.2 | 1.67 |
| Ti-6Al-4V, Annealed | NaCl | 16 | 120 | 3.8 | 10.0 | 1.64 |
| 17-4 PH, STA | $NaNO_3$ | 36 | 270 | 3.6 | 8.6 | 1.41 |
| 300M, Tempered | NaCl | 16 | 120 | 1.1 | 13.2 | 2.16 |
| 18%NI Marage, STA | NaCl | 16 | 120 | 1.0 | 13.2 | 2.16 |
| MAR-M509, As cast | NaCl | 16 | 120 | 1.2 | 9.6 | 1.57 |
| Astroloy | $NaNO_3$ | 32 | 240 | 4.0 | 12.5 | 2.05 |

SOURCE: J. A. Cross and A. U. Jollis, Electrochemical machining (ECM), Technical report AFML-TR-72-188, General Electric Company, 1972.

*Used for estimating current required in formula

$$I = \frac{A_w v_f}{k}$$

$I$ = current, amperes
$A_w$ = area being machined, $in^2$ [$cm^2$]
$v_f$ = feed rate, in/min [cm/min]
$k$ = metal removal constant, $in^3$/A-min [$cm^3$/A-min]

See also figure 11.5-7 and table 11.5-3.

†Can also be cut satisfactorily with 16 oz/gal NaCl.

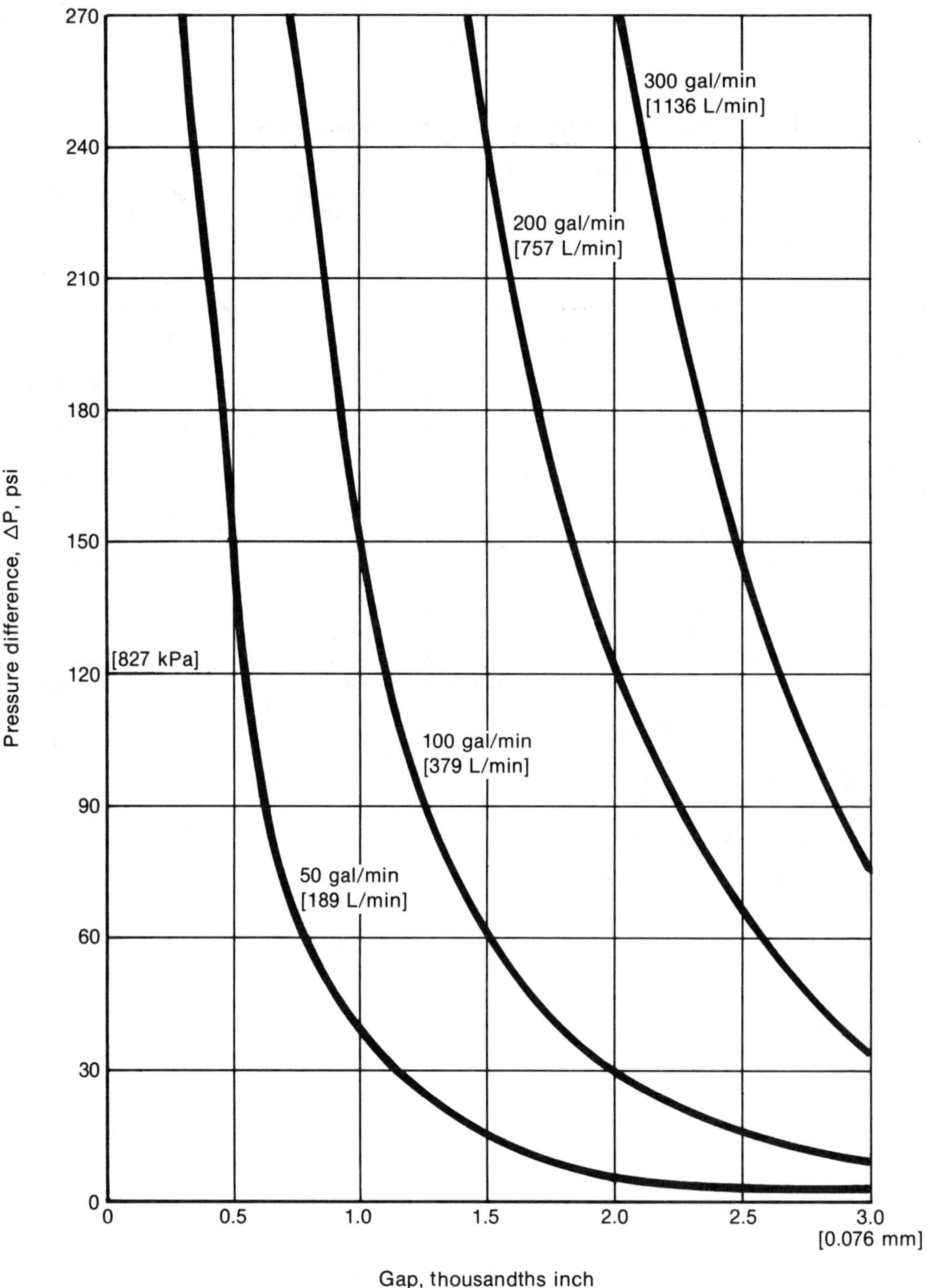


**Figure 11.5–19** Theoretical pressure drop for an idealized cathode as a function of flow rate and gap size. (Courtesy of Anocut Inc.)

**TABLE 11.5–10 Electrode and Tool Material Data for ECM**

| MATERIAL | ELECTRICAL RESISTIVITY | COST | TENSILE STRENGTH (RT) | | STIFFNESS (MODULUS ELASTICITY) | | THERMAL CONDUCTIVITY | MELTING POINT | | SPARK RESISTANCE | REPAIRABILITY | COMMENTS |
|---|---|---|---|---|---|---|---|---|---|---|---|---|
| | microhm-cm at 20°C | | ksi | MPa | psi x10⁶ | kPa x10⁶ | cal/cm²/cm/s/°C | °F | °C | | | |
| Copper (OF) as rolled | 1.71 | medium | 58 | 400 | 17.6 | 121 | 0.934 | 1,980 | 1,082 | poor | fair | Lowest resistivity. |
| Naval brass, (60Cu-39Zn) | 6.63 | low | 75 | 517 | 15.0 | 103 | 0.287 | 1,640 | 893 | poor | fair | Best general purpose electrode. |
| Bronze, commercial | 3.8 | medium | 88 | 606 | 15.0 | 103 | 0.479 | 1,625 | 885 | poor | fair | High silicon, aluminum or phosphor bronze can be used for additional strength. |
| Austenitic stainless steel, 316 or other | 73.0 | high | 85 | 586 | 29.0 | 200 | 0.039 | 2,525 | 1,385 | good | good | Most easily repaired (by arc welding). Avoid for long electrodes. |
| Beryllium copper (heat treated) | 7.8 (3.8 to 8.5) | high | 110 | 758 | 16.0 | 110 | 0.139 to 0.277 | 1,780 to 1,990 | 970 to 1,088 | poor | fair | High strength. Use for thin-wall electrodes when high electrolyte pressure required. |
| Copper tungsten | 5.0 (varies with composition) | very high | low | low | — | — | | | | very good | poor | Use when spark damage must be minimized. |
| Silver tungsten | 4.0 (varies with composition) | very high | 75 | 517 | — | — | | | | very good | poor | Use when spark damage must be minimized. |
| Tungsten, ductile | 13.1 | very high | 130 | 896 | 53 | 365 | 0.264 | 6,126 | 3,386 | excellent | poor | Difficult to fabricate. |
| Titanium alloys | 170 | high | 61 | 420 | 16.0 | 110 | 0.098 | 3,272 | 1,800 | fair | fair | Good corrosion resistance. |
| SAE 1045 carbon steel | 16.2 | low | 95 | 655 | 29.5 | 203 | 0.141 | 2,700 | 1,482 | poor | good | Not recommended for electrodes, corrodes easily. |

**TABLE 11.5–11 Insulating Materials for ECM**

| MATERIAL | TYPICAL SOURCE | METHOD OF APPLICATION |
|---|---|---|
| #260 Epoxy powder | 3M Company | Spray or fluid-bed dip |
| E-201 Epoxy powder | Armstrong Resins Inc. | Spray or fluid-bed dip |
| Microsol E-1003 | Michigan Chrome & Chemical | Liquid dip |
| Anocoat 6004 epoxy putty | Anocut Inc. | Knife |
| Anocoat 6005 | Anocut Inc. | Brush |
| Anocoat 6006 | Anocut Inc. | Spray or fluid-bed dip |
| Plexiglass | Various | Fabricate (for short life applications) |
| Epoxy-glass laminates | Various | Fabricate |
| Textolite 11558 (NEMA G10; sometimes called green glass) | General Electric Co. | Fabricate |

**TABLE 11.5–12 Organic Coatings to Protect Metals from Corrosion by ECM Electrolytes**

| COATING | CHEMICAL RESISTANCE | | | RESISTANCE TO ABRASION | REMARKS |
|---|---|---|---|---|---|
| | Dilute Hydrochloric and Sulphuric Acids | Brine Solution | Dilute Alkalis | | |
| Polyvinyl chloride (PVC) | S | S | S | Good | |
| Polythene (high density) | S | S | S | Good | Preferred to the low density material for electrical applications |
| Polythene (low density) | S | S | S | Fair | |
| Nylon | Limited | S | Limited | Very good | Very low water absorption. Electrical insulation excellent |
| PTFE | Good, but difficult to obtain coatings free of pores | | | Poor | |
| Chlorinated polyether (Penton) | S | S | S | Very good | |
| Solvent-based epoxy resin | Generally S, depending on formulation | S | As for acids | Good | Oven-drying preferred to air-drying |
| Solventless epoxy resin | S | S | S | Good | Better performance than solvent-based compositions |
| Polyurethane | S | S | S | Said to be similar to vitreous enamel | |
| Chlorinated rubber | S | S | S | Good | |
| Neoprene | Limited | S | S | Very good | |

SOURCE: A. E. De Barr and D. A. Oliver, *Electrochemical machining* (New York: American Elsevier, 1967), p. 181.
NOTE: S = Satisfactory

**TABLE 11.5–13 Volume Ratio and Specific Gravity for ECM Sludge**
(Ratio of volume of sludge to volume of metal removed for sodium chloride (NaCl) electrolytes)

| WORKPIECE MATERIAL | VOLUME RATIO | SPECIFIC GRAVITY |
|---|---|---|
| Nickel base alloy | 250:1 | 1.3 |
| Titanium base alloy | 500:1 | 1.1 |
| Iron base alloy | 150:1 | 1.3 |

SOURCE: J. A. Cross and A. U. Jollis, Electrochemical machining (ECM), Technical report AFML-TR-72-188, General Electric Company, 1972.

## 11.5 Electrochemical Machining—ECM

### SPECIFIC APPLICATIONS DATA

**TABLE 11.5–14 Specific Applications Data for Electrochemical Machining**

| Work material: | **SAE 1020 Steel, 60–75 $R_B$** | **52100 Steel, 65 $R_c$** | **T15 Tool Steel** |
|---|---|---|---|
| Workpiece configuration: | Friction disc  | 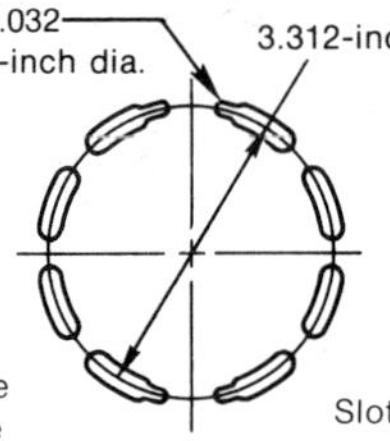 0.032 ±0.005-inch dia. 3.312-inch dia. Valve plate Slots 0.250 ±0.010 in 0.090 ±0.010 in | 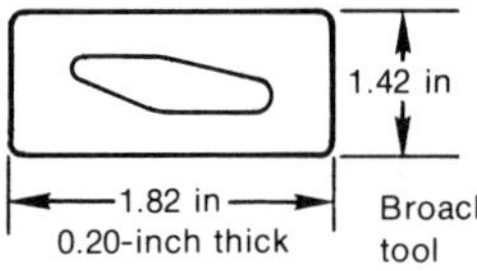 1.42 in 1.82 in 0.20-inch thick Broach tool |
| Electrolyte type: | NaCl | NaCl | NaCl |
| concentration, lb/gal: | 1.25 | 1.7 | 1.0 |
| inlet pressure, psi: | 60 to 100 | 100 to 150 | 200 |
| outlet pressure, psi: | 0 | 0 | 0 |
| flow rate, gal/min: | 10 | 18 to 20 | |
| temperature, °F: | 85 to 90 | 90 | 100 |
| Voltage, dc volts: | 20 | 13 | 13 |
| Current, A start: | 650 | 1,500 | 200 |
| end: | 600 | 2,200 | 250 |
| Feed rate, in/min: | zero (static tools) | 0.150 | 0.094 |
| Depth of cut, in: | 0.015 | 0.330 | 0.2 |
| Tolerance, in: | ±0.005 | ±0.007 | ±0.005 |
| Surface roughness, $\mu$in $R_a$: | burr free | 25 to 50 | 30 |
| Electrode material, filtration, benefits and remarks: | Brass electrode. 10 $\mu$m (micron) filters. 5-second dwell time. 72 grooves spaced at 5° intervals for 360° both sides. Groove width 0.040 inch; thickness 0.0693 inch ±0.0035 inch. One side machined at a time. | Brass electrode. 50 $\mu$m (micron) filters. All 8 kidney slots machined simultaneously. | Two pieces cut simultaneously. Piercing cut. |

**TABLE 11.5-14—Continued**

| Work material: | **A286, 25 $R_c$** | **Waspaloy** | **Udimet 700, forged** |
|---|---|---|---|
| Workpiece configuration: | 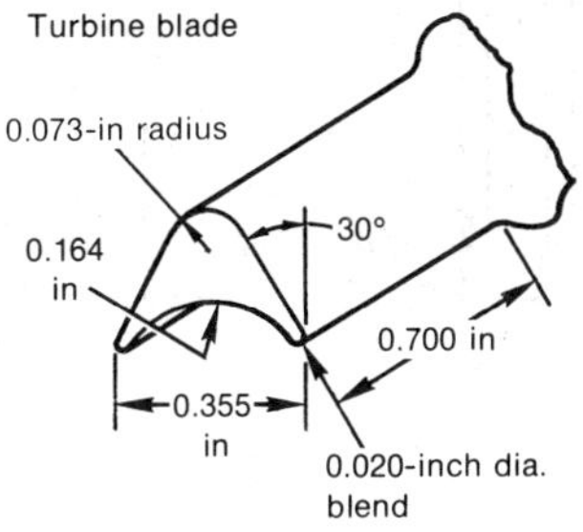 Turbine blade; 0.073-in radius; 0.164 in; 30°; 0.700 in; 0.355 in; 0.020-inch dia. blend | 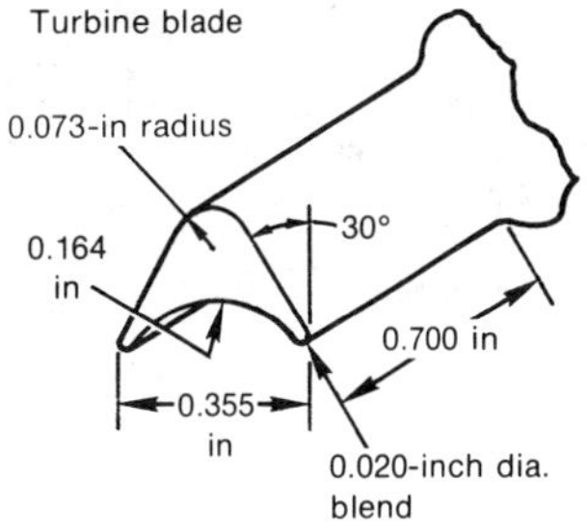 Turbine blade; 0.073-in radius; 0.164 in; 30°; 0.700 in; 0.355 in; 0.020-inch dia. blend | 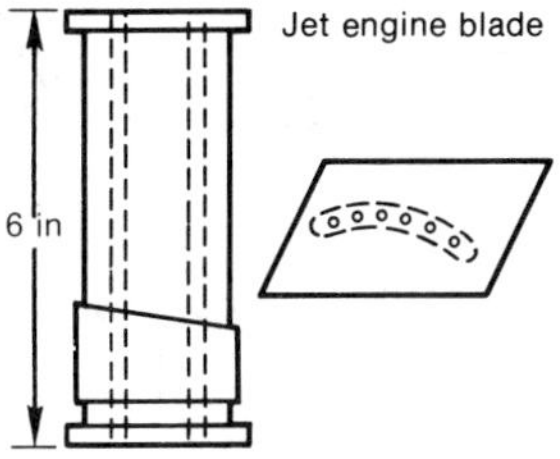 Jet engine blade; 6 in |
| Electrolyte type: | $NaNO_3$ | NaCl | $H_2SO_4$ |
| concentration, lb/gal: | 2.2 | 1.7 | 10% solution |
| inlet pressure, psi: | 130 to 200 | 130 to 200 | 15 |
| outlet pressure, psi: | 0 | 0 | 0 |
| flow rate, gal/min: | 2 | 2.5 | 0.6 |
| temperature, °F: | 110 | 90 | 95 |
| Voltage, dc volts: | 11 | 12 | 9 with periodic reversal |
| Current, A start: | 100 | 100 | 3 |
| end: | 150 | 150 | 10 |
| Feed rate, in/min: | 0.300 | 0.325 | 0.050 |
| Depth of cut, in: | 7.0 | 7.0 | 6.0 |
| Tolerance, in: | ±0.004 | ±0.004 | ±0.002 |
| Surface roughness, $\mu$in $R_a$: | 15 to 25 | 10 to 20 | 63 to 120 |
| Electrode material, filtration, benefits and remarks: | Copper-tungsten electrode. 50 $\mu$m (micron) filters. Constant section turbine blade. | Copper-tungsten electrodes. 50 $\mu$m (micron) filters. Constant section turbine blade. | Insulated titanium tubes as electrodes. 10 $\mu$m (micron) filters. 6 holes of slightly different sizes machined simultaneously. See also STEM process. |

**TABLE 11.5-14—Continued**

| Work material: | **Inconel Alloy 718, STA** | | **Tungsten** | **Uranium, 53–57 $R_A$** |
|---|---|---|---|---|
| Workpiece configuration: | 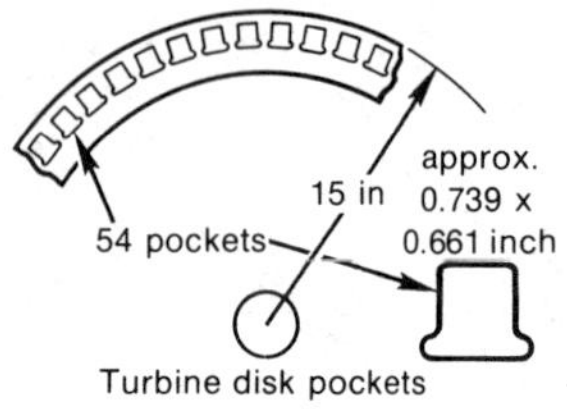 54 pockets; 15 in; approx. 0.739 x 0.661 inch; Turbine disk pockets | | 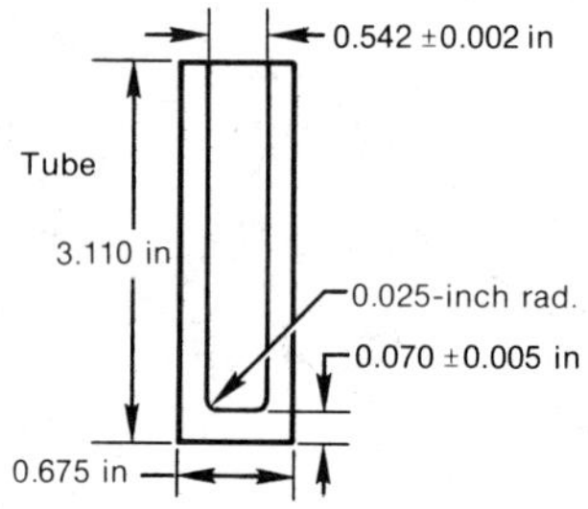 Tube; 0.542 ±0.002 in; 3.110 in; 0.025-inch rad.; 0.070 ±0.005 in; 0.675 in | 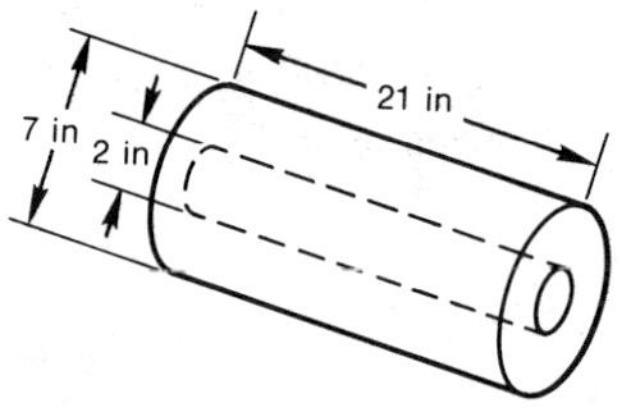 21 in; 7 in; 2 in |
| Electrolyte type: | NaCl | NaCl | NaOH | NaCl |
| concentration, lb/gal: | 1.0 ±0.1 | 1.0 | 1.5 | 1.0 |
| inlet pressure, psi: | 225 to 275 | 100, start; 160, end | 50 | 150 |
| outlet pressure, psi: | 25 to 75 | 60, start; 40, end | 0 | 0 |
| flow rate, gal/min: | 15 (est.) | 290, start; 235, end | 3 | |
| temperature, °F: | 105 ±5 | 100 | 80 | 110 |
| Voltage, dc volts: | 18 | 11 | 10 | 12.5 |
| Current, A start: | 130 (est.) | 2,800 to 3,500 | 450 | 400 (est.) |
| end: | 630 | 16,000 to 18,000 | 450 | 800 |
| Feed rate, in/min: | 0.060 | 0.032 | none | 0.10 |
| Depth of cut, in: | 0.2 | 0.3 | 3.04 | 11 |
| Tolerance, in: | ±0.003 | ±0.003 | ±0.002 | ±0.010 |
| Surface roughness, μin $R_a$: | 30 to 40, front; 75 to 150, side | 70 to 90 | 10 | |
| Electrode material, filtration, benefits and remarks: | Copper electrodes. 75 μm (micron) filter. Single pocket per stroke. | Brass electrodes (an assembly of 54 separate replaceable units). All 54 pockets per stroke. Nine-minute machining cycle. | Copper electrodes. 50 μm (micron) filter. Bore enlargement from 0.535 to 0.542 inch. Similar to ECP. | Copper tubing electrode. Trepanned 2-inch hole through 21-inch ingot. (Ingot turned end for end after penetration of the tool had reached one half the length of ingot.) |

**TABLE 11.5-14—Continued**

| Work material: | **Ti-6Al-4V** | **HC250 Alloy Steel, 54 $R_c$** | **316 Stainless Steel** |
|---|---|---|---|
| Workpiece configuration: | 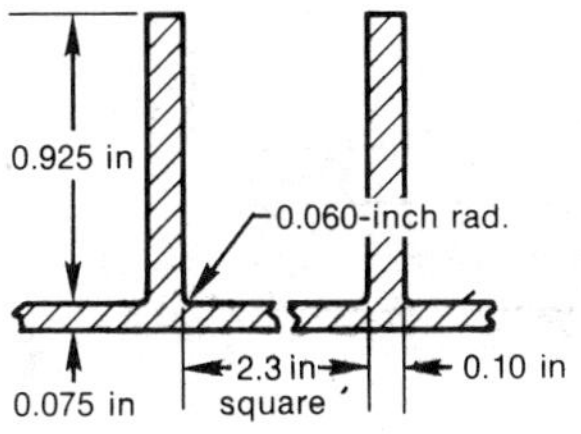 0.925 in; 0.060-inch rad.; 0.075 in; 2.3 in square; 0.10 in | 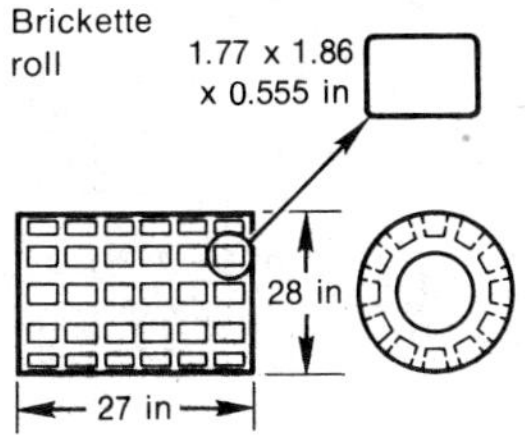 Brickette roll; 1.77 x 1.86 x 0.555 in; 28 in; 27 in | 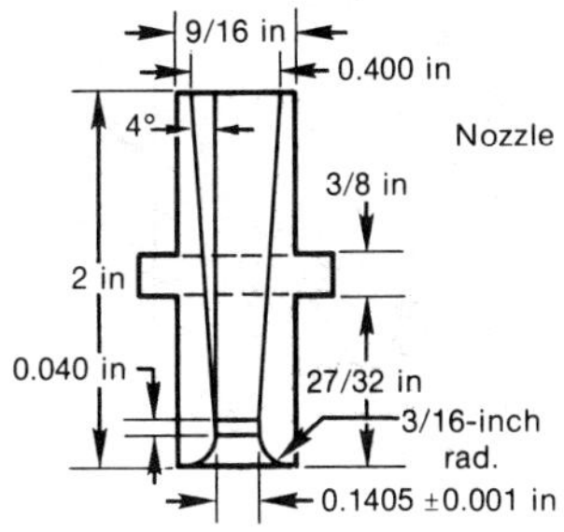 9/16 in; 0.400 in; 4°; Nozzle; 3/8 in; 2 in; 0.040 in; 27/32 in; 3/16-inch rad.; 0.1405 ±0.001 in |
| Electrolyte type: | NaCl | NaCl | NaCl |
| concentration, lb/gal: | 1.0 | 1.0 | 1.0 |
| inlet pressure, psi: | 190 ±10 | 210 | 80 |
| outlet pressure, psi: | 10 (est.) reverse flow | 150 | 0 |
| flow rate, gal/min: | 50 (est.) | | 2 |
| temperature, °F: | 115 to 125 | 100 | 80 |
| Voltage, dc volts: | 13 | 15 | 17 |
| Current, A start: | 2,000 (est.) | 3,000 | 20 |
| end: | 2,400 | 9,000 | 310 |
| Feed rate, in/min: | 0.050 | 0.060 | 0.230 |
| Depth of cut, in: | 0.925 | 0.555 | 1.74 |
| Tolerance, in: | ±0.002 | ±0.015 | 4° taper |
| Surface roughness, $\mu$in $R_a$: | 22 to 28, bottom; 120 to 240, side | 70 | 5 to 10 |
| Electrode material, filtration, benefits and remarks: | Copper-tungsten electrodes (side insulation). 0.008-inch front gap. 0.030-inch side gap (overcut). 0.060-inch corner radii. | 7 pockets cut simultaneously. 630 pockets in 45 rows of 4 each. One indexing vertically per row. | Copper electrode. 50 $\mu$m (micron) filters. |

**TABLE 11.5-14—Continued**

| Work material: | **18% Ni Maraging Steel** | **Inconel Alloy X750** | **4340 Steel** | |
|---|---|---|---|---|
| Workpiece configuration: | 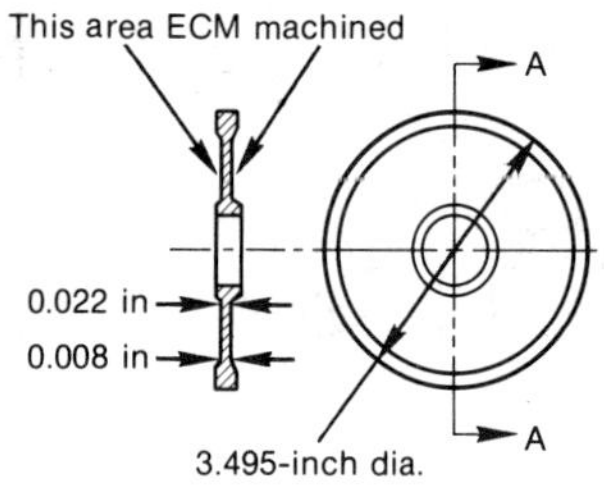 This area ECM machined<br>A<br>0.022 in<br>0.008 in<br>3.495-inch dia.<br>A | 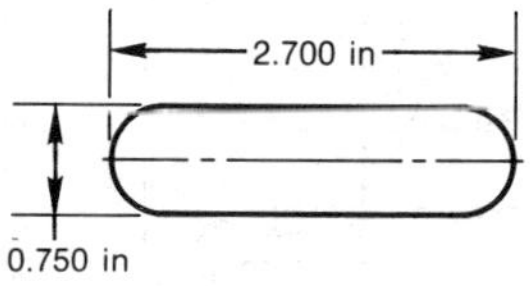 2.700 in<br>0.750 in | 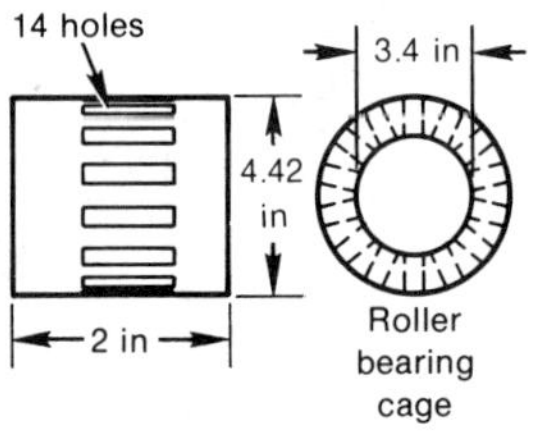 14 holes<br>4.42 in<br>2 in | 3.4 in<br>Roller bearing cage |
| Electrolyte type: | NaCl | NaCl + $NaNO_3$ | NaCl | NaCl |
| concentration, lb/gal: | 1.25 | 1.33 + 0.67 | 1.0 | 1.0 |
| inlet pressure, psi: | 60 to 100 | 80 (est.) | 200 | 200 |
| outlet pressure, psi: | 0 | 20 (est.) | 125 | 50 |
| flow rate, gal/min: | 17 | 10 (est.) | | |
| temperature, °F: | 110 | 100 (est.) | 100 | 100 |
| Voltage, dc volts: | 6.5 | 17 | 15 | 10.5 |
| Current, A start: | 1,000 | 400 | 5,000 | 1,000 |
| end: | 2,600 | 1400 | 6,000 | 1,200 |
| Feed rate, in/min: | 0.076 | 0.080 | 0.195 | 0.400 |
| Depth of cut, in: | 0.090 | 1.250 | 0.50 | 0.50 |
| Tolerance, in: | ±0.0003 on thickness | ±0.003 (est.) | Trepanning roughing cut | Finish cut ±0.002 |
| Surface roughness, $\mu$in $R_a$: | 4 to 10 | 35 to 50 | | 30 max. |
| Electrode material, filtration, benefits and remarks: | Copper electrode. 100 $\mu$m (micron) filter.<br>Workpiece rotated at 125 rpm (or ECT). One side machined at a time.<br>Waffle grid of pockets. | Copper-tungsten electrode. Through-hole slot. | Five parts cut simultaneously.<br>Trepanning roughing cut followed by finishing cut. 30 minute floor-to-floor time per part. | |

## PROCESS SUMMARY

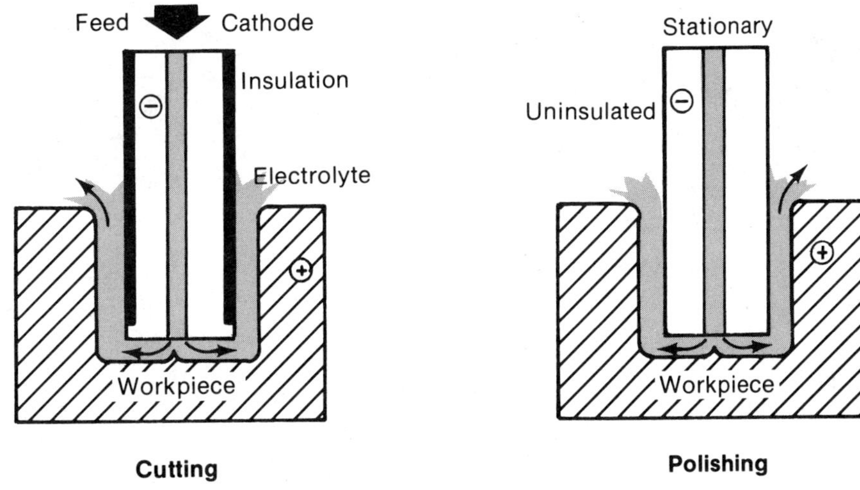


**Figure 11.6–1** ECP schematic.

### Principles

Electrochemical polishing (ECP) is a special form of electrochemical machining (ECM) arranged for cutting or polishing a workpiece (see figure 11.6–1). Polishing parameters are similar in range to those for cutting, but without the feed motion. ECP generally uses a larger gap and a lower current density than does ECM. This requires modestly higher voltages. (In contrast, electropolishing (ELP) uses still lower current densities, lower electrolyte flow and more remote electrodes.)

Sometimes, after ECM of a cavity, the ECM electrode is allowed to dwell at the end of the cut to accomplish a polishing action. Dwelling must be performed under carefully selected conditions and timing to prevent a bulge in the cut. Occasionally, the ECM electrode is retracted slightly (to double or triple the cutting gap) to promote better electrochemical polishing. An uninsulated electrode may also be inserted in place of the cutting electrode to polish the side walls of deep pockets.

### Practical Applications

The special efforts and electrodes required for ECP can be justified for high production or unusual applications on difficult-to-machine materials. Very fine finishes can be achieved with only a few seconds exposure. With proper insulation, only selected portions of the workpiece will receive the extra polishing. The rougher side walls on deep ECM cavities can be smoothed by the use of an uninsulated electrode.

### Operating Parameters

ECP operates in the same manner as does regular ECM; however, the values of the operating parameters are modified slightly. For ECP, the working gap is larger, typically 0.020 inch [0.50 mm]; the current density is much lower, less than or equal to 25 percent of cutting level; and the voltage is higher, up to 50 percent. The electrolyte flow rate is maintained at its usual high level.

### Material Removal Rates, Tolerances and Surface Technology

ECP removes insignificant amounts of material—only ten-thousandths inch. The accuracy of the timing of the "on" cycle controls the tolerances and the surface roughness. A surface roughness of 12 microinches $R_a$ [0.3 μm] can be attained easily in stainless steels or nickel alloys.

### Equipment and Tools

ECP requires modification of the electrodes and operating conditions of regular ECM equipment.

## PROCESS SUMMARY

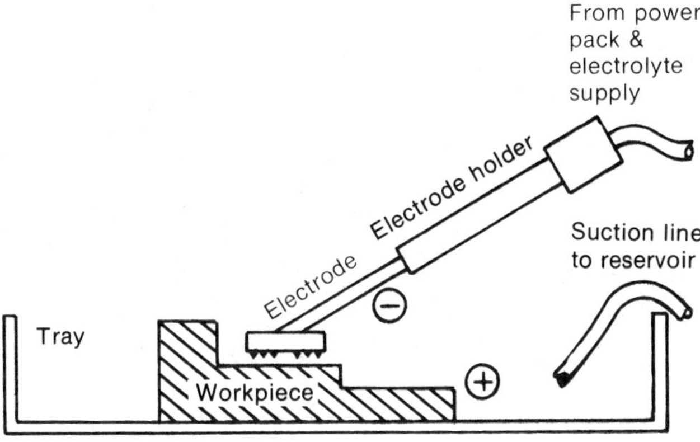


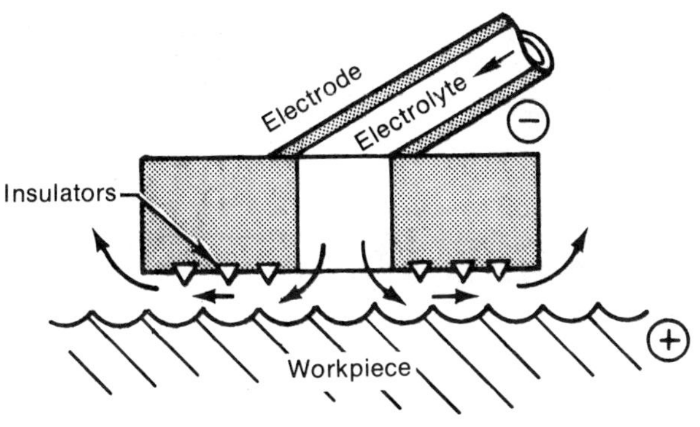


**Figure 11.7–1** ECS schematic.

### Principles

Electrochemical sharpening (ECS) is a special form of electrochemical machining arranged to accomplish sharpening or polishing by hand (see figure 11.7-1). A portable power pack and electrolyte reservoir supply a finger-held electrode with a small current and flow. The fixed gap incorporated on the several styles of shaped electrodes controls the flow rate. A suction tube picks up the used electrolyte for recirculation after filtration.

### Practical Applications

Rapid, localized finishing to low roughness levels can be achieved on any conductive material. Spot removal of unwanted recast layers or heat-affected areas is practical as part of a finishing bench operation.

### Operating Parameters

For ECS, only 10 amperes or less are required for the small electrodes and fractional gallon-per-minute flow rates. The voltage range is similar to that for ECM, 8 to 18 volts.

### Material Removal Rates, Tolerances and Surface Technology

ECS smooths without significant material removal. Roughness levels of 30 microinches $R_a$ [0.8 $\mu$m] from an initial 63 microinches $R_a$ [1.6 $\mu$m] are possible with a few seconds application.

### Equipment and Tools

ECS equipment has been packaged into a portable set; however, any small power pack, tray and pump arrangement suitable for use with the salt electrolytes can be adapted for use for this process.

## PROCESS SUMMARY

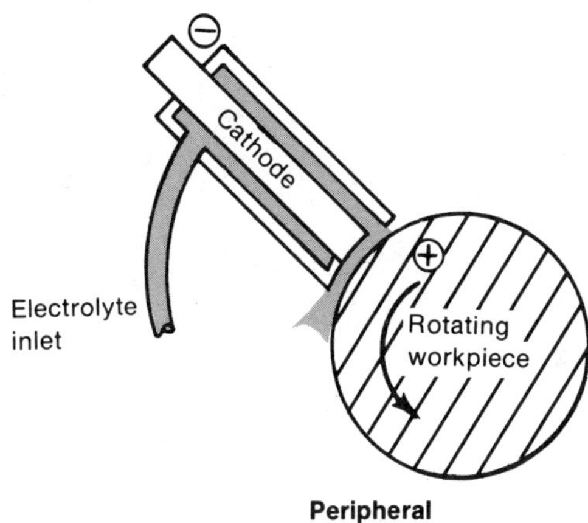


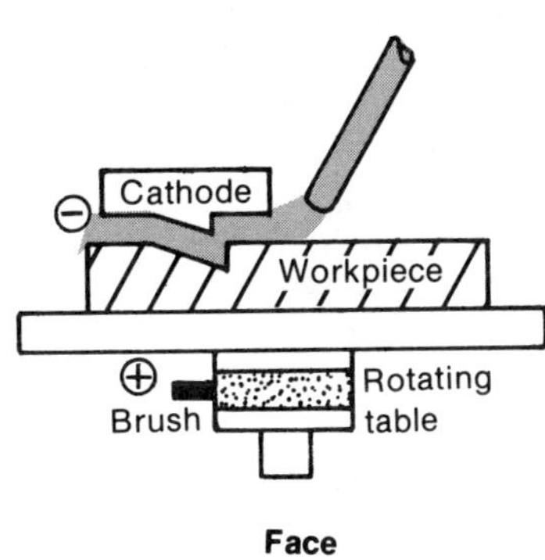


**Figure 11.8-1** ECT schematic.

### Principles

Electrochemical turning (ECT) is a special form of electrochemical machining designed to accommodate rotating workpieces (see figure 11.8-1). The rotation provides additional accuracy but complicates the equipment with the method of introducing the high currents to the rotating part. Electrolyte control may also be complicated because rotating seals are needed to properly direct the flow. Otherwise, the parameters and considerations of electrochemical machining apply equally to the turning mode.

### Practical Applications

Both peripheral and face cuts are practical on electrically conductive materials. Roughing of large disk forgings is practical with electrochemical turning. Some shops have plunged a full-face electrode into the face of the rotating disk. Turning of inside diameters has also been accomplished. ECT can be used to finish parts such as bearing races to close tolerances with surface roughness values better than 5 microinches $R_a$ [0.1 $\mu$m].

### Operating Parameters

The operating parameters and typical values are similar to those for electrochemical machining, section 11.5.

### Material Removal Rates, Tolerances and Surface Technology

ECT attains the same values as those for electrochemical machining.

### Equipment and Tools

ECT equipment is available by special order for face, internal or peripheral turning. Capacities range up to 20,000 amperes.

## PROCESS SUMMARY

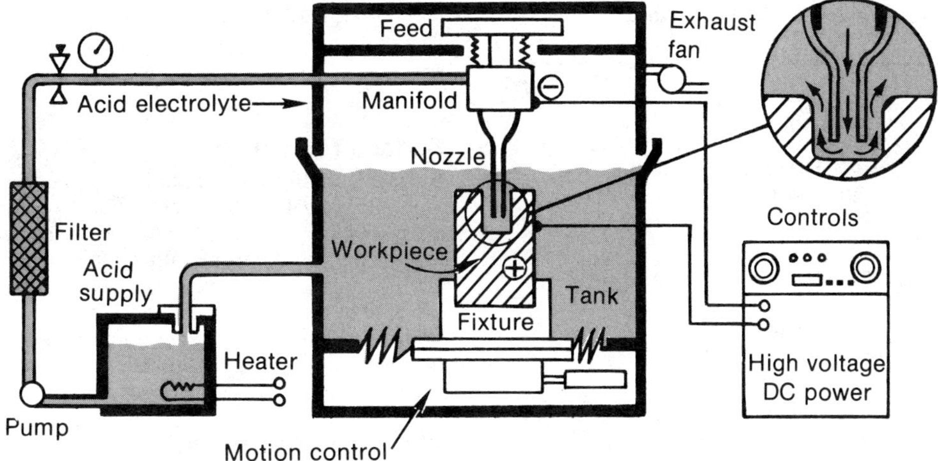


**Figure 11.9-1** ES schematic.

### Principles

Electro-stream™ (ES) is a special version of electrochemical machining adapted for drilling very small holes using high voltages and acid electrolytes (see figure 11.9-1). The voltages are more than 10 times those employed in ECM or STEM™, so special provisions for containment and protection are required. The feed must be controlled to match exactly the rate of dissolution of the workpiece material. Temperature, pressure, concentration and flow control are needed for the acid electrolyte, which is chosen to be chemically compatible with the workpiece metallurgical state. For corrosion protection, both the feed and the motion mechanisms are in pressurized chambers, separated from the acid electrolyte. The actual drilling takes place in a plastic chamber, suitably vented, with controls for automatic rapid advance of the tool, close programming of feed, dwell at hole breakthrough, if desired, and rapid retraction prior to indexing to the next location. The tool is a drawn-glass nozzle, one or two thousandths inch smaller than the desired hole size. An electrode inside the nozzle or the manifold ensures electrical contact with the acid. Multiple hole drilling predominates.

### Practical Applications

Holes ranging from 0.008- to 0.040-inch [0.2 to 1 mm] diameter with depth-to-diameter ratios up to 50:1 can be made in any conductive material. Table 11.9-1 lists materials that have been drilled with ES. Holes can be drilled at angles as shallow as 10 degrees. A principal application is the drilling of cooling holes in gas turbine components which are usually fabricated from nickel and cobalt alloys. Insertion of a nozzle with a formed tip permits right-angle drilling deep inside prior holes or cavities. Over 70 holes per machine stroke have been accomplished on multiple parts; automatic indexing permits the drilling of several hundred holes per part. One example is the drilling of 96 holes simultaneously in the leading edge of a superalloy gas turbine vane. Over 100,000 such parts have been drilled with a reject rate of 1-1/2 percent and an average drilling time of one minute per part. Up to 0.75-inch [19 mm] deep holes can be drilled.

**TABLE 11.9-1 Some Materials that Have Been Drilled Using ES**

| |
|---|
| With HCl: |
| Aluminum |
| IN-100 |
| Ti-6Al-4V |
| SEL 15 |
| With $H_2SO_4$: |
| 1010 Carbon steel |
| 304 Stainless steel |
| 316 Stainless steel |
| 321 Stainless steel |
| 8630 Alloy steel |
| Udimet 700 |
| Rene 41 |
| Rene 77 |
| Rene 80 |
| Rene 100 |
| Rene 120 |
| Rene 125 |
| IN-102 |
| Inconel alloy 625 |
| Inconel alloy 718 |
| IN-738 |
| Inconel alloy X-750 |
| Incoloy alloy 825 |
| Hastelloy alloy C |
| Hastelloy alloy X |
| HS-31 (X-40) |
| Haynes alloy 25 (L605) |
| Haynes alloy 188 |

# 11.9 Electro-stream™—ES

## Operating Parameters

**TABLE 11.9-2 Typical Values for ES Operating Parameters**

| | |
|---|---|
| Power supply | |
| type: | Direct current |
| voltage: | 150 to 850 V |
| current: | 20 to 200 milliamperes/hole (not independently adjustable) |
| Electrolyte | |
| type (s): | $H_2SO_4$, HCl, or proprietary mixtures |
| concentration: | 5 to 15% |
| temperature: | 80° to 120°F [27° to 49°C] |
| pressure: | 20 to 100 psi [138 to 689 kPa] |
| contamination: | 6 g/L max |
| Feed rate: | 0.030 to 0.180 in/min [0.76 to 4.6 mm/min] |
| Hole diameter: | 0.008 to 0.040 inch [0.2 to 1 mm] |
| Hole depth: | Up to 0.75 inch [19 mm] |
| Depth-to-diameter ratio: | Up to 50:1 |
| Holes per stroke: | Up to 100 |
| Drilling angle: | 10° to 170° |
| Diameter tolerance: | ±0.001 inch [0.025 mm] or ±5% of hole diameter |
| Surface roughness: | 16 to 63 μin $R_a$ [0.4 to 1.6 μm] |

## Material Removal Rates and Tolerances

Penetration rates of 1 to 2 thousandths inch per second [0.025 to 0.050 mm/s] are typical for the superalloys. The electrolytic dissolution follows Faraday's laws (see section 11.5), and current density is limited by the boiling of the electrolyte in the nozzle from the resistive heating. Diameter tolerance is typically ±0.001 inch [±0.025 mm] or plus or minus 5 percent for sizes above 0.020 inch [0.5 mm]. Straightness is 0.0005 inch per inch [0.0005 mm/mm].

## Surface Technology

Surface roughness in the holes ranges from 16 to 63 microinches $R_a$ [0.4 to 1.6 μm]. There are no metallurgical changes when the electrolyte and operating parameters are compatible with the metallurgical state of the workpiece, nor are there process-induced residual stresses in the surfaces. Thermal damage is nonexistent. Surface imperfections frequently are a reflection of material inhomogeneity rather than process variability.

## Equipment and Tools

ES equipment is available from the General Electric Company, Cincinnati, Ohio 45216. Equipment can be provided with all the ancillary devices needed to operate the system. Fixturing requires the careful selection of materials suitable for the process environment. The nozzles are drawn from special glasses and are designed for penetration either in the feed mode or in the stationary, "dwell" mode.

## Machining Characteristics

The highly corrosive environment can shorten the life of the equipment unless exceptionally good maintenance is rendered. Adjacent equipment must also be protected. Disposal of the metal-ion-contaminated electrolyte must be carried out in an environmentally satisfactory manner.

## PROCESS SUMMARY

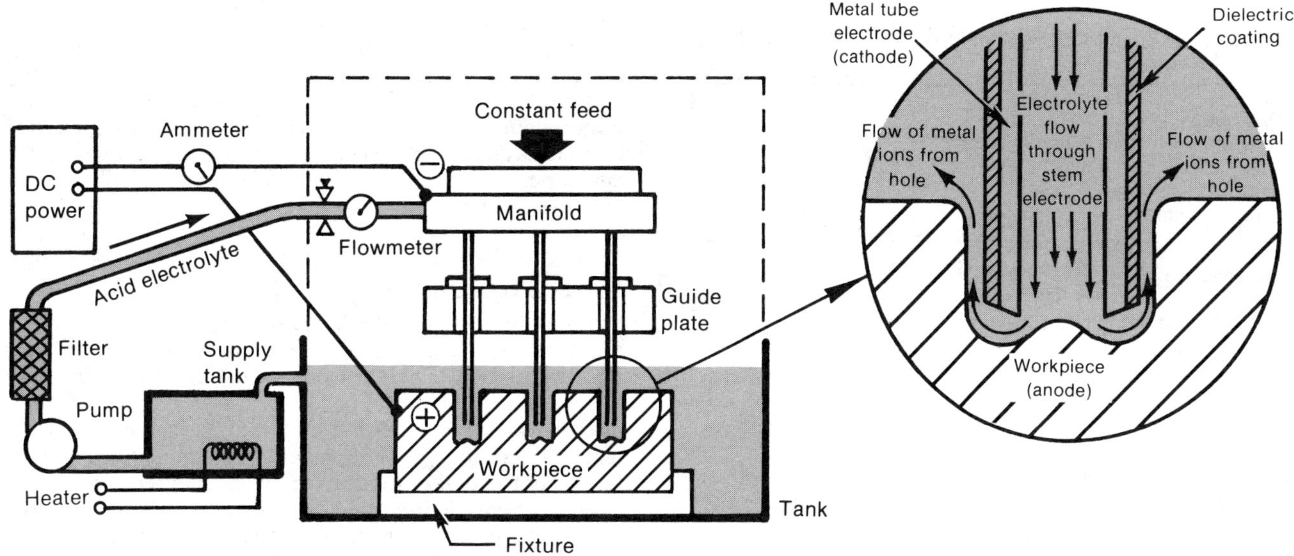


**Figure 11.10-1** STEM™ schematic.

### Principles

Shaped tube electrolytic machining (STEM™) is a specialized ECM technique for "drilling" small, deep holes utilizing acid electrolytes (see figure 11.10-1). Acid is used so that the dissolved metal will go into the solution rather than form a sludge, as is the case with the salt-type electrolytes of ECM. The electrode is a carefully straightened, acid-resistant, metal tube. The tube is coated with a film of enamel-type insulation. The acid is pressure fed through the tube and returns via a narrow gap between the tube insulation and the hole wall. The feed, constant within plus or minus one percent, advances the electrode into the workpiece at a rate exactly equal to the rate at which the workpiece material is dissolved. Multiple electrodes, even of varying diameters or shapes, may be used simultaneously. A guide plate is used to direct the electrodes. A solution of sulfuric acid is frequently used as the electrolyte when machining the nickel alloys. The electrolyte is heated and filtered, and flow monitors control the pressure. Tooling is frequently made of plastics, ceramics or titanium alloys to withstand the electrified, hot acid. The electrode can be shaped to drill odd cross sections, and, within the yield limits of the tooling material, divergent holes may be drilled simultaneously. The voltage used is modest and is reversed periodically for a fraction of a second to remove any film buildup on the electrode, but not long enough to remove substantial material that would change the electrode shape.

### Practical Applications

STEM™ is used to drill round or shaped holes in difficult-to-machine conductive materials. Table 11.10-1 lists materials that have been drilled using STEM™. Holes as deep as 24 inches [610 mm] can be produced with length-to-diameter ratios up to 300:1 and diameters ranging from 0.020 to 0.250 inch [0.5 to 6.4 mm]. Special and oval shapes should have a minimum width of 0.020 inch [0.5 mm] and a major-to-minor-axis ratio of 3:1. Over 100 holes per machine stroke are practical. The same tooling can produce different sizes of holes within a modest range by adjusting the feed, voltage, and other parameters (see figure 11.10-2). A typical case was the simultaneous drilling of sixteen 0.050-inch [1.27 mm] diameter holes in cast Udimet 700 nickel alloy. The holes were 9-inches [229 mm] deep in the center of 0.150-inch [3.8 mm] thick airfoil walls. The drilling rate was 0.001 inch per second [0.025 mm/s] using a 10 percent sulfuric acid electrolyte.

**TABLE 11.10-1 Some Materials that Have Been Drilled Using STEM™**

| | |
|---|---|
| 304 Stainless steel | Incoloy alloy 825 |
| 321 Stainless steel | Rene 41 |
| 414 Stainless steel | Rene 80 |
| 4130 Alloy steel | Rene 95 |
| 4340 Alloy steel | Rene 100 |
| M2 Tool steel | Ti-6Al-4V |
| DCM | Ti-6Al-2Sn-4Zr-2Mo |
| Stellite | Ti-8Al-1Mo-1V |
| Udimet 500 | Haynes alloy 25 (L605) |
| Udimet 700 | Haynes alloy 188 |
| Udimet 710 | HS-31 (X-40) |
| Inconel alloy 718 | Hastelloy alloy C |
| IN-100 | Hastelloy alloy X |
| IN-102 | Greek Ascoloy |
| Inconel alloy 625 | SEL 15 |
| IN-738 | TZM Molybdenum |
| Inconel alloy X-750 | |

# 11.10 Shaped Tube Electrolytic Machining—STEM™

## Operating Parameters

**TABLE 11.10–2 Typical Values for STEM™ Operating Parameters**

| | |
|---|---|
| Power supply | |
| type: | Direct current |
| voltage: | 5 to 15 V |
| forward on time: | 3 to 20 s |
| reverse time: | 0.2 to 0.4 s |
| Electrolyte | |
| type: | $H_2SO_4$ |
| concentration: | 6 to 15% volume |
| temperature: | 90° to 110°F [32° to 43°C] |
| pressure: | 10 to 45 psi [69 to 310 kPa] |
| metal content: | 8 oz/gal max. [6 g/L max.] |
| flow rate: | 1.2 to 9.1 in³/min [20 to 150 cm³/min] |
| Feed rate: | 0.030 to 0.080 in/min [0.76 to 2.0 mm/min] |
| Hole diameter: | 0.02 to 0.25 inch [0.50 to 6.4 mm] |
| Bare electrode, OD: | See figures 11.10–2 to 11.10–4 |
| Amperes per hole: | 1 to 40 A |
| Holes per stroke: | 1 to 100 |
| Hole length: | Up to 24 inches [up to 610 mm] |
| Hole length-to-diameter ratio: | Up to 300:1 |
| Hole tolerance on diameter: | ±5% |
| Hole tolerance, straightness: | 0.0015 in/in [0.0015 mm/mm] |
| Surface roughness, $R_a$: | 32 to 125 μin [0.8 to 3.2 μm] |

NOTES: Electrolyte concentration below 10% reduces attack on tube insulation.

Hole size and roughness increases with increasing electrolyte contamination levels.

Reverse polarity is frequently at lower voltages (typically 4 to 5 V) than the forward voltages.

## Material Removal Rates and Tolerances

Drilling penetration rates are typically 0.001 inch per second [0.025 mm/s] but can range from 0.0007 to 0.0025 inch per second [0.018 to 0.064 mm/s]. Several dozen holes have been drilled simultaneously, with tooling or setup costs being the practical limit. Runout is typically 0.0015 inch per inch [0.0015 mm/mm]; however, 0.001 inch per inch [0.001 mm/mm] can be achieved. Diameter tolerance is typically ±0.0015 inch [±0.038 mm] for hole diameters under 0.10 inch [2.5 mm] and plus or minus 10 percent for larger diameters. The material removal rates are governed by Faraday's laws (see section 11.5) and are limited by the current-carrying capacity of the tube and the boiling point of the electrolyte.

## Surface Technology

Surface roughness in the holes ranges from 32 to 125 microinches $R_a$ [0.8 to 3.2 μm]. There is no stress introduced into the surface. The electrolyte must be compatible with the metallurgical state of the workpiece to prevent intergranular attack or selective-etch pitting. The natural radiusing action of electrolytic machining eliminates all burrs. Close control of the electrolyte flow and degree of metal-ion contamination is required for quality assurance. Monitoring the current during operation can provide an early warning of size deviation.

## Equipment and Tools

The STEM™ process equipment is available from the General Electric Company, Cincinnati, Ohio 45216. The equipment includes controls, tanks, pumps, and so on, and enclosures, ventilation and plastic components to protect the operator from the electrified and corrosive environment. Several equipment sizes are available to accomplish multiple drilling up to 24-inches [610 mm] deep. The tubes or "drills" are usually made from high quality titanium with acid-resisting organic coatings to provide insulation. Carbide locators withstand the environment of machining.

## Machining Characteristics

Corrosion protection is required for equipment adjacent to STEM™ installations. Special tooling is required for through-hole drilling so that electrolyte pressure is not lost upon breakthrough. Careful straightening of the tubes is essential, as is end preparation and adhesion of the insulation.

## SELECTED DATA

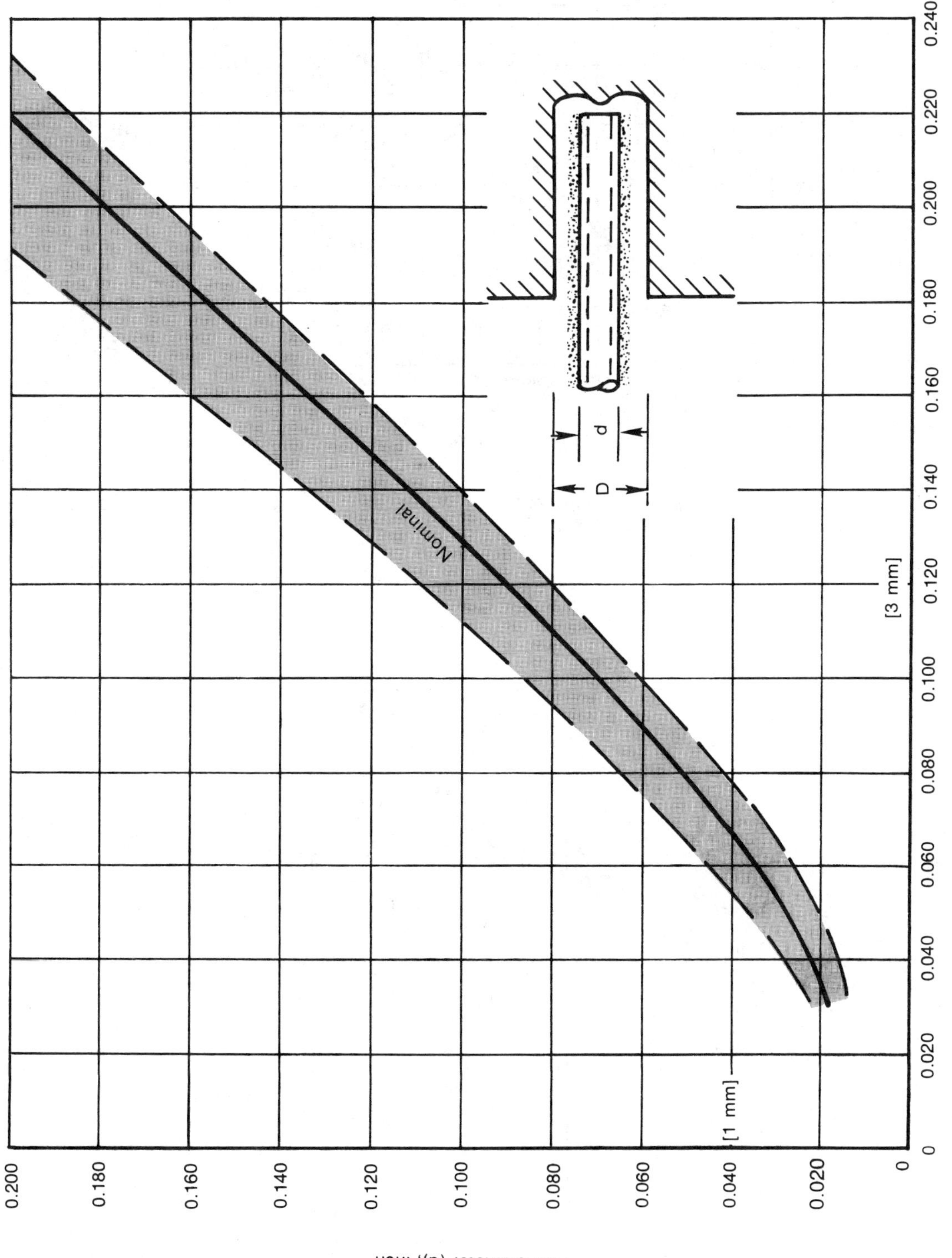


**Figure 11.10-2** Tool diameter versus hole diameter for STEM™ drilling. Shaded area indicates hole sizes readily attainable with a particular diameter tool by varying the other operating parameters.

# 11.10 Shaped Tube Electrolytic Machining—STEM™

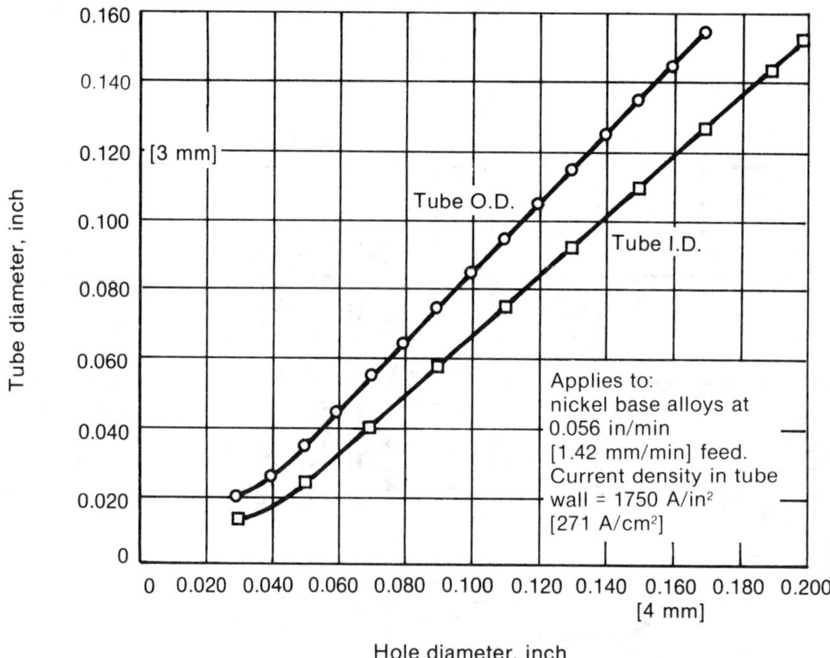


**Figure 11.10-3** Tube diameter versus hole diameter for STEM™ drilling nickel base alloys. (C. Jackson and R. D. Olson, Shaped tube electrolytic machining, Technical paper MR69-109, Society of Manufacturing Engineers, Dearborn, MI, 1969, p. 10)

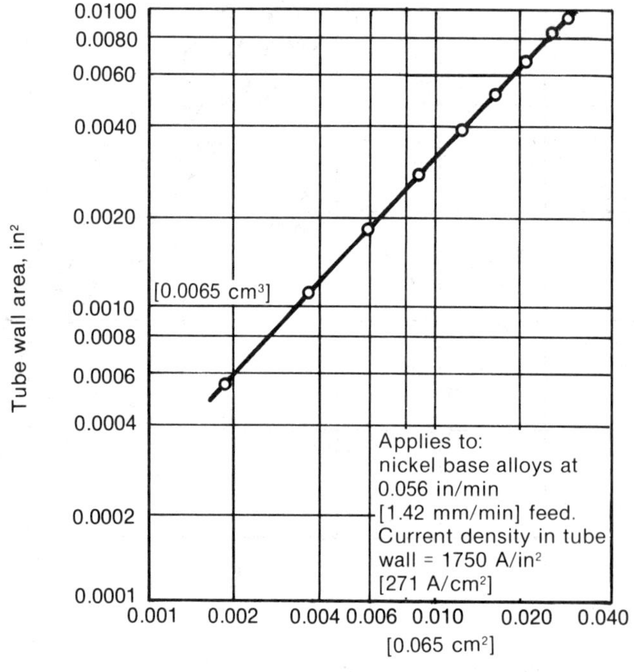


**Figure 11.10-4** Tube wall area versus hole area for STEM™ drilling nickel base alloys. (C. Jackson and R. D. Olson, p. 10)

## SPECIFIC APPLICATIONS DATA

**TABLE 11.10-3 Specific Applications Data for STEM™**

| OPERATING PARAMETER | WORK MATERIAL | | | | | | |
|---|---|---|---|---|---|---|---|
| | 304 stainless steel | Udimet 700 | Udimet 700 | Inconel alloy 718 | Rene 95 | Ti-8Al-1Mo-1V | L605 (Haynes alloy 25) |
| Electrolyte | | | | | | | |
| type | $H_2SO_4$ | $H_2SO_4$ | $H_2SO_4$ | $H_2SO_4$ | $H_2SO_4$ | HCl | $H_2SO_4$ |
| concentration, % vol. | 10 | 10 | 8 | 9–10 | 10 | 5.0 ±0.5 | 10 |
| temperature, °F | 90 | 100 ±2 | 95 | 100 ±2 | 100 | 100 ±5 | 100 |
| pressure, psi | 10 | 40 | 10 est. | 25 | 22 | 45 | 10 |
| metal content, g/L | | 2 | 2 | 3.0–6.5 | 4 | none | |
| flow rate, cm³/min | | 100 | 60 | 115 ±5 | 330 | 20 | |
| Voltage, dc volts | 10 | 6 | 10 | 7.4–7.8 | 6.5 | 13.5 | 6 |
| Forward on time, s | 8 | 7 | 8 | 10 | 10 | 10 | 3 |
| Reverse time, s | 0.3 | 0.1 | 0.3 | 0.3 | 0.3 | 0.3 | 0.3 |
| Feed rate, in³/min | 0.060 | 0.050 | 0.050 | 0.050 | 0.050 | 0.070 ±0.003 | 0.0042 |
| Hole diameter, in | 0.02 | 0.050 | 0.120 | 0.05 | 0.124 | 0.060 | 0.025 ±0.003 |
| Bare electrode, OD, in | 0.015 | 0.035 | 0.090 | 0.035 | 0.107 | 0.045 | 0.017 |
| Amperes per hole, A | 1 | 1 | 8 | 2 max. | 7.5 | 4.3 | 0.5 |
| Holes per stroke | 1 | 16 | 8 | 6 | 1 | 3 | 30 |
| Hole length, in | 6 | 9 | 2.5 | 4.0 | 1 | 8 | 0.030 |
| Hole length-to-diameter ratio | 300 | 180 | 21 | 80 | 20 | 134 | 1.2 |
| Hole tolerance on diameter, in | ±0.002 | ±0.0025 | | | ±0.002 | ±0.001 | |
| Hole straightness, in/in | | 0.001 | | | | | |
| Surface roughness, μin $R_a$ | | | | | 100–130 | 130–150 | |

## SECTION 12

# THERMAL NONTRADITIONAL MACHINING OPERATIONS


12.1 Electron Beam Machining—EBM ........ 12–3
12.2 Electrical Discharge Grinding—EDG ........ 12–11
12.3 Electrical Discharge Machining—EDM ........ 12–15
12.4 Electrical Discharge Sawing—EDS ........ 12–47
12.5 Electrical Discharge Wire Cutting—EDWC ........ 12–49
12.6 Laser Beam Machining—LBM ........ 12–55
12.7 Laser Beam Torch—LBT ........ 12–71
12.8 Plasma Beam Machining—PBM ........ 12–97

## PROCESS SUMMARY

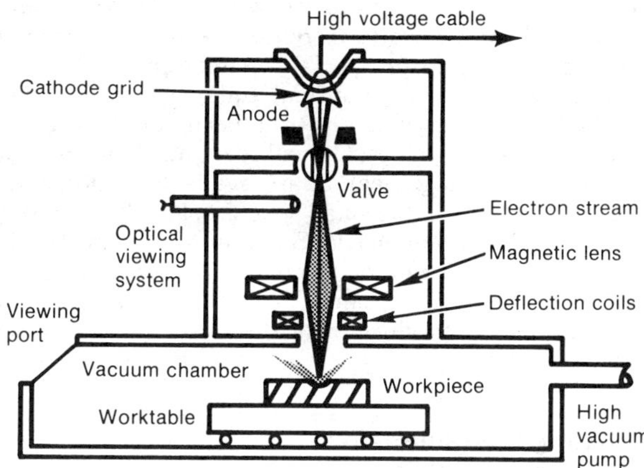


**Figure 12.1-1** EBM schematic.

### Principles

Electron beam machining (EBM) removes material by melting and vaporizing the workpiece at the point of impingement of a focused stream of high-velocity electrons (see figure 12.1-1). To eliminate scattering of the beam of electrons by contact with gas molecules, the work is done in a high-vacuum chamber. Electrons emanate from a triode electron beam gun and are accelerated to three-fourths the speed of light at the anode. The collision of the electrons with the workpiece immediately translates their kinetic energy into thermal energy. The low-inertia beam can be simply controlled by electromagnetic fields. Magnetic lenses focus the electron beam on the workpiece, where a 0.001-inch [0.025 mm] diameter spot can attain an energy density of up to $10^9$ watts per square inch [1.55 x $10^8$ W/cm$^2$] to melt and vaporize any material. The extremely fast response time of the beam is an excellent companion for three-dimensional computer control of beam deflection, beam focus, beam intensity and workpiece motion. View ports and optical tracking systems help guide the beam or the workpiece. Deflection coils permit magnetic control and programming of the beam to any desired pattern over a small area—about 1/4-inch [6.4 mm] square.

### Practical Applications

The power densities available in EBM will vaporize any known material (see examples, table 12.1-2). This permits a wide range of processing possibilities, as shown in figure 12.1-2. Micromachining of thin materials, engraving and hole drilling are the principal applications for EBM. The hole depth-to-diameter ratio can reach 100:1 with multiple pulses. The capability for machining microholes and narrow slots with high precision in a short time in any material also permits the production of unique designs. Figures 12.1-3 and 12.1-4 show perforation ranges and rates for EBM of steels and nickel alloys. The absence of mechanical contact and the suitability for automatic control enhance the process capabilities, but the necessity to work in a vacuum lengthens the floor-to-floor cycle time. One application placed 400,000 holes 80 microinches [2 μm] in diameter in each square inch of a 0.001-inch [0.025 mm] thick foil. Another application incorporates the drilling of holes approximately 0.020 inch [0.50 mm] in diameter in 0.250-inch [6.4 mm] thick material at speeds several times faster than for EDM or ECM. Metering orifices, rough wire dies and spinnerette holes are other applications.

### Operating Parameters

**TABLE 12.1-1 Typical Values for EBM Operating Parameters**

| | |
|---|---|
| Accelerating voltage: | 50 to 150 kV |
| Beam current: | 100 to 1,000 μA |
| Power: | 0.5 to 60 kW |
| Pulse time: | 4 to 64,000 μs |
| Pulse frequency: | 0.1 to 16,000 Hz |
| Vacuum: | $10^{-2}$ to $10^{-4}$ mm Mercury |
| Beam spot size minimum: | 0.0005 to 0.001 inch [0.013 to 0.025 mm] |
| Beam deflection range: | 0.25-inch [6.4 mm] square |
| Beam intensity: | $10^6$ to $10^{10}$ W/in$^2$ [1.55 x $10^5$ to 1.55 x $10^9$ W/cm$^2$] |
| Depth of cut: | Up to 0.25 inch [6.4 mm] |
| Narrowest cut: | 0.001 inch in 0.001-inch thick metal [0.025 mm in 0.025-mm thick metal] |
| Hole range, diameter: | 0.001 inch [0.025 mm] in 0.0007-inch [0.020 mm] thickness<br>0.040 inch [1 mm] in 0.20-inch [5 mm] thickness |
| Hole taper: | 1° to 2° typical |
| Hole angle to surface: | 20° to 90° |
| Cutting speed: | See tables 12.1-3 and 12.1-4 |
| Material removal rates: | Up to 0.0024 in$^3$/s [40 mm$^3$/s], see figure 12.1-5 |
| Penetration rates: | Up to 0.010 in/s [0.25 mm/s] |
| Perforation rates: | Up to 5,000 holes/second (if part can be traversed rapidly) |
| Tolerances: | ±0.001 inch [0.025 mm] on 0.125-inch [3.2 mm] diameter, see figure 12.1-6<br>±0.00005 inch [0.0013 mm] on 0.0005-inch [0.013 mm] diameter<br>±0.001 inch [0.025 mm] typical and ±0.0002 inch [0.005 mm] attainable, 10% otherwise |

# 12.1 Electron Beam Machining—EBM

## Material Removal Rates and Tolerances

The thermal properties of the work material and the power level of the EBM unit dictate the material removal rates. Tungsten can be machined at a rate of $0.92 \times 10^{-4}$ cubic inch per second [$15 \times 10^{-4}$ $cm^3/s$] with one kilowatt of power, while aluminum can be machined at a rate of $0.24 \times 10^{-3}$ cubic inch per second [$3.9 \times 10^{-3}$ $cm^3/s$] with the same power level, as shown in figure 12.1-5. Typical tolerances are about 10 percent of the slot width or hole diameter, as shown in figure 12.1-6. The taper present in slots and holes limits the depth-to-width ratio. With computer control of the pulsed beam and integrated workpiece motion, hole drilling rates up to 5,000 holes per second can be attained in thin materials, including ceramics.

## Surface Technology

Estimates of surface roughness for the small holes and cuts are in the neighborhood of 40 microinches $R_a$ [1 $\mu$m]. The short bursts of the beam limit the extent of the heat-affected zone; however, there is always a thin layer of recast or heat-affected material on the cut surface. Heat-affected material up to 0.010-inch [0.25 mm] thick has been observed. This material can be detrimental to the structural integrity of highly stressed components, and for such components, this material should be removed or modified.

## Equipment and Tools

Machines for EBM are produced with capacities ranging from a few hundred watts to 60,000 watts with vacuum chambers of nominal or special volume—up to 20-by-27-foot [6 by 9 m] capacity. Special equipment can include room-size chambers with electron beam guns traveling over NC positioning tables. Optical scans and controls with as much as 40:1 magnification are available to utilize the precision capability of the small spot size, high beam power densities and low beam inertia. Simultaneous control of the beam and the work motion is common. Proper programming of the deflection beam, "flying spot" scanners and pulse controls can produce intricate geometric cuts in the workpiece.

## Machining Characteristics

The "chips" from EBM are small dust particles or small beads formed from the solidified, expulsed molten material. Cleanliness efforts should be continuous. Cathode life declines at the higher power densities. Shielding of the secondary radiation is necessary for safety.

## SELECTED DATA

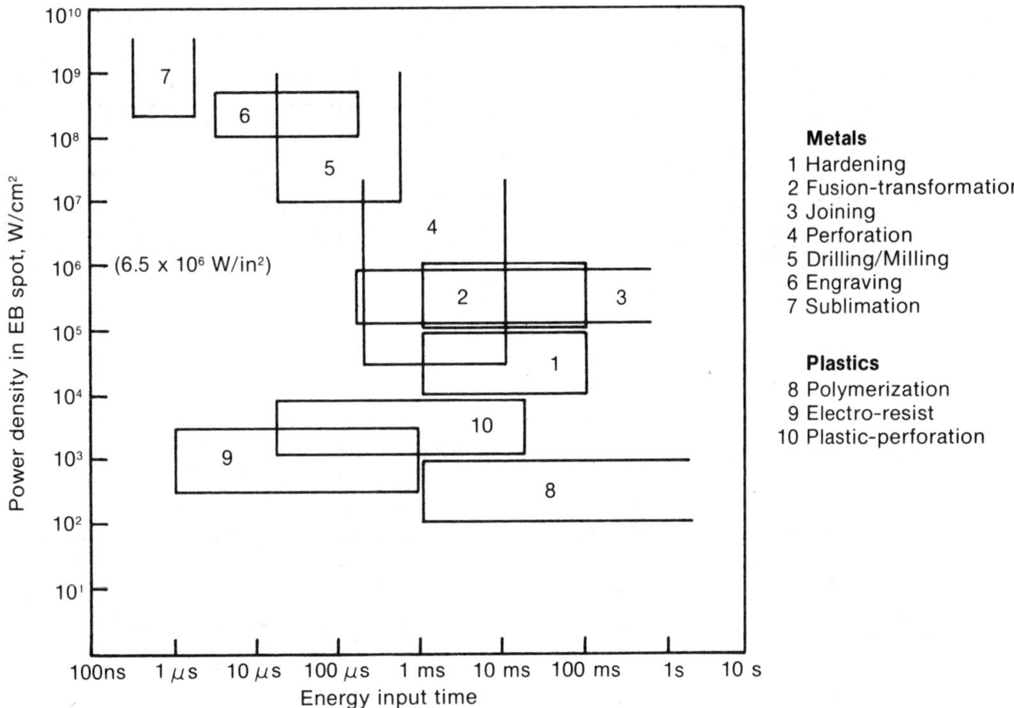


**Figure 12.1–2** Application range of electron beams. Power density and time of impact are typical for the various techniques. (J. Drew, Electron beam drilling, *Influence of metallurgy on hole making operations*, Metals Park, OH: American Society for Metals, 1978, p. 115)

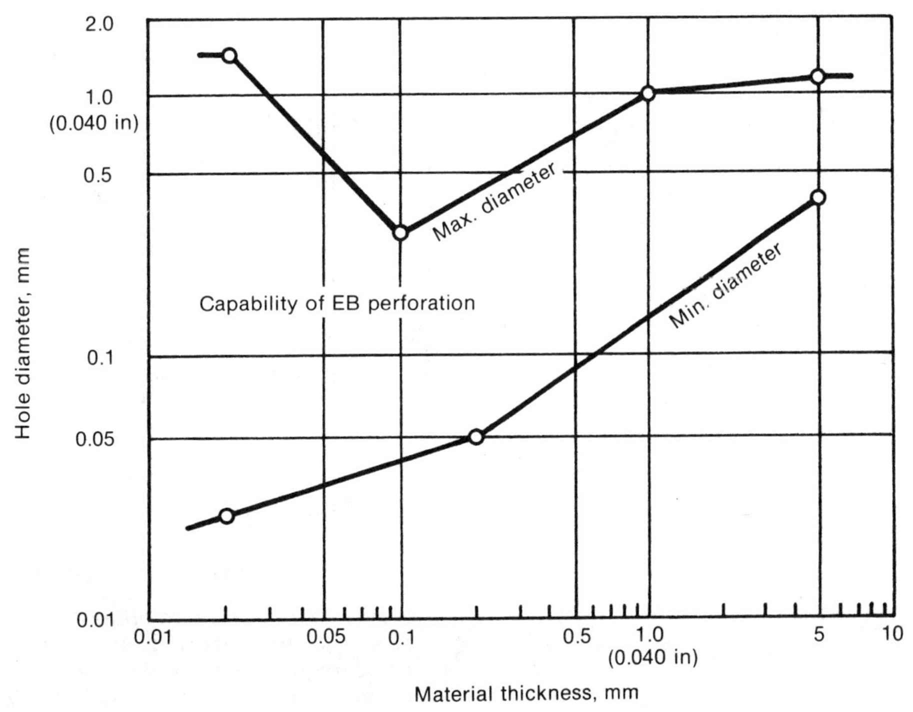


**Figure 12.1–3** Range of EB perforation for steel and nickel alloys. (J. Drew, p. 118)

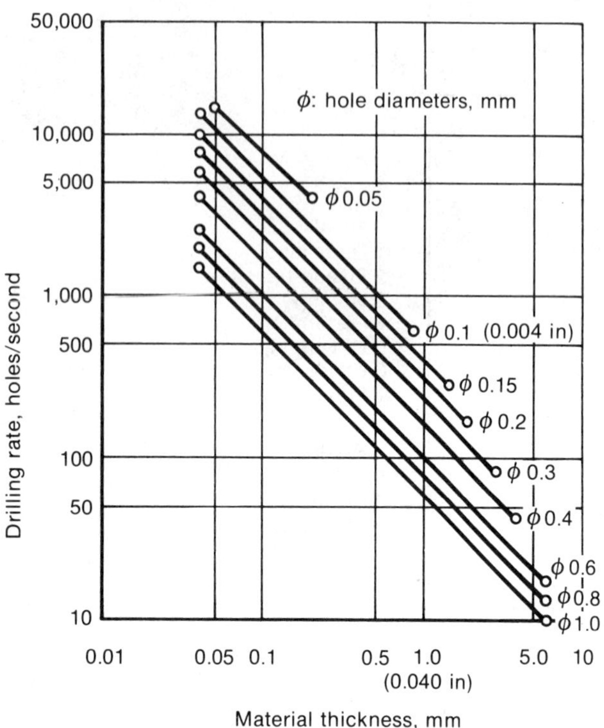


**Figure 12.1–4** EB perforation rate for steel and nickel alloys. (J. Drew, p. 119)

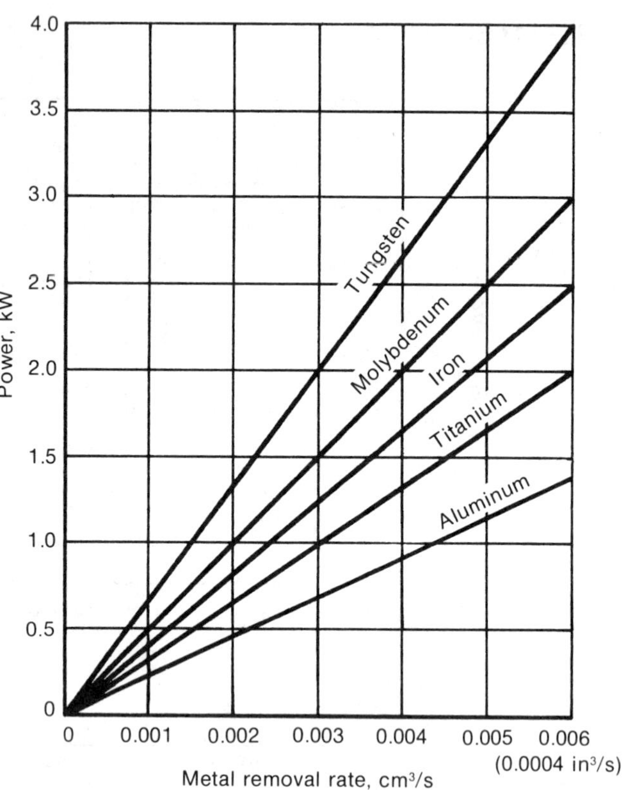


**Figure 12.1–5** Metal removal rates versus power, assuming 15 percent cutting efficiency. (R. K. Springborn, ed., *Non-traditional machining processes*, Dearborn, MI: American Society of Tool and Manufacturing Engineers, 1967, p. 141)

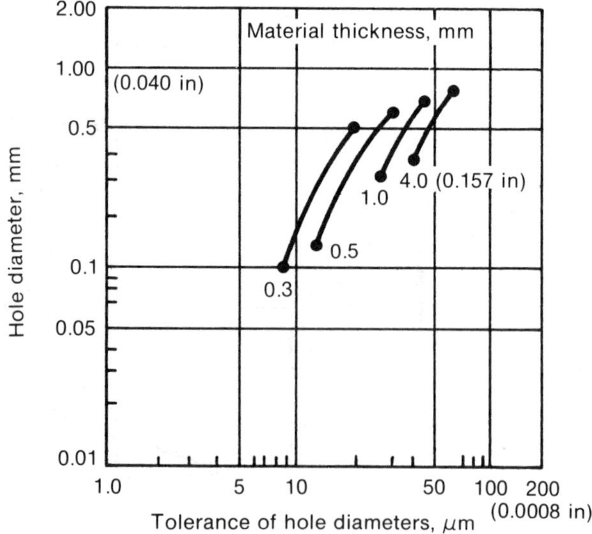


**Figure 12.1–6** Hole tolerances variation with hole size and material thickness. Generally, deeper holes require wider tolerances. (J. Drew, Electron beam tackles tough machining jobs, *Machine Design* 48 (February 26, 1976), p. 96)

**TABLE 12.1-2 Physical Thermal Properties of Various Materials**

| PROPERTY | MATERIAL | | | | |
|---|---|---|---|---|---|
| | Aluminum | Titanium | Molybdenum | Tungsten | Iron |
| Melting temperature, °C | 660 | 1,668 | 2,610 | 3,410 | 1,536 |
| Boiling temperature, °C | 2,450 | 3,260 | 5,560 | 5,930 | 3,000 |
| Specific heat, cal/g/°C | 0.215 | 0.126 | 0.061 | 0.032 | 0.11 |
| Heat of fusion, cal/g | 94.6 | 36.7 | 70.0 | 44.0 | 65.0 |
| Heat of vaporization, cal/g | 2,517.6 | 2,223.4 | 1,340 | 1,005.9 | 1,514.8 |
| Specific energy to vaporize, joules/cm$^3$ | $3.54 \times 10^4$ | $5.07 \times 10^4$ | $7.46 \times 10^4$ | $1.0 \times 10^5$ | $6.28 \times 10^4$ |
| Power to vaporize* (ratio to aluminum) | 100 | 150 | 220 | 290 | 180 |

*Relative power to remove equal volumes in equal time.

**TABLE 12.1-3 Holes Drilled by EBM in Various Materials**

| WORK MATERIAL | WORKPIECE THICKNESS | | HOLE DIAMETER | | DRILLING SPEED | ACCELERATING VOLTAGE | AVERAGE BEAM CURRENT | PULSE WIDTH | PULSE FREQUENCY |
|---|---|---|---|---|---|---|---|---|---|
| | in | mm | in | mm | s | kV | $\mu$A | $\mu$s | Hz |
| 400 Series stainless steel | 0.010 | 0.25 | 0.0005 | 0.013 | <1 | 130 | 60 | 4 | 3,000 |
| Alumina $Al_2O_3$ | 0.030 | 0.76 | 0.012 | 0.30 | 30 | 125 | 60 | 80 | 50 |
| Tungsten | 0.010 | 0.25 | 0.001 | 0.025 | <1 | 140 | 50 | 20 | 50 |
| 90-10 Tantalum-tungsten | 0.040 | 1.0 | 0.005 | 0.13 | <1 | 140 | 100 | 80 | 50 |
| 90-10 Tantalum-tungsten | 0.080 | 2.0 | 0.005 | 0.13 | 10 | 140 | 100 | 80 | 50 |
| 90-10 Tantalum-tungsten | 0.100 | 2.5 | 0.005 | 0.13 | 10 | 140 | 100 | 80 | 50 |
| Stainless steel | 0.040 | 1.0 | 0.005 | 0.13 | <1 | 140 | 100 | 80 | 50 |
| Stainless steel | 0.080 | 2.0 | 0.005 | 0.13 | 10 | 140 | 100 | 80 | 50 |
| Stainless steel | 0.100 | 2.5 | 0.005 | 0.13 | 10 | 140 | 100 | 80 | 50 |
| Aluminum | 0.100 | 2.5 | 0.005 | 0.13 | 10 | 140 | 100 | 80 | 50 |
| Tungsten | 0.016 | 0.41 | 0.003 | 0.076 | <1 | 130 | 100 | 80 | 50 |
| Quartz | 0.125 | 3.18 | 0.001 | 0.025 | <1 | 140 | 10 | 12 | 50 |

SOURCE: Adapted from R. K. Springborn, ed., p. 143.

NOTE: The main control parameters for shaping the hole are the pulse width for the depth of the hole, the beam current for the diameter of the hole and the power distribution within the beam as well as the position of the focus with respect to the workpiece.—J. Drew, Farrel Company.

**TABLE 12.1–4 Slots Cut by EBM in Various Materials**

| WORK MATERIAL | WORKPIECE THICKNESS | | SLOT DESCRIPTION AND/OR DIMENSIONS | | TIME OF CUT OR RATE | | ACCEL-ERATING VOLTAGE | AVERAGE BEAM | PULSE WIDTH | PULSE FRE-QUENCY |
|---|---|---|---|---|---|---|---|---|---|---|
| | in | mm | in | mm | | | kV | μA | μs | Hz |
| Stainless steel | 0.062 | 1.57 | Rectangle: 0.008 by 0.250 | Rectangle: 0.2 by 6.35 | 5 min | 5 min | 140 | 120 | 80 | 50 |
| Hardened steel | 0.125 | 3.18 | Rectangle: 0.018 by 0.072 | Rectangle: 0.46 by 1.83 | 10 min | 10 min | 140 | 150 | 80 | 50 |
| Stainless steel | 0.007 | 0.18 | 0.004 wide | 0.10 wide | 2 in/min | 50 mm/min | 130 | 50 | 80 | 50 |
| Brass | 0.010 | 0.25 | 0.004 wide | 0.10 wide | 2 in/min | 50 mm/min | 130 | 50 | 80 | 50 |
| Stainless steel | 0.002 | 0.050 | 0.002 wide | 0.050 wide | 4 in/min | 100 mm/min | 130 | 20 | 4 | 50 |
| Alumina $Al_2O_3$ | 0.030 | 0.75 | 0.004 wide | 0.10 wide | 24 in/min | 610 mm/min | 150 | 200 | 80 | 200 |
| Tungsten | 0.002 | 0.050 | 0.001 wide | 0.025 wide | 7 in/min | 175 mm/min | 150 | 30 | 80 | 50 |

SOURCE: Adapted from R. K. Springborn, ed., p. 145.

## SPECIFIC APPLICATIONS DATA

**TABLE 12.1–5 Specific Applications Data for Electron Beam Machining**

| MATERIAL | CONFIGURATION PRODUCED | CUTTING SPEED OR TIME | ACCELERATING VOLTAGE kV | AVERAGE BEAM CURRENT μA | PULSE WIDTH μs | PULSE FREQUENCY Hz |
|---|---|---|---|---|---|---|
| Alumina | | | | | | |
| 0.030-inch thick | 0.012-inch dia. hole | 30 s | 125 | 60 | 80 | 50 |
| Alumina wafers | | | | | | |
| 0.010-inch thick | 0.003-inch dia. hole | 10 s | 125 | 60 | 80 | 50 |
| 0.010-inch thick | 0.004-inch wide slot | 12 in/min | 90 | 150 | 80 | 150 |
| 0.030-inch thick | 0.004-inch wide slot | 24 in/min | 150 | 200 | 80 | 200 |
| Ferrite wafers | | | | | | |
| 0.010-inch thick | 0.001-inch dia. holes | <1 s | 140 | 25 | 5 | 50 |
| Microdiodes scribing | Scribed to approx. 1 mil depth | Approx. 60 in/min | 110 | 7 | 12 | 50 |
| Mylar tape | | | | | | |
| 0.0015-inch thick | Cut | 4,615 in/min | 110 | 600 | — | Continuous |
| Quartz crystal | | | | | | |
| 0.125-inch thick | 0.001-inch exit dia. | <1 s | 140 | 10 | 12 | 50 |
| Sapphire crystal | | | | | | |
| 0.026-inch thick | 0.0025-inch dia. hole | <30 s | 110 | 20 | 9 | 50 |
| 0.026-inch thick | 0.0015-inch dia. hole, hole taper approx. 2° | <30 s | 110 | 20 | 9 | 50 |
| Silicon wafers | | | | | | |
| 0.010-inch thick gold deposited | Scribed to 0.002-inch depth | 5 in/min | 130 | 70 | 4 | 3,000 |
| Aluminum | | | | | | |
| 0.100-inch thick | 0.005-inch dia. hole | 10 s | 140 | 100 | 80 | 50 |
| Aluminum, 5052 1/16-inch and 1/8-inch thick | Slotting | Manual | 130 | 15 | 500 | 300 |
| Copper (OFHC) | | | | | | |
| 0.002-inch thick traveling wave tube grid | Cut | Approx. 1 min | 130 | 50 | 2.5 | 25,000 (scanner) |
| Brass | | | | | | |
| 0.010-inch thick | 0.004-inch wide slot | 2 in/min | 130 | 50 | 80 | 50 |
| Brass, foil grid | | | | | | |
| 0.004-inch thick | Cut by flying spot scanner | Approx. 3 min | 120 | 50 | 4 | 3,000 |
| Steel, hardened | | | | | | |
| 0.125-inch thick | Rectangle—0.018 by 0.072 inch | 10 min | 140 | 150 | 80 | 50 |
| Steel, high speed tool | | | | | | |
| 0.1875-inch thick | 0.008-inch entrance width, 0.007-inch exit width, slot | — | 120 | 30,000 | 50,000 | 3.3 |
| Steel, drill | 0.014-inch dia. hole | Approx. 3 min | 140 | 200 | 80 | 50 |
| Steel, plate | | | | | | |
| 0.062-inch thick | Cut | 15 in/min | 150 | 9,000 | 2,100 | 330 |
| 0.040-inch thick | Cut | 30 in/min | 150 | 9,000 | 2,100 | 330 |
| Steel, 304 stainless ¼-inch thick | Cut (reopening of welded container) | 10 to 20 in/min | 130 | 5,000 | 5,300 | 35 |
| Steel, stainless (400 series) | | | | | | |
| 0.010-inch thick | 0.0005-inch dia. hole | <1 s | 130 | 60 | 4 | 3,000 |
| Steel, 410 stainless | | | | | | |
| 0.250-inch thick | 0.030-inch entrance width, 0.010-inch exit width, slot | — | 150 | 7,000 | 1,000 | 3.3 |

**TABLE 12.1-5—Continued**

| MATERIAL | CONFIGURATION PRODUCED | CUTTING SPEED OR TIME | ACCELERATING VOLTAGE kV | AVERAGE BEAM CURRENT $\mu$A | PULSE WIDTH $\mu$s | PULSE FREQUENCY Hz |
|---|---|---|---|---|---|---|
| Steel, stainless | | | | | | |
| 0.250-inch thick | 0.020- to 0.040-inch dia. holes | 3 min | 145 | 4,000 | 2,100 | 12.5 |
| 0.040-inch thick | 0.005-inch dia. hole | $<$1 s | 140 | 100 | 80 | 50 |
| 0.080-inch thick | 0.005-inch dia. hole | 10 s | 140 | 100 | 80 | 50 |
| 0.100-inch thick | 0.005-inch dia. hole | 10 s | 140 | 100 | 80 | 50 |
| 0.062-inch thick | Rectangle—0.008 by 0.250 inch $\pm$0.00025 inch | 5 min | 140 | 120 | 80 | 50 |
| 0.007-inch thick | 0.004-inch wide slot | 2 in/min | 130 | 50 | 80 | 50 |
| 0.002-inch thick | 0.002-inch wide slot | 4 in/min | 130 | 20 | 4 | 50 |
| Hastelloy | 0.200-inch dia. hole in 0.200-inch thick material | — | 130 | 5,000 | 5,300 | 100 |
| | 0.100-inch dia. hole in 0.450-inch thick material | 70 s | 130 | 5,000 | 5,300 | 100 |
| Molybdenum, shim | | | | | | |
| 0.010-inch thick | Less than 0.002-inch dia. holes on 0.003-inch centers | $<$1 s | 140 | 20 | 20 | 50 |
| Tantalum alloy (90Ta-10W) | | | | | | |
| 0.040-inch thick | 0.005-inch dia. hole | $<$1 s | 140 | 100 | 80 | 50 |
| 0.080-inch thick | 0.005-inch dia. hole | 10 s | 140 | 100 | 80 | 50 |
| 0.100-inch thick | 0.005-inch dia. hole | 10 s | 140 | 100 | 80 | 50 |
| Tantalum, 100A thin film resistor | Manual cut | — | 100 | 20 | 9 | 1,000 |
| Tungsten | | | | | | |
| 0.010-inch thick | 0.001-inch dia. hole | $<$1 s | 140 | 50 | 20 | 50 |
| 0.016-inch thick | 0.003-inch dia. hole | $<$1 s | 130 | 100 | 80 | 50 |
| 0.002-inch thick | 0.001-inch wide slot | 7 in/min | 150 | 30 | 80 | 50 |

## PROCESS SUMMARY

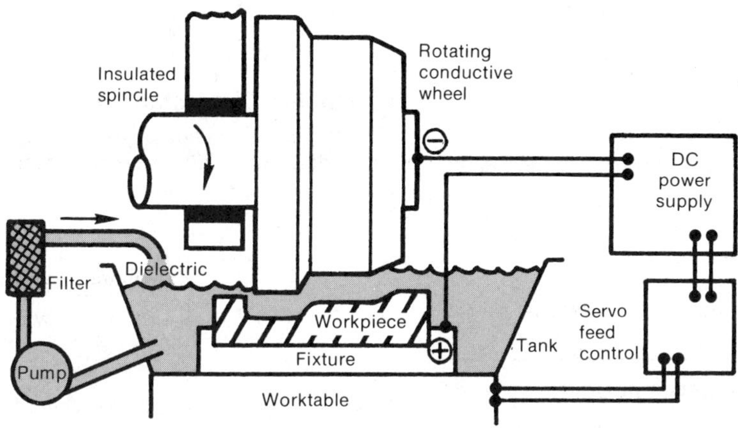


**Figure 12.2–1** EDG schematic.

### Principles

Electrical discharge grinding (EDG) is the removal of a conductive material by rapid, repetitive spark discharges between a rotating tool and the workpiece, which are separated by a flowing dielectric fluid (see figure 12.2–1). (EDG is similar to EDM except the electrode is in the form of a grinding wheel and the current is usually lower.) The spark gap is servo controlled. The insulated wheel and work table are connected to the direct-current pulse generator. A positive charge on the workpiece is "standard." Higher currents produce faster cutting, rougher finishes and deeper heat-affected zones in the workpiece. Wheel wear depends upon current density, work material, wheel material, dielectric and sharpness of corner details.

### Practical Applications

EDG provides greater accuracy in cutting hard materials such as carbide form tools, hardened gear racks or tungsten carbide throwaway bits, even though its cutting rates are low. Lamination die grinding in the hardened state is a frequent use of EDG. The absence of significant cutting forces permits the grinding of fragile shapes or closely spaced, thin slots in any conductive material. Cast irons usually are not electrical-discharge-ground because the sand inclusions can damage the grinding wheel.

### Operating Parameters

**TABLE 12.2–1 Typical Values for EDG Operating Parameters**

| | |
|---|---|
| Power supply | |
| type: | Direct current with pulse control |
| voltage: | 30 to 400 V |
| frequency: | 200 Hz to 500 kHz |
| current: | 0.5 to 200 A |
| Wheel | |
| type: | Graphite or sometimes brass |
| speed: | 100 to 600 fpm [30 to 180 m/min] |
| Dielectric: | Filtered hydrocarbon oil |
| Spark gap: | 0.0005 to 0.003 inch [0.013 to 0.076 mm] |

### Material Removal Rates and Tolerances

Material removal rates range from 0.01 to 0.15 cubic inch per hour [0.16 to 2.45 $cm^3/hr$], with the higher figures accompanied by surface roughness in the 63- to 125-microinch $R_a$ [1.6 to 3.2 $\mu m$] range. Corner radius depends upon overcut values used and ranges from 0.0005 to 0.005 inch

[0.013 to 0.13 mm]. Tolerances to ±0.0002 inch [±0.005 mm] are normal, with ±0.000050 inch [±0.0013 mm] achievable.

### Surface Technology

Surface roughness values decrease with an increase in spark frequency and are typically 16 to 125 microinches $R_a$ [0.4 to 3.2 μm]. The melting, vaporizing and resolidification of the surface of the workpiece leaves a heat-affected zone that can be from a few ten-thousandths- to a few thousandths-inch deep. Hardness alterations occur which also affect the material properties. Components intended for high-stress applications should have these heat-affected layers removed or modified to ensure the best surface integrity.

**TABLE 12.2-2 EDG Material Removal Rates versus Surface Roughness**

| MATERIAL REMOVAL RATE | | SURFACE ROUGHNESS, $R_a$ | |
|---|---|---|---|
| in³/hr | cm³/hr | μin | μm |
| 0.05 | 0.82 | 125 | 3.2 |
| 0.03 | 0.49 | 63 | 1.6 |
| 0.002 | 0.03 | 16 | 0.4 |

### Equipment

Equipment for EDG is available in a range of sizes. Systems are equipped with variable speed drive, insulated wheel spindles and a servomechanism to control the table speed.

### Machining Characteristics

Grinding action is nonexistent. Wheel wear occurs from the spark discharge. The wear ratio ranges from 100:1 to 0.1:1 with an average of 3:1.

$$\text{volumetric wear ratio} = \frac{\text{volume of work removed}}{\text{volume of wheel consumed}}$$

It is necessary to redress the wheel frequently if sharp radii are desired. Sharp radii also contribute to more rapid wheel wear than do round or flat shapes. About one-half of the wear ratio can be attributed to redressing. The dielectric fluid must be inserted between the wheel and the workpiece; consequently, the contact area is usually submerged in dielectric.

## SPECIFIC APPLICATIONS DATA

**TABLE 12.2-3 Specific Applications Data for Electrical Discharge Grinding of Stainless Steels**

| WORK | | | | | OPERATING CONDITIONS | | | | | RESULTS | | | | |
|---|---|---|---|---|---|---|---|---|---|---|---|---|---|---|
| Part No. | Type of Stainless Steel | Description of Workpiece | Shape Produced | Dimensions of Shape Produced | Speed* rpm | Volts | Current A | Capacitance mfd | Pulse Frequency kHz | Grinding Rate in³/hr | Cutting Time min | Volume Wear Ratio | Overcut in/side | Surface Roughness $\mu$in $R_a$ |
| 1 | 303 | Rod, 0.531-inch dia. by 1.219-inch long | 3 grooves | 0.188-inch wide, 0.218-inch deep; full-radius bottom | 20 | 70 | 10 | 6 | 32 | 0.470 | 6 | 3:1 | 0.0017 | 200 |
| 2 | 304 | Strip, 0.104 by 4.000 by 6.500 inch | Rib | 0.600-inch wide, 0.010-inch deep; central web 0.010-inch wide | 20 | 70 | 0.5 | 0.5 | 32 | 0.06 | 44 | 3:1 | 0.001 | 60 |
| 3 | 304 | Strip, 0.125 by 3 by 20 inch | 2 grooves | 2.800-inch wide, 0.020-inch deep | 20 | 70 | 2 | 2 | 130 | 0.156 | 400 | 6:1 | 0.001 | 60 |
| 4 | 304 | Hexagonal tubing, 2.25-by-0.050-inch wall | Face, step-cut end | Step, 0.187-inch deep | 34 | 80 | 8 | 2 | 32 | — | 30 | — | 0.001 | 125 |

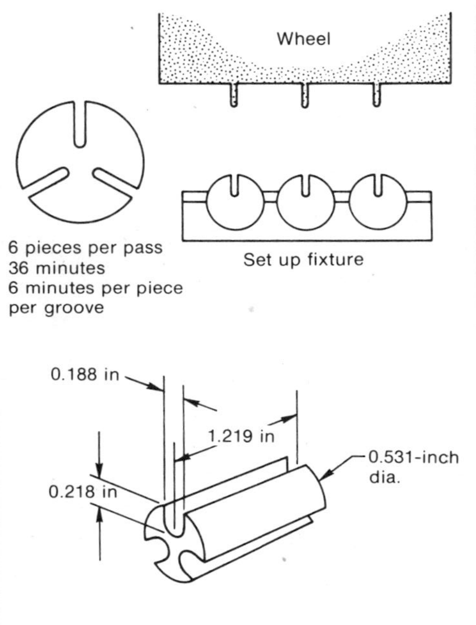

Part 1: Wheel; Set up fixture; 6 pieces per pass, 36 minutes, 6 minutes per piece per groove; 0.188 in; 1.219 in; 0.218 in; 0.531-inch dia.

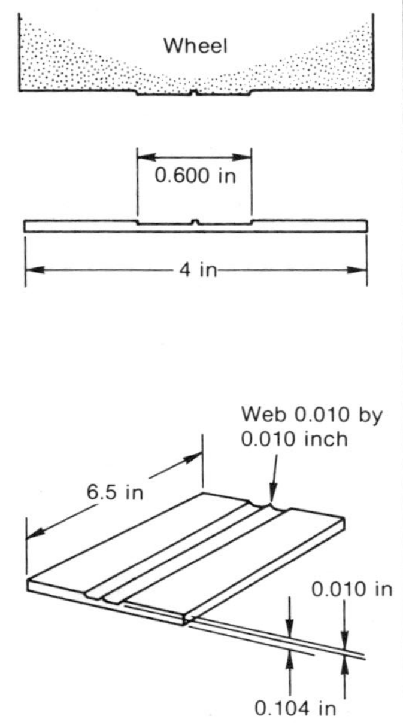

Part 2: Wheel; 0.600 in; 4 in; Web 0.010 by 0.010 inch; 6.5 in; 0.010 in; 0.104 in

Part 3: Wheei; 2.800 in; 3.000 in; 0.250 x 0.125-inch deep; 45°; Wheel; Herringbone pattern cut into periphery of wheel to improve wash action; 20 in; 0.125 in; 0.020 in; 0.020 in

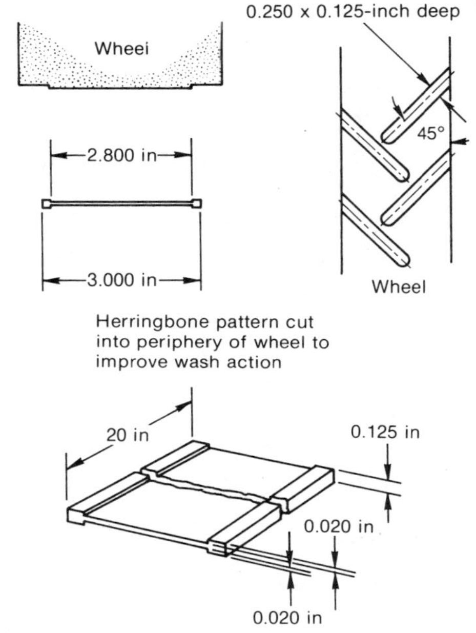

Part 4: Wheel; 0.187 in; 2.25 Hex tube x 0.050-inch wall; Surfaces machined one pass, 5 pieces 30 minutes, 6 minutes per piece; 2-1/4 in; 0.050 in; 0.187 in

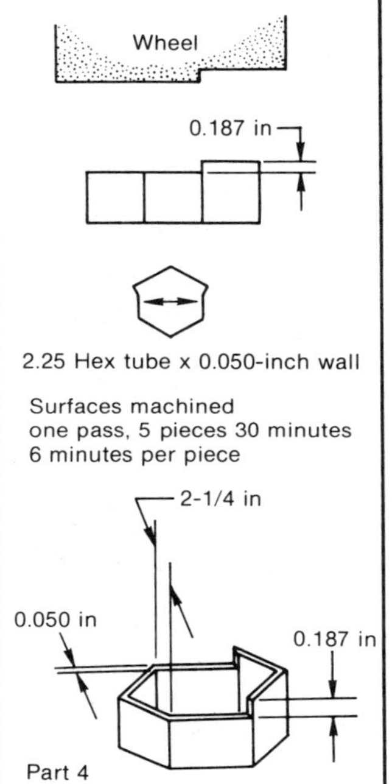

*Wheel diameter is 12 inches.

## PROCESS SUMMARY

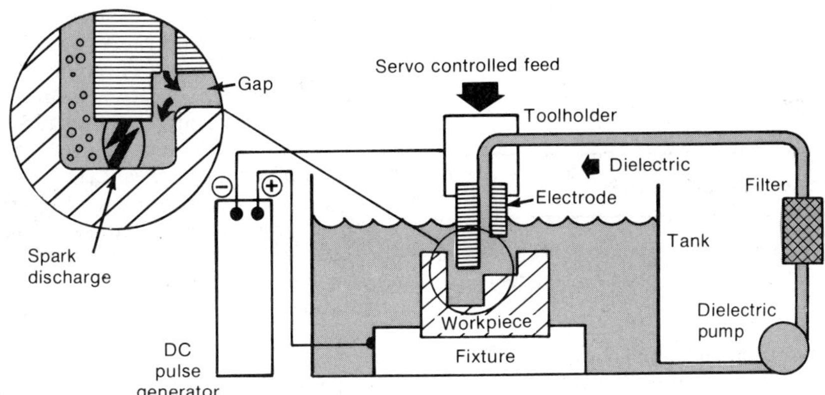


**Figure 12.3–1** EDM schematic.

### Principles

Electrical discharge machining (EDM) removes electrically conductive material by means of rapid, repetitive spark discharges from a pulsating direct-current power supply with dielectric flowing between the workpiece and the tool (see figure 12.3–1). The shaped tool (electrode) is fed into the workpiece under servo control. A spark discharge then breaks down the dielectric fluid. The frequency and energy per spark are set and controlled with a direct-current power source. The servo control maintains a constant gap between the tool and the workpiece while advancing the electrode. The dielectric oil cools and flushes out the vaporized and condensed material while reestablishing insulation in the gap. Surface roughness decreases with increased frequency and reduced current. Material removal rate, surface roughness and overcut all increase with a current increase or with a frequency decrease (or longer "on" time cycles). Electrode materials frequently used are brass, copper, copper-tungsten, tungsten wire and graphite. Small spark gaps are associated with closer tolerance control and slower cutting rates. Erosion occurs on the tool as well as on the workpiece, with wear ratios depending on spark wave-shape from the power source, electrode material and workpiece material. A nearly "no wear" combination of operating parameters can be found for electrical discharge machining of some steels when using reverse polarity, as opposed to "standard" polarity (positive on the workpiece).

### Practical Applications

EDM cuts any electrically conductive material regardless of its hardness and is particularly adapted for machining irregular slots or cavities. The scope of EDM applications extends from 0.002-inch [0.050 mm] diameter holes to 50 ton [45 metric tons] automotive die cavities. Because there is no physical contact, delicate structures can be cut successfully. Cutting is three-dimensional as the shaped electrode is fed into the workpiece. Because the sparks focus first on peaks and corners, burr-free cutting occurs. Multiple electrodes, automatic dressing, automatic positioners and NC motion control all contribute to electrical discharge machining's versatility. Tool and die work is a frequent EDM application, but mass production and even transfer-line applications exist. EDM production of small and/or shaped holes at shallow angles to the workpiece surface is commonplace. Narrow slots (0.002- to 0.012-inch [0.050 to 0.30 mm] wide), honeycomb cores and fragile parts can be cut in high-strength and high-hardness materials.

### Operating Parameters

See table 12.3–1.

### Material Removal Rates and Tolerances

Feed rates and material removal rates range from 0.00036 to 0.04 cubic inch per hour per ampere [0.0059 to 0.65 $cm^3/hr/A$]. Corner radii to 1/64 inch [0.4 mm] are common. Production tolerance to $\pm 0.001$ inch [$\pm 0.025$ mm] is normal; tolerances of $\pm 0.0002$ inch [$\pm 0.005$ mm] are repeatable with careful selection or development of cutting conditions. The recast layer can be controlled and is repeatable to a few ten-thousandths inch.

Material removal rates, surface roughness, recast layer and heat-affected zones all increase as spark intensity increases, see table 12.3–3.

Electrode erosion is expressed as wear ratio as follows (see also figure 12.3–19):

$$\text{volumetric wear ratio} = \frac{\text{volume of work removed}}{\text{volume of electrode consumed}}$$

or

$$\text{linear wear ratio} = \frac{\text{depth of cut}}{\text{length of electrode consumed}}$$

# 12.3 Electrical Discharge Machining—EDM

Corner radii are controlled by the spark length, while taper is controlled by the amount of cutting debris in the arc gap or fluid.

**TABLE 12.3–1 Typical Values for EDM Operating Parameters**

| INDEPENDENTLY CONTROLLABLE PARAMETERS | |
|---|---|
| Open circuit voltage: | 50 to 300 V |
| Frequency: | 50 Hz to 500 kHz or equivalent duty-cycle setting of on and off times per pulse |
| Dielectric type: | Hydrocarbon (petroleum) oils, deionized water, kerosene, gas (dry) (see table 12.3–10) |
| Dielectric flow pressure: | 28-inch [711 mm] vacuum to 70 psi [482 kPa] pressure |
| Electrode materials: | Graphite, copper, brass, zinc-tin, steel, copper-tungsten, copper-graphite, silver-tungsten, tungsten |
| Servo drive gap sensitivity control: | Gaps: 0.0005 to 0.005 inch [0.013 to 0.13 mm] |
| Capacitance: | (on some style machines) from trial cuts |
| Polarity: | "Standard" is positive on workpiece, negative on electrode (see table 12.3–8) |
| **DEPENDENT VARIABLES AND RESULTS** | |
| Average current: | 0.1 to 500 A (A few large machines use multiple 500 A power packs, with separate leads.) |
| Spark gap: | 0.0005 to 0.005 inch [0.013 to 0.13 mm] |
| Overcut: | 0.0002 to 0.020 inch/side [0.005 to 0.50 mm] |
| Material removal rate: | 0.003 to 1.5 in$^3$/hr [0.05 to 24.6 cm$^3$/hr] |
| Wear ratio (ratio of workpiece erosion to electrode erosion): | 0.5:1 to 100:1 (see table 12.3–12) |
| Surface roughness: | 8 to 250 microinches $R_a$ [0.2 to 6.3 $\mu$m] |
| Depth of recast plus heat-affected zone: | 0.0001 to 0.005 inch [0.0025 to 0.13 mm] |
| Corner radius: | 0.001 inch [0.025 mm] or equal to overcut |
| Taper | 0.0005 to 0.005 inch/inch/side [0.0005 to 0.005 mm/mm/side]. With proper tooling, taper can be eliminated. |

## Surface Technology

The surface texture is a series of overlapping, small craters that increase in size with increasing spark energy and/or lower spark frequency. Recast and heat-affected layers occur on all materials and range from 0.0001- to 0.005-inch [0.002 to 0.13 mm] deep. These layers should be removed or modified on critical or fatigue-sensitive surfaces. Surface roughness is typically in the 63- to 125-microinch $R_a$ [1.6 to 3.2 $\mu$m] range. Surface roughness from rough cuts can range up to 500 microinches $R_a$ [12.5 $\mu$m], and more costly, deluxe methods can attain 2 to 4 microinches $R_a$ [0.05 to 0.1 $\mu$m]. Surface residual stress is shallow (under 0.001 inch [0.025 mm]) but may be a high percent of tensile strength. The fatigue strength of EDM surfaces without post-treatment usually is severely reduced.

## Equipment

EDM equipment is available in a wide range of sizes from a bench model with a few amperes capacity to 13-by-18-foot [4 by 5.5 m] die sinkers with 5,000-ampere capacity (made up from multiple 500-ampere power packs). Automatic or NC controls are common. Automatic feed, interchangeable electrode holders and rotary-turret electrode holders are available to aid electrode changing and automation of the EDM process. Multiple electrodes and multi-lead power supplies enhance the productivity of many equipment types. Integrated systems are the usual order; thus, EDM machines can be placed almost anywhere in the normal shop. Fume vents are recommended, and tooling should provide a means for venting the liberated gases. The pulse power supply usually contains full control of "on and off" times for each discharge, as well as full control of the discharge energy.

The "grinding" mode of EDM is covered in section 12.2 on EDG. Section 12.5 on EDWC covers the wire cutting EDM operations. Orbital or planetary erosion is a recent addition to EDM capabilities. This method makes possible the almost constant enlargement of a die cavity of geometrically difficult contour with simple electrodes. Electrode motions range from simple orbital movement to complex combinations of vertical, eccentric and orbital motions, as shown in figure 12.3–2.

## Machining Characteristics

High quality output depends upon rigid machine tools and fixturing for close gap control via the servomechanism which controls the electrode movement.

Good safety practice makes it desirable to operate with the spark fully submerged in the dielectric. Gas and fumes generated should be carefully vented. Good hygiene, thorough washing by the machine operator and the use of a moisturizing cream can eliminate skin irritations caused by the dielectric oils.

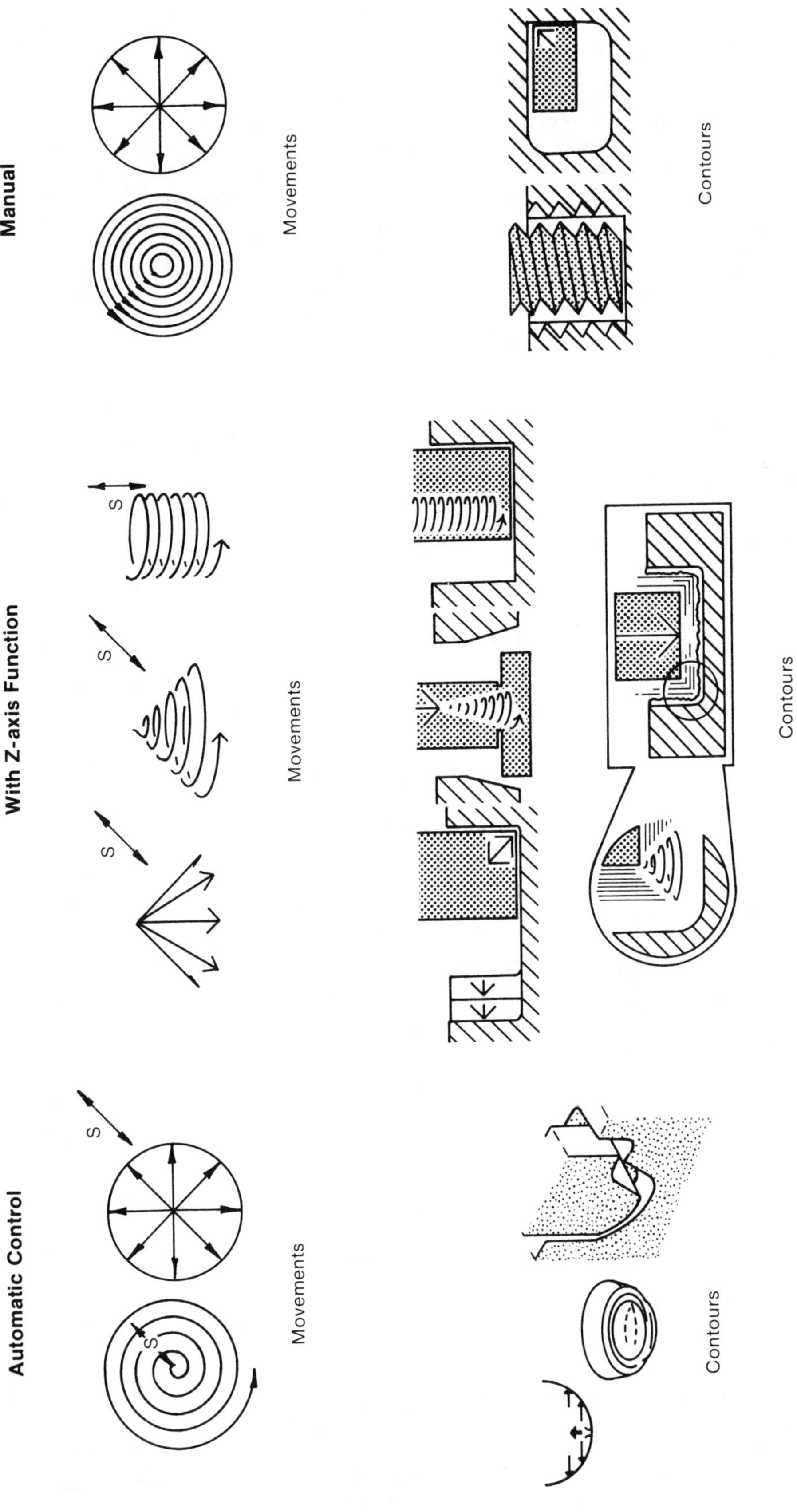


**Figure 12.3–2** Orbital or planetary erosion motions in EDM. **S** represents feed direction. (Courtesy of Agietron Corporation)

## 12.3 Electrical Discharge Machining—EDM

### SELECTED DATA

There are many combinations of EDM operating parameters that will remove material from conductive workpieces. Experience with particular equipment and/or systems plus a limited number of test points on particular electrode-material/workpiece-material combinations will provide the operator with reference points for improved estimating and operating efficiency. Most equipment manufacturers are supplying approved data for their installations. These data frequently are measured under "standard" conditions.[1] In practical application, deviations from "standard" might result from the use of different materials, dielectric fluids, or product forms or from different levels of operator experience.

The steps in table 12.3-2 are recommended as one logical sequence for selecting EDM operating parameters. Supporting data come from the other figures and tables in accordance with the terms as explained in figure 12.3-3.

[1]Swiss standard 37550 and a German recommendation VDI 3400 are in existence circa 1979.

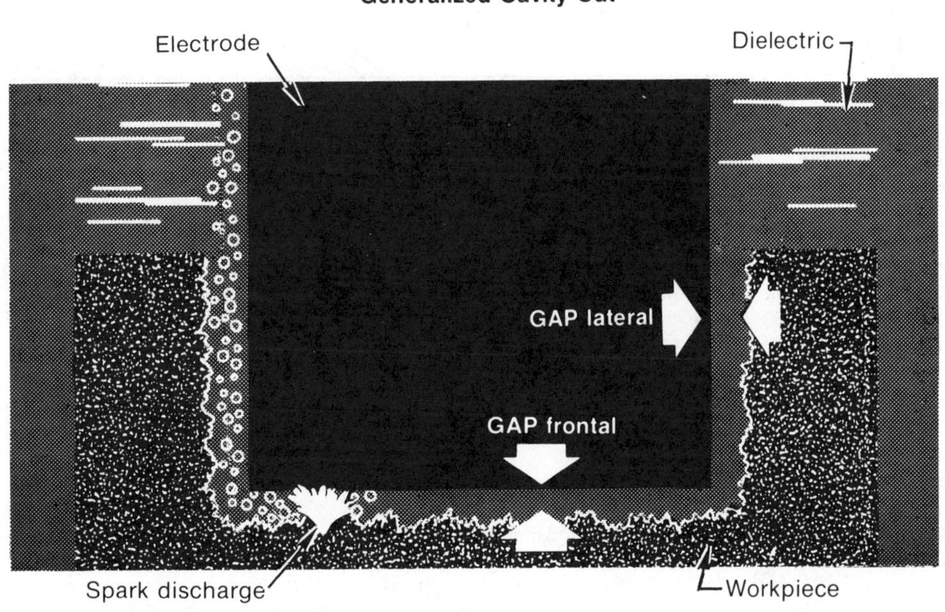


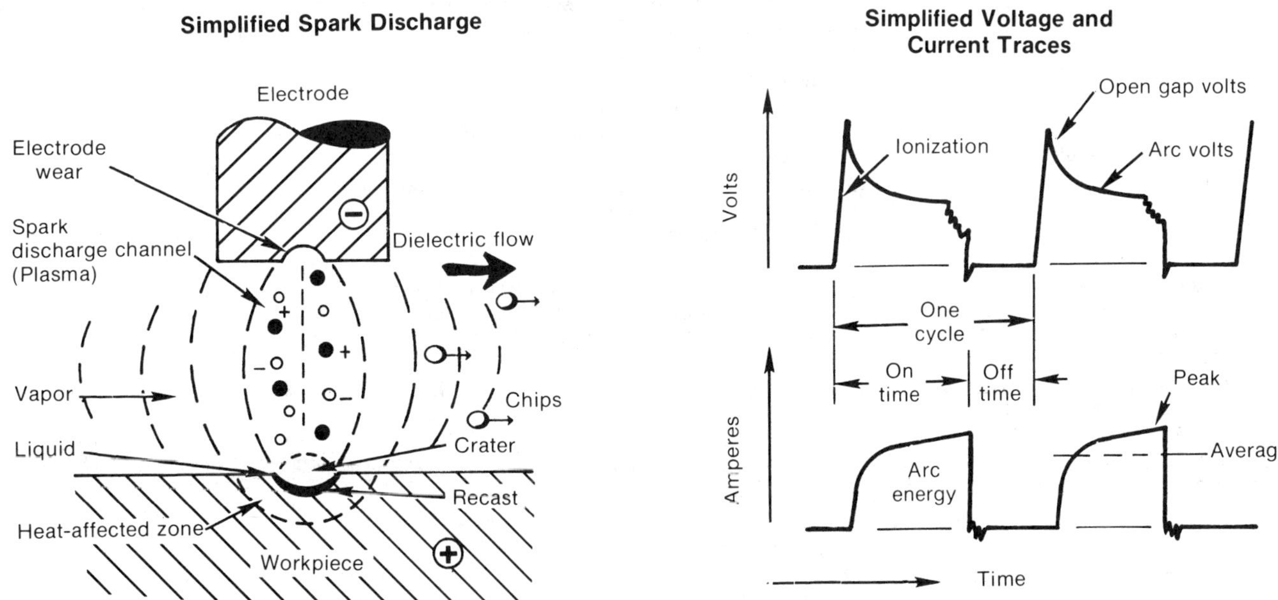


**Figure 12.3–3** EDM parameter terminology. (Drawing at top courtesy of Agietron Corporation)

**TABLE 12.3-2 Steps to Select EDM Operating Parameters**

| ITEM | DATA |
|---|---|
| 1. Type material, metallurgical state and melting point | Engineering drawing and material specifications. |
| 2. Surface roughness required (finishing operation allowance—if any) | Engineering drawing. (Consider recast and heat-affected zone removal *only* if application function requires it and then *only* on the critically loaded areas or surfaces. Recommend removal of 2 times recast layer shown on micrograph of cross section of trial cut.) |
| 3. Volume of material to be removed | From engineering drawing and raw stock shape, calculate $in^3$ [$cm^3$] to be removed. |
| 4. Material removal rate—preliminary | Figures 12.3–4 to 12.3–6; tables 12.3–3 and 12.3–4. |
| 5. Cutting time estimate | Material to be removed divided by material removal rate. |
| 6. Select type electrode material | Tables 12.3–5, 12.3–6 and 12.3–12. |
| 7. Make trade-offs among machining parameters | Table 12.3–7. According to work complexity, electrode, cost, desired precision, and quantity, a choice between high removal rate or normal or lowest wear must be fixed. |
| 8. Select operating polarity and open circuit voltage | Table 12.3–8 |
| 9. Select specific parameters:<br>On time—for roughing and finishing<br>Current<br>Capacitance<br>Material removal rate<br>Surface roughness<br>End wear<br>Corner wear | Figures 12.3–7 to 12.3–11 and table 12.3–9 are given for one specific electrode (graphite) and material (steel) combination. Similar curves on other material-electrode polarity combinations should be developed by the operator using his specific equipment, controls and materials. Many EDM builders have data for their own equipment, figures 12.3–12 to 12.3–15. For "drilling," see table 12.3–15 and figure 12.3–22. |
| 10. Dielectric selection | Table 12.3–10; figures 12.3–16 and 12.3–17. |
| 11. Flushing and filtration | Table 12.3–11. One of the most important parameters to secure high quality repeatability. Must be tailored to workpiece configuration, electrode shape and spark gap or overcuts. Consider use of orbiting electrodes, figure 12.3–2. Increasingly difficult flushing conditions, figure 12.3–18, can increase the machining cycle time.<br>Filtration degree for recirculated dielectric should increase as the spark gap decreases and the flushing becomes more difficult. Recommend 75 $\mu$m (micron) filtration maximum. |
| 12. Check availability of selected parameters on equipment to be used | Equipment ratings and control points. (Circuit type may require selection of capacitance.) |
| 13. Cutting time—refined | Based on revised material removal rate from item 9 and material to be removed item 3, revised by excluding provision for flushing holes—if any. On finish cuts, it is a good approximation to assume unworn corners on electrodes when calculating material to be removed. Allow a multiplying factor for flushing difficulties, figure 12.3–18. |
| 14. Overcut—the side wear between electrode and eroded workpiece—an allowance for electrode design | Figures 12.3–12 to 12.3–15. Increases with increase in spark intensity and material removal rate and voltage setting. Each workpiece material-electrode material combination has its own overcut, current, frequency charts. |
| 15. Wear—an allowance for electrode design, stroke setting and accuracy of fine details | Figure 12.3–19; table 12.3–12. Corner wear is usually the most critical or limiting wear. |
| 16. Number of electrodes | Use finishing electrode for fine details. It can frequently then serve as the roughing electrode on subsequent parts. Number to use depends on wear rate and depth of cut. |
| 17. Voltage setting | Figure 12.3–20; table 12.3–8. For maximum efficiency, use minimum volts to achieve average current. |
| 18. Select electrode manufacturing method | Often the most costly element for EDM. Many fabrication/machining methods available. Choice depends on electrode material production rate, shape and costs, tables 12.3–13 and 12.3–14. |
| 19. Electrode design | See items 2, 14, 15, 16 and 18. |
| 20. Assess extent of altered material zones on surface: recast, heat-affected zone (HAZ) | Figure 12.3–21. Depth of HAZ below recast approximately equal to depth of recast. (Note: If EDM'd surface is highly stressed, subject to stress corrosion or otherwise critically loaded, steps should be taken to ameliorate these surface integrity effects.) |
| 21. Trial cut | Several test cuts are recommended in the workpiece material using the selected conditions to check the resulting values for overcut, removal rate, surface roughness and wear ratio. These data will, in time, provide the operator with his own machinability tables for his particular conditions. |

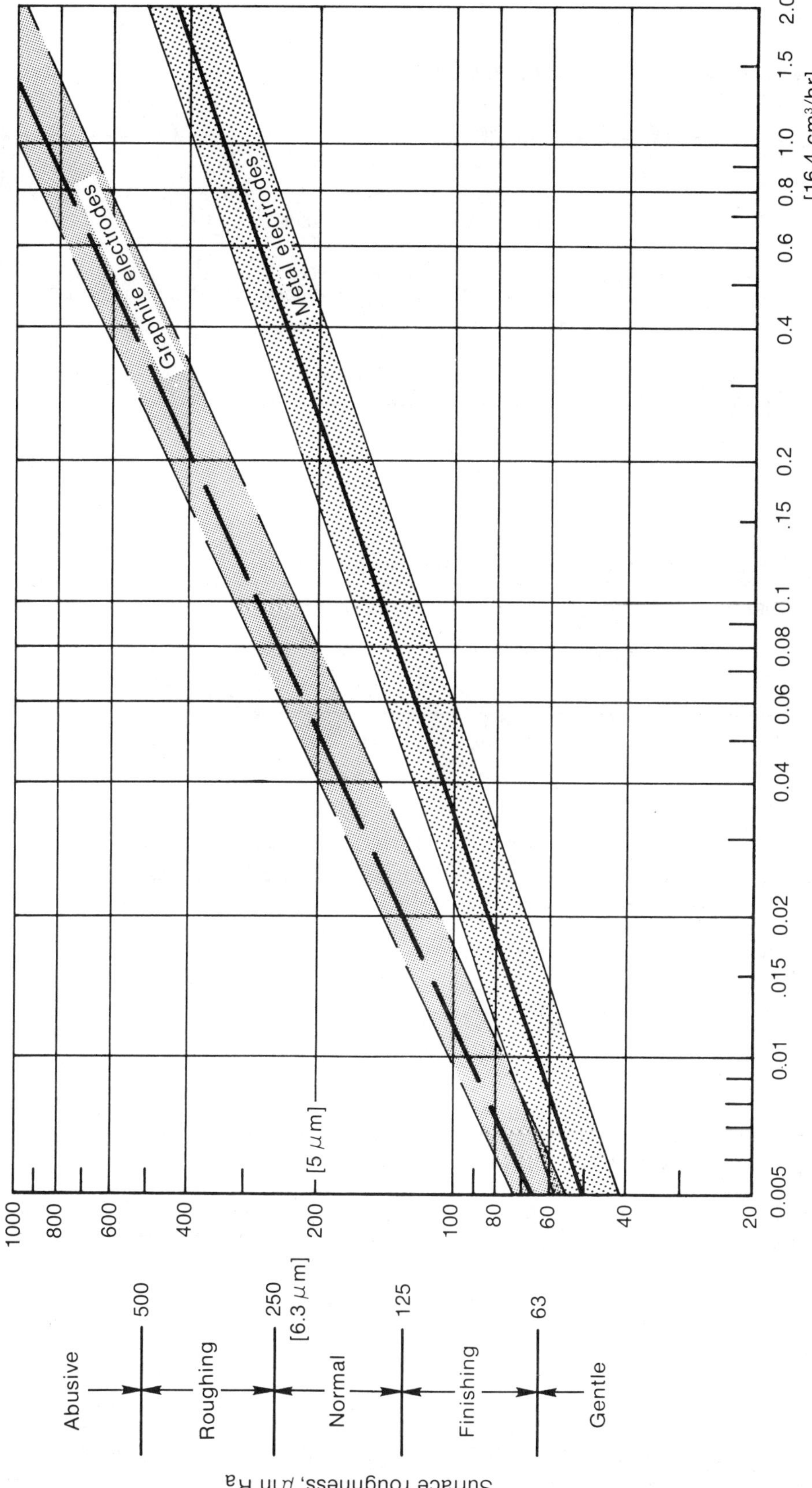


**Figure 12.3–4** **Material removal rate versus surface roughness for EDM with graphite or metal electrodes.** (Modified from G. Bellows, Surface integrity from electrical discharge machining of superalloys, Report R72AEG236, General Electric Company, 1972, p. 18)

**TABLE 12.3–3 Relationship of Metal Removal Rate in EDM to Surface Roughness and Expected Recast Material**

| PROCESSING LEVEL | SURFACE ROUGHNESS, $R_a$ | | METAL REMOVAL RATE | | EXPECTED RECAST MATERIAL | | | |
|---|---|---|---|---|---|---|---|---|
| | | | | | Predominant | | Maximum | |
| | μin | μm | in³/hr | cm³/hr | in | mm | in | mm |
| Gentle | <63 | <1.6 | <0.003 | <0.05 | <0.00022 | <0.0056 | 0.00054-0.00060 | 0.014-0.015 |
| Finishing | 63-125 | 1.6- 3.2 | 0.003-0.018 | 0.05-0.30 | 0.00022-0.00028 | 0.0056-0.0071 | 0.001-0.0013 | 0.025-0.033 |
| Normal | 125-250 | 3.2- 6.3 | 0.018-0.42 | 0.30-6.9 | 0.00028-0.00043 | 0.0071-0.011 | 0.0022-0.0033 | 0.056-0.084 |
| Roughing | 250-500 | 6.3-12.5 | 0.42-1.5 | 6.9-24.6 | 0.00043-0.00068 | 0.011-0.017 | 0.0055-0.0090 | 0.14-0.23 |
| Abusive | >500 | >12.5 | >1.5 | >24.6 | >0.00068 | >0.017 | to 0.014 | to 0.36 |

**TABLE 12.3–4 EDM Metal Removal Rates (Roughing Cuts)**

| CURRENT | METAL REMOVAL RATE* | | | |
|---|---|---|---|---|
| | Graphite Electrode | | Metal Electrode | |
| A | in³/hr | cm³/hr | in³/hr | cm³/hr |
| 25 | 1.5 | 24.6 | 0.75 | 12.3 |
| 50 | 3.0 | 49.2 | 1.5-1.8 | 24.6-29.5 |
| 100 | 5.5 | 90.1 | 2.7-4.4 | 44.2-72.1 |
| 200 | 9.0 | 147.5 | † | † |
| 400 | 16.0 | 262.0 | † | † |

NOTE: A common *rough* estimating guide is 1.0 in³/hr [16.4 cm³/hr] for 20 amperes
*Based on steel workpiece.
† Graphite is normally used for high-ampere applications.

**TABLE 12.3–5 Some Physical Properties of EDM Electrode Materials**

| ELECTRODE MATERIAL | MELTING POINT °C | BOILING POINT °C | HEAT TO VAPORIZE 1 cm³ FROM ROOM TEMPERATURE cal/cm³ | THERMAL CONDUCTIVITY Ag=100 percentage | ELECTRICAL CONDUCTIVITY Ag=100 percentage | THERMAL EXPANSION per °C x $10^{-6}$ | STRENGTH psi | MODULUS OF ELASTICITY psi x $10^6$ |
|---|---|---|---|---|---|---|---|---|
| Copper | 1,083 | 2,580 | 12,740 | 94.3 | 96.5 | 16.0 | 35,000 | 18.0 |
| Graphite | * | Over 4,000 | Approx. 20,000 | 30.0 | 0.1 | 4.5 | 5,000 | 0.86 |
| Tungsten | 3,395 | 5,930 | 22,680 | 29.6 | 48.1 | 4.6 | 600,000 | 51.0 |
| Iron | 1,535 | 2,800 | 16,900 | 16.2 | 16.2 | 15.0 | 40,000 | 27.0 |

SOURCE: E. I. Shobert, II, What happens in EDM, *Manufacturing Engineering* 77 (July 1976), p. 39.
*Sublimes or boils before melting at atmospheric pressure.

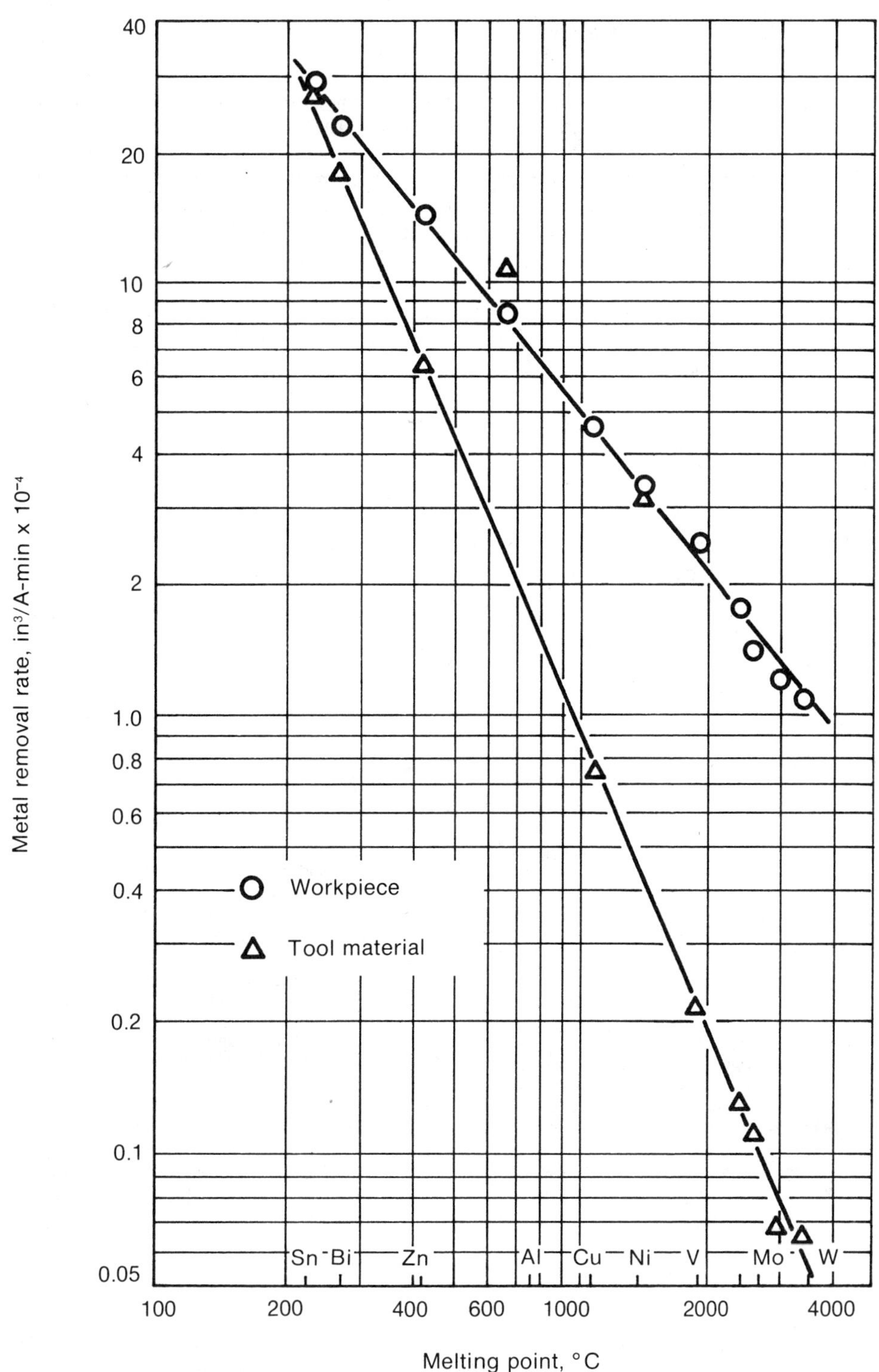


**Figure 12.3–5** Average metal removal rate versus melting point. (F. J. Demaine and B. Schneider, *A microscopic model of the EDM process,* TR 01.1178, IBM-Endicott Laboratories, January 1970, p. 19.)

**TABLE 12.3–6 Selection of EDM Electrode Material**

| ELECTRODE MATERIAL | GUIDE-LINES See Footnote | FORM | CORNER WEAR RATIO IN FINISHING | END WEAR RATIO IN ROUGHING | RELATIVE COST | MACHIN-ABILITY RATING | RECOM-MENDED USES | USES NOT RECOM-MENDED |
|---|---|---|---|---|---|---|---|---|
| Graphite | 1 | blocks, rod, tube, bar | 5:1 | to 100:1 | low | excellent | tooling | |
| Copper | 2 | bar, rod, sheet, wire, tube, forgings, stampings | 1:1 | 2:1 | medium | good | holes, slots | high accuracy and detail |
| Copper-graphite | 3 | blocks, rods | 2:1 | 4:1 | medium | fine | general purpose | |
| Brass | 4 | same as copper | 0.7:1 | 1:1 | low | good | holes and cavity sinking | high accuracy |
| Zinc alloys | 5 | cast, die casting | 0.7:1 | 2:1 | low | good | forging die cavities | holes |
| Steel | 6 | all forms | 1:1 | 2:1 | low | excellent | through holes | carbides |
| Copper-tungsten | 7 | bar, flats, shim stock, rod, wire, tube | 3:1 | 8:1 | medium | fair | slots, carbides | large areas |
| Silver-tungsten | 8 | sintered | 8:1 | 12:1 | high | fair | small slots, holes and intricate details | large areas |
| Tungsten | 9 | wire, rod, ribbon | 5:1 | 10:1 | high | poor | small holes | irregular holes |

NOTES:

1. GRAPHITE permits fast EDM rates and best wear ratio, is easy to machine, and offers excellent stability. Surface roughness produced on the workpiece is related to graphite particle size and density. Rotating graphite electrodes can produce roughness values less than 40 microinches $R_a$ [1.0 $\mu$m]. A number of different grades are available. Easily attached to tool holder with conductive cement. Needs good dielectric flow conditions.
2. COPPER produces fine finishes and is excellent for No-wear EDM. Easily machined or coined. Low wear ratio.
3. COPPER-GRAPHITE has best characteristics of both materials—more conductivity and strength than graphite. Will work well under poor flushing.
4. BRASS has a higher wear ratio than copper but can be economically machined for many operations. Not recommended for EDM of tungsten carbide.
5. ZINC-TIN ALLOYS are used for high production where many identical electrodes are needed because they are easily cast in metal molds. Can be pressure-cast and coined to an existing cavity. Complex shapes reproduced more readily than machining. Reusable by remelting. Not recommended for fine detail. Usually has poor wear ratio.
6. STEEL gives slow stock removal rates. Applicable when doing parting-line matching. Greater heat-affected zones. Readily machined, but wear ratio unsatisfactory for some types of steel.
7. COPPER-TUNGSTEN has good wear and finish characteristics. Can be brazed and has good rigidity. Used for close tolerances, fine detail, and low wear. Used on carbide. Used in slotting operations. Metal removal rates are lower than graphite, and the cost is higher. Use positive polarity on electrode when machining steels. Molded Cu-W electrodes are available for complex shapes or repetitive production.
8. SILVER-TUNGSTEN resists corner wear. Good for accurate detail or thinecuts. Provides low wear and faster removal rates than copper-tungsten. Is more expensive and generally limited to machining intricate carbide dies.
9. TUNGSTEN has high rigidity which makes it excellent for EDM of holes under 0.010-inch [0.25 mm] diameter. High cost but good wear ratio.

**TABLE 12.3–7 Trade-off Conditions Among EDM Parameters**

| SITUATION A | SITUATION B |
|---|---|
| High material removal rates | Low material removal rates |
| High surface roughness | Low surface roughness |
| Low frequency (long cycle time | High frequency (short cycle time) |
| High arc energy (high amperes and long on-time) | Low arc energy (low amperes and short on-time) |
| Large arc gap | Small arc gap |
| High overcut | Low overcut |
| Modest electrode wear | Low electrode wear |

**TABLE 12.3–8 EDM Polarity and Open Circuit Voltage Guidelines**

| ITEM | ELECTRODE (TOOL) POLARITY* | |
|---|---|---|
| | Negative | Positive |
| Metal removal rate | Medium to high | Medium to low |
| Work-to-wear ratios | Medium | High to "no wear"† |
| Surface roughness | | |
| steels | Good to rough | Good to excellent |
| other materials | Good (depends on other parameters) | Good (depends on other parameters) |

| WORKPIECE MATERIAL | ELECTRODE MATERIAL | ELECTRODE POLARITY* | OPEN CIRCUIT VOLTAGE V |
|---|---|---|---|
| Steel | Steel | Positive | Up to 270 |
| Steel | WCu | Positive | Up to 100 |
| Steel | Graphite | Positive | Up to 100 |
| Steel | Cu | Positive | Up to 150 |
| Steel | Cu | Negative | Up to 270 |
| Copper | WCu | Positive | Up to 100 |
| Copper | Cu | Negative | Up to 100 |
| Carbides | WCu | Negative | Up to 270 |

NOTE: These guidelines are subject to change depending upon power pack characteristics and voltage needed to attain desired average currents and cutting rates.

*Normal or "standard" polarity is usually considered to be positive on the workpiece and negative on the electrode tool. However, since the advent of solid state controls (circa 1970), some writings have reversed this, particularly for some materials where better wear ratios have been experienced.

†"No wear" EDM is the special combination of operating parameters that on some workpiece material-electrode material combinations yield wear ratios 100:1 or better, see table 12.3–12.

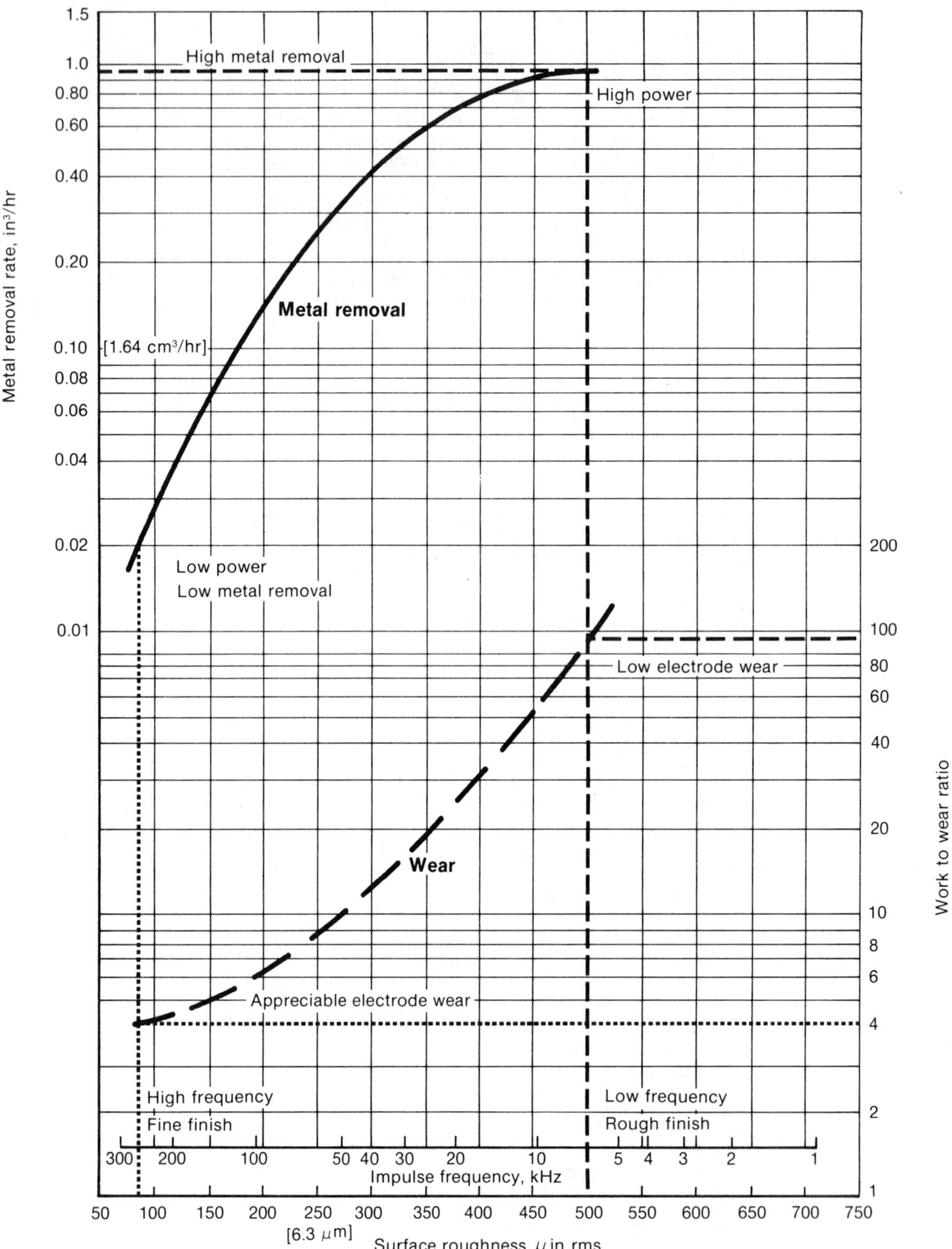


**Figure 12.3–6** Interrelations of EDM process variables. (Courtesy of Poco Graphite, Inc.)

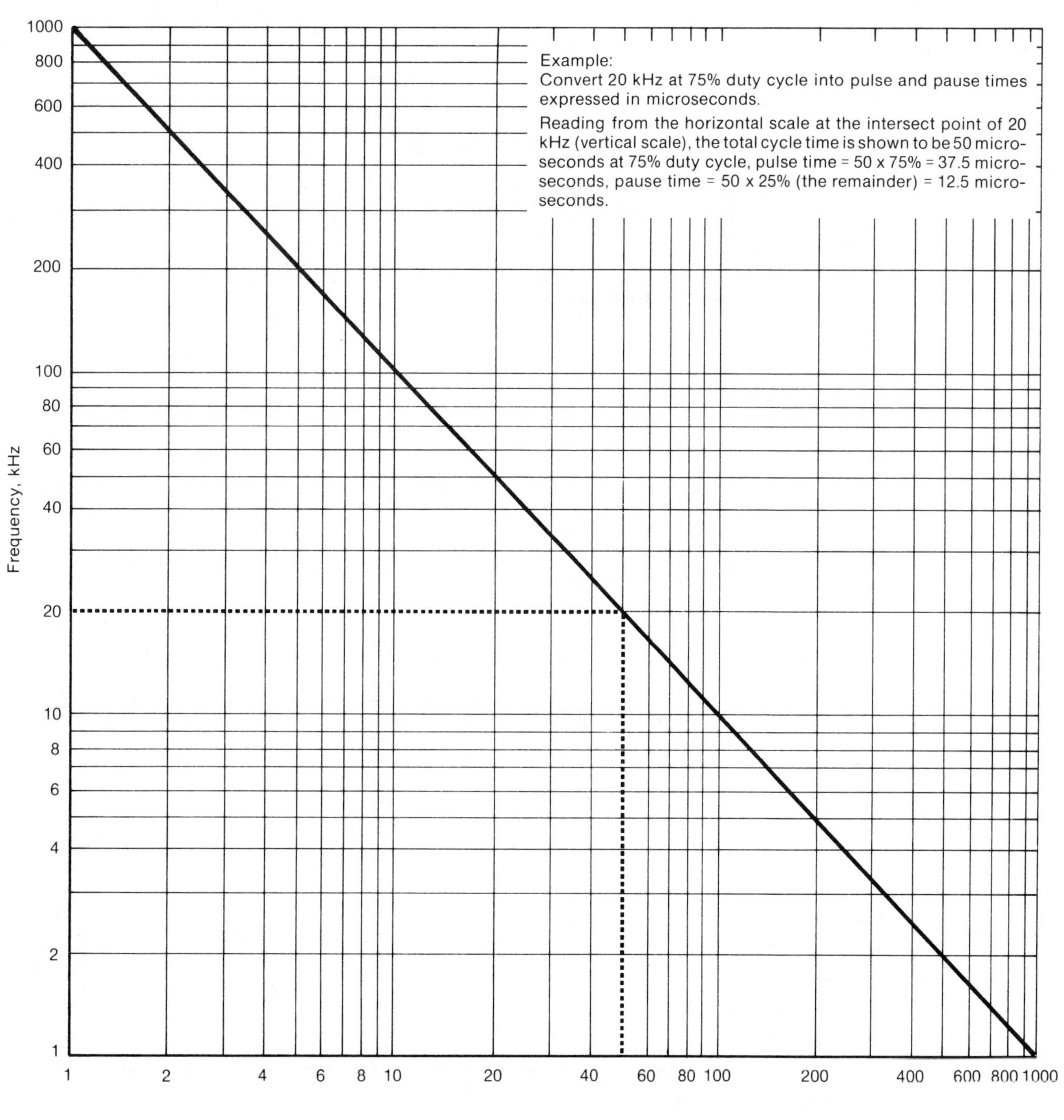


**Figure 12.3–7** Cycle time—frequency conversion. (Courtesy of Poco Graphite, Inc.)

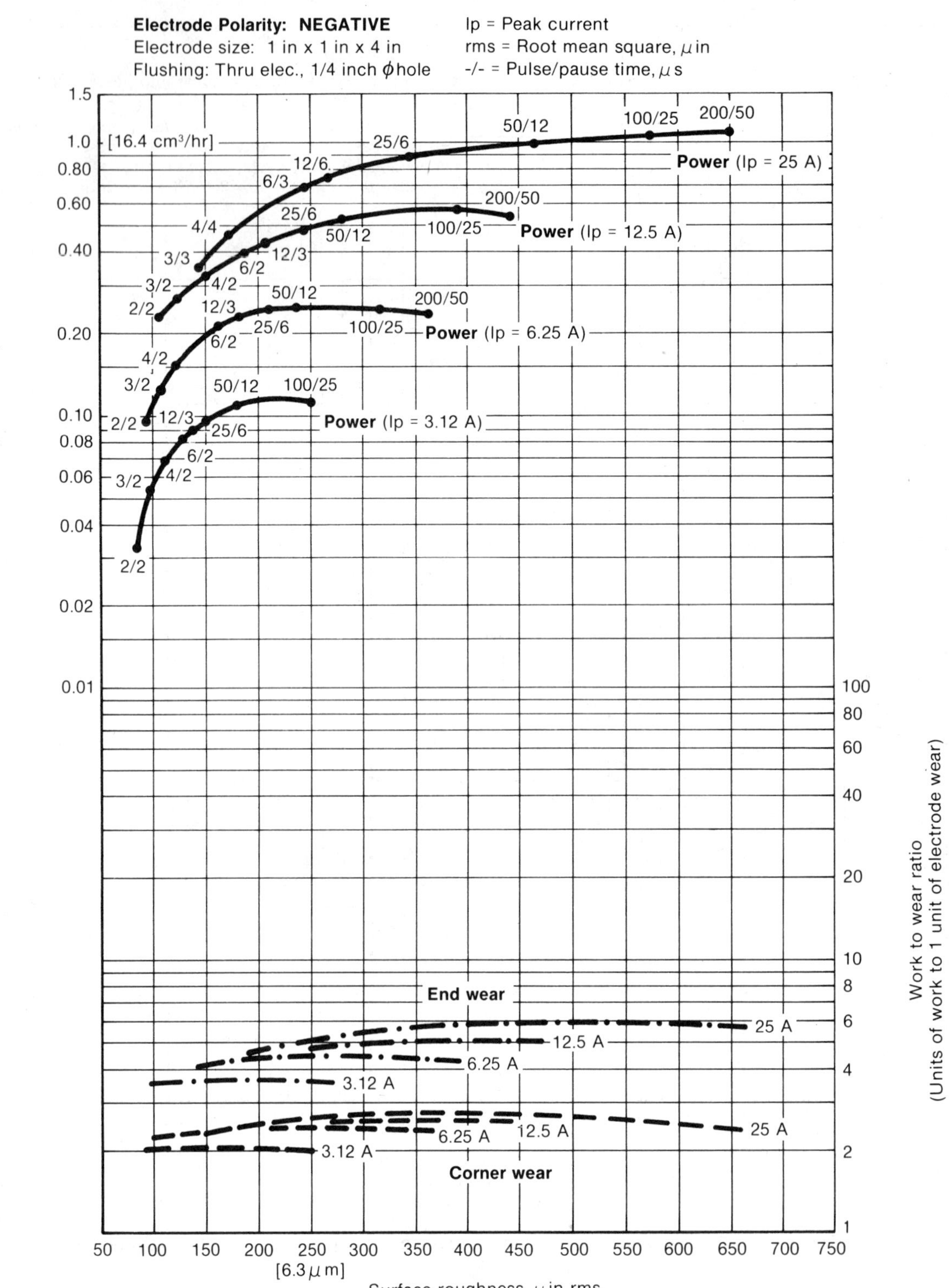


**Figure 12.3–8** Data for Poco EDM-1 electrode (negative polarity) cutting tool steel. (Courtesy of Poco Graphite, Inc.)

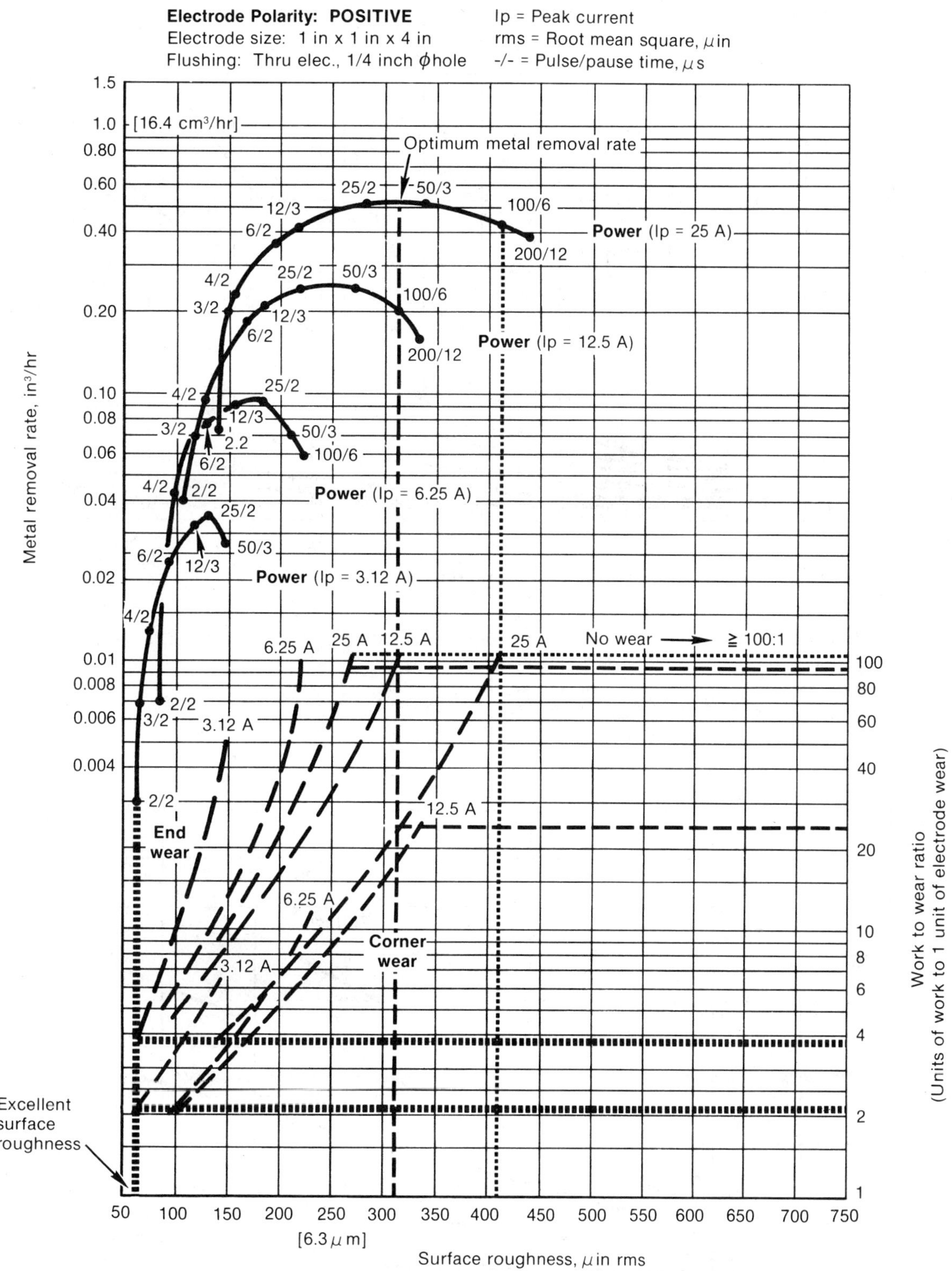


**Figure 12.3–9** Data for Poco EDM-3 electrode (positive polarity) cutting tool steel. (Courtesy of Poco Graphite, Inc.)

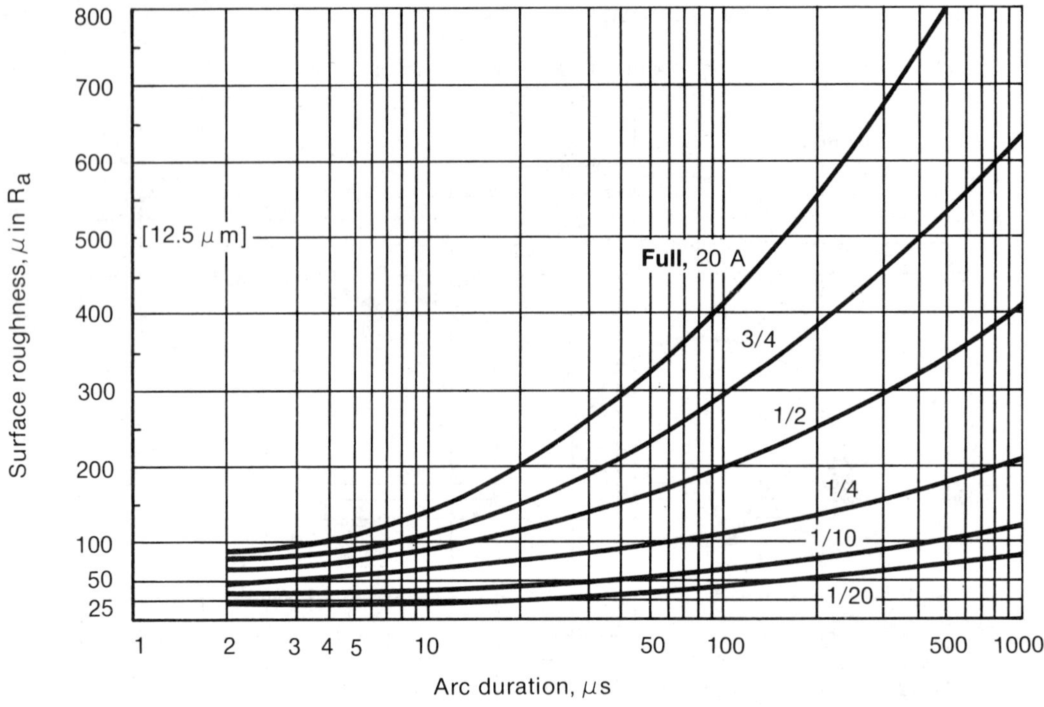


**Figure 12.3–10** Surface roughness versus arc duration at various currents for EDM of steel with graphite electrodes. (Courtesy of Hansvedt Engineering, Inc.)

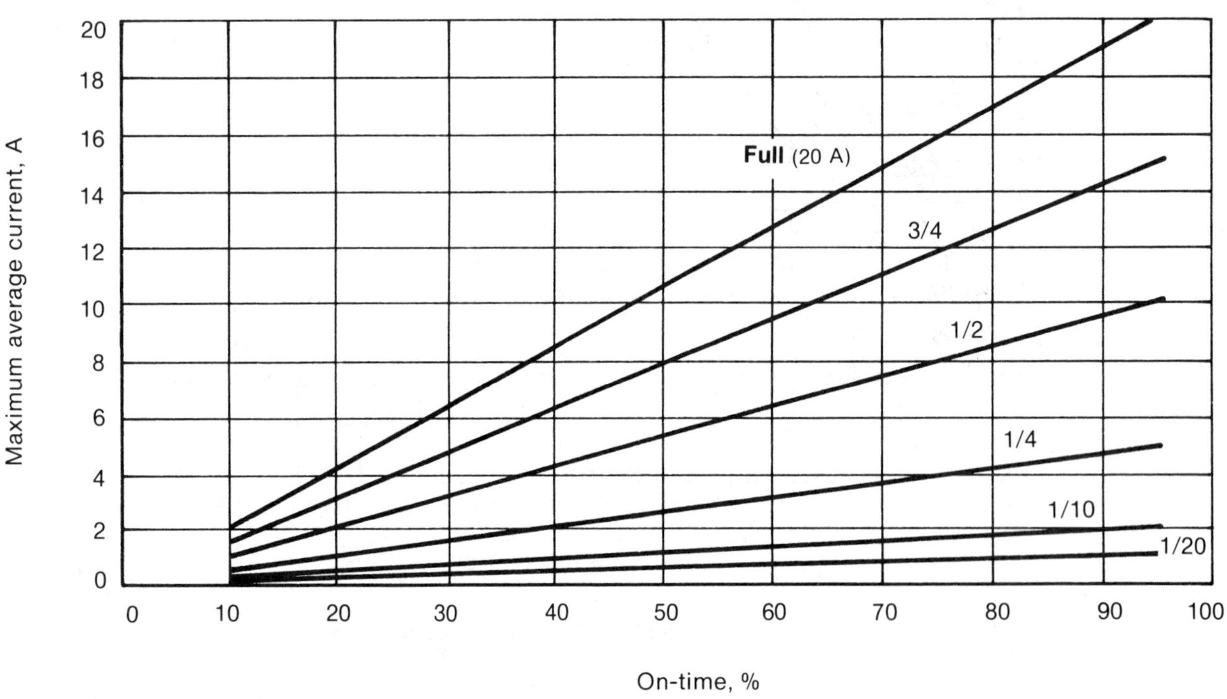


**Figure 12.3–11** Maximum average current versus percent age of on-time for EDM of steel with graphite electrodes. (Courtesy of Hansvedt Engineering, Inc.)

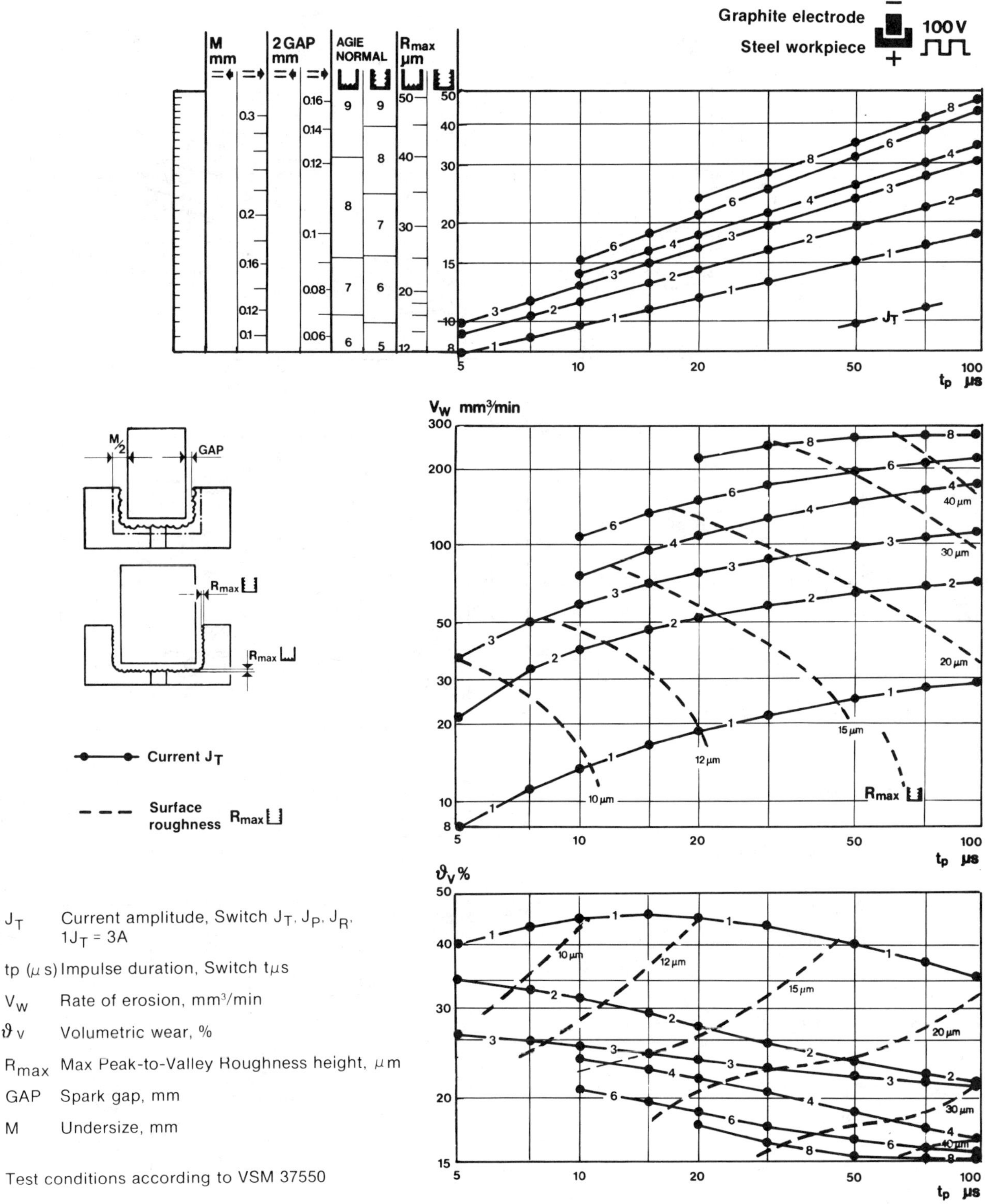


**Figure 12.3–12** Parameter selection for EDM. (Courtesy of Agietron Corporation)

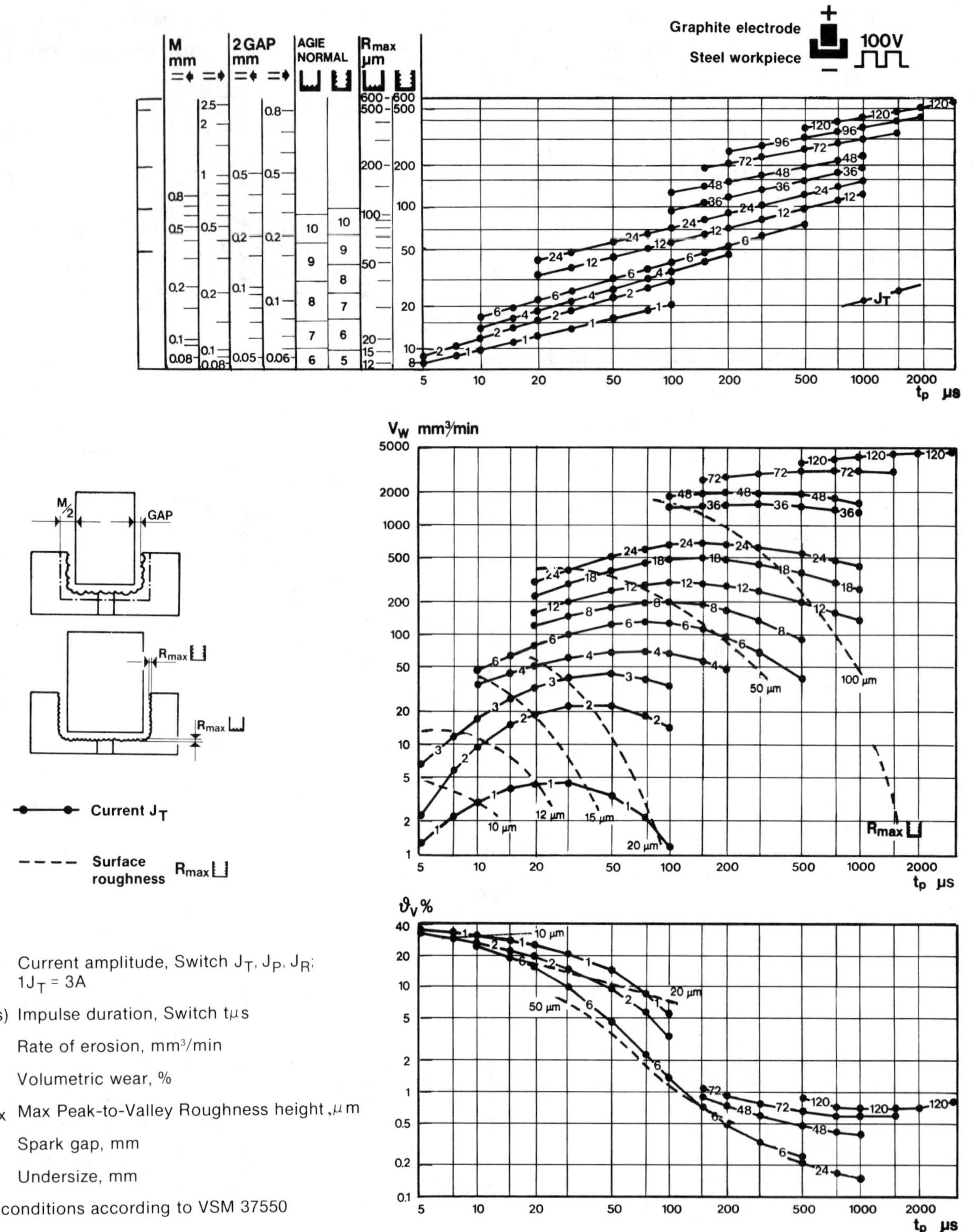


**Figure 12.3–13** Parameter selection for EDM. (Courtesy of Agietron Corporation)

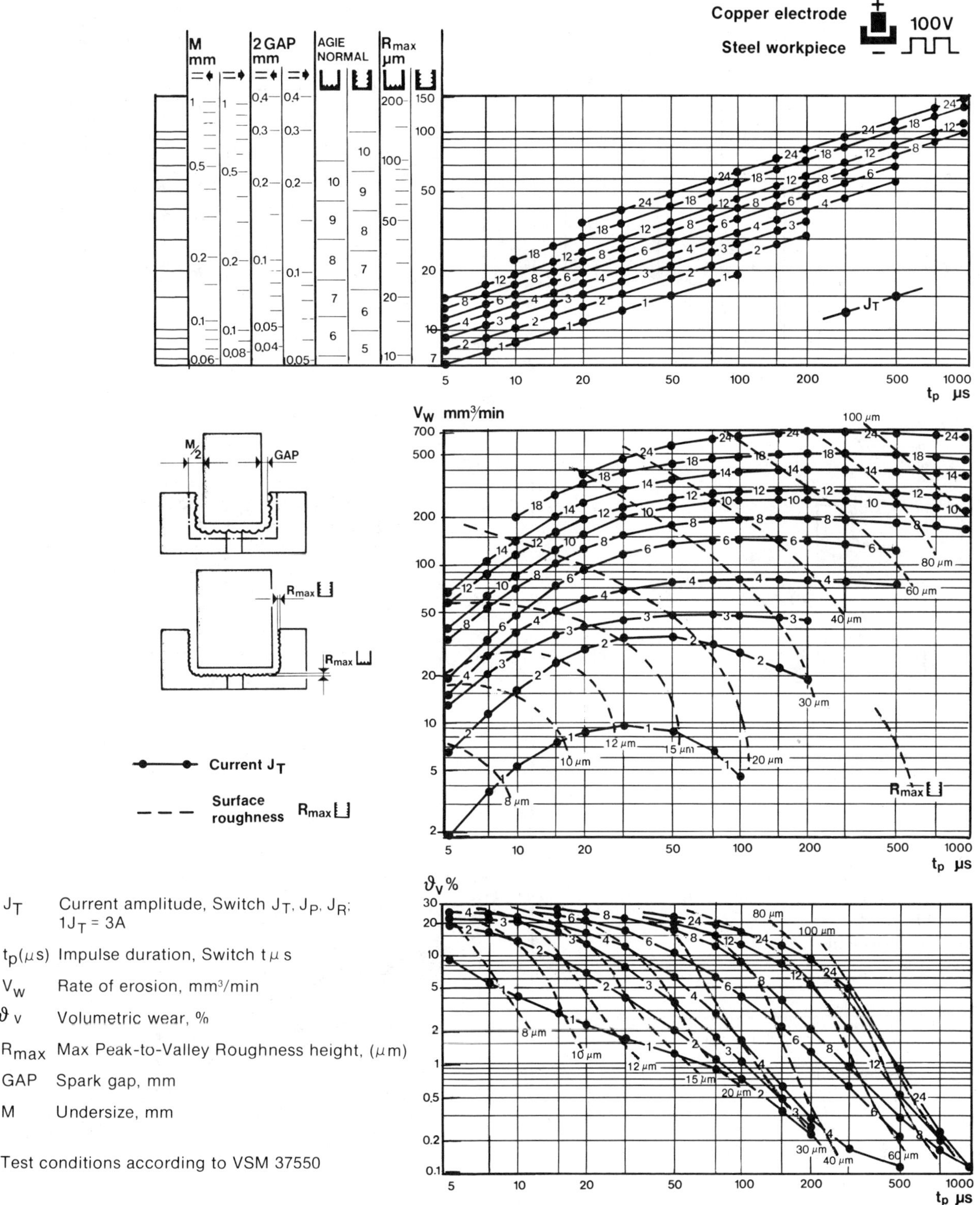


**Figure 12.3–14** Parameter selection for EDM. (Courtesy of Agietron Corporation)

# 12.3 Electrical Discharge Machining—EDM

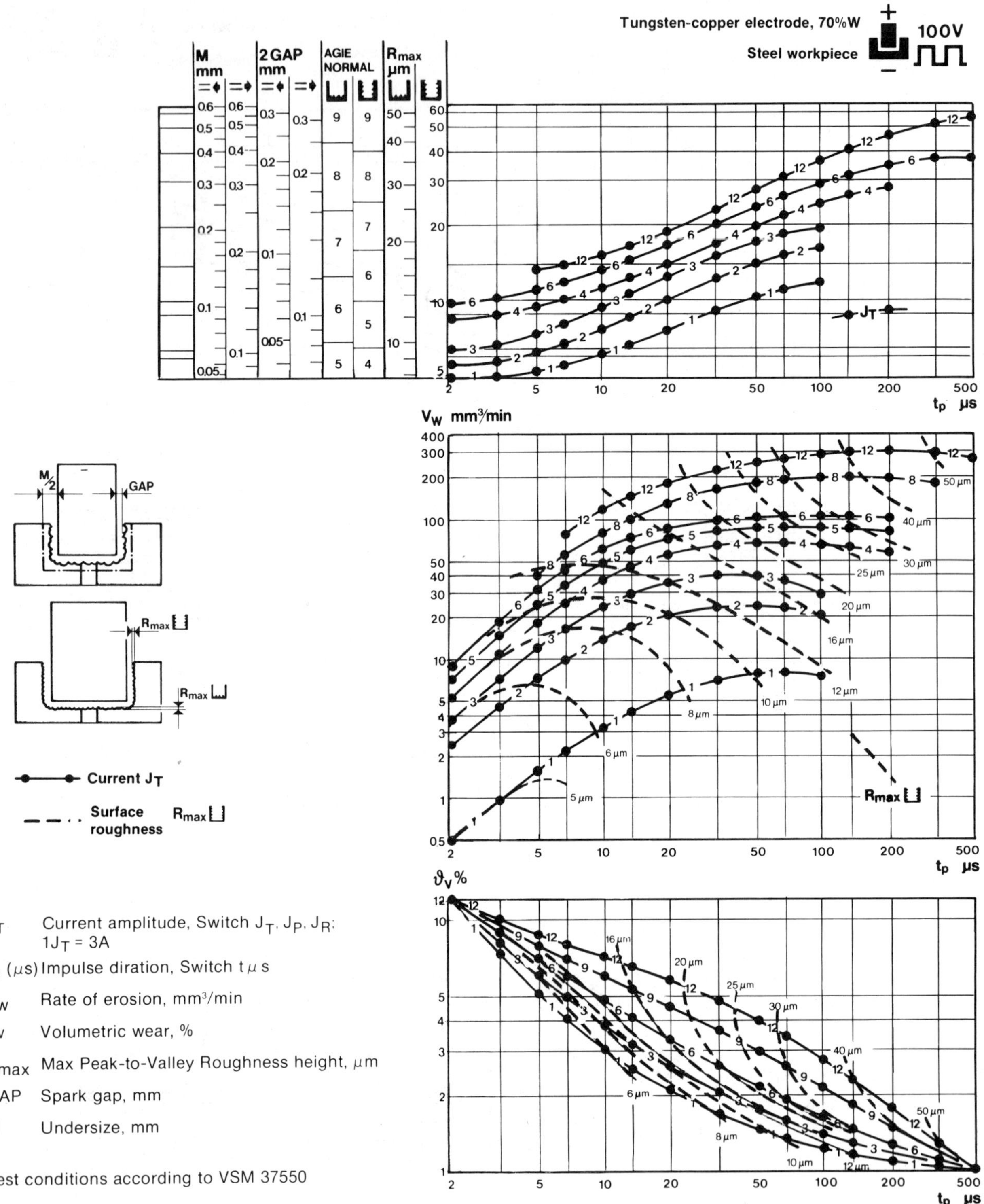


$J_T$ Current amplitude, Switch $J_T$, $J_P$, $J_R$: $1 J_T = 3A$

$t_p$ (µs) Impulse diration, Switch t µ s

$V_W$ Rate of erosion, mm³/min

$\vartheta_V$ Volumetric wear, %

$R_{max}$ Max Peak-to-Valley Roughness height, µm

GAP Spark gap, mm

M Undersize, mm

Test conditions according to VSM 37550

**Figure 12.3–15** Parameter selection for EDM. (Courtesy of Agietron Corporation)

**TABLE 12.3–9 Pulse/Pause Times versus Duty-Cycles: Conversion Chart**

| PULSE TIME μs | PAUSE TIME μs | | | |
|---|---|---|---|---|
| 2 | 2 | | | |
| 3 | 3 | 2 | | |
| 4 | 4 | 2 | | |
| 6 | 6 | 3 | 2 | |
| 12 | 12 | 6 | 3 | |
| 25 | 25 | 12 | 6 | 2 |
| 50 | 50 | 25 | 12 | 3 |
| 100 | 100 | 50 | 25 | 6 |
| 200 | 200 | 100 | 50 | 12 |
| 400 | 400 | 200 | 100 | 25 |
| 800 | 800 | 400 | 200 | 50 |
| 1,600 | 1,600 | 800 | 400 | 100 |
| | APPROXIMATE DUTY CYCLE | | | |
| | 50% | 66% | 80% | 95% |

SOURCE: Poco Graphite, Inc., Decatur, TX.

**TABLE 12.3–10 EDM Dielectrics**

Principal functions of the dielectric in EDM

- **Insulate** until required conditions are achieved.
- **Flush** particles out of spark gap.
- **Cool** machined particles, tool and workpiece.

Desired properties of the dielectric

- Low viscosity.
- High dielectric strength.
- High flash point.
- Freedom from acid, alkali and corrosion products (particularly sulfur compounds when machining sulfur-free steels).
- Low toxicity (particularly chlorine compounds).

Dielectrics used in EDM

- Hydrocarbon oils—the most widely used dielectric. Select type and viscosity to suit type cut, gap size and surface roughness (see figure 12.3–16). Generally cuts more smoothly after a few minutes use, or conditioning.
- Water—distilled and deionized. Used principally for micromachining and wire cutting machines (EDWC).
- Kerosene—good for superfinishing, but infrequently used. Deodorizing recommended and attention to safety precautions. Useful with tungsten electrodes.
- Silicone oils—infrequently used; more costly.
- Ethylene glycol solutions—rarely used.
- Gas—very specialized applications.

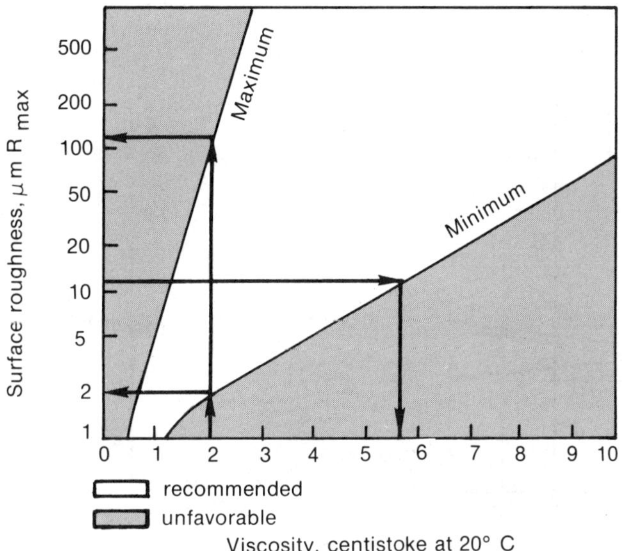


**Figure 12.3–16** Recommended dielectric viscosity for EDM as a function of peak to valley, $R_{max}$ surface roughness. (Courtesy of Agietron Corporation)

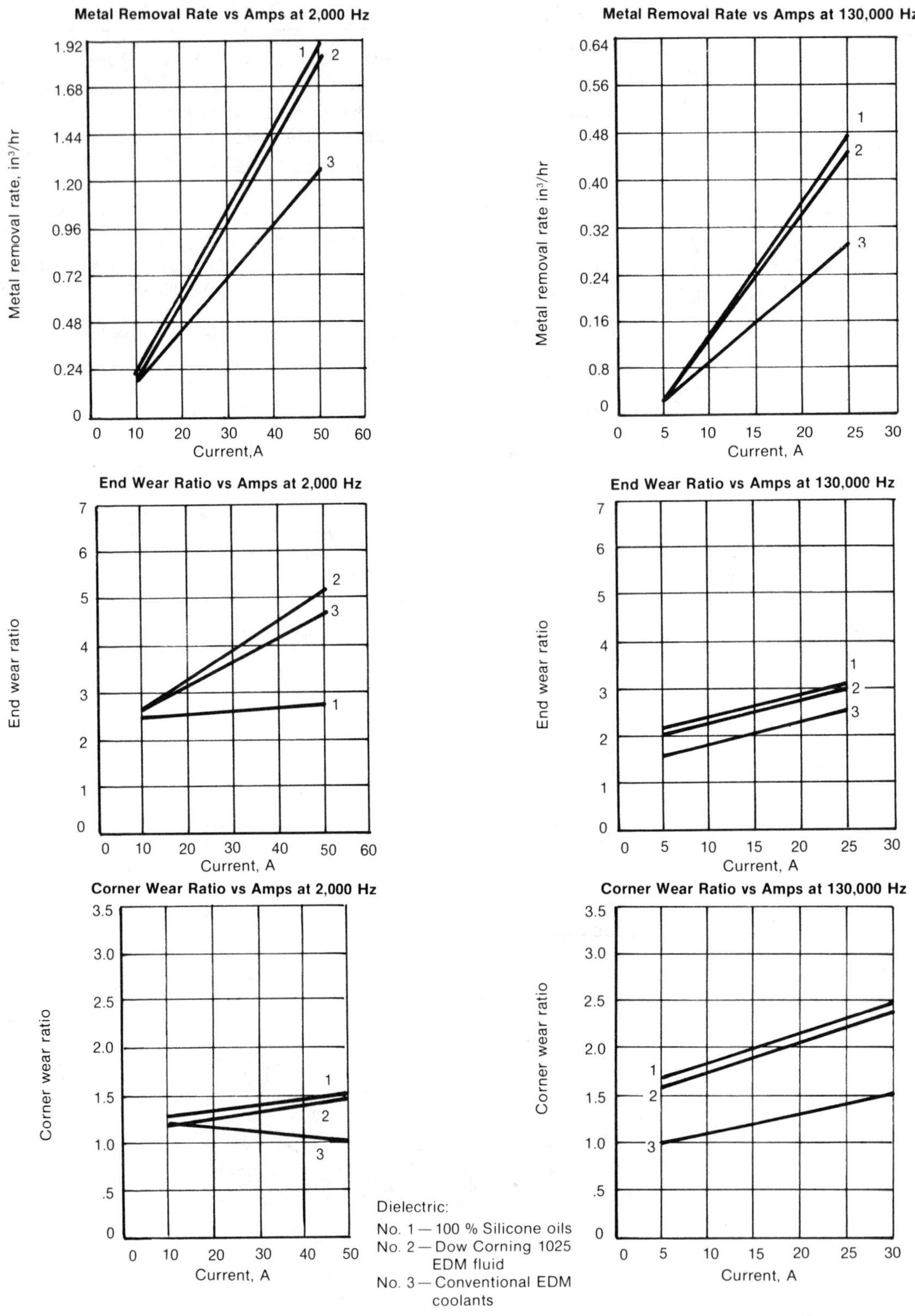


**Figure 12.3–17** Effect of dielectric on operating parameters. (Courtesy of Dow Corning Co.)

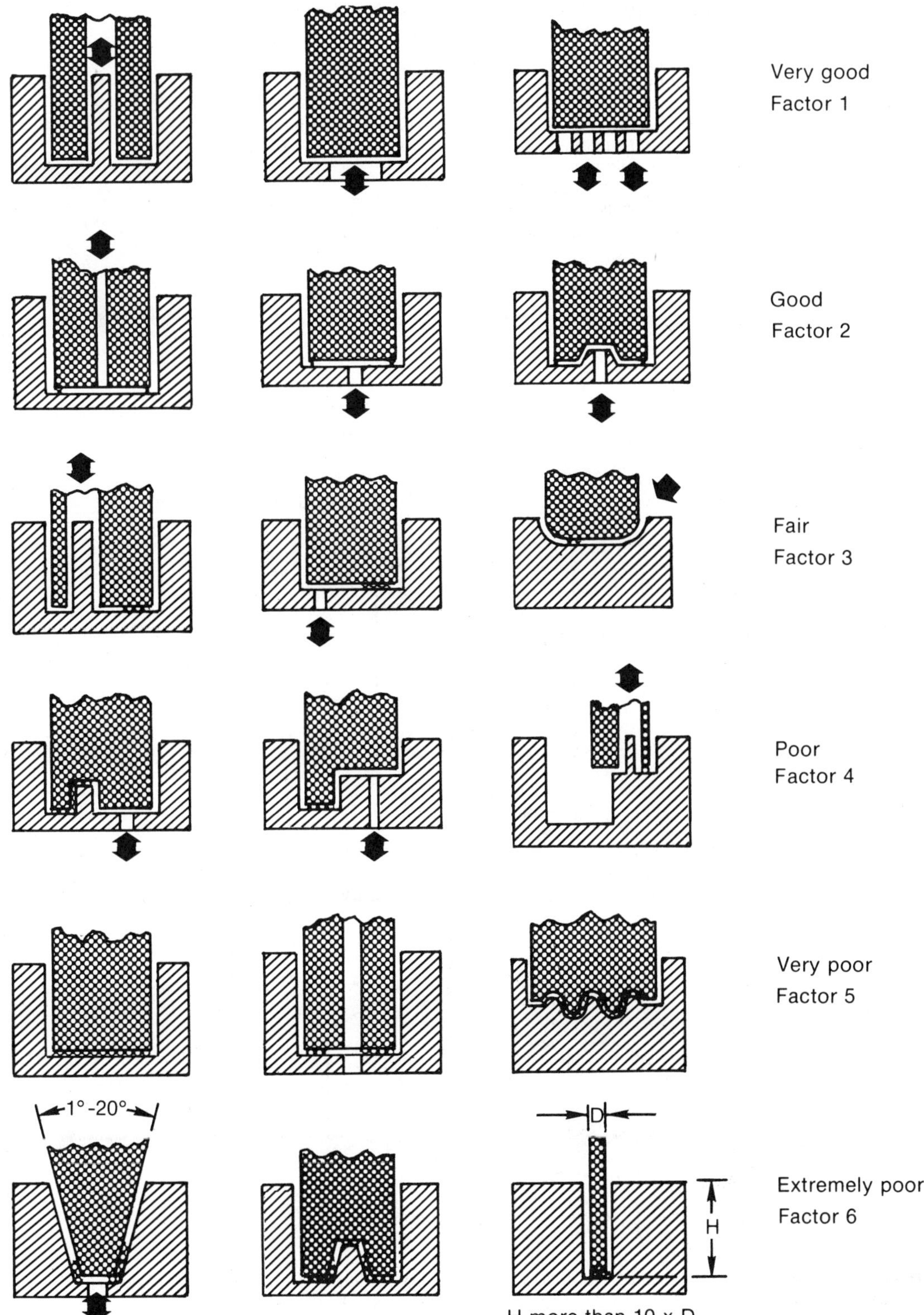


**Figure 12.3–18** Flushing evaluation chart. (L. Houman, How to estimate EDM time requirements, *Metal Stamping*, 11 (June 1977), p. 15)

**TABLE 12.3-11 Flushing Techniques Used in EDM**

- Flow over workpiece cutting area*
- Submerge work in tank
- Pressure through electrode or workpiece
- Suction through electrode or workpiece—or both pressure and suction
- Ram cycling
- Vibration on workpiece
- Invert cavity (workpiece on ram)

*Careful fire safety check recommended.

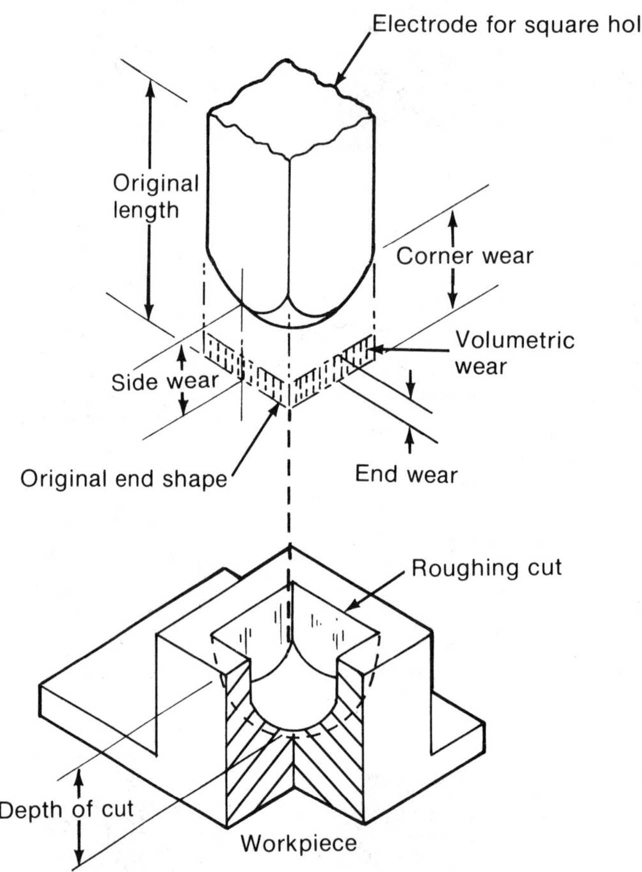


**Figure 12.3-19** Types of electrode wear in EDM. (Modified from H. W. Yankee, *Manufacturing processes*, Englewood Cliffs, NJ: Prentice-Hall, 1979, p. 311)

TABLE 12.3-12 Electrode Wear Ratio Chart

| PART 1 | | | |
|---|---|---|---|
| **ELECTRODE MATERIAL** | **WORKPIECE MATERIAL** | **POLARITY ON ELECTRODE** | **WEAR RATIO*** percentage |
| Aluminum | Steel | Positive | 700 |
| Brass | Brass | Negative | 50 |
| Brass | Carbide | Negative | 700 |
| Brass | Steel | Negative | 200 |
| Brass | Tungsten | Negative | 700 |
| Carbide | Carbide | Negative | 150 |
| Copper | Steel | Negative | 200 |
| Copper-tungsten | Carbide | Negative | 150 |
| Copper-tungsten | Copper-tungsten | Negative | 50 |
| Copper-tungsten | Steel | Positive | 30 |
| Copper-tungsten | Tungsten | Negative | 75 |
| Zinc-tin | Steel | Negative | 500 |
| Graphite | Steel | Negative | 50 |
| Graphite | Carbide | — | 150 (with caution) |
| Silver-tungsten | Steel | Negative | 20 |
| Steel | Steel | Positive | 150 |
| Graphite | Steel | Positive† | <1 |
| Copper | Steel | Positive† | <1 |

| PART 2 | | | |
|---|---|---|---|
| **ELECTRODE MATERIAL** | **WORKPIECE MATERIAL** | **POLARITY ON ELECTRODE** | **ELECTRODE CORNER WEAR§** percentage |
| Graphite | Steel | Positive | <1 |
| Graphite | Steel | Negative | 30 |
| Graphite | Inconel alloy 718 | Positive | <1 |
| Graphite | Inconel alloy 718 | Negative | 35 |
| Graphite | Aluminum | Positive | <1 |
| Graphite | Aluminum | Negative | 15 |
| Graphite | Copper | Negative | 40 |
| Copper-graphite | Carbide | Negative | 62 |
| Copper | Carbide | Negative | 50 |
| Copper | Steel | Positive | <1 |
| Copper | Copper | Negative | 40 |
| Copper | Graphite | Negative | 40 |
| Copper | Aluminum | Positive | <1 |
| Copper-tungsten | Steel | Positive | 6 |
| Copper-tungsten | Graphite | Negative | 40 |
| Copper-tungsten | Copper | Negative | 36 |
| Copper-tungsten | Copper-tungsten | Negative | 80 |
| Copper-tungsten | Graphite | Positive | 70 |
| Copper-tungsten | Carbide | Negative | 40 |
| Brass | Steel | Negative | 100 |
| Brass | Carbide | Negative | 500 |
| Steel | Steel | Positive or Negative | 100 |

SOURCE: Part 1, Hansvedt Engineering, Inc., Urbana, IL; Part 2, Elox Division, Colt Industries, Davidson, NC.

*Linear corner wear ratio $= \frac{\text{electrode length}}{\text{workpiece length}} \times 100$. Includes end wear in this total wear ratio so figure provides length of electrode required for a finished length (depth) of workpiece.

†Low-wear (pulse mode).

§Wear ratios are for 90° corner. Less than 1% electrode wear is achieved, between 450 and 50 microseconds on-time with copper, and 250 to 35 microseconds on-time with graphite. Settings vary with electrode material.

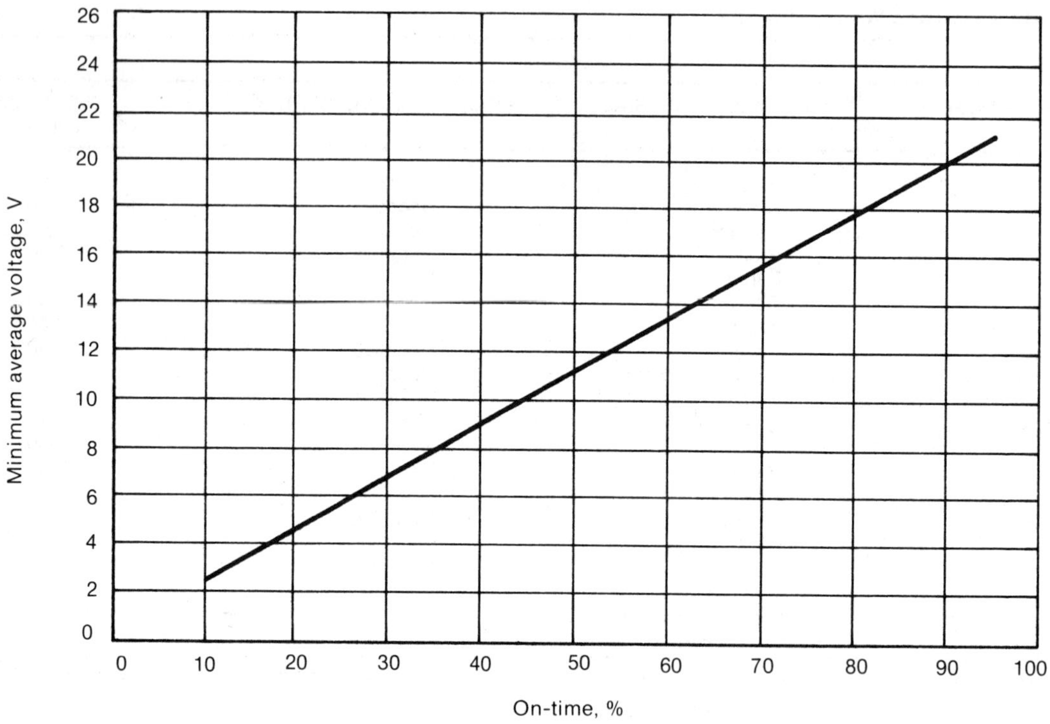


**Figure 12.3–20** Voltage setting for EDM. (Courtesy of Hansvedt Engineering, Inc.)

**TABLE 12.3–13 Electrode Manufacturing Methods**

- Conventional machining from solid block; mill, drill, turn, etc.
- Cut off from bar stock; rounds, squares, shapes, extrusions
- Tracer milling from plaster master
- Punching
- Total form machining (TFM, see section 10.6)
- Electroforming
- Casting (die, investment, plaster)
- Rotating discs
- Molding (cold, powder metal)
- Pressing, forging, coining
- Extruding
- Electrical discharge wire cutting (EDWC, see section 12.5)

**TABLE 12.3–14 Typical Equipment and Conditions for Machining Graphite**

| EQUIPMENT | ECONOMICAL TOLERANCES* (varying with size and volume) | | SURFACE ROUGHNESS $R_q$ | | TOOL TYPE | MACHINE SETTINGS Tool Speed | | FEED RATE |
|---|---|---|---|---|---|---|---|---|
| | in | mm | μin | μm | | ft/min | m/min | |
| Band saw | 0.06 to 0.19 | 1.5 to 4.8 | 350 to 600 | 8 to 15 | 0.5 inch [12.7 mm] blade width—4 teeth/inch [6.3 mm/pitch] | 5,000 | 1,525 | Feed by hand |
| 36-grit silicon wheel | 0.03 to 0.06 | 0.8 to 1.5 | 125† | 3.2 | Thickness 0.06 in [1.5 mm] | 7,000 | 2,130 | Feed by hand |
| Lathe | 0.003 to 0.125 | 0.08 to 3.2 | 50 to 250 | 1.27 to 6.3 | Radius tungsten-carbide tipped tools | 850 | 260 | 0.008 to 0.018 inch [0.2 to 0.5 mm] |
| Drill | 0.003 to 0.016 | 0.08 to 0.4 | 300 to 400 | 7.6 to 10 | Tungsten carbide | 110 | 34 | Feed by hand |
| Mill | 0.003 to 0.06 | 0.08 to 1.5 | 120 to 300 | 3.0 to 7.6 | Fly cutter with 2 tungsten-carbide bits, 6-inch [150 mm] diameter | 1,900 | 580 | 9 to 60 in/min [0.2 to 1.5 m/min] |
| Reamer | 0.002 to 0.005 | 0.05 to 0.13 | 35 to 65 | 0.89 to 1.7 | Tungsten-carbide reamer | 80 | 24 | 0.01 in [0.25 mm] cut |
| Hone | 0.001 to 0.003 | 0.03 to 0.08 | 20 to 32 | 0.5 to 0.8 | 400 to 800 grit | 200 | 60 | Force necessary |
| | | | | | | Pressure | | |
| | | | | | | psi | kPa | |
| Lapping machine | 0.001 to 0.003 | 0.03 to 0.08 | 15 to 30 | 0.4 to 0.8 | No. 320 or higher grit aluminum oxide wheels | 40 | 276 | Grind for 1 minute (variable) |

SOURCE: Airco Speer, St. Marys, PA.

NOTE: Graphite is readily machinable and can be machined using normal metalworking equipment. Power requirement for machining graphite is normally about half that required for steel. Good dust collection equipment is a "must" when machining graphite, since large quantities of abrasive dust that are generated will cause wear to the moving parts of machinery. Graphite is a brittle material and requires good-quality tools with sharp cutting edges to maintain tolerances. Sharp internal corners should be avoided if possible since these set up stress points that can cause cracking.

*Finer tolerances can be maintained by carefully controlling operations when only a few parts are being machined.

†Lower surface roughness values can be obtained by using finer-grit wheels.

**TABLE 12.3–15 Optimum Machining Parameters Used To Compare Mini-Rod® Electrodes with Metallic Electrodes**

| ELECTRODE MATERIAL | ELECTRODE POLARITY | PEAK CURRENT A | PULSE/PAUSE TIME μs | FREQUENCY kHz | DUTY CYCLE percentage | GAP VOLTAGE V |
|---|---|---|---|---|---|---|
| Poco Mini-Rod™ | Negative | 12.5 | 12/100 | 8.93 | 11 | 50 |
| Brass | Negative | 25.0 | 200/800 | 1.00 | 20 | 50 |
| Copper | Positive | 25.0 | 100/800 | 1.11 | 11 | 50 |
| Copper-tungsten | Positive | 25.0 | 25/100 | 8.00 | 20 | 50 |
| Tungsten | Positive | 25.0 | 50/1600 | 0.60 | 3 | 50 |

SOURCE: Poco Graphite, Inc., Decatur, TX.

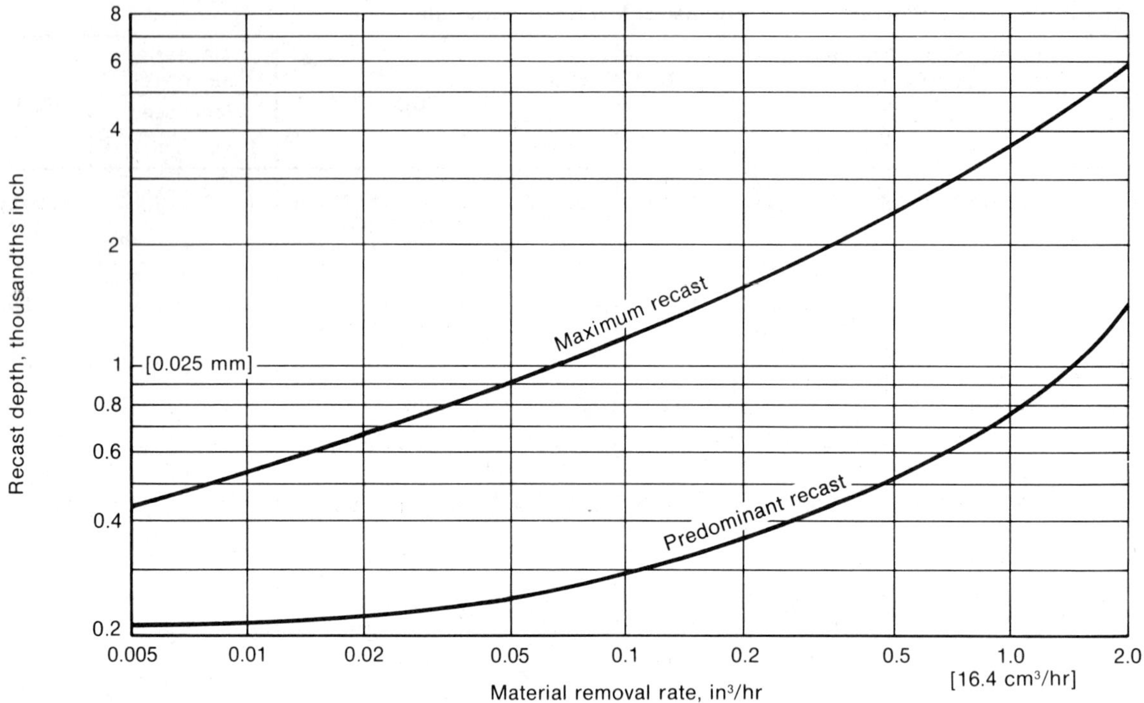


**Figure 12.3–21** Depth of recast versus process intensity (material removal rate) for EDM of a variety of materials. (Modified from G. Bellows, p. 19)

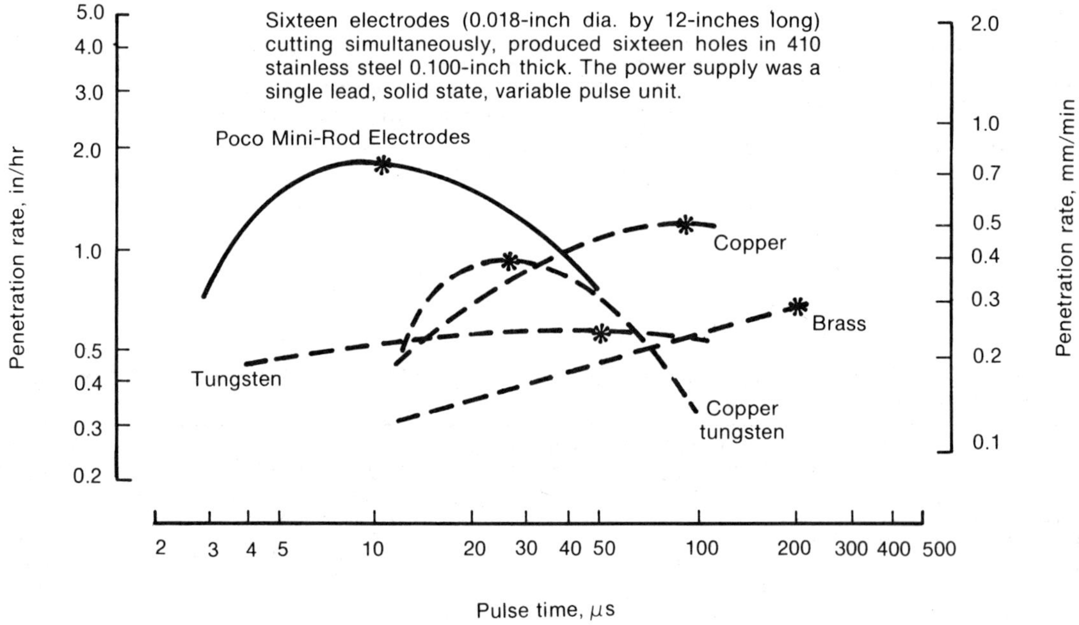


**Figure 12.3–22** EDM drilling performance data for various electrode materials. (Courtesy of Poco Graphite, Inc.)

## SPECIFIC APPLICATIONS DATA

**TABLE 12.3–16 Specific Applications Data for Electrical Discharge Machining**

| Work material: | **Rene 80, STA** | | **Hastelloy Alloy X** | **A6 Steel, 56 $R_c$** | |
|---|---|---|---|---|---|
| Workpiece configuration: | 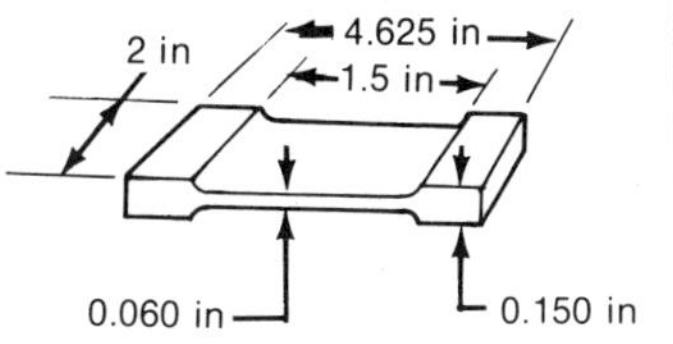 2 in; 4.625 in; 1.5 in; 0.060 in; 0.150 in; Fatigue test specimen | | 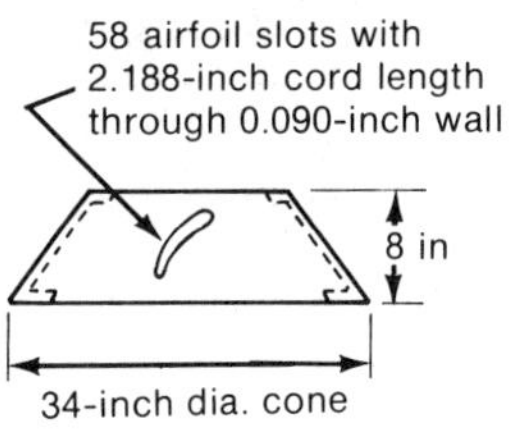 58 airfoil slots with 2.188-inch cord length through 0.090-inch wall; 8 in; 34-inch dia. cone | 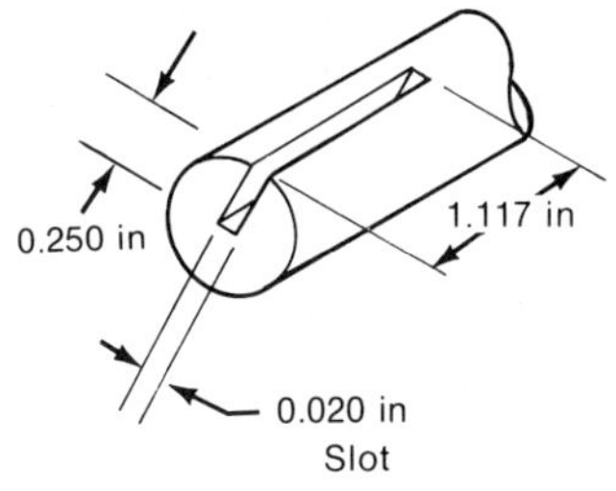 0.250 in; 1.117 in; 0.020 in Slot | |
| Power supply: | Elox HPR104 | Elox HPR104 | Ingersoll Special | Hansvedt SM-150B | Hansvedt SM-150B |
| Electrode material: | copper (rotating wheel) | copper (rotating wheel) | copper punching (used once) | graphite rotating disc | graphite rotating disc |
| Dielectric type: | Texaco 499 | Texaco 499 | Eloxal 13 oil | spray flushing Hansvedt SE-290 oil | spray flushing Hansvedt SE-290 oil |
| pressure, psi: | 0 | 0 | submerged | | |
| Operation type: | roughing | finishing | | roughing | finishing |
| Capacitance, $\mu$F: | 2 | 0 | 30 | 0 | 0 |
| Frequency, kHz: | 16 | 250 | 8 | 10 (70% on) | 250 (5% on) |
| Voltage (open circuit), V: | 80 | 70 | | 40 | 15 |
| Average current, A: | 4 to 5 | 1 | 25 to 30 | 4 | 2 |
| Polarity on electrode: | negative | negative | negative | positive | negative |
| Material removal rate, $in^3/hr$: | | | | | |
| Cutting time, min:sec: | | | | 24:00 | 9:00 |
| Surface roughness, $\mu$in $R_a$: | 205 | 90 | no burrs | | 32 |
| Tolerance, inch: | ±0.001 | ±0.001 | ±0.002 | | |
| Overcut, inch/side: | | | | 0.003 | 0.0005 |
| Taper, inch/side: | | | | | |
| Corner radii, inch/side: | | | | | |
| Linear wear ratio end: | | | | <1% | 0.002 |
| corner: | | | 15:1 by volume | | |
| Notes: | 45 strokes/min reciprocation. | 45 strokes/min reciprocation. | 4 electrodes simultaneously. Automatic index. | | |

**TABLE 12.3–16—Continued**

| Work material: | **420 Stainless Steel** | | | | **01 Tool Steel** | |
|---|---|---|---|---|---|---|
| Workpiece configuration: | 0.25-inch thick; 0.5 in; 0.5 in; Through hole | | | | 0.563-in. dia.; 1.312 in; 0.563-inch dia; Side view; Half of mold for fitting | |
| Power supply: | | | | | Hansvedt SM-150B | Hansvedt SM-150B |
| Electrode material: | graphite | brass | copper | zinc | copper | copper |
| Dielectric type: | oil | oil | oil | oil | Hansvedt SE-290 oil | Hansvedt SE-290 oil |
| pressure, psi: | | | | | | |
| Operation type | | | | | roughing | finishing |
| Capacitance, $\mu$F: | 24 | 24 | 24 | 16 | 0 | 0 |
| Frequency, kHz: | 16 | 16 | 16 | 16 | 2 (70% on) | |
| Voltage (open circuit), V: | | | | | 25 | 80 |
| Average current, A: | 25 | 25 | 25 | 15 | 10 | 1 |
| Polarity on electrode: | negative | negative | negative | negative | positive | positive |
| Material removal rate, $in^3/hr$: | 0.40 | 0.34 | 0.38 | 0.21 | | |
| Cutting time, min:sec: | | | | | 156:00 | 48:00 |
| Surface roughness, $\mu$in $R_a$: | 210 to 270 | 170 to 210 | 210 to 240 | 200 to 230 | | 24 |
| Tolerance, inch: | | | | | | |
| Overcut, inch/side: | 0.002 | 0.003 | 0.0024 | 0.0034 | | |
| Taper, inch/side: | | | | | | |
| Corner radii, inch/side: | | | | | | |
| Linear wear ratio end: | 5.45 | 2.08 | 3.55 | 6.45 | | 1% |
| corner: | | | | | | |
| Notes: | | | | | Ram cycling and pulse flush. One electrode for both roughing and finishing. | Ram cycling and pulse flush. One electrode for both roughing and finishing. |

**TABLE 12.3-16—Continued**

| Work material: | **Hastelloy Alloy X** | **H13 Tool Steel, 76 $R_A$** | | | |
|---|---|---|---|---|---|
| Workpiece configuration: | 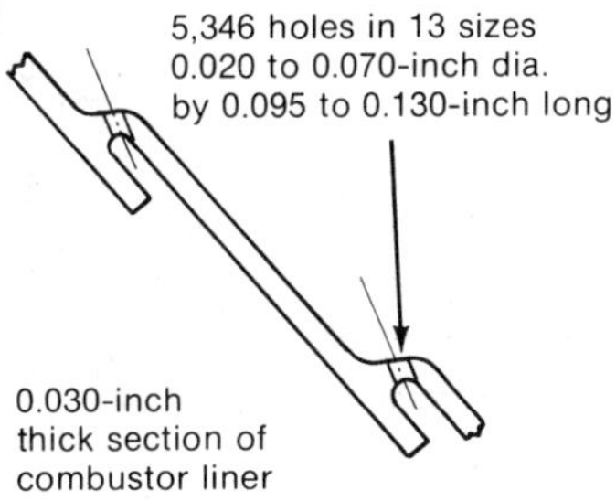 5,346 holes in 13 sizes 0.020 to 0.070-inch dia. by 0.095 to 0.130-inch long; 0.030-inch thick section of combustor liner | 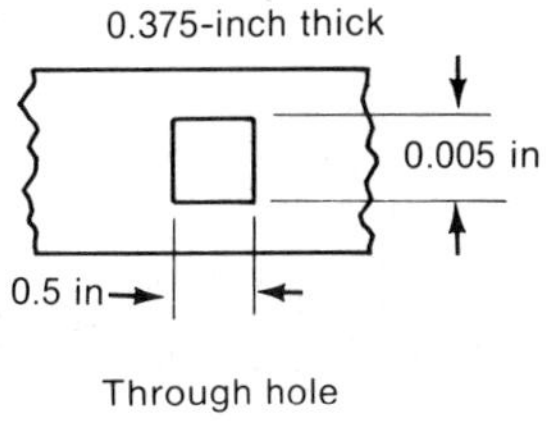 0.375-inch thick; 0.005 in; 0.5 in; Through hole | | | |
| Power supply: | Raycon solid state 40-ampere split lead with refeed control | Elox PS50 | Elox PS50 | Elox PS50 | Elox PS50 |
| Electrode material: | copper-graphite and Poco EDMC-3 | graphite (Gentrode 10) | graphite (Gentrode 10) | graphite (Gentrode 10) | graphite (Gentrode 10) |
| Dielectric type: | Texaco 499 | petroleum oil | petroleum oil | petroleum oil | petroleum oil |
| pressure, psi: | surface flow | 5 | 5 | 5 | 5 |
| Operation type: | | | | | |
| Capacitance, $\mu$F: | none | 56 | 28 | 14 | 5 |
| Frequency, kHz: | 10 (50 $\mu$s on-time; 50 $\mu$s off-time) | 20 | 20 | 20 | 20 |
| Voltage (open circuit), V: | | | | | |
| Current, A: | 3 | 38 | 38 | 15 | 5 |
| Polarity on electrode: | negative | negative | negative | negative | negative |
| Material removal rate, $in^3$/hr: | 0.11 in/min (7 to 26 holes/pass) | 1.98 | 1.84 | 0.70 | 0.16 |
| Cutting time, min:sec: | 0:50 | 1:40 | 1:50 | 4:45 | 19:40 |
| Surface roughness, $\mu$in $R_a$: | 150 (approx. 0.001-inch recast) | | | | |
| Tolerance, inch: | ±0.002 | | | | |
| Overcut, inch/side: | 0.002 | 0.0042 | 0.0040 | 0.0033 | 0.0019 |
| Taper, inch/side: | | 0.0018 | 0.0016 | 0.0015 | 0.0012 |
| Corner radii, inch/side: | | 0.013 | 0.010 | 0.007 | 0.005 |
| Linear wear ratio end: | 15% | 7.0 | 6.5 | 5.9 | 3.2 |
| corner: | | 1.2 | 1.1 | 1.1 | 1.2 |
| Notes: | Fully automatic multiple drilling with 3-axis NC control and multiple slides for electrode refeed cartridges. 11 holes average per pass. | ⅜-inch diameter predrilled hole for flushing. | ⅜-inch diameter predrilled hole for flushing. | ⅜-inch diameter predrilled hole for flushing. | ⅜-inch diameter predrilled hole for flushing. |

**TABLE 12.3-16—Continued**

| Work material: | **Low Carbon Steel** | **C11 Carbide** | **7520 Heat-treated Steel** |
|---|---|---|---|
| Workpiece configuration: | 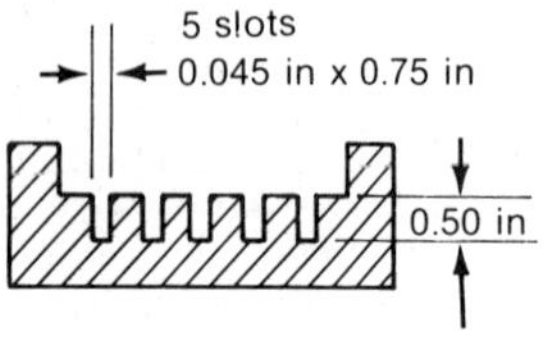 5 slots 0.045 in x 0.75 in<br>0.50 in | 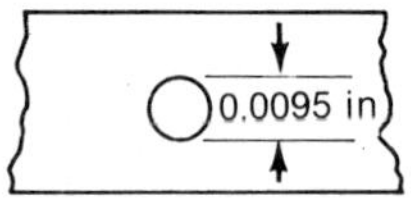 0.0095 in<br>0.125-inch thick | 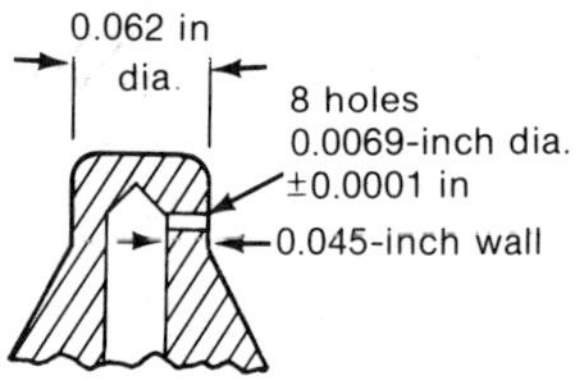 0.062 in dia.<br>8 holes 0.0069-inch dia. ±0.0001 in<br>0.045-inch wall<br>Fuel injector nozzle |
| Power supply: | | Elox PT5 | Raycon SH101 |
| Electrode material: | graphite | tungsten wire | tungsten |
| Dielectric<br>type:<br>pressure, psi: | <br>petroleum oil<br>10 | <br>mineral oil<br>0 | <br>oil |
| Operation type: | | | |
| Capacitance, $\mu$F: | 0.25 | 0.90 | |
| Frequency, kHz: | | 100 | |
| Voltage (open circuit), V: | 40 | | |
| Current, A: | 2 | 0.2 | |
| Polarity on electrode: | negative | negative | negative |
| Material removal rate, $in^3/hr$: | 0.030 | $9 \times 10^{-5}$ | |
| Cutting time, min:sec: | 240:00 | 5:25 | 0:30 per hole |
| Surface roughness, $\mu$in $R_a$: | 50 | | 16 to 18, no burrs |
| Tolerance, inch: | ±0.001 | | ±0.001 |
| Overcut, inch/side: | 0.0007 | 0.0007 | |
| Taper, inch/side: | ±0.001 | 0.0002 | >0.0002 |
| Corner, radii, inch/side: | | | |
| Linear wear ratio<br>end:<br>corner: | | | |
| Notes: | One slot per stroke. Electrode reground 3 times per slot. Quill recycling used. | | Automated fixture. |

## PROCESS SUMMARY

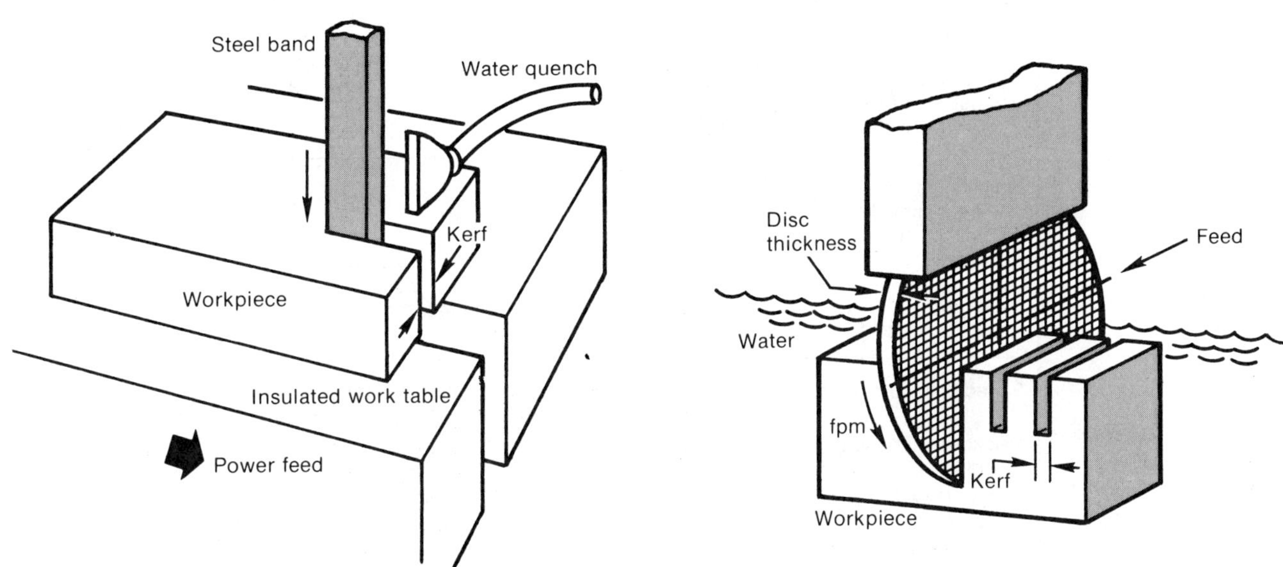


**Figure 12.4–1** EDS schematic. *Left*, band saw; *right*, disc saw.

### Principles

Electrical discharge sawing (EDS) is a variation of electrical discharge machining (EDM) that combines the motion of either a band saw or a circular disc saw with electrical erosion of the workpiece (see figure 12.4–1). The rapid-moving, untoothed, thin, special steel band or disc is guided into the workpiece by carbide-faced inserts. A kerf only 0.002- to 0.005-inch [0.050 to 0.13 mm] wider than the blade or disc is formed as they are fed into the workpiece. The feed can be (1) uniform with continuous arcing and constant feed or (2) servo-controlled to maintain a positive gap and arc control. The low-voltage, direct-current power supplies high currents to the gap, and water is used as a cooling quenchant for the tool, swarf and workpiece. Circular cutting is usually performed underwater, thereby reducing noise and fumes. While the work is power-fed into the band (or the disc into the work), it is not subjected to appreciable forces because the arc does the cutting, so fixturing can be minimal. Precision adjustment of the feed rate is made to be in exact balance with the arc erosion rate.

### Practical Applications

EDS will cut any electrically conductive material or, simultaneously, a mixture of such materials. It will cut about twice as fast as a conventional abrasive saw and has a significantly smaller kerf. The low cutting forces permit slicing of fragile or cellular structures like honeycomb, and because erosion rates are independent of workpiece hardness, it easily cuts aluminum, stainless steel or titanium. No-burr cutting produces few or no roll-over edges on thin materials. Cuts up to 40-inches [1 m] deep have been made on thin-walled heat exchanger tubular assemblies. The low forces involved result in few tracking problems, and the total indicator runout (TIR) is low. Billet and bar cutoff is a natural application.

### Operating Parameters

See table 12.4–1.

### Material Removal Rates and Tolerances

Cutting rates are approximately proportional to the melting point of the workpiece material. The rate is independent of material hardness, but the material must be electrically conductive. When cutting steels, band EDS provides cutting rates of 5 to 20 square inches per minute [32 to 129 $cm^2$/min], while circular EDS runs from 20 to 300 square inches per minute [129 to 1935 $cm^2$/min] (see table 12.4–2). Straightness runs from ±0.003 inch [±0.076 mm] TIR at the lower feed rates to ±0.016 inch [±0.40 mm] TIR at the higher feed rates.

## 12.4 Electrical Discharge Sawing—EDS

**TABLE 12.4–1 Typical Values for EDS Operating Parameters—Steel Workpieces**

| OPERATING PARAMETER | BAND EDS | DISC EDS |
|---|---|---|
| Power supply<br>type:<br>voltage:<br>current (varies with thickness): | <br>Direct current<br>6 to 24 V<br>400 to 1,500 A | <br>Direct current<br>18 to 35 V<br>1,000 to 15,000 A |
| Fluids<br>frequently<br>sometimes:<br>rarely:<br>Fluid flow: | <br>Water<br>Air<br>Water glass (sodium silicate)<br>5 to 13 gal/min [20 to 50 L/min] | <br>Water<br>Air<br>Water glass (sodium silicate)<br>Submerged |
| Speed of band or disc:<br>Thickness of band or disc:<br>Width of band or disc diameter: | 5,000 to 6,000 fpm [1,525 to 1,830 m/min]<br>0.020 to 0.030 inch [0.50 to 0.76 mm]<br>0.5 to 1.25 inch [12.7 to 31.8 mm] | 6,000 to 8,000 fpm [1,830 to 2,440 m/min]<br>0.025 to 0.20 inch [0.64 to 5.1 mm]<br>Up to 50-inch dia. [up to 1.25 m dia.] |
| Wear ratio, workpiece-to-band or workpiece-to-disc:<br>Cutting rates: | Approx. 3:1<br>5 to 20 in²/min [32 to 129 cm²/min] | Approx. 10:1<br>20 to 300 in²/min [129 to 1,935 cm²/min] |
| Surface roughness, $R_a$<br>coarse, high-ampere cutting:<br>fine, low-ampere cutting: | <br>400 to 500 μin [10 to 12.5 μm]<br>250 to 400 μin [6.3 to 10 μm] | <br>400 to 500 μin [10 to 12.5 μm]<br>250 to 400 μin [6.3 to 10 μm] |

**TABLE 12.4–2 Circular Arc Saw Cutting Rates**

| WORK MATERIAL | CUTTING RATE | |
|---|---|---|
| | in²/min | cm²/min |
| Aluminum | 775 | 5,000 |
| Tool steel | 280 | 1,800 |
| Stainless steel | 263 | 1,700 |
| Nickel alloys | 280 | 1,800 |
| Cobalt alloys | 280 | 1,800 |

### Surface Technology

Surface roughness values for EDS are usually greater than those for conventional band sawing or abrasive cutoff. Typically, values range from 400 to 500 microinches $R_a$ [10 to 12.5 μm] for high feed rates and 250 to 400 microinches $R_a$ [6.3 to 10 μm] for low feed rates. The arc discharges leave a recast and heat-affected zone on the surface that is 0.001-to 0.005-inch [0.025 to 0.13 mm] deep.

### Equipment and Tools

Tooling consists of simple holding devices because only low forces are involved. Good electrical contact is needed. Band arrangements are available with throats up to 48-inches [1.2 m] deep. Discs with 18- to 60-inch [0.5 to 1.5 m] blades are typical, with blades up to 72-inches [1.8 m] diameter available on special order. Current capacities up to 15,000 amperes have been used.

### Machining Characteristics

Fast, narrow-kerf, single-plane cutting is available. Both noise and fume pollution can be reduced when EDS is carried out underwater. Clean, easily recovered swarf aids recycling of expensive metals. Arc erosion also occurs on the blade or disc.

## PROCESS SUMMARY

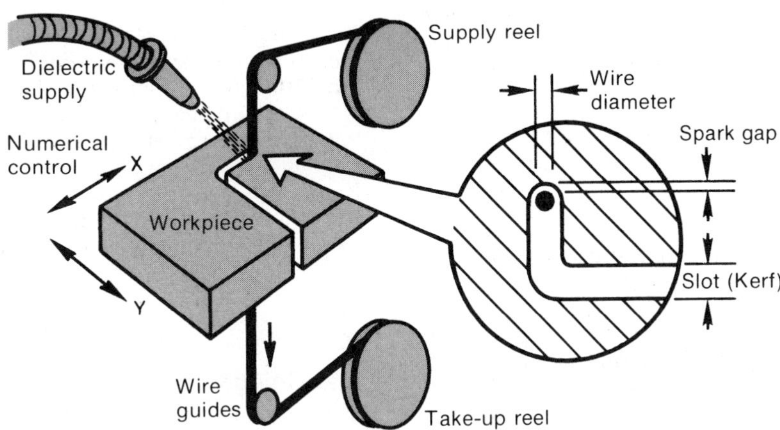


**Figure 12.5–1** EDWC schematic.

### Principles

Electrical discharge wire cutting (EDWC) is a special form of electrical discharge machining wherein the electrode is a continuously moving conductive wire (see figure 12.5–1). EDWC is often called traveling-wire EDM. A small-diameter, tensioned wire is guided to produce a straight, narrow-kerf cut. The slowly moving wire brings a fresh, constant-diameter electrode to the cutting gap, thereby enhancing kerf size control. Usually, a programmed or numerically controlled motion guides the cutting, while the width of the kerf is maintained by the wire size and discharge controls. The dielectric is oil or deionized water carried into the gap by motion of the wire. The wire is inexpensive enough to be used only once. The cutting is basically a two-axis process.

### Practical Applications

EDWC acts like an electrical band saw with the wire replacing the band and the spark discharge, the teeth. A straight cut perpendicular to the major axis of the workpiece has no flaring or bellmouth, and extremely tight corners can be cut with almost no radius. Punches, dies and stripper plates can be cut in any of the hardened, conductive tool materials. The same NC tape can be used repeatedly for short production runs. Mirror-image profile work and internal contours from a starting hole are frequent. Stacking of sheets for multiple cutting is possible. Conventional carbide dies can frequently be cut from a solid plate using EDWC, without the conventional segmenting and fitting. Operation without an attendant is a regular practice, and operating time has extended to as much as 50 hours of cutting with one reel of wire.

### Operating Parameters

**TABLE 12.5–1 Typical Values for EDWC Operating Parameters**

| | |
|---|---|
| Power supply | |
| type: | 55 to 60 V (open circuit volts to 300) |
| frequency: | Pulse time controlled 1 to 100 μs in time or 180 to 300 kHz with 3 kHz most frequent |
| current: | 1 to 32 A |
| Electrode wire | |
| types: | Brass, copper, tungsten, molybdenum |
| diameter: | 0.003 to 0.012 inch [0.076 to 0.30 mm]; most frequently used size is 0.008 inch [0.2 mm] |
| speed: | 0.1 to 6 in/s [2.5 to 150 mm/s] |
| Dielectric: | Deionized water, oil, or rarely, air, gas or plain water |
| Overcut (working gap): | 0.0008 to 0.0020 inch [0.02 to 0.05 mm]; usually 0.001 inch [0.025 mm] |

# 12.5 Electrical Discharge Wire Cutting—EDWC

### Material Removal Rates and Tolerances

Cutting of 0.001- to 6-inch [0.25 to 150 mm] thick materials can be done at a rate of approximately 4 square inches per hour [26 $cm^2$/hr], which on thin parts can yield cutting at 40 inches per minute [1 m/min] (see tables 12.5-2 and 12.5-3). Positioning accuracy up to ±0.0001 inch [±0.0025 mm] is normal in all metals. Workpiece accuracy usually is ±0.0005 inch [±0.013 mm], with special instances of ±0.0002 inch [±0.005 mm] and ±0.0001 inch [±0.0025 mm] in unique applications.

### Surface Technology

Surface roughness is typically in the 30- to 50-microinch $R_a$ [0.8 to 1.3 μm] range, and the surface has a matte or velvet-like texture. Special care and slower cutting can produce a surface roughness of 15 microinches $R_a$ [0.4 μm]. The recast and the heat-affected zone are very small and uniform with the low spark energy levels that are typically used. These layers should be removed or modified on critical or fatigue-sensitive surfaces.

### Equipment

Several manufacturers regularly build EDWC equipment with NC, tracer controls and all programming accessories. Die relief angle generators and offset controls are available on most machines along with kerf-width control via the gap setting and wire-diameter selection. Equipment is also available with cam or other mechanical programming for the wire motion as well as with standard EDM servo control for straight cutting.

Some machines have CNC or optically guided motion direct from an enlarged drawing. Table motions as much as 2 by 3 feet [0.6 by 1 m] have been made, with 1 by 1 foot [0.3 by 0.3 m] being more common. Workpiece thickness 3 to 6 inches [75 to 150 mm] can be accommodated.

Forces are low, so simple holding fixtures are practical. A 5-pound [2.25 kg] reel of 0.008-inch [0.2 mm] diameter copper wire can provide 36 hours of cutting.

### Machining Characteristics

The machine motions are similar to those for a band or a wire saw. Use of cutter offset control permits several sizes to be cut from the same NC tape. The cutting accuracy is high because the spark erosion emanates from a fresh, constant-size section of wire.

## SELECTED DATA

**TABLE 12.5-2 Average Performance Data For EDWC with Brass Wire Electrode**

| WORKPIECE MATERIAL | WIRE DIAMETER in [mm] | WORK THICKNESS, H in [mm] | CUTTING WIDTH, S in [μm] | SURFACE ROUGHNESS, $R_z$ (Ten-Point Height) μin [μm] | REMOVAL RATE, Vs $in^2/min$ [$cm^2/min$] |
|---|---|---|---|---|---|
| Steel, carbon chrome | 0.004 [0.1] | 0.08 to 0.80 [2 to 20] | 0.005 [130] | 8 to 12 [0.2 to 0.3] | 0.0109 [0.070] |
| | 0.006 [0.15] | 0.08 to 2.0 [2 to 50] | 0.0078 [198] | 14 to 20 [0.35 to 0.5] | 0.0186 [0.120] |
| | 0.008 [0.2] | 0.08 to 3.0 [2 to 75] | 0.0102 [259] | 14 to 28 [0.35 to 0.71] | 0.0388 [0.250] |
| | 0.010 [0.25] | 0.4 to 5.0 [10 to 125] | 0.0134 [340] | 14 to 28 [0.35 to 0.71] | 0.0388 [0.250] |
| | 0.012 [0.3] | 3.0 to 6.0 [75 to 150] | 0.0149 [378] | 14 to 20 [0.35 to 0.5] | 0.0388 [0.250] |
| Copper | 0.010 [0.25] | 0.08 to 1.6 [2 to 40] | 0.0126 [320] | 14 to 28 [0.35 to 0.7] | 0.03 [0.194] |

SOURCE: Agietron Corporation.
NOTE: To calculate cutting speed, divide removal rate (Vs) by work thickness (H).

# 12.5 Electrical Discharge Wire Cutting—EDWC

**TABLE 12.5-2 — Continued**

| WORKPIECE MATERIAL | WIRE DIAMETER in [mm] | WORK THICKNESS, H in [mm] | CUTTING WIDTH, S in [μm] | SURFACE ROUGHNESS, $R_z$ (Ten-Point Height) μin [μm] | REMOVAL RATE, Vs in²/min [cm²/min] |
|---|---|---|---|---|---|
| Carbide, 15% cobalt | 0.004 [0.1] | 0.08 to 0.8 [2 to 20] | 0.0075 [190] | 6 to 9 [0.15 to 0.24] | 0.0054 [0.035] |
| | 0.006 [0.15] | 0.08 to 1.2 [2 to 30] | 0.009 [229] | 9 to 11 [0.24 to 0.28] | 0.011 [0.071] |
| | 0.010 [0.25] | 0.08 to 2.0 [2 to 50] | 0.0142 [361] | 8 to 20 [0.2 to 0.5] | 0.019 [0.122] |
| Graphite | 0.010 [0.25] | 0.08 to 1.6 [2 to 40] | 0.0139 [351] | 14 to 24 [0.35 to 0.6] | 0.0186 [0.120] |
| Aluminum | 0.010 [0.25] | 0.08 to 1.6 [2 to 40] | 0.0134 [340] | 20 to 35 [0.5 to 35] | 0.093 [0.60] |

SOURCE: Agietron Corporation.
NOTE: To calculate cutting speed, divide removal rate (Vs) by work thickness (H).

TABLE 12.5-3 Average Performance Data for EDWC with Molybdenum Wire Electrode

| WORKPIECE MATERIAL | WIRE DIAMETER in [mm] | WORK THICKNESS, H in [mm] | CUTTING WIDTH, S in [μm] | SURFACE ROUGHNESS, $R_Z$ (Ten-Point Height) μin [μm] | REMOVAL RATE, Vs $in^2/min$ [$cm^2/min$] |
|---|---|---|---|---|---|
| Steel, carbon chrome | 0.003 [0.08] | 0.08 to 0.4 [2 to 10] 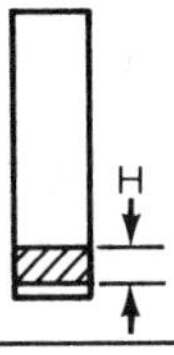 | 0.004 [105] 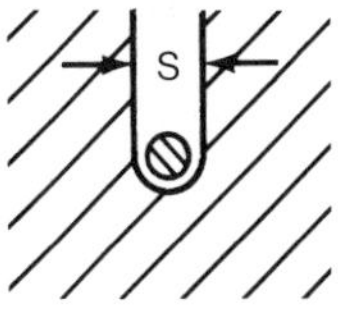 | 14 to 22 [0.35 to 0.55] 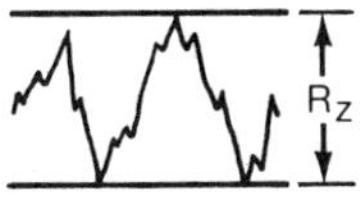 | 0.0078 [0.05]  |
| | 0.004 [0.1] | 0.08 to 0.4 [2 to 10] 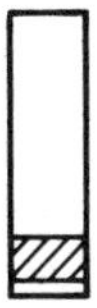 | 0.0049 [125] 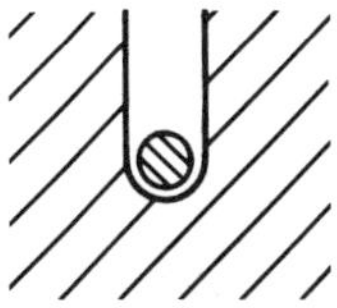 | 18 to 23 [0.47 to 0.59] 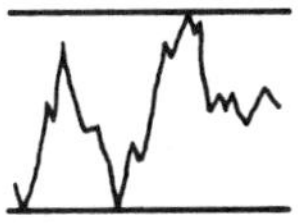 | 0.0109 [0.07]  |
| Carbide, 15% cobalt | 0.003 [0.08] | 0.08 to 0.5 [2 to 12.7]  | 0.004 [105] 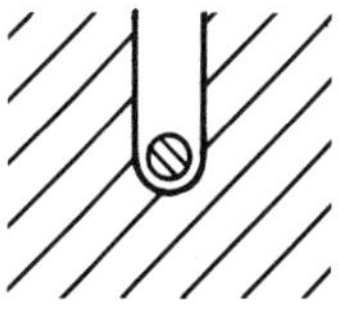 | 3 to 9 [0.078 to 0.23]  | 0.0062 [0.04] |
| | 0.004 [0.1] | 0.08 to 0.5 [2 to 12.7]  | 0.0053 [135] 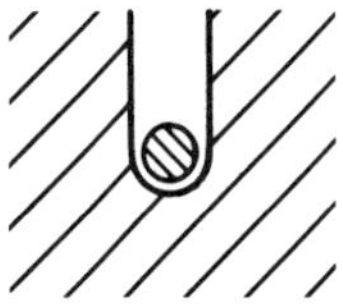 | 5 to 9 [0.118 to 0.23]  | 0.0093 [0.06]  |

SOURCE: Agietron Corporation.
NOTE: To calculate cutting speed, divide removal rate (Vs) by work thickness (H).

## PROCESS SUMMARY

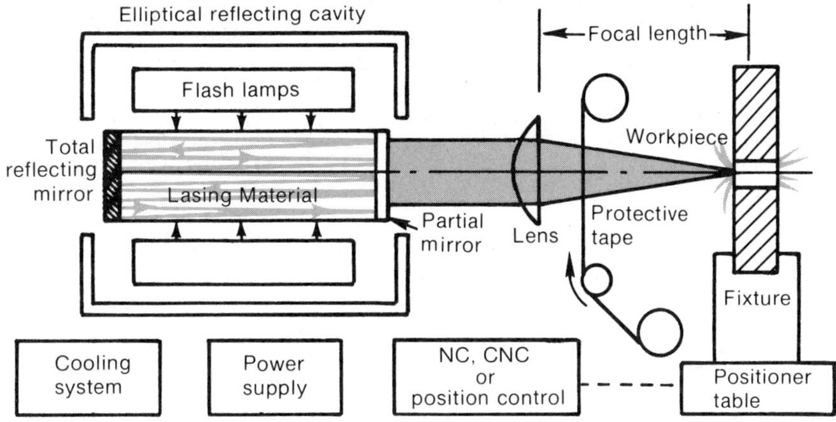


**Figure 12.6–1** LBM schematic.

### Principles

Laser beam machining (LBM) removes material by melting, ablating, and vaporizing the workpiece at the point of impingement of a highly focused beam of coherent monochromatic light (see figure 12.6–1). Laser is an acronym for "Light Amplification by Stimulated Emission of Radiation." The electromagnetic radiation operates at wavelengths from the visible to the infrared (see table 12.6–3). The principal lasers used for material removal are the neodymium-glass, the Nd-YAG (neodymium-yttrium aluminum garnet), the ruby and the carbon dioxide ($CO_2$). The last is a gas laser (most frequently used as a torch with an assisting gas—see LBT, laser beam torch), while the others are solid state lasing materials.

For pulsed operation, the power supply produces short, intense bursts of electricity into the flash lamps, which concentrate their light flux on the lasing material. The resulting energy from the excited atoms is released at a characteristic, constant frequency. The monochromatic light is amplified during successive reflections from the mirrors. The thoroughly collimated light exits through the partially reflecting mirror to the lens, which focuses it on or just below the surface of the workpiece. The small beam divergence, high peak power and single frequency provide excellent, small-diameter spots of light with energy densities up to $3 \times 10^{10}$ watts per square inch [$4.6 \times 10^9$ W/cm$^2$] which can sublime almost any material. Cutting requires energy densities of $10^7$ to $10^9$ watts per square inch [$1.55 \times 10^6$ to $1.55 \times 10^8$ W/cm$^2$], at which rate the thermal capacity of most materials cannot conduct energy into the body of the workpiece fast enough to prevent melting and vaporization. The initial or leading edge of the power spike is most important in $CO_2$ lasers to initiate vaporization in nonmetal drilling, which then promotes better absorption of the energy. Workpiece surface reflectivity, absorption coefficient, thermal conductivity, specific heat and heat of vaporization are all important when considering a laser application.

Cutting, welding, scribing and heat treating can all be done with the same laser by varying the energy density and by appropriately adjusting the focus (spot size) and the pulse duration. Sharp, short, Gaussian-mode repetitive pulses are best for cutting because they have high energy densities. Welding requires longer duration pulses and a broad spot. Welding energy densities are $10^5$ to $10^7$ watts per square inch [$1.55 \times 10^4$ to $1.55 \times 10^6$ W/cm$^2$]. Heat treating operates at under $10^6$ watts per square inch [$1.55 \times 10^5$ W/cm$^2$], and scribing requires very short cutting pulses (see figures 12.6–3 to 12.6–7). Heat treating and welding are not covered in this process summary.

During laser beam machining, the expulsed material forms metal gases and vapors, which subsequently solidify to dust or minute beads. In addition to protection for the lens from molten particles or splatter, a "chip" removal system—frequently a vacuum cleaner type—is desirable to keep the operating area clean and free from dust.

### Practical Applications[1]

LBM is not a mass material removal process—it is a fast, easily controlled process with a non-contact, non-wearing tool that imparts minimal heat and thermal distortion with minimal demands upon fixturing. The high energy density in a well-focused beam will melt and vaporize any material. Contour cuts and slitting of thin steels, aluminum, titanium alloys, and other metals are done at rapid rates with narrow kerfs; continuous-wave (CW) modes or high pulse rates are used. Texturing and etching are performed at lower energy levels. Small hole drilling can be done at shallow angles (15 degrees) to the surface. Multiple pulses permit

[1]Laser technology is advancing so rapidly that current limitations frequently are removed by more recent developments. A check with active laser-equipment builders is recommended for new applications.

hole drilling to 40:1 depth-to-diameter ratios on 0.005-inch [0.13 mm] diameter holes, while larger holes, 0.050-inch [1.27 mm] diameter, can be drilled in 0.50-inch [12.7 mm] thick material. Trepanning is done for still larger holes using NC or a circle-generator attachment. Plastics, rubber, cloth and similar organic-base materials cut more easily than the metals. Ceramics, glasses and composites of many types are drilled, scribed and diced. Microcircuit components, resistor trimming or circuit "deburring" make use of the laser's speed and precision. Viable applications for LBM range from diamond wire draw dies to catheter drilling to food sieves. The ability to direct and focus the laser beam into inaccessible locations permits unusual applications, such as balancing rotors dynamically while they are rotating.

For additional applications, see Laser Beam Torch, section 12.7—particularly for slitting and for contour cutting.

## Operating Parameters

**TABLE 12.6–1 Typical Values for LBM Operating Parameters and Process Capabilities**

| OPERATING PARAMETERS | | | | |
|---|---|---|---|---|
| Lasing material: | Ruby | Nd-YAG | Nd-glass | $CO_2$ |
| Type: | Solid state, smallest spot, high peak watts, most easily absorbed by metals | Solid state | Solid state | Gas, high efficiency |
| Composition: | 0.03 to 0.07% Cr | 1% Nd | Glass + 2 to 6% Nd | $CO_2 + He + N_2$ |
| Wavelength, micrometers (see figure 12.6–8): | 0.69 | 1.064 | 1.064 | 10.6 |
| Efficiency, %: | Up to 1 | 2 | 2 | 10 to 15 |
| Beam mode: | Pulse | Pulse or continuous | Pulse | Pulse or continuous |
| Spot size at focal point, in: | 0.0005 | 0.0005 | 0.001 | 0.003 |
| [mm]: | [0.013] | [0.013] | [0.025] | [0.076] |
| Pulse repetition rate, pps* (normal mode operation): | 1 to 10 | 1 to 300 or continuous wave | 1 to 3 | Continuous wave |
| Beam output, watts: | 10 to 100 | 10 to 1,000 | 10 to 100 | 250 to 10,000 |
| Peak power, kW (see figure 12.6–9): | 200 | 400 | 200 | 100 |
| Beam divergence, milliradians: | 5 to 7 | 1 to 5 | 5 to 7 | 0.1 to 10 |
| Excitation sources: | Krypton, xenon or tungsten-halogen lamps, electrical discharge | | | |

| PROCESS CAPABILITIES† | |
|---|---|
| Drilling diameter: | 0.00002 to 0.050 inch [0.005 to 1.27 mm] (larger with trepanning) |
| Drilling depth: | Up to 0.70 inch [17 mm] |
| Drilling angle to surface: | 15 to 90 degrees |
| Drilling taper: | 5 to 20% of hole diameter |
| Drilling length-to-diameter ratio: | Up to 50:1 |
| Drilling length, trepanned: | Up to 0.25 inch [6.4 mm] |
| Drilling tolerance: | ±5 to 20% of hole diameter |
| Minimum corner radius: | 0.010 inch [0.25 mm] |

*pps = pulses per second. Q-switched lasing can greatly increase all these repetition rates.

†See also table 12.6–3 and figures 12.6–10 to 12.6–14.

## Material Removal Rates and Tolerances

Material removal rates are slow, approximately 4 x $10^{-4}$ cubic inches per minute [6.5 x $10^{-3}$ $cm^3$/min]. Nevertheless, a laser can drill or slot thin materials faster than a workpiece can be repositioned by numerical control. Even if multiple laser pulses are necessary, each pulse typically takes only one-thousandth second. The removal rates are inversely proportional to the material thickness. Even deeper holes can be drilled rapidly, for example, 1 to 3 seconds for a 0.020-inch [0.50 mm] diameter hole in 0.10-inch [2.5 mm] thick nickel alloy. Because flash lamps have a long life, costs can be reduced to a few cents per slot, and for multiple arrays of holes, to tenths-of-a-cent per hole. Holes and slots will have irregular surfaces or roundness and will contain taper, yet location accuracies up to ±0.001 inch [±0.025 mm] are commonplace and limited primarily by the accuracies available in the positioning devices for the workpiece. Conventional, axicon or toric lenses are used to attain better productivity in unusual applications.

## Surface Technology

Laser machined surfaces are usually rough and have a recast texture. The rapidity of the application of energy reduces the heat-affected zone and recast to less than that from other thermal processes. The usual recast is about 0.001- to 0.002-inch [0.025 to 0.050 mm] thick, with some as much as 0.010 inch [0.25 mm], with a heat-affected zone below the recast, as shown in figure 12.6-13. In many materials, a hardness alteration will also occur below the surface. These surface effects can be detrimental to product integrity if the surface is highly stressed or if fatigue life is a concern. Removal or modification of such lasered surfaces is recommended (see section 18.3).

## Equipment and Tooling

The principal equipment emphasis is the control of the position of the workpiece under the focal point of the beam. Multiple-axis control is needed to allow the beam to follow undulations in the workpiece. Computer or numerical control makes use of the easy integration of the laser parameters—intensity, focus, pulse rate—with motion control or automation. Beam deflecting, beam splitting for multiple machining sites, circle generators for trepanning, special lenses (toric, axicon), rotating mirrors, and time sharing are all employed in special laser systems. The absence of mechanical contact, the low-inertia beam and the nearly forceless characteristics of the laser simplify the fixturing.

## Machining Characteristics

The ease with which the laser beam can be manipulated enhances its versatility. Adequate eye protection is needed for both direct and reflected laser light. The United States Bureau of Radiological Health has issued requirements that laser manufacturers should follow to assure personnel safety. The $CO_2$ laser wavelength is opaque to plastic, so shielding or enclosure is simplified. Dust control is necessary, and vacuum systems are the prevalent method for removing the "chips." The gases released by some materials during LBM should be checked with OSHA for safety effects.

# 12.6 Laser Beam Machining—LBM

## SELECTED DATA

There are several valid techniques for selecting operating parameters for LBM. The steps in table 12.6-2 are recommended as one logical sequence. Supporting data come from the other figures and tables in accordance with the terms as explained in figure 12.6-2.

**TABLE 12.6-2 Steps To Select LBM Operating Parameters**

| ITEM | DATA |
|---|---|
| 1. Workpiece material and its thermal characteristics | Engineering drawings, specifications, handbooks |
| 2. Select type laser, wavelength, and mode of operation | Tables 12.6-3 to 12.6-5 and figures 12.6-3 to 12.6-9. |
| 3. Type optics | Small spot size, easier with shorter wavelengths; better quality optics required with longer wavelengths |
| 4. Lens selection | Depth of focus (d), focal length (f), and spot size (s) determined on application need and configuration of workpiece |
| 5. Workpiece motion control | CNC, DNC, or NC depending on complexity of workpiece and volume of production |
| 6. Safety shields | Determined by material handling needs* |

*See also D. Sliney, Health hazards from laser material processing, Technical paper MR75-581, Society of Manufacturing Engineers, Dearborn, MI, 1975.

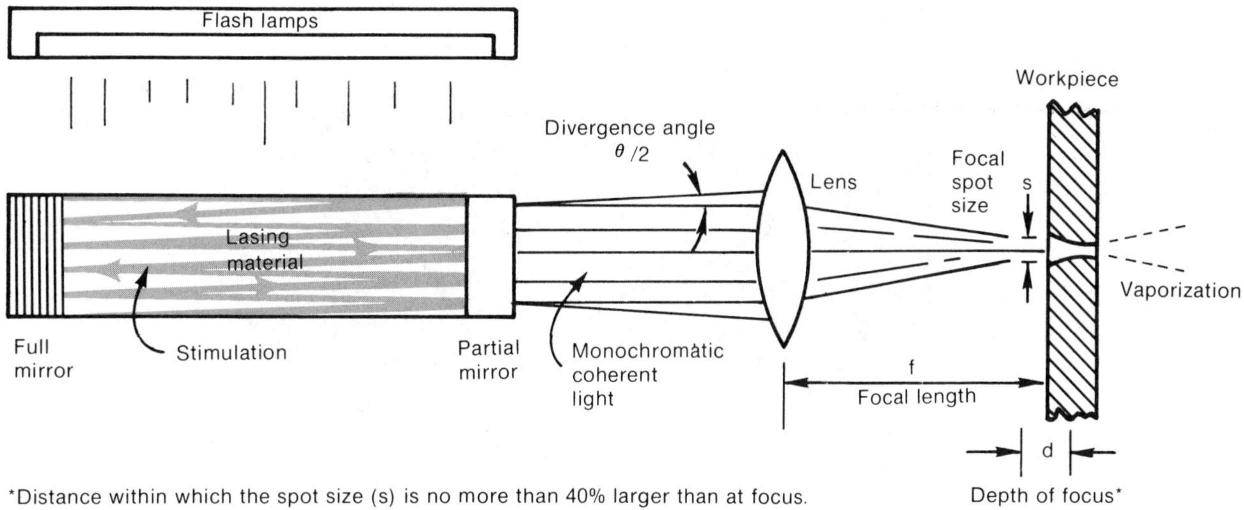


**Figure 12.6–2** LBM parameter terminology.

**TABLE 12.6–3 Commercial Lasers Suitable for Materials Processing**

| LASER TYPE | WAVELENGTH μm | MODE OF OPERATION | POWER W | PULSE REP. RATE pps | PULSE LENGTH | APPLICATION | COMMENTS |
|---|---|---|---|---|---|---|---|
| Argon | 0.4880<br>0.5145 | Repetitively pulsed | 20 peak; 0.005 average | 60 | 50 μs | Scribing thin films | Power low |
| Ruby | 0.6943 | Normal pulse | $2 \times 10^5$ peak | Low (5 to 10) | 0.2 to 7 ms | Large material removal in one pulse, drilling diamond dies, spot welding | Often uneconomical |
| Nd-Glass | 1.06 | Normal pulse | $2 \times 10^6$ peak | Low (0.2) | 0.5 to 10 ms | Large material removal in one pulse | Often uneconomical |
| Nd-YAG* | 1.06 | Continuous | 1,000 | — | — | Welding | Compact; economical at low powers |
| Nd-YAG | 1.06 | Repetitively Q-switched | $3 \times 10^5$ peak; 30 average | 1 to 24,000<br>300 | 50 to 250 ns<br>50 ns | Resistor trimming, electronic circuit fabrication | Compact and economical |
| Nd-YAG | 1.06 | Normal pulsed | 400 | 300 | 0.5 to 7 ms | Spot weld, drill | |
| $CO_2$† | 10.6 | Continuous | 15,000 | — | — | Cutting organic materials, oxygen-assisted metal cutting | Very bulky at high powers |
| $CO_2$ | 10.6 | Repetitively Q-switched | 75,000 peak 1.5 average | 400 | 50 to 200 ns | Resistor trimming | Bulky but economical |
| $CO_2$ | 10.6 | Superpulsed | 100 average | 100 | 100 μs and up | Welding, hole production, cutting | Bulky but economical |

SOURCE: Modified from J. F. Ready, Selecting a laser for material working, *Laser Focus* (March 1970), p. 40.

*Neodymium-yttrium aluminum garnet.

† $CO_2$ plus He plus $N_2$ mixture.

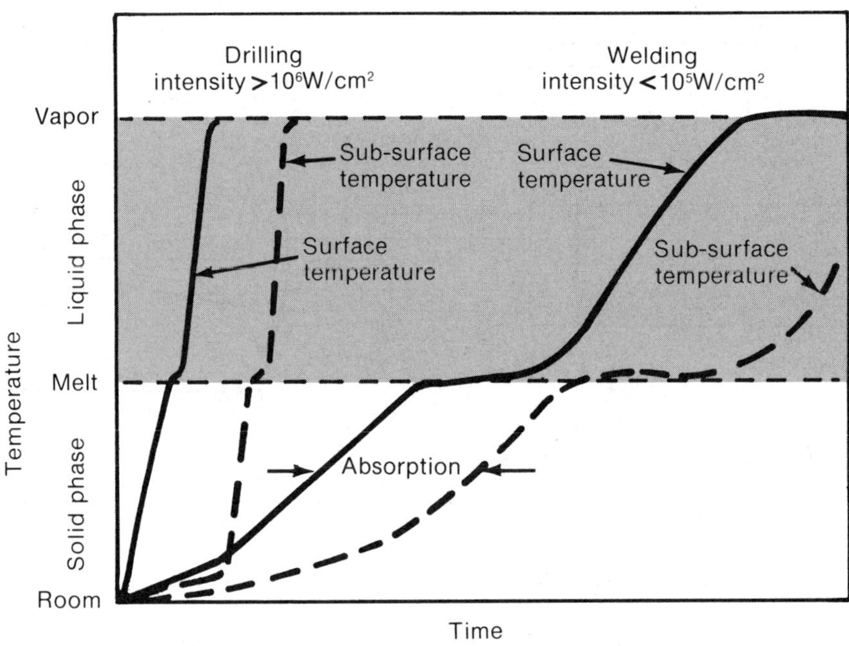


**Figure 12.6–3** Difference in energy and absorption between laser drilling and laser welding. (S. R. Bolin, Pulsed-laser metalworking, *American Machinist* 120 (October 1976) p. 124)

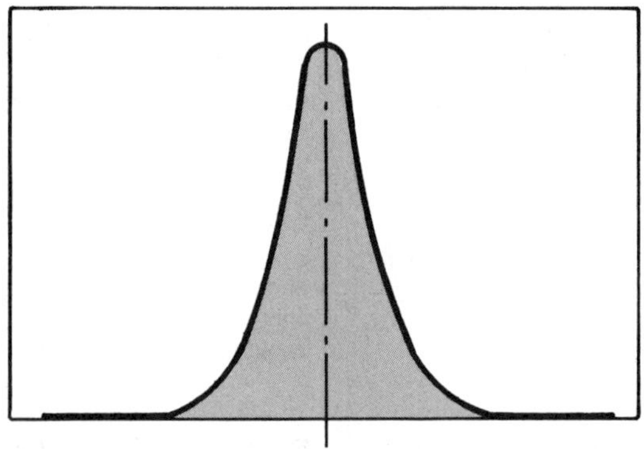

**Gaussian Mode** (or $TEM_{00}$)*. Ideally suited for high quality cutting. Power densities up to $10^9$ watts/inch² [1.55 x $10^8$ W/cm²]. Mode quality is maintained throughout the power range.

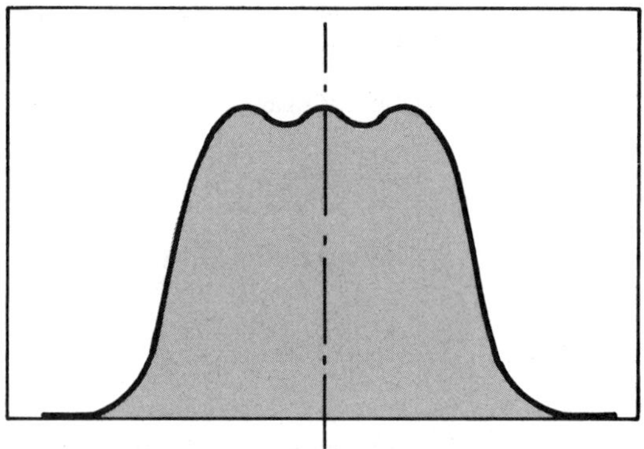

**Mixed Mode** (approximately 50% each of $TEM_{00}$ and $TEM_{01}$). Flat topped distribution makes this model ideal for welding and heat treating.

*TEM — transverse excitation mode

**Figure 12.6–4** Laser excitation modes for cutting versus welding. (Courtesy of GTE-Sylvania)

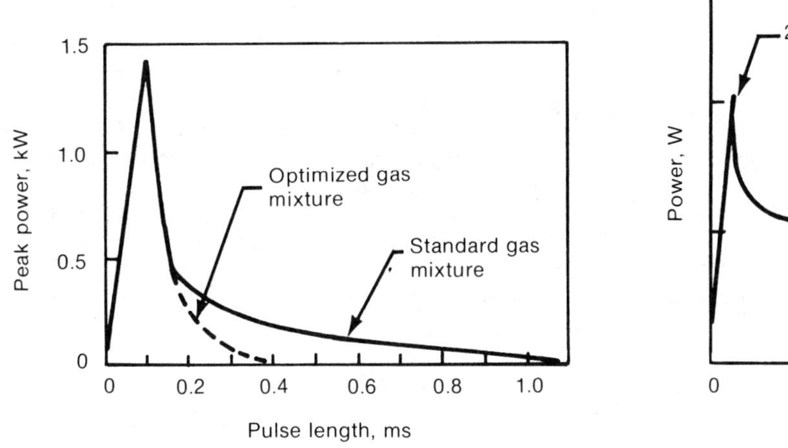


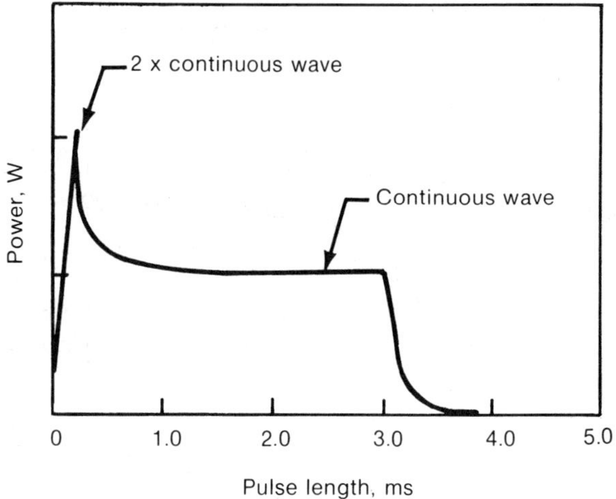


**Figure 12.6–5** Pulse configurations for $CO_2$ lasers. *Left*, short pulse length (less than 100 $\mu$s to peak); *right*, long pulse length. (Courtesy of Coherent, Inc.)

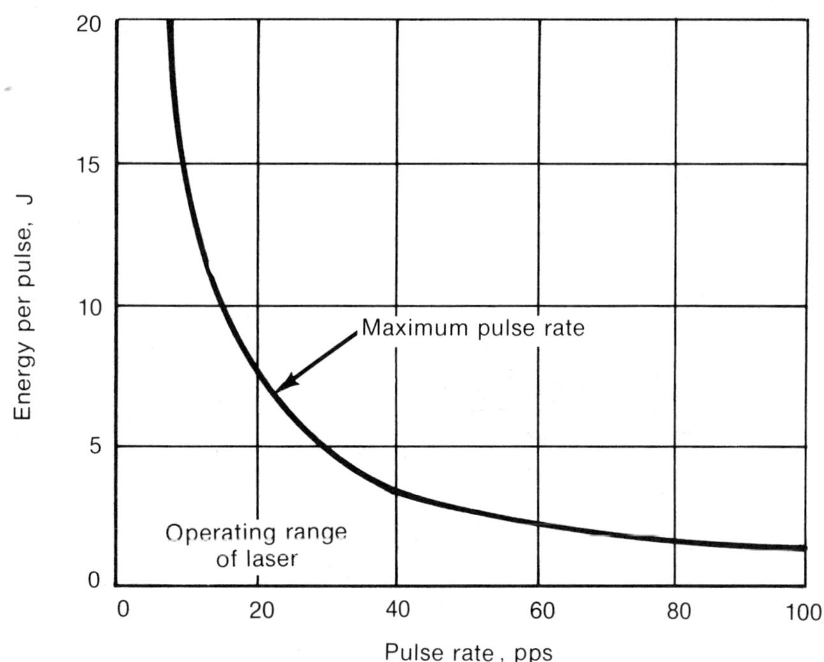


**Figure 12.6–6** Operating range of a flash-lamp pulsed Nd-YAG laser. (Courtesy of Coherent, Inc.)

# 12.6 Laser Beam Machining—LBM

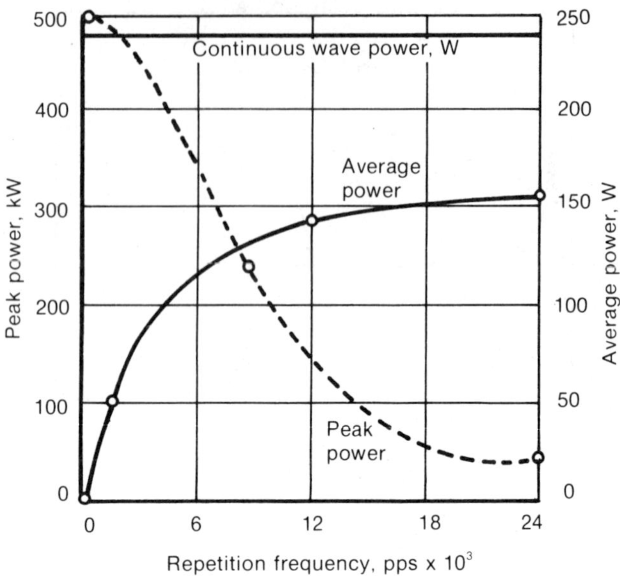


**Figure 12.6-7** Relationship of peak power and average power with variation in repetition frequency (YAG laser). (Courtesy of Holobeam Laser Company)

**TABLE 12.6-5 Selecting a Laser Type for Machining**

| APPLICATION | LASER TYPE |
|---|---|
| Large holes (to 0.060-inch [1.52 mm] diameter) | Ruby, Nd glass, Nd-YAG |
| Larger holes (trepanned) | Nd-YAG, $CO_2$ |
| Small holes (>0.010-inch [0.25 mm] diameter) | Ruby, Nd glass, Nd-YAG |
| Drilling (punching or percussion) | Nd-YAG, Ruby |
| Thick cutting | $CO_2$ + gas assist* |
| Thin slitting metals | Nd-YAG |
| Thin slitting plastics | $CO_2$ |
| Plastics | $CO_2$ |
| Metals | Nd-YAG, Ruby, Nd glass |
| Organics, nonmetals | Pulsed $CO_2$ |
| Ceramics | Pulsed $CO_2$, Nd-YAG |

*See also gas-assisted cutting under LBT, section 12.7.

**TABLE 12.6-4 Relative Power To Vaporize Equal Volumes in Equal Time**

| | |
|---|---|
| Aluminum | 1.0 |
| Titanium | 1.5 |
| Iron | 1.8 |
| Molybdenum | 2.2 |
| Tungsten | 2.9 |

SOURCE: *Metals Handbook: Vol. 3—Machining,* Metals Park, OH: American Society for Metals, 1967, p. 254.

This nomograph provides fast, convenient conversions among the various ways of describing light waves in energy, frequency and wavelength.

As an example, the left side of the nomograph shows that the 532-nanometer wavelength of a frequency-doubled neodymium-yag laser corresponds to an optical carrier frequency of $5.6 \times 10^{14}$ hertz, and a photon energy of 2.3 electronvolts or 18,800 wavenumbers ($cm^{-1}$).

The right side of the nomograph shows that the laser's spectral width of 0.5 $cm^{-1}$ is equivalent to a width of 15 gigahertz or 0.14 angstrom.

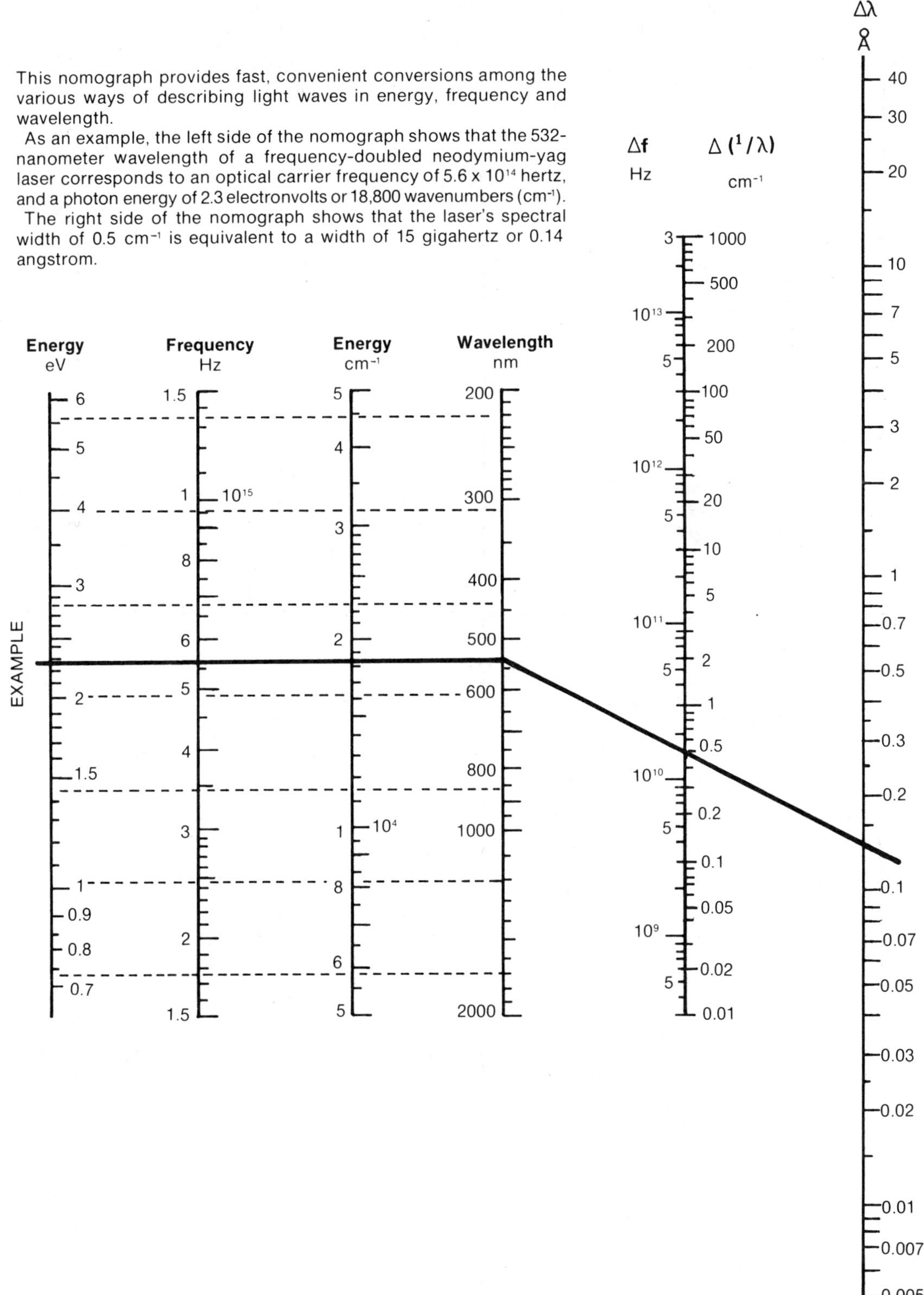


**Figure 12.6-8** Energy nomograph. (Courtesy of Photonetics Associates; reprinted from 1979 *Laser Focus Buyers' Guide*)

## 12.6 Laser Beam Machining—LBM

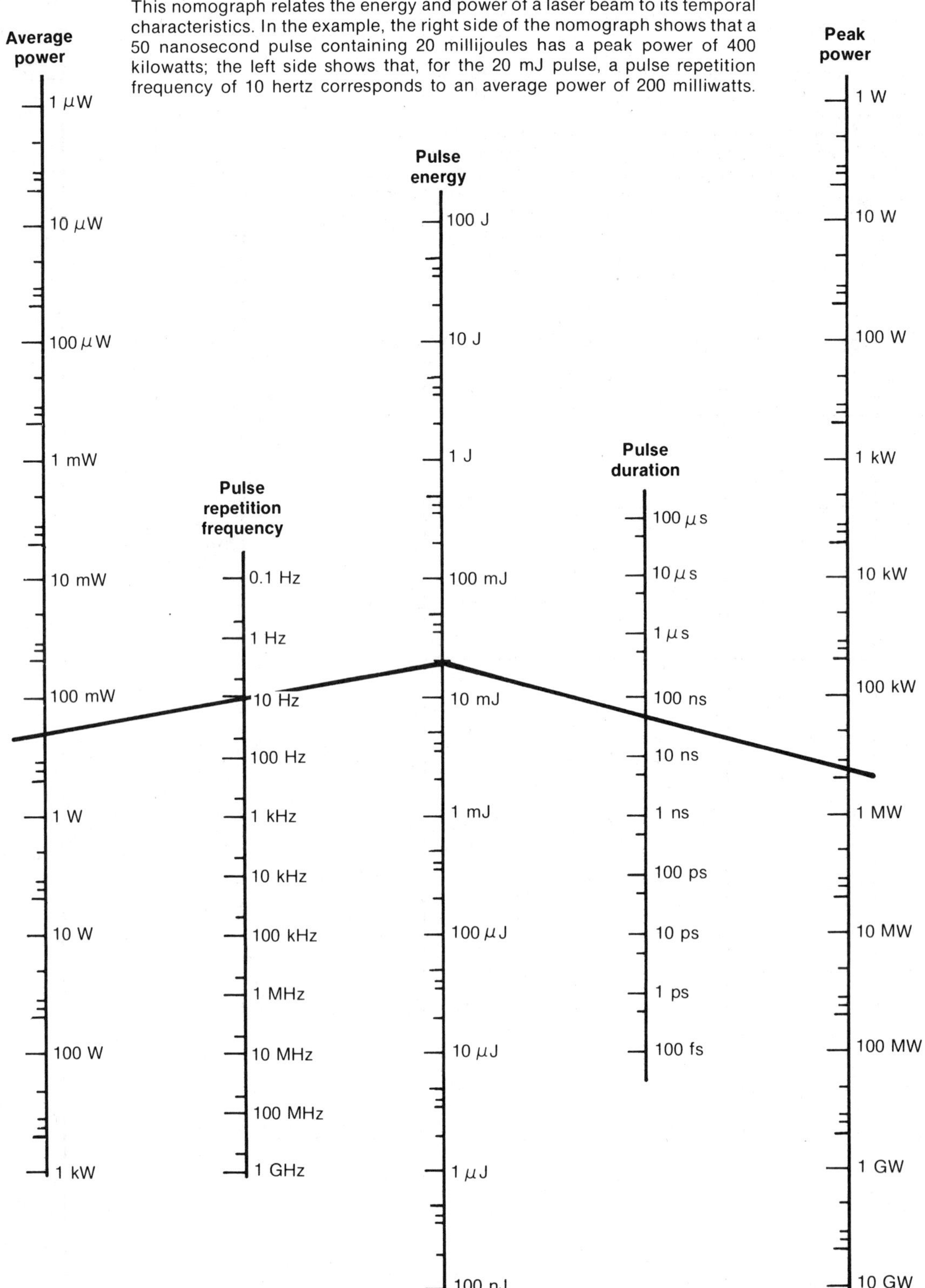


**Figure 12.6–9** Pulse energy and power nomograph. (Courtesy of Photonetics Associates; reprinted from 1979 *Laser Focus Buyers' Guide*)

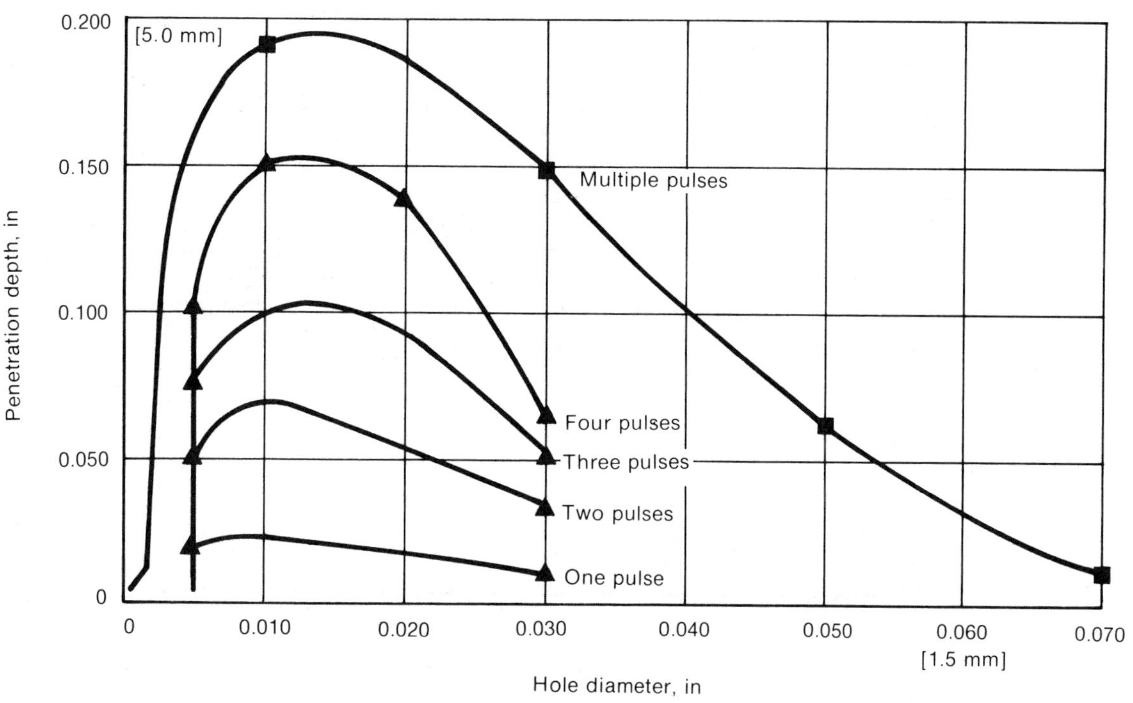


**Figure 12.6–10** Capabilities for laser drilling of superalloys.

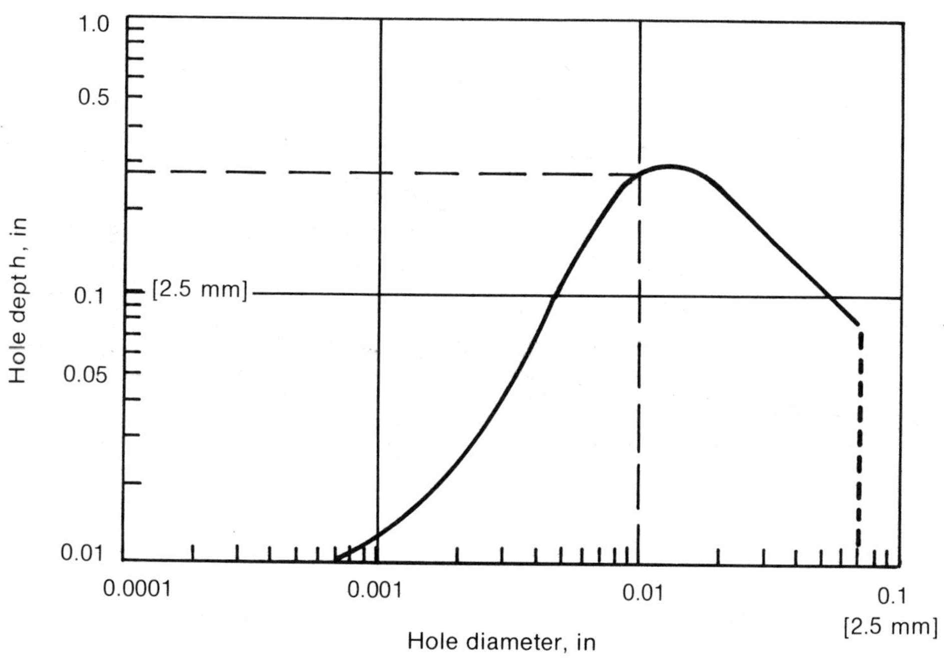


**Figure 12.6–11** Typical depth and diameter values for pulsed-laser drilling in ferrous alloys. (Adapted from S. R. Bolin, p. 126.)

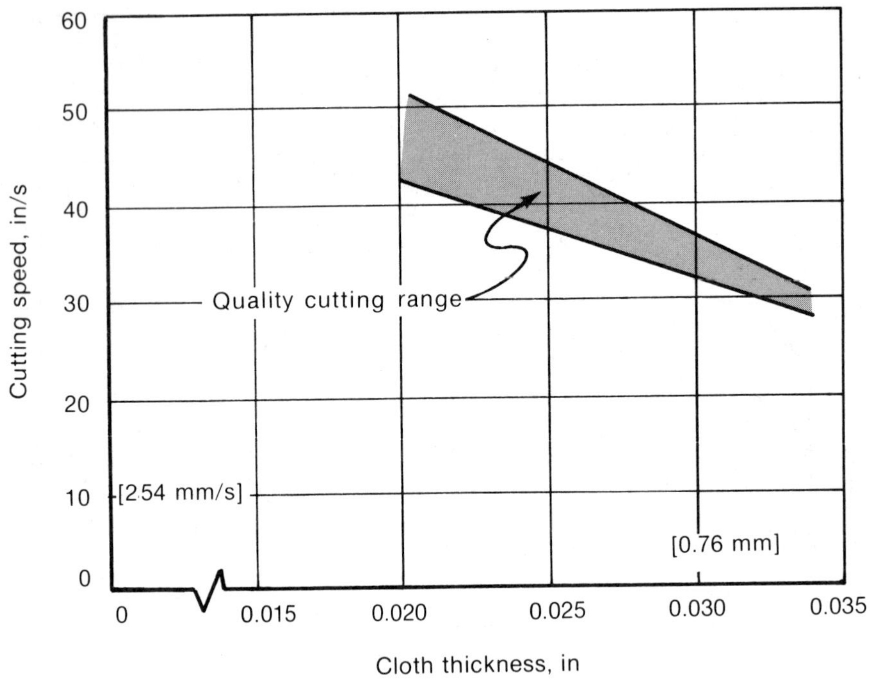


**Figure 12.6–12** $CO_2$-laser cloth cutting (250 watts $TEM_{00}$). (Courtesy of Coherent, Inc.)

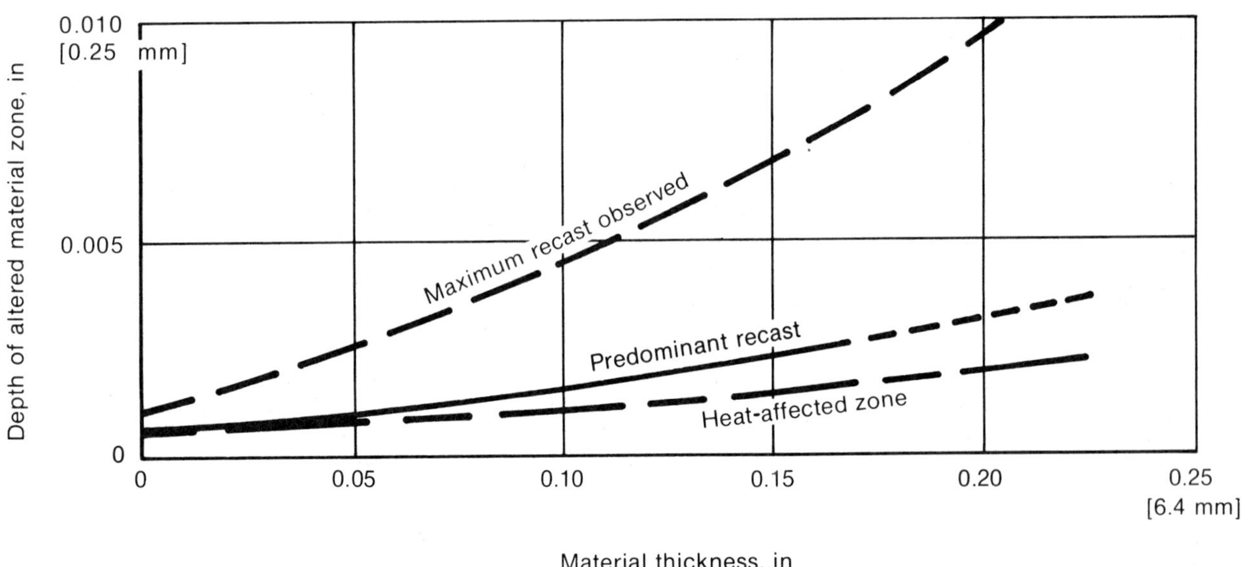


**Figure 12.6–13** Surface effects from LBM, circa 1978.

**TABLE 12.6-6 $CO_2$-Laser Drilling of Plastics**

| MATERIAL | THICKNESS | | HOLE DIAMETER | | | | PULSE LENGTH | POWER LEVEL |
|---|---|---|---|---|---|---|---|---|
| | | | Top | | Bottom | | | |
| | in | mm | in | mm | in | mm | ms | W |
| PVC | 0.025 | 0.64 | 0.010 | 0.25 | | | 5.0 | 50.0 |
| PVC | 0.025 | 0.64 | 0.030 | 0.76 | | | 40 | 50 |
| PVC | 0.020 | 0.51 | 0.012 | 0.30 | 0.010 | 0.25 | 10 | 50 |
| PVC | | | 0.023 | 0.58 | 0.020 | 0.51 | 30 | |
| Acrylic | 0.060 | 1.52 | 0.007 | 0.18 | 0.005 | 0.13 | 10 | 50 |
| Polyethylene | 0.006 | 0.15 | 0.014 | 0.36 | | | 4.5 | 150 |
| Polyethylene | 0.020 | 0.51 | 0.015 | 0.38 | | | 20 | 50 |
| Polystyrene | 0.013 | 0.33 | 0.010 | 0.25 | | | 5 | 50 |
| Polystyrene | 0.011 | 0.28 | 0.040 | 1.02 | | | 40 | 50 |
| ABS plastic | 0.030 | 0.76 | 0.040 | 1.02 | | | 100 | 50 |
| ABS plastic | 0.030 | 0.76 | 0.010 | 0.25 | | | 20 | 50 |
| Mylar | 0.0015 | 0.04 | 0.005 | 0.13 | | | 0.4 | 2.6 |
| Acrylic | 0.035 | 0.89 | 0.016 | 0.41 | | | 10 | 175 |
| Polycarbonate | 0.035 | 0.89 | 0.015 | 0.38 | | | 10 | 175 |
| Nylon | 0.035 | 0.89 | 0.011 | 0.28 | | | 10.0 | 175.0 |

SOURCE: Coherent, Inc.

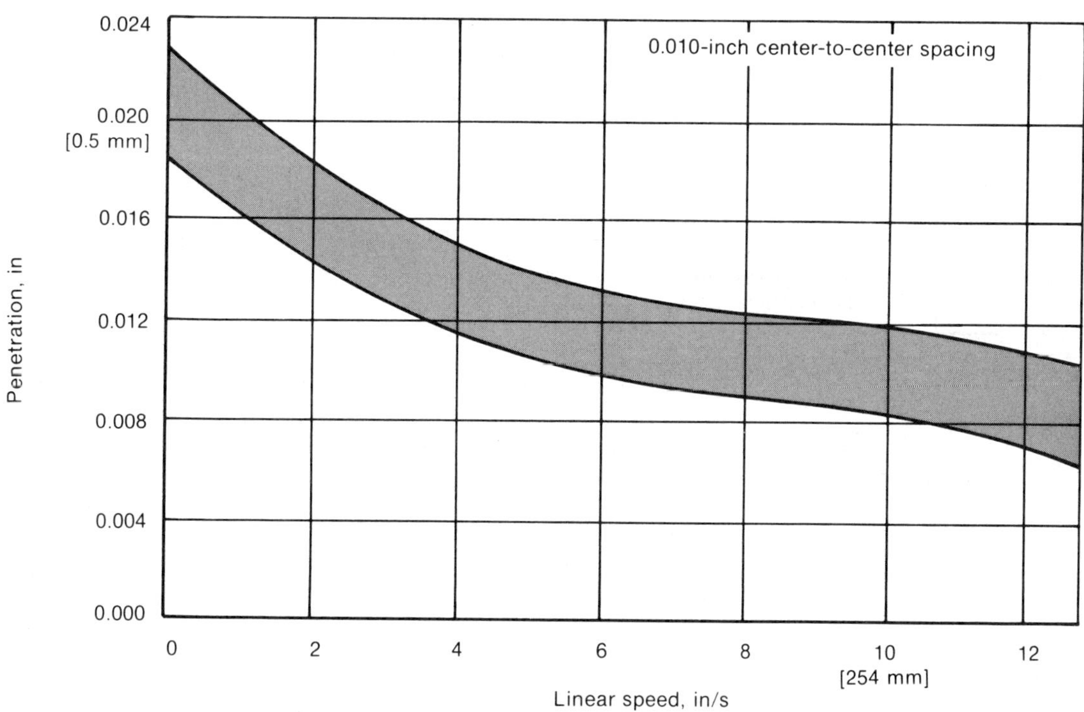


**Figure 12.6-14** Ceramic scribing chart—penetration depth versus speed using 185-watt continuous-wave $CO_2$ laser. Shaded area reflects operating domain expected in production environment with a 1.5-inch focal length lens. (Courtesy of Coherent, Inc.)

# 12.6 Laser Beam Machining—LBM

## SPECIFIC APPLICATIONS DATA

**TABLE 12.6–7 Specific Applications Data for Laser Beam Machining**

| Work material: | **ABS Plastic Strip** | **Cloth, Single Ply** | **Brass Caps** | |
|---|---|---|---|---|
| Workpiece configuration: | 0.010 in<br>0.030-inch thick | | 0.020-inch dia.<br>1-inch dia. | |
| Laser equipment type: | $CO_2$ | $CO_2$ | Nd-glass | Ruby |
| wavelength, $\mu$m: | 10.6 | 10.6 | 1.06 | 0.69 |
| power, W: | 250 | 500 | | |
| Operating conditions power level: | 50 W | | | 40 J |
| beam mode: | Pulsed | CW $TEM_{00}$ | Pulsed | Pulsed |
| pulses per second: | | | 1 | 1 |
| cycle time: | <1 s | Up to 40 in/s | 50 caps/min | |
| pulse length, ms: | 20 | | 0.150 | 0.60 |
| Tolerance, in: | | | ±0.0003 | |
| Remarks: | Drilled while strip moving. Laser deflection used to reduce relative motion. | Computer controlled nesting and cutting. Seals edge so no fraying. | Vibratory bowl feed for workpieces. | |

**TABLE 12.6-7—Continued**

| Work material: | **Stainless Steel Surgical Needle** | **Quartz Tubing, Cutoff** | **Rene 80** |
|---|---|---|---|
| Workpiece configuration: | 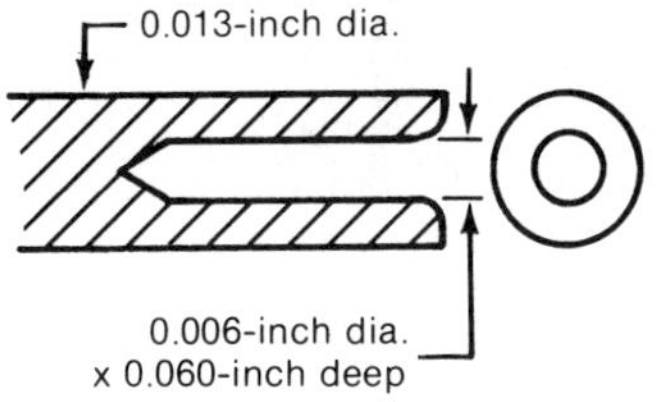 0.013-inch dia.<br>0.006-inch dia. x 0.060-inch deep | 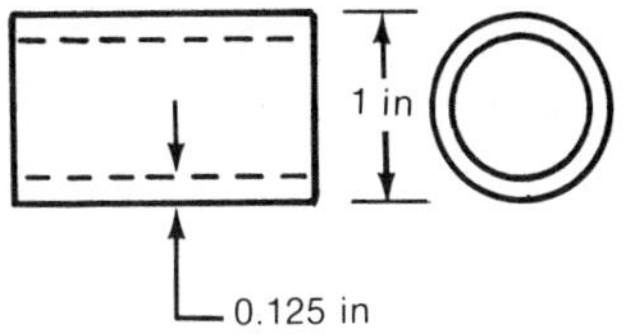 1 in<br>0.125 in | 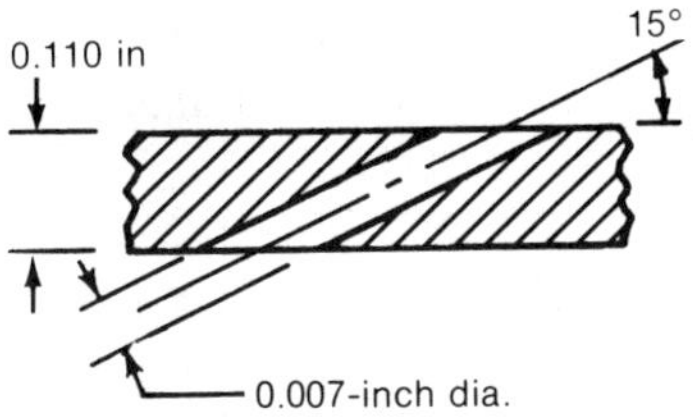 0.110 in<br>15°<br>0.007-inch dia. |
| Laser equipment type: | Ruby | $CO_2$ | Ruby |
| wavelength, μm: | 0.69 | 10.6 | 0.69 |
| power, W: | 150 J | 500 | 25 J |
| Operating conditions power level: | 3 J | 250 W | 14.5 J |
| beam mode: | Pulsed | CW | Pulsed |
| pulses per second: | 1 | | 1 |
| cycle time: | 10 s | 4 s | One pulse |
| pulse length, ms: | 1 | | 1 |
| Tolerance, in: | | | |
| Remarks: | | No cracks and ends smoothed. Room temperature cutting with tube rotating at 60 rpm. | 35:1 = length to diameter ratio.<br>0.002-inch recast. |

**TABLE 12.6-7—Continued**

| Work material: | **Kapton Insulated Wire Stripping (No. 26 to 1/0 Wire)** | **Rubber** | **Plastic** |
|---|---|---|---|
| Workpiece configuration: | 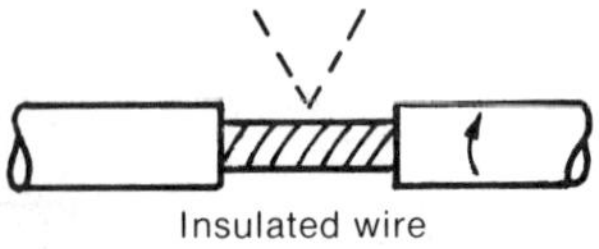 Insulated wire | 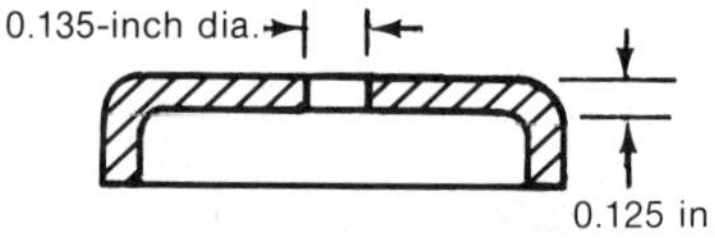 0.135-inch dia.; 0.125 in | 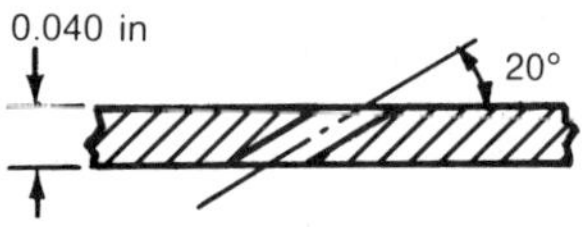 0.040 in; 20° |
| Laser equipment type: | $CO_2$ | Ruby | Ruby |
| wavelength, $\mu$m: | 10.6 | 0.69 | 0.69 |
| power, W: | 250 | 50 J | 50 J |
| Operating conditions power level: | 4 W | 27 J | 25 J |
| beam mode: | CW | Multi | Multi |
| pulses per second: | | 1 | 1 |
| cycle time: | 15 s for #18 wire | 30 shots | 3 shots |
| pulse length, ms: | | 5 | 6 |
| Tolerance, in: | No action on metal wire. | | |
| Remarks: | Laser beam rotated around wire. Spot size 0.005-inch dia. and $10^5$ W/in$^2$. | | |

## PROCESS SUMMARY

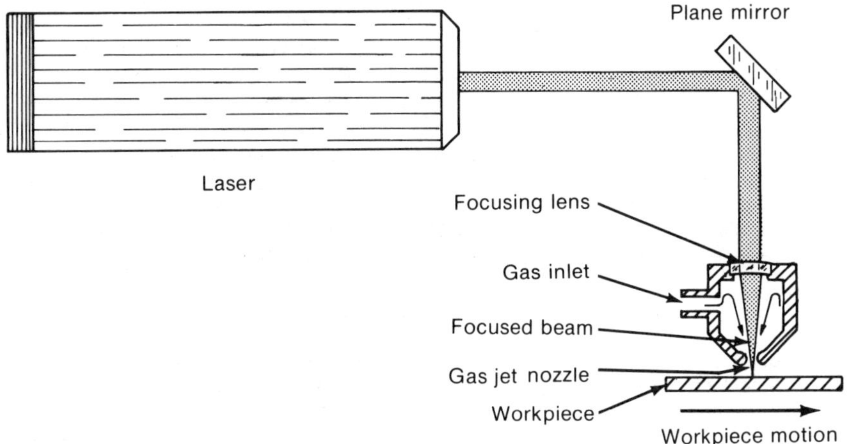


**Figure 12.7–1** LBT schematic.

### Principles

Laser beam torch (LBT) is a process in which material is removed by the simultaneous focusing of a laser beam and a gas stream on the workpiece (see figure 12.7–1). A continuous-wave (CW) laser or a pulsed laser with more than 100 pulses per second is focused on or slightly below the surface of the workpiece, and the absorbed energy causes localized melting. An oxygen gas stream promotes an exothermic reaction and purges the molten material from the cut. Argon or nitrogen gas is sometimes used to purge the molten material while also protecting the workpiece. Argon or nitrogen gas is often used when organic or ceramic materials are being cut. Close control of the spot size and the focus on the workpiece surface is required for uniform cutting. The type of gas used has only a modest effect on laser penetrating ability (see table 12.7–3 and figure 12.7–19). Typically, short laser pulses with high peak power are used for vaporizing materials (drilling and scribing) while long pulses are used for cutting and welding, as shown in figure 12.7–3. The $CO_2$ laser is the laser most often used for cutting.

### Practical Applications

The LBT cut is characterized by a narrow, tapered kerf and a shallow heat-affected zone. For a given power level, the highest cutting speed yields the best quality of cut and the minimum heat-affected zone. The assisting gas improves the quality and decreases the kerf. Complex contours can be cut in hard materials like silicon carbide, friable materials like glass, or sticky materials like confections. Thin materials are cut at high rates, with 1/8- to 3/8-inch [3.2 to 9.5 mm] thickness a practical limit. Mechanical, optical or numerical control of cuts is usual. Trimming of high strength alloy sheet products with LBT has reduced previous trimming costs as much as 50 percent. Plywood die-board slotting for steel-rule dies, contour-cut titanium plates for aircraft fabrications, stacks of cloth for suits, stripping of wire insulation, and drilling of contact lenses are all practical LBT applications. While most cutting is done on stock less than 0.10-inch [2.5 mm] thick, carbon steel up to 0.50-inch [12.7 mm] thick has been cut at 30 inches per minute [760 mm/min] with higher power lasers. The narrow heat-affected zone minimizes trimming and post-cutting cleanup operations. Before selecting laser as a cutting process, a comparison with other processes is desirable. Table 12.7–4 and figure 12.7–4 contain comparative data. The rapidity of the change in the state of the art of LBT makes it desirable to check with laser builders for the latest process capabilities.

### Operating Parameters

See table 12.7–1.

### Material Removal Rates and Tolerances

The cutting rates depend upon the thermal properties of the material (table 12.7–5), the material thickness and the power density in the focal spot (figure 12.7–3). A 0.002-inch [0.05 mm] diameter spot can attain a density of $7 \times 10^9$ watts per square inch [$1.1 \times 10^9$ W/mm$^2$], which is sufficient to vaporize any material. Cutting rates range from 30 inches per minute [750 mm/min] on 1/16-inch [1.6 mm] thick tool steel, to 240 inches per minute [6,100 mm/min] on 0.020-inch [0.5 mm] thick tinplate, to 500 inches per minute [12,700 mm/min] on 0.030-inch [0.75 mm] thick Lexan with a 1/2-kW continuous-wave $CO_2$ gas laser. This laser will cut 1/4-inch [6.4 mm] thick aluminum at 3 inches per minute [75 mm/min] or 1/4-inch [6.4 mm] thick titanium at 140 inches per minute [3,500 mm/min].

Accuracies up to ±0.004 inch [±0.1 mm] are attainable and depend upon the accuracies in the workpiece motion control. Taper is present in most kerfs, as shown in figure 12.7–20, and too high cutting speeds will impart a curved lay pattern in the cut.

# 12.7 Laser Beam Torch—LBT

**TABLE 12.7-1 Typical Values for LBT Operating Parameters**

| | |
|---|---|
| Laser types: | Continuous wave Nd-YAG (cutting)<br>Continuous wave $CO_2$ (cutting)<br>Q-switch or pulsed ND-YAG (drilling) |
| Power: | 0.25 to 16 kW |
| Gas | |
| types: | Oxygen, nitrogen, argon, air, helium |
| pressure: | 20 to 120 psi [140 to 825 kPa] |
| nozzle diameter: | 0.020 to 0.040 inch [0.5 to 1 mm] |
| nozzle gap: | 0.020 ±0.010 inch [0.5 ±0.25 mm] |
| Focus | |
| spot size: | 0.002 inch [0.05 mm] smallest diameter<br>0.003 to 0.010 inch [0.07 to 0.25 mm] usual |
| depth of focus*: | 0.030 to 0.33 inch [0.76 to 8.4 mm] |
| focal length: | 1.5 to 5 inches [40 to 130 mm] |
| focal point: | Surface or slightly below |
| Trim/cleanup setback: | 0.5 to 3 times stock thickness |
| Kerf: | 0.004 to 0.080 inch [0.1 to 2 mm]<br>(see table 12.7-7 and figures 12.7-20 to 12.7-24) |
| Heat-affected zone (HAZ): | 0.001 to 0.010 inch [0.025 to 0.25 mm] |
| Cutting rates (feed rate): | 10 to 300 in/min [250 to 7,500 mm/min]<br>(see table 12.7-7 and figures 12.7-6 to 12.7-19) |
| Cut thickness capability for | |
| nonferrous alloys: | Up to 0.040 inch [1 mm] |
| ferrous and titanium alloys: | Up to 0.200 inch [5 mm] |
| nonmetals: | Up to 0.10 inch [2.5 mm] |
| nickel alloys: | Up to 0.60 inch [1.5 mm] |
| plastics: | Up to 0.20 inch [5 mm] |
| Minimum outside trim radius: | 0.50 inch [12.7 mm] |

*Distance between which spot size is no more than 40 percent larger than at focus.

## Surface Technology

A heat-affected zone (HAZ) is present in all laser cuts but the rapidity of the metal removal can leave less HAZ than competing processes (see table 12.7-7 and LBM section, figure 12.6-13). Surface roughness is high—125 to 250 microinches $R_a$ [3.2 to 6.3 μm] being typical. The amount of stock removal needed to clean up the roughness or HAZ is minimal and ranges from one-fourth to three times the stock thickness. Dross attached to the underside of the cut usually is easy to remove.

## Equipment and Tools

LBT equipment is available in 1/4- to 16-kW sizes and in worktable sizes up to 4 by 6 feet [1.3 by 2 m]. NC, optically or mechanically guided torches or workpieces are available to cut straight or intricate contours. Multiple torch combinations and LBT combined with punch presses have been built. Laser cutting forces are almost nonexistent so fixturing is less complicated. Rotating mechanisms for the lens (or mirrors) are employed when trepanning larger-size holes.

## Machining Characteristics

It is essential to keep the laser beam focused accurately on the workpiece. Optical and mechanical contour followers are available to adjust the torch focus when working on sheets with undulations. Adequate fume and sight protection are needed for LBT.

The noncontacting tool in LBT increases tool life almost indefinitely. The choice of gas used can be influenced by the need for protection of the workpiece material (see table 12.7-6).

## SELECTED DATA

There are several valid techniques for selecting operating parameters for LBT. The steps in table 12.7-2 are recommended as one logical sequence. Supporting data come from the other figures and tables in accordance with the terms as explained in figure 12.7-2.

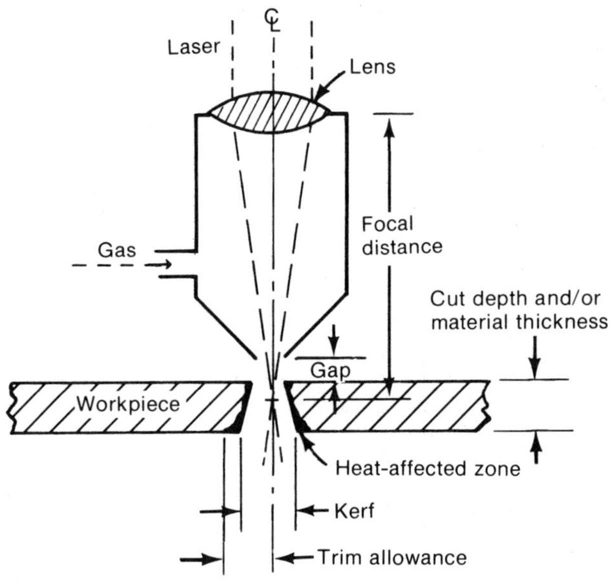


**Figure 12.7-2** LBT parameter terminology.

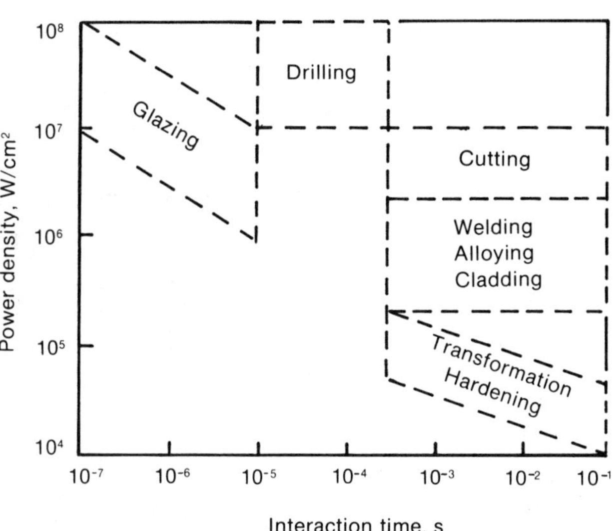


**Figure 12.7-3** Laser material processing applications in terms of power density and interaction time. (A. Schachrai and M. Castellani-Longo, Application of high power lasers in manufacturing, *Annals of the CIRP*, Vol. 28/2, 1979, p. 466)

**TABLE 12.7-2 Steps to Select LBT Operating Parameters**

| ITEM | DATA |
|---|---|
| 1. Workpiece material characteristics | Engineering drawing and table 12.7-5 |
| 2. Applicability of LBT | Comparison of saw, nibble, plasma- arc, oxyacetylene, etc., with laser, table 12.7-4 and figures 12.7-3 and 12.7-4 |
| 3. Select assisting gas | Tables 12.7-3 and 12.7-6 and figure 12.7-19 |
| 4. Material thickness (cut depth) | Engineering drawing |
| 5. Select power level and cutting rate | Figures 12.7-5 to 12.7-25 Table 12.7-7 |
| 6. Estimate kerf and heat-affected zone (HAZ) | Table 12.7-7 and figures 12.7-20 to 12.7-25 |
| 7. Template design with trim allowance | Figure 12.7-20 and item 6 |
| 8. Cutting gap setting | Trial cut to verify parameters and edge finish |

**TABLE 12.7-3 Penetration of $CO_2$ Laser in Perspex (Boxboard) with Various Gases**

| GAS | PENETRATION | |
|---|---|---|
| | in | mm |
| Helium | 0.93 | 23.5 |
| Nitrogen | 0.94 | 24.0 |
| Oxygen | 0.89 | 22.5 |
| Air | 0.96 | 24.5 |
| Argon | 1.00 | 25.5 |
| Carbon dioxide | 0.87 | 22.0 |
| Argon/hydrogen, 75/25 | 0.91 | 23.0 |

SOURCE: F. W. Lunau, Boc-Murex Welding Research and Development Laboratories, London.

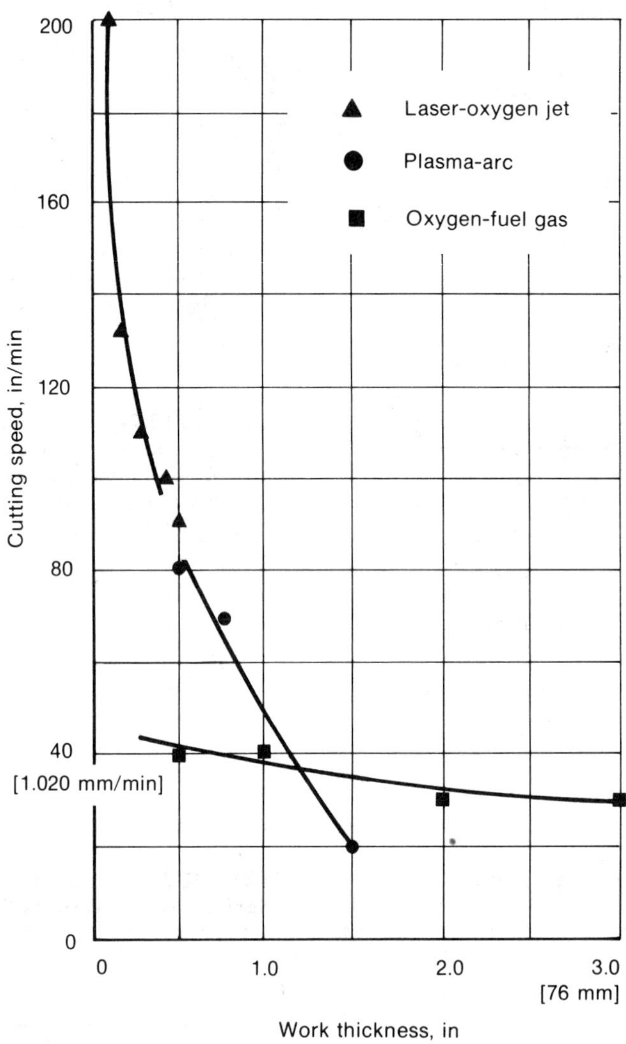


**Figure 12.7-4** Comparative cutting speeds of laser beam torch, plasma-arc torch and oxyacetylene torch for titanium alloys. (Courtesy of I. Slater, The Boeing Company)

**TABLE 12.7-4 Comparison of Cutting Rates Among Competing Processes**

| MATERIAL | TYPE OF CUT | WORK THICKNESS | | PUNCH PRESS NIBBLER | | BAND SAW | | OXYACETYLENE TORCH | | PLASMA-ARC TORCH | | LASER TORCH CW-$CO_2$ at 250 W plus $O_2$ gas assist | |
|---|---|---|---|---|---|---|---|---|---|---|---|---|---|
| | | | | Cutting Speed | | | | | | | | | |
| | | in | mm | in/min | m/min | in/min | m/min | in/min | m/min | in/min | m/min | in/min | m/min |
| Titanium alloy | Contour | 0.020 | 0.5 | | | 8.5 | 0.22 | | | | | 200 | 5 |
| | Contour | 0.062 | 1.5 | | | 5.8 | 0.15 | | | | | 160 | 4 |
| | Contour | 0.125 | 3 | 20 | 0.5 | 4.4 | 0.11 | 60 | 1.5 | 60 | 1.5 | 120 | 3 |
| | Straight | 0.250 | 6 | | | 23 | 0.58 | 100 | 2.54 | 167 | 4.2 | 200 | 5.1 |
| | Contour | 0.250 | 6 | | | 3 | 0.08 | 30 | 0.75 | 50 | 1.25 | 60 | 1.5 |
| | Contour | 0.50 | 12 | | | 1 | 0.03 | 20 | 0.5 | 35 | 0.9 | 40 | 1 |
| | Contour | 1 | 25 | | | | | 15 | 0.4 | 30 | 0.75 | 15* | 0.4 |
| | Contour | 2 | 50 | | | | | 16 | 0.4 | 22 | 0.6 | 5* | 0.13 |
| Steel | Contour | 0.020 | 0.5 | | | 8.5 | 0.22 | | | | | 80 | 2 |
| | Contour | 0.062 | 1.5 | | | 5.8 | 0.15 | | | | | 40 | 1 |
| | Contour | 0.125 | 3 | | | 4.4 | 0.11 | | | | | 20 | 0.5 |
| 4130 Alloy steel | Contour | 0.100 | 2.5 | 30 | 0.75 | 20 | 0.5 | 30 | 0.75 | 30 | 0.75 | 65 | 1.7 |
| 4340 Alloy steel | Contour | 0.375 | 8.6 | | | 14 | 0.36 | 80 | 2 | 88 | 2.2 | 35* | 0.9 |
| | Contour | 0.75 | 17 | | | 10 | 0.25 | 50 | 1.3 | 60 | 1.5 | 15* | 0.4 |
| Stainless steel | Contour | 0.050 | 1.3 | 30 | 0.75 | 25 | 0.6 | 10 | 0.25 | 10 | 0.25 | 18 | 0.46 |

*16 kw laser

**TABLE 12.7-5 Properties of Metals and Alloys Important for Oxygen-Laser Cutting**

| MATERIAL | ELECTRICAL RESISTIVITY at 68°F ohm-cm | THERMAL CONDUCTIVITY at 212°F BTU/hr/ft²/ft/°F | THERMAL DIFFUSIVITY ft²/hr | OXIDE | OXIDE MELTING POINT °F | HEAT OF COMBUSTION kcal/cm³metal |
|---|---|---|---|---|---|---|
| Ti-6Al-4V | 176 | 4.3 | 0.12 | $TiO_2$ | 3,360 | 12.9 |
| Titanium | 55 | 9.8 | 0.29 | $TiO_2$ | 3,360 | 12.9 |
| 4340 Alloy steel | | 22 | 0.43 | $Fe_2O_3$ | 2,840 | 14.1 |
| C1020 Carbon steel | 14.3 | 27 | 0.52 | $Fe_2O_3$ | 2,840 | 14.1 |
| Zircaloy-2 | 74 | 8.1 | 0.29 | $ZrO_2$ | 4,930 | 16.3 |
| 321 Stainless steel | 72 | 9.3 | 0.13 | $Cr_2O_4$ | 4,130 | 14.4 |
| 6061 Aluminum alloy | 3.8 | 99 | 2.5 | $Al_2O_3$ | 3,720 | 20.2 |
| Copper ETP | 1.71 | 226 | 4.4 | $Cu_2O$ | 2,260 | 2.8 |

**TABLE 12.7-6 Gas Selection for LBT**

| MATERIAL CLASS | BEST | OPTIONAL |
|---|---|---|
| Steels | Air | $O_2$, $N_2$ |
| Titanium | Helium | Helium/argon, $CO_2$ |
| Nonferrous alloys | $O_2$ | Air |
| Nickel alloys | $O_2$ | Air, $O_2$ |
| Nonmetals | $O_2$ | Air |
| Composites | $O_2$ | $N_2$, argon |
| Plastics | $O_2$ | Air, argon |
| Wood products | $N_2$ | Air, $O_2$ |

**TABLE 12.7-7 Feed Rates for $CO_2$ Laser Beam Torch**

| MATERIAL | THICKNESS | | CUTTING RATE | | POWER | ASSISTING GAS | | | KERF WIDTH | | | | HEAT-AFFECTED ZONE | |
|---|---|---|---|---|---|---|---|---|---|---|---|---|---|---|
| | | | | | | Type | Pressure | | Inlet | | Outlet | | | |
| | in | mm | in/min | mm/min | watts | | psi | kPa | in | mm | in | mm | in | mm |
| **STEELS** | | | | | | | | | | | | | | |
| Mild steel | 0.039 | 1.0 | 118 | 3,000 | 230* | $O_2$ | | | | | 0.004 | 0.1 | | |
| | 0.039 | 1.0 | 177 | 4,500 | 400 | $O_2$ | | | | | | | | |
| | 0.051 | 1.3 | 142 | 3,610 | 500 | air | | | | | | | | |
| | 0.063 | 1.6 | 98 | 2,500 | 500 | air | | | | | | | | |
| | 0.09 | 2.3 | 70 | 1,780 | 650 | $O_2$ | | | | | | | | |
| | 0.118 | 3.0 | 60 | 1,520 | 350 | $O_2$ | | | | | | | | |
| | 0.119 | 3.0 | 24 | 610 | 230* | $O_2$ | | | | | 0.008 | 0.2 | | |
| | 0.126 | 3.2 | 40 | 1,020 | 500 | $O_2$ | | | | | | | | |
| C1010 | 0.125 | 3.2 | 22 | 560 | 190 | $O_2$ | | | 0.04 | 1.0 | | | | |
| Low carbon | 0.125 | 3.2 | 100 | 2,540 | 4,000 | air | | | | | | | | |
| Mild steel | 0.252 | 6.4 | 20 | 510 | 500 | $O_2$ | | | | | | | | |
| Low carbon | 0.66 | 16.8 | 45 | 1,140 | 4,000 | air | | | | | | | | |
| Stainless steel | 0.004 | 0.1 | 197 | 5,000 | 500 | $O_2$ | | | | | | | | |
| | 0.012 | 0.3 | 146 | 3,710 | 500 | $O_2$ | | | | | | | | |
| 321 | 0.02 | 0.5 | 92 | 2,340 | 165 | $O_2$ | | | 0.012 | 0.3 | | | | |
| | 0.02 | 0.5 | 750 | 19,000 | 1,000 | $O_2$ | | | | | | | | |
| 17-7 PH, 321 | 0.025 | 0.6 | 180 | 4,570 | 250 | $O_2$ | | | | | | | | |
| | 0.032 | 0.8 | 650 | 16,500 | 1,000 | $O_2$ | | | | | | | | |
| | 0.039 | 1.0 | 65 | 1,650 | 500 | $O_2$ | | | | | | | | |
| 321 | 0.04 | 1.0 | 550 | 14,000 | 1,000 | $O_2$ | | | | | | | | |
| | 0.05 | 1.3 | 30 | 760 | 165 | $O_2$ | | | 0.02 | 0.5 | | | | |
| | 0.059 | 1.5 | 18 | 460 | 400 | $O_2$ | | | | | | | | |
| | 0.062 | 1.6 | 450 | 11,400 | 1,000 | $O_2$ | | | | | | | | |
| | 0.063 | 1.6 | 75 | 1,900 | 500 | $O_2$ | | | | | | | | |
| 321,410,17-7 PH | 0.063 | 1.6 | 60 | 1,520 | 250 | $O_2$ | 60 | 414 | | | | | | |
| | 0.08 | 2.0 | 325 | 8,260 | 1,000 | $O_2$ | | | | | | | | |
| 17-7 PH, 321 | 0.09 | 2.3 | 30 | 760 | 250 | $O_2$ | | | | | | | | |
| | 0.11 | 2.8 | 47 | 1,190 | 400 | $O_2$ | | | | | | | | |
| | 0.118 | 3.0 | 45 | 1,140 | 400 | $O_2$ | | | | | | | | |
| 321,410,17-7 PH | 0.125 | 3.2 | 10 | 250 | 250 | $O_2$ | 70 | 483 | | | | | | |
| 300 or 400 series | 0.125 | 3.2 | 100 | 2,540 | 3,000 | air | | | | | | | | |
| | 0.125 | 3.2 | 200 | 5,080 | 1,000 | $O_2$ | | | | | | | | |
| | 0.126 | 3.2 | 35 | 890 | 500 | $O_2$ | | | | | | | | |
| | 0.252 | 6.4 | 20 | 510 | 500 | $O_2$ | | | | | | | | |

NOTE: Cutting rates are not necessarily maximum for any given thickness. The rates consider other parameters for acceptable cuts, such as kerf width, gas consumption, depth of heat-affected zone (HAZ), etc.

*TEM 0.1.

TABLE 12.7-7—Continued

| MATERIAL | THICKNESS | | CUTTING RATE | | POWER | ASSISTING GAS | | | KERF WIDTH | | | | HEAT-AFFECTED ZONE | |
|---|---|---|---|---|---|---|---|---|---|---|---|---|---|---|
| | | | | | | Type | Pressure | | Inlet | | Outlet | | | |
| | in | mm | in/min | mm/min | watts | | psi | kPa | in | mm | in | mm | in | mm |
| Galvanized steel | 0.02 | 0.5 | 250 | 6,350 | 1,000 | $O_2$ | | | | | | | | |
| | 0.03 | 0.75 | 137 | 3,480 | 230* | $O_2$ | | | | | 0.004 | 0.1 | | |
| | 0.039 | 1.0 | 177 | 4,500 | 400 | $O_2$ | | | | | | | | |
| | 0.04 | 1.0 | 100 | 2,540 | 1,000 | $O_2$ | | | | | | | | |
| | 0.051 | 1.3 | 142 | 3,610 | 400 | $O_2$ | | | | | | | | |
| | 0.062 | 1.6 | 50 | 1,270 | 1,000 | $O_2$ | | | | | | | | |
| | 0.118 | 3.0 | 60 | 1,520 | 400 | $O_2$ | | | | | | | | |
| Tinplate steel | 0.02 | 0.5 | 240 | 6,100 | 500 | $O_2$ | | | | | | | | |
| High speed steel | 0.093 | 2.4 | 35 | 890 | 500 | $O_2$ | | | | | | | | |
| Hi C steel | 0.118 | 3.0 | 59 | 1,500 | 400 | $O_2$ | | | | | | | | |
| High speed steel | 0.134 | 3.4 | 24 | 610 | 500 | $O_2$ | | | | | | | | |
| | 0.165 | 4.2 | 16 | 410 | 500 | $O_2$ | | | | | | | | |
| | 0.205 | 5.2 | 27 | 690 | 500 | $O_2$ | | | | | | | | |
| | 0.22 | 5.6 | 23 | 580 | 500 | $O_2$ | | | | | | | | |
| | 0.28 | 7.1 | 3 | 75 | 500 | $O_2$ | | | | | | | | |
| Low alloy steel | 0.024 | 0.6 | 23 | 580 | 850 | $O_2$ | | | | | | | | |
| Maraging steel | 0.189 | 4.8 | 3 | 75 | 500 | $O_2$ | | | | | | | | |
| H25 tool steel | 0.063 | 1.6 | 31 | 790 | 500 | $O_2$ | | | | | | | | |
| Martensitic stainless steel | 0.2 | 5.1 | 30 | 760 | 650 | $O_2$ | | | | | | | | |
| | 0.31 | 7.9 | 15 | 380 | 650 | $O_2$ | | | | | | | | |
| 18-8 austenitic stainless steel | 0.039 | 1.0 | 59 | 1,500 | 230* | $O_2$ | | | | | 0.004 | 0.1 | | |
| | 0.125 | 3.2 | 30 | 760 | | | | | | | | | | |
| Nimonic | | | | | | | | | | | | | | |
| 90 | 0.039 | 1.0 | 96 | 2,440 | 500 | $O_2$ | | | | | | | | |
| 80 | 0.059 | 1.5 | 23 | 580 | 250 | $O_2$ | | | | | | | | |
| 90 | 0.059 | 1.5 | 90 | 2,290 | 850 | $O_2$ | | | | | | | | |
| 75 | 0.002 | 0.05 | 98 | 2,490 | 500 | $O_2$ | | | | | | | | |
| 80A | 0.028 | 0.7 | 180 | 4,570 | 650 | $O_2$ | | | | | | | | |
| 70 | 0.028 | 0.7 | 80 | 2,030 | 650 | $O_2$ | | | | | | | | |
| 75 | 0.047 | 1.2 | 51 | 1,300 | 500 | $O_2$ | | | | | | | | |
| 90 | 0.05 | 1.3 | 90 | 2,290 | 650 | $O_2$ | | | | | | | | |
| 75 | 0.079 | 2.0 | 31 | 780 | 500 | $O_2$ | | | | | | | | |
| 4340 steel | 0.025 | 0.6 | 120 | 3,050 | 250 | $O_2$ | 40 | 280 | | | | | | |
| | 0.063 | 1.6 | 65 | 1,650 | 250 | $O_2$ | 50† | 340† | | | | | | |
| | 0.09 | 2.3 | 60 | 1,520 | 250 | $O_2$ | 40 | 280 | | | | | | |

*TEM 0.1. †Nozzle diameter = 0.035 inch [0.9 mm].

**TABLE 12.7-7—Continued**

| MATERIAL | THICKNESS | | CUTTING RATE | | POWER | ASSISTING GAS | | | KERF WIDTH | | | | HEAT-AFFECTED ZONE | |
|---|---|---|---|---|---|---|---|---|---|---|---|---|---|---|
| | | | | | | Type | Pressure | | Inlet | | Outlet | | | |
| | in | mm | in/min | mm/min | watts | | psi | kPa | in | mm | in | mm | in | mm |
| 4340 steel (cont.) | 0.125 | 3.2 | 30 | 760 | 250 | $O_2$ | 60† | 410† | | | | | | |
| | 0.125 | 3.2 | 50 | 1,270 | 750 | $O_2$ | 75 | 520 | 0.032 | 0.81 | 0.052 | 1.32 | 0.013 | 0.33 |
| | 0.125 | 3.2 | 70 | 1,780 | 1,500 | $O_2$ | 120 | 830 | 0.035 | 0.89 | 0.041 | 1.04 | 0.011 | 0.28 |
| | 0.125 | 3.2 | 130 | 3,300 | 3,000 | $O_2$ | 120 | 830 | 0.027 | 0.69 | 0.042 | 1.07 | 0.025 | 0.64 |
| | 0.125 | 3.2 | 200 | 5,080 | 6,000 | $O_2$ | 200 | 1,380 | 0.021 | 0.53 | 0.024 | 0.61 | 0.014 | 0.36 |
| | 0.125 | 3.2 | 45 | 1,140 | 1,500 | $O_2$ | 200 | 1,380 | 0.04 | 1.02 | 0.043 | 1.09 | 0.012 | 0.30 |
| | 0.125 | 3.2 | 50 | 1,270 | 1,500 | $CO_2$ | 100 | 690 | 0.016 | 0.41 | 0.014 | 0.36 | 0.025 | 0.64 |
| | 0.125 | 3.2 | 80 | 2,030 | 3,000 | $CO_2$ | 120 | 830 | 0.013 | 0.33 | 0.015 | 0.38 | 0.026 | 0.66 |
| | 0.125 | 3.2 | 100 | 2,540 | 6,000 | $CO_2$ | 250 | 1,720 | 0.02 | 0.51 | 0.016 | 0.41 | 0.032 | 0.81 |
| | 0.125 | 3.2 | 80 | 2,030 | 3,000 | $N_2$ | 200 | 1,380 | 0.02 | 0.51 | 0.022 | 0.56 | 0.014 | 0.36 |
| | 0.125 | 3.2 | 80 | 2,030 | 3,000 | air | 200 | 1,380 | 0.024 | 0.61 | 0.014 | 0.36 | 0.021 | 0.53 |
| | 0.25 | 6.4 | 70 | 1,780 | 3,000 | $O_2$ | 150 | 1,030 | 0.047 | 1.19 | 0.058 | 1.47 | 0.014 | 0.36 |
| | 0.25 | 6.4 | 100 | 2,540 | 6,000 | $O_2$ | 180 | 1,240 | 0.061 | 1.55 | 0.063 | 1.60 | 0.045 | 1.14 |
| | 0.25 | 6.4 | 15 | 380 | 3,000 | $CO_2$ | 200 | 1,380 | 0.021 | 0.53 | 0.03 | 0.76 | 0.033 | 0.84 |
| | 0.25 | 6.4 | 30 | 760 | 6,000 | $CO_2$ | 200 | 1,380 | 0.025 | 0.64 | 0.025 | 0.64 | 0.033 | 0.84 |
| | 0.38 | 9.7 | 30 | 760 | 3,000 | $O_2$ | 200 | 1,380 | 0.055 | 1.40 | 0.129 | 3.27 | 0.038 | 0.97 |
| | 0.38 | 9.7 | 60 | 1,520 | 6,000 | $O_2$ | 200 | 1,380 | 0.054 | 1.37 | 0.125 | 3.18 | | |
| | 0.38 | 9.7 | 18 | 460 | 6,000 | $CO_2$ | 350 | 2,410 | 0.025 | 0.64 | 0.014 | 0.36 | 0.057 | 1.45 |
| | 0.75 | 19.1 | 45 | 1,140 | 10,000 | $O_2$ | | | | | | | | |
| | 0.75 | 19.1 | 9.4 | 240 | 10,000 | air | | | | | | | | |
| | 0.75 | 19.1 | 10 | 250 | 10,000 | He | | | 0.09 | 2.29 | 0.054 | 1.37 | 0.099 | 2.51 |
| | 0.75 | 19.1 | 15 | 380 | 15,000 | 75%He, 25%A | | | | | | | | |
| | 0.75 | 19.1 | 15 | 380 | 15,000 | He | | | 0.09 | 2.29 | 0.037 | 0.94 | 0.097 | 2.46 |
| **NONFERROUS** | | | | | | | | | | | | | | |
| Aluminum | 0.02 | 0.5 | 800 | 20,300 | 1,000 | $O_2$ | | | | | | | | |
| | 0.04 | 1.0 | 350 | 8,890 | 1,000 | $O_2$ | | | | | | | | |
| | 0.062 | 1.6 | 200 | 5,080 | 1,000 | $O_2$ | | | | | | | | |
| | 0.125 | 3.2 | 100 | 2,540 | 1,000 | $O_2$ | | | | | | | | |
| | 0.24 | 6.1 | 1.2 | 30 | 3,800 | $O_2$ | | | | | | | | |
| | 0.25 | 6.4 | 40 | 1,020 | 3,800 | air | | | | | | | | |
| | 0.5 | 12.7 | 40 | 1,020 | 10,000 | $O_2$ | | | | | | | | |
| | 0.5 | 12.7 | 90 | 2,290 | 15,000 | $O_2$ | | | | | | | | |
| | 0.5 | 12.7 | 30 | 760 | 5,700 | air | | | | | | | | |
| Brass | 0.005 | 0.1 | 156 | 3,960 | 275 | $O_2$ | | | | | | | | |
| Copper (dag coated) | 0.008 | 0.2 | 23 | 580 | 500 | $O_2$ | | | | | | | | |
| Niobium | 0.126 | 3.2 | 8 | 200 | 500 | $O_2$ | | | | | | | | |

†Nozzle diameter = 0.035 inch [0.9 mm].

**TABLE 12.7-7—Continued**

| MATERIAL | THICKNESS | | CUTTING RATE | | POWER | ASSISTING GAS | | | KERF WIDTH | | | | HEAT-AFFECTED ZONE | |
|---|---|---|---|---|---|---|---|---|---|---|---|---|---|---|
| | | | | | | Type | Pressure | | Inlet | | Outlet | | | |
| | in | mm | in/min | mm/min | watts | | psi | kPa | in | mm | in | mm | in | mm |
| Molybdenum | 0.002 | 0.05 | 16 | 400 | 500 | $O_2$ | | | | | | | | |
| Zircaloy | 0.018 | 0.5 | 600 | 15,200 | 230 | $O_2$ | | | 0.02 | 0.51 | | | | |
| Titanium, CP | 0.02 | 0.5 | 600 | 15,200 | 135 | $O_2$ | | | 0.015 | 0.38 | | | | |
| Ti-6Al-4V or Ti-6Al-6V-2Sn | 0.025 | 0.7 | 300 | 7,600 | 250 | $O_2$ | 40 | 280 | | | | | | |
| | 0.05 | 1.3 | 300 | 7,600 | 210 | $O_2$ | 50 | 340 | 0.03 | 0.76 | | | | |
| | 0.062 | 1.6 | 120 | 3,050 | 3,000 | N | 200 | 1,380 | 0.016 | 0.41 | 0.015 | 0.38 | 0.008 | 0.2 |
| | 0.062 | 1.6 | 120 | 3.050 | 3,000 | air | 200 | 1,380 | 0.036 | 0.91 | 0.03 | 0.76 | 0.021 | 0.53 |
| | 0.08 | 2 | 708 | 18,000 | 230* | $O_2$ | | | | | | 0.008 | 0.2 | |
| | 0.09 | 2.3 | 108 | 4,580 | 210 | $O_2$ | 40 | 280 | 0.03 | 0.76 | | | | |
| | 0.118 | 3 | 161 | 4,090 | 400 | air | | | | | | | | |
| | 0.125 | 3.2 | 200 | 5,000 | 250 | $O_2$ | 40 | 280 | | | | | | |
| | 0.125 | 3.2 | 60 | 1,520 | 1,500 | $CO_2$ | 200 | 1,380 | 0.019 | 0.48 | 0.02 | 0.51 | | |
| | 0.125 | 3.2 | 80 | 2,040 | 3,000 | air | 200 | 1,380 | 0.023 | 0.58 | 0.02 | 0.51 | 0.04 | 1 |
| | 0.125 | 3.2 | 80 | 2,040 | 3,000 | N | 200 | 1,380 | 0.022 | 0.56 | 0.022 | 0.56 | 0.023 | 0.58 |
| | 0.125 | 3.2 | 70 | 1,780 | 3,000 | N | 20 | 140 | 0.012 | 0.3 | 0.012 | 0.3 | 0.018 | 0.46 |
| | 0.18 | 4.6 | 170 | 1,780 | 250 | $O_2$ | 40 | 280 | | | | | | |
| | 0.200 | 5.1 | 130 | 3,300 | 650 | $O_2$ | | | | | | | | |
| | 0.212 | 5.4 | 160 | 4,000 | 250 | $O_2$ | 80 | 560 | | | | | | |
| | 0.25 | 6.4 | 80 | 2,030 | 250 | $O_2$ | 40 | 280 | | | | | | |
| | 0.25 | 6.4 | 108 | 2,740 | 250 | $O_2$ | | | 0.040 | 1.02 | 0.035 | 0.89 | 0.005 | 0.13 |
| | 0.25 | 6.4 | 80 | 2,030 | 10,000 | He | | | 0.061‡ | 1.55‡ | 0.019‡ | 0.48‡ | 0.040 | 1.02 |
| | 0.25 | 6.4 | 140 | 3,560 | 3,000 | air | | | | | | | | |
| | 0.25 | 6.4 | 60 | 1,520 | 750 | $O_2$ | 60 | 410 | 0.060 | 1.52 | 0.075 | 1.91 | 0.019 | 0.48 |
| | 0.25 | 6.4 | 100 | 2,540 | 1,500 | $O_2$ | 60 | 410 | 0.055 | 1.40 | 0.070 | 1.78 | 0.017 | 0.43 |
| | 0.25 | 6.4 | 140 | 3,560 | 3,000 | $O_2$ | 80 | 550 | 0.060 | 1.52 | 0.070 | 1.78 | 0.016 | 0.41 |
| | 0.25 | 6.4 | 180 | 4,570 | 6,000 | $O_2$ | 150 | 1,030 | 0.102 | 2.59 | 0.092 | 2.34 | 0.037 | 0.94 |
| | 0.25 | 6.4 | 14 | 360 | 3,000 | $CO_2$ | 300 | 2,070 | 0.022 | 0.56 | 0.032 | 0.81 | 0.061 | 1.55 |
| | 0.25 | 6.4 | 60 | 1,520 | 6,000 | $CO_2$ | 300 | 2,070 | 0.030 | 0.76 | 0.019 | 0.48 | 0.056 | 1.42 |
| | 0.25 | 6.4 | 142 | 3,610 | 500 | $O_2$ | | | | | | | | |
| | 0.35 | 8.9 | 60 | 1,520 | 250 | $O_2$ | 50 | 340 | | | | | | |
| | 0.39 | 9.9 | 100 | 2,540 | 260 | $O_2$ | | | 0.065 | 1.65 | | | 0.005 | 0.13 |
| | 0.39 | 9.9 | 110 | 2,800 | 230* | $O_2$ | | | | | 0.060 | 1.52 | | |
| | 0.40 | 10.2 | 60 | 1,520 | 250 | $O_2$ | | | 0.035 | 0.89 | 0.063 | 1.60 | | |
| | 0.50 | 12.7 | 40 | 1,020 | 250 | $O_2$ | 70 | 480 | | | | | | |
| | 0.75 | 19.1 | 50 | 1,270 | 1,500 | $O_2$ | 100 | 690 | 0.062 | 1.57 | 0.123 | 3.12 | 0.054 | 1.37 |
| | 0.75 | 19.1 | 60 | 1,520 | 3,000 | $O_2$ | 120 | 830 | 0.058 | 1.47 | 0.136 | 3.45 | 0.039 | 0.99 |

*TEM 0.1. ‡Smoother cut.

**TABLE 12.7-7—Continued**

| MATERIAL | THICKNESS | | CUTTING RATE | | POWER | ASSISTING GAS | | | KERF WIDTH | | | | HEAT-AFFECTED ZONE | |
|---|---|---|---|---|---|---|---|---|---|---|---|---|---|---|
| | | | | | | Type | Pressure | | Inlet | | Outlet | | | |
| | in | mm | in/min | mm/min | watts | | psi | kPa | in | mm | in | mm | in | mm |
| Ti-6Al-4V or Ti-6Al-6V-2Sn (cont.) | 0.75 | 19.1 | 80 | 2,030 | 6,000 | $O_2$ | 140 | 970 | 0.094 | 2.39 | 0.162 | 4.11 | 0.017 | 0.43 |
| | 0.75 | 19.1 | 15 | 380 | 10,000 | He | | | 0.067 | 1.70 | 0.036 | 0.91 | 0.105 | 2.67 |
| | 0.75 | 19.1 | 100 | 2,540 | 10,000 | $O_2$ | | | | | | | | |
| | 0.80 | 20.3 | 100 | 2,540 | 11,000 | $O_2$ | | | | | | | | |
| | 1.00 | 25.4 | 20 | 510 | 230* | $O_2$ | | | | | 0.137 | 3.48 | | |
| | 1.00 | 25.4 | 60 | 1,520 | 3,000 | $O_2$ | 150 | 1,030 | 0.085 | 2.16 | 0.190 | 4.83 | 0.046 | 1.17 |
| | 1.00 | 25.4 | 15 | 380 | 15,000 | He | | | 0.093 | 2.36 | 0.046‡ | 1.17‡ | 0.09 | 2.29 |
| | 1.25 | 31.8 | 50 | 1,270 | 3,000 | $O_2$ | | | | | | | | |
| | 2.00 | 50.8 | 20 | 510 | 3,000 | $O_2$ | | | | | | | | |
| | 2.00 | 50.8 | 40 | 1,020 | 6,000 | $O_2$ | 200 | 1,380 | 0.118 | 3.00 | 0.5‡ | 12.7‡ | | |
| | 2.00 | 50.8 | 5 | 130 | 15,000 | He | | | | | | | 0.165 | 4.19 |
| Titanium honeycomb | 0.25 | 6.4 | 100 | 2,540 | 3,000 | $O_2$ | | | | | | | | |
| **NICKEL ALLOYS** | | | | | | | | | | | | | | |
| Haynes alloy 188 | 0.02 | 0.5 | 90 | 2,290 | 250 | $O_2$ | 50 | 340 | | | | | | |
| | 0.063 | 1.6 | 10 | 250 | 250 | $O_2$ | 50 | 340 | | | | | | |
| TD-Ni Cr | 0.015 | 0.4 | 20 | 510 | 250 | $O_2$ | 70 | 480 | | | | | | |
| | 0.060 | 1.5 | 10 | 250 | 250 | $O_2$ | 60 | 410 | | | | | | |
| MN-nickel alloy | 0.003 | 0.1 | 67 | 1,700 | 500 | $O_2$ | | | | | | | | |
| Inconel alloy 718 | 0.500 | 12.7 | 50 | 1,270 | 11,000 | $O_2$ | | | | | | | | |
| Monel mesh | 2.000 | 50.8 | 4 | 100 | 500 | $O_2$ | | | | | | | | |
| Nickel-tungsten | 0.002 | 0.05 | 31 | 790 | 500 | $O_2$ | | | | | | | | |
| Waspaloy | 0.03 | 0.8 | 400 | 10,200 | 750 | $O_2$ | 100 | 690 | 0.01 | 0.25 | 0.01 | 0.25 | 0.005 | 0.13 |
| | 0.03 | 0.8 | 400 | 10,200 | 1,500 | $O_2$ | 100 | 690 | 0.01 | 0.25 | 0.008 | 0.20 | 0.004 | 0.10 |
| | 0.03 | 0.8 | 600 | 15,200 | 3,000 | $O_2$ | 100 | 690 | | | | | 0.002 | 0.05 |
| | 0.03 | 0.8 | 720 | 18,300 | 6,000 | $O_2$ | 175 | 1,210 | 0.022 | 0.56 | 0.018 | 0.46 | 0.002 | 0.05 |
| | 0.03 | 0.8 | 140 | 3,560 | 750 | $CO_2$ | 100 | 690 | 0.012 | 0.30 | 0.008 | 0.20 | 0.001 | 0.03 |
| | 0.03 | 0.8 | 280 | 7,110 | 1,500 | $CO_2$ | 100 | 690 | 0.007 | 0.20 | 0.006 | 0.15 | 0.001 | 0.03 |
| | 0.03 | 0.8 | 500 | 12,700 | 3,000 | $CO_2$ | 200 | 1,380 | 0.014 | 0.36 | 0.013 | 0.33 | 0.002 | 0.05 |
| | 0.03 | 0.8 | 650 | 16,500 | 6,000 | $CO_2$ | 250 | 1,720 | 0.016 | 0.41 | 0.015 | 0.38 | $<0.001$ | $<0.03$ |
| | 0.06 | 1.5 | 100 | 2,540 | 750 | $O_2$ | 100 | 690 | 0.011 | 0.28 | 0.011 | 0.28 | 0.005 | 0.13 |
| | 0.06 | 1.5 | 240 | 6,100 | 1,500 | $O_2$ | 200 | 1,380 | 0.010 | 0.25 | 0.010 | 0.25 | 0.003 | 0.08 |
| | 0.06 | 1.5 | 350 | 8,900 | 3,000 | $O_2$ | 100 | 690 | 0.018 | 0.46 | 0.018 | 0.46 | 0.004 | 0.10 |
| | 0.06 | 1.5 | 450 | 11,400 | 6,000 | $O_2$ | 150 | 1,030 | 0.017 | 0.43 | 0.030 | 0.76 | 0.045 | 1.14 |

*TEM 0.1. ‡Smoother cut.

TABLE 12.7-7—Continued

| MATERIAL | THICKNESS | | CUTTING RATE | | POWER | ASSISTING GAS | | | KERF WIDTH | | | | HEAT-AFFECTED ZONE | |
|---|---|---|---|---|---|---|---|---|---|---|---|---|---|---|
| | | | | | | Type | Pressure | | Inlet | | Outlet | | | |
| | in | mm | in/min | mm/min | watts | | psi | kPa | in | mm | in | mm | in | mm |
| Waspaloy (cont.) | 0.06 | 1.5 | 200 | 5,080 | 3,000 | $CO_2$ | 200 | 1,380 | 0.015 | 0.38 | 0.005 | 0.13 | 0.002 | 0.05 |
| | 0.06 | 1.5 | 350 | 8,900 | 6,000 | $CO_2$ | 200 | 1,380 | 0.023 | 0.58 | 0.022 | 0.56 | 0.003 | 0.08 |
| | 0.125 | 3.2 | 50 | 1,270 | 750 | $O_2$ | 200 | 1,380 | 0.011 | 0.28 | 0.013 | 0.33 | 0.004 | 0.10 |
| | 0.125 | 3.2 | 80 | 2,030 | 1,500 | $O_2$ | 100 | 690 | 0.001 | 0.03 | 0.012 | 0.30 | 0.006 | 0.15 |
| | 0.125 | 3.2 | 140 | 3,560 | 3,000 | $O_2$ | 120 | 830 | 0.019 | 0.48 | 0.023 | 0.58 | 0.006 | 0.15 |
| | 0.125 | 3.2 | 270 | 6,860 | 6,000 | $O_2$ | 150 | 1,030 | 0.017 | 0.43 | 0.030 | 0.76 | 0.003 | 0.08 |
| | 0.125 | 3.2 | 140 | 3,560 | 6,000 | $CO_2$ | 300 | 2,070 | 0.020 | 0.51 | 0.019 | 0.48 | 0.003 | 0.08 |
| Nickel | 0.005 | 0.1 | 156 | 3,960 | 275 | $O_2$ | | | | | | | | |
| Rene 41 | 0.02 | 0.5 | 80 | 2,030 | 250 | $O_2$ | 40 | 280 | | | | | | |
| | 0.05 | 1.3 | 20 | 510 | 250 | $O_2$ | 40 | 280 | | | | | | |
| | 0.063 | 1.6 | 15 | 380 | 250 | $O_2$ | 40 | 280 | | | | | | |
| Hastelloy alloy X | 0.02 | 0.5 | 70 | 1,780 | 250 | $O_2$ | 40 | 280 | | | | | | |
| | 0.02 | 0.5 | 52 | 1,320 | 165 | $O_2$ | | | | | | | | |
| | 0.05 | 1.3 | 30 | 760 | 250 | $O_2$ | 40 | 280 | | | | | | |
| **NONMETALS** | | | | | | | | | | | | | | |
| Alumina | 0.024 | 0.6 | 90 | 2,290 | 500 | $O_2$ | | | | | | | | |
| | 0.025 | 0.6 | 50 | 1,270 | 250 | $N_2$ | | | 0.012 | 0.3 | | | | |
| | 0.039 | 1.0 | 118 | 3,000 | 230* | $O_2$ | | | | | 0.004 | 0.1 | | |
| Asbestos cement | 0.150 | 3.8 | 0.6 | 15 | 200 | air | | | | | | | | |
| Carborundum | 0.040 | 1.0 | 19 | 480 | 200 | air | | | | | | | | |
| Ceramic | 0.025 | 0.6 | 500 | 12,700 | 1,000 | $O_2$ | | | | | | | | |
| Ceramic tile | 0.250 | 6.4 | 19 | 480 | 850 | $O_2$ | | | | | | | | |
| Glass | 0.010 | 0.3 | 42 | 1,070 | 400 | $O_2$ | | | | | | | | |
| Quartz | 0.125 | 3.2 | 29 | 740 | 500 | $O_2$ | | | | | | | | |
| Quartz glass | 0.075 | 1.9 | 24 | 610 | 230* | $O_2$ | | | | | 0.008 | 0.2 | | |
| Silica | 0.040 | 1.0 | 94 | 2,390 | 500 | $O_2$ | | | | | | | | |
| Soda glass | 0.005 | 0.1 | 125.0 | 3,180 | 200 | air | | | | | | | | |
| | 0.040 | 1.0 | 9.5 | 240 | 200 | air | | | | | | | | |
| | 0.100 | 2.5 | 2.5 | 64 | 200 | air | | | | | | | | |
| Tungsten carbide | 0.071 | 1.8 | 1.8 | 46 | 500 | $O_2$ | | | | | | | | |
| | 0.189 | 4.8 | 1.8 | 46 | 500 | $O_2$ | | | | | | | | |

*TEM 0.1.

TABLE 12.7-7—Continued

| MATERIAL | THICKNESS | | CUTTING RATE | | POWER | ASSISTING GAS | | | KERF WIDTH | | | | HEAT-AFFECTED ZONE | |
|---|---|---|---|---|---|---|---|---|---|---|---|---|---|---|
| | | | | | | Type | Pressure | | Inlet | | Outlet | | | |
| | in | mm | in/min | mm/min | watts | | psi | kPa | in | mm | in | mm | in | mm |
| **ORGANICS, PLASTICS, ETC.** | | | | | | | | | | | | | | |
| ABS Plastic | 0.035 | 0.9 | 324 | 8,230 | 250 | air | 60 | 413 | | | 0.025[§] | 0.64[§] | | |
| | 0.042 | 1.1 | 258 | 6,550 | 250 | air | 60 | 413 | | | 0.025 | 0.64 | | |
| | 0.075 | 1.9 | 90 | 2,290 | 250 | air | 60 | 413 | | | 0.030 | 0.76 | | |
| | 0.095 | 2.4 | 120 | 3,050 | 250 | air | 60 | 413 | | | [§] | [§] | | |
| | 0.100 | 2.5 | 150 | 3,810 | 240 | $N_2$ | | | 0.030 | 0.76 | | | | |
| (extruded) | 0.100 | 2.5 | 120 | 3,050 | 250 | air | 60 | 413 | | | 0.025 | 0.64 | | |
| | 0.160 | 4.1 | 160 | 4,060 | 240 | $O_2$ | | | | | | | | |
| Carpet, polyester | 0.250 | 6.4 | 120 | 3,050 | 200 | A | | | 0.020 | 0.51 | | | | |
| | 0.375 | 9.5 | 120 | 3,050 | 200 | A | | | 0.020 | 0.51 | | | | |
| | 0.390 | 9.9 | 102 | 2,590 | 230[§] | $N_2$ | | | | | 0.020 | 0.51 | | |
| Carpet, needle cord | 0.220 | 5.6 | 320 | 8,130 | 230* | $N_2$ | | | | | 0.012 | 0.30 | | |
| Cotton fabric, multilayer | 0.710 | 18.0 | 8 | 200 | 230* | $N_2$ | | | | | 0.027 | 0.69 | | |
| Delrin | 0.026 | 0.7 | 588 | 14,900 | 380 | $O_2$ | | | | | | | | |
| Fiberglass reinforced plastic | 0.130 | 3.3 | 24 | 610 | 230* | $N_2$ | | | | | 0.012 | 0.30 | | |
| Hardboard | 0.125 | 3.2 | 270 | 6,860 | 650 | $O_2$ | | | | | | | | |
| | 0.200 | 5.1 | 180 | 4,570 | 650 | $O_2$ | | | | | | | | |
| Leather | 0.080 | 2.0 | 16 | 410 | 200 | air | | | | | | | | |
| Lexan | 0.030 | 0.8 | 498 | 12,650 | 380 | $O_2$ | | | | | | | | |
| Mylar film | 0.001 | 0.03 | 48,000 | 1.22 km | 300 | $O_2$ | | | | | | | | |
| Nylon | 0.020 | 0.5 | 125 | 3,180 | 200 | air | | | | | | | | |
| PVC Sheet | 0.0012 | 0.03 | 3,000 | 76,000 | 300 | $O_2$ | | | | | | | | |
| | 0.012 | 0.3 | 118 | 3,000 | 100 | $O_2$ | | | | | | | | |
| | 0.280 | 7.0 | 47 | 1,200 | 230 | $N_2$ | | | | | 0.02 | 0.51 | | |
| | 0.393 | 10.0 | 189 | 4,800 | 340 | $O_2$ | | | | | | | | |
| Packageboard (Perspex) | 0.086 | 2.2 | 16 | 410 | 200 | air | | | | | | | | |
| | 0.250 | 6.4 | 5 | 130 | 200 | air | | | | | | | | |
| | 0.625 | 15.8 | 2.5 | 65 | 200 | air | | | | | | | | |
| | 1.250 | 31.2 | 12 | 300 | 650 | $O_2$ | | | | | | | | |
| Perspex, embossed | 0.120 | 3.0 | 177 | 4,500 | 230* | $N_2$ | | | | | 0.015 | 0.38 | | |
| Perspex, transparent | 0.390 | 10.0 | 32 | 810 | 230* | $N_2$ | | | | | 0.027 | 0.69 | | |

*TEM 0.1. [§]Smooth, clean cut.

**TABLE 12.7-7—Continued**

| MATERIAL | THICKNESS | | CUTTING RATE | | POWER | ASSISTING GAS | | | KERF WIDTH | | | | HEAT-AFFECTED ZONE | |
|---|---|---|---|---|---|---|---|---|---|---|---|---|---|---|
| | | | | | | Type | Pressure | | Inlet | | Outlet | | | |
| | in | mm | in/min | mm/min | watts | | psi | kPa | in | mm | in | mm | in | mm |
| Plexiglas | 0.125 | 3.2 | 150 | 3,810 | 250 | $O_2$ | 40 | 275 | | | | | | |
| Polycarbonate | 0.250 | 6.4 | 15 | 380 | 250 | air | 60 | 413 | | | 0.04# | 1.02# | | |
| Polyethylene | 0.060 | 1.5 | 78 | 1,980 | 250 | air | 60 | 413 | | | ** | ** | | |
| | 0.090 | 2.3 | 48 | 1,220 | 250 | air | 60 | 413 | | | ** | ** | | |
| | 0.115 | 2.9 | 39 | 990 | 250 | air | 60 | 413 | | | ** | ** | | |
| Polypropane | 0.500 | 12.7 | 9.6 | 240 | 250 | air | 60 | 413 | | | 0.055†† | 1.4†† | | |
| Polypropylene | 0.012 | 0.3 | 540 | 13,700 | 50 | air | 20 | 138 | | | 0.008†† | 0.20†† | | |
| | 0.220 | 5.6 | 27 | 680 | 230* | $N_2$ | | | | | 0.020 | 0.51 | | |
| | 0.045 | 1.1 | 180 | 4,570 | 250 | air | 60 | 413 | | | †† | †† | | |
| Polystyrene | 0.126 | 3.2 | 165 | 4,200 | 230* | $N_2$ | | | | | 0.015 | 0.38 | | |
| Polyurethane | 0.090 | 2.3 | 120 | 3,050 | 450 | air | 60 | 413 | | | § | § | | |
| Teflon | 0.002 | 0.05 | 600 | 15,200 | 380 | $O_2$ | | | | | | | | |
| Textiles, nylon | 0.004 | 0.1 | 7,900 | 200,000 | 230* | $N_2$ | | | | | 0.004 | 0.10 | | |
| Vinyl fabric | 0.030 | 0.8 | 300 | 7,600 | 250 | | | | | | ‡‡ | ‡‡ | | |
| **WOOD** | | | | | | | | | | | | | | |
| Plywood: | | | | | | | | | | | | | | |
| douglas fir comm. | 0.248 | 6.3 | 64 | 1,630 | 250 | $N_2$ | 50 | 340 | 0.018 | 0.46 | | | | |
| plywood | 0.250 | 6.4 | 180 | 4,570 | 650 | $O_2$ | | | | | | | | |
| plywood | 0.500 | 12.7 | 45 | 1,140 | | | | | | | | | | |
| douglas fir/urea | 0.615 | 15.6 | 26 | 660 | 250 | $N_2$ | 50 | 340 | 0.019 | 0.48 | | | | |
| douglas fir/ phenolic | 0.623 | 15.8 | 6 | 150 | 250 | $N_2$ | 50 | 340 | 0.025 | 0.64 | | | | |
| douglas fir comm. | 0.739 | 18.8 | 13 | 330 | 250 | $N_2$ | 50 | 340 | 0.021 | 0.53 | | | | |
| plywood | 0.750 | 19.1 | 200 | 5,080 | 3,000 | $O_2$ | | | | | | | | |
| Softwood | 0.50 | 12.7 | 60 | 1,520 | 650 | $O_2$ | | | | | | | | |
| Oak/Maple | 0.75 | 19.1 | 200 | 5,080 | 3,000 | $O_2$ | | | | | | | | |
| Spruce | 1.62 | 41.1 | 780 | 19,800 | 3,000 | $O_2$ | | | | | | | | |
| Oak | 0.450 | 11.4 | 5 | 130 | 200 | air | | | | | | | | |
| Teak | 0.625 | 15.9 | 1.8 | 46 | 200 | air | | | | | | | | |
| Deal | 1.250 | 31.8 | 2.5 | 64 | 200 | air | | | | | | | | |
| Soft maple | 0.506 | 12.9 | 36 | 910 | 250 | $N_2$ | 50 | 340 | 0.025 | 0.64 | | | | |
| Douglas fir | 0.856 | 21.7 | 15 | 380 | 250 | $N_2$ | 50 | 340 | 0.022 | 0.56 | | | | |
| White pine | 0.874 | 22.2 | 30 | 760 | 250 | $N_2$ | 50 | 340 | 0.021 | 0.53 | | | | |
| Southern pine | 0.784 | 19.9 | 11 | 280 | 250 | $N_2$ | 50 | 340 | 0.026 | 0.66 | | | | |
| Hard maple | 0.871 | 22.1 | 10 | 250 | 250 | $N_2$ | 50 | 340 | 0.022 | 0.56 | | | | |
| White oak | 0.866 | 22.0 | 10 | 250 | 250 | $N_2$ | 50 | 340 | 0.022 | 0.56 | | | | |

*TEM 0.1. #Some charring. ††Some beading.
§Smooth, clean cut. **Melt on bottom edge. ‡‡Clean cut.

**TABLE 12.7-7—Continued**

| MATERIAL | THICKNESS | | CUTTING RATE | | POWER | ASSISTING GAS | | | KERF WIDTH | | | | HEAT-AFFECTED ZONE | |
|---|---|---|---|---|---|---|---|---|---|---|---|---|---|---|
| | | | | | | Type | Pressure | | Inlet | | Outlet | | | |
| | in | mm | in/min | mm/min | watts | | psi | kPa | in | mm | in | mm | in | mm |
| Hickory | 0.722 | 18.3 | 11 | 280 | 250 | $N_2$ | 50 | 340 | 0.021 | 0.53 | | | | |
| Red oak | 0.989 | 25.1 | 13 | 330 | 250 | $N_2$ | 50 | 340 | 0.030 | 0.76 | | | | |
| Aspen | 0.749 | 19.0 | 33 | 840 | 250 | $N_2$ | 50 | 340 | 0.026 | 0.66 | | | | |
| Particleboard: | | | | | | | | | | | | | | |
| pine/phenolic | 0.522 | 13.3 | 2 | 50 | 250 | $N_2$ | 50 | 340 | 0.028 | 0.71 | | | | |
| pine/urea | 0.514 | 13.1 | 24 | 610 | 250 | $N_2$ | 50 | 340 | 0.019 | 0.48 | | | | |
| Knot: | | | | | | | | | | | | | | |
| southern pine | 1.035 | 26.3 | 3 | 75 | 250 | $N_2$ | 50 | 340 | 0.027 | 0.69 | | | | |
| red oak | 0.830 | 21.1 | 11 | 280 | 250 | $N_2$ | 50 | 340 | 0.018 | 0.46 | | | | |
| **COMPOSITES** | | | | | | | | | | | | | | |
| ABS/Polycarbonate; 50-50 | 0.140 | 3.6 | 40 | 1,020 | 250 | air | 60 | 410 | | | ## | ## | | |
| Boron aluminum[§§] | 0.008 | 0.2 | 30 | 762 | 250 | $O_2$ | 40 | 280 | | | | | | |
| Boron/aluminum tape | 0.010 | 0.3 | 450 | 11,430 | 3,000 | $O_2$ | | | | | | | | |
| | 0.125 | 3.2 | 50 | 1,270 | 1,000 | $O_2$ | | | | | | | | |
| Boron-clay filled | 0.015 | 0.4 | 270 | 6,860 | 250 | air | low flow | low flow | | | § | § | | |
| Boron/epoxy (cured) | 0.120 | 3.0 | 60 | 1,520 | 3,000 | $O_2$ | | | | | | | | |
| | 0.320 | 8.1 | 65 | 1,650 | 15,000 | $O_2$ | | | | | | | | |
| | 0.750 | 19.1 | 12 | 305 | 4,000 | $O_2$ | | | | | | | | |
| Boron fiber composite | 0.050 | 1.3 | 30 | 762 | 260 | $N_2$ | | | 0.03 | 0.76 | | | | |
| | 0.050 | 1.3 | 9 | 229 | 165 | A | | | 0.03 | 0.76 | | | | |
| Boron tape | 0.004 | 0.1 | 300 | 7,620 | 250 | $O_2$ | 40 | 280 | | | | | | |
| Fiberglass epoxy | 0.040 | 1.0 | 101 | 2,570 | 380 | $O_2$ | | | | | | | | |
| | 0.500 | 12.7 | 180 | 4,570 | 20,000 | $O_2$ | | | | | | | | |
| Glass-reinforced nylon | 0.028 | 0.7 | 347 | 8,810 | 380 | $O_2$ | | | | | | | | |
| Graphite epoxy | 0.102 | 2.6 | 75 | 1,910 | 1,000 | $O_2$ | | | | | | | | |
| Polyethylene-glass fiber | 0.125 | 3.2 | 3.6 | 91 | 250 | air | 60 | 410 | | | ## | ## | | |
| | 0.160 | 4.1 | 2.4 | 61 | 250 | air | 60 | 410 | | | ## | ## | | |
| | 0.190 | 4.8 | 1.8 | 46 | 250 | air | 60 | 410 | | | ## | ## | | |
| Polyester-glass fiber | 0.110 | 2.8 | 2.4 | 61 | 250 | air | 60 | 410 | | | *** | *** | | |

§Smooth, clean cut.
§§Uncured tape.
##Some edge discoloration.
***Small amount charring.

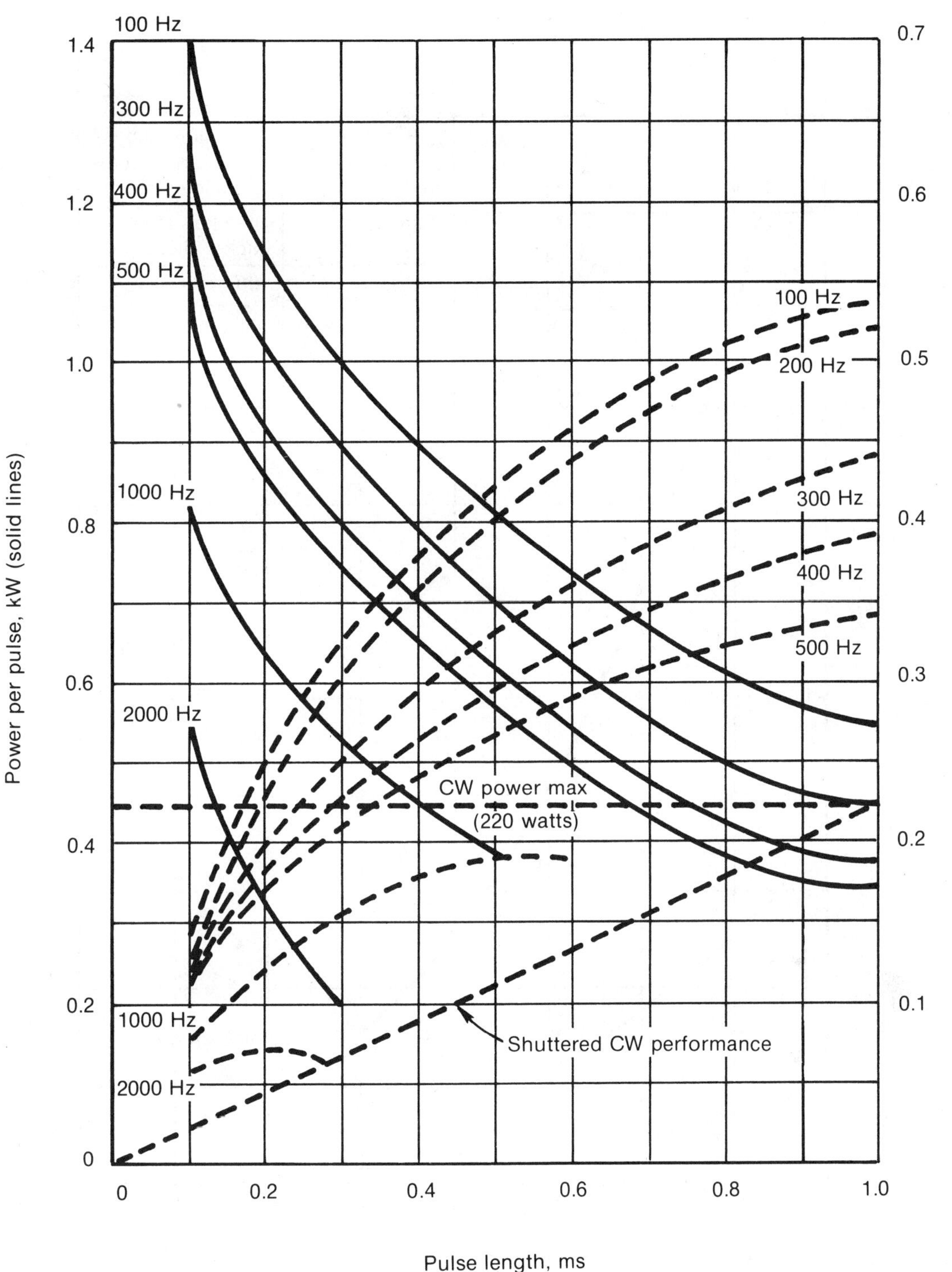


**Figure 12.7–5** Pulse power capabilities of 220 watt continuous-wave (CW) $CO_2$ laser. (Courtesy of Coherent, Inc.)

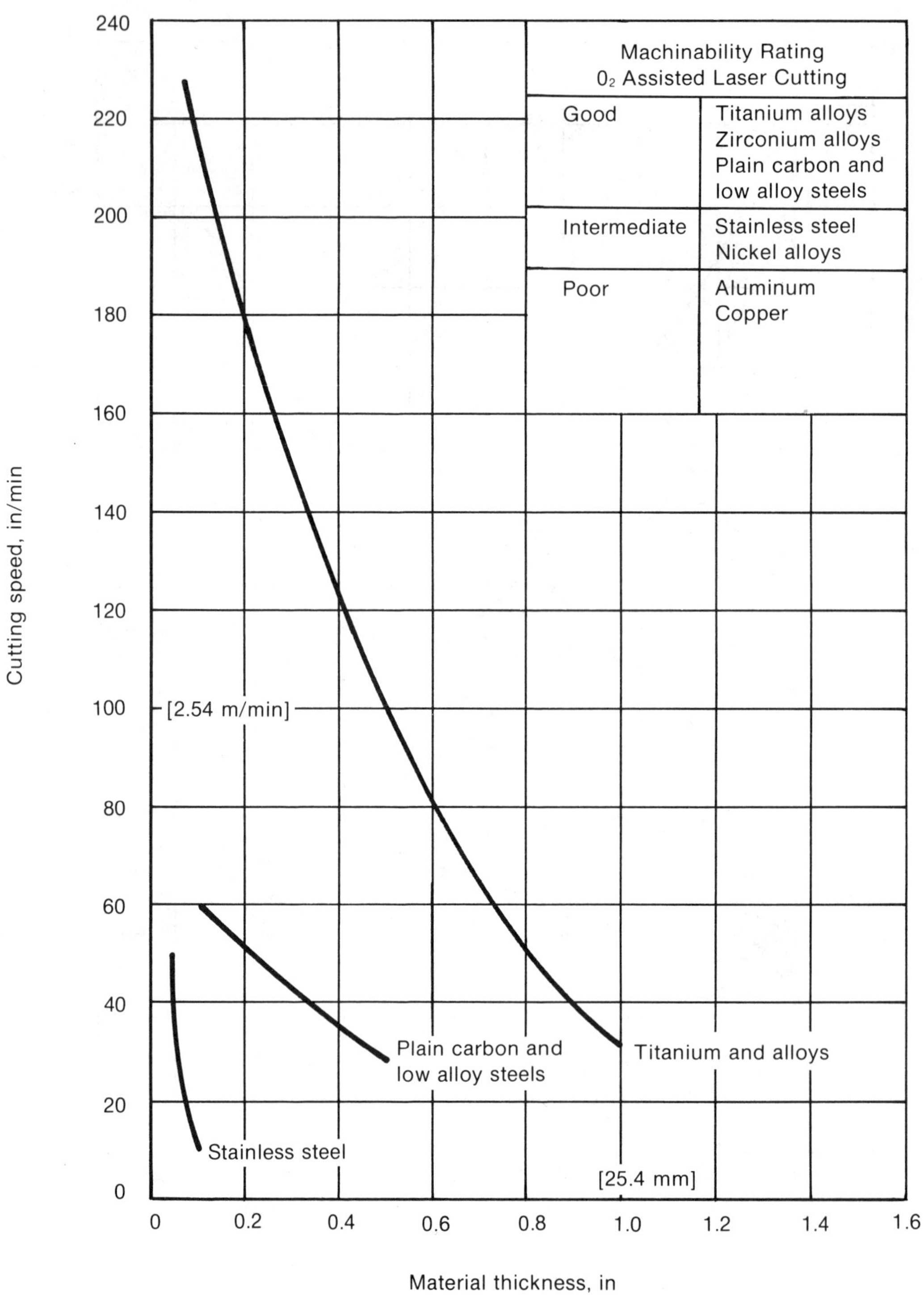


**Figure 12.7-6** Cutting rates for 325 watt $CO_2$ laser with $O_2$ assist in stainless steel, carbon and low alloy steel, and titanium alloys. (Courtesy of Coherent, Inc.)

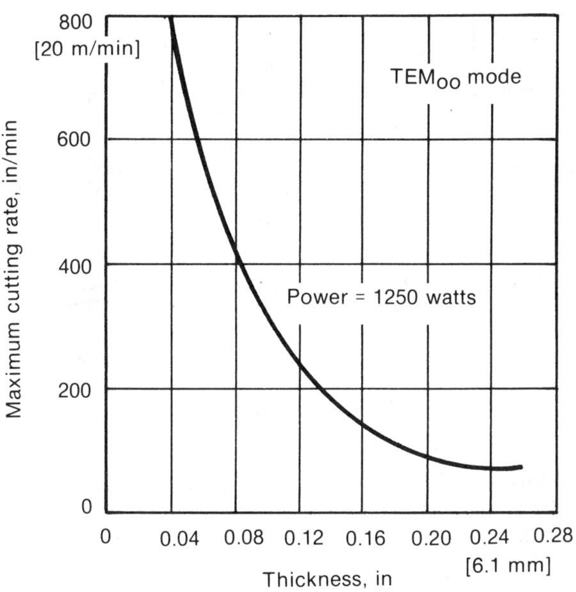


**Figure 12.7–7** Maximum material removal rate versus workpiece thickness for $CO_2$ laser cutting 302 stainless steel at 1250 watts with $O_2$ assist. Rate inversely proportional to thickness. (Courtesy of GTE-Sylvania)

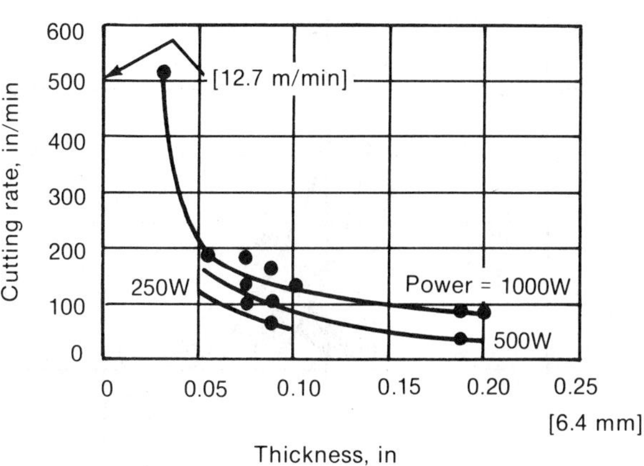


**Figure 12.7–9** LBT cutting of low carbon steel using a $CO_2$ laser with $O_2$ assist. (W. Hanson and P. Lawryk, Laser machining is now, *Modern Machine Shop* 50 (June 1977), p. 97)

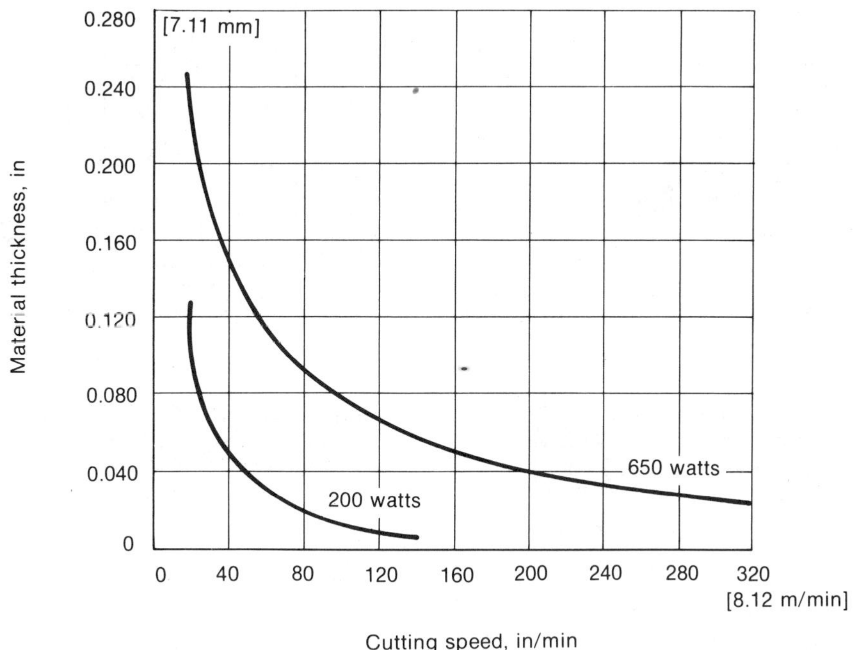


**Figure 12.7–8** Maximum material removal rate versus thickness for CW Nd-YAG laser cutting carbon steel with $O_2$ assist. (Courtesy of Holobeam Laser Inc.)

# 12.7 Laser Beam Torch—LBT

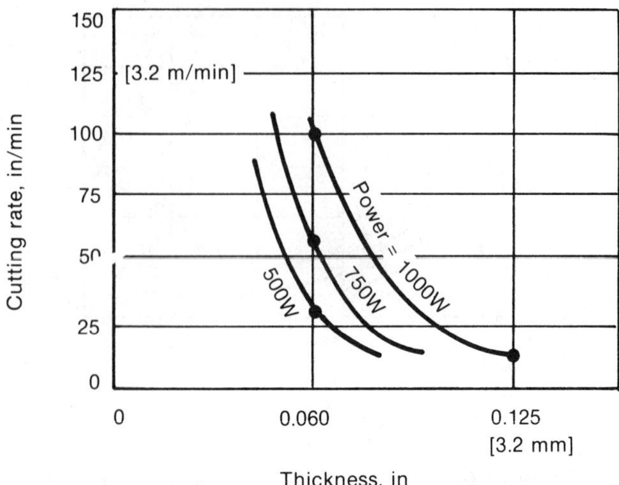


**Figure 12.7–10** LBT cutting of titanium using a $CO_2$ laser with argon gas assist. (W. Hanson and P. Lawryk, p. 97)

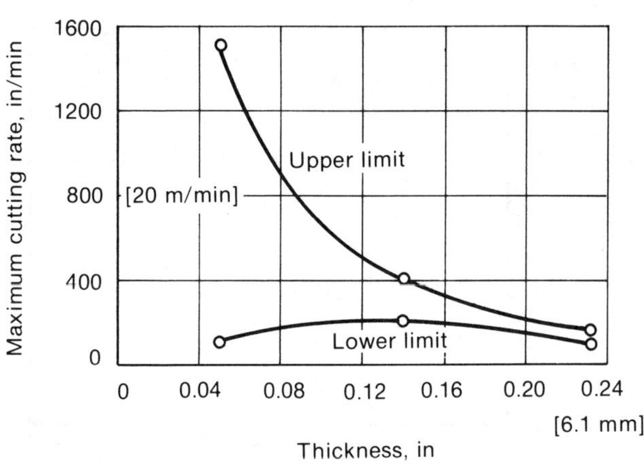


**Figure 12.7–12** $CO_2$ laser cutting rate for Ti-6Al-4V alloy. $TEM_{00}$ mode cutting at 1000 watts with no assisting gas. (Courtesy of GTE-Sylvania)

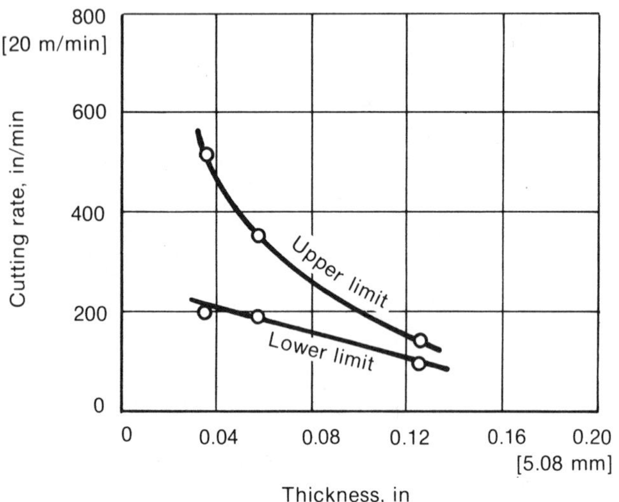


**Figure 12.7–11** $CO_2$ laser cutting rate for 1018 low carbon steel. $TEM_{00}$ mode cutting at 1000 watts with no assisting gas. (Courtesy of GTE-Sylvania)

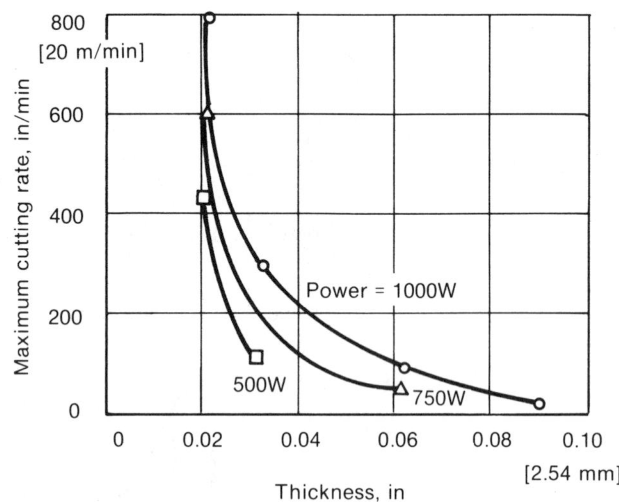


**Figure 12.7–13** $CO_2$ laser cutting rate for 6061 Aluminum alloy. $TEM_{00}$ mode cutting with $O_2$ assist. (Courtesy of GTE-Sylvania)

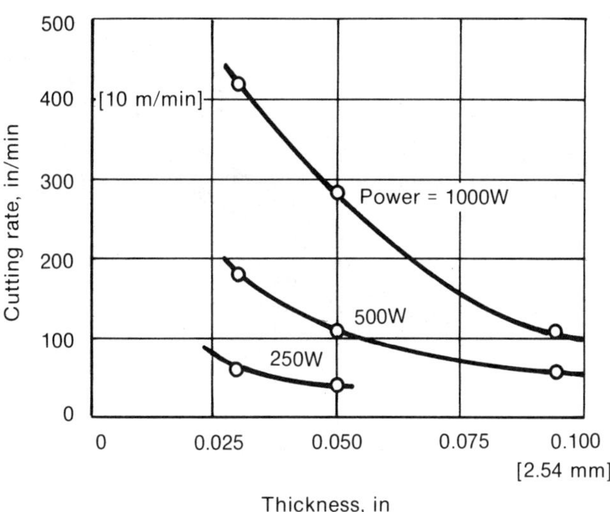


**Figure 12.7–14** $CO_2$ laser cutting rate for Hastelloy alloy X. $TEM_{00}$ mode cutting with $O_2$ assist. (Courtesy of GTE-Sylvania)

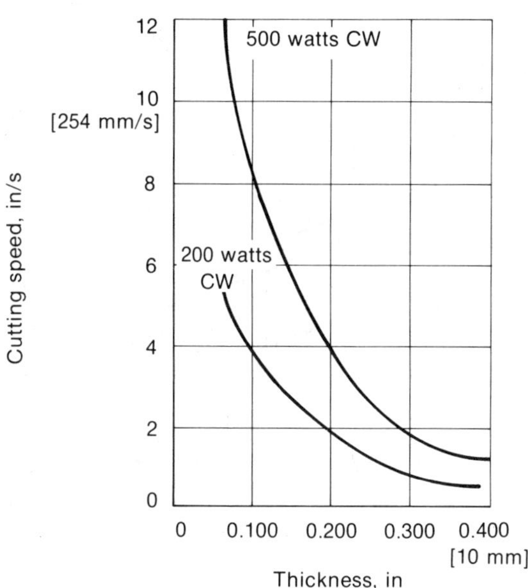


**Figure 12.7–16** Acrylic cutting rates with $CO_2$ laser with air assist. (R. J. Saunders and M. Pasturel, $CO_2$ laser beam machining of plastic and glass, Technical paper MR74-958, Society of Manufacturing Engineers, Dearborn, MI 1974, p. 8)

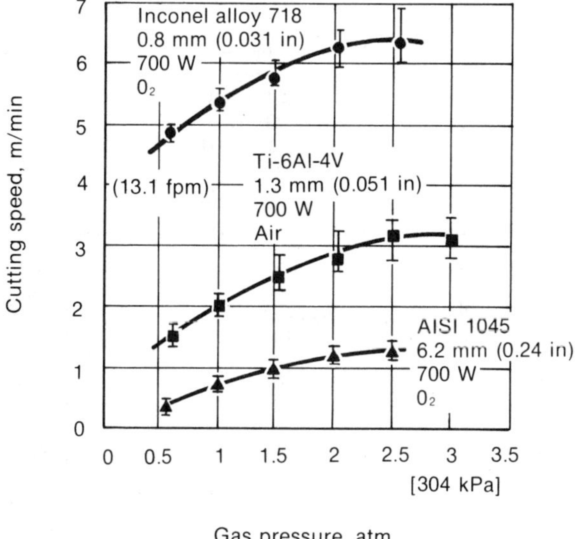


**Figure 12.7–15** Cutting speed as a function of gas pressure in cutting Inconel alloy 718, Ti-6Al-4V alloy and AISI 1045 steel. (A. Schachrai and M. Castellani-Longo, p. 466)

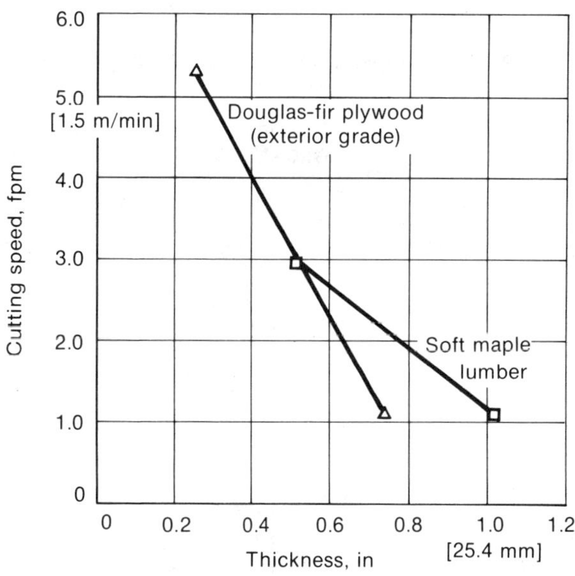


**Figure 12.7–17** LBT cutting of plywood and lumber with 250 watt $CO_2$ laser with 50 psi $N_2$ gas assist. (C. C. Peters and H. L. Marshall, Cutting wood materials by laser, FPL report 250, Forest Products Laboratory, 1975)

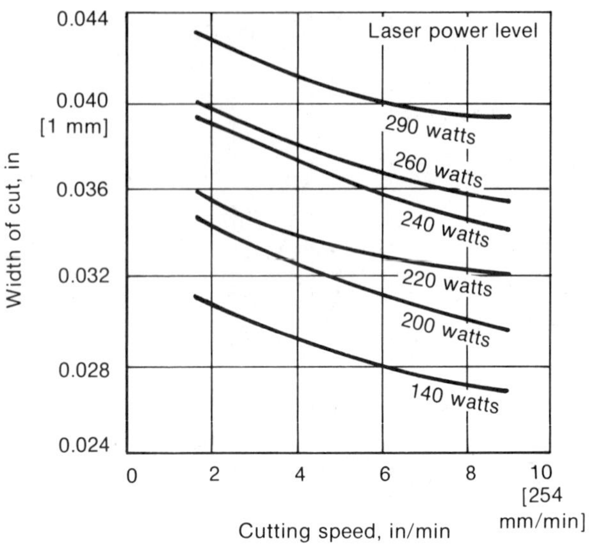


**Figure 12.7–18** LBT cutting of parallel-sided slots in 3/4-inch thick plywood for steel rule die manufacture with $CO_2$ laser using inert gas assist. (Courtesy of Coherent, Inc.)

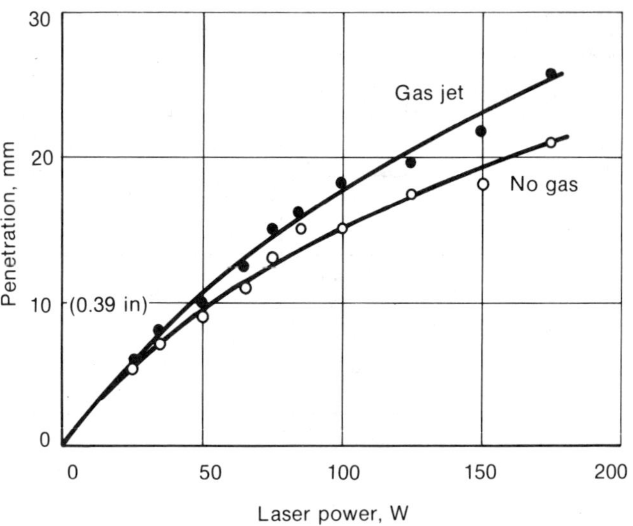


**Figure 12.7–19** Penetration of $CO_2$ laser in boxboard (Perspex) at 200 mm (7.87 in) focal length and 130 mm/min (5.1 in/min) cutting speed. (Courtesy of F. W. Lunau, BOC-Murex Welding Research and Development Laboratories, London)

**TABLE 12.7–8 Time Requirements for Drilling 0.004- to 0.005-inch [0.1 to 0.13 mm] Diameter Holes with Nd-YAG Laser**

| WORK MATERIAL | WORK THICKNESS | | DRILL TIME REQUIREMENT min | | | LAMP CURRENT USED | AVERAGE LASER POWER OF 1.06μm WAVELENGTH | MAXIMUM THICKNESS DRILLED | |
|---|---|---|---|---|---|---|---|---|---|
| | in | mm | Oxygen assist | Air | Argon assist | A | W | in | mm |
| 304 Stainless steel | 0.120 | 3.05 | 1.46 | 1.33* | 3.69 | 34 | 31 | 0.190 | 4.83 |
| Beryllium | 0.049 | 1.24 | 4.63 | 3.41 | 0.56* | 32 | 24 | 0.198 | 5.03 |
| 3003 Aluminum alloy | 0.065 | 1.65 | very long | 0.99 | 0.28* | 34 | 31 | 0.122 | 3.10 |
| Tungsten | 0.033 | 0.84 | 0.32 | 0.27* | 1.34 | 34 | 31 | 0.100 | 2.54 |
| Tantalum | 0.051 | 1.30 | 0.44 | 0.35* | 1.88 | 35 | 42 | 0.105 | 2.67 |
| Copper | 0.100 | 2.54 | 1.63 | 1.01* | very long | 34 | 31 | 0.125 | 3.18 |
| AZ 31 Magnesium alloy | 0.123 | 3.12 | very long | 12.50 | 0.36* | 35 | 42 | 0.123 | 3.12 |
| Uranium | 0.053 | 1.35 | 0.20* | 0.22 | 0.43 | 35 | 42 | 0.125 | 3.18 |

SOURCE: C. A. Pippin, Metal machining with neodymium laser, *Manufacturing Engineering* 76 (January 1976), p. 39.
*Optimum gas for drilling.

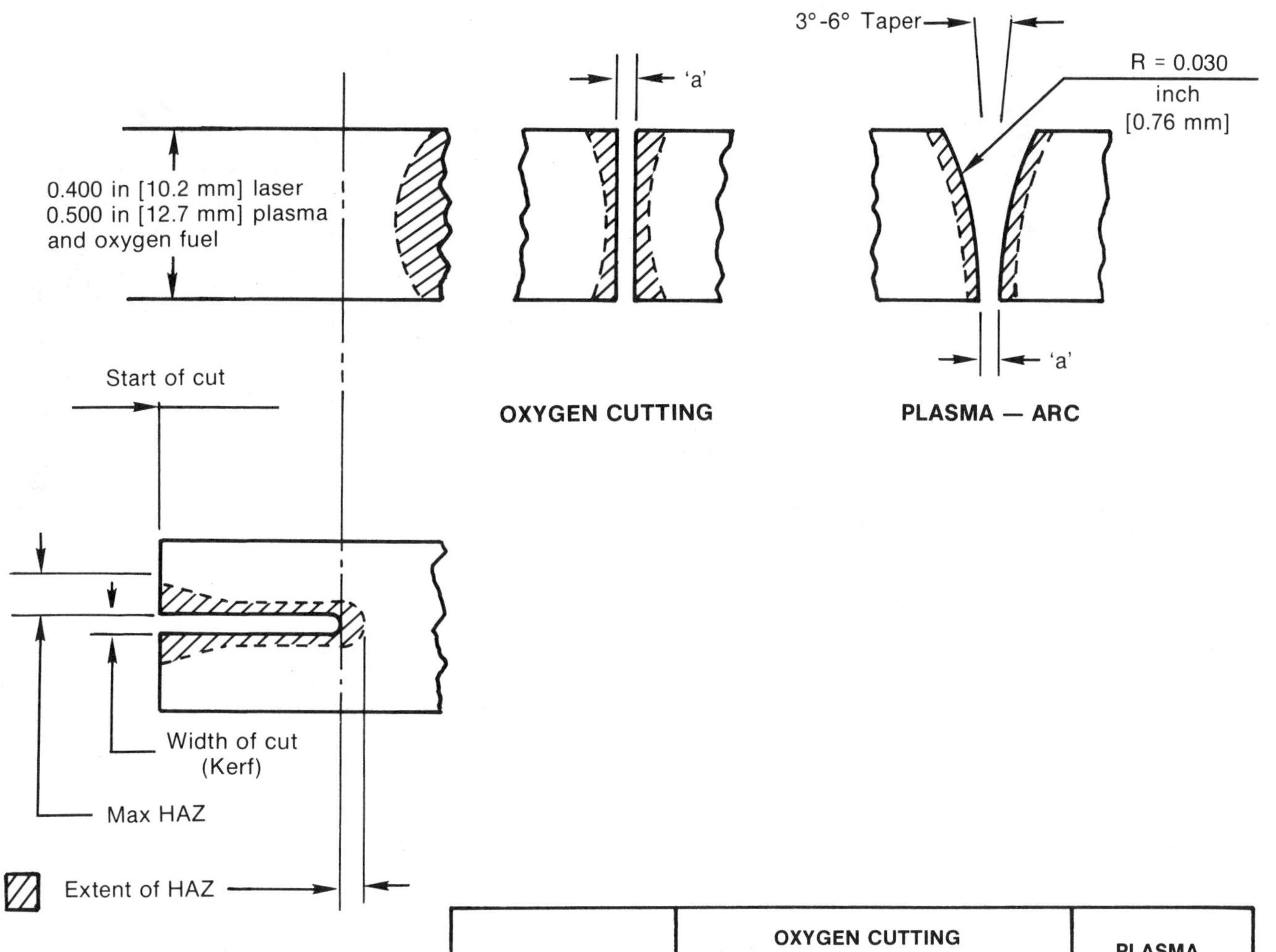


| | | OXYGEN CUTTING | | PLASMA ARC |
|---|---|---|---|---|
| | | Laser | Fuel Gas | |
| HAZ | in | 0.010 to 0.030 | 0.250 | 0.100 |
| | mm | 0.25 to 0.76 | 6.4 | 2.5 |
| Kerf 'a' | in | 0.035 to 0.055 | 0.080 | 0.312 |
| | mm | 0.89 to 1.4 | 2.0 | 7.9 |
| EMA | in | 0.050 | 0.375 | 0.200 |
| | mm | 1.27 | 9.5 | 5.1 |

HAZ = Heat Affected Zone
EMA = Edge Machining Allowance

**Figure 12.7–20** Comparison of kerfs produced by torch cutting techniques. (Courtesy of Coherent, Inc.)

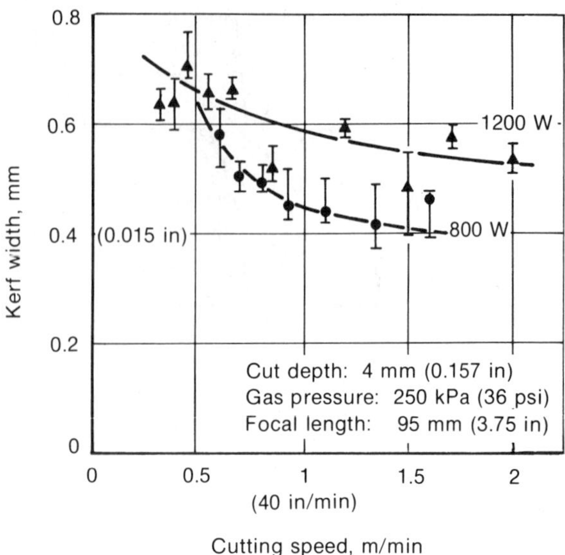


**Figure 12.7-21** Kerf width as a function of cutting speed in cutting AISI 316L stainless steel with $CO_2$ laser torch with $O_2$ assist. (A. Schachrai and M. Castellani-Longo, p. 466)

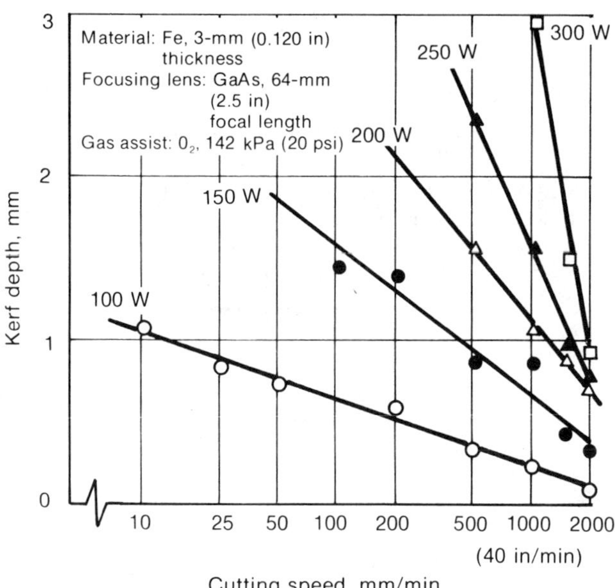


**Figure 12.7-23** Relationship between cutting speed and kerf depth with $CO_2$ laser. (A. Kobayashi, S. Shimakawa and Y. Nagano, Cutting of metals with a 300W $CO_2$ laser, *Annals of the CIRP*, Vol. 24/1, 1975, p. 122)

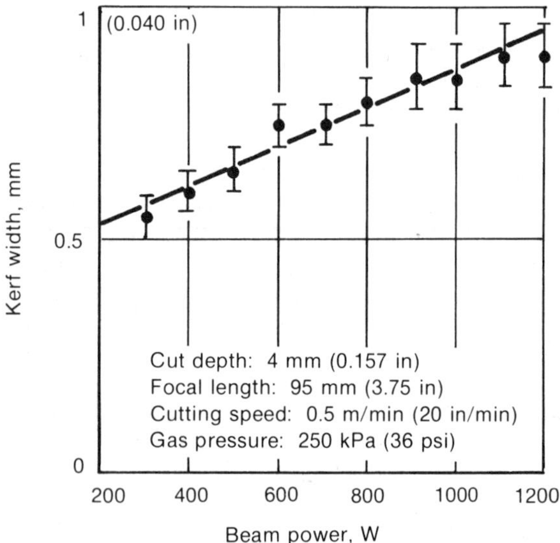


**Figure 12.7-22** Kerf width as a function of beam power in cutting AISI 1045 steel with $CO_2$ laser torch with $O_2$ assist. (A. Schachrai and M. Castellani-Longo, p. 466)

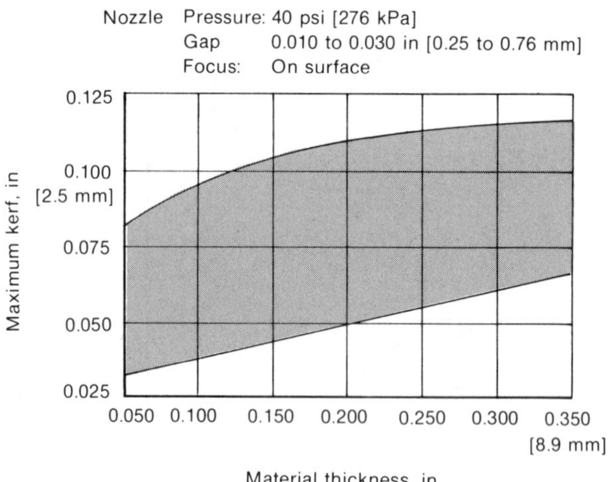


**Figure 12.7-24** Maximum kerf width versus material thickness when cutting a variety of metals with 250 watt $CO_2$ laser. (S. Trink, Laser cutting of aerospace materials, Technical paper EM73-214, Society of Manufacturing Engineers, Dearborn, MI, 1973, p. 18)

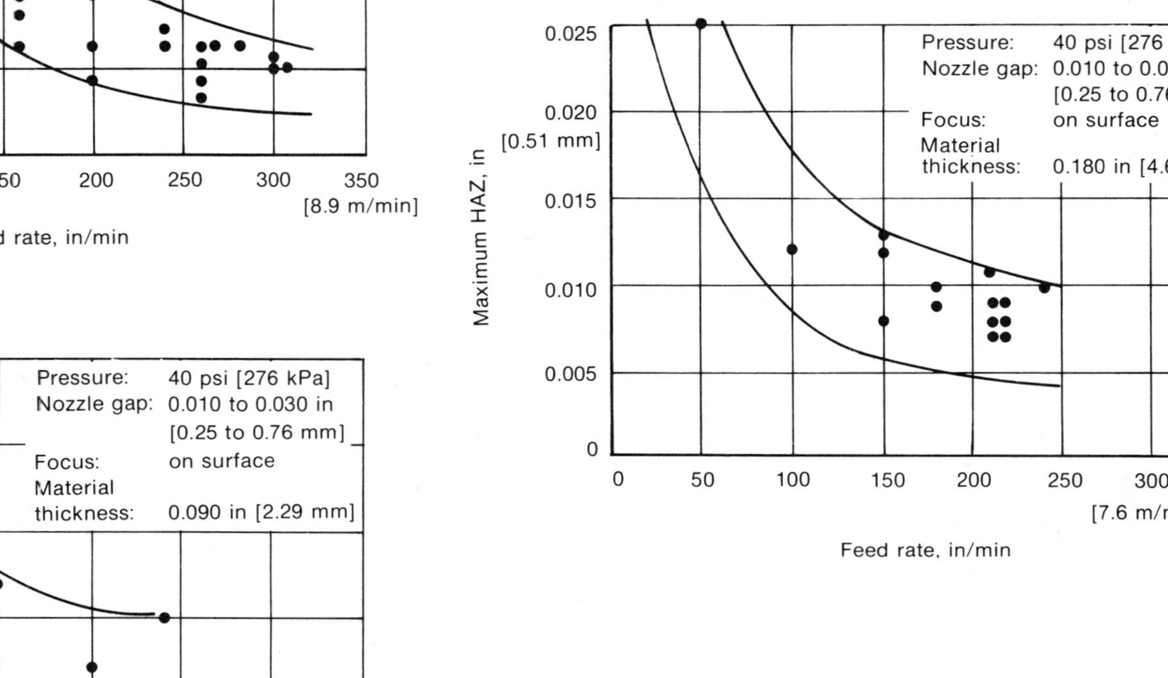


**Figure 12.7–25** Maximum heat-affected zone (HAZ) versus feed rate when cutting three different thicknesses of Ti-6Al-6V-2Sn alloy with 250 watt $CO_2$ laser with $O_2$ gas assist. (S. Trink, p. 18–19)

# 12.7 Laser Beam Torch—LBT

## SPECIFIC APPLICATIONS DATA

**TABLE 12.7–9 Specific Applications Data for Laser Beam Torch**

| Work material: | **Polyethylene Tubing (bi-wall)** | **Glass Contact Lens** | **Ti-6Al-6V-2Sn** |
|---|---|---|---|
| Workpiece configuration: | 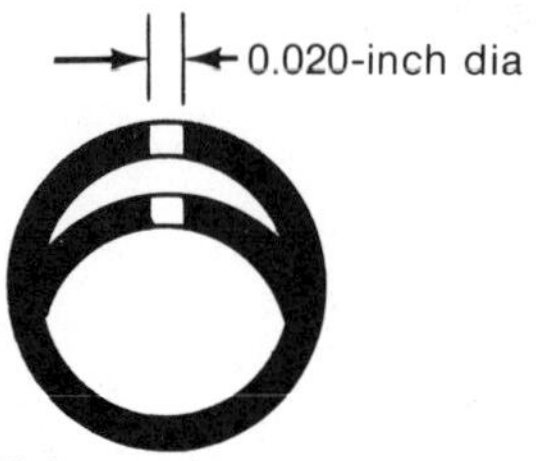 0.020-inch dia<br>Hole spacing on 6-to 24-inch centers | 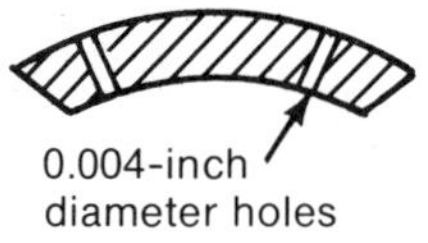 0.004-inch diameter holes | 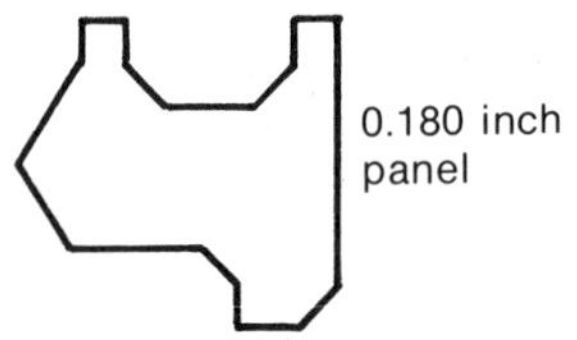 0.180 inch panel |
| Laser equipment type: | $CO_2$ | $CO_2$ | $CO_2$ |
| wavelength, $\mu$m: | 10.6 | 10.6 | 10.6 |
| power: | 250 W | 50 W | 250 W |
| gas: | | | $O_2$ |
| gas pressure, psi: | | | 40 |
| Operating conditions power level: | | | 250 W |
| beam mode: | CW | CW | CW |
| pulse length, ms: | 5 to 15 | | |
| cycle time: | | 1.5 s/lens | |
| feed rate, in/min: | | | 140 |
| Tolerance, in: | | | |
| Remarks: | Drill during extrusion using optical tracking. Extrusion up to 300 fpm.<br>No deburring.<br>No residue.<br>Wall thicknesses range from 0.010 to 0.040 inch. Pulse length depends on hole in outer wall; some holes through both walls. | 5 to 15 holes.<br>NC controlled. | Good saving by nesting parts. |

**TABLE 12.7-9—Continued**

| Work material: | **Aluminum Honeycomb** | **Fired Ceramic Alumina** | **Stainless Steel Heat Exchanger Component** |
|---|---|---|---|
| Workpiece configuration: | 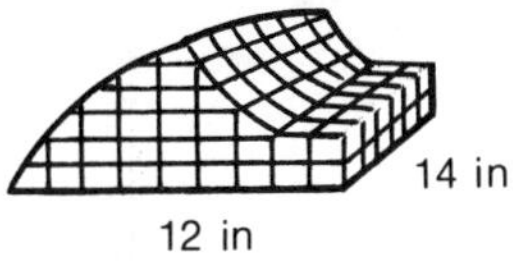 12 in; 14 in | 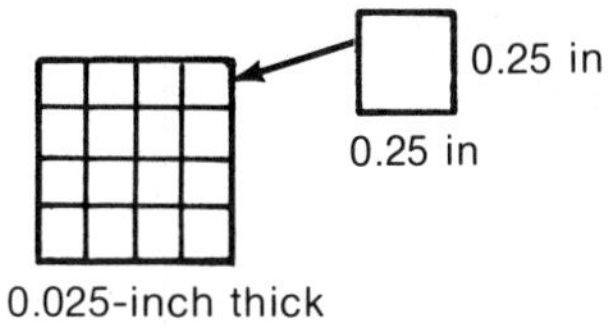 0.25 in; 0.25 in; 0.025-inch thick | 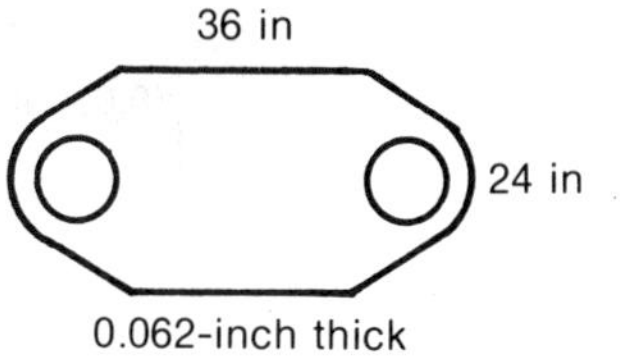 36 in; 24 in; 0.062-inch thick |
| Laser equipment type: | $CO_2$ | $CO_2$ | $CO_2$ |
| wavelength, $\mu$m: | 10.6 | 10.6 | 10.6 |
| power: | 1,000 W | 50 W | 2,500 W |
| gas: | $O_2$ | $N_2$ | $O_2$ |
| gas pressure, psi: | 40 | | |
| Operating conditions power level: | 1,000 W | 50 W | 1,000 W |
| beam mode: | CW | Pulsed | CW |
| pulse length, ms: | | 0.5 | |
| cycle time: | | | |
| feed rate, in/min: | 90 | 2 to 4 in/s | 450 |
| Tolerance, in: | | No HAZ | |
| Remarks: | In contour cuts, no burrs or bent combs. | Scribing with closely spaced holes, 0.003-inch dia. by 0.008-inch deep. | LBT substitute for large blanking dies. Vacuum-chuck worktable with NC motion. |

**TABLE 12.7-9—Continued**

| Work material: | **SAE 1008 Steel** | **Carbon Steel** | **Steel** |
|---|---|---|---|
| Workpiece configuration: | 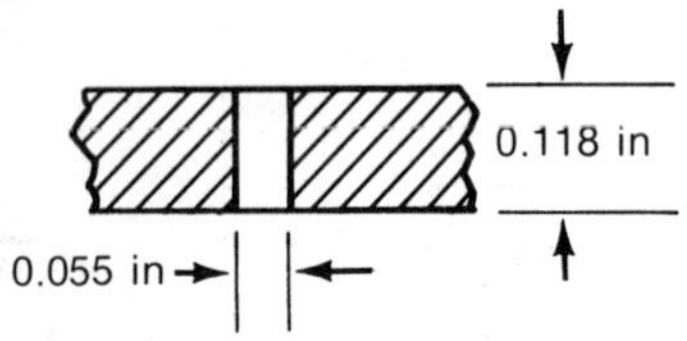 0.118 in; 0.055 in | 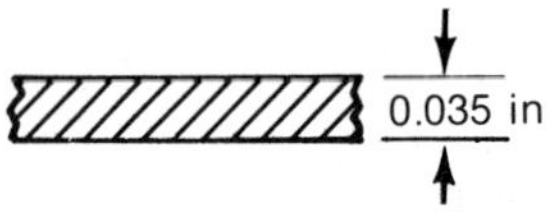 0.035 in | 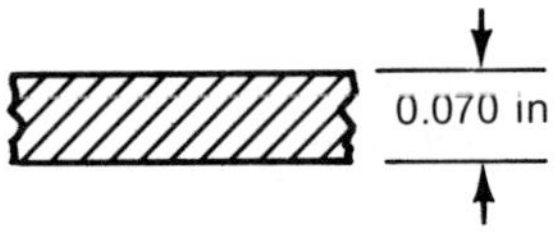 0.070 in |
| Laser equipment type: | YAG | YAG | YAG |
| wavelength, μm: | 1.06 | 1.06 | 1.06 |
| power: | 70 J | 200 W | 600 W |
| gas: | $O_2$ | yes | $O_2$ |
| gas pressure, psi: | | | |
| Operating conditions power level: | 55 J | 200 W | 600 W |
| beam mode: | Pulsed multi | CW multi | Multi |
| pulse length, ms: | 1 | | |
| cycle time: | 10 shots at 1 pps | | |
| feed rate, in/min: | | 50 | 50 |
| Tolerance, in: | | | |
| Remarks: | | Slitting | Slitting |

## PROCESS SUMMARY

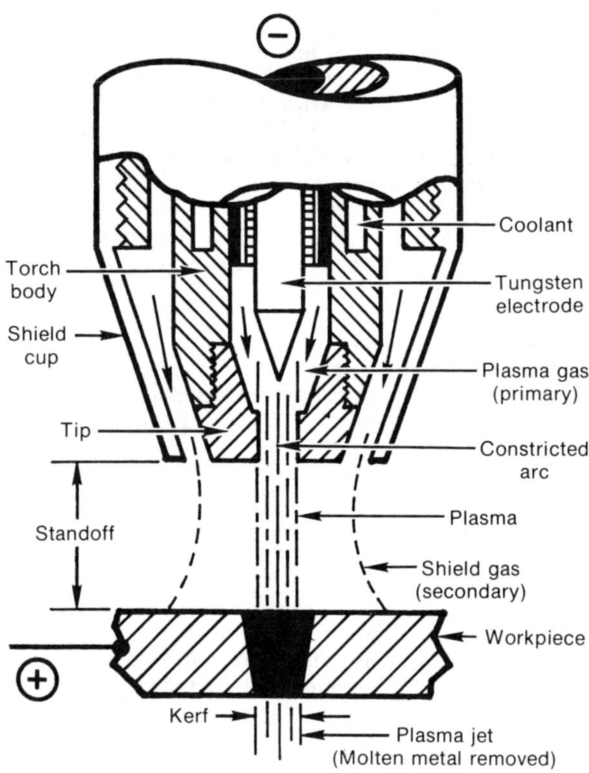


**Figure 12.8–1** PBM schematic

### Principles

Plasma beam machining (PBM) removes material by utilizing a superheated stream of electrically ionized gas (see figure 12.8–1). The 20,000° to 50,000°F [11,000° to 28,000°C] plasma is created inside a water-cooled nozzle by electrically ionizing a suitable gas such as nitrogen, hydrogen, argon, or mixtures of these gases. Since the process does not rely on the heat of combustion between the gas and the workpiece material, it can be used on almost any conductive metal. Generally, the arc is transferred to the workpiece which is made electrically positive. The plasma—a mixture of free electrons, positively-charged ions and neutral atoms—is initiated in a confined, gas-filled chamber by a high-frequency spark. The high-voltage direct-current power sustains the arc, which exits from the nozzle at near sonic velocity. The high-velocity gases blow away the molten metal "chips." Dual-flow torches use a secondary gas or water shield to assist in blowing the molten metal out of the kerf, giving a cleaner cut. Water shield or injection is sometimes used to assist in confining the arc, blasting away the scale and reducing smoke. Greater nozzle life and faster cutting speeds accompany use of water-injection-type torches. Control of nozzle standoff from the workpiece is important. One electrode size can be used to machine a wide range of materials and thicknesses by suitable adjustments to the power level, gas type, gas flow rate, traverse speed and flame angle. PBM is sometimes called plasma arc machining (PAM) or plasma arc cutting (PAC).

### Practical Applications

High cutting speed and the ability to cut exotic metals are the greatest operational advantages of PBM. Profile cutting of metals, particularly of stainless steel and aluminum, has been the most prominent commercial application; however, mild steel, alloy steel, titanium, bronze and most metals can be cut cleanly and rapidly. Multiple-torch cuts are possible on programmed or tracer-controlled cutting tables on plates up to 6-inches [150 mm] thick in stainless steel. Smoothness of cut with freedom from contaminants is an advantage. Well-attached dross on the underside of the cut can be a problem. The transfer arc is used for the thick cuts in metals. Applications for surface heat treatment exist, and the plasma arc is widely used for metal joining. The plasma torch also is used for "punching" holes, as an assist to turning, or for grooving. Plasma cutting has become popular as a maintenance tool since the development of LO-AMP™ portable plasma cutting equipment.

### Operating Parameters

**TABLE 12.8–1 Typical Values for PBM Operating Parameters**

| | |
|---|---|
| Power supply | |
| type: | Direct current |
| wattage: | Up to 200 kW |
| current: | 50 to 1,000 A |
| Primary gas | |
| type: | Argon-nitrogen (80%-20%), hydrogen |
| flow: | 15 to 200 $ft^3$/hr [0.42 to 5.66 $m^3$/hr] |
| Shielding gas | |
| type: | Nitrogen, oxygen, water, compressed air, $CO_2$ |
| flow: | Up to 400 $ft^3$/hr [11.3 $m^3$/hr] |
| Water shield flow (in place of shielding gas): | Up to 15 gal/hr [56.8 L/hr] |
| Plasma temperature: | 20,000° to 50,000°F [11,100° to 27,800°C] |
| Cutting speed: | 2 to 240 in/min [50 to 6,100 mm/min] |
| Standoff distance: | 0.25 to 3 inches [6.4 to 9.5 mm] |
| Nozzle orifice size: | 0.062- to 0.25-inch [1.59 to 6.35 mm] diameter (figure 12.8–2) |
| Kerf width: | 0.060- to 0.090-inch [1.52 to 2.29 mm] thin plate<br>0.188 inch [4.76] for 1-inch [25.4 mm] plate<br>0.75 inch [19 mm] for 6-inch [152 mm] plate |
| Kerf angle | |
| normal: | 2° to 7° |
| special: | 1° to 2° |
| Accuracy: | ±0.031 inch [0.81 mm] on 0.25- to 1.375-inch [6.4 to 35 mm] plate<br>±0.125 inch [3.2 mm] on 6- to 8-inch [152 to 203 mm] plate |
| Corner radii: | 0.156 inch [3.9 mm] minimum (increases with increasing cutting speeds to 1.50 inches [38 mm] at 240 in/min [6,100 mm/min] |

# 12.8 Plasma Beam Machining—PBM

## Material Removal Rates and Tolerances

Profile cutting can be done at 30 inches per minute [760 mm/min] on 1-inch [25.4 mm] thick aluminum plate, while 1/4-inch [6.4 mm] thick carbon steel can be cut at 160 inches per minute [4,060 mm/min]. The use of water injection can increase the carbon steel cutting rates to 240 inches per minute [6,100 mm/min] for 3/16-inch [4.76 mm] thick plate. Tables 12.8–2 and 12.8–5 through 12.8–7 show the range of cutting speeds for PBM of various materials and thicknesses. Table 12.8–9 and figure 12.8–4 contain feed and speed data for plasma arc turning.

The taper on the sides of the cut ranges from 2 to 7 degrees with the straight side determined by the direction of the cut. With special selection and controls, the taper can be held to 1 to 2 degrees. The kerf width usually is 3/32 to 3/16 inch [2.38 to 4.76 mm] for metals less than 1-inch [25.4 mm] thick and 3/8 to 3/4 inch [9.5 to 19 mm] for metals up to 6-inches [150 mm] thick. Corner radius is a minimum of 5/32 inch [3.97 mm] on thinner plates. Tolerances for slots and holes ordinarily range from ±1/32 inch [±0.81 mm] on 1/4-to 1-3/8-inch [6.4 to 35 mm] thick plates to ±1/8 inch [±3.18 mm] on 6-inch [152 mm] thick plates.

## Surface Technology

Grooving is not as severe as in gas torch cutting, and surface roughness can be as good as 32 microinches $R_a$ [0.8 μm]. The usual roughness is 63 to 125 microinches $R_a$ [1.6 to 3.2 μm]. A heat-affected zone can range from 1/32- to 3/16-inch [0.81 to 4.76 mm] deep, depending on the workpiece thermal properties, the work speed and the depth of cut. Sometimes an increase in hardness will accompany the heat-affected zone and will require an edge-finishing operation. In many applications, the cut edge may be adequate for the functional purpose; however, an allowance of 1/16 inch [1.6 mm] for finishing is typical for PBM.

## Equipment and Tools

Hand-held plasma torches and small portable equipment are available, as are NC or CNC programmable motion-controlled, multiple-torch arrangements. Optical followers and table sizes up to 44 by 82 feet [13.4 by 25 m] are used, with cutting speeds controllable from 2 to 240 inches per minute [50 to 6,100 mm/min]. Advanced systems with water injection, water shielding or water beds to catch the dross are available. In addition to flat contour cutting, attachments for lathe turning, grooving or gouging are made (see TAM).

Some advanced plasma cutting systems are designed for mounting on large NC shape-cutting machines and are equipped with torch standoff controls that automatically find and maintain the correct torch standoff from the plate.

Tooling is simple because there is little load reaction from the torch.

## Machining Characteristics

Eye shielding, and noise and fume control are needed to meet workplace environmental requirements. High-flow water shield (sometimes called a water muffler) may be used to recirculate cutting table water at a high flow rate (1,000 gallons per hour [3,780 L/hr]) through a ring mounted around the torch head. This provides a shield or curtain which reduces most of the light, smoke and noise created by the plasma cutting operation. Introduction of a dye into the water can reduce the glare of the arc.

## SELECTED DATA

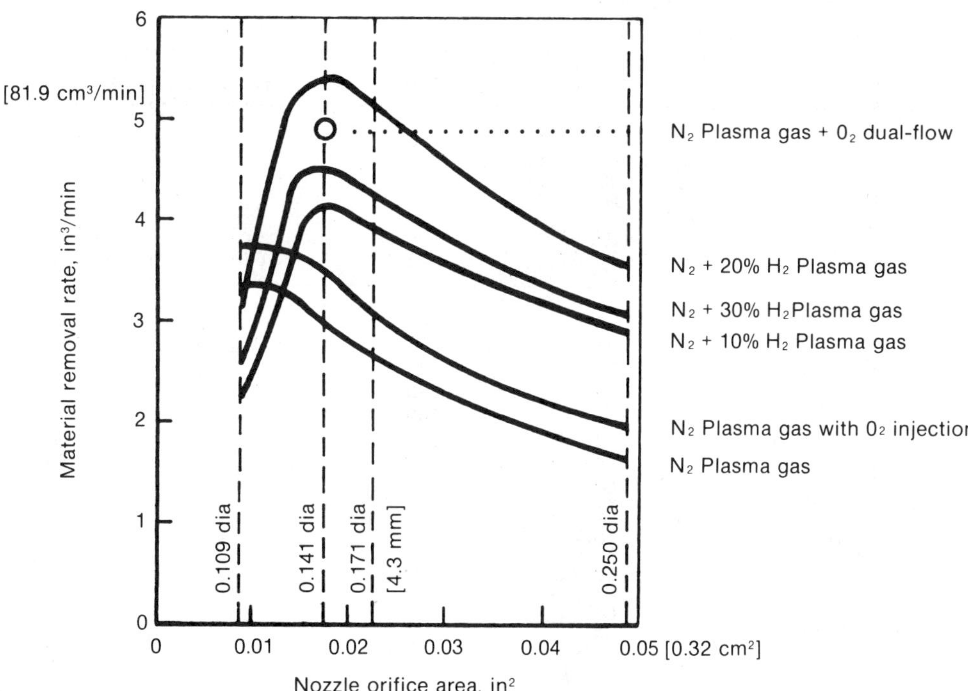


**Figure 12.8–2** Material removal rates for PBM with various gases and nozzle orifice sizes. (R. K. Springborn, ed., *Non-traditional machining processes*, Dearborn, MI: American Society of Tool and Manufacturing Engineers, 1967, p. 163)

**TABLE 12.8–2 Plasma Cutting Guide for Common Steel, Stainless Steel and Aluminum**

| WORKPIECE THICKNESS | | CUTTING SPEED | | POWER SELECTION Amperage, A | | |
|---|---|---|---|---|---|---|
| in | mm | in/min | cm/min | Steel | Stainless steel | Aluminum |
| 0.25 | 6.35 | 20 | 51 | 100 | | |
| | | 40 | 102 | 110 | | |
| | | 70 | 178 | 150 | 105 | |
| | | 100 | 254 | 200 | 140 | |
| | | 150 | 381 | | 210 | 135 |
| | | 160 | 406 | 280 | | |
| | | 200 | 508 | 420 | | |
| | | 430 | 1,090 | | | 550 |
| 0.50 | 12.70 | 20 | 51 | 170 | 135 | |
| | | 40 | 102 | | 190 | 120 |
| | | 70 | 178 | 260 | 250 | 180 |
| | | 100 | 254 | | 270 | 275 |
| | | 135 | 343 | 500 | | |
| | | 150 | 381 | | 700 | 450 |
| | | 175 | 445 | 1,000 | | |
| | | 210 | 533 | | 1,000 | |
| | | 280 | 711 | | | 1,000 |

SOURCE: J. A. Bagley, Plasma arc cutting, Technical paper MR69-578, Society of Manufacturing Engineers, Dearborn, MI, 1969, p. 23.

TABLE 12.8-2—Continued

| WORKPIECE THICKNESS | | CUTTING SPEED | | POWER SELECTION Amperage, A | | |
|---|---|---|---|---|---|---|
| in | mm | in/min | cm/min | Steel | Stainless steel | Aluminum |
| 0.75 | 19.05 | 10 | 25 | 180 | 130 | |
| | | 20 | 51 | 220 | 155 | 100 |
| | | 33 | 84 | 280 | | |
| | | 40 | 102 | 330 | 235 | 150 |
| | | 70 | 178 | 600 | 420 | 275 |
| | | 100 | 254 | 875 | 620 | 400 |
| 1 | 25.40 | 10 | 25 | 250 | 175 | 115 |
| | | 20 | 51 | 270 | 210 | 140 |
| | | 30 | 76 | | 270 | 270 |
| | | 40 | 102 | 530 | 350 | |
| | | 70 | 178 | 900 | | |
| | | 80 | 203 | 1,000 | 540 | 520 |
| | | 100 | 254 | | | 550 |
| | | 110 | 279 | | 1,000 | |
| | | 135 | 343 | | | 1,000 |
| 1.25 | 31.80 | 10 | 25 | 300 | 210 | 140 |
| | | 20 | 51 | 420 | 300 | 190 |
| | | 40 | 102 | 650 | 460 | 300 |
| | | 70 | 178 | 1,100 | 800 | 500 |
| | | 100 | 254 | | 1,140 | 620 |
| 1.50 | 38 | 10 | 25 | 400 | 280 | 180 |
| | | 20 | 51 | 590 | 420 | 270 |
| | | 40 | 102 | 875 | 620 | 400 |
| | | 70 | 178 | | 1,000 | 550 |
| | | 100 | 254 | | | 750 |
| 2 | 50 | 5 | 13 | 450 | 320 | 210 |
| | | 10 | 25 | 600 | 420 | 275 |
| | | 20 | 51 | 850 | 610 | 390 |
| | | 40 | 102 | | 950 | 600 |
| | | 70 | 178 | | | 800 |
| 2.5 | 63.5 | 5 | 13 | 585 | 410 | 235 |
| | | 10 | 25 | 785 | 550 | 320 |
| | | 20 | 51 | | 820 | 470 |
| 3 | 76 | 5 | 13 | 730 | 510 | 295 |
| | | 10 | 25 | 970 | 675 | 385 |
| | | 20 | 51 | | 1,020 | 585 |
| 3.5 | 89 | 5 | 13 | 790 | 550 | 320 |
| | | 10 | 25 | 1,050 | 730 | 420 |
| | | 20 | 51 | | 1,110 | 640 |
| 4 | 100 | 5 | 13 | 970 | 675 | 385 |
| | | 10 | 25 | | 900 | 520 |
| | | 20 | 51 | | | 800 |
| 4.5 | 114 | 5 | 13 | | 900 | 850 |
| 5 | 127 | 3 | 7.6 | | 1,100 | 900 |
| 5.5 | 140 | 3 | 7.6 | | 1,100 | 1,000 |

**TABLE 12.8–3 Nozzle Size and Gas Flow Selection**

| REQUIRED AMPERE RANGE | NOZZLE SIZE | | GAS FLOW* | | | | MAXIMUM AMPERAGE† |
|---|---|---|---|---|---|---|---|
| | | | $N_2$ | | $H_2$ | | |
| A | in | mm | ft³/hr | m³/hr | ft³/hr | m³/hr | A |
| 0 to 200 | 0.125 | 3.18 | 100 | 2.83 | 8 | 0.23 | 225 |
| 200 to 275 | 0.140 | 3.56 | 110 | 3.11 | 10 | 0.28 | 300 |
| 275 to 300 | 0.161 | 4.09 | 120 | 3.40 | 12 | 0.34 | 400 |
| 350 to 400 | 0.187 | 4.75 | 150 | 4.25 | 15 | 0.42 | 550 |
| 450 to 650 | 0.218 | 5.54 | 180 | 5.10 | 18 | 0.51 | 700 |
| over 650 | 0.250 | 6.35 | 225 | 6.37 | 20 | 0.57 | 1,100 |

SOURCE: J. A. Bagley, p. 23.

*Gas mix of $N_2$ and $H_2$ measured at 70 psi [480 kPa] input to 25 feet [7.6 m] of 3/8-inch [9.5 mm] diameter hose.

†Do not exceed this amperage for each nozzle size. To estimate power consumption multiply amperes by 200.

**TABLE 12.8–4 Plasma Arc Gouging Power and Speed in 5086 Aluminum Alloy**
(Constant 1/8-inch [3.2 mm] standoff, torch angle of 45°, and nitrogen gas)

| SPEED | | GROOVE SIZE | | | | MATERIAL REMOVAL RATE | |
|---|---|---|---|---|---|---|---|
| | | Width | | Depth | | | |
| in/min | mm/min | in | mm | in | mm | in³/min | cm³/min |
| 40 | 1,016 | 0.56 | 14.2 | 0.25 | 6.4 | 4.29 | 70.3 |
| 50 | 1,270 | 0.53 | 13.5 | 0.22 | 5.6 | 4.35 | 71.3 |
| 60 | 1,524 | 0.50 | 12.7 | 0.19 | 4.8 | 4.31 | 70.6 |
| 70 | 1,778 | 0.47 | 11.9 | 0.16 | 4.1 | 3.70 | 60.6 |
| 80 | 2,032 | 0.44 | 11.2 | 0.13 | 3.3 | 3.10 | 50.8 |
| 90 | 2,286 | 0.41 | 10.4 | 0.13 | 3.3 | 3.60 | 59.0 |
| **POWER** kVA | | **SPEED: 50 in/min [1270 mm/min]** | | | | | |
| 35.5 | | 0.47 | 11.9 | 0.22 | 5.6 | 3.90 | 63.9 |
| 38.9 | | 0.56 | 14.2 | 0.24 | 6.1 | 5.10 | 83.6 |
| 42.0 | | 0.59 | 15.0 | 0.26 | 6.6 | 5.80 | 95.1 |
| | | **SPEED: 60 in/min [1524 mm/min]** | | | | | |
| 35.8 | | 0.50 | 12.7 | 0.16 | 4.1 | 3.30 | 54.1 |
| 37.9 | | 0.53 | 13.5 | 0.16 | 4.1 | 3.50 | 57.4 |
| 42.6 | | 0.56 | 14.2 | 0.19 | 4.8 | 4.60 | 75.4 |

SOURCE: J. F. Alban, Jr., Plasma arc gouging of aluminum, *Welding Journal* 55 (November 1976), p. 956–57.

**TABLE 12.8-5 Cutting Speeds for Plasma Beam Machining**

| MACHINING CONDITIONS | MATERIAL THICKNESS in / mm | CUTTING SPEED in/min / m/min | | | | | |
|---|---|---|---|---|---|---|---|
| | | Carbon Steel | | Stainless Steel | | Aluminum | |
| | | Best | Max. | Best | Max. | Best | Max. |
| **100 amperes*** | 0.25 | 40 | 80 | 50 | 100 | 60 | 120 |
| | 0.50 | 16 | 25 | 20 | 30 | 28 | 43 |
| Primary gas: $N_2$ | 0.75 | NR | NR | 12 | 15 | 15 | 19 |
| at 55 cfh | 1.00 | NR | NR | 9 | 11 | 10 | 12 |
| 30 psi | | | | | | | |
| | 6.4 | 1.02 | 2.03 | 1.27 | 2.54 | 1.52 | 3.05 |
| Secondary gas: $CO_2$ | 12.7 | 0.41 | 0.64 | 0.51 | 0.76 | 0.71 | 1.09 |
| at 210 cfh | 19.0 | NR | NR | 0.30 | 0.38 | 0.38 | 0.48 |
| 40 psi | 25.4 | NR | NR | 0.23 | 0.28 | 0.25 | 0.30 |
| **200 amperes†** | 0.25 | 50 | 100 | 65 | 135 | 75 | 155 |
| | 0.50 | 30 | 40 | 50 | 70 | 60 | 85 |
| Primary gas: $N_2$ | 0.75 | 20 | 30 | 35 | 50 | 40 | 55 |
| at 70 cfh | 1.0 | 15 | 20 | 20 | 26 | 26 | 35 |
| 30 psi | 1.5 | NR | NR | 12 | 16 | 18 | 24 |
| Secondary gas: $CO_2$ | 6.4 | 1.27 | 2.54 | 1.65 | 3.43 | 1.90 | 3.94 |
| at 210 cfh | 12.7 | 0.76 | 1.07 | 1.27 | 1.78 | 1.52 | 2.16 |
| 40 psi | 19.0 | 0.51 | 0.76 | 0.89 | 1.27 | 1.02 | 1.40 |
| | 25.4 | 0.38 | 0.51 | 0.51 | 0.66 | 0.66 | 0.89 |
| | 38.1 | NR | NR | 0.30 | 0.40 | 0.46 | 0.61 |
| **400 amperes‡** | 0.5 | 40 | 65 | 75 | 120 | 90 | 150 |
| | 1.0 | 30 | 40 | 40 | 55 | 50 | 70 |
| Primary gas: $N_2$ | 1.5 | 20 | 30 | 25 | 38 | 35 | 50 |
| at 50 cfh | 2.0 | 14 | 23 | 17 | 28 | 28 | 45 |
| 20 psi | 2.5 | 10 | 13 | 12 | 15 | 16 | 20 |
| | 3.0 | 6 | 7 | 8 | 10 | 10 | 12 |
| Secondary gas: $CO_2$ | | | | | | | |
| at 210 cfh | 12.7 | 1.02 | 1.65 | 1.91 | 3.05 | 2.29 | 3.81 |
| 40 psi | 25.4 | 0.76 | 1.07 | 1.02 | 1.40 | 1.27 | 1.78 |
| | 38.1 | 0.51 | 0.76 | 0.64 | 0.97 | 0.89 | 1.27 |
| | 50.8 | 0.36 | 0.58 | 0.43 | 0.71 | 0.71 | 1.14 |
| | 63.5 | 0.25 | 0.33 | 0.30 | 0.38 | 0.41 | 0.51 |
| | 76.2 | 0.15 | 0.18 | 0.20 | 0.25 | 0.25 | 0.30 |

SOURCE: Data courtesy of Thermal-Dynamics Corp.

NOTES: NR = Not recommended.

*Minimum speed = 20 in/min [0.51 m/min] on 0.25-inch [6.4 mm] thick material. Maximum thickness for piercing applications is 0.375 inch [10 mm].

†Maximum thickness for piercing applications is 0.5 inch [13 mm].

‡Maximum thickness for piercing applications is 0.75 inch [19 mm].

**TABLE 12.8–6 Cutting Speeds with Water-Injection PBM**

| MACHINING CONDITIONS | MATERIAL THICKNESS in / mm | CUTTING SPEED in/min / m/min | | | | | |
|---|---|---|---|---|---|---|---|
| | | Carbon Steel | | Stainless Steel | | Aluminum | |
| | | Best | Max. | Best | Max. | Best | Max. |
| **300 amperes*** | 0.25 | 60 | 110 | 75 | 130 | 85 | 145 |
| | 0.50 | 40 | 60 | 50 | 70 | 55 | 80 |
| Primary gas: $N_2$ | 0.75 | 30 | 40 | 40 | 55 | 55 | 75 |
| at 75 cfh | 1.00 | 20 | 25 | 25 | 35 | 40 | 60 |
| 30 psi | 1.25 | 12 | 15 | 19 | 23 | 24 | 29 |
| | 1.50 | 12 | 15 | 15 | 18 | 20 | 25 |
| Water injection | | | | | | | |
| at 8 to 15 gal/hr | 6.4 | 1.52 | 2.79 | 1.91 | 3.30 | 2.16 | 3.68 |
| | 12.7 | 1.02 | 1.52 | 1.27 | 1.78 | 1.40 | 2.03 |
| | 19.0 | 0.76 | 1.02 | 1.02 | 1.40 | 1.40 | 1.91 |
| | 25.4 | 0.51 | 0.64 | 0.64 | 0.89 | 1.02 | 1.52 |
| | 31.8 | 0.30 | 0.38 | 0.48 | 0.58 | 0.61 | 0.74 |
| | 38.1 | 0.30 | 0.38 | 0.38 | 0.46 | 0.51 | 0.64 |

SOURCE: Data courtesy of Thermal-Dynamics Corp.

*Maximum thickness for piercing applications is 0.625 inch [16 mm].

**TABLE 12.8–7 Cutting Speeds for PBM of Sheet Metals**

| MACHINING MATERIAL CONDITIONS | THICKNESS | | CUTTING SPEED | | | |
|---|---|---|---|---|---|---|
| | | | Mild Steel | | Stainless Steel | |
| | in | mm | in/min | m/min | in/min | m/min |
| 50 to 100 amperes | 0.062 | 1.5 | 275 | 7.0 | 280 | 7.1 |
| | 0.125 | 3 | 190 | 4.8 | 195 | 5.0 |
| Primary gas: $N_2$ | 0.188 | 5 | 80 | 2.0 | 85 | 2.1 |
| at 15 cfh | | | | | | |
| 35 to 40 psi | 0.250 | 6 | 65 | 1.6 | 70 | 1.8 |
| | 0.375 | 10 | 50 | 1.3 | 55 | 1.4 |
| Secondary gas: $CO_2$ | 0.500 | 12 | 35 | 0.9 | 40 | 1.0 |
| at 150 cfh | | | | | | |
| 45 to 50 psi | | | | | | |

SOURCE: Data courtesy of W. A. Whitney Co.

**TABLE 12.8–8 Tolerances for PBM Hole Piercing**
(For 0.156-inch [4 mm], 9-gage steel)

| HOLE DIAMETER | | INCLUDED ANGLE | DIAMETER TOLERANCE | | OUT-OF-ROUND | |
|---|---|---|---|---|---|---|
| in | mm | degrees | in | mm | in | mm |
| 0.188 | 4.76 | 10 | ±0.01 | ±0.25 | 0.01 | 0.25 |
| 0.250 | 6.35 | 2 | ±0.01 | ±0.25 | 0.01 | 0.25 |
| 0.312 | 7.94 | 2 | ±0.01 | ±0.25 | 0.01 | 0.25 |
| 0.375 | 9.53 | 4 | ±0.01 | ±0.25 | 0.01 | 0.25 |
| 0.437 | 11.11 | 4 | ±0.02 | ±0.50 | 0.01 | 0.25 |

# 12.8 Plasma Beam Machining—PBM

**TABLE 12.8–9 Plasma Arc Turning Parameters**

| MATERIAL | TYPE OF CUT | CURRENT DCSP | ARC VOLTAGE | ROTATING SPEED | | FEED PER REVOLUTION | | METAL REMOVAL RATE | | POWER | |
|---|---|---|---|---|---|---|---|---|---|---|---|
| | | A | V | fpm | m/min | in | mm | in³/min | cm³/min | hp/in³/min | kW/cm³/min |
| Inconel | Rough<br>Smooth | 300<br>170 | 100<br>90 | 40<br>75 | 12<br>23 | 0.160<br>0.050 | 4.1<br>1.3 | 4.0<br>1.5 | 65.6<br>24.6 | 9.0<br>15.0 | 0.41<br>0.69 |
| Rene 41 | Rough<br>Smooth | 325<br>240 | 95<br>90 | 52<br>104 | 16<br>32 | 0.160<br>0.080 | 4.1<br>2.0 | 5.0<br>2.0 | 81.9<br>32.8 | 8.0<br>15.0 | 0.37<br>0.69 |
| Hastelloy alloy C | Rough<br>Smooth | 200<br>140 | 100<br>130 | 30<br>60 | 9<br>18 | 0.100<br>0.050 | 2.5<br>1.3 | 3.5<br>1.5 | 57.4<br>24.6 | 7.0<br>15.0 | 0.32<br>0.69 |
| Precipitation-hardening stainless steels | Rough<br>Smooth | 300<br>170 | 90<br>92 | 50<br>200 | 15<br>61 | 0.160<br>0.040 | 4.1<br>1.0 | 4.5<br>2.0 | 73.8<br>32.8 | 8.0<br>15.0 | 0.37<br>0.69 |

Torch
60°
Rotating workpiece
Metal removal

SOURCE: J. A. Bagley, p. 25.

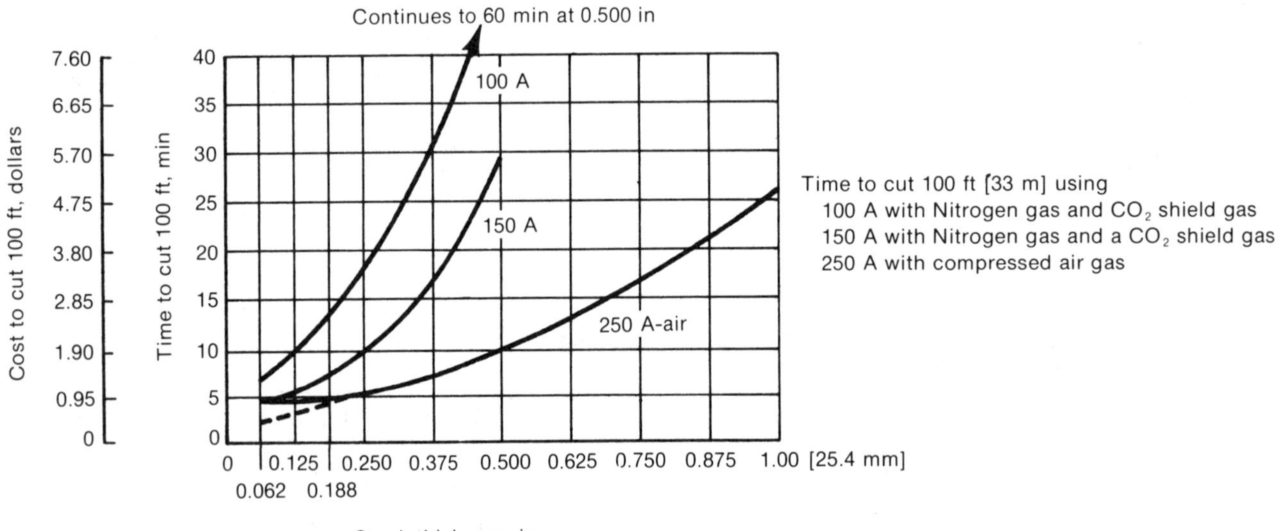


**Figure 12.8–3** Cost and time versus thickness in cutting mild steel with plasma arc. Cost includes gas, perishable parts and electricity, at prevailing rates in 1978, but does not include labor cost for time indicated. (Courtesy of W. A. Whitney Company)

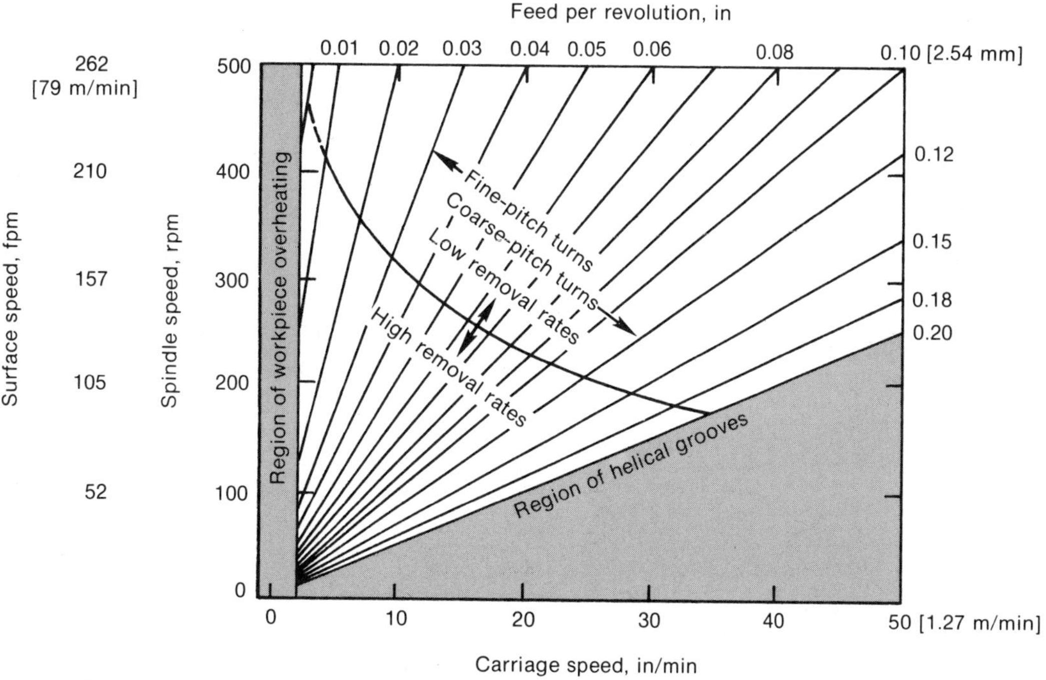


**Figure 12.8–4** Interrelationship among some of the factors involved in plasma arc turning. (R. K. Springborn, p. 165)

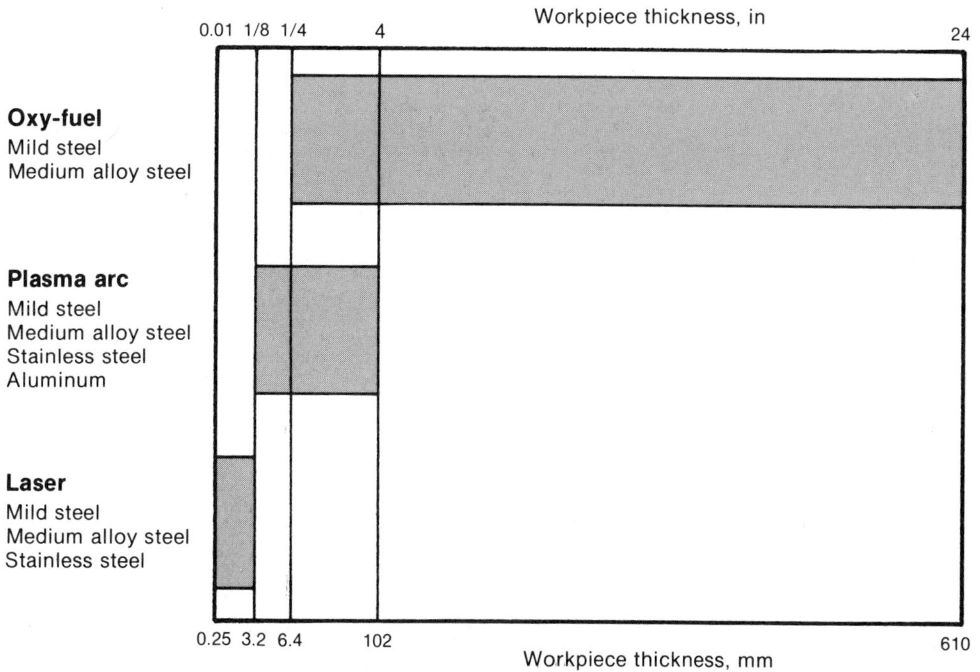


**Figure 12.8–5** Comparison of plasma arc, oxy-fuel and laser processes in cutting metals.

SECTION **13**

# CHEMICAL NONTRADITIONAL MACHINING OPERATIONS


13.1 Chemical Machining—CHM ........ 13–3
13.2 Electropolishing—ELP ........ 13–17
13.3 Photochemical Machining—PCM ........ 13–19
13.4 Thermochemical Machining—TCM or Thermal Energy Method—TEM ........ 13–25

## PROCESS SUMMARY

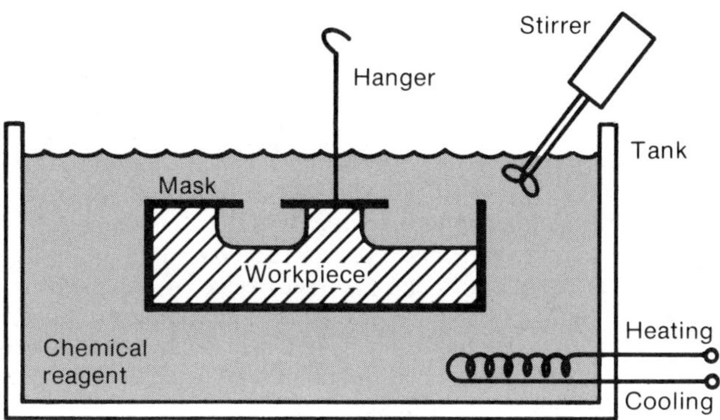


**Figure 13.1–1** CHM schematic.

### Principles

Chemical machining (CHM) is the controlled dissolution of a workpiece material by contact with a strong chemical reagent (see figure 13.1–1). The thoroughly cleaned workpiece is covered with a strippable, chemically resistant mask. Areas where chemical action is desired are outlined on the workpiece with the use of a template and then stripped of the mask. The workpiece is then submerged in the chemical reagent to remove material simultaneously from all exposed surfaces. The solution should be stirred or the workpiece should be agitated for more effective and more uniform action. Increasing the temperatures will also expedite the action. The machined workpiece is then washed and rinsed, and the remaining mask is removed. Multiple parts can be machined simultaneously in the same tank.

Contour machining is accomplished by successively stripping masks and resubmerging the workpiece in the chemical bath. Etching of the workpiece proceeds radially from the openings in the mask. This results in an undercut as well as a depth of cut. The ratio of distance etched beneath the mask to the distance etched into the workpiece (the etching factor) is typically 1:1 (figure 13.1–6). A controlled rate of immersion or withdrawal from the bath will produce tapered sections. It is preferred that the workpiece be oriented such that the grain is in the direction of the longest cut.

### Practical Applications

Nearly all materials can be chemically machined; however, the depth of cut has a practical limit of 0.25 to 0.5 inch [6.4 to 12.7 mm]. Large, shallow areas are especially suitable for CHM since removal is uniform and simultaneous. No burrs are produced and no workpiece surface stresses are generated by CHM. Short-run, quick-change, low-cost tooling offers process flexibility. Thin sheets, formed sheets and delicate cuts are particularly suitable for CHM. Sharp radii cannot be produced in the cutting direction.

Workpiece size is limited only by tank dimensions; aluminum wing spars have been chemically machined in 3-by-8-by-50-foot [1 by 2.5 by 15 m] tanks. Surface contours, scratches and waviness will be reproduced as CHM proceeds. Porous castings are poor candidates for CHM, and welds or welded areas are difficult to cut uniformly.

The maximum practical depth of cut varies with the material form as follows:

| | |
|---|---|
| Sheets and plates: | 0.50 inch [12.7 mm] |
| Forgings and castings: | 0.25 inch [6.4 mm] |
| Extrusions: | 0.15 inch [3.8 mm] |

See Photochemical Machining, section 13.3, for information on chemical machining of thin or small parts.

### Operating Parameters

**TABLE 13.1–1 Typical Values for CHM Operating Parameters**

| | |
|---|---|
| Etchant | |
| types: | Alkalines, acids, proprietary mixtures |
| concentration: | See table 13.1–3 |
| operating temperature: | 70° to 200°F [21° to 93°C] |
| Maskants: | Rubbers, PVC, polyethylene, neoprene and similar plastics or elastomers (tables 13.1–6 and 13.1–7) |
| Etching rate: | 0.0002 to 0.0025 in/min [0.005 to 0.064 mm/min] typical, 0.001 in/min [0.025 mm/min] |
| Etch factor: | 1:1 usual; up to 1:3 (table 13.1–3) (ratio of undercut to depth of cut) |
| Metal ion content: | A little is desirable but too much can stop etching. Uncontrolled etching begins at about 75 grams per liter [10 oz/gal]. |
| Tolerance: | Approximately ±10% of cut depth (tables 13.1–8 and 13.1–9, figures 13.1–7 to 13.1–13) |

### Material Removal Rates and Tolerances

The material removal rate is determined by the solution type, concentration and temperature, all of which are selected to be compatible with the particular metallurgical state of the work material (tables 13.1–3, 13.1–4, and 13.1–5). While the removal rate is low, about 0.001 inch per minute [0.025 mm/min], cutting occurs simultaneously on all exposed surfaces. The depth of cut is controlled by the time of immersion.

Uniformity of removal rates depends upon the chemical and metallurgical uniformity of the workpiece and the uniformity of the solution temperature, concentration and metal ion content (figures 13.1–2 to 13.1–5). Solution or workpiece agitation aids application of the etchant to the exposed surface. Too violent agitation or fluid directional constraint can lead to uneven cut times or grooving. With good time, temperature and solution control, accuracies up to ±0.0005 inch [±0.0127 mm] can be achieved for shallow depths of cut. Deep cuts yield tolerances up to ±0.005 inch [±0.13 mm]. Minimum width of cut or land between cuts should be twice the depth of cut plus the etch factor.

Scribe-and-peel masks can produce ±0.007-inch [±0.18 mm] lateral tolerance per edge. Screen masks can produce ±0.003-inch [±0.076 mm] lateral tolerance per edge, and photoresist masks can produce a tolerance of ±0.0005 inch [±0.013 mm]. (See tables 13.1-8 and 13.1-9 and figures 13.1-7 to 13.1-13.)

### Surface Technology

The chemical machining process does not introduce stress into the workpiece. The gentle chemical action of removing material by molecule-by-molecule dissolution results in surfaces free from residual stresses.

CHM surface roughness values are influenced by the roughness present in the initial workpiece surface and the degree of adherence to the preplanned operating conditions. Typically, very smooth surfaces (32 microinches $R_a$ [0.8 μm] or less) are slightly roughened, while rough surfaces (125 microinches $R_a$ [3.2 μm] or more) are slightly smoothed; see figure 13.1-14 and table 13.1-10. Surface roughness values increase as the metal ion concentration increases in the etchant.

Hydrogen embrittlement and intergranular attack can occur and can have detrimental effects on material properties unless post-treatment corrective action is taken.

### Equipment and Tooling

Equipment runs from small bench-top devices to outdoor tanks, 6 by 12 by 60 feet [1.8 by 3.7 by 18 m]. Both submerged and spray etching are used. Automation of sequence of wash and etchant immersion is commonplace. Agitation by air or by mechanical motion is frequently built into the material handling fixtures. Excessive air agitation can lead to excessive fuming. Multiple part handling racks are commonly used.

Masking, mask scribing and templates for scribing are of many types and shapes (see tables 13.1-6 and 13.1-7). The mask must be inert to the chemical reagent and the heat during etching. Adherence and ease of stripping are necessary to withstand handling. Dip-cut-and-peel masks are common, and silk-screen-applied resists and photoresists are used for higher production rates. Lateral accuracy, figures 13.1-9 and 13.1-10, is affected by type of mask:

| | |
|---|---|
| Cut-and-peel mask: | ±0.007 inch [±0.18 mm] |
| Screen resists: | ±0.003 inch [±0.076 mm] |
| Photoresists: | ±0.0005 inch [±0.013 mm] |

### Machining Characteristics

Chemically resistant floors, plumbing, ventilation and solution controls are required. Special environmental and safety precautions are required to meet local limitations. Automatic monitoring of the chemicals is commonplace. Chemical etchants, by their very nature, are very corrosive. Extreme care in handling is required to avoid spillage and splashing. Vigilance must be maintained at all times. The fumes are corrosive and may be toxic with health risks and risks to nearby equipment or building structures, see table 13.1-11.

## SELECTED DATA

There are several valid techniques for selecting operating parameters for CHM. The steps in table 13.1-2 are recommended as one logical sequence. Supporting data come from the other figures and tables.

**TABLE 13.1-2 Steps To Select CHM Operating Parameters**

| ITEM | DATA |
|---|---|
| 1. Workpiece material: type, heat treatment, grain size, metallurgical state, rolling direction | Engineering drawings |
| 2. Select compatible etchant: type, concentration, temperature | Tables 13.1-3, 13.1-4 and 13.1-5; figures 13.1-2 to 13.1-5 |
| 3. Estimate etching rate: | Tables 13.1-1, 13.1-3, 13.1-4 and 13.1-12; figures 13.1-2 to 13.1-5 |
| 4. Calculate time of immersion from depth of cut: | Drawing dimension divided by estimated etching rate |
| 5. Run test specimen to check estimates and establish final value for operating parameters: | Laboratory tests on specimen with exactly same finish and metallurgical state as the production workpiece |
| 6. Check surface roughness attained: | Specimens from item 5 using final operating parameters |
| 7. Quality control: | Deviations from surface roughness or etching rate during a production run signal the need for corrective changes in the etching bath conditions |
| 8. Tooling: | |
| maskant type | Tables 13.1-6 and 13.1-7 |
| template design and undercut allowances | Table 13.1-3; figures 13.1-6, 13.1-8, 13.1-9 and 13.1-10 |
| 9. Safety precautions: handling, ventilation, effluents | Local safety practices, OSHA guidelines, EPA guidelines |

# 13.1 Chemical Machining—CHM

**TABLE 13.1-3 Etchant Characteristics and General Applications**

| METAL THAT ETCHANT WILL ATTACK | ETCHANT | ETCHANT CONCENTRATION | OPERATING TEMPERATURE | | ETCH RATE WITH FRESH SOLUTION | | ETCH FACTOR* |
|---|---|---|---|---|---|---|---|
| | | | °F | °C | in/min | mm/min | |
| Alfenol | $FeCl_3$<br>$HNO_3:HCl:H_2O$ | 42° Be′†<br>1:1:2 | 120<br>100 to 120 | 49<br>38 to 49 | —<br>— | —<br>— | —<br>— |
| Aluminum alloys | $FeCl_3$<br>$HCl:HNO_3:H_2O$ | 12 to 18° Be′<br>10:1:9 | 120<br>120 | 49<br>49 | 0.001 +<br>0.001 to 0.002 | 0.025<br>0.025 to 0.050 | 1.5:1 to 2:1<br>2:1 (variable) |
| Cold rolled steels | $FeCl_3$<br>$HNO_3$ | 42° Be′<br>10 to 15% (vol.) | 120<br>120 | 49<br>49 | 0.001<br>0.001 | 0.025<br>0.025 | 2:1<br>1.5 to 2:1 |
| Copper and its alloys | $FeCl_3$<br>$(NH_4)_2S_2O_8$<br>Chromic acid<br>$CuCl_2$ | 42° Be′<br>2.2 lb/gal [263g/L] $H_2O$<br>commercially available<br>35° Be′ (regenerated) | 120<br>90 to 120<br>120<br>130 | 49<br>32 to 49<br>49<br>54 | 0.002<br>0.001<br>0.0015<br>0.00055 | 0.050<br>0.025<br>0.038<br>0.014 | 2.5 to 3:1<br>2 to 3:1<br>2 to 3:1<br>2.5 to 3:1 |
| Germanium | HF or $HF:HNO_3$ | various | — | — | — | — | — |
| Glass | HF or $HF:HNO_3$ | various | — | — | — | — | — |
| Gold | $HCl:HNO_3$ | 3:1 | 90 to 100 | 32 to 38 | 0.001 to 0.002 | 0.025 to 0.050 | — |
| Hardened tool steel | $HNO_3$ | 10 to 15% (vol.) | 100 to 120 | 38 to 49 | 0.0005 to 0.001 | 0.013 to 0.025 | 1:1 to 2:1 |
| Inconel | $FeCl_3$ | 42° Be′ | 130 | 54 | — | — | — |
| Inconel alloy X | $FeCl_3$ | 42° Be′ | 130 | 54 | — | — | — |
| Kovar | Chromic acid<br>$FeCl_3$ | commercially available<br>40° Be′ | 120<br>120 | 49<br>49 | 0.001<br>0.001 | 0.025<br>0.025 | 2 to 2.5:1<br>2:1 |
| Lead | $FeCl_3$ | 42° Be′ | 130 | 54 | — | — | — |
| Magnesium | $HNO_3$ | 12 to 15% (vol.) | 90 to 120 | 32 to 49 | 0.001 to 0.002 | 0.025 to 0.050 | — |
| Moly permalloy | $FeCl_3$ | 42° Be′ | 130 | 54 | — | — | — |
| Molybdenum | $H_2SO_4:HNO_3:H_2O$<br><br>$HNO_3:HCl:H_2O$ | 1:1:1 to 5<br><br>1:1:1 to 2 | 130<br><br>— | 54<br><br>— | 0.001 at 130°F varies with temp.<br>— | 0.025 at 54°C varies with temp.<br>— | —<br><br>— |
| Nickel | $FeCl_3$ | 42° Be′ | 120 | 49 | 0.0005 to 0.001 | 0.013 to 0.025 | 1:1 to 3:1 |
| Nickel-iron alloys | $FeCl_3$ | 42° Be′ | 120 | 49 | 0.0005 to 0.001 | 0.013 to 0.025 | 1:1 to 3:1 |
| Nickel-silver alloys | $FeCl_3$<br>Chromic acid<br>$(NH_4)_2S_2O_8$ | 42° Be′<br>commercially available<br>2.2 lb/gal [263g/L] $H_2O$ | 130<br>120<br>90 to 120 | 54<br>49<br>32 to 49 | —<br>—<br>— | —<br>—<br>— | —<br>—<br>— |
| Phosphor-bronze | $FeCl_3$<br>Chromic acid<br>$(NH_4)_2S_2O_8$ | 42° Be′<br>commercially available<br>2.2 lb/gal [263g/L] $H_2O$ | cold (80)<br>cold (80)<br>cold (80) | 27<br>27<br>27 | 0.0005<br>0.0005<br>0.0003 | 0.013<br>0.013<br>0.008 | 2:1<br>2:1<br>2:1 |
| Silicon | $HNO_3:HF:H_2O$ | various | 100 to 120 | 38 to 49 | slow | slow | — |
| Silicon steel | $FeCl_3$ | 42° Be′ | 130 | 54 | 0.001 | 0.025 | 1.5 to 2:1 |
| Silver | $HNO_3:H_2O$<br>$FeNO_3$ | 50 to 90% (vol.)<br>36° Be′ | 100 to 120<br>120 | 38 to 49<br>49 | 0.0005 to 0.001<br>0.0008 | 0.013 to 0.025<br>0.020 | —<br>— |
| Stainless steel | $FeCl_3$ | 42° Be′ | 130 | 54 | 0.0008 | 0.020 | 1.5 to 2:1 |
| Tin | $FeCl_3$ | 42° Be′ | 130 | 54 | — | — | — |
| Titanium | HF<br>$HF:HNO_3:H_2O$<br>$NH_4HF_2:HCl:H_2O$ | 10 to 50% (vol.)<br>various<br>various | 100 to 120<br>100 to 120<br>100 to 120 | 38 to 49<br>38 to 49<br>38 to 49 | —<br>—<br>— | —<br>—<br>— | —<br>—<br>— |
| Zinc | $HNO_3$ | 10 to 15% (vol.) | 100 to 120 | 38 to 49 | 0.001 | 0.025 | — |

*The ratio of depth of undercut to depth of cut.

†Baume′ specific gravity scale (Be′).

**TABLE 13.1–4 Comparison of Data and Characteristics of Systems for Chemical Milling Titanium, Steel, Aluminum, and Nickel- and Cobalt-Base Alloys**

| METAL | PRINCIPAL ETCHANT | ETCH RATE | | OPTIMUM ETCH DEPTH | | ETCHANT TEMPERATURE | | AVERAGE SURFACE ROUGHNESS, $R_a$ | |
|---|---|---|---|---|---|---|---|---|---|
| | | in/min | mm/min | in | mm | °F | °C | μin | μm |
| Titanium alloys | Hydrofluoric acid | 0.0006 to 0.0012 | 0.015 to 0.030 | 0.125 | 3.2 | 115 ±5 | 46 ±2.7 | 16 to 100 | 0.4 to 2.5 |
| Steels | Hydrochloric acid-nitric acid | 0.0006 to 0.0012 | 0.015 to 0.030 | 0.125 | 3.2 | 145 ±5 | 63 ±2.7 | 30 to 120 | 0.8 to 3.2 |
| Aluminum alloys | Sodium hydroxide | 0.0008 to 0.0012 | 0.020 to 0.030 | 0.125 | 3.2 | 195 ±5 | 90 ±2.7 | 80 to 120 | 2.0 to 3.2 |
| Nickel- and cobalt-base alloys | Nitric acid-hydrochloric acid-ferric chloride | 0.0004 to 0.0015 | 0.010 to 0.038 | 0.125 | 3.2 | 140 ±5 | 60 ±2.7 | 40 to 150 | 1.0 to 3.8 |

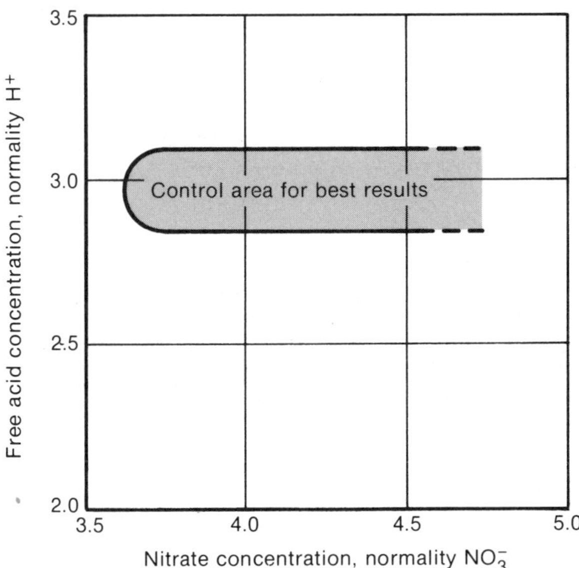


**Figure 13.1–2** Suitable working ranges for the free acid and nitrate concentrations of an aqua-regia-type etchant for 18/8 stainless steels. (W. T. Harris, *Chemical milling,* Oxford: Clarendon Press, 1976, p. 239)

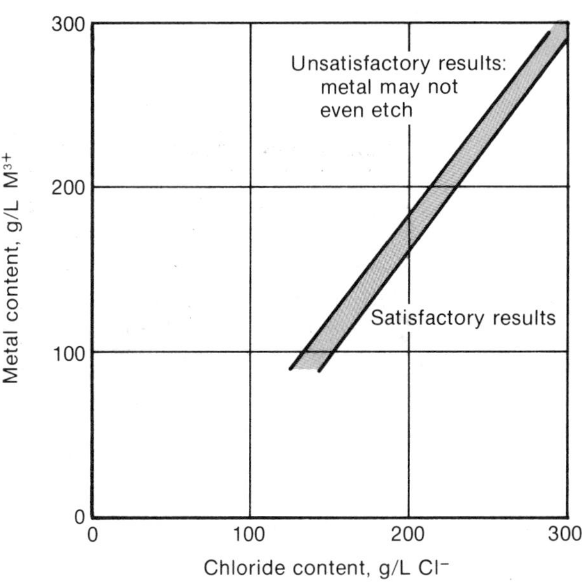


**Figure 13.1–3** Relationship between metal and chloride content for an aqua-regia-type etchant. (W. T. Harris, p. 238)

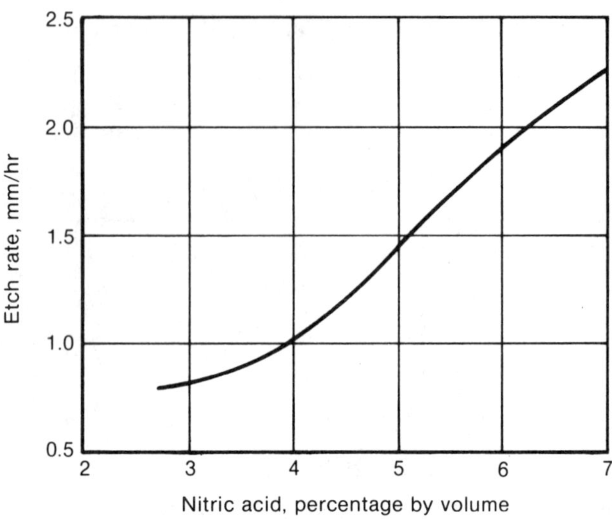


**Figure 13.1–4** The effect on etch rate of increasing the nitric acid concentration in an etchant containing 25 percent sulfuric acid at 70°C (158°F). (W. T. Harris, p. 234)

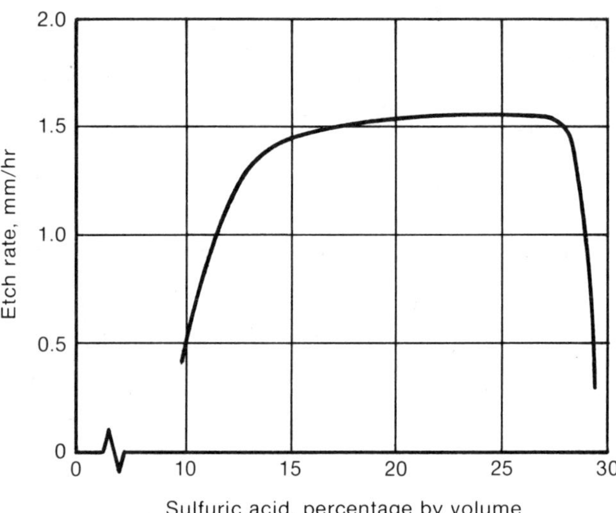


**Figure 13.1–5** The effect on etch rate of increasing the sulfuric acid concentration in an etchant containing 5 percent nitric acid at 70°C (158°F). (W. T. Harris, p. 234)

**TABLE 13.1–5 Etchants for Steel, Nickel and Cobalt Alloys**

| METAL ALLOY | $H^+$(N) | $NO_3^-$(N) | $PO^-$(N) | $F^-$(N) | METAL ION g/L |
|---|---|---|---|---|---|
| Mild steel | 2.0 to 3.5 | 0.2 to 4.1 | 1.5 to 3.0 | | 100 to 250 |
| Stainless steel | 2.0 to 3.5 | 0.2 to 4.1 | 1.5 to 3.0 | | 100 to 250 |
| Nickel alloys | 2.0 to 3.5 | 0.2 to 4.1 | 1.5 to 3.0 | | 100 to 250 |
| Cobalt alloys | 2.0 to 5.5 | 0.2 to 2.0 | 0 to 3.0 | 0.5 to 1.0 | 125 to 250 |

SOURCE: W. T. Harris, p. 235.

**TABLE 13.1-6 Typical Maskants for Various Metals**

| METAL | ETCHANT | TEMPERATURE °C | MASKANTS |
|---|---|---|---|
| Aluminum | Alkaline | 91 | Acrylonitrile rubber, butyl rubber, neoprene rubber |
| Ferrous metals | Acid | 54 | Polyvinyl chloride, polyethylene, butyl rubber |
| Magnesium | Acid | $<38$ | Polymers |
| Titanium | Acid | 21 to 35 | Translucent chlorinated polymers |
| Titanium | Acid | 54 | Vinyl polymers, elastomers |
| Beryllium | Acid | 21 to 54 | Vinyl, neoprene, butyl-based materials |
| Nickel | Acid | 45 to 50 | Neoprene |

**TABLE 13.1-7 Properties of Maskant Materials Widely Used in Chemical Milling**

| PROPERTY | MASKANT MATERIAL: Exposure to Oxidizing Acids | | | | |
|---|---|---|---|---|---|
| | | | Exposure to Acids or Alkalis | | |
| | Polyvinyl chloride | Polyethylene | Butyl rubber | Acrylonitrile rubber | Neoprene rubber |
| Ease of manufacture | Good | Good | Fair | Fair | Good |
| Shelf life, months | 6 to 12 | 6 to 12 | 4 to 6 | 3 to 6 | 6 to 8 |
| Solids, % | 25 to 55 | 60 to 100 | 20 to 25 | 15 to 25 | 25 to 35 |
| Ease of application: | | | | | |
| dipping | Good to fair | Poor to fair | Good | Good | Good |
| flow coating | Good to fair | Poor to fair | Good | Good | Good |
| air spraying | Good | Good | Poor | Poor | Fair |
| Type of cure | Air or heat | Heat | Heat | Air or heat | Air or heat |
| Tensile strength, $MN/m^2$ | | | | | |
| air-dried (24 hr) | — | — | — | 7 max. | 14 max. |
| heat cured | 7 to 17.5 | 6 to 11 | 6 to 11 | 11 to 17.5 | 11 to 21 |
| Resistance to etchant: | | | | | |
| deterioration | Good to excellent | Very good | Very good | Very good | Very good |
| permeability | Fair to good | Very good | Excellent | Very good | Very good |
| Heat limit °F* | 160 | 140 | 300 | 250 | 200 |

*Maximum useful temperature for intermittent exposure—also curing temperature.

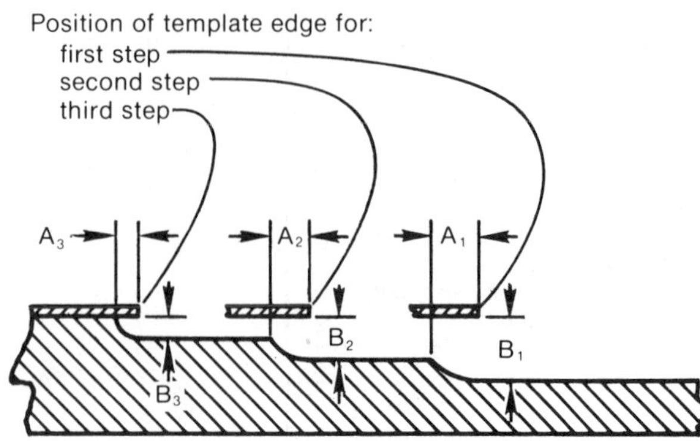


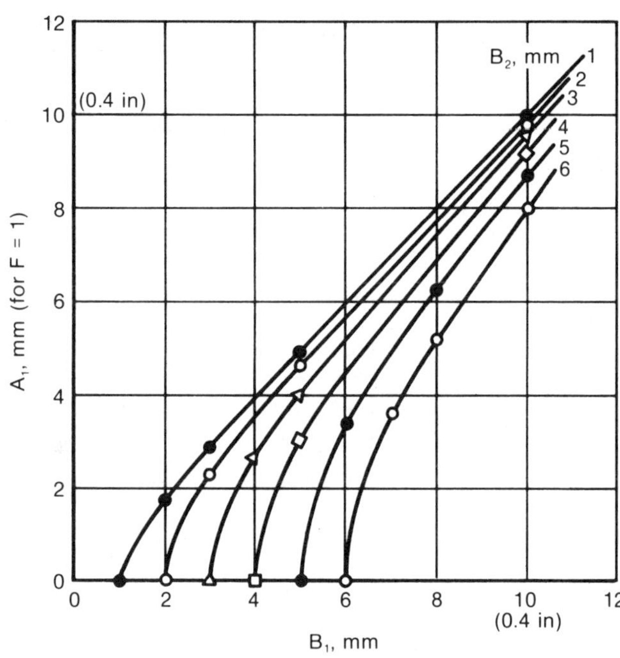


**Figure 13.1–6** Undercut allowance, A, for step chemical milling. Allowance is for adjacent steps $B_1$ and $B_2$ with various step depths. Note: Etch factor for CHM is ratio of undercut to depth of cut or F = A ÷ B. Caution: In photochemical machining, the factor is defined as the inverse ratio, B ÷ A. (W. T. Harris, p. 268–69)

**TABLE 13.1–8 Typical Surface Dimensional Tolerances**

| WORK MATERIAL | TOLERANCE, in | | | | | TOLERANCE, mm | | | | |
|---|---|---|---|---|---|---|---|---|---|---|
| | Workpiece Thickness, in | | | | | Workpiece Thickness, mm | | | | |
| | 0.002 | 0.005 | 0.010 | 0.020 | 0.030 | 0.05 | 0.13 | 0.25 | 0.50 | 0.75 |
| Aluminum alloys | ±0.0005 | ±0.0012 | ±0.0025 | ±0.005 | ±0.008 | ±0.013 | ±0.030 | ±0.063 | ±0.13 | ±0.20 |
| Copper and copper alloys | ±0.0004 | ±0.001 | ±0.002 | ±0.004 | ±0.006 | ±0.010 | ±0.025 | ±0.050 | ±0.10 | ±0.15 |
| Nickel | ±0.0005 | ±0.0012 | ±0.0025 | ±0.005 | ±0.008 | ±0.013 | ±0.030 | ±0.063 | ±0.13 | ±0.20 |
| Steel alloys | ±0.0004 | ±0.001 | ±0.002 | ±0.004 | ±0.006 | ±0.010 | ±0.025 | ±0.050 | ±0.10 | ±0.15 |
| Stainless steel | ±0.0005 | ±0.0012 | ±0.0025 | ±0.005 | ±0.008 | ±0.013 | ±0.030 | ±0.063 | ±0.13 | ±0.20 |

SOURCE: Chemcut Corporation, State College, PA.

**TABLE 13.1-9 Tolerance on Depth of Cut**

| MATERIAL | TOLERANCE, in | | | | TOLERANCE, mm | | | |
|---|---|---|---|---|---|---|---|---|
| | Depth of cut, in | | | | Depth of cut, mm | | | |
| | 0-0.050 | 0.051-0.100 | 0.101-0.250 | 0.251-0.500 | 0-1.25 | 1.25-2.50 | 2.5-6.4 | 6.4-12.7 |
| General | 0.002 | 0.003 | 0.004 | 0.006 | 0.05 | 0.075 | 0.10 | 0.15 |
| Aluminum and magnesium alloys | 0.001 | 0.0015 | 0.002 | 0.003 | 0.025 | 0.038 | 0.05 | 0.075 |
| Stainless steels, ferrous and nickel alloys | 0.002 | 0.003 | 0.004 | 0.006 | 0.05 | 0.075 | 0.10 | 0.15 |
| Titanium and beryllium alloys | 0.003 | 0.004 | 0.006 | 0.010 | 0.075 | 0.10 | 0.15 | 0.25 |

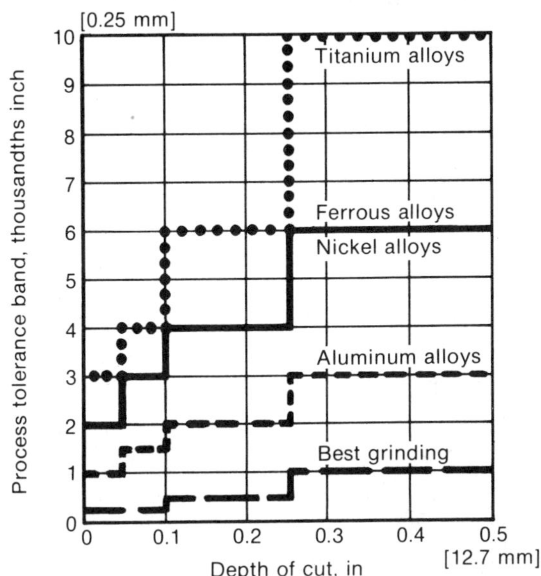


**Figure 13.1-7** CHM depth-of-cut tolerance for sheet materials.

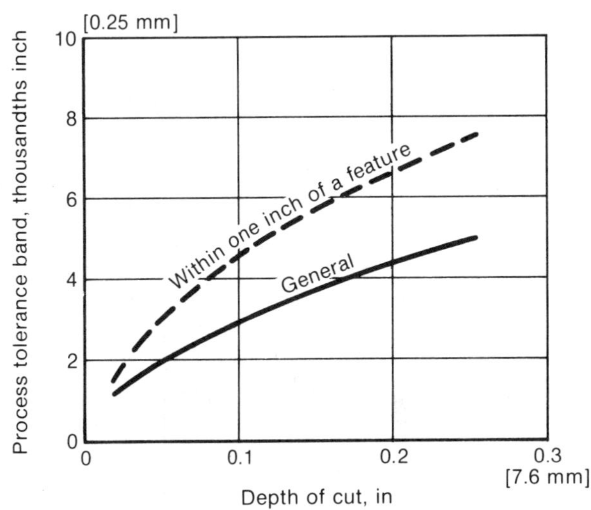


**Figure 13.1-8** Depth-of-cut tolerance capability of CHM.

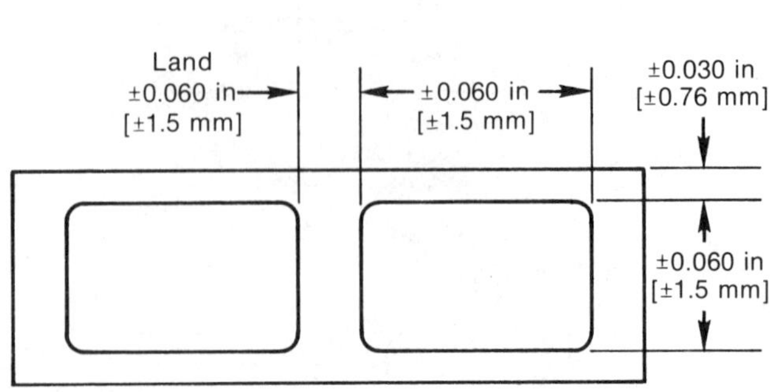


**Figure 13.1–9** Nominal lateral tolerance.

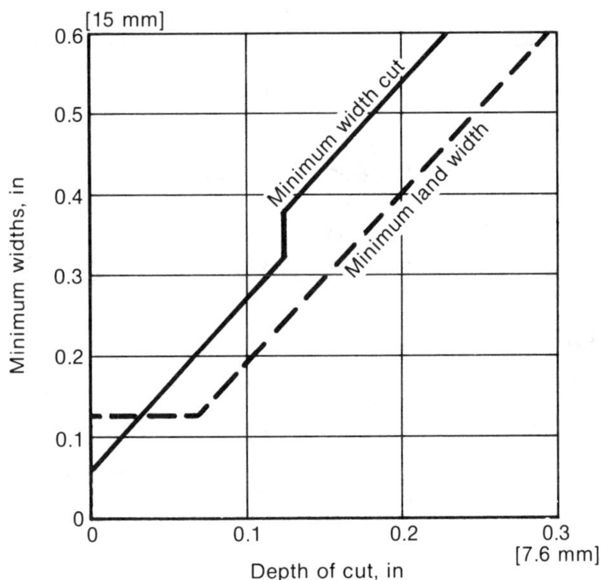


**Figure 13.1–10** Recommended minimum widths of cut and land for CHM.

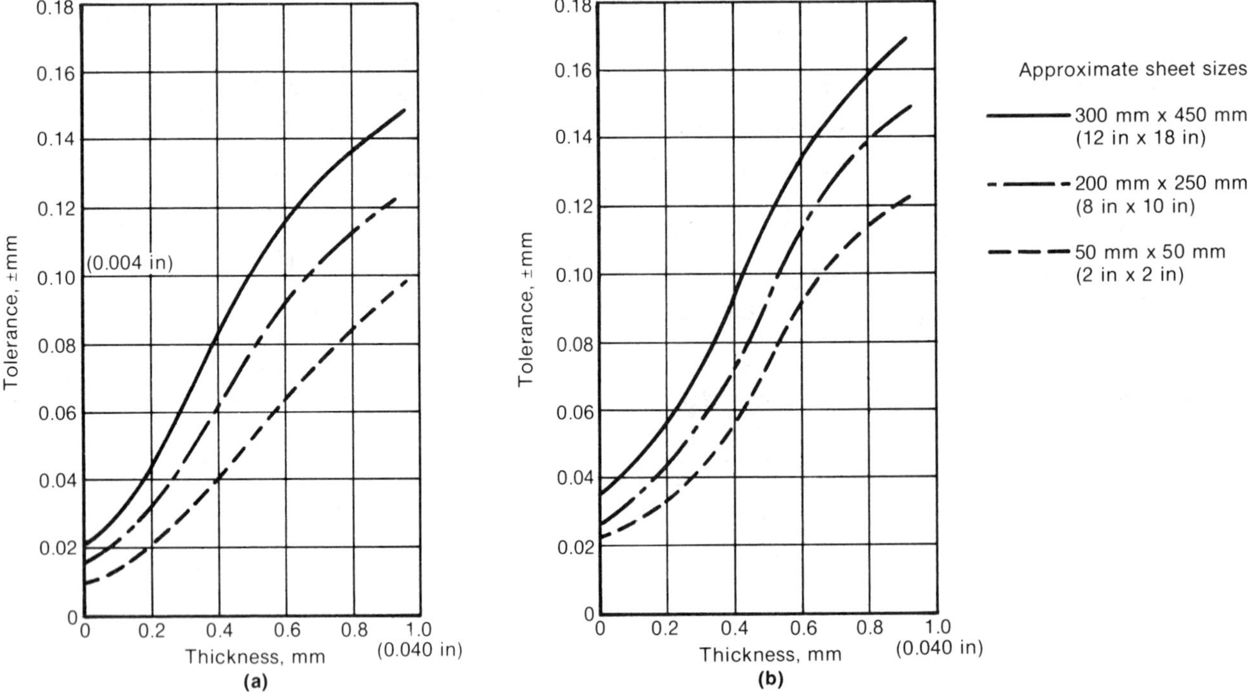


**Figure 13.1–11** Practical tolerances on etched dimensions, for a range of flat sheet sizes and material thickness. (a) For prototype and short runs; (b) for production runs. (W. T. Harris, p. 192)

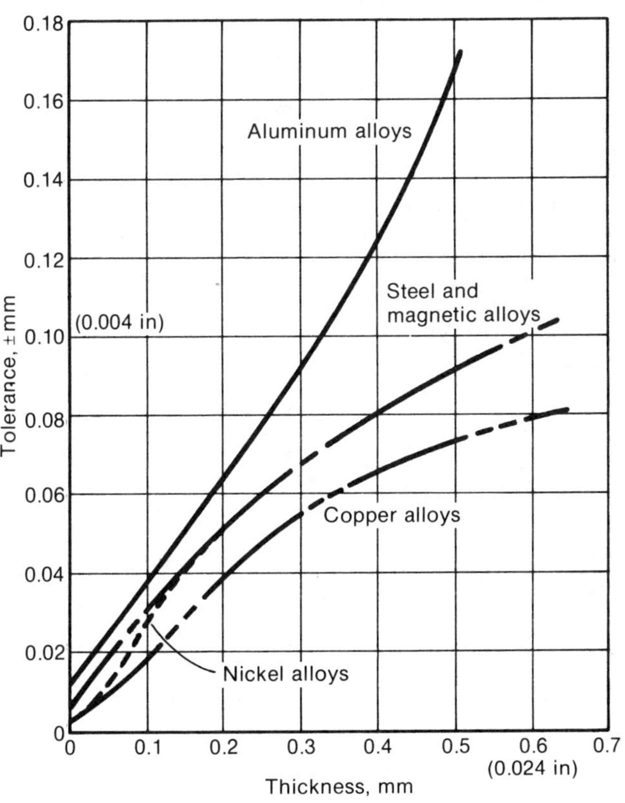


Figure 13.1–12 Normally attained dimensional tolerances for a range of materials. (W. T. Harris, p. 193)

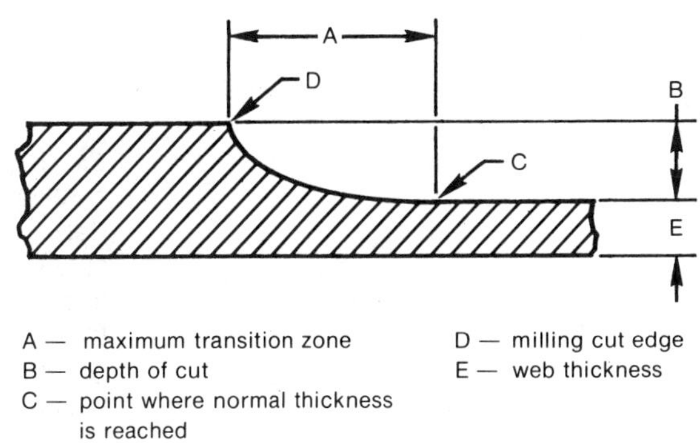


| Materials | Maximum Transition Zone |
|---|---|
| Aluminum alloys | 4 x depth of cut |
| Magnesium alloys | 6 x depth of cut |
| Ferrous alloys | 10 x depth of cut |
| Nickel and Cobalt alloys | 10 x depth of cut |
| Titanium alloys | 10 x depth of cut or 0.5 inch [12.7 mm] (whichever is greater) |

**Figure 13.1–13** Transition zones of CHM.

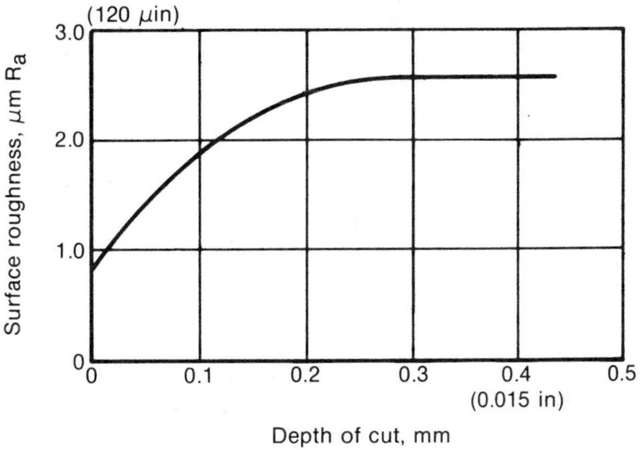


**Figure 13.1–14** The relationship between surface finish and depth of cut for chemical milling 7075 aluminum sheet. (W. T. Harris, p. 112)

**TABLE 13.1–10 Surface Roughness Achieved by Chemical Milling**

| MATERIAL | FORM | SURFACE ROUGHNESS AFTER 0.25 to 0.40 mm (0.010 to 0.015 in) REMOVED | |
|---|---|---|---|
| | | μm $R_a$ | μin $R_a$ |
| Aluminum alloys | Sheet | 2.0 to 3.8 | 80 to 150 |
| | Casting | 3.8 to 7.6 | 150 to 300 |
| | Forging | 2.5 to 6.4 | 100 to 250 |
| Magnesium alloys | Casting<br>Forging | 0.75 to 1.4 | 30 to 55 |
| Steel alloys | Sheet<br>Forging | 0.75 to 1.5 | 30 to 60 |
| Nickel alloys | Sheet | 0.75 to 1.0 | 30 to 40 |
| Titanium alloys | Sheet | 0.2 to 0.8 | 8 to 32 |
| | Casting | 0.75 to 1.5 | 30 to 60 |
| | Forging | 0.38 to 1.0 | 15 to 40 |
| Tungsten | Bar | 0.5 to 1.0 | 20 to 40 |
| Beryllium | Bar | 3.8 to 6.4 | 150 to 250 |
| Tantalum | Sheet | 0.25 to 0.5 | 10 to 20 |
| Columbium | Bar | 1.0 to 1.5 | 40 to 60 |
| Niobium (Columbium) | Sheet | 1.0 to 1.5 | 40 to 60 |
| Molybdenum | Sheet | 1.5 to 3.3 | 60 to 130 |

SOURCE: W. T. Harris, p. 114.

**TABLE 13.1–11 Summary of Health Risk Arising from the Commonly Used Etchants**

| MATERIAL BEING ETCHED | ETCHANT(S) | ETCHANT RISK(S) | FUME RISK(S) |
|---|---|---|---|
| Aluminum alloys | Sodium hydroxide | Splashes on to skin, mouth, or eyes | Fume not dangerous but hydrogen gas is explosive |
| Magnesium alloys | Sulphuric acid | Splashes on to skin, mouth, or eyes. Additions may cause spattering | Corrosive acidic fume. Hydrogen gas is explosive |
| Steel/nickel alloys | Hydrochloric, nitric, sulphuric and phosphoric acids | Highly corrosive splashes on to skin, mouth, or eyes | Corrosive and suffocating fume is also toxic |
| Titanium alloys | Hydrofluoric acid with nitric or chromic acid | Highly corrosive splashes on to skin, mouth, or eyes. Hydrofluoric acid is particularly dangerous | Corrosive and suffocating fume is also toxic |
| Copper alloys | Ferric chloride | Corrosive splashes on to skin, mouth, or eyes | Corrosive and suffocating fume |

SOURCE: W. T. Harris, p. 107.

**TABLE 13.1–12 Consolidated Chemical Machining Data**

| MATERIAL TYPE | ETCHANT | ETCHANT CONCENTRATION | OPERATING TEMPERATURE °F | OPERATING TEMPERATURE °C | ETCH RATE in x $10^{-3}$/min | ETCH RATE mm/min | MASKANT | ETCH FACTOR* | ROUTINE DEPTH OF CUT TOLERANCE ±in | ROUTINE DEPTH OF CUT TOLERANCE ±mm | SURFACE ROUGHNESS μin $R_a$ | SURFACE ROUGHNESS μm $R_a$ |
|---|---|---|---|---|---|---|---|---|---|---|---|---|
| Aluminum and aluminum alloys | $FeCl_3$ | 11–15°Be′[†] | 120 | 49 | 0.5–1.0 | 0.013–0.025 | Polymers | — | — | — | — | — |
| | NaOH | 150 g/L | 120 | 49 | 0.8–1.2 | 0.020–0.030 | Neoprene | 1.5–20 | 0.002 | 0.050 | 90–125 | 2.3–3.2 |
| Beryllium | $NH_4HF_2$ | 15% vol. | 80 | 27 | 0.4–0.6 | 0.010–0.015 | Neoprene | — | 0.001–0.003 | 0.025–0.076 | 125 | 3.2 |
| Brass/Bronze | $FeCl_3$ | 32°Be′ | 120 | 49 | 1.0 | 0.025 | — | — | 0.001 | 0.025 | 32 | 0.8 |
| Chromium | $K_3Fe(CN)$ | — | — | — | — | — | — | — | — | — | — | — |
| | HCl | 10–50% vol. | — | — | 0.5 | 0.013 | — | — | — | — | — | — |
| Cobalt alloys | $HNO_3$:HCl: $FeCl_3$ | — | 140 | 60 | 0.4–1.5 | 0.010–0.038 | — | — | — | — | 40–150 | 1.0–3.8 |
| Columbium | $HNO_3$:HF: $H_2O$ | 1:1:4 | Ambient | Ambient | 0.5–1.0 | 0.013–0.025 | — | — | — | — | — | — |
| Copper | $FeCl_3$ | 32°Be′ | 120 | 49 | 2.0 | 0.050 | — | 2.5–3.0 | 0.003 | 0.076 | — | — |
| | $CuCl_2$ | 42°Be′ | 120 | 49 | 1.2 | 0.030 | — | 2.5–3.0 | 0.003 | 0.076 | — | — |
| Gold | HCl:$HNO_3$ | 3:1 | 100 | 38 | 1.0–2.0 | 0.025–0.050 | — | 1.0 | — | — | — | — |
| | KI:$I_2$:$H_2O$ | 75 lb:25 lb: 15 gal | 100 | 38 | 1.0–2.0 | 0.025–0.050 | — | 1.0 | — | — | — | — |
| Inconel | $FeCl_3$ | 42°Be′ | 130 | 54 | 0.5–1.5 | 0.013–0.038 | Polyethylene | — | — | — | 40–150 | 1.0–3.8 |
| | HCl:$HNO_3$ | 42°Be′ | 130 | 54 | 0.5–1.5 | 0.013–0.038 | Polyethylene | — | — | — | 40–150 | 1.0–3.8 |
| Indium oxide | HCl:$HNO_3$ | 50:5% vol. | 120 | 49 | 0.5–1.0 | 0.013–0.025 | Polymers | — | — | — | — | — |
| Invar | $FeCl_3$ | 42°Be′ | 130 | 54 | — | — | — | — | — | — | — | — |
| Kovar | $FeCl_3$ | 42°Be′ | 130 | 54 | 0.75 | 0.019 | Polymers | 1.0–2.0 | — | — | — | — |
| Lead | $FeCl_3$ | 42°Be′ | 130 | 54 | 1.0 | 0.025 | — | 1.0 | — | — | — | — |
| Magnesium | $HNO_3$ | 12–15% vol. | 90 to 120 | 32 to 49 | 1.0–2.0 | 0.025–0.050 | Polymers | 1.0 | 0.002 | 0.051 | 50 | 1.3 |
| Molybdenum | $H_2SO_4$: $HNO_3$:$H_2O$ | 1:1:1 | 130 | 54 | 1.0 | 0.025 | — | 1.0 | — | — | — | — |
| Nickel, | $FeCl_3$ | 42°Be′ | 120 | 49 | 0.5–1.5 | 0.013–0.038 | Polyethylene | 1.0–3.0 | 0.002 | 0.051 | 40–150 | 1.0–3.8 |
| Nimonic | $FeCl_3$: $HNO_3$:HCl | — | 120 | 49 | 0.5–1.5 | 0.013–0.038 | Polyethylene | 1.0–3.0 | 0.002 | 0.051 | 40–150 | 1.0–3.8 |

*The ratio of depth of undercut to depth of cut.

[†] Baume specific gravity scale (Be′).

**TABLE 13.1-12—Continued**

| MATERIAL TYPE | ETCHANT | ETCHANT CONCEN-TRATION | OPERATING TEMPERATURE | | ETCH RATE | | MASKANT | ETCH FACTOR* | ROUTINE DEPTH OF CUT TOLERANCE | | SURFACE ROUGHNESS | |
|---|---|---|---|---|---|---|---|---|---|---|---|---|
| | | | °F | °C | in x $10^{-3}$/min | mm/min | | | ±in | ±mm | μin $R_a$ | μm $R_a$ |
| Phosphor-bronze | $FeCl_3$ | 42°Be′ | 80 | 26 | 0.5 | 0.013 | — | 2.0 | — | — | — | — |
| Silicon | HF | — | — | — | — | — | — | — | — | — | — | — |
| | $HF:HNO_3$ | — | — | — | — | — | — | — | — | — | — | — |
| Silver | $HNO_3:H_2O$ | 50–90% vol. | 120 | 49 | 0.5–1.0 | 0.013–0.025 | — | 1.0 | — | — | — | — |
| Stainless steel austenitic | $FeCl_3$ (or HCl: $HNO_3$) | 42°Be′ | 130 | 54 | 0.8–5.0 | 0.020–0.130 | Polyvinyl chloride | 1.5–2.0 | 0.004 | 0.1 | 63 | 1.6 |
| martensitic | $FeCl_3$ (or HCl: $HNO_3$) | 42°Be′ | 130 | 54 | 0.25 | 0.006 | Polyvinyl chloride | — | 0.004 | 0.1 | 125 | 3.2 |
| Mild steel | $FeCl_3$ (or HCl: $HNO_3$) | 32°Be′ | 120 | 49 | 1.0 | 0.025 | Polyvinyl chloride | 2.0 | 0.002 | 0.05 | 63 | 1.6 |
| Silicon steel | $FeCl_3$ (or HCl: $HNO_3$) | 32°Be′ | 130 | 54 | 1.0 | 0.025 | Polyvinyl chloride | 1.5–2.0 | 0.002 | 0.05 | 63 | 1.6 |
| Tool steel | $FeCl_3$ (or HCl: $HNO_3$) | 32°Be′ | 130 | 54 | 1.0 | 0.025 | Polyvinyl chloride | — | — | — | — | — |
| Tantalum | $HF:HNO_3$ | 10:5 | 120 | 49 | 0.2–0.5 | 0.005–0.013 | — | — | — | — | — | — |
| Tin | $FeCl_3$ (or HCL: $HNO_3$) | 32°Be′ | 130 | 54 | 1.0 | 0.025 | — | — | — | — | — | — |
| Titanium | HF | 10–50% vol. | 120 | 49 | 0.6–1.5 | 0.015–0.038 | Vinyl polymers | 1.0 | 0.002–0.0035 | 0.05–0.009 | 15–100 | 0.4–2.5 |
| | $HF:HNO_3$ (or HF: $CrO_3$) | — | 115±10 | 46±5.6 | 0.6–1.5 | 0.015–0.038 | Vinyl polymers | 1.0 | 0.002–0.005 | 0.05–0.13 | 15–100 | 0.4–2.5 |
| Tungsten | $K_3Fe(CN)_6$: NaOH | 20% wt.: 2–3% wt. | 70 to 130 | 21 to 54 | — | — | — | — | — | — | — | — |
| | $HF:HNO_3$ | 30–70% vol. | 80 | 27 | — | — | — | — | — | — | — | — |
| Uranium | $FeCl_3$ | 1,400 g/L | 75 | 24 | 0.04 | 0.001 | — | — | — | — | — | — |
| Zinc | $HNO_3$ | 10–15% vol. | 120 | 49 | 1.0 | 0.025 | — | — | — | — | — | — |
| Zirconium | $HNO_3$: $H_2SiF_6$: $NH_4HF_2$ | 4:2:1 | 90 | 32 | 1.8 | 0.046 | — | — | — | — | — | — |

*The ratio of depth of undercut to depth of cut.

† Baume specific gravity scale (Be′).

## PROCESS SUMMARY

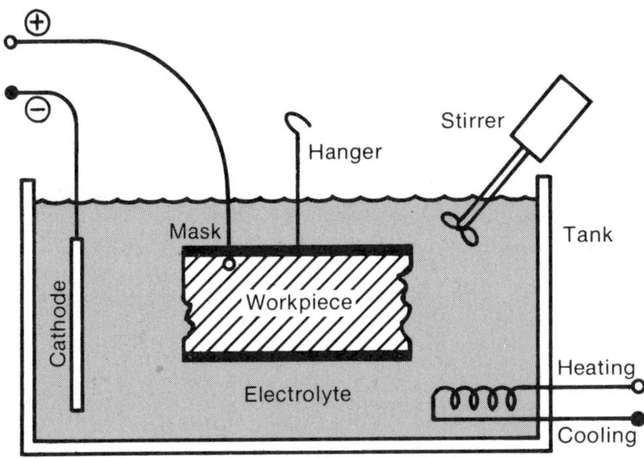


**Figure 13.2–1** ELP schematic.

### Principles

Electropolishing (ELP) is a specialized form of chemical machining which uses an electrical deplating action to enhance the chemical action (see figure 13.2–1). The chemical action from the concentrated heavy acids does most of the work, while the electrical action smooths or polishes the irregularities. A metal cathode is connected to a low-voltage, low-ampere, direct-current power source and is installed in the chemical bath near the workpiece. Usually, the cathode is not shaped or conformed to the surface being polished. The cutting action takes place over the entire exposed surface; therefore, a good flow of heated, fresh chemicals is needed in the cutting area to secure uniform finishes. The cutting action will concentrate first on burrs, fins and sharp corners. Masking, similar to that used with CHM, prevents cutting in unwanted areas. The proper choice of an electrolyte can enhance the "throwing power" of the direct current to polish recesses and speed the cutting action. Almost any electrically conductive material can be polished by this process.

### Practical Applications

Electropolishing's gentle removal action produces a workpiece surface of high quality and introduces no stresses. It is frequently used as a finishing process for appearance or smoothness, or for applications where the absence of process-induced residual stresses in the workpiece is desirable. Antifriction surfaces can be more consistent, and corrosion resistance can be improved. The pure or unaltered mechanical properties of a workpiece material remain after finishing with ELP.

On some alloys, for example, 17-4 PH, electropolishing is successful in removing tightly adhering scale after forging and as a preliminary process to passivation.

### Operating Parameters

**TABLE 13.2–1 Typical Values for ELP Operating Parameters**

| | |
|---|---|
| Bath | |
| types: | Acid mixtures, sodium hydroxide and proprietary solutions (see table 13.2–2) |
| temperature: | 70° to 200°F [21° to 93°C] |
| Current density: | 50 to several hundred amperes per square foot [0.05 to several times 0.1 A/cm²] |

### Material Removal Rates and Tolerances

The removal rates are very slow, on the order of 0.0005 to 0.0015 inch per minute [0.013 to 0.038 mm/min]; however, it usually requires only a few minutes exposure to achieve the desired surface improvement. The current density in the electrolyte concentrates around the surface asperities; consequently, erosion of the asperities takes place at a faster rate than erosion in the declivities. The surface, therefore, can be smoothed without considerable material removal.

### Surface Technology

Surface roughness depends upon the roughness of the initial surface and the time of immersion. The typical roughness values range from 4 to 32 microinches $R_a$ [0.1 to 0.8 μm], with 1 to 2 microinches $R_a$ [0.025 to 0.05 μm] achievable. When nonconducting masks are used, an additional depth of cutting is experienced adjacent to the mask. ELP produces a surface that is free from residual stress and essentially unblemished. The uniformity of surface conditions also is enhanced.

### Equipment and Tools

Commercial plating or chemical machining equipment can be readily adapted for ELP use.

### Machining Characteristics

Caution in handling the strong chemicals is essential. Suitable venting for fumes and environmentally proper disposal of spent fluids are necessary.

# 13.2 Electropolishing—ELP

## SELECTED DATA

**TABLE 13.2–2 Electropolishing Parameters for High Strength Alloys**

| MATERIAL | TYPE BATH | CURRENT DENSITY | | BATH TEMPERATURE | |
|---|---|---|---|---|---|
| | | A/ft² | A/cm² | °F | °C |
| 300 series stainless steel | Phosphoric acid | 100 | 0.10 | 122 | 50 |
| 300 & 400 series stainless steel | Phosphoric-Sulfuric acid | 30 | 0.03 | 80 to 280 | 27 to 138 |
| 300 & 400 series stainless steel | Phosphoric-Chromic acid | 100 to 1,000 | 0.10 to 1.0 | 80 to 175 | 27 to 80 |
| Waspaloy and Nimonic 80 | Phosphoric-Sulfuric acid | 860 | 0.9 | 82 to 90 | 28 to 32 |
| S-816 | Phosphoric-Sulfuric-Chromic acid | 500 | 0.5 | 200 to 260 | 93 to 126 |
| AMS 5615 | Phosphoric-Sulfuric acid | 400 | 0.4 | 200 | 93 |
| Beryllium (rapid agitation needed) | Perchloric-Ethanol-Butyl Cellosolve | 700 to 840 | 0.75 to 0.90 | <95 | <35 |
| | Phosphoric-Sulfuric-Glycerol-Ethanol | 1,900 to 3,700 | 2.0 to 4.0 | 68 | 20 |
| Titanium | Chromic-Hydrofluoric acid | 200 to 500 | 0.2 to 0.5 | 60 to 70 | 16 to 21 |
| | 10% Sulfamic acid in formamide (no water) | 6.5 to 8.1 | 0.007 to 0.0087 | 77 | 25 |
| | Phosphoric-Hydrofluoric-Methanol | 500 to 1,000 | 0.5 to 1.1 | 60 to 80 | 16 to 27 |
| Tungsten | Sodium Hydroxide | 30 to 50 | 0.03 to 0.05 | 68 | 20 |
| Rene 41 | Phosphoric-Sulfuric acid | 400 | 0.4 | 70 | 21 |
| Molybdenum | Ortho Phosphoric acid-Sulfuric acid-Molybdic Anhydride-Water | 3.9 to 5.8 | 0.004 to 0.006 | 160 | 71 |

## PROCESS SUMMARY

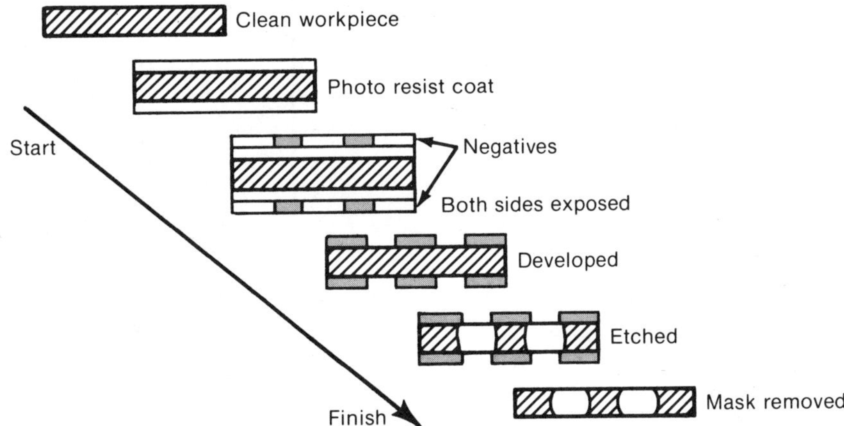


**Figure 13.3-1** PCM schematic.

### Principles

Photochemical machining (PCM) is a variation of CHM where the chemically resistant mask is applied to the workpiece by a photographic technique (see figure 13.3-1). A photographic negative, often a reduced image of an oversize master print, is applied to the workpiece and developed. Precise registry of duplicate negatives on each side of the sheet is essential for accurately blanked parts. Immersion or spray etching is used to remove the exposed material. The chemicals used must be active on the workpiece, but inactive against the photoresist mask. There will be some undercutting behind the mask, however, which limits the use of PCM to thin materials—up to 3/32 inch [2.4 mm] (see figure 13.3-2).

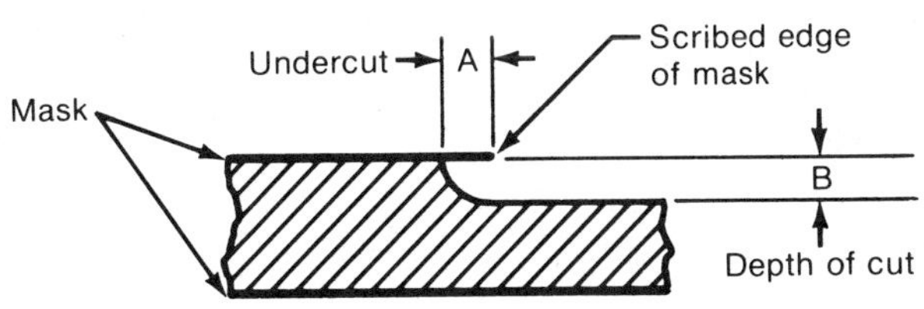


$$\text{Etch factor} = \frac{\text{Depth of cut}}{\text{Undercut}} = \frac{B}{A}$$

**Figure 13.3-2** Etch factor in photochemical machining. (Note: Inverse of that used in chemical machining.)

### Practical Applications

Photochemical blanking is burr free and is capable of producing intricate designs. A wide range of materials can be etched; brittle materials can be "blanked" since there is no mechanical strain. Small lot sizes can be produced at lower cost than with conventional mechanical presses, and there is a short cycle time from design to finished part. Cost of tooling is low and design changes can be quickly effected. Flat, thin-gauge, small complex parts ranging in thickness from 0.001 to 0.020 inch [0.025 to 0.50 mm] are most easily produced by PCM. Table 13.3-1 contains a list of work materials that can be processed with PCM.

### Operating Parameters

Operating parameters and typical values are similar to those for CHM; however, some metals that can be chemically machined do not easily respond to PCM because the photoresist masks are more fragile (for example, Waspaloy and Rene 41).

### Material Removal Rates and Tolerances

Cutting rates range from 0.0004 to 0.0020 inch per minute [0.01 to 0.05 mm/min], depending on the material and its metallurgical state (see tables in section 13.1). Tolerance increases with material thickness, as shown in tables 13.3-2 to 13.3-6 and figure 13.3-3. On surface dimensions, the typical tolerance is ±15 percent of the material thickness; however, any one dimension can be held to ±10 percent when other dimensions are allowed to vary by ±25 percent of the material thickness.

# 13.3 Photochemical Machining—PCM

**TABLE 13.3-1 Materials that Can Be Photochemically Machined**

**METALS**
Aluminum
Aluminum, anodized
Brass
Cold rolled steel
Chromium
Copper, oxygen free
Copper, rolled
Copper, electrolytic
Copper, beryllium
Copper (OFHC)
Elinvar Extra
Elgiloy
Gold
Havar
Inconel
Lead
Magnesium
Manganese
Molybdenum
Molybdenum, cold rolled
Molybdenum, arc cast
Monel
Nickel
Nickel silver
Nichrome
Ni Span C
Permanickel
Phospor bronze
Rehnium
Silver
Steels
- Stainless steel
  - 300 series
- PH 15-7 Steel
- PH 17-7 Steel
- Custom 455
- Spring steel

Tungsten
Udimet
Vanadium
Zinc
Zirconium
90-10 Copper nickel

GLASS TO METAL AND CERIUM TO METAL

ASTM F-15 Alloy
- Kovar
- Rodar
- Nicoseal
- Therlo
- Sealmet 29-17

ASTM F-30 Alloy
- Glass Sealing 42
- 142 Alloy
- Glass Sealing 46
- 146 Alloy
- Glass Sealing 49
- 4750
- Glass Sealing 52
- 152 Alloy

ASTM F-31 Alloy
- Glass Sealing 42-6
- Sylvania No. 4 Alloy
- Sealmet No. 4

SOFT MAGNETIC ALLOYS

44-50% Ni Fe Balance
- High Permeability 49
- 4750 Alloy
- Hipernik
- High Permeability 45

High Permeability Alloys
- Hymu 80
- 4-79 Permalloy
- Hipernom
- Mumetal
- Hymu 800
- Supermalloy
- 5-79 Permalloy

ELECTRICAL SILICON IRONS

Orthosil

CONTROLLED EXPANSION ALLOYS

Invar 36

**OTHERS**
Plastics
Polyesters
Polymides
Epoxy resins
Glass
Ceramics

SOURCE: Photo Chemical Machining Institute, Evanston, IL.

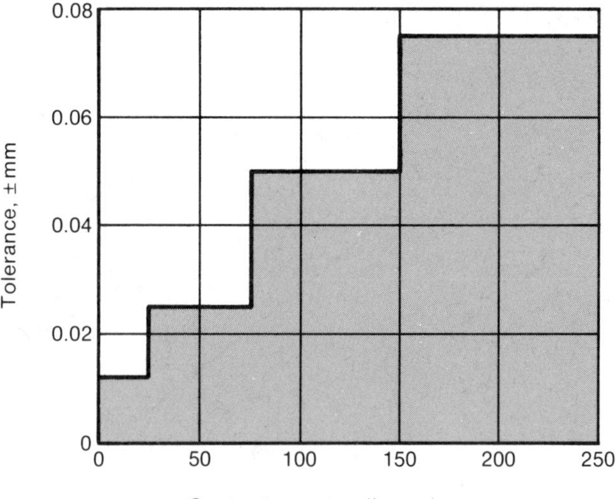


**Figure 13.3–3** Practical tolerances on center-to-center dimensions using photochemical milling. (W. T. Harris, *Chemical milling*, Oxford: Clarendon Press, 1976, p. 191)

### Surface Technology

Similar to that for CHM.

### Equipment and Tooling

Batch and automatic equipment are available for both dip and spray etching of strips up to 2-feet [0.6 m] wide. Some installations include an electrical assist to the light, low-concentration chemicals in order to obtain a straighter cut. Direct-current power, with low current density, is required. Photoreducing and photoreproducing equipment are readily available and can be integrated into continuous blanking lines.

### Machining Characteristics

Safety considerations in handling the corrosive chemicals and fumes are similar to those for CHM.

The etch factor for photochemical machining and the circuit board industry, figure 13.3–2, has been defined as the inverse of the ratio used in the chemical machining industry. The tool designer must exercise care to be sure which is applicable in his case.

## SELECTED DATA

**TABLE 13.3-2 Standard PCM Tolerances for Common Materials**

| WORKPIECE MATERIAL | TOLERANCE, inch | | | | | |
|---|---|---|---|---|---|---|
| | Workpiece Thickness, inch | | | | | |
| | 0.001 | 0.002 | 0.005 | 0.010 | 0.015 | 0.020 |
| Copper, copper alloys and glass sealing alloys (Nicoseal*) | ±0.0002 | ±0.0005 | ±0.001 | ±0.0015 | ±0.0025 | ±0.0035 |
| Nickel-silver | ±0.0005 | ±0.001 | ±0.001 | ±0.0015 | ±0.0025 | ±0.0035 |
| Magnetic Ni-Fe alloys (HyMu 80*) | ±0.0005 | ±0.001 | ±0.001 | ±0.0015 | ±0.0025 | ±0.0035 |
| Steel | ±0.0005 | ±0.001 | ±0.0015 | ±0.0015 | — | — |
| Nickel and stainless steel | ±0.0005 | ±0.001 | ±0.0015 | ±0.002 | ±0.003 | — |
| Aluminum and magnesium | ±0.001 | ±0.0015 | ±0.0025 | — | — | — |
| Plastics (Mylar†, Kapton†) | ±0.001 | ±0.0015 | ±0.0025 | ±0.005 | — | — |
| Molybdenum, titanium and exotics | ±0.0005 | ±0.001 | ±0.002 | — | — | — |

| WORKPIECE MATERIAL | TOLERANCE, mm | | | | | |
|---|---|---|---|---|---|---|
| | Workpiece Thickness, mm | | | | | |
| | 0.025 | 0.050 | 0.13 | 0.25 | 0.38 | 0.50 |
| Copper, copper alloys and glass sealing alloys (Nicoseal*) | ±0.005 | ±0.013 | ±0.025 | ±0.038 | ±0.063 | ±0.089 |
| Nickel-silver | ±0.013 | ±0.025 | ±0.025 | ±0.038 | ±0.063 | ±0.089 |
| Magnetic Ni-Fe alloys (HyMu 80*) | ±0.013 | ±0.025 | ±0.025 | ±0.038 | ±0.063 | ±0.089 |
| Steel | ±0.013 | ±0.025 | ±0.038 | ±0.038 | — | — |
| Nickel and stainless steel | ±0.013 | ±0.025 | ±0.038 | ±0.050 | ±0.076 | — |
| Aluminum and magnesium | ±0.025 | ±0.038 | ±0.063 | — | — | — |
| Plastics (Mylar†, Kapton†) | ±0.025 | ±0.038 | ±0.063 | ±0.13 | — | — |
| Molybdenum, titanium and exotics | ±0.013 | ±0.025 | ±0.050 | — | — | — |

SOURCE: ChemPar Corporation, Montgomeryville, PA.

*Registered Trademark—Carpenter Technology Corporation.

† Registered Trademark—E. I. duPont deNemours and Co. Inc.

**TABLE 13.3–3 Practical Tolerances Attainable for Prototype and Short PCM Runs**

| APPROXIMATE FLAT SIZE | TOLERANCE, inch | | | | | | |
|---|---|---|---|---|---|---|---|
| | Thickness, inch | | | | | | |
| in | 0.001 | 0.002 | 0.005 | 0.010 | 0.015 | 0.020 | 0.040 |
| 2 x 2 | Empirical | ±0.0005 | ±0.0007 | ±0.0010 | ±0.0015 | ±0.0020 | ±0.0040 |
| 8 x 10 | Empirical | ±0.0007 | ±0.0010 | ±0.0015 | ±0.0020 | ±0.0030 | ±0.0050 |
| 12 x 18 | Empirical | ±0.0010 | ±0.0015 | ±0.0020 | ±0.0030 | ±0.0040 | ±0.0060 |
| | TOLERANCE, mm | | | | | | |
| | Thickness, mm | | | | | | |
| mm | 0.025 | 0.050 | 0.13 | 0.25 | 0.38 | 0.50 | 1.0 |
| 50 x 50 | Empirical | ±0.013 | ±0.018 | ±0.025 | ±0.038 | ±0.051 | ±0.10 |
| 200 x 250 | Empirical | ±0.018 | ±0.025 | ±0.038 | ±0.050 | ±0.076 | ±0.13 |
| 300 x 450 | Empirical | ±0.025 | ±0.038 | ±0.050 | ±0.076 | ±0.10 | ±0.15 |

SOURCE: Data courtesy of Photo Chemical Machining Institute.

**TABLE 13.3–4 Practical Tolerances Attainable for PCM Production Runs**

| APPROXIMATE FLAT SIZE | TOLERANCE, inch | | | | | | |
|---|---|---|---|---|---|---|---|
| | Thickness, inch | | | | | | |
| in | 0.001 | 0.002 | 0.005 | 0.010 | 0.015 | 0.020 | 0.040 |
| 2 x 2 | Empirical | ±0.0010 | ±0.0010 | ±0.0015 | ±0.0020 | ±0.0030 | ±0.0050 |
| 8 x 10 | Empirical | ±0.0010 | ±0.0015 | ±0.0020 | ±0.0025 | ±0.0040 | ±0.0060 |
| 12 x 18 | Empirical | ±0.0015 | ±0.0020 | ±0.0025 | ±0.0035 | ±0.0045 | ±0.0070 |
| | TOLERANCE, mm | | | | | | |
| | Thickness, mm | | | | | | |
| mm | 0.025 | 0.050 | 0.13 | 0.25 | 0.38 | 0.50 | 1.0 |
| 50 x 50 | Empirical | ±0.025 | ±0.025 | ±0.038 | ±0.050 | ±0.076 | ±0.13 |
| 200 x 250 | Empirical | ±0.025 | ±0.038 | ±0.050 | ±0.063 | ±0.10 | ±0.15 |
| 300 x 450 | Empirical | ±0.038 | ±0.050 | ±0.063 | ±0.089 | ±0.11 | ±0.18 |

SOURCE: Data courtesy of Photo Chemical Machining Institute.

**TABLE 13.3–5 Practical Tolerances for Center-to-Center Dimensions in PCM**

| CENTER-TO-CENTER DIMENSIONS | | TOLERANCES | |
|---|---|---|---|
| in | mm | in | mm |
| 1.0 or less | 25 or less | ±0.0005 | ±0.013 |
| 1.0 to 3.0 | 25 to 75 | ±0.0010 | ±0.025 |
| 3.0 to 6.0 | 75 to 150 | ±0.0020 | ±0.050 |
| 6.0 to 10.0 | 150 to 250 | ±0.0030 | ±0.076 |

SOURCE: Data courtesy of Photo Chemical Machining Institute.

**TABLE 13.3–6 Practical Tolerances for Holes Made by PCM**

| MATERIAL THICKNESS | | PRACTICAL HOLE DIAMETER | | TOLERANCES ATTAINABLE | |
|---|---|---|---|---|---|
| in | mm | in | mm | in | mm |
| 0.0005 | 0.013 | Empirical | Empirical | Empirical | Empirical |
| 0.0010 | 0.025 | Empirical | Empirical | Empirical | Empirical |
| 0.0020 | 0.050 | 0.0030 | 0.076 | ±0.001 | ±0.025 |
| 0.0050 | 0.13 | 0.0060 | 0.15 | ±0.0015 | ±0.038 |
| 0.0070 | 0.18 | 0.0080 | 0.20 | ±0.0015 | ±0.038 |
| 0.0100 | 0.25 | 0.0120 | 0.30 | ±0.002 | ±0.051 |
| 0.0200 | 0.50 | 0.0260 | 0.66 | ±0.003 | ±0.076 |

SOURCE: Data courtesy of Photo Chemical Machining Institute.

**TABLE 13.3–7 Guidelines for Smallest Hole Sizes by PCM**

| METAL THICKNESS | | SMALLEST HOLE DIAMETER |
|---|---|---|
| in | mm | |
| less than 0.001 | less than 0.025 | Must be determined by test run |
| 0.001 to 0.005 | 0.025 to 0.13 | At least metal thickness |
| 0.005 or over | 0.13 or over | At least 110% metal thickness |

SOURCE: Data courtesy of Photo Chemical Machining Institute.

## PROCESS SUMMARY

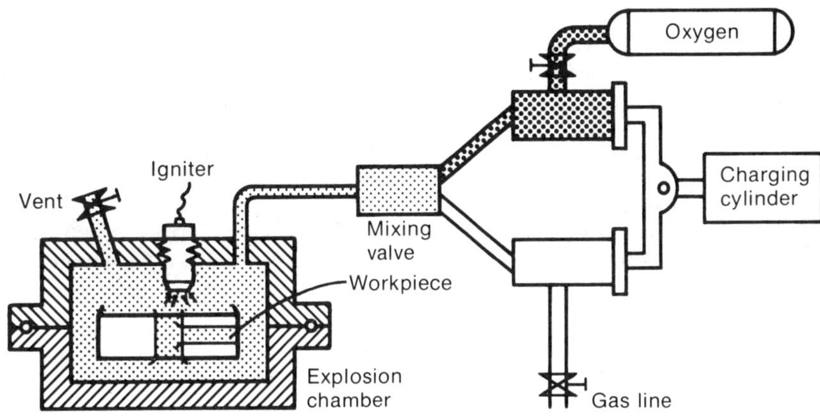


**Figure 13.4–1** TCM schematic.

### Principles

Thermochemical machining (TCM) removes workpiece material—usually only burrs and fins—by exposure of the workpiece to hot, corrosive gases (see figure 13.4–1). The process is sometimes called combustion machining, thermal deburring, or thermal energy method (TEM). The workpiece is exposed for a very short time to extremely hot gases, which are formed by detonating an explosive mixture. The ignition of the explosive—usually hydrogen or natural gas and oxygen—creates a transient thermal wave that vaporizes the burrs and fins. The main body of the workpiece remains unaffected and relatively cool because of its low surface-to-mass ratio and the brevity of the exposure to high temperatures.

### Practical Applications

TCM will remove burrs or fins from a wide range of materials, but it is particularly effective with materials of low thermal conductivity. It will deburr thermosetting plastics—but not thermoplastic materials. Any modest size workpiece requiring manual deburring or flash removal should be considered a candidate for thermal deburring. Die castings, gears, valves, rifle bolts and similar small parts are deburred readily, including blind, internal and intersecting holes in inaccessible locations. Carburetor parts are processed in automated equipment. Thin sections of fragile parts or those made from highly oxidation-resistant materials may be difficult for TCM. As a rule of thumb, the maximum burr thickness should be less than one-fifteenth of the thinnest feature on the workpiece. Uniformity of results and greater quality assurance over hand deburring is a special advantage of TCM.

### Operating Parameters

**TABLE 13.4–1 Typical Values for TCM Operating Parameters**

| | |
|---|---|
| Gas | |
| type: | Natural gas and oxygen or hydrogen and oxygen |
| mixture: | 4:1 to 9:1 (oxygen:gas) |
| pressure (initial): | 10 to 370 psi [69 to 255 kPa] (see table 13.4–2) |
| Thermal wave | |
| temperature: | Estimated 6,000°F [3,315°C] |
| time: | 1 to 2 microseconds: Mach 8 |

### Material Removal Rates and Tolerances

Burrs are removed in a few milliseconds and the total cycle time is typically 15 to 50 seconds. Handling time is the principal manufacturing rate determinant. Rifle bolts that formerly took 5 minutes each to deburr by hand are being processed with more consistent results at the rate of 2 pieces in 8 seconds in a batch processing setup. The bulk of the workpiece is generally not affected in TCM. With the application of extra energy, radii can be formed. Radiusing is limited by the conductivity of the workpiece material, for example:

| | |
|---|---|
| radii in steel | 0.002 to 0.030 inch [0.05 to 0.76 mm] |
| radii in aluminum and brass | 0.001 to 0.005 inch [0.025 to 0.13 mm] |

# 13.4 Thermochemical Machining—TCM
## or Thermal Energy Method—TEM

### Surface Technology

Other than the deburring or radiusing action, there is little effect of TCM on the surface of the workpiece. Thermal distortion is not a problem unless extremely close tolerances are involved. There can be a thin recast layer below thick burrs or fins. A thin oxide film is deposited on most parts. These films are easily removed with commercial solvents or washes.

### Equipment and Tools

Automatic equipment is commercially available with chambers up to 11 inches [280 mm] diameter and 9 inches [229 mm] high. A cycle time of 15 seconds is possible with this equiment. Fixturing is primarily oriented toward simplifying the movement of workpieces into and out of the processing chamber. Delicate parts may require holding fixtures to withstand buffeting during the thermal shock wave. Bulk loading baskets are used for handling small parts in lots up to hundreds per load.

### Machining Characteristics

Surface stains may result if the workpiece is not clean, dry and free of oil before TCM. The equipment has a stout chamber to contain the detonation. Preplanned handling of the pressurized gases is needed. Noise can range from a gentle "ping" to the sharp report like a .22-caliber rifle when the higher pressure settings are used. Even at full load, the average sound emission level generally remains below 85 decibels. Cleaning of the operating chamber is required at regular intervals.

## SELECTED DATA

**TABLE 13.4-2 Thermal Energy Deburring Parameters**

| MATERIAL | MIXTURE $O_2$ TO NATURAL GAS | INITIAL PRESSURE | |
|---|---|---|---|
| | | psi | kPa |
| Thermplastic | 9 to 1 | 10 to 25 | 69 to 172 |
| Zinc | 4.5 to 1 | 30 to 70 | 207 to 483 |
| Cast iron | 4.5 to 1 | 50 to 175 | 345 to 1,206 |
| Aluminum | 4.5 to 1 | 50 to 125 | 345 to 862 |
| Brass | 4.5 to 1 | 200 to 240 | 1,378 to 1,655 |
| Steel | 4.5 to 1 | 100 to 220 | 689 to 1,516 |
| Stainless steel | 4.5 to 1 | 175 to 370 | 1,206 to 2,551 |

SOURCE: SURF/TRAN Company, Madison Heights, MI.

NOTE: The pressure or amount of energy required for deburring increases with the heat conductivity of the material to be deburred.

## SECTION 14

# TOOL MATERIALS


14.1 General Guidelines for Selection of Tool Materials ..... 14–3
14.2 High Speed Steels ..... 14–4
14.3 Cast Alloys ..... 14–6
14.4 Carbides ..... 14–7
14.5 Micrograin Carbides ..... 14–15
14.6 Coated Carbides ..... 14–16
14.7 Ceramic Tool Materials ..... 14–17
14.8 Diamond Tools ..... 14–18

## 14.1 General Guidelines for Selection of Tool Materials

Selecting the optimum tool material is a major factor in realizing the full potential of a particular machine tool. While only one type of tool material is specified for some materials and operations in sections 1 through 7, there are a number of cases for which two or three material types are recommended. The following supplemental guidelines are offered as an approach for the logical selection of the best tool materials for machining a specific work material.

1. High speed steel tools are generally used for the following:

   - High volume, low cutting speed operations (for example, in screw machines)
   - Complex tool forms such as form tools, drills, cutoff tools, etc.
   - All sizes of end mills, drills, reamers, taps and gear cutters
   - Certain machining operations on problem materials, such as nickel base high temperature alloys
   - High positive rake requirements

2. Cast alloy cutting tool materials are selected as an intermediate between high speed steel and carbide tool materials. The high cobalt high speed steels also serve as intermediates, and there appears to be a trend for them to supersede the cast alloy tools.

3. Carbide tools are generally applicable when one or more of the following conditions exist:

   - Rigidity of the machine tool, tooling, and workpiece is acceptable.
   - Machine tool power is adequate for higher metal removal rates.
   - Workpiece configuration and machining operation permit higher cutting speeds.
   - High production rates are required.

4. Ceramic tools, high strength carbides, diamond tools, and the cast alloy tools referred to previously have rather specific application in contrast with the wide usage of high speed steel and carbide tools.

## 14.2 High Speed Steels

The high speed steels can be classified into three general types as follows:

1. Tungsten high speed steels
2. Molybdenum high speed steels
3. High speed steels containing cobalt

The chemical compositions of the AISI and the ISO high speed steels are listed in tables 14-1 and 14-2, respectively. Note that the type numbers and designations listed in table 14-2 are tentative, having been taken from the proposed draft international standard ISO/DIS 4957.

As a general rule, the tungsten grades such as T1, T2, etc., are not quite as tough as the molybdenum grades but are much simpler to heat treat. The molybdenum high speed steels such as M1, M2, M3, M7 and M10 are much more widely used than the tungsten grades. Both the tungsten and the molybdenum high speed steels can be hardened to 64 to 66 $R_C$ and are recommended for the machining of easy-to-machine materials. These include steels having a hardness up to 350 Bhn. While high speed steel types 1 and 2 can be used in machining steels having a higher hardness, the type 3 high speed steels containing cobalt are recommended. It should be noted that several of the high speed steels such as M3 and M4 contain more vanadium, thus providing increased wear resistance, but at the same time making the grinding of the tools more difficult.

**TABLE 14-1 Compositions and Applications of AISI-Type High Speed Tool Steels**

| TYPE | NOMINAL COMPOSITION, percentage | | | | | | APPLICATION |
|---|---|---|---|---|---|---|---|
| | C | W | Mo | Cr | V | Co | |
| 1. Tungsten High Speed Steels | | | | | | | |
| T1 | 0.75 | 18.00 | — | 4.00 | 1.00 | — | General purpose |
| T2 | 0.80 | 18.00 | — | 4.00 | 2.00 | — | General purpose—Higher strength |
| 2. Molybdenum High Speed Steels | | | | | | | |
| M1 | 0.80 | 1.50 | 8.00 | 4.00 | 1.00 | — | General purpose |
| M2 | 0.85 | 6.00 | 5.00 | 4.00 | 2.00 | — | General purpose |
| M3 Class 1 | 1.05 | 6.00 | 5.00 | 4.00 | 2.40 | — | Fine edge tools |
| M3 Class 2 | 1.20 | 6.00 | 5.00 | 4.00 | 3.00 | — | Fine edge tools |
| M4 | 1.30 | 5.50 | 4.50 | 4.00 | 4.00 | — | Abrasion resistant |
| M7 | 1.00 | 1.75 | 8.75 | 4.00 | 2.00 | — | Fine edge tools—Abrasion resistant |
| M10 | 0.85 | — | 8.00 | 4.00 | 2.00 | — | General purpose—High strength |
| 3. High Speed Steels Containing Cobalt | | | | | | | |
| M6 | 0.80 | 4.00 | 5.00 | 4.00 | 1.50 | 12.00 | Heavy cuts—Abrasion resistant |
| M30 | 0.80 | 2.00 | 8.00 | 4.00 | 1.25 | 5.00 | Heavy cuts—Abrasion resistant |
| M33 | 0.90 | 1.50 | 9.50 | 4.00 | 1.15 | 8.00 | Heavy cuts—Abrasion resistant |
| M34 | 0.90 | 2.00 | 8.00 | 4.00 | 2.00 | 8.00 | Heavy cuts—Abrasion resistant |
| M36 | 0.80 | 6.00 | 5.00 | 4.00 | 2.00 | 8.00 | Heavy cuts—Abrasion resistant |
| M41 | 1.10 | 6.75 | 3.75 | 4.25 | 2.00 | 5.00 | Heavy cuts—Abrasion resistant |
| M42 | 1.10 | 1.50 | 9.50 | 3.75 | 1.15 | 8.00 | Heavy cuts—Abrasion resistant |
| M43 | 1.20 | 2.75 | 8.00 | 3.75 | 1.60 | 8.25 | Heavy cuts—Abrasion resistant |
| M44 | 1.15 | 5.25 | 6.25 | 4.25 | 2.25 | 12.00 | Heavy cuts—Abrasion resistant |
| M46 | 1.25 | 2.00 | 8.25 | 4.00 | 3.20 | 8.25 | Heavy cuts—Abrasion resistant |
| M47 | 1.10 | 1.50 | 9.50 | 3.75 | 1.25 | 5.00 | Heavy cuts—Abrasion resistant |
| T4 | 0.75 | 18.00 | — | 4.00 | 1.00 | 5.00 | Heavy cuts |
| T5 | 0.80 | 18.00 | — | 4.00 | 2.00 | 8.00 | Heavy cuts—Abrasion resistant |
| T6 | 0.80 | 20.00 | — | 4.50 | 1.50 | 12.00 | Heavy cuts—Hard material |
| T8 | 0.75 | 14.00 | — | 4.00 | 2.00 | 5.00 | General purpose—Hard material |
| T15 | 1.50 | 12.00 | — | 4.00 | 5.00 | 5.00 | Extreme abrasion resistant |

The high speed steels with cobalt, type 3, contain 5 to 12 percent cobalt. The addition of cobalt provides greater hot hardness and wear resistance but results in a somewhat lower toughness. In general, it has been found that these grades are not particularly advantageous in the machining of the readily machinable materials. They are, however, most beneficial for machining steels having a hardness level above 350 Bhn and for the more difficult-to-machine metals, such as titanium and nickel base high temperature alloys. The T15 grade also has a higher percentage of vanadium; consequently, it is more difficult and more costly to grind than the other high speed steels containing cobalt.

In general, the high speed steels containing cobalt can be heat treated to a hardness level of 65 to 67 $R_C$. While the M40 series can be hardened to a level of 70 $R_C$, the hardness range usually recommended is 66 to 68 $R_C$. Above 68 $R_C$, high speed steel tools tend to be too brittle for most applications.

High speed steel cutters are widely used for the following machining operations:

- Operations using form tools
- Screw machine operations
- End milling
- Drilling
- Reaming (carbide-tipped reamers are also widely used)
- Tapping (nitrided taps are often used for difficult-to-machine alloys)
- Broaching
- Gear cutting

A recent development in high speed steels is the availability of indexable inserts made by the powder metal process. The sizes and shapes of the inserts correspond to those of some of the more popular carbide inserts. The current grade selection is limited, but in the future, selection is likely to expand.

**TABLE 14–2 Identification and Type Classification of High Speed Tool Steels**
(Per ISO/DIS 4957)

| TYPE OF STEEL | | | | CHEMICAL COMPOSITION, percentage§ | | | | | |
|---|---|---|---|---|---|---|---|---|---|
| Group | ISO No.* | AISI No.† | Designation*‡ | C | Co# | Cr | Mo# | V | W |
| Basic types | S1 | T1 | HS 18-0-1 | 0.73 to 0.83 | — | 3.50 to 4.50 | — | 0.90 to 1.20 | 17.2 to 18.7 |
| | S2 | M7 | HS 2-9-2 | 0.95 to 1.05 | — | 3.50 to 4.50 | 8.20 to 9.20 | 1.70 to 2.20 | 1.50 to 2.10 |
| | S3 | M1 | HS 1-8-1 | 0.77 to 0.87 | — | 3.50 to 4.50 | 8.00 to 9.00 | 0.90 to 1.40 | 1.40 to 2.00 |
| | S4 | M2 | HS 6-5-2 | 0.82 to 0.92 | — | 3.50 to 4.50 | 4.60 to 5.30 | 1.70 to 2.20 | 5.70 to 6.70 |
| Increased C + V-content | S5 | M3 | HS 6-5-3 | 1.15 to 1.30 | — | 3.50 to 4.50 | 4.60 to 5.30 | 2.70 to 3.20 | 5.70 to 6.70 |
| With Co-content | S6 | use T5 | HS 18-0-1-10 | 0.75 to 0.85 | 9.50 to 10.5 | 3.50 to 4.50 | — | 1.30 to 1.80 | 17.2 to 18.7 |
| | S7 | use T4 | HS 18-1-1-5 | 0.75 to 0.85 | 4.70 to 5.20 | 3.50 to 4.50 | 0.70 to 1.00 | 1.10 to 1.60 | 17.2 to 18.7 |
| | S8 | M35 | HS 6-5-2-5 | 0.85 to 0.95 | 4.70 to 5.20 | 3.50 to 4.50 | 4.60 to 5.30 | 1.70 to 2.20 | 5.70 to 6.70 |
| Increased C + V content + Co | S9 | T15 | HS 12-1-5-5 | 1.45 to 1.60 | 4.70 to 5.20 | 3.50 to 4.50 | 0.70 to 1.00 | 4.75 to 5.55 | 11.5 to 13.0 |
| | S10 | — | HS 10-4-3-10 | 1.20 to 1.35 | 9.50 to 10.5 | 3.50 to 4.50 | 3.20 to 3.90 | 3.00 to 3.50 | 9.00 to 10.0 |
| Increased C-content + Co | S11 | M42 | HS 2-9-1-8 | 1.05 to 1.20 | 7.50 to 8.50 | 3.50 to 4.50 | 9.00 to 10.0 | 0.90 to 1.40 | 1.30 to 1.90 |
| | S12 | M41 | HS 7-4-2-5 | 1.05 to 1.20 | 4.70 to 5.20 | 3.50 to 4.50 | 3.50 to 4.20 | 1.70 to 2.20 | 6.40 to 7.40 |

*The type numbers and designations are tentative and will be subject to alteration when the relevant International Standards have been established.

†Approximate AISI equivalent—See table 14–1.

‡Successively W-Mo-V-Co.

§For all steels: Silicon < 0.50%, Manganese < 0.40%, Phosphorus < 0.030%, and Sulphur < 0.030%.

#A dash in these columns indicates that these alloying elements should not be intentionally added to the heat and, that in the case of cobalt, the content of the cast should not exceed 1.00% and, in the case of molybedenum, it should not exceed 0.70%.

## 14.3 Cast Alloys

Cast alloys have been available for many years. These materials are generally cobalt-chrome-tungsten alloys with carbon and other alloy additions. They are not heat treatable, and the maximum hardness (55 to 65 $R_C$) occurs near the cast surface. As a result, cast alloy tools must be used as-cast with as little grinding as possible.

Cast alloy materials are not widely used. However, they find limited use as a compromise since they perform well at higher surface speeds than conventional high speed steels and are more resistant to chipping than standard carbide grades. A number of cast alloy tool materials and their compositions are shown in table 14-3.

**TABLE 14-3 Cast Alloy Tool Materials**

| TRADE NAME | NOMINAL COMPOSITION percentage | | | | | HARDNESS, $R_c$ | MANUFACTURER |
|---|---|---|---|---|---|---|---|
| | Co | Cr | W | C | OTHERS | | |
| Blackalloy 525 | 44.0 | 24.0 | 20.0 | 2.0 | | 61 to 63 | Blackalloy Company of America, Inc. Paterson, NJ |
| Blackalloy T.X. 90 | 42.0 | 24.0 | 22.0 | 2.0 | | 63 to 65 | |
| Haynes Stellite R 98M2 alloy | 38.0 | 30.0 | 18.5 | 2.0 | 1.2% boron, 4%V, 4% nickel | 63 | Co-Bo Cast Tool Co. Inc. Kokomo, IN |
| Haynes Stellite R Star J metal | 43.0 | 32.5 | 17.5 | 2.5 | | 61 | |
| Haynes Stellite R alloy No. 3 | 50.0 | 31.0 | 12.5 | 2.4 | | 59 | |
| Haynes Stellite R alloy No. 19 | 53.0 | 31.0 | 10.5 | 1.8 | | 55 | |
| Tantung G | 47.0 | 30.0 | 15.0 | 3.0 | 5% Ta or Cb | 60 to 63 | Fansteel, VR/Wesson Division Waukegan, IL |
| Tantung 144 | 45.0 | 28.0 | 18.0 | 3.0 | 6% Ta or Cb | 62 to 64.5 | |

NOTES: The procedure used for making cast alloy tools is to melt the particular analysis in an electric furnace and then cast the tools to shape with small stock allowance for finish grinding. The tools are at maximum hardness when removed from the molds and cannot be hot or cold worked and do not respond to heat treatment.

"Haynes Stellite" is a registered trademark of Cabot Corporation.

This chart is not to be considered an endorsement of any manufacturer's product or an approved list of any manufacturer's products.

## 14.4 Carbides

Each tool manufacturer listed in table 14-4 has a wide variety of carbide grades so that a specific grade can be selected for a given material and operation. Although most of the carbide grades are listed under the industry code (C-1, C-2, etc.) in table 14-5, this does not necessarily imply that the various manufacturers' grades under a specific code are equivalent.

The C-1 through C-4 grades are straight tungsten carbide bonded with cobalt and vary chiefly in cobalt content and grain size. They are widely used for machining cast irons, high temperature alloys, work hardening stainless steels, abrasive nonmetallic materials and nonferrous metals including titanium. Toughness decreases and hardness increases proceeding from C-1 to C-4. The C-2 grades provide a good compromise in properties and may be categorized as "general purpose" for the aforementioned materials.

The C-5 through C-8 grades include those which contain various combinations of tungsten carbide, tantalum carbide and titanium carbide bonded with cobalt. The C-8 category now also includes tools of straight titanium carbide bonded with molybdenum (or molybdenum carbide) and nickel. Other straight titanium carbide grades with properties in the C-5 through C-7 range are also available. Grades C-5 through C-8 are generally recommended for machining steels because they provide better crater resistance than the C-1 to C-4 grades. Toughness decreases and the hardness increases in going from C-5 to C-8. The C-6 grades are for general-purpose use.

For work materials such as highly alloyed cast irons, alloy steels over 50 $R_C$, ferritic or martensitic stainless steels and some high temperature alloys, the preferred carbide grades may come from either the C-1 to C-4 or the C-5 to C-8 group.

Table 14-6 has been used in classifying carbides according to use as specified by the International Organization for Standardization (ISO). Recommended grades produced by various manufacturers are listed according to ISO classifications in table 14-7.

**TABLE 14-4 Carbide Manufacturers (Carbide Grade Chart)**

| CARBIDE | MANUFACTURER |
|---|---|
| Adamas | ADAMAS CARBIDE CORPORATION, Kenilworth, NJ |
| Atrax | ATRAX RESEARCH & DEVELOPMENT CENTER, Wallace-Murray Corporation, Tuscaloosa, AL |
| Carboloy | CARBOLOY SYSTEMS DEPARTMENT, General Electric Company, Detroit, MI |
| Carmet | CARMET COMPANY, A Member Company of Allegheny Ludlum Industries, Madison Heights, MI |
| DoAll | DOALL COMPANY, Des Plaines, IL |
| Duramet | DURAMET CORPORATION, Warren, MI |
| Ex-Cell-O | EX-CELL-O CORPORATION, Detroit, MI |
| Greenleaf | GREENLEAF CORPORATION, Saegertown, PA |
| GTE (Walmet) | GTE SYLVANIA INCORPORATED, Walmet Division, Royal Oak, PA |
| Iscar | ISCAR METALS, INC., Edison, NJ |
| Kennametal | KENNAMETAL INC., Latrobe, PA |
| Newbide | NEWCOMER PRODUCTS, INC., Latrobe, PA |
| Sandvik (Coromant) | SANDVIK, INC., Fair Lawn, NJ |
| Seco | SECO TOOLS INC., Northbrook, IL |
| Talide | METAL CARBIDES CORPORATION, Youngstown, OH |
| Teledyne (Firthite) | TELEDYNE FIRTH STERLING, Pittsburgh, PA |
| TRW Wendt-Sonis | WENDT-SONIS DIVISION, TRW Inc., Rogers, AR |
| Tungsten Alloy | TUNGSTEN ALLOY MANUFACTURING CO., INC., Harrison, NJ |
| Ultra-Met | ULTRA-MET MANUFACTURING COMPANY, Farmington, MI |
| Valenite | THE VALERON CORPORATION, Oak Park, MI |
| VR/Wesson | FANSTEEL, VR/Wesson Division, Waukegan, IL |

**TABLE 14–5 Carbide Grade Chart, C Grade**

| TRADE NAME | INDUSTRY CODE | | | | | | | |
|---|---|---|---|---|---|---|---|---|
| | C-1 | C-2 | C-3 | C-4 | C-5 | C-6 | C-7 | C-8 |
| Adamas | B | A<br>AM<br>ACT-2† | PWX<br>ACT-2† | GU-2 | 499<br>434<br>ACT-5†<br>AG-5‡<br>474<br>Titan 50* | ACT-5†<br>AG-5‡<br>495<br>Titan 60* | 548<br>ACT-7†<br>Titan 80* | 490<br>ACT-7†<br>Titan 80*<br>Titan 100* |
| Atrax | M-22 | A6 | A7 | A7 | T3<br>T50<br>T56<br>T50C | T50<br>T64<br>T65 | T63<br>T76<br>T78<br>T76C | T8 |
| Carboloy | 518†<br>44A<br>820 | 570§<br>515†<br>523†<br>883 | 545§<br>570§<br>515†<br>523†<br>895 | 210<br>545§<br>999<br>570§ | 518†<br>370<br>375<br>390 | 570§<br>515†<br>516†<br>350<br>518†<br>370 | 545§<br>570§<br>515†<br>516†<br>350 | 210*<br>320 |
| Carmet | CA-12<br>CA-3 | CA-310<br>CA-4<br>CA-443<br>CA-8443†<br>CA-9443# | CA-7<br>CA-8443†<br>CA-9443#<br>R-03* | CA-8<br>R-03* | CA-610<br>CA-740<br>CA-721<br>CA-9740#<br>CA-9721# | CA-606<br>CA-720<br>CA-9720# | CA-711<br>CA-9711# | CA-704<br>R-03* |
| DoAll | DO-1<br>DO-30 | DO-2<br>DO-20<br>DO-46† | DO-3 | DO-4<br>DO-80 | DO-15<br>DO-35<br>DO-44† | DO-16<br>DO-36<br>DO-40† | DO-17<br>DO-34<br>DO-42† | DO-18<br>DO-80 |
| Duramet | DU1 | DU2<br>DU22 | DU3 | DU4 | DU5<br>DU55 | DU6 | DU7<br>DU78 | DU8 |
| Ex-Cell-o | E-8 | E-6<br>XLO-28<br>XLO-620<br>XL-202†† | E-5<br>XL-202†† | E-3 | 10A<br>8A<br>XL-602†† | 8A<br>606<br>6A<br>XL-85*<br>XL-602†† | 6AX<br>XLO-61<br>XL-86* | 509<br>XL-88* |
| Greenleaf | G-01<br>G-10 | G-02<br>G-20/G-30<br>G-1<br>Ti-2 | G-30<br>G-1<br>Ti-2 | G-1<br>G-40 | G-50<br>G-53<br>G-52<br>GA-5<br>Ti-5 | Ti-6/GA-6<br>G-60<br>G-54 | GA-6/G-1/<br>Ti-6<br>G-70/G-25 | Ti-6<br>GA-6/G-1<br>G-80/G-88 |
| G.T.E. (Walmet) | WA-110<br>WA-1<br>WA-59 | WA-2<br>P-2#<br>WA-35<br>WA-69<br>WA-63 | WA-3<br>P-3# | WA-4<br>P-4# | WA-54<br>P-54#<br>WA-5<br>P-5#<br>WA-57<br>P-57# | WA-47<br>P-47# | WA-7<br>WA-73<br>P-47# | WA-8 |
| Iscar | IC 28 | IC 2<br>IC 20<br>IC 24<br>IC 424 | IC 20<br>IC 424 | IC 4 | IC 50<br>IC 50M<br>IC 54<br>IC 656<br>IC 757 | IC 50M<br>IC 70<br>IC 656<br>IC 757 | IC 70<br>IC 78<br>IC 60T<br>IC 80T<br>IC 757 | IC 80T |
| Kennametal | K1<br>KC250†† | K6<br>K68<br>K8735<br>KC210††<br>KC250††<br>KC910§ | K68<br>KC210††<br>KC910§ | K11<br>KC910§ | K420<br>KC850†† | K420<br>K21<br>K2884<br>KC810††<br>KC850††<br>KC910§ | K4H<br>K45<br>KC810††<br>KC850††<br>KC910§ | K7H<br>K165*<br>KC910§ |
| Newbide | N-10 | N-22<br>NT-2<br>N25<br>NT25 | N-30<br>NT-2<br>N20<br>NT25 | N-40<br>NT25 | N-50/N-52<br>NT-5<br>N55<br>NT55 | N-60<br>NT55<br>NT6 | N-70/N-72<br>NT-6<br>NT55 | N-80<br>N-93/N-95<br>NT6 |

CAST IRON, NONFERROUS AND NONMETALLIC MATERIALS
- C-1 Roughing
- C-2 General Purpose
- C-3 Finishing
- C-4 Precision Finishing

STEEL AND STEEL ALLOYS
- C-5 Roughing
- C-6 General Purpose
- C-7 Finishing
- C-8 Precision Finishing

NOTE: Listings do not necessarily imply equivalency of various manufacturers' grades. This chart is not to be considered an endorsement of or an approved list of any manufacturer's products.

*Contains more than 50% titanium carbide.

†TiC coated.

‡TiN over TiC coated.

§$Al_2O_3$ coated.

#Tri-Phase titanium nitride coated.

**$Al_2O_3$ and TiC coated.

††Coated.

‡‡Triple-coated, Ti-C + Ti (C, N) + $Al_2O_3$.

**Table 14-5—Continued**

| TRADE NAME | INDUSTRY CODE | | | | | | | |
|---|---|---|---|---|---|---|---|---|
| | C-1 | C-2 | C-3 | C-4 | C-5 | C-6 | C-7 | C-8 |
| Sandvik (Coromant) | H20<br>HML | GC015**<br>GC315†<br>GC1025†<br>H1P<br>H20<br>HM<br>HML<br>GC310**<br>GC320**<br>GC135†<br>IRB | GC015**<br>GC315†<br>GC1025†<br>H1P<br>H10<br>HM<br>GC310**<br>GC320**<br>GC135†<br>IRB | GC015**<br>H05<br>H1P<br>H10<br>GC310** | S6<br>SM30<br>GC135†<br>S6/S4<br>GC1025†<br>S35 | GC1025†<br>GC015**<br>GC135†<br>S4<br>SM<br>SM30<br>S35<br>GC120# | GC1025†<br>GC015**<br>SIP<br>S2<br>SM<br>SM30<br>GC135†<br>GC120# | GC015**<br>SIP<br>F02*<br>GC1025†<br>GC120# |
| Seco | HX | SU41<br>HX<br>TP15‡‡<br>T15M‡‡ | H13<br>TX10‡‡<br>SU41<br>HX<br>TP15‡‡<br>T15M‡‡ | TX10‡‡<br>TP15‡‡<br>T15M‡‡ | G27<br>S4<br>S6<br>S60M<br>TP35†<br>S25M | S2<br>SU41<br>S25M<br>S10M<br>TP25†<br>TP35†<br>TP15‡‡ | S25M<br>S10M<br>S1F<br>TP15‡‡<br>TP25† | S1G<br>TP15‡‡ |
| Talide | C-89 | C-91 | C-93<br>C-936 | C-95 | S-880 | S-901 | S-92 | S-94 |
| Teledyne (Firthite) | H-91<br>H-17<br>H-6 | H-21<br>HTA<br>HA<br>HN + ††<br>TC + †† | HA<br>HTA<br>HN + ††<br>TC + ††<br>HF<br>SD-3* | HN + ††<br>TC + ††<br>HF<br>SD-3* | T-04<br>T-14<br>HN + 4††<br>TC + 4††<br>NTA<br>HN + ††<br>TC + †† | HN + ††<br>TC + ††<br>TXH<br>T-22<br>T-24 | HN + ††<br>T-25<br>TC + 1††<br>SD-3* | HN + ††<br>TC + 1††<br>SD-3* |
| TRW Wendt-Sonis | TRW-167‡<br>CQ-22T†<br>CQ-22<br>CQ-12 | TRW-027‡<br>U222†<br>CQ-2<br>CQ-23 | TRW-237‡<br>CQ-23T†<br>CQ-23<br>CQ-3 | TRW-237‡<br>CQ-23T†<br>CQ-4<br>CQ-23 | CY-12‡<br>CY-16‡<br>CY-17‡<br>U225†<br>TRW-715†<br>TRW-716†<br>CY-17<br>CY-12 | TRW-716‡<br>U225†<br>CY-16 | TRW-714‡<br>U227†<br>CY-14<br>Ti8* | TRW-731‡<br>CY-31T†<br>CY-31<br>Ti8* |
| Tungsten Alloy | A6 | 6B<br>6BH<br>10F | 4B | 3B | 11T<br>10T<br>9S | 10T<br>9S | 8T | 5S |
| Ultra-Met | Z-1<br>Z-10<br>Z-25<br>Z-9 | Z-2<br>Z-20 | Z-20<br>Z-3 | Z-4 | Z-5<br>Z-52<br>Z-517<br>Z-50 | Z-50<br>Z-6<br>Z-60 | Z-7<br>Z-70<br>Z-71<br>Z-8 | Z-8<br>Z-80* |
| Valenite | V-1N††<br>VC-1<br>VC-111<br>VC-101 | V-01**<br>V-91†<br>V-88†<br>VN-8‡<br>VN-2††<br>V-1N††<br>VC-2<br>VC-24<br>VC-27<br>VC-28<br>VC-29 | V-01**<br>V-91†<br>V-88†<br>VN-8‡<br>VN-2††<br>VC-3<br>VC-29 | V-01**<br>VC-3<br>VC-23 | V-88†<br>VN-8‡<br>V-99†<br>VN-5††<br>V-1N††<br>VC-5<br>VC-55<br>VC-56<br>VC-125 | V-01**<br>V-99†<br>V-90†<br>V-88†<br>VN-8‡<br>VN-5††<br>VC-6 | V-01**<br>V-90†<br>V-88†<br>VN-8‡<br>VN-5††<br>VC-7<br>VC-76 | V-01**<br>VC-83*<br>VC-8 |
| VR/Wesson | 2A3<br>2A68<br>VR-54<br>Ramet-1 | 2A5<br>VR-82<br>630†<br>Ramet-1 | VR-82<br>630†<br>2A7 | 2A7<br>630†<br>VR-65* | 650†<br>660†<br>VR-77 | 650†<br>660†<br>VR-75<br>VR-77 | 660†<br>670†<br>VR-73 | VR-65*<br>VR-71<br>670† |

CAST IRON, NONFERROUS AND NONMETALLIC MATERIALS
C-1 Roughing
C-2 General Purpose
C-3 Finishing
C-4 Precision Finishing

STEEL AND STEEL ALLOYS
C-5 Roughing
C-6 General Purpose
C-7 Finishing
C-8 Precision Finishing

*Contains more than 50% titanium carbide.
†TiC coated.
‡TiN over TiC coated.
§$Al_2O_3$ coated.
#Tri-Phase titanium nitride coated.
**$Al_2O_3$ and TiC coated.
††Coated.
‡‡Triple-coated, Ti-C + Ti (C, N) + $Al_2O_3$.

**TABLE 14–6 Classification of Carbides According to Use, per ISO 513-1975(E)**

| MAIN GROUPS OF CHIP REMOVAL | | | GROUPS OF APPLICATION | | | DIRECTION OF INCREASE IN CHARACTERISTIC | |
|---|---|---|---|---|---|---|---|
| Symbol | Broad Categories of Material To Be Machined | Distinguishing Colors | Designation | Material To Be Machined | Use and Working Conditions | Of Cut | Of Carbide |
| P | Ferrous metals with long chips | Blue | P 01 | Steel, steel castings | Finish turning and boring; high cutting speeds, small chip section, accuracy of dimensions and fine finish, vibration-free operation. | Increasing feed ↓ / Increasing speed ↑ | Toughness ↓ / Wear resistance ↑ |
| | | | P 10 | Steel, steel castings | Turning, copying, threading and milling, high cutting speeds, small or medium chip sections. | | |
| | | | P 20 | Steel, steel castings<br>Malleable cast iron with long chips | Turning, copying, milling, medium cutting speeds and chip sections, planing with small chip sections. | | |
| | | | P 30 | Steel, steel castings<br>Malleable cast iron with long chips | Turning, milling, planing, medium or low cutting speeds, medium or large chip sections, and machining in unfavorable conditions.* | | |
| | | | P 40 | Steel<br>Steel castings with sand inclusion and cavities | Turning, planing, slotting, low cutting speeds, large chip sections with the possibility of large cutting angles for machining in unfavorable conditions* and work on automatic machines. | | |
| | | | P 50 | Steel<br>Steel castings of medium or low tensile strength, with sand inclusion and cavities | For operations demanding very tough carbide: turning, planing, slotting, low cutting speeds, large chip sections, with the possibility of large cutting angles for machining in unfavorable conditions* and work on automatic machines. | | |
| M | Ferrous metals with long or short chips and nonferrous metals | Yellow | M 10 | Steel, steel castings, manganese steel<br>Gray cast iron, alloy cast iron | Turning, medium or high cutting speeds. Small or medium chip sections. | Increasing feed ↓ / Increasing speed ↑ | Toughness ↓ / Wear resistance ↑ |
| | | | M 20 | Steel, steel castings, austenitic or manganese steel, gray cast iron | Turning, milling. Medium cutting speeds and chip sections. | | |
| | | | M 30 | Steel, steel castings, austenitic steel, gray cast iron, high temperature resistant alloys | Turning, milling, planing. Medium cutting speeds, medium or large chip sections. | | |
| | | | M 40 | Mild free cutting steel, low tensile steel<br>Nonferrous metals and light alloys | Turning, parting off, particularly on automatic machines. | | |

*Raw material or components in shapes which are awkward to machine: casting or forging skins, variable hardness, etc., variable depth of cut, interrupted cut, work subject to vibrations.

TABLE 14-6—Continued

| MAIN GROUPS OF CHIP REMOVAL | | | GROUPS OF APPLICATION | | | DIRECTION OF INCREASE IN CHARACTERISTIC | |
|---|---|---|---|---|---|---|---|
| Symbol | Broad Categories of Material To Be Machined | Distinguishing Colors | Designation | Material To Be Machined | Use and Working Conditions | Of Cut | Of Carbide |
| K | Ferrous metals with short chips, nonferrous metals and nonmetallic materials | Red | K 01 | Very hard gray cast iron, chilled castings of over 85 Shore, high silicon aluminum alloys, hardened steel, highly abrasive plastics, hard cardboard, ceramics. | Turning, finish turning, boring, milling, scraping. | Increasing feed ↑ | Toughness ↑ |
| | | | K 10 | Gray cast iron over 220 Brinell, malleable cast iron with short chips, hardened steel, silicon aluminum alloys, copper alloys, plastics, glass, hard rubber, hard cardboard, porcelain, stone. | Turning, milling, drilling, boring, broaching, scraping. | | |
| | | | K 20 | Gray cast iron up to 220 Brinell, nonferrous metals: copper, brass, aluminum | Turning, milling, planing, boring, broaching, demanding very tough carbide. | | |
| | | | K 30 | Low hardness gray cast iron, low tensile steel, compressed wood | Turning, milling, planing, slotting, for machining in unfavorable conditions* and with the possibility of large cutting angles | | |
| | | | K 40 | Soft wood or hard wood<br>Nonferrous metals | Turning, milling, planing, slotting, for machining in unfavorable conditions* and with the possibility of large cutting angles. | ↓ Increasing speed | ↓ Wear resistance |

*Raw material or components in shapes which are awkward to machine: casting or forging skins, variable hardness, etc., variable depth of cut, interrupted cut, work subject to vibrations.

**TABLE 14-7 Carbide Grade Chart, ISO Grade**

| TRADE NAME | Group P<br>harder ← → tougher | | | | | | | | | | | Group M<br>harder ← → tougher | | | | | | Group K<br>harder ← → tougher | | | | | | |
|---|---|---|---|---|---|---|---|---|---|---|---|---|---|---|---|---|---|---|---|---|---|---|---|---|
| | P01 | P05 | P10 | P15 | P20 | P25 | P30 | P35 | P40 | P45 | P50 | M05 | M10 | M15 | M20 | M30 | M40 | K01 | K05 | K10 | K15 | K20 | K30 | K40 |
| Adamas | T100 | 490<br>ACT7<br>T80 | 548<br>ACT7<br>T80 | 495<br>ACT5<br>ACT7<br>T60 | 495<br>AG5<br>ACT5<br>T60 | 499<br>ACT5<br>T50<br>AG5 | 499<br>ACT5<br>AG5<br>T50 | ACT5<br>AG5<br>499<br>434 | 474<br>ACT5<br>AG5<br>434 | 474 | GG | | 548 | 548 | 548 | 434 | 474 | GU2 | GU2 | ACT2<br>PWX | ACT2<br>PWX | ACT2<br>AM | B | BB |
| Atrax | AT8<br>T76C | T76C<br>AT8 | T76<br>T76C | | T50C<br>T65 | T50C<br>T56 | T50<br>T50C<br>T56 | | T3 | | | AT8 | | | T50 | T56 | | A7 | A7 | A6 | | M22 | M30 | M31 |
| Carboloy | 210<br>320 | 545<br>320 | 545<br>570<br>515<br>516<br>350 | 570<br>515<br>516<br>350<br>370 | 570<br>515<br>516<br>518<br>370 | 570<br>515<br>516<br>518<br>370 | 570<br>515<br>516<br>518<br>370 | 515<br>516<br>518<br>370<br>390 | 516<br>518<br>370<br>375<br>390<br>395 | 518<br>370<br>375<br>390<br>395 | 370<br>390 | 545<br>570<br>523 | 545<br>570<br>523<br>895 | 570<br>523<br>895<br>883 | 570<br>523<br>883<br>518<br>370 | 523<br>883<br>518<br>370<br>820 | 883<br>44A<br>820 | 545<br>999<br>570 | 545<br>570<br>523<br>895 | 570<br>523<br>895<br>883 | 570<br>523<br>895<br>883<br>518 | 570<br>523<br>895<br>883<br>518 | 523<br>883<br>518<br>820 | 883<br>44A<br>820 |
| Carmet | CA<br>704<br>R03 | CA<br>711<br>CA<br>9711<br>R03 | CA<br>711<br>CA<br>9711 | CA<br>9720<br>CA<br>606 | CA<br>606<br>CA<br>720<br>CA<br>9720 | CA<br>721<br>CA<br>9721 | CA<br>721<br>CA<br>9721 | CA<br>721<br>CA<br>9721 | CA<br>610<br>CA<br>740<br>CA<br>9740 | CA<br>610<br>CA<br>740<br>CA<br>9740 | CA<br>610<br>CA<br>740 | CA<br>9443<br>R03 | CA<br>310<br>CA<br>711<br>CA<br>9443<br>CA<br>9721<br>CA<br>443 | CA<br>310<br>CA<br>711<br>CA<br>9443<br>CA<br>9721<br>CA<br>443 | CA<br>310<br>CA<br>721<br>CA<br>9721<br>CA<br>9740<br>CA<br>9443<br>CA<br>443 | CA<br>721<br>CA<br>740<br>CA<br>9740<br>CA<br>443 | CA<br>740 | CA<br>8<br>R03 | CA<br>7<br>CA<br>8443<br>CA<br>9443 | CA<br>310<br>CA<br>8443<br>CA<br>9443 | CA<br>310<br>CA<br>443<br>CA<br>8443<br>CA<br>9443 | CA<br>310<br>CA<br>4<br>CA<br>443 | CA<br>310<br>CA<br>3 | CA<br>12 |
| DoAll | D018 | D018 | D017<br>D042 | D042 | D016<br>D040<br>D044 | D015<br>D040<br>D044 | D015<br>D040<br>D044 | D035 | D035 | D035 | | D018 | D017 | D034 | D016 | D015 | D035 | D018 | | D02<br>D03<br>D020<br>D046 | D02<br>D046 | D02<br>D046 | D01 | D011 |
| Duramet | | DU6<br>DU7 | DU6 | DU55 | DU5<br>DU55 | DU55 | | | | | | | | | | | | | DU4 | DU3 | DU22 | DU2<br>DU22 | DU1 | |
| Ex-Cell-O | 6AX<br>509<br>XL88<br>XL602 | 6AX<br>XL88<br>XL602 | 6AX<br>XL061<br>XL88<br>XL602 | XL061<br>XL602 | 6A<br>606<br>XL061<br>XL86<br>XL602 | XL602 | 8A<br>XL85<br>XL602 | 8A | 8A | | 10A<br>XL502 | EA | EA | EA | EA | EA | EA | E3 | | E5<br>XL028<br>XL202 | XL202 | E6<br>XL202<br>XL620 | E6<br>E8 | E9 |
| G.T.E. (Walmet) | WA8 | WA73 | WA7<br>P-47 | WA7<br>P-47<br>P-5 | WA47<br>P-47<br>P-5 | WA5<br>WA47<br>P-47<br>WA57<br>P-5 | WA5<br>P-5<br>WA57<br>P-47 | WA5<br>P-5<br>P-54 | WA54<br>P-54<br>P-5 | WA54<br>P-54 | WA54 | | WA69 | | WA63 | | | WA4 | WA3 | P-2<br>WA3<br>WA2<br>WA35 | WA35<br>WA2<br>P-2 | WA2<br>P-2 | WA1<br>WA59<br>WA110 | |

NOTE: Listings do not necessarily imply equivalency of various manufacturers' grades. This chart is not to be considered an endorsement of or an approved list of any manufacturers' products.

**TABLE 14-7—Continued**

| TRADE NAME | Group P | | | | | | | | | | | Group M | | | | | | Group K | | | | | | |
|---|---|---|---|---|---|---|---|---|---|---|---|---|---|---|---|---|---|---|---|---|---|---|---|---|
| | harder ◀— —▶ tougher | | | | | | | | | | | harder ◀— —▶ tougher | | | | | | harder ◀— —▶ tougher | | | | | | |
| | P01 | P05 | P10 | P15 | P20 | P25 | P30 | P35 | P40 | P45 | P50 | M05 | M10 | M15 | M20 | M30 | M40 | K01 | K05 | K10 | K15 | K20 | K30 | K40 |
| Greenleaf | G80 | G1<br>G70<br>G80 | G1<br>G70<br>G74<br>GA6<br>Ti6 | G1<br>G70<br>G74<br>GA6<br>Ti6 | G1<br>G52<br>G53<br>G54<br>G74<br>GA5<br>GA6<br>Ti5<br>Ti6 | G1<br>G52<br>G53<br>G54<br>G74<br>GA5<br>GA6<br>Ti5<br>Ti6 | G1<br>G52<br>G52<br>G54<br>GA5<br>GA6<br>Ti5<br>Ti6 | G1<br>G50<br>GA5<br>G53<br>G54<br>GA5<br>GA6<br>Ti5<br>Ti6 | G50<br>G52<br>GA5<br>Ti5 | G01<br>G50<br>Ti5 | G01 | G02 | G1<br>G02<br>G74<br>GA6<br>Ti2<br>Ti6 | G1<br>G02<br>G10<br>G52<br>G74<br>GA5<br>GA6<br>Ti2<br>Ti5<br>Ti6 | G1<br>G10<br>G52<br>G74<br>GA5<br>GA6<br>Ti2<br>Ti5<br>Ti6 | G10<br>G50<br>G52<br>GA5<br>Ti5 | G01<br>G50 | G25<br>G40 | G23<br>G25<br>G30<br>G40<br>Ti2 | G02<br>G23<br>G25<br>G30<br>GA6<br>Ti2<br>Ti6 | G1<br>G02<br>G10<br>G23<br>G25<br>G30<br>Ti2 | G02<br>G10<br>G23<br>G25<br>Ti2 | G02<br>G10 | |
| Iscar | IC80T | IC60T<br>IC80T | IC60T<br>IC78<br>IC757 | IC757 | IC70<br>IC656<br>IC757 | IC50M<br>IC656<br>IC757 | IC50<br>IC50M<br>IC656<br>IC757 | IC54<br>IC656<br>IC50M | IC54<br>IC656<br>IC50M | IC54 | IC54 | | IC24 | IC24 | IC24 | IC28 | IC28 | IC4 | IC424 | IC20<br>IC424 | IC424 | IC2<br>IC20<br>IC424 | IC28 | IC12<br>IC28 |
| Kennametal | K165<br>KC910 | K7H<br>KC910 | K7H<br>K45<br>KC810<br>KC850<br>KC910 | K45<br>KC810<br>KC850<br>KC910 | K4H<br>K2884<br>KC810<br>KC850<br>KC910 | K21<br>K2884<br>KC810<br>KC850<br>KC910 | K21<br>K420<br>KC810<br>KC850 | K21<br>K420<br>KC810<br>KC850 | K420<br>K21<br>KC850 | K420<br>KC850 | K420 | K7H<br>KC910 | K45<br>KC810<br>KC850<br>KC910 | K4H<br>KC810<br>KC850<br>KC910 | K4H<br>K2884<br>KC810<br>KC850<br>KC910 | K420<br>K21<br>KC850 | K420<br>K21 | K11<br>KC910 | K11<br>KC210<br>KC910 | K68<br>KC210<br>KC910 | K68<br>KC210<br>KC910 | K6<br>KC210<br>KC250<br>KC910 | K1<br>KC250 | K1 |
| Newbide | N93<br>N95<br>NT6 | N80<br>NT6 | N70<br>N72<br>NT6<br>NT55 | N72<br>NT6<br>NT55 | N60<br>NT5<br>NT6<br>NT55 | N50<br>N55<br>NT55 | N50<br>N52<br>N55<br>NT55 | N52 | N52 | N52 | N52 | N40<br>N80<br>NT2 | N20<br>N60<br>NT25<br>NT55 | N20<br>N60<br>NT25<br>NT55 | N22<br>N25<br>N55<br>NT25<br>NT55 | N22<br>N25<br>N55<br>NT55 | N10<br>N52 | N40 | N20<br>NT25 | N20<br>NT25 | N20<br>NT25 | N22<br>NT2<br>N25<br>NT25 | N10 | N10 |
| Sandvik (Coromant) | F02<br>S1P | GC015<br>S1P | GC015<br>GC<br>1025<br>S1P<br>GC120 | GC015<br>GC<br>1025<br>S1P<br>SM<br>GC120 | GC015<br>GC<br>1025<br>S2<br>SM<br>SM30<br>GC135<br>GC120 | GC015<br>GC<br>135<br>GC<br>1025<br>S4<br>SM<br>SM30<br>GC120 | GC015<br>GC<br>135<br>GC<br>1025<br>S4<br>SM<br>SM30<br>S35<br>GC120 | GC135<br>GC<br>1025<br>S4<br>S6<br>SM30<br>S35 | GC135<br>S6<br>SM30<br>S35 | GC135<br>R4<br>S6 | R4 | | R1P<br>GC<br>015 | GC135<br>GC<br>315<br>SH<br>GC015 | GC135<br>GC<br>315<br>H20<br>SH<br>GC015 | H20<br>S6<br>GC<br>135 | R4<br>S6 | H1P<br>H05 | GC015<br>GC<br>315<br>GC<br>1025<br>H1P<br>H05<br>H10<br>GC310 | GC015<br>GC<br>315<br>GC<br>1025<br>H1P<br>H10<br>HM<br>GC310<br>GC320<br>IRB<br>GC135 | GC015<br>GC<br>315<br>GC<br>1025<br>H1P<br>H10<br>HM<br>GC310<br>GC320<br>IRB<br>GC135 | GC015<br>GC<br>315<br>GC<br>1025<br>H1P<br>H20<br>HM<br>HML<br>GC310<br>GC320<br>IRB<br>GC135 | H20<br>HML | |
| Seco | | S1G<br>TP15 | S1G<br>S1F<br>TP15<br>TP25<br>S10M | S1G<br>S1F<br>S2<br>TP15<br>TP25<br>TP35<br>S10M | S1F<br>S2<br>TP15<br>TP25<br>TP35<br>S10M<br>S25M | S2<br>S4<br>TP15<br>TP25<br>TP35<br>S10M<br>S25M | S4<br>TP15<br>TP25<br>TP35<br>S25M | S4<br>TP25<br>TP35<br>S25M<br>S60M | S6<br>TP35<br>S25M<br>S60M | S6<br>S60M | S6<br>S60M | TP15<br>TX10 | SU41<br>TP15<br>TP25<br>TX10 | SU41<br>TP15<br>TP25<br>TP35<br>TX10<br>S10M | SU41<br>G27<br>TP15<br>TP25<br>TP35<br>S10M<br>S25M<br>S60M | G27<br>TP15<br>TP25<br>TP35<br>S10M<br>S25M<br>S60M | S6<br>G27<br>TP35<br>S25M | | TP15<br>TX10<br>T15M | H13<br>TP15<br>TP25<br>TX10<br>T15M | H13<br>HX<br>TP15<br>TP25<br>TX10<br>T15M | H13<br>HX<br>TP15<br>TP25 | HX<br>TP25 | |
| Talide | S94 | | S92 | | S901 | | S880 | | | | | | | | | | | C95 | | C93<br>C936 | | C91 | C89<br>C99 | |

TABLE 14-6—Continued

| TRADE NAME | Group P | | | | | | | | | | | Group M | | | | | | Group K | | | | | | |
|---|---|---|---|---|---|---|---|---|---|---|---|---|---|---|---|---|---|---|---|---|---|---|---|---|
| | harder ◄— —► tougher | | | | | | | | | | | harder ◄— —► tougher | | | | | | harder ◄— —► tougher | | | | | | |
| | P01 | P05 | P10 | P15 | P20 | P25 | P30 | P35 | P40 | P45 | P50 | M05 | M10 | M15 | M20 | M30 | M40 | K01 | K05 | K10 | K15 | K20 | K30 | K40 |
| Teledyne (Firthite) | HN+<br>SD3<br>TC+1 | HN+<br>SD3<br>TC+1 | HN+<br>SD3<br>T25<br>TC+1 | HN+<br>T-24 | HN+<br>T-24<br>T22<br>TC+<br>TXH | HN+<br>NTA<br>TC+ | HN+<br>T-14<br>TC+ | HN+4<br>TC+4 | HN+4<br>TC+4<br>T04 | HN+4<br>TC+4 | H91<br>HN+4<br>TC+4 | SD3 | HN+<br>TC+1 | HN+<br>T-24 | HN+<br>TC+<br>TXH | HN+4<br>TC+4 | H91<br>HN+4<br>TC+4 | HF<br>HN+<br>TC+1 | HN+<br>HF<br>SD3<br>TC+1 | HA<br>HTA<br>HN+<br>TC+1 | HN+<br>HTA<br>TC+1 | H21<br>HN+<br>TC+1 | H17 | H91 |
| TRW Wendt-Sonis | 731<br>CY31T<br>Ti8<br>714<br>U227<br>CY31 | 714<br>U227<br>Ti8<br>CY31<br>731<br>CY31T | 714<br>U227<br>715<br>CY14<br>716<br>U225 | 714<br>U227<br>715<br>CY14<br>716<br>U225 | 714<br>U227<br>715<br>CY14<br>716<br>U225<br>CY16 | 715<br>716<br>U225<br>CY16<br>CY17T | 715<br>716<br>U225<br>CY16<br>CY17 | 715<br>716<br>U225<br>CY16<br>CY17 | 716<br>715<br>U225<br>CY16<br>CY17 | 717<br>715<br>CY17T<br>CY17 | CY17<br>717<br>CY17T | 731<br>CY31<br>027<br>U222<br>U227<br>CQ23 | 714<br>U227<br>715<br>716<br>U225<br>U222<br>CY14<br>CQ2 | U227<br>715<br>716<br>U225<br>U222<br>CY14<br>CQ2 | 716<br>U225<br>715<br>U227<br>U222<br>CY16<br>CY14<br>CQ2 | 716<br>U225<br>715<br>U222<br>CY16<br>CY14<br>CQ2 | CY17<br>717<br>CQ12<br>CY16<br>CQ2 | CQ4<br>CQ23<br>237<br>CQ23T | U222<br>CQ23<br>027<br>CQ4 | CQ2<br>CQ23<br>U222<br>237<br>027 | U222<br>CQ2<br>CQ23<br>027<br>237 | U222<br>CQ2<br>CQ23<br>027<br>CQ22 | CQ2<br>CQ12<br>CQ22<br>U222<br>027 | CQ22<br>CQ2<br>CQ12<br>CQ13 |
| Tungsten Alloy | | 5S | 8T | | 9S<br>10T | | 11T | | | | | | | | | | | 3B | | 4B | | 6B | A6 | A9 |
| Ultra-Met | Z-8<br>Z-80<br>ZT-70 | Z-8<br>ZT-70 | Z-70<br>Z-71<br>ZT-70<br>ZT-60<br>Z-7 | Z-70<br>ZT-70<br>ZT-60<br>ZT-50<br>Z-6 | Z-60<br>Z-50<br>Z-6<br>ZT-50<br>ZT-60<br>ZT-6 | ZT-52<br>ZT-50<br>Z-50 | ZT-50<br>ZT-52<br>Z-50<br>Z-52 | ZT-517<br>ZT-52<br>Z-52<br>Z-517<br>Z-50 | Z-517<br>Z-52<br>ZT-5<br>ZT-52 | Z-52<br>ZT-5<br>ZT-52 | Z-5<br>Z-52 | Z-8<br>ZT-7<br>ZT-2 | Z-71<br>Z-70<br>ZT-70<br>ZT-2<br>Z-2<br>Z-20 | Z-70<br>ZT-70<br>ZT-50<br>Z-2<br>Z-20<br>ZT-2 | Z-60<br>Z-50<br>ZT-60<br>ZT-50<br>Z-2<br>ZT-2 | Z-1<br>Z-10<br>Z-25<br>Z-50<br>ZT-52<br>Z-2 | Z-1<br>Z-25<br>Z-52 | Z-4<br>ZT-3 | Z-3<br>ZT-20<br>Z-20 | Z-20<br>ZT-2<br>ZT-20<br>Z-3 | Z-2<br>Z-20<br>ZT-2 | Z-2<br>Z-25<br>ZT-2<br>ZT-25 | Z-25<br>Z-10<br>Z-1 | Z-1<br>Z-25<br>Z-9 |
| Valenite | V-01<br>VC-83<br>VC-8 | V-01<br>V-88<br>VC-8<br>VC-76 | V-01<br>V-90<br>V-88<br>VN-8<br>VN-5<br>VC-7<br>VC-76 | V-01<br>V-90<br>V-88<br>VN-8<br>VN-5<br>VC-76<br>VC-6<br>VC-125 | V-01<br>V-90<br>V-88<br>VN-8<br>VN-5<br>V-99<br>VC-6<br>VC-55<br>VC-125 | V-01<br>V-90<br>V-88<br>VN-8<br>VN-5<br>V-99<br>VC-5<br>VC-55<br>VC-125 | V-01<br>V-88<br>VN-8<br>V-99<br>VC-5<br>VC-55 | V-88<br>VN-8<br>V-99<br>V-1N<br>VC-56 | V-99<br>V-1N<br>VC-56 | V-1N | | V-01 | V-01<br>V-88<br>VC-23 | V-01<br>V-88<br>VN-8<br>VN-5<br>VC-27<br>VC-23 | V-01<br>V-88<br>VN-8<br>VN-5<br>VC-5<br>VC-28<br>VC-27<br>VC-24<br>VC-23 | V-01<br>V-88<br>VN-8<br>V-1N<br>VC-5<br>VC-28<br>VC-24 | V-1N<br>VC-111<br>VC-101 | V-01<br>VC-3 | V-01<br>V-91<br>V-88<br>VC-3 | V-01<br>V-91<br>V-88<br>VN-8<br>VN-2 | V-01<br>V-91<br>V-88<br>VN-8<br>VN-2<br>VC-2<br>VC-24<br>VC-29 | V-01<br>V-91<br>V-88<br>VN-8<br>VN-2<br>VC-2<br>VC-29<br>VC-28<br>VC-24 | VN-8<br>V-1N<br>VC-2<br>VC-28<br>VC-24<br>VC-1<br>VC-111<br>VC-101 | V-1N<br>VC-111<br>VC-101 |
| VR/Wesson | VR65<br>VR71<br>670 | VR65<br>670 | 670<br>VR73<br>660 | 670<br>660 | 650<br>660<br>670<br>VR75 | 650<br>660<br>VR75<br>VR77 | 650<br>660<br>VR77 | VR77<br>650 | 650 | 650 | | VR65<br>630 | VR73<br>660<br>630 | VR75<br>660<br>630 | VR77<br>VR75<br>630<br>VR82<br>660 | VR77<br>630<br>VR82<br>650 | | | 630 | 2A5<br>2A7<br>630<br>VR82 | 2A5<br>630<br>VR82 | 2A5<br>630<br>VR54 | 2A68<br>Ramet<br>1<br>VR54 | 2A3<br>Ramet<br>1 |

## 14.5 Micrograin Carbides

A relatively new group of high strength carbides intended for machining applications is now available. These tungsten carbide-cobalt base materials are characterized by strength levels previously found only in impact and rock bit grades, coupled with much higher levels of hardness, as shown in figure 14-1. Their properties are achieved by various proprietary processes and depend in part on achieving and maintaining a very fine carbide grain size.

Manufacturers recommend these materials for improving metal removal rates and tool life over those possible with high speed steels for conditions where carbides with normal strength levels would chip or break. The major use of micrograin carbides at present is in cutoff and form tools, replacing high speed steel in low-surface-speed, high-production operations on screw machines and similar equipment. Some application is also being found for micrograin carbide tools in the machining of problem materials such as the high temperature alloys and for making punch and die sets intended for extended use.

The high edge-strength of micrograin carbides allows the use of high speed steel tool geometry and, when necessary, speeds as low as those used with high speed steels. Conversely, operation at normal carbide speeds may not be practical because the high cobalt binder level in these materials significantly reduces their resistance to cratering.

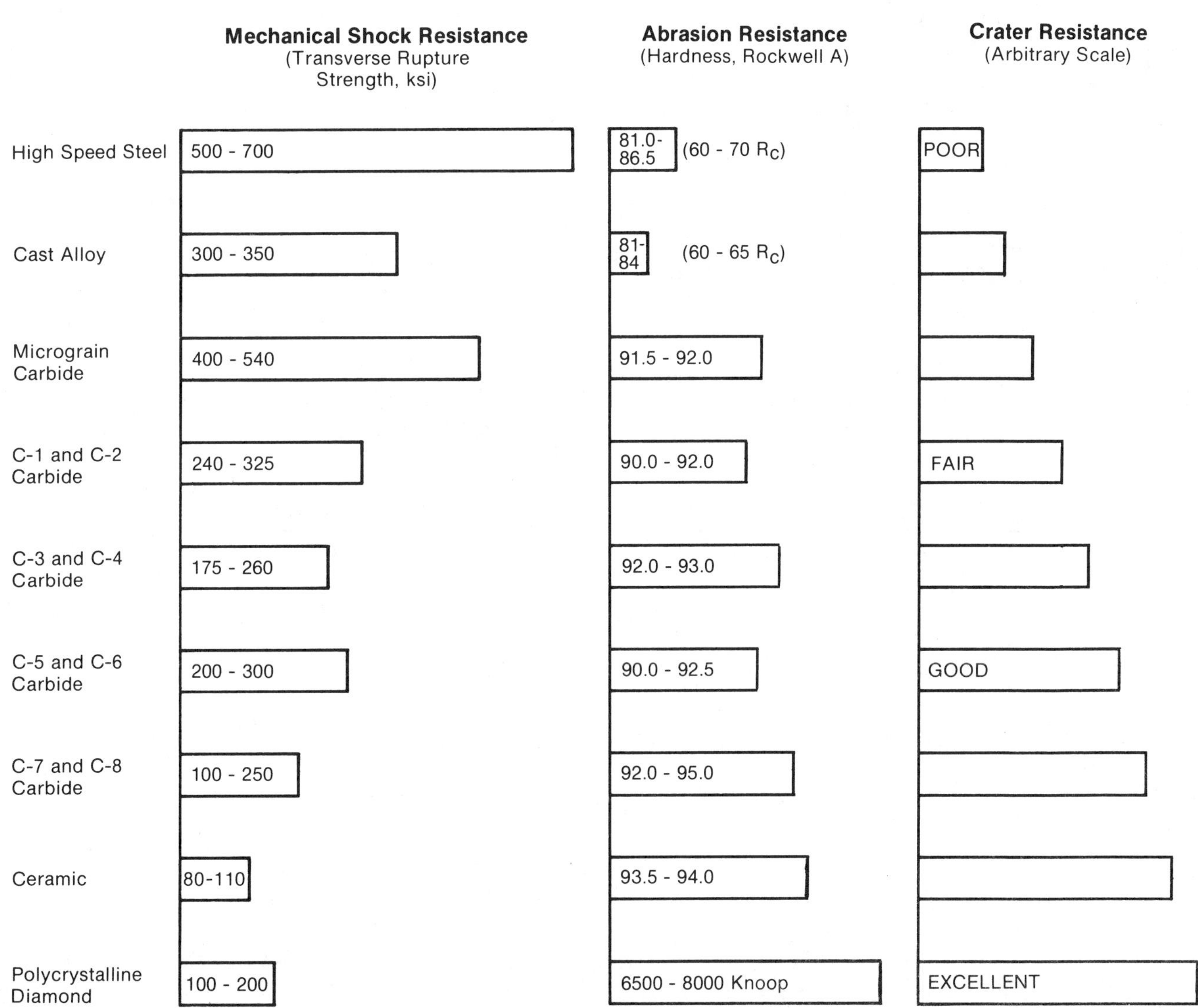


**Figure 14-1** Comparative properties of tool materials derived from published data for available grades. Crater resistance refers to ferrous metals.

## 14.6 Coated Carbides

Perhaps the most important advance in cutting tool technology during the past decade has been the development of coated carbide inserts. The impact of this improvement can be measured in terms of the major benefits of increased metal removal rate and tool life that have been found in turning and face milling of cast irons and steels. These two general types of work materials still represent the bulk of the work material being machined in the United States. The application of these tool materials overlaps much of the entire C-1 through C-8 range. The coatings are usually, but not necessarily, deposited on conventional carbide inserts. Special substrates are being developed which definitely are not suitable as metal cutting tools without the coatings.

Titanium carbide coatings are produced by reactive deposition from the gaseous phase, where titanium tetrachloride vapor is converted to fine grain titanium crystals. The process continues until the coating is about 0.0002-inch [5 μm] thick. Other coatings such as TiN, HfN, HfC, and $Al_2O_3$ are produced using the same technology with other gaseous components.

The improved machining performance of coated carbide inserts results from reduced friction, increased surface hardness and, most important of all, chemical inertness. The thin layer of coating provides a cobalt-free diffusion barrier between the workpiece and the insert. The reduction or elimination of diffusion results in an appropriate reduction of cratering and flank wear. For many applications, this results in either a longer tool life or a higher metal removal rate when compared to uncoated carbides. An exception to this advantage would probably occur in highly abrasive, lower cutting speed operations. Newer grades are being developed to successfully perform heavy roughing and interrupted cuts.

## 14.7 Ceramic Tool Materials

There are two main types of ceramic cutting tools, the pure alumina tools and the alloyed cermets (see table 14-8). The cermets are also aluminum-oxide-base materials containing various amounts of titanium carbide or other alloying ingredients. The alumina tools are white in appearance. Although these tools may be either hot pressed or cold pressed, they are usually cold pressed. The cold pressed alumina (CPA) inserts are normally applied in lighter-duty cuts, such as noninterrupted turning and boring.

**TABLE 14-8 Ceramic Inserts**

| IDENTIFICATION | MANUFACTURER |
|---|---|
| TRW-138<br>TRW-1322 | TRW Wendt-Sonis, Rogers, AR |
| VR-97<br>VR-100* | Fansteel, VR/Wesson Div., Waukegan, IL |
| Ceratip 30<br>Ceratip 42.<br>Cermet 61* | Ceratip—Division of Kyocera International Inc., San Diego, CA |
| C1 | NTK Division of NGK Spark Plug Co. Ltd., Harbor City, CA |
| GEM9<br>GEM1<br>GEM2*<br>GEM3* | Greenleaf Corporation, Saegertown, PA |
| K 090*<br>K 060 | Kennametal Inc., Latrobe, PA |
| V-32*<br>V-34 | Valenite Div., Valeron Corp., Detroit, MI |
| G-10*<br>G-20 | Babcock & Wilcox, Rochester, MI |
| CA-W<br>CA-B* | Carmet Company, Madison Heights, MI |

NOTE: Listings do not necessarily imply equivalency of various manufacturers' grades. This chart is not to be considered an endorsement of or an approved list of any manufacturer's products.
*Cermet

The cermet tools are hot pressed. Hot pressed cermet (HPC) tools are much tougher than the alumina inserts and are therefore applicable for roughing, for interrupted cuts, for some face milling and for machining materials of very high hardness (60 to 68 $R_C$).

Presently, the main uses for ceramic tool materials are in finish and semifinish turning. Typical applications are boring and facing of cast irons for the automotive industry, turning of steels for ordnance applications, and finish turning and facing of chilled iron and of hardened steel rolls. Where high rigidity has been designed into the machine tool and the tooling, roughing of cast iron and steel may be performed with ceramic materials.

In addition to the introduction of hot-pressed cermets, three recent developments in the preparation of ceramic inserts have expanded the usage of these tools. Those developments are as follows:

1. The grinding of K-lands on the cutting edges
2. The use of thicker inserts
3. The use of larger nose radii

The primary failure mode of ceramic inserts has always been either edge chipping or insert breakage. These three improvements have all helped reduce those problems.

A "K-land" is a precision-ground flat that removes the sharp cutting edge where the top and side faces of the insert intersect. The width of the flat varies from 0.002 to 0.012 inch [0.05 to 0.30 mm]. The angle of the bevel may vary from 30 to 45 degrees with respect to the top face.

A thicker insert will be less likely to break under the increased cutting forces produced by a chipped cutting edge. This increases insert life by preserving unused corners which would normally be lost if the insert cracks.

A larger nose radius will naturally present a stronger corner than will a smaller radius. This also helps to reduce edge chipping and nose breakage.

Even with the improved technology of adding K-lands and using thicker inserts, more care must be exercised in applying ceramic tools than is normally used in applying high speed steel or carbide inserts. In particular, it is necessary to avoid mechanical shock where possible and thermal shock at all times. Ceramic tools are rarely, if ever, used with cutting fluids. Their application in turning, boring or milling is always dry. Negative rake inserts are used almost exclusively.

## 14.8 Diamond Tools

Diamond tools are most generally used for machining nonferrous alloys and abrasive materials, such as presintered carbides and ceramics, graphite, fiberglass, rubber and high-silicon aluminum alloys. Man-made polycrystalline diamonds that are sintered under very high temperatures and pressures are now available as cutting tool inserts of various sizes and shapes. Polycrystalline diamonds are manufactured by companies such as the Specialty Materials Department of General Electric Company, Megadiamond Industries, Inc., and De Beers Industrial Diamond Division. These companies supply polycrystalline diamonds to the tool producers listed in table 14-9. Polycrystalline diamond tools have proved to be, in most cases, greatly superior to natural mined diamonds in normal machining operations on the work materials listed above. However, even though the polycrystalline diamonds are much tougher and able to withstand a great deal more abuse than natural mined diamonds, it is still good practice to observe as much care in handling the tools as possible. If natural mined diamonds are desired for use as the cutting tool material, then the following guidelines should be observed:

> Diamond tools are purchased from suppliers who are expert in the art of cutting and orienting diamonds and should be returned to them for resharpening. In-house grinding or reshaping should not be attempted. The diamond tool should be resharpened as soon as it becomes dull to minimize breakage, thereby increasing the number of possible resharpenings.

In general, diamond tools should be treated with the same care as would be used with any very hard, brittle tool material. However, even light interrupted cuts are possible with the stronger polycrystalline diamonds. Diamond tools are generally used with positive-rake tool geometry in order to reduce the forces on the tool, but there are other applications of negative-rake geometries that have proved successful. The new strength and reduced cost of polycrystalline diamond have opened up many new fields of machining, and new applications continue to be found.

The use of a cutting fluid will usually increase the tool life of the diamond, but where this is not practical, a stream of air can be used to cool the tool and keep the cutting edge free from chips.

**TABLE 14–9 Producers of Polycrystalline Diamond Tools**

Carboloy Division of General Electric Company
Detroit, MI

Citco, Inc.,
Chardon, OH

Crafts Company,
Waltham, MA

Diamonds Unlimited, Inc.,
Miami, FL

DICO Corporation,
Southfield, MI

Kennametal Inc.,
Latrobe, PA

E. C. Kitzel & Sons, Inc.,
Cleveland, OH

Megadiamond Industries, Inc.,
Chicago, IL

Neuber Industrial Diamond,
Boston, MA

New England Carbide Tool Company,
Peabody, MA

Precision Diamond Tool,
Elgin, IL

J. K. Smit & Sons,
Long Beach, CA

Tool Specialties,
Los Angeles, CA

Valeron Corporation,
Troy, MI

Wheel Trueing Tool Company,
Columbia, SC

NOTE: This is a partial listing. Contact the manufacturers of polycrystalline diamonds for a complete listing of companies supplying tools in your area.

# SECTION 15

# TOOL GEOMETRY

## TOOL GEOMETRY GUIDELINES

The following tables show the range of standard tool geometries that are used for various work materials. Negative rake is shown wherever applicable because it provides a stronger cutting edge and usually better tool life on roughing operations.

**Standard "off-the-shelf" tools** are satisfactory for most applications.

**"Modified" standard tools** for specific materials are offered by many tool manufacturers at costs considerably less than for special tools.

**Special tools** may be justified for high-production applications or for the machining of complex parts or difficult-to-machine alloys.

### SECTION CONTENTS


15.1 Turning and Boring Tools, Single Point ........ 15-2
15.2 Threading Tools, Single Point ........ 15-4
15.3 Die Threading Tools, Thread Chasers ........ 15-6
15.4 Face Mills ........ 15-8
15.5 Side and Slot Mills—Arbor Mounted ........ 15-10
15.6 End Mills—Peripheral and Slotting, High Speed Steel ........ 15-12
15.7 Drills, High Speed Steel Twist ........ 15-14
15.8 Oil-Hole or Pressurized-Coolant Drills, High Speed Steel ........ 15-16
15.9 Spade Drills, High Speed Steel ........ 15-16
15.10 Gun Drills, Carbide ........ 15-18
15.11 Reamers, High Speed Steel ........ 15-19
15.12 Reamers, Carbide ........ 15-20
15.13 Boring Tools, Carbide ........ 15-21
15.14 Taps, High Speed Steel ........ 15-23
15.15 Planing Tools ........ 15-25
15.16 Broaches, High Speed Steel ........ 15-27
15.17 Tool and Cutter Angles—Approximate Equivalents ........ 15-28

# 15.1 Turning and Boring Tools, Single Point

## TOOL GEOMETRY

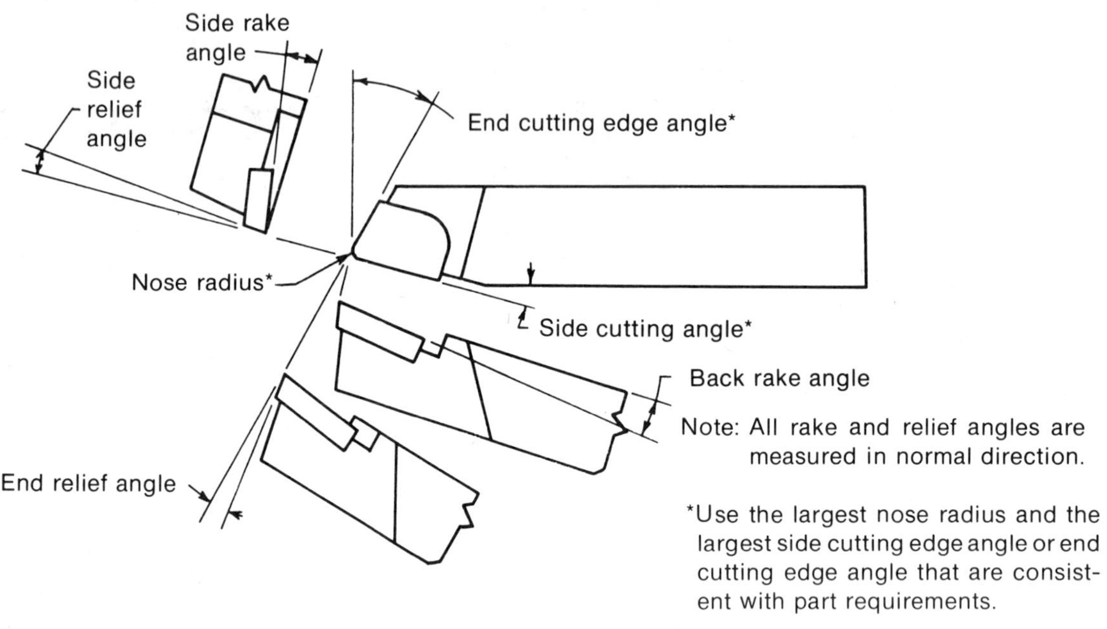


| MATERIAL | HARDNESS Bhn | HIGH SPEED STEEL Back Rake Angle degrees | HIGH SPEED STEEL Side Rake Angle degrees | HIGH SPEED STEEL End Relief Angle degrees | HIGH SPEED STEEL Side Relief Angle degrees | CARBIDE BRAZED Back Rake Angle degrees | CARBIDE BRAZED Side Rake Angle degrees | CARBIDE BRAZED Relief Angles degrees | CARBIDE INDEXABLE Back Rake Angle degrees | CARBIDE INDEXABLE Side Rake Angle degrees | CARBIDE INDEXABLE Relief Angles degrees |
|---|---|---|---|---|---|---|---|---|---|---|---|
| Free Machining Carbon Steels-Wrought<br>Carbon Steels-Wrought and Cast<br>Free Machining Alloy Steels-Wrought<br>Alloy Steels-Wrought and Cast<br>High Strength Steels-Wrought<br>Maraging Steels-Wrought<br>Tool Steels-Wrought<br>Nitriding Steels-Wrought<br>Armor Plate-Wrought<br>Structural Steels-Wrought | 85-225 | 10 | 12 | 5 | 5 | 0 | 6 | 7 | 0 | 5 | 5 |
| | 225-325 | 8 | 10 | 5 | 5 | 0 | 6 | 7 | 0 | 5 | 5 |
| | 325-52$R_c$ | 0 | 10 | 5 | 5 | 0 | 6 | 7 | -5 | -5 | 5 |
| | 52$R_c$-58$R_c$ | — | — | — | — | — | — | — | -5 | -5 | 5 |
| Free Machining Stainless Steels-Wrought | 135-275 | 5 | 8 | 5 | 5 | 0 | 6 | 7 | -5 | -5 | 5 |
| | 275-425 | 0 | 10 | 5 | 5 | 0 | 6 | 7 | -5 | -5 | 5 |
| Stainless Steels, Ferritic-Wrought and Cast | 135-185 | 5 | 8 | 5 | 5 | 0 | 6 | 7 | 0 | 5 | 5 |
| Stainless Steels, Austenitic-Wrought and Cast | 135-275 | 0 | 10 | 5 | 5 | 0 | 6 | 7 | 0 | 5 | 5 |
| Stainless Steels, Martensitic-Wrought and Cast | 135-325 | 0 | 10 | 5 | 5 | 0 | 6 | 7 | 0 | 5 | 5 |
| | 325-425<br>48$R_c$-52$R_c$ | 0 | 10 | 5 | 5 | 0 | 6 | 7 | -5 | -5 | 5 |

| MATERIAL | HARDNESS | HIGH SPEED STEEL | | | | CARBIDE | | | | | |
|---|---|---|---|---|---|---|---|---|---|---|---|
| | | | | | | BRAZED | | | INDEXABLE | | |
| | | Back Rake Angle | Side Rake Angle | End Relief Angle | Side Relief Angle | Back Rake Angle | Side Rake Angle | Relief Angles | Back Rake Angle | Side Rake Angle | Relief Angles |
| | Bhn | degrees | degrees | degrees | degrees | degrees | degrees | degrees | degrees | degrees | degrees |
| **Precipitation Hardening Stainless Steels-Wrought and Cast** | 150-450 | 0 | 10 | 5 | 5 | 0 | 6 | 7 | -5 | -5 | 5 |
| **Gray Cast Irons<br>Ductile Cast Irons<br>Malleable Cast Irons<br>Compacted Graphite Cast Irons<br>White Cast Irons** | 100-200 | 5 | 10 | 5 | 5 | 0 | 6 | 7 | -5 | -5 | 5 |
| | 200-300 | 5 | 8 | 5 | 5 | 0 | 6 | 7 | -5 | -5 | 5 |
| | 300-400 | 5 | 5 | 5 | 5 | -5 | -5 | 7 | -5 | -5 | 5 |
| **Aluminum Alloys-Wrought and Cast** | 30-150<br>500 kg | 20 | 15 | 12 | 10 | 3 | 15 | 7 | 0 | 5 | 5 |
| **Magnesium Alloys-Wrought and Cast** | 40-90<br>500 kg | 20 | 15 | 12 | 10 | 3 | 15 | 7 | 0 | 5 | 5 |
| **Titanium Alloys-Wrought and Cast** | 110-440 | 0 | 5 | 5 | 5 | 0 | 6 | 7 | -5 | -5 | 5 |
| **Copper Alloys-Wrought and Cast** | 40-200<br>500 kg | 5 | 10 | 8 | 8 | 0 | 8 | 7 | 0 | 5 | 5 |
| **Nickel Alloys-Wrought and Cast<br>Chrome-Nickel Alloys<br>Beryllium-Nickel Alloys** | 80-360 | 8 | 10 | 12 | 12 | 0 | 6 | 7 | -5 | -5 | 5 |
| **Nitinol Alloys-Wrought** | 210-340<br>$48R_c$-$52R_c$ | - | - | - | - | 0 | 5 | 7 | 0 | 5 | 5 |
| **High Temperature Alloys-Wrought and Cast** | 140-475 | 0 | 10 | 5 | 5 | 0 | 6 | 7 | 5 | 0 | 5 |
| **Columbium Alloys-Wrought, Cast, P/M<br>Molybdenum Alloys-Wrought, Cast, P/M<br>Tantalum Alloys-Wrought, Cast, P/M** | 170-290 | 0 | 20 | 5 | 5 | 0 | 20 | 7 | - | - | 5 |
| **Tungsten Alloys-Wrought, Cast, P/M** | 180-320 | - | - | - | - | -15 | 0 | 7 | - | - | - |
| **Zinc Alloys-Cast** | 80-100 | 10 | 10 | 12 | 4 | 5 | 5 | 7 | 0 | 5 | 5 |
| **Uranium Alloys-Wrought** | 190-210 | - | - | - | - | 0 | 0 | 7 | -5 | 0 | 7 |
| **Zirconium Alloys-Wrought** | 140-280 | 15 | 10 | 10 | 10 | 5 | 5 | 7 | 5 | 5 | 6 |
| **Thermoplastics** | All | 0 | 0 | 20 to 30 | 15 to 20 | 0 | 0 | 20 to 30 | 0 | 0 | 20 to 30 |
| **Thermosetting Plastics** | All | 0 | 0 | 20 to 30 | 15 to 20 | 0 | 15 | 7 | 0 | 15 | 5 |
| **Magnetic Alloys, Nickel- and Cobalt-Base<br>Controlled Expansion Alloys** | 125-250 | 10 | 8 | 8 | 8 | - | - | - | - | - | - |
| **Powder Metal Alloys — Copper** | All | 10 | 8 | 8 | 8 | 6 | 12 | 7 | - | - | - |
| **Powder Metal Alloys — Iron** | All | 0 | 0 | 8 | 8 | 6 | 16 | 7 | 0 | 0 | 5 |
| **Magnetic Core Iron** | 185-240 | 15 | 30 | 8 | 8 | 20 | 0 | 7 | 5 | 5 | 5 |
| **Carbon and Graphite** | All | 0 | 0 | 20 | 20 | 0 | 0 | 20 | 0 | 0 | 15 |
| **Machinable Carbide (Ferro-Tic)** | $40R_c$-$51R_c$ | -5 | -5 | 5 | 5 | -5 | -5 | 7 | -5 | -5 | 5 |
| **Machinable Glass Ceramic** | 250 Knoop<br>100 g | 0 | 15 | 5 | 5 | 1 | 0 | 7 | 0 | 0 | 5 |

# 15.2 Threading Tools, Single Point

## TOOL GEOMETRY

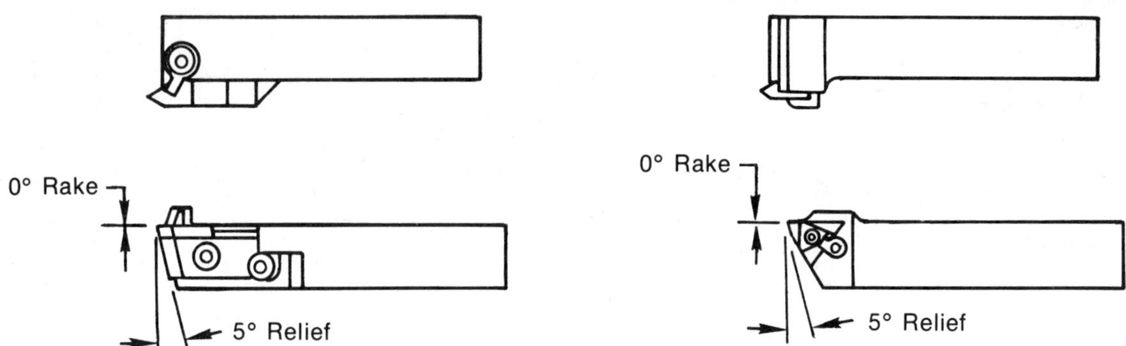


**Typical Carbide Threading Tools**

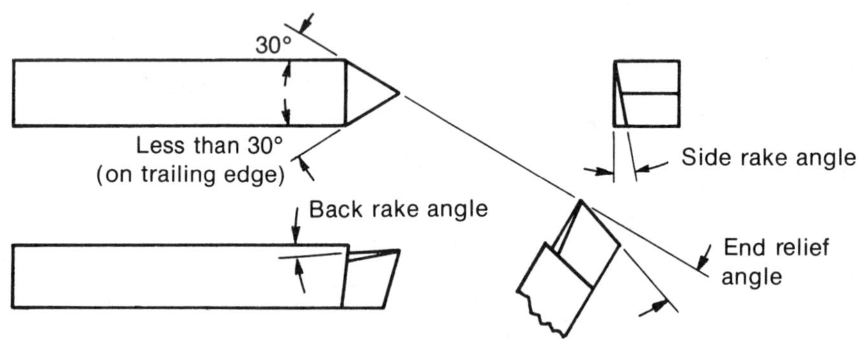


**Typical High Speed Steel Threading Tool**

| MATERIAL | | HARDNESS Bhn | HIGH SPEED STEEL Back Rake Angle degrees | Side Rake Angle degrees | End Relief Angle degrees |
|---|---|---|---|---|---|
| Free Machining Carbon Steels-Wrought<br>Carbon Steels-Wrought and Cast<br>Free Machining Alloy Steels-Wrought<br>Alloy Steels-Wrought and Cast<br>High Strength Steels-Wrought<br>Maraging Steels-Wrought<br>Tool Steels-Wrought<br>Nitriding Steels-Wrought<br>Armor Plate-Wrought<br>Structural Steels-Wrought | | 80-225 | 10 to 15 | 10 to 15 | 5 to 10 |
| | | 225-325 | 8 to 12 | 8 to 12 | 5 to 10 |
| | | 325-425 | 0 to 5 | 5 to 10 | 5 to 10 |
| Free Machining Stainless Steels-Wrought | | 135-275 | 5 to 10 | 5 to 8 | 5 to 10 |
| | | 275-425 | 0 to 5 | 10 to 15 | 5 to 10 |
| Stainless Steels, Ferritic-Wrought and Cast | | 135-185 | 5 to 10 | 5 to 8 | 5 to 10 |
| Stainless Steels, Austenitic-Wrought and Cast | | 135-275 | 0 to 5 | 10 to 15 | 5 to 10 |
| Stainless Steels, Martensitic and Precipitation Hardening-Wrought and Cast | | 135-425 | 0 to 5 | 10 to 15 | 3 to 10 |
| Gray, Ductile and Malleable Cast Irons<br>Compacted Graphite Cast Irons | | 100-200 | 5 to 8 | 8 to 12 | 5 to 10 |
| | | 200-400 | 0 to 5 | 5 to 10 | 5 to 10 |
| Aluminum Alloys-Wrought and Cast | | 30-150 500 kg | 20 to 30 | 10 to 15 | 5 to 10 |
| Magnesium Alloys-Wrought and Cast | | 40-90 500 kg | 20 to 30 | 10 to 15 | 5 to 10 |
| Titanium Alloys-Wrought and Cast | | 110-440 | 0 | 0 to 5 | 5 to 15 |
| Copper Alloys-Wrought and Cast* | Group 1 | 40-200 500 kg | 0 | 5 to 10 | 5 to 10 |
| | Group 2 | 40-200 500 kg | 3 to 7 | 5 to 10 | 5 to 10 |
| | Group 3 | 40-200 500 kg | 7 to 15 | 5 to 10 | 5 to 10 |
| Nickel Alloys-Wrought and Cast | | 80-360 | 6 to 9 | 9 to 12 | 2 to 12 |
| High Temperature Alloys-Wrought and Cast | | 140-475 | 0 to 10 | 10 to 15 | 5 to 12 |

*Copper alloy groups.

| Group 1 | | | | | | | Group 2 | | | | | | | Group 3 | | | | | | |
|---|---|---|---|---|---|---|---|---|---|---|---|---|---|---|---|---|---|---|---|---|
| 314 | 356 | 377 | 842 | 858 | 935 | 953 | 226 | 442 | 467 | 821 | 872 | 915 | 948 | 102 | 172 | 614 | 803 | 818 | 863 | 917 |
| 330 | 360 | 385 | 844 | 864 | 937 | 954 | 230 | 443 | 651 | 833 | 874 | 922 | 952 | 110 | 175 | 706 | 805 | 820 | 902 | 962 |
| 332 | 365 | 485 | 848 | 867 | 938 | 956 | 240 | 444 | 655 | 853 | 875 | 923 | 955 | 113 | 210 | 715 | 807 | 822 | 907 | 963 |
| 335 | 366 | 544 | 852 | 879 | 939 | 973 | 260 | 445 | 675 | 861 | 876 | 925 | 957 | 114 | 220 | 745 | 809 | 824 | 909 | 964 |
| 340 | 367 | 834 | 854 | 928 | 943 | 974 | 268 | 464 | 687 | 862 | 878 | 926 | 958 | 115 | 505 | 752 | 811 | 825 | 910 | 966 |
| 342 | 368 | 836 | 855 | 932 | 944 | 976 | 270 | 465 | 770 | 865 | 903 | 927 | | 116 | 510 | 754 | 813 | 826 | 911 | |
| 353 | 370 | 838 | 857 | 934 | 945 | 978 | 280 | 466 | 817 | 868 | 905 | 947 | | 122 | 521 | 757 | 814 | 827 | 913 | |
| | | | | | | | | | | | | | | 170 | 524 | 801 | 815 | 828 | 916 | |

# 15.3 Die Threading Tools, Thread Chasers

## TOOL GEOMETRY

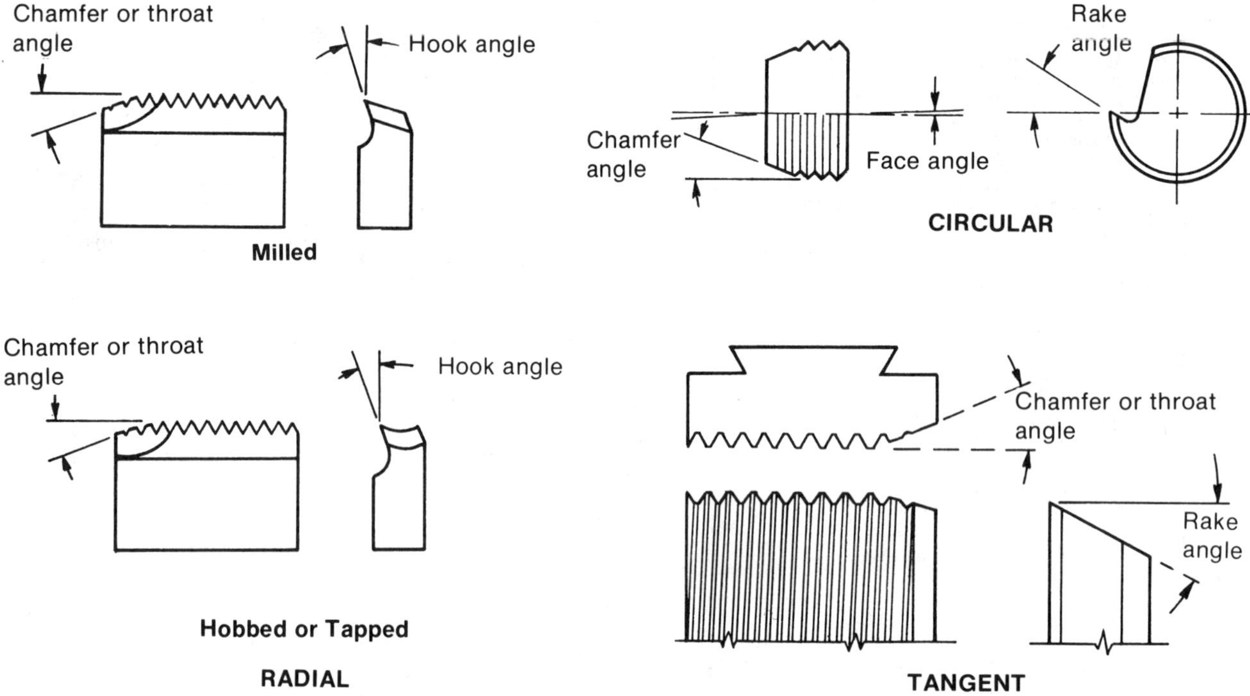

| MATERIAL | HARDNESS | HIGH SPEED STEEL | | | | | CARBIDE |
|---|---|---|---|---|---|---|---|
| | | CIRCULAR | | TANGENT | RADIAL | | RADIAL |
| | | | | | TAPPED | MILLED | |
| | Bhn | Rake Angle degrees | Face Angle degrees | Rake Angle degrees | Rake Angle degrees | Rake Angle degrees | Rake Angle degrees |
| **Free Machining Carbon Steels-Wrought**<br>**Carbon Steels-Wrought and Cast**<br>**Free Machining Alloy Steels-Wrought**<br>**Alloy Steels-Wrought and Cast**<br>**High Strength Steels-Wrought**<br>**Maraging Steels-Wrought**<br>**Tool Steels-Wrought**<br>**Nitriding Steels-Wrought**<br>**Armor Plate-Wrought**<br>**Structural Steels-Wrought** | 100-330 | 20 | 1-1/2 | 15 to 25 | 10 to 15 | 10 to 15 | 5 to 10 |
| | 85-225 | 20 | 1-1/2 | 25 to 35 | 8 to 15 | 8 to 15 | 5 to 10 |
| | 225-330 | 20 | 1-1/2 | 18 to 22 | 5 to 7 | 5 to 7 | 0 to 5 |
| **Free Machining Stainless Steels-Wrought** | 135-330 | 15 to 25 | 1-1/2 | 10 to 20 | 7 to 10 | 7 to 10 | 5 to 10 |
| **Stainless Steels, Ferritic-Wrought and Cast** | 135-185 | 20 to 25 | 1-1/2 | 20 to 28 | 10 to 15 | 10 to 12 | 5 to 10 |
| **Stainless Steels, Austenitic-Wrought and Cast** | 135-275 | 25 to 30 | 1-1/2 | 15 to 20 | 10 to 12 | 10 to 15 | 5 to 10 |
| **Stainless Steels, Martensitic-Wrought and Cast** | 135-200 | 15 to 20 | 1-1/2 | 20 to 28 | 10 to 12 | 7 to 10 | 5 to 10 |
| | 200-325 | 15 to 20 | 1-1/2 | 15 to 20 | 10 to 12 | 7 to 10 | 5 to 10 |
| **Gray Cast Irons**<br>**Compacted Graphite Cast Irons** | 110-330 | 10 | 2 | 15 to 25 | 0 to 5 | 0 to 5 | 0 to 5 |
| **Ductile and Malleable Cast Irons** | 110-330 | 10 | 1-1/2 | 18 | 7 to 10 | 7 to 10 | 5 to 10 |
| **Aluminum Alloys-Wrought and Cast** | 30-150 500 kg | 20 to 25 | 2 | 20 to 35 | 15 to 25 | 15 to 25 | 10 to 15 |
| **Magnesium Alloys-Wrought and Cast** | 40-90 500 kg | 20 to 25 | 2 | 20 to 35 | 15 | 15 | 10 |
| **Titanium Alloys-Wrought and Cast** | 110-340 | 5 to 20 | 0 to 1 | 5 to 20 | 0 to 10 | 0 to 10 | 5 to 10 |
| **Copper Alloys-Wrought and Cast*** — **Group 1** | 40-200 500 kg | -5 to 5 | 0 | 0 to 10 | 0 to-10 | 0 to -10 | 0 to 5 |
| **Group 2** | 40-200 500 kg | 10 to 20 | 1 to 2 | 20 to 25 | 0 to 10 | 0 to 10 | 0 to 5 |
| **Group 3** | 40-200 500 kg | 15 to 35 | 2 to 3 | 30 to 35 | 12 to 30 | 12 to 30 | 5 to 15 |
| **Nickel Alloys-Wrought and Cast** | 80-240 | 15 to 25 | 1 to 2 | 20 to 25 | 20 to 30 | 20 to 30 | 5 to 15 |
| | 240-320 | 10 to 20 | 1 to 2 | 15 to 20 | 8 to 15 | 8 to 15 | 5 to 10 |
| **High Temperature Alloys-Wrought and Cast** | 180-230 | 15 to 20 | 1 to 2 | 20 to 25 | 15 to 25 | 15 to 25 | 10 to 15 |
| | 230-320 | 5 to 15 | 1 to 2 | 15 to 20 | 10 to 15 | 10 to 15 | 10 |
| **Zinc Alloys-Cast** | 80-100 | 15 to 20 | 1 to 2 | 5 | 10 to 15 | 10 to 15 | 10 |
| **Thermoplastics** | $31R_R$-$103R_M$ | 5 to 10 | 0 | 5 to 10 | 0 to 5 | 0 to 5 | 5 |

*See section 15.2 for copper alloy groups.

# 15.4 Face Mills

## TOOL GEOMETRY

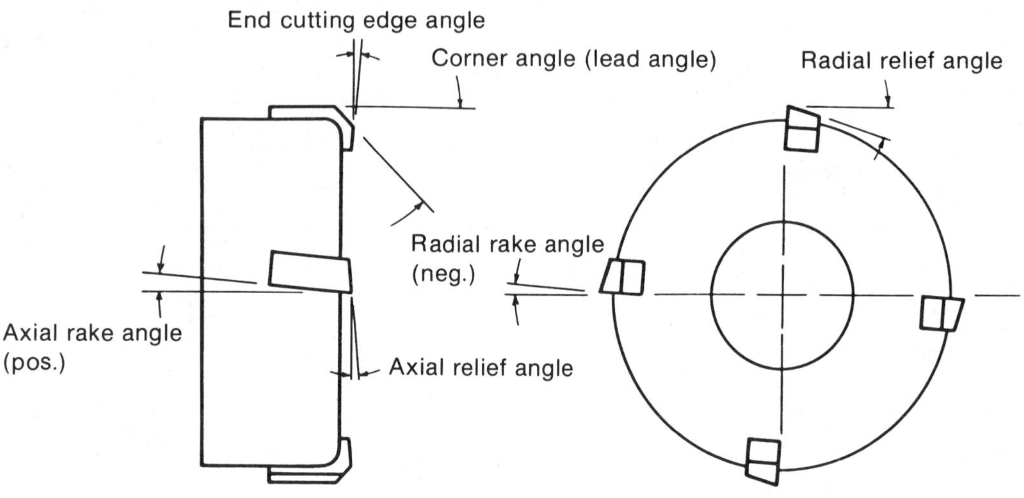


| MATERIAL | HARDNESS Bhn | HIGH SPEED STEEL Axial Rake Angle degrees | HIGH SPEED STEEL Radial Rake Angle degrees | INDEXABLE CARBIDE Axial Rake Angle degrees | INDEXABLE CARBIDE Radial Rake Angle degrees | BRAZED CARBIDE Axial Rake Angle degrees | BRAZED CARBIDE Radial Rake Angle degrees | CORNER ANGLE degrees | END CUTTING EDGE ANGLE degrees | AXIAL RELIEF ANGLE degrees | RADIAL RELIEF ANGLE degrees |
|---|---|---|---|---|---|---|---|---|---|---|---|
| **Free Machining Carbon Steels and Carbon Steels-Wrought and Cast Free Machining Alloy Steels and Alloy Steels-Wrought and Cast Maraging Steels-Wrought Tool Steels-Wrought Structural Steels-Wrought** | 85-270 | 10 to 15 | 10 to 15 | 5 to 7 | -5 to -14 | 0 to -7 | 0 to -7 | 30 | 5 to 10 | 5 to 7 | 3 to 7 |
| | 270-325 | 10 to 15 | 10 to 15 | -4 to -8 | -3 to -11 | 0 to -7 | 0 to -7 | 30 | 5 to 10 | 5 to 7 | 3 to 7 |
| | 325-425 | 10 to 12 | 10 to 12 | -4 to -8 | -3 to -11 | 0 to -10 | 0 to -10 | 30 | 5 to 10 | 5 to 7 | 3 to 7 |
| | $43R_c$-$50R_c$ | 5 to 10 | 5 to 10 | -4 to -8 | -3 to -11 | -5 to -15 | -5 to -15 | 45 | 4 to 7 | 5 to 7 | 3 to 7 |
| | $50R_c$-$56R_c$ | - | - | -4 to -8 | -3 to -11 | -5 to -15 | -5 to -15 | 45 | 4 to 7 | 8 | 8 |
| **High Strength Steels-Wrought** | 225-425 | 5 to 10 | 0 to 10 | -4 to -8 | -3 to -11 | -5 to -15 | -5 to -15 | 45 | 4 to 7 | 5 to 7 | 3 to 7 |
| | $45R_c$-$58R_c$ | - | - | -4 to -8 | -3 to -11 | -5 to -15 | -5 to -15 | 45 | 4 to 7 | 8 | 8 |
| **Nitiriding Steels-Wrought** | 200-350 | 5 to 10 | 5 to 10 | -4 to -8 | -3 to -11 | 0 to -10 | -5 to -15 | 45 | 5 to 10 | 5 to 7 | 3 to 5 |
| **Armor Plate-Wrought** | 250-320 | 0 to 5 | 0 to 5 | -4 to -8 | -3 to -11 | 0 to -10 | -5 to -15 | 45 | 4 to 7 | 5 to 7 | 3 to 5 |
| **Free Machining Stainless Steels-Wrought** | 135-275 | 10 to 15 | 10 to 12 | 5 to 11 | -5 to -11 | 0 | 0 to 5 | 45 | 5 | 8 to 10 | 8 to 10 |
| | 275-425 | 5 to 10 | 5 to 10 | 5 to 11 | -5 to -11 | 0 | 0 to -5 | 45 | 5 | 8 to 10 | 8 to 10 |

| MATERIAL | HARDNESS | HIGH SPEED STEEL | | INDEXABLE CARBIDE | | BRAZED CARBIDE | | CORNER ANGLE | END CUTTING EDGE ANGLE | AXIAL RELIEF ANGLE | RADIAL RELIEF ANGLE |
|---|---|---|---|---|---|---|---|---|---|---|---|
| | | Axial Rake Angle | Radial Rake Angle | Axial Rake Angle | Radial Rake Angle | Axial Rake Angle | Radial Rake Angle | | | | |
| | Bhn | degrees | degrees | degrees | degrees | degrees | degrees | degrees | degrees | degrees | degrees |
| **Stainless Steels, Ferritic and Austenitic-Wrought and Cast** | 135-275 | 10 to 15 | 10 to 12 | 5 to 11 | -5 to -11 | 0 to 5 | 0 to -5 | 45 | 5 | 8 to 10 | 8 to 10 |
| **Stainless Steels, Martensitic-Wrought and Cast** | 135-425 | 5 to 10 | 5 to 10 | 5 to 11 | -5 to -11 | 0 | 0 to -5 | 45 | 5 | 8 to 10 | 8 to 10 |
| **Precipitation Hardening Stainless Steels-Wrought and Cast** | 150-450 | 5 to 10 | 5 to 10 | 5 to 7 | 0 to 5 | 0 | 0 | 45 | 5 | 8 to 10 | 8 to 10 |
| **Gray Cast Irons<br>Ductile Cast Irons<br>Malleable Cast Irons<br>Compacted Graphite Cast Irons<br>White Cast Irons** | 100-400 | 20 to 30 | -5 to -10 | 5 to 11 | -5 to -11 | 5 to 10 | 5 to -10 | 45 | 5 to 10 | 4 to 7 | 4 to 7 |
| **Aluminum Alloys-Wrought and Cast** | 30-150<br>500 kg | 20 to 35 | 20 to 35 | 5 to 7 | 0 to 5 | 10 to 20 | 10 to 20 | 45 | 7 to 12 | 3 to 5 | 10 to 12 |
| **Magnesium Alloys-Wrought and Cast** | 40-90<br>500 kg | 20 to 35 | 20 to 35 | 5 to 7 | 0 to 5 | 10 to 20 | 10 to 20 | 45 | 7 to 12 | 3 to 5 | 10 to 12 |
| **Titanium Alloys-Wrought and Cast** | 110-440 | 5 | 5 | 0 to -5 | 0 to -5 | 0 to -5 | -10 | 45 | 6 to 12 | 10 to 12 | 10 to 12 |
| **Copper Alloys-Wrought and Cast** | 40-200<br>500 kg | 12 to 25 | 10 to 12 | 5 to 7 | 0 to 5 | 3 to 10 | 3 to 10 | 45 | 7 to 12 | 3 to 5 | 5 to 10 |
| **Nickel Alloys, Chrome-Nickel & Beryllium-Nickel Alloys-Wrought & Cast** | 80-360 | 7 | 15 | 5 to 11 | -5 to -14 | 5 to 10 | 0 to -5 | 45 | 5 | 7 to 9 | 7 to 9 |
| **Nitinol Alloys-Wrought** | 210-340<br>$48R_c$-$60R_c$ | - | - | 5 to 7 | 0 to 5 | 0 | 0 | 45 | 10 | 12 | 12 |
| **High Temperature Alloys-Wrought and Cast** | 200-475 | 5 to 10 | 5 to 10 | 0 to 5 | 0 to -5 | 0 to 5 | 0 to -5 | 45 | 5 | 7 to 10 | 7 to 10 |
| **Columbium Alloys-Wrought, Cast, P/M** | 170-225 | 0 | 20 | 5 to 7 | 0 to 5 | 0 | 10 | 45 | 5 to 10 | 10 | 10 |
| **Molybdenum Alloys-Wrought, Cast, P/M** | 220-290 | 0 | 20 | 5 to 7 | 0 to 5 | 0 | 0 | 45 | 5 to 10 | 10 | 10 |
| **Tantalum Alloys-Wrought, Cast, P/M** | 200-250 | 0 | 20 | 5 to 7 | 0 to 5 | 0 | 0 | 45 | 5 to 10 | 10 | 10 |
| **Tungsten Alloys-Wrought, Cast, P/M** | 180-320 | - | - | -4 to -8 | -3 to -11 | -15 | 0 | 45 | 5 to 10 | 15 | 15 |
| **Zinc Alloys-Cast** | 80-100 | 10 to 15 | 10 to 15 | 5 to 7 | 0 to 5 | 10 to 12 | 10 to 12 | 45 | 7 to 12 | 10 | 10 to 12 |

# 15.5 Side and Slot Mills—Arbor Mounted

## TOOL GEOMETRY

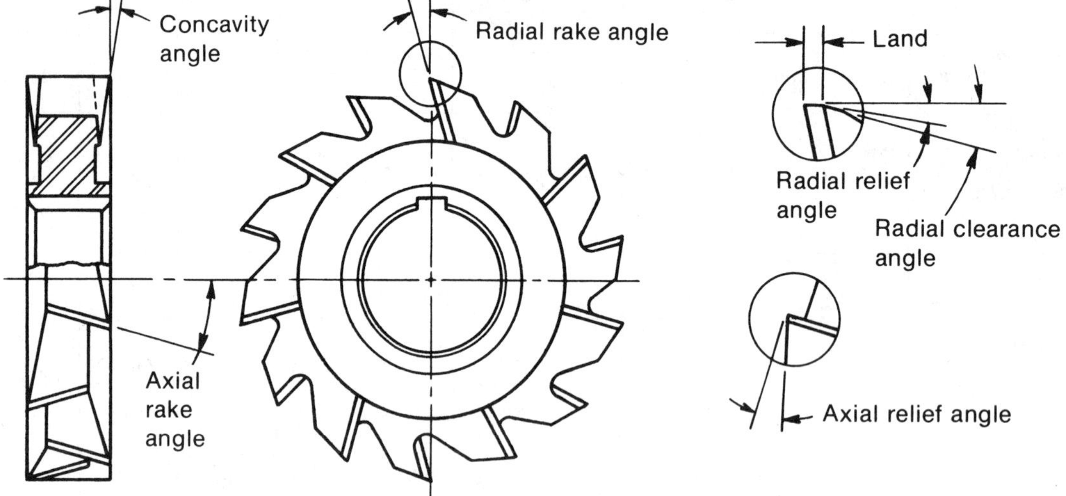

| MATERIAL | HARDNESS Bhn | HIGH SPEED STEEL Axial Rake Angle degrees | HIGH SPEED STEEL Radial Rake Angle degrees | HIGH SPEED STEEL Axial Relief Angle degrees | HIGH SPEED STEEL Radial Relief Angle degrees | CARBIDE Axial Rake Angle degrees | CARBIDE Radial Rake Angle degrees | CARBIDE Axial Relief Angle degrees | CARBIDE Radial Relief Angle degrees |
|---|---|---|---|---|---|---|---|---|---|
| Free Machining Carbon Steels-Wrought<br>Carbon Steels-Wrought and Cast<br>Free Machining Alloy Steels-Wrought | 85-325 | 10 to 15 | 10 to 15 | 3 to 5 | 4 to 8 | 0 to -5 | -5 to 5 | 2 to 4 | 5 to 8 |
| | 325-425 | 10 to 12 | 5 to 12 | 3 to 5 | 4 to 8 | 0 to -5 | -5 to 5 | 2 to 4 | 5 to 8 |
| | $45R_c$-$52R_c$ | 10 to 12 | 5 to 12 | 2 to 4 | 3 to 7 | -5 to -10 | 0 to -10 | 2 to 4 | 5 to 8 |
| Alloy Steels-Wrought and Cast | 125-425 | 10 to 12 | 5 to 12 | 3 to 5 | 4 to 8 | -5 to -10 | 0 to -10 | 2 to 5 | 5 to 8 |
| | $45R_c$-$52R_c$ | 10 to 12 | 5 to 12 | 2 to 4 | 3 to 7 | -5 to -10 | 0 to -10 | 2 to 4 | 3 to 6 |
| High Strength Steels, Maraging Steels and Tool Steels-Wrought | 100-$52R_c$ | 10 to 12 | 5 to 12 | 2 to 4 | 3 to 7 | -5 to -10 | 0 to -10 | 2 to 4 | 5 to 8 |
| Nitriding Steels-Wrought | 200-350 | 10 to 12 | 5 to 12 | 2 to 4 | 3 to 7 | -5 to -10 | 0 to -10 | 2 to 4 | 3 to 6 |
| Armor Plate-Wrought | 250-320 | 0 to 5 | 0 to 5 | 2 to 4 | 3 to 7 | -5 to -10 | -5 to -10 | 2 to 4 | 3 to 6 |
| Structural Steels-Wrought | 100-$50R_c$ | 10 to 12 | 5 to 12 | 3 to 5 | 4 to 8 | 0 to -5 | 0 to -10 | 2 to 4 | 5 to 8 |
| Free Machining Stainless Steels-Wrought | 135-425 | 10 to 12 | 5 to 12 | 3 to 5 | 4 to 8 | 0 to 5 | -5 to 5 | 2 to 4 | 5 to 8 |
| Stainless Steels, Ferritic-Wrought and Cast<br>Stainless Steels, Austenitic-Wrought and Cast | 135-$52R_c$ | 10 to 12 | 5 to 12 | 3 to 5 | 4 to 8 | 0 to 5 | -5 to 5 | 2 to 4 | 5 to 8 |
| Stainless Steels, Martensitic-Wrought and Cast | 135-$52R_c$ | 10 to 12 | 5 to 12 | 2 to 4 | 3 to 7 | -5 to -10 | 0 to -10 | 2 to 4 | 5 to 8 |
| Precipitation Hardening Stainless Steels-Wrought and Cast | 150-450 | 10 to 12 | 5 to 12 | 2 to 4 | 4 to 8 | 0 to -5 | 0 to -10 | 2 to 4 | 5 to 8 |
| Gray Cast Irons<br>Ductile Cast Irons<br>Malleable Cast Irons | 100-400 | 10 to 12 | 10 to 12 | 2 to 4 | 3 to 7 | 0 to -10 | 5 to -10 | 3 to 5 | 5 to 8 |
| Aluminum Alloys-Wrought and Cast | 30-150<br>500 kg | 12 to 25 | 10 to 20 | 5 to 7 | 5 to 11 | 10 to 20 | 5 to 15 | 5 to 7 | 7 to 10 |
| Magnesium Alloys-Wrought and Cast | 40-90<br>500 kg | 12 to 25 | 10 to 20 | 5 to 7 | 5 to 11 | 10 to 20 | 5 to 15 | 5 to 7 | 7 to 10 |
| Titanium Alloys-Wrought | 110-440 | 10 to 15 | 5 to 10 | 5 to 7 | 5 to 11 | 0 to -10 | 0 to -10 | 5 to 7 | 5 to 8 |
| Copper Alloys-Wrought and Cast | 40-200<br>500 kg | 12 to 25 | 10 to 20 | 5 to 7 | 5 to 11 | 10 to 20 | 5 to 10 | 4 to 7 | 5 to 8 |
| Nickel Alloys-Wrought and Cast | 80-360 | 10 to 20 | 10 to 15 | 3 to 5 | 4 to 8 | -5 to -10 | 0 to -10 | 3 to 5 | 5 to 8 |
| High Temperature Alloys-Wrought and Cast | 140-300 | 10 to 15 | 10 to 15 | 1 to 5 | 5 to 10 | -5 to -10 | 0 to -10 | 3 to 5 | 5 to 8 |
| | 300-475 | 10 to 12 | 5 to 12 | 1 to 5 | 4 to 8 | -5 to -10 | 0 to -10 | 3 to 5 | 5 to 8 |
| Columbium, Molybdenum Alloys-Wrought, Cast, P/M | 170-290 | 0 | 15 to 20 | 3 to 5 | 5 to 10 | 0 | 5 to 15 | 7 to 10 | 7 to 10 |
| Tantalum Alloys-Wrought, Cast, P/M | 200-250 | 0 | 15 to 20 | 3 to 5 | 5 to 10 | 0 | 5 to 15 | 7 to 10 | 7 to 10 |
| Tungsten Alloys-Wrought, Cast, P/M | 180-320 | - | - | - | - | -10 to -15 | 5 to 15 | 10 to 15 | 10 to 15 |
| Zinc Alloys-Cast | 80-100 | 10 to 20 | 10 to 20 | 5 to 7 | 8 to 11 | 10 to 15 | 10 to 15 | 7 to 10 | 7 to 10 |

### TOOL GEOMETRY

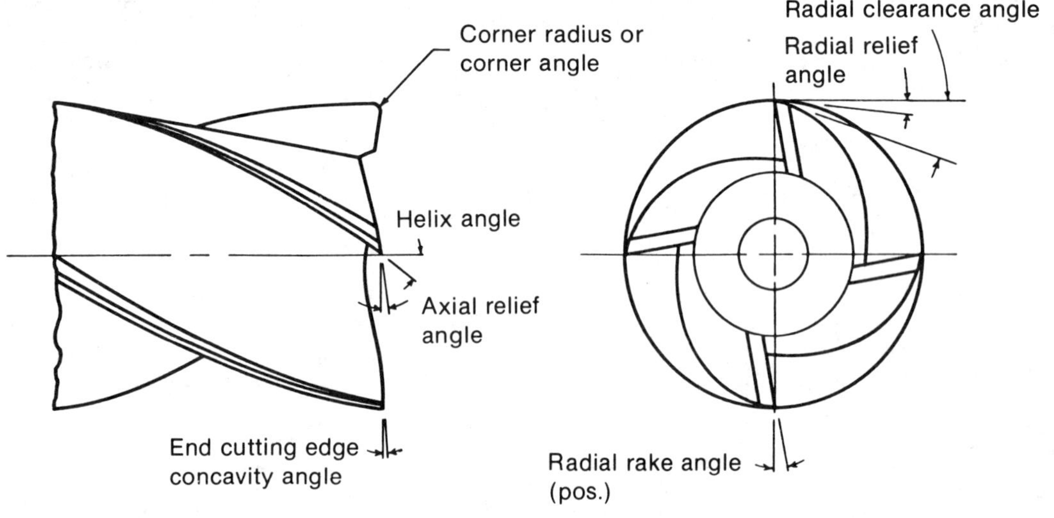

| NOMINAL CUTTER DIAMETER | GENERAL PURPOSE—30° to 35° HELIX Steels, Cast Irons, Copper Alloys, Titanium Alloys, Nickel Alloys, High Temperature Alloys and Zinc Alloys | | | 35° to 45° HELIX Aluminum and Magnesium Alloys | | |
|---|---|---|---|---|---|---|
| | Radial Primary Relief Angle | Primary Land Width | Radial Secondary Clearance Angle | Radial Primary Relief Angle | Primary Land Width | Radial Secondary Clearance Angle |
| in | degrees | in | degrees | degrees | in | degrees |
| 1/16 | 20 to 21 | 0.007-0.010 | 30 to 35 | 20 to 22 | 0.007-0.010 | 30 to 35 |
| 1/8 | 12 to 13 | 0.010-0.015 | 22 to 28 | 14 to 18 | 0.010-0.015 | 25 to 30 |
| 3/16 | 12 to 13 | 0.010-0.020 | 20 to 25 | 14 to 18 | 0.010-0.020 | 25 to 30 |
| 1/4 | 10 to 11 | 0.010-0.20 | 20 to 25 | 12 to 15 | 0.010-0.020 | 22 to 28 |
| 5/16 | 10 to 11 | 0.015-0.025 | 20 to 25 | 12 to 14 | 0.015-0.025 | 21 to 28 |
| 3/8 | 10 to 11 | 0.015-0.025 | 17 to 20 | 12 to 14 | 0.015-0.025 | 19 to 26 |
| 7/16 | 9 to 10 | 0.020-0.030 | 17 to 20 | 11 to 13 | 0.020-0.030 | 18 to 25 |
| 1/2 | 9 to 10 | 0.020-0.030 | 17 to 20 | 11 to 13 | 0.020-0.030 | 18 to 25 |
| 5/8 | 9 to 10 | 0.025-0.035 | 17 to 20 | 11 to 13 | 0.025-0.035 | 18 to 25 |
| 3/4 | 8 to 9 | 0.030-0.040 | 15 to 18 | 10 to 12 | 0.030-0.040 | 17 to 24 |
| 7/8 | 8 to 9 | 0.030-0.040 | 15 to 18 | 10 to 12 | 0.030-0.040 | 17 to 24 |
| 1 | 8 to 9 | 0.035-0.050 | 15 to 18 | 10 to 12 | 0.035-0.050 | 16 to 23 |
| 1-1/4 | 7 to 8 | 0.040-0.060 | 13 to 18 | 9 to 11 | 0.040-0.060 | 14 to 22 |
| 1-1/2 | 7 to 8 | 0.040-0.060 | 11 to 17 | 9 to 11 | 0.040-0.060 | 13 to 21 |
| 1-3/4 | 7 to 8 | 0.040-0.060 | 10 to 16 | 8 to 10 | 0.040-0.060 | 12 to 20 |
| 2 | 6 to 7 | 0.040-0.060 | 9 to 15 | 8 to 10 | 0.040-0.060 | 12 to 20 |
| mm | degrees | mm | degrees | degrees | mm | degrees |
| M1.6 | 20 to 21 | 0.200-0.250 | 30 to 35 | 20 to 22 | 0.200-0.250 | 30 to 35 |
| M3 | 14 to 15 | 0.250-0.350 | 24 to 30 | 15 to 20 | 0.250-0.350 | 25 to 30 |
| M4 | 12 to 13 | 0.250-0.400 | 22 to 28 | 14 to 18 | 0.250-0.400 | 25 to 30 |
| M6 | 10 to 11 | 0.250-0.500 | 20 to 25 | 12 to 15 | 0.250-0.500 | 22 to 28 |
| M7 | 10 to 11 | 0.250-0.500 | 20 to 25 | 12 to 15 | 0.250-0.500 | 22 to 28 |
| M8 | 10 to 11 | 0.400-0.650 | 20 to 25 | 12 to 14 | 0.400-0.650 | 19 to 26 |
| M10 | 10 to 11 | 0.400-0.650 | 17 to 20 | 12 to 14 | 0.400-0.650 | 19 to 26 |
| M12 | 9 to 10 | 0.500-0.750 | 17 to 20 | 11 to 13 | 0.500-0.750 | 18 to 25 |
| M14 | 9 to 10 | 0.500-0.750 | 17 to 20 | 11 to 13 | 0.500-0.750 | 18 to 25 |
| M16 | 9 to 10 | 0.650-0.900 | 17 to 20 | 11 to 13 | 0.650-0.900 | 18 to 25 |
| M20 | 8 to 9 | 0.750-1.000 | 15 to 18 | 10 to 12 | 0.750-1.000 | 17 to 24 |
| M22 | 8 to 9 | 0.75-1.000 | 15 to 18 | 10 to 12 | 0.75-1.000 | 17 to 24 |
| M25 | 8 to 9 | 0.900-1.250 | 15 to 18 | 10 to 12 | 0.900-1.250 | 16 to 23 |
| M32 | 7 to 8 | 1.000-1.500 | 13 to 18 | 9 to 11 | 1.000-1.500 | 14 to 22 |
| M40 | 7 to 8 | 1.000-1.500 | 11 to 17 | 9 to 11 | 1.000-1.500 | 13 to 21 |
| M45 | 7 to 8 | 1.000-1.500 | 10 to 16 | 8 to 10 | 1.000-1.500 | 12 to 20 |

# 15.7 Drills, High Speed Steel Twist
## TOOL GEOMETRY

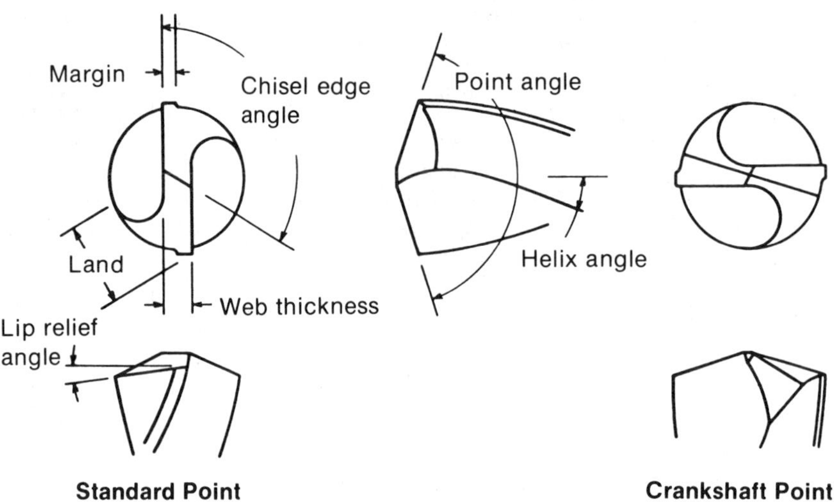


**Standard Point** **Crankshaft Point**

| MATERIAL | HARDNESS Bhn | DRILL TYPE | POINT ANGLE* degrees | LIP RELIEF ANGLE‡ degrees | HELIX ANGLE | POINT GRIND |
|---|---|---|---|---|---|---|
| **Free Machining Carbon Steels-Wrought<br>Carbon Steels-Wrought, Cast and P/M<br>Free Machining Alloy Steels-Wrought<br>Alloy Steels-Wrought, Cast and P/M<br>Maraging Steels-Wrought<br>Tool Steels-Wrought, Cast and P/M<br>Nitriding Steels-Wrought<br>Armor Plate-Wrought<br>Structural Steels-Wrought** | 85-225 | General Purpose | 118 | A | Standard | Standard |
| | 225-325 | General Purpose | 118 | A | Standard | Standard |
| | 325-425 | General Purpose | 118 to 135 | B | Standard | Crankshaft |
| | $45R_c$-$52R_c$ | Heavy Web | 118 to 135 | B | Standard | Crankshaft |
| **High Strength Steels-Wrought** | 175-325<br>325-$52R_c$ | Heavy Web | 118<br>118 to 135 | A<br>B | Standard<br>Standard | <br>Crankshaft |
| **Austenitic Manganese Steels-Cast** | 150-220 | Rail Drill | 135 | B | Low | Split |
| **Free Machining Stainless Steels-Wrought** | 135-425 | General Purpose | 118 | A | Standard | Crankshaft |
| **Stainless Steels, Ferritic-Wrought and Cast<br>Stainless Steels, Austenitic-Wrought, Cast and P/M<br>Stainless Steels, Martensitic-Wrought, Cast and P/M<br>Precipitation Hardening Stainless Steels-Wrought and Cast** | 135-200 | General Purpose | 118 to 135 | A | Standard | Standard |
| | 200-325 | General Purpose | 118 to 135 | A | Standard | Crankshaft |
| | 325-425<br>$48R_c$-$52R_c$ | Heavy Web | 118 to 135 | B | Standard | Crankshaft |
| **Gray Cast Irons<br>Ductile Cast Irons<br>Malleable Cast Irons<br>Compacted Graphite Cast Iron<br>White Cast Iron** | 110-220 | General Purpose | 118 | A | Standard | Standard |
| | 220-400 | Heavy Web | 118 | A | Standard | Standard |

NOTE: Use stub-length drills whenever possible on high strength materials.

*Chisel edge angle: 115° to 135°

‡See chart at end of table on next page.

| MATERIAL | HARDNESS Bhn | DRILL TYPE | POINT ANGLE* degrees | LIP RELIEF ANGLE‡ degrees | HELIX ANGLE | POINT GRIND |
|---|---|---|---|---|---|---|
| **Aluminum Alloys-Wrought and Cast** | 30-150 500 kg | Polished Flutes | 90 to 118 | C | High | Standard |
| **Magnesium Alloys-Wrought and Cast** | 40-90 500 kg | Polished Flutes | 70 to 118 | C | High | Standard |
| **Titanium Alloys-Wrought and Cast** | 110-275 | General Purpose | 118 to 135 | B | Standard | Crankshaft |
| | 275-440 | Heavy Web | 118 to 135 | B | Standard | Crankshaft |
| **Copper Alloys-Wrought, Cast and P/M** | 40-200 500 kg | Polished Flutes | 118 | C | Low | Standard |
| **Nickel Alloys-Wrought, Cast and P/M Chromium-Nickel Alloys-Cast** | 80-360 | General Purpose | 118 | B | Standard | Crankshaft |
| **Nitinol Alloys-Wrought** | 210-360 $48R_c$-$52R_c$ | General Purpose | 118 | B | Standard | Crankshaft |
| **High Temperature Alloys-Wrought and Cast** | 140-475 | Heavy Web | 118 to 135 | B | Standard | Crankshaft |
| **Columbium, Molybdenum and Tantalum Alloys-Wrought, Cast, P/M** | 170-290 | General Purpose | 118 | B | Standard | Standard |
| **Tungsten Alloys (Anviloy)†** | 290-320 | General Purpose | 118 | B | Standard | Standard |
| **Zinc Alloys-Cast** | 80-100 | General Purpose | 118 | C | Standard | Standard |
| **Uranium Alloys-Wrought§** | 190-210 | Special Carbide | 118 | 5-8 | 20° | -5° Land on Drill Lip |
| **Zirconium Alloys-Wrought** | 140-280 | General Purpose | 118 | A | Standard | Crankshaft |
| **Thermoplastics and Thermosetting Plastics†** | All | Special Polished | 60 to 90 | C | Low | Standard |
| **Magnetic Core Iron** | 185-240 | General Purpose | 100 to 118 | A | High | Standard |
| **Controlled Expansion Alloys** | 125-250 | General Purpose | 118 | A | Standard | Crankshaft |
| **Magnetic Alloys (Hi Perm 49, HyMu 80)** | 185-240 | General Purpose | 118 | A | Standard | Crankshaft |
| **Carbon and Graphite** | 8-100 Shore | General Purpose | 90 to 118 | C | Standard | Crankshaft |
| **Machinable Carbide (Ferro-Tic)** | $40R_c$-$51R_c$ | General Purpose | 118 | B | Standard | Crankshaft |

NOTE: Use stub-length drills whenever possible on high strength materials.
*Chisel edge angle: 115° to 135°
†For both hss and carbide drills.
§For carbide drills.
‡See following chart.

| Lip Relief Angles at Periphery | Drill Size | | | | | | | |
|---|---|---|---|---|---|---|---|---|
| | #80 to #61 | #60 to #41 | #40 to #31 | ⅛ to ¼ | ¼ to ⅜ | ⅜ to ½ | ½ to ¾ | 1 inch up |
| A | 24° | 21° | 18° | 16° | 14° | 12° | 10° | 8° |
| B | 20° | 18° | 16° | 14° | 12° | 10° | 8° | 7° |
| C | 26° | 24° | 22° | 20° | 18° | 16° | 14° | 12° |

## 15.8 Oil-Hole or Pressurized-Coolant Drills, High Speed Steel

### TOOL GEOMETRY

Drill geometry as shown in section 15.7 (HIGH SPEED STEEL TWIST DRILLS) may be used for Oil-Hole or Pressurized-Coolant Drilling.

Drills used for Oil-Hole or Pressurized-Coolant Drilling operations, however, have heavy or semi-heavy webs which require thinning of the point. Instructions for thinning are shown below.

| DRILL DIAMETER inches | WEB |
|---|---|
| Up to 1/4 | 12% |
| Over 1/4 to 1/2 | 10% |
| Over 1/2 to 3/4 | 8% |
| Over 3/4 to 1 | 8% |
| Over 1 to 1-1/4 | 7% |
| Over 1-1/4 to 1-1/2 | 7% |

**NOTCH THINNING OF HEAVY WEB OIL-HOLE DRILLS**

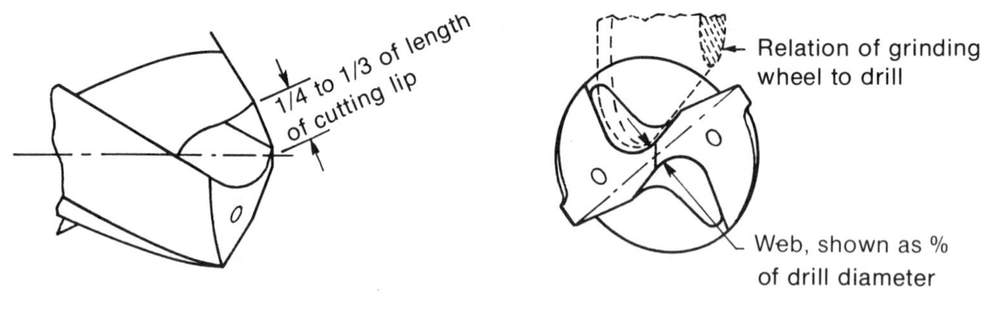


## 15.9 Spade Drills, High Speed Steel

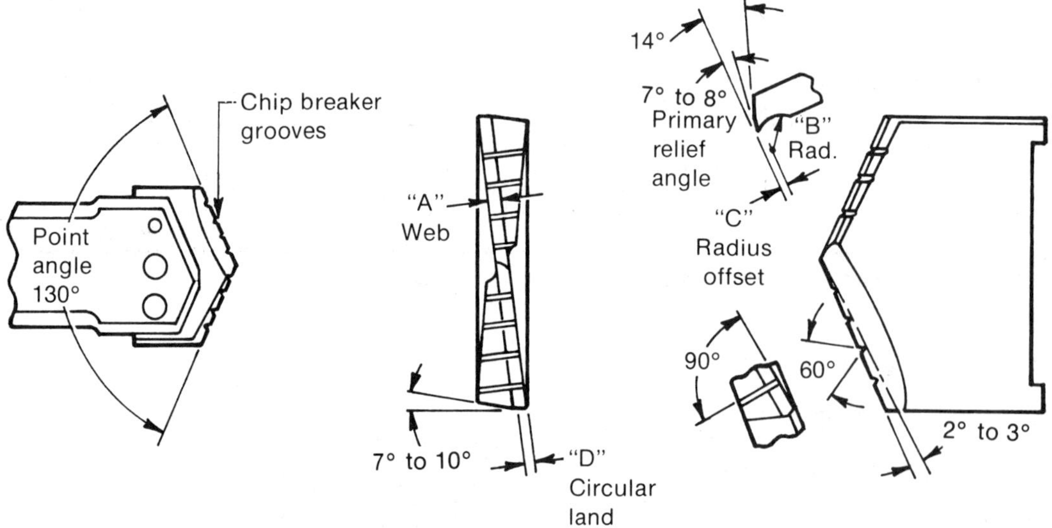

| DRILL SIZE RANGE inches | DIMENSIONS | MATERIAL* | | | | | | |
|---|---|---|---|---|---|---|---|---|
| | | STEELS, Wrought and Cast and CAST IRONS 100 to 175 Bhn inch | STEELS, Wrought and Cast and CAST IRONS 175 to 330 Bhn inch | STEELS, Wrought and Cast 330 Bhn to $52R_c$ inch | STAINLESS STEELS, Wrought and Cast inch | ALUMINUM and MAGNESIUM ALLOYS, Wrought and Cast inch | TITANIUM ALLOYS, Wrought and Cast inch | COPPER ALLOYS, Wrought and Cast inch |
| 1-1/32 to 1-1/4 | A | 1/16 | 5/64 | 5/64 | 1/16 | 5/64 | 1/16 | 5/64 |
| | B | 3/16 | 1/4 | 1/4 | 3/16 | 3/16 | 1/64 and 10° Rake | 5/32 |
| | C | 0 to 1/32 | 0 to 1/32 | 0 to 1/32 | 3/64 | 3/64 | 1/32 | 5/64 |
| | D | 1/16 | 1/16 | 1/16 | 3/64 | 1/16 | 1/16 | 1/16 |
| 1-9/32 to 1-1/2 | A | 5/64 | 3/32 | 5/64 | 5/64 | 3/32 | 5/64 | 3/32 |
| | B | 7/32 | 9/32 | 1/4 | 7/32 | 1/4 | 1/64 and 10° Rake | 5/32 |
| | C | 0 to 1/32 | 0 to 1/32 | 0 to 1/32 | 3/64 | 1/16 | 1/32 | 5/64 |
| | D | 5/64 | 5/64 | 5/64 | 3/64 | 1/16 | 5/64 | 5/64 |
| 1-9/16 to 2 | A | 3/32 | 7/64 | 3/32 | 5/64 | 3/32 | 3/32 | 7/64 |
| | B | 1/4 | 5/16 | 5/16 | 1/4 | 5/16 | 1/64 and 10° Rake | 3/16 |
| | C | 0 to 1/32 | 0 to 1/32 | 0 to 1/32 | 1/16 | 5/64 | 1/32 | 1/16 |
| | D | 5/64 | 5/64 | 5/64 | 1/16 | 1/16 | 1/16 | 5/64 |
| 2-1/16 to 2-1/2 | A | 7/64 | 1/8 | 7/64 | 7/64 | 7/64 | 7/64 | 9/64 |
| | B | 9/32 | 11/32 | 3/8 | 9/32 | 3/8 | 1/64 and 10° Rake | 7/32 |
| | C | 0 to 1/32 | 0 to 1/32 | 0 to 1/32 | 1/16 | 5/32 | 1/16 | 7/64 |
| | D | 5/64 | 3/32 | 5/64 | 1/16 | 5/64 | 5/64 | 5/64 |
| 2-9/16 to 3 | A | 1/8 | 9/64 | 1/8 | 1/8 | 7/32 | 5/32 | 11/64 |
| | B | 9/32 | 3/8 | 3/8 | 5/16 | 1/2 | 1/64 and 10° Rake | 1/4 |
| | C | 0 to 1/32 | 0 to 1/32 | 0 to 1/32 | 5/64 | 9/32 | 1/8 | 7/64 |
| | D | 3/32 | 3/32 | 3/32 | 5/64 | 3/32 | 3/32 | 3/32 |
| 3-1/16 to 3-1/2 | A | 5/32 | 11/64 | 5/32 | 5/32 | – | – | 13/64 |
| | B | 5/16 | 3/8 | 3/8 | 11/32 | – | – | 9/32 |
| | C | 0 to 1/32 | 0 to 1/32 | 0 to 1/32 | 3/32 | – | – | 1/16 |
| | D | 7/64 | 7/64 | 7/64 | 5/64 | – | – | 7/64 |
| 3-9/16 to 4 | A | 11/64 | 3/16 | 11/64 | 11/64 | – | – | 15/64 |
| | B | 5/16 | 7/16 | 3/8 | 3/8 | – | – | 5/16 |
| | C | 0 to 1/32 | 0 to 1/32 | 0 to 1/32 | 7/64 | – | – | 1/16 |
| | D | 1/8 | 1/8 | 1/8 | 3/32 | – | – | 1/8 |
| 4-1/16 to 5 | A | 3/16 | 7/32 | 3/16 | 3/16 | – | – | – |
| | B | 3/8 | 1/2 | 1/2 | 7/16 | – | – | – |
| | C | 0 to 1/32 | 0 to 1/32 | 0 to 1/32 | 1/8 | – | – | – |
| | D | 5/32 | 5/32 | 5/32 | 3/32 | – | – | – |

*For other alloys use recommendations for steel as starting geometry.

# 15.10 Gun Drills, Carbide
## TOOL GEOMETRY

**Slash or General Purpose Grind**

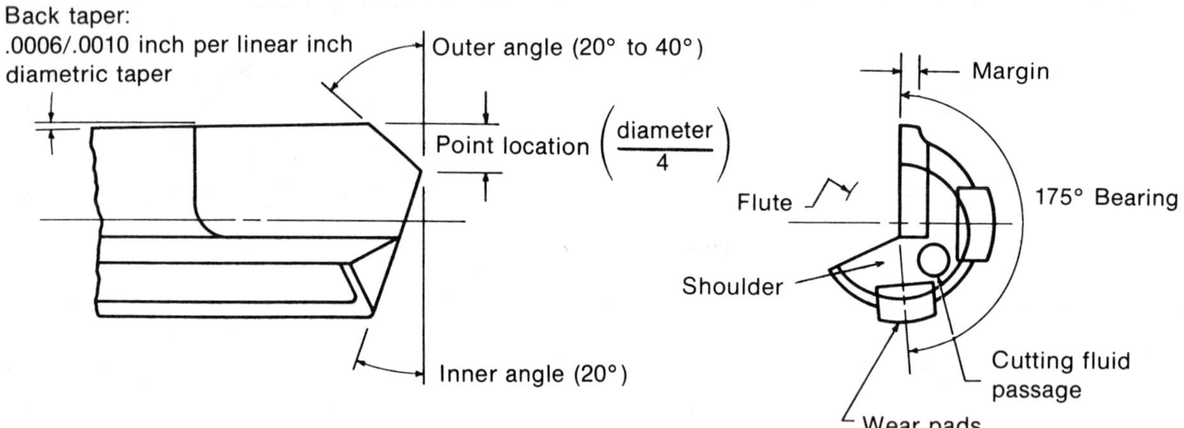


**Stack Grind**

(This grind is for difficult to machine alloys and stacked applications)

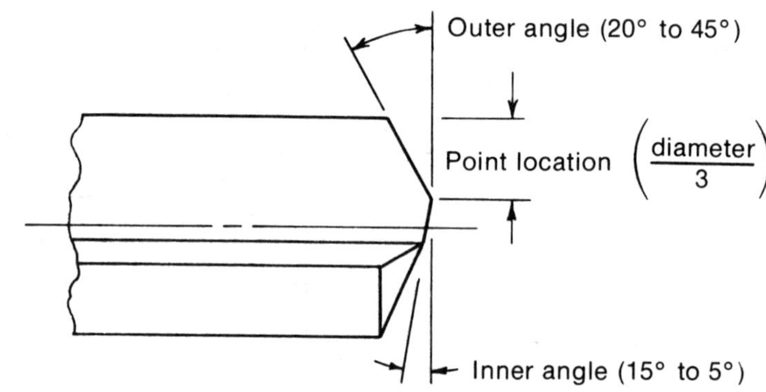

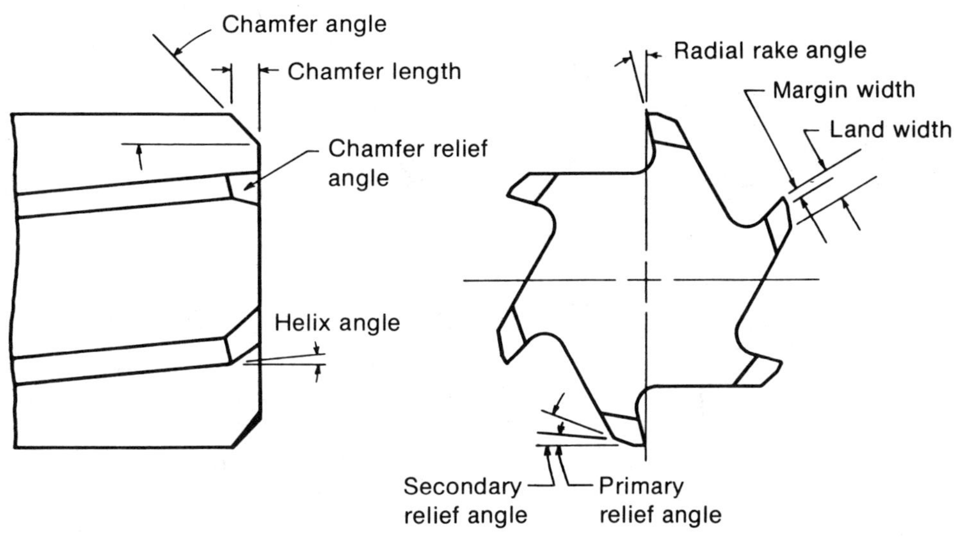


| DIAMETER inches | MARGIN WIDTH inch | PRIMARY RADIAL RELIEF ANGLE degrees | CHAMFER ANGLE degrees | CHAMFER RELIEF ANGLE degrees |
|---|---|---|---|---|
| Under 1/8 | 0.004-0.006 | 20 to 25 | 45 | 7 to 12 |
| 1/8 to 1/4 | 0.006-0.008 | 15 to 20 | 45 | 7 to 12 |
| 1/4 to 1/2 | 0.008-0.010 | 11 to 14 | 45 | 7 to 12 |
| 1/2 to 3/4 | 0.010-0.015 | 8 to 12 | 45 | 7 to 12 |
| 3/4 to 1 | 0.012-0.017 | 7 to 10 | 45 | 7 to 12 |
| 1 to 1-1/2 | 0.014-0.018 | 5 to 8 | 45 | 7 to 12 |
| 1-1/2 to 2 | 0.016-0.022 | 5 to 8 | 45 | 7 to 12 |
| Over 2 | 0.018-0.025 | 5 to 8 | 45 | 7 to 12 |

## Reamer Selection Guidelines

| STRAIGHT FLUTE | RIGHT HAND HELIX | LEFT HAND HELIX |
|---|---|---|
| • For general purpose use. | • Freer cutting.<br>• Improves finish.<br>• Reduces chatter.<br>• Requires rigid setups with no backlash. | • Reduces chatter on nonrigid setups.<br>• Use 30° to 45° helix for splines and keyways.<br>• Requires more thrust. |

# 15.12 Reamers, Carbide
## TOOL GEOMETRY

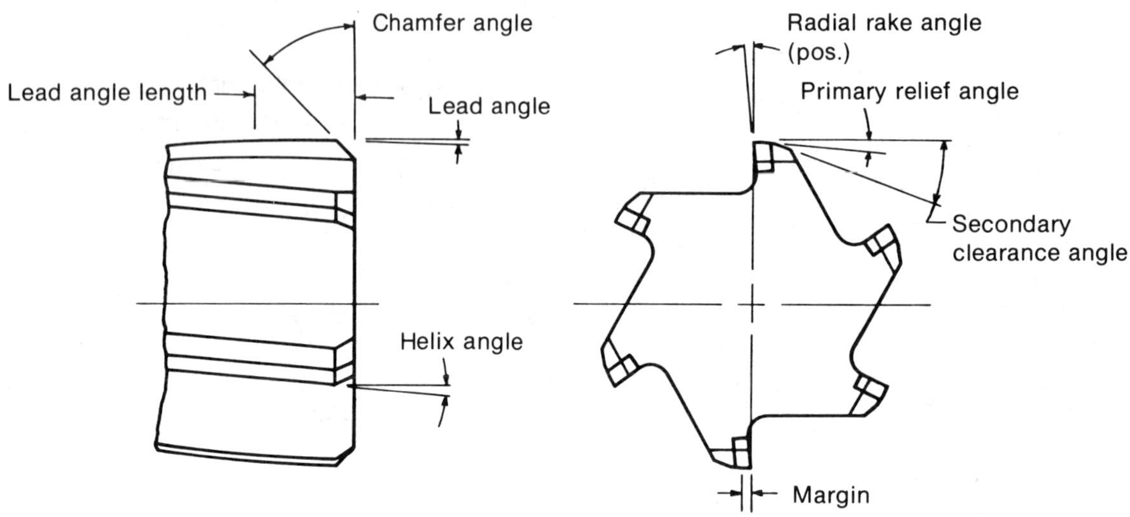


| DIAMETER inches | MARGIN OR CIRCULAR LAND inch | PRIMARY RADIAL RELIEF ANGLE degrees | SECONDARY CLEARANCE ANGLE degrees | LENGTH OF CHAMFER inch | CHAMFER PRIMARY RELIEF ANGLE degrees | CHAMFER SECONDARY CLEARANCE ANGLE degrees |
|---|---|---|---|---|---|---|
| 1/4 | 0.005-0.007 | 12 to 15 | 28 to 32 | 0.020-0.030 | 10 to 12 | 25 to 28 |
| 3/8 | 0.005-0.008 | 12 to 15 | 28 to 32 | 0.020-0.030 | 10 to 12 | 22 to 24 |
| 1/2 | 0.008-0.010 | 11 to 13 | 26 to 28 | 0.030-0.040 | 9 to 11 | 20 to 22 |
| 5/8 | 0.009-0.012 | 10 to 12 | 22 to 26 | 0.040-0.050 | 8 to 10 | 18 to 20 |
| 3/4 | 0.010-0.015 | 9 to 10 | 20 to 22 | 0.040-0.050 | 7 to 9 | 14 to 18 |
| 7/8 | 0.010-0.015 | 9 to 10 | 18 to 20 | 0.040-0.050 | 7 to 9 | 14 to 18 |
| 1 | 0.012-0.015 | 8 to 10 | 18 to 20 | 0.050-0.060 | 7 to 9 | 14 to 18 |
| 1-1/2 | 0.014-0.016 | 7 to 8 | 16 to 18 | 0.050-0.060 | 6 to 8 | 12 to 16 |
| 2 | 0.018-0.023 | 7 to 8 | 16 to 18 | 0.050-0.060 | 6 to 8 | 12 to 16 |

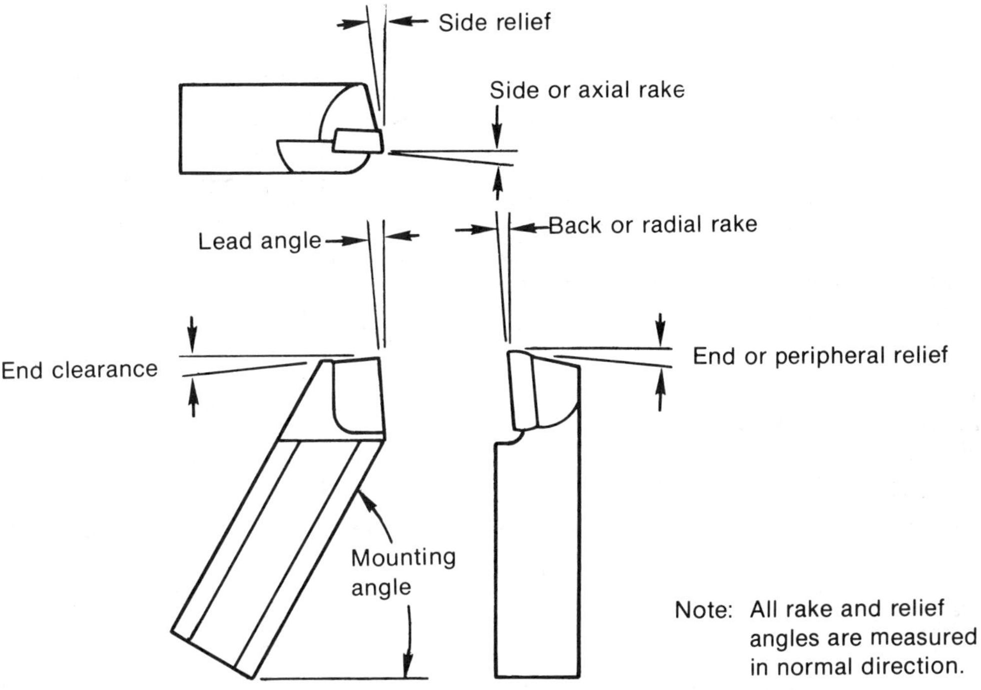


| MATERIAL | HARDNESS<br>Bhn | BACK OR RADIAL RAKE ANGLE*<br>degrees | SIDE OR AXIAL RAKE ANGLE<br>degrees | END OR PERIPHERAL RELIEF ANGLE†<br>degrees | SIDE RELIEF ANGLE<br>degrees |
|---|---|---|---|---|---|
| **Free Machining Carbon Steels-Wrought** | 100-425 | -3 to -10 | 0 to 15 | 5 to 10 | 2 to 3 |
| **Carbon Steels-Wrought and Cast** | 85-425 | -3 to -10 | 0 to 15 | 5 to 10 | 2 to 3 |
| **Free Machining Alloy Steels-Wrought** | 150-425<br>$45R_c$-$56R_c$ | -3 to -10 | 0 to 15 | 5 to 10 | 2 to 3 |
| **Alloy Steels-Wrought and Cast** | 125-425<br>$45R_c$-$56R_c$ | 0 to -6 | -3 to -8 | 5 to 10 | 2 to 3 |
| **High Strength Steels and Maraging Steels-Wrought** | 175-375<br>$43R_c$-$56R_c$ | 0 to -6 | -3 to -8 | 5 to 10 | 2 to 3 |
| **Tool Steels-Wrought** | 100-375<br>$48R_c$-$56R_c$ | 0 to -6 | -3 to -8 | 5 to 10 | 2 to 3 |
| **Nitriding Steels-Wrought** | 200-350 | 0 to -6 | -3 to -8 | 5 to 10 | 2 to 3 |
| **Armor Plate-Wrought** | 250-320 | -5 to -35 | 15 to 35 | 5 to 10 | 2 to 3 |

NOTE: The lead angle and end clearance are in many cases dictated by part shape and tool holder mounting angles. When a choice is possible, a larger lead angle and a smaller end clearance will usually decrease surface roughness and reduce the tendency to chatter.

*These values are based on boring tool being set on centerline. If tool is set above centerline the radial rake angle should be increased and the end relief angle should be decreased by a value of $\phi$ = arc tan H/R; where H = height above centerline and R = radius of bored hole.

†These end relief angles apply to bores 1/2 inch and larger. Smaller bores require greater relief angles.

# 15.13 Boring Tools, Carbide

## TOOL GEOMETRY

| MATERIAL | HARDNESS Bhn | BACK OR RADIAL RAKE ANGLE* degrees | SIDE OR AXIAL RAKE ANGLE degrees | END OR PERIPHERAL RELIEF ANGLE† degrees | SIDE RELIEF ANGLE degrees |
|---|---|---|---|---|---|
| Structural Steels-Wrought | 100-350 | -3 to -10 | 0 to 15 | 5 to 10 | 2 to 3 |
| | 350-500 | 0 to -6 | -3 to -8 | 5 to 10 | 2 to 3 |
| Free Machining Stainless Steels-Wrought | 135-425 | 3 to 10 | 0 to 15 | 5 to 10 | 2 to 3 |
| Stainless Steels, Ferritic-Wrought and Cast | 135-185 | 3 to 10 | 0 to 15 | 5 to 10 | 2 to 3 |
| Stainless Steels, Austenitic-Wrought and Cast | 135-275 | 3 to 10 | 0 to 15 | 5 to 10 | 2 to 3 |
| Stainless Steels, Martensitic-Wrought and Cast | 135-425 | 3 to 10 | 0 to 15 | 5 to 10 | 2 to 3 |
| Precipitation Hardening Stainless Steels-Wrought and Cast | 150-440 | 3 to 10 | 0 to 15 | 5 to 10 | 2 to 3 |
| Carbon and Alloy Steels-Cast | 120-400 | 0 to -6 | -3 to -8 | 5 to 10 | 2 to 3 |
| Gray Cast Irons | 110-320 | 0 | 0 | 5 to 10 | 5 to 8 |
| Ductile Cast Irons | 140-400 | 0 | 0 | 5 to 10 | 5 to 8 |
| Malleable Cast Irons | 110-280 | 0 | 0 | 5 to 10 | 5 to 8 |
| Aluminum Alloys-Wrought and Cast | 30-150 500 kg | 0 to 15 | 5 to 15 | 8 to 13 | 5 to 8 |
| Magnesium Alloys-Wrought and Cast | 40-90 500 kg | 0 to 15 | 5 to 15 | 8 to 13 | 5 to 8 |
| Titanium Alloys-Wrought | 110-440 | -3 to -10 | -3 to -8 | 8 to 12 | 3 to 5 |
| Copper Alloys-Wrought and Cast | 40-200 500 kg | 0 to 10 | 5 to 20 | 8 to 13 | 5 to 8 |
| Nickel Alloys-Wrought and Cast | 80-360 | 3 to 10 | 0 to 15 | 5 to 10 | 2 to 3 |
| High Temperature Alloys-Wrought and Cast | 140-475 | 3 to 10 | 0 to 15 | 5 to 10 | 2 to 5 |
| Columbium, Molybdenum and Tantalum Alloys-Wrought, Cast, P/M | 170-290 | 0 | 20 | 8 to 12 | 3 to 5 |
| Tungsten Alloys-Wrought, Cast, P/M | 180-320 | -15 | 0 | 5 | 5 |
| Zinc Alloys-Cast | 80-100 | 0 to 15 | 5 to 10 | 8 to 13 | 5 to 8 |

NOTE: The lead angle and end clearance are in many cases dictated by part shape and tool holder mounting angle. When a choice is possible, a larger lead angle and a smaller end clearance will usually decrease surface roughness and reduce the tendency to chatter.

*These values are based on boring tool being set on centerline. If tool is set above centerline the radial rake angle should be increased and the end relief angle should be decreased by a value of $\phi$ = arc tan H/R; where H = height above centerline and R = radius of bored hole.

†These end relief angles apply to bores 1/2 inch and larger. Smaller bores require greater relief angles.

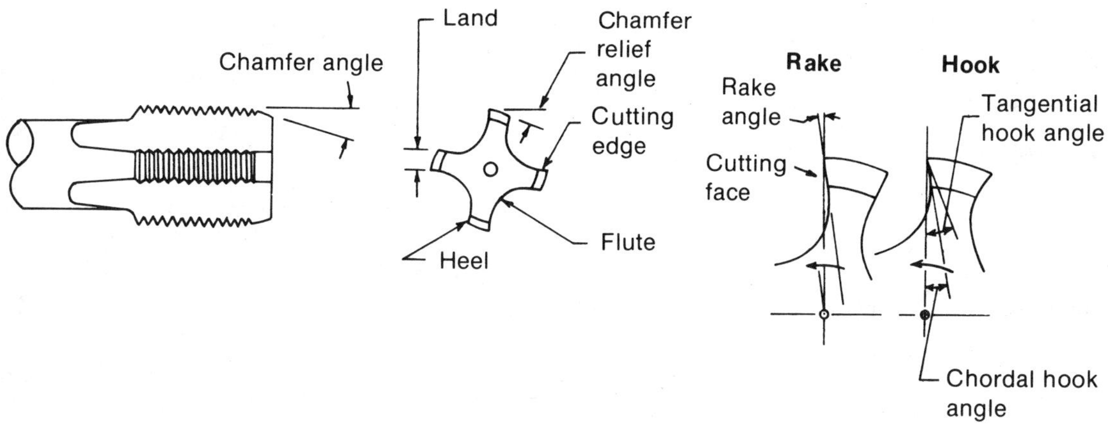


| MATERIAL | HARDNESS Bhn | HOOK OR RAKE ANGLE degrees | CHAMFER RELIEF ANGLE degrees | TYPE OF TAP | |
|---|---|---|---|---|---|
| | | | | Thru Hole | Blind Hole |
| **Steels-Wrought, Cast & P/M** | 85-200 | 7 to 10 | 8 | Spiral Point | Fast Spiral Flute |
| | 200-300 | 0 to 8 | 8 | Spiral Point | Fast Spiral Flute |
| | 300-375 | 0 | 6 | Modified 4 Flute Hand Tap | |
| | 375-425 | -3 to -6 | 6 | Modified 4 Flute Hand Tap | |
| | $48R_c$-$52R_c$ | -5 to -10 | 4 to 6 | Modified 4 Flute Hand Tap† | |
| **Stainless Steels, Ferritic, Martensitic & Precipitation Hardening-Wrought, Cast & P/M** | 135-275 | 8 to 12 | 8 | Heavy Duty Spiral Point | Heavy Duty Spiral Flute |
| | 275-325 | 0 to 5 | 8 | Heavy Duty Spiral Point | Heavy Duty Spiral Flute |
| | 325-425 | 0 | 6 to 8 | Modified 4 Flute Hand Tap | |
| **Stainless Steels, Austenitic-Wrought, Cast & P/M** | 135-275 | 15 to 20 | 10 | Heavy Duty Spiral Point | Heavy Duty Spiral Flute |
| **Gray Cast Irons Ductile Cast Irons Malleable Cast Irons** | 120-260 | 5 to 8 | 6 | 4 Flute Hand Tap | 4 Flute Hand Tap |
| | 260-330 | 0 to 3 | 6 | Modified 4 Flute Hand Tap | |
| **Aluminum and Magnesium Alloys-Wrought, Cast & P/M** | All | 10 to 20 | 12 | Spiral Point High Helix | Fast Spiral Flute |

†Special taps for these alloys are offered by some tap manufacturers.

# 15.14 Taps, High Speed Steel

## TOOL GEOMETRY

| MATERIAL | | HARDNESS Bhn | HOOK OR RAKE ANGLE degrees | CHAMFER RELIEF ANGLE degrees | TYPE OF TAP | |
|---|---|---|---|---|---|---|
| | | | | | Thru Hole | Blind Hole |
| **Titanium Alloys-Wrought and Cast** | | 110-275 | 10 to 15 | 12 | Modified Spiral Point | Modified 4 Flute Hand Tap |
| | | 275-440 | 6 to 10 | 12 | Modified Spiral Point | Modified 4 Flute Hand Tap |
| **Copper Alloys-Wrought, Cast & P/M*** | **Groups 1 & 2** | All | 0 to 8 | 10 | Spiral Point | Spiral Flute |
| | **Group 3** | All | 9 to 18 | 12 | Spiral Point | Fast Spiral Flute |
| **Nickel Alloys, Magnetic Alloys, Controlled Expansion Alloys-Wrought, Cast & P/M** | | 80-170 | 9 to 12 | 6 to 8 | Spiral Point | Spiral Flute |
| **High Temperature Alloys -Wrought and Cast** | | 140-425 | 0 to 10 | 4 to 6 | 2 Flute Spiral Point† | 3 Flute Hand Tap† Interrupted Thread |
| **Columbium, Molybdenum & Tantalum Alloys-Wrought, Cast & P/M** | | 170-290 | 10 to 12 | 6 to 8 | 2 Flute Spiral Point† | 4 Flute Hand Tap† |
| **Magnetic Core Iron** | | 185-240 | 12 to 15 | 6 to 8 | 2 Flute Spiral Point | Spiral Flute |
| **Zinc Alloys-Die Cast** | | 80-100 | 12 to 15 | 12 | Spiral Point High Hook | Fast Spiral Flute |
| **Thermoplastics** | | All | 5 to 8 | 12 | Spiral Point High Hook | Fast Spiral Flute |
| **Thermoset Plastics** | | All | 0 to 3 | 12 | Modified 4 Flute Hand Tap | |

*See section 15.2 for copper alloy groups.

†Special taps for these alloys are offered by some tap manufacturers.

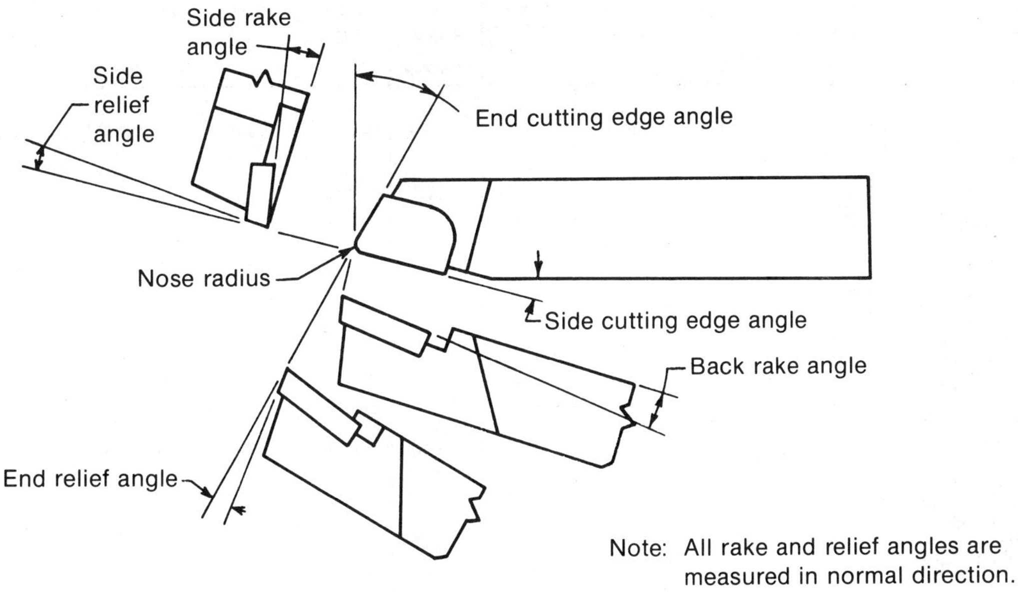
Side rake angle
Side relief angle
End cutting edge angle
Nose radius
Side cutting edge angle
Back rake angle
End relief angle
Note: All rake and relief angles are measured in normal direction.

# 15.15 Planing Tools

## TOOL GEOMETRY

| MATERIAL | HARDNESS Bhn | HIGH SPEED STEEL Back Rake Angle degrees | HIGH SPEED STEEL Side Rake Angle degrees | HIGH SPEED STEEL End Relief Angle degrees | HIGH SPEED STEEL Side Relief Angle degrees | HIGH SPEED STEEL End Cutting Edge Angle degrees | CARBIDE Back Rake Angle degrees | CARBIDE Side Rake Angle degrees | CARBIDE End Relief Angle degrees | CARBIDE Side Relief Angle degrees | CARBIDE End Cutting Edge Angle degrees |
|---|---|---|---|---|---|---|---|---|---|---|---|
| Free Machining Carbon Steels-Wrought<br>Carbon Steels-Wrought and Cast<br>Free Machining Alloy Steels-Wrought<br>Alloy Steels-Wrought and Cast<br>High Strength Steels-Wrought<br>Maraging Steels-Wrought<br>Tool Steels-Wrought<br>Nitriding Steels-Wrought<br>Armor Plate-Wrought<br>Structural Steels-Wrought | 85 to 375 | 5 to 10 | 6 to 10 | 10 to 15 | 6 to 10 | 5 to 15 | 0 to -5 | 15 to -15 | 3 to 5 | 3 to 5 | 4 to 15 |
| Free Machining Stainless Steels-Wrought<br>Stainless Steels, Ferritic-Wrought and Cast<br>Stainless Steels, Austenitic-Wrought and Cast<br>Stainless Steels, Martensitic-Wrought and Cast<br>Precipitation Hardening Stainless Steels-Wrought and Cast | 135 to 325 | 5 to 10 | 6 to 10 | 10 to 15 | 6 to 10 | 5 to 15 | 0 to -5 | 5 to -15 | 4 to 6 | 4 to 6 | 4 to 15 |
| Gray Cast Irons<br>Ductile Cast Irons<br>Malleable Cast Irons | 100 to 330 | 3 to 8 | 5 to 10 | 2 to 5 | 2 to 5 | 8 to 15 | 8 to -5 | 15 to -3 | 4 to 6 | 4 to 6 | 8 to 10 |
| Aluminum Alloys-Wrought and Cast | 30 to 150 500 kg | 10 to 20 | 30 to 40 | 8 to 15 | 8 to 15 | 7 to 9 | – | – | – | – | – |
| Magnesium Alloys-Wrought and Cast | 40 to 90 500 kg | 10 to 20 | 30 to 40 | 8 to 15 | 8 to 15 | 7 to 9 | – | – | – | – | – |
| Copper Alloys-Wrought and Cast | 40 to 200 500 kg | 5 to 10 | 5 to 10 | 8 to 12 | 8 to 12 | 5 to 15 | – | – | – | – | – |
| Nickel Alloys-Wrought and Cast | 80 to 320 | 0 to 5 | 5 to 10 | 4 to 10 | 4 to 10 | 10 to 15 | – | – | – | – | – |

NOTE: For all materials and for both high speed steel tools and carbide tools, use a 0°–30° side cutting edge angle. For best chip flow when planing mild and medium hard steels, try a 23° side cutting edge angle.

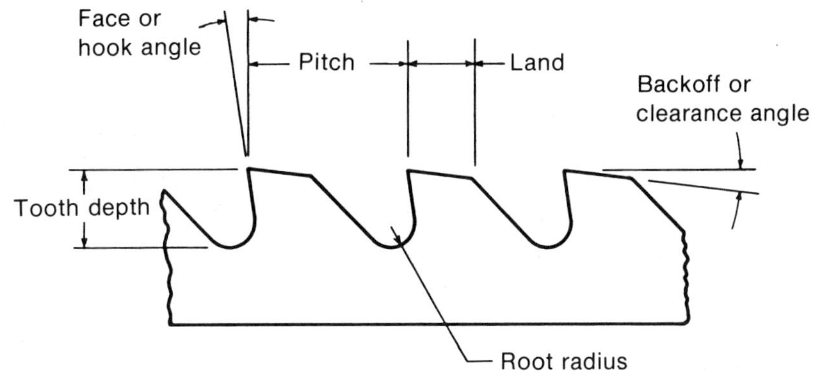


| MATERIAL | | HARDNESS<br>Bhn | HOOK ANGLE<br>degrees | CLEARANCE ANGLE<br>degrees |
|---|---|---|---|---|
| **Free Machining Carbon Steels-Wrought**<br>**Alloy Steels-Wrought and Cast** | | 100-375 | 15 to 20 | 2 to 3 |
| **Carbon Steels-Wrought** | | 85-375 | 15 to 20 | 1 to 2 |
| **Carbon Steels-Cast**<br>**Alloy Steels-Wrought and Cast**<br>**High Strength Steels-Wrought**<br>**Maraging Steels-Wrought** | | 120-375 | 8 to 15 | 1 to 3 |
| **Tool Steels-Wrought**<br>**Armor Plate-Wrought** | | 100-375 | 8 to 12 | 1 to 2 |
| **Nitriding Steels-Wrought** | | 200-350 | 8 to 15 | 1 to 2 |
| **Free Machining Stainless Steels-Wrought**<br>**Stainless Steels, Martensitic-Wrought and Cast** | | 135-425 | 8 to 12 | 1 to 2 |
| **Stainless Steels, Ferritic-Wrought and Cast** | | 135-185 | 12 to 18 | 2 to 3 |
| **Stainless Steels, Austenitic-Wrought and Cast** | | 135-275 | 12 to 18 | 1/2 to 2 |
| **Precipitation Hardening Stainless Steels-Wrought and Cast** | | 150-440 | 10 to 15 | 2 |
| **Gray Cast Irons** | | 110-320 | 6 to 8 | 2 to 3 |
| **Ductile and Malleable Cast Irons** | | 110-400 | 8 to 15 | 2 to 3 |
| **Aluminum Alloys-Wrought and Cast** | | 30-150<br>500 kg | 10 to 15 | 1 to 3 |
| **Magnesium Alloys-Wrought and Cast** | | 40-90<br>500 kg | 10 to 15 | 1 to 3 |
| **Titanium Alloys-Wrought** | | 110-440 | 8 to 20 | 2 to 8 |
| **Copper Alloys-Wrought and Cast*** | **Group 1** | 40-200<br>500 kg | -5 to 5 | 1 to 2 |
| | **Group 2** | 40-200<br>500 kg | 0 to 10 | 1 to 2 |
| | **Group 3** | 40-200<br>500 kg | 10 to 15 | 2 to 3 |
| **Nickel Alloys-Wrought and Cast** | | 80 to 360 | 12 to 18 | 1/2 to 2 |
| **High Temperature Alloys-Wrought and Cast** | | 140-475 | 15 to 20 | 2 to 8 |
| **Zinc Alloys-Cast** | | 80-100 | 10 to 15 | 2 to 5 |

*See section 15.2 for copper alloy groups.

# 15.17 Tool and Cutter Angles—Approximate Equivalents

## TOOL GEOMETRY

| LATHE AND PLANER | MILLING CUTTER | DRILL | BROACH |
|---|---|---|---|
| Back Rake Angle | Axial Rake Angle | Helix Angle | Hook Angle |
| Side Rake Angle | Radial Rake Angle | Radial Rake Angle | Side Rake Angle |
| Side Cutting Edge Angle | Corner Angle | 1/2 Point Angle | - |
| End Cutting Edge Angle | End Cutting Edge Concavity Angle | - | - |
| Side Relief Angle | Radial Relief Angle | - | Clearance Angle |
| End Relief Angle | Axial Relief Angle | Lip Relief Angle | Clearance Angle |

# SECTION 16

# CUTTING FLUIDS


16.1 Cutting Fluid Selection and Use .......... 16–3
Introduction .......... 16–3
Functions of Cutting Fluids .......... 16–3
Types of Cutting and Grinding Fluids .......... 16–3
Cutting Oils .......... 16–3
Water-Miscible (Water-Soluble) Fluids .......... 16–4
Gases .......... 16–7
Paste and Solid Lubricants .......... 16–7
Selection of a Cutting Fluid .......... 16–7
Machinability .......... 16–7
Other Considerations .......... 16–8
Application of Cutting Fluids .......... 16–9
Manual Application .......... 16–10
Flood Application .......... 16–10
High-Pressure Application .......... 16–12
Mist Systems .......... 16–12
Special Application Methods .......... 16–12
Maintenance of Cutting Fluids .......... 16–12
Cutting Oils .......... 16–12
Emulsifiable Oils .......... 16–13
Occupational Dermatitis .......... 16–13
Machine Cleaning Practice .......... 16–14
Disposal of Cutting Fluids .......... 16–14
Recycling Cutting Fluids .......... 16–14
Central Fluid System .......... 16–14
Economic Considerations .......... 16–15

16.2 Cutting Fluid Recommendations .......... 16–17

16.3 Cutting Fluid Key .......... 16–65

## INTRODUCTION

When properly applied, cutting fluids can increase productivity and reduce costs by making possible the use of higher cutting speeds, higher feed rates and greater depths of cut. Effective application of cutting fluids can also lengthen tool life, decrease surface roughness, increase dimensional accuracy and decrease the amount of power that is consumed when cutting dry. Knowledge of cutting fluid functions, types, physical limitations and compositions plays an important role in the selection and application of the proper fluid for a specific machining situation. Cutting fluid recommendations for various machining operations and work materials are given in section 16.2.

## FUNCTIONS OF CUTTING FLUIDS

Depending upon the machining operation being performed, a cutting fluid has one or more of the following functions:

- Cooling the tool, workpiece, and chip
- Lubricating (reducing friction and minimizing erosion on the tool)
- Controlling built-up edge (BUE) on the tool
- Flushing away chips
- Protecting the workpiece from corrosion

The relative importance of each function depends upon the work material, the cutting tool, the cutting conditions and the finish required on the part.

## TYPES OF CUTTING AND GRINDING FLUIDS

Many cutting fluids are available today to satisfy the requirements of modern machine tools. High speed cutting and grinding, increased metal removal rates, the emergence of new tool and work materials, and the demands for smoother finishes and closer tolerances have all contributed to the continual development of a wider variety of fluids that satisfy more specific requirements.

Although there is no all-purpose cutting fluid, some fluids offer considerable versatility while others are tailored for spefific applications. For economic reasons, most metalworking firms try to use as few different fluids as possible. Preferred fluids have long life and do not require constant changing or modifying.

Each of the basic types of fluids has distinctive features, advantages and limitations, although the dividing line is not always clearly identifiable. Understanding these basic differences is necessary to obtain optimum performance.

The four basic types of cutting fluids are as follows:

Cutting oils

- straight and compounded mineral oil (plus additives)

Water-miscible (water-soluble) fluids

- emulsifiable oils (soluble oils)
- chemical (synthetic) fluids
- semichemical (semisynthetic) fluids

Gases

Paste and solid lubricants

The cutting oils and the water-miscible fluids are the types most commonly used. Each of these has distinctive advantages over the other, depending on the type of machining operation and the material being machined. Cooling is best accomplished by water-miscible fluids, while oil-base fluids provide better lubrication.

### Cutting Oils

Cutting oils have a mineral oil base and may be used straight (uncompounded) or compounded (combined with polar additives and/or chemically active or inactive additives or compounds). They have excellent lubrication properties, good rust control and long life, but they do not cool as well as water-miscible fluids. Oils may be classified as inactive or active.

#### Inactive Cutting Oils

Inactive cutting oils (compounded with chemically inactive additives), have relatively high lubricity but low or no anti-welding properties. They have nonstaining properties which prevent discoloration of chemically sensitive materials.

**Straight mineral oils** do not have as good lubrication properties as do the compounded types, but they are lower in cost. Straight mineral oils are generally restricted to light-duty operations on metals that are easier to machine, such as aluminum, magnesium, brass and sulfurized or leaded free-machining steels, where the lubrication and cooling requirements are not severe. The oils are noncorrosive and stable and, if kept clean, can be used almost indefinitely. They lubricate all exposed moving parts, and minor leakage into or from gear boxes, bearings and hydraulic systems does not upset a machine's performance.

**Fatty oils** were once widely used as cutting fluids. The most common types are lard and rape seed oil. Their use has declined, partly because they are now difficult to obtain and are expensive, but mainly because modern additives blended with mineral oil are much more effective. Fatty oils are very polar and have high "oiliness"—anti-friction performance—but poor anti-weld characteristics. They oxidize readily and display a tendency to fume and to emit unpleasant odors.

**Compounded cutting oils** are made by blending polar additives and/or chemically active additives with mineral oil. Polar additives, such as certain fats, oils, waxes and synthetic materials, increase the load-carrying and cutting capability of mineral oil. Common polar additives are lard oil and castor oil. The function of any polar additive is to wet and to penetrate the chip/tool interface by reducing the interfacial tension between the carrier mineral oil and the metal. Such additives have been refined to minimize previous objections to the formation of disagreeable odors and the tendency to gum.

**Fatty-mineral oils** are combinations of one or more fatty oils blended in straight mineral oil. Lard oil is frequently used for this purpose. Fatty oils may comprise up to 40 percent of the blend depending on the application and are quite effective for many operations. The advantages are not great and are confined primarily to improvement of finish in the machining of mild steel, brass, copper and aluminum. Such blends are particularly suitable for machining the harder types of brass and copper, where straight mineral oil may not give the finish required and where the use of more active oils would cause staining. These oils are widely used in automatic screw machines where the operations are not exceptionally severe.

**Extreme pressure (EP) additives** are added to fluids used for machining operations where cutting forces are particularly high, such as tapping and broaching, or for operations performed with heavy feeds. Chemical or EP additives provide a tougher, more stable form of lubrication at the chip-tool interface. These additives include sulfur, chlorine or phosphorus compounds that react at high temperatures in the cutting zones to form metallic sulfides, chlorides and phosphides. In addition to providing extreme pressure lubrication, these additives provide a film on the tool surface with anti-weld properties that minimize the built-up edge. Sulfurized fatty-mineral oil blends have sulfur added in a strongly bonded, inactive form which may be totally nonstaining.

### Active Cutting Oils

Active cutting oils (compounded with chemically active additives) include sulfurized or phosphorized mineral oils, sulfo-chlorinated mineral oils, and sulfo-chlorinated fatty-mineral oil blends. These fluids have good anti-weld and extreme pressure lubrication qualities which improve tool life in high-temperature and high-pressure applications. They may, however, cause discoloration or staining of certain metals.

**Sulfur** can be added to a cutting oil in the form of sulfurized mineral oil or sulfurized fat. The sulfurized mineral oil is more active at lower temperatures and tends to severely stain aluminum, copper, brass, bronze and magnesium alloys. In comparison, the sulfurized fatty oil (which is sulfurized at a higher temperature) will not release sulfur as readily and, therefore, has less tendency to stain nonferrous metals or steel. Oils containing sulfur will form metallic sulfide films which will act as solid lubricants at temperatures up to about 1300°F [700°C].

**Chlorine** reacts and functions in essentially the same manner as sulfur. Inhibiting ingredients are added to prevent corrosion of ferrous surfaces, since chlorine is more reactive than sulfur. Chemical reaction is thus restricted to the chip-tool interface, where the temperatures are high. Chlorinated oils form an iron chloride film when reacting with ferrous workpiece materials or high speed steel tools. This film has a low shear strength and provides low friction up to about 750°F [400°C], beyond which it decomposes. Chlorinated oils usually will not stain nonferrous alloys.

When both chlorine and sulfur are added to cutting oils, anti-weld or EP characteristics are effective over a wider temperature range. Sulfo-chlorinated mineral oil and sulfo-chlorinated fatty oils are examples of such oils which are suitable for a wide range of applications.

**Phosphorus** performs as a mild EP lubricant or anti-friction additive when added to a cutting oil. The phosphide film breaks down at lower temperatures than do sulfide or chloride films and, therefore, is not as effective an anti-weld agent. Phosphorus is most effective in reducing friction and wear. These oils are nonstaining to most ferrous and nonferrous alloys.

These basic types of cutting oils can have various additive levels and may be used singly or in combination and at various viscosities to suit various applications. Some recent developments include oils in which the production of fume and oil-mist have been significantly reduced.

Applications of all compounded cutting oils (both active and inactive) are generally limited to machining operations on difficult-to-machine metals or form grinding from the solid. The high cost, the possible danger from smoke and fire, and the operator health problems often restrict the use of these oils to operations where other fluids do not provide satisfactory performance.

## Water-Miscible (Water-Soluble) Fluids

The water-miscible (water-soluble) cutting fluids are primarily used for high speed machining operations because they have better cooling capabilities. These fluids are also best for cooling machined parts to minimize thermal distortion.

Water-miscible cutting fluids are mixed with water at different ratios depending on the machining operation. For high-speed chip-making operations, they are normally mixed 1 part concentrate to 20 to 30 parts water. For many grinding operations where it is desirable to have a lighter fluid with more cooling action, the ratio is 1:40 or 1:50. Water-miscible fluids form mixtures ranging from emulsions to solutions when mixed with water. Because water has a high specific heat, high thermal conductivity and high heat of vaporization, it is one of the most effective cooling media known. Blended with water, the water-miscible fluids provide the combined cooling and moderate lubrication required by metal removal operations conducted at high speeds and lower pressures.

The water-miscible fluids can be classified as emulsifiable oils (soluble oils), chemical (synthetic) fluids or semi-chemical (semisynthetic) fluids. Fluids within these classes are available for light-, medium-, and heavy-duty performance.

### Emulsifiable Oils

Emulsifiable oils are commonly called soluble oils, emulsions, or emulsifiable cutting fluids. An emulsion is a suspension of oil droplets in water made by blending the oil with emulsifying agents and other materials. These emulsifiers (soap or soap-like materials) break the oil into minute particles and keep the particles dispersed in water for long periods of time. Bactericides—usually nonphenolic organic compounds specifically approved by the Environmental Protection Agency (EPA)—are added to control the growth of microorganisms such as bacteria, algae and fungi. If disposal is of no concern, phenolics may be used. The soaps, wetting agents, and couplers used as emulsifiers in water-miscible fluids reduce surface tension signficantly. As a result, the liquid has a greater tendency to foam when subjected to shear and turbulence. For this reason, water-miscible fluids sometimes cause foaming problems in operations such as gundrilling or flat-bed and double-disk grinding. With the use of special wetting agents and foam depressants, however, water-miscible fluids can be rendered sufficiently nonfoaming to be effective in almost all operations.

Emulsifiable oils combine the lubricating and rust-prevention properties of oil with water's excellent cooling properties. Emulsions, with their cooling-lubricating properties, are most effectively used for metalcutting operations with high cutting speeds and low cutting pressures accompanied by considerable heat generation.

Advantages of emulsifiable oils over straight or compounded cutting oils include greater reduction of heat, cleaner working conditions, economy resulting from dilution with water, better operator acceptance and improved health and safety benefits. They can be used for practically all light- and moderate-duty cutting operations, as well as for most heavy-duty applications except those involving extremely difficult-to-machine materials. Emulsifiable oils can be used for practically all grinding operations with the exception of severe grinding operations, such as form, thread and plunge grinding where wheel form is a critical factor. Extreme-pressure, compounded emulsifiable oils do not suffer from this limitation.

Cutting fluid manufacturers supply emulsifiable oils as concentrates that the user prepares by mixing with water. Mixtures range from 1 part oil in 100 parts water to a 1:5 oil-water ratio. The leaner emulsions are used for grinding or light-duty machining operations where cooling is the essential requirement. Lubricating properties and rust prevention increase with higher concentrations of oil. The four types of emulsifiable oils are summarized in table 16.1-1.

**General-purpose soluble oils** are milky fluids with mineral oil droplets of 0.0002-inch to 0.008-inch [0.005 mm to 0.2 mm] diameter. They are commonly used at dilutions of 1:10 to 1:40 for general purpose machining.

**Clear-type (or translucent) soluble oils** contain less oil (with higher proportions of corrosion inhibitors) and considerably more emulsifier than do milky emulsions. The clear type, therefore, consists of oil dispersions with smaller oil droplets which are more widely distributed. Since there is less dispersion of transmitted light, the fluid is less opaque, and the result is a translucent liquid. The translucency is not permanent, though, because often times the tiny oil droplets tend to coalesce and form larger droplets. These oils are generally used for grinding or for light-duty machining.

**Fatty soluble oils** have animal or vegetable fats or oils or other esters added to the mineral oil content to provide a range of fluids with enhanced lubricating properties.

**EP soluble oils** contain sulfur, chlorine or phosphorus additives to improve load-carrying performance. Since the EP concentrate is diluted 5 to 20 times when the emulsion is prepared, the lubricating capability is reduced. Where the lubricating capabilities of soluble-oil emulsions and the cooling properties of cutting oils are inadequate, EP soluble oils can satisfy both requirements in many cases. These fluids, commonly known as heavy-duty soluble oils, have in some cases replaced cutting oils for broaching, gear hobbing, gear shaping, and gear shaving.

### Chemical Fluids

Chemical (synthetic) fluids are chemical solutions consisting of inorganic and/or other materials dissolved in water and containing no mineral oil. All of these fluids are coolants; some are also lubricants. Chemical agents that go into these fluids include: amines and nitrites[1] for corrosion inhibitors; nitrates, for nitrite stabilization; phosphates and borates, for water softening; soaps and wetting agents, for lubrication and reduction of surface tension; phosphorus, chlorine and sulfur compounds, for chemical lubrication; glycols, as blending agents and humectants; and germicides, to control the growth of bacteria.

In general, the advantages of chemical fluids include economy, rapid heat dissipation, good size control, detergent properties which help keep the machine surfaces and the coolant systems clean, excellent workpiece visibility, easy mixing with little agitation, and high resistance to rancidity and rust. The increased cost of oils, plus OSHA and EPA requirements, may result in increased use of these and other water-miscible fluids.

A possible disadvantage of chemical fluids that may be encountered in some severe operations is insufficient lubricity, which may cause sticking and/or wear of moving machine tool parts. The dry, powdery residue or film left by chemical fluids is easy to remove. This residue can interfere with component movements if allowed to remain on the machine surfaces. Most of the stickiness is caused by the minerals in the water used. The mineral content in water can quadruple in six weeks as water is added to replace that lost by evaporation. Deionized water should be used to avoid this problem. Improved lubricating or EP properties, however, can be provided. Other limitations may include foaming problems in high agitation applications, and high detergency and alkalinity characteristics which may irritate sensitive hands when the concentration of the mix is not controlled.

Chemical fluids are usually classified into two general groups: true solutions and surface-active types (see table 16.1-1).

[1]CAUTION: The use of fluids containing nitrites may present a hazard and is presently under review by the National Institute of Occupational Safety and Health (NIOSH). Nitrites can react with amines to form nitrosamines which are carcinogenic. NIOSH may ban or control the use of nitrites in cutting fluids after completion of their review.

# 16.1 Cutting Fluid Selection and Use

**TABLE 16.1–1 Water-Miscible Cutting Fluids**

| CLASS | TYPE | GENERAL CHARACTERISTICS |
|---|---|---|
| Emulsifiable Oils | (1) General-Purpose Soluble Oils | Used at dilutions between 1:10 and 1:40 to give a milky emulsion. Used for general purpose machining. |
| | (2) Clear-Type Soluble Oils | Used at dilutions between 1:50 and 1:100. Their high emulsifier content results in emulsions which vary from translucent to clear. Used for grinding or light-duty machining. |
| | (3) Fatty Soluble Oils | Used at similar concentrations to (1) and of similar appearance. Their fat content makes them particularly good for general machining operations on nonferrous metals. |
| | (4) EP Soluble Oils | Generally contain sulfurized or chlorinated EP additives. Used at dilutions between 1:5 and 1:20 where a higher performance than that given by (1), (2) or (3) is required. |
| Chemical (Synthetic) Fluids | (1) True Solutions | Essentially solutions of chemical rust inhibitors in water. Used at dilutions between 1:50 and 1:100 for grinding operations on iron and steel. |
| | (2) Surface-Active Chemical Fluids | Contain mainly water-soluble rust inhibitors and surface-active load-carrying additives. Used at dilutions between 1:10 and 1:40 for cutting and at higher dilutions for grinding. Most are suitable for both ferrous and nonferrous metals. |
| | (3) EP Surface-Active Chemical Fluids | Similar in characteristics to (2) but containing EP additives to give higher machining performance when used with ferrous metals. Used at dilutions between 1:5 and 1:30. |
| Semichemical (Semisynthetic) Fluids | – | Essentially a combination of a chemical fluid and a small amount of emulsifiable oil in water forming a translucent, stable emulsion of small droplet size. EP additives are usually included permitting their use for moderate- and heavy-duty machining and grinding applications. |

**True-solution** fluids (without wetting agents), also called chemical solutions or chemical grinding fluids, primarily contain rust inhibitors (inorganic and organic nitrites[2]), sequestering agents, amines, phosphates, borates, glycols or ethylene or propylene oxide condensates. Some of these fluids contain highly developed corrosion inhibitors such as sodium nitrite[2] (for cast iron), triethanolamine (for both cast iron and steel) and sodium mercaptobenzothiazole (for reducing corrosion on brass, zinc and aluminum). True solutions, used at 1:50 or 1:100 ratios, are clear in appearance but are often colored with dyes to indicate their presence in water. These fluids are restricted to grinding operations where they prevent rust and permit rapid heat removal. True solutions have high surface tension (about equivalent to water). They have a tendency to leave a residue of hard or crystalline deposits formed by water evaporating. These chemicals can also be added to emulsifiable oils or to other chemical fluids to enhance their corrosion-inhibiting properties.

[2]CAUTION: The use of fluids containing nitrites may present a hazard and is presently under review by the National Institute of Occupational Safety and Health (NIOSH). Nitrites can react with amines to form nitrosamines which are carcinogenic. NIOSH may ban or control the use of nitrites in cutting fluids after completion of their review.

**Surface-active** chemical fluids are extremely fine colloidal solutions composed of inorganic and organic materials dissolved in water with the addition of wetting agents (surface active agents). The wetting agents improve the wetting action of the water and provide greater uniformity of both heat dissipation and anti-rust action. This type of fluid may include anti-foaming agents, humectants, mild lubricants (organic or inorganic) and water softeners. The lubricating (anti-wear) properties may be provided by a viscous polyglycol compound. Corrosion protection may be provided by a mixture of triethanolamine and caprylic acid, and the formulation may include a deactivator for copper alloys.

The surface-active type of chemical fluid has fair lubricity, low surface tension, and good rust-inhibiting properties and usually leaves a dry, hard or powdery residue that is easily removed. The slight tendency of these fluids to foam is usually not a serious problem in most operations. They are usually used at dilutions of 1 part concentrate in 10 to 40 parts water.

**EP surface-active** chemical fluids are similar to the plain (general-purpose) surface-active type but have chlorine, sulfur or phosphorus additives to provide extreme-pressure lubrication effects. These fluids are diluted at one part concentrate to between 5 and 30 parts water for tougher machining operations.

### Semichemical Fluids

Semichemical fluids or semisynthetic fluids are essentially a combination of chemical fluids and emulsifiable oils in water. These fluids are actually preformed chemical emulsions that contain only a small amount of emulsified mineral oil, about 5 to 30 percent of the base fluid, which has been added to form a translucent, stable emulsion of small droplet size. Since the usual EP additives can be incorporated (often more readily in the oil content than in the synthetic base), the lubricating performance can be varied to permit using such fluids for moderate- and heavy-duty machining and grinding applications.

Semichemical fluids combine some of the best qualities of chemical fluids and emulsifiable oils. The advantages and limitations are similar to those described for chemical fluids, except that semichemical fluids have better lubricating properties than do chemical fluids. They are also cleaner, with better rust and rancidity control than emulsifiable oils.

### Gases

Air is the most common gaseous fluid. It is present under atmospheric pressure for dry operations and also present when fluids are used. Air is sometimes compressed to provide better cooling, with a stream directed at the cutting zone to remove heat by forced convection. This also can be used to blow chips away, but safety precautions must be observed. Gases such as argon, helium and nitrogen are sometimes used to prevent oxidation of the workpiece and the chips, but the high cost of these gases generally makes them uneconomical for production applications.

Gases, such as Freon™ or $CO_2$, with boiling points below room temperature can be compressed and sprayed into the cutting zone to provide evaporative cooling. Use of liquid argon or nitrogen allows cooling to several hundred degrees below zero. Care is necessary, however, to prevent part warpage caused by large temperature differentials.

Advantages of inert gases include good cooling ability, increased tool life, a clear view of the operation, elimination of mist, and no contamination of the workpiece, chips or machine lubricants. The cost of some of these gases, however, can be extremely high.

### Paste and Solid Lubricants

There are also paste and solid lubricants that are usually applied manually by brush or by oilcan to the tool or workpiece in operations such as tapping and hand reaming. Grinding wheels are sometimes impregnated with solids possessing lubricating qualities. In special cases, such as knife grinding, wheels are treated with sulfur to produce a cooler action in wet grinding. Also, external application of grease sticks to grinding wheels can provide some lubrication, but this requires performing the operation dry. Solid waxes in stick form can be used on grinding wheels, sanding disks or belts, taps, and band- or circular-saw blades for lubrication to improve finish and tool life or to reduce burring or metal welding. Other solids most often used as heavy-duty lubricants include graphite, molybdenum disulfide, pastes, soaps and waxes.

## SELECTION OF A CUTTING FLUID

The proper choice of a cutting fluid depends on many complex interrelated factors. Of primary concern are machinability (or grindability) of the material, compatibility (metallurgical, chemical and human) and acceptability (fluid properties, reliability and stability).

### Machinability

The selection of the type of cutting fluid for use should be based on:

- Type of machining operation
- Material being machined
- Tool material
- Machining conditions—cutting speed, feed, and depth of cut

One of the most important factors in selecting a cutting fluid is the nature of the cutting operation itself. The various machining processes naturally differ in metal removal characteristics. The more difficult operations will place greater demands on a cutting fluid. Selection is, therefore, a matter of assessing the severity of the machining operation and marrying it to the appropriate cutting fluid. The approximate ratings for the machining processes in order of increasing severity, are as follows:

1. Grinding
2. Sawing
3. Turning, single-point tools
4. Planing and shaping
5. Milling
6. Drilling
7. Reaming
8. High-speed, light-feed screw machining
9. Screw machining with form tools
10. Boring
11. Deep-hole drilling
12. Gear cutting
13. Threading
14. Tapping
15. External broaching
16. Internal broaching

These ratings cannot be regarded as absolute, since variations in the tool geometry and workpiece material will change the severity of the machining operation.

For heavy-duty machining operations (such as tapping or broaching), medium- or heavy-duty cutting oils are generally used. Horizontal broaching of steel usually requires a heavier-bodied or more-chemically-active oil than does vertical surface broaching under comparable conditions. The heavier oil clings to the horizontal broach better than a water-miscible fluid, and chemical activity aids in efficient cutting. For vertical surface broaching of mild steels, emulsions or solutions may be used, but the usual choice is an oil.

Cutting oils can be ranked in order of increasing load-carrying capacity as follows:

1. Straight mineral oil
2. Mineral oil with fatty additives

3. Mineral oil with chlorinated additives
4. Mineral oil with sulfurized fatty additives
5. Mineral oil with free sulfur and sulfurized or chlorinated compounds

Slow speed operations require the lubricating properties of cutting oils. As a general rule, cutting oils should be used at cutting speeds below 100 feet per minute [30 m/min]. Cutting oils with EP lubricants are effective in a wide variety of machining operations on many materials up to speeds of 200 feet per minute [60 m/min]. At speeds of about 200 feet per minute [60 m/min], chemical action is much less effective than at the lower speeds. Chemical action diminishes quickly and becomes virtually nil as the cutting speed is increased to 400 feet per minute [120 m/min].

Both tapping and threading involve many small cutting edges in continuous contact with the work throughout the cut. The design of the tools and the nature of these operations shield the edges of the tools from the flow and the cooling effect of the cutting fluid, particularly in tapping.

Drilling can be a difficult operation when done on difficult-to-machine materials. Drilling speeds are generally slower than those used for other operations because the cutting edge is in continuous contact with the metal when cutting and the cutting edges are shielded from the flow and beneficial cooling action of the cutting fluid. The preferred fluids for conventional drilling operations are emulsifiable oils and sulfurized or chlorinated mineral oils. These fluids provide some lubricity to prevent chatter and friction-generated heat while carrying away the heat generated by chip formation. Oil-hole drills should be used wherever possible.

There are some operations for which oils are specially formulated. For example, honing requires the use of a thinner, paraffinic-base oil.

For many high speed operations, such as grinding and turning or milling with carbides, the most important benefit is provided by the superior cooling characteristics of the water-miscible fluids. In such instances, lubrication and antiweld properties are less important. At high speeds, time for reaction between fluid additives and the workpiece is reduced to a minimum. Further, the higher relative velocities of the work and the tool virtually preclude the fluid's reaching the work zone. Consequently, gross cooling is required to prevent catastrophic tool failure from crater wear, to prevent distortion in the workpiece from heat buildup and to control workpiece size.

Some emulsifiable oils and chemical fluids are formulated specifically for grinding operations and are used in concentrations of 1 part concentrate in 25 to 60 parts water. An increase in the richness of emulsifiable oil mixtures from 2.5 percent to 10 percent can improve the grinding ratio and the workpiece finish and reduce horsepower requirements. Note that the grinding ratio is a measure of the volume of material removed per unit volume of wheel wear.

Severe grinding operations, such as form, thread, and plunge grinding where wheel form is a critical factor, require the use of cutting oils or EP-compounded emulsifiable oils. It should be noted that for the low stress grinding technique, the use of an oil-base fluid results in less surface distortion. Grinding oils reduce friction, thus reducing the heat generated. They permit heavier cuts to be taken, produce smoother finishes, reduce wheel breakdown, and, hence, allow better form control. Oils suffer from the disadvantages of allowing heat buildup in the workpiece, tending to hold chips in suspension, and smoking or burning. The messiness of oil vapors is also a nuisance.

In summary, from a machinability viewpoint, cutting and grinding fluids should be selected to provide the following:

Greater Lubrication

- at relatively low speeds
- on the more difficult-to-machine materials
- for more difficult operations
- for better surface finish

Greater Cooling

- at relatively high speeds
- on the easier-to-machine materials
- for easier operations
- where heat buildup in the part is a problem

Types of cutting fluids recommended for various work materials and machining operations are listed in section 16.2.

### Other Considerations

A number of other important factors must be considered when selecting cutting and grinding fluids. These include compatibility of the fluid with the material being machined, the quality of the water in which it is mixed, tramp oil contamination and the machine tool. The acceptability of the fluid, in terms of its properties, reliability and stability, is also important. Other factors to be considered include the quantity of fluid used and the method of application, facilities for storing, handling, cleaning or recycling the fluid, the quality of water used in the aqueous emulsions or solutions, and the rust and rancidity control necessary. Economic considerations include any waste treatment facilities that may be required prior to disposal. The cutting fluid's effect on the total machining cost per piece must be considered, since the more expensive fluids may sometimes be more economical in the long run.

#### Compatibility with Workpiece Material

The cutting fluid selected for use must be compatible with the material being machined. The fluid should not cause staining or corrosion of the workpiece. It should prevent corrosion of the workpiece during machining, while as work-in-progress waiting for transfer, and possibly while in storage for a reasonable time. Care should be taken when selecting coolants from a corrosion standpoint, especially for nonferrous work materials.

Some copper alloys cannot be machined with highly sulfurized cutting oils since the active sulfur will stain the machined surface. Sulphur is often used in a form in which it is chemically combined with fatty oils, although for severe machining operations it is often used at low concentrations in its natural form as free sulfur. In this latter or active form, however, it is very reactive toward copper. For this reason, cutting fluids containing free sulfur should never be used in the machining of copper or its alloys.

Metals such as aluminum, magnesium and zinc (also galvanized steel) are attacked by acids or alkalies, thus causing corrosion. Selection of a cutting fluid for aluminum depends on the alloy. Active sulfur or chlorine additives may cause staining of some aluminum alloys. There are some fluids, however, with patented additives that improve finishes and help prevent staining. The use of highly alkaline grinding fluids on aluminum would probably produce badly corroded parts. Magnesium and its alloys, for example, *should never* be machined or ground with water-miscible fluids because the metal has an incendiary effect in water. Zinc and galvanized steel should not be machined with nitrite-containing fluids; oil-rich emulsions with a pH of approximately 8.0 have been found effective for machining these metals. Special fluids available for machining copper, aluminum and magnesium are listed in section 16.3.

Ferrous metals dissolve when attacked by acids and rust will form. Alkalies form a protective layer on most ferrous metals to prevent corrosion. Cutting fluids for ferrous metals are therefore kept on the alkaline side—to inhibit rusting of machines and machined parts.

Fluid concentration should be maintained; otherwise, the rust-inhibiting properties will be reduced. Some rust inhibitors are attacked by oxygen and acids in the air and in the water; eventually, the inhibitors will be used up.

The rust-inhibiting effect of soluble oils is rapidly lost when machining cast iron. Therefore, it is advisable to add some soluble oil concentrate regularly in order to maintain the proper concentration. To prevent corrosion, soluble oils must be 2 to 3 times more concentrated when machining cast iron than when machining steel.

Particular care should be exercised in selecting cutting fluids used for machining components which are highly stressed in service and which are exposed to unusual environments. For such parts, cutting fluids should be used which will not make the part vulnerable to stress corrosion attack in the event that surface films are retained on the workpiece. At present, the only way of selecting such cutting fluids is to choose them on the basis of extensive laboratory tests and/or service experience. An alternative is to provide unusual care in washing the cutting fluid away after machining with due consideration given to the difficulties of washing complex finished assemblies.

Cutting fluid selection can have a dramatic impact on the distortion and the residual stresses generated by grinding. Grinding is commonly done with water-miscible fluids, with oils or even dry. The differences among them relate to thermal cooling action, lubricity and possibly some chemical action. Water-miscible fluids have greater cooling capability and less lubricity than the oils; however, they also impose a greater heat shock on the workpiece. Dry grinding conditions are not usually compatible with low stress grinding or high surface integrity. Lubrication is the most important function of the grinding fluid relative to control of residual stresses. Occasionally, data will show little difference between water-miscible fluids and oils; however, most of these tests involve higher infeed rates or higher wheel speeds. The effect of infeed is more pronounced and tends to overwhelm the effects of the fluids. Grinding oils become increasingly effective at wheel speeds below 4,000 feet per minute [20 m/s]. It is apparent that grinding oils are superior for high integrity grinding.

#### Compatibility with Water

Water-miscible fluids must be compatible with the water in which they are mixed. Water quality is perhaps the most important single factor affecting fluid life. The chemistry and biology of the water have a strong impact on the cooling effectiveness and the life of the coolant. The best cooling action can be obtained from very fine emulsions, but such emulsions cannot be made with water that is high in either hardness or dissolved solids. Many things are affected by the hardness level or pH of the water, such as rate of biological growth, rust protection, evaporative effect, emulsion stability, and foaming. A high chloride or sulfate content can negate the rust protection that has been built into a coolant formulation. Hard water (over 200 ppm) can break some emulsions, resulting in scum formation that leads to many performance problems. Soft water (under 100 ppm) often produces excessive foam, especially under conditions of high agitation.

#### Compatibility with the Machine Tool

The cutting fluid must also be compatible with the machine tool in which it is being used. The use of water-miscible fluids with most machines causes few problems. Usually, the drive mechanisms and the feed mechanisms are some distance from the cutting zone, completely enclosed with their own lubrication system and adequately sealed against the ingress of water-containing cutting fluids. With some of the more complex machine tools, however, like automatic and semi-automatic lathes and gear cutting and broaching machines, water-miscible fluids cannot be used. Because many of these machines rely on a cutting fluid to lubricate moving parts near the cutting zone, the use of oils is mandatory. In other cases, problems have been experienced in preventing a water-miscible fluid from reaching drive mechanisms and, therefore, oils must also be used.

#### Acceptability

Another important consideration in selecting cutting fluids is acceptability to the operator. Fluids designed with the operator in mind will be clean to use, pleasant and safe to breathe, nontoxic and free from ingredients harsh to human skin. Bacterial growth in the coolant should be controllable before and during use. Fluid clarity makes it easier for the operator to see the work while it is being machined. A pleasant-smelling, generally clean, "non-sticky" fluid helps to make an operator's job more pleasant and productive.

## APPLICATION OF CUTTING FLUIDS

The way in which a cutting fluid is applied has a considerable influence on tool life and on the machining operation in general. Although there are many extremely effective devices and systems for supplying fluids to the cutting area, special equipment is not generally necessary for good results.

Even the best fluid, however, cannot perform its function unless it is effectively delivered to the cutting zone. Thus, a fluid chosen for its lubricating qualities must be directed so that it can form a film at the sliding surfaces. Likewise, a fluid used for cooling must gain reasonable access to the cutting edge of the tool. These conditions usually require that the fluid be forced into the cutting zone so that the heat can be removed as it is generated. Continuous application of the cutting fluid is preferable to intermittent application. Sporatic fluid application causes thermal cycling, which leads to the formation and propogation of microcracks in hard and brittle tool materials, such as carbides. Besides shortened tool life, intermittent fluid application can lead to irregular surface finishes.

A secondary advantage of proper fluid application is the efficient removal of chips. This also aids in prolonging tool life, since properly placed fluid nozzles can prevent blockage or packing of the chips in the flutes of milling cutters and drill bits.

## Manual Application

Paste and solid lubricants are applied manually by brush or oilcan to the tool and to the workpiece, mainly in tapping. More recent developments include pressurized aerosol dispersants and foams which cling to the tool and workpiece.

Manual application is an effective method where a small number of holes are to be drilled or tapped on a machine that is not equipped with a coolant system. When two different operations are performed on the same machine, manual application may be used in conjunction with the flood cooling system of the machine. The flood cooling may be used for a drilling operation with a moderately active fluid, whereas a highly active cutting oil can be applied manually for a tapping operation that follows.

## Flood Application

The most common method of applying fluids is to flood the tool and workpiece. A low-pressure pump delivers the cutting fluid through piping and valves to nozzles situated near the cutting zone. After flooding the cutting area, the fluid drains down over various parts of the machine into the chip pan where it flows to the sump of the coolant pump. The volume of the tank must be sufficient to allow time for cooling and for the settling of fine swarf and may require from five to fifty gallons or more depending on the type of machine. A course strainer on top of the collecting pan prevents larger chips from entering the tank and the fine strainer at the pump section. Important exceptions are grinding, honing, lapping and deep-hole boring machines where high-quality work depends upon removing the finer swarf and abrasive particles. Occasions arise when the inclusion of filtration equipment on other types of machines can avoid the gross contamination and overloading of the coolant with the metallic particles and help keep the fluid clean, thus prolonging its useful life.

Flood application of cutting fluids permits a continuous flow of fluid to the cutting zone and is efficient in flushing away chips. A copious stream of fluid should be applied so that the cutting tool edge and the work are completely enveloped. In addition to supplying an adequate amount of fluid to the cutting zone, a copious flow provides cooling action that prevents undue temperature rise. Proper cutting fluid application should not be neglected because of inadequate splash guards. Typical coolant flow requirements are given in table 16.1-2.

The geometry of flood application directly influences the effectiveness of cutting fluids. Nozzles should direct fluid flow so that fluid is not thrown off the workpiece or tool by centrifugal force. Two or more nozzles should be used—one for directing fluid into the cutting zone and the other for auxiliary cooling and flushing away chips.

### Turning and Boring

Turning and boring operations require the cutting fluid to be directed to the cutting zone where the chips form. This provides good cooling action by enveloping the cutting tool edge and the workpiece. An empirical rule is that the coolant nozzle should have an inside diameter of at least three-quarters the width of lathe-type cutting tools.

For heavy-duty turning and boring operations, a second nozzle supplying fluid from below and along the flank of the tool is desirable. The flow from the lower nozzle is not hindered by the chip and can be forced between the work and tool to help in providing lubrication at lower speeds.

### Drilling and Reaming

For horizontal drilling and reaming, fluid application through hollow tools is preferable to external application because it provides adequate flow at the cutting edges and flushes chips out of the hole. Even in vertical drilling, very little fluid gets down to the cutting zone because the spiral flutes (which are designed to pump out the chips) pump out the fluid. The use of oil-hole drills will solve this problem.

### Milling

Milling operations require the use of two nozzles directing a copious supply of fluid to both the incoming and the outgoing sides of the cutter. The fluid from one nozzle is pumped through the cutting zone by the cutter teeth, while fluid from the other nozzle washes away the chips as they emerge from the cutter. Standard round nozzles are sufficient for narrow cutters. Wide cutters require using fan-shaped nozzles at least three-quarters the width of the milling cutter to provide good coverage.

For face milling, use of a ring-type distributor consisting of a tube with many small holes can be beneficial. This directs the fluid at all cutting edges and keeps the cutter completely bathed in fluid to provide even cooling. When a particular size face mill is used often, the ring-type distributor can be supplemented with a special fan nozzle with a curved opening to match the cutter radius.

### Grinding

A copious flow of cutting fluid at low pressures will generally provide good results for grinding operations. Where application of a large volume of fluid results in undue splashing, it is better to install splash guards on the machine than to reduce the coolant flow.

The normal methods of applying fluids to grinding operations remove little heat before it has dissipated into the mass of the workpiece. Because high surface speeds are in-

**TABLE 16.1–2 Typical Coolant Flow Requirements**

| OPERATION | FLOW AT WORK | REMARKS |
|---|---|---|
| Turning | 5 gal/min [19 L/min]/tool | |
| Screw Machining | | |
| 1 inch [25 mm] dia. | 35 gal/min [132 L/min] | |
| 2 inch [50 mm] dia. | 45 gal/min [170 L/min] | |
| 3 inch [75 mm] dia. | 60 gal/min [227 L/min] | |
| Milling | | |
| Small cutters | 5 gal/min [19 L/min]/tool | |
| Large cutters | Up to 60 gal/min [227 L/min]/ tool | |
| Drilling, Reaming | | |
| 1 inch [25 mm] dia. | 2-3 gal/min [7.6-11 L/min] | |
| Drilling, Large | 2-3 gal/min x dia., inch [0.3-0.43 L/min x dia., mm] | |
| Gundrilling | | |
| External chip removal type | | |
| 0.18-0.37 inch [4.6-9.4 mm] dia. | 2-6 gal/min [7.6-23 L/min] | Fine filtration required. |
| 0.37-0.75 inch [9.4-19 mm] dia. | 5-17 gal/min [19-64 L/min] | |
| 0.75-1.25 inch [19-32 mm] dia. | 10-40 gal/min [38-151 L/min] | |
| 1.25-1.50 inch [32-38 mm] dia. | 17-50 gal/min [64-189 L/min] | Use higher flow rates for deeper holes and the largest diameters in each range. |
| Internal chip removal type | | |
| 0.31-0.37 inch [7.9-9.4 mm] dia. | 5-8 gal/min [19-30 L/min] | |
| 0.37-0.75 inch [9.4-19 mm] dia. | 8-26 gal/min [30-98 L/min] | |
| 0.75-1.18 inch [19-30 mm] dia. | 26-66 gal/min [98-250 L/min] | |
| 1.18-2.38 inch [30-60 mm] dia. | 66-130 gal/min [250-492 L/min] | |
| Trepanning | | |
| External chip removal heads | | |
| 2-3.5 inch [51-89 mm] dia. | 8-48 gal/min [30-182 L/min] | Fine filtration required. |
| 3.5-6 inch [89-152 mm] dia. | 16-80 gal/min [61-303 L/min] | |
| 6-8 inch [152-203 mm] dia. | 32-104 gal/min [121-394 L/min] | Use higher flow rates for deeper holes and the largest diameters in each range. If an emulsion is used instead of oil, increase the flow rate. |
| Internal chip removal heads | | |
| 2.37-6 inch [60-152 mm] dia. | 110-215 gal/min [416-814 L/min] | |
| 6-12 inch [152-305 mm] dia. | 215-340 gal/min [814-1,287 L/min] | |
| 12-18 inch [305-457 mm] dia. | 340-460 gal/min [1,287-1,741 L/min] | |
| 18-24 inch [457-610 mm] dia. | 460-570 gal/min [1,741-2,158 L/min] | |
| Honing | | |
| Small | 3 gal/min [11 L/min]/hole | Very fine filtration required. |
| Large | 5 gal/min [19 L/min]/hole | |
| Broaching | | |
| Small | 10 gal [38 L]/stroke | |
| Large | 3 gal/stroke x length of cut, inch [0.45 L/stroke x length of cut, mm] | |
| Centerless grinding | | |
| Small | 20 gal/min [76 L/min] | Fine filtration required. |
| Large | 40 gal/min [151 L/min] | |
| Other grinding | 5 gal/min/inch of wheel width [0.75 L/min/mm of wheel width] | |

volved, an entrained film of air usually encloses the grinding-wheel surface, and this prevents penetration of the fluid into the cutting zone. Special nozzles can be designed which will force the fluid through the air film and onto the wheel. These nozzles must be placed as close as possible to the workpiece to prevent complete loss of the fluid by the centrifugal forces of the wheel. Another method of overcoming the air film enclosing the grinding wheel is to use a close-fitting baffle that interrupts the air flow, creating a vacuum that sucks the grinding fluid into the wheel-work interface.

### High-Pressure Application

For some operations, such as gundrilling and trepanning, high-pressure fluid systems are normally used with fluids being applied at pressures ranging from 100 to 2,000 psi [690 to 13,790 kPa]. A gundrilling tool is essentially a single-point end cutter similar to a boring tool, except that it has an internal passage for fluid. Trepanning is a hole-making operation that cuts a cylindrical path into the metal, leaving a solid core. This core passes through the hollow cylindrical cutting head as a tool feeds into the metal. The cutting fluid is pumped around the outside of the tool under pressure, forcing chips back through the center. Cutting fluids for trepanning must have good EP and anti-weld properties, must be low enough in viscosity to flow freely around the tool, and must have good oiliness.

Deep-hole drilling presents the problem of maintaining a sufficient flow of cutting fluid to the cutting edges. One solution is the use of oil-groove, oil-hole, or oil-tube drills, which utilize drill-flute space for cutting fluid passages. The fluid, under 50 to 100 psi [345 to 690 kPa] pressure, is transfered to the drill by a rotating gland and is forced directly into the cutting zone. The fluid flowing from the hole assists in chip removal. Oil-hole drills have become very popular in recent years, particularly for deep holes. Their use represents a significant improvement over flooding as the method of getting the cutting fluid to the drill lips. Significant increases in tool life and productivity can also be achieved (that is, oil-hole cooling gives better tool life at higher speeds than flood cooling does).

High-pressure systems are sometimes used for other operations. The high pressure facilitates the fluid's reaching the chip tool interface. In grinding, a high-pressure jet also serves to clean the wheel.

### Mist Systems

Cutting fluids may also be applied in the form of an air-carried mist. Small jet equipment is used to disperse soluble oil or synthetic water-miscible cutting fluids as very fine droplets in a carrier such as air, at 10 to 80 psi [69 to 552 kPa], or occasionally as an aerosol. Water-miscible fluids are preferred over oil because oil presents possible health hazards and tends to clog. Mist application is best suited to operations where the cutting speed is high and the areas of cut are low, as in end milling. Cutting fluids normally chosen primarily for their cooling ability are used for mist application. The very small droplets come into contact with the hot tool, workpiece, or chip and evaporate and rapidly remove heat by vaporization. Mist cooling does not require the splash guards, the chip pans and the return hoses that are required for flood cooling. In addition, only small amounts of fluids are used, and these generally dry on the part or can be easily wiped away.

Mist systems are often advantageous because they:

- Provide better tool life than cutting dry
- Provide coolant when a flood system is not available or practical
- Apply fluids to otherwise inaccessible areas
- Provide higher fluid velocities at tool-workpiece interface than possible by flood cooling
- Reduce costs in some cases
- Give better visibility of the workpiece in cut

Disadvantages of mist systems include limited cooling capability and the need for venting.

Two types of mist generators are normally used—the aspirator type and the direct-pressure type. For the aspirator, a stream of air is blown over the open end of a tube which is immersed in the fluid. A partial vacuum is created and the fluid is drawn up the tube where it becomes entrained in the air stream. In the pressurized mist generator, either pressurized gas bottles or shop air may be used to force the fluid into the air stream.

### Special Application Methods

Chilled cutting fluids and highly pressurized gases have been shown effective in increasing tool life. These methods are more exotic than the conventional application methods. In special cases they may be economically justifiable. They may prove worthwhile to try in cases where nothing else works.

## MAINTENANCE OF CUTTING FLUIDS

Cutting fluids, like any other fluids that are used over and over again, must be cared for properly. There are several precautions that should be observed.

### Cutting Oils

Oil-type fluids perform satisfactorily if applied at full flow to the tools and the work and if sufficient volume is maintained in the system to hold the oil temperature around 70° to 75°F [21° to 24°C].

Cutting and grinding oils become contaminated rapidly during use. Extraneous materials, chips, dirt, etc., should be removed continously or at periodic intervals by filters, strainers, centrifuges, or settling tanks. The mechanical edge-type filter, incorporating metal strips or disks as the filter element, acts principally as a strainer. The absorbent-type filter uses paper disks, cotton waste, or cloth bags as the filtering element. Magnetic filters are suitable for separation of ferrous particles. Centrifuging is used for removing heavy contaminants and particles from oil. Centrifuging, together with a heating unit and settling tanks, is often used for extracting oil from chips.

All cutting oil systems should be drained at intervals, manually cleaned, flushed and replenished with filtered or new cutting oil. Frequency of cleaning depends upon individual conditions.

### Emulsifiable Oils

Emulsions generally require more maintenance and care than do cutting oils. When preparing emulsions, always add the oil to at least two or three times as much water. This initial mix should be agitated thoroughly while the oil is being added; otherwise, soap forms which combines with the water causing the mineral oil to separate out. If not enough water is used, or if the water is added to the oil, an invert emulsion will result (water particles are dispersed in the oil phase) which is undesirable for metalcutting. Premixed fluids, instead of plain water, should be used for preparing or for alterating the mix concentration.

Water used in preparing emulsions is very important. Hard water containing various minerals and salts often hinders or impedes emulsification. It is not uncommon for emulsions made with hard water to "break" readily, that is, to separate into a stratified condition with a layer of oil or creamy emulsion floating on the surface. Such separation is detrimental. On the other hand, water that is too soft will cause foaming. Use of a specially formulated hard-water soluble oil with soft water may lead to the formation of a bluish-black stain on freshly ground ferrous parts.

Water used for making cutting fluid mixtures should be as pure as possible for the most economical and trouble-free use. Boiler water condensate (when available) or deionized water (mineral free) should be used. The deionizer removes all minerals by chemical absorption so that the effluent is equivalent to distilled water. Consequently, no residues are left by evaporation of the water, and corrosion effects from minerals are eliminated.

Pretreatment of hard water is sometimes necessary. Water conditioning agents consisting usually of polyphosphate combinations are readily available, as is tri-sodium phosphate. The general rule for using these materials is to add about 1.5 ounces per 100 gallons [0.11 grams per liter] of water per grain of hardness. Chemical treatment of extremely hard water is far more economical than the purchase of a specially formulated soluble oil. Excessive use of polyphosphate water conditioners, however, will tend to increase bacteria, mold and fungi growth. Where water makeup rates are high, the use of deionized water is preferred.

Microorganisms in the water shorten the service life of a soluble-oil emulsion. Microorganisms of three types—bacteria, algae, and fungii—are often encountered in soluble oils, and all three have detrimental effects on emulsion stability. Many soluble oils are compounded with a bactericide, but the amount that can be added is limited by its solubility in the oil. When the emulsion is made, the bactericide is further diluted, reducing its effectiveness.

Rancidity, the term applied whenever a cutting fluid gives off a bad odor, is usually caused by bacterial growth. The rotten-egg stench emanating from the sump of a machine that has been shut down over a weekend is caused by bacteria that attack inorganic sulfates found in all natural waters. A quality cutting fluid and regular use of biocides where needed are the best insurance against rancidity.

Emulsion concentration is not always given the attention it deserves. In heavy-duty cutting operations, heat at the tool causes water to evaporate at a rate faster than the carryoff of oil on the machined parts. This results in an increased oil-to-water ratio, which if carried too far can cause an invert emulsion. The opposite applies to grinding operations where oil carryoff is higher and the emulsion becomes increasingly dilute with use. This may cause rusting unless the concentration is frequently checked and controlled.

The amount of time that an emulsion is kept in service varies widely—anywhere from one week to six months. Cooling an emulsion by mechanical circulation or refrigeration is useful in extending its service life and in producing better finish. Aeration, although an effective means of cooling an emulsion, will usually increase microorganism growth which results in reduced emulsion life. In use, emulsions should be held at a temperature between 55° and 70°F [13° and 21°C].

Before putting an emulsion into use, wash and flush the coolant system thoroughly. Deposits of all kinds must be removed. Emulsions are extremely susceptible to contamination. If there is any reason to suspect that bacteria are present in the system, flush it with a germicidal solution before putting in the new emulsion.

### Occupational Dermatitis

Dermatitis is not an affliction that must be tolerated. It can be closely controlled, or eliminated, by observing simple principles of cleanliness that the average person observes at home.

Dermatitis is frequently, though mistakenly, associated with the handling of petroleum products, particularly in the machine shop where the operator's hands or forearms are in prolonged contact with soluble oils and cutting oils. Combined with dirt, these fluids form grimy compounds that may become embedded in the skin, often blocking the pores and hair follicles. In many cases, these areas become infected and dermatitis sets in. This condition may occur with almost any material that is allowed to remain on the skin for a long period of time; it is not limited to petroleum products. A contributing factor may sometimes be the solvent action of the cutting fluid. Left on the skin, the fluid can dissolve the natural skin oils, inflaming the skin and causing it to crack.

The bactericides used in most cutting fluids generally have no effect on the incidence of dermatitis; however, formaldehyde-releasing bactericides may sensitize the skin of some people. People with thin skin are more susceptible to dermatitis. Fair, blond-complexioned people are usually more susceptible than are dark, oilier-skinned people. For hypersensitive workers, the only answer is a switch to another job—one in which they will not be exposed to cutting fluids.

Maintaining personal cleanliness, keeping the fluid clean, using commercially available hand creams, barrier creams or protective clothing, and installing splash guards on the machines—all can help to control dermatitis in the shop.

### Machine Cleaning Practice

For mineral-oil-base fluids, machine cleaning can generally be accomplished by periodic removal of chips, metal fines and sludge, followed by flushing with clean cutting oil.

No matter how effective a coolant clarifying system is on water-miscible fluids, machine cleaning must be performed eventually. Proper machine cleaning can extend coolant life 4 to 6 times, compared to simply removing the old coolant and replacing it with fresh coolant.

For machines not in operation, the proper cleaning procedure is as follows:

1. Pump out old coolant.
2. Clean out chips and oil residue.
3. Add cleaner mixed with water at a ratio of 1:50 to machine tool (fill reservoir).
4. Circulate cleaner for at least three hours until machine is clean and apply cleaning solution directly to all machine surfaces that are not in contact with circulating system.
5. When the machine is clean, pump out cleaner and remove all accumulated sediment from the sump.
6. Fill with enough plain water to circulate through all coolant lines and circulate for at least 15 minutes while rinsing surfaces previously cleaned in step 4.
7. Drain and refill with plain water, circulate, rinse and drain again.
8. Add new cutting fluid immediately to cover all exposed metal surfaces to prevent rusting.

By thoroughly cleaning a machine each time the coolant needs changing, all bacteria are removed and, consequently, are not present to immediately begin degrading the fresh coolant. Although a thorough cleaning takes considerably more time compared to pumping out and recharging, the coolant will last considerably longer; consequently, machines can be cleaned on a predetermined schedule instead of on a "catch-as-catch-can" program. The result is controllable machine downtime, which is always less expensive than emergency downtime.

### Disposal of Cutting Fluids

Disposal problems with straight oil products are minimized with proper batch-type recovery equipment. Straight oil fluids can be sterilized and water contaminants removed by heat. Settling and addition of base concentrates usually restore adequate quality for continued use. When necessary, oil-base fluids should be disposed of by burning (possibly as a fuel).

In order to meet federal, state, and local water-pollution-control laws, all water-miscible cutting fluids should undergo some sort of treatment before disposal into a lake, stream, or municipal sewer system. The chemicals considered as pollutants in water-miscible products are oil, nitrites, phenols, phosphates, PCB, and heavy metals. The oil content can be broken out of emulsion by an acid or aluminum sulfate treatment. The nitrite content of the effluent can then be destroyed by treatment with sulfamic acid.

In some states effluent containing more than 2 parts per billion of phenol or phenolic derivatives is prohibited. Where regulations are less stringent, phenols can be removed by the use of a carbon filter or slurry.

### Recycling Cutting Fluids

Recycling of cutting fluids can solve waste-disposal problems, reduce costs and ease pollution problems. A closed-loop system for cutting fluid recycling will have some loss, but 90 percent recovery is not unusual. Even lower rates of recovery can justify the expense of a closed-loop system.

One basic part of a recycling system is some kind of filter to remove metal chips and grinding swarf. Other elements are needed to remove tramp oils, to provide makeup fluid, and, unless the recycling system is part of a central fluid system, to haul, store, and pump fluid in and out of machine sumps.

A recycling system can be used for separate machines or used in a central fluid system as described in the following discussion.

## CENTRAL FLUID SYSTEM

Wherever possible, consideration should be given to the possible use of central systems for handling of recycled fluids from groups of machines. This is only practical if the machines are using the same cutting fluid throughout. A group of grinding machines may be linked together to handle the swarf by means of an integrated conveying system; however, all the machines must be grinding similar materials since mixed metals lose their scrap value. The central collection of swarf wetted by cutting fluids also reduces physical handling and improves working conditions.

Centralized coolant systems enable a plant to better maintain cutting fluids by providing a convenient checkpoint with ancillary equipment to better maintain fluid concentration. A central system also means that more controls or checks can be brought into the evaluation by minimizing the number of samples that need to be checked. Besides providing better control, a central system minimizes maintenance with a fluid recycling system. Centralized systems usually cost less than a number of individual machine units.

One major advantage of the central system is its efficient removal of metal fines from the cutting fluid with the feasibility of removing tramp oil by centrifuge or by other means. Centrifuging has an additional benefit for the fluid user. By centrifuging the coolant, both the tramp oil and the metallic fines are removed. With removal of these contaminants, about half of the bacteria in the coolant is also removed, since bacteria tend to grow at the interfaces of the coolant solution and the tramp oil droplets and metallic fines. The continuous removal of these contaminants, periodic quality control tests, and the systematic use of additives or product concentrates based on these tests—are all major factors which make the central system an effective means of realizing extended cutting fluid life. Increased fluid life reduces the disposal problems inherent in the use of water-miscible fluids.

Extending cutting- or grinding-fluid life in central-system applications requires greater attention to fluid quality control. The large number of machine tools and the productivity involved justify greater levels of laboratory effort—not only to determine concentration but also to determine rancidity-control performance, corrosion protection, volume of recirculating chips (dirt load) and degree of unwanted contaminants. The use of system log records (water, product input, concentration, charge dates) and fluid test records (concentration, pH, bacteria count, dirt load, tramp oil, corrosion protection) are vital to monitoring the condition of the cutting fluid and can be useful for identifying corrective action.

## ECONOMIC CONSIDERATIONS

The cost per piece produced is more important than the initial price of a cutting fluid because in many cases more expensive fluids may actually be more economical to use. Because the cutting fluid is part of a manufacturing system, an economic analysis of cutting fluids must take into consideration both the costs associated with the fluid itself and the costs affected by the fluid. These two cost divisions can be further subdivided into direct and indirect costs. In those cases where the influence of specific costs is great, it is possible to make a rough estimate of the savings without a complete cost analysis. In any event, the economic justification of cutting and grinding fluids should include not only the cost of the fluid but also its effect on the cost of tools, wheels, downtime, etc.

The selection of cutting fluids varies with the work material, the operations performed, etc. The benefits, such as reduced machining costs, increased production, decreased surface roughness, and greater accuracy, must be balanced against the costs of having too many different fluids. These include the cost of maintaining a large fluid inventory, fluid changing costs, fluid price differentials, and increased maintenance and disposal costs. Recycling of fluids can reduce the increasingly higher costs of fluid disposal.

The following tables contain Cutting Fluid Recommendations for the machining operations and work materials listed in sections 1 through 8. The code numbers in these tables refer to the following general types of fluids:

| Code No. | Type |
|---|---|
| 0 | Dry |
| 1 | Oils—Light Duty (General Purpose) |
| 2 | Oils—Medium Duty |
| 3 | Oils—Heavy Duty |
| 4 | Emulsifiable Oils—Light Duty (General Purpose) |
| 5 | Emulsifiable Oils—Heavy Duty |
| 6 | Chemicals and Synthetics—Light Duty (General Purpose) |
| 7 | Chemicals and Synthetics—Heavy Duty |
| 8 | Special—Light Duty |
| 9 | Special—Heavy Duty |

For each general type of cutting fluid, the Cutting Fluid Key, section 16.3, provides an extensive list of cutting fluids by code number, type and trade name, and manufacturer. For each type of fluid, there are a number of subtypes, the choice of which has been left to the user because of the lack of more specific machining data.

All cutting fluids in section 16.3 are identified by specific trade name and by the manufacturer of the fluid. The Cutting Fluid Key includes all those companies who answered inquiries regarding the classification of their fluids and does not imply recommendation of particular fluids or manufacturers.

Example: Selection of a cutting fluid for peripheral end milling of an alloy steel (48-50 $R_C$).

Step 1. The Cutting Fluid Recommendations tables, section 16.2, indicate a preference for a Type 3 fluid which is an Oil—Heavy Duty. See page 16-22.

Step 2. The Cutting Fluid Key, section 16.3, lists the following subtypes under Oils—Heavy Duty (see pages 16-72 through 16-77):

- Sulfurized Mineral Oil
- Compounded Sulfurized Mineral Oil
- Chlorinated Mineral Oil
- Sulfurized Mineral Oil + Sulfur, Chlorine, and Phosphorus Compounding
- Sulfo-Chlorinated Mineral Oil
- Sulfurized Mineral Oil + Fatty Oil
- Chlorinated Mineral Oil + Fatty Oil
- Sulfo-Chlorinated Mineral Oil + Fatty Oil
- Compounded Sulfo-Chlorinated Mineral Oil + Fatty Oil
- Sulfo-Chlorinated Fatty Oil
- Sulfurized Fatty Oil
- Mineral Oil + Added Fat, Sulfur, Chlorine, and Phosphorus Compounding

From the subtypes listed, select a specific product based on machine tool and workpiece characteristics, company policies concerning use of cutting fluids, availability of a fluid, preference for a particular cutting fluid supplier, economics, and other conditions.

# 16.2 Cutting Fluid Recommendations

| OPERATION | TOOL MATERIAL OR ABRASIVE | 1. Free Machining Carbon Steels, Wrought | | | 2. Carbon Steels, Wrought | | |
|---|---|---|---|---|---|---|---|
| | | 100 to 275 Bhn | 275 to 425 Bhn | 48 to 65 $R_C$ | 85 to 275 Bhn | 275 to 425 Bhn | 48 to 65 $R_C$ |
| 1.1 Turning, Single Point and Box Tools | HSS | 1, 4, 6 | 2, 5, 7 | 3, 5, 7 | 1, 4, 6 | 2, 5, 7 | 3, 5, 7 |
| | Carbide | 0, 4, 6 | 0, 4, 6 | 0, 4, 6 | 0, 4, 6 | 0, 4, 6 | 0, 4, 6 |
| 1.3 Turning | Diamond | — | — | — | — | — | — |
| 1.4 Turning, Cutoff and Form Tools | HSS | 2, 5, 7 | 3, 5, 7 | 3, 5, 7 | 2, 5, 7 | 3, 5, 7 | 3, 5, 7 |
| | Carbide | 0, 4, 6 | 0, 4, 6 | 5, 7 | 0, 4, 6 | 0, 4, 6 | 5, 7 |
| 1.5 Threading, Single Point | HSS | 2, 5 | 3, 6 | 3 | 2, 5 | 3, 6 | 3 |
| | Carbide | 1, 4, 6 | 1, 4, 6 | 1, 4, 6 | 1, 4, 6 | 1, 4, 6 | 1, 4, 6 |
| 1.6 Threading, Die | HSS | 2, 5 | 3, 6 | 3 | 2, 5 | 3, 6 | 3 |
| | Carbide | 1, 4, 6 | 1, 4, 6 | 1, 4, 6 | 1, 4, 6 | 1, 4, 6 | 1, 4, 6 |
| 1.7 Hollow Milling | HSS | 1, 4, 6 | 2, 5, 7 | 3 | 1, 4, 6 | 2, 5, 7 | 3 |
| | Carbide | 0, 4, 6 | 0, 4, 6 | 2, 5, 7 | 0, 4, 6 | 0, 4, 6 | 2, 5, 7 |
| 2.1 Face Milling | HSS | 1, 4, 6 | 2, 5, 7 | 3 | 1, 4, 6 | 2, 5, 7 | 3 |
| | Carbide | 0, 1, 4 | 0, 1, 4 | 0, 2 | 0, 1, 4 | 0, 1, 4 | 0, 2 |
| 2.2 Face Milling | Diamond | — | — | — | — | — | — |
| 2.4 Slab Milling | HSS | 1, 4, 6 | 2, 5, 7 | 3 | 1, 4, 6 | 2, 5, 7 | 3 |
| | Carbide | 0, 4, 6 | 0, 4, 6 | 2 | 0, 4, 6 | 0, 4, 6 | 2 |
| 2.5 Side and Slot Milling—Arbor Mounted Cutters | HSS | 1, 4, 6 | 2, 5, 7 | 3 | 1, 4, 6 | 2, 5, 7 | 3 |
| | Carbide | 0, 4, 6 | 0, 4, 6 | 2 | 0, 4, 6 | 0, 4, 6 | 2 |
| 2.6 End Milling—Peripheral | HSS | 1, 4, 6 | 2, 5, 7 | 3 | 1, 4, 6 | 2, 5, 7 | 3 |
| | Carbide | 0, 4, 6 | 0, 4, 6 | 3 | 0, 4, 6 | 0, 4, 6 | 3 |
| 2.8 End Milling—Slotting | HSS | 1, 4, 6 | 2, 5, 7 | 3 | 1, 4, 6 | 2, 5, 7 | 3 |
| | Carbide | 0, 4, 6 | 0, 4, 6 | 3 | 0, 4, 6 | 0, 4, 6 | 3 |
| 2.9 Thread Milling | HSS | 2, 5, 7 | 3, 5, 7 | 3 | 2, 5, 7 | 3, 5, 7 | 3 |
| 3.1 Drilling | HSS | 1, 4, 6 | 2, 5, 7 | 3 | 1, 4, 6 | 2, 5, 7 | 3 |
| | Carbide | 0, 4, 6 | 0, 4, 6 | 3 | 0, 4, 6 | 0, 4, 6 | 3 |
| 3.3 Oil-Hole or Pressurized-Coolant Drilling | HSS | 1, 4, 6 | 1, 4, 6 | 3 | 1, 4, 6 | 1, 4, 6 | 3 |
| | Carbide | 4, 6 | 4, 6 | 2 | 4, 6 | 4, 6 | 2 |
| 3.4 Spade Drilling | HSS | 1, 4, 6 | 1, 4, 6 | 3 | 1, 4, 6 | 1, 4, 6 | 3 |
| | Carbide | 0, 4, 6 | 0, 4, 6 | 2 | 0, 4, 6 | 0, 4, 6 | 2 |
| 3.5 Gundrilling | HSS | 1, 4, 6 | 2, 5, 7 | 3 | 1, 4, 6 | 2, 5, 7 | 3 |
| | Carbide | 1, 4, 6 | 1, 4, 6 | 2, 3 | 1, 4, 6 | 1, 4, 6 | 2, 3 |
| 3.6 Pressure Coolant Reaming (Gun Reaming) | HSS | 1, 5, 6 | 2, 6, 7 | 3 | 1, 5, 6 | 2, 6, 7 | 3 |
| | Carbide | 1, 4, 6 | 1, 4, 6 | 2, 5, 7 | 1, 4, 6 | 1, 4, 6 | 2, 5, 7 |
| 3.7 Reaming | HSS | 1, 5, 6 | 2, 6, 7 | 3 | 1, 5, 6 | 2, 6, 7 | 3 |
| | Carbide | 1, 4, 6 | 1, 4, 6 | 2, 3 | 1, 4, 6 | 1, 4, 6 | 2, 3 |
| 3.9 Boring | HSS | 1, 4, 6 | 2, 5, 7 | 3, 5, 7 | 1, 4, 6 | 2, 5, 7 | 3, 5, 7 |
| | Carbide | 0, 4, 6 | 0, 4, 6 | 2, 4, 6 | 0, 4, 6 | 0, 4, 6 | 2, 4, 6 |
| 3.11 Boring | Diamond | — | — | — | — | — | — |
| 3.12 Counterboring and Spotfacing | HSS | 1, 4, 6 | 2, 5, 7 | 3 | 1, 4, 6 | 2, 5, 7 | 3 |
| | Carbide | 0, 4, 6 | 0, 4, 6 | 3, 7 | 0, 4, 6 | 0, 4, 6 | 3, 7 |
| 3.13 Trepanning | HSS | 1, 4, 6 | 2, 5, 7 | 3 | 1, 4, 6 | 2, 5, 7 | 3 |
| | Carbide | 1, 4, 6 | 1, 4, 6 | 2, 3 | 1, 4, 6 | 1, 4, 6 | 2, 3 |
| 3.14 Honing | — | 80 | 80 | 90 | 80 | 80 | 90 |
| 3.15 Burnishing | — | 1, 5, 7 | 2, 5, 7 | — | 1, 5, 7 | 2, 5, 7 | — |
| 4.1 Tapping | HSS | 1, 5, 7 | 2, 5, 7 | 3 | 1, 5, 7 | 2, 5, 7 | 3 |

| OPERATION | TOOL MATERIAL OR ABRASIVE | 1. Free Machining Carbon Steels, Wrought | | | 2. Carbon Steels, Wrought | | |
|---|---|---|---|---|---|---|---|
| | | 100 to 275 Bhn | 275 to 425 Bhn | 48 to 65 $R_C$ | 85 to 275 Bhn | 275 to 425 Bhn | 48 to 65 $R_C$ |
| **5.1 Planing** | HSS | 1, 4, 6 | 2, 5, 7 | 3 | 1, 4, 6 | 2, 5, 7 | 3 |
| | Carbide | 0, 1, 4 | 0, 1, 4 | 2 | 0, 1, 4 | 0, 1, 4 | 2 |
| **5.2 Broaching** | HSS | 2, 5, 7 | 3, 5, 7 | 3 | 2, 5, 7 | 3, 5, 7 | 3 |
| | Carbide | 1, 4, 6 | 1, 4, 6 | 3 | 1, 4, 6 | 1, 4, 6 | 3 |
| **6.1 Power Hack Sawing** | HSS | 1, 4, 6 | 2, 5, 7 | 3 | 1, 4, 6 | 2, 5, 7 | 3 |
| **6.2 Power Band Sawing** | HSS | 1, 4, 6 | 2, 5, 7 | 3 | 1, 4, 6 | 2, 5, 7 | 3 |
| **6.4 Circular Sawing** | HSS | 1, 4, 6 | 2, 5, 7 | 3 | 2, 5, 7 | 3, 5, 7 | 3 |
| **6.5 Circular Sawing** | Carbide | 0, 1, 4 | 0, 1, 4 | 2, 3 | 1, 4, 6 | 2, 5, 7 | 2, 3 |
| **6.6 Abrasive Cutoff** | — | 1, 4, 6 | 1, 4, 6 | 3 | 1, 4, 6 | 2 | 3 |
| **7.1 Gear Hobbing** | HSS | 2, 5, 7 | 3, 5, 7 | 3 | 2, 5, 7 | 3, 5, 7 | 3 |
| **7.2 Gear Cutting, Straight and Spiral Bevel** | HSS | 2, 5, 7 | 3, 5, 7 | 3 | 2, 5, 7 | 3, 5, 7 | 3 |
| **7.3 Gear Shaping** | HSS | 2, 5, 7 | 3, 5, 7 | 3 | 2, 5, 7 | 3, 5, 7 | 3 |
| **7.4 Gear Shaving** | HSS | 2, 5, 7 | 3, 5, 7 | 3 | 2, 5, 7 | 3, 5, 7 | 3 |
| **7.5 Gear Grinding, Form** | — | 1, 4, 6 | 1, 4, 6 | 3 | 1, 4, 6 | 1, 4, 6 | 3 |
| **8.1 Surface Grinding—Horiz. Spdl., Recip. Table** | — | 4, 6 | 4, 6 | 3 | 1, 4, 6 | 2, 4, 6 | 3 |
| **8.2 Surface Grinding—Horiz. Spdl., Recip. Table** | CBN | — | — | 5, 4 | — | — | 5, 4 |
| **8.3 Surface Grinding—Horiz. Spdl., Recip. Table** | Diamond | — | — | — | — | — | — |
| **8.4 Surface Grinding—Vertical Spdl., Rotary Table** | — | 4, 6 | 4, 6 | 3 | 1, 4, 6 | 2, 4, 6 | 3 |
| **8.5 Cylindrical Grinding** | — | 4, 6 | 4, 6 | 3 | 1, 4, 6 | 2, 4, 6 | 3 |
| **8.6 Cylindrical Grinding** | CBN | — | — | 5, 4 | — | — | 5, 4 |
| **8.7 Cylindrical Grinding** | Diamond | — | — | — | — | — | — |
| **8.8 Internal Grinding** | — | 4, 6 | 4, 6 | 3 | 1, 4, 6 | 2, 4, 6 | 3 |
| **8.9 Internal Grinding** | CBN | — | — | 5, 4 | — | — | 5, 4 |
| **8.10 Internal Grinding** | Diamond | — | — | — | — | — | — |
| **8.11 Centerless Grinding** | — | 4, 6 | 4, 6 | 3 | 1, 4, 6 | 2, 4, 6 | 3 |
| **8.13 Abrasive Belt Grinding** | — | 0, 1, 4 | 0, 1, 4 | 1, 4, 6 | 0, 1, 4 | 0, 1, 4 | 1, 4, 6 |
| **8.14 Thread Grinding** | — | 1, 4, 6 | 1, 4, 6 | 3 | 1, 4, 6 | 2, 4, 6 | 3 |
| | | | | | | | |

# 16.2 Cutting Fluid Recommendations

| OPERATION | TOOL MATERIAL OR ABRASIVE | 3. Carbon and Ferritic Alloy Steels (High Temperature Service) | | | 4. Free Machining Alloy Steels, Wrought | | |
|---|---|---|---|---|---|---|---|
| | | — | 150 to 200 Bhn | — | 150 to 275 Bhn | 275 to 425 Bhn | 45 to 65 $R_C$ |
| **1.1 Turning, Single Point and Box Tools** | HSS | | 1, 4, 6 | | 1, 4, 6 | 2, 5, 7 | 3, 5, 7 |
| | Carbide | | 0, 4, 6 | | 0, 4, 6 | 0, 4, 6 | 0, 4, 6 |
| **1.3 Turning** | Diamond | | — | | — | — | — |
| **1.4 Turning, Cutoff and Form Tools** | HSS | | 2, 5, 7 | | 2, 5, 7 | 2, 5, 7 | 2, 5, 7 |
| | Carbide | | 0, 4, 6 | | 0, 4, 6 | 5, 7 | 5, 7 |
| **1.5 Threading, Single Point** | HSS | | 2, 5 | | 2, 5 | 2, 5 | 3 |
| | Carbide | | 1, 4, 6 | | 1, 4, 6 | 1, 4, 6 | 1, 4, 6 |
| **1.6 Threading, Die** | HSS | | 2, 5 | | 2, 5 | 2, 5 | 3 |
| | Carbide | | 1, 4, 6 | | 1, 4, 6 | 1, 4, 6 | 1, 4, 6 |
| **1.7 Hollow Milling** | HSS | | 1, 4, 6 | | 1, 5, 7 | 2, 5, 7 | 3 |
| | Carbide | | 0, 4, 6 | | 0, 4, 6 | 1, 4, 6 | 2, 5, 7 |
| **2.1 Face Milling** | HSS | | 1, 4, 6 | | 1, 5, 7 | 2, 5, 7 | 3 |
| | Carbide | | 0, 1, 4 | | 0, 4, 6 | 0, 4, 6 | 0, 2 |
| **2.2 Face Milling** | Diamond | | — | | — | — | — |
| **2.4 Slab Milling** | HSS | | 1, 4, 6 | | 1, 4, 6 | 2, 5, 7 | 3 |
| | Carbide | | 0, 4, 6 | | 0, 4, 6 | 1, 4, 6 | 2 |
| **2.5 Side and Slot Milling—Arbor Mounted Cutters** | HSS | | 1, 4, 6 | | 1, 5, 7 | 2, 5, 7 | 3 |
| | Carbide | | 0, 4, 6 | | 0, 4, 6 | 1, 4, 6 | 2 |
| **2.6 End Milling—Peripheral** | HSS | | 1, 4, 6 | | 1, 4, 6 | 2, 5, 7 | 3 |
| | Carbide | | 0, 4, 6 | | 0, 4, 6 | 0, 4, 6 | 3 |
| **2.8 End Milling—Slotting** | HSS | | 1, 4, 6 | | 1, 4, 6 | 2, 5, 7 | 3 |
| | Carbide | | 0, 4, 6 | | 0, 4, 6 | 0, 4, 6 | 3 |
| **2.9 Thread Milling** | HSS | | 2, 5, 7 | | 2, 5, 7 | 3, 5, 7 | 3 |
| **3.1 Drilling** | HSS | | 1, 4, 6 | | 1, 4, 6 | 2, 5, 7 | 3 |
| | Carbide | | 0, 4, 6 | | 0, 4, 6 | 1, 5, 7 | 3 |
| **3.3 Oil-Hole or Pressurized-Coolant Drilling** | HSS | | 1, 4, 6 | | 1, 4, 6 | 1, 4, 6 | 3 |
| | Carbide | | 4, 6 | | 4, 6 | 4, 6 | 2 |
| **3.4 Spade Drilling** | HSS | | 1, 4, 6 | | 1, 4, 6 | 2, 5, 7 | 3 |
| | Carbide | | 0, 4, 6 | | 0, 4, 6 | 1, 5, 7 | 2 |
| **3.5 Gundrilling** | HSS | | 1, 4, 6 | | 1, 4, 6 | 2, 5, 7 | 3 |
| | Carbide | | 1, 4, 6 | | 1, 4, 6 | 2, 5, 7 | 2, 3 |
| **3.6 Pressure Coolant Reaming (Gun Reaming)** | HSS | | 1, 5, 6 | | 1, 4, 6 | 2, 5, 7 | 3 |
| | Carbide | | 1, 4, 6 | | 1, 4, 6 | 1, 4, 6 | 2, 5, 7 |
| **3.7 Reaming** | HSS | | 1, 5, 6 | | 1, 4, 6 | 2, 5, 7 | 3 |
| | Carbide | | 1, 4, 6 | | 1, 4, 6 | 1, 4, 6 | 2, 3 |
| **3.9 Boring** | HSS | | 1, 4, 6 | | 1, 4, 6 | 2, 5, 7 | 3, 5, 7 |
| | Carbide | | 0, 4, 6 | | 0, 4, 6 | 4, 6 | 2, 4, 6 |
| **3.11 Boring** | Diamond | | — | | — | — | — |
| **3.12 Counterboring and Spotfacing** | HSS | | 1, 4, 6 | | 2, 5, 6 | 2, 5, 7 | 3 |
| | Carbide | | 0, 4, 6 | | 0, 4, 6 | 1, 4, 6 | 3, 7 |
| **3.13 Trepanning** | HSS | | 1, 4, 6 | | 1, 4, 6 | 2, 5, 7 | 3 |
| | Carbide | | 1, 4, 6 | | 1, 4, 6 | 1, 5, 7 | 2, 3 |
| **3.14 Honing** | — | | 80 | | 80 | 80 | 90 |
| **3.15 Burnishing** | — | | 1, 5, 7 | | 1, 5, 7 | 2, 5, 7 | — |
| **4.1 Tapping** | HSS | | 1, 5, 7 | | 2, 5, 7 | 3, 5, 7 | 3 |

| OPERATION | TOOL MATERIAL OR ABRASIVE | 3. Carbon and Ferritic Alloy Steels (High Temperature Service) | | | 4. Free Machining Alloy Steels, Wrought | | |
|---|---|---|---|---|---|---|---|
| | | — | 150 to 200 Bhn | — | 150 to 275 Bhn | 275 to 425 Bhn | 45 to 65 $R_C$ |
| **5.1 Planing** | HSS | | 1, 4, 6 | | 1, 4, 6 | 2, 5, 7 | 3 |
| | Carbide | | 0, 1, 4 | | 0, 4, 6 | 0, 4, 6 | 2 |
| **5.2 Broaching** | HSS | | 2, 5, 7 | | 2, 5, 7 | 3, 5, 7 | 3 |
| | Carbide | | 1, 4, 6 | | 1, 4, 6 | 1, 4, 6 | 3 |
| **6.1 Power Hack Sawing** | HSS | | 1, 4, 6 | | 1, 4, 6 | 2, 5, 7 | 3 |
| **6.2 Power Band Sawing** | HSS | | 1, 4, 6 | | 1, 4, 6 | 2, 5, 7 | 3 |
| **6.4 Circular Sawing** | HSS | | 1, 4, 6 | | 2, 5, 7 | 3, 5, 7 | 3 |
| **6.5 Circular Sawing** | Carbide | | 0, 1, 4 | | 1, 4, 6 | 2, 5, 7 | 2, 3 |
| **6.6 Abrasive Cutoff** | — | | 1, 4, 6 | | 1, 4, 6 | 2 | 3 |
| **7.1 Gear Hobbing** | HSS | | 2, 5, 7 | | 2, 5, 7 | 3, 5, 7 | 3 |
| **7.2 Gear Cutting, Straight and Spiral Bevel** | HSS | | 2, 5, 7 | | 2, 5, 7 | 3, 5, 7 | 3 |
| **7.3 Gear Shaping** | HSS | | 2, 5, 7 | | 2, 5, 7 | 3, 5, 7 | 3 |
| **7.4 Gear Shaving** | HSS | | 2, 5, 7 | | 2, 5, 7 | 3, 5, 7 | 3 |
| **7.5 Gear Grinding, Form** | — | | 1, 4, 6 | | 1, 4, 6 | 1, 4, 6 | 3 |
| **8.1 Surface Grinding—Horiz. Spdl., Recip. Table** | — | | 4, 6 | | 1, 4, 6 | 2, 4, 6 | 3 |
| **8.2 Surface Grinding—Horiz. Spdl., Recip. Table** | CBN | | — | | — | — | 5, 4 |
| **8.3 Surface Grinding—Horiz. Spdl., Recip. Table** | Diamond | | — | | — | — | — |
| **8.4 Surface Grinding—Vertical Spdl., Rotary Table** | — | | 4, 6 | | 1, 4, 6 | 2, 4, 6 | 3 |
| **8.5 Cylindrical Grinding** | — | | 4, 6 | | 1, 4, 6 | 2, 4, 6 | 3 |
| **8.6 Cylindrical Grinding** | CBN | | — | | — | — | 5, 4 |
| **8.7 Cylindrical Grinding** | Diamond | | — | | — | — | — |
| **8.8 Internal Grinding** | — | | 4, 6 | | 1, 4, 6 | 2, 4, 6 | 3 |
| **8.9 Internal Grinding** | CBN | | — | | — | — | 5, 4 |
| **8.10 Internal Grinding** | Diamond | | — | | — | — | — |
| **8.11 Centerless Grinding** | — | | 4, 6 | | 1, 4, 6 | 2, 4, 6 | 3 |
| **8.13 Abrasive Belt Grinding** | — | | 0, 1, 4 | | 0, 1, 4 | 0, 1, 4 | 1, 4, 6 |
| **8.14 Thread Grinding** | — | | 1, 4, 6 | | 1, 4, 6 | 2, 4, 6 | 3 |
| | | | | | | | |

## 16.2 Cutting Fluid Recommendations

| OPERATION | TOOL MATERIAL OR ABRASIVE | 5. Alloy Steels, Wrought | | | 6. High Strength Steels, Wrought | | |
|---|---|---|---|---|---|---|---|
| | | 125 to 275 Bhn | 275 to 425 Bhn | 45 to 65 $R_C$ | 225 to 275 Bhn | 275 to 425 Bhn | Over 45 $R_C$ |
| **1.1 Turning, Single Point and Box Tools** | HSS | 1, 4, 6 | 2, 5, 7 | 3, 5, 7 | 1, 4, 6 | 2, 5, 7 | 3, 5, 7 |
| | Carbide | 0, 4, 6 | 0, 4, 6 | 0, 4, 6 | 0, 4, 6 | 0, 4, 6 | 0, 4, 6 |
| **1.3 Turning** | Diamond | — | — | — | — | — | — |
| **1.4 Turning, Cutoff and Form Tools** | HSS | 2, 5, 7 | 2, 5, 7 | 2, 5, 7 | 2, 5, 7 | 2, 5, 7 | 2, 5, 7 |
| | Carbide | 0, 4, 6 | 5, 7 | 5, 7 | 0, 4, 6 | 5, 7 | 5, 7 |
| **1.5 Threading, Single Point** | HSS | 2, 5 | 2, 5 | 3 | 2, 5 | 2, 5 | 3 |
| | Carbide | 1, 4, 6 | 1, 4, 6 | 1, 4, 6 | 1, 4, 6 | 1, 4, 6 | 1, 4, 6 |
| **1.6 Threading, Die** | HSS | 2, 5 | 2, 5 | 3 | 2, 5 | 2, 5 | 3 |
| | Carbide | 1, 4, 6 | 1, 4, 6 | 1, 4, 6 | 1, 4, 6 | 1, 4, 6 | 1, 4, 6 |
| **1.7 Hollow Milling** | HSS | 1, 5, 7 | 2, 5, 7 | 3 | 1, 5, 7 | 2, 5, 7 | 3 |
| | Carbide | 0, 4, 6 | 1, 4, 6 | 2, 5, 7 | 0, 4, 6 | 1, 4, 6 | 2, 5, 7 |
| **2.1 Face Milling** | HSS | 1, 5, 7 | 2, 5, 7 | 3 | 1, 5, 7 | 2, 5, 7 | 3 |
| | Carbide | 0, 4, 6 | 0, 4, 6 | 0, 2 | 0, 4, 6 | 0, 4, 6 | 0, 2 |
| **2.2 Face Milling** | Diamond | — | — | — | — | — | — |
| **2.4 Slab Milling** | HSS | 1, 4, 6 | 2, 5, 7 | 3 | 1, 4, 6 | 2, 5, 7 | 3 |
| | Carbide | 0, 4, 6 | 1, 4, 6 | 2 | 0, 4, 6 | 1, 4, 6 | 2 |
| **2.5 Side and Slot Milling—Arbor Mounted Cutters** | HSS | 1, 5, 7 | 2, 5, 7 | 3 | 1, 5, 7 | 2, 5, 7 | 3 |
| | Carbide | 0, 4, 6 | 1, 4, 6 | 2 | 0, 4, 6 | 1, 4, 6 | 2 |
| **2.6 End Milling—Peripheral** | HSS | 1, 4, 6 | 2, 5, 7 | 3 | 1, 4, 6 | 2, 5, 7 | 3 |
| | Carbide | 0, 4, 6 | 0, 4, 6 | 3 | 0, 4, 6 | 0, 4, 6 | 3 |
| **2.8 End Milling—Slotting** | HSS | 1, 4, 6 | 2, 5, 7 | 3 | 1, 4, 6 | 2, 5, 7 | 3 |
| | Carbide | 0, 4, 6 | 0, 4, 6 | 3 | 0, 4, 6 | 0, 4, 6 | 3 |
| **2.9 Thread Milling** | HSS | 2, 5, 7 | 3, 5, 7 | 3 | 2, 5, 7 | 3, 5, 7 | 3 |
| **3.1 Drilling** | HSS | 1, 4, 6 | 2, 5, 7 | 3 | 1, 4, 6 | 2, 5, 7 | 3 |
| | Carbide | 0, 4, 6 | 1, 5, 7 | 3 | 0, 4, 6 | 1, 5, 7 | 3 |
| **3.3 Oil-Hole or Pressurized-Coolant Drilling** | HSS | 1, 4, 6 | 1, 4, 6 | 3 | 1, 4, 6 | 1, 4, 6 | 3 |
| | Carbide | 4, 6 | 4, 6 | 2 | 4, 6 | 4, 6 | 2 |
| **3.4 Spade Drilling** | HSS | 1, 4, 6 | 2, 5, 7 | 3 | 1, 4, 6 | 2, 5, 7 | 3 |
| | Carbide | 0, 4, 6 | 1, 5, 7 | 2 | 0, 4, 6 | 1, 5, 7 | 2 |
| **3.5 Gundrilling** | HSS | 1, 4, 6 | 2, 5, 7 | 3 | 1, 4, 6 | 2, 5, 7 | 3 |
| | Carbide | 1, 4, 6 | 2, 5, 7 | 2, 3 | 1, 4, 6 | 2, 5, 7 | 2, 3 |
| **3.6 Pressure Coolant Reaming (Gun Reaming)** | HSS | 1, 4, 6 | 2, 5, 7 | 3 | 1, 4, 6 | 2, 5, 7 | 3 |
| | Carbide | 1, 4, 6 | 1, 4, 6 | 2, 5, 7 | 1, 4, 6 | 1, 4, 6 | 2, 5, 7 |
| **3.7 Reaming** | HSS | 1, 4, 6 | 2, 5, 7 | 3 | 1, 4, 6 | 2, 5, 7 | 3 |
| | Carbide | 1, 4, 6 | 1, 4, 6 | 2, 3 | 1, 4, 6 | 1, 4, 6 | 2, 3 |
| **3.9 Boring** | HSS | 1, 4, 6 | 2, 5, 7 | 3, 5, 7 | 1, 4, 6 | 2, 5, 7 | 3, 5, 7 |
| | Carbide | 0, 4, 6 | 4, 6 | 2, 4, 6 | 0, 4, 6 | 4, 6 | 2, 4, 6 |
| **3.11 Boring** | Diamond | — | — | — | — | — | — |
| **3.12 Counterboring and Spotfacing** | HSS | 2, 5, 6 | 2, 5, 7 | 3 | 2, 5, 6 | 2, 5, 7 | 3 |
| | Carbide | 0, 4, 6 | 1, 4, 6 | 3, 7 | 0, 4, 6 | 1, 4, 6 | 3, 7 |
| **3.13 Trepanning** | HSS | 1, 4, 6 | 2, 5, 7 | 3 | 1, 4, 6 | 2, 5, 7 | 3 |
| | Carbide | 1, 4, 6 | 1, 5, 7 | 2, 3 | 1, 4, 6 | 1, 5, 7 | 2, 3 |
| **3.14 Honing** | — | 80 | 80 | 90 | 80 | 80 | 90 |
| **3.15 Burnishing** | — | 1, 5, 7 | 2, 5, 7 | — | 1, 5, 7 | 2, 5, 7 | — |
| **4.1 Tapping** | HSS | 2, 5, 7 | 3, 5, 7 | 3 | 2, 5, 7 | 3, 5, 7 | 3 |

| OPERATION | TOOL MATERIAL OR ABRASIVE | 5. Alloy Steels, Wrought | | | 6. High Strength Steels, Wrought | | |
|---|---|---|---|---|---|---|---|
| | | 125 to 275 Bhn | 275 to 425 Bhn | 45 to 65 $R_C$ | 225 to 275 Bhn | 275 to 425 Bhn | Over 45 $R_C$ |
| **5.1 Planing** | HSS | 1, 4, 6 | 2, 5, 7 | 3 | 1, 4, 6 | 2, 5, 7 | 3 |
| | Carbide | 0, 4, 6 | 0, 4, 6 | 2 | 0, 4, 6 | 0, 4, 6 | 2 |
| **5.2 Broaching** | HSS | 2, 5, 7 | 3, 5, 7 | 3 | 2, 5, 7 | 3, 5, 7 | 3 |
| | Carbide | 1, 4, 6 | 1, 4, 6 | 3 | 1, 4, 6 | 1, 4, 6 | 3 |
| **6.1 Power Hack Sawing** | HSS | 1, 4, 6 | 2, 5, 7 | 3 | 1, 4, 6 | 2, 5, 7 | 3 |
| **6.2 Power Band Sawing** | HSS | 1, 4, 6 | 2, 5, 7 | 3 | 1, 4, 6 | 2, 5, 7 | 3 |
| **6.4 Circular Sawing** | HSS | 2, 5, 7 | 3, 5, 7 | 3 | 2, 5, 7 | 3, 5, 7 | 3 |
| **6.5 Circular Sawing** | Carbide | 1, 4, 6 | 2, 5, 7 | 2, 3 | 1, 4, 6 | 2, 5, 7 | 2, 3 |
| **6.6 Abrasive Cutoff** | — | 1, 4, 6 | 2 | 3 | 1, 4, 6 | 2 | 3 |
| **7.1 Gear Hobbing** | HSS | 2, 5, 7 | 3, 5, 7 | 3 | 2, 5, 7 | 3, 5, 7 | 3 |
| **7.2 Gear Cutting, Straight and Spiral Bevel** | HSS | 2, 5, 7 | 3, 5, 7 | 3 | 2, 5, 7 | 3, 5, 7 | 3 |
| **7.3 Gear Shaping** | HSS | 2, 5, 7 | 3, 5, 7 | 3 | 2, 5, 7 | 3, 5, 7 | 3 |
| **7.4 Gear Shaving** | HSS | 2, 5, 7 | 3, 5, 7 | 3 | 2, 5, 7 | 3, 5, 7 | 3 |
| **7.5 Gear Grinding, Form** | — | 1, 4, 6 | 1, 4, 6 | 3 | 1, 4, 6 | 1, 4, 6 | 3 |
| **8.1 Surface Grinding—Horiz. Spdl., Recip. Table** | — | 1, 4, 6 | 2, 4, 6 | 3 | 1, 4, 6 | 2, 4, 6 | 3 |
| **8.2 Surface Grinding—Horiz. Spdl., Recip. Table** | CBN | — | — | 5, 4 | — | — | 5, 4 |
| **8.3 Surface Grinding—Horiz. Spdl., Recip. Table** | Diamond | — | — | — | — | — | — |
| **8.4 Surface Grinding—Vertical Spdl., Rotary Table** | — | 1, 4, 6 | 2, 4, 6 | 3 | 1, 4, 6 | 2, 4, 6 | 3 |
| **8.5 Cylindrical Grinding** | — | 1, 4, 6 | 2, 4, 6 | 3 | 1, 4, 6 | 2, 4, 6 | 3 |
| **8.6 Cylindrical Grinding** | CBN | — | — | 5, 4 | — | — | 5, 4 |
| **8.7 Cylindrical Grinding** | Diamond | — | — | — | — | — | — |
| **8.8 Internal Grinding** | — | 1, 4, 6 | 2, 4, 6 | 3 | 1, 4, 6 | 2, 4, 6 | 3 |
| **8.9 Internal Grinding** | CBN | — | — | 5, 4 | — | — | 5, 4 |
| **8.10 Internal Grinding** | Diamond | — | — | — | — | — | — |
| **8.11 Centerless Grinding** | — | 1, 4, 6 | 2, 4, 6 | 3 | 1, 4, 6 | 2, 4, 6 | 3 |
| **8.13 Abrasive Belt Grinding** | — | 0, 1, 4 | 0, 1, 4 | 1, 4, 6 | 0, 1, 4 | 0, 1, 4 | 1, 4, 6 |
| **8.14 Thread Grinding** | — | 1, 4, 6 | 2, 4, 6 | 3 | 1, 4, 6 | 2, 4, 6 | 3 |
| | | | | | | | |

# 16.2 Cutting Fluid Recommendations

| OPERATION | TOOL MATERIAL OR ABRASIVE | 7. Maraging Steels, Wrought | | 8. Tool Steels, Wrought | | | |
|---|---|---|---|---|---|---|---|
| | | | | Annealed | | Hardened | |
| | | 275 to 425 Bhn | 50 to 52 Rc | 100 to 200 Bhn | 200 to 325 Bhn | 325 to 425 Bhn | 45 to 65 Rc |
| 1.1 Turning, Single Point and Box Tools | HSS | 2, 5, 7 | 3, 5, 7 | 1, 4, 6 | 2, 5, 7 | 2, 5, 7 | 3, 5, 7 |
| | Carbide | 0, 4, 6 | 0, 4, 6 | 0, 4, 6 | 0, 4, 6 | 0, 4, 6 | 0, 4, 6 |
| 1.3 Turning | Diamond | — | — | — | — | — | — |
| 1.4 Turning, Cutoff and Form Tools | HSS | 2, 5, 7 | 2, 5, 7 | 2, 5, 7 | 2, 5, 7 | 2, 5, 7 | 3 |
| | Carbide | 5, 7 | 5, 7 | 2, 5, 7 | 5, 7 | 5, 7 | 5, 6, 7 |
| 1.5 Threading, Single Point | HSS | 2, 5 | 3 | 3 | 3 | 3 | 3 |
| | Carbide | 1, 4, 6 | 1, 4, 6 | 2, 5, 7 | 2, 5, 7 | 1, 4, 6 | 1, 4, 6 |
| 1.6 Threading, Die | HSS | 2, 5 | 3 | 3 | 3 | 3 | 3 |
| | Carbide | 1, 4, 6 | 1, 4, 6 | 2, 5, 7 | 2, 5, 7 | 1, 4, 6 | 1, 4, 6 |
| 1.7 Hollow Milling | HSS | 2, 5 | 3 | 3 | 3 | 3 | 3 |
| | Carbide | 1, 4, 6 | 2, 5, 7 | 2, 5, 7 | 2, 5, 7 | 2, 5, 7 | 2, 5, 7 |
| 2.1 Face Milling | HSS | 2, 5, 7 | 3 | 2, 5, 7 | 2, 5, 7 | 3 | 3 |
| | Carbide | 0, 4, 6 | 0, 2 | 0, 4, 6 | 0, 4, 6 | 0, 5, 7 | 0, 5, 7 |
| 2.2 Face Milling | Diamond | — | — | — | — | — | — |
| 2.4 Slab Milling | HSS | 2, 5, 7 | 3 | 2, 5, 7 | 2, 5, 7 | 3, 5, 7 | 3 |
| | Carbide | 1, 4, 6 | 2 | 1, 4, 6 | 1, 4, 6 | 2, 5, 7 | 3, 5, 7 |
| 2.5 Side and Slot Milling—Arbor Mounted Cutters | HSS | 2, 5, 7 | 3 | 2, 5, 7 | 2, 5, 7 | 3 | 3 |
| | Carbide | 1, 4, 6 | 2 | 1, 4, 6 | 1, 4, 6 | 2, 5, 7 | 2, 5, 7 |
| 2.6 End Milling—Peripheral | HSS | 2, 5, 7 | 3 | 2, 5, 7 | 2, 5, 7 | 3 | 3 |
| | Carbide | 0, 4, 6 | 3 | 1, 4, 6 | 1, 4, 6 | 2, 5, 7 | 3, 5, 7 |
| 2.8 End Milling—Slotting | HSS | 2, 5, 7 | 3 | 2, 5, 7 | 2, 5, 7 | 3 | 3 |
| | Carbide | 0, 4, 6 | 3 | 1, 4, 6 | 1, 4, 6 | 2, 5, 7 | 3, 5, 7 |
| 2.9 Thread Milling | HSS | 3, 5, 7 | 3 | 3 | 3 | 3 | 3 |
| 3.1 Drilling | HSS | 2, 5, 7 | 3 | 2, 5, 7 | 2, 5, 7 | 3 | 3 |
| | Carbide | 1, 5, 7 | 3 | 1, 4, 6 | 1, 4, 6 | 2, 5, 7 | 3, 5, 7 |
| 3.3 Oil-Hole or Pressurized-Coolant Drilling | HSS | 1, 4, 6 | 2 | 1, 4, 6 | 2, 5, 7 | 3 | 3 |
| | Carbide | 1, 4, 6 | 2 | 1, 4, 6 | 1, 4, 6 | 2, 5, 7 | 2, 5, 7 |
| 3.4 Spade Drilling | HSS | 2, 5, 7 | 3 | 2, 5, 7 | 2, 5, 7 | 2, 5, 7 | 3 |
| | Carbide | 1, 5, 7 | 2 | 1, 4, 6 | 1, 4, 6 | 2, 5, 7 | 2 |
| 3.5 Gundrilling | HSS | 2, 5, 7 | 3 | 3, 5, 7 | 2, 5, 7 | 2, 3 | 3 |
| | Carbide | 2, 5, 7 | 2, 3 | 2, 5, 7 | 2, 5, 7 | 2, 5, 7 | 3, 5, 7 |
| 3.6 Pressure Coolant Reaming (Gun Reaming) | HSS | 2, 5, 7 | 3 | 2 | 3 | 3 | 3 |
| | Carbide | 1, 4, 6 | 2, 3 | 2, 5, 7 | 2, 5, 7 | 2, 5, 7 | 3, 5, 7 |
| 3.7 Reaming | HSS | 2, 5, 7 | 3 | 3 | 3 | 3 | 3 |
| | Carbide | 1, 4, 6 | 2, 3 | 2, 4, 6 | 2, 4, 6 | 2, 5, 7 | 3, 5, 7 |
| 3.9 Boring | HSS | 2, 5, 7 | 3, 5, 7 | 2, 5, 7 | 2, 5, 7 | 2, 5, 7 | 3, 5, 7 |
| | Carbide | 4, 6 | 2, 4, 6 | 1, 4, 6 | 1, 4, 6 | 2, 5, 7 | 2, 5, 7 |
| 3.11 Boring | Diamond | — | — | — | — | — | — |
| 3.12 Counterboring and Spotfacing | HSS | 2, 5, 7 | 3 | 2, 5, 7 | 2, 5, 7 | 3, 5, 7 | 3 |
| | Carbide | 1, 4, 6 | 3, 7 | 1, 4, 6 | 1, 4, 6 | 2, 5, 7 | 3, 5, 7 |
| 3.13 Trepanning | HSS | 2, 5, 7 | 3 | 3, 5, 7 | 3, 5, 7 | 2, 3 | 3 |
| | Carbide | 1, 5, 7 | 2, 3 | 2, 5, 7 | 2, 5, 7 | 2, 5, 7 | 3, 5, 7 |
| 3.14 Honing | — | 90 | 90 | 90 | 90 | 90 | 90 |
| 3.15 Burnishing | — | 2, 5, 7 | — | 2 | 2 | — | — |
| 4.1 Tapping | HSS | 3, 5, 7 | 3 | 3 | 3 | 3 | 3 |

| OPERATION | TOOL MATERIAL OR ABRASIVE | 7. Maraging Steels, Wrought | | 8. Tool Steels, Wrought | | | |
|---|---|---|---|---|---|---|---|
| | | | | Annealed | | Hardened | |
| | | 275 to 425 Bhn | 50 to 52 $R_C$ | 100 to 200 Bhn | 200 to 325 Bhn | 325 to 425 Bhn | 45 to 65 $R_C$ |
| 5.1 Planing | HSS | 2, 5, 7 | 3 | 2, 5, 7 | 2, 5, 7 | 2, 5, 7 | 3 |
| | Carbide | 0, 4, 6 | 2 | 1, 4, 6 | 1, 4, 6 | 1, 5, 7 | 3, 5, 7 |
| 5.2 Broaching | HSS | 3, 5, 7 | 3 | 2, 5, 7 | 2, 5, 7 | 3 | 3 |
| | Carbide | 2, 5, 7 | 3 | 1, 4, 6 | 1, 4, 6 | 3 | 3 |
| 6.1 Power Hack Sawing | HSS | 2, 5, 7 | 3 | 1, 5, 7 | 1, 5, 7 | 3 | 3 |
| 6.2 Power Band Sawing | HSS | 2, 5, 7 | 3 | 1, 5, 7 | 1, 5, 7 | 3 | 3 |
| 6.4 Circular Sawing | HSS | 3, 5, 7 | 3 | 2, 5, 7 | 2, 5, 7 | 3 | 3 |
| 6.5 Circular Sawing | Carbide | 2, 5, 7 | 3 | 1, 4, 6 | 1, 4, 6 | 2, 5, 7 | 3, 5, 7 |
| 6.6 Abrasive Cutoff | — | 2 | 3 | 5, 7 | 5, 7 | 5, 7 | 5, 7 |
| 7.1 Gear Hobbing | HSS | 3, 5, 7 | 3 | 2, 3 | 2, 3 | 3 | 3 |
| 7.2 Gear Cutting, Straight and Spiral Bevel | HSS | 3, 5, 7 | 3 | 2, 3 | 2, 3 | 3 | 3 |
| 7.3 Gear Shaping | HSS | 3, 5, 7 | 3 | 2, 3 | 2, 3 | 3 | 3 |
| 7.4 Gear Shaving | HSS | 3, 5, 7 | 3 | 2, 3 | 2, 3 | 3 | 3 |
| 7.5 Gear Grinding, Form | — | 1, 4, 6 | 3 | 5, 7 | 5, 7 | 2, 5, 7 | 2, 5, 7 |
| 8.1 Surface Grinding—Horiz. Spdl., Recip. Table | — | 2, 4, 6 | 3 | 5, 7 | 5, 7 | 5, 7 | 5, 7 |
| 8.2 Surface Grinding—Horiz. Spdl., Recip. Table | CBN | — | 5, 4 | — | — | — | 5, 4 |
| 8.3 Surface Grinding—Horiz. Spdl., Recip. Table | Diamond | — | — | — | — | — | 5 |
| 8.4 Surface Grinding—Vertical Spdl., Rotary Table | — | 2, 4, 6 | 3 | 5, 7 | 5, 7 | 5, 7 | 5, 7 |
| 8.5 Cylindrical Grinding | — | 2, 4, 6 | 3 | 5, 7 | 5, 7 | 5, 7 | 5, 7 |
| 8.6 Cylindrical Grinding | CBN | — | 5, 4 | — | — | — | 5, 4 |
| 8.7 Cylindrical Grinding | Diamond | — | — | — | — | — | — |
| 8.8 Internal Grinding | — | 2, 4, 6 | 3 | 5, 7 | 5, 7 | 5, 7 | 5, 7 |
| 8.9 Internal Grinding | CBN | — | 5, 4 | — | — | — | 5, 4 |
| 8.10 Internal Grinding | Diamond | — | — | — | — | — | — |
| 8.11 Centerless Grinding | — | 2, 4, 6 | 3 | 5, 7 | 5, 7 | 5, 7 | 5, 7 |
| 8.13 Abrasive Belt Grinding | — | 1, 4, 6 | 2, 5, 7 | 1, 5, 7 | 1, 5, 7 | 1, 5, 7 | 1, 5, 7 |
| 8.14 Thread Grinding | — | 2, 4, 6 | 3 | 5, 7 | 5, 7 | 5, 7 | 5, 7 |
| | | | | | | | |

## 16.2 Cutting Fluid Recommendations

| OPERATION | TOOL MATERIAL OR ABRASIVE | 9. Nitriding Steels, Wrought | | 10. Armor Plate, Ship Plate & Aircraft Plate, Wrought | | 11. Structural Steels, Wrought | |
|---|---|---|---|---|---|---|---|
| | | 200 to 350 Bhn | 60 to 65 $R_C$ | 200 to 350 Bhn | 40 to 45 $R_C$ | 100 to 300 Bhn | Over 300 Bhn |
| 1.1 Turning, Single Point and Box Tools | HSS | 1, 4, 6 | — | 2, 5, 7 | 3, 5, 7 | 1, 4, 6 | 2, 5, 7 |
| | Carbide | 0, 4, 6 | 0, 4, 6 | 0, 4, 6 | 0, 4, 6 | 0, 4, 6 | 0, 4, 6 |
| 1.3 Turning | Diamond | — | — | — | — | — | — |
| 1.4 Turning, Cutoff and Form Tools | HSS | 2, 5, 7 | — | 2, 5, 7 | 2, 5, 7 | 2, 5, 7 | 2, 5, 7 |
| | Carbide | 4, 6 | 5, 7 | 5, 7 | 5, 7 | 0, 4, 6 | 5, 7 |
| 1.5 Threading, Single Point | HSS | 3 | — | 3 | 3 | 2, 5 | 2, 5 |
| | Carbide | 1, 4, 6 | 1, 4, 6 | 1, 4, 6 | 1, 4, 6 | 1, 4, 6 | 1, 4, 6 |
| 1.6 Threading, Die | HSS | 3 | — | 3 | 3 | 2, 5 | 2, 5 |
| | Carbide | 1, 4, 6 | 1, 4, 6 | 0, 4, 6 | 1, 4, 6 | 0, 4, 6 | 0, 4, 6 |
| 1.7 Hollow Milling | HSS | 3 | — | 3 | 3 | 2, 5 | 2, 5 |
| | Carbide | 2, 5, 7 | 2, 5, 7 | 0, 4, 6 | 2, 5, 7 | 0, 4, 6 | 0, 4, 6 |
| 2.1 Face Milling | HSS | 1, 4, 6 | — | 3, 5, 7 | 3 | 1, 5, 7 | 2, 5, 7 |
| | Carbide | 0, 4, 6 | 0, 2 | 0, 4, 6 | 0, 2 | 0, 4, 6 | 0, 4, 6 |
| 2.2 Face Milling | Diamond | — | — | — | — | — | — |
| 2.4 Slab Milling | HSS | 1, 4, 6 | — | 3, 5, 7 | 3 | 1, 4, 6 | 2, 5, 7 |
| | Carbide | 0, 4, 6 | 2 | 0, 4, 6 | 2 | 0, 4, 6 | 1, 4, 6 |
| 2.5 Side and Slot Milling—Arbor Mounted Cutters | HSS | 1, 4, 6 | — | 3, 5, 7 | 3 | 1, 4, 6 | 2, 5, 7 |
| | Carbide | 0, 4, 6 | 2 | 0, 4, 6 | 2 | 0, 4, 6 | 1, 4, 6 |
| 2.6 End Milling—Peripheral | HSS | 1, 4, 6 | — | 2, 5, 7 | 3 | 1, 4, 6 | 2, 5, 7 |
| | Carbide | 0, 4, 6 | 3 | 0, 4, 6 | 3 | 0, 4, 6 | 0, 4, 6 |
| 2.8 End Milling—Slotting | HSS | 1, 4, 6 | — | 2, 5, 7 | 3 | 1, 4, 6 | 2, 5, 7 |
| | Carbide | 0, 4, 6 | 3 | 0, 4, 6 | 3 | 0, 4, 6 | 0, 4, 6 |
| 2.9 Thread Milling | HSS | 2, 5, 7 | — | 7 | 3 | 2, 5, 7 | 3, 5, 7 |
| 3.1 Drilling | HSS | 1, 4, 6 | — | 2, 5, 7 | 3 | 1, 4, 6 | 2, 5, 7 |
| | Carbide | 0, 4, 6 | 3 | 0, 4, 6 | 3 | 0, 4, 6 | 2, 5, 7 |
| 3.3 Oil-Hole or Pressurized-Coolant Drilling | HSS | 1, 4, 6 | — | 2, 5, 7 | 3 | 1, 4, 6 | 2, 5, 7 |
| | Carbide | 4, 6 | 2 | 4, 6 | 2 | 4, 6 | 2, 5, 7 |
| 3.4 Spade Drilling | HSS | 1, 4, 6 | — | 2, 5, 7 | 3 | 1, 4, 6 | 1, 4, 6 |
| | Carbide | 0, 4, 6 | 2 | 0, 4, 6 | 2 | 0, 4, 6 | 0, 4, 6 |
| 3.5 Gundrilling | HSS | 1, 4, 6 | — | 2, 5, 7 | 3 | 1, 4, 6 | 2, 5, 7 |
| | Carbide | 1, 4, 6 | 2, 3 | 2, 5, 7 | 2, 3 | 1, 4, 6 | 2, 5, 7 |
| 3.6 Pressure Coolant Reaming (Gun Reaming) | HSS | 1, 5, 7 | — | 2, 5, 7 | 3 | 1, 4, 6 | 2, 5, 7 |
| | Carbide | 1, 4, 6 | 2, 5, 7 | 1, 4, 6 | 2, 5, 7 | 1, 4, 6 | 1, 4, 6 |
| 3.7 Reaming | HSS | 1, 4, 6 | — | 2, 5, 7 | 3 | 1, 4, 6 | 2, 5, 7 |
| | Carbide | 1, 4, 6 | 2, 3 | 1, 4, 6 | 2, 3 | 1, 4, 6 | 1, 4, 6 |
| 3.9 Boring | HSS | 1, 4, 6 | — | 1, 4, 6 | 3, 5, 7 | 1, 4, 6 | 2, 5, 7 |
| | Carbide | 2, 4, 6 | 2, 4, 6 | 2, 4, 6 | 2, 4, 6 | 0, 4, 6 | 4, 6 |
| 3.11 Boring | Diamond | — | — | — | — | — | — |
| 3.12 Counterboring and Spotfacing | HSS | 2, 5, 7 | — | 2, 5, 7 | 3 | 2, 5, 6 | 2, 5, 7 |
| | Carbide | 4, 6 | 3, 7 | 1, 4, 6 | 3, 7 | 0, 4, 6 | 1, 4, 6 |
| 3.13 Trepanning | HSS | 1, 4, 6 | — | 2, 5, 7 | 3 | 1, 4, 6 | 2, 5, 7 |
| | Carbide | 1, 4, 6 | 2, 3 | 2, 5, 7 | 2, 3 | 1, 4, 6 | 1, 5, 7 |
| 3.14 Honing | — | 90 | 90 | 90 | 90 | 90 | 90 |
| 3.15 Burnishing | — | 2 | — | 2 | — | 1, 4, 6 | 2, 5, 7 |
| 4.1 Tapping | HSS | 2, 5, 7 | — | 3, 5, 7 | 3 | 2, 5, 7 | 3, 5, 7 |

| OPERATION | TOOL MATERIAL OR ABRASIVE | 9. Nitriding Steels, Wrought | | 10. Armor Plate, Ship Plate & Aircraft Plate, Wrought | | 11. Structural Steels, Wrought | |
|---|---|---|---|---|---|---|---|
| | | 200 to 350 Bhn | 60 to 65 $R_C$ | 200 to 350 Bhn | 40 to 45 $R_C$ | 100 to 300 Bhn | Over 300 Bhn |
| 5.1 Planing | HSS | 2, 5 | — | 2, 5, 7 | 3 | 1, 4, 6 | 2, 5, 7 |
| | Carbide | 4, 6 | 2 | 0, 4, 6 | 2 | 0, 4, 6 | 0, 4, 6 |
| 5.2 Broaching | HSS | 2, 5, 7 | 3 | 3, 5, 7 | 3, 5, 7 | 2, 5, 7 | 2, 5, 7 |
| | Carbide | 4, 6 | 3 | 1, 5, 7 | 1, 5, 7 | 1, 5, 7 | 2, 5, 7 |
| 6.1 Power Hack Sawing | HSS | 1, 4, 6 | 3 | 2, 5, 7 | 3, 5, 7 | 1, 4, 6 | 2, 5, 7 |
| 6.2 Power Band Sawing | HSS | 1, 4, 6 | 3 | 2, 5, 7 | 3, 5, 7 | 1, 4, 6 | 2, 5, 7 |
| 6.4 Circular Sawing | HSS | 2, 5, 7 | 3 | 3, 5, 7 | 3, 5, 7 | 1, 4, 6 | 3, 5, 7 |
| 6.5 Circular Sawing | Carbide | 1, 4, 6 | 2, 3 | 1, 4, 6 | 1, 4, 6 | 1, 4, 6 | 2, 5, 7 |
| 6.6 Abrasive Cutoff | — | 1, 4, 6 | 3 | 4, 6 | 4, 6 | 1, 4, 6 | 2 |
| 7.1 Gear Hobbing | HSS | 2, 5, 7 | 3 | 7 | 7 | 2, 5, 7 | 3, 5, 7 |
| 7.2 Gear Cutting, Straight and Spiral Bevel | HSS | 2, 5, 7 | 3 | 7 | 7 | 2, 5, 7 | 3, 5, 7 |
| 7.3 Gear Shaping | HSS | 2, 5, 7 | 3 | 7 | 7 | 2, 5, 7 | 3, 5, 7 |
| 7.4 Gear Shaving | HSS | 2, 5, 7 | 3 | 7 | 7 | 2, 5, 7 | 3, 5, 7 |
| 7.5 Gear Grinding, Form | — | 2, 5, 7 | 3 | 2, 5 | 3, 5 | 1, 4, 6 | 1, 4, 6 |
| 8.1 Surface Grinding—Horiz. Spdl., Recip. Table | — | 4, 6 | 3 | 4, 6 | 4, 6 | 1, 4, 6 | 2, 4, 6 |
| 8.2 Surface Grinding—Horiz. Spdl., Recip. Table | CBN | — | 5, 4 | — | — | — | — |
| 8.3 Surface Grinding—Horiz. Spdl., Recip. Table | Diamond | — | — | — | — | — | — |
| 8.4 Surface Grinding—Vertical Spdl., Rotary Table | — | 4, 6 | 3 | 4, 6 | 4, 6 | 1, 4, 6 | 2, 4, 6 |
| 8.5 Cylindrical Grinding | — | 4, 6 | 3 | 4, 6 | 4, 6 | 1, 4, 6 | 2, 4, 6 |
| 8.6 Cylindrical Grinding | CBN | — | 5, 4 | — | — | — | — |
| 8.7 Cylindrical Grinding | Diamond | — | — | — | — | — | — |
| 8.8 Internal Grinding | — | 4, 6 | 3 | 4, 6 | 4, 6 | 1, 4, 6 | 2, 4, 6 |
| 8.9 Internal Grinding | CBN | — | 5, 4 | — | — | — | — |
| 8.10 Internal Grinding | Diamond | — | — | — | — | — | — |
| 8.11 Centerless Grinding | — | 4, 6 | 3 | 4, 6 | 4, 6 | 1, 4, 6 | 2, 4, 6 |
| 8.13 Abrasive Belt Grinding | — | 0, 1, 4 | 1, 4, 6 | 0, 1, 4 | 0, 1, 4 | 0, 1, 4 | 2, 5, 7 |
| 8.14 Thread Grinding | — | 4, 6 | 3 | 7 | 7 | 1, 4, 6 | 2, 4, 6 |
| | | | | | | | |

# 16.2 Cutting Fluid Recommendations

| OPERATION | TOOL MATERIAL OR ABRASIVE | 12. Free Machining Stainless Steels, Wrought | | | 13. Stainless Steels, Wrought — Ferritic and Austenitic | | |
|---|---|---|---|---|---|---|---|
| | | 135 to 275 Bhn | Over 275 Bhn | — | 135 to 275 Bhn | 275 to 375 Bhn | — |
| 1.1 Turning, Single Point and Box Tools | HSS | 2, 5, 7 | 2, 5, 7 | | 2, 5, 6 | 2, 5, 7 | |
| | Carbide | 0, 4, 6 | 0, 4, 6 | | 0, 4, 6 | 0, 4, 6 | |
| 1.3 Turning | Diamond | — | — | | — | — | |
| 1.4 Turning, Cutoff and Form Tools | HSS | 2, 5, 7 | 2, 5, 7 | | 2, 5, 7 | 2 | |
| | Carbide | 1, 4, 6 | 1, 4, 6 | | 4, 6 | 2, 5, 7 | |
| 1.5 Threading, Single Point | HSS | 2, 5 | 2, 5 | | 3 | 3 | |
| | Carbide | 1, 4, 6 | 1, 4, 6 | | 2, 5, 7 | 2, 5, 7 | |
| 1.6 Threading, Die | HSS | 2, 5 | 2, 5 | | 3 | 3 | |
| | Carbide | 1, 4, 6 | 1, 4, 6 | | 2, 5, 7 | 2, 5, 7 | |
| 1.7 Hollow Milling | HSS | 2, 5 | 2, 5 | | 3 | 3 | |
| | Carbide | 0, 4, 6 | 0, 4, 6 | | 2, 5, 7 | 2, 5, 7 | |
| 2.1 Face Milling | HSS | 1, 4, 6 | 1, 4, 6 | | 2, 5, 7 | 2, 5, 7 | |
| | Carbide | 0, 4, 6 | 0, 4, 6 | | 0, 4, 6 | 0, 4, 6 | |
| 2.2 Face Milling | Diamond | — | — | | — | — | |
| 2.4 Slab Milling | HSS | 1, 4, 6 | 1, 4, 6 | | 2, 5, 7 | 2, 5, 7 | |
| | Carbide | 1, 4, 6 | 1, 4, 6 | | 1, 4, 6 | 1, 5, 7 | |
| 2.5 Side and Slot Milling—Arbor Mounted Cutters | HSS | 1, 4, 6 | 1, 4, 6 | | 2, 5, 7 | 2, 5, 7 | |
| | Carbide | 1, 4, 6 | 1, 4, 6 | | 1, 4, 6 | 1, 5, 7 | |
| 2.6 End Milling—Peripheral | HSS | 1, 4, 6 | 1, 4, 6 | | 2, 5, 7 | 2, 5, 7 | |
| | Carbide | 0, 4, 6 | 0, 4, 6 | | 0, 4, 6 | 0, 4, 6 | |
| 2.8 End Milling—Slotting | HSS | 1, 4, 6 | 1, 4, 6 | | 2, 5, 7 | 2, 5, 7 | |
| | Carbide | 0, 4, 6 | 0, 4, 6 | | 0, 4, 6 | 0, 4, 6 | |
| 2.9 Thread Milling | HSS | 2 | 2 | | 3 | 3 | |
| 3.1 Drilling | HSS | 1, 4, 6 | 1, 4, 6 | | 2, 5, 7 | 2, 5, 7 | |
| | Carbide | 1, 4, 6 | 1, 4, 6 | | 1, 4, 6 | 1, 5, 7 | |
| 3.3 Oil-Hole or Pressurized-Coolant Drilling | HSS | 1, 4, 6 | 1, 4, 6 | | 2, 5, 7 | 2, 5, 7 | |
| | Carbide | 1, 4, 6 | 1, 4, 6 | | 1, 4, 6 | 1, 4, 6 | |
| 3.4 Spade Drilling | HSS | 1, 4, 6 | 1, 4, 6 | | 2, 5, 7 | 2, 5, 7 | |
| | Carbide | 1, 4, 6 | 1, 4, 6 | | 1, 4, 6 | 1, 4, 6 | |
| 3.5 Gundrilling | HSS | 2, 5, 7 | 2, 5, 7 | | 2, 5, 7 | 2, 5, 7 | |
| | Carbide | 1, 4, 6 | 1, 4, 6 | | 2, 5, 7 | 2, 5, 7 | |
| 3.6 Pressure Coolant Reaming (Gun Reaming) | HSS | 2 | 2 | | 3 | 3 | |
| | Carbide | 1, 4, 6 | 1, 4, 6 | | 2, 5, 7 | 2, 5, 7 | |
| 3.7 Reaming | HSS | 2 | 2 | | 3 | 3 | |
| | Carbide | 1, 4, 6 | 1, 4, 6 | | 2, 5, 7 | 2, 5, 7 | |
| 3.9 Boring | HSS | 1, 5, 6 | 1, 5, 6 | | 2, 5, 7 | 2, 5, 7 | |
| | Carbide | 0, 4, 6 | 0, 4, 6 | | 1, 4, 6 | 2, 5, 7 | |
| 3.11 Boring | Diamond | — | — | | — | — | |
| 3.12 Counterboring and Spotfacing | HSS | 1, 5, 6 | 1, 5, 6 | | 2, 5, 7 | 2, 5, 7 | |
| | Carbide | 1, 4, 6 | 1, 4, 6 | | 2, 5, 7 | 2, 5, 7 | |
| 3.13 Trepanning | HSS | 3, 5, 7 | 3, 5, 7 | | 3, 5, 7 | 3, 5, 7 | |
| | Carbide | 1, 4, 6 | 1, 4, 6 | | 2, 5, 7 | 2, 5, 7 | |
| 3.14 Honing | — | 90 | 90 | | 90 | 90 | |
| 3.15 Burnishing | — | 2 | 2 | | 2 | 2 | |
| 4.1 Tapping | HSS | 2 | 3 | | 3 | 3 | |

| OPERATION | TOOL MATERIAL OR ABRASIVE | 12. Free Machining Stainless Steels, Wrought | | | 13. Stainless Steels, Wrought Ferritic and Austenitic | | |
|---|---|---|---|---|---|---|---|
| | | 135 to 275 Bhn | Over 275 Bhn | — | 135 to 275 Bhn | 275 to 375 Bhn | — |
| **5.1 Planing** | HSS | 1, 4, 6 | 1, 4, 6 | | 2, 5, 7 | 2, 5, 7 | |
| | Carbide | 1, 4, 6 | 1, 4, 6 | | 1, 4, 6 | 1, 4, 6 | |
| **5.2 Broaching** | HSS | 2, 5, 7 | 3 | | 2, 5, 7 | 3 | |
| | Carbide | 1, 4, 6 | 2, 3 | | 1, 4, 6 | 2, 3 | |
| **6.1 Power Hack Sawing** | HSS | 1, 5, 7 | 3, 5, 7 | | 1, 5, 7 | 3, 5, 7 | |
| **6.2 Power Band Sawing** | HSS | 1, 5, 7 | 3, 5, 7 | | 1, 5, 7 | 3, 5, 7 | |
| **6.4 Circular Sawing** | HSS | 2, 5, 7 | 3, 5, 7 | | 2, 5, 7 | 3, 5, 7 | |
| **6.5 Circular Sawing** | Carbide | 1, 4, 6 | 4, 6 | | 1, 4, 6 | 4, 6 | |
| **6.6 Abrasive Cutoff** | — | 5, 7 | 2, 5, 7 | | 5, 7 | 2, 5, 7 | |
| **7.1 Gear Hobbing** | HSS | 2, 3 | 2 | | 2, 3 | 2 | |
| **7.2 Gear Cutting, Straight and Spiral Bevel** | HSS | 2, 3 | 2 | | 2, 3 | 2 | |
| **7.3 Gear Shaping** | HSS | 2, 3 | 2 | | 2, 3 | 2 | |
| **7.4 Gear Shaving** | HSS | 2, 3 | 2 | | 2, 3 | 2 | |
| **7.5 Gear Grinding, Form** | — | 2, 3 | 2 | | 5, 7 | 5, 7 | |
| **8.1 Surface Grinding—Horiz. Spdl., Recip. Table** | — | 5, 7 | 1, 4, 6 | | 5, 7 | 5, 7 | |
| **8.2 Surface Grinding—Horiz. Spdl., Recip. Table** | CBN | — | — | | — | — | |
| **8.3 Surface Grinding—Horiz. Spdl., Recip. Table** | Diamond | — | — | | — | — | |
| **8.4 Surface Grinding—Vertical Spdl., Rotary Table** | — | 5, 7 | 1, 4, 6 | | 5, 7 | 5, 7 | |
| **8.5 Cylindrical Grinding** | — | 5, 7 | 1, 4, 6 | | 5, 7 | 5, 7 | |
| **8.6 Cylindrical Grinding** | CBN | — | — | | — | — | |
| **8.7 Cylindrical Grinding** | Diamond | — | — | | — | — | |
| **8.8 Internal Grinding** | — | 5, 7 | 1, 4, 6 | | 5, 7 | 5, 7 | |
| **8.9 Internal Grinding** | CBN | — | — | | — | — | |
| **8.10 Internal Grinding** | Diamond | — | — | | — | — | |
| **8.11 Centerless Grinding** | — | 5, 7 | 1, 4, 6 | | 5, 7 | 5, 7 | |
| **8.13 Abrasive Belt Grinding** | — | 0, 2, 5 | 0, 1, 4 | | 0, 2, 5 | 0, 2, 5 | |
| **8.14 Thread Grinding** | — | 5, 7 | 5, 7 | | 5, 7 | 5, 7 | |
| | | | | | | | |

# 16.2 Cutting Fluid Recommendations

| OPERATION | | TOOL MATERIAL OR ABRASIVE | 13. Stainless Steels, Wrought | | | 14. Precipitation Hardening Stainless Steels, Wrought | | |
|---|---|---|---|---|---|---|---|---|
| | | | Martensitic | | | | | |
| | | | 135 to 275 Bhn | 275 to 425 Bhn | 48 to 56 $R_C$ | — | 150 to 440 Bhn | — |
| 1.1 | Turning, Single Point and Box Tools | HSS | 2, 5, 6 | 2, 5, 7 | 3, 5, 7 | | 2, 5, 7 | |
| | | Carbide | 0, 4, 6 | 0, 4, 6 | 0, 4, 6 | | 0, 4, 6 | |
| 1.3 | Turning | Diamond | — | — | — | | — | |
| 1.4 | Turning, Cutoff and Form Tools | HSS | 2, 5, 7 | 2 | 3 | | 2 | |
| | | Carbide | 4, 6 | 1, 5, 7 | 5, 6, 7 | | 1, 5, 7 | |
| 1.5 | Threading, Single Point | HSS | 3 | 3 | 3 | | 3 | |
| | | Carbide | 2, 5, 7 | 1, 5, 7 | 2, 5, 7 | | 1, 5, 7 | |
| 1.6 | Threading, Die | HSS | 3 | 3 | 3 | | 3 | |
| | | Carbide | 2, 5, 7 | 1, 5, 7 | 2, 5, 7 | | 1, 5, 7 | |
| 1.7 | Hollow Milling | HSS | 2 | 2 | 3 | | 2 | |
| | | Carbide | 0, 1, 4, 6 | 1, 5, 7 | 2, 5, 7 | | 1, 5, 7 | |
| 2.1 | Face Milling | HSS | 2, 5, 7 | 1, 5, 7 | 3 | | 1, 5, 7 | |
| | | Carbide | 0, 4, 6 | 0, 4, 6 | 0, 5, 7 | | 0, 4, 6 | |
| 2.2 | Face Milling | Diamond | — | — | — | | — | |
| 2.4 | Slab Milling | HSS | 2, 5, 7 | 2, 5, 7 | 3 | | 2, 5, 7 | |
| | | Carbide | 1, 4, 6 | 1, 5, 7 | 3, 5, 7 | | 1, 5, 7 | |
| 2.5 | Side and Slot Milling—Arbor Mounted Cutters | HSS | 2, 5, 7 | 2, 5, 7 | 3 | | 2, 5, 7 | |
| | | Carbide | 1, 4, 6 | 1, 5, 7 | 2, 5, 7 | | 1, 5, 7 | |
| 2.6 | End Milling—Peripheral | HSS | 2, 5, 7 | 2, 5, 7 | 3 | | 2, 5, 7 | |
| | | Carbide | 0, 4, 6 | 0, 4, 6 | 3, 5, 7 | | 0, 4, 6 | |
| 2.8 | End Milling—Slotting | HSS | 2, 5, 7 | 2, 5, 7 | 3 | | 2, 5, 7 | |
| | | Carbide | 0, 4, 6 | 0, 4, 6 | 3, 5, 7 | | 0, 4, 6 | |
| 2.9 | Thread Milling | HSS | 3 | 2 | 3 | | 2 | |
| 3.1 | Drilling | HSS | 2, 5, 7 | 2 | 3 | | 2 | |
| | | Carbide | 1, 4, 6 | 1, 5, 7 | 3, 5, 7 | | 1, 5, 7 | |
| 3.3 | Oil-Hole or Pressurized-Coolant Drilling | HSS | 2, 5, 7 | 2, 5, 7 | 3 | | 2 | |
| | | Carbide | 1, 4, 6 | 1, 4, 6 | 2, 5, 7 | | 2, 5, 7 | |
| 3.4 | Spade Drilling | HSS | 2, 5, 7 | 2 | 3 | | 2 | |
| | | Carbide | 1, 4, 6 | 2, 5, 7 | 2, 5, 7 | | 2, 5, 7 | |
| 3.5 | Gundrilling | HSS | 2, 5, 7 | 2 | 3 | | 2 | |
| | | Carbide | 2, 5, 7 | 2, 5, 7 | 3, 5, 7 | | 1, 5, 7 | |
| 3.6 | Pressure Coolant Reaming (Gun Reaming) | HSS | 3 | 3 | 3 | | 3 | |
| | | Carbide | 2, 5, 7 | 2, 5, 7 | 3, 5, 7 | | 2, 4, 6 | |
| 3.7 | Reaming | HSS | 3 | 3 | 3 | | 3 | |
| | | Carbide | 1, 5, 7 | 2, 5, 7 | 3, 5, 7 | | 2, 4, 6 | |
| 3.9 | Boring | HSS | 2, 5, 7 | 2, 5, 7 | 3, 5, 7 | | 2, 5, 7 | |
| | | Carbide | 1, 4, 6 | 2 | 2, 5, 7 | | 2 | |
| 3.11 | Boring | Diamond | — | — | — | | — | |
| 3.12 | Counterboring and Spotfacing | HSS | 2, 5, 7 | 2, 5, 7 | 3 | | 2, 5, 7 | |
| | | Carbide | 2, 5, 7 | 2, 5, 7 | 3, 5, 7 | | 2, 5, 7 | |
| 3.13 | Trepanning | HSS | 3, 5, 7 | 2 | 3 | | 2 | |
| | | Carbide | 2, 5, 7 | 2, 5, 7 | 3, 5, 7 | | 2, 5, 7 | |
| 3.14 | Honing | — | 90 | 90 | 90 | | 90 | |
| 3.15 | Burnishing | — | 2 | 2 | 2 | | 2 | |
| 4.1 | Tapping | HSS | 3 | 2 | 3 | | 2 | |

| OPERATION | TOOL MATERIAL OR ABRASIVE | 13. Stainless Steels, Wrought | | | 14. Precipitation Hardening Stainless Steels, Wrought | | |
|---|---|---|---|---|---|---|---|
| | | Martensitic | | | | | |
| | | 135 to 275 Bhn | 275 to 425 Bhn | 48 to 56 $R_C$ | — | 150 to 440 Bhn | — |
| 5.1 Planing | HSS | 2, 5, 7 | 2, 5, 7 | 3 | | 2, 5, 7 | |
| | Carbide | 1, 4, 6 | 0, 4, 6 | 3, 5, 7 | | 0, 4, 6 | |
| 5.2 Broaching | HSS | 2, 5, 7 | 3 | 3 | | 3 | |
| | Carbide | 1, 4, 6 | 2, 3 | 3 | | 2, 3 | |
| 6.1 Power Hack Sawing | HSS | 1, 5, 7 | 3, 5, 7 | 3 | | 3, 5, 7 | |
| 6.2 Power Band Sawing | HSS | 1, 5, 7 | 3, 5, 7 | 3 | | 3, 5, 7 | |
| 6.4 Circular Sawing | HSS | 2, 5, 7 | 3, 5, 7 | 3 | | 3, 5, 7 | |
| 6.5 Circular Sawing | Carbide | 1, 4, 6 | 4, 6 | 3, 5, 7 | | 4, 6 | |
| 6.6 Abrasive Cutoff | — | 5, 7 | 2, 5, 7 | 5, 7 | | 2, 5, 7 | |
| 7.1 Gear Hobbing | HSS | 2, 3 | 2 | 3 | | 2 | |
| 7.2 Gear Cutting, Straight and Spiral Bevel | HSS | 2, 3 | 2 | 3 | | 2 | |
| 7.3 Gear Shaping | HSS | 2, 3 | 2 | 3 | | 2 | |
| 7.4 Gear Shaving | HSS | 2, 3 | 2 | 3 | | 2 | |
| 7.5 Gear Grinding, Form | — | 5, 7 | 2 | 2, 5, 7 | | 2 | |
| 8.1 Surface Grinding—Horiz. Spdl., Recip. Table | — | 5, 7 | 1, 4, 6 | 5, 7 | | 1, 4, 6 | |
| 8.2 Surface Grinding—Horiz. Spdl., Recip. Table | CBN | — | — | 5, 4 | | — | |
| 8.3 Surface Grinding—Horiz. Spdl., Recip. Table | Diamond | — | — | — | | — | |
| 8.4 Surface Grinding—Vertical Spdl., Rotary Table | — | 5, 7 | 1, 4, 6 | 5, 7 | | 1, 4, 6 | |
| 8.5 Cylindrical Grinding | — | 5, 7 | 1, 4, 6 | 5, 7 | | 1, 4, 6 | |
| 8.6 Cylindrical Grinding | CBN | — | — | 5, 4 | | — | |
| 8.7 Cylindrical Grinding | Diamond | — | — | — | | — | |
| 8.8 Internal Grinding | — | 5, 7 | 1, 4, 6 | 5, 7 | | 1, 4, 6 | |
| 8.9 Internal Grinding | CBN | — | — | 5, 4 | | — | |
| 8.10 Internal Grinding | Diamond | — | — | — | | — | |
| 8.11 Centerless Grinding | — | 5, 7 | 1, 4, 6 | 5, 7 | | 1, 4, 6 | |
| 8.13 Abrasive Belt Grinding | — | 0, 1, 4 | 0, 1, 5 | 1, 5, 7 | | 1, 5, 7 | |
| 8.14 Thread Grinding | — | 5, 7 | 1, 4, 6 | 5, 7 | | 1, 4, 6 | |
| | | | | | | | |

## 16.2 Cutting Fluid Recommendations

| OPERATION | TOOL MATERIAL OR ABRASIVE | 15. Carbon Steels, Cast | | | 16. Alloy Steels, Cast | | |
|---|---|---|---|---|---|---|---|
| | | 100 to 300 Bhn | Over 48 $R_C$ | — | 150 to 250 Bhn | 250 to 400 Bhn | Over 48 $R_C$ |
| 1.1 Turning, Single Point and Box Tools | HSS | 1, 4, 6 | 2, 5, 7 | | 1, 4, 6 | 2, 5, 7 | 3, 5, 7 |
| | Carbide | 0, 4, 6 | 0, 4, 6 | | 0, 4, 6 | 0, 4, 6 | 0, 4, 6 |
| 1.3 Turning | Diamond | — | — | | — | — | — |
| 1.4 Turning, Cutoff and Form Tools | HSS | 2, 5, 7 | 3, 5, 7 | | 2, 5, 7 | 2, 5, 7 | 2, 5, 7 |
| | Carbide | 0, 4, 6 | 0, 4, 6 | | 0, 4, 6 | 5, 7 | 5, 7 |
| 1.5 Threading, Single Point | HSS | 2, 5 | 3, 6 | | 2, 5 | 2, 5 | 3 |
| | Carbide | 1, 4, 6 | 1, 4, 6 | | 1, 4, 6 | 1, 4, 6 | 1, 4, 6 |
| 1.6 Threading, Die | HSS | 2, 5 | 3, 6 | | 2, 5 | 2, 5 | 3 |
| | Carbide | 1, 4, 6 | 1, 4, 6 | | 1, 4, 6 | 1, 4, 6 | 1, 4, 6 |
| 1.7 Hollow Milling | HSS | 1, 4, 6 | 2, 5, 7 | | 1, 5, 7 | 2, 5, 7 | 3 |
| | Carbide | 0, 4, 6 | 0, 4, 6 | | 0, 4, 6 | 1, 4, 6 | 2, 5, 7 |
| 2.1 Face Milling | HSS | 1, 4, 6 | 2, 5, 7 | | 1, 5, 7 | 2, 5, 7 | 3 |
| | Carbide | 0, 1, 4 | 0, 1, 4 | | 0, 4, 6 | 0, 4, 6 | 0, 2 |
| 2.2 Face Milling | Diamond | — | — | | — | — | — |
| 2.4 Slab Milling | HSS | 1, 4, 6 | 2, 5, 7 | | 1, 4, 6 | 2, 5, 7 | 3 |
| | Carbide | 0, 4, 6 | 0, 4, 6 | | 0, 4, 6 | 1, 4, 6 | 2 |
| 2.5 Side and Slot Milling—Arbor Mounted Cutters | HSS | 1, 4, 6 | 2, 5, 7 | | 1, 5, 7 | 2, 5, 7 | 3 |
| | Carbide | 0, 4, 6 | 0, 4, 6 | | 0, 4, 6 | 1, 4, 6 | 2 |
| 2.6 End Milling—Peripheral | HSS | 1, 4, 6 | 2, 5, 7 | | 1, 4, 6 | 2, 5, 7 | 3 |
| | Carbide | 0, 4, 6 | 0, 4, 6 | | 0, 4, 6 | 0, 4, 6 | 3 |
| 2.8 End Milling—Slotting | HSS | 1, 4, 6 | 2, 5, 7 | | 1, 4, 6 | 2, 5, 7 | 3 |
| | Carbide | 0, 4, 6 | 0, 4, 6 | | 0, 4, 6 | 0, 4, 6 | 3 |
| 2.9 Thread Milling | HSS | 2, 5, 7 | 3, 5, 7 | | 2, 5, 7 | 2, 5, 7 | 3 |
| 3.1 Drilling | HSS | 1, 4, 6 | 2, 5, 7 | | 1, 4, 6 | 2, 5, 7 | 3 |
| | Carbide | 0, 4, 6 | 0, 4, 6 | | 0, 4, 6 | 1, 5, 7 | 3 |
| 3.3 Oil-Hole or Pressurized-Coolant Drilling | HSS | 1, 4, 6 | 1, 4, 6 | | 1, 4, 6 | 1, 4, 6 | 3 |
| | Carbide | 4, 6 | 4, 6 | | 4, 6 | 4, 6 | 2 |
| 3.4 Spade Drilling | HSS | 1, 4, 6 | 1, 4, 6 | | 1, 4, 6 | 2, 5, 7 | 3 |
| | Carbide | 0, 4, 6 | 0, 4, 6 | | 0, 4, 6 | 1, 5, 7 | 2 |
| 3.5 Gundrilling | HSS | 1, 4, 6 | 2, 5, 7 | | 1, 4, 6 | 2, 5, 7 | 3 |
| | Carbide | 1, 4, 6 | 1, 4, 6 | | 1, 4, 6 | 2, 5, 7 | 2, 3 |
| 3.6 Pressure Coolant Reaming (Gun Reaming) | HSS | 1, 5, 6 | 2, 6, 7 | | 1, 4, 6 | 2, 5, 7 | 3 |
| | Carbide | 1, 4, 6 | 1, 4, 6 | | 1, 4, 6 | 1, 4, 6 | 2, 5, 7 |
| 3.7 Reaming | HSS | 1, 5, 6 | 2, 6, 7 | | 1, 4, 6 | 2, 5, 7 | 3 |
| | Carbide | 1, 4, 6 | 1, 4, 6 | | 1, 4, 6 | 1, 4, 6 | 2, 3 |
| 3.9 Boring | HSS | 1, 4, 6 | 2, 5, 7 | | 1, 4, 6 | 2, 5, 7 | 3, 5, 7 |
| | Carbide | 0, 4, 6 | 0, 4, 6 | | 0, 4, 6 | 4, 6 | 2, 4, 6 |
| 3.11 Boring | Diamond | — | — | | — | — | — |
| 3.12 Counterboring and Spotfacing | HSS | 1, 4, 6 | 2, 5, 7 | | 2, 5, 6 | 2, 5, 7 | 3 |
| | Carbide | 0, 4, 6 | 0, 4, 6 | | 0, 4, 6 | 1, 4, 6 | 3, 7 |
| 3.13 Trepanning | HSS | 1, 4, 6 | 2, 5, 7 | | 1, 4, 6 | 2, 5, 7 | 3 |
| | Carbide | 1, 4, 6 | 1, 4, 6 | | 1, 4, 6 | 1, 5, 7 | 2, 3 |
| 3.14 Honing | — | 80 | 80 | | 80 | 80 | 90 |
| 3.15 Burnishing | — | 1, 5, 7 | 2, 5, 7 | | 1, 5, 7 | 2, 5, 7 | — |
| 4.1 Tapping | HSS | 1, 5, 7 | 2, 5, 7 | | 2, 5, 7 | 3, 5, 7 | 3 |

| OPERATION | TOOL MATERIAL OR ABRASIVE | 15. Carbon Steels, Cast | | | 16. Alloy Steels, Cast | | |
|---|---|---|---|---|---|---|---|
| | | 100 to 300 Bhn | Over 48 $R_C$ | — | 150 to 250 Bhn | 250 to 400 Bhn | Over 48 $R_C$ |
| 5.1 Planing | HSS | 1, 4, 6 | 2, 5, 7 | | 1, 4, 6 | 2, 5, 7 | 3 |
| | Carbide | 0, 1, 4 | 0, 1, 4 | | 0, 4, 6 | 0, 4, 6 | 2 |
| 5.2 Broaching | HSS | 2, 5, 7 | 3 | | 2, 5, 7 | 3, 5, 7 | 3 |
| | Carbide | 1, 4, 6 | 3 | | 1, 4, 6 | 1, 4, 6 | 3 |
| 6.1 Power Hack Sawing | HSS | 1, 4, 6 | 3 | | 1, 4, 6 | 2, 5, 7 | 3 |
| 6.2 Power Band Sawing | HSS | 1, 4, 6 | 3 | | 1, 4, 6 | 2, 5, 7 | 3 |
| 6.4 Circular Sawing | HSS | 2, 5, 7 | 3 | | 2, 5, 7 | 3, 5, 7 | 3 |
| 6.5 Circular Sawing | Carbide | 1, 4, 6 | 2, 3 | | 1, 4, 6 | 2, 5, 7 | 2, 3 |
| 6.6 Abrasive Cutoff | — | 1, 4, 6 | 3 | | 1, 4, 6 | 2 | 3 |
| 7.1 Gear Hobbing | HSS | 2, 5, 7 | 3 | | 2, 5, 7 | 3, 5, 7 | 3 |
| 7.2 Gear Cutting, Straight and Spiral Bevel | HSS | 2, 5, 7 | 3 | | 2, 5, 7 | 3, 5, 7 | 3 |
| 7.3 Gear Shaping | HSS | 2, 5, 7 | 3 | | 2, 5, 7 | 3, 5, 7 | 3 |
| 7.4 Gear Shaving | HSS | 2, 5, 7 | 3 | | 2, 5, 7 | 3, 5, 7 | 3 |
| 7.5 Gear Grinding, Form | — | 1, 4, 6 | 3 | | 1, 4, 6 | 1, 4, 6 | 3 |
| 8.1 Surface Grinding—Horiz. Spdl., Recip. Table | — | 1, 4, 6 | 3 | | 1, 4, 6 | 2, 4, 6 | 3 |
| 8.2 Surface Grinding—Horiz. Spdl., Recip. Table | CBN | — | 5, 4 | | — | — | 5, 4 |
| 8.3 Surface Grinding—Horiz. Spdl., Recip. Table | Diamond | — | — | | — | — | — |
| 8.4 Surface Grinding—Vertical Spdl., Rotary Table | — | 1, 4, 6 | 3 | | 1, 4, 6 | 2, 4, 6 | 3 |
| 8.5 Cylindrical Grinding | — | 1, 4, 6 | 3 | | 1, 4, 6 | 2, 4, 6 | 3 |
| 8.6 Cylindrical Grinding | CBN | — | 5, 4 | | — | — | 5, 4 |
| 8.7 Cylindrical Grinding | Diamond | — | — | | — | — | — |
| 8.8 Internal Grinding | — | 1, 4, 6 | 3 | | 1, 4, 6 | 2, 4, 6 | 3 |
| 8.9 Internal Grinding | CBN | — | 5, 4 | | — | — | 5, 4 |
| 8.10 Internal Grinding | Diamond | — | — | | — | — | — |
| 8.11 Centerless Grinding | — | 1, 4, 6 | 3 | | 1, 4, 6 | 2, 4, 6 | 3 |
| 8.13 Abrasive Belt Grinding | — | 0, 1, 4 | 1, 4, 6 | | 0, 1, 4 | 0, 1, 4 | 1, 4, 6 |
| 8.14 Thread Grinding | — | 1, 4, 6 | 3 | | 1, 4, 6 | 2, 4, 6 | 3 |
| | | | | | | | |

## 16.2 Cutting Fluid Recommendations

| OPERATION | TOOL MATERIAL OR ABRASIVE | 17. Tool Steels, Cast | | | 18. Stainless Steels, Cast — Austenitic and Ferritic | | |
|---|---|---|---|---|---|---|---|
| | | 150 to 250 Bhn | 325 to 375 Bhn | 48 to 65 $R_c$ | — | 135 to 210 Bhn | — |
| 1.1 Turning, Single Point and Box Tools | HSS | 1, 4, 6 | 2, 5, 7 | 3, 5, 7 | | 2, 5, 6 | |
| | Carbide | 0, 4, 6 | 0, 4, 6 | 0, 4, 6 | | 0, 4, 6 | |
| 1.3 Turning | Diamond | — | — | — | | — | |
| 1.4 Turning, Cutoff and Form Tools | HSS | 2, 5, 7 | 2, 5, 7 | 3 | | 2, 5, 7 | |
| | Carbide | | | | | 4, 6 | |
| 1.5 Threading, Single Point | HSS | 3 | 3 | 3 | | 3 | |
| | Carbide | 2, 5, 7 | 1, 4, 6 | 1, 4, 6 | | 2, 5, 7 | |
| 1.6 Threading, Die | HSS | 3 | 3 | 3 | | 3 | |
| | Carbide | 2, 5, 7 | 1, 4, 6 | 1, 4, 6 | | 2, 5, 7 | |
| 1.7 Hollow Milling | HSS | 3 | 3 | 3 | | 3 | |
| | Carbide | 2, 5, 7 | 2, 5, 7 | 2, 5, 7 | | 2, 5, 7 | |
| 2.1 Face Milling | HSS | 2, 5, 7 | 3 | 3 | | 2, 5, 7 | |
| | Carbide | 0, 4, 6 | 0, 5, 7 | 0, 5, 7 | | 0, 4, 6 | |
| 2.2 Face Milling | Diamond | — | — | — | | — | |
| 2.4 Slab Milling | HSS | 2, 5, 7 | 3, 5, 7 | 3 | | 2, 5, 7 | |
| | Carbide | 1, 4, 6 | 2, 5, 7 | 3, 5, 7 | | 1, 4, 6 | |
| 2.5 Side and Slot Milling—Arbor Mounted Cutters | HSS | 2, 5, 7 | 3 | 3 | | 2, 5, 7 | |
| | Carbide | 1, 4, 6 | 2, 5, 7 | 2, 5, 7 | | 1, 4, 6 | |
| 2.6 End Milling—Peripheral | HSS | 2, 5, 7 | 3 | 3 | | 2, 5, 7 | |
| | Carbide | 1, 4, 6 | 2, 5, 7 | 3, 5, 7 | | 0, 4, 6 | |
| 2.8 End Milling—Slotting | HSS | 2, 5, 7 | 3 | 3 | | 2, 5, 7 | |
| | Carbide | 1, 4, 6 | 2, 5, 7 | 3, 5, 7 | | 0, 4, 6 | |
| 2.9 Thread Milling | HSS | 3 | 3 | 3 | | 3 | |
| 3.1 Drilling | HSS | 2, 5, 7 | 3 | 3 | | 2, 5, 7 | |
| | Carbide | 1, 4, 6 | 2, 5, 7 | 3, 5, 7 | | 1, 4, 6 | |
| 3.3 Oil-Hole or Pressurized-Coolant Drilling | HSS | 1, 4, 6 | 2, 5, 7 | 3 | | 2, 5, 7 | |
| | Carbide | 1, 4, 6 | 2, 5, 7 | 2, 5, 7 | | 1, 4, 6 | |
| 3.4 Spade Drilling | HSS | 2, 5, 7 | 2, 5, 7 | 3 | | 2, 5, 7 | |
| | Carbide | 1, 4, 6 | 2, 5, 7 | 2 | | 1, 4, 6 | |
| 3.5 Gundrilling | HSS | 3, 5, 7 | 2, 5, 7 | 3 | | 2, 5, 7 | |
| | Carbide | 2, 5, 7 | 2, 5, 7 | 3, 5, 7 | | 2, 5, 7 | |
| 3.6 Pressure Coolant Reaming (Gun Reaming) | HSS | 2 | 3 | 3 | | 3 | |
| | Carbide | 2, 5, 7 | 2, 5, 7 | 3, 5, 7 | | 2, 5, 7 | |
| 3.7 Reaming | HSS | 3 | 3 | 3 | | 3 | |
| | Carbide | 2, 4, 6 | 2, 5, 7 | 3, 5, 7 | | 2, 5, 7 | |
| 3.9 Boring | HSS | 2, 5, 7 | 2, 5, 7 | 3, 5, 7 | | 2, 5, 7 | |
| | Carbide | 1, 4, 6 | 2, 5, 7 | 2, 5, 7 | | 1, 4, 6 | |
| 3.11 Boring | Diamond | — | — | — | | — | |
| 3.12 Counterboring and Spotfacing | HSS | 2, 5, 7 | 3, 5, 7 | 3 | | 2, 5, 7 | |
| | Carbide | 1, 4, 6 | 2, 5, 7 | 3, 5, 7 | | 2, 5, 7 | |
| 3.13 Trepanning | HSS | 3, 5, 7 | 2, 3 | 3 | | 3, 5, 7 | |
| | Carbide | 2, 5, 7 | 2, 5, 7 | 3, 5, 7 | | 2, 5, 7 | |
| 3.14 Honing | — | 90 | 90 | 90 | | 90 | |
| 3.15 Burnishing | — | 2 | 2 | — | | 2 | |
| 4.1 Tapping | HSS | 3 | 3 | 3 | | 3 | |

| OPERATION | TOOL MATERIAL OR ABRASIVE | 17. Tool Steels, Cast | | | 18. Stainless Steels, Cast Austenitic and Ferritic | | |
|---|---|---|---|---|---|---|---|
| | | 150 to 250 Bhn | 325 to 375 Bhn | 48 to 65 $R_C$ | — | 135 to 210 Bhn | — |
| 5.1 Planing | HSS | 2, 5, 7 | 2, 5, 7 | 3 | | 2, 5, 7 | |
| | Carbide | 1, 4, 6 | 1, 5, 7 | 3, 5, 7 | | 1, 4, 6 | |
| 5.2 Broaching | HSS | 2, 5, 7 | 3 | 3 | | 2, 5, 7 | |
| | Carbide | 1, 4, 6 | 3 | 3 | | 1, 4, 6 | |
| 6.1 Power Hack Sawing | HSS | 1, 5, 7 | 3 | 3 | | 1, 5, 7 | |
| 6.2 Power Band Sawing | HSS | 1, 5, 7 | 3 | 3 | | 1, 5, 7 | |
| 6.4 Circular Sawing | HSS | 2, 5, 7 | 3 | 3 | | 2, 5, 7 | |
| 6.5 Circular Sawing | Carbide | 1, 4, 6 | 2, 5, 7 | 3, 5, 7 | | 1, 4, 6 | |
| 6.6 Abrasive Cutoff | — | 5, 7 | 5, 7 | 5, 7 | | 5, 7 | |
| 7.1 Gear Hobbing | HSS | 2, 3 | 3 | 3 | | 2, 3 | |
| 7.2 Gear Cutting, Straight and Spiral Bevel | HSS | 2, 3 | 3 | 3 | | 2, 3 | |
| 7.3 Gear Shaping | HSS | 2, 3 | 3 | 3 | | 2, 3 | |
| 7.4 Gear Shaving | HSS | 2, 3 | 3 | 3 | | 2, 3 | |
| 7.5 Gear Grinding, Form | — | 5, 7 | 2, 5, 7 | 2, 5, 7 | | 5, 7 | |
| 8.1 Surface Grinding—Horiz. Spdl., Recip. Table | — | 5, 7 | 5, 7 | 5, 7 | | 5, 7 | |
| 8.2 Surface Grinding—Horiz. Spdl., Recip. Table | CBN | — | — | 5, 4 | | — | |
| 8.3 Surface Grinding—Horiz. Spdl., Recip. Table | Diamond | — | — | — | | — | |
| 8.4 Surface Grinding—Vertical Spdl., Rotary Table | — | 5, 7 | 5, 7 | 5, 7 | | 5, 7 | |
| 8.5 Cylindrical Grinding | — | 5, 7 | 5, 7 | 5, 7 | | 5, 7 | |
| 8.6 Cylindrical Grinding | CBN | — | — | 5, 4 | | — | |
| 8.7 Cylindrical Grinding | Diamond | — | — | — | | — | |
| 8.8 Internal Grinding | — | 5, 7 | 5, 7 | 5, 7 | | 5, 7 | |
| 8.9 Internal Grinding | CBN | — | — | 5, 4 | | — | |
| 8.10 Internal Grinding | Diamond | — | — | — | | — | |
| 8.11 Centerless Grinding | — | 5, 7 | 5, 7 | 5, 7 | | 5, 7 | |
| 8.13 Abrasive Belt Grinding | — | 1, 5, 7 | 1, 5, 7 | 1, 5, 7 | | 0, 2, 5 | |
| 8.14 Thread Grinding | — | 5, 7 | 5, 7 | 5, 7 | | 5, 7 | |
| | | | | | | | |

## 16.2 Cutting Fluid Recommendations

| OPERATION | TOOL MATERIAL OR ABRASIVE | 18. Stainless Steels, Cast — Martensitic | | | 19. Precipitation Hardening Stainless Steels, Cast | | |
|---|---|---|---|---|---|---|---|
| | | 135 to 275 Bhn | 275 to 425 Bhn | — | — | 325 to 450 Bhn | — |
| 1.1 Turning, Single Point and Box Tools | HSS | 2, 5, 6 | 2, 5, 7 | | | 2, 5, 7 | |
| | Carbide | 0, 4, 6 | 0, 4, 6 | | | 0, 4, 6 | |
| 1.3 Turning | Diamond | — | — | | | — | |
| 1.4 Turning, Cutoff and Form Tools | HSS | 2, 5, 7 | 2 | | | 2 | |
| | Carbide | 4, 6 | 1, 5, 7 | | | 1, 5, 7 | |
| 1.5 Threading, Single Point | HSS | 3 | 3 | | | 3 | |
| | Carbide | 2, 5, 7 | 1, 5, 7 | | | 1, 5, 7 | |
| 1.6 Threading, Die | HSS | 3 | 3 | | | 3 | |
| | Carbide | 2, 5, 7 | 1, 5, 7 | | | 1, 5, 7 | |
| 1.7 Hollow Milling | HSS | 2 | 2 | | | 2 | |
| | Carbide | 0, 1, 4, 6 | 1, 5, 7 | | | 1, 5, 7 | |
| 2.1 Face Milling | HSS | 2, 5, 7 | 1, 5, 7 | | | 1, 5, 7 | |
| | Carbide | 0, 4, 6 | 0, 4, 6 | | | 0, 4, 6 | |
| 2.2 Face Milling | Diamond | — | — | | | — | |
| 2.4 Slab Milling | HSS | 2, 5, 7 | 2, 5, 7 | | | 2, 5, 7 | |
| | Carbide | 1, 4, 6 | 1, 5, 7 | | | 1, 5, 7 | |
| 2.5 Side and Slot Milling—Arbor Mounted Cutters | HSS | 2, 5, 7 | 2, 5, 7 | | | 2, 5, 7 | |
| | Carbide | 1, 4, 6 | 1, 5, 7 | | | 1, 5, 7 | |
| 2.6 End Milling—Peripheral | HSS | 2, 5, 7 | 2, 5, 7 | | | 2, 5, 7 | |
| | Carbide | 0, 4, 6 | 0, 4, 6 | | | 0, 4, 6 | |
| 2.8 End Milling—Slotting | HSS | 2, 5, 7 | 2, 5, 7 | | | 2, 5, 7 | |
| | Carbide | 0, 4, 6 | 0, 4, 6 | | | 0, 4, 6 | |
| 2.9 Thread Milling | HSS | 3 | 2 | | | 2 | |
| 3.1 Drilling | HSS | 2, 5, 7 | 2 | | | 2 | |
| | Carbide | 1, 4, 6 | 1, 5, 7 | | | 1, 5, 7 | |
| 3.3 Oil-Hole or Pressurized-Coolant Drilling | HSS | 2, 5, 7 | 2, 5, 7 | | | 2 | |
| | Carbide | 1, 4, 6 | 1, 4, 6 | | | 2, 5, 7 | |
| 3.4 Spade Drilling | HSS | 2, 5, 7 | 2 | | | 2 | |
| | Carbide | 1, 4, 6 | 2, 5, 7 | | | 2, 5, 7 | |
| 3.5 Gundrilling | HSS | 2, 5, 7 | 2 | | | 2 | |
| | Carbide | 2, 5, 7 | 2, 5, 7 | | | 1, 5, 7 | |
| 3.6 Pressure Coolant Reaming (Gun Reaming) | HSS | 3 | 3 | | | 3 | |
| | Carbide | 2, 5, 7 | 2, 5, 7 | | | 2, 4, 6 | |
| 3.7 Reaming | HSS | 3 | 3 | | | 3 | |
| | Carbide | 1, 5, 7 | 2, 5, 7 | | | 2, 4, 6 | |
| 3.9 Boring | HSS | 2, 5, 7 | 2, 5, 7 | | | 2, 5, 7 | |
| | Carbide | 1, 4, 6 | 2 | | | 2 | |
| 3.11 Boring | Diamond | — | — | | | — | |
| 3.12 Counterboring and Spotfacing | HSS | 2, 5, 7 | 2, 5, 7 | | | 2, 5, 7 | |
| | Carbide | 2, 5, 7 | 2, 5, 7 | | | 2, 5, 7 | |
| 3.13 Trepanning | HSS | 3, 5, 7 | 2 | | | 2 | |
| | Carbide | 2, 5, 7 | 2, 5, 7 | | | 2, 5, 7 | |
| 3.14 Honing | — | 90 | 90 | | | 90 | |
| 3.15 Burnishing | — | 2 | 2 | | | 2 | |
| 4.1 Tapping | HSS | 3 | 2 | | | 2 | |

| OPERATION | TOOL MATERIAL OR ABRASIVE | 18. Stainless Steels, Cast | | | 19. Precipitation Hardening Stainless Steels, Cast | | |
|---|---|---|---|---|---|---|---|
| | | Martensitic | | | | | |
| | | 135 to 275 Bhn | 275 to 425 Bhn | — | — | 325 to 450 Bhn | — |
| 5.1 Planing | HSS | 2, 5, 7 | 2, 5, 7 | | | 2, 5, 7 | |
| | Carbide | 1, 4, 6 | 0, 4, 6 | | | 0, 4, 6 | |
| 5.2 Broaching | HSS | 2, 5, 7 | 3 | | | 3 | |
| | Carbide | 1, 4, 6 | 2, 3 | | | 2, 3 | |
| 6.1 Power Hack Sawing | HSS | 1, 5, 7 | 3, 5, 7 | | | 3, 5, 7 | |
| 6.2 Power Band Sawing | HSS | 1, 5, 7 | 3, 5, 7 | | | 3, 5, 7 | |
| 6.4 Circular Sawing | HSS | 2, 5, 7 | 3, 5, 7 | | | 3, 5, 7 | |
| 6.5 Circular Sawing | Carbide | 1, 4, 6 | 4, 6 | | | 4, 6 | |
| 6.6 Abrasive Cutoff | — | 5, 7 | 2, 5, 7 | | | 2, 5, 7 | |
| 7.1 Gear Hobbing | HSS | 2, 3 | 2 | | | 2 | |
| 7.2 Gear Cutting, Straight and Spiral Bevel | HSS | 2, 3 | 2 | | | 2 | |
| 7.3 Gear Shaping | HSS | 2, 3 | 2 | | | 2 | |
| 7.4 Gear Shaving | HSS | 2, 3 | 2 | | | 2 | |
| 7.5 Gear Grinding, Form | — | 5, 7 | 2 | | | 2 | |
| 8.1 Surface Grinding—Horiz. Spdl., Recip. Table | — | 5, 7 | 1, 4, 6 | | | 1, 4, 6 | |
| 8.2 Surface Grinding—Horiz. Spdl., Recip. Table | CBN | — | — | | | — | |
| 8.3 Surface Grinding—Horiz. Spdl., Recip. Table | Diamond | — | — | | | — | |
| 8.4 Surface Grinding—Vertical Spdl., Rotary Table | — | 5, 7 | 1, 4, 6 | | | 1, 4, 6 | |
| 8.5 Cylindrical Grinding | — | 5, 7 | 1, 4, 6 | | | 1, 4, 6 | |
| 8.6 Cylindrical Grinding | CBN | — | — | | | — | |
| 8.7 Cylindrical Grinding | Diamond | — | — | | | — | |
| 8.8 Internal Grinding | — | 5, 7 | 1, 4, 6 | | | 1, 4, 6 | |
| 8.9 Internal Grinding | CBN | — | — | | | — | |
| 8.10 Internal Grinding | Diamond | — | — | | | — | |
| 8.11 Centerless Grinding | — | 5, 7 | 1, 4, 6 | | | 1, 4, 6 | |
| 8.13 Abrasive Belt Grinding | — | 0, 1, 4 | 0, 1, 5 | | | 1, 5, 7 | |
| 8.14 Thread Grinding | — | 5, 7 | 1, 4, 6 | | | 1, 4, 6 | |
| | | | | | | | |

## 16.2 Cutting Fluid Recommendations

| OPERATION | TOOL MATERIAL OR ABRASIVE | 20. Austenitic Manganese Steels, Cast | | | 21. Gray Cast Irons | | |
|---|---|---|---|---|---|---|---|
| | | — | 150 to 220 Bhn | — | 120 to 320 Bhn | 45 to 60 $R_C$ | — |
| 1.1 Turning, Single Point and Box Tools | HSS | | — | | 4, 6 | — | |
| | Carbide | | 3, 5 | | 0, 4, 6 | — | |
| 1.3 Turning | Diamond | | — | | — | — | |
| 1.4 Turning, Cutoff and Form Tools | HSS | | 3, 5 | | 4, 6 | — | |
| | Carbide | | 3, 5 | | 0, 4, 6 | — | |
| 1.5 Threading, Single Point | HSS | | — | | 4, 6 | — | |
| | Carbide | | — | | 0, 4, 6 | — | |
| 1.6 Threading, Die | HSS | | — | | 4, 6 | — | |
| | Carbide | | — | | 0, 4, 6 | — | |
| 1.7 Hollow Milling | HSS | | — | | 5, 7 | — | |
| | Carbide | | 3, 5 | | 5, 7 | — | |
| 2.1 Face Milling | HSS | | — | | 4, 6 | — | |
| | Carbide | | 3, 5 | | 0, 4, 6 | — | |
| 2.2 Face Milling | Diamond | | — | | — | — | |
| 2.4 Slab Milling | HSS | | — | | 4, 6 | — | |
| | Carbide | | — | | 0, 4, 6 | — | |
| 2.5 Side and Slot Milling—Arbor Mounted Cutters | HSS | | — | | 4, 6 | — | |
| | Carbide | | 3, 5 | | 0, 4, 6 | — | |
| 2.6 End Milling—Peripheral | HSS | | — | | 4, 6 | — | |
| | Carbide | | — | | 0, 4, 6 | — | |
| 2.8 End Milling—Slotting | HSS | | — | | 4, 6 | — | |
| | Carbide | | — | | 0, 4, 6 | — | |
| 2.9 Thread Milling | HSS | | — | | 4, 6 | — | |
| 3.1 Drilling | HSS | | 3, 5 | | 4, 6 | — | |
| | Carbide | | 3, 5 | | 0, 4, 6 | — | |
| 3.3 Oil-Hole or Pressurized-Coolant Drilling | HSS | | — | | 4, 6 | — | |
| | Carbide | | 3, 5 | | 4, 6 | — | |
| 3.4 Spade Drilling | HSS | | — | | 4, 6 | — | |
| | Carbide | | 3, 5 | | 0, 4, 6 | — | |
| 3.5 Gundrilling | HSS | | — | | 1, 86 | — | |
| | Carbide | | 3, 5 | | 1, 86 | — | |
| 3.6 Pressure Coolant Reaming (Gun Reaming) | HSS | | — | | 4, 6 | — | |
| | Carbide | | — | | 4, 6 | — | |
| 3.7 Reaming | HSS | | — | | 4, 6 | — | |
| | Carbide | | — | | 4, 6 | — | |
| 3.9 Boring | HSS | | — | | 4, 6 | — | |
| | Carbide | | 3, 5 | | 0, 4, 6 | — | |
| 3.11 Boring | Diamond | | — | | — | — | |
| 3.12 Counterboring and Spotfacing | HSS | | — | | 4, 6 | — | |
| | Carbide | | 3, 5 | | 0, 4, 6 | — | |
| 3.13 Trepanning | HSS | | — | | 1, 86 | — | |
| | Carbide | | — | | 1, 86 | — | |
| 3.14 Honing | — | | — | | 80 | — | |
| 3.15 Burnishing | — | | — | | 5, 7 | — | |
| 4.1 Tapping | HSS | | — | | 4, 6 | — | |

| OPERATION | TOOL MATERIAL OR ABRASIVE | 20. Austenitic Manganese Steels, Cast | | | 21. Gray Cast Irons | | |
|---|---|---|---|---|---|---|---|
| | | — | 150 to 220 Bhn | — | 120 to 320 Bhn | 45 to 60 Rc | — |
| 5.1 Planing | HSS | | — | | 4, 6 | — | |
| | Carbide | | — | | 0, 4, 6 | — | |
| 5.2 Broaching | HSS | | — | | 5, 7 | — | |
| | Carbide | | — | | 0, 4, 6 | — | |
| 6.1 Power Hack Sawing | HSS | | 3, 5 | | 4, 6 | — | |
| 6.2 Power Band Sawing | HSS | | 3, 5 | | 4, 6 | — | |
| 6.4 Circular Sawing | HSS | | 3, 5 | | 4, 6 | — | |
| 6.5 Circular Sawing | Carbide | | 3, 5 | | 0, 4, 6 | — | |
| 6.6 Abrasive Cutoff | — | | 3, 5 | | 4, 6 | — | |
| 7.1 Gear Hobbing | HSS | | — | | 4, 6 | — | |
| 7.2 Gear Cutting, Straight and Spiral Bevel | HSS | | — | | 4, 6 | — | |
| 7.3 Gear Shaping | HSS | | — | | 4, 6 | — | |
| 7.4 Gear Shaving | HSS | | — | | 4, 6 | — | |
| 7.5 Gear Grinding, Form | — | | — | | 4, 6 | 5, 7 | |
| 8.1 Surface Grinding—Horiz. Spdl., Recip. Table | — | | 3, 5 | | 4, 6 | 5, 7 | |
| 8.2 Surface Grinding—Horiz. Spdl., Recip. Table | CBN | | — | | — | 5, 4 | |
| 8.3 Surface Grinding—Horiz. Spdl., Recip. Table | Diamond | | — | | — | 5 | |
| 8.4 Surface Grinding—Vertical Spdl., Rotary Table | — | | 3, 5 | | 4, 6 | 5, 7 | |
| 8.5 Cylindrical Grinding | — | | 3, 5 | | 4, 6 | 5, 7 | |
| 8.6 Cylindrical Grinding | CBN | | — | | — | 5, 4 | |
| 8.7 Cylindrical Grinding | Diamond | | — | | — | — | |
| 8.8 Internal Grinding | — | | 3, 5 | | 4, 6 | 5, 7 | |
| 8.9 Internal Grinding | CBN | | — | | — | 5, 4 | |
| 8.10 Internal Grinding | Diamond | | — | | — | — | |
| 8.11 Centerless Grinding | — | | 3, 5 | | 4, 6 | 5, 7 | |
| 8.13 Abrasive Belt Grinding | — | | 3, 5 | | 0, 4, 6 | 0, 5, 7 | |
| 8.14 Thread Grinding | — | | 3, 5 | | 4, 6 | 5, 7 | |
| | | | | | | | |

## 16.2 Cutting Fluid Recommendations

| OPERATION | TOOL MATERIAL OR ABRASIVE | 22. Compacted Graphite Cast Irons | | | 23. Ductile Cast Irons | | |
|---|---|---|---|---|---|---|---|
| | | — | 185 to 255 Bhn | — | 120 to 400 Bhn | 53 to 60 $R_C$ | — |
| 1.1 Turning, Single Point and Box Tools | HSS | | 4, 6 | | 4, 6 | — | |
| | Carbide | | 0, 4, 6 | | 0, 4, 6 | — | |
| 1.3 Turning | Diamond | | — | | — | — | |
| 1.4 Turning, Cutoff and Form Tools | HSS | | 4, 6 | | 4, 6 | — | |
| | Carbide | | 0, 4, 6 | | 0, 4, 6 | — | |
| 1.5 Threading, Single Point | HSS | | 4, 6 | | 4, 6 | — | |
| | Carbide | | 0, 4, 6 | | 0, 4, 6 | — | |
| 1.6 Threading, Die | HSS | | 4, 6 | | 4, 6 | — | |
| | Carbide | | 0, 4, 6 | | 0, 4, 6 | — | |
| 1.7 Hollow Milling | HSS | | 5, 7 | | 5, 7 | — | |
| | Carbide | | 5, 7 | | 5, 7 | — | |
| 2.1 Face Milling | HSS | | 4, 6 | | 4, 6 | — | |
| | Carbide | | 0, 4, 6 | | 0, 4, 6 | — | |
| 2.2 Face Milling | Diamond | | — | | — | — | |
| 2.4 Slab Milling | HSS | | 4, 6 | | 4, 6 | — | |
| | Carbide | | 0, 4, 6 | | 0, 4, 6 | — | |
| 2.5 Side and Slot Milling—Arbor Mounted Cutters | HSS | | 4, 6 | | 4, 6 | — | |
| | Carbide | | 0, 4, 6 | | 0, 4, 6 | — | |
| 2.6 End Milling—Peripheral | HSS | | 4, 6 | | 4, 6 | — | |
| | Carbide | | 0, 4, 6 | | 0, 4, 6 | — | |
| 2.8 End Milling—Slotting | HSS | | 4, 6 | | 4, 6 | — | |
| | Carbide | | 0, 4, 6 | | 0, 4, 6 | — | |
| 2.9 Thread Milling | HSS | | 4, 6 | | 4, 6 | — | |
| 3.1 Drilling | HSS | | 4, 6 | | 4, 6 | — | |
| | Carbide | | 0, 4, 6 | | 0, 4, 6 | — | |
| 3.3 Oil-Hole or Pressurized-Coolant Drilling | HSS | | 4, 6 | | 4, 6 | — | |
| | Carbide | | 4, 6 | | 4, 6 | — | |
| 3.4 Spade Drilling | HSS | | 4, 6 | | 4, 6 | — | |
| | Carbide | | 0, 4, 6 | | 0, 4, 6 | — | |
| 3.5 Gundrilling | HSS | | 1, 86 | | 1, 86 | — | |
| | Carbide | | 1, 86 | | 1, 86 | — | |
| 3.6 Pressure Coolant Reaming (Gun Reaming) | HSS | | 4, 6 | | 4, 6 | — | |
| | Carbide | | 4, 6 | | 4, 6 | — | |
| 3.7 Reaming | HSS | | 4, 6 | | 4, 6 | — | |
| | Carbide | | 4, 6 | | 4, 6 | — | |
| 3.9 Boring | HSS | | 4, 6 | | 4, 6 | — | |
| | Carbide | | 0, 4, 6 | | 0, 4, 6 | — | |
| 3.11 Boring | Diamond | | — | | — | — | |
| 3.12 Counterboring and Spotfacing | HSS | | 4, 6 | | 4, 6 | — | |
| | Carbide | | 0, 4, 6 | | 0, 4, 6 | — | |
| 3.13 Trepanning | HSS | | 1, 86 | | 1, 86 | — | |
| | Carbide | | 1, 86 | | 1, 86 | — | |
| 3.14 Honing | — | | 80 | | 80 | — | |
| 3.15 Burnishing | — | | 5, 7 | | 5, 7 | — | |
| 4.1 Tapping | HSS | | 4, 6 | | 4, 6 | — | |

| OPERATION | TOOL MATERIAL OR ABRASIVE | 22. Compacted Graphite Cast Irons | | | 23. Ductile Cast Irons | | |
|---|---|---|---|---|---|---|---|
| | | — | 185 to 255 Bhn | — | 120 to 400 Bhn | 53 to 60 $R_C$ | — |
| 5.1 Planing | HSS | | 4, 6 | | 4, 6 | — | |
| | Carbide | | 0, 4, 6 | | 0,4,6 | — | |
| 5.2 Broaching | HSS | | 5, 7 | | 5, 7 | — | |
| | Carbide | | 0, 4, 6 | | 0, 4, 6 | — | |
| 6.1 Power Hack Sawing | HSS | | 4, 6 | | 4, 6 | — | |
| 6.2 Power Band Sawing | HSS | | 4, 6 | | 4, 6 | — | |
| 6.4 Circular Sawing | HSS | | 4, 6 | | 4, 6 | — | |
| 6.5 Circular Sawing | Carbide | | 0, 4, 6 | | 0, 4, 6 | 0, 5, 7 | |
| 6.6 Abrasive Cutoff | — | | 4, 6 | | 4, 6 | 5, 7 | |
| 7.1 Gear Hobbing | HSS | | 4, 6 | | 4, 6 | — | |
| 7.2 Gear Cutting, Straight and Spiral Bevel | HSS | | 4, 6 | | 4, 6 | — | |
| 7.3 Gear Shaping | HSS | | 4, 6 | | 4, 6 | — | |
| 7.4 Gear Shaving | HSS | | 4, 6 | | 4, 6 | — | |
| 7.5 Gear Grinding, Form | — | | 4, 6 | | 4, 6 | 5, 7 | |
| 8.1 Surface Grinding—Horiz. Spdl., Recip. Table | — | | 4, 6 | | 4, 6 | 5, 7 | |
| 8.2 Surface Grinding—Horiz. Spdl., Recip. Table | CBN | | — | | — | 5, 4 | |
| 8.3 Surface Grinding—Horiz. Spdl., Recip. Table | Diamond | | — | | — | — | |
| 8.4 Surface Grinding—Vertical Spdl., Rotary Table | — | | 4, 6 | | 4, 6 | 5, 7 | |
| 8.5 Cylindrical Grinding | — | | 4, 6 | | 4, 6 | 5, 7 | |
| 8.6 Cylindrical Grinding | CBN | | — | | — | 5, 4 | |
| 8.7 Cylindrical Grinding | Diamond | | — | | — | — | |
| 8.8 Internal Grinding | — | | 4, 6 | | 4, 6 | 5, 7 | |
| 8.9 Internal Grinding | CBN | | — | | — | 5, 4 | |
| 8.10 Internal Grinding | Diamond | | — | | — | — | |
| 8.11 Centerless Grinding | — | | 4, 6 | | 4, 6 | 5, 7 | |
| 8.13 Abrasive Belt Grinding | — | | 0, 4, 6 | | 0, 4, 6 | 0, 5, 7 | |
| 8.14 Thread Grinding | — | | 4, 6 | | 4, 6 | 5, 7 | |
| | | | | | | | |

## 16.2 Cutting Fluid Recommendations

| OPERATION | TOOL MATERIAL OR ABRASIVE | 24. Malleable Cast Irons | | | 25. White Cast Irons (Abrasion Resistant) | | |
|---|---|---|---|---|---|---|---|
| | | 110 to 260 Bhn | Over 52 $R_C$ | — | — | 450 to 600 Bhn | — |
| 1.1 Turning, Single Point and Box Tools | HSS | 4, 6 | — | | | 4, 6 | |
| | Carbide | 0, 4, 6 | — | | | 0, 4, 6 | |
| 1.3 Turning | Diamond | — | — | | | — | |
| 1.4 Turning, Cutoff and Form Tools | HSS | 4, 6 | — | | | 4, 6 | |
| | Carbide | 0, 4, 6 | — | | | 0, 4, 6 | |
| 1.5 Threading, Single Point | HSS | 4, 6 | — | | | — | |
| | Carbide | 0, 4, 6 | — | | | — | |
| 1.6 Threading, Die | HSS | 4, 6 | — | | | — | |
| | Carbide | 0, 4, 6 | — | | | — | |
| 1.7 Hollow Milling | HSS | 5, 7 | — | | | — | |
| | Carbide | 5, 7 | — | | | — | |
| 2.1 Face Milling | HSS | 4, 6 | — | | | — | |
| | Carbide | 0, 4, 6 | — | | | 0, 4, 6 | |
| 2.2 Face Milling | Diamond | — | — | | | — | |
| 2.4 Slab Milling | HSS | 4, 6 | — | | | — | |
| | Carbide | 0, 4, 6 | — | | | — | |
| 2.5 Side and Slot Milling—Arbor Mounted Cutters | HSS | 4, 6 | — | | | — | |
| | Carbide | 0, 4, 6 | — | | | 0, 4, 6 | |
| 2.6 End Milling—Peripheral | HSS | 4, 6 | — | | | — | |
| | Carbide | 0, 4, 6 | — | | | 0, 4, 6 | |
| 2.8 End Milling—Slotting | HSS | 4, 6 | — | | | — | |
| | Carbide | 0, 4, 6 | — | | | — | |
| 2.9 Thread Milling | HSS | 4, 6 | — | | | — | |
| 3.1 Drilling | HSS | 4, 6 | — | | | 4, 6 | |
| | Carbide | 0, 4, 6 | — | | | 0, 4, 6 | |
| 3.3 Oil-Hole or Pressurized-Coolant Drilling | HSS | 4, 6 | — | | | 4, 6 | |
| | Carbide | 4, 6 | — | | | 0, 4, 6 | |
| 3.4 Spade Drilling | HSS | 4, 6 | — | | | 4, 6 | |
| | Carbide | 0, 4, 6 | — | | | 0, 4, 6 | |
| 3.5 Gundrilling | HSS | 1, 86 | — | | | — | |
| | Carbide | 1, 86 | — | | | — | |
| 3.6 Pressure Coolant Reaming (Gun Reaming) | HSS | 4, 6 | — | | | — | |
| | Carbide | 4, 6, | — | | | — | |
| 3.7 Reaming | HSS | 4, 6 | — | | | 4, 6 | |
| | Carbide | 4, 6 | — | | | 0, 4, 6 | |
| 3.9 Boring | HSS | 4, 6 | — | | | — | |
| | Carbide | 0, 4, 6 | — | | | 0, 4, 6 | |
| 3.11 Boring | Diamond | — | — | | | — | |
| 3.12 Counterboring and Spotfacing | HSS | 4, 6 | — | | | — | |
| | Carbide | 0, 4, 6 | — | | | 0, 4, 6 | |
| 3.13 Trepanning | HSS | 1, 86 | — | | | — | |
| | Carbide | 1, 86 | — | | | — | |
| 3.14 Honing | — | 80 | — | | | 80 | |
| 3.15 Burnishing | — | 5, 7 | — | | | — | |
| 4.1 Tapping | HSS | 4, 6 | — | | | 4, 6 | |

| OPERATION | TOOL MATERIAL OR ABRASIVE | 24. Malleable Cast Irons | | | 25. White Cast Irons (Abrasion Resistant) | | |
|---|---|---|---|---|---|---|---|
| | | 110 to 260 Bhn | Over 52 $R_C$ | — | — | 450 to 600 Bhn | — |
| 5.1 Planing | HSS | 4, 6 | — | | | — | |
| | Carbide | 0, 4, 6 | — | | | — | |
| 5.2 Broaching | HSS | 5, 7 | — | | | — | |
| | Carbide | 0, 4, 6 | — | | | — | |
| 6.1 Power Hack Sawing | HSS | 4, 6 | — | | | — | |
| 6.2 Power Band Sawing | HSS | 4, 6 | — | | | — | |
| 6.4 Circular Sawing | HSS | 4, 6 | — | | | — | |
| 6.5 Circular Sawing | Carbide | 0, 4, 6 | 0, 5, 7 | | | 0, 5, 7 | |
| 6.6 Abrasive Cutoff | — | 4, 6 | 5, 7 | | | 5, 7 | |
| 7.1 Gear Hobbing | HSS | 4, 6 | — | | | — | |
| 7.2 Gear Cutting, Straight and Spiral Bevel | HSS | 4, 6 | — | | | — | |
| 7.3 Gear Shaping | HSS | 4, 6 | — | | | — | |
| 7.4 Gear Shaving | HSS | 4, 6 | — | | | — | |
| 7.5 Gear Grinding, Form | — | 4, 6 | 5, 7 | | | — | |
| 8.1 Surface Grinding—Horiz. Spdl., Recip. Table | — | 4, 6 | 5, 7 | | | 5, 7 | |
| 8.2 Surface Grinding—Horiz. Spdl., Recip. Table | CBN | — | 5, 4 | | | — | |
| 8.3 Surface Grinding—Horiz. Spdl., Recip. Table | Diamond | — | — | | | — | |
| 8.4 Surface Grinding—Vertical Spdl., Rotary Table | — | 4, 6 | 5, 7 | | | 5, 7 | |
| 8.5 Cylindrical Grinding | — | 4, 6 | 5, 7 | | | 5, 7 | |
| 8.6 Cylindrical Grinding | CBN | — | 5, 4 | | | — | |
| 8.7 Cylindrical Grinding | Diamond | — | — | | | — | |
| 8.8 Internal Grinding | — | 4, 6 | 5, 7 | | | 5, 7 | |
| 8.9 Internal Grinding | CBN | — | 5, 4 | | | — | |
| 8.10 Internal Grinding | Diamond | — | — | | | — | |
| 8.11 Centerless Grinding | — | 4, 6 | 5, 7 | | | 5, 7 | |
| 8.13 Abrasive Belt Grinding | — | 0, 4, 6 | 0, 5, 7 | | | 0, 5, 7 | |
| 8.14 Thread Grinding | — | 4, 6 | 5, 7 | | | 5, 7 | |
| | | | | | | | |

## 16.2 Cutting Fluid Recommendations

| OPERATION | TOOL MATERIAL OR ABRASIVE | 26. High Silicon Cast Irons | | | 27. Chromium-Nickel Alloy Castings | | |
|---|---|---|---|---|---|---|---|
| | | — | 52 $R_C$ | — | — | 275 to 375 Bhn (500 kg) | — |
| 1.1 Turning, Single Point and Box Tools | HSS | | — | | | 3, 5, 7 | |
| | Carbide | | — | | | 5, 7 | |
| 1.3 Turning | Diamond | | — | | | — | |
| 1.4 Turning, Cutoff and Form Tools | HSS | | — | | | 3, 5, 7 | |
| | Carbide | | — | | | 5, 7 | |
| 1.5 Threading, Single Point | HSS | | — | | | — | |
| | Carbide | | — | | | — | |
| 1.6 Threading, Die | HSS | | — | | | — | |
| | Carbide | | — | | | — | |
| 1.7 Hollow Milling | HSS | | — | | | — | |
| | Carbide | | — | | | — | |
| 2.1 Face Milling | HSS | | — | | | 3, 5, 7 | |
| | Carbide | | — | | | 5, 7 | |
| 2.2 Face Milling | Diamond | | — | | | — | |
| 2.4 Slab Milling | HSS | | — | | | — | |
| | Carbide | | — | | | — | |
| 2.5 Side and Slot Milling—Arbor Mounted Cutters | HSS | | — | | | 3, 5, 7 | |
| | Carbide | | — | | | 5, 7 | |
| 2.6 End Milling—Peripheral | HSS | | — | | | 3, 5, 7 | |
| | Carbide | | — | | | 5, 7 | |
| 2.8 End Milling—Slotting | HSS | | — | | | 3, 5, 7 | |
| | Carbide | | — | | | — | |
| 2.9 Thread Milling | HSS | | — | | | 3, 5, 7 | |
| 3.1 Drilling | HSS | | — | | | 3, 5, 7 | |
| | Carbide | | — | | | — | |
| 3.3 Oil-Hole or Pressurized-Coolant Drilling | HSS | | — | | | 3, 5, 7 | |
| | Carbide | | — | | | 5, 7 | |
| 3.4 Spade Drilling | HSS | | — | | | 3, 5, 7 | |
| | Carbide | | — | | | — | |
| 3.5 Gundrilling | HSS | | — | | | — | |
| | Carbide | | — | | | — | |
| 3.6 Pressure Coolant Reaming (Gun Reaming) | HSS | | — | | | — | |
| | Carbide | | — | | | — | |
| 3.7 Reaming | HSS | | — | | | 3, 5, 7 | |
| | Carbide | | — | | | 5, 7 | |
| 3.9 Boring | HSS | | — | | | 3, 5, 7 | |
| | Carbide | | — | | | 5, 7 | |
| 3.11 Boring | Diamond | | — | | | — | |
| 3.12 Counterboring and Spotfacing | HSS | | — | | | 3, 5, 7 | |
| | Carbide | | — | | | 5, 7 | |
| 3.13 Trepanning | HSS | | — | | | — | |
| | Carbide | | — | | | — | |
| 3.14 Honing | — | | — | | | 90 | |
| 3.15 Burnishing | — | | — | | | — | |
| 4.1 Tapping | HSS | | — | | | 2, 3 | |

| OPERATION | TOOL MATERIAL OR ABRASIVE | 26. High Silicon Cast Irons | | | 27. Chromium-Nickel Alloy Castings | | |
|---|---|---|---|---|---|---|---|
| | | — | 52 $R_C$ | — | — | 275 to 375 Bhn (500 kg) | — |
| 5.1 Planing | HSS | | — | | | — | |
| | Carbide | | — | | | — | |
| 5.2 Broaching | HSS | | — | | | — | |
| | Carbide | | — | | | — | |
| 6.1 Power Hack Sawing | HSS | | — | | | 5, 7, 3 | |
| 6.2 Power Band Sawing | HSS | | — | | | 5, 7, 3 | |
| 6.4 Circular Sawing | HSS | | — | | | 5, 7, 3 | |
| 6.5 Circular Sawing | Carbide | | — | | | 5, 7, 2 | |
| 6.6 Abrasive Cutoff | — | | 5, 7 | | | 5, 7 | |
| 7.1 Gear Hobbing | HSS | | — | | | — | |
| 7.2 Gear Cutting, Straight and Spiral Bevel | HSS | | — | | | — | |
| 7.3 Gear Shaping | HSS | | — | | | — | |
| 7.4 Gear Shaving | HSS | | — | | | — | |
| 7.5 Gear Grinding, Form | — | | — | | | — | |
| 8.1 Surface Grinding—Horiz. Spdl., Recip. Table | — | | 5, 7 | | | 2, 3 | |
| 8.2 Surface Grinding—Horiz. Spdl., Recip. Table | CBN | | 5, 4 | | | — | |
| 8.3 Surface Grinding—Horiz. Spdl., Recip. Table | Diamond | | — | | | — | |
| 8.4 Surface Grinding—Vertical Spdl., Rotary Table | — | | 5, 7 | | | 2, 3 | |
| 8.5 Cylindrical Grinding | — | | 5, 7 | | | 2, 3 | |
| 8.6 Cylindrical Grinding | CBN | | 5, 4 | | | — | |
| 8.7 Cylindrical Grinding | Diamond | | — | | | — | |
| 8.8 Internal Grinding | — | | 5, 7 | | | 2, 3 | |
| 8.9 Internal Grinding | CBN | | 5, 4 | | | — | |
| 8.10 Internal Grinding | Diamond | | — | | | — | |
| 8.11 Centerless Grinding | — | | 5, 7 | | | 2, 3 | |
| 8.13 Abrasive Belt Grinding | — | | 0, 5, 7 | | | 3 | |
| 8.14 Thread Grinding | — | | 5, 7 | | | 2, 3 | |
| | | | | | | | |

## 16.2 Cutting Fluid Recommendations

| OPERATION | TOOL MATERIAL OR ABRASIVE | 28. Aluminum Alloys, Wrought | | 29. Aluminum Alloys, Cast | | 30. Magnesium Alloys, Wrought | |
|---|---|---|---|---|---|---|---|
| | | 30 to 150 Bhn (500 kg) | — | 40 to 125 Bhn (500 kg) | — | 50 to 90 Bhn (500 kg) | — |
| 1.1 Turning, Single Point and Box Tools | HSS | 4, 6, 82 | | 4, 6, 82 | | 1, 88 | |
| | Carbide | 0, 4, 6 | | 0, 4, 6 | | 1, 88 | |
| 1.3 Turning | Diamond | 0, 4, 6 | | 0, 4, 6 | | 1, 88 | |
| 1.4 Turning, Cutoff and Form Tools | HSS | 4, 6, 82 | | 4, 6, 82 | | 1, 88 | |
| | Carbide | 0, 4, 6 | | 0, 4, 6 | | 1, 88 | |
| 1.5 Threading, Single Point | HSS | 0, 1, 82 | | 0, 1, 82 | | 1, 88 | |
| | Carbide | 0, 1, 82 | | 0, 1, 82 | | 1, 88 | |
| 1.6 Threading, Die | HSS | 0, 1, 82 | | 0, 1, 82 | | 1, 88 | |
| | Carbide | 0, 1, 82 | | 0, 1, 82 | | 1, 88 | |
| 1.7 Hollow Milling | HSS | 0, 1, 82 | | 0, 1, 82 | | 1, 88 | |
| | Carbide | 0, 1, 82 | | 0, 1, 82 | | 1, 88 | |
| 2.1 Face Milling | HSS | 0, 1, 82 | | 0, 1, 82 | | 1, 88 | |
| | Carbide | 0, 1, 82 | | 0, 1, 82 | | 1, 88 | |
| 2.2 Face Milling | Diamond | 0, 4, 6 | | 0, 4, 6 | | 1, 88 | |
| 2.4 Slab Milling | HSS | 0, 1, 82 | | 0, 1, 82 | | 1, 88 | |
| | Carbide | 0, 1, 82 | | 0, 1, 82 | | 1, 88 | |
| 2.5 Side and Slot Milling—Arbor Mounted Cutters | HSS | 0, 1, 82 | | 0, 1, 82 | | 1, 88 | |
| | Carbide | 0, 1, 82 | | 0, 1, 82 | | 1, 88 | |
| 2.6 End Milling—Peripheral | HSS | 0, 1, 82 | | 0, 1, 82 | | 1, 88 | |
| | Carbide | 0, 1, 82 | | 0, 1, 82 | | 1, 88 | |
| 2.8 End Milling—Slotting | HSS | 0, 1, 82 | | 0, 1, 82 | | 1, 88 | |
| | Carbide | 0, 1, 82 | | 0, 1, 82 | | 1, 88 | |
| 2.9 Thread Milling | HSS | 0, 1, 82 | | 0, 1, 82 | | 1, 88 | |
| 3.1 Drilling | HSS | 0, 1, 82 | | 0, 1, 82 | | 1, 88 | |
| | Carbide | 0, 1, 82 | | 0, 1, 82 | | 1, 88 | |
| 3.3 Oil-Hole or Pressurized-Coolant Drilling | HSS | 1, 82 | | 1, 82 | | 1, 88 | |
| | Carbide | 1, 82 | | 1, 82 | | 1, 88 | |
| 3.4 Spade Drilling | HSS | 1, 82 | | 1, 82 | | 1, 88 | |
| | Carbide | 0, 82 | | 0, 82 | | 88 | |
| 3.5 Gundrilling | HSS | 1, 82 | | 1, 82 | | 1, 88 | |
| | Carbide | 1, 82 | | 1, 82 | | 1, 88 | |
| 3.6 Pressure Coolant Reaming (Gun Reaming) | HSS | 1, 82 | | 1, 82 | | 1, 88 | |
| | Carbide | 1, 82 | | 1, 82 | | 1, 88 | |
| 3.7 Reaming | HSS | 0, 1, 82 | | 0, 1, 82 | | 1, 88 | |
| | Carbide | 0, 1, 82 | | 0, 1, 82 | | 1, 88 | |
| 3.9 Boring | HSS | 4, 6, 82 | | 4, 6, 82 | | 1, 88 | |
| | Carbide | 0, 4, 6 | | 0, 4, 6 | | 1, 88 | |
| 3.11 Boring | Diamond | 0, 4, 6 | | 0, 4, 6 | | 1, 88 | |
| 3.12 Counterboring and Spotfacing | HSS | 4, 6, 82 | | 4, 6, 82 | | 1, 88 | |
| | Carbide | 0, 4, 6 | | 0, 4, 6 | | 1, 88 | |
| 3.13 Trepanning | HSS | 1, 82 | | 1, 82 | | 1, 88 | |
| | Carbide | 1, 82 | | 1, 82 | | 1, 88 | |
| 3.14 Honing | — | 80 | | 80 | | 80 | |
| 3.15 Burnishing | — | 1, 82 | | 1, 82 | | — | |
| 4.1 Tapping | HSS | 0, 1, 82 | | 0, 1, 82 | | 1, 88 | |

| OPERATION | TOOL MATERIAL OR ABRASIVE | 28. Aluminum Alloys, Wrought | | 29. Aluminum Alloys, Cast | | 30. Magnesium Alloys, Wrought | |
|---|---|---|---|---|---|---|---|
| | | 30 to 150 Bhn (500 kg) | — | 40 to 125 Bhn (500 kg) | — | 50 to 90 Bhn (500 kg) | — |
| 5.1 Planing | HSS | 4, 6, 82 | | 4, 6, 82 | | 1, 88 | |
| | Carbide | 0, 4, 6 | | 0, 4, 6 | | 1, 88 | |
| 5.2 Broaching | HSS | 4, 6, 82 | | 4, 6, 82 | | 0, 1, 88 | |
| | Carbide | 0, 4, 6 | | 0, 4, 6 | | 0, 1, 88 | |
| 6.1 Power Hack Sawing | HSS | 0, 1, 82 | | 0, 1, 82 | | 0, 1, 88 | |
| 6.2 Power Band Sawing | HSS | 0, 1, 82 | | 0, 1, 82 | | 0, 1, 88 | |
| 6.4 Circular Sawing | HSS | 1, 82 | | 1, 82 | | 1, 88 | |
| 6.5 Circular Sawing | Carbide | 0, 1, 82 | | 0, 1, 82 | | 0, 1, 88 | |
| 6.6 Abrasive Cutoff | — | 1, 82 | | 1, 82 | | 0, 1, 88 | |
| 7.1 Gear Hobbing | HSS | 1, 82 | | 1, 82 | | 0, 1, 88 | |
| 7.2 Gear Cutting, Straight and Spiral Bevel | HSS | 1, 82 | | 1, 82 | | 1, 88 | |
| 7.3 Gear Shaping | HSS | 1, 82 | | 1, 82 | | 0, 1, 88 | |
| 7.4 Gear Shaving | HSS | 1, 82 | | 1, 82 | | 0, 1, 88 | |
| 7.5 Gear Grinding, Form | — | 1, 82 | | 1, 82 | | 1, 88 | |
| 8.1 Surface Grinding—Horiz. Spdl., Recip. Table | — | 1, 82 | | 1, 82 | | 1, 88 | |
| 8.2 Surface Grinding—Horiz. Spdl., Recip. Table | CBN | — | | — | | — | |
| 8.3 Surface Grinding—Horiz. Spdl., Recip. Table | Diamond | — | | — | | — | |
| 8.4 Surface Grinding—Vertical Spdl., Rotary Table | — | 1, 82 | | 1, 82 | | 1, 88 | |
| 8.5 Cylindrical Grinding | — | 1, 82 | | 1, 82 | | 1, 88 | |
| 8.6 Cylindrical Grinding | CBN | — | | — | | — | |
| 8.7 Cylindrical Grinding | Diamond | — | | — | | — | |
| 8.8 Internal Grinding | — | 1, 82 | | 1, 82 | | 1, 88 | |
| 8.9 Internal Grinding | CBN | — | | — | | — | |
| 8.10 Internal Grinding | Diamond | — | | — | | — | |
| 8.11 Centerless Grinding | — | 1, 82 | | 1, 82 | | 1, 88 | |
| 8.13 Abrasive Belt Grinding | — | 1, 82 | | 1, 82 | | 1, 88 | |
| 8.14 Thread Grinding | — | 1, 82 | | 1, 82 | | 1, 88 | |
| | | | | | | | |

## 16.2 Cutting Fluid Recommendations

| OPERATION | TOOL MATERIAL OR ABRASIVE | 31. Magnesium Alloys, Cast | | 32. Titanium Alloys, Wrought | | 33. Titanium Alloys, Cast | |
|---|---|---|---|---|---|---|---|
| | | 50 to 90 Bhn (500 kg) | — | 110 to 440 Bhn | — | 150 to 350 Bhn | — |
| **1.1 Turning, Single Point and Box Tools** | HSS | 1, 88 | | 83 | | 83 | |
| | Carbide | 1, 88 | | 0, 83 | | 0, 83 | |
| **1.3 Turning** | Diamond | 1, 88 | | — | | — | |
| **1.4 Turning, Cutoff and Form Tools** | HSS | 1, 88 | | 83, 91 | | 83, 91 | |
| | Carbide | 1, 88 | | 83 | | 83 | |
| **1.5 Threading, Single Point** | HSS | 1, 88 | | 83, 91 | | 83, 91 | |
| | Carbide | 1, 88 | | 83 | | 83 | |
| **1.6 Threading, Die** | HSS | 1, 88 | | 83, 91 | | 83, 91 | |
| | Carbide | 1, 88 | | 83 | | 83 | |
| **1.7 Hollow Milling** | HSS | 1, 88 | | 83, 91 | | 83, 91 | |
| | Carbide | 1, 88 | | 83 | | 83 | |
| **2.1 Face Milling** | HSS | 1, 88 | | 83, 91 | | 83, 91 | |
| | Carbide | 1, 88 | | 0, 83 | | 0, 83 | |
| **2.2 Face Milling** | Diamond | 1, 88 | | — | | — | |
| **2.4 Slab Milling** | HSS | 1, 88 | | 83, 91 | | 83, 91 | |
| | Carbide | 1, 88 | | 83 | | 83 | |
| **2.5 Side and Slot Milling—Arbor Mounted Cutters** | HSS | 1, 88 | | 83, 91 | | 83, 91 | |
| | Carbide | 1, 88 | | 0, 83 | | 83 | |
| **2.6 End Milling—Peripheral** | HSS | 1, 88 | | 83, 91 | | 83, 91 | |
| | Carbide | 1, 88 | | 83 | | 83 | |
| **2.8 End Milling—Slotting** | HSS | 1, 88 | | 83, 91 | | 83, 91 | |
| | Carbide | 1, 88 | | 83 | | 83 | |
| **2.9 Thread Milling** | HSS | 1, 88 | | 83, 91 | | 83, 91 | |
| **3.1 Drilling** | HSS | 1, 88 | | 83, 91 | | 83, 91 | |
| | Carbide | 1, 88 | | 83 | | 83 | |
| **3.3 Oil-Hole or Pressurized-Coolant Drilling** | HSS | 1, 88 | | 83, 91 | | 83, 91 | |
| | Carbide | 1, 88 | | 83 | | 83 | |
| **3.4 Spade Drilling** | HSS | 1, 88 | | 83, 91 | | 83, 91 | |
| | Carbide | 88 | | 0, 83 | | 0, 83 | |
| **3.5 Gundrilling** | HSS | 1, 88 | | 83, 91 | | 83, 91 | |
| | Carbide | 1, 88 | | 83, 91 | | 83, 91 | |
| **3.6 Pressure Coolant Reaming (Gun Reaming)** | HSS | 1, 88 | | 83, 91 | | 83, 91 | |
| | Carbide | 1, 88 | | 83, 91 | | 83, 91 | |
| **3.7 Reaming** | HSS | 1, 88 | | 83, 91 | | 83, 91 | |
| | Carbide | 1, 88 | | 83, 91 | | 83, 91 | |
| **3.9 Boring** | HSS | 1, 88 | | 83, 91 | | 83, 91 | |
| | Carbide | 1, 88 | | 83, 91 | | 83, 91 | |
| **3.11 Boring** | Diamond | 1, 88 | | — | | — | |
| **3.12 Counterboring and Spotfacing** | HSS | 1, 88 | | 83, 91 | | 83, 91 | |
| | Carbide | 1, 88 | | 83 | | 83 | |
| **3.13 Trepanning** | HSS | 1, 88 | | 83, 91 | | 83, 91 | |
| | Carbide | 1, 88 | | 83 | | 83 | |
| **3.14 Honing** | — | 80 | | 90 | | 90 | |
| **3.15 Burnishing** | — | — | | — | | — | |
| **4.1 Tapping** | HSS | 1, 88 | | 1, 83 | | 1, 83 | |

| OPERATION | TOOL MATERIAL OR ABRASIVE | 31. Magnesium Alloys, Cast | | 32. Titanium Alloys, Wrought | | 33. Titanium Alloys, Cast | |
|---|---|---|---|---|---|---|---|
| | | 50 to 90 Bhn (500 kg) | — | 110 to 440 Bhn | — | 150 to 350 Bhn | — |
| 5.1 Planing | HSS | 1, 88 | | 83, 91 | | 83, 91 | |
| | Carbide | 1, 88 | | 83 | | 83 | |
| 5.2 Broaching | HSS | 0, 1, 88 | | 83, 91 | | 83, 91 | |
| | Carbide | 0, 1, 88 | | 83, 91 | | 83, 91 | |
| 6.1 Power Hack Sawing | HSS | 0, 1, 88 | | 83, 91 | | 83, 91 | |
| 6.2 Power Band Sawing | HSS | 0, 1, 88 | | 83, 91 | | 83, 91 | |
| 6.4 Circular Sawing | HSS | 1, 88 | | 83, 91 | | 83, 91 | |
| 6.5 Circular Sawing | Carbide | 0, 1, 88 | | 83, 91 | | 83, 91 | |
| 6.6 Abrasive Cutoff | — | 0, 1, 88 | | 81, 83 | | 81, 83 | |
| 7.1 Gear Hobbing | HSS | 0, 1, 88 | | 1, 83 | | 1, 83 | |
| 7.2 Gear Cutting, Straight and Spiral Bevel | HSS | 1, 88 | | 1, 83 | | 1, 83 | |
| 7.3 Gear Shaping | HSS | 0, 1, 88 | | 1, 83 | | 1, 83 | |
| 7.4 Gear Shaving | HSS | 0, 1, 88 | | 1, 83 | | 1, 83 | |
| 7.5 Gear Grinding, Form | — | 1, 88 | | 81, 83 | | 81, 83 | |
| 8.1 Surface Grinding—Horiz. Spdl., Recip. Table | — | 1, 88 | | 81, 83 | | 81, 83 | |
| 8.2 Surface Grinding—Horiz. Spdl., Recip. Table | CBN | — | | — | | — | |
| 8.3 Surface Grinding—Horiz. Spdl., Recip. Table | Diamond | — | | — | | — | |
| 8.4 Surface Grinding—Vertical Spdl., Rotary Table | — | 1, 88 | | 81, 83 | | 81, 83 | |
| 8.5 Cylindrical Grinding | — | 1, 88 | | 81, 83 | | 81, 83 | |
| 8.6 Cylindrical Grinding | CBN | — | | — | | — | |
| 8.7 Cylindrical Grinding | Diamond | — | | — | | — | |
| 8.8 Internal Grinding | — | 1, 88 | | 81, 83 | | 81, 83 | |
| 8.9 Internal Grinding | CBN | — | | — | | — | |
| 8.10 Internal Grinding | Diamond | — | | — | | — | |
| 8.11 Centerless Grinding | — | 1, 88 | | 81, 83 | | 81, 83 | |
| 8.13 Abrasive Belt Grinding | — | 1, 88 | | 81, 83 | | 81, 83 | |
| 8.14 Thread Grinding | — | 1, 88 | | 81, 83 | | 81, 83 | |
| | | | | | | | |

## 16.2 Cutting Fluid Recommendations

| OPERATION | TOOL MATERIAL OR ABRASIVE | 34. Copper Alloys, Wrought | | 35. Copper Alloys, Cast | | 36. Nickel Alloys, Wrought and Cast | |
|---|---|---|---|---|---|---|---|
| | | 10 to 100 $R_B$ | – | 40 to 425 Bhn | – | 80 to 360 Bhn | – |
| 1.1 Turning, Single Point and Box Tools | HSS | 4, 6, 87 | | 4, 6, 87 | | 3, 5, 7 | |
| | Carbide | 4, 6 | | 4, 6 | | 0, 5, 7 | |
| 1.3 Turning | Diamond | 4, 6, 87 | | 4, 6, 87 | | – | |
| 1.4 Turning, Cutoff and Form Tools | HSS | 4, 6, 87 | | 4, 6, 87 | | 3, 5, 7 | |
| | Carbide | 4, 6 | | 4, 6 | | 5, 7 | |
| 1.5 Threading, Single Point | HSS | 1, 87 | | 1, 87 | | 3, 5, 7 | |
| | Carbide | 0, 1, 87 | | 0, 1, 87 | | 5, 7 | |
| 1.6 Threading, Die | HSS | 1, 87 | | 1, 87 | | 3, 5, 7 | |
| | Carbide | 0, 1, 87 | | 0, 1, 87 | | 5, 7 | |
| 1.7 Hollow Milling | HSS | 1, 87 | | 1, 87 | | 3, 5, 7 | |
| | Carbide | 0, 1, 87 | | 0, 1, 87 | | 2, 5, 7 | |
| 2.1 Face Milling | HSS | 4, 6, 87 | | 4, 6, 87 | | 3, 5, 7 | |
| | Carbide | 4, 6 | | 4, 6 | | 0, 5, 7 | |
| 2.2 Face Milling | Diamond | 4, 6, 87 | | 4, 6, 87 | | – | |
| 2.4 Slab Milling | HSS | 4, 6, 87 | | 4, 6, 87 | | 3, 5, 7 | |
| | Carbide | 4, 6, 87 | | 4, 6, 87 | | 2, 5, 7 | |
| 2.5 Side and Slot Milling—Arbor Mounted Cutters | HSS | 4, 6, 87 | | 4, 6, 87 | | 3, 5, 7 | |
| | Carbide | 4, 6, 87 | | 4, 6, 87 | | 2, 5, 7 | |
| 2.6 End Milling—Peripheral | HSS | 4, 6, 87 | | 4, 6, 87 | | 3, 5, 7 | |
| | Carbide | 4, 6, 87 | | 4, 6, 87 | | 0, 5, 7 | |
| 2.8 End Milling—Slotting | HSS | 4, 6, 87 | | 4, 6, 87 | | 3, 5, 7 | |
| | Carbide | 4, 6, 87 | | 4, 6, 87 | | 0, 5, 7 | |
| 2.9 Thread Milling | HSS | 1, 87 | | 1, 87 | | 3, 5, 7 | |
| 3.1 Drilling | HSS | 4, 6, 87 | | 4, 6, 87 | | 3, 5, 7 | |
| | Carbide | 4, 6, 87 | | 4, 6, 87 | | 0, 5, 7 | |
| 3.3 Oil-Hole or Pressurized-Coolant Drilling | HSS | 4, 6, 87 | | 4, 6, 87 | | 3, 5, 7 | |
| | Carbide | 4, 6, 87 | | 4, 6, 87 | | 2, 5, 7 | |
| 3.4 Spade Drilling | HSS | 4, 6, 87 | | 4, 6, 87 | | 3, 5, 7 | |
| | Carbide | 4, 6, 87 | | 4, 6, 87 | | 0, 5, 7 | |
| 3.5 Gundrilling | HSS | 1, 86, 87 | | 1, 86, 87 | | 3, 5, 7 | |
| | Carbide | 1, 86, 87 | | 1, 86, 87 | | 2, 5, 7 | |
| 3.6 Pressure Coolant Reaming (Gun Reaming) | HSS | 1, 87 | | 1, 87 | | 5, 7, 3 | |
| | Carbide | 1, 87 | | 1, 87 | | 5, 7, 2 | |
| 3.7 Reaming | HSS | 1, 87 | | 1, 87 | | 5, 7, 3 | |
| | Carbide | 1, 87 | | 1, 87 | | 5, 7, 2 | |
| 3.9 Boring | HSS | 4, 6, 87 | | 4, 6, 87 | | 5, 7, 3 | |
| | Carbide | 4, 6 | | 4, 6 | | 0, 5, 7 | |
| 3.11 Boring | Diamond | 4, 6, 87 | | 4, 6, 87 | | – | |
| 3.12 Counterboring and Spotfacing | HSS | 4, 6, 87 | | 4, 6, 87 | | 5, 7, 3 | |
| | Carbide | 4, 6 | | 4, 6 | | 5, 7, 2 | |
| 3.13 Trepanning | HSS | 1, 86, 87 | | 1, 86, 87 | | 2, 86 | |
| | Carbide | 1, 86, 87 | | 1, 86, 87 | | 5, 7, 2 | |
| 3.14 Honing | – | 80 | | 80 | | 90 | |
| 3.15 Burnishing | – | 1, 87 | | 1, 87 | | – | |
| 4.1 Tapping | HSS | 1, 87 | | 1, 87 | | 2, 3 | |

| OPERATION | TOOL MATERIAL OR ABRASIVE | 34. Copper Alloys, Wrought | | 35. Copper Alloys, Cast | | 36. Nickel Alloys, Wrought and Cast | |
|---|---|---|---|---|---|---|---|
| | | 10 to 100 $R_B$ | — | 40 to 425 Bhn | — | 80 to 360 Bhn | — |
| 5.1 Planing | HSS | 4, 6, 87 | | 4, 6, 87 | | 5, 7, 3 | |
| | Carbide | 4, 6 | | 4, 6 | | 0, 5, 7 | |
| 5.2 Broaching | HSS | 4, 6, 87 | | 4, 6, 87 | | 5, 7, 3 | |
| | Carbide | 4, 6, 87 | | 4, 6, 87 | | 5, 7, 3 | |
| 6.1 Power Hack Sawing | HSS | 1, 87 | | 1, 87 | | 5, 7, 3 | |
| 6.2 Power Band Sawing | HSS | 1, 87 | | 1, 87 | | 5, 7, 3 | |
| 6.4 Circular Sawing | HSS | 4, 6, 87 | | 4, 6, 87 | | 5, 7, 3 | |
| 6.5 Circular Sawing | Carbide | 1, 87 | | 1, 87 | | 5, 7, 2 | |
| 6.6 Abrasive Cutoff | — | 1, 87 | | 1, 87 | | 5, 7 | |
| 7.1 Gear Hobbing | HSS | 1, 87 | | 1, 87 | | — | |
| 7.2 Gear Cutting, Straight and Spiral Bevel | HSS | 1, 87 | | 1, 87 | | — | |
| 7.3 Gear Shaping | HSS | 1, 87 | | 1, 87 | | — | |
| 7.4 Gear Shaving | HSS | 1, 87 | | 1, 87 | | — | |
| 7.5 Gear Grinding, Form | — | 1, 87 | | 1, 87 | | — | |
| 8.1 Surface Grinding—Horiz. Spdl., Recip. Table | — | 1, 87 | | 1, 87 | | 2, 3 | |
| 8.2 Surface Grinding—Horiz. Spdl., Recip. Table | CBN | — | | — | | — | |
| 8.3 Surface Grinding—Horiz. Spdl., Recip. Table | Diamond | 4, 87 | | 4, 87 | | — | |
| 8.4 Surface Grinding—Vertical Spdl., Rotary Table | — | 1, 87 | | 1, 87 | | 2, 3 | |
| 8.5 Cylindrical Grinding | — | 1, 87 | | 1, 87 | | 2, 3 | |
| 8.6 Cylindrical Grinding | CBN | — | | — | | — | |
| 8.7 Cylindrical Grinding | Diamond | 4, 87 | | 4, 87 | | — | |
| 8.8 Internal Grinding | — | 1, 87 | | 1, 87 | | 2, 3 | |
| 8.9 Internal Grinding | CBN | — | | — | | — | |
| 8.10 Internal Grinding | Diamond | 4, 87 | | 4, 87 | | — | |
| 8.11 Centerless Grinding | — | 1, 87 | | 1, 87 | | 2, 3 | |
| 8.13 Abrasive Belt Grinding | — | 1, 87 | | 1, 87 | | 3 | |
| 8.14 Thread Grinding | — | 1, 87 | | 1, 87 | | 2, 3 | |
| | | | | | | | |

# 16.2 Cutting Fluid Recommendations

| OPERATION | TOOL MATERIAL OR ABRASIVE | 37. Beryllium Nickel Alloys, Wrought and Cast | | 38. Nitinol Alloys, Wrought | | 39. High Temperature Alloys, Wrought and Cast | |
|---|---|---|---|---|---|---|---|
| | | 200 to 425 Bhn | 47 to 52 $R_C$ | 210 to 515 Bhn | — | 140 to 475 Bhn | — |
| 1.1 Turning, Single Point and Box Tools | HSS | 3, 5, 7 | — | 2, 5, 7 | | 3, 5, 7 | |
| | Carbide | 5, 7 | — | 1, 4, 6 | | 0, 4, 6 | |
| 1.3 Turning | Diamond | — | — | — | | — | |
| 1.4 Turning, Cutoff and Form Tools | HSS | 3, 5, 7 | — | 3, 5, 7 | | 3 | |
| | Carbide | 5, 7 | — | 2, 5, 7 | | 3 | |
| 1.5 Threading, Single Point | HSS | 3 | — | 3, 5, 7 | | 3 | |
| | Carbide | — | — | 2, 4, 6 | | 2, 5, 7 | |
| 1.6 Threading, Die | HSS | 3, 5, 7 | — | 3, 5, 7 | | 3 | |
| | Carbide | — | — | 2, 4, 6 | | 2, 5, 7 | |
| 1.7 Hollow Milling | HSS | — | — | 2, 3 | | 3 | |
| | Carbide | — | — | 2 | | 2, 5, 7 | |
| 2.1 Face Milling | HSS | 3, 5, 7 | — | 2, 3 | | 2, 5 | |
| | Carbide | 5, 7 | — | 2, 3 | | 0, 3 | |
| 2.2 Face Milling | Diamond | — | — | — | | — | |
| 2.4 Slab Milling | HSS | 3, 5, 7 | — | — | | 3 | |
| | Carbide | — | — | — | | 3 | |
| 2.5 Side and Slot Milling—Arbor Mounted Cutters | HSS | 3, 5, 7 | — | 2, 3 | | 3 | |
| | Carbide | 5, 7 | — | 2, 3 | | 3 | |
| 2.6 End Milling—Peripheral | HSS | 3, 5, 7 | — | 2, 3 | | 3 | |
| | Carbide | 5, 7 | — | 2, 3 | | 3 | |
| 2.8 End Milling—Slotting | HSS | 3, 5, 7 | — | 2, 3 | | 3 | |
| | Carbide | — | — | 2, 3 | | 3 | |
| 2.9 Thread Milling | HSS | 3, 5, 7 | — | 2, 3 | | 3 | |
| 3.1 Drilling | HSS | 3, 5, 7 | — | 2, 3 | | 3 | |
| | Carbide | — | — | 2, 3 | | 3 | |
| 3.3 Oil-Hole or Pressurized-Coolant Drilling | HSS | 3, 5, 7 | — | 2, 3 | | 3 | |
| | Carbide | 5, 7 | — | 2, 3 | | 3 | |
| 3.4 Spade Drilling | HSS | 3, 5, 7 | — | 2, 3 | | 3 | |
| | Carbide | — | — | 2, 3 | | 3 | |
| 3.5 Gundrilling | HSS | — | — | 2, 3, 86 | | 3 | |
| | Carbide | 5, 7 | — | 2, 3 | | 3 | |
| 3.6 Pressure Coolant Reaming (Gun Reaming) | HSS | — | — | 2, 3 | | 3 | |
| | Carbide | — | — | 2, 3 | | 3 | |
| 3.7 Reaming | HSS | 3, 5, 7 | — | 2, 3 | | 3 | |
| | Carbide | 5, 7 | — | 2, 3 | | 3 | |
| 3.9 Boring | HSS | 3, 5, 7 | — | 2, 5, 7 | | 3, 5, 7 | |
| | Carbide | 5, 7 | — | 2, 5, 7 | | 3 | |
| 3.11 Boring | Diamond | — | — | — | | — | |
| 3.12 Counterboring and Spotfacing | HSS | 3, 5, 7 | — | 2, 5, 7 | | 3 | |
| | Carbide | 5, 7 | — | 2, 5, 7 | | 3 | |
| 3.13 Trepanning | HSS | 3, 5, 7 | — | 2, 3, 86 | | 3 | |
| | Carbide | 5, 7 | — | 2, 86 | | 3 | |
| 3.14 Honing | — | 90 | — | 90 | | 90 | |
| 3.15 Burnishing | — | — | — | — | | — | |
| 4.1 Tapping | HSS | 3, 94 | — | 2, 3 | | 3 | |

| OPERATION | TOOL MATERIAL OR ABRASIVE | 37. Beryllium Nickel Alloys, Wrought and Cast | | 38. Nitinol Alloys, Wrought | | 39. High Temperature Alloys, Wrought and Cast | |
|---|---|---|---|---|---|---|---|
| | | 200 to 425 Bhn | 47 to 52 $R_C$ | 210 to 515 Bhn | — | 140 to 475 Bhn | — |
| 5.1 Planing | HSS | — | — | — | | — | |
| | Carbide | — | — | — | | — | |
| 5.2 Broaching | HSS | 5, 3 | — | — | | 3 | |
| | Carbide | 5, 7, 3 | — | — | | 3 | |
| 6.1 Power Hack Sawing | HSS | 5, 7, 3 | — | 2, 3 | | 3 | |
| 6.2 Power Band Sawing | HSS | 5, 7, 3 | — | 2, 3 | | 3 | |
| 6.4 Circular Sawing | HSS | 5, 7, 3 | — | 2, 3 | | 3 | |
| 6.5 Circular Sawing | Carbide | 5, 7, 2 | 5, 7, 2 | 2, 3 | | 3 | |
| 6.6 Abrasive Cutoff | — | 5, 7 | 5, 7 | 2, 3 | | 5, 7 | |
| 7.1 Gear Hobbing | HSS | — | — | — | | 3 | |
| 7.2 Gear Cutting, Straight and Spiral Bevel | HSS | — | — | — | | 3 | |
| 7.3 Gear Shaping | HSS | — | — | — | | 3 | |
| 7.4 Gear Shaving | HSS | — | — | — | | 3 | |
| 7.5 Gear Grinding, Form | — | — | — | — | | 3 | |
| 8.1 Surface Grinding—Horiz. Spdl., Recip. Table | — | 2, 3 | 2, 3 | 2, 3 | | 3 | |
| 8.2 Surface Grinding—Horiz. Spdl., Recip. Table | CBN | — | — | — | | 5, 3 | |
| 8.3 Surface Grinding—Horiz. Spdl., Recip. Table | Diamond | — | — | — | | — | |
| 8.4 Surface Grinding—Vertical Spdl., Rotary Table | — | 2, 3 | 2, 3 | 2, 3 | | 3 | |
| 8.5 Cylindrical Grinding | — | 2, 3 | 2, 3 | 2, 3 | | 3 | |
| 8.6 Cylindrical Grinding | CBN | — | — | — | | 5, 3 | |
| 8.7 Cylindrical Grinding | Diamond | — | — | — | | — | |
| 8.8 Internal Grinding | — | 2, 3 | 2, 3 | 2, 3 | | 3 | |
| 8.9 Internal Grinding | CBN | — | — | — | | 5, 3 | |
| 8.10 Internal Grinding | Diamond | — | — | — | | — | |
| 8.11 Centerless Grinding | — | 2, 3 | 2, 3 | 2, 3 | | 3 | |
| 8.13 Abrasive Belt Grinding | — | 3 | 3 | 2, 3 | | 3 | |
| 8.14 Thread Grinding | — | 2, 3 | 2, 3 | 2, 3 | | 3 | |
| | | | | | | | |

# 16.2 Cutting Fluid Recommendations

| OPERATION | TOOL MATERIAL OR ABRASIVE | 40. Refractory Metals, Wrought, Cast and P/M | | 41. Zinc Alloys, Cast | 42. Lead Alloys, Cast | 43. Tin Alloys, Cast | 44. Uranium, Wrought |
|---|---|---|---|---|---|---|---|
| | | 170 to 320 Bhn | — | 80 to 100 Bhn | 10 to 20 Bhn (500 kg) | 15 to 30 Bhn (500 kg) | 56 to 58 $R_A$ |
| 1.1 Turning, Single Point and Box Tools | HSS | 3, 5 | | 0, 4 | — | — | 4, 6 |
| | Carbide | 2, 5 | | 0 | — | — | 4, 6 |
| 1.3 Turning | Diamond | — | | — | — | — | — |
| 1.4 Turning, Cutoff and Form Tools | HSS | 3, 5 | | 0, 4 | — | — | 4, 6 |
| | Carbide | 2, 5 | | 0 | — | — | 4, 6 |
| 1.5 Threading, Single Point | HSS | — | | 0, 1, 4 | — | — | 4, 6 |
| | Carbide | — | | 0, 1 | — | — | 4, 6 |
| 1.6 Threading, Die | HSS | — | | 0, 1, 4 | — | — | 4, 6 |
| | Carbide | — | | 0, 1 | — | — | 4, 6 |
| 1.7 Hollow Milling | HSS | 3, 5, 7 | | 0, 1, 4 | — | — | 4, 6 |
| | Carbide | 2, 5, 7 | | 0, 1 | — | — | 4, 6 |
| 2.1 Face Milling | HSS | 2, 5 | | 0, 1 | — | — | 4, 6 |
| | Carbide | 2, 5 | | 0 | — | — | 4, 6 |
| 2.2 Face Milling | Diamond | — | | — | — | — | — |
| 2.4 Slab Milling | HSS | 3, 5 | | 0, 1, 4 | — | — | 4, 6 |
| | Carbide | — | | 0, 1 | — | — | 4, 6 |
| 2.5 Side and Slot Milling—Arbor Mounted Cutters | HSS | 3, 5 | | 0, 1, 4 | — | — | 4, 6 |
| | Carbide | 2, 5 | | 0 | — | — | 4, 6 |
| 2.6 End Milling—Peripheral | HSS | 2, 5 | | 0, 1, 4 | — | — | 4, 6 |
| | Carbide | 5 | | 0, 4 | — | — | 4, 6 |
| 2.8 End Milling—Slotting | HSS | 2, 5 | | 0, 1, 4 | — | — | 4, 6 |
| | Carbide | 5 | | 0, 4 | — | — | 4, 6 |
| 2.9 Thread Milling | HSS | 3 | | 0, 1, 4 | — | — | 4, 6 |
| 3.1 Drilling | HSS | 3 | | 0, 1, 4 | — | — | 4, 6 |
| | Carbide | 2 | | 0, 4 | — | — | 4, 6 |
| 3.3 Oil-Hole or Pressurized-Coolant Drilling | HSS | 3 | | 1, 4 | — | — | 4, 6 |
| | Carbide | 2 | | 1, 4 | — | — | 4, 6 |
| 3.4 Spade Drilling | HSS | 3 | | 0, 1, 4 | — | — | 4, 6 |
| | Carbide | 2 | | 0, 4 | — | — | 4, 6 |
| 3.5 Gundrilling | HSS | 3 | | 1, 86 | — | — | 4, 6 |
| | Carbide | 2 | | 1, 86 | — | — | 4, 6 |
| 3.6 Pressure Coolant Reaming (Gun Reaming) | HSS | 3 | | 1, 4 | — | — | 5, 7 |
| | Carbide | 3 | | 1, 4 | — | — | 4, 6 |
| 3.7 Reaming | HSS | 3 | | 0, 4 | — | — | 5, 7 |
| | Carbide | 3 | | 0, 4 | — | — | 4, 6 |
| 3.9 Boring | HSS | 3, 5 | | 0, 4 | — | — | 3 |
| | Carbide | 2, 5 | | 0 | — | — | 3 |
| 3.11 Boring | Diamond | — | | — | — | — | — |
| 3.12 Counterboring and Spotfacing | HSS | 3, 5 | | 0, 4 | — | — | 5, 7 |
| | Carbide | 2, 5 | | 0, 4 | — | — | 4, 6 |
| 3.13 Trepanning | HSS | 3 | | 4, 86 | — | — | 3 |
| | Carbide | 2 | | 1, 4 | — | — | 3 |
| 3.14 Honing | — | 90 | | 90 | — | — | 90 |
| 3.15 Burnishing | — | — | | — | — | — | — |
| 4.1 Tapping | HSS | 3 | | 1, 4 | — | — | 3 |

| OPERATION | TOOL MATERIAL OR ABRASIVE | 40. Refractory Metals, Wrought, Cast and P/M | | 41. Zinc Alloys, Cast | 42. Lead Alloys, Cast | 43. Tin Alloys, Cast | 44. Uranium, Wrought |
|---|---|---|---|---|---|---|---|
| | | 170 to 320 Bhn | — | 80 to 100 Bhn | 10 to 20 Bhn (500 kg) | 15 to 30 Bhn (500 kg) | 56 to 58 $R_A$ |
| **5.1 Planing** | HSS | — | | — | — | — | — |
| | Carbide | — | | — | — | — | — |
| **5.2 Broaching** | HSS | — | | 0, 1, 4 | — | — | — |
| | Carbide | — | | 0, 1 | — | — | — |
| **6.1 Power Hack Sawing** | HSS | 3 | | 1, 4 | 0* | 0* | 4, 6 |
| **6.2 Power Band Sawing** | HSS | 3 | | 1, 4 | 0* | 0* | 4, 6 |
| **6.4 Circular Sawing** | HSS | 3 | | 1, 4 | 0* | 0* | 5, 7 |
| **6.5 Circular Sawing** | Carbide | 3 | | 0, 4 | 0* | 0* | 5, 7 |
| **6.6 Abrasive Cutoff** | — | 1, 83 | | 0, 4 | — | — | 4, 6 |
| **7.1 Gear Hobbing** | HSS | — | | 1, 4 | — | — | — |
| **7.2 Gear Cutting, Straight and Spiral Bevel** | HSS | — | | 1, 4 | — | — | — |
| **7.3 Gear Shaping** | HSS | — | | 1, 4 | — | — | — |
| **7.4 Gear Shaving** | HSS | — | | 1, 4 | — | — | — |
| **7.5 Gear Grinding, Form** | — | — | | 1, 4 | — | — | — |
| **8.1 Surface Grinding—Horiz. Spdl., Recip. Table** | — | 1, 83 | | 4 | — | — | 4, 6 |
| **8.2 Surface Grinding—Horiz. Spdl., Recip. Table** | CBN | — | | — | — | — | — |
| **8.3 Surface Grinding—Horiz. Spdl., Recip. Table** | Diamond | — | | — | — | — | — |
| **8.4 Surface Grinding—Vertical Spdl., Rotary Table** | — | 1, 83 | | 4 | — | — | 4, 6 |
| **8.5 Cylindrical Grinding** | — | 1, 83 | | 4 | — | — | 4, 6 |
| **8.6 Cylindrical Grinding** | CBN | — | | — | — | — | — |
| **8.7 Cylindrical Grinding** | Diamond | — | | — | — | — | — |
| **8.8 Internal Grinding** | — | 1, 83 | | 4 | — | — | 4, 6 |
| **8.9 Internal Grinding** | CBN | — | | — | — | — | — |
| **8.10 Internal Grinding** | Diamond | — | | — | — | — | — |
| **8.11 Centerless Grinding** | — | 1, 83 | | 4 | — | — | 4, 6 |
| **8.13 Abrasive Belt Grinding** | — | 3, 91 | | — | — | — | — |
| **8.14 Thread Grinding** | — | 1, 83 | | 4 | — | — | 4, 6 |

*Compressed air may be used to facilitate chip removal.

## 16.2 Cutting Fluid Recommendations

| OPERATION | TOOL MATERIAL OR ABRASIVE | 45. Zirconium Alloys, Wrought | | 47. Powder Metal Alloys | 48. Machinable Carbides | | 49. Carbides |
|---|---|---|---|---|---|---|---|
| | | 140 to 280 Bhn | — | All Hardness | 40 to 51 $R_C$ | 68 to 70 $R_C$ | 89 to 94 $R_A$ |
| **1.1 Turning, Single Point and Box Tools** | HSS | 1, 4, 6 | | * | — | — | — |
| | Carbide | 4, 6 | | * | — | — | — |
| **1.3 Turning** | Diamond | — | | * | — | — | — |
| **1.4 Turning, Cutoff and Form Tools** | HSS | 1, 5, 7 | | * | — | — | — |
| | Carbide | 4, 6 | | * | — | — | — |
| **1.5 Threading, Single Point** | HSS | 1, 5, 7 | | * | — | — | — |
| | Carbide | 5, 7 | | * | — | — | — |
| **1.6 Threading, Die** | HSS | 1, 5, 7 | | * | — | — | — |
| | Carbide | 5, 7 | | * | — | — | — |
| **1.7 Hollow Milling** | HSS | 1, 5, 7 | | * | — | — | — |
| | Carbide | 4, 6 | | * | — | — | — |
| **2.1 Face Milling** | HSS | 1, 4, 6 | | * | — | — | — |
| | Carbide | 4, 6 | | * | — | — | — |
| **2.2 Face Milling** | Diamond | — | | * | — | — | — |
| **2.4 Slab Milling** | HSS | 1, 5, 7 | | * | — | — | — |
| | Carbide | 4, 6 | | * | — | — | — |
| **2.5 Side and Slot Milling—Arbor Mounted Cutters** | HSS | 1, 5, 7 | | * | — | — | — |
| | Carbide | 4, 6 | | * | — | — | — |
| **2.6 End Milling—Peripheral** | HSS | 1, 5, 7 | | * | — | — | — |
| | Carbide | 4, 6 | | * | — | — | — |
| **2.8 End Milling—Slotting** | HSS | 1, 5, 7 | | * | — | — | — |
| | Carbide | 4, 6 | | * | — | — | — |
| **2.9 Thread Milling** | HSS | 2, 5, 7 | | * | — | — | — |
| **3.1 Drilling** | HSS | 1, 5, 7 | | * | — | — | — |
| | Carbide | 4, 6 | | * | — | — | — |
| **3.3 Oil-Hole or Pressurized-Coolant Drilling** | HSS | 1, 5, 7 | | * | — | — | — |
| | Carbide | 4, 6 | | * | — | — | — |
| **3.4 Spade Drilling** | HSS | 1, 5, 7 | | * | — | — | — |
| | Carbide | 4, 6 | | * | — | — | — |
| **3.5 Gundrilling** | HSS | 1, 5, 7 | | * | — | — | — |
| | Carbide | 4, 6 | | * | — | — | — |
| **3.6 Pressure Coolant Reaming (Gun Reaming)** | HSS | 1, 5, 7 | | * | — | — | — |
| | Carbide | 4, 6 | | * | — | — | — |
| **3.7 Reaming** | HSS | 1, 5, 7 | | * | — | — | — |
| | Carbide | 4, 6 | | * | — | — | — |
| **3.9 Boring** | HSS | 1, 5, 7 | | * | — | — | — |
| | Carbide | 4, 6 | | * | — | — | — |
| **3.11 Boring** | Diamond | — | | * | — | — | — |
| **3.12 Counterboring and Spotfacing** | HSS | 1, 5, 7 | | * | — | — | — |
| | Carbide | 4, 6 | | * | — | — | — |
| **3.13 Trepanning** | HSS | 1, 5, 7 | | * | — | — | — |
| | Carbide | 4, 6 | | * | — | — | — |
| **3.14 Honing** | — | 90 | | * | 80 | 80 | 80 |
| **3.15 Burnishing** | — | — | | * | — | — | — |
| **4.1 Tapping** | HSS | — | | * | — | — | — |

*Use the same cutting fluid as for cast alloys unless prohibited by part requirements for cleanliness and porosity.

| OPERATION | TOOL MATERIAL OR ABRASIVE | 45. Zirconium Alloys, Wrought | | 47. Powder Metal Alloys | 48. Machinable Carbides | | 49. Carbides |
|---|---|---|---|---|---|---|---|
| | | 140 to 280 Bhn | — | All Hardness | 40 to 51 $R_C$ | 68 to 70 $R_C$ | 89 to 94 $R_A$ |
| **5.1 Planing** | HSS | — | | * | — | — | — |
| | Carbide | — | | * | — | — | — |
| **5.2 Broaching** | HSS | 5, 7 | | * | — | — | — |
| | Carbide | 4, 6 | | * | — | — | — |
| **6.1 Power Hack Sawing** | HSS | 4, 6 | | * | 0 | — | — |
| **6.2 Power Band Sawing** | HSS | 4, 6 | | * | 0 | — | — |
| **6.4 Circular Sawing** | HSS | 5, 7 | | * | 0 | — | — |
| **6.5 Circular Sawing** | Carbide | 4, 6 | | * | 0 | — | — |
| **6.6 Abrasive Cutoff** | — | 4, 6 | | * | — | — | — |
| **7.1 Gear Hobbing** | HSS | 2, 5, 7 | | * | — | — | — |
| **7.2 Gear Cutting, Straight and Spiral Bevel** | HSS | 2, 5, 7 | | * | — | — | — |
| **7.3 Gear Shaping** | HSS | 2, 5, 7 | | * | — | — | — |
| **7.4 Gear Shaving** | HSS | 2, 5, 7 | | * | — | — | — |
| **7.5 Gear Grinding, Form** | — | 2, 81 | | * | — | — | — |
| **8.1 Surface Grinding—Horiz. Spdl., Recip. Table** | — | 2, 81 | | * | 0, 4, 6 | 0, 4, 6 | 0, 4, 6 |
| **8.2 Surface Grinding—Horiz. Spdl., Recip. Table** | CBN | — | | * | — | — | — |
| **8.3 Surface Grinding—Horiz. Spdl., Recip. Table** | Diamond | — | | * | — | 4 | 4 |
| **8.4 Surface Grinding—Vertical Spdl., Rotary Table** | — | 2, 81 | | * | 0, 4, 6 | 0, 4, 6 | 0, 4, 6 |
| **8.5 Cylindrical Grinding** | — | 2, 81 | | * | 0, 4, 6 | 0, 4, 6 | 0, 4, 6 |
| **8.6 Cylindrical Grinding** | CBN | — | | * | — | — | — |
| **8.7 Cylindrical Grinding** | Diamond | — | | * | — | 4 | 4 |
| **8.8 Internal Grinding** | — | 2, 81 | | * | 0, 4, 6 | 0, 4, 6 | 0, 4, 6 |
| **8.9 Internal Grinding** | CBN | — | | * | — | — | — |
| **8.10 Internal Grinding** | Diamond | — | | * | — | 4 | 4 |
| **8.11 Centerless Grinding** | — | 2, 81 | | * | 0, 4, 6 | 0, 4, 6 | 0, 4, 6 |
| **8.13 Abrasive Belt Grinding** | — | 2, 3 | | * | 0, 4, 6 | 0, 4, 6 | 0, 4, 6 |
| **8.14 Thread Grinding** | — | 2, 81 | | * | 0, 4, 6 | 0, 4, 6 | 0, 4, 6 |
| | | | | | | | |

*Use the same cutting fluid as for cast alloys unless prohibited by part requirements for cleanliness and porosity.

## 16.2 Cutting Fluid Recommendations

| OPERATION | TOOL MATERIAL OR ABRASIVE | 50. Free Machining Magnetic Alloys | | 51. Magnetic Alloys | | 52. Free Machining Controlled Expansion Alloys | |
|---|---|---|---|---|---|---|---|
| | | 185 to 240 Bhn | — | 185 to 240 Bhn | 45 to 58 $R_C$ | 125 to 250 Bhn | — |
| **1.1 Turning, Single Point and Box Tools** | HSS | 1, 5, 7 | | 1, 5, 7 | — | 1, 5, 7 | |
| | Carbide | — | | — | — | — | |
| **1.3 Turning** | Diamond | — | | — | — | — | |
| **1.4 Turning, Cutoff and Form Tools** | HSS | 1, 5, 7 | | 1, 5, 7 | — | 1, 5, 7 | |
| | Carbide | 2, 5, 7 | | 2, 5, 7 | — | 2, 5, 7 | |
| **1.5 Threading, Single Point** | HSS | — | | — | — | — | |
| | Carbide | — | | — | — | — | |
| **1.6 Threading, Die** | HSS | — | | — | — | — | |
| | Carbide | — | | — | — | — | |
| **1.7 Hollow Milling** | HSS | — | | — | — | — | |
| | Carbide | — | | — | — | — | |
| **2.1 Face Milling** | HSS | 1, 5, 7 | | 1, 5, 7 | — | 1, 5, 7 | |
| | Carbide | — | | — | — | — | |
| **2.2 Face Milling** | Diamond | — | | — | — | — | |
| **2.4 Slab Milling** | HSS | 1, 5, 7 | | 1, 5, 7 | — | 1, 5, 7 | |
| | Carbide | — | | — | — | — | |
| **2.5 Side and Slot Milling—Arbor Mounted Cutters** | HSS | 1, 5, 7 | | 1, 5, 7 | — | 1, 5, 7 | |
| | Carbide | — | | 2, 5, 7 | — | — | |
| **2.6 End Milling—Peripheral** | HSS | 1, 5, 7 | | 1, 5, 7 | — | 1, 5, 7 | |
| | Carbide | — | | — | — | — | |
| **2.8 End Milling—Slotting** | HSS | 1, 5, 7 | | 1, 5, 7 | — | 1, 5, 7 | |
| | Carbide | — | | — | — | — | |
| **2.9 Thread Milling** | HSS | 1, 5, 7 | | 1, 5, 7 | — | 1, 5, 7 | |
| **3.1 Drilling** | HSS | 1, 5, 7 | | 1, 5, 7 | — | 1, 5, 7 | |
| | Carbide | — | | — | — | — | |
| **3.3 Oil-Hole or Pressurized-Coolant Drilling** | HSS | 1, 5, 7 | | 1, 5, 7 | — | 1, 5, 7 | |
| | Carbide | — | | — | — | — | |
| **3.4 Spade Drilling** | HSS | 1, 5, 7 | | 1, 5, 7 | — | — | |
| | Carbide | — | | — | — | — | |
| **3.5 Gundrilling** | HSS | — | | — | — | — | |
| | Carbide | — | | — | — | — | |
| **3.6 Pressure Coolant Reaming (Gun Reaming)** | HSS | — | | — | — | — | |
| | Carbide | — | | — | — | — | |
| **3.7 Reaming** | HSS | 1, 5, 7 | | 1, 5, 7 | — | 1, 5, 7 | |
| | Carbide | 2, 5, 7 | | 2, 5, 7 | — | 2, 5, 7 | |
| **3.9 Boring** | HSS | 1, 5, 7 | | 1, 5, 7 | — | 1, 5, 7 | |
| | Carbide | — | | — | — | — | |
| **3.11 Boring** | Diamond | — | | — | — | — | |
| **3.12 Counterboring and Spotfacing** | HSS | 1, 5, 7 | | 1, 5, 7 | — | 1, 5, 7 | |
| | Carbide | — | | 2, 5, 7 | — | — | |
| **3.13 Trepanning** | HSS | — | | — | — | — | |
| | Carbide | — | | — | — | — | |
| **3.14 Honing** | — | 90 | | 90 | — | 90 | |
| **3.15 Burnishing** | — | — | | — | — | — | |
| **4.1 Tapping** | HSS | 2, 3 | | 2, 3 | — | 2, 3 | |

| OPERATION | TOOL MATERIAL OR ABRASIVE | 50. Free Machining Magnetic Alloys | | 51. Magnetic Alloys | | 52. Free Machining Controlled Expansion Alloys | |
|---|---|---|---|---|---|---|---|
| | | 185 to 240 Bhn | — | 185 to 240 Bhn | 45 to 58 $R_C$ | 125 to 250 Bhn | — |
| 5.1 Planing | HSS | — | | — | — | — | |
| | Carbide | — | | — | — | — | |
| 5.2 Broaching | HSS | 5, 7, 3 | | 5, 7, 3 | — | 5, 7, 3 | |
| | Carbide | 5, 7, 3 | | 5, 7, 3 | — | 5, 7,3 | |
| 6.1 Power Hack Sawing | HSS | 5, 7, 3 | | 5, 7, 3 | — | 5, 7, 3 | |
| 6.2 Power Band Sawing | HSS | 5, 7, 3 | | 5, 7, 3 | — | 5, 7, 3 | |
| 6.4 Circular Sawing | HSS | 5, 7, 3 | | 5, 7, 3 | — | 5, 7, 3 | |
| 6.5 Circular Sawing | Carbide | 5, 7, 2 | | 5, 7, 2 | — | 5, 7, 2 | |
| 6.6 Abrasive Cutoff | — | 5, 7 | | 5, 7 | 2, 3 | 5, 7 | |
| 7.1 Gear Hobbing | HSS | — | | — | — | — | |
| 7.2 Gear Cutting, Straight and Spiral Bevel | HSS | — | | — | — | — | |
| 7.3 Gear Shaping | HSS | — | | — | — | — | |
| 7.4 Gear Shaving | HSS | — | | — | — | — | |
| 7.5 Gear Grinding, Form | — | — | | — | — | — | |
| 8.1 Surface Grinding—Horiz. Spdl., Recip. Table | — | 2, 3 | | 2, 3 | 2, 3 | 2, 3 | |
| 8.2 Surface Grinding—Horiz. Spdl., Recip. Table | CBN | — | | — | — | — | |
| 8.3 Surface Grinding—Horiz. Spdl., Recip. Table | Diamond | — | | — | — | — | |
| 8.4 Surface Grinding—Vertical Spdl., Rotary Table | — | 2, 3 | | 2, 3 | 2, 3 | 2, 3 | |
| 8.5 Cylindrical Grinding | — | 2, 3 | | 2, 3 | 2, 3 | 2, 3 | |
| 8.6 Cylindrical Grinding | CBN | — | | — | — | — | |
| 8.7 Cylindrical Grinding | Diamond | — | | — | — | — | |
| 8.8 Internal Grinding | — | 2, 3 | | 2, 3 | 2, 3 | 2, 3 | |
| 8.9 Internal Grinding | CBN | — | | — | — | — | |
| 8.10 Internal Grinding | Diamond | — | | — | — | — | |
| 8.11 Centerless Grinding | — | 2, 3 | | 2, 3 | 2, 3 | 2, 3 | |
| 8.13 Abrasive Belt Grinding | — | 3 | | 3 | 3 | 3 | |
| 8.14 Thread Grinding | — | 2, 3 | | 2, 3 | 2, 3 | 2, 3 | |
| | | | | | | | |

## 16.2 Cutting Fluid Recommendations

| OPERATION | TOOL MATERIAL OR ABRASIVE | 53. Controlled Expansion Alloys | | 54. Carbons & Graphites | | 55. Glasses & Ceramics | |
|---|---|---|---|---|---|---|---|
| | | | | | | Machinable Glass-Ceramic | Other |
| | | 125 to 250 Bhn | — | 40 to 100 Shore | — | 250 Knoop | Over 800 Knoop |
| 1.1 Turning, Single Point and Box Tools | HSS | 1, 5, 7 | | — | | 4, 93 | — |
| | Carbide | — | | — | | 4, 93 | — |
| 1.3 Turning | Diamond | — | | — | | 4, 93 | — |
| 1.4 Turning, Cutoff and Form Tools | HSS | 1, 5, 7 | | — | | 4, 93 | — |
| | Carbide | — | | — | | 4, 93 | — |
| 1.5 Threading, Single Point | HSS | — | | — | | 4, 93 | — |
| | Carbide | — | | — | | 4, 93 | — |
| 1.6 Threading, Die | HSS | — | | — | | — | — |
| | Carbide | — | | — | | — | — |
| 1.7 Hollow Milling | HSS | — | | — | | — | — |
| | Carbide | — | | — | | — | — |
| 2.1 Face Milling | HSS | 1, 5, 7 | | — | | — | — |
| | Carbide | — | | — | | — | — |
| 2.2 Face Milling | Diamond | — | | — | | 4, 93 | — |
| 2.4 Slab Milling | HSS | 1, 5, 7 | | — | | 4, 93 | — |
| | Carbide | — | | — | | — | — |
| 2.5 Side and Slot Milling—Arbor Mounted Cutters | HSS | 1, 5, 7 | | — | | 4, 93 | — |
| | Carbide | — | | — | | 4, 93 | — |
| 2.6 End Milling—Peripheral | HSS | 1, 5, 7 | | — | | 4, 93 | — |
| | Carbide | — | | — | | 4, 93 | — |
| 2.8 End Milling—Slotting | HSS | 1, 5, 7 | | — | | 4, 93 | — |
| | Carbide | — | | — | | — | — |
| 2.9 Thread Milling | HSS | 1, 5, 7 | | — | | 4, 93 | — |
| 3.1 Drilling | HSS | 1, 5, 7 | | — | | 4, 93 | — |
| | Carbide | — | | — | | — | — |
| 3.3 Oil-Hole or Pressurized-Coolant Drilling | HSS | 1, 5, 7 | | — | | 4, 93 | — |
| | Carbide | — | | — | | 4, 93 | — |
| 3.4 Spade Drilling | HSS | — | | — | | 4, 93 | — |
| | Carbide | — | | — | | 4, 93 | — |
| 3.5 Gundrilling | HSS | — | | — | | — | — |
| | Carbide | — | | — | | — | — |
| 3.6 Pressure Coolant Reaming (Gun Reaming) | HSS | — | | — | | — | — |
| | Carbide | — | | — | | — | — |
| 3.7 Reaming | HSS | 1, 5, 7 | | — | | — | — |
| | Carbide | 2, 5, 7 | | — | | — | — |
| 3.9 Boring | HSS | 1, 5, 7 | | — | | 4, 93 | — |
| | Carbide | — | | — | | 4, 93 | — |
| 3.11 Boring | Diamond | — | | — | | 4, 93 | — |
| 3.12 Counterboring and Spotfacing | HSS | 1, 5, 7 | | — | | 4, 93 | — |
| | Carbide | — | | — | | 4, 93 | — |
| 3.13 Trepanning | HSS | — | | — | | — | — |
| | Carbide | — | | — | | — | — |
| 3.14 Honing | — | 90 | | — | | 93 | — |
| 3.15 Burnishing | — | — | | — | | — | — |
| 4.1 Tapping | HSS | 2, 3 | | — | | — | — |

| OPERATION | TOOL MATERIAL OR ABRASIVE | 53. Controlled Expansion Alloys | | 54. Carbons & Graphites | | 55. Glasses & Ceramics | |
|---|---|---|---|---|---|---|---|
| | | | | | | Machinable Glass-Ceramic | Other |
| | | 125 to 250 Bhn | — | 40 to 100 Shore | — | 250 Knoop | Over 800 Knoop |
| **5.1 Planing** | HSS | — | | — | | — | — |
| | Carbide | — | | — | | — | — |
| **5.2 Broaching** | HSS | 5, 7, 3 | | 0* | | — | — |
| | Carbide | 5, 7, 3 | | 0* | | — | — |
| **6.1 Power Hack Sawing** | HSS | 5, 7, 3 | | 0* | | 4, 93 | — |
| **6.2 Power Band Sawing** | HSS | 5, 7, 3 | | 0* | | 4, 93 | — |
| **6.4 Circular Sawing** | HSS | 5, 7, 3 | | 0* | | 4, 93 | — |
| **6.5 Circular Sawing** | Carbide | 5, 7, 2 | | 0* | | 4, 93 | — |
| **6.6 Abrasive Cutoff** | — | 5, 7 | | 0* | | — | 4, 93 |
| **7.1 Gear Hobbing** | HSS | — | | — | | — | — |
| **7.2 Gear Cutting, Straight and Spiral Bevel** | HSS | — | | — | | — | — |
| **7.3 Gear Shaping** | HSS | — | | — | | — | — |
| **7.4 Gear Shaving** | HSS | — | | — | | — | — |
| **7.5 Gear Grinding, Form** | — | — | | — | | — | — |
| **8.1 Surface Grinding—Horiz. Spdl., Recip. Table** | — | 2, 3 | | 0* | | 4, 93 | 4, 93 |
| **8.2 Surface Grinding—Horiz. Spdl., Recip. Table** | CBN | — | | — | | — | — |
| **8.3 Surface Grinding—Horiz. Spdl., Recip. Table** | Diamond | — | | 0* | | 4, 93 | 4, 93 |
| **8.4 Surface Grinding—Vertical Spdl., Rotary Table** | — | 2, 3 | | 0* | | 4, 93 | 4, 93 |
| **8.5 Cylindrical Grinding** | — | 2, 3 | | 0* | | 4, 93 | 4, 93 |
| **8.6 Cylindrical Grinding** | CBN | — | | — | | — | — |
| **8.7 Cylindrical Grinding** | Diamond | — | | 0* | | 4, 93 | 4, 93 |
| **8.8 Internal Grinding** | — | 2, 3 | | 0* | | 4, 93 | 4, 93 |
| **8.9 Internal Grinding** | CBN | — | | — | | — | — |
| **8.10 Internal Grinding** | Diamond | — | | 0* | | 4, 93 | 4, 93 |
| **8.11 Centerless Grinding** | — | 2, 3 | | 0* | | 4, 93 | 4, 93 |
| **8.13 Abrasive Belt Grinding** | — | 3 | | 0* | | 4, 93 | 4, 93 |
| **8.14 Thread Grinding** | — | 2, 3 | | 0* | | 4, 93 | 4, 93 |

*Compressed air may be used to aid swarf removal.

## 16.2 Cutting Fluid Recommendations

| OPERATION | TOOL MATERIAL OR ABRASIVE | 56. Plastics | 58. Flame (Thermal) Sprayed Materials | | | 60. Precious Metals | 61. Rubber |
|---|---|---|---|---|---|---|---|
| | | — | — | — | — | — | — |
| 1.1 Turning, Single Point and Box Tools | HSS | 0, 4, 6 | | 1 | | — | — |
| | Carbide | 0, 4, 6 | | 1 | | — | — |
| 1.3 Turning | Diamond | 0, 4, 6 | | — | | — | — |
| 1.4 Turning, Cutoff and Form Tools | HSS | 0, 4, 6 | | — | | — | — |
| | Carbide | 0, 4, 6 | | — | | — | — |
| 1.5 Threading, Single Point | HSS | 0, 4, 6 | | — | | — | — |
| | Carbide | — | | — | | — | — |
| 1.6 Threading, Die | HSS | 0, 4, 6 | | — | | — | — |
| | Carbide | — | | — | | — | — |
| 1.7 Hollow Milling | HSS | 0, 4, 6 | | — | | — | — |
| | Carbide | — | | — | | — | — |
| 2.1 Face Milling | HSS | 0, 4, 6 | | — | | — | — |
| | Carbide | 0, 4, 6 | | — | | — | — |
| 2.2 Face Milling | Diamond | 0, 4, 6 | | — | | — | — |
| 2.4 Slab Milling | HSS | 0, 4, 6 | | — | | — | — |
| | Carbide | 0, 4, 6 | | — | | — | — |
| 2.5 Side and Slot Milling—Arbor Mounted Cutters | HSS | 0, 4, 6 | | — | | — | — |
| | Carbide | 0, 4, 6 | | — | | — | — |
| 2.6 End Milling—Peripheral | HSS | 0, 4, 6 | | — | | — | — |
| | Carbide | 0, 4, 6 | | — | | — | — |
| 2.8 End Milling—Slotting | HSS | 0, 4, 6 | | — | | — | — |
| | Carbide | 0, 4, 6 | | — | | — | — |
| 2.9 Thread Milling | HSS | 0, 4, 6 | | — | | — | — |
| 3.1 Drilling | HSS | 0, 4, 6 | | — | | — | — |
| | Carbide | 0, 4, 6 | | — | | — | — |
| 3.3 Oil-Hole or Pressurized-Coolant Drilling | HSS | — | | — | | — | — |
| | Carbide | — | | — | | — | — |
| 3.4 Spade Drilling | HSS | 0, 4, 6 | | — | | — | — |
| | Carbide | — | | — | | — | — |
| 3.5 Gundrilling | HSS | — | | — | | — | — |
| | Carbide | — | | — | | — | — |
| 3.6 Pressure Coolant Reaming (Gun Reaming) | HSS | — | | — | | — | — |
| | Carbide | — | | — | | — | — |
| 3.7 Reaming | HSS | 0, 4, 6 | | — | | — | — |
| | Carbide | 0, 4, 6 | | — | | — | — |
| 3.9 Boring | HSS | 0, 4, 6 | | — | | — | — |
| | Carbide | 0, 4, 6 | | — | | — | — |
| 3.11 Boring | Diamond | 0, 4, 6 | | — | | — | — |
| 3.12 Counterboring and Spotfacing | HSS | 0, 4, 6 | | — | | — | — |
| | Carbide | 0, 4, 6 | | — | | — | — |
| 3.13 Trepanning | HSS | — | | — | | — | — |
| | Carbide | — | | — | | — | — |
| 3.14 Honing | — | 80 | | — | | — | — |
| 3.15 Burnishing | — | — | | — | | — | — |
| 4.1 Tapping | HSS | 0, 4, 6 | | — | | — | — |

| OPERATION | TOOL MATERIAL OR ABRASIVE | 56. Plastics | 58. Flame (Thermal) Sprayed Materials | | | 60. Precious Metals<br>Silver | 61. Rubber |
|---|---|---|---|---|---|---|---|
| | | — | — | — | — | — | — |
| **5.1 Planing** | HSS | — | | — | | — | — |
| | Carbide | — | | — | | — | — |
| **5.2 Broaching** | HSS | 0, 1, 4, 6 | | — | | — | — |
| | Carbide | 0, 1, 4, 6 | | — | | — | — |
| **6.1 Power Hack Sawing** | HSS | 0, 1, 4, 6 | | — | | 4, 6 | 81 |
| **6.2 Power Band Sawing** | HSS | 0, 1, 4, 6 | | — | | 4, 6 | 81 |
| **6.4 Circular Sawing** | HSS | 0, 1, 4, 6 | | — | | 4, 6 | 81 |
| **6.5 Circular Sawing** | Carbide | 0, 1, 4, 6 | | — | | 4, 6 | 81 |
| **6.6 Abrasive Cutoff** | — | 0, 1, 4, 6 | | — | | — | — |
| **7.1 Gear Hobbing** | HSS | 0, 1, 4, 6 | | — | | — | — |
| **7.2 Gear Cutting, Straight and Spiral Bevel** | HSS | 0, 1, 4, 6 | | — | | — | — |
| **7.3 Gear Shaping** | HSS | 0, 1, 4, 6 | | — | | — | — |
| **7.4 Gear Shaving** | HSS | 0, 1, 4, 6 | | — | | — | — |
| **7.5 Gear Grinding, Form** | — | 0, 1, 4, 6 | | 0, 4, 6 | | — | — |
| **8.1 Surface Grinding—Horiz. Spdl., Recip. Table** | — | 0, 1, 4, 6 | | 0, 4, 6 | | 4, 6 | 4, 81 |
| **8.2 Surface Grinding—Horiz. Spdl., Recip. Table** | CBN | — | | — | | — | — |
| **8.3 Surface Grinding—Horiz. Spdl., Recip. Table** | Diamond | — | | — | | — | 0, 4, 81 |
| **8.4 Surface Grinding—Vertical Spdl., Rotary Table** | — | 0, 1, 4, 6 | | 0, 4, 6 | | 4, 6 | 4, 81 |
| **8.5 Cylindrical Grinding** | — | 0, 1, 4, 6 | | 0, 4, 6 | | 4, 6 | 4, 81 |
| **8.6 Cylindrical Grinding** | CBN | — | | — | | — | — |
| **8.7 Cylindrical Grinding** | Diamond | — | | — | | — | 0, 4, 81 |
| **8.8 Internal Grinding** | — | 0, 1, 4, 6 | | 0, 4, 6 | | 4, 6 | 4, 81 |
| **8.9 Internal Grinding** | CBN | — | | — | | — | — |
| **8.10 Internal Grinding** | Diamond | — | | — | | — | 0, 4, 81 |
| **8.11 Centerless Grinding** | — | 0, 1, 4, 6 | | 0, 4, 6 | | 4, 6 | 4, 81 |
| **8.13 Abrasive Belt Grinding** | — | 0, 1, 4, 6 | | 0, 4, 6 | | 4, 6 | 4, 81 |
| **8.14 Thread Grinding** | — | 0, 1, 4, 6 | | 0, 4, 6 | | 1 | 4, 81 |
| | | | | | | | |

| CODE NO. | TYPE AND TRADE NAME | MANUFACTURER |
|---|---|---|
| **0** | **DRY** | |
| **1** | **OILS—LIGHT DUTY (GENERAL PURPOSE)** | |
| | **Straight Mineral Oil** | |
| | Lafayette Blending Oil #1 | American Oil & Supply Co. |
| | Amoco Neutral Oil, AMOLITE Oil No. 10 thru 68 | Amoco Oil Co. |
| | Econ Oil 5, Econ Oil 10 | Ashland Oil, Inc. |
| | Diamond Oils, Topaz Oils, Gascon Oils | Atlantic Richfield Co. |
| | Mineral Oil H-5 | BP Oil, Inc. |
| | Bruko 1C, Bruko 2C, Bruko 3C, Bruko 4C | Bruce Products Corp. |
| | Chevron Utility Oil Grade 46 | Chevron U.S.A. Inc. |
| | CITGO Amplex 22, CITGO Amplex 32, CITGO Amplex 46 | Cities Service Co. |
| | Cook-Cut B-1, Cook-Cut B-5, Swiss-Cut 1816 | Cook's Industrial Lubricants, Inc. |
| | CORAY, FAXAM | Exxon Co., U.S.A. |
| | BLENDING OIL 3333 | Franklin Oil Corp. (Ohio) |
| | Gulf No. 372 Oil | Gulf Oil Corp. |
| | I.C. #1233, International Machine Oil #1 | International Chemical Co. |
| | IRMCO 248, IRMCO 249 | International Refining & Mfg. Co. |
| | Anza Oil | Lubrication Co. of America |
| | VANISHING OIL | Magnus Div., Economic Lab Inc. |
| | Kleer Kut Tri-Work | Pillsbury Chemical & Oil, Inc. |
| | Vehicle 6 | Process Research Products |
| | #3 OIL, #6 OIL | G. Whitfield Richards Co. |
| | SYN-KUT 100, SYZ-KUT 300 | Henry E. Sanson & Sons, Inc. |
| | Space Age Blending Oil | Space Age Chemlube |
| | Spindel Oil No. 2 and 4 | Specialty Products Co. |
| | Mineral Seal Oil, Factoil 39 | Standard Oil Co. (Ohio) |
| | Starcut-37 | Star Oil Company |
| | SUNICUT 113 | Sun Petroleum Products Co. |
| | 300 Oil 5, 522 Oil 19, Almag | Texaco Inc. |
| | NB BLENDING OIL | Tower Oil & Technology Co. |
| | W&B Cutting Oil 3016 | The White & Bagley Co. |
| | Magnacut, Macco Blending Oil | Witco Chemical Corp. |
| | R-100 CUTTING OIL | Arthur C. Withrow Co. |
| | ZURN BLENDING OILS | O F Zurn Co. |
| | **Straight Fatty Oil** | |
| | Lafayette Lard Oil #2 | American Oil & Supply Co. |
| | Brukut 11, Brukut 20 | Bruce Products Corp. |
| | Cook-Cut A-11 | Cook's Industrial Lubricants, Inc. |
| | W. S. LARD OIL | Franklin Oil Corp. (Ohio) |
| | I.C. #661-A, I.C. #1259 | International Chemical Co. |
| | IRMCO 040 | International Refining & Mfg. Co. |
| | Modoc Lard Oil | Lubrication Co. of America |
| | DO-12 | Magnus Div., Economics Lab Inc. |
| | Gipco Lard #1 | Pillsbury Chemical & Oil, Inc. |
| | Special Prime Lard Oil | Reilly-Whiteman, Inc. |
| | CUTZOL F.O. 512 | Rust-Lick, Inc. |
| | SYN-LARD | Henry E. Sanson & Sons, Inc. |
| | Toolife 254 | Specialty Products Co. |

| CODE NO. | TYPE AND TRADE NAME | MANUFACTURER |
|---|---|---|
| 1 | **OILS—LIGHT DUTY (GENERAL PURPOSE)** (Continued) | |
| | **Straight Fatty Oil** (Continued) | |
| | LARD OIL | Tower Oil & Technology Co. |
| | W&B Lard Oil 1948 | The White & Bagley Co. |
| | EWS LARD OIL (CODE 626) | Arthur C. Withrow Co. |
| | **Mineral Oil + Fatty Oil** | |
| | Lafayette SCF-20 | American Oil & Supply Co. |
| | AMOCUT Oil No. 144-F | Amoco Oil Co. |
| | ARCO ML Oils | Atlantic Richfield Co. |
| | Bezora 22 | BP Oil, Inc. |
| | Sparkut NF | Bruce Products Corp. |
| | Cook-Cut 25, Cook-Cut 30, Cook-Cut 1711 | Cook's Industrial Lubricants, Inc. |
| | FANOX N 33 | Exxon Co., U.S.A. |
| | B-650 CUTTING OIL | Franklin Oil Corp. (Ohio) |
| | Gulfcut 11A | Gulf Oil Corp. |
| | Habcool 344 | H & B Petroleum Co. |
| | Cut-Max 110, Cut-Max 135 | E. F. Houghton & Co. |
| | I.C. #152, I.C. #880 | International Chemical Co. |
| | IRMCO 240, IRMCO 242 | International Refining & Mfg. Co. |
| | MINERAL LARD OIL | Keystone Div., Pennwalt Corp. |
| | LPS 1, LPS 2, LPS 3 | LPS Research Laboratories, Inc. |
| | DO-5A, L-12 | Magnus Div., Economics Lab Inc. |
| | Mobilmet 33, Mobilmet 34, Mobilmet 35 | Mobil Oil Corp. |
| | Kleer Kut Dual Work | Pillsbury Chemical & Oil, Inc. |
| | Triple C 5500 | Pro-Chem Inc. |
| | Non-Ferrous Aid | Process Research Products |
| | Petrolard 116, Petrolard 112 | Reilly-Whiteman, Inc. |
| | NEAR-A-LARD OIL, SPECIAL NEAR-A-LARD OIL | G. Whitfield Richards Co. |
| | CUTZOL F.O. 506, CUTZOL F.O. 509 | Rust-Lick, Inc. |
| | SYN-KUT 115, SYN-KUT 135 | Henry E. Sanson & Sons, Inc. |
| | Space Age Blending Oil-Plus | Space Age Chemlube |
| | Toolife 250, Toolife 251, Toolife 256 | Specialty Products Co. |
| | Factokut L-2 | Standard Oil Co. (Ohio) |
| | Starcut-3950 | Star Oil Company |
| | SUNICUT 150 | Sun Petroleum Products Co. |
| | Extra Cutting Oil | Texaco Inc. |
| | TOWER E-933 | Tower Oil & Technology Co. |
| | W&B Minolard Cutting Oil | The White & Bagley Co. |
| | Magnacut 40 | Witco Chemical Corp. |
| | WITHROKUT 104 CUTTING OIL | Arthur C. Withrow Co. |
| | LARDOLEUM 5 | O F Zurn Co. |
| | **Mineral Oil + Additives** | |
| | Lafayette B-16 | American Oil & Supply Co. |
| | AMOCUT Tri-Purpose Cutting Oil | Amoco Oil Co. |
| | Promax 1110 | Ashland Oil, Inc. |
| | Clairo Oils | Atlantic Richfield Co. |
| | Sevora 37, Lumaro 25, Lumaro 36 | BP Oil, Inc. |
| | Belco 1727 Cutting Oil | Bel-Ray Co., Inc. |
| | Brukut 177-X | Bruce Products Corp. |

| CODE NO. | TYPE AND TRADE NAME | MANUFACTURER |
|---|---|---|
| 1 | **OILS—LIGHT DUTY (GENERAL PURPOSE)** (Continued) | |
| | **Mineral Oil + Additives** (Continued) | |
| | Chevron Metalworking Fluid 504 | Chevron U.S.A. Inc. |
| | Cook-Cut 70 | Cook's Industrial Lubricants, Inc. |
| | CURTIS 32-AS | Curtis Systems, Inc. |
| | EBONITE DUAL PURPOSE 702 | Franklin Oil Corp. (Ohio) |
| | Cindepol 150, Cindepol 300 | E. F. Houghton & Co. |
| | I.C. #852-A | International Chemical Co. |
| | IRMCO 200 Series | International Refining & Mfg. Co. |
| | 121 Wax Cut | S. C. Johnson & Son, Inc. |
| | LPS 1, LPS 2, LPS 3 | LPS Research Laboratories, Inc. |
| | Clear-Cut Ax | Lubrication Co. of America |
| | L-67, Magna-Draw 10 | Magnus Div., Economics Lab Inc. |
| | Translube, Metkut 310 | Metalloid Corp. |
| | Mobilmet 33, Mobilmet 34, Mobilmet 35, Mobilmet Omicron, Mobilmet Upsilon | Mobil Oil Corp. |
| | Magno Kut #11 | Pillsbury Chemical & Oil, Inc. |
| | PC Lapping Oil, LE-2 | Process Research Products |
| | NEAR-A-LARD #190-H OIL, #4-B OIL, #3-B OIL | G. Whitfield Richards Co. |
| | CUTZOL 101-C | Rust-Lick, Inc. |
| | SYZ-KUT 216, SYZ-KUT 236 | Henry E. Sanson & Sons Inc. |
| | Space Age Blending Oil–Plus | Space Age Chemlube |
| | Toolife VO | Specialty Products Co. |
| | Sulkleer 177X | Standard Oil Co. (Ohio) |
| | Starcut-3180 | Star Oil Company |
| | SUNICUT 330 | Sun Petroleum Products Co. |
| | Cleartex B | Texaco Inc. |
| | TOWERCUT F-901 | Tower Oil & Technology Co. |
| | VSCC 5468 | Van Straaten Chemical Co. |
| | W&B Cutting Oil 2169 | The White & Bagley Co. |
| | WITHROKUT 102 CUTTING OIL | Arthur C. Withrow Company |
| | **Mineral Oil + Sulfur and Chlorine Compounding** | |
| | Lafayette B–115 | American Oil & Supply Co. |
| | Promax 1220 | Ashland Oil, Inc. |
| | Chevron Metalworking Fluid 502 | Chevron U.S.A. Inc. |
| | CITGO Cutting Oil 400 | Cities Service Co. |
| | Habcool 301a | H & B Petroleum Co. |
| | IRMCO 200 Series | International Refining & Mfg. Co. |
| | Magno Kut #1 | Pillsbury Chemical & Oil, Inc. |
| | FRIGIDOL TRANSPARENT CUTTING OIL | G. Whitfield Richards Co. |
| | SYN-KUT 9-L | Henry E. Sanson & Sons, Inc. |
| | Space Age Blending Oil–Plus, Space Age T-61 Cutting Oil | Space Age Chemlube |
| | SUNICUT 350 | Sun Petroleum Products Co. |
| | Dualtex | Texaco Inc. |
| | W&B Cutting Oil 2169 | The White & Bagley Co. |
| | Macco Honing Oil, Macco Broach Oil 4-50, Epcut 31-BS-1, Maccut 16 | Witco Chemical Corp. |
| | WITHROKUT 114 CUTTING OIL | Arthur C. Withrow Co. |
| | **Mineral Oil + Fatty Oil + Additives** | |
| | Promax 5352 | Ashland Oil, Inc. |
| | Cook-Cut 70 | Cook's Industrial Lubricants, Inc. |

## 16.3 Cutting Fluid Key

| CODE NO. | TYPE AND TRADE NAME | MANUFACTURER |
|---|---|---|
| 1 | **OILS—LIGHT DUTY (GENERAL PURPOSE)** (Continued) | |
| | **Mineral Oil + Fatty Oil + Additives** (Continued) | |
| | DAUBERT MACRON M | Daubert Chemical Co. |
| | TF-460 | Metalloid Corp. |
| | CUTZOL 400-A | Rust-Lick, Inc. |
| | Dual Action Cutting Fluid #2 | Tapmatic Corp. |
| 2 | **OILS—MEDIUM DUTY** | |
| | **Sulfurized Mineral Oil** | |
| | Lafayette B-16 | American Oil & Supply Co. |
| | AMOCUT Oil No. 1195, AMOCUT Oil No. 159S | Amoco Oil Co. |
| | Promax 1004 | Ashland Oil, Inc. |
| | Cutting Oil 70, Cutting Oil 100, Cutting Oil 160 | Atlantic Richfield Co. |
| | Lumaro 24A | BP Oil, Inc. |
| | Brukut 4 | Bruce Products Corp. |
| | PETROCHEM 709 | Champions Choice, Inc. |
| | Cook-Cut 10, Cook-Cut 4985-Clear | Cook's Industrial Lubricants, Inc. |
| | DAUBERT GARIA S | Daubert Chemical Co. |
| | PEERLESS CUTTING OIL 3307 | Franklin Oil Corp. (Ohio) |
| | IRMCO 269 | International Refining & Mfg. Co. |
| | C-5, C-41, C-42, C-47 | Lubrication Co. of America |
| | Fine Kut 41 | Pillsbury Chemical & Oil, Inc. |
| | FRIGIDOL CUTTING OIL | G. Whitfield Richards Co. |
| | CUTZOL S.O. 501 | Rust-Lick, Inc. |
| | SYN-KUT 845 | Henry E. Sanson & Sons Inc. |
| | Toolife 100 | Specialty Products Co. |
| | Starcut-3719 | Star Oil Company |
| | SUNICUT 300 | Sun Petroleum Products Co. |
| | W&B Sulfurized Mineral Oil 3197 | The White & Bagley Co. |
| | **Compounded Sulfurized Mineral Oil** | |
| | Lafayette B-16 | American Oil & Supply Co. |
| | AMOCUT Oil No. 115FS, AMOCUT Oil No. 155FS | Amoco Oil Co. |
| | Promax 1220 | Ashland Oil, Inc. |
| | Tulkut 160, Honol Oils | Atlantic Richfield Co. |
| | Lumaro 34A, Sevora 26A, Sevora 41A, Sevora 53A | BP Oil, Inc. |
| | Brukut 4LV, Brukut 102SR | Bruce Products Corp. |
| | PETROCHEM 703 | Champions Choice, Inc. |
| | Chevron Metalworking Fluid 502 | Chevron U.S.A. Inc. |
| | L 2051, L 2052 | Chrysler Corp. |
| | DAUBERT GARIA C, DAUBERT GARIA H | Daubert Chemical Co. |
| | PEERLESS CUTTING OIL 967 | Franklin Oil Corp. (Ohio) |
| | IRMCO 259 | International Refining & Mfg. Co. |
| | RINCON 180 | Lubrication Co. of America |
| | CUTTING COMPOUND No. 6 | Magnus Div., Economics Lab Inc. |
| | Mobilmet 25, Mobilmet 26, Mobilmet 715, Mobilmet 735, Mobilmet Alpha, Mobilmet Sigma | Mobil Oil Corp. |
| | Fine Kut 41 | Pillsbury Chemical & Oil, Inc. |

| CODE NO. | TYPE AND TRADE NAME | MANUFACTURER |
|---|---|---|
| 2 | **OILS—MEDIUM DUTY** (Continued) | |
| | **Compounded Sulfurized Mineral Oil** (Continued) | |
| | FRIGIDOL L-6 CUTTING OIL | G. Whitfield Richards Co. |
| | SYZ-KUT 2023 | Henry E. Sanson & Sons, Inc. |
| | Space Age Z-33 Cutting Oil | Space Age Chemlube |
| | Sulkleer 19E, Sulkleer 30, Sulkleer 102S, Sulkleer 163A | Standard Oil Co. (Ohio) |
| | Starcut-3531 | Star Oil Company |
| | SUZICUT 400 | Sun Petroleum Products Co. |
| | TREFOLEX Cutting Compound | Trefolex Industries, Inc. |
| | VSCC 5601 | Van Straaten Chemical Co. |
| | W&B Cutting Oil 2151 | The White & Bagley Co. |
| | B-5 CUTTING OIL | Arthur C. Withrow Co. |
| | **Chlorinated Mineral Oil** | |
| | Lafayette CP-6535 | American Oil & Supply Co. |
| | AMOCUT Tri-Purpose Cutting Oil | Amoco Oil Co. |
| | TECTYL H-i40, PROMAX 1940 | Ashland Oil, Inc. |
| | Chevron Metalworking Fluid 504 | Chevron U.S.A. Inc. |
| | Cook-Cut 4938, Cook-Grind 4903 | Cook's Industrial Lubricants, Inc. |
| | Cindol 3106 | E. F. Houghton & Co. |
| | I.C. #484-M, I.C. #484-MS | International Chemical Co. |
| | Hon-E-Kut 740-3 | Lubrication Co. of America |
| | CUTTING COMPOUND No. 2, MAGNU DRAW 30 | Magnus Div., Economics Lab Inc. |
| | Taplube, Translube, Multi-slide 21 | Metalloid Corporation |
| | Spun Gold #1 | Pillsbury Chemical & Oil, Inc. |
| | Rapid Tap | Relton Corp. |
| | LUBE-A-TUBE "EP" #6 COMPOUND, LUBE-A-TUBE "EP" #214-P COMPOUND | G. Whitfield Richards Co. |
| | CUTZOL 101-C, CUTZOL 711 | Rust-Lick, Inc. |
| | CLARET-KUT 21 | Henry E. Sanson & Sons, Inc. |
| | Space Age Z-34 Cutting Oil | Space Age Chemlube |
| | Toolife 703, Toolife 705 | Specialty Products Co. |
| | Starcut-3260 | Star Oil Company |
| | SUNICUT 201 | Sun Petroleum Products Co. |
| | Dual Action Cutting Fluid #1 | Tapmatic Corp. |
| | VSCC 5487-G | Van Straaten Chemical Co. |
| | W&B Cutting 2981 | The White & Bagley Co. |
| | **Sulfo-Chlorinated Mineral Oil** | |
| | Lafayette A-13 | American Oil & Supply Co. |
| | AMOCUT Oil 117 BCS | Amoco Oil Co. |
| | Promax 1401, Promax 1455 | Ashland Oil, Inc. |
| | Brukut #7 | Bruce Products Corp. |
| | PETROCHEM 704 | Champions Choice, Inc. |
| | Chevron Metalworking Fluid 503 | Chevron U.S.A. Inc. |
| | L 402, L 405, L 407 | Chrysler Corp. |
| | MITEE Clear Thread Cutting Oil | DAP, Inc. |
| | I.C. #1280-A | International Chemical Co. |
| | IRMCO 220 | International Refining & Mfg. Co. |
| | LCA Speed-Cut 105, 102, 104, 104B, 103 | Lubrication Co. of America |
| | Flash Kut 101 | Pillsbury Chemical & Oil, Inc. |

| CODE NO. | TYPE AND TRADE NAME | MANUFACTURER |
|---|---|---|
| 2 | **OILS—MEDIUM DUTY** (Continued) | |
| | **Sulfo-Chlorinated Mineral Oil** (Continued) | |
| | FRIGIDOL #401-T COMPOUND | G. Whitfield Richards Co. |
| | SYN-KUT 99-L | Henry E. Sanson & Sons, Inc. |
| | Space Age Z-35 Cutting Oil | Space Age Chemlube |
| | Starcut-3531 | Star Oil Company |
| | SUNICUT 510 | Sun Petroleum Products Co. |
| | W&B Cutting Oil 2151 | The White & Bagley Co. |
| | "WADCO" Cutting and Threading Oil | J. C. Whitlam Manufacturing Co. |
| | **Sulfurized Mineral Oil + Fatty Oil** | |
| | Lafayette F-14 | American Oil & Supply Co. |
| | AMOCUT Oil No. 155FS | Amoco Oil Co. |
| | Promax 1023 | Ashland Oil, Inc. |
| | Autokut Oils | Atlantic Richfield Co. |
| | Sevora 26A, Sevora 41A, Sevora 53A | BP Oil, Inc. |
| | PETROCHEM 703 | Champions Choice, Inc. |
| | Cook-Cut 15, Cook-Cut 40, Cook-Cut 1880 | Cook's Industrial Lubricants, Inc. |
| | MITEE Dark Thread Cutting Oil | DAP, Inc. |
| | DAUBERT GARIA C, DAUBERT GARIA H | Daubert Chemical Co. |
| | PENNEX N 40 | Exxon Co., U.S.A. |
| | PEERLESS 930 | Franklin Oil Corp. (Ohio) |
| | Gulfcut 31A, Gulfcut 31C | Gulf Oil Corp. |
| | IRMCO 259 | International Refining & Mfg. Co. |
| | Clear Cutting Oil C-47 | Lubrication Co. of America |
| | Mobilmet 24, Mobilmet 25, Mobilmet 26, Mobilmet 37, Mobilmet 715, Mobilmet 735, Mobilmet Alpha, Mobilmet Sigma | Mobil Oil Corp. |
| | Flash Kut 101 | Pillsbury Chemical & Oil, Inc. |
| | FRIGIDOL BASE "A" COMPOUND | G. Whitfield Richards Co. |
| | CUTZOL C.F.O. 507 | Rust-Lick, Inc. |
| | SYZ-KUT 320 | Henry E. Sanson & Sons, Inc. |
| | Space Age Z-36 Cutting Oil | Space Age Chemlube |
| | Toolife 103 | Specialty Products Co. |
| | Sulkleer 30 | Standard Oil Co. (Ohio) |
| | Starcut-3570 | Star Oil Company |
| | SUNICUT 400 | Sun Petroleum Products Co. |
| | W&B Cutting Oil 1749 | The White & Bagley Co. |
| | WITHROKUT 132 CUTTING OIL | Arthur C. Withrow Co. |
| | **Chlorinated Mineral Oil + Fatty Oil** | |
| | Lumaro 36 | BP Oil, Inc. |
| | L 415 | Chrysler Corp. |
| | Cook-Cut 55, Swiss-Cut 51 | Cook's Industrial Lubricants, Inc. |
| | CURTIS 64-AS | Curtis Systems, Inc. |
| | DAUBERT MACRON M | Daubert Chemical Co. |
| | PEERLESS 1211 | Franklin Oil Corp. (Ohio) |
| | I.C. #152-S, I.C. #818 | International Chemical Co. |
| | IRMCO 754 | International Refining & Mfg. Co. |
| | Hon-E-Kut | Lubrication Co. of America |
| | EL CUT S | Magnus Div., Economics Lab Inc. |
| | Metkut #310, TF-308 | Metalloid Corp. |

| CODE NO. | TYPE AND TRADE NAME | MANUFACTURER |
|---|---|---|
| 2 | **OILS—MEDIUM DUTY** (Continued) | |
| | **Chlorinated Mineral Oil + Fatty Oil** (Continued) | |
| | Spun Gold #12 | Pillsbury Chemical & Oil, Inc. |
| | Petrosan EP43 | Reilly-Whiteman, Inc. |
| | NEAR-A-LARD #190-LV OIL, NEAR-A-LARD #13 | G. Whitfield Richards Co. |
| | CUTZOL C.F.O. 509, CUTZOL LC, HONZOL 11-B | Rust-Lick, Inc. |
| | SYN-KUT 3401, SYN-KUT M-2 | Henry E. Sanson & Sons Inc. |
| | Toolife 150, Toolife Swiss Screw Machine No. 3 | Specialty Products Co. |
| | Starcut-3360 | Star Oil Company |
| | W&B Cutting Oil 2871 | The White & Bagley Co. |
| | WITHROKUT 110 CUTTING OIL | Arthur C. Withrow Co. |
| | **Mineral Oil With Fatty Compounding + Sulfur and Chlorine Additives** | |
| | Promax 5352, Promax 1310 | Ashland Oil, Inc. |
| | Cilora 19A, Lumaro 40A | BP Oil, Inc. |
| | Brukut 9, Brukut 102, Brukut 102S, Bruko D-648 | Bruce Products Corp. |
| | Chevron Metalworking Fluid 503 | Chevron U.S.A. Inc. |
| | L 2045 | Chrysler Corp. |
| | CITGO Cutting Oil 425 | Cities Service Co. |
| | Cook's RTP Cutting Oil 4768 | Cook's Industrial Lubricants, Inc. |
| | DAUBERT GARIA D | Daubert Chemical Co. |
| | DORTAN N 14 | Exxon Co., U.S.A. |
| | Habcool 250 | H & B Petroleum Co. |
| | I.C. #634 | International Chemical Co. |
| | Mobilmet 715, Mobilmet 735 | Mobil Oil Corp. |
| | Triple C 53611 | Pro-Chem Inc. |
| | PRP-9112 | Process Research Products |
| | FRIGIDOL #401-P COMPOUND | G. Whitfield Richards Co. |
| | SYN-KUT 725, SYN-KUT 750 | Henry E. Sanson & Sons Inc. |
| | Space Age Z-37 Cutting Oil | Space Age Chemlube |
| | Sulkleer 177X | Standard Oil Co. (Ohio) |
| | SUNICUT 352 | Sun Petroleum Products Co. |
| | Transultex A, Transultex F | Texaco Inc. |
| | TOWERCUT F-932 | Tower Oil & Technology Co. |
| | W&B Cutting Oil 2688 | The White & Bagley Co. |
| | WITHROKUT 136 CUTTING OIL | Arthur C. Withrow Co. |
| | ZURNKUT 3 L X | O F Zurn Co. |
| | **Sulfo-Chlorinated Mineral Oil + Fatty Oil** | |
| | Lafayette C-115 | American Oil & Supply Co. |
| | AMOCUT Oil 139 BCS | Amoco Oil Co. |
| | Promax 1002 | Ashland Oil, Inc. |
| | Belco Cutting Oil 1735 | Bel-Ray Co., Inc. |
| | PETROCHEM 704 | Champions Choice, Inc. |
| | Chevron Metalworking Fluid 503 | Chevron U.S.A. Inc. |
| | Cook-Cut 4984-Dark, Cook-Cut 4889 | Cook's Industrial Lubricants, Inc. |
| | No. 80 Cutting Oil | DoALL Co. |
| | DORTAN N 34 | Exxon Co., U.S.A. |
| | Gulfcut 41M | Gulf Oil Corp. |
| | Cut-Max 206 | E. F. Houghton & Co. |
| | I.C. #283-C | International Chemical Co. |

| CODE NO. | TYPE AND TRADE NAME | MANUFACTURER |
|---|---|---|
| 2 | **OILS—MEDIUM DUTY** (Continued) | |
| | **Sulfo-Chlorinated Mineral Oil + Fatty Oil** (Continued) | |
| | IRMCO 268 | International Refining & Mfg. Co. |
| | Flash Kut 11 | Pillsbury Chemical & Oil, Inc. |
| | FRIGIDOL #401 COMPOUND | G. Whitfield Richards Co. |
| | SYZ-KUT 206, SYN-KUT 226, SYN-KUT 246 | Henry E. Sanson & Sons Inc. |
| | Space Age Z-38 Cutting Oil | Space Age Chemlube |
| | Toolife 202 | Specialty Products Co. |
| | Starcut-3531 | Star Oil Company |
| | SUZICUT 500 | Sun Petroleum Products Co. |
| | W&B Cutting Oil 2688 | The White & Bagley Co. |
| | Maccut 11 Concentrate, Maccut 16 BGA, Maccut 1652 DO, Maccut 19 BG | Witco Chemical Corp. |
| | ZURN ULTRAKUT 70 S 1 | O F Zurn Co. |
| | **Compounded Sulfo-Chlorinated Mineral Oil + Fatty Oil** | |
| | Lafayette C–75 | American Oil & Supply Co. |
| | AMOCUT Oil 168 BC | Amoco Oil Co. |
| | Promax 1002 | Ashland Oil, Inc. |
| | Transkut Oils | Atlantic Richfield Co. |
| | Cilora 39A, Cilora 59A | BP Oil, Inc. |
| | PETROCHEM 704 | Champions Choice, Inc. |
| | Chevron Metalworking Fluid 503 | Chevron U.S.A. Inc. |
| | Cook-Cut 20, Cook-Cut 1545 | Cook's Industrial Lubricants, Inc. |
| | DAUBERT GARIA D | Daubert Chemical Co. |
| | Gulfcut 41D, Gulfcut 41E | Gulf Oil Corp. |
| | I.C. #1884-T | International Chemical Co. |
| | IRMCO 215 | International Refining & Mfg. Co. |
| | LUBRICOOL 175 | Larson Chemical Specialties |
| | Magno Base 201 | Pillsbury Chemical & Oil, Inc. |
| | Gear Lap 2S | Process Research Products |
| | FRIGIDOL #401-PX COMPOUND | G. Whitfield Richards Co. |
| | BOOSTER-KUT 50 | Henry E. Sanson & Sons, Inc. |
| | Space Age Z-39–Cutting Oil | Space Age Chemlube |
| | Starcut-3671 | Star Oil Company |
| | SUNICUT 500 | Sun Petroleum Products Co. |
| | TOWERCUT D-904 | Tower Oil & Technology Co. |
| | VSCC 5299 | Van Straaten Chemical Co. |
| | W&B Cutting Oil 2688 | The White & Bagley Co. |
| | Maccut 1652 DO, Maccut 18-L | Witco Chemical Corp. |
| 3 | **OILS—HEAVY DUTY** | |
| | **Sulfurized Mineral Oil** | |
| | AMOCUT Oil No. 159S | Amoco Oil Co. |
| | PETROCHEM 709 | Champions Choice, Inc. |
| | CITGO Cutting Oil 110 | Cities Service Co. |
| | Cook-Cut 4985–Clear | Cook's Industrial Lubricants, Inc. |
| | PEERLESS 3307 | Franklin Oil Corp. (Ohio) |
| | Smooth Cut 111, 112, 113, 114 | Lubrication Co. of America |

| CODE NO. | TYPE AND TRADE NAME | MANUFACTURER |
|---|---|---|
| 3 | **OILS—HEAVY DUTY** (Continued) | |
| | **Sulfurized Mineral Oil** (Continued) | |
| | Komp Kut 5493 | Pillsbury Chemical and Oil, Inc. |
| | SYN-KUT 2023 AB | Henry E. Sanson & Sons Inc. |
| | Space Age 49-1-Cutting Oil | Space Age Chemlube |
| | Starcut-3068 | Star Oil Company |
| | SUNICUT 300 | Sun Petroleum Products Co. |
| | W&B Sulfurized Mineral Oil 3197 | The White & Bagley Co. |
| | **Compounded Sulfurized Mineral Oil** | |
| | AMOCUT Oil No. 189FS, AMOCUT Oil No. 155FS | Amoco Oil Co. |
| | Promax 1130-S | Ashland Oil, Inc. |
| | Cilora 39A, Cilora 59A, Lumaro 40A, Sevora 53A | BP Oil, Inc. |
| | Belco Cutting Oil S-1191 | Bel-Ray Co., Inc. |
| | Bruko D-269 | Bruce Products Corp. |
| | PETROCHEM 703 | Champions Choice, Inc. |
| | CITGO Cutting Oil 120 | Cities Service Co. |
| | Cook-Cut 4669, Cook-Cut 4770, Cook's Base 4876 | Cook's Industrial Lubricants, Inc. |
| | DAUBERT GARIA D, DAUBERT GARIA G | Daubert Chemical Co. |
| | PEERLESS 977, PEERLESS 967 | Franklin Oil Corp. (Ohio) |
| | Mobilmet 27, Mobilmet 45, Mobilmet 814, Mobilmet 826, Mobilmet 833, Mobilmet Gamma, Mobilmet Omega | Mobil Oil Corp. |
| | Gipco "A" | Pillsbury Chemical and Oil, Inc. |
| | FRIGIDOL CUTTING OIL | G. Whitfield Richards Co. |
| | SYN-KUT BASE 55 | Henry E. Sanson & Sons, Inc. |
| | Space Age 49-2 Cutting Oil | Space Age Chemlube |
| | Sulkleer 30, Sulkleer 102S | Standard Oil Co. (Ohio) |
| | Starcut-3068 | Star Oil Company |
| | SUNICUT 415 | Sun Petroleum Products Co. |
| | TOWER B-1136 | Tower Oil and Technology Co. |
| | VSCC 5299 | Van Straaten Chemical Co. |
| | W&B Cutting Oil 2555 | The White & Bagley Co. |
| | **Chlorinated Mineral Oil** | |
| | Lafayette CP-6535 | American Oil & Supply Co. |
| | Promax Q-1013 | Ashland Oil, Inc. |
| | Bruko D-337 | Bruce Products Corp. |
| | Cook's Base 1241, Cook's Base 4676 | Cook's Industrial Lubricants, Inc. |
| | PEERLESS 1211, TAPPING COMPOUND 1715 | Franklin Oil Corp. (Ohio) |
| | I.C. #871 | International Chemical Co. |
| | 122 Wax Cut | S. C. Johnson & Son, Inc. |
| | KOOLMIST FORMULA #88 TAP WAX | Kool Mist Corp. |
| | DO-91, DO-2A, MAGNU DRAW 40, MAGNU DRAW 50L, CHLOROLUBE 50H | Magnus Div., Economics Lab Inc. |
| | Hon-E-Cut 740, 741, 742 | Lubrication Co. of America |
| | Plastrol, Metdraw 407, Metdraw 446 | Metalloid Corp. |
| | Spun Gold #3 | Pillsbury Chemical and Oil, Inc. |
| | Die Gard 84 | Reilly-Whiteman, Inc. |
| | LUBE-A-TUBE #81 COMPOUND | G. Whitfield Richards Co. |
| | CUTZOL 711 | Rust-Lick, Inc. |
| | CLARET-KUT 35, CLARET-BASE | Henry E. Sanson & Sons, Inc. |

# 16.3 Cutting Fluid Key

| CODE NO. | TYPE AND TRADE NAME | MANUFACTURER |
|---|---|---|
| 3 | **OILS—HEAVY DUTY** (Continued) | |
| | **Chlorinated Mineral Oil** (Continued) | |
| | Space Age 49-3 Cutting Oil | Space Age Chemlube |
| | Toolife 720, Toolife 740, Toolife 770 | Specialty Products Co. |
| | Starcut-3260 | Star Oil Company |
| | SUNICUT 201 | Sun Petroleum Products Co. |
| | TOWER A-1123 | Tower Oil and Technology Co. |
| | VSCC 5487-G | Van Straaten Chemical Co. |
| | W&B Cutting Oil 2981 | The White & Bagley Co. |
| | WITHROBASE 434 CONCENTRATE | Arthur C. Withrow Co. |
| | **Sulfurized Mineral Oil + Sulfur, Chlorine, and Phosphorus Compounding** | |
| | Bruko D-643 | Bruce Products Corp. |
| | Cook-Cut 4749, Cook-Grind 4806, Cook-Cut 4895, Cook's Base 4601 | Cook's Industrial Lubricants, Inc. |
| | IRMCO 793 | International Refining & Mfg. Co. |
| | VISTA*KUT Heavy Duty | Lubrication Co. of America |
| | Triple C 5401 | Pro-Chem Inc. |
| | SYN-KUT FGO | Henry E. Sanson & Sons Inc. |
| | Toolife 203 | Specialty Products Co. |
| | **Sulfo-Chlorinated Mineral Oil** | |
| | Lafayette A–13 | American Oil & Supply Co. |
| | AMOCUT Oil No. 139 BCS, AMOCUT Oil No. 209 BCS | Amoco Oil Co. |
| | Promax 1455-S | Ashland Oil, Inc. |
| | PETROCHEM 707 | Champions Choice, Inc. |
| | Chevron Metalworking Fluid 503 | Chevron U.S.A. Inc. |
| | CITGO Cutting Oil 130 | Cities Service Co. |
| | Cook-Cut 4515 | Cook's Industrial Lubricants, Inc. |
| | I.C. #1226 | International Chemical Co. |
| | Thread Mate 101, 110, 202, 225 | Lubrication Co. of America |
| | Tapeze X-2, Thredwell | Metalloid Corp. |
| | Black Velvet | Pillsbury Chemical and Oil, Inc. |
| | FRIGIDOL TRANSPARENT CUTTING OIL, FRIGIDOL BASE "L" COMPOUND | G. Whitfield Richards Co. |
| | CLARET-KUT 570, CLARET-KUT 27 | Henry E. Sanson & Sons Inc. |
| | SWEPCO 705 Heavy Duty Cutting Oil | Southwestern Petroleum Corp. |
| | Space Age 49-4 Cutting Oil | Space Age Chemlube |
| | Toolife 201 | Specialty Products Co. |
| | Starcut-3033 | Star Oil Company |
| | SUNICUT 510 | Sun Petroleum Products Co. |
| | TOWERCUT D-908 | Tower Oil and Technology Co. |
| | W&B Cutting Oil 2151 | The White & Bagley Co. |
| | **Sulfurized Mineral Oil + Fatty Oil** | |
| | AMOCUT Oil No. 189FS, AMOCUT Oil No. 155FS | Amoco Oil Co. |
| | Promax 1023, Promax 1130-S, Promax 1160 | Ashland Oil, Inc. |
| | Autokut Oils | Atlantic Richfield Co. |
| | Cook's Honing Oil 1995, Cook-Cut 1998, Cook-Cut 4950 | Cook's Industrial Lubricants, Inc. |
| | No. 120 Cutting Oil | DoALL Co. |
| | PEERLESS 930 | Franklin Oil Corp. (Ohio) |
| | Gulfcut 31C | Gulf Oil Corp. |

| CODE NO. | TYPE AND TRADE NAME | MANUFACTURER |
|---|---|---|
| 3 | **OILS—HEAVY DUTY** (Continued) | |
| | **Sulfurized Mineral Oil + Fatty Oil** (Continued) | |
| | IRMCO 266 | International Refining & Mfg. Co. |
| | KEYCUT 101 | Keystone Division/Pennwalt Corp. |
| | Mobilmet 27, Mobilmet 29, Mobilmet 45, Mobilmet Gamma, Mobilmet Omega | Mobil Oil Corp. |
| | Spun Gold #14 | Pillsbury Chemical and Oil, Inc. |
| | FRIGIDOL BASE "A" COMPOUND | G. Whitfield Richards Co. |
| | CUTZOL SFO-507 | Rust-Lick, Inc. |
| | SYN-KUT 310 | Henry E. Sanson & Sons, Inc. |
| | Space Age 49-5 Cutting Oil | Space Age Chemlube |
| | Toolife 102 | Specialty Products Co. |
| | Sulkleer 30 | Standard Oil Co. (Ohio) |
| | Starcut-3161 | Star Oil Company |
| | SUNICUT 400 | Sun Petroleum Products Co. |
| | W&B Cutting Oil 2555 | The White & Bagley Co. |
| | WITHROBASE 424 CONCENTRATE | Arthur C. Withrow Co. |
| | **Chlorinated Mineral Oil + Fatty Oil** | |
| | TECTYL H-2, PROMAX 1002, TECTYL H-22, PROMAX 1022 | Ashland Oil, Inc. |
| | Bruko D-591, Bruko D-579 | Bruce Products Corp. |
| | Swiss-Cut 57, Swiss-Cut 71, Swiss Base 1956 | Cook's Industrial Lubricants, Inc. |
| | CURTIS 55-CUT | Curtis Systems, Inc. |
| | No. 150 Cutting Oil | DoALL Co. |
| | Cut-Max 568 | E. F. Houghton & Co. |
| | I.C. #484-EW, I.C. #845 | International Chemical Co. |
| | IRMCO 754 | International Refining & Mfg. Co. |
| | X-60 | Metalloid Corp. |
| | Die Gard 89 | Reilly-Whiteman, Inc. |
| | NEAR-A-LARD #20-A OIL | G. Whitfield Richards Co. |
| | CUTZOL LC, HONZOL 11-B | Rust-Lick, Inc. |
| | CLARET-KUT 10 | Henry E. Sanson & Sons, Inc. |
| | SWEPCO 725 EXTRA Heavy Duty Cutting Oil | Southwestern Petroleum Corp. |
| | Space Age 49-6 Cutting Oil | Space Age Chemlube |
| | Toolife 254, Toolife Swiss Screw Machine No. 1, Toolife Swiss Screw Machine No. 2 | Specialty Products Co. |
| | Starcut-3361 | Star Oil Company |
| | TOWER A-1123 | Tower Oil and Technology Co. |
| | W&B Cutting Oil 2872 | The White & Bagley Co. |
| | WITHROBASE 436 CONCENTRATE | Arthur C. Withrow Co. |
| | **Sulfo-Chlorinated Mineral Oil + Fatty Oil** | |
| | Lafayette C-13 | American Oil & Supply Co. |
| | AMOCUT Oil No. 139 BCS, AMOCUT Oil No. 209 BCS, AMOCUT Oil No. 309 BCS | Amoco Oil Co. |
| | Promax 1310 | Ashland Oil, Inc. |
| | Cilora 39A | BP Oil, Inc. |
| | Bruko D-621A | Bruce Products Corp. |
| | PETROCHEM THREAD KUT DARK | Champions Choice, Inc. |
| | Chevron Metalworking Fluid 503 | Chevron U.S.A. Inc. |
| | CITGO Cutting Oil 140 | Cities Service Co. |
| | Cook-Cut 1685, Cook-Cut 4949, Cook-Cut 4559 | Cook's Industrial Lubricants, Inc. |

| CODE NO. | TYPE AND TRADE NAME | MANUFACTURER |
|---|---|---|
| 3 | **OILS—HEAVY DUTY** (Continued) | |
| | **Sulfo-Chlorinated Mineral Oil + Fatty Oil** (Continued) | |
| | No. 240 Cutting Oil | DoALL Co. |
| | DORTAN N 36 | Exxon Co., U.S.A. |
| | Gulfcut 45A | Gulf Oil Corp. |
| | Cut-Max 236, Cut-Max Base 7 | E. F. Houghton & Co. |
| | I.C. #283-C, I.C. #629, I.C. #1220-A | International Chemical Co. |
| | IRMCO 229 | International Refining & Mfg. Co. |
| | Threadmate 110 | Lubrication Co. of America |
| | Black Velvet 77 | Pillsbury Chemical and Oil, Inc. |
| | LUBE-A-TUBE "EP-MR" #6 OIL, FRIGIDOL #401-P COMPOUND | G. Whitfield Richards Co. |
| | SYN-KUT SPECIAL 725, SYN-KUT SPECIAL 750 | Henry E. Sanson & Sons, Inc. |
| | Space Age 49-7 Cutting Oil | Space Age Chemlube |
| | Sulkleer 19E, Sulkleer 102S, Sulkleer 163A | Standard Oil Co. (Ohio) |
| | Starcut-3031 | Star Oil Company |
| | SUNICUT 501 | Sun Petroleum Products Co. |
| | TOWERCUT D-908 | Tower Oil and Technology Co. |
| | VSCC 5495 | Van Straaten Chemical Co. |
| | W&B Cutting Oil 2698 | The White & Bagley Co. |
| | Maccut 11 Concentrate, Maccut 16 BGA, Maccut 1652 DO | Witco Chemical Corp. |
| | WITHROKUT 158 CUTTING OIL | Arthur C. Withrow Co. |
| | **Compounded Sulfo-Chlorinated Mineral Oil + Fatty Oil** | |
| | Lafayette B-16 | American Oil & Supply Co. |
| | Promax 1022 | Ashland Oil, Inc. |
| | Ordnance Oils | Atlantic Richfield Co. |
| | Belco 730-S Cutting Oil | Bel-Ray Co., Inc. |
| | Bruko D-647, Bruko D-637 | Bruce Products Corp. |
| | PETROCHEM THREAD KUT DARK | Champions Choice, Inc. |
| | Chevron Metalworking Fluid 503 | Chevron U.S.A. Inc. |
| | Cook-Cut 20, Cook-Cut 1545, Cook-Tap #1, Cook-Tap 1157 | Cook's Industrial Lubricants, Inc. |
| | DAUBERT GARIA D, DAUBERT GARIA G | Daubert Chemical Co. |
| | PENNEX N 47 | Exxon Co., U.S.A. |
| | Gulfcut 45B, Gulfcut 21D | Gulf Oil Corp. |
| | Habcool 318, Habcool 277 | H & B Petroleum Co. |
| | I.C. #815-A, I.C. #1894 | International Chemical Co. |
| | IRMCO 790 | International Refining & Mfg. Co. |
| | M-60-3 | S. C. Johnson & Son, Inc. |
| | Exzact 750 | Lubrication Co. of America |
| | Black Velvet 25 | Pillsbury Chemical and Oil, Inc. |
| | LUBE-A-TUBE "EP-M" COMPOUND, FRIGIDOL #401-A COMPOUND | G. Whitfield Richards Co. |
| | BOOSTER-KUT PLV | Henry E. Sanson & Sons, Inc. |
| | Space Age 49-8 Cutting Oil | Space Age Chemlube |
| | Toolife 205 | Specialty Products Co. |
| | Starcut-3461 | Star Oil Company |
| | SUNICUT 500 | Sun Petroleum Products Co. |
| | TOWERCUT D-908 | Tower Oil and Technology Co. |
| | VSCC 5510-R, VSCC 5700 | Van Straaten Chemical Co. |
| | W&B Cutting Oil 2192 | The White & Bagley Co. |
| | Maccut 1652 DO, Maccut 18-L | Witco Chemical Corp. |
| | WITHROKUT 203 CUTTING OIL | Arthur C. Withrow Co. |

| CODE NO. | TYPE AND TRADE NAME | MANUFACTURER |
|---|---|---|

## 3 OILS—HEAVY DUTY (Continued)

### Sulfo-Chlorinated Fatty Oil

| | |
|---|---|
| Promax H-333 | Ashland Oil, Inc. |
| Cook's Compound 4710 | Cook's Industrial Lubricants, Inc. |
| IRMCO 252 | International Refining & Mfg. Co. |
| Elcoa Base 201 | Lubrication Co. of America |
| Magno Base 41 | Pillsbury Chemical and Oil, Inc. |
| LUBE-A-TUBE "EP-M" COMPOUND | G. Whitfield Richards Co. |
| SYN-KUT 99 | Henry E. Sanson & Sons Inc. |
| Space Age 49-9 Cutting Oil | Space Age Chemlube |
| Toolife 208 | Specialty Products Co. |
| W&B Base 2844 | The White & Bagley Co. |

### Sulfurized Fatty Oil

| | |
|---|---|
| Promax H-3 | Ashland Oil, Inc. |
| Cadrex #1, Safrex #1 | Cook's Industrial Lubricants, Inc. |
| IRMCO 055 | International Refining & Mfg. Co. |
| Nu-Mac 11 | Lubrication Co. of America |
| CUTTING BASE 66 | Magnus Div., Economics Lab Inc. |
| Gipco Base 11M | Pillsbury Chemical and Oil, Inc. |
| FRIGIDOL BASE "V" COMPOUND, FRIGIDOL BASE "VR" COMPOUND | G. Whitfield Richards Co. |
| CUTZOL SFO-507 | Rust-Lick, Inc. |
| SYN-KUT 340 | Henry E. Sanson & Sons Inc. |
| Toolife Base 17, Toolife 404 | Specialty Products Co. |
| W&B Base L | The White & Bagley Co. |
| "RUGGED CUT" Cutting and Threading Oil | J. C. Whitlam Manufacturing Co. |
| WITHROBASE 423 CONCENTRATE | Arthur C. Withrow Co. |

### Mineral Oil + Added Fat, Sulfur, Chlorine, and Phosphorus Compounding

| | |
|---|---|
| Cook's Base 4601, Cook-Cut 4749, Cook-Grind 4806 | Cook's Industrial Lubricants, Inc. |
| Gulfcut 41H | Gulf Oil Corp. |
| I.C. #1885 | International Chemical Co. |
| IRMCO 791 | International Refining & Mfg. Co. |
| LUBRICOOL 280 | Larson Chemical Specialties |
| Specialube H-104 | Lubrication Technology, Inc. |
| Sintolin X-2 | Process Research Products |
| SYN-KUT FGO | Henry E. Sanson & Sons Inc. |
| Sultex F | Texaco Inc. |
| ZURN ULTRAKUT 70 S | O F Zurn Co. |

## 4 EMULSIFIABLE OILS—LIGHT DUTY (GENERAL PURPOSE)

### Soluble Oil

| | |
|---|---|
| Lafayette Soluble Oil #1 | American Oil & Supply Co. |
| AMOCOOL Soluble Oil, Amocool Soluble Oil | Amoco Oil Co. |
| Adsol 1, Adsol 100 | Ashland Oil, Inc. |
| Crystex Sol. Oil, Emulsicool Sol. Oil, RISO Sol. Oil, Sol Oil DO, Sol. Oil Z, Tooltex Sol. Oil, No. 1 Sol. Oil, 943 Sol. Oil | Atlantic Richfield Co. |
| Soluble Oil #371, Fedaro CM, Fedara GS, Fedaro HW, Fedaro SB | BP Oil, Inc. |

| CODE NO. | TYPE AND TRADE NAME | MANUFACTURER |
|---|---|---|
| 4 | **EMULSIFIABLE OILS—LIGHT DUTY (GENERAL PURPOSE)** (Continued) | |
| | **Soluble Oil** (Continued) | |
| | Barco Soluble Oil 15 | Bel-Ray Co., Inc. |
| | Brusol A | Bruce Products Corp. |
| | PETROCHEM SOLUBLE OIL | Champions Choice, Inc. |
| | Chevron Soluble Oil | Chevron U.S.A. Inc. |
| | L 470, L 481, L 493, L 3022 HW, L 3033 NB | Chrysler Corp. |
| | CITGO Cutting Oil 205, CITGO Cutting Oil 210 | Cities Service Co. |
| | Cook-Cool 500, Cook-Cool 1235 | Cook's Industrial Lubricants, Inc. |
| | COOL SPEED 500, DAUBERT DROMUS B | Daubert Chemical Co. |
| | No. 470 Soluble Oil | DoALL Co. |
| | KUTWELL 30, KUTWELL 40 | Exxon Co., U.S.A. |
| | TOOL COOL 1854 | Franklin Oil Corp. (Ohio) |
| | Fremont-7040, Fremont-7042, Fremont-7043, Fremont-7045 | Fremont Industries, Inc. |
| | Gulfcut Soluble Oil | Gulf Oil Corp. |
| | Habcool 747 | H & B Petroleum Co. |
| | A.S.O. | E. F. Houghton & Co. |
| | I.C. #1651, Tooleze #2000 | International Chemical Co. |
| | KEYCUT SOLUBLE OIL | Keystone Division/Pennwalt Corp. |
| | 301, 303, 304, 333, 309, 322 Soluble Oils | Lubrication Co. of America |
| | CLEAN CUT, MAGNA COOL CC, EL SOL R | Magnus Div./Economics Lab Inc. |
| | TRIM SOL LC | Master Chemical Corp. |
| | Metsol, Oxitrol Air-O-Mist | Metalloid Corp. |
| | Hamikut | Harry Miller Corp. |
| | Mobilmet S-123, Mobilmet S-127, Mobilmet 104 | Mobil Oil Corp. |
| | Oakite Formula 59 | Oakite Products, Inc. |
| | Cutlass C 5500 | Pillsbury Chemical & Oil, Inc. |
| | Polar Sol 500 | Polar Chip, Inc. |
| | Triple C 3000 | Pro-Chem Inc. |
| | Rycosol 19, Cupromul 23 | Reilly-Whiteman, Inc. |
| | LUBE-WELL D-100 SPECIAL SOLUBLE OIL, LUBE-WELL S-42 SOLUBLE OIL | G. Whitfield Richards Co. |
| | CUTZOL WS-10, CUTZOL WS-11 | Rust-Lick, Inc. |
| | SYN-SOL D A, SYN-SOL 60, ANTISEPTIC SOLUBLE OIL | Henry E. Sanson & Sons Inc. |
| | Space Age T-61 Cutting Oil, Space Age Z-33 Cutting Oil, Space Age Z-34 Cutting Oil, Space Age Z-35 Cutting Oil, Space Age Z-36 Cutting Oil, Space Age Z-37 Cutting Oil, Space Age #78, Soluble Oil 20-1 | Space Age Chemlube |
| | Toolife 301 | Specialty Products Co. |
| | Staysol 4X, Staysol 77 | Standard Oil Co. (Ohio) |
| | Starsol-2950 | Star Oil Company |
| | SECO, SECO 11 | Sun Petroleum Products Co. |
| | Soluble Oil D | Texaco Inc. |
| | TOWERSOL H-978 | Tower Oil & Technology Co. |
| | VSCC 255-A, VSCC 653 | Van Straaten Chemical Co. |
| | W&B RP Soluble Oil 1999 | The White & Bagley Co. |
| | Macco 163, 230-JL-3, Grinding oil 445 | Witco Chemical Corp. |
| | WITHROSOL 310 SOLUBLE OIL, WITHROSOL 315 SOLUBLE OIL | Arthur C. Withrow Co. |
| | ZURNSOL 19 | O F Zurn Co. |

| CODE NO. | TYPE AND TRADE NAME | MANUFACTURER |
|---|---|---|
| 4 | **EMULSIFIABLE OILS—LIGHT DUTY (GENERAL PURPOSE)** (Continued) | |
| | **Semichemical** | |
| | AMOCOOL Transparent Coolant | Amoco Oil Co. |
| | Brukool 127 | Bruce Products Corp. |
| | COOL SPEED 515 | Daubert Chemical Co. |
| | TOOL COOL AR SOLUBLE | Franklin Oil Corp. (Ohio) |
| | I.C. 124 | International Chemical Co. |
| | IRMCO 302 | International Refining & Mfg. Co. |
| | Hamikleer | Harry Miller Corp. |
| | CHOICE-CUT | Monroe Chemical Co. Inc. |
| | Hy Kut 9737 | Pillsbury Chemical & Oil, Inc. |
| | Triple C 3020 | Pro-Chem Inc. |
| | Sintolin OP2, HD | Process Research Products |
| | Rycosol 60 SDG | Reilly-Whiteman, Inc. |
| | GWR COMPOUND | G. Whitfield Richards Co. |
| | CUTZOL WS-5100 | Rust-Lick, Inc. |
| | SYN-KUT 125-P | Henry E. Sanson & Sons, Inc. |
| | Ami-Lube X-15 | Tron-X Research Corp. |
| | VSCC 938 | Van Straaten Chemical Co. |
| | W&B Lubricant 1888 | The White & Bagley Co. |
| | Macco 158-C, Macco Syncrocut 167-DM-9, New Solube-D | Witco Chemical Corp. |
| | **Semisynthetic** | |
| | Brukool 82-C, Brukool 121 | Bruce Products Corp. |
| | 12 D 192 | Chrysler Corp. |
| | Cook-Coolex 40, Cook-Cool 1300, Cook-Cool 4893 | Cook's Industrial Lubricants, Inc. |
| | COOL SPEED 363 | Daubert Chemical Co. |
| | TOOL COOL 1369 | Franklin Oil Corp. (Ohio) |
| | I.C. #343-S | International Chemical Co. |
| | J-Cool, Trampol-X, Coldstream | S. C. Johnson & Son, Inc. |
| | CLEAR-CUT C | Larson Chemical Specialties |
| | Heavy Duty 302 Carbo-Flo, Florosol | Lubrication Co. of America |
| | Hamikleer | Harry Miller Corp. |
| | CHOICE-CUT | Monroe Chemical Co. Inc. |
| | Hy Kut 105 | Pillsbury Chemical & Oil, Inc. |
| | Triple C 2670 | Pro-Chem Inc. |
| | Sintolin OP2, HD | Process Research Products |
| | LUBE-WELL "CD" COMPOUND | G. Whitfield Richards Co. |
| | CUTZOL WS-15, CUTZOL WS-30 | Rust-Lick, Inc. |
| | SYZ-KUT H | Henry E. Sanson & Sons, Inc. |
| | Toolife 316 | Specialty Products Co. |
| | SUNICOOL 6 | Sun Petroleum Products Co. |
| | ME II Super Concentrate (Recirculating) | Tapmatic Corp. |
| | Ami-Lube X-20 | Tron-X Research Corp. |
| | VSCC 537, VSCC 545-E | Van Straaten Chemical Co. |
| | W&B Concentrate 3100 | The White & Bagley Co. |
| | Macco Syncrocut 980-W-1, Macco Syncrocut 980-MM-2 | Witco Chemical Corp. |

| CODE NO. | TYPE AND TRADE NAME | MANUFACTURER |
|---|---|---|
| 5 | **EMULSIFIABLE OILS—HEAVY DUTY** | |
| | **Sulfurized Fatty Soluble Oil** | |
| | Lafayette Soluble Oil #60 | American Oil & Supply Co. |
| | AMOCOOL HD Soluble Oil | Amoco Oil Co. |
| | PETROCHEM 100 | Champions Choice, Inc. |
| | L 485 HW | Chrysler Corp. |
| | Soluble Cadrex 1838 | Cook's Industrial Lubricants, Inc. |
| | KUTWELL EP 66 | Exxon Co., U.S.A. |
| | TOOL COOL 1342 | Franklin Oil Corp. (Ohio) |
| | Gulfcut Heavy Duty Soluble Oil | Gulf Oil Corp. |
| | All Purpose Base | E. F. Houghton & Co. |
| | I.C. #1648 | International Chemical Co. |
| | IRMCO 340 | International Refining & Mfg. Co. |
| | Hamikut | Harry Miller Corp. |
| | WHEELMATE 819 | Norton Co. |
| | Foil Kut 5350 | Pillsbury Chemical & Oil, Inc. |
| | Triple C 3002 | Pro-Chem Inc. |
| | FRIGIDOL #222 COMPOUND, FRIGIDOL #222-T COMPOUND | G. Whitfield Richards Co. |
| | ANTISEPTIC ALL PURPOSE BASE | Henry E. Sanson & Sons, Inc. |
| | Toolife 303 | Specialty Products Co. |
| | Solopex-2760 | Star Oil Company |
| | TOWER G-941 | Tower Oil & Technology Co. |
| | Macco Hi Sul 588-EM | Witco Chemical Corp. |
| | WITHROSOL 320 SOLUBLE OIL | Arthur C. Whitrow Co. |
| | **Chlorinated Soluble Oil** | |
| | Lafayette Soluble Oil #71 | American Oil & Supply Co. |
| | AMOCOOL SS Fluid | Amoco Oil Co. |
| | Adsol 2 | Ashland Oil, Inc. |
| | Trukut EP-C | Atlantic Richfield Co. |
| | 12 U 49, L 485 HW | Chrysler Corp. |
| | CIMPERIAL 18, CIMPERIAL 1010, CIMPERIAL 1011 | Cincinnati Milacron, Products Div. |
| | CITGO Cutting Oil 215, CITGO Cutting Oil 220 | Cities Service Co. |
| | Cook-Cool 2020, Cook-Cool 4782, Cook-Cool 4718 | Cook's Industrial Lubricants, Inc. |
| | CURTIS S-6, CURTIS S-7, CURTIS S-21, CURTIS S-DD-100 | Curtis Systems, Inc. |
| | COOL SPEED 2010 | Daubert Chemical Co. |
| | TOOL COOL 1385, TOOL COOL 1485 | Franklin Oil Corp. (Ohio) |
| | Hocut 733, Hocut 747 | E. F. Houghton & Co. |
| | I.C. #922 | International Chemical Co. |
| | MAGNU DRAW 10, MAGNU DRAW 30, MAGNU DRAW 40, MAGNU DRAW 50L, CHLOROLUBE 50H | Magnus Div./Economics Lab Inc. |
| | TRIM SOL | Master Chemical Corp. |
| | Plastrol, Metsol WOS | Metalloid Corp. |
| | Hamikut | Harry Miller Corp. |
| | PRIME-CUT | Monroe Chemical Co., Inc. |
| | WHEELMATE 607 | Norton Co. |
| | Sword Kut C5100 | Pillsbury Chemical & Oil, Inc. |
| | Sintolin E-1 | Process Research Products |
| | Rycosol 69 | Reilly-Whiteman Inc. |
| | LUBE-WHEEL S-47-H SOLUBLE OIL, LUBE-WELL S-47 SOLUBLE OIL | G. Whitfield Richards Co. |
| | CUTZOL WS-10, CUTZOL WS-20, CUTZOL WS-5050 | Rust-Lick, Inc. |

| CODE NO. | TYPE AND TRADE NAME | MANUFACTURER |
|---|---|---|
| 5 | **EMULSIFIABLE OILS—HEAVY DUTY** (Continued) | |
| | **Chlorinated Soluble Oil** (Continued) | |
| | TOOLMATE, SYN-SOL 1000 | Henry E. Sanson & Sons Inc. |
| | Space Age 49-3–Cutting Oil | Space Age Chemlube |
| | Toolife 315 | Specialty Products Co. |
| | Staysol 743 | Standard Oil Co. (Ohio) |
| | Starsol-2800 | Star Oil Company |
| | EMULSUZ 51 | Sun Petroleum Products Co. |
| | TOWERSOL G-966 | Tower Oil & Technology Co. |
| | VSCC 710 | Van Straaten Chemical Co. |
| | W&B HD Soluble Oil 2213 | The White & Bagley Co. |
| | Macco 472, Macco 472-MPA, Macco 478, Macco 768 MTS-3 | Witco Chemical Corp. |
| | WITHROSOL 323 SOLUBLE OIL | Arthur C. Whitrow Co. |
| | **Sulfo-Chlorinated Fatty Soluble Oil** | |
| | Lafayette Soluble Oil #71 | American Oil & Supply Co. |
| | Adsol 3 | Ashland Oil, Inc. |
| | Trukut EP-SC | Atlantic Richfield Co. |
| | Fedaro EPSC | BP Oil, Inc. |
| | Bel-Ray WS-6-HD | Bel-Ray Co., Inc. |
| | Brukool 100, Brusol C, Brukool 131 | Bruce Products Corp. |
| | PETROCHEM 101 | Champions Choice, Inc. |
| | CIMPERIAL 20 | Cincinnati Milacron, Products Div. |
| | Cook-Cool 1400 | Cook's Industrial Lubricants, Inc. |
| | DAUBERT DROMUS SD | Daubert Chemical Co. |
| | POWER-CUT No. 360 | DoALL Co. |
| | TOOL COOL 1397 | Franklin Oil Corp. (Ohio) |
| | Habcool 312 | H & B Petroleum Co. |
| | Hocut 3210-X, Permasol HD-20 | E. F. Houghton & Co. |
| | I.C. #939-NP | International Chemical Co. |
| | IRMCO 335 | International Refining & Mfg. Co. |
| | LUBRI-FLO 1000, LUBRI-FLO 440 | Larson Chemical Specialties |
| | EP COOLANT, MAGNA COOL EP, EL SOL-SC | Magnus Div./Economics Lab Inc. |
| | TRIM SOL S | Master Chemical Corp. |
| | Hamidraw | Harry Miller Corp. |
| | Mobilmet 140, Mobilmet 160 | Mobil Oil Corp. |
| | MONROE HD | Monroe Chemical Co., Inc. |
| | WHEELMATE 810, WHEELMATE 811 | Norton Co. |
| | Rapier Kut 5200 | Pillsbury Chemical & Oil, Inc. |
| | Triple C 3011 | Pro-Chem Inc. |
| | Sintolin E-100 | Process Research Products |
| | LUBE-WELL S-43 SOLUBLE OIL | G. Whitfield Richards Co. |
| | CUTZOL PB-10 | Rust-Lick, Inc. |
| | SYN-SOL 300 | Henry E. Sanson & Sons, Inc. |
| | SWEPCO 720 Water Soluble Cutting Oil | Southwestern Petroleum Corp. |
| | Space Age 49-4–Cutting Oil | Space Age Chemlube |
| | Staysol EPSC | Standard Oil Co. (Ohio) |
| | Solopex-2770 | Star Oil Company |
| | EMULSUN 54 | Sun Petroleum Products Co. |
| | Soluble Oil HD, TEXGOLD | Texaco Inc. |
| | TOWERSOL G919 | Tower Oil & Technology Co. |

# 16.3 Cutting Fluid Key

| CODE NO. | TYPE AND TRADE NAME | MANUFACTURER |
|---|---|---|
| 5 | **EMULSIFIABLE OILS—HEAVY DUTY** (Continued) | |
| | **Sulfo-Chlorinated Fatty Soluble Oil** (Continued) | |
| | VSCC 700, VSCC 708 | Van Straaten Chemical Co. |
| | Macco Extrudoil 51-DE, Spermasol EP, Kelcut 203-EP-Soluble | Witco Chemical Corp. |
| | WITHROSOL 322 SOLUBLE OIL | Arthur C. Whitrow Co. |
| | ZURNTURN 004 | O F Zurn Co. |
| | **Semichemical** | |
| | AMOCOOL SS Fluid | Amoco Oil Co. |
| | Brukool 129 | Bruce Products Corp. |
| | CIMPERIAL 15 | Cincinnati Milacron, Products Div. |
| | COOL SPEED 515 | Daubert Chemical Co. |
| | Libricoolant 930 Sol. Oil (Mineral with natural sulfur) | DuBois Chemicals |
| | I.C. #1667 | International Chemical Co. |
| | IRMCO 302 | International Refining & Mfg. Co. |
| | TRIM EP | Master Chemical Corp. |
| | Hamikleer | Harry Miller Corp. |
| | CHOICE-CUT | Monroe Chemical Co., Inc. |
| | Hy Kut 9737 | Pillsbury Chemical & Oil, Inc. |
| | POLAR SOL 510 | Polar Chip, Inc. |
| | Triple C 3010 | Pro-Chem Inc. |
| | LUBE-A-TUBE "XAM" COMPOUND | G. Whitfield Richards Co. |
| | CUTZOL WS-5100 | Rust-Lick, Inc. |
| | SYN-SOL | Henry E. Sanson & Sons, Inc. |
| | Space Age "Will-Cut" | Space Age Chemlube |
| | Ami-Lube SX-15 | Tron-X Research Corp. |
| | VSCC 707 | Van Straaten Chemical Co. |
| | W&B EP Concentrate 2828 | The White & Bagley Co. |
| | Syncrocut 167-DM-9, Macco 158-C | Witco Chemical Corp. |
| | ZURNSOL 253 | O F Zurn Co. |
| | **Semisynthetic** | |
| | Brukool 66, Brukool 128 | Bruce Products Corp. |
| | Cook-Coolex 40 | Cook's Industrial Lubricants, Inc. |
| | COOL SPEED 363 | Daubert Chemical Co. |
| | Fremont-7030, Fremont-7031, Fremont-7032, Fremont-7033, Fremont-7035, Fremont-7036 | Fremont Industries, Inc. |
| | I.C. 321K | International Chemical Co. |
| | IRMCO 303 | International Refining & Mfg. Co. |
| | Slipstream 54, Coldstream | S. C. Johnson & Son, Inc. |
| | Hon-E-Cool | Lubrication Co. of America |
| | Hamikleer | Harry Miller Corp. |
| | CHOICE-CUT | Monroe Chemical Co., Inc. |
| | Oakite Formula B-50 | Oakite Products, Inc. |
| | Hy Kut 9734 | Pillsbury Chemical & Oil, Inc. |
| | Triple C 2511 | Pro-Chem Inc. |
| | GWR COMPOUND | G. Whitfield Richards Co. |
| | CUTZOL WS-15, CUTZOL WS-30 | Rust-Lick, Inc. |
| | SYN-SOL DURO 55, P.O. CUTTING LIQUID | Henry E. Sanson & Sons Inc. |
| | Space Age "Will-Cut" | Space Age Chemlube |
| | Toolife 316 | Specialty Products Co. |

| CODE NO. | TYPE AND TRADE NAME | MANUFACTURER |
|---|---|---|
| **5** | **EMULSIFIABLE OILS—HEAVY DUTY** (Continued) | |
| | **Semisynthetic** (Continued) | |
| | ME II Super Concentrate (Recirculating) | Tapmatic Corp. |
| | Ami-Lube Regular | Tron-X Research Corp. |
| | VSCC 550-P | Van Straaten Chemical Co. |
| | W&B Concentrate 3100 | The White & Bagley Co. |
| | Macco 980-W-1, Macco Syncrocut 980-MM-2 | Witco Chemical Corp. |
| | ZURNSOL 254 | O F Zurn Co. |
| **6** | **CHEMICAL AND SYNTHETICS—LIGHT DUTY (GENERAL PURPOSE)** | |
| | **Solutions** | |
| | CAMPBELLENE, CAMPBELLENE NO-NITRITE | Allison Campbell Div./Acco Industries |
| | Lafayette Soluble #91 | American Oil & Supply Co. |
| | AMOCOOL Transparent Coolant | Amoco Oil Co. |
| | Adcool 1 | Ashland Oil, Inc. |
| | Brukool 130, Brukool 77, Brukool 78 | Bruce Products Corp. |
| | PETROCHEM CHEMICAL COOLANT | Champions Choice, Inc. |
| | L 1549, 12 D 192 | Chrysler Corp. |
| | CIMFREE 104, CIMFREE 106, CIMFREE 108 | Cincinnati Milacron, Products Div. |
| | Cook-Cool 1975 (Grinding), Cook-Cool 4988 (Grinding) | Cook's Industrial Lubricants, Inc. |
| | COOL SPEED 77-45, COOL SPEED 36 | Daubert Chemical Co. |
| | Kleen-Kool | DoALL Co. |
| | Lubricoolant 925, Lubricoolant 926 | DuBois Chemicals |
| | Fremont-5018 | Fremont Industries, Inc. |
| | Houghto-Grind 60 | E. F. Houghton & Co. |
| | I.C. 124 | International Chemical Co. |
| | IRMCO 120, IRMCO 110 | International Refining & Mfg. Co. |
| | 50 Cool, TL-131, Millstream | S. C. Johnson & Son, Inc. |
| | KEYCUT 106 | Keystone Division/Pennwalt Corp. |
| | KOOLMIST FORMULA #78 COOLANT | Kool Mist Corp. |
| | LPS 1, LPS 2, LPS 3 | LPS Research Laboratories, Inc. |
| | BLUE COOLANT NF | Magnus Div./Economics Lab Inc. |
| | TRIM 9106, TRIM 9106CS, TRIM REGULAR, TRIM 7030, TRIM 5050, TRIM 2575, TRIM MIST, TRIM 9106WA | Master Chemical Corp. |
| | Metgrind #432, Metgrind #468 | Metalloid Corp. |
| | Hamikleer | Harry Miller Corp. |
| | MONROE RI | Monroe Chemical Co., Inc. |
| | WHEELMATE 464, WHEELMATE 467 | Norton Co. |
| | Oakite Formula A, Oakite Formula C, Oakite Formula G | Oakite Products, Inc. |
| | Hone-Rite 306 | Pillsbury Chemical & Oil, Inc. |
| | Polar Chip 347 | Polar Chip, Inc. |
| | Sintolin R30, Aqualap 15C | Process Research Products |
| | Rycosol 66 | Reilly-Whiteman, Inc. |
| | GWR #110 | G. Whitfield Richards Co. |
| | RUST-LICK G-25-J, RUST-LICK G-25-AH, VYTRON | Rust-Lick, Inc. |
| | MIRROR-KOOL M-1, MIRROR-KOOL M-15 | Henry E. Sanson & Sons, Inc. |
| | Kool-Aide I | Space Age Chemlube |

# 16.3 Cutting Fluid Key

| CODE NO. | TYPE AND TRADE NAME | MANUFACTURER |
|---|---|---|
| 6 | **CHEMICAL AND SYNTHETICS—LIGHT DUTY (GENERAL PURPOSE)** (Continued) | |
| | **Solutions** (Continued) | |
| | Toolife 305 | Specialty Products Co. |
| | Molecular Edge (Squirt on), ME II Super Concentrate (Recirculating) | Tapmatic Corp. |
| | TOWERKEM W-2110 | Tower Oil & Technology Co. |
| | Ami-Lube Clear | Tron-X Research Corp. |
| | VSCC 990 | Van Straaten Chemical Co. |
| | Bio-Cool 500 | Westmont Industrial Products |
| | W&B Grinding Concentrate 1500 | The White & Bagley Co. |
| | Syncrocut 980-CE-3, Syncrocut 850-A-3 | Witco Chemical Corp. |
| | WITHROKOOL 334 CHEMICAL COOLANT | Arthur C. Withrow Co. |
| | ZURN KOOLKUT | O F Zurn Co. |
| | **Synthetic Emulsions** | |
| | Q-77 Coolant Concentrate | Associate Technical Sales Co. |
| | L 1542 | Chrysler Corp. |
| | CIMCOOL S2 | Cincinnati Milacron, Products Div. |
| | COOL SPEED 363 | Daubert Chemical Co. |
| | KUTWELL 82 | Exxon Co., U.S.A. |
| | Hocut 711 | E. F. Houghton & Co. |
| | I.C. #922 | International Chemical Co. |
| | Specialube 994 | Lubrication Technology, Inc. |
| | Hamikleer | Harry Miller Corp. |
| | MONROE NN | Monroe Chemical Co., Inc. |
| | WHEELMATE 405, WHEELMATE 414, WHEELMATE 419, WHEELMATE 613 | Norton Co. |
| | Hy Kut 105 | Pillsbury Chemical & Oil, Inc. |
| | Polar Cut 400 | Polar Chip, Inc. |
| | Triple C 2010 | Pro-Chem Inc. |
| | Rycosol 77, Rycosol 87 NF | Reilly-Whiteman, Inc. |
| | LUBE-WELL D-171 SOLUBLE OIL | G. Whitfield Richards Co. |
| | CUTZOL WS-15 | Rust-Lick, Inc. |
| | MIRROR-KOOL G-70 | Henry E. Sanson & Sons, Inc. |
| | Kool-Aide XII | Space Age Chemlube |
| | Toolife 316 | Specialty Products Co. |
| | SUNICOOL 6 | Sun Petroleum Products Co. |
| | Ami-Lube Syn-12 Series | Tron-X Research Corp. |
| | W&B Lubricant 1888 | The White & Bagley Co. |
| 7 | **CHEMICAL AND SYNTHETICS—HEAVY DUTY** | |
| | **Solutions** | |
| | CAMPBELLENE, CAMPBELLENE NO-NITRITE | Allison Campbell Div./Acco Industries |
| | Lafayette Synthetic #102 | American Oil & Supply Co. |
| | Adcool 2 | Ashland Oil, Inc. |
| | Brukool 106 | Bruce Products Corp. |
| | PETROCHEM 130Z | Champions Choice, Inc. |
| | 12 U 176, 12 U 182 B | Chrysler Corp. |

| CODE NO. | TYPE AND TRADE NAME | MANUFACTURER |
|---|---|---|
| 7 | **CHEMICAL AND SYNTHETICS—HEAVY DUTY** (Continued) | |
| | **Solutions** (Continued) | |
| | CIMFREE 234, CIMFREE 238, CIMFREE 305, CIMFREE 307 | Cincinnati Milacron, Products Div. |
| | Cook-Cool 1975 (Grinding), Cook-Cool 4988 (Grinding) | Cook's Industrial Lubricants, Inc. |
| | POWER-CUT HD-600 | DoALL Co. |
| | TOOL COOL 268 | Franklin Oil Corp. (Ohio) |
| | Fremont-7010, Fremont-7011 | Fremont Industries, Inc. |
| | Gulfcut Soluble CB-2 | Gulf Oil Corp. |
| | Hocut 702, Hocut 4206, Hydra-Cut 496 | E. F. Houghton & Co. |
| | I.C. 2614 | International Chemical Co. |
| | IRMCO 111 | International Refining & Mfg. Co. |
| | KOOLMIST FORMULA #77 COOLANT | Kool Mist Corp. |
| | CLEAR-CUT 44 | Larson Chemical Specialists |
| | Empactosol | Lubrication Co. of America |
| | Specialube 67D, Specialube 75 | Lubrication Technology, Inc. |
| | TRIM HD, TRIM EP | Master Chemical Corp. |
| | Metchem #429, Metchem #474 | Metalloid Corp. |
| | Hamikleer | Harry Miller Corp. |
| | WHEELMATE 658, WHEELMATE 689 | Norton Co. |
| | Kool Kut X350 | Pillsbury Chemical & Oil, Inc. |
| | POLAR CHIP 347L | Polar Chip, Inc. |
| | Sintolin ROD | Process Research Products |
| | GWR COMPOUND | G. Whitfield Richards Co. |
| | VYTRON, RUST-LICK G-52 | Rust-Lick, Inc. |
| | E-COOL, MIRROR-KUT G-7 | Henry E. Sanson & Sons Inc. |
| | Kool-Aide III | Space Age Chemlube |
| | Toolife 311 | Specialty Products Co. |
| | Molecular Edge (Squirt on), ME II Super Concentrate (Recirculating) | Tapmatic Corp. |
| | TOWERKEM W-2110 | Tower Oil & Technology Co. |
| | Ami-Lube 10X-Clear | Tron-X Research Corp. |
| | VSCC 951, VSCC 936 | Van Straaten Chemical Co. |
| | Bio-Cool 700 | Westmont Industrial Products |
| | W&B Concentrate 3100 | The White & Bagley Co. |
| | Syncrocut 105, Syncrocut 172 | Witco Chemical Corp. |
| | WITHROKOOL 340 CHEMICAL COOLANT | Arthur C. Withrow Co. |
| | ZURN ALL MET 6 | O F Zurn Co. |
| | **Synthetic Emulsions** | |
| | Adcool 3 | Ashland Oil, Inc. |
| | Brukool 132 | Bruce Products Corp. |
| | 12 U 41 | Chrysler Corp. |
| | CIMCOOL FIVE STAR 40, CIMCOOL S4, CIMCOOL S8 | Cincinnati Milacron, Products Div. |
| | COOL SPEED 363, COOL SPEED 515 | Daubert Chemical Co. |
| | I.C. #922 | International Chemical Co. |
| | Specialube 997A, Specialube 998 | Lubrication Technology, Inc. |
| | Hamikleer | Harry Miller Corp. |
| | MONROE NN | Monroe Chemical Co., Inc. |
| | WHEELMATE 603, WHEELMATE 613, WHEELMATE 674 | Norton Co. |
| | Hy Kut 9734 | Pillsbury Chemical & Oil, Inc. |
| | POLAR EZE 444 | Polar Chip, Inc. |
| | Triple C 2510 | Pro-Chem Inc. |

| CODE NO. | TYPE AND TRADE NAME | MANUFACTURER |
|---|---|---|
| 7 | **CHEMICAL AND SYNTHETICS—HEAVY DUTY** (Continued) | |
| | **Synthetic Emulsions** (Continued) | |
| | Rycosol 99 | Reilly-Whiteman, Inc. |
| | LUBE-WELL "X" COMPOUND | G. Whitfield Richards Co. |
| | CUTZOL WS-15 | Rust-Lick, Inc. |
| | KING KOOL | Henry E. Sanson & Sons, Inc. |
| | Toolife 314 | Specialty Products Co. |
| | Starsyn-5 | Star Oil Company |
| | Ami-Lube Syn-12X Series | Tron-X Research Corp. |
| | VSCC 950, VSCC 938 | Van Straaten Chemical Co. |
| | W&B Lubricant 2300 | The White & Bagley Co. |
| | WITHROKOOL 339 CHEMICAL COOLANT | Arthur C. Withrow Co. |
| | ZURN KOOLKUT 252 | O F Zurn Co. |
| 8 | **SPECIALS—LIGHT DUTY** | |
| 80 | **Honing Oil—Light Duty** | |
| | Lafayette KM-16 | American Oil & Supply Co. |
| | AMOCUT Oil No. 62FC | Amoco Oil Co. |
| | Econ 5, Promax 1074 | Ashland Oil, Inc. |
| | Honol Oils | Atlantic Richfield Co. |
| | Bezora 3 | BP Oil, Inc. |
| | Bruko D-666 (non-ferrous), Bruko D-667 (ferrous) | Bruce Products Corp. |
| | PETROCHEM HONING OIL | Champions Choice, Inc. |
| | CIMFREE 104, CIMFREE 200, CIMFREE 234, CIMFREE 238 | Cincinnati Milacron, Products Div. |
| | Cook's 250 Compound | Cook's Industrial Lubricants, Inc. |
| | DAUBERT GARIA H | Daubert Chemical Co. |
| | MICROCOOL IC-9S: 90-10 | Ex-Cell-O Tool & Abrasive Products |
| | MENTOR 28 | Exxon Co., U.S.A. |
| | Gulfcut 11A, Gulfcut 11D | Gulf Oil Corp. |
| | I.C. #155-M | International Chemical Co. |
| | IRMCO 278, IRMCO 229 | International Refining & Mfg. Co. |
| | LPS 1, LPS 2, LPS 3 | LPS Research Laboratories, Inc. |
| | LUBRICOOL H/O | Larson Chemical Specialties |
| | Elcoa Honing Oil 213 | Lubrication Co. of America |
| | L-12 | Magnus Div./Economics Lab Inc. |
| | TRIM REGULAR, TRIM 7030, TRIM SOL LC | Master Chemical Corp. |
| | TF 382 | Metalloid Corp. |
| | Hamilube | Harry Miller Corp. |
| | Alumicut | Mistic Metal Mover, Inc. |
| | Vacmul 3A | Mobil Oil Corp. |
| | Hone Rite 711 | Pillsbury Chemical & Oil, Inc. |
| | GRA Honing Oil | Process Research Products |
| | Petrosan 102 | Reilly-Whiteman, Inc. |
| | GWR HONING OIL | G. Whitfield Richards Co. |
| | HONZOL 11-B | Rust-Lick, Inc. |
| | SYN-HONE MM-LV | Henry E. Sanson & Sons Inc. |

| CODE NO. | TYPE AND TRADE NAME | MANUFACTURER |
|---|---|---|

## 8 SPECIALS—LIGHT DUTY (Continued)

### 80 Honing Oil—Light Duty (Continued)

| TYPE AND TRADE NAME | MANUFACTURER |
|---|---|
| Toolife Honing Oil | Specialty Products Co. |
| Factokut H | Standard Oil Co. (Ohio) |
| SUNICUT 102 | Sun Petroleum Products Co. |
| MAN-845, CK-50 (for cast iron and aluminum only) | Sunnen Products Co. |
| Almag | Texaco Inc. |
| TOWER D-204 | Tower Oil & Technology Co. |
| VSCC 5551-A | Van Straaten Chemical Co. |
| Bio-Hone 252 | Westmont Industrial Products |
| W&B Honing Oil 2088 | The White & Bagley Co. |
| Macco Honing Oil | Witco Chemical Corp. |
| WITHROKUT 104 CUTTING OIL | Arthur C. Withrow Co. |

### 81 Chemical Grinding Fluids: Amine Nitrite Solution, Sodium Nitrite Solution, etc.

(NOTE: Products with * are Nitrite Free. See Chemical Fluids in section 16.1.)

| TYPE AND TRADE NAME | MANUFACTURER |
|---|---|
| Bel-Ray Coolant 12X | Bel-Ray Co., Inc. |
| Brukool 77, Brukool 106 | Bruce Products Corp. |
| Cook-Cool 1975*, Cook-Cool 4988* | Cook's Industrial Lubricants, Inc. |
| COOL SPEED 77-45, COOL SPEED 363*, COOL SPEED 36* | Daubert Chemical Co. |
| DuBois 910 Synthetic (sodium nitride solution), DuBois 915 | DuBois Chemicals |
| TOOL COOL 1392 | Franklin Oil Corp. (Ohio) |
| Coolant Base 100 | E. F. Houghton & Co. |
| I.C. #45, I.C. 124* | International Chemical Co. |
| Electrisol 308 | Lubrication Co. of America |
| TRIM 9106*, TRIM REGULAR, TRIM 7030, TRIM 5050, TRIM 2575*, TRIM HD, TRIM EP, TRIM 9106WA*, TRIM CLEAR, TRIM HM, TRIM TI GRIND | Master Chemical Corp. |
| Metgrind #468* | Metalloid Corp. |
| Immunol, Hamikleer | Harry Miller Corp. |
| MONROE NN*, MONROE RI* | Monroe Chemical Co., Inc. |
| WHEELMATE 203, WHEELMATE 205 | Norton Co. |
| Hone Rite 3795 | Pillsbury Chemical & Oil, Inc. |
| Polar Chip 347*, GRINDEX 11B | Polar Chip, Inc. |
| X55-30A | Reilly-Whiteman, Inc. |
| VYTRON, RUST-LICK B, RUST-LICK G-25-J, RUST-LICK G-25-AH*, RUST-LICK G-1066D | Rust-Lick, Inc. |
| MIRROR-KOOL G-5, MIRROR-KUT G-7* | Henry E. Sanson & Sons, Inc. |
| Kool-Aide IV* | Space Age Chemlube |
| Starsyn-5 | Star Oil Company |
| VSCC 5551-A, Van Straaten 920*, Van Straaten 936* | Van Straaten Chemical Co. |
| W&B Grinding Coolant E-55 | The White & Bagley Co. |
| Syncrocut 172 | Witco Chemical Corp. |

### 82 Cutting Fluid for Aluminum

| TYPE AND TRADE NAME | MANUFACTURER |
|---|---|
| CAMPBELLENE NO-NITRITE | Allison Campbell Div./Acco Industies |
| Lafayette BM-112 | American Oil & Supply Co. |
| AMOCUT Oil No. 62FC | Amoco Oil Co. |
| Promax 1074, Promax 1075 | Ashland Oil, Inc. |
| ARCO ML Oils, Clairo Oils | Atlantic Richfield Co. |

| CODE NO. | TYPE AND TRADE NAME | MANUFACTURER |
|---|---|---|
| 8 | **SPECIALS—LIGHT DUTY** (Continued) | |
| | **82 Cutting Fluid for Aluminum** (Continued) | |
| | Fedora EPSC, Bezora 22 | BP Oil, Inc. |
| | Brukool 127, Bruko D-680 | Bruce Products Corp. |
| | PETROCHEM 709 | Champions Choice, Inc. |
| | Chevron Metalworking Fluid 504, Chevron Soluble Oil | Chevron U.S.A. Inc. |
| | CIMCOOL AL, CIMCOOL S2, CIMCOOL S4, CIMCOOL S8, CIMFREE 234, CIMFREE 305, CIMPERIAL 15, CIMPERIAL 16, CIMPERIAL 1010, CIMPERIAL 1011 | Cincinnati Milacron, Products Div. |
| | CITGO Cutting Oil 400 | Cities Service Co. |
| | Cook-Cut 1711, Cook-Cut 30, Cook-Cool 1235 (Water soluble) | Cook's Industrial Lubricants, Inc. |
| | COOL SPEED 515 | Daubert Chemical Co. |
| | No. 120 Cutting Oil, No. 470 Soluble Oil | DoALL Co. |
| | Freon T-B1, Freon HV-3 | E. I. DuPont Co. Inc. |
| | Gulfcut 11D | Gulf Oil Corp. |
| | Habcool 202 | H & B Petroleum Co. |
| | Cindol 3411, Cindol 3101, Cindol 3102, Cindol 3103 | E. F. Houghton & Co. |
| | I.C. #1231-A, I.C. #131-A | International Chemical Co. |
| | IRMCO 249, IRMCO 242, IRMCO 278 | International Refining & Mfg. Co. |
| | 121 Wax Cut, Slipstream 54, Coldstream, J-Cool | S. C. Johnson & Son, Inc. |
| | LPS 1, LPS 2, LPS 3 | LPS Research Laboratories, Inc. |
| | LUBRI-FLO 1000 | Larson Chemical Specialties |
| | Vista-Kut | Lubrication Co. of America |
| | DO-5A | Magnus Div./Economics Lab Inc. |
| | TRIM SOL LC, TRIM SOL, TRIM 9106 WA, TRIM HD | Master Chemical Corp. |
| | WOS, Metsol, TF-306 | Metalloid Corp. |
| | Hamikut | Harry Miller Corp. |
| | Alumicut (Al) | Mistic Metal Mover, Inc. |
| | PRIME-CUT | Monroe Chemical Co., Inc. |
| | WHEELMATE 607, WHEELMATE 674 | Norton Co. |
| | Sword Kut C5100, Sabre Kut 5400 | Pillsbury Chemical & Oil, Inc. |
| | Polar Cut 400 | Polar Chip, Inc. |
| | Triple C 3010 | Pro-Chem Inc. |
| | Sintolin E-1 | Process Research Products |
| | Petrosan 116, Rycosol 19 | Reilly-Whiteman, Inc. |
| | A-9 (A1) | Relton Corp. |
| | LUBE-WELL S-4A SOLUBLE OIL, LUBE-A-TUBE "XAM" COMPOUND | G. Whitfield Richards Co. |
| | CUTZOL WS-10, CUTZOL WS-11, CUTZOL WS-20, CUTZOL WS-5050, CUTZOL WS-500-A | Rust-Lick, Inc. |
| | SYN-KUT 105 | Henry E. Sanson & Sons, Inc. |
| | Kool-Aide I, Kool-Aide IV, Kool-Aide XII | Space Age Chemlube |
| | Toolife 250, Toolife 304 | Specialty Products Co. |
| | Factokut A-50-C, Factokut L-2, Staysol EPSC | Standard Oil Co. (Ohio) |
| | Starcut-3902 | Star Oil Company |
| | TAP MAGIC ALUMINUM | Steco Corp. |
| | SUNICUT 150 | Sun Petroleum Products Co. |
| | Molecular Edge (Squirt on), ME II Super Concentrate (Recirculating), Dual Action Cutting Fluid #2 | Tapmatic Corp. |
| | Almag | Texaco Inc. |
| | TOWERCUT E-915 | Tower Oil & Technology Co. |
| | Ami-Lube X-12 Series | Tron-X Research Corp. |

| CODE NO. | TYPE AND TRADE NAME | MANUFACTURER |
|---|---|---|
| **8** | **SPECIALS—LIGHT DUTY** (Continued) | |
| | **82 Cutting Fluid for Aluminum** (Continued) | |
| | VSCC 45, VSCC 653 | Van Straaten Chemical Co. |
| | Bio-Cool 700 | Westmont Industrial Products |
| | W&B Cutting Oil 3016 | The White & Bagley Co. |
| | ALUMTAP | Winfield Brooks Co., Inc. |
| | Macco 230-JL-3 | Witco Chemical Corp. |
| | WITHROKUT 101 CUTTING OIL | Arthur C. Withrow Co. |
| | **83 Cutting Fluid for Titanium—Light Duty** | |
| | CAMPBELLENE, CAMPBELLENE NO-NITRITE | Allison Campbell Div./Acco Industries |
| | AMOCOOL Soluble Oil | Amoco Oil Co. |
| | Promax 1002 | Ashland Oil, Inc. |
| | Crystex Sol. Oil, Emulsicool Sol. Oil, Sol. Oil DO, Sol. Oil Z, Tooltex Sol. Oil, No. 1 Sol. Oil, 943 Sol. Oil, Autokut Oils | Atlantic Richfield Co. |
| | Fedora EPSC | BP Oil, Inc. |
| | Brukool 4-37d | Bruce Products Corp. |
| | Chevron Metalworking Fluid 503 | Chevron U.S.A. Inc. |
| | CIMFREE 234, CIMFREE 238, CIMCOOL S4, CIMCOOL FIVE STAR 40 | Cincinnati Milacron, Products Div. |
| | Cook-Cool 1235, Cook-Coolex 40 | Cook's Industrial Lubricants, Inc. |
| | COOL SPEED 515 | Daubert Chemical Co. |
| | POWER-CUT HD-600 | DoALL Co. |
| | Freon T-B1, Freon HV-3 | E. I. DuPont Co. Inc. |
| | Hocut 711, Cindol 3212 | E. F. Houghton & Co. |
| | I.C. #48, I.C. 124 | International Chemical Co. |
| | IRMCO 278 | International Refining & Mfg. Co. |
| | TL-131 | S. C. Johnson & Son, Inc. |
| | LPS 1, LPS 2, LPS 3 | LPS Research Laboratories, Inc. |
| | Threadmate 2254 | Lubrication Co. of America |
| | TRIM TI GRIND, TRIM REGULAR, TRIM 7030, TRIM 9106WA, TRIM HD, TRIM SOL | Master Chemical Corp. |
| | Metkut #310 | Metalloid Corp. |
| | Hamikut | Harry Miller Corp. |
| | Alumicut | Mistic Metal Mover, Inc. |
| | CHOICE-CUT, PRIME-CUT, MONROE HD | Monroe Chemical Co., Inc. |
| | WHEELMATE 602 | Norton Co. |
| | Hone Rite TI-Cut | Pillsbury Chemical & Oil, Inc. |
| | Polar Chip 347 | Polar Chip, Inc. |
| | Sintolin OP2 | Process Research Products |
| | Rycosol 66 | Reilly-Whiteman, Inc. |
| | GWR COMPOUND | G. Whitfield Richards Co. |
| | CUTZOL WS-500-A, VYTRON, CUTZOL 400-A | Rust-Lick, Inc. |
| | MIRROR-KUT G-7, SYN-SOL 300 | Henry E. Sanson & Sons, Inc. |
| | Kool-Aide IV, Kool-Aide XII | Space Age Chemlube |
| | Staysol EPSC, Sulkleer 163A | Standard Oil Co. (Ohio) |
| | EMULSUN 31, EMULSUN 51 | Sun Petroleum Products Co. |
| | Transultex D, Sultex D | Texaco Inc. |
| | Ami-Lube X-12 Series | Tron-X Research Corp. |
| | VSCC 550 | Van Straaten Chemical Co. |
| | W&B Cutting Oil 2151 | The White & Bagley Co. |

| CODE NO. | TYPE AND TRADE NAME | MANUFACTURER |
|---|---|---|
| 8 | **SPECIALS—LIGHT DUTY** (Continued) | |
| | **83 Cutting Fluid for Titanium—Light Duty** (Continued) | |
| | Bio-Cool 500 | Westmont Industrial Products |
| | Macco 478 | Witco Chemical Corp. |
| | WITHROKUT 210 CUTTING OIL | Arthur C. Withrow Co. |
| | **84 Compressed Gas, Freon, Liquid $CO_2$** | |
| | Freon T-B1, Freon HV-3 | E. I. DuPont Co. Inc. |
| | **85 Grinding Fluid for Ferrous Metals** | |
| | CAMPBELLENE, CAMPBELLENE NO-NITRITE | Allison-Campbell Div./Acco Industries |
| | Promax 1310 | Ashland Oil, Inc. |
| | RISO Soluble Oil | Atlantic Richfield Co. |
| | PETROCHEM 108 | Champions Choice, Inc. |
| | Chevron Soluble Oil | Chevron U.S.A. Inc. |
| | CIMFREE 104, CIMFREE 234, CIMFREE 238, CIMFREE 305, CIMFREE S4, CIMCOOL S8, CIMCOOL FIVE STAR 40, CIMPERIAL 1010, CIMPERIAL 1011, CIMPERIAL 18, CIMPERIAL 20 | Cincinnati Milacron, Products Div. |
| | Cook-Cool 1975, Cook-Cool 4988 | Cook's Industrial Lubricants, Inc. |
| | COOL SPEED 36, COOL SPEED 77-45 | Daubert Chemical Co. |
| | Kleen-Kool, No. 470 Soluble Oil | DoALL Co. |
| | Fremont-5018M, Fremont-5019X, Fremont-7013 | Fremont Industries, Inc. |
| | I.C. #1676 | International Chemical Co. |
| | IRMCO 120 | International Refining & Mfg. Co. |
| | Millstream, TL-131 | S. C. Johnson & Son, Inc. |
| | LCA Trans Sol 303 | Lubrication Co. of America |
| | Specialube 67D | Lubrication Technology, Inc. |
| | MAGNA COOL CC, CLEAN CUT, EL SOL R, BLUE COOLANT NF | Magnus Div./Economics Lab Inc. |
| | TRIM REGULAR, TRIM 7030, TRIM 9106, TRIM 9106WA, TRIM SOL, TRIM SOL S, TRIM CE, TRIM HD, TRIM EP, TRIM 5050, TRIM 9106CS, TRIM HM, TRIM CLEAR | Master Chemical Corp. |
| | Metsol | Metalloid Corp. |
| | Hamikleer, Hamigrind | Harry Miller Corp. |
| | PRIME-CUT, CHOICE-CUT, MONROE HD, MONROE RI | Monroe Chemical Co., Inc. |
| | Kool Grind 9054 | Pillsbury Chemical & Oil, Inc. |
| | Polar Chip 347 | Polar Chip, Inc. |
| | Gear Lap RIG-12 | Process Research Products |
| | Rycosol 66 | Reilly-Whiteman, Inc. |
| | SEMI-FLUID VASTOLEIN COMPOUND, GRIND-WELL #110 GRINDING COMPOUND | G. Whitfield Richards Co. |
| | RUST-LICK G-25-J, RUST-LICK G-25-AH, VYTRON, RUST-LICK G-1066D, CUTZOL WS-15 | Rust-Lick, Inc. |
| | POWER'D OIL FOR GRINDING | Henry E. Sanson & Sons Inc. |
| | Kool-Aide I, Kool-Aide IV, Kool-Aide XII | Space Age Chemlube |
| | Toolife 100, Toolife 102, Toolife 103 | Specialty Products Co. |
| | Factocool 65 | Standard Oil Co. (Ohio) |
| | Star S.M. 3732 | Star Oil Company |
| | SECO 14, 3% EMULSION | Sun Petroleum Products Co. |
| | Ami-Lube Regular | Tron-X Research Corp. |
| | W&B Grinding Coolant E-55, W&B Grinding Concentrate 1500 | The White & Bagley Co. |

| CODE NO. | TYPE AND TRADE NAME | MANUFACTURER |
|---|---|---|
| **8** | **SPECIALS—LIGHT DUTY** (Continued) | |
| **85** | **Grinding Fluid for Ferrous Metals** (Continued) | |
| | Bio-Cool 500S | Westmont Industrial Products |
| | Syncrocut 980-W-1 | Witco Chemical Corp. |
| | WITHROKUT 117 CUTTING OIL | Arthur C. Withrow Co. |
| **86** | **Cutting Fluids for Gun Drilling and Trepanning** | |
| | AMOCUT Gundrill Oil | Amoco Oil Co. |
| | Promax 5352, Promax 1002, Promax 1074, Promax 1455-S | Ashland Oil, Inc. |
| | Ordnance Oils | Atlantic Richfield Co. |
| | Cilora 19A | BP Oil, Inc. |
| | Brusol A, Bruko D-666 | Bruce Products Corp. |
| | PETROCHEM 710 | Champions Choice, Inc. |
| | Chevron Metalworking Fluid 503 | Chevron U.S.A. Inc. |
| | CIMFREE 305, CIMPERIAL 1010, CIMPERIAL 1011, CIMPERIAL 20 | Cincinnati Milacron, Products Div. |
| | Cook-Cut 1940 | Cook's Industrial Lubricants, Inc. |
| | DAUBERT GARIA H, DAUBERT GARIA T | Daubert Chemical Co. |
| | Gulfcut 44A, Gulfcut 45A | Gulf Oil Corp. |
| | I.C. #814-A | International Chemical Co. |
| | IRMCO 229 | International Refining & Mfg. Co. |
| | LUBRI-FLO 1000 | Larson Chemical Specialties |
| | Threadmate Gun Drilling Oil 5110 | Lubrication Co. of America |
| | TRIM SOL S, TRIM EP, TRIM HD, TRIM SOL | Master Chemical Corp. |
| | Metkut #310 | Metalloid Corp. |
| | Hamikut | Harry Miller Corp. |
| | Mobilmet 25, Mobilmet Alpha | Mobil Oil Corp. |
| | PRIME-CUT, CHOICE-CUT, MONROE HD | Monroe Chemical Co., Inc. |
| | Thio Glo 162 | Pillsbury Chemical & Oil, Inc. |
| | Polar Chip 347 | Polar Chip, Inc. |
| | Triple C 5201 | Pro-Chem Inc. |
| | FRIGIDOL L-6 CUTTING OIL, NEAR-A-LARD #20-A | G. Whitfield Richards Co. |
| | CUTZOL SCFO-504 | Rust-Lick, Inc. |
| | SYN-KUT GUN DRILLING OIL C | Henry E. Sanson & Sons Inc. |
| | Kool-Aide IV | Space Age Chemlube |
| | Toolife 202, Toolife 203 | Specialty Products Co. |
| | Factocool 65 | Standard Oil Co. (Ohio) |
| | Starcut-3075 | Star Oil Company |
| | SUNICUT 425 | Sun Petroleum Products Co. |
| | Transultex A, Transultex F | Texaco Inc. |
| | TOWERCUT F-932, TOWERCUT B-825 | Tower Oil & Technology Co. |
| | Super Ami-Lube | Tron-X Research Corp. |
| | VSCC 5468 | Van Straaten Chemical Co. |
| | W&B Cutting Oil 2190 | The White & Bagley Co. |
| | Bio-Cool 500 | Westmont Industrial Products |
| | Macco GD 100-12 | Witco Chemical Corp. |
| | WITHROKUT 125 CUTTING OIL | Arthur C. Withrow Co. |
| | ZURNKUT 3 L X | O F Zurn Co. |

| CODE NO. | TYPE AND TRADE NAME | MANUFACTURER |
|---|---|---|
| 8 | **SPECIALS—LIGHT DUTY** (Continued) | |
| | **87 Cutting Fluids for Copper** | |
| | Lafayette Soluble Oil #70, Lafayette SCF-20 | American Oil & Supply Co. |
| | AMOCUT Oil No. 106BC, AMOCUT Oil 62FC, AMOCOOL SS Fluid | Amoco Oil Co. |
| | Promax 1074, Promax 1075, Promax 1710 | Ashland Oil, Inc. |
| | ARCO ML Oils, Clairo Oils | Atlantic Richfield Co. |
| | Bezora 22, Fedora EPSC | BP Oil, Inc. |
| | Belco 1727 Cutting Oil | Bel-Ray Co., Inc. |
| | PETROCHEM 712 | Champions Choice, Inc. |
| | Chevron Utility Oil Grade 46, Chevron Metalworking Fluid 504, Chevron Soluble Oil | Chevron U.S.A. Inc. |
| | CIMCOOL AL, CIMCOOL S2, CIMCOOL S4, CIMPERIAL 16, CIMPERIAL 1010, CIMPERIAL 1011 | Cincinnati Milacron, Products Div. |
| | Cook-Cut 30, Cook-Cut 25, Cook-Cool 1235 (Water Soluble) | Cook's Industrial Lubricants, Inc. |
| | COOL SPEED 500, DAUBERT MACRON M | Daubert Chemical Co. |
| | No. 120 Cutting Oil, No. 150 Cutting Oil, No. 470 Soluble Oil | DoALL Co. |
| | Gulfcut 44A, Gulfcut 45A | Gulf Oil Corp. |
| | I.C. #1440-M, I.C. #1873, Tooleze #2000 | International Chemical Co. |
| | IRMCO 242, IRMCO 282 | International Refining & Mfg. Co. |
| | J-Cool, Trampol-X | S. C. Johnson & Son, Inc. |
| | MINERAL LARD OIL | Keystone Division/Pennwalt Corp. |
| | LUBRI-FLO 1000 | Larson Chemical Specialties |
| | Elcoa Copper Cutting Oil 204N | Lubrication Co. of America |
| | MAGNA COOL CC, CLEAN CUT, CUTTING COMPOUND 6 | Magnus Div./Economics Lab Inc. |
| | TRIM SOL, TRIM CE, TRIM 9106, TRIM 9106CS, TRIM 9106WA | Master Chemical Corp. |
| | Alumicut | Mistic Metal Mover, Inc. |
| | Mobilmet 306 | Mobil Oil Corp. |
| | PRIME-CUT | Monroe Chemical Co., Inc. |
| | Sword Kut C5100 | Pillsbury Chemical & Oil, Inc. |
| | Polar Chip 347 BBC | Polar Chip, Inc. |
| | Triple C 5000 | Pro-Chem Inc. |
| | Sintolin E-1 | Process Research Products |
| | Petrosan 116, Cupromul 23 | Reilly-Whiteman, Inc. |
| | LUBE-WELL S-42-M SOLUBLE OIL, LUBE-WELL S-60-T SOLUBLE OIL, PERFECTED LUBE-WELL "AA" COMPOUND | G. Whitfield Richards Co. |
| | CUTZOL WS-11 | Rust-Lick, Inc. |
| | MIRROR-KUT G-7, SYN-KUT 125 | Henry E. Sanson & Sons, Inc. |
| | Kool-Aide IV, Kool-Aide XII | Space Age Chemlube |
| | Toolife 251, Toolife 304 | Specialty Products Co. |
| | Factokut A-50-C, Factokut L-2, Staysol EPSC | Standard Oil Co. (Ohio) |
| | Starcut-3902 | Star Oil Company |
| | TAP MAGIC ALUMINUM | Steco Corp. |
| | SUNICUT 201 | Sun Petroleum Products Co. |
| | Cleartex B, Soluble Oil D | Texaco Inc. |
| | TOWER E-933 | Tower Oil & Technology Co. |
| | Ami-Lube BR | Tron-X Research Corp. |
| | VSCC 545, VSCC 653 | Van Straaten Chemical Co. |
| | W&B Minolard Cutting Oil | The White & Bagley Co. |
| | Bio-Cool 500 | Westmont Industrial Products |
| | ALUMTAP | Winfield Brooks Co., Inc. |
| | Macco 768-1 | Witco Chemical Corp. |

| CODE NO. | TYPE AND TRADE NAME | MANUFACTURER |
|---|---|---|

## 8 SPECIALS—LIGHT DUTY (Continued)

### 87 Cutting Fluids for Copper (Continued)

| TYPE AND TRADE NAME | MANUFACTURER |
|---|---|
| WITHROKUT 103 CUTTING OIL | Arthur C. Withrow Co. |
| ZURNSOL 254 | O F Zurn Co. |

### 88 Cutting Fluids for Magnesium

(PRECAUTION: Consult with cutting fluid supplier or manufacturer to be sure fluids do not contain water or other chemicals which may support combustion of magnesium.)

| TYPE AND TRADE NAME | MANUFACTURER |
|---|---|
| Lafayette BM-112 | American Oil & Supply Co. |
| AMOCUT Oil 62FC | Amoco Oil Co. |
| Econ Oil 5 | Ashland Oil, Inc. |
| ARCO ML Oils, Clairo Oils | Atlantic Richfield Co. |
| TAP 99 | Beechem Laboratories |
| Mineral Oil H-5 | BP Oil, Inc. |
| PETROCHEM THREAD KUT LIGHT | Champions Choice, Inc. |
| Chevron Metalworking Fluid 504 | Chevron U.S.A. Inc. |
| CITGO Cutting Oil 400 | Cities Service Co. |
| Cook-Cut B-5, Cook-Cut 1711 | Cook's Industrial Lubricants, Inc. |
| No. 150 Cutting Oil | DoALL Co. |
| Gulfcut 11D | Gulf Oil Corp. |
| I.C. #1880 | International Chemical Co. |
| IRMCO 249, IRMCO 242, IRMCO 282 | International Refining & Mfg. Co. |
| KOOLMIST TAP SUPREME | Kool Mist Corp. |
| Elcomag 215 | Lubrication Co. of America |
| DO-5A | Magnus Div./Economics Lab Inc. |
| Alumicut | Mistic Metal Mover, Inc. |
| Kleensol 792 | Pillsbury Chemical & Oil, Inc. |
| Petrosan 112 | Reilly-Whiteman, Inc. |
| A-9 | Relton Corp. |
| NEAR-A-LARD OIL, #3-B OIL | G. Whitfield Richards Co. |
| SYN-KUT 105 | Henry E. Sanson & Sons, Inc. |
| Kool-Aide IV, Kool-Aide XII | Space Age Chemlube |
| Toolife 250 | Specialty Products Co. |
| Mineral Seal Oil | Standard Oil Co. (Ohio) |
| TAP MAGIC ALUMINUM | Steco Corp. |
| SUNICUT 102 | Sun Petroleum Products Co. |
| Almag | Texaco Inc. |
| TOWERCUT E-915 | Tower Oil & Technology Co. |
| Ami-Lube NC | Tron-X Research Corp. |
| W&B Cutting Oil 3016, W&B Minolard Cutting Oil | The White & Bagley Co. |
| ALUMTAP | Winfield Brooks Co., Inc. |
| Macco Magnacut | Witco Chemical Corp. |
| LARDOLEUM | O F Zurn Co. |

### 89 Tapping Fluids, Compounds and Pastes

| TYPE AND TRADE NAME | MANUFACTURER |
|---|---|
| Adsol 3, Promax H-10 | Ashland Oil, Inc. |
| CIMTAP, CIMTAP II | Cincinnati Milacron, Products Div. |
| Cook-Tap #1, Cook-Tap #2, Cook-Tap #3 | Cook's Industrial Lubricants, Inc. |
| Tapping Cream | DoALL Co. |
| Habcool 344 | H & B Petroleum Co. |

| CODE NO. | TYPE AND TRADE NAME | MANUFACTURER |
|---|---|---|
| **8** | **SPECIALS—LIGHT DUTY** (Continued) | |
| | **89 Tapping Fluids, Compounds and Pastes** (Continued) | |
| | TRIM TAP LIGHT, TRIM TAP HEAVY | Master Chemical Corp. |
| | Taps-All, Taplube | Metalloid Corp. |
| | Alumicut | Mistic Metal Mover, Inc. |
| | COOL-TOOL | Monroe Chemical Co., Inc. |
| | TAPZOL, CUTZOL WS-20 | Rust-Lick, Inc. |
| | THREADMATE | Henry E. Sanson & Sons, Inc. |
| | Space Age #20 Tapping Lube | Space Age Chemlube |
| | TAP MAGIC | The Steco Corp. |
| | Cutting Compound, Dual Action Cutting Fluid #1, Dual Action Cutting Fluid #2 | Tapmatic Corp. |
| | TREFOLEX Cutting Compound | Trefolex Industries, Inc. |
| | W&B Cutting Oil 2192 | The White & Bagley Co. |
| | TAPFREE | Winfield Brooks Co., Inc. |
| **9** | **SPECIALS—HEAVY DUTY** | |
| | **90 Honing Oil—Heavy Duty** | |
| | AMOCUT Gundrill Oil | Amoco Oil Co. |
| | Promax 1130 | Ashland Oil, Inc. |
| | Honol Oils | Atlantic Richfield Co. |
| | Bezora 3 | BP Oil, Inc. |
| | Bruko D-666 (non-ferrous), Bruko D-667 (ferrous) | Bruce Products Corp. |
| | PETROCHEM 708 | Champions Choice, Inc. |
| | Cook's Honing Oil 1995 | Cook's Industrial Lubricants, Inc. |
| | DAUBERT GARIA T | Daubert Chemical Co. |
| | MICROCOOL IC-9 HONING OIL (Concentrate), MICROCOOL IC-9S: 80-20 | Ex-Cell-O Tool & Abrasive Products |
| | I.C. #155 | International Chemical Co. |
| | IRMCO 229 | International Refining & Mfg. Co. |
| | LPS 1, LPS 2, LPS 3 | LPS Research Laboratories, Inc. |
| | Threadmate Honing Oil 228 | Lubrication Co. of America |
| | TRIM SOL S, TRIM EP, TRIM HD | Master Chemical Corp. |
| | Metcool 222, Metcool 999 | Metalloid Corp. |
| | Vacmul 3D, Vacmul 21C | Mobil Oil Corp. |
| | United Base | Pillsbury Chemical & Oil, Inc. |
| | PRP 9112 Honing Additive | Process Research Products |
| | Petrosan EP43 | Reilly-Whiteman, Inc. |
| | FRIGIDOL HONING OIL | G. Whitfield Richards Co. |
| | HONZOL 411-B | Rust-Lick, Inc. |
| | SYN-HONE MM | Henry E. Sanson & Sons, Inc. |
| | Factokut H | Standard Oil Co. (Ohio) |
| | Star-3701 | Star Oil Company |
| | MAN-852 (Sulfur Free), MB-30 | Sunnen Products Co. |
| | Transultex A | Texaco Inc. |
| | TOWER D-204 | Tower Oil & Technology Co. |
| | VSCC 5299-A | Van Straaten Chemical Co. |
| | W&B Honing Oil 2088 | The White & Bagley Co. |
| | Bio-Hone 253 | Westmont Industrial Products |
| | Macco Honing Oil | Witco Chemical Corp. |

| CODE NO. | TYPE AND TRADE NAME | MANUFACTURER |
|---|---|---|
| 9 | **SPECIALS—HEAVY DUTY** (Continued) | |
| | **91 Cutting Fluids for Titanium—Heavy Duty** | |
| | CAMPBELLENE, CAMPBELLENE NO-NITRITE | Allison Campbell Div./Acco Industries |
| | AMOCOOL HD Soluble Oil | Amoco Oil Co. |
| | Promax 1022 | Ashland Oil, Inc. |
| | Autokut Oils | Atlantic Richfield Co. |
| | Brukool 4-37d | Bruce Products Corp. |
| | CIMFREE 305, CIMPERIAL 1010, CIMPERIAL 1011, CIMPERIAL 20 | Cincinnati Milacron, Products Div. |
| | Cook-Coolex 40, Cook-Cool 2020, Cook-Cool 1400 | Cook's Industrial Lubricants, Inc. |
| | POWER-CUT HD-600, No. 470 Soluble Oil | DoALL Co. |
| | Freon T-B1, Freon HV-3 | E. I. DuPont Co. Inc. |
| | I.C. #48-SDW, I.C. 124 | International Chemical Co. |
| | IRMVO 282 | International Refining & Mfg. Co. |
| | TL-131 | S. C. Johnson & Son, Inc. |
| | LPS 1, LPS 2, LPS 3 | LPS Research Laboratories, Inc. |
| | Empact 51 | Lubrication Co. of America |
| | TRIM HD, TRIM SOL, TRIM REGULAR | Master Chemical Corp. |
| | X-60 | Metalloid Corp. |
| | MONROE HD, PRIME-CUT, CHOICE-CUT | Monroe Chemical Co., Inc. |
| | Rapier Kut 5200 | Pillsbury Chemical & Oil, Inc. |
| | Polar Chip 347 | Polar Chip, Inc. |
| | Rycosol 69 | Reilly-Whiteman, Inc. |
| | FRIGIDOL L-6 CUTTING OIL | G. Whitfield Richards Co. |
| | CUTZOL 400-A, CUTZOL WS-500-A, VYTRON | Rust-Lick, Inc. |
| | MIRROR-KUT G-7-M, BOOSTER KUT PLV, SYN-SOL 600 | Henry E. Sanson & Sons Inc. |
| | SUNICUT 520 | Sun Petroleum Products Co. |
| | Transultex D, Sultex D | Texaco Inc. |
| | Ami-Lube Super NC | Tron-X Research Corp. |
| | VSCC 708 | Van Straaten Chemical Co. |
| | Bio-Cool 500 | Westmont Industrial Products |
| | W&B HD Soluble Oil 2945 | The White & Bagley Co. |
| | Macco 478 | Witco Chemical Corp. |
| | ZURNAGENT BD | O F Zurn Co. |
| | **92 Biodegradable Soluble Oil** | |
| | Adsol 3 | Ashland Oil, Inc. |
| | Brukool 82-c, Bruko D-680 | Bruce Products Corp. |
| | Cook-Cool 4639D | Cook's Industrial Lubricants, Inc. |
| | IRMCO 303 | International Refining & Mfg. Co. |
| | Kool Kut X350 | Pillsbury Chemical & Oil, Inc. |
| | Triple C 2100 | Pro-Chem Inc. |
| | Rycosol 11 | Reilly-Whiteman, Inc. |
| | LUBE-WELL "CD" | G. Whitfield Richards Co. |
| | SYN-SOL 50, MIRROR-KUT G-7 | Henry E. Sanson & Sons Inc. |
| | ME II Super Concentrate (Recirculating) | Tapmatic Corp. |
| | Bio Ami-Lube | Tron-X Research Corp. |
| | VSCC 938 | Van Straaten Chemical Co. |
| | ZURNAGENT BD | O F Zurn Co. |

| CODE NO. | TYPE AND TRADE NAME | MANUFACTURER |
|---|---|---|
| 9 | **SPECIALS—HEAVY DUTY** (Continued) | |
| | **93 Ceramic and Glass Soluble Oil** | |
| | Soluble Oil #371 | BP Oil, Inc. |
| | PETROCHEM 101 | Champions Choice, Inc. |
| | CIMCOOL S2, CIMCOOL S4, CIMFREE 234, CIMFREE 238 | Cincinnati Milacron, Products Div. |
| | Cook-Coolex 40 | Cook's Industrial Lubricants, Inc. |
| | Hocut 219, Hocut 702, Hocut 714 | E. F. Houghton & Co. |
| | I.C. #993-G | International Chemical Co. |
| | IRMCO 189 | International Refining & Mfg. Co. |
| | 130 Wax Cool, Slipstream 54 | S. C. Johnson & Son, Inc. |
| | Ceramasol 3031 | Lubrication Co. of America |
| | PRIME-CUT | Monroe Chemical Co., Inc. |
| | Hy Kut 6638 | Pillsbury Chemical & Oil, Inc. |
| | Solap 30, MON-60 | Process Research Products |
| | Rycosol 77 | Reilly-Whiteman, Inc. |
| | LUBE-WELL "CD" | G. Whitfield Richards Co. |
| | CUTZOL WS-20, CUTZOL WS-15 | Rust-Lick, Inc. |
| | SYN-GRIND 60-P | Henry E. Sanson & Sons, Inc. |
| | Kool-Aide XII | Space Age Chemlube |
| | Glassgrind Oil | Standard Oil Co. (Ohio) |
| | VSCC S-212 | Van Straaten Chemical Co. |
| | W&B Lubricant 1888 | The White & Bagley Co. |
| | Grind Rite 174 | Witco Chemical Corp. |
| | **94 Tapping Fluids, Compounds and Pastes** | |
| | Promax H-7, Promax Q-1013, Promax H-333 | Ashland Oil, Inc. |
| | CIMTAP, CIMTAP II | Cincinnati Milacron, Products Div. |
| | Cook-Tap #1, Cook-Tap #2, Cook-Tap #3 | Cook's Industrial Lubricants, Inc. |
| | CURTIS DD-50, CURTIS DD-80, CURTIS DD-100, CURTIS DD-200, CURTIS SDD-100 (emulsifiable) | Curtis Systems, Inc. |
| | Tapping Compound | DoALL Co. |
| | TAPPING COMPOUND 1715 | Franklin Oil Corp. (Ohio) |
| | Habcool 318 | H & B Petroleum Co. |
| | Tapeze #1271 (Paste) | International Chemical Co. |
| | KOOL MIST FORMULA #88 TAP WAX | Kool Mist Corp. |
| | TRIM TAP LIGHT, TRIM TAP HEAVY | Master Chemical Corp. |
| | Tapeze X-2 | Metalloid Corp. |
| | Alumicut | Mistic Metal Mover, Inc. |
| | COOL-TOOL | Monroe Chemical Co., Inc. |
| | Rapid Tap, A-9 (for aluminum) | Relton Corp. |
| | TAPZOL, VYTRON TAPPING COMPOUND | Rust-Lick, Inc. |
| | TOOLMATE TAP | Henry E. Sanson & Sons, Inc. |
| | Space Age #23 Tapping Lube | Space Age Chemlube |
| | Startap-3202 | Star Oil Company |
| | Cutting Compound, Dual Action Cutting Fluid #1, Dual Action Cutting Fluid #2 | Tapmatic Corp. |
| | TREFOLEX Cutting Compound | Trefolex Industries, Inc. |
| | W&B Base 2844 | The White & Bagley Co. |
| | MOLY-DEE TAPPING FLUID | Arthur C. Withrow Co. |

SECTION 17

# POWER AND FORCE REQUIREMENTS IN MACHINING

17.1 Determining Forces in Machining .......... 17–3
17.2 Determining Power Requirements in Machining .......... 17–7
17.3 Estimating Forces in Turning .......... 17–23
17.4 Estimating Torque and Thrust in Drilling .......... 17–27

The forces produced in turning and milling can be determined by the use of dynamometers such as those shown in figures 17.1-1 and 17.1-2. The drill dynamometer shown in figure 17.1-3 measures the torque and the thrust force.

The equations given in these three figures provide the means for calculating power and torque from the dynamometer readings in both English and metric units.

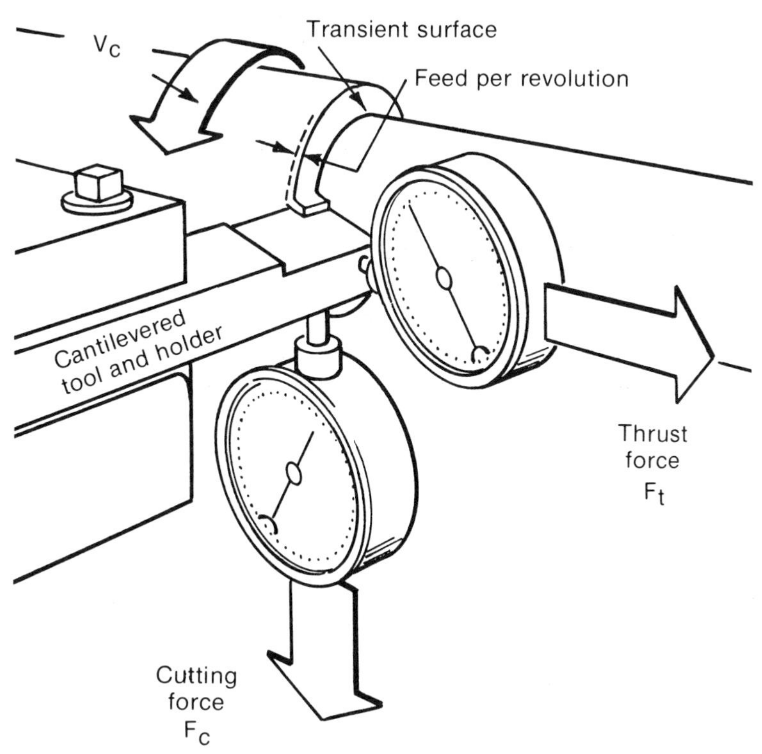


**Determination of English Unit Parameters Using Unit hp Values**

- d = Depth of cut, in
- $D_t$ = Workpiece diameter, in
- E = Efficiency of spindle drive
- $F_c$ = Cutting force, lb
- $f_r$ = Feed per revolution, in
- $hp_m$ = Horsepower at motor
- $hp_s$ = Horsepower at spindle
- P = Unit power, hp/in³/min
- Q = Metal removal rate, in³/min
- $T_s$ = Torque at spindle, in-lb
- $V_c$ = Cutting speed, fpm

$$hp_s = \frac{F_c \times V_c}{33{,}000}$$

$$hp_m = \frac{hp_s}{E}$$

$$Q = 12 \times d \times f_r \times V_c$$

$$P = \frac{hp_s}{Q}$$

$$T_s = F_c \times \frac{D_t}{2}$$

$$T_s = \frac{63{,}030 \times hp_s}{rpm}$$

**Determination of Metric Unit Parameters Using Unit kW Values**

- d = Depth of cut, mm
- $D_t$ = Workpiece diameter, mm
- E = Efficiency of spindle drive
- $F_c$ = Cutting force, N
- $f_r$ = Feed per revolution, mm
- $kW_m$ = Power at motor, kW
- $kW_s$ = Power at spindle, kW
- P = Unit power, kW/cm³/min
- Q = Metal removal rate, cm³/min
- $T_s$ = Torque at spindle, N·m
- $V_c$ = Cutting speed, m/min

$$kW_s = \frac{F_c \times V_c}{60{,}000}$$

$$kW_m = \frac{kW_s}{E}$$

$$Q = d \times f_r \times V_c$$

$$P = \frac{kW_s}{Q}$$

$$T_s = F_c \times \frac{D_t}{2000}$$

$$T_s = \frac{9549 \times kW_s}{rpm}$$

**Figure 17.1-1** Forces and power in turning.

## 17.1 Determining Forces in Machining

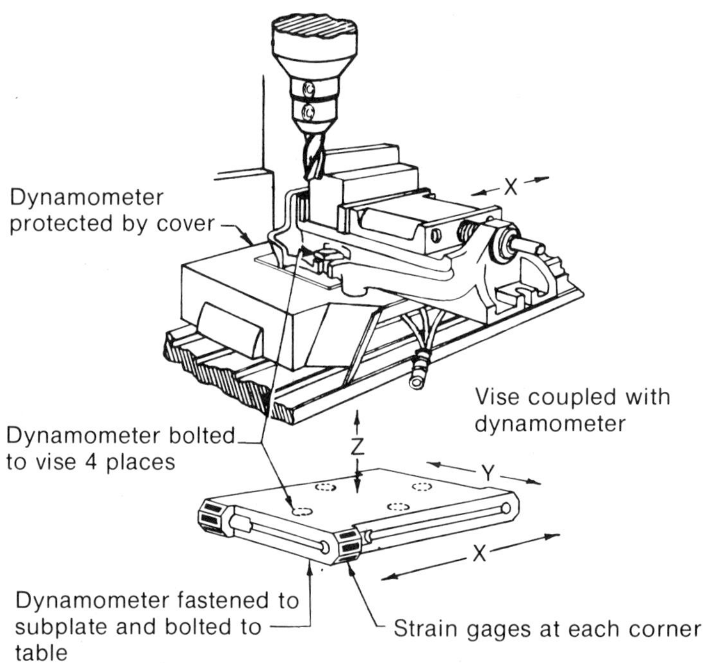


Three-force dynamometer designed especially for measuring machining forces during milling (end milling shown).

**Determination of English Unit Parameters Using Unit hp Values**

| | | |
|---|---|---|
| d | = | Depth of cut, in |
| E | = | Efficiency of spindle drive |
| $f_t$ | = | Feed, inch/tooth |
| n | = | Number of teeth in cutter |
| $hp_m$ | = | Horsepower at motor |
| $hp_s$ | = | Horsepower at spindle |
| P | = | Unit power, hp/in³/min |
| Q | = | Metal removal rate, in³/min |
| $T_s$ | = | Torque at spindle, in-lb |
| $V_c$ | = | Cutting speed, fpm |
| $F_c$ | = | Cutting force, lb |
| w | = | Width of cut, in |

$$hp_s = \frac{F_c \times V_c}{33{,}000}$$

$$hp_m = \frac{hp_s}{E}$$

$$Q = w \times d \times f_t \times n \times rpm$$

$$P = \frac{hp_s}{Q}$$

$$T_s = \frac{63{,}030 \times hp_s}{rpm}$$

**Determination of Metric Unit Parameters Using Unit kW Values**

| | | |
|---|---|---|
| d | = | Depth of cut, mm |
| E | = | Efficiency of spindle drive |
| $f_t$ | = | Feed, mm/tooth |
| n | = | Number of teeth in cutter |
| $kW_m$ | = | Power at motor, kW |
| $kW_s$ | = | Power at spindle, kW |
| P | = | Unit power, kW/cm³/min |
| Q | = | Metal removal rate, cm³/min |
| $T_s$ | = | Torque at spindle, N·m |
| $V_c$ | = | Cutting speed, m/min |
| $F_c$ | = | Cutting force, N |
| w | = | Width of cut, mm |

$$kW_s = \frac{F_c \times V_c}{60{,}000}$$

$$kW_m = \frac{kW_s}{E}$$

$$Q = \frac{w \times d \times f_t \times n \times rpm}{1000}$$

$$P = \frac{kW_s}{Q}$$

$$T_s = \frac{9549 \times kW_s}{rpm}$$

**Figure 17.1–2** Forces and power in milling.

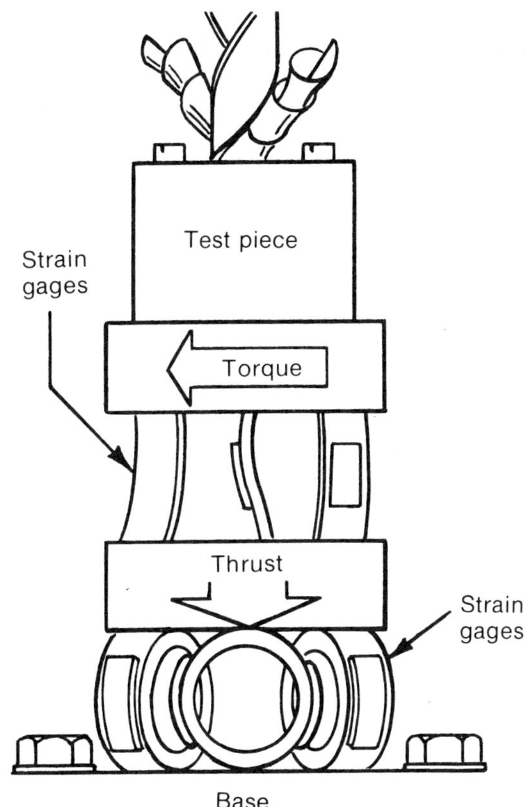


**Determination of English Unit Parameters Using Unit hp Values**

$D_d$ = Drill diameter, in
E = Efficiency of spindle drive
$f_r$ = Feed, inch/revolution
$hp_s$ = Horsepower at motor
$hp_m$ = Horsepower at spindle
P = Unit power, hp/in³/min
Q = Metal removal rate, in³/min
$T_s$ = Torque at spindle, in-lb
$V_c$ = Cutting speed

$$hp_s = \frac{T_s \times rpm}{63{,}030}$$

$$hp_m = \frac{hp_s}{E}$$

$$Q = \frac{\pi}{4} \times D_d^2 \times f_r \times rpm$$

$$P = \frac{hp_s}{Q}$$

$$T_s = 49{,}500 \times f_r \times D_d^2 \times P$$

**Determination of Metric Unit Parameters Using kW Values**

$D_d$ = Drill diameter, mm
E = Efficiency of spindle drive
$f_r$ = Feed, mm/revolution
$kW_s$ = Power at spindle, kW
$kW_m$ = Power at motor, kW
P = Unit power, kW/cm³/min
Q = Metal removal rate, cm³/min
$T_s$ = Torque at spindle, N•m
$V_c$ = Cutting speed, m/min

$$kW_s = \frac{T_s \times rpm}{9549}$$

$$kW_m = \frac{kW_s}{E}$$

$$Q = \frac{\pi \times D_d^2 \times f_r \times rpm}{4000}$$

$$P = \frac{kW_s}{Q}$$

$$T_s = 7.5 \times f_r \times D_d^2 \times P$$

**Figure 17.1–3** Forces and power in drilling.

The power required in machining can be determined by several different methods as follows:

**Use of a watt meter or ammeter at the spindle motor.** This is the simplest and most practical method of measuring power. The watt meter is more accurate and reaches greatest accuracy when the motor is loaded to full load. The power required at the cutter can be estimated by subtracting the idle power from the power produced while cutting.

**Calculation of power from the cutting force.** In order to utilize this method, it is necessary to measure the forces at the cutting tool using a dynamometer such as one shown in figure 17.1-1, 17.1-2 or 17.1-3. From the cutting force values, $F_c$, as measured with the turning dynamometer, the power at the spindle can be calculated. The power at the spindle in drilling can be calculated from the torque measured by the drill dynamometer in figure 17.1-3. To determine the power in milling, the torque must be measured with a torque dynamometer, and the power calculated from the torque values. The power at the motor is determined by dividing the power at the spindle by the efficiency of the spindle system. See figures 17.1-1 to 17.1-3 and the alignment chart, figure 17.2-11.

**Calculation of power from published values of unit power requirements.** An adequate estimate of power in machining can be made from the unit power requirements given in table 17.2-3 and figure 17.2-1 or table 17.2-4 and figure 17.2-2. The power is determined by multiplying the unit power requirements (hp/in$^3$/min or kW/cm$^3$/min) by the rate of metal removal (in$^3$/min or cm$^3$/min). The equations for performing these operations are given in tables 17.2-1 and 17.2-2. In addition, alignment charts are provided in figures 17.2-3 through 17.2-10 for determining the power required for turning, milling, drilling and broaching.

A list of the equations, data sheets, and alignment charts for determining the power requirements and other pertinent information on machining for both English and metric units is as follows:

Tables 17.2-1 and 17.2-2 – Shop Formulas for Turning, Milling, Drilling and Broaching

Tables 17.2-3 and 17.2-4 – Unit Power Requirements for Turning, Drilling and Milling

Figures 17.2-1 and 17.2-2 – Unit Power Requirements for Surface Broaching

Figures 17.2-3 and 17.2-4 – Alignment Chart for Determining Metal Removal Rate and Motor Power in Turning

Figures 17.2-5 and 17.2-6 – Alignment Chart for Determining Metal Removal Rate and Motor Power in Face Milling

Figures 17.2-7 and 17.2-8 – Alignment Chart for Determining Motor Power in Drilling

Figures 17.2-9 and 17.2-10 – Alignment Chart for Determining Metal Removal Rate and Motor Power in Surface Broaching

Figure 17.2-11 – Alignment Chart for Estimating Spindle Drive Efficiency

## 17.2 Determining Power Requirements in Machining

**TABLE 17.2-1 Shop Formulas for Turning, Milling, Drilling and Broaching—English Units**

| PARAMETER | TURNING | MILLING | DRILLING | BROACHING |
|---|---|---|---|---|
| Cutting speed, fpm | $V_c = .262 \times D_t \times rpm$ | $V_c = .262 \times D_m \times rpm$ | $V_c = .262 \times D_d \times rpm$ | $V_c$ |
| Revolutions per minute | $rpm = 3.82 \times \frac{V_c}{D_t}$ | $rpm = 3.82 \times \frac{V_c}{D_m}$ | $rpm = 3.82 \times \frac{V_c}{D_d}$ | — |
| Feed rate, in/min | $f_m = f_r \times rpm$ | $f_m = f_t \times n \times rpm$ | $f_m = f_r \times rpm$ | — |
| Feed per tooth, in | — | $f_t = \frac{f_m}{n \times rpm}$ | — | $f_t$ |
| Cutting time, min | $t = \frac{L}{f_m}$ | $t = \frac{L}{f_m}$ | $t = \frac{L}{f_m}$ | $t = \frac{L}{12\ V_c}$ |
| Rate of metal removal, $in^3/min$ | $Q = 12 \times d \times f_r \times V_c$ | $Q = w \times d \times f_m$ | $Q = \frac{\pi D^2}{4} \times f_m$ | $Q = 12 \times w \times d_t \times V_c$ |
| Horsepower required at spindle* | $hp_s = Q \times P$ | $hp_s = Q \times P$ | $hp_s = Q \times P$ | — |
| Horsepower required at motor* | $hp_m = \frac{Q \times P}{E}$ | $hp_m = \frac{Q \times P}{E}$ | $hp_m = \frac{Q \times P}{E}$ | $hp_m = \frac{Q \times P}{E}$ |
| Torque at spindle | $T_s = \frac{63030\ hp_s}{rpm}$ | $T_s = \frac{63030\ hp_s}{rpm}$ | $T_s = \frac{63030\ hp_s}{rpm}$ | — |

SYMBOLS:
- $D_t$ = Diameter of workpiece in turning, inches
- $D_m$ = Diameter of milling cutter, inches
- $D_d$ = Diameter of drill, inches
- $d$ = Depth of cut, inches
- $d_t$ = Total depth per stroke in broaching, inches
- $E$ = Efficiency of spindle drive
- $f_m$ = Feed rate, inches per minute
- $f_r$ = Feed, inches per revolution
- $f_t$ = Feed, inches per tooth
- $hp_m$ = Horsepower at motor
- $hp_s$ = Horsepower at spindle
- $L$ = Length of cut, inches
- $n$ = Number of teeth in cutter
- $P$ = Unit power, horsepower per cubic inch per minute
- $Q$ = Rate of metal removed, cubic inches per minute
- $rpm$ = Revolutions per minute of work or cutter
- $T_s$ = Torque at spindle, inch-pounds
- $t$ = Cutting time, minutes
- $V_c$ = Cutting speed, feet per minute
- $w$ = Width of cut, inches

*Unit power data are given in table 17.2-3 for turning, milling and drilling, and in figure 17.2-1 for broaching.

**TABLE 17.2–2 Shop Formulas for Turning, Milling, Drilling and Broaching—Metric Units**

| PARAMETER | TURNING | MILLING | DRILLING | BROACHING |
|---|---|---|---|---|
| Cutting speed, m/min | $V_c = \frac{\pi}{1000} \times D_t \times rpm$ | $V_c = \frac{\pi}{1000} \times D_m \times rpm$ | $V_c = \frac{\pi}{1000} \times D_d \times rpm$ | $V_c$ |
| Revolutions per minute | $rpm = \frac{1000}{\pi} \times \frac{V_c}{D_t}$ | $rpm = \frac{1000}{\pi} \times \frac{V_c}{D_m}$ | $rpm = \frac{1000}{\pi} \times \frac{V_c}{D_d}$ | — |
| Feed rate, mm/min | $f_m = f_r \times rpm$ | $f_m = f_t \times n \times rpm$ | $f_m = f_r \times rpm$ | — |
| Feed per tooth, mm | — | $f_t = \frac{f_m}{n \times rpm}$ | — | $f_t$ |
| Cutting time, min | $t = \frac{L}{f_m}$ | $t = \frac{L}{f_m}$ | $t = \frac{L}{f_m}$ | $t = \frac{L}{1000\ V_c}$ |
| Rate of metal removal, cm³/min | $Q = d \times f_r \times V_c$ | $Q = \frac{w \times d \times f_m}{1000}$ | $Q = \frac{\pi D_d^2}{4000} \times f_m$ | $Q = w \times d_t \times V_c$ |
| Power required at spindle* | $kW_s = Q \times P$ | $kW_s = Q \times P$ | $kW_s = Q \times P$ | — |
| Power required at motor* | $kW_m = \frac{Q \times P}{E}$ | $kW_m = \frac{Q \times P}{E}$ | $kW_m = \frac{Q \times P}{E}$ | $kW_m = \frac{Q \times P}{E}$ |
| Torque at spindle | $T_s = \frac{9549 kW_s}{rpm}$ | $T_s = \frac{9549 kW_s}{rpm}$ | $T_s = \frac{9549 kW_s}{rpm}$ | — |

SYMBOLS:
- $D_t$ = Diameter of workpiece in turning, millimeters
- $D_m$ = Diameter of milling cutter, millimeters
- $D_d$ = Diameter of drill, millimeters
- $d$ = Depth of cut, millimeters
- $d_t$ = Total depth per stroke in broaching, millimeters
- $E$ = Efficiency of spindle drive
- $f_m$ = Feed rate, millimeters per minute
- $f_r$ = Feed, millimeters per revolution
- $f_t$ = Feed, millimeters per tooth
- $kW_m$ = Power at motor, kW
- $kW_s$ = Power at spindle, kW
- $L$ = Length of cut, millimeters
- $n$ = Number of teeth in cutter
- $P$ = Unit power, kilowatts per cubic centimeter per minute
- $Q$ = Rate of metal removed, cubic centimeters per minute
- rpm = Revolutions per minute of work or cutter
- $T_s$ = Torque at spindle, newton-meters
- $t$ = Cutting time, minutes
- $V_c$ = Cutting speed, meters per minute
- $w$ = Width of cut, millimeters

*Unit power data are given in table 17.2-4 for turning, milling and drilling, and in figure 17.2-2 for broaching.

**TABLE 17.2–3 Average Unit Power Requirements for Turning, Drilling and Milling—English Units**

| MATERIAL | HARDNESS | UNIT POWER* hp/in³/min | | | | | |
|---|---|---|---|---|---|---|---|
| | | TURNING $P_t$ HSS AND CARBIDE TOOLS (feed .005-.020 ipr) | | DRILLING $P_d$ HSS DRILLS (feed .002-.008 ipr) | | MILLING $P_m$ HSS AND CARBIDE TOOLS (feed .005-.012 ipt) | |
| | Bhn | Sharp Tool | Dull Tool | Sharp Tool | Dull Tool | Sharp Tool | Dull Tool |
| STEELS, WROUGHT AND CAST<br>Plain Carbon<br>Alloy Steels<br>Tool Steels | 85-200 | 1.1 | 1.4 | 1.0 | 1.3 | 1.1 | 1.4 |
| | 35-40 $R_c$ | 1.4 | 1.7 | 1.4 | 1.7 | 1.5 | 1.9 |
| | 40-50 $R_c$ | 1.5 | 1.9 | 1.7 | 2.1 | 1.8 | 2.2 |
| | 50-55 $R_c$ | 2.0 | 2.5 | 2.1 | 2.6 | 2.1 | 2.6 |
| | 55-58 $R_c$ | 3.4 | 4.2 | 2.6 | 3.2† | 2.6 | 3.2 |
| CAST IRONS<br>Gray, Ductile and Malleable | 110-190 | 0.7 | 0.9 | 1.0 | 1.2 | 0.6 | 0.8 |
| | 190-320 | 1.4 | 1.7 | 1.6 | 2.0 | 1.1 | 1.4 |
| STAINLESS STEELS, WROUGHT AND CAST<br>Ferritic, Austenitic and Martensitic | 135-275 | 1.3 | 1.6 | 1.1 | 1.4 | 1.4 | 1.7 |
| | 30-45 $R_c$ | 1.4 | 1.7 | 1.2 | 1.5 | 1.5 | 1.9 |
| PRECIPITATION HARDENING STAINLESS STEELS | 150-450 | 1.4 | 1.7 | 1.2 | 1.5 | 1.5 | 1.9 |
| TITANIUM | 250-375 | 1.2 | 1.5 | 1.1 | 1.4 | 1.1 | 1.4 |
| HIGH TEMPERATURE ALLOYS<br>Nickel and Cobalt Base | 200-360 | 2.5 | 3.1 | 2.0 | 2.5 | 2.0 | 2.5 |
| Iron Base | 180-320 | 1.6 | 2.0 | 1.2 | 1.5 | 1.6 | 2.0 |
| REFRACTORY ALLOYS<br>Tungsten | 321 | 2.8 | 3.5 | 2.6 | 3.3† | 2.9 | 3.6 |
| Molybdenum | 229 | 2.0 | 2.5 | 1.6 | 2.0 | 1.6 | 2.0 |
| Columbium | 217 | 1.7 | 2.1 | 1.4 | 1.7 | 1.5 | 1.9 |
| Tantalum | 210 | 2.8 | 3.5 | 2.1 | 2.6 | 2.0 | 2.5 |
| NICKEL ALLOYS | 80-360 | 2.0 | 2.5 | 1.8 | 2.2 | 1.9 | 2.4 |
| ALUMINUM ALLOYS | 30-150<br>500 kg | 0.25 | 0.3 | 0.16 | 0.2 | 0.32 | 0.4 |
| MAGNESIUM ALLOYS | 40-90<br>500 kg | 0.16 | 0.2 | 0.16 | 0.2 | 0.16 | 0.2 |
| COPPER | 80 $R_B$ | 1.0 | 1.2 | 0.9 | 1.1 | 1.0 | 1.2 |
| COPPER ALLOYS | 10-80 $R_B$<br>80-100 $R_B$ | 0.64<br>1.0 | 0.8<br>1.2 | 0.48<br>0.8 | 0.6<br>1.0 | 0.64<br>1.0 | 0.8<br>1.2 |

*Power requirements at spindle drive motor, corrected for 80% spindle drive efficiency.

† Carbide

**TABLE 17.2-4 Average Unit Power Requirements for Turning, Drilling and Milling—Metric Units**

| MATERIAL | HARDNESS | UNIT POWER* kW/cm³/min | | | | | |
|---|---|---|---|---|---|---|---|
| | | TURNING $P_t$ HSS AND CARBIDE TOOLS (feed .12-.50 mm/r) | | DRILLING $P_d$ HSS DRILLS (feed .05-.20 mm/r) | | MILLING $P_m$ HSS AND CARBIDE TOOLS (feed .12-.30 mm/r) | |
| | Bhn | Sharp Tool | Dull Tool | Sharp Tool | Dull Tool | Sharp Tool | Dull Tool |
| STEELS, WROUGHT AND CAST<br>Plain Carbon<br>Alloy Steels<br>Tool Steels | 85-200 | 0.050 | 0.064 | 0.046 | 0.059 | 0.050 | 0.064 |
| | 35-40 $R_c$ | .064 | .077 | .064 | .077 | .068 | .086 |
| | 40-50 $R_c$ | .068 | .086 | .077 | .096 | .082 | .100 |
| | 50-55 $R_c$ | .091 | .114 | .096 | .118 | .096 | .118 |
| | 55-58 $R_c$ | .155 | .191 | .118 | .146† | .118 | .146 |
| CAST IRONS<br>Gray, Ductile and Malleable | 110-190 | .032 | .041 | .046 | .055 | .027 | .036 |
| | 190-320 | .064 | .077 | .073 | .091 | .050 | .064 |
| STAINLESS STEELS, WROUGHT AND CAST<br>Ferritic, Austenitic and Martensitic | 135-275 | .059 | .073 | .050 | .064 | .064 | .077 |
| | 30-45 $R_c$ | .064 | .077 | .055 | .068 | .068 | .086 |
| PRECIPITATION HARDENING STAINLESS STEELS | 150-450 | .064 | .077 | .055 | .068 | .068 | .086 |
| TITANIUM | 250-375 | .055 | .068 | .050 | .064 | .050 | .064 |
| HIGH TEMPERATURE ALLOYS<br>Nickel and Cobalt Base | 200-360 | .114 | .141 | .091 | .114 | .091 | .114 |
| Iron Base | 180-320 | .073 | .091 | .055 | .068 | .073 | .091 |
| REFRACTORY ALLOYS<br>Tungsten | 321 | .127 | .159 | .118 | .150† | .132 | .164 |
| Molybdenum | 229 | .091 | .114 | .073 | .091 | .073 | .091 |
| Columbium | 217 | .077 | .096 | .064 | .077 | .068 | .086 |
| Tantalum | 210 | .127 | .159 | .096 | .118 | .091 | .114 |
| NICKEL ALLOYS | 80-360 | .091 | .114 | .082 | .100 | .086 | .109 |
| ALUMINUM ALLOYS | 30-150<br>500 kg | .011 | .014 | .007 | .009 | .015 | .018 |
| MAGNESIUM ALLOYS | 40-90<br>500 kg | .007 | .009 | .007 | .009 | .007 | .009 |
| COPPER | 80 $R_B$ | .046 | .055 | .041 | .050 | .046 | .055 |
| COPPER ALLOYS | 10-80 $R_B$<br>80-100 $R_B$ | .029<br>.046 | .036<br>.055 | .022<br>.036 | .027<br>.046 | .029<br>.046 | .036<br>.055 |

*Power requirements at spindle drive motor, corrected for 80% spindle drive efficiency.

† Carbide

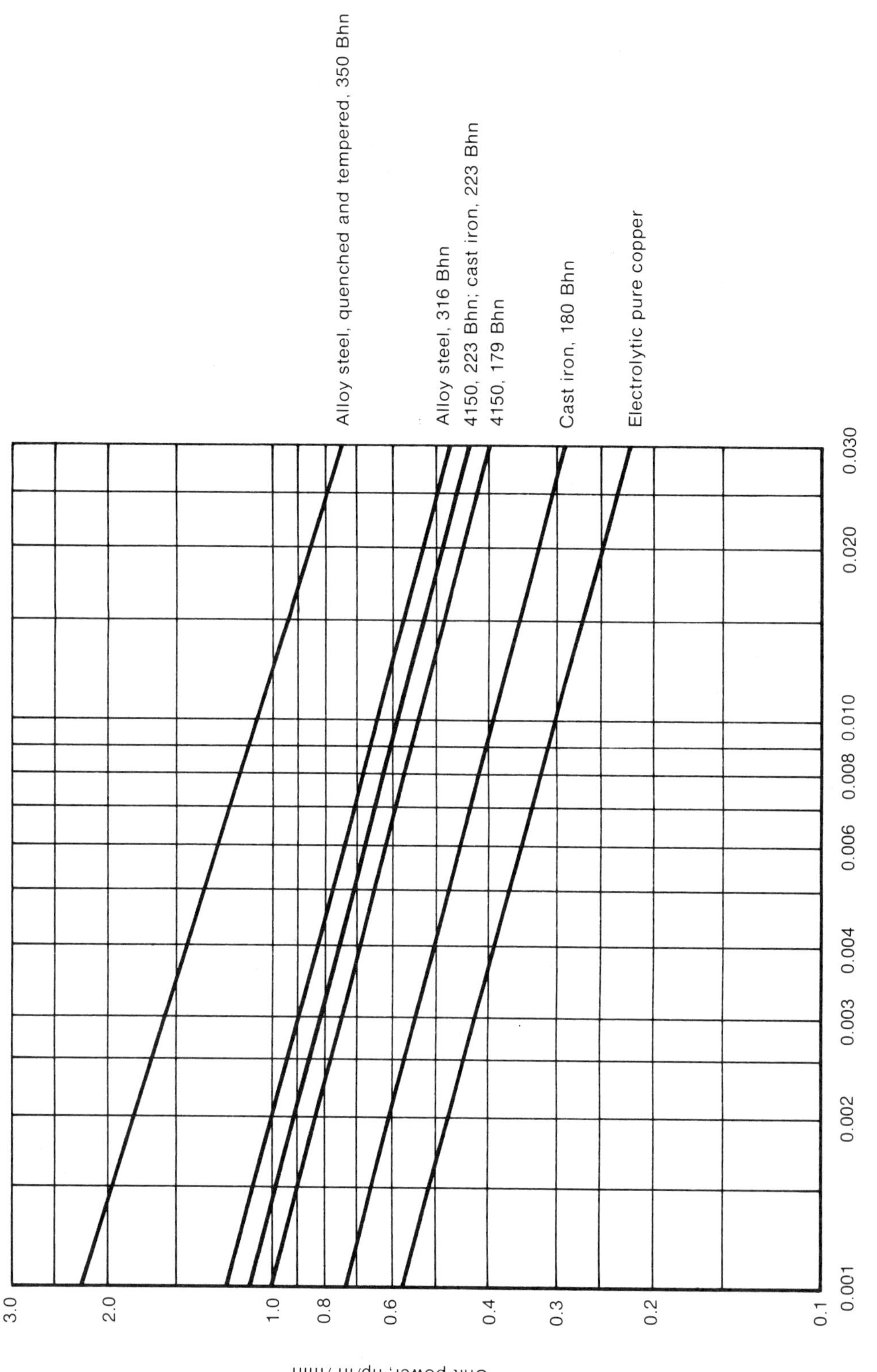


**Figure 17.2-1** Unit power requirements for surface broaching with HSS tools only—English units.

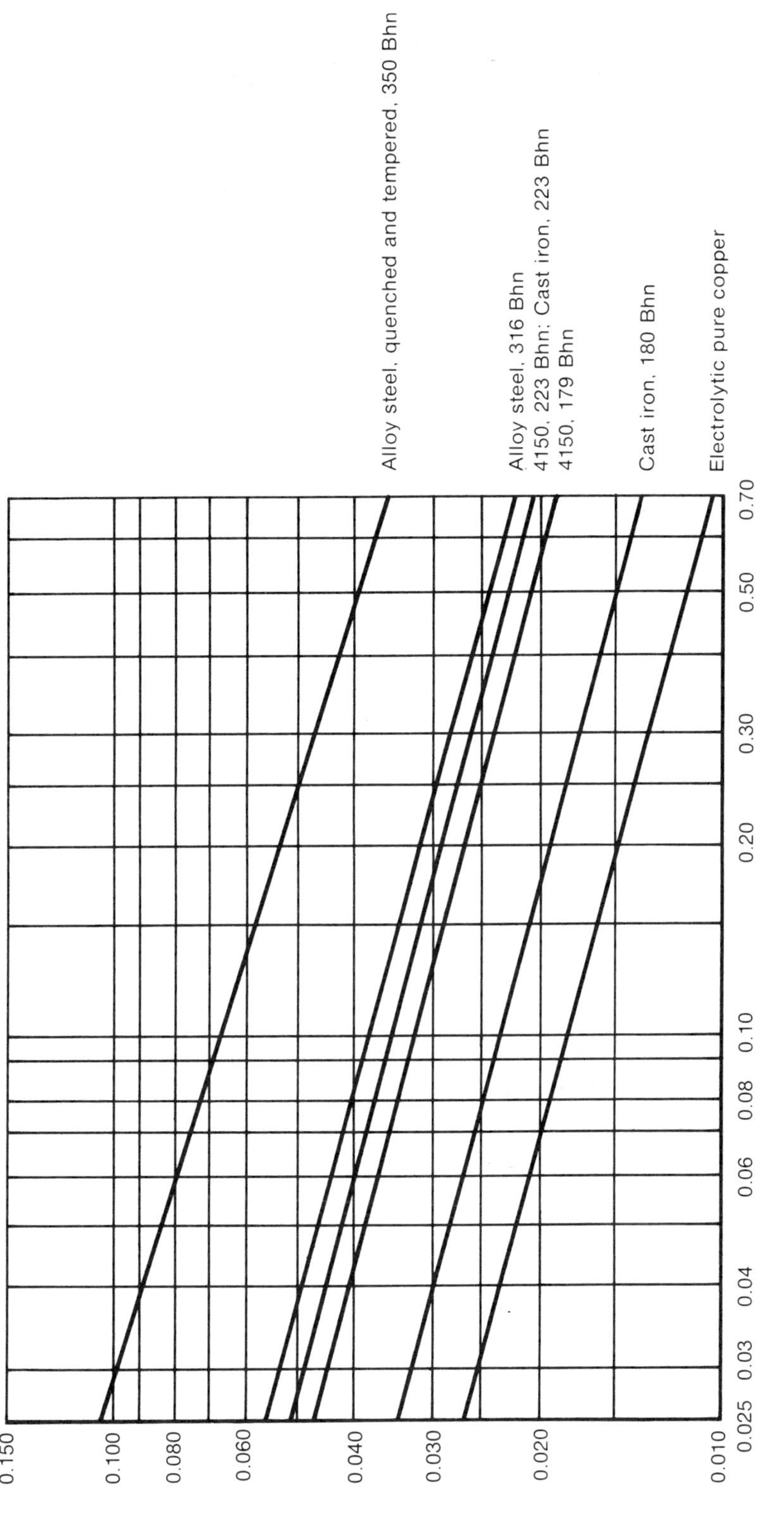


**Figure 17.2-2** Unit power requirements for surface broaching with HSS tools only—Metric units.

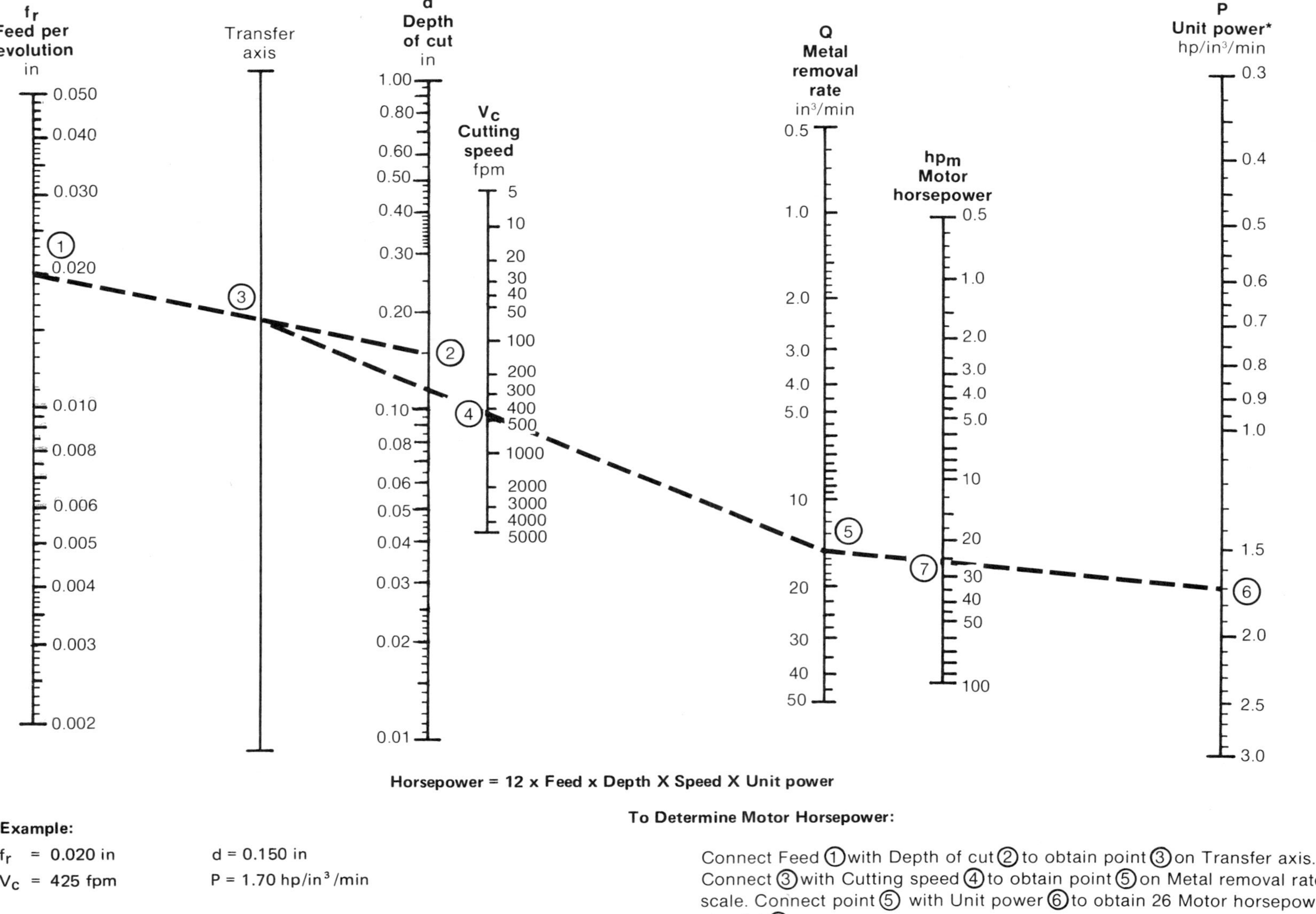


**Horsepower = 12 x Feed x Depth X Speed X Unit power**

**Example:**

$f_r$ = 0.020 in — d = 0.150 in

$V_c$ = 425 fpm — P = 1.70 hp/in³/min

* See table 17.2-3 for Unit power values. Corrected for dull cutter and 80% machine tool efficiency.

**To Determine Motor Horsepower:**

Connect Feed ① with Depth of cut ② to obtain point ③ on Transfer axis. Connect ③ with Cutting speed ④ to obtain point ⑤ on Metal removal rate scale. Connect point ⑤ with Unit power ⑥ to obtain 26 Motor horsepower at point ⑦

**Figure 17.2–3** Alignment chart for determining metal removal rate and motor horsepower in turning—English units.

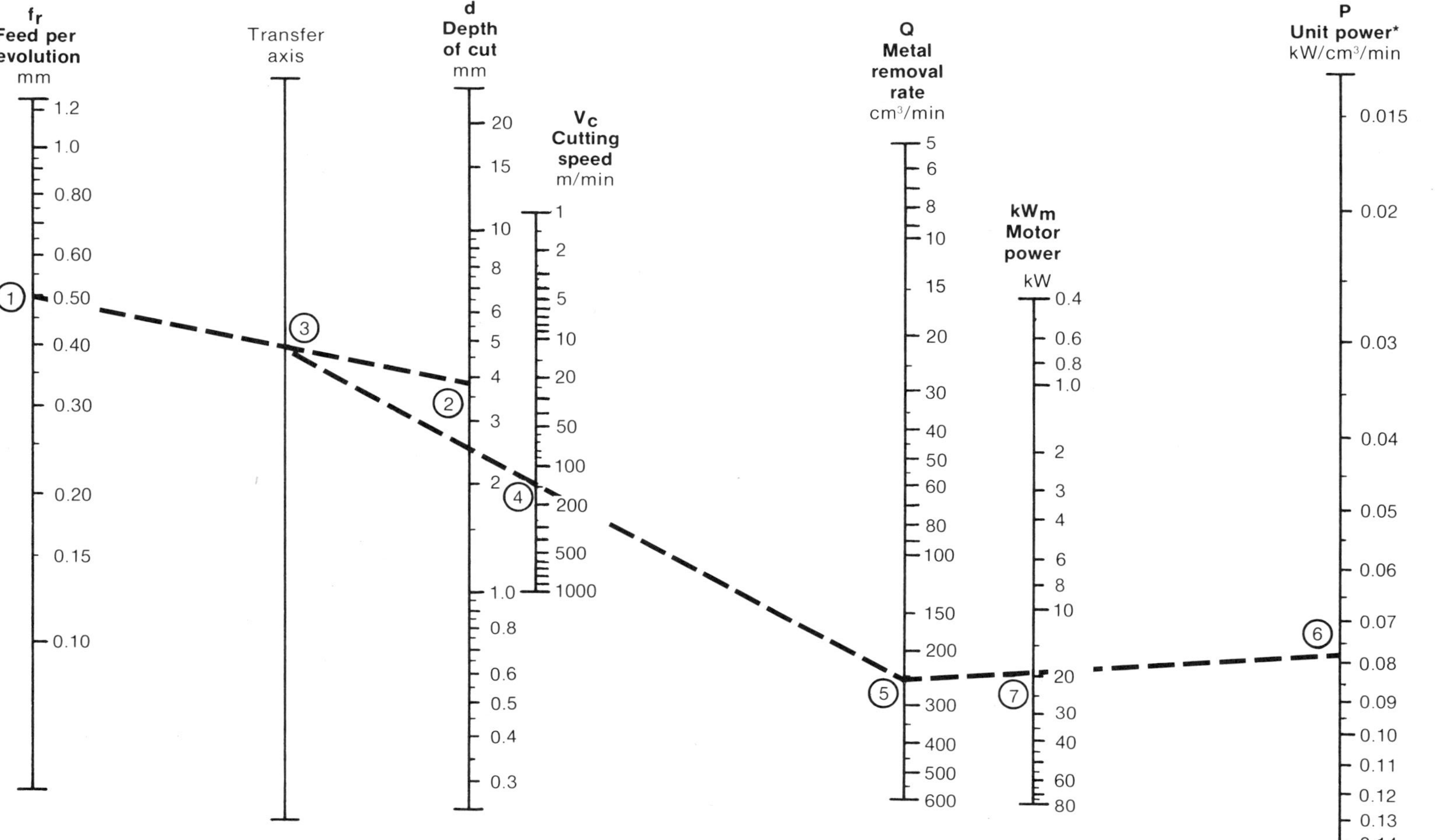


**Power = Feed x Depth x Speed x Unit power**

**To Determine Motor Power:**

Connect Feed ① with Depth of cut ② to obtain point ③ on Transfer axis. Connect ③ with Cutting speed ④ to obtain point ⑤ on Metal removal rate scale. Connect point ⑤ with Unit power ⑥ to obtain 19.02 kW at Motor, point ⑦

**Example:**

$f_r$ = 0.5 mm　　d = 3.8 mm

$V_c$ = 130 m/min　　P = 0.077 kW/cm³/min

*See table 17.2-4 for Unit power values. Corrected for dull cutter and 80% machine tool efficiency.

**Figure 17.2–4** Alignment chart for determining metal removal rate and motor power in turning—Metric units.

# 17.2 Determining Power Requirements in Machining

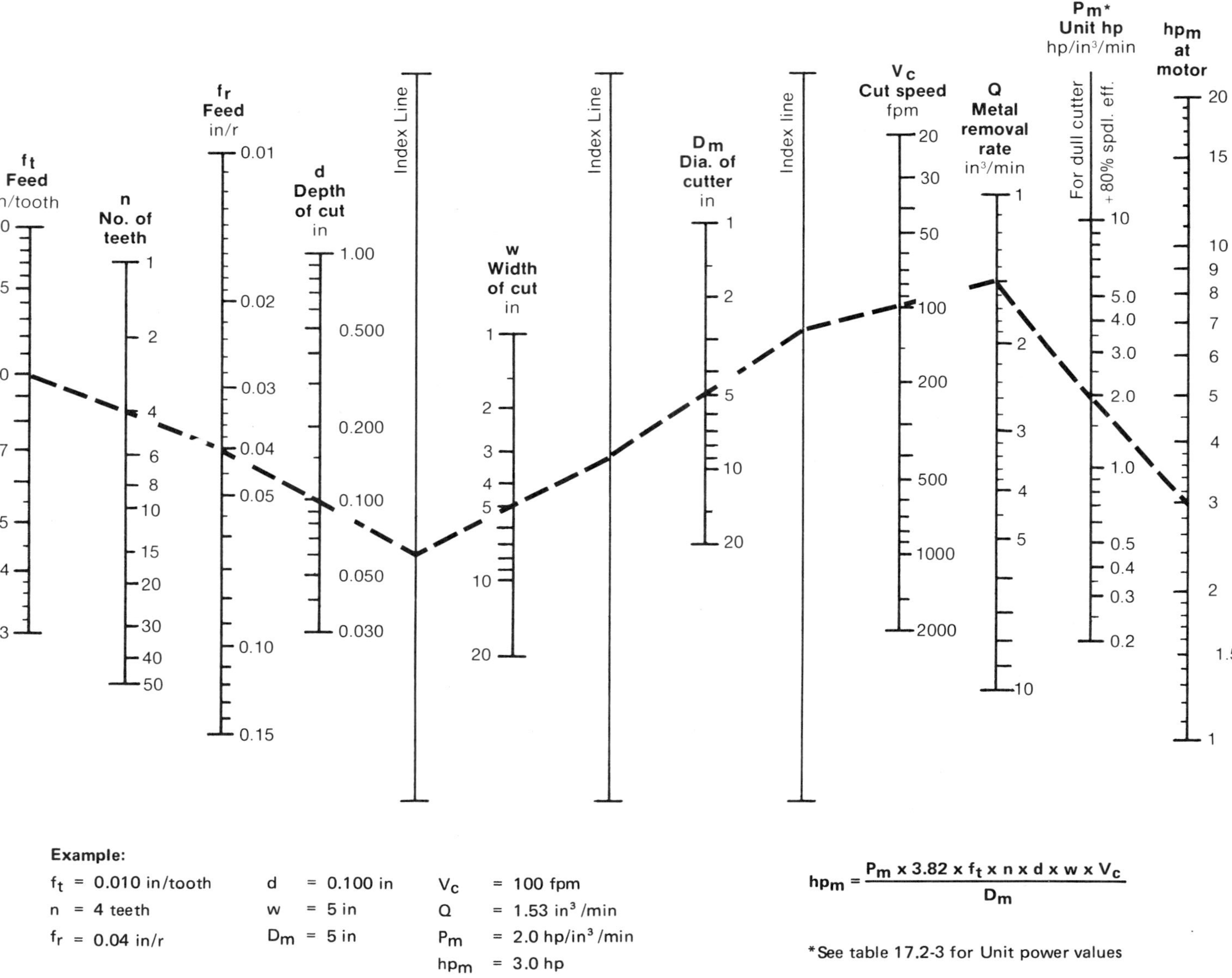


**Figure 17.2–5** Alignment chart for determining metal removal rate and motor horsepower in face milling—English units.

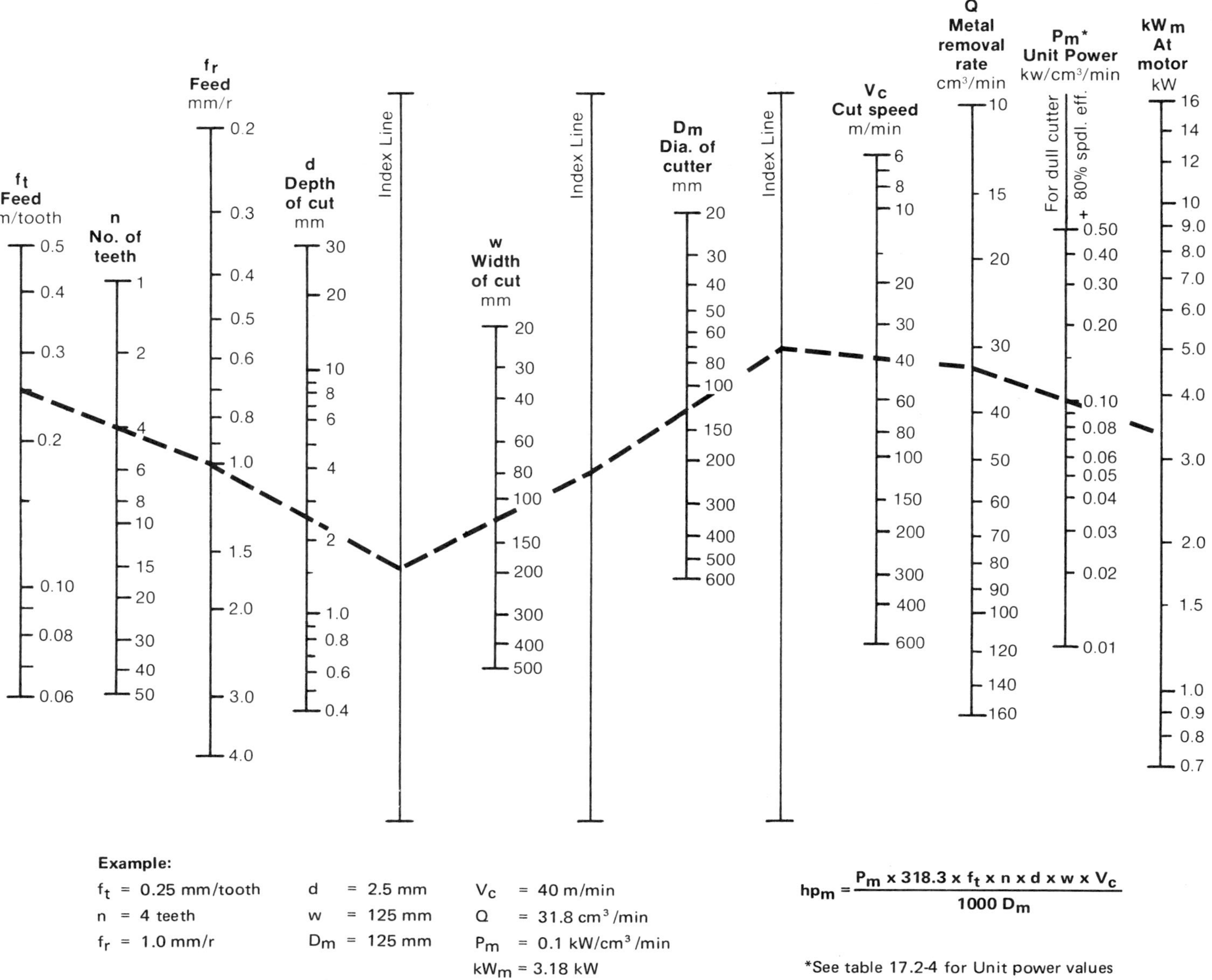


**Example:**

| | | |
|---|---|---|
| $f_t$ = 0.25 mm/tooth | d = 2.5 mm | $V_c$ = 40 m/min |
| n = 4 teeth | w = 125 mm | Q = 31.8 cm³/min |
| $f_r$ = 1.0 mm/r | $D_m$ = 125 mm | $P_m$ = 0.1 kW/cm³/min |
| | | $kW_m$ = 3.18 kW |

$$hp_m = \frac{P_m \times 318.3 \times f_t \times n \times d \times w \times V_c}{1000\ D_m}$$

*See table 17.2-4 for Unit power values

**Figure 17.2-6** Alignment chart for determining metal removal rate and motor power in face milling—Metric units.

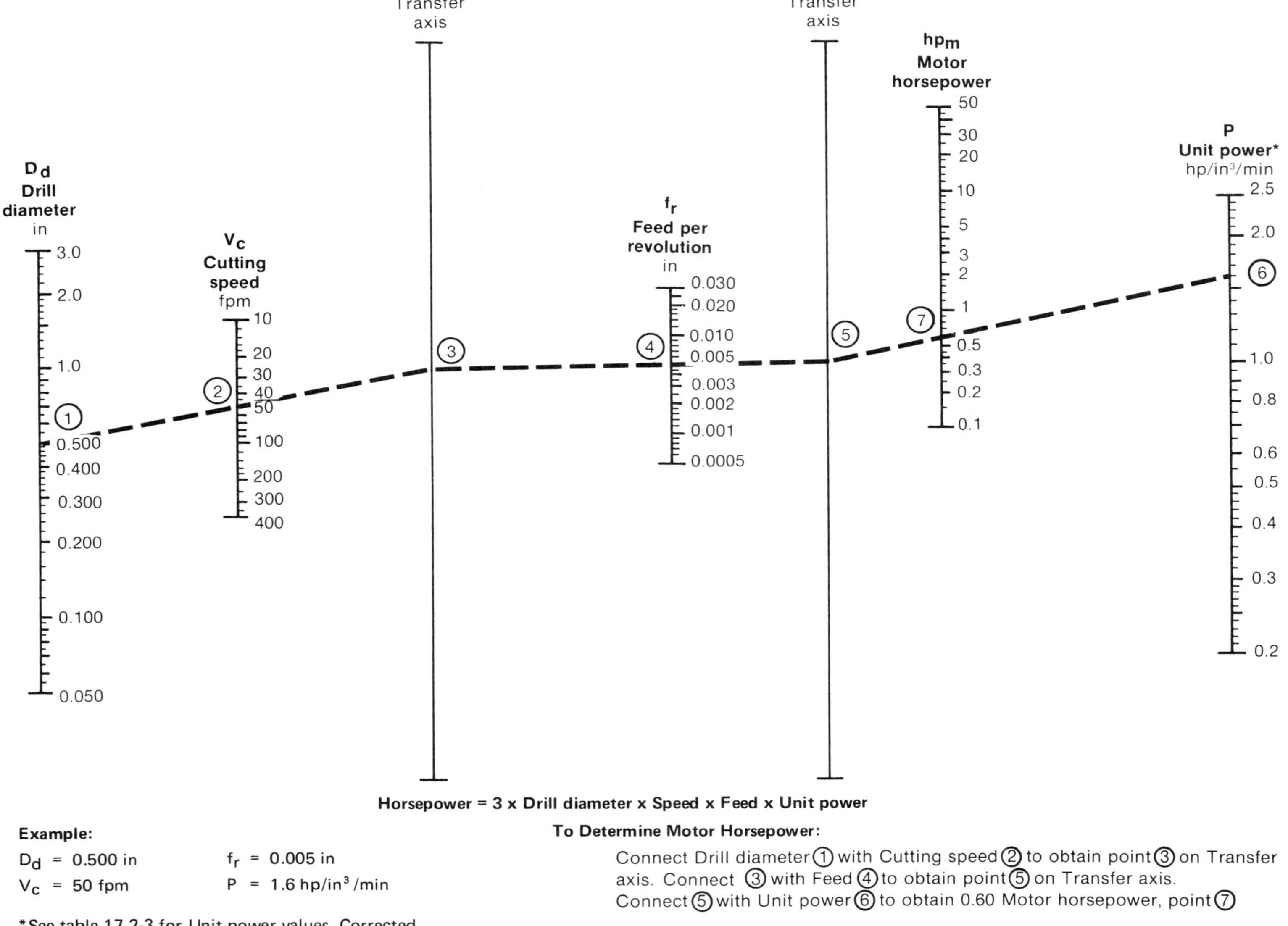


**Horsepower = 3 x Drill diameter x Speed x Feed x Unit power**

**Example:**

$D_d$ = 0.500 in $\quad$ $f_r$ = 0.005 in

$V_c$ = 50 fpm $\quad$ P = 1.6 hp/in³/min

*See table 17.2-3 for Unit power values. Corrected for dull drill and 80% machine tool efficiency.

**To Determine Motor Horsepower:**

Connect Drill diameter ① with Cutting speed ② to obtain point ③ on Transfer axis. Connect ③ with Feed ④ to obtain point ⑤ on Transfer axis. Connect ⑤ with Unit power ⑥ to obtain 0.60 Motor horsepower, point ⑦

**Figure 17.2–7** Alignment chart for determining motor horsepower in drilling—English units.

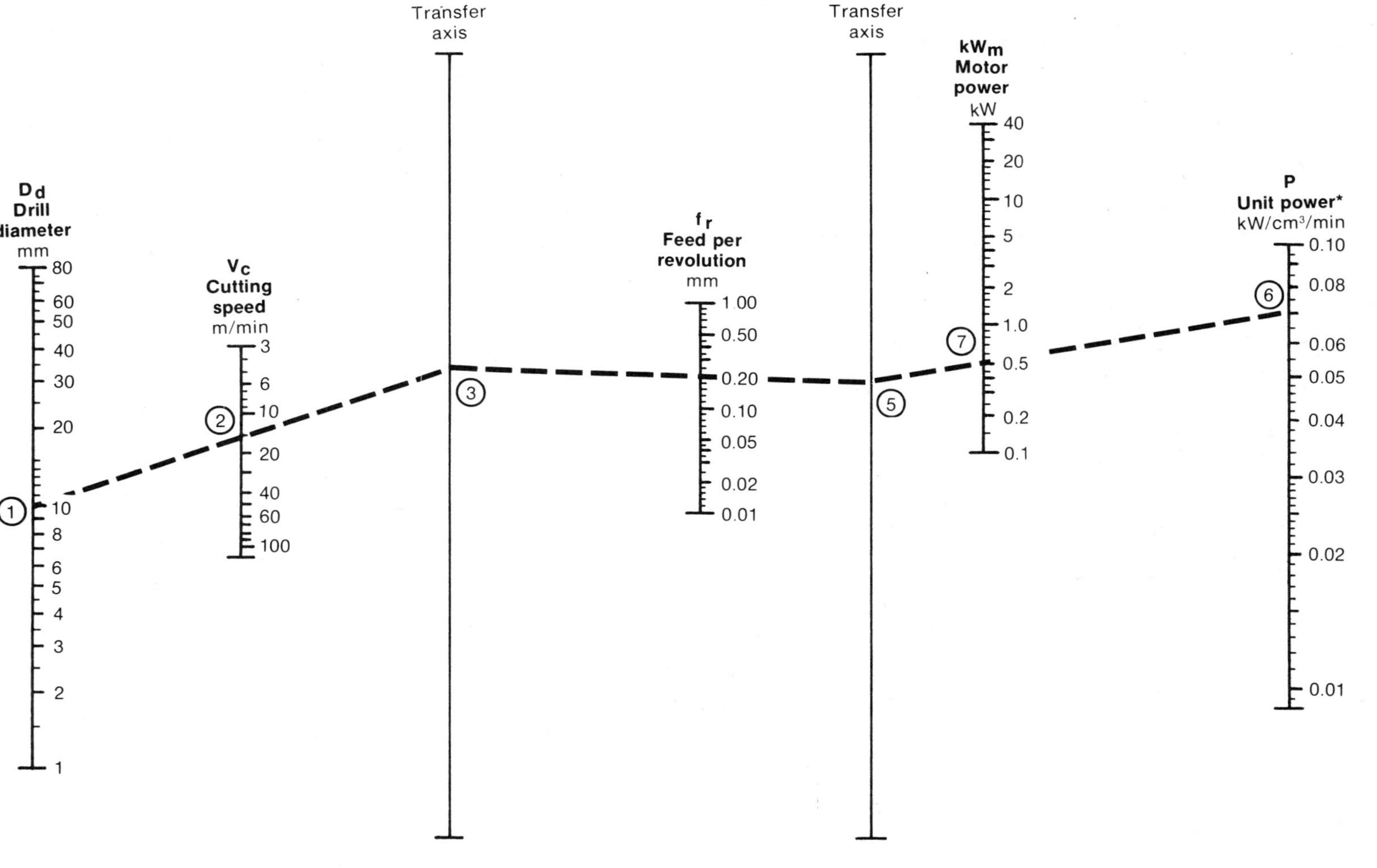


**Motor Power = 0.25 x Drill diameter x Speed x Feed x Unit Power**

**Example:**

$D_d$ = 10 mm  $f_r$ = 0.2 mm

$V_c$ = 15 m/min  P = 0.07 kW/cm³/min

**To Determine Motor Power:**

Connect Drill diameter ① with Cutting speed ② to obtain point ③ on Transfer axis. Connect ③ with Feed ④ to obtain point ⑤ on Transfer axis. Connect ⑤ with Unit power ⑥ to obtain 0.53 kW at motor power, point ⑦

*See table 17.2-4 for Unit power values. Corrected for dull drill and 80% machine tool efficiency.

**Figure 17.2–8** Alignment chart for determining motor power in drilling—Metric units.

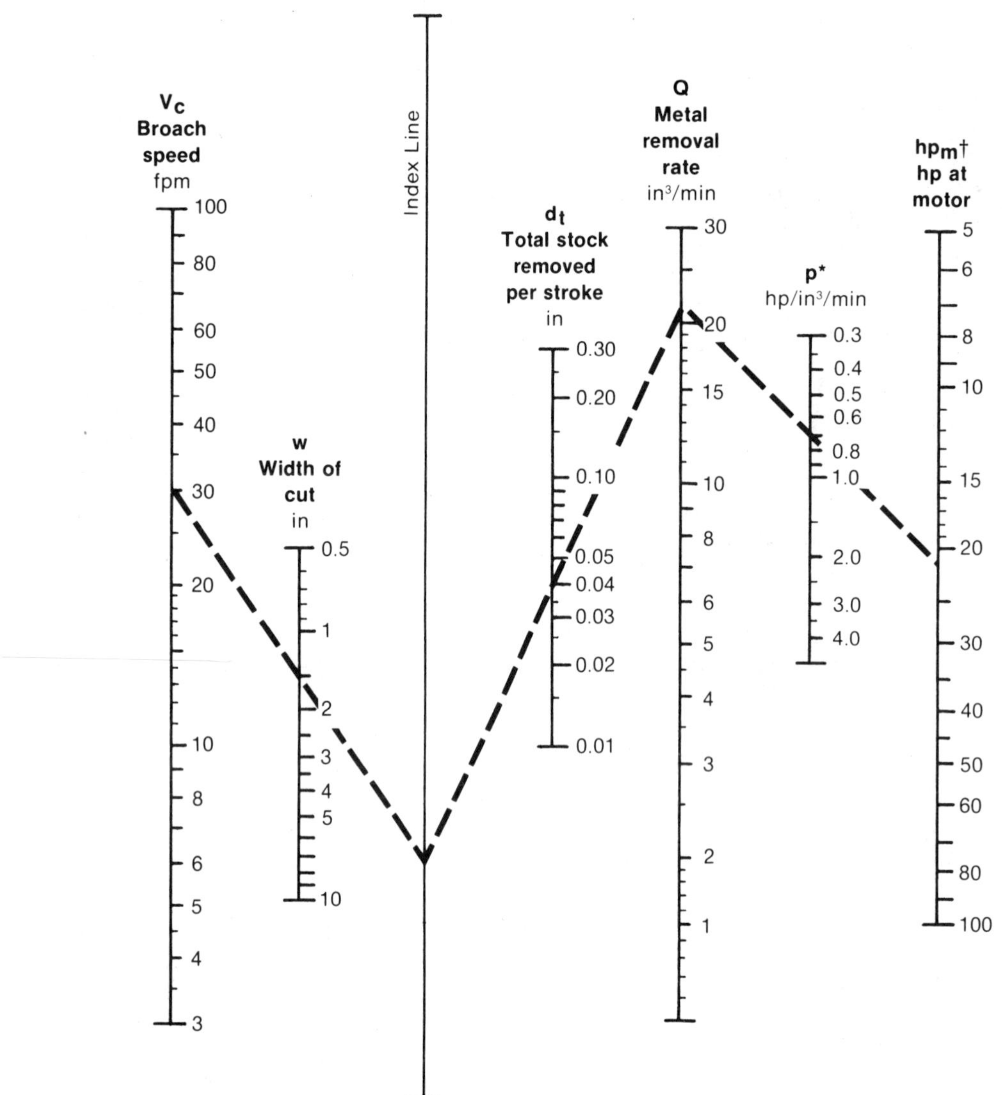


**Example:**

Material: Cast iron — HSS tools

Chipload 0.005 in/tooth

| | | | |
|---|---|---|---|
| $V_c$ | = 30 fpm | w | = 1.5 in |
| $d_t$ | = 0.040 in | Q | = 22 in³/min |
| P | = 0.7 hp/in³/min | $hp_m$ | = 22 hp |

$$Q = 12\, V_c \times w \times d_t \text{ in}^3/\text{min}$$

$$hp_m = \frac{Q \times P}{E} = \frac{Q \times P}{0.7}$$

*See figure 17.2-1 for Unit power values

†$hp_m$ at motor for 70% drive efficiency

**Figure 17.2–9** Alignment chart for determining metal removal rate and motor horsepower in surface broaching with high speed steel broaching tools—English units.

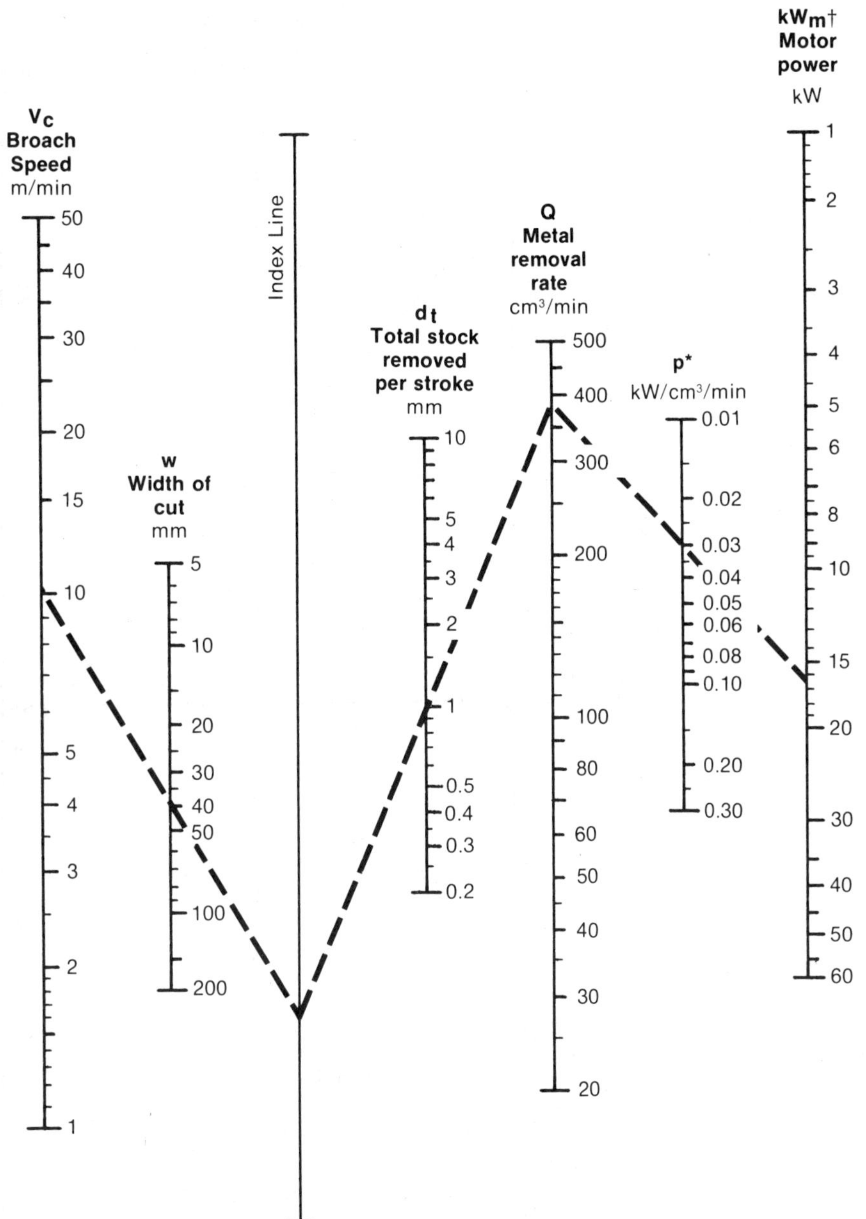


**Example:**

Material: Cast iron — HSS tools

Chipload 0.13 mm/tooth

| | |
|---|---|
| $V_c$ = 10 m/min | w = 38 mm |
| $d_t$ = 1 mm | Q = 380 cm³/min |
| P = 0.03 kW/cm³/min | $P_m$ = 16.3 kW |

$$Q = V_c \times w \times d_t \ \text{cm}^3/\text{min}$$

$$P_m = \frac{Q \times P}{E} = \frac{Q \times P}{0.7}$$

*See figure 17.2-2 for Unit power values

† $kW_m$ at motor for 70% Drive Efficiency

**Figure 17.2–10** Alignment chart for determining metal removal rate and motor power in surface broaching with high speed broaching tools—Metric units.

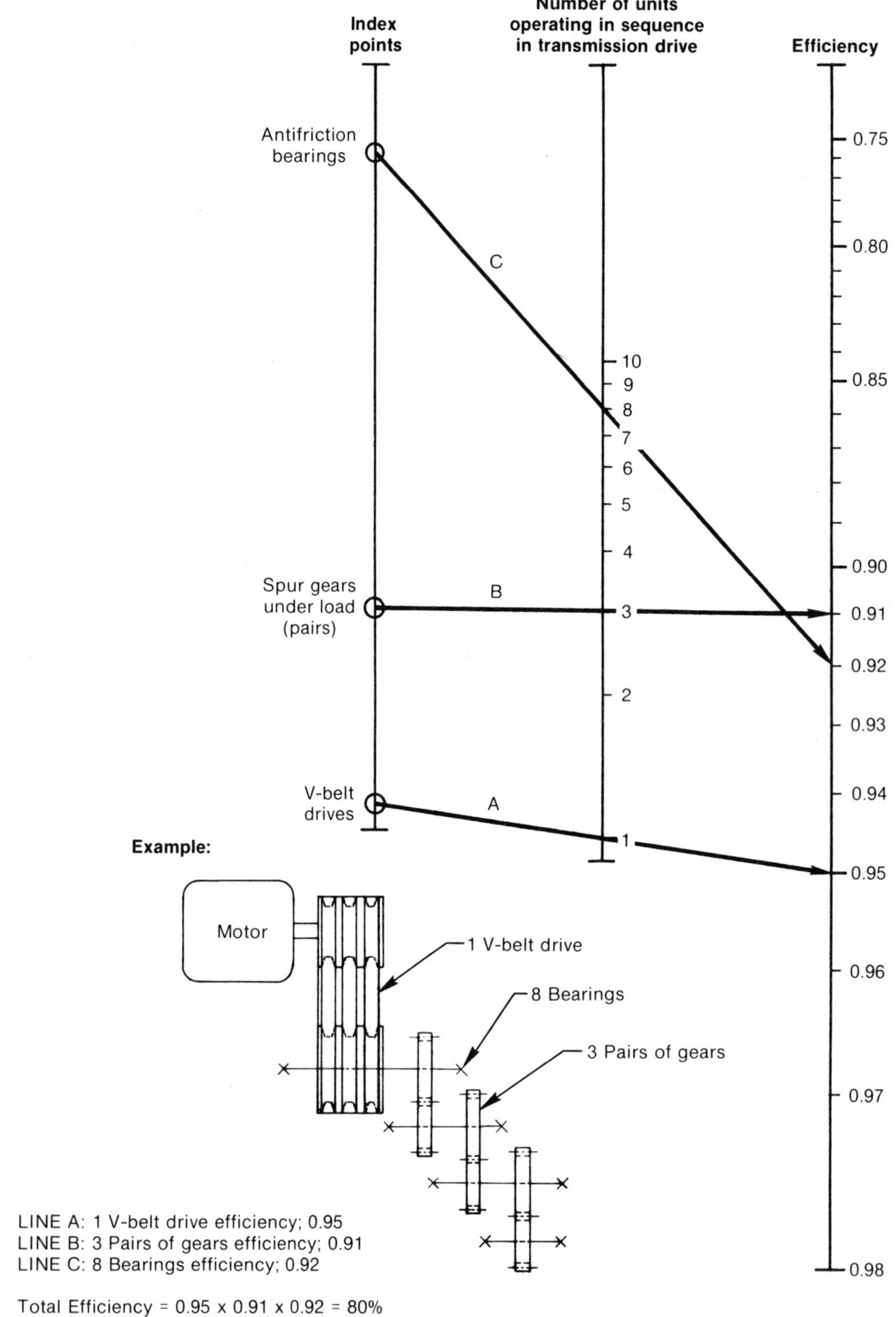


**Figure 17.2–11** Alignment chart for estimating spindle drive efficiency.

The turning forces listed in tables 17.3–1 to 17.3–3 are based upon a systematic analysis of empirical data; nevertheless, these values are to be applied only as starting points. Most of the data were derived from the following sources:

Kaczmarek, J. 1976. *Principles of machining by cutting, abrasion and erosion.* Stevenage, England: Peter Peregrinus Ltd.

König, W. and Essel, K. 1973. *Spezifische Schnittkraftwerte für die Zerspanung metallischer Werkstoffe* (Cutting force data for the machining of metallic materials). Düsseldorf, West Germany: Verlag Stahleisen.

Kronenberg, M. 1966. *Machining science and application.* Oxford, England: Pergamon Press.

Zorev, N. N. 1966. *Metal cutting mechanics.* Oxford, England: Pergamon Press.

Metcut Research Associates Inc. Internal reports.

The forces listed are as follows:

Tangential or cutting force, $F_t$
Radial force, $F_r$
Feed force, $F_f$

The values in tables 17.3–1 to 17.3–3 provide the turning force in pounds for a unit depth of 1.0 inch. In order to calculate the estimated cutting force in pounds, multiply the values given in the tables by the depth of cut in inches (to convert to newtons, multiply force in pounds by 4.48222).

Example: Material: Steel, 175 Bhn
Rake angle: −5°
Side cutting edge angle (SCEA): 15°
Feed: 0.015 inch per revolution
Depth of cut: 0.250 inch

Read the following values from the force tables: $F_t$ = 4890, $F_r$ = 1648, $F_f$ = 2500. Forces for this example are calculated as follows:

Tangential or cutting force = 4890 x 0.250 = 1223 pounds
Radial force = 1650 x 0.250 = 413 pounds
Feed force = 2500 x 0.250 = 625 pounds

**TABLE 17.3–1 Turning Force Data for a Unit Depth of Cut of 1 Inch**
(Data based on turning with carbide tool at 80 to 200 feet per minute)

| MATERIAL AND HARDNESS | TOOL RAKE ANGLE degrees | FEED PER REV. in | SCEA: 0° | | | SCEA: 15° | | | SCEA: 30° | | | SCEA: 45° | | |
|---|---|---|---|---|---|---|---|---|---|---|---|---|---|---|
| | | | $F_t$* lb. | $F_r$† lb. | $F_f$‡ lb. | $F_t$ lb. | $F_r$ lb. | $F_f$ lb. | $F_t$ lb. | $F_r$ lb. | $F_f$ lb. | $F_t$ lb. | $F_r$ lb. | $F_f$ lb. |
| Cast iron 260-320 Bhn | +5 to +10 | .005 | 2,061 | 598 | 1,383 | 2,078 | 602 | 1,407 | 2,133 | 614 | 1,487 | 2,238 | 638 | 1,647 |
| | | .010 | 3,498 | 1,050 | 1,947 | 3,527 | 1,057 | 1,980 | 3,619 | 1,078 | 2,094 | 3,798 | 1,120 | 2,320 |
| | | .015 | 4,766 | 1,460 | 2,378 | 4,806 | 1,470 | 2,420 | 4,932 | 1,500 | 2,558 | 5,175 | 1,558 | 2,835 |
| | | .020 | 5,936 | 1,844 | 2,742 | 5,985 | 1,856 | 2,790 | 6,142 | 1,895 | 2,950 | 6,445 | 1,968 | 3,268 |
| | | .025 | 7,038 | 2,211 | 3,062 | 7,096 | 2,225 | 3,116 | 7,282 | 2,272 | 3,293 | 7,640 | 2,360 | 3,648 |
| | | .030 | 8,088 | 2,564 | 3,350 | 8,155 | 2,580 | 3,410 | 8,370 | 2,634 | 3,603 | 8,780 | 2,736 | 3,992 |
| | | .040 | 10,074 | 3,240 | 3,862 | 10,157 | 3,260 | 3,930 | 10,424 | 3,328 | 4,153 | 10,937 | 3,457 | 4,602 |
| | | .050 | 11,944 | 3,883 | 4,312 | 12,043 | 3,908 | 4,388 | 12,358 | 3,990 | 4,637 | 12,966 | 4,144 | 5,138 |
| | −5 to −10 | .005 | 2,680 | 1,016 | 2,418 | 2,702 | 1,023 | 2,461 | 2,773 | 1,044 | 2,600 | 2,910 | 1,084 | 2,882 |
| | | .010 | 4,548 | 1,785 | 3,407 | 4,585 | 1,796 | 3,467 | 4,706 | 1,833 | 3,664 | 4,937 | 1,904 | 4,060 |
| | | .015 | 6,197 | 2,480 | 4,162 | 6,248 | 2,497 | 4,236 | 6,412 | 2,548 | 4,476 | 6,727 | 2,648 | 4,960 |
| | | .020 | 7,718 | 3,135 | 4,798 | 7,780 | 3,155 | 4,882 | 7,985 | 3,220 | 5,160 | 8,378 | 3,345 | 5,717 |
| | | .025 | 9,150 | 3,758 | 5,357 | 9,226 | 3,783 | 5,450 | 9,467 | 3,860 | 5,760 | 9,933 | 4,010 | 6,383 |
| | | .030 | 10,516 | 4,358 | 5,862 | 10,603 | 4,387 | 5,965 | 10,880 | 4,478 | 6,304 | 11,416 | 4,650 | 6,985 |
| | | .040 | 13,097 | 5,507 | 6,757 | 13,205 | 5,543 | 6,876 | 13,550 | 5,657 | 7,267 | 14,218 | 5,876 | 8,052 |
| | | .050 | 15,528 | 6,602 | 7,544 | 15,656 | 6,645 | 7,677 | 16,066 | 6,782 | 8,113 | 16,857 | 7,044 | 8,990 |

NOTE: To calculate force in pounds, multiply values given by depth of cut in inches. To convert to newtons, multiply force in pounds by 4.48222.
*$F_t$ = Tangential or Cutting force
†$F_r$ = Radial force
‡$F_f$ = Feed force

**TABLE 17.3-2 Turning Force Data for a Unit Depth of Cut of 1 Inch**
(Data based on turning with carbide tool at 200 to 600 feet per minute)

| MATERIAL AND HARDNESS | TOOL RAKE ANGLE degrees | FEED PER REV. in | SCEA: 0° | | | SCEA: 15° | | | SCEA: 30° | | | SCEA: 45° | | |
|---|---|---|---|---|---|---|---|---|---|---|---|---|---|---|
| | | | $F_t$* lb. | $F_r$† lb. | $F_f$‡ lb. | $F_t$ lb. | $F_r$ lb. | $F_f$ lb. | $F_t$ lb. | $F_r$ lb. | $F_f$ lb. | $F_t$ lb. | $F_r$ lb. | $F_f$ lb. |
| Cast iron 160-200 Bhn | +5 to +10 | .005 | 1,195 | 340 | 800 | 1,204 | 343 | 813 | 1,232 | 350 | 857 | 1,288 | 362 | 946 |
| | | .010 | 2,056 | 608 | 1,140 | 2,070 | 610 | 1,160 | 2,120 | 623 | 1,224 | 2,216 | 644 | 1,350 |
| | | .015 | 2,824 | 852 | 1,405 | 2,845 | 857 | 1,430 | 2,913 | 873 | 1,507 | 3,044 | 903 | 1,663 |
| | | .020 | 3,538 | 1,082 | 1,629 | 3,564 | 1,088 | 1,657 | 3,650 | 1,108 | 1,747 | 3,814 | 1,147 | 1,928 |
| | | .025 | 4,213 | 1,303 | 1,827 | 4,245 | 1,310 | 1,858 | 4,347 | 1,335 | 1,960 | 4,542 | 1,380 | 2,162 |
| | | .030 | 4,860 | 1,517 | 2,006 | 4,897 | 1,525 | 2,040 | 5,014 | 1,554 | 2,152 | 5,240 | 1,607 | 2,375 |
| | | .040 | 6,088 | 1,927 | 2,326 | 6,134 | 1,938 | 2,365 | 6,280 | 1,974 | 2,494 | 6,563 | 2,042 | 2,753 |
| | | .050 | 7,250 | 2,320 | 2,608 | 7,305 | 2,334 | 2,653 | 7,480 | 2,377 | 2,797 | 7,816 | 2,458 | 3,087 |
| | -5 to -10 | .005 | 1,555 | 590 | 1,400 | 1,566 | 593 | 1,424 | 1,604 | 604 | 1,502 | 1,676 | 625 | 1,658 |
| | | .010 | 2,675 | 1,050 | 2,000 | 2,695 | 1,056 | 2,034 | 2,760 | 1,075 | 2,145 | 2,884 | 1,112 | 2,367 |
| | | .015 | 3,675 | 1,470 | 2,463 | 3,703 | 1,480 | 2,505 | 3,790 | 1,507 | 2,640 | 3,962 | 1,558 | 2,915 |
| | | .020 | 4,603 | 1,868 | 2,855 | 4,638 | 1,880 | 2,903 | 4,750 | 1,914 | 3,062 | 4,963 | 1,980 | 3,378 |
| | | .025 | 5,482 | 2,250 | 3,202 | 5,523 | 2,263 | 3,256 | 5,656 | 2,304 | 3,434 | 5,910 | 2,385 | 3,790 |
| | | .030 | 6,323 | 2,618 | 3,516 | 6,371 | 2,634 | 3,576 | 6,524 | 2,683 | 3,770 | 6,817 | 2,775 | 4,160 |
| | | .040 | 7,920 | 3,327 | 4,076 | 7,980 | 3,346 | 4,145 | 8,172 | 3,408 | 4,370 | 8,540 | 3,526 | 4,824 |
| | | .050 | 9,433 | 4,006 | 4,570 | 9,504 | 4,030 | 4,648 | 9,732 | 4,104 | 4,902 | 10,170 | 4,246 | 5,410 |
| Cast iron 200-260 Bhn | +5 to +10 | .005 | 1,636 | 474 | 1,097 | 1,649 | 477 | 1,116 | 1,692 | 486 | 1,180 | 1,775 | 505 | 1,306 |
| | | .010 | 2,780 | 833 | 1,546 | 2,804 | 838 | 1,573 | 2,876 | 856 | 1,662 | 3,016 | 888 | 1,840 |
| | | .015 | 3,793 | 1,160 | 1,890 | 3,824 | 1,167 | 1,924 | 3,923 | 1,190 | 2,032 | 4,114 | 1,236 | 2,250 |
| | | .020 | 4,727 | 1,466 | 2,180 | 4,765 | 1,476 | 2,218 | 4,890 | 1,506 | 2,344 | 5,127 | 1,563 | 2,596 |
| | | .025 | 5,607 | 1,758 | 2,435 | 5,653 | 1,770 | 2,478 | 5,800 | 1,806 | 2,618 | 6,082 | 1,875 | 2,900 |
| | | .030 | 6,447 | 2,040 | 2,665 | 6,500 | 2,054 | 2,712 | 6,668 | 2,096 | 2,866 | 6,993 | 2,176 | 3,174 |
| | | .040 | 8,034 | 2,580 | 3,073 | 8,100 | 2,597 | 3,128 | 8,310 | 2,650 | 3,305 | 8,715 | 2,750 | 3,660 |
| | | .050 | 9,530 | 3,095 | 3,433 | 9,608 | 3,115 | 3,493 | 9,858 | 3,178 | 3,690 | 10,338 | 3,300 | 4,088 |
| | -5 to -10 | .005 | 2,126 | 806 | 1,918 | 2,143 | 811 | 1,952 | 2,200 | 828 | 2,062 | 2,306 | 860 | 2,284 |
| | | .010 | 3,614 | 1,418 | 2,704 | 3,644 | 1,427 | 2,752 | 3,738 | 1,457 | 2,908 | 3,920 | 1,512 | 3,220 |
| | | .015 | 4,930 | 1,974 | 3,306 | 4,970 | 1,987 | 3,364 | 5,098 | 2,027 | 3,555 | 5,347 | 2,104 | 3,938 |
| | | .020 | 6,144 | 2,495 | 3,813 | 6,194 | 2,511 | 3,880 | 6,355 | 2,563 | 4,100 | 6,664 | 2,660 | 4,540 |
| | | .025 | 7,288 | 2,993 | 4,258 | 7,348 | 3,012 | 4,334 | 7,538 | 3,074 | 4,578 | 7,905 | 3,190 | 5,072 |
| | | .030 | 8,380 | 3,473 | 4,660 | 8,448 | 3,495 | 4,743 | 8,668 | 3,566 | 5,012 | 9,090 | 3,703 | 5,552 |
| | | .040 | 10,444 | 4,390 | 5,375 | 10,530 | 4,418 | 5,470 | 10,803 | 4,508 | 5,780 | 11,328 | 4,680 | 6,402 |
| | | .050 | 12,390 | 5,266 | 6,004 | 12,490 | 5,300 | 6,110 | 12,815 | 5,408 | 6,456 | 13,438 | 5,615 | 7,150 |

NOTE: To calculate force in pounds, multiply values given by depth of cut in inches. To convert to newtons, multiply force in pounds by 4.48222.

*$F_t$ = Tangential or Cutting force

† $F_r$ = Radial force

‡ $F_f$ = Feed force

**TABLE 17.3-3 Turning Force Data for a Unit Depth of Cut of 1 Inch**
(Data based on turning with carbide tool at 300 to 600 feet per minute)

| MATERIAL AND HARDNESS | TOOL RAKE ANGLE degrees | FEED PER REV. in | SCEA: 0° | | | SCEA: 15° | | | SCEA: 30° | | | SCEA: 45° | | |
|---|---|---|---|---|---|---|---|---|---|---|---|---|---|---|
| | | | $F_t$* lb. | $F_r$† lb. | $F_f$‡ lb. | $F_t$ lb. | $F_r$ lb. | $F_f$ lb. | $F_t$ lb. | $F_r$ lb. | $F_f$ lb. | $F_t$ lb. | $F_r$ lb. | $F_f$ lb. |
| Steels 135-200 Bhn | +5 to +10 | .005 | 1,853 | 541 | 1,116 | 1,870 | 552 | 1,146 | 1,928 | 588 | 1,244 | 2,038 | 662 | 1,449 |
| | | .010 | 3,062 | 725 | 1,324 | 3,092 | 740 | 1,360 | 3,186 | 788 | 1,476 | 3,369 | 886 | 1,719 |
| | | .015 | 4,108 | 860 | 1,463 | 4,148 | 877 | 1,502 | 4,274 | 934 | 1,630 | 4,520 | 1,050 | 1,900 |
| | | .020 | 5,060 | 970 | 1,570 | 5,110 | 990 | 1,612 | 5,265 | 1,055 | 1,750 | 5,567 | 1,186 | 2,039 |
| | | .025 | 5,950 | 1,066 | 1,660 | 6,006 | 1,088 | 1,703 | 6,190 | 1,158 | 1,850 | 6,545 | 1,303 | 2,155 |
| | | .030 | 6,790 | 1,150 | 1,736 | 6,855 | 1,174 | 1,782 | 7,064 | 1,250 | 1,934 | 7,469 | 1,407 | 2,254 |
| | | .040 | 8,365 | 1,300 | 1,863 | 8,444 | 1,326 | 1,912 | 8,702 | 1,412 | 2,076 | 9,200 | 1,588 | 2,420 |
| | | .050 | 9,832 | 1,427 | 1,968 | 9,927 | 1,456 | 2,020 | 10,230 | 1,550 | 2,194 | 10,816 | 1,744 | 2,556 |
| | −5 to −10 | .005 | 2,136 | 860 | 1,646 | 2,155 | 873 | 1,682 | 2,215 | 914 | 1,804 | 2,333 | 996 | 2,054 |
| | | .010 | 3,582 | 1,285 | 2,112 | 3,614 | 1,304 | 2,160 | 3,715 | 1,365 | 2,316 | 3,911 | 1,488 | 2,037 |
| | | .015 | 4,846 | 1,624 | 2,444 | 4,890 | 1,648 | 2,500 | 5,027 | 1,726 | 2,680 | 5,293 | 1,880 | 3,050 |
| | | .020 | 6,006 | 1,918 | 2,711 | 6,060 | 1,946 | 2,772 | 6,230 | 2,038 | 2,973 | 6,560 | 2,220 | 3,384 |
| | | .025 | 7,093 | 2,182 | 2,938 | 7,156 | 2,215 | 3,004 | 7,358 | 2,318 | 3,222 | 7,747 | 2,526 | 3,668 |
| | | .030 | 8,127 | 2,425 | 3,138 | 8,198 | 2,460 | 3,208 | 8,430 | 2,577 | 3,440 | 8,875 | 2,807 | 3,917 |
| | | .040 | 10,070 | 2,864 | 3,480 | 10,160 | 2,906 | 3,558 | 10,446 | 3,043 | 3,816 | 11,000 | 3,315 | 4,344 |
| | | .050 | 11,895 | 3,258 | 3,772 | 12,000 | 3,306 | 3,856 | 12,338 | 3,462 | 4,135 | 12,990 | 3,771 | 4,708 |
| Steels 200-325 Bhn | +5 to +10 | .005 | 1,942 | 497 | 1,183 | 1,960 | 504 | 1,213 | 2,015 | 529 | 1,313 | 2,121 | 577 | 1,522 |
| | | .010 | 3,256 | 738 | 1,429 | 3,285 | 748 | 1,466 | 3,378 | 785 | 1,587 | 3,556 | 856 | 1,838 |
| | | .015 | 4,405 | 929 | 1,596 | 4,444 | 943 | 1,638 | 4,570 | 988 | 1,773 | 4,812 | 1,078 | 2,054 |
| | | .020 | 5,460 | 1,095 | 1,727 | 5,508 | 1,111 | 1,772 | 5,662 | 1,165 | 1,918 | 5,963 | 1,270 | 2,222 |
| | | .025 | 6,447 | 1,243 | 1,836 | 6,504 | 1,262 | 1,882 | 6,688 | 1,323 | 2,038 | 7,042 | 1,443 | 2,362 |
| | | .030 | 7,386 | 1,380 | 1,930 | 7,452 | 1,400 | 1,980 | 7,662 | 1,467 | 2,142 | 8,067 | 1,600 | 2,483 |
| | | .040 | 9,153 | 1,625 | 2,088 | 9,234 | 1,650 | 2,140 | 9,494 | 1,728 | 2,318 | 9,997 | 1,886 | 2,686 |
| | | .050 | 10,810 | 1,845 | 2,218 | 10,905 | 1,872 | 2,275 | 11,212 | 1,963 | 2,463 | 11,806 | 2,142 | 2,854 |
| | −5 to −10 | .005 | 2,192 | 1,040 | 1,842 | 2,214 | 1,058 | 1,888 | 2,285 | 1,117 | 2,038 | 2,422 | 1,235 | 2,350 |
| | | .010 | 3,592 | 1,478 | 2,262 | 3,628 | 1,504 | 2,318 | 3,743 | 1,587 | 2,503 | 3,968 | 1,754 | 2,886 |
| | | .015 | 4,795 | 1,815 | 2,551 | 4,842 | 1,846 | 2,614 | 4,997 | 1,948 | 2,823 | 5,297 | 2,154 | 3,256 |
| | | .020 | 5,885 | 2,100 | 2,779 | 5,945 | 2,136 | 2,848 | 6,134 | 2,254 | 3,075 | 6,502 | 2,491 | 3,546 |
| | | .025 | 6,900 | 2,350 | 2,969 | 6,969 | 2,390 | 3,042 | 7,191 | 2,523 | 3,285 | 7,623 | 2,790 | 3,788 |
| | | .030 | 7,857 | 2,578 | 3,134 | 7,936 | 2,622 | 3,212 | 8,189 | 2,767 | 3,468 | 8,680 | 3,058 | 4,000 |
| | | .040 | 9,644 | 2,982 | 3,414 | 9,740 | 3,033 | 3,498 | 10,052 | 3,200 | 3,777 | 10,655 | 3,538 | 4,356 |
| | | .050 | 10,306 | 3,338 | 3,647 | 11,420 | 3,395 | 3,737 | 11,784 | 3,584 | 4,036 | 12,490 | 3,962 | 4,654 |

NOTE: To calculate force in pounds, multiply values given by depth of cut in inches. To convert to newtons, multiply force in pounds by 4.48222.

*$F_t$ = Tangential or Cutting force

† $F_r$ = Radial force

‡ $F_f$ = Feed force

Tables 17.4–1 and 17.4–2 are tabulations of drilling torque and thrust values in English units derived from formulas given in the paper "On the drilling of metals, 2—The torque and thrust in drilling" by M. C. Shaw and C. J. Oxford, Jr., *Transactions of the ASME* 79 (January 1957), p. 139–148.

Example: Find the torque and thrust when drilling steel at 200 Bhn with a 1/2-inch diameter standard point drill using a feed rate of 0.006 inch per revolution.

In the table for standard point drills, locate 200 Bhn steel, read down the data column for a 1/2-inch drill to the 0.006-inch feed per revolution and read 124 inch-pounds torque and 735 pounds thrust.

Tables 17.4–3 and 17.4–4 give torque and thrust values for the same materials in metric units.

**TABLE 17.4–1 Torque and Thrust Values for Drilling with Standard Point Drills**

| MATERIAL AND HARDNESS | FEED PER REV. | TORQUE, inch-pounds | | | | | | | | THRUST, pounds | | | | | | | |
|---|---|---|---|---|---|---|---|---|---|---|---|---|---|---|---|---|---|
| | | Drill Diameter, in | | | | | | | | Drill Diameter, in | | | | | | | |
| | in | ⅛ | ¼ | ½ | ¾ | 1 | 1¼ | 1½ | 2 | ⅛ | ¼ | ½ | ¾ | 1 | 1¼ | 1½ | 2 |
| Steels 200 Bhn | 0.0005 | 1.4 | — | — | — | — | — | — | — | 44 | — | — | — | — | — | — | — |
| | .001 | 2.5 | 8.5 | — | — | — | — | — | — | 65 | 130 | — | — | — | — | — | — |
| | .002 | 4.4 | 15 | 52 | — | — | — | — | — | 102 | 194 | 382 | — | — | — | — | — |
| | .003 | 6 | 21 | 71 | — | — | — | — | — | 135 | 250 | 478 | — | — | — | — | — |
| | .004 | 8 | 26 | 90 | 183 | — | — | — | — | 166 | 304 | 568 | 832 | — | — | — | — |
| | .006 | 11 | 36 | 124 | 253 | 414 | 617 | 856 | 1,430 | 224 | 403 | 735 | 1,058 | 1,368 | 1,687 | 2,030 | 2,796 |
| | .008 | — | 45 | 156 | 318 | 521 | 776 | 1,078 | 1,800 | — | 496 | 891 | 1,269 | 1,629 | 1,997 | 2,388 | 3,246 |
| | .010 | — | 54 | 187 | 381 | 623 | 928 | 1,288 | 2,152 | — | 584 | 1,039 | 1,469 | 1,876 | 2,292 | 2,728 | 3,674 |
| | .012 | — | — | 216 | 440 | 721 | 1,074 | 1,491 | 2,490 | — | — | 1,182 | 1,662 | 2,115 | 2,575 | 3,054 | 4,085 |
| | .015 | — | — | 259 | 527 | 862 | 1,283 | 1,782 | 2,976 | — | — | 1,387 | 1,939 | 2,458 | 2,983 | 3,524 | 4,677 |
| | .020 | — | — | 325 | 663 | 1,085 | 1,615 | 2,243 | 3,746 | — | — | 1,711 | 2,378 | 3,001 | 3,629 | 4,269 | 5,614 |
| | .025 | — | — | — | — | 1,297 | 1,931 | 2,681 | 4,478 | — | — | — | — | 3,517 | 4,244 | 4,976 | 6,505 |
| | .030 | — | — | — | — | — | — | — | 5,782 | — | — | — | — | — | — | — | 7,360 |
| Steels 300 Bhn | 0.0005 | 1.9 | — | — | — | — | — | — | — | 56 | — | — | — | — | — | — | — |
| | .001 | 3.2 | 11 | — | — | — | — | — | — | 84 | 168 | — | — | — | — | — | — |
| | .002 | 5.6 | 19 | 67 | — | — | — | — | — | 132 | 251 | 494 | — | — | — | — | — |
| | .003 | 7.8 | 27 | 92 | — | — | — | — | — | 175 | 324 | 618 | — | — | — | — | — |
| | .004 | 9 | 33 | 116 | 236 | — | — | — | — | 215 | 293 | 733 | 1,075 | — | — | — | — |
| | .006 | 14 | 46 | 160 | 327 | 535 | 796 | 1,106 | 1,847 | 290 | 521 | 950 | 1,367 | 1,767 | 2,179 | 2,622 | 3,611 |
| | .008 | — | 58 | 202 | 411 | 673 | 1,003 | 1,392 | 2,325 | — | 640 | 1,151 | 1,639 | 2,104 | 2,580 | 3,084 | 4,193 |
| | .010 | — | 70 | 241 | 492 | 805 | 1,199 | 1,664 | 2,779 | — | 754 | 1,342 | 1,898 | 2,424 | 2,961 | 3,523 | 4,745 |
| | .012 | — | — | 279 | 569 | 932 | 1,386 | 1,925 | 3,216 | — | — | 1,526 | 2,146 | 2,732 | 3,327 | 3,945 | 5,276 |
| | .015 | — | — | 334 | 680 | 1,114 | 1,658 | 2,302 | 3,844 | — | — | 1,791 | 2,504 | 3,175 | 3,854 | 4,552 | 6,041 |
| | .020 | — | — | 420 | 856 | 1,402 | 2,087 | 2,897 | 4,839 | — | — | 2,211 | 3,071 | 3,876 | 4,688 | 5,514 | 7,251 |
| | .025 | — | — | — | — | 1,676 | 2,495 | 3,463 | 5,785 | — | — | — | — | 4,543 | 5,481 | 6,428 | 8,402 |
| | .030 | — | — | — | — | — | — | — | 6,693 | — | — | — | — | — | — | — | 9,507 |
| Steels 400 Bhn | .0005 | 2 | — | — | — | — | — | — | — | 62 | — | — | — | — | — | — | — |
| | .001 | 3.5 | 12 | — | — | — | — | — | — | 92 | 185 | — | — | — | — | — | — |
| | .002 | 6.2 | 21 | 73 | — | — | — | — | — | 144 | 274 | 542 | — | — | — | — | — |
| | .003 | 8.5 | 29 | 101 | — | — | — | — | — | 192 | 355 | 678 | — | — | — | — | — |
| | .004 | 11 | 37 | 127 | 259 | — | — | — | — | 236 | 431 | 805 | 1,179 | — | — | — | — |
| | .006 | 15 | 51 | 176 | 358 | 586 | 874 | 1,213 | 2,026 | 318 | 571 | 1,041 | 1,499 | 1,940 | 2,390 | 2,876 | 3,961 |
| | .008 | — | 64 | 222 | 451 | 739 | 1,100 | 1,527 | 2,550 | — | 702 | 1,262 | 1,797 | 2,307 | 2,829 | 3,383 | 4,598 |
| | .010 | — | 76 | 265 | 539 | 883 | 1,314 | 1,825 | 3,048 | — | 827 | 1,472 | 2,081 | 2,659 | 3,247 | 3,864 | 5,204 |
| | .012 | — | — | 306 | 624 | 1,022 | 1,521 | 2,112 | 3,527 | — | — | 1,674 | 2,354 | 2,996 | 3,648 | 4,327 | 5,787 |
| | .015 | — | — | 366 | 746 | 1,221 | 1,818 | 2,524 | 4,216 | — | — | 1,965 | 2,747 | 3,482 | 4,227 | 4,993 | 6,625 |
| | .020 | — | — | 461 | 939 | 1,537 | 2,289 | 3,178 | 5,307 | — | — | 2,425 | 3,368 | 4,251 | 5,142 | 6,047 | 7,953 |
| | .025 | — | — | — | — | 1,838 | 2,736 | 3,799 | 6,345 | — | — | — | — | 4,982 | 6,012 | 7,050 | 9,215 |
| | .030 | — | — | — | — | — | — | — | 7,341 | — | — | — | — | — | — | — | 10,430 |

**TABLE 17.4-1—Continued**

| MATERIAL AND HARDNESS | FEED PER REV. | TORQUE, inch-pounds<br>Drill Diameter, in | | | | | | | | THRUST, pounds<br>Drill Diameter, in | | | | | | | |
|---|---|---|---|---|---|---|---|---|---|---|---|---|---|---|---|---|---|
| | in | ⅛ | ¼ | ½ | ¾ | 1 | 1¼ | 1½ | 2 | ⅛ | ¼ | ½ | ¾ | 1 | 1¼ | 1½ | 2 |
| Aluminum Alloys | .0005 | 0.4 | — | — | — | — | — | — | — | 13 | — | — | — | — | — | — | — |
| | .001 | 0.7 | 2.5 | — | — | — | — | — | — | 19 | 38 | — | — | — | — | — | — |
| | .002 | 1.3 | 4.3 | 15 | — | — | — | — | — | 30 | 56 | 111 | — | — | — | — | — |
| | .003 | 1.8 | 6 | 21 | — | — | — | — | — | 39 | 73 | 139 | — | — | — | — | — |
| | .004 | 2.2 | 7.6 | 26 | 53 | — | — | — | — | 49 | 89 | 166 | 243 | — | — | — | — |
| | .006 | 3.1 | 11 | 36 | 74 | 121 | 180 | 250 | 417 | 66 | 118 | 214 | 308 | 399 | 492 | 592 | 815 |
| | .008 | — | 13 | 46 | 93 | 152 | 226 | 314 | 525 | — | 145 | 260 | 370 | 475 | 583 | 696 | 947 |
| | .010 | — | 16 | 55 | 111 | 182 | 271 | 376 | 628 | — | 170 | 303 | 428 | 547 | 669 | 796 | 1,072 |
| | .012 | — | — | 63 | 128 | 210 | 313 | 435 | 726 | — | — | 345 | 485 | 617 | 751 | 891 | 1,191 |
| | .015 | — | — | 75 | 154 | 251 | 324 | 520 | 868 | — | — | 404 | 566 | 717 | 870 | 1,028 | 1,364 |
| | .020 | — | — | 95 | 193 | 317 | 471 | 654 | 1,093 | — | — | 499 | 694 | 875 | 1,059 | 1,245 | 1,637 |
| | .025 | — | — | — | — | 378 | 563 | 782 | 1,306 | — | — | — | — | 1,026 | 1,238 | 1,451 | 1,897 |
| | .030 | — | — | — | — | — | — | — | 1,511 | — | — | — | — | — | — | — | 2,147 |
| Magnesium Alloys | .0005 | 0.2 | — | — | — | — | — | — | — | 7.3 | — | — | — | — | — | — | — |
| | .001 | 0.4 | 1.4 | — | — | — | — | — | — | 11 | 22 | — | — | — | — | — | — |
| | .002 | 0.7 | 2.5 | 8.6 | — | — | — | — | — | 17 | 32 | 64 | — | — | — | — | — |
| | .003 | 1 | 3.4 | 12 | — | — | — | — | — | 23 | 42 | 80 | — | — | — | — | — |
| | .004 | 1.3 | 4.3 | 15 | 30 | — | — | — | — | 28 | 51 | 95 | 139 | — | — | — | — |
| | .006 | 1.7 | 6 | 21 | 42 | 69 | 103 | 143 | 238 | 33 | 59 | 109 | 158 | 205 | 254 | 307 | 427 |
| | .008 | — | 7.5 | 26 | 53 | 87 | 129 | 180 | 300 | — | 83 | 148 | 211 | 272 | 333 | 398 | 541 |
| | .010 | — | 9 | 31 | 64 | 104 | 155 | 215 | 359 | — | 97 | 173 | 245 | 313 | 382 | 455 | 612 |
| | .012 | — | — | 36 | 73 | 120 | 179 | 250 | 415 | — | — | 197 | 277 | 352 | 429 | 509 | 681 |
| | .015 | — | — | 43 | 89 | 144 | 214 | 297 | 496 | — | — | 231 | 323 | 410 | 497 | 587 | 780 |
| | .020 | — | — | 54 | 110 | 181 | 269 | 374 | 624 | — | — | 285 | 396 | 500 | 605 | 711 | 936 |
| | .025 | — | — | — | — | 216 | 322 | 447 | 746 | — | — | — | — | 586 | 707 | 829 | 1,084 |
| | .030 | — | — | — | — | — | — | — | 864 | — | — | — | — | — | — | — | 1,227 |
| Copper Alloys | .0005 | 0.8 | — | — | — | — | — | — | — | 25 | — | — | — | — | — | — | — |
| | .001 | 1.5 | 5 | — | — | — | — | — | — | 38 | 76 | — | — | — | — | — | — |
| | .002 | 2.5 | 8.7 | 30 | — | — | — | — | — | 60 | 113 | 223 | — | — | — | — | — |
| | .003 | 3.5 | 12 | 42 | — | — | — | — | — | 79 | 146 | 279 | — | — | — | — | — |
| | .004 | 4.4 | 15 | 52 | 107 | — | — | — | — | 97 | 177 | 331 | 485 | — | — | — | — |
| | .006 | 6.1 | 21 | 73 | 148 | 242 | 360 | 499 | 834 | 131 | 235 | 429 | 617 | 798 | 983 | 1,184 | 1,630 |
| | .008 | — | 26 | 91 | 186 | 304 | 453 | 629 | 1,050 | — | 289 | 520 | 740 | 950 | 1,165 | 1,393 | 1,893 |
| | .010 | — | 31 | 109 | 222 | 364 | 541 | 751 | 1,255 | — | 341 | 606 | 857 | 1,095 | 1,337 | 1,591 | 2,143 |
| | .012 | — | — | 126 | 257 | 421 | 626 | 870 | 1,452 | — | — | 689 | 969 | 1,234 | 1,502 | 1,782 | 2,383 |
| | .015 | — | — | 151 | 307 | 503 | 749 | 1,039 | 1,736 | — | — | 809 | 1,131 | 1,434 | 1,740 | 2,056 | 2,728 |
| | .020 | — | — | 190 | 387 | 633 | 942 | 1,308 | 2,185 | — | — | 998 | 1,387 | 1,750 | 2,117 | 2,490 | 3,275 |
| | .025 | — | — | — | — | 757 | 1,127 | 1,564 | 2,612 | — | — | — | — | 2,052 | 2,475 | 2,903 | 3,795 |
| | .030 | — | — | — | — | — | — | — | 3,023 | — | — | — | — | — | — | — | 4,293 |
| Leaded Brass | .0005 | 0.4 | — | — | — | — | — | — | — | 13 | — | — | — | — | — | — | — |
| | .001 | 0.7 | 2.5 | — | — | — | — | — | — | 19 | 38 | — | — | — | — | — | — |
| | .002 | 1.3 | 4.3 | 15 | — | — | — | — | — | 30 | 56 | 111 | — | — | — | — | — |
| | .003 | 1.8 | 6 | 21 | — | — | — | — | — | 39 | 73 | 139 | — | — | — | — | — |
| | .004 | 2.2 | 7.6 | 26 | 53 | — | — | — | — | 49 | 89 | 166 | 243 | — | — | — | — |
| | .006 | 3.1 | 10 | 36 | 74 | 121 | 180 | 250 | 417 | 66 | 118 | 214 | 308 | 399 | 492 | 592 | 815 |
| | .008 | — | 13 | 46 | 93 | 152 | 226 | 314 | 525 | — | 145 | 259 | 370 | 475 | 583 | 696 | 947 |
| | .010 | — | 16 | 55 | 111 | 182 | 271 | 376 | 628 | — | 170 | 303 | 428 | 547 | 669 | 796 | 1,072 |
| | .012 | — | — | 63 | 128 | 210 | 313 | 435 | 726 | — | — | 344 | 485 | 616 | 751 | 891 | 1,191 |
| | .015 | — | — | 75 | 154 | 252 | 374 | 520 | 868 | — | — | 404 | 566 | 717 | 870 | 1,028 | 1,364 |
| | .020 | — | — | 95 | 193 | 317 | 471 | 654 | 1,093 | — | — | 499 | 694 | 875 | 1,059 | 1,245 | 1,637 |
| | .025 | — | — | — | — | 378 | 563 | 782 | 1,306 | — | — | — | — | 1,026 | 1,238 | 1,451 | 1,897 |
| | .030 | — | — | — | — | — | — | — | 1,511 | — | — | — | — | — | — | — | 2,147 |

**TABLE 17.4–2 Torque and Thrust Values for Drilling with Split or Crankshaft Point Drills**

| MATERIAL AND HARDNESS | FEED PER REV. | TORQUE, inch-pounds | | | | | | | | THRUST, pounds | | | | | | | |
|---|---|---|---|---|---|---|---|---|---|---|---|---|---|---|---|---|---|
| | | Drill Diameter, in | | | | | | | | Drill Diameter, in | | | | | | | |
| | in | ⅛ | ¼ | ½ | ¾ | 1 | 1¼ | 1½ | 2 | ⅛ | ¼ | ½ | ¾ | 1 | 1¼ | 1½ | 2 |
| Steels 200 Bhn | 0.0005 | 1.3 | — | — | — | — | — | — | — | 23 | — | — | — | — | — | — | — |
| | .001 | 2.3 | 7.9 | — | — | — | — | — | — | 40 | 71 | — | — | — | — | — | — |
| | .002 | 3.9 | 14 | 48 | — | — | — | — | — | 70 | 122 | 216 | — | — | — | — | — |
| | .003 | 5.4 | 19 | 66 | — | — | — | — | — | 96 | 169 | 297 | — | — | — | — | — |
| | .004 | 6.9 | 24 | 83 | 173 | — | — | — | — | 121 | 212 | 372 | 519 | — | — | — | — |
| | .006 | 9.5 | 33 | 115 | 239 | 401 | 599 | 831 | 1,395 | 167 | 292 | 512 | 714 | 905 | 1,091 | 1,273 | 1,630 |
| | .008 | — | 42 | 145 | 300 | 504 | 754 | 1,046 | 1,756 | — | 368 | 643 | 895 | 1,133 | 1,364 | 1,588 | 2,028 |
| | .010 | — | 50 | 173 | 359 | 603 | 901 | 1,251 | 2,099 | — | 439 | 768 | 1,067 | 1,350 | 1,623 | 1,888 | 2,405 |
| | .012 | — | — | 200 | 416 | 698 | 1,042 | 1,447 | 2,429 | — | — | 887 | 1,233 | 1,558 | 1,872 | 2,177 | 2,768 |
| | .015 | — | — | 239 | 497 | 834 | 1,246 | 1,730 | 2,904 | — | — | 1,060 | 1,471 | 1,858 | 2,230 | 2,591 | 3,290 |
| | .020 | — | — | 301 | 625 | 1,050 | 1,569 | 2,178 | 3,655 | — | — | 1,332 | 1,848 | 2,333 | 2,798 | 3,248 | 4,116 |
| | .025 | — | — | — | — | 1,255 | 1,875 | 2,603 | 4,369 | — | — | — | — | 2,784 | 3,337 | 3,872 | 4,902 |
| | .030 | — | — | — | — | — | — | — | 5,055 | — | — | — | — | — | — | — | 5,657 |
| Steels 300 Bhn | 0.0005 | 1.7 | — | — | — | — | — | — | — | 30 | — | — | — | — | — | — | — |
| | .001 | 2.9 | 10 | — | — | — | — | — | — | 52 | 91 | — | — | — | — | — | — |
| | .002 | 5.1 | 18 | 62 | — | — | — | — | — | 90 | 158 | 279 | — | — | — | — | — |
| | .003 | 7 | 24 | 85 | — | — | — | — | — | 124 | 218 | 383 | — | — | — | — | — |
| | .004 | 9 | 31 | 107 | 223 | — | — | — | — | 156 | 273 | 480 | 671 | — | — | — | — |
| | .006 | 12 | 43 | 149 | 308 | 517 | 773 | 1,074 | 1,802 | 216 | 378 | 662 | 922 | 1,169 | 1,409 | 1,644 | 2,106 |
| | .008 | — | 54 | 187 | 388 | 651 | 973 | 1,351 | 2,268 | — | 475 | 831 | 1,156 | 1,464 | 1,762 | 2,052 | 2,619 |
| | .010 | — | 64 | 224 | 464 | 779 | 1,164 | 1,615 | 2,711 | — | 567 | 992 | 1,378 | 1,744 | 2,096 | 2,439 | 2,107 |
| | .012 | — | — | 259 | 537 | 901 | 1,346 | 1,869 | 3,137 | — | — | 1,146 | 1,592 | 2,013 | 2,418 | 2,811 | 3,575 |
| | .015 | — | — | 309 | 642 | 1,077 | 1,609 | 2,235 | 3,751 | — | — | 1,369 | 1,900 | 2,400 | 2,881 | 3,347 | 4,249 |
| | .020 | — | — | 389 | 808 | 1,356 | 2,026 | 2,813 | 4,721 | — | — | 1,720 | 2,387 | 3,014 | 3,614 | 4,195 | 5,317 |
| | .025 | — | — | — | — | 1,621 | 2,422 | 3,363 | 5,644 | — | — | — | — | 3,597 | 4,311 | 5,002 | 6,332 |
| | .030 | — | — | — | — | — | — | — | 6,530 | — | — | — | — | — | — | — | 7,307 |
| Steels 400 Bhn | .0005 | 1.8 | — | — | — | — | — | — | — | 33 | — | — | — | — | — | — | — |
| | .001 | 3.2 | 11 | — | — | — | — | — | — | 57 | 100 | — | — | — | — | — | — |
| | .002 | 5.6 | 19 | 68 | — | — | — | — | — | 99 | 173 | 306 | — | — | — | — | — |
| | .003 | 7.7 | 27 | 94 | — | — | — | — | — | 136 | 239 | 420 | — | — | — | — | — |
| | .004 | 9.7 | 34 | 118 | 244 | — | — | — | — | 172 | 300 | 527 | 736 | — | — | — | — |
| | .006 | 13 | 47 | 163 | 338 | 568 | 848 | 1,178 | 1,976 | 237 | 414 | 726 | 1,011 | 1,283 | 1,546 | 1,803 | 2,310 |
| | .008 | — | 59 | 205 | 426 | 714 | 1,068 | 1,482 | 2,488 | — | 521 | 911 | 1,268 | 1,606 | 1,932 | 2,250 | 2,873 |
| | .010 | — | 70 | 245 | 509 | 854 | 1,276 | 1,772 | 2,974 | — | 622 | 1,088 | 1,512 | 1,913 | 2,299 | 2,675 | 3,407 |
| | .012 | — | — | 284 | 589 | 988 | 1,477 | 2,050 | 3,441 | — | — | 1,257 | 1,746 | 2,208 | 2,652 | 3,083 | 3,921 |
| | .015 | — | — | 339 | 704 | 1,181 | 1,765 | 2,451 | 4,113 | — | — | 1,501 | 2,084 | 2,633 | 3,160 | 3,671 | 4,661 |
| | .020 | — | — | 427 | 886 | 1,487 | 2,222 | 3,085 | 5,178 | — | — | 1,887 | 2,618 | 3,305 | 3,964 | 4,601 | 5,832 |
| | .025 | — | — | — | — | 1,778 | 2,656 | 3,688 | 6,190 | — | — | — | — | 3,945 | 4,728 | 5,486 | 6,945 |
| | .030 | — | — | — | — | — | — | — | 7,162 | — | — | — | — | — | — | — | 8,014 |
| Aluminum Alloys | .0005 | 0.4 | — | — | — | — | — | — | — | 6.8 | — | — | — | — | — | — | — |
| | .001 | 0.7 | 2.3 | — | — | — | — | — | — | 12 | 21 | — | — | — | — | — | — |
| | .002 | 1.1 | 4 | 14 | — | — | — | — | — | 20 | 36 | 63 | — | — | — | — | — |
| | .003 | 1.6 | 5.5 | 19 | — | — | — | — | — | 28 | 49 | 87 | — | — | — | — | — |
| | .004 | 2 | 7 | 24 | 50 | — | — | — | — | 35 | 62 | 108 | 152 | — | — | — | — |
| | .006 | 2.8 | 9.6 | 34 | 70 | 117 | 175 | 242 | 407 | 49 | 85 | 149 | 208 | 264 | 318 | 371 | 476 |
| | .008 | — | 12 | 42 | 88 | 147 | 220 | 305 | 512 | — | 107 | 188 | 261 | 331 | 398 | 463 | 591 |
| | .010 | — | 14 | 50 | 105 | 176 | 263 | 365 | 612 | — | 128 | 224 | 311 | 394 | 473 | 551 | 701 |
| | .012 | — | — | 58 | 121 | 203 | 304 | 422 | 708 | — | — | 259 | 359 | 455 | 546 | 635 | 807 |
| | .015 | — | — | 70 | 145 | 243 | 363 | 505 | 847 | — | — | 309 | 429 | 542 | 651 | 756 | 960 |
| | .020 | — | — | 88 | 182 | 306 | 457 | 635 | 1,066 | — | — | 389 | 539 | 681 | 816 | 947 | 1,201 |
| | .025 | — | — | — | — | 366 | 547 | 759 | 1,274 | — | — | — | — | 812 | 973 | 1,129 | 1,430 |
| | .030 | — | — | — | — | — | — | — | 1,475 | — | — | — | — | — | — | — | 1,650 |

**TABLE 17.4-2—Continued**

| MATERIAL AND HARDNESS | FEED PER REV. | TORQUE, inch-pounds | | | | | | | | THRUST, pounds | | | | | | | |
|---|---|---|---|---|---|---|---|---|---|---|---|---|---|---|---|---|---|
| | | Drill Diameter, in | | | | | | | | Drill Diameter, in | | | | | | | |
| | in | ⅛ | ¼ | ½ | ¾ | 1 | 1¼ | 1½ | 2 | ⅛ | ¼ | ½ | ¾ | 1 | 1¼ | 1½ | 2 |
| Magnesium Alloys | .0005 | 0.2 | — | — | — | — | — | — | — | 3.9 | — | — | — | — | — | — | — |
| | .001 | 0.4 | 1.3 | — | — | — | — | — | — | 6.7 | 12 | — | — | — | — | — | — |
| | .002 | 0.7 | 2.3 | 8 | — | — | — | — | — | 12 | 20 | 36 | — | — | — | — | — |
| | .003 | 0.9 | 3.2 | 11 | — | — | — | — | — | 16 | 28 | 49 | — | — | — | — | — |
| | .004 | 1.1 | 4 | 14 | 29 | — | — | — | — | 20 | 35 | 62 | 87 | — | — | — | — |
| | .006 | 1.6 | 5.5 | 19 | 40 | 67 | 100 | 138 | 232 | 28 | 49 | 85 | 119 | 151 | 182 | 212 | 272 |
| | .008 | — | 6.9 | 24 | 50 | 84 | 126 | 174 | 293 | — | 61 | 107 | 149 | 189 | 227 | 265 | 338 |
| | .010 | — | 8.3 | 29 | 60 | 100 | 150 | 208 | 350 | — | 73 | 128 | 178 | 225 | 270 | 315 | 401 |
| | .012 | — | — | 33 | 69 | 116 | 174 | 241 | 405 | — | — | 148 | 205 | 260 | 312 | 363 | 461 |
| | .015 | — | — | 40 | 83 | 139 | 208 | 288 | 484 | — | — | 177 | 245 | 310 | 372 | 432 | 548 |
| | .020 | — | — | 50 | 104 | 175 | 261 | 363 | 609 | — | — | 222 | 308 | 389 | 466 | 541 | 686 |
| | .025 | — | — | — | — | 209 | 312 | 434 | 728 | — | — | — | — | 464 | 556 | 645 | 817 |
| | .030 | — | — | — | — | — | — | — | 843 | — | — | — | — | — | — | — | 943 |
| Copper Alloys | .0005 | 0.8 | — | — | — | — | — | — | — | 14 | — | — | — | — | — | — | — |
| | .001 | 1.3 | 4.6 | — | — | — | — | — | — | 23 | 41 | — | — | — | — | — | |
| | .002 | 2.3 | 8 | 28 | — | — | — | — | — | 41 | 71 | 126 | — | — | — | — | — |
| | .003 | 3.2 | 11 | 39 | — | — | — | — | — | 56 | 98 | 173 | — | — | — | — | — |
| | .004 | 4 | 14 | 49 | 101 | — | — | — | — | 71 | 123 | 217 | 303 | — | — | — | — |
| | .006 | 5.5 | 19 | 67 | 139 | 234 | 349 | 485 | 814 | 98 | 171 | 299 | 416 | 528 | 637 | 743 | 951 |
| | .008 | — | 24 | 84 | 175 | 294 | 440 | 610 | 1,024 | — | 214 | 275 | 522 | 661 | 796 | 927 | 1,183 |
| | .010 | — | 29 | 101 | 209 | 352 | 525 | 730 | 1,225 | — | 256 | 448 | 622 | 788 | 947 | 1,102 | 1,403 |
| | .012 | — | — | 117 | 242 | 407 | 608 | 844 | 1,417 | — | — | 518 | 719 | 909 | 1,093 | 1,270 | 1,614 |
| | .015 | — | — | 140 | 290 | 486 | 727 | 1,009 | 1,694 | — | — | 618 | 858 | 1,089 | 1,301 | 1,512 | 1,919 |
| | .020 | — | — | 176 | 365 | 612 | 915 | 1,270 | 2,132 | — | — | 777 | 1,078 | 1,361 | 1,632 | 1,895 | 2,401 |
| | ,025 | — | — | — | — | 732 | 1,094 | 1,519 | 2,549 | — | — | — | — | 1,624 | 1,947 | 2,259 | 2,860 |
| | .030 | — | — | — | — | — | — | — | 2,949 | — | — | — | — | — | — | — | 3,330 |
| Leaded Brass | .0005 | 0.4 | — | — | — | — | — | — | — | 6.8 | — | — | — | — | — | — | — |
| | .001 | 0.7 | 2.3 | — | — | — | — | — | — | 12 | 21 | — | — | — | — | — | — |
| | .002 | 1.1 | 4 | 14 | — | — | — | — | — | 20 | 36 | 63 | — | — | — | — | — |
| | .003 | 1.6 | 5 | 19 | — | — | — | — | — | 28 | 49 | 87 | — | — | — | — | — |
| | .004 | 2 | 7 | 24 | 50 | — | — | — | — | 35 | 62 | 108 | 152 | — | — | — | — |
| | .006 | 2.8 | 9.6 | 34 | 70 | 117 | 175 | 242 | 407 | 50 | 85 | 149 | 208 | 264 | 318 | 371 | 476 |
| | .008 | — | 12 | 42 | 88 | 147 | 220 | 305 | 512 | — | 107 | 188 | 261 | 331 | 398 | 463 | 591 |
| | .010 | — | 14 | 50 | 105 | 176 | 263 | 365 | 612 | — | 128 | 224 | 311 | 394 | 473 | 551 | 701 |
| | .012 | — | — | 58 | 121 | 203 | 304 | 422 | 708 | — | — | 259 | 359 | 455 | 546 | 635 | 807 |
| | .015 | — | — | 70 | 145 | 243 | 363 | 505 | 847 | — | — | 309 | 429 | 542 | 651 | 756 | 960 |
| | .020 | — | — | 88 | 182 | 306 | 457 | 635 | 1,066 | — | — | 389 | 539 | 681 | 816 | 947 | 1,201 |
| | .025 | — | — | — | — | 366 | 547 | 759 | 1,274 | — | — | — | — | 812 | 973 | 1,129 | 1,430 |
| | .030 | — | — | — | — | — | — | — | 1,475 | — | — | — | — | — | — | — | 1,650 |

**TABLE 17.4–3 Torque and Thrust Values for Drilling with Metric Standard Point Drills**

| MATERIAL AND HARDNESS | FEED PER REV. | TORQUE, newton-meters | | | | | | | | THRUST, newtons | | | | | | | |
|---|---|---|---|---|---|---|---|---|---|---|---|---|---|---|---|---|---|
| | | Drill Diameter, mm | | | | | | | | Drill Diameter, mm | | | | | | | |
| | mm | 3 | 6 | 12 | 18 | 25 | 30 | 35 | 50 | 3 | 6 | 12 | 18 | 25 | 30 | 35 | 50 |
| Steels 200 Bhn | .013 | 0.16 | – | – | – | – | – | – | – | 196 | – | – | – | – | – | – | – |
| | .025 | 0.28 | 0.96 | – | – | – | – | – | – | 289 | 578 | – | – | – | – | – | – |
| | .050 | 0.50 | 1.70 | 5.88 | – | – | – | – | – | 454 | 863 | 1,700 | – | – | – | – | – |
| | .075 | 0.68 | 2.37 | 8.02 | – | – | – | – | – | 602 | 1,139 | 2,126 | – | – | – | – | – |
| | .102 | 0.90 | 2.94 | 10.17 | 20.68 | – | – | – | – | 7,406 | 1,352 | 2,527 | 3,700 | – | – | – | – |
| | .15 | 1.24 | 4.07 | 14.00 | 28.59 | 47 | 70 | 97 | 162 | 1,000 | 1,793 | 3,270 | 4,706 | 6,085 | 7,504 | 9,030 | 12,437 |
| | .20 | – | 5.08 | 17.6 | 36 | 59 | 88 | 122 | 203 | – | 2,206 | 3,963 | 5,636 | 7,246 | 8,883 | 10,622 | 14,439 |
| | .25 | – | 6.10 | 21 | 43 | 70 | 105 | 146 | 243 | – | 2,598 | 4,622 | 6,534 | 8,345 | 10,195 | 12,135 | 16,343 |
| | .36 | – | – | 24 | 50 | 81 | 121 | 168 | 281 | – | – | 5,258 | 7,393 | 9,408 | 11,454 | 13,585 | 18,171 |
| | .40 | – | – | 29 | 60 | 97 | 145 | 201 | 336 | – | – | 6,170 | 8,625 | 10,934 | 13,271 | 15,676 | 20,804 |
| | .50 | – | – | 37 | 75 | 123 | 182 | 253 | 423 | – | – | 7,613 | 10,578 | 13,349 | 16,143 | 18,989 | 24,972 |
| | .65 | – | – | – | – | 147 | 218 | 303 | 506 | – | – | – | – | 15,644 | 18,876 | 21,692 | 28,936 |
| | .75 | – | – | – | – | – | – | – | 653 | – | – | – | – | – | – | – | 32,739 |
| Steels 300 Bhn | .013 | 0.21 | – | – | – | – | – | – | – | 249 | – | – | – | – | – | – | – |
| | .025 | 0.36 | 1.24 | – | – | – | – | – | – | 374 | 750 | – | – | – | – | – | – |
| | .050 | 0.63 | 2.15 | 7.57 | – | – | – | – | – | 587 | 1,117 | 2,197 | – | – | – | – | – |
| | .075 | 0.88 | 3.05 | 10.4 | – | – | – | – | – | 778 | 1,441 | 2,749 | – | – | – | – | – |
| | .102 | 1.01 | 3.73 | 13.1 | 26.7 | – | – | – | – | 946 | 1,303 | 3,263 | 4,782 | – | – | – | – |
| | .15 | 1.53 | 5.20 | 18.1 | 36.9 | 60.4 | 89.9 | 125 | 208 | 1,290 | 2,318 | 4,226 | 6,081 | 7,860 | 9,693 | 11,663 | 16,063 |
| | .20 | – | 6.55 | 22.8 | 46.4 | 76.0 | 113 | 157 | 263 | – | 2,847 | 5,120 | 7,291 | 9,359 | 11,476 | 13,718 | 18,651 |
| | .25 | – | 7.90 | 27.2 | 55.6 | 91 | 135 | 188 | 314 | – | 3,354 | 5,970 | 8,443 | 10,782 | 13,171 | 15,671 | 21,107 |
| | .36 | – | – | 31.5 | 64.3 | 105 | 157 | 217 | 363 | – | – | 6,788 | 9,546 | 12,153 | 14,800 | 17,548 | 23,469 |
| | .40 | – | – | 37.7 | 76.8 | 126 | 187 | 299 | 434 | – | – | 7,967 | 11,138 | 14,123 | 17,143 | 20,248 | 26,872 |
| | .50 | – | – | 47.5 | 96.7 | 158 | 236 | 327 | 547 | – | – | 9,835 | 13,660 | 17,241 | 20,853 | 24,527 | 32,254 |
| | .65 | – | – | – | – | 189 | 282 | 391 | 654 | – | – | – | – | 20,208 | 24,380 | 28,593 | 37,374 |
| | .75 | – | – | – | – | – | – | – | 756 | – | – | – | – | – | – | – | 42,289 |
| Steels 400 Bhn | .013 | 0.23 | – | – | – | – | – | – | – | 276 | – | – | – | – | – | – | – |
| | .025 | 0.40 | 1.36 | – | – | – | – | – | – | 409 | 823 | – | – | – | – | – | – |
| | .050 | 0.70 | 2.37 | 8.24 | – | – | – | – | – | 643 | 1,219 | 2,411 | – | – | – | – | – |
| | .075 | 0.96 | 3.28 | 11.4 | – | – | – | – | – | 854 | 1,579 | 3,016 | – | – | – | – | – |
| | .102 | 1.24 | 4.18 | 14.4 | 29.3 | – | – | – | – | 1,050 | 1,918 | 3,581 | 5,245 | – | – | – | – |
| | .15 | 1.69 | 5.76 | 19.9 | 40.4 | 66.2 | 98.7 | 137 | 229 | 1,415 | 2,540 | 4,630 | 6,668 | 8,630 | 10,631 | 12,793 | 17,619 |
| | .20 | –– | 7.23 | 25.1 | 50.9 | 83.5 | 124 | 173 | 288 | – | 3,123 | 5,614 | 7,993 | 10,262 | 12,584 | 15,048 | 20,453 |
| | .25 | – | 8.58 | 29.9 | 60.9 | 99.8 | 148 | 206 | 344 | – | 3,679 | 6,548 | 9,257 | 11,828 | 14,443 | 17,188 | 23,149 |
| | .36 | – | – | 34.6 | 70.5 | 115 | 172 | 239 | 398 | – | – | 7,446 | 10,471 | 13,327 | 16,227 | 19,247 | 25,742 |
| | .40 | – | – | 41.3 | 84.2 | 138 | 205 | 285 | 476 | – | – | 8,741 | 12,219 | 15,489 | 18,803 | 22,210 | 29,469 |
| | .50 | – | – | 52.0 | 106 | 174 | 259 | 359 | 600 | – | – | 10,787 | 14,982 | 18,910 | 22,873 | 26,898 | 35,377 |
| | .65 | – | – | – | – | 208 | 309 | 429 | 717 | – | – | – | – | 22,161 | 26,743 | 31,360 | 40,990 |
| | .75 | – | – | – | – | – | – | – | 829 | – | – | – | – | – | – | – | 46,395 |
| Aluminum Alloys | .013 | 0.05 | – | – | – | – | – | – | – | 57.8 | – | – | – | – | – | – | – |
| | .025 | 0.08 | 0.28 | – | – | – | – | – | – | 84.5 | 169 | – | – | – | – | – | – |
| | .050 | 0.15 | 0.49 | 1.69 | – | – | – | – | – | 133 | 251 | 494 | – | – | – | – | – |
| | .075 | 0.20 | 0.68 | 2.37 | – | – | – | – | – | 173 | 325 | 618 | – | – | – | – | – |
| | .102 | 0.25 | 0.86 | 2.94 | 5.99 | – | – | – | – | 218 | 396 | 738 | 1,080 | – | – | – | – |
| | .15 | 0.35 | 1.19 | 4.07 | 8.36 | 13.7 | 20.3 | 28.2 | 47.0 | 294 | 525 | 952 | 1,370 | 1,775 | 2,188 | 2,633 | 3,625 |
| | .20 | – | 1.47 | 5.20 | 10.5 | 17.2 | 25.5 | 35.5 | 59.3 | – | 645 | 1,157 | 1,646 | 2,113 | 2,593 | 3,096 | 4,212 |
| | .25 | – | 1.81 | 6.21 | 12.5 | 20.6 | 30.6 | 42.5 | 71 | – | 756 | 1,348 | 1,904 | 2,433 | 2,976 | 3,541 | 4,768 |
| | .36 | – | – | 7.11 | 14.5 | 23.7 | 35.4 | 49.1 | 82 | – | – | 1,535 | 2,157 | 2,745 | 3,341 | 3,963 | 5,298 |
| | .40 | – | – | 8.47 | 17.4 | 28.3 | 42.2 | 58.8 | 98 | – | – | 1,797 | 2,518 | 3,189 | 3,870 | 4,573 | 6,097 |
| | .50 | – | – | 10.7 | 21.8 | 35.8 | 53.2 | 73.9 | 124 | – | – | 2,220 | 3,087 | 3,892 | 4,710 | 5,538 | 7,282 |
| | .65 | – | – | – | – | 42.7 | 63.6 | 88.4 | 148 | – | – | – | – | 4,564 | 5,507 | 6,454 | 8,438 |
| | .75 | – | – | – | – | – | – | – | 171 | – | – | – | – | – | – | – | 9,550 |

## 17.4 Estimating Torque and Thrust in Drilling

**TABLE 17.4-3—Continued**

| MATERIAL AND HARDNESS | FEED PER REV. | TORQUE, newton-meters | | | | | | | | THRUST, newtons | | | | | | | |
|---|---|---|---|---|---|---|---|---|---|---|---|---|---|---|---|---|---|
| | | Drill Diameter, mm | | | | | | | | Drill Diameter, mm | | | | | | | |
| | mm | 3 | 6 | 12 | 18 | 25 | 30 | 35 | 50 | 3 | 6 | 12 | 18 | 25 | 30 | 35 | 50 |
| Magnesium Alloys | .013 | 0.023 | — | — | — | — | — | — | — | 32 | — | — | — | — | — | — | — |
| | .025 | 0.045 | 0.158 | — | — | — | — | — | — | 49 | 98 | — | — | — | — | — | — |
| | .050 | 0.079 | 0.282 | 0.971 | — | — | — | — | — | 75 | 142 | 285 | — | — | — | — | — |
| | .075 | 0.113 | 0.384 | 1.34 | — | — | — | — | — | 102 | 187 | 356 | — | — | — | — | — |
| | .102 | 0.147 | 0.486 | 1.69 | 3.45 | — | — | — | — | 125 | 227 | 423 | 618 | — | — | — | — |
| | .15 | 0.192 | 0.678 | 2.37 | 4.75 | 7.80 | 11.64 | 16.16 | 26.9 | 147 | 262 | 485 | 703 | 912 | 1,130 | 1,366 | 1,900 |
| | .20 | — | 0.874 | 2.94 | 5.99 | 9.83 | 14.6 | 20.3 | 33.9 | — | 369 | 658 | 939 | 1,210 | 1,481 | 1,770 | 2,406 |
| | .25 | — | 1.02 | 3.50 | 7.23 | 11.8 | 17.5 | 24.3 | 40 | — | 431 | 770 | 1,090 | 1,392 | 1,700 | 2,024 | 2,722 |
| | .36 | — | — | 4.07 | 8.25 | 13.6 | 20.2 | 28.2 | 46.9 | — | — | 876 | 1,232 | 1,568 | 1,908 | 2,264 | 3,029 |
| | .40 | — | — | 4.86 | 10.1 | 16.3 | 24.2 | 33.6 | 56 | — | — | 1,028 | 1,437 | 1,824 | 2,210 | 2,610 | 3,470 |
| | .50 | — | — | 6.10 | 12.4 | 20.5 | 30.4 | 42.3 | 70.5 | — | — | 1,268 | 1,761 | 2,224 | 2,691 | 3,163 | 4,164 |
| | .65 | — | — | — | — | 24.4 | 36.4 | 50.5 | 84.3 | — | — | — | — | 2,607 | 3,145 | 3,688 | 4,822 |
| | .75 | — | — | — | — | — | — | — | 97.6 | — | — | — | — | — | — | — | 5,458 |
| Copper Alloys | .013 | 0.090 | — | — | — | — | — | — | — | 113 | — | — | — | — | — | — | — |
| | .025 | 0.169 | 0.565 | — | — | — | — | — | — | 169 | 338 | — | — | — | — | — | — |
| | .050 | 0.282 | 0.983 | 3.39 | — | — | — | — | — | 267 | 503 | 992 | — | — | — | — | — |
| | .075 | 0.395 | 1.36 | 4.75 | — | — | — | — | — | 351 | 649 | 1,241 | — | — | — | — | — |
| | .102 | 0.497 | 1.71 | 5.88 | 12.1 | — | — | — | — | 431 | 787 | 1,472 | 2,157 | — | — | — | — |
| | .15 | 0.689 | 2.37 | 8.25 | 16.7 | 27.3 | 40.6 | 56.4 | 94.2 | 583 | 1,045 | 1,908 | 2,745 | 3,550 | 4,373 | 5,267 | 7,250 |
| | .20 | — | 2.94 | 10.3 | 21 | 34.3 | 51.2 | 71.1 | 119 | — | 1,286 | 2,313 | 3,292 | 4,226 | 5,182 | 6,196 | 8,420 |
| | .25 | — | 3.50 | 12.3 | 25 | 41 | 61.1 | 84.9 | 142 | — | 1,517 | 2,696 | 3,812 | 4,870 | 5,947 | 7,077 | 9,533 |
| | .36 | — | — | 14.2 | 29 | 47.6 | 70.7 | 98.3 | 164 | — | — | 3,065 | 4,310 | 5,489 | 6,681 | 7,927 | 10,600 |
| | .40 | — | — | 17.1 | 34.7 | 56.8 | 84.6 | 117 | 196 | — | — | 3,600 | 5,030 | 6,379 | 7,740 | 9,146 | 12,135 |
| | .50 | — | — | 21.5 | 43.7 | 71.5 | 106 | 148 | 247 | — | — | 4,439 | 6,170 | 7,784 | 9,417 | 11,076 | 14,568 |
| | .65 | — | — | — | — | 85.5 | 127 | 177 | 295 | — | — | — | — | 9,128 | 11,009 | 12,913 | 16,881 |
| | .75 | — | — | — | — | — | — | — | 342 | — | — | — | — | — | — | — | 19,096 |
| Leaded Brass | .013 | 0.045 | — | — | — | — | — | — | — | 56.5 | — | — | — | — | — | — | — |
| | .025 | 0.079 | 0.282 | — | — | — | — | — | — | 84.5 | 169 | — | — | — | — | — | — |
| | .050 | 0.147 | 0.486 | 1.69 | — | — | — | — | — | 133 | 251 | 494 | — | — | — | — | — |
| | .075 | 0.203 | 0.678 | 2.37 | — | — | — | — | — | 173 | 325 | 618 | — | — | — | — | — |
| | .102 | 0.249 | 0.859 | 2.94 | 5.99 | — | — | — | — | 218 | 396 | 738 | 1,081 | — | — | — | — |
| | .15 | 0.350 | 1.19 | 4.07 | 8.36 | 13.7 | 20.3 | 28.2 | 47 | 294 | 525 | 952 | 1,370 | 1,775 | 2,188 | 2,633 | 3,625 |
| | .20 | — | 1.47 | 5.20 | 10.5 | 17.2 | 25.5 | 35.5 | 59.3 | — | 645 | 1,152 | 1,646 | 2,113 | 2,593 | 3,096 | 4,212 |
| | .25 | — | 1.81 | 6.21 | 12.5 | 20.6 | 30.6 | 42.5 | 70.9 | — | 756 | 1,348 | 1,904 | 2,433 | 2,976 | 3,541 | 4,768 |
| | .36 | — | — | 7.11 | 14.5 | 23.7 | 35.4 | 49 | 82 | — | — | 1,530 | 2,157 | 2,740 | 3,340 | 3,963 | 5,298 |
| | .40 | — | — | 8.47 | 17.4 | 28.5 | 42.3 | 58.8 | 98 | — | — | 1,797 | 2,518 | 3,189 | 3,870 | 4,573 | 6,067 |
| | .50 | — | — | 10.7 | 21.8 | 35.8 | 53.2 | 73.9 | 123 | — | — | 2,220 | 3,087 | 3,892 | 4,710 | 5,538 | 7,282 |
| | .65 | — | — | — | — | 42.7 | 63.6 | 88.3 | 147 | — | — | — | — | 4,564 | 5,507 | 6,454 | 8,438 |
| | .75 | — | — | — | — | — | — | — | 171 | — | — | — | — | — | — | — | 9,550 |

**TABLE 17.4-4 Torque and Thrust Values for Drilling with Metric Split or Crankshaft Point Drills**

| MATERIAL AND HARDNESS | FEED PER REV. | TORQUE, newton-meters | | | | | | | | THRUST, newtons | | | | | | | |
|---|---|---|---|---|---|---|---|---|---|---|---|---|---|---|---|---|---|
| | | Drill Diameter, mm | | | | | | | | Drill Diameter, mm | | | | | | | |
| | mm | 3 | 6 | 12 | 18 | 25 | 30 | 35 | 50 | 3 | 6 | 12 | 18 | 25 | 30 | 35 | 50 |
| Steels 200 Bhn | .013 | 0.15 | — | — | — | — | — | — | — | 102 | — | — | — | — | — | — | — |
| | .025 | 0.26 | 0.89 | — | — | — | — | — | — | 178 | 316 | — | — | — | — | — | — |
| | .050 | 0.44 | 1.58 | 5.42 | — | — | — | — | — | 311 | 543 | 961 | — | — | — | — | — |
| | .075 | 0.61 | 2.15 | 7.46 | — | — | — | — | — | 427 | 752 | 1,321 | — | — | — | — | — |
| | .102 | 0.78 | 2.71 | 9.38 | 20 | — | — | — | — | 538 | 943 | 1,655 | 2,309 | — | — | — | — |
| | .15 | 1.07 | 3.73 | 13 | 27 | 45 | 68 | 94 | 158 | 743 | 1,300 | 2,277 | 3,176 | 4,026 | 4,853 | 5,663 | 7,250 |
| | .20 | — | 4.75 | 16.4 | 34 | 57 | 85 | 118 | 198 | — | 1,637 | 2,860 | 3,981 | 5,040 | 6,067 | 7,064 | 9,021 |
| | .25 | — | 5.64 | 19.5 | 41 | 68 | 102 | 141 | 237 | — | 1,953 | 3,416 | 4,746 | 6,005 | 7,219 | 8,398 | 10,698 |
| | .36 | — | — | 22.5 | 47 | 79 | 118 | 163 | 274 | — | — | 3,946 | 5,485 | 6,930 | 8,327 | 9,684 | 12,313 |
| | .40 | — | — | 27 | 56 | 94 | 141 | 195 | 328 | — | — | 4,715 | 6,543 | 8,265 | 9,920 | 11,525 | 14,635 |
| | .50 | — | — | 34 | 71 | 119 | 177 | 246 | 413 | — | — | 5,925 | 8,220 | 10,378 | 12,446 | 14,448 | 18,309 |
| | .65 | — | — | — | — | 142 | 212 | 294 | 494 | — | — | — | — | 12,384 | 14,844 | 17,224 | 21,805 |
| | .75 | — | — | — | — | — | — | — | 571 | — | — | — | — | — | — | — | 25,164 |
| Steels 300 Bhn | .013 | 0.19 | — | — | — | — | — | — | — | 133 | — | — | — | — | — | — | — |
| | .025 | 0.33 | 1.13 | — | — | — | — | — | — | 231 | 405 | — | — | — | — | — | — |
| | .050 | 0.58 | 2.03 | 7 | — | — | — | — | — | 400 | 703 | 1,241 | — | — | — | — | — |
| | .075 | 0.79 | 2.71 | 9.6 | — | — | — | — | — | 552 | 970 | 1,704 | — | — | — | — | — |
| | .102 | 1.01 | 3.50 | 12.1 | 25.2 | — | — | — | — | 694 | 1,214 | 2,135 | 2,985 | — | — | — | — |
| | .15 | 1.36 | 4.86 | 16.8 | 34.8 | 58.4 | 87.3 | 121 | 204 | 961 | 1,681 | 2,945 | 4,100 | 5,200 | 6,268 | 7,313 | 9,368 |
| | .20 | — | 6.10 | 21.1 | 43.8 | 73.6 | 110 | 153 | 256 | — | 2,113 | 3,696 | 5,142 | 6,512 | 7,838 | 9,128 | 11,650 |
| | .25 | — | 7.23 | 25.3 | 52.4 | 88 | 132 | 182 | 306 | — | 2,522 | 4,413 | 6,130 | 7,758 | 9,323 | 10,849 | 13,820 |
| | .36 | — | — | 29.2 | 60.6 | 102 | 152 | 211 | 354 | — | — | 5,098 | 7,082 | 8,954 | 10,756 | 12,504 | 15,902 |
| | .40 | — | — | 34.9 | 72.5 | 122 | 182 | 253 | 424 | — | — | 6,090 | 8,452 | 10,676 | 12,815 | 14,888 | 18,900 |
| | .50 | — | — | 44 | 91.3 | 153 | 229 | 318 | 533 | — | — | 7,650 | 10,618 | 13,407 | 16,076 | 18,660 | 23,651 |
| | .65 | — | — | — | — | 183 | 274 | 380 | 638 | — | — | — | — | 16,000 | 19,176 | 22,250 | 28,166 |
| | .75 | — | — | — | — | — | — | — | 738 | — | — | — | — | — | — | — | 32,503 |
| Steels 400 Bhn | .013 | 0.20 | — | — | — | — | — | — | — | 147 | — | — | — | — | — | — | — |
| | .025 | 0.36 | 1.24 | — | — | — | — | — | — | 254 | 445 | — | — | — | — | — | — |
| | .050 | 0.63 | 2.15 | 7.68 | — | — | — | — | — | 440 | 770 | 1,361 | — | — | — | — | — |
| | .075 | 0.87 | 3.05 | 10.6 | — | — | — | — | — | 605 | 1,063 | 1,868 | — | — | — | — | — |
| | .102 | 1.10 | 3.84 | 13.3 | 27.6 | — | — | — | — | 765 | 1,334 | 2,344 | 3,274 | — | — | — | — |
| | .15 | 1.47 | 5.31 | 18.4 | 38.2 | 64.2 | 95.8 | 133 | 223 | 1,054 | 1,842 | 3,229 | 4,497 | 5,707 | 6,877 | 8,020 | 10,275 |
| | .20 | — | 6.66 | 23.2 | 48.1 | 80.7 | 121 | 167 | 281 | — | 2,318 | 4,052 | 5,640 | 7,144 | 8,594 | 10,008 | 12,780 |
| | .25 | — | 7.91 | 27.7 | 57.5 | 96.5 | 144 | 200 | 336 | — | 2,767 | 4,840 | 6,726 | 8,510 | 10,226 | 11,900 | 15,155 |
| | .36 | — | — | 32.0 | 66.5 | 112 | 167 | 232 | 389 | — | — | 5,591 | 7,767 | 9,822 | 11,797 | 13,714 | 17,441 |
| | .40 | — | — | 38.3 | 79.5 | 133 | 199 | 277 | 465 | — | — | 6,677 | 9,270 | 11,712 | 14,056 | 16,329 | 20,733 |
| | .50 | — | — | 48.2 | 100 | 168 | 251 | 349 | 585 | — | — | 8,394 | 11,645 | 14,700 | 17,633 | 20,466 | 25,942 |
| | .65 | — | — | — | — | 200 | 300 | 417 | 700 | — | — | — | — | 17,548 | 21,031 | 24,403 | 30,893 |
| | .75 | — | — | — | — | — | — | — | 810 | — | — | — | — | — | — | — | 35,648 |
| Aluminum Alloys | .013 | 0.05 | — | — | — | — | — | — | — | 30.2 | — | — | — | — | — | — | — |
| | .025 | 0.08 | 0.26 | — | — | — | — | — | — | 53.4 | 93.4 | — | — | — | — | — | — |
| | .050 | 0.12 | 0.45 | 1.58 | — | — | — | — | — | 88.9 | 160 | 280 | — | — | — | — | — |
| | .075 | 0.18 | 0.62 | 2.15 | — | — | — | — | — | 125 | 218 | 387 | — | — | — | — | — |
| | .102 | 0.23 | 0.79 | 2.71 | 5.65 | — | — | — | — | 156 | 276 | 480 | 676 | — | — | — | — |
| | .15 | 0.36 | 1.08 | 3.84 | 7.91 | 13.2 | 19.8 | 27.4 | 46.0 | 218 | 378 | 663 | 925 | 1,174 | 1,415 | 1,650 | 2,117 |
| | .20 | — | 1.36 | 4.75 | 9.9 | 16.6 | 24.8 | 34.5 | 57.8 | — | 476 | 836 | 1,161 | 1,472 | 1,770 | 2,060 | 2,629 |
| | .25 | — | 1.58 | 5.65 | 11.9 | 19.9 | 29.7 | 41.2 | 69.1 | — | 569 | 996 | 1,383 | 1,753 | 2,104 | 2,451 | 3,118 |
| | .36 | — | — | 6.55 | 13.7 | 22.9 | 34.3 | 47.7 | 80.0 | — | — | 1,152 | 1,597 | 2,024 | 2,429 | 2,825 | 3,590 |
| | .40 | — | — | 7.90 | 16.4 | 27.5 | 41.0 | 57.1 | 95.7 | — | — | 1,375 | 1,908 | 2,410 | 2,896 | 3,363 | 4,270 |
| | .50 | — | — | 9.94 | 20.6 | 34.6 | 51.6 | 71.7 | 120 | — | — | 1,730 | 2,398 | 3,029 | 3,630 | 4,212 | 5,342 |
| | .65 | — | — | — | — | 41.4 | 61.8 | 85.8 | 144 | — | — | — | — | 3,612 | 4,337 | 5,022 | 6,360 |
| | .75 | — | — | — | — | — | — | — | 167 | — | — | — | — | — | — | — | 7,340 |

**TABLE 17.4-4—Continued**

| MATERIAL AND HARDNESS | FEED PER REV. | TORQUE, newton-meters | | | | | | | | THRUST, newtons | | | | | | | |
|---|---|---|---|---|---|---|---|---|---|---|---|---|---|---|---|---|---|
| | | Drill Diameter, mm | | | | | | | | Drill Diameter, mm | | | | | | | |
| | mm | 3 | 6 | 12 | 18 | 25 | 30 | 35 | 50 | 3 | 6 | 12 | 18 | 25 | 30 | 35 | 50 |
| Magnesium Alloys | .013 | 0.023 | — | — | — | — | — | — | — | 17.3 | — | — | — | — | — | — | — |
| | .025 | 0.045 | 0.147 | — | — | — | — | — | — | 29.8 | 53.4 | — | — | — | — | — | — |
| | .050 | 0.079 | 0.260 | 0.903 | — | — | — | — | — | 53.3 | 88.9 | 160 | — | — | — | — | — |
| | .075 | 0.102 | 0.362 | 1.24 | — | — | — | — | — | 71.1 | 125 | 218 | — | — | — | — | — |
| | .102 | 0.124 | 0.452 | 1.58 | 3.28 | — | — | — | — | 89 | 156 | 276 | 387 | — | — | — | — |
| | .15 | 0.181 | 0.621 | 2.15 | 4.52 | 7.57 | 11.3 | 15.6 | 26.2 | 125 | 218 | 378 | 529 | 672 | 809 | 943 | 1,210 |
| | .20 | — | 0.780 | 2.71 | 5.65 | 9.49 | 14.2 | 19.7 | 33.1 | — | 271 | 476 | 663 | 841 | 1,010 | 1,179 | 1,503 |
| | .25 | — | 0.938 | 3.28 | 6.78 | 11.3 | 16.9 | 22.6 | 39.5 | — | 325 | 569 | 792 | 1,000 | 1,200 | 1,400 | 1,784 |
| | .36 | — | — | 3.73 | 7.80 | 13.1 | 19.7 | 27.2 | 45.8 | — | — | 658 | 912 | 1,157 | 1,388 | 1,615 | 2,050 |
| | .40 | — | — | 4.52 | 9.38 | 15.7 | 23.5 | 32.5 | 54.7 | — | — | 787 | 1,090 | 1,379 | 1,655 | 1,922 | 2,438 |
| | .50 | — | — | 5.65 | 11.8 | 19.8 | 29.5 | 41 | 68.8 | — | — | 988 | 1,370 | 1,730 | 2,073 | 2,406 | 3,051 |
| | .65 | — | — | — | — | 23.6 | 35.3 | 49 | 82.3 | — | — | — | — | 2,064 | 2,473 | 2,869 | 3,634 |
| | .75 | — | — | — | — | — | — | — | 95.2 | — | — | — | — | — | — | — | 4,195 |
| Copper Alloys | .013 | 0.090 | — | — | — | — | — | — | — | 62 | — | — | — | — | — | — | — |
| | .025 | 0.147 | 0.520 | — | — | — | — | — | — | 102 | 182 | — | — | — | — | — | — |
| | .050 | 0.260 | 0.904 | 3.16 | — | — | — | — | — | 182 | 316 | 560 | — | — | — | — | — |
| | .075 | 0.362 | 1.24 | 4.41 | — | — | — | — | — | 249 | 436 | 770 | — | — | — | — | — |
| | .102 | 0.451 | 1.58 | 5.54 | 11.4 | — | — | — | — | 315 | 547 | 965 | 1,348 | — | — | — | — |
| | .15 | 0.621 | 2.15 | 7.57 | 15.7 | 26.4 | 39.4 | 54.8 | 92.0 | 436 | 761 | 1,330 | 1,850 | 2,349 | 2,834 | 3,305 | 4,230 |
| | .20 | — | 2.71 | 9.49 | 19.8 | 32.2 | 49.7 | 68.9 | 116 | — | 952 | 1,223 | 2,321 | 2,940 | 3,541 | 4,124 | 5,262 |
| | .25 | — | 3.28 | 11.4 | 23.6 | 39.8 | 59.3 | 82.5 | 138 | — | 1,139 | 1,993 | 2,767 | 3,505 | 4,212 | 4,902 | 6,240 |
| | .36 | — | — | 13.2 | 27.3 | 46 | 68.7 | 95.4 | 160 | — | — | 2,304 | 3,198 | 4,043 | 4,862 | 5,649 | 7,180 |
| | .40 | — | — | 15.8 | 32.8 | 54.9 | 82.1 | 114 | 191 | — | — | 2,749 | 3,817 | 4,844 | 5,787 | 6,726 | 8,536 |
| | .50 | — | — | 19.9 | 41.2 | 69.1 | 103 | 143 | 241 | — | — | 3,456 | 4,795 | 6,054 | 7,259 | 8,429 | 10,680 |
| | .65 | — | — | — | — | 82.7 | 124 | 172 | 288 | — | — | — | — | 7,224 | 8,638 | 10,049 | 12,722 |
| | .75 | — | — | — | — | — | — | — | 333 | — | — | — | — | — | — | — | 14,679 |
| Leaded Brass | .013 | 0.045 | — | — | — | — | — | — | — | 30.2 | — | — | — | — | — | — | — |
| | .025 | 0.079 | 0.260 | — | — | — | — | — | — | 53.4 | 93.4 | — | — | — | — | — | — |
| | .050 | 0.124 | 0.452 | 1.58 | — | — | — | — | — | 89.0 | 160 | 280 | — | — | — | — | — |
| | .075 | 0.181 | 0.621 | 2.15 | — | — | — | — | — | 125 | 218 | 387 | — | — | — | — | — |
| | .102 | 0.226 | 0.791 | 2.71 | 5.65 | — | — | — | — | 156 | 276 | 480 | 676 | — | — | — | — |
| | .15 | 0.316 | 1.08 | 3.84 | 7.91 | 13.2 | 19.8 | 27.3 | 46.0 | 222 | 378 | 663 | 925 | 1,174 | 1,415 | 1,650 | 2,117 |
| | .20 | — | 1.36 | 4.75 | 9.94 | 16.6 | 24.9 | 34.5 | 57.8 | — | 476 | 836 | 1,161 | 1,472 | 1,770 | 2,060 | 2,629 |
| | .25 | — | 1.58 | 5.65 | 11.9 | 19.9 | 29.7 | 41.2 | 69.0 | — | 569 | 996 | 1,383 | 1,753 | 2,104 | 2,450 | 3,118 |
| | .36 | — | — | 6.55 | 13.7 | 22.9 | 34.3 | 47.7 | 80.0 | — | — | 1,152 | 1,597 | 2,024 | 2,429 | 2,825 | 3,590 |
| | .40 | — | — | 7.90 | 16.4 | 27.5 | 41.0 | 57.0 | 95.7 | — | — | 1,375 | 1,908 | 2,410 | 2,896 | 3,363 | 4,270 |
| | .50 | — | — | 9.94 | 20.6 | 34.6 | 51.6 | 71.7 | 120 | — | — | 1,730 | 2,398 | 3,029 | 3,629 | 4,212 | 5,342 |
| | .65 | — | — | — | — | 41.3 | 61.8 | 85.8 | 144 | — | — | — | — | 3,612 | 4,328 | 5,022 | 6,360 |
| | .75 | — | — | — | — | — | — | — | 167 | — | — | — | — | — | — | — | 7,340 |

# SECTION 18

# SURFACE TECHNOLOGY


18.1 Introduction to Surface Technology ... 18–3
Trends and Concerns ... 18–3
Relationship of Surface Texture and Surface Integrity ... 18–3
Economics ... 18–3
Quality Assurance ... 18–3

18.2 Surface Texture ... 18–5
Introduction ... 18–5
Surface Texture Defined ... 18–5
Surface Texture Symbols ... 18–6
Surface Texture Measurement ... 18–9
Surface Texture Produced by the Machining Process ... 18–11
Machining Cost and Surface Texture ... 18–12
Dimensional Tolerance Versus Surface Texture ... 18–12
Surface Texture and Quality Assurance ... 18–12
Preferred Roughness Values ... 18–14
Theoretical Roughness Produced by Milling Cutters and by Lathe Tools ... 18–15

18.3 Surface Integrity ... 18–39
Introduction ... 18–39
Definitions ... 18–39
Benefits from the Application of Surface Integrity Practices ... 18–39
Problems Amenable to Surface Integrity Discipline ... 18–39
Altered Material Zones (AMZ) ... 18–40
AMZ Definitions ... 18–42
Evaluation Techniques ... 18–45
Process Intensities and Surface Integrity ... 18–47
Metallographic Sectioning and Special Preparation Techniques ... 18–47
Microhardness Determinations ... 18–47
Residual Stress Determination ... 18–49
Fatigue Strength Testing ... 18–51
Stress Corrosion Cracking ... 18–54
Combined Effects ... 18–54
General Guidelines ... 18–56
Precautions in Use of Guidelines ... 18–56
Key Indicators of Need for Surface Integrity Control ... 18–56
Guidelines for Material Removal Process in General ... 18–56
Comparison of Altered Material Zones from Material Removal Processes ... 18–57
Implementing Surface Integrity Programs ... 18–57

# 18 SURFACE TECHNOLOGY

Specific Guidelines ........ 18–59
Specific Guidelines and Data for Mechanical Material Removal Processes ........ 18–59
Specific Guidelines and Data for Abrasive Material Removal Processes ........ 18–74
Specific Guidelines and Data for Electrical Material Removal Processes ........ 18–91
Specific Guidelines and Data for Thermal Material Removal Processes ........ 18–98
Specific Guidelines and Data for Chemical Material Removal Processes ........ 18–111
Specific Guidelines and Data for Post-Treatment Processes to Aid Surface Integrity ........ 18–116
Surface Integrity by Materials—High Cycle Fatigue Results ........ 18–123
Quality Assurance for Surface Integrity ........ 18–127
Specific Guidelines and Special Comments on Quality Assurance for Surface Integrity ........ 18–132
Economics and Applications of Surface Integrity ........ 18–133
Selected Sources of Surface Integrity Data ........ 18–134
References ........ 18–134
Bibliography ........ 18–135

One of the principal design considerations for highly stressed components should be the surface condition produced during manufacturing. This consideration must include a thorough review of the surface alterations resulting from specific machining operations. Surface technology is the activity that describes, details, and evaluates the condition of both the surface and the surface layers of manufactured components. The historical emphasis on surface texture (roughness, waviness, and lay) has been extended to include the assessment of effects just below the surface, that is, surface integrity, thereby including the influence of the outermost boundary of a component as well as those outermost layers which differ measurably from the base material.

## TRENDS AND CONCERNS

In recent years, the demand for increased safety, reliability, and service life has caused considerable attention to be directed to the pronounced influence of the types of surface conditions produced by various manufacturing methods on the resulting properties and performance of components once in service. New and unusual service requirements and conservation of energy and/or materials have brought about the use of thinner sections of improved (and more difficult to fabricate) materials and the introduction of design criteria that employ these materials much closer to their full potential. While part drawings continue to call out configuration, dimensions, tolerances and material properties, surface quality specifications must be added to complete the component definition. All the design elements noted above are intimately interrelated and directly affect component cost and producibility.

The manufacturer, recognizing the above design trends and characteristics, also must be concerned with the attainment of accuracy in thinner parts where distortion from processing stresses is a frequent result. Some of the newer and higher performance alloys used in the thinner and lighter parts also are likely to be altered or more sensitive to thermal gradients or the heat generated by some processes. The emphasis on quality assurance and product reliability increases the importance of recognizing the impact that manufacturing processes may have on material properties. As always, in addition to these concerns and trends the fabrication of a product or component must be accomplished within the constraints of both economics and environmental effects.

## RELATIONSHIP OF SURFACE TEXTURE AND SURFACE INTEGRITY

Traditionally, surface texture (roughness, waviness and lay) has been accepted as the criterion which controls the quality of a surface. Direct relationships are widely assumed to exist between surface roughness and fatigue strength as well as other properties. Ample data have now been accumulated to indicate that surface texture is only part of the consideration. Metallurgical and other alterations below the surface also have a major influence on material performance. This becomes particularly important where high stresses or severe environments are encountered by the workpiece.

While most alterations occur in the first 0.015 inch [0.38 mm] below the surface, this distance can be a significant part of the total thickness of the thinner parts being designed. In addition, it should be noted that many component failures have originated within the first few thousandths inch below the geometrical surface. These facts alone support the need for design, manufacturing, and quality assurance engineers to exhibit interest in and understanding of both aspects of surface technology—surface texture and surface integrity—in order to produce safe, reliable and long life components.

The influence of manufactured surfaces on component properties has been investigated most extensively for the material removal processes. Figure 18.1-1 shows the surface technology effects studied to date. During machining, process energies blend with work material properties to produce a wide variety of these surface effects. Surface effects produced during the machining process affect the properties of the work material and in turn of the component reliability, as shown in figure 18.1-2. Studies have demonstrated that each material removal process has its own effect on each work material. It has also been shown that for each process the intensity or energy level may have a distinctive effect on the specific metallurgical state of the work material.

## ECONOMICS

The cost of producing a machined component increases as surface specifications increase. Sophisticated and extensive surface texture or surface integrity requirements are justified only if they are necessary for the component part to fulfill its function. Many parts require only a roughing operation to attain their desired shape and tolerance. Finishing operations may improve quality but they add cost, and while emphasis on surface integrity considerations increases component reliability, cost also may increase. To eliminate unnecessary costs it is important to apply texture and integrity specifications selectively to only those areas of a part where they are needed, rather than to impose these requirements across the entire surface of the part. The relationship between surface texture and machining costs is discussed further in section 21.

The application of surface integrity principles and guidelines can increase the number of good parts produced or, conversely, reduce manufacturing losses. Surface integrity evaluations should consider the entire part processing sequence and emphasize the achievement of high quality in the "as shipped" surface produced. The price for inattention to surface integrity can be premature component failure or loss of reliability.

## QUALITY ASSURANCE

Surface technology ranks with dimensional configuration and tolerance as visual evidence of quality. Deterioration of surface quality with time or with number of parts produced is a handy clue to deteriorating tools or process settings. A 10-percent rule of thumb has been found to be a practical limit within which normal manufacturing operations can be conducted without jeopardizing surface quality. This rule states that if surface roughness increases by 10 percent (in microinches $R_a$ or rms) or if any one of the operating parameters has been reset or "adjusted" by plus or minus 10 percent from the planned values, immediate questions should be raised as to the control status of the process. If two or more parameters are "adjusted" by 10 percent or more, the process is no longer controlled and should be shut down until corrective action is effective.

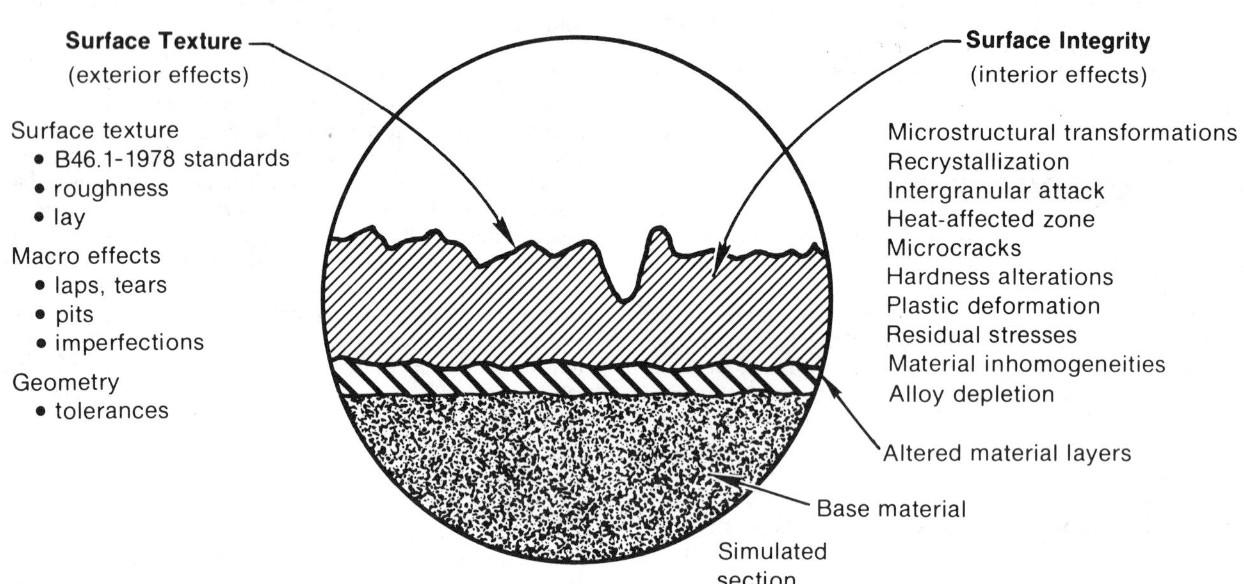


**Figure 18.1–1** Surface technology effects.

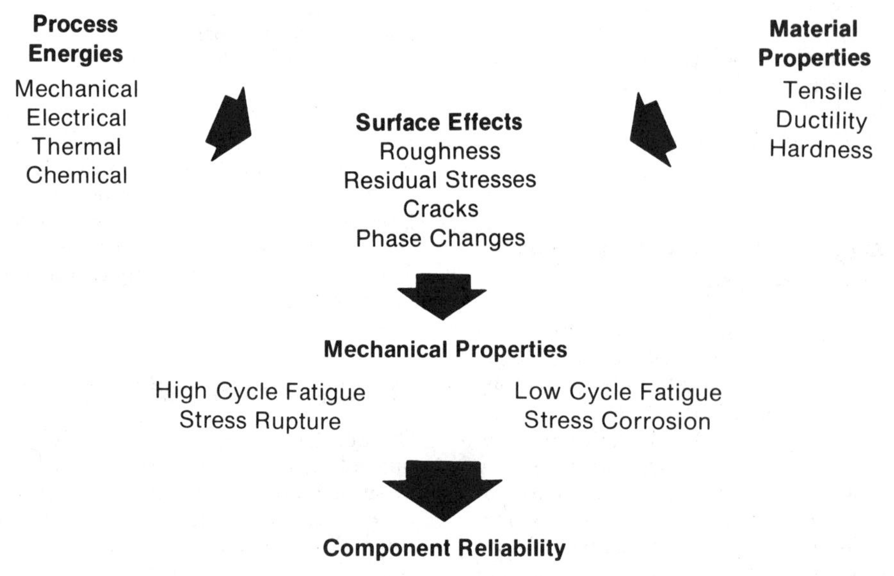


**Figure 18.1–2** Interaction of processes with materials.

## INTRODUCTION

The surfaces produced by machining generally are irregular and complex. Despite this, the majority of machined parts can satisfactorily perform their functions with general, uncomplicated surface texture specifications. Many machining processes can meet these general surface texture requirements with ordinary process control and a minimum of quality control. Dimensional checks and visual examination for macro imperfections usually are sufficient. When surface roughness requirements exceed about 63 microinches $R_a$ [1.6 μm] most companies will use a visual check rather than a measurement of the roughness profile. Often the visual check is aided by using sets of sample roughness specimens. Any more exacting specifications for most surfaces would serve no practical purpose and would result in needless expense.

Parts that are highly engineered, heavily stressed or subjected to unusual environments usually have more specific and detailed surface quality requirements. Tolerance requirements for these parts may also need to be closely allied to the roughness allowed. For some of these parts, history or testing has demonstrated a direct relationship between performance and surface texture. Quality control to measure surface texture is, therefore, necessary for these parts. The following are examples of parts that require enhanced surface texture specifications in order to improve fatigue strength, corrosion resistance, cleanliness, appearance, coatability, sealing or product performance:

1. Antifriction bearings and airfoils
2. Objects operating in corrosive environments
3. Food preparation devices
4. Telescope lense, plug gages and rolling mill rolls
5. Journal bearings
6. Painted or coated surfaces
7. Sealing surfaces or friction clamped assemblies

Parts that are critically stressed, product-life-limiting or safety controlling, or those specific areas on such parts, should have additional specifications that include surface integrity considerations, as described in section 18.3.

## SURFACE TEXTURE DEFINED

Surface texture is the repetitive or random deviations from the nominal surface which form the three-dimensional surface topography. A variety of mechanical, electronic and optical devices are available to measure these deviations by sampling the profile of the workpiece. Figure 18.2-1 shows a comparison between the measured and the nominal profiles for a given surface.

American National Standard ANSI B46.1-1978 describes, standardizes and calls out acceptable measuring instrumentation for surface texture.[1] (Internationally, ISO R468 compares in content.) According to the standard, surface texture includes four elements—roughness, waviness, lay and flaws. Figure 18.2-2 shows these characteristics in relation to a unidirectional lay surface.

[1]ANSI B46.1-1978, "Surface Texture" (which includes ANSI Y14.36-1978 "Surface Texture Symbols") should be available to and be read by everyone concerned with specifying or measuring surface texture. Copies are available from the American Society of Mechanical Engineers, United Engineering Center, 345 East 47th Street, New York, NY 10017.

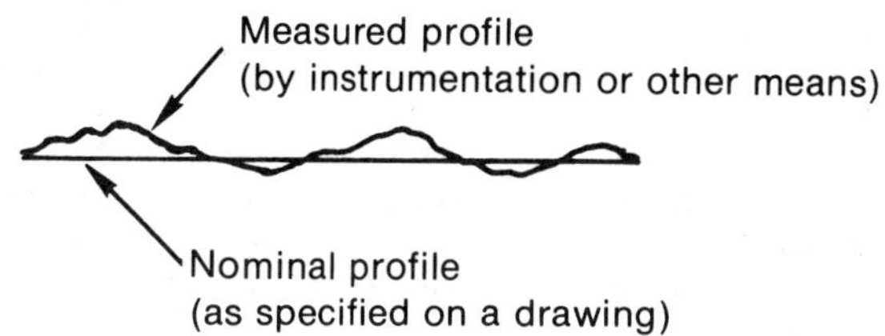


**Figure 18.2-1** Surface texture profile. (Based on ANSI B46.1-1978)

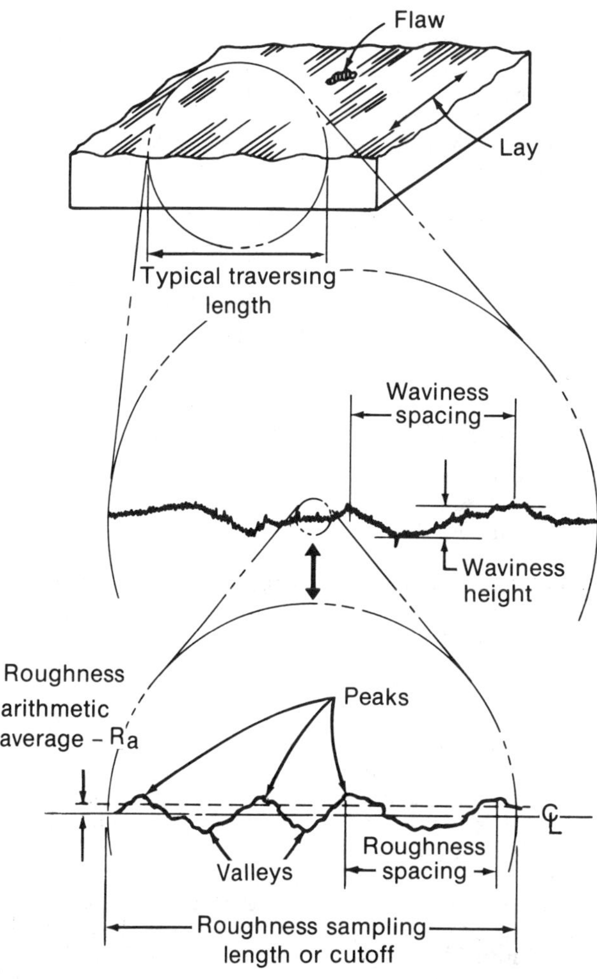


**Figure 18.2-2** Pictorial display of unidirectional lay surface characteristics. (Based on ANSI B46.1-1978)

# 18.2 Surface Texture

Definitions for the four elements of surface texture are as follows:

*Roughness* consists of the finer irregularities which generally result from the inherent action of the production process. These include transverse feed marks and other irregularities within the limits of the sampling length.

*Waviness* includes all irregularities whose spacing is greater than the roughness sampling length and less than the waviness sampling length. Waviness may result from machine or work deflections, chatter, vibration, heat treatment or cutting tool runout. Roughness may be considered superimposed on a "wavy" surface.

*Lay* is the direction of the predominant surface pattern, ordinarily determined by the production method used.

*Flaws* are unintentional irregularities which occur at one place or at relatively infrequent or widely varying intervals on the surface. Flaws include cracks, blow holes, inclusions, checks, ridges, scratches, etc. Unless otherwise specified, the effect of flaws shall not be included in the roughness average measurements. Where flaws are to be restricted or controlled, a special note as to the method of inspection should be included on the drawing or in the specifications.

*Surface finish* is a colloquial term widely used to denote the general quality of a surface. Surface finish is not specifically tied to the texture or characteristic pattern of the surface, nor is it tied to specific roughness values; however, a "good" finish implies low roughness values and vice versa. The term surface finish is not as precisely defined as are the terminologies used in the American National Standard, nor is it necessarily expressed numerically.

## SURFACE TEXTURE SYMBOLS

The description and specification of surface texture is accomplished by means of symbols and conventions as presented in the ANSI Standard Y14.36-1978 (almost in complete agreement with ISO 1302). The symbol used to designate surface roughness is the check mark with a horizontal extension as shown in figure 18.2-3. Roughness, which is the most commonly used surface parameter, is specified by placing the height rating in microinches (one microinch = 0.000001 inch) or micrometers (one micrometer = 0.000001 meter) to the left of the check mark. If there is a maximum and a minimum rating, the two numbers are placed one above the other. Roughness is defined as the arithmetic average (AA) deviation of the surface expressed in microinches from a mean line or centerline (see figures 18.2-2 and 18.2-4). $R_a$ is a symbol for roughness that has been adopted internationally. Surface roughness is still sometimes displayed with the symbols AA, or CLA or c.l.a. Many instruments still in use employ an average deviation from the roughness centerline which is the root mean square (rms), also expressed in microinches ($R_q$). While still used frequently, rms actually has been obsolete since about 1950. Roughness measuring instruments calibrated in rms read approximately 25 percent higher on a given surface than those instruments calibrated for $R_a$ (see table 18.2-1). The difference is usually much less than the point-to-point variations on any given machined surface.

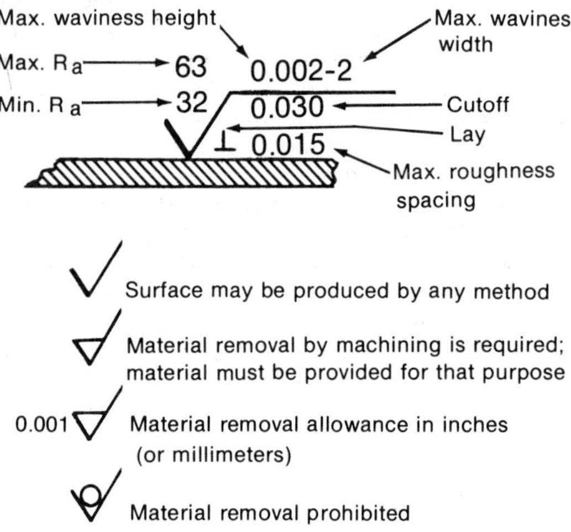


**Figure 18.2-3** Surface texture symbols used for drawings or specifications. In this example, all values are in inches except $R_a$ values, which are in microinches (millionths of an inch). Metric values (millimeters and micrometers, respectively) are used on metric drawings. (Based on ANSI Y14.36–1978)

**TABLE 18.2-1 Ratio of Root Mean Square Roughness to Arithmetic Average Roughness Values**

| | |
|---|---|
| Root mean square roughness: | $R_q$ |
| Arithmetic average roughness: | $R_a$ |
| Theoretical ratio of sine waves, $R_q/R_a$: | 1.11 |
| Actual ratios of $R_q/R_a$ for various processes: | |
| turning | 1.17 to 1.26 |
| milling | 1.16 to 1.40 |
| surface grinding | 1.22 to 1.27 |
| plunge grinding | 1.26 to 1.28 |
| soft honing | 1.29 to 1.48 |
| hard honing | 1.50 to 2.10 |
| electrical discharge machining | 1.24 to 1.27 |
| shot peening | 1.24 to 1.28 |
| Practical first approximation of $R_q/R_a$: | |
| for most processes | 1.25 |
| for honing | 1.45 |

SOURCE: J. Peters; P. Vanherck; and M. Sastrodinoto, Assessment of surface typology analysis techniques, *Annals of the CIRP* 28/2, 1979.

The roughness width (or sampling) cutoff is the greatest spacing of repetitive surface irregularities to be included in the measurement of the average roughness height (figure 18.2-2). It is specified in inches and is placed below the horizontal extension of the check mark. A waviness height specification is rated in inches as the peak-to-valley distance; and waviness spacing or width, also rated in inches, is the spacing of successive peaks or valleys (figure 18.2-2). Waviness height and waviness width values are placed above the horizontal extension of the check mark. The symbol designating the lay is placed under the extension of

## $R_a$ $R_q$

$R_a$ is the universally recognized parameter of roughness. It is the arithmetic mean of the departures y of the profile from the mean line. It is normally determined as the mean results of several consecutive sampling lengths L.

$R_q$ is the corresponding rms parameter.

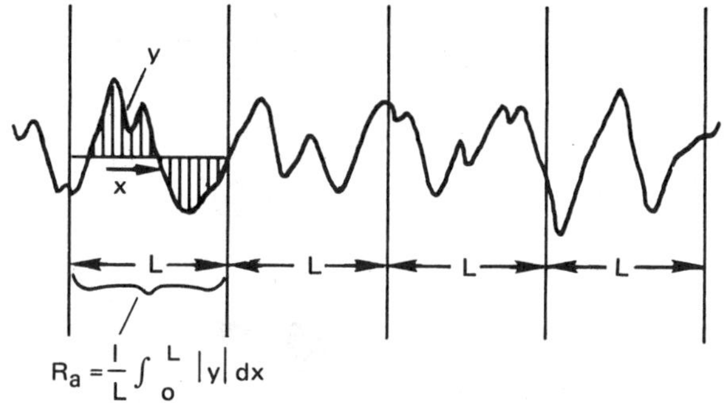


$$R_a = \frac{1}{L} \int_0^L |y|\, dx$$

$$R_q = \sqrt{\frac{1}{L} \int_0^L y^2 (x) dx}$$

## $R_z$

Ten-Point Height is the average distance between the five highest peaks and the five deepest valleys within the sampling length and measured perpendicular to it.

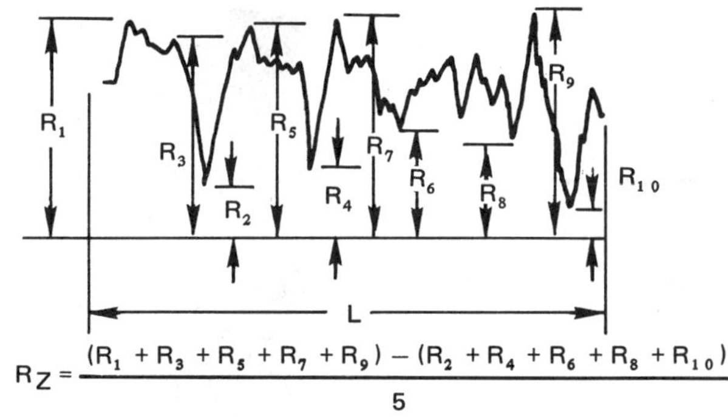


$$R_z = \frac{(R_1 + R_3 + R_5 + R_7 + R_9) - (R_2 + R_4 + R_6 + R_8 + R_{10})}{5}$$

## $R_t$ $R_{max}$ $R_{tm}$

$R_t$ is the maximum peak-to-valley height within the assessment length. $R_{max}$ is the maximum peak-to-valley height within a sampling length L. But because the value can be greatly affected by a spurious scratch or particle of dirt on the surface, it is more usual to use the average ($R_{tm}$) of five consecutive sampling lengths.

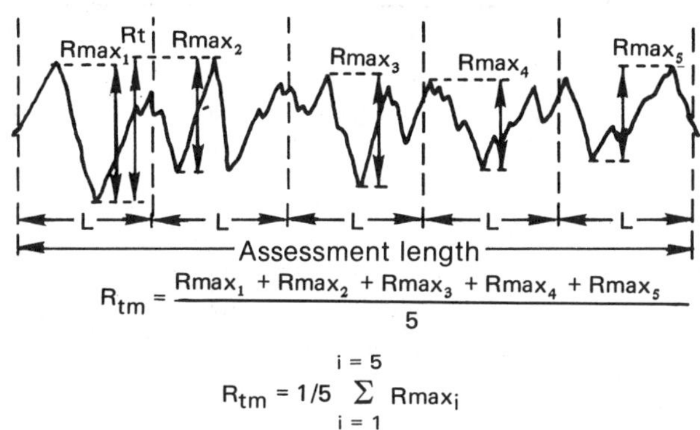


$$R_{tm} = \frac{Rmax_1 + Rmax_2 + Rmax_3 + Rmax_4 + Rmax_5}{5}$$

$$R_{tm} = 1/5 \sum_{i=1}^{i=5} Rmax_i$$

## $R_p$ $R_{pm}$

$R_p$ is the maximum profile height from the mean line within the sampling length. $R_{pm}$ is the mean value of $R_p$ determined over 5 sampling lengths.

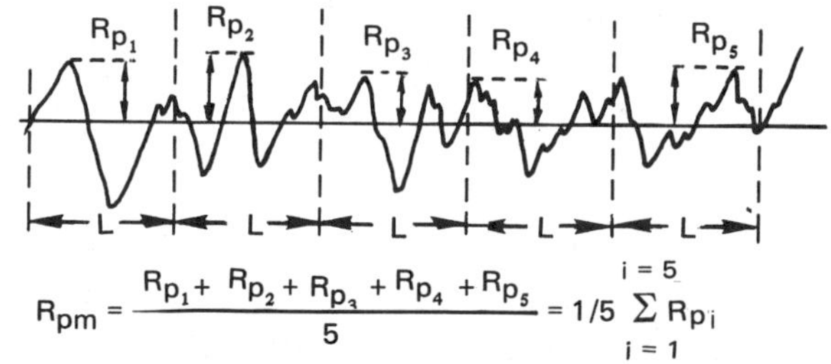


$$R_{pm} = \frac{R_{p_1} + R_{p_2} + R_{p_3} + R_{p_4} + R_{p_5}}{5} = 1/5 \sum_{i=1}^{i=5} R_{p_i}$$

## $P_c$

Peak count is the number of peak/valley pairs per inch projecting through a band of width b centered about the mean line.

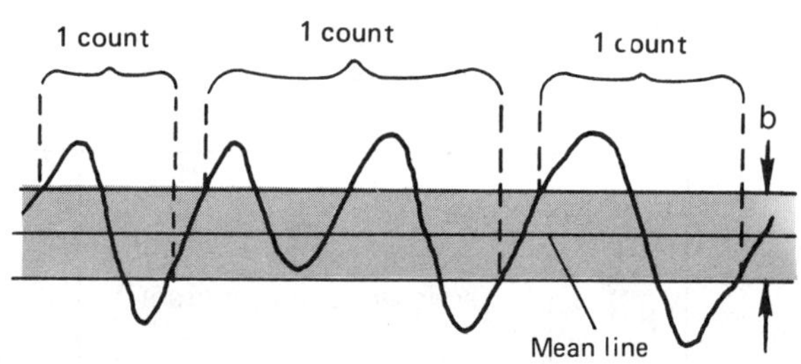


**Figure 18.2–4** Some commonly used surface texture symbols and their definitions. (Based on ANSI B46.1–1978)

the check mark. This symbol indicates the direction of the lay relative to the nominal surface. Figure 18.2-5 explains the various symbols used to designate lay. The pitted non-directional or protuberant designation of lay, P, is useful to describe surfaces produced by some of the nontraditional machining operations. The surface texture symbol also can designate the extent of material removal desired—from none to any amount, as shown in figure 18.2-3.

| Lay Symbol | Meaning | Example Showing Direction of Tool Marks | Photographs of Examples |
|---|---|---|---|
| = | Lay approximately parallel to the line representing the surface to which ,the symbol is applied. | 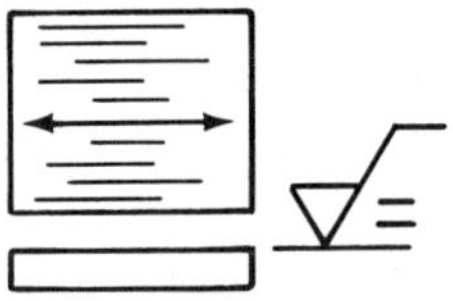 = | |
| ⊥ | Lay approximately perpendicular to the line representing the surface to which the symbol is applied. | 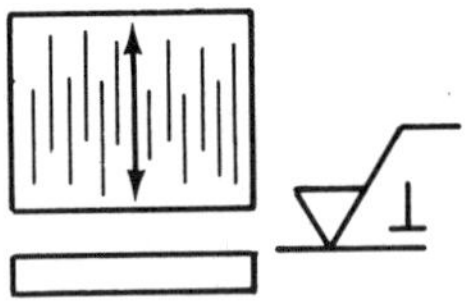 ⊥ | 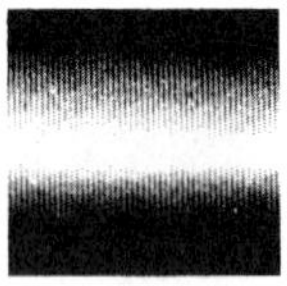 |
| X | Lay anglular in both directions to line representing the surface to which the symbol is applied. | 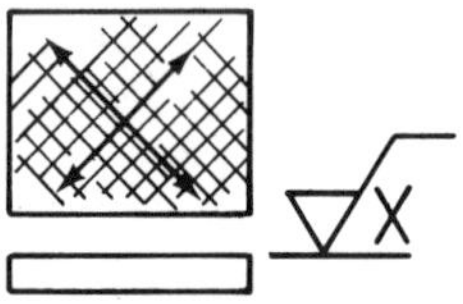 X | |
| M | Lay multidirectional. | 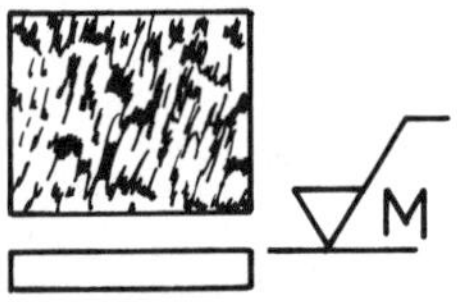 M |  |
| C | Lay approximately circular relative to the center of the surface to which the symbol is applied. | 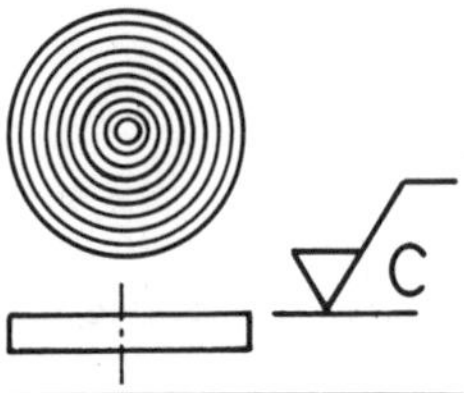 C | 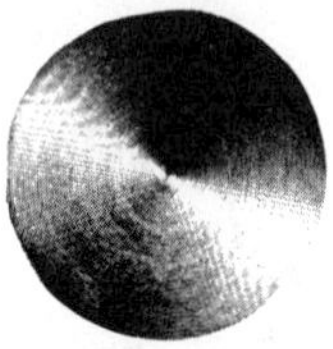 |
| R | Lay approximately radial relative to the center of the surface to which the symbol is applied. | 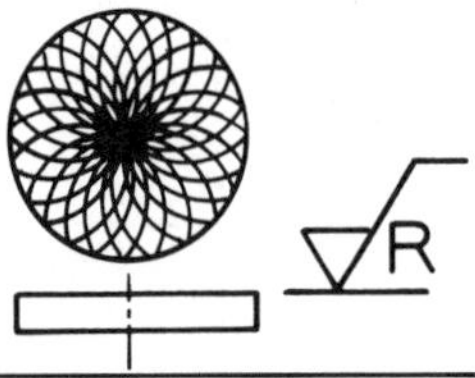 R | 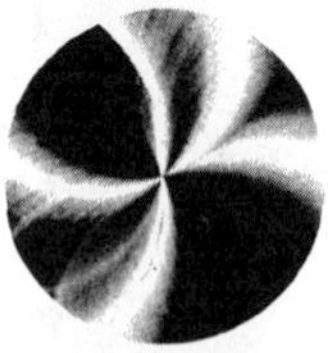 |
| P | Lay particulate, non-directional, or protuberant. | 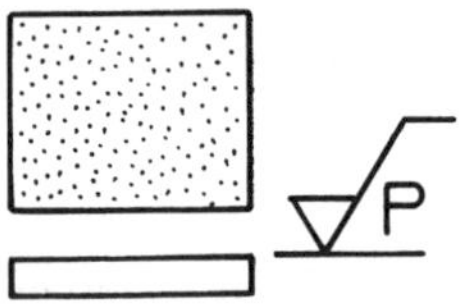 P | |

**Figure 18.2-5** Lay symbols for surface texture designation. (Based on ANSI Y14.36-1978)

## SURFACE TEXTURE MEASUREMENT

The most prevalent measuring technique for surface texture employs a mechanical-electronic device whose readout indicates the roughness of the surface profile taken during the passage of a small radius stylus over a short straight line path on the surface. The most common diamond stylus has a 0.0004 inch [10 μm] radius and usually is used with a 0.030 inch [0.8 mm] cutoff width. The total stylus travel is usually 20 to 60 times the cutoff width with the electronic circuitry continuously averaging the readings over the set cutoff width. These instruments can read average roughness, $R_a$, peak count or other roughness designations depending on the particular instrument design.

It should be kept in mind that electronic surface measuring devices generally indicate the roughness but do not indicate the physical character of the surface. Several surfaces can, in effect, be quite different in appearance and still yield similar roughness values, as shown in figure 18.2-6. One approach to more exactly describing the surface roughness is the microtopographic map, as shown in figure 18.2-7. These maps are made by successive, closely spaced stylus generated traces. A still more detailed examination is sometimes made using the scanning electron microscope (SEM). The series of views with SEM shown in figure 18.2-8 utilizes increasing magnifications of 20X, 200X, 1000X, 2000X and 5000X which are easy to relate to the view with a 10X hand-held magnifier.

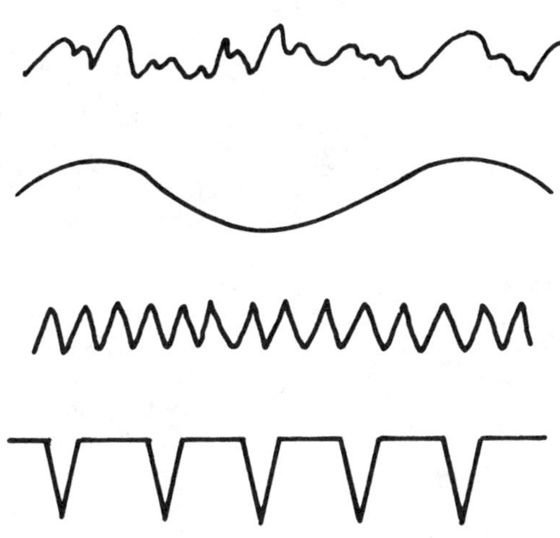

**Figure 18.2-6** Sketches of cross-sections of surfaces greatly different in character but having approximately the same surface finish level.

**Figure 18.2-7** Microtopographic map of Blanchard-ground surface of stainless steel. 200X in X-Y plane, 2000X in Z direction. (Courtesy of Gould Inc.)

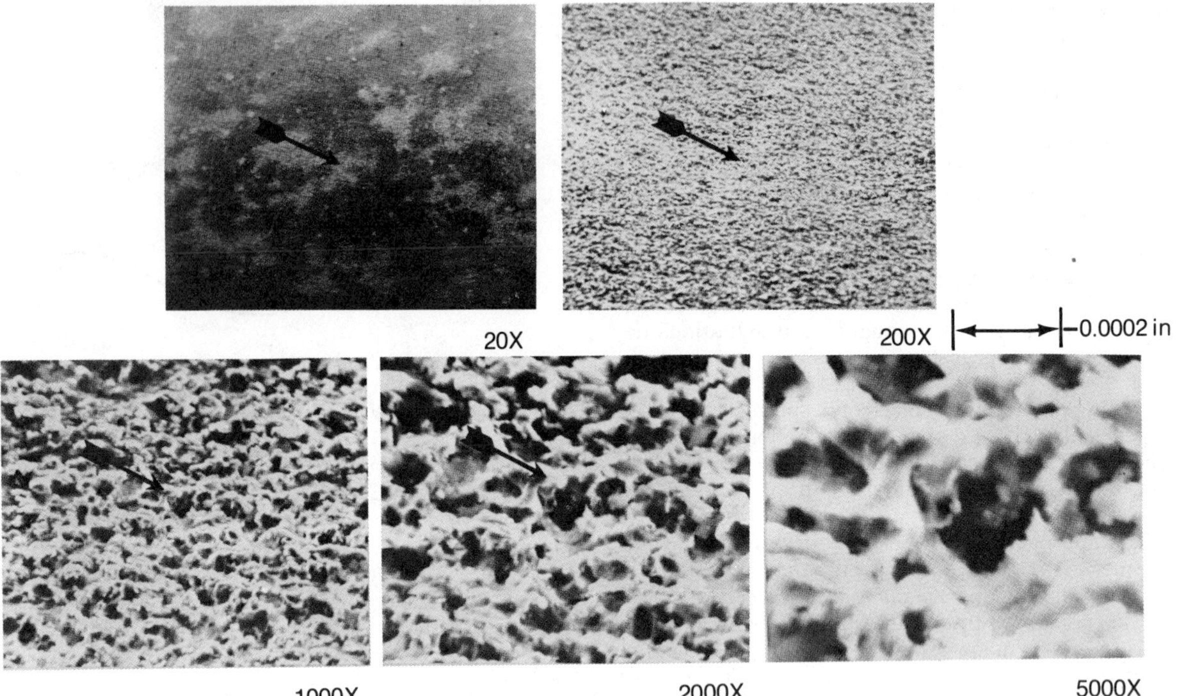


**Figure 18.2–8** Scanning electron microscope (SEM) views taken at a 55° angle of a typical surface of 17-4 PH stainless steel produced by electrochemical machining (ECM). Arithmetic average roughness ($R_a$) was 29 microinches [0.74 $\mu$m], and maximum peak-to-valley roughness height ($R_t$) was 152 microinches [3.86 $\mu$m].

## SURFACE TEXTURE PRODUCED BY THE MACHINING PROCESS

To a large degree the surface texture produced by a material removal process is characteristic for that particular process. The range of roughnesses typically obtained for a variety of manufacturing processes is shown in figure 18.2-9. This chart also indicates that it is possible to exceed the usual range under unusual or specially controlled conditions.

The selection of surface texture values involves more than merely designating a particular process. The ability of a processing operation to produce a specific surface roughness depends on many factors. In turning, for example, the surface roughness is geometrically related to the nose radius of the tool and the feed per revolution. For surface grinding, the final surface depends on the type of grinding wheel, the method of wheel dressing, the wheel speed, the table speed, cross feed and down feed, and the grinding fluid. For electrical discharge machining (EDM), the roughness level is related directly to the individual spark discharge energy level. A change in any of the process operating parameters may have a significant effect on the final surface produced.

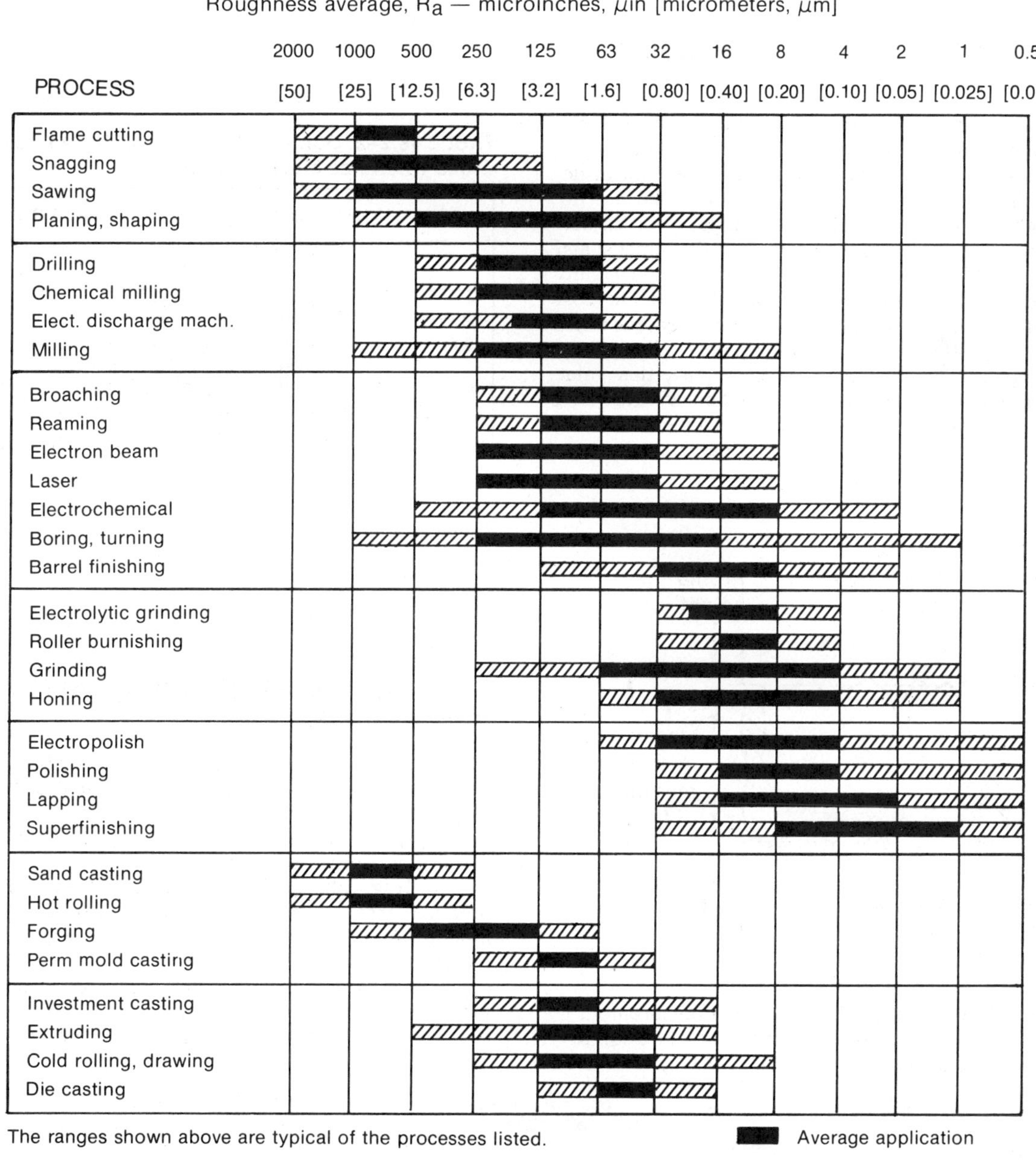


**Figure 18.2-9** Surface roughness produced by common production methods. (From ANSI B46.1-1978).

## MACHINING COST AND SURFACE TEXTURE

The cost of producing a machined surface increases with increasing requirements for finer finishes. Certain machining operations, such as rough turning and milling, are necessary to shape a component to its required dimensions. Additional operations to refine the surface are needed only to permit the surface to perform functions which it could not otherwise perform. A surface roughness of 63 microinches [1.6 μm] or coarser can be obtained at a reasonable cost by general roughing and semi-finishing operations. The relationship of surface texture to the cost of machining is discussed further in section 21.

## DIMENSIONAL TOLERANCE VERSUS SURFACE TEXTURE

There is a direct relationship between the dimensional tolerance of a part and its permissible surface roughness since the roughness measurement involves the average linear deviation of the actual surface from the nominal surface defined by the dimension. If the deviations induced by the surface roughness exceed those permitted by the dimensional tolerance, the dimension will be subject to an uncertainty beyond the tolerance, as shown in figure 18.2-10. On most surfaces the total profile height is approximately four times the measured (arithmetic average) roughness. When measurements are made on a diameter of a part, this value would be doubled. It follows that the roughness value on a diameter should not exceed one-eighth the dimensional tolerance on the diameter if useful dimensional controls are to be maintained.

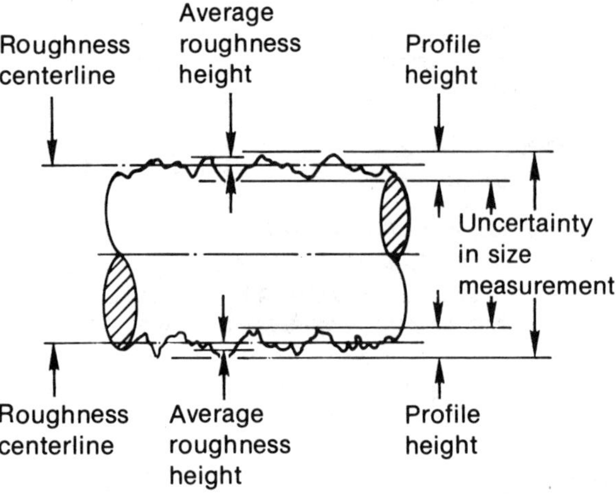


**Figure 18.2-10** Uncertainty in size measurement in relation to surface roughness parameters for a cylindrical surface. (From ANSI B46.1-1978).

Each application must be evaluated on its own merit. Table 18.2-2, if used with discretion, may serve as a guide to the surface roughness values that may be necessary where machine work must be held to close tolerance for other reasons than merely surface texture. It must be further realized that the practical control of tolerance is also influenced by the size of the part, the overall size of the surface being cut, and the material removal operations involved. A chart showing how attainable dimensional tolerances tend to vary as a function of part size is shown in figure 18.2-11.

**TABLE 18.2-2 Guide to Surface Roughness Values for Close Tolerance Machine Work**

| DIMENSIONAL TOLERANCES | | SURFACE ROUGHNESS | |
|---|---|---|---|
| in | mm | μin | μm |
| Below 0.0002 | Below 0.005 | Below 8 | Below 0.2 |
| 0.0002 to 0.0005 | 0.05 to 0.012 | 8 to 16 | 0.2 to 0.4 |
| 0.0005 to 0.0010 | 0.012 to 0.025 | 16 to 32 | 0.4 to 0.8 |
| 0.0010 to 0.0020 | 0.025 to 0.05 | 32 to 63 | 0.8 to 1.6 |
| 0.0020 to 0.0100 | 0.05 to 0.25 | 63 to 250 | 1.6 to 6.3 |

## SURFACE TEXTURE AND QUALITY ASSURANCE

Quality assurance for the surfaces produced by material removal processes should include assessment of the surface roughness and all other surface texture factors. This does not imply that the effect of scratches, tool marks, sharp corners and other geometric considerations can be overlooked. It is well established that all of these elements can produce stress concentrations that can lead to premature fatigue failures. Historically, surface roughness has been the prime criterion for surface quality and a guide to acceptable fatigue strength. Some recent data indicate that for some alloys surface roughness is not the critical criterion for high cycle fatigue strength. Section 18.3 discusses this matter in greater detail.

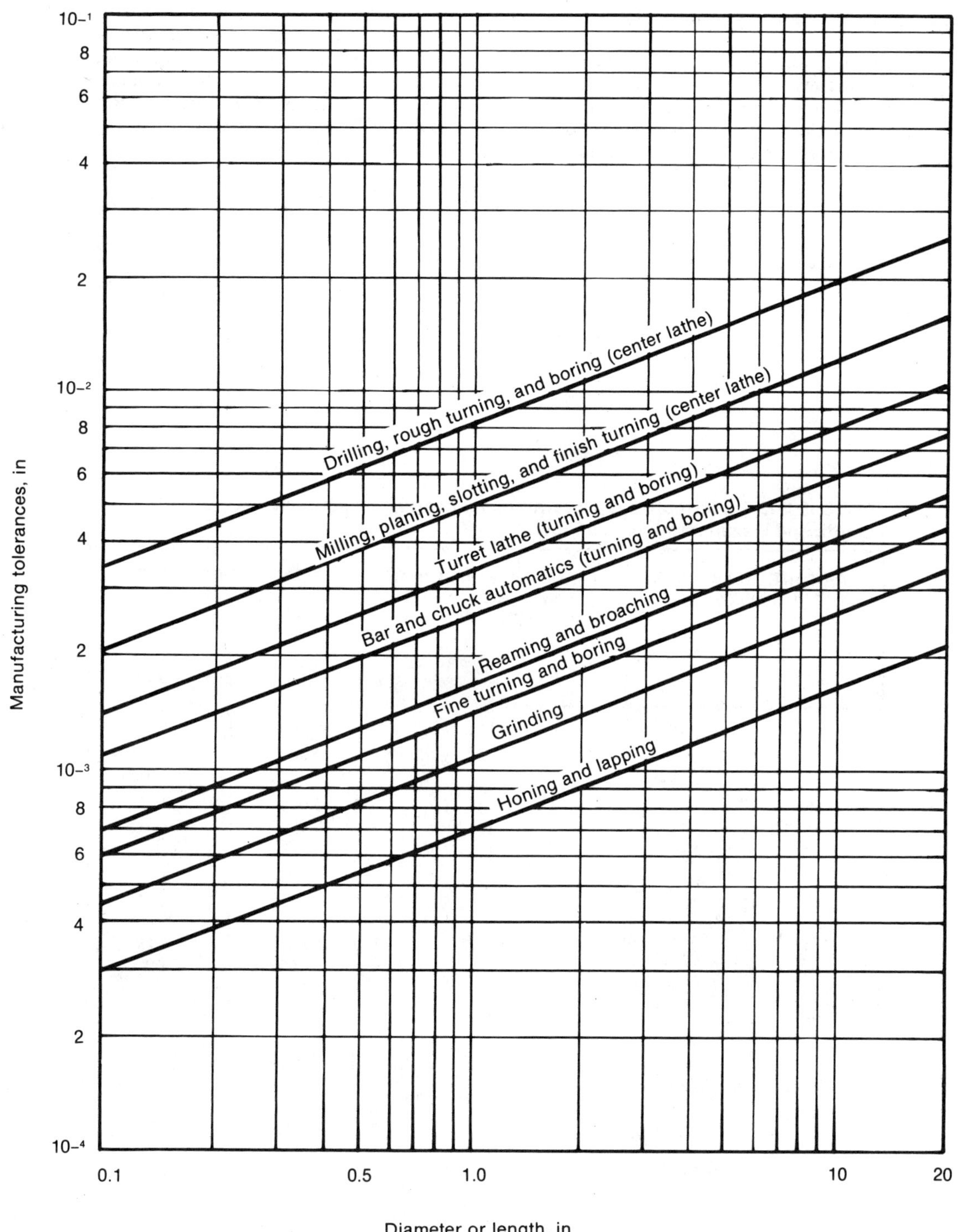


**Figure 18.2-11** General effect of part size on manufacturing tolerances. (Adapted from *Manufacturing planning and estimating handbook*, New York: McGraw-Hill, 1963, p. 20-23)

# 18.2 Surface Texture

## PREFERRED ROUGHNESS VALUES

With all the factors that can influence the generation of a surface roughness value, with the dearth of data linking a specific roughness value to the function performed by a specific surface, and with the variability introduced by the minute sample used in most measuring techniques, it is impractical to place undue emphasis on achieving a specific roughness number. To minimize the variety of drawing callouts, the American National Standards Institute in the standard, ANSI Y14.36-1978, promotes the use of preferred values for average roughess ($R_a$), roughness cutoff length and maximum waviness height. These values are listed in table 18.2-3.

**TABLE 18.2-3 Preferred Values for Arithmetic Average Roughness ($R_a$), Cutoff Length, and Maximum Waviness Height**

| ARITHMETIC AVERAGE ROUGHNESS ($R_a$) | | STANDARD ROUGHNESS SAMPLING LENGTH (CUTOFF)† | | MAXIMUM WAVINESS HEIGHT | |
|---|---|---|---|---|---|
| μin | μm | in | mm | in | mm |
| 0.5 | 0.012 | 0.003 | 0.08 | 0.00002 | 0.0005 |
| 1* | 0.025* | 0.010 | 0.25 | 0.00003 | 0.0008 |
| 2* | 0.050* | 0.030 | 0.80 | 0.00005 | 0.0012 |
| 3 | 0.075 | 0.1 | 2.5 | 0.00008 | 0.0020 |
| 4* | 0.10* | 0.3 | 8.0 | 0.0001 | 0.0025 |
| 5 | 0.125 | 1.0 | 25.0 | 0.0002 | 0.005 |
| 6 | 0.15 | | | 0.0003 | 0.008 |
| 8* | 0.20* | | | 0.0005 | 0.012 |
| 10 | 0.25 | | | 0.0008 | 0.020 |
| 13 | 0.32 | | | 0.001 | 0.025 |
| 16* | 0.40* | | | 0.002 | 0.05 |
| 20 | 0.50 | | | 0.003 | 0.08 |
| 25 | 0.63 | | | 0.005 | 0.12 |
| 32* | 0.80* | | | 0.008 | 0.20 |
| 40 | 1.00 | | | 0.010 | 0.25 |
| 50 | 1.25 | | | 0.015 | 0.38 |
| 63* | 1.60* | | | 0.020 | 0.50 |
| 80 | 2.0 | | | 0.030 | 0.80 |
| 100 | 2.5 | | | | |
| 125* | 3.2* | | | | |
| 160 | 4.0 | | | | |
| 200 | 5.0 | | | | |
| 250* | 6.3* | | | | |
| 320 | 8.0 | | | | |
| 400 | 10.0 | | | | |
| 500* | 12.5* | | | | |
| 600 | 15 | | | | |
| 800 | 20 | | | | |
| 1000* | 25* | | | | |

SOURCE: ANSI Y14.36-1978.

*Recommended.

†When no value is specified, the value of 0.030 inch [0.8 mm] applies.

## THEORETICAL ROUGHNESS PRODUCED BY MILLING CUTTERS AND BY LATHE TOOLS

It is possible to calculate the theoretical surface roughness produced by face milling cutters in a milling operation and by lathe tools in turning. Calculation of the theoretical surface roughness is important because it represents the approximate finish that can be obtained with a sharp tool. The actual surface roughness may be different because built-up edge or burnishing may occur during the cutting process.

For face milling, the cutter tooth can be classified according to one of the following common configurations:[2]

Type A: Zero nose radius
Type B: Round
Type C: Nose radius, end cutting edge angle and side cutting edge angle

These configurations are shown in figure 18.2-12. Formulas have been developed which determine the arithmetic average roughness, $R_a$, and the peak-to-valley roughness height, $R_t$, for these configurations.[3]

The theoretical surface roughness produced by a Type-A tooth in face milling is a function of the side cutting edge angle (SCEA), the end cutting edge angle (ECEA) and the feed per tooth. Calculated values of $R_a$ and $R_t$ for this tooth configuration are contained in tables 18.2-4 through 18.2-8.

The theoretical roughness produced by the round Type-B tooth in milling is a function of the tooth radius and the feed per tooth. Table 18.2-9 contains calculated values of $R_a$ and $R_t$ for this configuration.

The Type-C tooth produces theoretical roughness values which are a function of the nose radius, the ECEA and the feed per tooth. Values $R_a$ and $R_t$ that were calculated for this configuration are given in tables 18.2-10 through 18.2-19.

For turning operations, lathe tools can be characterized as either Type B or Type C (see figure 18.2-13). For common Type-C tools, the theoretical surface roughness is determined by the nose radius, the ECEA and the feed per revolution. When a round Type-B tool is used, theoretical roughness is a function of the tool radius and the feed per revolution. The similarity between the roughness determinations for the Type-B and Type-C configurations for both the turning tools and the milling cutters results in identical sets of theoretical roughness values for each tool type in each operation. Hence, the table of theoretical values of $R_a$ and $R_t$ values for the Type-B tooth in milling (table 18.2-9) applies equally to the Type-B tool in turning. Similarly, the values in tables 18.2-10 through 18.2-19 for the Type-C tooth in milling apply equally to the Type-C tool in turning.

A series of graphs depicts the arithmetic average surface roughness for the various tool configurations as follows:

Type A: Zero-radius face milling cutter; figure 18.2-14
Type B: Round tool; figure 18.2-15
Type C: Tool containing a nose radius plus end cutting edge angle; figures 18.2-16, 18.2-17, and 18.2-18

As previously mentioned, the actual surface roughness generally will be poorer than that calculated from geometric considerations such as shown in tables 18.2-4 through 18.2-19. (The theoretical calculations all assume that the cutting conditions are such that the tool profile is duplicated exactly in the workpiece surface.) During the process of chip formation, a built-up edge or plastically deformed material may be produced at the cutting edge. Most of this built-up edge breaks off and flows away with the chip. However, part of this built-up edge can remain welded to the finished workpiece surface and adds to the surface roughness, as shown in figure 18.2-19. The presence of a built-up edge can increase the surface roughness by a factor of 1.2 to 2.5. The added roughness, or the amount of built-up edge, is a function of the cutting tool material and the work material. The amount of built-up edge present on the surface can be minimized by using higher cutting speeds and by using chemically active cutting fluids. Additional factors which can add to the surface roughness include chatter, vibration and tool wear.

Special tool shapes can create finer- or micro-finishes. A "wiper" blade, figure 18.2-20, is one such tool shape used. Its broad, convex face reduces feed marks. Normally only one wiper blade is required in multiblade milling cutters. Higher cutting speeds do not appear to reduce the efficiency of wiper blades.

[2]Type A, B and C tools are the same designations used by G. R. Dickinson in his paper, "A survey of factors affecting surface finish," *Proceedings, Institution of Mechanical Engineers* 182 (Part 3K, 1968): 135-147.

[3]H. L. Fischer and J. T. Elrod, "Surface finish as a function of tool geometry and feed—A theoretical approach," *Microtecnic* 25 (April 1971): 175-178.

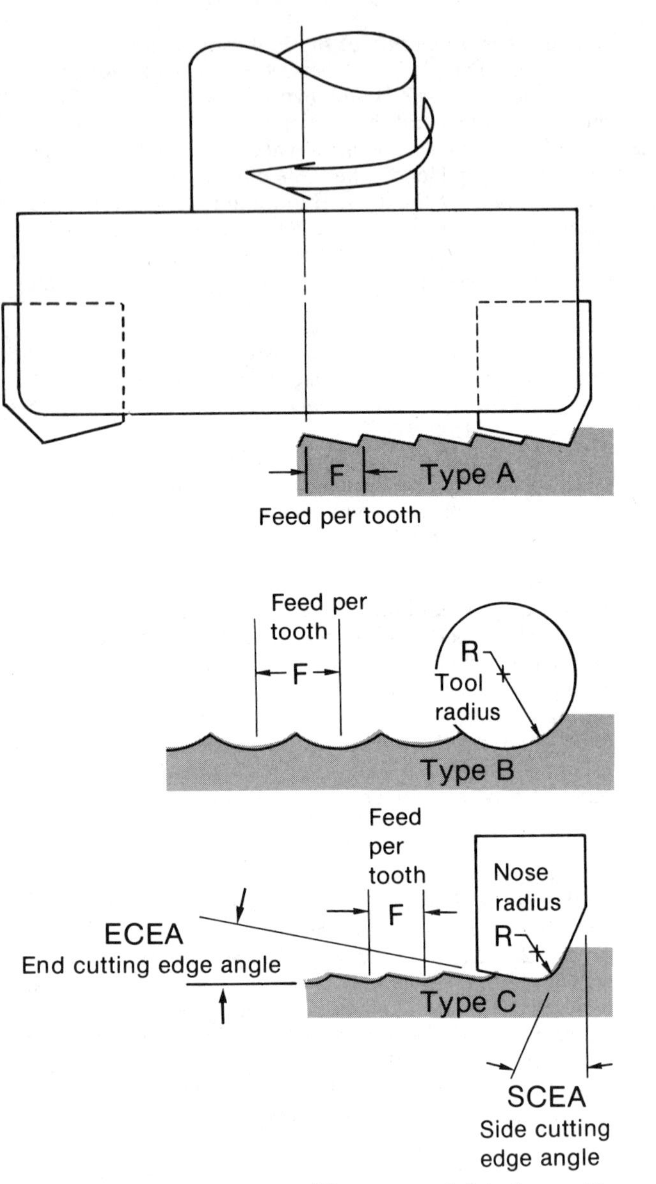


**Figure 18.2–12** Surface produced by face milling with the types of tools used in the theoretical surface finish calculations.

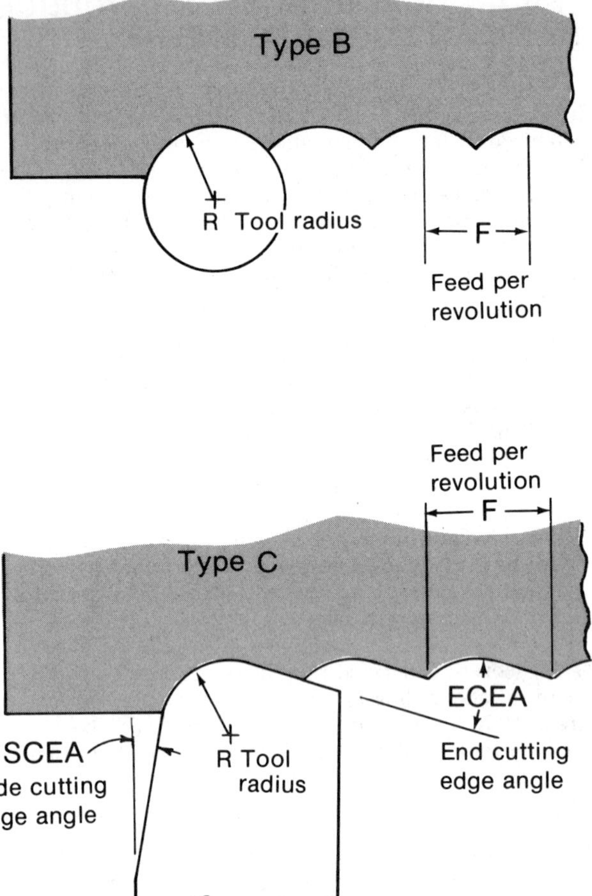


**Figure 18.2–13** Surfaces produced by turning with the types of tools used in the theoretical surface finish calculations.

**TABLE 18.2-4 THEORETICAL Values for Arithmetic-Average Roughness ($R_a$) and Maximum Peak-to-Valley Roughness Height ($R_t$) for Milling—Tool Type A**

| ZERO NOSE RADIUS | | | | | | | | | | | | | | | | SCEA=0° |
|---|---|---|---|---|---|---|---|---|---|---|---|---|---|---|---|---|
| FEED PER TOOTH | 1° ECEA | | 2° ECEA | | 3° ECEA | | 4° ECEA | | 5° ECEA | | 6° ECEA | | 7° ECEA | | 8° ECEA | |
| | $R_a$ | $R_t$ | $R_a$ | $R_t$ | $R_a$ | $R_t$ | $R_a$ | $R_t$ | $R_a$ | $R_t$ | $R_a$ | $R_t$ | $R_a$ | $R_t$ | $R_a$ | $R_t$ |
| in | microinches | | | | | | | | | | | | | | | |
| 0.001 | 4.4 | 17. | 8.7 | 35. | 13. | 52. | 17. | 70. | 22. | 87. | 26. | 105. | 31. | 123. | 35. | 141. |
| 0.002 | 8.7 | 35. | 17. | 70. | 26. | 105. | 35. | 140. | 44. | 175. | 53. | 210. | 61. | 246. | 70. | 281. |
| 0.003 | 13. | 52. | 26. | 105. | 39. | 157. | 52. | 210. | 66. | 262. | 79. | 315. | 92. | 368. | 105. | 422. |
| 0.004 | 17. | 70. | 35. | 140. | 52. | 210. | 70. | 280. | 87. | 350. | 105. | 420. | 123. | 491. | 141. | 562. |
| 0.005 | 22. | 87. | 44. | 175. | 66. | 262. | 87. | 350. | 109. | 437. | 131. | 526. | 153. | 614. | 176. | 703. |
| 0.006 | 26. | 105. | 52. | 210. | 79. | 314. | 105. | 420. | 131. | 525. | 158. | 631. | 184. | 737. | 211. | 843. |
| 0.007 | 31. | 122. | 61. | 244. | 92. | 367. | 122. | 489. | 153. | 612. | 184. | 736. | 215. | 859. | 246. | 984. |
| 0.008 | 35. | 140. | 70. | 279. | 105. | 419. | 140. | 559. | 175. | 700. | 210. | 841. | 246. | 982. | 281. | 1124. |
| 0.009 | 39. | 157. | 79. | 314. | 118. | 472. | 157. | 629. | 197. | 787. | 236. | 946. | 276. | 1105. | 316. | 1265. |
| 0.010 | 44. | 175. | 87. | 349. | 131. | 524. | 175. | 699. | 219. | 875. | 263. | 1051. | 307. | 1228. | 351. | 1405. |
| 0.012 | 52. | 209. | 105. | 419. | 157. | 629. | 210. | 839. | 262. | 1050. | 315. | 1261. | 368. | 1473. | 422. | 1686. |
| 0.014 | 61. | 244. | 122. | 489. | 183. | 734. | 245. | 979. | 306. | 1225. | 368. | 1471. | 430. | 1719. | 492. | 1968. |
| 0.016 | 70. | 279. | 140. | 559. | 210. | 839. | 280. | 1119. | 350. | 1400. | 420. | 1682. | 491. | 1965. | 562. | 2249. |
| 0.018 | 79. | 314. | 157. | 629. | 236. | 943. | 315. | 1259. | 394. | 1575. | 473. | 1892. | 553. | 2210. | 632. | 2530. |
| 0.020 | 87. | 349. | 175. | 698. | 262. | 1048. | 350. | 1399. | 437. | 1750. | 526. | 2102. | 614. | 2456. | 703. | 2811. |
| 0.025 | 109. | 436. | 218. | 873. | 328. | 1310. | 437. | 1748. | 547. | 2187. | 657. | 2628. | 767. | 3070. | 878. | 3514. |
| 0.030 | 131. | 524. | 262. | 1048. | 393. | 1572. | 524. | 2098. | 656. | 2625. | 788. | 3153. | 921. | 3684. | 1054. | 4216. |
| 0.035 | 153. | 611. | 306. | 1222. | 459. | 1834. | 612. | 2447. | 766. | 3062. | 920. | 3679. | 1074. | 4297. | 1230. | 4919. |
| 0.040 | 175. | 698. | 349. | 1397. | 524. | 2096. | 699. | 2797. | 875. | 3500. | 1051. | 4204. | 1228. | 4911. | 1405. | 5622. |
| 0.045 | 196. | 785. | 393. | 1571. | 590. | 2358. | 787. | 3147. | 984. | 3937. | 1182. | 4730. | 1381. | 5525. | 1581. | 6324. |
| 0.050 | 218. | 873. | 437. | 1746. | 655. | 2620. | 874. | 3496. | 1094. | 4374. | 1314. | 5255. | 1535. | 6139. | 1757. | 7027. |
| mm | micrometers | | | | | | | | | | | | | | | |
| 0.020 | 0.09 | 0.35 | 0.17 | 0.70 | 0.26 | 1.0 | 0.35 | 1.4 | 0.44 | 1.7 | 0.53 | 2.1 | 0.61 | 2.5 | 0.70 | 2.8 |
| 0.040 | 0.17 | 0.70 | 0.35 | 1.4 | 0.52 | 2.1 | 0.70 | 2.8 | 0.87 | 3.5 | 1.1 | 4.2 | 1.2 | 4.9 | 1.4 | 5.6 |
| 0.060 | 0.26 | 1.0 | 0.52 | 2.1 | 0.79 | 3.1 | 1.0 | 4.2 | 1.3 | 5.2 | 1.6 | 6.3 | 1.8 | 7.4 | 2.1 | 8.4 |
| 0.080 | 0.35 | 1.4 | 0.70 | 2.8 | 1.0 | 4.2 | 1.4 | 5.6 | 1.7 | 7.0 | 2.1 | 8.4 | 2.5 | 9.8 | 2.8 | 11. |
| 0.100 | 0.44 | 1.7 | 0.87 | 3.5 | 1.3 | 5.2 | 1.7 | 7.0 | 2.2 | 8.7 | 2.6 | 11. | 3.1 | 12. | 3.5 | 14. |
| 0.120 | 0.52 | 2.1 | 1.0 | 4.2 | 1.6 | 6.3 | 2.1 | 8.4 | 2.6 | 10. | 3.2 | 13. | 3.7 | 15. | 4.2 | 17. |
| 0.140 | 0.61 | 2.4 | 1.2 | 4.9 | 1.8 | 7.3 | 2.4 | 9.8 | 3.1 | 12. | 3.7 | 15. | 4.3 | 17. | 4.9 | 20. |
| 0.160 | 0.70 | 2.8 | 1.4 | 5.6 | 2.1 | 8.4 | 2.8 | 11. | 3.5 | 14. | 4.2 | 17. | 4.9 | 20. | 5.6 | 22. |
| 0.180 | 0.79 | 3.1 | 1.6 | 6.3 | 2.4 | 9.4 | 3.1 | 13. | 3.9 | 16. | 4.7 | 19. | 5.5 | 22. | 6.3 | 25. |
| 0.200 | 0.87 | 3.5 | 1.7 | 7.0 | 2.6 | 10. | 3.5 | 14. | 4.4 | 17. | 5.3 | 21. | 6.1 | 25. | 7.0 | 28. |
| 0.250 | 1.1 | 4.4 | 2.2 | 8.7 | 3.3 | 13. | 4.4 | 17. | 5.5 | 22. | 6.6 | 26. | 7.7 | 31. | 8.8 | 35. |
| 0.300 | 1.3 | 5.2 | 2.6 | 10. | 3.9 | 16. | 5.2 | 21. | 6.6 | 26. | 7.9 | 32. | 9.2 | 37. | 11. | 42. |
| 0.350 | 1.5 | 6.1 | 3.1 | 12. | 4.6 | 18. | 6.1 | 24. | 7.7 | 31. | 9.2 | 37. | 11. | 43. | 12. | 49. |
| 0.400 | 1.7 | 7.0 | 3.5 | 14. | 5.2 | 21. | 7.0 | 28. | 8.7 | 35. | 11. | 42. | 12. | 49. | 14. | 56. |
| 0.450 | 2.0 | 7.9 | 3.9 | 16. | 5.9 | 24. | 7.9 | 31. | 9.8 | 39. | 12. | 47. | 14. | 55. | 16. | 63. |
| 0.500 | 2.2 | 8.7 | 4.4 | 17. | 6.6 | 26. | 8.7 | 35. | 11. | 44. | 13. | 53. | 15. | 61. | 18. | 70. |
| 0.600 | 2.6 | 10. | 5.2 | 21. | 7.9 | 31. | 10. | 42. | 13. | 52. | 16. | 63. | 18. | 74. | 21. | 84. |
| 0.700 | 3.1 | 12. | 6.1 | 24. | 9.2 | 37. | 12. | 49. | 15. | 61. | 18. | 74. | 21. | 86. | 25. | 98. |
| 0.800 | 3.5 | 14. | 7.0 | 28. | 10. | 42. | 14. | 56. | 17. | 70. | 21. | 84. | 25. | 98. | 28. | 112. |
| 1.000 | 4.4 | 17. | 8.7 | 35. | 13. | 52. | 17. | 70. | 22. | 87. | 26. | 105. | 31. | 123. | 35. | 141. |
| 1.200 | 5.2 | 21. | 10. | 42. | 16. | 63. | 21. | 84. | 26. | 105. | 32. | 126. | 37. | 147. | 42. | 169. |

# 18.2 Surface Texture

**TABLE 18.2-5 THEORETICAL Values for Arithmetic-Average Roughness ($R_a$) and Maximum Peak-to-Valley Roughness Height ($R_t$) for Milling—Tool Type A**

| ZERO NOSE RADIUS | | | | | | | | | | | | | | | | SCEA=15° |
|---|---|---|---|---|---|---|---|---|---|---|---|---|---|---|---|---|
| FEED PER TOOTH | 1° ECEA | | 2° ECEA | | 3° ECEA | | 4° ECEA | | 5° ECEA | | 6° ECEA | | 7° ECEA | | 8° ECEA | |
| | $R_a$ | $R_t$ | $R_a$ | $R_t$ | $R_a$ | $R_t$ | $R_a$ | $R_t$ | $R_a$ | $R_t$ | $R_a$ | $R_t$ | $R_a$ | $R_t$ | $R_a$ | $R_t$ |
| in | microinches | | | | | | | | | | | | | | | |
| 0.001 | 4.3 | 17. | 8.6 | 35. | 13. | 52. | 17. | 69. | 21. | 85. | 26. | 102. | 30. | 119. | 34. | 135. |
| 0.002 | 8.7 | 35. | 17. | 69. | 26. | 103. | 34. | 137. | 43. | 171. | 51. | 204. | 59. | 238. | 68. | 271. |
| 0.003 | 13. | 52. | 26. | 104. | 39. | 155. | 51. | 206. | 64. | 256. | 77. | 307. | 89. | 357. | 102. | 406. |
| 0.004 | 17. | 69. | 35. | 138. | 52. | 207. | 69. | 275. | 85. | 342. | 102. | 409. | 119. | 475. | 135. | 542. |
| 0.005 | 22. | 87. | 43. | 173. | 65. | 258. | 86. | 343. | 107. | 427. | 128. | 511. | 149. | 594. | 169. | 677. |
| 0.006 | 26. | 104. | 52. | 208. | 78. | 310. | 103. | 412. | 128. | 513. | 153. | 613. | 178. | 713. | 203. | 813. |
| 0.007 | 30. | 122. | 61. | 242. | 90. | 362. | 120. | 480. | 150. | 598. | 179. | 716. | 208. | 832. | 237. | 948. |
| 0.008 | 35. | 139. | 69. | 277. | 103. | 413. | 137. | 549. | 171. | 684. | 204. | 818. | 238. | 951. | 271. | 1084. |
| 0.009 | 39. | 156. | 78. | 311. | 116. | 465. | 154. | 618. | 192. | 769. | 230. | 920. | 267. | 1070. | 305. | 1219. |
| 0.010 | 43. | 174. | 86. | 346. | 129. | 517. | 172. | 686. | 214. | 855. | 256. | 1022. | 297. | 1189. | 339. | 1354. |
| 0.012 | 52. | 208. | 104. | 415. | 155. | 620. | 206. | 824. | 256. | 1026. | 307. | 1227. | 357. | 1426. | 406. | 1625. |
| 0.014 | 61. | 243. | 121. | 484. | 181. | 724. | 240. | 961. | 299. | 1197. | 358. | 1431. | 416. | 1664. | 474. | 1896. |
| 0.016 | 69. | 278. | 138. | 554. | 207. | 827. | 275. | 1098. | 342. | 1368. | 409. | 1636. | 475. | 1902. | 542. | 2167. |
| 0.018 | 78. | 313. | 156. | 623. | 233. | 930. | 309. | 1236. | 385. | 1539. | 460. | 1840. | 535. | 2140. | 609. | 2438. |
| 0.020 | 87. | 347. | 173. | 692. | 258. | 1034. | 343. | 1373. | 427. | 1710. | 511. | 2045. | 594. | 2377. | 677. | 2709. |
| 0.025 | 109. | 434. | 216. | 865. | 323. | 1292. | 429. | 1716. | 534. | 2137. | 639. | 2556. | 743. | 2972. | 847. | 3386. |
| 0.030 | 130. | 521. | 259. | 1038. | 388. | 1550. | 515. | 2059. | 641. | 2565. | 767. | 3067. | 892. | 3566. | 1016. | 4063. |
| 0.035 | 152. | 608. | 303. | 1211. | 452. | 1809. | 601. | 2402. | 748. | 2992. | 894. | 3578. | 1040. | 4161. | 1185. | 4740. |
| 0.040 | 174. | 695. | 346. | 1384. | 517. | 2067. | 686. | 2746. | 855. | 3419. | 1022. | 4089. | 1189. | 4755. | 1354. | 5418. |
| 0.045 | 195. | 782. | 389. | 1557. | 581. | 2326. | 772. | 3089. | 962. | 3847. | 1150. | 4600. | 1337. | 5349. | 1524. | 6095. |
| 0.050 | 217. | 869. | 432. | 1730. | 646. | 2584. | 858. | 3432. | 1069. | 4274. | 1278. | 5111. | 1486. | 5944. | 1693. | 6772. |
| mm | micrometers | | | | | | | | | | | | | | | |
| 0.020 | 0.09 | 0.35 | 0.17 | 0.69 | 0.26 | 1.0 | 0.34 | 1.4 | 0.43 | 1.7 | 0.51 | 2.0 | 0.59 | 2.4 | 0.68 | 2.7 |
| 0.040 | 0.17 | 0.69 | 0.35 | 1.4 | 0.52 | 2.1 | 0.69 | 2.7 | 0.85 | 3.4 | 1.0 | 4.1 | 1.2 | 4.8 | 1.4 | 5.4 |
| 0.060 | 0.26 | 1.0 | 0.52 | 2.1 | 0.78 | 3.1 | 1.0 | 4.1 | 1.3 | 5.1 | 1.5 | 6.1 | 1.8 | 7.1 | 2.0 | 8.1 |
| 0.080 | 0.35 | 1.4 | 0.69 | 2.8 | 1.0 | 4.1 | 1.4 | 5.5 | 1.7 | 6.8 | 2.0 | 8.2 | 2.4 | 9.5 | 2.7 | 11. |
| 0.100 | 0.43 | 1.7 | 0.86 | 3.5 | 1.3 | 5.2 | 1.7 | 6.9 | 2.1 | 8.5 | 2.6 | 10. | 3.0 | 12. | 3.4 | 14. |
| 0.120 | 0.52 | 2.1 | 1.0 | 4.2 | 1.6 | 6.2 | 2.1 | 8.2 | 2.6 | 10. | 3.1 | 12. | 3.6 | 14. | 4.1 | 16. |
| 0.140 | 0.61 | 2.4 | 1.2 | 4.8 | 1.8 | 7.2 | 2.4 | 9.6 | 3.0 | 12. | 3.6 | 14. | 4.2 | 17. | 4.7 | 19. |
| 0.160 | 0.69 | 2.8 | 1.4 | 5.5 | 2.1 | 8.3 | 2.7 | 11. | 3.4 | 14. | 4.1 | 16. | 4.8 | 19. | 5.4 | 22. |
| 0.180 | 0.78 | 3.1 | 1.6 | 6.2 | 2.3 | 9.3 | 3.1 | 12. | 3.8 | 15. | 4.6 | 18. | 5.3 | 21. | 6.1 | 24. |
| 0.200 | 0.87 | 3.5 | 1.7 | 6.9 | 2.6 | 10. | 3.4 | 14. | 4.3 | 17. | 5.1 | 20. | 5.9 | 24. | 6.8 | 27. |
| 0.250 | 1.1 | 4.3 | 2.2 | 8.6 | 3.2 | 13. | 4.3 | 17. | 5.3 | 21. | 6.4 | 26. | 7.4 | 30. | 8.5 | 34. |
| 0.300 | 1.3 | 5.2 | 2.6 | 10. | 3.9 | 16. | 5.1 | 21. | 6.4 | 26. | 7.7 | 31. | 8.9 | 36. | 10. | 41. |
| 0.350 | 1.5 | 6.1 | 3.0 | 12. | 4.5 | 18. | 6.0 | 24. | 7.5 | 30. | 8.9 | 36. | 10. | 42. | 12. | 47. |
| 0.400 | 1.7 | 6.9 | 3.5 | 14. | 5.2 | 21. | 6.9 | 27. | 8.5 | 34. | 10. | 41. | 12. | 48. | 14. | 54. |
| 0.450 | 2.0 | 7.8 | 3.9 | 16. | 5.8 | 23. | 7.7 | 31. | 9.6 | 38. | 12. | 46. | 13. | 53. | 15. | 61. |
| 0.500 | 2.2 | 8.7 | 4.3 | 17. | 6.5 | 26. | 8.6 | 34. | 11. | 43. | 13. | 51. | 15. | 59. | 17. | 68. |
| 0.600 | 2.6 | 10. | 5.2 | 21. | 7.8 | 31. | 10. | 41. | 13. | 51. | 15. | 61. | 18. | 71. | 20. | 81. |
| 0.700 | 3.0 | 12. | 6.1 | 24. | 9.0 | 36. | 12. | 48. | 15. | 60. | 18. | 72. | 21. | 83. | 24. | 95. |
| 0.800 | 3.5 | 14. | 6.9 | 28. | 10. | 41. | 14. | 55. | 17. | 68. | 20. | 82. | 24. | 95. | 27. | 108. |
| 1.000 | 4.3 | 17. | 8.6 | 35. | 13. | 52. | 17. | 69. | 21. | 85. | 26. | 102. | 30. | 119. | 34. | 135. |
| 1.200 | 5.2 | 21. | 10. | 42. | 16. | 62. | 21. | 82. | 26. | 103. | 31. | 123. | 36. | 143. | 41. | 163. |

**TABLE 18.2-6 THEORETICAL Values for Arithmetic-Average Roughness ($R_a$) and Maximum Peak-to-Valley Roughness Height ($R_t$) for Milling—Tool Type A**

ZERO NOSE RADIUS — SCEA=30°

| FEED PER TOOTH | 1° ECEA | | 2° ECEA | | 3° ECEA | | 4° ECEA | | 5° ECEA | | 6° ECEA | | 7° ECEA | | 8° ECEA | |
|---|---|---|---|---|---|---|---|---|---|---|---|---|---|---|---|---|
| | $R_a$ | $R_t$ | $R_a$ | $R_t$ | $R_a$ | $R_t$ | $R_a$ | $R_t$ | $R_a$ | $R_t$ | $R_a$ | $R_t$ | $R_a$ | $R_t$ | $R_a$ | $R_t$ |
| in | microinches | | | | | | | | | | | | | | | |
| 0.001 | 4.3 | 17. | 8.6 | 34. | 13. | 51. | 17. | 67. | 21. | 83. | 25. | 99. | 29. | 115. | 32. | 130. |
| 0.002 | 8.6 | 35. | 17. | 68. | 25. | 102. | 34. | 134. | 42. | 167. | 50. | 198. | 57. | 229. | 65. | 260. |
| 0.003 | 13. | 52. | 26. | 103. | 38. | 153. | 50. | 202. | 62. | 250. | 74. | 297. | 86. | 344. | 97. | 390. |
| 0.004 | 17. | 69. | 34. | 137. | 51. | 203. | 67. | 269. | 83. | 333. | 99. | 396. | 115. | 459. | 130. | 520. |
| 0.005 | 22. | 86. | 43. | 171. | 64. | 254. | 84. | 336. | 104. | 416. | 124. | 495. | 143. | 573. | 162. | 650. |
| 0.006 | 26. | 104. | 51. | 205. | 76. | 305. | 101. | 403. | 125. | 500. | 149. | 595. | 172. | 688. | 195. | 780. |
| 0.007 | 30. | 121. | 60. | 240. | 89. | 356. | 118. | 470. | 146. | 583. | 173. | 694. | 201. | 803. | 227. | 910. |
| 0.008 | 35. | 138. | 68. | 274. | 102. | 407. | 134. | 538. | 167. | 666. | 198. | 793. | 229. | 917. | 260. | 1040. |
| 0.009 | 39. | 156. | 77. | 308. | 114. | 458. | 151. | 605. | 187. | 750. | 223. | 892. | 258. | 1032. | 292. | 1170. |
| 0.010 | 43. | 173. | 86. | 342. | 127. | 509. | 168. | 672. | 208. | 833. | 248. | 991. | 287. | 1147. | 325. | 1300. |
| 0.012 | 52. | 207. | 103. | 411. | 153. | 610. | 202. | 807. | 250. | 999. | 297. | 1189. | 344. | 1376. | 390. | 1560. |
| 0.014 | 60. | 242. | 120. | 479. | 178. | 712. | 235. | 941. | 291. | 1166. | 347. | 1387. | 401. | 1605. | 455. | 1820. |
| 0.016 | 69. | 276. | 137. | 548. | 203. | 814. | 269. | 1075. | 333. | 1333. | 396. | 1585. | 459. | 1835. | 520. | 2080. |
| 0.018 | 78. | 311. | 154. | 616. | 229. | 916. | 302. | 1210. | 375. | 1499. | 446. | 1784. | 516. | 2064. | 585. | 2340. |
| 0.020 | 86. | 346. | 171. | 685. | 254. | 1017. | 336. | 1344. | 416. | 1666. | 495. | 1982. | 573. | 2293. | 650. | 2600. |
| 0.025 | 108. | 432. | 214. | 856. | 318. | 1272. | 420. | 1680. | 521. | 2082. | 619. | 2477. | 717. | 2866. | 812. | 3250. |
| 0.030 | 130. | 518. | 257. | 1027. | 382. | 1526. | 504. | 2016. | 625. | 2498. | 743. | 2973. | 860. | 3440. | 975. | 3900. |
| 0.035 | 151. | 605. | 300. | 1198. | 445. | 1780. | 588. | 2352. | 729. | 2915. | 867. | 3468. | 1003. | 4013. | 1137. | 4550. |
| 0.040 | 173. | 691. | 342. | 1369. | 509. | 2035. | 672. | 2689. | 833. | 3331. | 991. | 3964. | 1147. | 4586. | 1300. | 5200. |
| 0.045 | 194. | 778. | 385. | 1540. | 572. | 2289. | 756. | 3025. | 937. | 3748. | 1115. | 4459. | 1290. | 5160. | 1462. | 5850. |
| 0.050 | 216. | 864. | 428. | 1712. | 636. | 2543. | 840. | 3361. | 1041. | 4164. | 1239. | 4955. | 1433. | 5733. | 1625. | 6500. |
| mm | micrometers | | | | | | | | | | | | | | | |
| 0.020 | 0.09 | 0.35 | 0.17 | 0.68 | 0.25 | 1.0 | 0.34 | 1.3 | 0.42 | 1.7 | 0.50 | 2.0 | 0.57 | 2.3 | 0.65 | 2.6 |
| 0.040 | 0.17 | 0.69 | 0.34 | 1.4 | 0.51 | 2.0 | 0.67 | 2.7 | 0.83 | 3.3 | 0.99 | 4.0 | 1.1 | 4.6 | 1.3 | 5.2 |
| 0.060 | 0.26 | 1.0 | 0.51 | 2.1 | 0.76 | 3.1 | 1.0 | 4.0 | 1.2 | 5.0 | 1.5 | 5.9 | 1.7 | 6.9 | 1.9 | 7.8 |
| 0.080 | 0.35 | 1.4 | 0.68 | 2.7 | 1.0 | 4.1 | 1.3 | 5.4 | 1.7 | 6.7 | 2.0 | 7.9 | 2.3 | 9.2 | 2.6 | 10. |
| 0.100 | 0.43 | 1.7 | 0.86 | 3.4 | 1.3 | 5.1 | 1.7 | 6.7 | 2.1 | 8.3 | 2.5 | 9.9 | 2.9 | 11. | 3.2 | 13. |
| 0.120 | 0.52 | 2.1 | 1.0 | 4.1 | 1.5 | 6.1 | 2.0 | 8.1 | 2.5 | 10.0 | 3.0 | 12. | 3.4 | 14. | 3.9 | 16. |
| 0.140 | 0.60 | 2.4 | 1.2 | 4.8 | 1.8 | 7.1 | 2.4 | 9.4 | 2.9 | 12. | 3.5 | 14. | 4.0 | 16. | 4.5 | 18. |
| 0.160 | 0.69 | 2.8 | 1.4 | 5.5 | 2.0 | 8.1 | 2.7 | 11. | 3.3 | 13. | 4.0 | 16. | 4.6 | 18. | 5.2 | 21. |
| 0.180 | 0.78 | 3.1 | 1.5 | 6.2 | 2.3 | 9.2 | 3.0 | 12. | 3.7 | 15. | 4.5 | 18. | 5.2 | 21. | 5.8 | 23. |
| 0.200 | 0.86 | 3.5 | 1.7 | 6.8 | 2.5 | 10. | 3.4 | 13. | 4.2 | 17. | 5.0 | 20. | 5.7 | 23. | 6.5 | 26. |
| 0.250 | 1.1 | 4.3 | 2.1 | 8.6 | 3.2 | 13. | 4.2 | 17. | 5.2 | 21. | 6.2 | 25. | 7.2 | 29. | 8.1 | 32. |
| 0.300 | 1.3 | 5.2 | 2.6 | 10. | 3.8 | 15. | 5.0 | 20. | 6.2 | 25. | 7.4 | 30. | 8.6 | 34. | 9.7 | 39. |
| 0.350 | 1.5 | 6.0 | 3.0 | 12. | 4.5 | 18. | 5.9 | 24. | 7.3 | 29. | 8.7 | 35. | 10. | 40. | 11. | 45. |
| 0.400 | 1.7 | 6.9 | 3.4 | 14. | 5.1 | 20. | 6.7 | 27. | 8.3 | 33. | 9.9 | 40. | 11. | 46. | 13. | 52. |
| 0.450 | 1.9 | 7.8 | 3.9 | 15. | 5.7 | 23. | 7.6 | 30. | 9.4 | 37. | 11. | 45. | 13. | 52. | 15. | 58. |
| 0.500 | 2.2 | 8.6 | 4.3 | 17. | 6.4 | 25. | 8.4 | 34. | 10. | 42. | 12. | 50. | 14. | 57. | 16. | 65. |
| 0.600 | 2.6 | 10. | 5.1 | 21. | 7.6 | 31. | 10. | 40. | 12. | 50. | 15. | 59. | 17. | 69. | 19. | 78. |
| 0.700 | 3.0 | 12. | 6.0 | 24. | 8.9 | 36. | 12. | 47. | 15. | 58. | 17. | 69. | 20. | 80. | 23. | 91. |
| 0.800 | 3.5 | 14. | 6.8 | 27. | 10. | 41. | 13. | 54. | 17. | 67. | 20. | 79. | 23. | 92. | 26. | 104. |
| 1.000 | 4.3 | 17. | 8.6 | 34. | 13. | 51. | 17. | 67. | 21. | 83. | 25. | 99. | 29. | 115. | 32. | 130. |
| 1.200 | 5.2 | 21. | 10. | 41. | 15. | 61. | 20. | 81. | 25. | 100. | 30. | 119. | 34. | 138. | 39. | 156. |

## 18.2 Surface Texture

**TABLE 18.2-7 THEORETICAL Values for Arithmetic-Average Roughness ($R_a$) and Maximum Peak-to-Valley Roughness Height ($R_t$) for Milling—Tool Type A**

| ZERO NOSE RADIUS | | | | | | | | | | | | | | | | SCEA=45° |
|---|---|---|---|---|---|---|---|---|---|---|---|---|---|---|---|---|
| FEED PER TOOTH | 1° ECEA | | 2° ECEA | | 3° ECEA | | 4° ECEA | | 5° ECEA | | 6° ECEA | | 7° ECEA | | 8° ECEA | |
| | $R_a$ | $R_t$ | $R_a$ | $R_t$ | $R_a$ | $R_t$ | $R_a$ | $R_t$ | $R_a$ | $R_t$ | $R_a$ | $R_t$ | $R_a$ | $R_t$ | $R_a$ | $R_t$ |
| in | microinches | | | | | | | | | | | | | | | |
| 0.001 | 4.3 | 17. | 8.4 | 34. | 12. | 50. | 16. | 65. | 20. | 80. | 24. | 95. | 27. | 109. | 31. | 123. |
| 0.002 | 8.6 | 34. | 17. | 67. | 25. | 100. | 33. | 131. | 40. | 161. | 48. | 190. | 55. | 219. | 62. | 246. |
| 0.003 | 13. | 51. | 25. | 101. | 37. | 149. | 49. | 196. | 60. | 241. | 71. | 285. | 82. | 328. | 92. | 370. |
| 0.004 | 17. | 69. | 34. | 135. | 50. | 199. | 65. | 261. | 80. | 322. | 95. | 380. | 109. | 437. | 123. | 493. |
| 0.005 | 21. | 86. | 42. | 169. | 62. | 249. | 82. | 327. | 101. | 402. | 119. | 476. | 137. | 547. | 154. | 616. |
| 0.006 | 26. | 103. | 51. | 202. | 75. | 299. | 98. | 392. | 121. | 483. | 143. | 571. | 164. | 656. | 185. | 739. |
| 0.007 | 30. | 120. | 59. | 236. | 87. | 349. | 114. | 457. | 141. | 563. | 166. | 666. | 191. | 766. | 216. | 863. |
| 0.008 | 34. | 137. | 67. | 270. | 100. | 398. | 131. | 523. | 161. | 644. | 190. | 761. | 219. | 875. | 246. | 986. |
| 0.009 | 39. | 154. | 76. | 304. | 112. | 448. | 147. | 588. | 181. | 724. | 214. | 856. | 246. | 984. | 277. | 1109. |
| 0.010 | 43. | 172. | 84. | 337. | 124. | 498. | 163. | 654. | 201. | 805. | 238. | 951. | 273. | 1094. | 308. | 1232. |
| 0.012 | 51. | 206. | 101. | 405. | 149. | 598. | 196. | 784. | 241. | 965. | 285. | 1141. | 328. | 1312. | 370. | 1479. |
| 0.014 | 60. | 240. | 118. | 472. | 174. | 697. | 229. | 915. | 282. | 1126. | 333. | 1332. | 383. | 1531. | 431. | 1725. |
| 0.016 | 69. | 274. | 135. | 540. | 199. | 797. | 261. | 1046. | 322. | 1287. | 380. | 1522. | 437. | 1750. | 493. | 1972. |
| 0.018 | 77. | 309. | 152. | 607. | 224. | 896. | 294. | 1176. | 362. | 1448. | 428. | 1712. | 492. | 1968. | 555. | 2218. |
| 0.020 | 86. | 343. | 169. | 675. | 249. | 996. | 327. | 1307. | 402. | 1609. | 476. | 1902. | 547. | 2187. | 616. | 2464. |
| 0.025 | 107. | 429. | 211. | 844. | 311. | 1245. | 408. | 1634. | 503. | 2011. | 594. | 2378. | 683. | 2734. | 770. | 3081. |
| 0.030 | 129. | 515. | 253. | 1012. | 373. | 1494. | 490. | 1961. | 603. | 2414. | 713. | 2853. | 820. | 3281. | 924. | 3697. |
| 0.035 | 150. | 600. | 295. | 1181. | 436. | 1743. | 572. | 2287. | 704. | 2816. | 832. | 3329. | 957. | 3828. | 1078. | 4313. |
| 0.040 | 172. | 686. | 337. | 1350. | 498. | 1992. | 654. | 2614. | 805. | 3218. | 951. | 3804. | 1094. | 4374. | 1232. | 4929. |
| 0.045 | 193. | 772. | 380. | 1518. | 560. | 2241. | 735. | 2941. | 905. | 3620. | 1070. | 4280. | 1230. | 4921. | 1386. | 5545. |
| 0.050 | 214. | 858. | 422. | 1687. | 622. | 2490. | 817. | 3268. | 1006. | 4023. | 1189. | 4755. | 1367. | 5468. | 1540. | 6161. |
| mm | micrometers | | | | | | | | | | | | | | | |
| 0.020 | 0.09 | 0.34 | 0.17 | 0.67 | 0.25 | 1.00 | 0.33 | 1.3 | 0.40 | 1.6 | 0.48 | 1.9 | 0.55 | 2.2 | 0.62 | 2.5 |
| 0.040 | 0.17 | 0.69 | 0.34 | 1.3 | 0.50 | 2.0 | 0.65 | 2.6 | 0.80 | 3.2 | 0.95 | 3.8 | 1.1 | 4.4 | 1.2 | 4.9 |
| 0.060 | 0.26 | 1.0 | 0.51 | 2.0 | 0.75 | 3.0 | 0.98 | 3.9 | 1.2 | 4.8 | 1.4 | 5.7 | 1.6 | 6.6 | 1.8 | 7.4 |
| 0.080 | 0.34 | 1.4 | 0.67 | 2.7 | 1.00 | 4.0 | 1.3 | 5.2 | 1.6 | 6.4 | 1.9 | 7.6 | 2.2 | 8.7 | 2.5 | 9.9 |
| 0.100 | 0.43 | 1.7 | 0.84 | 3.4 | 1.2 | 5.0 | 1.6 | 6.5 | 2.0 | 8.0 | 2.4 | 9.5 | 2.7 | 11. | 3.1 | 12. |
| 0.120 | 0.51 | 2.1 | 1.0 | 4.0 | 1.5 | 6.0 | 2.0 | 7.8 | 2.4 | 9.7 | 2.9 | 11. | 3.3 | 13. | 3.7 | 15. |
| 0.140 | 0.60 | 2.4 | 1.2 | 4.7 | 1.7 | 7.0 | 2.3 | 9.1 | 2.8 | 11. | 3.3 | 13. | 3.8 | 15. | 4.3 | 17. |
| 0.160 | 0.69 | 2.7 | 1.3 | 5.4 | 2.0 | 8.0 | 2.6 | 10. | 3.2 | 13. | 3.8 | 15. | 4.4 | 17. | 4.9 | 20. |
| 0.180 | 0.77 | 3.1 | 1.5 | 6.1 | 2.2 | 9.0 | 2.9 | 12. | 3.6 | 14. | 4.3 | 17. | 4.9 | 20. | 5.5 | 22. |
| 0.200 | 0.86 | 3.4 | 1.7 | 6.7 | 2.5 | 10.0 | 3.3 | 13. | 4.0 | 16. | 4.8 | 19. | 5.5 | 22. | 6.2 | 25. |
| 0.250 | 1.1 | 4.3 | 2.1 | 8.4 | 3.1 | 12. | 4.1 | 16. | 5.0 | 20. | 5.9 | 24. | 6.8 | 27. | 7.7 | 31. |
| 0.300 | 1.3 | 5.1 | 2.5 | 10. | 3.7 | 15. | 4.9 | 20. | 6.0 | 24. | 7.1 | 29. | 8.2 | 33. | 9.2 | 37. |
| 0.350 | 1.5 | 6.0 | 3.0 | 12. | 4.4 | 17. | 5.7 | 23. | 7.0 | 28. | 8.3 | 33. | 9.6 | 38. | 11. | 43. |
| 0.400 | 1.7 | 6.9 | 3.4 | 13. | 5.0 | 20. | 6.5 | 26. | 8.0 | 32. | 9.5 | 38. | 11. | 44. | 12. | 49. |
| 0.450 | 1.9 | 7.7 | 3.8 | 15. | 5.6 | 22. | 7.4 | 29. | 9.1 | 36. | 11. | 43. | 12. | 49. | 14. | 55. |
| 0.500 | 2.1 | 8.6 | 4.2 | 17. | 6.2 | 25. | 8.2 | 33. | 10. | 40. | 12. | 48. | 14. | 55. | 15. | 62. |
| 0.600 | 2.6 | 10. | 5.1 | 20. | 7.5 | 30. | 9.8 | 39. | 12. | 48. | 14. | 57. | 16. | 66. | 18. | 74. |
| 0.700 | 3.0 | 12. | 5.9 | 24. | 8.7 | 35. | 11. | 46. | 14. | 56. | 17. | 67. | 19. | 77. | 22. | 86. |
| 0.800 | 3.4 | 14. | 6.7 | 27. | 10.0 | 40. | 13. | 52. | 16. | 64. | 19. | 76. | 22. | 87. | 25. | 99. |
| 1.000 | 4.3 | 17. | 8.4 | 34. | 12. | 50. | 16. | 65. | 20. | 80. | 24. | 95. | 27. | 109. | 31. | 123. |
| 1.200 | 5.1 | 21. | 10. | 40. | 15. | 60. | 20. | 78. | 24. | 97. | 29. | 114. | 33. | 131. | 37. | 148. |

**TABLE 18.2-8 THEORETICAL Values for Arithmetic-Average Roughness ($R_a$) and Maximum Peak-to-Valley Roughness Height ($R_t$) for Milling—Tool Type A**

ZERO NOSE RADIUS — SCEA=60°

| FEED PER TOOTH | 1° ECEA | | 2° ECEA | | 3° ECEA | | 4° ECEA | | 5° ECEA | | 6° ECEA | | 7° ECEA | | 8° ECEA | |
|---|---|---|---|---|---|---|---|---|---|---|---|---|---|---|---|---|
| | $R_a$ | $R_t$ | $R_a$ | $R_t$ | $R_a$ | $R_t$ | $R_a$ | $R_t$ | $R_a$ | $R_t$ | $R_a$ | $R_t$ | $R_a$ | $R_t$ | $R_a$ | $R_t$ |
| in | microinches | | | | | | | | | | | | | | | |
| 0.001 | 4.2 | 17. | 8.2 | 33. | 12. | 48. | 16. | 62. | 19. | 76. | 22. | 89. | 25. | 101. | 28. | 113. |
| 0.002 | 8.5 | 34. | 16. | 66. | 24. | 96. | 31. | 125. | 38. | 152. | 44. | 178. | 51. | 203. | 57. | 226. |
| 0.003 | 13. | 51. | 25. | 99. | 36. | 144. | 47. | 187. | 57. | 228. | 67. | 267. | 76. | 304. | 85. | 339. |
| 0.004 | 17. | 68. | 33. | 132. | 48. | 192. | 62. | 249. | 76. | 304. | 89. | 356. | 101. | 405. | 113. | 452. |
| 0.005 | 21. | 85. | 41. | 165. | 60. | 240. | 78. | 312. | 95. | 380. | 111. | 445. | 127. | 506. | 141. | 565. |
| 0.006 | 25. | 102. | 49. | 198. | 72. | 288. | 94. | 374. | 114. | 456. | 133. | 534. | 152. | 608. | 170. | 678. |
| 0.007 | 30. | 119. | 58. | 231. | 84. | 336. | 109. | 437. | 133. | 532. | 156. | 622. | 177. | 709. | 198. | 791. |
| 0.008 | 34. | 136. | 66. | 263. | 96. | 384. | 125. | 499. | 152. | 608. | 178. | 711. | 203. | 810. | 226. | 904. |
| 0.009 | 38. | 152. | 74. | 296. | 108. | 432. | 140. | 561. | 171. | 684. | 200. | 800. | 228. | 911. | 254. | 1017. |
| 0.010 | 42. | 169. | 82. | 329. | 120. | 480. | 156. | 624. | 190. | 760. | 222. | 889. | 253. | 1013. | 283. | 1130. |
| 0.012 | 51. | 203. | 99. | 395. | 144. | 577. | 187. | 748. | 228. | 912. | 267. | 1067. | 304. | 1215. | 339. | 1356. |
| 0.014 | 59. | 237. | 115. | 461. | 168. | 673. | 218. | 873. | 266. | 1064. | 311. | 1245. | 354. | 1418. | 396. | 1582. |
| 0.016 | 68. | 271. | 132. | 527. | 192. | 769. | 249. | 998. | 304. | 1216. | 356. | 1423. | 405. | 1620. | 452. | 1808. |
| 0.018 | 76. | 305. | 148. | 593. | 216. | 865. | 281. | 1123. | 342. | 1368. | 400. | 1601. | 456. | 1823. | 509. | 2034. |
| 0.020 | 85. | 339. | 165. | 659. | 240. | 961. | 312. | 1247. | 380. | 1520. | 445. | 1778. | 506. | 2025. | 565. | 2261. |
| 0.025 | 106. | 424. | 206. | 823. | 300. | 1201. | 390. | 1559. | 475. | 1899. | 556. | 2223. | 633. | 2531. | 706. | 2826. |
| 0.030 | 127. | 508. | 247. | 988. | 360. | 1441. | 468. | 1871. | 570. | 2279. | 667. | 2668. | 759. | 3038. | 848. | 3391. |
| 0.035 | 148. | 593. | 288. | 1153. | 420. | 1682. | 546. | 2183. | 665. | 2659. | 778. | 3112. | 886. | 3544. | 989. | 3956. |
| 0.040 | 169. | 678. | 329. | 1317. | 480. | 1922. | 624. | 2495. | 760. | 3039. | 889. | 3557. | 1013. | 4050. | 1130. | 4521. |
| 0.045 | 191. | 762. | 370. | 1482. | 541. | 2162. | 702. | 2807. | 855. | 3419. | 1000. | 4001. | 1139. | 4556. | 1272. | 5086. |
| 0.050 | 212. | 847. | 412. | 1646. | 601. | 2402. | 780. | 3119. | 950. | 3799. | 1111. | 4446. | 1266. | 5063. | 1413. | 5651. |
| mm | micrometers | | | | | | | | | | | | | | | |
| 0.020 | 0.08 | 0.34 | 0.16 | 0.66 | 0.24 | 0.96 | 0.31 | 1.2 | 0.38 | 1.5 | 0.44 | 1.8 | 0.51 | 2.0 | 0.57 | 2.3 |
| 0.040 | 0.17 | 0.68 | 0.33 | 1.3 | 0.48 | 1.9 | 0.62 | 2.5 | 0.76 | 3.0 | 0.89 | 3.6 | 1.0 | 4.1 | 1.1 | 4.5 |
| 0.060 | 0.25 | 1.0 | 0.49 | 2.0 | 0.72 | 2.9 | 0.94 | 3.7 | 1.1 | 4.6 | 1.3 | 5.3 | 1.5 | 6.1 | 1.7 | 6.8 |
| 0.080 | 0.34 | 1.4 | 0.66 | 2.6 | 0.96 | 3.8 | 1.2 | 5.0 | 1.5 | 6.1 | 1.8 | 7.1 | 2.0 | 8.1 | 2.3 | 9.0 |
| 0.100 | 0.42 | 1.7 | 0.82 | 3.3 | 1.2 | 4.8 | 1.6 | 6.2 | 1.9 | 7.6 | 2.2 | 8.9 | 2.5 | 10. | 2.8 | 11. |
| 0.120 | 0.51 | 2.0 | 0.99 | 4.0 | 1.4 | 5.8 | 1.9 | 7.5 | 2.3 | 9.1 | 2.7 | 11. | 3.0 | 12. | 3.4 | 14. |
| 0.140 | 0.59 | 2.4 | 1.2 | 4.6 | 1.7 | 6.7 | 2.2 | 8.7 | 2.7 | 11. | 3.1 | 12. | 3.5 | 14. | 4.0 | 16. |
| 0.160 | 0.68 | 2.7 | 1.3 | 5.3 | 1.9 | 7.7 | 2.5 | 10.0 | 3.0 | 12. | 3.6 | 14. | 4.1 | 16. | 4.5 | 18. |
| 0.180 | 0.76 | 3.0 | 1.5 | 5.9 | 2.2 | 8.6 | 2.8 | 11. | 3.4 | 14. | 4.0 | 16. | 4.6 | 18. | 5.1 | 20. |
| 0.200 | 0.85 | 3.4 | 1.6 | 6.6 | 2.4 | 9.6 | 3.1 | 12. | 3.8 | 15. | 4.4 | 18. | 5.1 | 20. | 5.7 | 23. |
| 0.250 | 1.1 | 4.2 | 2.1 | 8.2 | 3.0 | 12. | 3.9 | 16. | 4.7 | 19. | 5.6 | 22. | 6.3 | 25. | 7.1 | 28. |
| 0.300 | 1.3 | 5.1 | 2.5 | 9.9 | 3.6 | 14. | 4.7 | 19. | 5.7 | 23. | 6.7 | 27. | 7.6 | 30. | 8.5 | 34. |
| 0.350 | 1.5 | 5.9 | 2.9 | 12. | 4.2 | 17. | 5.5 | 22. | 6.6 | 27. | 7.8 | 31. | 8.9 | 35. | 9.9 | 40. |
| 0.400 | 1.7 | 6.8 | 3.3 | 13. | 4.8 | 19. | 6.2 | 25. | 7.6 | 30. | 8.9 | 36. | 10. | 41. | 11. | 45. |
| 0.450 | 1.9 | 7.6 | 3.7 | 15. | 5.4 | 22. | 7.0 | 28. | 8.5 | 34. | 10. | 40. | 11. | 46. | 13. | 51. |
| 0.500 | 2.1 | 8.5 | 4.1 | 16. | 6.0 | 24. | 7.8 | 31. | 9.5 | 38. | 11. | 44. | 13. | 51. | 14. | 57. |
| 0.600 | 2.5 | 10. | 4.9 | 20. | 7.2 | 29. | 9.4 | 37. | 11. | 46. | 13. | 53. | 15. | 61. | 17. | 68. |
| 0.700 | 3.0 | 12. | 5.8 | 23. | 8.4 | 34. | 11. | 44. | 13. | 53. | 16. | 62. | 18. | 71. | 20. | 79. |
| 0.800 | 3.4 | 14. | 6.6 | 26. | 9.6 | 38. | 12. | 50. | 15. | 61. | 18. | 71. | 20. | 81. | 23. | 90. |
| 1.000 | 4.2 | 17. | 8.2 | 33. | 12. | 48. | 16. | 62. | 19. | 76. | 22. | 89. | 25. | 101. | 28. | 113. |
| 1.200 | 5.1 | 20. | 9.9 | 40. | 14. | 58. | 19. | 75. | 23. | 91. | 27. | 107. | 30. | 122. | 34. | 136. |

**TABLE 18.2-9 THEORETICAL Values for Arithmetic-Average Roughness ($R_a$) and Maximum Peak-to-Valley Roughness Height ($R_t$) for Turning or Milling—Tool Type B**

| FEED/REV. (Turning) Or FEED/TOOTH (Milling) | DIAMETER | | | | | | | | | | | | | | | |
|---|---|---|---|---|---|---|---|---|---|---|---|---|---|---|---|---|
| | 0.25 in 8 mm | | 0.375 in 10 mm | | 0.50 in 12 mm | | 0.625 in 16 mm | | 0.75 in 20 mm | | 1.0 in 25 mm | | 1.25 in 32 mm | | 1.5 in 40 mm | |
| | $R_a$ | $R_t$ | $R_a$ | $R_t$ | $R_a$ | $R_t$ | $R_a$ | $R_t$ | $R_a$ | $R_t$ | $R_a$ | $R_t$ | $R_a$ | $R_t$ | $R_a$ | $R_t$ |
| in | microinches | | | | | | | | | | | | | | | |
| 0.001 | 0.31 | 1.0 | 0.16 | 0.67 | 0.12 | 0.50 | 0.10 | 0.40 | 0.08 | 0.33 | 0.08 | 0.25 | 0.05 | 0.20 | 0.05 | 0.17 |
| 0.002 | 1.0 | 4.0 | 0.75 | 2.7 | 0.50 | 2.0 | 0.40 | 1.6 | 0.32 | 1.3 | 0.25 | 1.0 | 0.20 | 0.80 | 0.20 | 0.67 |
| 0.003 | 2.3 | 9.0 | 1.6 | 6.0 | 1.2 | 4.5 | 0.90 | 3.6 | 0.80 | 3.0 | 0.60 | 2.3 | 0.45 | 1.8 | 0.40 | 1.5 |
| 0.004 | 4.1 | 16. | 2.7 | 11. | 2.1 | 8.0 | 1.7 | 6.4 | 1.5 | 5.3 | 1.0 | 4.0 | 0.80 | 3.2 | 0.70 | 2.7 |
| 0.005 | 6.4 | 25. | 4.3 | 17. | 3.2 | 13. | 2.6 | 10. | 2.2 | 8.3 | 1.7 | 6.3 | 1.5 | 5.0 | 1.1 | 4.2 |
| 0.006 | 9.2 | 36. | 6.2 | 24. | 4.6 | 18. | 3.7 | 14. | 3.2 | 12. | 2.4 | 9.0 | 1.9 | 7.2 | 1.5 | 6.0 |
| 0.007 | 13. | 49. | 8.4 | 33. | 6.3 | 25. | 5.0 | 20. | 4.2 | 16. | 3.2 | 12. | 2.6 | 9.8 | 2.5 | 8.2 |
| 0.008 | 16. | 64. | 11. | 43. | 8.2 | 32. | 6.6 | 26. | 5.5 | 21. | 4.2 | 16. | 3.5 | 13. | 3.0 | 11. |
| 0.009 | 21. | 81. | 14. | 54. | 10. | 41. | 8.3 | 32. | 7.0 | 27. | 5.2 | 20. | 4.2 | 16. | 3.7 | 14. |
| 0.010 | 26. | 100. | 17. | 67. | 13. | 50. | 10. | 40. | 8.6 | 33. | 6.4 | 25. | 5.2 | 20. | 4.4 | 17. |
| 0.012 | 37. | 144. | 25. | 96. | 18. | 72. | 15. | 58. | 12. | 48. | 9.3 | 36. | 7.4 | 29. | 6.3 | 24. |
| 0.014 | 50. | 196. | 34. | 131. | 25. | 98. | 20. | 78. | 17. | 65. | 13. | 49. | 10. | 39. | 8.5 | 33. |
| 0.016 | 66. | 256. | 44. | 171. | 33. | 128. | 26. | 102. | 22. | 85. | 16. | 64. | 13. | 51. | 11. | 43. |
| 0.018 | 83. | 324. | 55. | 216. | 42. | 162. | 33. | 130. | 28. | 108. | 21. | 81. | 17. | 65. | 14. | 54. |
| 0.020 | 103. | 401. | 68. | 267. | 51. | 200. | 41. | 160. | 34. | 133. | 26. | 100. | 21. | 80. | 17. | 67. |
| 0.025 | 161. | 627. | 107. | 417. | 80. | 313. | 64. | 250. | 53. | 208. | 40. | 156. | 32. | 125. | 27. | 104. |
| 0.030 | 232. | 903. | 154. | 601. | 116. | 450. | 92. | 360. | 77. | 300. | 58. | 225. | 46. | 180. | 39. | 150. |
| 0.035 | 316. | 1231. | 210. | 818. | 157. | 613. | 126. | 490. | 105. | 409. | 79. | 306. | 63. | 245. | 52. | 204. |
| 0.040 | 413. | 1610. | 274. | 1070. | 206. | 801. | 164. | 641. | 137. | 534. | 103. | 400. | 82. | 320. | 68. | 267. |
| 0.045 | 523. | 2042. | 347. | 1355. | 260. | 1015. | 208. | 811. | 173. | 676. | 130. | 507. | 104. | 405. | 87. | 338. |
| 0.050 | 647. | 2526. | 429. | 1674. | 321. | 1253. | 257. | 1002. | 214. | 834. | 160. | 625. | 128. | 500. | 107. | 417. |
| 0.060 | 935. | 3653. | 619. | 2416. | 463. | 1807. | 370. | 1443. | 308. | 1202. | 231. | 901. | 185. | 720. | 154. | 600. |
| 0.070 | 1278. | 5000. | 844. | 3296. | 631. | 2462. | 504. | 1966. | 420. | 1637. | 315. | 1227. | 252. | 981. | 210. | 817. |
| 0.080 | 1677. | 6573. | 1105. | 4316. | 825. | 3221. | 659. | 2571. | 549. | 2139. | 411. | 1603. | 329. | 1281. | 274. | 1067. |
| 0.090 | 2135. | 8381. | 1402. | 5480. | 1046. | 4083. | 835. | 3257. | 695. | 2710. | 520. | 2029. | 416. | 1622. | 347. | 1351. |
| 0.100 | 2654. | 10436. | 1736. | 6790. | 1293. | 5051. | 1032. | 4026. | 858. | 3348. | 643. | 2506. | 514. | 2003. | 428. | 1669. |
| mm | micrometers | | | | | | | | | | | | | | | |
| 0.020 | 0.00 | 0.01 | 0.00 | 0.01 | 0.00 | 0.01 | 0.00 | 0.01 | 0.00 | 0.01 | 0.00 | 0.00 | 0.00 | 0.00 | 0.00 | 0.00 |
| 0.040 | 0.01 | 0.05 | 0.01 | 0.04 | 0.01 | 0.03 | 0.01 | 0.03 | 0.00 | 0.02 | 0.00 | 0.02 | 0.00 | 0.01 | 0.00 | 0.01 |
| 0.060 | 0.03 | 0.11 | 0.02 | 0.09 | 0.02 | 0.08 | 0.01 | 0.06 | 0.01 | 0.05 | 0.00 | 0.04 | 0.01 | 0.03 | 0.00 | 0.02 |
| 0.080 | 0.05 | 0.20 | 0.04 | 0.16 | 0.03 | 0.13 | 0.03 | 0.10 | 0.02 | 0.08 | 0.02 | 0.06 | 0.01 | 0.05 | 0.00 | 0.04 |
| 0.100 | 0.08 | 0.31 | 0.06 | 0.25 | 0.05 | 0.21 | 0.04 | 0.16 | 0.03 | 0.13 | 0.03 | 0.10 | 0.02 | 0.08 | 0.02 | 0.06 |
| 0.120 | 0.12 | 0.45 | 0.09 | 0.36 | 0.08 | 0.30 | 0.06 | 0.23 | 0.05 | 0.18 | 0.04 | 0.14 | 0.03 | 0.11 | 0.02 | 0.09 |
| 0.140 | 0.16 | 0.61 | 0.13 | 0.49 | 0.10 | 0.41 | 0.08 | 0.31 | 0.06 | 0.25 | 0.05 | 0.20 | 0.04 | 0.15 | 0.03 | 0.12 |
| 0.160 | 0.21 | 0.80 | 0.16 | 0.64 | 0.14 | 0.53 | 0.10 | 0.40 | 0.08 | 0.32 | 0.07 | 0.26 | 0.05 | 0.20 | 0.05 | 0.16 |
| 0.180 | 0.26 | 1.0 | 0.21 | 0.81 | 0.17 | 0.68 | 0.13 | 0.51 | 0.10 | 0.41 | 0.08 | 0.32 | 0.06 | 0.25 | 0.05 | 0.20 |
| 0.200 | 0.32 | 1.3 | 0.26 | 1.0 | 0.21 | 0.83 | 0.16 | 0.63 | 0.13 | 0.50 | 0.10 | 0.40 | 0.08 | 0.31 | 0.06 | 0.25 |
| 0.250 | 0.50 | 2.0 | 0.40 | 1.6 | 0.33 | 1.3 | 0.25 | 0.98 | 0.20 | 0.78 | 0.16 | 0.63 | 0.13 | 0.49 | 0.10 | 0.39 |
| 0.300 | 0.72 | 2.8 | 0.58 | 2.3 | 0.48 | 1.9 | 0.36 | 1.4 | 0.29 | 1.1 | 0.23 | 0.90 | 0.18 | 0.70 | 0.14 | 0.56 |
| 0.350 | 0.98 | 3.8 | 0.79 | 3.1 | 0.65 | 2.6 | 0.49 | 1.9 | 0.39 | 1.5 | 0.31 | 1.2 | 0.25 | 0.96 | 0.20 | 0.77 |
| 0.400 | 1.3 | 5.0 | 1.0 | 4.0 | 0.86 | 3.3 | 0.64 | 2.5 | 0.51 | 2.0 | 0.41 | 1.6 | 0.32 | 1.3 | 0.26 | 1.0 |
| 0.450 | 1.6 | 6.3 | 1.3 | 5.1 | 1.1 | 4.2 | 0.81 | 3.2 | 0.65 | 2.5 | 0.52 | 2.0 | 0.41 | 1.6 | 0.32 | 1.3 |
| 0.500 | 2.0 | 7.8 | 1.6 | 6.3 | 1.3 | 5.2 | 1.0 | 3.9 | 0.80 | 3.1 | 0.64 | 2.5 | 0.50 | 2.0 | 0.40 | 1.6 |
| 0.600 | 2.9 | 11. | 2.3 | 9.0 | 1.9 | 7.5 | 1.4 | 5.6 | 1.2 | 4.5 | 0.92 | 3.6 | 0.72 | 2.8 | 0.58 | 2.3 |
| 0.700 | 3.9 | 15. | 3.1 | 12. | 2.6 | 10. | 2.0 | 7.7 | 1.6 | 6.1 | 1.3 | 4.9 | 0.98 | 3.8 | 0.79 | 3.1 |
| 0.800 | 5.1 | 20. | 4.1 | 16. | 3.4 | 13. | 2.6 | 10. | 2.1 | 8.0 | 1.6 | 6.4 | 1.3 | 5.0 | 1.0 | 4.0 |
| 1.000 | 8.0 | 31. | 6.4 | 25. | 5.4 | 21. | 4.0 | 16. | 3.2 | 13. | 2.6 | 10. | 2.0 | 7.8 | 1.6 | 6.3 |
| 1.200 | 12. | 45. | 9.3 | 36. | 7.7 | 30. | 5.8 | 23. | 4.6 | 18. | 3.7 | 14. | 2.9 | 11. | 2.3 | 9.0 |
| 1.500 | 18. | 71. | 14. | 57. | 12. | 47. | 9.0 | 35. | 7.2 | 28. | 5.8 | 23. | 4.5 | 18. | 3.6 | 14. |
| 2.000 | 32. | 127. | 26. | 101. | 22. | 84. | 16. | 63. | 13. | 50. | 10. | 40. | 8.0 | 31. | 6.4 | 25. |
| 2.500 | 51. | 200. | 41. | 159. | 34. | 132. | 25. | 98. | 20. | 78. | 16. | 63. | 13. | 49. | 10. | 39. |

**TABLE 18.2–10 THEORETICAL Values for Arithmetic-Average Roughness ($R_a$) and Maximum Peak-to-Valley Roughness Height ($R_t$) for Turning or Milling—Tool Type C**

**TOOL NOSE RADIUS = 0.016 in [0.20 mm]**

| FEED/REV. (Turning) Or FEED/TOOTH (Milling) | 3° ECEA | | 5° ECEA | | 6° ECEA | | 10° ECEA | | 15° ECEA | | 30° ECEA | | 40° ECEA | | 45° ECEA | |
|---|---|---|---|---|---|---|---|---|---|---|---|---|---|---|---|---|
| | $R_a$ | $R_t$ | $R_a$ | $R_t$ | $R_a$ | $R_t$ | $R_a$ | $R_t$ | $R_a$ | $R_t$ | $R_a$ | $R_t$ | $R_a$ | $R_t$ | $R_a$ | $R_t$ |
| in | microinches | | | | | | | | | | | | | | | |
| 0.001 | 2.1 | 8.0 | 2.1 | 8.0 | 2.1 | 8.0 | 2.1 | 8.0 | 2.1 | 8.0 | 2.1 | 8.0 | 2.1 | 8.0 | 2.1 | 8.0 |
| 0.002 | 8.2 | 31. | 8.2 | 32. | 8.2 | 32. | 8.2 | 32. | 8.2 | 32. | 8.2 | 32. | 8.2 | 32. | 8.2 | 32. |
| 0.003 | 17. | 63. | 19. | 72. | 19. | 72. | 19. | 72. | 19. | 72. | 19. | 72. | 19. | 72. | 19. | 72. |
| 0.004 | 27. | 97. | 32. | 121. | 33. | 126. | 33. | 129. | 33. | 129. | 33. | 129. | 33. | 129. | 33. | 129. |
| 0.005 | 37. | 134. | 47. | 174. | 50. | 187. | 52. | 202. | 52. | 202. | 52. | 202. | 52. | 202. | 52. | 202. |
| 0.006 | 47. | 172. | 63. | 231. | 68. | 251. | 75. | 290. | 75. | 291. | 75. | 291. | 75. | 291. | 75. | 291. |
| 0.007 | 57. | 211. | 79. | 291. | 87. | 320. | 101. | 386. | 102. | 398. | 102. | 398. | 102. | 398. | 102. | 398. |
| 0.008 | 68. | 252. | 96. | 352. | 107. | 390. | 129. | 488. | 133. | 522. | 133. | 522. | 133. | 522. | 133. | 522. |
| 0.009 | 79. | 293. | 113. | 415. | 127. | 463. | 159. | 594. | 169. | 659. | 169. | 663. | 169. | 663. | 169. | 663. |
| 0.010 | 90. | 334. | 131. | 480. | 147. | 538. | 191. | 705. | 209. | 804. | 210. | 823. | 210. | 823. | 210. | 823. |
| 0.012 | 112. | 419. | 166. | 612. | 189. | 692. | 255. | 937. | 295. | 1114. | 305. | 1200. | 305. | 1200. | 305. | 1200. |
| 0.014 | 135. | 506. | 202. | 748. | 231. | 851. | 322. | 1179. | 388. | 1444. | 421. | 1659. | 421. | 1659. | 421. | 1659. |
| 0.016 | 158. | 594. | 239. | 887. | 275. | 1014. | 390. | 1430. | 485. | 1792. | 557. | 2206. | 557. | 2207. | 557. | 2207. |
| 0.018 | 181. | 683. | 276. | 1028. | 319. | 1180. | 460. | 1689. | 585. | 2154. | 716. | 2817. | 718. | 2858. | 718. | 2858. |
| 0.020 | 204. | 773. | 314. | 1171. | 363. | 1349. | 531. | 1953. | 688. | 2528. | 891. | 3470. | 904. | 3627. | 904. | 3627. |
| 0.025 | 263. | 1001. | 409. | 1537. | 477. | 1781. | 714. | 2637. | 955. | 3508. | 1389. | 5271. | 1492. | 5937. | 1509. | 6137. |
| 0.030 | 323. | 1232. | 507. | 1911. | 592. | 2224. | 903. | 3347. | 1233. | 4540. | 1945. | 7286. | 2224. | 8678. | 2314. | 9277. |
| mm | micrometers | | | | | | | | | | | | | | | |
| 0.020 | 0.06 | 0.25 | 0.06 | 0.25 | 0.06 | 0.25 | 0.06 | 0.25 | 0.06 | 0.25 | 0.06 | 0.25 | 0.06 | 0.25 | 0.06 | 0.25 |
| 0.040 | 0.23 | 0.85 | 0.26 | 0.99 | 0.26 | 1.0 | 0.26 | 1.0 | 0.26 | 1.0 | 0.26 | 1.0 | 0.26 | 1.0 | 0.26 | 1.0 |
| 0.060 | 0.43 | 1.6 | 0.54 | 2.0 | 0.57 | 2.1 | 0.58 | 2.3 | 0.58 | 2.3 | 0.58 | 2.3 | 0.58 | 2.3 | 0.58 | 2.3 |
| 0.080 | 0.63 | 2.3 | 0.86 | 3.1 | 0.93 | 3.4 | 1.0 | 4.0 | 1.0 | 4.0 | 1.0 | 4.0 | 1.0 | 4.0 | 1.0 | 4.0 |
| 0.100 | 0.85 | 3.1 | 1.2 | 4.4 | 1.3 | 4.8 | 1.6 | 6.0 | 1.6 | 6.4 | 1.6 | 6.4 | 1.6 | 6.4 | 1.6 | 6.4 |
| 0.120 | 1.1 | 3.9 | 1.5 | 5.6 | 1.7 | 6.3 | 2.2 | 8.1 | 2.3 | 9.1 | 2.4 | 9.2 | 2.4 | 9.2 | 2.4 | 9.2 |
| 0.140 | 1.3 | 4.8 | 1.9 | 6.9 | 2.1 | 7.8 | 2.8 | 10. | 3.2 | 12. | 3.2 | 13. | 3.2 | 13. | 3.2 | 13. |
| 0.160 | 1.5 | 5.6 | 2.2 | 8.3 | 2.5 | 9.4 | 3.5 | 13. | 4.1 | 15. | 4.2 | 17. | 4.2 | 17. | 4.2 | 17. |
| 0.180 | 1.7 | 6.5 | 2.6 | 9.6 | 3.0 | 11. | 4.1 | 15. | 5.0 | 19. | 5.4 | 21. | 5.4 | 21. | 5.4 | 21. |
| 0.200 | 2.0 | 7.4 | 3.0 | 11. | 3.4 | 13. | 4.8 | 18. | 6.0 | 22. | 6.8 | 27. | 6.8 | 27. | 6.8 | 27. |
| 0.250 | 2.5 | 9.6 | 3.9 | 15. | 4.5 | 17. | 6.6 | 24. | 8.5 | 31. | 11. | 42. | 11. | 44. | 11. | 44. |
| 0.300 | 3.1 | 12. | 4.9 | 18. | 5.6 | 21. | 8.4 | 31. | 11. | 41. | 16. | 60. | 16. | 66. | 17. | 67. |
| 0.350 | 3.7 | 14. | 5.8 | 22. | 6.8 | 25. | 10. | 38. | 14. | 51. | 21. | 79. | 23. | 92. | 24. | 96. |

**TABLE 18.2–11 THEORETICAL Values for Arithmetic-Average Roughness ($R_a$) and Maximum Peak-to-Valley Roughness Height ($R_t$) for Turning or Milling—Tool Type C**

| TOOL NOSE RADIUS = 0.031 in [0.40 mm] | | | | | | | | | | | | | | | | |
|---|---|---|---|---|---|---|---|---|---|---|---|---|---|---|---|---|
| FEED/REV. (Turning) Or FEED/TOOTH (Milling) | 3° ECEA | | 5° ECEA | | 6° ECEA | | 10° ECEA | | 15° ECEA | | 30° ECEA | | 40° ECEA | | 45° ECEA | |
| | $R_a$ | $R_t$ | $R_a$ | $R_t$ | $R_a$ | $R_t$ | $R_a$ | $R_t$ | $R_a$ | $R_t$ | $R_a$ | $R_t$ | $R_a$ | $R_t$ | $R_a$ | $R_t$ |
| in | microinches | | | | | | | | | | | | | | | |
| 0.001 | 1.0 | 4.0 | 1.0 | 4.0 | 1.0 | 4.0 | 1.0 | 4.0 | 1.0 | 4.0 | 1.0 | 4.0 | 1.0 | 4.0 | 1.0 | 4.0 |
| 0.002 | 4.1 | 16. | 4.1 | 16. | 4.1 | 16. | 4.1 | 16. | 4.1 | 16. | 4.1 | 16. | 4.1 | 16. | 4.1 | 16. |
| 0.003 | 9.3 | 36. | 9.3 | 36. | 9.3 | 36. | 9.3 | 36. | 9.3 | 36. | 9.3 | 36. | 9.3 | 36. | 9.3 | 36. |
| 0.004 | 16. | 63. | 16. | 64. | 16. | 64. | 16. | 64. | 16. | 64. | 16. | 64. | 16. | 64. | 16. | 64. |
| 0.005 | 25. | 93. | 26. | 100. | 26. | 100. | 26. | 100. | 26. | 100. | 26. | 100. | 26. | 100. | 26. | 100. |
| 0.006 | 34. | 125. | 37. | 144. | 37. | 145. | 37. | 145. | 37. | 145. | 37. | 145. | 37. | 145. | 37. | 145. |
| 0.007 | 43. | 159. | 50. | 191. | 50. | 196. | 51. | 197. | 51. | 197. | 51. | 197. | 51. | 197. | 51. | 197. |
| 0.008 | 53. | 194. | 64. | 242. | 66. | 253. | 66. | 257. | 66. | 257. | 66. | 257. | 66. | 257. | 66. | 257. |
| 0.009 | 63. | 230. | 79. | 294. | 82. | 312. | 84. | 326. | 84. | 326. | 84. | 326. | 84. | 326. | 84. | 326. |
| 0.010 | 73. | 268. | 94. | 349. | 100. | 373. | 103. | 403. | 103. | 403. | 103. | 403. | 103. | 403. | 103. | 403. |
| 0.012 | 94. | 344. | 126. | 462. | 136. | 503. | 149. | 579. | 149. | 582. | 149. | 582. | 149. | 582. | 149. | 582. |
| 0.014 | 115. | 423. | 159. | 581. | 174. | 639. | 202. | 771. | 204. | 795. | 204. | 795. | 204. | 795. | 204. | 795. |
| 0.016 | 136. | 503. | 192. | 704. | 213. | 781. | 259. | 975. | 267. | 1043. | 267. | 1043. | 267. | 1043. | 267. | 1043. |
| 0.018 | 158. | 585. | 226. | 830. | 253. | 927. | 319. | 1189. | 339. | 1318. | 339. | 1326. | 339. | 1326. | 339. | 1326. |
| 0.020 | 180. | 669. | 261. | 959. | 294. | 1076. | 381. | 1410. | 417. | 1609. | 420. | 1646. | 420. | 1646. | 420. | 1646. |
| 0.025 | 236. | 881. | 350. | 1291. | 398. | 1463. | 543. | 1992. | 636. | 2389. | 665. | 2613. | 665. | 2613. | 665. | 2613. |
| 0.030 | 293. | 1099. | 441. | 1634. | 506. | 1865. | 711. | 2607. | 873. | 3232. | 972. | 3842. | 972. | 3842. | 972. | 3842. |
| 0.035 | 351. | 1321. | 534. | 1985. | 615. | 2277. | 885. | 3247. | 1120. | 4124. | 1348. | 5320. | 1350. | 5370. | 1350. | 5370. |
| 0.040 | 409. | 1545. | 628. | 2343. | 726. | 2698. | 1063. | 3906. | 1377. | 5055. | 1783. | 6940. | 1808. | 7253. | 1808. | 7253. |
| 0.045 | 468. | 1772. | 723. | 2706. | 839. | 3127. | 1244. | 4583. | 1640. | 6020. | 2263. | 8684. | 2355. | 9456. | 2360. | 9581. |
| 0.050 | 527. | 2001. | 819. | 3074. | 953. | 3562. | 1429. | 5274. | 1910. | 7015. | 2778. | 10542. | 2983. | 11874. | 3018. | 12273. |
| 0.060 | 646. | 2465. | 1013. | 3821. | 1185. | 4447. | 1805. | 6693. | 2466. | 9080. | 3889. | 14572. | 4447. | 17356. | 4629. | 18554. |
| mm | micrometers | | | | | | | | | | | | | | | |
| 0.020 | 0.03 | 0.13 | 0.03 | 0.13 | 0.03 | 0.13 | 0.03 | 0.13 | 0.03 | 0.13 | 0.03 | 0.13 | 0.03 | 0.13 | 0.03 | 0.13 |
| 0.040 | 0.13 | 0.50 | 0.13 | 0.50 | 0.13 | 0.50 | 0.13 | 0.50 | 0.13 | 0.50 | 0.13 | 0.50 | 0.13 | 0.50 | 0.13 | 0.50 |
| 0.060 | 0.28 | 1.1 | 0.29 | 1.1 | 0.29 | 1.1 | 0.29 | 1.1 | 0.29 | 1.1 | 0.29 | 1.1 | 0.29 | 1.1 | 0.29 | 1.1 |
| 0.080 | 0.46 | 1.7 | 0.51 | 2.0 | 0.51 | 2.0 | 0.51 | 2.0 | 0.51 | 2.0 | 0.51 | 2.0 | 0.51 | 2.0 | 0.51 | 2.0 |
| 0.100 | 0.65 | 2.4 | 0.79 | 3.0 | 0.80 | 3.1 | 0.80 | 3.1 | 0.80 | 3.1 | 0.80 | 3.1 | 0.80 | 3.1 | 0.80 | 3.1 |
| 0.120 | 0.85 | 3.1 | 1.1 | 4.0 | 1.1 | 4.3 | 1.2 | 4.5 | 1.2 | 4.5 | 1.2 | 4.5 | 1.2 | 4.5 | 1.2 | 4.5 |
| 0.140 | 1.1 | 3.9 | 1.4 | 5.1 | 1.5 | 5.5 | 1.6 | 6.2 | 1.6 | 6.2 | 1.6 | 6.2 | 1.6 | 6.2 | 1.6 | 6.2 |
| 0.160 | 1.3 | 4.6 | 1.7 | 6.3 | 1.9 | 6.9 | 2.1 | 8.0 | 2.1 | 8.1 | 2.1 | 8.1 | 2.1 | 8.1 | 2.1 | 8.1 |
| 0.180 | 1.5 | 5.4 | 2.0 | 7.5 | 2.2 | 8.2 | 2.6 | 9.9 | 2.6 | 10. | 2.6 | 10. | 2.6 | 10. | 2.6 | 10. |
| 0.200 | 1.7 | 6.2 | 2.4 | 8.7 | 2.6 | 9.6 | 3.2 | 12. | 3.2 | 13. | 3.2 | 13. | 3.2 | 13. | 3.2 | 13. |
| 0.250 | 2.2 | 8.3 | 3.2 | 12. | 3.6 | 13. | 4.7 | 17. | 5.1 | 20. | 5.1 | 20. | 5.1 | 20. | 5.1 | 20. |
| 0.300 | 2.8 | 10. | 4.1 | 15. | 4.7 | 17. | 6.3 | 23. | 7.2 | 27. | 7.4 | 29. | 7.4 | 29. | 7.4 | 29. |
| 0.350 | 3.4 | 13. | 5.0 | 19. | 5.7 | 21. | 7.9 | 29. | 9.5 | 36. | 10. | 40. | 10. | 40. | 10. | 40. |
| 0.400 | 3.9 | 15. | 5.9 | 22. | 6.8 | 25. | 9.6 | 35. | 12. | 44. | 14. | 54. | 14. | 54. | 14. | 54. |
| 0.450 | 4.5 | 17. | 6.9 | 26. | 7.9 | 29. | 11. | 42. | 14. | 53. | 17. | 69. | 17. | 69. | 17. | 69. |
| 0.500 | 5.1 | 19. | 7.8 | 29. | 9.0 | 34. | 13. | 48. | 17. | 62. | 22. | 85. | 22. | 88. | 22. | 88. |
| 0.600 | 6.3 | 24. | 9.7 | 36. | 11. | 42. | 17. | 62. | 22. | 82. | 31. | 120. | 33. | 132. | 33. | 135. |
| 0.700 | 7.5 | 28. | 12. | 44. | 14. | 51. | 21. | 76. | 28. | 102. | 42. | 158. | 46. | 183. | 48. | 192. |

**TABLE 18.2–12 THEORETICAL Values for Arithmetic-Average Roughness ($R_a$) and Maximum Peak-to-Valley Roughness Height ($R_t$) for Turning or Milling—Tool Type C**

| TOOL NOSE RADIUS = 0.047 in [0.60 mm] | | | | | | | | | | | | | | | | |
|---|---|---|---|---|---|---|---|---|---|---|---|---|---|---|---|---|
| FEED/REV. (Turning) Or FEED/TOOTH (Milling) | 3° ECEA | | 5° ECEA | | 6° ECEA | | 10° ECEA | | 15° ECEA | | 30° ECEA | | 40° ECEA | | 45° ECEA | |
| | $R_a$ | $R_t$ | $R_a$ | $R_t$ | $R_a$ | $R_t$ | $R_a$ | $R_t$ | $R_a$ | $R_t$ | $R_a$ | $R_t$ | $R_a$ | $R_t$ | $R_a$ | $R_t$ |
| in | microinches | | | | | | | | | | | | | | | |
| 0.001 | 0.69 | 2.7 | 0.69 | 2.7 | 0.69 | 2.7 | 0.69 | 2.7 | 0.69 | 2.7 | 0.69 | 2.7 | 0.69 | 2.7 | 0.69 | 2.7 |
| 0.002 | 2.7 | 11. | 2.7 | 11. | 2.7 | 11. | 2.7 | 11. | 2.7 | 11. | 2.7 | 11. | 2.7 | 11. | 2.7 | 11. |
| 0.003 | 6.2 | 24. | 6.2 | 24. | 6.2 | 24. | 6.2 | 24. | 6.2 | 24. | 6.2 | 24. | 6.2 | 24. | 6.2 | 24. |
| 0.004 | 11. | 43. | 11. | 43. | 11. | 43. | 11. | 43. | 11. | 43. | 11. | 43. | 11. | 43. | 11. | 43. |
| 0.005 | 17. | 67. | 17. | 67. | 17. | 67. | 17. | 67. | 17. | 67. | 17. | 67. | 17. | 67. | 17. | 67. |
| 0.006 | 25. | 94. | 25. | 96. | 25. | 96. | 25. | 96. | 25. | 96. | 25. | 96. | 25. | 96. | 25. | 96. |
| 0.007 | 33. | 124. | 34. | 131. | 34. | 131. | 34. | 131. | 34. | 131. | 34. | 131. | 34. | 131. | 34. | 131. |
| 0.008 | 42. | 155. | 44. | 171. | 44. | 171. | 44. | 171. | 44. | 171. | 44. | 171. | 44. | 171. | 44. | 171. |
| 0.009 | 51. | 188. | 56. | 216. | 56. | 217. | 56. | 217. | 56. | 217. | 56. | 217. | 56. | 217. | 56. | 217. |
| 0.010 | 60. | 222. | 68. | 263. | 69. | 268. | 69. | 268. | 69. | 268. | 69. | 268. | 69. | 268. | 69. | 268. |
| 0.012 | 80. | 291. | 96. | 362. | 99. | 379. | 99. | 386. | 99. | 386. | 99. | 386. | 99. | 386. | 99. | 386. |
| 0.014 | 99. | 364. | 126. | 468. | 132. | 498. | 135. | 526. | 135. | 526. | 135. | 526. | 135. | 526. | 135. | 526. |
| 0.016 | 120. | 439. | 157. | 579. | 167. | 624. | 176. | 689. | 176. | 689. | 176. | 689. | 176. | 689. | 176. | 689. |
| 0.018 | 141. | 516. | 189. | 694. | 204. | 754. | 224. | 869. | 224. | 874. | 224. | 874. | 224. | 874. | 224. | 874. |
| 0.020 | 162. | 594. | 222. | 812. | 242. | 890. | 275. | 1059. | 277. | 1081. | 277. | 1081. | 277. | 1081. | 277. | 1081. |
| 0.025 | 215. | 796. | 305. | 1119. | 339. | 1243. | 418. | 1568. | 435. | 1699. | 435. | 1700. | 435. | 1700. | 435. | 1700. |
| 0.030 | 270. | 1003. | 392. | 1439. | 441. | 1614. | 572. | 2115. | 626. | 2413. | 630. | 2469. | 630. | 2469. | 630. | 2469. |
| 0.035 | 326. | 1215. | 480. | 1768. | 545. | 1999. | 732. | 2691. | 840. | 3182. | 864. | 3395. | 864. | 3395. | 864. | 3395. |
| 0.040 | 382. | 1430. | 570. | 2106. | 651. | 2393. | 898. | 3291. | 1070. | 3996. | 1140. | 4489. | 1140. | 4489. | 1140. | 4489. |
| 0.045 | 439. | 1649. | 662. | 2451. | 758. | 2797. | 1067. | 3911. | 1309. | 4848. | 1459. | 5764. | 1459. | 5764. | 1459. | 5764. |
| 0.050 | 497. | 1870. | 754. | 2801. | 867. | 3208. | 1240. | 4546. | 1555. | 5733. | 1824. | 7214. | 1824. | 7237. | 1824. | 7237. |
| 0.060 | 613. | 2318. | 942. | 3514. | 1090. | 4047. | 1594. | 5860. | 2065. | 7583. | 2674. | 10410. | 2712. | 10880. | 2712. | 10880. |
| 0.070 | 731. | 2773. | 1132. | 4242. | 1316. | 4907. | 1958. | 7218. | 2594. | 9523. | 3647. | 13937. | 3834. | 15357. | 3852. | 15670. |
| 0.080 | 850. | 3233. | 1325. | 4982. | 1545. | 5782. | 2330. | 8613. | 3140. | 11540. | 4707. | 17759. | 5163. | 20406. | 5270. | 21344. |
| 0.090 | 969. | 3697. | 1520. | 5732. | 1777. | 6671. | 2708. | 10040. | 3699. | 13621. | 5834. | 21858. | 6671. | 26034. | 6943. | 27831. |
| mm | micrometers | | | | | | | | | | | | | | | |
| 0.020 | 0.02 | 0.08 | 0.02 | 0.08 | 0.02 | 0.08 | 0.02 | 0.08 | 0.02 | 0.08 | 0.02 | 0.08 | 0.02 | 0.08 | 0.02 | 0.08 |
| 0.040 | 0.09 | 0.33 | 0.09 | 0.33 | 0.09 | 0.33 | 0.09 | 0.33 | 0.09 | 0.33 | 0.09 | 0.33 | 0.09 | 0.33 | 0.09 | 0.33 |
| 0.060 | 0.19 | 0.75 | 0.19 | 0.75 | 0.19 | 0.75 | 0.19 | 0.75 | 0.19 | 0.75 | 0.19 | 0.75 | 0.19 | 0.75 | 0.19 | 0.75 |
| 0.080 | 0.34 | 1.3 | 0.34 | 1.3 | 0.34 | 1.3 | 0.34 | 1.3 | 0.34 | 1.3 | 0.34 | 1.3 | 0.34 | 1.3 | 0.34 | 1.3 |
| 0.100 | 0.51 | 1.9 | 0.54 | 2.1 | 0.54 | 2.1 | 0.54 | 2.1 | 0.54 | 2.1 | 0.54 | 2.1 | 0.54 | 2.1 | 0.54 | 2.1 |
| 0.120 | 0.69 | 2.6 | 0.77 | 3.0 | 0.77 | 3.0 | 0.77 | 3.0 | 0.77 | 3.0 | 0.77 | 3.0 | 0.77 | 3.0 | 0.77 | 3.0 |
| 0.140 | 0.89 | 3.2 | 1.0 | 3.9 | 1.0 | 4.1 | 1.1 | 4.1 | 1.1 | 4.1 | 1.1 | 4.1 | 1.1 | 4.1 | 1.1 | 4.1 |
| 0.160 | 1.1 | 4.0 | 1.3 | 5.0 | 1.4 | 5.2 | 1.4 | 5.4 | 1.4 | 5.4 | 1.4 | 5.4 | 1.4 | 5.4 | 1.4 | 5.4 |
| 0.180 | 1.3 | 4.7 | 1.6 | 6.0 | 1.7 | 6.4 | 1.7 | 6.8 | 1.7 | 6.8 | 1.7 | 6.8 | 1.7 | 6.8 | 1.7 | 6.8 |
| 0.200 | 1.5 | 5.4 | 1.9 | 7.1 | 2.1 | 7.7 | 2.2 | 8.4 | 2.2 | 8.4 | 2.2 | 8.4 | 2.2 | 8.4 | 2.2 | 8.4 |
| 0.250 | 2.0 | 7.4 | 2.7 | 10. | 3.0 | 11. | 3.4 | 13. | 3.4 | 13. | 3.4 | 13. | 3.4 | 13. | 3.4 | 13. |
| 0.300 | 2.5 | 9.4 | 3.6 | 13. | 3.9 | 14. | 4.7 | 18. | 4.9 | 19. | 4.9 | 19. | 4.9 | 19. | 4.9 | 19. |
| 0.350 | 3.1 | 11. | 4.4 | 16. | 4.9 | 18. | 6.3 | 23. | 6.7 | 26. | 6.7 | 26. | 6.7 | 26. | 6.7 | 26. |
| 0.400 | 3.6 | 14. | 5.3 | 19. | 6.0 | 22. | 7.8 | 29. | 8.7 | 33. | 8.8 | 34. | 8.8 | 34. | 8.8 | 34. |
| 0.450 | 4.2 | 16. | 6.2 | 23. | 7.0 | 26. | 9.4 | 35. | 11. | 41. | 11. | 44. | 11. | 44. | 11. | 44. |
| 0.500 | 4.8 | 18. | 7.1 | 26. | 8.1 | 30. | 11. | 41. | 13. | 49. | 14. | 55. | 14. | 55. | 14. | 55. |
| 0.600 | 5.9 | 22. | 8.9 | 33. | 10. | 38. | 14. | 53. | 18. | 66. | 20. | 80. | 20. | 80. | 20. | 80. |
| 0.700 | 7.1 | 27. | 11. | 40. | 12. | 46. | 18. | 66. | 23. | 84. | 28. | 111. | 28. | 113. | 28. | 113. |
| 0.800 | 8.2 | 31. | 13. | 47. | 15. | 55. | 22. | 79. | 28. | 103. | 37. | 144. | 38. | 153. | 38. | 153. |
| 1.000 | 11. | 40. | 16. | 62. | 19. | 72. | 29. | 107. | 39. | 143. | 58. | 218. | 63. | 248. | 64. | 258. |

**TABLE 18.2–13 THEORETICAL Values for Arithmetic-Average Roughness ($R_a$) and Maximum Peak-to-Valley Roughness Height ($R_t$) for Turning or Milling—Tool Type C**

TOOL NOSE RADIUS = 0.063 in [0.80 mm]

| FEED/REV. (Turning) Or FEED/TOOTH (Milling) | 3° ECEA | | 5° ECEA | | 6° ECEA | | 10° ECEA | | 15° ECEA | | 30° ECEA | | 40° ECEA | | 45° ECEA | |
|---|---|---|---|---|---|---|---|---|---|---|---|---|---|---|---|---|
| | $R_a$ | $R_t$ | $R_a$ | $R_t$ | $R_a$ | $R_t$ | $R_a$ | $R_t$ | $R_a$ | $R_t$ | $R_a$ | $R_t$ | $R_a$ | $R_t$ | $R_a$ | $R_t$ |
| in | microinches | | | | | | | | | | | | | | | |
| 0.001 | 0.51 | 2.0 | 0.51 | 2.0 | 0.51 | 2.0 | 0.51 | 2.0 | 0.51 | 2.0 | 0.51 | 2.0 | 0.51 | 2.0 | 0.51 | 2.0 |
| 0.002 | 2.1 | 8.0 | 2.1 | 8.0 | 2.1 | 8.0 | 2.1 | 8.0 | 2.1 | 8.0 | 2.1 | 8.0 | 2.1 | 8.0 | 2.1 | 8.0 |
| 0.003 | 4.6 | 18. | 4.6 | 18. | 4.6 | 18. | 4.6 | 18. | 4.6 | 18. | 4.6 | 18. | 4.6 | 18. | 4.6 | 18. |
| 0.004 | 8.2 | 32. | 8.2 | 32. | 8.2 | 32. | 8.2 | 32. | 8.2 | 32. | 8.2 | 32. | 8.2 | 32. | 8.2 | 32. |
| 0.005 | 13. | 50. | 13. | 50. | 13. | 50. | 13. | 50. | 13. | 50. | 13. | 50. | 13. | 50. | 13. | 50. |
| 0.006 | 18. | 72. | 18. | 72. | 18. | 72. | 18. | 72. | 18. | 72. | 18. | 72. | 18. | 72. | 18. | 72. |
| 0.007 | 25. | 98. | 25. | 98. | 25. | 98. | 25. | 98. | 25. | 98. | 25. | 98. | 25. | 98. | 25. | 98. |
| 0.008 | 33. | 126. | 33. | 128. | 33. | 128. | 33. | 128. | 33. | 128. | 33. | 128. | 33. | 128. | 33. | 128. |
| 0.009 | 41. | 155. | 42. | 162. | 42. | 162. | 42. | 162. | 42. | 162. | 42. | 162. | 42. | 162. | 42. | 162. |
| 0.010 | 50. | 186. | 51. | 200. | 51. | 200. | 51. | 200. | 51. | 200. | 51. | 200. | 51. | 200. | 51. | 200. |
| 0.012 | 68. | 250. | 74. | 287. | 74. | 289. | 74. | 289. | 74. | 289. | 74. | 289. | 74. | 289. | 74. | 289. |
| 0.014 | 87. | 318. | 100. | 382. | 101. | 392. | 101. | 393. | 101. | 393. | 101. | 393. | 101. | 393. | 101. | 393. |
| 0.016 | 106. | 388. | 128. | 483. | 131. | 504. | 132. | 514. | 132. | 514. | 132. | 514. | 132. | 514. | 132. | 514. |
| 0.018 | 126. | 461. | 158. | 588. | 164. | 623. | 167. | 651. | 167. | 651. | 167. | 651. | 167. | 651. | 167. | 651. |
| 0.020 | 146. | 535. | 188. | 697. | 199. | 746. | 206. | 805. | 206. | 805. | 206. | 805. | 206. | 805. | 206. | 805. |
| 0.025 | 198. | 726. | 268. | 983. | 291. | 1072. | 323. | 1251. | 323. | 1263. | 323. | 1263. | 323. | 1263. | 323. | 1263. |
| 0.030 | 251. | 925. | 351. | 1283. | 387. | 1418. | 459. | 1742. | 467. | 1827. | 467. | 1827. | 467. | 1827. | 467. | 1827. |
| 0.035 | 305. | 1129. | 435. | 1596. | 486. | 1778. | 606. | 2266. | 639. | 2492. | 639. | 2500. | 639. | 2500. | 639. | 2500. |
| 0.040 | 360. | 1337. | 522. | 1917. | 587. | 2151. | 761. | 2817. | 833. | 3213. | 839. | 3286. | 839. | 3286. | 839. | 3286. |
| 0.045 | 416. | 1548. | 610. | 2246. | 691. | 2534. | 921. | 3390. | 1045. | 3976. | 1068. | 4190. | 1068. | 4190. | 1068. | 4190. |
| 0.050 | 472. | 1762. | 700. | 2581. | 796. | 2925. | 1085. | 3982. | 1270. | 4774. | 1327. | 5218. | 1327. | 5218. | 1327. | 5218. |
| 0.060 | 586. | 2198. | 882. | 3266. | 1011. | 3727. | 1422. | 5210. | 1743. | 6458. | 1941. | 7671. | 1941. | 7671. | 1941. | 7671. |
| 0.070 | 701. | 2641. | 1067. | 3968. | 1230. | 4552. | 1769. | 6489. | 2239. | 8240. | 2692. | 10622. | 2696. | 10719. | 2696. | 10719. |
| 0.080 | 817. | 3090. | 1255. | 4684. | 1452. | 5394. | 2125. | 7808. | 2751. | 10102. | 3560. | 13860. | 3609. | 14477. | 3609. | 14477. |
| 0.090 | 935. | 3544. | 1445. | 5410. | 1678. | 6251. | 2487. | 9161. | 3278. | 12032. | 4519. | 17345. | 4700. | 18876. | 4710. | 19119. |
| 0.100 | 1053. | 4002. | 1637. | 6146. | 1906. | 7121. | 2856. | 10543. | 3818. | 14021. | 5549. | 21056. | 5954. | 23705. | 6023. | 24495. |
| mm | micrometers | | | | | | | | | | | | | | | |
| 0.020 | 0.02 | 0.06 | 0.02 | 0.06 | 0.02 | 0.06 | 0.02 | 0.06 | 0.02 | 0.06 | 0.02 | 0.06 | 0.02 | 0.06 | 0.02 | 0.06 |
| 0.040 | 0.06 | 0.25 | 0.06 | 0.25 | 0.06 | 0.25 | 0.06 | 0.25 | 0.06 | 0.25 | 0.06 | 0.25 | 0.06 | 0.25 | 0.06 | 0.25 |
| 0.060 | 0.14 | 0.56 | 0.14 | 0.56 | 0.14 | 0.56 | 0.14 | 0.56 | 0.14 | 0.56 | 0.14 | 0.56 | 0.14 | 0.56 | 0.14 | 0.56 |
| 0.080 | 0.26 | 1.0 | 0.26 | 1.0 | 0.26 | 1.0 | 0.26 | 1.0 | 0.26 | 1.0 | 0.26 | 1.0 | 0.26 | 1.0 | 0.26 | 1.0 |
| 0.100 | 0.40 | 1.5 | 0.40 | 1.6 | 0.40 | 1.6 | 0.40 | 1.6 | 0.40 | 1.6 | 0.40 | 1.6 | 0.40 | 1.6 | 0.40 | 1.6 |
| 0.120 | 0.56 | 2.1 | 0.58 | 2.3 | 0.58 | 2.3 | 0.58 | 2.3 | 0.58 | 2.3 | 0.58 | 2.3 | 0.58 | 2.3 | 0.58 | 2.3 |
| 0.140 | 0.74 | 2.8 | 0.79 | 3.1 | 0.79 | 3.1 | 0.79 | 3.1 | 0.79 | 3.1 | 0.79 | 3.1 | 0.79 | 3.1 | 0.79 | 3.1 |
| 0.160 | 0.93 | 3.4 | 1.0 | 4.0 | 1.0 | 4.0 | 1.0 | 4.0 | 1.0 | 4.0 | 1.0 | 4.0 | 1.0 | 4.0 | 1.0 | 4.0 |
| 0.180 | 1.1 | 4.1 | 1.3 | 4.9 | 1.3 | 5.1 | 1.3 | 5.1 | 1.3 | 5.1 | 1.3 | 5.1 | 1.3 | 5.1 | 1.3 | 5.1 |
| 0.200 | 1.3 | 4.8 | 1.6 | 5.9 | 1.6 | 6.2 | 1.6 | 6.3 | 1.6 | 6.3 | 1.6 | 6.3 | 1.6 | 6.3 | 1.6 | 6.3 |
| 0.250 | 1.8 | 6.6 | 2.3 | 8.6 | 2.4 | 9.2 | 2.5 | 9.8 | 2.5 | 9.8 | 2.5 | 9.8 | 2.5 | 9.8 | 2.5 | 9.8 |
| 0.300 | 2.3 | 8.5 | 3.1 | 11. | 3.3 | 12. | 3.6 | 14. | 3.6 | 14. | 3.6 | 14. | 3.6 | 14. | 3.6 | 14. |
| 0.350 | 2.8 | 10. | 3.9 | 14. | 4.3 | 16. | 4.9 | 19. | 5.0 | 19. | 5.0 | 19. | 5.0 | 19. | 5.0 | 19. |
| 0.400 | 3.4 | 12. | 4.8 | 17. | 5.3 | 19. | 6.3 | 24. | 6.5 | 25. | 6.5 | 25. | 6.5 | 25. | 6.5 | 25. |
| 0.450 | 3.9 | 15. | 5.6 | 21. | 6.3 | 23. | 7.8 | 29. | 8.2 | 32. | 8.3 | 32. | 8.3 | 32. | 8.3 | 32. |
| 0.500 | 4.5 | 17. | 6.5 | 24. | 7.3 | 27. | 9.4 | 35. | 10. | 39. | 10. | 40. | 10. | 40. | 10. | 40. |
| 0.600 | 5.6 | 21. | 8.2 | 30. | 9.3 | 34. | 13. | 46. | 14. | 55. | 15. | 58. | 15. | 58. | 15. | 58. |
| 0.700 | 6.7 | 25. | 10. | 37. | 11. | 42. | 16. | 58. | 19. | 71. | 20. | 81. | 20. | 81. | 20. | 81. |
| 0.800 | 7.9 | 30. | 12. | 44. | 14. | 50. | 19. | 71. | 24. | 88. | 27. | 107. | 27. | 107. | 27. | 107. |
| 1.000 | 10. | 38. | 16. | 58. | 18. | 67. | 26. | 97. | 34. | 125. | 43. | 169. | 44. | 176. | 44. | 176. |
| 1.200 | 13. | 48. | 19. | 73. | 23. | 84. | 34. | 124. | 45. | 163. | 63. | 239. | 66. | 264. | 66. | 270. |
| 1.500 | 16. | 61. | 25. | 95. | 29. | 111. | 45. | 166. | 61. | 225. | 95. | 357. | 108. | 422. | 112. | 449. |

**TABLE 18.2–14 THEORETICAL Values for Arithmetic-Average Roughness ($R_a$) and Maximum Peak-to-Valley Roughness Height ($R_t$) for Turning or Milling—Tool Type C**

| TOOL NOSE RADIUS = 0.125 in [1.2 mm] | | | | | | | | | | | | | | | | |
|---|---|---|---|---|---|---|---|---|---|---|---|---|---|---|---|---|
| FEED/REV. (Turning) Or FEED/TOOTH (Milling) | 3° ECEA | | 5° ECEA | | 6° ECEA | | 10° ECEA | | 15° ECEA | | 30° ECEA | | 40° ECEA | | 45° ECEA | |
| | $R_a$ | $R_t$ | $R_a$ | $R_t$ | $R_a$ | $R_t$ | $R_a$ | $R_t$ | $R_a$ | $R_t$ | $R_a$ | $R_t$ | $R_a$ | $R_t$ | $R_a$ | $R_t$ |
| in | microinches | | | | | | | | | | | | | | | |
| 0.001 | 0.26 | 1.0 | 0.26 | 1.0 | 0.26 | 1.0 | 0.26 | 1.0 | 0.26 | 1.0 | 0.26 | 1.0 | 0.26 | 1.0 | 0.26 | 1.0 |
| 0.002 | 1.0 | 4.0 | 1.0 | 4.0 | 1.0 | 4.0 | 1.0 | 4.0 | 1.0 | 4.0 | 1.0 | 4.0 | 1.0 | 4.0 | 1.0 | 4.0 |
| 0.003 | 2.3 | 9.0 | 2.3 | 9.0 | 2.3 | 9.0 | 2.3 | 9.0 | 2.3 | 9.0 | 2.3 | 9.0 | 2.3 | 9.0 | 2.3 | 9.0 |
| 0.004 | 4.1 | 16. | 4.1 | 16. | 4.1 | 16. | 4.1 | 16. | 4.1 | 16. | 4.1 | 16. | 4.1 | 16. | 4.1 | 16. |
| 0.005 | 6.4 | 25. | 6.4 | 25. | 6.4 | 25. | 6.4 | 25. | 6.4 | 25. | 6.4 | 25. | 6.4 | 25. | 6.4 | 25. |
| 0.006 | 9.2 | 36. | 9.2 | 36. | 9.2 | 36. | 9.2 | 36. | 9.2 | 36. | 9.2 | 36. | 9.2 | 36. | 9.2 | 36. |
| 0.007 | 13. | 49. | 13. | 49. | 13. | 49. | 13. | 49. | 13. | 49. | 13. | 49. | 13. | 49. | 13. | 49. |
| 0.008 | 16. | 64. | 16. | 64. | 16. | 64. | 16. | 64. | 16. | 64. | 16. | 64. | 16. | 64. | 16. | 64. |
| 0.009 | 21. | 81. | 21. | 81. | 21. | 81. | 21. | 81. | 21. | 81. | 21. | 81. | 21. | 81. | 21. | 81. |
| 0.010 | 26. | 100. | 26. | 100. | 26. | 100. | 26. | 100. | 26. | 100. | 26. | 100. | 26. | 100. | 26. | 100. |
| 0.012 | 37. | 144. | 37. | 144. | 37. | 144. | 37. | 144. | 37. | 144. | 37. | 144. | 37. | 144. | 37. | 144. |
| 0.014 | 50. | 196. | 50. | 196. | 50. | 196. | 50. | 196. | 50. | 196. | 50. | 196. | 50. | 196. | 50. | 196. |
| 0.016 | 65. | 252. | 66. | 256. | 66. | 256. | 66. | 256. | 66. | 256. | 66. | 256. | 66. | 256. | 66. | 256. |
| 0.018 | 82. | 310. | 83. | 324. | 83. | 324. | 83. | 324. | 83. | 324. | 83. | 324. | 83. | 324. | 83. | 324. |
| 0.020 | 99. | 372. | 103. | 401. | 103. | 401. | 103. | 401. | 103. | 401. | 103. | 401. | 103. | 401. | 103. | 401. |
| 0.025 | 145. | 534. | 160. | 621. | 161. | 627. | 161. | 627. | 161. | 627. | 161. | 627. | 161. | 627. | 161. | 627. |
| 0.030 | 192. | 706. | 228. | 864. | 231. | 895. | 232. | 903. | 232. | 903. | 232. | 903. | 232. | 903. | 232. | 903. |
| 0.035 | 242. | 885. | 301. | 1122. | 311. | 1185. | 316. | 1231. | 316. | 1231. | 316. | 1231. | 316. | 1231. | 316. | 1231. |
| 0.040 | 292. | 1070. | 377. | 1394. | 398. | 1492. | 413. | 1610. | 413. | 1610. | 413. | 1610. | 413. | 1610. | 413. | 1610. |
| 0.045 | 343. | 1259. | 456. | 1675. | 488. | 1812. | 523. | 2040. | 523. | 2042. | 523. | 2042. | 523. | 2042. | 523. | 2042. |
| 0.050 | 395. | 1453. | 536. | 1965. | 581. | 2144. | 645. | 2501. | 647. | 2526. | 647. | 2526. | 647. | 2526. | 647. | 2526. |
| 0.060 | 502. | 1850. | 701. | 2567. | 773. | 2835. | 917. | 3484. | 935. | 3653. | 935. | 3653. | 935. | 3653. | 935. | 3653. |
| 0.070 | 610. | 2258. | 871. | 3191. | 971. | 3557. | 1213. | 4533. | 1277. | 4984. | 1278. | 5000. | 1278. | 5000. | 1278. | 5000. |
| 0.080 | 720. | 2673. | 1044. | 3834. | 1174. | 4303. | 1523. | 5635. | 1667. | 6426. | 1677. | 6573. | 1677. | 6573. | 1677. | 6573. |
| 0.090 | 831. | 3096. | 1221. | 4491. | 1381. | 5068. | 1843. | 6781. | 2091. | 7951. | 2135. | 8381. | 2135. | 8381. | 2135. | 8381. |
| 0.100 | 944. | 3525. | 1399. | 5161. | 1592. | 5850. | 2170. | 7963. | 2539. | 9547. | 2654. | 10436. | 2654. | 10436. | 2654. | 10436. |
| mm | micrometers | | | | | | | | | | | | | | | |
| 0.020 | 0.01 | 0.04 | 0.01 | 0.04 | 0.01 | 0.04 | 0.01 | 0.04 | 0.01 | 0.04 | 0.01 | 0.04 | 0.01 | 0.04 | 0.01 | 0.04 |
| 0.040 | 0.04 | 0.17 | 0.04 | 0.17 | 0.04 | 0.17 | 0.04 | 0.17 | 0.04 | 0.17 | 0.04 | 0.17 | 0.04 | 0.17 | 0.04 | 0.17 |
| 0.060 | 0.10 | 0.38 | 0.10 | 0.38 | 0.10 | 0.38 | 0.10 | 0.38 | 0.10 | 0.38 | 0.10 | 0.38 | 0.10 | 0.38 | 0.10 | 0.38 |
| 0.080 | 0.17 | 0.67 | 0.17 | 0.67 | 0.17 | 0.67 | 0.17 | 0.67 | 0.17 | 0.67 | 0.17 | 0.67 | 0.17 | 0.67 | 0.17 | 0.67 |
| 0.100 | 0.27 | 1.0 | 0.27 | 1.0 | 0.27 | 1.0 | 0.27 | 1.0 | 0.27 | 1.0 | 0.27 | 1.0 | 0.27 | 1.0 | 0.27 | 1.0 |
| 0.120 | 0.39 | 1.5 | 0.39 | 1.5 | 0.39 | 1.5 | 0.39 | 1.5 | 0.39 | 1.5 | 0.39 | 1.5 | 0.39 | 1.5 | 0.39 | 1.5 |
| 0.140 | 0.52 | 2.0 | 0.52 | 2.0 | 0.52 | 2.0 | 0.52 | 2.0 | 0.52 | 2.0 | 0.52 | 2.0 | 0.52 | 2.0 | 0.52 | 2.0 |
| 0.160 | 0.68 | 2.6 | 0.68 | 2.7 | 0.68 | 2.7 | 0.68 | 2.7 | 0.68 | 2.7 | 0.68 | 2.7 | 0.68 | 2.7 | 0.68 | 2.7 |
| 0.180 | 0.85 | 3.2 | 0.87 | 3.4 | 0.87 | 3.4 | 0.87 | 3.4 | 0.87 | 3.4 | 0.87 | 3.4 | 0.87 | 3.4 | 0.87 | 3.4 |
| 0.200 | 1.0 | 3.8 | 1.1 | 4.2 | 1.1 | 4.2 | 1.1 | 4.2 | 1.1 | 4.2 | 1.1 | 4.2 | 1.1 | 4.2 | 1.1 | 4.2 |
| 0.250 | 1.5 | 5.5 | 1.7 | 6.4 | 1.7 | 6.5 | 1.7 | 6.5 | 1.7 | 6.5 | 1.7 | 6.5 | 1.7 | 6.5 | 1.7 | 6.5 |
| 0.300 | 2.0 | 7.2 | 2.4 | 8.9 | 2.4 | 9.3 | 2.4 | 9.4 | 2.4 | 9.4 | 2.4 | 9.4 | 2.4 | 9.4 | 2.4 | 9.4 |
| 0.350 | 2.5 | 9.0 | 3.1 | 12. | 3.2 | 12. | 3.3 | 13. | 3.3 | 13. | 3.3 | 13. | 3.3 | 13. | 3.3 | 13. |
| 0.400 | 3.0 | 11. | 3.9 | 14. | 4.1 | 15. | 4.3 | 17. | 4.3 | 17. | 4.3 | 17. | 4.3 | 17. | 4.3 | 17. |
| 0.450 | 3.5 | 13. | 4.7 | 17. | 5.0 | 19. | 5.4 | 21. | 5.5 | 21. | 5.5 | 21. | 5.5 | 21. | 5.5 | 21. |
| 0.500 | 4.0 | 15. | 5.5 | 20. | 6.0 | 22. | 6.7 | 26. | 6.7 | 26. | 6.7 | 26. | 6.7 | 26. | 6.7 | 26. |
| 0.600 | 5.1 | 19. | 7.1 | 26. | 7.9 | 29. | 9.5 | 36. | 9.7 | 38. | 9.7 | 38. | 9.7 | 38. | 9.7 | 38. |
| 0.700 | 6.2 | 23. | 8.8 | 32. | 9.9 | 36. | 13. | 47. | 13. | 52. | 13. | 52. | 13. | 52. | 13. | 52. |
| 0.800 | 7.3 | 27. | 11. | 39. | 12. | 44. | 16. | 58. | 17. | 66. | 18. | 69. | 18. | 69. | 18. | 69. |
| 1.000 | 9.5 | 36. | 14. | 52. | 16. | 59. | 22. | 81. | 26. | 98. | 28. | 109. | 28. | 109. | 28. | 109. |
| 1.200 | 12. | 44. | 18. | 66. | 20. | 75. | 29. | 106. | 36. | 132. | 41. | 161. | 41. | 161. | 41. | 161. |
| 1.500 | 15. | 58. | 23. | 87. | 27. | 101. | 40. | 145. | 51. | 187. | 65. | 254. | 66. | 263. | 66. | 263. |
| 2.000 | 21. | 81. | 33. | 124. | 38. | 144. | 58. | 214. | 78. | 285. | 115. | 435. | 125. | 496. | 127. | 517. |

# 18.2 Surface Texture

**TABLE 18.2–15 THEORETICAL Values for Arithmetic-Average Roughness ($R_a$) and Maximum Peak-to-Valley Roughness Height ($R_t$) for Turning or Milling—Tool Type C**

**TOOL NOSE RADIUS = 0.188 in [1.6 mm]**

| FEED/REV. (Turning) Or FEED/TOOTH (Milling) | 3° ECEA | | 5° ECEA | | 6° ECEA | | 10° ECEA | | 15° ECEA | | 30° ECEA | | 40° ECEA | | 45° ECEA | |
|---|---|---|---|---|---|---|---|---|---|---|---|---|---|---|---|---|
| | $R_a$ | $R_t$ | $R_a$ | $R_t$ | $R_a$ | $R_t$ | $R_a$ | $R_t$ | $R_a$ | $R_t$ | $R_a$ | $R_t$ | $R_a$ | $R_t$ | $R_a$ | $R_t$ |
| in | microinches | | | | | | | | | | | | | | | |
| 0.001 | 0.06 | 0.67 | 0.06 | 0.67 | 0.06 | 0.67 | 0.06 | 0.67 | 0.06 | 0.67 | 0.06 | 0.67 | 0.06 | 0.67 | 0.06 | 0.67 |
| 0.002 | 0.62 | 2.7 | 0.62 | 2.7 | 0.62 | 2.7 | 0.62 | 2.7 | 0.62 | 2.7 | 0.62 | 2.7 | 0.62 | 2.7 | 0.62 | 2.7 |
| 0.003 | 1.5 | 6.0 | 1.5 | 6.0 | 1.5 | 6.0 | 1.5 | 6.0 | 1.5 | 6.0 | 1.5 | 6.0 | 1.5 | 6.0 | 1.5 | 6.0 |
| 0.004 | 2.7 | 11. | 2.7 | 11. | 2.7 | 11. | 2.7 | 11. | 2.7 | 11. | 2.7 | 11. | 2.7 | 11. | 2.7 | 11. |
| 0.005 | 4.3 | 17. | 4.3 | 17. | 4.3 | 17. | 4.3 | 17. | 4.3 | 17. | 4.3 | 17. | 4.3 | 17. | 4.3 | 17. |
| 0.006 | 6.2 | 24. | 6.2 | 24. | 6.2 | 24. | 6.2 | 24. | 6.2 | 24. | 6.2 | 24. | 6.2 | 24. | 6.2 | 24. |
| 0.007 | 8.4 | 33. | 8.4 | 33. | 8.4 | 33. | 8.4 | 33. | 8.4 | 33. | 8.4 | 33. | 8.4 | 33. | 8.4 | 33. |
| 0.008 | 11. | 43. | 11. | 43. | 11. | 43. | 11. | 43. | 11. | 43. | 11. | 43. | 11. | 43. | 11. | 43. |
| 0.009 | 14. | 54. | 14. | 54. | 14. | 54. | 14. | 54. | 14. | 54. | 14. | 54. | 14. | 54. | 14. | 54. |
| 0.010 | 17. | 67. | 17. | 67. | 17. | 67. | 17. | 67. | 17. | 67. | 17. | 67. | 17. | 67. | 17. | 67. |
| 0.012 | 25. | 96. | 25. | 96. | 25. | 96. | 25. | 96. | 25. | 96. | 25. | 96. | 25. | 96. | 25. | 96. |
| 0.014 | 34. | 131. | 34. | 131. | 34. | 131. | 34. | 131. | 34. | 131. | 34. | 131. | 34. | 131. | 34. | 131. |
| 0.016 | 44. | 171. | 44. | 171. | 44. | 171. | 44. | 171. | 44. | 171. | 44. | 171. | 44. | 171. | 44. | 171. |
| 0.018 | 55. | 216. | 55. | 216. | 55. | 216. | 55. | 216. | 55. | 216. | 55. | 216. | 55. | 216. | 55. | 216. |
| 0.020 | 68. | 267. | 68. | 267. | 68. | 267. | 68. | 267. | 68. | 267. | 68. | 267. | 68. | 267. | 68. | 267. |
| 0.025 | 106. | 406. | 107. | 417. | 107. | 417. | 107. | 417. | 107. | 417. | 107. | 417. | 107. | 417. | 107. | 417. |
| 0.030 | 149. | 558. | 154. | 601. | 154. | 601. | 154. | 601. | 154. | 601. | 154. | 601. | 154. | 601. | 154. | 601. |
| 0.035 | 194. | 718. | 210. | 817. | 210. | 818. | 210. | 818. | 210. | 818. | 210. | 818. | 210. | 818. | 210. | 818. |
| 0.040 | 241. | 885. | 273. | 1050. | 274. | 1069. | 274. | 1070. | 274. | 1070. | 274. | 1070. | 274. | 1070. | 274. | 1070. |
| 0.045 | 289. | 1058. | 341. | 1296. | 347. | 1343. | 347. | 1355. | 347. | 1355. | 347. | 1355. | 347. | 1355. | 347. | 1355. |
| 0.050 | 338. | 1236. | 414. | 1552. | 426. | 1630. | 429. | 1674. | 429. | 1674. | 429. | 1674. | 429. | 1674. | 429. | 1674. |
| 0.060 | 438. | 1604. | 565. | 2090. | 596. | 2238. | 619. | 2416. | 619. | 2416. | 619. | 2416. | 619. | 2416. | 619. | 2416. |
| 0.070 | 541. | 1985. | 723. | 2656. | 778. | 2882. | 844. | 3287. | 844. | 3296. | 844. | 3296. | 844. | 3296. | 844. | 3296. |
| 0.080 | 646. | 2376. | 886. | 3244. | 966. | 3556. | 1099. | 4230. | 1105. | 4316. | 1105. | 4316. | 1105. | 4316. | 1105. | 4316. |
| 0.090 | 753. | 2775. | 1052. | 3850. | 1160. | 4253. | 1376. | 5226. | 1402. | 5480. | 1402. | 5480. | 1402. | 5480. | 1402. | 5480. |
| 0.100 | 861. | 3181. | 1221. | 4472. | 1357. | 4970. | 1669. | 6265. | 1736. | 6786. | 1736. | 6790. | 1736. | 6790. | 1736. | 6790. |
| mm | micrometers | | | | | | | | | | | | | | | |
| 0.020 | 0.01 | 0.03 | 0.01 | 0.03 | 0.01 | 0.03 | 0.01 | 0.03 | 0.01 | 0.03 | 0.01 | 0.03 | 0.01 | 0.03 | 0.01 | 0.03 |
| 0.040 | 0.03 | 0.13 | 0.03 | 0.13 | 0.03 | 0.13 | 0.03 | 0.13 | 0.03 | 0.13 | 0.03 | 0.13 | 0.03 | 0.13 | 0.03 | 0.13 |
| 0.060 | 0.07 | 0.28 | 0.07 | 0.28 | 0.07 | 0.28 | 0.07 | 0.28 | 0.07 | 0.28 | 0.07 | 0.28 | 0.07 | 0.28 | 0.07 | 0.28 |
| 0.080 | 0.13 | 0.50 | 0.13 | 0.50 | 0.13 | 0.50 | 0.13 | 0.50 | 0.13 | 0.50 | 0.13 | 0.50 | 0.13 | 0.50 | 0.13 | 0.50 |
| 0.100 | 0.20 | 0.78 | 0.20 | 0.78 | 0.20 | 0.78 | 0.20 | 0.78 | 0.20 | 0.78 | 0.20 | 0.78 | 0.20 | 0.78 | 0.20 | 0.78 |
| 0.120 | 0.29 | 1.1 | 0.29 | 1.1 | 0.29 | 1.1 | 0.29 | 1.1 | 0.29 | 1.1 | 0.29 | 1.1 | 0.29 | 1.1 | 0.29 | 1.1 |
| 0.140 | 0.39 | 1.5 | 0.39 | 1.5 | 0.39 | 1.5 | 0.39 | 1.5 | 0.39 | 1.5 | 0.39 | 1.5 | 0.39 | 1.5 | 0.39 | 1.5 |
| 0.160 | 0.51 | 2.0 | 0.51 | 2.0 | 0.51 | 2.0 | 0.51 | 2.0 | 0.51 | 2.0 | 0.51 | 2.0 | 0.51 | 2.0 | 0.51 | 2.0 |
| 0.180 | 0.65 | 2.5 | 0.65 | 2.5 | 0.65 | 2.5 | 0.65 | 2.5 | 0.65 | 2.5 | 0.65 | 2.5 | 0.65 | 2.5 | 0.65 | 2.5 |
| 0.200 | 0.80 | 3.1 | 0.80 | 3.1 | 0.80 | 3.1 | 0.80 | 3.1 | 0.80 | 3.1 | 0.80 | 3.1 | 0.80 | 3.1 | 0.80 | 3.1 |
| 0.250 | 1.2 | 4.6 | 1.3 | 4.9 | 1.3 | 4.9 | 1.3 | 4.9 | 1.3 | 4.9 | 1.3 | 4.9 | 1.3 | 4.9 | 1.3 | 4.9 |
| 0.300 | 1.7 | 6.2 | 1.8 | 7.0 | 1.8 | 7.0 | 1.8 | 7.0 | 1.8 | 7.0 | 1.8 | 7.0 | 1.8 | 7.0 | 1.8 | 7.0 |
| 0.350 | 2.1 | 7.9 | 2.4 | 9.4 | 2.5 | 9.6 | 2.5 | 9.6 | 2.5 | 9.6 | 2.5 | 9.6 | 2.5 | 9.6 | 2.5 | 9.6 |
| 0.400 | 2.6 | 9.6 | 3.1 | 12. | 3.2 | 12. | 3.2 | 13. | 3.2 | 13. | 3.2 | 13. | 3.2 | 13. | 3.2 | 13. |
| 0.450 | 3.1 | 11. | 3.9 | 14. | 4.0 | 15. | 4.1 | 16. | 4.1 | 16. | 4.1 | 16. | 4.1 | 16. | 4.1 | 16. |
| 0.500 | 3.6 | 13. | 4.6 | 17. | 4.9 | 18. | 5.0 | 20. | 5.0 | 20. | 5.0 | 20. | 5.0 | 20. | 5.0 | 20. |
| 0.600 | 4.6 | 17. | 6.2 | 23. | 6.7 | 25. | 7.3 | 28. | 7.3 | 28. | 7.3 | 28. | 7.3 | 28. | 7.3 | 28. |
| 0.700 | 5.7 | 21. | 7.8 | 29. | 8.6 | 32. | 9.8 | 38. | 9.9 | 39. | 9.9 | 39. | 9.9 | 39. | 9.9 | 39. |
| 0.800 | 6.8 | 25. | 9.5 | 35. | 11. | 39. | 13. | 48. | 13. | 51. | 13. | 51. | 13. | 51. | 13. | 51. |
| 1.000 | 8.9 | 33. | 13. | 48. | 15. | 53. | 19. | 69. | 20. | 79. | 20. | 80. | 20. | 80. | 20. | 80. |
| 1.200 | 11. | 42. | 16. | 61. | 19. | 69. | 25. | 92. | 29. | 109. | 30. | 117. | 30. | 117. | 30. | 117. |
| 1.500 | 15. | 55. | 22. | 81. | 25. | 92. | 35. | 129. | 43. | 159. | 47. | 187. | 47. | 187. | 47. | 187. |
| 2.000 | 20. | 77. | 31. | 116. | 36. | 134. | 53. | 193. | 68. | 250. | 87. | 339. | 88. | 351. | 88. | 351. |
| 2.500 | 26. | 100. | 41. | 153. | 47. | 177. | 71. | 262. | 94. | 347. | 136. | 516. | 145. | 577. | 146. | 593. |

**TABLE 18.2–16 THEORETICAL Values for Arithmetic-Average Roughness ($R_a$) and Maximum Peak-to-Valley Roughness Height ($R_t$) for Turning or Milling—Tool Type C**

TOOL NOSE RADIUS = 0.250 in [2.4 mm]

| FEED/REV. (turning) Or FEED/TOOTH (milling) | 3° ECEA | | 5° ECEA | | 6° ECEA | | 10° ECEA | | 15° ECEA | | 30° ECEA | | 40° ECEA | | 45° ECEA | |
|---|---|---|---|---|---|---|---|---|---|---|---|---|---|---|---|---|
| | $R_a$ | $R_t$ | $R_a$ | $R_t$ | $R_a$ | $R_t$ | $R_a$ | $R_t$ | $R_a$ | $R_t$ | $R_a$ | $R_t$ | $R_a$ | $R_t$ | $R_a$ | $R_t$ |
| in | microinches | | | | | | | | | | | | | | | |
| 0.001 | 0.13 | 0.50 | 0.13 | 0.50 | 0.13 | 0.50 | 0.13 | 0.50 | 0.13 | 0.50 | 0.13 | 0.50 | 0.13 | 0.50 | 0.13 | 0.50 |
| 0.002 | 0.51 | 2.0 | 0.51 | 2.0 | 0.51 | 2.0 | 0.51 | 2.0 | 0.51 | 2.0 | 0.51 | 2.0 | 0.51 | 2.0 | 0.51 | 2.0 |
| 0.003 | 1.2 | 4.5 | 1.2 | 4.5 | 1.2 | 4.5 | 1.2 | 4.5 | 1.2 | 4.5 | 1.2 | 4.5 | 1.2 | 4.5 | 1.2 | 4.5 |
| 0.004 | 2.1 | 8.0 | 2.1 | 8.0 | 2.1 | 8.0 | 2.1 | 8.0 | 2.1 | 8.0 | 2.1 | 8.0 | 2.1 | 8.0 | 2.1 | 8.0 |
| 0.005 | 3.2 | 13. | 3.2 | 13. | 3.2 | 13. | 3.2 | 13. | 3.2 | 13. | 3.2 | 13. | 3.2 | 13. | 3.2 | 13. |
| 0.006 | 4.6 | 18. | 4.6 | 18. | 4.6 | 18. | 4.6 | 18. | 4.6 | 18. | 4.6 | 18. | 4.6 | 18. | 4.6 | 18. |
| 0.007 | 6.3 | 25. | 6.3 | 25. | 6.3 | 25. | 6.3 | 25. | 6.3 | 25. | 6.3 | 25. | 6.3 | 25. | 6.3 | 25. |
| 0.008 | 8.2 | 32. | 8.2 | 32. | 8.2 | 32. | 8.2 | 32. | 8.2 | 32. | 8.2 | 32. | 8.2 | 32. | 8.2 | 32. |
| 0.009 | 10. | 41. | 10. | 41. | 10. | 41. | 10. | 41. | 10. | 41. | 10. | 41. | 10. | 41. | 10. | 41. |
| 0.010 | 13. | 50. | 13. | 50. | 13. | 50. | 13. | 50. | 13. | 50. | 13. | 50. | 13. | 50. | 13. | 50. |
| 0.012 | 18. | 72. | 18. | 72. | 18. | 72. | 18. | 72. | 18. | 72. | 18. | 72. | 18. | 72. | 18. | 72. |
| 0.014 | 25. | 98. | 25. | 98. | 25. | 98. | 25. | 98. | 25. | 98. | 25. | 98. | 25. | 98. | 25. | 98. |
| 0.016 | 33. | 128. | 33. | 128. | 33. | 128. | 33. | 128. | 33. | 128. | 33. | 128. | 33. | 128. | 33. | 128. |
| 0.018 | 42. | 162. | 42. | 162. | 42. | 162. | 42. | 162. | 42. | 162. | 42. | 162. | 42. | 162. | 42. | 162. |
| 0.020 | 51. | 200. | 51. | 200. | 51. | 200. | 51. | 200. | 51. | 200. | 51. | 200. | 51. | 200. | 51. | 200. |
| 0.025 | 80. | 313. | 80. | 313. | 80. | 313. | 80. | 313. | 80. | 313. | 80. | 313. | 80. | 313. | 80. | 313. |
| 0.030 | 115. | 446. | 116. | 450. | 116. | 450. | 116. | 450. | 116. | 450. | 116. | 450. | 116. | 450. | 116. | 450. |
| 0.035 | 155. | 591. | 157. | 613. | 157. | 613. | 157. | 613. | 157. | 613. | 157. | 613. | 157. | 613. | 157. | 613. |
| 0.040 | 198. | 744. | 206. | 801. | 206. | 801. | 206. | 801. | 206. | 801. | 206. | 801. | 206. | 801. | 206. | 801. |
| 0.045 | 243. | 903. | 260. | 1014. | 260. | 1015. | 260. | 1015. | 260. | 1015. | 260. | 1015. | 260. | 1015. | 260. | 1015. |
| 0.050 | 289. | 1068. | 321. | 1242. | 321. | 1253. | 321. | 1253. | 321. | 1253. | 321. | 1253. | 321. | 1253. | 321. | 1253. |
| 0.060 | 385. | 1411. | 455. | 1728. | 462. | 1790. | 463. | 1807. | 463. | 1807. | 463. | 1807. | 463. | 1807. | 463. | 1807. |
| 0.070 | 483. | 1769. | 601. | 2245. | 622. | 2370. | 631. | 2462. | 631. | 2462. | 631. | 2462. | 631. | 2462. | 631. | 2462. |
| 0.080 | 584. | 2139. | 754. | 2787. | 795. | 2984. | 825. | 3221. | 825. | 3221. | 825. | 3221. | 825. | 3221. | 825. | 3221. |
| 0.090 | 687. | 2519. | 911. | 3350. | 976. | 3624. | 1046. | 4081. | 1046. | 4083. | 1046. | 4083. | 1046. | 4083. | 1046. | 4083. |
| 0.100 | 791. | 2906. | 1072. | 3931. | 1162. | 4288. | 1291. | 5002. | 1293. | 5051. | 1293. | 5051. | 1293. | 5051. | 1293. | 5051. |
| mm | micrometers | | | | | | | | | | | | | | | |
| 0.020 | 0.01 | 0.02 | 0.01 | 0.02 | 0.01 | 0.02 | 0.01 | 0.02 | 0.01 | 0.02 | 0.01 | 0.02 | 0.01 | 0.02 | 0.01 | 0.02 |
| 0.040 | 0.02 | 0.08 | 0.02 | 0.08 | 0.02 | 0.08 | 0.02 | 0.08 | 0.02 | 0.08 | 0.02 | 0.08 | 0.02 | 0.08 | 0.02 | 0.08 |
| 0.060 | 0.05 | 0.19 | 0.05 | 0.19 | 0.05 | 0.19 | 0.05 | 0.19 | 0.05 | 0.19 | 0.05 | 0.19 | 0.05 | 0.19 | 0.05 | 0.19 |
| 0.080 | 0.09 | 0.33 | 0.09 | 0.33 | 0.09 | 0.33 | 0.09 | 0.33 | 0.09 | 0.33 | 0.09 | 0.33 | 0.09 | 0.33 | 0.09 | 0.33 |
| 0.100 | 0.13 | 0.52 | 0.13 | 0.52 | 0.13 | 0.52 | 0.13 | 0.52 | 0.13 | 0.52 | 0.13 | 0.52 | 0.13 | 0.52 | 0.13 | 0.52 |
| 0.120 | 0.19 | 0.75 | 0.19 | 0.75 | 0.19 | 0.75 | 0.19 | 0.75 | 0.19 | 0.75 | 0.19 | 0.75 | 0.19 | 0.75 | 0.19 | 0.75 |
| 0.140 | 0.26 | 1.0 | 0.26 | 1.0 | 0.26 | 1.0 | 0.26 | 1.0 | 0.26 | 1.0 | 0.26 | 1.0 | 0.26 | 1.0 | 0.26 | 1.0 |
| 0.160 | 0.34 | 1.3 | 0.34 | 1.3 | 0.34 | 1.3 | 0.34 | 1.3 | 0.34 | 1.3 | 0.34 | 1.3 | 0.34 | 1.3 | 0.34 | 1.3 |
| 0.180 | 0.43 | 1.7 | 0.43 | 1.7 | 0.43 | 1.7 | 0.43 | 1.7 | 0.43 | 1.7 | 0.43 | 1.7 | 0.43 | 1.7 | 0.43 | 1.7 |
| 0.200 | 0.53 | 2.1 | 0.53 | 2.1 | 0.53 | 2.1 | 0.53 | 2.1 | 0.53 | 2.1 | 0.53 | 2.1 | 0.53 | 2.1 | 0.53 | 2.1 |
| 0.250 | 0.84 | 3.3 | 0.84 | 3.3 | 0.84 | 3.3 | 0.84 | 3.3 | 0.84 | 3.3 | 0.84 | 3.3 | 0.84 | 3.3 | 0.84 | 3.3 |
| 0.300 | 1.2 | 4.6 | 1.2 | 4.7 | 1.2 | 4.7 | 1.2 | 4.7 | 1.2 | 4.7 | 1.2 | 4.7 | 1.2 | 4.7 | 1.2 | 4.7 |
| 0.350 | 1.6 | 6.1 | 1.6 | 6.4 | 1.6 | 6.4 | 1.6 | 6.4 | 1.6 | 6.4 | 1.6 | 6.4 | 1.6 | 6.4 | 1.6 | 6.4 |
| 0.400 | 2.0 | 7.6 | 2.1 | 8.3 | 2.1 | 8.3 | 2.1 | 8.3 | 2.1 | 8.3 | 2.1 | 8.3 | 2.1 | 8.3 | 2.1 | 8.3 |
| 0.450 | 2.5 | 9.3 | 2.7 | 11. | 2.7 | 11. | 2.7 | 11. | 2.7 | 11. | 2.7 | 11. | 2.7 | 11. | 2.7 | 11. |
| 0.500 | 3.0 | 11. | 3.3 | 13. | 3.3 | 13. | 3.3 | 13. | 3.3 | 13. | 3.3 | 13. | 3.3 | 13. | 3.3 | 13. |
| 0.600 | 3.9 | 14. | 4.7 | 18. | 4.8 | 19. | 4.8 | 19. | 4.8 | 19. | 4.8 | 19. | 4.8 | 19. | 4.8 | 19. |
| 0.700 | 4.9 | 18. | 6.2 | 23. | 6.4 | 24. | 6.6 | 26. | 6.6 | 26. | 6.6 | 26. | 6.6 | 26. | 6.6 | 26. |
| 0.800 | 5.9 | 22. | 7.7 | 29. | 8.2 | 31. | 8.6 | 34. | 8.6 | 34. | 8.6 | 34. | 8.6 | 34. | 8.6 | 34. |
| 1.000 | 8.0 | 29. | 11. | 40. | 12. | 44. | 13. | 52. | 13. | 53. | 13. | 53. | 13. | 53. | 13. | 53. |
| 1.200 | 10. | 37. | 14. | 52. | 16. | 58. | 19. | 72. | 19. | 76. | 19. | 76. | 19. | 76. | 19. | 76. |
| 1.500 | 13. | 50. | 19. | 71. | 22. | 80. | 28. | 104. | 31. | 118. | 31. | 120. | 31. | 120. | 31. | 120. |
| 2.000 | 19. | 71. | 28. | 105. | 32. | 119. | 44. | 163. | 52. | 196. | 55. | 218. | 55. | 218. | 55. | 218. |
| 2.500 | 25. | 93. | 37. | 139. | 43. | 159. | 61. | 225. | 77. | 283. | 89. | 351. | 89. | 351. | 89. | 351. |

**TABLE 18.2–17 THEORETICAL Values for Arithmetic-Average Roughness ($R_a$) and Maximum Peak-to-Valley Roughness Height ($R_t$) for Turning or Milling—Tool Type C**

| TOOL NOSE RADIUS = 3.0 mm | | | | | | | | | | | | | | | | |
|---|---|---|---|---|---|---|---|---|---|---|---|---|---|---|---|---|
| FEED/REV. (Turning) Or FEED/TOOTH (Milling) | 3° ECEA | | 5° ECEA | | 6° ECEA | | 10° ECEA | | 15° ECEA | | 30° ECEA | | 40° ECEA | | 45° ECEA | |
| | $R_a$ | $R_t$ | $R_a$ | $R_t$ | $R_a$ | $R_t$ | $R_a$ | $R_t$ | $R_a$ | $R_t$ | $R_a$ | $R_t$ | $R_a$ | $R_t$ | $R_a$ | $R_t$ |
| mm | micrometers | | | | | | | | | | | | | | | |
| 0.020 | 0.01 | 0.02 | 0.01 | 0.02 | 0.01 | 0.02 | 0.01 | 0.02 | 0.01 | 0.02 | 0.01 | 0.02 | 0.01 | 0.02 | 0.01 | 0.02 |
| 0.040 | 0.02 | 0.07 | 0.02 | 0.07 | 0.02 | 0.07 | 0.02 | 0.07 | 0.02 | 0.07 | 0.02 | 0.07 | 0.02 | 0.07 | 0.02 | 0.07 |
| 0.060 | 0.04 | 0.15 | 0.04 | 0.15 | 0.04 | 0.15 | 0.04 | 0.15 | 0.04 | 0.15 | 0.04 | 0.15 | 0.04 | 0.15 | 0.04 | 0.15 |
| 0.080 | 0.07 | 0.27 | 0.07 | 0.27 | 0.07 | 0.27 | 0.07 | 0.27 | 0.07 | 0.27 | 0.07 | 0.27 | 0.07 | 0.27 | 0.07 | 0.27 |
| 0.100 | 0.11 | 0.42 | 0.11 | 0.42 | 0.11 | 0.42 | 0.11 | 0.42 | 0.11 | 0.42 | 0.11 | 0.42 | 0.11 | 0.42 | 0.11 | 0.42 |
| 0.120 | 0.15 | 0.60 | 0.15 | 0.60 | 0.15 | 0.60 | 0.15 | 0.60 | 0.15 | 0.60 | 0.15 | 0.60 | 0.15 | 0.60 | 0.15 | 0.60 |
| 0.140 | 0.21 | 0.82 | 0.21 | 0.82 | 0.21 | 0.82 | 0.21 | 0.82 | 0.21 | 0.82 | 0.21 | 0.82 | 0.21 | 0.82 | 0.21 | 0.82 |
| 0.160 | 0.27 | 1.1 | 0.27 | 1.1 | 0.27 | 1.1 | 0.27 | 1.1 | 0.27 | 1.1 | 0.27 | 1.1 | 0.27 | 1.1 | 0.27 | 1.1 |
| 0.180 | 0.35 | 1.4 | 0.35 | 1.4 | 0.35 | 1.4 | 0.35 | 1.4 | 0.35 | 1.4 | 0.35 | 1.4 | 0.35 | 1.4 | 0.35 | 1.4 |
| 0.200 | 0.43 | 1.7 | 0.43 | 1.7 | 0.43 | 1.7 | 0.43 | 1.7 | 0.43 | 1.7 | 0.43 | 1.7 | 0.43 | 1.7 | 0.43 | 1.7 |
| 0.250 | 0.67 | 2.6 | 0.67 | 2.6 | 0.67 | 2.6 | 0.67 | 2.6 | 0.67 | 2.6 | 0.67 | 2.6 | 0.67 | 2.6 | 0.67 | 2.6 |
| 0.300 | 0.96 | 3.8 | 0.96 | 3.8 | 0.96 | 3.8 | 0.96 | 3.8 | 0.96 | 3.8 | 0.96 | 3.8 | 0.96 | 3.8 | 0.96 | 3.8 |
| 0.350 | 1.3 | 5.1 | 1.3 | 5.1 | 1.3 | 5.1 | 1.3 | 5.1 | 1.3 | 5.1 | 1.3 | 5.1 | 1.3 | 5.1 | 1.3 | 5.1 |
| 0.400 | 1.7 | 6.5 | 1.7 | 6.7 | 1.7 | 6.7 | 1.7 | 6.7 | 1.7 | 6.7 | 1.7 | 6.7 | 1.7 | 6.7 | 1.7 | 6.7 |
| 0.450 | 2.1 | 8.0 | 2.2 | 8.4 | 2.2 | 8.4 | 2.2 | 8.4 | 2.2 | 8.4 | 2.2 | 8.4 | 2.2 | 8.4 | 2.2 | 8.4 |
| 0.500 | 2.6 | 9.6 | 2.7 | 10. | 2.7 | 10. | 2.7 | 10. | 2.7 | 10. | 2.7 | 10. | 2.7 | 10. | 2.7 | 10. |
| 0.600 | 3.5 | 13. | 3.9 | 15. | 3.9 | 15. | 3.9 | 15. | 3.9 | 15. | 3.9 | 15. | 3.9 | 15. | 3.9 | 15. |
| 0.700 | 4.4 | 16. | 5.2 | 20. | 5.2 | 20. | 5.3 | 20. | 5.3 | 20. | 5.3 | 20. | 5.3 | 20. | 5.3 | 20. |
| 0.800 | 5.4 | 20. | 6.6 | 25. | 6.8 | 26. | 6.9 | 27. | 6.9 | 27. | 6.9 | 27. | 6.9 | 27. | 6.9 | 27. |
| 1.000 | 7.4 | 27. | 9.7 | 36. | 10. | 38. | 11. | 42. | 11. | 42. | 11. | 42. | 11. | 42. | 11. | 42. |
| 1.200 | 9.5 | 35. | 13. | 47. | 14. | 51. | 15. | 60. | 16. | 61. | 16. | 61. | 16. | 61. | 16. | 61. |
| 1.500 | 13. | 47. | 18. | 65. | 20. | 72. | 24. | 90. | 24. | 95. | 24. | 95. | 24. | 95. | 24. | 95. |
| 2.000 | 18. | 68. | 26. | 97. | 30. | 109. | 39. | 144. | 43. | 166. | 44. | 172. | 44. | 172. | 44. | 172. |
| 2.500 | 24. | 89. | 35. | 131. | 40. | 148. | 55. | 203. | 66. | 246. | 69. | 273. | 69. | 273. | 69. | 273. |

**TABLE 18.2–18 THEORETICAL Values for Arithmetic-Average Roughness ($R_a$) and Maximum Peak-to-Valley Roughness Height ($R_t$) for Turning or Milling—Tool Type C**

| TOOL NOSE RADIUS = 5.0 mm | | | | | | | | | | | | | | | | |
|---|---|---|---|---|---|---|---|---|---|---|---|---|---|---|---|---|
| FEED/REV. (Turning) Or FEED/TOOTH (Milling) | 3° ECEA | | 5° ECEA | | 6° ECEA | | 10° ECEA | | 15° ECEA | | 30° ECEA | | 40° ECEA | | 45° ECEA | |
| | $R_a$ | $R_t$ | $R_a$ | $R_t$ | $R_a$ | $R_t$ | $R_a$ | $R_t$ | $R_a$ | $R_t$ | $R_a$ | $R_t$ | $R_a$ | $R_t$ | $R_a$ | $R_t$ |
| mm | micrometers | | | | | | | | | | | | | | | |
| 0.020 | 0.01 | 0.01 | 0.01 | 0.01 | 0.01 | 0.01 | 0.01 | 0.01 | 0.01 | 0.01 | 0.01 | 0.01 | 0.01 | 0.01 | 0.01 | 0.01 |
| 0.040 | 0.01 | 0.04 | 0.01 | 0.04 | 0.01 | 0.04 | 0.01 | 0.04 | 0.01 | 0.04 | 0.01 | 0.04 | 0.01 | 0.04 | 0.01 | 0.04 |
| 0.060 | 0.02 | 0.09 | 0.02 | 0.09 | 0.02 | 0.09 | 0.02 | 0.09 | 0.02 | 0.09 | 0.02 | 0.09 | 0.02 | 0.09 | 0.02 | 0.09 |
| 0.080 | 0.04 | 0.16 | 0.04 | 0.16 | 0.04 | 0.16 | 0.04 | 0.16 | 0.04 | 0.16 | 0.04 | 0.16 | 0.04 | 0.16 | 0.04 | 0.16 |
| 0.100 | 0.06 | 0.25 | 0.06 | 0.25 | 0.06 | 0.25 | 0.06 | 0.25 | 0.06 | 0.25 | 0.06 | 0.25 | 0.06 | 0.25 | 0.06 | 0.25 |
| 0.120 | 0.09 | 0.36 | 0.09 | 0.36 | 0.09 | 0.36 | 0.09 | 0.36 | 0.09 | 0.36 | 0.09 | 0.36 | 0.09 | 0.36 | 0.09 | 0.36 |
| 0.140 | 0.13 | 0.49 | 0.13 | 0.49 | 0.13 | 0.49 | 0.13 | 0.49 | 0.13 | 0.49 | 0.13 | 0.49 | 0.13 | 0.49 | 0.13 | 0.49 |
| 0.160 | 0.16 | 0.64 | 0.16 | 0.64 | 0.16 | 0.64 | 0.16 | 0.64 | 0.16 | 0.64 | 0.16 | 0.64 | 0.16 | 0.64 | 0.16 | 0.64 |
| 0.180 | 0.21 | 0.81 | 0.21 | 0.81 | 0.21 | 0.81 | 0.21 | 0.81 | 0.21 | 0.81 | 0.21 | 0.81 | 0.21 | 0.81 | 0.21 | 0.81 |
| 0.200 | 0.26 | 1.0 | 0.26 | 1.0 | 0.26 | 1.0 | 0.26 | 1.0 | 0.26 | 1.0 | 0.26 | 1.0 | 0.26 | 1.0 | 0.26 | 1.0 |
| 0.250 | 0.40 | 1.6 | 0.40 | 1.6 | 0.40 | 1.6 | 0.40 | 1.6 | 0.40 | 1.6 | 0.40 | 1.6 | 0.40 | 1.6 | 0.40 | 1.6 |
| 0.300 | 0.58 | 2.3 | 0.58 | 2.3 | 0.58 | 2.3 | 0.58 | 2.3 | 0.58 | 2.3 | 0.58 | 2.3 | 0.58 | 2.3 | 0.58 | 2.3 |
| 0.350 | 0.79 | 3.1 | 0.79 | 3.1 | 0.79 | 3.1 | 0.79 | 3.1 | 0.79 | 3.1 | 0.79 | 3.1 | 0.79 | 3.1 | 0.79 | 3.1 |
| 0.400 | 1.0 | 4.0 | 1.0 | 4.0 | 1.0 | 4.0 | 1.0 | 4.0 | 1.0 | 4.0 | 1.0 | 4.0 | 1.0 | 4.0 | 1.0 | 4.0 |
| 0.450 | 1.3 | 5.1 | 1.3 | 5.1 | 1.3 | 5.1 | 1.3 | 5.1 | 1.3 | 5.1 | 1.3 | 5.1 | 1.3 | 5.1 | 1.3 | 5.1 |
| 0.500 | 1.6 | 6.3 | 1.6 | 6.3 | 1.6 | 6.3 | 1.6 | 6.3 | 1.6 | 6.3 | 1.6 | 6.3 | 1.6 | 6.3 | 1.6 | 6.3 |
| 0.600 | 2.3 | 8.9 | 2.3 | 9.0 | 2.3 | 9.0 | 2.3 | 9.0 | 2.3 | 9.0 | 2.3 | 9.0 | 2.3 | 9.0 | 2.3 | 9.0 |
| 0.700 | 3.1 | 12. | 3.1 | 12. | 3.1 | 12. | 3.1 | 12. | 3.1 | 12. | 3.1 | 12. | 3.1 | 12. | 3.1 | 12. |
| 0.800 | 4.0 | 15. | 4.1 | 16. | 4.1 | 16. | 4.1 | 16. | 4.1 | 16. | 4.1 | 16. | 4.1 | 16. | 4.1 | 16. |
| 1.000 | 5.8 | 21. | 6.4 | 25. | 6.4 | 25. | 6.4 | 25. | 6.4 | 25. | 6.4 | 25. | 6.4 | 25. | 6.4 | 25. |
| 1.200 | 7.7 | 28. | 9.1 | 35. | 9.2 | 36. | 9.3 | 36. | 9.3 | 36. | 9.3 | 36. | 9.3 | 36. | 9.3 | 36. |
| 1.500 | 11. | 39. | 14. | 50. | 14. | 53. | 14. | 57. | 14. | 57. | 14. | 57. | 14. | 57. | 14. | 57. |
| 2.000 | 16. | 58. | 21. | 79. | 23. | 86. | 26. | 100. | 26. | 101. | 26. | 101. | 26. | 101. | 26. | 101. |
| 2.500 | 21. | 78. | 30. | 109. | 33. | 121. | 40. | 150. | 41. | 159. | 41. | 159. | 41. | 159. | 41. | 159. |

**TABLE 18.2–19 THEORETICAL Values for Arithmetic-Average Roughness ($R_a$) and Maximum Peak-to-Valley Roughness Height ($R_t$) for Turning or Milling—Tool Type C**

| TOOL NOSE RADIUS = 10.0 mm | | | | | | | | | | | | | | | | |
|---|---|---|---|---|---|---|---|---|---|---|---|---|---|---|---|---|
| FEED/REV. (Turning) Or FEED/TOOTH (Milling) | 3° ECEA | | 5° ECEA | | 6° ECEA | | 10° ECEA | | 15° ECEA | | 30° ECEA | | 40° ECEA | | 45° ECEA | |
| | $R_a$ | $R_t$ | $R_a$ | $R_t$ | $R_a$ | $R_t$ | $R_a$ | $R_t$ | $R_a$ | $R_t$ | $R_a$ | $R_t$ | $R_a$ | $R_t$ | $R_a$ | $R_t$ |
| mm | micrometers | | | | | | | | | | | | | | | |
| 0.020 | 0.00 | 0.01 | 0.00 | 0.01 | 0.00 | 0.01 | 0.00 | 0.01 | 0.00 | 0.01 | 0.00 | 0.01 | 0.00 | 0.01 | 0.00 | 0.01 |
| 0.040 | 0.01 | 0.02 | 0.01 | 0.02 | 0.01 | 0.02 | 0.01 | 0.02 | 0.01 | 0.02 | 0.01 | 0.02 | 0.01 | 0.02 | 0.01 | 0.02 |
| 0.060 | 0.02 | 0.05 | 0.02 | 0.05 | 0.02 | 0.05 | 0.02 | 0.05 | 0.02 | 0.05 | 0.02 | 0.05 | 0.02 | 0.05 | 0.02 | 0.05 |
| 0.080 | 0.03 | 0.08 | 0.03 | 0.08 | 0.03 | 0.08 | 0.03 | 0.08 | 0.03 | 0.08 | 0.03 | 0.08 | 0.03 | 0.08 | 0.03 | 0.08 |
| 0.100 | 0.03 | 0.13 | 0.03 | 0.13 | 0.03 | 0.13 | 0.03 | 0.13 | 0.03 | 0.13 | 0.03 | 0.13 | 0.03 | 0.13 | 0.03 | 0.13 |
| 0.120 | 0.05 | 0.18 | 0.05 | 0.18 | 0.05 | 0.18 | 0.05 | 0.18 | 0.05 | 0.18 | 0.05 | 0.18 | 0.05 | 0.18 | 0.05 | 0.18 |
| 0.140 | 0.07 | 0.25 | 0.07 | 0.25 | 0.07 | 0.25 | 0.07 | 0.25 | 0.07 | 0.25 | 0.07 | 0.25 | 0.07 | 0.25 | 0.07 | 0.25 |
| 0.160 | 0.08 | 0.32 | 0.08 | 0.32 | 0.08 | 0.32 | 0.08 | 0.32 | 0.08 | 0.32 | 0.08 | 0.32 | 0.08 | 0.32 | 0.08 | 0.32 |
| 0.180 | 0.11 | 0.41 | 0.11 | 0.41 | 0.11 | 0.41 | 0.11 | 0.41 | 0.11 | 0.41 | 0.11 | 0.41 | 0.11 | 0.41 | 0.11 | 0.41 |
| 0.200 | 0.13 | 0.50 | 0.13 | 0.50 | 0.13 | 0.50 | 0.13 | 0.50 | 0.13 | 0.50 | 0.13 | 0.50 | 0.13 | 0.50 | 0.13 | 0.50 |
| 0.250 | 0.20 | 0.78 | 0.20 | 0.78 | 0.20 | 0.78 | 0.20 | 0.78 | 0.20 | 0.78 | 0.20 | 0.78 | 0.20 | 0.78 | 0.20 | 0.78 |
| 0.300 | 0.29 | 1.1 | 0.29 | 1.1 | 0.29 | 1.1 | 0.29 | 1.1 | 0.29 | 1.1 | 0.29 | 1.1 | 0.29 | 1.1 | 0.29 | 1.1 |
| 0.350 | 0.39 | 1.5 | 0.39 | 1.5 | 0.39 | 1.5 | 0.39 | 1.5 | 0.39 | 1.5 | 0.39 | 1.5 | 0.39 | 1.5 | 0.39 | 1.5 |
| 0.400 | 0.51 | 2.0 | 0.51 | 2.0 | 0.51 | 2.0 | 0.51 | 2.0 | 0.51 | 2.0 | 0.51 | 2.0 | 0.51 | 2.0 | 0.51 | 2.0 |
| 0.450 | 0.65 | 2.5 | 0.65 | 2.5 | 0.65 | 2.5 | 0.65 | 2.5 | 0.65 | 2.5 | 0.65 | 2.5 | 0.65 | 2.5 | 0.65 | 2.5 |
| 0.500 | 0.80 | 3.1 | 0.80 | 3.1 | 0.80 | 3.1 | 0.80 | 3.1 | 0.80 | 3.1 | 0.80 | 3.1 | 0.80 | 3.1 | 0.80 | 3.1 |
| 0.600 | 1.2 | 4.5 | 1.2 | 4.5 | 1.2 | 4.5 | 1.2 | 4.5 | 1.2 | 4.5 | 1.2 | 4.5 | 1.2 | 4.5 | 1.2 | 4.5 |
| 0.700 | 1.6 | 6.1 | 1.6 | 6.1 | 1.6 | 6.1 | 1.6 | 6.1 | 1.6 | 6.1 | 1.6 | 6.1 | 1.6 | 6.1 | 1.6 | 6.1 |
| 0.800 | 2.1 | 8.0 | 2.1 | 8.0 | 2.1 | 8.0 | 2.1 | 8.0 | 2.1 | 8.0 | 2.1 | 8.0 | 2.1 | 8.0 | 2.1 | 8.0 |
| 1.000 | 3.2 | 13. | 3.2 | 13. | 3.2 | 13. | 3.2 | 13. | 3.2 | 13. | 3.2 | 13. | 3.2 | 13. | 3.2 | 13. |
| 1.200 | 4.6 | 18. | 4.6 | 18. | 4.6 | 18. | 4.6 | 18. | 4.6 | 18. | 4.6 | 18. | 4.6 | 18. | 4.6 | 18. |
| 1.500 | 7.1 | 27. | 7.2 | 28. | 7.2 | 28. | 7.2 | 28. | 7.2 | 28. | 7.2 | 28. | 7.2 | 28. | 7.2 | 28. |
| 2.000 | 12. | 43. | 13. | 50. | 13. | 50. | 13. | 50. | 13. | 50. | 13. | 50. | 13. | 50. | 13. | 50. |
| 2.500 | 16. | 60. | 20. | 74. | 20. | 77. | 20. | 78. | 20. | 78. | 20. | 78. | 20. | 78. | 20. | 78. |

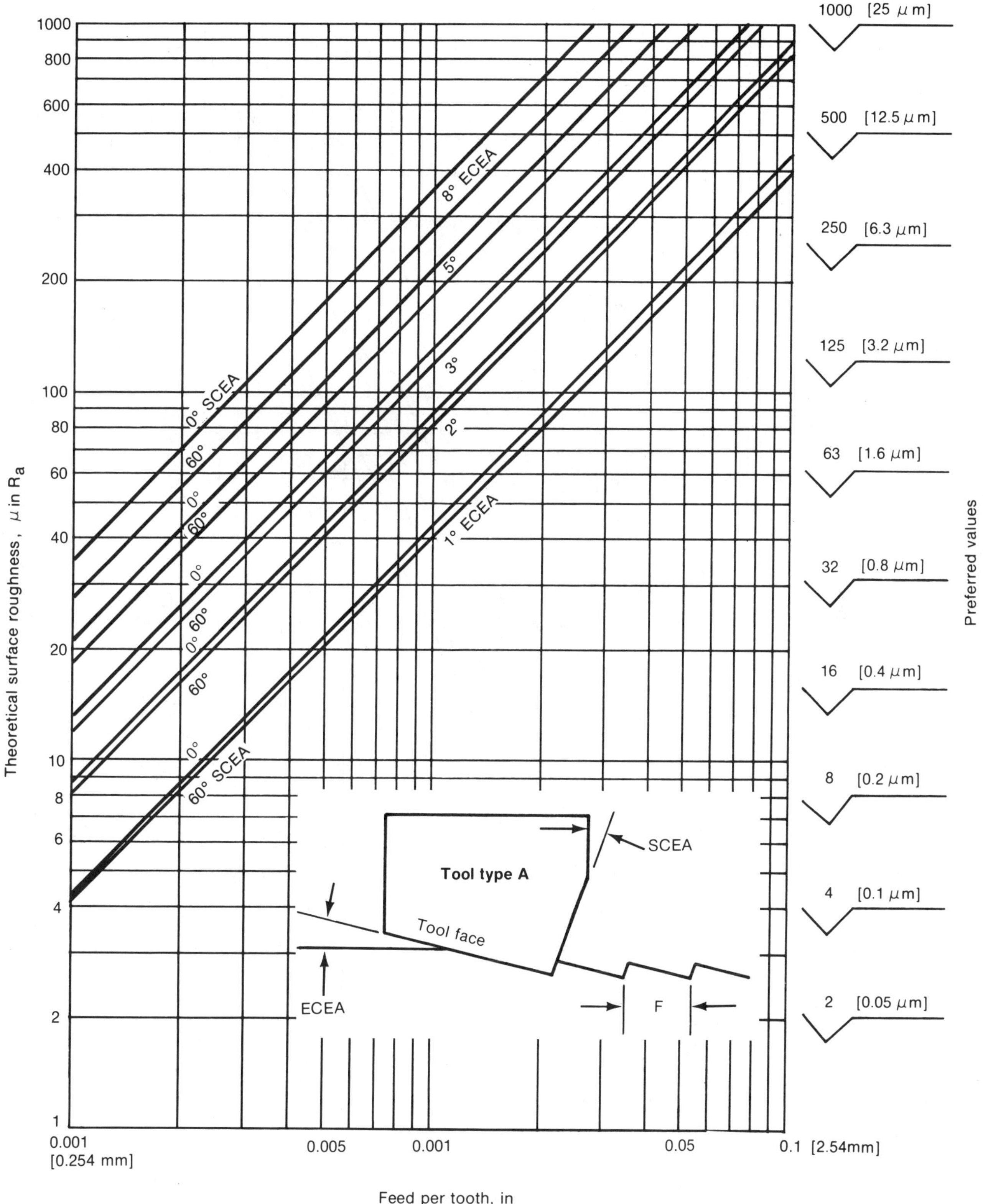


**Figure 18.2–14** Theoretical surface roughness for a face milling cutter containing teeth with a zero nose radius.

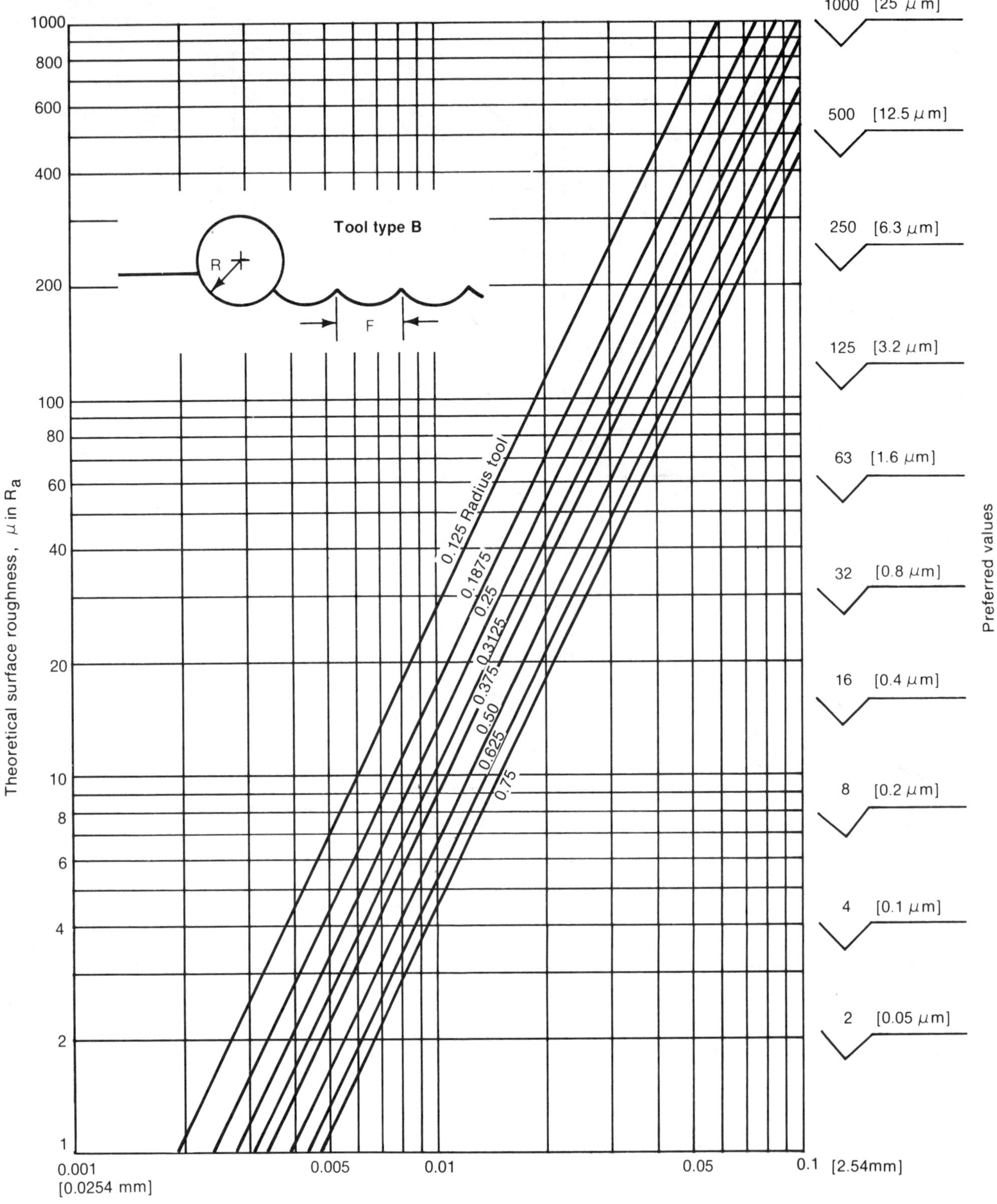


**Figure 18.2–15** Theoretical surface roughness for turning or face milling tools with round cutting edges.

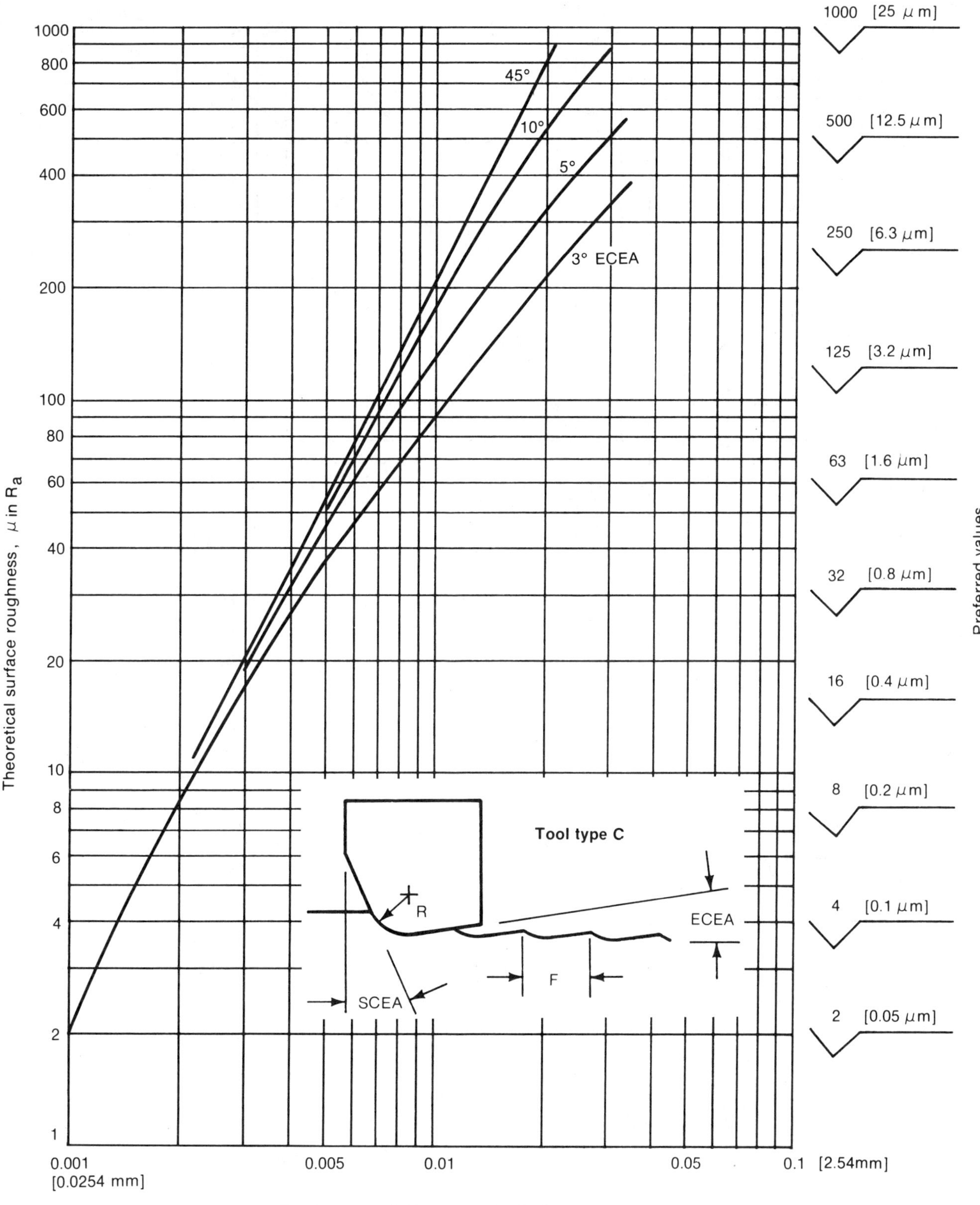


**Figure 18.2-16** Theoretical surface roughness for turning or face milling tools with a radius of 0.0156 inch [0.39 mm] and various ECEA.

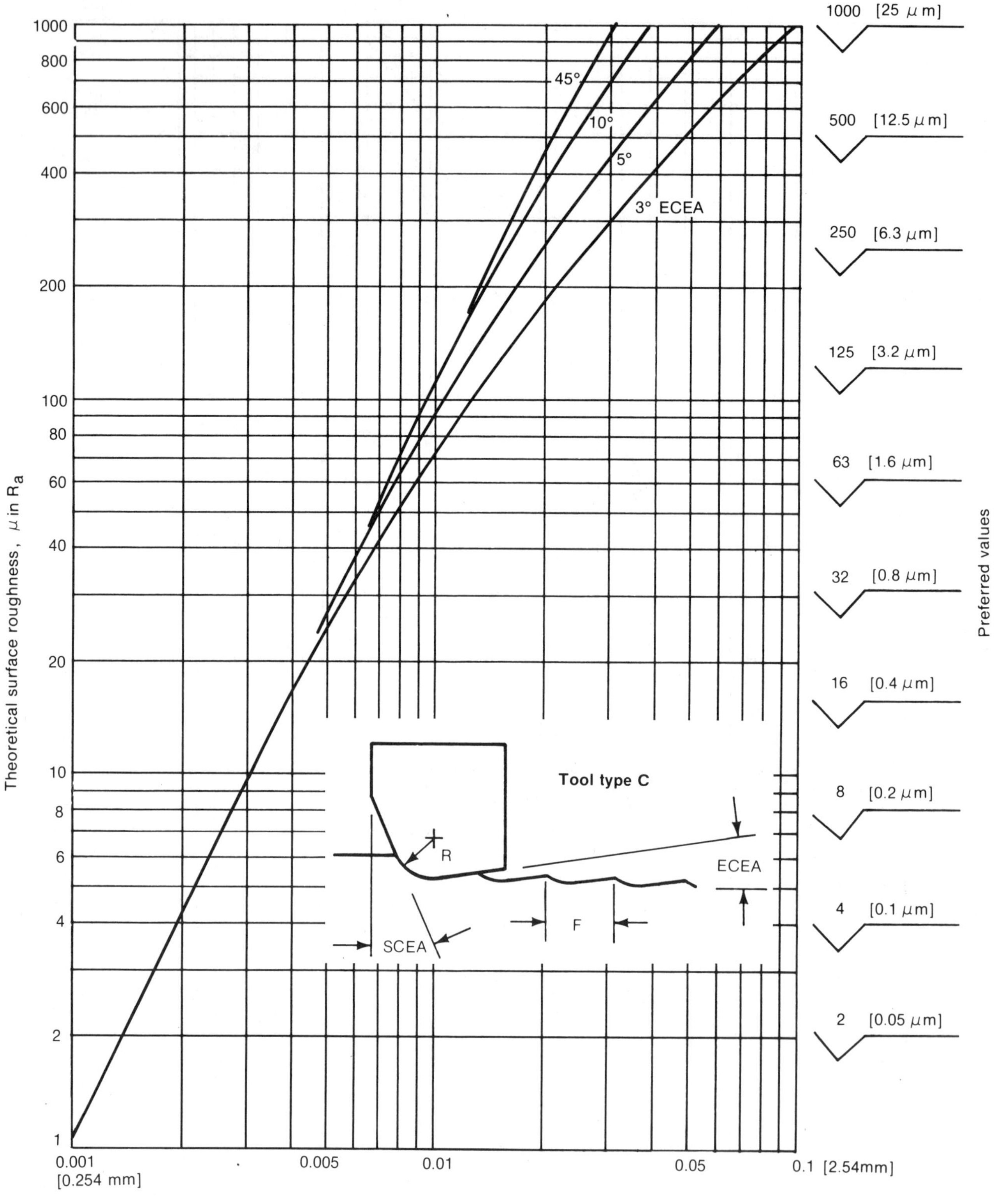


**Figure 18.2-17** Theoretical surface roughness for turning or face milling tools with a radius of 0.0312 inch [0.79 mm] and various ECEA.

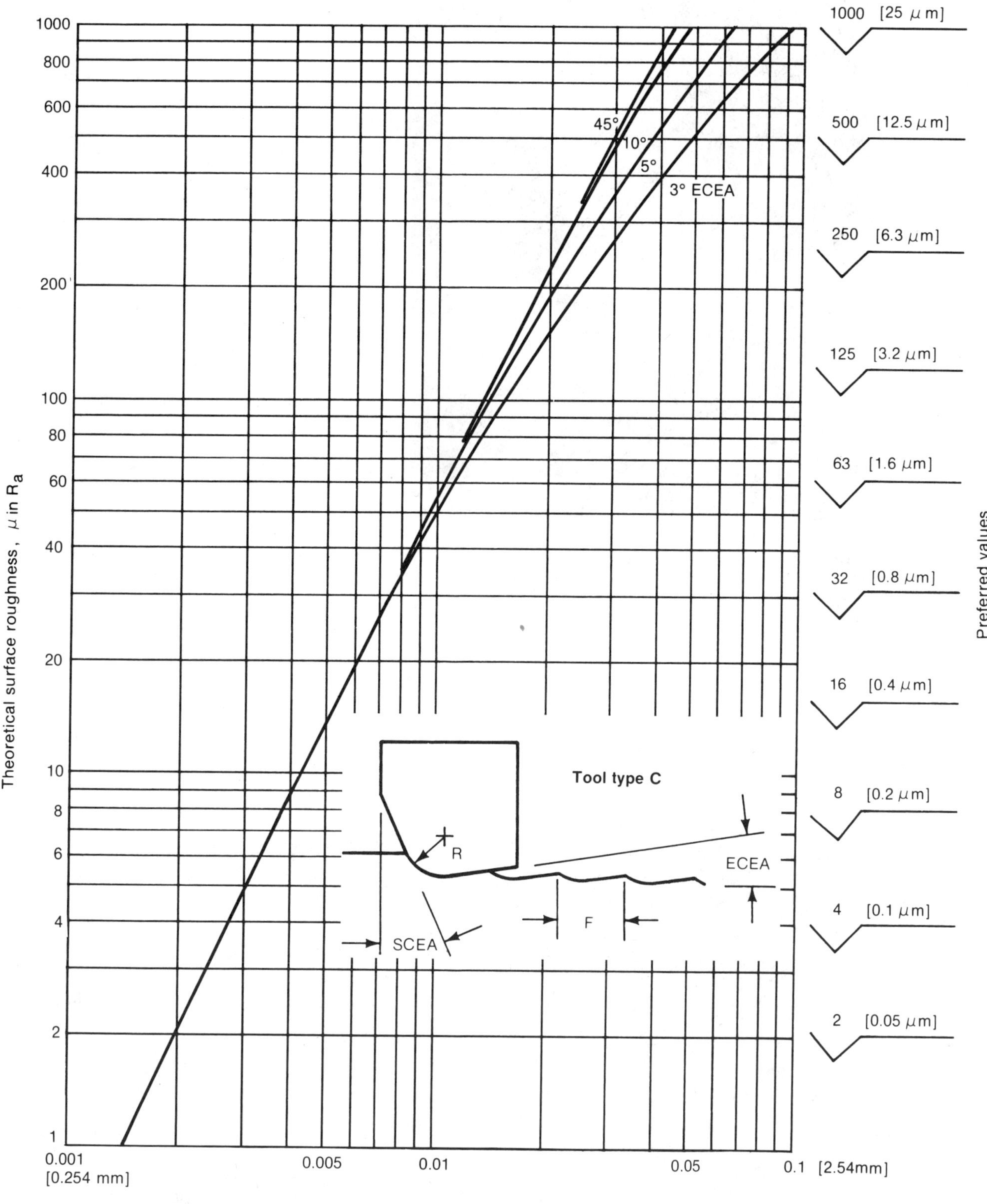


**Figure 18.2–18** Theoretical surface roughness for turning or face milling tools with a radius of 0.0625 inch [1.59 mm] and various ECEA.

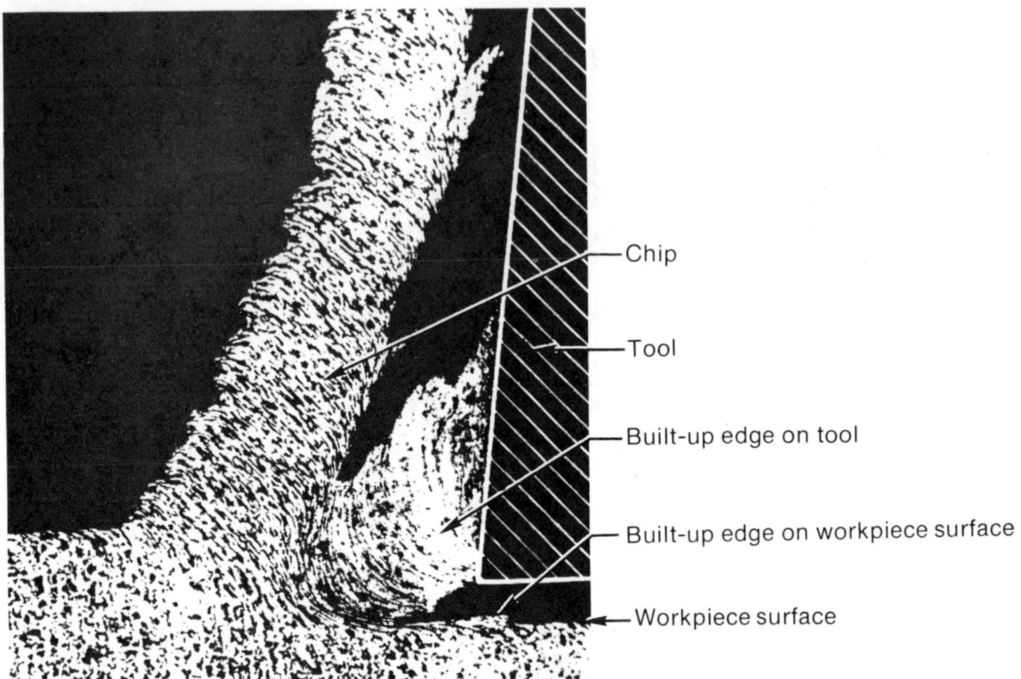


**Figure 18.2–19** Built-up edge retained on workpiece surface is detrimental to good surface texture. (Photo from A. J. Pekelharing, *Annals of the CIRP* 23(2): 207)

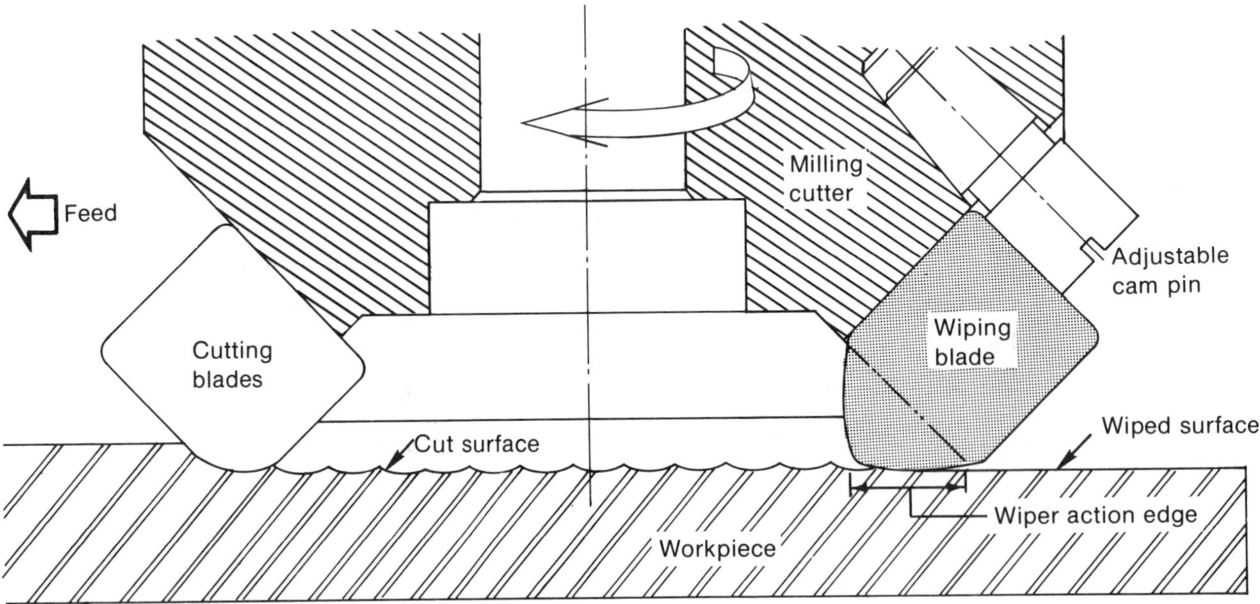


**Figure 18.2–20** Wiper blade or micro-finish blade for smoothing feed marks from face milling.

# INTRODUCTION

## Definitions

The specification and manufacture of unimpaired or enhanced surfaces requires an understanding of the interrelationship of metallurgy, machinability and mechanical testing. To satisfy this requirement, an encompassing discipline known as surface integrity was introduced and has gained worldwide acceptance. Surface integrity is the description and control of the many possible alterations produced in a surface layer during manufacturing, including their effects on the material properties and the performance of the surface in service. Surface integrity is achieved by the selection and control of manufacturing processes according to the evaluation of the process effects on significant engineering properties of work materials.

For highly stressed, critically loaded or specially engineered surfaces, surface integrity is an added requirement. It is the assessment of the overall engineering quality of a surface and its ability to perform required service functions. It joins dimensions, tolerances and surface texture as an essential specification for highly engineered components in order for them to satisfy modern concerns with safety, product reliability and life cycle costs.

Dynamic loading is recognized as an important factor in the design of many engineering structures in all types of applications. Where dynamic loading is present, operating stresses frequently are limited by the fatigue characteristics of the structural material to be used. Service histories and failure analyses of dynamically loaded components illustrate clearly that fatigue failures often initiate at or just below the surface of a component. Thus, it may be concluded that fatigue behavior is sensitive to surface conditions. Also, in considering stress corrosion resistance, it is recognized that the surface condition of a component is a primary factor in determining susceptibility to attack and possible subsequent part failure.

While all manufacturing processes may have significant effects on the workpiece material and its properties, the bulk of currently available data relates to conventional and nontraditional material removal processes.

## Benefits from the Application of Surface Integrity Practices

The disciplined use of surface integrity practices for the manufacture of surfaces which are highly stressed and/or critically loaded can yield significant benefits as follows:

- Enhanced component integrity with resultant increase in safety and service life and decrease in maintenance costs
- Better understanding of the effects of manufacturing process parameters as they relate to reliability of components
- Decreased costs achieved by limiting surface integrity processing only to highly stressed surfaces and simultaneously relaxing costly specification requirements for generating the other surfaces of a component
- Reduction in manufacturing scrap or rework
- Improved process control and quality control through understanding of effects from process control limits
- Better producibility definition
- Strengthened value analysis
- Better defined manufacturing and/or design leeways
- Guidance to advanced process design or application
- More disciplined process selection

## Problems Amenable to Surface Integrity Discipline

Before applying surface integrity technology, one must first decide whether a surface integrity problem actually exists. The control of surface integrity generally adds cost to the manufacturing process; therefore, surface integrity should not be a consideration unless a problem exists or may be considered as a possibility. Surface integrity must be considered in the manufacture of highly stressed components used in applications involving human safety, high cost, and predictable component life. If a surface integrity problem exists, then the pertinent surface characteristics and the affected engineering properties of the material must be monitored. Some examples of the types of problems for which the application of surface integrity principles can be helpful are as follows:

- *overheating or burning*; grinding burns on high strength alloys; for example, gears, landing gear cylinders, bearings, turbine blades—see figure 18.3-1
- *microcracks* and *surface irregularities* as initiation sites for failures
- *distortion* and loss of dimensional quality, particularly in thin components
- *residual stresses* lowering fatigue endurance strength
- *residual stresses* combined with severe environments leading to early stress corrosion failures
- *reduction in fatigue strength* from metallurgically altered surfaces; for example, untempered or overtempered martensite, recast or resolidifed areas from thermal processes
- *metallurgical or mechanical alterations* produced as a result of excessively high removal rates or process energy
- *thin sections* of components operating at high stress levels
- applications where *life or safety reliability concerns* are particularly stringent
- holes with *high depth-to-diameter ratios* where maintaining drill sharpness may be difficult

# 18.3 Surface Integrity

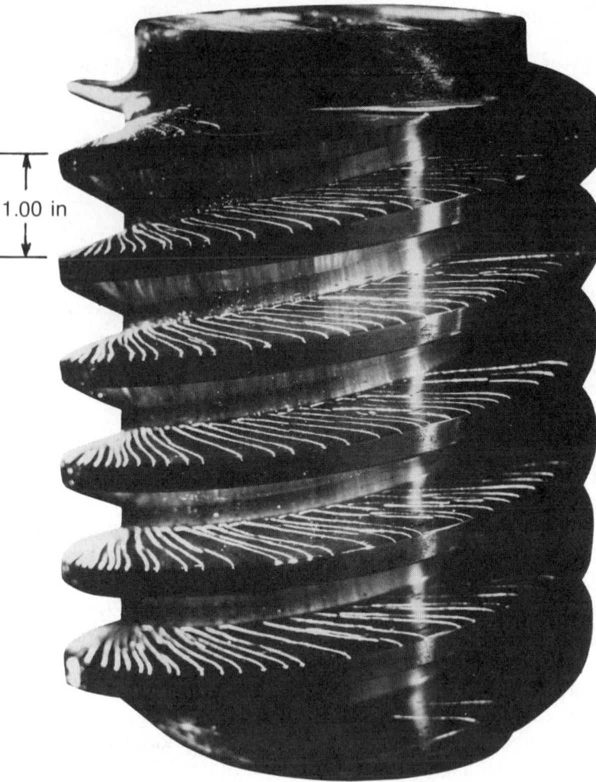


**Figure 18.3-1** Failure in grinding of a carburized 8620 steel worm gear; an absence of surface integrity from leading edge dullness on plunge gear grinding. Ultraviolet light photograph of fluorescent penetrant indications found during inspection after grinding. (G. Bellows[13])

A practical balance must be found between over-controlling the manufacturing process with its attendant costs and unneeded benefits and under-controlling with its possible decrease in quality, increase in scrapped components, or loss of reliability. Good application of surface integrity practices can result in cost improvement and quality enhancement by insisting on greater surface quality only on surface areas where needed and relaxing specifications for the bulk of the component surfaces.

## ALTERED MATERIAL ZONES (AMZ)

Surface integrity is concerned primarily with the host of effects a manufacturing process produces below the visible surface. The subsurface characteristics occur in various layers or zones. The subsurface altered material zones (AMZ) can be as simple as a stress condition different from that in the body of the material or as complex as a microstructure change interlaced with intergranular attack (IGA). Changes can be caused by chemical, thermal, electrical, or mechanical energy and may affect both the physical and the metallurgical properties of the material. The principal causes of alterations to a surface during material removal operations are as follows:

- high temperatures and high temperature gradients
- plastic deformation or mechanical strain
- chemical reactions or absorption on nascent machined surfaces
- excessive electrical currents
- excessive energy densities during processing

The subsurface altered material zones can be grouped by the principal energy modes that produce them, as shown in table 18.3-1. Table 18.3-2 lists alterations that have been observed in several classes of materials as a result of various machining processes.

**TABLE 18.3-1 Altered Material Zones by Principal Energy Mode**

MECHANICAL:
- Plastic deformations (as result of hot or cold working)
- Tears and laps and crevice-like defects (associated with "built-up edge" produced in machining)
- Hardness alterations
- Cracks (macroscopic and microscopic)
- Residual stress distribution in surface layer
- Processing inclusions introduced
- Plastically deformed debris as a result of grinding
- Voids, pits, burrs or foreign material inclusions in surface

METALLURGICAL:
- Transformation of phases
- Grain size and distribution
- Precipitate size and distribution
- Foreign inclusions in material
- Twinning
- Recrystallization
- Untempered martensite (UTM) or overtempered martensite (OTM)
- Resolutioning or austenite reversion

CHEMICAL:
- Intergranular attack (IGA)
- Intergranular corrosion (IGC)
- Intergranular oxidation (IGO)
- Preferential dissolution of microconstituents
- Contamination
- Embrittlement—by chemical absorption of elements such as hydrogen, chlorine, etc.
- Pits or selective etch
- Corrosion
- Stress corrosion

THERMAL:
- Heat-affected zone (HAZ)
- Recast or redeposited material
- Resolidified material
- Splattered particles or remelted metal deposited on surface

ELECTRICAL:
- Conductivity change
- Magnetic change
- Resistive heating or overheating

SOURCE: G. Bellows and D. N. Tishler.[25]

**TABLE 18.3-2 Summary of Possible Surface Alterations Resulting From Various Metal Removal Processes**

| MATERIAL | CONVENTIONAL METAL REMOVAL METHODS | | NONTRADITIONAL REMOVAL METHODS | | |
|---|---|---|---|---|---|
| | Milling, Drilling, or Turning | Grinding | EDM | ECM | CHM |
| Steels:<br>Nonhardenable<br>1018 | R,<br>PD,<br>L & T | R,<br>PD | R,<br>MCK,<br>RC | R,<br>SE,<br>IGA | R,<br>SE,<br>IGA |
| Hardenable<br>4340<br>D6ac | R,<br>PD,<br>L & T<br>MCK,<br>UTM,<br>OTM | R,<br>PD,<br>MCK,<br>UTM,<br>OTM | R,<br>MCK,<br>RC,<br>UTM,<br>OTM | R,<br>SE,<br>IGA | R,<br>SE,<br>IGA |
| Tool Steel<br>D2 | R,<br>PD,<br>L & T<br>MCK,<br>UTM,<br>OTM | R,<br>PD,<br>MCK,<br>UTM,<br>OTM | R,<br>MCK,<br>RC,<br>UTM,<br>OTM | R,<br>SE,<br>IGA | R,<br>SE,<br>IGA |
| Stainless (martensitic)<br>410 | R,<br>PD,<br>L & T<br>MCK,<br>UTM,<br>OTM | R,<br>PD,<br>MCK,<br>UTM,<br>OTM | R,<br>MCK,<br>RC,<br>UTM,<br>OTM | R,<br>SE,<br>IGA | R,<br>SE,<br>IGA |
| Stainless (austenitic)<br>302 | R,<br>PD,<br>L & T | R,<br>PD | R,<br>MCK,<br>RC | R,<br>SE,<br>IGA | R,<br>SE,<br>IGA |
| Precipitation hardening<br>17-4 PH | R,<br>PD,<br>L & T,<br>OA | R,<br>PD,<br>OA | R,<br>MCK,<br>RC,<br>OA | R,<br>SE,<br>IGA | R,<br>SE,<br>IGA |
| Maraging (18% Ni)<br>250 Grade | R,<br>PD,<br>L & T,<br>RS,<br>OA | R,<br>PD,<br>RS,<br>OA | R,<br>RC,<br>RS,<br>OA | R,<br>SE,<br>IGA | R,<br>SE,<br>IGA |
| Nickel and Cobalt Base Alloys:<br>Inconel alloy 718,<br>Rene 41<br>HS 31<br>IN-100 | HAZ<br>R,<br>PD,<br>L & T,<br>MCK | HAZ<br>R,<br>PD,<br>MCK | R,<br>MCK,<br>RC | R,<br>SE,<br>IGA | R,<br>SE,<br>IGA |
| Titanium Alloy:<br>Ti-6Al-4V | HAZ<br>R,<br>PD,<br>L & T | HAZ<br>R,<br>PD,<br>MCK | R,<br>MCK,<br>RC | R,<br>SE,<br>IGA | R,<br>SE |
| Refractory Alloys:<br>TZM | R,<br>L & T,<br>MCK | R,<br>MCK | R,<br>MCK | R,<br>SE,<br>IGA | R,<br>SE |
| Tungsten (pressed and sintered) | R,<br>L & T,<br>MCK | R,<br>MCK | R,<br>MCK | R,<br>SE,<br>MCK,<br>IGA | R,<br>SE,<br>MCK,<br>IGA |

SOURCE: Revised from M. Field, J. F. Kahles and J. T. Cammett.[12]

Key:
- R — Roughness of surface
- PD — Plastic deformation and plastically deformed debris
- L & T — Laps and tears and crevice-like defects
- MCK — Microcracks
- SE — Selective etch
- IGA — Intergranular attack
- UTM — Untempered martensite
- OTM — Overtempered martensite
- OA — Overaging
- RS — Resolution or austenite reversion
- RC — Recast, respattered metal, or vapor deposited metal
- HAZ — Heat affected zone

# 18.3 Surface Integrity

## AMZ Definitions

*Cracks* are narrow ruptures or separations that alter the continuity of a surface. They are usually tight and characterized by sharp edges or sharp changes in direction with depth-to-width ratio of 4:1 or greater and discernible with the unaided eye or with 10X or less magnification.

*Microcracks* are those cracks requiring greater than 10X magnification for discernment, as shown in figures 18.3-2 and 18.3-4.

*Craters* are surface depressions with rough edges, approximately round or oval and shallow, usually with a depth-to-width ratio of less than 4:1. This term is frequently applied to the impressions left by the individual spark discharges in electrical discharge machining (EDM), as shown in figure 18.3-3, but is also used to describe massive depressions resulting from short circuit accidents in EDM or ECM. See also *Pits*.

*Hardness alterations* are changes in hardness of a surface layer as a result of heat, mechanical deformation or chemical change during processing. (Alterations of less than ±2 points Rockwell C or equivalent from the hardness of the bulk of the material are not considered significant.) (See also figure 18.3-15.)

*Heat-affected zones* (HAZ) are those portions of a material not melted, yet subjected to sufficient thermal energy to produce microstructural alterations or microhardness alterations, as shown in figure 18.3-4.

*Inclusions* are small particles in the surface layer of an object and may be either foreign or a part of the normal composition of the material, as shown in figure 18.3-4.

*Intergranular attack* (IGA) is a form of corrosion or attack in which preferential reactions are concentrated at the surface grain boundaries, usually in the form of sharp notches or discontinuities, as shown in figure 18.3-5. These effects are sometimes supplemented by intergranular oxidation (IGO) derived from exposure to elevated temperatures or intergranular corrosion (IGC) derived from exposure to active chemical reagents.

*Laps, folds or seams* are defects in a surface from continued plastic working of overlapping surfaces, as shown in figure 18.3-6.

*Low stress surface* is one containing a residual stress less than 20 ksi [138 MPa] or 10 percent of tensile strength, whichever is greater, at depths below the surface greater than 0.001 inch [0.025 mm]. Sometimes called a stress-free surface.

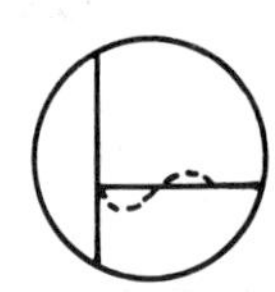

*Metallurgical transformations* are those microstructural changes resulting from external influences. They include recrystallization, alloy depletion, chemical reactions, and resolidified or redeposited or recast layers. Figure 18.3-7 shows an example of metallurgical transformation.

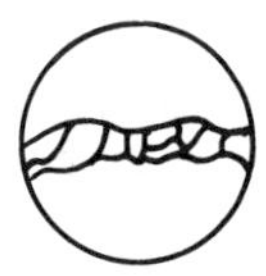

*Microcracks*—See *Cracks*.

*Pits* are shallow depressions resembling a small crater with rounded edges and less than 4:1 depth-to-width ratio. Also describes a specialized form of localized or selective etching or corrosion that results in holes or pockets left by the mechanical removal of small particles or inclusions from the surface. Pits also are the result of dents from the impingement of foreign particles against the surface. Pits sometimes are associated from electrochemical action into regions slightly removed from the high current density regions. See figure 18.3-8. See also *Craters*.

*Plastic deformation* is a microstructural change as a result of exceeding the yield point of the material and generally includes elongation of the grain structure and increased hardness. See figure 18.3-9.

*Recrystallization* is the formation of new, strain-free grain or crystal structure from that existing in the material prior to processing, usually as a result of plastic deformation and subsequent heating or from a phase change occurring during heating. See figure 18.3-10.

*Recast material* is a general term applied to surfaces that have at some point in the processing become molten and then resolidified. Frequently includes a conglomerate mixture of redeposited or remelted material, as shown in figure 18.3-11.

*Redeposited material* is that material, which in the material removal process, is removed from the surface in a molten state, and then prior to solidification is reattached to the surface. This is sometimes called splattered metal and is also shown in figure 18.3-11.

*Remelted or resolidified material* is that portion of the surface which during the metal removal process become molten, but is not removed from the surface prior to resolidification. See figure 18.3-11.

*Residual stresses* are those stresses which are present in a material after all external influences (forces, thermal gradients, or external energy) have been removed. See figure 18.3-16.

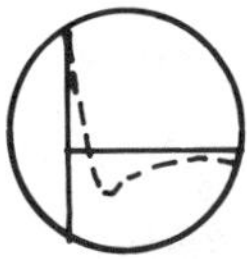

*Selective etch* is a form of in-process corrosion or attack in which preferential reactions are concentrated within and through the grains or concentrated on certain constituents in the base material. See figure 18.3-12.

*Splattered metal*—See *Redeposited material* and figure 18.3-11.

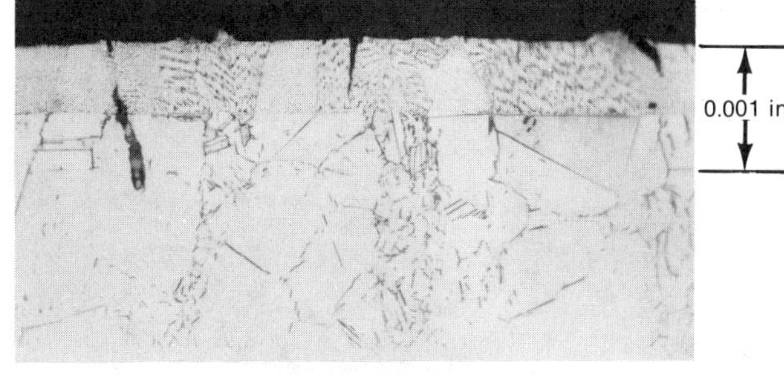


**Figure 18.3-4** Example of heat-affected zone. Laser machining of Inconel Alloy 718 (solution treated and aged), 1000X. Note epitaxial growth of grains nucleating from base metal grains. Also shows microcracks and an inclusion. (Surface Integrity Encyclopedia[1])

**Figure 18.3-2** Example of microcrack from grinding of Udimet 700C, 250X. (Surface Integrity Encyclopedia[1])

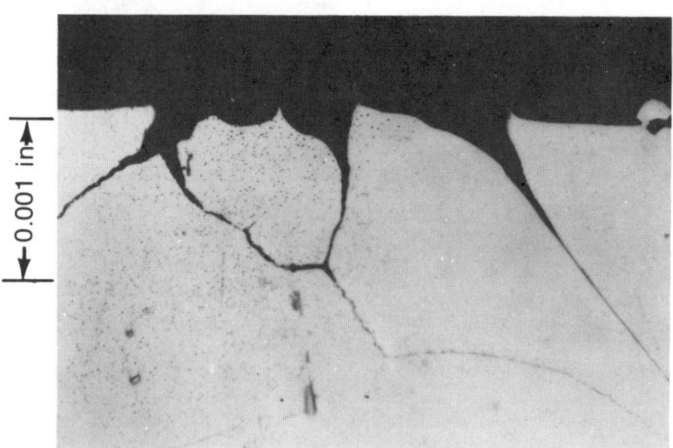


**Figure 18.3-5** Example of intergranular attack, ECM of Waspaloy (aged, 40 $R_C$), 1000X. (W. P. Koster et al[8])

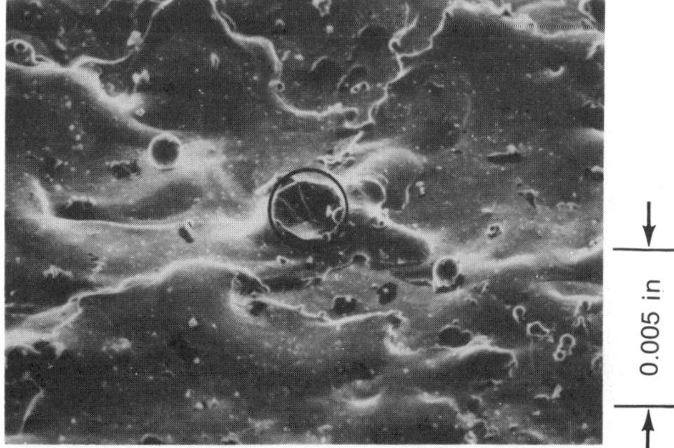


**Figure 18.3-3** Example of craters, EDM of Inconel Alloy 718 (solution treated and aged), SEM at 90° and 215X. (Surface Integrity Encyclopedia[1])

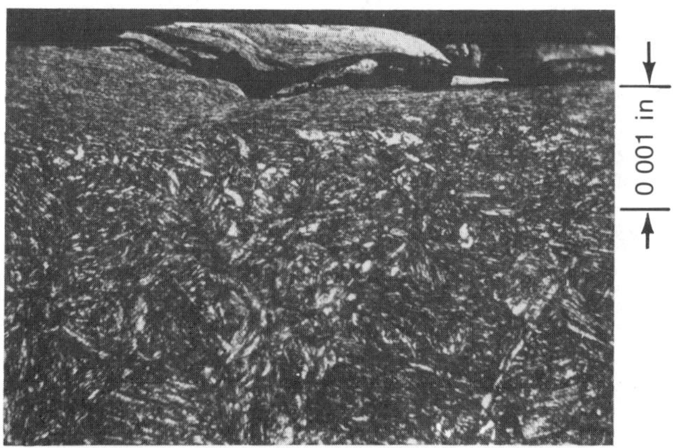


**Figure 18.3-6** Example of lap, fold and tears. Drilling of 4340 steel (quenched and tempered, 50 $R_C$), 1000X perpendicular to lay.

# 18.3 Surface Integrity

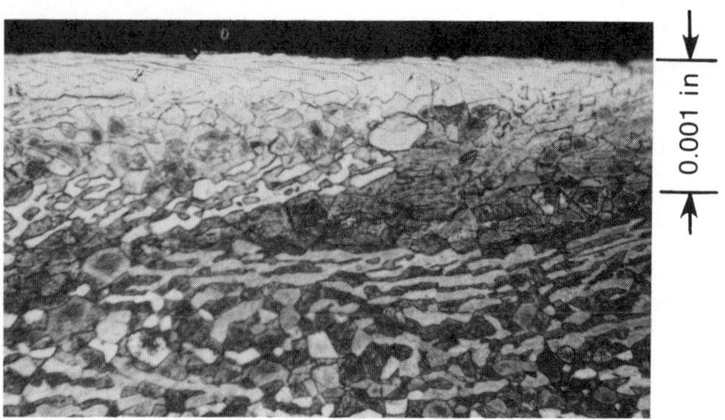


**Figure 18.3-7** Example of metallurgical transformation. Reformed alpha case crystals of Ti-6Al-4V during grinding, 1000X. Note also the plastic deformation.

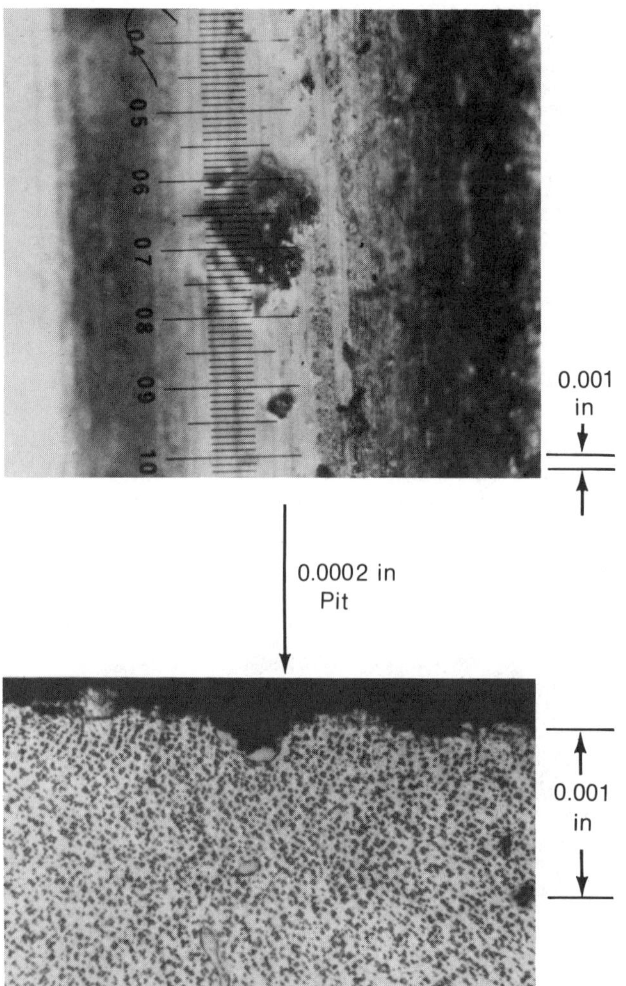


**Figure 18.3-8** Example of pit from grinding of Rene 80 (solution treated). *Top*, surface view at 40X; *bottom*, section at 1000X. (Surface Integrity Encyclopedia[1])

**Figure 18.3-9** Example of plastic deformation. Conventional turning of Inconel Alloy 718 (solution treated and aged) with dull tool, 500X, approximately 0.0036-inch deep. (Surface Integrity Encyclopedia[1])

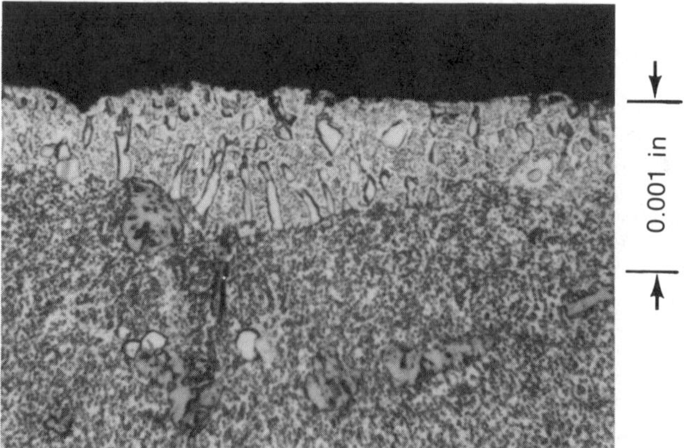


**Figure 18.3-10** Example of recrystallization. Recrystallized layer from age heat treatment following grinding of Rene 80, 1000X. (Surface Integrity Encyclopedia[1])

**Figure 18.3–11** Example of recast, redeposited and splattered material. EDM of 4340 steel (quenched and tempered, 50 $R_C$), 1000X. Note underlying untempered martensite in the 0.002 inch-deep heat-affected zone. (W. P. Koster et al[8])

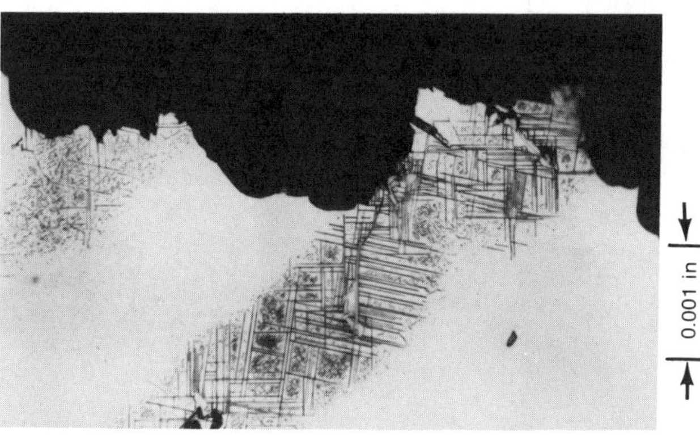


**Figure 18.3–12** Example of selective etching. Chemical machining of Inconel Alloy 718C (solution treated and aged, 40 $R_C$), 1000X. (W. P. Koster et al[8])

## EVALUATION TECHNIQUES

Experimental procedures have been developed to provide three increasingly deeper levels of surface integrity evaluation. The *minimum data set* is the least expensive approach and should be considered only as a screening test for the analysis of surface effects. It is essentially metallographic information supplemented with microhardness and surface texture measurements. Table 18.3–3 describes the components of the minimum data set.

The *standard data set* provides more in-depth data for more critical applications. It includes the results from the minimum data set along with residual stress (or distortion) profiles and high cycle fatigue screening tests, as outlined in table 18.3–4. Generally, the standard data set has information at two or more diverse levels of process intensity to represent finishing and roughing (or abusive) machining conditions. Figure 18.3–13 is an example of the standard data set produced in grinding of 4340 steel, quenched and tempered to 50 $R_C$.

The *extended data set* provides data gathered from statistically designed fatigue programs and yields data suitable for detailed designing. It also includes specialized data such as stress corrosion life with specific types of environmental exposures. Components of the extended data set are given in table 18.3–5.

**TABLE 18.3-3 Minimum Surface Integrity Data Set**

1. Surface texture
   - roughness measurement (per ANSI B46.1–1978) or microtopographic map (figure 18.2-7, section 18.2)
   - lay designation or photo
   - occasionally includes scanning electron microscope (SEM) series of photos at increasing magnifications (recommended series 20, 200, 1000, and 2000X) (figure 18.2-8, section 18.2)
2. Macrostructure (10X or less)
   - macrocracks or surface imperfections (pits, laps, etc.)
   - macroetch indications (e.g., fluorescent penetrant or magnetic flux or similar)
   - chemical etchant tests*
3. Microstructure (cross section examination at 1000X preferred)
   - microcracks
   - plastic deformation (section parallel to lay suggested)
   - phase transformations
   - intergranular attack
   - micro defects (laps, inclusions, etc.)
   - built-up edge or deposits of debris
   - recast layers
   - selective etching
   - metallurgical transformations
4. Microhardness alterations
   - heat-affected zones

SOURCE: G. Bellows and D. N. Tishler.[25]
*Nital for UTM in hardened steels, 1% HF for alpha case in titanium alloys.

**TABLE 18.3-4 Standard Surface Integrity Data Set**

1. Minimum data set
2. Residual stress profile—or distortion measurements
3. Fatigue tests (screening only—recommend using full-reverse bending at room temperature using tapered area flat specimens with results to $10^7$ cycles)

SOURCE: G. Bellows and D. N. Tishler.[25]

## 18.3 Surface Integrity

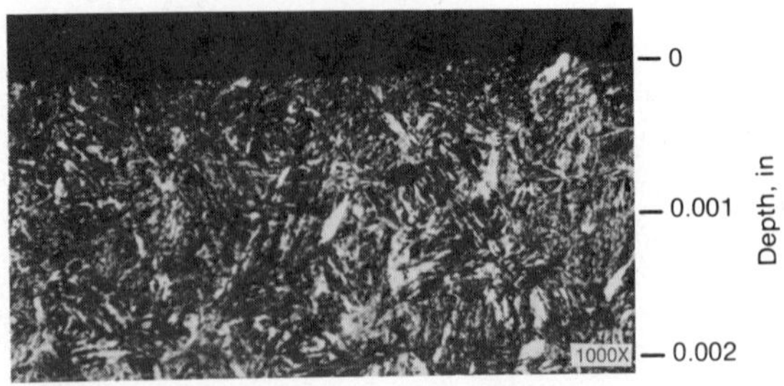


LOW STRESS CONDITIONS — No visible surface alterations in microstructure were detected.

ABUSIVE CONDITIONS — A total heat affected zone of 0.013 inch was produced. The white layer shown above is 65 $R_C$ and approximately 0.001-0.002 inch deep but in some sections was up to 0.005 inch deep.

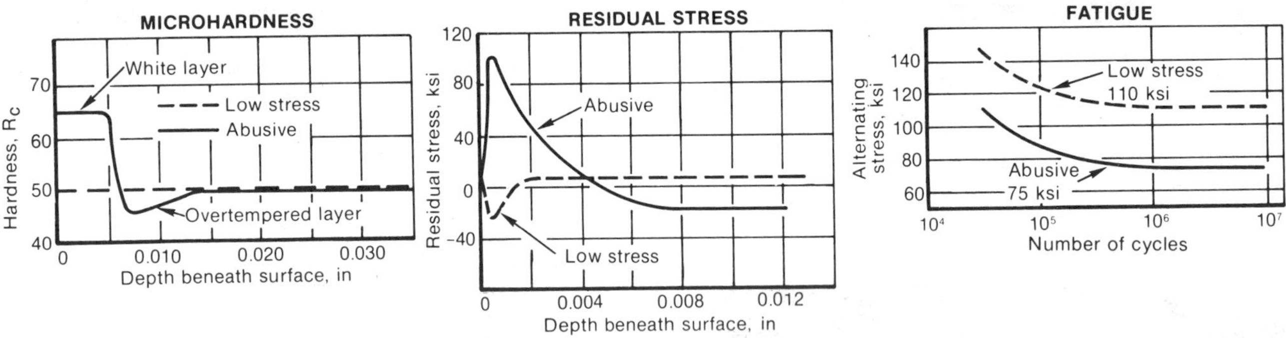


**Figure 18.3–13** Surface integrity standard data set for grinding of 4340 steel (quenched and tempered, 50 $R_C$). Data are for two levels of grinding intensity: low stress and abusive. (W. P. Koster et al[9])

**TABLE 18.3-5 Extended Surface Integrity Data Set**

1. Standard data set
2. Fatigue tests—statistical data to establish design data
3. Stress corrosion tests at selected environmental conditions
4. Additional mechanical testing
   - tensile
   - stress rupture
   - creep
   - specialized: friction, wear, sealing, bearing performance
   - fracture toughness
   - low cycle fatigue
   - elevated or cryogenic temperature
   - crack propagation
   - surface chemistry

SOURCE: G. Bellows and D. N. Tishler.[25]

### Process Intensities and Surface Integrity

Processes are usually operated over a range of intensities. The roughing and finishing modes are one reflection of these differing intensities or energy densities. From a surface integrity and quality assurance standpoint, it is necessary to consider the change in surface effects over the full range of energy levels expected to be used. These data can then be used to establish reasonable quality control limits. This range of process intensity levels is often implied in the machining terminology that indicates increasingly intense processing, such as the following:

- low stress surface
- finish machining (or "gentle")
- conventional (or "standard")
- roughing (or "off standard")
- abusive (or accidental conditions)

The surface integrity of components produced with roughing or off-standard operating conditions is frequently less than that produced under standard conditions. The standard surface integrity data set evaluations should include at least two levels of process intensity. It should be emphasized that ultimately abusive and gentle processing can be differentiated only by mechanical testing (fatigue, stress corrosion, etc.) and/or service performance.

**Figure 18.3–14** A 1-1/4 inch diameter metallographic mount prepared by the vacuum epoxy method. The white layer contains aluminum oxide pellets. (M. Field, J. F. Kahles, and J. T. Cammett[12])

### Metallographic Sectioning and Special Preparation Techniques

Microscopic examination is an important and inexpensive means for rapid initial surface integrity evaluation. However, special metallographic techniques are necessary for studying surface phenomena. Surface microstructure alterations are generally very shallow, usually of the order of 0.001 to 0.003 inch [0.025 to 0.076 mm]. Under very abusive conditions, alterations as deep as 0.005 to 0.015 inch [0.127 to 0.38 mm] have been noted. Often significant microstructure alterations, microcracks, or flaws as shallow as 0.0001 inch [0.0025 mm] in depth are developed during material removal processing. Thus, it is necessary to employ sectioning, mounting and other metallographic techniques which do not alter or destroy the pertinent surface and which maintain high edge retention. One successful technique consists of the application of vacuum epoxy mounting as outlined in table 18.3–6 and shown in figure 18.3–14. It should be noted that the frequently employed edge retention technique of plating can alter some of the surface effects that might be present.

Various types of microscopy are available for the evaluation of surface metallurgy including optical microscopy, scanning electron microscopy (SEM), and transmission electron microscopy (TEM). Optical microscopy is the least expensive and the most widely applicable method. It is advisable to examine metallurgical mounts first in the unetched condition, thereby accentuating the profile, microcracks, inclusions, voids, and crevice-like defects. After etching, specimens are reexamined for microstructural identification and grain boundary conditions.

### Microhardness Determinations

Microhardness determinations can be made on the previously described metallurgical mounts using a microhardness tester with either a Knoop or a Vickers indenter. Microhardness studies are important for identifying the possible overall effects of heat-affected zones caused by specific material removal operations. Examples of surface effects revealed by microhardness variations include the following:

- areas of untempered or overtempered martensite
- softening from resolutioned austenite
- chemical softening (sometimes associated with chemical machining or electrochemical machining)
- heat-affected zones
- plastic deformation or work hardened zones

Figure 18.3–15 illustrates some of the types of hardness alterations observed.

**TABLE 18.3-6 Mounting Technique and Metallographic Preparation for Edge Retention**

1. Samples are sectioned from the workpiece in a manner which leads to the least possible distortion or burring. Bandsawing or hacksawing is preferred. A minimum of 0.030 inch [0.76 mm] is then removed from the cut surface using a 120-grit silicon carbide paper on a low-speed polisher.
2. Copper molds (or tubes), 1¼ inches [31.8 mm] inside diameter by 2¾ inches [70 mm] high, are placed on a pallet approximately 5 inches [125 mm] in diameter. The inner surface of the molds and surface of the pallet are previously sprayed with a silicone releasing agent.
3. After placing a metallurgical specimen in a mold, a mixture of epoxy resin, hardener, and pelletized aluminum oxide, sufficient to produce a layer of ¼ to ⅜ inch [6.4 to 9.5 mm] in depth, is poured over the specimen. The ratio of resin to hardener is 4 to 1. The amount of pellets added is in the range of 10 to 15 grams. The hardness or abrasive level of the pelletized material used (low, medium, or high fired) is strictly a function of the alloy to be prepared and its hardness characteristics.
4. The pallet containing the molds is placed in a vacuum chamber (at a vacuum of $1 \times 10^{-2}$ to $1 \times 10^{-3}$ torr) in order to degas the mixture, thereby improving the adherence of the epoxy and pellets to the surface of the specimen. When vigorous bubbling of the mixture decreases after vacuum impregnation, sufficient resin and hardener (4 to 1 ratio) is added to produce a mount approximately 1 inch [25 mm] high.
5. The mounts are cured at a temperature not greater than 70°F [21°C] for approximately 10 hours. Casting of the mounts is accomplished during the latter portion of the laboratory workday so that curing occurs overnight.
6. After curing, they are placed in an oven at a temperature of 150°F [66°C] for a period of one hour after which they are removed from the molds.
7. Approximately 0.020 inch [0.50 mm] of stock is then removed from the as-mounted metal surface on a positive positioning automatic polishing unit, using the side of a 1-by-13-inch [25 x 330 mm] aluminum oxide 320-grit grinding wheel as the grinding medium. Water is used as a coolant.
8. Subsequent rough grinding is performed wet on silicon carbide papers or equivalent ranging from 240 to 600 grit.
9. For steels and nickel- and cobalt-base superalloys, the intermediate polish is performed on an automatic polisher using a polishing cloth with a soft nap texture and 6-micrometer diamond paste. The final polish is achieved using deep nap or pile cloth similar to billiard cloth with a suspension of 0.1 micrometer or finer aluminum oxide in water.

   Titanium and refractory alloys require an etch-polish cycle (using a slurry of hydrogen peroxide, water, and 0.1 micrometer or finer aluminum oxide) which is accomplished between a diamond polish and a final polish procedure. The final polish for titanium and refractory alloys is accomplished on a vibratory polisher using a deep pile cloth with a suspension of 0.1 micrometer or finer aluminum oxide in water.
10. Samples are etched by swabbing. Examples of some typical etchants used are given below:

| Material | Etchant |
|---|---|
| Steels | 2% $HNO_3$ and 98% Denatured Anhydrous Alcohol |
| Nickel Base Alloys | 100 ml HCl, 5 g $CuCl_2 \cdot 2H_2O$, and 100 ml Denatured Anhydrous Alcohol |
| Titanium Alloys | 2% HF and 98% $H_2O$ or 2% HF, 3% $HNO_3$, and 95% $H_2O$ |

SOURCE: L. R. Gatto and T. D. DiLullo.[26]

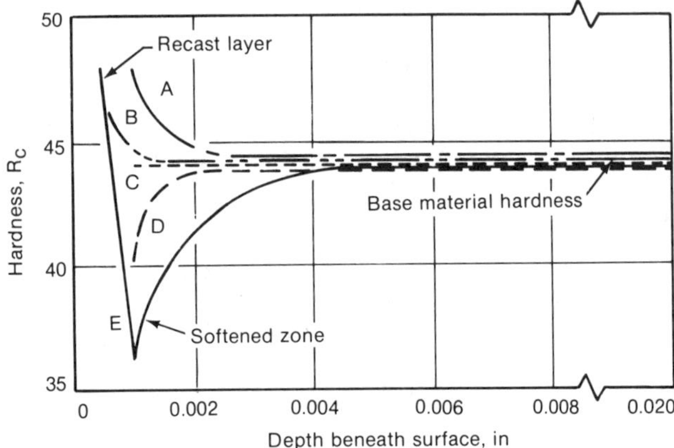


**Figure 18.3-15** Microhardness traverses on Inconel Alloy 718 (solution treated and aged) from a variety of machining processes. Key: A, roughing EDM and shot peening; B, gentle ECM and glass bead peen; C, gentle grinding; D, ECM; E, EDM. (G. Bellows and D. N. Tishler[14])

When determining the microhardness using a Knoop indenter, a 100-gram load is generally employed. With this load and carefully performed readings, accurate hardness readings can be made to within 0.001 inch [0.025 mm] of the surface without producing edge yielding. Surface hardness readings to within 0.0005 inch [0.0127 mm] of the surface may be made using 25- to 50-gram loads, but extreme care must be taken even with these light loads. It is recommended that if the light load is used, hardness readings also be taken at a considerable distance from the edge and compared with 100- and 500-gram load Knoop readings. Knoop readings are generally converted to Rockwell C readings for a ready appreciation of the hardness relationships which exist near and below the affected surface layer.

### Residual Stress Determination

Distortion may occur in the manufacture of components as a result of heat treating or material removal processes which can leave substantial residual stresses in the surface layers. There are several ways of determining the residual stress profile. The two most common are X-ray diffraction and layer removal-deflection techniques. Most of the machining-induced residual stress occurs in the first 0.0005 to 0.010 inch [0.0127 to 0.254 mm] below the surface. Some of the descriptors used in residual stress discussions are shown in figure 18.3-16. Figures 18.3-17 and 18.3-18 illustrate results of residual stress profiles determined from two materials and are typical of the range of effects observed. It should be noted that single surface residual stress measurements can be quite misleading, and a profile for the first few thousandths of an inch below the surface is essential. The peak value and its depth below the surface are related to the high cycle fatigue strength and magnitude of distortion. For example, note in figure 18.3-18 that the surface residual stress is zero, yet substantial stress is present less than 0.001 inch [0.025 mm] below the surface.

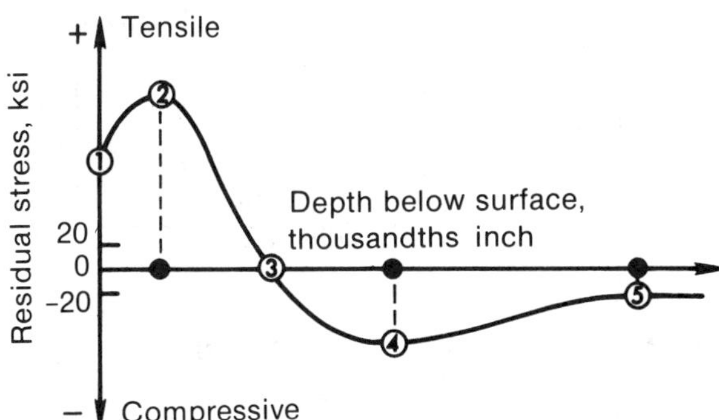


**Figure 18.3-16** Residual stress profile descriptors. Key: 1, surface residual stress—value at zero depth; 2, peak stress (tensile)—value at defined depth; 3, cross over—depth in thousandths inch; 4, peak stress (compressive)—value at defined depth; 5, depth of residual stress—depth below the surface where stress declines to and remains less than 20,000 psi or an inconsequential value or less than 10% of tensile strength. (G. Bellows[13])

Carefully measured distortion on strip test specimens as shown in figures 18.3-19 and 18.3-20 can also be used as an overall residual stress indicator. In these measurements the initially flat and unstressed blank, figure 18.3-19, has one surface machined by the process being investigated. The distortion after the test cut is measured in the fixture shown in figure 18.3-20. The distortion variation with different process intensities is illustrated in the grinding of D6ac as shown in figure 18.3-21.

The use of X-ray diffraction for measuring residual stress has by and large superseded the deflection test strip method. Much smaller areas can be measured with greater accuracy. Accuracy of plus or minus 5 percent can be obtained.

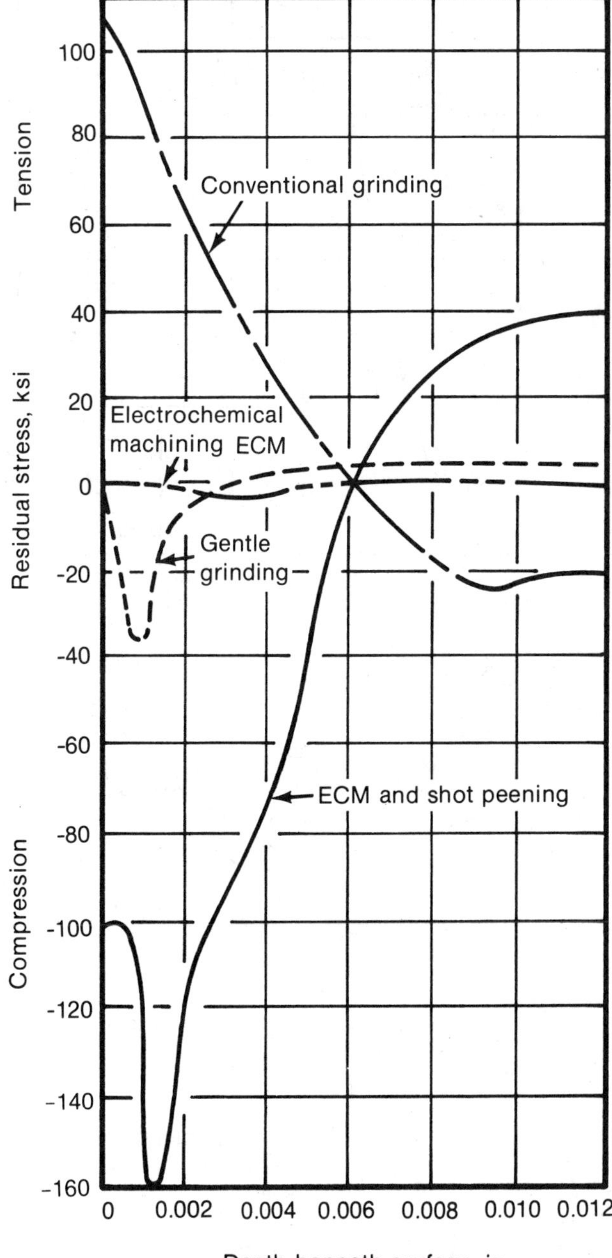


**Figure 18.3-17** Residual stress produced in Inconel Alloy 718 (solution treated and aged) by several machining processes. (G. Bellows and D. N. Tishler[14])

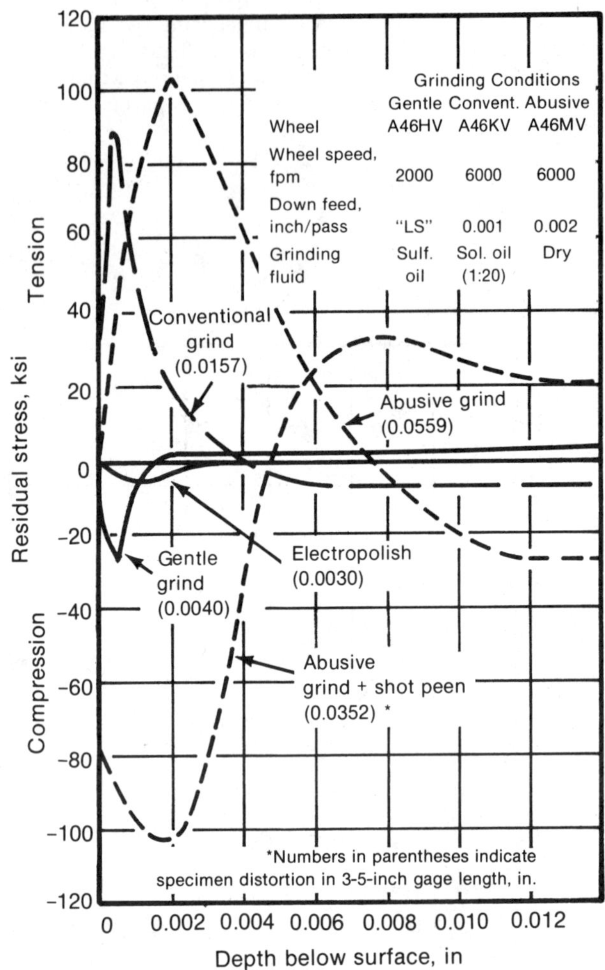


**Figure 18.3–18** Residual surface stress in 4340 steel (quenched and tempered, 50 $R_C$) produced by surface grinding. (M. Field, J. F. Kahles, and J. T. Cammett[12])

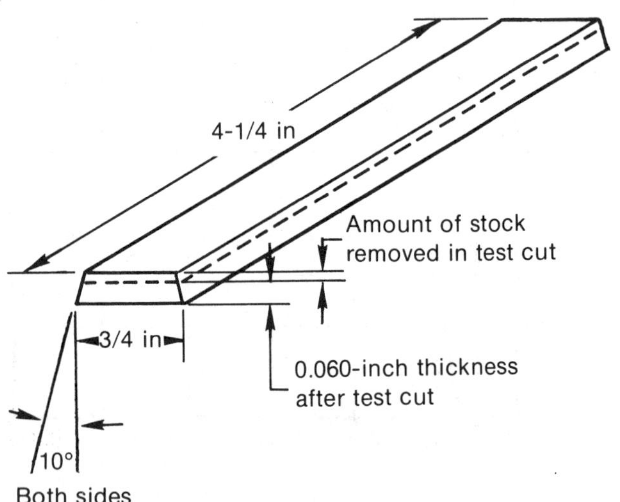


**Figure 18.3–19** Distortion and residual stress test specimen. (M. Field, J. F. Kahles and J. T. Cammett[12])

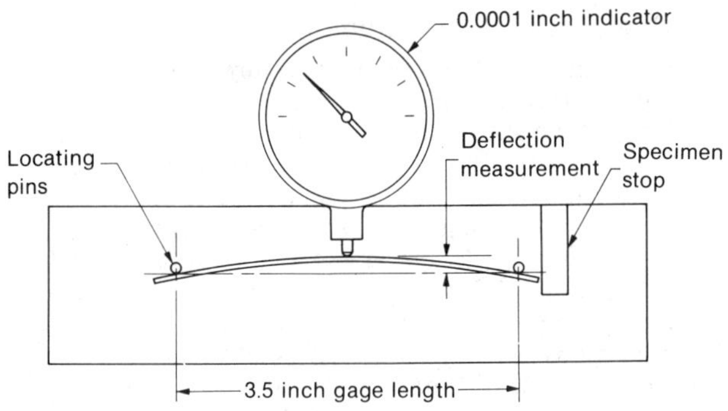


**Figure 18.3–20** Deflection measurement fixture. (M. Field, J. F. Kahles and J. T. Cammett[12])

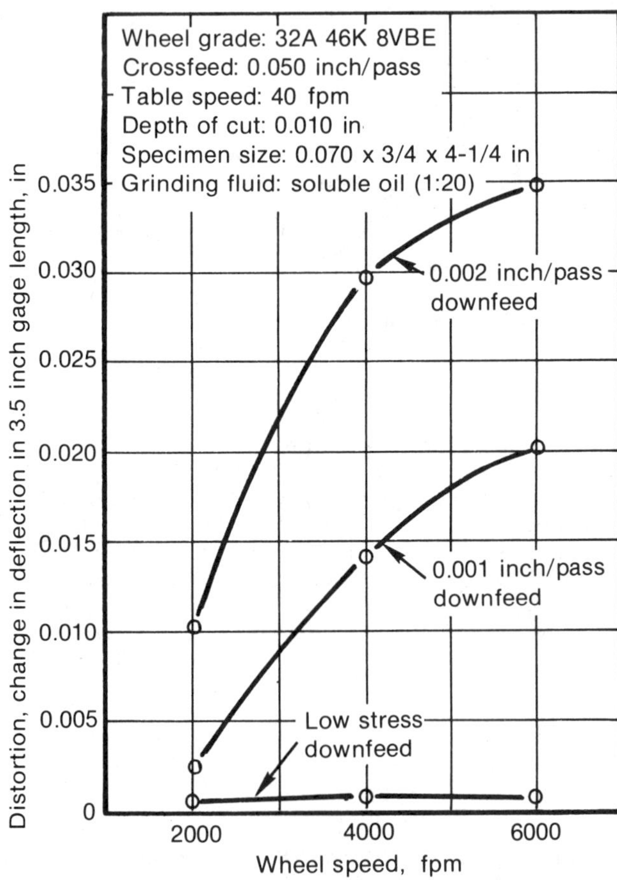


**Figure 18.3–21** Change in deflection versus wheel speed for surface grinding D6ac steel (quenched and tempered, 56 $R_C$). Effect of down feed. (M. Field, W. P. Koster and J. B. Kohls[15])

### Fatigue Strength Testing

It is well established that variations in material removal operations can produce great differences in the mechanical performance of the workpiece material. Fatigue strength is perhaps the most sensitive property. The extensive range of process influence on high cycle fatigue strength is shown in table 18.3-7 for Ti-5Al-2.5Sn alloy and in table 18.3-8 for Inconel Alloy 718. High cycle and low cycle fatigue strengths are dynamic properties that are surface dependent and are linked on a simple continuum for AF 95 (Rene 95) in figure 18.3-22. Similar effects are found at elevated temperatures. For the simple screening tests that are part of the standard data set, however, it is sufficient to collate or compare results from room temperature, full-reverse cantilever bending high cycle fatigue tests, the setup for which is shown in figure 18.3-23. A test specimen that produces repeatable results and exposes an ample test surface to the test stress level is shown in figure 18.3-24. The results are frequently displayed on an S-N (stress versus number of cycles to failure) curve, figure 18.3-25, or in a bar graph using the stress level at $10^7$ cycles as representative of the endurance strength.

**TABLE 18.3-7 Comparative High Cycle Fatigue Strengths and Surface Roughnesses of Ti-5Al-2.5Sn Alloy (formerly A110 Alloy) from Several Processes**

| PROCESS | FATIGUE STRENGTH* | | SURFACE ROUGHNESS | |
|---|---|---|---|---|
| | ksi | MPa | $\mu$in $R_a$ | $\mu$m $R_a$ |
| Ultrasonic machining (USM) | 98 | 676 | 20 | 0.5 |
| Conventional slab milling (CML) | 86 | 593 | 6 to 8 | 0.15 to 0.20 |
| Chemical machining plus vacuum anneal (CHM) | 77 | 531 | 30 to 40 | 0.8 to 1.0 |
| Shot peened (CPE)† | 76 | 524 | — | — |
| As rolled | 61 | 421 | — | — |
| Chemical machining (CHM) | 59 | 407 | 30 to 40 | 0.8 to 1.0 |
| Conventional traverse grinding (CGS) | 53 | 365 | 16 to 18 | 0.40 to 0.45 |
| Conventional longitudinal grinding (CGS) | 52 | 359 | 16 to 18 | 0.40 to 0.45 |
| Electrical discharge machining (EDM) | 22 | 152 | 30 to 40 | 0.8 to 1.0 |

SOURCE: Data from R. J. Rooney.[27]

*Room temperature, $10^7$ cycles full-reverse bending.

† P28 steel shot; 0.010 to 0.012 $A_2$ Almen level.

**TABLE 18.3-8 Effect of Machining and Peening on Fatigue Strength of Inconel 718 (Solution Treated and Aged 44 $R_c$)**

| OPERATION | FATIGUE STRENGTH* | | |
|---|---|---|---|
| | ksi | MPa | Percent of gentle grind |
| Gentle surface grinding | 60 | 414 | 100 |
| Conventional surface grinding | 24 | 165 | 40 |
| Gentle turning | 60 | 414 | 100 |
| Abusive turning | 60 | 414 | 100 |
| Standard ECM | 39 | 269 | 65 |
| Off-standard ECM | 39 | 269 | 65 |
| Standard ECM plus Peen | 78 | 538 | 130 |
| Off-standard ECM plus Peen | 67 | 462 | 112 |
| Finish EDM | 22 | 152 | 37 |
| Rough EDM | 22 | 152 | 37 |
| Finish EDM plus Peen | 66 | 455 | 110 |
| Rough EDM plus Peen | 75 | 517 | 125 |
| Electropolishing (ELP) | 42 | 290 | 70 |
| ELP plus Peen | 78 | 538 | 130 |

*Room temperature, $10^7$ cycles full-reverse bending.

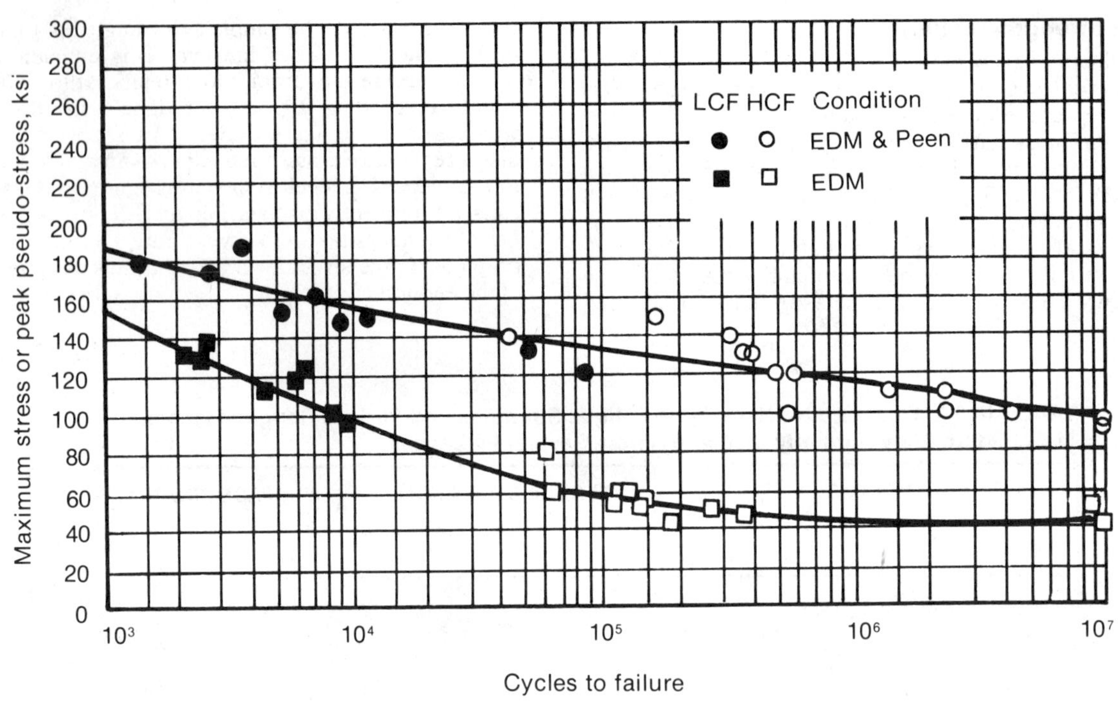


**Figure 18.3-22** Example of high cycle and low cycle fatigue strength continuum in test data from electrical discharge machining of AF 95 (Rene 95) at 1000°F. (G. Bellows and W. P. Koster[16])

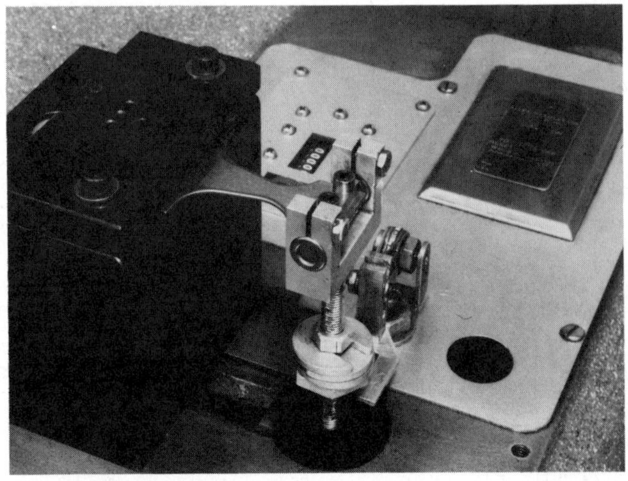

**Figure 18.3-23** High cycle fatigue test setup. (M. Field, J. F. Kahles and J. T. Cammett[12])

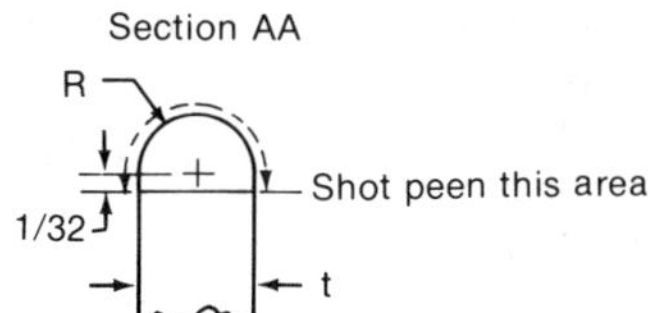


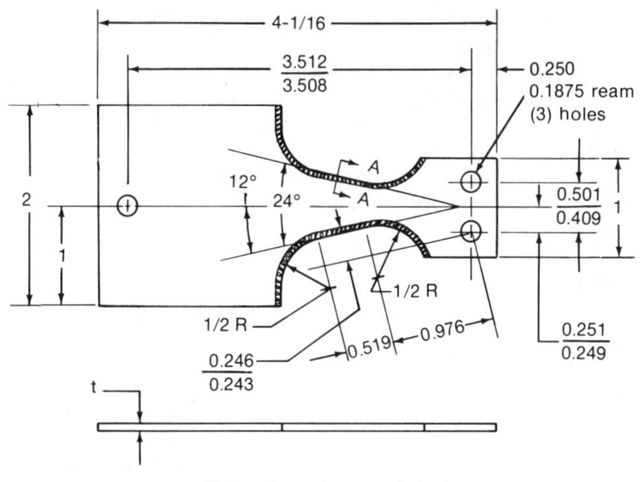


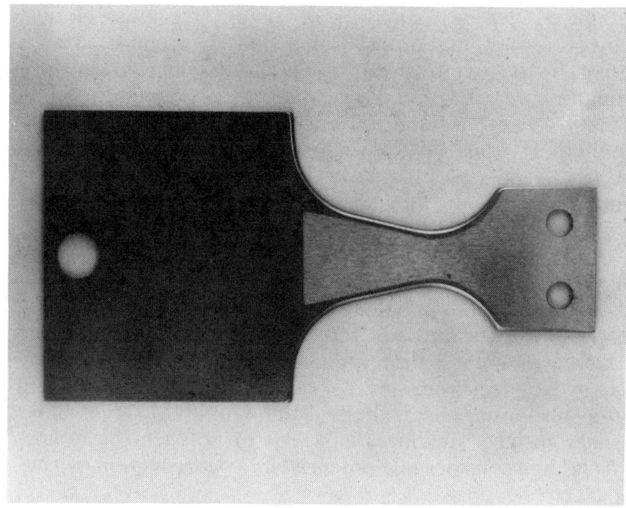

**Figure 18.3–24** Fatigue test specimen with tapered test section to secure a test area subjected to test stress. (M. Field, J. F. Kahles and J. T. Cammett[12])

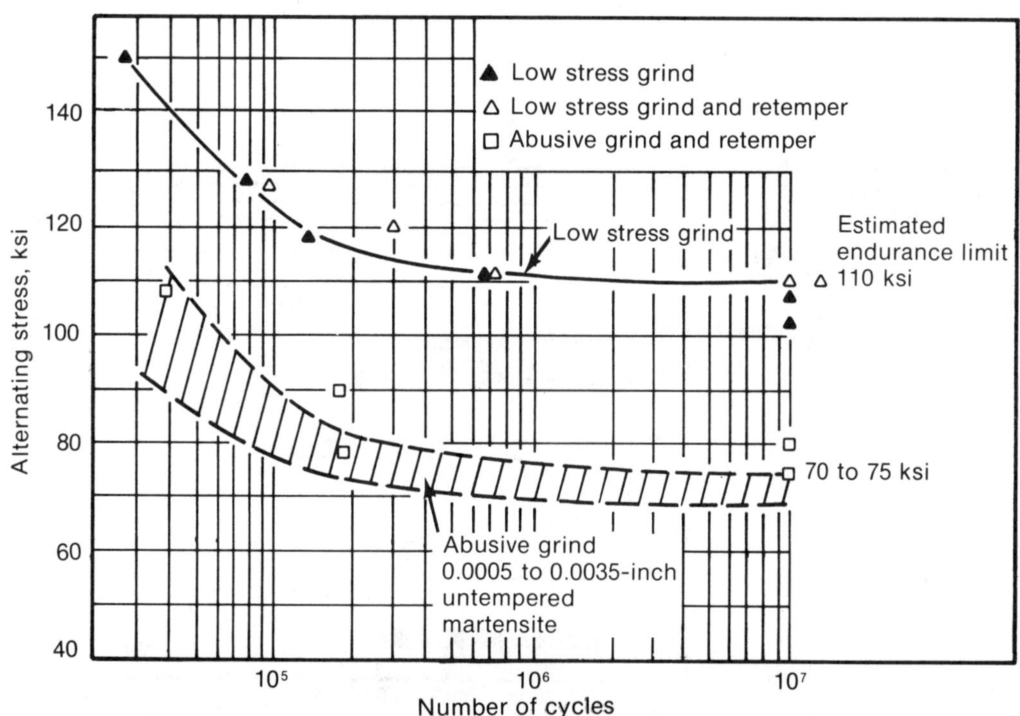


**Figure 18.3–25** S-N curve of high cycle fatigue test data of ground 4340 steel (quenched and tempered, 50 $R_C$). (M. Field, J. F. Kahles and J. T. Cammett[12])

### Surface Roughness and Fatigue Strength

In the testing of several alloys, it has been observed that surface roughness is not the critical factor that it has traditionally been assumed to be. It appears that the effects of roughness can be overshadowed by the effects of residual stresses, the presence of metallurgically changed phases, plastically deformed surfaces or other effects. As indicated in figure 18.3-26, the gently ground surfaces on 4340 steel produced the best high cycle fatigue strengths. Notice, however, that in the case of abusively ground surfaces other factors were at work within the range of surface roughnesses tried, and all roughnesses developed the same 65 ksi [448 MPa] fatigue strength. Observe also that in the milling of Ti-6Al-6V-2Sn and in the face turning of Inconel Alloy 718, the fatigue strength did not vary with surface roughness.

Further studies of the effects of surface roughness on fatigue strength have been made in which the data represent conditions where residual stresses from the machining process were essentially zero. Figure 18.3-27 shows the differences among three materials that range from practically no effect from surface roughness (Inconel Alloy 718) to a substantial effect (the two titanium alloys). For alloys where surface roughness has little or no effect on fatigue strength, a considerable potential for cost reduction exists.

In analyzing surface roughness, it is not implied that the effect of scratches, tool marks, sharp corners and other geometric considerations can be overlooked. The effect of these factors as producers of stress concentrations leading to premature fatigue failure is well established.

In summary, an important point to consider is to test and evaluate the influence of finish and other factors affecting the endurance strength of particular material and process combinations. Caution must be exercised before drawing general conclusions from the limited data available at this time concerning surface roughness and surface integrity interrelationships.

### Stress Corrosion Cracking

The two surface integrity effects having the most direct bearing on design and application considerations are fatigue strength and stress corrosion cracking propensity. The former lends itself to a fairly simple comparison, as has already been noted. The latter is more complex and no simple evaluation procedure has evolved that yields comparable data. The number of possible combinations of materials, processes, metallurgical states, energy levels, service stresses, environmental fluids or gases is beyond counting. Therefore, each application should be analyzed for its own situation and evaluated by carefully structured testing.

### Combined Effects

It is rare that the dynamic properties results can be attributed to a single surface integrity effect, such as residual stress, finish, hardness, etc. One of the more pronounced relationships which has been observed exists between the peak residual stress value and the fatigue endurance strength, as demonstrated in figure 18.3-28. The tri-axis graph, figure 18.3-29, illustrates the relationships among fatigue endurance strength, residual stress and surface roughness for Inconel Alloy 718. These results are for specific materials and process combinations, and other combinations may demonstrate different relationships. Caution should be taken not to draw general conclusions from the limited data available.

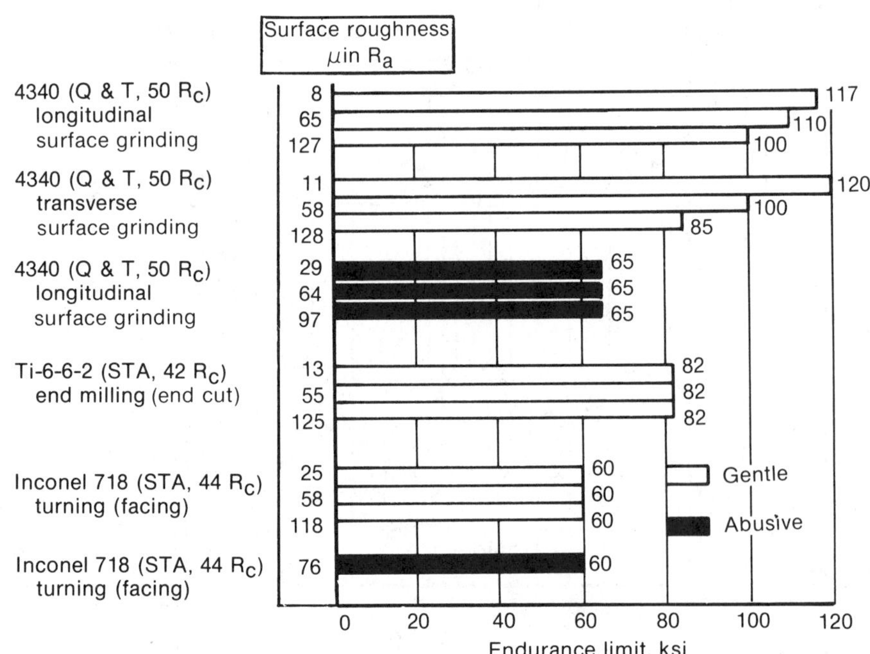


**Figure 18.3-26** Summary of high cycle fatigue strengths—surface roughness study. Fatigue strength in full-reverse bending at $10^7$ cycles at room temperature. (W. P.Koster et al[9])

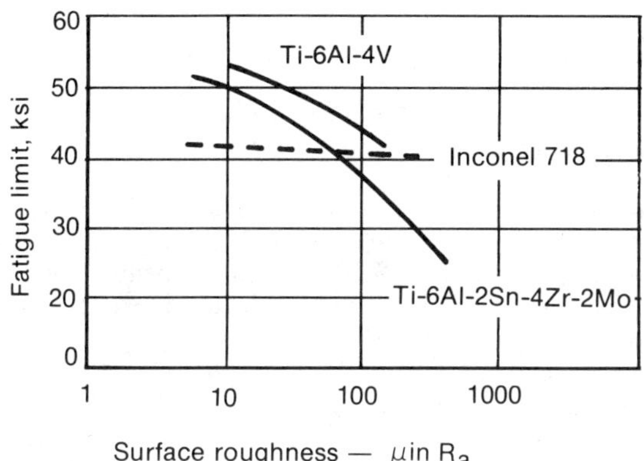


**Figure 18.3–27** Fatigue strength versus surface roughness with negligible residual surface stress (full-reverse bending at $10^7$ cycles at room temperature). (G. Bellows and W. P. Koster[16])

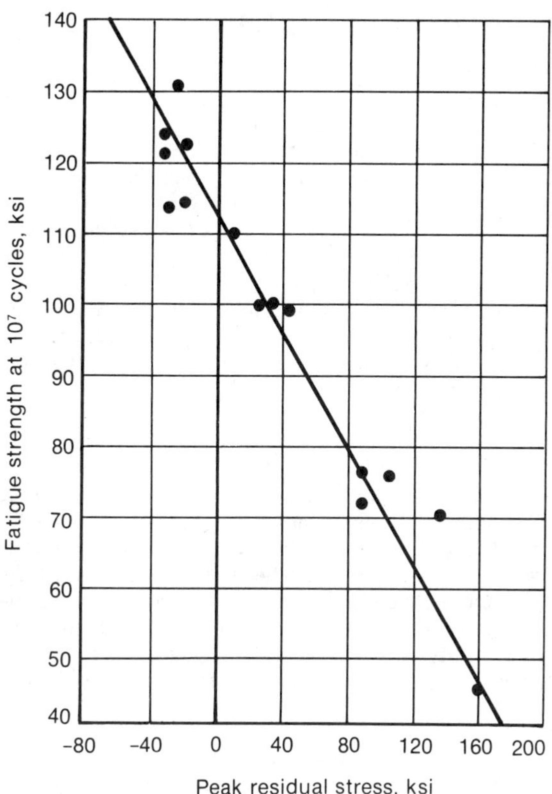


**Figure 18.3–28** Correlation between peak residual stress and fatigue strength in 4340 steel at various levels of grinding intensity. Flat tapered specimens were tested in full-reverse bending. (W. P. Koster[11])

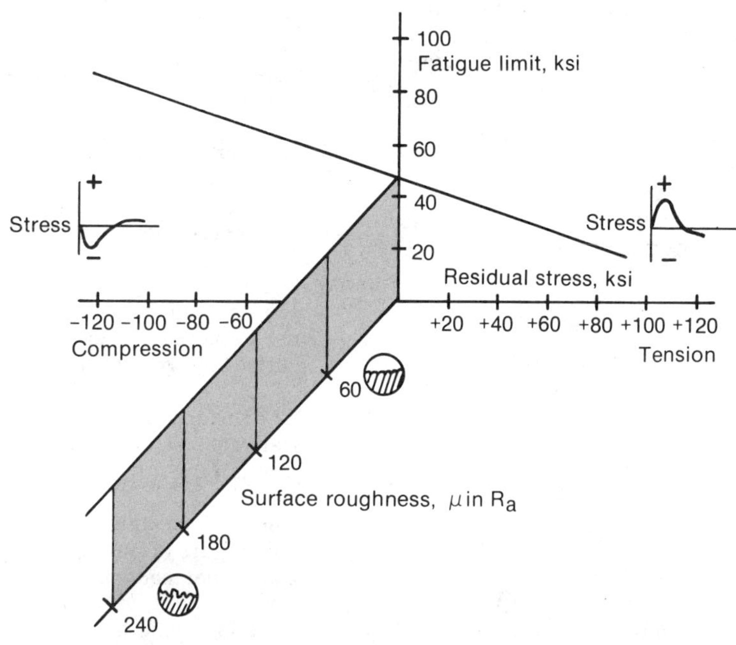


**Figure 18.3–29** Relationships among surface roughness, residual stress and high cycle fatigue endurance strength at room temperature in Inconel Alloy 718 (solution treated and aged, 44 $R_C$). (G. Bellows and W. P. Koster[16])

# 18.3 Surface Integrity

## GENERAL GUIDELINES

There is an increasing awareness that surface integrity effects can significantly impact product integrity, longevity and reliability. Specific data on every material and process combination are not available. Reliable data must also consider the several processing intensities that may be used and the several metallurgical states that may be presented for processing. Until more data can be accumulated, reliance must be placed on sets of guidelines. Detailed data supporting the guidelines are contained in the references. Basically, each level of processing intensity for each manufacturing process has a distinct set of surface integrity effects when it is applied to a specific metallurgical state of a workpiece material.

### Precautions in Use of Guidelines

The following guidelines can only serve as general or starting recommendations. Data and experience gathered to date indicate that these practices lead to the production of unimpaired or enhanced surfaces. However, knowledge of the surface technology at this time is such that general recommendations are not always applicable to all specific surface integrity situations. For highly critical or highly stressed surfaces, it is mandatory to make specific individual evaluations.

Surface integrity control generally results in increased costs and decreased production rates. Therefore, surface integrity practices should be implemented only where a definite need exists. Process parameters which enhance surface integrity should be applied selectively to critical parts or critical areas of given parts to help minimize costs.

The guidelines are intended primarily for processes used in producing the final or finished surface of a workpiece. It is important to know the depth of surface alterations produced during the roughing or prior operations so that adequate provision can be made for removing any damaged layers. In maintaining surface integrity, the sequence of operations is as important as the specific processing level.

Ample recognition must be given to the necessity for relentless process control once the desired surface integrity processing parameters have been selected. Equal consideration for control of the metallurgical state and condition of the material being presented for processing is essential.

### Key Indicators of Need for Surface Integrity Control

- Hazards to life and/or the possibility of high economic property loss.
- Distortion in thin parts.
- Cracking in processing or in service.
- Short service life.
- Requirements for manufacturing parts using sensitive alloys, such as high strength steels, nickel and cobalt base high temperature alloys, titanium alloys, beryllium, and refractory alloys.
- New or indeterminate environmental conditions including stress, temperature, and atmosphere.
- Requirements for designs which approach more complete utilization of material properties.
- Manufacturing requirements calling for use of material removal processes, such as rough conventional machining, heavy grinding, EDM and laser beam machining, which characteristically cause surface alternations.

### Guidelines for Material Removal Processes in General

While most of the surface integrity investigations to date have involved material removal processes, this does not mean that forming, joining, coating or other processes are exempt from surface integrity considerations. It only indicates that insufficient data are available for these processes.

Some general guidelines for all material removal processes are as follows:

1. Thorough component or product testing is one of the best assurances for and checks of surface integrity. The tests should be run with surfaces produced by the complete and exact sequence of production operations.
2. Apply surface integrity requirements only to the critical or highly stressed zones of the component part. Do not apply "all over".
3. Highly stressed areas of critical components should be evaluated carefully to assess the impact of the full sequence of processes that generate the "as-shipped" surface. The control of the sequence of processes is as important to surface integrity as is the selection and maintenance of the process operating parameters.
4. Control of the material metallurgical state is as important as control of the process parameters.
5. Material inhomogeneities or anomalies (sometimes even those within specification limits) can have component integrity effects as serious as the surface integrity effects from different process intensities.
6. Designers, shop supervisors, and quality control and process engineers must be educated and trained in order to increase their appreciation of the magnitude of surface integrity effects from manufacturing processes.
7. The surface integrity from processes other than material removal or machining processes should also be evaluated.
8. The surface integrity effects from conventional machining are of the same magnitude as those from nontraditional machining.
9. Metallographic sections at high magnification, taken parallel and perpendicular to the lay pattern, provide an effective early alert to potential surface integrity problems.

10. Post-processing treatments, such as heat treatment, shot peening, roller burnishing, low stress grinding, etc., may offset some, but not necessarily all, of the otherwise detrimental surface integrity effects.

11. Hand-controlled operations have a tendency to produce variability in surface effects and should be considered suspect.

12. Low process energy intensities and low material removal rates are characteristic of most, but not all, material removal processes which provide acceptable surface integrity.

13. Rigid, high quality machine tools and fixtures are desirable.

14. Cutting fluids should be fresh or well controlled and carefully, completely and quickly removed from the workpiece when the operation is completed.

15. Deburring of all machined edges is desirable.

16. Parts stored for extended periods should be covered with a protective coating to prevent corrosion.

## Comparison of Altered Material Zones from Material Removal Processes

Examination of several thousand section photomicrographs of surfaces generated by a wide variety of material removal processes has provided data for table 18.3-9, a summary of the depth of surface integrity effects. These data show the maximum depth of alterations observed to date for two particular operating levels of each listed process. The two process intensity levels displayed represent (a) finishing, "gentle" or low stress conditions or (b) roughing, "off-standard" or "abusive" conditions. The former represent what might be expected from well controlled, good surface integrity operations, while the latter represents high intensity or even accidental conditions, such as a delayed application of coolants. The data cover the full range of material types.

## Implementing Surface Integrity Programs

The following steps are recommended for instituting an in-house surface integrity program:

### Education and Training

Manufacturing and product engineering personnel are generally not sufficiently acquainted with the many types of surface alterations which can be produced during material removal operations. Product or design engineers generally do not realize that high temperatures prevail in conventional machining at the workpiece-tool interface, while manufacturing people often do not realize the extent to which metallurgical changes occur and how seriously they lower mechanical properties. This situation points out the advisability of educational efforts to help increase general cognizance of the importance of surface integrity control. In addition to general educational programs, specific training is required for machine operators and quality control personnel. Some companies, for example, require qualification of personnel in order to perform drilling and reaming of holes. Other plants give short courses on how to inspect for rehardened and overtempered martensite in the manufacture of high strength steel components.

### Specifying Surface Integrity Quality

Overspecification or too extensive use of surface integrity quality specifications can significantly increase manufacturing costs. Component part drawings should call for the surface integrity controls only on the areas or zones that are highly stressed or critical in preference to notes calling out "apply all over". Good value engineering and coordination with manufacturing personnel will produce design details with the best component integrity.

### Manufacturing Engineering

Carefully integrated machine, operation and inspection planning will improve the surface integrity results and contain the manufacturing costs. Simple metallographic examination of test pieces can provide a preview of the surface integrity results to be expected during production. These checks can prevent expensive and time consuming surprises during initial production.

### Maintaining Operating Parameters

Particular attention should be given to the necessity for maintaining specified operating parameters. Machine operators should be thoroughly trained to follow the proper procedures outlined by manufacturing engineering. All parameters for material removal should be specified by manufacturing engineering. For chip removal operations, these include tool material, tool geometry, cutting speed, feed, cutting fluid, and tool life (maximum number of parts per tool grind). Similarly, appropriate parameters must be set up for operations such as grinding, EDM and ECM. Machinists should be informed as to the reason and necessity for maintaining good surface integrity and surface quality and should be made aware that it is not currently possible to guarantee maintenance of surface integrity by final part inspection. Management needs to be ever alert to the desire on the part of operators to try "just one more notch" on the control. Without requalification, this relaxation of standards may result in loss of surface integrity.

### Quality Audits

Since no instrument can measure the surface integrity produced, reliance is frequently placed on process control. Good quality control practice includes randomly spaced, periodic audits of the operating conditions actually being used as compared to the preplanned values. Some companies use the manufacturing engineer most knowledgeable in a process to conduct the audit.

### Guidelines and Specifications

The guideline concept has been used by some organizations to assist in the implementation of surface integrity processing. Booklets can be designed to fit the specific needs of a company, division, or department. Guidelines are particularly useful prior to the development of practical and meaningful specifications.

Preparation of processing specifications is highly recommended for critical parts. In fact, some of the aerospace producers have written detailed manufacturing specifications incorporating machining parameters and procedures in an effort to maintain surface integrity. These specifica-

# 18.3 Surface Integrity

**TABLE 18.3-9 Comparison of Depth of Surface Integrity Effects Observed in Material Removal Processes**

| PROPERTY AND TYPE OF EFFECT | CONDITION | MAXIMUM OBSERVED DEPTH OF EFFECT* inch / millimeter | | | | | | | |
|---|---|---|---|---|---|---|---|---|---|
| | | Turning or Milling | Drilling | Grinding | Chemical Machining | Electro-chemical Machining | Electro-chemical Grinding | Electrical Discharge Machining | Laser Beam Machining |
| **MECHANICAL ALTERED MATERIAL ZONES** | | | | | | | | | |
| Plastic deformation (PD) | finishing§ | 0.0017 | 0.0008 | 0.0003 | *** | *** | *** | *** | *** |
| | roughing # | 0.0030 | 0.0047 | 0.0035 | *** | *** | *** | *** | *** |
| | finishing | 0.043 | 0.020 | 0.008 | *** | *** | *** | *** | *** |
| | roughing | 0.076 | 0.119 | 0.089 | *** | *** | *** | *** | *** |
| Plastically deformed debris ($PD^2$) | finishing | *** | *** | 0.0005 | *** | *** | *** | *** | *** |
| | roughing | *** | *** | 0.0013 | *** | *** | *** | *** | *** |
| | finishing | *** | *** | 0.013 | *** | *** | *** | *** | *** |
| | roughing | *** | *** | 0.033 | *** | *** | *** | *** | *** |
| Hardness alteration† | finishing | 0.0005 | 0.0010 | 0.0015 | 0.0010 | 0.0014 | 0.0007 | 0.0010 | — |
| | roughing | 0.0050 | 0.0200 | 0.0100 | 0.0031 | 0.0020 | 0.0015 | 0.0080 | — |
| | finishing | 0.013 | 0.025 | 0.038 | 0.025 | 0.036 | 0.018 | 0.025 | — |
| | roughing | 0.127 | 0.508 | 0.254 | 0.079 | 0.051 | 0.038 | 0.203 | — |
| Microcracks or Macrocracks | finishing | 0.0005 | 0.0005 | 0.0005 | *** | 0.0003 | 0.0000 | 0.0005 | 0.0006 |
| | roughing | 0.0015 | 0.0015 | 0.0090 | *** | 0.0015 | 0.0010 | 0.0070 | 0.0040 |
| | finishing | 0.013 | 0.013 | 0.013 | *** | 0.008 | 0.000 | 0.013 | 0.015 |
| | roughing | 0.038 | 0.038 | 0.229 | *** | 0.038 | 0.025 | 0.178 | 0.102 |
| Residual stress‡ | finishing | 0.0060 | — | 0.0005 | 0.0010 | 0.0000 | 0.0000 | 0.0020 | 0.0002 |
| | roughing | 0.0140 | — | 0.0125 | 0.0010 | 0.0000 | 0.0000 | 0.0030 | — |
| | finishing | 0.152 | — | 0.013 | 0.025 | 0.000 | 0.000 | 0.051 | 0.005 |
| | roughing | 0.356 | — | 0.318 | 0.025 | 0.000 | 0.000 | 0.076 | — |
| **METALLURGICAL ALTERED MATERIAL ZONES** | | | | | | | | | |
| Recrystallization | finishing | — | — | 0.005 | *** | *** | *** | *** | *** |
| | roughing | — | — | — | *** | *** | *** | *** | *** |
| | finishing | — | — | 0.013 | *** | *** | *** | *** | *** |
| | roughing | — | — | — | *** | *** | *** | *** | *** |
| Intergranular attack (IGA) | finishing | *** | *** | *** | 0.0003 | 0.0003 | 0.000 | *** | *** |
| | roughing | *** | *** | *** | 0.0060 | 0.0015 | — | *** | *** |
| | finishing | *** | *** | *** | 0.008 | 0.008 | 0.000 | *** | *** |
| | roughing | *** | *** | *** | 0.152 | 0.038 | — | *** | *** |
| Selective etch, pits, protuberances | finishing | 0.0004 | — | 0.0002 | 0.0006 | 0.0004 | 0.0001 | 0.0005 | — |
| | roughing | 0.0010 | 0.3000 | 0.0004 | 0.0015 | 0.0025 | 0.0005 | 0.0016 | — |
| | finishing | 0.010 | — | 0.005 | 0.015 | 0.010 | 0.003 | 0.013 | — |
| | roughing | 0.025 | 0.076 | 0.010 | 0.038 | 0.064 | 0.013 | 0.041 | — |
| Metallurgical transformations | finishing | 0.0004 | 0.0015 | 0.0005 | — | 0.0000 | 0.0001 | 0.0006 | 0.0006 |
| | roughing | 0.0030 | 0.0200 | 0.0060 | — | 0.0002 | 0.0003 | 0.0050 | 0.0015 |
| | finishing | 0.010 | 0.038 | 0.013 | — | 0.000 | 0.003 | 0.015 | 0.015 |
| | roughing | 0.076 | 0.508 | 0.152 | — | 0.005 | 0.008 | 0.127 | 0.038 |
| Heat-affected zone (HAZ) or Recast layers | finishing | 0.0001 | — | 0.0007 | *** | *** | *** | 0.0006 | 0.0006 |
| | roughing | 0.0010 | 0.0030 | 0.0125 | *** | *** | *** | 0.0050 | 0.0015 |
| | finishing | 0.003 | — | 0.018 | *** | *** | *** | 0.015 | 0.015 |
| | roughing | 0.025 | 0.076 | 0.318 | *** | *** | *** | 0.127 | 0.038 |

SOURCE: G. Bellows and W. P. Koster.[16]

NOTE: A dash (—) in the table indicates no or insufficient data.

A triple asterisk (***) in the table indicates no occurrences or not expected.

*Normal to the surface.

†Depth to point where hardness becomes less than ±2 points $R_c$ (or equivalent) of bulk material hardness (hardness converted from Knoop microhardness measurements).

‡Depth to point where residual stress becomes and remains less than 20 ksi [138 MPa] or 10% of tensile strength, whichever is greater.

§Finishing, "gentle" or low stress conditions.

# Roughing, "off-standard" or abusive conditions.

tions, even though widely used in subcontract work, are generally considered proprietary and under the control of the prime contractor. It is suggested that the manufacturing or engineering departments of the large producers be contacted directly concerning the availability of their specifications.

**Testing Specific Applications**

Thorough component or product testing is one of the best assurances for checking of surface integrity. Tests should be conducted on surfaces of parts that have completed the full sequence of processing steps. While metallographic examination and some nondestructive tests can be helpful and can reveal clues to the surface integrity, the most critical applications should preferably be supported by specific specialized testing where the test conditions reflect the expected operating environment.

## SPECIFIC GUIDELINES

While several thousand bits of data have been used to develop specific surface integrity guidelines, only the summary tabulations or statements are included in this handbook.[1] These summaries are accompanied by a representative selection of specific data to illustrate the types of effects observed.

The guidelines are organized and presented according to the process groups that follow:

- Mechanical Processes, including milling, turning and drilling
- Abrasive Processes, including grinding, sanding and low stress grinding
- Electrical Processes, including electrochemical machining and electrochemical grinding
- Thermal Processes, including electrical discharge machining, laser beam machining and electron beam machining
- Chemical Processes, including chemical machining and photochemical machining

For each process group, the following information is provided:

- Depth of surface integrity effects observed
- Photomicrographs of good and poor surfaces
- Representative residual stress data
- High cycle fatigue strength values for several materials
- Minimum or standard data sets where available
- Comments by major material groups
- Comments on special situations
- List of guidelines

These data emphasize the importance of the machining process; however, equal consideration should be given to the metallurgical state of the workpiece material.[2]

### Specific Guidelines and Data for MECHANICAL Material Removal Processes

Conventional mechanical machining surface integrity guidelines have been developed principally from data generated by single- or multiple-point tools operated in the turning, milling (end, peripheral or face) or drilling mode. Table 18.3–10 combines and summarizes the surface integrity data observed from all of these. Figures 18.3–30 through 18.3–44 illustrate both good and poor surface conditions as seen in cross section, high magnification photomicrographs. Figures 18.3–45 and 18.3–46 show residual stress profiles found from the surface inward a few thousandths of an inch perpendicular to the surface. Figures 18.3–47 and 18.3–48 compare the high cycle fatigue strengths of a variety of materials. These strengths are the values at $10^7$ cycles of failure from full-reverse bending cantilever tests run on tapered section specimens at room temperature. The reference base or "handbook" value in most cases was data generated from low-stress-ground specimens or hand-polished specimens. Other mechanical material removal illustrations are figures 18.3–6, 18.3–7, 18.3–9 and 18.3–26.

Conventional machining processes have been used so long that it is easy to forget that they are characterized by mechanical plastic deformation, regular lay patterns and sometimes substantial heat affects from too forceful a passage of the cutter point over the surface. The degree of dullness of the tool is a major factor in determining the surface effects. With the approach of the end of the tool life, extensive roughness, tears, laps, or deposited built-up edge can result in abused surfaces. (See section 18.2, figure 18.2–19.)

Surface integrity data for the mechanical nontraditional processes are not extensive and should be investigated in critical applications using techniques suggested previously. Generally the low energy levels and gentle material removal rates result in a minimum of altered material zones.

**Chip Cutting Process Guidelines that Promote Good Surface Integrity Results**

1. Sharp tools are essential for establishing surface integrity in turning, milling, and similar single-point tool cutting processes. Plastic flow is frequently present and dull tools can produce laps, tears and roughness which can be initiating sites for fatigue failure.
2. Maximum flank wearlands should be limited to 0.005 to 0.008 inch [0.127 to 0.203 mm]. This is the point at which the wearland becomes visible to the naked eye.
3. Cutting parameters should be selected which provide long tool life in order to assist in keeping tools sharp.
4. Form cutters tend to produce surface damage more readily than do finishing tools which generate a form.
5. The heat from the passage of a cutting tool, especially a dull one, can produce metallurgical transformations.

6. Evidence of burning on the surface should initiate a check of the depth, which can extend to 0.010 inch [0.254 mm].

7. Residual stresses from dull cutting tools are frequently compressive, below the surface.

8. Sharp drills should be used to help avoid serious surface layer alterations.

9. Dwelling during drilling should be avoided; galling, torn or discolored surfaces should be cause for rejection or correction.

10. Drill fixturing rigidity is desirable.

11. Deburring is imperative on both the entrance and the exit of holes.

12. Drilled and reamed holes should be chamfered on both the exit and the entrance.

13. Reaming stock allowances should be controlled.

14. A maximum number of holes per reamer should be specified and maintained despite the visual appearance of reamer.

15. Hand feeding of straight reamed holes should be avoided.

16. Alignment, tool geometry and tool condition are important controls in hand reaming.

17. Finish boring operations should maintain roughness limits and avoid laps and tears. Only very small wearlands can be tolerated with about 0.005 inch [0.127 mm] as a limit. One finish pass per cutting edge is a frequently set limit.

18. Honing is an excellent finishing operation for developing surface integrity.

19. Corrosion potential of some cutting fluids should be checked. Old fluids should be checked because old fluids often corrode workpieces more rapidly.

20. Finish roughness values should not serve as the only criteria for development of fatigue strength.

### Special Comments Concerning Surface Integrity Practices for Chip Cutting Processes

**Rigid, high quality machine tools are essential.** They must be designed with the ranges of speeds and feeds necessary to meet surface integrity requirements. See comments below on the need for selecting speeds and feeds which give long tool life.

**Cutting tools should be inspected carefully prior to use.** Cutting tools must be carefully ground, and the most rigid tool design should be employed. For example, stub length drills should be used instead of a jobber's length wherever possible. Cobalt or premium grade high speed steel should be used wherever carbide is not applicable so as to do everything possible to provide a low rate of tool wear. All tools should be inspected after grinding to insure that previous wear, chipping, galling, etc., have been corrected to meet tool specifications. After grinding, the cutting edges of all tools should be protected to prevent accidental damage from transit or handling. Tools should be double checked by the machine tool operator for obvious defects.

**Sharp tools help establish surface integrity in turning and milling.** For chip removal operations, as for grinding operations, it is important to produce a surface which has a minimum or preferably an absence of surface alterations; that is, the surface layers should be similar to the base material below the surface.

For turning and milling, there are two very important steps which will improve surface integrity. First, select machining conditions which will give long tool life and produce good surface finish. Second, machine with sharp tools. Sharp tools minimize distortion and generally lead to better control during machining. The maximum flank wear when turning or milling should be limited to approximately 0.005 to 0.008 inch [0.127 to 0.203 mm]. A good rule of thumb is to remove the tool when the wearland becomes visible to the naked eye since the aforementioned wearland is just barely visible to the naked eye. Dull tools develop high compressive stresses which cause distortion, and very often they produce tears, laps and metallurgical alterations of the surface, including untempered and overtempered martensite in steels. For turning most alloys, carbide tools tend to have the lowest wear rate and at the same time make it possible to meet reasonable production rates. Indexable carbide tools, in particular, should be used whenever possible. This expedites the changing of cutting edges whenever tool wear reaches its specified limit.

In order to determine machining conditions which will provide long tool life, it is necessary to refer to tool life data for specific materials and material removal operations. References 3 through 6 supply data for tool materials, tool geometry, feeds, depth of cut, cutting fluids, etc., for the purpose of achieving long tool life. If tool life data are not available, the material closest in characteristics to the actual work material can be used for initial consideration of the machining parameters. In addition, it is recommended that tool life data be studied carefully in order to develop the relationship between machining parameters and tool life. In this way, one can get an idea of the effects of changes in speeds, feeds, tool materials, etc., on tool life and be in a better position to decide on departures from the recommended machining parameters.

Any evidence of burning as a result of a tool or cutter breakdown should be reported to supervision. Care must be taken to remove sufficient stock after tool breakdown to completely remove the effects of the burning which may extend as much as 0.005 to 0.010 inch [0.127 to 0.254 mm] below the machined surface.

There is a tendency for more surface damage to occur when a long cutting edge produces the final surface. This situation usually occurs when turning to a shoulder or when milling a radius into a component. The use of especially sharp cutters for finishing is desirable in order to minimize the alterations. For turning, whenever possible a shoulder or large radius should be generated by the finishing tool rather than being formed by the long cutting edge of a forming tool.

**Sharp drills should be used to avoid serious surface alterations.** Holes in highly stressed components should be free of tears, laps, and untempered or overtempered martensite. To minimize defects, the wearland on drills should be limited to 0.005 to 0.008 inch [0.127 to 0.203 mm]. Wherever possible, all hand feeding during drilling should be avoided. When drilling assemblies, a rigid machine tool or gantry type drill should be employed in preference to portable drilling equipment. Dwelling should also be avoided because it produces damage and may even friction weld the drill to the workpiece. The operator should visually check the hole and drill after each operation. If average or localized wear exceeds specifications, the drill should be replaced. If abnormal conditions develop in the hole, it should be marked and inspected thoroughly before assembly. Galling, torn surfaces, or discoloration from overheating are causes for rejection. When a drill wears excessively or actually breaks during the drilling of sensitive alloys, the operator should notify appropriate personnel so that remedial steps can be taken. Accidents of this type indicate that machining conditions have not been selected and/or used properly. Coolant-fed drills may help minimize surface damage, but no supporting data are available.

**Proper drill fixturing assists in minimizing damage during drilling.** When drilling holes 1/4 inch [6.4 mm] or larger, a drill fixture or bushing should be used. Where accessibility permits, a drill backup should be used to minimize burring.

**Finishing of drilled holes is imperative.** The entrance and exit of all holes should be carefully deburred and chamfered. All holes should be finish reamed after drilling whenever possible to insure better surface finish and surface integrity. (See comments that follow.)

**Bushing clearance is needed on deep holes.** For holes with depths 3 times the diameter, or more, ample clearance should be provided between the drill bushing and the workpiece for the chips to flow.

**Special precautions should be taken when reaming holes in sensitive alloys.** Since reaming often serves as a final hole finishing operation, all machining parameters must be controlled. Stock allowances must also be controlled. Power feeding of power driven machines should be employed for reaming of straight holes. When using power driven machines for tapered holes, hand feeding is permissible, but power feeding is preferred. Hand reaming of tapered holes using a tap wrench may be permissible after power reaming. If hand reaming is performed, special attention should be given to the selection of tool material, the reamer geometry, and the accuracy in grinding the reamer. In addition, it may be advisable to provide for alignment of the reamer to insure accuracy during the hand reaming operation.

Double ream all straight holes 5/16 inch [7.9 mm] or larger with a minimum metal removal of 3/64 inch [1.2 mm] on the diameter. On smaller holes, the minimum metal removal should be 1/64 inch [0.40 mm] on the diameter. The operator should visually check the reamer for sharpness after each operation. At the first sign of chipping, localized wear or average flank wear beyond specification, the reamer should be replaced, and the hole should be inspected. Also, regardless of the hole and reamer condition, a maximum number of holes should be specified for reamer replacement. Each reamed hole should be carefully inspected for surface roughness, galling, smearing, scratches, etc. The entrance and exit of all holes should be carefully deburred and chamfered.

**Deburring and chamfering should be used to remove all sharp edges.** Drilled and reamed holes should be countersunk or chamfered at the entrance and exit to remove the entire burr because sharp edges and burrs are common sources of component failure. To countersink, use power feed units if possible and use a countersink which completely avoids chattering. Generally, low spindle speeds are desirable and chamfering tools should be kept sharp. The operator should visually inspect the tool after each cut, and it should be replaced if there is any visual evidence of wear.

The breaking of edges or radiusing may be done by abrasive deburring using a low speed, power hand drill. When chamfering a part, a minimum of 0.010 inch [0.254 mm] chamfer is advisable. The break edges and chamfers should be carefully examined for compliance with surface roughness requirements. This is especially important on the entrance and exit of holes where fatigue failures have a tendency to originate. The chamfers should be carefully blended into the adjacent surfaces.

**Honing is an excellent finishing operation for developing surface integrity.** Honing is usually used only when finish requirements or tolerances are too close for practical use of other finishing operations such as reaming, grinding, etc. A multi-stone head is preferred; heads with steel shoes and/or steel wipers are not recommended. Honing produces less surface deformation and fewer surface integrity effects than any other conventional mechanical hole-finishing process.

**Boring may be used as a finish machining operation if roughness is within the manufacturing engineering limits.** When boring, extra precautions should be taken in the preparation of the cutting edge to provide extremely low surface roughness values. The tool wearland in finish boring should be limited to 0.005 inch [0.127 mm], but often it should be far less than this in order to achieve the desired accuracy and surface roughness.

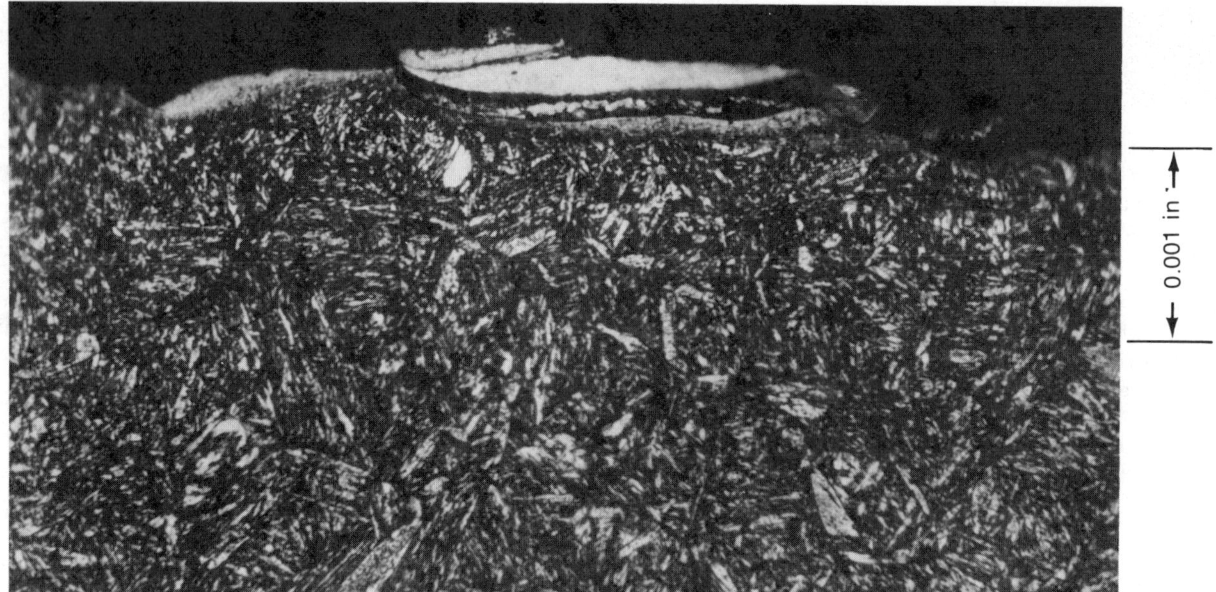


**Figure 18.3–30** Built-up edge from face milling 4340 steel (quenched and tempered, 50 $R_C$). Machining operations such as turning, drilling, milling, etc., especially under poor machining conditions, are likely to produce surfaces with a built-up edge. This photomicrograph shows that portion of the built-up edge which was left on the workpiece. Fracture analyses of failed parts have shown that cracks associated with the built-up edge have served as origins of the fracture.

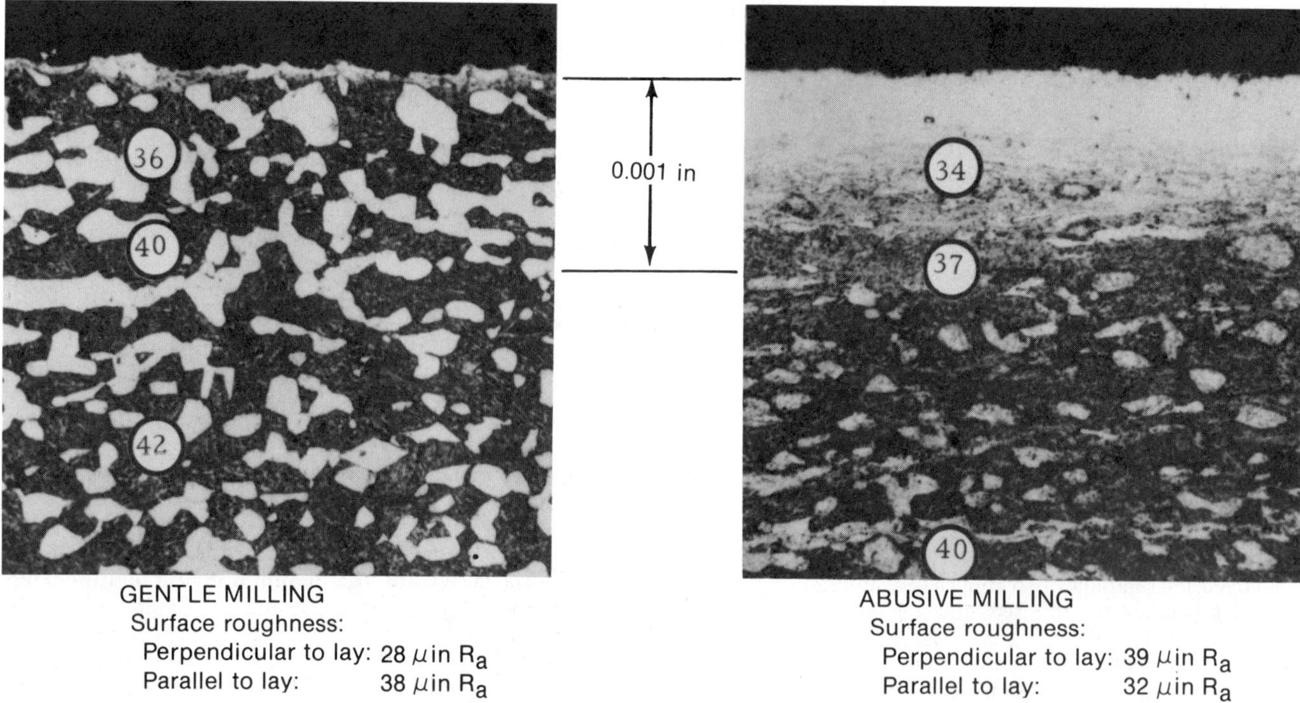


GENTLE MILLING
Surface roughness:
Perpendicular to lay: 28 $\mu$in $R_a$
Parallel to lay: 38 $\mu$in $R_a$

ABUSIVE MILLING
Surface roughness:
Perpendicular to lay: 39 $\mu$in $R_a$
Parallel to lay: 32 $\mu$in $R_a$

**Figure 18.3–31** Surface characteristics of Ti-6Al-6V-2Sn (solution treated and aged, 42 $R_C$) produced by peripheral end milling. Orientation: longitudinal mill, surface sections perpendicular to lay. Slight evidence of distortion can be seen on the gently milled sample. Very pronounced alterations are associated with abusive milling. A maximum hardness loss of 8 points $R_C$ and a total affected depth of 0.004 inch can be attributed to abusive milling. Approximately 6 points $R_C$ surface hardness loss was measured on the gently milled samples. In this case, however, the effect was much shallower, not exceeding 0.002 inch. Indicated hardness data on the photomicrographs are $R_C$ values converted from Knoop hardness measurements. Surface roughness measurements are averages of readings made on all specimens from each group. (W. P. Koster et al[9])

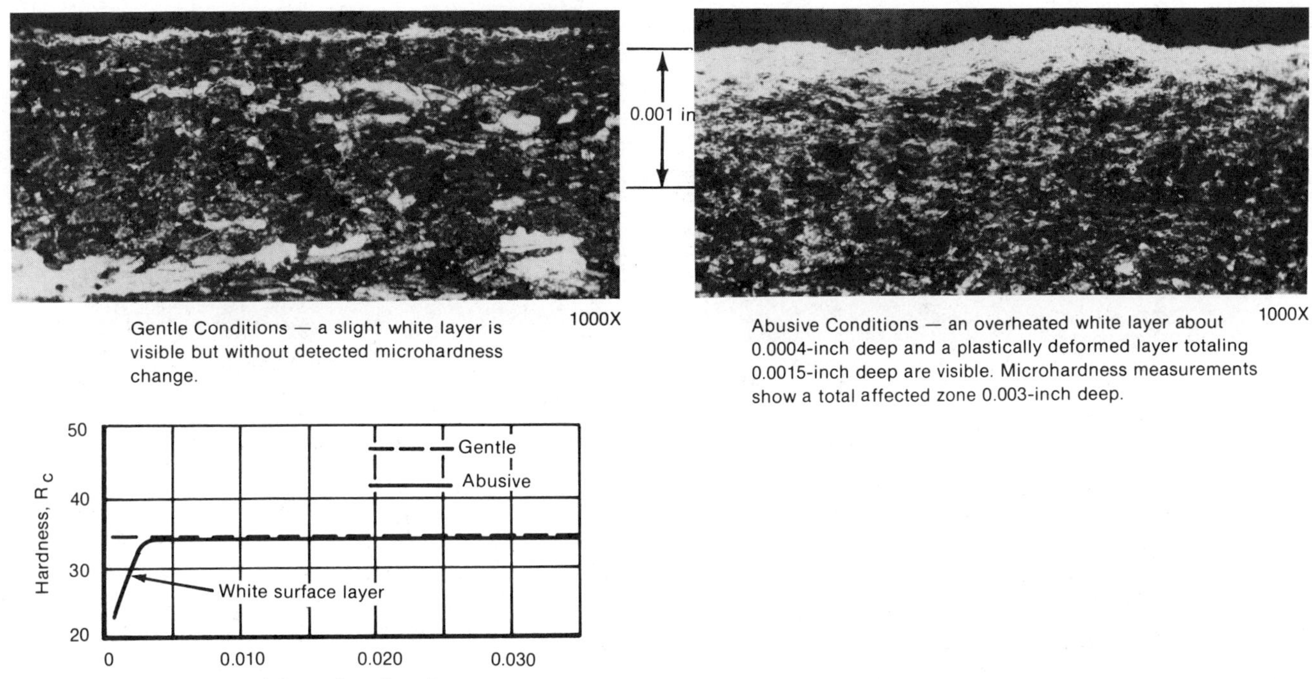
0.001 in
1000X
Gentle Conditions — a slight white layer is visible but without detected microhardness change.
1000X
Abusive Conditions — an overheated white layer about 0.0004-inch deep and a plastically deformed layer totaling 0.0015-inch deep are visible. Microhardness measurements show a total affected zone 0.003-inch deep.
Hardness, R c
50
40
30
20
Gentle
Abusive
White surface layer
0
0.010
0.020
0.030
Depth beneath surface, in

**Figure 18.3–32** Surface characteristics of Ti-6Al-4V (aged, 33 $R_C$) produced by face milling. (Surface Integrity Encyclopedia[1])

0.001 in

Built-up edge produced by sharp tool. Note microcrack originating at base of edge.

Tearout of untempered martensite in flat surface produced by dull tool.

**Figure 18.3–33** Surfaces produced by carbide face milling 4340 steel (quenched and tempered, 40 $R_C$). (M. Field, W. P. Koster and J. B. Kohls[15])

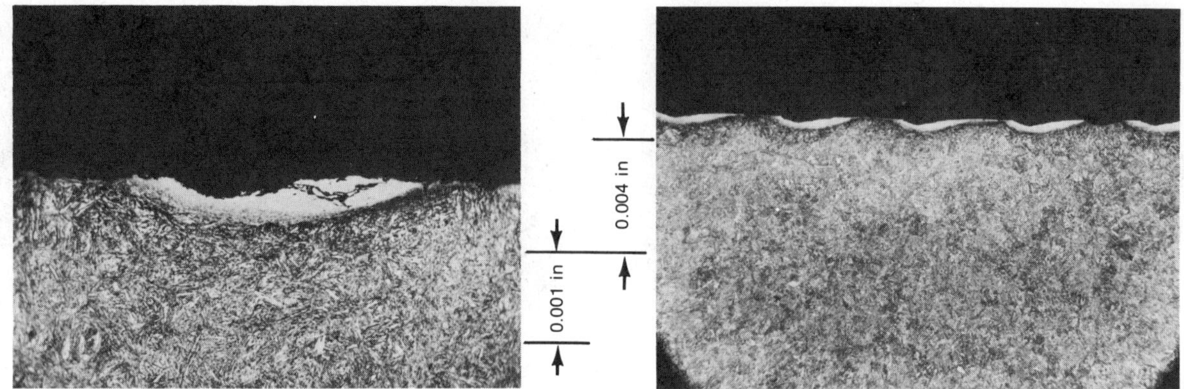


ABUSIVE CONDITIONS — Photomicrographs at 1000X (left) and 250X (right) showing white rehardened patches of martensite. The interval of the patches corresponds to the feed of the cutting tool. Thin zones of overtempered martensite 0.001-inch deep with hardness as low as 46 $R_C$ are found beneath each patch.

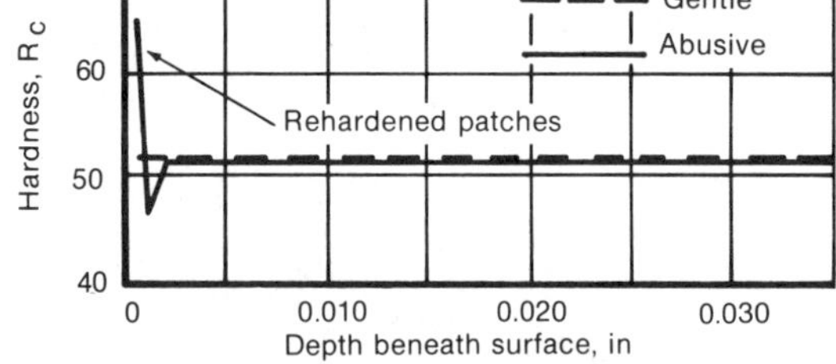


**Figure 18.3-34** Surface characteristics of 4340 steel (quenched and tempered, 52 $R_C$) produced by face milling. (M. Field, W. P. Koster and J. B. Kohls[15])

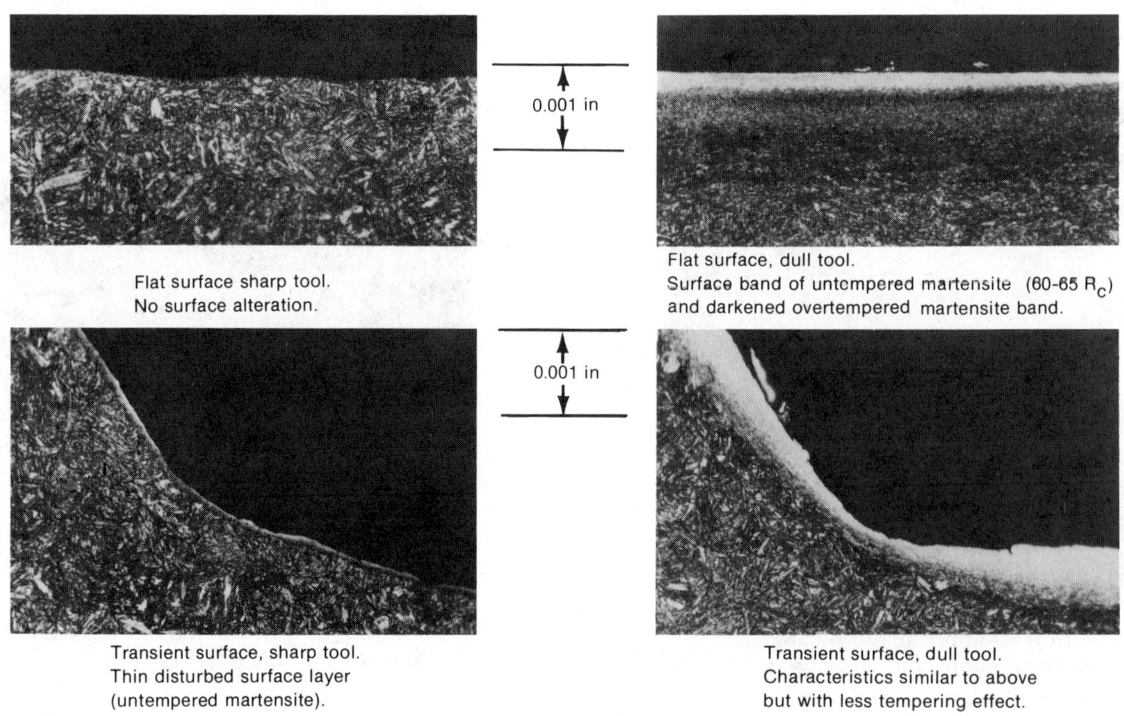

0.001 in
Flat surface sharp tool.
No surface alteration.
Flat surface, dull tool.
Surface band of untempered martensite (60-65 $R_C$)
and darkened overtempered martensite band.
0.001 in
Transient surface, sharp tool.
Thin disturbed surface layer
(untempered martensite).
Transient surface, dull tool.
Characteristics similar to above
but with less tempering effect.


**Figure 18.3-35** Surfaces produced by carbide face milling D6ac (quenched and tempered, 48 $R_C$). (M. Field, W. P. Koster and J. B. Kohls[15])

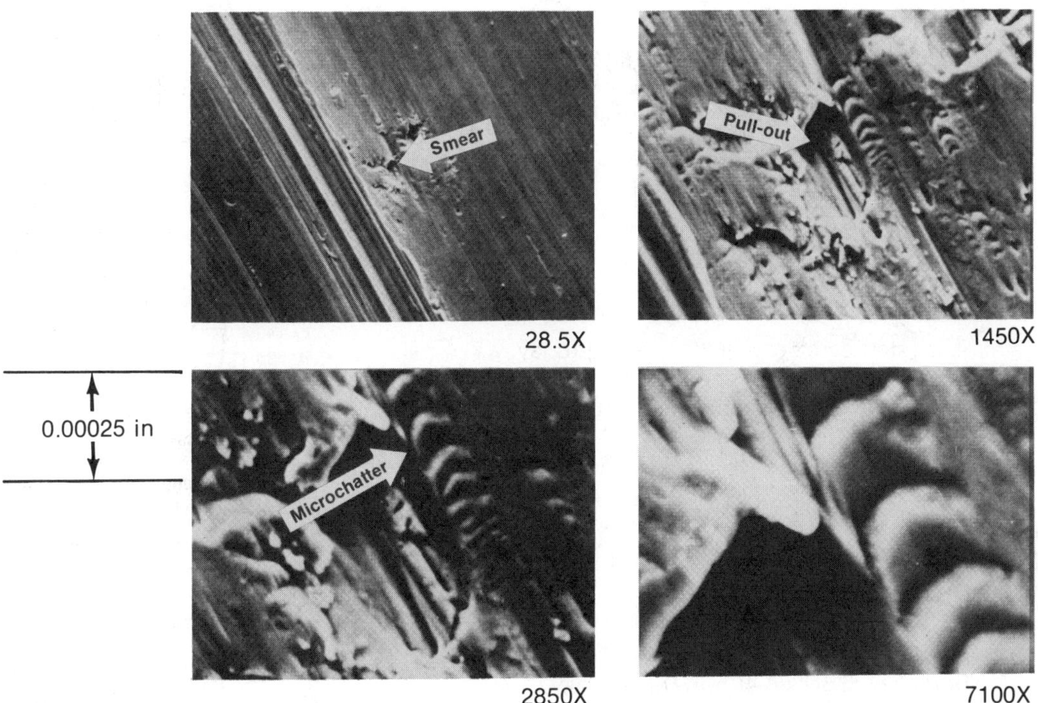


**Figure 18.3-36** Turning Inconel Alloy 718 (solution treated and aged) with sharp tungsten carbide showing anomaly areas of smeared metal, material pull-outs and microchatter. Views at successively greater magnification from scanning electron microscope taken at 15° from perpendicular. Plastically deformed surfaces as seen in figure 18.3-37. (G. Bellows[17])

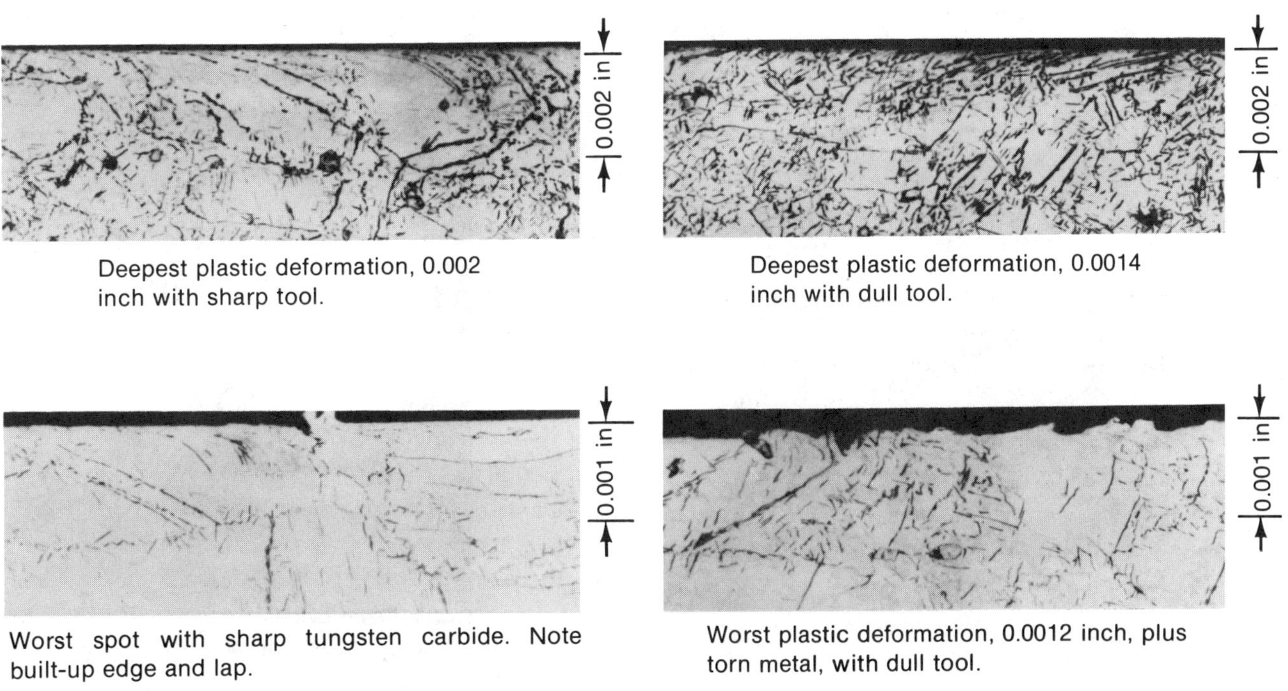


Deepest plastic deformation, 0.002 inch with sharp tool.

Deepest plastic deformation, 0.0014 inch with dull tool.

Worst spot with sharp tungsten carbide. Note built-up edge and lap.

Worst plastic deformation, 0.0012 inch, plus torn metal, with dull tool.

**Figure 18.3-37** Surfaces produced by turning Inconel Alloy 718 (solution treated and aged) with sharp and dull tungsten carbide tools. Sharp tool had 0.005-inch wearland and dull tool had 0.013 inch. Surface roughness was 63 microinches $R_a$ in both cases. (G. Bellows[17])

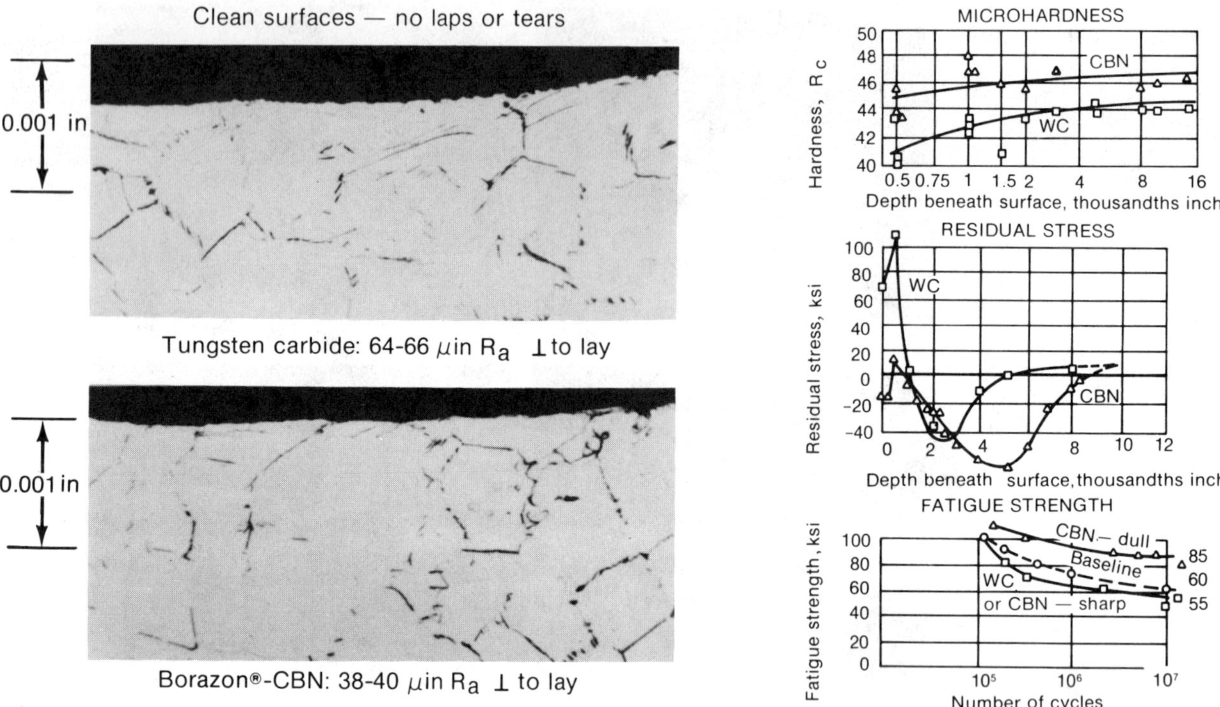


**Figure 18.3–38** Turning Inconel Alloy 718 (solution treated and aged) standard surface integrity data set comparing performance of tungsten carbide tools with Borazon®-CBN. Baseline fatigue curve made with low-stress-ground specimens. (G. Bellows[17])

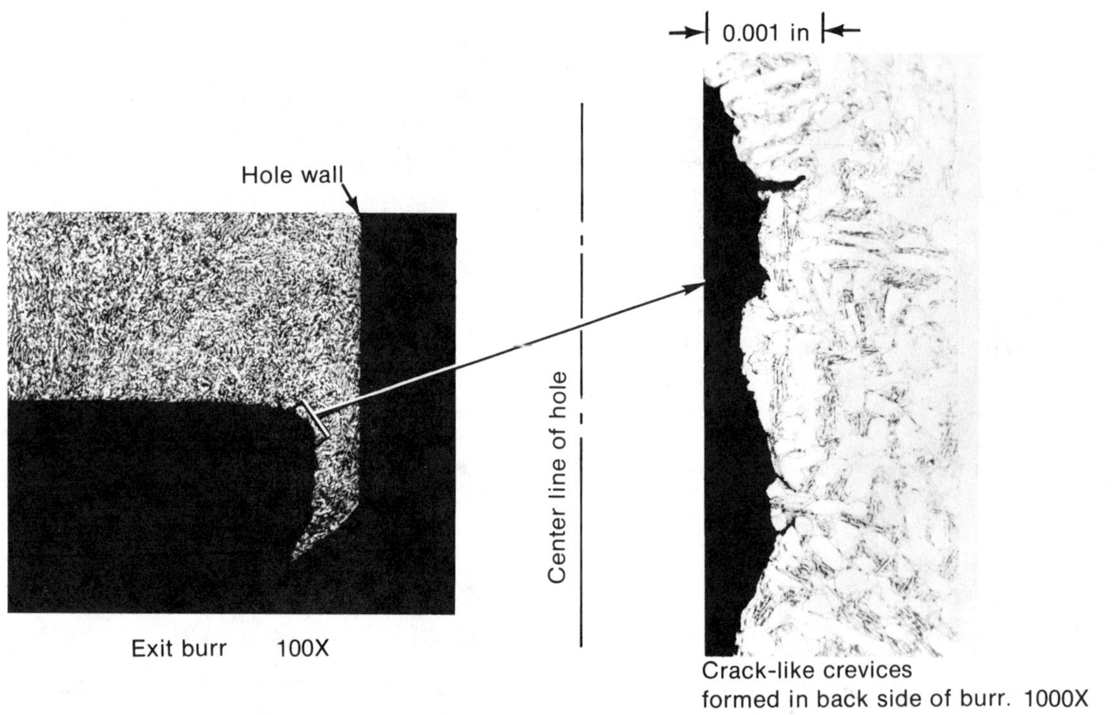


**Figure 18.3–39** Drilling of Ti-6Al-6V-2Sn. *Right*, crack-like crevices commonly found in drilled and reamed holes. *Left*, the section shown is a burr on the workpiece surface. Burrs have been shown to be responsible for appreciable reduction of fatigue strength in steels and in titanium alloys.

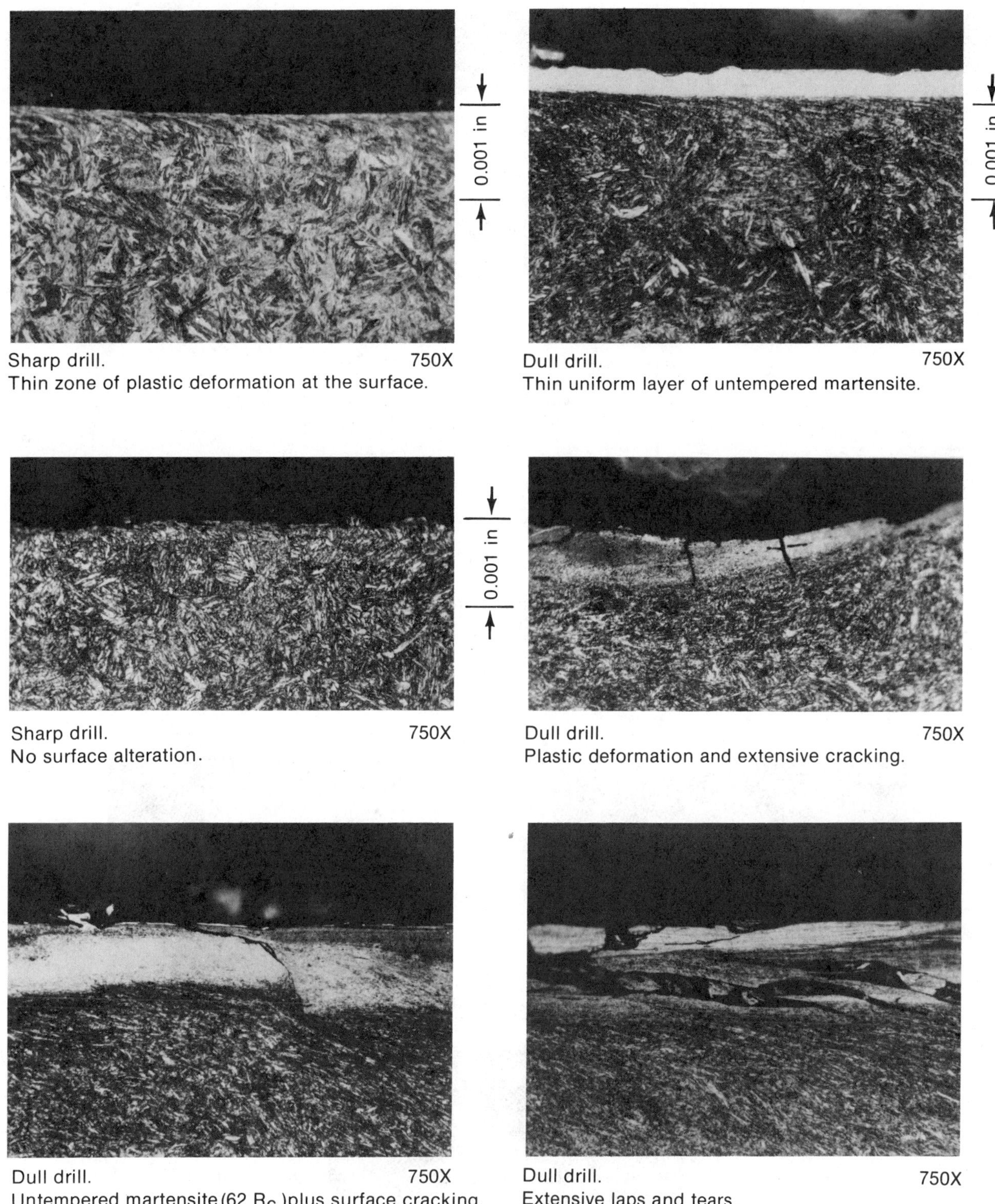


Sharp drill. 750X
Thin zone of plastic deformation at the surface.

Dull drill. 750X
Thin uniform layer of untempered martensite.

Sharp drill. 750X
No surface alteration.

Dull drill. 750X
Plastic deformation and extensive cracking.

Dull drill. 750X
Untempered martensite (62 $R_C$) plus surface cracking.

Dull drill. 750X
Extensive laps and tears.

**Figure 18.3–40** Surfaces produced by drilling 4340 steel. Top two photos represent carbide drilling 4340 (quenched and tempered, 54 $R_C$); remaining photos represent high speed steel drilling 4340 (quenched and tempered, 48 $R_C$). (M. Field, W. P. Koster and J. B. Kohls[15])

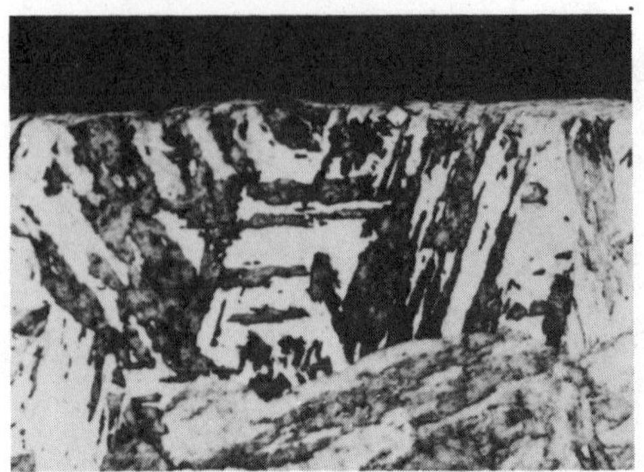

Gentle conditions — very thin trace of cold work may be seen on the surface. 500X

Abusive conditions — an overaged or resolutioned layer 0.001-inch deep at 37 $R_C$ is found on the surface. Total affected depth is approximately 0.002 inch. 500X

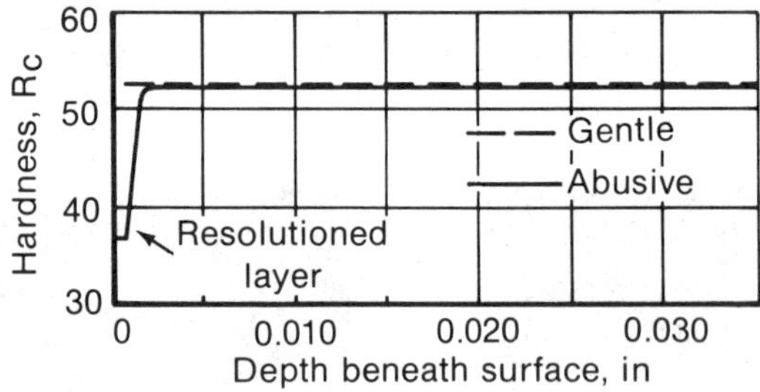


**Figure 18.3–41** Surface characteristics of 18% nickel maraging steel Grade 250 (aged, 52 $R_C$) produced by drilling. (M. Field, W. P. Koster and J. B. Kohls[15])

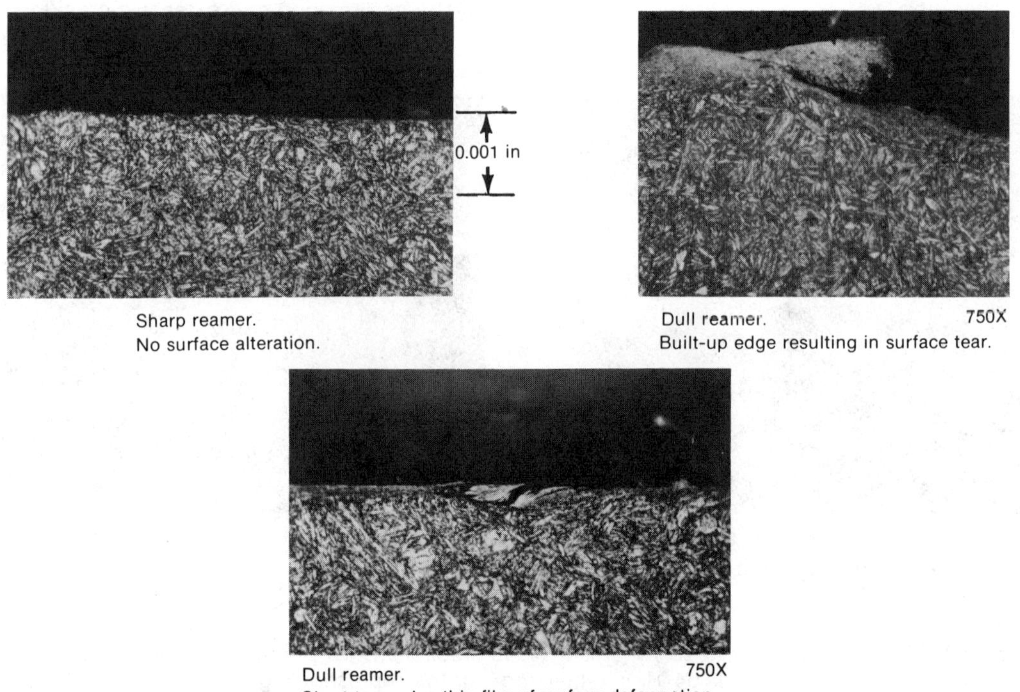


Sharp reamer.
No surface alteration.

Dull reamer. 750X
Built-up edge resulting in surface tear.

Dull reamer. 750X
Short tears plus thin film of surface deformation.

**Figure 18.3–42** Surfaces produced by high speed steel power taper reaming 4340 steel (quenched and tempered, 40 $R_C$). (M. Field, W. P. Koster and J. B. Kohls[15])

**Figure 18.3–43** Surfaces produced by high speed steel power taper reaming 4340 steel (quenched and tempered, 48 $R_C$). (M. Field, W. P. Koster and J. B. Kohls[15])

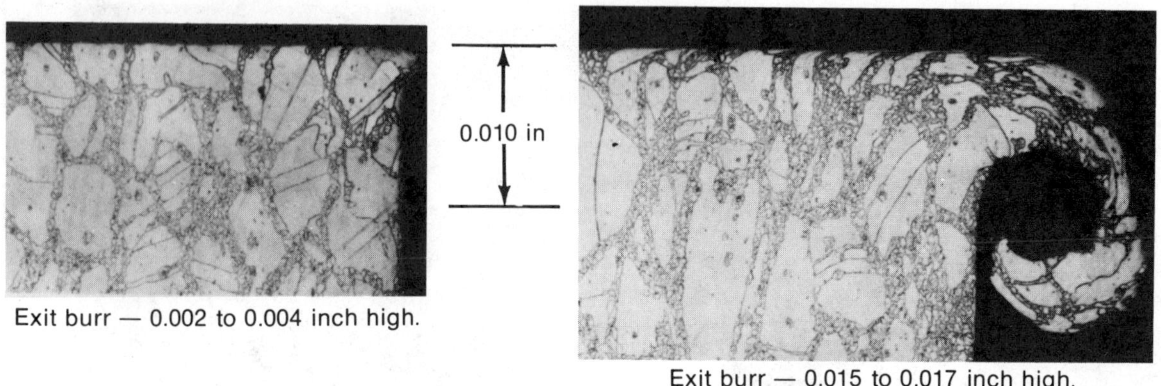


Exit burr — 0.002 to 0.004 inch high.

Exit burr — 0.015 to 0.017 inch high.

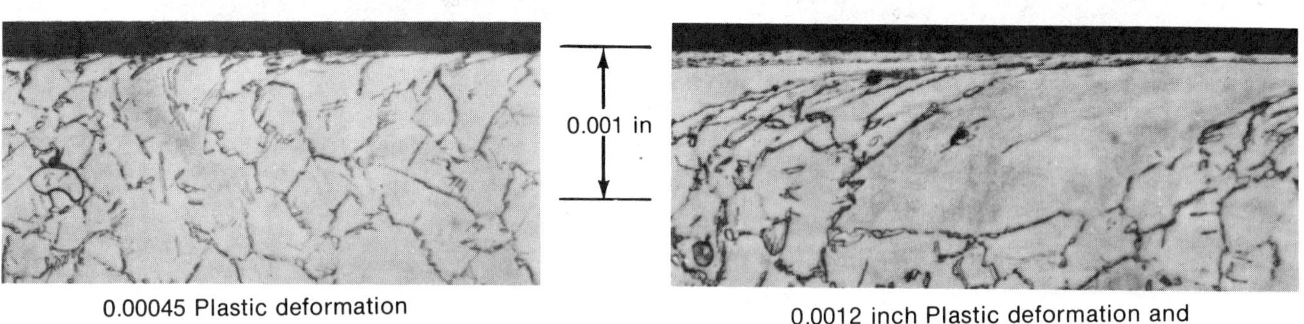


0.00045 Plastic deformation

0.0012 inch Plastic deformation and 0.0001 inch smeared metal.

INITIAL BROACHING PASS

LAST PASS, #254

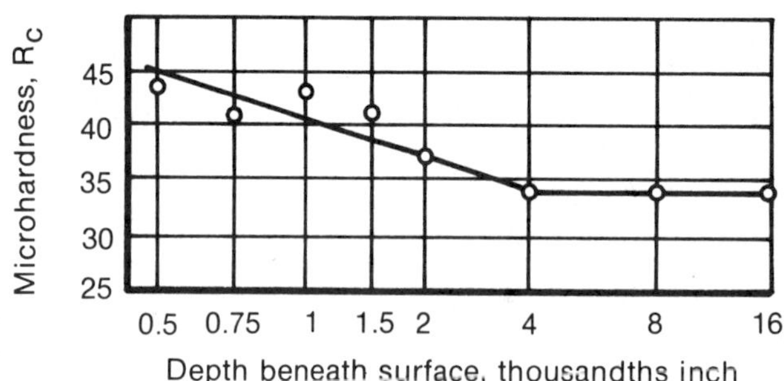


**Figure 18.3–44** Broaching of Inconel Alloy 718 (solution treated and aged) with T15 high speed steel tools. Wearland increased to 0.006 to 0.007 inch during 254 cuts for a total of 158 inches of cut. Microhardness traverse taken after 254 passes. (Surface Integrity Encyclopedia[1])

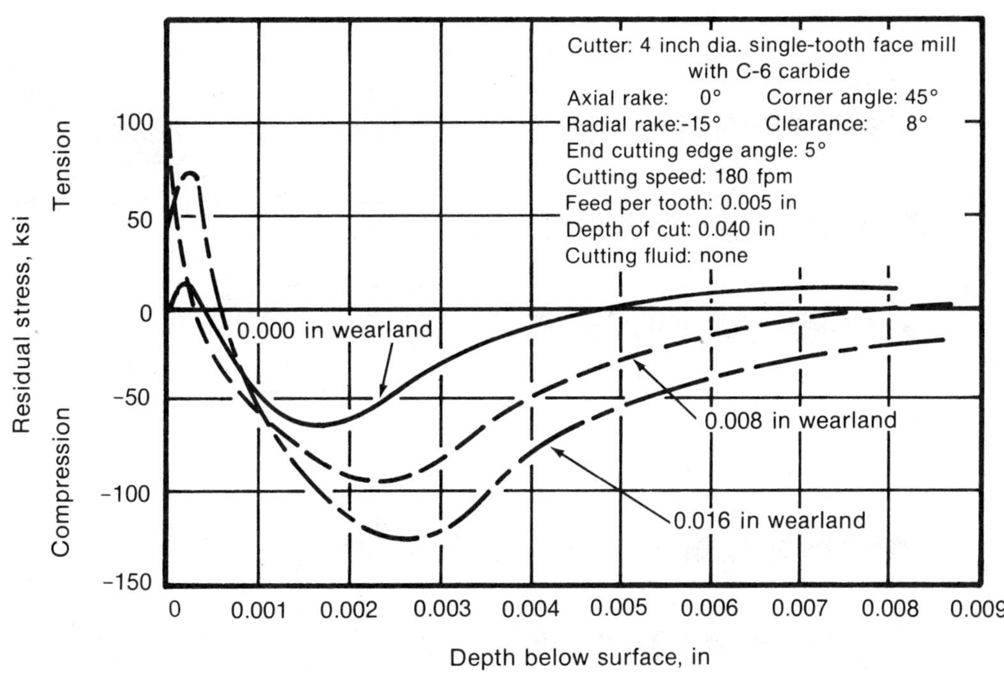


**Figure 18.3–45** Effect of tool wear on residual stress in milled surface, face milling of 4340 steel (52 $R_C$). Note that the amount of residual compressive stress produced in this specimen is related to the tool sharpness. The worn tool produced a deeper compressive layer. (W. P. Koster and J. B. Kohls[18])

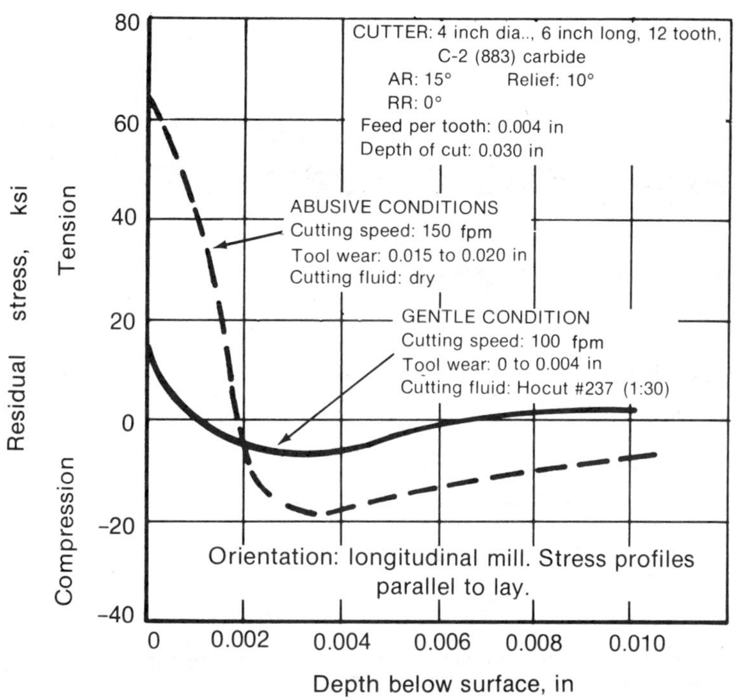


**Figure 18.3–46** Residual surface stress profiles of Ti-6Al-6V-2Sn (solution treated and aged, 42 $R_C$) produced by peripheral end milling. (W. P. Koster and J. B. Kohls[18])

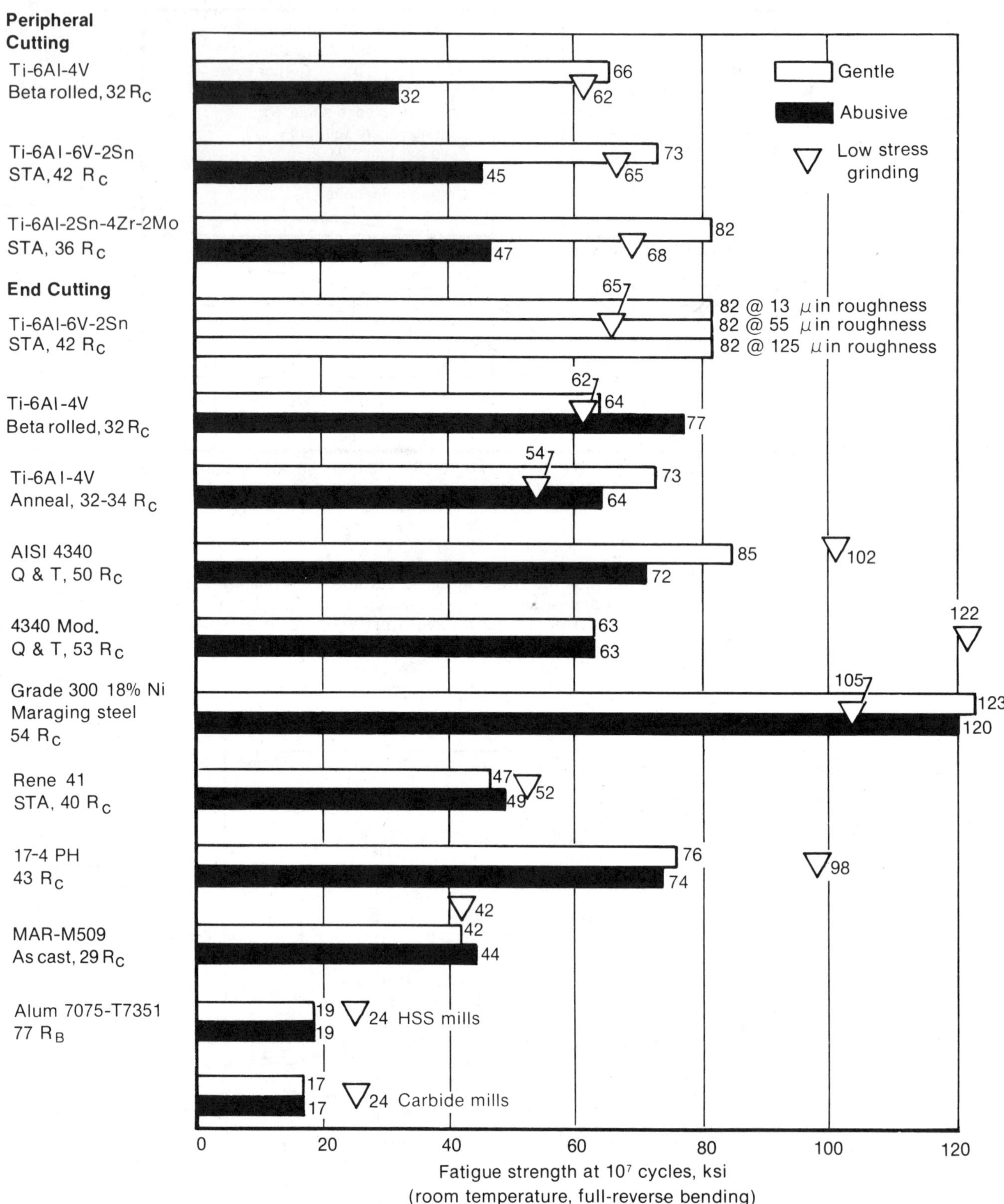


**Figure 18.3–47** Summary of high cycle fatigue strength—end milling.

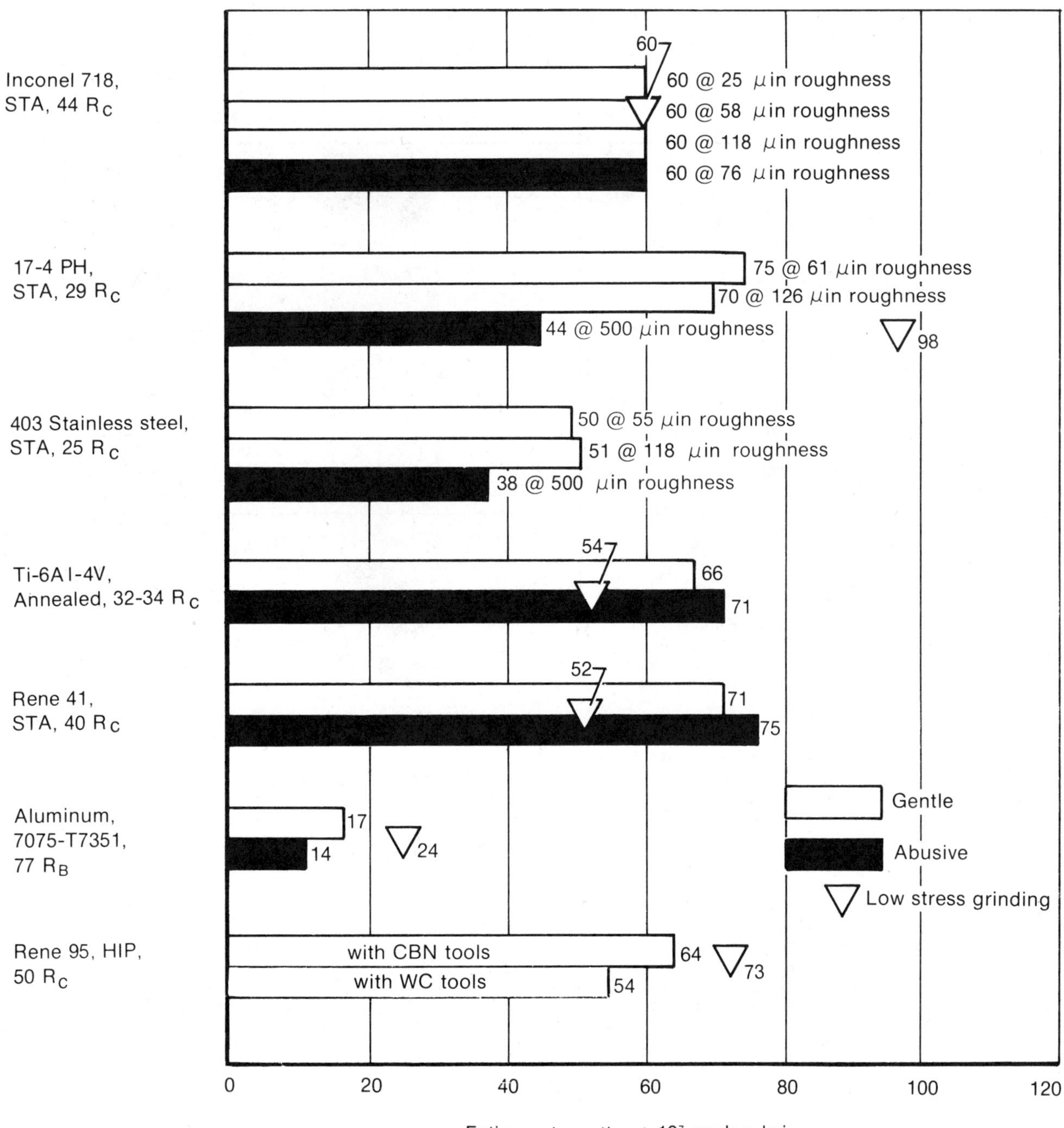


**Figure 18.3–48** Summary of high cycle fatigue strength—face turning.

**TABLE 18.3-10 Surface Integrity Effects Observed in Mechanical Material Removal Processes**

| PROPERTY AND TYPE OF EFFECT | FINISHING, "GENTLE" OR LOW STRESS CONDITIONS | | ROUGHING, "OFF-STANDARD" OR ABUSIVE CONDITIONS | |
|---|---|---|---|---|
| Surface roughness: | μin $R_a$ | μm $R_a$ | μin $R_a$ | μm $R_a$ |
| Average range | 8 to 63 | 0.2 to 1.6 | 125 to 250 | 3.2 to 6.3 |
| Less frequent range | 1 to 125 | 0.025 to 3.2 | 63 to 1,000 | 1.6 to 25 |
| Mechanical altered material zones: | inch* | mm* | inch* | mm* |
| Plastic deformation (PD) | 0.0017 | 0.04 | 0.0047 | 0.12 |
| Plastically deformed debris ($PD^2$) | *** | *** | *** | *** |
| Hardness alteration† | 0.0010 | 0.025 | 0.0200 | 0.50 |
| Microcracks or Macrocracks | 0.0005 | 0.013 | 0.0015 | 0.04 |
| Residual stress‡ | 0.0060 | 0.15 | 0.0140 | 0.36 |
| Metallurgical altered material zones: | inch* | mm* | inch* | mm* |
| Recrystallization | — | — | — | — |
| Intergranular attack (IGA) | *** | *** | *** | *** |
| Selective etch, pits, protuberances | 0.0004 | 0.01 | 0.0010 | 0.025 |
| Metallurgical transformations | 0.0015 | 0.04 | 0.0200 | 0.50 |
| Heat-affected zone (HAZ) or Recast layer | 0.0001 | 0.003 | 0.0030 | 0.076 |
| High cycle fatigue (HCF): | percent | | percent | |
| Variation from "handbook" values at room temperature§ | −50 to +22 | | −48 to +2 | |

SOURCE: G. Bellows and J. B. Kohls.[20]

NOTE: A dash (—) in the table indicates no or insufficient data. A triple asterisk (***) in the table indicates no occurrences or not expected.

*Maximum observed depth normal to the surface.

†Depth to point where hardness becomes less than ±2 points $R_c$ (or equivalent) of bulk material hardness (hardness converted from Knoop microhardness measurements).

‡Depth to point where residual stress becomes and remains less than 20 ksi [138 MPa] or 10% of tensile strength, whichever is greater.

§Handbook values from HCF testing frequently are generated from specimens that have been processed using low stress grinding, hand or gentle machine polishing or occasionally electropolishing. For this table, "handbook" values represent HCF strength values generated from low-stress-ground specimens (with their minor amount of retained but enhancing compressive residual stress). The values are the $10^7$ cycle strengths from full-reverse bending tests at room temperature.

## Specific Guidelines and Data for ABRASIVE Material Removal Processes

The heat in the workpiece surface resulting from the use of any of the varieties of abrasive machining is the principal ingredient affecting surface integrity. The amount of heat is proportional to the intensity with which the process is applied. The degree of surface effects is a function of the thermal properties of the workpiece material and its metallurgical response to the heating level. Residual stress and the attendant distortion can be problems, as can smeared metal or the debris from plastic deformation.

Grinding produces the greatest range of fatigue endurance sensitivity among the several material removal processes and alloys studied to date. Titanium alloys are the most drastically affected by grinding variables, although some cast nickel base materials are close behind. The ferrous materials, except the high strength varieties, are less sensitive to grinding variables than these other alloys groups, although very significant losses in fatigue strength are associated with both conventional and rough grinding levels of intensity. Even shallow, isolated spots of untempered or overtempered martensite can be detrimental to fatigue endurance strength. Hand sanding or deburring operations need strict control to reduce the human variables that can influence surface reliability. Abrasive cutoff operations are particularly susceptible to surface damage. Table 18.3-11 summarizes the surface integrity effects observed during the examination of several thousand pieces of data.

The charts and guidelines that follow have baselines the same as those for the mechanical material removal processes. Figures 18.3–49 through 18.3–54 illustrate good and poor surface effects from grinding. Figure 18.3–55 shows typical microhardness alterations, and figures 18.3–56 to 18.3–61 show variations of the residual stress or distortion patterns. High cycle fatigue responses are charted in figures 18.3–62 and 18.3–63. Other grinding illustrations are figures 18.3–1, 18.3–2, 18.3–8, 18.3–10, 18.3–13, 18.3–15, 18.3–17, 18.3–18, 18.3–21, 18.3–25, 18.3–26, and 18.3–28.

**Abrasive Process Guidelines that Promote Good Surface Integrity Results**

1. Low stress grinding techniques can reduce distortion and surface damage. Frequent coarse dressings of softer grade wheels, lower infeeds and reduced grinding wheel speeds with lubricating-type cutting fluids are the principal elements of low stress grinding. A shallow depth of low value tensile or compressive residual stress can be attained.

2. Higher work material speeds can aid both productivity and surface integrity. Lower wheel surface speeds aid surface integrity without sacrificing productivity.

3. Frequent coarse dressing of grinding wheels can reduce surface damage by keeping the wheels open and sharp, thus helping to reduce temperatures at the wheel-workpiece interface.

4. Modifications to established low stress grinding procedures should not be attempted unless testing confirms that they can be tolerated.

5. If low stress grinding is specified for finishing operations, then conventional grinding can be used to within 0.010 inch [0.254 mm] of the finished size, provided that the work materials are not sensitive to cracking.

6. Lubricating-type fluids and coolants should have a copious and positively directed flow and should be checked for their chemical action on the specific material being ground.

7. Conventional grinding conditions should not be used for finish grinding of highly sensitive alloys such as high strength steels, high temperature nickel and cobalt base alloys, and titanium or molybdenum alloys.

8. The heat-affected zones from rough grinding can be much deeper than the surface discolorations might indicate. Microhardness traverses can generally detect the depth of surface alterations.

9. High stress grinding of martensitic steels can create untempered martensitic zones with overtempered martensitic zones below them, both of which will limit fatigue strength.

10. Crack detection tests should be made when rough grinding materials. Existing surface cracks can propagate ahead of the grinding when finishing some thermally sensitive materials. Low stress grinding should be used to remove more than the altered material zones from any prior conventional grinding operation.

11. Hand grinding of sensitive alloys should be discouraged, unless done under very careful control.

12. Strict instructions and controls over power hand sanders should be maintained.

13. Abrasive cutoff frequently has a harsh and deep surface effect, and these altered layers should be removed by more gentle processes.

14. Surface X-ray residual stress measurements can be misleading. A full profile is best, or enough checks should be made below the surface to correct for penetration of the X-rays and the slope of the profile.

15. Good dressing action on the grinding wheel is promoted by cleanliness and sharpness of the dressing tool. Pauses or dwells in the dressing tool action can dull the wheel surface, particularly with the roll form, crush roll or form block dressing tools.

**Special Comments Concerning Surface Integrity Practices for Abrasive Machining Processes**

**Reduce grinding distortion and surface damage by using "low stress" conditions during finish grinding.** Conventional finish grinding practices should be replaced by low stress grinding procedures in order to minimize grinding distortion and to reduce the possibility of producing extensive surface metallurgical alterations and cracking. Low stress grinding procedures are described in table 18.3–12. Some of the metallurgical alterations caused by abusive grinding do not produce cracking immediately, but they develop conditions leading to delayed cracking, which may occur later "on the shelf" or prematurely in service, thereby seriously complicating inspection practices. Low stress grinding in comparison with conventional practices employs softer grade grinding wheels, reduced grinding wheel speeds, reduced infeed rates, and chemically active cutting fluids. Low stress conditions should be used for removing the last 0.010 inch [0.254 mm]. See tables 18.3–13 to 18.3–17.

**Modification of low stress grinding procedures should not be attemped unless service experience or testing programs in the shop and/or laboratory confirm that compromises can be tolerated.** For certain type parts made of sensitive alloys, no compromises should be permitted in the specification of low stress grinding parameters. Experience has shown that it is sometimes possible to relax low stress procedures in order to accommodate equipment limitations or to increase production rates. Table 18.3–17 is an example of a set of grinding conditions, case A, which produced scrap as a result of grinding cracks in the fir-tree section of IN-100 turbine blades. Substitution conditions, case B, are essentially low stress except for the infeed. Experience had indicated that in this instance an infeed as high as 0.002 inch [0.050 mm] could be tolerated.

An almost identical cracking problem was encountered in the production of Inconel Alloy 713C turbine blades. This problem was also solved by implementation of the recommendations shown for case B of table 18.3–17. Parts made of cast HS-31 and cast WI-52 have also cracked in conventional grinding. Because of these experiences and the lack of other data, it is suggested that all cast nickel and cast cobalt base high temperature alloys be processed using low stress grinding conditions for the highly stressed or critically loaded surfaces.

**If low stress grinding is required for finish grinding, then conventional grinding can be used to within 0.010 inch [0.254 mm] of finish size if the materials being ground are not sensitive to cracking.** During *conventional grinding*, metallurgical alterations, including stresses, are usually confined to within approximately 0.005 inch [0.127 mm] of the surface or less. This makes it possible to remove most of the stock by conventional practices and then finish grind using low stress methods. However, crack detection tests should be made to see that conventional grinding of the material in question does not create cracks of greater depth. Stock allowances of the order of 0.010 inch [0.254 mm] or greater, if necessary, are suggested to compensate for location inaccuracies in holding fixtures.

**Finish grinding of critical and/or highly stressed surfaces of sensitive alloys exposed to high service stresses should be performed using low stress grinding instead of conventional grinding.** Some high performance, high strength alloys are also sensitive to the thermal shocks accompanying grinding. A partial list of these alloys is as follows:

High Strength Steels, Wrought
- 4340 or 4340 Mod. at 40–54 $R_C$
- 300M or D6ac at 48–56 $R_C$
- Maraging Steels
- HP9-4-45

High Temperature Nickel and Cobalt Base Alloys, Wrought
- Rene 41
- Rene 80
- Udimet 700
- AF 95 (Rene 95)
- Waspaloy
- AF2-1DA
- Inconel Alloy 718
- Rene 125

Titanium Alloys, Wrought
- Ti-6Al-4V
- Ti-5Al-2.5Sn
- Ti-6Al-2Sn-4Zr-2Mo
- Ti-6Al-6V-2Sn

Table 18.3–18 and figure 18.3–62 provide high cycle, room temperature fatigue data showing the adverse effects of abusive and of conventional grinding compared with gentle or low stress grinding for several classes of aerospace materials.

Fatigue strength reduction of martensitic steels as a result of abusive grinding has been determined to be approximately equal, regardless of the quantity of untempered martensite (UTM) or overtempered martensite (OTM) present. Traces of UTM or OTM as well as 0.004-inch [0.10 mm] deep layers were associated with the same fatigue strength reduction, figure 18.3–25. Reduction of the fatigue strength was found to be about 30 to 35 percent. For quenched and tempered steels, OTM is frequently produced under the UTM. Even if the UTM is removed by gentle grinding, the fatigue strength is still reduced 25 to 30 percent. This indicates that overtempered martensite may be as important as untempered martensite in influencing fatigue strength.

**Conventional grinding conditions should not be employed for either roughing or finishing alloys which are ultra-sensitive to the grinding process.** From experience, certain alloys have been found to crack easily during conventional grinding. For such alloys, it is likely that the cracks created will not be eliminated by finish grinding. The high temperature nickel and cobalt base *cast* alloys (such as IN-100, Inconel Alloy 713C, MAR-M509, HS-31, WI-52, Udimet 500, Udimet 700, Inconel Alloy 738, etc.), in contrast with the *wrought* nickel and cobalt base high temperature alloys, are examples of alloys which should never be ground conventionally. See table 18.3–17 for a specific case history involving cracking of a cast nickel base high temperature alloy. Metals and alloys which are brittle, such as beryllium and tungsten, tend to respond similarly.

**Frequent dressing of grinding wheels can reduce surface damage by keeping the wheels open and sharp, thus helping to reduce temperatures at the wheel-workpiece interface.** Automatic dressing and wheel compensation contribute to the economic feasibility of frequent dressing. Crush dressing and diamond roll dressing can also be used to minimize the cost of frequent dressing.

**Cutting fluids, properly applied, help promote surface integrity.** In order to get fluid to the wheel-workpiece interface, position the fluid nozzle as close to the wheel as possible. A general rule of thumb, regarding the quantity of fluid needed, is to use at least two gallons of fluid per horsepower per minute.

**Hand grinding of sensitive alloys should be discouraged.** The inherent lack of control has been responsible for creating many surface defects.

**Abrasive cutoff requires special surface integrity considerations.** Abrasive cutoff operations generally cause deeper surface alterations than grinding, sometimes as much as 0.050 inch [1.27 mm], and capability for minimizing damage from cutoff operations varies from plant to plant. It is recommended, therefore, that when abrasive cutoff is used that steps be taken to determine the extent of the disturbed layers and proper stock allowances be made for subsequent cleanup by suitable machining. Also, the entire cut surface should be examined because the temperatures generated are subject to extreme variation across the cut.

**Controls for hand power sanders should be maintained.** Test data reported in figure 18.3–63 show that good fatigue results can be obtained for such alloys as 4340 at 50 $R_C$ and Ti-6Al-4V provided that controlled conditions are maintained. The conditions are as follows:

Use a low belt speed (2000 feet per minute [10 m/s] maximum, for example).

Use a flexible (rubber) support for sanding disks or belts.

Use an abrasive grit generally no coarser than 80, and use coolant if feasible.

**At present, the relatively new high speed grinding processes should not be used for finishing of highly stressed structural parts unless an extended data set is developed.** Presently, insufficient mechanical property data are available to justify the recommendation of high speed grinding for critical parts even though increased production rates are attainable. High speed grinding increases the difficulty of the effective application of the grinding fluid, especially on complex parts, thereby making it more difficult to avoid surface damage from overheating.

**Abrasive processing and especially finish grinding must be accomplished under strict process control when employed for the manufacture of aerospace components.** Tests have shown that uncontrolled abusive as well as conventional grinding can decrease substantially the low and high cycle fatigue strengths at room and at elevated temperatures for titanium alloys, nickel base high temperature alloys and high strength steels. In addition to the room temperature high cycle fatigue data shown in table 18.3-18 and figure 18.3-62, data have been developed for elevated temperature high cycle fatigue and for low cycle fatigue. Tests on AF 95 (Rene 95) show that at 1000F [538C] the endurance limit ($10^7$ cycles, full-reverse bending) is lowered from 98 ksi [676 MPa] for gentle grinding to 48 ksi [331 MPa] for abusive grinding. Low cycle fatigue properties are affected adversely by abusive grinding, as shown in table 18.3-19.

**Adjustment of the grinding conditions can substantially alter the surface conditions.** When it is not possible or practical to control one of the operating parameters to low stress values, alteration of all of the other parameters to the best values can frequently achieve the desired surface quality, as shown in figure 18.3-64.

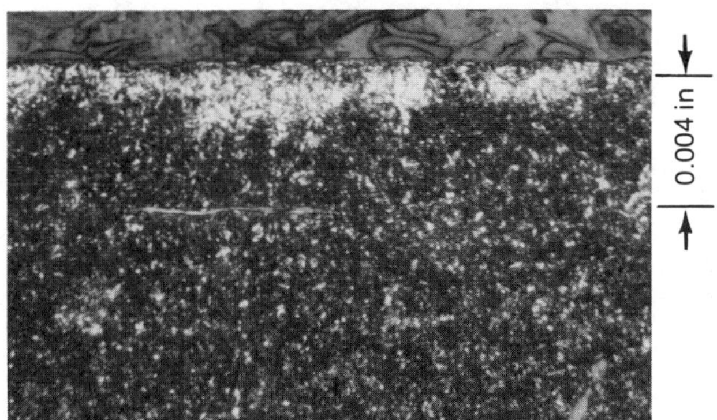


**Figure 18.3-49** Surface produced by abusive grinding of 4340 steel (50 $R_C$). The thin white martensitic layer in the photomicrograph is often found in the surfaces of hardened alloy steels after grinding. It has been found that even isolated patches of untempered or overtempered martensite as thin as 0.0002 inch can reduce fatigue strength of high strength steels as much as 35% to 40%.

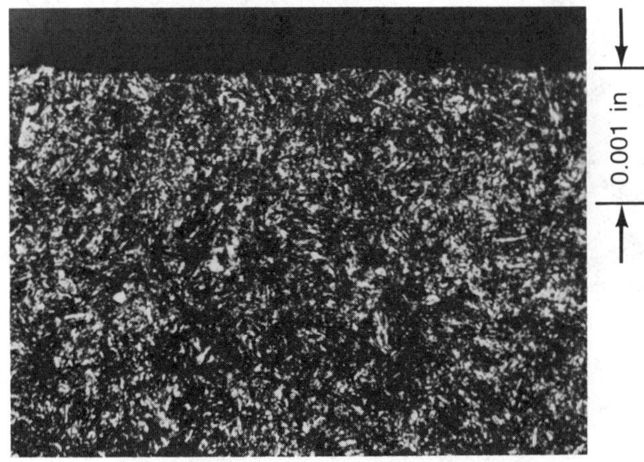


Infeed: 0.010 in/min
Surface roughness: 11 rms 1000X

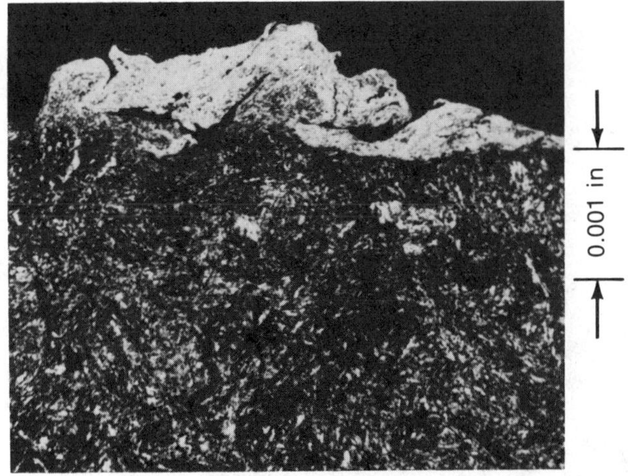


Infeed: 4.9 in/min
Surface roughness: 160 rms 1000X

**Figure 18.3-50** Surfaces produced by plunge grinding 4340 steel (45 $R_C$). Grinding Wheel: 32A54MVBEX113. Wheel Speed: 11,500 fpm. (M. Field[19])

Gentle conditions
Surface roughness: 45 $\mu$in $R_a$

Conventional conditions
Surface roughness: 40 $\mu$in $R_a$

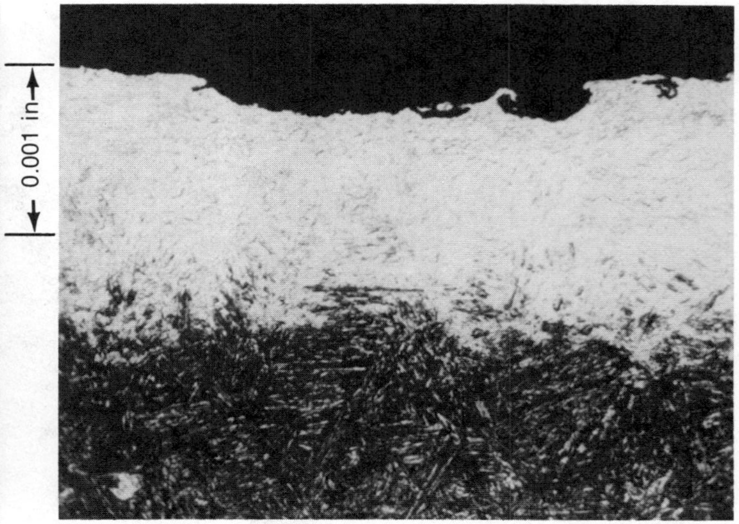

0.001 in

Abusive conditions
Surface roughness: 50 $\mu$in $R_a$

Gentle grinding produced no visible surface alterations. Conventional grinding shows evidence of spotty surface rehardening and underlying overtempering or softening. Abusive grinding produced a rehardened surface layer averaging 0.001 inch deep and an underlying overtempered zone approximately 0.004 inch deep. 1000X

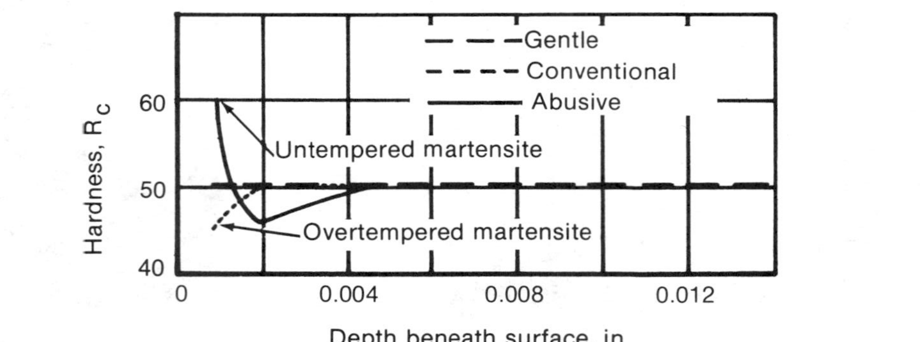


**Figure 18.3–51** Surface characteristics of 4340 steel (quenched and tempered, 50 $R_C$) produced by surface grinding. (See figure 18.3–18 for residual stress values corresponding to these section views and figure 18.3–13 or 18.3–62 for fatigue test results.) (W. P. Koster et al[8])

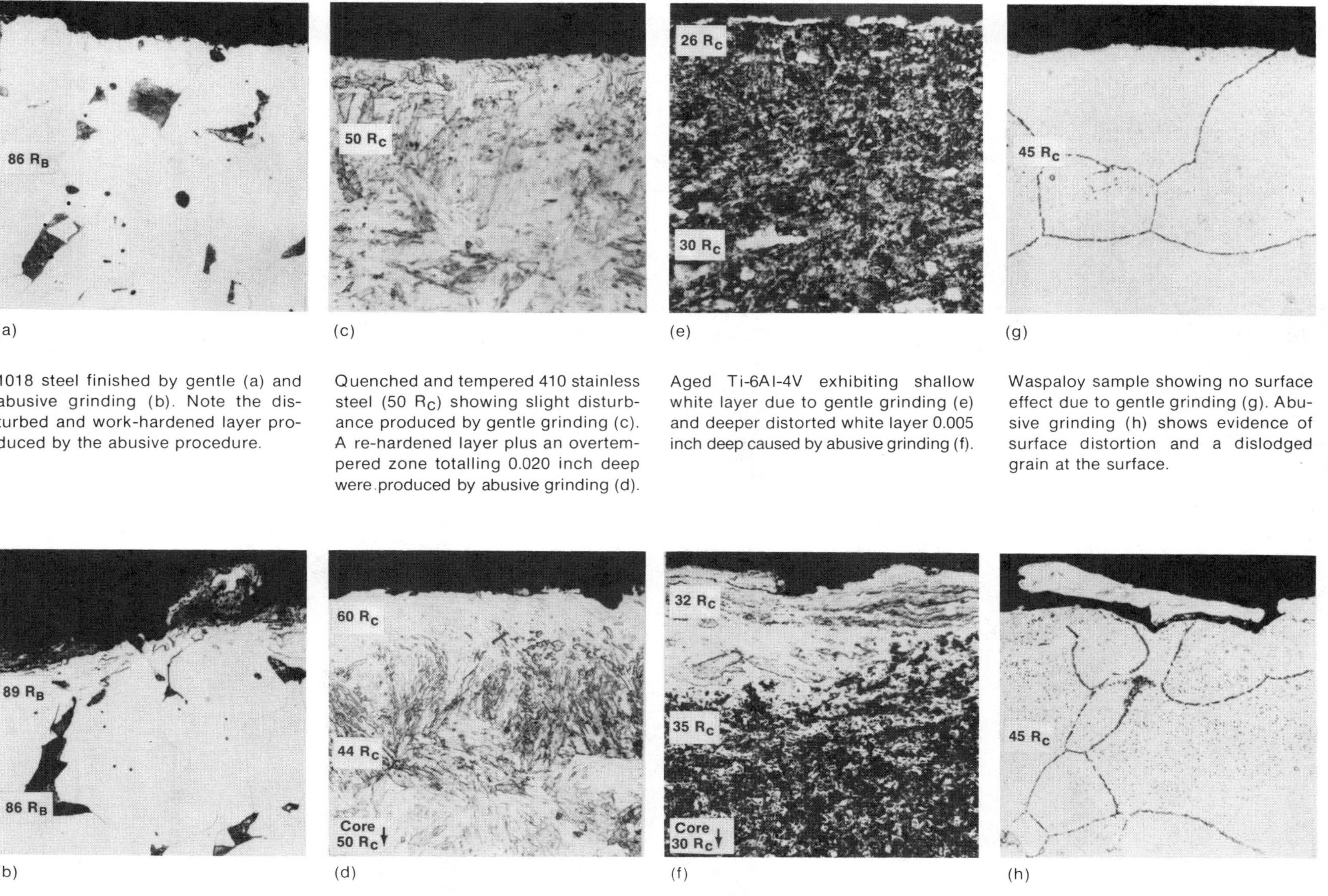


(a) (c) (e) (g)

1018 steel finished by gentle (a) and abusive grinding (b). Note the disturbed and work-hardened layer produced by the abusive procedure.

Quenched and tempered 410 stainless steel (50 $R_C$) showing slight disturbance produced by gentle grinding (c). A re-hardened layer plus an overtempered zone totalling 0.020 inch deep were produced by abusive grinding (d).

Aged Ti-6Al-4V exhibiting shallow white layer due to gentle grinding (e) and deeper distorted white layer 0.005 inch deep caused by abusive grinding (f).

Waspaloy sample showing no surface effect due to gentle grinding (g). Abusive grinding (h) shows evidence of surface distortion and a dislodged grain at the surface.

(b) (d) (f) (h)

**Figure 18.3–52** Surfaces of alloys produced by surface grinding.

Abusively ground — no subsequent heat treatment.

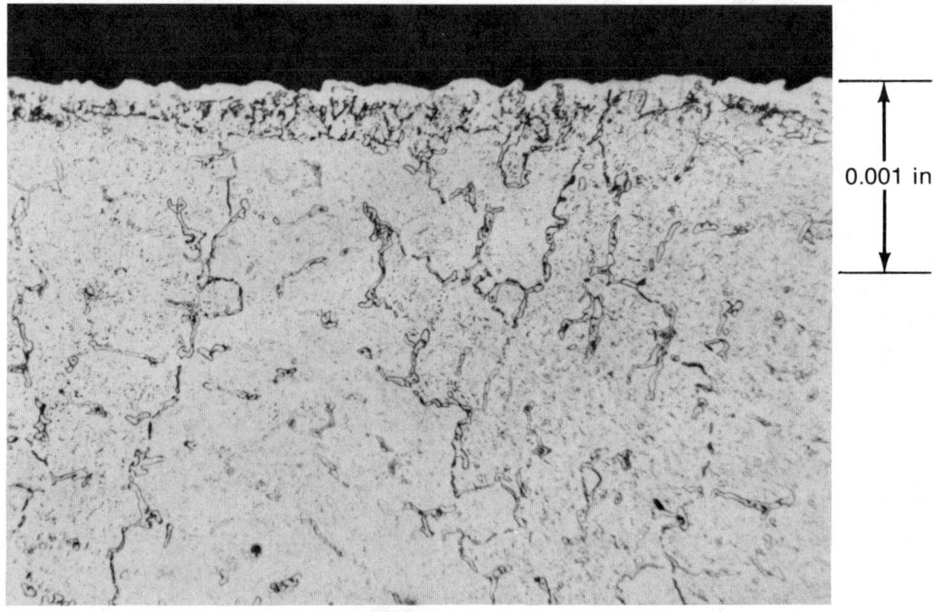


Abusively ground — heat treated, after grinding, at 2000°F for one vacuum. Recrystallized layer at surface.

**Figure 18.3-53** Recrystallization resulting from abusive grinding of HS-31 alloy.

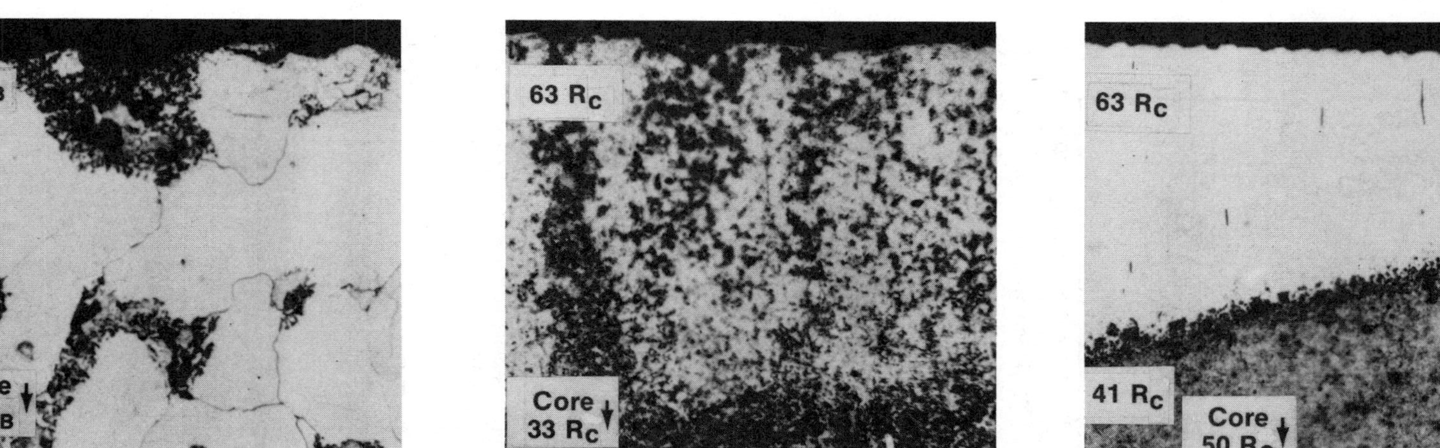


1000X
1018 steel showing partially re-austenitized pearlite which extends to a depth of about 0.005 inch.

100X
Annealed 4340 steel showing re-hardened layer 0.003 inch deep.

100X
Hardened 4340 steel (50 $R_C$) showing re-hardened surface zone and overtempered layer beneath. Total effect reached 0.050 inch.

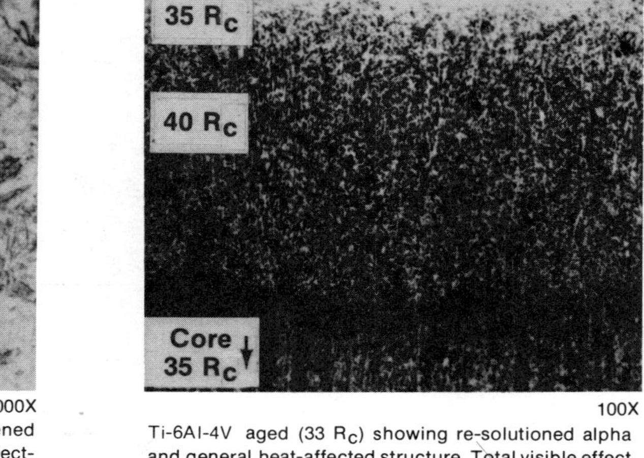


1000X
Hardened 410 stainless (50 $R_C$) showing re-hardened surface and overtempered sub-zone. Total depth affected approached 0.090 inch.

100X
Ti-6Al-4V aged (33 $R_C$) showing re-solutioned alpha and general heat-affected structure. Total visible effect was 0.030 inch.

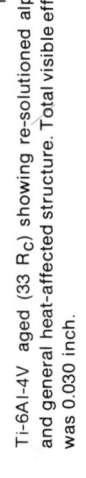

1000X
Waspaloy, which showed little or no microhardness or structural change.

**Figure 18.3–54** Surfaces of alloys produced by an abrasive cutoff operation.

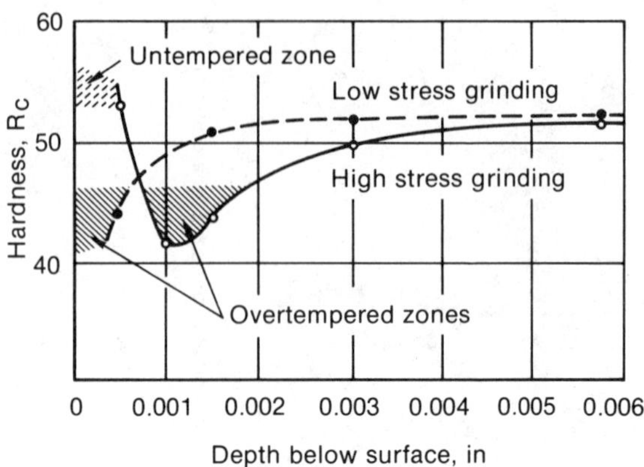


**Figure 18.3–55** Typical hardness alteration traverses of subsurfaces as a result of grinding 4340 steel (quenched and tempered, 53 $R_C$) at low stress and high stress levels. (G. Bellows[13])

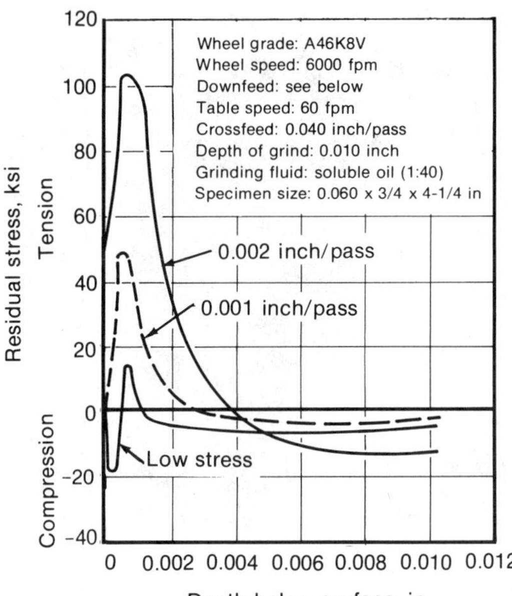


**Figure 18.3–57** Effect of downfeed on residual stress in ground surface of 4340 steel (52 $R_C$). It should be noted that high residual tensile stresses are reduced by grinding at lower wheel speeds and at lower downfeeds. The corresponding distortions (measured at the center of a 0.060 inch by 3/4 inch by 4-1/4 inch specimen) are shown in figures 18.3–59 and 18.3–60. (M. Field, W. P. Koster and J. B. Kohls[15])

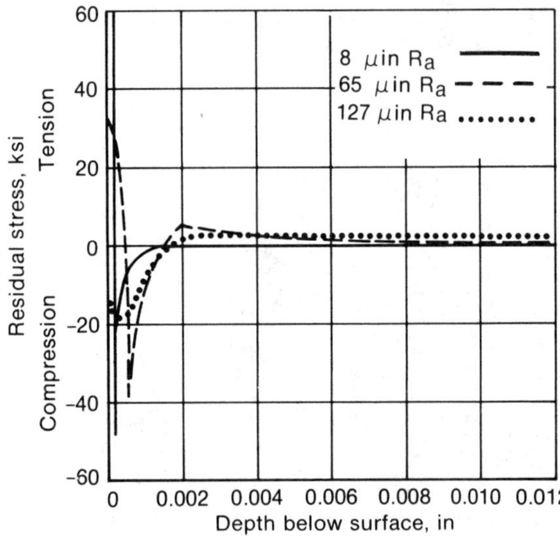


**Figure 18.3–56** Residual stress profiles of 4340 steel (quenched and tempered, 50 $R_C$) for various values of surface roughness. Stress profiles and grinding lay parallel to longitudinal axis of specimens. (G. Bellows[13])

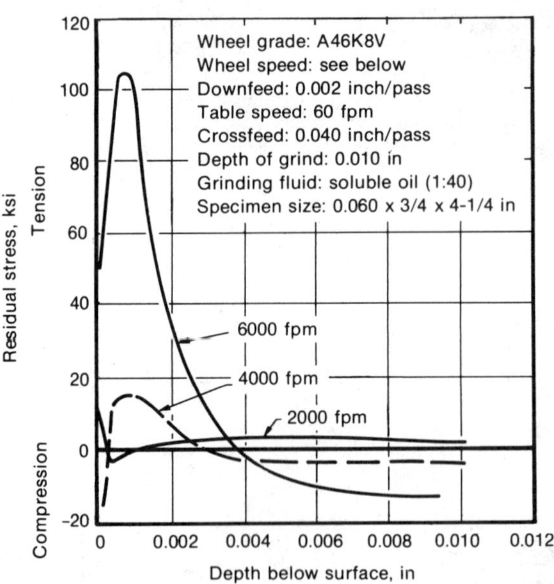


**Figure 18.3–58** Effect of wheel speed on residual stress in ground surface 4340 steel (52 $R_C$). It should be noted that high residual tensile stresses are reduced by grinding at lower wheel speeds and at lower downfeeds. The corresponding distortions (measured at the center of a 0.060 inch by 3/4 inch by 4-1/4 inch specimen) are shown in figures 18.3–59 and 18.3–60. (M. Field, W. P. Koster and J. B. Kohls[15])

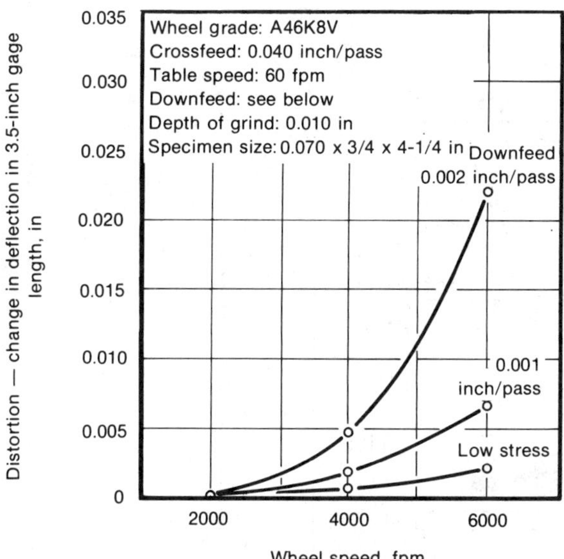


**Figure 18.3-59** Effect of downfeed on change in deflection versus wheel speed when surface grinding 4340 steel (52 $R_C$). Note the low distortion produced by low wheel speed and low stress downfeed. A typical low stress downfeed calls for careful control in removal of the last 0.010 inch in finishing. The first 0.008 inch is taken at 0.0005 inch/pass and the last 0.002 inch at 0.0002 inch/pass. Residual stress values for these tests are shown on figures 18.3-57 and 18.3-58. (M. Field, W. P. Koster and J. B. Kohls[15])

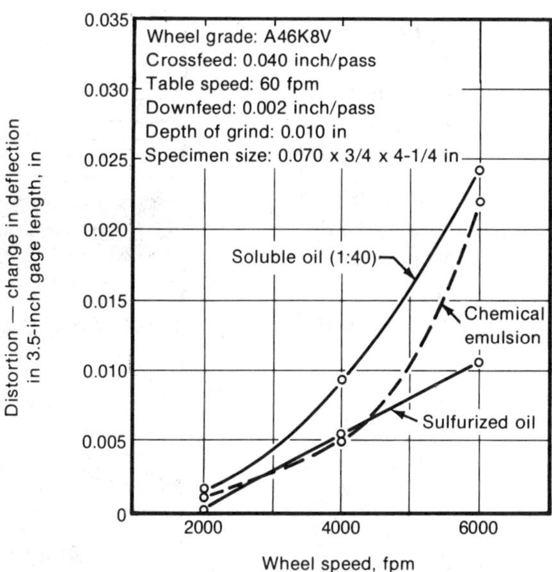


**Figure 18.3-61** Effect of grinding fluid on change in deflection versus wheel speed when surface grinding 4340 steel (52 $R_C$). Distortion is minimized by use of low wheel speeds and also by using sulfurized oil. (M. Field, W. P. Koster and J. B. Kohls[15])

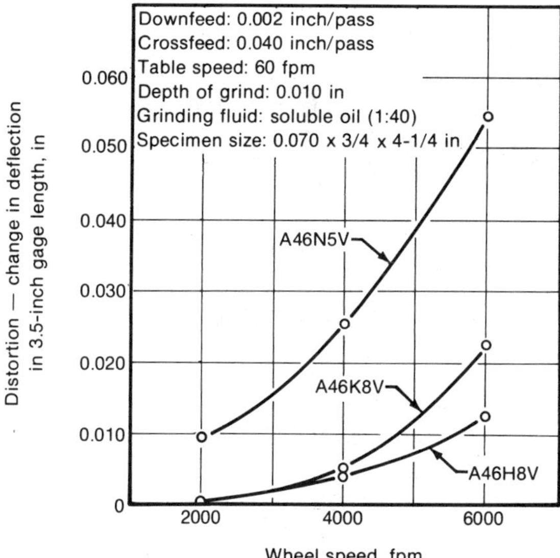


**Figure 18.3-60** Effect of wheel grade on change in deflection versus wheel speed when surface grinding 4340 steel (52 $R_C$). Note that low wheel speeds and soft wheels tend to minimize distortion. Residual stress values for these tests are shown on figures 18.3-57 and 18.3-58. (M. Field, W. P. Koster and J. B. Kohls[15])

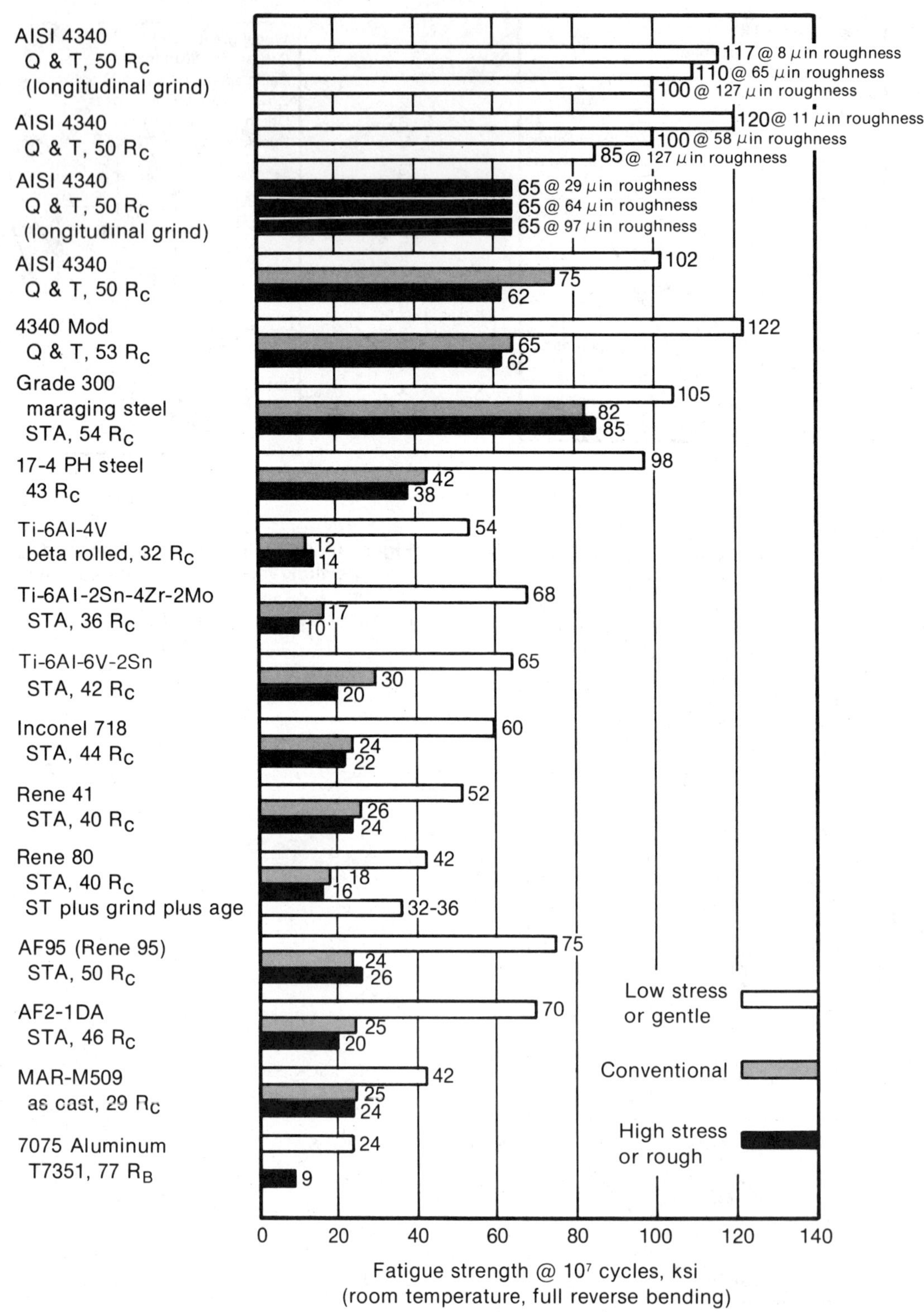


**Figure 18.3–62** Summary of high cycle fatigue strength—surface traverse grinding.

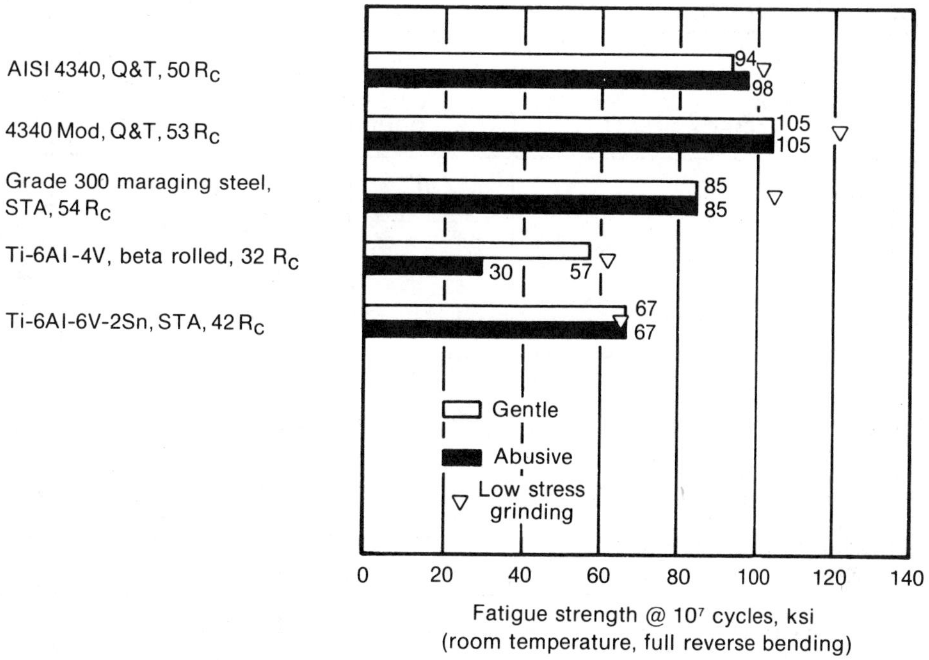

AISI 4340, Q&T, 50 $R_C$

4340 Mod, Q&T, 53 $R_C$

Grade 300 maraging steel, STA, 54 $R_C$

Ti-6Al-4V, beta rolled, 32 $R_C$

Ti-6Al-6V-2Sn, STA, 42 $R_C$

94 98 105 105 85 85 30 57 67 67

Gentle

Abusive

▽ Low stress grinding

0 20 40 60 80 100 120 140

Fatigue strength @ $10^7$ cycles, ksi (room temperature, full reverse bending)

**Figure 18.3–63** Summary of high cycle fatigue strength—hand sanding. Controlled hand grinding or hand sanding generally results in fatigue strength levels only slightly lower than those produced by gentle or low stress grinding. With one exception, the materials studied were not sensitive to variations in hand sanding processes over the range studied. (G. Bellows and W. P. Koster[16])

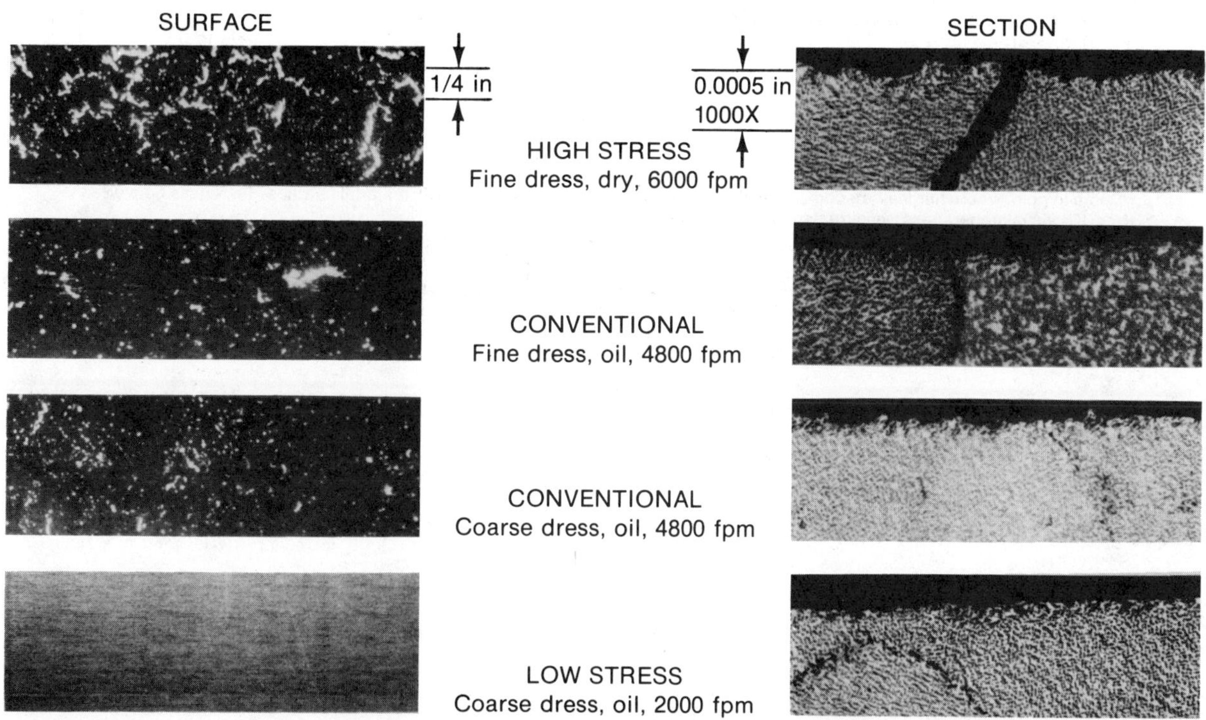


**Figure 18.3–64** Effect of different grinding intensities. *Left*, ultraviolet photographs of fluorescent penetrant inspections of the four surfaces of a bar of cast nickel base alloy ground under four different conditions of surface grinding. *Right*, corresponding section photomicrographs at 1000X. These photos reveal the presence of cracks and pull-out. The low-stress-ground surface is free of microcracks. (Surface Integrity Encyclopedia[1])

**TABLE 18.3-11 Surface Integrity Effects Observed in Abrasive Material Removal Processes**

| PROPERTY AND TYPE OF EFFECT | FINISHING, "GENTLE" OR LOW STRESS CONDITIONS | | ROUGHING, "OFF-STANDARD" OR ABUSIVE CONDITIONS | |
|---|---|---|---|---|
| Surface roughness: | μin $R_a$ | μm $R_a$ | μin $R_a$ | μm $R_a$ |
| Average range | 4 to 63 | 0.1 to 1.6 | 32 to 150 | 0.8 to 3.8 |
| Less frequent range | 1 to 125 | 0.025 to 3.2 | 8 to 250 | 0.2 to 6.3 |
| Mechanical altered material zones: | inch* | mm* | inch* | mm* |
| Plastic deformation (PD) | 0.0003 | 0.0076 | 0.0035 | 0.09 |
| Plastically deformed debris ($PD^2$) | 0.0005 | 0.013 | 0.0013 | 0.03 |
| Hardness alteration† | 0.0015 | 0.04 | 0.0100 | 0.25 |
| Microcracks or Macrocracks | 0.0005 | 0.013 | 0.0090 | 0.23 |
| Residual stress† | 0.0005 | 0.013 | 0.0125 | 0.32 |
| Metallurgical altered material zones: | inch* | mm* | inch* | mm* |
| Recrystallization | 0.0005 | 0.013 | — | — |
| Intergranular attack (IGA) | *** | *** | *** | *** |
| Selective etch, pits, protuberances | 0.0002 | 0.005 | 0.0004 | 0.010 |
| Metallurgical transformations | 0.0005 | 0.013 | 0.0060 | 0.15 |
| Heat-affected zone (HAZ) or Recast layer | 0.0007 | 0.018 | 0.0500 | 1.27 |
| High cycle fatigue (HCF): | percent | | percent | |
| Variation from "handbook" values at room temperature§ | 0 to +62 | | —95 to 0 | |

SOURCE: G. Bellows and J. B. Kohls.[20]

NOTE: A dash (—) in the table indicates no or insufficient data. A triple asterisk (***) in the table indicates no occurrences or not expected.

*Maximum observed depth normal to the surface.

†Depth to point where hardness becomes less than ±2 points $R_c$ (or equivalent) of bulk material hardness (hardness converted from Knoop microhardness measurements).

‡Depth to point where residual stress becomes and remains less than 20 ksi [138 MPa] or 10% of tensile strength, whichever is greater.

§Handbook values from HCF testing frequently are generated from specimens that have been processed using low stress grinding, hand or gentle machine polishing or occasionally electropolishing. For this table, "handbook" values represent HCF strength values generated from low-stress-ground specimens (with their minor amount of retained but enhancing compressive residual stress). The values are the $10^7$ cycle strengths from full-reverse bending tests at room temperature.

**TABLE 18.3-12 Low Stress Grinding Procedures**

| GRINDING PARAMETERS | STEELS AND NICKEL BASE HIGH TEMPERATURE ALLOYS | | TITANIUM | |
|---|---|---|---|---|
| | English | Metric | English | Metric |
| **SURFACE GRINDING** | | | | |
| Wheel | A46HV | A46HV | C60HV | C60HV |
| Wheel speed | 2500 to 3000 fpm* | 13 to 15 m/s* | 2000 to 3000 fpm* | 10 to 15 m/s* |
| Downfeed per pass | 0.0002 to 0.0005 inch† | 0.005 to 0.013 mm† | 0.0002 to 0.0005 inch† | 0.005 to 0.013 mm† |
| Table speed | 40 to 100 fpm‡ | 12 to 30 m/min‡ | 40 to 100 fpm‡ | 12 to 30 m/min‡ |
| Crossfeed per pass | 0.040 to 0.050 inch | 1 to 1.25 mm | 0.040 to 0.050 inch | 1 to 1.25 mm |
| Grinding fluid | Highly sulfurized oil | Highly sulfurized oil | See section 16.3, Codes 81 and 83 | See section 16.3 Codes 81 and 83 |
| **TRAVERSE CYLINDRICAL GRINDING** | | | | |
| Wheel | A60IV | A60IV | C60HV | C60HV |
| Wheel speed | 2500 to 3000 fpm* | 13 to 15 m/s* | 2000 to 3000 fpm* | 10 to 15 m/s* |
| Infeed per pass | 0.0002 to 0.0005 inch† | 0.005 to 0.13 mm† | 0.0002 to 0.0005 inch† | 0.005 to 0.013 mm† |
| Work speed | 70 to 100 fpm‡ | 20 to 30 m/min‡ | 70 to 100 fpm‡ | 20 to 30 m/min‡ |
| Grinding fluid | Highly sulfurized oil | Highly sulfurized oil | See section 16.3, Codes 81 and 83 | See section 16.3, Codes 81 and 83 |

NOTE: For a wide variety of metals (including high strength steels, high temperature alloys, titanium and refractory alloys), low stress grinding practices develop very low residual tensile stresses. In some materials the residual stress produced near the surface is actually in compression instead of tension.

*Low stress grinding requires wheel speeds lower than the conventional 6000 feet per minute [30 meters per second]. In order to apply low stress grinding, it would be preferable to have a variable speed grinder. Since most grinding machines do not have wheel speed control, it is necessary to add a variable speed drive or make pulley modifications.

†Downfeeds or infeeds in the range of 0.0002 to 0.0005 inch per pass [0.005 to 0.013 mm/pass] have been found satisfactory for steels, nickel base high temperature alloys, and titanium alloys. A typical feed schedule calls for removing the last 0.010 inch [0.254 mm] of stock as follows: remove 0.008 inch [0.2 mm] at 0.0005 inch per pass [0.013 mm/pass] and remove the last 0.002 inch [0.05 mm] at 0.0002 inch per pass [0.005 mm/pass].

‡Increased work speeds even above those indicated are considered to be advantageous toward improving surface integrity.

**TABLE 18.3-13 Process Parameter Guidelines for Low Stress Grinding Results**

| PROCESS PARAMETERS | GUIDELINES |
|---|---|
| 1. Grinding wheel dressing technique | Frequent and coarse to maintain sharpness. Maintain sharpness of dressing tool. Avoid dwell in using crush, roll, or single-point dressing tools. |
| 2. Wheel speed | Low, under 3,500 fpm [18 m/s]. |
| 3. Downfeed (or infeed) rate | 0.0002 to 0.0005 inch/pass [0.005 to 0.013 mm/pass] with programmed reduction from conventional rates. |
| 4. Grinding fluid | Oil-base fluid is preferred. |
| 5. Wheel classification | Soft grade (G, H or I*). Open structure (6 or more). Grain size (60 or coarser). |
| 6. Table (workpiece) speed | High, 50 fpm [15 m/min] and up. Crossfeed is preferred to plunge motion. |
| 7. Grinding fluid flow control | Adequate to high fluid flow. Assure placement of fluid between wheel and workpiece. Flow controlling nozzle design. Reduce air film on wheel. |

SOURCE: G. Bellows.[13]

NOTE: Process parameter guidelines are listed in descending order of significance to low residual stress generation in the workpiece surface. A machine and setup with good rigidity, freedom from vibration or chatter and well maintained with fine cleanliness are also an aid to grinding performance.

*Cylindrical grinding frequently requires use of harder wheels (with J grade prevalent); however, the other parameters must be selected to compensate for this extra hardness.

**TABLE 18.3-14 Typical Infeed Schedule for Attaining Low Stress Surfaces by Grinding**

| STEP | WORKPIECE STOCK REMOVAL* | INFEED RATE | |
|---|---|---|---|
| | | inch/pass | mm/pass |
| Rough Grinding | From raw material dimension to within 0.010 inch [0.25 mm] of desired final dimension | up to 0.002† | up to 0.050† |
| Dress wheel | | | |
| Semi-finish Grinding | From 0.010 inch [0.25 mm] to 0.001 inch [0.025 mm] of final size | 0.0003 to 0.0006 | 0.008 to 0.015 |
| Dress wheel | Optional | | |
| Finish Grinding (to low stress levels) | From 0.001 inch [0.025 mm] to final size | 0.0001 to 0.0002 | 0.0025 to 0.005 |
| | Optional: at final size, with continued application of grinding fluid | 2 to 4 spark-out passes | |

SOURCE: G. Bellows.[13]

*Data applies to the radius for cylindrical grinding and to the depth for surface grinding.

†This roughing rate should be such that roughness, tears, cracks, or microcracks in the surface are of lesser magnitude than the material remaining to be removed by the finishing and semi-finishing cuts.

**TABLE 18.3-15 Influence of LSG Parameters on Production Rates**

| RANKING IN INFLUENCE ON RESIDUAL STRESS | DIRECT INFLUENCE ON MATERIAL REMOVAL RATE AND PRODUCTION TIME | MINOR INFLUENCE ON PRODUCTIVITY | NO INFLUENCE ON PRODUCTION RATE |
|---|---|---|---|
| 1 (most)<br>2<br>3 | Infeed rate | Dressing | Wheel speed* |
| 4<br>5<br>6 | Table speed | Wheel grade | Lubricant |
| 7<br>8<br>9 (least) | | | Machine type and quality<br>Rigidity of setup<br>Coolant nozzle design |

SOURCE: G. Bellows.[13]

NOTE: It is possible to achieve the "best of both worlds"—high surface integrity *and* good productivity—if a careful selection among the LSG parameters is made.

*At very low sheel speeds, a wheel acts like a softer wheel and needs more frequent dressing and thereby has a minor influence on production rates.

**TABLE 18.3-16 Low Stress Grinding Troubleshooter Checklist**

1. Check that **planned machine settings** are being used.
2. Is **wheel grade** correct?
3. Check for correctness and freshness of wheel **dressing.**
4. Check for changes in **workpiece material** or its metallurgical state.
5. Are set values of **feed and speed** being attained and is **infeed program** of reductions being followed?
6. Is **grinding fluid** correct, clean and inserted properly?
7. Check fixtures and machine rigidity.
8. Is workpiece clean and initially free from cracks?
9. Check inspection techniques.

SOURCE: G. Bellows.[13]

**TABLE 18.3-17 Grinding Parameters for IN-100 Turbine Blades (Case A resulted in cracks; Case B resulted in no cracks)**

| GRINDING PARAMETERS | CASE A (Cracks) | CASE B (No Cracks) |
|---|---|---|
| Wheel | 38A10018VBE | 38A8018VBE |
| Wheel speed | 5,500 fpm | 2,800 fpm |
| Table speed | 20 fpm | 20 fpm |
| Infeed per pass | 0.004 inch | 0.002 inch |
| Fluid | Sulfo-chlorinated oil | Highly sulfurized oil |
| Grinding cycle | Rough: 0.060 inch at 0.004 inch/pass, dress. Leave 0.100 inch/side for finish operation.<br><br>Finish: 0.012 inch from finish size (3 passes at 0.004 inch/pass; 2 sparkout passes). | Dress: feed 0.060 inch at 0.002 inch/pass;<br><br>Dress: feed 0.060 inch at 0.002 inch/pass;<br><br>Dress: feed 0.010 inch at 0.002 inch/pass to finish size. |

**TABLE 18.3-18 Fatigue Strength Variation from Different Grinding Conditions**

| MATERIAL | FATIGUE STRENGTH* | | | | | |
|---|---|---|---|---|---|---|
| | Gentle or Low Stress Grinding | | Conventional Grinding | | Abusive Grinding | |
| | ksi | MPa | ksi | MPa | ksi | MPa |
| High Strength Steels: | | | | | | |
| 4340 (Q & T, 50 $R_c$) | 102 | 703 | 70 | 483 | 62 | 427 |
| 4340 Mod. (Q & T, 53 $R_c$) | 122 | 841 | 65 | 448 | 62 | 427 |
| Maraging steel, 300 Grade (STA, 54 $R_c$) | 105 | 724 | 82 | 565 | 85 | 586 |
| High Temperature Nickel Base Alloys: | | | | | | |
| Inconel Alloy 718 (STA, 44 $R_c$) | 60 | 414 | 24 | 165 | – | – |
| AF 95 (STA, 50 $R_c$) | 75 | 517 | 24 | 165 | 26 | 179 |
| AF2-1DA (STA, 46 $R_c$) | 70 | 483 | 25 | 172 | 20 | 138 |
| Rene 80 (STA, 40 $R_c$) | 42 | 290 | 18 | 124 | 16 | 110 |
| Titanium Alloys: | | | | | | |
| Ti-6Al-4V (Beta rolled, 32 $R_c$) | 62 | 427 | – | – | 13 | 90 |
| Ti-6Al-6V-2Sn (STA, 42 $R_c$) | 65 | 448 | 30 | 207 | 20 | 138 |
| Ti-6Al-2Sn-4Zr-2Mo (STA, 36 $R_c$) | 68 | 469 | 17 | 117 | 10 | 69 |

*Room temperature, $10^7$ cycles full-reverse bending.

**TABLE 18.3-19 Low Cycle Fatigue Test Data From Surface-Ground Materials**

| MATERIAL | GRINDING CONDITION | STRESS FOR FAILURE AT 20,000 CYCLES* | |
|---|---|---|---|
| | | ksi | MPa |
| 4340 (Q & T, 50 $R_c$) | Gentle surface grind | 165 | 1,138 |
| | Abusive surface grind | 142 | 979 |
| Ti-6Al-4V (Beta rolled, 32 $R_c$) | Gentle surface grind | 92 | 634 |
| | Abusive surface grind | 71 | 490 |
| Inconel alloy 718 (STA, 44 $R_c$) | Gentle surface grind | 160 | 1,103 |
| | Abusive surface grind | 127 | 876 |
| Rene 80 (STA, 40 $R_c$) | Gentle surface grind | 55 | 379 |
| | Abusive surface grind | 45 | 310 |

*At room temperature.

### Specific Guidelines and Data for ELECTRICAL Material Removal Processes

The molecule by molecule dissolution of the electrically conductive workpiece in electrochemical machining (ECM) is a forceless material removal that introduces no residual stresses into the workpiece. The surface texture is smooth, has no lay pattern, and can approach a metallograph-polish caliber even to revealing individual grain patterns. The absence of beneficial residual stress and/or cold working in the surface frequently results in high cycle fatigue strengths less than that produced with conventionally prepared surfaces. If needed, the addition of a post-ECM process such as shot peening or rolling can add a beneficial compressive residual stress. Some slight softening of the surface has been observed in some alloy-electrolyte combinations. The key to good ECM surfaces is a careful match of the electrolyte composition with the metallurgical state of the workpiece material plus careful control of the operating parameters, especially the current density. Selective etching, intergranular attack (IGA) or pitting may occur in the low current density areas adjacent to the main electrode high current density areas. Static material properties have not been found to be affected by ECM. Hydrogen embrittlement has also not been attributed to ECM (the hydrogen ions form at the cathode and are in the molecular state before reaching the anodic workpiece surface).

The charts and guidelines that follow have baselines the same as those for the mechanical material removal processes. Figures 18.3-65 to 18.3-73 illustrate good or poor surface effects from the electrochemical processes. Microhardness examples are shown in figures 18.3-67, 18.3-71 and 18.3-72, and residual stress patterns are shown in figures 18.3-66 and 18.3-72. High cycle fatigue strength summary data are shown in figure 18.3-74. Table 18.3-20 is a summary of the surface integrity effects observed during the examination of several thousand pieces of data. Other electrical machining illustrations are figures 18.3-5, 18.3-15, and 18.3-17.

**Electrical Process Guidelines that Promote Good Surface Integrity Results**

1. Surface roughness standards should be reassessed when applying ECM because ECM produces unusual surface textures with no lay pattern.

2. Well controlled ECM does not induce any stresses into the surface.

3. High current densities are desirable in the cutting gap for good finishes and rapid metal removal.

4. The current density in the workpiece should be carefully planned to prevent overheating in the material.

5. Relentless process controls should be supplemented with periodic metallographic checks from representative surfaces.

6. The work material heat treat state should be precisely known and controlled to assure the best repeatability of ECM surfaces.

7. Contact between the electrolyte and workpiece without current flow should be minimized.

8. Localized overheating of the workpiece as a result of poor connections or short circuits should be carefully examined for the extent of any damage. Removal of the surface discoloration is insufficient, as demonstrated in figure 18.3-75.

9. Careful and complete rinsing of the electrolyte from the workpiece is essential. Ultrasonic agitation of the cleaning bath is helpful.

10. Selective etching, excessive roughness, intergranular attack or pitting are evidences of less than optimum operating conditions or tooling practices. These conditions can occur on surfaces subjected to low current density in the presence of the electrolyte or in areas adjacent to the main electrode cutting face.

11. Room temperature fatigue endurance strength is generally lower as compared to that of low stress conventionally ground specimens and represents the unblemished or unaffected material properties state. The lack of any residual stress in electrochemically machined surfaces can account for the differences observed in fatigue strength.

12. The use of a post-ECM shot peening or rolling treatment may be desirable to enhance component fatigue strength.

13. Anomalies in the workpiece surface or microvariations in material composition can influence the surface quality and may be exposed by ECM.

14. Some surface softening will occur in many, but not all, materials.

15. An increase in roughness of the ECM surface can be a significant alert to the departure from optimum or preplanned operating conditions.

16. Hydrogen embrittlement is not attributable to ECM.

**Special Comments Concerning Surface Integrity Practices for Electrical Machining Processes**

**High current densities produce improved finishes and high material removal rates.** In sharp contrast to conventional machining, ECM finishes improve with the higher material removal rates that accompany high current densities in the main electrode working gap. Rates of electrode advance of 0.001 to 0.005 inch per second [0.025 to 0.127 mm/s] and current densities of 500 to 1500 amperes per square inch [78 to 232 A/cm$^2$] are frequent. The upper limit for current densities is usually the internal, resistive heating in the workpiece. Inaccuracies or distorting forces can accompany this internal heating. Poor electrical connections or clamps can produce highly localized hot spots that can involve temperatures to the point that microstructural alteration occurs, as shown in figure 18.3-75.

**Surface roughness can be a quality control guideline.** The usual fine ECM finish, if tracked or plotted, sequencially and frequently can be a good quality control technique. A gradually changing roughness usually accompanies a slow change in ECM operating parameters; for example, rising sludge content, electrolyte temperature decline, etc. A sudden change in roughness frequently can be traced to a change in the metallurgical state of the workpiece material; for example, a new lot of stock.

**Areas of low current density should be suspects for intergranular attack or selective etching.** At increasing distances from the principal electrode cutting face, the voltage, current and current density all decline and with them the cutting rates. At specific low values, pitting, intergranular attack, selective etching or anodic films can occur. Insulation of the sides of the electrode can concentrate the electrolytic action. Masks on the workpiece surface are only partially effective in reducing these undesirable effects because a groove of removed material will frequently appear at the edge of such masks. These stray etching areas usually have so little material removal activity that dimensional problems are nonexistent and the cosmetic appearance effects can be confined to the anodic film or "blush" that is readily removed.

**Random pitting in workpiece surfaces adjacent to ECM'd areas can result from material anomalies.** Occasionally groups or single pits will appear on the workpiece in areas adjacent to the ECM'd area. The pitted areas have been exposed to low current density from free flowing electrolyte or electrified mists. The electrolytic action concentrates in the anomalies in the work material, thus generating a pit or ring- shaped depression (around a less conductive inhomogeneity in the material). Some use has been made of this selective cutting tendency as an inspection aid to assess the surface uniformity of material stock.

**Control of the metallurgical state of the work material is needed.** The proper selection of electrolyte, current density and the other operating parameters must include consideration of the heat treat or metallurgical state of the material. While electrolytic machining will continue at almost the same rate regardless of the material hardness, the presence or absence of IGA can depend on the heat treat state. Many alloys will be free of IGA when ECM'd in the hardened state; some, however, must be ECM'd in the annealed state to be free of IGA. A simple metallographic section of a test piece or coupon is good quality assurance. For example, the massive IGA shown in figure 18.3-68 disappeared completely when the M252 alloy was electrochemically machined in the annealed state.

**Accidental short circuits should be carefully examined for subsurface damage.** Modern short circuit protection devices have rapid enough action to protect surfaces from severe damage (and in some cases show only barely detectable pin pricks). Older, slower devices have been known to produce microcracking as deep as 0.10 inch [2.54 mm]. Careful examination, removal of damaged areas or reworking can protect the workpiece surface integrity. Occasionally, overheated areas will show only as a light surface discoloration. While the appearance is easily restored, cases have been recorded of nuggets of metallurgically altered material of considerable extent. Figure 18.3-75 illustrates one such nugget that occurred from overheating a Ti-6Al-4V alloy by concentrating the current flow under a connector as a result of a loose adjacent connector.

**ECM produces unblemished surfaces.** With the lack of mechanical contact with the workpiece, the gentle molecule-by-molecule dissolution of the material, and the minute surface chemistry effects, ECM produces a nearly unblemished surface. From the standpoint of assessment of the pure, uncontaminated or unaffected material properties, specimens prepared by well controlled ECM are unexcelled. Even polishing with worn crocus cloth can have measurable impact on high cycle fatigue values.

**Mechanical material removal action in electrochemical grinding (ECG) should be controlled to minimum values.** The abrasive action of the electrically conductive wheel used in ECG is intended for removal of the anodic films formed on workpiece surfaces. It is *not* used for significant material removal. The principal material removal action is still electrolytic. Too heavy abrasive action can result in short circuits, abrasive burns or too much residual stress, as in conventional grinding. Less than 5 percent, and frequently as little as 1 percent, of the material removed comes from the mechanical action. One rule of thumb states that no more than two light scratches from abrasive action should be seen in each square inch of surface. The primary purpose of the abrasive action is to provide a clean nascent surface so that the electrolytic action can proceed at the most rapid pace consistent with overheating the workpiece, wheel or electrolyte. With good electrolyte flow control and good operating parameters, ECG will result in stress free and unexcelled surface integrity, as is demonstrated in figure 18.3-73.

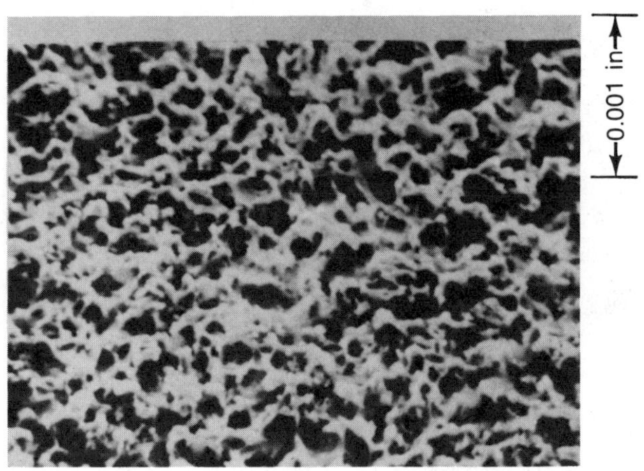


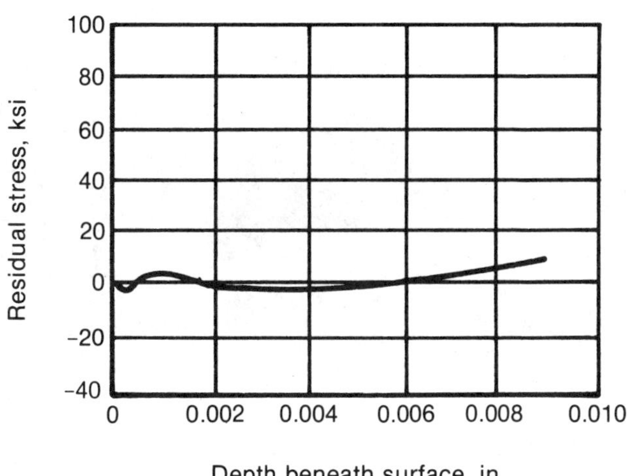


**Figure 18.3-66** Typical residual stress profile from ECM (17-4 PH). (G. Bellows and J. B. Kohls[20])

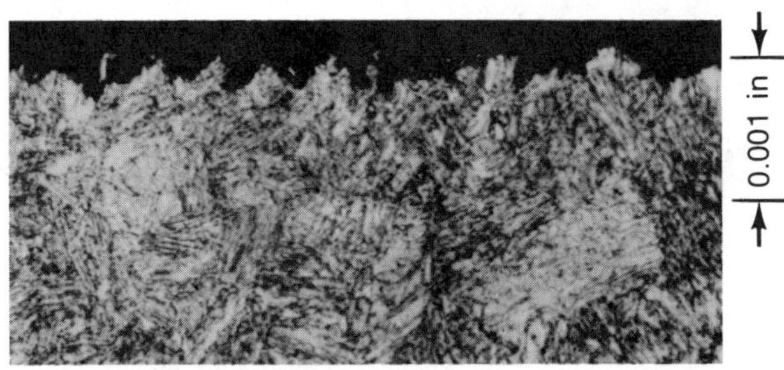


**Figure 18.3-65** Typical surface characteristics of 17-4 PH produced by electrochemical machining. *Top*, SEM of surface at 45° and 1000X; *bottom*, section at 1000X. Surface roughness is 29 microinches $R_a$; $R_t$ = 152 microinches. (Surface Integrity Encyclopedia[1])

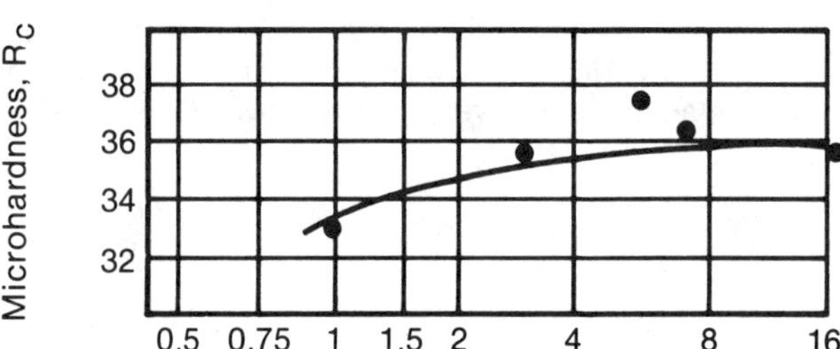


**Figure 18.3-67** Typical microhardness traverse from ECM showing surface softening or Rebinder effect (17-4 PH). (G. Bellows and J. B. Kohls[20])

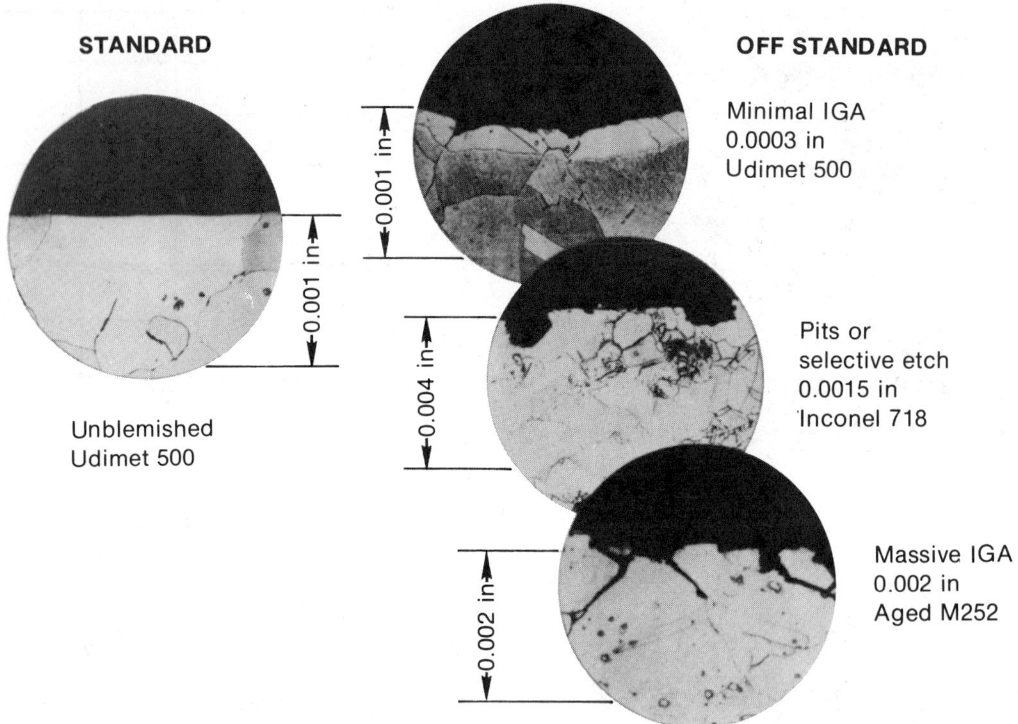


**Figure 18.3–68** Comparison of surface sections of "standard" and "off-standard" electrochemical machining of several nickel alloys. (G. Bellows[21])

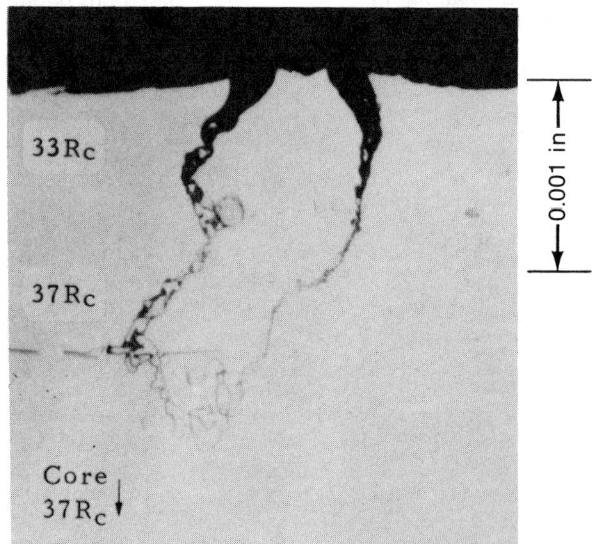


**Figure 18.3–69** Electrochemical machining of cast Rene 41 (aged, 47 $R_C$). Photomicrograph shows a surface having a pronounced tendency toward intergranular attack during an ECM process. Surface alterations, such as shown, are difficult to detect, but they are detrimental and can be avoided through the use of proper process parameters and exacting control of ECM conditions. (W. P. Koster et al[8])

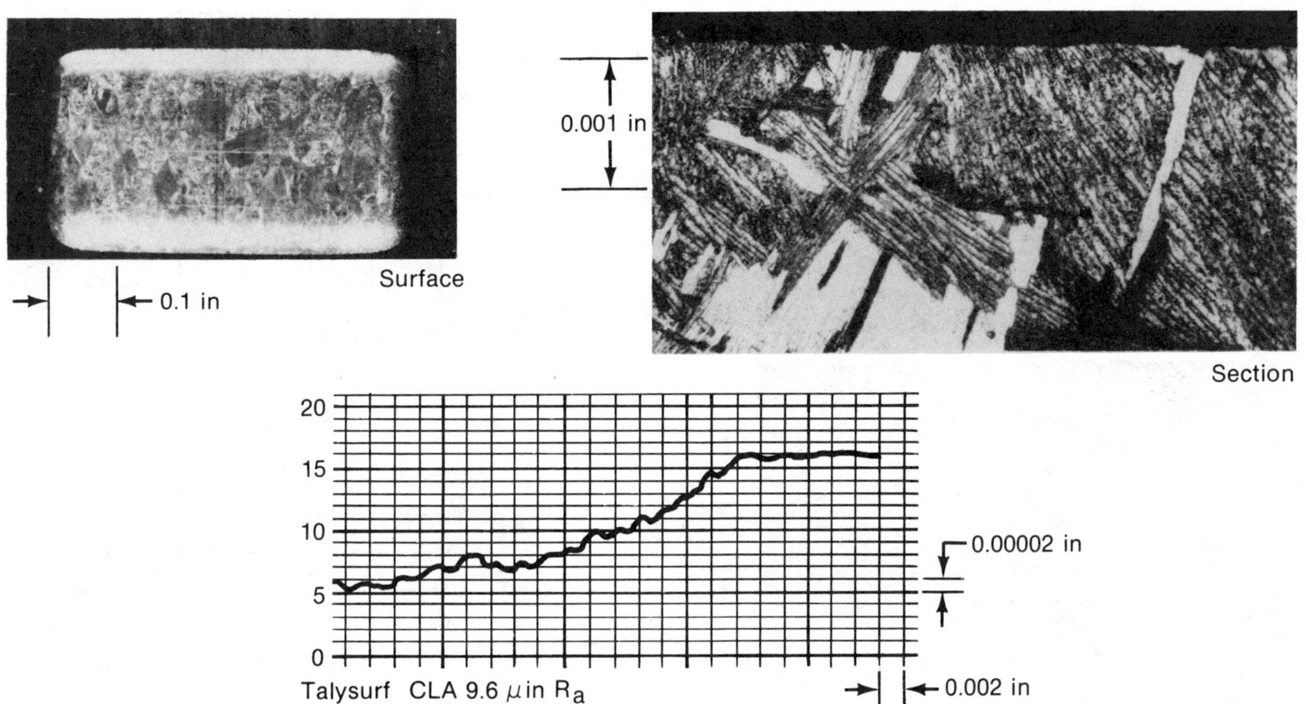


**Figure 18.3–70** ECM of Ti-6Al-4V (beta processed). Surface view at 5X, section at 1000X and stylus trace of well performed ECM with NaCl electrolyte. (Surface Integrity Encyclopedia[1])

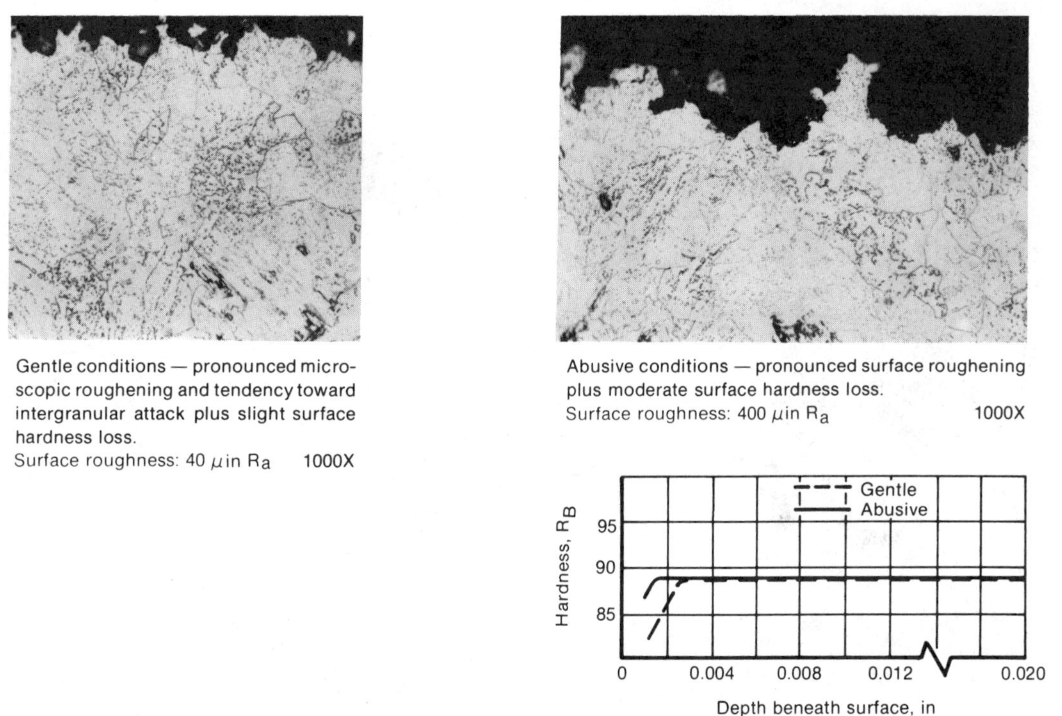

Gentle conditions — pronounced microscopic roughening and tendency toward intergranular attack plus slight surface hardness loss.
Surface roughness: 40 μin Ra 1000X
Abusive conditions — pronounced surface roughening plus moderate surface hardness loss.
Surface roughness: 400 μin Ra 1000X
Gentle
Abusive
Hardness, RB
95
90
85
0
0.004
0.008
0.012
0.020
Depth beneath surface, in


**Figure 18.3–71** Surface characteristics of type 410 stainless steel (annealed, 89 $R_B$) produced by ECM. (W. P. Koster et al[8])

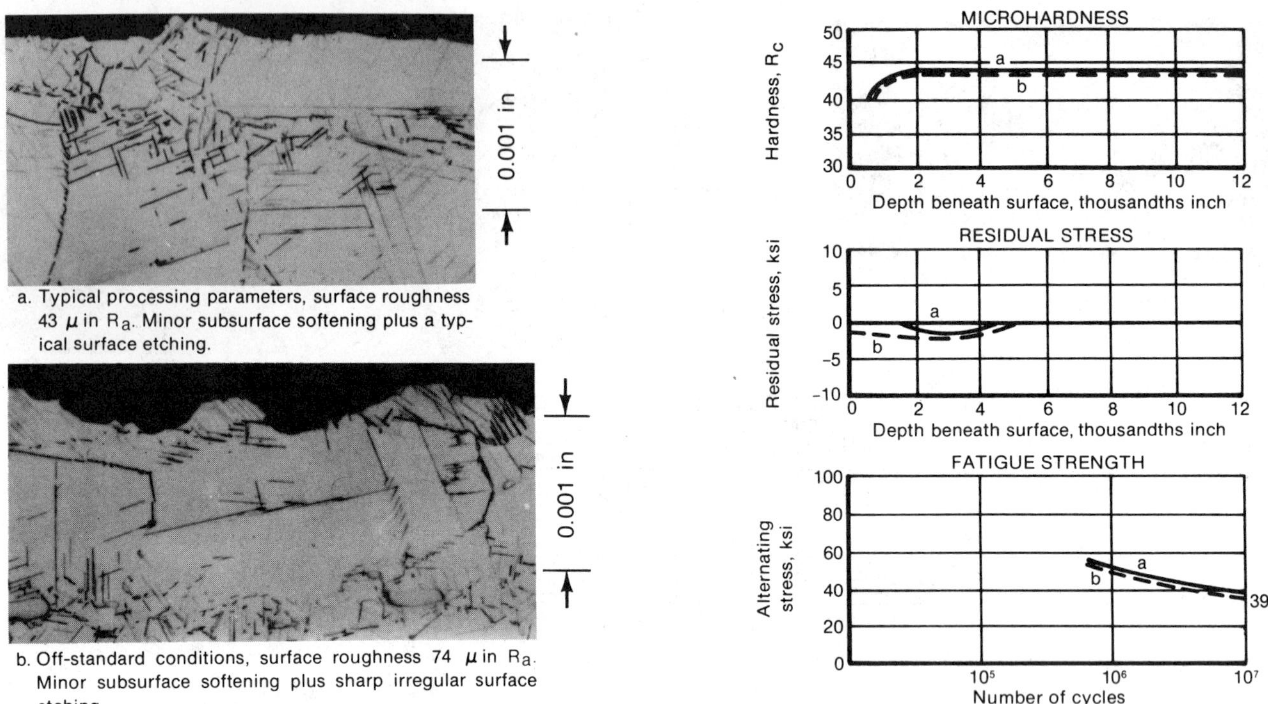


**Figure 18.3–72** Standard surface integrity data set for ECM of Inconel Alloy 718 (solution treated and aged, 44 $R_C$). (Surface Integrity Encyclopedia[1])

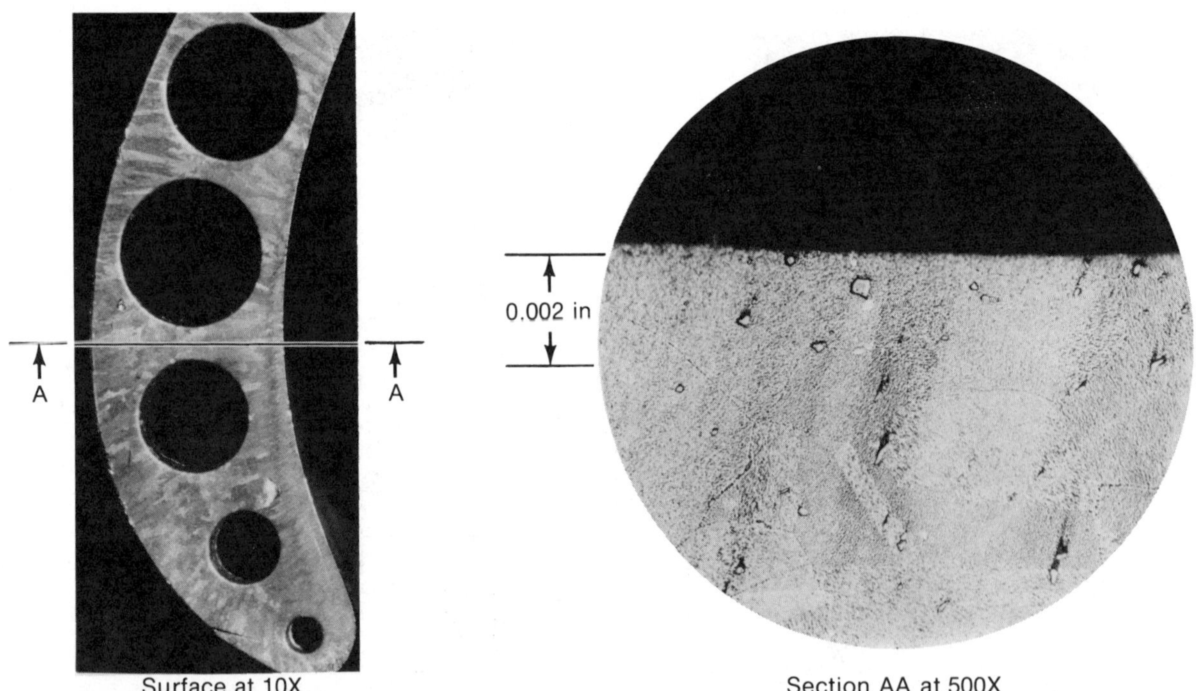


**Figure 18.3–73** Surface characteristics of wrought Udimet 700 turbine airfoil tip produced by ECG. Note the nearly metallographic polish and the absence of scratches in the surface view and unblemished cross section. (Surface Integrity Encyclopedia[1])

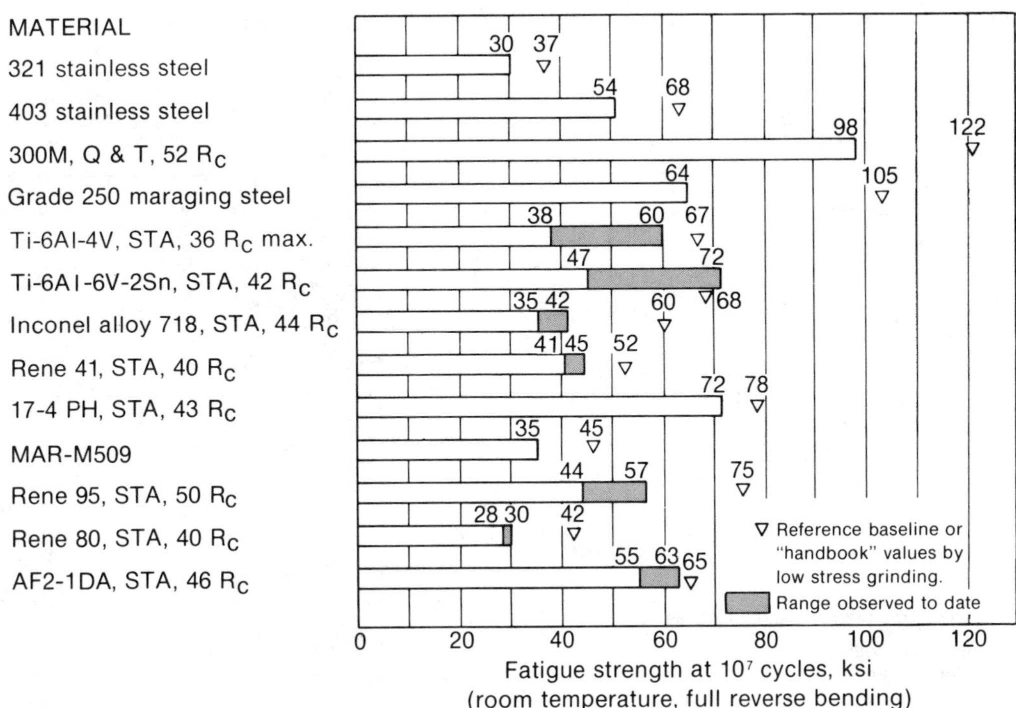


**Figure 18.3–74** Summary of high cycle fatigue strengths from electrochemical machining. Note: In the observed ranges, the lower values most generally represent the "off-standard" operating conditions, while the upper values are representative of "standard" or preplanned conditions.

**Figure 18.3–75** Metallurgical alteration below the ECM surface of Ti-6Al-4V caused by current concentration under an electrical connection adjacent to a poor (or high resistance) electrical connection, 40X. (G. Bellows[21])

**TABLE 18.3-20 Surface Integrity Effects Observed in Electrical Material Removal Processes**

| PROPERTY AND TYPE OF EFFECT | FINISHING, "GENTLE" OR LOW STRESS CONDITIONS | | ROUGHING, "OFF-STANDARD" OR ABUSIVE CONDITIONS | |
|---|---|---|---|---|
| Surface roughness: | μin $R_a$ | μm $R_a$ | μin $R_a$ | μm $R_a$ |
| Average range | 8 to 32 | 0.2 to 0.8 | 63 to 250 | 1.6 to 6.3 |
| Less frequent range | 2 to 125 | 0.05 to 3.2 | 8 to 500 | 0.2 to 12.7 |
| Mechanical altered material zones: | inch* | mm* | inch* | mm* |
| Plastic deformation (PD) | — | — | — | — |
| Plastically deformed debris ($PD^2$) | — | — | — | — |
| Hardness alteration† | 0.0014 | 0.036 | 0.0020 | 0.050 |
| Microcracks or Macrocracks | 0.0003 | 0.008 | 0.0015 # | 0.038 |
| Residual stress‡ | 0 | 0 | 0 | 0 |
| Metallurgical altered material zones: | inch* | mm* | inch* | mm* |
| Recrystallization | — | — | — | — |
| Intergranular attack (IGA) | 0.0003 | 0.008 | 0.0015 | 0.038 |
| Selective etch, pits, protuberances | 0.0004 | 0.010 | 0.0025 | 0.064 |
| Metallurgical transformations | — | — | — | — |
| Heat-affected zone (HAZ) or Recast layer | — | — | — | — |
| High cycle fatigue (HCF): | percent | | percent | |
| Variation from "handbook" values at room temperatures§ | +6 to −33 | | −20 to −45 | |

SOURCE: G. Bellows and J. B. Kohls.[20]

NOTE: A dash (—) in the table indicates no or insufficient data.

*Maximum observed depth normal to the surface.

†Depth to point where hardness becomes less than ±2 points $R_c$ (or equivalent) of bulk material has (hardness converted from Knoop microhardness measurements).

‡Depth to point where residual stress becomes and remains less than 20 ksi [138 MPa] or 10% of tensile strength, whichever is greater.

§Handbook values from HCF testing frequently are generated from specimens that have been processed using low stress grinding, hand or gentle machine polishing or occasionally electropolishing. For this table, "handbook" values represent HCF strength values generated from low-stress-ground specimens (with their minor amount of retained but enhancing compressive residual stress). The values are the $10^7$ cycle strengths from full-reverse bending tests at room temperature.

# Short circuit accidents have caused microcracking as deep as 0.10 inch [0.25 mm].

## Specific Guidelines and Data for THERMAL Material Removal Processes

The thermal material removal processes utilize a variety of heat sources to melt, vaporize or sublime the workpiece surface. The rate of material removal is a function of the energy density which can vary from 100 watts per square inch [15.5 w/cm²] with plasma arc or electrical discharge machining (EDM) to 3 x $10^{10}$ watts per square inch [0.47 x $10^{10}$ w/cm²] for laser machining. The surface texture reflects the impingement of the heat source and the molten state that occurred, as is shown in figures 18.3-3 and 18.3-76. Surface roughness measurements of reasonable accuracy can be made on these surfaces if stylus traces are averaged over several readings taken in different directions. Roughness increases with increase in process energy level which is directly proportional to the material removal rate, as shown in figure 18.3-77.

The shallow thermal impact of these processes is accompanied by the very rapid quench rate from the heat sink of the bulk of the material. These transient thermal waves produce a recast and/or resolidified layer(s) on the surface with a heat-affected zone (HAZ) below the recast layer(s). Lightly attached splatter and globules appear at higher energy levels. Figure 18.3-78 illustrates most of the effects found. The thickness of the recast layers increases with an increase in the material removal rate, as shown in figure 18.3-79. When EDM is well controlled, only a few ten thousandths inch of recast is found, as shown in figure 18.3-80. The recast layer has a heat-affected zone immediately beneath it and of a thickness approximately the same as the maximum recast. Figure 18.3-81 shows typical microhardness traverses with the much harder recast and the usually softened zone below.

Since most of the thermal machining data has been generated using EDM, the temptation to extend these data to other thermal processes should be resisted. The significantly different energy rates of laser, for example, produce significantly different effects, as demonstrated in figure 18.3-4. The charts and guidelines that follow have baselines the same as those for the mechanical material removal processes. Figures 18.3-76 to 18.3-92 illustrate good or poor surface effects from thermal material removal processes. Microhardness examples are shown in figures 18.3-81, 18.3-82, 18.3-84, 18.3-86, 18.3-87, and 18.3-89 and typical residual stress patterns in figures 18.3-88 and 18.3-89. A summary of high cycle fatigue strengths from EDM is contained in figure 18.3-94 and table 18.3-22. Table 18.3-21 is a summary of the surface integrity effects observed during the examination of several thousand pieces of data. Other thermal material removal illustrations are figures 18.3-3, 18.3-4, 18.3-11, and 18.3-15.

### Thermal Process Guidelines that Promote Good Surface Integrity Results

1. The surface texture is composed of a random array of overlapping craters or cusps, sometimes with macrocracks at the roughing level of processing, as shown in figures 18.3-3 and 18.3-76.
2. Surface roughness standards should be reassessed when using these processes because they produce unusual surface textures with no lay patterns and require different methods of checking.
3. The prior metallurgical condition of the workpiece is as important to generating surface quality as is the selection of operating parameters.
4. Careful monitoring of the preselected operating parameters is necessary. Some older types of equipment do not maintain their initial settings throughout a long day or run.
5. The depth of the heat-affected zone (HAZ) or recast structure on the surface is approximately proportional to the magnitude of the energy impinging on that surface. It is always present to some degree.
6. The HAZ can induce hardness variations and a substantial tensile residual stress in the surface layers.
7. The depth of the HAZ below the recast layers is approximately equal to the depth of the maximum recast. The recast layer in EDM can be controlled precisely and is usually only a few ten-thousandths inch thick.
8. Fatigue strength is frequently severely reduced by the HAZ(s).
9. Highly stressed or critical surfaces should have the HAZ produced by thermal material removal processes removed or modified by a post-processing treatment.
10. Removal of the altered material zones (AMZ) may not be necessary if component or laboratory tests shown that design requirements are being satisfied.
11. Microholes (those less than one millimeter in diameter) produced by thermal processes may not be detrimental to fatigue endurance strength. A check by facture mechanics of the critical hole size for a specific material is desirable.
12. On thin components, high current densities may overheat the workpiece.
13. Thorough cleaning to remove dielectric fluids, beads and vapor residue is desirable.
14. The microcracks per inch of cross section can be a valuable clue to the relative thermal sensitivity of materials, table 18.3-23.

### Special Comments Concerning Surface Integrity Practices for Thermal Machining Processes

**Whenever EDM is used in the manufacture of highly stressed structural parts, the heat affected layer which is produced should be removed or given a remedial post treatment.** The altered surface layer(s) which is produced during EDM lowers fatigue strength of alloys significantly. The altered layer consists of a recast layer with or without microcracks, some of which may extend into the base metal, plus metallurgical alterations such as rehardened and tempered layers, heat-affected zones, intergranular precipitates, etc. Concern over the lowered fatigue strength is in reference to highly stressed structures. For many tool and die applications, the altered layer has not caused problems, and there have been reports of improved die life in special applications. Generally, during EDM roughing, the layer showing microstructural changes, including a melted and resolidified layer, is less than 0.005-inch [0.127 mm] deep; while during EDM finishing, it is less than 0.001 inch [0.025 mm].

The detrimental high cycle fatigue effects of the surface alterations caused by EDM are shown in comparison with low stress grinding in table 18.3-22 and figure 18.3-94.

**Post treatment to restore fatigue strength is recommended to follow EDM of critical or highly stressed surfaces.** Since all alloys examined to date show a significant decline in fatigue strength when their surfaces are produced by EDM, it is necessary to apply a modifying post-EDM process in order to restore the surface integrity. There are several effective processes to accomplish restoration or even enhancement. It is recommended that a special surface integrity evaluation be made for critically loaded surfaces. The evaluation should be made on test workpieces or coupons to which the *full* sequence of operations generating the "as-shipped" surface has been applied. Among the frequently used post-treatment processes are the following:

- removal of layers by low stress grinding
- removal of layers by chemical machining
- addition of a metallurgical type coating
- reheat treatment
- application of shot peening

See figure 18.3-93 and the Specific Guidelines for Post-Treatment Processes.

**The thermal nature of these processes always produces a recast and underlying heat-affected zone (HAZ) on the surface being machined.** The intensity of the process energy does affect the amount of recast and microcracking. However, the thermal sensitivity or chemical complexity of the alloy can also affect the microcracking propensity. Using a constant EDM energy level, the number of microcracks per inch was measured on several nickel alloys. Table 18.3-23 shows the increasing rate of microcrack appearances as the complexity of the alloy increases. These values are the count found in an examination at 1000X of 2 inches [50 mm] of EDM cut surface.

This same study found no significant surface effects differences between cast and rolled states of Hastelloy Alloy X or Rene 41 when processed using EDM. It is probable that the energy density in EDM is such as to overwhelm the differences in the energy state between casting and rolling in these alloys.

The degree of microcracking can be a valuable comparison technique to assess the relative thermal sensitivity of alloys.

**Laser beam machining (LBM) develops surfaces showing the effects of melting and vaporization.** Applications for LBM are not yet common. Recently, however, LBM with oxygen assist has been investigated for cutting instead of shearing various alloys including titanium. Also, LBM has been used for the production of small holes. Wherever LBM is used in manufacturing highly stressed structural members, it should be remembered that in the application of this method the surface is subjected to melting and vaporization. Indications are that such surfaces should be removed or modified. It becomes difficult, however, to apply secondary processing to parts containing very small diameter holes or narrow slots. As a first step, it is suggested that critical parts made by electron beam machining and laser beam machining be tested to see if surface alterations lower the critical mechanical properties.

**The concern for EDM recast should be confined only to critically loaded surfaces.** Too frequently the reaction to the microcracking seen in the recast layers from EDM or other thermal machining methods is either to reject the processes or apply a drawing note to finish by mechanical means if such a process is used. This approach is costly, eliminates useful alternatives and ignores the altered material zones from mechanical processing (see milling "recast," figure 18.3-34). A more fruitful approach is to apply surface integrity evaluations *only* to those areas of a workpiece that are fatigue or stress corrosion or critically limited in some manner peculiar to the specific applications. Concentrated effort to improve these areas by the method used for their generation or by the application of post-processing treatments is eminently worthy to produce the best surface and component integrity. It is for these areas that the decision must be made to remove or to modify EDM. For all other less critical or noncritical surfaces, specifications can be relaxed with the potential for considerable cost improvements. These less critical areas are usually several times as extensive as the critical areas.

**The recast layer is *not* a crack propogation barrier.** Many section photomicrographs will show microcracks only in the recast layer. It is wrong to assume that the sharp metallurgical structural change at the interface between the recast and the base metal will act as a barrier for the propagation of microcracks. The single metallographic section is too small a sample to assure that no microcrack extends deeper than the recast, and no data are available on such microcrack propagation characteristics.

**Very minor fatigue strength differences exist between finishing and roughing EDM.** Finishing and roughing EDM are characterized by a wide difference in surface roughness levels (50 to 200 microinches $R_a$ [1.25 to 5.0 μm $R_a$]) and wide differences in recast layer thicknesses (0.0002 to 0.005 inch [0.005 to 0.127 mm]). High cycle fatigue strength, however, is nearly the same for finishing as for roughing EDM, as shown in figures 18.3-93 and 18.3-94. The roughness values are not as significant as it has classically been assumed to be. Not only the recast layer but also the underlying heat affected zone can be detrimental. For critically loaded surfaces, it is desirable to remove the HAZ or to modify the surface preparation or to conduct extensive testing to assess the reliability of the surface generated for particular applications.

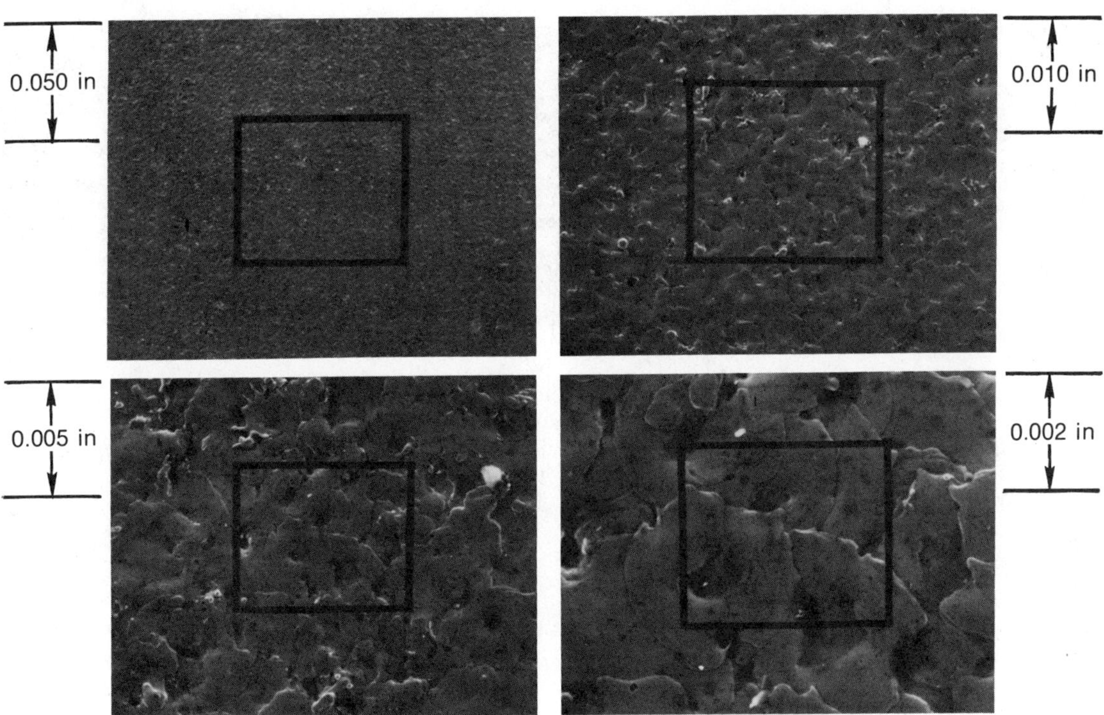


**Figure 18.3–76** Scanning electron microscope photomicrographs at successively greater magnifications of finish electrical discharge machining of Inconel Alloy 718 (solution treated and aged). Views taken at 84° to the surface. Surface roughness is 62 microinches $R_a$. (Surface Integrity Encyclopedia[1])

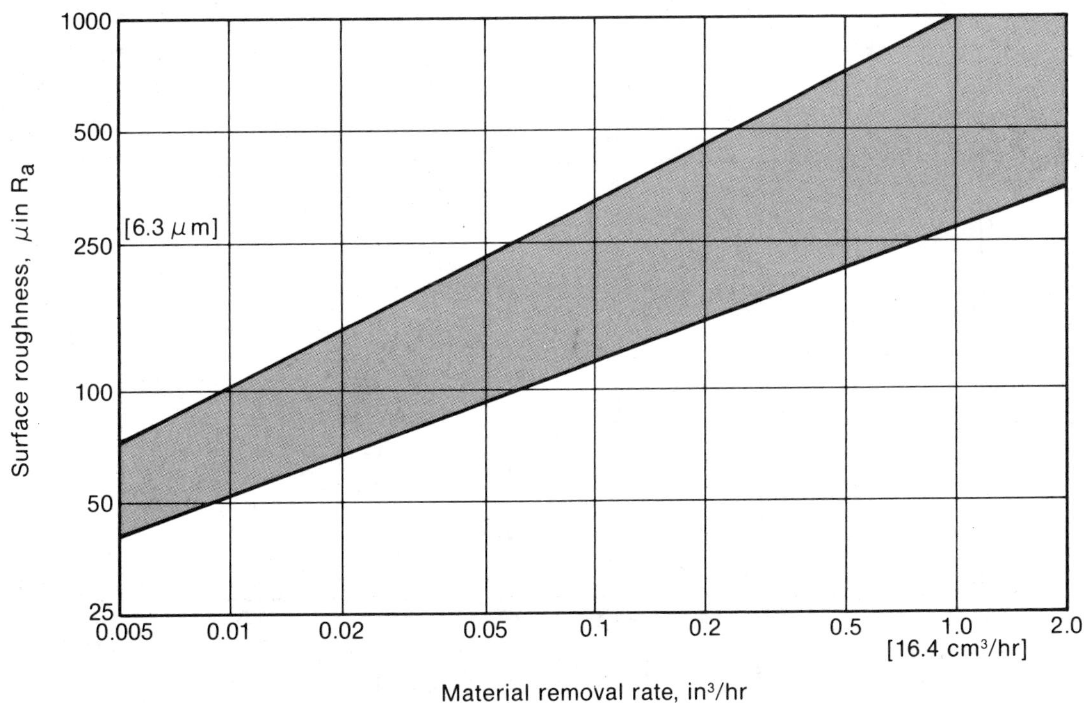


**Figure 18.3–77** Approximate relation of surface roughness to material removal rate for electrical discharge machining. (G. Bellows and J. B. Kohls[20])

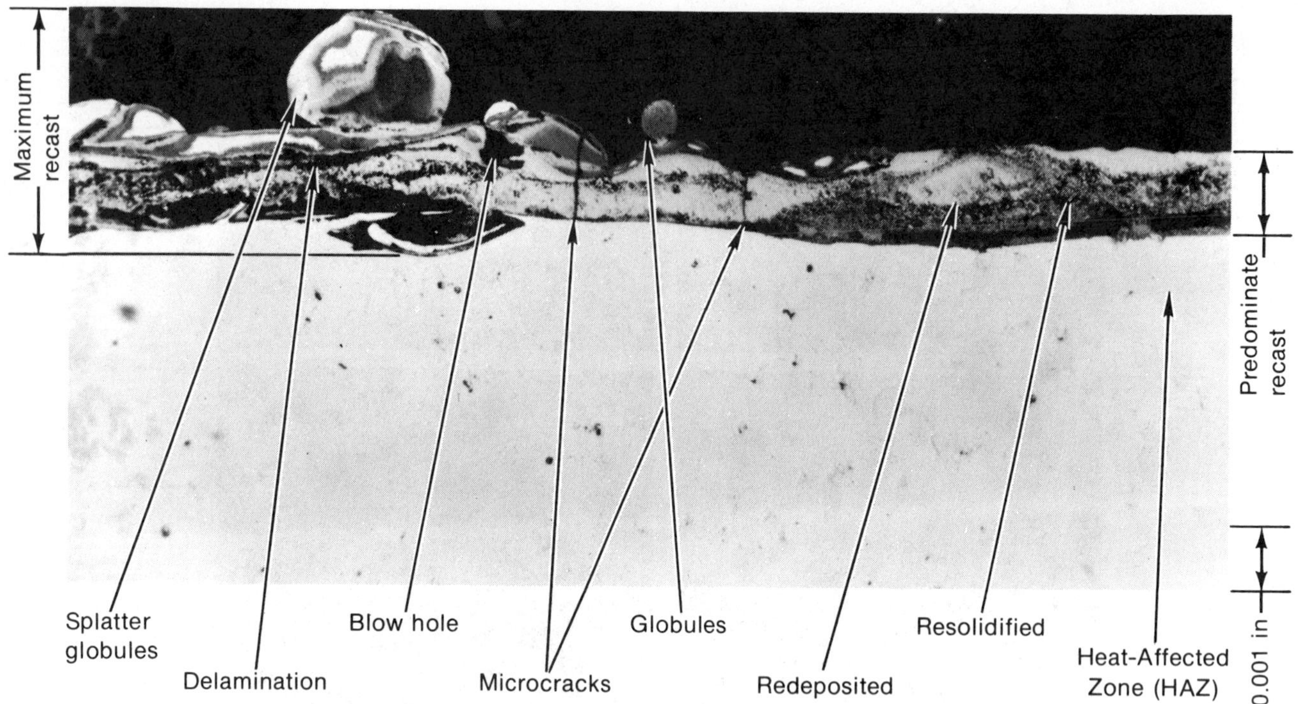


**Figure 18.3–78** Recast surface with types of alterations and their descriptors (Hastelloy Alloy X at 300X). (G. Bellows and J. B. Kohls[20])

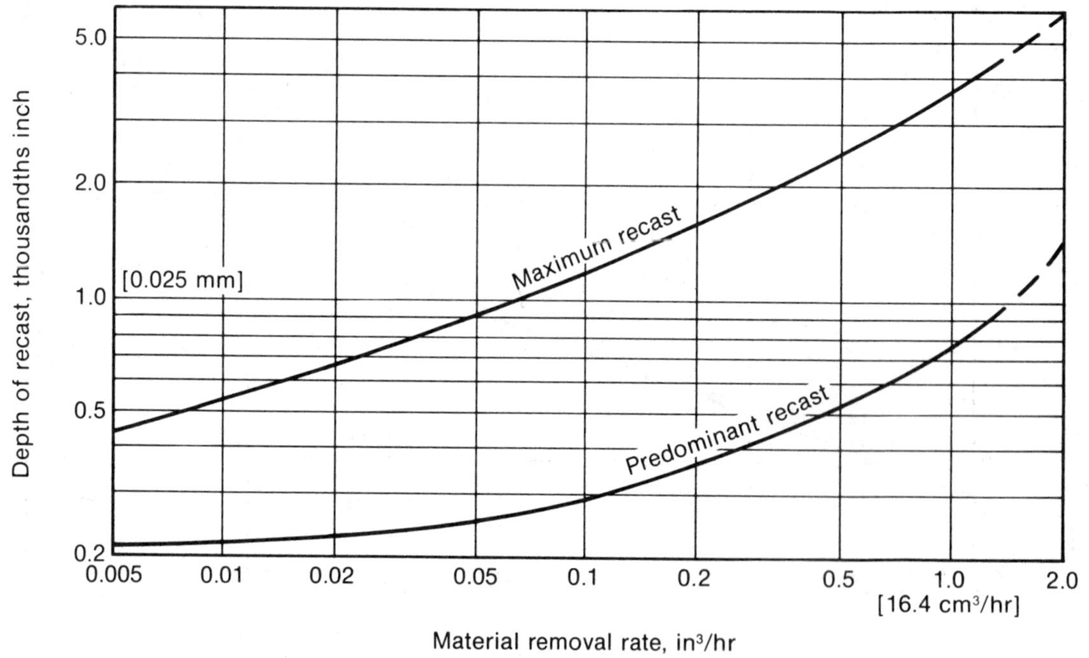


**Figure 18.3–79** Depth of recast from electrical discharge machining versus process intensity (material removal rate) for a wide variety of materials. (G. Bellows and J. B. Kohls[20])

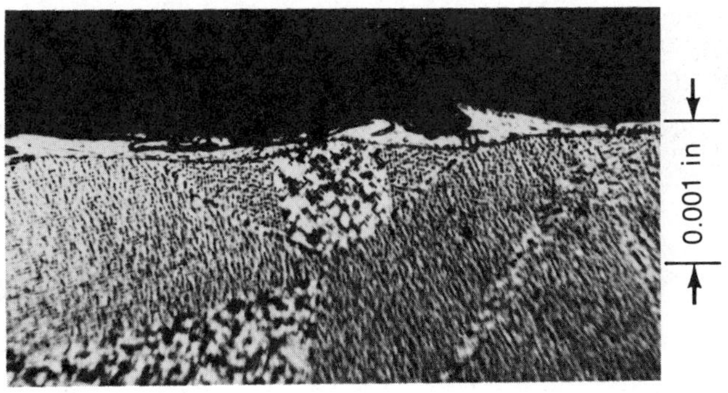


**Figure 18.3–80** Typical well controlled EDM recast and heat-affected zone (0.0002 inch predominant recast from EDM of Udimet 700). (Surface Integrity Encyclopedia[1])

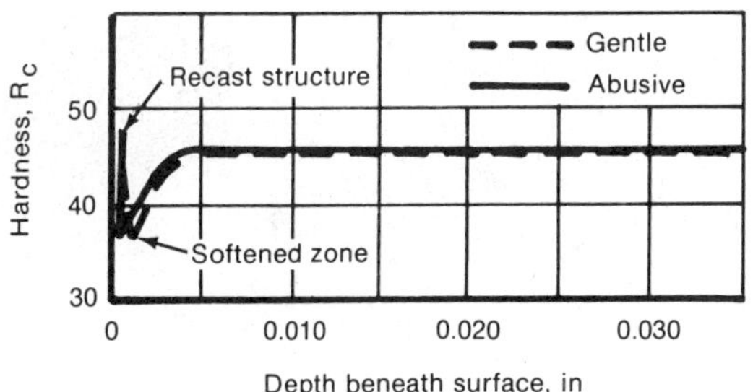


**Figure 18.3–81** Typical microhardness traverse on EDM surface of Inconel Alloy 718 (solution treated and aged). (G. Bellows and J. B. Kohls[20])

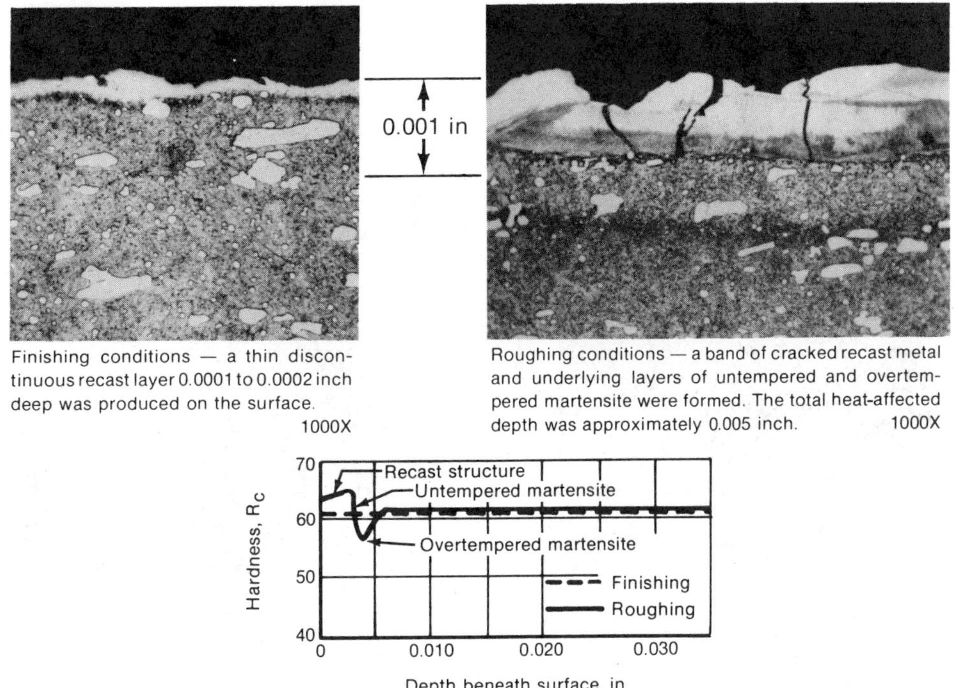


**Figure 18.3–82** Surface characteristics of type D2 tool steel (quenched and tempered, 61 $R_C$) produced by EDM.

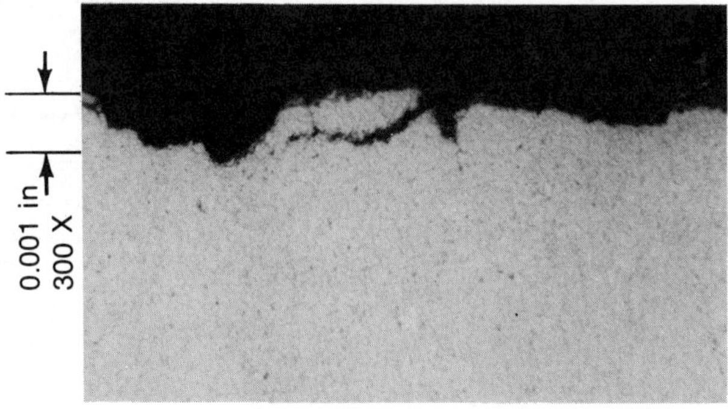


**Figure 18.3-83** Surface characteristics of tungsten carbide (883) produced by EDM. (Surface Integrity Encyclopedia[1])

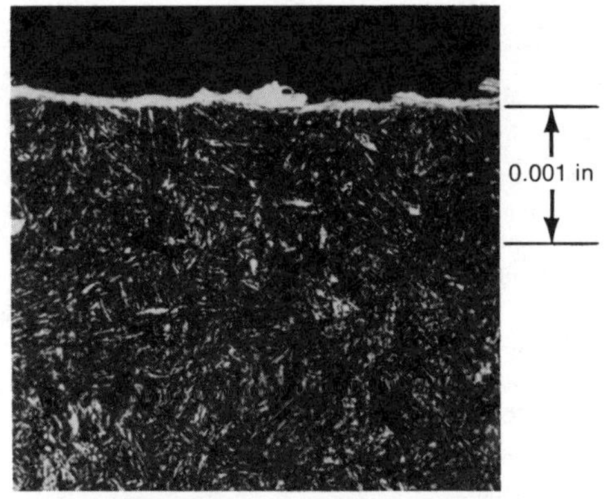


Finishing conditions — discontinuous patches of recast metal plus a thin layer of rehardened martensite (0.0001 inch) are seen. No hardness change was detected. 1000X

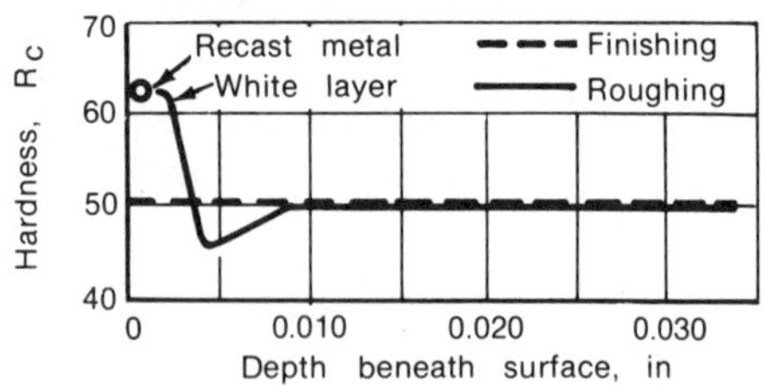

Roughing conditions — globs of recast metal (62 $R_C$) are spattered on a white layer of rehardened martensite totaling 0.003 inch deep, also at 62 $R_C$. An overtempered zone as soft as 46 $R_C$ is found beneath the surface. Total depth of layer affected approaches 0.010 inch. 1000X

**Figure 18.3-84** Surface characteristics of 4340 steel (quenched and tempered, 50 $R_C$) produced by electrical discharge machining.

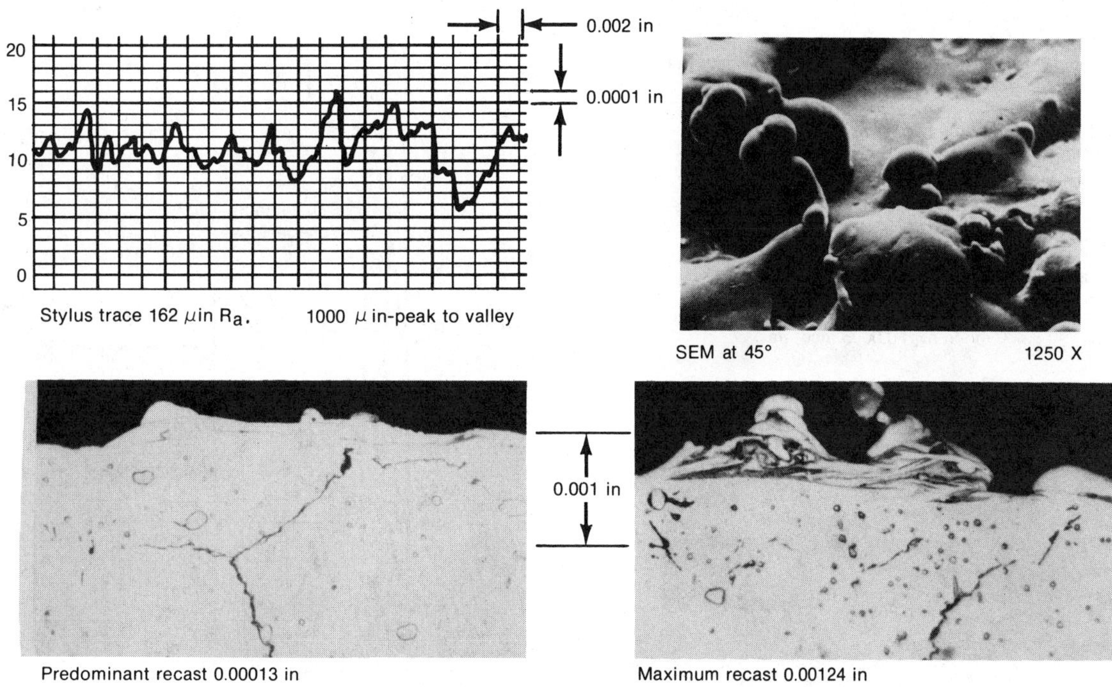


**Figure 18.3–85** Typical surface characteristics of rolled Hastelloy Alloy X from "normal" EDM. (Surface Integrity Encyclopedia[1])

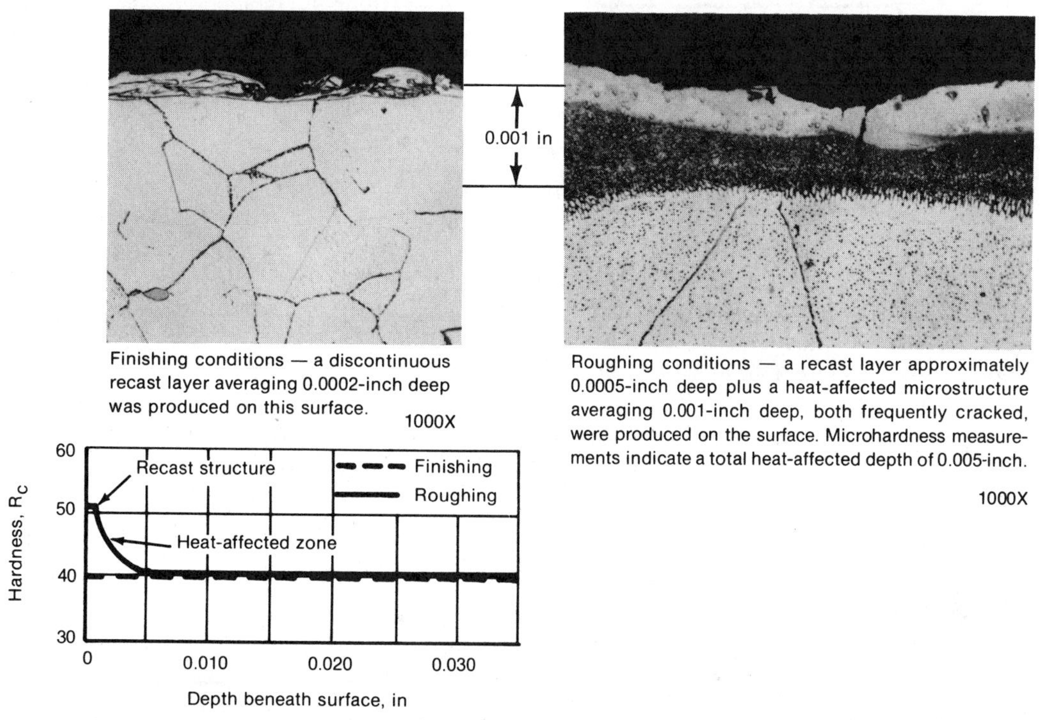

0.001 in
Finishing conditions — a discontinuous recast layer averaging 0.0002-inch deep was produced on this surface.
1000X
Roughing conditions — a recast layer approximately 0.0005-inch deep plus a heat-affected microstructure averaging 0.001-inch deep, both frequently cracked, were produced on the surface. Microhardness measurements indicate a total heat-affected depth of 0.005-inch.
1000X
Recast structure
Finishing
Roughing
Heat-affected zone
Hardness, Rc
60
50
40
30
0
0.010
0.020
0.030
Depth beneath surface, in


**Figure 18.3–86** Surface characteristics of Waspaloy (aged, 40 $R_C$) produced by EDM. (W. P. Koster et al[8])

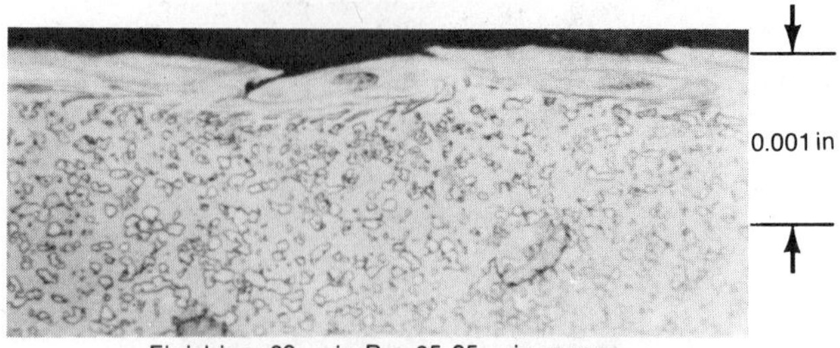

0.001 in
Finishing: 89 μin Ra; 85-95 μin range
Recast: 0.0003 inch avg; 0.0005 inch max.


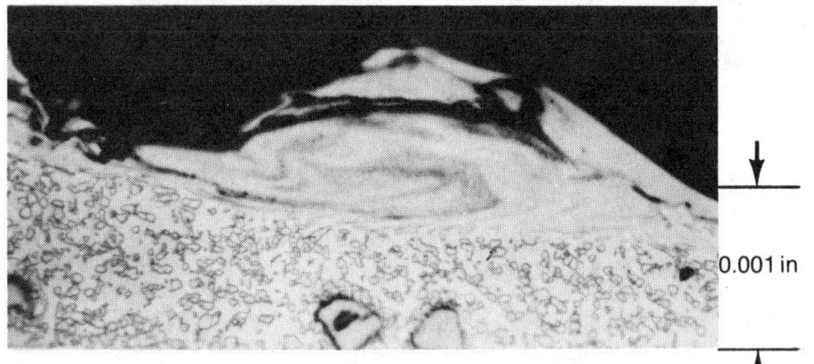


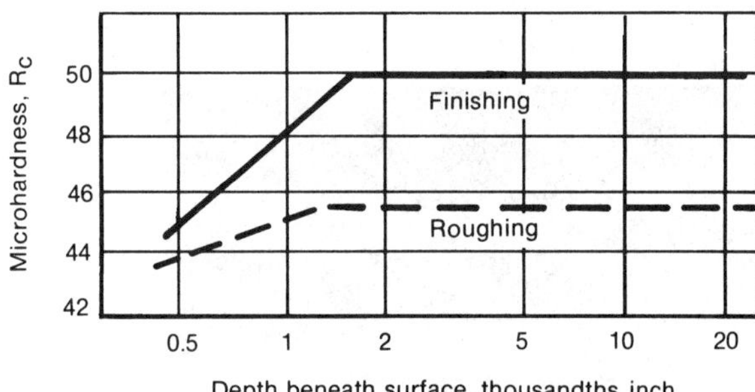


**Figure 18.3–87** Surface characteristics of nickel alloy AF2-1DA (solution treated and aged) produced by EDM. (Surface Integrity Encyclopedia[1])

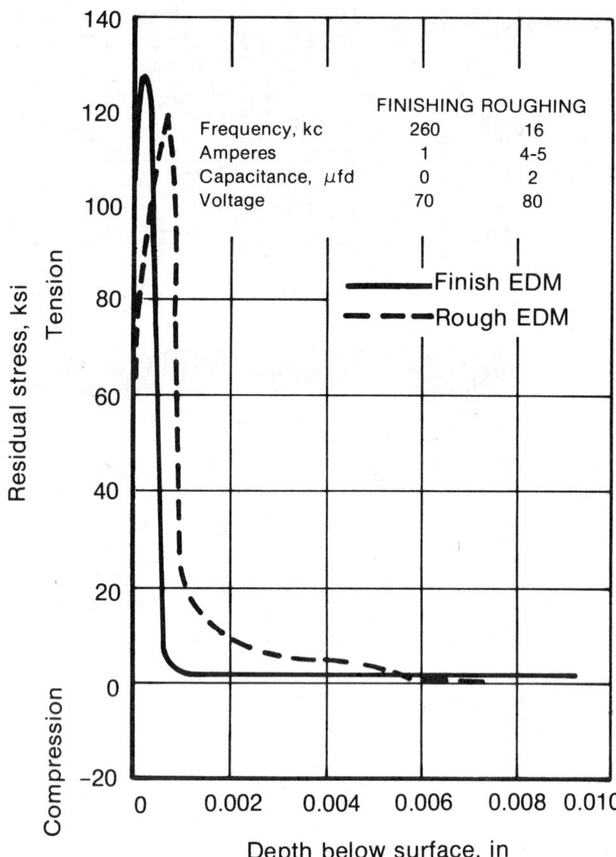


**Figure 18.3–88** Residual surface stress profiles in AF 95 (Rene 95) (solution treated and aged, 50 $R_C$) produced by EDM. See also figure 18.3–93. (W. P. Koster and J. B. Kohls[18])

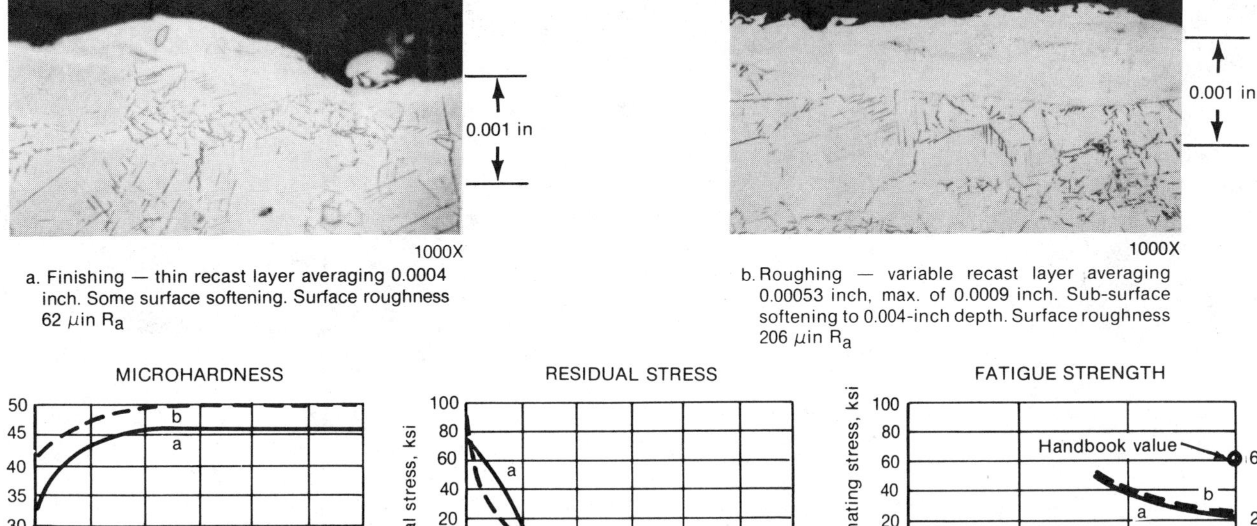


**Figure 18.3-89** Surface integrity standard data set of EDM of Inconel Alloy 718 (solution treated and aged). Note that thin layers of recast (or the HAZ) can be as detrimental to fatigue strength as layers many times as thick. In this alloy, the reduction is 63% from the "handbook" high cycle fatigue strength of 60 ksi. (W. P. Koster et al[9])

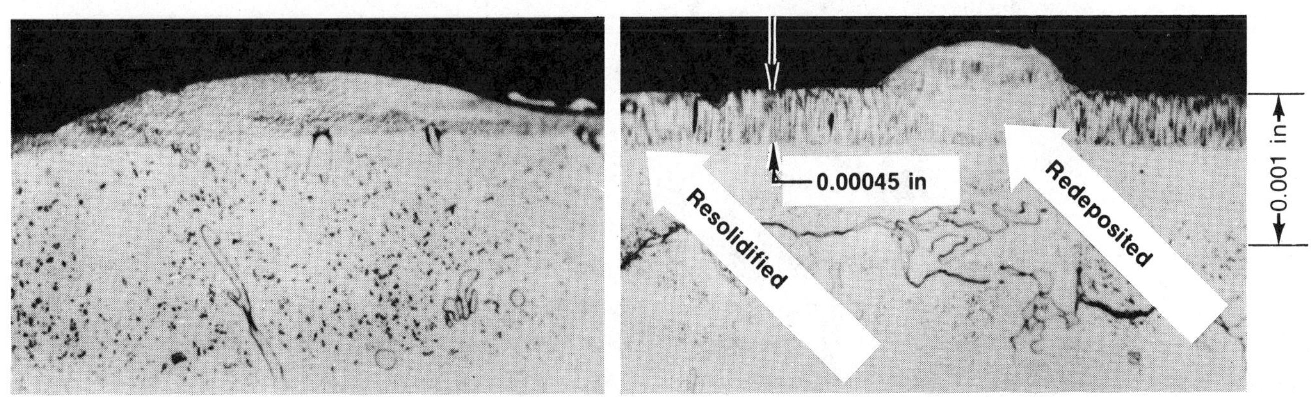


**Figure 18.3-90** Typical surface characteristics of hole drilled with laser in cast cobalt alloy X40, 1000X. (Surface Integrity Encyclopedia[1])

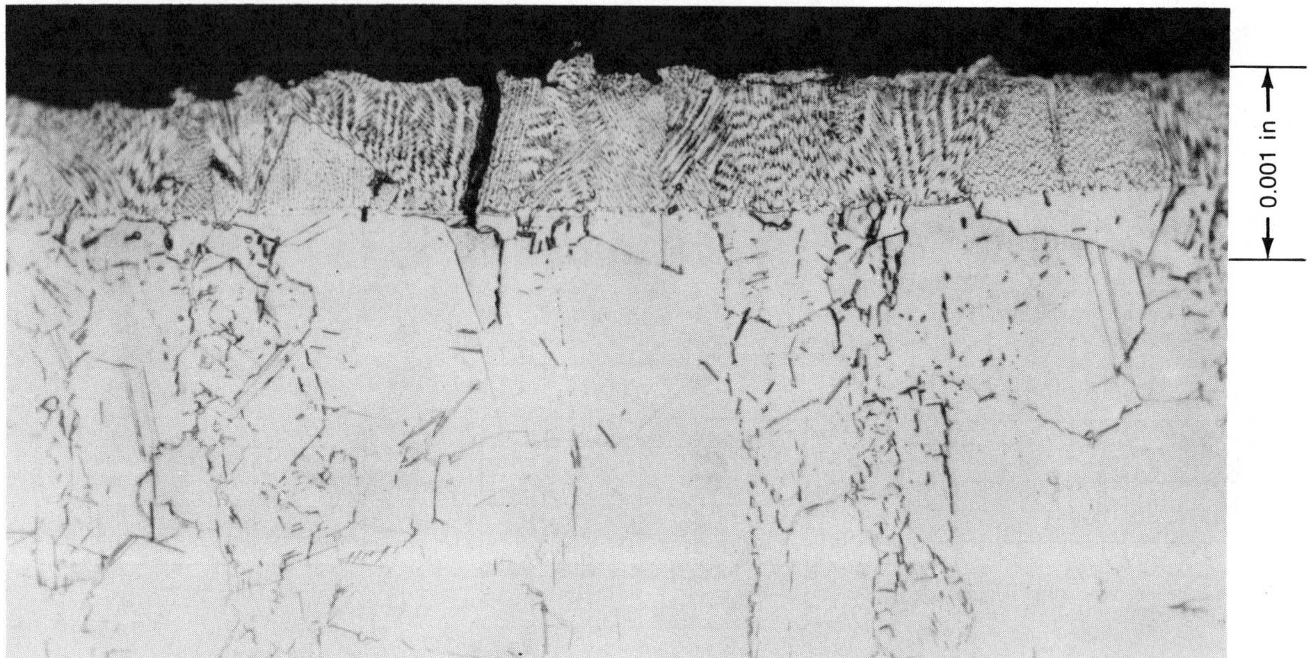


**Figure 18.3–91** Surface characteristics of hole in Inconel Alloy 718 produced by LBM showing worst microcrack observed, 0.0008-inch long. Predominant recast 0.0007 inch, maximum recast 0.0029 inch. Note epitaxial growth in deposited layer nucleated on base metal grains. (Surface Integrity Encyclopedia[1])

**Figure 18.3–92** Surface characteristics of hole drilled in Rene 80 (solution treated and aged) with ruby laser. Note the two distinct recast layers from vapor deposited during the two laser pulses.

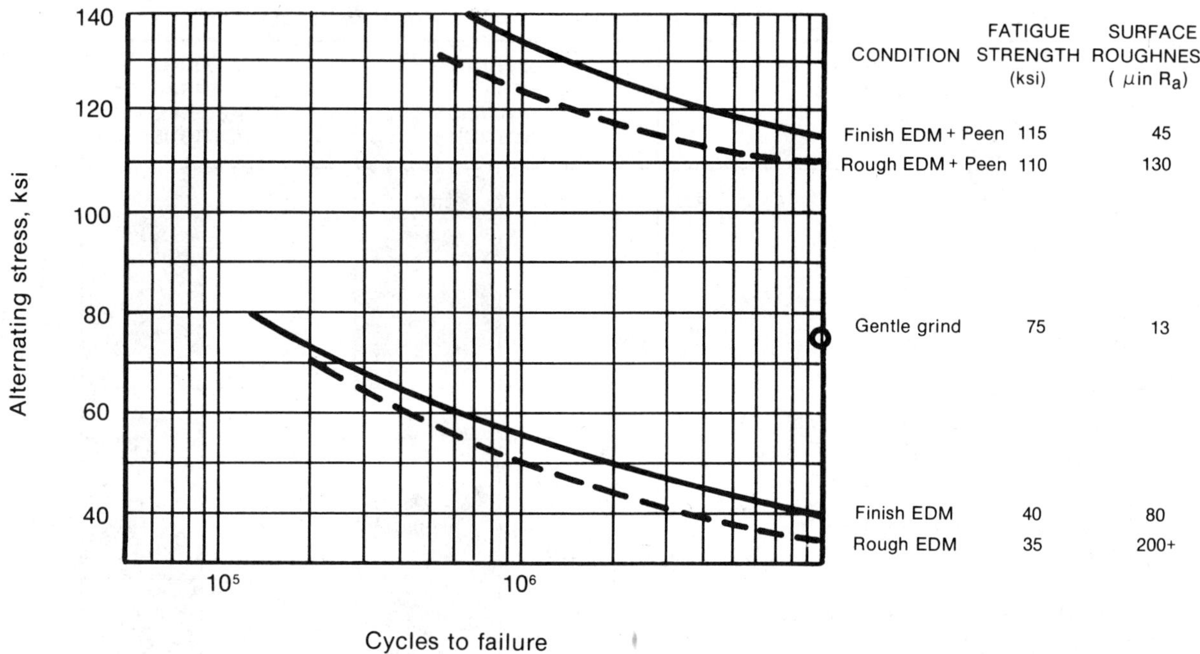


**Figure 18.3–93** High cycle fatigue characteristics of AF 95 (Rene 95) (solution treated and aged, 50 $R_C$) produced by EDM. Mode: cantilever bending, zero mean stress, temperature 75°F. See also figure 18.3–88. Note that finish EDM has about the same reduction in fatigue strength as rough EDM as compared to gentle low stress grinding. Note also the recovery and enhancement of fatigue strength with a post-EDM shot peen treatment. (W. P. Koster and J. B. Kohls[18])

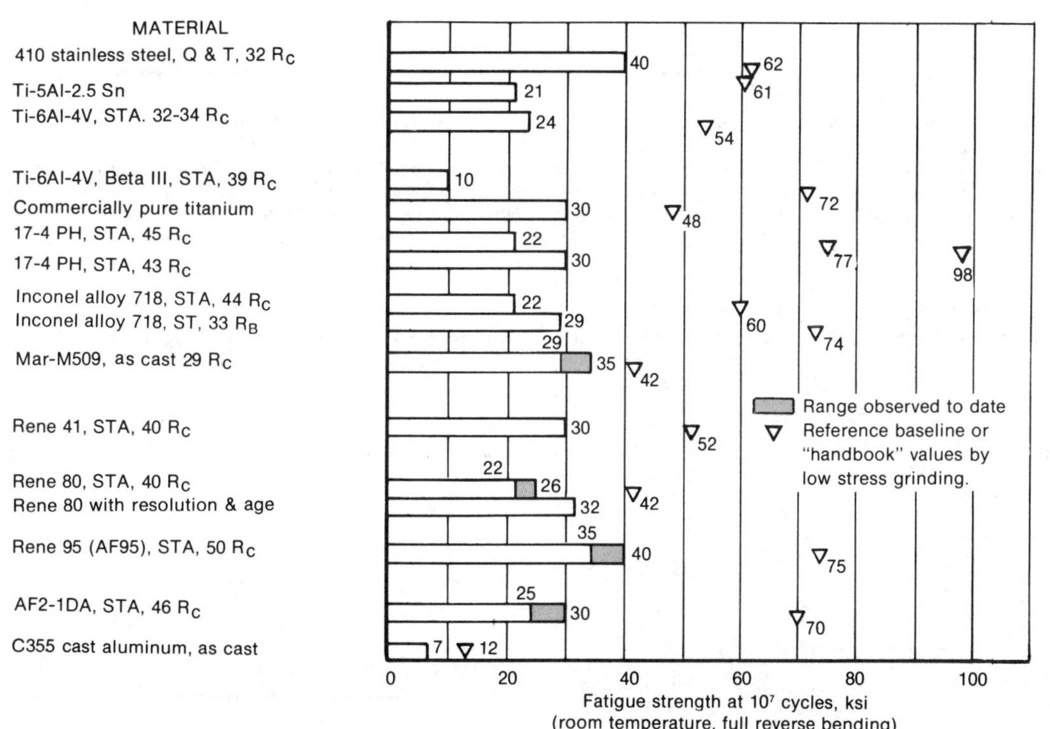


**Figure 18.3–94** Summary of high cycle fatigue strength from electrical discharge machining.

**TABLE 18.3-21 Surface Integrity Effects Observed in Thermal Material Removal Processes**

| PROPERTY AND TYPE OF EFFECT | FINISHING, "GENTLE" OR LOW STRESS CONDITIONS | | ROUGHING, "OFF-STANDARD" OR ABUSIVE CONDITIONS | |
|---|---|---|---|---|
| Surface roughness: | μin $R_a$ | μm $R_a$ | μin $R_a$ | μm $R_a$ |
| Average range | 32 to 125 | 0.8 to 3.2 | 125 to 500 | 3.2 to 12.7 |
| Less frequent range | 2 to 250 | 0.05 to 6.3 | 32 to 1,000 | 0.8 to 25 |
| Mechanical altered material zones: | inch* | mm* | inch* | mm* |
| Plastic deformation (PD) | *** | *** | *** | *** |
| Plastically deformed debris ($PD^2$) | *** | *** | *** | *** |
| Hardness alteration† | 0.0011 | 0.028 | 0.0080 | 0.203 |
| Microcracks or Macrocracks | 0.0003 | 0.008 | 0.0070 | 0.188 |
| Residual stress‡ | 0.0020 | 0.050 | 0.0030 | 0.076 |
| Metallurgical altered material zones: | inch* | mm* | inch* | mm* |
| Recrystallization | *** | *** | *** | *** |
| Intergranular attack (IGA) | *** | *** | *** | *** |
| Selective etch, pits, protuberances | — | — | — | — |
| Metallurgical transformations | — | — | — | — |
| Heat-affected zone (HAZ) or Recast layer | 0.0006 | 0.015 | 0.0050 | 0.127 |
| High cycle fatigue (HCF): | percent | | percent | |
| Variation from "handbook" values at room temperature§ | −17 to −96 | | −48 to −64 | |

SOURCE: G. Bellows and J. B. Kohls.[20]

NOTE: A dash (—) in the table indicates no or insufficient data. A triple asterisk (***) in the table indicates no occurrences or not expected.

*Maximum observed depth normal to the surface.

†Depth to point where hardness becomes less than ±2 points $R_c$ (or equivalent) of bulk material has (hardness converted from Knoop microhardness measurements).

‡Depth to point where residual stress becomes and remains less than 20 ksi [138 MPa] or 10% of tensile strength, whichever is greater.

§Handbook values from HCF testing frequently are generated from specimens that have been processed using low stress grinding, hand or gentle machine polishing or occasionally electropolishing. For this table, "handbook" values represent HCF strength values generated from low-stress-ground specimens (with their minor amount of retained but enhancing compressive residual stress). The values are the $10^7$ cycle strengths from full-reverse bending tests at room temperature.

**TABLE 18.3-22 Fatigue Strength Comparison for Materials Processed Using EDM or Low Stress Grinding**

| MATERIAL | FATIGUE STRENGTH* | | | |
|---|---|---|---|---|
| | Gentle or Low Stress Grinding | | EDM (Finishing Conditions) | |
| | ksi | MPa | ksi | MPa |
| Inconel 718 (STA, 44 $R_c$) | 60 | 414 | 22 | 152 |
| AF 95 (STA, 50 $R_c$) | 75 | 517 | 40 | 276 |
| | 98† | 676† | 50† | 345† |
| Rene 80 (STA, 40 $R_c$) | 42 | 290 | 26 | 179 |
| AF2-1DA (STA, 46 $R_c$) | 70 | 483 | 30 | 207 |

NOTE: Room temperature unless noted otherwise.

*$10^7$ cycles full-reverse bending.

†At 1000°F [538°C].

**TABLE 18.3-23 Cracking Sensitivity of Nickel Alloys from EDM**

| MATERIAL | NUMBER OF MICROCRACKS PER INCH [25 mm]* |
|---|---|
| Hastelloy X (least sensitive) | 1.0 |
| Rene 41 | 1.5 |
| Rene 77 | 2.5 |
| Rene 80 | 3.5 |
| Rene 120 (most sensitive) | 4.6 |

SOURCE: G. Bellows.[28]

*Number of microcracks observed in two inches [50 millimeters] of cross section after the specimen was electrical discharge machined at a normal level equivalent to a material removal rate of 0.18 to 0.27 in³/hour. Observations were made at a magnification of 1000X.

### Specific Guidelines and Data for CHEMICAL Material Removal Processes

The chemical family of material removal processes is characterized by an absence of stress introduced into the workpiece as a result of the process. The gentle chemical action of dissolution, molecule-by-molecule or grain-face by grain-face, is typically smooth; however, very smooth surfaces (32 microinches $R_a$ or less [0.8 $\mu$m $R_a$ or less]) are slightly roughened, while rough surfaces (125 microinches $R_a$ or more [3.2 $\mu$m $R_a$ or more]) are usually smoothed. The uniformity and fineness of the grain structure in the original workpiece is a principal factor in the final roughness. A slight softening of the surface sometimes attributed to the Rebinder effect* is frequently, but not always, found, as demonstrated in figure 18.3-100. Surface discontinuities, such as a weld joint, frequently result in different surface texture and cutting rates.

Achieving good surface integrity from the chemical material removal processes begins with a careful matching of the reagents to the metallurgical state and composition of the work material. Proper process control, particularly the chemical bath "freshness," is essential.

Selective etching, pitting or IGA can occur with improper reagents or overly contaminated or stagnant chemicals. Hydrogen absorption and possible embrittlement can occur with some alloys of steel, titanium or nickel. Post-processing bakes are recommended for this condition.

Static mechanical properties have not been found to be effected by chemical machining (CHM). The absence of beneficial compressive residual stress from chemical machining has the effect of reducing the high cycle fatigue strength when compared to conventionally prepared test specimens.

The charts and guidelines that follow have baselines the same as those for the mechanical material removal processes. Figures 18.3-12 and 18.3-95 through 18.3-100 illustrate some of the good and poor surface effects from chemical material removal processes. A summary of the high cycle fatigue strengths from chemical machining and electropolishing (ELP) is shown in figure 18.3-101. Table 18.3-24 is a summary of the surface integrity effects observed during the examination of several hundred pieces of data.

**Chemical Process Guidelines that Promote Good Surface Integrity Results**

1. Chemical machining does not induce any significant stress in the machined surfaces.
2. Surface roughness standards should be reassessed when applying CHM because CHM produces surfaces with unusual textures and no lay pattern.
3. The chemical reagents must be matched to the expected metallurgical state of the work material.
4. Surface roughness variations during processing can be a good indicator of changes in the processing conditions, or a change in the metallurgical state of the work material.
5. Careful rinsing of the solutions from the finished workpiece is essential.
6. The control of the metallurgical and heat treat state of the work material is as essential as the control of the process parameters.
7. Selective etching, intergranular attack and pitting can result from off-standard conditions such as high etchant temperatures, incomplete stirring, depleted or unbalanced solutions or contaminated solutions, and variations in the metallurgical state of the work material.
8. Weld areas usually show a different rate of cutting than the base material and an increase in surface roughness.
9. Steel, titanium and nickel base alloys which are susceptible to hydrogen embrittlement should have a post heat treatment of a few hours at a temperature of about 375° to 400°F [191° to 204°C]. Treatment should be applied immediately after chemical processing.
10. Room-temperature, high cycle fatigue strength generally is lower when compared to conventionally prepared, low-stress-ground specimens. This is probably caused by the absence of residual surface stresses.
11. The use of post treatments to add a compressively stressed surface layer may be desirable for enhancing a component's fatigue strength.
12. A test coupon for metallurgical evaluation should be made at least at 90- to 180-day intervals, or whenever the chemical solution is changed.
13. For quality assurance, a deviation of one operating parameter by more than 10 percent from the preplanned value should be cause for alarm and checking. If more than one parameter has deviated by more than 10 percent, the operation should be shut down until corrections have returned the process to a controlled state.

**Special Comments Concerning Surface Integrity Practices for Chemical Machining Processes**

**Special cognizance should be taken of the surface softening that occurs in CHM and ECM of aerospace materials.** Microhardness studies on aerospace alloys have shown that chemical machining and electrochemical machining produce a soft layer in a majority of the aerospace alloys.[8] Hardness reductions for chemical machining and electrochemical machining range from 3 to 6 Rockwell C points to a depth of 0.001 inch [0.025 mm] for chemical machining and 0.002 inch [0.050 mm] for electrochemical machining. Shot peening or other suitable post processing should be used on such surfaces to enhance mechanical properties. This surface softening is sometimes labeled the Rebinder effect.

*The Rebinder effect is the modest softening of surface layers of an object upon application of selected chemical solutions that vary with the material of the object.[7]

**Surface integrity evaluations should be made when chemical (CHM) and electrochemical (ECM) processes are used for finishing critical parts.** The fatigue strength of surfaces produced by controlled chemical and electrochemical processes has often been found to be lower than that produced by some of the more commonly used material removal processes. This is generally attributed to the unworked, stress-free surface produced by processes such as electrochemical machining (ECM), electropolishing (ELP), and chemical machining (CHM). There is evidence that processes such as milling and polishing may sometimes provide beneficial fatigue resistance as a result of cold working or compressive residual stresses; therefore, when substituting CHM for other machining processes, it may be necessary to add post-processing operations such as steel shot or glass bead peening or mechanical polishing. Some companies require peening of all chemically or electrochemically machined surfaces of highly stressed structural parts.

**Heat treat state is a principal factor in good chemical machining results.** The chemical machining of Rene 41, for example, is very sensitive to the heat treat state of the material. Material removal is at the rate of 0.0005 inch per minute [0.0127 mm/min], and the surface roughens slightly when in the full solution-treated state. When aged, however, massive IGA appears. Figure 18.3-99 illustrates the differences. These differences occur with identical processing parameters. The material quenching rate is the principal variable. The best chemical machining of Rene 41 occurs when the material is in a fine-grained structure with the best uniformity. The quick quench in hot rolling, plus the well-worked structure of thin (under 0.100 inch [2.54 mm]) sheet stock yields the best material condition for chemical machining. Increased thickness and slower quench rates tend to produce increased roughness and slower material removal rates. The aged condition usually produces a surface that is rough (150 to 400 microinches $R_a$ [3.75 to 10 $\mu$m $R_a$]) and filled with IGA. In the worst case, a film forms on the surface exposed to the chemicals, inhibiting the action. This film must be mechanically removed before chemical machining can be continued.

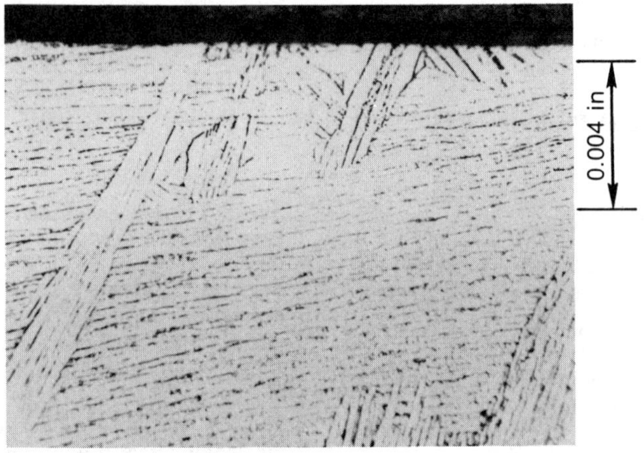


**Figure 18.3–95** Unblemished surface section of cast Ti-5Al-2.5Sn (A110 alloy) produced by well controlled chemical machining, 250X. (Surface Integrity Encyclopedia[1])

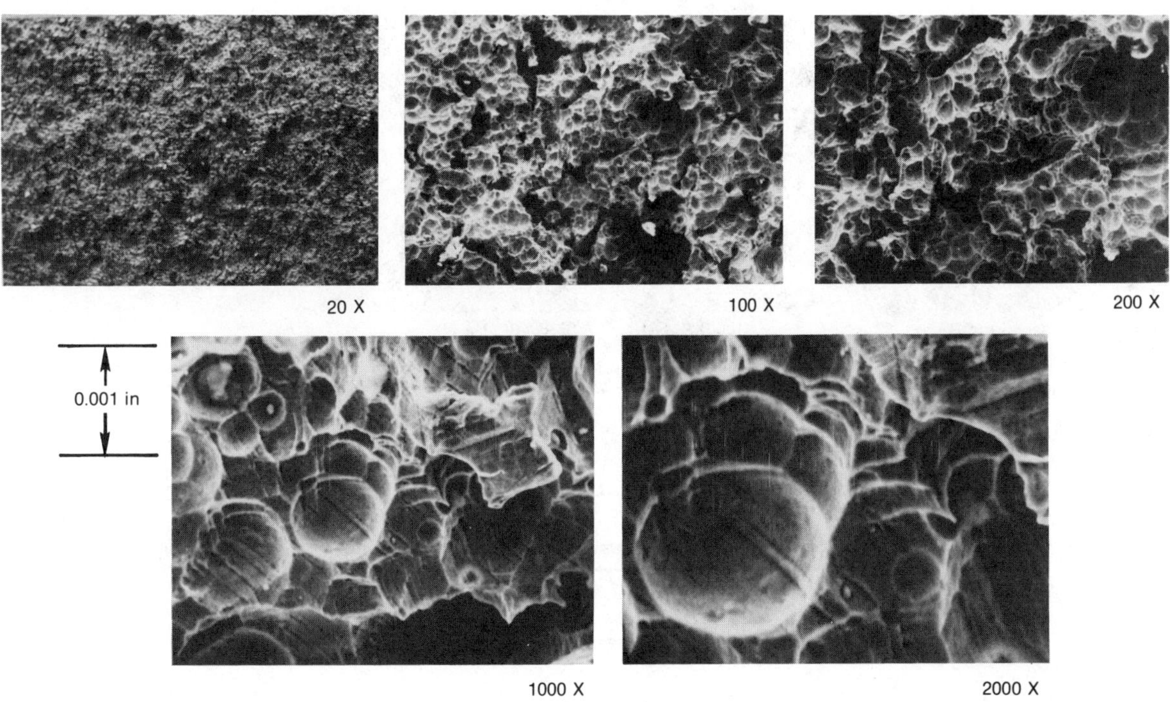


**Figure 18.3–96** Surface texture of chemically machined Rene 41. SEM photos at 20, 100, 200, 1000, and 2000X taken at 84° to surface. Surface roughness 500 microinches $R_a$ from off-standard conditions. (G. Bellows[22])

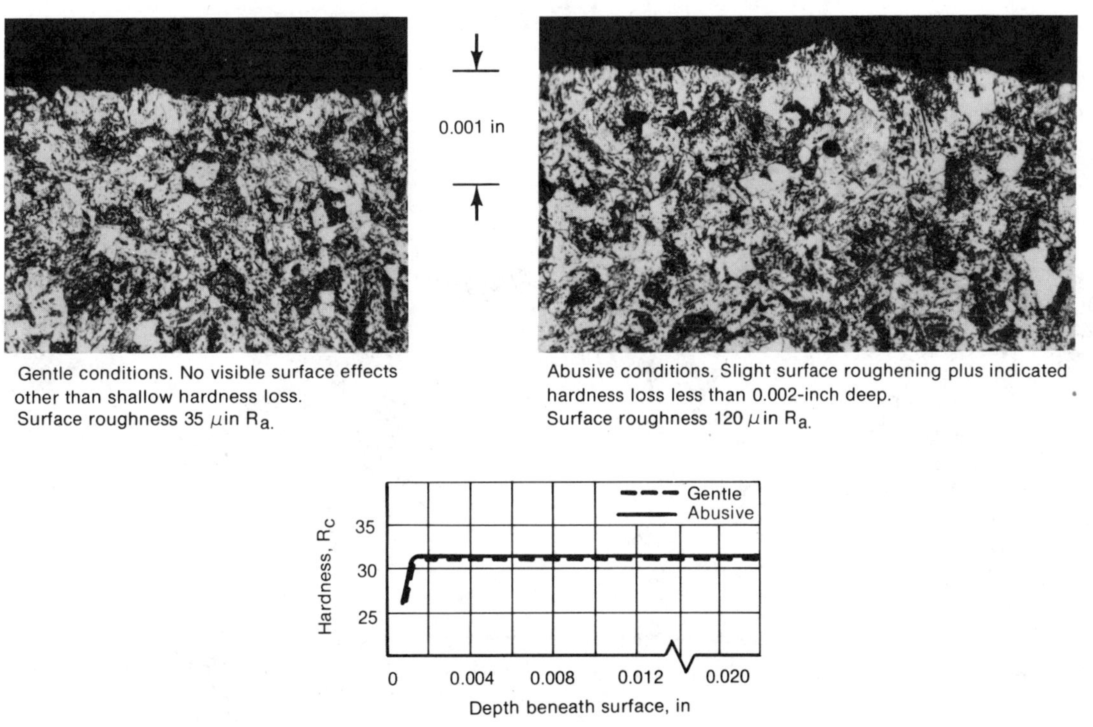


**Figure 18.3–97** Surface characteristics of 4340 steel (annealed, 31-36 $R_C$) produced by CHM.

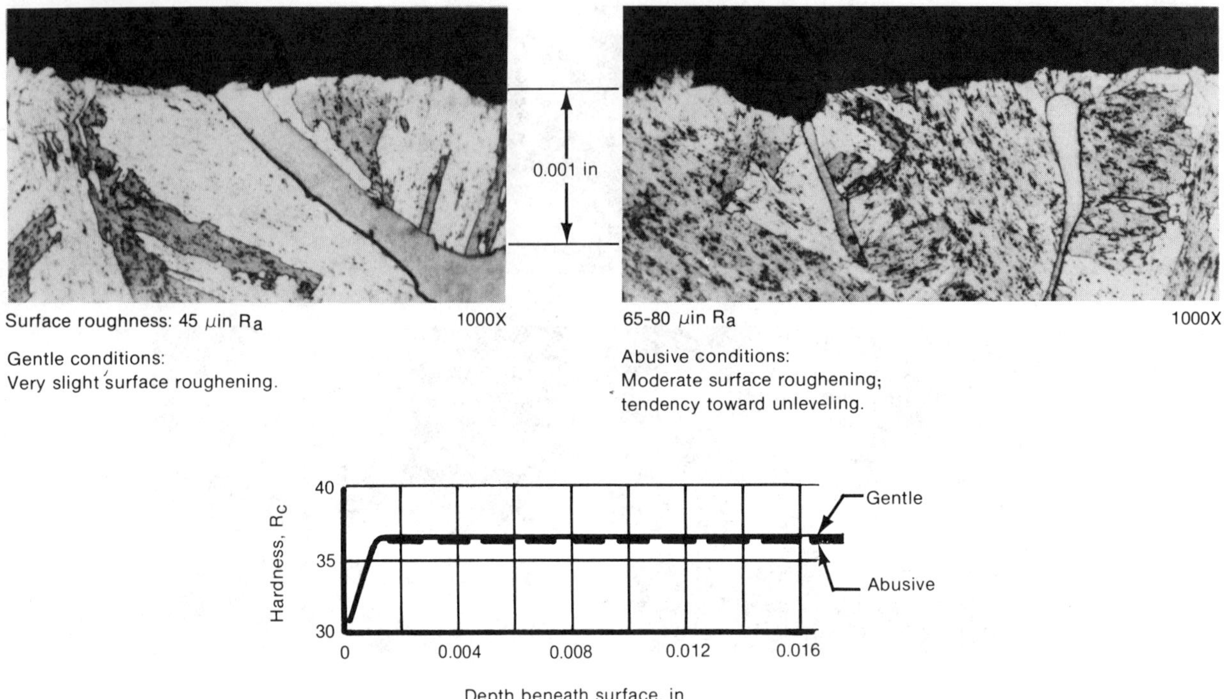


**Figure 18.3–98** Surface characteristics of 17-4 PH stainless steel produced by CHM. (W. P. Koster et al[8])

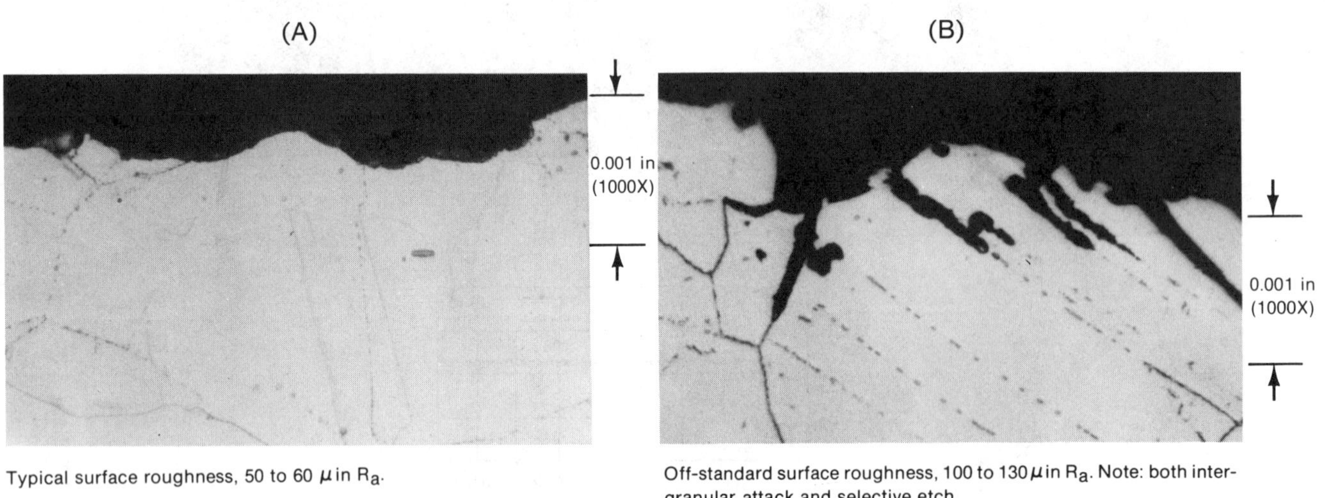


**Figure 18.3–99** Comparison of surfaces of Rene 41 (45 $R_C$) when chemically machined in (A) solution treated state and (B) solution treated and aged state. (G. Bellows[23])

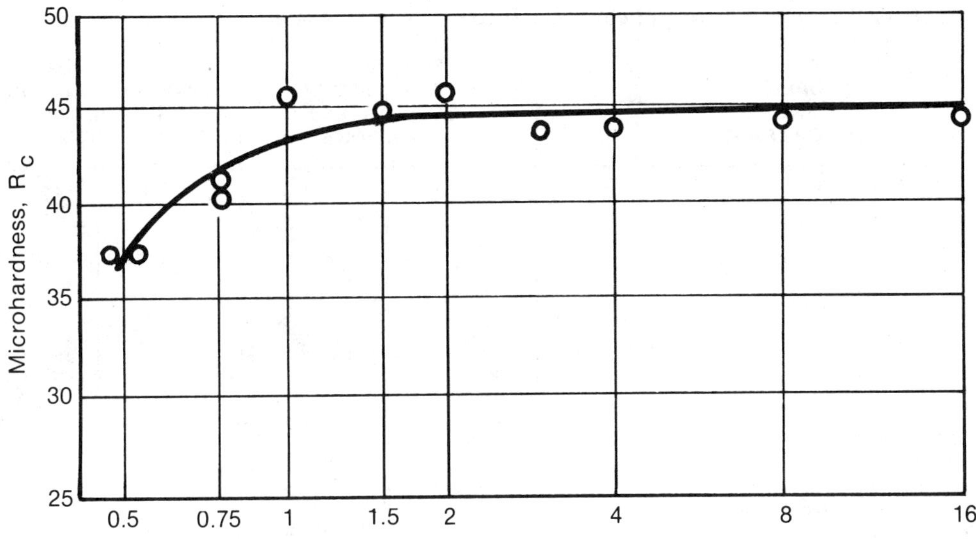


**Figure 18.3–100** Rebinder (softening) effect on chemically machined Rene 41 (solution treated and aged), surface roughness 190 to 290 microinches $R_a$. (G. Bellows[23])

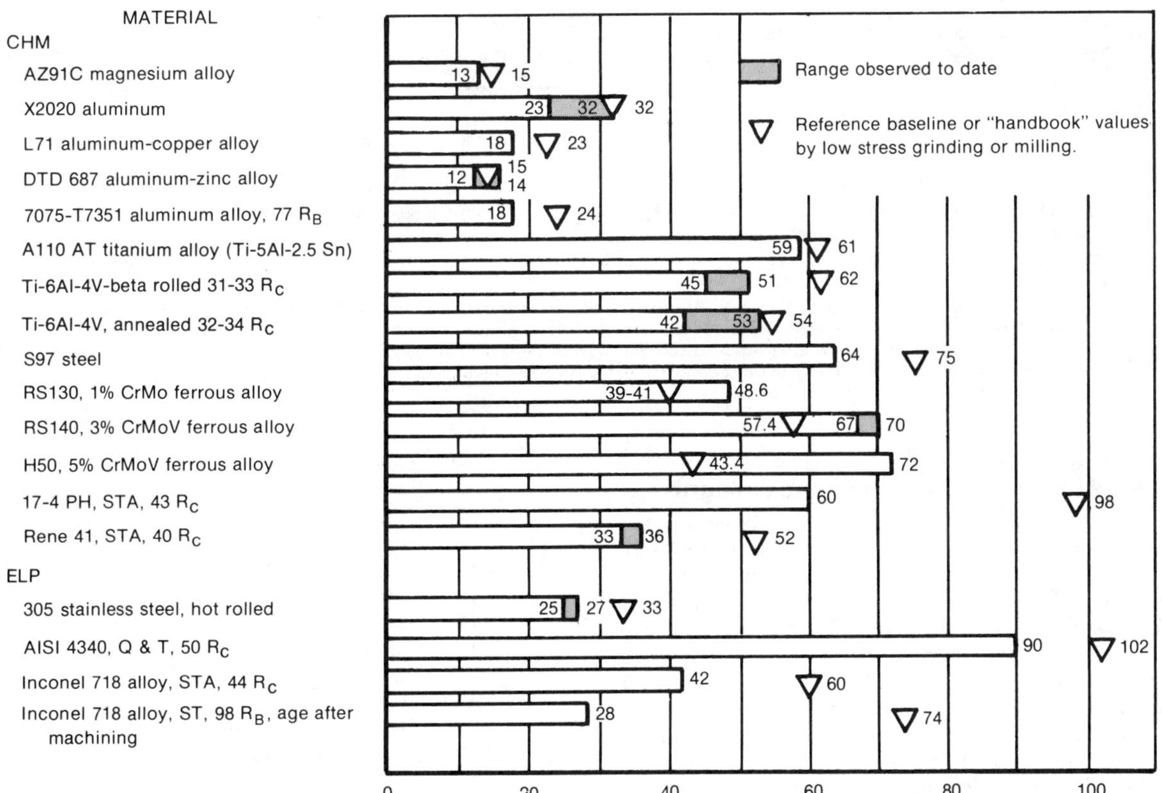


**Figure 18.3–101** Summary of high cycle fatigue strength from chemical machining (CHM) and electropolishing (ELP). (G. Bellows[23])

**TABLE 18.3-24 Surface Integrity Effects Observed in Chemical Material Removal Processes**

| PROPERTY AND TYPE OF EFFECT | FINISHING, "GENTLE" OR LOW STRESS CONDITIONS | | ROUGHING, "OFF-STANDARD" OR ABUSIVE CONDITIONS | |
|---|---|---|---|---|
| Surface roughness: | μin $R_a$ | μm $R_a$ | μin $R_a$ | μm $R_a$ |
| Average range | 63 to 250 | 1.6 to 6.3 | 125 to 500 | 3.2 to 12.5 |
| Less frequent range | 8 to 500 | 0.2 to 12.5 | 63 to 500 | 1.6 to 12.5 |
| Mechanical altered material zones: | inch* | mm* | inch* | mm* |
| Plastic deformation (PD) | *** | *** | *** | *** |
| Plastically deformed debris ($PD^2$) | *** | *** | *** | *** |
| Hardness alteration† | 0.0010 | 0.025 | 0.0031 | 0.079 |
| Microcracks or Macrocracks | — | — | — | — |
| Residual stress‡ | 0.0010 | 0.025 | 0.0010 | 0.025 |
| Metallurgical altered material zones: | inch* | mm* | inch* | mm* |
| Recrystallization | *** | *** | *** | *** |
| Intergranular attack (IGA) | 0.0003 | 0.008 | 0.0060 | 0.152 |
| Selective etch, pits, protuberances | 0.0006 | 0.015 | 0.0015 | 0.038 |
| Metallurgical transformations | — | — | — | — |
| Heat-affected zone (HAZ) or Recast layer | *** | *** | *** | *** |
| High cycle fatigue (HCF): | percent | | percent | |
| Variation from "handbook" values at room temperature§ | +18 to −39 | | −22 to −37 | |

SOURCE: G. Bellows and J. B. Kohls.[20]

NOTE: A dash (—) in the table indicates no or insufficient data. A triple asterisk (***) in the table indicates no occurrences or not expected.

*Maximum observed depth normal to the surface.

†Depth to point where hardness becomes less than ±2 points $R_c$ (or equivalent) of bulk material has (hardness converted from Knoop microhardness measurements).

‡Depth to point where residual stress becomes and remains less than 20 ksi [138 MPa] or 10% of tensile strength, whichever is greater.

§Handbook values from HCF testing frequently are generated from specimens that have been processed using low stress grinding, hand or gentle machine polishing or occasionally electropolishing. For this table, "handbook" values represent HCF strength values generated from low-stress-ground specimens (with their minor amount of retained but enhancing compressive residual stress). The values are the $10^7$ cycle strengths from full-reverse bending tests at room temperature.

## Specific Guidelines and Data for POST-TREATMENT Processes to Aid Surface Integrity

Finishing operations or post-treatment processing have long been used in manufacturing. These operations are used to improve the surface roughness or appearance, to remove undesirable surface layers, or to enhance the beneficial compressive layers. The surface integrity produced by these finishing processes, or sequence of processes, should be assessed with the same diligence as that used for the prime fabricating process. While shot peening has been the most extensively used post treatment, consideration should be given to other techniques to achieve the desired effects. Often, other treatments may be used that result in less cost or greater productivity or better quality assurance. Some of the effective post-treatment processes are as follows:

AFM Abrasive flow machining
CBN Roller burnishing
CBS Belt sanding
CGH Hand grinding
CHM Chemical machining
CHN Honing
CPE Peening—shot, sand
CPG Glass bead peening
CPO Buffing
CTB Abrasive barrel tumbling
CVD Chemical vapor deposition
ECD Electrochemical deburring
ECP Electrochemical polishing
ELP Electropolishing
HT Heat treatment
LSG Low stress grinding
LST Laser shock treatment
USC Ultrasonic cleaning
USP Ultrasonic peening
VSR Vibration stress relief

Very gentle, low energy intensity conventional turning, milling or reaming can be quite effective as finishing operations; however, on stress-critical areas these processes should be carefully evaluated as previously detailed. Process and/or quality control must be rigorous and continuous with much insistence on the use of sharp tools. Chemical and electrochemical processes with their absences of residual stress need special evaluation if it is desirable to add residual stress or cold working.

Figure 18.3-102 shows the shot peened surface of 18% nickel maraging steel Grade 250. Figure 18.3-103 illustrates the surface section of solution treated and aged Inconel Alloy 718 after shot peening, with off-standard ECM as the primary material removal method. Figure 18.3-103 depicts two levels of peening: level 1, using S110 shot of 50 to 55 $R_C$ hardness with 300 percent coverage to Almen strip intensity of 0.006 to 0.008 A; and level 2, which was the same peening as level 1 except coverage was 125 percent.

Figure 18.3-104 charts the residual stress pattern induced in solution treated and aged Inconel Alloy 718. The low-stress-ground specimens, 0.076 inch by 0.75 inch by 4 inches [approx. 2 mm by 20 mm by 100 mm], were shot peened with S110 steel shot to Almen intensity level 0.006 to 0.008 A. The residual stress profiles were prepared on specimens after exposure to the indicated temperatures for 100 hours in a vacuum. The shot peening pattern is essentially retained through 1300°F [704°C] for Inconel Alloy 718. This results in a 6 to 16 percent improvement in high cycle fatigue strength at 1000°F [538°C]. On such specimens, an overpeened (that is, damaged) condition can occur when a 0.012- to 0.014-A intensity level is used on thin specimens such as these.

Figure 18.3-105 summarizes the high cycle fatigue strengths found for various materials subsequent to shot peening. These data are comparable to those illustrated for the material removal processes. Most of the tests were run on specimens peened to 0.006 to 0.008 $A_2$ or 0.004 to 0.006 $A_2$ on the Almen strip scale with 125 to 300 percent coverage.

Figure 18.3-106 summarizes the high cycle fatigue strength found with several types of post-processing treatments on one alloy, Inconel Alloy 718.

**Post-Treatment Guidelines that Promote Good Surface Integrity Results**

1. Careful washing should be employed to remove all traces of fluids used during machining that could contribute to stress corrosion.

2. Low temperature heat treatment will remove any hydrogen picked up during processing.

3. Heat treatments following material removal are of limited usefulness.

4. Abrasive tumbling can be effective for improving surface texture and for adding a modest compressive residual stress to aid fatigue strength.

5. Peening with steel shot or sand or glass beads can be useful to add substantial beneficial compressive residual stress, improve surface cleanliness and enhance fatigue strength.

6. Mechanically gentle processing can be effective with specific evaluation of each situation.

7. The benefits of shot peening can be nullified or severely decreased by exposure to high local stresses or temperatures or by vibration which allows the surface stress to relax by plastic deformation either from creep or yielding during "post-peen" processing or during operation.

8. The possibility that peening, burnishing, lapping and similar processes can cover up, but not overcome, the existing microcracks should be assessed.

9. Human variability means that hand finishing should be avoided if possible or controlled rigorously.

10. It is possible to overpeen surfaces and create excessive core stresses, microcracks, untempered martensite, laps or similar abuses.

11. The junction between peened and unpeened surfaces should be "feathered" to avoid stress concentrations.

**Special Comments Concerning Surface Integrity Practices for Post-Treating Processes**

**Cracks, heat-affected and other detrimental layers created during material removal processes should be removed (or altered) from critically stressed areas of component parts.** Microstructural examination and microhardness testing can frequently establish the depth of adversely altered layers and other defects. Certain critical situations may require mechanical testing to be certain that no alterations have been produced which cannot be detected by microexamination. Removal of these defective layers by gentle processes should be accomplished. Some post-treatment processes can satisfactorily alter these layers or even enhance the fatigue strength results.

**Steel shot and glass bead peening as well as burnishing can be used to improve surface integrity.** A considerable number of studies have confirmed that the fatigue life determined in laboratory tests and verified by field performance is measurably enhanced by peening.[8] Laboratory tests have shown that shot peening is effective in substantially increasing high cycle and low cycle fatigue strength at both room and elevated temperatures for typical titanium alloys, high temperature alloys, and high strength steels, regardless of the type of prior material removal processing. For example, electrical discharge machining and abusive grinding are two processes which are among the most detrimental to these alloy groups. But when added as a post-processing treatment, shot peening has been shown capable of fully restoring the room and elevated temperature low cycle and high cycle fatigue strengths for these alloy groups. Since the above comments are based upon laboratory tests of specimens, component evaluations are recommended. Peening, which puts the surface layer into compression and cold works the surface, must be performed under controlled conditions. Specifications for controlled peening should include consideration of factors such as the cleanliness and surface roughness of the part being machined; the type, the chemistry, the geometry and the hardness of the shot and its fluid carrier; and peening time, intensity and coverage. Reports also indicate that peening improves resistance to corrosion and reduction in stress corrosion susceptibility. There are indications that cold working of drilled holes by burnishing is also beneficial as is roller burnishing. Some reports contain precautions against overpeening in order to avoid fatigue damage and to reduce the possibility of masking flaws such as fine cracks.

**Evaluate the use of controlled shot peening practices to restore fatigue life of components processed by electrical, chemical, and thermal removal processes.** Shot peening has been shown to be extremely effective in improving the fatigue life of *specimens* processed by electrochemical machining (ECM), electrical discharge machining (EDM), and electropolishing (ELP).[9] Component tests are recommended to confirm the favorable trends shown in tests on laboratory specimens. Examples of shot peening improvement in the fatigue strength of specimens are shown in table 18.3-25.

**Post heat treatments following material removal are of limited usefulness.** Stress relief treatments, used to soften hardened layers produced during grinding of steels, do not restore the hardness of overtempered layers which are present immediately below the damaged surface layers. Also, heat treatment does not heal any cracks produced during material removal. Some companies have advised the use of tempering operations to relieve stresses after electrical discharge machining (EDM) of steels. Stresses may be relieved but fatigue properties are not improved sufficiently. Evidence has also been presented to show that annealing treatments following EDM on nickel base alloys, such as Rene 41, Inconel Alloy 625, Inconel Alloy 718 and Monel Alloy K-500, seriously lower tensile strength and ductility as a result of carbon diffusion.[10] Experience has shown that heat treatments in some cases have improved surface integrity of damaged surfaces but not nearly as effectively as shot peening. Low temperature heat treatments are helpful in eliminating embrittlement for operations where hydrogen may be picked up during processing, such as in plating. Elimination of hydrogen is time- and temperature-dependent and also depends upon the alloy being treated. Steels, for example, are often treated at about 375° to 400°F [191° to 204°C] for periods of 8 hours or more.

**Abrasive tumbling is an effective process for improving surface properties including fatigue.** This process is less applicable than shot peening for many of the very large parts required for aerospace applications. Both abrasive tumbling and shot peening usually require an added polishing operation when very high finish requirements must be satisfied. Care must be taken not to remove the thin favorable surface layer established by peening or tumbling. Abrasive tumbling can be used to reverse unfavorable tensile stresses by inducing a compression stressed surface layer.

**Washing procedures should be employed for critical parts and assemblies to remove all traces of cutting fluids which may cause corrosion.** Typical suspect compounds are sulfur compounds on aluminum and nickel base alloys and chlorine compounds on titanium alloys.[11] Currently, some companies do not allow any chlorine-containing cutting fluids to be used in processing titanium parts which are to be used at room or elevated temperatures. Other companies use this precaution only for parts which are subjected to temperatures over 500°F [260°C]. For applications at less than 500°F [260°C], carefully controlled washing procedures are often used to remove the chlorinated and sulfurized cutting oils. These fluids are particularly effective for chip removal operations such as drilling, tapping and broaching. Since complete agreement does not exist among manufacturers regarding cutting fluid practices, subcontractors are obliged to follow the policies and procedures established by the prime contractor.

**Protection of parts.** Parts should not be stored for extended periods without being carefully washed and then covered with a coating of oil for corrosion prevention.

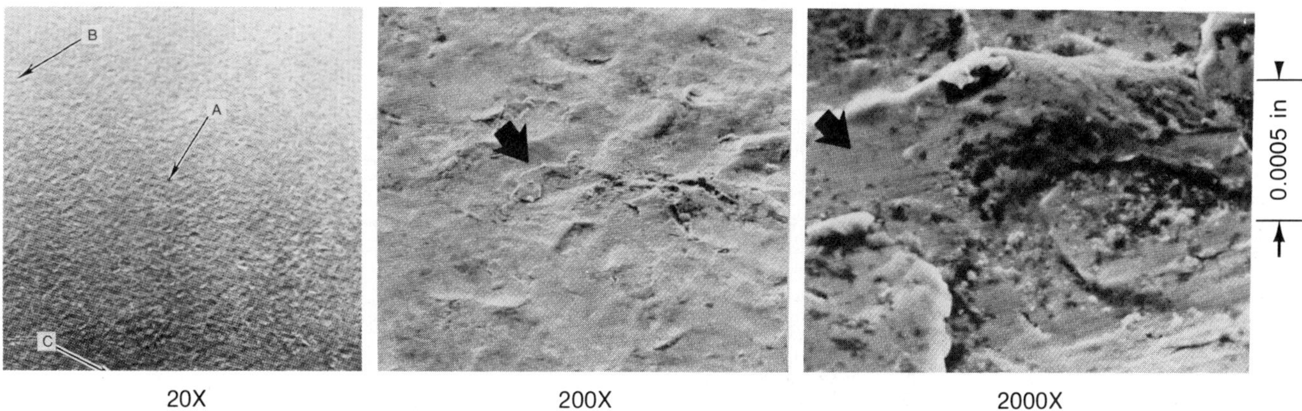


**Figure 18.3–102** Typical surface characteristics of shot-peened Grade 250 Maraging steel. Surface machined with ECM, then 300% coverage shot peened with S110 shot of 45 to 50 $R_C$ hardness to Almen level 0.004 to 0.006 $A_2$. View by SEM at 45° at increasing magnifications of 20, 200 and 2000X. Note impression of single shot in highest magnification view. Surface roughness of 38 microinches $R_a$. (Surface Integrity Encyclopedia[1])

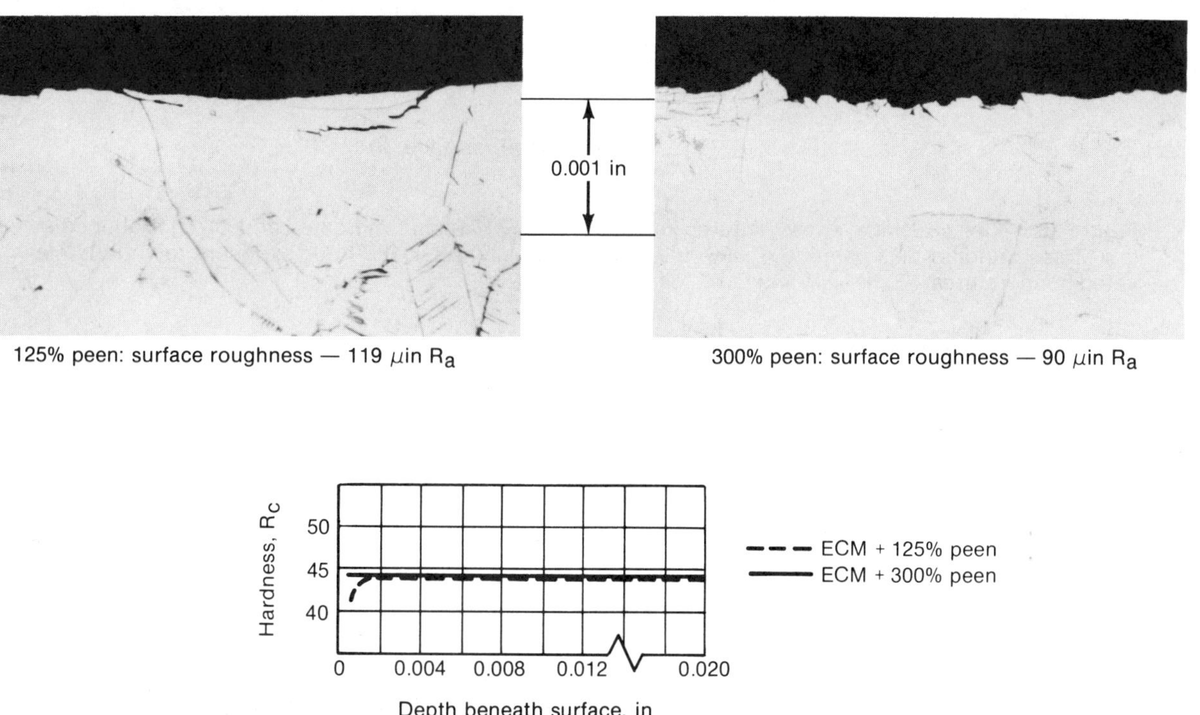


**Figure 18.3–103** Surface characteristics of Inconel Alloy 718 (solution treated and aged, 44 $R_C$) produced by ECM plus shot peening. Note roughness similar to ECM but with additional tears and cracks, perhaps from peening. Subsurface strain lines from peening also visible. (W. P. Koster et al[8])

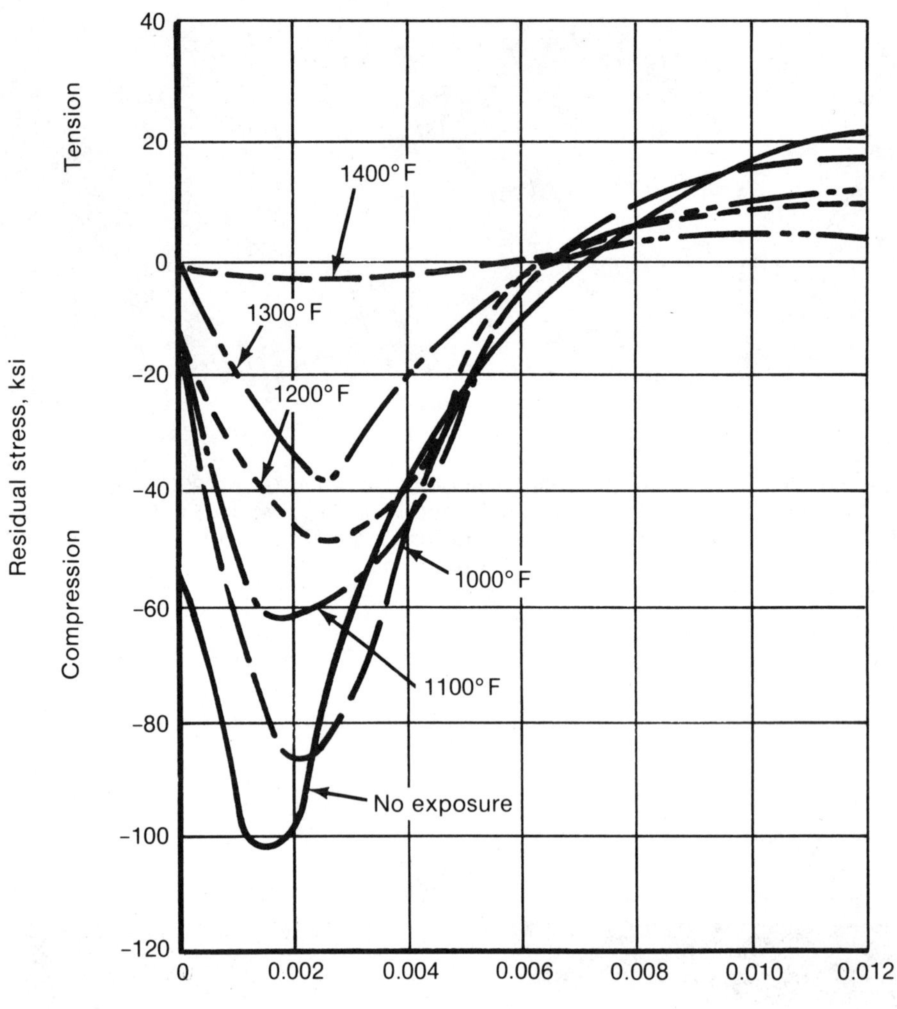


**Figure 18.3–104** Residual stress patterns in Inconel Alloy 718 (solution treated and aged) produced by low stress grinding plus peening (0.006- to 0.008-A intensity) after 100-hour exposure to various elevated temperatures. (G. Bellows and R. M. Niemi[24])

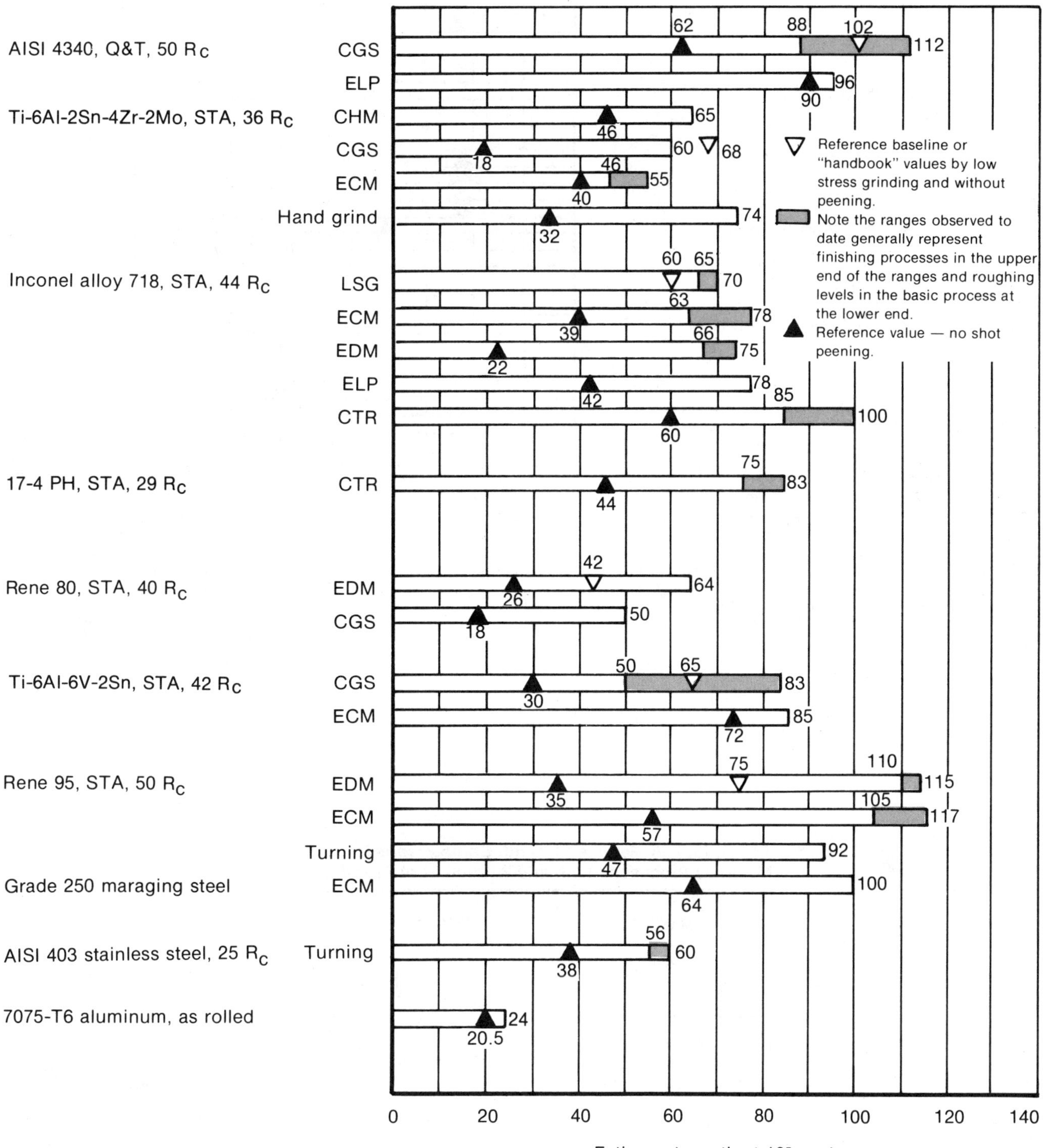


**Figure 18.3–105** Summary of high cycle fatigue strength of materials shot peened subsequent to basic material removal process.

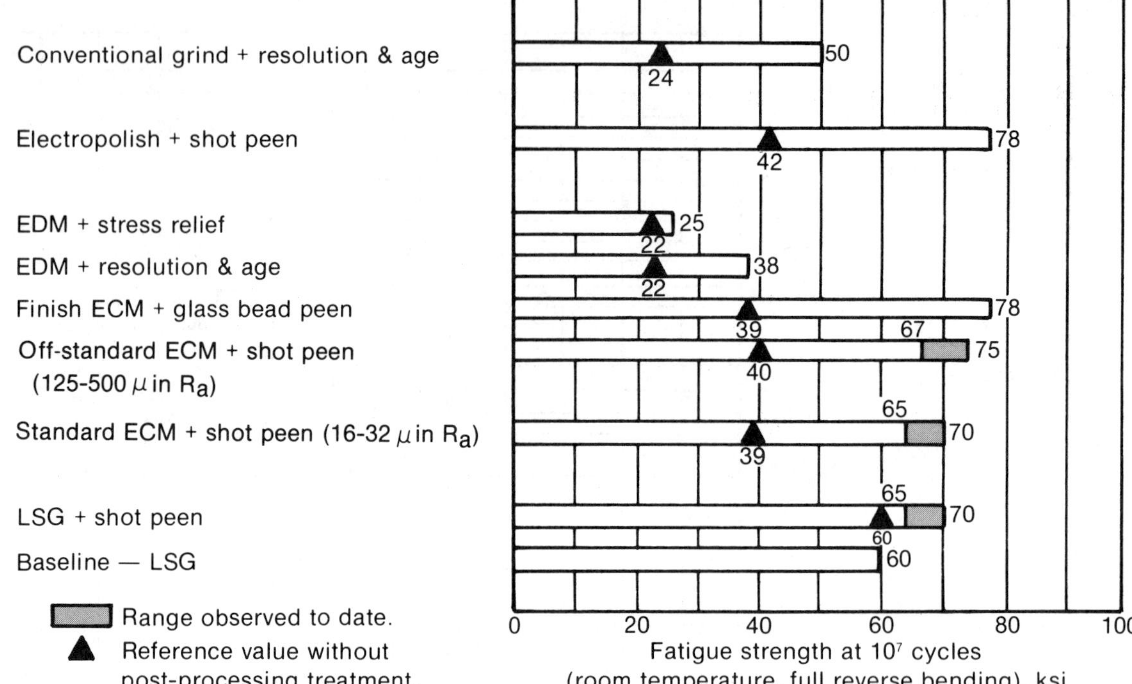


**Figure 18.3–106** Summary of high cycle fatigue strength of Inconel Alloy 718 (solution treated and aged) with several types of post-processing treatments. Baseline reference, or "handbook" value, from low stress grinding (LSG).

**TABLE 18.3-25 Fatigue Strength Comparisons for Materials Processed With and Without Shot Peening**

| MATERIAL AND PROCESS | FATIGUE STRENGTH* | | | |
|---|---|---|---|---|
| | Without Peen | | With Peen | |
| | ksi | MPa | ksi | MPa |
| AF 95 (STA, 50 $R_c$) | | | | |
| ECM | 57 | 393 | 105 | 724 |
| EDM | 40 | 276 | 115 | 793 |
| Low stress or Gentle grinding | 75 | 517 | — | — |
| Inconel 718 (STA, 44 $R_c$) | | | | |
| ECM | 39 | 269 | 78 | 538 |
| ELP | 42 | 290 | 78 | 538 |
| Low stress or Gentle grinding | 60 | 414 | — | — |
| Ti-6Al-6V-2Sn (STA, 42 $R_c$) | | | | |
| ECM | 72 | 496 | 85 | 586 |
| Low stress or Gentle grinding | 65 | 448 | — | — |

*Room temperature, $10^7$ cycles full-reverse bending.

## SURFACE INTEGRITY BY MATERIALS—HIGH CYCLE FATIGUE RESULTS

It is helpful in selecting the processes to be used in fabricating a component part to compare the relative fatigue strengths from a variety of processes. Value engineering can be strengthened by use of surface integrity data to assure a more complete and disciplined approach to process selection. Tables 18.3-7, 18.3-8, and 18.3-26 to 18.3-30 provide such data for a number of alloys. Caution and restraint should be exercised before transferring these data to similar alloys. It has been found that each intensity level of a process will have a unique surface integrity response for each metallurgical state of the work material. The data presented are the room temperature high cycle fatigue strength at $10^7$ cycles taken from full-reverse bending tests.

**TABLE 18.3-26 Comparative Fatigue Strengths and Surface Roughnesses of 4340 Steel (Quenched and Tempered 50 $R_c$) from Several Processes**

| PROCESS | FATIGUE STRENGTH* | | SURFACE ROUGHNESS | |
|---|---|---|---|---|
| | ksi | MPa | μin $R_a$ | μm $R_a$ |
| Grinding: | | | | |
| Gentle longitudinal surface grinding | 117 | 807 | 8 | 0.20 |
| Gentle longitudinal surface grinding | 110 | 758 | 65 | 1.60 |
| Gentle longitudinal surface grinding | 100 | 689 | 127 | 3.20 |
| Gentle traverse surface grinding | 120 | 827 | 11 | 0.28 |
| Gentle traverse surface grinding | 100 | 689 | 58 | 1.50 |
| Gentle traverse surface grinding | 85 | 586 | 128 | 3.30 |
| Low stress surface grinding (LSG) | 102 | 703 | 45 | 1.10 |
| Conventional surface grinding | 70 | 483 | 40 | 1.00 |
| Abusive grinding | 62 | 427 | 50 | 1.25 |
| Abusive longitudinal surface grinding | 65 | 448 | 29 | 0.70 |
| Abusive longitudinal surface grinding | 65 | 448 | 64 | 1.60 |
| Abusive longitudinal surface grinding | 65 | 448 | 97 | 2.50 |
| Low stress grinding plus shot peen | 112 | 772 | 25 | 0.63 |
| Abusive grinding plus shot peen | 88-92 | 607-634 | 40-76 | 1.00-1.90 |
| Gentle hand sanding | 94 | 648 | 115 | 2.90 |
| Abusive hand sanding | 98 | 676 | 105 | 2.70 |
| Abusive hand sanding plus shot peen | 118 | 558 | 56 | 1.40 |
| Milling: | | | | |
| Finish end milling—end cut | 85 | 586 | 75 | 1.90 |
| Rough end milling—end cut | 72 | 496 | 71 | 1.80 |
| Chemical NTM: | | | | |
| Electropolishing (ELP) | 90 | 621 | 15 | 0.40 |
| ELP plus shot peen | 96 | 662 | 46 | 1.20 |

*Room temperature, $10^7$ cycles full-reverse bending.

**TABLE 18.3-27 Comparative Fatigue Strengths and Surface Roughnesses of Ti-6Al-4V from Several Processes**

| MATERIAL CONDITION AND PROCESS | FATIGUE STRENGTH* | | SURFACE ROUGHNESS | |
|---|---|---|---|---|
| | ksi | MPa | μin $R_a$ | μm $R_a$ |
| BETA ROLLED 32 $R_c$ | | | | |
| Grinding: | | | | |
| Low stress surface grinding | 62 | 427 | 35 | 0.90 |
| Conventional surface grinding | 12 | 83 | 45 | 1.1 |
| Abusive surface grinding | 13 | 90 | 65 | 1.60 |
| Gentle hand sanding | 57 | 393 | 80 | 2.0 |
| Abusive hand sanding | 30 | 207 | 80 | 2.0 |
| Milling: | | | | |
| End milling—sharp cutter | 64 | 441 | 67 | 1.70 |
| End milling—dull cutter | 77 | 531 | 84 | 2.10 |
| Peripheral milling—sharp cutter | 70 | 483 | 17 | 0.40 |
| Peripheral milling—dull cutter | 32 | 221 | 59 | 1.50 |
| Chemical NTM: | | | | |
| Standard CHM | 51 | 352 | 20 | 0.50 |
| Off-standard CHM | 45 | 310 | 165 | 4.20 |
| SOLUTION TREATED AND AGED | | | | |
| Electrical NTM: | | | | |
| Standard ECM | 60 | 414 | 14 | 0.35 |
| Off-standard ECM | 40 | 276 | 165 | 4.20 |
| ANNEALED 32-34 $R_c$ | | | | |
| Grinding: | | | | |
| Low stress surface grinding | 54 | 372 | 24 | 0.60 |
| Conventional surface grinding | 12 | 83 | 30 | 0.75 |
| Abusive surface grinding | 14 | 97 | 48 | 1.20 |
| Turning: | | | | |
| Gentle turning | 66 | 455 | 22 | 0.55 |
| Rough turning | 71 | 490 | 34 | 0.85 |
| Milling: | | | | |
| Gentle end milling—end cut | 73 | 503 | 16 | 0.40 |
| Rough end milling—end cut | 64 | 441 | 20 | 0.50 |
| Chemical NTM: | | | | |
| Standard CHM | 53 | 365 | 18 | 0.45 |
| Off-standard CHM | 42 | 290 | 157 | 3.90 |
| Thermal NTM: | | | | |
| EDM | 24 | 165 | 69 | 1.75 |

*Room temperature, $10^7$ cycles full-reverse bending.

**TABLE 18.3-28 Comparative Fatigue Strengths and Surface Roughnesses of Ti-6Al-2Sn-4Zr-2Mo (Solution Treated and Aged 36 $R_c$) from Several Processes**

| PROCESS | FATIGUE STRENGTH* | | SURFACE ROUGHNESS | |
|---|---|---|---|---|
| | ksi | MPa | μin $R_a$ | μm $R_a$ |
| Grinding: | | | | |
| Low stress surface grinding (LSG) | 68 | 469 | 39 | 1.00 |
| Conventional surface grinding | 17 | 117 | 41 | 1.00 |
| Abusive surface grinding | 10 | 69 | | |
| Hand grinding | 32 | 221 | 120 | 3.00 |
| LSG plus age | 50 | 345 | | |
| Conventional surface grinding plus shot peen | 60 | 414 | | |
| Hand grinding plus stress relief | 44 | 303 | | |
| Hand grinding plus shot peen | 74 | 510 | | |
| Milling: | | | | |
| Gentle milling—peripheral cut | 82 | 565 | 36 | 0.90 |
| Rough milling—peripheral cut | 47 | 324 | 77 | 1.90 |
| Electrical NTM: | | | | |
| Standard ECM | 40 | 276 | 32-125 | 0.80- 3.20 |
| Off-standard ECM | 25 | 172 | 125-500 | 3.20-12.50 |
| STEM | 35 | 241 | | |
| Standard ECM plus shot peen | 55 | 379 | | |
| Off-standard ECM plus shot peen | 46-50 | 317-345 | | |
| Chemical NTM: | | | | |
| ELP | 50 | 345 | | |
| CHM | 46 | 317 | | |
| CHM plus shot peen | 65 | 448 | | |

*Room temperature, $10^7$ cycles full-reverse bending.

**TABLE 18.3-29 Comparative Fatigue Strengths and Surface Roughnesses of Inconel Alloy 718 (Solution Treated and Aged 44 $R_c$) from Several Processes**

| PROCESS | FATIGUE STRENGTH* | | SURFACE ROUGHNESS | |
|---|---|---|---|---|
| | ksi | MPa | μin $R_a$ | μm $R_a$ |
| Grinding: | | | | |
| Low stress surface grinding (LSG) | 60 | 414 | 15 | 0.38 |
| Conventional surface grinding | 24 | 165 | 26 | 0.65 |
| Abusive surface grinding | 22 | 152 | 70 | 1.75 |
| Hand grinding | 40 | 276 | 64 | 1.60 |
| LSG plus resolution and age | 58 | 400 | 15 | 0.38 |
| LSG plus superfinish and age | 49 | 338 | <5 | <0.125 |
| LSG plus vacuum age | 74 | 510 | 17 | 0.43 |
| LSG plus shot peen | 65-70 | 448-483 | | |
| LSG plus shot peen | 70 | 483 | — | — |
| Conventional surface grinding plus resolution and age | 50 | 345 | 26 | 0.65 |
| Conventional surface grinding plus stress relief | 58 | 400 | — | — |
| Hand grinding plus stress relief | 40 | 276 | 32 | 0.80 |
| Turning: | | | | |
| Gentle face turning | 60 | 414 | 25 | 0.63 |
| Gentle face turning | 60 | 414 | 58 | 1.45 |
| Gentle face turning | 60 | 414 | 188 | 4.70 |
| Abusive face turning | 60 | 414 | 76 | 1.90 |
| Conventional turning—Borazon® CBN tool | 85 | 586 | 39 | 1.00 |
| Conventional turning—Carbide | 55 | 379 | 65 | 1.60 |
| Electrical NTM: | | | | |
| Standard ECM | 39-40 | 269-276 | 43 | 1.10 |
| Off-standard ECM | 39-40 | 269-276 | 74-500 | 1.9-12.5 |
| STEM | 39 | 269 | 125 | 3.20 |
| Standard ECM plus vacuum age | 42 | 290 | 19 | 0.48 |
| Off-standard ECM plus vacuum age | 42 | 290 | 55 | 1.38 |
| Standard ECM plus glass bead peen | 78 | 538 | 69 | 1.70 |
| Standard ECM plus shot peen | 65-78 | 448-538 | 74 | 1.90 |
| Off-standard ECM plus shot peen | 67-75 | 462-517 | 90-119 | 2.25-3.00 |
| Thermal NTM: | | | | |
| Finish EDM | 22 | 152 | 60 | 1.50 |
| Rough EDM | 22 | 152 | 170 | 4.30 |
| Finish EDM plus vacuum age | 29 | 200 | 65 | 1.60 |
| Rough EDM plus vacuum age | 29 | 200 | 155 | 3.90 |
| Rough EDM plus resolution and age | 26-38 | 179-262 | 175 | 4.40 |
| Rough EDM plus stress relief | 25 | 172 | 221 | 5.50 |
| Finish EDM plus glass bead peen | 66 | 455 | 43 | 1.10 |
| Rough EDM plus shot peen | 75 | 517 | 125 | 3.20 |
| Chemical NTM: | | | | |
| CHM | 40 | 276 | 16 | 0.40 |
| ELP | 40-42 | 276-290 | 15 | 0.40 |
| ELP plus shot peen | 78 | 538 | 43 | 1.10 |
| ELP plus vacuum age | 28 | 193 | 15 | 0.40 |

*Room temperature, $10^7$ cycles full-reverse bending.

**TABLE 18.3-30 Comparative Fatigue Strengths and Surface Roughnesses of Rene 80 (Solution Treated and Aged) from Several Processes**

| PROCESS | FATIGUE STRENGTH* | | SURFACE ROUGHNESS | |
|---|---|---|---|---|
| | ksi | MPa | μin $R_a$ | μm $R_a$ |
| **Grinding:** | | | | |
| Low stress grinding (LSG) | 42-43 | 290-296 | 16 | 0.40 |
| Conventional grinding | 18 | 124 | | |
| Abusive grinding | 16 | 110 | | |
| Low stress grinding—finishing plus age heat treat | 33-40 | 228-276 | | |
| Low stress grinding—finishing plus full heat treat† | 35 | 241 | | |
| Low stress grinding—finishing plus full heat treat and shot peen† | 53-57 | 365-393 | | |
| Conventional grinding—roughing plus age heat treat | 32-35 | 221-241 | | |
| Conventional grinding—roughing plus full heat treat† | 33-36 | 228-248 | | |
| Conventional grinding—roughing plus full heat treat and shot peen† | 54 | 372 | | |
| Ultra-high speed grinding plus vacuum heat treat | 40 | 276 | | |
| Ultra-high speed grinding plus shot peen | 50 | 345 | | |
| **Electrical NTM:** | | | | |
| Standard ECM conditions | 28 | 193 | 20 | 0.50 |
| Off-standard ECM conditions | 28 | 193 | 40 | 1.00 |
| Finish ECG | 30 | 207 | | |
| Rough ECG | 30 | 207 | | |
| **Thermal NTM:** | | | | |
| Finish EDM | 24-26 | 165-179 | 70 | 1.75 |
| Rough EDM | 22-26 | 152-179 | 150 | 3.75 |
| Solution treat plus EDM | 25 | 172 | | |
| EDM plus resolution and age | 32 | 221 | | |
| EDM with recast removed by LSG | 34 | 234 | | |
| EDM plus shot peen | 64 | 441 | | |

*Room temperature, $10^7$ cycles full-reverse bending.
† Full heat treatment includes aging plus equivalent of brazing and coating cycles.

## QUALITY ASSURANCE FOR SURFACE INTEGRITY

Quality control practices should be reviewed and amplified in order to satisfy high surface quality requirements. No instrument exists that will measure surface integrity. Properties which may require investigation in order to establish surface integrity include surface texture, surface metallurgy, mechanical properties, surface chemistry and other engineering properties. Table 18.3-31 is a summary of standard and specialized techniques generally employed for these measurements.

It is advisable to become acquainted with the possible types of surface alterations which result from the combination of a metal removal process with a specific material. Information of this type is provided in tables 18.3-2 and 18.3-32. Nondestructive testing is the most desirable approach to evaluating surface integrity effects. The principal nondestructive testing methods for evaluating surface integrity and subsurface characteristics are listed in tables 18.3-33 and 18.3-34. Destructive techniques are also shown in table 18.3-34 for comparison.

**TABLE 18.3-31 Techniques for Surface Integrity Measurement**

| PROPERTY | STANDARD TECHNIQUES | SPECIALIZED TECHNIQUES |
|---|---|---|
| Surface Texture | Contact (tracer point or stylus measurement)—Linear traverse or area traverse | Noncontact measurements—<br>Optical Microscopy<br>Interference microscopy, etc.<br>Three dimensional surface profile by combination of roughness measurement and computer plotting<br>Scanning electron microscopy |
| Surface Metallurgy | | |
| Microstructure | Metallurgical sectioning—Optical microscopy | Transmission electron microscopy |
| Microhardness | Microhardness testing—Knoop or Vickers indenter | |
| Microcracks and Crevice-like defects | Metallurgical sectioning—Optical microscopy<br>Nondestructive—Macroetching<br>Penetrant inspection<br>etc. | Scanning electron microscopy<br>Nondestructive—Eddy Current<br>Ultrasonic<br>etc. |
| Static Mechanical Properties | | |
| Tensile strength & Ductility | Tensile testing | |
| Stress rupture & Creep | Creep testing | |
| Residual stress | X-ray diffraction<br>Layer removal—Curvature measurement | |
| Stress corrosion | *Steels & Aluminum:* Constant prestress with alternate immersion in salt solution and air<br>*Titanium:* Specimen coated with salt or halide and exposed to high temperature and stress | Stress corrosion crack propagation testing |
| Fracture toughness | | Fracture toughness testing |
| Dynamic Mechanical Properties | | |
| High cycle fatigue | Bending fatigue testing, $A^* = \infty$ | Axial fatigue testing, $A^* = 0.95$ |
| Low cycle fatigue | Bending fatigue testing, $A = \infty$ | Axial fatigue testing, $A = 0.95$ |
| Surface Chemistry | Electron microprobe analysis<br>Dissolution of the surface layer by chemical or electrochemical etching followed by spectroscopy, spectrophotometry, etc. | Electron spectroscopy—Electron spectroscopy for chemical analysis (ESCA)<br>Auger spectroscopy<br>Ion spectroscopy—<br>Ion scattering spectrometry<br>Ion probe mass spectroscopy |
| Other Engineering Properties<br>Friction<br>Wear<br>Fretting<br>Galling & Seizing<br>Corrosion<br>Reflectivity<br>Electrical properties<br>Etc. | A variety of standard and specialized techniques have been developed to evaluate these properties for specific applications. | |

SOURCE: M. Field, J. F. Kahles and J. T. Cammett.[12]

*A = ratio of dynamic/static stress.

**TABLE 18.3-32 Surface Hardness Changes That May Result from Various Metal Removal Processes**

| MATERIAL | CONVENTIONAL PROCESSES Milling, Drilling, Turning, or Grinding | | NONTRADITIONAL PROCESSES EDM | | NONTRADITIONAL PROCESSES ECM or CHM | |
|---|---|---|---|---|---|---|
| | Surface Alteration | Hardness Change | Surface Alteration | Hardness Change | Surface Alteration | Hardness Change |
| Steels | PD<br>UTM<br>OTM<br>RS<br>OA | Increase<br>Increase<br>Decrease<br>Decrease<br>Decrease | RC<br>UTM<br>OTM<br>RS<br>OA | Increase<br>Increase<br>Decrease<br>Decrease<br>Decrease | None | Decrease |
| Nickel- and Cobalt-base superalloys | PD | Increase | RC | Increase | None | Decrease |
| Titanium alloys | PD | Increase | RC | Increase | None | Decrease |
| Refractory alloys (TZM, Tungsten) | | | RC | No change | None | Decrease |

SOURCE: M. Field, J. F. Kahles and J. T. Cammett.[12]

Key: PD — Plastic deformation and plastically deformed debris
UTM — Untempered martensite
OTM — Overtempered martensite
RS — Resolution or austenite reversion
OA — Overaging
RC — Recast, resplattered metal or vapor deposited metal

**TABLE 18.3-33 Nondestructive Techniques for Detecting Surface Inhomogeneities in Metals**

| TECHNIQUE | APPLICATION | STATUS |
|---|---|---|
| 1. Visual Inspection | Surface defects | Widely employed |
| 2. Automatic Optical Scanning | Surface defects | Specialized |
| 3. Dye Penetrant | Surface defects | Widely employed |
| 4. Magnetic Particle | Surface defects | Widely employed |
| 5. Acid Macroetch | Crack detection | Widely employed |
| | Surface phase transformations | Widely employed |
| | Redeposited & resolidified metal | Widely employed |
| 6. Eddy Current | Surface defects | Widely employed |
| | Inclusion & subsurface defects | Widely employed |
| | Surface hardness changes | Specialized |
| | Plastic deformation | Specialized |
| | Surface phase transformation | Needs development |
| | Subsurface fatigue | Needs development |
| | Residual stress analysis | Needs development |
| 7. Ultrasonic Pulse Echo | Inclusions & subsurface defects | Widely employed |
| | Surface defects | Specialized |
| 8. Ultrasonic Velocity | Residual stress analysis | Specialized |
| | Surface phase transformation | Needs development |
| | Plastic deformation | Needs development |
| 9. Ultrasonic Attenuation | Recrystallization & grain growth | Specialized |
| | Residual stress analysis | Needs development |
| 10. High Frequency Ultrasonic | High resolution of surface defects | Specialized |
| 11. X-Ray Radiography | Subsurface defects | Widely employed |
| 12. X-Ray Diffraction | Residual stress analysis | Widely employed |
| | Surface phase transformation | Specialized |
| 13. X-Ray Spectroscopy | Composition changes | Needs development |
| 14. Neutron Radiography | Subsurface defects | Specialized |
| 15. Gamma Radiography | Subsurface defects | Specialized |
| 16. Radioactive Gas Penetrant | Surface defects | Needs development |
| | Recrystallization & grain growth | Needs development |
| | Plastic deformation | Needs development |
| | Subsurface defects | Needs development |
| 17. Surface Electrical Resistance | Surface defects | Needs development |
| | Surface phase transformation | Needs development |
| 18. Beta Backscatter | Surface hardness changes | Needs development |
| | Plastic deformation | Needs development |
| 19. Electrochemical Potential | Residual stress analysis | Needs development |
| 20. Laser Probe Mass Spectrometry | Surface composition | Needs development |

SOURCE: M. Field, J. F. Kahles and J. T. Cammett.[12]

**TABLE 18.3-34 Testing Techniques Used To Detect and Locate Surface Inhomogeneities in Metals**

| METALLURGICAL INHOMOGENEITY | NONDESTRUCTIVE TECHNIQUES | | | DESTRUCTIVE TECHNIQUES |
|---|---|---|---|---|
| | Commonly Employed | Specialized | Possible With Further Developments | |
| Macrocracks | Visual inspection<br>Binocular inspection<br>Magnetic particle<br>Penetrant<br>Eddy current<br>Acid macroetch | Ultrasonic pulse echo<br><br>Automatic optical scanning | Acoustic impact | Optical metallography |
| Microcracks | Binocular inspection<br>High sensitivity fluorescent penetrant<br>Magnetic particle | Ultrasonic pulse echo, surface waves and lamb waves | Radioactive gas penetrant<br>High frequency ultrasonic<br>Acoustic impact<br>Surface electrical resistance | Optical metallography<br>Scanning electron microscopy<br>Transmission electron microscopy |
| Tears, laps & pits | Visual with etch<br>Magnetic particle<br>Eddy current<br>Penetrant | Automatic optical scanning | Radioactive gas penetrant | Optical metallography<br>Scanning electron microscopy |
| IGA & selective etch | | High sensitivity fluorescent penetrant | | Macroetch<br>Optical electron microscopy<br>Transmission electron microscopy |
| Surface phase transformation (OTM, UTM, resolutioning, etc.) | Macroetch | X-Ray diffraction<br>Magnetic particle | Ultrasonic velocity<br>Surface electrical resistance<br>Eddy current | Optical metallography |
| Composition changes (oxidation, decarb, etc.) | | | Laser probe<br>Mass spectrometry<br>X-Ray spectroscopy | Wet chemical analysis<br>Electron microprobe |
| Surface hardness changes | Superficial hardness testing<br>Ultrasonic hardness testing | Eddy current<br>X-Ray diffraction | Beta backscatter | Microhardness traverse |
| Redeposited & resolidified metal | Macroetch<br>Visual inspection | | | Optical metallography |
| Recrystallization & grain growth | | Ultrasonic attenuation | Radioactive gas penetrant | Optical metallography |
| Plastic deformation (cold work, hot work) | Superficial hardness testing | Eddy current<br>Magnetic particle | Beta backscatter<br>Radioactive gas penetrant<br>Ultrasonic velocity | Optical metallography<br>Microhardness traverse |
| Inclusions & voids | Ultrasonic pulse echo shear wave, surface wave, and lamb wave<br>Penetrant<br>X-Ray radiography<br>Eddy current<br>Magnetic particle | Gamma radiography<br>Neutron radiography | Radioactive gas penetrant | Optical metallography |
| Residual stresses | X-Ray diffraction | Ultrasonic velocity | Eddy current<br>Electrochemical potential<br>Ultrasonic attentuation<br>Magneto-absorption | Parting-out<br>Layer removal<br>X-Ray diffraction |
| Distortion | Visual inspection | Metrology | | |

SOURCE: M. Field, J. F. Kahles and J. T. Cammett.[12]

After exhausting the nondestructive testing methods, reliance is then placed on destructive testing of extra workpieces or coupons by simulating the process factors. In some situations, it is necessary to resort to detailed process parameter planning followed up by operator education and periodic audits for compliance to the preplanned conditions. Few quality assurance engineers are fully satisfied with a process control approach; however, it can be effective—if production supervision exercises a relentless insistence upon performance according to standards.

It is vital to surface integrity that as much attention be placed on the quality of the material presented for processing as is placed upon process parameter control. The metallurgical state will affect the final "as-shipped" surface quality as much as the intensity level at which the process is operated. Similarly the *full* sequence of operations applied to a surface must be evaluated, and the evaluation must be made with the *exact* sequence used during production.

The best quality assurance is the evaluation of surfaces of specific parts when exposed to the full environmental effects expected; namely, heat, chemicals, stress, vibration, etc. The minimum data set can provide a better evaluation, but the extended data set is recommended for the most critically loaded or exposed surfaces. This controlled, systematic testing technique is recommended for safety critical areas of component parts.

## Specific Guidelines and Special Comments on Quality Assurance for Surface Integrity

### 1. Detection of Untempered and Overtempered Martensite

Parts manufactured from martensitic high strength steel can be visually inspected after an acid etch for evidence of untempered or overtempered martensite (frequently found from grinding and some forms of mechanical cutting). Typically, untempered martensite appears white and overtempered areas appear darker than the background material. A specific etching technique for detecting grinding damage in hardened steel is given in table 18.3-35.

### 2. Conventional Versus Nontraditional Processes

The original surface integrity investigations concentrated heavily on grinding and the nontraditional material removal processes like electrical discharge machining and electrochemical machining. More recently, the older, conventional, generally mechanical material removal processes have been shown to have an equal range of surface integrity effects. It is vitally important that the application of surface integrity practices include an assessment of *every* process that effects the final, "as-shipped" surface.

### 3. Suggested Experimental Programs

Systematic metallurgical and mechanical testing programs for establishing and controlling surface integrity are essen-

**TABLE 18.3-35 Etching Technique for Detecting Grinding Damage in Hardened Steel**

| OPERATION | SOLUTION USED | DESCRIPTION, TIME, OR FUNCTION |
|---|---|---|
| **Double Etch Method** | | |
| (1) Etch No. 1 | 4 to 5% Nitric acid in water | Until black, 5 to 10 seconds. Do not over etch. |
| (2) Rinse | Warm water | To remove acid |
| (3) Rinse | Methanol (or acetone*) | To remove water |
| (4) Etch No. 2 | 5 to 10% Hydrochloric acid in methanol (or acetone*) | Until black smut is removed, 5 to 10 seconds |
| (5) Rinse | Running warm water | To remove acid |
| (6) Neutralize | 2% Sodium carbonate + phenolpthalien indicator in water | To neutralize any remaining acid |
| (7) Rinse | Methanol | To remove water |
| (8) Dry | Warm air blast | |
| (9) Oil dip | Low viscosity mineral oil with rust inhibitor | Enhance contrast, prevent corrosion |
| **Nital Etch Method** | | |
| (1) Etch | 5 to 10% Nitric acid in ethanol or methanol | Until contrast is evident |
| (2) Repeat steps 5.9 above | | |

SOURCE: M. Field, J. F. Kahles and J. T. Cammett.[12]

NOTE: Dark areas on etched surfaces show overtempering, light areas show rehardening, and uniform gray indicates no injury.

*4% $HNO_3$ in water for Etch No. 1 used with 2% HCl in acetone for Etch No. 2 sometimes gives greater sensitivity on high carbon hardened steel. It is important that appropriate precautions be taken to avoid fire hazards, and good ventilation must be provided.

tial for highly critical parts and have been detailed previously. Functional testing of the final product of field testing under actual service conditions is a desirable step for the most critical applications.

**4. Comparative Photomicrographs**

Comparison of current results during production with sets of photomicrographs that display both the "standard" and "off-standard" conditions to be expected are helpful. The growing data accumulations in surface integrity encyclopedias also are useful. The Machinability Data Center has the most extensive collection of this information. These data also are useful in appraising or establishing quality control limits or process parameter limits.

**5. Specifications**

Several companies now have special specifications for processing to secure the best surface integrity. These are very helpful to the vendors for guidance and to quality control for measurement. These special specifications should be used with care and only for those portions or surfaces of a component part deemed critical, otherwise costs will rise unnecessarily. Standards for surface integrity are being prepared both by the American National Standards Institute (ANSI) and by the International Organization for Standardization (ISO).

**6. Accidents Which Occur During Processing Should Be Reported to Quality Control**

Operators should be instructed to notify supervisors of all accidents or damage including all visual evidence of damage. Grinding burn, breaking of drills, reamers or taps in holes, or shorting of electrodes in both electrical discharge machining and electrochemical machining are examples of the types of mishaps which should be reported. Cleanup of surface discoloration is not necessarily sufficient to remove all of the damage which may extend several thousandths inch below the surface, as demonstrated in figure 18.3-75. In case of damage, parts should be systematically reviewed.

**7. Inspection Practices Should Be Reviewed and Amplified in order To Satisfy Surface Integrity Requirements**

There are several nondestructive and destructive testing practices which are being employed to test surface conditions. Brief discussions of these practices are presented in the following:

a. Microscopic examination including microhardness testing is used on a sampling basis in order to determine the kind of surface layer being produced and its depth. This method may be used to check for microcracks, pits, folds, tears, laps, built-up edge, intergranular attack, sparking, etc. Since many surface alterations are shallow, it is essential to use good edge retention techniques for microscopic examination. See table 18.3-6.

b. The white layer or overtempered martensite produced during grinding of steels can be detected by immersion etching using a 3 to 5 percent aqueous nitric acid solution. The procedures used vary among producers and are quite detailed and very exacting, consisting essentially of appropriate and specific precleaning, etching, and post-etching procedures, including cleaning, drying, and application of rust inhibitors, as outlined in table 18.3-35.

c. Magnetic particle inspection, penetrant inspection, ultrasonic testing, and eddy current techniques are recommended for detecting macrocracks. Most of the inspection techniques currently being used should be further refined by using more care in their application. Direct visual examination should be supplemented by macroscopic examination at low and medium magnifications (5 to 20X). Some of these methods require low levels of surface roughness on the surface being examined in order to reduce background "noise" to the instrument.

d. X-ray diffraction methods are available to detect residual surface stresses. This method may be helpful in process development as well as in spot checking of finished parts. The entire residual stress profile in the surface layer can be destructively determined by X-ray diffraction plus etch techniques or by a deflection-etch technique.

**8. Process Control Practices**

Where process control is relied upon to maintain surface integrity, deterioration of the process control can frequently be detected by observing the change in surface finish. A gradual drift to rougher values can signal a gradual drift in process parameters. One practical rule of thumb states that if, as a corrective measure, any one of the operating parameters of a process is reset or "adjusted" by plus or minus 10 percent or more from the planned values, immediate questions should be raised as to the control status of the process. If two or more parameters are "adjusted" by 10 percent or more, the process is out of control and should be shut down until corrective action is effective.

## ECONOMICS AND APPLICATIONS OF SURFACE INTEGRITY

A frequent initial reaction to surface integrity is a concern for productivity and costs. The generally gentle, low speed, lower energy levels employed in machining unimpaired or enhanced surfaces do increase costs and lower production rates. However, this is only part of the overall cost picture. Consideration should be given to the added costs of defective or scrapped parts or the cost of rework or replacement during the course of manufacture when component parts fail to pass quality control inspections. Subsequent to manufacture, the costs of early product failures, complaints, liability or its insurance also must be included. Many of these costs can be ameliorated by attention to surface integrity and its contribution to product reliability and product integrity.

The more disciplined approach of surface integrity can cope with the more stringent designs that involve increasingly thinner sections of stronger materials. The principal saving can come from applying surface integrity techniques selectively. Specific surface integrity specifications should not be applied all over a component part but only to the critically stressed areas. Consultation with the design engineer will frequently reveal that only a very small area, a radii, a shoulder or a section will be exposed to high stress or severe working environment. The balance of the surfaces are lightly loaded and a relaxation of specifications can be justified without compromising the product integrity.

By eliminating "all-over"-type specifications and concentrating on specific critical areas, substantial cost reductions have been made. Indeed, some of the "all-over" specifications have been found to be carry-over from previous generations of designers with little or no data to support the specific required levels. One set of gas turbine discs of Inconel Alloy 718 had for years been specified to have a 63-microinch $R_a$ [1.6 μm] roughness all over. When both high cycle and low cycle fatigue tests showed that the particular alloy was not sensitive to roughness level (see table 18.3-29 and figure 18.3-48), the roughness level was changed to 125 microinches $R_a$ [3.1 μm] on all but one critical radii which was changed to 32 microinches $R_a$ [0.8 μm]. A 7 to 12 percent reduction in machining time was effected on the six discs while increasing the quality assurance. The savings was several hundred thousand dollars per year at the then current level of production.

## SELECTED SOURCES OF SURFACE INTEGRITY DATA

The following References and Bibliography provide much detailed information on specific surface integrity situations as well as general background information. Information may also be obtained on a continuing basis from the Machinability Data Center (MDC). MDC is collecting information and data on surface integrity as they become available and is evaluating the data for application to the material removal industry. These data are included in an encyclopedic collection of photomicrographs, tables and graphs of surface integrity effects from a wide variety of material and process combinations and are available to aid discussions or comparisons. Specific inquiry services are available to U. S. Government agencies and to private industry. Inquiries should be directed to Supervisor, Technical Inquiries, Machinability Data Center, 3980 Rosslyn Drive, Cincinnati, Ohio 45209; Telephone: 513-271-9510; TWX: 810-461-2840.

### References


1. Surface integrity encyclopedia. (Special unpublished collection of data and effects from specific material-process combinations.) Maintained for reference by the Machinability Data Center, Cincinnati, OH.

2. *Influence of metallurgy on machinability*, 1975, and *Influence of metallurgy on hole making operations*, 1977. Metals Park, OH: American Society for Metals.

3. Zlatin, N. and Field, M. 1971. Machinability parameters on new and selective aerospace materials. U. S. Air Force Technical Report AFML-TR-71-95, Metcut Research Associates Inc., Cincinnati, OH.

4. Zlatin, N. and Field, M. 1969. Machinability parameters on new and selective aerospace materials. U. S. Air Force Technical Report AFML-TR-69-144, Metcut Research Associates Inc., Cincinnati, OH.

5. Field, M.; Mehl, C. L.; and Kahles, J. F. 1966. *Machining data for numerical control.* AFMDC 66-1. Cincinnati, OH: Air Force Machinability Data Center.

6. Field, M.; Mehl, C. L.; and Kahles, J. F. 1968. *Supplement to machining data for numerical control.* AFMDC 68-2. Cincinnati, OH: Air Force Machinability Data Center.

7. Glaeser, W. A. 1969. Surface effects in metal deformation. Technical Paper MF69-101, American Society of Tool and Manufacturing Engineers, Dearborn, MI.

8. Koster, W. P. et al. 1970. Surface integrity of machined structural components. U. S. Air Force Technical Report AFML-TR-70-11, Metcut Research Associates Inc., Cincinnati, OH.

9. Koster, W. P. et al. 1972. Manufacturing methods for surface integrity of machined structural components. U. S. Air Force Technical Report AFML-TR-71-258, Metcut Research Associates Inc., Cincinnati, OH.

10. Werner, A. R. and Olson, P. C. 1968. EDM—A metal removal process. Technical Paper MR68-710, American Society of Tool and Manufacturing Engineers, Dearborn, MI.

11. Koster, W. P. 1974. Surface integrity of machined materials. U. S. Air Force Technical Report AFML-TR-74-60, Metcut Research Associates Inc., Cincinnati, OH.

12. Field, M.; Kahles, J. F.; and Cammett, J. T. 1972. A review of measuring methods for surface integrity. *Annals of the CIRP* 21(2): 219–238.

13. Bellows, G. 1978. *Low stress grinding: For quality production.* MDC 78-103. Cincinnati, OH: Machinability Data Center, Metcut Research Associates Inc.

14. Bellows, G. and Tishler, D. N. 1970. Impact of surface integrity on material properties. Report TM70-975, General Electric Company, Cincinnati, OH.

15. Field, M.; Koster, W. P.; Kohls, J. B. 1970. *Machining of high strength steels with emphasis on surface integrity.* AFMDC 70-1. Cincinnati, OH: Air Force Machinability Data Center.

16. Bellows, G. and Koster, W. P. 1972. Surface integrity—Update '72. Report TM72-384, General Electric Company, Cincinnati, OH.

17. Bellows, G. 1974. Surface integrity from finish turning of Inconel 718. Paper presented at 5th Annual Carbide Cutting and Forming Seminar, held 25-27 June 1974, at Purdue University, West Lafayette, IN.

18. Koster, W. P. and Kohls, J. B. 1972. Relation of surface integrity to cost and reliability of structural components. Technical Paper IQ72-207, Society of Manufacturing Engineers, Dearborn, MI.

19. Field, M. 1974. Plastically deformed debris and built-up edge produced on surfaces by chip removal and

abrasive machining processes. *Annals of the CIRP* 23(1): 191–192.

20. Bellows, G. and Kohls, J. B. 1976. Surface integrity of nontraditional material removal processes. Technical Paper MRR76-12, Society of Manufacturing Engineers, Dearborn, MI.

21. Bellows, G. 1970. Surface integrity of electrochemical machining. Technical Paper 70-GT-111, American Society of Mechanical Engineers, New York, NY.

22. Bellows, G. 1973. Surface integrity of nontraditional machining processes. Technical Paper IQ73-633, Society of Manufacturing Engineers, Dearborn, MI.

23. Bellows, G. 1977. *Chemical machining: Production with chemistry*. MDC 77-102. Cincinnati, OH: Machinability Data Center, Metcut Research Associates Inc.

24. Bellows, G. and Niemi, R. M. 1971. Surface integrity of machined Inconel 718 alloy. Technical Paper IQ71-239, Society of Manufacturing Engineers, Dearborn, MI.

25. Bellows, G. and Tishler, D. N. 1970. Introduction to surface integrity. Report TM70-974, General Electric Company, Cincinnati, OH.

26. Gatto, L. R. and DiLullo, T. D. 1971. Metallographic techniques for determining surface alterations in machining. Technical Paper IQ71-225, Society of Manufacturing Engineers, Dearborn, MI.

27. Rooney, R. J. 1957. The effects of various machining processes on the reversed-bending fatigue strength of A-110AT titanium alloy sheet. U. S. Air Force Technical Report WADC-TR-57-310, Wright Air Development Center, Wright-Patterson Air Force Base, OH.

28. Bellows, G. 1972. Surface integrity from electrical discharge machining of superalloys. Report R72AEG236 (not fully released at this time), General Electric Company, Cincinnati, OH.

### Bibliography

American Society for Testing and Materials. 1978. Standard methods of tension testing of metallic materials. ASTM Standard E8. Philadelphia, PA.

Anderson, A. F.; Bellows, G.; and Burgbacher, D. E. 1973. Production processing effects in ground Rene 80. Report R73AEG269, General Electric Company, Cincinnati, OH.

Arzt, P. R. 1968. Machining ultra-high strength steels for aerospace applications. Technical Paper MR68-803, American Society of Tool and Manufacturing Engineers, Dearborn, MI.

Bellows, G. 1973. Specimen preparation and processing for USAF surface integrity program MMP721-2. Report TM73-765, General Electric Company, Cincinnati, OH.

——. 1972. Application of surface integrity to production of jet engines. *Metals Engineering Quarterly* 12(4): 55–58.

——. 1972. Advanced material removal processes. Technical Paper MRR72-02, Society of Manufacturing Engineers, Dearborn, MI.

——. 1969. Surface integrity of electrochemical machining. Report R69AEG172, General Electric Company, Cincinnati, OH.

——. 1968. Impact of nonconventional material removal processes on the surface integrity of materials. Technical Paper MR68-518, American Society of Tool and Manufacturing Engienering, Dearborn, MI.

——. 1968. Surface integrity guidelines. Report R68AEG192, General Electric Company, Cincinnati, OH.

Bellows, G. and Tishler, D. N. 1970. Manufacturing process effects on surface integrity. Report TM70-976, General Electric Company, Cincinnati, OH.

——. 1970. Surface integrity processing guidelines. Report TM70-977, General Electric Company, Cincinnati, OH.

Christenson, A. L. 1961. The measurement of stress by x-ray. Information Report TR 182, Society of Automotive Engineers, New York, NY.

Clorite, P. A. and Reed, E. C. 1958. Influence of various grinding conditions upon residual stresses in titanium. *Transactions of the ASME* 80 Series B (1): 297–301.

Cross, J. R. 1972. Electrochemical machining. U. S. Air Force Technical Report AFML-TR-72-188, Cross Technical Sales, Cincinnati, OH.

Decneut, A. and Peters, J. 1973. Continuous measurement of residual stress in thin cylindrical pieces using deflection-etching techniques. In *Proceedings of the international conference on surface technology*, pp. 262–281. Dearborn, MI: Society of Manufacturing Engineers.

Donachie, Jr., M. J. and Sprague, R. A. 1968. Mechanical properties, residual stress and surfaces. Technical Paper EM68-519, American Society of Tool and Manufacaturing Engineers, Dearborn, MI.

Field, M. 1973. Surface integrity—A new requirement for improving reliability of aerospace hardware. Paper presented at 18th National SAMPE Symposium, held 3–5 April 1973, at Los Angeles, CA.

Field, M. and Kahles, J. F. 1971. Review of surface integrity of machined components. *Annals of the CIRP* 20(2): 153–163.

——. 1964. The surface integrity of machined-and-ground high strength steels. In *Problems in the load-carrying application of high-strength steels*, pp. 54–77. DMIC Report 210. Columbus OH: Defense Metals Information Center.

Field, M.; Kahles, J. F.; and Koster, W. P. 1966. The surface effects produced in nonconventional metal removal—Comparison with conventional machining techniques. *Metals Engineering Quarterly* 6(3): 32–45.

## 18.3 Surface Integrity


Fleming, C. M. and Arzt, P. R. 1971. Integrity control procedures for machining 300M steel and titanium aircraft structures. Technical Paper IQ71-238, Society of Manufacturing Engineers, Dearborn, MI.

Gurklis, J. A. 1965. Metal removal by electrochemical methods and its effects on mechanical properties of metals. DMIC Report 213, Defense Metals Information Center, Columbus, OH.

Kahles, J. F.; Bellows, G.; and Field, M. 1969. Surface integrity guidelines for machining. Technical Paper MR69-730, American Society of Tool and Manufacturing Engineers, Dearborn, MI.

Kahles, J. F. and Field, M. 1973. Impact of surface integrity on product manufacturing and performance. In *Proceedings of the international conference on surface technology*, pp. 652–668. Dearborn, MI: Society of Manufacturing Engineers.

——. 1968. Surface integrity—A new requirement for surfaces generated by material removal methods. In *Properties and metrology of surfaces, Proceedings 1967-1968*, Vol. 182, Part 3K, pp. 31–45. London, England: The Institution of Mechanical Engineers.

Koster, W. P. 1976. Surface integrity: An emerging criterion for quality assurance. Technical Paper IQ76-926, Society of Manufacturing Engineers, Dearborn, MI.

Koster, W. P. and Fritz, L. J. 1970. Surface integrity in conventional machining. Technical Paper 70-GT-100, American Society of Mechanical Engineers, New York, NY.

Koster, W. P.; Fritz, L. J.; and Kohls, J. B. 1971. Surface integrity in machining of 4340 steel and Ti-6Al-4V. Technical Paper IQ71-237, Society of Manufacturing Engineers, Dearborn, MI.

Kroll, R. J.; Westermann, F. E.; and Cuddeback, J. K. 1970. Expansion of derivation of Stablein's equation for calculating residual stress in a machined surface. Metcut Research Associates Inc., Cincinnati, OH.

Littman, W. E. 1967. Control of residual stress in metal surfaces. In *International conference on manufacturing technology, Proceedings*, pp. 1303–1317. Dearborn, MI: American Society of Tool and Manufacturing Engineers.

——. 1967. The influence of grinding on workpiece technology. Technical Paper MR67-593, American Society of Tool and Manufacturing Engineers, Dearborn, MI.

Mattson, R. L. 1956. Fatigue, residual stresses and surface cold working. In *International conference on fatigue of metals*, Session 7, Paper 5, pp. 6–12. London, England: The Institution of Mechanical Engineers.

McAdams, H. T. and Reese, P. A. 1970. Surface topography and metal fatigue (Project STOP). CAL Report No. KB-2952-D-1. Final report. Cornell Aeronautical Laboratory, Inc., Buffalo, NY.

Niemi, R. M. 1971. Surface integrity prediction. Technical Paper IQ71-226, Society of Manufaturing Engineers, Dearborn, MI.

National Aeronautics and Space Administration. 1967. Nondestructive testing: trends and techniques. NASA SP-5082. Washington, DC.

Paxton, H. W. and Proctor, R. P. M. 1968. The effects of machining and grinding on the stress-corrosion cracking susceptibility of metals and alloys. Technical Paper EM68-520, American Society of Tool and Manufacturing Engineers, Dearborn, MI.

Phillips, J. L. 1967. Effect of machining on the service life of high-strength low-alloy steels. In *Conference on machinability*, pp. 93–101. ISI Special Report 94. London, England: The Iron and Steel Institute.

Prevey, P. S. and Koster, W. P. 1972. Effect of surface integrity on fatigue of structural alloys at elevated temperatures. Metcut Research Associates Inc., Cincinnati, OH (Presented at the University of Connecticut, June 18–22, 1972).

Society of Manufacturing Engineers. 1973. *Proceedings of international conference on surface technology.* Dearborn, MI.

Schroeer, R. 1968. Research on exploratory development of nondestructive methods for crack detection. U. S. Air Force Technical Report AFML-TR-67-167 (Part II), Arvin Systems Inc., Dayton, OH.

Slack, R. B. 1967. The effect of several variables on fatigue of a nickel-base alloy. Master of Science Dissertation, Rensselaer Polytechnic Institute, Hartford Graduate Center, Troy, NY.

Society of Automotive Engineers. 1971. *Residual stress measurement by x-ray diffraction.* SAE J784a. New York, NY.

Tarasov, L. P. 1946. Detection causes and prevention of injury in ground surfaces. *Transactions of American Society for Metals* 36: 389–439.

Tarasov, L. P.; Hyler, W. S.; and Letner, H. R. 1958. Effects of grinding direction and abrasive tumbling on the endurance limit of hardened steel. *American Society for Testing Materials Proceedings* 58: 528–539.

Westermann, F. E. 1971. Determining the distortion and residual stresses produced by metal removal operations. Technical Paper IQ71-224, Society of Manufacturing Engineers, Dearborn, MI.

Whitehouse, D. J. 1972. Modern methods of assessing the quality and function of surface texture. Technical Paper IQ72-206, Society of Manufacturing, Dearborn, MI.

Williamson, J. B. P. 1968. Physical aspects of a surface. Technical Paper EM68-513, American Society of Tool and Manufacturing Engineers, Dearborn, MI.

# SECTION 19

# MACHINING GUIDELINES


19.1 General Machining Guidelines ..... 19–3

19.2 Guidelines for Drilling ..... 19–5
- Introduction ..... 19–5
- Types of Drills ..... 19–5
- Drill Geometry ..... 19–8
- Hole Sizes ..... 19–8
- Accuracy of Drilled Holes ..... 19–9
- Tool Materials for Drills ..... 19–9
- Coolant Systems for Drilling ..... 19–10
- Thin Sheet Drilling ..... 19–10
- Drilling Nonmetallics ..... 19–10

19.3 Tool Life ..... 19–11

19.4 Types of Machinability Data ..... 19–15
- Tables of Recommended Machining Conditions ..... 19–15
- Wide Range Machining Data ..... 19–15
- Cost and Production Rate Determination and Analysis ..... 19–15
- Mathematical Correlation of Machining Parameters with Tool Life and Subsequent Optimization of Cost and Production Rates ..... 19–15
- Machinability Ratings ..... 19–15

1. Good machining practice requires a *rigid setup* in addition to the selection of proper cutting speed, feed, tool material, tool geometry and cutting fluid. Rigidity in itself does not ensure a successful machining operation; however, without it, tools are subject to premature failure.

2. The machine tool must be capable of providing the rigidity required for the machining conditions used. If the size of the machine tool is not adequate or if looseness exists in the moving parts, such as spindle bearings or gibs, chatter will occur and poor tool life will result. When a rigid setup cannot be made, the feed and/or depth of cut must be reduced accordingly.

3. Excessive tool overhang is a source of trouble in a machining operation. When this condition exists, poor tool life and surface finish result, and dimensional accuracy is difficult to maintain. Stub-length drills should be used instead of jobbers-length drills where the depth of hole permits. Milling cutters should be mounted as close to the spindle as the job will allow. The length of end mills should be kept at a minimum. Climb milling usually gives better tool life and surface finish than does conventional milling if the machine tool and setup have sufficient rigidity and the feed mechanism is free from backlash.

4. Misalignment and tool runout cause other machining problems, such as oversize and bellmouthed holes when reaming and rough and torn threads when tapping and die threading.

5. Maintenance of cutting tools must be given careful consideration in the development of good machining practice. One common problem is the failure to change tools at the proper time. Tool costs become exorbitant on those jobs where tool failure is the criterion for tool change. Usually this approach results in excessive regrinding time or, in the case of a milling cutter, the replacement of broken teeth with new ones. Tooth replacement requires considerable time in addition to regrinding time.

6. When dimensional accuracy and surface integrity are not critical, high speed steel tools should be removed when the wearland on the flank of the tool reaches approximately 0.060 inch [1.5 mm] width. In the case of carbide tools, the maximum width should not be allowed to exceed 0.030 inch [0.75 mm], otherwise complete tool failure may occur. On components where dimensional accuracy and surface integrity are critical, the tool wear must be carefully limited, see section 18.3.

7. Wearland measurement is not always possible on the tool; however, it is practical to instruct the operator to change tools after a predetermined number of pieces have been machined. The number of parts to be machined per tool should be set conservatively so that the cutter will not fail. Occasionally by this procedure, a cutter may be removed before it is dull; therefore, the resharpening time will be short, but catastrophic failure will have been avoided.

8. The cutting fluid system should provide a copious flow of cutting fluid to the area where the chip is being formed. In the case of machining operations where cutting fluids are used with carbide tools, a continuous flow of the cutting fluid is imperative. Interrupted or intermittent flow can cause thermal shock and breakage of the carbide tool.

9. The concept of good machining practice involves consideration of all factors associated with the machining operation. Each detail—workpiece, fixturing, speed, feed, tool material, tool geometry, cutting fluid, and the machine tool itself—must be given careful attention to ensure success for the machining operation under consideration.

10. The *machinability* of a work material must be determined in order to select the proper machining conditions. The machinability of a material can be defined in terms of three major factors: surface integrity, tool life, and power or force requirements.

## INTRODUCTION

Drills are among the most widely used tools in the shop, but their selection, grinding, and usage often receive the least care. A drilled hole is usually either a clearance hole for a fastener or a starting point for tapping, reaming, or boring. Drilling usually can be done satisfactorily with standard off-the-shelf twist drills. Where large quantities of precision holes must be produced, certain refinements in practice are necessary as follows:

- The drills should be machine ground in order to develop accurate geometry of the point.
- Point angle, web thickness, and lip clearance should be suitable for the material being drilled.
- The work should be properly supported and clamped and the drill should be guided in a drill bushing of suitable fit and length.
- Adequately powered and rigid machine tools should be utilized.

## TYPES OF DRILLS

### Twist Drills

A *twist drill* is an end cutting tool having one or more cutting edges and having helical and sometimes straight flutes for the passage of chips and the admission of a cutting fluid. See figure 19.2-1 and table 19.2-1 for the description and application of twist drills.

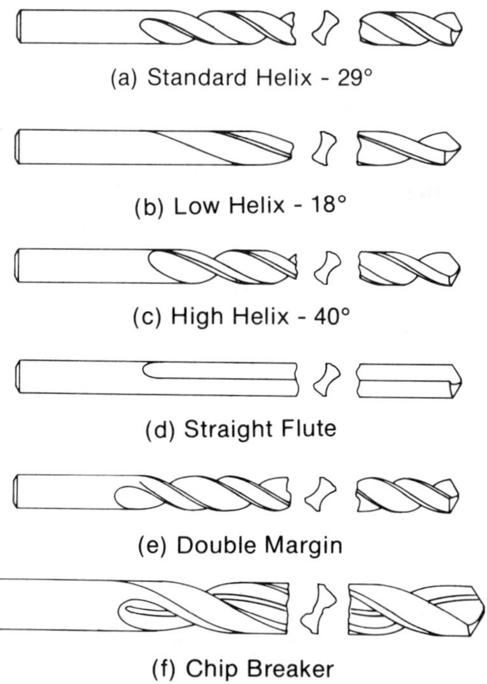


**Figure 19.2-1** Twist drills.

A *standard twist drill* has two cutting edges, two helical flutes, and a straight or tapered shank.

*Straight shank drills* have cylindrical shanks which may be of the same or of a different diameter than the body of the drill.

*Taper shank drills* are preferable to the straight shank type for drilling medium and large diameter holes. The taper on the shank conforms to one of the tapers in the ANSI Standard Series. Taper shank drills generally have a driving tang and are directly fitted into tapered holes in drilling machine spindles or into driving sockets.

*Double margin drills* have a second pair of margins at the rear of the lands which act as steadying elements. Double margin drills may produce more accurate holes. Their size range is from 1/8 to 2 inches [3.2 to 50 mm].

*Chip breaker drills* have features built into the drill design to prevent long, stringy chips. They have a diameter range from 3/8 to 2 inches [9.5 to 50 mm].

### Alternate Drills

When applications arise that cannot be met by using a twist drill, one of the following alternate drills can be considered:

Step Drills, figure 19.2-2
Subland Drills, figure 19.2-3
Flat Drills, figure 19.2-4
Spade Drills, figure 19.2-5
Core Drills, figure 19.2-6
Oil-Hole or Pressurized-Coolant Drills, figure 19.2-7
Gun Drills, figure 19.2-8

See tables 19.2-1 through 19.2-3 for the descriptions and applications of alternate drills.

**Figure 19.2-2** Step drills.

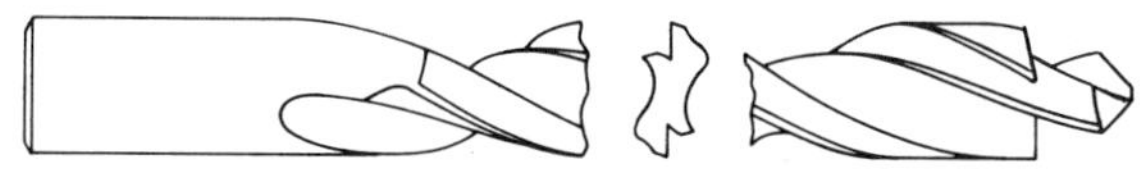

**Figure 19.2-3** Subland drills.

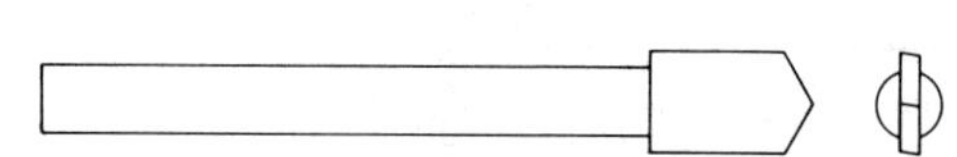

**Figure 19.2-4** Flat drills.

## 19.2 Guidelines for Drilling

**TABLE 19.2-1 Characteristics of Twist, Step and Subland Drills**

| | TWIST DRILLS | STEP DRILLS | SUBLAND DRILLS |
|---|---|---|---|
| Description | Twist drills are the most common. Most of these drills are made with two helical flutes which allow the removal of chips and the admission of a cutting fluid to the tip of the drill. See figure 19.2-1. | Step drills have two or more diameters produced by grinding various steps on the diameter of the drill. The steps or different diameters are ground on the same land. See figure 19.2-2. | Subland drills are combination tools having separate lands or margins which extend the full length of the flutes for each of the two or more diameters. See figure 19.2-3. |
| Application | Twist drills are used to produce the majority of clearance holes for fasteners and pilot holes for subsequent reaming or tapping. | Step drills produce two or more diameters in one pass of the drill, and also serve as combined drills and countersinks, combined drills and counterbores, and combined drills and reamers. | Two or more diameters can be drilled simultaneously. The different lands can be ground to permit drilling and countersinking for flathead screws or drilling and counterboring for socket head screws. Two or more diameters can be maintained constant throughout the life of the drill even after resharpening. |
| Tool Material | High speed steel, solid carbide, or carbide-tipped for special applications. | High speed steel | High speed steel |
| Size range | Standard range: 1/8 to 1-1/2 inches [3 to 35 mm]<br>Micro range: 0.0059 to 1/8 inch [.15 to 3 mm]<br>Large range: 1-1/2 to 6 inches [38 to 150 mm] | Standard range: 1/4 to 2 inches [6 to 50 mm] | Standard range: 5/32 to 3/4 inch [4 to 18 mm]<br>For special sizes, the largest diameter should be no greater than twice the smallest diameter because of the variation in cutting speeds. |

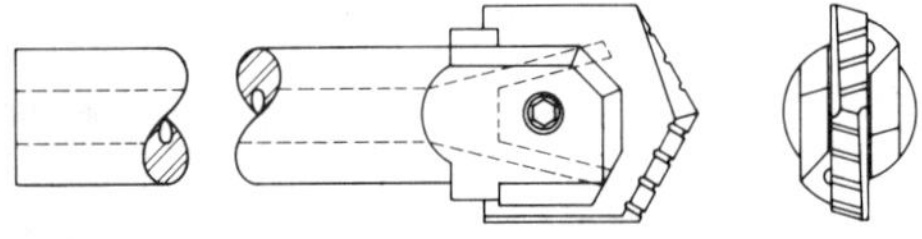

**Figure 19.2-5** Spade drills.

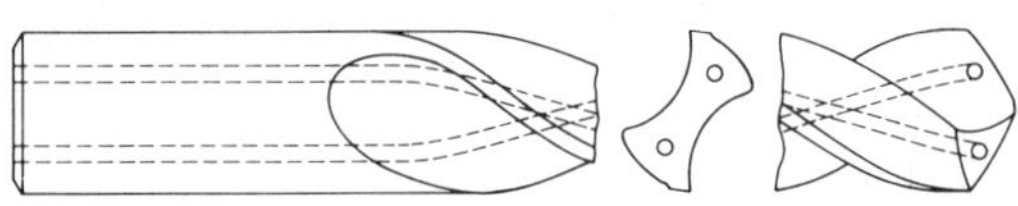

**Figure 19.2-7** Oil-hole or pressurized-coolant drills.

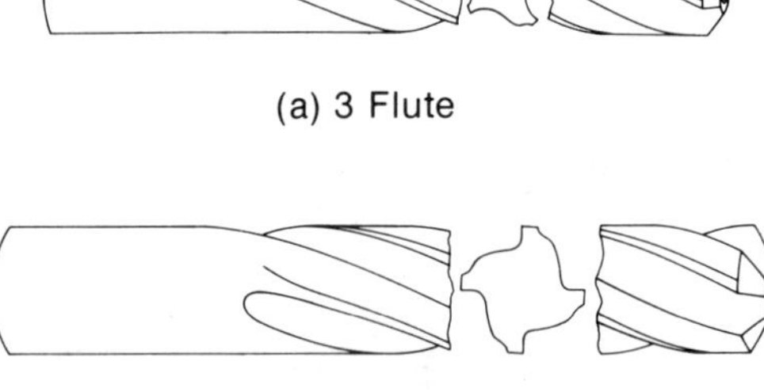

(a) 3 Flute

(b) 4 Flute

**Figure 19.2-6** Core drills.

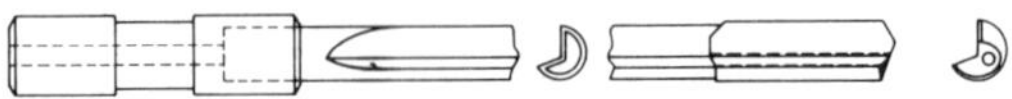

**Figure 19.2-8** Gun drills.

**TABLE 19.2–2 Characteristics of Flat, Spade and Core Drills**

| | FLAT DRILLS | SPADE DRILLS | CORE DRILLS |
|---|---|---|---|
| Description | Flat drills derive their name from the shape of the drill body which is flat rather than the normal round configuration. The drills are produced by grinding tapered opposite flats on the drill body. The flats are usually not parallel, creating a web which is thicker at the shank. A slight back taper of 0.001 inch [0.025 mm] or less is ground on the drill body. See figure 19.2–4. | A spade drill has a removable tip or bit clamped in a special holder attached on the drill shank. See figure 19.2–5. Most holders have an axial hole through the shank to allow the passage of cutting fluid under pressure to the drill point. Replacement blades cost about 1/6 as much as a twist drill. | Core drills are multi-fluted, and have no point, similar in design to a reamer. See figure 19.2–6. |
| Application | Flat drills have a low production capacity, but because of their simplicity in design and low cost are occasionally used for drilling hard forgings and castings. | Spade drills smaller than 1-3/4 inch [44 mm] should be used only for shallow holes, one diameter deep vertical, two diameters deep horizontal. Do not use a center drill for starting a spade drill. Use instead a short spade drill having the same point angle as the drill to be started. | Core drills are used to enlarge cored, forged or previously drilled holes. They do not produce the original holes. Core drills are extremely rigid, and because of the greater number of flutes (3 to 6), they can be used at higher feed rates than twist drills. |
| Tool Material | Standard size: HSS or carbide<br>Micro size: HSS | Blade: HSS or carbide<br>Holder and shank: Alloy steel | High speed steel or carbide tipped |
| Size Range | Standard range: 3/32 to 1/2 in. [2.4 to 13 mm]<br>Micro range: 0.001 to 3/32 inch [0.025 to 2.4 mm] | Standard range: 1 to 5 inches [25 to 125 mm]<br>Special range: 3/4 to 1 inch [18 to 25 mm] and 5 to 15 inches [125 to 380 mm] | Standard range:<br>3 flute: 1/4 to 3/4 inch [6 to 18 mm]<br>4 flute: 1/2 to 1-1/4 inches [13 to 32 mm] |

**TABLE 19.2–3 Characteristics of Oil-Hole or Pressurized-Coolant and Gun Drills**

| | OIL-HOLE OR PRESSURIZED-COOLANT DRILLS | GUN DRILLS |
|---|---|---|
| Description | An oil-hole drill is either a twist drill or a straight fluted drill which has one or more continuous holes through its body and shank to permit the passage of cutting fluid under pressure. The holes can be through either the lands or the web of the drill. See figure 19.2–7. Chip-breaking grooves are sometimes ground in the drill point to break and curl the chips to aid ejection. | A gun drill is a special straight flute drill for producing deep holes. It has a straight fluted tip, cutting to center, mounted in a tubing shank. The tips are often brazed to the shanks, but in the larger sizes, the tips are keyed or pinned to the shanks for convenience in removal for regrinding. Both the tip and the shank allow for ample volume of coolant under high pressure to reach the cutting edges and also to wash away the chips. See figure 19.2–8. |
| Application | Oil-hole drills are very suitable for making deep holes where flood applications of a cutting fluid would be ineffective. Increased metal removal rates are possible in both normal and deep holes by using oil-hole drills instead of flood cooling. Both constant pressure and pulsating pressures from 50 to 300 psi [345 to 2068 kPa] are employed. | Gun drills are used for making very deep holes that require extremely close tolerance in straightness and diameter. |
| Tool Material | High speed steel or carbide-tipped steel for special applications. | The tip is usually carbide, and the shank is alloy steel tubing. |
| Size Range | Standard range: 5/32 to 1-1/2 inches [4 to 38 mm] | Standard range: 1/8 to 2 inches [3 to 50 mm] |

# 19.2 Guidelines for Drilling

## DRILL GEOMETRY

### Drill Point

**Point Angle**

Standard drills have point angles of 118 degrees. The point angle may be increased up to 150 degrees for very hard or tough materials. For softer materials such as aluminum or copper alloys, the point angle may be 90 to 100 degrees.

**Double Point Angle**

When drilling very hard or abrasive materials, double angles are sometimes ground on the point to reduce chipping of the corners of the lips. This is done by first grinding the larger included angle (118 to 140 degrees) and then regrinding the smaller included angle (60 to 90 degrees) on the corners, giving the effect of a chamfer at the corner.

**Thinned Web**

The central web of the drill increases in thickness toward the shank. Hence, the point must be thinned as the drill is ground back so as to prevent excessive end pressure. This is usually required when the web thickness exceeds one-fifth of the drill diameter. Techniques used to reduce the web thickness include the undercut thinned point, the notched point, and the crankshaft point.

**Radius Corner**

A radial lip twist drill has curved outer lips. These curves allow stress to be uniformly distributed over the entire cutting area. This design may produce a smoother finish in the hole and may reduce the burrs at breakthrough. This point can be produced on a special machine designed for that purpose or on general-purpose cutter grinding machines.

**Spiral Point**

Reducing the point from a regular chisel edge to a spiral point gives the drill a better self-centering effect. The spiral point must be ground on a spiral-point-grinding machine.

**Multiple Point and Relief Angle**

Drills having points with more than one included angle and more than one angle of relief (faceted) can be self-centering and chip-breaking for use in hard materials.

### Helix Angle

**Standard Helix**

A standard helix (approximately 29 degrees) high speed steel drill is used for drilling most cast irons and steels. See figure 19.2-1 (a).

**Low Helix**

Low helix drills (approximately 18 degrees) have reduced axial rakes which facilitate the penetration of brass, plastics, and other soft materials. See figure 19.2-1 (b).

**High Helix**

High helix drills are made with wide flutes and narrow lands, for drilling deep holes in mild ferrous and nonferrous materials. The higher helix (approximately 40 degrees) removes chips more effectively. See figure 19.2-1 (c).

**Straight Flute**

Straight flute drills (zero helix) are especially adapted for drilling brass and other soft or very hard metals. See figure 19.2-1 (d).

### Lip Clearance

For most drilling, the lip clearance varies from 7 to 25 degrees depending on the drill size. Smaller drills require more clearance, while larger drills require less. In general, for harder materials, the clearance should be as small as possible to allow maximum strength to the cutting edge.

## HOLE SIZES

**Micro Drilling**

For holes from 0.001- to 0.020-inch [0.025 to 0.5 mm] diameter, flat drills are used over this entire size range. Twist drills are used for hole sizes over 0.006 inch [0.15 mm]. Speeds must be much lower and feed rates higher than are used for larger drills so as not to produce a powder that will pack in the flutes.

**Normal Drilling**

For holes from 0.014- to 3-inch [0.35 to 75 mm] diameter, standard twist drills are generally used.

**Large Holes**

For holes larger than 3 inches [75 mm] in diameter, spade drills, gun drills, or trepanning tools are needed.

**Deep Holes**

When drilling deep holes with twist drills, the recommended speeds and feeds should be reduced as the holes become deeper, as shown in table 19.2-4. The drill should also be retracted periodically to clear the chips and lubricate the points.

**TABLE 19.2-4 Reduction of Speed and Feed for Drilling Deep Holes with Twist Drills**

| HOLE DEPTH drill diameters | REDUCE SPEED BY | REDUCE FEED BY |
|---|---|---|
| 3 | 10% | 10% |
| 4 | 20% | 10% |
| 5 | 30% | 20% |
| 6 | 35% | 20% |
| 8 | 40% | 20% |

Oil-hole drills are helpful in deep-hole drilling. Hole depths up to 30 diameters are sometimes possible. The feed rate should be reduced as the holes become deeper, as shown in table 19.2-5.

**TABLE 19.2-5 Reduction of Feed for Deep-Hole Drilling with Oil-Hole Drills**

| HOLE DEPTH drill diameters | REDUCE FEED BY |
|---|---|
| 5 to 7 | 10% |
| 8 to 10 | 20% |
| over 10 | 30% |

## ACCURACY OF DRILLED HOLES

### Accuracy of Holes Drilled with Twist Drills

There are two elements of hole accuracy—location and oversize. Table 19.2-6 indicates the accuracy of holes drilled with twist drills with and without center-drilled holes and drill bushings. Holes drilled without center-drilled holes and drill bushings tend to be bellmouthed.

### Effect of Drill Point

The point angle affects hole location and size. Point angles which are more acute than standard point angles (smaller included angle) have more of a self-centering effect on the drill. The blunter the point angle, the greater the care and skill which must be exercised to control hole location and size. Special point grinds, such as spiral points, also improve the centering effect of the drill.

### Point Grinding Faults

Cutting lips of nonuniform length or angles will produce oversize holes and excessive margin wear. Insufficient clearance will result in excessive drill thrust and will produce excessive heating and burning of the work material and drill point. A too large lip clearance, leaving insufficient support behind the cutting edges, results in chipping and crumbling of the cutting edges.

### Hole Location

The machine tool and the setup affect the accuracy of the location of the hole. Approximate placement capabilities for three general categories of machining with respect to true location are as follows:

| | |
|---|---|
| Jig borer | ±0.0002 inch [±0.005 mm] |
| NC drill | ±0.0025 inch [±0.064 mm] |
| Drill press (with bushings in drill jig) | ±0.002 inch [±0.05 mm] |
| Drill press without bushing | ±0.015 inch [±0.08 mm] |

## TOOL MATERIALS FOR DRILLS

### High Speed Steel

Several common grades such as M1, M7 and M10 [S2 and S3] are used in the majority of drilling operations. See section 14.2 for complete listing of HSS compositions.

### Cobalt High Speed Steel

Grades containing combinations of high cobalt with high vanadium or high carbon such as T15, M33, and M41 through M47 [S9, S11 and S12] are very useful in drilling hard steels, titanium alloys and nickel-base alloys. The cobalt grades usually give higher tool life or faster production than regular high speed steel. The cobalt grades are more expensive and in some cases are more difficult to grind than the common high speed steels.

### Surface Treated HSS Drills

Various surface treatments such as cyaniding and nitriding can be applied to high speed steel drills to increase the hardness of the outer layer of materials. Cyanided and nitrided HSS drills are considered when drilling hard and abrasive materials which allow very short tool life of untreated HSS drills. Flash chrome plating is used on occasion when drilling small-diameter holes in printed circuit boards, carbon and graphite, and some nonferrous metals. Other treatments such as special polishing and black oxiding are used to minimize friction between the drill and the workpiece, galling of chips in the flutes, and excessive work material buildup on the drill lips.

**TABLE 19.2-6 Average Accuracy of Holes Drilled with Twist Drills**

| CONDITION | DIAMETER | | | | | |
|---|---|---|---|---|---|---|
| | 1/8 to 1/4 inch 3 to 6 mm | | 1/4 to 3/4 inch 6 to 18 mm | | 3/4 to 1-1/2 inch 18 to 35 mm | |
| | Oversize | Location | Oversize | Location | Oversize | Location |
| No center drilled hole—no bushing | 0.003 | ±0.007 | 0.006 | ±0.008 | 0.008 | ±0.009 |
| Center drilled hole—no bushing | 0.003 | ±0.004 | 0.003 | ±0.004 | 0.004 | ±0.005 |
| Drill bushing | 0.002 | ±0.002 | 0.003 | ±0.002 | 0.004 | ±0.003 |
| No center drilled hole—no bushing | 0.075 | ±0.18 | 0.15 | ±0.20 | 0.20 | ±0.23 |
| Center drilled hole—no bushing | 0.075 | ±0.10 | 0.075 | ±0.10 | 0.10 | ±0.13 |
| Drill bushing | 0.05 | ±0.05 | 0.075 | ±0.05 | 0.10 | ±0.075 |

### Carbide Drills

Twist or straight-flute drills can be tipped with carbide cutting edges or be made of solid tungsten carbide.

- Carbide-tipped drills are usually preferred because they are less expensive than solid carbide drills. Standard carbide-tipped drill diameter range is usually from 3/32 to 1-1/2 inches [2.4 to 38 mm].
- Solid carbide drill diameter range is from 1/16 to 1/2 inch [1.6 to 13 mm] (or No. 80 to No. 13).
- Applications for carbide drilling:

  a. For steels harder than 50 $R_C$ to avoid surface damage to the steel

  b. High production drilling on cast iron and aluminum

  c. Drilling of reinforced plastics including circuit boards

## COOLANT SYSTEMS FOR DRILLING

### Flood

The most commonly used method of applying coolant to the drill is the overhead flood, where a constant stream of liquid is directed at the hole.

### Mist

Application of a mist of coolant to the drill is very useful in situations where a recirculating coolant system would be impractical or impossible, such as on large sheets or airframes.

### Oil-Hole or Pressurized-Coolant Drilling System

This application is used often in screw machines and turret lathes for drilling deep holes. The coolant is pumped under constant or pulsating pressure (see section 3.3).

Recent availability of relatively inexpensive pump and rotating seal equipment will permit the use of oil-hole drills in almost any drill press, old or new, provided that it is in reasonably good condition. Oil-hole or coolant-fed HSS drills are now available in a wide range of sizes in both high speed steel and carbide-tipped styles. Oil-hole usage may, in many cases, result in increased metal removal rates over flood cooling for normal depth holes as well as for deep holes.

## THIN SHEET DRILLING

Very thin sheets, either metallic or nonmetallic, should be backed up with a solid block of wood or soft steel to prevent cracking in brittle materials, deformation in metals, or oversize holes. The formation of burrs, both at the entrance and the exit of the drill, is a common occurrence in drilling "sandwich" layers of sheets, and delamination of pressed and sintered alloys may occur under heavy feeds. The use of a backup minimizes both burring and delamination. Burrs and delamination can also be minimized by controlling the operation to ensure that drills are sharp at all times.

Very large assemblies of sheet material, usually aluminum, titanium or steel alloys, are frequently drilled in the aircraft industry. The larger assembly size usually dictates the use of portable drilling equipment. This equipment may be electric- or pneumatic-powered. The drills may be hand-fed when drilling aluminum alloys, but are power-fed when drilling titanium alloys or steel.

## DRILLING NONMETALLICS

### Plastics

Use low helix drills with deep-polished, large-radius flutes to facilitate chip removal. Point angles should be 60 to 90 degrees; use 60 degrees on sheet plastics.

Some plastics produce undersize holes. To overcome this problem, use drills 0.002- to 0.003-inch [0.05 to 0.08 mm] oversize. Before using a cutting fluid, check the possible damaging reaction to the plastic; or use an air-cooling jet.

### Composites

Composite materials containing glass, graphite, boron and Kevlar fibers are finding increasing applications in aerospace, automotive, marine and commercial items. Basically, these materials are composed of high-strength fibers impregnated in a plastic matrix. These materials are very hard and abrasive by nature; therefore, conventional drills made of both high speed steel and carbide find limited usage because their tool life is short and they tend to delaminate the material on exiting the drilled hole.

These problems have practically eliminated the use of HSS drills and have led to the development of a family of new drill configurations and drilling technology utilizing both carbide and diamond. According to present technology, the drilling of small-diameter holes, up to 3/16-inch [4.75 mm] diameter, is best performed using solid carbide drills having a low flute helix and a smaller than normal included point angle. These drills have less tendency to delaminate the material when a back-up material is impractical to use.

For larger diameter holes, industry has turned to a widespread use of diamond-impregnated core drills. These drills have brought vast improvements in hole quality, hole size and drill life. Running at much higher speeds than conventional drilling speeds, diamond core drills grind the hard fibers, thus eliminating fiber whiskers in the hole.

A new development in the use of diamond-impregnated core drills employs ultrasonic vibration to produce faster penetration rates. However used, diamond core drills do have a tendency to delaminate upon exiting the hole; therefore, it is recommended that wherever possible the underside of the hole be backed up with wood, aluminum or plastic. A newer family of composites composed of fibers in a metal matrix, such as aluminum, are posing problems in drilling which require additional process developments, such as utilizing rotary ultrasonic core drilling, electrolytic diamond drilling and lasers to produce acceptable holes.

Tool life is one of the most important factors in the evaluation of machinability. Specifically, the manufacturing engineer needs to know the relation of tool life to speeds, feeds and the other pertinent machining parameters. For production operations, tool life is usually expressed as the number of pieces machined per tool grind. In machinability testing, tool life is generally defined as the cutting time in minutes to produce a given wearland for a set of machining conditions. This cutting time can be converted to cubic inches of metal removed for a given depth of cut.

Turning tests usually are used for evaluating the machinability of a material in terms of tool life. This operation is used because of the simplicity of the cutting tool. In addition, all of the machining conditions, such as speed, feed, tool geometry, tool material, and cutting fluid, can be readily controlled. By varying one of the machining conditions and keeping the others constant, it is possible to determine the effect of such a change on tool life. The relationships among the various machining variables are obtained in this manner.

In a machining test such as turning, the tool is removed after a given amount of wear is produced on the flank of the tool. The wearland is usually 0.015 or 0.030 inch [0.38 or 0.75 mm] for carbide and 0.060 inch [1.5 mm] for high speed steel. It is general practice in these tests to stop the machining at frequent intervals and examine the wearland on the tool. The time to produce the wearland is then recorded. A typical data sheet for turning tests is shown in figure 19.3-1. This procedure is continued until the wearland reaches a predetermined width, which in these tests was 0.015 inch [0.38 mm]. It should be noted that intermittently measured tool wear rates may not fully indicate rates of continuous cutting wear. Hence, wear tests should be run both ways.

Initially, a tool life curve is obtained with several different speeds, keeping the other variables constant, thus obtaining a relationship between cutting speed and tool life. The recommended cutting speed for a high speed steel tool is generally the one which produces a 60-minute tool life. With carbide, the tool life can be much shorter, particularly if a throwaway-type tool blank is used. In this case, a 30-minute tool life may be satisfactory. When a tool life curve is available for a range of cutting speeds, the tool engineer can select the cutting speed which will give him the best compromise between tool life and production rate. Tool life curves for various feeds which show the effect of cutting speed are presented in figures 19.3-2 and 19.3-3. Figure 19.3-2 shows that for tests on 8640 steel using a carbide tool, the longest tool life was obtained at 300 feet per minute [90 m/min] with a feed per revolution of 0.010 inch [0.25 mm]. Figure 19.3-3 shows the tool life in cubic inches of metal removed for the same test conditions. In this case, the greatest tool life was obtained at 300 feet per minute [90 m/min] with a feed per revolution of 0.020 inch [0.5 mm], even though the tool life in minutes was less than that obtained when using a feed per revolution of 0.010 inch [0.25 mm].

Once the tool life curves have been established showing the effect of cutting speed, other tests can be run to develop the relationships between tool life and other machining variables. Tool life curves on the effect of feed for a constant cutting speed are shown in figures 19.3-4 and 19.3-5. Tool life versus feed is given for three different cutting speeds. The curves in figure 19.3-4 show tool life in minutes. The longest tool life was obtained using a feed per revolution of 0.010 inch [0.25 mm] with a cutting speed of 300 feet per minute [90 m/min]. Figure 19.3-5 shows tool life in cubic inches of metal removed. At a cutting speed of 300 feet per minute [90 m/min], tool life was greatest when using a feed per revolution of 0.020 inch [0.5 mm]. For complete results, a three-dimensional graph of speed and feed versus tool life is necessary.

Other tool life tests can be performed to evaluate the effect of the other machining variables, such as carbide grade, tool geometry, and cutting fluid. Figure 19.3-6 shows a set of tool life curves comparing the effect of a carbide grade on tool life. Tool life curves for various materials can be grouped on the same chart to compare the relative machinability of each of the materials when machined under similar conditions. A group of tool life curves for turning various materials is shown in figure 19.3-7.

This presentation on tool life curves has dealt mainly with turning tests. While turning tests are most widely used, it is preferable whenever possible to test using the particular machining operation in question, such as turning, milling, drilling, reaming, tapping, etc. The tool life curve is a valuable source of machining information; from it the tool engineer, designer, or planner can obtain machining conditions which will produce maximum production at a minimum cost for the machining operation. A further discussion of cost is presented in section 21.

PROJECT NUMBER 18C    TEST NUMBER 11    MACHINABILITY TEST DATA SHEET    OBSERVED BY HWH    DATE 5/15/59

MATERIAL 8640 STEEL    TOOL CARBIDE    MATERIAL C-6

HEAT TREATMENT NORMALIZED    MICROSTRUCTURE 75% PEARLITE, 25% FERRITE

BACK RAKE 0°    RESULTANT RAKE —    NOSE RADIUS .040"

SIDE RAKE 6°    INCLINATION —    CHIP BREAKER —

SCEA 0°    SIDE RELIEF 5°

HARDNESS 190 - 192 BHN    ECEA 5°    END RELIEF 5°

MACHINE AMERICAN PACEMAKER    CUTTING FLUID DRY

| BAR No. | SPEED fpm | BAR Dia. in. | rpm | DEPTH of CUT, in. | FEED in./rev. | TOOL WEAR, in. | | TIME min. | POWER kw. input | SURFACE FINISH, in. | METAL REMOVED | |
|---|---|---|---|---|---|---|---|---|---|---|---|---|
| | | | | | | uniform | localized | | | | cu. in. | per in. edge |
| S-3 | 600 | 3.910 | 587 | .100 | .010 | .003 | — | 2'40" | | | | |
| (192) | | 3.510 | 654 | | | .009 | — | 7'42" | | | | |
| | | 3.310 | 694 | | | .013 | .014 | 9'55" | | | | |
| | | 3.110 | 739 | | | .015 | .017 | 11' | | | | |
| S-3 | 300 | 3.110 | 370 | .100 | .010 | .001 | — | 2'10" | | | | |
| | | 2.510 | 458 | | | .003 | — | 13'16" | | | | |
| | | 2.110 | 545 | | | .004 | — | 19'20" | | | | |
| S-5 | 300 | 3.800 | 301 | | | .006 | — | 24'38" | | | | |
| (190) | | 3.400 | 337 | | | .010 | .011 | 34'24" | | | | |
| | | 3.200 | 358 | | | .012 | .0125 | 38'50" | | | | |
| | | 3.000 | 382 | | | .013 | .014 | 41' | | | | |
| | | 2.800 | 408 | | | .0145 | .015 | 43' | | | | |
| | | 2.800 | 408 | | | .0155 | .016 | 44' | | | | |
| S-5 | 400 | 2.800 | 527 | .100 | .010 | .002 | — | 2'13" | | | | |
| | | 2.200 | 694 | | | .007 | — | 9'48" | | | | |
| | | 2.000 | 763 | | | .008 | — | 11'55" | | | | |
| S-6 | | 3.970 | 385 | | | .010 | — | 16' | | | | |
| (190) | | 3.570 | 428 | | | .013 | .014 | 23'44" | | | | |
| | | 3.370 | 453 | | | .0145 | .016 | 26' | | | | |
| | | 3.370 | 453 | | | .015 | .0165 | 27' | | | | |

**Figure 19.3–1** Typical data sheet for turning tests.

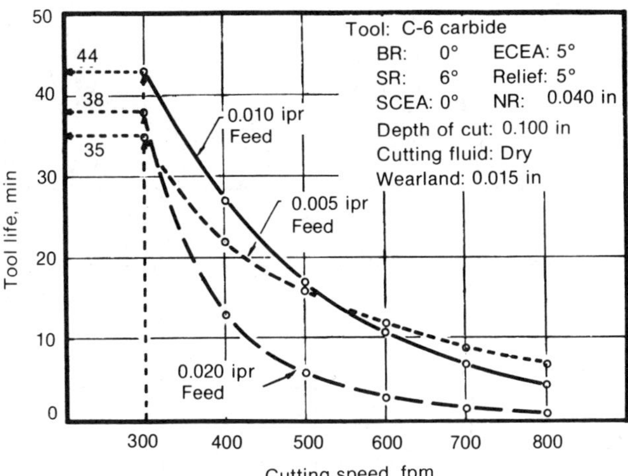


**Figure 19.3-2** Effect of cutting speed on tool life for turning 8640 steel, 190 Bhn.

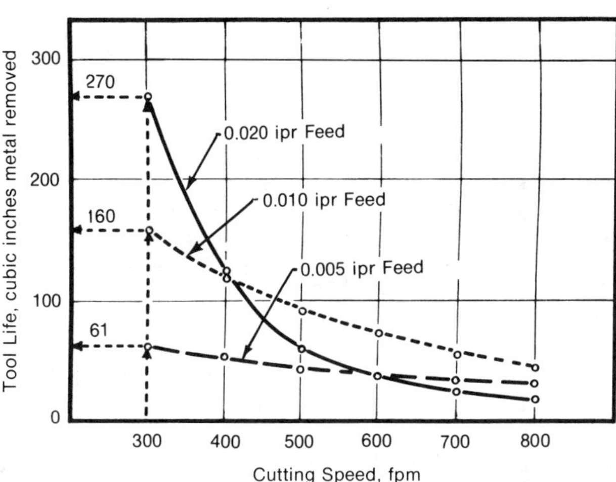


**Figure 19.3-3** Effect of cutting speed on tool life for turning 8640 steel, 190 Bhn. See figure 19.3-2 for test conditions.

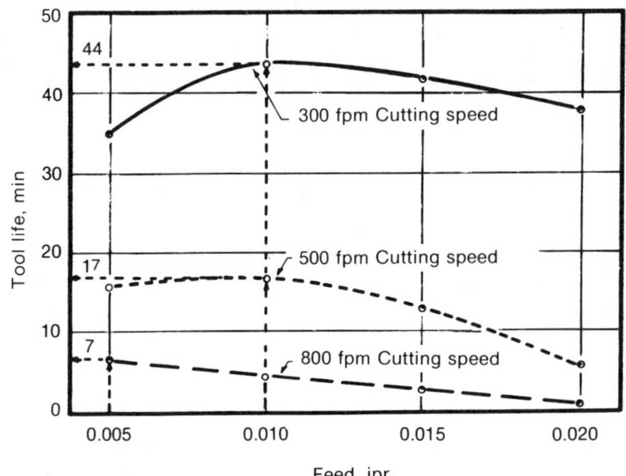


**Figure 19.3-4** Effect of feed on tool life for turning 8640 steel, 190 Bhn. See figure 19.3-2 for test conditions.

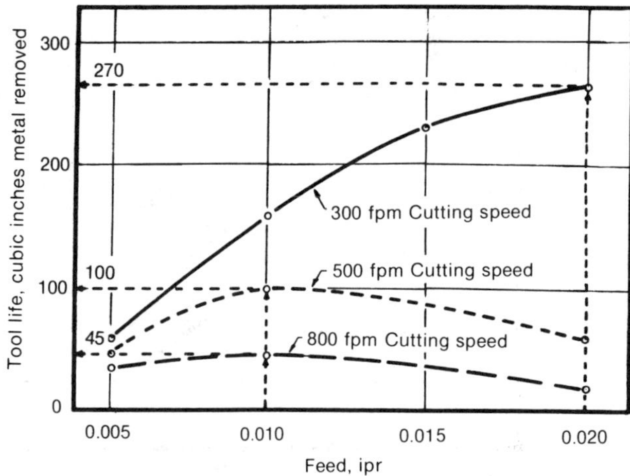


**Figure 19.3-5** Effect of feed on tool life for turning 8640 steel, 190 Bhn. See figure 19.3-2 for test conditions.

# 19.3 Tool Life

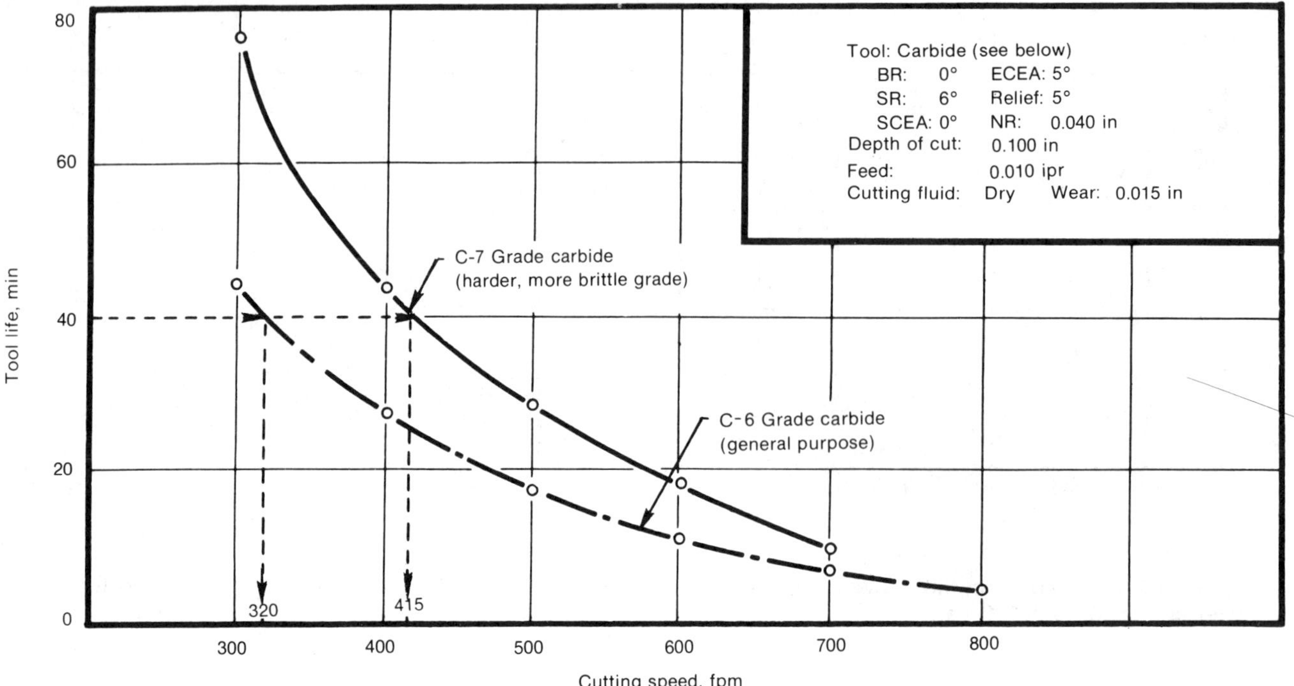


**Figure 19.3–6** Effect of carbide grade on tool life for turning 8640 steel, 190 Bhn. Best tool life was obtained at 300 fpm using C-7 grade carbide. For a 40-minute tool life, AISI 8640 steel can be turned at 415 fpm using C-7 grade carbide, compared to 320 fpm with C-6 grade carbide.

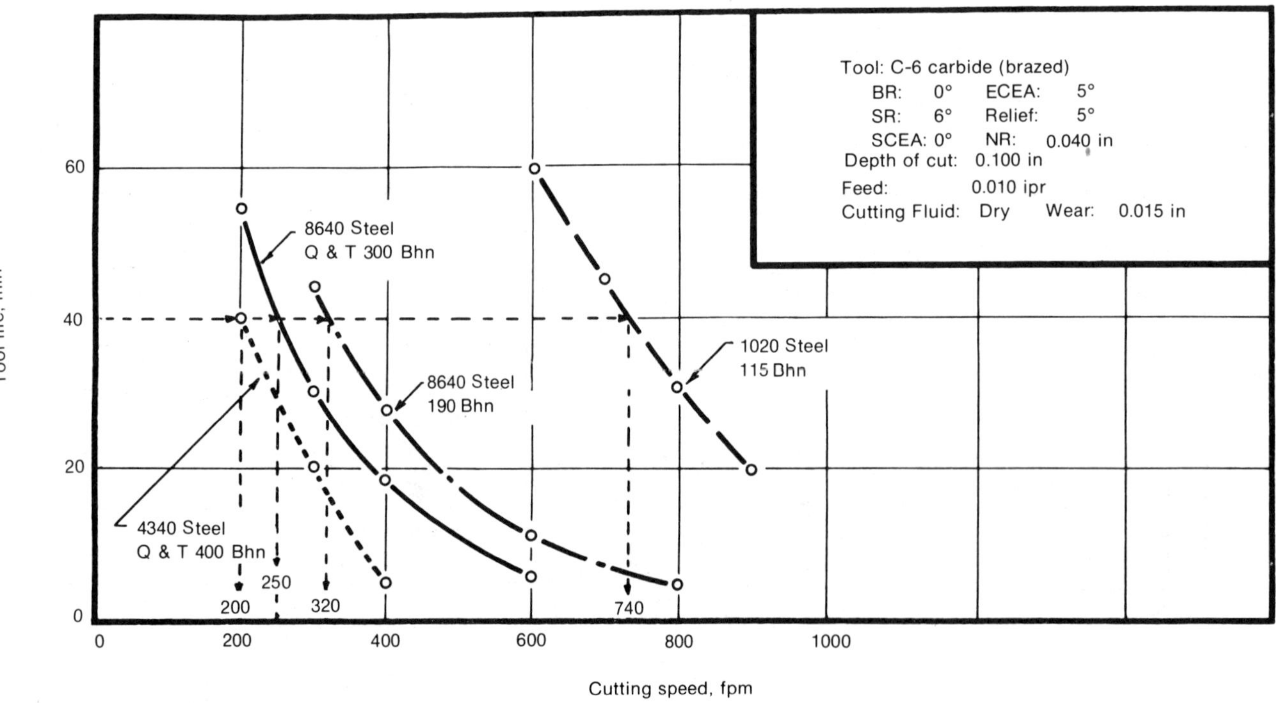


**Figure 19.3–7** Effect of cutting speed on tool life for turning various materials. For a 40-minute tool life turn the above steels at: 740 fpm for 1020 steel, 115 Bhn; 320 fpm for 8640 steel, 190 Bhn; 250 fpm for 8640 steel, quenched and tempered to 300 Bhn; and 200 fpm for 4340 steel, quenched and tempered to 400 Bhn.

There are various degrees of sophistication of machinability data available for use by industry. A general discussion of the types of data will be given.

### Tables of Recommended Machining Conditions

This *Machining Data Handbook* contains the practical starting conditions for most of the significant work materials and machining operations. It supplies the recommended feed, speed, tool material, tool geometry, and cutting fluid as well as data for determining horsepower requirements for various machining operations. It also contains the most practical and readily accessible data for general shop application.

### Wide Range Machining Data

Detailed machining data covering the tool life obtained under a wide range of machining parameters such as speed, feed, tool material, tool geometry, and cutting fluids are often useful in selecting a set of conditions for machining at lower costs or higher production rates. The detailed machining parameters can be obtained from laboratory studies or from shop studies by varying the pertinent parameters and observing their effect on tool life. Data formats for presentation of wide range machining data are shown in section 21.

### Cost and Production Rate Determination and Analysis

It is possible and desirable to calculate the total cost and production rate on any machining operation. Section 21 supplies the necessary procedures for making these calculations based on the relation of tool life to the machining parameters.

### Mathematical Correlation of Machining Parameters with Tool Life and Subsequent Optimization of Cost and Production Rates

Attempts have been and are being made to correlate tool life with the other pertinent machining parameters. The usual approach is to obtain empirical data relating tool life to speed, feed, and depth and width of cut. In some cases, these data can be put into the fairly simplified equation:

$$vT^nf^ad^bw^c = K$$

where: v = cutting speed
T = tool life
f = feed
d = depth
w = width

and n, a, b, c, and K are constants.

Unfortunately, it has been found that most work materials when machined by the various operations do not follow this simple relationship except for restricted regions of cutting speeds and feeds. Despite this, certain combinations of work materials and operations that obey this simple law permit one to obtain this mathematical relationship between tool life and the various machining parameters. For these situations, it is possible to calculate the machining conditions which provide minimum cost or maximum production rate or maximum profit. Whenever such calculations are made, it is wise to verify the calculation by shop trial.

### Machinability Ratings

There appears to be an endless desire to provide machinability ratings for materials. The original machinability rating index was based upon turning B1112 steel at 180 feet per minute [55 m/min] with a high speed steel tool for an index of 100 percent. This index was developed years ago under strong influence from the Independent Research Committee on Cutting Fluids.

Since that time, high speed steels have lost ground to carbides in turning. It is tempting to try to use a single numerical value to designate the degree of difficulty (or ease) with which a particular material can be machined. Unfortunately, it has been found that this machinability rating of a given work material changes with the type of operation involved and with the tool material selected. The inconsistencies and futility of applying machinability ratings can be seen in table 19.4–1.

Machinability ratings are given for four work materials and three operations—turning, milling and drilling. In addition, the rating is given for both high speed steel and carbides for the turning and milling. In machining 4340 steel, the rating is seen to vary from 25 to 88 depending upon the operation and tool material.

**TABLE 19.4–1 Machinability Index Inconsistencies**

| OPERATION | WORK MATERIAL | | | |
|---|---|---|---|---|
| | 4340 Steel 340 Bhn | Ti-3Al-13V-11Cr 285 Bhn | 17-7 PH Steel 170 Bhn | 250 Maraging Steel 330 Bhn |
| Turning | | | | |
| HSS | 25 | — | 32 | 54 |
| Carbide | 31 | 16 | 18 | 59 |
| Milling | | | | |
| HSS | 25 | 16 | 37 | 74 |
| Carbide | 88 | 20 | 53 | 83 |
| Drilling | | | | |
| HSS | 42 | 17 | 42 | 92 |

The application of the machinability index should be restricted to very special situations where the ratings have meaningful and consistent values. Another application of the machinability rating system involves the use of relative machining time, an example of which is given. A more precise and logical approach to the selection of machining data is to use the data contained in the *Machining Data Handbook*.

**SECTION 20**

# GRINDING AND ABRASIVE MACHINING


20.1 Grinding Wheels .......... 20–3
Conventional Wheels .......... 20–3
Cubic Boron Nitride (CBN) Grinding Wheels .......... 20–8
Diamond Grinding .......... 20–11

20.2 Grinding Guidelines .......... 20–13
Grinding Recommendations .......... 20–13
Modifying Wheel Recommendations .......... 20–13
Modifying Operating Conditions to Change Wheel Grade Action .......... 20–14
Precautions .......... 20–15

20.3 Surface Roughness and Tolerances .......... 20–17

20.4 Grinding Formulas and Charts .......... 20–21

20.5 Abrasive Machining .......... 20–33
Vertical Spindle, Rotary Table Surface Grinding .......... 20–33
Reciprocating Table Surface Grinding .......... 20–33
Abrasive Belt Machining .......... 20–33
High-Efficiency or High-Speed Grinding .......... 20–34
Cylindrical Crush Dress Grinding .......... 20–35
Creep Feed Grinding .......... 20–35

## CONVENTIONAL WHEELS

### Wheel Identification

Table 20.1-1 shows the standard marking system for identifying grinding wheels and other bonded abrasives (per ANSI B74.13-1977 and ISO 525-1975E).

Tables 20.1-2 and 20.1-3 list the codes that are used by the various grinding wheel manufacturers for aluminum oxide wheels and silicon carbide wheels, respectively.

#### Abrasive Types

The two most common abrasives used in industrial applications are aluminum oxide and silicon carbide. Aluminum oxide is a synthetic material produced in an electric furnace. Aluminum oxide abrasives have a wide variety of properties, such as toughness and friability, which can be tailored to specific job requirements. Tough aluminum oxide is used for rough grinding operations. Semifriable aluminum oxide is a general-purpose type. White friable aluminum oxide is used for tool grinding and for heat sensitive materials, while pink friable aluminum oxide is used for difficult-to-grind alloys.

Silicon carbide, also a product of the electric furnace, is harder than aluminum oxide and is used primarily for grinding nonmetallic, nonferrous and low-tensile-strength materials. Green friable silicon carbide is used for general-purpose grinding of tungsten carbide. Various combinations of aluminum oxide and silicon carbide abrasives are available only in organic bonds—resinoid, rubber and shellac.

#### Grain Size

Grain size influences surface roughness, stock removal rate, chip size and corner holding ability. Surface roughness depends on a number of factors, such as wheel hardness, structure and dressing methods, but finer-grain abrasives tend to produce surfaces with lower roughness values.

#### Wheel Grade

The grade or hardness of a wheel is a measure of the strength of the bond holding the grains in the wheel. The major factor affecting bond strength is the amount of bond holding an individual grain. Wheel grades range from A to Z in order of increasing bond content.

Wheel grade requirements vary inversely with the workpiece hardness. Harder materials require softer-grade wheels, while softer materials require harder-grade wheels.

#### Wheel Structure

The wheel structure number indicates the porosity or density of the wheel. Low numbers indicate a dense structure, while high numbers indicate an open structure.

Wheels having an open structure are recommended for surface grinding, while wheels having a dense structure are recommended for cylindrical, centerless and form grinding. Dense wheel structures are recommended when low surface roughness values are required.

#### Bond

Of the several bond types listed in table 20.1-1, resinoid, rubber and vitrified are the most widely used types.

Vitrified bonds are more brittle than resinoid bonds but are superior for holding form. Normal speeds for vitrified wheels are 6,500 or 8,500 feet per minute [33 or 43 m/s]. Higher speeds of 10,000 to 16,500 feet per minute [51 to 64 m/s] require special wheels, tested and approved for the particular speed and machine.

Normal speeds for resinoid-bond wheels range from 6,500 to 9,500 feet per minute [33 to 48 m/s] depending on the actual grading. In some cases, cutoff wheels and snagging wheels can be run at speeds up to 16,500 feet per minute [64 m/s]. In any case, the speed marked on the blotter is the maximum permissible speed. Resinoid-bond wheels are used for wet grinding and for cutoff operations.

### Truing and Dressing

Before truing a grinding wheel, the spindle should be run until it is up to operating temperature. Then the diamond truing tool should be placed at the center of the wheel and fed out to the wheel edge, and then from edge to edge. A copious flow of coolant should be directed at the wheel-tool interface.

The face of the wheel must be prepared for the job it is expected to do by dressing. When dressing wheels for finishing operations, the diamond tool is fed slowly across the face; when dressing wheels for roughing operations, the diamond tool is fed more rapidly.

**TABLE 20.1–1 Markings for Identifying Grinding Wheels and Other Bonded Abrasives**
(Standard marking system chart per ANSI B74.13-1977 and ISO 525-1975E)

| 51 | A | 36 | | | | L | | | 5 | | V | 23 |
|---|---|---|---|---|---|---|---|---|---|---|---|---|
| **PREFIX** | **ABRASIVE TYPE** | **ABRASIVE (GRAIN) SIZE** | | | | **GRADE** | | | **STRUCTURE** | | **BOND TYPE*** | **MANUFACTURER'S RECORD** |
| | | **Coarse** | **Medium** | **Fine** | **Very Fine** | **Soft** | **Medium** | **Hard** | **Dense to Open** | | | |
| Manufac-turer's symbol indicating exact kind of abrasive *(use optional)* | A - Aluminum Oxide<br><br>C - Silicon Carbide | 8<br>10<br>12<br>14<br>16<br>20<br>24 | 30<br>36<br>46<br>54<br>60 | 70<br>80<br>90<br>100<br>120<br>150<br>180 | 220<br>240<br>280<br>320<br>400<br>500<br>600 | A E<br>B F<br>C G<br>D H | I M<br>J N<br>K O<br>L P | Q V<br>R W<br>S X<br>T Y<br>U Z | 1<br>2<br>3<br>4<br>5<br>6<br>7<br>8 | 9<br>10<br>11<br>12<br>13<br>14<br>15<br>16<br>etc. | B - Resinoid<br>BF - Resinoid Reinforced<br>E - Shellac<br>O - Oxy-chloride<br>R - Rubber<br>RF - Rubber Reinforced<br>S - Silicate<br>V - Vitrified<br>Mg - Magnesia | Manufacturer's private marking to identify wheel *(use optional)* |
| | | | | | | | | | *(use optional)* | | | |

*ISO 525-1975(E) specifies a Mg-Magnesia bond not contained in ANSI B74.13-1977 and does not include bond types O-Oxychloride and S-Silicate.

**TABLE 20.1–2 Comparative Aluminum Oxide Wheel Marking Symbols**
(Includes both manufacturers' prefixes and abrasive symbols)

| MANUFACTURER | HEAVY DUTY | | | REGULAR | SEMI-FRIABLE | FRIABLE | MIXTURE OF ALUMINUM OXIDE | MIXTURE OF Al. Ox. & SiC |
|---|---|---|---|---|---|---|---|---|
| | Aluminum Oxide | Zirconia | Sintered | | | | | |
| Acme | A | 3XT | 7XT | A | FA | — | WA | CA |
| | — | 8XT | — | RA | — | — | — | C1A |
| | — | — | — | — | — | — | — | C2A |
| Bancroft | NA | ZA | VA | A | 1A | 2A | 5A | 7A |
| Bay State | — | — | 1A | A | 3A | 9A | 8A | CA |
| | RA | — | 15A | — | — | — | 2A | MA |
| | — | 27A | 25A | — | — | 17A | — | GA |
| | — | 33A | — | — | — | — | 16A | — |
| Bendix | RA | — | KA | A | FA | WA | GA | CA |
| | — | — | 6A | — | BA | — | EA | — |
| | — | 31A | JA | — | — | — | DA | — |
| | — | 2A | 22A | A | 35 | 48 | 15 | CA |
| | — | — | 88A | — | 29 | — | 18 | — |
| | — | — | — | — | — | — | 115A | — |
| | — | — | — | A | — | — | FBA | AC |
| | — | — | — | — | — | — | FWA | — |
| | — | — | — | — | — | — | BWA | — |
| Blanchard | — | — | — | A | 5A | 9A | 59A | CA |
| | — | — | — | — | 6A | — | 91A | — |
| Brightboy | — | — | — | ALO | — | — | — | — |
| | — | — | — | F | — | — | — | — |
| Buckeye | A | ZA | SA | A | FA | WA | MA | CA |
| Bullard | 3A | — | A | FA | WA | 6A | 6TA | CA |
| Carborundum | A | VA | WA | A | BA | AA | DA | CA |
| | — | XA | — | — | GA | — | BGA | — |
| | — | — | — | — | PA | — | FA | — |
| | — | — | — | — | SA | — | — | — |
| Chicago | TA | ZA | — | 51A | 12A | 53A | 50A | CA |
| | — | ZX | — | — | 52A | — | 54A | — |
| | — | AZ | — | — | 55A | — | — | — |
| Cincinnati | — | — | — | A | 2A | 9A | 97A | CA |
| | — | — | — | — | 4A | — | — | C2A |
| | — | — | — | — | — | 12A | — | C4A |
| | — | — | — | — | — | — | — | C12A |
| Colonial | HA | — | — | A | TA | 20A | — | 5TAC |
| | — | — | — | — | — | 21A | — | 10TAC |
| | — | — | — | — | — | 22A | — | 5CTA |
| | — | — | — | — | — | 25A | — | 10CTA |
| | — | — | — | — | — | 24A | 5TA | — |
| | — | — | — | — | — | 26A | 10TA | — |
| | — | — | — | — | — | — | 15TA | — |
| de Sanno | A | XA | — | A | 7A | 9A | 97A | CA |
| | — | YA | — | AT | 7AT | 10A | JA | DA |
| | — | ZA | — | — | 8A | RA | 7RA | FA |
| | — | — | — | — | — | 12A | 4A | — |
| | — | XZ | — | — | — | — | KA | GA |
| | — | — | — | — | — | — | KTA | HA |
| | — | — | — | — | — | — | SA | — |
| | — | — | — | — | — | — | PTA | — |
| Edmar | HA | ZA | SA | A | FA | WA | MA | CA |
| Ferro (FMR Electro) | TA | ZT | EA | A | FA | 13A | 2A | BA |
| | YA | BZT | XA | — | JA | 36A | 15A | CA |
| | — | CZT | — | — | TFA | WA | HA | DA |
| | — | — | — | — | — | — | LA | — |
| Fisher | M | ZR | — | — | A | B | AB | AC |

SOURCE: Grinding Wheel Institute.

# 20.1 Grinding Wheels

TABLE 20.1-2—Continued

| MANUFACTURER | HEAVY DUTY | | | REGULAR | SEMI-FRIABLE | FRIABLE | MIXTURE OF ALUMINUM OXIDE | MIXTURE OF Al. Ox. & SiC |
|---|---|---|---|---|---|---|---|---|
| | Aluminum Oxide | Zirconia | Sintered | | | | | |
| Gardner | —<br>—<br>—<br>—<br>—<br>— | —<br>—<br>—<br>—<br>—<br>— | —<br>—<br>—<br>—<br>—<br>— | A<br>84A<br>87A<br>—<br>—<br>— | 82A<br>—<br>—<br>—<br>—<br>— | 80A<br>—<br>—<br>—<br>—<br>— | 81A<br>83A<br>85A<br>86A<br>88A<br>91A | AC<br>—<br>—<br>—<br>—<br>— |
| Grier | —<br>—<br>— | —<br>—<br>— | —<br>—<br>— | A<br>—<br>— | 27A<br>—<br>— | 28A<br>22A<br>29A | 78A<br>—<br>— | CA<br>—<br>— |
| Hanson | TA<br>— | ZA<br>— | MA<br>— | —<br>— | BA<br>FA | RA<br>WA | HA<br>W2A | CA<br>AC |
| ITT Abrasive | HA<br>— | XA<br>— | MA<br>— | A<br>— | KA<br>PA | WA<br>— | ZA<br>EA | JA<br>— |
| Jowitt & Rodgers | — | — | — | G | B | W | BW | BX |
| MWA | 5A<br>— | 54A<br>— | 63A<br>— | 1A<br>— | 8A<br>11A | 2A<br>— | SFA<br>S11A | 8AC<br>11AC |
| National | JA<br>—<br>—<br>—<br>—<br>— | 14A<br>ZA<br>24A<br>34A<br>ZMA<br>— | 6A<br>—<br>—<br>—<br>—<br>— | A<br>—<br>—<br>—<br>—<br>— | PA<br>BA<br>HA<br>—<br>LA<br>BKA | SA<br>WA<br>—<br>—<br>—<br>— | MA<br>—<br>—<br>—<br>—<br>— | CA<br>CKA<br>—<br>—<br>—<br>— |
| Ney Abrasives | — | — | — | RW | A | WA | HA | CA |
| Norton | —<br>—<br>—<br>—<br>— | 66A<br>68A<br>ZF<br>ZS<br>NZ | —<br>76A<br>—<br>—<br>— | A<br>—<br>—<br>—<br>— | 57<br>—<br>—<br>—<br>— | 32<br>38<br>—<br>—<br>— | 16<br>19<br>23<br>53<br>— | AC<br>—<br>—<br>—<br>— |
| Osborn | 3A | — | — | 7A | 2A | 1A | — | AC |
| Pacific | —<br>—<br>—<br>—<br>—<br>— | 19A<br>20A<br>21A<br>22A<br>—<br>— | —<br>—<br>—<br>—<br>—<br>— | —<br>A<br>—<br>—<br>—<br>— | —<br>35A<br>—<br>—<br>—<br>— | —<br>50A<br>—<br>47A<br>49A<br>— | —<br>40A<br>—<br>42A<br>45A<br>— | 69A<br>70A<br>—<br>71A<br>72A<br>73A |
| Red Hill | TA<br>NA | ZA<br>XA | —<br>— | A<br>— | FA<br>— | WA<br>— | HA<br>— | AC<br>— |
| SGL Abrasives | 3A<br>—<br>— | 15A<br>18A<br>20A | —<br>—<br>— | A<br>—<br>— | 14A<br>4A<br>— | 10A<br>—<br>— | HA<br>KA | RA<br>EA<br>— |
| Universal | —<br>—<br>—<br>— | 4A<br>4B<br>4D<br>— | —<br>—<br>—<br>— | A<br>—<br>—<br>— | FA<br>—<br>—<br>— | WA<br>SA<br>RA (Ruby)<br>EA (Em-Rald) | WAA<br>WRA<br>EAA<br>RFA | CA<br>FCA<br>—<br>— |
| Waltham | A<br>—<br>—<br>21A<br>—<br>—<br>—<br>— | 28A<br>29A<br>48A<br>51A<br>54A<br>30A<br>49A<br>53A | 46A<br>47A<br>—<br>—<br>—<br>—<br>—<br>— | A<br>—<br>—<br>—<br>—<br>—<br>—<br>— | —<br>—<br>—<br>2168A<br>68A<br>—<br>—<br>— | BA<br>—<br>—<br>25A<br>WA<br>11A<br>12A<br>3A (Pink) | DA<br>—<br>—<br>—<br>14A<br>24A<br>—<br>— | —<br>—<br>—<br>—<br>CA<br>16A<br>—<br>— |
| Westfield | TA<br>—<br>— | —<br>—<br>— | —<br>—<br>— | A<br>—<br>— | FA<br>77A<br>— | 75A<br>—<br>— | 72A<br>73A<br>74A | CA<br>76A<br>— |

**TABLE 20.1–3 Comparative Silicon Carbide Wheel Marking Symbols**
(Includes both manufacturers' prefixes and abrasive symbols)

| MANUFACTURER | HEAVY DUTY | REGULAR | SEMI-FRIABLE | FRIABLE | SILICON CARBIDE MIXTURE |
|---|---|---|---|---|---|
| Acme | — | C | — | GC | — |
| Bancroft | C | C | — | GC | — |
| Bay State | 7C | C | — | 1C | — |
| Bendix | SC<br>— | C<br>C | —<br>— | GC<br>GC | MC<br>HC |
| Blanchard | — | C | — | 9C | — |
| Brightboy | —<br>— | S1C<br>S | —<br>— | —<br>— | —<br>— |
| Buckeye | HC | C | — | GC | CA |
| Carborundum | TC<br>WC | C<br>— | SC<br>BC | GC<br>— | RC<br>— |
| Chicago | —<br>— | 49C<br>— | —<br>— | 49CG<br>48C | 47C<br>— |
| Cincinnati | — | 6C | — | 5C | 7C |
| Colonial | C | C | — | GC | — |
| de Sanno | HC | C | — | GC | GCC |
| Edmar | HC | C | FC | GC | MC |
| Ferro (FMR Electro) | TC | C | — | GC | RC |
| Fisher | — | C | — | GC | — |
| Gardner | —<br>— | C<br>— | —<br>— | GC<br>SC | AC<br>— |
| Hanson | HC | — | — | GC | 4C |
| ITT Abrasive | C | C | — | GC | SC |
| Jowitt & Rodgers | — | X | — | GC | — |
| MWA | C | 3C | — | 7C | 3C |
| National | — | C | — | GC | RC |
| Ney Abrasives | — | C | — | GC | — |
| Norton | — | 37C | — | 39C | 74C |
| Osborn | 3C | — | — | 1C | — |
| Pacific | — | 75C | — | 80C | 77C |
| Red Hill | — | C | — | GC | — |
| SGL Abrasives | — | C | — | 7C | — |
| Universal | — | C | BC | GC | — |
| Waltham | 21C<br>— | C<br>C | —<br>— | GC<br>— | LC<br>MC |
| Westfield | — | C | — | GC | — |
| Bullard | C | FC | GC | 6C | — |

SOURCE: Grinding Wheel Institute.

## CUBIC BORON NITRIDE (CBN) GRINDING WHEELS

### Applications

CBN grinding wheels are recommended for alloys that are difficult to grind with conventional abrasives; that is, steels and cast irons with hardnesses above 50 $R_C$ and nickel and cobalt high temperature alloys with hardnesses above 35 $R_C$.

### Wheel Identification

Table 20.1-4 shows the standard marking system for diamond and cubic boron nitride abrasives (per ANSI B74.13-1977). Table 20.1-5 lists the codes that are used by the various wheel manufacturers for these abrasives.

#### Bond Types

CBN grinding wheels are available with resinoid, vitrified and metal bonds. Resinoid-bond wheels produce a good finish at high metal removal rates and provide fairly good wheel life. Vitrified-bond wheels are better for holding form and provide longer wheel life. Metal-bond wheels are free cutting and give high metal removal rates.

Metal-bond, electroplated wheels contain a single layer of CBN grains and cannot be trued or dressed. Therefore, it is imperative that the spindle be running true and that all mounting surfaces be clean and free of burrs. When these wheels are mounted, the runout should not exceed 0.0005 inch [0.013 mm] TIR in either the radial or the lateral directions.

#### Grit Size

Because of the sharpness of CBN, it is necessary to use finer grain sizes than are used with conventional abrasives to obtain the same surface roughness level. Extremely fine finishes are difficult to obtain with CBN abrasives because there is little or no burnishing of the surface since CBN abrasives wear so little.

### Grinding Fluids

Hardened steels and cast irons can be ground dry with CBN wheels; however, using a grinding fluid will improve both the surface roughness and the wheel life. A light-duty soluble oil will suffice for these materials.

When grinding the superalloys, a heavy-duty, active grinding fluid, that is, sulfo-chlorinated oil or sulfo-chlorinted soluble oil, is a must. These heavy-duty grinding fluids will also improve grinding ratios when grinding hardened steels with CBN wheels.

### Truing and Dressing

It is of utmost importance that a CBN wheel be properly conditioned before attempting to grind with it. Both resinoid-bond and vitrified-bond wheels can be trued using a rotary diamond dresser, a brake-controlled truing device or a metal-bonded, 100 to 180 mesh size diamond truing tool. A single-point or cluster diamond should not be used. The wheel should be flooded with a copious supply of coolant while truing.

After truing, resinoid-bond CBN wheels must be dressed to open up the glazed surface left by the truing process. A conditioning stick containing approximately 200-mesh aluminum oxide or silicon carbide abrasives in a soft vitrified bond can be used. It is best to jog the wheel and press the conditioning stick into the wheel while it is slowing down. This removes the bond, exposing the sharp cutting edges. When the conditioning stick starts to wear rapidly, the wheel is properly dressed. Vitrified-bond CBN wheels usually do not need to be dressed or conditioned. They can be used immediately after truing.

### Machine Requirements

To realize maximum benefits from using CBN wheels, the machine must have sufficient rigidity, horsepower and spindle speed. Machine vibrations that can be tolerated with conventional abrasives because of their more rapid wheel breakdown will cause CBN wheels to chatter.

### Surface Integrity

Because CBN wheels remain sharp longer, they produce less heat and deformation in the workpiece, thus less residual stress. This permits the use of conventional speeds and feeds on critical parts that would otherwise require "low stress" grinding procedures if conventional abrasives were used.

**TABLE 20.1–4 Markings for Identifying Diamond and Cubic Boron Nitride Grinding Wheels and Other Bonded Abrasives**
(Standard marking system chart per ANSI B74.13-1977)

| M | D | 120 | N | 100 | B | 77 | 1/8 | ___ |
|---|---|---|---|---|---|---|---|---|
| PREFIX | ABRASIVE TYPE | ABRASIVE (GRAIN) SIZE | GRADE | CONCENTRATION | BOND TYPE | BOND MODIFICATION | DEPTH OF ABRASIVE | MANUFACTURER'S RECORD |
| Manufacturer's symbol indicating exact kind of abrasive *(use optional)* | B- CBN<br>D- Diamond | see below | see below | Manufacturer's designation. May be number or symbol. | B - Resin<br>M - Metal<br>V - Vitrified | Manufacturer's notation of special bond type or modification | Working depth of abrasive section in inches or millimeters *(inches illustrated)* | Manufacturer's identification symbol *(use optional)* |

**ABRASIVE (GRAIN) SIZE**

| Coarse | | to | Very Fine |
|---|---|---|---|
| 8 | 30 | 90 | 280 |
| 10 | 36 | 100 | 320 |
| 12 | 46 | 120 | 400 |
| 14 | 54 | 150 | 500 |
| 16 | 60 | 180 | 600 |
| 20 | 70 | 220 | etc. |
| 24 | 80 | 240 | |

**GRADE**

| Soft | | Medium | | Hard | |
|---|---|---|---|---|---|
| A | E | I | M | Q | V |
| B | F | J | N | R | W |
| C | G | K | O | S | X |
| D | H | L | P | T | Y |
| | | | | U | Z |

**ABRASIVE TYPE**

B- CBN
D- Diamond

**BOND TYPE**

B - Resin
M - Metal
V - Vitrified

**TABLE 20.1–5 Comparative Diamond and Cubic Boron Nitride Wheel Marking Symbols**

| MANUFACTURER | CUBIC BORON NITRIDE | DIAMOND |
|---|---|---|
| Bay State | 1B (Borazon) | D, 1D, 3D, 4D, 6D, 9D |
| Bendix | B (Cubic boron nitride) | D (Manufactured) |
| Carborundum | CB (Nickel-coated resin-bond cubic boron nitride)<br>B (uncoated cubic boron nitride) | D<br>MD (Manufactured)<br>CMD (Coated manufactured)<br>SND (Selected natural) |
| Cincinnati | 1BN (Metal-coated cubic boron nitride)<br>2BN (Cubic boron nitride) | MD (Manufactured)<br>CMD (Coated manufactured)<br>AMD (Coated manufactured)<br>EMD (Coated manufactured)<br>DMD (Coated manufactured) |
| Norton | CB (Cubic boron nitride) | D<br>SD (Manmade)<br>ASD (Coated manufactured)<br>ASDC (Coated manufactured)<br>AZD (Coated manufactured) |
| Universal | BZ (Borazon) | D (Natural)<br>CD (Coated natural)<br>MD (Manufactured)<br>WD (Metal coated manufactured)<br>DD (Metal coated manufactured)<br>SD (Metal coated manufactured) |
| Waltham | BN (Borazon) | MD (Manufactured)<br>D (Natural)<br>ND (Manufactured nickel coated)<br>CD (Manufactured copper coated) |

SOURCE: Grinding Wheel Institute.

## DIAMOND GRINDING

Diamond grinding technology has developed considerably during the past 30 years with respect to both grinding wheels with natural grit and, more recently, grinding wheels with synthetic grit. Today, diamond wheels are available in a large variety of mesh sizes, diamond concentrations, and bond types. Table 20.1-4 lists the standard markings used for diamond wheels. The designations for abrasive types used by various wheel manufacturers are listed in table 20.1-5.

Grain types have been developed for specific grinding operations and materials. More recent developments have opened up new applications for diamond wheels, especially in the field of steel and carbide grinding utilizing metal-clad, diamond-grain wheels.

Diamond wheels are available in several bonds: metal, vitrified and resinoid. Each bond type provides different results, depending on the type of grinding operation and the material being ground.

**Metal-bond** wheels are extremely strong, long-life wheels having diamonds securely held in a metal matrix. Chief uses are cutoff operations with thin wheels, offhand grinding, chip-breaker grinding and electrolytically assisted grinding. Metal-bond wheels should always be used with a coolant.

**Vitrified-bond** wheels are diamonds bonded together with ceramic materials. This bond type provides long wheel life at good grinding rates. The principal applications for vitrified-bond diamond wheels are surface grinding, cylindrical grinding, offhand grinding and internal grinding.

**Resinoid-bond** wheels are the most popular type because they provide excellent grinding speed and relatively long life. A resinoid diamond wheel is ideally suited for grinding carbide cutting tools where maximum protection from heating and cracking are a consideration. This type of bond is best for machine grinding operations, such as cutter grinding, surface and cylindrical grinding. Resinoid bond has excellent form holding qualities, is easily dressed, and is most suitable for diamond wheels with formed faces.

### Wet Grinding

Wherever possible, diamond wheels should be used wet. This promotes longer wheel life, decreases workpiece surface roughness and reduces wheel dressing frequency. For machines not equipped with flood coolant systems or where the operator must observe the work at all times, the mist or wick method can be used. The primary reason for recommending wet grinding with diamond wheels is that the diamond itself has a relatively low thermal threshold and will "burn" at relatively low (1,100° to 1,300°F [600° to 700°C]) temperatures. The heat at the crystal-work interface can easily reach the low temperature range in dry grinding.

Wet grinding with diamond wheels eliminates some of the common problems associated with dry grinding. Since workpiece temperatures are kept at a low level, the potential for surface cracks or possible catastrophic cracking of the workpiece is greatly reduced. Dust problems also are eliminated. The flushing action of a copious supply of coolant helps to keep the diamond wheel cutting free, thus preventing loading of the wheel.

### Dry Grinding

If dry grinding is absolutely necessary, wheel speed must be kept below 4,500 feet per minute [23 m/s], with 3,500 to 4,000 feet per minute [18 to 20 m/s] preferable. With care and practice, an experienced operator can become proficient at dry grinding of hard materials such as carbide and ceramics. One problem associated with dry grinding is dust; therefore, adequate dust-collecting systems need to be employed. Wheel wear will be greater than with wet grinding because softer wheels must be used to help keep the grinding temperatures as low as possible. Although resinoid-bond wheels are recommended for dry diamond grinding of hard, brittle materials such as carbide, glass and ceramics, high grinding temperatures can burn or crack the grinding wheel. Metal- and vitrified-bond diamond wheels, on the other hand, if run dry on hard, brittle materials may cause cracking of the workpiece material.

### Machines for Grinding with Diamonds

Diamond grinding requires a higher level of machine tool integrity than might be required for grinding with a silicon carbide or an aluminum oxide wheel. Spindles must run true, bearings should be in good condition and good maintenance procedures should be followed.

### Wheel Mounting

When mounting a diamond wheel, extra care should be taken to be sure that back plates, flanges and spindles are clean, free of burrs and running true. Straight wheels should be tightened gently between the flanges and then lightly tapped into position so that they run within 0.0005 inch [0.013 mm] TIR. The flanges should then be completely tightened.

If the grinding machine has a tapered spindle nose, it is common practice to mount each diamond wheel on a separate collet or adaptor. If left as a unit, the mounted wheel and adaptor can then be removed and replaced as needed, and the time and abrasive lost in truing can be avoided.

## GRINDING RECOMMENDATIONS

The grinding conditions listed in section 8 of this handbook are provided as starting recommendations. These starting conditions may have to be adjusted to reflect (1) workpiece characteristics, (2) requirements for the ground surface, (3) economic objectives and (4) operational conditions. Obviously, these grinding recommendations represent only an approximation of the optimum conditions achievable by testing and experimenting.

These recommendations should handle most grinding situations. In a job shop, it is often necessary to avoid wheel changes; a few general-purpose wheels usually are used to grind almost every type of material, even if at less than optimum efficiency. Where large production quantities are involved, testing and modifications may be necessary to obtain optimum results.

Despite the use of standard grinding wheel markings (table 20.1-1), wheels with identical markings made by various manufacturers may vary in actual performance. Individual wheel makers have their own modifications of both abrasive and bond which cannot be included in the data tables. Likewise, the structure number is omitted from the wheel identification because its significance differs among various makes of wheels. The structure provides chip clearance and is also a result of grain size and proportion of bond. The best or standard structure is usually derived for most grain size and grade combinations as a result of experience and testing. The manufacturers of grinding wheels publish their own tables of wheel recommendations which include grain and bond modifications as well as structure numbers.

Recommendations in section 8 are predicated on average shop conditions. Deviations from such basic conditions will generally affect the manner in which grinding wheels perform. Some of the variables discussed later have opposing effects and may balance each other when present concurrently. Other variables can have similar effects which can be expected to work additively. An understanding of the directions in which process variables interact with grinding wheel performance permits modification of the controllable variables to improve overall operating conditions or economics.

## MODIFYING WHEEL RECOMMENDATIONS

The following list of guidelines are provided for use in modifying the grinding wheel recommendations provided in section 8 to improve production or to meet the specific requirements of an application. No application, however, should be so finely tuned that normal variations would affect the output rate or the quality. The best approach is to accommodate any variations likely to occur. The best wheel for any application is the one that compromises the ability to cut rapidly with the ability to hold form, maintain surface roughness requirements and last longer.

**Wheels recommended in section 8 are suitable for both rough and finish grinding in one setup. Where two setups are used, use a wheel one grade harder for roughing and a wheel one grade softer for finishing.**

### General Guidelines

- To remove a substantial amount of stock—wet grinding is recommended over dry grinding.
- To remove stock faster—use a coarser grain wheel with a more open structure and a less friable abrasive.
- To produce a smoother finish—use a finer grain wheel with a denser structure and a less friable abrasive. (See table 20.3-1.)
- To generate form—use a finer grain wheel with a more dense structure. (See table 20.3-3.)
- To grind large areas—use a softer grade and coarser grain wheel.
- To grind small areas—use a harder grade and a finer grain wheel.
- To grind soft metals—use a harder grade and a coarser grain wheels.
- To grind hard metals—use a softer grade and a finer grain wheel.
- To improve workpiece finish—dress the wheel to a fine finish.
- To minimize heat, warpage and surface damage in the workpiece—maintain wheel sharpness (see section 18.3, Surface Integrity).
- If the grinding wheel breaks down too fast—use a wheel with a less friable abrasive, a harder grade and a denser structure.
- If the grinding wheel glazes and burns—use a wheel with a more friable abrasive, a softer grade and a more open structure.

### Guidelines for Surface Grinding

- To minimize heat and warpage in the workpiece—use fast table speeds, light downfeeds, and dress the wheel before final size.
- To improve workpiece finish—dress the wheel to a fine finish.
- When using wheels larger than 14-inch [356 mm] diameter—use one grade softer bond.

### Guidelines for Surface Grinding, Vertical Spindle, Rotary Table

- To produce smoother workpiece finish and closer tolerances—use faster table speed, lighter downfeed and proper sparkout.
- To increase material removal rate—increase table speed and downfeed rate.

# 20.2 Grinding Guidelines

### Guidelines for Cylindrical Grinding

- When using wheels larger than 14-inch [356 mm] diameter—use one grade softer bond.
- For heavier stock removal—use a faster traverse speed and a slower work speed and/or increase the depth of cut.
- To improve workpiece finish—use a slower traverse speed and a faster work speed and/or decrease the depth of cut.
- When using wider wheels—use a softer grade wheel (infeed grinding).
- When using narrower wheels—use a harder grade wheel (infeed grinding).

### Guidelines for Centerless Grinding

- When using wheels smaller than 20-inch [500 mm] diameter—use a one grade harder wheel.
- To improve workpiece finish—use a smaller angle of draw and a faster regulating wheel speed.
- For heavier stock removal—use a larger angle of draw and a slower regulating wheel speed and grind closer to center.
- To produce roundness—grind as high above center as possible.
- To grind long bars—grind below center on a flat blade.

### Guidelines for Internal Grinding

- For long bores—use softer grade wheels and wider wheels, if possible.
- For low powered machines—use softer grade wheels.
- For light spindle machines—use softer grade wheels.

## MODIFYING OPERATING CONDITIONS TO CHANGE WHEEL GRADE ACTION

Most grinding problems (other than those related to machine condition) arise from the action of the wheel grade (or hardness) which is a direct function of the wheel sharpness or wear rate. Indicators of too little wear (that is, the wheel is acting hard) are: the wheel glazing, loading or not cutting freely; workpieces with burning, heat checking (grinding cracks) or out of roundness; finishes getting progressively better (due to glazing); and finely spaced chatter marks or squealing. Indicators of too much wear (that is, the wheel is acting soft) are: the wheel breaking down too fast; workpieces with finishes getting progressively worse; widely spaced chatter marks; and scratches, fishtails, taper or lack of accuracy.

The action of the wheel grade can be altered by adjusting the other grinding conditions to achieve proper grinding action when it is not desirable or possible to change the wheel. Table 20.2-1 contains guidelines for modifying wheel action.

**TABLE 20.2-1 Guidelines for Modifying Wheel Action**

| OPERATION | To Make Wheel Act HARDER | To Make Wheel Act SOFTER |
|---|---|---|
| Internal grinding | Decrease work rpm<br>Decrease wheel reciprocation<br>Increase wheel rpm<br>Dress wheel at slower rate | Increase work rpm<br>Increase wheel reciprocation<br>Decrease wheel rpm<br>Dress wheel at faster rate |
| Surface grinding | Decrease table speed<br>Decrease crossfeed<br>Increase wheel rpm<br>Dress wheel at slower rate | Increase table speed<br>Increase crossfeed<br>Decrease wheel rpm<br>Dress wheel at faster rate |
| Cylindrical grinding (traverse) | Decrease traverse speed<br>Increase wheel rpm<br>Dress wheel at slower rate | Increase traverse speed<br>Decrease wheel rpm<br>Dress wheel at faster rate |
| Centerless grinding | Decrease feed wheel rpm<br>Dress grinding wheel at slower rate | Increase feed wheel rpm<br>Dress grinding wheel at faster rate |

## PRECAUTIONS

### Safety

Follow all requirements for safe wheel use. Do not exceed the maximum speed marked on the wheel or shown in ANSI B7.1-1978, *Safety Requirements for the Use, Care, and Protection of Abrasive Wheels.* See table 20.2-2.

### Cutting Fluids

Use the proper cutting fluid, use the proper flow rates, and direct the fluid so that is reaches the cutting zone. See section 16, Cutting Fluids.

### Surface Integrity

To avoid damage to the ground surface in the form of grinding burn, heat checks (grinding cracks), warpage, residual stress or other surface alterations, see section 10.4, Low Stress Grinding, for proper grinding procedures. Also see section 18.3, Surface Integrity.

**TABLE 20.2–2 Standard Maximum Wheel Speeds**

| CLASSIFICATION NUMBER | TYPES OF WHEELS | MAXIMUM WHEEL SPEED, fpm | | | | | |
|---|---|---|---|---|---|---|---|
| | | INORGANIC BONDS | | | ORGANIC BONDS | | |
| | | Low Strength | Medium Strength | High Strength | Low Strength | Medium Strength | High Strength |
| 1 | Type 1—Straight Wheels—except classifications 6, 7, 9, 10, 11, 12, 13 and 14 below.<br>Type 4*—Taper Side Wheels<br>Types 5, 7, 20, 21, 22, 23, 24, 25, 26 Recessed, Dovetailed and/or relieved wheels. (Except Classification 7 and 14 below.)<br>Type 12—Dish Wheels<br>Type 13—Saucer Wheels<br>Types 16, 17, 18, 19—Cones and Plugs | 5,500 | 6,000 | 6,500 | 6,500 | 8,000 | 9,500 |
| 2 | Type 2—Cylinder Wheels including plate mounted, inserted nut and projecting stud—Segments | 5,000 | 5,500 | 6,000 | 5,000 | 6,000 | 7,000 |
| 3 | Cup Shape Tool Grinding Wheels (For Fixed Base Machines)<br>Type 6—Straight Side Cups<br>Type 11—Flaring Cups | 4,500 | 5,000 | 6,000 | 6,000 | 7,500 | 8,500 |
| 4 | Cup Shape Snagging Wheels (For Portable Machines)<br>Type 6—Straight Side Cups<br>Type 11—Flaring Cups | 4,500 | 5,500 | 6,500 | 6,000 | 8,000 | 9,500 |
| 5 | Abrasive Discs: Plate Mounted<br>Inserted Nut and Projecting Stud: Solid or Segmental | 5,500 | 6,000 | 6,500 | 5,500 | 7,000 | 8,500 |
| 6† | Reinforced Wheels Type 1<br>Max. Dia. 4″<br>Max. Thickness 1/4″ | NA‡ | NA | NA | 9,500 | 12,500 | 16,000 |
| | Max. Dia. 10″<br>Max. Thickness 1/2″ | NA | NA | NA | 9,500 | 12,500 | 14,200 |
| | All Other Diameters and Thicknesses | NA | NA | NA | 9,500 | 12,500 | 12,500 |
| | Reinforced Wheels—Types 27 and 28<br>Max. Dia. 9″<br>Max. Thickness 3/8″ | NA | NA | NA | 9,500 | 12,500 | 14,200 |
| | Max. Dia. 9″—Over 3/8″ Thick | NA | NA | NA | NA | 9,500 | 11,000 |
| 7 | Type 1 Wheels for Bench and Pedestal Grinders<br>Types 1 and 5 for Surface Grinders only in sizes<br>7″ dia. up to 2″ thick and up to 2″ hole | 5,500 | 6,325 | 6,600 | 6,500 | 8,000 | 9,500 |
| | 8″ dia. up to 2″ thick and up to 2″ hole | 5,500 | 6,325 | 7,550 | 6,500 | 8,000 | 9,500 |
| 8 | Diamond and Cubic Boron Nitride Wheels | NA | NA | 6,500 | NA | NA | 9,500 |
| | Exceptions: 1. Metal Bond | NA | NA | 12,000 | NA | NA | NA |
| | 2. Steel Centered Cutting-off Wheels | NA | NA | 16,000 | NA | NA | 16,000 |
| 9 | Cutting-off Wheels Type 1 and 27A<br>Larger than 16″ dia., Including Reinforced Organic | NA | NA | NA | 9,500 | 12,000 | 14,200 |
| 10 | Cutting-off Wheels Type 1 and 27A<br>16″ dia., and Smaller—Including Reinforced Organic | NA | NA | NA | 9,500 | 12,000 | 16,000 |
| 11 | Thread and Flute Grinding Wheels | 8,000 | 10,000 | 12,000 | 8,000 | 10,000 | 12,000 |
| 12 | Crankshaft and Camshaft Grinding Wheels | 5,500 | 8,000 | 8,500 | 6,500 | 8,000 | 9,500 |
| 13 | Type 1 Snagging Wheels 16″ diameter and larger, Organic bond including reinforced:<br>Used on swing frame grinders, designed for this speed | NA | NA | NA | NA | NA | 12,500 |
| | Used on semi-automatic snagging grinders, designed for this speed | NA | NA | NA | NA | NA | 16,500 |
| 14 | Internal Wheels—Type 1 and 5<br>Maximum dia. 6″ | 5,500 | 8,000 | 8,500 | 6,500 | 8,500 | 9,500 |
| 15 | Mounted Wheels<br>(See standard for limitations) | NA | NA | 10,000 | NA | NA | 10,000 |

SOURCE: ANSI B7.1-1978, p. 65

NOTE: Abrasive wheels are produced in a wide range of specifications to perform satisfactorily in a wide range of grinding applications. As a general rule, hard material requires a soft grade wheel and soft material is best ground by a hard grade wheel. Different wheel grades have different strengths; harder grade wheels are generally stronger than softer grade wheels, although other ingredients such as grit size, structure and bond type play a part in the overall strength of an abrasive wheel specification. Therefore, the maximum allowable operating speeds vary depending on the strength of the wheel. It is for this reason that speeds shown are listed in columns labeled Low, Medium, and High Strength.

*Non-standard shape.

†Classification 6 excludes cutoff wheels.

‡NA = not applicable.

## SURFACE ROUGHNESS

The grain size of a grinding wheel will determine the approximate surface roughness that can be obtained on the workpiece, as given in table 20.3-1. The structure of the wheel and the dressing procedure will also affect the surface roughness to some degree. A finer-grain-size wheel will usually produce surfaces with lower roughness values at some sacrifice of stock removal capability.

Obtaining surface roughness values of less than 10 to 15 microinches $R_a$ [0.25 to 0.38 μm] requires special attention to work speeds and crossfeed or traverse rates. Very smooth surfaces require abrasive grain sizes of 220 and finer.

## TOLERANCES

The tolerances achievable with various grinding operations are given in table 20.3-2. The production tolerances can be held without difficulty; however, larger tolerances where acceptable will be more economical. The precision tolerances can be held with care but will be more costly.

### Form Requirements

Grinding of forms and fillets usually requires the use of 80 grain size and finer wheels, as indicated in table 20.3-3.

### Thread Grinding Requirements

In conventional grinding, the standard practice is to use the coarsest practical grain size for fastest stock removal. In thread grinding, however, the coarsest practical grain size is limited by the maximum allowable radius at the bottom of the thread. This radius is usually given in terms of root width, which is the distance between the points of tangency of the arc and the sides of the teeth. The root width varies inversely with the number of threads per inch [or metric pitch]; the greater the number of threads per inch, the smaller the maximum allowable root width.

To alter the wheel recommendations in section 8.14, table 20.3-4 may be used as a guide for selecting the coarsest allowable grain size for a thread grinding wheel. If a particular standard calls for a narrower root width, the grain size must be chosen for the maximum allowable root width. Another factor to be considered is the quality of the finish. Where a very low surface roughness value is required on threads of relatively coarse pitch, it is advisable to use a finer grain wheel than would be necessary to hold the proper root width.

**TABLE 20.3-1 Grinding Wheel Grain Sizes for Producing Various Surface Roughnesses**

| GRAIN SIZE | SURFACE ROUGHNESS | |
|---|---|---|
| | μin | μm |
| 46 | 32 | 0.80 |
| 54 | 20 to 32 | 0.50 to 0.80 |
| 60 | 15 to 20 | 0.38 to 0.50 |
| 80 | 10 to 15 | 0.25 to 0.38 |
| 120 | 8 to 10 | 0.20 to 0.25 |

**TABLE 20.3–2 Production and Precision Grinding Tolerances**

| GRINDING OPERATION | TOLERANCES (Plus or Minus) | | | |
|---|---|---|---|---|
| | inch | | mm | |
| | Production | Precision | Production | Precision |
| CYLINDRICAL GRINDING | | | | |
| Diameters | 0.00025 | 0.00001 | 0.0064 | 0.00025 |
| Shoulders | | | | |
| Shoulder to Shoulder | 0.00025 | 0.0005 | 0.0064 | 0.0127 |
| Traverse Grinding to a Shoulder | 0.002 | 0.001 | 0.050 | 0.025 |
| Corners and Radii | | | | |
| External Corners | Sharp | Sharp | | |
| Internal Corner Radii | 0.005 | 0.0025 | 0.13 | 0.063 |
| Spherical Sections (oscillating grinders) | | | | |
| Diameters | 0.00015 | | 0.0038 | |
| Location of Centers | 0.001 | | 0.025 | |
| CENTERLESS GRINDING | | | | |
| Diameters and Parallelism | 0.0001 | 0.000025 | 0.0025 | 0.00064 |
| Roundness | 0.000012 | | 0.0003 | |
| Concentricity of Stepped Diameters | 0.00025 | 0.0001 | 0.0064 | 0.0025 |
| THREAD GRINDING | | | | |
| Lead Error (inch per inch) | 0.00025 | 0.00001 | 0.0064 | 0.00025 |
| Pitch Diameter | 0.0005 | 0.0002 | 0.0127 | 0.0050 |
| Roundness | 0.00025 | | 0.0064 | |
| Concentricity (thread form with OD) | 0.0005 | | 0.0127 | |
| Grooves (width) | 0.001 | | 0.025 | |
| SURFACE GRINDING | | | | |
| Reciprocating Table Grinder | | | | |
| Flatness | 0.0002 | 0.00015 | 0.0050 | 0.0038 |
| Thickness | 0.0003 | 0.00015 | 0.0076 | 0.0038 |
| Rotary Table Grinder | | | | |
| Flatness | 0.0002 | 0.0001 | 0.0050 | 0.0025 |
| Parallelism | 0.0002 | 0.00005 | 0.0050 | 0.0013 |
| Thickness | 0.001 | 0.0002 | 0.025 | 0.0050 |
| INTERNAL GRINDING | | | | |
| Holes (using automatic sizing devices) | 0.00025 | 0.00005 | 0.0064 | 0.0013 |
| Face Runout (squareness of shoulder to bore) | 0.00025 | 0.00005 | 0.0064 | 0.0013 |

SOURCE: Adapted from H. E. Trucks, *Designing for economical production,* Dearborn, MI: Society of Manufacturing Engineers, 1974, p. 34.

**TABLE 20.3–3 Grinding Wheel Grain Sizes for the Radius of Forms or Fillets**

| GRAIN SIZE | MINIMUM RADIUS | |
|---|---|---|
| | inch | mm |
| 80 | 0.010 | 0.254 |
| 120 | 0.007 | 0.178 |
| 180 | 0.005 | 0.127 |
| 220 | 0.004 | 0.102 |
| 280 | 0.003 | 0.076 |
| 320 | 0.002 | 0.051 |
| 500 | 0.001 | 0.025 |

TABLE 20.3-4 Coarsest Allowable Grain Sizes for Thread Grinding Wheels

| AMERICAN STANDARD UNIFIED THREADS | | COARSEST ALLOWABLE GRAIN SIZE | | IFI-500 TRIAL METRIC THREADS* | |
|---|---|---|---|---|---|
| Threads Per Inch | Root Width in | Vitrified Wheels | Resinoid Wheels | Pitch | Root Width mm |
| 80 | 0.0016 | 240 | 220 | 0.35 | 0.044 |
| 72 | 0.0017 | 240 | 220 | — | — |
| 64 | 0.0019 | 220 | 180 | — | — |
| — | (0.0020) | 220 | 180 | 0.40 | 0.050 |
| 56 | 0.0022 | 220 | 180 | 0.45 | 0.056 |
| — | (0.0024) | 220 | 180 | 0.50 | 0.062 |
| 48 | 0.0026 | 220 | 180 | — | — |
| 44 | 0.0028 | 220 | 180 | — | — |
| — | (0.0029) | 220 | 180 | 0.60 | 0.075 |
| 40 | 0.0031 | 220 | 180 | — | — |
| 36 | 0.0034 | 180 | 180 | — | — |
| — | (0.0035) | 180 | 180 | 0.70 | 0.088 |
| 32 | 0.0039 | 180 | 150 | 0.80 | 0.100 |
| 28 | 0.0045 | 150 | 120 | — | — |
| — | (0.0049) | 150 | 120 | 1.00 | 0.125 |
| 24 | 0.0052 | 150 | 120 | — | — |
| — | (0.0061) | 120 | 120 | 1.25 | 0.156 |
| 20 | 0.0062 | 120 | 120 | — | — |
| 18 | 0.0069 | 120 | 100 | — | — |
| — | 0.0074 | 120 | 100 | 1.50 | 0.188 |
| 16 | 0.0078 | 120 | 100 | — | — |
| — | 0.0086 | 120 | 100 | 1.75 | 0.219 |
| 14 | 0.0089 | 100 | 100 | — | — |
| 13 | 0.0096 | 100 | 90 | — | — |
| — | 0.0100 | 100 | 90 | 2.0 | 0.250 |
| 12 | 0.0104 | 90 | 90 | — | — |
| 11 | 0.0113 | 90 | 90 | — | — |
| — | (0.0123) | 90 | 90 | 2.5 | 0.312 |
| 10 | 0.0125 | 90 | 90 | — | — |
| 9 | 0.0139 | 90 | 80 | — | — |
| — | (0.0148) | 90 | 80 | 3.0 | 0.375 |
| 8 | 0.0156 | 80 | 80 | — | — |
| — | (0.0172) | 80 | 80 | 3.5 | 0.438 |
| 7 | 0.0178 | 80 | 80 | — | — |
| — | (0.0197) | 80 | 80 | 4.0 | 0.500 |
| 6 | 0.0208 | 80 | 80 | — | — |
| — | (0.0221) | 80 | 80 | 4.5 | 0.562 |
| — | (0.0246) | 80 | 70 | 5.0 | 0.625 |
| 5 | 0.0250 | 80 | 70 | — | — |
| — | (0.0271) | 80 | 70 | 5.5 | 0.688 |
| 4.5 | 0.0278 | 80 | 70 | — | — |
| — | (0.0295) | 80 | 70 | 6 | 0.750 |
| 4 | 0.0312 | 70 | 70 | — | — |

SOURCE: Adapted from Thread Grinding, 18th Edition, 1963, published by Norton Company.

NOTES: Pitch = 1 ÷ number of threads per inch.
Root width = 0.125 × Pitch.

*Also applies to British Standard ISO Metric Threads.

Shop equations and data for determining pertinent information on various types of grinding operations are given in tables 20.4-1 through 20.4-4 and figures 20.4-1 through 20.4-7 as follows:

| | |
|---|---|
| Table 20.4-1 | Definitions of symbols used in grinding formulas and sketches, English units. |
| Table 20.4-2 | Definitions of symbols used in grinding formulas, Metric units. |
| Figure 20.4-1 | Sketches of various grinding operations. |
| Table 20.4-3 | Grinding formulas, English units. |
| Table 20.4-4 | Grinding formulas, Metric units. |
| Figure 20.4-2 | Unit power for surface traverse grinding: horizontal spindle—reciprocating table. |
| Figure 20.4-3 | Unit power for centertype cylindrical plunge grinding. |
| Figure 20.4-4 | Alignment chart for determining metal removal rate and horsepower required at motor: traverse grinding (reciprocating table) for horizontal and vertical spindle surface grinding. |
| Figure 20.4-5 | Alignment chart for determining metal removal rate and horsepower required at motor: plunge grinding (reciprocating table) for horizontal and vertical spindle surface grinding. |
| Figure 20.4-6 | Alignment chart for determining metal removal rate and horsepower required at motor: traverse grinding for centertype cylindrical, centerless and internal grinding. |
| Figure 20.4-7 | Alignment chart for determining metal removal rate and horsepower required at motor: plunge grinding for centertype cylindrical, centerless and internal grinding. |

# 20.4 Grinding Formulas and Charts

**TABLE 20.4–1 Definitions of Symbols Used in Grinding Formulas and Sketches, English Units**

| Symbol | Definition |
|---|---|
| $a$ | = Depth of grind, inches per pass |
| $a_t$ | = Total depth of grind, inches |
| $b$ | = Width of cut in plunge grinding, inches |
| $b_c$ | = Crossfeed, inches per pass |
| $b_s$ | = Original width of grinding wheel, inches |
| $b_w$ | = Width of workpiece, inches |
| $d_o$ | = OD of grinding wheel, inches (vertical spindle surface grinder) |
| $d_i$ | = ID of grinding wheel, inches (vertical spindle surface grinder) |
| $d_r$ | = Regulating wheel diameter, inches (for centerless grinding) |
| $d_{s1}$ | = Initial grinding wheel diameter, inches |
| $d_{s2}$ | = Final grinding wheel diameter, inches |
| $d_t$ | = Average diameter of workpiece path on rotary table, inches (vertical spindle surface grinder) |
| $d_{w1}$ | = Original work diameter, inches |
| $d_{w2}$ | = Final work diameter, inches |
| $E$ | = Efficiency of grinding wheel drive |
| $f_t$ | = Table traverse feed rate or throughfeed of work in centerless grinding, inches per minute |
| $f_p$ | = Plunge infeed rate, inches per minute |
| $G$ | = Grinding ratio = $\frac{\text{Volume metal removed, in}^3}{\text{Volume wheel wear, in}^3}$ |
| $L$ | = Length of workpiece ground, inches |
| $hp_g$ | = Horsepower required at motor, hp |
| $n_r$ | = Regulating wheel rpm |
| $n_s$ | = Grinding wheel rpm |
| $n_t$ | = Rotary table rpm (vertical spindle surface grinder) |
| $n_w$ | = Work rpm |
| $P_g$ | = Unit horsepower, horsepower per cubic inch per minute |
| $s$ | = Feed per revolution of work, inches |
| $t$ | = Axial wear from face of grinding wheel (vertical spindle surface grinder) |
| $V_s$ | = Peripheral speed of grinding wheel, feet per minute |
| $V_w$ | = Peripheral speed of work, feet per minute |
| $w_m$ | = Maximum contact width of grinding wheel on work (vertical spindle surface grinder) |
| $Z$ | = Metal removal rate, cubic inches per minute |
| $\alpha$ | = Regulating wheel inclination angle, degrees (centerless grinder) |

**TABLE 20.4–2 Definitions of Symbols Used in Grinding Formulas and Sketches, Metric Units**

| Symbol | Definition |
|---|---|
| $a$ | = Depth of grind, millimeters per pass |
| $a_t$ | = Total depth of grind, millimeters |
| $b$ | = Width of cut in plunge grinding, millimeters |
| $b_c$ | = Crossfeed, millimeters per pass |
| $b_s$ | = Original width of grinding wheel, millimeters |
| $b_w$ | = Width of workpiece, millimeters |
| $d_o$ | = OD of grinding wheel, millimeters (vertical spindle surface grinder) |
| $d_i$ | = ID of grinding wheel, millimeters (vertical spindle surface grinder) |
| $d_r$ | = Regulating wheel diameter, millimeters (for centerless grinding) |
| $d_{s1}$ | = Initial grinding wheel diameter, millimeters |
| $d_{s2}$ | = Final grinding wheel diameter, millimeters |
| $d_t$ | = Average diameter of workpiece path on rotary table, millimeters (vertical spindle surface grinder) |
| $d_{w1}$ | = Original work diameter, millimeters |
| $d_{w2}$ | = Final work diameter, millimeters |
| $E$ | = Efficiency of grinding wheel drive |
| $f_t$ | = Table traverse feed rate or throughfeed of work in centerless grinding, millimeters per minute |
| $f_p$ | = Plunge infeed rate, millimeters per minute |
| $G$ | = Grinding ratio = $\frac{\text{Volume metal removed, mm}^3}{\text{Volume wheel wear, mm}^3}$ |
| $L$ | = Length of workpiece ground, millimeters |
| $kW_g$ | = Power required at motor, kilowatts |
| $n_r$ | = Regulating wheel rpm |
| $n_s$ | = Grinding wheel rpm |
| $n_t$ | = Rotary table rpm (vertical spindle surface grinder) |
| $n_w$ | = Work rpm |
| $P_g$ | = Unit power, kilowatts per cubic centimeter per minute |
| $s$ | = Feed per revolution of work, millimeters |
| $t$ | = Axial wear from face of grinding wheel, millimeters (vertical spindle surface grinder) |
| $V_s$ | = Peripheral speed of grinding wheel, meters per second |
| $V_w$ | = Peripheral speed of work, meters per minute |
| $w_m$ | = Maximum contact width of grinding wheel on work, millimeters (vertical spindle surface grinder) |
| $Z$ | = Metal removal rate, cubic centimeters per minute |
| $\alpha$ | = Regulating wheel inclination angle, degrees (centerless grinder) |

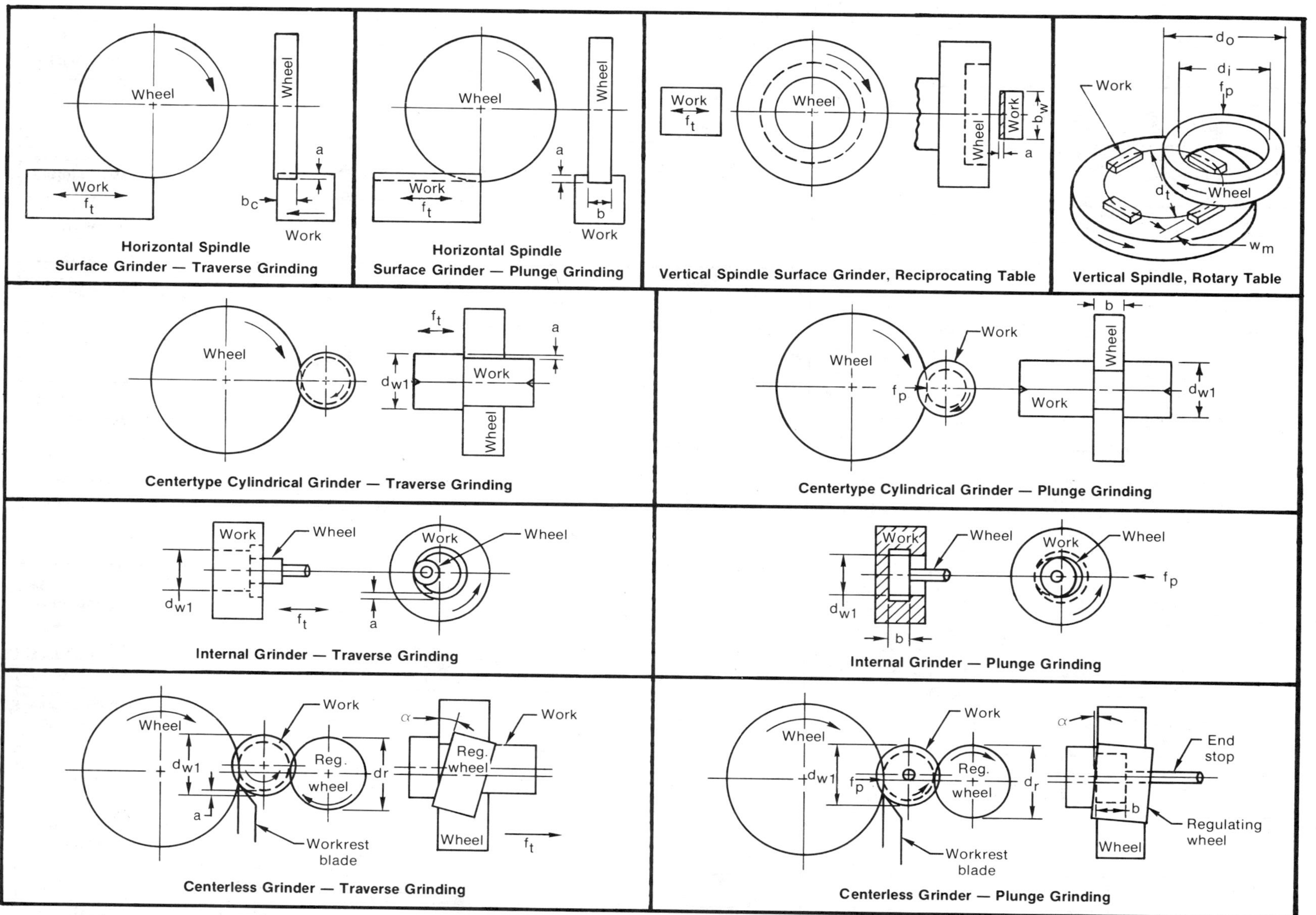


**Figure 20.4-1** Sketches of various grinding operations. See tables 20.4-1 and 20.4-2 for definitions of symbols.

TABLE 20.4-3 Grinding Formulas, English Units

| PARAMETER | SURFACE GRINDING | | | CENTERTYPE CYLINDRICAL GRINDING | CENTERLESS GRINDING | INTERNAL GRINDING |
|---|---|---|---|---|---|---|
| | HORIZONTAL SPINDLE | VERTICAL SPINDLE | | | | |
| | Reciprocating Table | Reciprocating Table | Rotary Table | | | |
| **Peripheral wheel speed,** ft/min | $v_s = 0.262 \times d_{s1} \times n_s$ | $v_s = 0.262 \times d_{s1} \times n_s$ | $v_s = 0.262 \times d_{s1} \times n_s$ | $v_s = 0.262 \times d_{s1} \times n_s$ | $v_s = 0.262 \times d_{s1} \times n_s$ | $v_s = 0.262 \times d_{s1} \times n_s$ |
| **Wheel rpm** | $n_s = 3.82 \frac{v_s}{d_{s1}}$ | $n_s = 3.82 \frac{v_s}{d_{s1}}$ | $n_s = 3.82 \frac{v_s}{d_{s1}}$ | $n_s = 3.82 \frac{v_s}{d_{s1}}$ | $n_s = 3.82 \frac{v_s}{d_{s1}}$ | $n_s = 3.82 \frac{v_s}{d_{s1}}$ |
| **Peripheral work speed,** ft/min | $v_w$ | $v_w$ | $v_w = 0.262 \times d_t \times n_t$ | $v_w = 0.262 \times d_{w1} \times n_w$ | $v_w = 0.262 \times d_{w1} \times n_w$ | $v_w = 0.262 \times d_{w1} \times n_w$ |
| **Work rpm** | — | — | $n_t = \frac{3.82 \times v_w}{d_t}$ | $n_w = 3.82 \times \frac{v_w}{d_{w1}}$ | $n_w = 3.82 \times \frac{v_w}{d_{w1}}$ | $n_w = 3.82 \times \frac{v_w}{d_{w1}}$ |
| **Horsepower at motor** | $hp_g = \frac{P_g \times Z}{E}$ | $hp_g = \frac{P_g \times Z}{E}$ | $hp_g = \frac{P_g \times Z}{E}$ | $hp_g = \frac{P_g \times Z}{E}$ | $hp_g = \frac{P_g \times Z}{E}$ | $hp_g = \frac{P_g \times Z}{E}$ |
| **TRAVERSE GRINDING** | | | | | | |
| **Table traverse feed rate,** in/min | $f_t$ | $f_t$ | — | $f_t = s \times n_w$ | $f_t = \pi \times d_r \times n_r \times \sin \alpha$ | $f_t = s \times n_w$ |
| **Metal removal rate,** in³/min | $Z = a \times b_c \times f_t$ | $Z = a \times b_w \times f_t$ | — | $Z = \pi \times a \times f_t \times d_{w1}$ | $Z = \pi \times a \times f_t \times d_{w1}$ | $Z = \pi \times a \times f_t \times d_{w1}$ |
| **G ratio** | $G = \frac{1.273 \times L \times b_w \times a_t}{b_s (d_{s1}^2 - d_{s2}^2)}$ | $G = \frac{1.273 \times L \times b_w \times a_t}{t (d_o^2 - d_i^2)}$ | — | $G = \frac{L (d_{w1}^2 - d_{w2}^2)}{b_s (d_{s1}^2 - d_{s2}^2)}$ | $G = \frac{L (d_{w1}^2 - d_{w2}^2)}{b_s (d_{s1}^2 - d_{s2}^2)}$ | $G = \frac{L (d_{w1}^2 - d_{w2}^2)}{b_s (d_{s1}^2 - d_{s2}^2)}$ |
| **PLUNGE GRINDING** | | | | | | |
| **Feed rate,** in/min | $f_t$ | $f_t$ | $f_p = s \times n_t$ | $f_p$ | $f_p$ | $f_p$ |
| **Metal removal rate,** in³/min | $Z = a \times b \times f_t$ | $Z = a \times b_w \times f_t$ | $Z = \pi \times w_m \times f_p \times d_t$ | $Z = \pi \times b \times f_p \times d_{w1}$ | $Z = \pi \times b \times f_p \times d_{w1}$ | $Z = \pi \times b \times f_p \times d_{w1}$ |
| **G ratio** | $G = \frac{1.273 \times L \times b \times a_t}{b_s (d_{s1}^2 - d_{s2}^2)}$ | $G = \frac{1.273 \times L \times b_w \times a_t}{t (d_o^2 - d_i^2)}$ | $G = \frac{1.273 \times L \times b_w \times a_t}{t (d_o^2 - d_i^2)}$ | $G = \frac{b (d_{w1}^2 - d_{w2}^2)}{b_s (d_{s1}^2 - d_{s2}^2)}$ | $G = \frac{b (d_{w1}^2 - d_{w2}^2)}{b_s (d_{s1}^2 - d_{s2}^2)}$ | $G = \frac{b (d_{w1}^2 - d_{w2}^2)}{b_s (d_{s1}^2 - d_{s2}^2)}$ |

See table 20.4-1 for definition of symbols.

**TABLE 20.4–4 Grinding Formulas, Metric Units**

| PARAMETER | SURFACE GRINDING | | | CENTERTYPE CYLINDRICAL GRINDING | CENTERLESS GRINDING | INTERNAL GRINDING |
|---|---|---|---|---|---|---|
| | HORIZONTAL SPINDLE | VERTICAL SPINDLE | | | | |
| | Reciprocating Table | Reciprocating Table | Rotary Table | | | |
| **Peripheral wheel speed,** m/s | $v_s = 0.000052\ d_{sI} \times n_s$ | $v_s = 0.000052\ d_{sI} \times n_s$ | $v_s = 0.000052\ d_{sI} \times n_s$ | $v_s = 0.000052\ d_{sI} \times n_s$ | $v_s = 0.000052\ d_{sI} \times n_s$ | $v_s = 0.000052\ d_{sI} \times n_s$ |
| **Wheel rpm** | $n_s = 19231 \frac{v_s}{d_{sI}}$ | $n_s = 19231 \frac{v_s}{d_{sI}}$ | $n_s = 19231 \frac{v_s}{d_{sI}}$ | $n_s = 19231 \frac{v_s}{d_{sI}}$ | $n_s = 19231 \frac{v_s}{d_{sI}}$ | $n_s = 19231 \frac{v_s}{d_{sI}}$ |
| **Peripheral work speed,** m/min | $v_w$ | $v_w$ | $v_w = 0.0032\ d_t \times n_t$ | $v_w = 0.0032\ d_{wI} \times n_w$ | $v_w = 0.0032\ d_{wI} \times n_w$ | $v_w = 0.0032\ d_{wI} \times n_w$ |
| **Work rpm** | — | — | $n_t = 318 \frac{v_w}{d_t}$ | $n_w = 318 \frac{v_w}{d_{wI}}$ | $n_w = 318 \frac{v_w}{d_{wI}}$ | $n_w = 318 \frac{v_w}{d_{wI}}$ |
| **Power at motor,** kW | $kW_g = \frac{P_g \times Z}{E}$ | $kW_g = \frac{P_g \times Z}{E}$ | $kW_g = \frac{P_g \times Z}{E}$ | $kW_g = \frac{P_g \times Z}{E}$ | $kW_g = \frac{P_g \times Z}{E}$ | $kW_g = \frac{P_g \times Z}{E}$ |
| **TRAVERSE GRINDING** | | | | | | |
| **Table traverse feed rate,** mm/min | $f_t$ | $f_t$ | — | $f_t = s \times n_w$ | $f_t = \pi \times d_r \times n_r \times \sin\alpha$ | $f_t = s \times n_w$ |
| **Metal removal rate,** cm³/min | $Z = \frac{a \times b_c \times f_t}{1000}$ | $Z = \frac{a \times b_w \times f_t}{1000}$ | — | $Z = \frac{\pi \times a \times f_t \times d_{wI}}{1000}$ | $Z = \frac{\pi \times a \times f_t \times d_{wI}}{1000}$ | $Z = \frac{\pi \times a \times f_t \times d_{wI}}{1000}$ |
| **G ratio** | $G = \frac{1.273 \times L \times b_w \times a_t}{t\,(d_o^2 - d_i^2)}$ | $G = \frac{1.273 \times L \times b_w \times a_t}{t\,(d_o^2 - d_i^2)}$ | — | $G = \frac{L\,(d_{wI}^2 - d_{w2}^2)}{b_s\,(d_{sI}^2 - d_{s2}^2)}$ | $G = \frac{L\,(d_{wI}^2 - d_{w2}^2)}{b_s\,(d_{sI}^2 - d_{s2}^2)}$ | $G = \frac{L\,(d_{wI}^2 - d_{w2}^2)}{b_s\,(d_{sI}^2 - d_{s2}^2)}$ |
| **PLUNGE GRINDING** | | | | | | |
| **Feed rate,** mm/min | $f_t$ | $f_t$ | $f_p = s \times n_t$ | $f_p$ | $f_p$ | $f_p$ |
| **Metal removal rate,** cm³/min | $Z = \frac{a \times b \times f_t}{1000}$ | $Z = \frac{a \times b_w \times f_t}{1000}$ | $Z = \frac{\pi \times w_m \times f_p \times d_t}{1000}$ | $Z = \frac{\pi \times b \times f_p \times d_{wI}}{1000}$ | $Z = \frac{\pi \times b \times f_p \times d_{wI}}{1000}$ | $Z = \frac{\pi \times b \times f_p \times d_{wI}}{1000}$ |
| **G ratio** | $G = \frac{1.273 \times L \times b \times a_t}{t\,(d_{sI}^2 - d_i^2)}$ | $G = \frac{1.273 \times L \times b_w \times a_t}{b_s\,(d_{sI}^2 - d_{s2}^2)}$ | $G = \frac{1.273 \times L \times b_w \times a_t}{t\,(d_o^2 - d_i^2)}$ | $G = \frac{b\,(d_{wI}^2 - d_{w2}^2)}{b_s\,(d_{sI}^2 - d_{s2}^2)}$ | $G = \frac{b\,(d_{wI}^2 - d_{w2}^2)}{b_s\,(d_{sI}^2 - d_{s2}^2)}$ | $G = \frac{b\,(d_{wI}^2 - d_{w2}^2)}{b_s\,(d_{sI}^2 - d_{s2}^2)}$ |

See table 20.4-2 for definition of symbols.

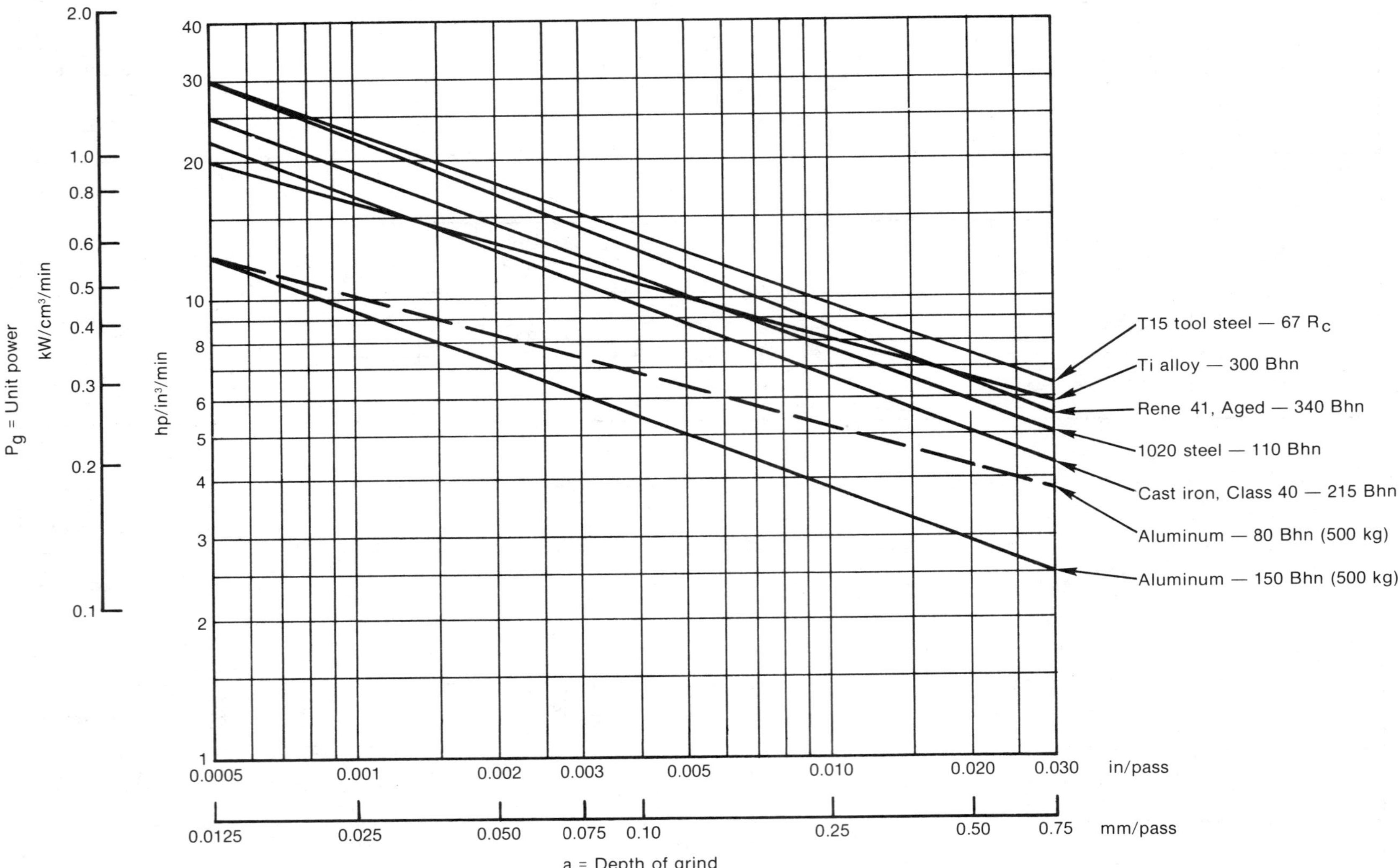


**Figure 20.4–2** Unit power for surface traverse grinding: horizontal spindle—reciprocating table.

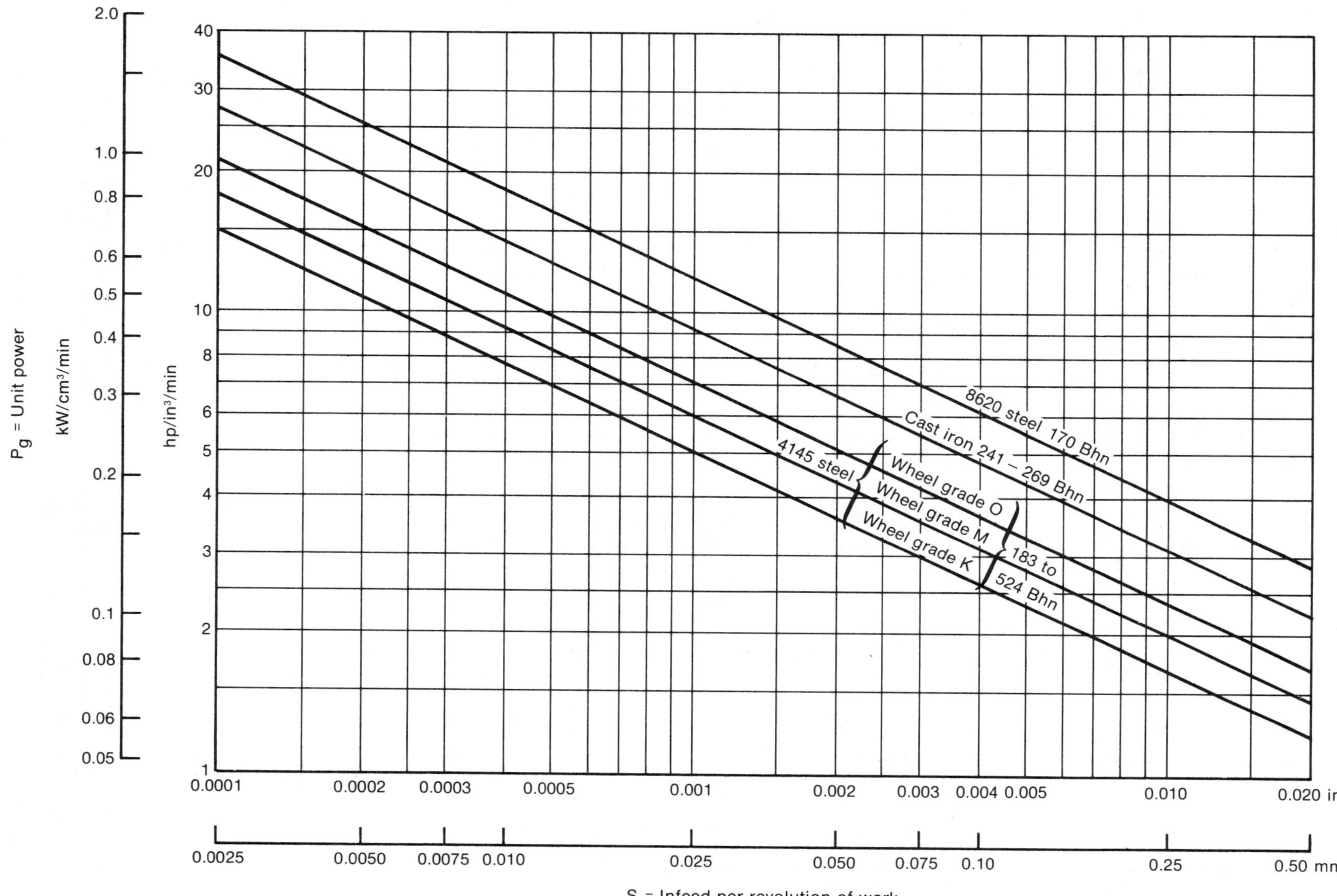


**Figure 20.4–3** Unit power for centertype cylindrical plunge grinding.

## 20.4 Grinding Formulas and Charts

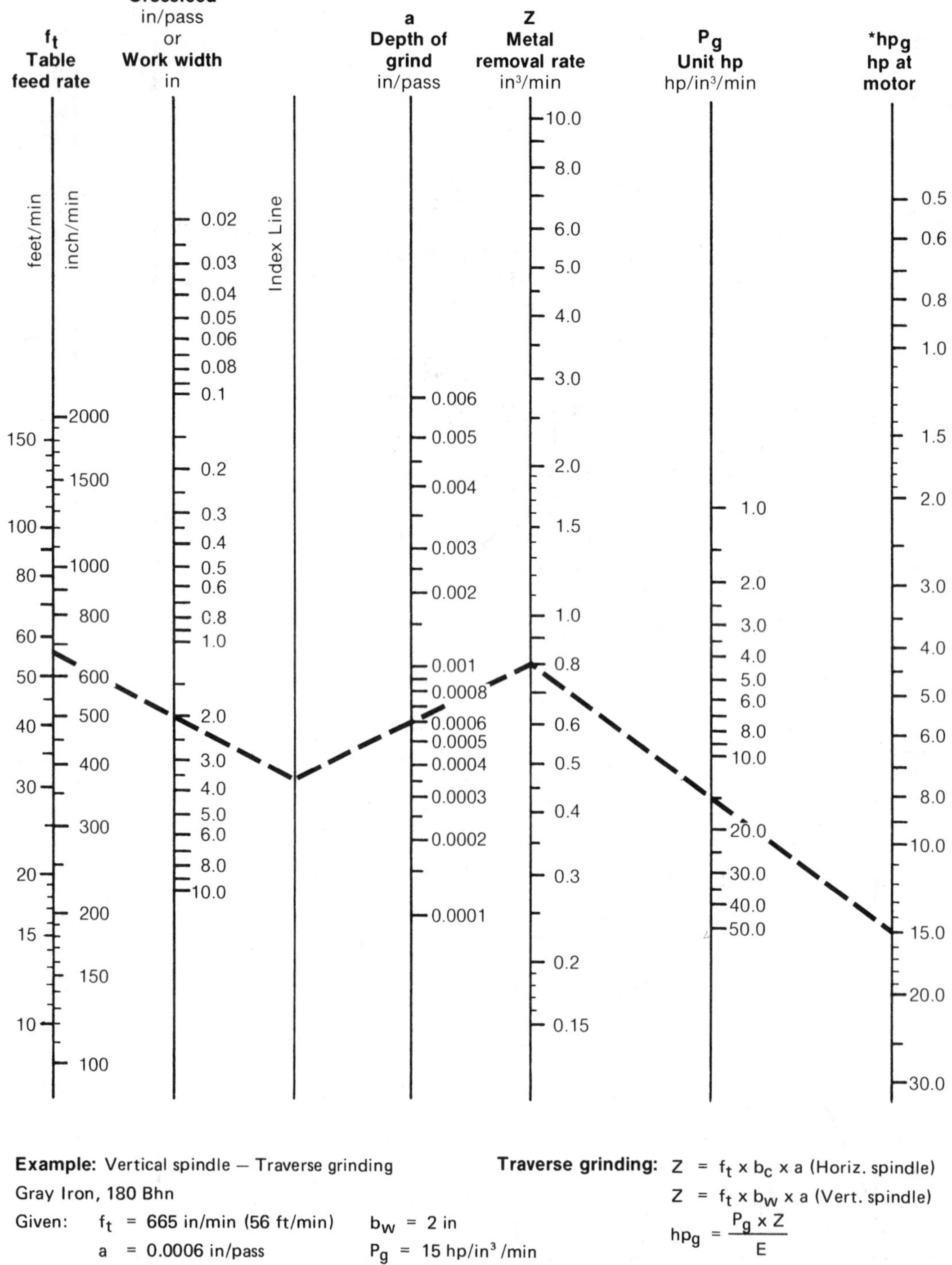


**Example:** Vertical spindle — Traverse grinding

Gray Iron, 180 Bhn

| | | | |
|---|---|---|---|
| Given: | $f_t$ = 665 in/min (56 ft/min) | | $b_w$ = 2 in |
| | $a$ = 0.0006 in/pass | | $P_g$ = 15 hp/in³/min |
| Then: | $Z$ = 0.8 in³/min | | $hp_g$ = 15 hp |

**Traverse grinding:** $Z = f_t \times b_c \times a$ (Horiz. spindle)

$Z = f_t \times b_w \times a$ (Vert. spindle)

$$hp_g = \frac{P_g \times Z}{E}$$

*$hp_g$ = Horsepower at spindle motor for 80% spindle drive efficiency

**Figure 20.4–4** Alignment chart for determining metal removal rate and horsepower required at motor: traverse grinding (reciprocating table) for horizontal and vertical spindle surface grinding.

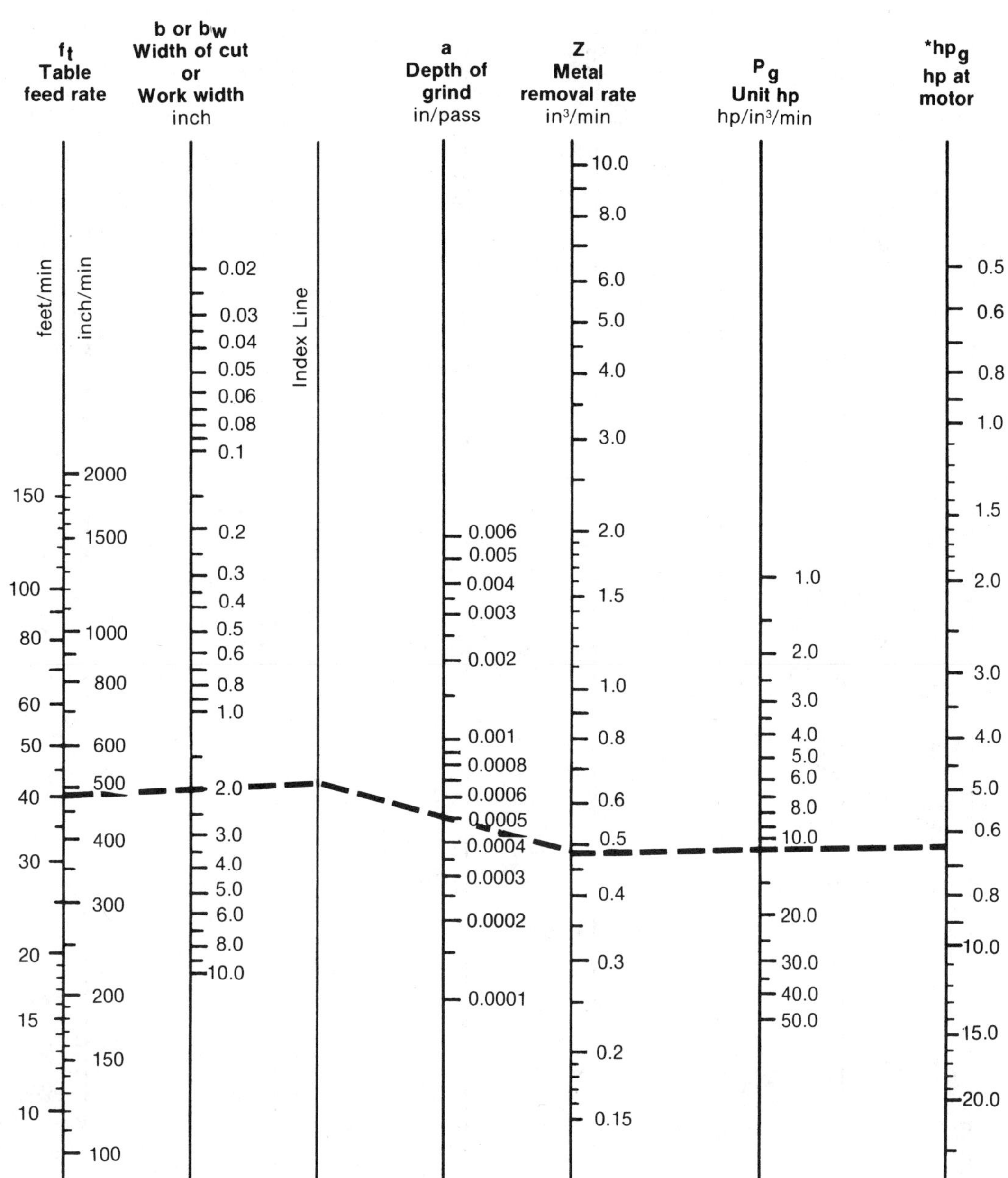


**Example:** Vertical spindle — Plunge grinding

1020 steel, 120 Bhn

Given: $f_t$ = 40 ft/min  $b_w$ = 2 in  a = 0.0005 in/pass

$P_g$ = 10.5 hp/in³/min

Then: Z = 0.48 in³/min  $hp_g$ = 6.4 hp

**Plunge grinding:** $Z = f_t \times b \times a$ (Horiz. spindle)

$Z = f_t \times b_w \times a$ (Vert. spindle)

$$hp_g = \frac{P_g \times Z}{E}$$

*$hp_g$ = Horsepower at spindle motor for 80% spindle drive efficiency

**Figure 20.4–5** Alignment chart for determining metal removal rate and horsepower required at motor: plunge grinding (reciprocating table) for horizontal and vertical spindle surface grinding.

# 20.4 Grinding Formulas and Charts

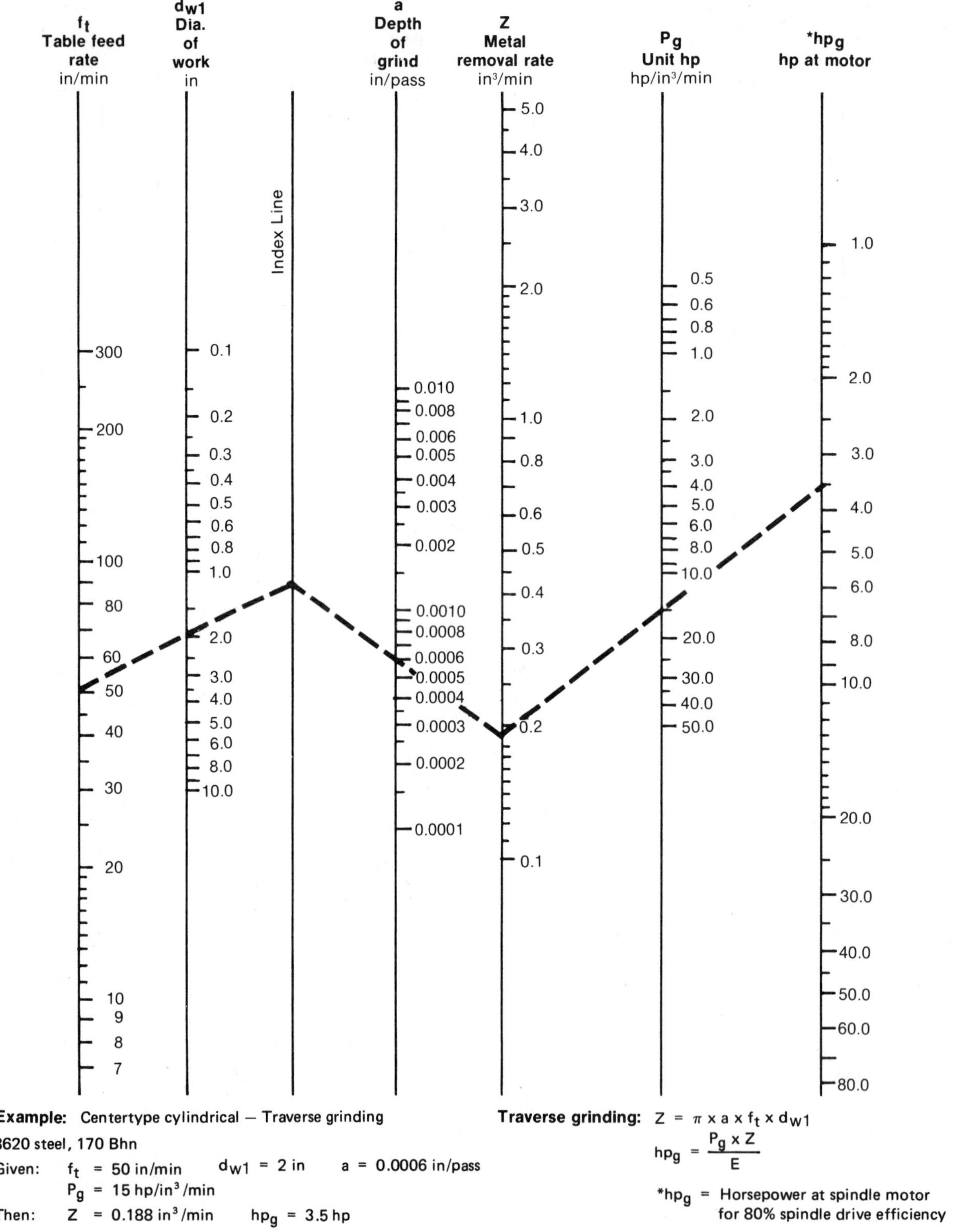


**Figure 20.4–6** Alignment chart for determining metal removal rate and horsepower required at motor: traverse grinding for centertype cylindrical, centerless and internal grinding.

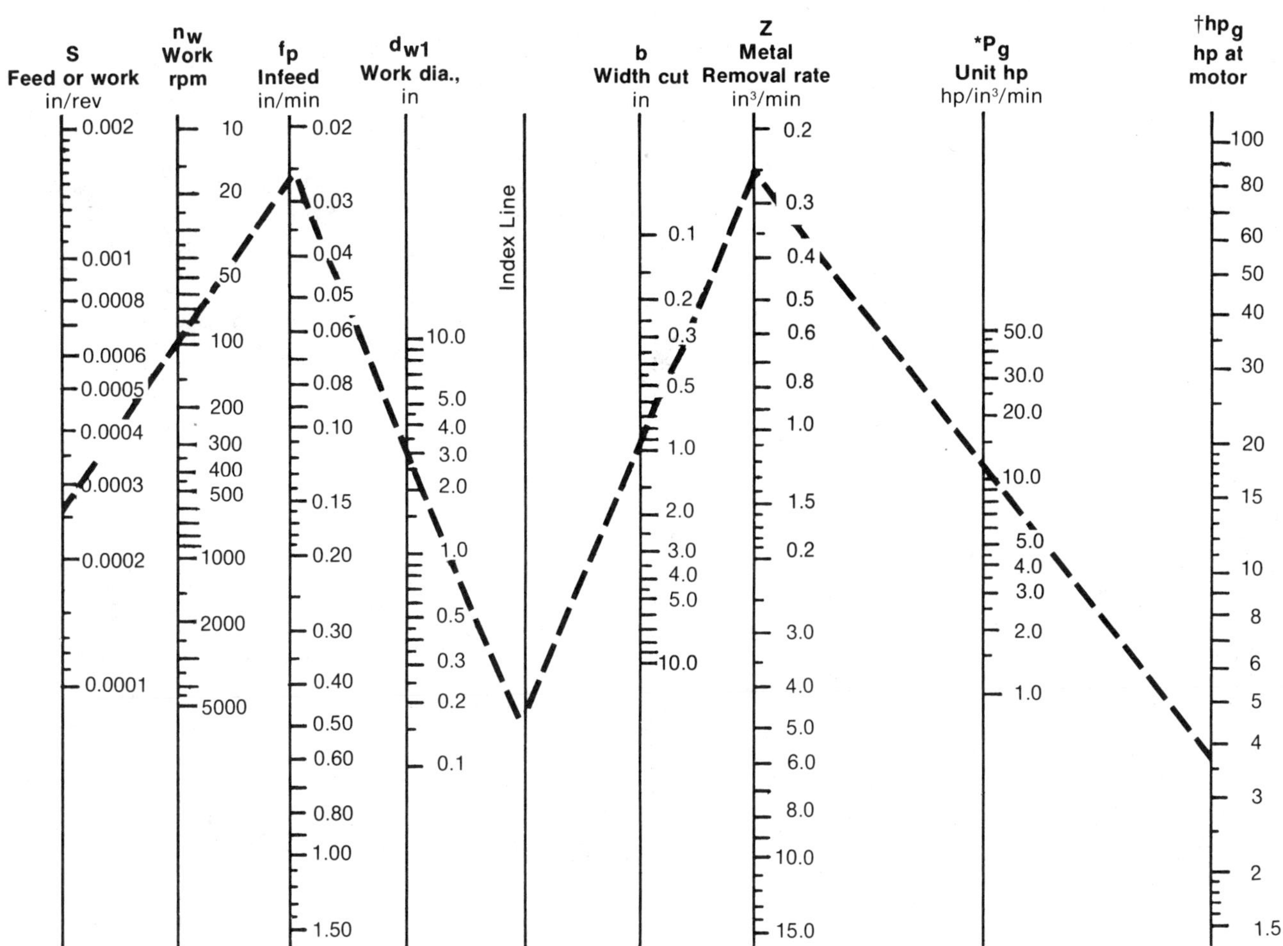


**Example:** Centertype cylindrical — Plunge grinding

4145 steel, 316 Bhn

Given: $S = 0.00025$ in/rev $\quad n_w = 100$ rpm $\quad d_{w1} = 3$ in

$b = 1$ in $\quad P_g = 11.6$ hp/in³/min

Then: $fp = 0.026$ in/min $\quad Z = 0.25$ in³/min $\quad hp_g = 3.6$ hp

**Plunge grinding:** $Z = \pi \times b \times f_p \times d_{w1}$

$$hp_g = \frac{P_g \times Z}{E}$$

*See figure 20.4-3 for unit power values for centertype cylindrical grinding

†$hp_g$ = Horsepower at spindle motor for 85% spindle drive efficiency

**Figure 20.4–7** Alignment chart for determining metal removal rate and horsepower required at motor: plunge grinding for centertype cylindrical, centerless and internal grinding.

Grinding has traditionally been used to provide final workpiece finish and dimensional accuracy. About 1962, the term "abrasive machining" appeared in the literature to cover those grinding operations used for stock removal as well as for the production of the final finish and accuracy.

There are various types of abrasive grinding operations; the prominent ones are described below.

## VERTICAL SPINDLE, ROTARY TABLE SURFACE GRINDING

Vertical spindle surface grinding has been employed for many years as a stock removal operation on many common commercial alloys. It has been used in the automotive industry for grinding cast iron and steel components in lieu of milling. The normal stock removal of 0.125 to 0.250 inch [3.2 to 6.4 mm] on castings and forgings has been accomplished on parts which have essentially flat, unencumbered surfaces. In recent years, there has been an extension of the application of vertical spindle surface grinding to many larger parts, including machine tool beds and tables, cylinder heads, gear housings, motor cases, etc. Machines are being built with table sizes up to 14-foot [4 m] diameter and up to 300 horsepower [225 kW]. It may be advantageous to redesign parts to take advantage of the benefits of this operation.

The recommended grinding conditions for this operation are contained in section 8.4, where it may be noted that wheel speeds in most common use vary between 3,400 and 5,000 feet per minute [17 and 25 m/s].

## RECIPROCATING TABLE SURFACE GRINDING

Abrasive grinding is being applied for heavy stock removal using reciprocating table machines with both vertical and horizontal grinding wheel spindles. The application of the horizontal spindle, reciprocating table abrasive grinding is being augmented by the application of crush dressing to the grinding wheel, which minimizes idle time in the grinding operation. Data for surface grinding with horizontal spindle, reciprocating table machines are given in section 8.1. Horizontal spindle, reciprocating table surface grinding competes with milling on certain metals such as high temperature alloys which are extremely difficult to mill.

## ABRASIVE BELT MACHINING

### The Process

Abrasive belt machining, once confined to polishing and minor metal removal, has become a high-metal-removal-rate process. The advent of stronger and sharper grains, better bonds, and stronger belts accounts in part for the increasing popularity of abrasive belt machining. It can take the place of conventional grinding, polishing and similar operations. For high production, one machine can be equipped with several heads, each with a different belt so that roughing and finishing can be performed in one pass. This reduces floor space and capital investment requirements. Further economies may result from the use of worn, coarse-grain belts for finishing passes.

### The Abrasive Belt Machine

Machines used for belt grinding have two common features: all are very rigid and all utilize high horsepower. Rigidity is at least as important as horsepower in obtaining good abrasive economies. The common types of machines in belt machining are centerless belt grinder, surface belt grinder, cylindrical belt grinder, and roll belt grinder.

### The Abrasive Belt

Grinding belts used for abrasive machining are resin bond coated with either aluminum oxide, silicon carbide, or diamond grains. The grain is electrostatically coated to the belt in order to control orientation, dispersal and thickness. There is up to three times more chip clearance in a coarse grinding belt than in a comparable grinding wheel. This permits high metal removal rates at low pressures. Most failures of belts are caused by loading, glazing and stripping (shedding) which are discussed as follows:

- Loading is the filling of areas between abrasive grains. The use of lubricants and a weaker bond can retard loading.
- Glazing is the dulling of the abrasive grains resulting from the use of strong bonds. It can be eliminated by changing to a weaker bond and using a more "aggressive" contact wheel.
- Stripping is the shedding or breaking away of the bond and grains from the belt backing. It can be corrected by changing to a bond that has greater adhesion.

### Contact Wheels and Platens

When the correct contact wheel or platen is used behind the belt, belt sharpness and cutting efficiency are maintained by breakdown of the dulled abrasive grains, thus producing and thereby presenting sharp grains to the work during the useful life of the belt. Stock removal rates are approximately in direct proportion to the pressure applied, within the capabilities of the grain size and bond type used.

The ability of the contact wheel or platen to make the belt cut is termed "aggressiveness." For optimum stock removal, a hard rubber (90 durometer) contact wheel with 45° serrations, 0.25-inch [6.4 mm] land, and 0.50-inch [12.7 mm] wide grooves provides aggressive cutting action and sheds dulled abrasive grains from the belt to eliminate glazing. Hard contact wheels are used both for heavy stock removal and for high accuracy. It is not uncommon to hold accuracy between ±0.0005 inch [±0.013 mm]. Good abrasive belt operations consume only 4.5 to 7 horsepower per cubic inch per minute [0.20 to 0.32 kW/cm$^3$/min] when grinding low carbon steel, compared to 6 to 10 horsepower per cubic inch per minute [0.27 to 0.46 kW/cm$^3$/min] for comparable bonded wheel operations. Centerless belt grinding removal rates vary from 0.25 cubic inch per inch [0.16 cm$^3$/mm] of belt width on small bars to 1 cubic inch per inch [0.65 cm$^3$/mm] of belt width on larger bars. In comparison, the new wide-wheel centerless grinders remove only 0.6 to 0.75 cubic inch per inch [0.39 to 0.48 $cm^3/mm$] width on large bars and require 8.3 horsepower per inch [6.2 kW] width of wheel compared to 7.5 horsepower per inch [5.6 kW] width for a belt.

There are six types of belts as follows:

- Paper-backed with two glue coatings.
- Paper-backed with a glue making-coat and a heat-resistant resin sizing-coat.
- Cloth-backed with a glue making-coat and a heat-resistant resin sizing-coat.
- Cloth-backed with two resin coats.
- Cloth-backed, specially waterproofed with two resin coats.
- Belts containing dry lubricants.

### Abrasive Belt Machining Conditions

Abrasive belt speeds range from 600 feet per minute [3 m/s] on certain ceramics to 14,000 feet per minute [70 m/s] on nonferrous snag grinding. Most abrasive machining is done between 2,500 and 5,500 feet per minute [13 and 28 m/s], with the most commonly used speed being 4,500 feet per minute [23 m/s]. Feeds vary from one inch per minute [25.4 mm/min] for heavy slabbing cuts to 240 feet per minute [73 m/min] for precision centerless belt grinding applications.

## HIGH-EFFICIENCY OR HIGH-SPEED GRINDING

The name high-efficiency grinding is being used for abrasive machining employing very high grinding wheel speeds. Wheel speeds of 12,000 to 18,000 feet per minute [60 to 90 m/s] are being used today. By using high wheel speeds, it has been found that the rate of metal removal can be increased.

### Types of High-Efficiency Grinding Machines

**Shoe-Type Centerless Grinders** – These appear to be the most numerous machines in production today and are used primarily for abrasive grinding of bearing races.

**Cylindrical Plunge Grinders** – There are many of these machines out in the field for grinding of pinion gears, cam shafts, steering knuckles and other components as will be described below.

**Internal Grinders** – These machines are being used primarily for the grinding of bearing races.

**Surface Grinders and Centers Grinders** – Development work on the high spindle speed application of these grinders is under way, but few of these grinders are being used in production.

### Machine Design Considerations

Grinding machines designed for high-efficiency grinding require special considerations which prohibit the conversion of existing grinders into high-efficiency grinders. The design requirements for high-efficiency grinders include the following:

- High spindle speeds to provide grinding wheel speeds of 12,000 to 18,000 feet per minute [60 to 90 m/s].
- High work speeds, at least double that of normal grinders.
- High horsepower to accommodate the high rates of metal removal.
- High rigidity and freedom from vibration.
- Automatic wheel balancing on the spindle to minimize grinding wheel vibration.
- Provision for the safety of the operator, the passerby and the machine tool, including use of special guard systems to contain the wheel in case of wheel breakage.
- Special coolant application systems to ensure that the grinding fluid reaches the grinding zone at the high wheel speeds.
- Filtration systems to remove the large quantities of swarf from the coolant system.
- Cooling systems to prevent overheating of the grinding fluid.
- The use of live centers rather than dead centers for cylindrical grinders to accommodate the increased work speed.

### Grinding Wheels

Special grinding wheels have to be designed to withstand the high wheel speeds. Some of the design requirements include the following:

- Providing a composite wheel with a hard, fine section toward the center, or a single-graded wheel impregnated with resinoid in the center section for added strength.
- Wheel design involving clamping of wheel segments.

### Grinding Wheel Dressing

Diamond roll wheel dressing is being used to reduce the time for dressing the wheel.

### Types of Materials and Components

Low carbon carburizing steel: stems of gears
Medium carbon alloy steel: steering knuckles, gear stems, Universal spiders
Bearing steels—hardened: bearing races
Cast iron: crankshafts, slip yoke

### Applications

Most of the high-efficiency cylindrical grinding applications are plunge grinding operations. Cylindrical plunge grinding may economically replace turning under the following conditions where:

- The product of the diameter times the length of the grind is greater than 4.
- The ground length does not usually exceed 7 inches [178 mm].
- The stock removal is limited to about 0.25 inch [6.4 mm] on the diameter.
- The tolerance is less than ±0.003 inch [±0.075 mm].
- Special problems (for example, interrupted cuts) which might be more restrictive on turning than on grinding.
- The surface roughness is 30 microinches $R_a$ [0.76 μm] or greater.
- The workpieces are mass produced, are manufactured in large batches, or are similar parts produced in short runs.

Table 20.5-1 illustrates some workpieces which were successfully high-efficiency machined and compares them with the above list of economic guidelines.

## CYLINDRICAL CRUSH DRESS GRINDING

A recent addition to abrasive machining has been the design of cylindrical grinders for removing large amounts of stock by plunge grinding both straight and formed parts. The grinding wheels are crush dressed. The width of the wheel or the length of the ground workpiece can be as much as 12 inches [305 mm]. A necessary requirement for this grinding system includes a high pressure coolant jet spray which is traversed back and forth across the grinding wheel. This jet spray blasts metal and swarf loose from the grinding wheel and maintains a clean wheel. This method of grinding is being used on steels and high temperature alloys. It is being applied to the grinding of jet engine components and mechanical test specimens instead of turning and finish grinding.

## CREEP FEED GRINDING

Creep feed grinding is a process whereby the grinding wheel takes the full depth of cut in a single pass at an extremely slow feed rate. This process was developed in Europe some twenty years ago but did not gain wide acceptance until recently when specially designed grinding machines with high stability, increased power and improved drive systems became available. With such machines it is possible to remove as much as 1 inch [25 mm] of metal from flat or cylindrical surfaces in a single pass, which rivals the capacity of milling machines.

Metal removal rates as high as 1 cubic inch per second [16.4 $cm^3/s$] with wheel speeds of 20,000 feet per minute [102 m/s] have been reported.

With the present state of the art, it is imperative that the process parameters be evaluated for each new application. This is especially true if the surface integrity of the part is a consideration.

**TABLE 20.5-1 Mass Production Applications of High-Efficiency Grinding**

| WORKPIECE | MATERIAL | CRITICAL TOLERANCE inch | PRODUCT OF d x L | DEPTH OF STOCK (on diameter) inch | SURFACE ROUGHNESS μin $R_a$ |
|---|---|---|---|---|---|
| Stem pinion | 8620 | ±.00025 or ±.0015 | 7.5 | varied 1/8 to 1 | none |
| Slip yoke | Cast iron | ±.0005 | 5.6 to 9.0 | 3/8 | 15 to 30 or 30 to 60 |
| Steering knuckle | Steel | ±.0035 | 5.5 | 1/4 | none |
| Sun gear | Steel | ±.001 | 7 | 3/8 | none |
| Universal spider | Steel | ±.001 | 4.3 | 1/4 | 63 max. |
| Compressor cage | Cast iron | ±.00075 | 25 | 1/4 | 40 max. |
| Crankshaft | Cast iron | ±.0005 | 4.5 | 1/4 | 50 max. |

SOURCE: C. B. Matson, High effiency centertype grinding, Technical paper MR71-147, Society of Manufacturing Engineers, Dearborn, MI, 1971.

SECTION **21**

# ECONOMICS IN MACHINING AND GRINDING


Introduction .......... 21–3
Effect of Engineering Design on Cost .......... 21–3
Effect of Surface Roughness and Dimensional Tolerance on Cost .......... 21–3
Machining Cost and Production Rate .......... 21–3
Procedure for Calculating Cost and Production Rate .......... 21–4
Determining Optimum Cutting Conditions .......... 21–5
Shop Procedures for Optimization of Machining Conditions .......... 21–6
Technique for Calculating Cost and Production Rate .......... 21–6

## INTRODUCTION

The annual labor and overhead cost for operating metal-cutting machine tools in the United States has been estimated to be approximately $115 billion, as shown in table 21-1. The total shipments of metalcutting machinery in the United States have exceeded $2 billion per year, and in 1979 almost $2 billion was spent for small cutting tools. From these data, it is obvious that metalcutting is an enormous industry in the U.S.A. and that even a modest reduction in the cost of operating machine tools could very well result in nationwide savings of billions of dollars per year. It is apparent that similarly high proportionate savings also can be achieved by individual companies in industry.

**TABLE 21-1 Estimated Annual Labor and Overhead Cost for Operating Metalcutting Machine Tools in Industries in the United States**

| | |
|---|---|
| Total number of metalcutting machine tools | = 2,301,500* |
| Average labor + overhead cost | = $25 per hour |
| Average working day | = 8 hours |
| Number of working days per year | = 250 |
| Average number of direct labor personnel per machine | = 1 |
| Total cost of labor + overhead: 2,301,500 x $25 x 8 x 250 x 1 | = $115,075,000,000 |

It appears reasonable to conclude that the cost of labor + overhead for machining required for manufacturing in the U.S.A. is of the order of:

$115,000,000,000 Annually

*Based on *12th American Machinist Inventory* (1976–1978). (See summary in *American Machinist,* December 1978, p. 133–148.)

There are two general approaches for studying the cost of machining in metalcutting plants. The first is to examine the subject in a general or qualitative way, and the second is to make a quantitative analysis of cost and production rate. Both of these approaches are discussed in this section.

## EFFECT OF ENGINEERING DESIGN ON COST

When examining costs in a qualitative way, it should be pointed out that the basic cost in machining a component is determined by the design of the component. The basic cost in machining is first determined by the material selected for the part, that is, by the material per se, its heat treatment, and resultant metallurgical condition and hardness. Figure 21-1 indicates the relative cost of machining a given component made from a wide variety of alloys. Based upon a part cost of $10 for turning 7075-T6 aluminum alloy, it can be seen that turning 1020 steel would cost $25, turning 4340 steel, quenched and tempered to 52 $R_C$, would cost $100, and turning Inconel 700 would cost over $300. A detailed table of the relative time for machining the same group of alloys using various machining operations and tool materials is given in table 21-2.

The basic cost of machining also is determined by the size, rigidity and geometry of the component. A rigid part with simple surfaces can obviously be machined quicker and at lower costs than a flimsy part with complex surfaces. In addition, the engineering design specifications for accuracy and surface roughness play a major role in determining the machining cost.

## EFFECT OF SURFACE ROUGHNESS AND DIMENSIONAL TOLERANCE ON COST

The relative cost of machining is closely associated with both dimensional tolerance and surface roughness. The percentage machining cost in machining steel parts is exemplified in figure 21-2. As the surface roughness requirement becomes more stringent, the cost increases because more than one cut on a given machine tool is required to achieve that level of roughness. For example, a surface might have to be rough turned, then semi-finish turned, then finally finish turned to achieve the required surface roughness. The surface roughness specification may be even finer than that readily obtained by a chip removal process, and in this case, a second operation such as grinding or even a third operation such as honing may be required to achieve the surface roughness as well as the dimensional accuracy. Figure 21-2 illustrates the great increase in cost that occurs in order to achieve the desired finer surface roughnesses.

The overall machining costs also are associated with dimensional tolerance. As a matter of fact, close tolerance requirements usually dictate finer surface roughnesses. The interrelationship of dimensional tolerances, surface roughness, and their effect on machining costs are illustrated in the charts of figure 21-3.

## MACHINING COST AND PRODUCTION RATE

It is possible to quantitatively determine the cost and production rate for a given machining operation on a specific work material. Equations have been developed to determine the cost per piece and the production rate in pieces per hour for specific machining operations. The total cost for machining comprises costs associated with operating the machine tool and costs associated with the cutter and its reconditioning. The machine tool cost can be broken down into idle cost, tool reconditioning cost, and feed or cutting cost. The idle cost, which consists of rapid traverse, load and unload, and tool change costs, remains constant with change in cutting speed, while the feed or cutting cost decreases with increasing cutting speed. The tool reconditioning cost generally increases with increasing cutting speed because greater cutter wear rate is associated with the higher speeds. The total cost is the sum of all of the above cost elements. This total cost is seen to go through a minimum at some intermediate cutting speed and then sharply increase. The production rate in pieces per hour increases with increasing cutting speed, goes through a maximum and then decreases. Curves demonstrating these relationships of machining cost and production rate to cutting speed are contained in figure 21-4.

Although the previous discussion has centered on the relationship of cost to cutting speed, it should be pointed out that cost also is a function of other machining parameters, such as feed, depth of cut, width of cut, tool material, cutting fluid, etc.

## PROCEDURE FOR CALCULATING COST AND PRODUCTION RATE

The following steps describe a method for calculating machining cost and production rate.

### 1. Machining Data and Machining Parameters

Data relating the tool life to machining parameters must be obtained for those work materials that are to be machined. Data can be obtained from handbooks and may also be obtained from historical shop experiences. It is important to have a well-defined format for recording and storing significant data. Typical data formats for turning, milling, drilling, reaming and tapping are shown in tables 21-3 through 21-5.

### 2. Cost and Production Rate Equations

Equations for determining costs, operating times, and production rates for various machining operations are listed in tables 21-6 and 21-8 for English units and tables 21-7 and 21-9 for metric units. The first bracket of terms in the cost equations determines the time and cost for the machine tool, and the second bracket of terms determines the cost for the cutter and its reconditioning. In tables 21-6 and 21-7, the cost equations are given for turning, milling, drilling or reaming, tapping, and center drilling or chamfering. These equations apply to both conventional and numerical control machine tools. It should be noted that a separate equation is provided for calculating the handling and setup costs. In the case of conventional machining, the two terms in this equation should be added into the appropriate operation equation in order to calculate the handling and setup costs of a specific operation such as turning or milling, etc. In the case of NC machining, the individual equations are used for each machining operation in sequence, while the handling and setup equation is used separately, as will be illustrated in an example to follow. The equations for determining the operating time and the production rate for each operation are given in tables 21-8 and 21-9. It is desirable to program all of the cost and production rate equations on a computer for ease of calculations.

### 3. Supporting Data Required to Complete Calculations

In addition to tool life data, time study data and tool cost data are required for a cost and production rate determination. Tables 21-10 and 21-11 list and define the variable factors that appear in the cost and production rate equations. Figure 21-5 illustrates the setup factors for various machining operations. Note that "R" equals the total rapid traverse distance for a tool or cutter in one operation, whereas "e" equals the extra travel in feed.

The derivation of the terms in the cost equations for turning and milling are given in table 21-12.

The procedure described above has been applied in several situations to generate the examples that follow.

## Examples

### 1. Example of a Cost and Production Rate Determination in Turning on a Conventional Engine Lathe

Table 21-13 contains the time study and cost data for turning a shaft 3.5 inches in diameter by 19 inches long on a conventional engine lathe. The material was 4340 steel, quenched and tempered to 300 Bhn. Three types of tools (that is, brazed carbide, throwaway carbide, and high speed steel) were used for the cost investigation. The tool life data are given in table 21-14. The tool life versus cutting speed data sets are denoted by the circled numbers 1, 2, 3, and 4 for the C-7 carbide tool material, and 5, 6, 7, and 8 for the high speed steel tool material.

The lathe tool setup for turning using the brazed carbide, the throwaway carbide and the solid high speed steel tools was the same as that illustrated in figure 21-5.

Using the above data and the equations for turning from tables 21-6 and 21-8, the machining cost and production rate were determined. The calculations were performed on an IBM 1130 computer, and a printout of the results is shown in figure 21-6. It should be noted that there are three sets of calculations: one for brazed carbide tools, one for throwaway carbide tools, and one for solid high speed steel tools. Using the tool life data for the C-7 carbide tool material and the time study data for brazed carbide tools, cost calculations were made at each of four cutting speeds.

Notice that the printout contains not only the total cost per piece and the production rate in pieces per hour but also all the cost factors that make up the total cost. In the case of the brazed carbide tool, there were ten cost factors; for the throwaway carbide tools, seven; and for the solid high speed steel, eight. A quick glance at the cost factors in figure 21-6 indicates which are significant and which are insignificant. For example, with the brazed carbide tool, when cutting at 470 feet per minute, the total cost was \$5.33 per piece. Of this, the feeding cost was \$1.50, the load and unload cost \$0.92, the setup cost \$0.42, the tool change cost \$0.49, and the tool sharpening cost \$1.48. The less significant factors were as follows: rapid traverse cost \$0.11, tool depreciation cost \$0.13, rebrazing cost \$0.16, tip cost \$0.10, and the grinding wheel cost \$0.02.

For the throwaway carbide tools, figure 21-6, it can be seen that all the tool costs were virtually nil. On the other hand, when turning with solid high speed steel, the tool cost was an appreciable portion of the total cost. Thus, when turning with high speed steel at 45 feet per minute, the total cost was \$21.29 per piece, the tool change cost was \$1.29, and the tool sharpening cost \$2.58.

The cost and production rates for turning 4340 steel at 300 Bhn were plotted against cutting speed. Figure 21-7 contains these plots for brazed carbide, throwaway carbide and high speed steel tools. The data show that the cost per piece decreased as the cutting speed increased when using throwaway carbide tools. The cost per piece was at a minimum at approximately 360 feet per minute for the brazed carbide, and the cost per piece was at a minimum at approximately 60 feet per minute for the high speed steel.

The minimum costs for turning within the range of experimental data were about $3.09 with the throwaway carbide, $4.44 with the brazed carbide and $18.62 with the high speed steel. The maximum production rate, also within the range of experimental data, was 8 pieces per hour for the throwaway carbide tools, 7 pieces per hour with the brazed carbide tools, and 1.8 pieces per hour with the high speed steel tools.

For those who would like to calculate the cost manually, the cost calculations for the example just described are given in table 21-15 for turning with the brazed carbide tool.

**2. Example of a Cost and Production Rate Determination in Face Milling on a Conventional Milling Machine**

Table 21-16 contains the time study and cost data for face milling a block 2 inches wide by 8 inches long. The work material was 4340 steel, quenched and tempered to 341 Bhn. A two-inch wide cut was taken on the eight-inch long workpiece with 4-inch diameter milling cutters. A solid high speed steel cutter and two types of carbide cutters (inserted tooth and throwaway insert) were used. The setup used for this example was the same as that for milling illustrated in figure 21-5.

The tool life data for the alloy are given in table 21-17. In this table, the sets of data relating tool life in inches per tooth to cutting speed in feet per minute for the alloy are identified by the circled numbers 1 through 6.

The IBM 1130 computer was used to calculate the cost per piece using the milling equation from table 21-6 and the production rate using the milling equation from table 21-8. The computer printout of the results is shown in figure 21-8.

Examination of the computer printout reveals the significant cost factors and the comparative costs per piece and production rates among the types of tools used. For instance, the printout shows that the use of a throwaway carbide cutter resulted in higher production and lower costs. This is more evident from examining figure 21-9, which indicates that the cost per piece in face milling the 4340 steel did not vary as sharply for the throwaway carbide cutter as it did for the other cutters. The production rate curve for the throwaway cutter lies above the other two types. Of specific interest is that for the data used in this example, the mimimum cost and maximum production rate occurred for the case of throwaway carbide cutters at 550 feet per minute.

**3. Example of a Cost and Production Rate Determination in Drilling Using a Conventional Drill**

Table 21-18 contains the time study and cost data for drilling holes in 4340 steel, annealed to 212 Bhn. High speed steel drills were used to drill five holes, one-half-inch deep in each part. The drill diameter used was one-quarter inch. The geometric setup used was the same as that illustrated for drilling in figure 21-5.

The tool life data is listed in table 21-19. The sets of data giving the drill life in number of holes are identified by the circled numbers 1 through 8. The cost and production equations required that the drill life ($T_t$) be expressed in inches. Therefore, it was necessary to multiply the drill life, in number of holes drilled, by the hole length to get the drill life in inches of travel required to dull the drill. Using data set 1, the drill life was 30 holes. The length of each hole was 0.5 inch. Therefore, the drill life was $T_t$ = 15 inches (30 x 0.5 inch).

Figure 21-10 is the computer printout of the results from using the drilling equation from table 21-6 to calculate the cost per piece and the drilling equation from table 21-8 to determine the production rate.

Figure 21-11 shows the cost and production rate curves for this drilling operation. The data show that the 0.005 inch per revolution feed can achieve higher production rates and lower costs than the 0.002 inch per revolution feed, provided that proper cutting speeds are used.

**4. Example of Machining a Part on an NC Machining Center**

An example of a 4340 steel forging machined on an NC machining center is shown in figure 21-12. The operations were to face mill the top surface, peripheral end mill the sides and ends, mill two slots and drill two holes. The time study and tool cost data are given in table 21-20, and the pertinent machining parameter and tool life data for the various operations are given in table 21-21. The cost and production rates for the individual operations were calculated on the computer and a printout is given in figure 21-13. Again, the computer prints not only the total cost but also the individual cost factors. It should be noted that calculations were made for each of the individual operations, and for each operation a line of data shows the cost corresponding to the feed, speed and tool life combination. The minimum cost for each of the operations in the printout of figure 21-13 is circled, and a summary of the best machining conditions is given in figure 21-14.

## DETERMINING OPTIMUM CUTTING CONDITIONS

As shown in the previous examples, there are cases where the optimum cutting conditions lie beyond the available data. In order to optimize, it is necessary to determine the mathematical relationship between tool life and the pertinent cutting parameters such as speed, feed, and depth and width of cut. Probably the most common approach is to assume that Taylor's equation relating tool life to cutting speed applies:

for turning

$$vT^n = S$$

for milling, drilling, reaming and tapping

$$vT_t^n = S_t$$

where:

v = cutting speed
T and $T_t$ = tool life
S and $S_t$ = reference cutting speed
n = tool life exponent

(See table 21-10 or 21-11 for detailed definition of variables.)

With these Taylor equations, the cost for the various machining operations can be minimized. This has been done by substituting the appropriate Taylor equation into the cost equation, differentiating the cost with respect to cutting speed, and setting the derivative equal to zero. The cutting speed calculations for minimum cost ($v_{min\ cost}$) and for maximum production ($v_{max\ prod}$) for the conventional as well as the NC machining operations are given in the equations in tables 21-22 and 21-23. The corresponding tool life equations for minimum cost and maximum production are given in table 21-24. The values of $v_{min\ cost}$ and $v_{max\ prod}$ in the denominators must be calculated using the appropriate equation(s) from table 21-22 or 21-23.

The validity of the optimized determination depends, in turn, on the validity of the constants n and S or $S_t$ in the Taylor equation. The most accurate values of these constants can be determined by actual experimentation (that is, by actual cutting tests). In so doing, it is also possible to determine the range of cutting conditions over which the Taylor equation applies. The least reliable procedure is to pick an average value for the Taylor constant from a handbook. It is suggested that if one assumes the Taylor relationship and proceeds to determine optimum cutting conditions, these conditions should be verified by actual shop tests.

The relationship of cutting conditions to tool life as determined by actual cutting tests may be too complex to be represented by the simple Taylor equation. Techniques are now available to mathematically model the more complex relationship of tool life to cutting speed, feed, and depth of cut. The mathematical model can, in turn, be used to determine cost and production rate over the wide range of machining conditions with subsequent determination of minimum cost, minimum cutting time and maximum production rates. These techniques are too complex for inclusion in this handbook.

## SHOP PROCEDURES FOR OPTIMIZATION OF MACHINING CONDITIONS

When it is felt necessary to decrease cost or increase productivity of a machining operation, it is possible to experiment on the shop floor by changing machining conditions. It is necessary to keep careful records of the changes in cutting conditions and the effect of these changes on tool life as well as on production rates.

The first step is usually to increase the feed. The feed may be increased until either the specified surface roughness is no longer obtained or the tool life starts to decrease. With the best feed, the next step is to increase and then decrease the cutting speed and observe the change in tool life. By this manner, the combination of feed and speed can be selected to achieve maximum productivity. Additional factors which may be investigated for their effect on tool life include cutting tool material, tool geometry, and cutting fluid.

Although the feed rate as a result of shop trials can be readily observed, it is not possible to directly observe the effect of these trials on the overall machining cost since the machining cost involves not only machine tool time but also the cutter cost. It is necessary, therefore, to calculate the overall machining cost and production rate as a function of the machining parameters using the procedure outlined previously.

## TECHNIQUE FOR CALCULATING COST AND PRODUCTION RATE

Although the equations used for the cost and production rate calculations involve only simple mathematics, the detailed analysis of a given machining setup will become very time consuming because many factors must be included. With this in mind, techniques have been developed that allow the rapid analysis of a machining operation through the use of computers and programmable calculators.

The cost and production rate equations given in tables 21-6 and 21-8 have been programmed in FORTRAN IV for use on a computer with as little as 8K memory, such as an IBM 1130. This program, called NCECO, is available through the Machinability Data Center and is supplied in card-deck form and includes full documentation. With minor modifications for input-output, this program will operate on any digital computer supporting the FORTRAN IV language. The program will accept any combination of machining operations required for a part setup including: turning, face milling, end milling, drilling, reaming, tapping, center drilling and chamfering. Capability exists to determine the effects of using all types of tools, that is, high speed steel, brazed carbide, or throwaway carbide. To run this program, data such as that in table 21-13 is compiled for the machining operation. These data are then punched on cards which are input to the computer. The computer then prints out a detailed cost breakdown such as that shown in figure 21-13.

With the advent of programmable calculators, it is now possible to perform these calculations quickly without utilizing a large computer. A calculator generally uses semiconductor electronics that allow the solving of a mathematical problem using a series of keystrokes on the machine. A programmable calculator has the ability to store and automatically execute the series of keystrokes necessary to solve a particular problem. Using this feature, a programmable calculator can be instructed to perform all the calculations necessary for cost and production rate analysis with the user supplying only the data necessary for the calculations. Two such programmable calculators are the Hewlett-Packard HP-67 and the Texas Instrument TI-59.

The Hewlett-Packard HP-67 is a pocket-sized programmable calculator powered by alternating current or rechargeable batteries. All output is by digital display. The programs are stored on small magnetic strips and are read by the machine simply by inserting a strip in a slot on the side of the machine. The small memory size of this machine makes it necessary to use a separate memory strip for each type of machining operation. The magnetic strips are stored in a 3-by-5-inch card which gives all the necessary instructions for running a complete analysis. Since this calculator is pocket-sized, it is well suited for doing these calculations right on the shop floor. An enlarged version of the instructions and sample inputs and outputs are shown in figure 21-15.

The Texas Instruments TI-59 programmable calculator is similar to the Hewlett-Packard calculator. When attached to a TI-PC100A printer, cost and time analysis can be accomplished in an interactive mode. Instructions are printed out to the user describing the necessary inputs. The results of the calculations are also printed with appropriate labels. Also included is the capability to save the input data on magnetic strips for later analysis without having to re-input all the data by hand. An example of the cost and production time program run on the Texas Instruments TI-59 calculator with printer is shown in figure 21-16.

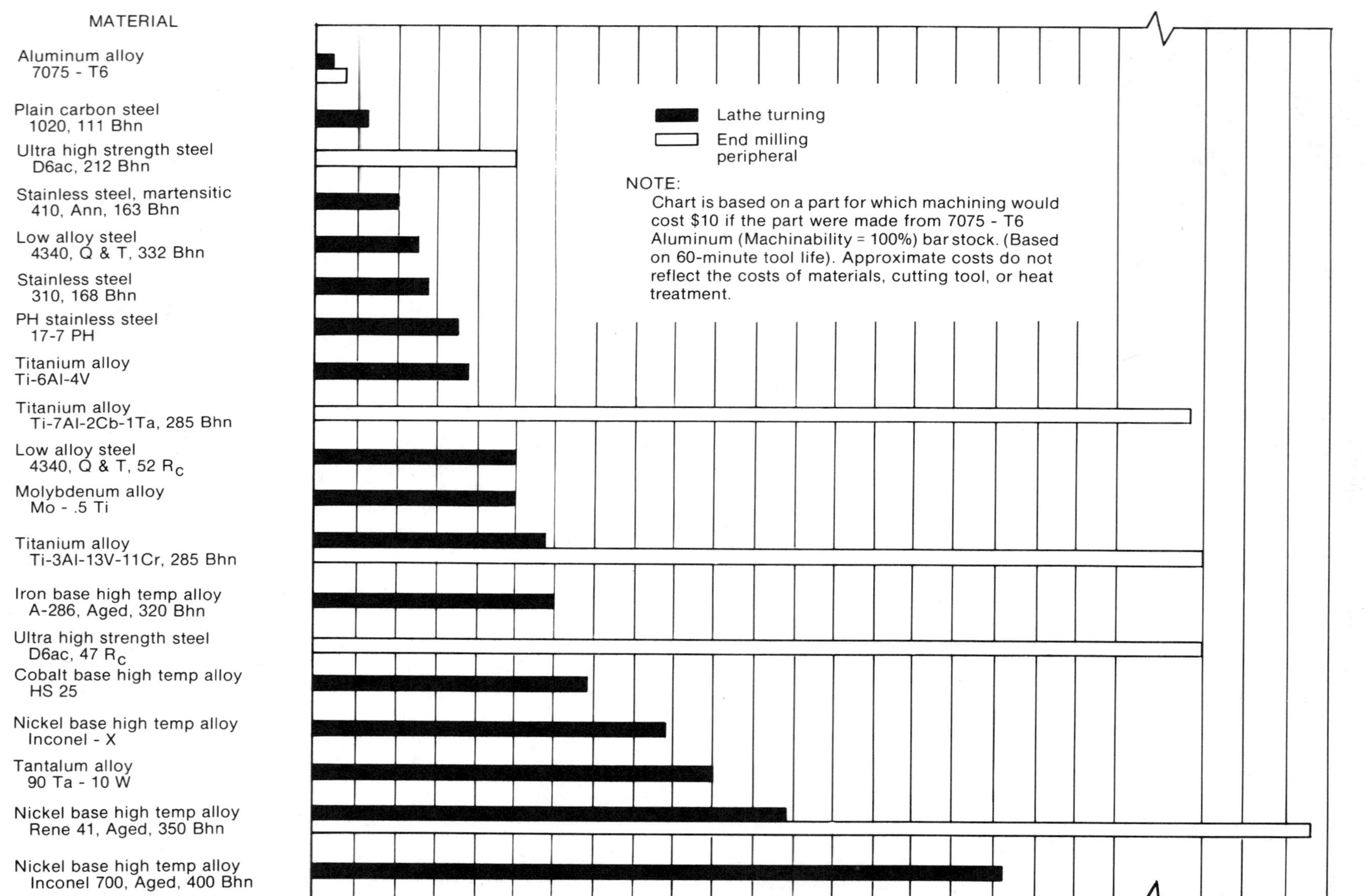


**Figure 21–1** **Relative machining costs for hard and soft aerospace materials.** (Adapted from Profile milling requirements for the hard metals 1965–1970, Report of the Ad Hoc Machine Tool Advisory Committee to the Department of the Air Force, May 1965)

**TABLE 21-2 Relative Machining Time of Various Alloys Compared to 4340 Steel, 300 Bhn**

| MATERIAL GROUP | MATERIAL | CONDI-TION* | HARD-NESS Bhn | RELATIVE MACHINING TIME | | | | | | | |
|---|---|---|---|---|---|---|---|---|---|---|---|
| | | | | Turning | | Face Milling | | End Milling | | Drilling | Tapping |
| | | | | HSS | Carbide | HSS | Carbide | HSS | Carbide | HSS | HSS |
| Alloy steels | | Q & T | 300 | 1.0 | 1.0 | 1.0 | 1.0 | 1.0 | 1.0 | 1.0 | 1.0 |
| | | Q & T | 400 | 2.5 | 1.7 | 5.0 | 2.5 | 2.0 | 2.5 | 1.7 | 2.5 |
| | 4340 | Q & T | 500 | 3.3 | 3.3 | 10.0 | 3.3 | 5.0 | 10.0 | 5.0 | 5.0 |
| | | Q & T | 520 | 5.0 | 5.0 | 25.0 | 10.0 | 7.0 | 11.1 | 10.0 | 10.0 |
| | | Ann | 210 | 0.8 | 0.8 | 0.4 | 0.5 | 0.7 | 0.6 | 0.8 | 0.7 |
| | | Norm | 320 | 1.3 | 1.0 | 1.7 | 1.7 | 1.0 | 1.0 | 1.0 | 1.0 |
| High strength steels | D6ac | Ann | 220 | 0.8 | 0.8 | 0.6 | 0.7 | 0.5 | 0.7 | 0.8 | 1.0 |
| | | Q & T | 560 | — | 14.0 | — | 10.0 | — | 25.0 | 1.1[‡] | — |
| | | Ann | 200 | 0.8 | 0.9 | 0.5 | 0.5 | 0.9 | 1.0 | 0.8 | 1.3 |
| | H11 | Q & T | 350 | 1.7 | 2.0 | 1.7 | 2.0 | 3.3 | 5.0 | 1.4 | 1.7 |
| | | Q & T | 520 | — | 5.0 | 25.0 | 10.0 | — | 10.0 | 10.0 | 10.0 |
| Maraging steels | 200, 250, | Ann | 320 | 1.7 | 0.8 | 1.0 | 1.4 | 0.9 | 0.9 | 0.9 | 1.3 |
| | 300 Grades | Mar | 500 | 3.3 | 2.5 | 2.5 | 5.0 | 5.0 | 5.0 | 3.0 | 5.0 |
| Stainless steels, austenitic | 302, 304 317, 321 | Ann | 180 | 0.8 | 0.9 | 1.1 | 1.0 | 1.0 | 1.1 | 0.8 | 1.0 |
| Stainless steels, martensitic | 410, 420 | Ann | 180 | 0.9 | 0.8 | 0.8 | 0.7 | 1.4 | 0.8 | 0.7 | 0.8 |
| | | Q & T | 300 | 1.7 | 1.3 | 2.0 | 2.0 | 1.7 | 1.3 | 0.8 | 1.3 |
| | 420 | Q & T | 400 | 2.5 | 3.3 | 5.0 | 5.0 | 3.3 | 5.0 | 1.7 | 2.5 |
| Stainless steels, precipitation hardening | 17-4 PH 17-7 PH | Ann | 170 | 1.0 | 0.9 | 1.1 | 0.9 | 1.1 | 0.8 | 1.0 | 1.0 |
| | AM-350 | Ann | 300 | 1.4 | 1.4 | 1.4 | 1.4 | 1.4 | 1.0 | 1.4 | 1.3 |
| | 17-4 PH 17-7 PH AM-350 | Q & T | 400 | 2.5 | 2.0 | 3.3 | 3.3 | 2.5 | 5.0 | 2.5 | 2.5 |
| Titanium alloys | Ti-100A | — | 175 | 0.7 | 0.7 | 0.8 | 1.4 | 0.4 | 0.8 | 0.5 | 0.8 |
| | Ti-5Al-2.5Sn | Ann | 300 | 1.4 | 1.4 | 1.4 | 2.5 | 0.5 | 1.3 | 0.7 | 1.0 |
| | Ti-6Al-4V | Ann | 310 | 1.7 | 2.0 | 2.0 | 3.3 | 1.0 | 2.0 | 1.2 | 1.3 |
| | | STA | 365 | 2.0 | 2.5 | 2.0 | 3.3 | 2.0 | 2.0 | 1.7 | 1.7 |
| | Ti-7Al-4Mo Ti-6Al-6V-2Sn Ti-8Al-1Mo-1V | Ann | 320 | 2.0 | 2.0 | 2.5 | 2.0 | 1.1 | 1.7 | 0.9 | 1.4 |
| | Ti-6Al-6V-2Sn Ti-7Al-4Mo | STA | 420 | 2.5 | 3.3 | 2.5 | 5.0 | 1.4 | 2.5 | 1.4 | 2.5 |
| | Ti-13V-11Cr-3Al | ST | 310 | 2.5 | 5.0 | 3.3 | 3.3 | 2.0 | 3.3 | 3.3 | 2.5 |
| | | STA | 400 | 2.5 | 5.0 | 3.3 | 10.0 | 3.3 | 4.0 | 10.0 | 3.3 |
| High temperature alloys | Reńe 41 | ST | 280 | 5.0 | 5.0 | 10.0 | 16.7 | 5.0 | 5.0 | 3.3 | 2.5 |
| | Inconel 700 U-500 | STA | 365 | 10.0 | 5.0 | 10.0 | 16.7 | 10.0 | 12.5 | 10.0 | 5.0 |
| Nickel base | Inconel 718 | ST | 270 | 5.0 | 5.0 | 3.3 | 10.0 | 3.3 | 10.0 | 3.3 | 2.5 |
| | | STA | 370 | 5.0 | 5.0 | 5.0 | 10.0 | 12.5 | 12.5 | 5.0 | 5.0 |
| Cobalt base | HS25 | ST | 200 | 3.3 | 5.0 | 5.0 | 10.0 | 5.0 | 5.0 | 3.3 | 2.5 |
| | S-816 | STA | 300 | 5.0 | 10.0 | 10.0 | 12.5 | 10.0 | 10.0 | 5.0 | 3.3 |
| Iron base | A-286 | ST | 200 | 1.7 | 2.5 | 1.4 | 3.3 | 2.0 | 3.3 | 1.7 | 1.7 |
| | 19-9DL | STA | 300 | 2.0 | 3.3 | 3.3 | 5.0 | 3.3 | 12.5 | 3.3 | 2.5 |
| Aluminum alloys | 7075-T6 | STA | 75 to 150[†] | 0.12 | 0.3 | 0.06 | 0.1 | 0.1 | 0.3 | 0.1 | 0.5 |

Source: R. L. Vaughn and N. Zlatin, "Producibility Aspects of Aerospace Products with Regard to Machinability," ASTME Paper WE S7-29, 1967.

*Condition: Ann—Annealed; Norm—Normalized; Mar—Maraged; Q & T—Quenched and Tempered; ST—Solution Treated; STA—Solution Treated and Aged

[†] Brinell Hardness for 500 kg load

[‡] Carbide Drill

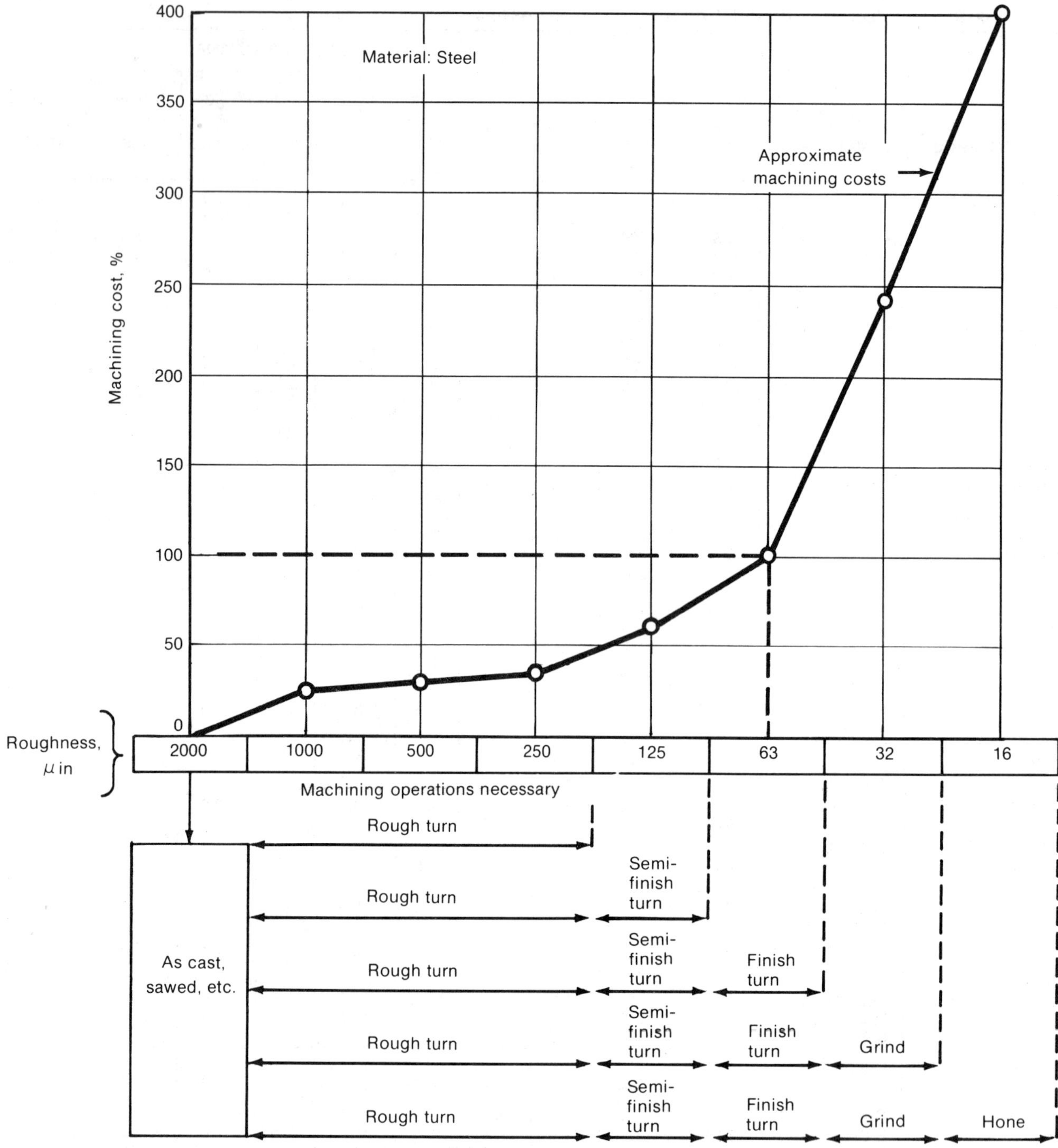


**Figure 21–2** Relative machining costs and surface roughness for steel parts. (Courtesy of General Electric Company)

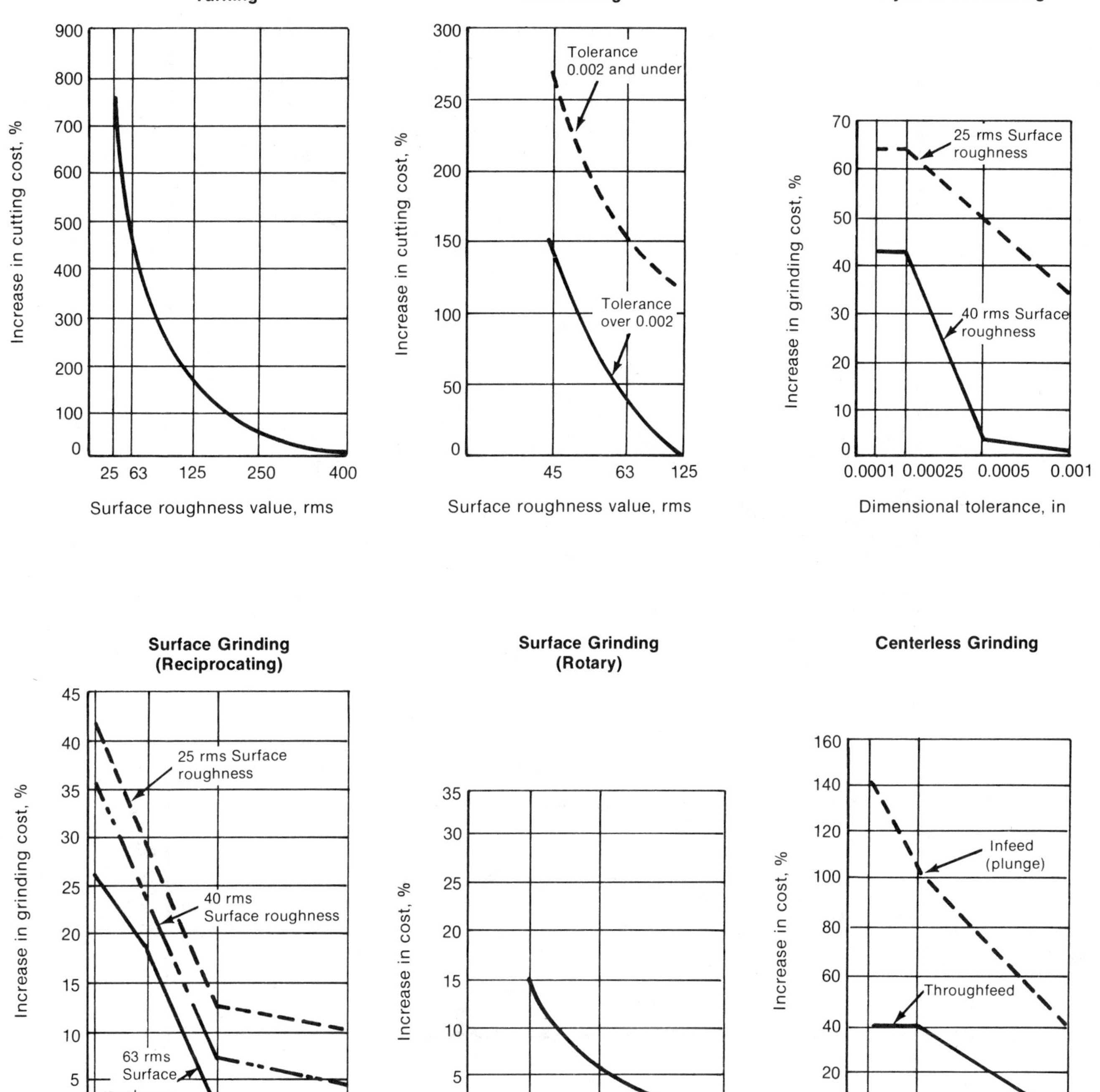


**Figure 21–3** Effect of tolerance and surface roughness on machining costs for several operations. (L. J. Bayer, Analysis of manufacturing costs relative to product design, Paper No. 56-A-9, American Society of Mechanical Engineers, New York, NY, 1956)

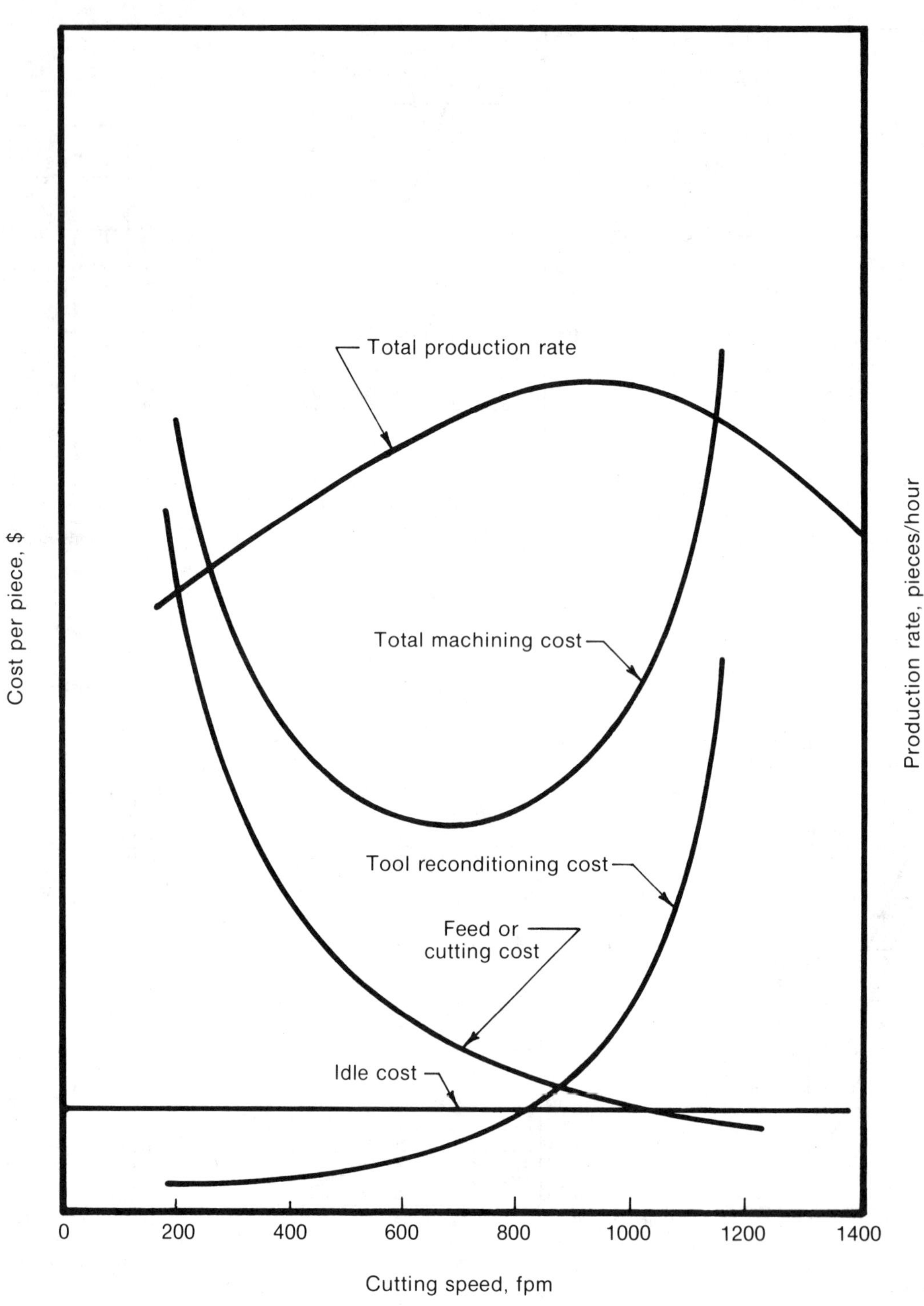


**Figure 21–4** Machining cost and production rate versus cutting speed.

**TABLE 21-3 Tool Life Data for Turning**

| MATERIAL | CONDITION AND MICROSTRUCTURE | HARD-NESS Bhn | TOOL MATERIAL | | TOOL GEOMETRY | | | | | | CUTTING FLUID CODE * | DEPTH OF CUT in | FEED ipr | TOOL LIFE END POINT in | TOOL LIFE, min vs CUTTING SPEED, fpm | | | |
|---|---|---|---|---|---|---|---|---|---|---|---|---|---|---|---|---|---|---|
| | | | Trade Name | Industry Grade | BR° | SR° | SCEA° | ECEA° | Relief° | Nose Radius in | | | | | | | | |
| Alloy Steels 8640 | Quenched and tempered | 400 | — | T1 HSS | 0 | 15 | 0 | 5 | 5 | 0.005 | 11 1:20 | 0.060 | 0.009 | 0.060 | 5 | 15 | 30 | |
| | Tempered martensite | | | | | | | | | | | | | | 80 | 67 | 58 | |
| 8640 | Annealed | 170 | 78 | C-7 | 0 | 6 | 0 | 6 | 6 | 0.040 | 00 | 0.100 | 0.010 | 0.015 | 15 | 30 | 45 | 60 |
| | 50%P-50%F | | | | | | | | | | | | | | 610 | 490 | 420 | 373 |
| 8640 | Spheroidized | 180 | 78 | C-7 | 0 | 6 | 0 | 6 | 6 | 0.040 | 00 | 0.100 | 0.010 | 0.015 | 15 | 30 | 45 | 60 |
| | Spheroidized carbides + ferrite | | | | | | | | | | | | | | 695 | 580 | 525 | 485 |
| 8640 | Annealed | 190 | 78 | C-7 | 0 | 6 | 0 | 6 | 6 | 0.040 | 00 | 0.100 | 0.010 | 0.015 | 15 | 30 | 45 | 60 |
| | 75%P-25%F | | | | | | | | | | | | | | 590 | 450 | 380 | 335 |
| 8640 | Annealed | 250 | 78B | C-6 | 0 | 6 | 0 | 6 | 6 | 0.040 | 00 | 0.100 | 0.010 | 0.015 | 1 | 6 | 20 | 66 |
| | Widmanstätten | | | | | | | | | | | | | | 910 | 600 | 440 | 300 |
| 8640 | Annealed | 250 | 78 | C-7 | 0 | 6 | 0 | 6 | 6 | 0.040 | 00 | 0.100 | 0.010 | 0.015 | 15 | 30 | 45 | 60 |
| | Widmanstätten | | | | | | | | | | | | | | 640 | 520 | 440 | 375 |
| 8640 | Quenched and tempered | 300 | 78 | C-7 | 0 | 6 | 0 | 6 | 6 | 0.040 | 00 | 0.100 | 0.010 | 0.015 | 5 | 15 | 20 | |
| | Tempered martensite | | | | | | | | | | | | | | 660 | 400 | 315 | |
| 8640 | Quenched and tempered | 400 | 78 | C-7 | 0 | 6 | 0 | 6 | 6 | 0.040 | 00 | 0.100 | 0.010 | 0.015 | 5 | 15 | 20 | |
| | Tempered martensite | | | | | | | | | | | | | | 480 | 365 | 318 | |
| 52100 | Spheroidized | 190 | — | T1 HSS | 0 | 15 | 0 | 5 | 5 | 0.005 | 11 1:20 | 0.060 | 0.009 | 0.060 | 15 | 30 | 35 | |
| | Spheroidized carbides + ferrite | | | | | | | | | | | | | | 137 | 123 | 120 | |
| 52100 | Spheroidized | 190 | 78B | C-6 | 0 | 6 | 0 | 6 | 6 | 0.040 | 00 | 0.100 | 0.010 | 0.015 | 15 | 30 | 45 | |
| | Spheroidized carbides + ferrite | | | | | | | | | | | | | | 430 | 340 | 300 | |

*Cutting Fluid Code
00 Dry
11 Soluble Oil

**TABLE 21–4 Tool Life Data for Milling**

**FACE MILLING**

| MATERIAL | CONDITION AND MICROSTRUCTURE | HARDNESS Bhn | TOOL MATL. Trade Name | TOOL MATL. Industry Grade | UP OR DOWN MILLING | TOOL GEOMETRY AR° | RR° | CA° | TR° | Incl° | ECEA° | End Rel.° / Cor. Rel.° | CUTTING FLUID CODE * | DEPTH OF CUT in | WIDTH OF CUT in | FEED PER TOOTH in | TOOL LIFE END POINT in | TOOL LIFE/TOOTH inches work travel vs CUTTING SPEED, fpm | | | |
|---|---|---|---|---|---|---|---|---|---|---|---|---|---|---|---|---|---|---|---|---|---|
| Stainless steels Martensitic 410 | Quenched & tempered | 353 | 370 | C-6 | Up | 0 | −7 | 45 | −5 | 5 | 5 | 8 | 00 | 0.100 | 2.0 | 0.010 | 0.016 | 15 | 45 | 60 | 90 |
| | Tempered martensite | | | | | | | | | | | 8 | | | | | | 540 | 420 | 365 | 230 |
| 410 | Quenched & tempered | 45 $R_c$ | — | T15 HSS | Up | 0 | 0 | 45 | 0 | 0 | 5 | 8 | 11 1:20 | 0.060 | 2.0 | 0.005 | 0.060 | 15 | 30 | 50 | |
| | Tempered martensite | | | | | | | | | | | 8 | | | | | | 100 | 79 | 68 | |

**END MILL SLOTTING**

| MATERIAL | CONDITION AND MICROSTRUCTURE | HARDNESS Bhn | TOOL MATL. Trade Name | TOOL MATL. Industry Grade | CUTTER TYPE | DIA. in | NO. TEETH | FLUTE LENGTH in | UP OR DOWN MILLING | TOOL GEOMETRY Helix Angle° | RR° | Chamfer | ECEA° | End Rel.° / Periph. Rel.° | CUTTING FLUID CODE * | DEPTH OF CUT in | WIDTH OF CUT in | FEED PER TOOTH in | TOOL LIFE END POINT in | TOOL LIFE/CUTTER inches work travel vs CUTTING SPEED, fpm | | | |
|---|---|---|---|---|---|---|---|---|---|---|---|---|---|---|---|---|---|---|---|---|---|---|---|
| Alloy steel 4340 | Annealed | 213 | — | M2 HSS | Solid HSS | 0.750 | 4 | 2 | — | 30 | 10 | 45° x 0.060 in | 1 | 3 | 11 1:20 | 0.250 | 0.750 | 0.002 | 0.012 | 50 | 120 | 240 | |
| | Spheroidized carbides ferrite | | | | | | | | | | | | | 7 | | | | | | 190 | 153 | 125 | |

**PERIPHERAL END MILLING**

| MATERIAL | CONDITION AND MICROSTRUCTURE | HARDNESS Bhn | TOOL MATL. Trade Name | TOOL MATL. Industry Grade | CUTTER TYPE | DIA. in | NO. TEETH | FLUTE LENGTH in | UP OR DOWN MILLING | TOOL GEOMETRY Helix Angle° | RR° | Chamfer | ECEA° | End Rel.° / Periph. Rel.° | CUTTING FLUID CODE * | DEPTH OF CUT in | WIDTH OF CUT in | FEED PER TOOTH in | TOOL LIFE END POINT in | TOOL LIFE/CUTTER inches work travel vs CUTTING SPEED, fpm | | | |
|---|---|---|---|---|---|---|---|---|---|---|---|---|---|---|---|---|---|---|---|---|---|---|---|
| High strength steel D6ac | Annealed | 223 | — | M2 HSS | Solid HSS | 0.750 | 4 | 2 | Down | 30 | 10 | 45° x 0.060 in | 1 | 3 | 11 1:20 Mist | 0.250 | 0.750 | 0.004 | 0.012 | 75 | 130 | 260 | |
| | 60% P-40% F | | | | | | | | | | | | | 7 | | | | | | 290 | 240 | 190 | |

*Cutting Fluid Code
00 Dry
11 Soluble Oil

**TABLE 21-5 Tool Life Data for Drilling, Reaming and Tapping**

**DRILLING**

| MATERIAL | CONDITION AND MICROSTRUCTURE | HARDNESS Bhn | DRILL MATL. Trade Name | DRILL MATL. Industry Grade | TYPE DRILL | DRILL SIZE Dia. in | DRILL SIZE Length in | DRILL SIZE Flute Length in | DRILL GEOMETRY Type Point | DRILL GEOMETRY Helix Angle° | DRILL GEOMETRY Point Angle° | DRILL GEOMETRY Lip Relief° | CUTTING FLUID CODE * | DEPTH OF HOLE in | FEED ipr | DRILL LIFE END POINT in | DRILL LIFE no. of holes vs CUTTING SPEED, fpm | | | | |
|---|---|---|---|---|---|---|---|---|---|---|---|---|---|---|---|---|---|---|---|---|---|
| Alloy steel 4340 | Quenched & tempered / Tempered martensite | 341 | — | M2 HSS | Twist | 0.250 | 4.0 | 2.75 | Standard | 29 | 118 | 7 | 31 | 0.5 thru | 0.002 | 0.015 | 25 | 50 | 75 | 100 | |
| | | | | | | | | | | | | | | | | | 98 | 84 | 76 | 70 | |
| " | " / " | " | — | " | " | " | " | " | " | " | " | " | " | " | 0.005 | " | 25 | 50 | 75 | 100 | |
| | | | | | | | | | | | | | | | | | 80 | 65 | 56 | 50 | |

**REAMING**

| MATERIAL | CONDITION AND MICROSTRUCTURE | HARDNESS Bhn | REAMER DESCRIPTION: TOOL MATL. Trade Name | TOOL MATL. Industry Grade | Dia. in | No. of Flutes | Style | TOOL GEOMETRY: Helix & Hand | Chamfer | Rel.° | CUTTING FLUID CODE * | STOCK ALLOW. ON DIA. in | LENGTH OF HOLE in | FEED ipr | TOOL LIFE END POINT in | REAMER LIFE no. of holes vs CUTTING SPEED, fpm | | | | |
|---|---|---|---|---|---|---|---|---|---|---|---|---|---|---|---|---|---|---|---|---|
| High strength steel 250 Grade maraging steel | Annealed & maraged / Martensite | 50 $R_c$ | — | M2 HSS | 0.272 | 6 | Chucking | 0° RH | 45° x 0.060 in | 7 | 52 | 0.022 | 0.5 thru | 0.005 | 0.006 | 20 | 35 | 170 | | |
| | | | | | | | | | | | | | | | | 90 | 80 | 50 | | |

**TAPPING**

| MATERIAL | CONDITION AND MICROSTRUCTURE | HARDNESS Bhn | TAP MATERIAL | TAP SIZE | NO. OF FLUTES | TAP STYLE | PERCENT OF THREAD | CUTTING FLUID CODE * | DEPTH OF HOLE in | TAP LIFE END POINT in | TAP LIFE no. of holes vs CUTTING SPEED, fpm | | | | |
|---|---|---|---|---|---|---|---|---|---|---|---|---|---|---|---|
| High temperature alloy—iron base, wrought A-286 | Solution treated & aged / Austenitic | 320 | M10 HSS | 5/16-18NC | 2 | Plug spiral point | 75 | 53 | 0.5 thru | Tap breakage | 12 | 31 | 132 | | |
| | | | | | | | | | | | 50 | 40 | 30 | | |

*Cutting Fluid Code
31 Sulfurized mineral oil + fatty oil—light duty
52 Sulfo-chlorinated mineral oil + fatty oil—medium duty
53 Highly chlorinated mineral oil—heavy duty

**TABLE 21–6 Generalized Machining Cost Equations for Use with English Units**

**Turning**

$$C = M\left[\frac{D(L+e)}{3.82 f_r v} + \frac{R}{r} + t_i + \frac{DLt_d}{3.82 f_r vT} + \frac{DL}{3.82 f_r vT}\left[\frac{C_p}{(K_1 + 1)} + Gt_s + \frac{Gt_b}{k_2} + \frac{C_c}{k_3} + C_w + Gt_p\right]\right]$$

**Milling**

$$C = M\left[\frac{D(L+e)}{3.82 Z f_t v} + \frac{R}{r} + t_i + \frac{Lt_d}{ZT_t} + \frac{L}{ZT_t}\left[\frac{C_p}{(K_1 + 1)} + Gt_s + \frac{Gt_b}{k_2} + \frac{ZC_c}{k_3} + C_w + Gt_p\right]\right]$$

**Drilling or Reaming**

$$C = M\left[\frac{D(L+e)}{3.82 f_r v} + \frac{R}{r} + t_i + \frac{Lt_d}{T_t} + \frac{L}{T_t}\left[\frac{C_p}{(K_1 + 1)} + Gt_s + Gt_p\right]\right]$$

**Tapping**

$$C = M\left[\frac{mD(L+e)}{1.91v} + \frac{R}{r} + t_i + \frac{Lt_d}{T_t} + \frac{L}{T_t}\left[\frac{C_p}{(K_1 + 1)} + Gt_s + Gt_p\right]\right]$$

**Center Drilling or Chamfering**

$$C = M\left[\frac{D(L+e)}{3.82 f_r v} + \frac{R}{r} + t_i + \frac{u_c t_d}{T_h} + \frac{u_c}{T_h}\left[\frac{C_p}{(K_1 + 1)} + Gt_s + Gt_p\right]\right]$$

**Handling and Setup**

$$C = M\left[t_L + \frac{t_o}{N_L}\right]$$

Terms, left to right:

- $\frac{\$}{Min}$
- Feeding Time
- Rapid Traverse Time
- Load & Unload Time
- Setup Time
- Cutter Index Time
- Dull Tool Replacement Time
- Tool Depreciation Cost
- Tool Resharpening Cost
- Rebrazing or Blade Reset Cost
- Insert or Blade Cost
- Grinding Wheel Cost
- Tool Presetting Cost

**TABLE 21–7 Generalized Machining Cost Equations for Use with Metric Units**

**Turning**

$$C = M\left[\frac{D(L+e)}{318f_r\,v} + \frac{R}{r} + t_i + \frac{DLt_d}{318f_r\,vT}\right] + \frac{DL}{318f_r\,vT}\left[\frac{C_p}{(K_1+1)} + Gt_s + \frac{Gt_b}{k_2} + \frac{C_c}{k_3} + C_w + Gt_p\right]$$

**Milling**

$$C = M\left[\frac{D(L+e)}{318Zf_t\,v} + \frac{R}{r} + t_i + \frac{Lt_d}{ZT_t}\right] + \frac{L}{ZT_t}\left[\frac{C_p}{(K_1+1)} + Gt_s + \frac{Gt_b}{k_2} + \frac{ZC_c}{k_3} + C_w + Gt_p\right]$$

**Drilling or Reaming**

$$C = M\left[\frac{D(L+e)}{318f_r\,v} + \frac{R}{r} + t_i + \frac{Lt_d}{T_t}\right] + \frac{L}{T_t}\left[\frac{C_p}{(K_1+1)} + Gt_s + Gt_p\right]$$

**Tapping**

$$C = M\left[\frac{D(L+e)}{159pv} + \frac{R}{r} + t_i + \frac{Lt_d}{T_t}\right] + \frac{L}{T_t}\left[\frac{C_p}{(K_1+1)} + Gt_s + Gt_p\right]$$

**Center Drilling or Chamfering**

$$C = M\left[\frac{D(L+e)}{318f_r\,v} + \frac{R}{r} + t_i + \frac{u_c t_d}{T_h}\right] + \frac{u_c}{T_h}\left[\frac{C_p}{(K_1+1)} + Gt_s + Gt_p\right]$$

**Handling and Setup**

$$C = M\left[t_L + \frac{t_o}{N_L}\right]$$

| $\frac{\$}{Min}$ | Feeding Time | Rapid Traverse Time | Load & Unload Time | Setup Time | Cutter Index Time | Dull Tool Replacement Time | Tool Depreciation Cost | Tool Resharpening Cost | Rebrazing or Blade Reset Cost | Insert or Blade Cost | Grinding Wheel Cost | Tool Presetting Cost |
|---|---|---|---|---|---|---|---|---|---|---|---|---|

**TABLE 21–8 Equations for Calculating Operating Time Per Piece and Production Rate When Using English Units**

**EQUATIONS FOR OPERATING TIME PER PIECE**

**Turning**

$$t_m = \frac{D(L+e)}{3.82f_r v} + \frac{R}{r} + t_i + \frac{DLt_d}{3.82f_r vT}$$

**Milling**

$$t_m = \frac{D(L+e)}{3.82Zf_t v} + \frac{R}{r} + t_i + \frac{Lt_d}{ZT_t}$$

**Drilling and Reaming**

$$t_m = \frac{D(L+e)}{3.82f_r v} + \frac{R}{r} + t_i + \frac{Lt_d}{T_t}$$

**Tapping**

$$t_m = \frac{mD(L+e)}{1.91v} + \frac{R}{r} + t_i + \frac{Lt_d}{T_t}$$

**Center Drilling or Chambering**

$$t_m = \frac{D(L+e)}{3.82f_r v} + \frac{R}{r} + t_i + \frac{u_c t_d}{T_h}$$

**EQUATION FOR PRODUCTION RATE**

$$P = \frac{60}{\left(\Sigma t_m + t_L + \frac{t_o}{N_L}\right)}$$

**TABLE 21–9 Equations for Calculating Operating Time Per Piece and Production Rate When Using Metric Units**

**EQUATIONS FOR OPERATING TIME PER PIECE**

**Turning**

$$t_m = \frac{D(L+e)}{318f_r v} + \frac{R}{r} + t_i + \frac{DLt_d}{318f_r vT}$$

**Milling**

$$t_m = \frac{D(L+e)}{318Zf_t v} + \frac{R}{r} + t_i + \frac{Lt_d}{ZT_t}$$

**Drilling and Reaming**

$$t_m = \frac{D(L+e)}{318f_r v} + \frac{R}{r} + t_i + \frac{Lt_d}{T_t}$$

**Tapping**

$$t_m = \frac{D(L+e)}{159pv} + \frac{R}{r} + t_i + \frac{Lt_d}{T_t}$$

**Center Drilling or Chamfering**

$$t_m = \frac{D(L+e)}{318f_r v} + \frac{R}{r} + t_i + \frac{u_c t_d}{T_h}$$

**EQUATION FOR PRODUCTION RATE**

$$P = \frac{60}{\left(\Sigma t_m + t_L + \frac{t_o}{N_L}\right)}$$

**TABLE 21-10 Cost and Production Rate Equation Variables**
(Definitions and applications for variables in English units)

| VARIABLE | DEFINITION | APPLIES TO OPERATION | | | | |
|---|---|---|---|---|---|---|
| | | Turn | Mill | Drill and Ream | Tap | Center Drill |
| $C$ | Cost for machining one workpiece; \$/workpiece | ✓ | ✓ | ✓ | ✓ | ✓ |
| $C_c$ | Cost of each insert or inserted blade; \$/blade | ✓ | ✓ | No | No | No |
| $C_p$ | Purchase cost of tool or cutter; \$/cutter | ✓ | ✓ | ✓ | ✓ | ✓ |
| $C_w$ | Cost of grinding wheel for resharpening tool or cutter; \$/cutter | ✓ | ✓ | No | No | No |
| $d$ | Depth of cut; in | ✓ | ✓ | No | No | No |
| $D$ | Dia. of work in turning, of tool in milling, drilling, reaming, tapping; in | ✓ | ✓ | ✓ | ✓ | ✓ |
| $e$ | Extra travel at feedrate ($f_r$ or $f_t$) including approach, overtravel, and all positioning moves; in | ✓ | ✓ | ✓ | ✓ | ✓ |
| $f_r$ | Feed per revolution; in | ✓ | No | ✓ | No | ✓ |
| $f_t$ | Feed per tooth; in | No | ✓ | No | No | No |
| $G$ | Labor + overhead in tool reconditioning department; \$/min | ✓ | ✓ | ✓ | ✓ | ✓ |
| $k_l$ | No. of times lathe tool, or milling cutter, or drill, or reamer or tap is resharpened before being discarded | ✓ | ✓ | ✓ | ✓ | ✓ |
| $k_2$ | No. of times lathe tool or milling cutter is resharpened before inserts or blades are rebrazed or reset | ✓ | ✓ | No | No | No |
| $k_3$ | No. of times blades (or inserts) are resharpened (or indexed) before blades (or inserts) are discarded | ✓ | ✓ | No | No | No |
| $L$ | Length of workpiece in turning and milling or sum of length of all holes of same diameter in drilling, reaming, tapping; in | ✓ | ✓ | ✓ | ✓ | ✓ |
| $m$ | No. of threads per inch | No | No | No | ✓ | No |
| $M$ | Labor + overhead cost on lathe, milling machine or drilling machine; \$/min | ✓ | ✓ | ✓ | ✓ | ✓ |
| $n$ | Tool life exponent in Taylor's Equation | ✓ | ✓ | ✓ | ✓ | No |
| $N_L$ | No. of workpieces in lot | ✓ | ✓ | ✓ | ✓ | ✓ |
| $N_s$ | No. of pieces between sharpenings | ✓ | ✓ | ✓ | ✓ | ✓ |
| $P$ | Production rate per hour; workpieces/hour | ✓ | ✓ | ✓ | ✓ | ✓ |
| $r$ | Rapid traverse rate; in/min | ✓ | ✓ | ✓ | ✓ | ✓ |
| $R$ | Total rapid traverse distance for a tool or cutter on one part; in | ✓ | ✓ | ✓ | ✓ | ✓ |
| $S$ | Reference cutting speed for a tool life of $T = 1$ min; fpm | ✓ | No | No | No | No |
| $S_t$ | Reference cutting speed for a tool life of $T_t = 1$ inch; fpm | No | ✓ | ✓ | ✓ | No |
| $t_b$ | Time to rebraze lathe tool or cutter teeth or reset blades; min | ✓ | ✓ | No | No | No |
| $t_d$ | Time to replace dull cutter in tool changer storage unit; min | ✓ | ✓ | ✓ | ✓ | ✓ |
| $t_i$ | Time to index from one type cutter to another between operations (automatic or manual); min | ✓ | ✓ | ✓ | ✓ | ✓ |
| $t_L$ | Time to load and unload workpiece; min | ✓ | ✓ | ✓ | ✓ | ✓ |
| $t_m$ | Time (average) to complete one operation; min | ✓ | ✓ | ✓ | ✓ | ✓ |
| $t_o$ | Time to setup machine tool for operation; min | ✓ | ✓ | ✓ | ✓ | ✓ |
| $t_p$ | Time to preset tools away from machine (in toolroom); min | ✓ | ✓ | ✓ | ✓ | ✓ |
| $t_s$ | Time to resharpen lathe tool, milling cutter, drill, reamer or tap; min/tool | ✓ | ✓ | ✓ | ✓ | ✓ |
| $T$ | Tool life measured in minutes to dull a lathe tool; min | ✓ | No | No | No | No |
| $T_h$ | No. of holes per resharpening | No | No | No | No | ✓ |
| $T_t$ | Tool life measured in inches travel of work or tool to dull a drill, reamer, tap or one milling cutter tooth; in | No | ✓ | ✓ | ✓ | No |
| $u_c$ | No. of holes center drilled or chamfered in workpiece | No | No | No | No | ✓ |
| $v$ | Cutting speed; fpm | ✓ | ✓ | ✓ | ✓ | ✓ |
| $w$ | Width of cut; in | No | ✓ | No | No | No |
| $Z$ | No. of teeth in milling cutter or no. of flutes in a tap | No | ✓ | No | ✓ | No |

**TABLE 21–11 Cost and Production Rate Equation Variables**
(Definitions and applications for variables in metric units)

| VARIABLE | DEFINITION | APPLIES TO OPERATION | | | | |
|---|---|---|---|---|---|---|
| | | Turn | Mill | Drill and Ream | Tap | Center Drill |
| $C$ | Cost for machining one workpiece; \$/workpiece | ✓ | ✓ | ✓ | ✓ | ✓ |
| $C_c$ | Cost of each insert or inserted blade; \$/blade | ✓ | ✓ | No | No | No |
| $C_p$ | Purchase cost of tool or cutter; \$/cutter | ✓ | ✓ | ✓ | ✓ | ✓ |
| $C_w$ | Cost of grinding wheel for resharpening tool or cutter; \$/cutter | ✓ | ✓ | No | No | No |
| $d$ | Depth of cut; mm | ✓ | ✓ | No | No | No |
| $D$ | Dia. of work in turning, of tool in milling, drilling, reaming, tapping; mm | ✓ | ✓ | ✓ | ✓ | ✓ |
| $e$ | Extra travel at feedrate ($f_r$ or $f_t$) including approach, overtravel, and all positioning moves; mm | ✓ | ✓ | ✓ | ✓ | ✓ |
| $f_r$ | Feed per revolution; mm | ✓ | No | ✓ | No | ✓ |
| $f_t$ | Feed per tooth; mm | No | ✓ | No | No | No |
| $G$ | Labor + overhead in tool reconditioning department; \$/min | ✓ | ✓ | ✓ | ✓ | ✓ |
| $k_1$ | No. of times lathe tool, or milling cutter, or drill, or reamer or tap is resharpened before being discarded | ✓ | ✓ | ✓ | ✓ | ✓ |
| $k_2$ | No. of times lathe tool or milling cutter is resharpened before inserts or blades are rebrazed or reset | ✓ | ✓ | No | No | No |
| $k_3$ | No. of times blades (or inserts) are resharpened (or indexed) before blades (or inserts) are discarded | ✓ | ✓ | No | No | No |
| $L$ | Length of workpiece in turning and milling or sum of length of all holes of same diameter in drilling, reaming, tapping; mm | ✓ | ✓ | ✓ | ✓ | ✓ |
| $M$ | Labor + overhead cost on lathe, milling machine or drilling machine; \$/min | ✓ | ✓ | ✓ | ✓ | ✓ |
| $n$ | Tool life exponent in Taylor's Equation | ✓ | ✓ | ✓ | ✓ | No |
| $N_L$ | No. of workpieces in lot | ✓ | ✓ | ✓ | ✓ | ✓ |
| $N_s$ | No. of pieces between sharpenings | ✓ | ✓ | ✓ | ✓ | ✓ |
| $p$ | Pitch of thread, mm | No | No | No | ✓ | No |
| $P$ | Production rate per hour; workpieces/hour | ✓ | ✓ | ✓ | ✓ | ✓ |
| $r$ | Rapid traverse rate; mm/min | ✓ | ✓ | ✓ | ✓ | ✓ |
| $R$ | Total rapid traverse distance for a tool or cutter on one part; mm | ✓ | ✓ | ✓ | ✓ | ✓ |
| $S$ | Reference cutting speed for a tool life of $T = 1$ min; m/min | ✓ | No | No | No | No |
| $S_t$ | Reference cutting speed for a tool life of $T_t = 1$ mm; m/min | No | ✓ | ✓ | ✓ | No |
| $t_b$ | Time to rebraze lathe tool or cutter teeth or reset blades; min | ✓ | ✓ | No | No | No |
| $t_d$ | Time to replace dull cutter in tool changer storage unit; min | ✓ | ✓ | ✓ | ✓ | ✓ |
| $t_i$ | Time to index from one type cutter to another between operations (automatic or manual); min | ✓ | ✓ | ✓ | ✓ | ✓ |
| $t_L$ | Time to load and unload workpiece; min | ✓ | ✓ | ✓ | ✓ | ✓ |
| $t_m$ | Time (average) to complete one operation; min | ✓ | ✓ | ✓ | ✓ | ✓ |
| $t_o$ | Time to setup machine tool for operation; min | ✓ | ✓ | ✓ | ✓ | ✓ |
| $t_p$ | Time to preset tools away from machine (in toolroom); min | ✓ | ✓ | ✓ | ✓ | ✓ |
| $t_s$ | Time to resharpen lathe tool, milling cutter, drill, reamer or tap; min/tool | ✓ | ✓ | ✓ | ✓ | ✓ |
| $T$ | Tool life measured in minutes to dull a lathe tool; min | ✓ | No | No | No | No |
| $T_h$ | No. of holes per resharpening | No | No | No | No | ✓ |
| $T_t$ | Tool life measured in mm travel of work or tool to dull a drill, reamer, tap or one milling cutter tooth; mm | No | ✓ | ✓ | ✓ | No |
| $u_c$ | No. of holes center drilled or chamfered in workpiece | No | No | No | No | ✓ |
| $v$ | Cutting speed; m/min | ✓ | ✓ | ✓ | ✓ | ✓ |
| $w$ | Width of cut; mm | No | ✓ | No | No | No |
| $Z$ | No. of teeth in milling cutter or no. of flutes in a tap | No | ✓ | No | ✓ | No |

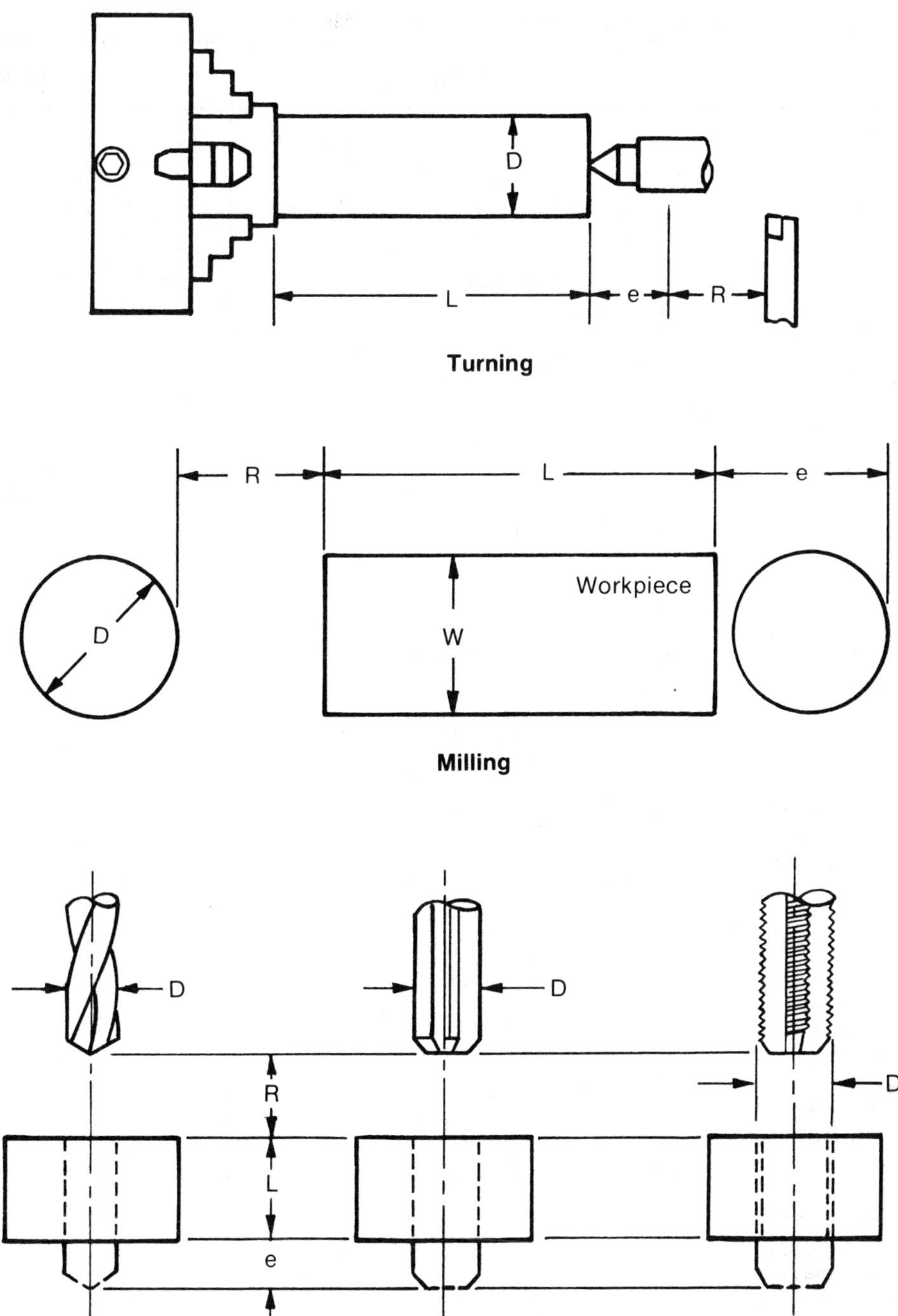


**Figure 21–5** Setups for various machining operations.

**TABLE 21-12 Derivation of Terms in Cost Equations for Turning and Milling**

| TERMS | TURNING | MILLING |
|---|---|---|
| rpm to Cut speed | $v = \frac{\pi D \times rpm}{12}$; $rpm = \frac{3.82 v}{D}$ | $rpm = \frac{3.82 v}{D}$ |
| Feed rate, in/min | $\frac{in}{rev} \times \frac{rev}{min} = f_r \times \frac{3.82 v}{D}$ | $\frac{in}{tooth} \times \text{no. teeth} \times \frac{rev}{min} = f_t \times Z \times \frac{3.82 v}{D}$ |
| $N_s$ = No. pieces between sharpenings | $\frac{\text{Tool life, min}}{\text{Feed time, min}} = \frac{3.82 f_r v T}{DL}$ | $\frac{\text{Tool life (Total length cut, in)}}{\text{Length each piece, in}} = \frac{Z T_t}{L}$ |
| Feed time = $\frac{\text{Distance in feed}}{\text{in/min}}$ | $\frac{L+e}{\frac{3.82 f_r v}{D}} = \frac{D(L+e)}{3.82 f_r v}$ | $\frac{(e + L) D}{3.82 Z f_t v}$ |
| Rapid traverse time | $\frac{R}{r}$ | $\frac{R}{r}$ |
| Setup time = $\frac{\text{Time to set up}}{\text{No. pieces in lot}}$ | $\frac{t_o}{N_L}$ | $\frac{t_o}{N_L}$ |
| Tool change time = $\frac{\text{Time to change tool}}{N_s = \text{No. pieces between sharpenings}}$ | $\frac{DL t_d}{3.82 f_r v T}$ | $\frac{L t_d}{Z T_t}$ |
| Tool deprec. cost = $\frac{1}{N_s} \times \frac{\text{Purchased cost of cutter}}{\text{No. times cutter is resharpened before discarding}}$ | $\frac{1}{N_s} \times \frac{C_p}{(k_1 + 1)}$ | $\frac{1}{N_s} \times \frac{C_p}{(k_1 + 1)}$ |
| Tool resharp. cost = $\frac{1}{N_s} \times G \times$ Time to resharpen cutter | $\frac{1}{N_s} \times G t_s$ | $\frac{1}{N_s} \times G t_s$ |
| Rebraze (or blade reset) cost = $\frac{1}{N_s} \times \frac{\text{Time to rebraze}}{\text{No. times tool is resharpened before blades are rebrazed}}$ | $\frac{1}{N_s} \times G \frac{t_b}{k_2}$ | $\frac{1}{N_s} \times G \frac{t_b}{k_2}$ |
| Insert or blade cost = $\frac{1}{N_s} \times \frac{\text{Cost of each insert or blade}}{\text{No. times blades are resharpened before blades are discarded}}$ | $\frac{1}{N_s} \times \frac{C_c}{k_3}$ | $\frac{1}{N_s} \times \frac{Z C_c}{k_3}$ |
| Grinding wheel cost = $\frac{1}{N_s} \times$ Cost of grinding wheel for resharpening tool or cutter | $\frac{1}{N_s} \times C_w$ | $\frac{1}{N_s} \times C_w$ |

**TABLE 21–13 Time Study and Cost Data for Turning 4340 Steel Shaft**

| VARIABLE | DEFINITION | BRAZED CARBIDE TOOL | THROWAWAY CARBIDE TOOL | SOLID HSS TOOL |
|---|---|---|---|---|
| $R$ | Total rapid traverse distance for a tool on one part, in | 27.2 | 27.2 | 27.2 |
| $C_c$ | Cost of each carbide tip or insert, $ | 5.00 | 3.15 | — |
| $C_p$ | Purchase cost of tool, $ | 6.70 | 28.30 | 18.30 |
| $C_w$ | Cost of grinding wheel for resharpening tool, $ | 0.07 | — | 0.02 |
| $d$ | Depth of cut, in | 0.1 | 0.1 | 0.1 |
| $D$ | Diameter of work in turning, in | 3.5 | 3.5 | 3.5 |
| $f_r$ | Feed per revolution, in | * | * | * |
| $G$ | Labor and overhead cost on tool grinder, $/min | 0.40 | — | 0.40 |
| $k_1$ | No. of times lathe tool is resharpened before discarding (or no. times insert is indexed before throwaway holder is discarded | 12 | 2000 | 36 |
| $k_2$ | No. of times lathe tool is resharpened before rebrazing or resetting | 6 | — | — |
| $k_3$ | No. of times insert is resharpened (or indexed) before insert is discarded | 12 | 8 | — |
| $L$ | Length of workpiece in turning, in | 19 | 19 | 19 |
| $M$ | Labor and overhead cost on lathe, $/min | 0.15 | 0.15 | 0.15 |
| $N_L$ | No. of pieces in lot | 20 | 20 | 20 |
| $r$ | Rapid traverse rate, in/min | 100 | 100 | 100 |
| $t_b$ | Time to rebraze lathe tool, min | 10 | — | — |
| $t_d$ | Time to change and reset tool or time to index throwaway insert, min | 5 | 4 | 5 |
| $t_L$ | Time to load and unload workpiece, min | 2.3 | 2.3 | 2.3 |
| $t_o$ | Time to set up lathe for operation, min | 21 | 21 | 21 |
| $t_s$ | Time to resharpen tool, min | 15 | — | 10 |
| $T$ | Tool life, total time to dull tool, min | * | * | * |
| $v$ | Cutting speed, fpm | * | * | * |
| $e$ | Extra travel of tool in feed (includes approach and overtravel in feed) | 0.5 | 0.5 | 0.5 |

*These values are taken from Tool Life Data, table 21–14.

**TABLE 21–14 Tool Life Data for Turning 4340 Steel**

| MATERIAL | CONDITION AND MICROSTRUCTURE | HARDNESS Bhn | TOOL MATERIAL | | TOOL GEOMETRY | | | | | | CUTTING FLUID CODE * | DEPTH OF CUT in | FEED ipr | TOOL LIFE END POINT in | TOOL LIFE, min vs CUTTING SPEED, fpm | | | |
|---|---|---|---|---|---|---|---|---|---|---|---|---|---|---|---|---|---|---|
| | | | Trade Name | Industry Grade | BR° | SR° | SCEA° | ECEA° | Relief° | Nose Radius in | | | | | | | | |
| 4340 | Quenched and tempered | 300 | 78 | C-7 | 0 | 6 | 0 | 6 | 6 | 0.040 | 00 | 0.100 | 0.010 | 0.015 | ① 15 | ② 30 | ③ 45 | ④ 60 |
| | Tempered martensite | | | | | | | | | | | | | | 470 | 400 | 360 | 325 |
| 4340 | Quenched and tempered | 300 | — | T1 HSS | 0 | 15 | 0 | 5 | 5 | 0.005 | 11 1:20 | 0.060 | 0.010 | 0.060 | ⑤ 15 | ⑥ 30 | ⑦ 45 | ⑧ 60 |
| | Tempered martensite | | | | | | | | | | | | | | 77 | 63 | 54 | 45 |

*Cutting Fluid Code
00 Dry
11 Soluble Oil

COST AND PRODUCTION RATE FOR TURNING

BRAZED CARBIDE TOOLS

| DATA* SET* NO.* | WORK MATERIAL | *HARD* NESS* | *TOOL* MATL* | *CUT* SPD* F/M* | FEED IN/REV* | *TOOL* LIFE* MIN* | *FEED* COST* $ | *RAPD* TRAV* $ | *LOAD* UNLD* $ | SET* UP $ | *TOOL* CHNG* $ | *TOOL* DEPR* $ | *TOOL* SHPN* $ | RE* BRAZ* $ | TIP* COST* $ | *GRIND* WHEEL* $ | **TOTAL **COST **$/PC | **PROD **RATE **PC/HR |
|---|---|---|---|---|---|---|---|---|---|---|---|---|---|---|---|---|---|---|
| 1 | AISI 4340 | 300 | C-7 | 470 | 0.0100 | 15 | 1.50 | 0.11 | 0.92 | 0.42 | 0.49 | 0.13 | 1.48 | 0.16 | 0.10 | 0.02 | 5.33 | 7.0 |
| 2 | AISI 4340 | 300 | C-7 | 400 | 0.0100 | 30 | 1.76 | 0.11 | 0.92 | 0.42 | 0.29 | 0.07 | 0.87 | 0.10 | 0.06 | 0.01 | 4.61 | 6.9 |
| 3 | AISI 4340 | 300 | C-7 | 360 | 0.0100 | 45 | 1.95 | 0.11 | 0.92 | 0.42 | 0.21 | 0.06 | 0.64 | 0.07 | 0.04 | 0.01 | 4.44 | 6.6 |
| 4 | AISI 4340 | 300 | C-7 | 325 | 0.0100 | 60 | 2.17 | 0.11 | 0.92 | 0.42 | 0.18 | 0.05 | 0.54 | 0.06 | 0.04 | 0.01 | 4.48 | 6.3 |

COST AND PRODUCTION RATE FOR TURNING

THROWAWAY CARBIDE TOOLS

| DATA* SET* NO.* | WORK MATERIAL | *HARD* NESS* | *TOOL* MATL* | *CUT* SPD* F/M* | FEED IN/REV* | *TOOL* LIFE* MIN* | *FEED* COST* $ | *RAPD* TRAV* $ | *LOAD* UNLD* $ | SET* UP $ | *INDX* INST* $ | *HLDR* DEPR* $ | *INSERT* COST* $ | **TOTAL **COST **$/PC | **PROD **RATE **PC/HR |
|---|---|---|---|---|---|---|---|---|---|---|---|---|---|---|---|
| 1 | AISI 4340 | 300 | C-7 | 470 | 0.0100 | 15 | 1.50 | 0.11 | 0.92 | 0.42 | 0.04 | 0.00 | 0.10 | 3.09 | 8.0 |
| 2 | AISI 4340 | 300 | C-7 | 400 | 0.0100 | 30 | 1.76 | 0.11 | 0.92 | 0.42 | 0.02 | 0.00 | 0.06 | 3.29 | 7.4 |
| 3 | AISI 4340 | 300 | C-7 | 360 | 0.0100 | 45 | 1.95 | 0.11 | 0.92 | 0.42 | 0.02 | 0.00 | 0.04 | 3.46 | 7.0 |
| 4 | AISI 4340 | 300 | C-7 | 325 | 0.0100 | 60 | 2.17 | 0.11 | 0.92 | 0.42 | 0.01 | 0.00 | 0.04 | 3.66 | 6.6 |

COST AND PRODUCTION RATE FOR TURNING

SOLID HIGH SPEED STEEL TOOLS

| DATA* SET* NO.* | WORK MATERIAL | *HARD* NESS* | *TOOL* MATL* | *CUT* SPD* F/M* | FEED IN/REV* | *TOOL* LIFE* MIN * | *FEED* COST* $ | *RAPD* TRAV* $ | *LOAD* UNLD* $ | SET* UP $ | *TOOL* CHNG* $ | *TOOL* DEPR* $ | *TOOL* SHPN* $ | *GRIND* WHEEL* $ | **TOTAL **COST **$/PC | **PROD **RATE **PC/HR |
|---|---|---|---|---|---|---|---|---|---|---|---|---|---|---|---|---|
| 5 | AISI 4340 | 300 | T-1 | 77 | 0.0100 | 15 | 9.14 | 0.11 | 0.92 | 0.42 | 3.01 | 0.75 | 6.03 | 0.03 | 20.41 | 1.8 |
| 6 | AISI 4340 | 300 | T-1 | 63 | 0.0100 | 30 | 11.17 | 0.11 | 0.92 | 0.42 | 1.84 | 0.46 | 3.68 | 0.02 | 18.62 | 1.7 |
| 7 | AISI 4340 | 300 | T-1 | 54 | 0.0100 | 45 | 13.03 | 0.11 | 0.92 | 0.42 | 1.43 | 0.35 | 2.87 | 0.01 | 19.15 | 1.5 |
| 8 | AISI 4340 | 300 | T-1 | 45 | 0.0100 | 60 | 15.64 | 0.11 | 0.92 | 0.42 | 1.29 | 0.32 | 2.58 | 0.01 | 21.29 | 1.3 |

**Figure 21-6** Printout of cost and production rate results for turning 4340 steel with three tool materials.

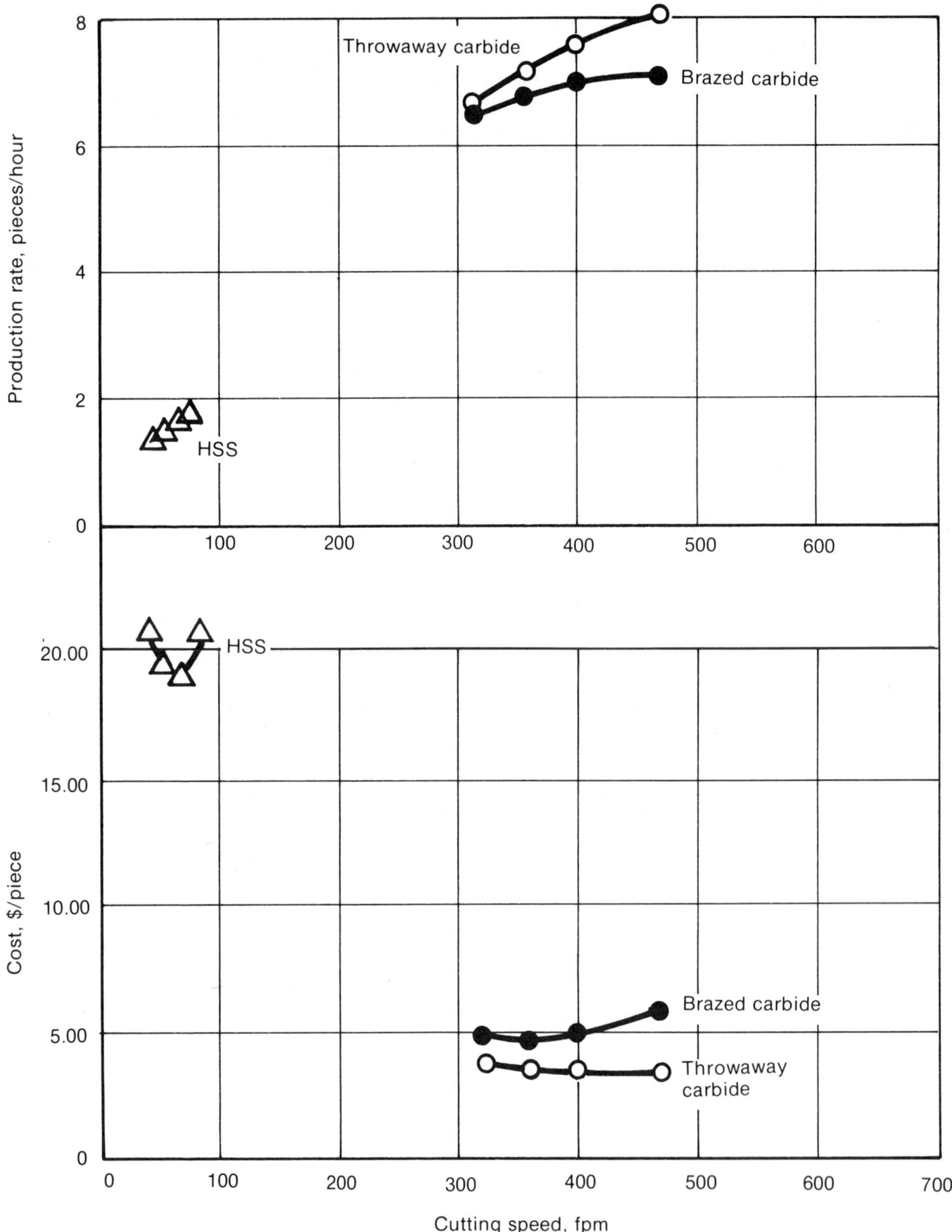


**Figure 21–7** Cutting speed versus cost and production rate for turning 4340 steel (quenched and tempered, 300 Bhn) with three tool materials.

**TABLE 21–15 Manual Cost Calculations for Turning 4340 Steel Shaft with Brazed Carbide Tool**

**Operation** Turn, shaft, 3.5-inch diameter by 19 inches long

**Material:** 4340 Steel, Q&T, 300 Bhn

**References:** Table 21-13, Brazed carbide tools
Table 21-14, Data set 1
Figure 21-6, Brazed carbide tools, data set 1

$$\text{Feed cost} = M \times \frac{DL}{3.82\, f_r\, v} = \frac{0.40 \times 3.5 \times 19}{3.82 \times 0.010 \times 470} = \$1.48$$

$$\text{Rapid traverse cost} = M \times \frac{2a + L}{r} = 0.40 \times \frac{2 \times 4 + 19}{100} = \$0.11$$

$$\text{Load \& unload cost} = M \times t_L = 0.40 \times 2.3 = \$0.92$$

$$\text{Setup cost} = M \times \frac{t_o}{N_L} = 0.40 \times \frac{21}{20} = \$0.43$$

$$\text{Tool change cost} = M \times \frac{DL\, t_c}{3.82\, f_r\, v\, T} = \frac{0.40 \times 3.5 \times 19 \times 5}{3.82 \times 0.010 \times 470 \times 15} = \$0.49$$

$$\text{Tool depreciation cost} = \frac{1}{N_s} \times \frac{C_p}{k_1 + 1} = \frac{DL}{3.82\, f_r\, vt} \times \frac{C_p}{k_1 + 1}$$

$$= \frac{3.5 \times 19}{3.82 \times 0.010 \times 470 \times 15} \times \frac{6.70}{12 + 1} = \frac{1}{4.04} \times \frac{6.70}{13} = \$0.13$$

$$\text{Tool resharpening cost} = \frac{1}{N_s} \times 0.40 \times t_s = \frac{1}{4.04} \times 0.40 \times 15 = \$1.48$$

**TABLE 21-16 Time Study and Cost Data for Face Milling, 2 Inches Wide by 8 Inches Long**
(Material: 4340 steel, quenched and tempered, 341 Bhn)

| VARIABLE | DEFINITION | INSERTED TOOTH CARBIDE TIP | THROWAWAY INSERT | SOLID HSS CUTTER |
|---|---|---|---|---|
| a | Approach of cutter to work, in | 9.0 | 9.0 | 9.0 |
| $C_c$ | Cost of each inserted tooth, throwaway insert, or carbide tip, $ | 2.50 | 2.35 | — |
| $C_p$ | Purchase cost of cutter, $ | 137.00 | 248.00 | 310.00 |
| $C_w$ | Cost of grinding wheel for resharpening cutter, $/cutter | 0.30 | — | 0.35 |
| d | Depth of cut, in | * | * | * |
| D | Diameter of milling cutter, in | 4.0 | 4.0 | 4.0 |
| e | Overtravel of milling cutter past workpiece, in. | 5.0 | 5.0 | 5.0 |
| $f_t$ | Feed per tooth, in | * | * | * |
| G | Labor and overhead cost on cutter grinder, $/min | 0.40 | — | 0.40 |
| $k_1$ | No. of times cutter is resharpened before being discarded | 9000 | 9000 | 20 |
| $k_2$ | No. of times cutter is resharpened before inserts (or blades) are reset (or rebrazed) | 4 | — | — |
| $k_3$ | No. of times blades (or inserts) are resharpened (or indexed) before blades (or inserts) are discarded | 12 | 8 | — |
| L | Length of workpiece, in | 8.0 | 8.0 | 8.0 |
| M | Labor and overhead cost on milling machine, $/min | 0.40 | 0.40 | 0.40 |
| N | No. of workpieces in lot | 100 | 100 | 100 |
| r | Rapid traverse rate, in/min | 150 | 150 | 150 |
| $t_b$ | Time to reset blades or to rebraze cutter teeth, min | 30 | — | — |
| $t_c$ | Time to change cutter or index all inserts in cutter, min | 10.0 | 6.0 | 10.0 |
| $t_L$ | Time to load and unload workpiece, min | 3 | 3 | 3 |
| $t_o$ | Time to set up milling machine for operation, min | 60 | 60 | 60 |
| $t_s$ | Time to resharpen cutter, min/cutter | 80 | — | 80 |
| $T_t$ | Tool life measured in inches travel of work to dull one cutter tooth, in | * | * | * |
| v | Cutting speed, fpm | * | * | * |
| w | Width of cut, in | 2.0 | 2.0 | 2.0 |
| Z | No. of teeth in milling cutter | 6 | 6 | 14 |

*These values are taken from Tool Life Data, table 21-17.

TABLE 21-17 Tool Life Data for Face Milling 4340 Steel

| MATERIAL | CONDITION AND MICROSTRUCTURE | HARD-NESS Bhn | TOOL MATL. Trade Name | TOOL MATL. Industry Grade | UP OR DOWN MILL-ING | TOOL GEOMETRY AR° | RR° | CA° | TR° | Incl° | ECEA° | End Rel.° / Cor. Rel.° | CUT-TING FLUID CODE * | DEPTH OF CUT in | WIDTH OF CUT in | FEED PER TOOTH in | TOOL LIFE END POINT in | TOOL LIFE/TOOTH inches work travel vs CUTTING SPEED, fpm | | |
|---|---|---|---|---|---|---|---|---|---|---|---|---|---|---|---|---|---|---|---|---|
| 4340 | Quenched & tempered | 341 | — | T15 HSS | Up | 0 | 0 | 30 | 0 | 0 | 5 | 6 | 11 1:20 | 0.100 | 2.0 | 0.010 | 0.060 | ① 17 | ② 22 | ③ 32 |
| | Tempered martensite | | | | | | | | | | | 6 | | | | | | 93 | 76 | 62 |
| 4340 | Quenched & tempered | 341 | 370 | C-6 | Up | 0 | −7 | 45 | −5 | 5 | 5 | 6 | 00 | 0.100 | 2.0 | 0.005 | 0.016 | ④ 20 | ⑤ 50 | ⑥ 80 |
| | Tempered martensite | | | | | | | | | | | 6 | | | | | | 680 | 550 | 340 |

*Cutting Fluid Code
00 Dry
11 Soluble Oil

COST AND PRODUCTION RATE FOR MILLING

INSERTED TOOTH - CARBIDE TIP

| DATA SET NO. | WORK MATERIAL | HARDNESS | TOOL MATL | CUT SPD F/M | FEED/TOOTH IN | TOOL LIFE IN/TH | FEED COST $ | RAPD TRAV $ | LOAD UNLD $ | SET-UP $ | CUTR CHNG $ | BODY DEPR $ | CUTR SHPN $ | BLAD RSET $ | BLAD COST $ | GRND WHL. $ | TOTAL COST $/PC. | PROD RATE PC/HR |
|---|---|---|---|---|---|---|---|---|---|---|---|---|---|---|---|---|---|---|
| 4 | AISI 4340 | 341 | C-6 | 680 | 0.005 | 20.0 | 0.27 | 0.08 | 1.20 | 0.24 | 0.27 | 0.00 | 2.13 | 0.20 | 0.08 | 0.02 | 4.49 | 11.7 |
| 5 | AISI 4340 | 341 | C-6 | 550 | 0.005 | 50.0 | 0.33 | 0.08 | 1.20 | 0.24 | 0.11 | 0.00 | 0.85 | 0.08 | 0.03 | 0.01 | 2.93 | 12.2 |
| 6 | AISI 4340 | 341 | C-6 | 340 | 0.005 | 80.0 | 0.53 | 0.08 | 1.20 | 0.24 | 0.07 | 0.00 | 0.53 | 0.05 | 0.02 | 0.00 | 2.73 | 11.3 |

COST AND PRODUCTION RATE FOR MILLING

THROWAWAY INSERT

| DATA SET NO. | WORK MATERIAL | HARDNESS | TOOL MATL | CUT SPD F/M | FEED/TOOTH IN | TOOL LIFE IN/TH | FEED COST $ | RAPD TRAV $ | LOAD UNLD $ | SET-UP $ | INDX INST $ | BODY DEPR $ | INSERT COST $ | TOTAL COST $/PC. | PROD RATE PC/HR |
|---|---|---|---|---|---|---|---|---|---|---|---|---|---|---|---|
| 4 | AISI 4340 | 341 | C-6 | 680 | 0.005 | 20.0 | 0.27 | 0.08 | 1.20 | 0.24 | 0.16 | 0.00 | 0.12 | 2.07 | 12.3 |
| 5 | AISI 4340 | 341 | C-6 | 550 | 0.005 | 50.0 | 0.33 | 0.08 | 1.20 | 0.24 | 0.06 | 0.00 | 0.05 | 1.96 | 12.5 |
| 6 | AISI 4340 | 341 | C-6 | 340 | 0.005 | 80.0 | 0.53 | 0.08 | 1.20 | 0.24 | 0.04 | 0.00 | 0.03 | 2.13 | 11.4 |

COST AND PRODUCTION RATE FOR MILLING

SOLID HIGH SPEED STEEL CUTTER

| DATA SET NO. | WORK MATERIAL | HARDNESS | TOOL MATL | CUT SPD F/M | FEED/TOOTH IN | TOOL LIFE IN/TH | FEED COST $ | RAPD TRAV $ | LOAD UNLD $ | SET-UP $ | CUTR CHNG $ | CUTR DEPR $ | CUTR SHPN $ | GRND WHL. $ | TOTAL COST $/PC. | PROD RATE PC/HR |
|---|---|---|---|---|---|---|---|---|---|---|---|---|---|---|---|---|
| 1 | AISI 4340 | 341 | T15 | 93 | 0.010 | 17.0 | 0.42 | 0.08 | 1.20 | 0.24 | 0.13 | 0.50 | 1.08 | 0.01 | 3.66 | 11.6 |
| 2 | AISI 4340 | 341 | T15 | 76 | 0.010 | 22.0 | 0.51 | 0.08 | 1.20 | 0.24 | 0.10 | 0.38 | 0.83 | 0.01 | 3.36 | 11.2 |
| 3 | AISI 4340 | 341 | T15 | 62 | 0.010 | 32.0 | 0.63 | 0.08 | 1.20 | 0.24 | 0.07 | 0.26 | 0.57 | 0.01 | 3.06 | 10.8 |

**Figure 21–8** Printout of cost and production rate results for face milling 4340 steel with three cutter materials.

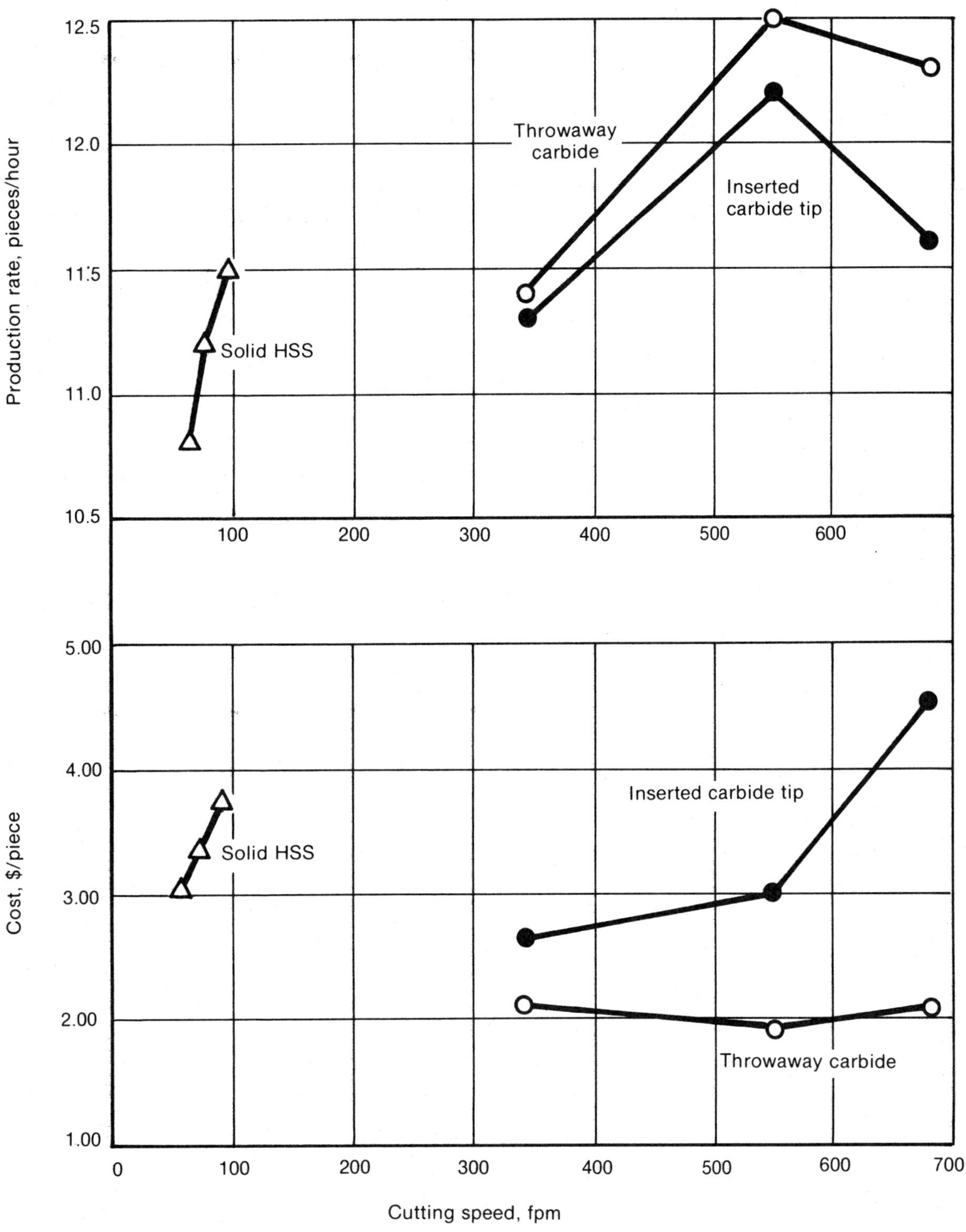


**Figure 21–9** Cutting speed versus cost and production rate for face milling 4340 steel (quenched and tempered, 341 Bhn) with three tool materials.

**TABLE 21–18 Time Study and Cost Data for Drilling Five Holes, 0.25-Inch Diameter by 0.5-Inch Deep**
(Material: 4340 steel, annealed, 212 Bhn)

| VARIABLE | DEFINITION | DATA |
|---|---|---|
| $a$ | Approach of drill to work, in | 3.0 |
| $C_p$ | Purchase cost of drill, $/drill | 0.92 |
| $D$ | Diameter of drill, in | 0.25 |
| $f_r$ | Feed per revolution, in | * |
| $G$ | Labor and overhead cost on tool grinder, $/min | 0.40 |
| $k_1$ | No. of times drill is resharpened before being discarded | 12 |
| $L$ | Sum of lengths of all holes of same diameter, in | 2.5 |
| $M$ | Labor and overhead cost on drilling machine, $/min | 0.40 |
| $N_L$ | No. of workpieces in lot | 70 |
| $r$ | Rapid traverse rate, in/min | 100 |
| $t_c$ | Time to change drill, min | 0.5 |
| $t_L$ | Time to load and unload workpiece, min | 1.0 |
| $t_o$ | Time to set up drill for operation, min | 25 |
| $t_s$ | Time to resharpen drill, min/drill | 5 |
| $T_t$ | Tool life in inches travel of drill to dull drill, in | † |
| $u$ | No. of holes of same diameter in workpiece | 5 |
| $v$ | Cutting speed, fpm | * |
| $e$ | Extra travel in drilling, in | 2.5 |

*These values are taken from Tool Life Data, table 21–19.
†Drill life data given in table 21–19 are in number of holes to dull drill, and equation requires these data in inches. Therefore, to obtain $T_t$, the number of holes is multiplied by the hole length (0.5 inch).

**TABLE 21-19 Tool Life Data for Drilling Holes in 4340 Steel**

| MATERIAL | CONDITION AND MICROSTRUCTURE | HARDNESS Bhn | DRILL MATL. Trade Name | DRILL MATL. Industry Grade | TYPE DRILL | DRILL SIZE Dia. in | DRILL SIZE Length in | DRILL SIZE Flute Length in | DRILL GEOMETRY Type Point | DRILL GEOMETRY Helix Angle° | DRILL GEOMETRY Point Angle° | DRILL GEOMETRY Lip Relief° | CUTTING FLUID CODE * | LENGTH OF HOLE in | FEED ipr | DRILL LIFE END POINT in | DRILL LIFE no. of holes vs CUTTING SPEED, fpm | | | |
|---|---|---|---|---|---|---|---|---|---|---|---|---|---|---|---|---|---|---|---|---|
| 4340 | Annealed | 212 | — | M1 HSS | Twist | 0.250 | 2.5 | 1.375 | Standard | 29 | 118 | 7 | 11 1:20 | 0.5 thru | 0.002 | 0.015 | ① 30 | ② 50 | ③ 100 | ④ 220 |
| | Spheroidized carbides + ferrite | | | | | | | | | | | | | | | | 174 | 158 | 142 | 125 |
| ″ | ″ | ″ | — | ″ | ″ | ″ | ″ | ″ | ″ | ″ | ″ | ″ | ″ | ″ | 0.005 | ″ | ⑤ 10 | ⑥ 100 | ⑦ 200 | ⑧ 280 |
| | ″ | | | | | | | | | | | | | | | | 140 | 97 | 82 | 75 |

*Cutting Fluid Code
11 Soluble Oil

```
                         COST AND PRODUCTION RATE FOR DRILLING

DATA*    WORK     *HARD*TOOL*    *CUT * FEED *DRILL*    *FEED*RAPD*LOAD*SET-*DRLL*DRLL*DRLL*    **TOTAL**PROD*
SET *             *NESS*MATL*    *SPD *       *LIFE *    *COST*TRAV*UNLD* UP *CHNG*DEPR*SHPN*    **COST **RATE*
NO. * MATERIAL *    *    *       *F/M *IN/REV* IN  *    * $  * $  * $  * $  * $  * $  * $  *    ** $/PC**PC/HR

 1  AISI 4340   212  M-1     174  0.0020  15.0    0.38 0.04 0.40 0.14 0.03 0.01 0.33      1.34  24.1
 2  AISI 4340   212  M-1     158  0.0020  25.0    0.41 0.04 0.40 0.14 0.02 0.01 0.20      1.23  23.5
 3  AISI 4340   212  M-1     142  0.0020  50.0    0.46 0.04 0.40 0.14 0.01 0.00 0.10      1.16  22.7
 4  AISI 4340   212  M-1     125  0.0020 110.0    0.52 0.04 0.40 0.14 0.00 0.00 0.05      1.16  21.5
 5  AISI 4340   212  M-1     140  0.0050   5.0    0.19 0.04 0.40 0.14 0.10 0.04 1.00      1.91  27.5
 6  AISI 4340   212  M-1      97  0.0050  50.0    0.27 0.04 0.40 0.14 0.01 0.00 0.10      0.97  27.7
 7  AISI 4340   212  M-1      82  0.0050 100.0    0.32 0.04 0.40 0.14 0.00 0.00 0.05      0.96  26.3
 8  AISI 4340   212  M-1      75  0.0050 140.0    0.35 0.04 0.40 0.14 0.00 0.00 0.04      0.98  25.5
```

**Figure 21-10** Printout of cost and production rate results for drilling 4340 steel with M-1 high speed steel tool.

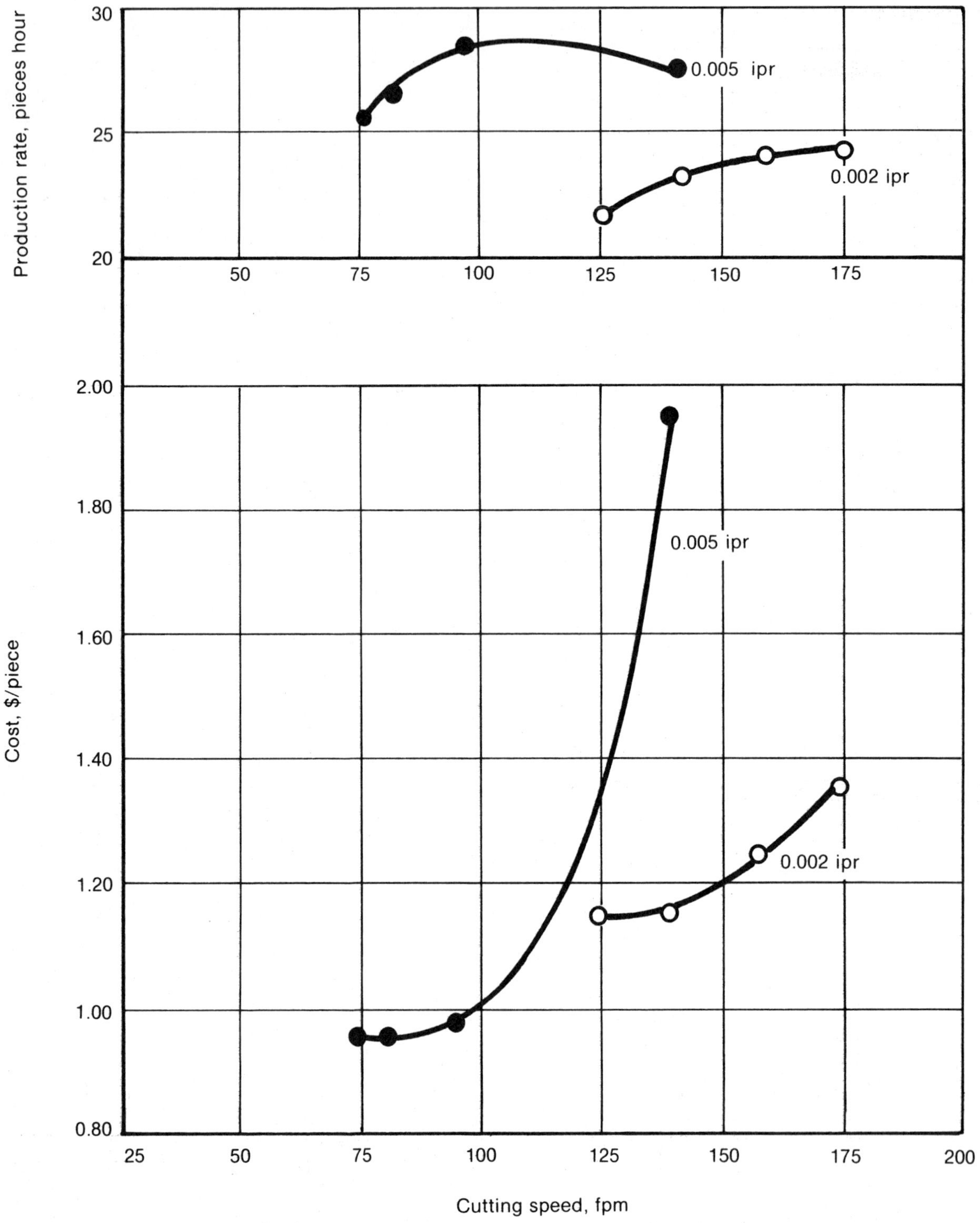


**Figure 21–11** Cutting speed versus cost and production rate for drilling holes in 4340 steel (annealed to 212 Bhn) using two feed rates.

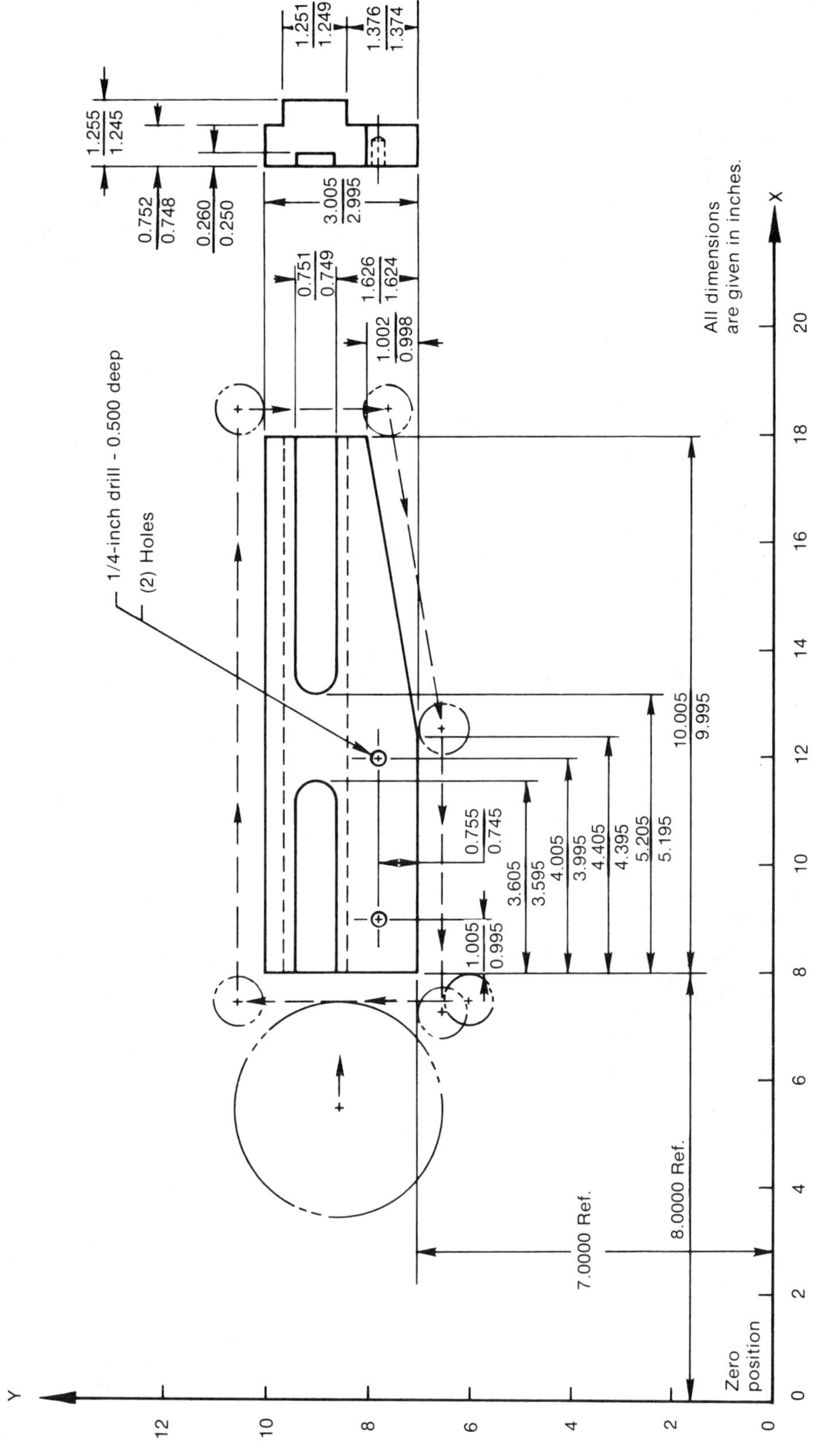


**Figure 21-12** Part machined on an NC machining center.

**TABLE 21–20 Time Study and Cost Data for Machining 4340 Steel Forging on an NC Machining Center**

| OPER. NO. | OPERATION | | |
|---|---|---|---|
| N 010 | Face mill top surface | Part No. | 456987-003 |
| N 020 | Peripheral end mill sides and ends | Part Name | Bracket—Slider |
| N 030 | End mill two slots in top surface | Material | AISI 4340 Steel Forg. |
| N 040 | Drill (2) 0.250-inch dia. holes | Condition | Normalized |
| | | Hardness | 331 Bhn |

| VARIABLE | DEFINITION | OPERATION NO. | | | |
|---|---|---|---|---|---|
| | | N 010 | N 020 | N 030 | N 040 |
| $C_c$ | Cost of each insert or inserted blade; $/blade | 3.40 | — | — | — |
| $C_p$ | Purchase cost of tool or cutter; $/cutter | 245.00 | 15.00 | 10.00 | 0.80 |
| $C_w$ | Cost of grinding wheel for resharpening tool or cutter; $/cutter | — | 0.10 | 0.10 | — |
| d | Depth of cut; in | 0.100 | 0.250 | 0.250 | — |
| D | Dia. of tool in milling and drilling; in | 4.000 | 1.000 | 0.750 | 0.250 |
| e | Extra travel at feedrate ($f_r$ or $f_t$) including approach, overtravel, and all positioning moves; in | 5.00 | 5.50 | 2.25 | 0.50 |
| $f_r$ | Feed per revolution; in | — | — | — | * |
| $f_t$ | Feed per tooth; in | * | * | * | — |
| G | Labor + overhead in tool reconditioning department; $/min | 0.40 | 0.40 | 0.40 | 0.40 |
| $k_1$ | No. of times milling cutter or drill is reused before being discarded | 9000 | 7 | 7 | 12 |
| $k_2$ | No. of times milling cutter is resharpened before inserts (or blades) are rebrazed (or reset) | — | — | — | — |
| $k_3$ | No. of times blades (or inserts) are resharpened (or indexed) before blade (or inserts) are discarded | 8 | — | — | — |
| L | Length of workpiece in milling or sum of length of all holes of same diameter in drilling; in | 10.00 | 26.25 | 8.40 | 1.00 |
| M | Labor + overhead cost of machining center | 0.83 | 0.83 | 0.83 | 0.83 |
| $N_L$ | No. of workpieces in lot | 20 | 20 | 20 | 20 |
| r | Rapid traverse rate; | 150 | 150 | 150 | 150 |
| R | Total rapid traverse distance for a cutter on one part; in | 43.00 | 27.25 | 46.35 | 41.00 |
| $t_b$ | Time to rebraze cutter teeth or reset blades; min | — | — | — | — |
| $t_d$ | Time to replace dull cutter in tool changer storage unit; min | 3.0 | 3.0 | 3.0 | 3.0 |
| $t_i$ | Time to index from one type cutter to another between operations (automatic or manual); min | 0.1 | 0.1 | 0.1 | 0.1 |
| $t_L$ | Time to load and unload workpiece; min | 2.0 | 2.0 | 2.0 | 2.0 |
| $t_o$ | Time to setup machine tool for operation; min | 60 | 60 | 60 | 60 |
| $t_p$ | Time to preset tools away from machine (in toolroom); min | 20 | 15 | 15 | 12 |
| $t_s$ | Time to resharpen lathe tool, milling cutter, drill, reamer or tap; min/tool | — | 15 | 15 | 5 |
| $T_t$ | Tool life measured in inches travel of work or tool to dull a drill, reamer, tap or one milling cutter tooth; in | * | * | * | * |
| V | Cutting speed; fpm | * | * | * | * |
| w | Width of cut; in | 3.50 | 0.75 | 0.75 | — |
| Z | No. of teeth in milling cutter or no. of flutes in a tap | 6 | 4 | 4 | — |

*These values taken from Tool Life Data, table 21–21.

**TABLE 21–21 Tool Life Data for Machining 4340 Steel on an NC Machining Center**

**FACE MILLING***

| MATERIAL | CONDITION AND MICROSTRUCTURE | HARDNESS Bhn | TOOL MATL. Trade Name | TOOL MATL. Industry Grade | UP OR DOWN MILLING | TOOL GEOMETRY AR° | RR° | CA° | TR° | Incl° | ECEA° | End Rel.° / Cor. Rel.° | CUTTING FLUID CODE † | DEPTH OF CUT in | WIDTH OF CUT in | FEED PER TOOTH in | TOOL LIFE END POINT in | TOOL LIFE/TOOTH inches work travel vs CUTTING SPEED, fpm | | | |
|---|---|---|---|---|---|---|---|---|---|---|---|---|---|---|---|---|---|---|---|---|---|
| Alloy steel 4340 | Normalized | 320 | 370 | C-6 | Up | 0 | −7 | 45 | −5 | 5 | 5 | 6 | 00 | 0.100 | 4.0 | 0.005 | 0.016 | ① 47 | ② 70 | ③ 83 | ④ 95 |
| | Acicular | | | | | | | | | | | 6 | | | | | | 670 | 540 | 445 | 345 |

**PERIPHERAL END MILLING**

| MATERIAL | CONDITION AND MICROSTRUCTURE | HARDNESS Bhn | TOOL MATL. Trade Name | TOOL MATL. Industry Grade | CUTTER TYPE | DIA. in | NO. TEETH | FLUTE LENGTH in | UP OR DOWN MILLING | TOOL GEOMETRY Helix Angle° | RR° | Chamfer | ECEA° | End Rel.° / Periph. Rel.° | CUTTING FLUID CODE † | DEPTH OF CUT in | WIDTH OF CUT in | FEED PER TOOTH in | TOOL LIFE END POINT in | TOOL LIFE/CUTTER inches work travel vs CUTTING SPEED, fpm | | | |
|---|---|---|---|---|---|---|---|---|---|---|---|---|---|---|---|---|---|---|---|---|---|---|---|
| Alloy steel 4340 | Normalized | 331 | — | M2 HSS | Solid HSS Double end | 0.750 | 4 | 2 | Down | 30 | 10 | 45° x 0.060 in | 1 | 3 | 11 1:20 Flood | 0.250 | 0.750 | 0.004 | 0.012 | ⑤ 50 | ⑥ 100 | ⑦ 140 | ⑧ 220 |
| | Acicular | | | | | | | | | | | | | 7 | | | | | | 155 | 100 | 80 | 70 |

**END MILL SLOTTING**

| MATERIAL | CONDITION AND MICROSTRUCTURE | HARDNESS Bhn | TOOL MATL. Trade Name | TOOL MATL. Industry Grade | CUTTER TYPE | DIA. in | NO. TEETH | FLUTE LENGTH in | UP OR DOWN MILLING | TOOL GEOMETRY Helix Angle° | RR° | Chamfer | ECEA° | End Rel.° / Periph. Rel.° | CUTTING FLUID CODE † | DEPTH OF CUT in | WIDTH OF CUT in | FEED PER TOOTH in | TOOL LIFE END POINT in | TOOL LIFE/CUTTER inches work travel vs CUTTING SPEED, fpm | | | |
|---|---|---|---|---|---|---|---|---|---|---|---|---|---|---|---|---|---|---|---|---|---|---|---|
| Alloy steel 4340 | Normalized | 331 | — | M2 HSS | Solid HSS Double end | 0.750 | 4 | 2 | — | 30 | 10 | 45° x 0.060 in | 1 | 3 | 11 1:20 | 0.250 | 0.750 | 0.002 | 0.012 | ⑨ 50 | ⑩ 110 | ⑪ 190 | |
| | Acicular | | | | | | | | | | | | | 7 | | | | | | 65 | 54 | 43 | |

**DRILLING**

| MATERIAL | CONDITION AND MICROSTRUCTURE | HARDNESS Bhn | DRILL MATL. Trade Name | DRILL MATL. Industry Grade | TYPE DRILL | DRILL SIZE Dia. in | Length in | Flute Length in | DRILL GEOMETRY Type Point | Helix Angle° | Point Angle° | Lip Relief° | CUTTING FLUID CODE † | DEPTH OF HOLE in | FEED ipr | DRILL LIFE END POINT in | DRILL LIFE no. of holes vs CUTTING SPEED, fpm | | | |
|---|---|---|---|---|---|---|---|---|---|---|---|---|---|---|---|---|---|---|---|---|
| Alloy steel 4340 | Normalized | 331 | — | M1 HSS | Twist | 0.250 | 2.5 | 1.375 | Standard | 29 | 118 | 7 | 11 1:20 | 0.5 thru | 0.002 | 0.015 | ⑫ 20 | ⑬ 50 | ⑭ 100 | ⑮ 160 |
| | Acicular | | | | | | | | | | | | | | | | 90 | 74 | 60 | 50 |
| " | " | " | — | " | " | " | " | " | " | " | " | " | " | " | 0.005 | " | ⑯ 10 | ⑰ 50 | ⑱ 100 | ⑲ 180 |
| | " | | | | | | | | | | | | | | | | 74 | 45 | 36 | 30 |

*Single tooth cutter except as noted.

† Cutting Fluid Code
00 Dry
11 Soluble Oil

N/C MACHINING COSTS AND OPERATION TIMES
FOR
PART NO.-456987-003 * PART NAME-BRACKET-SLIDER * LOT SIZE- 20
MATERIAL-AISI 4340 STEEL,NORMALIZED,ACICULAR * HARDNESS- 331

| DATA SET NO. | OPER. NO. | MACHG OPER. | TOOL MATL | SPEED SFM | FEED | TOOL LIFE | FEED COST $ | RAPD TRAV $ | LOAD UNLD $ | SET-UP $ | TOOL INDX $ | TOOL REPL $ | TOOL DEPR $ | TOOL SHPN $ | BLAD RSET $ | INST COST $ | GRND WHL $ | TOOL PRST $ | TOTAL COST $ | OPER. TIME MIN. |
|---|---|---|---|---|---|---|---|---|---|---|---|---|---|---|---|---|---|---|---|---|
| 30001 | N010 | FM-TA | C-6 | 670. | 0.005 | 47. | 0.65 | 0.24 | | | 0.08 | 0.09 | 0.00 | | | 0.09 | | 0.28 | 1.43 | 1.27 |
| 30002 | N010 | FM-TA | C-6 | 540. | 0.005 | 70. | 0.80 | 0.24 | | | 0.08 | 0.06 | 0.00 | | | 0.06 | | 0.19 | 1.44 | 1.43 |
| 30003 | N010 | FM-TA | C-6 | 445. | 0.005 | 83. | 0.98 | 0.24 | | | 0.08 | 0.05 | 0.00 | | | 0.05 | | 0.16 | 1.56 | 1.62 |
| 30004 | N010 | FM-TA | C-6 | 345. | 0.005 | 95. | 1.26 | 0.24 | | | 0.08 | 0.04 | 0.00 | | | 0.04 | | 0.14 | 1.81 | 1.96 |
| 30005 | N020 | EM-HSS | M2 | 155. | 0.004 | 50. | 2.78 | 0.15 | | | 0.08 | 0.33 | 0.25 | 0.79 | | | 0.01 | 0.79 | 5.18 | 4.03 |
| 30006 | N020 | EM-HSS | M2 | 100. | 0.004 | 100. | 4.31 | 0.15 | | | 0.08 | 0.16 | 0.12 | 0.39 | | | 0.01 | 0.39 | 5.63 | 5.67 |
| 30007 | N020 | EM-HSS | M2 | 80. | 0.004 | 140. | 5.39 | 0.15 | | | 0.08 | 0.12 | 0.09 | 0.28 | | | 0.00 | 0.28 | 6.40 | 6.92 |
| 30008 | N020 | EM-HSS | M2 | 70. | 0.004 | 220. | 6.16 | 0.15 | | | 0.08 | 0.07 | 0.06 | 0.18 | | | 0.00 | 0.18 | 6.88 | 7.79 |
| 30009 | N030 | EM-HSS | M2 | 65. | 0.002 | 50. | 3.34 | 0.26 | | | 0.08 | 0.10 | 0.05 | 0.25 | | | 0.00 | 0.25 | 4.34 | 4.56 |
| 30010 | N030 | EM-HSS | M2 | 54. | 0.002 | 110. | 4.02 | 0.26 | | | 0.08 | 0.05 | 0.02 | 0.11 | | | 0.00 | 0.11 | 4.66 | 5.31 |
| 30011 | N030 | EM-HSS | M2 | 43. | 0.002 | 190. | 5.05 | 0.26 | | | 0.08 | 0.03 | 0.01 | 0.07 | | | 0.00 | 0.07 | 5.56 | 6.52 |
| 30012 | N040 | DRILL | M1 | 90. | 0.002 | 10. | 0.45 | 0.23 | | | 0.08 | 0.25 | 0.01 | 0.20 | | | | 0.48 | 1.70 | 1.22 |
| 30013 | N040 | DRILL | M1 | 74. | 0.002 | 25. | 0.55 | 0.23 | | | 0.08 | 0.10 | 0.00 | 0.08 | | | | 0.19 | 1.23 | 1.16 |
| 30014 | N040 | DRILL | M1 | 60. | 0.002 | 50. | 0.68 | 0.23 | | | 0.08 | 0.05 | 0.00 | 0.04 | | | | 0.10 | 1.18 | 1.25 |
| 30015 | N040 | DRILL | M1 | 50. | 0.002 | 80. | 0.81 | 0.23 | | | 0.08 | 0.03 | 0.00 | 0.02 | | | | 0.06 | 1.24 | 1.39 |
| 30016 | N040 | DRILL | M1 | 74. | 0.005 | 5. | 0.22 | 0.23 | | | 0.08 | 0.50 | 0.01 | 0.40 | | | | 0.96 | 2.40 | 1.24 |
| 30017 | N040 | DRILL | M1 | 45. | 0.005 | 25. | 0.36 | 0.23 | | | 0.08 | 0.10 | 0.00 | 0.08 | | | | 0.19 | 1.05 | 0.93 |
| 30018 | N040 | DRILL | M1 | 36. | 0.005 | 50. | 0.45 | 0.23 | | | 0.08 | 0.05 | 0.00 | 0.04 | | | | 0.10 | 0.95 | 0.98 |
| 30019 | N040 | DRILL | M1 | 30. | 0.005 | 90. | 0.54 | 0.23 | | | 0.08 | 0.03 | 0.00 | 0.02 | | | | 0.05 | 0.96 | 1.06 |
| LOAD, UNLOAD AND SETUP TIME AND COST | | | | | | | | | 1.66 | 2.49 | | | | | | | | | 4.15 | 5.00 |

NOTE - FEED UNITS ARE IPR FOR TURNING, DRILLING,REAMING, AND CENTERDRILLING OR CHAMFERING
FEED UNITS ARE IPT FOR MILLING
TOOL LIFE UNITS ARE IN, FOR MILLING, DRILLING, REAMING,AND TAPPING
TOOL LIFE UNITS ARE MIN. FOR TURNING
TOOL LIFE UNITS ARE NO. OF HOLES FOR CENTERDRILLING OR CHAMFERING

**Figure 21-13** Printout of cost and production rate results for machining 4340 steel forging on an NC machining center.

N/C MACHINING COSTS AND OPERATION TIMES
FOR
PART NO.-456987-003 * PART NAME-BRACKET-SLIDER * LOT SIZE- 20
MATERIAL-AISI 4340 STEEL,NORMALIZED,ACICULAR * HARDNESS- 331

| DATA SET NO. | OPER NO. | MACHG OPER. | TOOL MATL | SPEED SFM | FEED | TOOL LIFE | FEED COST $ | RAPD TRAV $ | LOAD UNLD $ | SET-UP $ | TOOL INDX $ | TOOL REPL $ | TOOL DEPR $ | TOOL SHPN $ | BLAD RSET $ | INST COST $ | GRND WHL $ | TOOL PRST $ | TOTAL COST $ | OPER. TIME MIN. |
|---|---|---|---|---|---|---|---|---|---|---|---|---|---|---|---|---|---|---|---|---|
| 30001 | N010 | FM-TA | C-6 | 670. | 0.005 | 47. | 0.65 | 0.24 | | | 0.08 | 0.09 | 0.00 | | | 0.09 | | 0.28 | 1.43 | 1.27 |
| 30005 | N020 | EM-HSS | M2 | 155. | 0.004 | 50. | 2.78 | 0.15 | | | 0.08 | 0.33 | 0.25 | 0.79 | | | 0.01 | 0.79 | 5.18 | 4.03 |
| 30009 | N030 | EM-HSS | M2 | 65. | 0.002 | 50. | 3.34 | 0.26 | | | 0.08 | 0.10 | 0.05 | 0.25 | | | 0.00 | 0.25 | 4.34 | 4.56 |
| 30018 | N040 | DRILL | M1 | 36. | 0.005 | 50. | 0.45 | 0.23 | | | 0.08 | 0.05 | 0.00 | 0.04 | | | | 0.10 | 0.95 | 0.98 |
| LOAD, UNLOAD AND SETUP TIME AND COST | | | | | | | | | 1.66 | 2.49 | | | | | | | | | 4.15 | 5.00 |
| | | | | | | | 7.22 | .88 | 1.66 | 2.49 | .32 | .57 | .30 | 1.08 | | .09 | .01 | 1.42 | 16.05 | 15.84 |

NOTE - FEED UNITS ARE IPR FOR TURNING, DRILLING,REAMING, AND CENTERDRILLING OR CHAMFERING
FEED UNITS ARE IPT FOR MILLING
TOOL LIFE UNITS ARE IN, FOR MILLING, DRILLING, REAMING,AND TAPPING
TOOL LIFE UNITS ARE MIN. FOR TURNING
TOOL LIFE UNITS ARE NO. OF HOLES FOR CENTERDRILLING OR CHAMFERING

**Figure 21–14** Best conditions for machining 4340 steel forging on an NC machining center.

**TABLE 21–22 Optimized Cutting Speed Equations for Use with English Units**

**For Turning**

$$V_{min.\ cost} = S\left[\frac{nM(L+e)}{L(1\text{-}n)\left(Mt_d + \dfrac{C_p}{k_1+1} + Gt_s + \dfrac{Gt_b}{k_2} + \dfrac{C_c}{k_3} + C_w + Gt_p\right)}\right]^n$$

$$V_{max.\ prod.} = S\left[\frac{n(L+e)}{(1\text{-}n)Lt_d}\right]^n$$

**For Milling**

$$V_{min.\ cost} = (S_t)^{\frac{1}{n+1}}\left[\frac{nMD(L+e)}{3.82f_t\,L\left(Mt_d + \dfrac{C_p}{k_1+1} + Gt_s + \dfrac{Gt_b}{k_2} + \dfrac{ZC_c}{k_3} + C_w + Gt_p\right)}\right]^{\frac{n}{n+1}}$$

$$V_{max.\ prod.} = (S_t)^{\frac{1}{n+1}}\left[\frac{nD(L+e)}{3.82f_t\,Lt_d}\right]^{\frac{n}{n+1}}$$

**For Drilling or Reaming**

$$V_{min.\ cost} = (S_t)^{\frac{1}{n+1}}\left[\frac{nMD(L+e)}{3.82f_r\,L\left(Mt_d + \dfrac{C_p}{k_1+1} + Gt_s + Gt_p\right)}\right]^{\frac{n}{n+1}}$$

$$V_{max.\ prod.} = (S_t)^{\frac{1}{n+1}}\left[\frac{nD(L+e)}{3.82f_r\,Lt_d}\right]^{\frac{n}{n+1}}$$

**For Tapping**

$$V_{min.\ cost} = (S_t)^{\frac{1}{n+1}}\left[\frac{mnMD(L+e)}{1.91L\left(Mt_u + \dfrac{C_p}{k_1+1} + Gt_s + Gt_p\right)}\right]^{\frac{n}{n+1}}$$

$$V_{max.\ prod.} = (S_t)^{\frac{1}{n+1}}\left[\frac{mnD(L+e)}{1.91Lt_d}\right]^{\frac{n}{n+1}}$$

**TABLE 21–23 Optimized Cutting Speed Equations for Use with Metric Units**

**For Turning**

$$V_{min.\,cost} = S\left[\frac{nM(L+e)}{L(1-n)\left(Mt_d + \dfrac{C_p}{k_1+1} + Gt_s + \dfrac{Gt_b}{k_2} + \dfrac{C_c}{k_3} + C_w + Gt_p\right)}\right]^n$$

$$V_{max.\,prod.} = S\left[\frac{n(L+e)}{(1-n)Lt_d}\right]^n$$

**For Milling**

$$V_{min.\,cost} = (S_t)^{\frac{1}{n+1}}\left[\frac{nMD(L+e)}{318f_t\,L\left(Mt_d + \dfrac{C_p}{k_1+1} + Gt_s + \dfrac{Gt_b}{k_2} + \dfrac{ZC_c}{k_3} + C_w + Gt_p\right)}\right]^{\frac{n}{n+1}}$$

$$V_{max.\,prod.} = (S_t)^{\frac{1}{n+1}}\left[\frac{nD(L+e)}{318f_t\,Lt_d}\right]^{\frac{n}{n+1}}$$

**For Drilling or Reaming**

$$V_{min.\,cost} = (S_t)^{\frac{1}{n+1}}\left[\frac{nMD(L+e)}{318f_r\,L\left(Mt_d + \dfrac{C_p}{k_1+1} + Gt_s + Gt_p\right)}\right]^{\frac{n}{n+1}}$$

$$V_{max.\,prod.} = (S_t)^{\frac{1}{n+1}}\left[\frac{nD(L+e)}{318f_r\,Lt_d}\right]^{\frac{n}{n+1}}$$

**For Tapping**

$$V_{min.\,cost} = (S_t)^{\frac{1}{n+1}}\left[\frac{nMD(L+e)}{159pL\left(Mt_d + \dfrac{C_p}{k_1+1} + Gt_s + Gt_p\right)}\right]^{\frac{n}{n+1}}$$

$$V_{max.\,prod.} = (S_t)^{\frac{1}{n+1}}\left[\frac{nD(L+e)}{159pLt_d}\right]^{\frac{n}{n+1}}$$

**TABLE 21–24 Optimized Tool Life Equations for Use with English or Metric Units**

**For Turning**

$$T_{min.\ cost} = \left[\frac{S}{V_{min.\ cost}}\right]^{\frac{1}{n}}$$

$$T_{max.\ prod.} = \left[\frac{S}{V_{max.\ prod.}}\right]^{\frac{1}{n}}$$

**For Milling, Drilling, Reaming, or Tapping**

$$T_{t\ min.\ cost} = \left[\frac{S_t}{V_{min.\ cost}}\right]^{\frac{1}{n}}$$

$$T_{t\ max.\ prod.} = \left[\frac{S_t}{V_{max.\ prod.}}\right]^{\frac{1}{n}}$$

**TURNING INSERTS — COST & PRODUCTION**
*MACHINABILITY DATA CENTER PROGRAM*

| STEP | HP-67/97 | | KEYS | OUTPUT |
|---|---|---|---|---|
| 1 | Load Strip | | | |
| 2 | Initialize | | RTN R/S | 3.00† |
| 3 | Load-Unload Time<br>Index Time<br>Rapid Trav Dist<br>Rapid Trav Rate | min<br>min<br>in<br>in/min | ENTER<br>ENTER<br>ENTER<br>R/S | 4.00† |
| 4 | Insert Cost<br># Cutting Edges<br>Set-up Time<br># Pieces in Lot | $<br><br>min<br> | ENTER<br>ENTER<br>ENTER<br>R/S | 5.00† |
| 5 | Preset Cost<br>Overhead<br>Dull Tool Rep Time | $<br>$/min<br>min | ENTER<br>ENTER<br>R/S | 6.00† |
| 6 | Diam Work<br>Extra Travel<br>Length of Cut | in<br>in<br>in | ENTER<br>ENTER<br>R/S | 7.00† |
| 7 | Tool Life<br>Speed<br>Feed | min<br>fpm<br>ipr | ENTER<br>ENTER<br>R/S | PTS/TL§ |
| 8 | for Times<br>or for Costs | | CF 1<br>SF 1 | |
| 9 | Constant (cost or time)<br>Feeding (cost or time)<br>Tool (cost or time)<br>Total (cost or time) | | A<br>B<br>C<br>D | $orMin<br>$or Min<br>$orMin<br>$orMin |
| * | for New Tool Life, V&F | | E | 7.00† |
| * | for New Geometry | | GTO 6 R/S | 6.00† |

†These prompting outputs give the numbers of the next step to be executed.

§This is the output of the number of parts that can be machined with one edge of one insert.

(1) *Select the proper magnetic strip and insert it into the calculator.*

(2) *Initialize*

(3) *Enter data and press appropriate keys*

| |
|---|
| 2.3<br>0<br>27.2<br>100. |
| 3.15<br>8.<br>21.<br>20. |
| 0<br>.40<br>.4 |
| 3.5<br>.2<br>19. |
| 60.<br>325.<br>.01 |

**11.2 Parts per edge**

(4) *Execute steps 8 and 9 for output*

| Cost ($) | Prod. Time (min.) |
|---|---|
| 1.45 | 3.62 |
| 2.17 | 5.41 |
| .05 | .04 |
| 3.67 | 9.07 |

(5) *Loop back using * steps if desired*

**Figure 21–15** Instructions and sample inputs and outputs for cost and production calculations using the HP-67 calculator.

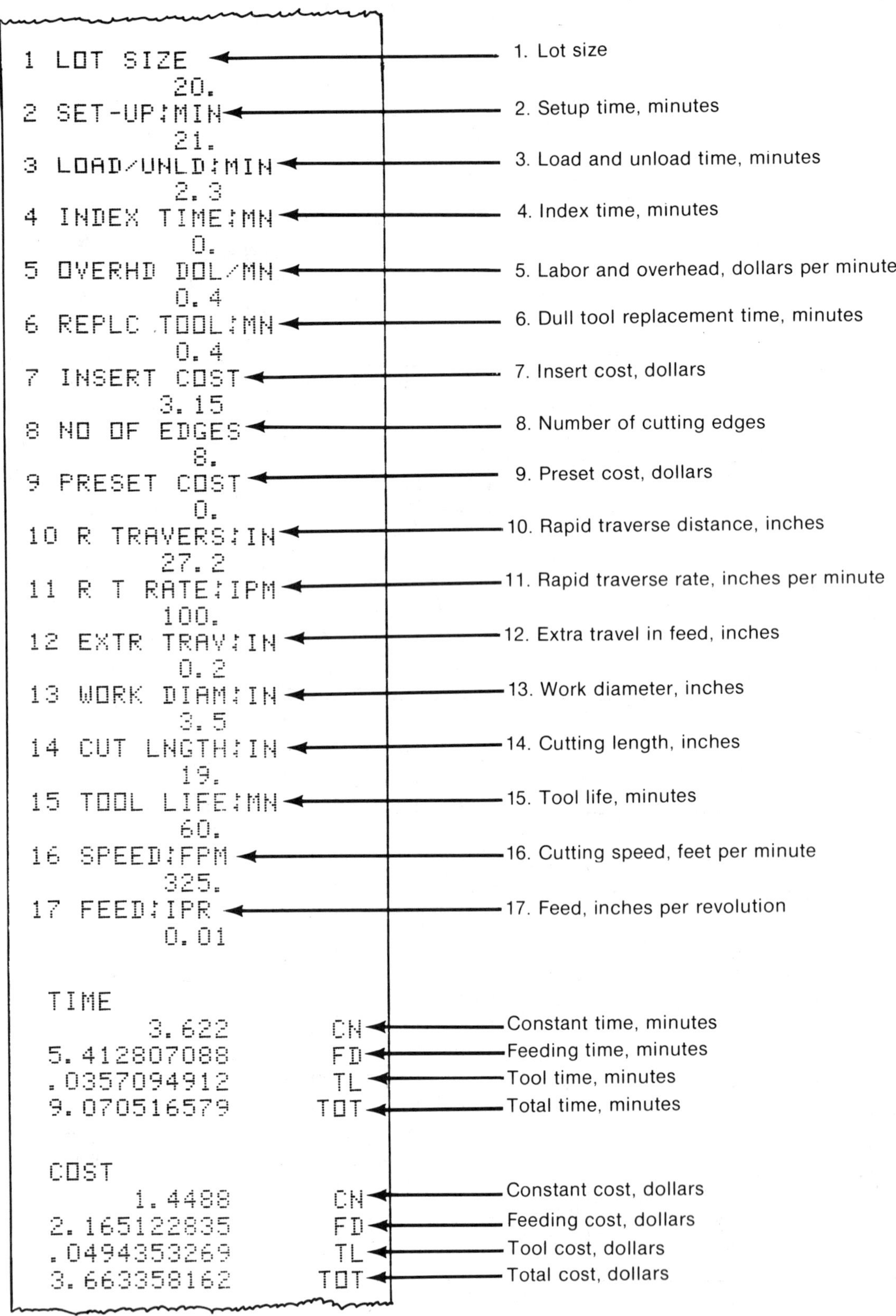
1 LOT SIZE
20.
2 SET-UP:MIN
21.
3 LOAD/UNLD:MIN
2.3
4 INDEX TIME:MN
0.
5 OVERHD DOL/MN
0.4
6 REPLC TOOL:MN
0.4
7 INSERT COST
3.15
8 NO OF EDGES
8.
9 PRESET COST
0.
10 R TRAVERS:IN
27.2
11 R T RATE:IPM
100.
12 EXTR TRAV:IN
0.2
13 WORK DIAM:IN
3.5
14 CUT LNGTH:IN
19.
15 TOOL LIFE:MN
60.
16 SPEED:FPM
325.
17 FEED:IPR
0.01
TIME
3.622 CN
5.412807088 FD
.0357094912 TL
9.070516579 TOT
COST
1.4488 CN
2.165122835 FD
.0494353269 TL
3.663358162 TOT
1. Lot size
2. Setup time, minutes
3. Load and unload time, minutes
4. Index time, minutes
5. Labor and overhead, dollars per minute
6. Dull tool replacement time, minutes
7. Insert cost, dollars
8. Number of cutting edges
9. Preset cost, dollars
10. Rapid traverse distance, inches
11. Rapid traverse rate, inches per minute
12. Extra travel in feed, inches
13. Work diameter, inches
14. Cutting length, inches
15. Tool life, minutes
16. Cutting speed, feet per minute
17. Feed, inches per revolution
Constant time, minutes
Feeding time, minutes
Tool time, minutes
Total time, minutes
Constant cost, dollars
Feeding cost, dollars
Tool cost, dollars
Total cost, dollars

**Figure 21–16** Example of the cost and production program as run on the TI-59 calculator.

# SECTION 22

# MACHINE CHATTER AND VIBRATION


Introduction .......... 22–3
Causes of Vibration .......... 22–3
Damping .......... 22–5
Responses of Structures to Periodic Forces .......... 22–6
Self-Excited Chatter .......... 22–12
Troubleshooting Chatter .......... 22–17
Regenerative Chatter Checklist .......... 22–22

## INTRODUCTION

Problems arising from machine tool-workpiece vibration are among the major obstacles to greater productivity in metal cutting. The vibratory motion between a cutting tool and a workpiece is recorded on the workpiece surface and, depending upon the severity of the oscillatory motion and the surface finish specifications, may result in parts with unsatisfactory surface quality. Excessive vibration or chatter may also seriously decrease tool life. In addition, chatter affects the life of bearings and other machine components, although this is not well understood.[1-4]

Higher production rates and more severe cutting conditions encountered in modern shops reveal another vibration problem—excessive noise. This problem usually stems from the high frequency vibrational condition known as chatter, and while it may have only a slight physical effect on surface finish, recent government regulations have focused sharply upon the noise environment of the machine tool operator.

It is necessary to solve a vibration problem if unsatisfactory tool life, surface finish, or noise result. Many experienced shop people are aware of chatter and vibration, but few have the training to identify the types of vibration problems and the appropriate methods for their solution. The discussion that follows provides an understanding of the causes of machine tool-workpiece vibration problems and alternative approaches for their elimination.

## CAUSES OF VIBRATION

### Deflection of Machine Elements

In order to analyze vibration problems, it is important first to understand the deflection characteristics of materials. The most important material property governing deflection is the modulus of elasticity. For most structures, the amount of elastic, nonpermanent, deflection for a given load is inversely proportional to the modulus of elasticity. Thus, a higher modulus will result in a lower deflection. Table 22-1 gives the modulus of elasticity of materials commonly used in machine tools, tooling, and fixturing.

**TABLE 22-1 Modulus of Elasticity of Some Common Materials**

| MATERIAL | MODULUS OF ELASTICITY, E (psi) |
|---|---|
| Aluminum | 9,000,000 |
| Cast iron | 14,000,000 (varies widely) |
| Steel | 29,000,000 |
| Tungsten carbide | 85,000,000 |

It is important to note that all steels, regardless of alloy content, heat treatment or hardness, have approximately the same modulus of elasticity. The elastic modulus depends on the fundamental atomic structure of the material, while strength and hardness are very dependent on small percentages of elements which do not affect the modulus. Of the materials listed in table 22-1, aluminum is the poorest material for minimizing the deflection of machine structures. Cast iron is more resistant to deflection than aluminum, steel is even better, and tungsten carbide is superior. Since cast iron is an easily obtainable material and less expensive to machine than steel, it is often used in machine tool structures in spite of its lower modulus of elasticity. Tungsten carbide is very expensive and costly to machine; hence, it is used where performance is more important than cost.

For a given material, the deflection of a structure is a function of the applied load and the geometry. Equation 22-1 can be used to predict the deflection of the cantilever beam with a round cross section, illustrated in figure 22-1.

$$d = \frac{64PL^3}{3\pi ED^4} \qquad \text{(Eq. 22-1)}$$

where:

d = beam deflection, inches
P = force on the end of the beam, pounds
L = length of the beam, inches
D = beam diameter, inches
E = modulus of elasticity, pounds per square inch

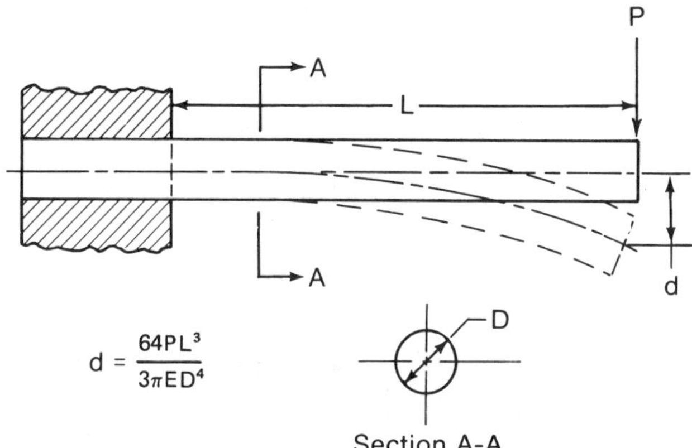


**Figure 22-1** Deflection of a cantilever beam.

Equation 22-1 only applies "exactly" to a slender beam solidly held on one end by a massive base. However, it can be used to approximate the deflection of a boring bar extending from a tool post or tool holder, a slender workpiece sticking out of a vise, a spindle nose extending beyond its front bearings, or any "slender" beam or "thin" plate that projects for some unsupported distance. In this case, "slender" and "thin" are relative terms and may refer to some very heavy cross sections that have an unsupported length several times their thickness. In equation 22-1, the beam diameter (D) is raised to the fourth power and the beam length (L) to the third power, implying that both are very important quantities influencing beam deflection.

For a given beam, a given percentage increase in overhang length (L) produces a much greater percentage increase in deflection. For example, a doubling of beam length results in an eight-fold increase in deflection. Thus, in tooling and fixture design, an absolute minimum of overhang should be used. Boring bar design is an example where a length-to-diameter ratio of greater than four or five to one is considered impractical. A given percentage increase in diameter (D) has an even greater effect on reducing deflection than length had on increasing deflection; hence, it is important to have the diameter of overhanging members as large as possible.

### Structural Stiffness

The stiffness of a structure can be interpreted as the amount of force required to produce an inch of deflection. Equation 22-2 can be used to calculate the stiffness of a structure in pounds per inch.

$$K = \frac{P}{d} \qquad \text{(Eq. 22-2)}$$

where:

K = stiffness, pounds per inch
P = force, pounds
d = deflection, inches

For example, if a machine deflects 0.001 inch under a load of 800 pounds, the stiffness is:

$$K = \frac{800}{0.001} = 800{,}000 \text{ lb/in}$$

If the stiffness of a structure is known, its deflection under load can be calculated from a modified form of equation 22-2:

$$d = \frac{P}{K}$$

The term "directional stiffness" is used to describe the stiffness of a structure when loaded in different directions. This is an important concept because many machines, tools, and workpieces have both stiff and weak directions. If a machining setup is made so that cutting forces are aimed toward the weak direction, excessive deflection and possible chatter will occur. If the setup is changed so that the forces are in the stiff direction, the problem will be minimized or eliminated. As an example, consider figure 22-2; if this beam is loaded in the Z or the X directions, it will be very rigid. Loading the beam in the Y direction produces the greatest deformation and, therefore, the lowest directional stiffness.

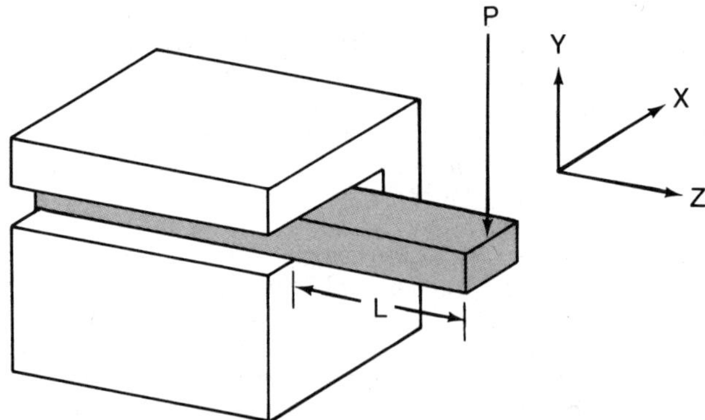


**Figure 22-2** Illustration of the concept of directional stiffness. For this beam the stiff directions are X and Z; the weak direction is Y.

### Forcing of Structures

The important terminology of vibration analysis is illustrated in figure 22-3 and is used extensively in the following discussion. Included are:

*Amplitude*, A, is the maximum displacement from the equilibrium position. It is one-half of the peak-to-valley distance and is *not* the total displacement from the topmost peak to the lowest valley.

*Period*, T, is the time required to complete one full cycle of vibration.

*Frequency*, f, is the number of cycles which are completed in a given amount of time, usually one second.

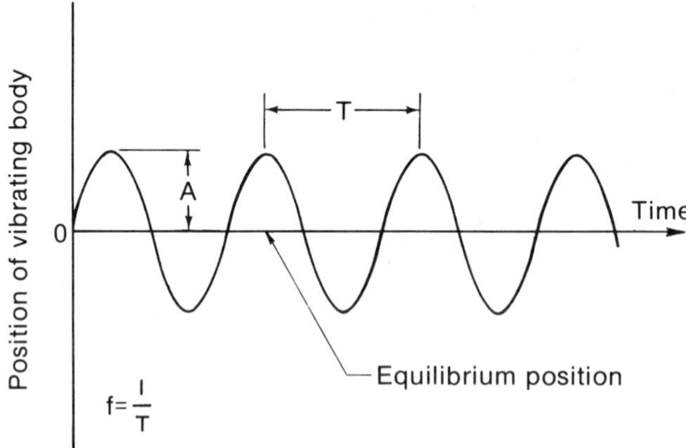


**Figure 22-3** Periodic motion of a vibrating structure.

Equation 22-3 is an important relationship between frequency and period:

$$f\,(\text{cycles per second}) = \frac{1}{T\,(\text{sec})} \qquad \text{(Eq. 22-3)}$$

Another name for "cycles per second" is Hertz. For example, 20 cycles per second is the same as 20 Hertz. While the term Hertz is the official standard for frequency in the U. S. A., many prefer to use the more descriptive term cycles per second, or cps. The term cps will be used in this discussion.

Most people are familiar with things that start vibrating by giving them a push and letting go. If the "structure" vibrates, it does so at a frequency determined by the structure itself, not by the force or displacement given it. The initiation of vibration in machine tools is often by forces that fluctuate periodically, that is, by forces that vary with time. The structure is said to be "excited" by the fluctuating force. Motor unbalance is a common method by which periodic forces in machinery are introduced. Another example of periodic forces in machining is the variation in cutting forces introduced by cutting through a keyway contained in a shaft. The cutting force will drop to zero each time the tool tip enters the keyway and return to its nominal value after the tool re-enters the workpiece.

The occurrence of forced vibrations in a milling operation can be expected for three major reasons. First, the number of teeth in contact with the workpiece is changing constantly. Second, the angular orientation of the force changes as the tooth rotates. Third, the chip thickness changes as the tooth moves through the arc of contact. Figure 22-4 is an illustration of an actual recording of milling forces.[14] Forced vibration is obviously present in all three cutting force components, $F_x$, $F_y$, and $F_z$.

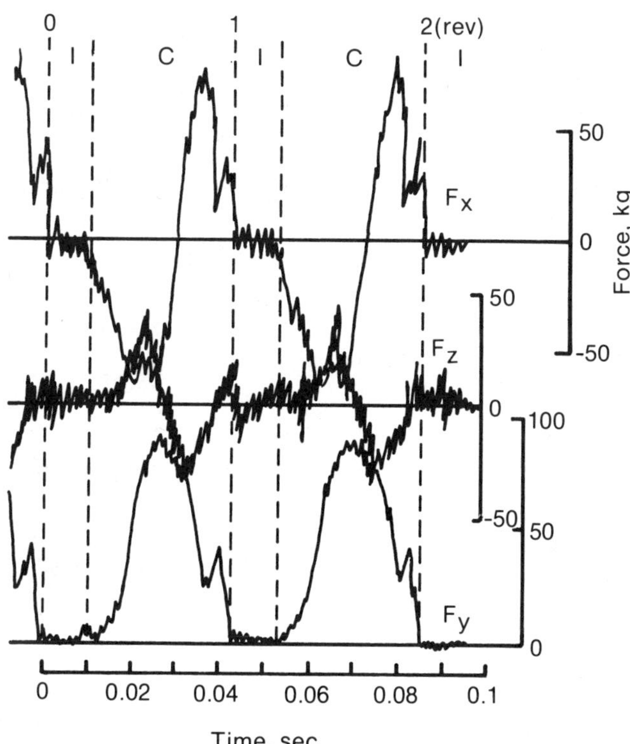


**Figure 22-4** Fluctuation of the cutting force components in a milling operation. (Adapted from Hosoi[14])

## DAMPING

Damping is the behavior of a system that tends to resist rapid movements. If a beam is struck with a hammer and vibrates for a long time, it is said to be lightly damped. If the beam stops vibrating almost immediately, it is said to be highly damped. Theoretically, a system that had no damping would go on vibrating forever, at the same initial amplitude. This can not occur in practice because all materials have some amount of internal damping. Because of the basic way in which materials deform, all of the work that is done to deflect an elastic member can not be recovered when the member is released. Most of the energy is recovered, but a small amount remaining in the member is converted into heat. This small energy loss causes the vibration amplitude to decrease very slightly each cycle.

Different materials have different amounts of internal damping. Figure 22-5 illustrates the different damping characteristics of cast iron and steel and shows clearly that cast iron has much higher internal damping. This figure might lead to the conclusion that cast iron machine structures have a much better vibration resistance than those built from steel.

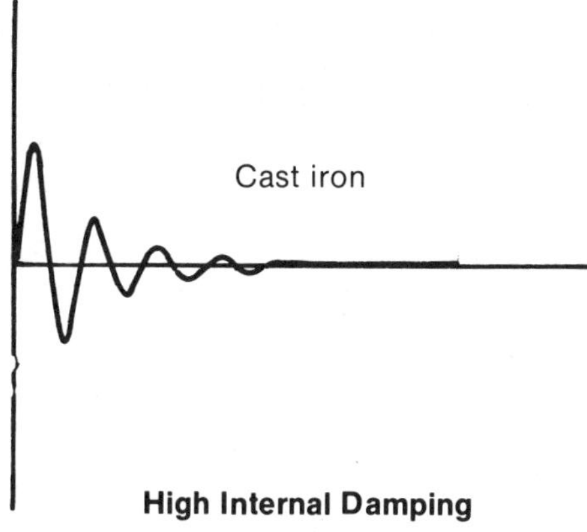


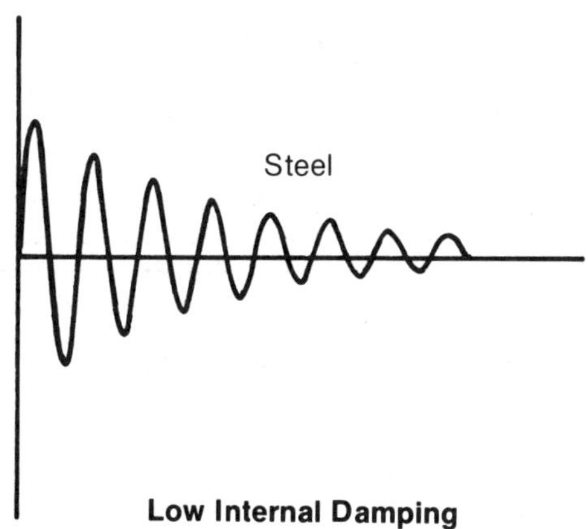


**Figure 22-5** Diagrams showing the difference in internal damping of cast iron and steel.

Material damping, however, is not the whole story because joint damping, or "squeeze film" damping, is also very important in machinery. Oil becomes trapped between mating surfaces that move apart during vibration. Oil is sucked in and fills the gap as surfaces move apart and is forced out when they close. This motion of the oil heats it, absorbing energy. While this is the most powerful source of damping in joints, some damping takes place in dry joints as a result of friction and the squeezing of air in and out. This damping is generally much less significant, however.

Figure 22-6 illustrates that the damping of a machine increases as components are added to it. Most of the damping in the completed machine results from the joints between the various components, rather than from the internal damping of the structural components.

The third major category of damping is that intentionally added to a machine to improve its resistance to vibration. A common, non-metalcutting example of this is the automobile shock absorber. Similarly, tuned damped vibration absorbers provide a convenient way to improve the vibration resistance of many machine tools.

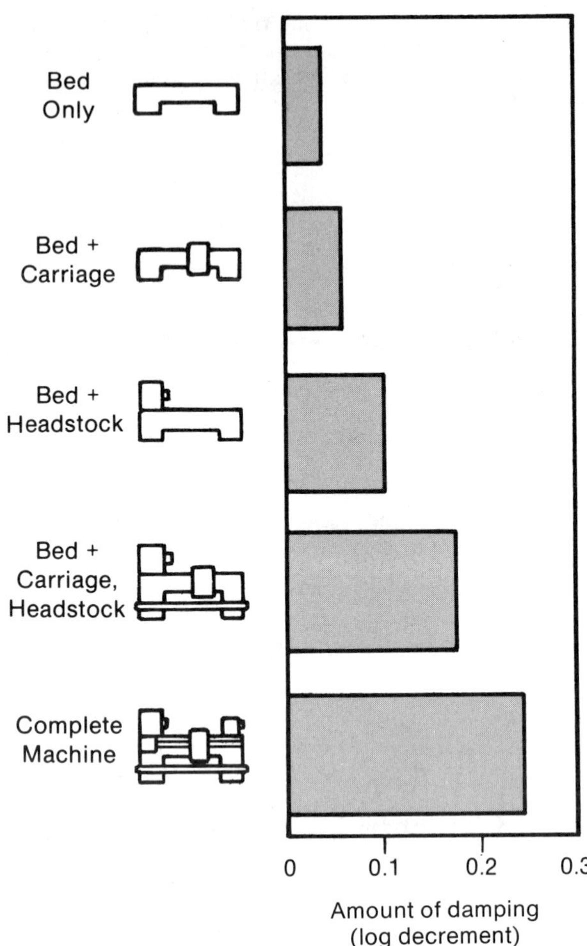


**Figure 22-6** Influence of material and of joints on damping. (Adapted from Peters[5])

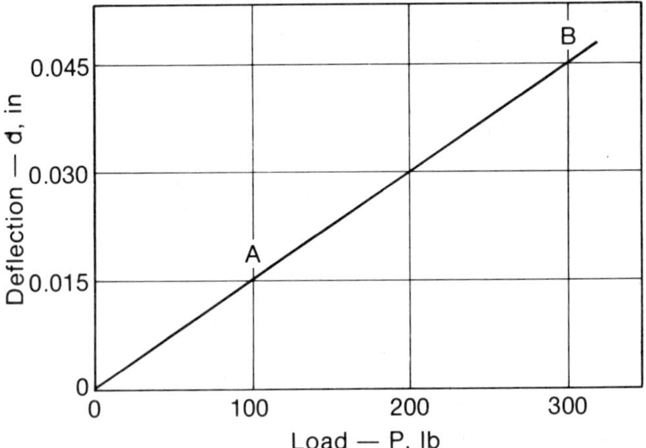


**Figure 22-7** Deflection versus static load behavior of a cantilever beam.

## RESPONSES OF STRUCTURES TO PERIODIC FORCES

### Static and Dynamic Stiffness

The analysis of the behavior of structures under dynamic loads is far more complicated than that of structural behavior under steady loads. The basic principle is that at very low frequencies the structure responds to dynamic loads the same as it does to static loads. To emphasize this, consider figure 22-7 which shows the deflection versus static load behavior of a cantilever beam.

This graph could be computed from the deflection equation for a specific beam at various loads. At 100 pounds, the deflection is 0.015 inch; at 200 pounds, it is 0.030 inch, etc. The static stiffness, K, is the value of steady force acting on the structure divided by the deflection it produces. The static stiffness of the beam producing the data of figure 22-7 is 6,670 pounds per inch as shown by the following calculation:

$$K = \frac{300}{0.045} = 6670 \text{ lb/in}$$

Note that any load along the line in figure 22-7 divided by the corresponding deflection gives this same value of static stiffness.

Suppose a load that varies from 100 to 300 pounds is applied to the beam. (The peak-to-valley variation in load is 300 − 100 = 200 pounds. The force amplitude is half of the peak-to-valley value, or 100 pounds.) Suppose the frequency of load application is one cycle every two hours, or 0.00014 cycles per second. This is so slow that vibration analysis is not necessary to predict what will happen. The force and deflection will creep slowly up and down the static deflection curve between points A and B in figure 22-7. At any point in time, the value of the deflection for the varying force acting at that instant will be the same as the value of the deflection if the force were static.

If the force variation is increased to one cycle per minute, the speed is still so slow that the bar does not know that it is being vibrated. If the force variation is increased to one cycle per second, or even faster, it is not obvious that the results will be the same. In order to understand what happens at higher frequencies of force variation, it is necessary to introduce the concept of dynamic stiffness. Dynamic stiffness, $K_d$, is the ratio of the force amplitude, F, to the vibration amplitude, A, at a certain vibration frequency, $K_d = K$; that is, the dynamic and static stiffnesses are the same.

The concept of dynamic stiffness, $K_D$, can be understood from the curve of figure 22-7 showing load versus deflection. For the example illustrated in figure 22-7, where a peak-to-valley force of 200 pounds (F = 100 pounds) gave a peak-to-valley displacement of 0.045 − 0.015 = 0.030 inch (A = 0.015 inch), the dynamic stiffness is

$$K_d = \frac{100}{0.015} = 6670 \text{ lb/in}$$

where:

F = force amplitude (the maximum increase in the oscillatory force from its average value, that is one half of its peak-to-valley value)

A = vibration amplitude.

Note that this is exactly the same as the value of K, illustrating the fact that as long as the force and displacement data fall on the same curve, their ratios remain the same and the dynamic stiffness is the same as the static stiffness.

Figure 22-8 shows the force pattern acting on a structure at a low frequency and the resulting motion of the structure. This illustrates a very basic point of vibration troubleshooting; namely, *the structure almost always vibrates at the same frequency as the periodic force that is causing it to vibrate.* In figure 22-8 there is a "phase difference" between the two waves; that is, the peaks of the displacement trace occur slightly later than those of the force trace. In spite of this, the period, T, is the same for both waves and consequently the frequency must be the same.

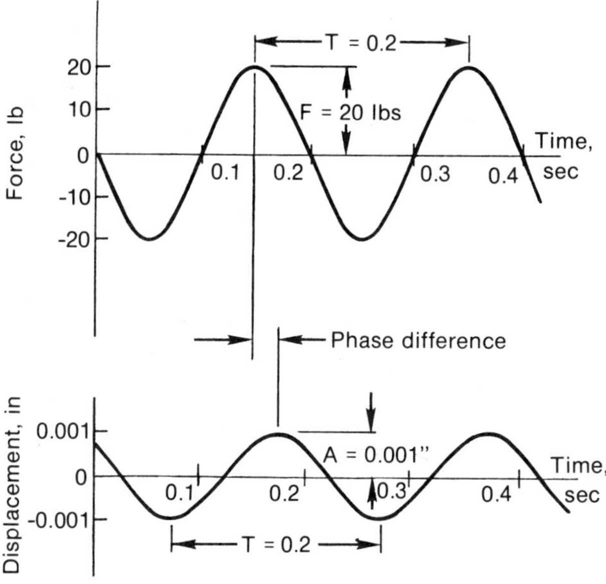


**Figure 22-8** Representative force and displacement patterns of a structure vibrating at a low frequency.

There are a few rare cases of unusual structures and unusual force wave shapes where the above relationship between displacement and force is not true, but these cases require an expert to analyze. For purposes of investigating common vibration problems, always proceed as though this rule (italicized above) were true, because it will lead to the best approach to the problem; namely, measure the vibration frequency and look for possible sources of forcing at that same frequency.

## Behavior of Simple Structures

Suppose a heavy object is suspended from a weak spring and the upper end of the spring is held in someone's hand. Further suppose the hand is moved up and down, very slowly at first, then at higher and higher frequencies. Let us examine what happens to the heavy object. It would be observed first that the heavy object always moves at the same frequency as the hand. This is elementary, but important. It suggests that when there is a suspected forced vibration problem the cause of the problem can be tracked down by knowing its frequency.

At very low frequencies the heavy object moves exactly the same distance as the hand. Thus, the amplitude of the hand and object are the same and there is no additional stretch in the spring. This behavior is typical of all structures. As the frequency increases, the spring begins to stretch and the amplitude of the object becomes larger than that of the hand. Eventually a frequency is reached where a given hand movement results in a very large object amplitude. This condition is called resonance. The frequency where resonance occurs is called the natural frequency, $f_n$, which is also the frequency at which the dynamic stiffness is the lowest. As the frequency is increased beyond the natural frequency, the amplitude of the object continually decreases until finally the hand can shake very rapidly while the object appears to remain motionless.

It may be concluded from these observations that a machine structure will vibrate easily if it is forced at frequencies near its natural frequency. It will hardly vibrate if forced at a frequency well above the natural frequency. At frequencies well below the natural frequency, the structure will vibrate at some intermediate amplitude. The natural frequency of a simple structure, such as a heavy object suspended on a spring, can be calculated from equation 22-4.

$$f_n = \frac{9.8}{\pi}\sqrt{\frac{K}{W}} \qquad \text{(Eq. 22-4)}$$

where:

$f_n$ = natural frequency, cycles per second
K = spring stiffness, pounds per inch
W = object weight, pounds

When troubleshooting vibration problems, equation 22-4 is seldom used to estimate the natural frequency; however, it is useful for estimating the effects of changes in stiffness and weight.

## Behavior of Complex Structures

Figures 22-9 and 22-10 show respectively a motor on its mounting bracket and its frequency response. The system has a static stiffness of 8,330 pounds per inch and a natural frequency of 20 cycles per second. In frequency analysis, it is common to plot the compliance, which is the reciprocal of the static stiffness. Figure 22-11 represents the frequency response of a machine column having a stiffness of 50,000 pounds per inch and a natural frequency of 50. If the bracket of figure 22-9 is attached to this column, the frequency response of the bracket mounted on the column will contain the natural frequency of both the bracket and the column. A composite frequency response can be constructed using simple logic. At low frequencies the motion of the weak bracket is much greater than that of the stiff column. Therefore, up to the lowest natural frequency, only bracket deflection occurs and the composite response

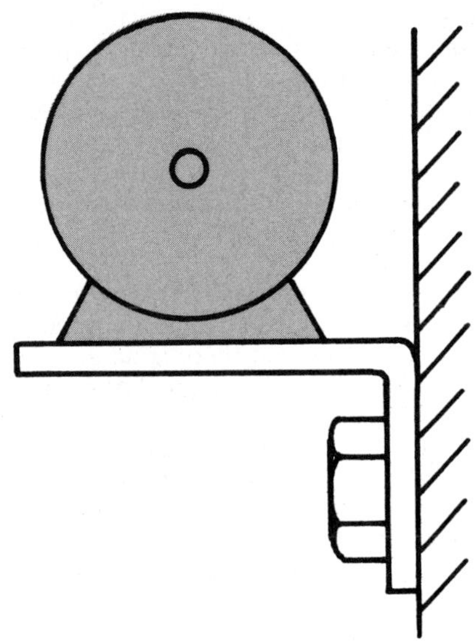

Figure 22-9 Schematic diagram of a motor mounted on a bracket.

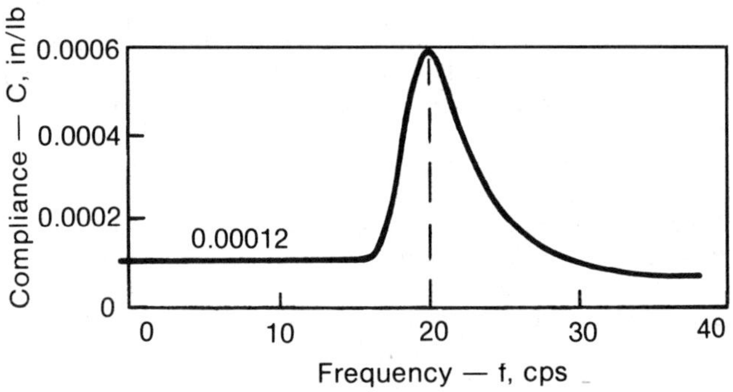

Compliance — C, in/lb
0.0006
0.0004
0.0002
0
0.00012
0 10 20 30 40
Frequency — f, cps

**Figure 22-10** Frequency response of the system shown in figure 22-9.

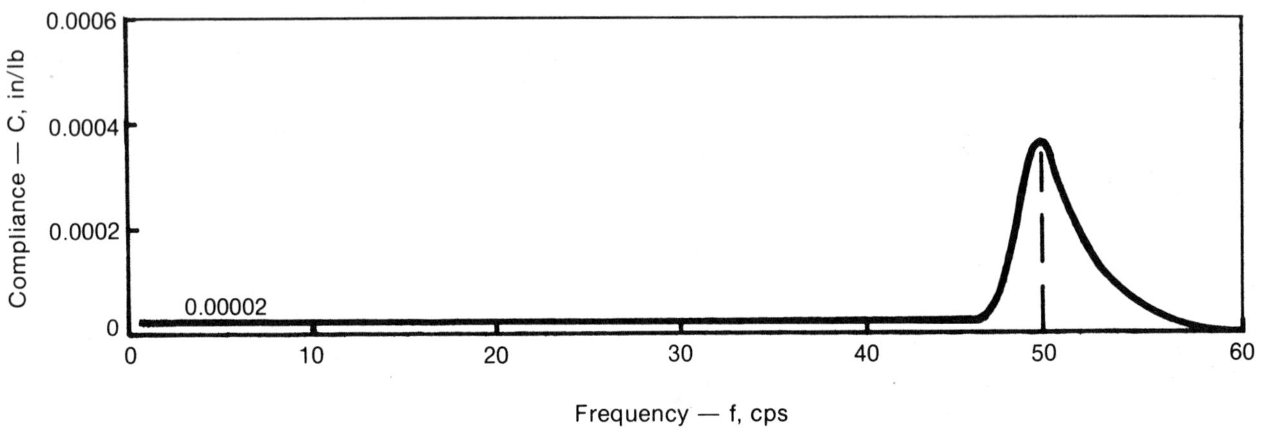


**Figure 22-11** Frequency response of a machine column.

curve will be identical to the bracket response curve. At the resonance of the bracket, both the motor and bracket will be vibrating and forcing the column. However, since the motion of the column is still very low at this frequency, the combined frequency response will still be the same as the bracket response through its resonance. The frequency response of the bracket-column will continue to be equal to the frequency response of the bracket alone, until the forcing frequency is close to the column resonance. At this frequency, and at higher frequencies, the column response is much greater than the bracket motion; therefore, the composite response is equal to the column response.

The composite frequency response for this system is illustrated in figure 22-12. This is the response curve that would be observed if the motor and bracket were tested while mounted on this column. Under these conditions there are two natural frequencies, namely, at 20 and at 50 cycles per second.

Most real machine structures are much more complicated than the above example; for instance, figure 22-13, taken from reference 15, shows the frequency response of a drill press. Logarithmic scales are used as graph axes in figure 22-13 to compress it to a reasonable size. The peaks in this figure represent resonances of the structure at 23, 33, 38, 70, 100, 110, 170, 230, 300, 380 and 500 cycles per second. Each peak is associated with a different vibration pattern of the various components that make up the drill press. Although the frequency response of most machine tools is more like figure 22-13 than the system response described by figure 22-12, vibration problems are generally not this complex because they often involve only one frequency. Consequently, the other resonances can be ignored

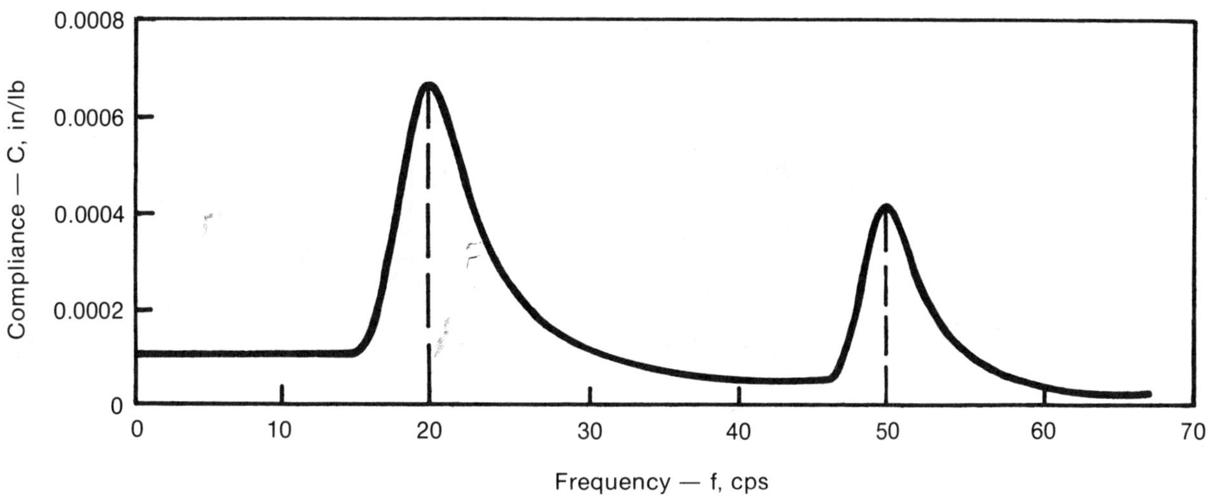


**Figure 22–12** Composite frequency response of a mounted motor and column.

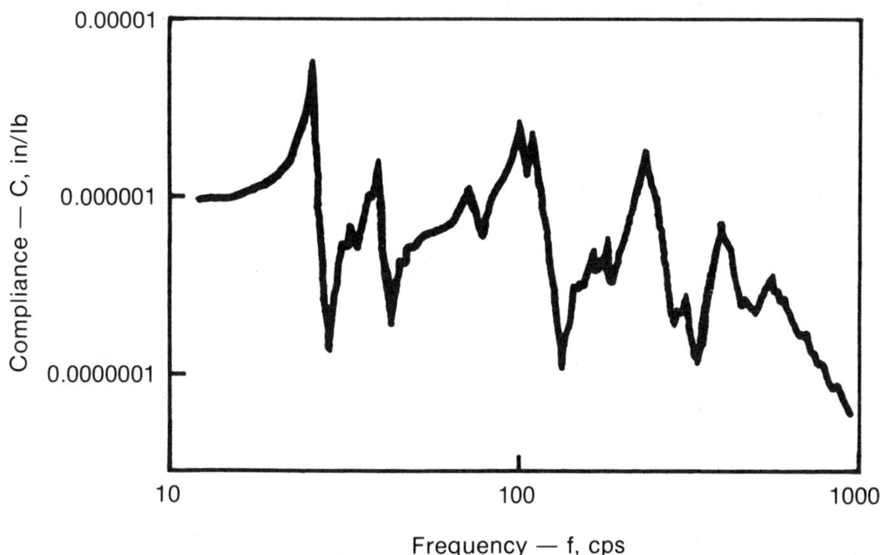


**Figure 22–13** Frequency response diagram of a drill press.

and the real machine considered as a complex system with only one natural frequency.

### Influence of Damping

It was noted that damping has a strong influence on the amplitude of free vibration when a system is started vibrating and allowed to die out. Damping also has a very strong influence on the amplitude of forced vibration, particularly at resonance. The family of curves in figure 22–14 shows the frequency response of a simple vibrating system with varying amounts of damping. In this figure the amplitude of the force is F, the vibration amplitude is A, and the frequency is f.

The degree to which a vibrating system is damped can be measured by its "resonance amplification" which is the vibration amplitude at resonance divided by the vibration amplitude at low frequencies for equal force amplitudes. The resonance amplification is also the ratio of the static stiffness, K, to the dynamic stiffness, $K_d$, at resonance. In figure 22–15 the static compliance is 0.00001 inches per pound while the maximum compliance at resonance is 0.00005 inches per pound. Therefore, the resonance amplification is (0.00005 ÷ 0.00001 =) 5.

For a lightly damped system the resonance amplification might be ten to thirty, while for heavy damping it might be one-and-one-half to four. These ranges point out the importance of damping in machinery. It is sometimes fairly easy to reduce the vibration amplitude by adding damping to a lightly damped structure vibrating near resonance. Almost any damping added to a lightly damped structure improves dynamic stiffness by a factor of two or three which, in turn,

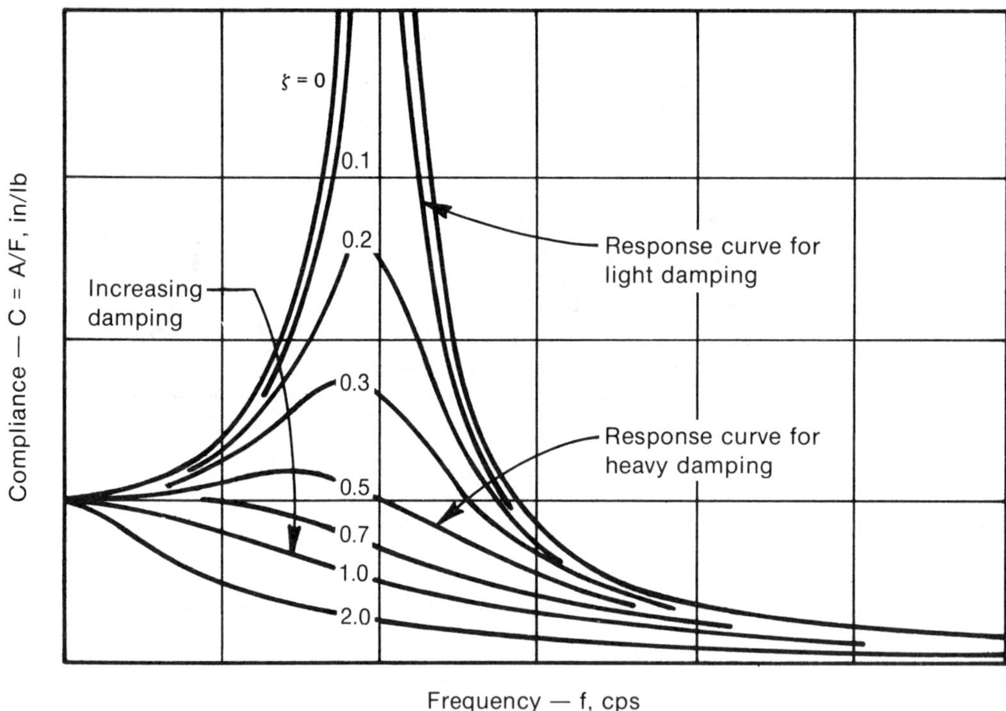


**Figure 22–14** Influence of damping on the frequency response of a vibrating system.

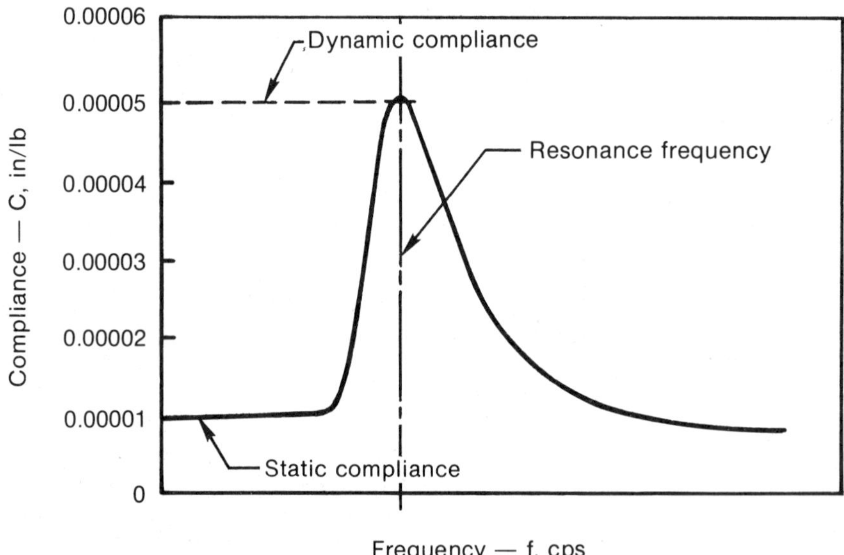


**Figure 22–15** Determination of resonance amplification.

reduces vibration amplitudes by one-half to one-third of their original value. The same reduction in vibration amplitude can be obtained by increasing the static stiffness, K, (that is, decreasing the static compliance) which may be both difficult and expensive.

### Data From Machine Structures

Figure 22-16 presents typical values of major natural frequencies on 13 different machine tools.[6-10] The 40 natural frequencies from these machines were combined into groups. Three natural frequencies were observed between

0 and 25 cycles per second, seven between 25 and 50 cycles per second, eight between 50 and 75 cycles per second; etc. The data of figure 22-16 show that natural frequencies often occur in the range of 0 to 100 cycles per second, that a fair number are widely spread over the range of 100 to 300 cycles per second, and that there were no important natural frequencies over 300 cycles per second for this particular group of machine tools.

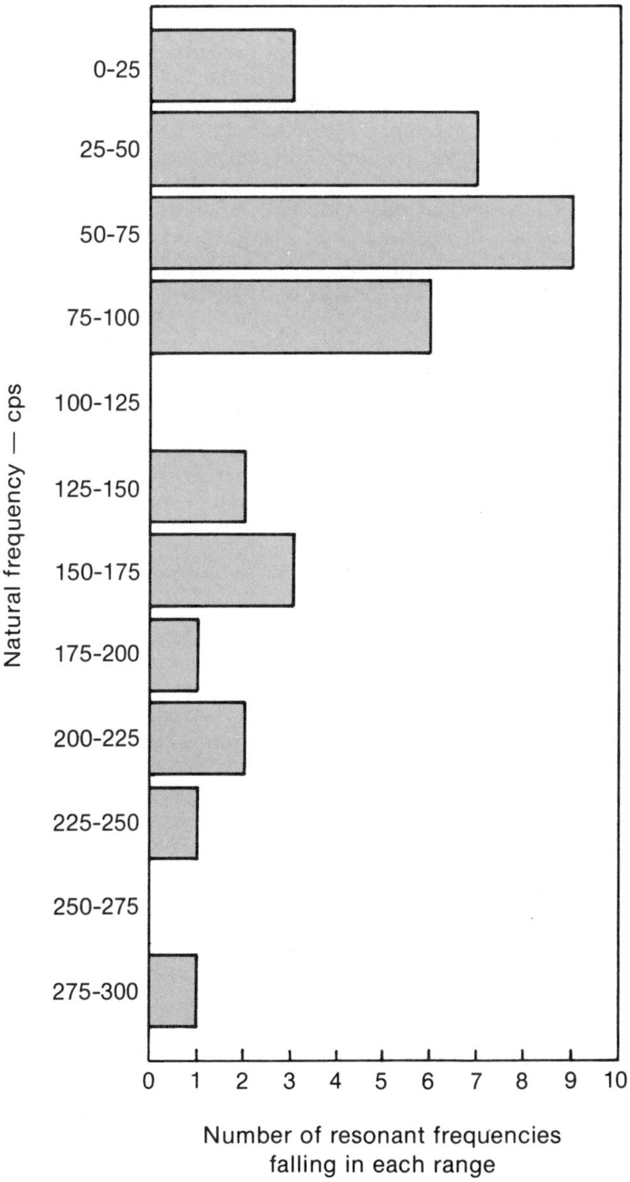

Natural frequency — cps

0-25
25-50
50-75
75-100
100-125
125-150
150-175
175-200
200-225
225-250
250-275
275-300

0 1 2 3 4 5 6 7 8 9 10

Number of resonant frequencies falling in each range

**Figure 22-16** Typical values of natural frequencies on thirteen machine tools. (Compiled from Long,[6] Koenigsberger,[7] Gaudissart,[8] Tobias,[9] and Tlusty[10])

Low natural frequencies are typical of the vibration of heavy components, such as a whole machine rocking on its mounts. The 100 to 300 cycles per second range corresponds to vibrations of good, stiff, lower weight components, for example, a lathe bed. In practice, vibration problems occasionally occur at natural frequencies well above 300 cycles per second but, in most cases, these are associated with tool, not machine, vibration. Equation 22-4 demonstrates the characteristics necessary to obtain high natural frequency values. If a component has a stiffness of 500,000 pounds per inch, then its effective weight can be only about 19.5 pounds to give a natural frequency as high as 500 cycles per second, since:

$$f_n = \frac{9.8}{\pi}\sqrt{\frac{500{,}000}{19.5}} = 500 \text{ cps}$$

Any weight greater than 19.5 pounds will result in a frequency lower than 500 cycles per second. Boring bars and slender end mills are typical components with natural frequencies in the range of 500 cycles per second and above.

Another useful number is the resonance amplification, that is, the ratio of static to dynamic stiffness. Figure 22-17 shows this quantity for seven different machine tools.[6,7,9] These values normally fall in the range of 5 to 30. Lower values are obtained when the vibration involves sliding joints, such as the rocking of a lathe saddle on its ways. Higher values correspond to the vibration of solid components such as the twisting of a milling machine column.

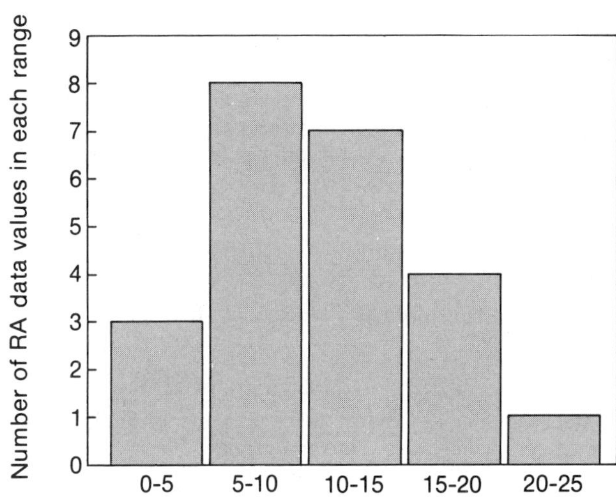


**Figure 22-17** Resonance amplification data from seven machine tools. (Compiled from Long,[6] Koenigsberger,[7] and Tobias[9])

The final value needed from the testing of real machines is the dynamic stiffness at resonance. Listed in table 22-2 are values obtained from nine machine tools of all sizes and shapes.

**TABLE 22-2 Dynamic Stiffness at Resonance**

| MACHINE | DYNAMIC STIFFNESS, lb/in |
|---|---|
| A | 15,400 |
| B | 27,500 |
| C | 33,300 |
| D | 33,500 |
| E | 58,900 |
| F | 71,500 |
| G | 122,000 |
| H | 410,000 |
| I | 478,000 |

Source: Data compiled from Long,[6] Koenigsberger,[7] Gandissart,[8] Tobias,[9] and Tlusty.[10]

A dynamic stiffness of 15,000 pounds per inch represents a light duty machine tool, not meant for heavy stock removal. Stiffness values of 400,000 and up represent machines that are capable of very heavy stock removal. Productivity of heavier duty machine tools will often be limited by the rigidity of the tooling, workpiece, or fixtures put on them.

## SELF-EXCITED CHATTER

The vibrations discussed to this point were caused by forces resulting from factors including unbalance, eccentricity, and misalignments. There is, however, a totally different type of vibration known as self-excited chatter. When there is no external forcing, the structure itself determines whether there will be self-excited chatter, based on its rigidity and the cutting conditions. There are two types of self-excited chatter, high frequency chatter and regenerative chatter.

### High Frequency Chatter

High frequency chatter is the result of high frequency vibration that sometimes occurs when tools become very dull. The tool vibrates in the direction of the cutting speed and results in high-pitched noise and a workpiece surface that is not deeply scalloped. This type of chatter is caused by the decrease in the rubbing force on the tool flank as the cutting speed increases. During cutting, a hard spot in the work, or a small vibration from a very small unbalanced force, may result in a very slight increase in cutting force. This, in turn, causes a very slight increase in the deflection of the cutting tool in the direction of the cutting speed. While the tool is deflecting in this direction, the speed is decreased slightly, causing a slight increase in cutting force, which causes more deflection, etc. This cycle continues until the cutting force has built up so high that the tool snaps back to its original position, overshoots, and the cycle repeats itself. This is not an important type of self-excited chatter and can always be solved by resharpening the tool. It is described here only to keep it from being confused with other types of machine tool vibration.

### Regenerative Chatter

The important type of self-excited chatter is regenerative chatter. Regenerative chatter is difficult to troubleshoot since it is almost impossible to predict why the machine is chattering. Nearly every machining variable affects it sometimes, but only a few affect it all the time. This is the source of much of the shop lore about chatter, some of which is usually true and some of which is seldom true.

In regenerative chatter, a small disturbance, as a result of small forced vibrations, cutting force fluctuations, etc., causes the tool to vibrate relative to the workpiece. Small bumps are left on the machined surface. Since most turning cuts normally overlap the previous cut to some degree, these bumps make the cut a little deeper when they show up on the next revolution and have the same effect as a vibratory force. This causes the tool to vibrate even though the original disturbance has disappeared. For a well-behaved cut, the bumps left by the forced vibration on the second revolution will be a little smaller than the original bumps, and after a number of revolutions, they will totally disappear.

In a situation leading to chatter, the bumps will become greater each revolution and eventually become so great that visible, audible chatter occurs. The mathematical study of whether these bumps will grow or die out is called stability theory. This study is described in several different ways in some excellent technical publications.[9-12] The most important practical findings of stability theory are as follows:

1. In some ranges of machine speed, the tendency to chatter increases or decreases as the speed is changed slightly. This is because severe chatter can only occur when the wave that is feeding back from the previous revolution is in a certain relationship to the present motion of the tool. For example, if the bump in the workpiece surface is increasing the force, and at the same time the tool is moving out of the work decreasing the force, the net result might be zero force variation, and no tendency to chatter. This phenomenon relates to the cutting speed and the natural frequency of the machine. As speed is changed, the machine might go from heavy chatter to no chatter and back again to heavy chatter.

2. The chatter frequency is very close to that machine natural frequency having the highest resonance peak (lowest dynamic stiffness).

3. Chatter never occurs until the severity of the cut reaches a certain level, dependent on the minimum dynamic stiffness of the machine structure at resonance. Above a certain level of cut severity, chatter always occurs. The threshold level depends on many different cutting and structural parameters and is very difficult to predict accurately.

4. The maximum severity of cut which can be taken without chatter is closely related to the value of the min-

imum dynamic stiffness at the highest resonance peak. The frequency response that is most important is not the response of the machine but the response to the forcing between the tool and the workpiece, which includes the tool, tool holder, machine, fixturing, and workpiece. Doubling the least dynamic stiffness permits taking a cut that is approximately twice as heavy.

5. The tendency to chatter increases in proportion to the total length cutting edge engaged in the cut. The total cutting force and power are not important. How large a chip is being produced is not important.

6. The tendency of various workpiece materials to chatter is in proportion to the cutting forces they produce. Materials which produce high cutting forces are capable of only light chatter-free cuts. Materials producing low machining forces are capable of heavy cuts without chatter.

Chatter occurs when the cutting severity is equal to or greater than the structural vibration resistance. Mathematically, chatter occurs if:

$$QL_c \geq CK_{dm} \qquad \text{(Eq. 22-5)}$$

where:

$Q$ = workpiece material tendency to chatter
$L_c$ = total length of cutting edge engaged
$C$ = factor depending on cutting speed and resonant frequency
$K_{dm}$ = minimum dynamic stiffness at the highest resonance peak of the machine tool-workpiece system

Table 22-3 contains a compilation of Q values for a wide range of workpiece materials and hardness ranges.

**TABLE 22-3 Chatter Tendency for Common Work Materials**

| MATERIAL GROUP | HARDNESS Bhn or $R_c$ | CHATTER TENDENCY Q |
|---|---|---|
| Free machining carbon steels, wrought | 100-150 | 0.9 |
| | 150-200 | 1.0 |
| | 200-275 | 1.2 |
| | 275-325 | 1.4 |
| | 325-375 | 1.6 |
| | 375-425 | 1.8 |
| Carbon steels, wrought and cast | 85-125 | 0.8 |
| | 125-175 | 1.0 |
| | 175-225 | 1.2 |
| | 225-275 | 1.4 |
| | 275-325 | 1.6 |
| | 325-375 | 1.8 |
| | 375-425 | 2.0 |
| Free machining alloy steels, wrought | 150-200 | 1.1 |
| | 200-275 | 1.4 |
| | 275-325 | 1.6 |
| | 325-375 | 1.8 |
| | 375-425 | 2.0 |
| | 45- 48 $R_c$ | 2.2 |
| | 48- 52 $R_c$ | 2.5 |
| Alloy steels, wrought and cast; High strength steels, wrought; and Maraging steels, wrought | 125-175 | 1.1 |
| | 175-225 | 1.3 |
| | 225-275 | 1.5 |
| | 275-325 | 1.7 |
| | 325-375 | 1.9 |
| | 375-425 | 2.1 |
| | 45- 48 $R_c$ | 2.4 |
| | 48- 52 $R_c$ | 2.9 |
| Tool steels, wrought and cast | 150-200 | 1.3 |
| | 200-275 | 1.6 |
| | 275-325 | 1.8 |
| | 325-375 | 2.0 |
| | 375-425 | 2.2 |
| | 45- 48 $R_c$ | 2.7 |
| | 48- 52 $R_c$ | 3.2 |

**TABLE 22-3—Continued**

| MATERIAL GROUP | | HARDNESS Bhn or $R_c$ | CHATTER TENDENCY Q |
|---|---|---|---|
| Nitriding steels, wrought | | 200-250 | 1.4 |
| | | 300-350 | 1.8 |
| Armor plate | | 200-250 | 1.6 |
| | | 250-300 | 1.8 |
| | | 300-350 | 2.0 |
| | | 350-400 | 2.3 |
| | | 400-450 | 2.6 |
| Structural steels, wrought | | 100-150 | 1.0 |
| | | 150-200 | 1.2 |
| | | 200-250 | 1.4 |
| | | 300-350 | 1.6 |
| | | 350-400 | 1.8 |
| | | 400-450 | 2.0 |
| Free machining stainless steels, wrought | austenitic | 135-185 | 1.5 |
| | | 185-225 | 1.7 |
| | | 225-275 | 1.9 |
| | | 275-325 | 2.1 |
| | | 325-375 | 2.3 |
| | ferritic & martensitic | 135-185 | 1.5 |
| | | 185-225 | 1.7 |
| | | 225-275 | 1.9 |
| | | 275-325 | 2.3 |
| | | 375-425 | 2.7 |
| Stainless steels, wrought and cast | austenitic | 135-185 | 1.8 |
| | | 185-225 | 1.9 |
| | | 225-275 | 2.1 |
| | | 275-325 | 2.3 |
| | | 325-375 | 2.6 |
| | ferritic & martensitic | 135-185 | 1.8 |
| | | 185-225 | 1.9 |
| | | 225-275 | 2.1 |
| | | 275-325 | 2.3 |
| | | 325-375 | 2.6 |
| | | 375-425 | 2.9 |
| | | 42- 45 $R_c$ | 3.2 |
| | | 45- 48 $R_c$ | 3.4 |
| | | 48- 52 $R_c$ | 4.0 |
| Precipitation hardening stainless steels, wrought and cast | | 150-200 | 2.4 |
| | | 275-325 | 2.7 |
| | | 325-375 | 3.0 |
| | | 375-440 | 3.4 |
| Gray cast iron | | 120-150 | 0.6 |
| | | 150-200 | 0.8 |
| | | 200-220 | 1.0 |
| | | 220-260 | 1.2 |
| Ductile cast iron | | 140-190 | 0.9 |
| | | 190-225 | 1.1 |
| | | 225-260 | 1.3 |
| | | 260-300 | 1.5 |

**TABLE 22-3—Continued**

| MATERIAL GROUP | HARDNESS Bhn or $R_c$ | CHATTER TENDENCY Q |
|---|---|---|
| Malleable cast iron | 110-160 | 0.9 |
| | 160-200 | 1.1 |
| | 200-240 | 1.3 |
| | 240-280 | 1.5 |
| | 280-320 | 1.7 |
| Aluminum alloys, wrought and cast | 30- 80 | 0.5 |
| | 80-125 | 0.6 |
| | (500 kg load) | |
| Magnesium alloys, wrought and cast | 50- 90 | 0.4 |
| | (500 kg load) | |
| Titanium alloys, wrought and cast | 150-200 | 1.2 |
| | 200-250 | 1.4 |
| | 250-300 | 1.6 |
| | 300-350 | 1.8 |
| | 350-400 | 2.0 |
| | 400-450 | 2.2 |
| Copper alloys, wrought | 10- 60 $R_B$ | 0.6-0.9 |
| | 60-100 $R_B$ | 1.0-1.2 |
| Copper alloys, cast | 40-100 | 0.6-0.8 |
| | 100-200 | 0.8-1.6 |
| | (500 kg load) | |
| Nickel alloys, wrought and cast | 80-150 | 1.5 |
| | 150-200 | 2.2 |
| | 200-250 | 2.8 |
| | 300-320 | 3.5 |
| High temperature alloys | 200-300 | 2.0-3.0 |
| | 300-375 | 2.0-5.0 |
| | 375-425 | 3.0-6.0 |
| Refractory alloys, wrought, cast, and P/M | 175-250 | 2.0-6.0 |
| | 250-300 | 3.0-7.0 |
| | 300-350 | 4.0-7.0 |
| Zinc alloys, cast | 80-100 | 0.5 |
| Lead alloys | 5- 30 | 0.3 |
| Tin alloys | (500 kg load) | |

### Machining Tests

A number of researchers have made cutting tests to study self-excited chatter. Figure 22-18 shows what happens during the turning of a bar if the depth of cut is gradually increased. The vibration amplitude is measured and plotted as a function of the depth of cut. The low vibration level at light cuts represents the background vibration level of the machine resulting from unbalance and irregularities in the drive train. The sharp increase in vibration at a certain "critical width of cut" is characteristic of regenerative chatter. The actual amplitude during chatter is likely to vary greatly. Not much is known about how the amplitude is determined, but once chatter occurs, the amplitude is almost always too great. The solution is to eliminate chatter rather than to determine why the machine vibrates at a certain level.

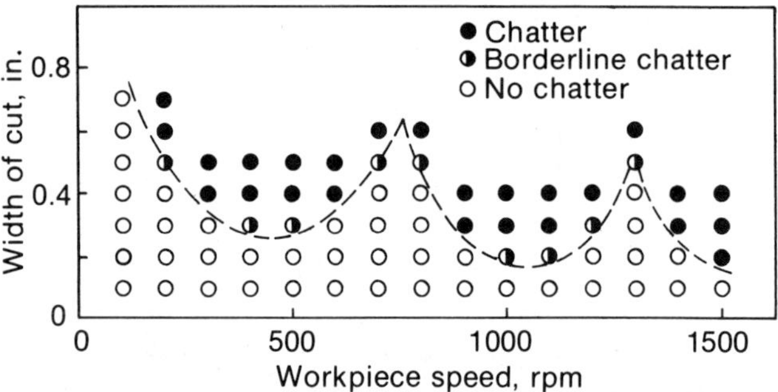


**Figure 22-19** Experimental results of a study on regenerative chatter. Lobed line denotes borderline of chatter.

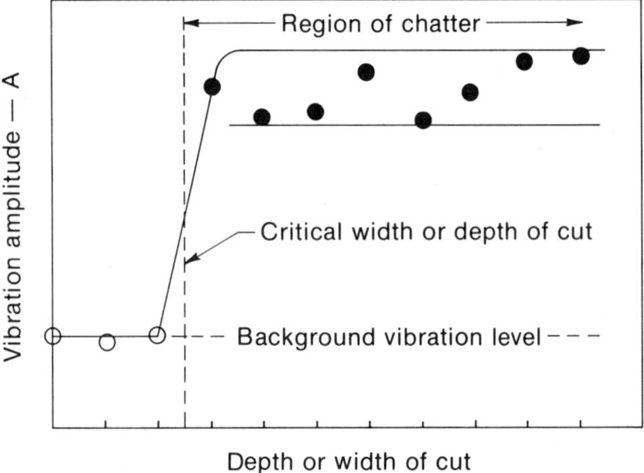


**Figure 22-18** Experimental procedure to determine the critical width of cut for regenerative chatter.

Repeating this experiment at various cutting speeds demonstrates how this "critical width" is affected. Tests were run at the speeds and widths of cut denoted by the circles in figure 22-19. The solid circles represent chatter, the open circles represent chatter-free cutting, and the half-solid circles represent borderline chatter (more than background vibration but less than violent chatter). The dashed line is an attempt to sketch the borderline of chatter. Useful concepts to be learned from this figure include the following:

1. Below a certain width of cut, there is no chatter at any speed.
2. At slightly greater widths of cut, it is possible to get into or out of chatter by changing workpiece speed.
3. Chatter occurs at large widths of cut except at very low speeds.
4. Very large chatter-free widths of cut can be taken only if the cutting speed is low enough.

The reason for the "lobes" in the figure where certain speeds enable increased chatter-free widths of cut is the interrelationship between cutting speed and regenerative chatter. The lack of chatter at very low cutting speeds is called "low-speed stability." This phenomenon is not completely understood, but it is likely the result of small areas of tool-workpiece contact just behind the cutting edge. Experiments have been performed to show that dull tools and low tool flank clearance angles increase low-speed stability.

While chatter theory is very good at explaining experimental results (other than the enhanced stability at low cutting speeds), it is difficult to use in solving chatter problems. Chatter theory is not good at predicting just when a machine will start to chatter. More theory development is necessary before this can be done.

### Factors Influencing Chatter

It seems logical that the more power consumed in a cut, and the higher the cutting forces, the greater the likelihood of chatter. This is wrong. Chatter theory says that the severity of the cut is the chatter tendency of the material multiplied by the length of tool edge in the cut. This relationship was given as equation 22-5. More power may or may not mean greater tendency to chatter. The chatter tendency of a workpiece material is closely related to the cutting forces and power that occur. The theoretical chatter tendency of a variety of workpiece materials is listed in table 22-3. The tendency to chatter is generally proportional to the expected cutting power. Low values are associated with aluminum, intermediate values with ordinary cast irons and steels, and high values with the difficult-to-machine alloys often found in aerospace applications. The chatter tendency numbers in the table show that, all other factors being equal, a cut three times as heavy can be taken with a 175 Bhn free machining carbon steel as with a 350 Bhn cast precipitation hardening stainless steel.

Another major factor influencing cutting severity is the length of the cutting edge. In rough turning, the length of the cutting edge in the workpiece is approximately equal to the radial depth of cut. Thus, if one-half inch is being taken off the diameter of a bar, the cut is twice as severe as if one-quarter inch were being removed. Note the lack of importance of the feed rate. According to chatter theory and supporting test data, light feeds are just as likely to cause chatter as heavy feeds. This implies that a radial depth of cut of 0.300 inch with a feed per revolution of 0.001 inch chatters sooner than a 0.250-inch radial depth of cut with a feed per revolution of 0.030 inch. While the heavier cut requires much more power, chatter is less likely to occur.

The important dimension affecting chatter in plunge cutting with a cutoff or form tool is the width of the tool. When cutting off a flimsy workpiece, a 1/8-inch wide cutoff tool is twice as likely to chatter as is a 1/16-inch wide tool. Again, the feed rate is an unimportant variable from the viewpoint of chatter. In drilling, the length of the cutting edges engaged is approximately equal to the drill diameter. The chisel edge has a strong stabilizing influence not predicted by theory. When a hole is enlarged and the chisel edge is not cutting, the tendency to chatter is proportional to the difference between the diameters.

Chatter in milling is a more complicated process to understand because its characteristics are three-dimensional. However, it can be simplified by the following procedure. The length of cut of a single tooth is first identified as illustrated in figure 22-20. The number of teeth in contact is then determined using the method shown in figure 22-21. The cutter should be "rotated" to produce the maximum number of teeth in contact, namely, the situation when one tooth (for example, tooth 1) is about to leave the cut. The total length of cutting edge in contact is the length per tooth times the number of teeth in contact. This number determines the cut severity. As in turning, feed rate is not considered since the chatter tendency in face milling is independent of the table feed rate. Consequently, for the same metal removal rate, a shallow cut at high feeds is preferable to a deep cut at light feeds.

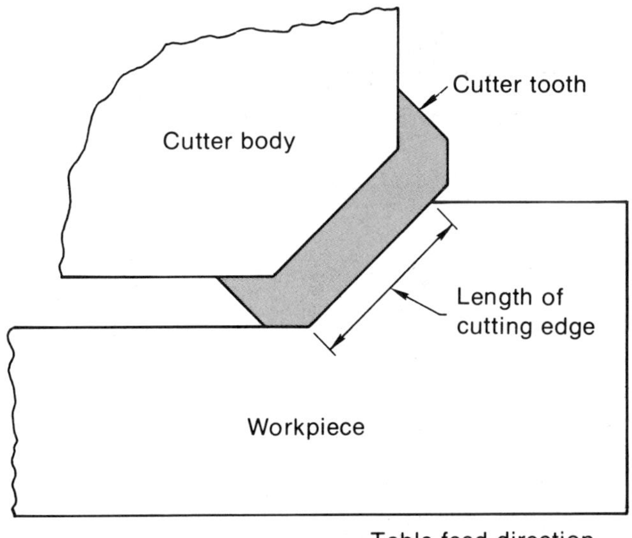


**Figure 22-20** Schematic diagram showing length of cutting edge in milling.

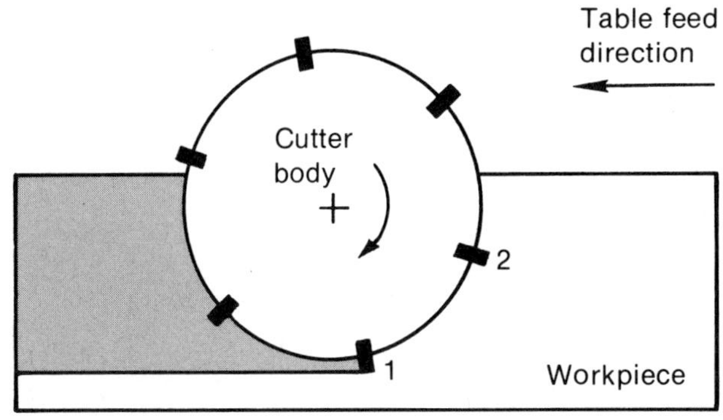


Number of teeth in contact, N = 2

**Figure 22-21** Schematic diagram illustrating the method that can be used to determine the number of teeth in contact with the workpiece during milling.

## TROUBLESHOOTING CHATTER

### Problem Identification

There are two types of vibration in machine tools: forced and self-excited. Multi-tooth cutter impact, gear drive irregularities and unbalance, and motion of the machine foundation contribute to forced vibration which results in surface irregularities in machined parts. In cutting operations free from these factors, vibration enters as self-excited chatter, a basic instability of the cutting mechanism itself. Figure 22-22 contains a flow chart that may be used to identify which type of vibration problem is present. The following discussion amplifies the steps in this process and provides guidelines relative to dealing with both types of vibration.

The vast majority of chatter problems are caused by self-excited vibration. A less frequent problem is chatter resulting from interrupted cuts forcing the structure at its resonant frequency. To solve these problems, the first step is to be able to distinguish between the two. Alternative fixes can then be logically accomplished.

The key to troubleshooting chatter problems is determining the chatter frequency. While special sensors and electronic instruments can be used, it is often possible to determine frequency without instruments by measuring the spacing between chatter marks on the workpiece surface. Figure 22-23 shows chatter marks on a typical turned surface. In the enlargement, the marks are numbered for easy reference. The correct measurement of chatter "wave length" is between corresponding points of mark 1 and mark 2, or of mark 3 and mark 4, or of mark 5 and mark 6. Note that the spiral, marks 1-3-5-7, has no significance, since the tool path is to be followed in making measurements. The tool makes mark 1, mark 2, etc., as it moves along the feed lines. After a complete revolution of the workpiece, it passes adjacent to mark 1 making mark 3, followed by mark 4, and another revolution later with marks 5 and 6, etc.

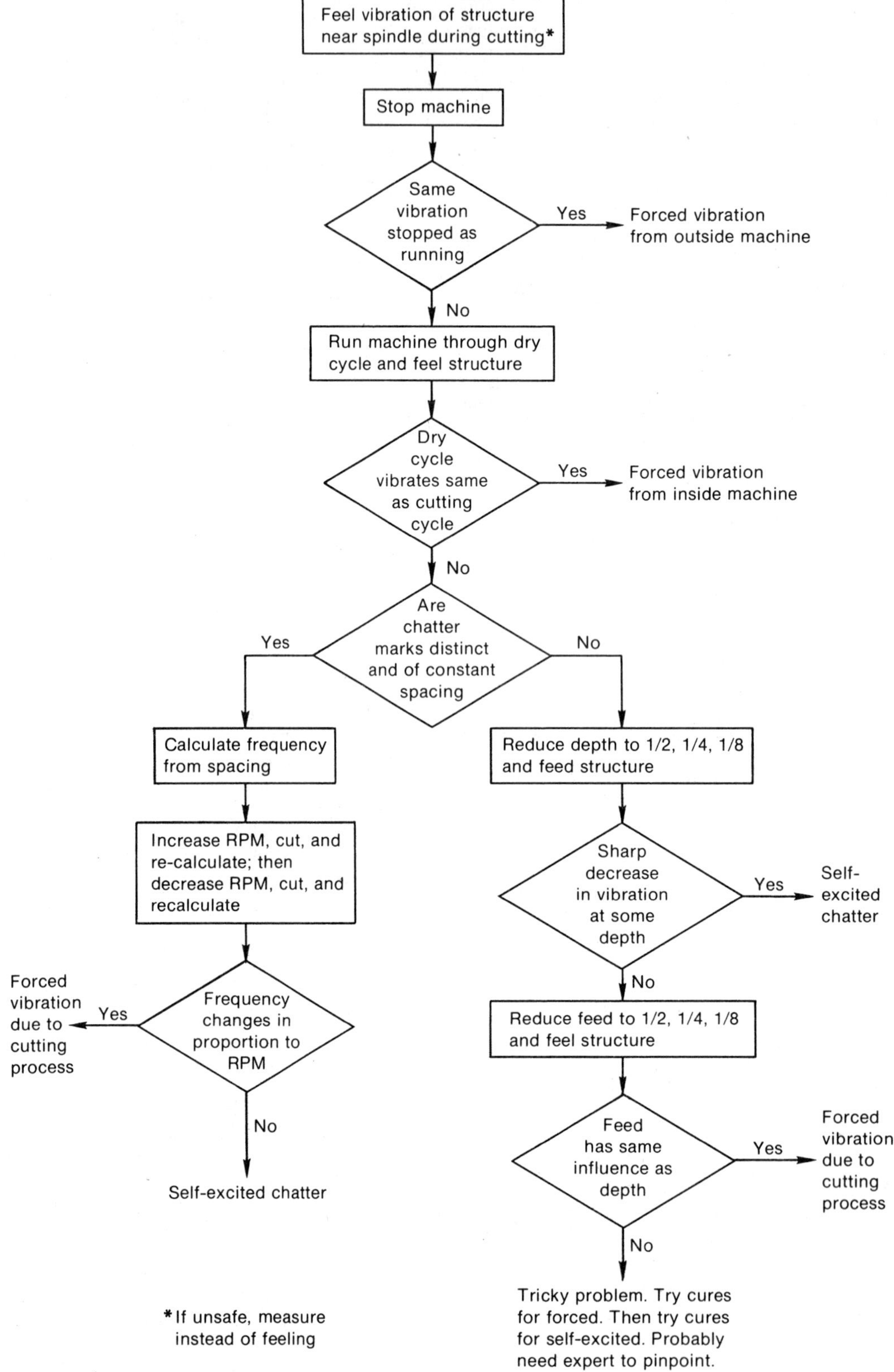


**Figure 22–22** Flow chart for identifying vibration problems.

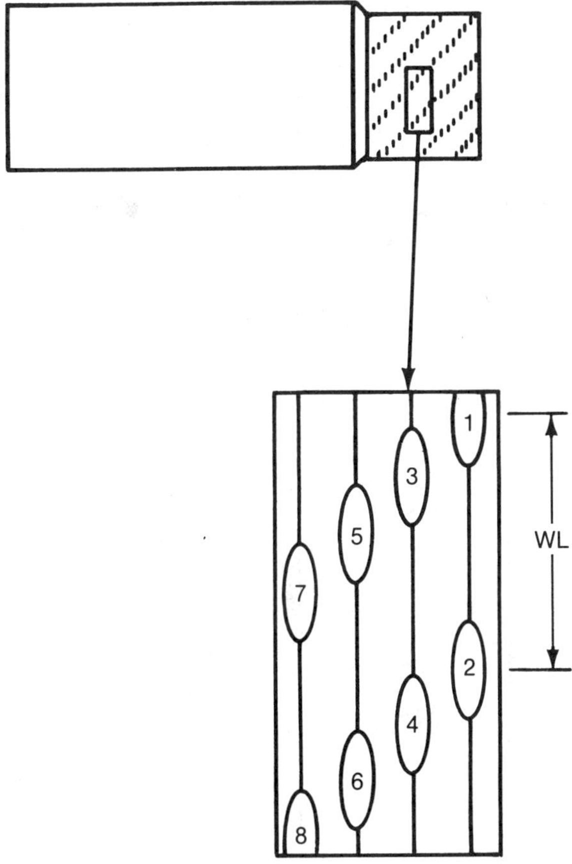


WL = Chatter wave length

**Figure 22-23** Chatter marks on a turned surface.

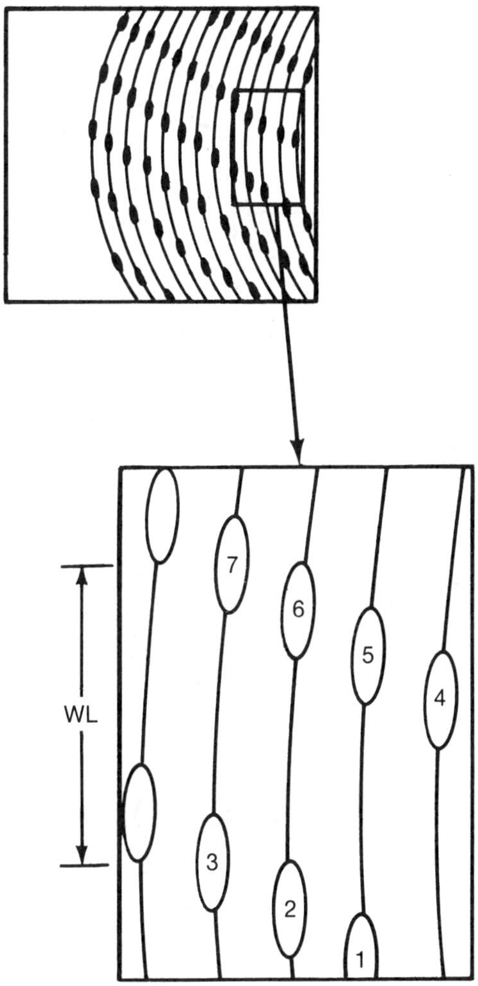

WL = Chatter wave length

**Figure 22-24** Chatter marks on a milled surface.

Figure 22-24 shows a similar situation for face milling. Again, no importance should be given to the spiral patterns, marks 1-2-3, or marks 4-5-6. By following the feed lines, the marks were found to be made in the following sequence:

Tooth 1: Mark 4; off the workpiece.
Tooth 2: Mark 5; mark 1; off the workpiece.
Tooth 3: Mark 6; mark 2; off the workpiece.

The chatter wave length is again measured by moving along, not across, feed lines.

In practice, the chatter wave length should be measured several times. Unequal measurements are usually caused by continuing vibration during the disengagment of the tool from the work. These extra vibration marks overlay the true pattern and make the spacing of marks meaningless. This problem can be minimized or avoided by disengaging the feed as quickly as possible or by rapid traversing away from the workpiece in the direction perpendicular to the chatter-marked surface.

Once the chatter wave length or spacing is measured, the frequency of vibration may be calculated as follows:

$$f = \frac{V}{5\,(WL)} \qquad \text{(Eq. 22-6)}$$

where:

f = vibration frequency in cycles per second
V = cutting speed in feet per minute
WL = length (or chatter mark spacing) in inches
5 = constant introduced because of the units used

For example, if the chatter wave length spacing in the previous milling example is 0.1 inch, and if the cutting speed is 200 feet per minute, then the chatter frequency is:

$$f = \frac{200}{5\,(0.1)} = 400 \text{ cps}$$

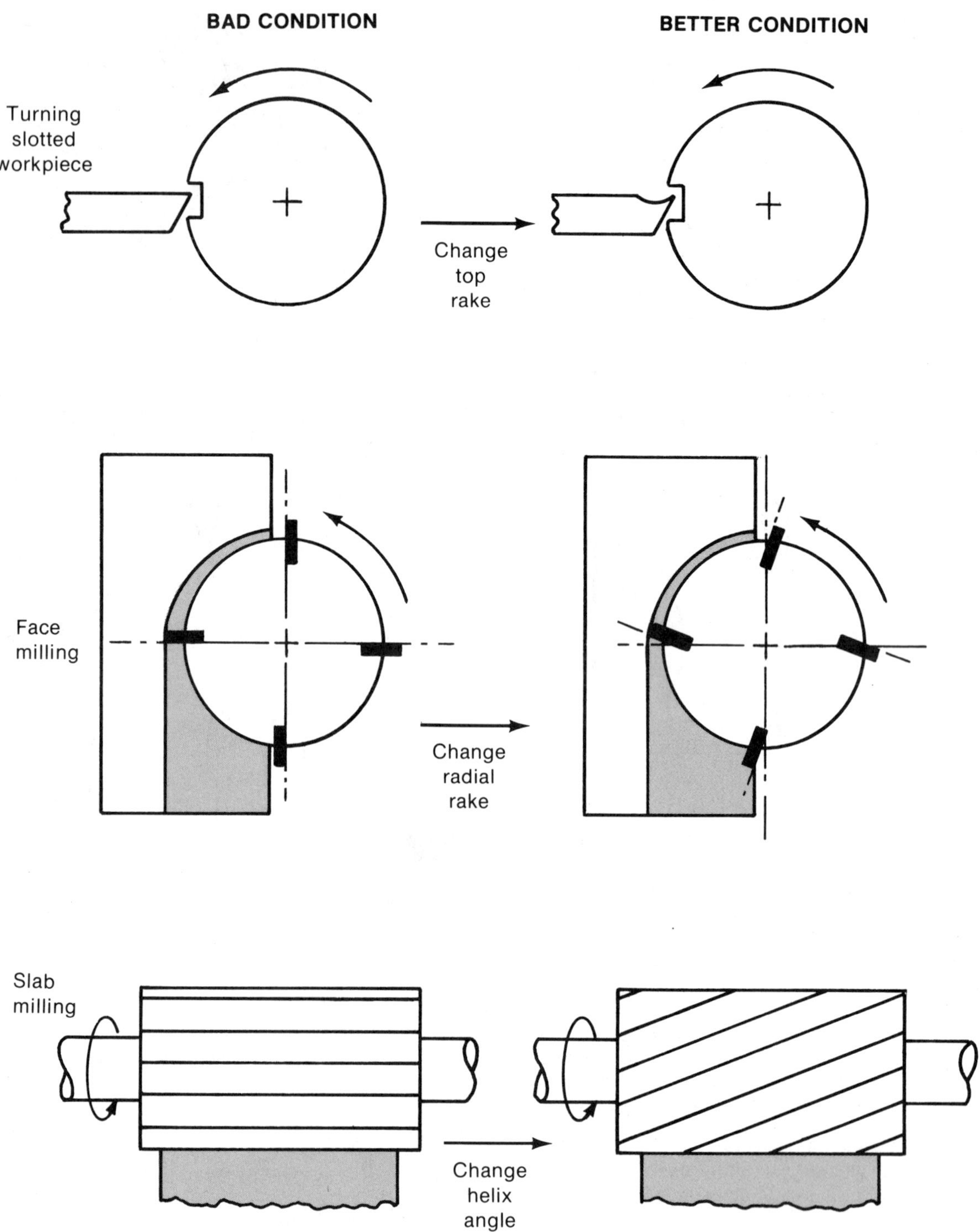


**Figure 22–25** Suggested methods for reducing the magnitude of forced vibrations.

Whether the vibration problem is the result of interrupted cutting can be determined quickly by increasing or decreasing the cutting speed by 50 percent. After each cut at a different speed, measure the chatter marks and recalculate the frequency. The chatter is the result of interrupted cutting if the frequency changes in proportion to the cutting speed. The fundamental principle here is that self-excited chatter occurs at a frequency near the natural frequency, regardless of the spindle speed, but forced vibrations always occur at the forcing frequency. A 50-percent change in cutting speed will make very little change in the frequency of self-excited chatter. The same change in speed should change the frequency of forced vibration by 50 percent.

Another sure way to identify self-excited chatter is to reduce the severity of the cut. As the length of cutting edge engaged is decreased, a point is reached where a small decrease in cutting edge length causes a dramatic decrease in vibration. The decrease may be associated with "width of cut" or "depth of cut" depending on the particular cutting operation; however, it is never associated with the feed rate.

### Elimination of Forced Vibration

If the problem is identified as being forced vibration, try either to reduce the magnitude of the forcing vibration or to increase the stiffness of the machine tool system. One way to reduce the magnitude of the forced vibration is to change tool angles so that the cutting load is picked up gradually. Figure 22-25 illustrates how bad situations can be changed into better ones. Another possible way to reduce the forced vibration is to reduce the feed rate. The vibration amplitude should change in proportion to the feed rate; therefore, if only a slight reduction is needed, decreasing the feed rate is an appropriate solution.

A change in the cutting speed often has dramatic effects on reducing forced vibration. Either an increase or a decrease in cutting speed, depending on which side of the resonance peak the machine is operating, may reduce the vibration enough to solve the problem. If the machine tool is being forced at a frequency just below the natural frequency, a small increase in speed will increase vibration amplitude, and further increases in speed will reduce vibration amplitude. On the other hand, a reduction in speed will immediately reduce amplitude. The opposite response is true if the forced vibration started out above the natural frequency. Since there is no way to tell in advance which situation applies, try either an increase or a decrease in cutting speed of at least 20 percent, whichever is easiest. If the results are not positive, try changing the speed in the other direction.

A technique that sometimes works in milling is to change the direction of the cutting force away from a weak direction of the structure toward a strong direction. This cannot always be done scientifically because force direction or machine directional rigidity may not be known. Changes such as feeding a face mill in the Y-direction rather than in the X-direction or changing the table feed direction in slab milling have been known to work.

### Elimination of Self-Excited Vibration

Decreasing the length of the cutting edge in the cut will eventually eliminate self-excited chatter. In turning, this means reducing the radial depth of cut. In face milling, the axial depth can be reduced or every other tooth can be removed from the cutter with a corresponding reduction in feed rate. In cutting with an end mill, either the radial depth or the axial depth may be reduced, or a cutter with fewer teeth may be used.

Dulling or stoning the cutting edge is a method that often reduces regenerative self-excited chatter, particularly at low cutting speeds. Increasing the feed rate is another helpful method of reducing self-excited chatter. Self-excited chatter occurs less frequently for a heavier chip load of 0.010 inch or greater than for a light chip load of 0.005 inch or less. The heavier feed rate increases the cutting force but not the length of cutting edge engaged in the workpiece and, hence, slightly reduces the tendency of the material to chatter.

If the cutting speed is low (less than 200 feet per minute) or if the chatter frequency is high (1,000 cycles per second or more), a reduction in the cutting speed tends to reduce chatter. Almost any size cut can be taken without chatter if the cutting speed is lowered sufficiently. This added stability at low speeds is associated with springback of the machined surface just behind the cutting edge. A small contact force is created that tends to push the tool out of the work and resist chatter. The effect is much more pronounced for tools with low clearance angles. Tool wear stabilizes the cut since the worn edge acts like a low clearance angle generating a force which opposes tool penetration into the work. If the clearance angle and the speed can be reduced, chances of solving the problem are improved.

One final technique which applies to both self-excited and forced vibration is redirecting the cutting force toward a stiffer direction of the structure. For example, the turning tool with a 45-degree lead angle on the top in figure 22-26 will chatter much sooner than the one shown on the bottom with a zero lead angle. Because the cutting force is perpendicular to the cutting edge, a major portion of the force is aimed in the weak bending direction of the bar in the first case (a). However, in the second case (b) the force is aimed along the very rigid axis of the bar. There is no simple rule for aiming the force in a more rigid direction except in the simplest of cases. A combination of analysis and judgment is required.

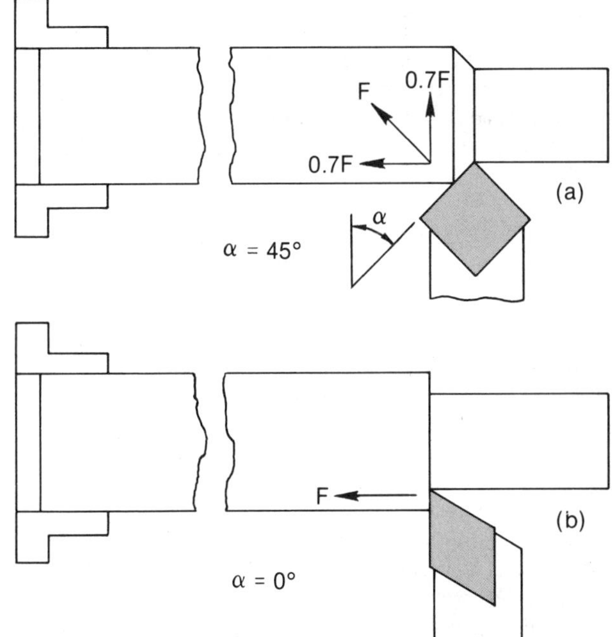


**Figure 22-26** Influence of lead angle, $\alpha$, on direction of cutting force, F.

## REGENERATIVE CHATTER CHECKLIST

By applying the principles outlined above, many difficult vibration and chatter problems can be solved. Others require experts who have powerful instruments to give them data not available from a simple approach. Often the measurements lead to the conclusion that the machine has a weakness that can be corrected. Further analysis leads to suggestions for stiffening the machine, or adding a vibration damper. Even if it is eventually necessary to go to an expert for a solution to a particular vibration problem, the suggestions contained in table 22-4 should always be tried first. There is a good chance that one of these simple techniques will yield a solution.

**TABLE 22-4 Checklist of Methods To Eliminate Vibration and Chatter Problems**

**TURNING**

- Lower the stock removal per pass.
- Try a heavier feed if presently feeding below 0.010 inch per rev.
- Reduce the width of cut of the form tool.
- Use a steadyrest on slender bars.
- Chuck slender bars at their center and turn end-for-end.
- Use a smaller nose radius on the tool to keep the cutting forces away from the radial direction.
- Modify the cutting edge angle to be more radial keeping the forces away from the radial direction.
- Consider using the Kennametal Roller Devibrator® (or equivalent) on slender bars.
- Minimize the tool overhang.

**BORING**

- Use the largest possible boring bar diameter.
- Reduce the boring bar length to as short as possible.
- Make the tool nose radius as small as possible to minimize the radial force.
- Use a side cutting edge angle as radial as possible to reduce the radial force.
- Reduce the stock removal in roughing.
- Consider a damped boring bar for L/D's greater than 6.
- Consider solid carbide boring bars for L/D's over 4.
- Consider reaming after semi-finish boring for finish boring problems.
- Consider multi-tooth boring bars for roughing operations.

**MILLING**

- Reduce the stock removal per pass.
- Reduce the width of cut.
- Try heavier feeds if presently feeding below 0.010 inch per tooth.
- Try reducing the RPM when using high speed steel cutters.
- Try dulling the cutting angles of high speed steel cutters.
- Avoid high wrap-around angles in end mill cutting.
- Generate an interior radius rather than plunging the cutter.
- Try using serrated edge end mills.
- Use a cutter with half as many teeth.
- Remove half of the inserts in a face milling cutter.
- Try a face mill with a random tooth space design.
- In face milling, use more of a corner angle to get cutting edge more toward the finished surface and reduce x and y horizontal force components.
- In face milling, try feeding at 90° to the present feed direction.
- Try an axial stepped milling cutter design.
- Try changing from climb to up milling (or vice versa) to change the force direction.

**GRINDING**

- Reduce the width of grind.
- Reduce the stock removal rate.
- Dress the grinding wheel more frequently.
- Use a sharper dress (more lead, greater depth, sharper diamond) if possible.
- Use a softer grinding wheel.
- Reduce excessively high work speeds.
- Use a backrest on slender workpieces.

## REFERENCES


1. Ostwald, P. F. and Shamblin, J. E. 1968. Effects of dynamic chip breaking upon surface microgeometry and free chip dimension. *Transactions of the ASME* 90 (Series B): 71-78.

2. Kristoffy, I. et al. 1965. Influence of vibrational energy on metalworking processes. U. S. Air Force Technical Report AFML-TR-65-211, Wright-Patterson Air Force Base, OH.

3. Langenecker, B. et al. 1964. Effect of ultrasound on deformation characteristics of structural metals. U. S. Navy NAVWEPS Report 8482, NOTS TP 3447, U. S. Naval Ordnance Test Station, China Lake, CA.

4. Andrew, C. 1965. Chatter in horizontal milling. *Proceedings of the Institution of Mechanical Engineers* 179 (Part 1, No. 28): 877-898.

5. Peters, J. 1965. Survey of methods for increasing the damping of machine tools. Report to the General Assembly of CIRP.

6. Long, G. W. et al. 1968. The effect and control of chatter vibrations in machine tool processes. U. S. Air Force Technical Report AFML-TR-68-35, Wright-Patterson Air Force Base, OH.

7. Koenigsberger, F. and Tlusty, J. 1969. *Machine tool structures*, Volume 1. London: Pergamon Press.

8. Gaudissart, Ph. and Van Herck, P. 1964. Influence des vibrations sur le fini de surface et l'erreur de forme dans les passes de finition. CRIF Report, University of Louvain, Louvain, Belgium.

9. Tobias, S. A. 1961. *Schwingungen an Werkzeugmaschinen.* Munich: Carl Hanser Verlag.

10. Tlusty, J. et al. 1962. *Selbsterregte Schwingungen an Werkzeugmaschinen.* Berlin: VEB Verlag Technik.

11. Hahn, R. 1954. On the theory of regenerative chatter in precision grinding operations. *Transactions of the ASME* 76 (May): 593-597.

12. Merritt, H. 1965. Theory of self-excited machine tool chatter. *Transactions of the ASME* 87 (Series B): 447-454.

13. Kegg, R. L. and Sisson, T. R. 1968. Trouble shooting chatter by improving cutting conditions. Technical Paper MR68-615, Society of Manufacturing Engineers, Dearborn, MI.

14. Hosoi, T. and Hoshi, T. 1977. Cutting action of ball end mill with a spiral edge. *Annals of the CIRP* 26(1): 49-53.

15. How to troubleshoot machine chatter and vibration. 1978. Manufacturing Education, P. O. Box 36050, Cincinnati, OH 45236.

# SECTION 23

# NUMERICAL CONTROL MACHINING


23.1 NC Machining Guidelines ........ 23-3
- Introduction ........ 23-3
- Selection of Parts for NC Machining ........ 23-3
- Selection of Machining Conditions for NC Machining ........ 23-5
- Tooling for NC Machining ........ 23-6

23.2 NC Vocabulary ........ 23-9

## INTRODUCTION

Numerical control (NC) is a sophisticated and versatile form of automation that continues to significantly impact the manufacturing industry. One of the more commonly accepted definitions of NC is that given by the Electronics Industries Association (EIA): "A system in which motions are controlled by the direct insertion of numerical data at some point. The system must automatically interpret at least some portion of this data."

The numerical control system consists of the five basic, interrelated components as follows:

1. Data input devices
2. Director (or machine control unit)
3. Machine tool or other controlled equipment
4. Servo-drives for each axis of motion
5. Feedback devices for each axis of motion

These components operate together to provide an integrated system that performs nearly all of the machine operator's functions. While there are many varieties of these components available, they all can be categorized into one of the five groups listed above. The major components of a typical NC machine tool system are shown in figure 23.1-1.

Numerically controlled machine tools are commonly classified as being either point-to-point or continuous path. The simplest form of numerical control is the point-to-point machine tool used for operations such as drilling, tapping, boring, punching, spot welding, or other operations which can be completed at a fixed coordinate position with respect to the workpiece. The tool does not contact the workpiece until the desired coordinate position has been reached; consequently, the exact path by which this position is reached is not important. The programming of these systems is elementary and can be done by anyone who has both the ability to read engineering drawings and a basic understanding of machining principles and practices.

With continuous path (that is, contouring) NC systems, there is contact between the workpiece and the tool as the relative movements are made. Continuous path NC systems are used primarily for milling and turning operations that can profile and sculpture workpieces. Other NC continuous path operations include: flame cutting, sawing, grinding, welding, and even operations such as the application of adhesives. We should note that continuous path systems can be programmed to perform point-to-point operations, although the reverse (while technically possible) is infrequently done. One important concept of continuous path operation is that the positional relationship between the tool and the workpiece must be provided for at all times because the path the cutter follows in moving from one position to another ultimately determines the workpiece shape.

## SELECTION OF PARTS FOR NC MACHINING

NC is not the answer to all manufacturing problems. Conventional machine tools or automatic transfer-line machinery may provide better solutions to particular problems. Usually, the answer is based on lot size. NC is often advantageous for small or moderate lot sizes, for example, from 1 to 100. It is important to note that there is a predominance of lot sizes of this magnitude in industry today.

As manufacturing costs continue to rise, the economic necessity for cost-reduction programs and increased machine-tool utilization requires that careful consideration be given to the selection of numerical control applications. Since various philosophies exist among companies as to when NC should be used, there currently is no general rule that can be applied.

Parts selection for NC should be based on an economic evaluation, including scheduling and machine availability. Economic considerations affecting NC part selection include alternative methods, tooling, machine loadings, manual versus computer-assisted part programming, and other applicable factors.

Thus, numerical control should be used only where it is more economical or does the work better or faster, or where it is more accurate than other methods. The selec-

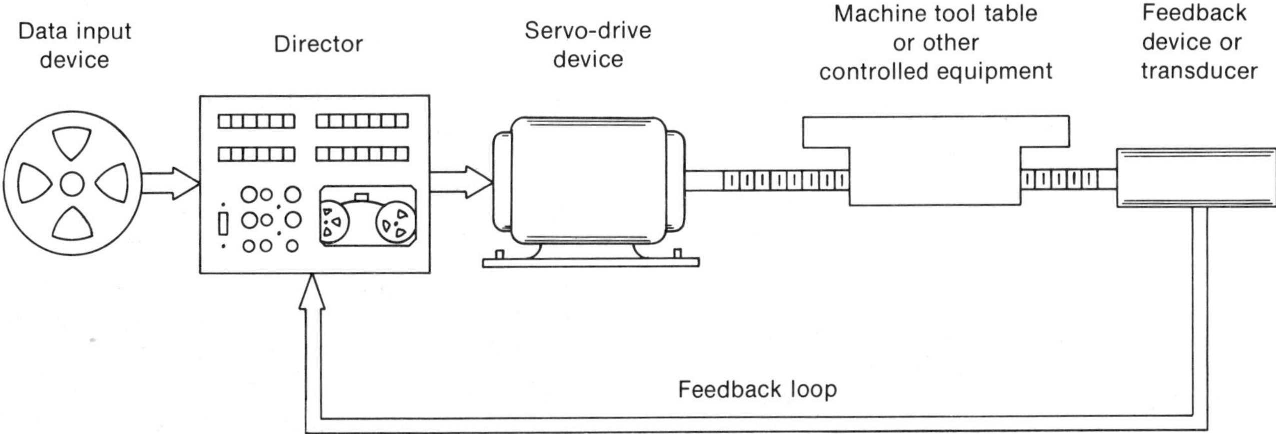


**Figure 23.1-1** Simplified numerical control system.

tion of parts to be assigned to NC has a significant effect on its payoff. The following guidelines which may be used for parts selection describe those parts for which NC may be applicable:

1. **Parts which require SUBSTANTIAL TOOLING COSTS in relation to the total manufacturing costs by conventional methods**

   For example, NC should be used where expensive drill jigs, templates, or models are otherwise required. Drill bushings are not needed for NC; consequently, the initial or replacement costs associated with drill jigs in conventional drilling do not apply to NC.

2. **Parts which require LENGTHY SETUP TIMES compared to the machine run time in conventional machining**

   Rapid setup of NC jobs is common because part fixturing is simple. This makes NC economical for jobs where the machine run time is short.

3. **Parts which are machined in SMALL OR VARIABLE LOTS**

   Usually, the savings in setup time and handling using NC machine tools outweighs the higher operating costs of NC compared to conventional equipment.

4. **A WIDE DIVERSITY OF PARTS requiring frequent changes of the machine setup and a large tooling inventory if conventionally machined**

   This is a common situation in the aerospace, the machine tool, and the job shop industries. The quantities are not large enough to justify transfer machines as used in the automotive industry. The part dimensions and configurations may cover such a broad range that adequate inventory may be prohibitively large. NC may be used to provide spare part capability without storing the parts or bulky fixtures and tooling after the product is no longer marketed.

5. **Parts which are PRODUCED AT INTERMITTENT TIMES because demand for them is cyclic**

   This occurs where part requirements are seasonal or subject to production scheduling for assembly. Whereas conventional machining requires the costly storage of templates, tools and fixtures, NC requires only tape storage.

6. **Parts which have COMPLEX CONFIGURATIONS requiring close tolerances and machined relationships**

   NC with multiaxis capability makes simultaneous machining motions possible, and parts can be produced which previously were impossible to produce by conventional methods.

7. **Parts which have MATHEMATICALLY DEFINED COMPLEX CONTOURS**

   In general, if a part surface can be defined mathematically, it can be machined by NC.

8. **Parts which require REPEATABILITY from part to part and lot to lot**

   Data stored on tape or other input media will not permit an operation to be forgotten or skipped.

9. **VERY EXPENSIVE parts where human error would be very costly and increasingly so as the part nears completion**

   Even prior to machining, the cost of the forging or casting required may be high or the material itself may be expensive.

10. **HIGH PRIORITY parts where lead time and flow time are serious considerations**

    Usually, the flow time from blueprint to the finish-machined part is considerably less with NC machining, provided that the programming can be expedited through any backlog. Special tooling, if required, is usually simpler for NC than for conventional machining and can be designed and built faster.

11. **Parts with ANTICIPATED DESIGN CHANGES**

    Changes can be made on the tape or other data input media with relative ease compared to changes in hard tooling.

12. **Parts which involve a LARGE NUMBER OF OPERATIONS or MACHINE SETUPS**

    Part operations may be complicated, tedious, and difficult to follow. NC machining allows data on the tape to be utilized repeatedly once it is satisfactorily programmed. This is a considerable advantage over conventional machining where the complexity of following the instructions continues for each part. NC can also utilize a variety of successive, interchangeable tools without time losses for machine stoppage, setups, and transfers. For example, parts requiring straight-cut milling combined with many drilling, reaming, and tapping operations are excellent choices for the less sophisticated machining centers.

13. **Parts where NONUNIFORM CUTTING CONDITIONS are required**

    The cutting speed, feed, or depth of cut may need to be varied because of the part configuration. Such variations may be incorporated into the NC program.

14. **Parts which require 100 percent INSPECTION or require measuring many checkpoints resulting in high inspection costs**

    With the inherent dimensional repeatability of NC, it is often necessary to inspect only the first and last pieces completely and to spot check the critical dimensions on the other pieces.

**15. FAMILY OF PARTS**

These are parts that are in the same general class with respect to configuration or manufacturing sequence but vary in dimensions. This situation is particularly common in the machine tool industry. Common setups may be used for these parts, and programming time may be reduced by using the same basic format or macro's for all the parts in the family.

**16. MIRROR-IMAGE PARTS**

In the aerospace industry, it is quite common to have dimensionally identical parts which are mirror images of each other. These are usually referred to as left- and right-hand parts. Many similar situations exist outside the aerospace industry. For example, a housing with internally and externally machined surfaces may be impossible to fabricate in one piece. For manufacturing purposes, it could be made in two symmetrical halves (that is, left- and right-hand parts). With NC, both parts may be made utilizing one program tape by actuating the mirror-image switch on the control and changing to opposite hand cutting tools if required. The mirror-image capability is provided as an option on most NC systems.

**17. NEW PARTS for which conventional tooling does not already exist**

Production of a new product alone is seldom sufficient justification for using NC. Significant cost savings do not begin to accrue until new parts are planned for NC. However, in the first stages of the NC installation, a proper selection balance should be maintained between old and new parts. In this manner, during the learning period the NC machine can be properly loaded while allowing conventional machining of the old parts should any problem arise. This dual planning eliminates needless part delays and possible stock outs.

**18. Parts which are suitable for MAXIMUM MACHINING on NC machine tools**

Although it is best if parts can be machined completely on NC machine tools, allied conventional machining operations often are required. An exception is the manufacturing cell or work center concept common to group technology. In this situation, the operator of an NC machine may also do part deburring, drilling or tapping of holes on peripheral equipment manually while waiting for the NC machine to finish its cycle. By balancing operations timewise between the NC machine and manual equipment, a reduction in the overall processing time can be achieved. Usually, however, if only the most time-consuming, conventional operations are replaced by NC, many of the benefits of NC will be minimized or eliminated.

Cost studies should be performed for parts that are selected when one or more of the above guidelines apply. Sometimes even simple configuration parts will show savings by NC production, so cost studies are advised for these parts as well. Various cutting conditions should be investigated and analyzed to determine the most economic operating conditions. Computer and calculator programs are available to carry out the simple, but long and tedious, calculations.[1,2] Machining data relating tool life with cutting speed and other parameters are required for these calculations. Such data are contained in this and in other MDC publications. Section 21 of this handbook and reference 3 at the end of this section outline and demonstrate this approach.

## SELECTION OF MACHINING CONDITIONS FOR NC MACHINING

The importance of selecting proper machining conditions is an essential step in good machining practice. To be successful, every machining operation using NC equipment must be based on a sound foundation of planning, coordination and feedback. The most widely used criteria for machinability include tool life, surface roughness, accuracy, and power consumption. Of these, tool life is the most commonly used and has been extensively related to the cutting conditions. Machinability data are needed to predict the life of a cutting tool in relation to some known reference value. It is important to be aware of all the variables that may affect the resulting tool life whenever data are accumulated for future use. Some of the more important variables, in broad groupings, include the following:

- Cutting tool design including material
- Tool characteristics
- Workpiece properties
- Characteristics of the machine tool
- Dimensions of the cut
- Presence or absence of a cutting fluid and its properties

With a knowledge of these variables and the degree of variability within each grouping, it is possible to achieve predictable tool life.

Machinability data in various degrees of sophistication are available for use by industry for NC machining. These data include (1) recommended starting conditions and (2) wide range machining data.

This *Machining Data Handbook* contains recommended starting conditions for most of the significant work materials and machining operations. It supplies the recommended feed, speed, tool material, tool geometry, and cutting fluid, as well as data for determining the horsepower requirements for various machining operations. These recommended starting conditions are currently the most practical and readily accessible data for general shop application.

Detailed machining data covering the tool life obtained under a wide range of machining parameters, such as speed, feed, tool material, tool geometry, and cutting fluids, are very useful in selecting a set of conditions for machining at lower costs or higher production rates. Detailed machining parameters can be obtained from laboratory studies or from shop studies by varying the pertinent parameters and observing their effect on tool life.

Machinability data that are accurate, reliable and appropriate are essential to the success of NC machining. The philosophy of NC manufacturing, involving its high level of automation, demands the use of the best available data. The high cost and productive capacity of NC are frequently such critical factors that every effort must be made to find the best set of operating conditions.

Since data for conventional machining are often conservative, they are not particularly suitable for NC machining operations. Further, such data are often too simple or based on information that is too generalized. NC machining conditions selected from these data may result in significant machine underutilization and reduced profits.

For NC machining, the feeds and speeds are most often computed quantities rather than being dependent solely on operator judgment. The effect of this is more consistent results because the operator-induced variables are minimized or eliminated. Another approach sometimes employed with NC feed and speed selection is to program the cutting conditions on the high side and then use override controls to suit actual conditions.

Specialized machinability data systems are being developed for NC because simple data systems are inadequate. Reasons for this inadequacy are as follows:

- The higher cost of NC machine tools and machining time
- Programming costs
- Variations in NC machine tools and NC production requirements
- The shift of responsibility for NC operations from operator to programmer

The computer has an integral role in both the development and the maintenance of specialized machinability data systems for NC. The number of variables, including materials, cutting tools, etc., suggests that for a system to be both practical and efficient, and at the same time accurate and realiable, use of the computer is essential. Also, in a computerized machinability system, data can be easily updated to incorporate the current conditions existing for a given operation.

Work is in progress in the United States and abroad to develop computerized machinability data systems. These systems vary in their final purpose and application, but many are being devised to faciltate both numerical control programming and machining. Some are being designed for numerical control as well as for conventional machining. They all have as a common purpose the systematic and rapid selection of data for the ultimate user.

## TOOLING FOR NC MACHINING

NC machine tools provide more cutting time than conventional machines. Conventional machines may use some 30 percent of their total available time actually cutting, while for NC machines, cutting time may be on the order of 70 percent. A major contributing factor in this reduction of nonproductive time has been the development of quick-change and preset tooling. Because there is high initial investment in the capital equipment used, it is essential that every avenue be followed that increases the percentage of machining time. The primary objective of tooling for NC is to assure that there is complete compatibility between it and all the engineering and economic criteria by which numerically controlled machines are justified.

Tooling systems for NC must incorporate the principles of eliminating operator error and maximizing productivity. These goals can be achieved using the strategies that follow:

1. Automatic tool changing for sequential operations
2. Off-machine presetting of tools
3. Speeding up the manual replacement and changeover of tools
4. Reducing machine adjustments caused by tool deflection and wear
5. Using the NC program to facilitate tool selection and changing

For NC applications, it is typical that less hard tooling is required than for comparable operations on conventional machines. Numerical control, with its relatively inexpensive tooling, eliminates the need for expensive jigs, fixtures, cams, master parts, form tools, etc. Lower tooling costs enhance the justifiability of parts for production with NC equipment. The inherent accuracy and repeatability of parts produced using NC result in a need for less assembly tooling. It should be emphasized that while less tooling is required for NC machining, the tooling that is used must be more precise so that the inherent precision of the NC equipment can be fully utilized.

Because numerical control machining is flexible, it is important to select tooling that is simple in design so that it functions with a high degree of dependability. The tooling should be designed so that the cost of maintenance is low and components may be easily replaced or interchanged. Complex tool designs are more likely to cause machine downtime. While tool costs are a very small part of the finished product cost when things are going properly, improper tool design can cause excessive difficulty and high manufacturing costs. For some applications, there may be a need for heavy stock removal; for others, thin wall sections may have to be machined. If the tooling system is carefully designed, both applications can be handled with a single tool holder; only the geometry or grade of the insert need be changed.

NC machining has introduced a number of new concepts for the operation and organization of modern machine shops. One of the more important concepts has been the introduction of tool standardization. Tool standardization becomes increasingly important as the number of NC machines in a plant increases. Tool standardization means finding the minimum number of types of tools, adapted to the NC system, that will do the majority of the jobs on as wide a variety of machines as possible.

The advantages of tool standardization can be signficant. With standardization, substantial reductions can be expected in the amounts of both initial and backup tooling. In addition, the inventory of tools can be maintained at a reasonable level, and the interchangeability of tooling among various machines can provide a decided intangible operational advantage. The specific method by which tools can be standardized obviously will vary. For single point tools, these methods might include control dimensions, insert styles and size, and tool holder style.

Tooling which can be quickly inserted into a machine tool spindle and easily locked in place with a minimum of effort and which will cut to a predetermined size without further adjustment while on the machine is referred to as quick-change, preset tooling. The term preset tooling applies to a wide variety of tooling from simple indexable-insert, single-point tools to coded, preset tool holders for use in automatic tool changers. Preset tooling is not a new idea; however, with the increasing use of NC machine tools, its application is rapidly growing. Because NC machines are costly to operate, the changeover time between jobs must be minimized. Preset tooling can sharply reduce this downtime.

The actual presetting of tools is not difficult, as there are mechanical and optical tool-setting devices readily available. Most commonly, the tools will be set in toolrooms; although, in some cases where machine operators have both the time and the ability, they may make up preset tools at the machine tool location. Presetting the tools in the toolroom is most desirable, especially when considering the potential costs of operator mistakes and machine downtime for trial cuts or interruptions to adjust tooling. Many shops maintain a file of tool-setting drawings used by both programmers and toolroom employees. These drawings describe the cutter, identify it by a tool number, and establish a setting length. With this approach, all parties can be confident that the job will be done correctly.

A number of numerical control machines have automatic tool-changing capabilities which further enhance the application and use of preset tooling. While the actual mechanism by which the tools are interchanged between the spindle and tool-changer magazine varies from machine to machine, it is important to note that the operation is both programmable and fast (on the order of a few seconds), which reduces nonproductive time. Preset tools for automatic tool-changer systems are either positioned in their numerical machining sequence or separately identified and coded to respond to specific tape commands.

Although most presetting of tools is done on equipment in an area away from the machine tool, there is a different arrangement on some NC machine tools that allows presetting the tools right on the machine. On these machines, control circuits are used which enable "zeroing" on each tool in relationship to the tape and the workpiece. The operator mounts a tool in the spindle and advances it to the workpiece until it contacts a feeler gage. While in contact with the feeler gage, tool length compensation controls are adjusted to establish a zero starting point for that tool. This procedure is repeated for each tool used on that job.

## REFERENCES

1. NCECO–NC ECOnomics Computer Program. 1973. Cincinnati, OH: Machinability Data Center.

2. Machining Cost and Production Time Analysis Programs for Hewlett-Packard calculators HP-65, HP-67 and HP-97 for turning, end milling, drilling, reaming, tapping, chamfering and center drilling. Cincinnati, OH: Machinability Data Center.

3. Field, M. and Ackenhausen, A. F. 1968. *Determination and analysis of machining costs and production rates using computer techniques.* AFMDC 68-1. Cincinnati, OH: Air Force Machinability Data Center.

**A-Axis:** An angle defining rotary motion around the X-axis such that positive *a* is in the direction to advance a right-handed screw in the +X direction.

**Absolute Accuracy:** Accuracy measured from a reference zero which must be specified.

**Absolute Address:** The actual address of a computer memory location, as contrasted to a relative address which may not be determined until it is used during the execution of an instruction.

**Absolute Coordinate:** The values of the X, Y, and Z coordinates with respect to the origin of the coordinate system.

**Absolute Dimension:** A dimension expressed with respect to the origin of a coordinate axis not necessarily coincident with the absolute zero point. Also called *Coordinate Dimension.*

**Absolute Reference Point:** An arbitrarily fixed zero location on the machine table from which part coordinates are dimensioned. For antithesis see *Floating Zero.*

**Absolute System:** A system in which all coordinate locations are measured from a fixed location on the machine table or from an absolute zero point. The zero point on some machines is established by the manufacturer at a fixed location, while on other machines it may be established by the NC programmer.

**Absolute Zero Point:** The point of absolute zero for all machine axes.

**ACC:** See *Adaptive Control Constrained.*

**Acceptance Test:** A test for evaluating the performance, capabilities and conformity of software or hardware to predefined specifications.

**Accuracy:** Conformity of an indicated value to the true value, that is, an actual or an accepted standard value (see *Precision* and *Reproducibility*). The accuracy of an NC system is measured by the difference between the actual position taken by a machine slide and the position called for. Accuracy is the joint effect of method, observer, apparatus and environment.

**ACO:** See *Adaptive Control Optimization.*

**Acoustic Coupler:** An electronic device that sends and receives digital data through a standard telephone handset. To transmit data, the digital signals are converted to audible tones that are acoustically coupled to a telephone handset. To receive data, the acoustically coupled audible signals are converted to digital signals.

**Active Storage:** Data storage locations in an NC control system which hold the data actively being transformed into motion. Contrast with *Buffer Storage.*

**ADAPT:** A computer-aided NC parts programming language comparable to APT but with more limited capabilities. Developed for medium-scale computers and essentially used for two-axis contouring.

**Adaptive Control:** A control method using sensors for real time measurement of process variables with calculation and adjustment of control parameters as a means of achieving near optimum process performance. The purpose of adaptive control is the optimization of the independent parameters (for example, speeds and feeds) consistent with process constraints such as quality of surface finish and cutter life.

**Adaptive Control Constrained (ACC):** A control system involving in-process measurement used to obtain maximum machine productivity by using limiting values for certain machine parameters such as torque or spindle deflection.

**Adaptive Control Optimization (ACO):** A control system involving in-process measurement and adjustment of variable cutting parameters to achieve optimum production conditions.

**Address:** In NC, a character or group of characters initiating a word and used to identify the data which follow in the word.

**AIA:** Aerospace Industries Association.

**Algorithm:** A procedure for accomplishing a given result by proceeding on a logical step-by-step basis. NC and computer programs are developed by this method.

**Alphanumeric Code:** A code consisting of letters, digits and associated special characters.

**Analog:** The use of physical variables, such as distance, rotation, or voltage, to represent and correspond with numerical variables that occur in a computation; contrasts with *Digital.* In NC, a system which utilizes electrical voltage magnitudes or ratios to represent physical axis positions.

**Analog-to-Digital (A/D) Converter:** A hardware device which senses an analog signal and converts it to a representation in digital form.

**ANSI:** American National Standards Institute.

**APT:** **A**utomatically **P**rogrammed **T**ools. A computer-aided programming language which describes the part illustrated on an engineering drawing and takes the form of a sequence of statements that define the part geometry, cutter operations, and machine tool capabilities. Initially developed for 3-, 4-, and 5-axis milling machines but presently is capable of a wider range of applications including turning and point-to-point work.

**ARELEM:** Acronym for **AR**ithmetic **ELEM**ent. A part of the APT processor that calculates the cutter locations based on the cutter motion commands and the canonical forms of the geometrical surfaces.

**ASCII:** American Standard Code for Information Interchange. One of two major standard codes used for information interchange among data processing systems, communication systems, and associated equipment. It uses a coded character set consisting of 128 unique, 7-bit binary coded characters.

## 23.2 NC Vocabulary

**Assembler:** A computer program which converts a symbolic assembly language program into an executable object (binary-coded) program.

**Assembly Language:** An operation language composed of brief expressions which is translated by an assembler into a machine language.

**Assembly Program:** A program which translates the programmer's symbolic language to machine language, assigns absolute addresses to instruction and data words, and integrates subroutines into the main routine.

**Automated Process Planning:** Creation of process plans, with computer assistance either partially or totally, for items in a given family.

**Automatic Acceleration and Deceleration:** A control system feature providing for smooth changes in the velocity of the machine tool slide.

**Automatic Programming:** Any technique whereby a digital computer transforms programming from a form that is easy for a human being to produce into a form that is efficient for the computer to carry out.

**Automatic Tape Rewind:** A control system feature that causes the input tape to be rewound to the initial starting block after reaching the end of the program.

**AUTOSPOT:** Acronym for **AUTO**matic **S**ystem for **PO**sitioning **T**ools. A general-purpose computer program designed to aid the parts programmer in preparing instructions for NC machine tools. Point-to-point capability only.

**Auxiliary Function:** A function of a machine control other than the control of the coordinates of a workpiece or tool. Usually on/off type operations.

**Axis:** A general direction along which the relative movements of the tool and workpiece occur. One of the reference lines of a coordinate system.

**Axis Inversion:** See *Mirror-Image Programming.*

**B-Axis:** An angle defining rotary motion around the Y-axis such that positive *b* is in the direction to advance a right-handed screw in the +Y direction.

**Backlash:** A relative movement between interacting mechanical parts resulting from looseness and deflections.

**Base:** A quantity used implicitly to define systems of representing numbers by positional notation. See *Radix.*

**Batch Processing:** A method of processing jobs on a computer where they are organized and handled one at a time. Contrast with *Time Sharing.*

**Baud:** A unit of signaling speed equal to the number of discrete conditions of signal events per second. Used as a measure of serial data flow between a computer and/or communication devices.

**Binary:** (1) Pertaining to a characteristic or property involving a selection, choice, or condition in which there are two possibilities. (2) Pertaining to the number system with a radix of two. (3) A system for describing numbers using only two digits (zero and one).

**Binary-Coded-Decimal (BCD) Notation:** A number code in which a four-digit binary number is used to represent each digit in the decimal numbering system.

**Binary Digit:** In binary notation, either of the characters zero or one. It is equivalent to an "on" or "off" condition, or to a "yes" or "no."

**Bit:** (1) An abbreviation of **B**inary dig**IT**. (2) A single character of a language employing exactly two distinct kinds of characters. A bit, for instance, can be the absence or presence of a hole in paper tape.

**Block:** A group of words that contains all the instructions for a single operation. In a positioning system, a block will include the coordinates of the position, together with all the instructions for auxiliary functions necessary to complete an operation. The blocks are separated by an end-of-block character.

**Block Address Format:** A tape programming method for NC systems in which only the instructions that need changing are punched into the tape.

**Block Delete:** A feature providing for the skipping of certain blocks by programming a slash (/) code immediately ahead of the block. Useful when the operator desires to omit certain cuts on a particular part configuration.

**Block Length:** The total number of records, words or characters contained in one block.

**BTR (Behind the Tape Reader):** A method of inputting data directly into a machine control unit from an external source other than a tape reader.

**Buffer Storage:** A place where information from a block of tape is electronically saved or stored so that it can be transferred to active storage instantly. The next block is then put into the buffer storage while the previous command is being performed. This avoids waiting for the tape to go through the tape reader. Contrast with *Active Storage.*

**Bug:** An error or flaw in a program making it incapable of performing the objectives for which it was written.

**Bulk Memory:** A memory device for storing large quantities of data, for example, disk, drum, or magnetic tape. This type of memory is not programmable, as is core memory.

**Byte:** A series of binary bits which are operated on as a unit in a computer. A byte may be comprised of eight, twelve, or sixteen bits, for example. See *K.*

**CAD:** See *Computer-Aided Design.*

**CAM:** See *Computer-Aided Manufacturing.*

**C-Axis:** An angle defining rotary motion around the Z-axis such that positive *c* is in the direction to advance a right-handed screw in the +Z direction.

**Canned Cycle:** In NC, a fixed function in hardware or software in which several operations are performed in a predetermined sequence and ending with a return to the beginning condition. See *Macro.*

**Cartesian Coordinates:** A system of two or three mutually perpendicular axes along which any point can be located in terms of distance and direction from any other point.

**Cathode Ray Tube (CRT):** An electronic device containing a screen on which graphic or printed information may be displayed.

**Central Processing Unit (CPU):** The basic memory or logic of a computer that includes the circuits controlling the processing and execution of instructions.

**Chad:** The pieces of material that are removed when punching holes in tape or cards.

**Channel:** A path parallel to the edge of a tape along which information may be stored by means of the presence or absence of magnetic areas or of holes, singly or in sets. Also known as *Level.*

**Character:** One of a set of elementary marks, such as numerals or alphabetic letters, which may be combined to express information. A character includes all the marks, such as a group of holes punched in one row of a tape, which are necessary to completely identify it.

**Chip:** Small piece of semiconductor material on which electronic components are formed. Integrated circuits are formed on chips.

**CIMS:** See *Computer Integrated Manufacturing System.*

**Circular Interpolation:** The simultaneous and coordinated control of two axes of motion such that the resulting cutter path describes the arc of a circle.

**CL Information:** Stands for **C**utter **L**ocation and describes the coordinates of the path of the center of the cutter resulting from a basic computer program. This information is common to all machine-tool-system combinations and is the input to the postprocessor.

**Closed Loop System:** A control system in which the output, or some result of the output, is measured and fed back for comparison with the input for the purpose of reducing any difference between input command and output functional response. Contrast with *Open Loop System.*

**CNC:** See *Computer Numerical Control.*

**Code:** The system of organized information that can be understood and acted upon by a control system, computer or data processing equipment.

**Code Conversion:** A process for changing the bit grouping for a character in one code into the corresponding bit group for a character in a second code.

**Command:** A signal from a machine control unit initiating one step in the execution of a complete program.

**Communication Link:** The physical means of connecting, transmitting and receiving data.

**Compatibility:** The degree to which tapes, languages, and programming can be interchanged among various machine tools and NC systems.

**Compensation:** When contouring, compensation is a displacement, always normal to the cutting path and workpiece programmed surface, to account for the difference between the actual and the programmed radius or diameter of the cutting tool.

**Compile:** To convert a computer program written in a high-level language such as *FORTRAN* into the binary-coded instructions that the computer can interpret.

**Computer-Aided Design (CAD):** The use of computers to aid in the design of products.

**Computer-Aided Manufacturing (CAM):** The use of computers to assist in any or all phases of manufacturing. NC is a subset of CAM.

**Computer Graphics:** A man-oriented system which uses the capabilities of a computer to create, transform and display pictorial and symbolic data.

**Computer Integrated Manufacturing System (CIMS):** Also known by several other names such as Flexible Manufacturing Systems and Computer Managed Parts Manufacturing. A multi-machine manufacturing complex linked by a material handling system and including features such as tool changers, load/unload stations, etc. Dissimilar workpieces are typically introduced into the system, then transported randomly and simultaneously to the NC machine tools and other processing stations, all under the control of a supervisory computer.

**Computer (Computerized) Numerical Control (CNC):** A self-contained NC system for a single machine tool utilizing a dedicated computer controlled by stored instructions to perform some or all of the basic NC functions. The usage of punched tape and tape readers is eliminated, except for possible use as a backup capability in the event of computer failure. It may become part of a DNC system through a direct link to a central computer.

**Console:** Part of a computer or NC machine tool system used for communication between the operator and the computer or machine tool.

**Continuous Path:** An NC machine type in which the rate and direction of relative movement of machine members is under continuous numerical control such that the machine travels through the designated path at the specified rate.

**Contour Control System:** A system in which the cutting

tool path is continuously controlled by the coordinated and simultaneous motion of two or more axes. Also called contouring.

**Control System:** An arrangement of interconnected elements interacting to maintain some condition of a machine or to modify it in a prescribed manner.

**Controller:** An apparatus of unitized or sectional design, through which commands are introduced and manipulated. A program controller has the following functions: data computation, encoding, storage, readout, process computation and output. See *Director.*

**Conversational Mode:** Communication between a terminal and the computer in which each entry from the terminal elicits a response from the computer and vice versa.

**Coordinate Dimension:** See *Absolute Dimension.*

**Coordinate Dimension Word:** A word defining an absolute dimension with respect to a specified reference zero.

**Coordinates:** See *Cartesian Coordinates.*

**Cornering:** Suddenly changing the direction of tool travel from one straight path to another.

**CPU:** See *Central Processing Unit.*

**CRT:** See *Cathode Ray Tube.*

**Cursor:** Visual movable pointer used on a CRT.

**Cutter Compensation:** A feature on some NC machines which gives the operator freedom to modify in a direction normal to the programmed tool path for changes in cutter radius, length or deflection.

**Cutter Offset:** The distance from the part surface to the axial center of a rotary cutter or to the center of a radius. The difference between the part surface and the cutter path during a machine operation.

**Cutter Path:** The path described by the center of a cutter in order to generate the desired part configuration.

**Cycle:** (1) A space or time interval during which a set of events or phenomena is completed. (2) A set of operations completely performed in a predetermined order.

**Data:** A representation of facts, concepts or instructions in a formalized manner suitable for communication, interpretation or processing by humans or by automatic means.

**Data Acquisition:** (1) Acquisition, evaluation and recording of data. (2) Identifying, isolating and gathering source data in a usable form to be centrally processed.

**Data Bank:** A comprehensive collection of libraries of data for computer processing, and stored on drum, disk or other storage media.

**Data Base:** A collection of data fundamental to an enterprise. Comprised of comprehensive files of information having predetermined structure and organization, and suitable for communication, interpretation or processing by humans or automatic means.

**Data Conversion:** Changing the format, method of storage, language, coding, symbolism or general form of information without changing the intelligence or logic content.

**Data Link:** The communication lines, modems and communication controls of all stations connected to the line used in the transmission of information between two or more stations.

**Data Set:** An electronic device providing an interface for the transmission of data to remote stations. See also *Acoustic Coupler.* The terms Modem and Data Set are often used interchangeably.

**Dead Band:** The range through which an input can be varied to the servo portion without initiating response at the machine tool. Generally, the narrower the dead band, the better the response of a machine-tool-system combination. Analogous to mechanical backlash. Also known as *Dead Zone.*

**Dead Zone:** See *Dead Band.*

**Debug:** Debugging is the process of detecting, locating and eliminating programming errors. Most programs of any complexity have to go through a debugging process.

**Deceleration Distance:** The calculated distance for deceleration of an axis of motion to avoid position overshoot.

**Decimal Code:** A code in which each allowable position has one of ten possible states.

**Decimal Digit:** One of the integers 0 through 9 when used in the scale of ten.

**Decoder:** A device which translates data from a coded form to an analog or other easily recognized form without significant loss of information.

**Dedicated Computer:** A computer devoted exclusively to a singular machine or application.

**Delete Character:** A character used primarily to obliterate any erroneous or unwanted characters.

**Diagnostic Routine:** A preventive maintenance check of key NC system components by use of a special programmed tape and/or electronic troubleshooting instruments. Specifically designed to locate and identify errors and malfunctions when they occur.

**Digit:** (1) One of the integers 0 through 9 when used for numbering in the scale of ten, regardless of position in the type of code in which they appear. (2) A character in any number system that may be used singly or in combination with other digits.

**Digital:** Information and values expressed in discrete terms. In a digital computer such terms are generated by a combination of binary on/off or positive/negative signals. This is the opposite of analog, where a fluctuating signal strength determines the fluctuations of values.

**Digital Input Data:** Information supplied to a machine control unit or director in the form of digits, quantized pulses, or some other coding elements.

**Digital-to-Analog (D/A) Converter:** A device which transforms digital data into analog data.

**Digitizer:** A device which gives a digital representation of the value of an analog quantity.

**Direct Numerical Control (DNC):** The use of a shared computer for distribution of part program data via data lines to remote machine tools. DNC systems will typically contain provisions for collection, display, or editing of part programs, operator instructions, or other data related to the NC system.

**Director:** That portion of an NC system which receives the input data signals, and converts and amplifies these signals so as to make them usable by the power servos or control elements in the machine. Also known as *Machine Control Unit* or *Controller.*

**Discrete Jog:** Movement of a selected axis for some predetermined distance and direction. Used for repositioning the machine setup and for precise movement of the tool for checking purposes.

**Disk Storage:** The storing of data in the form of magnetic spots on thin, circular, metal plates that have been coated on one or both sides with magnetizable material. The data are stored and retrieved by read-write heads positioned over the face of the disk.

**DNC:** See *Direct Numerical Control.*

**Downtime:** Time interval during which a device is malfunctioning or is in the process of being debugged.

**Dwell:** A timed delay of programmed or established duration.

**Edit:** To modify the form or format of data; for example, to add or delete characters.

**Editor:** A program permitting a user to create new or to modify existing data files.

**EIA:** Electronic Industries Association. An organization that has promulgated many standards pertaining to NC.

**Encoder:** An electromagnetic transducer commonly used to convert angular or linear position to digital data. See also *Resolver.*

**End of Block (EOB):** A symbol or character punched on the tape that denotes the end of one block of data. Used to stop the tape reader after a block of information has been read.

**End of Program:** A miscellaneous function (m02) indicating the last block in a program and the completion of a workpiece. Stops spindle, coolant, and feed after completion of all commands in the block. Used to reset control and/or machine.

**End of Tape:** A miscellaneous function (m03) which stops spindle, coolant, and feed after completion of all commands in the block. Used to reset control and/or machine.

**EOB:** See *End of Block.*

**Error:** The difference between an indicated and the desired values of a measured signal.

**Error Register or Counter:** A device for accumulating and signaling the algebraic difference between the quantized signal representing the desired machine position and the quantized signal representing the instantaneous position of the machine.

**EXAPT:** Acronym for **EX**tended subset of **APT**. An APT language processor developed in Europe by a German group. The most commonly used EXAPT processors are for point-to-point and lathe work.

**External Storage:** A storage unit such as punched tape or a floppy disk which is external to the computer. Also known as Auxiliary Storage.

**Feed Holes:** Holes which are punched in paper tape that enable the tape to be driven by a sprocket wheel.

**Feed Rate:** A term used to quantify the rate of movement of a machine element, expressed as a unit of distance relative to time or other machine function such as spindle rotation, table stroke, etc.

**Feed Rate Number (FRN):** A coded number which describes the feed rate. The coding method may differ among machines and control systems.

**Feed Rate Override:** A manual function which allows the operator to modify the programmed feed rate by a selected multiplier. This can be done on a percentage basis on some machines or a completely new feed rate on others.

**Feedback:** Information returned from the output of a machine or process for use as input in subsequent operations.

**Feedback Device:** A control system element which converts linear or rotary motion to an electrical signal for comparison to the input signal, for example, *Resolver, Encoder, Inductosyn.*

**Feedback Loop:** The part of a closed-loop system which enables the comparison of response with command.

**Field:** A set of one or more characters, not necessarily in the same word, which is treated as a unit of information.

**Fixed Block Format:** An NC tape format in which the number and sequence of words and characters appearing in successive blocks are constant. Note that the number of characters is always the same; thus, all blocks are the same length. Contrast with *Variable Block Format*.

**Fixed Cycle:** A preset series of operations which direct machine axis movement and/or cause spindle operation to complete such actions as boring, drilling, tapping, or combinations thereof.

**Fixed Sequential Format:** An NC tape format whereby each word in the format is identified by its position. Words must be presented in a specific order and all possible words preceding the last desired word must be present in the block.

**Fixed Zero:** A reference position of the origin of the coordinate system; typically characteristic of machines having absolute feedback elements.

**Fixture Compensation:** A control feature that permits the alignment of a machine tool to a fixture in lieu of changing the fixture location to align with a machine tool. The control automatically compensates for changing alignment values on machine tools with rotary axis motion.

**Flexowriter:** An automatic typewriter that incorporates an eight-track tape reader and punch. Made by the Friden Corporation for preparation of punched tape.

**Flip-Flop:** A physical element or circuit having two stable static states, either of which may be induced by means of suitable input signals.

**Floating Zero:** (Not to be confused with *Zero Offset*) A characteristic of an NC machine tool control permitting the zero reference point of an axis to be established readily at any point in the full travel of the machine tool. The control retains no information on the location of any previously established zeros. Contrast with *Absolute Zero Point*.

**Floppy Disk:** A magnetic based disk used to store information. Used for data input to NC machine control units.

**Following Error:** The distance lag between the actual position and the command position in a contouring machine at any specific time.

**Format:** The physical arrangement of data on a program tape and the overall pattern in which it is organized and presented.

**Format Classification:** An abbreviated notation of the codes used to identify particular systems; it consists of the three following parts: Format Classification Shorthand, Format Classification Detailed Shorthand, and Itemized Data of the Format Contents and Machine Specifications.

**FORTRAN:** Acronym for **FOR**mula **TRAN**slation, a universal computer language in which many NC master programs are written.

**Fourth Generation:** The change in NC control logic technology to include computer architecture.

**FRN:** See *Feed Rate Number*.

**Full Range Zero Offset:** A feature on some NC systems which gives the ability to introduce a dimension which will always be added to the tape command dimension which in turn results in the effect of moving the programmed zero point. Also known as Full Range Floating Zero.

**g-Function:** A preparatory code on a program tape addressed by the letter *g* that indicates a special function or cycle in an NC system.

**Gain:** The amount of increase in a signal as it passes through a control system or a control element. If a signal becomes smaller, it is said to be attenuated. The gain in a control system refers to its sensitivity and its ability to raise the power of a signal to a required output.

**Garbage:** Erroneous, faulty, unwanted or extraneous data in a computer or NC program.

**Gate:** A logic circuit with one output and one or more inputs, designed so that the output is energized only when certain input conditions are met.

**General Processor:** Also referred to as master programs. This is the program containing the basic intelligence for NC work that is stored in computer memory before individual workpiece programs can be processed. APT and ADAPT are examples of general processors. Many machine tool builders, machine control unit manufacturers, and service organizations have prepared and make available general processors.

**Graphical Input:** Input to NC systems which comes directly from a graphic description such as lines drawn on a cathode ray tube or information obtained by a scanner from drawings.

**Group Technology:** (1) The classification and coding of parts on the basis of design similarities. (2) The grouping of parts into production families based on processing similarities so that the parts in a particular "family" could then be processed together. (3) The grouping of diverse machines together to produce a particular family of parts.

**Hardware:** Physical equipment, as opposed to the program or method of use, for example, mechanical, magnetic, electrical or electronic devices. Contrast with *Software*.

**Hardwired:** An NC system which handles data input, sequencing, control functions and formats by fixed and committed circuit connections. Changes can only be made by changing physical components or interconnections.

**Hexadecimal:** A numbering system with a radix or base of sixteen. Valid digits range from 0 (zero) through F, where F represents the highest units position. Useful in computers because of the economy of memory space and simple representation in binary form. *Binary* and *Octal* are the other most frequently used systems.

**Hierarchy:** A group or series arranged in a rank order.

**Hold:** An untimed delay in the program, terminated by an operator or interlock action.

**Hollerith:** The set of codes used in punch cards utilizing 12 rows per column and usually 80 columns per card.

**Home Position:** A fixed location in the basic coordinate axes of the machine tool. Usually the point at which everything is fully retracted for a tool or pallet change.

**Hunting:** A condition in the control system in which a machine component moves slightly back and forth seeking to null out to an absolute stop condition. Generally, considered an unsatisfactory situation.

**Hysteresis:** The difference between the response of a unit or system to an increasing signal and the response to a decreasing signal.

**IC:** See *Integrated Circuit.*

**IEEE:** Institute of Electrical and Electronic Engineers.

**Inch/Metric Input:** The capability of selecting inch or metric programming on an NC system without the need to change the feedback devices. Selection is by switch or by program controls.

**Incremental Coordinates:** Coordinates measured from the preceding value in a sequence of values.

**Incremental Dimension:** A dimension defined with respect to the preceding point in a sequence of points.

**Incremental System:** An NC system in which each coordinate or positional dimension, both input and feedback, is taken from the last position rather than from a common datum point, as in an absolute system.

**Indexing:** One-axis movement of an NC system component to a precise point through numeric commands.

**Inductosyn:** Trademark for the Farrand Controls resolver, in which an output signal is produced by inductive coupling between metallic patterns in two glass members separated by a small air space. Produced in both rotary and linear configurations.

**Input:** The actual transfer of information into a computer or machine control unit by an appropriate medium such as punched tape.

**Input/Output (I/O) Device:** Input/output equipment used to communicate with a computer or control system.

**Input Resolution:** The smallest dimension increment that can be programmed as an input to the system.

**Input Translator:** A phase whereby the source statements are read one at a time, checked for errors in punctuation, ordering, syntax and incomplete statements.

**Instruction:** A statement which tells the director to cause some operation to be performed.

**Instruction Set:** The list of machine language instructions that can be performed by a computer.

**Integrated Circuit (IC):** An electronic circuit that is packaged as a single small unit, ranging in size from about 0.3 inches square to about 2 inches square. The circuits vary in complexity and function from simple logic gates to microprocessors, amplifiers and special devices such as A/D and D/A converters. If the integrated circuit is constructed on a single semiconductor substrate, it is called monolithic. If it is constructed of several independently fabricated circuits that are interconnected, it is called hybrid. The sizes of the circuits are further classified by the terms Small-Scale Integration (SSI), Medium-Scale Integration (MSI), Large-Scale Integration (LSI), dependent on the complexity of the circuit. Slang terms for an integrated circuit are "Bug" and "Chip."

**Interface:** The medium by which two separate elements are joined. There is an interface between a machine control unit and a machine tool comprised of electrical circuitry and drive motors. If a computer directly controls a machine tool, there is an appropriate interface.

**Interlock:** An arrangement in the control of a machine to prevent further operations until the operation in process is completed.

**Interpolation:** The insertion of intermediate information based on assumed order of computation. For example, cutter paths are controlled by interpolation between fixed points by assuming the intermediate points are on a line, a circle or a parabola.

**Interpolator:** A device for defining the path to be followed, and the rate of travel of a cutting tool or of a machine-tool slide or element, when supplied with a coded mathematical description (circular, linear, parabolic, etc.) of the same. It provides the intermediate points between programmed end points to produce smooth curves or straight lines.

**I/O:** See *Input/Output Device.*

**ISO:** International Organization for Standardization.

**Iteration:** Repetitive computations in which the output of each step is the input to the subsequent step.

**Jog:** The manual movement of a selected axis in either direction at varying feed rates for the purpose of accomplishing a small movement of the axis.

**K:** K comes from the French "Kilo" meaning thousand. In data processing, K is used when defining memory locations to mean 1,024 locations in which information is stored and processed. The memory capacity of computers is normally a power of two, thus an 8K computer memory has 8 times 1,024, or 8,192 locations.

**Keypunch:** A keyboard actuated device used to punch holes in cards to represent data.

## 23.2 NC Vocabulary

**Language:** A defined group of representative characters and symbols with rules necessary for their interpretation and application. When applied according to the rules, a language is meaningful to a computer, NC system, or data processing devices.

**Language Translator:** A general term for any assembler, compiler or other routine that accepts statements in one language and produces equivalent statements in another language.

**Lead Screw Compensation:** Automatic compensation for measured lead screw errors.

**Leader:** Blank tape which appears ahead of and after a section of coded tape.

**Leading Zero:** Redundant zeros added to the left side of a number to fix the position of the decimal point.

**Level:** Same as *Channel.*

**Linear Interpolation:** The approximation of curves of circular arcs by a series of straight lines or chords which combined produce a smooth curve (within tolerances) along the contour machine part. Linear interpolation may be performed by the contouring control itself, or it may be externally performed by manual or computer-assisted means.

**Loop Tape:** A short length of punched tape joined end-to-end in the form of a loop for continuous or repetitive reading.

**m-Function:** Similar to a g-function except that the m-functions control miscellaneous actions of the machine tool, such as turning coolant on and off or operating power clamps.

**Machine Control Unit (MCU):** See *Director.*

**Machine Datum:** The point in a machine system used as a basis for establishing a coordinate system.

**Machine Language:** A set of symbols, characters or signs (and the rules for combining them) which convey to a computer the instructions or information to be processed. This language is unintelligible to people unless the symbols and the rules for their use are understood. Special equipment is usually needed to convert this language from the form in which it is stored in the computer to a form perceptible to people.

**Machine Zero:** The origin of the coordinates in the machine system.

**Machining Center:** A machine tool, often numerically controlled, that can automatically drill, ream, tap, bore and mill workpieces and is frequently equipped with an automatic tool-changing system.

**Macro:** A group of instructions that are organized as a group and can be called as a unit.

**Magic 3:** A method of coding speeds and feeds specified in EIA standards.

**Magnetic Tape:** Tape constructed from metal or plastic and coated with a magnetic material that can store information at a much higher density than other media such as paper tape.

**Maintenance:** Any activity intended to eliminate faults or to keep hardware or programs in satisfactory working condition, including tests, measurements, replacements, adjustments and repairs.

**Management Information Systems:** Management performed with the aid of automatic data processing.

**Manual Data Input:** A means for the manual insertion of commands into the NC control system by means other than program tape.

**Manual Feed Rate Override:** See *Feed Rate Override.*

**Manual Mode:** A mode of operation in which NC machine tools can be completely controlled by manually initiating movements and/or cycles.

**Manual Part Programming:** See *Part Programming, Hand or Manual.*

**Manuscript:** An ordered list of NC program instructions. See also *Part Program.*

**Memory:** A term referring to the equipment and media used for holding information in machine language in electrical or magnetic form. See *K.*

**Microcomputer:** A computer that is constructed using a microprocessor as the basic element.

**Microinstruction:** A basic or elementary machine instruction.

**Microprocessor:** A basic element of a central processing unit that is a single integrated circuit. A microprocessor has a limited instruction set which is usually expanded by microprogramming. A microprocessor requires additional circuits to become a suitable central processing unit. See *Microcomputer.*

**Microprogramming:** A method of operation of the CPU in which each complete instruction starts the execution of a sequence of instructions, called microinstructions, which are generally at a more elementary level.

**Microsecond:** One-millionth of a second.

**Millisecond:** One-thousandth of a second.

**Mirror-Image Programming:** A feature of some machine control units that provides for the reversing of all instructions programmed for a specific axis (usually X) by means of a switch. This means that a single program tape can produce two mirror-image parts.

**Miscellaneous Function:** One of a group of special or auxiliary functions of a machine such as spindle stop, coolant on, clamp. See *g*, *m*, and *t* functions.

**Mnemonic Symbol:** An alphanumeric designation chosen to assist the programmer in remembering, such as mpy for multiply.

**Modem:** In data transmission, a modulator/demodulator device which converts data from one form to another for equipment compatibility.

**Mylar:** A DuPont trademark for polyester film, often used as a base for magnetically coated or perforated information media.

**Nanosecond:** One-billionth of a second.

**NC:** See *Numerical Control System.*

**Noise:** An unwanted stray signal in a control system which can interfere with normal operation of a machine control unit or computer.

**NMTBA:** National Machine Tool Builders Association.

**Number System:** A code for expressing numerical data in which the symbols and rules for their use are defined.

**Numerical Control (NC) System:** A system in which actions are controlled by the direct insertion of numerical data at some point. The system must automatically interpret at least some portion of this data.

**Numerical Data:** Data in which information is expressed by a set of numbers or symbols that can only assume definite discrete values.

**Object Deck:** A deck of tabulating cards containing the condensed computer instructions for handling a specific general processor. The programmer usually writes only one instruction per card. These instructions are then processed by the computer and condensed to an object deck where a multitude of instructions may be included on a single card. It is the customary practice for program writing services to furnish only the object deck to the customer. In this manner the service is furnishing a specific capability but is not releasing its proprietary methods and logic.

**Octal:** Pertaining to a number system with a radix or base of 8. The system is used in computers because of its economy of memory, ease of conversion to other number bases, and simple application to binary logic circuitry. *Binary* and *Hexadecimal* are the other frequently used systems with computers.

**Offline Operation:** Operation of peripheral equipment independent of the central processor of a computer system.

**Offset:** A displacement in the axial direction of the tool which is the difference between the actual tool length and that established by a programmer.

**Ones Complement:** The radix-minus-one complement in binary notation.

**Online Operation:** Applies to computer operations and calculations which are performed by the computer itself or the major portion of the computer.

**Open Loop System:** A control system that has no means for comparing the output with the input for control purposes, that is, there is no feedback. Contrast with *Closed Loop System.*

**Operation Number:** A number indicating the position of an operation in a sequence of operations.

**Optimize:** The rearrangement of instructions of data in NC or computer applications to obtain the best balance between operating efficiency and best use of hardware capacity.

**Optional Stop:** A miscellaneous function command similar to a program stop, except that the control ignores the command unless the operator has previously pushed a button to recognize the command.

**Origin:** A reference point whose coordinate values are zero.

**Output:** The opposite of input. The actual transfer of processed information out of a computer. It may be output onto tabulating cards, punched tape, magnetic tape, printed sheets, drum storage or even to another computer.

**Overshoot:** The amount of overtravel that takes place when a cornering maneuver is made at high feed rates by the machine tool traveling member. The overshoot may be caused by the inertia of the mechanical drive system. Contrast with *Undercut.*

**Paper Tape:** See *Punched Tape.*

**Paper Tape Reader:** A device that senses and translates the holes in perforated tape into electrical signals.

**Parabolic Interpolation:** A mode of contouring control utilizing parabolic arcs to approximate curves by automatic means within the control system. The arcs are also blended automatically. This method of interpolation is particularly applicable to mold or die work where free form shapes can be generated with a minimum number of program steps.

**Parallel Entry:** All characters displayed simultaneously as opposed to serial.

**Parity Check:** A method of automatically checking to reduce possibility of tape errors caused by a malfunctioning tape punch. In the EIA method of character coding, the parity column will have a punch if the character itself is comprised of an even number of punches. Thus any row of tape will always have an odd number of holes (not counting the sprocket holes). ASCII characters will always have an even number when the parity check hole is included. The incorporation of a parity check feature makes it possible to automatically stop the machine when a punched tape error is detected.

**Park:** A programmed instruction for moving the tool to a location safe for conducting tool and workpiece inspection.

**Part Program:** An ordered set of instructions in a language and form required to cause operations to be effected under automatic control, which then is either written in the form of a manuscript on an input medium or stored as input data for processing in a computer to obtain a manuscript.

**Part Programming, Computer:** The preparation of a manuscript in computer language and format required to accomplish a given task. The necessary calculations are performed by the computer.

**Part Programming, Hand or Manual:** The preparation of a manuscript in machine control language and format required to accomplish a given task. The necessary calculations are performed manually.

**Pattern Manipulation:** A part programming feature which permits the programmer to reuse a previously programmed pattern at other positions on the part.

**Peripheral Equipment:** Any unit of equipment, distinct from the central processing unit, which may provide the system with outside communication.

**Phase Shift:** A time difference between the input and output signal of a servo loop.

**Picosecond:** One-millionth of one microsecond.

**Plotter:** A device which will draw a facsimile of coded data input such as the cutter path of an NC program. Some plotters make "hard" paper copy and others use a CRT scope.

**Point-to-Point Control System:** See *Positioning Control System.*

**Position Readout:** A visual display of the absolute position of a machine axis. Derived from a position feedback device normally attached to the lead screw of the machine.

**Position Sensor:** A device for measuring a position, and for converting this measurement into a form convenient for transmission, as a source of feedback. See *Transducer.*

**Positioning Control System:** Discrete or point-to-point control, in which the controlled motion is required only to reach a given end point, with no path control during the transition from one end point to the next.

**Positioning Time:** The time required to rapid traverse the tool from one coordinate to another.

**Postprocessor:** A computer routine for translating the output of a general programming routine into a machine language suitable for a specific NC machine.

**Precision:** Closeness of agreement among repeated measurements of the same characteristics by the same method under the same condition. See *Accuracy* and *Reproducibility.*

**Preparatory Function:** A command changing the mode of operation of the control, such as from positioning to contouring, or calling for a "canned cycle" of the machine. A command to establish the mode of operation, which is generally noted at the beginning of a block and consists of a letter character "g" plus a 2-digit number.

**Preset Tool:** A cutting tool placed in a holder so that a predetermined geometrical relationship will exist with a gage point. Tooling is adjusted offline to be compatible with a programmed tool path.

**Preventive Maintenance:** Maintenance specifically intended to prevent faults from occurring during subsequent operation.

**Printer:** An output mechanism which prints or typewrites characters in either parallel or serial entry.

**Printout:** A printed sheet giving all the data of a program that has either been manually or computer processed. The printout is used for reference and visual checking. Normally the printout will be made at the same time a tape is punched.

**Processor:** (1) That portion of a computer which controls the operation input and output devices and operates on the received, stored and transmitted data. Its circuitry includes the functions of memory, logic, arithmetic and control. (2) The software required to implement some function of a computer. Generally used in connection with the implementation of a source language.

**Program:** A set of instructions that defines a desired sequence of conditions for a process or function, and the operations required between these conditions. See *Part Program* and *Manuscript.*

**Program Stop:** A miscellaneous function command to stop the spindle, coolant, and feed after completion of other commands in the block. It is necessary for the operator to push a button in order to continue with the remainder of the program.

**Programmer:** In NC work there are two important types of programmers. A workpiece programmer writes the instructions for the computer to act upon to develop a specific program tape. A computer programmer develops the routines that give the computer the basic intelligence to act upon instructions when they are prepared by the workpiece programmer.

**Programming:** Preparing a detailed sequence of operating instructions for a particular problem.

**Programming Language:** A language used by the programmer to describe the problem in terms associated with the problem itself.

**PROM:** Programmable **R**ead **O**nly **M**emory. A memory that can be programmed only by special routines. Once programmed with permanent data, such as a mathematical formula, it becomes a ROM or Read Only Memory.

**Pulse:** A pattern of variation of a quantity, such as voltage or current, consisting of an abrupt change from one level to another, followed by an abrupt change to the original level.

**Punched Card:** A card of constant size and shape, suitable for punching in a meaningful pattern and for mechanical handling. The punched holes are usually sensed electrically or mechanically.

**Punched Tape:** An input medium comprising a paper or plastic tape into which a pattern of holes is punched so as to represent data.

**Quadrant:** One of the quarters of the rectangular or Cartesian coordinate dimensioning system.

**Radix:** Base or root; a number that is arbitrarily made the fundamental number of a number system. The quantity of different symbols required to express any integer in a number system.

**Random Access:** Indicates equality of access time to all memory locations, independent of the location of the previous memory reference.

**Random Tool Selection:** A feature permitting the next tool to be loaded from any position in an automatic tool changer as contrasted to loading from the next available location in the changer.

**Rapid Traverse:** Tool movement at maximum feed rate from one cutting operation to another.

**Read:** To sense the characters involved in numerical input information.

**Read-In:** Numerical display of input data, that is, what is presented by the coded tape to the control and not the actual position assumed by the machine. Commonly referred to, erroneously, as *Read-Out.*

**Read-Out:** A numerical display of the actual position of, for instance, a machine slide, as provided by some feedback device. May be digital visual display, punched tape, punched cards, automatic typewriter, etc. See *Read-In.*

**Real Time:** The ability of a computer to function and control a process as that process occurs. For example: A computer directly controlling the operation of a machine tool would be a real time application. If a computer generates a program tape and the tape is utilized at some later time, the computer is used in an "offline" application.

**Rectangular Coordinates:** A system of two or three mutually perpendicular axes along which any point can be located in terms of distance and direction from any other point.

**Reference Block:** A block within the program identified by an "o" (letter o) or "h" (letter h) in place of the word address "n" and containing sufficient data to enable resumption of the program following an interruption.

**Reference Point:** A point defined within the limits of travel to locate the spindle in relation to the part.

**Relative Coordinates:** See *Incremental Coordinates.*

**Reproducibility:** (1) The exactness with which measurement of a given value can be duplicated. (2) The ability of a system or element to maintain its output/input precision over a relatively long period of time. Also called repeatability. See *Accuracy* and *Precision.*

**Rerun:** A repeat of a machine run, usually because of a correction, an interrupt, or a false start.

**Resolution:** The shortest distance between two adjacent cutter locations which can be distinguished from each other by the NC system. Commonly available resolutions are 0.001 or 0.0002 or 0.0001 inch.

**Resolver:** A rotary or linear electrical feedback device used to feed back a signal to the machine control unit error detector. The resolver is usually on the end of a leadscrew of an NC machine tool. See *Encoder.*

**Retrofit:** Modification of a machine originally operated by manual or tracer control to one that operates by NC controls. The machine may require considerable alteration and repair so that the NC controls will function properly and accurately.

**Routine:** A set of instructions arranged in the correct sequence to direct the control computer to perform one of a series of operations.

**Row:** A path perpendicular to the edge of a tape, along which information may be stored by means of the presence or absence of magnetized areas, or of holes, singly or in sets.

**s-Word:** An NC program code preceded by the word "s" which determines spindle speed.

**Sampled Data:** Data in which the information content is ascertainable only at discrete intervals of time. Sampled data can be analog or digital.

**Scanner:** Equipment used to digitize coordinate information from a master and convert it into punched tape for later re-creation of the master shape on an NC machine.

**SCR:** See *Silicon Controlled Rectifier.*

**Sculptured Surface:** (1) A mathematically described surface consisting of a composite of interconnected, bounded, parametric surface patches, each patch representing an image of a unit square in parametric space. (2) Any free form, nonplanar surface often nonanalytic; the surfaces of complex shapes usually obtained by hand or sculpturing techniques.

# 23.2 NC Vocabulary

**Second Generation:** In NC, the period of technology associated with transistors.

**Segment:** That portion of a path joining two successive points stored in the position-storage medium.

**Sensor:** A transducer or other device whose input is a quantitative measure of some external physical phenomenon and whose output can be read by a computer.

**Sequence Number:** A series of numerals programmed on a tape and displayed as a Read-In. It is normally used to identify blocks or groups of blocks in a manuscript and is identified by the letter *n* as the address character.

**Sequential:** Items arranged in some predetermined logical order.

**Serial:** Items handled one after another rather than simultaneously.

**Servomechanism:** A power device for directly effecting machine motion. It embodies a closed-loop system in which the controlled variable is mechanical position. Usually some amplification is necessary between the relatively weak feedback signal and the strong command signal.

**Setpoint:** The final or target value of a controlled variable which is to be reached.

**Set-up Time:** Time required to adjust a machine and attach the proper tooling to make a particular product.

**Sign Digit:** A plus or a minus sign for arbitrarily designating the positive or negative characteristic of a coordinate.

**Signal:** Information transmittal from one point in a control system to another.

**Silicon Controlled Rectifier (SCR):** An electronic device that is generally used in control systems for high-power loads, such as for an electric heating element.

**Skip:** An instruction to proceed to the next instruction.

**Software:** All of the program manuscripts, tapes, decks of tabulating cards, methods sheets, flow charts and other programming documentation associated with computers and NC. Its counterpart is the physical hardware comprising the computer or NC system. Contrast with *Hardware.*

**Softwired Numerical Control:** See *Computer Numerical Control.*

**Source Language:** A computer input language designed for easy or convenient use which can be translated, interpreted and executed by a computer, using an appropriate processor. The output is in the language of the NC machine tool, usually as a punched tape.

**Spindle Speed Override:** A feature which allows the operator to modify the programmed spindle speed by a selected multiplier.

**Spindle Speed Selection:** A control function that provides control of spindle speed through input tape commands.

**Static Behavior:** The functioning of a control system under fixed or steady-state conditions as contrasted with dynamic behavior, which refers to behavior under changing conditions.

**Steady State:** A characteristic or condition exhibiting only negligible change over an arbitrarily long period of time.

**Step-Change:** A change from one value to another by a stepped increment as opposed to a smooth variation.

**Stepping Motor:** An electric motor in which the rotor moves in small fixed angular steps controlled by a sequence of electrical pulses to the stator windings. The angle through which the rotor turns depends on the number of applied pulses while the rotational speed depends on the rate of pulse application.

**Stiction:** The combination of adhesion (sticking) and friction which must be overcome to move a machine member from rest.

**Storage Medium:** A device into which information can be introduced, held and then extracted at a later time.

**Stored Program Numerical Control:** Usually the same as CNC. The distinguishing feature is an internal memory that can be altered by receiving new instructions. The physical configuration is not altered by programming.

**Straight-Cut Control System:** A system of control in which controlled cutting action occurs only along a path parallel to linear, circular or other machine ways.

**Subroutine:** A portion of the total NC program, stored in the computer's memory, and available upon call to accomplish a particular operation, usually a mathematical calculation. At its conclusion, control reverts to the master routine.

**Swarf Cut:** The removal of a large section of material such as clamping lugs or tooling pads from a part by cutting the material off the part with a profiling cutter pass.

**Symmetry Switching:** See *Mirror-Image Programming.*

**Synchro:** A wound-rotor magnetic induction transducer, capable of electrically transmitting or receiving angular positional information.

**System:** An organized collection of interdependent and interactive personnel, machines, and methods required to accomplish a set of specific functions.

**t-Function:** A code identifying a tool select command on a program tape. As with *m* and *g* codes, the *t* also appears as a lower case letter on a printout.

**Tab:** A nonprinting spacing action on tape preparation

devices, whose code is necessary to the *Tab Sequential Format.*

**Tab Sequential Format:** An NC tape format in which words are identified by the number of tab characters preceding the word in the block. The first character in each word is a tab character. Words must be presented in a special order, but all characters in a word, except the tab character, may be omitted when the command represented by that word is not desired.

**Tape Editor:** A means that allows the operator to replace, delete, or otherwise change one or more blocks of data.

**Tape Feed:** A mechanism for feeding tape to be read or sensed by the machine.

**Tape Rewind:** A feature that automatically returns the tape to the initial block of information. It will always be located in the last block if it is part of the tape input information.

**Temporary Storage:** Internal storage locations reserved for intermediate and partial results.

**Third Generation:** In NC, the period of technology associated with integrated circuits.

**Time Sharing:** The use of a computer memory for two or more purposes during the same time period. Contrast with *Batch Processing.*

**Tool-Change Time:** The time to change from one type of tool to another. Used especially to describe automatic tool changers.

**Tool Function:** A command identifying a tool and calling for its selection either automatically or manually. The actual changing of the tool may be accomplished by a separate tool-change command.

**Tool Offset:** An incremental displacement correction for tool position parallel to a controlled axis.

**Tooling:** A set of required standard or special tools for the production of a particular part, including jigs, fixtures, gages, cutting tools, etc. Specifically excludes machine tools.

**Track:** A patch parallel to the edge of a tape along which information may be stored by means of the presence or absence of magnetic areas or of holes, singly or in sets.

**Transducer:** A device for converting one form of energy into another form; for example, a pneumatic signal into an electric signal.

**Transistor:** A device employing a semiconductor with three or more electrodes for controlling and regulating the flow of electrons; performs much the same functions as a vacuum tube.

**Transistor-Transistor-Logic (TTL):** A common logic configuration used in integrated circuits characterized by high-speed and noise immunity.

**Translation:** The changing of information from one form to another.

**TTL:** See *Transistor-Transistor-Logic.*

**Turn Key System:** An NC or computer system installed by a vendor who has total responsibility for building, installing and testing the system.

**Turning Center:** A lathe-type NC machine tool capable of automatically boring, turning, threading and facing parts; often equipped with a system for automatically changing or indexing cutting tools.

**Undercut:** A machine cut where the cutter failed to arrive at a programmed point following a command change in direction. Also called Undershoot. Contrast with *Overshoot.*

**Universal Fixturing:** A set of building-block-type components from which a fixture or fixture setup may be constructed.

**Unwind:** To reverse the direction of a screw or a gear for a period sufficient to reduce the torsional stresses due to windup.

**Uptime:** The percentage of total working time in which the NC machine is in satisfactory operation condition. Opposite to *Downtime.*

**USASCII:** Same as *ASCII.*

**Variable Block Format:** An NC tape format which allows the number of words in successive blocks to vary. This applies to both *Tab Sequential* and *Word Address* programs. Contrast with *Fixed Block Format.*

**Verify:** To check, usually by automatic means, one typing or recording of data against another in order to minimize the number of human and machine errors in the data transcription.

**Windup:** Lost motion in a mechanical system proportional to the applied force or torque.

**Word:** An ordered set of characters referring to a unit of information. The normal unit in which information may be stored, transmitted, or acted upon.

**Word Address Format:** An NC tape format in which the addressing of each word in a block is done by one or more characters which identify the meaning of the word.

**Workpiece Program:** A program which gives the instructions for machining a specific workpiece. This should not be confused with computer master programs or processor languages which equip a computer to handle the individual workpiece programs.

**Write:** The imparting or introduction of information into the memory or data input devices of a computer or control system.

**X-Axis:** The axis of motion that is horizontal and parallel to the work holding surface. If Z is horizontal, positive X is to the right looking from the spindle toward the workpiece. If the Z-axis is vertical, when looking from the spindle toward its supporting column(s) the positive X-axis is to the right on single column machines or forward on dual column or gantry machines. On machines generating a surface of revolution, such as lathes, X motions shall be radial, and normally the positive direction of motion shall be away from the center of revolution. Where the linear motion can cross the centerline of rotation, positive motion shall be in the direction of maximum displacement from the center of rotation.

**Y-Axis:** The axis of motion that is perpendicular to both X and Z. Positive Y is in the direction to make a right-handed set of coordinates, that is, +X rotated into +Y advances a right-handed screw in the +Z direction.

**Z-Axis:** The axis of motion that is parallel to the principal spindle of the machine. If there are several spindles, one shall be selected as the principal one. If there is no spindle, the Z-axis is perpendicular to the work holding surface. If the principal spindle can be swiveled or gimbaled, the Z-axis is parallel to the spindle axis when the spindle is in its zero position. Positive Z is in the direction from the work holding means toward the tool holding means. Positive Z motion increases the distance between the work and the tool.

**Zero:** The point from which all coordinate dimensions are programmed in an absolute system.

**Zero Offset:** (Not to be confused with *Floating Zero*) A machine control feature permitting the fixed zero point on an axis to be shifted readily over a specified range. The control retains information on the location of the "permanent" zero.

**Zero Reset:** See *Zero Synchronization*.

**Zero Shift:** Manually orienting the slides of an NC machine so that the zero point is shifted to any point within the range of the measuring system.

**Zero Suppression:** The elimination of nonsignificant zeros either before or after the significant figures in a tape command. The purpose is to reduce the number of characters that are required to be read.

**Zero Synchronization:** A technique which permits automatic recovery of a precise position after the machine axis has been approximately positioned by manual control.

# SECTION 24

# COMPUTER-AIDED MANUFACTURING TECHNOLOGIES


Introduction .................... 24–3
Computers in Manufacturing Planning .................... 24–4
  Group Technology .................... 24–4
  Computer-Aided Process Planning .................... 24–6
  Manufacturing Data Bases .................... 24–7
Computers in Manufacturing Control .................... 24–10
  DNC and CNC .................... 24–10
  Adaptive Control .................... 24–10

## INTRODUCTION

The introduction of computers into the manufacturing environment has revolutionized manufacturing technology. As computers have become more powerful with their increasing ability to perform complex calculations quickly and manage vasts amounts of data, they have found their way into numerous manufacturing applications including part design and analysis, production scheduling and control, inventory and cost accounting, process control, and optimization and machinability data analyses. Time-sharing computer networks have brought the computer to design engineers, process planners, production control personnel and, in many cases, out to the machine operators at the NC machine tools. The days when the computer was simply an accounting tool or a scientific research tool are gone. Every phase of the production process has been affected, and the future promises that even more work will be taken over by the computer. This should hopefully lead to design engineers being more creative, manufacturing engineers tackling more problems, and products being better engineered and produced more efficiently.

The earliest applications of computers in manufacturing came in the general accounting area (job control, inventory control, shipping/receiving, payroll, etc.). Engineering applications emerged in three general areas: design and drafting, planning and scheduling, and fabrication (see figure 24-1). The first applications were in computer-assisted drafting. Design analysis routines then emerged for evaluating the mechanical properties of the part. Today, this area is referred to as CAD, computer-aided design, and has seen the largest growth in commercial systems. The introduction of highly advanced interactive graphics (IAG) systems accelerated this growth. The designer using the IAG cathode ray tube (CRT) can change his design by trying a number of new design modifications. The interactive loop gives him almost instantaneous review of the effect of the modifications on his design, and the process can be conducted until a satisfactory design is obtained. The computer can then analyze the mechanical, structural and dynamic properties of the design before directing graphic plotters to make finished drawings. To date, the status of CAD has been restricted to special applications primarily in the aerospace and automotive industries where complex parts with contoured surfaces are designed. However, CAD has brought an enormous capability to the hands of the design engineer, and its expansion will continue with the emergence of new software, hardware and data bases.

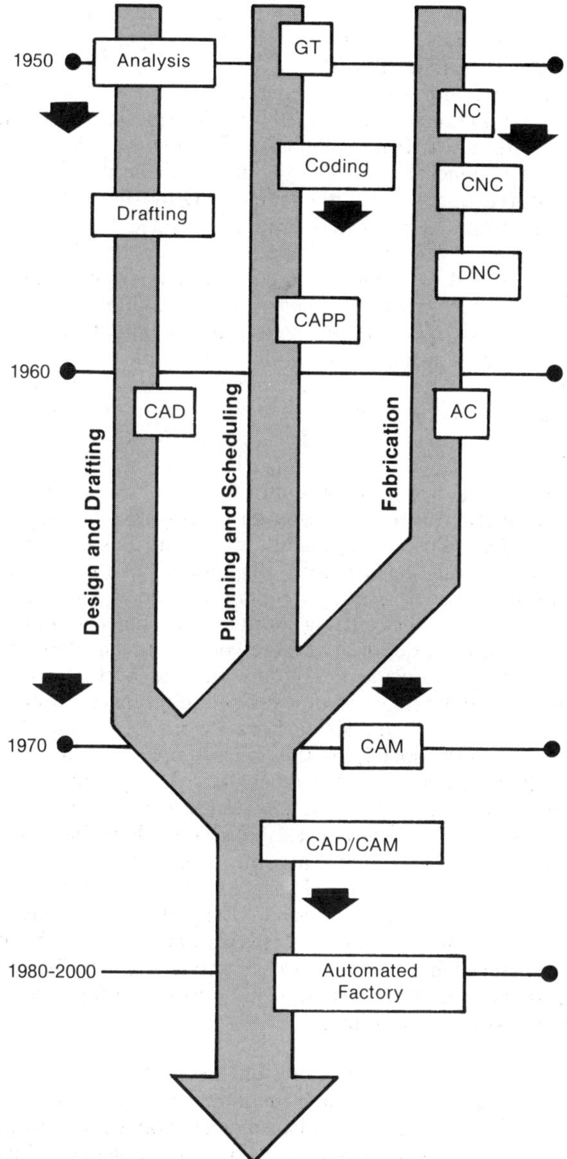


**Figure 24–1** Historical development of CAD/CAM software concepts.

The influence of computers in production has been in two areas: manufacturing planning and manufacturing control. These two areas comprise what has come to be referred to as CAM, computer-aided manufacturing. Earliest CAM developments were in the planning area since production planning was chronologically the next step after part design and did not rely on the retrofitting or redesign of any hardware. The growth of computers in manufacturing control was dependent upon the growth of NC and NC machine tools. Early work in the planning area dealt with group technology (GT). Group technology classifies parts having similar design features and/or similar manufacturing sequences and has become one of the first formal disciplines to bring together part design and part manufacturing. Out of group technology have grown the concepts of part families, cellular manufacturing, part coding for manufacturing, computer-assisted process planning and centralized manufacturing data bases. Computer-assisted process planning generates the production operation sequence, tool layout, fixturing requirements and recommended cutting speeds and feed rates. Manufacturing data bases provide the raw information on tooling, machine tools and cutting performance to run process planning programs. Group technology, computer-assisted process planning and manufacturing data bases are discussed in more detail in the section "Computers in Manufacturing Planning."

The use of computers in manufacturing control grew out of NC technology. The preparation of NC tapes is one area which was enhanced with the introduction of computer-assisted NC programming languages (APT, CINTURN, TOOLPATH, etc.) and with the work done in interfacing CAD with CAM. It is now possible with some CAD/CAM systems to generate the NC tape directly from the designed part stored in the computer memory. Only slight revisions of the computer-generated tape are needed for inserting cutting speed, feed rate, coolant, etc. NC controllers have been modified to incorporate mini- and micro-computers. These control systems are now called computer numerical control (CNC). Computer networks for monitoring and controlling several machine tools, namely, direct numerical control (DNC), have been implemented. CNC and DNC are discussed in more detail in the section "Computers in Manufacturing Control."

A more advanced process control concept which has recently emerged from the research laboratories is adaptive control (AC). The concept of adaptive control is to monitor critical process performance responses, such as cutting forces, horsepower, cutting temperature and tool wear, and then to adjust the cutting parameters (that is, cutting speed, feed rate, depth of cut, etc.) on-line to maintain optimal cutting performance. Numerous research activities have been undertaken to develop reliable sensing systems, appropriate performance critera, workable controlling logic and high speed, on-the-floor computing capability. Not all the elements of the envisioned AC system are sufficiently developed for commercial use (for example, tool wear sensing), but a few workable systems have been demonstrated. These systems are the constraint-control type which monitors one critical response, such as horsepower or cutting forces, and adjusts one parameter (commonly the feed rate) when the critical (maximum) level of the response is reached. In constraint-control AC, feasible cutting is assured; however, optimal cutting performance is not necessarily maintained.

Research work is being done on the performance index-control type system which monitors not only the constraints of the process but also some economical or productivity performance index, and then continually adjusts several cutting parameters to maintain optimal performance. Adaptive control is discussed in more detail in the section "Computers in Manufacturing Control."

## COMPUTERS IN MANUFACTURING PLANNING

### Group Technology

#### Introduction

Group technology is a manufacturing philosophy whose basis is the recognition that similarities occur in the design and manufacture of discrete parts. Despite the tendency to consider each manufactured part as being unique and despite the fact that a casual visual examination of a total part population normally will not reveal commonalities, parts can be categorized into groups or families if their fundamental attributes are identified. Through the systematic classification and grouping of parts into families based on their design or production similarities, significant cost reductions can be achieved. Parts classification is a necessary activity in group technology systems and can be accomplished using one of a variety of classification and coding schemes.

The scope of group technology is broad. Since its inception, it has been associated with part family manufacturing and has been considered to affect all areas of a manufacturing company including design, process planning, manufacturing, assembly, etc. It should also be noted that group technology applies both to mass production where the approach of fixed automation (for example, transfer-line machines) is used and to small lot, discrete-parts manufacturing. In the typical manufacturing plant, the excessive setup time caused by the product mix and small lot sizes may be the most significant part of the total production time. Furthermore, plants typically have a functional layout of equipment; consequently, jobs take a nearly unpredictable path through the plant in order to reach all the necessary processing locations. Production scheduling and production control become very complicated, and actual information on the status of any particular job is nearly impossible to obtain.

Group technology alleviates the situation of mass confusion by first grouping parts into families having manufacturing similarities. In this way, different parts requiring similar machines and tooling may be processed in a sequence that increases the quantity per setup, thereby significantly reducing setup times and costs. Machines used in the production of similar part families are grouped together forming a machine group or cell (hence, the name cellular manufacturing). This layout has the effect of reducing the scope of the problems of production scheduling, production control, material handling, etc., and at the same time tends to improve the operators' morale. Problems related to tooling, for example, can be simplified through the use of fixtures common to an entire part family.

#### Benefits and Problems

The implementation of group technology will yield advantages in areas such as component standardization, reliability of estimates, effective machine operation, productivity, costing accuracy, customer service and order potential. At the same time, implementation of group technology can be expected to bring reductions in planning effort, paperwork, setup time, downtime, work-in-process, work movement, overall production times, finished parts stock and overall costs. There are numerous case studies in the technical literature that cite specific benefits obtained through the implementation of group technology. Representative of these benefits are reductions in the following categories:

- 52 percent in new part design
- 10 percent in number of drawings through standardization
- 30 percent in new shop drawings
- 60 percent in industrial engineering time
- 20 percent in production floor space required

- 42 percent in raw material stocks
- 69 percent in setup time
- 70 percent in throughput time
- 62 percent in work-in-process inventory
- 82 percent in overdue orders

In addition to these advantages, group technology enhances the humanization of work and improves the quality of life in a manufacturing environment. Because machines are grouped to accommodate part families, the workers are able to clearly see their contribution to the final product. It has also been found that setup changes from one part to another similar one require less skill from employees than was formerly demanded. This creates possibilities for job enlargement. Indeed, one of the major benefits of group technology is increased worker satisfaction.

As may be expected, some problems in implementing group technology have been noted. These include the following:

- Group technology requires a steady workload to assure worker satisfaction since the work-in-process is low.
- Group technology may require major changes in the organization of the production system involving such items as redesign of products, standardization and modularization of subassemblies and regrouping of equipment.
- Since group technology typically involves fewer people than machines, some of the machines will have a low utilization level.
- Group technology is particularly applicable to situations where there are a large number of small batches.
- Supervision of a manufacturing cell organized according to the principles of group technology may be more difficult than supervision of a conventional functional layout.

### Classification and Coding

Classification and coding for group technology applications is a very complex problem. Although many systems have been developed throughout the world and countless efforts have been made to improve them, there is not, as yet, any universally accepted and acclaimed system.

A well-designed classification and coding system provides many benefits and facilitates group technology applications in many areas of company operations. The major benefits of a well-designed classification and coding system can be summarized as follows:

- facilitates formation of part families and machine groups
- allows quick retrieval of designs, drawings and production plans
- brings about a reduction in design duplication
- permits design rationalization and reduction of design costs
- secures reliable workpiece statistics
- facilitates accurate estimation of machine tool requirements, logical machine loadings and optimized capital expenditures
- permits rationalization of tooling setups and reductions in setup time and overall production time
- brings about rationalization and improvement in tool design
- allows for rationalization of production planning procedures and scheduling
- improves accurate cost accounting and estimation procedures
- causes better utilization of machine tools, workholding devices and manpower
- improves the effective use of NC machines

Although all departments may use a classification and coding system in one way or another, the two major users of the system are the design/engineering and manufacturing areas. Therefore, a system should be evaluated to determine whether it fulfills all the requirements and provides all the information needed by each of these two departments. Table 24-1 is representative of some of the information and data that should be included in a well-designed classification and coding system.

**TABLE 24-1 Basic Information Requirements of Classification and Coding Systems**

**DESIGN/ENGINEERING AREA**

main shape
shape element
material
rough shape and size
major dimensions
minor dimensions
tolerances
functions
assembly
drawing size, etc.

**MANUFACTURING AREA**

major operation
minor operation
major dimension and size ratio
rough shape and size
machine tool
workholding devices
cutting tools
lot size
setup time
production time
operation sequence
surface roughness
special treatment
inspection
assembly, etc.

While a classification system for group technology applications should be based on classes which are themselves rational, explicit and mutually exclusive, a coding system is a symbolic form of shorthand which can take a number of different forms in communicating specific information or data. There are two basic coding system structures for part classification. These are the hierarchical structure and the chain-type structure. In a *hierarchical* code, the information represented by each subsequent digit is dependent on the preceding digit, and the code number is assigned by a step-by-step examination of part characteristics as they relate to a series of coding charts. A *chain-type* structure is a fixed-digit-significance code where a certain digit value always represents the same feature. Most systems are designed with a combination of these two basic forms, although some systems use a hierarchical code for primary codes and a chain-type code for secondary codes.

### Relationship of Group Technology to Numerical Control

Although both group technology and numerical control are manufacturing-related technologies that have been in existence for several decades, attention has only recently focused on their interrelationship. Many advantages of using NC machine tools have been cited, and a number of these advantages take on even greater importance in a group technology environment. Often, the combination of group technology and numerical control produces synergistic results. NC machine tools are capable of producing parts economically in smaller lot sizes than can ordinarily be justified when conventional machine tools are used. Since group technology is a manufacturing philosophy particularly applicable to small batch production, it should not be surprising that the two technologies have many things in common. These similarities, however, should not lead one to assume that they are completely compatible. It is necessary to understand the interrelationship of group technology and NC so that in situations where they are both applied, they can be combined to achieve maximum effectiveness.

The greatest potential for the application of group technology to NC machine shops is in the area of *cellular manufacturing*. This aspect, while offering more potential benefits specifically for NC shops than the other aspects of group technology, should not cause one to ignore the benefits from the other aspects discussed previously. Often overlooked is the position which a numerically controlled machine tool occupies in a manufacturing cycle and the effect that this has on the work-in-process. An NC drilling machine, for example, is commonly used for operations near the end of a manufacturing cycle and consequently is very effective in reducing work-in-process. Conversely, NC lathes are frequently the first machines in a production cycle; and since they are efficient producers of partly finished work, they can substantially increase work-in-process unless they are effectively supported by subsequent operations. Cellular manufacturing is an appropriate method for providing the dedicated support of other machines that is necessary to fully utilize a machining center.

Group technology can also make economically feasible the use of NC equipment which might otherwise be too expensive for an ordinary job shop. To a small job shop owner, the advantages of numerical control often seem outweighed by the many problems and costs of production planning and underutilization of equipment. Appropriate implementation of group technology can reduce these problems to a point where it can become economically beneficial to introduce NC equipment into the job shop.

## Computer-Aided Process Planning

Process planning refers to the selection of the procedures required to convert a part design into a piece of hardware. The steps in process planning are shown in figure 24-2. They include raw material selection, development of a sequence of manufacturing operations, selection of fixtures, development of tooling and tool configuration, development of cut sequence and NC program, and selection of detailed machining technology, such as feed, speed and tool-change frequency. Typically, the end product of process planning is a set of detailed operation instructions and NC tapes which are used by the machinist in making the part. Process planning may be performed by more than one person. Typically, these people are referred to as planners, methods engineers, manufacturing engineers, process planners, value engineers, or NC programmers.

Process planning is one of the active areas of CAD/CAM implementation. Currently, the computer is being utilized to facilitate certain steps in the process planning sequence. These are: (1) the development of operation sequences, (2) the selection of tooling and the preparation of NC programs and graphical operator instructions, and (3) the selection of machining conditions. In each of these areas, research in the U. S. and abroad is bringing about substantial improvements.

### Development of Operation Sequences

The sequence of machining and nonmachining operations that are required to produce a part are generally chosen on the basis of past experience with similar parts. The planner uses his memory or a stored document to produce a sequence that is typically used for the particular part type that he is planning. Along with the standard sequence, the processing time and work station associated with the operation may be retrieved. The standard sequence is then altered to fit the specific part being planned. Additional operations may be required, work stations or planned time may be altered, or operations may be deleted. This approach is referred to as the *variant* method of process planning. The final plan is a variation of a standard plan.

The variant approach lends itself to computerization with the incorporation of a group technology classification and coding scheme. First, all parts which have been processed previously are coded using a group technology scheme. Second, the parts are divided into part families, each part family requiring a unique sequence of operations. Standard sequences are stored on the computer for each part family. When a new part arrives for planning, the group technology code for the new part is used to retrieve the standard operation sequence for the appropriate part family. Finally, an interactive editor is used to alter the standard sequence to fit the part, and the final sequence is stored in the computer data base by the part number to which it applies. The variant approach is currently exemplified by the CAM-I CAPP system (Computer Aided Manufacturing-International, Inc., Automated Process Planning).

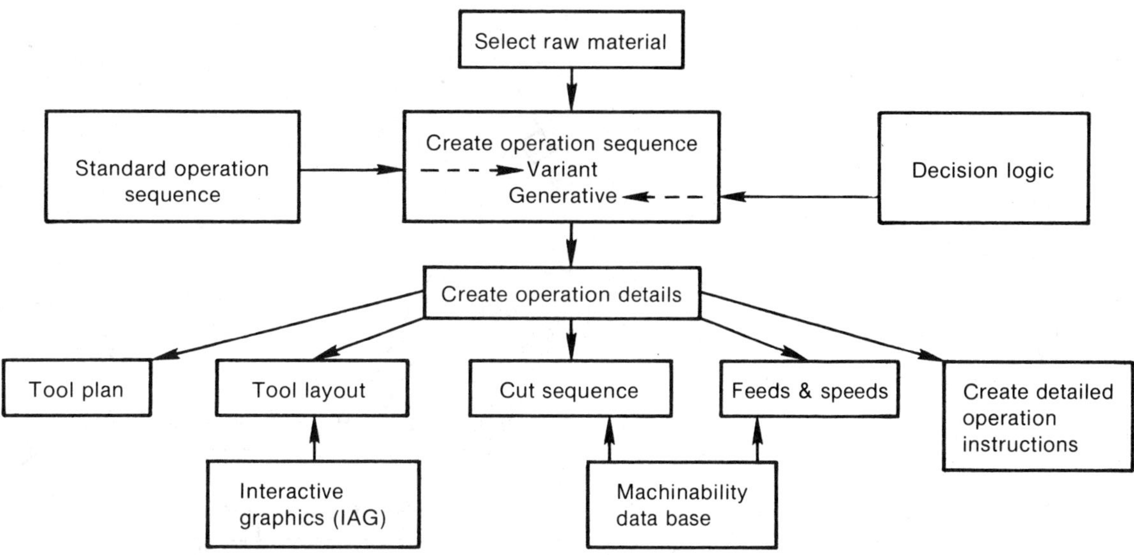


**Figure 24–2** Process planning overview.

Another approach to developing a sequence of operations is termed the *generative* approach. In this approach, the parts are again broken into part families, but a rigorous analysis is made for each part family to determine what features of the part require which operations. This results in the process planning logic for the part family, which is stored as a decision model. When a new part requires planning, it first must be analyzed to determine what features that are incorporated in the decision logic are present on the part. The decision model is retrieved and an operation sequence is generated by processing the decision model with the part features. Essentially, the computer executes the thought process of the planner.

### Interactive Graphics in Process Planning

An important computerized tool that is playing an increasing role in the task of the process planner is interactive graphics (IAG). There are about a dozen IAG systems that are commercially available which can be used by a planner to develop tool layouts, assist in NC programming and compose graphical operation instructions. This graphical composition and analysis of tool layout and NC geometry provides an extremely powerful verification capability which previously required hours of time-consuming drafting and analysis. Now, a tooling layout can be composed and driven by an NC tape to simulate the cutting process in a matter of minutes. Figure 24–3 shows a time exposure of an IAG display of a turret layout driven through its NC cutter path.

### Selection of Machining Conditions

Often the specific machining conditions, such as tool material, feed, speed and coolant are not specified in a process plan. The machinist is left to make these decisions when the part arrives for machining. This is changing, however, because both the cost of sophisticated NC equipment and the complexity of available tool materials and grades are increasing. The computer is being used to select the appropriate grade of cutting tool and to evaluate the feeds and speeds to reduce cost and machining time. In some NC programming systems, such as ADAPT and GETURN, these machining conditions are automatically specified in the NC program. As an alternative, computerized files of machining data are being developed which can be used to evaluate the cost, time, cutting forces, surface roughness, and chatter that result over a range of feeds and speeds. This analysis enables an NC programmer or planner to specify cutting conditions which meet the manufacturing requirements of the part while attempting to minimize cost and production time.

It is apparent that process planning is enjoying the benefits of computer automation. The planner can look forward to more and more automated assistance through interactive graphics, NC programming, group technology and manufacturing data bases. The computer is also moving the process planning function from an experience-based procedure to a data-based procedure. The experience of manufacturing engineers and planners is being put into the computer, and this stored data is providing the decision logic to support existing and future process planning systems.

## Manufacturing Data Bases

At the heart of computer-assisted process planning, group technology or CAD systems is the manufacturing data base. The manufacturing data base provides the ability to store the vast amounts of data that are necessary to support the engineering analyses performed by these systems in formats that can be easily and quickly searched to retrieve individual data elements. Without this ability, engineering analyses performed by process planning, group technology or CAD systems would be extremely slow or impossible.

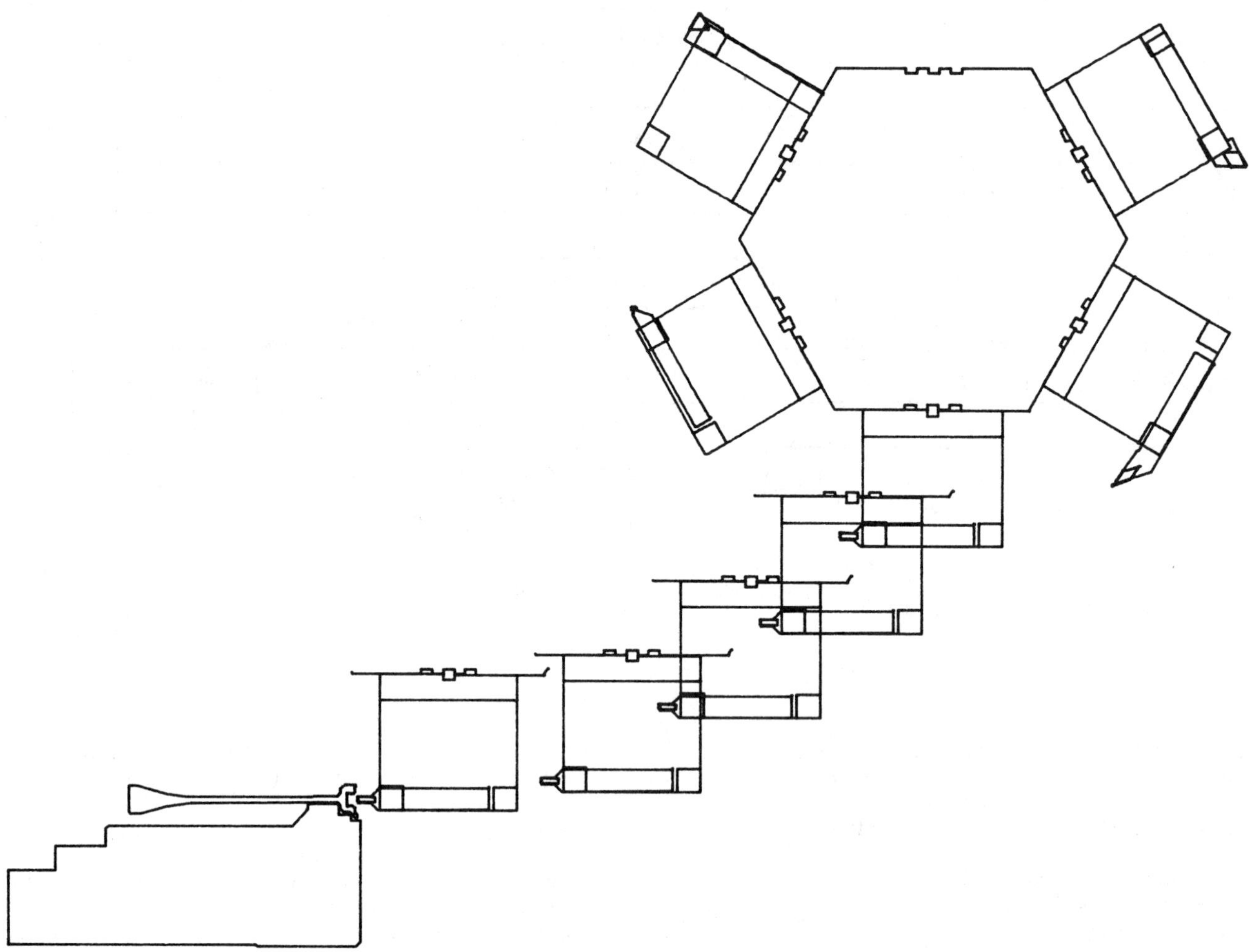

**Figure 24–3** Interactive graphics display of a tooling layout.

The growth of computer data bases in engineering analysis has been relatively recent. The first applications of data bases were inventory control, personnel/payroll and customer order processing. Early inventory control systems were primarily punched-card systems. Each item in inventory had a corresponding computer card. When the item was removed, the card was removed and read by the computer. The system generally operated in batch mode and generally provided only daily summaries of inventory status. Eventually, inventory systems expanded to include other tasks such as shipping/receiving, billing, customer order processing and other accounting functions. Very quickly, the demand for computer information processing systems grew in all fields, including manufacturing. Two major computer developments accelerated computer information system growth: time-sharing, real-time computers and the development of data base managers.

The time-sharing computer allows multiple users to use the computer simultaneously from remote locations. Cathode ray tubes (CRT) are used to communicate with the computer, eliminating much of the hard-copy input (cards) and output. Users are able to obtain immediate answers to their questions rather than waiting hours or days until their job is processed. A data base manager (DBM) is software which efficiently stores and retrieves data using specialized cataloging methods. DBM's allow the data to be structured and cross-referenced in many ways that tremendously increase the speed of data retrieval. Numerous DBM's and strategies for data structuring are available. Manufacturing data base systems incorporate the DBM to perform the routine storage, retrieval, updating and searching tasks. Figure 24–4 shows how a typical manufacturing data base might be structured. The structuring of the data base and the processing of the retrieved data are areas

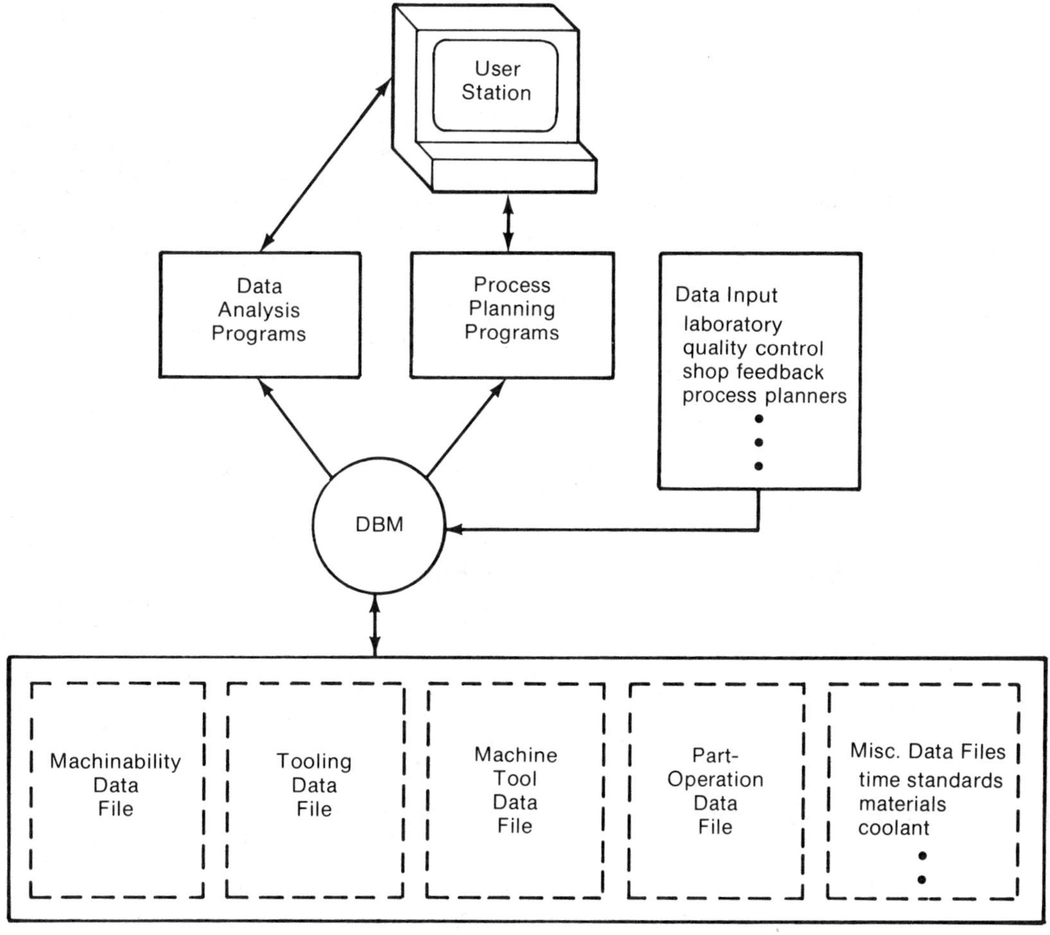


**Figure 24–4** Typical manufacturing data base structure.

where new manufacturing data base developments are forthcoming.

Manufacturing data bases consist of many types of data files, such as the following: geometrical description of part design, workpiece material properties, tooling, machine tool, raw material geometrical description, part/operation description and machinability data. All of these files are related to each other in varying degrees, and, depending on the engineering analysis performed, one or several of the data files may be used. All of the files except the machinability data file simply store and retrieve the description of the major characteristics of each data item. For example, the tooling file may store the following data for a given cutting tool: company code number, tool material, size and shape, rake, clearance and/or helix angles, quantity in stock, inventory location, regrinding information, price, vendor and recommended applications. The machine tool, workpiece material and part/operation files would be similar, but with different data items (for example, the machine tool file would store horsepower, table size, available speeds and feeds, etc.). The geometrical description files store in a special coded form the size and shape of a part. Using Cartesian coordinates and codes for holes, fillets, cylinders, etc., the part is completely described. The coded sequences of numbers are stored and used for interactive graphics and computer drafting routines.

The machinability data file is unique in many respects. Early machinability data files operated in a fashion similar to the other data files; that is, a series of recommended cutting speeds and feed rates were stored for numerous combinations of work material, tool material, cutting fluid and cut size. The structure of the data file was similar to a machinability data handbook. Given the required inputs (tool, cut size, material), the computer would look up the recommended speed and feed. The advantages over a printed handbook were that the recommended cutting conditions could be easily updated, more combinations of tools and cut sizes (depths of cut) could be entered, and automatic interpolation between recommended values could be incorporated. The major disadvantage was that significant engineering analysis was required not only to set up the original recommended cutting speeds and feeds but also to update and maintain current recommendations. The lack of adequate data base maintenance is the most common reason given for the failure of the early machinability data bases.

More recent machinability data base systems use a different approach. Rather than storing recommended cutting conditions and simply retrieving them, the new systems actually perform the engineering analysis necessary to determine the recommended cutting conditions. The recommendations are computed when they are needed and use a base set of raw machinability data. The raw machinability data may be machinability test results, shop-adjusted speeds and feeds, and/or published machinability data. The data may be tool life, cutting forces, surface roughness, horsepower or tool breakage data. The data bank comprises the most basic data elements; no preprocessing of the data is required. The engineering logic used to generate speeds and feeds is programmed into the computer system. Rather than the engineer performing his analysis every time new data become available, he establishes the procedure once and lets the computer perform the analysis when needed. The advantages are that the maintenance is minimal, that the engineer can concentrate his effort on the procedure or logic to use rather than on many repetitive analyses, and that more sophisticated analyses can be performed which are customized to the individual situation encountered. As a result, cutting speeds and feed rates are recommended which are closer to the optimum and may result in higher productivity.

## COMPUTERS IN MANUFACTURING CONTROL

### DNC and CNC

The primary advantage of using computers for NC operations is better control of the supporting functions which include parts program generation, data transmission, storage and fit techniques, and possibly the elimination of paper tapes and associated equipment. In addition, computer-assisted NC does indeed offer significant advantages with respect to control functions and can be justified as a direct replacement for conventional NC.

There are two different concepts of computer control of machine tools: CNC (computerized numerical control) and DNC (direct numerical control).

With CNC, the computer—usually a general purpose minicomputer—is dedicated to the control of only one machine. This minicomputer receives all pertinent machine information, calculates or performs logic functions to update machine instructions, then sends the instructions to the machine. Figure 24-5 shows a block diagram of a CNC system having both the central processing unit (CPU) which takes over the data handling and the control hardware which manipulates the NC machine tool. In addition, the CPU can perform data processing similar to that performed by a tape preparation system.

Conversely, DNC utilizes a central computer, multiplexed (time shared) to control a number of machines simultaneously (see figure 24-6). The main function of the computer is to police each machine and to make certain that it is supplied with sufficient data. DNC systems are usually open loop, which means that the computer updates the data when requested by each machine but does not follow up during machine operation.

Although CNC costs about the same as NC for the same level of sophistication, CNC has some advantages. The flexibility of CNC permits ready addition or reconfiguration of functions by simple software modifications rather than by the physical modifications required for hard-wired NC. Programming, including program modification, is simplified with CNC. CNC also permits the use of error message displays, the logging of management data and the use of diagnostic programs for detecting malfunctions.

DNC offers several operating advantages. First, the tape reader, which usually is the most downtime-prone component of a machine control unit, is bypassed. Second, a program in computer storage is much easier to access for operational use, for revision or editing, or for quick and easy interaction between programmer and machine tool. Third, the same computer that directs the operation of a machine tool can also be used for auxiliary purposes such as downtime recording, performance tabulation, real-time machine status and other operational items of interest to management. Gathered information may either be kept in the controlling computer's memory or it may be transmitted to some other type of data gathering unit where it may be organized for use by the manufacturing management staff.

An advanced-design DNC unit could also be used to sense operating conditions and even to make modifications to programmed instructions based upon the gathered data. With all operational instructions coded in terms of unseen computer software, DNC does require programmers and supervisors who have a very clear and thorough understanding of production operations in order to exercise full and optimum control. DNC systems can be extremely effective when combined with first-rate systems know-how, but the initial cost will probably be high and the support must always be of the highest caliber.

Even though CNC (by definition) has its own dedicated computer, it can also communicate readily with other computers as part of a DNC system. As production DNC systems develop, CNC machine tools can be readily integrated. This integration has greatly reduced the dependence on a single computer. In the case of a failure in the DNC computer, the CNC systems can continue to operate the machine tools satisfactorily.

### Adaptive Control

Adaptive control (AC) is a process control method that uses sensors for real-time measurement of process variables and calculates and adjusts control parameters to achieve near-optimum process performance. Adaptive control of the metalcutting process was introduced in the early 1960's and has since demonstrated that its use can increase productivity and reduce costs. While some machine tools can be purchased today with AC as an option, it is more common to retrofit AC units onto existing machine tools. The metalcutting applications of AC often cite increases in metal removal rate on the order of 50 percent; yet, AC has had a broader impact on the metalcutting field than the

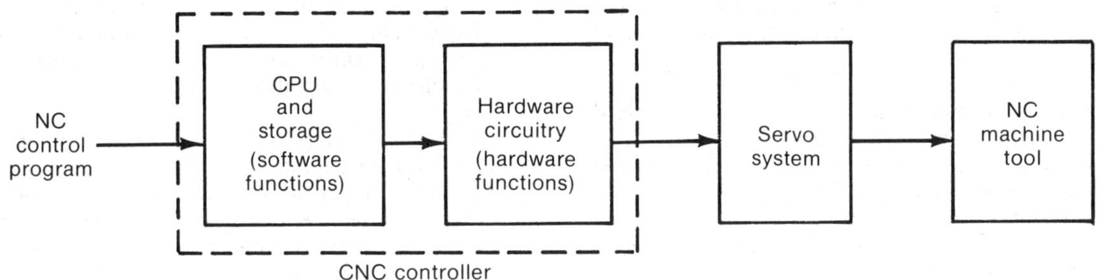


**Figure 24–5** Typical CNC system diagram.

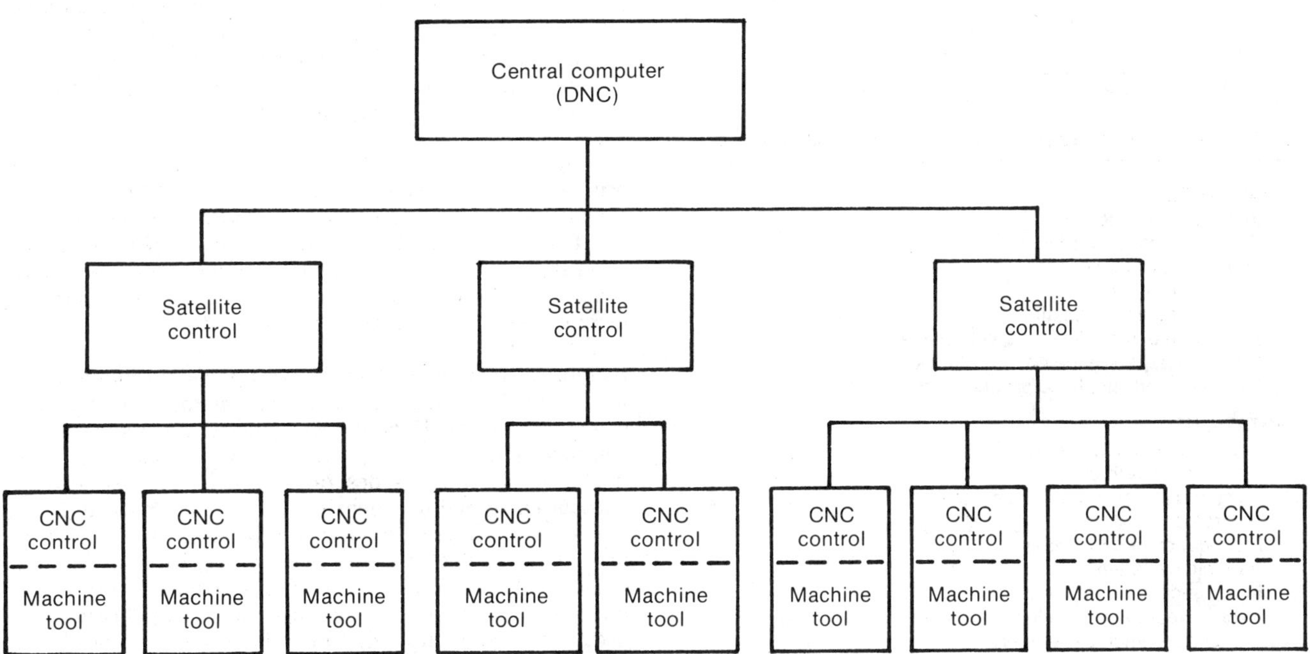


**Figure 24–6** DNC system configuration with satellite and CNC computer controls.

simple optimization of machining rates. Since it requires an understanding of the total machining operation as a system, AC has fostered in-depth studies of the system components and their interactions. Metalcutting applications of AC are still rather limited in number, primarily because system costs are high and it is difficult to obtain sensors suitable for a process affected by many difficult-to-predict process parameters.

For conventional and numerical control, machining speed and feed data such as that contained in this handbook as well as tool life data from either curves or computerized data banks are extremely helpful for selecting machining parameters. These data, however, are not completely adequate for AC. In order to optimize, demands are now being made for data on tool forces developed during machining and on the effect on these forces of variables such as cutter design, tooth design and geometry, type of work material and hardness, and extent of tool wear. With these data, reasonable constraints can be selected to maximize feeds and speeds without the risk of cutter breakage. The advantages and limitations of AC are listed in table 24-2.

**TABLE 24-2 Advantages and Limitations of AC Systems**

**ADVANTAGES**
- Increased productivity
- Reduced cost of production
- Improved product quality
- Lower scrap rates
- Higher machine utilization
- Protection of machine, tool, workpiece from damage
- Reduced dependence on operator skill
- Improved and simplified programming

**LIMITATIONS**
- High cost and complexity
- Poor system reliability
- Existing systems do not guarantee maximum metal removal rates
- Inadequacy of data banks
- Variable cycle times
- Increased maintenance costs
- Expensive and unreliable sensors for key AC process performance parameters

The various types of AC systems can be classified as follows, recognizing that there is no simple definition of AC.

*Feedback AC* In these systems, some process parameter, such as motor current or horsepower, is predetermined to be optimal. The machine tool system has the capability of monitoring the level of the process parameter and compensating for differences in its level by adjusting the cutting speed or feed rate.

*Constraint AC* In these systems, the cutting speed and/or feed is maximized within specified limits for the machine or cutting tool. Common constraints include maximum torque, force, or horsepower. These are the most prevalent systems currently in use since they are relatively simple and inexpensive. Most are based only on the adjustment of the feed to reach to optimum condition within the constraint region.

*Optimization AC* In these systems, an index of performance or a figure of merit (usually based on some economic function involving minimum cost or maximum production rate) is established. During operation, the process performance is compared with the index or merit figure. The system has the capability of deciding how and when to use the results of the comparison to improve the performance through implementation of a change in operating parameters. Most systems of this type are still laboratory systems used for research. Development of indices of performance that are realistic yet simple and of sensors that are reliable and inexpensive are the two necessary steps toward production usage of these systems.

The area of sensor development for AC systems is appropriately receiving much attention. The R&D efforts are focusing on cost reduction and on improved sensor reliability. The state of the art of sensor technology is such that several of the key metalcutting parameters including tool wear and surface quality cannot be easily or accurately monitored in the production environment.

It appears that interest in AC as applied to material removal was first generated by development work on rather complex systems. However, it appears that industry will continue to initiate more extensive use of AC with relatively simple systems designed to achieve cost savings without using highly expensive or delicate sensor and control systems. Interest is currently being shown in sensors and controls which can easily be adapted to existing machine tools. As further economic advantages are recognized, more complex AC systems will find application, especially as computer-aided manufacturing becomes more widely used.

SECTION **25**

# MACHINING STANDARDS

25.1 Machining Standards by Subject ........ 25–3
25.2 Alphabetical List of Machining Standards ........ 25–9
25.3 Sources of Machining Standards ........ 25–23

**Abrasives, Abrasive Wheels and Belts:**
GGG-W-290B
GGG-W-301B
MIL-W-16102C(1)
MIL-W-17929B(1)
Abrasive Grains,
ANSI B74.4-1977
ANSI B74.5-1964 (R1976)
ANSI B74.6-1964 (R1976)
ANSI B74.8-1977
Diamond,
ANSI B74.1-1966
ANSI B74.3-1974
ANSI B74.16-1971 (R1976)
ANSI B74.17-1973 (R1978)
DWMI 101
DMWI 102
DWMI 103
DWMI 104
DWMI 105
DWMI 106
Dressing,
ANSI B67.1-1975
Identification Codes and Dimensions,
ANSI B74.1-1966
ANSI B74.2-1974
ANSI B74.3-1974
ANSI B74.9-1970 (R1976)
ANSI B74.10-1977
ANSI B74.11-1967 (R1978)
ANSI B74.12-1977
ANSI B74.13-1977
ANSI B74.18-1977
ISO 525-1975
ISO/R 603-1967
ISO 1117-1975
ISO 1929-1974
ISO 2933-1974
ISO 2976-1973
ISO 3919-1977
ISO 5429-1977
Mounting,
ANSI B5.35-1969 (R1975)
Safety,
ANSI B7.1-1978

**Computer Software:**
ANSI X3.37-1977

**Cutting Fluids:**
ANSI Z11.142-1964 (R1974)
ASTM D88-56 (1973)
ASTM D92-72 (1977)
ASTM D94-71 (1976)
ASTM D129-64 (1973)
ASTM D130-75
ASTM D156-64 (1977)
ASTM D287-67 (1977)
ASTM D445-74
ASTM D808-63 (1976)
ASTM D892-74
ASTM D1479-64 (1973)
ASTM D1662-69 (1974)
ASTM D2670-67 (1977)
ASTM D2881-73 (1978)
MIL-C-46113B
MIL-C-46149
VV-C-846A(1)
VV-C-850A(1)

**Cutting Tools:**
ISO 2584-1972
ISO 2585-1972
ISO 3337-1975
ISO 3859-1977
ISO 3940-1977
ISO 5421-1977
Bur Blanks,
ANSI B94.13-1976
Carbide Tipped Turning Tools,
ISO 242-1975
ISO 243-1975
ISO 504-1975
ISO 514-1975
Counterbores,
ISO 4206-1977
ISO 4207-1977
MIL-C-936
Countersinks,
ISO 3293-1975
ISO 3294-1975
ISO 4204-1977
ISO 4205-1977
Cut-off Blades,
ANSI B94.3-1965 (R1972)
Designations,
ISO 513-1975
Drills,
ANSI B94.12-1977
ANSI B94.49-1975
ANSI B104.1-1964
ISO 235/2-1972
ISO 2306-1972
ISO 3314-1975
NAS 932
NAS 937
NAS 965
See also **Cutting Tools:** Twist Drills
End Mills,
ANSI B94.19-1977
ANSI B94.20-1977
ISO 1641/1-1978
ISO 1641/2-1978
ISO 1641/3-1978
ISO 2324-1972
MS-15704B
MS-15705A
MS-15706A
MS-15707B
MS-15708B
MS-15709B
MS-15710B
MS-15711B
MS-15713A
MS-15714A
MS-15715A
MS-15716B
NAS 986

Files,
GGG-F-00340(2)
Forming Tools,
ANSI B94.32-1954 (R1971)
Geometry,
ISO 3002/1-1977
Hobs,
AGMA 120.01-1975
ANSI B94.7-1966 (R1972)
Honing Stones,
ISO 3920-1976
ISO 3921-1976
Inserts,
ANSI B94.4-1976
ANSI B94.24-1976
ANSI B94.25-1975
ISO 883-1976
ISO 1832-1977
ISO 3364-1977
ISO 3365/1-1977
Knurling,
ANSI B94.6-1966 (R1972)
Milling Cutters,
ANSI B94.8-1967 (R1972)
ANSI B94.19-1977
GGG-C-755E
ISO 240-1975
ISO 532-1974
ISO 2587-1972
ISO 2780-1973
ISO 2940/1-1974
ISO 2940/2-1974
ISO 3365/1-1977
ISO 3855-1977
ISO 3860-1976
MS-15687C
MS-15688B
MS-15689A
MS-15690B
MS-15691A
MS-15693A
MS-15694B
MS-15695B
MS-15696B
MS-15698B
MS-15699B
MS-15700B
MS-15701A
MS-15702B
MS-15717B
MS-15718B
MS-15719B
See also **Cutting Tools:** End Mills
Reamers,
ANSI B94.2-1977
ANSI B94.20-1977
ISO 236/1-1976
ISO 236/2-1976
ISO 521-1975
ISO 522-1975
ISO 2250-1972
ISO 3465-1975
ISO 3466-1975
ISO 3467-1975
Saw Blades,
ANSI B94.42-1972 (R1978)
ANSI B94.51-1976 (1977)
ANSI B94.52-1977
ANSI B122.1-1970
GGG-B-421C
GGG-B-431A(1)
GGG-S-66B(SH)
ISO 2296-1972
ISO 2336-1972
ISO 2924-1973
ISO 4875/1-1978
MIL-H-938
MS-17008
MS-17010
MS-17011
MS-17012
Shaper Cutters,
ANSI B94.21-1968 (R1974)
Single-Point Tools,
ANSI B94.5-1974
ANSI B94.10-1967 (R1972)
ANSI B94.37-1972
ANSI B94.50-1975
ISO 3286-1976
Taps,
ANSI B94.1-1977
ANSI B94.9-1971
ISO 529-1975
ISO 2283-1972
ISO 2857-1973
Twist Drills,
ANSI B94.11-1967 (R1972)
ANSI B94.20-1977
GGG-D-751C(2)
ISO 235/1-1975
ISO 494-1975
ISO 3291-1975
ISO 3292-1975
ISO 3438-1975
ISO 3439-1975
NAS 907

**Grinding of Steel:**
MIL-STD-866A
MIL-STD-1571

**Industrial Engineering:**
ANSI Z94.10-1972
ANSI Z94.12-1972
MIL-STD-1567

**Inspection Machines:**
NAS 971

**Limits, Fits and Tolerances:**
ANSI Y14.5-1973
Cylindrical Parts,
ANSI B4.1-1967 (R1974)
Metric,
ANSI B4.2-1978
ANSI B4.3-1978

**Machine Tool Components, Accessories and Attachments:**
- Boring Machines,
  - MIL-STD-950A
- Broaching Machines,
  - MIL-STD-945A
- Bushings,
  - ANSI B94.33-1974
- Chucks,
  - ANSI B5.8-1972 (R1979)
  - GGG-C-350B
  - ISO 3089-1974
  - ISO 3442-1975
  - MIL-C-45576
  - MIL-C-45833C
- Clamps, Vises and Wedges,
  - GGG-V191D
  - GGG-V-433A
  - GGG-W-220B(1)
- Drilling and Tapping Machines,
  - MIL-STD-932
  - OO-M-340B
- Drivers,
  - ANSI B94.35-1972
- Electrical and Ultrasonic Erosion Machines,
  - MIL-STD-955A
- Gear Cutting and Finishing Machines,
  - MIL-STD-958A
- Grinding Machines,
  - MIL-STD-948A
  - OO-G-673A
  - OO-M-340B
- Holders, Tool Insert and Bit,
  - ANSI B94.26-1969
  - ANSI B94.45-1973
  - ANSI B94.46-1973
  - ANSI B94.47-1973
  - ANSI B94.48-1976
  - MIL-H-4081B(2)
  - NAS 970
  - OO-C-00845
  - OO-C-851A(1)
  - OO-H-581A
- Lapping Plate,
  - GGG-P-446A
- Lathes,
  - MIL-STD-931
  - OO-D-565A
- Machining Centers,
  - MIL-STD-938A
- Milling Machines,
  - MIL-STD-956A
  - OO-M-340B
- Miscellaneous Machine Tools,
  - MIL-STD-957A
- Pilots,
  - ISO 4208-1977
- Planers and Shapers,
  - MIL-STD-959A
- Quill Flanges,
  - ANSI B5.41-1968 (R1973)
- Saws and Filing Machines,
  - MIL-STD-941A
- Socket,
  - OO-S-550A
- Spindle Noses,
  - ANSI B5.9-1967 (R1972)
  - ANSI B5.11-1964 (R1973)
  - ANSI B5.18-1972
  - ANSI B5.40-1977
- Step Blocks,
  - GGG-B-461A
- Tables,
  - MIL-T-80218
- Tapers,
  - ANSI B5.10-1963 (R1972)
- T-Slots,
  - ANSI B5.1-1975

**Machine Tools:**
  - ISO 229-1973
  - ISO/R230-1961
  - NAS 938
- Accuracy,
  - ANSI B5.16-1952 (R1979)
  - ISO 1701-1974
  - ISO 1708-1975
  - ISO 1984-197A
  - ISO 1985-1974
  - ISO 1986-1974
  - ISO 2407-1973
  - ISO 2423-1974
  - ISO 2433-1973
  - ISO 2772/1-1973
  - ISO 2772/2-1974
  - ISO 2773/1-1973
  - ISO 2773/2-1973
  - ISO 3070/0-1975
  - ISO 3070/1-1975
  - ISO 3070/2-1978
  - ISO 3190-1975
  - ISO 3655/0-1976
  - ISO 3655/1-1976
  - ISO 3686-1976
  - ISO 4703-1977
  - NAS 979
  - NMTBA (1968)
  - See also standards for specific machine tools
- Belt Grinders,
  - MIL-G-80176A
- Boring Machines,
  - ISO 3070/0-1975
  - ISO 3070/1-1975
  - ISO 3070/2-1978
  - ISO 3686-1976
  - MIL-B-4554B
  - MIL-B-9925C
  - MIL-B-80012B
  - NAS 910
  - *Numerically Controlled,*
  - MIL-B-80004B
  - MIL-B-80094B
  - MIL-B-80202(1)
  - MIL-B-80210A
  - MIL-B-80248(1)
  - NAS 954

NAS 963
NAS 978
Broaching Machines,
MIL-B-80203
Buffing-grinding-polishing Machines,
OO-B-791(2)
W-G-656D(3)
Center Hole Grinders,
MIL-G-80179
Computerized Numerical Control,
NAS 995
Cutoff Machines,
MIL-C-80021A(1)
MIL-C-80071A
Cylindrical Grinders,
ANSI B5.33-1970
ANSI B5.37-1970
ANSI B5.42-1970
ISO 2407-1973
ISO 2433-1973
MIL-G-23368B(1)
MIL-G-23823C
MIL-G-80048B(1)
Direct Numerical Control,
NAS 993
Disc Grinders,
MIL-G-80126A
Drilling Machines,
ISO 2423-1974
ISO 2772/1-1973
ISO 2772/2-1974
ISO 2773/1-1973
ISO 2773/2-1973
ISO 3190-1975
ISO 3686-1976
MIL-B-4554B
MIL-D-3141B
MIL-D-17674D
MIL-D-18072C
MIL-D-80010C(1)
MIL-D-80076B
MIL-D-80150A
MIL-D-80193A
MIL-D-80225
NAS 910
OO-D-696A
*Numerically Controlled,*
MIL-B-80094B
MIL-D-80077B
MIL-D-80078B
MIL-D-80084A
NAS 948
NAS 954
NAS 960
NAS 978
Electrochemical Machines,
MIL-E-80100A
MIL-E-80238
Electrical Discharge Machines,
MIL-E-80023B
MIL-E-80130A
MIL-E-80140B
MIL-E-80157A
MIL-E-80158A
Electrolytic Grinders,
MIL-E-80022A
Filing Machines,
MIL-F-12490B
MIL-F-23808
Gear Hobbing Machines,
MIL-G-80194
MIL-G-80195
MIL-G-80196
Grinding Machines,
OO-G-00661B(2)
OO-G-669F
See also under specific types
Honing Machines,
MIL-H-80003B
MIL-H-80009B
MIL-H-80216
Internal Grinders,
MIL-G-80066B
MIL-G-80073B
MIL-G-80180A
MIL-G-80228
Jig Grinders,
MIL-G-23681B
Keyseating Machine,
MIL-K-80221
Lapping Machines,
MIL-L-80005A
MIL-L-80199
Lathes,
ISO 1708-1975
ISO 3655/0-1976
ISO 3655/1-1976
MIL-B-80012B
MIL-L-4370B
MIL-L-4533D
MIL-L-11285B
MIL-L-13896D(2)
MIL-L-17966D
MIL-L-19368B
MIL-L-23225B
MIL-L-23251B(2)
MIL-L-23257B(1)
MIL-L-23400B(1)
MIL-L-45100B(1)
MIL-L-45943
MIL-L-45958
MIL-L-80006B
MIL-L-80007B
MIL-L-80018A
MIL-L-80032B
MIL-L-80061A
MIL-L-80175A
NAS 931
OO-L-125C(1)
*Numerically Controlled,*
MIL-B-80202(1)
MIL-B-80240(1)
MIL-L-80053B
MIL-L-80123A
MIL-L-80219A
MIL-L-80245(1)
NAS 966

Lubrication,
- ISO/TR 3498-1974
- ISO 5169-1977
- ISO 5170-1977

Machining Centers,
- MIL-M-80169A
- MIL-M-80182(1)
- MIL-M-80255

Metric Units,
- ANSI B5.51M-1979

Milling Machines,
- ANSI B5.45-1972
- ISO 1701-1974
- ISO 1984-1974
- ISO 3070/0-1975
- ISO 3070/1-1975
- ISO 3070/2-1978
- MIL-B-4554B
- MIL-M-13934D
- MIL-M-23589B
- MIL-M-23855B
- MIL-M-45577A
- MIL-M-80016B
- MIL-M-80033B
- MIL-M-80044A
- MIL-M-80051A(2)
- MIL-M-80058A(1)
- MIL-M-80059A(1)
- MIL-M-80063A(1)
- MIL-M-80064A
- MIL-M-80097A
- MIL-M-80099A
- MIL-M-80127A
- MIL-M-80129A
- MIL-M-80155A
- MIL-M-80177A
- MIL-M-80206A
- MIL-M-80213A
- MIL-M-80215
- NAS 909
- NAS 910
- NAS 914
- NAS 929
- NAS 936
- OO-M-59C

*Adaptive Control,*
- NAS 994

*Numerically Controlled,*
- MIL-B-80004B
- MIL-B-80094B
- MIL-M-80168A
- MIL-M-80184A
- MIL-M-80204A
- MIL-M-80208A
- MIL-M-80209A
- MIL-M-80212(1)
- NAS 911
- NAS 913
- NAS 954
- NAS 977
- NAS 978
- NAS 984

Multi-Purpose Machines,
- MIL-M-80096B
- MIL-M-80139A
- MIL-M-80142A

Pipe Cutting Machines,
- MIL-C-45044

Planers or Shapers,
- MIL-P-80136A
- MIL-S-19240A(1)
- MIL-D-23375B
- MIL-S-23800A(1)
- MIL-S-23910A(2)
- MIL-S-80001A

Plasma or Flame Cutting Machines,
- MIL-C-80036B
- MIL-C-80090B

Plunge Grinders,
- MIL-G-80143A

Retoothing Machine,
- MIL-R-19169B

Safety,
- ANSI B11.6-1975
- ANSI B11.8-1974
- ANSI B11.9-1975
- ANSI B11.13-1975

Saws,
- GGG-S-66B(SH)
- MIL-S-13875E
- MIL-S-23807
- MIL-S-23847B(1)
- MIL-S-24034B
- MIL-S-24035C
- MIL-S-24036B
- MIL-S-26789B(1)
- MIL-S-45814C
- MIL-S-45957
- MIL-S-80201A(1)
- OO-S-240E

Screw Machines,
- MIL-S-23378B

Surface Grinders,
- ANSI B5.32-1977
- ANSI B5.32.1-1977
- ANSI B5.44-1971
- ISO 1985-1974
- ISO 1986-1974
- ISO 4703-1977
- MIL-G-80088B
- MIL-G-80125A
- MIL-G-80145A
- MIL-G-80163A
- MIL-G-80239

Thread Grinders,
- MIL-G-80162A

Threading Machines,
- MIL-T-80166(2)
- OO-T-290C

Tool and Cutter Grinders,
- MIL-G-4515B
- MIL-G-19120B
- MIL-G-23900C
- MIL-G-45072E
- MIL-G-80025A
- MIL-G-80065B
- MIL-G-80189

NAS 973
NAS 998
OO-G-692C(1)
Utilization,
NMTBA (1975)

**Machining Tests and Standards:**
ANSI B94.34-1946 (R1971)
ANSI B94.36-1956 (R1971)
ASTM D1666-64 (1976)
ASTM E618-77T
ISO 3685-1977
MIL-STD-1242
NAS 979

**Materials:**
Carbides,
ANSI H9.7-1973
ANSI H9.10-1973
ANSI H9.24-1973
ANSI H9.26-1969 (R1974)
ASTM B276-54 (1972)
ASTM B294-76
ASTM B311-58 (1972)
ASTM B390-64 (1976)
ASTM B406-76
ASTM B421-76
ASTM B437-67 (1973)
ASTM B485-76
ASTM B611-76
for Tools, Fixtures and Gages,
ANSI B52.1-1962
Tool Steel,
ASTM A597-75
ASTM A600-76
ASTM E352-76a

**Numerical Control:**
ANSI C85.1-1963
ANSI X8.1-1968
ANSI X8.2-1968
ANSI X8.3-1968
EIA Automation Bulletin No. 3-C
EIA Automation Bulletin No. 4
EIA RS-227A
EIA RS-244B
EIA RS-267A
EIA RS-274C
EIA RS-281A
EIA RS-358A
EIA RS-408
EIA RS-431
EIA RS-447
NAS 938
NAS 943
NAS 953
NAS 955
NAS 968
NAS 969
See also under specific machine tools

**Peening:**
Glass Bead,
ANSI/SAE J1173-1977
MIL-STD-852
Shot,
MIL-S-13165B(1)

**Plasma, Oxygen, Air and Electron Beam Cutting:**
AWS A6.3-69
AWS C4.1-77
AWS C5.2-73
AWS C5.3-74
AWS F2.1-78

**Safety:**
ANSI Z43.1-1966
ANSI Z49.1-1973

**Surface Technology:**
AGMA 118.01-1973
ANSI B46.1-1978
ANSI Y14.36-1978
ISO/R 468-1966
ISO 1302-1974
ISO 1878-1974
ISO 1879-1974
ISO 1880-1979
ISO 2632/1-1975
ISO 2632/2-1977
ISO 3274-1975

| DESIGNATION | TITLE |
|---|---|
| AGMA 118.01-1973 | Information Sheet—Gear Tooth Surface Texture for Aerospace Gearing (Surface Roughness, Waviness, Form and Lay) |
| AGMA 120.01-1975 | Fine- and Coarse-Pitch Hobs |
| ANSI B4.1-1967 (R1974) | Preferred Limits and Fits for Cylindrical Parts |
| ANSI B4.2-1978 | Preferred Metric Limits and Fits |
| ANSI B4.3-1978 | Tolerance for Metric Dimensioned Products, General |
| ANSI B5.1-1975 | T-Slots—Their Bolts, Nuts, Tongues and Cutters |
| ANSI B5.8-1972 (R1979) | Chucks and Chuck Jaws |
| ANSI B5.9-1967 (R1972) | Spindle Noses for Tool Room Lathes, Engine Lathes, Turret Lathes, and Automatic Lathes |
| ANSI B5.10-1963 (R1972) | Machine Tapers |
| ANSI B5.11-1964 (R1973) | Spindle Noses and Adjustable Adapters for Multiple Head Spindle Drilling Heads |
| ANSI B5.16-1952 (R1979) | Accuracy of Engine and Tool Room Lathes |
| ANSI B5.18-1972 | Spindle Noses and Tool Shanks for Milling Machines |
| ANSI B5.32-1977 | Grinding Machines, Surface, Reciprocating Table—Horizontal Spindle |
| ANSI B5.32.1-1977 | Grinding Machines, Surface, Reciprocating Table—Verticle Spindle |
| ANSI B5.33-1970 | External Cylindrical Grinding Machines—Plain |
| ANSI B5.35-1969 (R1975) | Machine Mounting Specifications for Abrasive Discs and Plate Mounted Wheels |
| ANSI B5.37-1970 | External Cylindrical Grinding Machines—Centerless |
| ANSI B5.40-1977 | Spindle Noses and Tool Shanks for Horizontal Boring Machines |
| ANSI B5.41-1968 (R1973) | Quill Flanges and Spindle Ends for Upright and Horizontal Drill Spindles for Production Type Drilling Machines |
| ANSI B5.42-1970 | External Cylindrical Grinding Machines—Universal |
| ANSI B5.43M-1979 | Modular Machine Tool Standards |
| ANSI B5.44-1971 | Rotary Table Surface Grinding Machines |
| ANSI B5.45-1972 | Milling Machines |
| ANSI B5.50-1978 | V Flange Tool Shanks for Machining Centers with Automatic Tool Changers |
| ANSI B5.51M-1979 | Preferred SI Units for Machine Tools |
| ANSI B7.1-1978 | Safety Requirements for the Use, Care, and Protection of Abrasive Wheels |
| ANSI B11.6-1975 | Safety Requirements for the Construction and Use of Lathes |
| ANSI B11.8-1974 | Safety Requirements for the Construction, Care and Use of Drilling, Milling and Boring Machines |
| ANSI B11.9-1975 | Safety Requirements for the Construction, Care and Use of Grinding Machines |
| ANSI B11.13-1975 | Safety Requirements for the Construction, Care and Use of Single and Multiple-Spindle Automatic Screw/Bar and Chucking Machines |
| ANSI B46.1-1978 | Surface Texture, Surface Roughness, Waviness and Lay |
| ANSI B52.1-1962 | Classification of Materials for Tools, Fixtures, and Gages |
| ANSI B67.1-1975 | Diamond Dressing Tools |
| ANSI B74.1-1966 | Identification Code for Diamond Wheel Shapes |
| ANSI B74.2-1974 | Specifications for Shapes and Sizes of Grinding Wheels and Shapes, Sizes and Identification of Mounted Wheels |
| ANSI B74.3-1974 | Specifications for Shapes and Sizes of Diamond Grinding Wheels, Hand Hones and Mounted Wheels |
| ANSI B74.4-1977 | Test for Bulk Density of Abrasive Grains |
| ANSI B74.5-1964 (R1976) | Test for Capillarity of Abrasive Grains |
| ANSI B74.6-1964 (R1976) | Procedure for Sampling of Abrasive Grains |
| ANSI B74.8-1977 | Ball Mill Test for Friability of Abrasive Grain |
| ANSI B74.9-1970 (R1976) | Specifications for Shapes, Sizes and Identification of Mounted Wheel Mandrels |
| ANSI B74.10-1977 | Specifications for Grading of Abrasive Microgrits |
| ANSI B74.11-1967 (R1978) | Specifications for Tumbling Chip Abrasives |
| ANSI B74.12-1977 | Specifications for Size of Abrasive Grain—Grinding Wheels, Polishing and General Industrial Uses |
| ANSI B74.13-1977 | Markings for Identifying Grinding Wheels and Other Bonded Abrasives |
| ANSI B74.16-1971 (R1976) | Checking the Size of Diamond Abrasive Grain |

## 25.2 Alphabetical List of Machining Standards

| DESIGNATION | TITLE |
|---|---|
| ANSI B74.17-1973 (R1978) | Bulk Density of Diamond Abrasive Grains, Test for |
| ANSI B74.18-1977 | Specifications for Grading of Certain Abrasive Grain on Coated Abrasive Products |
| ANSI B94.1-1977 | Blanks and Semi-Finished Blanks for Solid Carbide Taps |
| ANSI B94.2-1977 | Reamers |
| ANSI B94.3-1965 (R1972) | Straight Cut-Off Blades for Lathes and Screw Machines |
| ANSI B94.4-1976 | Identification System for Indexable Inserts for Cutting Tools |
| ANSI B94.5-1974 | Carbide Blanks and Single-Point Brazed Tools |
| ANSI B94.6-1966 (R1972) | Knurling |
| ANSI B94.7-1966 (R1972) | Hobs |
| ANSI B94.8-1967 (R1972) | Inserted Blade Milling Cutter Bodies |
| ANSI B94.9-1979 | Taps—Cut and Ground Threads |
| ANSI B94.10-1967 (R1972) | High-Speed Steel and Cast Nonferrous Single-Point Tools and Tool Holders |
| ANSI B94.11-1967 (R1972) | Twist Drills |
| ANSI B94.12-1977 | Carbide-Tipped Masonry Drills and Blanks for Carbide-Tipped Masonry Drills |
| ANSI B94.13-1976 | Blanks for Carbide Burs |
| ANSI B94.19-1977 | Milling Cutters and End Mills |
| ANSI B94.20-1977 | Specifications for Carbide Blanks for Twist Drills, Reamers, End Mills, and Random Rod |
| ANSI B94.21-1968 (R1974) | Gear Shaper Cutters |
| ANSI B94.24-1976 | Heavy Duty Carbide Inserts for Cutting Tools |
| ANSI B94.25-1975 | Indexable Inserts for Cutting Tools |
| ANSI B94.26-1969 | Indexable (Throw-Away) Insert Holders |
| ANSI B94.32-1954 (R1971) | Circular and Dove Tailed Forming Tool Blanks |
| ANSI B94.33-1974 | Jig Bushings |
| ANSI B94.34-1946 (R1971) | Life Tests of Single-Point Tools Made of Materials Other Than Sintered Carbides |
| ANSI B94.35-1972 | Drill Drivers—Split-Sleeve, Collet-Type |
| ANSI B94.36-1956 (R1971) | Life Tests for Single-Point Tools of Sintered Carbide |
| ANSI B94.37-1972 | Carbide Blanks and Cutting Tools, Single-Point, Carbide-Tipped, Roller Turner Type |
| ANSI B94.42-1972 (R1978) | Carbide Blanks for Tipping Circular Saws |
| ANSI B94.45-1973 | Precision Indexable Insert Holders |
| ANSI B94.46-1973 | Carbide Seats Used with Indexable Inserts for Clamp Type Holders |
| ANSI B94.47-1973 | Carbide Chip Breakers Used with Indexable Inserts for Clamp Type Holders |
| ANSI B94.48-1976 | Precision Indexable Insert Cartridges |
| ANSI B94.49-1975 | Spade Drill Blades and Spade Drill Holders |
| ANSI B94.50-1975 | Single-Point Cutting Tools, Basic Nomenclature and Definitions for |
| ANSI B94.51-1976 (1977) | Specifications for Band Saw Blades (Metal Cutting) |
| ANSI B94.52-1977 | Hack Saw Blades |
| ANSI B104.1-1964 | Dimensions for Diamond Core Drill (DCDMA Bulletin No. 2) |
| ANSI B122.1-1970 | Band Saw Blades |
| ANSI C85.1-1963 | Terminology for Automatic Control (includes 1966, 1972 supplements) |
| ANSI H9.7-1973 | Recommended Practice for Evaluating Cemented Carbides for Apparent Porosity (ASTM B276-54, reaffirmed 1972) |
| ANSI H9.10-1973 | Method for Hardness Testing of Cemented Carbides |
| ANSI H9.24-1973 | Test for Electrical Resistivity of Cemented Tungsten Carbides (ASTM B421-76) |
| ANSI H9.26-1969 (R1974) | Method for Tension Testing of Cemented Carbides (ASTM B437-67) |
| ANSI/SAE J1173-1977 | Size Classification and Characteristics of Glass Beads for Peening |
| ANSI X3.37-1977 | Programming Language APT |
| ANSI X8.1-1968 | Axis and Motion Nomenclature for Numerically Controlled Machines (EIA RS-267-A) |
| ANSI X8.2-1968 | Interchangeable Perforated Tape Vairable Block Format for Positioning, Contouring and Contouring/Positioning Numerically Controlled Machine Tools (EIA RS-274-C) |
| ANSI X8.3-1968 | Interchangeable Perforated Tape Variable Block Format for Positioning and Straight Cut Numerically Controlled Machine Tools (EIA RS-273-A) |
| ANSI Y14.5-1973 | Dimensioning and Tolerancing |

| DESIGNATION | TITLE |
|---|---|
| ANSI Y14.36-1978 | Surface Texture Symbols |
| ANSI Z11.142-1964 (R1974) | Method of Test for Emulsion Stability of Soluble Cutting Oils (ASTM D1479-64-1968) |
| ANSI Z43.1-1966 | Ventilation Control of Grinding, Polishing, and Buffing Operations |
| ANSI Z49.1-1973 | Safety in Welding and Cutting |
| ANSI Z94.10-1972 | Production Planning and Control |
| ANSI Z94.12-1972 | Work Measurement and Methods |
| ASTM A 597-75 | Specification for Cast Tool Steel |
| ASTM A 600-76 | Specification for High-Speed Tool Steel |
| ASTM B 276-54 (1972) | Recommended Practice for Evaluating Cemented Carbides for Apparent Porosity (ANSI H9.7-1973) |
| ASTM B 294-76 | Standard Method for Hardness Testing of Cemented Carbides (ANSI H9.10-1973) |
| ASTM B 311-58 (1972) | Test for Density of Cemented Carbides (Approved by ANSI) |
| ASTM B 390-64 (1976) | Evaluating Apparent Grain Size and Distribution of Cemented Tungsten Carbides (Approved by ANSI) |
| ASTM B 406-76 | Test for Determination of Transverse Rupture Strength of Cemented Carbides (Approved by ANSI) |
| ASTM B 421-76 | Test for Electrical Resistivity of Cemented Tungsten Carbides (ANSI H9.24-1973) |
| ASTM B 437-67 (1973) | Tension Testing of Cemented Carbides (ANSI H9.26-1969) |
| ASTM B 485-76 | Method for Diametral Compression Testing of Cemented Carbides |
| ASTM B 611-76 | Test for Abrasive Wear Resistance of Cemented Carbides |
| ASTM D 88-56 (1973) | Test for Saybolt Viscosity (Approved by ANSI) |
| ASTM D 92-72 (1977) | Test for Flash and Fire Points by Cleveland Open Cup (Approved by ANSI) |
| ASTM D 94-71 (1976) | Test for Saponification Number by Color-Indicator Tritration (Approved by ANSI) |
| ASTM D 129-64 (1973) | Test for Sulfur in Petroleum Products by the Bomb Method (Approved by ANSI) |
| ASTM D 130-75 | Test for Detection of Copper Corrosion from Petroleum Products by the Copper Strip Tarnish Test (Approved by ANSI) |
| ASTM D 156-64 (1977) | Test for Saybolt Color of Petroleum Products, Saybolt Chromometer Method (Approved by ANSI) |
| ASTM D 287-67 (1977) | Test for API Gravity of Crude Petroleum and Petroleum Products, Hydrometer Method (Approved by ANSI) |
| ASTM D 445-74 | Test for Viscosity of Transparent and Opaque Liquids, Kinematic and Dynamic Viscosities (Approved by ANSI) |
| ASTM D 808-63 (1976) | Test for Chlorine in New and Used Petroleum Products—Bomb Method (Approved by ANSI) |
| ASTM D 892-74 | Test for Foaming Characteristics of Lubricating Oils (Approved by ANSI) |
| ASTM D 1479-64 (1973) | Test for Emulsion Stability of Soluble Cutting Oils (ANSI Z11.142-1964, reaffirmed 1972) |
| ASTM D 1662-69 (1974) | Test for Active Sulfur in Cutting Fluids (Approved by ANSI) |
| ASTM D 1666-64 (1976) | Conducting Machining Tests of Wood and Wood-Base Materials |
| ASTM D 2670-67 (1977) | Measuring Wear Properties of Fluid Lubricants—Falex Method (Approved by ANSI) |
| ASTM D 2881-73 (1978) | Metal Working Fluids and Related Materials, Classification of |
| ASTM E 352-76a | Methods for Chemical Analysis of Tool Steel and Other Similar Medium- and High-Alloy Steels |
| ASTM E 618-77T | Evaluating Machining Performance Using an Automatic Screw/Bar Machine |
| AWS A6.3-69 | Recommended Safe Practices for Plasma Arc Cutting |
| AWS C4.1-77 | Criteria for Describing Oxygen-Cut Surfaces |
| AWS C5.2-73 | Recommended Practices for Plasma-Arc Cutting |
| AWS C5.3-74 | Recommended Practices for Air Carbon-Arc Gouging and Cutting |
| AWS F2.1-78 | Recommended Safe Practices for Electron Beam Welding and Cutting |
| DWMI 101 | Variables Affecting Diamond Wheels in Carbide Grinding |
| DWMI 102 | Test Procedures for Grinding Carbides with Coolants |
| DWMI 103 | Test Procedures for Dry Grinding Tungsten Carbide with Resinoid Bonded Diamond Wheels |
| DWMI 104 | Test Procedures for Evaluating Diamond Circular Saw Blades Cutting Non-Metallic Materials |

## 25.2 Alphabetical List of Machining Standards

| DESIGNATION | TITLE |
|---|---|
| DWMI 105 | Diamond Grinding of Carbides |
| DWMI 106 | Techniques for Consumer Evaluation of Resinoid Grinding Wheels |
| EIA Automation Bulletin No. 3-C | Glossary of Terms for Numerically Controlled Machines |
| EIA Automation Bulletin No. 4 | Recommended Interchangeable Perforated Tape Variable Block Format for Contouring and Positioning Numerically Controlled Stored Program Machines |
| EIA RS-227A | One Inch Perforated Tape |
| EIA RS-244B | Character Code for Numerical Machine Control Perforated Tape |
| EIA RS-267A | Axis and Motion Nomenclature for Numerically Controlled Machines (ANSI X8.1-1968) |
| EIA RS-274C | Interchangeable Perforated Tape Variable Block Format for Positioning, Contouring and Contouring/Positioning Numerically Controlled Machines (ANSI X8.2-1968) |
| EIA RS-281A | Electrical and Construction Standards for Numerical Machine Control |
| EIA RS-358A | Subset of American National Standard Code for Information Interchange (ASCII) for Numerical Machine Control Perforated Tape |
| EIA RS-408 | Interface between Numerical Control Equipment and Data Terminal Equipment Employing Parallel Binary Data Interchange |
| EIA RS-431 | Electrical Interface between Numerical Control and Machine Tools |
| EIA RS-447 | Operational Command and Data Format for Numerically Controlled Machines |
| GGG-B-421C | Blades, Band-Saw Wood Cutting, Metal Cutting, and Meat Cutting |
| GGG-B-431A(1) | Blades, Saw, Round (Machine and Hand) |
| GGG-B-461A | Blocks, Step, Machinists' |
| GGG-C-350B | Chucks, Drills; Adapters, Arbors, and Keys, Drill Chuck |
| GGG-C-755E | Cutters, Milling |
| GGG-D-751C(2) | Drills, Twist (Taper-Square, Straight, and Taper-Round Shanks) |
| GGG-F-00340(2) | Files, Rotary |
| GGG-P-446A | Plate, Lapping |
| GGG-S-66B (SH) | Saws, Hole: Blades, Hole Saw and Arbors, Hole Saw |
| GGG-V-191D | V-Block and Clamp |
| GGG-V-443A | Vises, Machine Table: Bench and Machine Table: Contour Gripping |
| GGG-W-220B(1) | Wedges, Steel |
| GGG-W-290B | Wheels, Abrasive, Snagging |
| GGG-W-301B | Wheel, Buffing and Polishing |
| ISO 229-1973 | Machine Tools—Speeds and Feeds |
| ISO/R 230-1961 | Machine Tool Test Code |
| ISO 235/1-1975 | Parallel Shank Twist Drills, Jobber and Stub Series, Morse Taper Shank Twist Drills and Core Drills |
| ISO 235/2-1972 | Core Drills with Parallel Shanks and Morse Taper Shanks—Recommended Stocked Sizes |
| ISO 236/1-1976 | Hand Reamers |
| ISO 236/2-1976 | Long Fluted Machine Reamers, Morse Taper Shanks |
| ISO 240-1975 | Milling Cutters—Interchangeability Dimensions for Cutter Arbors or Cutter Mandrels—Metric Series and Inch Series |
| ISO 242-1975 | Carbide Tips for Brazing on Turning Tools |
| ISO 243-1975 | Turning Tools with Carbide Tips—External Tools |
| ISO/R 468-1966 | Surface Roughness |
| ISO 494-1975 | Parallel Shank Twist Drills—Long Series |
| ISO 504-1975 | Turning Tools with Carbide Tips—Designation and Marking |
| ISO 513-1975 | Application of Carbides for Machining by Chip Removal—Designation of the Main Groups of Chip Removal and Groups of Application |
| ISO 514-1975 | Turning Tools with Carbide Tips—Internal Tools |
| ISO 521-1975 | Machine Chucking Reamers with Parallel Shanks or Morse Taper Shanks |
| ISO 522-1975 | Special Tolerances for Reamers |
| ISO 523-1974 | Milling Cutters—Recommended Range of Outside Diameters |
| ISO 525-1975 | Bonded Abrasive Products—General Features—Designation, Ranges of Dimensions, and Profiles |

| DESIGNATION | TITLE |
|---|---|
| ISO 529-1975 | Short Machine Taps and Hand Taps |
| ISO/R 603-1967 | Bonded Abrasive Products—Grinding-Wheel Dimensions (Part I) |
| ISO 883-1976 | Indexable (Throwaway) Carbide Inserts without Fixation Hole—Dimensions |
| ISO 1117-1975 | Bonded Abrasive Products—Grinding-Wheel Dimensions (Part II) |
| ISO 1302-1974 | Method of Indicating Surface Texture on Drawings |
| ISO 1641/1-1978 | End Mills and Slot Drills—Part I: Milling Cutters with Parallel Shanks |
| ISO 1641/2-1978 | End Mills and Slot Drills—Part II: Milling Cutters with Morse Taper Shanks |
| ISO 1641/3-1978 | End Mills and Slot Drills—Part III: Milling Cutters with 7/24 Taper Shanks |
| ISO 1701-1974 | Test Conditions for Milling Machines with Table of Variable Height, with Horizontal or Vertical Spindle—Testing of the Accuracy |
| ISO 1708-1975 | Test Conditions for General Purpose Parallel Lathes—Testing of the Accuracy |
| ISO 1832-1977 | Indexable (Throwaway) Inserts for Cutting Tools—Designation—Code of Symbolization |
| ISO 1878-1974 | Classification of Instruments and Devices for Measurement and Evaluation of the Geometrical Parameters of Surface Finish |
| ISO 1879-1974 | Instruments for the Measurement of Surface Roughness by the Profile Method—Vocabulary |
| ISO 1880-1979 | Instruments for the Measurement of Surface Roughness by the Profile Method—Contact (Stylus) Instruments of Progressive Profile Transformation—Profile Recording Instruments |
| ISO 1929-1974 | Abrasive Belts—Designation, Dimensions and Tolerances |
| ISO 1984-1974 | Test Conditions for Milling Machines with Table of Fixed Height with Horizontal or Verticle Spindle—Testing of Accuracy |
| ISO 1985-1974 | Test Conditions for Surface Grinding Machines with Vertical Grinding Wheel Spindle and Reciprocating Table—Testing of Accuracy |
| ISO 1986-1974 | Test Conditions for Surface Grinding Machines with Horizontal Grinding Wheel Spindle and Reciprocating Table—Testing of Accuracy |
| ISO 2250-1972 | Finishing Reamers for Morse and Metric Tapers, with Parallel Shanks and Morse Taper Shanks |
| ISO 2283-1972 | Long Shank Machine Taps with Nominal Diameters from 3 to 24 mm and 1/8 to 1 inch, with Amendment 1-1977 |
| ISO 2296-1972 | Metal Slitting Saws with Fine and Coarse Teeth—Metric Series |
| ISO 2306-1972 | Drills for Use Prior to Tapping Screw Threads |
| ISO 2324-1972 | End Mills with 7/24 Taper Shanks—Standard Series and Long Series |
| ISO 2336-1972 | Hand and Machine Hacksaw Blades—Dimensions for Lengths up to 450 mm and Pitches up to 6.3 mm |
| ISO 2407-1973 | Test Conditions for Internal Cylindrical Grinding Machines with Horizontal Spindle—Testing of Accuracy |
| ISO 2423-1974 | Test Conditions for Radial Drilling Machines with the Arm Adjustable in Height—Testing of the Accuracy |
| ISO 2433-1973 | Test Conditions for External Cylindrical Grinding Machines with a Movable Table—Testing of Accuracy |
| ISO 2584-1972 | Cylindrical Cutters with Plain Bore and Key Drive—Metric Series |
| ISO 2585-1972 | Slotting Cutters with Plain Bore and Key Drive—Metric Series |
| ISO 2587-1972 | Side and Face Milling Cutters with Plain Bore and Key Drive—Metric Series |
| ISO 2632/1-1975 | Roughness Comparison Specimens—Part I: Turned, Ground, Bored, Milled, Shaped and Planed |
| ISO 2632/2-1977 | Roughness Comparison Specimens—Part II: Spark-Eroded, Shot Blasted and Grit Blasted, and Polished |
| ISO 2772/1-1973 | Test Conditions for Box Type Vertical Drilling Machines—Testing of the Accuracy—Part I: Geometrical Tests |
| ISO 2772/2-1974 | Test Conditions for Box Type Vertical Drilling Machines—Testing of the Accuracy—Part II: Practical Test |
| ISO 2773/1-1973 | Test Conditions for Pillar Type Vertical Drilling Machines—Testing of the Accuracy—Part I: Geometrical Tests |
| ISO 2773/2-1973 | Test Conditions for Pillar Type Vertical Drilling Machines—Testing of the Accuracy—Part II: Practical Test |

## 25.2 Alphabetical List of Machining Standards

| DESIGNATION | TITLE |
|---|---|
| ISO 2780-1973 | Milling Cutters with Tenon Drive—Interchangeability Dimensions with Cutter Arbors—Metric Series |
| ISO 2857-1973 | Ground Thread Taps for ISO Metric Threads of Tolerances 4H to 8H and 4G to 6G Coarse and Fine Pitches—Manufacturing Tolerances on the Threaded Portion |
| ISO 2924-1973 | Solid and Segmental Circular Saws for Cold Cutting of Metals—Interchangeability Dimensions of the Drive—Saw Diameter Range 224 to 2240 mm |
| ISO 2933-1974 | Bonded Abrasive Products—Grinding Wheel Dimensions (Part III) |
| ISO 2940/1-1974 | Milling Cutters Mounted on Centring Arbors having a 7/24 Taper—Part I: Fitting Dimensions—Centring Arbors |
| ISO 2940/2-1974 | Milling Cutters Mounted on Centring Arbors having a 7/24 Taper—Part II: Inserted Tooth Cutters |
| ISO 2976-1973 | Abrasive Belts—Selection of Width/Length Combinations |
| ISO 3002/1-1977 | Geometry of the Active Part of Cutting Tools—Part I: General Terms, Reference Systems, Tool and Working Angles |
| ISO 3070/0-1975 | Test Conditions for Boring and Milling Machines with Horizontal Spindle—Testing of the Accuracy—Part 0: General Introduction |
| ISO 3070/1-1975 | Test Conditions for Boring and Milling Machines with Horizontal Spindle—Testing of the Accuracy—Part I: Table Type Machines |
| ISO 3274-1975 | Instruments for the Measurement of Surface Roughness by the Profile Method—Contact (Stylus) Instruments of Consecutive Profile Transformation—Contact Profile Meters, System M |
| ISO 3293-1975 | Morse Taper Shank Countersinks for Angles 60 degrees, 90 degrees and 120 degrees inclusive |
| ISO 3294-1975 | Parallel Shank Countersinks for Angles 60, 90 and 120 degrees inclusive |
| ISO 3314-1975 | Shell Drills with Taper Bore (Taper Bore 1:30 (included)) with Slot Drive |
| ISO 3337-1975 | T-Slot Cutters with Plain or Flatted Parallel Shanks and with Morse Taper Shanks having Tapped Hole—Metric Series |
| ISO 3364-1977 | Indexable (Throwaway) Carbide Inserts with Cylindrical Fixation Hole—Dimensions |
| ISO 3365/1-1977 | Indexable (Throwaway) Carbide Inserts for Milling Cutters—Dimensions—Part I—Square Inserts |
| ISO 3438-1975 | Subland Twist Drills with Morse Taper Shanks for Holes Prior to Tapping Screw Threads |
| ISO 3439-1975 | Subland Twist Drills with Parallel Shanks for Holes Prior to Tapping Screw Threads |
| ISO 3442-1975 | Self-Centring Chucks for Machine Tools with Two-Piece Jaws (Tongue and Groove Type)—Sizes for Interchangeability and Acceptance Test Specifications |
| ISO 3465-1975 | Hand Taper Pin Reamers |
| ISO 3466-1975 | Machine Taper Pin Reamers with Parallel Shanks |
| ISO 3467-1975 | Machine Taper Pin Reamers with Morse Taper Shanks |
| ISO/TR 3498-1974 | Lubricants for Machine Tools |
| ISO 3655/0-1976 | Test Conditions for Vertical Turning and Boring Lathes with One or Two Columns—Testing of the Accuracy—Part 0: General Introduction |
| ISO 3655/1-1976 | Test Conditions for Vertical Turning and Boring Lathes with One or Two Columns—Testing of the Accuracy—Part I: Lathes with a Single Fixed or Movable Table |
| ISO 3685-1977 | Tool-Life Testing with Single-Point Turning Tools |
| ISO 3686-1976 | Test Conditions for Turret and Single Spindle Co-ordinate Drilling and Boring Machines with Table of Fixed Height with Vertical Spindle—High Accuracy Machines—Testing of the Accuracy |
| ISO 3855-1977 | Milling Cutters—Nomenclature Bilingual Edition |
| ISO 3859-1977 | Inverse Dovetail Cutters and Dovetail Cutters with Parallel Shanks |
| ISO 3860-1976 | Bore Cutters with Key Drive—Form Milling Cutters with Constant Profile |
| ISO 3919-1977 | Coated Abrasives—Flap Wheels with Shafts—Designation and Dimensions |
| ISO 3920-1976 | Honing Stones of Square Section—Designation and Dimensions |
| ISO 3921-1976 | Honing Stones of Rectangular Section—Designation and Dimensions |
| ISO 3940-1977 | Tapered Die-Sinking Cutters with Parallel Shanks |
| ISO 4204-1977 | Countersinks, 90 degrees, with Morse Taper Shanks and Detachable Pilots |
| ISO 4205-1977 | Countersinks, 90 degrees, with Parallel Shanks and Solid Pilot |
| ISO 4206-1977 | Counterbores with Parallel Shanks and Solid Pilots |

| DESIGNATION | TITLE |
|---|---|
| ISO 4207-1977 | Counterbores with Morse Taper Shanks and Detachable Pilots |
| ISO 4208-1977 | Detachable Pilots for Use with Counterbores and 90 degrees Countersinks—Dimensions |
| ISO 4703-1977 | Test Conditions for Surface Grinding Machines with Two Columns—Machines for Grinding Slideways—Testing of Accuracy |
| ISO 4875/1-1978 | Metal Cutting Band Saw Blades—Part I: Definitions and Terminology |
| ISO 5169-1977 | Machine Tools—Presentation of Lubrication Instructions |
| ISO 5170-1977 | Machine Tools—Lubrication Systems |
| ISO 5421-1977 | Ground High Speed Steel Tool Bits |
| ISO 5429-1977 | Coated Abrasives—Flap Wheels with Incorporated Flanges or Separate Flanges—Designation and Dimensions |
| MIL-B-4554B | Boring, Drilling and Milling Machine, Horizontal, Table Type, Heavy Duty, 3, 4 and 5 Inch Spindle |
| MIL-B-9925C | Boring Machine, Jig, Vertical, Manual Control |
| MIL-B-80004B | Boring and Milling Machine, Precision, Horizontal Spindle, Table Type, Manually or Numerically Controlled |
| MIL-B-80012B | Boring and Turning Machines, Vertical, and Lathes, Vertical, Turret, Manual |
| MIL-B-80094B | Boring, Drilling and Milling Machines, Horizontal Spindle, Table Type, Numerically Controlled, Perforated Tape |
| MIL-B-80202(1) | Boring and Turning Machines, Vertical, Numerically Controlled, Perforated Tape |
| MIL-B-80203 | Broaching Machine, Hydraulic, Horizontal, Internal, Single Ram, Manual Control |
| MIL-B-80210A | Boring Machine, Jig, Vertical, Numerically Controlled Perforated Tape |
| MIL-B-80248(1) | Boring and Turning Machines, Vertical, Numerical Control by Manual Data Input 36 Through 96 Inch Tables |
| MIL-C-936 | Counterbore and Spot-facer, Adjustable |
| MIL-C-45044 | Cutting-grooving-beveling Machine, Pipe, Gasoline Engine Driven, Skid Mounted, 8-Inch Capacity |
| MIL-C-45576 | Chuck, Centering, Lathe(L) |
| MIL-C-45833C | Chuck, Independent Jaw (Medium Duty) |
| MIL-C-46113B | Cutting-fluid Concentrates, Transparent |
| MIL-C-46149 | Cutting Fluid, Sulfur and Chlorine Additive |
| MIL-C-80021A(1) | Cut-off Machine, Abrasive Disk or Circular Saw Blade, Metalcutting, Chop Stroke, Dry Cut, Floor Mounted |
| MIL-C-80036B | Cutting Machine, Oxygen Flame and Plasma Arc, Automatic, Stationary |
| MIL-C-80071A | Cutoff Machine, Abrasive Disk, Metal-Cutting, Wet Cut |
| MIL-C-80090B | Cutting System, Metal, Plasma Arc |
| MIL-D-3141B | Drilling Machine, Upright, Round Column, Floor Type, 21 Inch Swing,(L) Superseded by MIL-D-80002 |
| MIL-D-17674D | Drilling Machine, Upright, Back-geared, Fixed Head, 21-Inch Swing, 1-1/2 Inch Drill(L) |
| MIL-D-18072C | Drilling Machine, Upright, Sensitive, Bench Type, Single Spindle, 5/16 Inch and Under, Hand Feed |
| MIL-D-45815C | Drilling Machine, Radial, Bench Type, Sliding Arm, Swivel Head, Hand Feed |
| MIL-D-80000C | Drilling Machine, Upright, Round Column, Floor and Pedestal, Hand Feed |
| MIL-D-80002C | Drilling Machine, Upright, Round and Box Column, Sliding Head, Plain Table, Power Feed |
| MIL-D-80010C(1) | Drilling Machine, Upright, Turret Head, Automatic, 3/4 to 1-1/2 Inch Drill Capacity |
| MIL-D-80017B | Drilling Machine, Upright, Round and Box Column Sliding Head, Floor Mounted, Geared Drive, Power Feed |
| MIL-D-80038B(1) | Drilling Machine, Upright, Bench Type, Round Column, Electric Motor Driven, Hand Feed |
| MIL-D-80039A(1) | Drilling Machine, Upright, Bench Type, Box Column, Sliding Head, Electric Motor Driven, Hand Feed |
| MIL-D-80076B | Drilling Machine, Upright, Box Column, Floor, Plain Table, Hand Feed |

L = Limited coordination document

## 25.2 Alphabetical List of Machining Standards

| DESIGNATION | TITLE |
|---|---|
| MIL-D-80077B | Drilling Machine, Upright, Turret Type, Sliding Head, Bed Type, Numerically Controlled |
| MIL-D-80078B | Drilling Machine, Upright, Turret Head, Traveling Column, Numerically Controlled, Perforated Tape |
| MIL-D-80084A | Drilling Machine, Upright, Turret Type, Fixed Clearance, Sliding Turret Carrier, Numerically Controlled |
| MIL-D-80150A | Drilling Machine, Upright, Turret Head, Sliding Carrier, Six Spindles, Bench Mounted |
| MIL-D-80193A | Drilling Machine, Vertical, Multi-Spindle, Adjustable Joint, Standard |
| MIL-D-80225 | Drilling Machine, Deep Hole, Horizontal Spindle, Floor Mounted, Gun Type |
| MIL-E-80022A | Electro Erosion Machine, Electrolytic, Carbide Tool Bit and Chip Breaker Grinder |
| MIL-E-80023B | Electrical Discharge Machine, Vertical, Electro-Erosion Machine, Electrical Discharge, Vertical, Ram and Quill Type (Non-rotating, Non-tilting Head) |
| MIL-E-80100A | Electro-erosion Machine, Metal, Electro-chemical, Vertical Ram Type, Cavity Sinking |
| MIL-E-80130A | Electro-erosion Machine, Metal Disintegrating, Electric Arc, Vibrating Head, Drill Type |
| MIL-E-80140B | Electro-erosion Machine, Electrical Discharge, Horizontally Moving Head, Bridge Type |
| MIL-E-80157A | Electro-erosion Machine, Electrical Discharge, Angular Tilting Quill Head, Rotating Ram, Knee Type Bed |
| MIL-E-80158A | Electro-erosion Machine, Electrical Discharge, (C) Frame, Angular Tilting Head, Quill Type, Rotating Spindle |
| MIL-E-80222 | End Finishing Machine, Tube and Rod, Deburring, Chamfering, and Facing, Bench and Floor Mounted, 2, 3 and 5 Inch Diameter Workpiece Capacity |
| MIL-E-80238 | Electro-erosion Machine, Electro-chemical, Horizontal Ram Type, Cavity Sinking |
| MIL-F-12490B | Filing Machine, Saw Blade(L) |
| MIL-F-23808 | Filing Machine, Saw Blade, Band and Circular, Automatic(L) |
| MIL-G-4515B | Grinding Machine, Knife, Horizontal Spindle, Electric Motor Driven |
| MIL-G-19120B | Grinding Machines, Saw Tooth, Hand Operated, Electric Motor Driven(L) |
| MIL-G-23368B(1) | Grinding Machine, Cylindrical, External, Plain, Power |
| MIL-G-23681B | Grinding Machine, Jig, Vertical |
| MIL-G-23823C | Grinding Machine, Cylindrical, External, Center Type, Universal, Traveling Table |
| MIL-G-23900C | Grinding Machine, Universal and Tool |
| MIL-G-45072E | Grinding Machine, Drill, Manual Operation |
| MIL-G-80025A | Grinding Machine, Carbide Tool Bit, Floor Type, Electric Motor Driven |
| MIL-G-80048B(1) | Grinding Machine, Centerless, External, Cylindrical |
| MIL-G-80065B | Grinding Machine, Drill Point, Floor Mounted, Vertical Spindle, Automatic Cycle |
| MIL-G-80066B | Grinding Machine, Internal Chucking, Plain, Horizontal, Electric Motor Driven |
| MIL-G-80073B | Grinding Machine, Internal, Chucking, Universal |
| MIL-G-80088B | Grinding Machine, Surface, Refiprocating Table, Horizontal Spindle, Power Feed |
| MIL-G-80125A | Grinding Machine, Surface, Rotary, Vertical Spindle, Single Table |
| MIL-G-80126A | Grinding Machine, Disc, Horizontal, Double End, Single Spindle |
| MIL-G-80143A | Grinding Machine, Plunge, External, Multi-Form |
| MIL-G-80145A | Grinding Machine, Surface, Reciprocating Table, Horizontal Spindle, Hand Feed |
| MIL-G-80162A | Grinding Machine, Thread, External, Automatic |
| MIL-G-80163A | Grinding Machine, Surface, Rotary Table, Horizontal Spindle |
| MIL-G-80176A | Grinding Machine, Abrasive Belt, Platen Type, Light, Medium and Heavy Duty, 4, 6, 8, and 10 Inch Belt Widths |
| MIL-G-80179 | Grinding Machine, Center Hole, Vertical, Floor Mounted, Electric Motor Driven |
| MIL-G-80180A | Grinding Machine, Internal, Chucking, Horizontal, Electric Motor Driven |
| MIL-G-80189 | Grinding Machine, Circular Saw Blade, Automatic |
| MIL-G-80194 | Gear Hobbing Machines, Vertical |
| MIL-G-80195 | Gear Hobbing Machines, Horizontal, Mechanical |

L = Limited coordination document

| DESIGNATION | TITLE |
|---|---|
| MIL-G-80196 | Gear Hobbing Machines, Horizontal, Hydraulic |
| MIL-G-80228 | Grinding Machine, Internal, Planetary Head, Horizontal Spindle, Electric Motor Driven |
| MIL-G-80239 | Grinding Machine Inside/Outside, Surface, Rotary Table Vertical Spindle, Universal |
| MIL-H-938 | Hole-cutters and Hole-saws |
| MIL-H-4081B(2) | Holder, Countersink, Adjustable Micrometer Stop and Cutters, Countersink |
| MIL-H-80003B | Honing Machine, Vertical, Single Spindle, 1-1/2 to 12 Inch Bore, 12 to 80 Inch Stroke |
| MIL-H-80009B | Honing Machines, Horizontal, Single Spindle, Mandrel Type, Wet Application |
| MIL-H-80216 | Honing Machine, Horizontal, Internal, Single Spindle, Long Stroke (8 Feet to 25 Feet) |
| MIL-K-80221 | Keyseating Machine, Mechanical and Hydraulic, Reciprocating Cutter Type |
| MIL-L-4370B | Lathe, Engine, Heavy Duty, 4025, and 40-Inch Swing, Raised and Hollow Spindle |
| MIL-L-4533D | Lathe, Turret, Horizontal, Universal, Ram Type |
| MIL-L-11285B | Lathe, Engine, 9 by 22 Inch Capacity, Bench Type(L) |
| MIL-L-13896D(2) | Lathe, Engine, Gap, Sliding Bed |
| MIL-L-17966D | Lathe, Engine, Geared Head, 21 and 26 Inch |
| MIL-L-19368B | Lathe, Engine, Motor Driven, Bench, 16 Inch Swing, Nodular Iron Construction, Submarine Use(L) |
| MIL-L-23225B | Lathe, Precision, Jewelers and Instrument |
| MIL-L-23249C(1) | Lathe, Engine and Toolroom, Light Duty, 13 and 15 Inch |
| MIL-L-23250B | Lathe, Engine or Toolroom, Back Gear Belt Drive, 12 Inch Through 17 Inch |
| MIL-L-23251B(2) | Lathe, Engine and Tool Room 2516 and 3220 |
| MIL-L-23257B(1) | Lathe, Engine and Tool Room, Heavy Duty 1610 and 2013 |
| MIL-L-23400B(1) | Lathe, Engine and Tool Room, 14, 16 and 20 Inch |
| MIL-L-45100B(1) (WC) | Lathe, Turret, Horizontal(L) |
| MIL-L-45943 (WC) | Lathe, Engine, Floor Mounted, 18 Inch Swing(L) |
| MIL-L-45958 | Lathe, Engine, Floor Mounted, 25 Inch Swing(L) |
| MIL-L-80005A | Lapping Machine, Metalworking, High Precision, Flat Surfaces, Single Face, Free Abrasive Type |
| MIL-L-80006B | Lathe, Turret, Horizontal, Universal, Saddle Type |
| MIL-L-80007B | Lathe, Toolroom, Variable Speed, 10 Inch |
| MIL-L-80018A | Lathes, Turret, Horizontal, Chucking, and Bar and Chucking, Automatic, Saddle Type |
| MIL-L-80032B | Lathe, Chucking and Lathe, Bar, Single Spindle, Horizontal, Automatic, Turret-in-column Type |
| MIL-L-80053B | Lathe, Engine, Heavy Duty, Numerically Controlled, Perforated Tape |
| MIL-L-80061A | Lathe, Bench, Variable Speed, Plain and Turret, 9 Inch |
| MIL-L-80123A | Lathe, Chucking, and Lathes, Bar, Single Spindle, Horizontal, Turret-in-column Type, Automatic and Numerically Controlled by Perforated Tape |
| MIL-L-80175A | Lathe, Speed, Horizontal Spindle, Single End, Bench and Floor Mounted |
| MIL-L-80199 | Lapping Machine, Flat Surfaces, Gauge Lapping Type, 7 Inch Wheel |
| MIL-L-80219A | Lathe, Turret, Horizontal, Vertical Bed and Slant Bed, Numerically Controlled, Perforated Tape |
| MIL-L-80245(1) | Lathe, Multiple Purpose, Chucking, Horizontal Spindle, Numerically Controlled, Perforated Tape |
| MIL-M-13934D | Milling Machine, Horizontal, Universal(L) |
| MIL-M-23589B | Milling Machine, Vertical, Rotary |
| MIL-M-23855B | Milling Machine, Vertical, Knee Type, Hydraulic, Three Dimensional, Duplicating, Profiling and Duplicating |
| MIL-M-45577A | Milling Machine, Swivel Head, Ram Type, Universal(L) |
| MIL-M-80016B | Milling Machine, Vertical, Swivel Head, Knee and Column, Ram and Turret Type |
| MIL-M-80033B | Milling Machine, Combination Horizontal and Vertical, Knee Type |
| MIL-M-80044A | Milling Machine, Horizontal, Knee Type, Number 2 |

L = Limited coordination document

## 25.2 Alphabetical List of Machining Standards

| DESIGNATION | TITLE |
|---|---|
| MIL-M-80051A(2) | Milling Machine, Vertical, Knee Type, Size Number 2, 28 Inches Mininum Longitudinal Travel |
| MIL-M-80058A(1) | Milling Machine, Vertical, Knee Type, Size No. 3 and 4, 34 and 43 Inches Minimum Longitudinal Table Travel |
| MIL-M-80059A(1) | Milling Machine, Vertical, Knee Type, Size No. 5 and 6, 50 and 60 Inches Minimum Longitudinal Table Travel |
| MIL-M-80063A(1) | Milling Machine, Horizontal, Knee Type, Size Numbers 3 and 4 |
| MIL-M-80064A | Milling Machine, Horizontal, Knee Type, Size Numbers 5 and 6 |
| MIL-M-80096B | Multi-Purpose Machine, Drilling, Boring, Milling and Tapping, Vertical, Single Spindle, Fixed Height Compound Table, Numerically Controlled, Perforated Tape, Manual Tool Change |
| MIL-M-80097A | Milling Machine, Bed Type, Vertical, 3-Dimensional Power Feed |
| MIL-M-80099A | Milling Machine, Bed Type, Horizontal, 3-Axis Power Feed |
| MIL-M-80127A | Milling Machine, Bed Type, Ram Mounted Vertical Spindle, Manual and Hydraulic Tracer Controlled Profiling and Duplicating |
| MIL-M-80129A | Milling Machine, Portable, Ram and Columnar Types |
| MIL-M-80139A | Multi-Purpose Machines, Drilling, Boring, Milling and Tapping, Vertical, Single Spindle, Compound Worktable, Numerically Controlled, Perforated Tape, Automatic Tool Changer. Superseded by MIL-M-80255. |
| MIL-M-80142A | Multi-Purpose Machine, Drilling, Boring, Milling and Tapping, Horizontal, Single Spindle, Fixed Height Compound Table, Numerically Controlled, Perforated Tape, Manual Tool Change |
| MIL-M-80155A | Milling Machines, Bed Type, Vertical Spindle, Manual and Hydraulic or Electronic Tracer Controlled |
| MIL-M-80168A | Milling Machine, Bed Type, Vertical, Double Housing, Numerically Controlled, Profiling, 30 to 40 Horsepower Per Spindle, Single and Multi-Spindle Machines |
| MIL-M-80169A | Machining Center, Horizontal, Single Spindle, Integral Rotary Indexing Type Worktable, Numerically Controlled, Perforated Tape, Automatic Tool Change |
| MIL-M-80177A | Milling Machine, Planer Type, Double Housing |
| MIL-M-80182(1) | Machining Centers, Horizontal Spindle, Automatic Tool Changer, 3 Linear Axes, Rotary Machine Table, 3 and 4 Axis Numerical Controls |
| MIL-M-80184A | Milling Machine, Bed Type, Vertical, Double Housing, Numerically Controlled Profiling, 5 to 25 Horsepower per Spindle, Single and Multi-Spindle Machines |
| MIL-M-80204A | Milling Machine, Horizontal Profilers, Traveling Table Type, Three Axes, Numerically Controlled, Single and Multi-Spindle Machine |
| MIL-M-80206A | Milling Machine, Bed Type, Horizontal Spindle, Automatic Cycle, 14 Inch Minimum Table Width, Plain and Duplex Machines |
| MIL-M-80208A | Milling Machine, Horizontal Profiler, Traveling Column Type, 3 Axis, Numerically Controlled, Single and Multi-Spindle Machines |
| MIL-M-80209A | Milling Machines, Knee Type, Numerically Controlled, Positioning or Contouring, 1 to 4 Horsepower Spindle Drive |
| MIL-M-80212(1) | Milling Machine, Bed Type, Ram Mounted Vertical Spindle, Numerically Controlled Profiler |
| MIL-M-80213A | Milling Machine, Bed Type, Horizontal Spindle, Automatic Cycle, 18 and 20 Inch Minimum Table Width, Plain and Duplex Machines |
| MIL-M-80215 | Milling Machine, Bed Type, Horizontal Spindle, Automatic Cycle, 24, 26 and 30 Inch Minimum Table Width, Plain and Duplex Machines |
| MIL-M-80255 | Machining Centers, Vertical, Single Spindle, Computer Numerical Control (CNC) |
| MIL-P-80136A | Planers, Metal Cutting, Openside, Hydraulic, Shaper Planer Type |
| MIL-R-19169B | Retoothing Machine, Handsaw |
| MIL-S-13165B(1) | Shot Peening of Metal Parts |
| MIL-S-13875E | Saw, Band, Metal Cutting, 14 Inch |
| MIL-S-19240A(1) | Shaper, Metal-cutting Horizontal, Bench, Mechanical, 7-Inch Stroke |
| MIL-S-23375B | Shaper, Metal Cutting, Horizontal Ram, Hydraulic, Plain and Universal Table, Sizes 24 and 28 |
| MIL-S-23378B | Screw Machine, Automatic, Single Spindle |
| MIL-S-23800A(1) | Shaper, Metal Cutting, Horizontal, Plain and Universal, Mechanical, Medium Duty 14, 16, 20 and 24 Inch Stroke |

| DESIGNATION | TITLE |
|---|---|
| MIL-S-23807 | Saw, Circular, Traverse Cut, 8 Foot Capacity(L) |
| MIL-S-23847B(1) | Saw, Band, Cut Off, Horizontal, Manual and Automatic |
| MIL-S-23910A(2) | Shaper, Metal Cutting, Horizontal, Plain and Universal, Mechanical, Heavy Duty, 16, 20, 24, 28, 32 and 36 Inch Stroke |
| MIL-S-24034B | Saw, Band, Metal Cutting, Contour, Power Feed Table (3 and 7-1/2 Hp), and Non-Feeding Table (2 and 3 Hp), 16, 26, 36, and 60 Inch Throat |
| MIL-S-24035C | Saws, Band, Cutoff, Traveling, Tilting, Vertical Column |
| MIL-S-24036B | Saw, Band, Cutoff, Horizontal, Heavy Duty |
| MIL-S-26789B(1) | Saw, Band, Vertical, Contour, Friction Sawing, 36 Inch Throat Width, 20 Inch Throat Height |
| MIL-S-45814C | Saw, Band, Cutoff, Horizontal, and Combination Horizontal and Vertical |
| MIL-S-45957 | Saw, Band, Metal Cutting, Contour, 24-Inch Throat, 12-Inch Vertical Clearance(L) |
| MIL-S-80001A | Shapers, Metal Cutting, Vertical, Hydraulic |
| MIL-S-80201A(1) | Saw, Band, Metal Cutting, Vertical, Contour |
| MIL-STD-852 | Glass Bead Peening Procedures(L) |
| MIL-STD-866A | Grinding of Chrome Plated Steel and Steel Parts Heat Treated to 180,000 psi or over(L) |
| MIL-STD-931 | Attachments and Accessories for Lathes, Metal Working |
| MIL-STD-932 | Attachments and Accessories for Drilling and Tapping Machines |
| MIL-STD-938A | Attachments and Accessories for Machining Centers |
| MIL-STD-941A | Attachments and Accessories for Sawing and Filing Machines |
| MIL-STD-945A | Attachment and Accessories for Broaching Machines |
| MIL-STD-948A | Attachment and Accessories for Grinding Machines |
| MIL-STD-950A | Attachments and Accessories for Boring Machines |
| MIL-STD-955A | Attachments and Accessories for Electrical and Ultrasonic Erosion Machines |
| MIL-STD-956A | Attachments and Accessories for Milling Machines |
| MIL-STD-957A | Attachments and Accessories for Miscellaneous Machine Tools |
| MIL-STD-958A | Attachments and Accessories for Gear Cutting and Finishing Machines |
| MIL-STD-959A | Attachments and Accessories for Planers and Shapers |
| MIL-STD-1242 | Machining Standards for Fire Control Material(L) |
| MIL-STD-1567 | Work Measurement(L) |
| MIL-STD-1571 | Steel, High Strength, Grinding of(L) |
| MIL-T-80166(2) | Threading Machine, Single Point, Lathe Type, Semi-Automatic (4 Inch External Thread Diameter, 4 Inch Thread Length, 4 to 40 Threads per Inch) |
| MIL-T-80218 | Table, Rotary, Horizontal, Manual and Power Feed, Primary and Auxiliary Use, Round: 24 to 72 Inch; Rectangular: 24 x 24 to 60 x 84 Inch |
| MIL-W-16102C(1) | Wheel, Abrasive (Cutting-off)(Q) |
| MIL-W-17929B(1) | Wheel, Abrasive, Precision(Q) |
| MS-15687C | Cutter, Milling, Plain, Light Duty |
| MS-15688B | Cutter, Milling, Plain, Heavy Duty |
| MS-15689A | Cutter, Milling, Plain, Helical |
| MS-15690B | Cutter, Milling, Half-Side |
| MS-15691A | Cutter, Milling, Side Staggered-tooth |
| MS-15693A | Cutter, Milling, Metal Slitting Saw, Side-Chip Clearance |
| MS-15694B | Cutter, Milling, Metal Slitting Saw, Side-Chip Clearance, Staggered-tooth |
| MS-15695B | Cutter, Milling, Angle, Single, Plain Hole |
| MS-15696B | Cutter, Milling, Angle, Double |
| MS-15698B | Cutter, Milling, Concave |
| MS-15699B | Cutter, Milling, Convex |
| MS-15700B | Cutter, Milling, Corner Rounding |
| MS-15701A | Cutter, Milling, Involute Gear, 14-1/2 Degree Pressure Angle |
| MS-15702B | Cutter, Milling, Miter and Bevel Gear, 14-1/2 Degree Pressure Angle |

L = Limited coordination document

Q = Qualified products list issued

## 25.2 Alphabetical List of Machining Standards

| DESIGNATION | TITLE |
|---|---|
| MS-15704B | Cutter, Milling, End, Single-end Multiple Flute, ASA Taper Shank |
| MS-15705A | Cutter, Milling, Single End, Two-Flute, Brown and Sharpe Taper Shank |
| MS-15706A | Cutter, Milling, Single End, Multiple Flute, Brown and Sharpe Taper Shank |
| MS-15707B | Cutter, Milling, End, Single-end, Two Flute, Straight Shank |
| MS-15708B | Cutter, Milling, End, Single-end, Multiple-flute, Straight-shank |
| MS-15709B | Cutter, Milling, End, Single-end, Two Flute, Flatted Straight-shank |
| MS-15710B | Cutter, Milling, End, Single-end, Multiple-flute, Flatted Straight-shank |
| MS-15711B | Cutter, Milling, End, Single-end, Ball End, Two Flute, Flatted Straight-shank |
| MS-15713A | Cutter, Milling, End, Double-end, Multiple-Flute, Straight-shank |
| MS-15714A | Cutter, Milling, End, Double-end, Two Flute, Flatted Straight-shank |
| MS-15715A | Cutter, Milling, End, Double-end, Multiple-Flute, Flatted Straight-shank |
| MS-15716B | Cutter, Milling, Shell End, Removable Shank, Spiral-flute |
| MS-15717B | Cutter, Milling, T-slot, Staggered Tooth, Weldon Shank |
| MS-15718B | Cutter, Milling, Woodruff Key-seat, Straight-shank |
| MS-15719B | Cutter, Milling, Screw Slotting |
| MS-17008 | Blade, Band Saw, Metal Cutting, Flexible Back, Conventional Tooth(L) |
| MS-17010 | Blade, Band Saw, Wood Cutting, Narrow, Spring Tempered |
| MS-17011 | Blade, Band Saw, Wood Cutting, Wide, Spring Tempered |
| MS-17012 | Blade, Band Saw, Metal Cutting, Flexible Back, Skip Tooth General Purpose |
| NAS 907 | High Speed Steel Twist Drills 1/16 Inch thru 1/2 Inch |
| NAS 909 | Milling Machine—General Purpose, Knee and Bed Types, Horizontal and Vertical |
| NAS 910 | Horizontal Boring, Drilling and Milling Machine |
| NAS 911 | Milling Machine—Numerically Controlled Skin/Profile |
| NAS 912 | Spar Mill, Numerically Controlled |
| NAS 913 | Milling Machine—Numerically Controlled, Profiling and Contouring |
| NAS 914 | Milling Machine, Tracer Controlled, Profiling and Contouring—Vertical |
| NAS 929 | Performance Requirement—Work Holder Airframe Spar Milling Machine |
| NAS 931 | Specification for Dual Turning Machines |
| NAS 932 | Drills—Threaded Hexagonal Shank Sizes #40 thru Letter "F" |
| NAS 936 | Milling Machine—Precision Bed-Type, Horizontal and Vertical Spindles |
| NAS 937 | Drills—Procurement Specification Stepped Double Margin Sizes 5/32 Inch thru 1/2 Inch |
| NAS 938 | Machine Axis and Motion Nomenclature |
| NAS 943 | Interchangeable Perforated Tape Standards Variable Block Format for Positioning and Straight Cut Numerically Controlled Equipment (EIA RS-273) (Inactive for design after 30 August 1974, see NAS 955) |
| NAS 948 | Printed Circuit Board Drilling Machine—Numerically Controlled (Inactive for design after 15 September 1975) |
| NAS 953 | Numerical Control System |
| NAS 954 | Numerically Controlled Horizontal Boring, Drilling and Milling Machine |
| NAS 955 | Interchangeable Perforated Tape Standards Variable Block Format for Numerically Controlled Equipment (EIA RS-274) |
| NAS 960 | Drilling Machines—Numerically Controlled |
| NAS 963 | Numerically Controlled Horizontal and Vertical Jig Boring Machines |
| NAS 965 | Drills, High Speed Steel Threaded Shank 1/16 Inch thru "F" |
| NAS 966 | Lathe, Precision—Numerically Controlled |
| NAS 968 | Interchangeable Digital Magnetic Tape Standards, Variable Block, Word Address Format, Contouring and Contouring/Positioning Equipment (Inactive for Design after 31 October 1974) |
| NAS 969 | Interchangeable Digital Magnetic Tape Cartridge—For Numerically Controlled Machines (Inactive for Design after 31 October 1974) |
| NAS 970 | Tool Holders—Basic for Numerically Controlled Machine Tools |

L = Limited coordination document

| DESIGNATION | TITLE |
|---|---|
| NAS 971 | Measuring Inspection Machine—Precision—Numerically Controlled (Inactive for Design after 15 April 1978) |
| NAS 973 | Tool and Cutter Grinding Machine |
| NAS 977 | Milling Machine, Contour, Numerical Control, Multiaxis, Traveling Gantry/Table, Combination High and Low Speed |
| NAS 978 | Numerically Controlled Combination Drilling, Boring, Milling and Tapping Machines (Inactive for Design after 31 October 1974) |
| NAS 979 | Uniform Cutting Tests—NAS Series Metal Cutting Equipment Specifications |
| NAS 984 | Numerically Controlled Milling Machines Knee and Bed Types—Horizontal and Vertical |
| NAS 986 | End Mills, High Speed Steel .125 Inch thru 3.00 Inch Diameter |
| NAS 993 | Direct Numerical Control System |
| NAS 994 | Adaptive Control System for Numerically Controlled Milling Machines |
| NAS 995 | Computerized Numerical Control (CNC) System for Numerically Controlled Machine Tools |
| NAS 998 | Grinder—End Mill, Numerically Controlled |
| NMTBA(1968) | Definition and Evaluation of Accuracy and Repeatability for Numerically Controlled Machine Tools |
| NMTBA(1975) | Guidelines for the Measurement of Machine Tool Utilization |
| OO-B-791(2) | Buffing-grinding-polishing Machines (Bench and Pedestal, Electric) |
| OO-C-00845 | Cutter Bits, Tool Holder (Carbide Tip) |
| OO-C-851A(1) | Cutter Bit, Tool Holder (Solid High-speed Steel) |
| OO-D-565A | Dogs, Lathe |
| OO-D-696A | Drilling Machines, Radial, Floor Mounted |
| OO-G-00661B(2) | Grinding Machine, Bench, Hand-Operated |
| OO-G-669F | Grinder, Pneumatic, Horizontal and Vertical |
| OO-G-673A | Grinding Machine, Machine Tool Attachment; (Precision, Tool Post, and Angle Plate) |
| OO-G-692C(1) | Grinding Machine, Tool and Cutter, and Grinding and Gashing Machines, Tool, Cutter, Flute and Radius |
| OO-H-581A | Holders, Cutting Tool; Engine Lathe, Shaper and Planer |
| OO-L-125C(1) | Lathe, Engine and Toolroom, 10 Inch |
| OO-M-59C | Machine, Milling, Bench Type, Horizontal, Vertical, General Purpose |
| OO-M-340B | Milling, Grinding, Drilling, Slotting Attachment |
| OO-S-240E | Saw, Power, Hack (6 Inch thru 24 Inch Workpiece Capacity) |
| OO-S-550A | Socket, Taper-Shank Tool (Machine) |
| OO-T-290C | Threading Machines, Pipe and Bolt, Power Driven (Bench and Floor Type) |
| VV-C-846A(1) | Cutting Fluids: Emulsifiable Oils |
| VV-C-850A(1) | Cutting Fluids: Sulfurized Fatty and Mineral Oils |
| W-G-656D(3) | Grinding Machine, and Grinding and Buffing Machine Utility, Bench and Pedestal, Electric |

| SERIES CODE | PUBLISHER |
|---|---|
| AGMA | American Gear Manufacturers Association<br>Standards Department<br>1901 North Fort Myer Drive<br>Arlington, Virginia 22209 |
| ANSI | American National Standards Institute, Inc.<br>1430 Broadway<br>New York, New York 10018<br><br>(Many of the standards listed may also be obtained at a discount to members from:<br><br>The American Society of Mechanical Engineers<br>United Engineering Center<br>345 E. 47th Street<br>New York, New York 10017) |
| ASTM | American Society for Testing and Materials<br>1916 Race Street<br>Philadelphia, Pennsylvania 19103 |
| AWS | American Welding Society Inc.<br>2501 N.W. 7th Street<br>Miami, Florida 33125 |
| DWMI | Diamond Wheel Manufacturers' Institute<br>2130 Keith Building<br>Cleveland, Ohio 44115 |
| EIA | Electronic Industries Association<br>Engineering Department<br>2001 Eye Street, N.W.<br>Washington, DC 20006 |
| GGG, OO, VV, W | General Services Administration<br>Federal Supply Service<br>Washington, DC 20406 |
| ISO | International Organization for Standardization<br>Case postale 56<br>1211 GENEVE 20<br>Switzerland<br><br>(ISO standards may be purchased in the U.S. from the American National Standards Institute.) |
| MIL, MS | Commanding Officer<br>Naval Publications and Forms Center<br>5801 Tabor Avenue<br>Philadelphia, Pennsylvania 19120 |
| NAS | National Standards Association Inc.<br>427 Rugby Avenue<br>Washington, DC 20014 |
| NMTBA | National Machine Tool Builders' Association<br>7901 Westpark Drive<br>McLean, Virginia 22101 |

SECTION 26

# MATERIALS INDEX


26.1 Numerical List of Materials ........ 26–3
26.2 Alphabetical List of Materials ........ 26–17
26.3 Chemical Composition by Material Group ........ 26–37

# SECTION 26

# MATERIALS INDEX


26.1 Numerical List of Materials ........ 26–3
26.2 Alphabetical List of Materials ........ 26–17
26.3 Chemical Composition by Material Group ........ 26–37

| MATERIAL | UNS DESIGNATION | MATERIAL GROUP |
|---|---|---|
| 1Sb | | Lead Alloys, Cast—Lead Antimony Alloys |
| 4Sb | | Lead Alloys, Cast—Lead Antimony Alloys |
| 6Sb | | Lead Alloys, Cast—Lead Antimony Alloys |
| 8Sb | | Lead Alloys, Cast—Lead Antimony Alloys |
| 9Sb | | Lead Alloys, Cast—Lead Antimony Alloys |
| 15-5 PH | S15500 | Precipitation Hardening Stainless Steel, Wrought |
| 16-6 PH | | Precipitation Hardening Stainless Steel, Wrought |
| 16-25-6 | | High Temperature Alloy—Iron Base, Wrought |
| 17-4 PH | S17400 | Precipitation Hardening Stainless Steel, Wrought |
| 17-4 PH | S17400 | Precipitation Hardening Stainless Steel, Cast |
| 17-7 PH | S17700 | Precipitation Hardening Stainless Steel, Wrought |
| 17-4 Cu Mo | | Precipitation Hardening Stainless Steel, Wrought |
| 19-9 DL | K63198 | High Temperature Alloy—Iron Base, Wrought |
| 30 | | Structural Steel, Wrought |
| 35 | | Structural Steel, Wrought |
| 42 | | Structural Steel, Wrought |
| 45 | | Structural Steel, Wrought |
| 50 | | Structural Steel, Wrought |
| 50Mo-50Ag | | Powder Metal Alloy—Refractory Metal Base |
| 50W + C-50Ag | | Powder Metal Alloy—Refractory Metal Base |
| 51W-49Ag | | Powder Metal Alloy—Refractory Metal Base |
| 55 | | Structural Steel, Wrought |
| 55W-45Cu | | Powder Metal Alloy—Refractory Metal Base |
| 56W + C-44Cu | | Powder Metal Alloy—Refractory Metal Base |
| 60 | | Structural Steel, Wrought |
| 61Mo-39Ag | | Powder Metal Alloy—Refractory Metal Base |
| 62Cu-18Ni-18Zn-2Sn | | Powder Metal Alloy—Copper-Nickel Alloy |
| 65 | | Structural Steel, Wrought |
| 65W-35Ag | | Powder Metal Alloy—Refractory Metal Base |
| 67Ni-30Cu-3Fe | | Powder Metal Alloy—Nickel and Nickel Alloy |
| 68.5Cu-30Zn-1.5Pb | | Powder Metal Alloy—Brass |
| 70 | | Structural Steel, Wrought |
| 70Cu-30Zn | | Powder Metal Alloy—Brass |
| 72.5W-27.5Ag | | Powder Metal Alloy—Refractory Metal Base |
| 74W-26Cu | | Powder Metal Alloy—Refractory Metal Base |
| 75 | | Structural Steel, Wrought |
| 77Cu-15Pb-7Sn-1Fe-1C | | Powder Metal Alloy—Bronze |
| 80 | | Structural Steel, Wrought |
| 85 | | Structural Steel, Wrought |
| 85W-15Ag | | Powder Metal Alloy—Refractory Metal Base |
| 87W-13Cu | | Powder Metal Alloy—Refractory Metal Base |
| 88Al-5Sn-4Pb-3Cu | | Powder Metal Alloy—Aluminum Alloy |
| 90 | | Structural Steel, Wrought |
| 90Cu-10Ni | | Powder Metal Alloy—Copper-Nickel Alloy |
| 90Cu-10Zn | | Powder Metal Alloy—Brass |
| 90Cu-10Zn-0.5Pb | | Powder Metal Alloy—Brass |
| 90.5Al-5Sn-4Cu | | Powder Metal Alloy—Aluminum Alloy |
| 95Cu-5Al | | Powder Metal Alloy—Bronze |
| 100 | | Structural Steel, Wrought |
| 101 | C10100 | Copper Alloy, Wrought |
| 102 | C10200 | Copper Alloy, Wrought (Oxygen Free) |
| 104 | C10400 | Copper Alloy, Wrought |

# 26.1 Numerical List of Materials

| MATERIAL | UNS DESIGNATION | MATERIAL GROUP |
|---|---|---|
| 105 | C10500 | Copper Alloy, Wrought |
| 107 | C10700 | Copper Alloy, Wrought |
| 109 | C10900 | Copper Alloy, Wrought |
| 110 | C11000 | Copper Alloy, Wrought (Electrolytic Tough Pitch) |
| 110 | | Structural Steel, Wrought |
| 111 | C11100 | Copper Alloy, Wrought |
| 113 | C11300 | Copper Alloy, Wrought (Tough Pitch With Age) |
| 114 | C11400 | Copper Alloy, Wrought (Tough Pitch With Age) |
| 115 | C11500 | Copper Alloy, Wrought (Tough Pitch With Age) |
| 116 | C11600 | Copper Alloy, Wrought (Tough Pitch With Age) |
| 119 | C11900 | Copper Alloy, Wrought |
| 120 | C12000 | Copper Alloy, Wrought |
| 120 Grade | | Maraging Steel, Wrought |
| 121 | C12100 | Copper Alloy, Wrought |
| 122 | C12200 | Copper Alloy, Wrought (Phosphorus Deoxidized) |
| 125 | C12500 | Copper Alloy, Wrought |
| 127 | C12700 | Copper Alloy, Wrought |
| 128 | C12800 | Copper Alloy, Wrought |
| 129 | C12900 | Copper Alloy, Wrought |
| 130 | C13000 | Copper Alloy, Wrought |
| 135 | | Structural Steel, Wrought |
| 140 | | Structural Steel, Wrought |
| 142 | C14200 | Copper Alloy, Wrought |
| 143 | C14300 | Copper Alloy, Wrought |
| 145 | | Structural Steel, Wrought |
| 145 | C14500 | Copper Alloy, Wrought |
| 147 | C14700 | Copper Alloy, Wrought |
| 150 | C15000 | Copper Alloy, Wrought |
| 150 | | Structural Steel, Wrought |
| 155 | C15500 | Copper Alloy, Wrought |
| 160 | | Structural Steel, Wrought |
| 162 | C16200 | Copper Alloy, Wrought |
| 165 | C16500 | Copper Alloy, Wrought |
| 165 | | Structural Steel, Wrought |
| 170 | C17000 | Copper Alloy, Wrought (Beryllium Copper) |
| 172 | C17200 | Copper Alloy, Wrought (Beryllium Copper) |
| 173 | C17300 | Copper Alloy, Wrought |
| 175 | C17500 | Copper Alloy, Wrought (Beryllium Copper) |
| 180 Grade | K91940 | Maraging Steel, Wrought |
| 182 | C18200 | Copper Alloy, Wrought |
| 184 | C18400 | Copper Alloy, Wrought |
| 185 | C18500 | Copper Alloy, Wrought |
| 185 | | Structural Steel, Wrought |
| 187 | C18700 | Copper Alloy, Wrought |
| 189 | C18900 | Copper Alloy, Wrought |
| 190 | C19000 | Copper Alloy, Wrought |
| 191 | C19100 | Copper Alloy, Wrought |
| 192 | C19200 | Copper Alloy, Wrought |
| 194 | C19400 | Copper Alloy, Wrought |
| 195 | C19500 | Copper Alloy, Wrought |
| 200 Grade | | Maraging Steel, Wrought |
| 201 | S20100 | Stainless Steel, Wrought—Austenitic |

| MATERIAL | UNS DESIGNATION | MATERIAL GROUP |
|---|---|---|
| 201.0 | A02010 | Aluminum Alloy, Cast–Sand and Permanent Mold Cast |
| 202 | S20200 | Stainless Steel, Wrought–Austenitic |
| 203EZ | S20300 | Free Machining Stainless Steel, Wrought–Austenitic |
| 208.0 | A02080 | Aluminum Alloy, Cast–Sand Cast |
| 210 | C21000 | Copper Alloy, Wrought (Gilding, 95%) |
| 210 | | Structural Steel, Wrought |
| 213.0 | A02130 | Aluminum Alloy, Cast–Sand and Permanent Mold Cast |
| 220 | C22000 | Copper Alloy, Wrought (Commercial Bronze 90%) |
| 222.0 | A02220 | Aluminum Alloy, Cast–Sand and Permanent Mold Cast |
| 224.0 | A02240 | Aluminum Alloy, Cast–Sand Cast |
| 226 | C22600 | Copper Alloy, Wrought (Jewelry Bronze 87.5%) |
| 230 | C23000 | Copper Alloy, Wrought (Red Brass, 85%) |
| 240 | C24000 | Copper Alloy, Wrought (Low Brass, 80%) |
| 242.0 | A02420 | Aluminum Alloy, Cast–Sand and Permanent Mold Cast |
| 250 Grade | K92890 | Maraging Steel, Wrought |
| 260 | C26000 | Copper Alloy, Wrought (Cartridge Brass 70%) |
| 268 | C26800 | Copper Alloy, Wrought (Yellow Brass) |
| 270 | C27000 | Copper Alloy, Wrought (Yellow Brass) |
| 280 | C28000 | Copper Alloy, Wrought (Muntz Metal) |
| 295.0 | A02950 | Aluminum Alloy, Cast–Sand Cast |
| 300 Grade | K93120 | Maraging Steel, Wrought |
| 300M | K44220 | High Strength Steel, Wrought |
| 301 | S30100 | Stainless Steel, Wrought–Austenitic |
| 302 | S30200 | Stainless Steel, Wrought–Austenitic |
| 302B | S30215 | Stainless Steel, Wrought–Austenitic |
| 303 | S30300 | Free Machining Stainless Steel, Wrought–Austenitic |
| 303MA | S30345 | Free Machining Stainless Steel, Wrought–Austenitic |
| 303Pb | S30360 | Free Machining Stainless Steel, Wrought–Austenitic |
| 303 Plus X | S30310 | Free Machining Stainless Steel, Wrought–Austenitic |
| 303Se | S30323 | Free Machining Stainless Steel, Wrought–Austenitic |
| 304 | S30400 | Stainless Steel, Wrought–Austenitic |
| 304L | S30403 | Stainless Steel, Wrought–Austenitic |
| 305 | S30500 | Stainless Steel, Wrought–Austenitic |
| 308 | S30800 | Stainless Steel, Wrought–Austenitic |
| 308.0 | A03080 | Aluminum Alloy, Cast–Sand and Permanent Mold Cast |
| 309 | S30900 | Stainless Steel, Wrought–Austenitic |
| 309S | S30908 | Stainless Steel, Wrought–Austenitic |
| 310 | S31000 | Stainless Steel, Wrought–Austenitic |
| 310S | S31008 | Stainless Steel, Wrought–Austenitic |
| 314 | S31008 | Copper Alloy, Wrought (Leaded Commercial Bronze) |
| 314 | S31400 | Stainless Steel, Wrought–Austenitic |
| 316 | S31600 | Stainless Steel, Wrought–Austenitic |
| 316 | C31600 | Copper Alloy, Wrought |
| 316L | S31603 | Stainless Steel, Wrought–Austenitic |
| 317 | S31700 | Stainless Steel, Wrought–Austenitic |
| 319.0 | A03190 | Aluminum Alloy, Cast–Sand and Permanent Mold Cast |
| 321 | S32100 | Stainless Steel, Wrought–Austenitic |
| 328.0 | A03280 | Aluminum Alloy, Cast–Sand Cast |
| 330 | C33000 | Copper Alloy, Wrought (Low Leaded Brass) |
| 330 | N08330 | Stainless Steel, Wrought–Austenitic |
| 332 | C33200 | Copper Alloy, Wrought (High Lead Brass) |
| 333.0 | A03330 | Aluminum Alloy, Cast–Permanent Mold Cast |

# 26.1 Numerical List of Materials

| MATERIAL | UNS DESIGNATION | MATERIAL GROUP |
|---|---|---|
| 335 | C33500 | Copper Alloy, Wrought (Low Leaded Brass) |
| 340 | C34000 | Copper Alloy, Wrought (Medium Leaded Brass) |
| 342 | C34200 | Copper Alloy, Wrought (High Leaded Brass) |
| 347 | S34700 | Stainless Steel, Wrought—Austenitic |
| 348 | S34800 | Stainless Steel, Wrought—Austenitic |
| 349 | C34900 | Copper Alloy, Wrought |
| 350 | C35000 | Copper Alloy, Wrought |
| 350 Grade | | Maraging Steel, Wrought |
| 353 | C35300 | Copper Alloy, Wrought (High Leaded Brass) |
| 354.0 | A03540 | Aluminum Alloy, Cast—Permanent Mold Cast |
| 355.0 | A03550 | Aluminum Alloy, Cast—Sand and Permanent Mold Cast |
| 356 | C35600 | Copper Alloy, Wrought (Extra High Leaded Brass) |
| 356.0 | A03560 | Aluminum Alloy, Cast—Sand and Permanent Mold Cast |
| 357.0 | A03570 | Aluminum Alloy, Cast—Sand and Permanent Mold Cast |
| 359.0 | A03590 | Aluminum Alloy, Cast—Sand and Permanent Mold Cast |
| 360.0 | A03600 | Aluminum Alloy, Cast—Die Casting |
| 360 | C36000 | Copper Alloy, Wrought (Free Cutting Brass) |
| 365 | C36500 | Copper Alloy, Wrought (Leaded Muntz Metal) |
| 366 | C36600 | Copper Alloy, Wrought (Leaded Muntz Metal + As) |
| 367 | C36700 | Copper Alloy, Wrought (Leaded Muntz Metal + Sb) |
| 368 | C36800 | Copper Alloy, Wrought (Leaded Muntz Metal + P) |
| 370 | C37000 | Copper Alloy, Wrought (Free Cutting Muntz Metal) |
| 377 | C37700 | Copper Alloy, Wrought (Forging Brass) |
| 380.0 | A03800 | Aluminum Alloy, Cast—Die Casting |
| 383.0 | A03830 | Aluminum Alloy, Cast—Die Casting |
| 384 | S38400 | Stainless Steel, Wrought—Austenitic |
| 385 | S38500 | Stainless Steel, Wrought—Austenitic |
| 385 | C38500 | Copper Alloy, Wrought (Architectural Bronze) |
| 390.0 | A03900 | Aluminum Alloy, Cast—Die Casting |
| 392.0 | A03920 | Aluminum Alloy, Cast—Die Casting |
| 403 | S40300 | Stainless Steel, Wrought—Martensitic |
| 405 | S40500 | Stainless Steel, Wrought—Ferritic |
| 409 | S40900 | Stainless Steel, Wrought—Ferritic |
| 410 | S41000 | Stainless Steel, Wrought—Martensitic |
| 411 | C41100 | Copper Alloy, Wrought |
| 413 | C41300 | Copper Alloy, Wrought |
| 413.0 | A04130 | Aluminum Alloy, Cast—Die Casting |
| 414 | S41400 | Stainless Steel, Wrought—Martensitic |
| 416 | S41600 | Free Machining Stainless Steel, Wrought—Martensitic |
| 416Se | S41623 | Free Machining Stainless Steel, Wrought—Martensitic |
| 416 Plus X | S41610 | Free Machining Stainless Steel, Wrought—Martensitic |
| 420 | S42000 | Stainless Steel, Wrought—Martensitic |
| 420F | S42020 | Free Machining Stainless Steel, Wrought—Martensitic |
| 420F Se | S42023 | Free Machining Stainless Steel, Wrought—Martensitic |
| 422 | S42200 | Stainless Steel, Wrought—Martensitic |
| 425 | C42500 | Copper Alloy, Wrought |
| 429 | S42900 | Stainless Steel, Wrought—Ferritic |
| 430 | S43000 | Stainless Steel, Wrought—Ferritic |
| 430F | S43020 | Free Machining Stainless Steel, Wrought—Ferritic |
| 430F Se | S43023 | Free Machining Stainless Steel, Wrought—Ferritic |
| 431 | S43100 | Stainless Steel, Wrought—Martensitic |
| 434 | S43400 | Stainless Steel, Wrought—Ferritic |

| MATERIAL | UNS DESIGNATION | MATERIAL GROUP |
|---|---|---|
| 435 | C43500 | Copper Alloy, Wrought |
| 436 | S43600 | Stainless Steel, Wrought—Ferritic |
| 440A | S44002 | Stainless Steel, Wrought—Martensitic |
| 440B | S44003 | Stainless Steel, Wrought—Martensitic |
| 440C | S44004 | Stainless Steel, Wrought—Martensitic |
| 440F | S44020 | Free Machining Stainless Steel, Wrought—Martensitic |
| 440F Se | S44023 | Free Machining Stainless Steel, Wrought—Martensitic |
| 442 | C44200 | Copper Alloy, Wrought (Admiralty, Uninhibited) |
| 442 | S44200 | Stainless Steel, Wrought—Ferritic |
| 443 | C44300 | Copper Alloy, Wrought (Admiralty + As) |
| 444 | C44400 | Copper Alloy, Wrought (Admiralty + Sb) |
| 445 | C44500 | Copper Alloy, Wrought (Admiralty + P) |
| 446 | S44600 | Stainless Steel, Wrought—Ferritic |
| 464 | C46400 | Copper Alloy, Wrought (Naval Brass) |
| 465 | C46500 | Copper Alloy, Wrought (Naval Brass + As) |
| 466 | C46600 | Copper Alloy, Wrought (Naval Brass + Sb) |
| 467 | C46700 | Copper Alloy, Wrought (Naval Brass + P) |
| 482 | C48200 | Copper Alloy, Wrought |
| 485 | C48500 | Copper Alloy, Wrought (Leaded Naval Brass) |
| 501 | S50100 | Stainless Steel, Wrought—Martensitic |
| 502 | S50200 | Stainless Steel, Wrought—Martensitic |
| 505 | C50500 | Copper Alloy, Wrought (Phosphor Bronze) |
| 510 | C51000 | Copper Alloy, Wrought (Phosphor Bronze) |
| 511 | C51100 | Copper Alloy, Wrought |
| 514.0 | A05140 | Aluminum Alloy, Cast—Sand Cast |
| 518.0 | A05180 | Aluminum Alloy, Cast—Die Casting |
| 520.0 | A05200 | Aluminum Alloy, Cast—Sand Cast |
| 521 | C52100 | Copper Alloy, Wrought (Phosphor Bronze) |
| 524 | C52400 | Copper Alloy, Wrought (Phosphor Bronze) |
| 535.0 | A05350 | Aluminum Alloy, Cast—Sand Cast |
| 544 | C54400 | Copper Alloy, Wrought (Free Cutting Phosphor Bronze) |
| 608 | C60800 | Copper Alloy, Wrought |
| 610 | C61000 | Copper Alloy, Wrought |
| 613 | C61300 | Copper Alloy, Wrought |
| 614 | C61400 | Copper Alloy, Wrought (Aluminum Bronze) |
| 618 | C61800 | Copper Alloy, Wrought |
| 619 | C61900 | Copper Alloy, Wrought |
| 623 | C62300 | Copper Alloy, Wrought |
| 624 | C62400 | Copper Alloy, Wrought |
| 625 | C62500 | Copper Alloy, Wrought |
| 630 | C63000 | Copper Alloy, Wrought |
| 632 | C63200 | Copper Alloy, Wrought |
| 638 | C63800 | Copper Alloy, Wrought |
| 642 | C64200 | Copper Alloy, Wrought |
| 651 | C65100 | Copper Alloy, Wrought (Low Silicon Bronze) |
| 655 | C65500 | Copper Alloy, Wrought (High Silicon Bronze) |
| 667 | C66700 | Copper Alloy, Wrought |
| 674 | C67400 | Copper Alloy, Wrought |
| 675 | C67500 | Copper Alloy, Wrought (Manganese Bronze) |
| 687 | C68700 | Copper Alloy, Wrought (Aluminum Brass Arsenical) |
| 688 | C68800 | Copper Alloy, Wrought |
| 694 | C69400 | Copper Alloy, Wrought |

# 26.1 Numerical List of Materials

| MATERIAL | UNS DESIGNATION | MATERIAL GROUP |
|---|---|---|
| 705.0 | A07050 | Aluminum Alloy, Cast—Sand and Permanent Mold Cast |
| 706 | C70600 | Copper Alloy, Wrought (Copper Nickel) |
| 707.0 | A07070 | Aluminum Alloy, Cast—Sand and Permanent Mold Cast |
| 710 | C71000 | Copper Alloy, Wrought |
| 713.0 | A07130 | Aluminum Alloy, Cast—Sand and Permanent Mold Cast |
| 715 | C71500 | Copper Alloy, Wrought (Copper Nickel, 30%) |
| 725 | C72500 | Copper Alloy, Wrought |
| 745 | C74500 | Copper Alloy, Wrought (Nickel Silver, 65-10) |
| 752 | C75200 | Copper Alloy, Wrought (Nickel Silver, 65-18) |
| 754 | C75400 | Copper Alloy, Wrought (Nickel Silver, 65-15) |
| 757 | C75700 | Copper Alloy, Wrought (Nickel Silver, 65-12) |
| 770 | C77000 | Copper Alloy, Wrought (Nickel Silver, 58-18) |
| 771.0 | A07710 | Aluminum Alloy, Cast—Sand Cast |
| 782 | C78200 | Copper Alloy, Wrought |
| 801 | C80100 | Copper Alloy, Cast |
| 803 | C80300 | Copper Alloy, Cast |
| 805 | C80500 | Copper Alloy, Cast |
| 807 | C80700 | Copper Alloy, Cast |
| 809 | C80900 | Copper Alloy, Cast |
| 811 | C81100 | Copper Alloy, Cast |
| 813 | C81300 | Copper Alloy, Cast (High) |
| 814 | C81400 | Copper Alloy, Cast (High) |
| 815 | C81500 | Copper Alloy, Cast (High) |
| 817 | C81700 | Copper Alloy, Cast (High) |
| 818 | C81800 | Copper Alloy, Cast (High) |
| 820 | C82000 | Copper Alloy, Cast (High) |
| 821 | C82100 | Copper Alloy, Cast (High) |
| 822 | C82200 | Copper Alloy, Cast (High) |
| 824 | C82400 | Copper Alloy, Cast (High) |
| 825 | C82500 | Copper Alloy, Cast (High) |
| 826 | C82600 | Copper Alloy, Cast (High) |
| 827 | C82700 | Copper Alloy, Cast (High) |
| 828 | C82800 | Copper Alloy, Cast (High) |
| 833 | C83300 | Copper Alloy, Cast (Red Brass) |
| 834 | C83400 | Copper Alloy, Cast (Red Brass) |
| 836 | C83600 | Copper Alloy, Cast (Red Brass) |
| 838 | C83800 | Copper Alloy, Cast (Red Brass) |
| 842 | C84200 | Copper Alloy, Cast (Semi-Red Brass) |
| 844 | C84400 | Copper Alloy, Cast (Semi-Red Brass) |
| 848 | C84800 | Copper Alloy, Cast (Semi-Red Brass) |
| 850.0 | A08500 | Aluminum Alloy, Cast—Sand and Permanent Mold Cast |
| 852 | C85200 | Copper Alloy, Cast (Yellow Brass) |
| 853 | C85300 | Copper Alloy, Cast (Yellow Brass) |
| 854 | C85400 | Copper Alloy, Cast (Yellow Brass) |
| 855 | C85500 | Copper Alloy, Cast (Yellow Brass) |
| 857 | C85700 | Copper Alloy, Cast (Yellow Brass) |
| 858 | C85800 | Copper Alloy, Cast (Yellow Brass) |
| 861 | C86100 | Copper Alloy, Cast (High Strength Yellow Brass) |
| 862 | C86200 | Copper Alloy, Cast (High Strength Yellow Brass) |
| 863 | C86300 | Copper Alloy, Cast (High Strength Yellow Brass) |
| 864 | C86400 | Copper Alloy, Cast (High Strength Yellow Brass) |
| 865 | C86500 | Copper Alloy, Cast (High Strength Yellow Brass) |

| MATERIAL | UNS DESIGNATION | MATERIAL GROUP |
|---|---|---|
| 867 | C86700 | Copper Alloy, Cast (High Strength Yellow Brass) |
| 868 | C86800 | Copper Alloy, Cast (High Strength Yellow Brass) |
| 872 | C87200 | Copper Alloy, Cast (Silicon Bronze) |
| 874 | C87400 | Copper Alloy, Cast (Silicon Brass) |
| 875 | C87500 | Copper Alloy, Cast (Silicon Brass) |
| 876 | C87600 | Copper Alloy, Cast (Silicon Brass) |
| 878 | C87800 | Copper Alloy, Cast (Silicon Brass) |
| 879 | C87900 | Copper Alloy, Cast (Silicon Brass) |
| 902 | C90200 | Copper Alloy, Cast (Tin Bronze) |
| 903 | C90300 | Copper Alloy, Cast (Tin Bronze) |
| 905 | C90500 | Copper Alloy, Cast (Tin Bronze) |
| 907 | C90700 | Copper Alloy, Cast (Tin Bronze) |
| 909 | C90900 | Copper Alloy, Cast (Tin Bronze) |
| 910 | C91000 | Copper Alloy, Cast (Tin Bronze) |
| 911 | C91100 | Copper Alloy, Cast (Tin Bronze) |
| 913 | C91300 | Copper Alloy, Cast (Tin Bronze) |
| 915 | C91500 | Copper Alloy, Cast (Tin Bronze) |
| 916 | C91600 | Copper Alloy, Cast (Tin Bronze) |
| 917 | C91700 | Copper Alloy, Cast (Tin Bronze) |
| 922 | C92200 | Copper Alloy, Cast (Leaded Tin Bronze) |
| 923 | C92300 | Copper Alloy, Cast (Leaded Tin Bronze) |
| 925 | C92500 | Copper Alloy, Cast (Leaded Tin Bronze) |
| 926 | C92600 | Copper Alloy, Cast (Leaded Tin Bronze) |
| 927 | C92700 | Copper Alloy, Cast (Leaded Tin Bronze) |
| 928 | C92800 | Copper Alloy, Cast (Leaded Tin Bronze) |
| 932 | C93200 | Copper Alloy, Cast (High Leaded Tin Bronze) |
| 934 | C93400 | Copper Alloy, Cast (High Leaded Tin Bronze) |
| 935 | C93500 | Copper Alloy, Cast (High Leaded Tin Bronze) |
| 937 | C93700 | Copper Alloy, Cast (High Leaded Tin Bronze) |
| 938 | C93800 | Copper Alloy, Cast (High Leaded Tin Bronze) |
| 939 | C93900 | Copper Alloy, Cast (High Leaded Tin Bronze) |
| 943 | C94300 | Copper Alloy, Cast (High Leaded Tin Bronze) |
| 944 | C94400 | Copper Alloy, Cast (High Leaded Tin Bronze) |
| 945 | C94500 | Copper Alloy, Cast (High Leaded Tin Bronze) |
| 947 | C94700 | Copper Alloy, Cast (Nickel Tin Bronze) |
| 948 | C94800 | Copper Alloy, Cast (Nickel Tin Bronze) |
| 952 | C95200 | Copper Alloy, Cast (Aluminum Bronze) |
| 953 | C95300 | Copper Alloy, Cast (Aluminum Bronze) |
| 954 | C95400 | Copper Alloy, Cast (Aluminum Bronze) |
| 955 | C95500 | Copper Alloy, Cast (Aluminum Bronze) |
| 956 | C95600 | Copper Alloy, Cast (Aluminum Bronze) |
| 957 | C95700 | Copper Alloy, Cast (Aluminum Bronze) |
| 958 | C95800 | Copper Alloy, Cast (Aluminum Bronze) |
| 962 | C96200 | Copper Alloy, Cast (Copper Nickel) |
| 963 | C96300 | Copper Alloy, Cast (Copper Nickel) |
| 964 | C96400 | Copper Alloy, Cast (Copper Nickel) |
| 966 | C96600 | Copper Alloy, Cast (Copper Nickel) |
| 973 | C97300 | Copper Alloy, Cast (Nickel Silver) |
| 974 | C97400 | Copper Alloy, Cast (Nickel Silver) |
| 976 | C97600 | Copper Alloy, Cast (Nickel Silver) |
| 978 | C97800 | Copper Alloy, Cast (Nickel Silver) |
| 993 | C99300 | Copper Alloy, Cast |

# 26.1 Numerical List of Materials

| MATERIAL | UNS DESIGNATION | MATERIAL GROUP |
|---|---|---|
| 1005 | G10050 | Carbon Steel, Wrought–Low Carbon |
| 1006 | G10060 | Carbon Steel, Wrought–Low Carbon |
| 1008 | G10080 | Carbon Steel, Wrought–Low Carbon |
| 1009 | G10090 | Carbon Steel, Wrought–Low Carbon |
| 1010 | G10100 | Carbon Steel, Wrought–Low Carbon |
| 1010 | G10100 | Carbon Steel, Cast–Low Carbon |
| 1011 | G10110 | Carbon Steel, Wrought–Low Carbon |
| 1012 | G10120 | Carbon Steel, Wrought–Low Carbon |
| 1013 | G10130 | Carbon Steel, Wrought–Low Carbon |
| 1015 | G10150 | Carbon Steel, Wrought–Low Carbon |
| 1016 | G10160 | Carbon Steel, Wrought–Low Carbon |
| 1017 | G10170 | Carbon Steel, Wrought–Low Carbon |
| 1018 | G10180 | Carbon Steel, Wrought–Low Carbon |
| 10L18 | | Free Machining Carbon Steel, Wrought–Low Carbon Leaded |
| 1019 | G10190 | Carbon Steel, Wrought–Low Carbon |
| 1020 | G10200 | Carbon Steel, Wrought–Low Carbon |
| 1020 | J02000 | Carbon Steel, Cast–Low Carbon |
| 1021 | G10210 | Carbon Steel, Wrought–Low Carbon |
| 1022 | G10220 | Carbon Steel, Wrought–Low Carbon |
| 1023 | G10230 | Carbon Steel, Wrought–Low Carbon |
| 1025 | G10250 | Carbon Steel, Wrought–Low Carbon |
| 1026 | G10260 | Carbon Steel, Wrought–Low Carbon |
| 1029 | G10290 | Carbon Steel, Wrought–Low Carbon |
| 1030 | G10300 | Carbon Steel, Wrought–Medium Carbon |
| 1030 | J03005 | Carbon Steel, Cast–Medium Carbon |
| 1033 | C10330 | Carbon Steel, Wrought–Medium Carbon |
| 1035 | G10350 | Carbon Steel, Wrought–Medium Carbon |
| 1037 | G10370 | Carbon Steel, Wrought–Medium Carbon |
| 1038 | G10380 | Carbon Steel, Wrought–Medium Carbon |
| 1039 | G10390 | Carbon Steel, Wrought–Medium Carbon |
| 1040 | G10400 | Carbon Steel, Wrought–Medium Carbon |
| 1040 | J04000 | Carbon Steel, Cast–Medium Carbon |
| 1042 | G10420 | Carbon Steel, Wrought–Medium Carbon |
| 1043 | G10430 | Carbon Steel, Wrought–Medium Carbon |
| 1044 | G10440 | Carbon Steel, Wrought–Medium Carbon |
| 1045 | G10450 | Carbon Steel, Wrought–Medium Carbon |
| 10L45 | | Free Machining Carbon Steel, Wrought–Medium Carbon Leaded |
| 1046 | G10460 | Carbon Steel, Wrought–Medium Carbon |
| 1049 | G10490 | Carbon Steel, Wrought–Medium Carbon |
| 1050 | G10500 | Carbon Steel, Wrought–Medium Carbon |
| 1050 | J05000 | Carbon Steel, Cast–Medium Carbon |
| 10L50 | | Free Machining Carbon Steel, Wrought–Medium Carbon Leaded |
| 1053 | G10530 | Carbon Steel, Wrought–Medium Carbon |
| 1055 | G10550 | Carbon Steel, Wrought–Medium Carbon |
| 1060 | A91060 | Aluminum Alloy, Wrought |
| 1060 | G10600 | Carbon Steel, Wrought–High Carbon |
| 1064 | G10640 | Carbon Steel, Wrought–High Carbon |
| 1065 | G10650 | Carbon Steel, Wrought–High Carbon |
| 1069 | G10690 | Carbon Steel, Wrought–High Carbon |
| 1070 | G10700 | Carbon Steel, Wrought–High Carbon |
| 1074 | G10740 | Carbon Steel, Wrought–High Carbon |
| 1075 | G10750 | Carbon Steel, Wrought–High Carbon |

| MATERIAL | UNS DESIGNATION | MATERIAL GROUP |
|---|---|---|
| 1078 | G10780 | Carbon Steel, Wrought—High Carbon |
| 1080 | G10800 | Carbon Steel, Wrought—High Carbon |
| 1084 | G10840 | Carbon Steel, Wrought—High Carbon |
| 1085 | G10850 | Carbon Steel, Wrought—High Carbon |
| 1086 | G10860 | Carbon Steel, Wrought—High Carbon |
| 1090 | G10900 | Carbon Steel, Wrought—High Carbon |
| 1095 | G10950 | Carbon Steel, Wrought—High Carbon |
| 1100 | A91100 | Aluminum Alloy, Wrought |
| 1108 | G11080 | Free Machining Carbon Steel, Wrought—Low Carbon Resulfurized |
| 1109 | G11090 | Free Machining Carbon Steel, Wrought—Low Carbon Resulfurized |
| 1110 | G11100 | Free Machining Carbon Steel, Wrought—Low Carbon Resulfurized |
| 1115 | | Free Machining Carbon Steel, Wrought—Low Carbon Resulfurized |
| 1116 | G11160 | Free Machining Carbon Steel, Wrought—Low Carbon Resulfurized |
| 1117 | G11170 | Free Machining Carbon Steel, Wrought—Low Carbon Resulfurized |
| 11L17 | | Free Machining Carbon Steel, Wrought—Low Carbon Leaded |
| 1118 | G11180 | Free Machining Carbon Steel, Wrought—Low Carbon Resulfurized |
| 1119 | G11190 | Free Machining Carbon Steel, Wrought—Low Carbon Resulfurized |
| 1132 | G11320 | Free Machining Carbon Steel, Wrought—Medium Carbon Resulfurized |
| 1137 | G11370 | Free Machining Carbon Steel, Wrought—Medium Carbon Resulfurized |
| 11L37 | | Free Machining Carbon Steel, Wrought—Medium Carbon Leaded |
| 1139 | G11390 | Free Machining Carbon Steel, Wrought—Medium Carbon Resulfurized |
| 1140 | G11400 | Free Machining Carbon Steel, Wrought—Medium Carbon Resulfurized |
| 1141 | G11410 | Free Machining Carbon Steel, Wrought—Medium Carbon Resulfurized |
| 11L41 | | Free Machining Carbon Steel, Wrought—Medium Carbon Leaded |
| 1144 | G11440 | Free Machining Carbon Steel, Wrought—Medium Carbon Resulfurized |
| 11L44 | | Free Machining Carbon Steel, Wrought—Medium Carbon Leaded |
| 1145 | A91145 | Aluminum Alloy, Wrought |
| 1145 | G11450 | Free Machining Carbon Steel, Wrought—Medium Carbon Resulfurized |
| 1146 | G11460 | Free Machining Carbon Steel, Wrought—Medium Carbon Resulfurized |
| 1151 | G11510 | Free Machining Carbon Steel, Wrought—Medium Carbon Resulfurized |
| 1175 | A91175 | Aluminum Alloy, Wrought |
| 1211 | G12110 | Free Machining Carbon Steel, Wrought—Low Carbon Resulfurized |
| 1212 | G12120 | Free Machining Carbon Steel, Wrought—Low Carbon Resulfurized |
| 1213 | G12130 | Free Machining Carbon Steel, Wrought—Low Carbon Resulfurized |
| 12L13 | G12134 | Free Machining Carbon Steel, Wrought—Low Carbon Leaded |
| 12L14 | G12144 | Free Machining Carbon Steel, Wrought—Low Carbon Leaded |
| 1215 | G12150 | Free Machining Carbon Steel, Wrought—Low Carbon Resulfurized |
| 12L15 | | Free Machining Carbon Steel, Wrought—Low Carbon Leaded |
| 1235 | A91235 | Aluminum Alloy, Wrought |
| 1320 | A91235 | Alloy Steel, Cast—Low Carbon |
| 1330 | G13300 | Alloy Steel, Wrought—Medium Carbon |
| 1330 | | Alloy Steel, Cast—Medium Carbon |
| 1335 | G13350 | Alloy Steel, Wrought—Medium Carbon |
| 1340 | G13400 | Alloy Steel, Wrought—Medium Carbon |
| 1340 | | Alloy Steel, Cast—Medium Carbon |
| 1345 | G13450 | Alloy Steel, Wrought—Medium Carbon |
| 1513 | G15130 | Carbon Steel, Wrought—Low Carbon |
| 1518 | G15180 | Carbon Steel, Wrought—Low Carbon |
| 1522 | G15220 | Carbon Steel, Wrought—Low Carbon |
| 1524 | G15240 | Carbon Steel, Wrought—Medium Carbon |
| 1525 | G15250 | Carbon Steel, Wrought—Medium Carbon |
| 1526 | G15260 | Carbon Steel, Wrought—Medium Carbon |

## 26.1 Numerical List of Materials

| MATERIAL | UNS DESIGNATION | MATERIAL GROUP |
|---|---|---|
| 1527 | G15270 | Carbon Steel, Wrought—Medium Carbon |
| 1536 | G15360 | Carbon Steel, Wrought—Medium Carbon |
| 1541 | G15410 | Carbon Steel, Wrought—Medium Carbon |
| 1547 | G15470 | Carbon Steel, Wrought—Medium Carbon |
| 1548 | G15480 | Carbon Steel, Wrought—Medium Carbon |
| 1551 | G15510 | Carbon Steel, Wrought—Medium Carbon |
| 1552 | G15520 | Carbon Steel, Wrought—Medium Carbon |
| 1561 | G15610 | Carbon Steel, Wrought—High Carbon |
| 1566 | G15660 | Carbon Steel, Wrought—High Carbon |
| 1572 | G15720 | Carbon Steel, Wrought—High Carbon |
| 2011 | A92011 | Aluminum Alloy, Wrought |
| 2014 | A92014 | Aluminum Alloy, Wrought |
| 2017 | A92017 | Aluminum Alloy, Wrought |
| 2018 | A92018 | Aluminum Alloy, Wrought |
| 2021 | A92021 | Aluminum Alloy, Wrought |
| 2024 | A92024 | Aluminum Alloy, Wrought |
| 2025 | A92025 | Aluminum Alloy, Wrought |
| 2117 | A92117 | Aluminum Alloy, Wrought |
| 2218 | A92218 | Aluminum Alloy, Wrought |
| 2219 | A92219 | Aluminum Alloy, Wrought |
| 2315 | | Alloy Steel, Cast—Low Carbon |
| 2320 | | Alloy Steel, Cast—Low Carbon |
| 2325 | | Alloy Steel, Cast—Medium Carbon |
| 2330 | | Alloy Steel, Cast—Medium Carbon |
| 2618 | A92618 | Aluminum Alloy, Wrought |
| 3003 | A93003 | Aluminum Alloy, Wrought |
| 3004 | A93004 | Aluminum Alloy, Wrought |
| 3005 | A93005 | Aluminum Alloy, Wrought |
| 4012 | G40120 | Alloy Steel, Wrought—Low Carbon |
| 4023 | G40230 | Alloy Steel, Wrought—Low Carbon |
| 4024 | G40240 | Alloy Steel, Wrought—Low Carbon |
| 4027 | G40270 | Alloy Steel, Wrought—Medium Carbon |
| 4028 | G40280 | Alloy Steel, Wrought—Medium Carbon |
| 4032 | G40320 | Alloy Steel, Wrought—Medium Carbon |
| 4032 | A94032 | Aluminum Alloy, Wrought |
| 4037 | G40370 | Alloy Steel, Wrought—Medium Carbon |
| 4042 | G40420 | Alloy Steel, Wrought—Medium Carbon |
| 4047 | G40470 | Alloy Steel, Wrought—Medium Carbon |
| 4110 | | Alloy Steel, Cast—Low Carbon |
| 4118 | G41180 | Alloy Steel, Wrought—Low Carbon |
| 4120 | | Alloy Steel, Cast—Low Carbon |
| 4125 | | Alloy Steel, Cast—Medium Carbon |
| 4130 | G41300 | Alloy Steel, Wrought—Medium Carbon |
| 4130 | J13408 | Alloy Steel, Cast—Medium Carbon |
| 41L30 | | Free Machining Alloy Steel, Wrought—Medium Carbon Leaded |
| 4135 | G41350 | Alloy Steel, Wrought—Medium Carbon |
| 4137 | G41370 | Alloy Steel, Wrought—Medium Carbon |
| 4140 | G41400 | Alloy Steel, Wrought—Medium Carbon |
| 4140 | J14047 | Alloy Steel, Cast—Medium Carbon |
| 4140 | | Free Machining Alloy Steel, Wrought—Medium Carbon Resulfurized |
| 4140Se | | Free Machining Alloy Steel, Wrought—Medium Carbon Resulfurized |
| 41L40 | | Free Machining Alloy Steel, Wrought—Medium Carbon Leaded |

| MATERIAL | UNS DESIGNATION | MATERIAL GROUP |
|---|---|---|
| 4142 | G41420 | Alloy Steel, Wrought–Medium Carbon |
| 4142Te | | Free Machining Alloy Steel, Wrought–Medium Carbon Resulfurized |
| 4145 | G41450 | Alloy Steel, Wrought–Medium Carbon |
| 4145Se | | Free Machining Alloy Steel, Wrought–Medium Carbon Resulfurized |
| 41L45 | | Free Machining Alloy Steel, Wrought–Medium and High Carbon Leaded |
| 4147 | G41470 | Alloy Steel, Wrought–Medium Carbon |
| 4147Te | | Free Machining Alloy Steel, Wrought–Medium Carbon Resulfurized |
| 41L47 | | Free Machining Alloy Steel, Wrought–Medium Carbon Leaded |
| 4150 | G41500 | Alloy Steel, Wrought–Medium Carbon |
| 4150 | | Free Machining Alloy Steel, Wrought–Medium Carbon Resulfurized |
| 41L50 | | Free Machining Alloy Steel, Wrought–Medium Carbon Leaded |
| 4161 | G41610 | Alloy Steel, Wrought–Medium Carbon |
| 4320 | G43200 | Alloy Steel, Wrought–Low Carbon |
| 4320 | | Alloy Steel, Cast–Low Carbon |
| 4330 | | Alloy Steel, Cast–Medium Carbon |
| 4330V | K23080 | High Strength Steel, Wrought |
| 4340 | G43400 | Alloy Steel, Wrought–Medium Carbon |
| 4340 | J24055 | Alloy Steel, Cast–Medium Carbon |
| 4340 | | High Strength Steel, Wrought |
| 4340Si | J24060 | High Strength Steel, Wrought |
| 43L40 | | Free Machining Alloy Steel, Wrought–Medium and High Carbon Leaded |
| 4419 | G44190 | Alloy Steel, Wrought–Low Carbon |
| 4422 | G44220 | Alloy Steel, Wrought–Low Carbon |
| 4427 | G44270 | Alloy Steel, Wrought–Medium Carbon |
| 4615 | G46150 | Alloy Steel, Wrought–Low Carbon |
| 4617 | G46170 | Alloy Steel, Wrought–Low Carbon |
| 4620 | G46200 | Alloy Steel, Wrought–Low Carbon |
| 4621 | G46210 | Alloy Steel, Wrought–Low Carbon |
| 4626 | G46260 | Alloy Steel, Wrought–Medium Carbon |
| 4718 | G47180 | Alloy Steel, Wrought–Low Carbon |
| 4720 | G47200 | Alloy Steel, Wrought–Low Carbon |
| 4815 | G48150 | Alloy Steel, Wrought–Low Carbon |
| 4817 | G48170 | Alloy Steel, Wrought–Low Carbon |
| 4820 | G48200 | Alloy Steel, Wrought–Low Carbon |
| 5005 | A95005 | Aluminum Alloy, Wrought |
| 5015 | G50150 | Alloy Steel, Wrought–Low Carbon |
| 50B40 | G50401 | Alloy Steel, Wrought–Medium Carbon |
| 50B44 | G50441 | Alloy Steel, Wrought–Medium Carbon |
| 5046 | G50460 | Alloy Steel, Wrought–Medium Carbon |
| 50B46 | G50461 | Alloy Steel, Wrought–Medium Carbon |
| 5050 | A95050 | Aluminum Alloy, Wrought |
| 50B50 | G50501 | Alloy Steel, Wrought–Medium Carbon |
| 5052 | A95052 | Aluminum Alloy, Wrought |
| 5056 | A95056 | Aluminum Alloy, Wrought |
| 5060 | G50600 | Alloy Steel, Wrought–Medium Carbon |
| 50B60 | G50601 | Alloy Steel, Wrought–Medium Carbon |
| 5083 | A95083 | Aluminum Alloy, Wrought |
| 5086 | A95086 | Aluminum Alloy, Wrought |
| 5115 | G51150 | Alloy Steel, Wrought–Low Carbon |
| 5120 | G51200 | Alloy Steel, Wrought–Low Carbon |
| 5130 | G51300 | Alloy Steel, Wrought–Medium Carbon |
| 5132 | G51320 | Alloy Steel, Wrought–Medium Carbon |

## 26.1 Numerical List of Materials

| MATERIAL | UNS DESIGNATION | MATERIAL GROUP |
|---|---|---|
| 51L32 | | Free Machining Alloy Steel, Wrought–Medium Carbon Leaded |
| 5135 | G51350 | Alloy Steel, Wrought–Medium Carbon |
| 5140 | G51400 | Alloy Steel, Wrought–Medium Carbon |
| 5145 | G51450 | Alloy Steel, Wrought–Medium Carbon |
| 5147 | G51470 | Alloy Steel, Wrought–Medium Carbon |
| 5150 | G51500 | Alloy Steel, Wrought–Medium Carbon |
| 5154 | A95154 | Aluminum Alloy, Wrought |
| 5155 | G51550 | Alloy Steel, Wrought–Medium Carbon |
| 5160 | G51600 | Alloy Steel, Wrought–Medium Carbon |
| 51B60 | G51601 | Alloy Steel, Wrought–Medium Carbon |
| 5252 | A95252 | Aluminum Alloy, Wrought |
| 5254 | A95254 | Aluminum Alloy, Wrought |
| 5454 | A95454 | Aluminum Alloy, Wrought |
| 5456 | A95456 | Aluminum Alloy, Wrought |
| 5457 | A95457 | Aluminum Alloy, Wrought |
| 5652 | A95652 | Aluminum Alloy, Wrought |
| 5657 | A95657 | Aluminum Alloy, Wrought |
| 6053 | A96053 | Aluminum Alloy, Wrought |
| 6061 | A96061 | Aluminum Alloy, Wrought |
| 6063 | A96063 | Aluminum Alloy, Wrought |
| 6066 | A96066 | Aluminum Alloy, Wrought |
| 6070 | A96070 | Aluminum Alloy, Wrought |
| 6101 | A96101 | Aluminum Alloy, Wrought |
| 6118 | G61180 | Alloy Steel, Wrought–Low Carbon |
| 6150 | G61500 | Alloy Steel, Wrought–Medium Carbon |
| 6151 | A96151 | Aluminum Alloy, Wrought |
| 6253 | A96253 | Aluminum Alloy, Wrought |
| 6262 | A96262 | Aluminum Alloy, Wrought |
| 6463 | A96463 | Aluminum Alloy, Wrought |
| 6951 | A96951 | Aluminum Alloy, Wrought |
| 7001 | A97001 | Aluminum Alloy, Wrought |
| 7004 | A97004 | Aluminum Alloy, Wrought |
| 7005 | A97005 | Aluminum Alloy, Wrought |
| 7039 | A97039 | Aluminum Alloy, Wrought |
| 7049 | A97049 | Aluminum Alloy, Wrought |
| 7050 | A97050 | Aluminum Alloy, Wrought |
| 7075 | A97075 | Aluminum Alloy, Wrought |
| 7079 | A97079 | Aluminum Alloy, Wrought |
| 7175 | A97175 | Aluminum Alloy, Wrought |
| 7178 | A97178 | Aluminum Alloy, Wrought |
| 8020 | | Alloy Steel, Cast–Low Carbon |
| 8030 | | Alloy Steel, Cast–Medium Carbon |
| 80B30 | | Alloy Steel, Cast–Medium Carbon |
| 8040 | | Alloy Steel, Cast–Medium Carbon |
| 8115 | G81150 | Alloy Steel, Wrought–Low Carbon |
| 81B45 | G81451 | Alloy Steel, Wrought–Medium Carbon |
| 8430 | | Alloy Steel, Cast–Medium Carbon |
| 8440 | | Alloy Steel, Cast–Medium Carbon |
| 8617 | G86170 | Alloy Steel, Wrought–Low Carbon |
| 8620 | G86200 | Alloy Steel, Wrought–Low Carbon |
| 8620 | | Alloy Steel, Cast–Low Carbon |
| 86L20 | | Free Machining Alloy Steel, Wrought–Medium Carbon Leaded |

| MATERIAL | UNS DESIGNATION | MATERIAL GROUP |
|---|---|---|
| 8622 | G86220 | Alloy Steel, Wrought–Low Carbon |
| 8625 | G86250 | Alloy Steel, Wrought–Medium Carbon |
| 8627 | G86270 | Alloy Steel, Wrought–Medium Carbon |
| 8630 | G86300 | Alloy Steel, Wrought–Medium Carbon |
| 8630 | | Alloy Steel, Cast–Medium Carbon |
| 8637 | G86370 | Alloy Steel, Wrought–Medium Carbon |
| 8640 | G86400 | Alloy Steel, Wrought–Medium Carbon |
| 8640 | | Alloy Steel, Cast–Medium Carbon |
| 86L40 | | Free Machining Alloy Steel, Wrought–Medium Carbon Leaded |
| 8642 | G86420 | Alloy Steel, Wrought–Medium Carbon |
| 8645 | G86450 | Alloy Steel, Wrought–Medium Carbon |
| 86B45 | G86451 | Alloy Steel, Wrought–Medium Carbon |
| 8650 | G86500 | Alloy Steel, Wrought–Medium Carbon |
| 8655 | G86550 | Alloy Steel, Wrought–Medium Carbon |
| 8660 | G86600 | Alloy Steel, Wrought–Medium Carbon |
| 8740 | G87400 | Alloy Steel, Wrought–Medium Carbon |
| 8742 | G87420 | Alloy Steel, Wrought–Medium Carbon |
| 8822 | G88220 | Alloy Steel, Wrought–Low Carbon |
| 9254 | G92540 | Alloy Steel, Wrought–Medium Carbon |
| 9255 | G92550 | Alloy Steel, Wrought–Medium Carbon |
| 9260 | G92600 | Alloy Steel, Wrought–Medium Carbon |
| 9310 | G93106 | Alloy Steel, Wrought–Low Carbon |
| 94B15 | G94151 | Alloy Steel, Wrought–Low Carbon |
| 94B17 | G94171 | Alloy Steel, Wrought–Low Carbon |
| 94B30 | G94301 | Alloy Steel, Wrought–Medium Carbon |
| 9525 | | Alloy Steel, Cast–Medium Carbon |
| 9530 | | Alloy Steel, Cast–Medium Carbon |
| 9535 | | Alloy Steel, Cast–Medium Carbon |
| 98BV40 | | High Strength Steel, Wrought |
| 50100 | G50986 | Alloy Steel, Wrought–High Carbon |
| 51100 | G51986 | Alloy Steel, Wrought–High Carbon |
| 52100 | G52986 | Alloy Steel, Wrought–High Carbon |
| 52L100 | | Free Machining Alloy Steel, Wrought–High Carbon Leaded |

| MATERIAL | UNS DESIGNATION | MATERIAL GROUP |
|---|---|---|
| A2 | T30102 | Tool Steel, Wrought—Cold Work |
| A3 | T30103 | Tool Steel, Wrought—Cold Work |
| A4 | T30104 | Tool Steel, Wrought—Cold Work |
| A6 | T30106 | Tool Steel, Wrought—Cold Work |
| A7 | T30107 | Tool Steel, Wrought—Cold Work |
| A8 | T30108 | Tool Steel, Wrought—Cold Work |
| A9 | T30109 | Tool Steel, Wrought—Cold Work |
| A10 | T30110 | Tool Steel, Wrought—Cold Work |
| A140 | | Aluminum Alloy, Cast—Sand Cast |
| A-286 | K66286 | High Temperature Alloy—Iron Base, Wrought |
| A332.0 | A13320 | Aluminum Alloy, Cast—Permanent Mold Cast |
| A356.0 | A13560 | Aluminum Alloy, Cast—Sand and Permanent Mold Cast |
| A360.0 | A13600 | Aluminum Alloy, Cast—Die Casting |
| A380.0 | A13800 | Aluminum Alloy, Cast—Die Casting |
| A384.0 | A03840 | Aluminum Alloy, Cast—Die Casting |
| A413.0 | A14130 | Aluminum Alloy, Cast—Die Casting |
| A514.0 | A15140 | Aluminum Alloy, Cast—Permanent Mold Cast |
| A712.0 | A17120 | Aluminum Alloy, Cast—Sand Cast |
| A850.0 | A18500 | Aluminum Alloy, Cast—Sand and Permanent Mold Cast |
| ABS | | Plastic—Thermoplastic |
| AC41A | Z35531 | Zinc Alloy, Cast |
| Acetal | | Plastic—Thermoplastic |
| ACI Grade CB-7Cu | J92170 | Precipitation Hardening Stainless Steel, Cast |
| ACI Grade CD-4MCu | J93370 | Precipitation Hardening Stainless Steel, Cast |
| Acrylic | | Plastic—Thermoplastic |
| Acrylonitrile-Butadiene-Styrene | | Plastic—Thermoplastic |
| AF2-1DA | | High Temperature Alloy—Nickel Base, Wrought |
| AF-71 | | Precipitation Hardening Stainless Steel, Wrought |
| AFC-77 | K65770 | Precipitation Hardening Stainless Steel, Wrought |
| AG40A | Z33520 | Zinc Alloy, Cast |
| AiResist 13 | | High Temperature Alloy—Cobalt Base, Cast |
| AiResist 213 | | High Temperature Alloy—Cobalt Base, Wrought |
| AiResist 215 | | High Temperature Alloy—Cobalt Base, Cast |
| Al-1Mg-0.6Si-0.25Cu | | Powder Metal Alloy—Aluminum Alloy |
| Al-0.6Mg-0.4Si | | Powder Metal Alloy—Aluminum Alloy |
| Al-4.4Cu-0.8Si-0.4Mg | | Powder Metal Alloy—Aluminum Alloy |
| Allyl (DAP) | | Plastic—Thermosetting Plastic |
| Allyl-Glass filled | | Plastic—Thermosetting Plastic |
| Allyl-Fiber filled | | Plastic—Thermosetting Plastic |
| Almar 362 | S36200 | Precipitation Hardening Stainless Steel, Wrought |
| Alnico I | | Magnetic Alloy |
| Alnico II | | Magnetic Alloy |
| Alnico III | | Magnetic Alloy |
| Alnico IV | | Magnetic Alloy |
| Alnico V | | Magnetic Alloy |
| Alnico V-7 | | Magnetic Alloy |
| Alnico XII | | Magnetic Alloy |
| Alumina | | Glasses & Ceramics—Ceramic |
| Alumina (Pure) | | Flame (Thermal) Sprayed Material—Inorganic Coating Material |
| Alumina (Grey) containing Titania | | Flame (Thermal) Sprayed Material—Inorganic Coating Material |

## 26.2 Alphabetical List of Materials

| MATERIAL | UNS DESIGNATION | MATERIAL GROUP |
|---|---|---|
| Alumina, Nickel Aluminide Blends | | Flame (Thermal) Sprayed Material—Inorganic Coating Material |
| Alumina-Mullite | | Glasses & Ceramics—Ceramic |
| Aluminum Alloys | | Flame (Thermal) Sprayed Material |
| Aluminum Oxide | | Glasses & Ceramics—Ceramic |
| Aluminum Silicate | | Glasses & Ceramics—Ceramic |
| AM60A | M10600 | Magnesium Alloy, Cast |
| AM100A | M10100 | Magnesium Alloy, Cast |
| AM-350 | S35000 | Precipitation Hardening Stainless Steel, Wrought |
| AM-355 | S35500 | Precipitation Hardening Stainless Steel, Wrought |
| AM-355 | S35500 | Precipitation Hardening Stainless Steel, Cast |
| AM-362 | S36200 | Precipitation Hardening Stainless Steel, Wrought |
| AM-363 | | Precipitation Hardening Stainless Steel, Wrought |
| AMS 4803 | Z33520 | Zinc Alloy, Cast |
| Anviloy 1100 | | Refractory Alloy, Cast, P/M—Tungsten Alloy |
| Anviloy 1150 | | Refractory Alloy, Cast, P/M—Tungsten Alloy |
| Anviloy 1200 | | Refractory Alloy, Cast, P/M—Tungsten Alloy |
| AS41A | M10410 | Magnesium Alloy, Cast |
| ASTAR 811C | | Refractory Alloy, Wrought, Cast, P/M—Tantalum |
| ASTM A27: | | |
| Grade N1 | J02500 | Alloy Steel, Cast—Medium Carbon |
| Grade N2 | J03500 | Alloy Steel, Cast—Medium Carbon |
| Grade U-60-30 | J02500 | Alloy Steel, Cast—Medium Carbon |
| Grade 60-30 | J03000 | Alloy Steel, Cast—Medium Carbon |
| Grade 65-35 | J03001 | Alloy Steel, Cast—Medium Carbon |
| Grade 70-36 | J03501 | Alloy Steel, Cast—Medium Carbon |
| Grade 70-40 | J02501 | Alloy Steel, Cast—Medium Carbon |
| ASTM A47: | | |
| Grade 32510 | F22200 | Malleable Cast Iron—Ferritic |
| Grade 35018 | F22400 | Malleable Cast Iron—Ferritic |
| ASTM A48: | | |
| Class 20 | F11401 | Gray Cast Iron—Ferritic |
| Class 25 | F11701 | Gray Cast Iron—Pearlitic-Ferritic |
| Class 30 | F12101 | Gray Cast Iron—Pearlitic |
| Class 35 | F12401 | Gray Cast Iron—Pearlitic |
| Class 40 | F12801 | Gray Cast Iron—Pearlitic |
| Class 45 | F13101 | Gray Cast Iron—Pearlitic + Free Carbides |
| Class 50 | F13501 | Gray Cast Iron—Pearlitic + Free Carbides |
| Class 55 | F13801 | Gray Cast Iron—Pearlitic or Acicular + Free Carbides |
| Class 60 | F14101 | Gray Cast Iron—Pearlitic or Acicular + Free Carbides |
| ASTM A128: | | |
| Grade A | J91101 | Austenitic Manganese Steel, Cast |
| Grade B-1 | J91119 | Austenitic Manganese Steel, Cast |
| Grade B-2 | J91129 | Austenitic Manganese Steel, Cast |
| Grade B-3 | J91139 | Austenitic Manganese Steel, Cast |
| Grade B-4 | J91149 | Austenitic Manganese Steel, Cast |
| Grade C | J91309 | Austenitic Manganese Steel, Cast |
| Grade D | J91459 | Austenitic Manganese Steel, Cast |
| Grade E-1 | J91249 | Austenitic Manganese Steel, Cast |
| Grade E-2 | J91339 | Austenitic Manganese Steel, Cast |
| Grade F | J69990 | Austenitic Manganese Steel, Cast |

| MATERIAL | UNS DESIGNATION | MATERIAL GROUP |
|---|---|---|
| ASTM A148: | | |
| Grade 80-40 | D50400 | Alloy Steel, Cast–Medium Carbon |
| Grade 80-50 | D50500 | Alloy Steel, Cast–Medium Carbon |
| Grade 90-60 | D50600 | Alloy Steel, Cast–Medium Carbon |
| Grade 105-85 | D50850 | Alloy Steel, Cast–Medium Carbon |
| Grade 120-95 | D50950 | Alloy Steel, Cast–Medium Carbon |
| Grade 150-125 | D51250 | Alloy Steel, Cast–Medium Carbon |
| Grade 175-145 | D51450 | Alloy Steel, Cast–Medium Carbon |
| ASTM A216: | | |
| Grade WCA | J02502 | Alloy Steel, Cast–Medium Carbon |
| Grade WCB | J03002 | Alloy Steel, Cast–Medium Carbon |
| Grade WCC | J02503 | Alloy Steel, Cast–Medium Carbon |
| ASTM A217: | | |
| Grade C5 | J42045 | Stainless Steel, Cast–Ferritic |
| Grade C12 | J82090 | Stainless Steel, Cast–Ferritic |
| Grade CA-15 | J91150 | Stainless Steel, Cast–Martensitic |
| Grade WC1 | J12522 | Alloy Steel, Cast–Medium Carbon |
| Grade WC4 | J12082 | Alloy Steel, Cast–Medium Carbon |
| Grade WC5 | J22000 | Alloy Steel, Cast–Medium Carbon |
| Grade WC6 | J12072 | Alloy Steel, Cast–Medium Carbon |
| Grade WC9 | J21890 | Alloy Steel, Cast–Low Carbon |
| ASTM A220: | | |
| Grade 40010 | F22830 | Malleable Cast Iron–Pearlitic |
| Grade 45006 | F23131 | Malleable Cast Iron–Pearlitic |
| Grade 45008 | F23130 | Malleable Cast Iron–Pearlitic |
| Grade 50005 | F23530 | Malleable Cast Iron–Pearlitic |
| Grade 60004 | F24130 | Malleable Cast Iron–Tempered Martensite |
| Grade 70003 | F24830 | Malleable Cast Iron–Tempered Martensite |
| Grade 80002 | F25530 | Malleable Cast Iron–Tempered Martensite |
| Grade 90001 | F26230 | Malleable Cast Iron–Tempered Martensite |
| ASTM A296: | | |
| Grade CA6N | J91650 | Stainless Steel, Cast–Ferritic |
| Grade CA-6NM | J91540 | Stainless Steel, Cast–Ferritic |
| Grade CA-15 | J91150 | Stainless Steel, Cast–Martensitic |
| Grade CA15M | J91151 | Stainless Steel, Cast–Martensitic |
| Grade CA-40 | J91153 | Stainless Steel, Cast–Martensitic |
| Grade CB-30 | J91803 | Stainless Steel, Cast–Ferritic |
| Grade CC-50 | J92615 | Stainless Steel, Cast–Ferritic |
| Grade CD4MCu | J93370 | Stainless Steel, Cast–Ferritic |
| Grade CE-30 | J93423 | Stainless Steel, Cast–Ferritic |
| Grade CF-3 | J92700 | Stainless Steel, Cast–Austenitic |
| Grade CF-3M | J92800 | Stainless Steel, Cast–Austenitic |
| Grade CF-8 | J92600 | Stainless Steel, Cast–Austenitic |
| Grade CF-8C | J92710 | Stainless Steel, Cast–Austenitic |
| Grade CF-8M | J92900 | Stainless Steel, Cast–Austenitic |
| Grade CF-16F | J92701 | Stainless Steel, Cast–Austenitic |
| Grade CF-20 | J92602 | Stainless Steel, Cast–Austenitic |
| Grade CG-8M | J93000 | Stainless Steel, Cast–Austenitic |
| Grade CG-12 | J93001 | Stainless Steel, Cast–Austenitic |
| Grade CH-20 | J93402 | Stainless Steel, Cast–Austenitic |
| Grade CK-20 | J94202 | Stainless Steel, Cast–Austenitic |

| MATERIAL | UNS DESIGNATION | MATERIAL GROUP |
|---|---|---|
| ASTM A296 *(continued)* | | |
| Grade CN-7M | J95150 | Stainless Steel, Cast–Austenitic |
| Grade CN-7MS | J94650 | Stainless Steel, Cast–Austenitic |
| Grade CW-12M | | High Temperature Alloy–Nickel Base, Cast |
| Grade CY-40 | N07750 | High Temperature Alloy–Nickel Base, Cast |
| Grade CZ-100 | N02100 | Nickel Alloy, Cast |
| Grade M-35 | | Nickel Alloy, Cast |
| Grade N-12M | | High Temperature Alloy–Nickel Base, Cast |
| ASTM A297: | | |
| Grade HC | J92605 | Stainless Steel, Cast–Ferritic |
| Grade HD | J93005 | Stainless Steel, Cast–Austenitic |
| Grade HE | J93403 | Stainless Steel, Cast–Austenitic |
| Grade HF | J92603 | Stainless Steel, Cast–Austenitic |
| Grade HH | J93503 | Stainless Steel, Cast–Austenitic |
| Grade HI | J94003 | Stainless Steel, Cast–Austenitic |
| Grade HK | J94224 | Stainless Steel, Cast–Austenitic |
| Grade HL | J94604 | Stainless Steel, Cast–Austenitic |
| Grade HN | J94213 | Stainless Steel, Cast–Austenitic |
| Grade HP | J95705 | Stainless Steel, Cast–Austenitic |
| Grade HT | J94605 | Stainless Steel, Cast–Austenitic |
| Grade HU | J95405 | Stainless Steel, Cast–Austenitic |
| Grade HW | N08001 | High Temperature Alloy–Nickel Base, Cast |
| Grade HX | N06006 | High Temperature Alloy–Nickel Base, Cast |
| ASTM A351: | | |
| Grade CD-4MCu | J93370 | Precipitation Hardening Stainless Steel, Cast |
| Grade CF-3 | J92500 | Stainless Steel, Cast–Austenitic |
| Grade CF-3A | J92500 | Stainless Steel, Cast–Austenitic |
| Grade CF-3M | J92800 | Stainless Steel, Cast–Austenitic |
| Grade CF-3MA | J92800 | Stainless Steel, Cast–Austenitic |
| Grade CF-8 | J92600 | Stainless Steel, Cast–Austenitic |
| Grade CF-8A | | Stainless Steel, Cast–Austenitic |
| Grade CF-8C | J92710 | Stainless Steel, Cast–Austenitic |
| Grade CF-8M | J92900 | Stainless Steel, Cast–Austenitic |
| Grade CF-10MC | J92971 | Stainless Steel, Cast–Austenitic |
| Grade CH-8 | J93400 | Stainless Steel, Cast–Austenitic |
| Grade CH-10 | J93401 | Stainless Steel, Cast–Austenitic |
| Grade CH-20 | J93402 | Stainless Steel, Cast–Austenitic |
| Grade CK-20 | J94202 | Stainless Steel, Cast–Austenitic |
| Grade CN-7M | J95150 | Stainless Steel, Cast–Austenitic |
| Grade HK-30 | J94203 | Stainless Steel, Cast–Austenitic |
| Grade HK-40 | J94204 | Stainless Steel, Cast–Austenitic |
| Grade HT-30 | J94603 | Stainless Steel, Cast–Austenitic |
| ASTM A352: | | |
| Grade LC1 | J12522 | Alloy Steel, Cast–Medium Carbon |
| Grade LC2 | J22500 | Alloy Steel, Cast–Medium Carbon |
| Grade LC2-1 | J42215 | Alloy Steel, Cast–Medium Carbon |
| Grade LC3 | J31550 | Alloy Steel, Cast–Low Carbon |
| Grade LC4 | J41500 | Alloy Steel, Cast–Low Carbon |
| Grade LCA | J02504 | Carbon Steel, Cast–Medium Carbon |
| Grade LCB | J03003 | Carbon Steel, Cast–Medium Carbon |
| Grade LCC | J02505 | Carbon Steel, Cast–Medium Carbon |

| MATERIAL | UNS DESIGNATION | MATERIAL GROUP |
|---|---|---|
| ASTM A356: | | |
| Grade 1 | J03502 | Carbon Steel, Cast–Medium Carbon |
| Grade 2 | J12523 | Alloy Steel, Cast–Medium Carbon |
| Grade 5 | J12540 | Alloy Steel, Cast–Medium Carbon |
| Grade 6 | J12073 | Alloy Steel, Cast–Medium Carbon |
| Grade 8 | J11691 | Alloy Steel, Cast–Medium Carbon |
| Grade 9 | J21610 | Alloy Steel, Cast–Medium Carbon |
| Grade 10 | J22090 | Alloy Steel, Cast–Medium Carbon |
| ASTM A369: | | |
| Grade FPA | K02501 | Carbon and Ferritic Alloy Steel (High Temperature Service) |
| Grade FPB | K03006 | Carbon and Ferritic Alloy Steel (High Temperature Service) |
| Grade FP1 | K11522 | Carbon and Ferritic Alloy Steel (High Temperature Service) |
| Grade FP2 | K11547 | Carbon and Ferritic Alloy Steel (High Temperature Service) |
| Grade FP3b | K21509 | Carbon and Ferritic Alloy Steel (High Temperature Service) |
| Grade FP5 | K41545 | Carbon and Ferritic Alloy Steel (High Temperature Service) |
| Grade FP7 | K61595 | Carbon and Ferritic Alloy Steel (High Temperature Service) |
| Grade FP9 | K90941 | Carbon and Ferritic Alloy Steel (High Temperature Service) |
| Grade FP11 | K11597 | Carbon and Ferritic Alloy Steel (High Temperature Service) |
| Grade FP12 | K11562 | Carbon and Ferritic Alloy Steel (High Temperature Service) |
| Grade FP21 | K31545 | Carbon and Ferritic Alloy Steel (High Temperature Service) |
| Grade FP22 | K21590 | Carbon and Ferritic Alloy Steel (High Temperature Service) |
| ASTM A389: | | |
| Grade C23 | J12080 | Alloy Steel, Cast–Medium Carbon |
| Grade C24 | J12092 | Alloy Steel, Cast–Medium Carbon |
| ASTM A426: | | |
| Grade CP1 | J02508 | Carbon Steel, Cast–Low Carbon |
| Grade CP2 | J11542 | Alloy Steel, Cast–Low Carbon |
| Grade CP5 | J42045 | Alloy Steel, Cast–Low Carbon |
| Grade CP5b | J41545 | Alloy Steel, Cast–Low Carbon |
| Grade CP7 | J61544 | Stainless Steel, Cast–Martensitic |
| Grade CP9 | J82091 | Stainless Steel, Cast–Martensitic |
| Grade CP11 | J12073 | Alloy Steel, Cast–Low Carbon |
| Grade CP12 | J11562 | Alloy Steel, Cast–Low Carbon |
| Grade CP15 | J11522 | Alloy Steel, Cast–Low Carbon |
| Grade CP21 | J31545 | Alloy Steel, Cast–Low Carbon |
| Grade CP22 | J21891 | Alloy Steel, Cast–Low Carbon |
| Grade CPCA15 | J91151 | Stainless Steel, Cast–Martensitic |
| ASTM A436: | | |
| Type 1 | F41000 | Gray Cast Iron–Austenitic (NI-RESIST) |
| Type 1b | F41001 | Gray Cast Iron–Austenitic (NI-RESIST) |
| Type 2 | F41002 | Gray Cast Iron–Austenitic (NI-RESIST) |
| Type 2b | F41003 | Gray Cast Iron–Austenitic (NI-RESIST) |
| Type 3 | F41004 | Gray Cast Iron–Austenitic (NI-RESIST) |
| Type 4 | F41005 | Gray Cast Iron–Austenitic (NI-RESIST) |
| Type 5 | F41006 | Gray Cast Iron–Austenitic (NI-RESIST) |
| Type 6 | F41007 | Gray Cast Iron–Austenitic (NI-RESIST) |
| ASTM A439: | | |
| Type D-2 | F43000 | Ductile Cast Iron–Austenitic (NI-RESIST Ductile) |
| Type D-2B | F43001 | Ductile Cast Iron–Austenitic (NI-RESIST Ductile) |
| Type D-2C | F43002 | Ductile Cast Iron–Austenitic (NI-RESIST Ductile) |
| Type D-3 | F43003 | Ductile Cast Iron–Austenitic (NI-RESIST Ductile) |

## 26.2 Alphabetical List of Materials

| MATERIAL | UNS DESIGNATION | MATERIAL GROUP |
|---|---|---|
| ASTM A439 *(continued)* | | |
| Type D-3A | F43004 | Ductile Cast Iron—Austenitic (NI-RESIST Ductile) |
| Type D-4 | F43005 | Ductile Cast Iron—Austenitic (NI-RESIST Ductile) |
| Type D-5 | F43006 | Ductile Cast Iron—Austenitic (NI-RESIST Ductile) |
| Type D-5B | F43007 | Ductile Cast Iron—Austenitic (NI-RESIST Ductile) |
| ASTM A451: | | |
| Grade CPF3 | | Stainless Steel, Cast—Austenitic |
| Grade CPF3A | | Stainless Steel, Cast—Austenitic |
| Grade CPF3M | | Stainless Steel, Cast—Austenitic |
| Grade CPF8 | | Stainless Steel, Cast—Austenitic |
| Grade CPF8A | J92600 | Stainless Steel, Cast—Austenitic |
| Grade CPF8C | J92710 | Stainless Steel, Cast—Austenitic |
| Grade CPF8C (Ta Max.) | J92710 | Stainless Steel, Cast—Austenitic |
| Grade CPF8M | J92900 | Stainless Steel, Cast—Austenitic |
| Grade CPF10MC | J92971 | Stainless Steel, Cast—Austenitic |
| Grade CPH8 | J93400 | Stainless Steel, Cast—Austenitic |
| Grade CPH10 | J93402 | Stainless Steel, Cast—Austenitic |
| Grade CPH20 | J93402 | Stainless Steel, Cast—Austenitic |
| Grade CPK20 | J94202 | Stainless Steel, Cast—Austenitic |
| ASTM A452: | | |
| Grade TP 304H | S30409 | Stainless Steel, Cast—Austenitic |
| Grade TP 316H | S31609 | Stainless Steel, Cast—Austenitic |
| Grade TP 347H | S34709 | Stainless Steel, Cast—Austenitic |
| ASTM A486: | | |
| Class 70 | J03503 | Alloy Steel, Cast—Medium Carbon |
| Class 90 | J03504 | Alloy Steel, Cast—Medium Carbon |
| Class 120 | J03504 | Alloy Steel, Cast—Medium Carbon |
| ASTM A487: | | |
| Class 1N | J03004 | Alloy Steel, Cast—Medium Carbon |
| Class 2N | J13005 | Alloy Steel, Cast—Medium Carbon |
| Class 4N | J13047 | Alloy Steel, Cast—Medium Carbon |
| Class 6N | J13855 | Alloy Steel, Cast—Medium Carbon |
| Class 8N | J22091 | Alloy Steel, Cast—Medium Carbon |
| Class 9N | J13345 | Alloy Steel, Cast—Medium Carbon |
| Class 10N | J23015 | Alloy Steel, Cast—Medium Carbon |
| Class 1Q | J03004 | Alloy Steel, Cast—Medium Carbon |
| Class 2Q | J13005 | Alloy Steel, Cast—Medium Carbon |
| Class 4Q | J13047 | Alloy Steel, Cast—Medium Carbon |
| Class 4QA | J13047 | Alloy Steel, Cast—Medium Carbon |
| Class 6Q | J13855 | Alloy Steel, Cast—Medium Carbon |
| Class 7Q | J12084 | Alloy Steel, Cast—Medium Carbon |
| Class 8Q | J22091 | Alloy Steel, Cast—Medium Carbon |
| Class 9Q | J13345 | Alloy Steel, Cast—Medium Carbon |
| Class 10Q | J23015 | Alloy Steel, Cast—Medium Carbon |
| Class CA6NM | J91540 | Stainless Steel, Cast—Ferritic |
| Class CA15a | J91150 | Stainless Steel, Cast—Martensitic |
| Class CA15M | J91151 | Stainless Steel, Cast—Martensitic |
| Class DN | J04500 | Alloy Steel, Cast—Medium Carbon |
| ASTM A494: | | |
| Grade CW-12M-1 | N10002 | High Temperature Alloy—Nickel Base, Cast |
| Grade CW-12M-2 | N10002 | High Temperature Alloy—Nickel Base, Cast |
| Grade CZ-100 | N02100 | Nickel Alloy, Wrought and Cast |

| MATERIAL | UNS DESIGNATION | MATERIAL GROUP |
|---|---|---|
| Grade CY-40 | N07750 | High Temperature Alloy—Nickel Base, Cast |
| Grade M-35 | | Nickel Alloy, Wrought and Cast |
| Grade N-12M-1 | N10001 | High Temperature Alloy—Nickel Base, Cast |
| Grade N-12M-2 | N10001 | High Temperature Alloy—Nickel Base, Cast |
| ASTM A518 | F47003 | High Silicon Cast Iron |
| ASTM A532: | | |
| Class I, Type A | F45000 | White Cast Iron (Abrasion Resistant) |
| Class I, Type B | F45001 | White Cast Iron (Abrasion Resistant) |
| Class I, Type C | F45002 | White Cast Iron (Abrasion Resistant) |
| Class I, Type D | F45003 | White Cast Iron (Abrasion Resistant) |
| Class II, Type A | F45004 | White Cast Iron (Abrasion Resistant) |
| Class II, Type B | F45005 | White Cast Iron (Abrasion Resistant) |
| Class II, Type C | F45006 | White Cast Iron (Abrasion Resistant) |
| Class II, Type D | F45007 | White Cast Iron (Abrasion Resistant) |
| Class II, Type E | F45008 | White Cast Iron (Abrasion Resistant) |
| Class III, Type A | F45009 | White Cast Iron (Abrasion Resistant) |
| ASTM A536: | | |
| Grade 60-40-18 | F32800 | Ductile Cast Iron—Ferritic |
| Grade 65-45-12 | F33100 | Ductile Cast Iron—Ferritic |
| Grade 80-55-06 | F33800 | Ductile Cast Iron—Ferritic-Pearlitic |
| Grade 100-70-03 | F34800 | Ductile Cast Iron—Pearlitic-Martensitic |
| Grade 120-90-02 | F36200 | Ductile Cast Iron—Martensitic |
| ASTM A538: | | |
| Grade A | | Maraging Steel, Wrought |
| Grade B | | Maraging Steel, Wrought |
| Grade C | | Maraging Steel, Wrought |
| ASTM A560: | | |
| Grade 50Cr-50Ni | R20500 | Chromium-Nickel Alloy Casting |
| Grade 60Cr-40Ni | R20600 | Chromium-Nickel Alloy Casting |
| ASTM A571: | | |
| Type D-2M | F43010 | Ductile Cast Iron—Austenitic (NI-RESIST Ductile) |
| ASTM A597: | | |
| Grade CA-2 | T90102 | Tool Steel, Cast—Cold Work |
| Grade CD-2 | T90402 | Tool Steel, Cast—Cold Work |
| Grade CD-5 | T90405 | Tool Steel, Cast—Cold Work |
| Grade CH-12 | T90812 | Tool Steel, Cast—Hot Work |
| Grade CH-13 | T90813 | Tool Steel, Cast—Hot Work |
| Grade CO-1 | T91501 | Tool Steel, Cast—Cold Work |
| Grade CS-5 | T91905 | Tool Steel, Cast—Shock Resisting |
| ASTM A602: | | |
| Grade M3210 | F20000 | Malleable Cast Iron—Ferritic |
| Grade M4504 | F20001 | Malleable Cast Iron—Pearlitic |
| Grade M5003 | F20002 | Malleable Cast Iron—Pearlitic |
| Grade M5503 | F20003 | Malleable Cast Iron—Tempered Martensite |
| Grade M7002 | F20004 | Malleable Cast Iron—Tempered Martensite |
| Grade M8501 | F20005 | Malleable Cast Iron—Tempered Martensite |
| ASTM A608: | | |
| Grade HC30 | J92613 | Stainless Steel, Cast—Ferritic |
| Grade HD50 | J93015 | Stainless Steel, Cast—Austenitic |
| Grade HE35 | J93413 | Stainless Steel, Cast—Austenitic |
| Grade HF30 | J92803 | Stainless Steel, Cast—Austenitic |
| Grade HH30 | J93513 | Stainless Steel, Cast—Austenitic |

# 26.2 Alphabetical List of Materials

| MATERIAL | UNS DESIGNATION | MATERIAL GROUP |
|---|---|---|
| ASTM A608 *(continued)* | | |
| Grade HH33 | J93633 | Stainless Steel, Cast—Austenitic |
| Grade HI35 | J94013 | Stainless Steel, Cast—Austenitic |
| Grade HK30 | J94203 | Stainless Steel, Cast—Austenitic |
| Grade HK40 | J94204 | Stainless Steel, Cast—Austenitic |
| Grade HL30 | J94613 | Stainless Steel, Cast—Austenitic |
| Grade HL40 | J94614 | Stainless Steel, Cast—Austenitic |
| Grade HN40 | J94214 | Stainless Steel, Cast—Austenitic |
| Grade HT50 | J94805 | Stainless Steel, Cast—Austenitic |
| Grade HU50 | J95404 | Stainless Steel, Cast—Austenitic |
| Grade HW50 | N08005 | High Temperature Alloy—Nickel Base, Cast |
| Grade HX50 | N06050 | High Temperature Alloy—Nickel Base, Cast |
| ASTM B23: | | |
| Alloy 1 | L13910 | Tin Alloy, Cast—Tin Babbit Alloy |
| Alloy 2 | L13890 | Tin Alloy, Cast—Tin Babbit Alloy |
| Alloy 3 | L13840 | Tin Alloy, Cast—Tin Babbit Alloy |
| Alloy 7 | L05237 | Lead Alloy, Cast—Lead Babbit Alloy |
| Alloy 8 | L05188 | Lead Alloy, Cast—Lead Babbit Alloy |
| Alloy 11 | L13870 | Tin Alloy, Cast—Tin Babbit Alloy |
| Alloy 13 | L05153 | Lead Alloy, Cast—Lead Babbit Alloy |
| Alloy 15 | L05155 | Lead Alloy, Cast—Lead Babbit Alloy |
| ASTM B367: | | |
| Grade C-1 | R50250 | Titanium Alloy, Cast—Commercially Pure |
| Grade C-2 | R50400 | Titanium Alloy, Cast—Commercially Pure |
| Grade C-3 | R50550 | Titanium Alloy, Cast—Commercially Pure |
| Grade C-4 | R50700 | Titanium Alloy, Cast—Commercially Pure |
| Grade C-5 | R56400 | Titanium Alloy, Cast—Alpha and Alpha-Beta Alloy |
| Grade C-6 | R54520 | Titanium Alloy, Cast—Alpha and Alpha-Beta Alloy |
| Grade C-7A | R52250 | Titanium Alloy, Cast—Commercially Pure |
| Grade C-7B | R52400 | Titanium Alloy, Cast—Commercially Pure |
| Grade C-8A | R52550 | Titanium Alloy, Cast—Commercially Pure |
| Grade C-8B | | Titanium Alloy, Cast—Commercially Pure |
| Astroloy | | High Temperature Alloy—Nickel Base, Wrought |
| AZ21A | M11210 | Magnesium Alloy, Wrought |
| AZ31B | M11311 | Magnesium Alloy, Wrought |
| AZ31C | M11312 | Magnesium Alloy, Wrought |
| AZ61A | M11610 | Magnesium Alloy, Wrought |
| AZ63A | M11630 | Magnesium Alloy, Cast |
| AZ80A | M11800 | Magnesium Alloy, Wrought |
| AZ81A | M11810 | Magnesium Alloy, Cast |
| AZ91A | M11910 | Magnesium Alloy, Cast |
| AZ91B | M11912 | Magnesium Alloy, Cast |
| AZ91C | M11914 | Magnesium Alloy, Cast |
| AZ92A | M11920 | Magnesium Alloy, Cast |
| B295.0 | A22950 | Aluminum Alloy, Cast—Permanent Mold Cast |
| B443.0 | A24430 | Aluminum Alloy, Cast—Sand and Permanent Mold Cast |
| B514.0 | A25140 | Aluminum Alloy, Cast—Sand Cast |
| B850.0 | A28500 | Aluminum Alloy, Cast—Sand and Permanent Mold Cast |
| B-1900 | | High Temperature Alloy—Nickel Base, Cast |
| Babbit (High Arsenic) | | Lead Alloy, Cast—Lead Babbit Alloy |

| MATERIAL | UNS DESIGNATION | MATERIAL GROUP |
|---|---|---|
| Barium Titanate | | Flame (Thermal) Sprayed Material—Inorganic Coating Material |
| Berylco 41C | | Beryllium Nickel Alloy, Wrought and Cast |
| Berylco 42C | | Beryllium Nickel Alloy, Wrought and Cast |
| Berylco 43C | | Beryllium Nickel Alloy, Wrought and Cast |
| Berylco 440 | | Beryllium Nickel Alloy, Wrought and Cast |
| Beryllia | | Glasses & Ceramics—Ceramic |
| Beryllium Oxide | | Glasses & Ceramics—Ceramic |
| Boron | | Flame (Thermal) Sprayed Material—Inorganic Coating Material |
| Boron Epoxy | | Composite |
| Bronze | | Flame (Thermal) Sprayed Material—Sprayed Metals (Group II) |
| Brush Alloy 200C | | Beryllium Nickel Alloy, Wrought and Cast |
| Brush Alloy 220C | | Beryllium Nickel Alloy, Wrought and Cast |
| Brush Alloy 260C | | Beryllium Nickel Alloy, Wrought and Cast |
| Brush Grade Carbon | | Carbons & Graphites |
| C103 | | Refractory Alloy, Wrought, Cast, P/M—Columbium |
| C129Y | | Refractory Alloy, Wrought, Cast, P/M—Columbium |
| C355.0 | A33500 | Aluminum Alloy, Cast—Sand and Permanent Mold Cast |
| C443.0 | A04430 | Aluminum Alloy, Cast—Die Casting |
| Calcium Titanate | | Flame (Thermal) Sprayed Material—Inorganic Coating Material |
| Calcium Zirconate | | Flame (Thermal) Sprayed Material—Inorganic Coating Material |
| Carbon Steel | | Flame (Thermal) Sprayed Material—Sprayed Metals (Group III) |
| Cb-1Zr | | Refractory Alloy, Wrought, Cast, P/M—Columbium |
| Cb-752 | | Refractory Alloy, Wrought, Cast, P/M—Columbium |
| Cellulose Acetate | | Plastic—Thermoplastic |
| Chlorotrifluoroethylene | | Plastic—Thermoplastic—Fluorocarbon |
| Chromium | | Flame (Thermal) Sprayed Material—Sprayed Metals (Group II) |
| Chromium Carbide | | Flame (Thermal) Sprayed Material—Sprayed Carbide |
| Chromium Carbide-Cobalt Blend | | Flame (Thermal) Sprayed Material—Sprayed Carbide |
| Chromium Disilicide | | Flame (Thermal) Sprayed Material—Inorganic Coating Material |
| Chromium Oxide | | Flame (Thermal) Sprayed Material—Inorganic Coating Material |
| Chromium Plate | | Plated Material |
| Co-Cr-B | | Flame (Thermal) Sprayed Material—Corrosion and Oxidation Resistant Coating |
| Co-Cr-B Alloy (Self Fluxing) | | Flame (Thermal) Sprayed Material—Sprayed Metals (Group I) |
| Cobalt | | Flame (Thermal) Sprayed Material—Sprayed Metals (Group II) |
| Cobalt (40%), Zirconia Blend | | Flame (Thermal) Sprayed Material—Inorganic Coating Material |
| Columax-5 | | Magnetic Alloy, Cast |
| Columbium | | Flame (Thermal) Sprayed Material—Inorganic Coating Material |
| Columbium Carbide | | Flame (Thermal) Sprayed Material—Sprayed Carbide |
| Copper | C38600 | Powder Metal Alloy |
| Copper Alloys | | Flame (Thermal) Sprayed Material |
| CT-0010-N | | Powder Metal Alloy—Bronze |
| CT-0010-R | | Powder Metal Alloy—Bronze |
| CT-0010-S | | Powder Metal Alloy—Bronze |
| CTFE | | Plastic—Thermoplastic—Fluorocarbon |
| Custom 450 | S45000 | Precipitation Hardening Stainless Steel, Wrought |
| Custom 455 | S45500 | Precipitation Hardening Stainless Steel, Wrought |
| CZN-1818-T | | Powder Metal Alloy—Copper-Nickel Alloy |
| CZN-1818-U | | Powder Metal Alloy—Copper-Nickel Alloy |

## 26.2 Alphabetical List of Materials

| MATERIAL | UNS DESIGNATION | MATERIAL GROUP |
|---|---|---|
| CZN-1818-W | | Powder Metal Alloy—Copper-Nickel Alloy |
| CZNP-1618-U | | Powder Metal Alloy—Copper-Nickel Alloy |
| CZNP-1618-W | | Powder Metal Alloy—Copper-Nickel Alloy |
| CZP-0218-T | | Powder Metal Alloy—Brass |
| CZP-0218-U | | Powder Metal Alloy—Brass |
| CZP-0218-W | | Powder Metal Alloy—Brass |
| D2 | T30402 | Tool Steel, Wrought—Cold Work |
| D3 | T30403 | Tool Steel, Wrought—Cold Work |
| D4 | T30404 | Tool Steel, Wrought—Cold Work |
| D5 | T30405 | Tool Steel, Wrought—Cold Work |
| D6ac | | High Strength Steel, Wrought |
| D7 | T30407 | Tool Steel, Wrought—Cold Work |
| D712.0 | A47120 | Aluminum Alloy, Cast—Sand Cast |
| DAP | | Plastic—Thermosetting Plastic |
| Diallyl Phthalate | | Plastic—Thermosetting Plastic |
| Discaloy | K66220 | High Temperature Alloy—Iron Base, Wrought |
| DURANICKEL Alloy 301 | N03301 | Nickel Alloy, Wrought |
| Duriclor | | High Silicon Cast Iron |
| Duriron | | High Silicon Cast Iron |
| EC | | Aluminum Alloy, Wrought |
| EZ33A | M12330 | Magnesium Alloy, Cast |
| Epoxy | | Plastic—Thermosetting Plastic |
| F-0000-N | | Powder Metal Alloy—Iron |
| F-0000-P | | Powder Metal Alloy—Iron |
| F-0000-R | | Powder Metal Alloy—Iron |
| F-0000-S | | Powder Metal Alloy—Iron |
| F-0000-T | | Powder Metal Alloy—Iron |
| F-0005-S | | Powder Metal Alloy—Steel |
| F-0008-P | | Powder Metal Alloy—Steel |
| F-0008-S | | Powder Metal Alloy—Steel |
| F1 | T60601 | Tool Steel, Wrought—Special Purpose |
| F2 | T60602 | Tool Steel, Wrought—Special Purpose |
| F332.0 | A63320 | Aluminum Alloy, Cast—Permanent Mold Cast |
| FC-0205-S | | Powder Metal Alloy—Steel |
| FC-0208-P | | Powder Metal Alloy—Steel |
| FC-0208-S | | Powder Metal Alloy—Steel |
| FC-0508-P | | Powder Metal Alloy—Steel |
| FC-1000-N | | Powder Metal Alloy—Steel |
| Ferro-Tic | | Machinable Carbide |
| Fiberglass Epoxy | | Composite |
| FN-0205-S | | Powder Metal Alloy—Steel |
| FN-0205-T | | Powder Metal Alloy—Steel |
| FN-0405-R | | Powder Metal Alloy—Steel |
| FN-0405-S | | Powder Metal Alloy—Steel |
| FN-0405-T | | Powder Metal Alloy—Steel |
| FS-85 | | Refractory Alloy, Wrought, Cast, P/M—Columbium |

| MATERIAL | UNS DESIGNATION | MATERIAL GROUP |
|---|---|---|
| FS-291 | | Refractory Alloy, Wrought, Cast, P/M—Columbium |
| FSX-414 | | High Temperature Alloy—Cobalt Base, Cast |
| Furan | | Plastic—Thermosetting Plastic |
| FX-1005-T | | Powder Metal Alloy—Steel |
| FX-2008-T | | Powder Metal Alloy—Steel |
| GE-218 | | Refractory Alloy, Wrought, Cast, P/M—Tungsten Alloy |
| Glass | | Glasses & Ceramics |
| Glass (Kovar sealing) | | Flame (Thermal) Sprayed Material—Inorganic Coating Material |
| GMR-235 | | High Temperature Alloy—Nickel Base, Cast |
| GMR-235D | | High Temperature Alloy—Nickel Base, Cast |
| Gold | P00010 | Precious Metal |
| Graphite | | Carbons & Graphites |
| Graphite Epoxy | | Composite |
| Greek Ascoloy | S41800 | Stainless Steel, Wrought—Martensitic |
| Gyromet | | Refractory Alloy, Wrought, Cast, P/M—Tungsten Alloy |
| H10 | T20810 | Tool Steel, Wrought—Hot Work |
| H11 | | High Strength Steel, Wrought |
| H11 | T20811 | Tool Steel, Wrought—Hot Work |
| H12 | T20812 | Tool Steel, Wrought—Hot Work |
| H13 | | High Strength Steel, Wrought |
| H13 | T20813 | Tool Steel, Wrought—Hot Work |
| H14 | T20814 | Tool Steel, Wrought—Hot Work |
| H19 | T20819 | Tool Steel, Wrought—Hot Work |
| H21 | T20821 | Tool Steel, Wrought—Hot Work |
| H22 | T20822 | Tool Steel, Wrought—Hot Work |
| H23 | T20823 | Tool Steel, Wrought—Hot Work |
| H24 | T20824 | Tool Steel, Wrought—Hot Work |
| H25 | T20825 | Tool Steel, Wrought—Hot Work |
| H26 | T20826 | Tool Steel, Wrought—Hot Work |
| H42 | T20842 | Tool Steel, Wrought—Hot Work |
| Hastelloy Alloy B | N10001 | High Temperature Alloy—Nickel Base, Wrought or Cast |
| Hastelloy Alloy B-2 | N10665 | High Temperature Alloy—Nickel Base, Wrought |
| Hastelloy Alloy C | N10002 | High Temperature Alloy—Nickel Base, Wrought or Cast |
| Hastelloy Alloy C-276 | N10276 | High Temperature Alloy—Nickel Base, Wrought |
| Hastelloy Alloy D | | High Temperature Alloy—Nickel Base, Cast |
| Hastelloy Alloy G | N06007 | High Temperature Alloy—Nickel Base, Wrought |
| Hastelloy Alloy S | | High Temperature Alloy—Nickel Base, Wrought |
| Hastelloy Alloy X | N06002 | High Temperature Alloy—Nickel Base, Wrought |
| Haynes Alloy 25 | R30605 | High Temperature Alloy—Cobalt Base, Wrought |
| Haynes Alloy 188 | R30188 | High Temperature Alloy—Cobalt Base, Wrought |
| Haynes Alloy 263 | N07263 | High Temperature Alloy—Nickel Base, Wrought |
| Hexaboron Silicide | | Flame (Thermal) Sprayed Material—Inorganic Coating Material |
| Hi Perm 49 | | Magnetic Alloy |
| Hi Perm 49-FM | | Free Machining Magnetic Alloy |
| Hiduminium RR-350 | | Aluminum Alloy, Cast—Sand Cast |
| HK31A | M13310 | Magnesium Alloy, Cast |
| HK31A | M13310 | Magnesium Alloy, Wrought |
| HM21A | M13210 | Magnesium Alloy, Wrought |

## 26.2 Alphabetical List of Materials

| MATERIAL | UNS DESIGNATION | MATERIAL GROUP |
|---|---|---|
| HM31A | M13312 | Magnesium Alloy, Wrought |
| HNM | | Precipitation Hardening Stainless Steel, Wrought |
| HOWMET #3 | | High Temperature Alloy—Cobalt Base, Cast |
| HP9-4-20 | | High Strength Steel, Wrought |
| HP9-4-25 | | High Strength Steel, Wrought |
| HP9-4-30 | | High Strength Steel, Wrought |
| HP9-4-45 | | High Strength Steel, Wrought |
| HS-6 | R30006 | High Temperature Alloy—Cobalt Base, Cast |
| HS-21 | R30021 | High Temperature Alloy—Cobalt Base, Cast |
| HS-31 | R30031 | High Temperature Alloy—Cobalt Base, Cast |
| HY80 | J42015 | Armor Plate, Ship Plate, Aircraft Plate, Wrought |
| HY100 | J42240 | Armor Plate, Ship Plate, Aircraft Plate, Wrought |
| HY180 | | Armor Plate, Ship Plate, Aircraft Plate, Wrought |
| HY230 | | Maraging Steel, Wrought |
| HyMu 80 | | Magnetic Alloy |
| Hyflux Alnico V-7 | | Magnetic Alloy, Cast |
| ILZRO 12 | | Zinc Alloy, Cast |
| IN-100 | N13100 | High Temperature Alloy—Nickel Base, Cast |
| IN-102 | N06102 | High Temperature Alloy—Nickel Base, Wrought |
| IN-738 | | High Temperature Alloy—Nickel Base, Cast |
| IN-792 | | High Temperature Alloy—Nickel Base, Cast |
| Incoloy Alloy 800 | N08800 | High Temperature Alloy—Iron Base, Wrought |
| Incoloy Alloy 800H | N08810 | High Temperature Alloy—Iron Base, Wrought |
| Incoloy Alloy 801 | N08801 | High Temperature Alloy—Iron Base, Wrought |
| Incoloy Alloy 802 | N08802 | High Temperature Alloy—Iron Base, Wrought |
| Incoloy Alloy 804 | N06804 | High Temperature Alloy—Nickel Base, Wrought |
| Incoloy Alloy 825 | N08825 | High Temperature Alloy—Nickel Base, Wrought |
| Incoloy Alloy 901 | N09901 | High Temperature Alloy—Nickel Base, Wrought |
| Incoloy Alloy 903 | N19903 | High Temperature Alloy—Nickel Base, Wrought |
| Inconel Alloy 600 | N06600 | High Temperature Alloy—Nickel Base, Wrought |
| Inconel Alloy 601 | N06601 | High Temperature Alloy—Nickel Base, Wrought |
| Inconel Alloy 617 | N06617 | High Temperature Alloy—Nickel Base, Wrought |
| Inconel Alloy 625 | N06625 | High Temperature Alloy—Nickel Base, Wrought |
| Inconel Alloy 700 | | High Temperature Alloy—Nickel Base, Wrought |
| Inconel Alloy 702 | N07702 | High Temperature Alloy—Nickel Base, Wrought |
| Inconel Alloy 706 | N09706 | High Temperature Alloy—Nickel Base, Wrought |
| Inconel Alloy 713C | N07713 | High Temperature Alloy—Nickel Base, Cast |
| Inconel Alloy 718 | N07718 | High Temperature Alloy—Nickel Base, Wrought |
| Inconel Alloy 718 | N07718 | High Temperature Alloy—Nickel Base, Cast |
| Inconel Alloy 721 | N07721 | High Temperature Alloy—Nickel Base, Wrought |
| Inconel Alloy 722 | N07722 | High Temperature Alloy—Nickel Base, Wrought |
| Inconel Alloy X-750 | N07750 | High Temperature Alloy—Nickel Base, Wrought |
| Inconel Alloy 751 | N07751 | High Temperature Alloy—Nickel Base, Wrought |
| Invar | K93600 | Controlled Expansion Alloy |
| Invar 36 | K93602 | Free Machining Controlled Expansion Alloy |
| Iron | | Flame (Thermal) Sprayed Material |
| Iron | | Flame (Thermal) Sprayed Material—Sprayed Metals (Group III) |
| J-1570 | | High Temperature Alloy—Cobalt Base, Wrought |

| MATERIAL | UNS DESIGNATION | MATERIAL GROUP |
|---|---|---|
| K1A | M18010 | Magnesium Alloy, Cast |
| Kevlar 49 | | Composite |
| Kovar | K94610 | Controlled Expansion Alloy |
| | | |
| L2 | T61202 | Tool Steel, Wrought–Special Purpose |
| L6 | T61206 | Tool Steel, Wrought–Special Purpose |
| L7 | T61207 | Tool Steel, Wrought–Special Purpose |
| L605 | R30605 | High Temperature Alloy–Cobalt Base, Wrought |
| Lead | | Flame (Thermal) Sprayed Material |
| | | |
| M1 | T11301 | Tool Steel, Wrought–High Speed |
| M2 | T11302 | Tool Steel, Wrought–High Speed |
| M3 Class 1 | T11312 | Tool Steel, Wrought–High Speed |
| M3 Class 2 | T11323 | Tool Steel, Wrought–High Speed |
| M4 | T11304 | Tool Steel, Wrought–High Speed |
| M6 | T11306 | Tool Steel, Wrought–High Speed |
| M7 | T11307 | Tool Steel, Wrought–High Speed |
| M10 | T11310 | Tool Steel, Wrought–High Speed |
| M30 | T11330 | Tool Steel, Wrought–High Speed |
| M33 | T11333 | Tool Steel, Wrought–High Speed |
| M34 | T11334 | Tool Steel, Wrought–High Speed |
| M36 | T11336 | Tool Steel, Wrought–High Speed |
| M41 | T11341 | Tool Steel, Wrought–High Speed |
| M42 | T11342 | Tool Steel, Wrought–High Speed |
| M43 | T11343 | Tool Steel, Wrought–High Speed |
| M44 | T11344 | Tool Steel, Wrought–High Speed |
| M46 | T11346 | Tool Steel, Wrought–High Speed |
| M47 | T11347 | Tool Steel, Wrought–High Speed |
| M-50 | T11350 | Alloy Steel, Wrought–High Carbon |
| M252 | N07252 | High Temperature Alloy–Nickel Base, Wrought |
| MACOR | | Glasses & Ceramics–Machinable Glass-Ceramic |
| Magnesia | | Glasses & Ceramics–Ceramic |
| Magnesia Alumina Spinel | | Flame (Thermal) Sprayed Material–Inorganic Coating Material |
| Magnesium Oxide | | Glasses & Ceramics–Ceramic |
| Magnesium Zirconate | | Flame (Thermal) Sprayed Material–Inorganic Coating Material |
| Magnetic Core Iron-FM (up to 2.5% Si) | | Free Machining Magnetic Alloy |
| Magnetic Core Iron (up to 4% Si) | | Magnetic Alloy |
| Mallory 2000 | | Refractory Alloy, Wrought, Cast, P/M–Tungsten Alloy |
| Manganese | | Manganese, Wrought |
| MAR-M200 | | High Temperature Alloy–Nickel Base, Cast |
| MAR-M246 | | High Temperature Alloy–Nickel Base, Cast |
| MAR-M302 | | High Temperature Alloy–Cobalt Base, Cast |
| MAR-M322 | | High Temperature Alloy–Cobalt Base, Cast |
| MAR-M421 | | High Temperature Alloy–Nickel Base, Cast |
| MAR-M432 | | High Temperature Alloy–Nickel Base, Cast |
| MAR-M509 | | High Temperature Alloy–Cobalt Base, Cast |
| MAR-M905 | | High Temperature Alloy–Cobalt Base, Wrought |
| MAR-M918 | | High Temperature Alloy–Cobalt Base, Wrought |

## 26.2 Alphabetical List of Materials

| MATERIAL | UNS DESIGNATION | MATERIAL GROUP |
|---|---|---|
| Mechanical Grades | | Carbons & Graphites |
| Melamine | | Plastic–Thermosetting Plastic |
| Mica | | Glasses & Ceramics–Mineral Silicate |
| MIL-S-12560 | | Armor Plate, Ship Plate, Aircraft Plate, Wrought |
| MIL-S-16216 | J42015 | Armor Plate, Ship Plate, Aircraft Plate, Wrought |
| MIL-S-16216 | J42240 | Armor Plate, Ship Plate, Aircraft Plate, Wrought |
| Mo | | Refractory Alloy, Wrought, Cast, P/M–Molybdenum |
| Mo-50Re | | Refractory Alloy, Wrought, Cast, P/M–Molybdenum |
| Molybdenum | | Flame (Thermal) Sprayed Material–Sprayed Metals (Group II) |
| Molybdenum Disilicide | | Flame (Thermal) Sprayed Material–Inorganic Coating Material |
| Monel | | Flame (Thermal) Sprayed Material–Sprayed Metals (Group II) |
| MONEL Alloy 400 | N04400 | Nickel Alloy, Wrought |
| MONEL Alloy 401 | N04401 | Nickel Alloy, Wrought |
| MONEL Alloy 404 | N04404 | Nickel Alloy, Wrought |
| MONEL Alloy 502 | N05502 | Nickel Alloy, Wrought |
| MONEL Alloy K500 | N05500 | Nickel Alloy, Wrought |
| MONEL Alloy R405 | N04405 | Nickel Alloy, Wrought |
| Mullite | | Flame (Thermal) Sprayed Material–Inorganic Coating Material |
| Mullite | | Glasses & Ceramics–Ceramic |
| N-155 | R30155 | High Temperature Alloy–Iron Base, Wrought |
| NASA Co-W-Re | | High Temperature Alloy–Cobalt Base, Cast |
| Ni-Cr-B | | Flame (Thermal) Sprayed Material–Corrosion and Oxidation Resistant Coating |
| Ni-Cr-B Alloys (Self Fluxing) | | Flame (Thermal) Sprayed Material–Sprayed Metals (Group I) |
| Nickel | | Flame (Thermal) Sprayed Material–Sprayed Metals (Group II) |
| Nickel | | Powder Metal Alloy–Nickel and Nickel Alloy |
| Nickel 200 | N02200 | Nickel Alloy, Wrought and Cast |
| Nickel 201 | N02201 | Nickel Alloy, Wrought |
| Nickel 205 | N02205 | Nickel Alloy, Wrought |
| Nickel 211 | N02211 | Nickel Alloy, Wrought |
| Nickel 220 | N02220 | Nickel Alloy, Wrought |
| Nickel 230 | N02230 | Nickel Alloy, Wrought |
| Nickel (40%), Alumina Blend | | Flame (Thermal) Sprayed Material–Inorganic Coating Material |
| Nickel Alloys | | Flame (Thermal) Sprayed Material |
| Nickel Chrome Steel (Special) | | Flame (Thermal) Sprayed Material–Sprayed Metals (Group I) |
| Nickel Oxide | | Flame (Thermal) Sprayed Material–Inorganic Coating Material |
| Nickel Plate | | Plated Material |
| Nimonic 75 | N06075 | High Temperature Alloy–Nickel Base, Wrought |
| Nimonic 80 | N07080 | High Temperature Alloy–Nickel Base, Wrought |
| Nimonic 90 | N07090 | High Temperature Alloy–Nickel Base, Wrought |
| Nimonic 95 | | High Temperature Alloy–Nickel Base, Wrought |
| Niobium Alloy | | Flame (Thermal) Sprayed Material–Inorganic Coating Material |
| NI-SPAN-C 902 | N09902 | Nickel Alloy, Wrought |
| Nitinol 55Ni-45Ti | | Nitinol Alloy, Wrought |
| Nitinol 56Ni-44Ti | | Nitinol Alloy, Wrought |
| Nitinol 60Ni-40Ti | | Nitinol Alloy, Wrought |
| Nitralloy 125 | | Nitriding Steel, Wrought |
| Nitralloy 135 | | Nitriding Steel, Wrought |
| Nitralloy 135 Mod. | | Nitriding Steel, Wrought |

| MATERIAL | UNS DESIGNATION | MATERIAL GROUP |
|---|---|---|
| Nitralloy 225 | | Nitriding Steel, Wrought |
| Nitralloy 230 | | Nitriding Steel, Wrought |
| Nitralloy EZ | | Nitriding Steel, Wrought |
| Nitralloy N | | Nitriding Steel, Wrought |
| Nitrex 1 | | Nitriding Steel, Wrought |
| Nitronic 32 | S24100 | Stainless Steel, Wrought—Austenitic |
| Nitronic 33 | S24000 | Stainless Steel, Wrought—Austenitic |
| Nitronic 40 | S21900 | Stainless Steel, Wrought—Austenitic |
| Nitronic 50 | S20910 | Stainless Steel, Wrought—Austenitic |
| Nitronic 60 | S21800 | Stainless Steel, Wrought—Austenitic |
| Nylons, Unfilled | | Plastics—Thermoplastics |
| Nylons, 35% Glass reinforced | | Plastics—Thermoplastics |
| O1 | T31501 | Tool Steel, Wrought—Cold Work |
| O2 | T31502 | Tool Steel, Wrought—Cold Work |
| O6 | T31506 | Tool Steel, Wrought—Cold Work |
| O7 | T31507 | Tool Steel, Wrought—Cold Work |
| P2 | T51602 | Tool Steel, Wrought—Mold |
| P4 | T51604 | Tool Steel, Wrought—Mold |
| P5 | T51605 | Tool Steel, Wrought—Mold |
| P6 | T51606 | Tool Steel, Wrought—Mold |
| P20 | T51620 | Tool Steel, Wrought—Mold |
| P21 | T51621 | Tool Steel, Wrought—Mold |
| PERMANICKEL Alloy 300 | N03300 | Nickel Alloy, Wrought |
| PH 13-8Mo | S13800 | Precipitation Hardening Stainless Steel, Wrought |
| PH 14-8Mo | S14800 | Precipitation Hardening Stainless Steel, Wrought |
| PH 15-7Mo | S15700 | Precipitation Hardening Stainless Steel, Wrought |
| Phenolic | | Plastic—Thermosetting Plastic |
| Platinum | | Flame (Thermal) Sprayed Material |
| Platinum | P04995 | Precious Metal |
| Polyamides—Unfilled | | Plastic—Thermoplastic |
| Polyamides—35% Glass reinforced | | Plastic—Thermoplastic |
| Polyimide | | Plastic—Thermosetting Plastic |
| Polyimide—Glass filled | | Plastic—Thermosetting Plastic |
| Polyarylether | | Plastic—Thermoplastic |
| Polybutadiene | | Plastic—Thermosetting Plastic |
| Polycarbonate | | Plastic—Thermoplastic |
| Polyethylene | | Plastic—Thermoplastic |
| Polypropylene | | Plastic—Thermoplastic |
| Polysulfone | | Plastic—Thermoplastic |
| Polystyrene | | Plastic—Thermoplastic |
| Polyurethane | | Plastic—Thermosetting Plastic |
| Porcelain Enamel | | Glasses & Ceramics—Ceramic |
| Precipitation Hardening Steel | | Flame (Thermal) Sprayed Material—Sprayed Metals (Group III) |
| PYROCERAM | | Glasses & Ceramics—Polycrystalline Glass |

## 26.2 Alphabetical List of Materials

| MATERIAL | UNS DESIGNATION | MATERIAL GROUP |
|---|---|---|
| QE22A | M18220 | Magnesium Alloy, Cast |
| | | |
| Rare Earth Oxides | | Flame (Thermal) Sprayed Material—Inorganic Coating Material |
| Refractaloy 26 | | High Temperature Alloy—Nickel Base, Wrought |
| Rene 41 | N07041 | High Temperature Alloy—Nickel Base, Wrought |
| Rene 63 | | High Temperature Alloy—Nickel Base, Wrought |
| Rene 77 | | High Temperature Alloy—Nickel Base, Wrought |
| Rene 80 | | High Temperature Alloy—Nickel Base, Cast |
| Rene 95 | | High Temperature Alloy—Nickel Base, Wrought |
| Rene 100 | N13100 | High Temperature Alloy—Nickel Base, Cast |
| Rene 125 | | High Temperature Alloy—Nickel Base, Cast |
| Rubber | | Rubber |
| | | |
| S1 | T41901 | Tool Steel, Wrought—Shock Resisting |
| S2 | T41902 | Tool Steel, Wrought—Shock Resisting |
| S5 | T41905 | Tool Steel, Wrought—Shock Resisting |
| S6 | T41906 | Tool Steel, Wrought—Shock Resisting |
| S7 | T41907 | Tool Steel, Wrought—Shock Resisting |
| S-816 | R30816 | High Temperature Alloy—Cobalt Base, Wrought |
| SAE J158: | | |
| Grade M3210 | F20000 | Malleable Cast Iron—Ferritic |
| Grade M4504 | F20001 | Malleable Cast Iron—Pearlitic |
| Grade M5003 | F20002 | Malleable Cast Iron—Pearlitic |
| Grade M5503 | F20003 | Malleable Cast Iron—Tempered Martensite |
| Grade M7002 | F20004 | Malleable Cast Iron—Tempered Martensite |
| Grade M8501 | F20005 | Malleable Cast Iron—Tempered Martensite |
| SAE J431c: | | |
| Grade G1800 | F10004 | Gray Cast Iron—Ferritic |
| Grade G2500 | F10005 | Gray Cast Iron—Pearlitic-Ferritic |
| Grade G3000 | F10006 | Gray Cast Iron—Pearlitic |
| Grade G3500 | F10007 | Gray Cast Iron—Pearlitic + Free Carbides |
| Grade G4000 | F10008 | Gray Cast Iron—Pearlitic + Free Carbides |
| SAE J434c: | | |
| Grade D4018 | F32800 | Ductile Cast Iron—Ferritic |
| Grade D4512 | F33100 | Ductile Cast Iron—Ferritic |
| Grade D5506 | F33800 | Ductile Cast Iron—Ferritic-Pearlitic |
| Grade D7003 | F34800 | Ductile Cast Iron—Pearlitic-Martensitic |
| Grade DQ&T | F30000 | Ductile Cast Iron—Martensitic |
| SAE J438b: | | |
| Type W108 | T72301 | Tool Steel, Wrought—Water Hardening |
| Type W109 | T72301 | Tool Steel, Wrought—Water Hardening |
| Type W110 | T72301 | Tool Steel, Wrought—Water Hardening |
| Type W112 | T72301 | Tool Steel, Wrought—Water Hardening |
| Type W209 | T72302 | Tool Steel, Wrought—Water Hardening |
| Type W210 | T72302 | Tool Steel, Wrought—Water Hardening |
| Type W310 | T72303 | Tool Steel, Wrought—Water Hardening |
| SAE J469a: | | |
| Grade 903 (AG40A) | Z33520 | Zinc Alloy, Cast |
| Grade 925 (AC41A) | Z35531 | Zinc Alloy, Cast |
| SEL | | High Temperature Alloy—Nickel Base, Cast |

| MATERIAL | UNS DESIGNATION | MATERIAL GROUP |
|---|---|---|
| SEL 15 | | High Temperature Alloy—Nickel Base, Cast |
| Silicon Carbide | | Glasses & Ceramics—Ceramic |
| Silicon Nitride | | Glasses & Ceramics—Ceramic |
| Silicone | | Plastic—Thermosetting Plastic |
| Silicone-Glass filled | | Plastic—Thermosetting Plastic |
| Silver | | Flame (Thermal) Sprayed Material |
| Silver | P07010 | Precious Metal |
| Silver Plate | | Plated Material |
| SS-303-R | | Powder Metal Alloy—Stainless Steel |
| SS-304-R | | Powder Metal Alloy—Stainless Steel |
| SS-316-R | | Powder Metal Alloy—Stainless Steel |
| SS-410-R | | Powder Metal Alloy—Stainless Steel |
| Stainless Steels | | Flame (Thermal) Sprayed Material—Sprayed Metals (Group I) |
| Stainless Steels—Austenitic | | Flame (Thermal) Sprayed Material |
| Stainless Steels—Ferritic | | Flame (Thermal) Sprayed Material |
| Stainless W | S17600 | Precipitation Hardening Stainless Steel, Wrought |
| Steels | | Flame (Thermal) Sprayed Material |
| Structural Steels | | See Numerical Listing According to Tensile Yield Strength |
| T1 | T12001 | Tool Steel, Wrought—High Speed |
| T2 | T12002 | Tool Steel, Wrought—High Speed |
| T4 | T12004 | Tool Steel, Wrought—High Speed |
| T5 | T12005 | Tool Steel, Wrought—High Speed |
| T6 | T12006 | Tool Steel, Wrought—High Speed |
| T8 | T12008 | Tool Steel, Wrought—High Speed |
| T15 | T12015 | Tool Steel, Wrought—High Speed |
| T-111 | | Refractory Alloy, Wrought, Cast, P/M—Tantalum |
| T-222 | | Refractory Alloy, Wrought, Cast, P/M—Tantalum |
| Ta-10W | | Refractory Alloy, Wrought, Cast, P/M—Tantalum |
| Ta63 | | Refractory Alloy, Wrought, Cast, P/M—Tantalum |
| Ta-Hf | | Refractory Alloy, Wrought, Cast, P/M—Tantalum |
| Tantalum | | Flame (Thermal) Sprayed Material—Inorganic Coating Material |
| Tantalum Carbide | | Flame (Thermal) Sprayed Material—Sprayed Carbide |
| TD-Nickel | | High Temperature Alloy—Nickel Base, Wrought |
| TD-Ni-Cr | | High Temperature Alloy—Nickel Base, Wrought |
| Tetrafluoroethylene | | Plastic—Thermoplastic—Fluorocarbon |
| TFE | | Plastic—Thermoplastic—Fluorocarbon |
| Thermoplastics | | Plastics |
| Thermosetting Plastics | | Plastics |
| Thoria | | Glasses & Ceramics—Ceramic |
| Thorium Oxide | | Glasses & Ceramics—Ceramic |
| TiCODE-12 | | Titanium Alloy, Wrought—Commercially Pure |
| Ti-0.2Pd | R52400 | Titanium Alloy, Wrought or Cast—Commercially Pure |
| Ti-17 | | Titanium Alloy, Wrought—Alpha-Beta Alloy |
| Ti 98.9 | | Titanium Alloy, Wrought—Commercially Pure |
| Ti 99.0 | | Titanium Alloy, Wrought or Cast—Commercially Pure |
| Ti 99.2 | | Titanium Alloy, Wrought—Commercially Pure |
| Ti 99.5 | | Titanium Alloy, Wrought—Commercially Pure |
| Ti-1Al-8V-5Fe | | Titanium Alloy, Wrought—Alpha and Alpha-Beta Alloy |
| Ti-2Al-11Sn-5Zr-1Mo | R54790 | Titanium Alloy, Wrought—Alpha and Alpha-Beta Alloy |
| Ti-3Al-2.5V | R56320 | Titanium Alloy, Wrought—Alpha-Beta Alloy |

## 26.2 Alphabetical List of Materials

| MATERIAL | UNS DESIGNATION | MATERIAL GROUP |
|---|---|---|
| Ti-3Al-8V-6Cr-4Mo-4Zr | R58640 | Titanium Alloy, Wrought—Beta Alloy |
| Ti-4.5Sn-6Zr-11.5Mo | R58030 | Titanium Alloy, Wrought—Beta Alloy |
| Ti-5Al-2Sn-2Zr-4Mo-4Cr | | Titanium Alloy, Wrought—Alpha-Beta Alloy |
| Ti-5Al-2.5Sn | R54521 | Titanium Alloy, Wrought or Cast—Alpha Alloy |
| Ti-5Al-2.5Sn ELI | R54520 | Titanium Alloy, Wrought—Alpha Alloy |
| Ti-5Al-6Sn-2Zr-1Mo | R54560 | Titanium Alloy, Wrought—Alpha and Alpha-Beta Alloy |
| Ti-6Al-2Cb-1Ta-0.8Mo | R56210 | Titanium Alloy, Wrought—Alpha-Beta Alloy |
| Ti-6Al-2Sn-4Zr-2Mo | R54620 | Titanium Alloy, Wrought or Cast—Alpha-Beta Alloy |
| Ti-6Al-2Sn-4Zr-2Mo-.25Si | | Titanium Alloy, Wrought—Alpha-Beta Alloy |
| Ti-6Al-2Sn-4Zr-6Mo | R56260 | Titanium Alloy, Wrought—Alpha-Beta Alloy |
| Ti-6Al-4V | R56401 | Titanium Alloy, Wrought or Cast—Alpha-Beta Alloy |
| Ti-6Al-4V ELI | R56400 | Titanium Alloy, Wrought—Alpha-Beta Alloy |
| Ti-6Al-6V-2Sn | R56620 | Titanium Alloy, Wrought—Alpha-Beta Alloy |
| Ti-7Al-4Mo | R56740 | Titanium Alloy, Wrought—Alpha-Beta Alloy |
| Ti-8Al-1Mo-1V | R54810 | Titanium Alloy, Wrought or Cast—Alpha and Alpha-Beta Alloy |
| Ti-8Mn | R56080 | Titanium Alloy, Wrought—Alpha-Beta Alloy |
| Ti-8Mo-8V-2Fe-3Al | R58820 | Titanium Alloy, Wrought—Beta Alloy |
| Ti-10V-2Fe-3Al | | Titanium Alloy, Wrought—Beta Alloy |
| Ti-13V-11Cr-3Al | R58010 | Titanium Alloy, Wrought—Beta Alloy |
| Tin | | Flame (Thermal) Sprayed Material |
| Titania | | Glasses & Ceramics—Ceramic |
| Titania (50%), Alumina Blend | | Flame (Thermal) Sprayed Material—Inorganic Coating Material |
| Titanium Carbide | | Carbide |
| Titanium Carbide | | Flame (Thermal) Sprayed Material—Sprayed Carbide |
| Titanium Diboride | | Glasses & Ceramics—Ceramic |
| Titanium Oxide | | Flame (Thermal) Sprayed Material—Inorganic Coating Material |
| Titanium Oxide | | Glasses & Ceramics—Ceramic |
| TRW VI A | | High Temperature Alloy—Nickel Base, Cast |
| Tungsten | | Flame (Thermal) Sprayed Material—Inorganic Coating Material |
| Tungsten, 85% Density | | Refractory Alloy, Wrought, Cast, P/M—Tungsten |
| Tungsten, 93% Density | | Refractory Alloy, Wrought, Cast, P/M—Tungsten |
| Tungsten, 96% Density | | Refractory Alloy, Wrought, Cast, P/M—Tungsten |
| Tungsten, 100% Density | | Refractory Alloy, Wrought, Cast, P/M—Tungsten |
| Tungsten-2 Thoria | | Refractory Alloy, Wrought, Cast, P/M—Tungsten |
| Tungsten Carbide | | Carbide |
| Tungsten Carbide | | Flame (Thermal) Sprayed Material—Sprayed Carbide |
| Tungsten Carbide-Cobalt | | Flame (Thermal) Sprayed Material—Sprayed Carbide |
| Tungsten Carbide (Cobalt)-Nickel Alloy Blend | | Flame (Thermal) Sprayed Material—Sprayed Carbide |
| TZC | | Refractory Alloy, Wrought, Cast, P/M—Molybdenum |
| TZM | | Refractory Alloy, Wrought, Cast, P/M—Molybdenum |
| Udimet 500 | N07500 | High Temperature Alloy—Nickel Base, Wrought or Cast |
| Udimet 630 | | High Temperature Alloy—Nickel Base, Wrought |
| Udimet 700 | | High Temperature Alloy—Nickel Base, Wrought or Cast |
| Udimet 710 | | High Temperature Alloy—Nickel Base, Wrought |
| Unitemp 1753 | | High Temperature Alloy—Nickel Base, Wrought |
| Uranium | M08990 | Uranium, Wrought |

| MATERIAL | UNS DESIGNATION | MATERIAL GROUP |
|---|---|---|
| V-36 | | High Temperature Alloy–Cobalt Base, Wrought |
| V-57 | | High Temperature Alloy–Iron Base, Wrought |
| | | |
| W1 | T72301 | Tool Steel, Wrought–Water Hardening |
| W2 | T72302 | Tool Steel, Wrought–Water Hardening |
| W5 | T72305 | Tool Steel, Wrought–Water Hardening |
| W-5Re | | Refractory Alloy, Wrought, Cast, P/M–Tungsten Alloy |
| W-7Ni-4Cu | | Refractory Alloy, Wrought, Cast, P/M–Tungsten Alloy |
| W-10Ag | | Refractory Alloy, Wrought, Cast, P/M–Tungsten Alloy |
| W-15Mo | | Refractory Alloy, Wrought, Cast, P/M–Tungsten Alloy |
| W-25Re | | Refractory Alloy, Wrought, Cast, P/M–Tungsten Alloy |
| W-25Re-30Mo | | Refractory Alloy, Wrought, Cast, P/M–Tungsten Alloy |
| W-545 | K66545 | High Temperature Alloy–Iron Base, Wrought |
| Waspaloy | N07001 | High Temperature Alloy–Nickel Base, Wrought |
| WC-3015 | | Refractory Alloy, Wrought, Cast, P/M–Columbium |
| WI-52 | | High Temperature Alloy–Cobalt Base, Cast |
| | | |
| X-40 | R30031 | High Temperature Alloy–Cobalt Base, Cast |
| X-45 | | High Temperature Alloy–Cobalt Base, Cast |
| | | |
| Yttrium Zirconate | | Flame (Thermal) Sprayed Material–Inorganic Coating Material |
| | | |
| ZDC No. 7 | Z33520 | Zinc Alloy, Cast |
| ZE41A | M16410 | Magnesium Alloy, Cast |
| ZE63A | M16630 | Magnesium Alloy, Cast |
| ZH62A | M16620 | Magnesium Alloy, Cast |
| Zinc | | Flame (Thermal) Sprayed Material |
| Zircaloy 2 (Grade 32) | R60802 | Zirconium Alloy, Wrought |
| Zircaloy 4 (Grade 34) | R60804 | Zirconium Alloy, Wrought |
| Zircon | | Glasses & Ceramics–Ceramic |
| Zirconia | | Glasses & Ceramics–Ceramic |
| Zirconia (Lime Stabilized) | | Flame (Thermal) Sprayed Material–Inorganic Coating Material |
| Zirconia, Nickel-Aluminide Blends | | Flame (Thermal) Sprayed Material–Inorganic Coating Material |
| Zirconium Oxide | | Glasses & Ceramics–Ceramic |
| Zirconium Oxide (Hafnia Free, Lime Stabilized) | | Flame (Thermal) Sprayed Material–Inorganic Coating Material |
| Zirconium Silicate | | Flame (Thermal) Sprayed Material–Inorganic Coating Material |
| Zirconium Silicate | | Glasses & Ceramics–Ceramic |
| ZK40A | M16400 | Magnesium Alloy, Wrought |
| ZK51A | M16510 | Magnesium Alloy, Cast |
| ZK60A | M16600 | Magnesium Alloy, Wrought |
| ZK61A | M16610 | Magnesium Alloy, Cast |
| Zr-0.001%Hf (Grade 21) | | Zirconium Alloy, Wrought |
| Zr-2%Hf (Grade 11) | | Zirconium Alloy, Wrought |

## CHEMICAL COMPOSITION

The following tables contain approximate chemical compositions for materials included in this edition of the *Machining Data Handbook*. Chemistries are presented in the same material groupings as are specified in the machining recommendations, sections 1 through 8.

These tables are designed to assist in locating machining recommendations for material designations not listed in the handbook, but for which a chemical composition is known. Scanning the tables, the known composition can be matched with an identical or similar material designation and handbook group. Once this has been accomplished, machining recommendations can be selected for the material using the material group and designation identified.

Please note that chemical compositions were not provided for the following material groups:

22. Compacted Graphite Cast Irons
44. Uranium, Wrought
46. Manganese, Wrought
48. Machinable Carbides
49. Carbides
54. Carbons & Graphites
55. Glasses & Ceramics
56. Plastics
57. Composites
58. Flame (Thermal) Sprayed Materials
59. Plated Materials
60. Precious Metals
61. Rubber

The chemical elements and their symbols are listed below.

| | |
|---|---|
| Actinium | Ac |
| Aluminum | Al |
| Americium | Am |
| Antimony | Sb |
| Argon | A |
| Arsenic | As |
| Astatine | At |
| Barium | Ba |
| Berkelium | Bk |
| Beryllium | Be |
| Bismuth | Bi |
| Boron | B |
| Bromine | Br |
| Cadmium | Cd |
| Calcium | Ca |
| Californium | Cf |
| Carbon | C |
| Cerium | Ce |
| Cesium | Cs |
| Chlorine | Cl |
| Chromium | Cr |
| Cobalt | Co |
| Columbium | Cb |
| Copper | Cu |
| Curium | Cm |
| Dysprosium | Dy |
| Einsteinium | E |
| Erbium | Er |
| Europium | Eu |
| Fermium | Fm |
| Fluorine | F |
| Francium | Fr |
| Gadolinium | Gd |
| Gallium | Ga |
| Germanium | Ge |
| Gold | Au |
| Hafnium | Hf |
| Helium | He |
| Holmium | Ho |
| Hydrogen | H |
| Indium | In |
| Iodine | I |
| Iridium | Ir |
| Iron | Fe |
| Krypton | Kr |
| Lanthanum | La |
| Lead | Pb |
| Lithium | Li |
| Lutetium | Lu |
| Magnesium | Mg |
| Manganese | Mn |
| Mendelevium | Mv |
| Mercury | Hg |
| Molybdenum | Mo |
| Neodymium | Nd |
| Neon | Ne |
| Neptunium | Np |
| Nickel | Ni |
| Nitrogen | N |
| Nobelium | No |
| Osmium | Os |
| Oxygen | O |
| Palladium | Pd |
| Phosphorus | P |
| Platinum | Pt |
| Plutonium | Pu |
| Polonium | Po |
| Potassium | K |
| Praseodymium | Pr |
| Promethium | Pm |
| Protactinium | Pa |
| Radium | Ra |
| Radon | Rn |
| Rhenium | Re |
| Rhodium | Rh |
| Rubidium | Rb |
| Ruthenium | Ru |
| Samarium | Sm |
| Scandium | Sc |
| Selenium | Se |
| Silicon | Si |
| Silver | Ag |
| Sodium | Na |
| Strontium | Sr |
| Sulfur | S |
| Tantalum | Ta |
| Technetium | Tc |
| Tellurium | Te |
| Terbium | Tb |
| Thallium | Tl |
| Thorium | Th |
| Thulium | Tm |
| Tin | Sn |
| Titanium | Ti |
| Tungsten | W |
| Uranium | U |
| Vanadium | V |
| Xenon | Xe |
| Ytterbium | Yb |
| Yttrium | Y |
| Zinc | Zn |
| Zirconium | Zr |

# 26.3 Chemical Composition by Material Group

| Fe | C | Ni | Cr | Mo | V | Mn | Si | P | S | Pb | Co | Ti | Al | — | — | Other | MATERIAL DESIGNATION |
|---|---|---|---|---|---|---|---|---|---|---|---|---|---|---|---|---|---|
| COMPOSITION, % | | | | | | | | | | | | | | | | | |
| **1. FREE MACHINING CARBON STEELS, WROUGHT — Low Carbon Resulfurized** | | | | | | | | | | | | | | | | | |
| bal | .08 | — | — | — | — | .65 | — | .04 | .10 | — | — | — | — | | | | 1108 |
| bal | .09 | — | — | — | — | .75 | — | .04 | .10 | — | — | — | — | | | | 1109 |
| bal | .10 | — | — | — | — | .45 | — | .04 | .10 | — | — | — | — | | | | 1110 |
| bal | .11 | — | — | — | — | .75 | — | .10 | .12 | — | — | — | — | | | | 1211 |
| bal | .12 | — | — | — | — | .85 | — | .10 | .19 | — | — | — | — | | | | 1212 |
| bal | .13 | — | — | — | — | .85 | — | .10 | .28 | — | — | — | — | | | | 1213 |
| bal | .15 | — | — | — | — | .75 | — | .04 | .10 | — | — | — | — | | | | 1115 |
| bal | .15 | — | — | — | — | .90 | — | .06 | .31 | — | — | — | — | | | | 1215 |
| bal | .16 | — | — | — | — | 1.30 | — | .04 | .20 | — | — | — | — | | | | 1116 |
| bal | .17 | — | — | — | — | 1.20 | — | .04 | .10 | — | — | — | — | | | | 1117 |
| bal | .18 | — | — | — | — | 1.50 | — | .04 | .10 | — | — | — | — | | | | 1118 |
| bal | .19 | — | — | — | — | 1.20 | — | .04 | .30 | — | — | — | — | | | | 1119 |
| **1. FREE MACHINING CARBON STEELS, WROUGHT — Medium Carbon Resulfurized** | | | | | | | | | | | | | | | | | |
| bal | .32 | — | — | — | — | 1.50 | — | .04 | .10 | — | — | — | — | | | | 1132 |
| bal | .37 | — | — | — | — | 1.50 | — | .04 | .10 | — | — | — | — | | | | 1137 |
| bal | .39 | — | — | — | — | 1.50 | — | .04 | .20 | — | — | — | — | | | | 1139 |
| bal | .40 | — | — | — | — | .85 | — | .04 | .10 | — | — | — | — | | | | 1140 |
| bal | .41 | — | — | — | — | 1.50 | — | .04 | .10 | — | — | — | — | | | | 1141 |
| bal | .44 | — | — | — | — | 1.50 | — | .04 | .30 | — | — | — | — | | | | 1144 |
| bal | .45 | — | — | — | — | .85 | — | .04 | .05 | — | — | — | — | | | | 1145 |
| bal | .46 | — | — | — | — | .85 | — | .04 | .10 | — | — | — | — | | | | 1146 |
| bal | .51 | — | — | — | — | .85 | — | .04 | .10 | — | — | — | — | | | | 1151 |
| **1. FREE MACHINING CARBON STEELS, WROUGHT — Low Carbon Leaded** | | | | | | | | | | | | | | | | | |
| bal | .09 | — | — | — | — | .90 | — | .06 | .30 | .25 | — | — | — | | | | 12L15 |
| bal | .13 | — | — | — | — | .85 | — | .10 | .29 | .25 | — | — | — | | | | 12L13 |
| bal | .15 | — | — | — | — | 1.00 | — | .07 | .31 | .25 | — | — | — | | | | 12L14 |
| bal | .17 | — | — | — | — | 1.20 | — | .04 | .10 | .25 | — | — | — | | | | 11L17 |
| bal | .18 | — | — | — | — | .75 | — | .04 | .05 | .25 | — | — | — | | | | 10L18 |

| COMPOSITION, % | | | | | | | | | | | | | | | | | MATERIAL DESIGNATION |
|---|---|---|---|---|---|---|---|---|---|---|---|---|---|---|---|---|---|
| Fe | C | Ni | Cr | Mo | V | Mn | Si | P | S | Pb | Co | Ti | Al | — | — | Other | |
| **1. FREE MACHINING CARBON STEELS, WROUGHT — Medium Carbon Leaded** | | | | | | | | | | | | | | | | | |
| bal | .37 | — | — | — | — | 1.50 | — | .04 | .10 | .25 | — | — | — | | | | 11L37 |
| bal | .41 | — | — | — | — | 1.50 | — | .04 | .10 | .25 | — | — | — | | | | 11L41 |
| bal | .44 | — | — | — | — | 1.50 | — | .04 | .28 | .25 | — | — | — | | | | 11L44 |
| bal | .45 | — | — | — | — | .75 | — | .04 | .05 | .25 | — | — | — | | | | 10L45 |
| bal | .50 | — | — | — | — | .75 | — | .04 | .05 | .25 | — | — | — | | | | 10L50 |
| **2. CARBON STEELS, WROUGHT — Low Carbon** | | | | | | | | | | | | | | | | | |
| bal | .05 | — | — | — | — | .30 | — | .04 | .05 | — | — | — | — | | | | 1005 |
| bal | .06 | — | — | — | — | .35 | — | .04 | .05 | — | — | — | — | | | | 1006 |
| bal | .08 | — | — | — | — | .40 | — | .04 | .05 | — | — | — | — | | | | 1008 |
| bal | .09 | — | — | — | — | .60 | — | .04 | .05 | — | — | — | — | | | | 1009 |
| bal | .10 | — | — | — | — | .45 | — | .04 | .05 | — | — | — | — | | | | 1010 |
| bal | .11 | — | — | — | — | .75 | — | .04 | .05 | — | — | — | — | | | | 1011 |
| bal | .12 | — | — | — | — | .45 | — | .04 | .05 | — | — | — | — | | | | 1012 |
| bal | .13 | — | — | — | — | .65 | — | .04 | .05 | — | — | — | — | | | | 1013 |
| bal | .13 | — | — | — | — | 1.30 | — | .04 | .05 | — | — | — | — | | | | 1513 |
| bal | .15 | — | — | — | — | .45 | — | .04 | .05 | — | — | — | — | | | | 1015 |
| bal | .16 | — | — | — | — | .75 | — | .04 | .05 | — | — | — | — | | | | 1016 |
| bal | .17 | — | — | — | — | .45 | — | .04 | .05 | — | — | — | — | | | | 1017 |
| bal | .18 | — | — | — | — | .75 | — | .04 | .05 | — | — | — | — | | | | 1018 |
| bal | .18 | — | — | — | — | 1.30 | — | .04 | .05 | — | — | — | — | | | | 1518 |
| bal | .19 | — | — | — | — | .85 | — | .04 | .05 | — | — | — | — | | | | 1019 |
| bal | .20 | — | — | — | — | .45 | — | .04 | .05 | — | — | — | — | | | | 1020 |
| bal | .21 | — | — | — | — | .75 | — | .04 | .05 | — | — | — | — | | | | 1021 |
| bal | .22 | — | — | — | — | .85 | — | .04 | .05 | — | — | — | — | | | | 1022 |
| bal | .22 | — | — | — | — | 1.30 | — | .04 | .05 | — | — | — | — | | | | 1522 |
| bal | .23 | — | — | — | — | .45 | — | .04 | .05 | — | — | — | — | | | | 1023 |
| bal | .25 | — | — | — | — | .45 | — | .04 | .05 | — | — | — | — | | | | 1025 |
| bal | .26 | — | — | — | — | .75 | — | .04 | .05 | — | — | — | — | | | | 1026 |
| bal | .29 | — | — | — | — | .75 | — | .04 | .05 | — | — | — | — | | | | 1029 |

| COMPOSITION, % | | | | | | | | | | | | | | | | | MATERIAL DESIGNATION |
|---|---|---|---|---|---|---|---|---|---|---|---|---|---|---|---|---|---|
| Fe | C | Ni | Cr | Mo | V | Mn | Si | P | S | Pb | Co | Ti | Al | — | — | Other | |
| **2. CARBON STEELS, WROUGHT — Medium Carbon** | | | | | | | | | | | | | | | | | |
| bal | .24 | — | — | — | — | 1.50 | — | .04 | .05 | — | — | — | — | | | | 1524 |
| bal | .25 | — | — | — | — | .95 | — | .04 | .05 | — | — | — | — | | | | 1525 |
| bal | .26 | — | — | — | — | 1.30 | — | .04 | .05 | — | — | — | — | | | | 1526 |
| bal | .27 | — | — | — | — | 1.40 | — | .04 | .05 | — | — | — | — | | | | 1527 |
| bal | .30 | — | — | — | — | .75 | — | .04 | .05 | — | — | — | — | | | | 1030 |
| bal | .33 | — | — | — | — | .85 | — | .04 | .05 | — | — | — | — | | | | 1033 |
| bal | .35 | — | — | — | — | .75 | — | .04 | .05 | — | — | — | — | | | | 1035 |
| bal | .36 | — | — | — | — | 1.40 | — | .04 | .05 | — | — | — | — | | | | 1536 |
| bal | .37 | — | — | — | — | .85 | — | .04 | .05 | — | — | — | — | | | | 1037 |
| bal | .38 | — | — | — | — | .75 | — | .04 | .05 | — | — | — | — | | | | 1038 |
| bal | .39 | — | — | — | — | .85 | — | .04 | .05 | — | — | — | — | | | | 1039 |
| bal | .40 | — | — | — | — | .75 | — | .04 | .05 | — | — | — | — | | | | 1040 |
| bal | .41 | — | — | — | — | 1.50 | — | .04 | .05 | — | — | — | — | | | | 1541 |
| bal | .42 | — | — | — | — | .75 | — | .04 | .05 | — | — | — | — | | | | 1042 |
| bal | .43 | — | — | — | — | .85 | — | .04 | .05 | — | — | — | — | | | | 1043 |
| bal | .44 | — | — | — | — | .45 | — | .04 | .05 | — | — | — | — | | | | 1044 |
| bal | .45 | — | — | — | — | .75 | — | .04 | .05 | — | — | — | — | | | | 1045 |
| bal | .46 | — | — | — | — | .85 | — | .04 | .05 | — | — | — | — | | | | 1046 |
| bal | .47 | — | — | — | — | 1.50 | — | .04 | .05 | — | — | — | — | | | | 1547 |
| bal | .48 | — | — | — | — | 1.30 | — | .04 | .05 | — | — | — | — | | | | 1548 |
| bal | .49 | — | — | — | — | .75 | — | .04 | .05 | — | — | — | — | | | | 1049 |
| bal | .50 | — | — | — | — | .75 | — | .04 | .05 | — | — | — | — | | | | 1050 |
| bal | .51 | — | — | — | — | 1.00 | — | .04 | .05 | — | — | — | — | | | | 1551 |
| bal | .52 | — | — | — | — | 1.40 | — | .04 | .05 | — | — | — | — | | | | 1552 |
| bal | .53 | — | — | — | — | .85 | — | .04 | .05 | — | — | — | — | | | | 1053 |
| bal | .55 | — | — | — | — | .75 | — | .04 | .05 | — | — | — | — | | | | 1055 |

| COMPOSITION, % | | | | | | | | | | | | | | | | MATERIAL DESIGNATION |
|---|---|---|---|---|---|---|---|---|---|---|---|---|---|---|---|---|
| Fe | C | Ni | Cr | Mo | V | Mn | Si | P | S | Pb | Co | Ti | Al | — | — | Other | |
| **2. CARBON STEELS, WROUGHT — High Carbon** | | | | | | | | | | | | | | | | | |
| bal | .60 | — | — | — | — | .75 | — | .04 | .05 | — | — | — | — | | | | 1060 |
| bal | .61 | — | — | — | — | .90 | — | .04 | .05 | — | — | — | — | | | | 1561 |
| bal | .64 | — | — | — | — | .65 | — | .04 | .05 | — | — | — | — | | | | 1064 |
| bal | .65 | — | — | — | — | .75 | — | .04 | .05 | — | — | — | — | | | | 1065 |
| bal | .66 | — | — | — | — | 1.00 | — | .04 | .05 | — | — | — | — | | | | 1566 |
| bal | .69 | — | — | — | — | .55 | — | .04 | .05 | — | — | — | — | | | | 1069 |
| bal | .70 | — | — | — | — | .75 | — | .04 | .05 | — | — | — | — | | | | 1070 |
| bal | .72 | — | — | — | — | 1.20 | — | .04 | .05 | — | — | — | — | | | | 1572 |
| bal | .74 | — | — | — | — | .65 | — | .04 | .05 | — | — | — | — | | | | 1074 |
| bal | .75 | — | — | — | — | .55 | — | .04 | .05 | — | — | — | — | | | | 1075 |
| bal | .78 | — | — | — | — | .45 | — | .04 | .05 | — | — | — | — | | | | 1078 |
| bal | .80 | — | — | — | — | .75 | — | .04 | .05 | — | — | — | — | | | | 1080 |
| bal | .84 | — | — | — | — | .75 | — | .04 | .05 | — | — | — | — | | | | 1084 |
| bal | .85 | — | — | — | — | .85 | — | .04 | .05 | — | — | — | — | | | | 1085 |
| bal | .86 | — | — | — | — | .45 | — | .04 | .05 | — | — | — | — | | | | 1086 |
| bal | .90 | — | — | — | — | .75 | — | .04 | .05 | — | — | — | — | | | | 1090 |
| bal | .95 | — | — | — | — | .45 | — | .04 | .05 | — | — | — | — | | | | 1095 |
| **3. CARBON AND FERRITIC ALLOY STEELS (HIGH TEMPERATURE SERVICE)** | | | | | | | | | | | | | | | | | |
| bal | .20 | — | — | — | — | .60 | .20 | .04 | .05 | — | — | — | — | | | | ASTM A369 Grade FPA |
| bal | .25 | — | — | — | — | .65 | .20 | .04 | .05 | — | — | — | — | | | | ASTM A369 Grade FPB |
| bal | .15 | — | — | .55 | — | .55 | .30 | .04 | .04 | — | — | — | — | | | | ASTM A369 Grade FPI |
| bal | .15 | — | .65 | .55 | — | .45 | .20 | .04 | .04 | — | — | — | — | | | | ASTM A369 Grade PF2 |
| bal | .10 | — | 1.00 | .55 | — | .45 | .40 | .04 | .04 | — | — | — | — | | | | ASTM A369 Grade FP12 |
| bal | .10 | — | 1.25 | .55 | — | .45 | .75 | .03 | .03 | — | — | — | — | | | | ASTM A369 Grade FP11 |
| bal | .10 | — | 2.00 | .55 | — | .45 | .40 | .03 | .03 | — | — | — | — | | | | ASTM A369 Grade FP3b |

# 26.3 Chemical Composition by Material Group

| COMPOSITION, % | | | | | | | | | | | | | | | | | MATERIAL DESIGNATION |
|---|---|---|---|---|---|---|---|---|---|---|---|---|---|---|---|---|---|
| Fe | C | Ni | Cr | Mo | V | Mn | Si | P | S | Pb | Co | Ti | Al | — | — | Other | |
| **3. CARBON AND FERRITIC ALLOY STEELS (HIGH TEMPERATURE SERVICE) (continued)** | | | | | | | | | | | | | | | | | |
| bal | .10 | — | 2.30 | 1.00 | — | .45 | .40 | .03 | .03 | — | — | — | — | | | | ASTM A369 Grade FP22 |
| bal | .10 | — | 3.00 | 1.00 | — | .45 | .40 | .03 | .03 | — | — | — | — | | | | ASTM A369 Grade FP21 |
| bal | .10 | — | 5.00 | .55 | — | .45 | .40 | .03 | .03 | — | — | — | — | | | | ASTM A369 Grade FP5 |
| bal | .10 | — | 7.00 | .55 | — | .45 | .75 | .03 | .03 | — | — | — | — | | | | ASTM A369 Grade FP7 |
| bal | .10 | — | 9.00 | 1.00 | — | .45 | .75 | .03 | .03 | — | — | — | — | | | | ASTM A369 Grade FP9 |
| **4. FREE MACHINING ALLOY STEELS, WROUGHT — Medium Carbon Resulfurized** | | | | | | | | | | | | | | | | | |
| bal | .40 | — | 1.00 | .20 | — | .90 | .25 | .03 | .10 | — | — | — | — | | | | 4140 |
| bal | .40 | — | 1.00 | .20 | — | .90 | .25 | .03 | .10 | — | — | — | — | | | .04Se | 4140Se |
| bal | .42 | — | 1.00 | .20 | — | .90 | .25 | .03 | .10 | — | — | — | — | | | .04Te | 4142Te |
| bal | .45 | — | 1.00 | .20 | — | .90 | .25 | .03 | .10 | — | — | — | — | | | .04Se | 4145Se |
| bal | .47 | — | 1.00 | .20 | — | .90 | .25 | .03 | .10 | — | — | — | — | | | .04Te | 4147Te |
| bal | .50 | — | 1.00 | .20 | — | .90 | .25 | .03 | .10 | — | — | — | — | | | | 4150 |
| **4. FREE MACHINING ALLOY STEELS, WROUGHT — Medium and High Carbon Leaded** | | | | | | | | | | | | | | | | | |
| bal | .30 | — | .95 | .20 | — | .50 | .25 | .03 | .04 | .25 | — | — | — | | | | 41L30 |
| bal | .40 | — | 1.00 | .20 | — | .85 | .25 | .03 | .10 | .25 | — | — | — | | | | 41L40 |
| bal | .45 | — | 1.00 | .20 | — | .85 | .25 | .03 | .04 | .25 | — | — | — | | | | 41L45 |
| bal | .47 | — | 1.00 | .20 | — | .85 | .25 | .03 | .04 | .25 | — | — | — | | | | 41L47 |
| bal | .50 | — | 1.00 | .20 | — | .85 | .25 | .03 | .04 | .30 | — | — | — | | | | 41L50 |
| bal | .32 | — | .90 | — | — | .70 | .25 | .03 | .04 | .25 | — | — | — | | | | 51L32 |
| bal | 1.00 | — | 1.50 | — | — | .35 | .25 | .02 | .02 | .25 | — | — | — | | | | 52L100 |
| bal | .20 | .60 | .50 | .20 | — | .80 | .25 | .03 | .04 | .25 | — | — | — | | | | 86L20 |
| bal | .40 | .60 | .50 | .20 | — | .85 | .25 | .03 | .04 | .25 | — | — | — | | | | 86L40 |
| bal | .40 | 1.85 | .80 | .25 | — | .70 | .25 | .03 | .04 | .25 | — | — | — | | | | 43L40 |

| COMPOSITION, % | | | | | | | | | | | | | | | | | MATERIAL DESIGNATION |
|---|---|---|---|---|---|---|---|---|---|---|---|---|---|---|---|---|---|
| Fe | C | Ni | Cr | Mo | V | Mn | Si | P | S | Pb | Co | Ti | Al | — | — | Other | |
| **5. ALLOY STEELS, WROUGHT — Low Carbon** | | | | | | | | | | | | | | | | | |
| bal | .15 | — | .40 | — | — | .40 | .25 | .03 | .04 | — | — | — | — | | | | 5015 |
| bal | .15 | — | .80 | — | — | .80 | .25 | .03 | .04 | — | — | — | — | | | | 5115 |
| bal | .18 | — | .50 | .10 | — | .80 | .25 | .03 | .04 | — | — | — | — | | | | 4118 |
| bal | .18 | — | .60 | — | .10 | .60 | .25 | .03 | .04 | — | — | — | — | | | | 6118 |
| bal | .19 | — | .55 | — | — | .55 | .25 | .03 | .04 | — | — | — | — | | | | 4419 |
| bal | .20 | — | .80 | — | — | .80 | .25 | .03 | .04 | — | — | — | — | | | | 5120 |
| bal | .12 | — | — | .20 | — | .90 | .25 | .03 | .04 | — | — | — | — | | | | 4012 |
| bal | .22 | — | — | .40 | — | .80 | .25 | .03 | .04 | — | — | — | — | | | | 4422 |
| bal | .23 | — | — | .25 | — | .80 | .25 | .03 | .04 | — | — | — | — | | | | 4023 |
| bal | .24 | — | — | .25 | — | .80 | .25 | .03 | .04 | — | — | — | — | | | | 4024 |
| bal | .15 | 1.80 | — | .25 | — | .55 | .25 | .03 | .04 | — | — | — | — | | | | 4615 |
| bal | .15 | 3.50 | — | .25 | — | .50 | .25 | .03 | .04 | — | — | — | — | | | | 4815 |
| bal | .17 | 1.80 | — | .25 | — | .55 | .25 | .03 | .04 | — | — | — | — | | | | 4617 |
| bal | .17 | 3.50 | — | .25 | — | .50 | .25 | .03 | .04 | — | — | — | — | | | | 4817 |
| bal | .20 | 1.80 | — | .25 | — | .55 | .25 | .03 | .04 | — | — | — | — | | | | 4620 |
| bal | .20 | 3.50 | — | .25 | — | .60 | .25 | .03 | .04 | — | — | — | — | | | | 4820 |
| bal | .21 | 1.80 | — | .25 | — | .80 | .25 | .03 | .04 | — | — | — | — | | | | 4621 |
| bal | .10 | 3.30 | 1.20 | .10 | — | .55 | .25 | .02 | .02 | — | — | — | — | | | | 9310 |
| bal | .15 | .30 | .40 | .10 | — | .80 | .25 | .03 | .04 | — | — | — | — | | | | 8115 |
| bal | .15 | .50 | .50 | .10 | — | .90 | .25 | .03 | .04 | — | — | — | — | | | .0005B min. | 94B15 |
| bal | .17 | .50 | .50 | .10 | — | .90 | .25 | .03 | .04 | — | — | — | — | | | .0005B min. | 94B17 |
| bal | .17 | .60 | .50 | .20 | — | .80 | .25 | .03 | .04 | — | — | — | — | | | | 8617 |
| bal | .18 | 1.10 | .50 | .35 | — | .80 | — | — | — | — | — | — | — | | | | 4718 |
| bal | .20 | .60 | .50 | .20 | — | .80 | .25 | .03 | .04 | — | — | — | — | | | | 8620 |
| bal | .20 | 1.10 | .50 | .20 | — | .60 | .25 | .03 | .04 | — | — | — | — | | | | 4720 |
| bal | .20 | 1.80 | .50 | .25 | — | .55 | .25 | .03 | .04 | — | — | — | — | | | | 4320 |
| bal | .22 | .60 | .50 | .20 | — | .80 | .25 | .03 | .04 | — | — | — | — | | | | 8622 |
| bal | .22 | .60 | .50 | .35 | — | .90 | .25 | .03 | .04 | — | — | — | — | | | | 8822 |

| COMPOSITION, % | | | | | | | | | | | | | | | | | MATERIAL DESIGNATION |
|---|---|---|---|---|---|---|---|---|---|---|---|---|---|---|---|---|---|
| Fe | C | Ni | Cr | Mo | V | Mn | Si | P | S | Pb | Co | Ti | Al | — | — | Other | |
| 5. ALLOY STEELS, WROUGHT — Medium Carbon | | | | | | | | | | | | | | | | | |
| bal | .30 | — | — | — | — | 1.75 | .25 | .03 | .04 | — | — | — | — | | | | 1330 |
| bal | .35 | — | — | — | — | 1.80 | .25 | .03 | .04 | — | — | — | — | | | | 1335 |
| bal | .40 | — | — | — | — | 1.80 | .25 | .03 | .04 | — | — | — | — | | | | 1340 |
| bal | .45 | — | — | — | — | 1.80 | .25 | .03 | .04 | — | — | — | — | | | | 1345 |
| bal | .55 | — | — | — | — | .85 | 2.00 | .03 | .04 | — | — | — | — | | | | 9255 |
| bal | .60 | — | — | — | — | .90 | 2.00 | .03 | .04 | — | — | — | — | | | | 9260 |
| bal | .30 | — | 1.00 | — | — | .70 | .25 | .03 | .04 | — | — | — | — | | | | 5130 |
| bal | .32 | — | .90 | — | — | .70 | .25 | .03 | .04 | — | — | — | — | | | | 5132 |
| bal | .35 | — | 1.00 | — | — | .70 | .25 | .03 | .04 | — | — | — | — | | | | 5135 |
| bal | .40 | — | .50 | — | — | .90 | .25 | .03 | .04 | — | — | — | — | | | .0005B min. | 50B40 |
| bal | .40 | — | .80 | — | — | .80 | .25 | .03 | .04 | — | — | — | — | | | | 5140 |
| bal | .44 | — | .50 | — | — | .90 | .25 | .03 | .04 | — | — | — | — | | | .0005B min. | 50B44 |
| bal | .45 | — | .80 | — | — | .80 | .25 | .03 | .04 | — | — | — | — | | | | 5145 |
| bal | .46 | — | .30 | — | — | .90 | .25 | .03 | .04 | — | — | — | — | | | | 5046 |
| bal | .46 | — | .30 | — | — | .90 | .25 | .03 | .04 | — | — | — | — | | | .0005B min. | 50B46 |
| bal | .47 | — | 1.00 | — | — | .85 | .25 | .03 | .04 | — | — | — | — | | | | 5147 |
| bal | .50 | — | .50 | — | — | .90 | .25 | .03 | .04 | — | — | — | — | | | .0005B min. | 50B50 |
| bal | .50 | — | .80 | — | — | .80 | .25 | .03 | .04 | — | — | — | — | | | | 5150 |
| bal | .50 | — | 1.00 | — | .15 | .80 | .25 | .03 | .04 | — | — | — | — | | | | 6150 |
| bal | .54 | — | .70 | — | — | .70 | 1.40 | .03 | .04 | — | — | — | — | | | | 9254 |
| bal | .55 | — | .80 | — | — | .80 | .25 | .03 | .04 | — | — | — | — | | | | 5155 |
| bal | .60 | — | .50 | — | — | .90 | .25 | .03 | .04 | — | — | — | — | | | | 5060 |
| bal | .60 | — | .50 | — | — | .90 | .25 | .03 | .04 | — | — | — | — | | | .0005B min. | 50B60 |
| bal | .60 | — | .80 | — | — | .90 | .25 | .03 | .04 | — | — | — | — | | | | 5160 |
| bal | .60 | — | .80 | — | — | .90 | .25 | .03 | .04 | — | — | — | — | | | .0005B min. | 51B60 |
| bal | .27 | — | — | .25 | — | .80 | .25 | .03 | .04 | — | — | — | — | | | | 4027 |
| bal | .27 | — | — | .40 | — | .80 | .25 | .03 | .04 | — | — | — | — | | | | 4427 |
| bal | .28 | — | — | .25 | — | .80 | .25 | .03 | .04 | — | — | — | — | | | | 4028 |
| bal | .32 | — | — | .25 | — | .80 | .25 | .03 | .04 | — | — | — | — | | | | 4032 |

| COMPOSITION, % | | | | | | | | | | | | | | | | | MATERIAL DESIGNATION |
|---|---|---|---|---|---|---|---|---|---|---|---|---|---|---|---|---|---|
| Fe | C | Ni | Cr | Mo | V | Mn | Si | P | S | Pb | Co | Ti | Al | — | — | Other | |
| **5. ALLOY STEELS, WROUGHT — Medium Carbon (continued)** | | | | | | | | | | | | | | | | | |
| bal | .37 | — | — | .25 | — | .80 | .25 | .03 | .04 | — | — | — | — | | | | 4037 |
| bal | .42 | — | — | .25 | — | .80 | .25 | .03 | .04 | — | — | — | — | | | | 4042 |
| bal | .47 | — | — | .25 | — | .80 | .25 | .03 | .04 | — | — | — | — | | | | 4047 |
| bal | .26 | .90 | — | .20 | — | .55 | .25 | .03 | .04 | — | — | — | — | | | | 4626 |
| bal | .30 | — | .95 | .20 | — | .50 | .25 | .03 | .04 | — | — | — | — | | | | 4130 |
| bal | .35 | — | 1.00 | .20 | — | .80 | .25 | .03 | .04 | — | — | — | — | | | | 4135 |
| bal | .37 | — | 1.00 | .20 | — | .80 | .25 | .03 | .04 | — | — | — | — | | | | 4137 |
| bal | .40 | — | 1.00 | .20 | — | .90 | .25 | .03 | .04 | — | — | — | — | | | | 4140 |
| bal | .42 | — | 1.00 | .20 | — | .90 | .25 | .03 | .04 | — | — | — | — | | | | 4142 |
| bal | .45 | — | 1.00 | .20 | — | .90 | .25 | .03 | .04 | — | — | — | — | | | | 4145 |
| bal | .47 | — | 1.00 | .20 | — | .90 | .25 | .03 | .04 | — | — | — | — | | | | 4147 |
| bal | .50 | — | 1.00 | .20 | — | .90 | .25 | .03 | .04 | — | — | — | — | | | | 4150 |
| bal | .61 | — | .80 | .30 | — | .90 | .25 | .03 | .04 | — | — | — | — | | | | 4161 |
| bal | .25 | .60 | .50 | .20 | — | .80 | .25 | .03 | .04 | — | — | — | — | | | | 8625 |
| bal | .27 | .60 | .50 | .20 | — | .80 | .25 | .03 | .04 | — | — | — | — | | | | 8627 |
| bal | .30 | .50 | .40 | .12 | — | .90 | .25 | .03 | .04 | — | — | — | — | | | .0005B min. | 94B30 |
| bal | .30 | .60 | .50 | .20 | — | .80 | .25 | .03 | .04 | — | — | — | — | | | | 8630 |
| bal | .37 | .60 | .50 | .20 | — | .90 | .25 | .03 | .04 | — | — | — | — | | | | 8637 |
| bal | .40 | .60 | .50 | .20 | — | .90 | .25 | .03 | .04 | — | — | — | — | | | | 8640 |
| bal | .40 | .60 | .50 | .30 | — | .90 | .25 | .03 | .04 | — | — | — | — | | | | 8740 |
| bal | .40 | 1.80 | .80 | .20 | — | .70 | .25 | .03 | .04 | — | — | — | — | | | | 4340 |
| bal | .42 | .60 | .50 | .20 | — | .90 | .25 | .03 | .04 | — | — | — | — | | | | 8642 |
| bal | .42 | .60 | .50 | .25 | — | .90 | .25 | .03 | .04 | — | — | — | — | | | | 8742 |
| bal | .45 | .30 | .45 | .12 | — | 1.00 | .25 | .03 | .04 | — | — | — | — | | | .0005B min. | 81B45 |
| bal | .45 | .60 | .50 | .20 | — | .90 | .25 | .03 | .04 | — | — | — | — | | | | 8645 |
| bal | .45 | .60 | .50 | .20 | — | .90 | .25 | .03 | .04 | — | — | — | — | | | .0005B min. | 86B45 |
| bal | .50 | .60 | .50 | .20 | — | .90 | .25 | .03 | .04 | — | — | — | — | | | | 8650 |
| bal | .55 | .60 | .50 | .20 | — | .90 | .25 | .03 | .04 | — | — | — | — | | | | 8655 |
| bal | .60 | .60 | .50 | .20 | — | .90 | .25 | .03 | .04 | — | — | — | — | | | | 8660 |

# 26.3 Chemical Composition by Material Group

| COMPOSITION, % | | | | | | | | | | | | | | | | MATERIAL DESIGNATION |
|---|---|---|---|---|---|---|---|---|---|---|---|---|---|---|---|---|
| Fe | C | Ni | Cr | Mo | V | Mn | Si | P | S | Pb | Co | Ti | Al | — | — | Other | |
| **5. ALLOY STEELS, WROUGHT — High Carbon** | | | | | | | | | | | | | | | | | |
| bal | 1.00 | — | .50 | — | — | .35 | .25 | .02 | .02 | — | — | — | — | | | | 50100 |
| bal | 1.00 | — | 1.00 | — | — | .35 | .25 | .02 | .02 | — | — | — | — | | | | 51100 |
| bal | 1.00 | — | 1.50 | — | — | .35 | .25 | .02 | .02 | — | — | — | — | | | | 52100 |
| bal | .80 | — | 4.10 | 4.25 | 1.00 | .30 | .20 | .01 | .01 | — | — | — | — | | | | M-50 |
| **6. HIGH STRENGTH STEELS, WROUGHT** | | | | | | | | | | | | | | | | | |
| bal | .20 | 9.00 | .75 | 1.00 | .10 | .30 | .10 | — | — | — | 4.00 | — | — | | | | HP9-4-20 |
| bal | .25 | 9.00 | .45 | .45 | — | .20 | .10 | — | — | — | 4.00 | — | — | | | | HP9-4-25 |
| bal | .30 | 9.00 | 1.00 | 1.00 | .10 | .20 | .10 | — | — | — | 4.00 | — | — | | | | HP9-4-30 |
| bal | .32 | 1.80 | .90 | .43 | .08 | .90 | .30 | — | — | — | — | — | — | | | | 4330V |
| bal | .35 | — | 5.00 | 1.50 | .40 | .30 | 1.00 | — | — | — | — | — | — | | | | H11 |
| bal | .35 | — | 5.00 | 1.50 | 1.00 | .30 | 1.00 | — | — | — | — | — | — | | | | H13 |
| bal | .40 | 1.80 | .80 | .20 | — | .70 | .30 | — | — | — | — | — | — | | | | 4340 |
| bal | .40 | .75 | .90 | .60 | .02 | .90 | .60 | — | — | — | — | — | — | | | .0007B | 98BV40 |
| bal | .40 | 1.80 | .80 | .40 | .08 | — | 1.60 | — | — | — | — | — | — | | | | 4340Si |
| bal | .45 | 9.00 | .30 | .30 | — | .20 | .10 | — | — | — | 4.00 | — | — | | | | HP9-4-45 |
| bal | .45 | .55 | 1.05 | 1.00 | .07 | .75 | .25 | — | — | — | — | — | — | | | | D6ac |
| bal | .45 | 1.80 | .80 | .40 | .08 | .80 | 1.60 | — | — | — | — | — | — | | | | 300M |
| **7. MARAGING STEELS, WROUGHT** | | | | | | | | | | | | | | | | | |
| bal | .13 | 9.00 | — | — | — | .90 | .23 | — | — | — | — | — | — | | | | 120 Grade |
| bal | .02 | 12.00 | 4.80 | 3.00 | — | — | — | — | — | — | — | — | — | | | | 180 Grade |
| bal | .01 | 17.50 | — | 4.60 | — | .10 | .10 | — | — | — | 11.80 | 1.80 | .15 | | | | 350 Grade |
| bal | .03 | 18.00 | — | 4.80 | — | .10 | .10 | — | — | — | 7.50 | .40 | .10 | | | | 250 Grade |
| bal | .03 | 18.00 | — | 4.30 | — | .10 | .10 | .01 | .01 | — | 8.00 | .20 | .10 | | | | ASTM A538 Grade A |
| bal | .03 | 18.00 | — | 4.80 | — | .10 | .10 | .01 | .01 | — | 8.00 | .40 | .10 | | | | ASTM A538 Grade B |
| bal | .03 | 18.00 | — | 3.25 | — | .10 | .10 | — | — | — | 8.50 | .20 | .10 | | | | 200 Grade |
| bal | .03 | 18.50 | — | 4.80 | — | .10 | .10 | — | — | — | 9.00 | .65 | .10 | | | | 300 Grade |
| bal | .03 | 18.50 | — | 5.00 | — | .10 | .10 | .01 | .01 | — | 9.00 | .75 | .10 | | | | ASTM A538 Grade C |

| COMPOSITION, % | | | | | | | | | | | | | | | | | MATERIAL DESIGNATION |
|---|---|---|---|---|---|---|---|---|---|---|---|---|---|---|---|---|---|
| Fe | C | W | Mo | Cr | Co | V | Mn | Ni | Si | — | — | — | — | — | — | Other | |
| 8. TOOL STEELS, WROUGHT — High Speed | | | | | | | | | | | | | | | | | |
| bal | .85 | — | 8.00 | 4.00 | — | 2.00 | — | — | — | | | | | | | | M10 |
| bal | .85 | 1.50 | 8.50 | 4.00 | — | 1.00 | — | — | — | | | | | | | | M1 |
| bal | .85 | 6.00 | 5.00 | 4.00 | — | 2.00 | — | — | — | | | | | | | | M2 |
| bal | 1.00 | 1.75 | 8.75 | 4.00 | — | 2.00 | — | — | — | | | | | | | | M7 |
| bal | 1.05 | 6.00 | 5.00 | 4.00 | — | 2.40 | — | — | — | | | | | | | | M3 Class 1 |
| bal | 1.20 | 6.00 | 5.00 | 4.00 | — | 3.00 | — | — | — | | | | | | | | M3 Class 2 |
| bal | 1.30 | 5.50 | 4.50 | 4.00 | — | 4.00 | — | — | — | | | | | | | | M4 |
| bal | .75 | 18.00 | — | 4.00 | — | 1.00 | — | — | — | | | | | | | | T1 |
| bal | .80 | 18.00 | — | 4.00 | — | 2.00 | — | — | — | | | | | | | | T2 |
| bal | .75 | 14.00 | — | 4.00 | 5.00 | 2.00 | — | — | — | | | | | | | | T8 |
| bal | .75 | 18.00 | — | 4.00 | 5.00 | 1.00 | — | — | — | | | | | | | | T4 |
| bal | .80 | 18.00 | — | 4.00 | 8.00 | 2.00 | — | — | — | | | | | | | | T5 |
| bal | .80 | 20.00 | — | 4.50 | 12.00 | 1.50 | — | — | — | | | | | | | | T6 |
| bal | 1.50 | 12.00 | — | 4.00 | 5.00 | 5.00 | — | — | — | | | | | | | | T15 |
| bal | .80 | 2.00 | 8.00 | 4.00 | 5.00 | 1.25 | — | — | — | | | | | | | | M30 |
| bal | .80 | 4.00 | 5.00 | 4.00 | 12.00 | 1.50 | — | — | — | | | | | | | | M6 |
| bal | .80 | 6.00 | 5.00 | 4.00 | 8.00 | 2.00 | — | — | — | | | | | | | | M36 |
| bal | .90 | 1.50 | 9.50 | 4.00 | 8.00 | 1.15 | — | — | — | | | | | | | | M33 |
| bal | .90 | 2.00 | 8.00 | 4.00 | 8.00 | 2.00 | — | — | — | | | | | | | | M34 |
| bal | 1.10 | 1.50 | 9.50 | 3.75 | 8.00 | 1.15 | — | — | — | | | | | | | | M42 |
| bal | 1.10 | 1.50 | 9.50 | 3.75 | 5.00 | 1.25 | — | — | — | | | | | | | | M47 |
| bal | 1.10 | 6.75 | 3.75 | 4.25 | 5.00 | 2.00 | — | — | — | | | | | | | | M41 |
| bal | 1.15 | 5.25 | 6.25 | 4.25 | 12.00 | 2.00 | — | — | — | | | | | | | | M44 |
| bal | 1.20 | 2.75 | 8.00 | 3.75 | 8.25 | 1.60 | — | — | — | | | | | | | | M43 |
| bal | 1.25 | 2.00 | 8.25 | 4.00 | 8.25 | 3.20 | — | — | — | | | | | | | | M46 |

# 26.3 Chemical Composition by Material Group

| COMPOSITION, % | | | | | | | | | | | | | | | | | MATERIAL DESIGNATION |
|---|---|---|---|---|---|---|---|---|---|---|---|---|---|---|---|---|---|
| Fe | C | W | Mo | Cr | Co | V | Mn | Ni | Si | — | — | — | — | — | — | Other | |
| **8. TOOL STEELS, WROUGHT — Hot Work** | | | | | | | | | | | | | | | | | |
| bal | .35 | — | 1.50 | 5.00 | — | .40 | — | — | — | | | | | | | | H11 |
| bal | .35 | — | 1.50 | 5.00 | — | 1.00 | — | — | — | | | | | | | | H13 |
| bal | .40 | — | 2.50 | 3.25 | — | .40 | — | — | — | | | | | | | | H10 |
| bal | .25 | 15.00 | — | 4.00 | — | — | — | — | — | | | | | | | | H25 |
| bal | .30 | 12.00 | — | 12.00 | — | — | — | — | — | | | | | | | | H23 |
| bal | .35 | 11.00 | — | 2.00 | — | — | — | — | — | | | | | | | | H22 |
| bal | .35 | 9.00 | — | 3.50 | — | — | — | — | — | | | | | | | | H21 |
| bal | .40 | 5.00 | — | 5.00 | — | — | — | — | — | | | | | | | | H14 |
| bal | .45 | 15.00 | — | 3.00 | — | — | — | — | — | | | | | | | | H24 |
| bal | .50 | 18.00 | — | 4.00 | — | 1.00 | — | — | — | | | | | | | | H26 |
| bal | .40 | 4.25 | — | 4.25 | 4.25 | 2.00 | — | — | — | | | | | | | | H19 |
| bal | .35 | 1.50 | 1.50 | 5.00 | — | .40 | — | — | — | | | | | | | | H12 |
| bal | .60 | 6.00 | 5.00 | 4.00 | — | 2.00 | — | — | — | | | | | | | | H42 |
| **8. TOOL STEELS, WROUGHT — Cold Work** | | | | | | | | | | | | | | | | | |
| bal | .50 | — | 1.40 | 5.00 | — | 1.00 | — | 1.50 | — | | | | | | | | A9 |
| bal | .55 | 1.25 | 1.25 | 5.00 | — | — | — | — | — | | | | | | | | A8 |
| bal | .70 | — | 1.25 | 1.00 | — | — | 2.00 | — | — | | | | | | | | A6 |
| bal | .90 | .50 | — | .50 | — | — | 1.00 | — | — | | | | | | | | O1 |
| bal | .90 | — | — | — | — | — | 1.60 | — | — | | | | | | | | O2 |
| bal | 1.00 | — | 1.00 | 1.00 | — | — | 2.00 | — | — | | | | | | | | A4 |
| bal | 1.00 | — | 1.00 | 5.00 | — | — | — | — | — | | | | | | | | A2 |
| bal | 1.20 | 1.75 | — | .75 | — | — | — | — | — | | | | | | | | O7 |
| bal | 1.25 | — | 1.00 | 5.00 | — | 1.00 | — | — | — | | | | | | | | A3 |
| bal | 1.35 | — | 1.50 | — | — | — | 1.80 | 1.80 | 1.25 | | | | | | | | A10 |
| bal | 1.45 | — | .25 | — | — | — | .80 | — | 1.00 | | | | | | | | O6 |
| bal | 1.50 | — | 1.00 | 12.00 | — | 1.00 | — | — | — | | | | | | | | D2 |
| bal | 1.50 | — | 1.00 | 12.00 | 3.00 | — | — | — | — | | | | | | | | D5 |
| bal | 2.25 | — | — | 12.00 | — | — | — | — | — | | | | | | | | D3 |

| COMPOSITION, % | | | | | | | | | | | | | | | | | MATERIAL DESIGNATION |
|---|---|---|---|---|---|---|---|---|---|---|---|---|---|---|---|---|---|
| Fe | C | W | Mo | Cr | Co | V | Mn | Ni | Si | — | — | — | — | — | — | Other | |
| **8. TOOL STEELS, WROUGHT — Cold Work (continued)** | | | | | | | | | | | | | | | | | |
| bal | 2.25 | — | 1.00 | 12.00 | — | — | — | — | — | | | | | | | | D4 |
| bal | 2.25 | 1.00 | 1.00 | 5.25 | — | 4.75 | — | — | — | | | | | | | | A4 |
| bal | 2.35 | — | 1.00 | 12.00 | — | 4.00 | — | — | — | | | | | | | | D7 |
| **8. TOOL STEELS, WROUGHT — Shock Resisting** | | | | | | | | | | | | | | | | | |
| bal | .45 | — | .40 | 1.50 | — | — | 1.40 | — | 2.25 | | | | | | | | S6 |
| bal | .50 | — | 1.40 | 3.25 | — | — | — | — | — | | | | | | | | S7 |
| bal | .50 | 2.50 | — | 1.50 | — | — | — | — | — | | | | | | | | S1 |
| bal | .50 | — | .50 | — | — | — | — | — | 1.00 | | | | | | | | S2 |
| bal | .55 | — | .40 | — | — | — | .80 | — | 2.00 | | | | | | | | S5 |
| **8. TOOL STEELS, WROUGHT — Mold** | | | | | | | | | | | | | | | | | |
| bal | .07 | — | .20 | 2.00 | — | — | — | .50 | — | | | | | | | | P2 |
| bal | .07 | — | .75 | 5.00 | — | — | — | — | — | | | | | | | | P4 |
| bal | .10 | — | — | 2.25 | — | — | — | — | — | | | | | | | | P5 |
| bal | .10 | — | — | 1.50 | — | — | — | 3.50 | — | | | | | | | | P6 |
| bal | .20 | — | — | — | — | — | — | 4.00 | — | | | | | | | 1.20Al | P21 |
| bal | .35 | — | .40 | 1.70 | — | — | — | — | — | | | | | | | | P20 |
| **8. TOOL STEELS, WROUGHT — Special Purpose** | | | | | | | | | | | | | | | | | |
| bal | .70 | — | — | .70 | — | — | — | 1.50 | — | | | | | | | | L6 |
| bal | .80 | — | — | 1.00 | — | .20 | — | — | — | | | | | | | | L2 |
| bal | 1.00 | — | .40 | 1.50 | — | — | .35 | — | .30 | | | | | | | | L7 |
| bal | 1.00 | 1.25 | — | — | — | — | — | — | — | | | | | | | | F1 |
| bal | 1.25 | 3.50 | — | — | — | — | — | — | — | | | | | | | | F2 |
| **8. TOOL STEELS, WROUGHT — Water Hardening** | | | | | | | | | | | | | | | | | |
| bal | .60 | — | — | — | — | — | — | — | — | | | | | | | | W1 |
| bal | .60 | — | — | — | — | .25 | — | — | — | | | | | | | | W2 |
| bal | .80 | — | — | .15 | — | — | .35 | — | .35 | | | | | | | | SAE J438b Type W108 |
| bal | .90 | — | — | .15 | — | — | .35 | — | .35 | | | | | | | | SAE J438b Type W109 |

| COMPOSITION, % | | | | | | | | | | | | | | | | | MATERIAL DESIGNATION |
|---|---|---|---|---|---|---|---|---|---|---|---|---|---|---|---|---|---|
| Fe | C | W | Mo | Cr | Co | V | Mn | Ni | Si | — | — | — | — | — | — | Other | |
| **8. TOOL STEELS, WROUGHT — Water Hardening (continued)** | | | | | | | | | | | | | | | | | |
| bal | .90 | — | — | .15 | — | .25 | .35 | — | .35 | | | | | | | | SAE J438b Type W209 |
| bal | 1.05 | — | — | .15 | — | — | .35 | — | .35 | | | | | | | | SAE J438b Type W110 |
| bal | 1.05 | — | — | .15 | — | .25 | .35 | — | .35 | | | | | | | | SAE J438b Type W210 |
| bal | 1.05 | — | — | .15 | — | .45 | .35 | — | .35 | | | | | | | | SAE J438b Type W310 |
| bal | 1.10 | — | — | .50 | — | — | — | — | — | | | | | | | | W5 |
| bal | 1.20 | — | — | .15 | — | — | .35 | — | .35 | | | | | | | | SAE J438b Type W112 |

| COMPOSITION, % | | | | | | | | | | | | | | | | | MATERIAL DESIGNATION |
|---|---|---|---|---|---|---|---|---|---|---|---|---|---|---|---|---|---|
| Fe | C | Ni | Cr | Mo | V | Mn | Si | P | S | Al | Cu | — | — | — | — | Other | |
| **9. NITRIDING STEELS, WROUGHT** | | | | | | | | | | | | | | | | | |
| bal | .23 | 3.50 | 1.25 | .25 | — | .55 | .30 | — | — | 1.00 | — | | | | | | Nitralloy N |
| bal | .25 | — | — | .80 | — | .55 | .30 | — | — | 1.00 | — | | | | | | Nitralloy 225 |
| bal | .25 | — | 1.15 | .20 | — | .55 | .30 | — | — | .90 | — | | | | | | Nitralloy 125 |
| bal | .30 | — | — | .80 | — | .50 | .30 | — | — | 1.25 | — | | | | | | Nitralloy 230 |
| bal | .35 | — | 1.15 | .20 | — | .55 | .30 | — | — | .90 | — | | | | | | Nitralloy 135 |
| bal | .35 | — | 1.25 | .20 | — | .80 | .30 | — | — | 1.00 | — | | | | | .20Se | Nitralloy EZ |
| bal | .40 | — | 1.60 | .38 | — | .55 | .30 | — | — | 1.05 | — | | | | | | Nitrex 1 |
| bal | .42 | — | 1.60 | .37 | — | .55 | .30 | — | — | 1.00 | — | | | | | | Nitralloy 135 Mod. |
| **10. ARMOR PLATE, SHIP PLATE, AIRCRAFT PLATE, WROUGHT** | | | | | | | | | | | | | | | | | |
| bal | .10 | .50 | .30 | .07 | .10 | .30 | .20 | .02 | .02 | — | — | | | | | | MIL-S-12560 |
| bal | .11 | 10.00 | 2.00 | 1.00 | — | — | — | — | — | — | — | | | | | 10.00Co | HY180 |
| bal | .15 | 3.00 | 1.40 | .40 | .03 | .25 | .25 | .02 | .02 | — | .25 | | | | | .02Ti | MIL-S-16216 (HY80) |
| bal | .18 | 3.00 | 1.40 | .40 | — | .25 | .25 | .02 | .02 | — | — | | | | | | HY80 |
| bal | .16 | 3.25 | 1.40 | .40 | .03 | .25 | .25 | .02 | .02 | — | .25 | | | | | .02Ti | MIL-S-16216 (HY100) |
| bal | .20 | 3.25 | 1.40 | .40 | — | .25 | .25 | .02 | .02 | — | — | | | | | | HY100 |

| COMPOSITION, % | | | | | | | | | | | | | | | | | MATERIAL DESIGNATION |
|---|---|---|---|---|---|---|---|---|---|---|---|---|---|---|---|---|---|
| Fe | C | Ni | Cr | Mo | V | Mn | Si | P | S | Al | Cu | — | — | — | — | Other | |
| **11. STRUCTURAL STEELS, WROUGHT*** | | | | | | | | | | | | | | | | | |
| bal | .06 | 1.00 | — | — | — | .35 | — | — | — | — | 1.50 | | | | | V | 85 |
| bal | .12 | — | — | — | — | .75 | — | — | — | — | — | | | | | Cb-V | 35 |
| bal | .15 | — | .60 | .30 | — | 1.20 | — | — | — | — | — | | | | | B-Zr | 90 |
| bal | .15 | — | — | .30 | — | 1.30 | — | — | — | — | — | | | | | B | 150 |
| bal | .16 | .80 | .50 | .30 | — | .80 | — | — | — | — | .20 | | | | | V-B | 145 |
| bal | .16 | 3.00 | .50 | .50 | — | .80 | — | — | — | — | .20 | | | | | V-B | 140 |
| bal | .18 | 1.30 | 1.00 | .30 | — | .60 | — | — | — | — | — | | | | | V-B-Zr | 100 |
| bal | .18 | 1.50 | .70 | .30 | — | 1.00 | — | — | — | — | — | | | | | V | 80 |
| bal | .18 | — | .70 | .30 | — | 1.00 | — | — | — | — | — | | | | | B-Zr | 110 |
| bal | .20 | — | .70 | .20 | — | .80 | — | — | — | — | — | | | | | Zr | 165 |
| bal | .20 | — | — | — | — | .90 | — | — | — | — | — | | | | | | 30 |
| bal | .21 | — | — | — | — | 1.25 | — | — | — | — | — | | | | | Cb-V | 42 |
| bal | .22 | — | — | — | — | 1.25 | — | — | — | — | — | | | | | Cb-V | 45 |
| bal | .23 | — | .50 | .20 | — | 1.20 | — | — | — | — | .20 | | | | | V-B | 135 |
| bal | .23 | .60 | .60 | — | — | 1.25 | — | — | — | — | .30 | | | | | | 50 |
| bal | .25 | — | — | — | — | 1.35 | — | — | — | — | — | | | | | Cb-V | 55 |
| bal | .26 | — | — | — | — | 1.35 | — | — | — | — | — | | | | | Cb-V | 60 |
| bal | .26 | — | — | — | — | 1.35 | — | — | — | — | — | | | | | Cb-V | 65 |
| bal | .26 | — | — | — | — | 1.40 | — | — | — | — | — | | | | | Cb-V | 70 |
| bal | .26 | .31 | — | .50 | — | 1.00 | — | — | — | — | — | | | | | V | 75 |
| bal | .28 | — | — | .20 | — | 1.50 | — | — | — | — | .20 | | | | | B | 160 |
| bal | .28 | — | — | .20 | — | 1.50 | — | — | — | — | .20 | | | | | B | 185 |
| bal | .28 | — | — | .20 | — | 1.50 | — | — | — | — | .20 | | | | | B | 210 |

*In this handbook, Structural Steels are designated by yield strength in units of 1000 psi. For example, 50 means a steel with 50,000 psi yield strength. Chemical compositions are therefore approximate so as to apply to a variety of trade name steels.

# 26.3 Chemical Composition by Material Group

| COMPOSITION, % | | | | | | | | | | | | | | | | | MATERIAL DESIGNATION |
|---|---|---|---|---|---|---|---|---|---|---|---|---|---|---|---|---|---|
| Fe | C | Cr | Ni | Mo | Mn | Si | P | S | Al | Cu | — | — | — | — | — | Other | |
| **12. FREE MACHINING STAINLESS STEELS, WROUGHT — Ferritic** | | | | | | | | | | | | | | | | | |
| bal | .12 | 17.00 | — | .60* | 1.25 | 1.00 | .06 | .15 | — | — | | | | | | | 430F |
| bal | .12 | 17.00 | — | — | 1.25 | 1.00 | .06 | .06 | — | — | | | | | | .15Se min. | 430F Se |
| **12. FREE MACHINING STAINLESS STEELS, WROUGHT — Austenitic** | | | | | | | | | | | | | | | | | |
| bal | .08 | 17.00 | 6.00 | .50 | 6.50 | 1.00 | .04 | .35 | — | 2.00 | | | | | | | 203EZ |
| bal | .15 | 18.00 | 9.00 | .60* | 2.00 | 1.00 | .20 | .15 | — | — | | | | | | | 303 |
| bal | .15 | 18.00 | 9.00 | — | 2.00 | 1.00 | .20 | .06 | — | — | | | | | | .15Se min. | 303Se |
| bal | .15 | 18.00 | 9.00 | .50 | 2.00 | 1.00 | .05 | .16 | .80 | — | | | | | | | 303MA |
| bal | .15 | 18.00 | 9.00 | .60 | 2.00 | 1.00 | .04 | .25 | — | — | | | | | | .25Pb | 303Pb |
| bal | .15 | 13.00 | 8.50 | .60 | 4.50 | 1.00 | .20 | .15 | — | — | | | | | | | 303 Plus X |
| **12. FREE MACHINING STAINLESS STEELS, WROUGHT — Martensitic** | | | | | | | | | | | | | | | | | |
| bal | .15 | 13.00 | — | .60* | 1.25 | 1.00 | .06 | .15 | — | — | | | | | | | 416 |
| bal | .15 | 13.00 | — | — | 1.25 | 1.00 | .06 | .06 | — | — | | | | | | .15Se min. | 416Se |
| bal | .15 | 13.00 | — | .60 | 2.50 | 1.00 | .06 | .15 | — | — | | | | | | | 416 Plus X |
| bal | .15 | 13.00 | — | .60* | 1.25 | 1.00 | .06 | .15 | — | — | | | | | | | 420F |
| bal | .40 | 13.00 | — | — | 1.25 | 1.00 | .06 | .06 | — | — | | | | | | .15Se min. | 420F Se |
| bal | 1.20 | 17.00 | — | .75* | 1.25 | 1.00 | .06 | .15 | — | — | | | | | | | 440F |
| bal | 1.20 | 17.00 | — | — | 1.25 | 1.00 | .06 | .06 | — | — | | | | | | .15Se min. | 440F Se |
| **13. STAINLESS STEELS, WROUGHT — Ferritic** | | | | | | | | | | | | | | | | | |
| bal | .08 | 11.00 | .50 | — | 1.00 | 1.00 | .04 | .04 | — | — | | | | | | .75Ti | 409 |
| bal | .08 | 13.00 | — | — | 1.00 | 1.00 | .04 | .03 | .20 | — | | | | | | | 405 |
| bal | .12 | 15.00 | — | — | 1.00 | 1.00 | .04 | .03 | — | — | | | | | | | 429 |
| bal | .12 | 17.00 | — | — | 1.00 | 1.00 | .04 | .03 | — | — | | | | | | | 430 |
| bal | .12 | 17.00 | — | 1.00 | 1.00 | 1.00 | .04 | .03 | — | — | | | | | | | 434 |
| bal | .12 | 17.00 | — | 1.00 | 1.00 | 1.00 | .04 | .03 | — | — | | | | | | .70Cb + Ta | 436 |
| bal | .20 | 20.00 | — | — | 1.00 | 1.00 | .04 | .03 | — | — | | | | | | | 442 |
| bal | .20 | 25.00 | — | — | 1.50 | 1.00 | .04 | .03 | — | — | | | | | | .25N | 446 |

*Optional.

| COMPOSITION, % | | | | | | | | | | | | | | | | | MATERIAL DESIGNATION |
|---|---|---|---|---|---|---|---|---|---|---|---|---|---|---|---|---|---|
| Fe | C | Cr | Ni | Mo | Mn | Si | P | S | Al | Cu | — | — | — | — | — | Other | |
| 13. STAINLESS STEELS, WROUGHT — Austenitic | | | | | | | | | | | | | | | | | |
| bal | .03 | 17.00 | 12.00 | 2.50 | 2.00 | 1.00 | .04 | .03 | — | — | | | | | | | 316L |
| bal | .03 | 19.00 | 10.00 | — | 2.00 | 1.00 | .04 | .03 | — | — | | | | | | | 304L |
| bal | .06 | 22.00 | 12.50 | 2.50 | 6.00 | 1.00 | .04 | .03 | — | — | | | | | | .30N, .20Cb, .20V | Nitronic 50 |
| bal | .08 | 12.50 | 15.00 | — | 2.00 | 1.00 | .04 | .03 | — | — | | | | | | | 385 |
| bal | .08 | 16.00 | 18.00 | — | 2.00 | 1.00 | .04 | .03 | — | — | | | | | | | 384 |
| bal | .08 | 17.00 | 12.00 | 2.50 | 2.00 | 1.00 | .04 | .03 | — | — | | | | | | | 316 |
| bal | .08 | 18.00 | 10.50 | — | 2.00 | 1.00 | .04 | .03 | — | — | | | | | | Ti 5xC min. | 321 |
| bal | .08 | 18.00 | 11.00 | — | 2.00 | 1.00 | .04 | .03 | — | — | | | | | | Cb + Ta 10xC min. | 347 |
| bal | .08 | 18.00 | 11.00 | — | 2.00 | 1.00 | .04 | .03 | — | — | | | | | | Cb + Ta 10xC min., .10Ta, .20Co | 348 |
| bal | .08 | 18.00 | 3.00 | — | 14.50 | 1.00 | .06 | .03 | — | — | | | | | | .30N | Nitronic 33 |
| bal | .08 | 19.00 | 13.00 | 3.50 | 2.00 | 1.00 | .04 | .03 | — | — | | | | | | | 317 |
| bal | .08 | 19.00 | 9.00 | — | 2.00 | 1.00 | .04 | .03 | — | — | | | | | | | 304 |
| bal | .08 | 20.00 | 11.00 | — | 2.00 | 1.00 | .04 | .03 | — | — | | | | | | | 308 |
| bal | .08 | 20.00 | 6.50 | — | 10.00 | 1.00 | .06 | .03 | — | — | | | | | | .30N | Nitronic 40 |
| bal | .08 | 23.00 | 13.50 | — | 2.00 | 1.00 | .04 | .03 | — | — | | | | | | | 309S |
| bal | .08 | 25.00 | 21.00 | — | 2.00 | 1.50 | .04 | .03 | — | — | | | | | | | 310S |
| bal | .10 | 17.00 | 8.50 | — | 9.00 | 4.50 | — | — | — | — | | | | | | .10N | Nitronic 60 |
| bal | .12 | 18.00 | 12.00 | — | 2.00 | 1.00 | .04 | .03 | — | — | | | | | | | 305 |
| bal | .15 | 15.50 | 35.00 | — | 2.00 | 1.50 | .04 | .04 | — | — | | | | | | | 330 |
| bal | .15 | 18.00 | 1.50 | — | 14.00 | 1.00 | .06 | .03 | — | — | | | | | | .35N | Nitronic 32 |
| bal | .15 | 17.00 | 4.50 | — | 7.50 | 1.00 | .06 | .03 | — | — | | | | | | | 201 |
| bal | .15 | 17.00 | 7.00 | — | 2.00 | 1.00 | .04 | .03 | — | — | | | | | | | 301 |
| bal | .15 | 18.00 | 5.00 | — | 10.00 | 1.00 | .06 | .03 | — | — | | | | | | .25N | 202 |
| bal | .15 | 18.00 | 9.00 | — | 2.00 | 1.00 | .04 | .03 | — | — | | | | | | | 302 |
| bal | .15 | 18.00 | 9.00 | — | 2.00 | 3.00 | .04 | .03 | — | — | | | | | | | 302B |
| bal | .20 | 23.00 | 13.50 | — | 2.00 | 1.00 | .04 | .03 | — | — | | | | | | | 309 |
| bal | .25 | 25.00 | 21.00 | — | 2.00 | 3.00 | .04 | .03 | — | — | | | | | | | 314 |
| bal | .25 | 25.00 | 21.00 | — | 2.00 | 1.50 | .04 | .03 | — | — | | | | | | | 310 |

# 26.3 Chemical Composition by Material Group

| COMPOSITION, % | | | | | | | | | | | | | | | | | MATERIAL DESIGNATION |
|---|---|---|---|---|---|---|---|---|---|---|---|---|---|---|---|---|---|
| Fe | C | Cr | Ni | Mo | Mn | Si | P | S | Al | Cu | — | — | — | — | — | Other | |
| **13. STAINLESS STEELS, WROUGHT — Martensitic** | | | | | | | | | | | | | | | | | |
| bal | .10 | 5.00 | — | .55 | 1.00 | 1.00 | .04 | .03 | — | — | | | | | | | 502 |
| bal | .10 | 5.00 | — | .55 | 1.00 | 1.00 | .04 | .03 | — | — | | | | | | | 501 |
| bal | .15 | 12.25 | — | — | 1.00 | .50 | .04 | .03 | — | — | | | | | | | 403 |
| bal | .15 | 12.50 | — | — | 1.00 | 1.00 | .04 | .03 | — | — | | | | | | | 410 |
| bal | .15 | 12.50 | 2.00 | — | 1.00 | 1.00 | .04 | .03 | — | — | | | | | | | 414 |
| bal | .15 | 13.00 | — | — | 1.00 | 1.00 | .04 | .03 | — | — | | | | | | | 420 |
| bal | .20 | 13.00 | 2.00 | .50 | .50 | .50 | — | — | — | — | | | | | | 3.00W | Greek Ascoloy |
| bal | .20 | 16.00 | 2.00 | — | 1.00 | 1.00 | .04 | .03 | — | — | | | | | | | 431 |
| bal | .25 | 13.00 | .75 | 1.00 | 1.00 | 1.00 | .04 | .03 | — | — | | | | | | 1.00W, .35V | 422 |
| bal | .75 | 17.00 | — | .75 | 1.00 | 1.00 | .04 | .03 | — | — | | | | | | | 440A |
| bal | .95 | 17.00 | — | .75 | 1.00 | 1.00 | .04 | .03 | — | — | | | | | | | 440B |
| bal | 1.20 | 17.00 | — | .75 | 1.00 | 1.00 | .04 | .03 | — | — | | | | | | | 440C |
| **14. PRECIPITATION HARDENING STAINLESS STEELS, WROUGHT** | | | | | | | | | | | | | | | | | |
| bal | .03 | 12.00 | 8.50 | — | .25 | .25 | — | — | — | 2.00 | | | | | | 1.30Ti, .30Co | Custom 455 |
| bal | .05 | 11.50 | 4.50 | — | .20 | .15 | — | — | — | — | | | | | | .30Ti | AM-363 |
| bal | .05 | 12.75 | 8.00 | 2.25 | .10 | .10 | .01 | .008 | 1.10 | — | | | | | | | PH 13-8Mo |
| bal | .05 | 14.25 | 6.50 | — | .05 | .30 | — | — | — | — | | | | | | .75Ti | Almar 362 (AM-362) |
| bal | .05 | 14.50 | 8.00 | 2.50 | 1.00 | 1.00 | .01 | .01 | 1.10 | — | | | | | | | PH 14-8Mo |
| bal | .05 | 15.00 | 4.50 | — | 1.00 | 1.00 | .04 | .03 | — | 3.50 | | | | | | | 15-5 PH |
| bal | .05 | 15.50 | 6.50 | .75 | .50 | .50 | .03 | .03 | — | 1.50 | | | | | | Cb 8xC min. | Custom 450 |
| bal | .05 | 16.50 | 4.50 | — | 1.00 | 1.00 | .04 | .03 | — | 4.00 | | | | | | | 17-4 PH |
| bal | .07 | 17.00 | 7.00 | — | 1.00 | 1.00 | .04 | .03 | .20 | — | | | | | | .70Ti | Stainless W |
| bal | .08 | 16.00 | 7.00 | — | 1.00 | 1.00 | — | — | 1.00 | — | | | | | | .70Ti | 16-6 PH |
| bal | .09 | 15.00 | 7.00 | 2.50 | 1.00 | 1.00 | .04 | .03 | 1.10 | — | | | | | | | PH 15-7Mo |
| bal | .09 | 17.00 | 7.00 | — | 1.00 | 1.00 | .04 | .03 | 1.10 | — | | | | | | | 17-7 PH |
| bal | .10 | 17.00 | 4.50 | 3.00 | 1.00 | .50 | .04 | .03 | — | — | | | | | | | AM-350 |
| bal | .12 | 16.00 | 14.00 | 2.50 | .75 | .50 | .02 | .01 | — | 3.00 | | | | | | | 17-14 Cu Mo |
| bal | .13 | 15.00 | 4.00 | 2.50 | 1.10 | .75 | .04 | .03 | — | — | | | | | | | AM-355 |

| COMPOSITION, % | | | | | | | | | | | | | | | | MATERIAL DESIGNATION |
|---|---|---|---|---|---|---|---|---|---|---|---|---|---|---|---|---|
| Fe | C | Cr | Ni | Mo | Mn | Si | P | S | Al | Cu | — | — | — | — | Other | |
| **14. PRECIPITATION HARDENING STAINLESS STEELS, WROUGHT (continued)** | | | | | | | | | | | | | | | | |
| bal | .15 | 14.50 | — | 5.00 | .02 | .10 | .005 | .01 | — | — | | | | | 13.50Co, .50V | AFC-77 |
| bal | .25 | 12.00 | — | 3.00 | 18.00 | .25 | .02 | .02 | — | — | | | | | .20B, .80V | AF-71 |
| bal | .30 | 18.50 | 9.50 | — | 3.50 | .50 | .25 | .25 | — | — | | | | | | HNM |

| COMPOSITION, % | | | | | | | | | | | | | | | | MATERIAL DESIGNATION |
|---|---|---|---|---|---|---|---|---|---|---|---|---|---|---|---|---|
| Fe | C | Ni | Cr | Mo | Mn | Si | P | S | — | — | — | — | — | — | Other | |
| **15. CARBON STEELS, CAST — Low Carbon** | | | | | | | | | | | | | | | | |
| bal | .10 | — | — | — | .20 | .55 | .04 | .05 | | | | | | | | 1010 |
| bal | .15 | — | — | .55 | .55 | .30 | .04 | .04 | | | | | | | | ASTM A426 Grade CP1 |
| bal | .20 | — | — | — | .67 | .38 | .04 | .05 | | | | | | | | 1020 |
| **15. CARBON STEELS, CAST — Medium Carbon** | | | | | | | | | | | | | | | | |
| bal | .25 | — | — | — | .70 | .60 | .04 | .04 | | | | | | | | ASTM A352 Grade LCA |
| bal | .25 | — | — | — | 1.20 | .60 | .04 | .04 | | | | | | | | ASTM A352 Grade LCC |
| bal | .30 | — | — | — | 1.00 | .60 | .04 | .04 | | | | | | | | ASTM A352 Grade LCB |
| bal | .30 | — | — | — | .70 | .45 | .04 | .05 | | | | | | | | 1030 |
| bal | .35 | — | — | — | .70 | .60 | .03 | .03 | | | | | | | | ASTM A356 Grade 1 |
| bal | .40 | — | — | — | .75 | .50 | .04 | .05 | | | | | | | | 1040 |
| bal | .50 | — | — | — | .70 | .42 | .04 | .05 | | | | | | | | 1050 |
| **16. ALLOY STEELS, CAST — Low Carbon** | | | | | | | | | | | | | | | | |
| bal | .20 | — | — | — | 1.35 | .45 | .04 | .04 | | | | | | | | 1320 |
| bal | .15 | 3.50 | — | — | .65 | .60 | .04 | .04 | | | | | | | | ASTM A352 Grade LC3 |
| bal | .15 | 3.50 | — | — | .75 | .50 | .04 | .04 | | | | | | | | 2315 |
| bal | .15 | 4.50 | — | — | .65 | .60 | .04 | .04 | | | | | | | | ASTM A352 Grade LC4 |
| bal | .20 | 3.40 | — | — | .75 | .50 | .04 | .04 | | | | | | | | 2320 |
| bal | .15 | — | — | .55 | .40 | 1.00 | .03 | .03 | | | | | | | | ASTM A426 Grade CP15 |
| bal | .20 | — | — | .20 | 1.30 | .35 | .04 | .04 | | | | | | | | 8020 |

| COMPOSITION, % | | | | | | | | | | | | | | | | | MATERIAL DESIGNATION |
|---|---|---|---|---|---|---|---|---|---|---|---|---|---|---|---|---|---|
| Fe | C | Ni | Cr | Mo | Mn | Si | P | S | — | — | — | — | — | — | — | Other | |
| **16. ALLOY STEELS, CAST — Low Carbon (continued)** | | | | | | | | | | | | | | | | | |
| bal | .10 | — | .90 | .30 | .75 | .50 | .04 | .04 | | | | | | | | | 4110 |
| bal | .15 | — | .70 | .55 | .50 | .30 | .04 | .04 | | | | | | | | | ASTM A426 Grade CP2 |
| bal | .15 | — | 1.00 | .55 | .50 | .50 | .04 | .04 | | | | | | | | | ASTM A426 Grade CP12 |
| bal | .15 | — | 1.25 | .55 | .60 | .60 | .03 | .03 | | | | | | | | | ASTM A426 Grade CP11 |
| bal | .15 | — | 2.50 | 1.10 | .50 | .60 | .03 | .03 | | | | | | | | | ASTM A426 Grade CP22 |
| bal | .15 | — | 3.00 | .95 | .50 | .50 | .03 | .03 | | | | | | | | | ASTM A426 Grade CP21 |
| bal | .15 | — | 5.00 | .55 | .50 | 1.50 | .03 | .03 | | | | | | | | | ASTM A426 Grade CP5b |
| bal | .15 | — | 5.50 | .55 | .50 | .60 | .03 | .03 | | | | | | | | | ASTM A426 Grade CP5 |
| bal | .18 | — | 2.50 | 1.10 | .60 | .60 | .04 | .04 | | | | | | | | | ASTM A217 Grade WC9 |
| bal | .20 | — | .98 | .35 | .75 | .50 | .04 | .04 | | | | | | | | | 4120 |
| bal | .20 | 1.80 | .50 | .25 | .60 | .50 | .03 | .04 | | | | | | | | | 4320 |
| bal | .20 | .60 | .50 | .20 | .60 | .50 | .03 | .04 | | | | | | | | | 8620 |
| **16. ALLOY STEELS, CAST — Medium Carbon** | | | | | | | | | | | | | | | | | |
| bal | * | * | * | * | * | * | .05 | .06 | | | | | | | | | ASTM A148 Grade 80-40 |
| bal | * | * | * | * | * | * | .05 | .06 | | | | | | | | | ASTM A148 Grade 80-50 |
| bal | * | * | * | * | * | * | .05 | .06 | | | | | | | | | ASTM A148 Grade 90-60 |
| bal | * | * | * | * | * | * | .05 | .06 | | | | | | | | | ASTM A148 Grade 105-85 |
| bal | * | * | * | * | * | * | .05 | .06 | | | | | | | | | ASTM A148 Grade 120-95 |
| bal | * | * | * | * | * | * | .05 | .06 | | | | | | | | | ASTM A148 Grade 150-125 |
| bal | * | * | * | * | * | * | .05 | .06 | | | | | | | | | ASTM A148 Grade 175-145 |
| bal | .35 | * | * | * | * | * | .05 | .06 | | | | | | | | | ASTM A486 Class 90 |
| bal | .35 | * | * | * | * | * | .05 | .06 | | | | | | | | | ASTM A486 Class 120 |
| bal | .35 | — | — | — | .90 | .80 | .05 | .06 | | | | | | | | | ASTM A486 Class 70 |
| bal | .25 | — | — | — | .75 | .80 | .05 | .06 | | | | | | | | | ASTM A27 Grade N1 |
| bal | .35 | — | — | — | .60 | .80 | .05 | .06 | | | | | | | | | ASTM A27 Grade N2 |
| bal | .30 | — | — | — | .60 | .80 | .05 | .06 | | | | | | | | | ASTM A27 Grade U-60-30 |
| bal | .30 | — | — | — | .60 | .80 | .05 | .06 | | | | | | | | | ASTM A27 Grade 60-30 |
| bal | .30 | — | — | — | .70 | .80 | .05 | .06 | | | | | | | | | ASTM A27 Grade 65-35 |

*Composition selected by manufacturer. Data listed represent chemical requirements per ASTM reference.

| COMPOSITION, % | | | | | | | | | | | | | | | | | MATERIAL DESIGNATION |
|---|---|---|---|---|---|---|---|---|---|---|---|---|---|---|---|---|---|
| Fe | C | Ni | Cr | Mo | Mn | Si | P | S | — | — | — | — | — | — | — | Other | |
| **16. ALLOY STEELS, CAST — Medium Carbon (continued)** | | | | | | | | | | | | | | | | | |
| bal | .35 | — | — | — | .70 | .80 | .05 | .06 | | | | | | | | | ASTM A27 Grade 70-36 |
| bal | .25 | — | — | — | 1.20 | .80 | .05 | .06 | | | | | | | | | ASTM A27 Grade 70-40 |
| bal | .25 | — | — | — | .70 | .60 | .04 | .04 | | | | | | | | | ASTM A216 Grade WCA |
| bal | .30 | — | — | — | 1.00 | .60 | .04 | .04 | | | | | | | | | ASTM A216 Grade WCB |
| bal | .25 | — | — | — | 1.20 | .60 | .04 | .04 | | | | | | | | | ASTM A216 Grade WCC |
| bal | .30 | — | — | — | 1.50 | .50 | .04 | .04 | | | | | | | | | 1330 |
| bal | .40 | — | — | — | 1.55 | .50 | .04 | .04 | | | | | | | | | 1340 |
| bal | .25 | 2.50 | — | — | .70 | .60 | .04 | .04 | | | | | | | | | ASTM A352 Grade LC2 |
| bal | .25 | 3.45 | — | — | .75 | .45 | .04 | .04 | | | | | | | | | 2325 |
| bal | .30 | 3.50 | — | — | .75 | .45 | .04 | .04 | | | | | | | | | 2330 |
| bal | .30 | — | — | .20 | 1.30 | .35 | .04 | .04 | | | | | | | | | 8030 |
| bal | .30 | — | — | .20 | 1.30 | .35 | .04 | .04 | | | | | | | | .003B | 80B30 |
| bal | .40 | — | — | .20 | 1.20 | .35 | .04 | .04 | | | | | | | | | 8040 |
| bal | .30 | — | — | .40 | 1.35 | .35 | .04 | .04 | | | | | | | | | 8430 |
| bal | .40 | — | — | .40 | 1.60 | .35 | .04 | .04 | | | | | | | | | 8440 |
| bal | .25 | — | — | .55 | .70 | .60 | .04 | .04 | | | | | | | | | ASTM A217 Grade WC1 |
| bal | .25 | — | — | .55 | .70 | .60 | .04 | .04 | | | | | | | | | ASTM A352 Grade LC1 |
| bal | .25 | — | — | .55 | .70 | .60 | .03 | .03 | | | | | | | | | ASTM A356 Grade 2 |
| bal | .45 | — | — | — | .70 | .80 | .04 | .04 | | | | | | | | .03V | ASTM A487 Class DN |
| bal | .30 | — | — | — | 1.00 | .80 | .04 | .04 | | | | | | | | .08V | ASTM A487 Class 1N |
| bal | .30 | — | — | — | 1.00 | .80 | .04 | .04 | | | | | | | | .08V | ASTM A487 Class 1Q |
| bal | .30 | — | — | .20 | 1.20 | .80 | .04 | .04 | | | | | | | | .03V | ASTM A487 Class 2N |
| bal | .30 | — | — | .20 | 1.20 | .80 | .04 | .04 | | | | | | | | .03V | ASTM A487 Class 2Q |
| bal | .25 | — | .60 | .50 | .70 | .60 | .03 | .03 | | | | | | | | | ASTM A356 Grade 5 |
| bal | .25 | — | .75 | .30 | .65 | .50 | .04 | .04 | | | | | | | | | 4125 |
| bal | .30 | — | .75 | .20 | .65 | .50 | .04 | .04 | | | | | | | | | 4130 |
| bal | .40 | — | 1.00 | .20 | .65 | .50 | .04 | .04 | | | | | | | | | 4140 |
| bal | .20 | — | 1.25 | .55 | .70 | .60 | .04 | .04 | | | | | | | | | ASTM A217 Grade WC6 |

| COMPOSITION, % | | | | | | | | | | | | | | | | | MATERIAL DESIGNATION |
|---|---|---|---|---|---|---|---|---|---|---|---|---|---|---|---|---|---|
| Fe | C | Ni | Cr | Mo | Mn | Si | P | S | — | — | — | — | — | — | — | Other | |
| **16. ALLOY STEELS, CAST — Medium Carbon (continued)** | | | | | | | | | | | | | | | | | |
| bal | .20 | — | 1.25 | .55 | .70 | .60 | .03 | .03 | | | | | | | | | ASTM A356 Grade 6 |
| bal | .20 | — | 2.25 | 1.00 | .70 | .60 | .03 | .03 | | | | | | | | | ASTM A356 Grade 10 |
| bal | .33 | — | 1.00 | .25 | .80 | .80 | .04 | .04 | | | | | | | | .03V | ASTM A487 Class 9N |
| bal | .33 | — | 1.00 | .25 | .80 | .80 | .04 | .04 | | | | | | | | .03V | ASTM A487 Class 9Q |
| bal | .20 | — | 1.00 | 1.10 | .60 | .60 | .04 | .04 | | | | | | | | .20V | ASTM A389 Grade C24 |
| bal | .20 | — | 1.25 | .55 | .60 | .60 | .04 | .04 | | | | | | | | .20V | ASTM A389 Grade C23 |
| bal | .20 | — | 1.25 | 1.10 | .80 | .50 | .03 | .03 | | | | | | | | .10V | ASTM A356 Grade 8 |
| bal | .20 | — | 1.25 | 1.10 | .80 | .50 | .03 | .03 | | | | | | | | .30V | ASTM A356 Grade 9 |
| bal | .20 | — | 2.25 | 1.00 | .80 | .80 | .04 | .04 | | | | | | | | .03V | ASTM A487 Class 8N |
| bal | .20 | — | 2.25 | 1.00 | .80 | .80 | .04 | .04 | | | | | | | | .03V | ASTM A487 Class 8Q |
| bal | .25 | .50 | .60 | .35 | 1.40 | .50 | .04 | .04 | | | | | | | | | 9525 |
| bal | .30 | .65 | .60 | .40 | 1.40 | .50 | .04 | .04 | | | | | | | | | 9530 |
| bal | .30 | .60 | .50 | .20 | .85 | .40 | .04 | .04 | | | | | | | | | 8630 |
| bal | .40 | .60 | .50 | .20 | .85 | .40 | .04 | .04 | | | | | | | | | 8640 |
| bal | .35 | .55 | .60 | .35 | 1.50 | .50 | .04 | .04 | | | | | | | | | 9535 |
| bal | .20 | .90 | .80 | 1.10 | .60 | .60 | .04 | .04 | | | | | | | | | ASTM A217 Grade WC5 |
| bal | .20 | 1.00 | .70 | .55 | .70 | .60 | .04 | .04 | | | | | | | | | ASTM A217 Grade WC4 |
| bal | .30 | 1.80 | .65 | .20 | .75 | .50 | .04 | .04 | | | | | | | | | 4330 |
| bal | .40 | 1.80 | .80 | .30 | .75 | .50 | .04 | .04 | | | | | | | | | 4340 |
| bal | .22 | 3.00 | 1.60 | .50 | .65 | .50 | .04 | .04 | | | | | | | | | ASTM A352 Grade LC2-1 |
| bal | .30 | .60 | .60 | .25 | 1.00 | .80 | .04 | .04 | | | | | | | | .03V | ASTM A487 Class 4N |
| bal | .30 | .60 | .60 | .25 | 1.00 | .80 | .04 | .04 | | | | | | | | .03V | ASTM A487 Class 4Q |
| bal | .30 | .60 | .60 | .25 | 1.00 | .80 | .04 | .04 | | | | | | | | .03V | ASTM A487 Class 4QA |
| bal | .38 | .60 | .60 | .35 | 1.50 | .80 | .04 | .04 | | | | | | | | .03V | ASTM A487 Class 6N |
| bal | .38 | .60 | .60 | .35 | 1.50 | .80 | .04 | .04 | | | | | | | | .03V | ASTM A487 Class 6Q |
| bal | .20 | .90 | .60 | .50 | .80 | .80 | .04 | .04 | | | | | | | | .08V, .004B, .35Cu | ASTM A487 Class 7Q |
| bal | .30 | 1.80 | .75 | .30 | — | .80 | .04 | .04 | | | | | | | | .03V | ASTM A487 Class 10N |
| bal | .30 | 1.80 | .75 | .30 | — | .80 | .04 | .04 | | | | | | | | .03V | ASTM A487 Class 10Q |

| COMPOSITION, % | | | | | | | | | | | | | | | | | MATERIAL DESIGNATION |
|---|---|---|---|---|---|---|---|---|---|---|---|---|---|---|---|---|---|
| Fe | C | W | Mo | Cr | V | Si | Mn | — | — | — | — | — | — | — | — | Other | |
| **17. TOOL STEELS, CAST** | | | | | | | | | | | | | | | | | |
| bal | .35 | 1.35 | 1.50 | 5.25 | .35 | 1.50 | .75 | | | | | | | | | | ASTM A597 Grade CH-12 |
| bal | .36 | — | 1.50 | 5.25 | 1.00 | 1.50 | .75 | | | | | | | | | | ASTM A597 Grade CH-13 |
| bal | .60 | — | .60 | .35 | .35 | 2.25 | 1.00 | | | | | | | | | | ASTM A597 Grade CS-5 |
| bal | .95 | .50 | — | .70 | .30 | 1.50 | 1.30 | | | | | | | | | | ASTM A597 Grade CO-1 |
| bal | 1.00 | — | 1.20 | 5.25 | .35 | 1.50 | .75 | | | | | | | | | | ASTM A597 Grade CA-2 |
| bal | 1.45 | — | .95 | 12.00 | .45 | 1.50 | .75 | | | | | | | | | 3.00Co, .50Ni | ASTM A597 Grade CD-5 |
| bal | 1.50 | — | .95 | 12.00 | .70 | 1.50 | 1.00 | | | | | | | | | .85Co | ASTM A597 Grade CD-2 |

| COMPOSITION, % | | | | | | | | | | | | | | | | | MATERIAL DESIGNATION |
|---|---|---|---|---|---|---|---|---|---|---|---|---|---|---|---|---|---|
| Fe | C | Cr | Ni | Mo | Mn | Si | P | S | Cu | — | — | — | — | — | — | Other | |
| **18. STAINLESS STEELS, CAST — Ferritic** | | | | | | | | | | | | | | | | | |
| bal | .20 | 6.00 | — | .55 | .60 | .75 | .04 | .04 | — | | | | | | | | ASTM A217 Grade C5 |
| bal | .20 | 9.00 | — | 1.00 | .50 | 1.00 | .04 | .04 | — | | | | | | | | ASTM A217 Grade C12 |
| bal | .06 | 11.50 | 7.00 | — | .50 | 1.00 | .02 | .02 | — | | | | | | | | ASTM A296 Grade CA6N |
| bal | .30 | 20.00 | 2.00 | — | 1.00 | 1.50 | .04 | .04 | — | | | | | | | | ASTM A296 Grade CB-30 |
| bal | .50 | 28.00 | 4.00 | — | 1.00 | 1.50 | .04 | .04 | — | | | | | | | | ASTM A296 Grade CC-50 |
| bal | .30 | 29.00 | 9.00 | — | 1.50 | 2.00 | .04 | .04 | — | | | | | | | | ASTM A296 Grade CE-30 |
| bal | .06 | 13.00 | 4.00 | .80 | 1.00 | 1.00 | .04 | .03 | — | | | | | | | | ASTM A296 Grade CA-6NM |
| bal | .06 | 13.00 | 4.00 | .80 | 1.00 | 1.00 | .04 | .03 | — | | | | | | | | ASTM A487 Class CA6NM |
| bal | .04 | 26.00 | 5.50 | 2.00 | 1.00 | 1.00 | .04 | .04 | 3.00 | | | | | | | | ASTM A296 Grade CD4MCu |
| bal | .30 | 28.00 | 4.00 | .50 | 1.00 | 1.50 | .04 | .04 | — | | | | | | | | ASTM A608 Grade HC30 |
| bal | .50 | 28.00 | 4.00 | .50 | 1.00 | 2.00 | .04 | .04 | — | | | | | | | | ASTM A297 Grade HC |
| **18. STAINLESS STEELS, CAST — Austenitic** | | | | | | | | | | | | | | | | | |
| bal | .03 | 19.00 | 9.00 | — | 1.50 | 2.00 | .04 | .04 | — | | | | | | | | ASTM A296 Grade CF-3 |
| bal | .03 | 19.00 | 9.00 | — | 1.50 | 2.00 | .04 | .04 | — | | | | | | | | ASTM A351 Grade CF-3 |
| bal | .03 | 19.00 | 9.00 | — | 1.50 | 2.00 | .04 | .04 | — | | | | | | | | ASTM A351 Grade CF-3A |
| bal | .03 | 19.00 | 9.00 | — | 1.50 | 2.00 | .04 | .04 | — | | | | | | | | ASTM A451 Grade CPF3 |

| COMPOSITION, % | | | | | | | | | | | | | | | | | MATERIAL DESIGNATION |
|---|---|---|---|---|---|---|---|---|---|---|---|---|---|---|---|---|---|
| Fe | C | Cr | Ni | Mo | Mn | Si | P | S | Cu | — | — | — | — | — | — | Other | |
| **18. STAINLESS STEELS, CAST — Austenitic (continued)** | | | | | | | | | | | | | | | | | |
| bal | .03 | 19.00 | 9.00 | — | 1.50 | 2.00 | .04 | .04 | — | | | | | | | | ASTM A451 Grade CPF3A |
| bal | .08 | 19.00 | 9.00 | — | 1.50 | 2.00 | .04 | .04 | — | | | | | | | | ASTM A296 Grade CF-8 |
| bal | .08 | 19.00 | 9.00 | — | 1.50 | 2.00 | .04 | .04 | — | | | | | | | | ASTM A351 Grade CF-8 |
| bal | .08 | 19.00 | 9.00 | — | 1.50 | 2.00 | .04 | .04 | — | | | | | | | | ASTM A351 Grade CF-8A |
| bal | .08 | 19.00 | 9.00 | — | 1.50 | 2.00 | .04 | .04 | — | | | | | | | | ASTM A451 Grade CPF8 |
| bal | .08 | 19.00 | 9.00 | — | 1.50 | 1.00 | .04 | .04 | — | | | | | | | | ASTM A451 Grade CPF8A |
| bal | .08 | 19.00 | 9.00 | — | 2.00 | .75 | .04 | .03 | — | | | | | | | | ASTM A452 Grade TP 304H |
| bal | .20 | 19.00 | 9.00 | — | 1.50 | 2.00 | .04 | .04 | — | | | | | | | | ASTM A296 Grade CF-20 |
| bal | .08 | 19.00 | 10.00 | — | 1.50 | 2.00 | .04 | .04 | — | | | | | | | Cb 8xC, 1.0 max. | ASTM A296 Grade CF-8C |
| bal | .08 | 19.00 | 10.00 | — | 1.50 | 2.00 | .04 | .04 | — | | | | | | | Cb 8xC, 1.0 max. | ASTM A351 Grade CF-8C |
| bal | .08 | 19.00 | 10.00 | — | 2.00 | .75 | .04 | .03 | — | | | | | | | Cb + Ta 8xC, 1.0 max. | ASTM A452 Grade TP 347H |
| bal | .08 | 19.00 | 10.00 | — | 1.50 | 1.00 | .04 | .04 | — | | | | | | | Cb 8xC, 1.0 max. | ASTM A451 Grade CPF8C |
| bal | .08 | 19.00 | 10.00 | — | 1.50 | 1.00 | .04 | .04 | — | | | | | | | Ta .10 max.; Cb 8xC, 1.0 max. | ASTM A451 Grade CPF8C Ta max. |
| bal | .12 | 22.00 | 12.00 | — | 1.50 | 2.00 | .04 | .04 | — | | | | | | | | ASTM A296 Grade CG-12 |
| bal | .08 | 25.00 | 12.00 | — | 1.50 | 1.50 | .04 | .04 | — | | | | | | | | ASTM A351 Grade CH-8 |
| bal | .08 | 25.00 | 12.00 | — | 1.50 | 1.00 | .04 | .04 | — | | | | | | | | ASTM A451 Grade CPH8 |
| bal | .10 | 25.00 | 12.00 | — | 1.50 | 2.00 | .04 | .04 | — | | | | | | | | ASTM A351 Grade CH-10 |
| bal | .10 | 25.00 | 12.00 | — | 1.50 | 1.00 | .04 | .04 | — | | | | | | | | ASTM A451 Grade CPH10 |
| bal | .20 | 25.00 | 12.00 | — | 1.50 | 1.00 | .04 | .04 | — | | | | | | | | ASTM A451 Grade CPH20 |
| bal | .20 | 25.00 | 12.00 | — | 1.50 | 2.00 | .04 | .04 | — | | | | | | | | ASTM A296 Grade CH-20 |
| bal | .20 | 25.00 | 12.00 | — | 1.50 | 2.00 | .04 | .04 | — | | | | | | | | ASTM A351 Grade CH-20 |
| bal | .20 | 25.00 | 20.00 | — | 1.50 | 1.00 | .04 | .04 | — | | | | | | | | ASTM A451 Grade CPK20 |
| bal | .20 | 25.00 | 20.00 | — | 1.50 | 1.75 | .04 | .04 | — | | | | | | | | ASTM A351 Grade CK-20 |
| bal | .20 | 25.00 | 20.00 | — | 2.00 | 2.00 | .04 | .04 | — | | | | | | | | ASTM A296 Grade CK-20 |
| bal | .30 | 25.00 | 20.00 | — | 1.50 | 1.75 | .04 | .04 | — | | | | | | | | ASTM A351 Grade HK-30 |
| bal | .40 | 25.00 | 20.00 | — | 1.50 | 1.75 | .04 | .04 | — | | | | | | | | ASTM A351 Grade HK-40 |

| COMPOSITION, % | | | | | | | | | | | | | | | | | MATERIAL DESIGNATION |
|---|---|---|---|---|---|---|---|---|---|---|---|---|---|---|---|---|---|
| Fe | C | Cr | Ni | Mo | Mn | Si | P | S | Cu | — | — | — | — | — | — | Other | |
| **18. STAINLESS STEELS, CAST — Austenitic (continued)** | | | | | | | | | | | | | | | | | |
| bal | .30 | 15.00 | 35.00 | .50 | 2.00 | 2.50 | .04 | .04 | — | | | | | | | | ASTM A351 Grade HT-30 |
| bal | .60 | 15.00 | 35.00 | .50 | 2.00 | 2.50 | .04 | .04 | — | | | | | | | | ASTM A297 Grade HT |
| bal | .10 | 17.00 | 13.00 | 3.00 | 2.00 | .75 | .04 | .03 | — | | | | | | | | ASTM A452 Grade TP 316H |
| bal | .10 | 17.00 | 15.00 | 2.00 | 1.50 | 1.00 | .04 | .04 | — | | | | | | | Cb 10xC, 1.2 max. | ASTM A451 Grade CPF10MC |
| bal | .10 | 17.00 | 15.00 | 2.00 | 1.50 | 1.50 | .04 | .04 | — | | | | | | | Cb 10xC, 1.2 max. | ASTM A351 Grade CF-10MC |
| bal | .50 | 17.00 | 35.00 | .50 | 1.50 | 2.00 | .04 | .04 | — | | | | | | | | ASTM A608 Grade HT50 |
| bal | .03 | 19.00 | 10.00 | 3.00 | 1.50 | 1.50 | .04 | .04 | — | | | | | | | | ASTM A296 Grade CF-3M |
| bal | .03 | 19.00 | 10.00 | 3.00 | 1.50 | 1.50 | .04 | .04 | — | | | | | | | | ASTM A351 Grade CF-3M |
| bal | .03 | 19.00 | 10.00 | 3.00 | 1.50 | 1.50 | .04 | .04 | — | | | | | | | | ASTM A351 Grade CF-3MA |
| bal | .03 | 19.00 | 10.00 | 3.00 | 1.50 | 1.50 | .04 | .04 | — | | | | | | | | ASTM A451 Grade CPF3M |
| bal | .07 | 19.00 | 24.00 | 3.00 | 1.00 | 3.00 | .04 | .03 | 2.00 | | | | | | | | ASTM A296 Grade CN-7MS |
| bal | .08 | 19.00 | 10.00 | 3.00 | 1.50 | 2.00 | .04 | .04 | — | | | | | | | | ASTM A296 Grade CF-8M |
| bal | .08 | 19.00 | 10.00 | 3.00 | 1.50 | 1.50 | .04 | .04 | — | | | | | | | | ASTM A351 Grade CF-8M |
| bal | .08 | 19.00 | 10.00 | 3.00 | 1.50 | 1.00 | .04 | .04 | — | | | | | | | | ASTM A451 Grade CPF8M |
| bal | .08 | 19.00 | 11.00 | 4.00 | 1.50 | 1.50 | .04 | .04 | — | | | | | | | | ASTM A296 Grade CG-8M |
| bal | .16 | 19.00 | 9.00 | .80 | 1.50 | 2.00 | .04 | .04 | — | | | | | | | | ASTM A296 Grade CF-16F |
| bal | .30 | 19.00 | 9.00 | .50 | 1.50 | 2.00 | .04 | .04 | — | | | | | | | | ASTM A608 Grade HF30 |
| bal | .30 | 19.00 | 9.00 | .50 | 2.00 | 2.00 | .04 | .04 | — | | | | | | | | ASTM A297 Grade HF |
| bal | .50 | 19.00 | 39.00 | .50 | 1.50 | 2.00 | .04 | .04 | — | | | | | | | | ASTM A608 Grade HU50 |
| bal | .60 | 19.00 | 39.00 | .50 | 2.00 | 2.50 | .04 | .04 | — | | | | | | | | ASTM A297 Grade HU |
| bal | .07 | 20.00 | 29.00 | 3.00 | 1.50 | 1.50 | .04 | .04 | — | | | | | | | | ASTM A351 Grade CN-7M |
| bal | .07 | 20.00 | 29.00 | 3.00 | 1.50 | 1.50 | .04 | .04 | 4.00 | | | | | | | | ASTM A296 Grade CN-7M |
| bal | .40 | 20.00 | 25.00 | .50 | 1.50 | 2.00 | .04 | .04 | — | | | | | | | | ASTM A608 Grade HN40 |
| bal | .40 | 20.00 | 25.00 | .50 | 2.00 | 2.00 | .04 | .04 | — | | | | | | | | ASTM A297 Grade HN |
| bal | .30 | 25.00 | 12.00 | .50 | 1.50 | 2.00 | .04 | .04 | — | | | | | | | | ASTM A608 Grade HH30 |
| bal | .30 | 25.00 | 20.00 | .50 | 1.50 | 2.00 | .04 | .04 | — | | | | | | | | ASTM A608 Grade HK30 |
| bal | .33 | 25.00 | 12.00 | .50 | 1.50 | 2.00 | .04 | .04 | — | | | | | | | | ASTM A608 Grade HH33 |
| bal | .40 | 25.00 | 20.00 | .50 | 1.50 | 2.00 | .04 | .04 | — | | | | | | | | ASTM A608 Grade HK40 |

# 26.3 Chemical Composition by Material Group

| COMPOSITION, % | | | | | | | | | | | | | | | | | MATERIAL DESIGNATION |
|---|---|---|---|---|---|---|---|---|---|---|---|---|---|---|---|---|---|
| Fe | C | Cr | Ni | Mo | Mn | Si | P | S | Cu | — | — | — | — | — | — | Other | |
| **18. STAINLESS STEELS, CAST — Austenitic (continued)** | | | | | | | | | | | | | | | | | |
| bal | .50 | 25.00 | 12.00 | .50 | 2.00 | 2.00 | .04 | .04 | — | | | | | | | | ASTM A297 Grade HH |
| bal | .50 | 25.00 | 20.00 | .50 | 2.00 | 2.00 | .04 | .04 | — | | | | | | | | ASTM A297 Grade HK |
| bal | .60 | 26.00 | 35.00 | .50 | 2.00 | 2.50 | .04 | .04 | — | | | | | | | | ASTM A297 Grade HP |
| bal | .40 | 28.00 | 15.00 | .50 | 2.00 | 2.00 | .04 | .04 | — | | | | | | | | ASTM A297 Grade HI |
| bal | .40 | 28.00 | 15.00 | .50 | 1.50 | 2.00 | .04 | .04 | — | | | | | | | | ASTM A608 Grade HI35 |
| bal | .50 | 28.00 | 5.00 | .50 | 1.50 | 2.00 | .04 | .04 | — | | | | | | | | ASTM A297 Grade HD |
| bal | .50 | 28.00 | 5.00 | .50 | 1.50 | 2.00 | .04 | .04 | — | | | | | | | | ASTM A608 Grade HD50 |
| bal | .30 | 29.00 | 20.00 | .50 | 1.50 | 2.00 | .04 | .04 | — | | | | | | | | ASTM A608 Grade HL30 |
| bal | .40 | 29.00 | 9.00 | .50 | 1.50 | 2.00 | .04 | .04 | — | | | | | | | | ASTM A608 Grade HE35 |
| bal | .40 | 29.00 | 9.00 | .50 | 2.00 | 2.00 | .04 | .04 | — | | | | | | | | ASTM A297 Grade HE |
| bal | .40 | 29.00 | 20.00 | .50 | 1.50 | 2.00 | .04 | .04 | — | | | | | | | | ASTM A608 Grade HL40 |
| bal | .50 | 29.00 | 20.00 | .50 | 2.00 | 2.00 | .04 | .04 | — | | | | | | | | ASTM A297 Grade HL |
| **18. STAINLESS STEELS, CAST — Martensitic** | | | | | | | | | | | | | | | | | |
| bal | .15 | 7.00 | — | .60 | .50 | 1.00 | .03 | .03 | — | | | | | | | | ASTM A426 Grade CP7 |
| bal | .20 | 9.00 | — | 1.00 | .50 | 1.00 | .03 | .03 | — | | | | | | | | ASTM A426 Grade CP9 |
| bal | .15 | 12.00 | — | .50 | 1.00 | 1.50 | .04 | .04 | — | | | | | | | | ASTM A426 Grade CPCA15 |
| bal | .15 | 12.00 | 1.00 | .50 | 1.00 | 1.50 | .04 | .04 | — | | | | | | | | ASTM A217 Grade CA-15 |
| bal | .15 | 12.00 | 1.00 | .50 | 1.00 | 1.50 | .04 | .04 | — | | | | | | | | ASTM A296 Grade CA-15 |
| bal | .15 | 12.00 | 1.00 | .50 | 1.00 | 1.50 | .04 | .04 | — | | | | | | | | ASTM A487 Class CA15a |
| bal | .15 | 12.00 | 1.00 | 1.00 | 1.00 | .65 | .04 | .04 | — | | | | | | | | ASTM A296 Grade CA15M |
| bal | .15 | 12.00 | 1.00 | 1.00 | 1.00 | .65 | .04 | .04 | — | | | | | | | .03V | ASTM A487 Class CA15M |
| bal | .40 | 12.00 | 1.00 | .50 | 1.00 | 1.50 | .04 | .04 | — | | | | | | | | ASTM A296 Grade CA-40 |
| **19. PRECIPITATION HARDENING STAINLESS STEELS, CAST** | | | | | | | | | | | | | | | | | |
| bal | .10 | 16.00 | 4.00 | 3.00 | 1.00 | .50 | — | — | — | | | | | | | .10N | AM-355 |
| bal | .07 | 17.00 | 4.00 | — | 1.00 | 1.00 | — | — | 2.50 | | | | | | | | ACI Grade CB-7Cu |
| bal | .07 | 17.00 | 4.00 | — | 1.00 | 1.00 | — | — | 2.50 | | | | | | | | 17-4 PH |
| bal | .04 | 25.00 | 5.00 | 2.00 | 1.00 | 1.00 | — | — | 3.00 | | | | | | | | ACI Grade CD-4MCu |
| bal | .04 | 25.00 | 5.00 | 2.00 | 1.00 | 1.00 | .04 | .04 | 3.00 | | | | | | | | ASTM A351 Grade CD-4MCu |

| Fe | C | Mn | Cr | Ni | Mo | Si | P | — | — | — | — | — | — | — | — | Other | MATERIAL DESIGNATION |
|---|---|---|---|---|---|---|---|---|---|---|---|---|---|---|---|---|---|
| **20. AUSTENITIC MANGANESE STEELS, CAST** | | | | | | | | | | | | | | | | | |
| bal | 1.30 | 8.00 | — | — | 1.00 | 1.00 | .07 | | | | | | | | | | ASTM A128 Grade F |
| bal | 1.25 | 11.00 | — | — | — | 1.00 | .07 | | | | | | | | | | ASTM A128 Grade A |
| bal | 1.00 | 14.00 | — | — | — | 1.00 | .07 | | | | | | | | | | ASTM A128 Grade B-1 |
| bal | 1.15 | 14.00 | — | — | — | 1.00 | .07 | | | | | | | | | | ASTM A128 Grade B-2 |
| bal | 1.25 | 14.00 | — | — | — | 1.00 | .07 | | | | | | | | | | ASTM A128 Grade B-3 |
| bal | 1.30 | 14.00 | — | — | — | 1.00 | .07 | | | | | | | | | | ASTM A128 Grade B-4 |
| bal | 1.20 | 14.00 | 2.00 | — | — | 1.00 | .07 | | | | | | | | | | ASTM A128 Grade C |
| bal | 1.20 | 14.00 | — | 3.50 | — | 1.00 | .07 | | | | | | | | | | ASTM A128 Grade D |
| bal | 1.20 | 14.00 | — | — | 1.20 | 1.00 | .07 | | | | | | | | | | ASTM A128 Grade E-1 |
| bal | 1.40 | 14.00 | — | — | 2.00 | 1.00 | .07 | | | | | | | | | | ASTM A128 Grade E-2 |

| Fe | C | Si | Mn | Cr | Ni | Cu | P | S | Mo | — | — | — | — | — | — | Other | MATERIAL DESIGNATION |
|---|---|---|---|---|---|---|---|---|---|---|---|---|---|---|---|---|---|
| **21. GRAY CAST IRONS — Ferritic** | | | | | | | | | | | | | | | | | |
| bal | 3.50 | 2.40 | — | — | — | — | — | — | — | | | | | | | | ASTM A48 Class 20 |
| bal | 3.50 | 2.60 | .80 | — | — | — | .25 | .15 | — | | | | | | | | SAE J431c Grade G1800 |
| **21. GRAY CAST IRONS — Pearlitic-Ferritic** | | | | | | | | | | | | | | | | | |
| bal | 3.30 | 2.20 | — | — | — | — | — | — | — | | | | | | | | ASTM A48 Class 25 |
| bal | 3.30 | 2.20 | .90 | — | — | — | .20 | .15 | — | | | | | | | | SAE J431c Grade G2500 |
| **21. GRAY CAST IRONS — Pearlitic** | | | | | | | | | | | | | | | | | |
| bal | 3.00 | 1.90 | .60 | — | — | — | .20 | .06 | — | | | | | | | | ASTM A48 Class 40 |
| bal | 3.10 | 2.00 | — | — | — | — | — | — | — | | | | | | | | ASTM A48 Class 35 |
| bal | 3.20 | 2.20 | .50 | — | — | — | .20 | .06 | — | | | | | | | | ASTM A48 Class 30 |
| bal | 3.30 | 2.10 | .90 | — | — | — | .15 | .15 | — | | | | | | | | SAE J431c Grade G3000 |

# 26.3 Chemical Composition by Material Group

| COMPOSITION, % | | | | | | | | | | | | | | | | | MATERIAL DESIGNATION |
|---|---|---|---|---|---|---|---|---|---|---|---|---|---|---|---|---|---|
| Fe | C | Si | Mn | Cr | Ni | Cu | P | S | Mo | — | — | — | — | — | — | Other | |
| 21. GRAY CAST IRONS — Pearlitic plus Free Carbides | | | | | | | | | | | | | | | | | |
| bal | 3.15 | 2.00 | — | — | — | — | .20 | .10 | — | | | | | | | | ASTM A48 Class 45 |
| bal | 3.15 | 2.00 | — | — | — | — | .20 | .10 | — | | | | | | | | ASTM A48 Class 50 |
| bal | 3.15 | 2.20 | .90 | — | — | — | .12 | .15 | — | | | | | | | | SAE J431c Grade G3500 |
| bal | 3.15 | 2.00 | 1.00 | — | — | — | .10 | .15 | — | | | | | | | | SAE J431c Grade G4000 |
| 21. GRAY CAST IRONS — Pearlitic or Acicular plus Free Carbides | | | | | | | | | | | | | | | | | |
| bal | 3.50 | 1.80 | .90 | — | — | — | .20 | .06 | — | | | | | | | | ASTM A48 Class 55 |
| bal | 3.50 | 1.80 | — | — | — | — | .20 | .06 | — | | | | | | | | ASTM A48 Class 60 |
| 21. GRAY CAST IRONS — Austenitic (NI-RESIST) | | | | | | | | | | | | | | | | | |
| bal | 2.40 | 1.50 | 1.00 | .10 | 35.00 | .50 | — | .12 | — | | | | | | | | ASTM A436 Type 5 |
| bal | 2.60 | 1.50 | 1.00 | 3.00 | 30.00 | .50 | — | .12 | — | | | | | | | | ASTM A436 Type 3 |
| bal | 2.60 | 5.50 | 1.00 | 5.00 | 31.00 | .50 | — | .12 | — | | | | | | | | ASTM A436 Type 4 |
| bal | 3.00 | 2.50 | 1.00 | 2.00 | 16.00 | 6.50 | — | .12 | — | | | | | | | | ASTM A436 Type 1 |
| bal | 3.00 | 2.50 | 1.00 | 3.00 | 16.00 | 6.50 | — | .12 | — | | | | | | | | ASTM A436 Type 1b |
| bal | 3.00 | 2.50 | 1.00 | 2.00 | 20.00 | .50 | — | .12 | — | | | | | | | | ASTM A436 Type 2 |
| bal | 3.00 | 2.50 | 1.00 | 4.00 | 20.00 | .50 | — | .12 | — | | | | | | | | ASTM A436 Type 2b |
| bal | 3.00 | 2.00 | 1.00 | 1.50 | 20.00 | 4.50 | — | .12 | 1.00 | | | | | | | | ASTM A436 Type 6 |
| 23. DUCTILE CAST IRONS — Ferritic | | | | | | | | | | | | | | | | | |
| bal | 3.50 | 2.50 | .60 | — | — | — | .10 | .03 | — | | | | | | | | SAE J434c Grade D4018 |
| bal | 3.50 | 2.50 | .60 | — | — | — | .10 | .03 | — | | | | | | | | SAE J434c Grade D4512 |
| bal | 3.60 | 2.30 | .60 | — | — | — | .08 | .03 | — | | | | | | | | ASTM A536 Grade 60-40-18 |
| bal | 3.60 | 2.30 | .60 | — | — | — | .08 | .03 | — | | | | | | | | ASTM A536 Grade 65-45-12 |
| 23. DUCTILE CAST IRONS — Ferritic-Pearlitic | | | | | | | | | | | | | | | | | |
| bal | 3.50 | 2.50 | .60 | — | — | — | .10 | .03 | — | | | | | | | | SAE J434c Grade D5506 |
| bal | 3.60 | 2.30 | .60 | — | — | — | .08 | .03 | — | | | | | | | | ASTM A536 Grade 80-55-06 |
| 23. DUCTILE CAST IRONS — Pearlitic-Martensitic | | | | | | | | | | | | | | | | | |
| bal | 3.50 | 2.50 | .60 | — | — | — | .10 | .03 | — | | | | | | | | SAE J434c Grade D7003 |
| bal | 3.60 | 2.30 | .60 | — | — | — | .10 | .03 | — | | | | | | | | ASTM A536 Grade 100-70-03 |

| COMPOSITION, % | | | | | | | | | | | | | | | | | MATERIAL DESIGNATION |
|---|---|---|---|---|---|---|---|---|---|---|---|---|---|---|---|---|---|
| Fe | C | Si | Mn | Cr | Ni | Cu | P | S | Mo | — | — | — | — | — | — | Other | |
| **23. DUCTILE CAST IRONS — Martensitic** | | | | | | | | | | | | | | | | | |
| bal | 3.50 | 2.50 | .60 | — | — | — | .10 | .03 | — | | | | | | | | SAE J434c Grade DQ&T |
| bal | 3.60 | 2.30 | .60 | — | — | — | .10 | .03 | — | | | | | | | | ASTM A536 Grade 120-90-02 |
| **23. DUCTILE CAST IRONS — Austenitic (NI-RESIST Ductile)** | | | | | | | | | | | | | | | | | |
| bal | 2.40 | 2.80 | 1.00 | .10 | 36.00 | — | .08 | — | — | | | | | | | | ASTM A439 Type D-5 |
| bal | 2.40 | 2.80 | 1.00 | 3.00 | 36.00 | — | .08 | — | — | | | | | | | | ASTM A439 Type D-5B |
| bal | 2.60 | 2.80 | 1.00 | 1.50 | 32.00 | — | .08 | — | — | | | | | | | | ASTM A439 Type D-3A |
| bal | 2.60 | 2.80 | 1.00 | 3.00 | 32.00 | — | .08 | — | — | | | | | | | | ASTM A439 Type D-3 |
| bal | 2.60 | 6.00 | 1.00 | 5.00 | 32.00 | — | .08 | — | — | | | | | | | | ASTM A439 Type D-4 |
| bal | 2.70 | 2.00 | 4.00 | .20 | 24.00 | — | .08 | — | — | | | | | | | | ASTM A571 Type D-2M |
| bal | 2.90 | 3.00 | 2.20 | .50 | 24.00 | — | .08 | — | — | | | | | | | | ASTM A439 Type D-2C |
| bal | 3.00 | 3.00 | 1.00 | 2.50 | 22.00 | — | .08 | — | — | | | | | | | | ASTM A439 Type D-2 |
| bal | 3.00 | 3.00 | 1.00 | 3.50 | 22.00 | — | .08 | — | — | | | | | | | | ASTM A439 Type D-2B |
| **24. MALLEABLE CAST IRONS — Ferritic** | | | | | | | | | | | | | | | | | |
| bal | 2.20 | 1.15 | .35 | — | — | — | .16 | .08 | — | | | | | | | | ASTM A47 Grade 35018 |
| bal | 2.50 | 1.30 | .40 | — | — | — | .02 | .16 | — | | | | | | | | ASTM A47 Grade 32510 |
| bal | 2.50 | 1.50 | 1.00 | — | — | — | .10 | .15 | — | | | | | | | | ASTM A602 Grade M3210 |
| bal | 2.50 | 1.50 | 1.00 | — | — | — | .10 | .15 | — | | | | | | | | SAE J158 Grade M3210 |
| **24. MALLEABLE CAST IRONS — Pearlitic** | | | | | | | | | | | | | | | | | |
| bal | 2.20 | 1.15 | .60 | — | — | — | .10 | .08 | — | | | | | | | | ASTM A220 Grade 40010 |
| bal | 2.20 | 1.15 | .60 | — | — | — | .10 | .08 | — | | | | | | | | ASTM A220 Grade 45006 |
| bal | 2.20 | 1.15 | .70 | — | — | — | .10 | .08 | — | | | | | | | | ASTM A220 Grade 45008 |
| bal | 2.20 | 1.20 | .80 | — | — | — | .08 | .10 | — | | | | | | | | ASTM A220 Grade 50005 |
| bal | 2.20 | 1.30 | .60 | — | — | — | .08 | .10 | — | | | | | | | | ASTM A602 Grade M4504 |
| bal | 2.20 | 1.30 | .60 | — | — | — | .08 | .10 | — | | | | | | | | SAE J158 Grade M4504 |
| bal | 2.20 | 1.30 | .80 | — | — | — | .08 | .10 | — | | | | | | | | SAE J158 Grade M5003 |

## 26.3 Chemical Composition by Material Group

| COMPOSITION, % | | | | | | | | | | | | | | | | | MATERIAL DESIGNATION |
|---|---|---|---|---|---|---|---|---|---|---|---|---|---|---|---|---|---|
| Fe | C | Si | Mn | Cr | Ni | Cu | P | S | Mo | — | — | — | — | — | — | Other | |
| **24. MALLEABLE CAST IRONS — Tempered Martensite** | | | | | | | | | | | | | | | | | |
| bal | 2.20 | 1.10 | .80 | — | — | — | .10 | .10 | — | | | | | | | | ASTM A220 Grade 80002 |
| bal | 2.20 | 1 30 | .80 | — | — | — | .08 | .10 | — | | | | | | | | ASTM A602 Grade M5503 |
| bal | 2.20 | 1.30 | .80 | — | — | — | .08 | .10 | — | | | | | | | | SAE J158 Grade M5503 |
| bal | 2.30 | 1.20 | .50 | — | — | — | .08 | .10 | — | | | | | | | | ASTM A220 Grade 60004 |
| bal | 2.30 | 1.30 | 1.00 | — | — | — | .08 | .10 | — | | | | | | | | ASTM A602 Grade M7002 |
| bal | 2.30 | 1.30 | 1.00 | — | — | — | .08 | .10 | — | | | | | | | | SAE J158 Grade M7002 |
| bal | 2.30 | 1.30 | 1.00 | — | — | — | .08 | .10 | — | | | | | | | | ASTM A220 Grade 70003 |
| bal | 2.30 | 1.30 | 1.00 | — | — | — | .10 | .10 | — | | | | | | | | ASTM A602 Grade M8501 |
| bal | 2.30 | 1.30 | 1.00 | — | — | — | .10 | .10 | — | | | | | | | | SAE J158 Grade M8501 |
| bal | 2.30 | 1.30 | 1.00 | — | — | — | .10 | .10 | — | | | | | | | | ASTM A220 Grade 90001 |
| **25. WHITE CAST IRONS (ABRASION RESISTANT)** | | | | | | | | | | | | | | | | | |
| bal | 2.30 | 1.00 | 1.00 | 21.00 | 1.50 | 1.20 | .10 | .06 | 1.50 | | | | | | | | ASTM A532 Class II, Type D |
| bal | 2.60 | 1.00 | 1.00 | 13.00 | .50 | 1.20 | .10 | .06 | .75 | | | | | | | | ASTM A532 Class II, Type A |
| bal | 2.60 | 1.00 | 1.00 | 16.00 | .50 | 1.20 | .10 | .06 | 2.00 | | | | | | | | ASTM A532 Class II, Type B |
| bal | 2.70 | 1.00 | 1.00 | 25.00 | 1.50 | 1.20 | .10 | .06 | 1.50 | | | | | | | | ASTM A532 Class III, Type A |
| bal | 2.75 | .80 | 1.30 | 3.00 | 4.00 | — | .30 | .15 | 1.00 | | | | | | | | ASTM A532 Class I, Type B |
| bal | 3.00 | 1.00 | 1.00 | 21.00 | 1.50 | 1.20 | .10 | .06 | 1.50 | | | | | | | | ASTM A532 Class II, Type E |
| bal | 3.00 | 1.75 | 1.30 | 9.00 | 6.00 | — | .10 | .15 | 1.00 | | | | | | | | ASTM A532 Class I, Type D |
| bal | 3.20 | 1.00 | 1.00 | 16.00 | .50 | 1.20 | .10 | .06 | 3.00 | | | | | | | | ASTM A532 Class II, Type C |
| bal | 3.30 | .80 | 1.30 | 3.00 | 4.00 | — | .30 | .15 | 1.00 | | | | | | | | ASTM A532 Class I, Type A |
| bal | 3.50 | .80 | 1.30 | 1.30 | 3.50 | — | .30 | .15 | 1.00 | | | | | | | | ASTM A532 Class I, Type C |
| **26. HIGH SILICON CAST IRONS** | | | | | | | | | | | | | | | | | |
| bal | .85 | 14.50 | .50 | — | — | — | .07 | .08 | — | | | | | | | | Duriron |
| bal | .85 | 14.50 | .65 | — | — | — | — | — | 3.00 | | | | | | | | Duriclor |
| bal | 1.00 | 14.50 | 1.50 | .50 | — | .50 | — | — | .50 | | | | | | | | ASTM A518 |

| Ni | Cr | C | Si | Mn | Al | Ti | Fe | N | P | S | — | — | — | — | — | Other | MATERIAL DESIGNATION |
|---|---|---|---|---|---|---|---|---|---|---|---|---|---|---|---|---|---|
| **27. CHROMIUM-NICKEL ALLOY CASTINGS** | | | | | | | | | | | | | | | | | |
| bal | 50.00 | .10 | 1.00 | .30 | .25 | .50 | 1.00 | .30 | .02 | .02 | | | | | | | ASTM A560 Grade 50Cr-50Ni |
| bal | 60.00 | .10 | 1.00 | .30 | .25 | .50 | 1.00 | .30 | .02 | .02 | | | | | | | ASTM A560 Grade 60Cr-40Ni |

| Al | Si | Fe | Cu | Mn | Mg | Cr | Ni | Zn | Ti | Sn | — | — | — | — | — | Other | MATERIAL DESIGNATION |
|---|---|---|---|---|---|---|---|---|---|---|---|---|---|---|---|---|---|
| **28. ALUMINUM ALLOYS, WROUGHT** | | | | | | | | | | | | | | | | | |
| 99.45 | — | — | — | — | — | — | — | — | — | — | | | | | | | EC |
| 99.60 | — | .25 | — | .03 | .03 | — | — | .05 | .03 | — | | | | | | | 1060 |
| 99.75 | .15 | * | .10 | .02 | .02 | — | — | .04 | .02 | — | | | | | | | 1175 |
| 99.45 | .55 | * | .05 | .05 | .05 | — | — | .05 | .03 | — | | | | | | | 1145 |
| 99.35 | .65 | * | .05 | .05 | .05 | — | — | .10 | .06 | — | | | | | | | 1235 |
| 99.00 | 1.00 | * | .12 | .05 | — | — | — | .10 | — | — | | | | | | | 1100 |
| bal | .08 | .10 | .10 | .03 | 1.00 | — | — | .05 | — | — | | | | | | | 5657 |
| bal | .08 | .10 | .10 | .10 | 2.80 | — | — | .05 | — | — | | | | | | | 5252 |
| bal | .08 | .10 | .20 | .45 | 1.20 | — | — | .05 | — | — | | | | | | | 5457 |
| bal | .12 | .15 | 2.30 | .10 | 2.30 | .04 | — | 6.50 | .06 | — | | | | | | | 7050 |
| bal | .15 | .20 | 1.60 | .10 | 2.50 | .25 | — | 5.60 | .10 | — | | | | | | | 7175 |
| bal | .20 | .30 | 6.30 | .30 | — | — | — | — | — | — | | | | | | | 2021 |
| bal | .20 | .30 | 6.80 | .40 | .02 | — | — | .10 | .10 | — | | | | | | | 2219 |
| bal | .25 | .35 | .05 | .50 | 1.50 | .05 | — | 4.00 | .05 | — | | | | | | .15Zr | 7004 |
| bal | .25 | .35 | 1.60 | .20 | 2.50 | .15 | — | 7.70 | .10 | — | | | | | | | 7049 |
| bal | .25 | .40 | .10 | .10 | 2.80 | .35 | — | .10 | — | — | | | | | | | 5052 |
| bal | .25 | .40 | .10 | 1.00 | 3.00 | .20 | — | .25 | .20 | — | | | | | | | 5454 |
| bal | .25 | .40 | .10 | .10 | 3.90 | .35 | — | .20 | .20 | — | | | | | | | 5154 |
| bal | .25 | .40 | .10 | 1.00 | 5.50 | .20 | — | .25 | .20 | — | | | | | | | 5456 |
| bal | .25 | 1.30 | 2.70 | — | 1.80 | — | 1.00 | .10 | .10 | — | | | | | | | 2618 |

*Included with silicon.

| COMPOSITION, % | | | | | | | | | | | | | | | | | MATERIAL DESIGNATION |
|---|---|---|---|---|---|---|---|---|---|---|---|---|---|---|---|---|---|
| Al | Si | Fe | Cu | Mn | Mg | Cr | Ni | Zn | Ti | Sn | — | — | — | — | — | Other | |
| **28. ALUMINUM ALLOYS, WROUGHT (continued)** | | | | | | | | | | | | | | | | | |
| bal | .30 | .40 | .10 | .20 | 2.80 | .20 | — | 4.00 | — | — | | | | | | | 7039 |
| bal | .30 | .40 | .10 | .20 | 5.00 | .20 | — | .10 | — | — | | | | | | | 5056 |
| bal | .30 | .40 | .60 | .20 | 3.30 | .20 | — | 4.30 | — | — | | | | | | | 7079 |
| bal | .30 | .70 | .20 | .20 | 1.00 | .10 | — | .25 | — | — | | | | | | | 5005 |
| bal | .30 | .70 | .25 | 1.50 | 1.30 | — | — | .25 | — | — | | | | | | | 3004 |
| bal | .35 | .40 | .10 | .70 | 1.80 | .20 | — | 5.00 | .06 | — | | | | | | .14Zr | 7005 |
| bal | .35 | .40 | 2.60 | .20 | 3.40 | .35 | — | 8.00 | .20 | — | | | | | | | 7001 |
| bal | .40 | * | .04 | .01 | 2.80 | .35 | — | .10 | — | — | | | | | | | 5652 |
| bal | .40 | .40 | .10 | 1.00 | 4.90 | .25 | — | .25 | .15 | — | | | | | | | 5083 |
| bal | .40 | .50 | .10 | .70 | 4.50 | .25 | — | .25 | .15 | — | | | | | | | 5086 |
| bal | .40 | .50 | 2.00 | .30 | 2.90 | .25 | — | 6.00 | .20 | — | | | | | | | 7075 |
| bal | .40 | .50 | 2.40 | .30 | 3.10 | .35 | — | 7.00 | .20 | — | | | | | | | 7178 |
| bal | .40 | .70 | .20 | .10 | 1.80 | .10 | — | .25 | — | — | | | | | | | 5050 |
| bal | .40 | .70 | 6.00 | — | — | — | — | .30 | — | — | | | | | | | 2011 |
| bal | .45 | * | .05 | .01 | 3.90 | .35 | — | .20 | .05 | — | | | | | | | 5254 |
| bal | .50 | .50 | 4.90 | .90 | 1.50 | .10 | — | .25 | .15 | — | | | | | | | 2024 |
| bal | .50 | .80 | .40 | .10 | .80 | — | — | .20 | — | — | | | | | | | 6951 |
| bal | .55 | .35 | .10 | — | 1.40 | .35 | — | .10 | — | — | | | | | | | 6053 |
| bal | .60 | .15 | .20 | .05 | .90 | — | — | — | — | — | | | | | | | 6463 |
| bal | .60 | .35 | .10 | .10 | .90 | .10 | — | .10 | .10 | — | | | | | | | 6063 |
| bal | .60 | .70 | .20 | 1.50 | — | — | — | .10 | — | — | | | | | | | 3003 |
| bal | .60 | .70 | .30 | 1.50 | .60 | .10 | — | .25 | .10 | — | | | | | | | 3005 |
| bal | .65 | .50 | .10 | — | 1.50 | .35 | — | 2.40 | — | — | | | | | | | 6253 |
| bal | .70 | .50 | .10 | .03 | .80 | .03 | — | .10 | — | — | | | | | | | 6101 |
| bal | .80 | .70 | .40 | .15 | 1.20 | .15 | — | .25 | .15 | — | | | | | | | 6262 |
| bal | .80 | .70 | .40 | .15 | 1.20 | .35 | — | .25 | .15 | — | | | | | | | 6061 |
| bal | .80 | .70 | 3.00 | .20 | .50 | .10 | — | .25 | — | — | | | | | | | 2117 |

*Included with silicon.

| COMPOSITION, % | | | | | | | | | | | | | | | | | MATERIAL DESIGNATION |
|---|---|---|---|---|---|---|---|---|---|---|---|---|---|---|---|---|---|
| Al | Si | Fe | Cu | Mn | Mg | Cr | Ni | Zn | Ti | Sn | — | — | — | — | — | Other | |
| **28. ALUMINUM ALLOYS, WROUGHT (continued)** | | | | | | | | | | | | | | | | | |
| bal | .80 | .70 | 4.50 | .80 | .80 | .10 | — | .25 | .15 | — | | | | | | | 2017 |
| bal | .90 | 1.00 | 4.50 | .20 | .90 | .10 | 2.00 | .25 | — | — | | | | | | | 2018 |
| bal | .90 | 1.00 | 4.50 | .20 | 1.80 | .10 | 2.00 | .25 | — | — | | | | | | | 2218 |
| bal | 1.00 | .70 | 5.00 | 1.00 | .80 | .10 | — | .25 | .15 | — | | | | | | | 2014 |
| bal | 1.00 | 1.00 | 5.00 | .80 | .05 | .10 | — | .25 | .15 | — | | | | | | | 2025 |
| bal | 1.20 | 1.00 | .35 | .20 | .80 | .35 | — | .25 | .15 | — | | | | | | | 6151 |
| bal | 1.70 | .50 | .40 | 1.00 | 1.20 | .10 | — | .25 | .15 | — | | | | | | | 6070 |
| bal | 1.80 | .50 | 1.20 | 1.10 | 1.40 | .40 | — | .25 | .20 | — | | | | | | | 6066 |
| bal | 13.50 | 1.00 | 1.30 | — | 1.30 | .10 | 1.30 | .25 | — | — | | | | | | | 4032 |
| **29. ALUMINUM ALLOYS, CAST — Sand and Permanent Mold** | | | | | | | | | | | | | | | | | |
| bal | — | — | 8.00 | .50 | 6.00 | — | .50 | — | — | — | | | | | | | A140 |
| bal | .06 | .10 | 5.00 | .35 | — | — | — | — | .35 | — | | | | | | .15V, .25Zr | 224.0 |
| bal | .10 | .15 | 4.80 | .35 | .35 | — | — | — | .25 | — | | | | | | | 201.0 |
| bal | .15 | .15 | .05 | .25 | 7.00 | — | — | — | .25 | — | | | | | | | 535.0 |
| bal | .15 | .15 | .10 | .10 | 1.00 | .20 | — | 7.00 | .20 | — | | | | | | | 771.0 |
| bal | .15 | .50 | .65 | .05 | .80 | — | — | 7.00 | .25 | — | | | | | | | A712.0 |
| bal | .20 | .80 | .20 | .60 | 1.80 | .40 | — | 3.00 | .25 | — | | | | | | | 705.0 |
| bal | .20 | .80 | .20 | .60 | 2.40 | .40 | — | 4.50 | .25 | — | | | | | | | 707.0 |
| bal | .20 | 1.50 | 5.00 | .25 | — | — | 1.50 | — | .20 | — | | | | | | .20Zr,.25Sb,.25Co | Hiduminium RR-350 |
| bal | .25 | .30 | .25 | .15 | 10.00 | — | — | .15 | .25 | — | | | | | | | 520.0 |
| bal | .25 | 1.10 | 1.00 | .60 | .50 | .35 | .15 | 8.00 | .25 | — | | | | | | | 713.0 |
| bal | .30 | .40 | .10 | .30 | 4.00 | — | — | 2.00 | .20 | — | | | | | | | A514.0 |
| bal | .30 | .50 | .25 | .10 | .65 | .60 | — | 6.00 | .20 | — | | | | | | | D712.0 |
| bal | .35 | .50 | .15 | .35 | 4.00 | — | — | .15 | .25 | — | | | | | | | 514.0 |
| bal | .40 | .70 | 2.00 | .10 | .90 | — | 1.50 | — | .20 | 7.00 | | | | | | | B850.0 |

| COMPOSITION, % | | | | | | | | | | | | | | | | | MATERIAL DESIGNATION |
|---|---|---|---|---|---|---|---|---|---|---|---|---|---|---|---|---|---|
| Al | Si | Fe | Cu | Mn | Mg | Cr | Ni | Zn | Ti | Sn | — | — | — | — | — | Other | |
| **28. ALUMINUM ALLOYS, CAST — Sand and Permanent Mold (continued)** | | | | | | | | | | | | | | | | | |
| bal | .70 | .70 | 1.00 | .10 | .10 | — | 1.00 | — | .20 | 7.00 | | | | | | | 850.0 |
| bal | .70 | 1.00 | 4.50 | .35 | 1.80 | .25 | 2.30 | .35 | .25 | — | | | | | | | 242.0 |
| bal | 1.50 | 1.00 | 5.00 | .35 | .03 | — | — | .35 | .25 | — | | | | | | | 295.0 |
| bal | 2.00 | .60 | .35 | .80 | 4.00 | .25 | — | .35 | .25 | — | | | | | | | B514.0 |
| bal | 2.00 | 1.50 | 10.00 | .50 | .35 | — | .50 | .80 | .25 | — | | | | | | | 222.0 |
| bal | 2.50 | .70 | 1.00 | .10 | .10 | — | .70 | — | .20 | 7.00 | | | | | | | A850.0 |
| bal | 3.00 | 1.20 | 5.00 | .35 | .05 | — | .35 | .50 | .25 | — | | | | | | | B295.0 |
| bal | 3.00 | 1.20 | 8.00 | .60 | .10 | — | .35 | 2.50 | .25 | — | | | | | | | 213.0 |
| bal | 3.50 | 1.20 | 4.00 | .50 | .10 | — | .35 | 1.00 | .25 | — | | | | | | | 208.0 |
| bal | 5.50 | .20 | 1.25 | .10 | .60 | — | — | .10 | .20 | — | | | | | | | C355.0 |
| bal | 5.50 | .60 | 1.25 | .50 | .60 | .20 | — | .35 | .25 | — | | | | | | | 355.0 |
| bal | 6.00 | .80 | .15 | .35 | .05 | — | — | .35 | .25 | — | | | | | | | B443.0 |
| bal | 6.00 | 1.00 | 5.00 | .50 | .10 | — | — | 1.00 | .25 | — | | | | | | | 308.0 |
| bal | 6.50 | 1.00 | 4.00 | .50 | .10 | — | .35 | 1.00 | .25 | — | | | | | | | 319.0 |
| bal | 7.50 | .15 | .05 | .03 | .60 | — | — | .05 | .20 | — | | | | | | | 357.0 |
| bal | 7.50 | .20 | .20 | .10 | .40 | — | — | .10 | .20 | — | | | | | | | A356.0 |
| bal | 7.50 | .60 | .25 | .35 | .40 | — | — | .35 | .25 | — | | | | | | | 356.0 |
| bal | 8.50 | 1.00 | 2.00 | .60 | .60 | .35 | .25 | 1.50 | .25 | — | | | | | | | 328.0 |
| bal | 9.00 | .20 | .20 | .10 | .60 | — | — | .10 | .20 | — | | | | | | | 359.0 |
| bal | 9.00 | .20 | 1.80 | .10 | .50 | — | — | .10 | .20 | — | | | | | | | 354.0 |
| bal | 10.00 | 1.00 | 4.00 | .50 | .50 | — | .50 | 1.00 | .25 | — | | | | | | | 333.0 |
| bal | 10.50 | 1.20 | 4.00 | .50 | 1.00 | — | .50 | 1.00 | .25 | — | | | | | | | F332.0 |
| bal | 13.00 | 1.20 | 1.50 | .35 | 1.00 | — | 2.50 | .35 | .25 | — | | | | | | | A332.0 |

| COMPOSITION, % | | | | | | | | | | | | | | | | | MATERIAL DESIGNATION |
|---|---|---|---|---|---|---|---|---|---|---|---|---|---|---|---|---|---|
| Al | Si | Fe | Cu | Mn | Mg | Cr | Ni | Zn | Ti | Sn | — | — | — | — | — | Other | |
| **29. ALUMINUM ALLOYS, CAST — Die Castings** | | | | | | | | | | | | | | | | | |
| bal | .35 | 1.80 | .25 | .35 | 8.00 | — | .15 | .15 | — | .15 | | | | | | | 518.0 |
| bal | 5.50 | 2.00 | .60 | .35 | .10 | — | .50 | .50 | — | .15 | | | | | | | C443.0 |
| bal | 8.50 | 1.30 | 3.50 | .50 | .10 | — | .50 | 3.00 | — | .35 | | | | | | | A380.0 |
| bal | 8.50 | 2.00 | 3.50 | .50 | .10 | — | .50 | 3.00 | — | .35 | | | | | | | 380.0 |
| bal | 9.50 | 1.30 | .60 | .35 | .50 | — | .50 | .50 | — | .15 | | | | | | | A360.0 |
| bal | 9.50 | 2.00 | .60 | .35 | .50 | — | .50 | .50 | — | .15 | | | | | | | 360.0 |
| bal | 10.50 | 1.30 | 2.50 | .50 | .10 | — | .30 | 3.00 | — | .15 | | | | | | | 383.0 |
| bal | 11.50 | 1.30 | 4.00 | .50 | .10 | — | .50 | 1.00 | — | .35 | | | | | | | A384.0 |
| bal | 12.00 | 1.30 | 1.00 | .35 | .10 | — | .50 | .50 | — | .15 | | | | | | | A413.0 |
| bal | 12.00 | 2.00 | 1.00 | .35 | .10 | — | .50 | .50 | — | .15 | | | | | | | 413.0 |
| bal | 17.00 | 1.00 | 4.50 | .10 | .60 | — | — | .10 | .20 | — | | | | | | | 390.0 |
| bal | 19.00 | 1.50 | .60 | .50 | 1.00 | — | .50 | .50 | .20 | .30 | | | | | | | 392.0 |

| COMPOSITION, % | | | | | | | | | | | | | | | | | MATERIAL DESIGNATION |
|---|---|---|---|---|---|---|---|---|---|---|---|---|---|---|---|---|---|
| Mg | Al | Mn | Zn | Zr | Cu | Ni | Si | Fe | — | — | — | — | — | — | — | Other | |
| **30. MAGNESIUM ALLOYS, WROUGHT** | | | | | | | | | | | | | | | | | |
| bal | — | — | .30 | .70 | .10 | .01 | — | — | | | | | | | | 3.20Th | HK31A |
| bal | — | .80 | — | — | — | — | — | — | | | | | | | | 2.00Th | HM21A |
| bal | — | 1.20 | — | — | — | — | — | — | | | | | | | | 3.00Th | HM31A |
| bal | — | — | 4.00 | .45 | .10 | .01 | — | — | | | | | | | | | ZK40A |
| bal | — | — | 5.50 | .45 | — | — | — | — | | | | | | | | | ZK60A |
| bal | 2.00 | .15 | 1.20 | — | .05 | .002 | .05 | .005 | | | | | | | | .20Ca | AZ21A |
| bal | 3.00 | .20 | 1.00 | — | — | — | — | — | | | | | | | | .04Ca | AZ31B |
| bal | 3.00 | .15 | 1.00 | — | .10 | .03 | .10 | — | | | | | | | | | AZ31C |
| bal | 6.50 | .15 | 1.00 | — | .05 | .005 | .10 | .005 | | | | | | | | | AZ61A |
| bal | 8.50 | .12 | .50 | — | .05 | .005 | .10 | .005 | | | | | | | | | AZ80A |

| COMPOSITION, % | | | | | | | | | | | | | | | | | MATERIAL DESIGNATION |
|---|---|---|---|---|---|---|---|---|---|---|---|---|---|---|---|---|---|
| Mg | Al | Mn | Zn | Zr | Cu | Ni | Si | Fe | — | — | — | — | — | — | — | Other | |
| 31. MAGNESIUM ALLOYS, CAST | | | | | | | | | | | | | | | | | |
| bal | — | — | 2.60 | .70 | .10 | .01 | — | — | | | | | | | | 3.20 Rare Earths | EZ33A |
| bal | — | — | .30 | .70 | .10 | .01 | — | — | | | | | | | | 3.20Th | HK31A |
| bal | — | — | 2.10 | .70 | .10 | .01 | — | — | | | | | | | | 3.20Th | HZ32A |
| bal | — | — | — | .60 | — | — | — | — | | | | | | | | | K1A |
| bal | — | — | — | .70 | .10 | .01 | — | — | | | | | | | | 2.20 Rare Earths, 2.50 Ag | QE22A |
| bal | — | .15 | 4.20 | .70 | .10 | .01 | — | — | | | | | | | | 1.20 Rare Earths | ZE41A |
| bal | — | — | 5.70 | .70 | — | — | — | — | | | | | | | | 2.50 Rare Earths | ZE63A |
| bal | — | — | 5.70 | .70 | .10 | .01 | — | — | | | | | | | | 1.80Th | ZH62A |
| bal | — | — | 4.60 | .70 | .10 | .01 | — | — | | | | | | | | | ZK51A |
| bal | — | — | 6.00 | .80 | — | — | — | — | | | | | | | | | ZK61A |
| bal | 4.20 | .35 | .12 | — | .06 | .03 | 1.00 | — | | | | | | | | | AS41A |
| bal | 6.00 | .13 | .20 | — | .35 | .03 | .50 | — | | | | | | | | | AM60A |
| bal | 6.00 | .15 | 3.00 | — | .15 | .01 | .25 | — | | | | | | | | | AZ63A |
| bal | 7.60 | .13 | .70 | — | .10 | .01 | .30 | — | | | | | | | | | AZ81A |
| bal | 8.70 | .13 | .70 | — | .10 | .01 | .30 | — | | | | | | | | | AZ91C |
| bal | 9.00 | .10 | 2.00 | — | .25 | .01 | .30 | — | | | | | | | | | AZ92A |
| bal | 9.00 | .13 | .70 | — | .10 | — | .50 | — | | | | | | | | .03Mo | AZ91A |
| bal | 9.00 | .15 | .70 | — | .25 | .01 | .20 | — | | | | | | | | | AZ91B |
| bal | 10.00 | .10 | .30 | — | .10 | .01 | .30 | — | | | | | | | | | AM100A |

| COMPOSITION, % | | | | | | | | | | | | | | | | | MATERIAL DESIGNATION |
|---|---|---|---|---|---|---|---|---|---|---|---|---|---|---|---|---|---|
| Ti | Al | Sn | Zr | Mo | V | Cr | Fe | Si | Pd | O | — | — | — | — | — | Other | |
| 32. TITANIUM ALLOYS, WROUGHT | | | | | | | | | | | | | | | | | |
| 99.5 | — | — | — | — | — | — | — | — | — | — | | | | | | | Ti 99.5 |
| 99.2 | — | — | — | — | — | — | — | — | — | — | | | | | | | Ti 99.2 |

| COMPOSITION, % | | | | | | | | | | | | | | | | | MATERIAL DESIGNATION |
|---|---|---|---|---|---|---|---|---|---|---|---|---|---|---|---|---|---|
| Ti | Al | Sn | Zr | Mo | V | Cr | Fe | Si | Pd | O | — | — | — | — | — | Other | |
| 32. TITANIUM ALLOYS, WROUGHT (continued) | | | | | | | | | | | | | | | | | |
| 99.0 | — | — | — | — | — | — | — | — | — | — | | | | | | | Ti 99.0 |
| bal | — | — | — | — | — | — | — | — | .20 | — | | | | | | | Ti-0.2 Pd |
| 98.9 | — | — | — | — | — | — | — | — | — | — | | | | | | | Ti 98.9 |
| bal | — | — | — | .30 | — | — | .30 | — | — | — | | | | | | .80 Ni | Ti CODE-12 |
| bal | — | — | — | — | — | — | — | — | — | — | | | | | | 8.00 Mn | Ti-8 Mn |
| bal | 1.00 | — | — | — | 8.00 | — | 5.00 | — | — | — | | | | | | | Ti-1Al-8V-5Fe |
| bal | 2.50 | 11.00 | 5.00 | 1.00 | — | — | .10 | .25 | — | — | | | | | | | Ti-2Al-11Sn-5Zr-1Mo |
| bal | 3.00 | — | 4.00 | 4.00 | 8.00 | 6.00 | .30 | — | — | — | | | | | | | Ti-3Al-8V-6Cr-4Mo-4Zr |
| bal | 3.00 | — | — | — | 2.50 | — | — | — | — | — | | | | | | | Ti-3Al-2.5V |
| bal | 3.00 | — | — | 8.00 | 8.00 | — | 2.00 | — | — | — | | | | | | | Ti-8Mo-8V-2Fe-3Al |
| bal | 3.00 | — | — | — | 10.00 | — | 2.00 | — | — | — | | | | | | | Ti-10V-2Fe-3Al |
| bal | 3.00 | — | — | — | 13.00 | 11.00 | .35 | — | — | — | | | | | | | Ti-13V-11Cr-3Al |
| bal | 5.00 | 2.50 | — | — | — | — | — | — | — | — | | | | | | | Ti-5Al-2.5Sn |
| bal | 5.00 | 2.50 | — | — | — | — | — | — | — | — | | | | | | Extra Low Interstitials | Ti-5Al-2.5Sn ELI |
| bal | 5.00 | 6.00 | 2.00 | 1.00 | — | — | — | — | — | — | | | | | | | Ti-5Al-6Sn-2Zr-1Mo |
| bal | 5.00 | 2.00 | 2.00 | 4.00 | — | 4.00 | .30 | — | — | — | | | | | | | Ti-5Al-2Sn-2Zr-4Mo-4Cr |
| bal | 5.00 | 2.00 | 2.00 | 4.00 | — | 4.00 | .30 | — | — | — | | | | | | | Ti-17 |
| bal | 6.00 | — | — | .80 | — | — | .20 | — | — | — | | | | | | 2.00Cb 1.00Ta | Ti-6Al-2Cb-1Ta-0.8Mo |
| bal | 6.00 | 2.00 | 4.00 | 2.00 | — | — | .25 | — | — | — | | | | | | | Ti-6Al-2Sn-4Zr-2Mo |
| bal | 6.00 | 2.00 | 4.00 | 2.00 | — | — | — | .25 | — | — | | | | | | | Ti-6Al-2Sn-4Zr-2Mo-.25Si |
| bal | 6.00 | 2.00 | 4.00 | 6.00 | — | — | .15 | — | — | — | | | | | | | Ti-6Al-2Sn-4Zr-6Mo |
| bal | 6.00 | — | — | — | 4.00 | — | .30 | — | — | — | | | | | | | Ti-6Al-4V |
| bal | 6.00 | — | — | — | 4.00 | — | — | — | — | — | | | | | | Extra Low Interstitials | Ti-6Al-4V ELI |
| bal | 6.00 | 2.00 | — | — | 6.00 | — | .70 | — | — | — | | | | | | .70Cu | Ti-6Al-6V-2Sn |
| bal | 7.00 | — | — | 4.00 | — | — | .30 | — | — | — | | | | | | | Ti-7Al-4Mo |
| bal | 8.00 | — | — | 1.00 | 1.00 | — | .30 | — | — | — | | | | | | | Ti-8Al-1Mo-1V |
| bal | — | 4.50 | 6.00 | 11.50 | — | — | .35 | — | — | — | | | | | | | Ti-11.5Mo-6Zr-4.5Sn |

| COMPOSITION, % | | | | | | | | | | | | | | | | | MATERIAL DESIGNATION |
|---|---|---|---|---|---|---|---|---|---|---|---|---|---|---|---|---|---|
| Ti | Al | Sn | Zr | Mo | V | Cr | Fe | Si | Pd | O | — | — | — | — | — | Other | |
| **33. TITANIUM ALLOYS, CAST** | | | | | | | | | | | | | | | | | |
| 99.0 | — | — | — | — | — | — | — | — | — | — | | | | | | | Ti 99.0 |
| bal | — | — | — | — | — | — | — | — | .20 | — | | | | | | | Ti-0.2Pd |
| bal | — | — | — | — | — | — | .20 | — | — | .18 | | | | | | | ASTM B367 Grade C-1 |
| bal | — | — | — | — | — | — | .30 | — | — | .25 | | | | | | | ASTM B367 Grade C-2 |
| bal | — | — | — | — | — | — | .20 | — | .12 | .18 | | | | | | | ASTM B367 Grade C-7A |
| bal | — | — | — | — | — | — | .30 | — | .12 | .25 | | | | | | | ASTM B367 Grade C-7B |
| bal | — | — | — | — | — | — | .30 | — | — | .35 | | | | | | | ASTM B367 Grade C-3 |
| bal | — | — | — | — | — | — | .50 | — | — | .40 | | | | | | | ASTM B367 Grade C-4 |
| bal | — | — | — | — | — | — | .30 | — | .12 | .35 | | | | | | | ASTM B367 Grade C-8A |
| bal | — | — | — | — | — | — | .50 | — | .12 | .40 | | | | | | | ASTM B367 Grade C-8B |
| bal | 5.00 | 2.50 | — | — | — | — | — | — | — | — | | | | | | | Ti-5Al-2.5Sn |
| bal | 5.00 | 2.50 | — | — | — | — | .50 | — | — | .20 | | | | | | | ASTM B367 Grade C-6 |
| bal | 6.00 | — | — | — | 4.00 | — | .30 | — | — | .20 | | | | | | | Ti-6Al-4V |
| bal | 6.00 | — | — | — | 4.00 | — | .50 | — | — | .20 | | | | | | | ASTM B367 Grade C-5 |
| bal | 6.00 | 2.00 | 4.00 | 2.00 | — | — | — | — | — | — | | | | | | | Ti-6Al-2Sn-4Zr-2Mo |
| bal | 8.00 | — | — | 1.00 | 1.00 | — | — | — | — | — | | | | | | | Ti-8Al-1Mo-1V |

| COMPOSITION, % | | | | | | | | | | | | | | | | | MATERIAL DESIGNATION |
|---|---|---|---|---|---|---|---|---|---|---|---|---|---|---|---|---|---|
| Cu | Ag | As | Sb | P | Te | Fe | Sn | Zn | Pb | Al | Si | Mn | Ni | Co | Cr | Other | |
| **34. COPPER ALLOYS, WROUGHT** | | | | | | | | | | | | | | | | | |
| 99.99 | — | — | — | .003 | .001 | — | — | — | — | — | — | — | — | — | — | | 101 |
| 99.95 | — | — | — | — | — | — | — | — | — | — | — | — | — | — | — | | 102 |
| 99.95 | .03 | — | — | — | — | — | — | — | — | — | — | — | — | — | — | | 104 |
| 99.95 | .03 | — | — | — | — | — | — | — | — | — | — | — | — | — | — | | 105 |
| 99.95 | .08 | — | — | — | — | — | — | — | — | — | — | — | — | — | — | | 107 |
| 99.90 | .04 | — | — | — | — | — | — | — | — | — | — | — | — | — | — | | 109 |
| 99.90 | — | — | — | — | — | — | — | — | — | — | — | — | — | — | — | | 110 |
| 99.90 | — | — | — | — | — | — | — | — | — | — | — | — | — | — | — | | 111 |

| COMPOSITION, % | | | | | | | | | | | | | | | | | MATERIAL DESIGNATION |
|---|---|---|---|---|---|---|---|---|---|---|---|---|---|---|---|---|---|
| Cu | Ag | As | Sb | P | Te | Fe | Sn | Zn | Pb | Al | Si | Mn | Ni | Co | Cr | Other | |
| **34. COPPER ALLOYS, WROUGHT (continued)** | | | | | | | | | | | | | | | | | |
| 99.90 | .03 | — | — | — | — | — | — | — | — | — | — | — | — | — | — | | 113 |
| 99.90 | .03 | — | — | — | — | — | — | — | — | — | — | — | — | — | — | | 114 |
| 99.90 | .05 | — | — | — | — | — | — | — | — | — | — | — | — | — | — | | 115 |
| 99.90 | .08 | — | — | — | — | — | — | — | — | — | — | — | — | — | — | | 116 |
| 99.93 | — | — | — | .01 | — | — | — | — | — | — | — | — | — | — | — | | 119 |
| 99.90 | — | — | — | .01 | — | — | — | — | — | — | — | — | — | — | — | | 120 |
| 99.90 | .01 | — | — | .01 | — | — | — | — | — | — | — | — | — | — | — | | 121 |
| 99.90 | — | — | — | .02 | — | — | — | — | — | — | — | — | — | — | — | | 122 |
| 99.88 | — | .01 | .003 | — | .02 | — | — | — | .004 | — | — | — | .05 | — | — | .003Bi | 125 |
| 99.88 | .03 | .01 | .003 | — | .02 | — | — | — | .004 | — | — | — | .05 | — | — | .003Bi | 127 |
| 99.88 | .03 | .01 | .003 | — | .02 | — | — | — | .004 | — | — | — | .05 | — | — | .003Bi | 128 |
| 99.88 | .05 | .01 | .003 | — | .02 | — | — | — | .004 | — | — | — | .05 | — | — | .003Bi | 129 |
| 99.88 | .08 | .01 | .003 | — | .02 | — | — | — | .004 | — | — | — | .05 | — | — | .003Bi | 130 |
| 99.60 | — | .30 | — | .02 | — | — | — | — | — | — | — | — | — | — | — | | 142 |
| 99.90 | — | — | — | — | — | — | — | — | — | — | — | — | — | — | — | .15Cd | 143 |
| 99.50 | — | — | — | .01 | .50 | — | — | — | — | — | — | — | — | — | — | | 145 |
| 99.60 | — | — | — | — | — | — | — | — | — | — | — | — | — | — | — | .40S | 147 |
| 99.80 | — | — | — | — | — | — | — | — | — | — | — | — | — | — | — | .15Zr | 150 |
| 99.75 | .07 | — | — | .06 | — | — | — | — | — | — | — | — | — | — | — | .11Mg | 155 |
| 99.00 | — | — | — | — | — | — | — | — | — | — | — | — | — | — | — | 1.00Cd | 162 |
| 98.60 | — | — | — | — | — | — | .60 | — | — | — | — | — | — | — | — | .80Cd | 165 |
| 98.30 | — | — | — | — | — | .10 | — | — | — | — | — | — | .10 | .10 | — | 1.70Be | 170 |
| 98.10 | — | — | — | — | — | .10 | — | — | — | — | — | — | .10 | .10 | — | 1.90Be | 172 |
| 97.70 | — | — | — | — | — | .10 | — | — | .40 | — | — | — | .10 | .10 | — | 1.90Be | 173 |
| 96.90 | — | — | — | — | — | .08 | — | — | — | — | — | — | — | 2.50 | — | .60Be | 175 |
| 99.10 | — | — | — | — | — | .08 | — | — | .05 | — | .08 | — | — | — | .90 | | 182 |
| 99.20 | — | — | — | — | — | .10 | — | .50 | — | — | .08 | — | — | — | .80 | | 184 |
| 99.20 | .10 | — | — | — | — | — | — | — | .01 | — | — | — | — | — | .70 | | 185 |
| 99.00 | — | — | — | — | — | — | — | — | 1.00 | — | — | — | — | — | — | | 187 |

| COMPOSITION, % | | | | | | | | | | | | | | | | MATERIAL DESIGNATION |
|---|---|---|---|---|---|---|---|---|---|---|---|---|---|---|---|---|
| Cu | Ag | As | Sb | P | Te | Fe | Sn | Zn | Pb | Al | Si | Mn | Ni | Co | Cr | Other | |
| **34. COPPER ALLOYS, WROUGHT (continued)** | | | | | | | | | | | | | | | | | |
| 98.70 | — | — | — | .05 | — | — | .80 | .10 | .02 | .01 | .30 | .20 | — | — | — | | 189 |
| 98.60 | — | — | — | .30 | — | — | — | — | — | — | — | — | 1.10 | — | — | | 190 |
| 98.20 | — | — | — | .20 | .50 | — | — | — | — | — | — | — | 1.10 | — | — | | 191 |
| 98.97 | — | — | — | .03 | — | 1.00 | — | — | — | — | — | — | — | — | — | | 192 |
| 97.40 | — | — | — | .04 | — | 2.40 | .03 | .10 | .03 | — | — | — | — | — | — | | 194 |
| 97.00 | — | — | — | .10 | — | 1.50 | .60 | — | — | — | — | — | — | .80 | — | | 195 |
| bal | — | — | — | — | — | .05 | — | 4.90 | .05 | — | — | — | — | — | — | | 210 |
| bal | — | — | — | — | — | .05 | — | 9.90 | .05 | — | — | — | — | — | — | | 220 |
| bal | — | — | — | — | — | .05 | — | 12.40 | .05 | — | — | — | — | — | — | | 226 |
| bal | — | — | — | — | — | .05 | — | 14.90 | .05 | — | — | — | — | — | — | | 230 |
| bal | — | — | — | — | — | .05 | — | 19.90 | .05 | — | — | — | — | — | — | | 240 |
| bal | — | — | — | — | — | .05 | — | 29.90 | .05 | — | — | — | — | — | — | | 260 |
| bal | — | — | — | — | — | .05 | — | 33.80 | .15 | — | — | — | — | — | — | | 268 |
| bal | — | — | — | — | — | .05 | — | 34.85 | .10 | — | — | — | — | — | — | | 270 |
| bal | — | — | — | — | — | .05 | — | 39.70 | .25 | — | — | — | — | — | — | | 280 |
| bal | — | — | — | — | — | .10 | — | 9.00 | 1.90 | — | — | — | — | — | — | | 314 |
| bal | — | — | — | .10 | — | .10 | — | 8.00 | 1.80 | — | — | — | 1.00 | — | — | | 316 |
| bal | — | — | — | — | — | .07 | — | 33.40 | .50 | — | — | — | — | — | — | | 330 |
| bal | — | — | — | — | — | .07 | — | 32.30 | 1.60 | — | — | — | — | — | — | | 332 |
| bal | — | — | — | — | — | .10 | — | 34.40 | .50 | — | — | — | — | — | — | | 335 |
| bal | — | — | — | — | — | .10 | — | 33.90 | 1.00 | — | — | — | — | — | — | | 340 |
| bal | — | — | — | — | — | .10 | — | 33.40 | 2.00 | — | — | — | — | — | — | | 342 |
| bal | — | — | — | — | — | .10 | — | 37.40 | .30 | — | — | — | — | — | — | | 349 |
| bal | — | — | — | — | — | .10 | — | 36.40 | 1.00 | — | — | — | — | — | — | | 350 |
| bal | — | — | — | — | — | .10 | — | 36.10 | 1.80 | — | — | — | — | — | — | | 353 |
| bal | — | — | — | — | — | .10 | — | 35.40 | 2.50 | — | — | — | — | — | — | | 356 |
| bal | — | — | — | — | — | .25 | — | 35.25 | 3.00 | — | — | — | — | — | — | | 360 |
| bal | — | — | — | — | — | .15 | .25 | 39.00 | .60 | — | — | — | — | — | — | | 365 |

| COMPOSITION, % | | | | | | | | | | | | | | | | | MATERIAL DESIGNATION |
|---|---|---|---|---|---|---|---|---|---|---|---|---|---|---|---|---|---|
| Cu | Ag | As | Sb | P | Te | Fe | Sn | Zn | Pb | Al | Si | Mn | Ni | Co | Cr | Other | |
| **34. COPPER ALLOYS, WROUGHT (continued)** | | | | | | | | | | | | | | | | | |
| bal | — | .06 | — | — | — | .10 | .20 | 39.00 | .60 | — | — | — | — | — | — | | 366 |
| bal | — | — | .06 | — | — | .10 | .20 | 39.00 | .60 | — | — | — | — | — | — | | 367 |
| bal | — | — | — | .06 | — | .10 | .20 | 39.00 | .60 | — | — | — | — | — | — | | 368 |
| bal | — | — | — | — | — | .10 | — | 38.50 | 1.00 | — | — | — | — | — | — | | 370 |
| bal | — | — | — | — | — | .25 | — | 37.75 | 2.00 | — | — | — | — | — | — | | 377 |
| bal | — | — | — | — | — | .30 | — | 39.70 | 3.00 | — | — | — | — | — | — | | 385 |
| 90.00 | — | — | — | — | — | .05 | .50 | 9.35 | .10 | — | — | — | — | — | — | | 411 |
| 90.00 | — | — | — | — | — | .05 | 1.00 | .85 | .10 | — | — | — | — | — | — | | 413 |
| 88.50 | — | — | — | .20 | — | .05 | 2.00 | 9.20 | .05 | — | — | — | — | — | — | | 425 |
| 81.00 | — | — | — | — | — | .05 | .90 | 17.95 | .10 | — | — | — | — | — | — | | 435 |
| 72.00 | — | — | — | — | — | .05 | 1.00 | 26.90 | .07 | — | — | — | — | — | — | | 442 |
| 71.00 | — | .07 | — | — | — | .06 | 1.00 | 27.80 | .07 | — | — | — | — | — | — | | 443 |
| 71.00 | — | — | .07 | — | — | .06 | 1.00 | 27.80 | .07 | — | — | — | — | — | — | | 444 |
| 71.00 | — | — | — | .07 | — | .06 | 1.00 | 27.80 | .07 | — | — | — | — | — | — | | 445 |
| 60.00 | — | — | — | — | — | .10 | .80 | 38.90 | .20 | — | — | — | — | — | — | | 464 |
| 60.00 | — | .07 | — | — | — | .10 | .80 | 38.80 | .20 | — | — | — | — | — | — | | 465 |
| 60.00 | — | — | .07 | — | — | .10 | .80 | 38.80 | .20 | — | — | — | — | — | — | | 466 |
| 60.00 | — | — | — | .07 | — | .10 | .80 | 38.80 | .20 | — | — | — | — | — | — | | 467 |
| 60.50 | — | — | — | — | — | .10 | .80 | 37.90 | .70 | — | — | — | — | — | — | | 482 |
| 60.00 | — | — | — | — | — | .10 | .70 | 37.40 | 1.80 | — | — | — | — | — | — | | 485 |
| bal | — | — | — | .25 | — | .10 | 1.10 | .20 | .05 | — | — | — | — | — | — | | 505 |
| bal | — | — | — | .25 | — | .10 | 5.00 | .20 | .05 | — | — | — | — | — | — | | 510 |
| bal | — | — | — | .25 | — | .10 | 4.20 | .20 | .05 | — | — | — | — | — | — | | 511 |
| bal | — | — | — | .25 | — | .10 | 8.00 | .20 | .05 | — | — | — | — | — | — | | 521 |
| bal | — | — | — | .25 | — | .10 | 10.00 | .20 | .05 | — | — | — | — | — | — | | 524 |
| bal | — | — | — | .40 | — | .10 | 4.00 | 4.00 | 4.00 | — | — | — | — | — | — | | 544 |
| bal | — | .25 | — | — | — | .10 | — | — | .10 | 5.00 | — | — | — | — | — | | 608 |
| bal | — | — | — | — | — | .40 | — | .20 | .02 | 8.00 | .10 | — | — | — | — | | 610 |

| COMPOSITION, % | | | | | | | | | | | | | | | | | MATERIAL DESIGNATION |
|---|---|---|---|---|---|---|---|---|---|---|---|---|---|---|---|---|---|
| Cu | Ag | As | Sb | P | Te | Fe | Sn | Zn | Pb | Al | Si | Mn | Ni | Co | Cr | Other | |
| **34. COPPER ALLOYS, WROUGHT (continued)** | | | | | | | | | | | | | | | | | |
| bal | — | — | — | — | — | 2.00 | .30 | — | — | 7.00 | — | .50 | .50 | — | — | | 613 |
| bal | — | — | — | .01 | — | 2.00 | — | .20 | .01 | 7.00 | — | .75 | — | — | — | | 614 |
| bal | — | — | — | — | — | 1.00 | — | .02 | .02 | 10.00 | .10 | — | — | — | — | | 618 |
| bal | — | — | — | — | — | 4.00 | .40 | .60 | .02 | 9.00 | — | — | — | — | — | | 619 |
| bal | — | — | — | — | — | 3.00 | .40 | — | — | 10.00 | .20 | .50 | 1.00 | — | — | | 623 |
| bal | — | — | — | — | — | 3.00 | .20 | — | — | 11.00 | .20 | .30 | — | — | — | | 624 |
| bal | — | — | — | — | — | 4.00 | — | — | — | 13.00 | — | 2.00 | — | — | — | | 625 |
| bal | — | — | — | — | — | 3.00 | .20 | .20 | — | 10.00 | .20 | 1.00 | 4.00 | — | — | | 630 |
| bal | — | — | — | — | — | 4.00 | — | — | .02 | 9.00 | .10 | 3.00 | 5.00 | — | — | | 632 |
| bal | — | — | — | — | — | .05 | — | .40 | .05 | 2.80 | 1.80 | .10 | .10 | .40 | — | | 638 |
| bal | — | .10 | — | — | — | .25 | .20 | .40 | .05 | 7.00 | 1.80 | .10 | .25 | — | — | | 642 |
| bal | — | — | — | — | — | .60 | — | 1.00 | .05 | — | 1.50 | .70 | — | — | — | | 651 |
| bal | — | — | — | — | — | .60 | — | 1.00 | .05 | — | 3.00 | 1.00 | .60 | — | — | | 655 |
| bal | — | — | — | — | — | .10 | — | 28.00 | .07 | — | — | 1.20 | — | — | — | | 667 |
| bal | — | — | — | — | — | .25 | .20 | 35.00 | .40 | 1.20 | 1.00 | 2.80 | .20 | — | — | | 674 |
| bal | — | — | — | — | — | 1.40 | 1.00 | 38.00 | .20 | .20 | — | .10 | — | — | — | | 675 |
| bal | — | .10 | — | — | — | .06 | — | 20.00 | .07 | 2.00 | — | — | — | — | — | | 687 |
| bal | — | — | — | — | — | .05 | — | 22.00 | .05 | 3.40 | — | — | — | .40 | — | | 688 |
| bal | — | — | — | — | — | .20 | — | 14.00 | .30 | — | 4.00 | — | — | — | — | | 694 |
| bal | — | — | — | — | — | 1.40 | — | .80 | .05 | — | — | .80 | 10.00 | — | — | | 706 |
| bal | — | — | — | — | — | .80 | — | .80 | .05 | — | — | .80 | 20.00 | — | — | | 710 |
| bal | — | — | — | — | — | .50 | — | .80 | .05 | — | — | .80 | 30.00 | — | — | | 715 |
| bal | — | — | — | — | — | .40 | 2.30 | .50 | .05 | — | — | .20 | 9.50 | — | — | | 725 |
| bal | — | — | — | — | — | .20 | — | 24.00 | .05 | — | — | .50 | 10.00 | — | — | | 745 |
| bal | — | — | — | — | — | .20 | — | 16.00 | .05 | — | — | .50 | 18.00 | — | — | | 752 |
| bal | — | — | — | — | — | .20 | — | 19.00 | .10 | — | — | .50 | 15.00 | — | — | | 754 |
| bal | — | — | — | — | — | .20 | — | 22.00 | .05 | — | — | .50 | 12.00 | — | — | | 757 |
| bal | — | — | — | — | — | .20 | — | 26.00 | .05 | — | — | .50 | 18.00 | — | — | | 770 |
| bal | — | — | — | — | — | .30 | — | 24.00 | 2.00 | — | — | .50 | 8.00 | — | — | | 782 |

| COMPOSITION, % | | | | | | | | | | | | | | | | | MATERIAL DESIGNATION |
|---|---|---|---|---|---|---|---|---|---|---|---|---|---|---|---|---|---|
| Cu | Ag | As | Sb | P | S | Fe | Sn | Zn | Pb | Al | Si | Mn | Ni | Co | Cr | Other | |
| **35. COPPER ALLOYS, CAST** | | | | | | | | | | | | | | | | | |
| 99.95 | — | — | — | — | — | — | — | — | — | — | — | — | — | — | — | | 801 |
| 99.75 | .03 | — | — | — | — | — | — | — | — | — | — | — | — | — | — | | 803 |
| 99.75 | .03 | — | — | — | — | — | — | — | — | — | — | — | — | — | — | .01B | 805 |
| 99.75 | — | — | — | — | — | — | — | — | — | — | — | — | — | — | — | .01B | 807 |
| 99.70 | .03 | — | — | — | — | — | — | — | — | — | — | — | — | — | — | | 809 |
| 99.70 | — | — | — | — | — | — | — | — | — | — | — | — | — | — | — | | 811 |
| bal | — | — | — | — | — | — | — | — | — | — | — | — | — | .80 | — | .06Be | 813 |
| bal | — | — | — | — | — | — | — | — | — | — | — | — | — | — | .80 | .06Be | 814 |
| bal | — | — | — | — | — | .05 | .05 | .05 | .02 | .05 | .10 | — | — | — | 1.00 | | 815 |
| bal | 1.00 | — | — | — | — | — | — | — | — | — | — | — | .90 | .90 | — | .40Be | 817 |
| bal | 1.00 | — | — | — | — | — | — | — | — | — | — | — | — | 1.60 | — | .40Be | 818 |
| bal | — | — | — | — | — | .05 | .05 | .05 | .02 | .05 | .10 | — | .10 | 2.40 | .05 | .60Be | 820 |
| bal | — | — | — | — | — | — | — | — | — | — | — | — | .90 | .90 | — | .50Be | 821 |
| bal | — | — | — | — | — | — | — | — | — | — | — | — | 1.50 | — | — | .60Be | 822 |
| bal | — | — | — | — | — | .10 | .05 | .05 | .02 | .10 | — | — | .10 | .25 | .05 | 1.70Be | 824 |
| bal | — | — | — | — | — | .15 | .05 | .05 | .02 | .10 | .25 | — | .10 | .45 | .05 | 2.00Be | 825 |
| bal | — | — | — | — | — | .15 | .05 | .05 | .02 | .10 | .25 | — | .10 | .50 | .05 | 2.30Be | 826 |
| bal | — | — | — | — | — | .15 | .05 | .05 | .02 | .10 | .10 | — | 1.25 | — | .05 | 2.45Be | 827 |
| bal | — | — | — | — | — | .15 | .05 | .05 | .02 | .10 | .25 | — | .10 | .45 | .05 | 2.60Be | 828 |
| bal | — | — | — | — | — | — | 1.50 | 4.00 | 1.50 | — | — | — | — | — | — | | 833 |
| bal | — | — | — | — | — | — | .10 | 10.00 | .25 | — | — | — | — | — | — | | 834 |
| bal | — | — | .20 | .05 | .04 | .20 | 5.00 | 5.00 | 5.00 | .005 | .005 | — | .50 | — | — | | 836 |
| bal | — | — | .20 | .03 | .04 | .20 | 4.00 | 7.00 | 6.00 | .005 | .005 | — | .50 | — | — | | 838 |
| bal | — | — | .20 | .03 | .04 | .30 | 5.00 | 12.50 | 2.50 | .005 | .005 | — | .40 | — | — | | 842 |
| bal | — | — | .20 | .02 | .04 | .30 | 3.00 | 9.00 | 7.00 | .005 | .005 | — | .50 | — | — | | 844 |
| bal | — | — | .20 | .02 | .04 | .30 | 3.00 | 15.00 | 6.00 | .005 | .005 | — | .50 | — | — | | 848 |
| bal | — | — | .15 | .02 | .03 | .40 | 1.00 | 24.00 | 3.00 | .005 | .05 | — | .50 | — | — | | 852 |
| bal | — | — | — | — | — | — | .40 | 28.00 | .40 | — | — | — | — | — | — | | 853 |
| bal | — | — | — | — | — | .50 | 1.00 | 27.00 | 3.00 | .20 | .05 | — | — | — | — | | 854 |

| COMPOSITION, % | | | | | | | | | | | | | | | | | MATERIAL DESIGNATION |
|---|---|---|---|---|---|---|---|---|---|---|---|---|---|---|---|---|---|
| Cu | Ag | As | Sb | P | S | Fe | Sn | Zn | Pb | Al | Si | Mn | Ni | Co | Cr | Other | |
| 35. COPPER ALLOYS, CAST (continued) | | | | | | | | | | | | | | | | | |
| bal | — | — | — | — | — | .10 | .10 | 34.00 | .10 | .80 | — | .10 | .10 | — | — | | 855 |
| bal | — | — | — | — | — | .60 | 1.00 | 32.00 | 1.00 | .30 | .05 | — | .50 | — | — | | 857 |
| bal | — | .05 | .05 | .01 | .05 | .40 | 1.00 | 39.00 | 1.00 | .25 | .20 | .20 | — | — | — | | 858 |
| bal | — | — | — | — | — | 3.00 | .20 | 20.00 | .20 | 5.00 | — | 4.00 | — | — | — | | 861 |
| bal | — | — | — | — | — | 3.00 | .20 | 25.00 | .20 | 4.00 | — | 3.00 | .50 | — | — | | 862 |
| bal | — | — | — | — | — | 3.00 | .20 | 23.00 | .20 | 6.00 | — | 3.00 | — | — | — | | 863 |
| bal | — | — | — | — | — | 1.00 | .75 | 37.00 | 1.00 | .75 | — | .75 | — | — | — | | 864 |
| bal | — | — | — | — | — | 1.00 | .50 | 38.00 | .25 | 1.00 | — | 1.00 | .25 | — | — | | 865 |
| bal | — | — | — | — | — | 1.50 | 1.00 | 36.00 | 1.00 | 1.50 | — | 2.00 | .50 | — | — | | 867 |
| bal | — | — | — | — | — | 2.00 | .50 | 36.00 | .10 | 1.00 | — | 3.00 | 3.00 | — | — | | 868 |
| bal | — | — | — | — | — | 2.00 | .50 | 4.00 | .25 | 1.00 | 4.00 | 1.00 | — | — | — | | 872 |
| bal | — | — | — | — | — | — | — | 14.00 | .50 | .40 | 3.00 | — | — | — | — | | 874 |
| bal | — | — | — | — | — | — | — | 14.00 | .25 | .25 | 4.00 | — | — | — | — | | 875 |
| bal | — | — | — | — | — | — | — | 6.00 | .25 | — | 5.00 | — | — | — | — | | 876 |
| bal | — | .05 | .05 | .01 | .05 | .10 | .20 | 14.00 | .10 | .10 | 4.00 | .10 | .10 | — | — | | 878 |
| bal | — | .05 | .05 | .01 | .05 | .20 | .20 | 33.00 | .20 | .10 | 1.00 | .10 | .25 | — | — | | 879 |
| bal | — | — | .10 | .05 | .05 | .10 | 7.00 | .30 | .20 | .005 | .005 | — | .25 | — | — | | 902 |
| bal | — | — | .10 | .05 | .05 | .10 | 8.00 | 4.00 | .20 | .005 | .005 | — | .50 | — | — | | 903 |
| bal | — | — | .10 | .05 | .05 | .10 | 10.00 | 2.00 | .20 | .005 | .005 | — | .50 | — | — | | 905 |
| bal | — | — | .10 | .20 | .05 | .10 | 11.00 | .30 | .40 | .005 | .005 | — | .25 | — | — | | 907 |
| bal | — | — | .10 | .05 | .05 | .10 | 13.00 | .20 | .20 | .005 | .005 | — | .25 | — | — | | 909 |
| bal | — | — | .10 | .05 | .05 | .10 | 15.00 | 1.00 | .10 | .005 | .005 | — | .40 | — | — | | 910 |
| bal | — | — | .10 | .50 | .05 | .20 | 16.00 | .20 | .15 | .005 | .005 | — | .25 | — | — | | 911 |
| bal | — | — | .10 | .50 | .05 | .20 | 19.00 | .20 | .15 | .005 | .005 | — | .25 | — | — | | 913 |
| bal | — | — | .10 | .25 | .05 | .10 | 10.00 | .20 | 2.50 | .005 | .005 | — | 3.50 | — | — | | 915 |
| bal | — | — | .15 | .20 | .05 | .15 | 10.50 | .20 | .20 | .005 | .005 | — | 1.50 | — | — | | 916 |
| bal | — | — | .10 | .25 | .05 | .10 | 12.00 | .20 | .20 | .005 | .005 | — | 1.50 | — | — | | 917 |
| bal | — | — | .15 | .05 | .05 | .20 | 6.00 | 4.50 | 1.50 | .005 | .005 | — | .50 | — | — | | 922 |

| COMPOSITION, % | | | | | | | | | | | | | | | | | MATERIAL DESIGNATION |
|---|---|---|---|---|---|---|---|---|---|---|---|---|---|---|---|---|---|
| Cu | Ag | As | Sb | P | S | Fe | Sn | Zn | Pb | Al | Si | Mn | Ni | Co | Cr | Other | |
| **35. COPPER ALLOYS, CAST (continued)** | | | | | | | | | | | | | | | | | |
| bal | — | — | .15 | .05 | .05 | .20 | 8.00 | 4.00 | .50 | .005 | .005 | — | .50 | — | — | | 923 |
| bal | — | — | .15 | .25 | .05 | .10 | 11.00 | .25 | 1.00 | .005 | .005 | — | 1.00 | — | — | | 925 |
| bal | — | — | .15 | .03 | .05 | .10 | 10.00 | 2.00 | 1.00 | .005 | .005 | — | .40 | — | — | | 926 |
| bal | — | — | .15 | .20 | .05 | .10 | 10.00 | .50 | 2.00 | .005 | .005 | — | .50 | — | — | | 927 |
| bal | — | — | .15 | .05 | .05 | .10 | 16.00 | .40 | 5.00 | .005 | .005 | — | .40 | — | — | | 928 |
| bal | — | — | .25 | .10 | .04 | .10 | 7.00 | 3.00 | 7.00 | .005 | .005 | — | .50 | — | — | | 932 |
| bal | — | — | .35 | .30 | .04 | .10 | 8.00 | .40 | 8.00 | .005 | .005 | — | .50 | — | — | | 934 |
| bal | — | — | .20 | .05 | .04 | .10 | 5.00 | 1.00 | 9.00 | .005 | .005 | — | .50 | — | — | | 935 |
| bal | — | — | .40 | .10 | .04 | .10 | 10.00 | .40 | 10.00 | .005 | .005 | — | .50 | — | — | | 937 |
| bal | — | — | .40 | .05 | .04 | .10 | 7.00 | .40 | 15.00 | .005 | .005 | — | .50 | — | — | | 938 |
| bal | — | — | .35 | .75 | .04 | .20 | 6.00 | .80 | 15.00 | .005 | .005 | — | .40 | — | — | | 939 |
| bal | — | — | .40 | .05 | .04 | .10 | 5.00 | .40 | 25.00 | .005 | .005 | — | .50 | — | — | | 943 |
| bal | — | — | .40 | .35 | .04 | .10 | 8.00 | .40 | 11.00 | .005 | .005 | — | .50 | — | — | | 944 |
| bal | — | — | .40 | .05 | .04 | .10 | 7.00 | .60 | 20.00 | .005 | .005 | — | .50 | — | — | | 945 |
| bal | — | — | .10 | .05 | .05 | .15 | 5.00 | 2.00 | .05 | .005 | .005 | .10 | 5.00 | — | — | | 947 |
| bal | — | — | .10 | .05 | .05 | .15 | 5.00 | 2.00 | .65 | .005 | .005 | .10 | 5.00 | — | — | | 948 |
| bal | — | — | — | — | — | 3.00 | — | — | — | 9.00 | — | — | — | — | — | | 952 |
| bal | — | — | — | — | — | 1.00 | — | — | — | 10.00 | — | — | — | — | — | | 953 |
| bal | — | — | — | — | — | 4.00 | — | — | — | 11.00 | — | .25 | 1.50 | — | — | | 954 |
| bal | — | — | — | — | — | 4.00 | — | — | — | 11.00 | — | 2.50 | 4.00 | — | — | | 955 |
| bal | — | — | — | — | — | — | — | — | — | 7.00 | 2.00 | — | .15 | — | — | | 956 |
| bal | — | — | — | — | — | 3.00 | — | — | .03 | 8.00 | .10 | 12.00 | 2.00 | — | — | | 957 |
| bal | — | — | — | — | — | 4.00 | — | — | .03 | 9.00 | .10 | 1.00 | 4.00 | — | — | | 958 |
| bal | — | — | — | — | — | 1.40 | — | — | .02 | — | .15 | 1.00 | 10.00 | — | — | .50Cb | 962 |
| bal | — | — | — | — | — | .70 | — | — | .02 | — | .40 | .50 | 20.00 | — | — | .50Cb | 963 |
| bal | — | — | — | — | — | .90 | — | — | .02 | — | .40 | 1.00 | 30.00 | — | — | .50Cb | 964 |
| bal | — | — | — | — | — | 1.00 | — | — | .01 | — | .10 | .50 | 30.00 | — | — | .50Be | 966 |
| bal | — | — | .20 | .05 | .04 | .75 | 2.00 | 20.00 | 10.00 | .005 | .10 | .25 | 12.00 | — | — | | 973 |

| Cu | Ag | As | Sb | P | S | Fe | Sn | Zn | Pb | Al | Si | Mn | Ni | Co | Cr | Other | MATERIAL DESIGNATION |
|---|---|---|---|---|---|---|---|---|---|---|---|---|---|---|---|---|---|
| COMPOSITION, % | | | | | | | | | | | | | | | | | |
| **35. COPPER ALLOYS, CAST (continued)** | | | | | | | | | | | | | | | | | |
| bal | — | — | — | — | — | .75 | 3.00 | 16.00 | 5.00 | — | — | .25 | 17.00 | — | — | | 974 |
| bal | — | — | .15 | .05 | .04 | .75 | 4.00 | 8.00 | 4.00 | .005 | .10 | .50 | 20.00 | — | — | | 976 |
| bal | — | — | .10 | .05 | .04 | .75 | 5.00 | 2.00 | 2.00 | .005 | .10 | .50 | 25.00 | — | — | | 978 |
| bal | — | — | — | — | — | .70 | .05 | — | .02 | 11.00 | .02 | — | 15.00 | 1.50 | — | | 993 |

| Ni | C | Mn | Fe | Si | Cu | Mg | Ti | Al | P | S | — | — | — | — | — | Other | MATERIAL DESIGNATION |
|---|---|---|---|---|---|---|---|---|---|---|---|---|---|---|---|---|---|
| COMPOSITION, % | | | | | | | | | | | | | | | | | |
| **36. NICKEL ALLOYS, WROUGHT AND CAST** | | | | | | | | | | | | | | | | | |
| bal | — | .40 | 48.50 | .50 | — | — | 2.55 | .55 | — | — | | | | | | 5.30Cr | NI-SPAN-C 902 |
| bal | — | .50 | .60 | 1.00 | — | — | .60 | 4.35 | — | — | | | | | | | DURANICKEL Alloy 301 |
| bal | — | .75 | 1.00 | — | 28.00 | — | — | 3.00 | — | — | | | | | | | MONEL Alloy 502 |
| bal | — | 1.00 | 1.25 | .25 | 31.50 | — | — | — | — | — | | | | | | | MONEL Alloy 400 |
| bal | .01 | .18 | .20 | .18 | .13 | — | — | — | — | — | | | | | | | Nickel 201 |
| bal | .04 | .10 | .05 | .03 | .05 | .05 | .03 | — | — | — | | | | | | | Nickel 220 |
| bal | .05 | .08 | — | .02 | .05 | .06 | — | — | — | — | | | | | | | Nickel 230 |
| bal | .05 | 1.60 | .38 | .13 | 55.00 | — | — | — | — | | | | | | | | MONEL Alloy 401 |
| bal | .08 | — | .25 | .05 | 44.00 | — | — | .03 | — | — | | | | | | | MONEL Alloy 404 |
| bal | .08 | .18 | .10 | .08 | .08 | .05 | .03 | — | — | — | | | | | | | Nickel 205 |
| bal | .08 | .18 | .20 | .18 | .13 | — | — | — | — | — | | | | | | | Nickel 200 |
| bal | .10 | 4.75 | .38 | .08 | .13 | — | — | — | — | — | | | | | | | Nickel 211 |
| bal | .13 | .75 | 1.00 | — | 29.50 | — | .60 | 2.70 | — | — | | | | | | | MONEL Alloy K500 |
| bal | .15 | 1.00 | 1.25 | .25 | 31.50 | — | — | — | — | — | | | | | | | MONEL Alloy R405 |
| bal | .20 | .25 | .30 | .18 | .13 | .35 | .40 | — | — | — | | | | | | | PERMANICKEL Alloy 300 |
| bal | .35 | 1.50 | 3.50 | 2.00 | 30.00 | — | — | — | .03 | .03 | | | | | | | ASTM A296 Grade M-35 |
| bal | .35 | 1.50 | 3.50 | 2.00 | 30.00 | — | — | — | .03 | .03 | | | | | | | ASTM A494 Grade M-35 |
| bal | 1.00 | 1.50 | 3.00 | 2.00 | 1.25 | — | — | — | .03 | .03 | | | | | | | ASTM A296 Grade CZ-100 |
| bal | 1.00 | 1.50 | 3.00 | 2.00 | 1.25 | — | — | — | .03 | .03 | | | | | | | ASTM A494 Grade CZ-100 |

| COMPOSITION, % | | | | | | | | | | | | | | | | MATERIAL DESIGNATION |
|---|---|---|---|---|---|---|---|---|---|---|---|---|---|---|---|---|
| Ni | C | Be | Ti | Cr | — | — | — | — | — | — | — | — | — | — | Other | |
| **37. BERYLLIUM NICKEL ALLOYS, WROUGHT AND CAST** | | | | | | | | | | | | | | | | |
| bal | — | 2.20 | — | — | | | | | | | | | | | | Brush Alloy 200C |
| bal | — | 1.95 | .50 | — | | | | | | | | | | | | Berylco 440 |
| bal | .60 | 2.20 | — | — | | | | | | | | | | | | Brush Alloy 220C |
| bal | .30 | 2.70 | — | — | | | | | | | | | | | | Brush Alloy 260C |
| bal | .10 | 2.75 | — | .40 | | | | | | | | | | | | Berylco 41C |
| bal | .10 | 2.75 | — | 6.00 | | | | | | | | | | | | Berylco 43C |
| bal | .10 | 2.75 | — | 12.00 | | | | | | | | | | | | Berylco 42C |

| COMPOSITION, % | | | | | | | | | | | | | | | | MATERIAL DESIGNATION |
|---|---|---|---|---|---|---|---|---|---|---|---|---|---|---|---|---|
| Ni | Ti | — | — | — | — | — | — | — | — | — | — | — | — | — | Other | |
| **38. NITINOL ALLOYS, WROUGHT** | | | | | | | | | | | | | | | | |
| bal | 40.00 | | | | | | | | | | | | | | | Nitinol 60Ni-40Ti |
| bal | 44.00 | | | | | | | | | | | | | | | Nitinol 56Ni-44Ti |
| bal | 45.00 | | | | | | | | | | | | | | | Nitinol 55Ni-45Ti |

| COMPOSITION, % | | | | | | | | | | | | | | | | MATERIAL DESIGNATION |
|---|---|---|---|---|---|---|---|---|---|---|---|---|---|---|---|---|
| Ni | Cr | Co | Mo | W | Ta | Cb | Al | Ti | Fe | Mn | Si | C | B | Zr | Cu | Other | |
| **39. HIGH TEMPERATURE ALLOYS — Nickel Base, Wrought** | | | | | | | | | | | | | | | | | |
| bal | 1.00 | 1.00 | 28.00 | — | — | — | — | — | 2.00 | .50 | .05 | .02 | — | — | — | | Hastelloy Alloy B-2 |
| bal | 1.50 | 2.00 | 28.00 | — | — | — | — | — | 5.00 | .50 | .50 | .05 | — | — | — | .40V | Hastelloy Alloy B |
| bal | 12.00 | 10.00 | 3.00 | 6.00 | 1.50 | — | 4.60 | 3.00 | .30 | .05 | .05 | .35 | .01 | .10 | — | | AF2-1DA |
| bal | 13.00 | — | 6.00 | — | — | — | .20 | 2.80 | 34.00 | .40 | .40 | .05 | .01 | — | — | | Incoloy Alloy 901 |
| bal | 14.00 | 8.00 | 3.50 | 3.50 | — | 3.50 | 3.50 | 2.50 | — | — | — | .15 | .01 | .05 | — | | Rene 95 |
| bal | 14.00 | 15.00 | 6.00 | 3.50 | — | — | 3.80 | 2.50 | .50 | .10 | .20 | .05 | .01 | — | — | | Rene 63 |

| COMPOSITION, % | | | | | | | | | | | | | | | | | MATERIAL DESIGNATION |
|---|---|---|---|---|---|---|---|---|---|---|---|---|---|---|---|---|---|
| Ni | Cr | Co | Mo | W | Ta | Cb | Al | Ti | Fe | Mn | Si | C | B | Zr | Cu | Other | |
| **39. HIGH TEMPERATURE ALLOYS — Nickel Base, Wrought (continued)** | | | | | | | | | | | | | | | | | |
| bal | 15.00 | — | 3.00 | 3.00 | 3.00 | * | .50 | .60 | 7.00 | — | — | .04 | .01 | .03 | — | .04Mg | IN-102 |
| bal | 15.00 | 2.00 | 17.00 | 5.00 | — | — | — | — | 6.00 | — | — | .04 | — | — | — | | Hastelloy Alloy C |
| bal | 15.00 | 15.00 | 4.20 | — | — | — | 4.30 | 3.30 | .40 | .10 | .10 | .07 | .01 | .04 | — | | Rene 77 |
| bal | 15.00 | 17.00 | 5.25 | — | — | — | 4.40 | 3.50 | — | — | — | .06 | .03 | — | — | | Astroloy |
| bal | 15.00 | 18.50 | 5.20 | — | — | — | 4.30 | 3.50 | — | — | — | .08 | .03 | — | — | | Udimet 700 |
| bal | 15.00 | 28.50 | 3.70 | — | — | — | 3.00 | 2.20 | .70 | .10 | .30 | .12 | — | — | — | | Inconel Alloy 700 |
| bal | 15.50 | — | — | — | — | — | — | — | 8.00 | .50 | .20 | .08 | — | — | — | | Inconel Alloy 600 |
| bal | 15.50 | — | — | — | — | — | .70 | 2.35 | 7.00 | .50 | .35 | .04 | — | — | — | | Inconel Alloy 722 |
| bal | 15.50 | — | — | — | — | — | 1.20 | 2.30 | 7.00 | .50 | .25 | .05 | — | — | .25 | | Inconel Alloy 751 |
| bal | 15.50 | — | — | — | — | 1.00 | .70 | 2.50 | 7.00 | .50 | .20 | .04 | — | — | — | | Inconel Alloy X-750 |
| bal | 15.50 | 1.50 | 14.50 | — | — | — | .20 | — | 1.00 | .50 | .40 | .01 | .01 | — | — | .02La | Hastelloy Alloy S |
| bal | 15.50 | 2.00 | 16.00 | 4.00 | — | — | — | — | 6.00 | .50 | .05 | .02 | — | — | — | .35V | Hastelloy Alloy C-276 |
| bal | 15.60 | — | — | — | — | — | 3.40 | .70 | .40 | .05 | .20 | .04 | — | — | — | | Inconel Alloy 702 |
| bal | 16.00 | — | — | — | — | — | .10 | 3.00 | 8.00 | 2.25 | .15 | .07 | — | — | .20 | | Inconel Alloy 721 |
| bal | 16.00 | — | — | — | — | 2.90 | .20 | 1.80 | 40.00 | .20 | .20 | .03 | — | — | — | | Inconel Alloy 706 |
| bal | 16.25 | 7.20 | 1.60 | 8.40 | — | — | 1.90 | 3.20 | 9.50 | .05 | .10 | .24 | .01 | .06 | — | | Unitemp 1753 |
| bal | 18.00 | — | 3.00 | 3.00 | — | 6.50 | .50 | 1.00 | 18.00 | — | — | .03 | — | — | — | | Udimet 630 |
| bal | 18.00 | 15.00 | 3.00 | 1.50 | — | — | 2.50 | 5.00 | — | — | — | .07 | .02 | — | — | | Udimet 710 |
| bal | 18.00 | 18.50 | 4.00 | — | — | — | 2.90 | 2.90 | — | — | — | .08 | .01 | .05 | — | | Udimet 500 |
| bal | 18.00 | 20.00 | 3.20 | — | — | — | .20 | 2.60 | 18.00 | .80 | 1.00 | .03 | — | — | — | | Refractaloy 26 |
| bal | 19.00 | — | 3.00 | — | — | 5.00 | .50 | .90 | 18.50 | .20 | .20 | .04 | — | — | — | | Inconel Alloy 718 |
| bal | 19.00 | 11.00 | 10.00 | — | — | — | 1.50 | 3.10 | — | — | — | .09 | .01 | — | — | | Rene 41 |
| bal | 19.50 | — | — | — | — | — | — | .40 | 8.00 | .50 | .50 | .10 | — | — | — | | Nimonic 75 |
| bal | 19.50 | 2.00 | — | — | — | — | .65 | 2.25 | 5.00 | .50 | .50 | .06 | — | — | — | | Nimonic 80 |
| bal | 19.50 | 13.50 | 4.30 | — | — | — | 1.30 | 3.00 | — | — | — | .08 | .01 | .06 | — | | Waspaloy |

*Included with tantalum.

| COMPOSITION, % | | | | | | | | | | | | | | | | | MATERIAL DESIGNATION |
|---|---|---|---|---|---|---|---|---|---|---|---|---|---|---|---|---|---|
| Ni | Cr | Co | Mo | W | Ta | Cb | Al | Ti | Fe | Mn | Si | C | B | Zr | Cu | Other | |
| **39. HIGH TEMPERATURE ALLOYS — Nickel Base, Wrought (continued)** | | | | | | | | | | | | | | | | | |
| bal | 19.50 | 16.50 | — | — | — | — | 1.45 | 2.45 | 5.00 | .50 | .75 | .07 | .003 | .06 | — | | Nimonic 90 |
| bal | 19.50 | 18.00 | — | — | — | — | 2.00 | 3.50 | 5.00 | 1.00 | 1.00 | .10 | — | — | — | | Nimonic 95 |
| bal | 20.00 | 10.00 | 10.00 | — | — | — | 1.00 | 2.60 | — | .50 | .50 | .15 | .01 | — | — | | M252 |
| bal | 20.00 | 20.00 | 6.00 | — | — | — | .50 | 2.00 | .60 | .50 | .30 | .06 | — | — | .20 | | Haynes Alloy 263 |
| bal | 21.50 | — | — | 3.00 | — | — | .20 | .90 | 30.00 | .65 | .50 | .03 | — | — | 2.25 | | Incoloy Alloy 825 |
| bal | 21.50 | — | 9.00 | — | — | 3.60 | .20 | .20 | 2.50 | .20 | .20 | .05 | — | — | — | | Inconel Alloy 625 |
| bal | 21.80 | 1.50 | 9.00 | .60 | — | — | — | — | 18.50 | .50 | .50 | .10 | — | — | — | | Hastelloy Alloy X |
| bal | 22.00 | 2.00 | 6.50 | .50 | 2.10 | * | — | — | 19.50 | 1.50 | .50 | .03 | — | — | 2.00 | | Hastelloy Alloy G |
| bal | 22.00 | 12.50 | 9.00 | — | — | — | 1.00 | — | — | — | — | .07 | — | — | — | | Inconel Alloy 617 |
| bal | 23.00 | — | — | — | — | — | 1.40 | — | 14.10 | .50 | .20 | .05 | — | — | — | | Inconel Alloy 601 |
| bal | 29.50 | — | — | — | — | — | .25 | .60 | 25.40 | .75 | .50 | .06 | — | — | .40 | | Incoloy Alloy 804 |
| bal | — | 15.00 | — | — | — | 3.00 | .70 | 1.40 | 42.00 | — | — | — | — | — | — | | Incoloy Alloy 903 |
| bal | — | — | — | — | — | — | — | — | — | — | — | — | — | — | — | 2.00ThO$_2$ | TD-Nickel |
| bal | 20.00 | — | — | — | — | — | — | — | — | — | — | .10 | — | — | — | 2.00ThO$_2$ | TD-Ni-Cr |
| **39. HIGH TEMPERATURE ALLOYS — Nickel Base, Cast** | | | | | | | | | | | | | | | | | |
| bal | .70 | — | 28.00 | — | — | — | — | — | 5.00 | .70 | .70 | .10 | — | — | — | .50V | ASTM A494 Grade N-12M-1 |
| bal | .70 | — | 30.00 | — | — | — | — | — | 5.00 | .70 | .70 | .10 | — | — | — | .50V | ASTM A296 Grade N-12M |
| bal | .70 | — | 32.00 | — | — | — | — | — | 2.50 | .70 | .70 | .05 | — | — | — | | ASTM A494 Grade N-12M-2 |
| bal | .70 | 2.00 | 28.00 | — | — | — | — | — | 5.00 | .70 | .70 | .10 | — | — | — | .50V | Hastelloy Alloy B |
| bal | 1.00 | — | — | — | — | — | — | — | 2.00 | 1.00 | 9.00 | .12 | — | — | 3.00 | | Hastelloy Alloy D |
| bal | 6.00 | 7.50 | 2.00 | 5.80 | 9.00 | .50 | 5.40 | 1.00 | — | — | — | .13 | .02 | .13 | — | .50Re, .43Hf | TRW VI A |
| bal | 7.00 | .10 | 16.50 | .40 | — | — | .40 | — | 4.00 | .60 | .70 | .06 | .01 | — | .30 | | Hastelloy Alloy N |
| bal | 8.00 | 10.00 | 6.00 | .05 | — | .05 | 6.00 | 1.00 | .30 | .15 | .20 | .10 | .02 | .08 | — | | B-1900 |
| bal | 8.90 | 10.00 | 2.00 | 7.00 | 3.80 | — | 4.75 | 2.50 | — | — | — | .10 | .02 | .05 | — | 1.50Hf | Rene 125 |
| bal | 9.00 | 10.00 | — | 12.50 | — | 1.00 | 5.00 | 2.00 | — | — | — | .15 | .02 | .05 | — | | MAR-M200 |

*Included with tantalum.

| COMPOSITION, % | | | | | | | | | | | | | | | | MATERIAL DESIGNATION |
|---|---|---|---|---|---|---|---|---|---|---|---|---|---|---|---|---|
| Ni | Cr | Co | Mo | W | Ta | Cb | Al | Ti | Fe | Mn | Si | C | B | Zr | Cu | Other | |
| **39. HIGH TEMPERATURE ALLOYS — Nickel Base, Cast (continued)** | | | | | | | | | | | | | | | | | |
| bal | 9.00 | 10.00 | 2.50 | 10.00 | 1.50 | — | 5.50 | 1.50 | — | — | — | .15 | .02 | .05 | — | | MAR-M246 |
| bal | 10.00 | 15.00 | 3.00 | — | — | — | 5.50 | 4.70 | — | — | — | .18 | .01 | .06 | — | 1.00V | IN-100 (Rene 100) |
| bal | 10.50 | 13.50 | 6.15 | 1.50 | .50 | — | 5.50 | 2.50 | .40 | — | — | .08 | .02 | — | — | | SEL 15 |
| bal | 12.00 | — | .50 | — | — | — | — | — | — | 1.00 | 1.50 | .55 | — | — | — | | ASTM A297 Grade HW |
| bal | 12.00 | — | .50 | — | — | — | — | — | — | 1.00 | 1.50 | .50 | — | — | — | | ASTM A608 Grade HW50 |
| bal | 12.40 | 9.00 | 1.90 | 3.80 | 3.90 | — | 3.10 | 4.50 | — | — | — | .12 | .02 | .10 | — | | IN-792 |
| bal | 12.50 | — | 4.50 | — | 2.00 | * | 6.00 | .75 | 5.00 | 1.00 | 1.00 | .20 | — | — | — | | Inconel Alloy 713C |
| bal | 14.00 | 9.50 | 4.00 | 4.00 | — | — | 3.00 | 5.00 | — | — | — | .17 | .02 | .03 | — | | Rene 80 |
| bal | 15.00 | — | 17.00 | 5.00 | — | — | — | — | 6.00 | .70 | .70 | .08 | — | — | — | | Hastelloy Alloy C |
| bal | 15.00 | 18.50 | 5.20 | — | — | — | 4.30 | 3.50 | — | — | — | .08 | .03 | — | — | | Udimet 700 |
| bal | 15.00 | 22.00 | 4.40 | — | — | — | 4.35 | 2.35 | .70 | — | — | .08 | .02 | — | — | | SEL |
| bal | 15.50 | — | — | — | — | — | — | — | 10.00 | 1.00 | 2.00 | .30 | — | — | — | | ASTM A296 Grade CY-40 |
| bal | 15.50 | — | — | — | — | — | — | — | 10.00 | 1.00 | 2.00 | .30 | — | — | — | | ASTM A494 Grade CY-40 |
| bal | 15.50 | — | 5.25 | — | — | — | 3.00 | 2.00 | 10.00 | .20 | .40 | .15 | .06 | — | — | | GMR-235 |
| bal | 15.50 | — | 5.00 | — | — | — | 3.50 | 2.50 | 4.50 | .05 | .20 | .15 | .05 | — | — | | GMR-235D |
| bal | 15.50 | 10.00 | 1.75 | 3.50 | — | 1.75 | 4.25 | 1.75 | — | — | — | .15 | .02 | .05 | — | | MAR-M421 |
| bal | 15.50 | 20.00 | — | 3.00 | 2.00 | 2.00 | 2.80 | 4.30 | — | — | — | .15 | .02 | .05 | — | | MAR-M432 |
| bal | 16.00 | 8.50 | 1.75 | 2.60 | 1.75 | .90 | 3.40 | 3.40 | .40 | .10 | .15 | .17 | .01 | .10 | — | | IN-738 |
| bal | 16.50 | — | 17.00 | 5.00 | — | — | — | — | 6.00 | .70 | .70 | .10 | — | — | — | .30V | ASTM A494 Grade CW-12M-1 |
| bal | 17.00 | — | .50 | — | — | — | — | — | — | 1.00 | 1.50 | .55 | — | — | — | | ASTM A297 Grade HX |
| bal | 17.00 | — | .50 | — | — | — | — | — | — | 1.00 | 1.50 | .50 | — | — | — | | ASTM A608 Grade HX50 |
| bal | 18.00 | — | 18.00 | 5.00 | — | — | — | — | 7.00 | .70 | 1.00 | .10 | — | — | — | .40V | ASTM A296 Grade CW-12M |
| bal | 18.60 | — | 3.10 | — | — | 5.00 | .40 | .90 | 18.50 | .20 | .30 | .04 | — | — | — | | Inconel Alloy 718 |
| bal | 19.00 | — | 19.00 | — | — | — | — | — | 2.50 | .70 | .70 | .05 | — | — | — | | ASTM A494 Grade CW-12M-2 |
| bal | 19.00 | 10.00 | 10.00 | — | — | — | 1.00 | 2.60 | — | .40 | .40 | .15 | .01 | — | — | | M252 |
| bal | 19.00 | 18.00 | 4.20 | — | — | — | 3.00 | 3.00 | — | — | — | .08 | .01 | .05 | — | | Udimet 500 |

*Included with tantalum.

| COMPOSITION, % | | | | | | | | | | | | | | | | | MATERIAL DESIGNATION |
|---|---|---|---|---|---|---|---|---|---|---|---|---|---|---|---|---|---|
| Ni | Cr | Co | Mo | W | Ta | Cb | Al | Ti | Fe | Mn | Si | C | B | Zr | Cu | Other | |
| **39. HIGH TEMPERATURE ALLOYS — Cobalt Base, Wrought** | | | | | | | | | | | | | | | | | |
| — | 19.00 | bal | — | 4.70 | 6.50 | — | 3.50 | — | — | — | — | .18 | — | .15 | — | .10Y | AiResist 213 |
| 10.00 | 20.00 | bal | — | 15.00 | — | — | — | — | — | 1.50 | .50 | .10 | — | — | — | | Haynes Alloy 25 (L605) |
| 20.00 | 20.00 | bal | — | — | 7.50 | — | — | .50 | — | — | — | .05 | — | .10 | — | | MAR-M905 |
| 20.00 | 20.00 | bal | — | — | 7.50 | — | — | — | .40 | .10 | .10 | .05 | — | .10 | — | | MAR-M918 |
| 20.00 | 20.00 | bal | 4.00 | 4.00 | — | — | — | — | 4.00 | 1.20 | .40 | .38 | — | — | — | | S-816 |
| 20.00 | 25.00 | bal | 4.00 | 2.00 | — | 2.00 | — | — | 3.00 | 1.00 | .40 | .26 | — | — | — | | V-36 |
| 22.00 | 22.00 | bal | — | 14.00 | — | — | — | — | 2.50 | 1.00 | .40 | .10 | — | — | — | .08La | Haynes Alloy 188 |
| 28.00 | 20.00 | bal | — | 7.00 | — | — | — | 4.00 | 2.00 | — | — | — | — | — | — | | J-1570 |
| **39. HIGH TEMPERATURE ALLOYS — Cobalt Base, Cast** | | | | | | | | | | | | | | | | | |
| — | 3.00 | bal | — | 25.00 | — | — | — | 1.00 | — | — | — | .40 | — | 1.00 | — | 2.00Re | NASA Co-W-Re |
| — | 19.00 | bal | — | 4.50 | 7.50 | — | 4.30 | — | — | — | — | .35 | — | .13 | — | .17Y | AiResist 215 |
| — | 21.50 | bal | — | 10.00 | 9.00 | — | — | — | — | — | — | .85 | .01 | .20 | — | | MAR-M302 |
| — | 21.50 | bal | — | 9.00 | 4.50 | — | — | .75 | — | .10 | .10 | 1.00 | — | 2.25 | — | | MAR-M322 |
| .70 | 21.00 | bal | — | 11.00 | — | 2.00 | 3.50 | — | 2.00 | .40 | — | .45 | — | — | — | .10Y | AiResist 13 |
| 1.00 | 21.00 | bal | — | 11.00 | — | 2.00 | — | — | 2.00 | .25 | .25 | .45 | — | — | — | | WI-52 |
| 2.50 | 28.00 | bal | — | 5.00 | — | — | — | — | 3.00 | — | — | 1.00 | — | — | — | | HS-6 |
| 3.00 | 27.00 | bal | 5.00 | — | — | — | — | — | 1.00 | .60 | .60 | .25 | .01 | — | — | | HS-21 |
| 3.00 | 31.00 | bal | — | 12.50 | — | — | — | — | 3.00 | 1.00 | 1.00 | 2.45 | — | — | — | | HOWMET #3 |
| 10.00 | 23.50 | bal | — | 7.00 | 3.50 | — | — | .20 | 1.00 | .10 | .10 | .60 | .01 | .50 | — | | MAR-M509 |
| 10.00 | 25.00 | bal | — | 7.50 | — | — | — | — | 1.50 | .75 | .75 | .50 | — | — | — | | HS-31 (X-40) |
| 10.00 | 29.00 | bal | — | 7.00 | — | — | — | — | 1.50 | .70 | .70 | .25 | .01 | — | — | | FSX-414 |
| 10.50 | 25.50 | bal | — | 7.00 | — | — | — | — | 2.00 | .70 | — | .25 | .01 | — | — | | X-45 |

| COMPOSITION, % | | | | | | | | | | | | | | | | | MATERIAL DESIGNATION |
|---|---|---|---|---|---|---|---|---|---|---|---|---|---|---|---|---|---|
| Ni | Cr | Co | Mo | W | Ta | Cb | Al | Ti | Fe | Mn | Si | C | B | Zr | Cu | Other | |
| **39. HIGH TEMPERATURE ALLOYS — Iron Base, Wrought** | | | | | | | | | | | | | | | | | |
| 9.00 | 19.00 | — | 1.00 | 1.00 | — | — | — | — | bal | 1.10 | .60 | .30 | — | — | — | | 19-9DL |
| 20.00 | 21.00 | 20.00 | 3.00 | 2.50 | — | 1.00 | — | — | bal | 1.50 | .50 | .15 | — | — | — | .15N | N-155 |
| 25.00 | 16.00 | — | 6.00 | — | — | .40 | — | .30 | bal | 1.35 | .70 | .12 | — | — | — | | 16-25-6 |
| 26.00 | 13.50 | — | 1.50 | — | — | — | .20 | 2.85 | bal | 1.50 | .40 | .08 | .08 | — | — | | W-545 |
| 26.00 | 13.50 | — | 2.70 | — | — | — | .10 | 1.70 | bal | .90 | .80 | .04 | .01 | — | — | | Discaloy |
| 26.00 | 15.00 | — | 1.30 | — | — | — | .20 | 2.00 | bal | 1.35 | .50 | .05 | .02 | — | — | | A-286 |
| 27.00 | 14.80 | — | 1.25 | — | — | — | .25 | 3.00 | bal | .35 | .75 | .08 | .01 | — | — | .50V | V-57 |
| 32.00 | 20.50 | — | — | — | — | — | — | 1.10 | bal | .80 | .50 | .05 | — | — | — | | Incoloy Alloy 801 |
| 32.50 | 21.00 | — | — | — | — | — | .40 | .40 | bal | .80 | .50 | .05 | — | — | — | | Incoloy Alloy 800 |
| 32.50 | 21.00 | — | — | — | — | — | .40 | .40 | bal | .80 | .50 | .08 | — | — | .40 | .008S | Incoloy Alloy 800H |
| 32.50 | 21.50 | — | — | — | — | — | — | — | bal | .80 | .40 | .40 | — | — | — | | Incoloy Alloy 802 |

| COMPOSITION, % | | | | | | | | | | | | | | | | | MATERIAL DESIGNATION |
|---|---|---|---|---|---|---|---|---|---|---|---|---|---|---|---|---|---|
| W | Mo | Cb | Ta | Hf | Zr | Ti | Ni | Fe | Cu | Re | C | — | — | — | — | Other | |
| **40. REFRACTORY ALLOYS, WROUGHT, CAST, P/M — Columbium** | | | | | | | | | | | | | | | | | |
| — | — | bal | — | 10.00 | .50 | 1.00 | — | — | — | — | — | | | | | | C103 |
| 10.00 | — | bal | — | 10.00 | — | — | — | — | — | — | — | | | | | .10Y | C129Y |
| — | — | bal | — | — | 1.00 | — | — | — | — | — | — | | | | | | Cb-1Zr |
| 10.00 | — | bal | — | — | 2.50 | — | — | — | — | — | — | | | | | | Cb-752 |
| 11.00 | — | bal | 28.00 | — | 1.00 | — | — | — | — | — | — | | | | | | FS-85 |
| 10.00 | — | bal | 10.00 | — | — | — | — | — | — | — | .01 | | | | | .01$O_2$, .01N | FS-291 |
| 13.50 | — | bal | — | 29.00 | 1.00 | — | — | — | — | — | .30 | | | | | | WC-3015 |

| COMPOSITION, % | | | | | | | | | | | | | | | | | MATERIAL DESIGNATION |
|---|---|---|---|---|---|---|---|---|---|---|---|---|---|---|---|---|---|
| W | Mo | Cb | Ta | Hf | Zr | Ti | Ni | Fe | Cu | Re | C | — | — | — | — | Other | |
| **40. REFRACTORY ALLOYS, WROUGHT, CAST, P/M — Molybdenum** | | | | | | | | | | | | | | | | | |
| — | 99.00 | — | — | — | — | — | — | — | — | — | — | | | | | | Mo |
| — | bal | — | — | — | — | — | — | — | — | 50.00 | — | | | | | | Mo-50Re |
| — | bal | — | — | — | .30 | 1.25 | — | — | — | — | — | | | | | | TZC |
| — | bal | — | — | — | .08 | .50 | — | — | — | — | — | | | | | | TZM |
| **40. REFRACTORY ALLOYS, WROUGHT, CAST, P/M — Tantalum** | | | | | | | | | | | | | | | | | |
| 8.00 | — | — | bal | 1.00 | — | — | — | — | — | 1.00 | .02 | | | | | | ASTAR 811C |
| 8.00 | — | — | bal | 2.40 | — | — | — | — | — | — | — | | | | | | T-111 |
| 9.60 | — | — | bal | 2.40 | — | — | — | — | — | — | .01 | | | | | | T-222 |
| 10.00 | — | — | bal | — | — | — | — | — | — | — | — | | | | | | Ta-10W |
| — | — | — | bal | 1.00 | — | — | — | — | — | — | — | | | | | | Ta-Hf |
| 2.50 | — | .15 | bal | — | — | — | — | — | — | — | — | | | | | | Ta63 |
| **40. REFRACTORY ALLOYS, WROUGHT, CAST, P/M — Tungsten Alloys** | | | | | | | | | | | | | | | | | |
| bal | — | — | — | — | — | — | — | — | — | — | — | | | | | doped | GE-218 |
| bal | 4.00 | — | — | — | — | — | 4.00 | 2.00 | — | — | — | | | | | | Anviloy 1100 |
| bal | 4.00 | — | — | — | — | — | 4.00 | 2.00 | — | — | — | | | | | | Anviloy 1150 |
| bal | 4.00 | — | — | — | — | — | 4.00 | 2.00 | — | — | — | | | | | | Anviloy 1200 |
| bal | * | — | — | — | — | — | 3.00 | * | 7.00* | — | — | | | | | | Gyromet |
| bal | — | — | — | — | — | — | 5.00† | — | † | — | — | | | | | | Mallory 2000 |
| bal | — | — | — | — | — | — | — | — | — | — | — | | | | | 10.00Ag | W-10Ag |
| bal | 15.00 | — | — | — | — | — | — | — | — | — | — | | | | | | W-15Mo |
| bal | — | — | — | — | — | — | 7.00 | — | 4.00 | — | — | | | | | | W-7Ni-4Cu |
| bal | — | — | — | — | — | — | — | — | — | 5.00 | — | | | | | | W-5Re |
| bal | — | — | — | — | — | — | — | — | — | 25.00 | — | | | | | | W-25Re |
| bal | 30.00 | — | — | — | — | — | — | — | — | 25.00 | — | | | | | | W-25Re-30Mo |

*7.00 Cu + Mo + Fe.
†5.00 Ni + Cu.

# 26.3 Chemical Composition by Material Group

| Zn | Al | Mg | Cu | Fe | Pb | Cd | Sn | — | — | — | — | — | — | — | — | Other | MATERIAL DESIGNATION |
|---|---|---|---|---|---|---|---|---|---|---|---|---|---|---|---|---|---|
| **41. ZINC ALLOYS, CAST** | | | | | | | | | | | | | | | | | |
| bal | 3.90 | — | .25 | — | — | — | — | | | | | | | | | | ZDC No. 7 |
| bal | 12.00 | .02 | .90 | — | — | — | — | | | | | | | | | | ILZRO 12 |
| bal | 3.90 | .04 | .08 | .08 | .01 | .01 | .01 | | | | | | | | | | AMS 4803 |
| bal | 3.90 | .04 | .20 | .08 | .01 | .01 | .01 | | | | | | | | | | SAE J469a Grade 903 (AG40A) |
| bal | 3.90 | .05 | 1.00 | .10 | .01 | .01 | .01 | | | | | | | | | | SAE J469a Grade 925 (AC41A) |

(COMPOSITION, % spans the columns Zn through Other.)

| Pb | Sn | Sb | Cu | Fe | As | Bi | Zn | Al | Cd | — | — | — | — | — | — | Other | MATERIAL DESIGNATION |
|---|---|---|---|---|---|---|---|---|---|---|---|---|---|---|---|---|---|
| **42. LEAD ALLOYS, CAST** | | | | | | | | | | | | | | | | | |
| bal | — | 1.00 | — | — | — | — | — | — | — | | | | | | | | 1Sb |
| bal | — | 4.00 | — | — | — | — | — | — | — | | | | | | | | 4Sb |
| bal | — | 6.00 | — | — | — | — | — | — | — | | | | | | | | 6Sb |
| bal | — | 8.00 | — | — | — | — | — | — | — | | | | | | | | 8Sb |
| bal | — | 9.00 | — | — | — | — | — | — | — | | | | | | | | 9Sb |
| bal | 1.00 | 16.00 | .60 | .10 | 1.10 | .10 | .01 | .01 | .05 | | | | | | | | ASTM B23 Alloy 15 |
| bal | 5.00 | 15.00 | .50 | .10 | .50 | .10 | .01 | .01 | .05 | | | | | | | | ASTM B23 Alloy 8 |
| bal | 6.00 | 10.00 | .50 | .10 | .25 | .10 | .01 | .01 | .05 | | | | | | | | ASTM B23 Alloy 13 |
| bal | 10.00 | 15.00 | .50 | .10 | .50 | .10 | .01 | .01 | .05 | | | | | | | | ASTM B23 Alloy 7 |

(COMPOSITION, % spans the columns Pb through Other.)

| Sn | Sb | Pb | Cu | Fe | As | Bi | Zn | Al | Cd | — | — | — | — | — | — | Other | MATERIAL DESIGNATION |
|---|---|---|---|---|---|---|---|---|---|---|---|---|---|---|---|---|---|
| **43. TIN ALLOYS, CAST** | | | | | | | | | | | | | | | | | |
| bal | 4.50 | .35 | 4.50 | .08 | .10 | .08 | .01 | .01 | .05 | | | | | | | | ASTM B23 Alloy 1 |
| bal | 7.50 | .35 | 3.50 | .08 | .10 | .08 | .01 | .01 | .05 | | | | | | | | ASTM B23 Alloy 2 |
| bal | 8.00 | .35 | 8.00 | .08 | .10 | .08 | .01 | .01 | .05 | | | | | | | | ASTM B23 Alloy 3 |
| bal | 7.00 | .50 | 6.00 | .08 | .10 | .08 | .01 | .01 | .05 | | | | | | | | ASTM B23 Alloy 11 |

(COMPOSITION, % spans the columns Sn through Other.)

| COMPOSITION, % | | | | | | | | | | | | | | | | | MATERIAL DESIGNATION |
|---|---|---|---|---|---|---|---|---|---|---|---|---|---|---|---|---|---|
| Zr | Hf | Sn | Cr | Fe | — | — | — | — | — | — | — | — | — | — | — | Other | |
| **45. ZIRCONIUM ALLOYS, WROUGHT** | | | | | | | | | | | | | | | | | |
| 99.50 | * | — | — | — | | | | | | | | | | | | | Zr-0.001%Hf (Grade 21) |
| 99.20 | * | — | .20 | † | | | | | | | | | | | | | Zr-2%Hf (Grade 11) |
| bal | — | 1.45 | .10 | .14 | | | | | | | | | | | | | Zircaloy 2 (Grade 32) |
| bal | — | 1.45 | .10 | .21 | | | | | | | | | | | | | Zircaloy 4 (Grade 34) |

*Included with zirconium.
†Included with chromium.

| COMPOSITION, % | | | | | | | | | | | | | | | | | MATERIAL DESIGNATION |
|---|---|---|---|---|---|---|---|---|---|---|---|---|---|---|---|---|---|
| Al | Cu | Mg | Sn | Si | Pb | — | — | — | — | — | — | — | — | — | — | Other | |
| **47. POWDER METAL ALLOYS — Aluminum Alloys** | | | | | | | | | | | | | | | | | |
| bal | .25 | 1.00 | — | .60 | — | | | | | | | | | | | | A1-1Mg-0.6Si-0.25cu |
| bal | — | .60 | — | .40 | — | | | | | | | | | | | | Al-0.6Mg-0.4Si |
| bal | 4.40 | .40 | — | .80 | — | | | | | | | | | | | | Al-4.4Cu-0.8Si-0.4Mg |
| 90.50 | 4.00 | — | 5.00 | — | — | | | | | | | | | | | | 90.5Al-5Sn-4Cu |
| 88.00 | 3.00 | — | 5.00 | — | 4.00 | | | | | | | | | | | | 88Al-5Sn-4Pb-3Cu |

| COMPOSITION, % | | | | | | | | | | | | | | | | | MATERIAL DESIGNATION |
|---|---|---|---|---|---|---|---|---|---|---|---|---|---|---|---|---|---|
| Fe | C | Cu | Ni | Cr | Mn | Si | P | S | — | — | — | — | — | — | — | Other | |
| **47. POWDER METAL ALLOYS — Irons** | | | | | | | | | | | | | | | | | |
| 97.70 | .30 | — | — | — | — | — | — | — | | | | | | | | 2.00 max. | F-0000-N |
| 97.70 | .30 | — | — | — | — | — | — | — | | | | | | | | 2.00 max. | F-0000-P |
| 97.70 | .30 | — | — | — | — | — | — | — | | | | | | | | 2.00 max. | F-0000-R |
| 97.70 | .30 | — | — | — | — | — | — | — | | | | | | | | 2.00 max. | F-0000-S |
| 97.70 | .30 | — | — | — | — | — | — | — | | | | | | | | 2.00 max. | F-0000-T |

| COMPOSITION, % | | | | | | | | | | | | | | | | | MATERIAL DESIGNATION |
|---|---|---|---|---|---|---|---|---|---|---|---|---|---|---|---|---|---|
| Fe | C | Cu | Ni | Cr | Mn | Si | P | S | — | — | — | — | — | — | — | Other | |
| **47. POWDER METAL ALLOYS — Steels** | | | | | | | | | | | | | | | | | |
| 97.00 | .80 | — | — | — | — | — | — | — | | | | | | | | 2.00 max. | F-0008-P |
| 97.00 | .80 | — | — | — | — | — | — | — | | | | | | | | 2.00 max. | F-0008-S |
| 97.50 | .50 | — | — | — | — | — | — | — | | | | | | | | 2.00 max. | F-0005-S |
| 95.00 | .50 | 2.50 | — | — | — | — | — | — | | | | | | | | 2.00 max. | FC-0205-S |
| 94.50 | .80 | 2.70 | — | — | — | — | — | — | | | | | | | | 2.00 max. | FC-0208-P |
| 94.50 | .80 | 2.70 | — | — | — | — | — | — | | | | | | | | 2.00 max. | FC-0208-S |
| 93.20 | .80 | 5.00 | — | — | — | — | — | — | | | | | | | | 2.00 max. | FC-0508-P |
| 89.00 | .20 | 10.00 | — | — | — | — | — | — | | | | | | | | 2.00 max. | FC-1000-N |
| 95.00 | .50 | 1.50 | 2.00 | — | — | — | — | — | | | | | | | | 2.00 max. | FN-0205-S |
| 95.00 | .50 | 1.50 | 2.00 | — | — | — | — | — | | | | | | | | 2.00 max. | FN-0205-T |
| 93.00 | .50 | 1.00 | 4.50 | — | — | — | — | — | | | | | | | | 2.00 max. | FN-0405-R |
| 93.00 | .50 | 1.00 | 4.50 | — | — | — | — | — | | | | | | | | 2.00 max. | FN-0405-S |
| 93.00 | .50 | 1.00 | 4.50 | — | — | — | — | — | | | | | | | | 2.00 max. | FN-0405-T |
| 85.00 | .50 | 11.50 | — | — | — | — | — | — | | | | | | | | 4.00 max. | FX-1005-T |
| 78.00 | .50 | 20.00 | — | — | — | — | — | — | | | | | | | | 4.00 max. | FX-2008-T |
| **47. POWDER METAL ALLOYS — Stainless Steels** | | | | | | | | | | | | | | | | | |
| bal | .08 | — | 9.00 | 19.00 | 2.00 | 1.00 | .04 | .03 | | | | | | | | | SS-304-R |
| bal | .08 | — | 12.00 | 17.00 | 2.00 | 1.00 | .04 | .03 | | | | | | | | 2.50Mo | SS-316-R |
| bal | .15 | — | — | 12.50 | 1.00 | 1.00 | .04 | .03 | | | | | | | | | SS-410-R |
| bal | .15 | — | 9.00 | 18.00 | 2.00 | 1.00 | .20 | .15 | | | | | | | | .50Mo | SS-303-R |

| COMPOSITION, % | | | | | | | | | | | | | | | | | MATERIAL DESIGNATION |
|---|---|---|---|---|---|---|---|---|---|---|---|---|---|---|---|---|---|
| W | C | Mo | Ag | Cu | — | — | — | — | — | — | — | — | — | — | — | Other | |
| **47. POWDER METAL ALLOYS — Refractory Metal Base** | | | | | | | | | | | | | | | | | |
| — | — | 61.00 | 39.00 | — | | | | | | | | | | | | | 61Mo-39Ag |
| — | — | 50.00 | 50.00 | — | | | | | | | | | | | | | 50Mo-50Ag |

| COMPOSITION, % | | | | | | | | | | | | | | | | | MATERIAL DESIGNATION |
|---|---|---|---|---|---|---|---|---|---|---|---|---|---|---|---|---|---|
| W | C | Mo | Ag | Cu | — | — | — | — | — | — | — | — | — | — | — | Other | |
| **47. POWDER METAL ALLOYS — Refractory Metal Base (continued)** | | | | | | | | | | | | | | | | | |
| 50.00 | * | — | 50.00 | — | | | | | | | | | | | | | 50W + C-50Ag |
| 51.00 | — | — | 49.00 | — | | | | | | | | | | | | | 51W-49Ag |
| 55.00 | — | — | — | 45.00 | | | | | | | | | | | | | 55W-45Cu |
| 56.00 | * | — | — | 44.00 | | | | | | | | | | | | | 56W + C-44Cu |
| 65.00 | — | — | 35.00 | — | | | | | | | | | | | | | 65W-35Ag |
| 72.50 | — | — | 27.50 | — | | | | | | | | | | | | | 72.5W-27.5Ag |
| 74.00 | — | — | — | 26.00 | | | | | | | | | | | | | 74W-26Cu |
| 85.00 | — | — | 15.00 | — | | | | | | | | | | | | | 85W-15Ag |
| 87.00 | — | — | — | 13.00 | | | | | | | | | | | | | 87W-13Cu |

*Included with tungsten.

| COMPOSITION, % | | | | | | | | | | | | | | | | | MATERIAL DESIGNATION |
|---|---|---|---|---|---|---|---|---|---|---|---|---|---|---|---|---|---|
| Cu | Zn | Sn | Pb | Ni | Fe | C | Al | — | — | — | — | — | — | — | — | Other | |
| **47. POWDER METAL ALLOYS — Brasses** | | | | | | | | | | | | | | | | | |
| 78.00 | 20.00 | .10 | 1.60 | — | .30 | — | — | | | | | | | | | | CZP-0218-T |
| 78.00 | 20.00 | .10 | 1.60 | — | .30 | — | — | | | | | | | | | | CZP-0218-U |
| 78.00 | 20.00 | .10 | 1.60 | — | .30 | — | — | | | | | | | | | | CZP-0218-W |
| 90.00 | 10.00 | — | — | — | — | — | — | | | | | | | | | | 90Cu-10Zn |
| 90.00 | 10.00 | — | .50 | — | — | — | — | | | | | | | | | | 90Cu-10Zn-0.5Pb |
| 70.00 | 30.00 | — | — | — | — | — | — | | | | | | | | | | 70Cu-30Zn |
| 68.50 | 30.00 | — | 1.50 | — | — | — | — | | | | | | | | | | 68.5Cu-30Zn-1.5Pb |
| **47. POWDER METAL ALLOYS — Bronzes** | | | | | | | | | | | | | | | | | |
| 88.00 | — | 10.00 | — | — | .50 | 1.50 | — | | | | | | | | | | CT-0010-N |
| 88.00 | — | 10.00 | — | — | .50 | 1.50 | — | | | | | | | | | | CT-0010-R |
| 88.00 | — | 10.00 | — | — | .50 | 1.50 | — | | | | | | | | | | CT-0010-S |
| 95.00 | — | — | — | — | — | — | 5.00 | | | | | | | | | | 95Cu-5Al |
| 77.00 | — | 7.00 | 15.00 | — | 1.00 | 1.00 | — | | | | | | | | | | 77Cu-15Pb-7Sn-1Fe-1C |

| COMPOSITION, % | | | | | | | | | | | | | | | | | MATERIAL DESIGNATION |
|---|---|---|---|---|---|---|---|---|---|---|---|---|---|---|---|---|---|
| Cu | Zn | Sn | Pb | Ni | Fe | C | Al | — | — | — | — | — | — | — | — | Other | |
| **47. POWDER METAL ALLOYS — Copper-Nickel Alloys** | | | | | | | | | | | | | | | | | |
| 64.00 | 18.00 | — | — | 18.00 | — | — | — | | | | | | | | | | CZN-1818-T |
| 64.00 | 18.00 | — | — | 18.00 | — | — | — | | | | | | | | | | CZN-1818-U |
| 64.00 | 18.00 | — | — | 18.00 | — | — | — | | | | | | | | | | CZN-1818-W |
| 64.00 | 16.50 | — | 1.50 | 18.00 | — | — | — | | | | | | | | | | CZNP-1618-U |
| 64.00 | 16.50 | — | 1.50 | 18.00 | — | — | — | | | | | | | | | | CZNP-1618-W |
| 90.00 | — | — | — | 10.00 | — | — | — | | | | | | | | | | 90Cu-10Ni |
| 62.00 | 18.00 | 2.00 | — | 18.00 | — | — | — | | | | | | | | | | 62Cu-18Ni-18Zn-2Sn |
| **47. POWDER METAL ALLOYS — Nickel Alloys** | | | | | | | | | | | | | | | | | |
| 30.00 | — | — | — | 67.00 | 3.00 | — | — | | | | | | | | | | 67Ni-30Cu-3Fe |

| COMPOSITION, % | | | | | | | | | | | | | | | | | MATERIAL DESIGNATION |
|---|---|---|---|---|---|---|---|---|---|---|---|---|---|---|---|---|---|
| Fe | Al | Ni | Co | Mo | Ti | Si | Mn | C | P | Cb | Cu | — | — | — | — | Other | |
| **50. FREE MACHINING MAGNETIC ALLOYS** | | | | | | | | | | | | | | | | | |
| bal | — | — | — | — | — | 2.50 | .40 | .05 | .12 | — | — | | | | | | Magnetic Core Iron-FM |
| bal | — | 49.00 | — | — | — | .35 | .50 | .05 | — | — | — | | | | | | Hi Perm 49-FM |
| **51. MAGNETIC ALLOYS** | | | | | | | | | | | | | | | | | |
| bal | — | — | — | — | — | 4.00 | .15 | .05 | — | — | — | | | | | | Magnetic Core Iron |
| bal | — | 49.00 | — | — | — | .35 | .50 | .05 | — | — | — | | | | | | Hi Perm 49 |
| bal | — | 80.00 | — | 4.20 | — | .35 | .50 | .02 | — | — | — | | | | | | HyMu 80 |
| bal | 6.00 | 18.00 | 35.00 | — | 8.00 | — | — | — | — | — | — | | | | | | Alnico XII |
| bal | 7.80 | 13.50 | 25.00 | — | — | — | — | — | — | .80 | 3.00 | | | | | | Columax-5 |
| bal | 8.00 | 14.00 | 24.00 | — | — | — | — | — | — | — | 3.00 | | | | | | Alnico V-7 |
| bal | 8.00 | 14.00 | 24.00 | — | — | — | — | — | — | — | 3.00 | | | | | | Hyflux Alnico V-7 |

| COMPOSITION, % | | | | | | | | | | | | | | | | | MATERIAL DESIGNATION |
|---|---|---|---|---|---|---|---|---|---|---|---|---|---|---|---|---|---|
| Fe | Al | Ni | Co | Mo | Ti | Si | Mn | C | P | Cb | Cu | — | — | — | — | Other | |
| **51. MAGNETIC ALLOYS (continued)** | | | | | | | | | | | | | | | | | |
| bal | 8.00 | 14.00 | 24.00 | — | 1.00 | — | — | — | — | — | 3.00 | | | | | | Alnico V |
| bal | 10.00 | 17.00 | 2.50 | — | — | — | — | — | — | — | — | | | | | | Alnico II |
| bal | 12.00 | 20.00 | 5.00 | — | — | — | — | — | — | — | — | | | | | | Alnico I |
| bal | 12.00 | 25.00 | — | — | — | — | — | — | — | — | 6.00 | | | | | | Alnico III |
| bal | 12.00 | 28.00 | 5.00 | — | — | — | — | — | — | — | — | | | | | | Alnico IV |

| COMPOSITION, % | | | | | | | | | | | | | | | | | MATERIAL DESIGNATION |
|---|---|---|---|---|---|---|---|---|---|---|---|---|---|---|---|---|---|
| Fe | Ni | Co | C | Mn | Si | — | — | — | — | — | — | — | — | — | — | Other | |
| **52. FREE MACHINING CONTROLLED EXPANSION ALLOYS** | | | | | | | | | | | | | | | | | |
| bal | 36.00 | — | .12 | .35 | .30 | | | | | | | | | | | | Invar 36 |
| **53. CONTROLLED EXPANSION ALLOYS** | | | | | | | | | | | | | | | | | |
| bal | 35.00 | — | .12 | .50 | .50 | | | | | | | | | | | | Invar |
| bal | 29.00 | 17.00 | .02 | .30 | .20 | | | | | | | | | | | | Kovar |

SECTION 27

# GLOSSARY

**Abrasive Belt Grinding:** Removal of material from a workpiece, roughing and/or finishing, with a power-driven, abrasive-coated belt.

**Abrasive Belt Polishing:** Finishing a workpiece with a power-driven, abrasive-coated belt in order to develop a very good finish.

**Abrasive Cutoff:** Severing a workpiece by means of a thin, abrasive wheel.

**Abrasive Flow Machining:** See section 10.1.

**Abrasive Jet Machining:** See section 10.2.

**Abrasive Machining:** Used to accomplish heavy stock removal at high rates by use of a free-cutting grinding wheel.

**ACI:** Alloy Casting Institute, as in ACI Grade CB-7Cu Cast Stainless Steel.

**Acicular:** Refers to alloyed cast irons which develop high strength on cooling and which have a needlelike microstructure containing martensite mixed with other microconstituents.

**Active Oil:** An oil which contains chemically active ingredients to prevent metal-to-metal friction (by boundary lubrication) when the pressures become too great for the natural film strength of the oil.

**Aged:** A heat treatment of a previously solution heat treated alloy at temperatures which permit solid state precipitation to occur. (This treatment usually develops an increase in the hardness and strength of an alloy but may decrease machinability.)

**AISI:** American Iron and Steel Institute. Issues general classification standards for irons and steels.

**Alloy Steels:** Steels with carbon from 0.1 to 1.1 percent and containing alloying elements such as nickel, chromium, molybdenum, and vanadium. (The total of all such alloying elements in these type steels is usually less than 5 percent.)

**Alpha-Beta Titanium Alloys:** See *Titanium Alloys.*

**Alpha Titanium Alloys:** See *Titanium Alloys.*

**AMS:** Aerospace Material Specification published by the Society of Automotive Engineers (SAE), as in AMS 4803A.

**Annealed:** A heat treatment of an alloy above its critical or recrystallization temperature, usually followed by furnace cooling. (The treatment results in low hardness, high ductility, and usually improves tool life.)

**Arc Cast:** The as-cast condition produced by melting an alloy with an electrical arc.

**As Cast:** See *Cast.*

**ASTM:** American Society for Testing and Materials, as in standards such as ASTM A27, A148, etc.

**Austempering:** Quenching a ferrous alloy from a temperature above the transformation range, in a medium having a rate of heat abstraction fast enough to prevent the formation of high-temperature transformation products, and then holding the alloy, until transformation is complete, at a temperature below that of pearlite formation and above that of martensite formation.

**Austenite:** A solid solution of one or more elements in face-centered-cubic iron. (In most alloy steels, austenite is stable at heat treating temperatures and changes to other constituents such as pearlite, ferrite, martensite, etc. In the austenitic stainless steels, it is stable at room temperature as a result of alloying.)

**Austenitic:** Of or pertaining to *Austenite.*

**B:** See *Boron Steels.*

**Beta Titanium Alloys:** See *Titanium Alloys.*

**Bhn:** See *Brinell Hardness.*

**Boring:** Enlarging a hole by removing metal with a single- or occasionally a multiple-point cutting tool moving parallel to the axis of rotation of the work or tool.

> **Single-Point Boring:** Cutting with a single-point tool.
>
> **Precision Boring:** Cutting to tolerances held within narrow limits.
>
> **Gun Boring:** Cutting of deep holes.
>
> **Jig Boring:** Cutting of high-precision and accurate-location holes.
>
> **Groove Boring:** Cutting accurate recesses in hole walls.

**Boron Steels:** Steels with boron additions in amounts of 0.0005 to 0.003 percent to increase hardenability, as in 86B45, 94B30, etc.

**Boundary Lubrication:** Lubrication by a solid lubricant (a material which has relatively low shear strength). This lubricant may be introduced in solid form, or it may be formed on fresh metal surfaces by chemical reaction.

**Brinell Hardness:** A test for determining the hardness of a material by forcing a hard steel or carbide ball of specified diameter into it under a specified load. (The result is expressed as the Brinell hardness number, which is the value obtained by dividing the applied load in kilograms by the surface area of the resulting impression in square millimeters.)

**Broaching:** Cutting with a tool which consists of a bar having a single edge of a series of cutting edges (that is, teeth) on its surface. The cutting edges of multiple-tooth or successive, single-tooth broaches increase in size and/or change in shape. The broach cuts in a straight line or axial direction when relative motion is produced in relation to the workpiece, which may also be rotating. The entire cut is made in single or multiple passes over the workpiece to shape the required surface contour.

**Pull Broaching:** Tool pulled through or over workpiece.

**Push Broaching:** Tool pushed over or through workpiece.

**Chain Broaching:** A continuous, high-production surface broach.

**Tunnel Broaching:** Work travels through an enclosed area containing broach inserts.

**BUE:** See *Built-up Edge*.

**Buffing:** A two-stage operation: 1) cutting down and 2) coloring. Cutting down removes scratch marks from rough polishing, stretch marks from forming, die marks or other surface imperfections. It makes a relatively smooth surface smoother. Coloring refines the cut-down surface and brings out maximum luster.

**Built-up Edge (BUE):** A piece of work material which has been strain hardened and pressure welded to the cutting edge of a tool or to the surface of a workpiece.

**Burnishing:** Finish sizing and smooth finishing of surfaces (previously machined or ground) by displacement, rather than removal, of minute surface irregularities with smooth, point-of-line contact, fixed or rotating tools.

**Carbides:** A compound of carbon with one or more metallic elements. Often refers to one of the general class of pressed and sintered cutting tools which contain tungsten carbide plus smaller amounts of titanium and tantalum carbides along with cobalt which acts as a binder. (It is also used to describe hard compounds in steels and cast irons.)

**Carbon Steels:** Steels with carbon plus small amounts of manganese, silicon, sulfur, phosphorus, or other elements.

**Carburized:** A condition attained by using carburizing heat treatments to increase the carbon content at the surface of steels. (This is done by placing the parts in contact with carbonaceous solids, liquids, or gases at approximately 1600° to 1700°F [870° to 927°C]. The depth of penetration and amount of carbon pickup are a function of the time, temperature and carburizing medium. Hardening treatments of carburized surfaces develop hardnesses up to 65 $R_C$.)

**Case Hardening:** A heat treating method by which the surface layer of alloys is made substantially harder than the interior. (Carburizing and nitriding are common ways of case hardening steels.)

**Cast:** A state of the alloy after solidification of the molten alloy.

**Cast Steels:** Steels which are cast to shape and used without having been hot rolled or forged but which are usually heat treated to produce appropriate material properties.

**Cemented Carbides:** Carbides that are mixed with cobalt metal powder, milled in ball mills and then pressed into desired shapes and sintered. (Used primarily for cutting tools, wire-drawing dies and parts subject to heavy abrasive wear.)

**Ceramics:** The general class of hard, brittle, and high-melting nonmetallic materials such as aluminum oxide, zirconium oxide, and beryllium oxide. (This same term also is used to designate aluminum oxide and other similar type ceramic cutting tools.)

**Chatter Marks:** Imperfections on the work surface usually caused by vibrations of the tool and/or workpiece.

**Chemical Coolant:** A cutting fluid which does not contain any mineral oil; usually a true solution in water or a fine colloidal solution.

**Chemical Machining:** See section 13.1.

**Chlorine:** A common extreme pressure (EP) additive used to promote lubrication.

**Cold Drawn:** A condition produced by reducing hot rolled rod to size by pulling the rod through a die of the desired diameter. (This is frequently done at room temperature but, in any event, must be carried out below the recrystallization temperature of the material. This process often improves machinability by providing better surface finish and chip control.)

**Cold Form Tapping:** Producing internal threads by displacing material rather than removing it as either the tap or the workpiece is rotated. The thread form is produced by a tool, which has neither flutes nor cutting edges. The tool resembles a simple screw when viewed from the side, but the end view shows that both the major and minor diameters have irregular contours for displacing the work material.

**Cold Rolled:** An alloy condition developed by reducing hot rolled stock by passing it between rolls to the desired thickness. Temperature limitations are the same as for cold drawing.

**Compacted Graphite Cast Irons:** Class of cast irons characterized by graphite that is interconnected within eutectic cells, as is the flake graphite in gray iron, but is coarser and more rounded. (Well suited for applications where neither gray iron nor ductile iron is completely satisfactory.)

**Controlled Expansion Alloys:** Special group of iron alloys with low thermal expansion coefficients recommended for applications that require very low dimensional changes with variations in temperature.

**Coolant:** Liquid used to cool the work and tool and to prevent rusting or corrosion. See also *Cutting Fluid*.

**Corrosion Resistant Steels:** A term often used to describe the cast stainless steels and to differentiate those cast steels used primarily for corrosion applications.

**Counterboring:** Removal of material to enlarge a hole for part of its depth with a rotary, pilot-guided, end-cutting tool having two or more cutting lips and usually having straight or helical flutes for the passage of chips and the admission of a cutting fluid.

**Countersinking:** Beveling or tapering the work material around the periphery of a hole creating a concentric surface at an angle less than 90° with the centerline of the hole for the purpose of chamfering holes or recessing screw and rivet heads.

**Coupling Agent:** A mutual solvent; an emulsifier.

**Creaming:** A concentrating of oil droplets of an emulsion near the surface when the emulsion stands quiescent for a sufficient period of time.

**Cutting Fluid:** Fluid (liquid, gas, or mist) applied to the working part of a tool or cutter to promote more efficient machining. See also *Coolant.*

**Cutting Oil:** A cutting fluid used where lubrication is the prime consideration and the cooling factor is of less importance.

**Deburring:** Removal of burrs, sharp edges or fins from parts by filing, grinding or rolling the work in a barrel with abrasives suspended in a suitable liquid medium. Sometimes called "burring."

**Die Cast:** Alloy condition resulting from casting in metallic dies.

**Drawing:** A term often used to mean the same as tempering, which is the preferred designation. ("Drawing" should be reserved for hot and cold mechanical working operations.)

**Drilling:** Hole making with a rotary, end-cutting tool having one or more cutting lips and one or more helical or straight flutes or tubes for the ejection of chips and the passage of a cutting fluid.

> **Center Drilling:** Drilling a conical hole in the end of workpiece.
>
> **Core Drilling:** Enlarging a hole with a chamfer-edged, multiple-flute drill.
>
> **Spade Drilling:** Drilling with a flat blade drill tip.
>
> **Step Drilling:** Using a multiple-diameter drill.
>
> **Gun Drilling:** Drilling a deep hole using special straight-flute drills (often with a single lip) and cutting fluid at high pressures.
>
> **Oil-Hole or Pressurized-Coolant Drilling:** Using a drill with one or more continuous holes through its body and shank to permit the passage of a high-pressure cutting fluid which emerges at the drill point and ejects chips.

**Ductile Cast Irons:** Irons made by inoculating liquid iron from a cupola with magnesium-nickel or other inoculants to produce graphite in a nodular form. (These cast irons are sometimes called "nodular irons.")

**Electrical Discharge Grinding:** See section 12.2.

**Electrical Discharge Machining:** See section 12.3.

**Electrical Discharge Sawing:** See section 12.4.

**Electrical Discharge Wire Cutting:** See section 12.5.

**Electrochemical Deburring:** See section 11.1.

**Electrochemical Discharge Machining:** See section 11.2.

**Electrochemical Grinding:** See section 11.3.

**Electrochemical Honing:** See section 11.4.

**Electrochemical Machining:** See section 11.5.

**Electrochemical Polishing:** See section 11.6.

**Electrochemical Sharpening:** See section 11.7.

**Electrochemical Turning:** See section 11.8.

**Electron Beam Machining:** See section 12.1.

**Electropolishing:** See section 13.2.

**Electro-stream™:** See section 11.9.

**Emulsifiable Oil:** A straight oil or blend which contains an emulsifier or coupling agent so it will form a stable emulsion in water.

**Emulsion:** An oily mass in suspension in a watery liquid or vice versa.

**EP Additive:** See *Extreme Pressure Additive.*

**EPA:** United States Environmental Protection Agency.

**Extreme Pressure Additive:** A compound which reacts with the surface of the metal (or tool) to form thin films of metallic compounds (usually, a chloride, sulfide, or phosphate) which have relatively low shear-strength.

**Extruded:** Wrought material produced by forcing an alloy or other material through a die.

**Fatty Acid:** Any of the series of saturated or unsaturated acids ($C_nH_{2n}O_2$) such as stearic, oleic, and palmitic acids which occur in natural fats and natural oils.

**Fatty Oils:** Organic oils; the most common are lard oil and sperm oil.

**Ferrite:** A solid solution of one or more elements in body-centered-cubic iron. (Ferritic microstructure is commonly found in steels and cast irons and generally allows for good tool life.)

**Ferritic:** Of or pertaining to *Ferrite.*

**Filled Plastics:** Plastics containing a relatively inert material that has been added to modify strength, permanence, working properties, or other qualities, or to lower cost.

**Flame Hardened:** Surface hardened by controlled torch heating followed by quenching with water or air.

**Flame (Thermal) Sprayed Materials:** Hard, wear resistant materials deposited as coatings on the surfaces of metals by a blast-gas-stream-ignited mixture of oxygen and acetylene.

**Forged:** A cold or hot mechanical working process performed by presses or hammers to shape metals. (Treatments such as annealing or normalizing should generally be applied after forging to improve machinability.)

**Gear Cutting:** Producing tooth profiles of equal spacing on the periphery, internal surface, or face of a workpiece by means of an alternate shear gear-form cutter or a gear generator.

**Gear Hobbing:** Gear cutting by use of a tool resembling a worm gear in appearance, having helically spaced cutting teeth. In a single-thread hob, the rows of teeth advance exactly one pitch as the hob makes one revolution. With only one hob, it is possible to cut interchangeable gears of a given pitch of any number of teeth within the range of the hobbing machine.

**Gear Milling:** Gear cutting with a milling cutter that has been formed to the shape of the tooth space to be cut. The tooth spaces are machined one at a time.

**Gear Shaping:** Gear cutting with a reciprocating gear-shaped cutter rotating in mesh with the work blank.

**Gear Shaving:** A finishing operation performed with a serrated rack or gear-like cutter in mesh with the gear, but with their axis skewed.

**Germicide:** Any agent which destroys germs or micro-organisms.

**Glasses:** Inorganic products of fusion which have cooled to a rigid condition without crystallizing. Typically hard and relatively brittle.

**Gray Cast Irons:** Alloys primarily of iron, carbon, and silicon along with other alloying elements in which the graphite is in flake form. (These irons are characterized by low ductility but have many other desirable properties, such as good castability and good damping capacity.)

**Grinding:** Material removal by use of abrasive grains held by a binder.

**Surface Grinding:** Producing a flat surface with a rotating grinding wheel as the workpiece passes under the wheel.

**Cylindrical Grinding:** Grinding the outside diameters of cylindrical workpieces held between centers.

**Internal Grinding:** Grinding the inside of a rotating workpiece by use of a wheel spindle which rotates and reciprocates through the length of depth of the hole being ground.

**Centerless Grinding:** Grinding cylindrical surfaces without use of fixed centers to rotate the work. The work is supported and rotates between three fundamental machine components: the grinding wheel, the regulating wheel, and the work guide blade.

**Gear Grinding:** Removal of material to obtain correct gear tooth form by grinding. This is one of the more exact methods of finishing gears.

**Thread Grinding:** Thread cutting by use of a suitably formed grinding wheel.

**H:** A letter listed along with the composition of aluminum and magnesium alloys indicating that the alloy has been cold worked. (H12, H14, etc., are used to indicate various degrees of cold working.)

**Halogen:** The group of elements: chlorine, fluorine, bromine, and iodine.

**Hardened:** Designates condition produced by various heat treatments, such as quench hardening, age hardening, and precipitation hardening.

**Hardness:** The relative ability of a material to resist indentation.

**Heat Resistant Steels:** Cast steels which are highly alloyed and used for high temperature applications such as furnace conveyor parts, etc.

**Heat Treatment:** Heating and cooling of metal or an alloy in the solid state for the purpose of obtaining certain desirable conditions or properties.

**High Speed Steels:** Tool steels which contain tungsten, molybdenum, vanadium, cobalt, carbon, and other elements.

**High Strength Steels:** Class of steels capable of heat treatment to develop yield strengths of 200,000 psi [138 MPa] or higher. (Originally developed to obtain improved strength-to-weight ratios in materials for transportation equipment.)

**High Temperature Alloys:** Generally alloys of nickel, cobalt, or iron which are used at temperatures in excess of

1200°F [650°C] in applications such as rockets, jet engine blades, and compressor and turbine discs, etc.

**Hollow Milling:** Using a special end-cutting mill so designed to leave a core after feeding into or through the workpiece.

**Honing:** A finishing operation using fine grit abrasive stones to produce accurate dimensions and excellent finish.

**Hot Machining:** Machining in which the workpiece shear zone is heated by auxiliary means to reduce the shear strength and increase the machinability of the material.

**Hot Rolled:** A term used to describe alloys which are rolled at temperatures above the recrystallization temperature. (Many alloys are hot rolled, and machinability of such alloys may vary because cooling conditions differ from lot to lot.)

**Hydrodynamic Lubrication:** Lubrication where the viscosity of the lubricant keeps the surfaces separated by a fluid film.

**Hydrodynamic Machining:** See section 10.3.

**Induction Hardened:** Surface or through hardened using induction heating followed by quenching with water or air.

**Inverted Emulsion:** A dispersion of droplets of water in oil which is produced when a small quantity of water is mixed with a relatively large quantity of oil.

**ISO:** International Organization for Standardization. International standardizing body comprising the national standardizing bodies of several countries.

**Knoop Hardness:** Microhardness measured using a diamond indenter ground to pyramidal form and expressed as the ratio of the load applied to the indenter (kilograms) to the unrecovered projected area ($mm^2$).

**L:** See *Leaded Alloys.*

**Lapping:** A finishing operation using fine abrasive grits loaded into a lapping material such as cast iron. Lapping provides major refinements in the workpiece including: 1) extreme accuracy of dimension, 2) correction of minor imperfections of shape, 3) refinement of surface finish, and 4) close fit between mating surfaces.

**Laser Beam Machining:** See section 12.6.

**Laser Beam Torch:** See section 12.7.

**Leaded Alloys:** Alloys to which lead has been added to improve machinability, as in 41L30, 86L20, etc.

**Low Stress Grinding:** See section 10.4.

**Machinability:** The relative difficulty of a machining operation with regard to tool life, surface roughness, and power consumption.

**Machinable Carbides:** Carbide grains in a steel matrix readily machinable in the annealed state and hardenable by heat treatment to 64 to 70 $R_C$.

**Magnetic Alloys:** Alloys composed primarily or totally of nonferrous metals that can be used for making permanent magnets.

**Malleable Cast Irons:** Irons made by malleablizing white cast iron. See also *Malleablizing.*

**Malleablizing:** Process of annealing brittle white cast iron in such a way that the combined carbon is wholly or partly transformed to graphitic or temper-carbon nodules in a ferritic or pearlitic microstructure, thus providing a ductile and machinable material.

**Maraged:** Describes a series of heat treatments used to treat high strength steels of complex composition (maraging steels) by aging of martensite.

**Maraging Steels:** Special class of steels that are hardened by the precipitation of intermetallic compounds at temperatures of about 900°F [480°C].

**Martensite:** An acicular or needlelike microstructure that is formed in quenched steels. (It is very hard and brittle in the as-quenched form and therefore is usually tempered before being placed into service. The harder forms of tempered martensite have poor machinability.)

**Martensitic:** Of or pertaining to *Martensite.*

**Metal Slitting:** An operation using a thin circular saw blade to produce a narrow slit in the workpiece. The workpiece is fed into the saw usually on a setup similar to peripheral milling.

**Metallic Soap:** Any oil of mineral origin, such as petroleum.

**Milling:** Using a rotary tool with one or more teeth which engage the workpiece and remove material as the workpiece moves past the rotating cutter.

> **Face Milling:** Milling a surface perpendicular to the axis of the cutter. Peripheral cutting edges remove the bulk of the material while the face cutting edges generate the workpiece finish.
>
> **End Milling:** Milling accomplished with a tool having cutting edges on its cylindrical surfaces as well as on its end. In peripheral end milling, the peripheral cutting edges on the cylindrical surface are used; while in end milling–slotting, both end and peripheral cutting edges remove metal.
>
> **Side and Slot Milling:** Milling of the side or slot of a workpiece using a peripheral cutter.
>
> **Slab Milling:** Milling of a surface parallel to the axis

of a helical, multiple-toothed cutter mounted on an arbor.

**Straddle Milling:** Peripheral milling a workpiece on both sides at once using two cutters spaced as required.

**Miscible:** Capable of being mixed.

**NIOSH:** United States National Institute for Occupational Safety and Health.

**Nitrided:** Surface hardened by absorption of nitrogen from a gaseous or salt bath medium.

**Nitriding:** A heat treating method in which nitrogen is diffused into the surface of a solid ferrous alloy. (This is done by heating the metal at a temperature of about 950°F [510°C] in contact with ammonia gas or other suitable nitrogenous materials. Because nitrides form, the surface becomes much harder than the interior. Depth of the nitrided surface is a function of the length of time of exposure and can vary from 0.0005- to 0.032-inch [0.013 to 0.81 mm] thick. Hardness is generally 65 to 70 $R_C$; consequently, these structures are almost always ground.)

**Nitriding Steels:** Steels which are selected because they form good case hardened structures during the nitriding process. (In these steels, elements such as aluminum and chromium are important for producing a good case.)

**Nodular Cast Irons:** See *Ductile Cast Irons.*

**Normalized:** Heat treatment of ferrous alloys above the transformation range, followed by cooling in still air to a temperature below the transformation range. (This is often done to refine or homogenize the grain structure of castings, forgings, and wrought steel products.)

**O:** A letter often listed along with the composition of aluminum and magnesium alloys indicating that the alloy has been annealed.

**OSHA:** United States Occupational Safety and Health Administration.

**Pearlite:** A microconstituent found in iron-base alloys consisting of a lamellar (platelike) composite of ferrite and iron carbide. (This structure results from the decomposition of austenite and is very common in cast irons and annealed steels.)

**Pearlitic:** Of or pertaining to *Pearlite.*

**Permanent Mold:** A mold, usually made of metal, which is used repeatedly for the production of a number of castings of the same form.

**PH:** Letters used with the composition of precipitation hardening stainless steels to designate that special heat treatments using in part a refrigeration cycle have been performed on the material to develop specific properties.

**Photochemical Machining:** See section 13.3.

**Planing:** Producing flat surfaces by linear reciprocal motion of the work and the table to which it is attached relative to a stationary, single-point cutting tool.

**Plasma Beam Machining:** See section 12.8.

**Plastics:** Materials that contain as an essential ingredient an organic substance of large molecular weight, are solid in the finished state, and can be shaped by flow at some stage in their manufacture or processing into finished articles.

**Plated Materials:** Metals applied in thin films to the surface of a material by dipping or by electrodeposition.

**P/M:** See *Powder Metallurgy.*

**Polishing:** Removal of metal by the action of abrasive grains carried to the work by a flexible support, generally either a wheel or a coated abrasive belt.

**Powder Metal Alloys:** Alloyed metal powders that have been pressed or rolled into compacts that are sintered to produce shaped components or bar stock.

**Powder Metallurgy:** A process involving pressing of powdered metal followed by heating (sintering) to produce bearings, friction materials, refractory alloys, etc.

**Precipitation Hardening Stainless Steels:** Certain stainless steel compositions which respond to precipitation hardening or aging treatment.

**Pressed** (with reference to *Powder Metallurgy*): The "as pressed" condition of alloys when compacted from powders. (Pressing of powdered metals is generally followed by sintering. Some difficult- or impossible-to-machine materials such as tungsten and carbides are shaped by various machining operations in the "green" condition, generally produced by low temperature presintering.)

**Quenched:** Rapid cooling of alloys by immersion in liquids or gases after heating.

**Reaming:** An operation in which a previously formed hole is sized and contoured accurately by using a rotary cutting tool (reamer) with one or more cutting elements (teeth). The principal support for the reamer during the cutting action is obtained from the workpiece.

**Form Reaming:** Reaming to a contour shape.

**Taper Reaming:** Using a special reamer for taper pins.

**Hand Reaming:** Using a long lead reamer which permits reaming by hand.

**Pressure-Coolant Reaming (or Gun Reaming):** Using a multiple-lip, end-cutting tool through which coolant is forced at high pressure to flush chips ahead of the tool or back through the flutes for finishing of deep holes.

**Refractory Alloys:** Alloys or elements which have melting temperatures above 4000°F [2200°C]. (The most common refractory alloys of current industrial interest are tungsten, molybdenum, columbium, and tantalum.)

**Resulfurized Alloys:** Alloys which have sulfur added to improve machinability.

**Rockwell Hardness:** A measure of the difference in depth of penetration of an indenter between a major and a minor load. (Normally a minor load of 10 kilograms is applied, followed by the application and release of a major load which will vary from 60 to 150 kilograms. The more commonly used Rockwell scales are Rockwell "C" [$R_C$], using a diamond sphero-conical penetrator and a 150 kilogram major load, and Rockwell "B" [$R_B$], using a 1/16 inch diameter steel ball penetrator and a major load of 100 kilograms. Other scales used in this data book are $R_A$, $R_E$, $R_F$, $R_H$, $R_M$, and $R_R$.

**Rolled:** See *Cold Rolled, Hot Rolled.*

**Rotary Filing and Burring:** Machining or smoothing surfaces with contour-fitting rotary tools where only a minimum amount of material is to be removed.

**Routing:** Cutting out and contouring edges of various shapes in a relatively thin material using a small diameter rotating cutter which is operated at fairly high speeds.

**rpm:** Revolutions per minute.

**Sawing:** Using a toothed blade or disc to sever parts or cut contours.

**Circular Sawing:** Using a circular saw fed into the work by motion of either the workpiece or the blade.

**Power Band Sawing:** Using a long, multiple-tooth continuous band resulting in a uniform cutting action as the workpiece is fed into the saw.

**Power Hack Sawing:** Sawing in which a reciprocating saw blade is fed into the workpiece.

**SAE:** Society of Automotive Engineers, as in SAE G1800.

**Sand Casting:** Non-precision process for producing castings using molds made from natural or synthetic sand.

**Shaped Tube Electrolytic Machining:** See section 11.10.

**Shaping:** Using single-point tools fixed to a ram reciprocated in a linear motion past the work.

**Form Shaping:** Shaping with a tool ground to provide a specified shape.

**Contour Shaping:** Shaping of an irregular surface, usually with the aid of a tracing mechanism.

**Internal Shaping:** Shaping of internal forms such as keyways and guides.

**Shore Hardness** (A and D): Indentation hardness measured using a durometer having an indenter formed from hardened steel rod. (Type A durometers measure softer materials, while Type D durometers are used for harder materials.)

**Sintered:** The sintered condition results from a heating of pressed powdered materials for specified times at elevated temperature. Improved strength and other benefits result, but generally machinability is decreased.

**Skiving:** Generating cylindrical forms by moving a form tool laterally through a rotating workpiece.

**Snagging:** Heavy stock removal of superfluous material from a workpiece by using a portable or swing grinder mounted with a coarse grain abrasive wheel.

**Soluble Oil:** See *Emulsifiable Oil.*

**Solution Treated:** A process by which it is possible to dissolve microconstituents by taking certain alloys to an elevated temperature and then keeping them in solution after quenching. (Often a solution treatment is followed by a precipitation or aging treatment to improve the mechanical properties. Most high temperature alloys which are solution treated and aged machine better in the solution treated state just before they are aged.)

**Spotfacing:** Using a rotary, hole-piloted end-facing tool to produce a flat surface normal to the axis of rotation of the tool on or slightly below the workpiece surface.

**Spray Lubrication:** Describes the application of a lubricant by means of spraying equipment in much the same manner as paint is sprayed.

**ST:** See *Solution Treated.*

**STA:** Solution treated and aged. See *Solution Treated* and *Aged.*

**Stainless Steels:** Steels which have good or excellent corrosion resistance. There are three broad classes of stainless steels—ferritic, austenitic, and martensitic. These various classes are produced through the use of various alloying elements in differing quantities.

**Strain Hardening:** See *Work Hardening.*

**Stress Corrosion:** Corrosion facilitated by high residual surface stress imposed by machining or grinding operations.

**Stressed Relieved:** The treatment used to relieve the internal stresses induced by forming or heat treating operations. (It consists of heating a part uniformly, followed by cooling slow enough so as not to reintroduce stresses. To obtain low stress levels in steels and cast irons, temperatures as high as 1250°F [575°C] may be required.)

**Structural Steels:** Steels covering a wide range of strengths and used for structural purposes. These steels are sometimes called high strength steels or constructional steels. In this handbook, the term high strength steels is reserved primarily for those steels which are used at strength levels of 200,000 psi [138 MPa] or greater.

**Sub-zero Machining:** Using refrigerant or other means for cooling the workpiece during, or before, machining.

**Sulfo-chlorinated Oil:** Cutting oil containing sulfur and chlorine.

**Sulfur:** A common extreme pressure (EP) additive used to promote boundary lubrication.

**Superfinishing:** An abrasive process utilizing either a bonded stock for a cylindrical workpiece or a cup wheel for flat and spherical work. A large contact area, approximately 30 percent, exists between workpiece and abrasive. The object of superfinishing is to remove surface fragmentation and to correct inequalities in geometry, such as grinding feed marks and chatter marks.

**Surface Active Agent:** Materials capable of lowering surface and interfacial tensions. See *Wettability*.

**Synthetic Fluids:** Products which do not contain any mineral oil and usually form a true solution in water.

**T:** A letter often listed along with the numbers used to designate aluminum and magnesium alloys indicating that the alloy has been subjected to heat treatment. (T4, T5, T6, etc., indicate various specific heat treatments.)

**Tapping:** Producing internal threads with a cylindrical cutting tool having two or more peripheral cutting elements shaped to cut threads of the desired size and form. By a combination of rotary and axial motions, the leading end of the tap cuts the thread while the tap is supported mainly by the thread it produces.

**Tempered:** Reheating of hardened steels or hardened cast irons to reduce hardness and to lower internal stress. (The temperature used depends upon the mechanical properties specified but generally ranges from 300° to 1200°F [150° to 650°C].)

**Tempered Martensite:** A two-phase microstructure containing carbide particles in a ferrite matrix which results from reheating martensite to improve toughness.

**Thermally Assisted Machining:** See section 10.5.

**Thermochemical Machining:** See section 13.4.

**Thermoplastic:** A plastic which can be softened repeatedly by an increase and hardened by a decrease in temperature. (Heating causes physical changes rather than chemical.)

**Thermosetting Plastic:** A plastic which upon heating is changed chemically into a material having new properties.

**Threading:** Producing external threads on a cylindrical surface.

**Die Threading:** A process for cutting external threads on cylindrical or tapered surfaces by the use of solid or self-opening dies.

**Single-Point Threading:** Turning threads on a lathe.

**Thread Grinding:** See definition under *Grinding*.

**Thread Milling:** A method of cutting screw threads with a milling cutter.

**Titanium Alloys:** Alloys of the element titanium. (They combine light weight, excellent strength, moderately good high temperature strength, and excellent corrosion resistance. Various alloying elements and heat treatments provide differing crystal structures. Depending upon the crystal structure of the finished alloy, they are classified as alpha, beta, or alpha-beta alloys.)

**Titanium Carbide:** Carbides, used as tool materials, in which the greatest amount of carbide is titanium instead of tungsten and which also contain other carbides plus nickel and molybdenum as binders.

**Tool Steels:** Steels used to make cutting tools and dies. Many of these steels have considerable quantities of alloying elements such as chromium, carbon, tungsten, molybdenum, and other elements. These form hard carbides which provide good wearing qualities but at the same time decrease machinability. Tool steels in the trade are classified for the most part by their applications, such as hot work die (type H), cold work die (types D, A and O), high speed (types T and M), shock resisting (type S), mold (type P), special purpose (type L), and water hardening (type W).

**Total Form Machining:** See section 10.6.

**Tramp Oil:** Leakage into the cutting fluid system from the hydraulic or the lubrication systems of machine tools.

**Transformation Temperature:** The temperature at which a change in phase occurs. In a metal, phase transformation occurs at a specific temperature.

**Trepanning:** Cutting with a boring tool so designed as to leave an unmachined core when the operation is completed.

**Tumble Grinding:** Various surfacing operations ranging from deburring and polishing to honing and microfinishing of metallic parts before and after electroplating.

**Turning:** Generating cylindrical forms by removing metal with a single-point cutting tool moving parallel to the axis of rotation of the work.

**Single-Point Turning:** Using a tool with one cutting edge.

**Face Turning:** Turning a surface perpendicular to the axis of the workpiece.

**Form Turning:** Using a tool with a special shape.

**Turning Cutoff:** Severing the workpiece with a special lathe tool.

**Box Tool Turning:** Turning the end of a workpiece with one or more cutters mounted in a box-like frame, primarily for finish cuts.

**Ultimate Strength:** The maximum stress expressed in pounds per square inch or pascals which a material will carry before breaking under a slowly applied, continually increasing load.

**Ultrasonic Machining:** See section 10.7.

**Water Jet Machining:** See section 10.8.

**Wettability:** The relative ease with which a liquid spreads over a surface.

**Wetting Agent:** An additive which reduces surface and interfacial tension and, therefore, facilitates spreading of a fluid over a surface.

**White Cast Iron:** Hard, brittle, difficult-to-machine irons containing 2 to 6 percent carbon, the majority of which is present as iron carbides ($Fe_3C$). (Excellent material for certain wear- and abrasion-resistant applications.)

**Work Hardening:** A hardening process which may occur during cold working or machining.

**Wrought:** Condition of a material which has been worked mechanically as in forging, rolling, drawing, etc.

# SECTION 28

# SUBJECT INDEX

Abrasive belt grinding, 8-177 to 8-178
Abrasive belt machining, 20-33 to 20-34
Abrasive cutoff, 6-129 to 6-148
  surface integrity data, 18-81
Abrasive flow machining, 10-3 to 10-14
  applications data, 10-12 to 10-14
  hole enlargement, 10-11
  stock removal and flow rates, 10-7 to 10-10
Abrasive jet machining, 10-15 to 10-20
Abrasive machining, 20-33 to 20-35
Adaptive control, 24-10 to 24-12

Ballizing, 3-399 to 3-401
Band sawing, *see* Power band sawing
Bearingizing, 3-396 to 3-398
Boring, 3-223 to 3-267
  with ceramic tools, 3-269 to 3-294
  with diamond tools, 3-295 to 3-298
  tool geometry for carbide tools, 15-21 to 15-22
  tool geometry for single point tools, 15-2 to 15-3
Broaching, 5-25 to 5-46
  tool geometry for high speed steel broaches, 15-27
Broaching, surface:
  power requirements, 17-7 to 17-9, 17-12 to 17-13, 17-20 to 17-21
  unit power requirements, 17-12 to 17-13
Burnishing, 3-393 to 3-408

CAM, *see* Computer-aided manufacturing
Carbide grade chart:
  C grades, 14-8 to 14-9
  ISO grades, 14-12 to 14-14
Carbide tools, *see* Tool materials, carbide
Cast alloy tools, *see* Tool materials, cast alloys
Centerless grinding, 8-143 to 8-173
  guidelines for, 20-14
  horsepower and metal removal rate alignment chart, 20-30 to 20-31
  illustrated, 20-23
  parameter formulas, 20-24 to 20-25
  tolerances for, 20-18
  work traverse rates, 8-175 to 8-176
Ceramic tools, *see* Tool materials, ceramics
Cermet tools, *see* Tool materials, cermets
Chatter, 22-1 to 22-23
  regenerative, 22-12 to 22-16, 22-22
  self-excited, 22-12 to 22-17, 22-21
  troubleshooting, 22-17 to 22-21
Chatter tendency for common work materials, 22-13 to 22-15
Chemical compositions, 26-37 to 26-95
Chemical elements, list of, 26-37
Chemical machining, 13-3 to 13-16
  consolidated data, table, 13-15 to 13-16
  operating parameters, steps to selection of, 13-5
  surface integrity data, 18-111 to 18-115
Chemical milling, *see* Chemical machining
Circular sawing:
  with carbide tipped blade (performance data), 6-121 to 6-128
  with high speed steel blade, 6-75 to 6-119
CNC (computerized numerical control), 24-10
Cold form tapping, 4-43 to 4-46
Computer-aided manufacturing, 24-1 to 24-12
Computer-aided process planning, 24-6 to 24-7
Core drilling, diamond tools, 3-51
Costs, machining, *see* Economics, machining and grinding
Counterboring, 3-299 to 3-345
Creep feed grinding, 20-35
Cutoff, *see* Abrasive cutoff; Turning, cutoff tools
Cutting fluids, 16-1 to 16-96
  application, methods of, 16-9 to 16-12
  central fluid systems, 16-14 to 16-15
  coolant flow requirements, table of, 16-11
  economic considerations, 16-15
  maintenance of, 16-12 to 16-14
  recommendations by machining operation and material group, table of, 16-17 to 16-63
    key to types and trade names, 16-65 to 16-96
  selection of, 16-7 to 16-9
  types of, 16-3 to 16-7
Cylindrical grinding, 8-61 to 8-92
  with cubic boron nitride wheels, 8-93 to 8-96
  with diamond wheels, 8-97 to 8-100
  guidelines for, 20-14
  horsepower and metal removal rate alignment chart, 20-30 to 20-31
  illustrated, 20-23
  parameter formulas, 20-24 to 20-25
  tolerances for, 20-18
  unit power chart, 20-27

Damping, 22-5 to 22-6
  influence on forced vibration amplitude, 22-9 to 22-10
Deburring:
  abrasive flow, 10-3 to 10-14
  abrasive jet, 10-15 to 10-17
  electrochemical, 11-3 to 11-4
Deep rolling, 3-394 to 3-395
Diamond tools, *see* Tool materials, diamond
DNC (direct numerical control), 24-10
Drilling, 3-3 to 3-50
  accuracy of holes, 19-9
  coolant systems for, 19-10
  guidelines for, 19-5 to 19-10
  hole sizes, 19-8 to 19-9
  of nonmetallics, 19-10
  oil hole, *see* Oil hole drilling
  power requirements, 17-5, 17-7 to 17-11, 17-18 to 17-19
  pressurized-coolant, *see* Pressurized-coolant drilling
  of thin sheet, 19-10
  tool geometry, 15-14 to 15-18
  tool geometry for high speed steel twist drills, 15-14 to 15-15
  torque and thrust, estimating, 17-27 to 17-34
  unit power requirements, 17-10 to 17-11
  *See also* Core drilling, diamond tools; Spade drilling; Gun drilling
Drills:
  geometry, 19-8, *See also* Tool geometry *for specific drill type*
  tool materials for, 19-9 to 19-10
  types of, 19-5 to 19-7

Economics, machining and grinding, 21-1 to 21-44
  cost and production rate calculations, 21-4 to 21-7, 21-13 to 21-39, 21-43 to 21-44
  cost and production rate versus cutting speed, 21-3 to 21-4, 21-12
  dimensional tolerance versus cost, 21-3, 21-11
  engineering design versus cost, 21-3, 21-8 to 21-9
  surface roughness versus cost, 21-3, 21-10 to 21-11
Electrical discharge grinding, 12-11 to 12-13
  applications data, 12-13
Electrical discharge machining, 12-15 to 12-46
  applications data, 12-43 to 12-46
  operating parameters, steps to selection of, 12-18 to 12-20
  surface integrity data, 18-98 to 18-107, 18-109 to 18-110
Electrical discharge sawing, 12-47 to 12-48
Electrical discharge wire cutting, 12-49 to 12-53
  performance data, 12-51 to 12-53
Electrochemical deburring, 11-3 to 11-4
Electrochemical discharge grinding, 11-5 to 11-8
Electrochemical discharge machining, *see* Electrochemical discharge grinding

Electrochemical grinding, **11**-9 to **11**-22
  applications data, **11**-21 to **11**-22
  operating parameters, steps to selection of, **11**-12
Electrochemical honing, **11**-23 to **11**-24
Electrochemical machining, **11**-25 to **11**-62
  applications data, **11**-58 to **11**-62
  electrolyte selection guide, table, **11**-38 to **11**-41
  operating parameters, steps to selection of, **11**-29 to **11**-30
  surface integrity data, **18**-91 to **18**-97
Electrochemical polishing, **11**-63
Electrochemical sharpening, **11**-65
Electrochemical turning, **11**-67
Electropolishing, **13**-17 to **13**-18
  of high strength alloys, **13**-18
Electro-stream™, **11**-69 to **11**-70
Electron beam machining, **12**-3 to **12**-10
  applications data, **12**-9 to **12**-10
End milling:
  peripheral, **2**-127 to **2**-178
    with diamond tools, **2**-179
    tool geometry for high speed steel end mills, **15**-12 to **15**-13
  slotting, **2**-181 to **2**-231
    tool geometry for high speed steel end mills, **15**-12 to **15**-13

Face milling, **2**-3 to **2**-48
  with diamond impregnated cup wheel, **2**-53
  with diamond tools, **2**-49 to **2**-52
  power requirements, **17**-16 to **17**-17
  tool geometry, **15**-8 to **15**-9
Force requirements, **17**-1 to **17**-5, **17**-23 to **17**-34

Gear cutting:
  straight and spiral bevel, **7**-31 to **7**-46
  with planer-type generators, *see* Gear shaping
Gear grinding, form, **7**-89 to **7**-92
Gear hobbing, **7**-3 to **7**-29
Gear shaping, **7**-47 to **7**-67
Gear shaving, **7**-69 to **7**-88
Gentle grinding, *see* Low stress grinding
Glossary, **27**-1 to **27**-11
Grinding:
  formulas, **20**-24 to **20**-25
    definitions of symbols, **20**-22
  guidelines for, **20**-13 to **20**-16
  maximum wheel speeds, **20**-16
  metal removal rate, **20**-28 to **20**-31
  power requirements, **20**-28 to **20**-31
  precautions, **20**-15 to **20**-16
  surface integrity data, **18**-43 to **18**-44, **18**-46, **18**-50 to **18**-51, **18**-53 to **18**-55, **18**-58, **18**-74 to **18**-80, **18**-82 to **18**-90
  surface roughness, **20**-17
  tolerances, **20**-17 to **20**-19
  unit power requirements, **20**-26 to **20**-27
  *See also* Abrasive belt grinding; Abrasive cutoff; Centerless grinding; Cylindrical grinding; Electrical discharge grinding; Electrochemical discharge grinding; Electrochemical grinding; Internal grinding; Low stress grinding; Surface grinding; Thread grinding
Grinding fluids, *see* Cutting fluids
Grinding wheels, **20**-3 to **20**-11
  conventional, **20**-3 to **20**-7
  cubic boron nitride, **20**-8 to **20**-10
  diamond, **20**-9 to **20**-11
  manufacturers' markings, **20**-5 to **20**-7, **20**-10
  recommendations, modifying of, **20**-13 to **20**-14
  standard marking system charts, **20**-4, **20**-9
Group technology, **24**-4 to **24**-6
  classification and coding, **24**-5 to **24**-6
  relationship to numerical control, **24**-6

Gun drilling, **3**-137 to **3**-170
  tool geometry for carbide drills, **15**-18
Gun reaming, *see* Pressure-coolant reaming

Hack sawing, *see* Power hack sawing
High-efficiency grinding, **20**-34 to **20**-35
High-speed grinding, *see* High-efficiency grinding
Hollow milling, **1**-147 to **1**-180
Hone-forming, **11**-24
Honing, **3**-377 to **3**-392
  with cubic boron nitride stones, **3**-384 to **3**-387
  with diamond tools, **3**-388
  general guidelines for, **3**-373 to **3**-375
  recommendations, **3**-376 to **3**-383
Horsepower requirements, *see* Power requirements
  *See also specific operation*
Hydrodynamic machining, **10**-21 to **10**-36
  operating parameters, steps to selection of, **10**-23

Internal grinding, **8**-101 to **8**-132
  with cubic boron nitride wheels, **8**-133 to **8**-137
  with diamond wheels, **8**-139 to **8**-141
  guidelines for, **20**-14
  horsepower and metal removal rate alignment chart, **20**-30 to **20**-31
  illustrated, **20**-23
  parameter formulas, **20**-24 to **20**-25
  tolerances for, **20**-18

Laser beam machining, **12**-55 to **12**-70
  applications data, **12**-68 to **12**-70
  operating parameters, steps to selection of, **12**-58
  surface integrity data, **18**-100, **18**-107 to **18**-108
Laser beam torch, **12**-71 to **12**-96
  applications data, **12**-94 to **12**-96
  operating parameters, steps to selection of, **12**-73
Laser drilling, *see* Laser beam machining
Low stress grinding, **10**-37 to **10**-38, **18**-74 to **18**-90

Machinability ratings, **19**-15
Machine structures:
  frequency response analysis, **22**-7 to **22**-10
  response to periodic forces, **22**-6 to **22**-12
  stiffness of, **22**-4, **22**-6 to **22**-12
Machining guidelines, **19**-1 to **19**-15
  for drilling, **19**-5 to **19**-10
  general, **19**-3
  tool life, **19**-11 to **19**-14
Manufacturing data bases, **24**-7 to **24**-10
Materials index, **26**-1 to **26**-95
  alphabetical, **26**-17 to **26**-35
  chemical compositions, **26**-37 to **26**-95
  numerical, **26**-3 to **26**-15
Milling:
  power requirements, **17**-4, **17**-7 to **17**-11, **17**-16 to **17**-17
  unit power requirements, **17**-10 to **17**-11
  *See also* End milling; Face milling; Side and slot milling; Slab milling; Thread milling

NC, *see* Numerical control machining
Nontraditional machining, **9**-1 to **9**-5, **10**-1 to **10**-66, **11**-1 to **11**-75, **12**-1 to **12**-105, **13**-1 to **13**-26
  material removal rates for, compared to conventional processes, **9**-5
  surface roughnesses for, compared to conventional processes, **9**-4
  tolerances for, compared to conventional processes, **9**-4 to **9**-5
  *See also listing for specific process*

Numerical control machining, **23-1** to **23-22**
machining conditions selection, **23-5** to **23-6**
part selection guidelines, **23-3** to **23-5**
tooling, **23-6** to **23-7**
vocabulary, **23-9** to **23-22**

Oil-hole drilling, **3-53** to **3-94**
tool geometry for high speed steel drills, **15-16**
Optimum cutting conditions, **21-5** to **21-6**, **21-40** to **21-42**

Peening:
surface integrity data, **18-116** to **18-122**
Photochemical machining, **13-19** to **13-23**
Planing, **5-3** to **5-24**
tool geometry, **15-25** to **15-26**
Plasma arc machining, *see* Plasma beam machining
Plasma beam machining, **12-97** to **12-105**
Power band sawing:
with diamond coated band, **6-73**
with high speed steel blade, **6-35** to **6-72**
Power hack sawing:
with high speed steel blade, **6-3** to **6-34**
Power requirements, **17-1** to **17-22**
equations for calculating, **17-3** to **17-9**
spindle drive efficiency, alignment chart, **17-22**
*See also specific material removal process*
Pressure-coolant reaming (gun reaming), **3-171** to **3-172**
Pressurized-coolant drilling, **3-53** to **3-94**
tool geometry for high speed steel drills, **15-16**
Process planning, *see* Computer-aided process planning

Quality assurance:
and surface integrity, **18-127** to **18-133**
and surface texture, **18-12**

Reaming, **3-173** to **3-219**
with diamond tools, **3-221**
tool geometry:
for carbide reamers, **15-20**
for high speed steel reamers, **15-19**
*See also* Pressure-coolant reaming (gun reaming)
Roll peening, *see* Bearingizing
Roller burnishing, **3-389** to **3-393**
*See also* Skiving plus roller burnishing
Rotary ultrasonic machining, **10-43** to **10-63**

Shaped tube electrolytic machining, **11-71** to **11-75**
applications data, **11-75**
Shot peening, *see* Peening
Side and slot milling (arbor mounted cutters), **2-81** to **2-125**
tool geometry, **15-10** to **15-11**
Skiving plus roller burnishing, **3-402** to **3-404**
Slab milling, **2-55** to **2-80**
Slotting, *see* End milling, slotting; Side and slot milling (arbor mounted cutters)
Spade drilling, **3-95** to **3-136**
tool geometry for high speed steel spade drills, **15-16** to **15-17**
Specifications, *see* Standards, machining
Spotfacing, **3-299** to **3-345**
Standards, machining, **25-1** to **25-23**
sources of, **25-23**
STEM™, *see* Shaped tube electrolytic machining
Surface alterations, **18-40** to **18-45**
Surface finish, *see* Surface texture
Surface grinding:
horizontal spindle, reciprocating table, **8-3** to **8-28**
with cubic boron nitride wheels, **8-29** to **8-38**
with diamond wheels, **8-39** to **8-42**
guidelines for, **20-13** to **20-14**, **20-33**
horsepower and metal removal rate alignment charts, **20-28** to **20-29**
illustrated, **20-23**
parameter formulas, **20-25** to **20-26**
tolerances for, **20-18**
unit power chart, **20-26**
vertical spindle, rotary table, **8-43** to **8-60**
guidelines for, **20-13** to **20-14**, **20-33**
horsepower and metal removal rate alignment charts, **20-28** to **20-29**
illustrated, **20-23**
parameter formulas, **20-25** to **20-26**
tolerances for, **20-18**
Surface integrity, **18-39** to **18-136**
altered material zones (AMZ), **18-40** to **18-45**
definition of, **18-39**
economic considerations, **18-133** to **18-134**
evaluation techniques, **18-45** to **18-55**
for combined effects, **18-54** to **18-55**
extended data set, **18-45** to **18-46**
fatigue strength testing, **18-51** to **18-55**
metallographic sectioning, **18-47**
microhardness determination, **18-47** to **18-48**
minimum data set, **18-45**
residual stress determination, **18-49** to **18-50**
standard data set, **18-45** to **18-46**
for stress corrosion cracking, **18-54**
fatigue strengths and surface roughnesses from various processes:
for 4340 steel, **18-54**, **18-123**
for Inconel alloy 718, **18-54**, **18-126**
for Rene 80, **18-127**
for Ti-5Al-2.5Sn (A110), **18-51**
for Ti-6Al-4V, **18-124**
for Ti-6Al-6V-2Sn, **18-54**
for Ti-6Al-2Sn-4Zr-2Mo, **18-125**
guidelines:
for abrasive material removal processes, **18-74** to **18-90**
for chemical material removal processes, **18-111** to **18-116**
for electrical material removal processes, **18-91** to **18-98**
general, **18-56** to **18-59**
for mechanical material removal processes, **18-59** to **18-74**
for post-treatment processes, **18-116** to **18-122**
specific, **18-59** to **18-122**
for thermal material removal processes, **18-98** to **18-110**
measurement techniques, table of, **18-128**
and process intensity, **18-47**
and quality assurance, **18-127** to **18-133**
nondestructive tests, **18-130** to **18-131**
relationship to surface texture, **18-3**
selected sources of data, list of, **18-134** to **18-136**
Surface roughness, *see* Surface texture
Surface texture, **18-3** to **18-38**
definition of, **18-5** to **18-6**
and dimensional tolerance, **18-12**
effect on machining cost, **18-3**, **18-12**, **21-3**, **21-10** to **21-11**
measurement of, **18-9** to **18-10**
preferred roughness values, **18-14**
and quality assurance, **18-12**
relationship to surface integrity, **18-3**
roughness values for machining processes, **18-11**
symbols, **18-6** to **18-8**
theoretical roughness values, **18-15** to **18-38**

Tapping, **4-3** to **4-41**
cold form, *see* Cold form tapping
tool geometry, high speed steel taps, **15-23** to **15-24**

Thermal energy method, *see* Thermochemical machining
Thermally assisted machining, **10**-39 to **10**-40
  comparison with conventional turning, table, **10**-40
Thermochemical machining, **13**-25 to **13**-26
Thread grinding, **8**-179 to **8**-194
  tolerances for, **20**-17 to **20**-19
Thread milling, **2**-233 to **2**-269
Threading:
  die, **1**-139 to **1**-145
    tool geometry for die threading tools, thread chasers, **15**-6 to **15**-7
  single point, **1**-135 to **1**-138
    tool geometry for single point tools, **15**-4 to **15**-5
Tool geometry, **15**-1 to **15**-28
  boring tools, carbide, **15**-21 to **15**-22
  boring tools, single point, **15**-2 to **15**-3
  broaches, high speed steel, **15**-27
  die threading tools, **15**-6 to **15**-7
  drills, high speed steel twist, **15**-14 to **15**-15
  end mills, peripheral and slotting, high speed steel, **15**-12 to **15**-13
  face mills, **15**-8 to **15**-9
  gun drills, carbide, **15**-18
  oil-hole drills, high speed steel, **15**-16
  planing tools, **15**-25 to **15**-26
  pressurized-coolant drills, high speed steel, **15**-16
  reamers, carbide, **15**-20
  reamers, high speed steel, **15**-19
  side and slot mills, arbor mounted, **15**-10 to **15**-11
  spade drills, high speed steel, **15**-16 to **15**-17
  taps, high speed steel, **15**-23 to **15**-24
  thread chasers, **15**-6 to **15**-7
  threading tools, single point, **15**-4 to **15**-5
  tool and cutter angle equivalents, **15**-28
  turning tools, single point, **15**-2 to **15**-3
Tool life, **19**-11 to **19**-14
Tool materials, **14**-1 to **14**-18
  carbides, **14**-7 to **14**-16
    ISO classification, **14**-10 to **14**-11
  cast alloys, **14**-6
  ceramics, **14**-17
  cermets, **14**-17
  coated carbides, **14**-16
  diamond, **14**-18
  high speed steels, **14**-4 to **14**-5
    AISI-type, **14**-4
    ISO-type, **14**-5
  micrograin carbides, **14**-15
  selection of, **14**-3
Tool steels, *see* Tool materials, high speed steels
Total form machining, **10**-41 to **10**-42
Trepanning, **3**-347 to **3**-375
Turning:
  with box tools, **1**-3 to **1**-57
  with ceramic tools, **1**-59 to **1**-84
  with cutoff tools, **1**-89 to **1**-134
  with diamond tools, **1**-85 to **1**-88
  forces, estimating of, **17**-23 to **17**-25
  with form tools, **1**-89 to **1**-134
  power requirements, **17**-3, **17**-8 to **17**-11, **17**-14 to **17**-15
  with single point tools, **1**-3 to **1**-57
  tool geometry for single point tools, **15**-2 to **15**-3
  unit power requirements, **17**-10 to **17**-11

Ultrasonic machining, **10**-43 to **10**-63
  applications data, **10**-61 to **10**-63
  operating parameters, steps to selection of, **10**-46
Unit power requirements, **17**-10 to **17**-13
  for surface broaching, **17**-12 to **17**-13
  for drilling, **17**-10 to **17**-11
  for grinding, **20**-26 to **20**-27
  for milling, **17**-10 to **17**-11
  for turning, **17**-10 to **17**-11

Vibration, **22**-1 to **22**-23
  causes of, **22**-3 to **22**-5
    deflection, **22**-3 to **22**-4
    forcing, **22**-4 to **22**-5
  *See also* Chatter

Water jet machining, **10**-65 to **10**-66

## NOTES

## NOTES